WILSON AND GISVOLD'S

Textbook of Organic Medicinal and Pharmaceutical Chemistry

WILSON AND GISVOLD'S

Textbook of Organic Medicinal and Pharmaceutical Chemistry

Ninth Edition

EDITED BY

Jaime N. Delgado, Ph.D.

Division of Medicinal Chemistry, College of Pharmacy
University of Texas at Austin, Austin, Texas

AND

William A. Remers, Ph.D.

Department of Pharmaceutical Sciences, College of Pharmacy
University of Arizona, Tucson, Arizona

17 Contributors

J. B. Lippincott Company Philadelphia
New York London Hagerstown

Production Manager: Janet Greenwood
Acquisitions Editor: Lisa McAllister
Manuscript Editor: Marguerite Hague
Production: Till & Till, Inc.
Compositor: Science Typographers, Inc.
Printer/Binder: The Murray Printing Company

Ninth Edition

6 5 4 3 2 1

Library of Congress Cataloging in Publication Data

Wilson and Gisvold's textbook of organic medicinal and pharmaceutical chemistry. — 9th ed. / edited by Jaime N. Delgado and William A. Remers : 17 contributors.
p. cm.
Includes bibliographical references.
Includes index.
ISBN 0-397-50877-8
1. Chemistry, Pharmaceutical. 2. Chemistry, Organic. I. Wilson, Charles Owens, 1911– . II. Gisvold, Ole, 1904– . III. Delgado, Jaime N. IV. Remers, William A. (William Alan), 1932– . V. Title: Textbook of organic medicinal and pharmaceutical chemistry.
[DNLM: 1. Chemistry, Pharmaceutical. QV 744 W754]
RS403.T43 1991
615'.19—dc20
DNLM/DLC
for Library of Congress

90-13652
CIP

The authors and publisher have exerted every effort to ensure that drug selection and dosage set forth in this text are in accord with current recommendations and practice at the time of publication. However, in view of ongoing research, changes in government regulations, and the constant flow of information relating to drug therapy and drug reactions, the reader is urged to check the package insert for each drug for any change in indications and dosage and for added warnings and precautions. This is particularly important when the recommended agent is a new or infrequently employed drug.

CONTENTS

CONTRIBUTORS

John H. Block, Ph.D.
Professor of Medicinal Chemistry
College of Pharmacy
Oregon State University, Corvallis, Oregon

Neal Castagnoli, Jr., Ph.D.
Professor of Chemistry and Pharmaceutical Chemistry
School of Pharmacy, University of California
San Francisco, California

George H. Cocolas, Ph.D.
Professor of Medicinal Chemistry and Associate Dean
School of Pharmacy
University of North Carolina at Chapel Hill
Chapel Hill, North Carolina

Charles M. Darling, Ph.D.
Professor of Pharmaceutical Chemistry, School of Pharmacy
Auburn University, Auburn, Alabama

Jaime N. Delgado, Ph.D.
Professor of Medicinal Chemistry
Jacques P. Servier Regents Professorship in Pharmacy
College of Pharmacy, The University of Texas at Austin
Austin, Texas

Dwight S. Fullerton, Ph.D.
Dean, College of Pharmacy
Utah State University, Salt Lake City, Utah

Patrick E. Hanna, Ph.D.
Associate Professor of Medicinal Chemistry
College of Pharmacy and School of Medicine
University of Minnesota, Minneapolis, Minnesota

Thomas J. Holmes, Jr., Ph.D.
Associate Professor of Medicinal Chemistry and Associate Dean
School of Pharmacy
Campbell University, Buies Creek, North Carolina

Pedro L. Huerta, Jr., Ph.D.
Professor of Medicinal Chemistry, School of Pharmacy
Southwestern Oklahoma State University
Weatherford, Oklahoma

Eugene I. Isaacson, Ph.D.
Professor of Medicinal Chemistry
College of Pharmacy
Idaho State University, Pocatello, Idaho

Daniel A. Koechel, Ph.D.
Professor of Pharmacology
Medical College of Ohio, Toledo, Ohio

Lawrence K. Low, Ph.D.
Research Associate in Medicinal Chemistry
Department of Pharmaceutical Sciences
School of Pharmacy, University of Washington
Seattle, Washington

Arnold R. Martin, Ph.D.
Professor of Medicinal Chemistry, College of Pharmacy
Head, Department of Pharmaceutical Sciences
The University of Arizona, Tucson, Arizona

Gustavo R. Ortega, Ph.D.
Associate Professor of Medicinal Chemistry, School of Pharmacy
Southwestern Oklahoma State University
Weatherford, Oklahoma

Vilas A. Prabhu, Ph.D.
Associate Professor of Medicinal Chemistry, School of Pharmacy
Southwestern Oklahoma State University
Weatherford, Oklahoma

William A. Remers, Ph.D.
Professor of Medicinal Chemistry
College of Pharmacy, The University of Arizona
Tucson, Arizona

Robert E. Willette, Ph.D.
Director, Duo Research Division
Research Designs, Inc., Annapolis, Maryland

PREFACE

For more than four decades "Wilson and Gisvold's Textbook" has been a standard in the literature of medicinal chemistry, and the recent editions edited by Professor Robert F. Doerge have continued to receive national and international acceptance. Generations of students and faculty have depended on this textbook not only for undergraduate courses in medicinal chemistry but also to supplement graduate studies. Moreover, students in other health sciences have found certain chapters useful at one time or another. It was therefore decided to revise the book, and the current editors and authors worked on the ninth edition with the objective of continuing the tradition of a modern textbook for undergraduate students and also for graduate students who need a general review of medicinal chemistry. Since the chapters include a blend of chemical and pharmacological principles necessary for understanding structure-activity relationships and molecular mechanisms of drug action, the book should be useful in supporting courses in medicinal chemistry and in complementing pharmacology courses.

The nature and breadth of medicinal chemistry are discussed in the introduction, and the chapters on physico-chemical properties and on drug metabolism provide the fundamental theme of the conceptual approach to the study of medicinal chemistry. This theme continues as the basis for all the subsequent chapters on the classical types of therapeutic agents. The chapters include summaries of current research trends that lead the reader to the original literature via bibliographies. Documentation and references continue to be an important feature of the book.

An important area of medicinal chemistry is the chemistry and biology of pharmaceutically important natural products, hence chapters include antibiotics, glycosides, and the study of alkaloids as prototypes that led to synthetic medicinals. The table of contents reflects the important pharmacological types of therapeutic agents. To the extent possible each type is structurally characterized with mechanisms of action, therapeutic and pharmaceutical applications and limitations.

The editors extend thanks to all the authors who have effectively cooperated in the preparation of the current edition. Collectively, the authors represent many years of teaching and research experience in medicinal chemistry. Their students also deserve acknowledgement of their critiques which directly and indirectly contributed to the quality of this edition.

Finally, our gratitude is extended to Professors Charles O. Wilson and Ole Gisvold the originators of the book and the editors and authors who worked on, and stimulated, the continuation of the many revisions that led to the current one. They significantly contributed to the education of countless pharmacists, medicinal chemists, and other pharmaceutical scientists.

Jaime N. Delgado
William A. Remers

CHAPTER 1

Introduction

Jaime N. Delgado
William A. Remers

The practice of medicinal chemistry is devoted to the discovery and development of new agents for treating diseases. Most of the activity in this discipline is directed to new natural or synthetic organic compounds. Inorganic compounds continue to have their place in therapy, for example, as antacids, mineral supplements, or radiopharmaceuticals, but organic compounds with increasingly specific pharmacologic activities are clearly the dominant force. This volume treats many aspects of organic medicinals: how they are discovered, how they act, and how they are developed into clinical agents. The process of establishing a new drug is exceedingly complex and involves the talents of people from a variety of disciplines, including chemistry, biochemistry, physiology, pharmacology, pharmaceutics, and medicine. Medicinal chemistry is concerned mainly with the organic, analytical, and biologic chemical aspects of this process, but its people must interact productively with those in the other disciplines. Thus, it occupies a strategic position at the interface of chemistry and biology. There is considerable overlap with other disciplines; for example, medicinal chemistry and pharmacology both are concerned with mode of action and structure–activity relationships of drugs. However, this kind of overlap facilitates productive interactions in research.

The earliest drug discoveries were made by the presumably random sampling of higher plants. Herbal remedies have been important throughout human history, and they still are. Crude plant products, such as opium and belladonna, have been valuable drugs for centuries. Knowledge of crude drugs was codified into the discipline of pharmacognosy, which has held a distinguished place in the pharmaceutical sciences. This field has changed with time, embracing aspects of microbiology, when the antibiotics were discovered, and venturing into drug biosynthesis, as biochemistry developed. However, in recent years the introduction of new synthetic pharmaceuticals has outpaced that of natural products. Furthermore, the isolated and purified active principles have largely superseded preparations of the parent crude drugs, making it unnecessary for pharmacists to verify the identity of natural materials. These factors, together with the approach of teaching drugs by therapeutic categories, has led to deemphasis on pharmacognosy in the pharmacy curriculum and often to its combination with medicinal chemistry. This textbook, although it does not presume to teach pharmacognosy, contains much valuable information and insight into the sources, isolation, and standardization of natural-product drugs.

Although many natural products are used in pharmaceuticals in their original chemical structures, successful efforts have been made to improve their pharmaceutic and therapeutic properties by structural modifications. Some of these modifications are relatively simple, such as the formation of the phosphate ester of hydrocortisone. This derivative increases water solubility of the drug. Other modifications may be more substantial, as in the replacement of penicillin and cephalosporin side chains with new ones that modify the antibacterial spectrum. Another approach to improving therapeutic properties is to identify that portion of a natural molecule responsible for its biologic activity and synthesize new molecules that are based on it. This active portion is known as the *essential structural unit*. The development of local anesthetics from cocaine is a prime example of this approach.

Hundreds of thousands of new organic chemicals are prepared annually throughout the world, and many of them are entered into pharmacologic screens to determine if they have useful biologic activity. This process of random screening is inefficient, but it has resulted in the identification of new lead compounds not produced naturally or imagined by chemists. Such lead compounds form the basis of a series of analogues intended to optimize the therapeutic activity. The antitubercular drug, ethambutol, was developed in this way. More recently, emphasis has been placed on rational design of new pharmaceuticals. This is a difficult thing to do because detailed knowledge of the macromolecular receptor is important, and considerations of bioavailability and pharmacokinetics must be addressed. Some notable successes have resulted from the design of specific inhibitors of important enzymes. For example, the antihypertensive drug captopril was designed to inhibit peptidyl dipeptidase, an enzyme that cleaves angiotensin I to the potent vasoconstrictor angiotensin II. Another example is the design of the anticancer drug 5-fluorouracil as an inhibitor of thymidylate synthetase. Even if the receptor is not known in detail, rational design can be applied to selective modification of the structure of a neurotransmitter or hormone to limit its spectrum of activity (decrease side effects) or convert it into an antagonist. Thus, new adrenergic agents, such as terbutaline, provide relief of bronchoconstriction in asthmatics, without unduly stimulating the heart. The development of cimetidine as an antiulcer drug was based on antagonizing H_2 receptors for histamine. It involved a combination of chemical intuition and careful attention to changes in pharmacologic activity induced by varying the physicochemical properties of lead antagonists, based on the histamine structure.

Once a new pharmaceutical lead compound has been discovered, extensive and costly efforts usually are made to prepare a series of analogues in the hope that even better activity will be found. In an effort to improve the efficiency of analogue development, a variety of statistical methods have been introduced. They range from the Hansch approach, in which analysis of variance is used to derive an equation expressing the quantitative relationships between functional group changes and biologic activity, to pattern recognition and factor analysis methods. Nonquantitative methods, such as the Topliss approach, also are popular. Computer-aided design, including quantitative energy calculations and graphical methods, has been rapidly introduced in the pharmaceutical industry. It is too early to evaluate its effect on drug discovery.

The metabolism of drugs is an important topic in medicinal chemistry and considerable effort is spent on detailed analysis of the bioconversions that a new drug undergoes. Modern analytical methods, such as mass spectrometry, permit the identification of minute amounts of metabolites. The usual effect of metabolism is to convert drugs into compounds that are less active, less toxic, and more readily excreted. However, there are examples in which metabolism sustains or increases the activity. Thus, the *N*-demethyl metabolite, desipramine, contributes significantly to the antidepressant activity of imipramine. Metabolism of drugs and other xenobiotics can also increase toxicity. For example, the conversion of benzene into its diol epoxide results in liver damage. Bioactivating transformations can be used to advantage in designing inactive prodrugs that are converted into active species in the body. A representative process is hydrolysis of the ester functionality of clofibrate to give the corresponding acid, which has hypolipidemic activity.

Once a new compound shows potential for clinical development, extensive studies on its pharmacology and toxicology are undertaken. Dosage forms are optimized to provide appropriate solubility and bioavailability, and the pharmacokinetics are determined. When adequate information has been obtained, application can be made to the Food and Drug Administration (FDA) for investigational new drug (IND) status. When this status is granted, clinical trials may be undertaken. These activities are indirect, but important, considerations for medicinal chemistry. If the new drug is approved, medicinal chemistry again becomes directly involved in the large-scale synthesis, if appropriate, and in the analytical techniques of quality assurance. Many other skills, such as packaging, distribution, and marketing, are important in bringing a new drug to the patient. All of the operations in this complex and sophisticated process are concerns of pharmacy.

Future trends in drug development will include an increasing role for genetically engineered proteins, following on the successful introduction of insulin, clotting factor VIII, and interferon. However, organic medicinals, designed with increasing attention to receptor mapping and enzyme active site geometry and functionality, should constitute almost all new drugs for several years.

CHAPTER 2

Physicochemical Properties in Relation to Biologic Action

John H. Block

INTRODUCTION

Modern drug design, as compared with, *Let's make a change on an existing compound or synthesize a new structure and see what happens*, is a fairly recent discipline still in its infancy. It is based on modern chemical techniques utilizing recent knowledge of disease mechanisms and receptor properties. A good understanding of how the drug is transported into the body, distributed throughout the body compartments, metabolically altered by the liver and other organs, and excreted from the patient is required, along with the structural characteristics of the receptor. Acid–base chemistry is utilized to aid in formulation and biodistribution. Those structural attributes and substituent patterns responsible for optimum pharmacologic activity can be predicted many times by statistical techniques, such as regression analysis. Conformational analysis permits the medicinal chemist to predict the drug's three-dimensional shape that is *seen* by the receptor. With the isolation and structural determination of specific receptors and the availability of computer software that can estimate the three-dimensional shape of the receptor, it is possible to design molecules that will show an optimum fit to the receptor.

HISTORY

Initially, drugs were extracted from plant sources to obtain agents such as digitalis, quinine, and morphine, medicinal agents that are still in use today. Specific plants are selected by the chemist because the crude preparations were being used for treatment of medical conditions by the local population where the plant grew.

Early drug design started with elucidation of the structure of the natural product, followed by selective changes in the molecule. The latter was done for many reasons, including the reduction of an undesirable pharmacologic response (side effect); obtaining a better pharmacokinetic response; altering the drug's metabolism; securing a more plentiful, less costly supply; and producing a competing product. Let us use the morphine alkaloid as an example. Literally thousands of compounds have been synthesized in an attempt to separate the desired analgesia from the undesirable addiction liability. This tremendous effort in numerous research laboratories, over many years and involving many scientists, had been minimally successful until a better understanding of the opiate receptor developed (see Chap. 9).

In other examples, there has been good success at this empirical approach. Alteration of the cocaine structure has led to the very successful local anesthetics that lack cocaine's undesirable central effects (see Chap. 15). In contrast with this success story, there have been no significant commercial synthetic replacements for digitalis or colchicine.

Synthetic medicinal chemistry as a discipline became more intense in the 1900s, but many of the so-called principal compounds still were based on a natural product, a fortuitous observation, or an unsuspected chemical reaction. The phenothiazines (see Chap. 9) were first synthesized as antihistamines, but a careful pharmacologic evaluation led to their use as major tranquilizing agents that revolutionized the care of the severely mentally ill patient. The benzodiazepines (see Chap. 9) originated from an unexpected ring enlargement and resulted in a very important group of central nervous system relaxants.

Economic factors have stimulated the type of scientific investigations required to carry out focused new drug development. It has become increasingly costly to develop a new drug that will be approved by the U. S. Food and Drug Administration (FDA). At one time, safety was the main criterion for FDA approval. Today, demonstration of efficacy is an essential requirement, along with safety considerations. This has led to (1) increased basic research on the disease process for which a drug treatment is sought, (2) mathematically modeling the pharmacokinetics of the drug's distribution, (3) elucidation of the biochemistry of the pharmacologic response from the drug, (4) learning the metabolic fate of the drug, (5) defining those specific structural characteristics of the drug responsible for the desired pharmacologic response, and (6), where possible, visualizing the structural characteristics of the receptor. Although the number of new compounds introduced annually has decreased from earlier years, the products now coming into use are showing dramatic effects in the treatment of disease. More importantly, because of the intensive background investigation that has led to the design of today's new agents, a better understanding of the drug's mechanism of action is known. Indeed, this is an exciting time to practice pharmacy.

OVERVIEW

A drug is a chemical molecule. Following introduction into the body, a drug must pass through many barriers, survive alternate sites of attachment and storage, and avoid significant metabolic destruction before it reaches the site of action, usually a receptor on or in a cell (Fig. 2-1). At the receptor, the following equilibrium usually holds.

$$\text{Drug} + \text{Receptor} \rightleftharpoons \text{Drug–Receptor Complex} \rightarrow \text{Pharmacologic Response} \quad \text{(Rx. 2-1)}$$

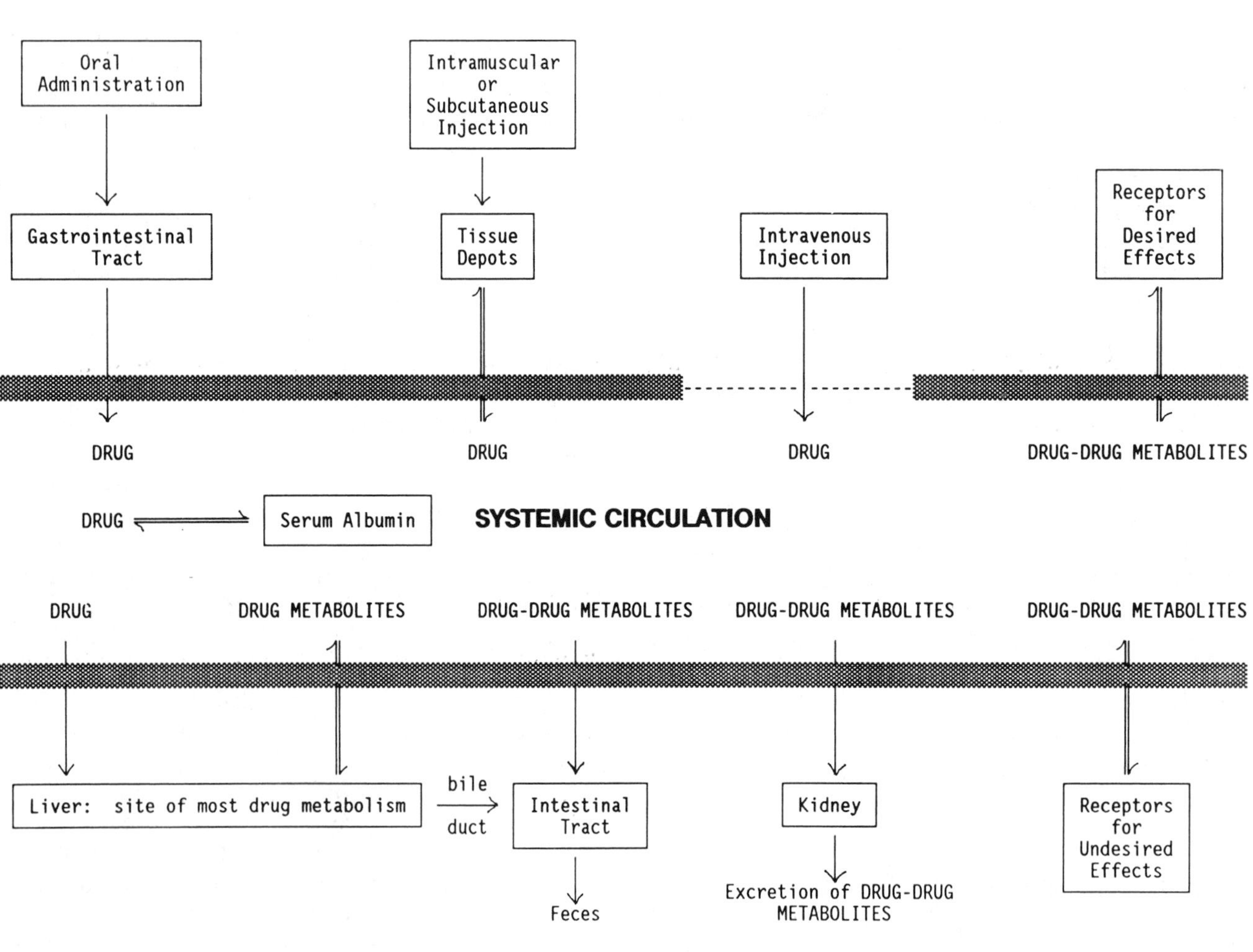

FIG. 2-1. Summary of drug distribution: Solid bars: Drug must pass through membranes. Broken lines: Drug administered directly into systemic circulation.

The ideal drug molecule will show favorable-binding characteristics to the receptor, such that the equilibrium lies to the right. At the same time the drug will be expected to dissociate from the receptor and reenter systemic circulation to be excreted. The major exceptions include the alkylating agents used in cancer chemotherapy (see Chap. 8) and a few inhibitors of the enzyme, acetylcholinesterase (see Chap. 12). Both of these subclasses of pharmacologic agents form covalent bonds with the receptor. In these cases the cell must destroy the receptor or, as with the alkylating agents, the cell would be replaced, ideally with a normal cell. In other words, the usual use of drugs in medical treatment call for the drug's effect to last for only a finite period. Then, if it is to be repeated, the drug will be administered again. If the patient does not tolerate the drug well, it is even more important that the agent dissociate from the receptor and be excreted from the body.

DRUG DISTRIBUTION

ORAL ADMINISTRATION

An examination of the *obstacle course* (see Fig. 2-1) faced by the drug will give a better understanding of what is involved in developing a commercially feasible product. Assume that the drug is administered orally. The drug must go into solution for it to pass through the gastrointestinal mucosa. Even drugs administered as true solutions may not remain in solution as they enter the acidic stomach and then pass into the alkaline intestinal tract. (This will be explained further in the discussion on acid–base chemistry.) The ability of the drug to dissolve is governed by several factors, including its chemical structure, variation in particle size and particle surface area, nature of the crystal form, type of coating, and type of tablet matrix. By varying the formulation containing the drug and physical characteristics of the drug, it is possible to have a drug dissolve quickly or slowly, the latter being the situation for many of the sustained-action products. An example is orally administered sodium phenytoin; for which variation of both the crystal form and tablet adjuvants can significantly alter the bioavailability of this drug, which is widely used in the treatment of epilepsy.

Chemical modification is also used to a limited extent. For example, sulfasalazine, used in the treatment of ulcerative colitis, passes through a substantial portion of the intestinal tract before being metabolized to sulfapyridine and 5-aminosalicylic acid. The latter compound is believed to be the active agent for the treatment of ulcerative colitis.

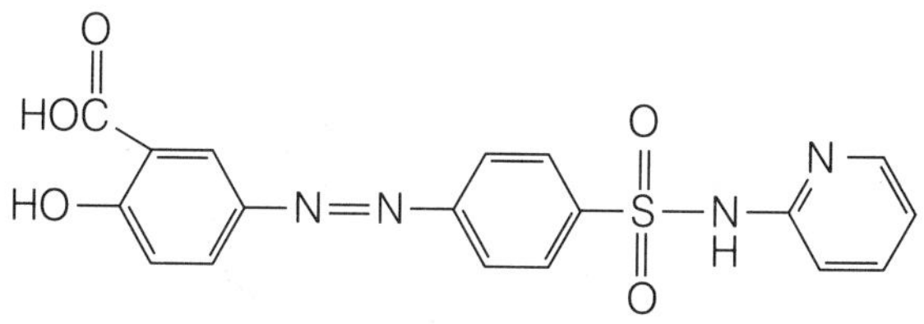

Sulfasalazine

Any compound passing through the gastrointestinal tract will encounter the many and various digestive enzymes that, in theory, can degrade the drug molecule. In practice, when a new drug entity is under investigation, it will probably be dropped from further consideration if it is found unable to survive in the intestinal tract. An exception would be a drug for which there is no other effective product available, or one that provides a more effective treatment over existing products and can be administered by an alternate route, usually parenteral.

In contrast, these same digestive enzymes can be used to advantage. Chloramphenicol is water-soluble enough that it comes in contact with the taste receptors on the tongue, producing an unpalatable bitterness. To mask this intense bitter taste, the palmitic acid moiety is added as an ester of the chloramphenicol's primary alcohol. This reduces the parent drug's water solubility so much that it can be formulated as a suspension that passes over the bitter taste receptors on the tongue. Once in the intestinal tract, the ester linkage is hydrolyzed by the digestive esterases to the active antibiotic, chloramphenicol, and the very common dietary fatty acid, palmitic acid.

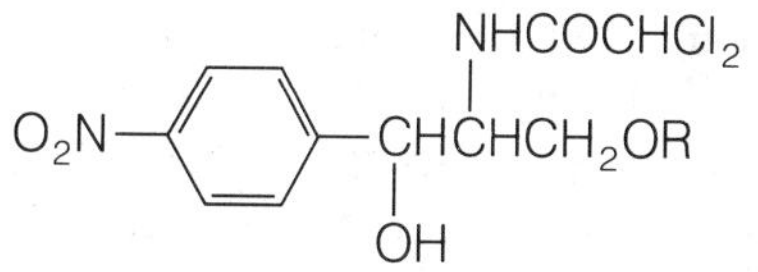

$O_2N-C_6H_4-CH(OH)CH(NHCOCHCl_2)CH_2OR$

Chloramphenicol: R = H;

Chloramphenicol Palmitate: $R = C(=O)(CH_2)_{14}CH_3$

Sulfasalazine and chloramphenicol palmitate are examples of *prodrugs*. Most prodrugs are compounds that are inactive in their native form, but are easily metabolized to the active agent. Sulfasalazine and chloramphenicol palmitate are examples of prodrugs that are cleaved to smaller compounds, one of which will be the active drug. Others are metabolic precursors to the active form. An example of this type of prodrug is menadione, a simple naphthoquinone, which is converted in the liver to vitamin $K_{2(20)}$.

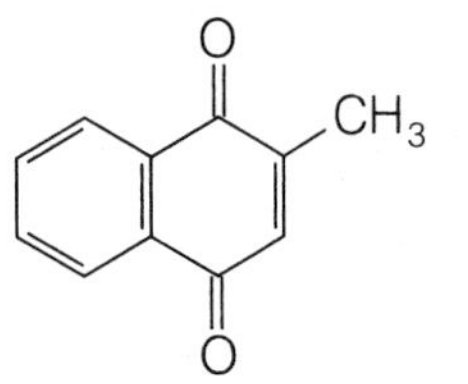

Menadione

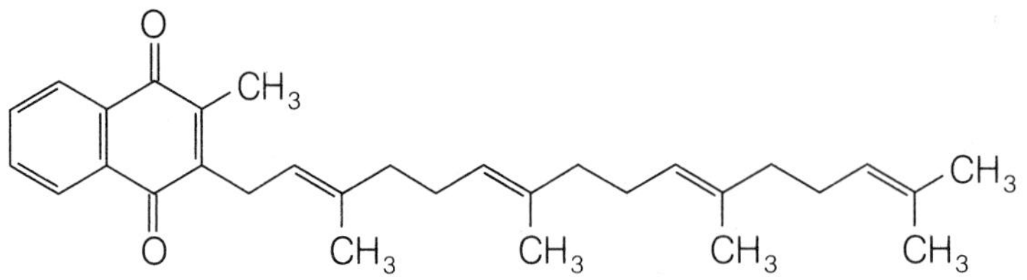

Vitamin $K_{2(20)}$

Occasionally the prodrug approach is used to enhance the absorption of a drug that is poorly absorbed from the gastrointestinal tract. Enalapril is the ethyl ester of enalaprilic acid, an active inhibitor of angiotensin-converting enzyme. The ester prodrug is much more readily absorbed orally than the carboxylic acid.

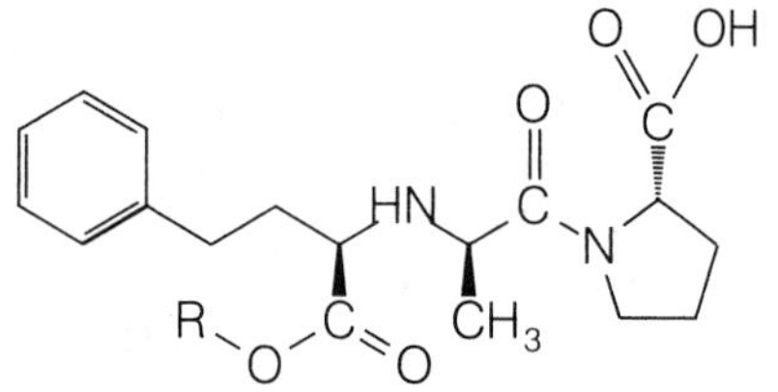

Enalapril: R = C_2H_5; Enalaprilic Acid: R = H

Unless the drug is intended to act locally in the gastrointestinal tract, it will have to pass through the gastrointestinal mucosa barrier into the venous circulation to reach the receptor site. This involves distribution, or partitioning, between the aqueous environment of the gastrointestinal tract, the lipid bilayer cell membrane of the mucosa cells, possibly the aqueous interior of the mucosa cells, the lipid bilayer membranes on the venous side of the gastrointestinal tract, and the aqueous environment of venous circulation. Some very lipid-soluble drugs may follow the route of dietary lipids by becoming part of the mixed micelles, passing through the mucosa cells into the thoracic duct of the lymphatic system and then into the venous circulation.

The drug's passage through the mucosa cells can be passive or active. As will be discussed later in this chapter, the lipid membranes are very complex, with a highly ordered structure. Part of this membrane is a series of channels or tunnels that form, disappear, and reform. There are receptors that move compounds into the cell by a process called pinocytosis. Most drug molecules are too large to enter the cell by an active transport mechanism through the passages. On the other hand, some drugs do resemble a normal metabolic precursor or intermediate, and they can be actively transported into the cell.

PARENTERAL ADMINISTRATION

Many times there will be therapeutic advantages to bypass the intestinal barrier by using parenteral (injectable) dosage forms; for example, because of illness, the patient cannot tolerate or is incapable of accepting drugs orally. Some drugs are so rapidly and completely metabolized to inactive products in the liver (first-pass effect) that oral administration is precluded. But that does not mean that the drug administered by injection is not confronted by obstacles (see Fig. 2-1). Intravenous administration places the drug directly into the circulatory system, from which it will be rapidly distributed throughout the body, including tissue depots and the liver, in which most biotransformations occur, in addition to the receptors.

It is possible to inject the drug directly into specific organs or areas of the body. Intraspinal and intracerebral routes will place the drug directly into the spinal fluid or brain, respectively, bypassing a specialized tissue, the blood–brain barrier, which protects the brain from exposure to diverse metabolites and chemicals. The blood–brain barrier is composed of membranes of tightly joined epithelial cells lining the cerebral capillaries. The net result is that the brain is not exposed to the same variety of compounds that other organs are. Local anesthetics are examples of administration of a drug directly on the desired nerve. A spinal block is a form of anesthesia performed by injecting a local anesthetic directly into the spinal cord at a specific location to block transmission along specific neurons.

Many of the injections a patient will experience in a lifetime will be subcutaneous or intramuscular. These parenteral routes produce a depot in the tissues (see Fig. 2.1) from which the drug must reach the blood or lymph to produce systematic effects. Once in systemic circulation, the drug will undergo the same distributive phenomena as orally and intravenously administered agents. In general, the same factors that control the drug's passage through the gastrointestinal mucosa will also determine the rate of movement out of the tissue depot.

The prodrug approach also can be used to alter the solubility characteristics which, in turn, can increase the flexibility in possible dosage forms. The solubility of methylprednisolone can be altered from essentially water-insoluble methylprednisolone acetate, to slightly water-insoluble methylprednisolone, to water-soluble methylprednisolone sodium succinate. The water-soluble sodium succinate salt is used in oral, intravenous, and intramuscular dosage forms. Methylprednisolone, itself, is normally found in tablets. The acetate ester is found in topical ointments and sterile aqueous suspensions for intramuscular injection. Both the succinate and acetate esters are hydrolyzed to the active methylprednisolone by the patient's own systemic hydrolytic enzymes.

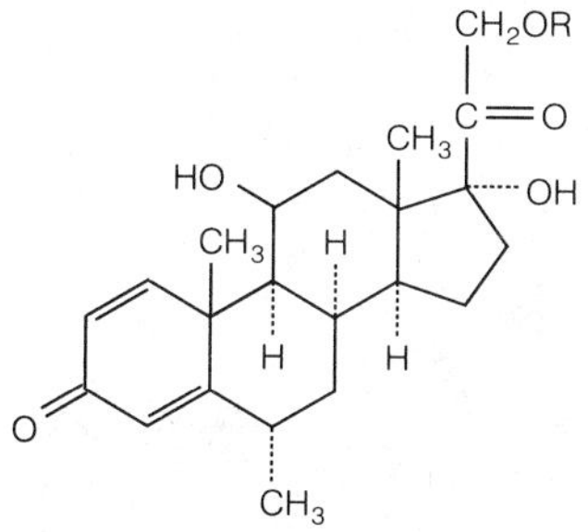

Methylprednisolone (R = H)

Ester Available:

Methylprednisolone Acetate: R = $COCH_3$

Salt Available:

Methylprednisolone Sodium Succinate:

R = $COCH_2CH_2COO^-Na^+$

PROTEIN BINDING

Once the drug enters the systemic circulation (see Fig. 2-1) it can undergo several events. It may stay in solution, but many drugs will be bound to the serum proteins, usually albumin. Thus, a new equilibrium must be considered. Depending on the equilibrium constant, the drug can remain in systemic circulation bound to albumin for a considerable period and be unavailable to the sites of biotransformation, the pharmacologic receptors, and excretion.

$$\text{Drug} + \text{Albumin} \rightleftharpoons \text{Drug–Albumin Complex} \qquad \text{(Rx. 2-2)}$$

The effect of protein binding can have a profound result on the drug's effective solubility, biodistribution, half-life in the body, and interaction with other drugs. A drug with such poor water solubility that therapeutic concentrations of the unbound (active) drug normally cannot be maintained, still can be a very effective agent. The albumin–drug complex acts as a reservoir by providing concentrations of free drug large enough to cause a pharmacologic response.

Protein binding may also limit access to certain body compartments. The placenta is able to block passage of proteins from maternal to fetal circulation. Consequently, drugs that normally would be expected to cross the placenta barrier and possibly harm the fetus are retained in the maternal circulation bound to the mother's serum proteins.

Protein binding also can prolong the drug's duration of action. The drug–protein complex is too large to pass through the renal glomerular membranes, preventing rapid excretion of the drug. Protein binding limits the amount of drug available for biotransformation (see later and Chap. 3) and for interaction with specific receptor sites. For example the trypanocide, suramin, remains in the body in the protein-bound form as long as three months. The maintenance dose for this drug is based on weekly administration. At first this might seem to be an advantage to the patient. It can be, but it also means that should the patient have serious adverse reactions, it will require a substantial length of time before the concentration of drug falls below toxic levels.

The drug–protein-binding phenomenon can lead to some interesting drug–drug interactions, resulting when one drug displaces another from the binding site on albumin. Diverse drugs can displace the anticoagulant, warfarin, from its albumin-binding sites. This increases the effective concentration of warfarin at the receptor, leading to an increased prothrombin time (increased time for clot formation) and potential hemorrhage.

TISSUE DEPOTS

The drug also can be stored in tissue depots. Neutral fat constitutes some 20% to 50% of body weight and constitutes a depot of considerable importance. The more lipophilic the drug is, the more likely it will concentrate in these pharmacologically inert depots. The short-acting, lipophilic barbiturate, thiopental, reportedly disappears into tissue protein, redistributes into body fat and, then, slowly diffuses back out of the tissue depots, but in concentrations too low for any pharmacologic response. Hence, only the initially administered thiopental is present in high enough concentrations to combine with its receptors. In general, structural changes in the barbiturate series (see Chap. 9) that favor partitioning into the lipid tissue stores decreases the duration of action, but increases central nervous system depression. Conversely, the barbiturates with the slowest onset and longest duration of action contain the more polar side chains. This latter group of barbiturates both enters and leaves the central nervous system very slowly, as compared with the more lipophilic thiopental.

DRUG METABOLISM

All substances, including drugs, metabolites, and nutrients that are in the circulatory system, will pass through the liver. Most molecules absorbed from the gastrointestinal tract will enter the portal vein and be transported to the liver. A large proportion of a drug will partition or be transported into the hepatocyte, where it may be metabolized by hepatic enzymes to inactive chemicals during the initial trip through the liver by what is known as the first-pass effect. Over 60% of the local anesthetic–antiarrhythmic agent, lidocaine, is metabolized during its initial passage through the liver, resulting in it being impractical to administer orally. When used for cardiac arrhythmias, it is administered intravenously. This rapid metabolism of lidocaine is used to advantage when stabilizing a patient with cardiac arrhythmias. Should too much lidocaine be adminis-

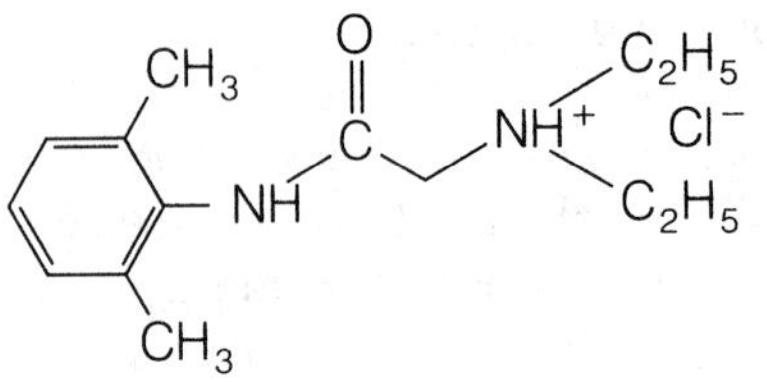

Lidocaine

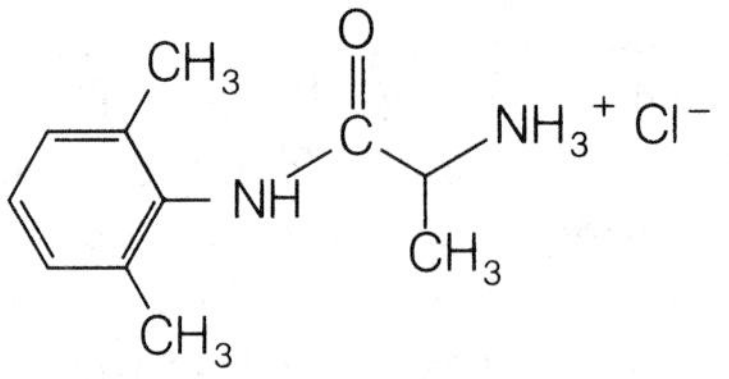

Tocainide

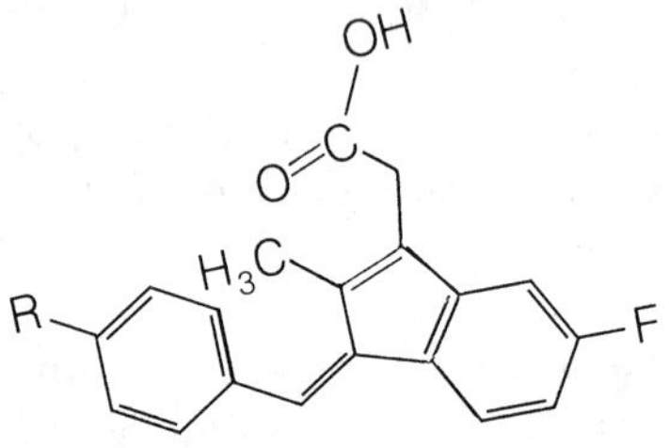

Sulindac: R = CH_3SO^-; Active Sulfide Metabolite: R = CH_3S^-

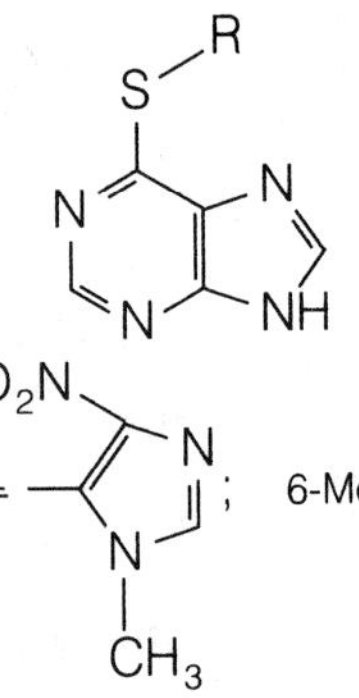

Azathioprene R = ; 6-Mercaptopurine: R = H

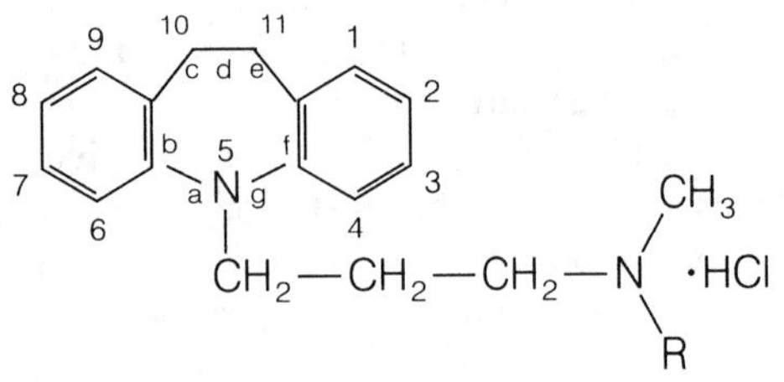

Imipramine: R = CH_3
Desipramine: R = H

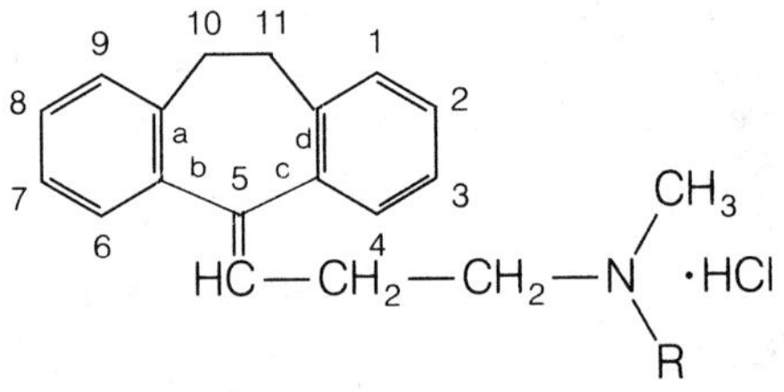

Amitriptyline: R = CH_3
Nortriptyline: R = H

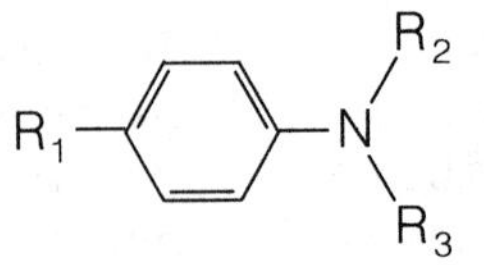

Phenacetin: $R_1 = C_2H_5O^-$, $R_2 = CH_3CO^-$, $R_3 = H$;
Acetaminophen: $R_1 = HO^-$, $R_2 = CH_3CO^-$, $R_3 = H$;

tered intravenously, toxic responses will tend to abate because of the rapid biotransformation to inactive metabolites. An understanding of the metabolic labile site on lidocaine led to the development of the primary amine analogue, tocainide. In contrast with lidocaine's half-life of less than two hours, tocainide's half-life is about 15 hours with 40% of the drug excreted unchanged. The development of orally active antiarrhythmic agents is discussed in more detail in Chapter 14.

A study of the metabolic fate of a drug is a requirement for all new products. Many times it is found that the metabolites also are active. Indeed, sometimes the metabolite is the pharmacologically active molecule. These drug metabolites can provide the leads for additional investigations of potentially new products. Examples of where an inactive parent drug is converted to an active metabolite include the nonsteroidal anti-inflammatory agent, sulindac, being reduced to the active sulfide metabolite; the immunosuppressant, azathioprine, being cleaved to the purine antimetabolite, 6-mercaptopurine; and purine and pyrimidine antimetabolites and antiviral agents being conjugated to their nucleotide form. Many times both the parent drug and its metabolite are active, which has led to two commercial products, instead of just one, being marketed. About 75% to 80% of phenacetin (now withdrawn from the U. S. market) is converted to acetaminophen. In the tricyclic antidepressant series (see Chap. 10) imipramine and amitriptyline, are *N*-demethylated to desipramine and nortriptyline, respectively. All four compounds have been marketed in the United States. The topic of drug metabolism is more fully discussed in Chapter 3.

Although a drug's metabolism can be a source of frustration for the medicinal chemist, pharmacist, and physician, and can lead to inconvenience and compliance problems for the patient, it is fortunate that the body has the ability to metabolize foreign molecules (xenobiotics). Otherwise, many of these substances could remain in the body for years. This has been the complaint against certain lipophilic chemical pollutants, including the once, very popular insecticide, DDT. After entering the body, these chemicals sit in body tissues, slowly diffusing out of the depots and potentially harming the individual on a prolonged basis for several years.

EXCRETION

The main route of excretion of a drug and its metabolites is through the kidney. For some drugs, enterohepatic circulation (see Fig. 2-1), in which the drug reenters the intestinal tract from the liver through the bile duct, can be an important part of the agent's distribution in the body and route of excretion. The drug or drug metabolite can reenter systemic circulation by passing once again through the intestinal mucosa. A portion of it also may be excreted in the feces. Nursing mothers must be concerned because drugs and their metabolites can be excreted in human milk and be ingested by the nursing infant. Usually, the end products of drug metabolism are very water-soluble relative to the parent molecule. Obviously, drugs that are bound to serum protein or show favorable partitioning into tissue depots are going to be excreted more slowly for the reasons already discussed.

This does not mean that for those drugs that remain in the body for longer periods, lower doses can be administered or the drug can be taken fewer times per day by the patient. Several variables determine dosing regimens, of which the affinity of the drug for the receptor is crucial. Reexamine Reaction 2-1 and Figure 2-1. If the equilibrium does not favor formation of the drug–receptor complex, higher and usually more frequent doses will have to be administered. If the partitioning into tissue stores, metabolic degradation, or excretion are favored, it will take more drug, and usually more frequent administration, to maintain therapeutic concentrations at the receptor.

RECEPTOR

With the exception of general anesthetics (see Chap. 9), the working model for a pharmacologic response consists of a drug binding to a specific receptor. Many drug receptors actually are used by endogenously produced ligands. Cholinergic agents interact with the same receptors as the neurotransmitter acetylcholine. Synthetic corticosteroids bind to the same receptors as cortisone and hydrocortisone. Many times, receptors for the same ligand will be found in a variety of tissues throughout the body. The nonsteroidal anti-inflammatory agents (see Chap. 17) inhibit the prostaglandin-forming enzyme cyclooxygenase which is found in nearly every tissue. This class of drugs has a long list of side effects, with many patients' complaints. Note in Figure 2-1 that, depending on which receptors contain bound drug, there may be desired or undesired effects. This is because there are a variety of receptors with similar structural requirements found in several organs and tissues. Thus, the nonsteroidal anti-inflammatory drugs combine with the desired cyclooxygenase receptors at the site of the inflammation and the receptors in the gastrointestinal mucosa causing severe discomfort and sometimes ulceration. One of the newer antihistamines, terfenadine, is claimed to cause less sedation because it does not readily penetrate the blood–brain barrier. The rationale is that less of this antihistamine is available for the receptors in the central nervous system that are responsible for the sedation response characteristic of antihistamines. In contrast, some antihistamines are used for their central nervous system depressant activity, which would imply that a considerable amount of the administered dose is crossing the blood–brain barrier relative to binding to the histamine-1 (H_1) receptors in the periphery.

Although it is normal to think of side effects as undesirable, they sometimes can be beneficial and lead to new products. The successful development of oral hypoglycemic agents, used in the treatment of diabetes, began when it was found that certain sulfonamides had a hypoglycemic effect (see Chap. 5). Nevertheless, a real problem in drug therapy is patient compliance to take the drug a directed. Drugs that cause serious problems and discomfort tend to be avoided by the patient.

SUMMARY

One of the goals is to design drugs that will interact with receptors at specific tissues. There are several ways to do this, including (1) altering the molecule which, in turn, can change the biodistribution, (2) increasing the specificity for the desired receptor (desired pharmacologic response) while decreasing the affinity for undesired receptor (producer of side effects), and (3) the still experimental approach of attaching the drug to a monoclonal antibody that will bind to a specific tissue that is antigenic for the antibody. Alteration of biodistribution can be done by changing the drug's solubility, enhancing its ability to resist being metabolized (usually in the liver), altering the formulation or physical characteristics of the drug, and changing the route of administration. If a drug molecule can be designed in such a way that its binding to the desired receptor is enhanced, relative to the undesired receptor, and biodistribution remains favorable, smaller doses of the drug can be administered. This, in turn, reduces the amount of drug available for binding to those receptors responsible for its side effects.

Thus, the medicinal chemist is confronted with several challenges in designing a bioactive molecule. A good fit to a specific receptor is desirable, but the drug would normally be expected to eventually dissociate from the receptor. The specificity for the receptor would be such that side effects would be minimal. The drug would be expected to clear the body within a reasonable time. Its rate of metabolic degradation should allow reasonable dosing schedules and, ideally, oral administration. Many times the drug chosen for commercial sales has been selected from the hundreds of compounds that have been screened. It usually is a compromise product that meets a medical need, while demonstrating good patient acceptance.

ACID–BASE PROPERTIES

INTRODUCTION

Most drugs used today can be classified as acids or bases. As will be noted shortly, many drugs can behave as either acids or bases as they begin their journey into the patient in different dosage forms and end up in systemic circulation. A drug's acid–base properties can influence greatly its biodistribution and partitioning characteristics.

Over the years, at least four major definitions of acids or bases have been developed. The model commonly used in pharmacy and biochemistry was developed independently by Lowry and Brönsted. In their definition an *acid* is defined as a proton donor and a *base* as a proton acceptor. Notice that for a base, *there is no mention of the hydroxide ion.*

ACID–CONJUGATE BASE

Representative examples of pharmaceutically important acidic drugs are listed in Table 2-1. Each acid, or proton donor, yields a *conjugate base*. The latter is the product produced after the proton is lost from the acid. Notice the diversity in structure of these proton donors. They include the classic hydrochloric acid, the weakly acidic dihydrogen phosphate anion, the ammonium cation such as is found in ammonium chloride, the carboxylic acetic acid, the enolic form of phenobarbital, the carboxylic acid moiety of indomethacin, the imide structure of saccharin, and the protonated amine of ephedrine. Because all are proton donors, they must be treated as acids when calculating *p*Hs of a solution or percentage ionization of the drug. At the same time, it will be noted shortly that there are important differences in the pharmaceutic properties of ephedrine hydrochloride (an acid salt of an amine) as compared with indomethacin, phenobarbital, or saccharin.

TABLE 2-1

EXAMPLES OF ACIDS

	Acid	⟶	H^+	+	Conjugate Base
(a)	Hydrochloric acid HCl	⟶	H^+	+	Cl^{-}*
(b)	Sodium dihydrogen phosphate (monobasic sodium phosphate) NaH_2PO_4 (Na^+, H_2PO_4)*	⟶	H^+	+	$NaHPO_4^{2-}$ (Na^{+}*, HPO_4^{2-})
(c)	Ammonium chloride NH_4Cl (NH_4^+, Cl^-)*	⟶	H^+	+	NH_3 (Cl^-)*
(d)	Acetic acid CH_3COOH	⟶	H^+	+	CH_3COO^-
(e)	Phenobarbital 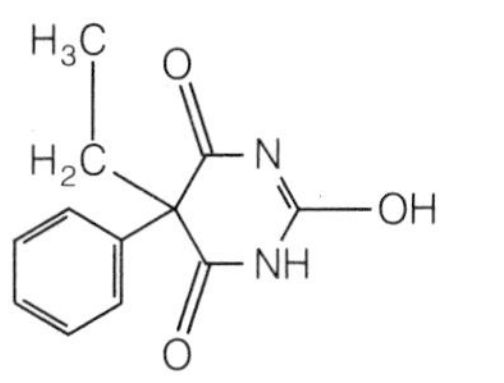	⟶	H^+	+	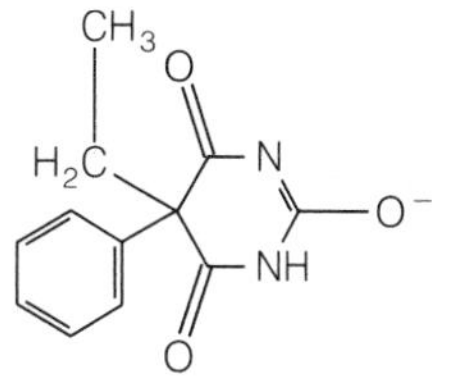
(f)	Indomethacin 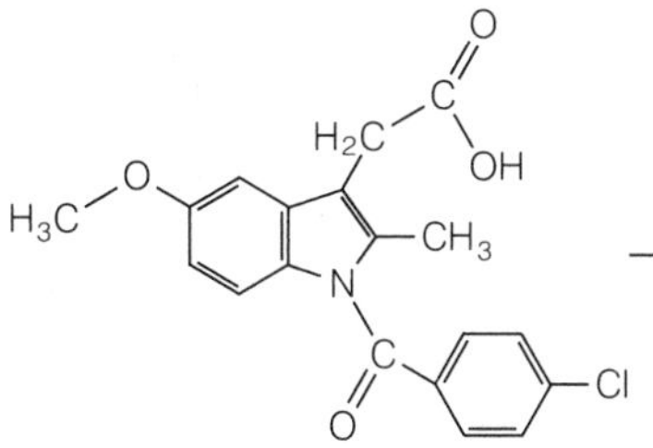	⟶	H^+	+	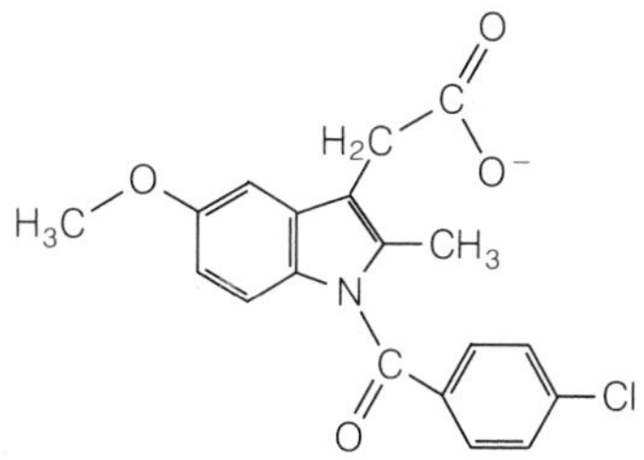
(g)	Saccharin 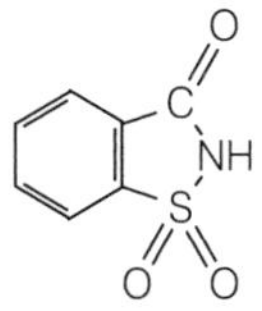	⟶	H^+	+	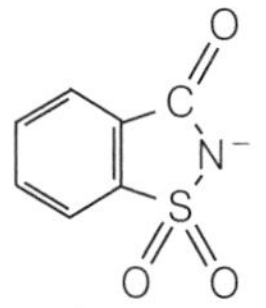
(h)	Ephedrine hydrochloride 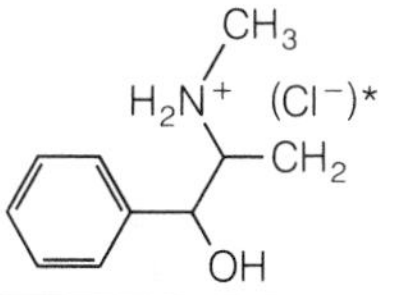 (Cl^-)*	⟶	H^+ (Cl^-*)	+	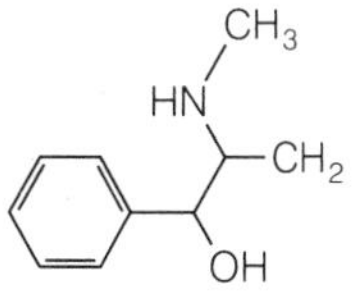

*The sodium cation and chloride anion do not take part in these reactions.

BASE–CONJUGATE ACID

The Brönsted-Lowry theory defines a *base* as a molecule that accepts a proton. The product resulting from the addition of a proton to the base is the *conjugate acid*. Pharmaceutically important bases are listed in Table 2-2. Again there are a variety of structures, including the easily recognizable base, sodium hydroxide, the basic component of an important physiologic buffer, sodium monohydrogen phosphate (conjugate base of dihydrogen phosphate), sodium acetate (conjugate base of acetic acid), ammonia (conjugate base of the ammonium cation), the carboxylate form of acetic acid (conjugate base of acetic acid), the enolate form of phenobarbital (conjugate base of phenobarbital), the carboxylate form

TABLE 2-2

EXAMPLES OF BASES

	Base	+	H^+	⟶	Conjugate Acid		
(a)	Sodium hydroxide NaOH (Na^{+}*, OH^-)	+	H^+	⟶	H_2O	+	Na^{+}*
(b)	Sodium monohydrogen phosphate (dibasic sodium phosphate) Na_2HPO_4 (2 Na^{+}*, HPO_4^{2-})	+	H^+	⟶	$H_2PO_4^-$	+	2 Na^{+}*
(c)	Ammonia NH_3	+	H^+	⟶	NH_4^+		
(d)	Sodium acetate CH_3COONa (CH_3COO^-, Na^{+}*)	+	H^+	⟶	CH_3COOH	+	Na^{+}*
(e)	Phenobarbital sodium 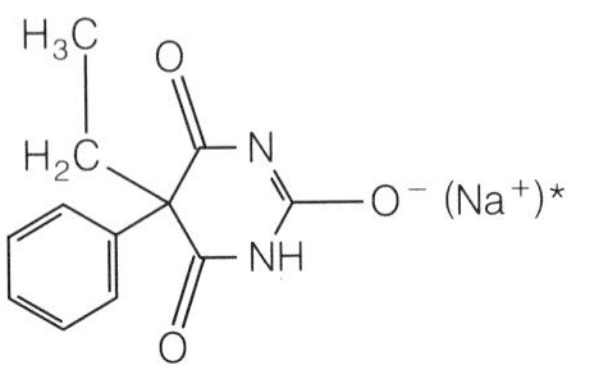 (Na^+)*	+	H^+	⟶	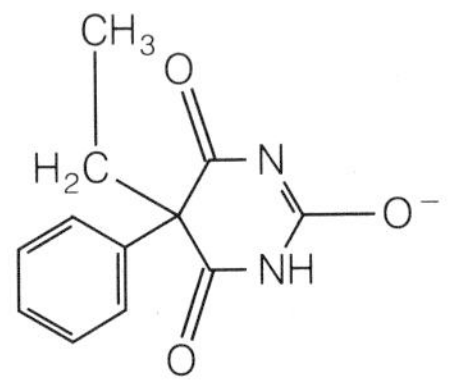	+	Na^{+}*
(f)	Indomethacin sodium 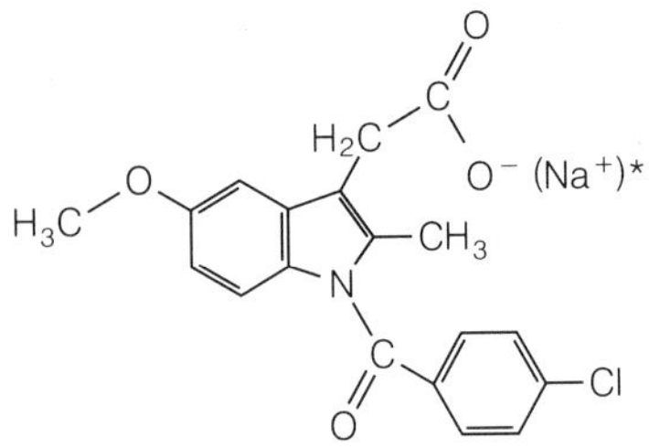 (Na^+)*	+	H^+	⟶	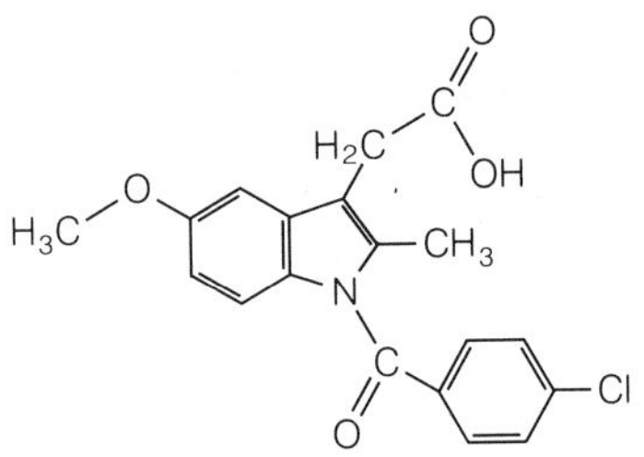	+	Na^{+}*
(g)	Saccharin sodium 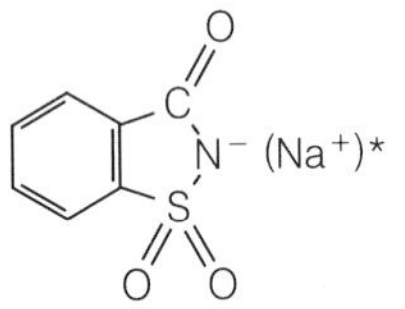 (Na^+)*	+	H^+	⟶	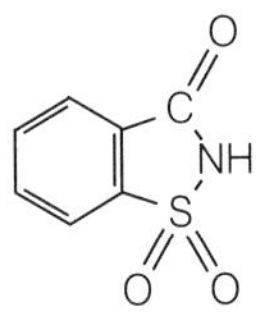	+	Na^{+}*
(h)	Ephedrine 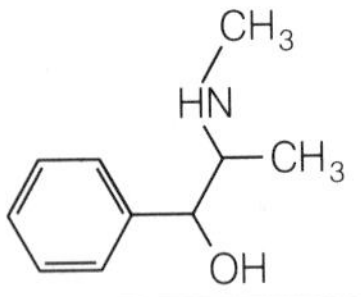	+	H^+	⟶	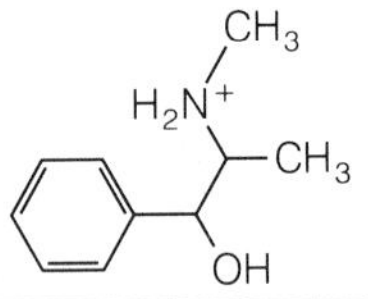		

*The sodium cation is present only to maintain charge balance. It plays no direct acid–base role.

of indomethacin (conjugate base of indomethacin), the imidate form of saccharin (conjugate base of saccharin), and the amine ephedrine (conjugate base of ephedrine hydrochloride). Notice, the conjugate acid products in Table 2-2 are the reactant acids in Table 2-1. Also notice that, whereas phenobarbital, indomethacin, and saccharin are un-ionized in the protonated form, the protonated (acidic) forms of ammonia and ephedrine are ionized salts (see Table 2-1). The opposite is true when examining the basic forms of these drugs. The basic forms of phenobarbital, indomethacin, and saccharin are anions, whereas ammonia and ephedrine are electrically neutral (see Table 2-2). It is important, then, to realize that each of the examples in Tables 2-1 and 2-2 can function as either a proton donor (acid) or proton acceptor (base). This can best be understood by emphasizing the concept of conjugate acid–conjugate base. Complicated as it first may seem, conjugate acids and conjugate bases are nothing more

TABLE 2-3

EXAMPLES OF ACID–BASE REACTIONS (WITH THE EXCEPTION OF HYDROCHLORIC ACID WHOSE CONJUGATE BASE (Cl^-) HAS NO BASIC PROPERTIES IN WATER AND SODIUM HYDROXIDE WHICH GENERATES HYDROXIDE, THE REACTION OF THE CONJUGATE BASE IN WATER IS SHOWN FOR EACH ACID)

Acid	+	Base	⇌	Conjugate Acid	+	Conjugate Base
Hydrochloric acid						
(a) HCl	+	H_2O	⟶	H_3O^+	+	Cl^-
Sodium hydroxide						
(b) H_2O	+	NaOH	⟶	H_2O	+	$OH^-(Na^+)$*
Sodium dihydrogen phosphate and its conjugate base, sodium monohydrogen phosphate						
(c) $H_2PO_4^-(Na^+)$*	+	H_2O	⇌	H_3O^+	+	$HPO_4^{2-}(Na^+)$*
(d) H_2O	+	$HPO_4^{2-}(2Na^+)$*	⇌	$H_2PO_4^{2-}(Na^+)$*	+	$OH^-(Na^+)$*
Ammonium chloride and its conjugate base, ammonia						
(e) $NH_4^+(Cl^-)$*	+	H_2O	⇌	$H_3O^+(Cl^-)$*	+	NH_3
(f) H_2O	+	NH_3	⇌	NH_4^+	+	OH^-
Acetic acid and its conjugate base, sodium acetate						
(g) CH_3COOH	+	H_2O	⇌	H_3O^+	+	CH_3COO^-
(h) H_2O	+	$CH_3COO^-(Na^+)$*	⇌	CH_3COOH	+	$OH^-(Na^+)$*

Indomethacin and its conjugate base, indomethacin sodium, will show the identical acid–base chemistry.

Phenobarbital and its conjugate base, phenobarbital sodium

(i) [phenobarbital structure, –OH form] + H_2O ⇌ H_3O^+ + [phenobarbital anion, –O^- form]

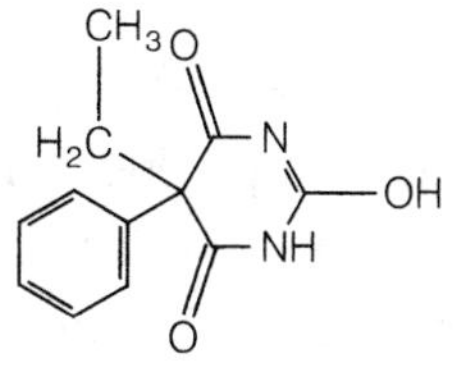

(j) H_2O + [phenobarbital anion, –O^- (Na^+)*] ⇌ [phenobarbital, –OH form] + $OH^-(Na^+)$*

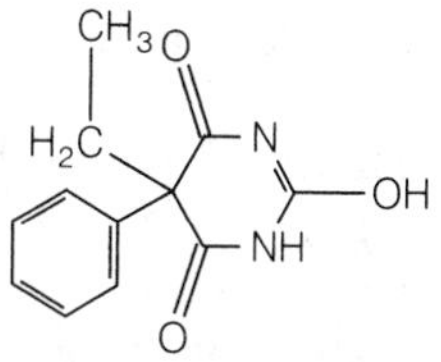

Saccharin and its conjugate base, saccharin sodium

(k) [saccharin, NH] + H_2O ⇌ H_3O^+ + [saccharin anion, N^-]

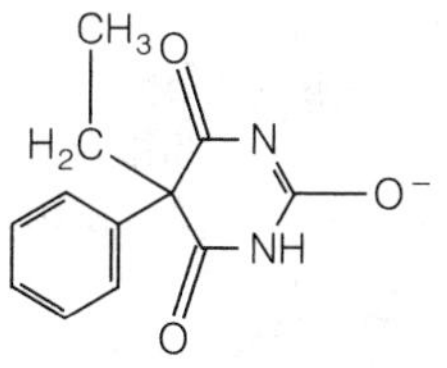

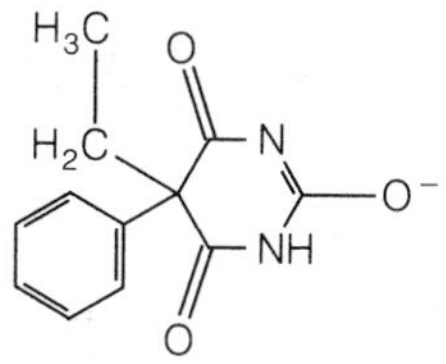

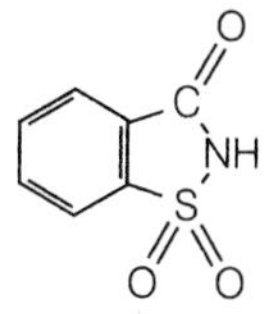

(l) H_2O + [saccharin anion, N^- (Na^+)*] ⇌ [saccharin, NH] + $OH^-(Na^+)$*

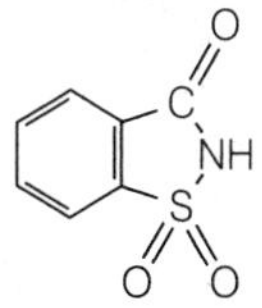

Ephedrine HCl and its conjugate base, ephedrine

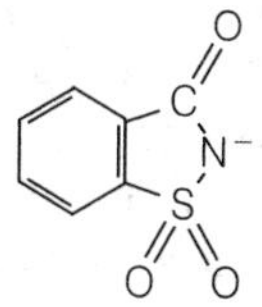

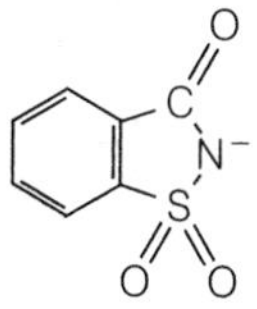

(m) [ephedrine HCl, H_2N^+ (Cl^-)*] + H_2O ⇌ $H_3O^+(Cl^-)$* + [ephedrine, HN]

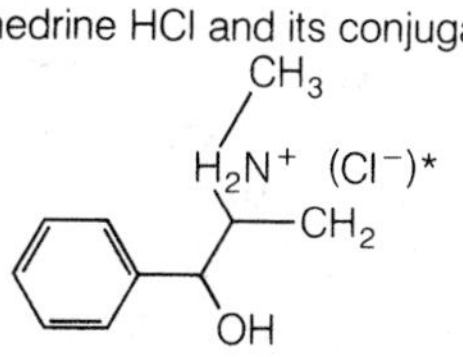

(n) H_2O + [ephedrine, HN] ⇌ [ephedrine cation, H_2N^+] + OH^-

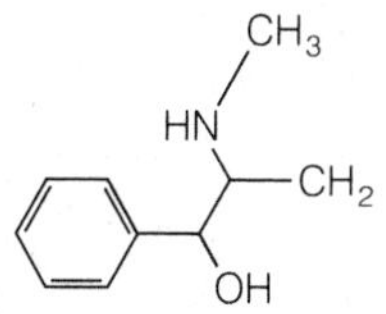

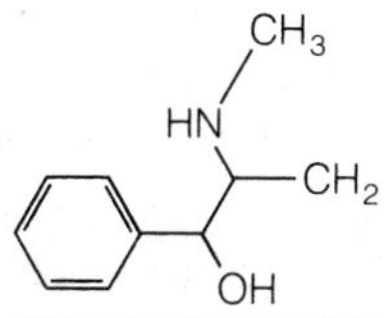

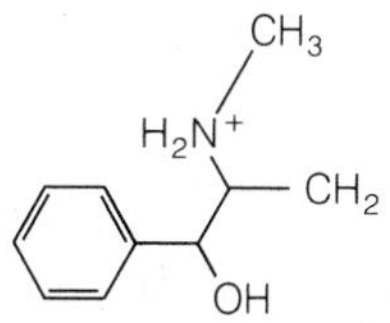

*The chloride anion and sodium cation are present only to maintain charge balance. These anions play no other acid–base role.

than the products of an acid–base reaction. In other words, they will appear to the right of the reaction arrows. The examples from Tables 2-1 and 2-2 are rewritten in Table 2-3 as complete acid–base reactions.

Careful study of Table 2-3 will show water functioning as a proton acceptor (base) in reactions a, c, e, g, i, k, m and proton donor (base) in reactions b, d, f, h, j, l, n. Hence, water is known as an *amphoteric* substance. Water either can be a weak base accepting a proton to form the strongly acidic hydrated proton or hydronium ion, H_3O^+, (reactions a, c, e, g, i, k, m) or a weak acid donating a proton to form the strongly basic hydroxide anion, OH^-, (reactions b, d, f, h, j, l, n).

ACID STRENGTH

Although any acid–base reaction can be written as an equilibrium reaction, an attempt has been made in Table 2-3 to indicate which sequences are unidirectional or show only a small reversal. For hydrochloric acid, the conjugate base, Cl^-, is so weak a base that it essentially does not function as a proton acceptor. That is why the chloride anion was not included as a base in Table 2-2. In a similar manner, water is such a weak conjugate acid that there is little reverse reaction involving water donating a proton to the hydroxide anion of sodium hydroxide.

A logical question to ask, at this point, is how does one predict in which direction an acid–base reaction lies and to what extent does the reaction go to completion. The common physicochemical measurement that contains this information is known as the pK_a. The pK_a is the negative logarithm of the modified equilibrium constant for an acid–base reaction written such that water will be the base or proton acceptor. It can be derived as follows.

Assume that a weak acid, HA, reacts with water.

$$\underset{\text{Acid}}{\mathrm{HA}} + \underset{\text{Base}}{\mathrm{H_2O}} \rightleftharpoons \underset{\text{Conj. Acid}}{\mathrm{H_3O^+}} + \underset{\text{Conj. Base}}{\mathrm{A^-}} \qquad \text{(Rx. 2-3)}$$

The equilibrium constant for reaction 2-3 is

$$K_{eq} = \frac{[H_3O^+][A^-]}{[HA][H_2O]} = \frac{[\text{conj. acid}][\text{conj. base}]}{[\text{acid}][\text{base}]} \qquad \text{(Eq. 2-1)}$$

It turns out that in a dilute solution of the weak acid, the molar concentration of water can be treated as a constant 55.5 *M*. Thus, with $[H_2O] = 55.5$, Equation 2-1 can be simplified to

$$K_a = K_{eq}[H_2O] = K_{eq}(55.5) = \frac{[H_3O^+][A^-]}{[HA]} = \frac{[\text{conj. acid}][\text{conj. base}]}{[\text{acid}]} \qquad \text{(Eq. 2-2)}$$

By definition

$$pK_a = -\log K_a \qquad \text{(Eq. 2-3)}$$

and

$$pH = -\log[H_3O^+] \qquad \text{(Eq. 2-4)}$$

The modified equilibrium constant, K_a, is customarily converted to pK_a (the negative logarithm) to put it on the same scale as pH. Therefore, rewriting Equation 2-2 in logarithmic form produces

$$\log K_a = \log[H_3O^+] + \log[A^-] - \log[HA] = \log[H_3O^+] + \log[\text{conj. base}] - \log[\text{acid}] \qquad \text{(Eq. 2-5)}$$

Rearranging Equation 2-5 gives

$$-\log[H_3O^+] = -\log K_a + \log[A^-] - \log[HA] = -\log K_a + \log[\text{conj. base}] - \log[\text{acid}] \qquad \text{(Eq. 2-6)}$$

Substituting Equations 2-3 and 2-4 into 2-6 produces

$$pH = pK_a + \log\frac{[A^-]}{[HA]} = pK_a + \log\frac{[\text{conj. base}]}{[\text{acid}]} \qquad \text{(Eq. 2-7)}$$

Equation 2-7 is more commonly called the Henderson–Hasselbalch equation and is the basis for most calculations involving weak acids and bases. It is the basis for calculating the pH of solutions of weak acids, weak bases, and used to calculate the pH of buffers consisting of weak acids and their conjugate bases or weak bases and their conjugate acids. Because the pK_a is a modified equilibrium constant, it corrects for the fact that weak acids do *not* completely react with water.

A very similar set of equations would be obtained from the reaction of a protonated amine, BH^+, in water. The reaction would be

Acid	Base	Conj. Acid	Conj. Base

$$BH^+ + H_2O \rightleftharpoons H_3O^+ + B \qquad \text{(Rx. 2-4)}$$

$$K_{eq} = \frac{[H_3O^+][B]}{[BH^+][H_2O]} = \frac{[\text{conj. acid}][\text{conj. base}]}{[\text{acid}][\text{base}]} \qquad \text{(Eq. 2-8)}$$

Notice that Equation 2-8 is identical with Equation 2-1 when the general [conj. acid] [conj. base] representation is used. Therefore, with the same simplifying assumption that water will remain at a constant concentration of 55.5 *M*, Equation 2-8 can be rewritten as

$$K_a = K_{eq}(55.5) = \frac{[H_3O^+][B]}{[BH^+]} = \frac{[\text{conj. acid}][\text{conj. base}]}{[\text{acid}]} \qquad \text{(Eq. 2-9)}$$

Rearranging Equation 2-9 into logarithmic form and substituting the relationships expressed in Equations 2-3 and 2-4 yields the same Henderson–Hasselbalch equation (Eq. 2-10).

$$pH = pK_a + \log\frac{[B]}{[BH^+]} = pK_a + \log\frac{[\text{conj. base}]}{[\text{acid}]} \qquad \text{(Eq. 2-10)}$$

Rather than trying to remember the specific form of the Henderson–Hasselbalch equation for an HA or BH^+ acid, it is simpler to use the general form of the equation (Eq. 2-11) expressed in both Equations 2-7 and 2-10. With this version of the equation,

$$pH = pK_a + \log\frac{[\text{conj. base}]}{[\text{acid}]} \qquad \text{(Eq. 2-11)}$$

there is no need to remember whether the species in the numerator/denominator is ionized or unionized. The molar concentration of the proton acceptor is the term in the numerator and the molar concentration of the proton donator is the denominator term.

What about weak bases such as amines? In aqueous solutions, water will function as the proton donor or acid (Rx. 2-5) producing the familiar hydroxide anion (conjugate base). Traditionally, a modified equilibrium constant called the pK_b was derived following the same steps that produced Equation 2-2. It is now more common to express the basicity of a chemical in terms of the pK_a using the relationship

Acid	Base	Conj. Acid	Conj. Base

$$H_2O + B \longrightarrow BH^+ + OH^- \qquad \text{(Rx. 2-5)}$$

$$pK_a = pK_b - 14 \qquad \text{(Eq. 2-12)}$$

WARNING! It is **IMPORTANT** to recognize that a pK_a for a base is in reality the pK_a of the conjugate acid (acid donor or protonated form) of the base. The pK_a for ephedrine is listed in the Appendix as 9.6 and for ammonia 9.3. In reality this is the pK_a of the protonated form, such as ephedrine hydrochloride (Rx. *m* in Table 2-3) and ammonium chloride (Rx. *e* in Table 2-3), respectively. This easily is confusing. It is crucial that the chemistry of the drug be understood when interpreting a pK_a value. When reading tables of pK_a values, such as those found in the Appendix, it is essential to realize that the listed value is for the proton donor form of the molecule, no matter what form is indicated by the name. See Table 2-4 for several worked examples of how the pK_a is used to calculate *p*Hs of solutions, required ratios of [conjugate base]/[acid], and percentage ionization at specific *p*Hs.

Just how strong, or weak, are the acids the reactions of which in water are illustrated in Table 2-3? Bear in mind that the K_as or pK_as are modified equilibrium constants that will indicate whether the acid's reaction in water tends to go to completion resulting in the formation of large amounts of conjugate acid and conjugate base or whether the base accepts few protons from water forming little conjugate acid and conjugate base.

Refer back to Equation 2-2 and using the K_a values in Table 2-5, substitute the K_a term for each of the foregoing acids listed. For hydrochloric acid it should be obvious that a K_a of 1.26×10^6 means that the numerator term containing the concentrations of the conjugate acid and conjugate base products is huge relative to the denominator term representing the concentration of the reactants. In other words, there essentially is no unreacted HCl left in an aqueous solution of hydrochloric acid. At the other extreme is ephedrine HCl with a pK_a of 9.6 or K_a of 2.51×10^{-10}. Here, the denominator representing the concentration of ephedrine HCl greatly predominates over that of the products which, in this example, would be ephedrine (conjugate base) and H_3O^+ (conjugate acid). In other words, the protonated form of ephedrine is a very poor proton donor. It holds onto the proton. Indeed, free ephedrine (the conjugate base in this reaction) is an excellent proton acceptor.

TABLE 2-4

EXAMPLES OF CALCULATIONS REQUIRING THE pK_a

1. What is the ratio of ephedrine to ephedrine HCl (pK_a 9.6) in the intestinal tract at pH 8.0? Use Equation 2-11.

$$8.0 = 9.6 + \log \frac{[\text{ephedrine}]}{[\text{ephedrine HCl}]} = -1.6$$

$$\frac{[\text{ephedrine}]}{[\text{ephedrine HCl}]} = 0.025$$

The number whose log is −1.6 is 0.025, meaning that there are 25 parts ephedrine for every 1000 parts ephedrine HCl in the intestinal tract whose environment is *p*H 8.0.

2. What is the *p*H of a buffer containing 0.1 *M* acetic acid (pK_a 4.8) and 0.08 *M* sodium acetate? Use Equation 2-11.

$$pH = 4.8 + \log \frac{0.08}{0.1} = 4.7$$

3. What is the *p*H of a 0.1 *M* acetic acid solution? Use the following equation for calculating the *p*H of a solution containing either an HA or BH^+ acid.

$$pH = \frac{pK_a - \log [\text{acid}]}{2} = 2.9$$

4. What is the *p*H of a 0.08 *M* sodium acetate solution? Remember, even though this is the conjugate base of acetic acid, the pK_a will still be used. The pK_w term in the following equation corrects for the fact that a proton acceptor (acetate anion) is present in the solution. The equation for calculating the *p*H of a solution containing either an A^- or B base is

$$pH = \frac{pK_w + pK_a + \log [\text{base}]}{2} = 8.9$$

5. What is the *p*H of an ammonium acetate solution? The pK_a of the ammonium (NH_4^+) cation is 9.3. Always bear in mind that the pK_a refers to the ability of the proton donor form to release the proton into water to form H_3O^+. Since this is the salt of a weak acid (NH_4^+) and the conjugate base of a weak acid (acetate anion), the following equation is used. Note that molar concentration is not a variable in this calculation.

$$pH = \frac{pK_{a_1} + pK_{a_2}}{2} = 7.1$$

6. What is the percentage ionization of ephedrine HCl (pK_a 9.6) in an intestinal tract buffered at *p*H 8.0 (see example 1). Use Equation 2-14 because this is a BH^+ acid.

$$\%\ \text{ionization} = \frac{100}{1 + 10^{(8.0-9.6)}} = 97.6\%$$

Only 2.4% of ephedrine is present as the un-ionized conjugate base.

7. What is the percentage ionization of indomethacin (pK_a 4.5) in an intestinal tract buffered at *p*H 8.0. Use Equation 2-13 because this is an HA acid.

$$\%\ \text{ionization} = \frac{100}{1 + 10^{(4.5-8.0)}} = 99.97\%$$

For all practical purposes indomethacin is present only as the anionic conjugate base in that region of the intestine buffered at *p*H 8.0.

TABLE 2-5

REPRESENTATIVE K_a AND pK_a VALUES FROM THE REACTIONS LISTED IN TABLE 2-3 (SEE THE APPENDIX).

	K_a	pK_a
Hydrochloric acid	1.26×10^{6}	−6.1
Dihydrogen phosphate	6.31×10^{-8}	7.2
Ammonia (ammonium)	5.01×10^{-10}	9.3
Acetic acid	1.58×10^{-5}	4.8
Phenobarbital	3.16×10^{-8}	7.5
Saccharin	2.51×10^{-2}	1.6
Indomethacin	3.16×10^{-5}	4.5
Ephedrine (as the HCl salt)	2.51×10^{-10}	9.6

A general rule for determining if a chemical is strong or weak acid or base is the following:

pK_a <2	strong acid; essentially no basic properties in water
pK_a 4–6	weak acid; very weak conjugate base
pK_a 8–10	very weak acid; weak conjugate base
pK_a >12	essentially no acidic properties in water; strong conjugate base

It is important to realize that this delineation is only approximate. Also other properties become important when considering cautions in handling acids and bases. Phenol has a pK_a of 9.9, slightly less than that of ephedrine HCl. Why is phenol considered corrosive to the skin whereas ephedrine HCl or free ephedrine considered innocuous when applied to the skin? Phenol has the ability to partition through the normally protective lipid layers of the skin. Because of this property, this extremely weak acid has carried the name carbolic acid. Thus, the pK_a simply tells a person the acid properties of the protonated form of the chemical. It does not represent anything else concerning potential toxicities.

PERCENTAGE IONIZATION

By using the drug's pK_a, the formulation pharmacist can adjust the *p*H to ensure maximum water solubility (ionic form of the drug) or maximum solubility in nonpolar media (nonionic form). This is where the understanding of the drug's acid–base chemistry becomes important.

Acids can be divided into two types, HA and BH^+, on the basis of the ionic form of the acid (or conjugate base). The HA acids go from nonionized acids to ionized (polar) conjugate bases (Rx. 2-6). An ionized BH^+ (polar) acid goes to a nonionized (nonpolar) conjugate base (Rx. 2-7). In general, pharmaceutically important HA acids include the inorganic acids (e.g., HCl, H_2SO_4), enols (e.g., barbiturates, hydantoins), carboxylic acids (e.g., low-molecular-weight

$pK_{a_3} = 9.6$ HO–(C$_6$H$_4$)–CH(NH$_3^+$)–CONH–CH–CH–S–C(CH$_3$)$_2$; CO—N——CH—COO$^-$; $pK_{a_2} = 7.4$; $pK_{a_1} = 2.4$

Amoxicillin

organic acids, arylacetic acids, *N*-aryl anthranilic acids, salicylic acids), and imides and amides (e.g., saccharin, sulfonamides). The chemistry is simpler for the pharmaceutically important BH^+ acids: They are all protonated amines. A polyfunctional drug can have several pK_as (e.g., amoxicillin).

Acid	Base	Conj. Acid	Conj. Base
HA (un-ionized) +	H_2O ⟶	H_3O^+ (ionized) +	A^-

(Rx. 2-6)

BH^+ (ionized) +	H_2O ⟶	H_3O^+ +	B (un-ionized)

(Rx. 2-7)

It is possible to calculate the percentage ionization of a drug by using Equations 2-13 for HA acids and 2-14 for BH^+ acids.

HA acids

$$\%\ \text{ionization} = \frac{100}{1 + 10^{(pK_a - pH)}} \qquad \text{(Eq. 2-13)}$$

BH^+ acids

$$\%\ \text{ionization} = \frac{100}{1 + 10^{(pH - pK_a)}} \qquad \text{(Eq. 2-14)}$$

A plot of percentage ionization versus *p*H is illustrative of how the degree of ionization can be shifted significantly with small changes in *p*H. The curves for an HA acid (indomethacin) and BH^+ (protonated ephedrine) are shown in Figure 2-2. First, note that when $pH = pK_a$, the compound is 50% ionized (or 50% unionized). In other words, when the pK_a is equal to the *p*H, the molar concentration of the acid equals the molar concentration of its conjugate base. In the Henderson–Hasselbalch equation, the $pK_a = pH$ when log[conj. base]/[acid] = 1. An increase of 1 *p*H unit from the pK_a (increase in alkalinity) causes an HA acid (indomethacin) to become 90.9% in the ionized conjugate base form, but results in a BH^+ acid (ephedrine HCl) decreasing its percentage ionization to only 9.1%. An increase of 2 *p*H units essentially shifts an HA acid to complete ionization (99%) and a BH^+ acid to the nonionic conjugate base form (0.99%).

Just the opposite is seen when the medium is made more acidic relative to the drug's pK_a value. Increasing the hydrogen ion concentration (decreasing the *p*H) will shift the equilibrium to the left, thereby increasing the concentration of the acid and decreasing the concentration of conjugate base. For indomethacin, a decrease of 1 *p*H unit below the pK_a will increase the concentration of un-ionized (protonated) indomethacin to 90.9%. Similarly, a decrease of 2 *p*H units results in only 0.99% of the indomethacin being present in the ionized conjugate base form. The opposite is seen for the BH^+ acids. The percentage of ephedrine present as the ionized (protonated) acid at 1 *p*H unit below the pK_a is 90.9% and 2 *p*H units below the pK_a is 99.0%. These results can be summarized in Table 2-6.

With this knowledge in mind, it is easier to explain and predict why problems and discomfort caused by *p*H extremes can occur with the use of some preparations. Phenytoin (pK_a 8.3) injection must be adjusted to *p*H 12 with sodium hydroxide to ensure complete ionization and maximize water solubility. In theory, a *p*H of 10.3 will result in 99.0% of the drug being in the anionic water-soluble conjugate base. To reduce even further that insoluble 1% in the acid form and maintain excess alkalin-

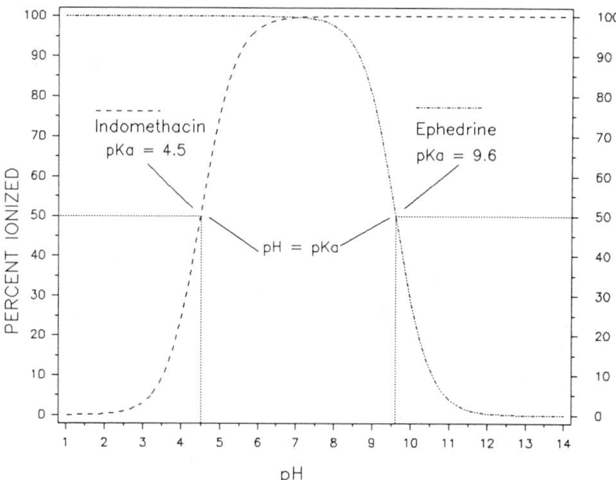

FIG. 2-2. Percentage ionized versus pH for indomethacin (pK_a 4.5) and ephedrine (pK_a 9.6).

TABLE 2-6

PERCENTAGE IONIZATION RELATIVE TO THE pK_a

	Ionization (%)	
	HA Acids	BH^+ Acids
pK_a − 2 pH units	0.99	99.0
pK_a − 1 pH unit	9.1	90.9
pK_a = pH	50.0	50.0
pK_a + 1 pH unit	90.9	9.1
pK_a + 2 pH units	99.0	0.99

ity, the pH is raised to 12 to obtain 99.98% of the drug in the ionized form. Even then, a cosolvent system of 40% propylene glycol, 10% ethyl alcohol, and 50% water for injection is used to ensure a complete solution. This highly alkaline solution is irritating to the patient and generally cannot be administered as an admixture with other intravenous fluids that are buffered more closely at physiologic pH (7.4). This decrease in pH will result in the parent un-ionizedphenytoin precipitating out of solution.

Phenytoin Sodium

Adjustments in pH to maintain water solubility can sometimes lead to chemical stability problems. An example is indomethacin, which is unstable in an alkaline medium. Therefore, the preferred oral liquid dosage form is a suspension buffered at pH 4 to 5. Because this is near the drug's pK_a of 4.5, only 50% will be in the water-soluble form. There is a medical indication requiring the intravenous administration of indomethacin to premature infants. The intravenous dosage form is the lyophilized (freeze-dried) sodium salt, which is reconstituted just before use.

DRUG DISTRIBUTION AND pK_a

The pK_a can have a pronounced effect on the pharmacokinetics of the drug. Drugs will be transported in the aqueous environment of the blood. Those drugs that are in an ionized form will tend to distribute throughout the body more rapidly than un-ionized (nonpolar) molecules. With few exceptions, the drug must leave the polar environment of the plasma to reach the site of action. In general, drugs pass through the nonpolar membranes of capillary walls, cell membranes, and blood–brain barrier in

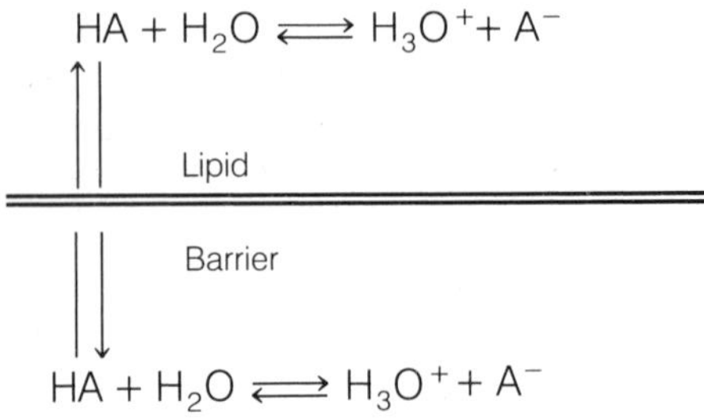

FIG. 2-3. *Passage of HA acids through lipid barriers.*

the un-ionized (nonpolar form). For HA acids, it will be the parent acid that will readily cross these membranes (Fig. 2-3). Just the opposite is the situation for the BH^+ acids. The un-ionized conjugate base will be the species most readily crossing the nonpolar membranes (Fig. 2-4).

Consider the changing pH environment experienced by the drug molecule when orally administered. The drug will first encounter the acidic stomach where the pH can range from 2 to 6 depending on the presence of food. The HA acids with pK_as of 4 to 5 will tend to be nonionic and be absorbed partially through the gastric mucosa. (The main reason why most acidic drugs are absorbed from the intestinal tract, rather than the stomach, is that the microvilli of the intestinal mucosa provide a huge surface area relative to what is found in the gastric mucosa of the stomach.) In contrast, amines (pK_a 9 to 10) will be protonated (BH^+ acids) and usually will not be absorbed until reaching the mildly alkaline intestinal tract (pH ~ 8). Even here, only a portion of the amine-containing drugs will be in their nonpolar conjugate base form. Whenever a nonpolar form of either an HA acid (as the acid) or a BH^+ acid (as the conjugate base) passes across the lipid barrier, the equilibrium ratio of conjugate base/acid must be maintained. The actual conjugate base/acid ratio on each side of the membrane will be dependent on the pH of the medium in which the drug is at any given location.

For example, once in systemic circulation, the plasma pH of 7.4 will be one of the determinants of whether the drug will tend to remain in the aqueous

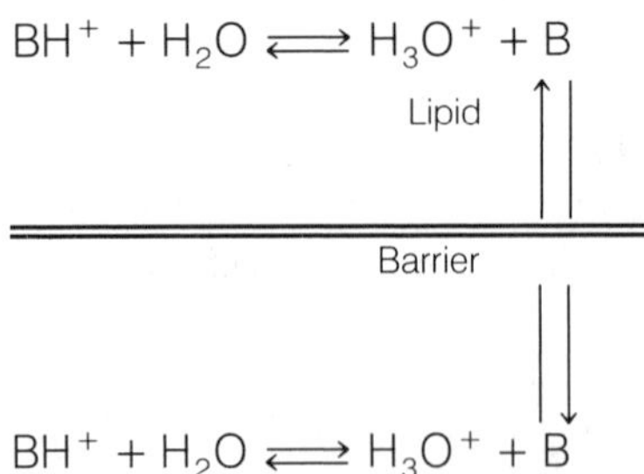

FIG. 2-4. *Passage of BH^+ acids through the lipid barrier.*

environment of the blood or to partition across lipid membranes into hepatic tissue to be metabolized, into the kidney for excretion, into tissue depots for storage, or to the receptor tissue, resulting in a pharmacological response. Examine the structure of amoxicillin with its three pK_as. At pH 7.4, the carboxyl group will be in its anionic conjugate base form; the primary amine will be 50% in the cationic acid form and 50% un-ionized conjugate base; and the phenolic moiety will be in the un-ionized acid form. Accordingly, the drawing is a reasonable representation of amoxicillin's ionic state in the body. A useful exercise is to calculate the [conj. base]/[acid] ratio using the Henderson–Hasselbalch equation (see Eq. 2-11) or percentage ionization for ephedrine (pK_a 9.6; see Eq. 2-14) and indomethacin (pK_a 4.5; see Eq. 2-13) at pH 3.5 (stomach), pH 8.0 (intestine), and pH 7.4 (plasma) (see examples 1, 6, and 7 in Table 2-4.) Of course, the previously discussed effect of protein binding can greatly alter any prediction of biodistribution that is based solely on pK_a.

STATISTICAL PREDICTION OF PHARMACOLOGIC ACTIVITY

INTRODUCTION

Just as mathematical modeling is used to explain and model many chemical processes, it has been the goal of medicinal chemists to quantify the effect of a structural change on a defined pharmacologic response. This would meet three goals in drug design.

1. Predict the biologic activity in untested compounds
2. Define the structural requirements required for a good fit between the drug molecule and the receptor
3. Design a test set of compounds to maximize the amount of information concerning structural requirements for activity from a minimum number of compounds tested

This aspect of medicinal chemistry is commonly referred to as quantitative structure–activity relationships, or QSAR.

The goals of QSAR studies were first proposed around 1865–1870 by Crum-Brown and Fraser, who showed that the gradual chemical modification in the molecular structure of a series of poisons produced some important differences in their action.[1] They postulated that the physiologic action, θ, of a molecule is some function of its chemical constitution, C. This can be expressed as

$$\theta = f(C) \qquad \text{(Eq. 2-15)}$$

Equation 2-15 states that a defined change in chemical structure would result in a predictable change in physiologic action. The problem now becomes one of numerically defining chemical structure. It is still a fertile area of research. What has been found is that biologic response can be predicted from physicochemical properties such as vapor pressure, water solubility, electronic parameters, steric descriptors, and partition coefficient (Eq. 2-16). Today, the partition coefficient has become the single most imporant physicochemical measurement for QSAR studies. Note that equation 2-16 is the equation for a straight line ($Y = mx + b$).

$$\log BR = a(\text{physicochemical property}) + c, \qquad \text{(Eq. 2-16)}$$

where

BR = a defined pharmacologic response usually expressed in millimoles, such as the effective dose in 50% of the subjects (ED_{50}), the lethal dose in 50% of the subjects (LD_{50}), or minimum inhibitory concentration (MIC). It is common to express the biologic response as a reciprocal, $\log 1/BR$ or $\log 1/C$.

a = the regression coefficient or slope of the straight line.

c = the intercept term on the Y axis (when the physicochemical property equals zero).

As has previously been emphasized, the drug will go through a series of partitioning steps, leaving the aqueous extracellular fluids, pass through lipid membranes, and enter other aqueous environments before reaching the receptor. In this sense, a drug is undergoing the same partitioning phenomena that happens to any chemical in a separatory funnel containing water and a nonpolar solvent, such as hexane, chloroform, or ether. The difference between the separatory funnel model and what actually occurs in the body is that the partitioning in the funnel will reach an equilibrium at which the rate of chemical leaving the aqueous phase and entering the organic phase will equal the rate of chemical moving from the organic phase to the aqueous phase. This is not the physiologic situation. Refer to Figure 2-1, and note that the dynamic changes are occurring to the drug, such as it being metabolized, bound to serum albumin, excreted from the body, and bound to receptors. The circumstance for the drug is not static. Upon administration, the drug will be *pushed* through the membranes because of the high concentration of drug in the extracellular fluids relative to the concentration in the intracellular compartments. In an attempt to maintain equilibrium ratios, the flow of the drug will be from systemic circulation, through the membranes, onto the receptors. As the drug is metabolized and excreted from the body, it will be *pulled* back across the membranes, and the concentration of drug at the receptors will decrease.

The question that now must be asked is: What immiscible nonpolar solvent system best mimics the aqueous–lipid membrane barriers found in the body? It is now realized that the *n*-octanol–water system is an excellent estimator of drug partitioning in biologic systems. To appreciate why this is so, it is necessary to understand the chemical nature of the lipid membranes.

These membranes are not anhydrous fatty or oily structures. As a first approximation they can be considered as bilayers composed of lipids consisting of a polar cap and large hydrophobic tail. Phosphoglycerides are major components of lipid bilayers. Other groups of bifunctional lipids include the sphinogomyelins, galactocerebrosides, and plasmologens. The hydrophobic portion is composed largely of unsaturated fatty acids, mostly with *cis* double bonds. In addition there is a considerable amount of cholesterol esters, protein, and charged mucopolysaccharides in the lipid membranes. The final result is that these membranes are highly organized structures comprising channels for transport of important molecules, such as metabolites, chemical regulators (hormones), amino acids, glucose, and fatty acids, into the cell and removal of waste products and biochemically produced products from the cell. The cellular membranes are dynamic with the channels, forming and disappearing depending on the cell's and body's needs.

In addition, the membranes on the surface of nucleated cells have specific antigenic markers by which the immune system monitors the cell's status. There are receptors on the cell surface at which hormones, such as epinephrine and insulin, bind, thereby setting off a series of biochemical events within the cell. Some of these receptors are used by viruses to gain entrance into the cells where the virus reproduces. As newer instrumental techniques are developed, and genetic cloning permits isolation of the genetic material responsible for forming and regulating the structures on the cell surface, the

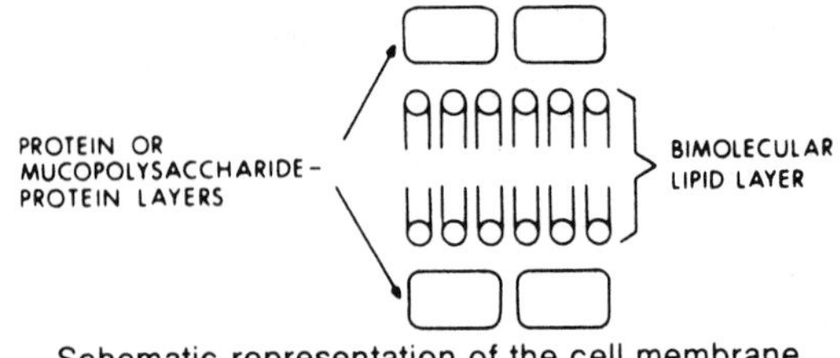

FIG. 2-5. *Schematic representation of the cell membrane.*

image of a passive lipid membrane has disappeared to be replaced by a very complex, highly organized dynamically functioning structure.

For purposes of the partitioning phenomenon, picture the cellular membranes as two layers of lipids (Fig. 2-5). The two outer layers, one facing the interior of the cell and the other the exterior, will consist of the polar ends of the bifunctional lipids. Keep in mind that these surfaces are exposed to an aqueous polar environment. The polar ends of the charged phospholipids and other bifunctional lipids will be solvated by the water molecules. There also will be considerable amounts of charged proteins and mucopolysaccharides present on the surface. In contrast, the interior of the membrane will be populated by the hydrophobic aliphatic chains from the fatty acid esters.

PARTITION COEFFICIENT

With this representation in mind, a partial explanation can be presented for why the *n*-octanol–water partitioning system seems to mimic the lipid membranes–water systems found in the body. It turns out that *n*-octanol is not as nonpolar as initially might be predicted. Water-saturated octanol contains 2.3 *M* water, whereas *n*-octanol-saturated water contains little of the organic phase. The water in the *n*-octanol phase apparently approximates the polar properties of the lipid bilayer, whereas the lack of octanol in the water phase mimics the physiologic aqueous compartments that are relatively free of nonpolar components. In contrast, partitioning systems such as hexane–water and chloroform–water contain so little water in the organic phase that they

---

Hydrophobic Tail — [Fatty Acid Ester—OCH_2 ; Fatty Acid Ester—OCH]

$H_2C—O—P(=O)(O^-)—R$ — Hydrophilic Head

Lecithin: $R = —OCH_2CH_2N^+(CH_3)_3$
Cephalin: $R = —OCH_2CH_2NH_3^+$

are poor models for the lipid bilayer–water system found in the body. At the same time, it must be remembered that the n-octanol–water system is only an approximation of the actual environment found in the interface between the cellular membranes and the extracellular–intracellular fluids.

The basic procedure for obtaining a partition coefficient is to shake a weighed amount of chemical in a flask containing a measured amount of water-saturated octanol and octanol-saturated water. Many times the aqueous phase will be buffered with a phosphate buffer at pH 7.4 to reflect physiologic pH. This corrects for the ratio of [conjugate base]/[acid] found in vivo. The amount of chemical in one or both of the phases is determined by an appropriate analytical technique and the partition coefficient calculated from Equation 2-17. The octanol–water partition coefficient has been determined for thousands of compounds, including drugs, agricultural chemicals, biochemical intermediates and metabolites, and common chemicals. Many of these determinations have been obtained in several other organic solvent–aqueous systems, such as ether, chloroform, triolein, and hexane.

$$P = \frac{[\text{chemical}]_{\text{oct}}}{[\text{chemical}]_{\text{aq}}} \qquad \text{(Eq. 2-17)}$$

The determination of partition coefficients is tedious and time-consuming. Some chemicals are too unstable and either degrade during the procedure, which can take several hours, or cannot be obtained in sufficient purity for an accurate determination. This has led to attempts at approximating the partition coefficient. Perhaps the most popular approach has been high-performance liquid chromatography (HPLC) or thin layer chromatography (TLC). In each case, the support phase is nonpolar, either by permanent bonding (usually octadecylsilane) or a coating of octanol, mineral oils, or related materials. The mobile phase usually has some water-miscible organic solvent to hold enough of the chemical, the partition coefficient of which is being determined, in solution. Sometimes, the partition coefficient is calculated from the retention data by regression analysis with Equation 2-18. The a and c terms have the same uses as in Equation 2-16.

$$\log P = a(\log \text{retention}) + c \quad \text{(Eq. 2-18)}$$

This model has at least two limitations. First, it usually only works when determining the retention times of chemicals of the same chemical class and similar substitution patterns. Second, there should be a set of partition coefficients previously determined by the classic shake-flask method for a set of compounds exhibiting very similar chemistry in terms of class and substituent pattern. Because of these limitations it is common to use directly the retention data in the prediction of biologic response (Eq. 2-19).

$$\log BR = a(\log \text{retention}) + c, \quad \text{(Eq. 2-19)}$$

A chemical's retention, on a chromatographic support is the result of a combination of its partitioning, steric, and electronic properties. Because these same physicochemical properties are important variables in determining a drug's biologic response, excellent correlations have been obtained between chromatographic retention parameters and biologic response. Although the model represented by Equation 2-19 is useful in predicting biologic response, it is not as definitive as the models presented later (Eqs. 2-20 to 2-22), because the precise physicochemical properties are combined into one chromatographic retention term. In other words, it is not possible to determine the relative importance of lipophilicity, electronic effects, or steric influence on the biologic response when using Equation 2-19.

Most recently, there has been a concentrated effort to calculate the partition coefficient on the basis of the atomic components of the molecule. Each atom type is assumed to contribute a fixed amount to the chemical's partition coefficient. Because this assumption breaks down quickly, several correction factors are used. Cyclohexene will serve as an example. For purposes of comparison, the observed octanol–water partition coefficient (expressed as a logarithm) is 2.96.

$$\begin{aligned}\log P &= 6(\text{carbon atoms}) + 12(\text{hydrogen atoms})\\ &\quad + (n - 1)\text{bonds} + \text{double-bond correction}\\ \log P &= 6(0.20) + 12(0.23) + 5(-0.09)\\ &\quad + (-0.55) = 2.96\end{aligned}$$

Because of the correction factors, these calculations become so complex that they must be done by a computer. Convenient as the calculation method may be, it must be realized that their accuracy is dependent on first determining experimental partition coefficients of chemicals exhibiting very similar chemistry. It is from these experimentally determined partition coefficients that the correction factors are derived.

There are simpler methods, which will give reasonably correct results, for estimating lipophilicity. These are based on the additive effect on the partition coefficient that is seen when varying a series of substituents on the same molecule. Over the years, fairly extensive tables have been developed that contain the contribution (π) of a wide variety of substituents to the partition coefficient. The method can be illustrated for chlorobenzene. The log P for benzene is 2.13 and for chlorobenzene is 2.84. The π value for the chlorine substituent is obtained by subtracting the log P values for benzene and chlorobenzene.

$$\pi_{Cl} = \log P_{chlorobenzene} - \log P_{benzene}$$

$$\pi_{Cl} = 2.84 - 2.13 = 0.71$$

OTHER PHYSICOCHEMICAL PARAMETERS

There are a series of other constants that measure the contribution by substituents to the molecule's total physicochemical properties. These include Hammett's σ constant; Taft's steric parameter, E_s; Charton's steric parameter, ν; Verloop's multidimensional steric parameters, L, B_1, B_5; and molar refractivity, MR. The latter has become the second most useful physicochemical parameter used in QSAR modeling. It is a complex term that is based on the molecule's refractive index, molecular weight, and density and can be considered a measure of the molecule's bulk and electronic character. One reason for its popularity is that it is easy to calculate from tables of atoms by using a minimum of correction factors. Of this group, it is most easy to locate values for π, σ, E_s, and MR. A representative list can be found in Table 2-7.

Table 2-7 is illustrative of several items that must be remembered when selecting substituents to be evaluated in terms of the type of factors that influence a biologic response. For electronic parameters such as σ, the location on an aromatic ring is important because of resonance versus inductive effects. Notice the twofold differences seen between σ_{meta} and σ_{para} for the three aliphatic substituents, iodo, and severalfold difference for methoxy, amino, fluoro, and phenolic hydroxyl.

Selection of substituents from a certain chemical class may not really test the influence of a parameter on biologic activity. There is little numerical difference in the σ_{meta} or σ_{para} values of the four aliphatic groups or the four halogens. It is not uncommon to go to the tables and find missing parameters, such as the E_s values for acetyl and N-acyl.

Nevertheless, the medicinal chemists can use information from extensive tables of physicochemical parameters, to minimize the number of substituents required to find out if the biologic response is sensitive to electronic, steric, or partitioning effects. This is done by selecting substituents in each of the numerical ranges for the different parameters. In Table 2-7 there are three ranges of π values: (−1.23 to −0.55), (−0.28 to 0.56), and (0.71 to 1.55); three ranges of MR values: (0.92 to 2.85), (5.02 to 8.88), and (10.30 to 14.96); two main clusters of σ values: one for the aliphatic substituents and the other for the halogens. Depending on the biologic responses obtained from testing the new compounds, it is possible to determine if lipophilicity (partitioning), steric bulk (molar refraction), or electron withdrawing or donating properties are important determinants of the desired biologic response.

TABLE 2-7

SAMPLING OF PHYSICOCHEMICAL PARAMETERS USED IN QUANTITATIVE STRUCTURE ACTIVITY RELATIONSHIP INVESTIGATIONS

Substituent Group	π	σ_{meta}	σ_{para}	E_s	MR
$-H$	0.00	0.00	0.00	0.00	1.03
$-CH_3$	0.56	−0.07	−0.17	−1.24	5.65
$-CH_2CH_3$	1.02	−0.07	−0.15	−1.31	10.30
$-CH_2CH_2CH_3$	1.55	−0.07	−0.13	−1.60	14.96
$-C(CH_3)_2$	1.53	−0.07	−0.15	−1.71	14.96
$-OCH_3$	−0.02	0.12	−0.27	−0.55	7.87
$-NH_2$	−1.23	−0.16	−0.66	−0.61	5.42
$-F$	0.14	0.34	0.06	−0.46	0.92
$-Cl$	0.71	0.37	0.23	−0.97	6.03
$-Br$	0.86	0.39	0.23	−1.16	8.88
$-I$	1.12	0.35	0.18	−1.40	13.94
$-CF_3$	0.88	0.43	0.54	−2.40	5.02
$-OH$	−0.67	0.12	−0.37	−0.55	2.85
$-COCH_3$	−0.55	0.38	0.50		11.18
$-NHCOCH_3$	−0.97	0.21	0.00		14.93
$-NO_2$	−0.8	0.71	0.78	−2.52	7.36
$-CN$	−0.57	0.56	0.66	−0.51	6.33

(From Hansch, C., Leo, A. J.: Substituent Constants for Correlation Analysis in Chemistry and Biology. New York, John Wiley & Sons, 1979.)[2]

QSAR MODELS

Currently there are three models or equations seen in QSAR analysis that use physicochemical parameters. These are illustrated (Fig. 2-6) by using the logarithm of the partition coefficient (log P) as the physicochemical parameter. First there is the linear model (Eq. 2-20). Because several plots indicated a nonlinear relationship between biologic response and the partition coefficient, a parabolic model was developed (Eq. 2-21). Examination of Figure 2-6 shows an optimum log P (log P_0) at which maximum biologic activity will be obtained before a decrease in activity is seen. One explanation for this phenomenon is that hydrophilic drugs will tend to stay in the aqueous phase, whereas lipophilic chemicals will prefer the lipid bilayer. In both cases, less drug is being transported to the receptor, resulting in the actual concentration of receptor-bound drug being decreased. In other words, the equilibrium seen in Reaction 2-1 shifts to the left. There will be a group of drugs whose log P will place them near the top of the parabola, such that there lipophilic–hydrophilic

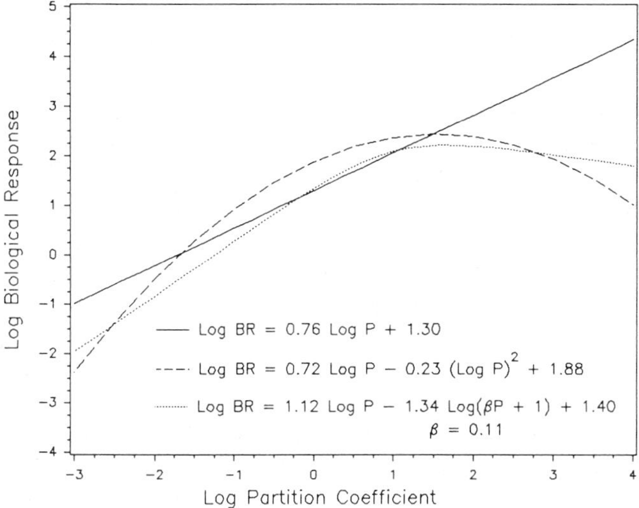

FIG. 2-6. *Log biologic response versus log partition coefficient using linear, parabolic, and bilinear models.*

balance will permit them to penetrate both aqueous and lipid barriers and reach the receptor.

$$\log 1/C = a(\log P) + c \qquad \text{(Eq. 2-20)}$$

$$\log 1/C = a(\log P) - b(\log P)^2 + c \qquad \text{(Eq. 2-21)}$$

The third QSAR equation in current use is the bilinear model (Eq. 2-22). There are several interpretations for the β term. One explanation is based on the ratio of the rate constant for diffusion out of the octanol layer into the aqueous environment being different from the rate of diffusion out of the aqueous layer into the octanol. In other words, what may be simulated with the bilinear model is recognition that the rate of diffusion from the extracellular fluids into the lipid bilayer is different from the rate of diffusion out of the lipid bilayer into the intracellular environment. Another interpretation is recognition that the kinetics of partitioning through the lipid bilayer will be different from the kinetics of binding to the receptor. A third explanation takes into account the different volumes of the aqueous and lipid bilayers in the biologic system.

$$\log 1/C = a(\log P) - b\log(\beta P + 1) + c \qquad \text{(Eq. 2-22)}$$

With this background in mind, three examples of QSAR equations taken from the medicinal chemistry literature will be presented. One will show a linear relationship (Eq. 2-23) and the others a parabolic (Eq. 2-24) and a bilinear (Eq. 2-25) correlation. A study of a group of griseofulvin analogues showed that there is a linear relationship (Eq. 2-23) between the biologic response and both lipophilicity (log P) and electronic character (σ).[3] It was suggested that the antibiotic activity may depend on the enone system facilitating the addition of griseofulvin to a nucleophilic group such as the SH moiety in a fungal enzyme.

$$\log BR = (0.56)\log P + (2.19)\sigma_x - 1.32 \qquad \text{(Eq. 2-23)}$$

Griseofulvin: R = R_1 = R_2 = OCH_3, R_3 = Cl

A parabolic relationship (Eq. 2-24) was reported for a series of substituted acetylated salicylates (substituted aspirins) tested for anti-inflammatory activity.[4] There is a nonlinear relationship between the biologic response and lipophilicity and a significant detrimental steric effect seen with substituents at position 4. The two STERIMOL parameters used in this equation were L, defined as the length of the substituent along the axis of the bond between the first atom of the substituent and the parent molecule and B_2, defined as a width parameter. Steric effects were not considered significant at position 3 because the STERIMOL parameters for substituents at position 3 are not part of Equation 2-23. The optimal partition coefficient (log P_0) for the substituted aspirins in this assay was 2.6.

$$\log 1/ED_{50} = 1.03\log P - 0.20(\log P)^2 - 0.05L_{(4)} - 0.24B_{2(4)} + 2.29 \qquad \text{(Eq. 2-24)}$$

Aspirin: X = Y = H

Related to these mathematical QSAR models, based on biologic responses, is the QSAR used to analyze pharmacokinetic activity. One example of this (Eq. 2-25) is a simulation of barbiturate absorption that leads to the bilinear model.[5,6]

$$\log k_{\text{DIFF}} = 0.949 \log P - 1.238 \log(\beta P + 1) - 3.131 \quad \text{(Eq. 2-25)}$$

At this point, it is appropriate to ask the question: Are all the determinations of partition coefficients and compilation of physical chemical parameters useful only when a statistically valid QSAR model is obtained? The answer is a firm, "**No.**" One of the most useful spinoffs from the field of QSAR has been the application of experimental design to the selection of new compounds to be synthesized and tested. Let us assume that a new series of drug molecules are to be synthesized that are based on the following structure. The goal is to test the effect of the 16 substituents in Table 2-7 at each of the three position on our new series. The number of possible analogues is equal to 16^3 or 4096 compounds, assuming all three position will always be substituted with one of the substituents from Table 2-7. If hydrogen is included when a position is not substituted, there are 17^3 or 4913 different combinations. The problem is to select a few substituents that will represent the different ranges or clusters of values for lipophilicity, electronic influence, and bulk. An initial design set could include the methyl and propyl from the aliphatic, fluorine and chlorine from the halogens, *N*-acetyl and phenol from the substituents showing hydrophilicity, and a range of electronic and bulk values. Including hydrogen, there will be 7^3 or 343 different combinations. Obviously that is too many for an initial evaluation. Instead certain rules have been devised to maximize the

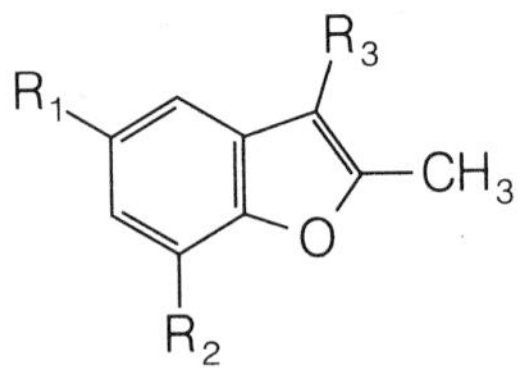

information obtained from a minimum number of compounds. These include

1. Each substituent must occur more than once at each position in which it is found.
2. The number of times that each substituent at a particular position appears should be approximately equal.
3. No two substituents should be present in a constant combination.
4. When combinations of substituents are a necessity, they should not occur more frequently than any other combination.

Following these guidelines, the initial test set can be reduced to 24 to 26 compounds. Dependent on the precision of the biologic tests, it will be possible to see if the data will fit a QSAR model. Even an approximate model usually will indicate the types of substituents to test further, what positions on the molecules are sensitive to substitution, and, if sensitive, to what degree variation in lipophilic, electronic, or bulk character is important. Just to be assured that the model is valid, it is a good idea to synthesize a couple of compounds that the model predicts would be inactive. As each group of new compounds are tested, the QSAR model is refined until the investigators have a fairly good idea about what substituent patterns are important for desired activity. It must be emphasized that these same techniques that are used to develop potent compounds with desired activity also can be used to evaluate the influence of substituent patterns on undesired toxic effects and pharmacokinetic properties.

CLASSIFICATION METHODS

There are other statistical techniques besides regression analysis that are used in drug design. These fit under the classification of multivariate statistics and include discriminant analysis, principal component analysis, and pattern recognition. The latter can consist of a mixture of statistical and nonstatistical methodologies. Usually, the goal is to try to ascertain those physicochemical parameters and structural attributes that contribute to a class or type of biologic activity. Then the chemicals are classified into groupings, such as carcinogenic or noncarcinogenic, sweet or bitter, active or inactive, and depressant or stimulant.

The term *multivariate* is used because of the wide variety and number of independent or descriptor variables that may be used. The same physicochemical parameters seen in QSAR analyses are used, but, in addition, the software in the computer programs "breaks" the molecule down into substructures. These structural fragments also become variables. Examples of the typical substructures used include carbonyls, enones, conjugation, rings of different sizes and types, *N*-substitution patterns, and aliphatic substitution patterns such as 1,3- or 1,2-disubstituted. The end result is that for even a moderately sized molecule, typical of most drugs, there can be 50 to 100 variables.

The technique is to develop a large set of chemicals well characterized in terms of the biologic activity that is going to be predicted. This is known as the training set. It should contain hundreds, if not thousands of compounds, divided into active and inactive types. The multivariate technique is now carried out. The algorithms are designed to group the underlying commonalities and select those variables having the greatest influence on biologic activity. The predictive ability is then measured with a test set of compounds that have been put through the same evaluations as the training set.

There are several examples of successful applications of this technique.[7] One study consisted of a diverse group of 140 tranquilizers and 79 sedatives subjected to a two-way classification study (tranquilizers versus sedatives). The ring types included phenothiazines, indoles, benzodiazepines, barbiturates, diphenylmethanes, and a variety of heterocycles. Sixty-nine descriptors were used initially to characterize the molecules. Eleven of these descriptors were crucial to the classification, 54 had intermediate use and were dependent on the composition of the training set, and four descriptors were of little use. The overall range of prediction accuracy was 88% to 92%. The results with the 54 descriptors indicates an important pitfall seen when large numbers of descriptors are used. The inclusion or exclusion of descriptors and parameters can be dependent on the composition of the training set. It is important that the training set be representative of the population of chemicals that are going to be evaluated.

Classification techniques lend themselves to studies lacking quantitative data. An interesting classification problem involved olfactory stimulants in which the goal was to select chemicals that had a musk odor. A group 300 unique compounds were selected from a group of odorants that included 60 musk odorants plus 49 camphor, 44 floral, 32 ethereal, 41 mint, 51 pungent, and 23 putrid odorants. Initially 68 descriptors were evaluated. Dependent on the approach, the descriptors were reduced in number down to 11–16, consisting mostly of bond types. With use of this small number, the 60 musk odorants could be selected from the remaining 240 compounds, with an accuracy of 95% to 97%.

The use of classification techniques in medicinal chemistry is still in its early stages. Although the statistical and pattern recognition methodologies have been in use for a long time, there still needs to be considerable research into their proper use, and further testing of their predictive power is needed. The goal of scanning data bases of already synthesize compounds to select compounds for pharmacologic evaluation will require considerable additional development in the various multivariate techniques.

DRUG–RECEPTOR INTERACTIONS

At this point, let us assume that the drug has entered systemic circulation (see Fig. 2-1), passed through the lipid barriers, and is now going to make contact with the receptor. As was illustrated in Reaction 2-1, this is an equilibrium process. A good ability to fit the receptor will favor binding and the desired pharmacologic response. In contrast, a poor fit will favor the reverse reaction. With only a small amount of drug bound to the receptor, there will be a much smaller pharmacologic effect. Indeed, if the amount of drug bound to the receptor is too small, there may be no discernible response observed. There are many variables that contribute to a drug's binding to the receptor. These include the structural class, the three-dimensional shape of the molecule, and types of chemical bonding involved in the binding of the drug to the receptor.

Most drugs that belong to the same pharmacologic class have certain structural features in common. The barbiturates act on specific central nervous system receptors, causing depressant effects; hydantoins act on central nervous system receptors producing an anticonvulsant response; benzodiazepines combine with the γ-aminobutyric acid (GABA) receptors, with resulting anxiolytic activity; steroids can be divided into such classes as corticosteroids, anabolic steroids, progestogens, and estrogens, each acting on specific receptors; nonsteroidal anti-inflammatory agents inhibit enzymes required for the prostaglandin cascade; penicillins and cephalosporins inhibit enzymes required to construct the bacterial cell wall; and tetracyclines act on bacterial ribosomes.

RECEPTOR

With the isolation and characterization of receptors becoming a common occurrence, it is hard to realize that the concept of receptors began as a postulation. It had been realized early that molecules with certain structural features would elucidate a specific biologic response. Very slight changes in structure could cause significant changes in biologic activity. These structural variations could increase or decrease activity or change an agonist into an antagonist. This early and fundamentally correct interpretation called for the drug (ligand) to fit onto some surface (the receptor) that had fairly strict structural requirements if proper binding of the drug was to occur. The initial receptor model was based on a rigid lock-and-key concept of the drug (key) fitting onto a receptor (lock). It has been used to explain why certain structural attributes produce a predictable pharmacologic action. This model is still useful, although it is important to realize that both the drug and, especially, the receptor can have considerable flexibility. Molecular graphics, with programs that calculate the preferred conformations of drug and receptor, show that the receptor can undergo an adjustment in three-dimensional structure when the drug makes contact. To use current space age language, the drug docks with the receptor.

More complex receptors are now being isolated and characterized. The first receptors to be isolated and characterized were the reactive and regulatory sites on enzymes. Acetylcholinesterase, dihydrofolate reductase, and angiotensin-converting enzyme are examples of enzyme receptors. Most drug receptors probably are receptors for natural ligands used to regulate cellular biochemistry and function and to communicate between cells. Receptors include a relatively small region of a macromolecule that may be an isolatable enzyme, a structural and functional component of a cell membrane, or a specific intracellular substance, such as a protein or a nucleic acid. Specific regions of these macromolecules are visualized as being oriented in space in a manner that permits their functional groups to interact with the complementary functional groups of the drug. This interaction initiates changes in structure and function of the macromolecule that ultimately lead to the observable biologic response. The concept of specifically oriented functional areas forming a receptor leads directly to specific structural requirement for functional groups of a drug that must be complementary to the receptor.

It is now possible to isolate membrane-bound receptors, although it still is difficult to elucidate their structural chemistry because, once separated from the cell membranes, these receptors lose their native shape. This is because the membrane is required to hold the receptor in its correct shape. One method of receptor isolation is affinity chromatography. In this technique, a ligand, many times an altered drug molecule known to combine with the receptor, is attached to a chromatographic support phase. A solution containing the desired receptor is passed over this column. The receptor will combine with the ligand. It is common to add a chemically reactive grouping to the drug, resulting in the receptor and drug covalently binding with each other. The drug–receptor complex now is washed from the column, after which it is then characterized further.

A more recent technique uses recombinant DNA. The gene for the receptor is located and cloned. It is transferred into a bacterium or some other animal that then produces the receptor in large enough quantities to permit further study. Furthermore, it sometimes is possible to determine the DNA sequence of the cloned gene. With use of the genetic code for amino acids, the amino acid sequence of the protein component of the receptor can be determined.

The preceding discussion in this chapter emphasized that the cell membrane is a highly organized, dynamic structure that interacts with small molecules in specific ways. The previous focus was on the lipid bilayer component of this complex structure. The receptor component of the membranes appear to be mainly protein. They constitute a potentially highly organized region of the cell membrane. The same type of molecular specificity seen in such proteins as enzymes and antibodies also is a property of the drug receptor. The nature of the amide link in proteins provides a unique opportunity for the formation of multiple internal hydrogen bonds, as well as internal formation of hydrophobic, van der Waals, and ionic bonds by side-chain groups, leading to such organized structures as the α-helix, which contains about four amino acid residues for each turn of the helix. An organized protein structure would hold the amino acid side chains at relatively fixed positions in space and available for specific interactions with a small molecule.

Proteins have the potential to adopt many different conformations in space without breaking their covalent amide linkages. They may shift from highly coiled structures to partially disorganized structures, with parts of the molecule existing in random chain, or to folded sheet structures, contingent on the environment. In the monolayer of a cell membrane, the interaction of a foreign small molecule with an organized protein may lead to a drastic change in the structural and physical properties of the membrane. Such changes could well be the initiating events in the production of a tissue or organ response to a drug, such as the ion-translocating effects produced by interaction of acetylcholine and the cholinergic receptor.[8]

The large body of information now available on relationships between chemical structure and biologic activity strongly supports the concept of flexible receptors. The fit of drugs onto or into macromolecules is only rarely an all-or-none process as pictured by the earlier lock-and-key concept of a receptor. Rather, the binding or partial insertion of groups of moderate size onto or into a macromolecular pouch appears to be a continuous process, at least over a limited range, as indicated by the frequently occurring regular increase and decrease in biologic activity as one ascends a homologous series of drugs. A range of productive associations between drug and receptor may be pictured, which lead to agonist responses, such as those produced by cholinergic drugs. Similarly, strong associations may lead to unproductive changes in the configuration of the macromolecule, leading to an antagonistic or blocking response, such as that produced by anticholinergic agents. The fundamental structural unit of the drug receptor is generally considered to be protein, although this may be supplemented by its associations with other units, such as mucopolysaccharides and nucleic acids.

In the maximally extended protein, the distance between peptide bonds ("identity distance") is 3.61 Å. For many types of biologic activity, the distance between functional groups leading to maximal activity approximates this identity distance or some whole-number multiple of it. Many parasympathomimetic (acetylcholinelike) and parasympatholytic (cholinergic-blocking) agents have a separation of 7.2 Å (2 × 3.6) between the ester carbonyl group and nitrogen.[9] This distance is doubled between quaternary nitrogens of curarelike drugs; 14.5 Å (4 × 3.61).[10] The preferred separation of hydrogen-bonding groups in estrogenic compounds (e.g., hydroxyls of diethylstilbestrol) is 14.5 Å (4 × 3.61).[11]

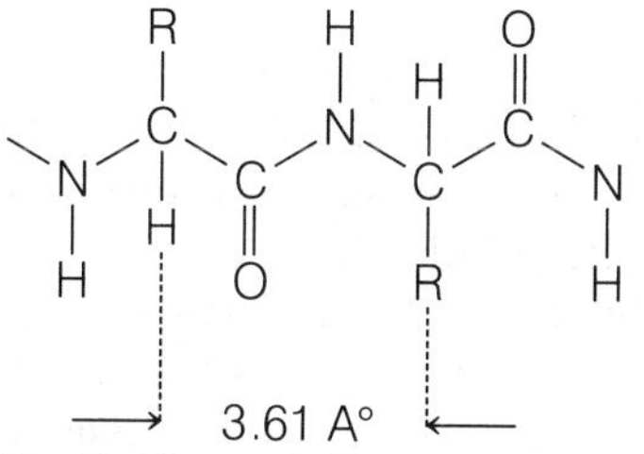

3.61 A°

Identity Distance in Extended Protein

A related spacing of 5.5 Å, which corresponds to two turns of the α-helical structure common to proteins, is found between functional groups of many drugs. The most frequently occurring of these is the $R—X—CH_2—CH_2—NR'_2$ (X = N; X = O, or X = C) structure that is present in local anesthetics, antihistamines, adrenergic-blocking agents, and others.[12]

Studies involving the relative effectiveness of various molecules of well-defined structural and functional types have contributed to an understanding of the stereochemical and physicochemical properties of their biologic receptors. Pfeiffer[9] concluded that parasympathomimetic stimulant action depends on two adjacent oxygen atoms at distances of approximately 5.0 Å and 7.0 Å from a methyl group or groups attached to nitrogen. Because these compounds (acetylcholine, methacholine, urecholine, and others) do not have rigid structures, the actual distance between the oxygen and the methyl groups varies; however, the more extended conformations would be favored in solution.

Welsh and Taub[13] have concluded that a carbonyl group at a maximum distance of 7 Å from the quaternary nitrogen is an important linking group with the acetylcholine receptor protein of the *Venus* heart. They suggest that some type of bond forms between the carbonyl carbon or ketone oxygen and an appropriate group in the protein molecule.

The nature of the acetylcholinesterase receptor site probably has been investigated more thoroughly than the reactive site of any other enzyme. On the basis of studies with enzyme inhibitors, Nachmansohn and Wilson[14] suggested two functional sites: a center of high electron density that binds the cationic nitrogen, and an esteratic site that interacts with the carbonyl carbon atom. Friess and his coworkers[15] attempted to define the distance between the anionic and the esteratic sites by studying enzyme inhibition with cyclic aminoalcohols (e.g., *cis*-2-dimethylaminocyclohexanol) and their esters. The *cis*-isomers were more active than the *trans*, and a distance of about 2.5 Å was indicated as separating the nitrogen and the oxygen and, by inference, the receptors that bind these on the enzyme. Krupka and Laidler[16] correlated previous stereochemical studies with kinetic data and described a complex esteratic site made up of three components: a basic

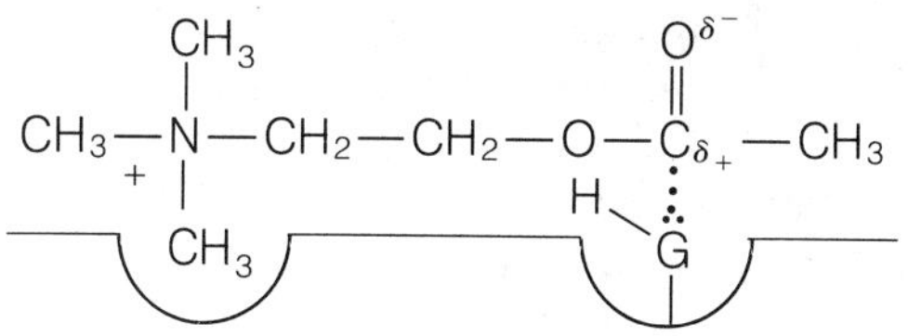

Anionic Site Acetylcholinesterase

Esteratic Site (Nachmansohn and Wilson)[14]

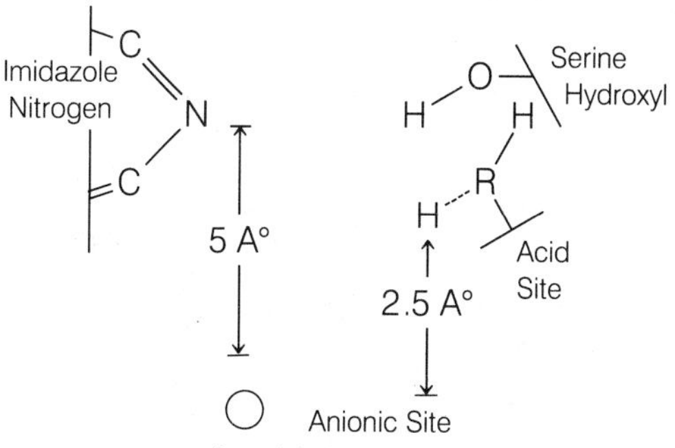

Imidazole Nitrogen

Serine Hydroxyl

5 A°

Acid Site

2.5 A°

Anionic Site Acetylcholinesterase (Krupka and Laidler)[16]

site (imidazole nitrogen, 5 Å from the anionic site), an acid site, 2.5 Å from the anionic site, and a serine hydroxyl group. Following stereospecific binding of acetylcholine, the serine hydroxyl is acetylated to effect ester cleavage. Subsequently, a water molecule is held in the proper position through hydrogen bonds with imidazole, serine is deacetylated (hydrolyzed), and the reactive enzyme is regenerated.

THE DRUG–RECEPTOR INTERACTION: FORCES INVOLVED

A biologic response is produced by the interaction of a drug with a functional or organized group of molecules, which may be called the biologic receptor site. This interaction would be expected to take place by utilizing the same bonding forces involved as those when simple molecules interact. These, together with typical examples, are collected in Table 2-8.

TABLE 2-8

TYPES OF CHEMICAL BONDS

Bond Type	Bond Strength (kcal / mol)	Example
Covalent	40–140	$CH_3—OH$
Reinforced ionic	10	$R—NH_2^{\oplus}—H\cdots O^{\ominus}—C(=O)—R'$
Ionic	5	$R_4N^{\oplus}\cdots{}^{\ominus}I$
Hydrogen	1–7	$—OH\cdots O=$; $—OH\cdots C=C$
Ion-dipole	1–7	$R_4N^{\oplus}\cdots :NR_3$
Dipole-dipole	1–7	$O=C\cdots :NR_3$ ($\delta^-\delta^+$)
van der Waals'	0.5–1	$C\cdots C$
Hydrophobic	1	See Text

(Adapted from a table *in* Albert, A. Selective Toxicity. New York, John Wiley & Sons, 1986, p. 183.)

Most drugs do not possess functional groups of a type that would lead to ready formation of the strong and essentially irreversible covalent bonds between drug and biologic receptors. In most cases, it is desirable that the drug leave the receptor site when the concentration decreases in the extracellular fluids; therefore, most useful drugs are held to their receptors by ionic or weaker bonds. However, occasionally, when relatively long-lasting or irreversible effects are desired (e.g., antibacterial, anticancer), drugs that form covalent bonds are effective and useful.

The alkylating agents, such as the nitrogen mustards (e.g., mechlorethamine) used in cancer chemotherapy, furnish an example of drugs that act by formation of covalent bonds (Fig. 2-7). These are believed to form the reactive immonium ion intermediates, which alkylate and thereby link together proteins or nucleic acids, preventing their normal participation in cell division.

Covalent bond formation between drug and receptor is the basis of Baker's[17] concept of *active-site-directed irreversible inhibition*. Considerable experimental evidence on the nature of enzyme inhibitors

$$CH_3—N(CH_2CH_2Cl)_2 \longrightarrow CH_3—{}^{+}N(CH_2CH_2)(CH_2CH_2Cl) \xrightarrow{RH} CH_3—N(CH_2CH_2Cl)(CH_2CH_2R)$$

Mechlorethamine — Immonium Ion — Alkylated Protein or Nucleic Acid

$$\longrightarrow CH_3—{}^{+}N(CH_2CH_2)(CH_2CH_2R) \xrightarrow{R'H} CH_3—N(CH_2CH_2R')(CH_2CH_2R)$$

Cross-linked Protein or Nucleic Acid

FIG. 2-7. Formation of the immonium cation and its alkylation of protein or nucleic acid: R,R′ = free amino groups of proteins, adenyl, or phosphate groups of nucleic acids.

has supported this concept. Compounds studied possess appropriate structural features for reversible and highly selective association with an enzyme. If, in addition, the compounds carry reactive groups capable of forming covalent bonds, the substrate may be irreversibly bound to the drug–receptor complex by covalent bond formation with reactive groups adjacent to the active site. In studies with reversibly binding antimetabolites that carried additional alkylating and acylating groups of varying reactivities, selective irreversible binding by the related enzymes, lactic dehydrogenase and glutamic dehydrogenase, has been demonstrated. The selectivity of response has been attributed to the formation of a covalent bond between the carbophenoxyamino substituent of 5-(carbophenoxyamino)salicylic acid and a primary amino group in glutamic dehydrogenase[18] and between the maleamyl substituent of 4-(maleamyl)salicylic acid and a sulfhydryl group in lactic dehydrogenase.[19] Assignments of covalent bond formation with specific groups in the enzymes are based on the fact that the α,β-unsaturated carbonyl system of maleamyl groups reacts most rapidly with sulfhydryl groups, much more slowly with amino groups, and extremely slowly with hydroxyl groups. In contrast, the carbophenoxy group will react only with a primary amino group on a protein. The diuretic drug, ethacrynic acid (see Chap. 13), is an α,β-unsaturated ketone, thought to act by covalent bond formation with sulfhydryl groups of ion transport systems in the renal tubules.

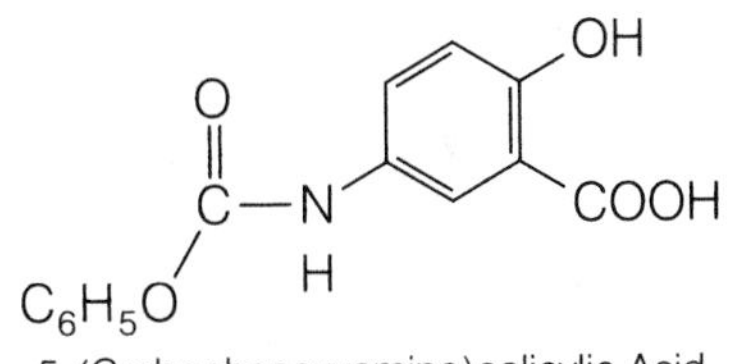

C_6H_5O–C(=O)–NH–(OH, COOH substituted benzene ring)

5-(Carbophenoxyamino)salicylic Acid

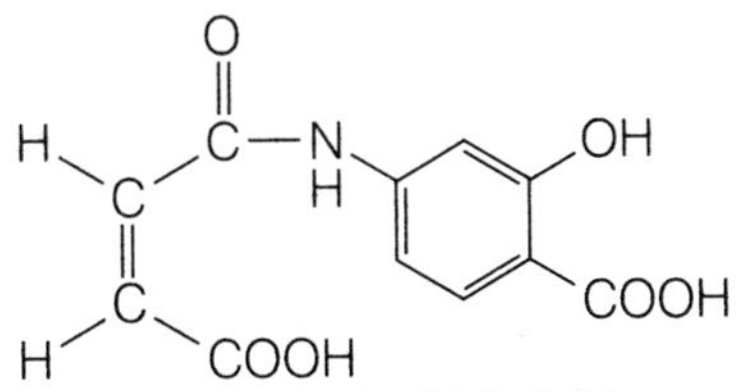

HOOC–CH=CH–C(=O)–NH–(OH, COOH substituted benzene ring)

4-(Maleamyl)salicylic Acid

Other examples of covalent bond formation between drug and biologic receptor site include the reaction of arsenicals and mercurials with essential sulfhydryl groups, the acylation of bacterial cell wall constituents by penicillin, and the inhibition of cholinesterase by the organic phosphates.

It is desirable that most drug effects be reversible. For this to occur, relatively weak forces must be involved in the drug–receptor complex, yet be strong enough that other binding sites will not competitively deplete the site of action. Compounds with a high degree of structural specificity may orient several weak-binding groups, such that the summation of their interactions with specifically oriented complementary groups on the receptor will provide a total bond strength sufficient for a stable combination.

Consequently, for drugs acting by virtue of their structural specificity, binding to the receptor site will be carried out by hydrogen bonds, ionic bonds, ion–dipole and dipole–dipole interactions, van der Waals and hydrophobic forces. Ionization at physiologic pH would normally occur with the carboxyl, sulfonamido, and aliphatic amino groups, as well as the quaternary ammonium group at any pH. These sources of potential ionic bonds are frequently found in active drugs. Differences in electronegativity between carbon and other atoms, such as oxygen and nitrogen, lead to an unsymmetric distribution of electrons (dipoles) that are also capable of forming weak bonds with regions of high or low electron density, such as ions or other dipoles. Carbonyl, ester, amide, ether, nitrile, and related groups that contain such dipolar functions are frequently found in equivalent locations in structurally specific drugs. Many examples may be found among the potent analgesics, the cholinergic-blocking agents, and local anesthetics.

The relative importance of the *hydrogen bond* in the formation of a drug–receptor complex is difficult to assess. Many drugs possess groups, such as carbonyl, hydroxyl, amino, and imino, with the structural capabilities of acting as acceptors or donors in the formation of hydrogen bonds. However, such groups would usually be solvated by water, as would the corresponding groups on a biologic receptor. Relatively little net change in free energy would be expected in exchanging a hydrogen bond with a water molecule for one between drug and receptor. However, in a drug–receptor combination, several forces could be involved, including the hydrogen bond, which would contribute to the stability of the interaction. Where multiple hydrogen bonds may be formed, the total effect may be sizeable, such as that demonstrated by the stability of the protein α-helix, and by the stabilizing influence of hydrogen bonds between specific base pairs in the double helical structure of DNA.

van der Waals forces are attractive forces created by the polarizability of molecules and are exerted when any two uncharged atoms approach very closely. Their strength is inversely proportional to the seventh power of the distance. Although individually weak, the summation of their forces provides a significant bonding factor in higher molecular weight compounds. For example, it is not possible to distill normal alkanes with more than 80 carbon atoms, because the energy of about 80 kcal/mol required to separate the molecules is approximately equal to the energy required to break a carbon–carbon covalent bond. Flat structures, such as aromatic rings, permit close approach of atoms. With van der Waals force approximately 0.5 to 1.0 kcal/mol for each atom, about six carbons (a benzene ring) would be necessary to match the strength of a hydrogen bond. The aromatic ring is frequently found in active drugs, and a reasonable explanation for its requirement for many types of biologic activity may be derived from the contributions of this flat surface to van der Waals binding to a correspondingly flat receptor area.

The *hydrophobic bond* is a concept used to explain attractive interactions between nonpolar regions of the receptor and the drug. Explanations such as the "isopropyl moiety of the drug fits into a hydrophobic cleft on the receptor composed of the hydrocarbon side chains of the amino acids valine, isoleucine, and leucine" are commonly used to explain why a nonpolar substituent at a particular position on the drug molecule is important for activity. Over the years, the concept of hydrophobic bonds has developed. There has been considerable controversy over whether or not the bond actually exists. Thermodynamic arguments on the gain in entropy (decrease in ordered state) when hydrophobic groups cause a partial collapse of the ordered water structure on the surface of the receptor have been proposed to validate a hydrophobic bonding model. There are two problems with this concept. First, the term *hydrophobic* implies repulsion. The term for attraction is *hydrophilicity*. Second and, perhaps, more important, there is no truly water-free region on the receptor. This is true, even in the areas populated by the nonpolar side amino acid side chains. An alternate approach is to consider only the concept of hydrophilicity and lipophilicity. The predominating water molecules solvate polar moieties, effectively squeezing the nonpolar residues toward each other.

Isosterism

The term *isosterism* has been widely used to describe the selection of structural components, the steric, electronic, and solubility characteristics of which make them interchangeable in drugs of the same pharmacologic class. The concept of isosterism has evolved and changed significantly in the years since its introduction by Langmuir[20] in 1919. Langmuir, while seeking a correlation that would explain similarities in physical properties for nonisomeric molecules, defined *isosteres* as compounds or groups of atoms having the same number and arrangement of electrons. Those isosteres that were isoelectric (i.e., with the same total charge as well as same number of electrons) would possess similar physical properties. For example, the molecules N_2 and CO, both possessing 14 total electrons and no charge, show similar physical properties. Related examples described by Langmuir were CO_2 and N_2O, and N_3^- and NCO^-.

With increased understanding of the structures of molecules, less emphasis has been placed on the number of electrons involved, for variations in hybridization during bond formation may lead to considerable differences in the angles, the lengths, and the polarities of bonds formed by atoms with the same number of peripheral electrons. Even the same atom may vary widely in its structural and electronic characteristics when it forms a part of a different functional group. Thus, nitrogen is part of a planar structure in the nitro group, but forms the apex of a pyramidal structure in ammonia and the amines.

Groups of atoms that impart similar physical or chemical properties to a molecule, because of similarities in size, electronegativity, or stereochemistry, are now frequently referred to under the general term of *isostere*. The early recognition that benzene and thiophene were alike in many of their properties led to the term "ring equivalents" for the vinylene group (—CH=CH—) and divalent sulfur (—S—). This concept has led to replacement of the sulfur atom in the phenothiazine ring system of tranquilizing agents with the vinylene group to produce the dibenzazepine class of antidepressant drugs (see Chap. 10). The vinylene group in an aromatic ring system may be replaced by other atoms isosteric to sulfur, such as oxygen (furan) or NH (pyrrole); however, in such cases, aromatic character is significantly decreased.

Examples of isosteric pairs that possess similar steric and electronic configurations are the carboxylate (COO^-) and sulfonamido (SO_2NR^-) ions, ketone (CO) and sulfone (SO_2) groups, chloride (Cl) and trifluoromethyl (CF_3) groups. Divalent ether (—O—), sulfide (—S—), amine (—NH—), and methylene ($—CH_2—$) groups, although dissimilar electronically, are sufficiently alike in their steric nature to be frequently interchangeable in drugs.

Compounds may be altered by isosteric replacements of atoms or groups, to develop analogues with select biologic effects, or to act as antagonists to normal metabolites. Each series of compounds showing a specific biologic effect must be considered separately, for there are no general rules that will predict whether biologic activity will be increased or decreased. It appears that when isosteric replacement involves the bridge connecting groups necessary for a given response, a gradation of like effects results, with steric factors (bond angles) and relative polar character being important. Some examples of this type are as follows:

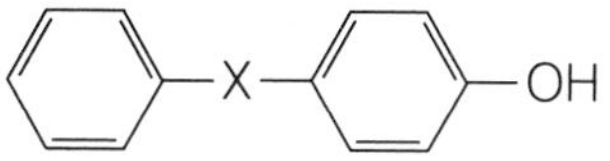

Antibacterial: X = S, Se, O, NH, CH_2

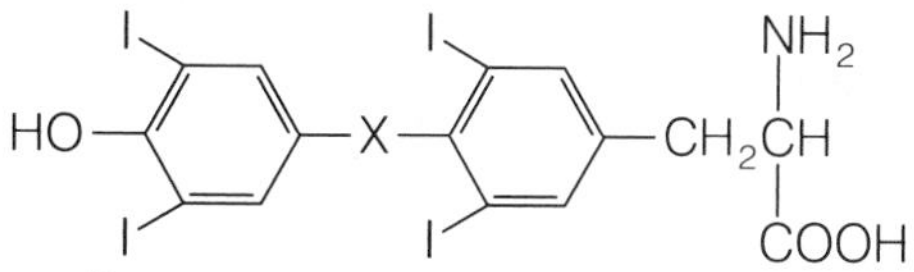

Thyroid Hormone Analogs: X = O, S, CH_2

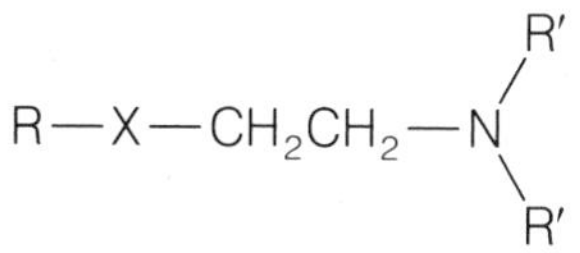

$R—X—CH_2CH_2—NR'_2$

Antihistamines: X = O, NH, CH_2
Cholinergic Blocking Agents: X = —COO—, —CONH—, —COS—

When a group is present in a part of a molecule in which it may be involved in an essential interaction or may influence the reactions of neighboring groups, isosteric replacement sometimes produces analogues that act as antagonists. Some examples from the field of cancer chemotherapy are:

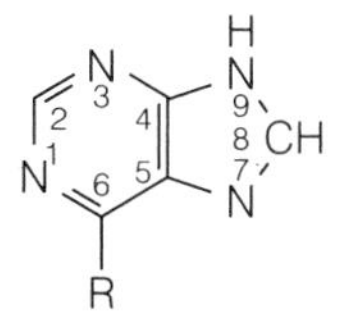

	R	
Adenine	NH_2	Metabolites
Hypoxanthine	OH	
6-Mercaptopurine	SH	Antimetabolite

The 6-NH_2 and 6-OH groups appear to play essential roles in the hydrogen-bonding interactions of base pairs during nucleic acid replication in cells. The substitution of the significantly weaker hydrogen-bonding isosteric sulfhydryl groups results in a partial blockage of this interaction and a decrease in the rate of cellular synthesis.

In a similar fashion, replacement of the hydroxyl group of pteroylglutamic acid (folic acid) by the amino group leads to methotrexate, an antagonist useful in the treatment of certain types of cancer.

As a better understanding develops of the nature of the interactions between drug, metabolizing enzymes, and biologic receptor, selection of isosteric groups with particular electronic, solubility, and steric properties should permit the rational preparation of more selectively acting drugs. But, in the meanwhile, results obtained by the systematic application of the principles of isosteric replacement are aiding in the understanding of the nature of these receptors.

Steric Features of Drugs

Regardless of the ultimate mechanism by which the drug and the receptor interact, the drug must approach the receptor and fit closely to its surface. Steric factors determined by the stereochemistry of the receptor site surface and that of the drug molecules are, therefore, of primary importance in determining the nature and the efficiency of the drug–receptor interaction. With the possible exception of the general anesthetics, such drugs must possess a high degree of structural specificity to initiate a response at a particular receptor.

Some structural features contribute a high degree of structural rigidity to the molecule. For example, aromatic rings are planar, and the atoms attached directly to these rings are held in the plane of the aromatic ring. Hence, the quaternary nitrogen and carbamate oxygen attached directly to the benzene ring in the cholinesterase inhibitor, neostigmine, are restricted to the plane of the ring and, consequently, the spatial arrangement of at least these atoms is established.

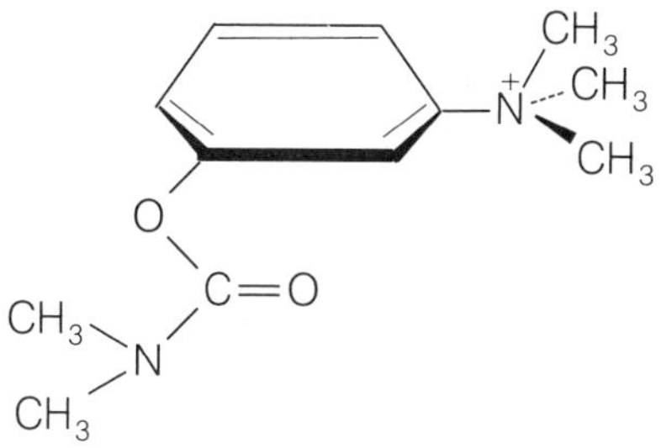

Neostigmine

The relative positions of atoms attached directly to multiple bonds are also fixed. For the double bond, *cis* and *trans* isomers result. For example, diethylstilbestrol exists in two fixed stereoisomeric forms. *trans*-Diethylstilbestrol is estrogenic, whereas the *cis*-isomer is only 7% as active. In *trans*-diethylstilbestrol, resonance interactions and minimal steric interference tend to hold the two aromatic rings and connecting ethylene carbon atoms in the same plane.

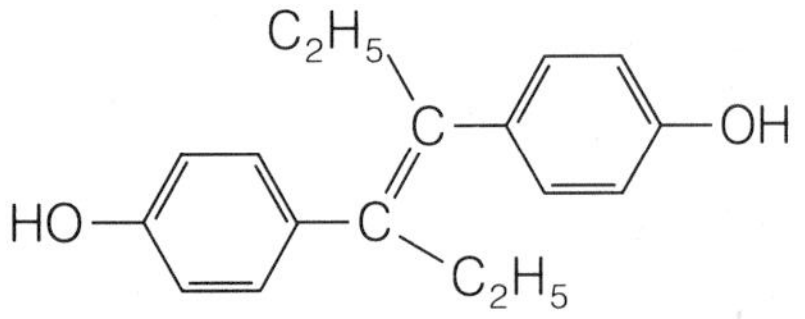

trans-Diethylstilbestrol

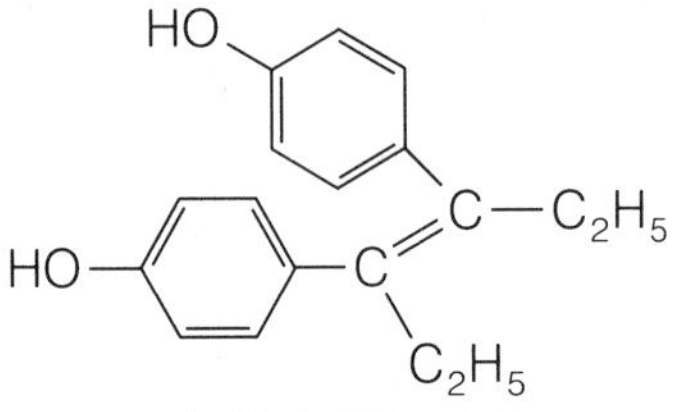

cis-Diethylstilbestrol

Geometric isomers, such as the *cis* and the *trans* isomers, hold structural features at different relative positions in space. These isomers also have significantly different physical and chemical properties. Therefore, their distributions in the biologic medium are different, as well as their capabilities for interacting with a biologic receptor in a structurally specific manner.

More subtle differences exist for *conformational* isomers. Similarly to geometric isomers, these exist as different arrangements in space for the atoms or groups in a single classic structure. Rotation about bonds allows interconversion of conformational isomers; however, an energy barrier between isomers is often sufficiently high for their independent existence and reaction. Differences in reactivity of functional groups, or interaction with biologic receptors, may be due to differences in steric requirements. In certain semirigid ring systems, such as the steroids, conformational isomers show significant differences in biologic activities (see Chap. 18). Methods for calculating these energy barriers will be discussed later.

The principles of conformational analysis have established some generalizations about the more stable structures for reduced (nonaromatic) ring systems. In the cyclohexane derivatives, bulky groups tend to be held approximately in the plane of the ring, the *equatorial* position. Substituents attached to bonds perpendicular to the general plane of the ring (*axial* position) are particularly susceptible to steric crowding. Thus, 1,3-diaxial substituents larger than hydrogen may repel each other, twisting the flexible ring and placing the substituents in the less crowded equatorial conformation.

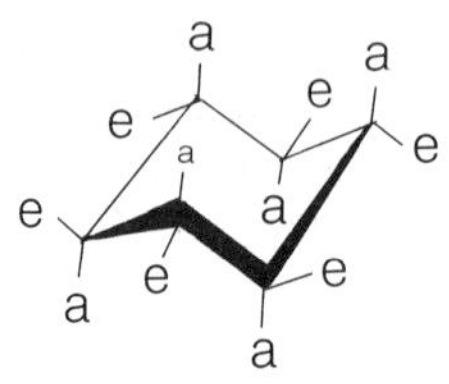

Equatorial (e) and *axial (a)* substitution in the chair form of cyclohexane.

Similar calculations may be made for reduced heterocyclic ring systems, such as substituted piperidines. Generally, an equilibrium mixture of conformers may exist. For example, the potent analgesic trimeperidine (see Chap. 17) has been calculated to exist largely in the form in which the bulky phenyl group is in the equatorial position, this form being favored by 7 kcal/mol over the axial species. The ability of a molecule to produce potent analgesia has been related to the relative spatial positioning of a flat aromatic nucleus, a connecting aliphatic or alicyclic chain, and a nitrogen atom, which exists largely in the ionized form at physiologic pH.[21] It might be expected that one of the conformers would be responsible for the analgesic activity; however, here, it appears that both the axially and the equatorially oriented phenyl group may contribute. In structurally related isomers the conformations of which are fixed by the fusion of an additional ring, both compounds in which the phenyl group is the axial and those in which it is in the equatorial position have equal analgesic potency.[22]

In a related study of conformationally rigid diastereoisomeric analogues of meperidine, the *endo*-phenyl epimer was more potent than was the *exo*-isomer.[23] However, the *endo*-isomer penetrated brain tissue more effectively because of slight differences in pK_a values and partition coefficients between the isomers. This emphasizes the importance of considering differences in physical properties of closely related compounds before interpreting differences in biologic activities solely on steric grounds and relative spatial positioning of functional groups.

Open chains of atoms, which form an important part of many drug molecules, are not equally free to assume all possible conformations, there being some that are sterically preferred.[24] Energy barriers to free rotation of the chains are present, owing to interactions of nonbonded atoms. For example, the atoms tend to position themselves in space such that they occupy staggered positions, with no two atoms directly facing (eclipsed). Thus, for butane at 37°, the calculated relative probabilities for four possible conformations show that the maximally extended *trans* form is favored 2 : 1 over the two equivalent bent (skew) forms. The *cis* form, in which all of the atoms are facing or *eclipsed*, is much hindered, and

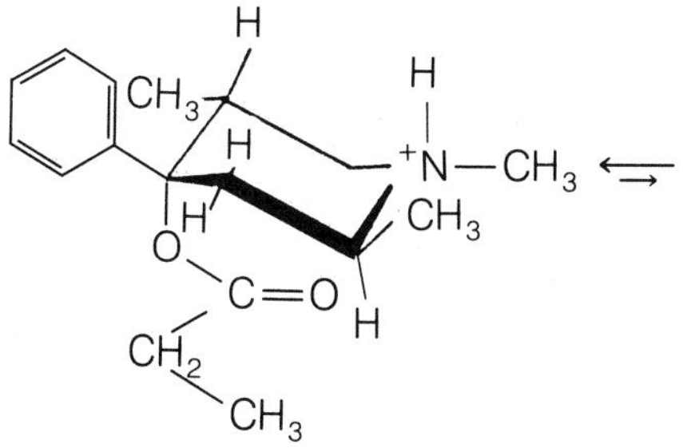

Trimeperidine (*equatorial*-phenyl)

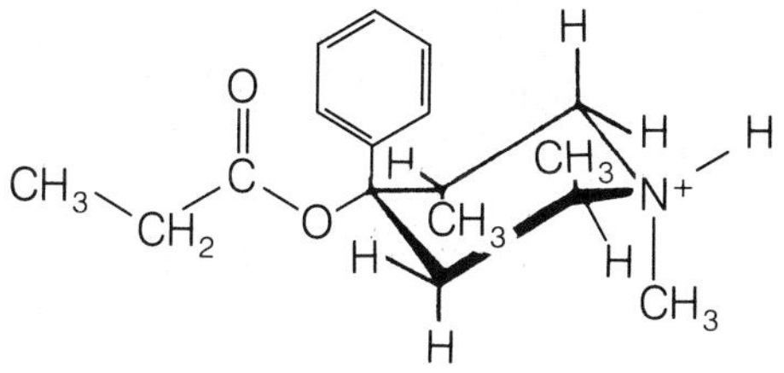

Trimeperidine (*axial*-phenyl)

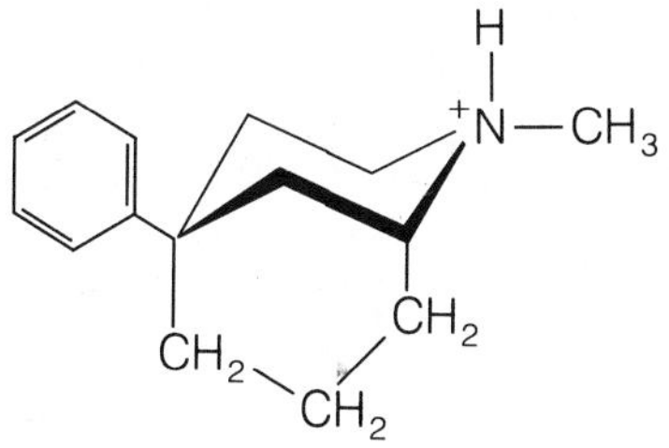

Equatorial-phenyl (analgesic ED_{50} 18.4 mg/kg)

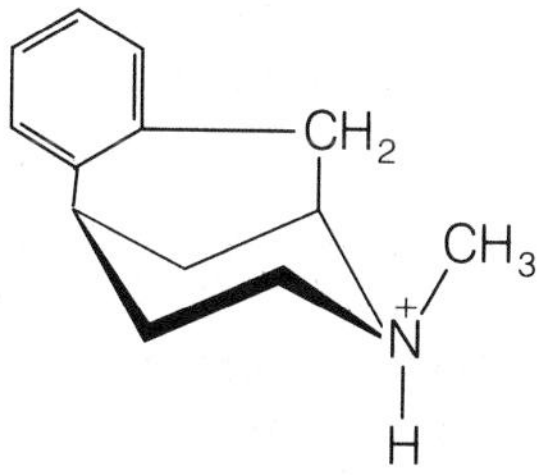

Axial-phenyl analgesic ED_{50} 18.7 mg/kg)

Ring-fused Analgesics

only about 1:1000 molecules may be expected to be in this conformation at normal temperatures.

Nonbonded interactions in polymethylene chains tend to favor the most extended *trans* conformations, although some of the partially extended *skew* conformations also exist. A branched methyl group reduces somewhat the preference for the *trans* form in that portion of the chain and, therefore, the probability distribution for the length of the chain is shifted toward the shorter distances. This situation is present in substituted chains that contain the elements of many drugs, such as the β-phenylethylamines. It should be noted that such amines are largely protonated at physiologic pH, and exist in a charged tetra-covalent form. Accordingly, their stereochemistry closely resembles that of carbon, although in the following diagrams, the hydrogen atoms attached to nitrogen are not shown. As may be expected, the fully extended *trans* form, with maximal separation of the phenyl ring and the nitrogen atom, is favored, and a smaller population of the two equivalent *skew* forms, in which the ring and the nitrogen are closer together, exists in solution. Introduction of an α-methyl group alters the favored position of the *trans* form, as positioning of the bulky methyl group away from the phenyl group (*skew* form 2) also results in a decrease in nonbonded interactions. Clearly, *skew* form 1 with both the methyl and the amine group close to phenyl is less favorable. The overall result is a reduction in the average distance between the aromatic group and the basic nitrogen atom in α-methyl-substituted β-arylethylamines. This steric factor influences the strength of the binding interaction with a biologic receptor required to produce a given pharmacologic effect. It is possible that the altered stereochemistry of α-methyl-β-arylethylamines may partially account for their slow rate of metabolic deamination (see Chap. 11).

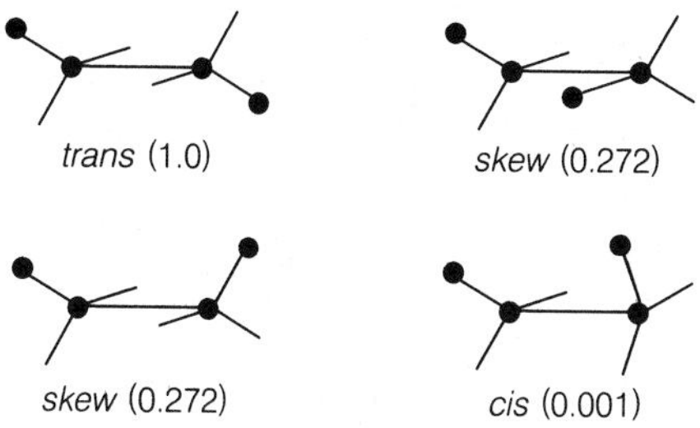

trans (1.0) *skew* (0.272) *skew* (0.272) *cis* (0.001)

Relative probabilities for the existence of conformations of butane

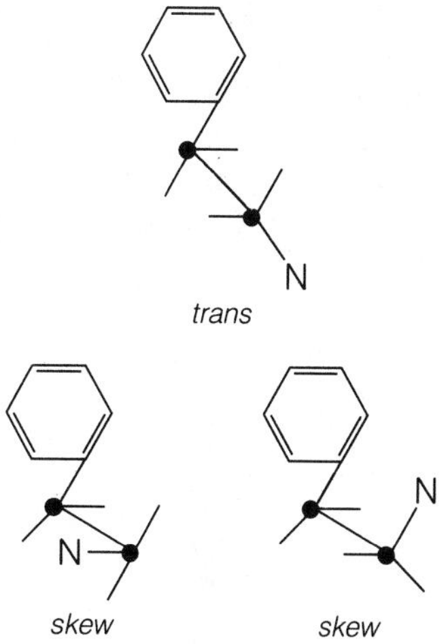

trans

skew *skew*

Conformantions of β-phenylethylamines

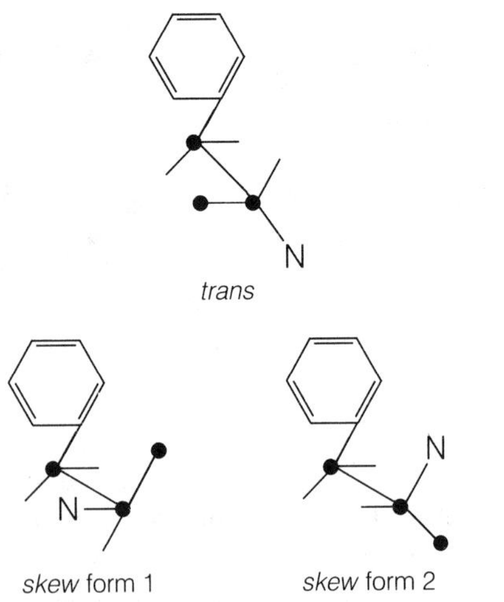

trans

skew form 1 *skew* form 2

Conformations of α-Methyl-β-phenylethylamines

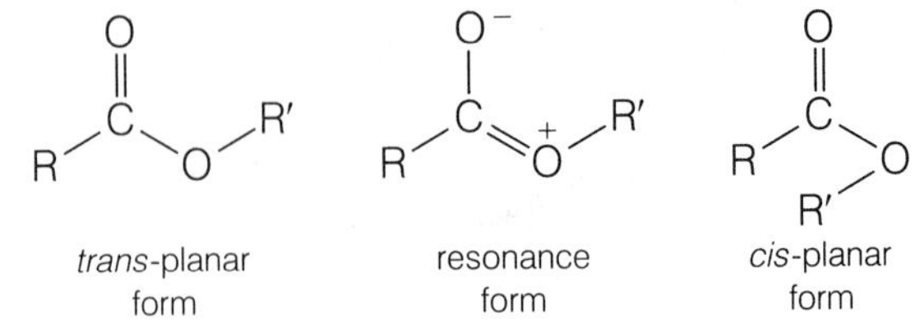

trans-planar form resonance form *cis*-planar form

Stabilizing planar structure of esters

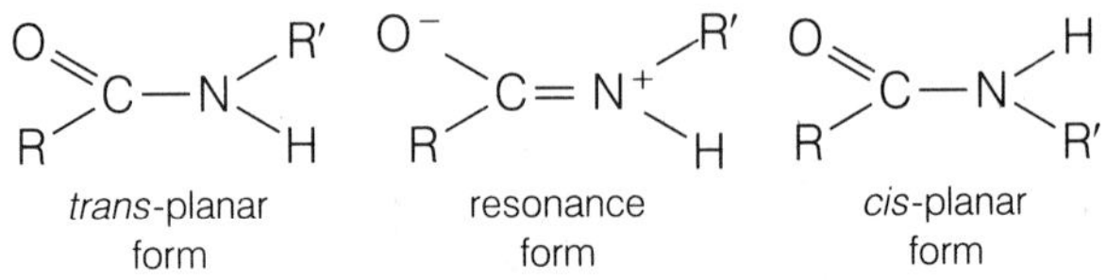

trans-planar form resonance form *cis*-planar form

Stabilizing planar structure of amides

The introduction of atoms other than carbon into a chain strongly influences the conformation of the chain. Because of resonance contributions of forms in which a double bond occupies the central bonds of esters and amides, a planar configuration is favored, in which minimal steric interference of bulky substituents occurs. Hence, an ester is mainly in the *trans*, rather than the *cis* form. For the same reason, the amide linkage is essentially planar, with the more bulky substituents occupying the *trans* position. Therefore, ester and amide linkages in a chain tend to hold bulky groups in a plane and to separate them as far as possible. As components of the side chains of drugs, ester and amide groups favor fully extended chains and, also, add polar character to that segment of the chain.

The foregoing considerations make it clear that the ester linkages in succinyl choline provide both a polar segment, which is readily hydrolyzed by plasma cholinesterase (see Chap. 12), and additional stabilization to the fully extended form. This form is also favored by repulsion of the positive charges at the ends of the chain.

The conformations favored by stereochemical considerations may be further influenced by *intramolecular interactions* between specific groups in the molecule. *Electrostatic forces*, involving attractions by groups of opposite charge, or repulsion by groups of like charge, may alter molecular size and shape. Consequently, the terminal positive charges on the polymethylene bis-quaternary ganglionic blocking agent, hexamethonium, and the neuromuscular-blocking agent, decamethonium, make it most likely that the ends of these molecules are maximally separated in solution.

$$(CH_3)_3\overset{+}{N}—(CH_2)_n—\overset{+}{N}(CH_3)_3$$

Hexamethonium $n = 6$

Decamethonium $n = 10$

In some cases *dipole–dipole interactions* appear to influence structure in solution. Methadone may exist partially in a cyclic form in solution, because of dipolar attractive forces between the basic nitrogen and carbonyl group.[25] In such a conformation, it closely resembles the conformationally more rigid potent analgesics, morphine, meperidine, and their analogues (see Chap. 17), and it may be this form that interacts with the analgesic receptor.

An intramolecular *hydrogen bond*, usually formed between donor —OH and =NH groups, and acceptor oxygen (:Ö=) and nitrogen (:N≡) atoms,

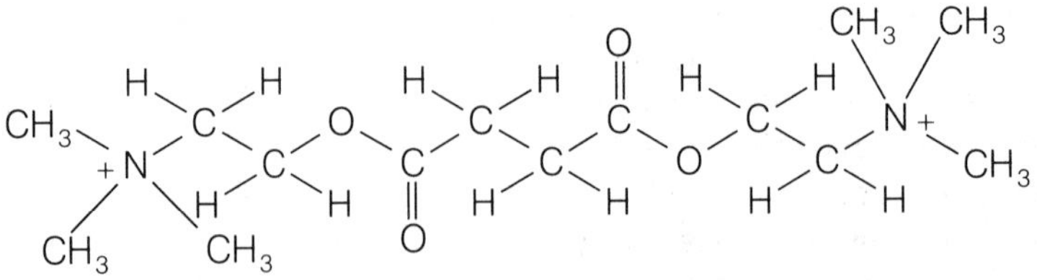

Extended form of succinyl choline

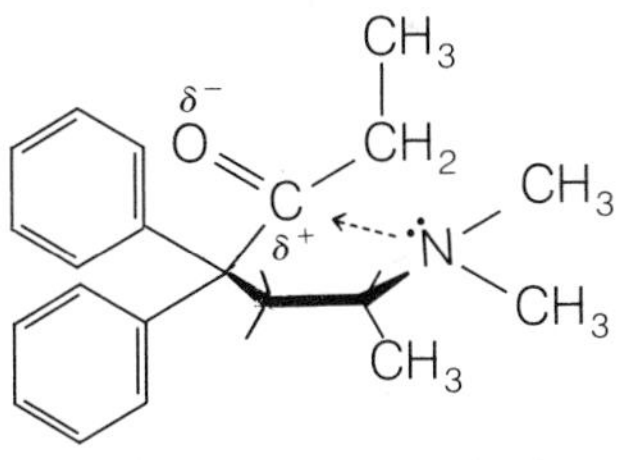

Ring conformation of methadone by dipolar interactions

might be expected to add stability to a particular conformation of a drug in solution. However, in aqueous solution donor and acceptor groups tend to be bonded to water, and little gain in free energy would be achieved by the formation of an intramolecular hydrogen bond, particularly if unfavorable steric factors involving nonbonded interactions were introduced in the process. Therefore, it is likely that internal hydrogen bonds play only a secondary role to steric factors in determining the conformational distribution of flexible drug molecules.

Conformational Flexibility and Multiple Modes of Action

It has been proposed that the conformational flexibility of most open-chain neurohormones, such as acetylcholine, epinephrine, serotonin, and related physiologically active biomolecules, such as histamine, permits multiple biologic effects to be produced by each molecule, by virtue of the ability to interact in a different and unique conformation with different biologic receptors. Thus, it has been suggested that acetylcholine may interact with the muscarinic receptor of postganglionic parasympathetic nerves and with acetylcholinesterase in the fully extended conformation and, in a different, more-folded structure, with the nicotinic receptors at ganglia and at neuromuscular junctions.[26,27] Acetylcholine bromide exists in a quasi-ring form in the crystal, with an *N*-methyl hydrogen atom close to, and perhaps forming a hydrogen bond with, the backbone oxygen.[28] In solution, however, it is able to assume a continuous series of conformations, some of which are energetically favored over others.[27]

Conformationally rigid acetylcholinelike molecules have been used to study the relationships between these various possible conformations of acetylcholine and their biologic effects. (+)-*trans*-2-Acetoxycyclopropyl trimethylammonium iodide, in which the quaternary nitrogen atom and acetoxyl groups are held apart in a conformation approximating that of the extended conformation of acetylcholine, was about five times more active than acetylcholine in its muscarinic effect on dog blood pressure, and equiactive to acetylcholine in its muscarinic effect on the guinea pig ileum.[29] The (+)-*trans*-isomer was hydrolyzed by acetylcholinesterase at a rate equal to the rate of hydrolysis of acetylcholine. It was inactive as a nicotinic agonist. In contrast, the (−)-*trans*-isomer and the mixed (+),(−)-*cis*-isomers were 1/500 and 1/10,000 as active as acetylcholine in muscarinic tests on guinea-pig ileum and were inactive as nicotinic agonists. Similarly, the *trans*-diaxial relationship between the quaternary nitrogen and acetoxyl group led to maximal muscarinic response and rate of hydrolysis by true acetylcholinesterase in a series of isomeric 3-trimethylammonium-2-acetoxyldecalins.[30] These results could be interpreted that either acetylcholine was acting in a *trans* conformation at the muscarinic receptor, and was not acting in a *cisoid* conformation at the nicotinic receptor, or that the nicotinic response is highly sensitive to steric effects of substituents being used to orient the molecule.

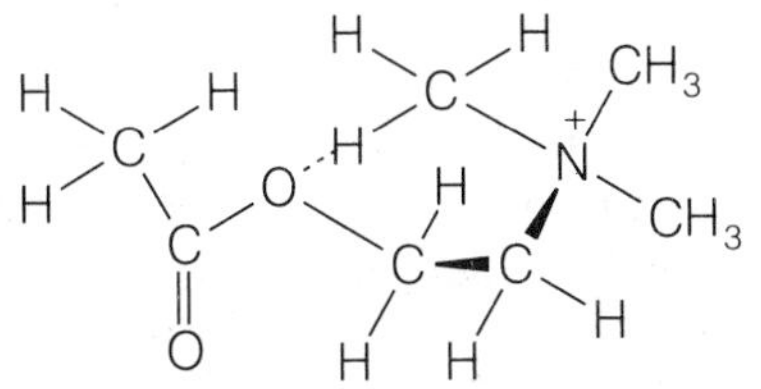

Quasi-ring form of acetylcholine

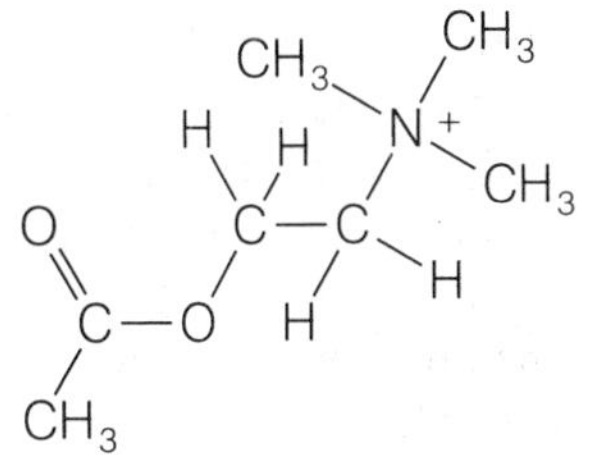

Extended conformation of acetylcholine

In contrast with the concept of acetylcholine reacting with muscarinic and nicotinic receptors in different conformations, Chothia[31] has proposed that acetylcholine interacts in the same conformation, but in a different manner, with each receptor. The conformations of acetylcholine (Fig. 2-8) are primarily defined by rotations about the $C_\alpha—C_\beta$ and $C_\beta—O_1$ bonds, because the $C_1—N—C_\alpha—C_\beta$ sequence and the $O_2—C_4—O_1—C_\beta$ ester group exist largely in planar conformations owing to steric and resonance factors. Acetylcholine and several selective muscarinic and nicotinic agents have been shown to be in closely similar conformations in the

trans-2-Acetoxycyclopropyl Trimethylammonium Iodide

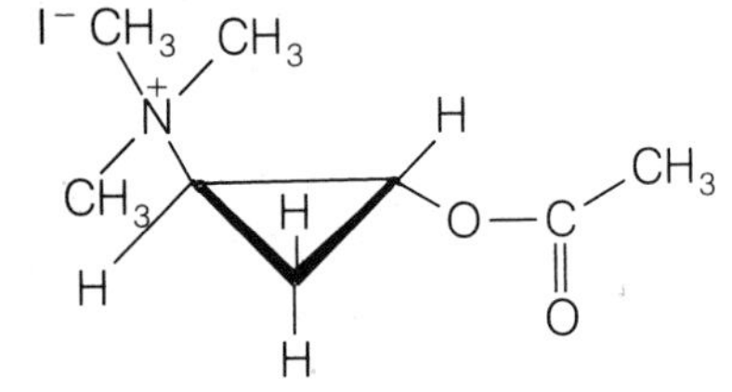

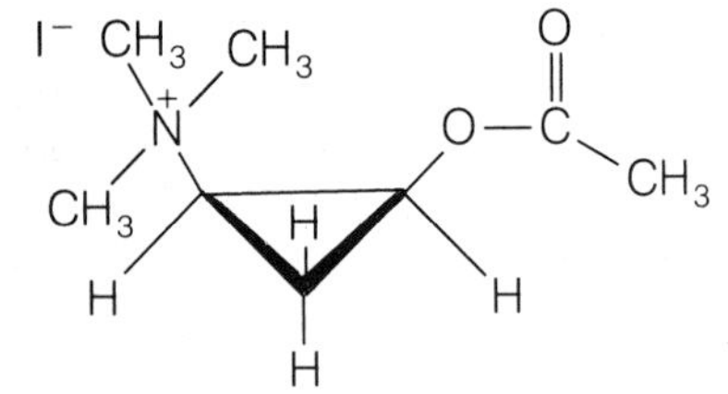

cis-2-Acetoxycyclopropyl Trimethylammonium Iodide

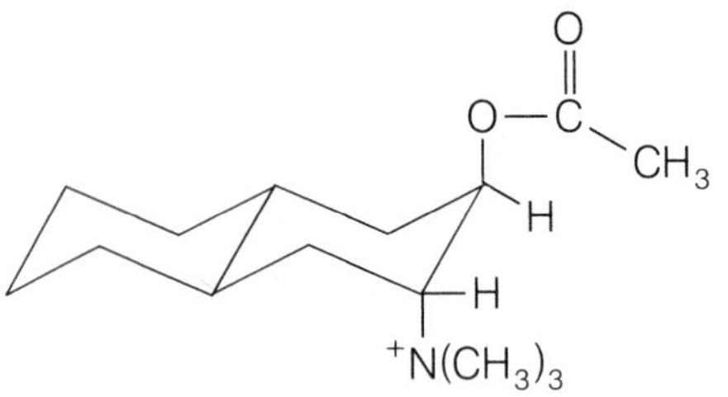

trans-diaxial 3-Trimethylammonium-2-acetoxydecalin

FIG. 2-8. *Acetylcholine conformation and receptor specificity.*

crystal state.[32] In these compounds the $C_\alpha—C_\beta$ bond, or its equivalent, is rotated so that the N and O_1 (ether oxygen) are about 60° to 75° from the *cis* coplanar conformation. The $C_\alpha—C_\beta—O_1—C_4$ atoms are essentially in a *trans* planar extended chain. This conformation presents a methyl side, defined by a plane close to C_2, O_1, and C_5 (methyl carbon), and a carbonyl side, defined by a plane close to C_3, C_β, and O_2 (carbonyl oxygen). Compounds with high muscarinic and low nicotinic activity, such as *trans*-2-acetoxycyclopropyl trimethylammonium iodide, L-(+)-acetyl-β-methylcholine, and muscarine, show structures in the crystal state that have free access to their methyl sides, whereas their carbonyl sides are blocked by the spatial position occupied by the extra methyl or methylene groups. In preferential nicotinic agonists, such as L-(+)-acetyl-α-methylcholine, the carbonyl side is exposed, and access to the methyl side is blocked. Chothia[31] has proposed that the methyl sides of acetylcholine and its predominantly muscarinic analogues interact with the muscarine receptor, whereas it is the interaction with groups on the carbonyl side of acetylcholine and its nicotinic analogues that activates the nicotinic receptor.

With use of an approach that focuses on the parent molecule, rather than on conformationally fixed analogues, molecular orbital calculations have indicated that histamine may exist in two extended conformations (A, B) of nearly equal and minimal energy[33] rather than the earlier predicted coiled form (C) involving intramolecular hydrogen bonds.[34] In one extended conformation (A), one imidazole ring nitrogen atom is about 4.55 Å from the side chain nitrogen, whereas in conformation B this distance is about 3.60 Å. Histamine receptors have been differentiated into at least two classes, there being different structural requirements for stimulation of smooth muscle, such as the guinea pig ileum (histamine H_1-receptor, blocked by classic antihistamines), and for the stimulation of secretion of gastric acid (histamine H_2-receptor, not blocked by classic antihistamines). It is proposed, on the basis of the internitrogen distance of closest approach of 4.8 ± 0.2 Å for the relatively rigid antihistamine triprolidine, that histamine acts on smooth muscle (H_1-receptor) in conformation A, in which the intranitrogen distance of 4.55 Å closely approximates the spacing found in the specific antagonist. It is further presumed that the histamine-induced release of gastric acid may be brought about by a histamine H_2-receptor interaction in an alternate conformation of closer internitrogen spacing, such as conformation B.

The histamine H_2-receptor antagonist,[35] cimetidine (see Chap. 16) contains uncharged polar residues on the side chain, such as the thio-

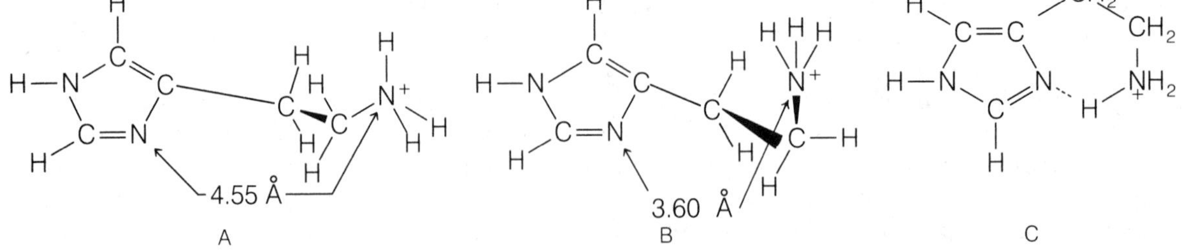

4.55 Å — A; 3.60 Å — B; C

Conformations of Histamine

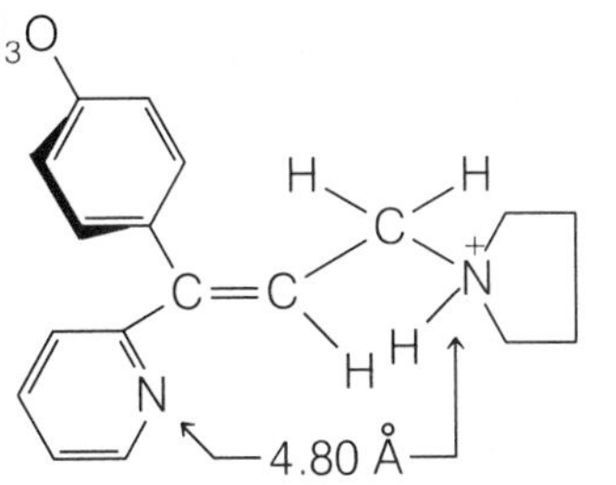

CH_3O ... H, C, H, N, C=C, H, H, N, 4.80 Å

Triprolidine (antihistamine)

urea (—NH—CS—NH—) or *N*-cyanoguanidine [—NH—C(=NCN)NH—] groups. These polar residues are separated from the imidazole ring by chains four-atoms long, two-atoms longer than the dimethylene side chain of histamine. It appears likely that H_2-antagonist activity results from the interaction of the side chain and its polar residues with a receptor region distinct from that with which the positively charged side chain of histamine interacts.

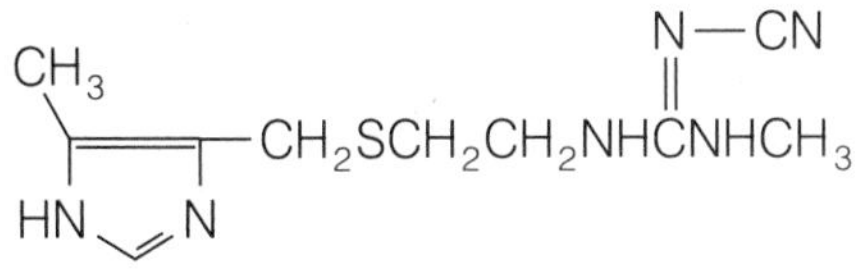

CH_3, HN, N, $CH_2SCH_2CH_2NHCNHCH_3$, N—CN

Cimetidine

Optical Isomerism and Biologic Activity

The widespread occurrence of differences in biologic activities for *optical isomers* has been of particular importance in the development of theories on the nature of drug–receptor interactions. *Diastereoisomers*, compounds with two or more asymmetric centers, have the same functional groups and, therefore, can undergo the same types of chemical reactions. However, the diastereoisomers (e.g., ephedrine, *pseudo*ephedrine, see Chap. 11) have different physical properties, undergo different rates of reactions, have substituent groups that occupy different relative positions in space, and the different biologic properties shown by such isomers may be accounted for by the influence of any of these factors on drug distribution, metabolism, or interaction with the drug receptor.

However, *optical enantiomers*, also called *optical antipodes* (mirror images) present a very different situation, for they are compounds the physical and chemical properties of which are usually considered identical, except for their ability to rotate the plane of polarized light. Here one might expect the compounds to have the same biologic activity. However, such is not representative of many of the enantiomers that have been investigated.

As examples of compounds the optical isomers of which show different activities, the following may be cited: (−)-hyoscyamine is 15 to 20 times more active as a mydriatic than (+)-hyoscyamine; (−)-hyoscine is 16 to 18 times as active as (+)-hyoscine; (−)-epinephrine is 12 to 15 times more active as a vasoconstrictor than (+)-epinephrine; (+)-norhomoepinephrine is 160 times more active as a pressor than (−)-norhomoepinephrine; (−)-synephrine has 60 times the pressor activity of (+)-synephrine; (−)-amino acids are either tasteless or bitter, whereas (+)-amino acids are sweet; (+)-ascorbic acid has good antiscorbutic properties, whereas (−)-ascorbic acid has none.

Although it is well established that optical antipodes have different physiologic activities, there are different interpretations of why this is so. Differences in distribution of isomers, without considering differences in action at the receptor site, could account for different activities for optical isomers. Diastereoisomer formation with optically active components of the body fluids (e.g., plasma proteins) could lead to differences in absorption, distribution, and metabolism. Distribution could also be affected by preferential metabolism of one of the optical antipodes by a stereospecific enzyme (e.g., D-amino acid oxidase). Preferential adsorption could also occur at a stereospecific site of loss (e.g., protein binding). Cushny[36] accounted for this difference by assuming that the optical antipodes reacted with an optically active receptor site to produce diastereoisomers with different physical and chemical properties. Easson and Stedman,[37] taking a somewhat different view, point out that optical antipodes can, in theory, have different physiologic effects for the same reason that structural isomers can have different effects (i.e., because of different molecular arrangements, one antipode can react with a hypothetical receptor, whereas the other cannot). Assuming a receptor in tissues to which a drug can be attached and have activity only if the complementary parts B, D, C are superimposed, it is apparent that of the two enantiomers, only I can be so superimposed. Under these conditions, I therefore would be active, and II would show no activity. This interpretation,

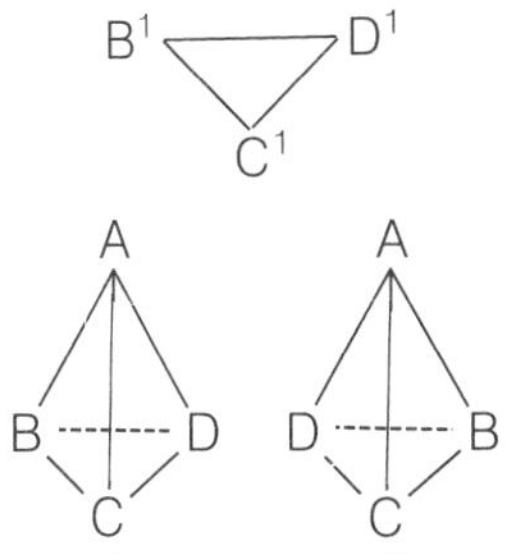

B^1 D^1 C^1

A, B, D, C — I

A, D, B, C — II

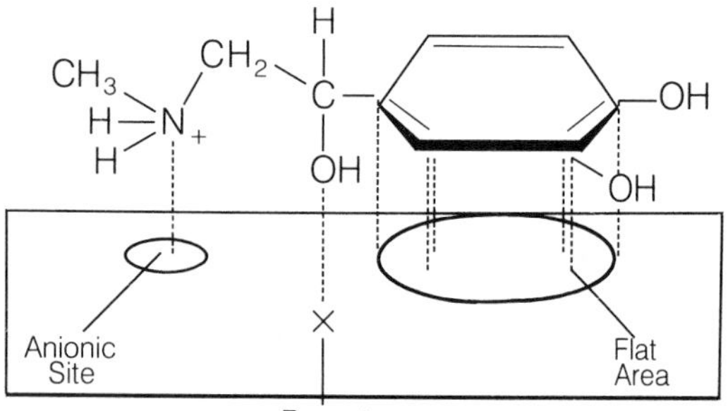

(−)-Epinephrine — more active

in a sense, is not greatly different from that given by Cushny, because the receptor has a unique configuration, not much different from that of an optically active compound. Both theories demand a structure of unique configuration in the body, but in the one theory, only one enantiomer reacts, whereas, in the other, they both react, with one combination having greater biologic activity than the other.

Easson and Stedman[37] have also postulated that the optical antipodes of epinephrine owe their differences in activity to a difference in ease of attachment to the receptor surface. This is illustrated below for the pressor activity of (−)- and (+)-epinephrine.[38]

Thus, only in (−)-epinephrine can the three groups essential for maximal pressor activity in sympathomimetic amines—the positively charged nitrogen, the aromatic ring, and the alcoholic hydroxyl group—attach to the complementary receptor surface. In the (+)-isomer, any two binding groups may orient to attach, but not all three. This is consistent with the observation[39] that deoxyepinephrine, which lacks the alcoholic hydroxyl and, therefore, may only bind in two positions, has about the same pressor effect as (+)-epinephrine.

CALCULATED CONFORMATIONS

It should now be obvious that it is important for medicinal chemists to obtain an accurate understanding of the conformation of the drug molecule.

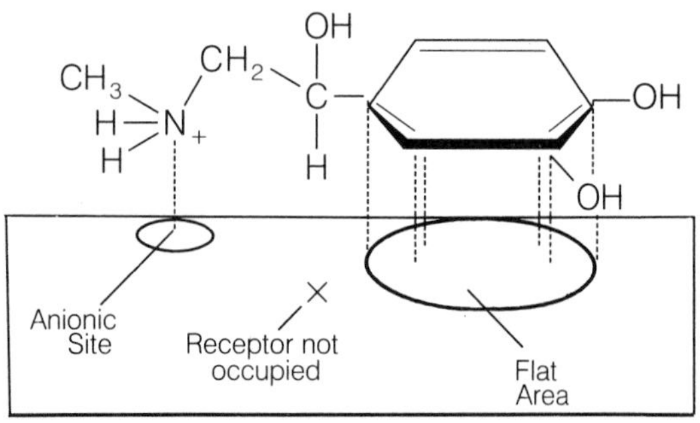

(+)-Epinephrine — less active

Originally molecular models were constructed from kits containing a variety of atoms of different valence and oxidation states. Hence, there would be carbons suitable for carbon–carbon single, double, and triple bonds; carbon–oxygen bonds for alcohols or ethers and the carbonyl moiety; carbon–nitrogen bonds for amines, amides, imines, and nitriles; and carbons for three-, four-, five-, and larger-membered rings. More complete sets include a variety of heteroatoms, including nitrogen, oxygen, and sulfur of various oxidation states. These kits might be ball and stick, stick or wire only, or space filling. The latter contain attempts at realistically visualizing the effect of a larger atom, such as sulfur, relative to the smaller oxygen. The diameters of the atoms in these kits are proportional to the van der Waal radii, usually corrected for overlap effects. In contrast the wire models usually depict accurate intra-atomic distances between distances. A skilled chemist using these kits usually can obtain a reasonably accurate three-dimensional representation. This is particularly true if it is a moderately simple molecule, with considerable rigidity. An extreme example is a steroid with the relatively inflexible fused ring system. In contrast, molecules with chains consisting of several atoms can assume many shapes. Yet, only one shape or conformation can be expected to fit onto the receptor.

There are now three quantitative ways to obtain estimations of preferred molecular shapes required for a good fit at the receptor. The first, which is the oldest and is considered to be the most accurate, is x-ray crystallography. When properly done, resolution down to a few angstrom units can be obtained. This permits an accurate mathematical description of the molecule, providing atomic coordinates in three-dimensional space that can be drawn with a chemical graphics program. A serious limitation of this technique is the requirement for a carefully grown crystal. Some chemicals will not form crystals. Others form crystals with mixed symmetries. Nevertheless, with the newer computational techniques, including high-speed computers, large data bases of x-ray crystallographic data are now available. These data bases can be searched looking for structures, including substructures, similar to the molecule of interest. Depending on how close the match is, it is possible to obtain a fairly good idea of a low-energy conformation of the drug molecule.

Because of the drawbacks to x-ray crystallography, two purely computational methods that require only a knowledge of the molecular structure are utilized. The two approaches are known as quantum mechanics and molecular mechanics. Both are based on assumptions that (1) a molecule's three-dimensional geometry is a function of the forces acting on the molecule and (2) that these forces can be expressed by a set of equations that pertain to all molecules. For the most part, both computational techniques assume that the molecule is in an isolated system in a vacuum. Solvation effects from water, which is common to any biologic system, tend to be ignored, although this is changing. Calculations now can include limited numbers of water molecules, the number dependent on the length of available computer time. Interestingly, many crystals grown for x-ray analysis can contain water in the crystal lattice.

There are fundamental differences between the approaches of quantum and molecular mechanics. They illustrate the dilemma that can confront a scientist. Quantum mechanics is derived from basic theoretical principles. The model, itself, is exact, but the equations used in the technique are only approximate. The molecular properties are derived from the electronic structure of the molecule. The assumption is made that the distribution of electrons within a molecule can be described by a linear sum of functions that represent an atomic orbital. (For carbon this would be s, p_x, p_y, etc). Quantum mechanics is computation-intensive, with the calculation time for obtaining an approximate solution increasing by about N^4 times, where N is the number of such functions. Until the advent of the high-speed super computers, quantum mechanics in its *pure* form was restricted to small molecules. In other words, it was not practical to conduct a quantum mechanical analysis of a drug molecule.

To make the technique more practical, simplifying techniques have been developed. Although the computing time is decreased, the accuracy of the outcome is also lessened. In general, the quantum mechanics type of calculations in medicinal chemistry is a method that is *still waiting to happen*. It is being used by those laboratories with access to large-scale computing, but there is considerable debate on its usefulness because so many simplifying approximations need to be made for larger molecules.

In contrast, medicinal chemists are embracing molecular mechanics. This approach is derived from empirical observations. In contrast with quantum mechanics, the equations in molecular mechanics have exact solutions. At the same time, it must be realized that the parameters that are used in these equations are adjusted to ensure that the outcome fits experimental observations. In place of the fundamental electronic structure used in quantum mechanics, molecular mechanics uses a model consisting of balls (the atoms) connected by springs (the bonds). The total energy of a molecule consists of the sum of the following energy terms:

E_c: stretching and compressing of the bonds (springs)
E_b: bending about a central atom
E_t: rotation about bonds
E_v: van der Waals interactions
E_u: electrostatic interactions

Each atom is defined (parametrized) in terms of these energy terms. What this means is that the validity of molecular mechanics is dependent on the accuracy of the parameterization process. From a historical point, saturated hydrocarbons have proved easy to parameterize, followed by selective heteroatoms, such as ether oxygens, amines, and such. Unsaturated systems, including aromaticity, caused problems because of the delocalization of the electrons, but this seems to have been solved. Charged atoms, such as the carboxylate anion and protonated amine, can prove to be a real problem particularly if the charge is delocalized. Nevertheless, molecular mechanics increasingly is being used by medicinal chemists to gain a better understanding of the preferred conformation of drug molecules and the macromolecules that compose a receptor. The computer programs are readily available and run on relatively inexpensive computers.

The only way to test the validity of the outcome from either quantum or molecular mechanics calculations is to compare the calculated solution with actual experimental data. Obviously, crystallographic data would provide a reliable measure of the accuracy of least one of the low-energy conformers. As that is not always feasible, other physicochemical measurements are used for comparison. These include comparing calculated vibrational energies, heats of formation, dipole moments, and relative conformational energies, with measured values. When results are inconsistent, the parameter values are adjusted. This readjustment of the parameters is analogous to the fragment approach for calculating octanol–water partition coefficients. The values for the fragments and the accompanying correction factors are determined by comparing calculated partition coefficients with a large population of experimentally determined partition coefficients.

A couple of examples will help explain how these computational methods are used to obtain an understanding of the drugs conformation. Examine the conformations of butane and β-phenethylamines. As the carbons rotate, the substituents will become eclipsed as they pass in front of each other. These are such high-energy conformations that few of the molecules will normally exist in this form. The situation becomes much more complicated with increasing numbers of atoms in the chains. Notice, the two conformations of acetylcholine. Because of intramolecular interactions, a quasi-ring conformation is suggested. A similar situation is shown in the illustration for methadone. Both quantum mechanics and molecular mechanics are used to calculate the energy required to move from one conformation to another. Figure 2-9 is a hypothetical plot for a

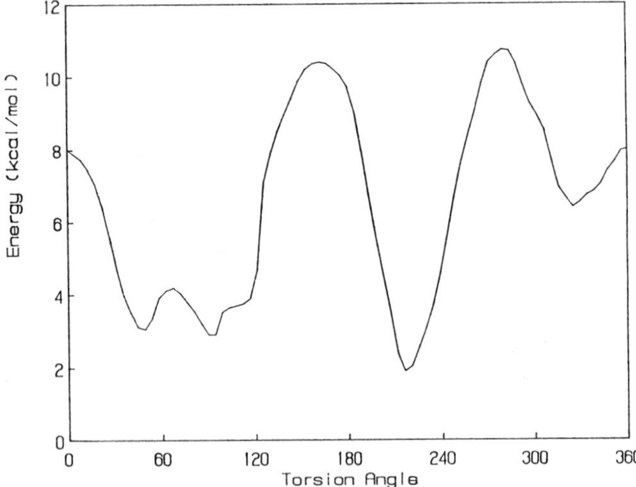

FIG. 2-9. *Diagram showing the energy maxima and minima as two substituted carbons connected by a single bond are rotated 360° relative to each other.*

complex molecule containing a chain with several single bonds made up of substituted carbons. The plot is for only one of the bonds. A similar type of drawing would be possible for each bond. The programs are designed to rotate about only one single bond at a time, holding the others in a constant conformation. What is the best conformation for the other bonds when carrying out the rotation on the bond in question? This will be based on information available for related substructures and commonsense. If a thorough analysis is to be done, the low-energy conformation should be determined for each bond. Contour-type plots on which the energy of the torsion angle for one bond is plotted against the energy of the torsion angle of a second bond are common. It is not uncommon to see several minimums, which are shown in Figure 2-9. Which one does the receptor prefer? This can be difficult to answer because occasionally there is a relatively low-energy barrier separating one conformation from another. Sometimes increased activity is seen by introduction of a small substituent, such as a methyl group, that can *lock* a chain into a preferred conformation. This has been suggested for a series of phenylpropionic acids with anti-inflammatory activity.[40]

THE DRUG–RECEPTOR INTERACTION AND SUBSEQUENT EVENTS

Once bound at the receptor site, drugs may act, either to initiate a response (*stimulant* or *agonist* action), or to decrease the activity potential of that receptor (*antagonist* action) by blocking access to it by active molecules. The chain of events leading to an observable biologic response must be initiated in some fashion by either the process of formation or the nature of the drug–receptor complex. Current theories on the mechanism of action of drugs at the receptor level are based primarily on the studies of Clark[41] and Gaddum,[42] whose work supports the assumption that the tissue response is proportional to the number of receptors occupied. The "occupancy theory" of drug action has been modified by Ariëns[43] and Stephenson,[44] who have divided the drug–receptor interaction into two steps: (1) combination of drug and receptor and (2) production of effect. Thus, any drug may have structural features that contribute independently to the *affinity* for the receptor and to the efficiency with which the drug–receptor combination initiates the response (*intrinsic activity* or *efficacy*). The Ariëns–Stephenson concept retains the assumption that the response is related to the number of drug–receptor complexes.

In the Ariëns–Stephenson theory, both agonist and antagonist molecules possess structural features that would enable formation of a drug–receptor complex (strong affinity). However, only the agonist possesses the ability to cause a stimulant action (i.e., possesses intrinsic activity). The affinity of a drug may be estimated by comparison of the dose required to produce a pharmacologic response with the dose required by a standard drug. Thus, acetylcholine produces a normal *S*-shaped curve if the logarithm of the dose is plotted against the percentage contraction of the rat jejunum (a segment of the small intestine). A series of related alkyl trimethylammonium salts (ethoxyethyl trimethylammonium, pentyl trimethylammonium, propyl trimethylammonium; Fig. 2-10) are able to produce the same degree of contraction of the tissue as does acetylcholine, but higher doses are required. The

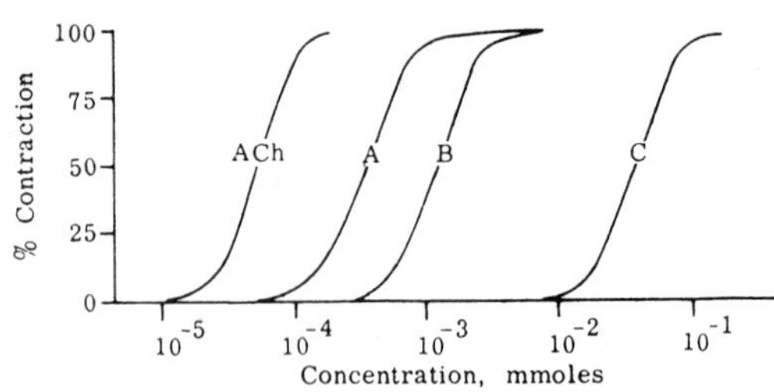

FIG. 2-10. *Dose–response curves for contraction produced by acetylcholine (ACh) and alkyltrimethylammonium salts on the rat jejunum.*

A. $CH_3CH_2OCH_2CH_2\overset{+}{N}Me_3$

B. $CH_3CH_2CH_2CH_2CH_2\overset{+}{N}Me_3$

C. $CH_3CH_2CH_2\overset{+}{N}Me_3$.

(Modified from Ariëns, E. J., Simonis, A. M.: J. Pharm. Pharmacol. 16:137, 289, 1964.)

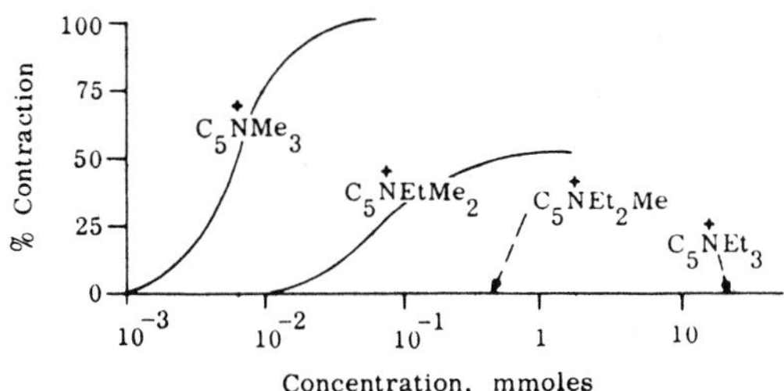

FIG. 2-11. *Dose-response curves for contraction produced by pentyl trialkylammonium salts on the rat jejunum. (Modified from Ariëns, E. J., Simonis, A. M.: J. Pharm. Pharmacol. 16:137, 289, 1964.)*

shape of the dose-response curve is the same, but the series of parallel curves are shifted to higher dose levels. Therefore, the alkyl trimethylammonium compounds are said to possess the same intrinsic activity as acetylcholine, being able to produce the same maximal response, but to show a lower affinity for the receptor, as larger amounts of drug are required.

By contrast, structural change of a molecule can lead to a gradual decline in the maximal height and slope of the log dose-response curves (Fig. 2-11) in which situation the loss in activity may be attributed to a decline in intrinsic activity. For example, pentyl trimethylammonium ion is able to produce a full acetylcholinelike contraction. Successive substitution of methyl by ethyl groups (pentylethyl dimethylammonium, pentyldiethyl methylammonium, pentyl triethylammonium) leads to successive decreases in the maximal effect obtainable, with pentyl triethylammonium ion producing no observable contraction. The loss in acetylcholinelike activity for pentyl triethylammonium ion is apparently caused by a loss in intrinsic activity, without a significant decrease in the affinity for the receptor because the compound acts as a competitive inhibitor (antagonist) for active derivatives of the same series.

For an antagonist, it is desirable to have high affinity and low or zero intrinsic activity—that is, to bind firmly to the receptor, but to be devoid of activity. Many examples are available for which structural modifications of an agonist molecule lead successively to compounds with decreasing agonist and increasing antagonist activity. Such modifications on acetylcholinelike structures, usually by addition of bulky nonpolar groups to either end (or both ends) of the molecule, may lead to the complete antagonistic activity found in the parasympatholytic compounds (e.g., atropine) discussed in Chapter 12.

In contrast with the occupancy theory, Croxatto[45] and Paton[46] have proposed that excitation by a stimulant drug is proportional to the *rate* of drug–receptor combination, rather than to the number of receptors occupied. The *rate theory* of drug action proposes that the rate of association and dissociation of an agonist is rapid, and this leads to the production of numerous impulses per unit time. An antagonist, with strong receptor-binding properties, would have a high rate of association but a low rate of dissociation. The occupancy of receptors by antagonists, assumed to be a nonproductive situation, prevents the productive events of association by other molecules. This concept is supported by the fact that even blocking molecules are known to cause a brief stimulatory effect before blocking action develops. During the initial period of drug–receptor contact when few receptors are occupied, the rate of association would be at a maximum. When a substantial number of sites are occupied, the rate of association would fall below the level necessary to evoke a biologic response.

The *occupation* and the *rate* theories of drug action do not provide specific models at the molecular level to account for a drug acting as agonist or antagonist. The *induced-fit* theory of enzyme–substrate interaction,[47] in which combination with the substrate induces a change in conformation of the enzyme, leading to an enzymatically active orientation of groups, provides the basis for similar explanations of mechanisms of drug action at receptors. Assuming that protein constituents of membranes play a role in regulating ion flow, it has been proposed[48] that acetylcholine may interact with the protein and alter the normal forces that stabilize the structure of the protein, thereby producing a transient rearrangement in the membrane structure and a consequent change in its ion-regulating properties. If the structural change of the protein led to a configuration in which the stimulant drug was bound less firmly and dissociated, the conditions of the *rate* theory would be satisfied. A drug–protein combination that did not lead to a structural change would result in a stable binding of the drug and a blocking action.

A related hypothesis (the *macromolecular perturbation theory*) of the mode of acetylcholine action at the muscarinic (postganglionic parasympathetic) receptor has been advanced by Belleau.[49] It is proposed that interaction of small molecules (substrate or drug) with a macromolecule (such as the protein of a drug receptor) may lead either to *specific conformational perturbations* (SCP) or to *nonspecific conformational perturbations* (NSCP). A SCP (specific change in structure or conformation of a protein molecule) would result in the specific response of an agonist (i.e., the drug receptor would possess intrinsic activity). If a NSCP occurs, no stimulant response would be obtained, and an an-

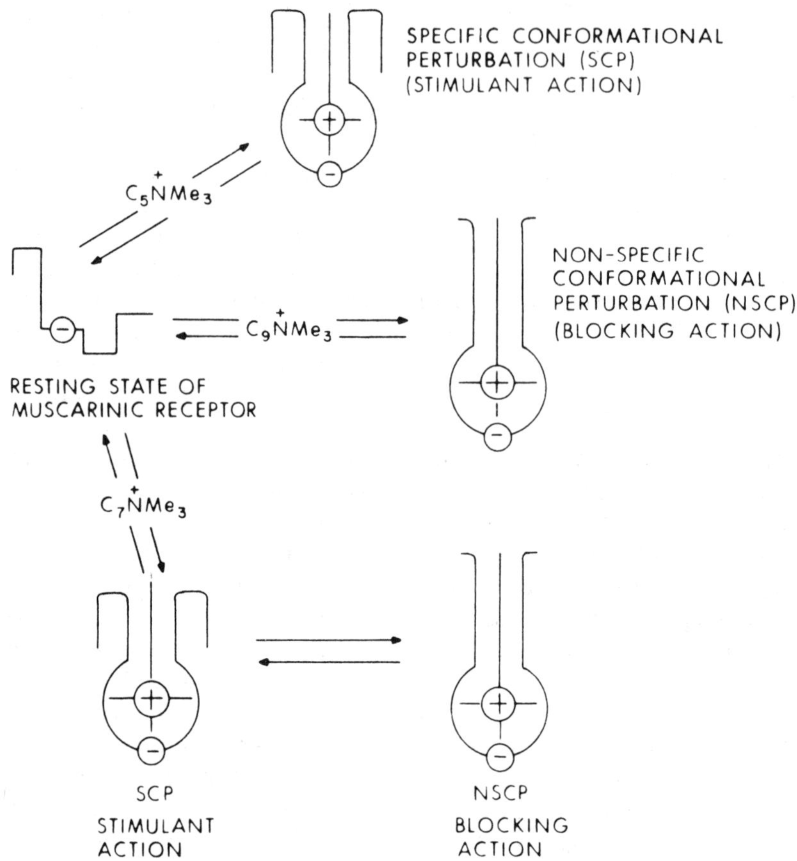

FIG. 2-12. Schematic representation of alkyl trimethylammonium ions reacting with the muscarinic receptor. (Modified from Belleau, B.: J. Med. Chem. 7:776, 1964.)

tagonistic or blocking action may be produced. If a drug possesses features that contribute to formation of both a SCP and a NSCP, an equilibrium mixture of the two complexes may result, which would account for a partial stimulant action.

The alkyl trimethylammonium ions ($R—\overset{+}{N}Me_3$), in which the alkyl group, R, is varied from 1 to 12 carbon atoms, provide a homologous series of muscarinic drugs that serve as models for the macromolecular perturbation theory of events that may occur at the drug receptor. With these simple analogues, hydrophobic forces, in addition to ion-pair formation, are considered to be the most important in contributing to receptor binding. Lower alkyl trimethylammonium ions (C_1 to C_6) stimulate the muscarinic receptor and are considered to possess a chain length that is able to form a hydrophobic bond with nonpolar regions of the receptor, altering receptor structure in a specific perturbation (Fig. 2-12; e.g., $C_5\overset{+}{N}Me_3$). With a chain of 8 to 12 carbon atoms, the antagonistic action observed is considered to result from a nonspecific conformational perturbation (NSCP) of a network of nonpolar residues at the periphery of the catalytic surface (see Fig. 2-12; e.g., $C_9\overset{+}{N}Me_3$). The intermediate heptyl and octyl derivatives act as partial agonists, and it is considered that they may form an equilibrium mixture of drug–receptor combinations, with both active SCP forms and inactive NSCP forms present (see Fig. 2-12; e.g., $C_7\overset{+}{N}Me_3$).

REFERENCES

1. Crum-Brown, A., Fraser, T.: R. Soc. Edinburgh 25:151, 1868–1869.
2. Hansch, C., Leo, A. J.: Substituent Constants for Correlation Analysis in Chemistry and Biology. New York, John Wiley & Sons, 1979.
3. Hansch, C., Lien, E. J.: J. Med. Chem. 14:653, 1971.
4. Dearden, J. C., George, E.: J. Pharm. Pharmacol. 31(Suppl.):45P, 1979.
5. Kubinyi, H.: The bilinear model. In Kuchar, M. (ed.). QSAR in Design of Bioactive Molecules. Barcelona, J. R. Prous, 1984.
6. Kubinyi, H. J.: Med. Chem. 20:625, 1971.
7. Stuper, A. J., Brügger, W. E., Jurs, P. C.: Computer Assisted Studies of Chemical Structure and Biological Function. New York, John Wiley & Sons, 1979.

8. Michaelson, D. M., Raftery, M. A.: Proc. Natl. Acad. Sci. USA 71:4768, 1974.
9. Pfeiffer, C.: Science 107:94, 1948.
10. Barlow, R. B., Ing, H. R.: Br. J. Pharmacol. 3:298, 1948.
11. Fisher, A., Keasling, A., Schueler, F.: Proc. Soc. Exp. Biol. Med. 81:439, 1952.
12. Gero, A., Reese, V. J.: Science, 123:100, 1956.
13. Welsh, J. H., Taub, R.: J. Pharmacol. Exp. Ther. 103:62, 1951.
14. Nachmansohn, D., Wilson, I. B.: Adv. Enzymol. 12:259, 1951.
15. Fries, S. L., et al.: J. Am. Chem. Soc. 76:1363, 1954; 78:199, 1956; 79:3269, 1957; 80:5687, 1958.
16. Krupka, R. M., Laidler, K. J.: J. Am. Chem. Soc. 83:1458, 1961.
17. Baker, B. R.: J. Pharm. Sci. 53:347, 1964.
18. Baker, B. R., Patel, R. P.: J. Pharm. Sci. 52:927, 1963.
19. Baker, B. R., Alumaula, P. I.: J. Pharm. Sci. 52:915, 1963.
20. Langmuir, I.: J. Am. Chem. Soc. 41:1543, 1919.
21. Beckett, A. H., Casy, A. F.: J. Pharm. Pharmacol. 6:986, 1954.
22. Eddy, N. B.: Chem. Ind. p. 1462, 1959.
23. Portoghese, P. S., Mikhail, A. A., Kupferberg, H. J.: J. Med. Chem. 11:219, 1968.
24. Gill, E. W.: Prog. Med. Chem. 4:39, 1965.
25. Beckett, A. H.: J. Pharm. Pharmacol. 8:848, 1956.
26. Martin-Smith, M., Smail, G. A., Stenlake, J. B.: J. Pharm. Pharmacol. 19:561, 1967.
27. Kier, L. B.: Mol. Pharmacol. 3:487, 1967; 4:70, 1968.
28. Chothia, C., Pauling, P.: Nature 219:1156, 1968.
29. Chiou, C. Y., Long, J. P., Cannon, J. G., Armstrong, P. D.: J. Pharmacol. Exp. Ther. 166:243, 1969.
30. Smissman, E., Nelson, W., Day, J., LaPidus, J.: J. Med. Chem. 9:458, 1966.
31. Chothia, C.: Nature 225:36, 1970.
32. Chothia, C., Pauling, P.: Nature 226:541, 1970.
33. Kier, L. B.: J. Med. Chem. 11:441, 1968.
34. Niemann, C. C., Hayes, J. T.: J. Am. Chem. Soc. 64:2288, 1942.
35. Black, J. W., et al.: Nature 236:385, 1972.
36. Cushy, A. R.: Biological Relations of Optically Active Isomeric Substances. Baltimore, Williams & Wilkins, 1926.
37. Easson, L. H., Steadman, E.: Biochem. J. 27:1257, 1933.
38. Beckett, A.: Prog. Drug Res. 1:455–530, 1959.
39. Blaschko, H.: Proc. R. Soc. Lond. (Ser. B) 137:307, 1950.
40. Smeyers, Y. G., Cuellare-Rodriguez, S., Galvez-Ruano, E., Arias-Perez, M. S.: J. Pharm. Sci. 74:47, 1985.
41. Clark, A. J.: J. Physiol. 61:530, 547, 1926.
42. Gaddum, J. H.: J. Physiol. 61:141, 1926; 89:7P, 1937.
43. Ariens, E. J., Simonis, A. M.: J. Pharm. Pharmacol. 16:137, 289, 1964.
44. Stephenson, R. P.: Br. J. Pharmacol. 11:379, 1956.
45. Croxatto, R., Huidobro, F.: Arch. Int. Pharmacodyn. 106:207, 1956.
46. Paton, W. D. M.: Proc. R. Soc. Lond. (Ser. B.) 154:21, 1961.
47. Koshland, D. E.: Proc. Natl. Acad. Sci. USA 44:98, 1958.
48. Nachmansohn, D.: Chemical and Molecular Basis of Nerve Activity. New York, Academic Press, 1959.
49. Belleau, B.: J. Med. Chem. 7:776, 1964.

SELECTED READINGS

Albert, A.: Selective Toxicity, 7th ed. New York, Chapman & Hall, 1985.

Keverling Buisman, J. A.: Biological activity and chemical structure. In Nauta, W. T., Rekker, R. F. (eds.), Pharmacochemistry Library, Vol. 2, New York, Elsevier, 1977.

Franke, R.: Theoretical drug design methods. In Nauta, W. T., Rekker, R. F. (eds.), Pharmacochemistry Library, Vol. 7. New York, Elsevier, 1984.

Martin, Y. C.: Quantitative Drug Design. In Grunewald, G. (ed.), Medicinal Research, Vol. 8. New York, Marcel Dekker, 1978.

Mutschler, E., Windterfeldt, E. (eds.): Trends in Medicinal Chemistry. Berlin, VCH Publishers, 1987.

Olson, E. C., Christoffersen, R. E.: Computer assisted drug design. In Comstock, M. J. (ed.). ACS Symp. Ser. Vol. 112, Washington, DC, American Chemical Society, 1979.

Topliss, J. G.: Quantitative structure–activity relationships of drugs. In Medicinal Chemistry, A Series of Monographs, Vol. 19, New York, Academic Press, 1983.

CHAPTER 3

Metabolic Changes of Drugs and Related Organic Compounds

Lawrence K. Low
Neal Castagnoli, Jr.

Metabolism plays a central role in the elimination of drugs and other foreign compounds (xenobiotic) from the body. Most organic compounds entering the body are relatively lipid-soluble (lipophilic). Therefore, to be absorbed, they must traverse the lipoprotein membranes of the lumen walls of the gastrointestinal (GI) tract. Once in the bloodstream, these molecules can diffuse passively through other membranes to reach various target organs to effect their pharmacologic actions. Owing to their reabsorption in the renal tubules, lipophilic compounds are not excreted to any substantial extent in the urine.

If lipophilic drugs or xenobiotics were not metabolized to polar, water-soluble products that are readily excretable, they would remain indefinitely in the body, eliciting their biologic effects. Thus, the formation of water-soluble metabolites not only enhances drug elimination, but also leads to compounds that are generally pharmacologically inactive and relatively nontoxic. Consequently, drug metabolism reactions have been traditionally regarded as *detoxication* (or *detoxification*) processes.[1] However, it is incorrect to assume that drug metabolism reactions are always detoxifying. Many drugs are biotransformed to pharmacologically active metabolites. Some metabolites have significant activity that contributes substantially to the pharmacologic effect ascribed to the parent drug. Occasionally, the parent compound is inactive and must be converted to a biologically active metabolite.[2,3] In addition, it is becoming increasingly clear that not all metabolites are nontoxic. Indeed many toxic side effects (e.g., tissue necrosis, carcinogenicity, teratogenicity) of drugs and environmental contaminants can be directly attributable to the formation of chemically reactive metabolites that are highly detrimental to the body.[4-6]

GENERAL PATHWAYS OF DRUG METABOLISM

Drug metabolism reactions have been divided into two categories: *phase I* (*functionalization*) and *phase II* (*conjugation*) reactions.[1,7] Phase I, or functionalization reactions, includes oxidative, reductive, and hydrolytic biotransformations (Box 3-1).[8] The purpose of these reactions is to introduce a polar functional group (e.g., OH, COOH, NH_2, SH) into the xenobiotic molecule. This can be achieved by direct introduction of the functional group (e.g., aromatic and aliphatic hydroxylation) or by modifying or "unmasking" existing functionalities (e.g., reduction of ketones and aldehydes to alcohols; oxidation of alcohols to acids; hydrolysis of ester and amides to yield COOH, NH_2, and OH groups; reduction of azo and nitro compounds to give NH_2 moieties; oxidative *N*-, *O*-, and *S*-dealkylation to give NH_2, OH, and SH groups). Although phase I reactions may not produce sufficiently hydrophilic or inactive metabolites, they generally tend to provide a functional group or "handle" in the molecule that can undergo subsequent phase II reactions.

The purpose of phase II reactions is to attach small, polar, and ionizable endogenous compounds such as glucuronic acid, sulfate, glycine, and other amino acids to the functional handles of phase I metabolites to form water-soluble conjugated products. Parent compounds that already have existing functional groups, such as OH, COOH, and NH_2, are often directly conjugated by phase II enzymes. Conjugated metabolites are readily excreted in the urine and are generally devoid of pharmacologic activity and toxicity. Other phase II pathways, such

BOX 3-1. GENERAL SUMMARY OF PHASE I AND PHASE II METABOLIC PATHWAYS

PHASE I OR FUNCTIONALIZATION REACTIONS

Oxidative Reactions

Oxidation of aromatic moieties
Oxidation of olefins
Oxidation at benzylic, allylic carbon atoms, and carbon atoms α to carbonyl and imines
Oxidation at aliphatic and alicyclic carbon atoms
Oxidation involving carbon–heteroatom systems:
- Carbon – nitrogen systems (aliphatic and aromatic amines; includes *N*-dealkylation, oxidative deamination, *N*-oxide formation, *N*-hydroxylation).
- Carbon–oxygen systems (*O*-dealkylation)
- Carbon–sulfur systems (*S*-dealkylation, *S*-oxidation, and desulfuration)

Oxidation of alcohols and aldehydes
Other miscellaneous oxidative reactions

Reductive Reactions

Reduction of aldehydes and ketones
Reduction of nitro and azo compounds
Miscellaneous reductive reactions

Hydrolytic Reactions

Hydrolysis of esters and amides
Hydration of epoxides and arene oxides by epoxide hydrase

PHASE II OR CONJUGATION REACTIONS

Glucuronic acid conjugation
Sulfate conjugation
Conjugation with glycine, glutamine and other amino acids
Glutathione or mercapturic acid conjugation
Acetylation
Methylation

as methylation and acetylation, serve to terminate or attenuate biologic activity, whereas glutathione conjugation serves to protect the body against chemically reactive compounds or metabolites. Thus, it is apparent that phase I and phase II reactions complement one another in detoxifying and facilitating the elimination of drugs and xenobiotics.

To illustrate, consider the principal psychoactive constituent of marijuana, Δ^1-tetrahydrocannabinol (Δ^1-THC). This lipophilic molecule (octanol/water partition coefficient ~ 6000)[9] undergoes allylic hydroxylation to give 7-hydroxy-Δ^1-THC in humans.[10] More polar than its parent compound, the 7-hydroxy metabolite is further oxidized to the corresponding carboxylic acid derivative Δ^1-THC-7-oic acid, which is ionized (pK_a COOH ~ 5) at physiologic pH. Subsequent conjugation of this metabolite (either at the COOH or phenolic OH) with glucuronic acid leads to water-soluble products that are readily eliminated in the urine.[11]

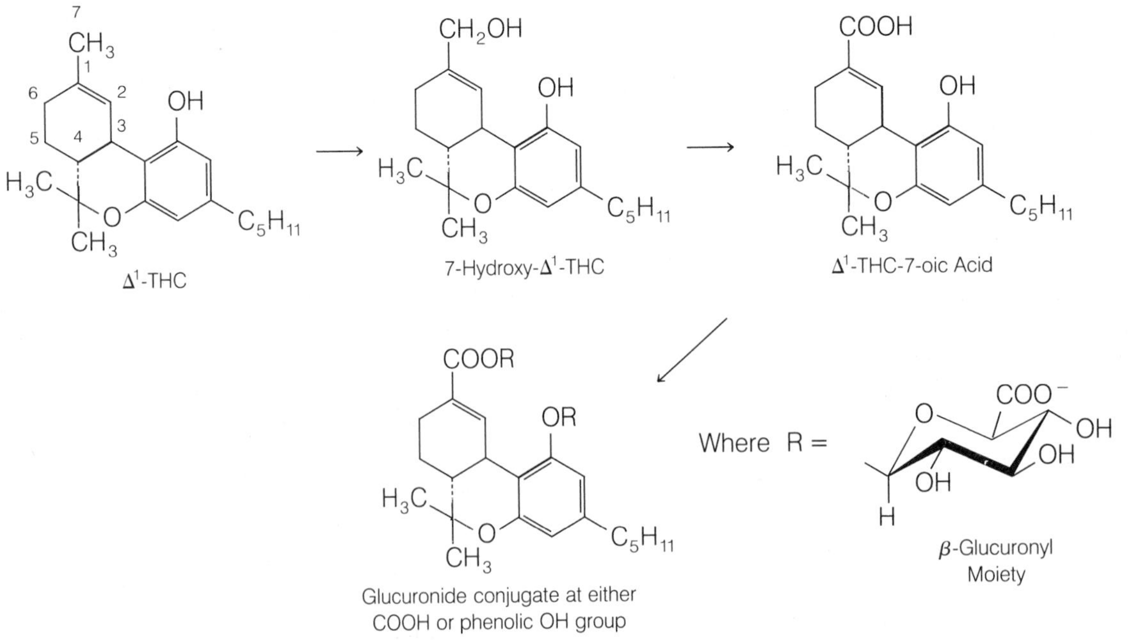

In the foregoing series of biotransformations, the parent Δ^1-THC molecule is made increasingly more polar, ionizable, and hydrophilic. The attachment of the glucuronyl moiety (with its ionized carboxylate group and three polar hydroxyl groups; see structure) to the Δ^1-THC metabolites notably favors partitioning of the conjugated metabolites into an aqueous medium.

The purpose of this chapter is to provide the student with a broad overview of drug metabolism. Various phase I and phase II biotransformation pathways (see Box 3-1) will be outlined. Representative drug examples for each pathway will be presented. Drug metabolism examples in humans will be emphasized, although discussion of metabolism in other mammalian systems is necessary. The central role of the cytochrome P-450 monooxygenase system in oxidative drug biotransformation will be elaborated. Discussion of other enzyme systems involved in phase I and phase II reactions will be presented in their respective sections. In addition to stereochemical factors that may affect drug metabolism, biologic factors such as age, sex, heredity, disease state, and species variation will be considered. The effects of enzyme induction and inhibition on drug metabolism, as well as a section on pharmacologically active metabolites, will be included.

SITES OF DRUG BIOTRANSFORMATION

Although biotransformation reactions may occur in many tissues, the liver is, by far, the most important organ in drug metabolism.[12] It is particularly rich in almost all the drug metabolizing enzymes to be discussed in this chapter. The liver is a well-perfused organ and plays a paramount role in the detoxification and metabolism of endogenous and exogenous compounds present in the bloodstream. Orally administered drugs that are absorbed through the gastrointestinal tract must first pass through the liver. Therefore, they are susceptible to hepatic metabolism (first-pass effect) before reaching the systemic circulation. Dependent on the drug, this metabolism can sometimes be quite significant and, as a result, decrease oral bioavailability. For example, in humans several drugs are extensively metabolized by the first-pass effect.[13] The following list includes some of those drugs:

Isoproterenol
Lidocaine
Meperidine
Morphine
Nitroglycerin
Pentazocine
Propoxyphene
Propranolol
Salicylamide

Some drugs (e.g., lidocaine) are so effectively removed by first-pass metabolism that they are ineffective when given orally.[14] Thus, one can appreciate the enormous metabolizing capability of the liver.

Because most drugs are administered orally, the intestine appears to play an important role in the extrahepatic metabolism of xenobiotics. For example, in humans orally administered isoproterenol undergoes considerable sulfate conjugation in the intestinal wall.[15] Several other drugs (e.g., levodopa, chlorpromazine, and diethylstilbestrol)[16] have also been reported to be metabolized in the gastrointestinal tract. Esterases and lipases present in the intestine may be particularly important in carrying out hydrolysis of many ester prodrugs (see later section on hydrolysis).[17] Bacterial flora present in the intestine and colon appear to play an important role in the reduction of many aromatic azo and nitro drugs (e.g., sulfasalazine).[18] Intestinal β-glucuronidase enzymes are capable of hydrolyzing glucuronide conjugates excreted in the bile, thereby liberating the free drug or its metabolite for possible reabsorption (enterohepatic circulation or recycling).[19]

Although other tissues, such as kidney, lungs, adrenal glands, placenta, brain, and skin, have some degree of drug-metabolizing capability, the biotransformations that they carry out are oftentimes more substrate-selective and more limited to particular types of reactions (e.g., oxidation, glucuronidation).[20] In many instances, the full metabolic capabilities of these tissues have not been fully explored.

ROLE OF CYTOCHROME P-450 MONOOXYGENASES IN OXIDATIVE BIOTRANSFORMATIONS

Of the various phase I reactions that will be considered, oxidative biotransformation processes are, by far, the most common and important in drug metabolism. The general stoichiometry that describes the oxidation of many xenobiotics (R—H) to their corresponding oxidized metabolites (R—OH) is given by the following equation:[21]

$$RH + NADPH + O_2 + H^+ \rightarrow ROH + NADP^+ + H_2O$$

The enzyme systems carrying out this biotransformation are referred to as *mixed function oxidases* or *monooxygenases*.[22,23] The reaction requires both molecular oxygen and the reducing agent NADPH (reduced form of nicotinamide adenosine dinucleotide phosphate). It should be emphasized that during this oxidative process, one atom of molecular oxygen (O_2) is introduced into the substrate R — H to form R — OH, and the other oxygen atom is incorporated into water. The mixed function oxidase system[24] is actually made up of several components, the most important being an enzyme called cytochrome P-450, which is responsible for transferring *an oxygen atom* to the substrate R — H. Other important components of this system include the NADPH-dependent cytochrome P-450 reductase and a NADH-linked cytochrome b_5. The latter two components, along with the cofactors NADPH and NADH supply the reducing equivalents (namely electrons) needed in the overall metabolic oxidation of foreign compounds. The proposed mechanistic scheme by which the cytochrome P-450 monooxygenase system catalyzes the conversion of molecular oxygen to an "activated oxygen" species will be elaborated later.

The cytochrome P-450 enzyme is a heme-protein.[25] The heme portion is an iron-containing porphyrin called protoporphyrin IX, and the protein portion is called the apoprotein. Cytochrome P-450 is found in high concentrations in the liver, the major organ involved in the metabolism of xenobiotics. The presence of this enzyme in many other tissues (e.g., lung, kidney, intestine, skin, placenta, adrenal cortex) is a reflection that these tissues have drug-oxidizing capability too. The name cytochrome P-450 is derived from the fact that the reduced (Fe^{2+}) form of this enzyme binds with carbon monoxide to form a complex that has a distinguishing spectroscopic absorption maximum at 450 nm.[26]

An important feature of the hepatic cytochrome P-450 mixed function oxidase system is its ability to metabolize an almost unlimited number of diverse substrates by a variety of oxidative transformations.[27] This versatility is believed to be attributable to the substrate nonspecificity of cytochrome P-450 as well as to the presence of multiple forms of the enzyme.[28] Some of these P-450 enzymes are selectively inducible by various chemicals (e.g., phenobarbital, benzo[*a*]pyrene, 3-methylcholanthrene).[29] One of these inducible forms of the enzyme (cytochrome P-448)[30] is of particular interest and will be discussed later.

The cytochrome P-450 monooxygenases are located in the endoplasmic reticulum, a highly organized and complex network of intracellular membranes that is particularly abundant in tissues such as the liver.[30] When these tissues are disrupted by homogenization, the endoplasmic reticulum loses its structure and is converted into small vesicular bodies known as microsomes.

Microsomes isolated from hepatic tissue appear to retain all of the mixed function oxidase capabilities of intact hepatocytes; because of this, microsomal preparations (with the necessary cofactors, e.g., NADPH, Mg^{2+}) are frequently utilized for in vitro drug metabolism studies. Because of its membrane-bound nature, the cytochrome P-450 monooxygenase system appears to be housed in a lipoidal environment. This may explain, in part, why lipophilic xenobiotics are generally good substrates for the monooxygenase system.[31]

The catalytic role that the cytochrome P-450 monooxygenase system plays in the oxidation of xenobiotics is summarized in the cycle shown in Figure 3-1.[32,33] The initial step of this catalytic reaction cycle starts with the binding of the substrate to the oxidized (Fe^{3+}) resting state of cytochrome P-450 to form a P-450–substrate complex. The next step involves the transfer of one electron from NADPH-dependent cytochrome P-450 reductase to the P-450–substrate complex. This one-electron transfer reduces Fe^{3+} to Fe^{2+}. It is this reduced (Fe^{2+}) P-450–substrate complex that is capable of binding dioxygen (O_2). The dioxygen–P-450-substrate complex which is formed then undergoes another one-electron reduction (by cytochrome P-450 reductase–NADPH and/or cytochrome b_5 reductase–NADH) to yield what is believed to be a peroxide dianion–P-450 (Fe^{3+})–substrate complex. Water (containing one of the oxygen atoms from the original dioxygen molecule) is released from the latter intermediate to form an activated oxygen–P-450–substrate complex (Fig. 3-2). The activated oxygen $[FeO]^{3+}$ in this complex is highly electron-deficient and is a potent oxidizing agent. The activated oxygen is transferred to the substrate (RH) and the oxidized substrate product (ROH) released from the enzyme complex to regenerate the oxidized form of cytochrome P-450.

It is important to recognize that the key sequence of events appears to center around the alteration of a dioxygen–P-450–substrate complex to an activated oxygen–P-450–substrate complex, which is then capable of effecting the critical transfer of oxygen from P-450 to the substrate.[34,35] In view of the potent oxidizing nature of the activated oxygen being transferred, it is not surprising that numerous substrates are capable of being oxidized by cytochrome P-450. The mechanistic details of oxygen activation and transfer in cytochrome P-450-catalyzed reactions continue to be an active area of research in drug metabolism.[32]

The many types of oxidative reactions carried out by cytochrome P-450 will be enumerated in the sections to follow. Many of these oxidative pathways are summarized schematically in Figure 3-3 (see also Box 3-1).[34]

The versatility of cytochrome P-450 in carrying out a variety of oxidation reactions on a multitude of substrates may be attributable to the multiple forms of the enzyme. Consequently, it is important for the student to realize that the biotransformation of a parent xenobiotic to several oxidized metabolites is carried out not just by one form of P-450, but, more

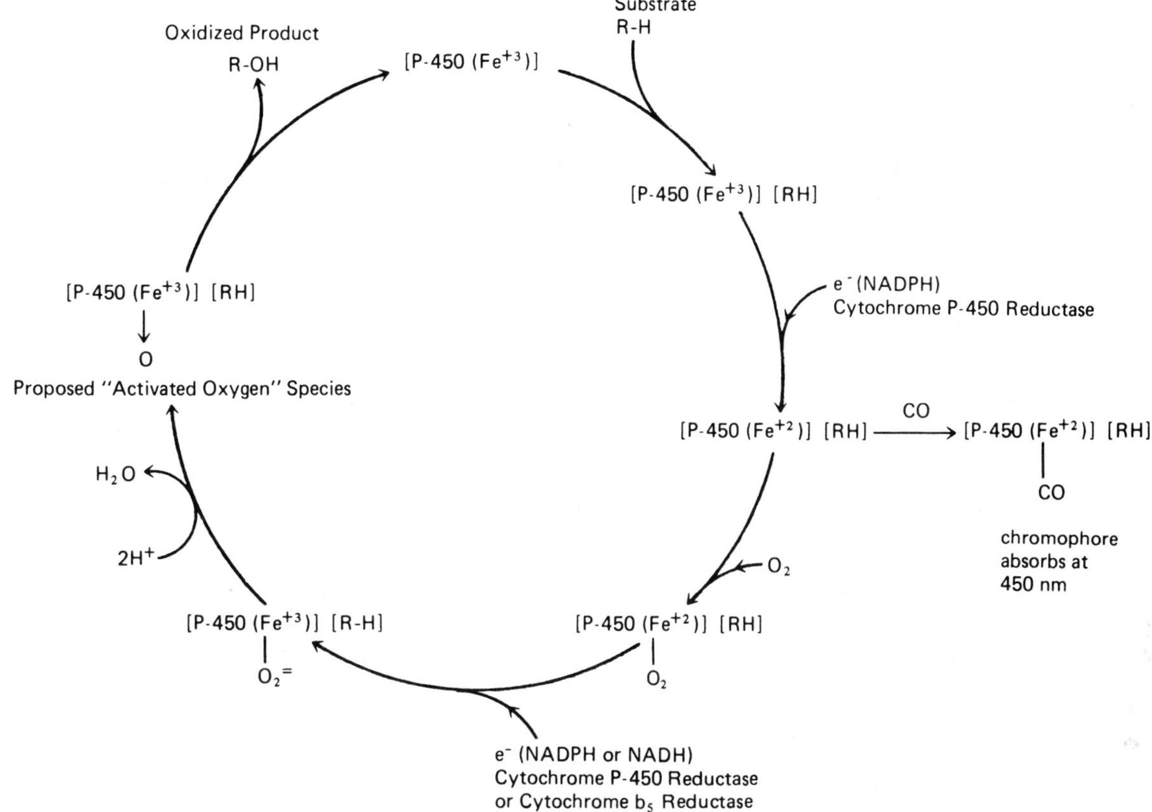

FIG. 3-1. *Proposed catalytic reaction cycle involving cytochrome P-450 in the oxidation of xenobiotics.*

likely, by several different forms.[36] Extensive studies indicate that the apoprotein portions of various cytochrome P-450s appear to differ from one another in their tertiary three-dimensional structure (owing to differences in amino acid sequence or the makeup of the polypeptide chain).[25,28] Because the apoprotein portion is important in substrate binding and catalytic transfer of activated oxygen, these structural differences may account for some substrates being preferentially or more efficiently oxidized by one particular form of cytochrome P-450.

OXIDATIVE REACTIONS

OXIDATION OF AROMATIC MOIETIES

Aromatic hydroxylation refers to the mixed function oxidation of aromatic compounds (arenes) to their corresponding phenolic metabolites (arenols).[37] Almost all aromatic hydroxylation reactions are believed to proceed initially through an epoxide intermediate called an arene oxide, which rearranges rapidly and spontaneously to the arenol product in

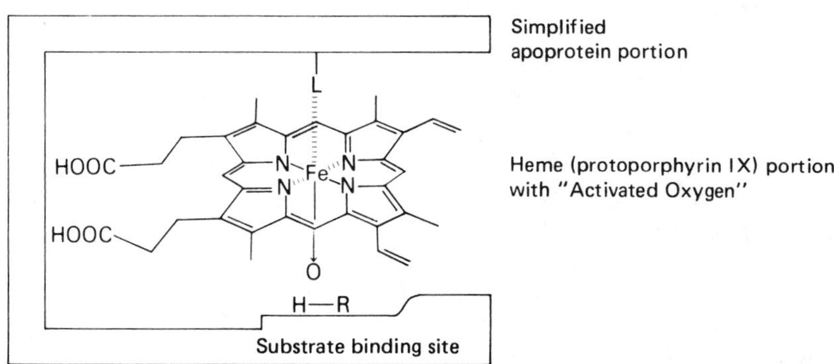

FIG. 3-2. *A simplified depiction of the proposed activated oxygen–cytochrome P-450–substrate complex. Note the simplified apoprotein portion and the heme (protoporphyrin IX) portion of cytochrome P-450 and the close proximity of the substrate RH undergoing oxidation.*

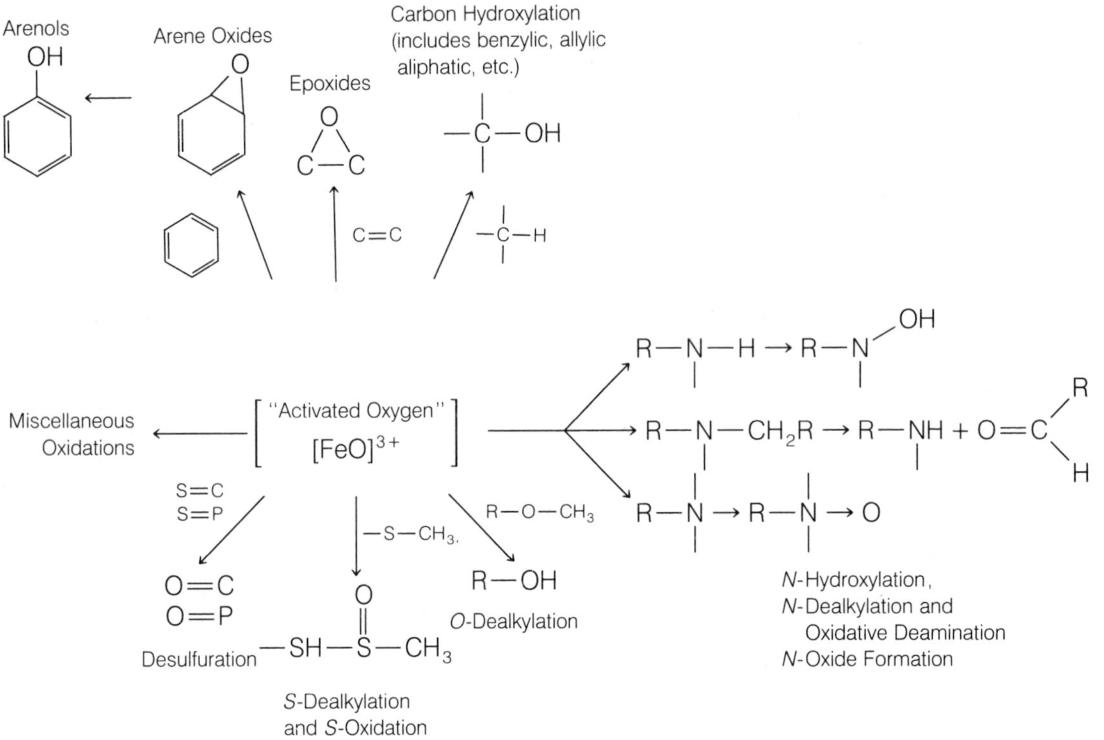

FIG. 3-3. *Schematic summary of cytochrome P-450-catalyzed oxidation reactions. (Adapted from Ullrich, v.: Top. Curr. Chem. 83:68, 1979.)*

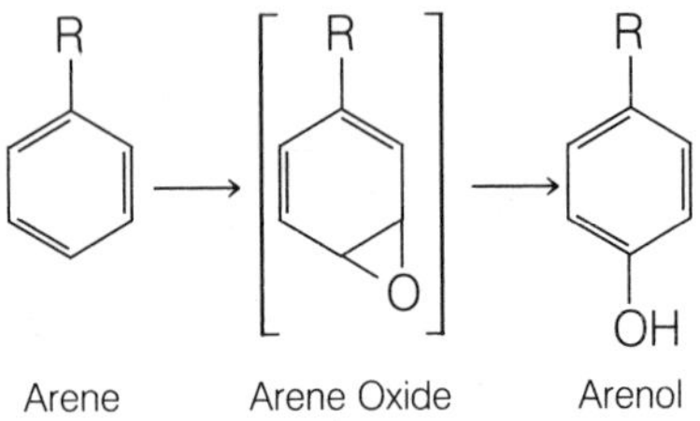

most instances. The importance of arene oxides in the formation of arenols and in other metabolic and toxicologic reactions will be discussed shortly.[38] Attention will now focus on the aromatic hydroxylation of several drugs and xenobiotics.

Most foreign compounds containing aromatic moieties are susceptible to aromatic oxidation. In humans, aromatic hydroxylation is a major route of metabolism for many drugs containing phenyl groups. Important therapeutic agents such as propranolol,[39] phenobarbital,[40] phenytoin,[41] phenylbutazone,[42] phenformin,[43] 17α-ethinylestradiol,[44] (*S*)(−)-warfarin,[45] among others undergo extensive aromatic oxidation (Figure 3-4 shows structure and site of hydroxylation). In most of the drugs just mentioned, hydroxylation occurs at the *para*-position.[47] Most phenolic metabolites formed from aromatic oxidation undergo further conversion to polar and water-soluble glucuronide or sulfate conjugates, which are readily excreted in the urine. For example, the major urinary metabolite of phenytoin found in humans is the *O*-glucuronide conjugate of *p*-hydroxyphenytoin.[41] Interestingly, the *para*-hydroxylated metabolite of phenylbutazone, oxyphenbutazone, is pharmacologically active and has been marketed, itself, as an anti-inflammatory

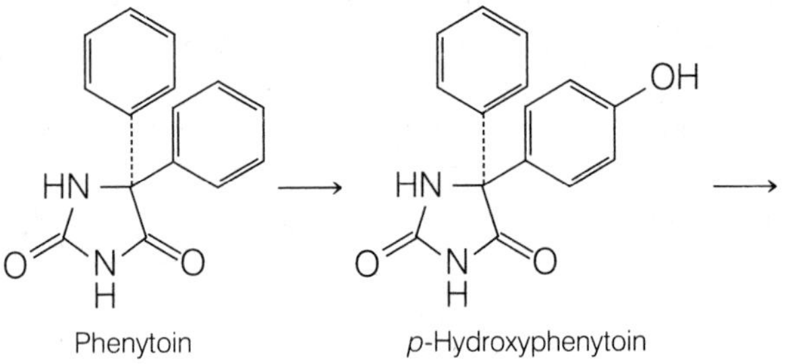

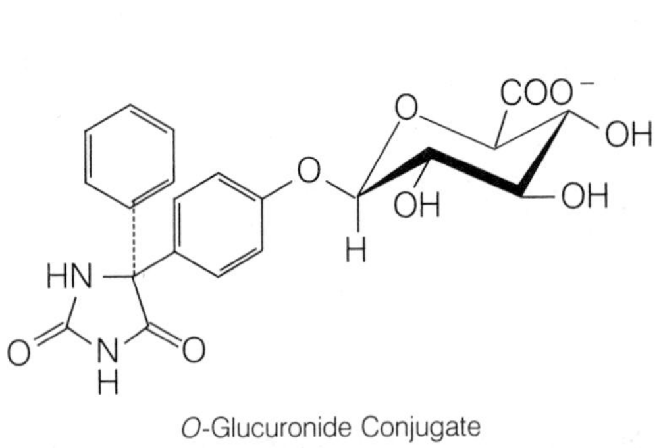

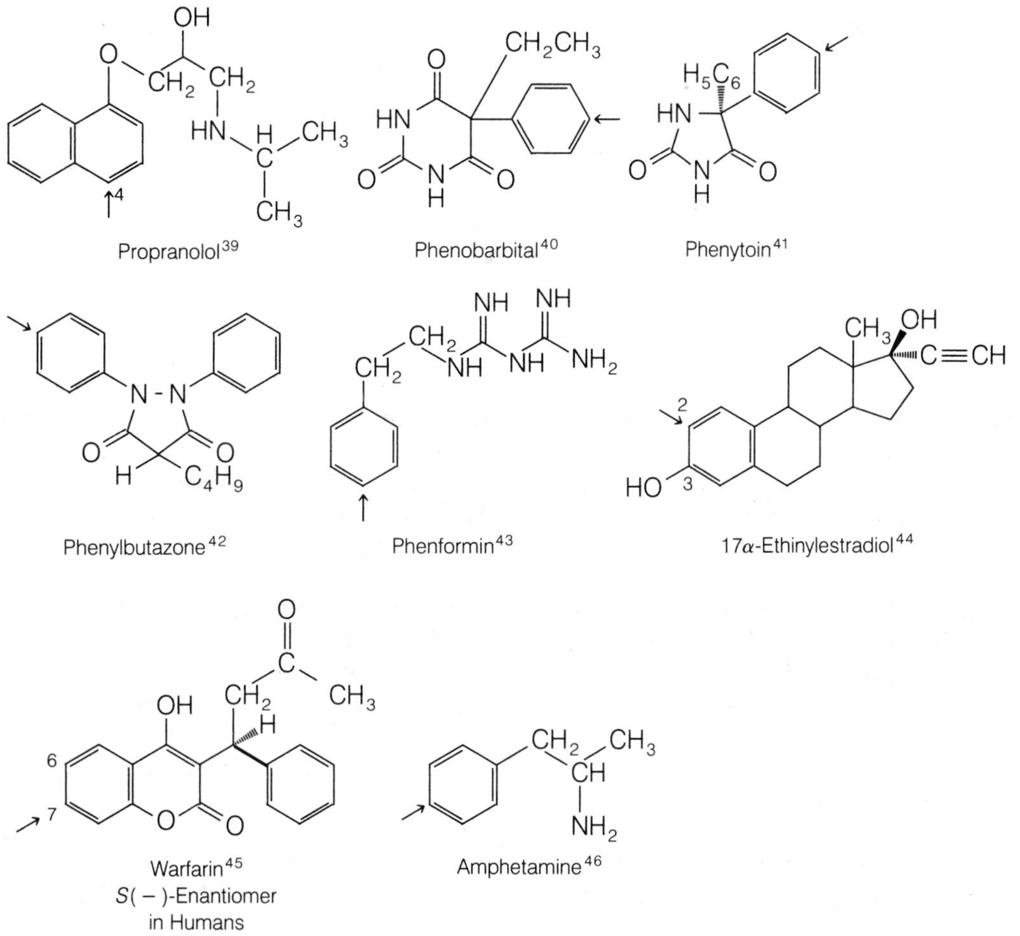

Propranolol[39]

Phenobarbital[40]

Phenytoin[41]

Phenylbutazone[42]

Phenformin[43]

17α-Ethinylestradiol[44]

Warfarin[45]
S(−)-Enantiomer
in Humans

Amphetamine[46]

FIG. 3-4. *Examples of drugs and xenobiotics that undergo aromatic hydroxylation in humans. Arrow indicates site of aromatic hydroxylation.*

agent (Tandearil, Oxalid).[42] Of the two enantiomeric forms of the oral anticoagulant warfarin (Coumadin), only the more active $S(-)$-enantiomer has been shown to undergo substantial aromatic hydroxylation to 7-hydroxywarfarin in humans.[45] In contrast, the $(R)(+)$-enantiomer is metabolized by keto reduction (see later section on stereochemical aspects of drug metabolism).[46]

Often the substituents attached to the aromatic ring may influence the ease of hydroxylation.[47] As a general rule, microsomal aromatic hydroxylation reactions appear to proceed most readily in activated (electron-rich) rings, whereas deactivated aromatic rings (e.g., those containing electron-withdrawing groups Cl, $-\overset{+}{N}R_3$, COOH, SO_2NHR) are generally slow or resistant to hydroxylation. The deactivating groups (Cl, $-\overset{+}{N}H=C$) present in the antihypertensive clonidine (Catapres) may explain why this rug undergoes little aromatic hydroxylation in humans.[48] The uricosuric agent probenecid (Benemid), with its electron-withdrawing carboxy and sulfamido groups, has not been reported to undergo any aromatic hydroxylation.[49]

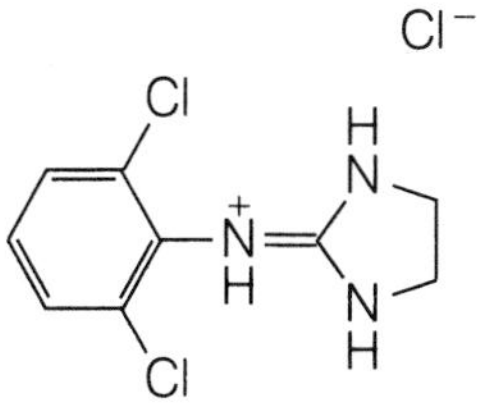

Clonidine Hydrochloride

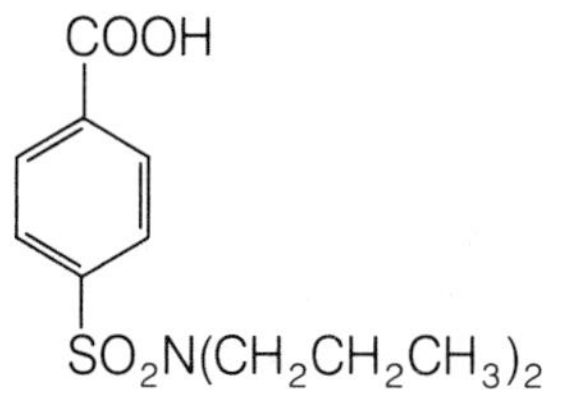

Probenecid

For compounds in which two aromatic rings are present, hydroxylation occurs preferentially in the more electron-rich ring. For example, aromatic hydroxylation of diazepam (Valium) occurs primarily in the more-activated ring to yield 4′-hydroxydiazepam.[50] A similar situation is seen in the 7-hydroxylation of the antipsychotic agent chlorpromazine (Thorazine)[51] and in the *para*-hydroxylation of *p*-chlorobiphenyl to *p*′-chloro-*p*′-hydroxybiphenyl.[52]

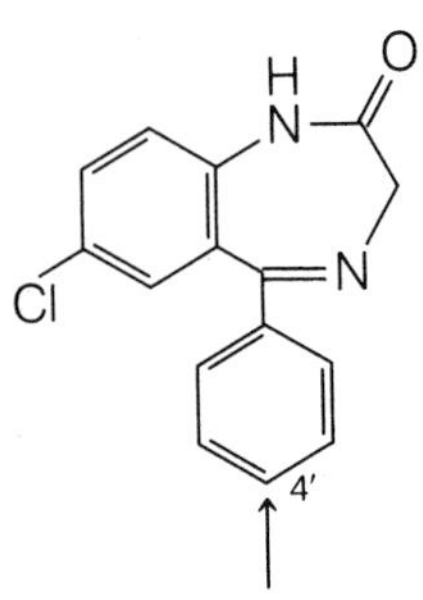

Diazepam

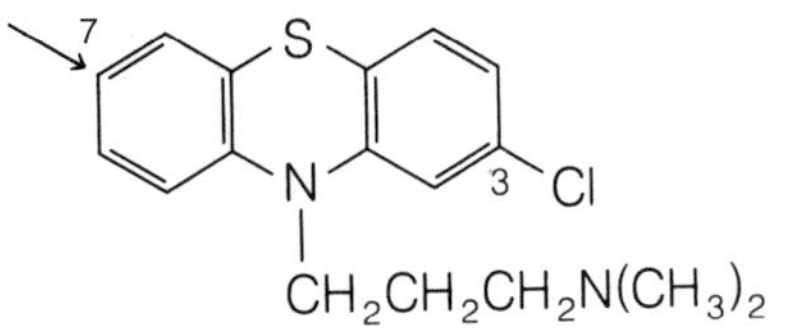

Chlorpromazine

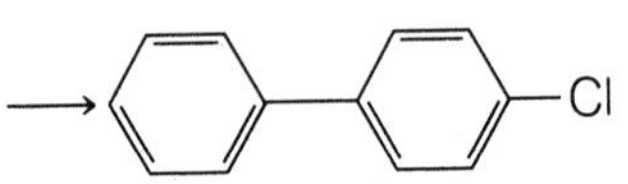

$CH_2CH_2CH_2N(CH_3)_2$

p-Chlorobiphenyl

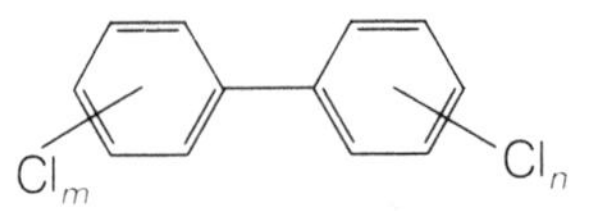

Polychlorinated Biphenyl Mixtures

The number of chlorine atoms (*m*, *n*) present in two aromatic rings varies considerably

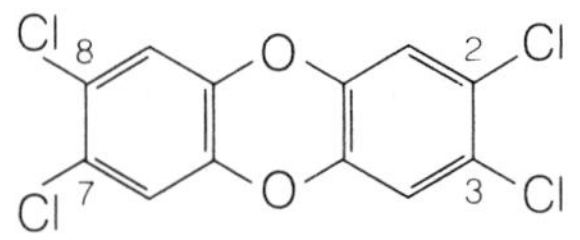

2,3,7,8-Tetrachlorodibenzo-*p*-dioxin (TCDD)

Recent environmental pollutants, such as polychlorinated biphenyls (PCBs) and 2,3,7,8-tetrachlorodibenzo-*p*-dioxin (TCDD), have attracted considerable public concern over their toxicity and health hazards. These compounds appear to be resistant to aromatic oxidation because of the numerous electronegative chlorine atoms present in the aromatic rings composing their structures. The metabolic stability coupled to the lipophilicity of these environmental contaminants probably explains their long persistence in the body once absorbed.[53,54]

Arene oxide intermediates are formed when a double bond in aromatic moieties is epoxidized. Arene oxides are of significant toxicologic concern, because these intermediates are electrophilic and chemically reactive (owing to the strained three-membered epoxide ring). Detoxification of arene oxides occurs mainly by spontaneous rearrangement to arenols, but enzymatic hydration to *trans*-dihydrodiols and enzymatic conjugation with glutathione (GSH) also play very important roles (Fig. 3-5).[37,38] If not effectively detoxified by the first three pathways in Figure 3-5, arene oxides will covalently bind with nucleophilic groups present on proteins, DNA, and RNA, thereby leading to serious cellular toxicity.[5,37] A few examples of reactive arene oxides causing carcinogenicity and cytotoxicity are examined in Figure 3-5.

Quantitatively, the most important detoxification reaction for arene oxides is the spontaneous rearrangement to the corresponding arenols. Many times, this rearrangement is accompanied by a novel intramolecular hydride (deuteride) migration called

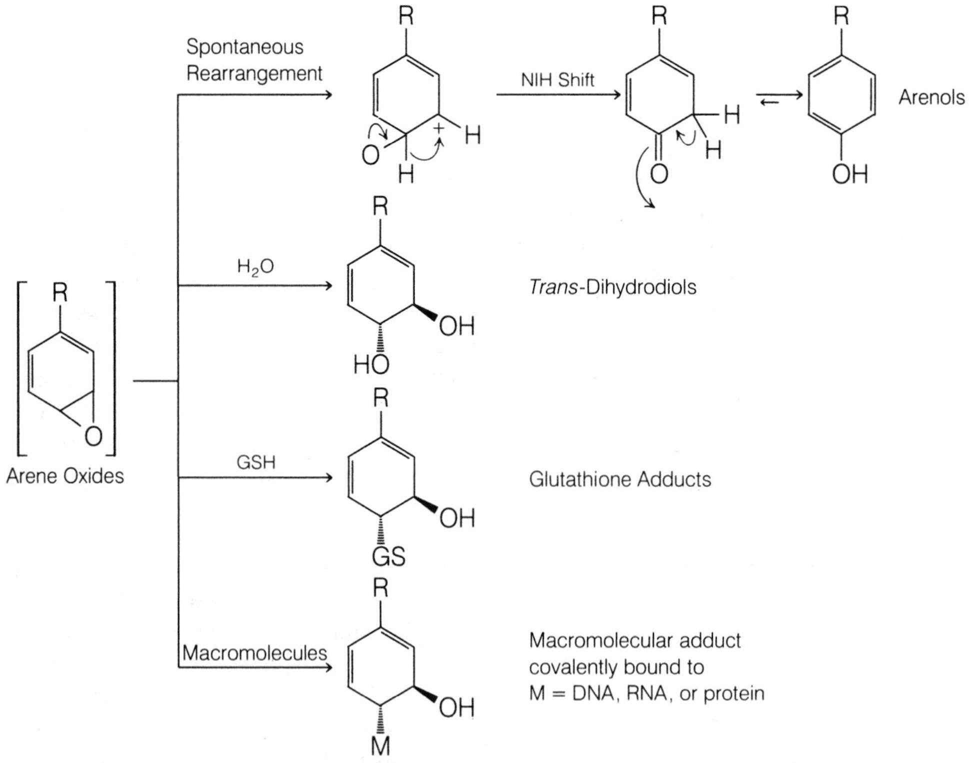

Spontaneous Rearrangement

NIH Shift

Arenols

H_2O

Trans-Dihydrodiols

Arene Oxides

GSH

Glutathione Adducts

Macromolecules

Macromolecular adduct covalently bound to M = DNA, RNA, or protein

FIG. 3-5. *Possible reaction pathways for arene oxides.*[37,38]

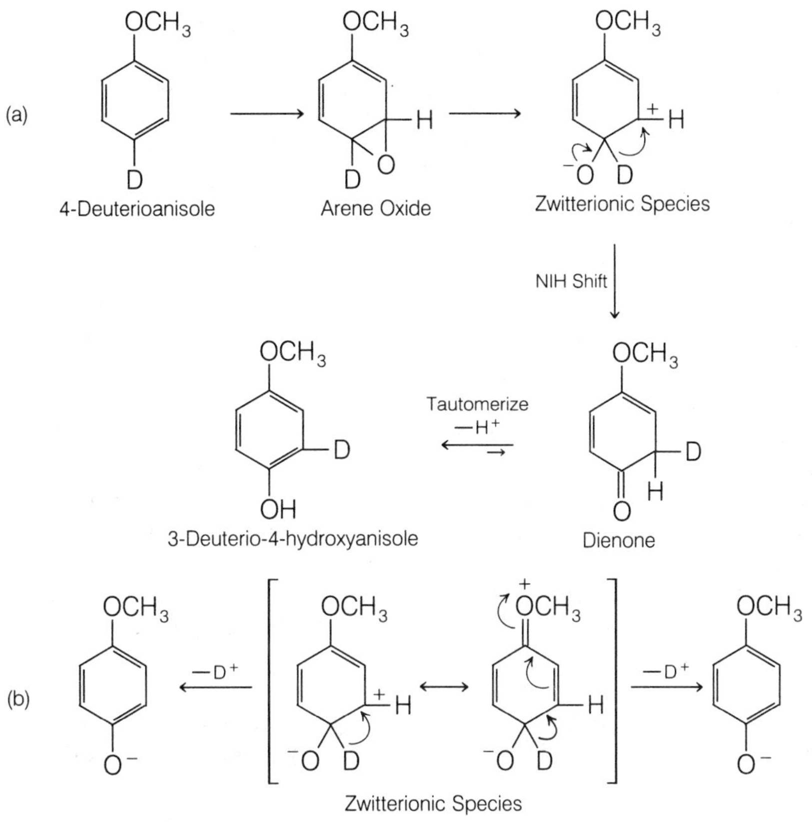

FIG. 3-6. *(a) General features of the NIH shift or 1,2-hydride (deuteride) shift in the mixed function oxidation of 4-deuterioanisole to 3-deuterio-4-hydroxyanisole; (b) Direct loss of D^+ from zwitterionic species leading to no retention of deuterium in 4-hydroxyanisole.*

the "NIH shift."[55] It was named after the National Institute of Health (NIH) laboratory in Bethesda, Maryland, where this process was discovered. The general features of the NIH shift are illustrated with the mixed function aromatic oxidation of 4-deuterioanisole to 3-deuterio-4-hydroxyanisole, as shown in Figure 3-6.[56]

After its metabolic formation, the arene oxide ring opens in the direction that generates the most resonance-stabilized carbocation (positive charge on C-3 carbon is resonance-stabilized by the OCH_3 group). The zwitterionic species (positive charge on the C-3 carbon atom and negative charge on the oxygen atom) then undergoes a 1,2-deuteride shift (NIH shift) to form the dienone. Final transformation of the dienone to 3-deuterio-4-hydroxyanisole occurs with the preferential loss of a proton because of the weaker bond energy of the C — H bond (as compared with the C — D bond). Thus, the deuterium is retained in the molecule by undergoing this intramolecular NIH shift. The experimental observation of an NIH shift for aromatic hydroxylation of a drug or xenobiotic is taken as indirect evidence for the involvement of an arene oxide.

In addition to the NIH shift, the zwitterionic species may also undergo direct loss of D^+ to generate 4-hydroxyanisole, in which there is no retention of deuterium (see Fig. 3-6). The alternative pathway (direct loss of D^+) may be more favorable than the NIH shift in some aromatic oxidation reactions. Therefore, depending on the substituent group on the arene, some aromatic hydroxylation reactions do not display any NIH shift.

Two extremely important enzymatic reactions also aid in neutralizing the reactivity of arene oxides. The first of these involves the hydration (i.e., nucleophilic attack of water on the epoxide) of arene oxides to yield inactive *trans*-dihydrodiol metabolites (see Fig. 3-5). This reaction is catalyzed by microsomal enzymes called epoxide hydrases.[57] Often epoxide hydrase inhibitors such as cyclohexene oxide and 1,1,1-trichloropropene-2,3-oxide have been utilized to demonstrate the detoxification role of these enzymes. Addition of these inhibitors is frequently accompanied by an increase in toxicity of the arene oxide being tested, because formation of nontoxic dihydrodiols is blocked. For example, the mutagenicity of benzo[*a*]pyrene-4,5-oxide, as measured by the Ames *Salmonella typhimurium* test system, is potentiated when cyclohexene oxide is added.[58] Dihydrodiol metabolites have been reported in the metabolism of several aromatic hydrocarbons

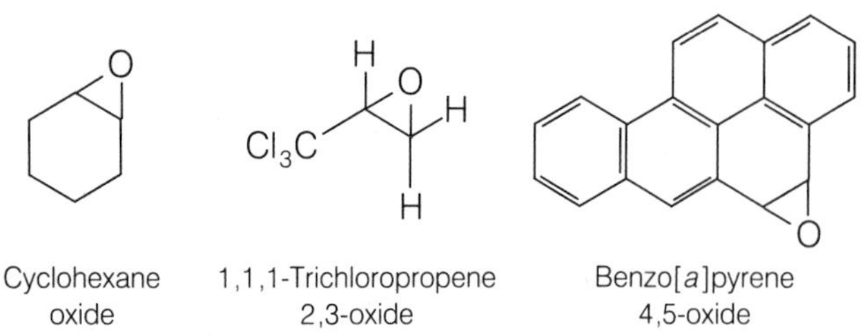

Cyclohexane oxide

1,1,1-Trichloropropene 2,3-oxide

Benzo[*a*]pyrene 4,5-oxide

(e.g., naphthalene, benzo[*a*]pyrene, and other related polycyclic aromatic hydrocarbons).[37] A few drugs (e.g., phenytoin,[59] phenobarbital,[60] glutethimide[61]) have also been observed to yield dihydrodiol products as minor metabolites in humans. Dihydrodiol products are susceptible to conjugation with glucuronic acid, as well as an enzymatic dehydrogenation to the corresponding catechol metabolite, as exemplified by the metabolism of phenytoin.[59]

A second enzymatic reaction involves nucleophilic ring opening of the arene oxide by the sulfhydryl group present in glutathione to yield the corresponding *trans*-1,2-dihydro-1-*S*-glutathionyl-2-hydroxy adduct or glutathione adduct (see Fig. 3-5).[37] The reaction is catalyzed by various glutatione *S*-transferases.[62] Because glutathione is found in practically all mammalian tissues, it plays an important role not only in the detoxification of arene oxides, but also in the detoxification of a variety of other chemically reactive and potentially toxic intermediates. Initially, formed glutathione adducts from arene oxides are modified in a series of reactions to yield "premercapturic acid" or mercapturic acid metabolites.[63] Since it is classified as a phase II pathway, glutathione conjugation will be covered in greater detail in a later section.

Because of their electrophilic and reactive nature, arene oxides may also undergo spontaneous reactions with nucleophilic functionalities present on biomacromolecules.[38] Such reactions lead to modified protein, DNA, and RNA structures, and often cause dramatic alterations in how these macromolecules function. Much of the cytotoxicity and irreversible lesions caused by arene oxides are presumed to be the result of their covalent binding to cellular components. Several well-established examples of reactive arene oxides that cause serious toxicity are presented in the following.

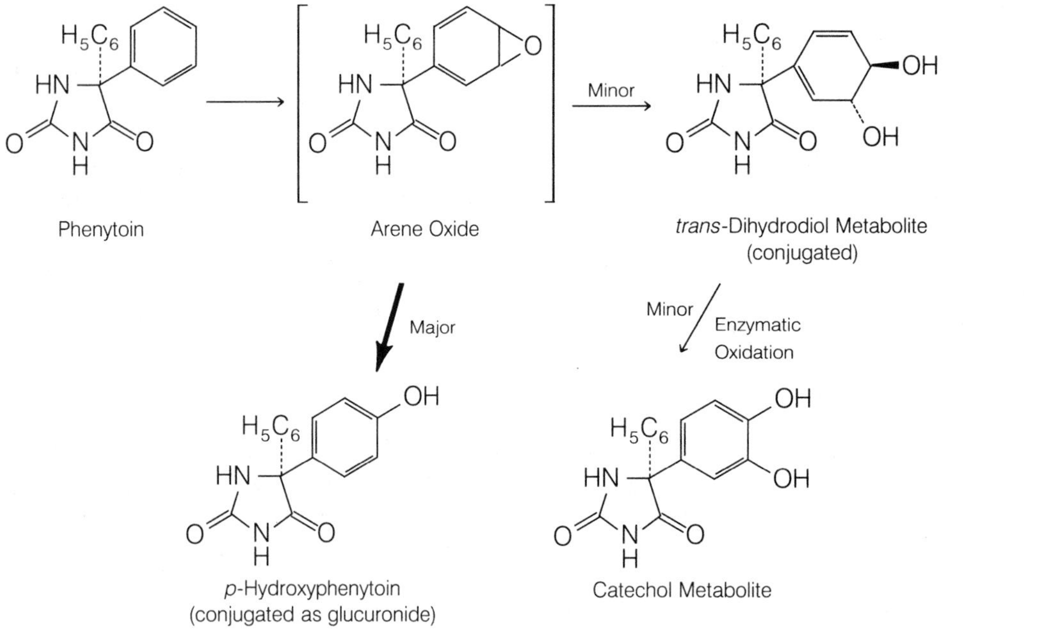

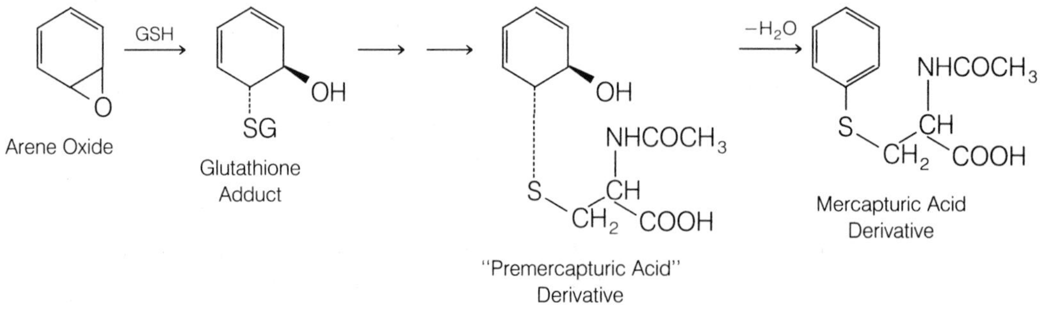

Phenytoin → [Arene Oxide] —Minor→ *trans*-Dihydrodiol Metabolite (conjugated)

Arene Oxide —Major→ *p*-Hydroxyphenytoin (conjugated as glucuronide)

trans-Dihydrodiol Metabolite —Minor, Enzymatic Oxidation→ Catechol Metabolite

Arene Oxide —GSH→ Glutathione Adduct → → "Premercapturic Acid" Derivative —$-H_2O$→ Mercapturic Acid Derivative

The administration of bromobenzene to rats causes severe liver necrosis.[64] Extensive in vivo and in vitro studies indicate that the liver damage results from the interaction of a chemically reactive metabolite, namely 4-bromobenzene oxide, with hepatocytes.[65] Extensive covalent binding to hepatic tissue was confirmed by using radiolabeled bromobenzene. The severity of necrosis correlated well with the amount of covalent binding to hepatic tissue. It was demonstrated, with diethyl maleate or large doses of bromobenzene in rats, that the depletion of hepatic glutatione led to more severe liver necrosis.

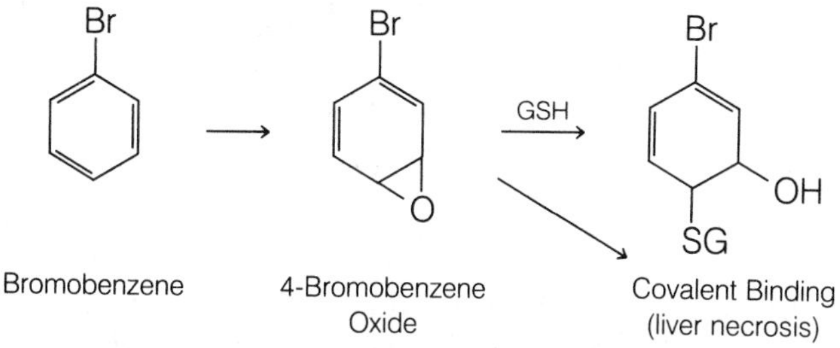

Br Br GSH Br OH O SG

Bromobenzene 4-Bromobenzene Oxide Covalent Binding (liver necrosis)

Polycyclic aromatic hydrocarbons are ubiquitous environmental contaminants that are formed from auto emission, refuse burning, industrial processes, cigarette smoke, and other combustion processes. Benzo[*a*]pyrene, a potent carcinogenic agent, is perhaps the most extensively studied of these polycyclic aromatic hydrocarbons.[66] From inspection of its structure, aromatic hydroxylation of benzo[*a*]pyrene can obviously occur at a number of positions. The identification of several dihydrodiol metabolites is viewed as indirect evidence for the formation and involvement of arene oxides in the metabolism of benzo[*a*]pyrene. Although certain arene oxides of benzo[*a*]pyrene (e.g., 4,5-oxide, 7,8-oxide, 9,10-oxide) appear to display some mutagenic and tumorigenic activity, it does not appear that they represent the ultimate reactive species responsible for benzo[*a*]pyrene's carcinogenicity. In recent years, extensive studies have led to the characterization of a specific sequence of metabolic reactions (Fig. 3-7) that give rise to a highly reactive intermediate that covalently binds to DNA. Metabolic activation of benzo[*a*]pyrene to the ultimate carcinogenic species involves an initial epoxidation reaction to form the 7,8-oxide, which is then converted by epoxide hydrase to (−)-7(*R*),8(*R*)-dihydroxy-7,8-dihydrobenzo[*a*]pyrene.[67] The two-step enzymatic formation of this *trans*-dihydrodiol is stereospecific. Subsequent epoxidation at the 9,10-double bond of the latter metabolite generates predominantly (+)-7(*R*),8(*S*)-dihydroxy-9(*R*),10(*R*)-oxy-7,8,9,10-tetrahydrobenzo[*a*]pyrene or (+)7,8-diol-9,10-epoxide. It is this key electrophilic diol epoxide metabolite that readily reacts with DNA to form many covalently bound adducts.[68,69] Careful degradation studies have shown that the principal adduct involves attack of the C-2 amino group of deoxyguanosine at C-10 of the diol epoxide. Clearly, these reactions are responsible for genetic code alterations that ultimately lead to the malignant transformations. Covalent binding of the diol epoxide metabolite to deoxyadenosine and to deoxycytidine have also been established.[69]

Another carcinogenic polycyclic aromatic hydrocarbon, 7,12-dimethylbenz[*a*]anthracene, also forms

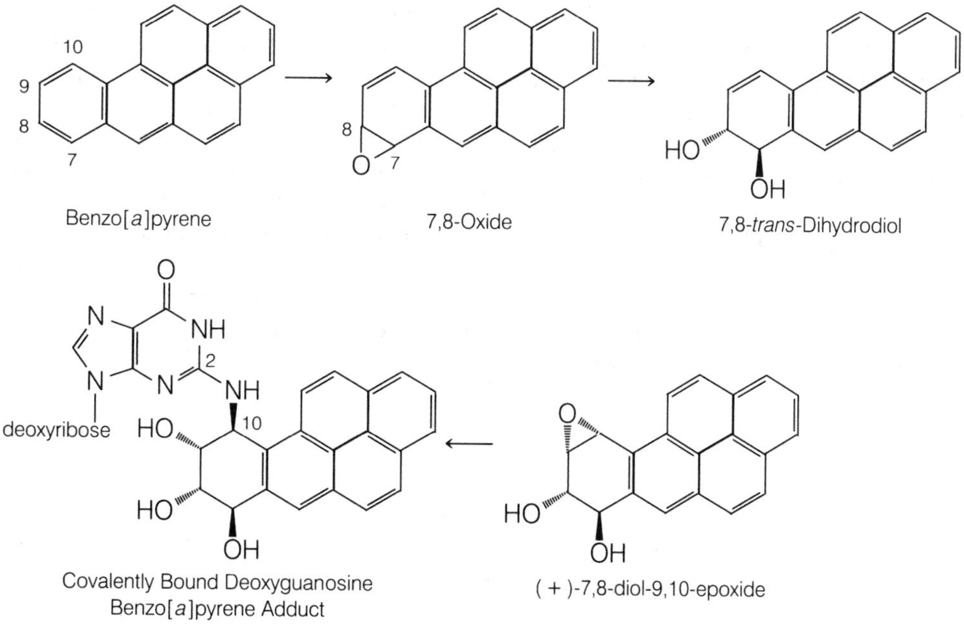

10 9 8 7 8 O 7 HO OH

Benzo[*a*]pyrene 7,8-Oxide 7,8-*trans*-Dihydrodiol

O N NH 2 N N NH deoxyribose HO 10 HO OH O HO OH

Covalently Bound Deoxyguanosine Benzo[*a*]pyrene Adduct

(+)-7,8-diol-9,10-epoxide

FIG. 3-7. *Metabolic sequence leading to the formation of the ultimate carcinogenic species of benzo[a]pyrene, (+)-7R,8S-dihydroxy-9R,10R-oxy-7,8,9,10-tetrahydrobenzo[a]pyrene or (+)-7.8-diol-9,10-epoxide.*

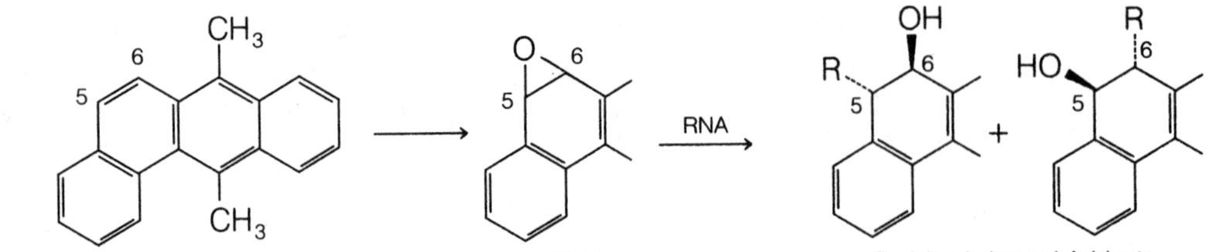

7,12-Dimethylbenz[*a*]anthracene → 5,6-Oxide —RNA→ Covalently bound Adducts to Guanosine

Where R =

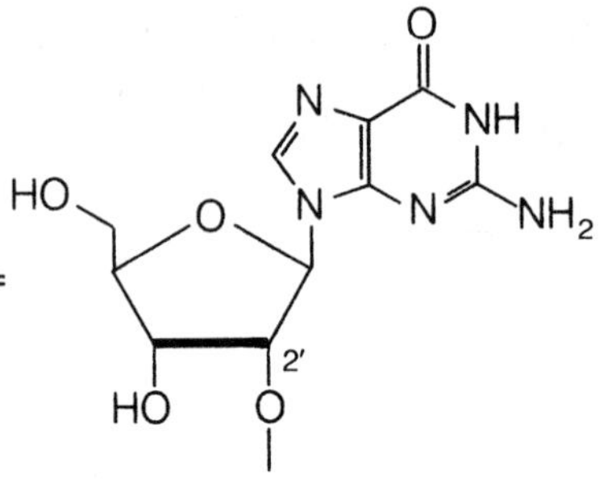

covalent adducts to nucleic acids (RNA).[70] The ultimate carcinogenic reactive species apparently is the 5,6-oxide that results from epoxidation of the 5,6-double bond in this aromatic hydrocarbon. The arene oxide intermediate binds covalently to guanosine residues of RNA to yield the two adducts shown below.

OXIDATION OF OLEFINS

The metablic oxidation of olefinic carbon–carbon double bonds leads to the corresponding epoxide (or oxirane). Epoxides derived from olefins generally tend to be somewhat more stable than the arene oxides formed from aromatic compounds. A few epoxides are stable enough that they are directly measurable in biologic fluids (e.g., plasma, urine). As with their arene oxide counterparts, epoxides are susceptible to enzymatic hydration by epoxide hydrase to form *trans*-1,2-dihydrodiols (also called 1,2-diols or 1,2-dihydroxy compounds).[57] In addition, several epoxides undergo glutathione conjugation.[71]

A well-known example of olefinic epoxidation is the metabolism, in humans, of the anticonvulsant drug carbamazepine (Tegretol) to carbamazepine-10,11-epoxide.[72] The epoxide is reasonably stable and can be quantitatively measured in the plasma of patients receiving the parent drug. Recent evidence suggests that the epoxide metabolite has marked anticonvulsant activity and, therefore, may contribute substantially to the therapeutic effect of the parent drug.[73] Subsequent hydration of the epoxide produces 10,11-dihydroxycarbamazepine, an important urinary metabolite (10–30%) in humans.[72]

Epoxidation of the olefinic 10,11-double bond in the antipsychotic agent protriptyline (Vivactl)[74] and in the H_1-histamine antagonist cyproheptadine (Periactin)[75] has also been demonstrated. Frequently, the epoxides formed from the biotransformation of an olefinic compound are minor products, owing to their further conversion to the corresponding 1,2-diols. For example, dihydroxyalcofenac is a major human urinary metabolite of the anti-inflammatory agent alclofenac.[76] However, the epoxide metabolite from which it is derived is present in minute amounts. The presence of the dihydroxy metabolite (called secodiol) of secobarbital but not the epoxide product has also been reported in humans.[77]

Indirect evidence for the formation of epoxides comes also from the isolation of glutathione or mercapturic acid metabolites. After administration of styrene to rats, two urinary metabolites have been

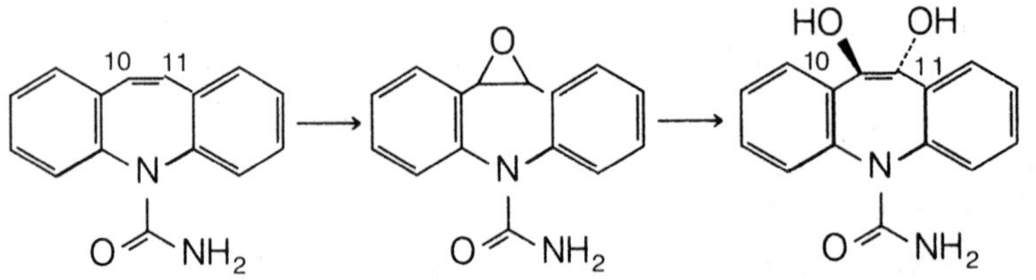

Carbamazepine → Carbamazepine-10,11-epoxide → *trans*-10,11-Dihydroxy-carbamazepine

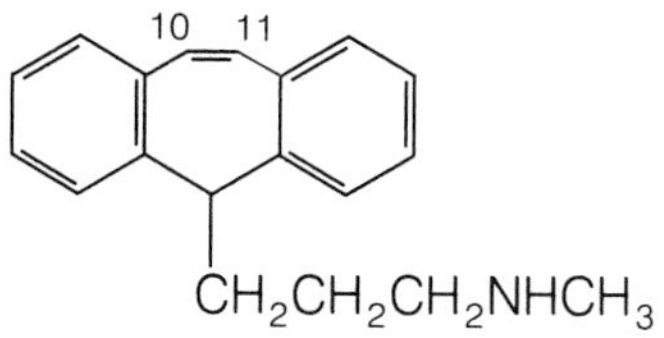

$CH_2CH_2CH_2NHCH_3$

Protriptyline

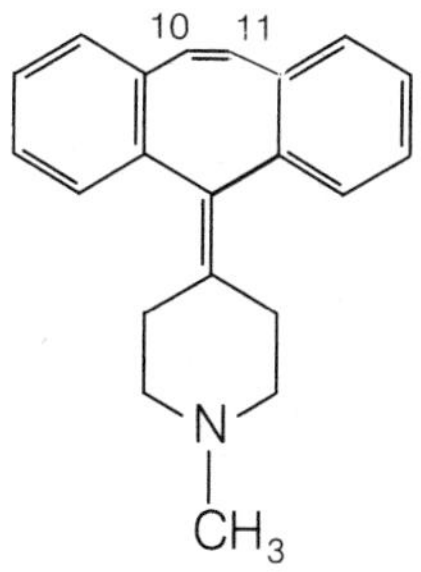

CH_3

Cyproheptadine

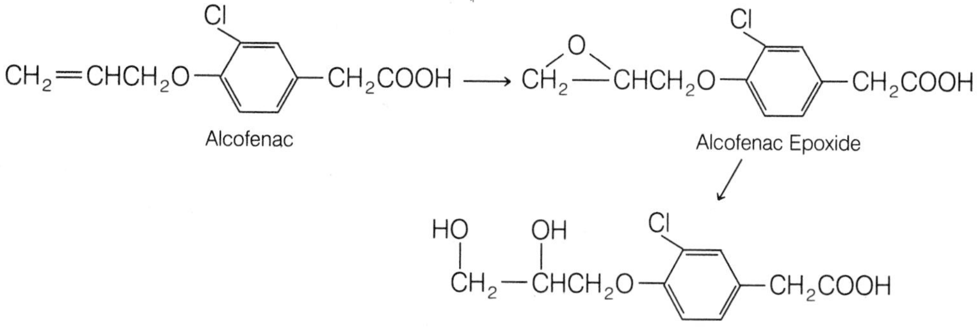

Alcofenac

Alcofenac Epoxide

Dihydroxyalcofenac

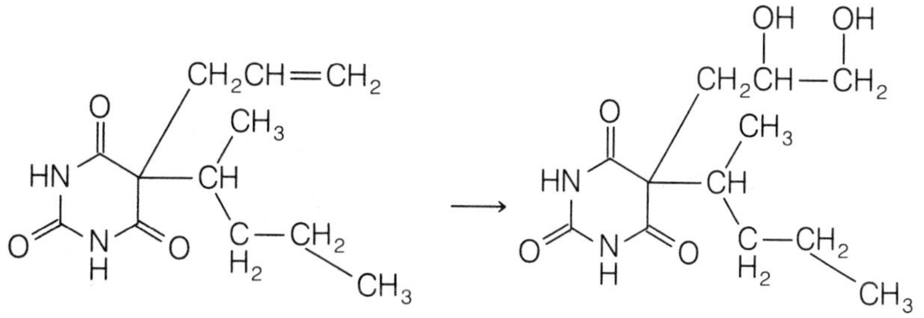

Secobarbital

Secodiol
5-(2,3-Dihydroxypropyl)-5-(1-methylbutyl)-barbituric Acid

identified as the isomeric mercapturic acid derivatives resulting from nucleophilic attack of glutathione on the intermediate epoxide.[78] In addition, styrene oxide covalently binds to rat liver microsomal proteins and nucleic acids.[79] These results indicate that styrene oxide is relatively reactive toward nucleophiles (e.g., glutathione and nucleophilic groups on protein and nucleic acids).

There apparently are divers metabolically generated epoxides that display similar chemical reactivity toward nucleophilic functionalities. Accordingly, it has been suggested that the toxicity of some olefinic compounds may be a consequence of their metabolic conversion to chemically reactive epoxides.[80] One example that clearly links metabolic epoxidation as a biotoxification pathway involves aflatoxin B_1. This naturally occurring carcinogenic agent contains an olefinic (C2–C3) double bond adjacent to a cyclic ether oxygen. The hepatocarcinogenicity of aflatoxin B_1 has been clearly linked to its metabolic oxidation to the corresponding 2,3-oxide, which is extremely reactive.[81] Extensive in vitro and in vivo metabolic studies indicate that this 2,3-oxide binds covalently to DNA, RNA, and proteins. A major DNA adduct has been isolated and characterized as 2,3-dihydro-2-(N^7-guanyl)-3-hydroxyaflatoxin B_1.[82]

Other olefinic compounds, such as vinyl chloride,[83] stilbene,[84] and the carcinogenic estrogenic agent diethylstilbestrol (DES),[85] have been observed to undergo metabolic epoxidation. It has been suggested that the corresponding epoxide metabolites may be the reactive species responsible for mediating the cellular toxicity seen with these compounds.

An interesting group of olefin-containing compounds is known to cause the destruction of cy-

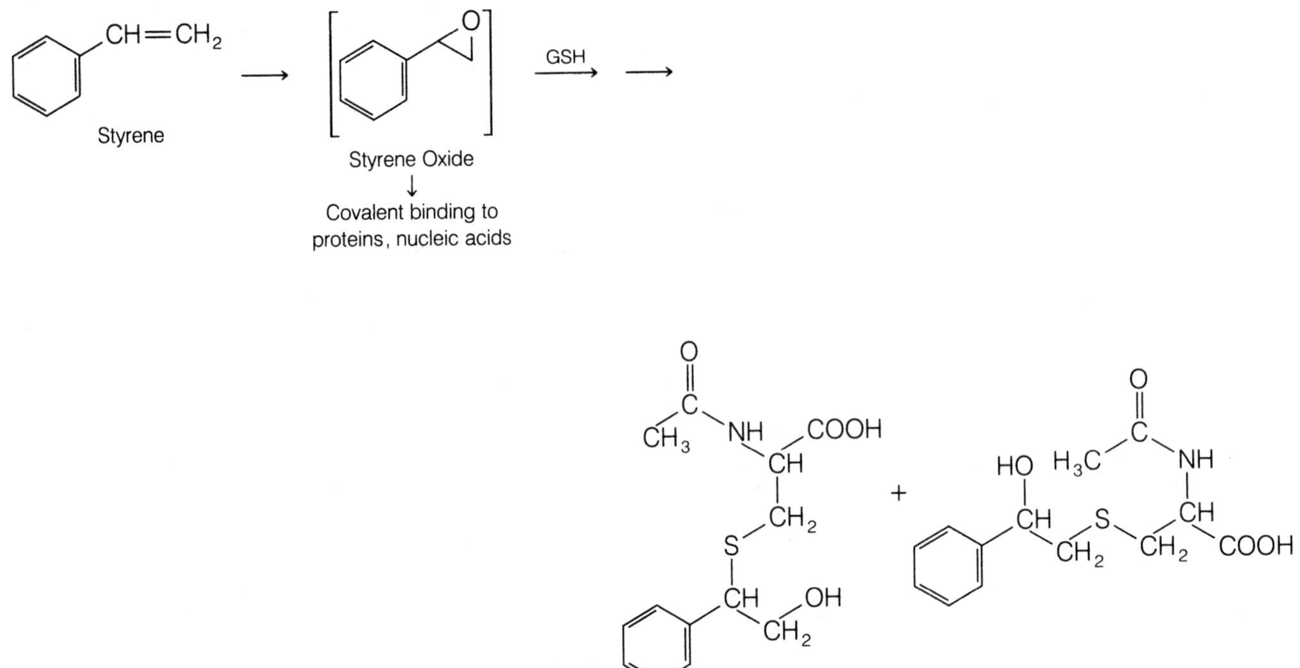

Styrene → [Styrene Oxide] —GSH→ →

Styrene Oxide ↓ Covalent binding to proteins, nucleic acids

Mercapturic Acid Derivative (major) + Mercapturic Acid Derivative (minor)

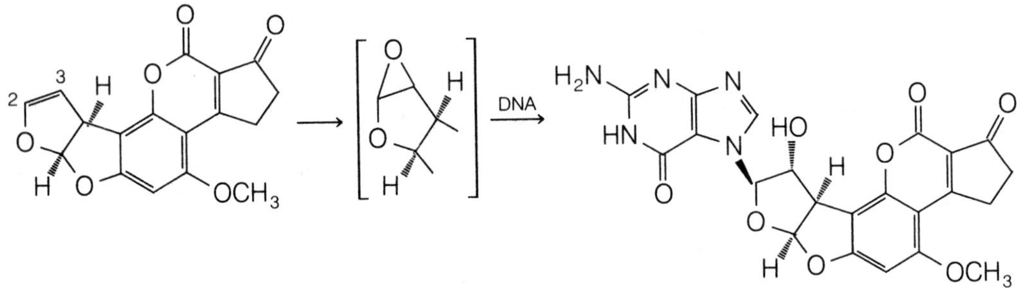

Aflatoxin B_1 → 2,3-Expoxide —DNA→ 2,3-Dihydro-2-(N^7-guanyl)-3-hydroxyaflatoxin B_1

tochrome P-450.[86] Compounds belonging to this group include allylisopropylacetamide,[87] secobarbital,[88] and the volatile anesthetic agent, fluroxene.[89] It is believed that the olefinic moiety present in these compounds is metabolically activated by cytochrome P-450 to form a very reactive intermediate that covalently binds to the heme portion of cytochrome P-450.[90] The abnormal heme derivatives or "green pigments" that result from this covalent interaction have been characterized as *N*-alkylated protoporphyrins in which the *N*-alkyl moiety is directly derived from the olefin administered.[86,90] Chronic administration of the foregoing three agents is expected to lead to inhibition of oxidative drug metabolism and to potential drug interactions and prolonged pharmacologic effects.

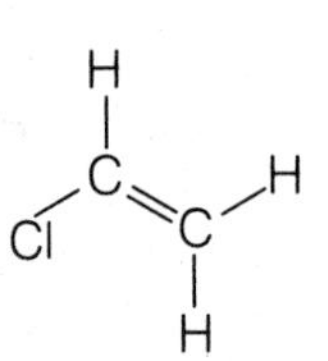

Vinyl chloride

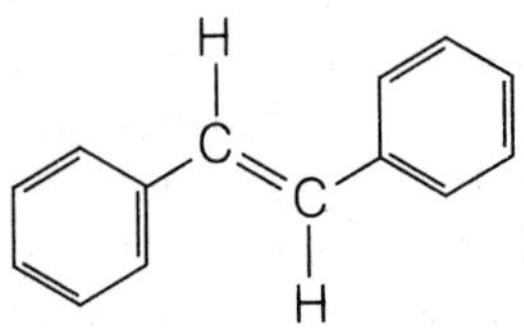

Stilbene

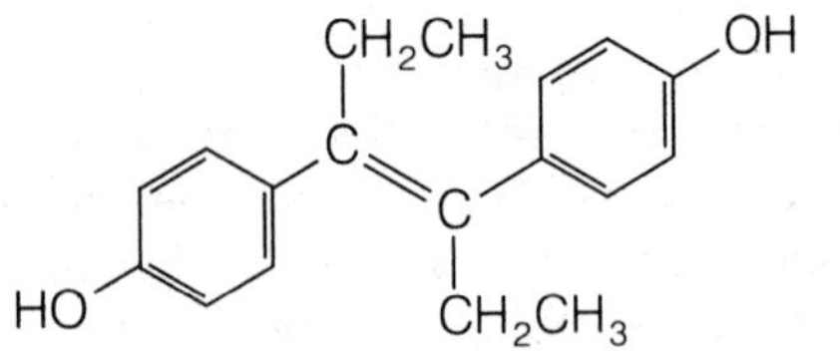

Diethylstilbestrol (DES)

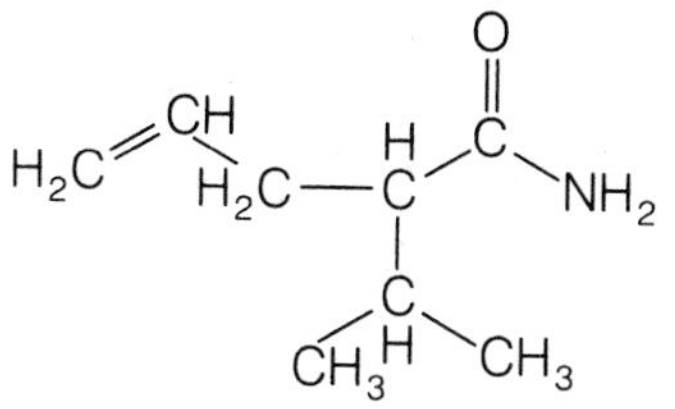

Allylisopropylacetamide

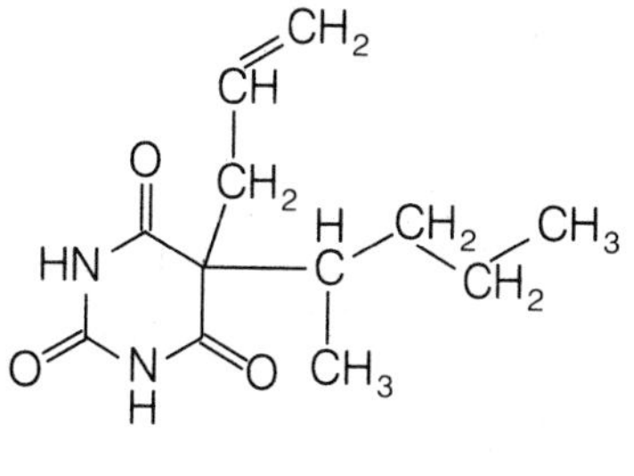

Secobarbital

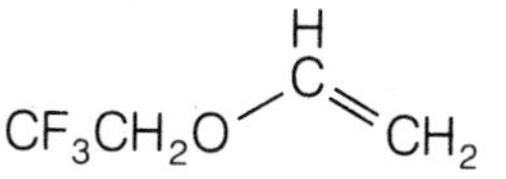

Fluroxene

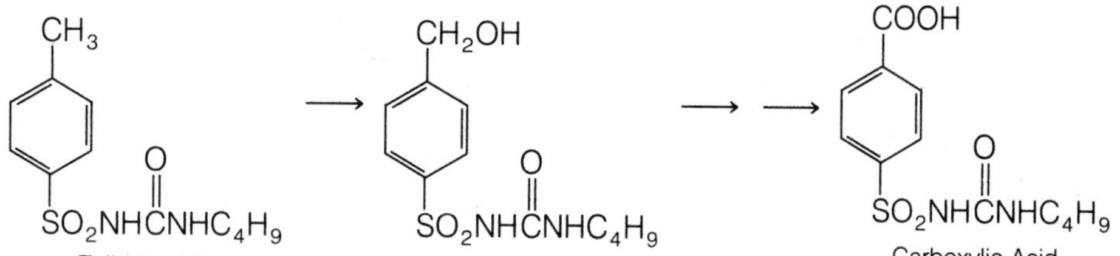

Tolbutamide

Alcohol Metabolite

Carboxylic Acid Metabolite

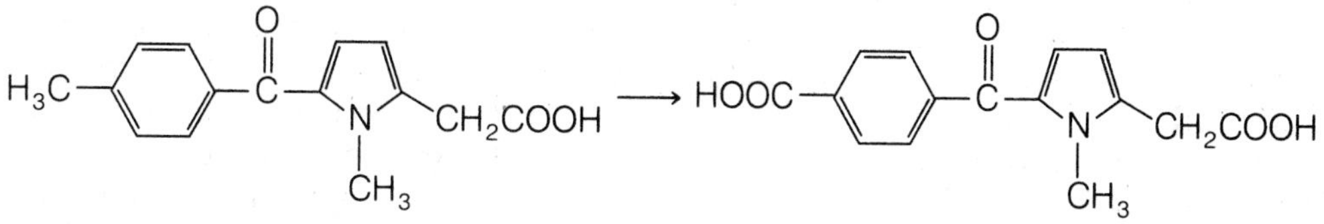

Tolmetin

Dicarboxylic Acid Metabolite

OXIDATION AT BENZYLIC CARBON ATOMS

Carbon atoms attached to aromatic rings (benzylic position) are susceptible to oxidation, thereby forming the corresponding alcohol (or carbinol) metabolite.[47] Primary alcohol metabolites are often oxidized further to aldehydes and carboxylic acids ($CH_2OH \rightarrow CHO \rightarrow COOH$), and secondary alcohols are converted to ketones by soluble alcohol and aldehyde dehydrogenases.[91] Alternatively, the alcohol may be directly conjugated with glucuronic acid.[92] The benzylic carbon atom present in the oral hypoglycemic agent tolbutamide (Orinase) is extensively oxidized to the corresponding alcohol and carboxylic acid. Both metabolites have been isolated in the urine of humans.[93] Similarly, the "benzylic" methyl group in the anti-inflammatory agent tolmetin (Tolectin) undergoes oxidation to yield the dicarboxylic acid product as the major metabolite in humans.[94] The sedative hypnotic agent methaqualone has been observed to undergo benzylic oxidation at its C-2′ methyl group to give 2′-hydroxymethylmethaqualone as a minor metabolite.[95] Benzylic hydroxylation occurs to a significant extent in the metabolism of the β-adrenergic blocker metoprolol (Lopressor) to yield α-hydroxymetoprolol.[96] Additional examples of drugs and xenobiotics undergoing benzylic oxidation are shown in Figure 3-8.

OXIDATION AT ALLYLIC CARBON ATOMS

Microsomal hydroxylation at allylic carbon atoms is commonly observed in drug metabolism. An illustrative example of allylic oxidation is given by the psychoactive component of marijuana, Δ^1-tetrahydrocannabinol (Δ^1-THC). This molecule contains three allylic carbon centers (C-7, C-6, and C-3). Allylic hydroxylation occurs extensively at C-7 to yield 7-hydroxy-Δ^1-THC as the major plasma metabolite in humans.[10] Pharmacologic studies show that this

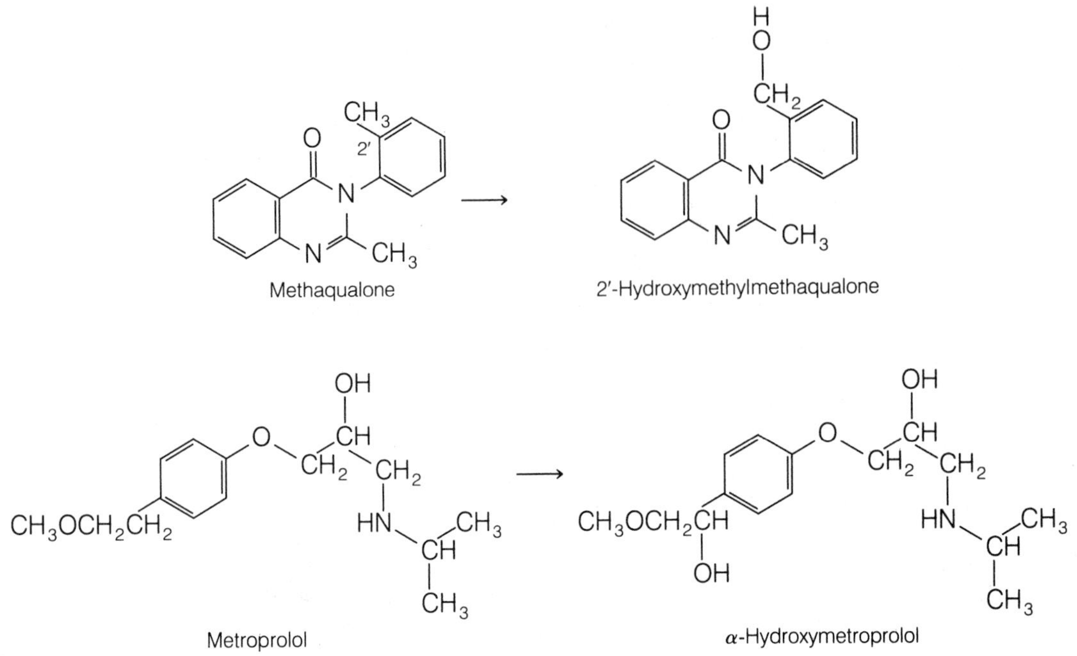

Methaqualone → 2′-Hydroxymethylmethaqualone

Metroprolol → α-Hydroxymetroprolol

7-hydroxy metabolite is as active or even more active than Δ^1-THC, itself, and may contribute significantly to the overall control nervous system psychotomimetic effects of the parent compound.[103] Hydroxylation also occurs to a minor extent at the allylic C-6 position to give both the epimeric 6α- and 6β-hydroxy metabolites.[10] Metabolism does not occur at C-3, presumably because of steric hindrance.

The antiarrhythmic agent, quinidine, is metabolized by allylic hydroxylation to 3-hydroxyquinidine, the principal plasma metabolite found in humans.[104] Recent reports indicate that this metabolite shows significant antiarrhythmic activity in animals and possibly also in humans.[105]

Other examples of allylic oxidation include the sedative hypnotic hexobarbital (Sombulex) and the analgesic pentazocine (Talwin). The 3′-hydroxylated metabolite formed from hexobarbital is susceptible to glucuronide conjugation as well as further oxidation to the 3′-oxo compound.[106] Hexobarbital is a

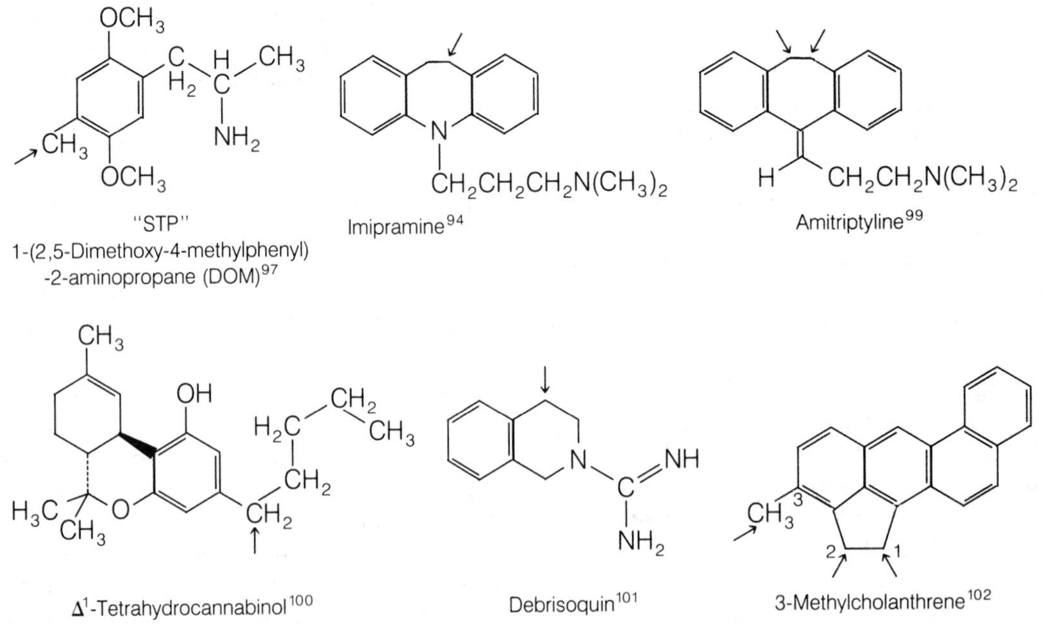

"STP" 1-(2,5-Dimethoxy-4-methylphenyl) -2-aminopropane (DOM)[97]

Imipramine[94]

Amitriptyline[99]

Δ^1-Tetrahydrocannabinol[100]

Debrisoquin[101]

3-Methylcholanthrene[102]

FIG. 3-8. Examples of drugs and xenobiotics undergoing benzylic hydroxylation. Arrow indicates site of hydroxylation.

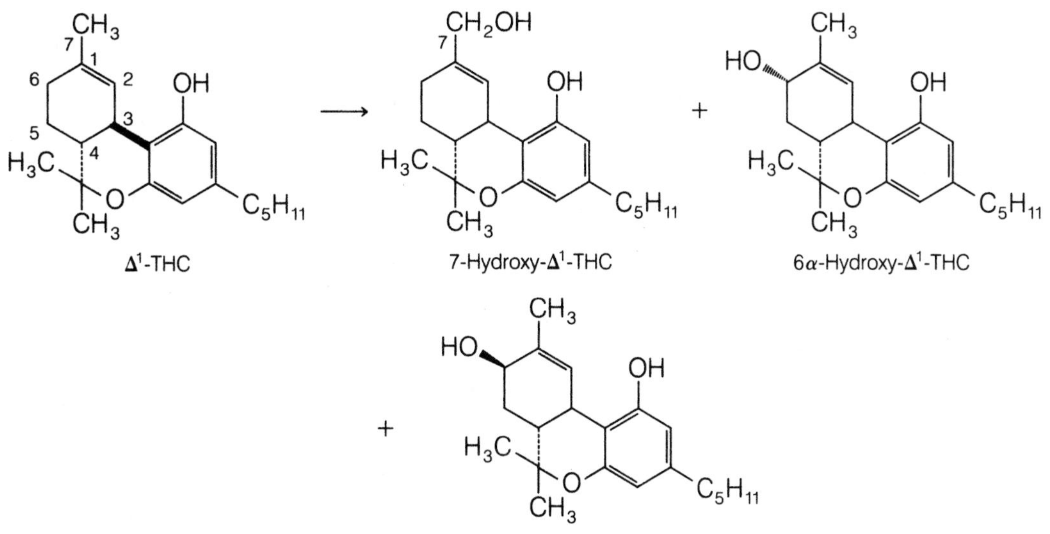

Δ^1-THC → 7-Hydroxy-Δ^1-THC + 6α-Hydroxy-Δ^1-THC + 6β-Hydroxy-Δ^1-THC

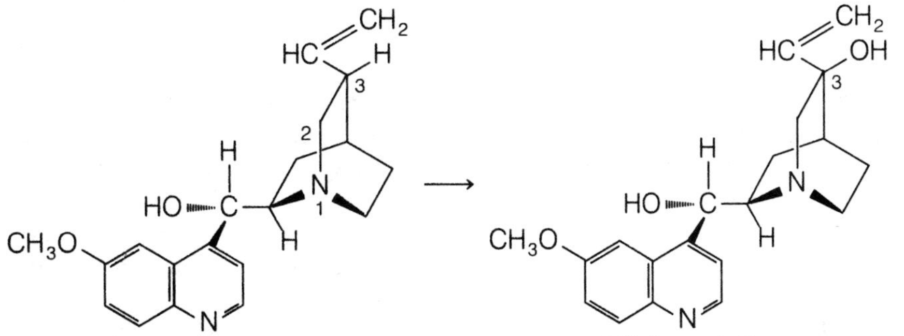

Quinidine → 3-Hydroxyquinidine

chiral barbiturate derivative that exists in two enantiomeric forms. Studies in humans indicate that the pharmacologically less active (*R*)(−)-enantiomer is more rapidly metabolized than its (*S*)(+)-isomer.[107] Pentazocine undergoes allylic hydroxylation at the two terminal methyl groups of its *N*-butenyl side chain to yield either the *cis* or *trans* alcohol metabolites shown in the diagrams. In humans, greater amounts of the *trans* alcohol are formed.[108]

For the hepatocarcinogenic agent, safrole, allylic hydroxylation is involved in a bioactivation pathway leading to the formation of chemically reactive metabolites.[109] This process involves initial hydroxylation at the C-1′ carbon of safrole. It should be noted that this center is both allylic and benzylic. The hydroxylated metabolite then undergoes further conjugation to form a sulfate ester. This chemically reactive ester intermediate presumably undergoes nucleophilic displacement reactions with DNA or RNA in vitro to form covalently bound adducts.[110] As shown in the following scheme (see Fig. 3-8), nucleophilic attack by DNA, RNA, or other nucleophiles is facilitated by a good leaving group (e.g., SO_4^{2-}) at the C-1′ position. The leaving group tendency of the alcohol OH group, itself, is not great enough to facilitate displacement reactions. Impor-

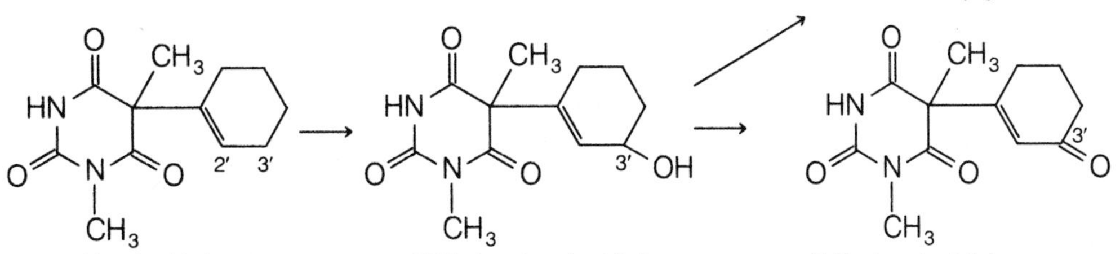

Hexobarbital → 3′-Hydroxyhexobarbital → 3′-Oxohexobarbital; 3′-Hydroxyhexobarbital → *O*-Glucuronide conjugate

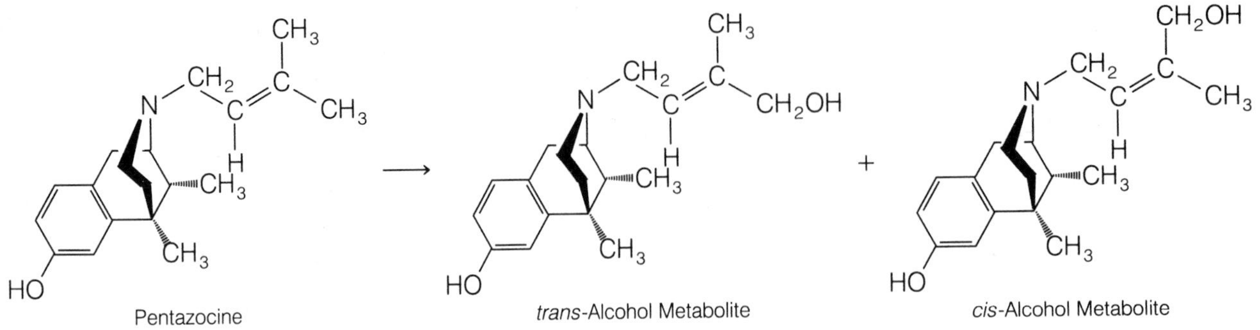

Pentazocine → *trans*-Alcohol Metabolite + *cis*-Alcohol Metabolite

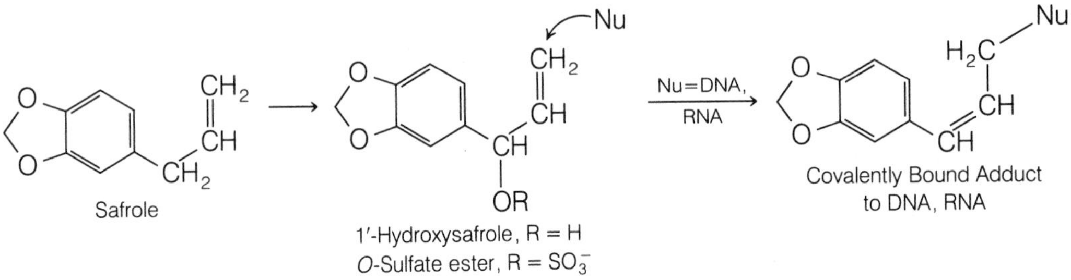

Safrole → 1′-Hydroxysafrole, R = H; *O*-Sulfate ester, R = SO_3^- → (Nu = DNA, RNA) Covalently Bound Adduct to DNA, RNA

tantly, allylic hydroxylation generally is not a pathway that leads to the generation of reactive intermediates. Its involvement in the biotoxification of safrole appears to be an exception.

OXIDATION AT CARBON ATOMS ALPHA TO CARBONYLS AND IMINES

The mixed function oxidase system also oxidizes carbon atoms adjacent (i.e., alpha) to carbonyl and imino (C=N) functionalities. An important class of drugs undergoing this type of oxidation is the benzodiazepines. For example, diazepam (Valium), flurazepam (Dalmane), and nimetazepam, all are oxidized to their corresponding 3-hydroxy metabolites.[111] The C-3 carbon atom undergoing hydroxylation is α to both a lactam carbonyl and an immino functionality.

For diazepam, the hydroxylation reaction proceeds with remarkable stereoselectivity to form primarily (90%) 3-hydroxydiazepam (also called *N*-methyloxazepam) having the (*S*) absolute configuration at C-3.[112] Further *N*-demethylation of the

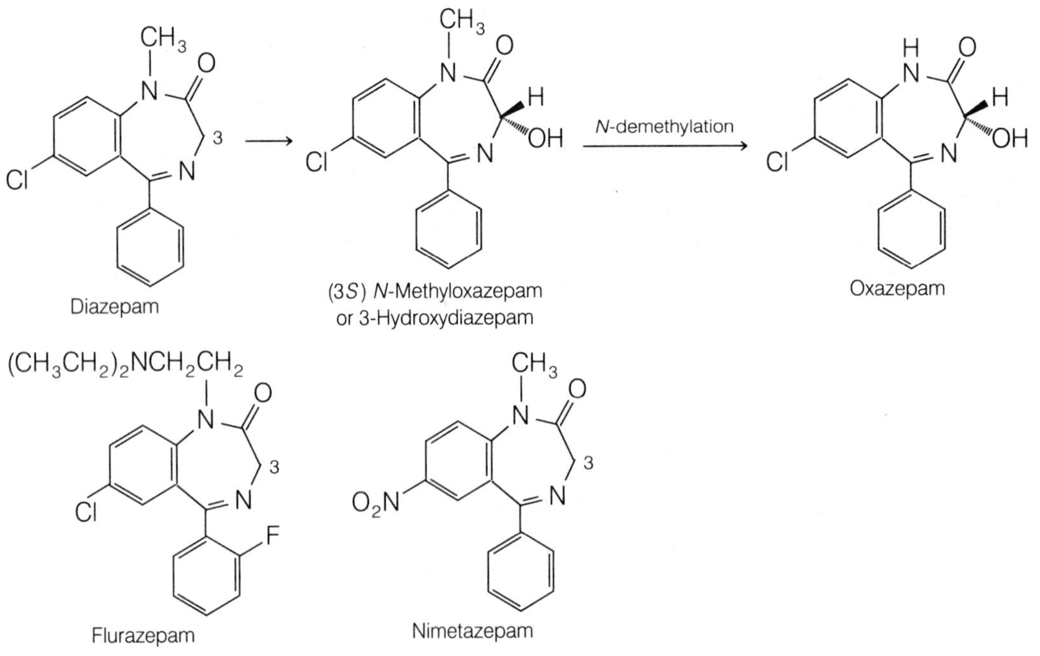

Diazepam → (3*S*) *N*-Methyloxazepam or 3-Hydroxydiazepam → (*N*-demethylation) Oxazepam

Flurazepam

Nimetazepam

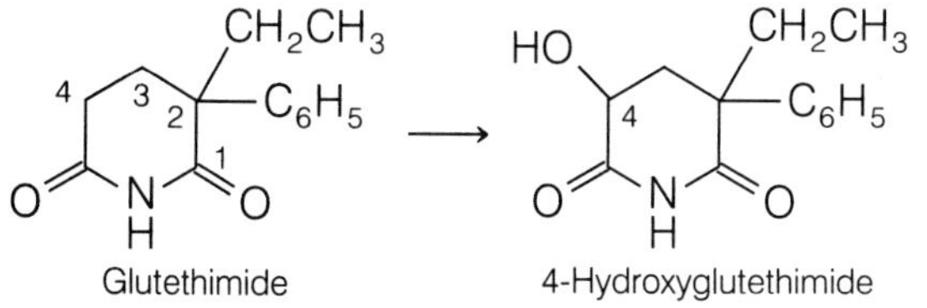

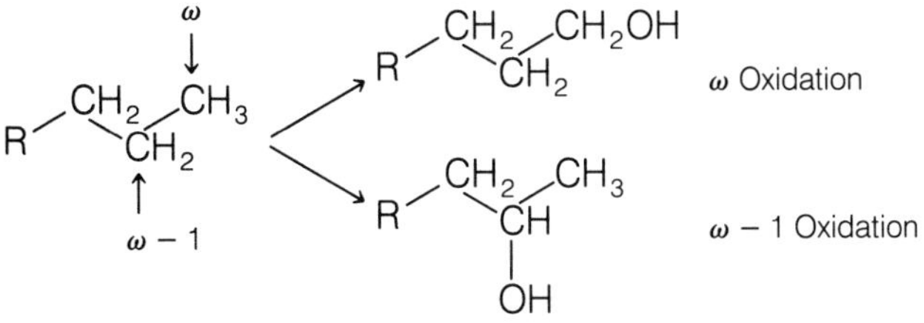

Glutethimide → 4-Hydroxyglutethimide

ω Oxidation

ω − 1 Oxidation

latter metabolite gives rise to the pharmacologically active 3(*S*)(+)-oxazepam.

Hydroxylation of the carbon atom α to carbonyl functionalities generally occurs only to a limited extent in drug metabolism. An illustrative example involves the hydroxylation of the sedative hypnotic, glutethimide (Doriden), to 4-hydroxyglutethimide.[113]

OXIDATION AT ALIPHATIC AND ALICYCLIC CARBON ATOMS

Alkyl or aliphatic carbon centers are subject to mixed function oxidation. Metabolic oxidation at the terminal methyl group is often referred to as *ω-oxidation*, and oxidation of the penultimate carbon atom (i.e., next-to-the-last carbon) is called *ω − 1 oxidation*.[47] The initial alcohol metabolites formed from these enzymatic ω- and ω − 1 oxidations are susceptible to further oxidation to yield aldehyde, ketones, or carboxylic acids. Alternatively, the alcohol metabolites may undergo glucuronide conjugation.

Aliphatic ω- and ω − 1 hydroxylations commonly take place in drug molecules having straight or branched alkyl chains. For example, the antiepileptic agent valproic acid (Depakene) undergoes both ω- and ω − 1 oxidation to the 5-hydroxy and 4-hydroxy metabolites, respectively.[114] Further oxidation of the 5-hydroxy metabolite yields 2-*n*-propylglutaric acid.

Numerous barbiturates and oral hypoglycemic sulfonylureas also have aliphatic side chains that are susceptible to oxidation. For example, the sedative hypnotic amobarbital (Amytal) undergoes extensive ω − 1 oxidation to the corresponding 3′-hydroxylated metabolite.[115] Other barbiturates, such as pentobarbital,[116] thiamylal,[117] and secobarbital,[77] have also been reported to be metabolized by way of ω and ω − 1 oxidation.

The *n*-propyl side chain attached to the oral hypoglycemic agent chlorpropamide (Diabinese) undergoes extensive ω − 1 hydroxylation to yield the secondary alcohol 2′-hydroxychlorpropamide as a major urinary metabolite in humans.[118]

Omega and ω − 1 oxidation of the isobutyl moiety present in the anti-inflammatory agent ibuprofen (Motrin) yield the corresponding carboxylic acid and tertiary alcohol metabolites.[119] Additional examples of drugs reported to undergo aliphatic hydroxylation include meprobamate,[120] glutethimide,[113] ethosuximide,[121] and phenylbutazone.[122]

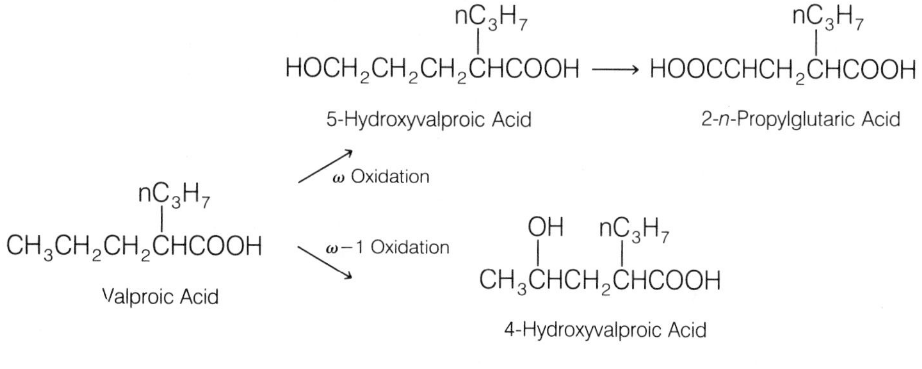

5-Hydroxyvalproic Acid → 2-*n*-Propylglutaric Acid

ω Oxidation

ω−1 Oxidation

Valproic Acid

4-Hydroxyvalproic Acid

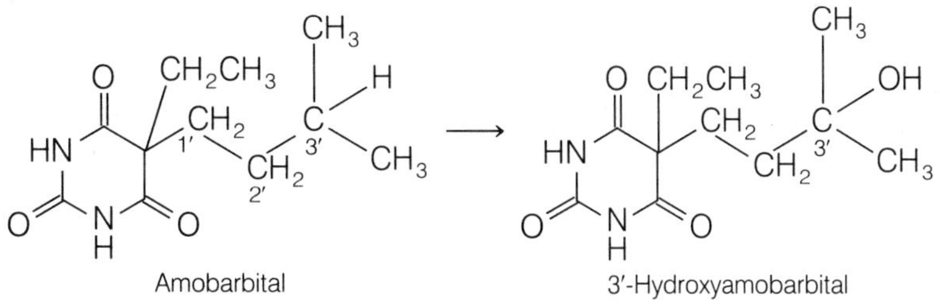

Amobarbital → 3′-Hydroxyamobarbital

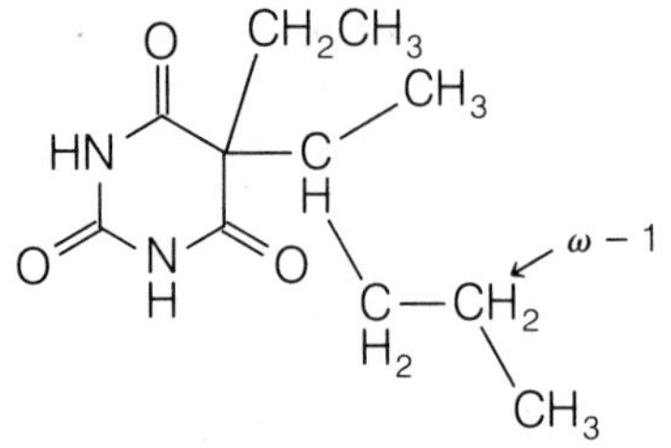

Pentobarbital

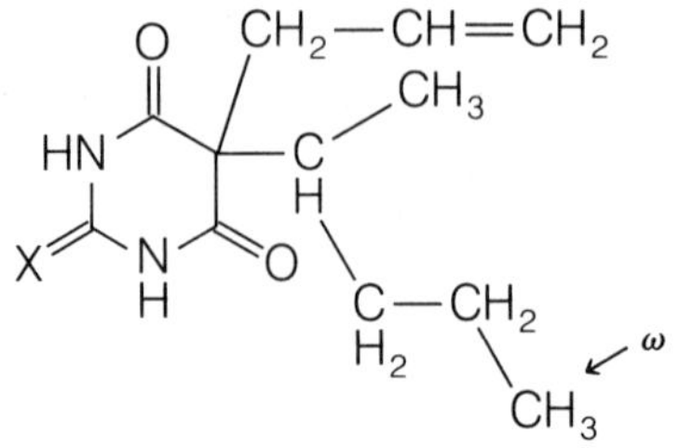

Thiamylal X = S
Secobarbital X = O

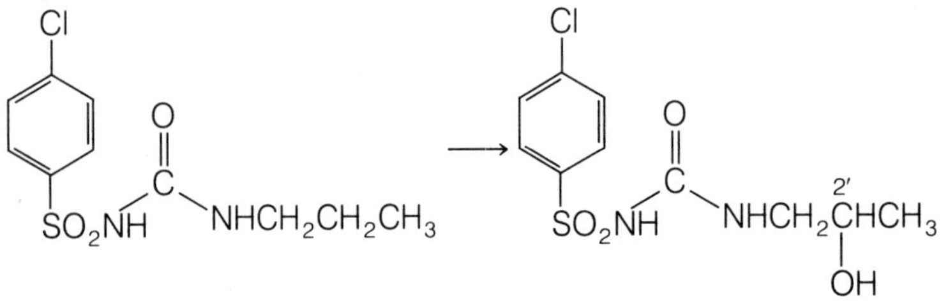

Chlorpropamide

2′-Hydroxychlorpropamide

The cyclohexyl group is commonly found in many medicinal agents and is also susceptible to mixed function oxidation (alicyclic hydroxylation).[47] Enzymatic introduction of a hydroxyl group into a monosubstituted cyclohexane ring generally occurs at C-3 or C-4 and can lead to *cis* and *trans* conformational stereo-isomers, as shown in the diagrammed scheme.

An example illustrating this hydroxylation pathway is seen in the metabolism of the oral hypoglycemic agent acetohexamide (Dymelor). In humans, the *trans*-4-hydroxycyclohexyl product has been reported as a major metabolite.[123] Small amounts of the other possible stereoisomers, namely, the *cis*-4-, *cis*-3-, and *trans*-3-hydroxycyclohexyl derivatives, have also been detected. Another related oral hypoglycemic agent, glipizide, is oxidized in humans to the *trans*-4- and *cis*-3-hydroxylcyclohexyl metabolites in about a 6 : 1 ratio.[124]

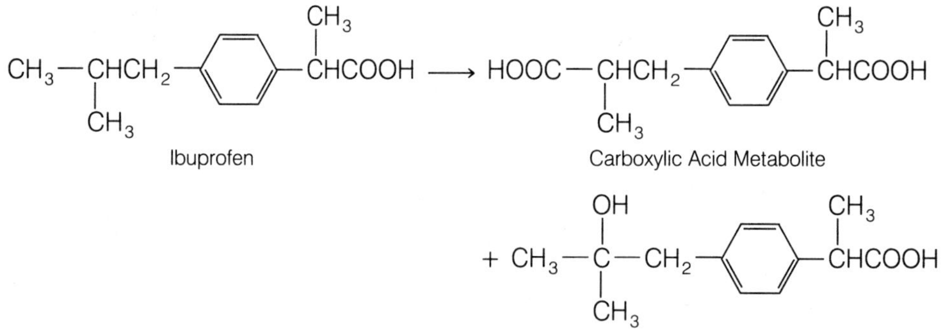

Ibuprofen

Carboxylic Acid Metabolite

Tertiary Alcohol Metabolite

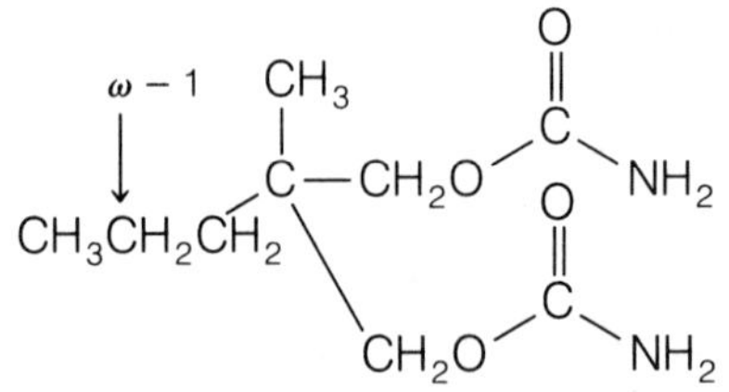

Meprobamate

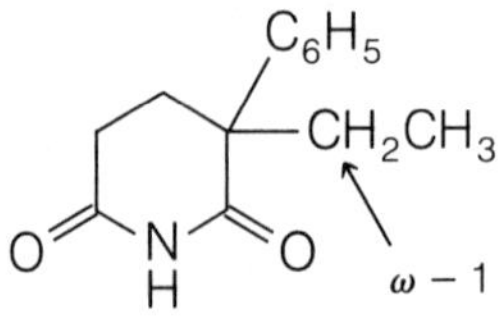

Glutethimide

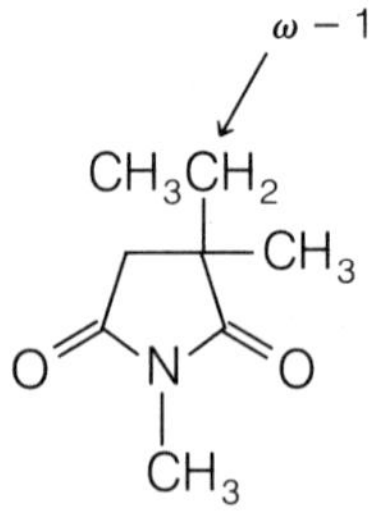

Ethosuximide

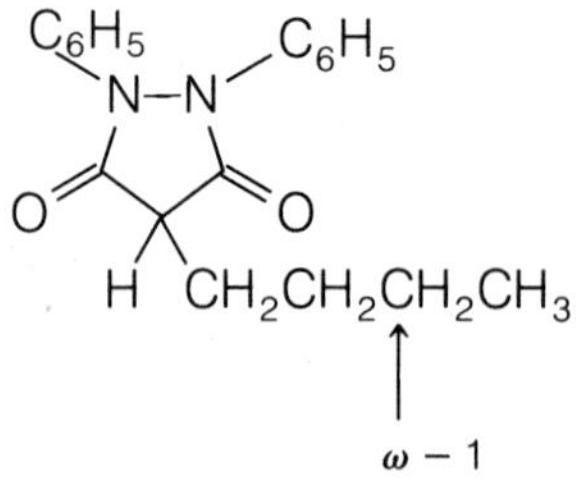

Phenylbutazone

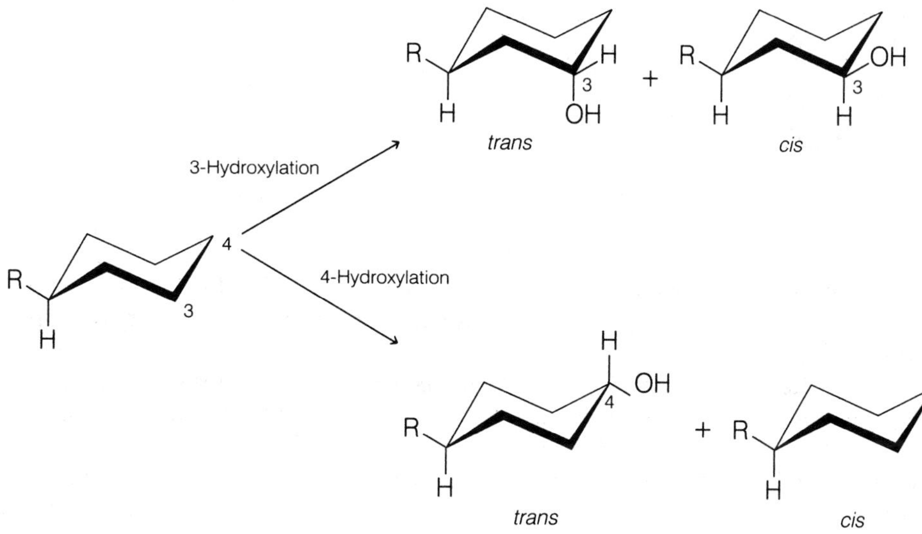

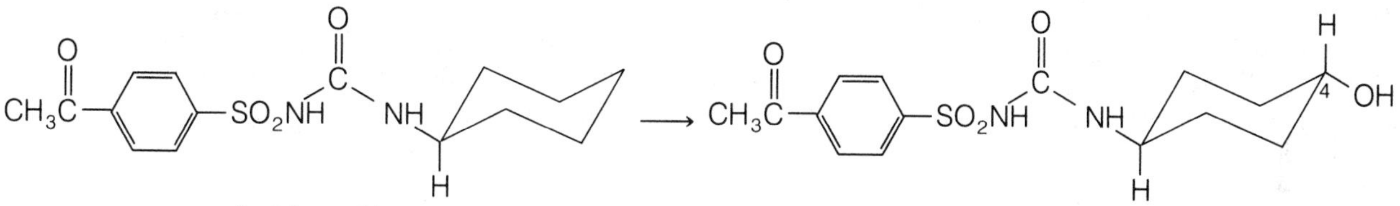

3-Hydroxylation

4-Hydroxylation

trans

cis

trans

cis

Acetohexamide

trans-4-Hydroxyacetohexamide

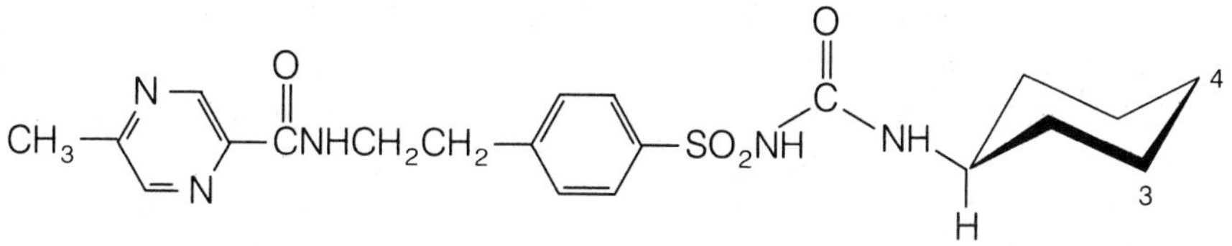

Glipizide

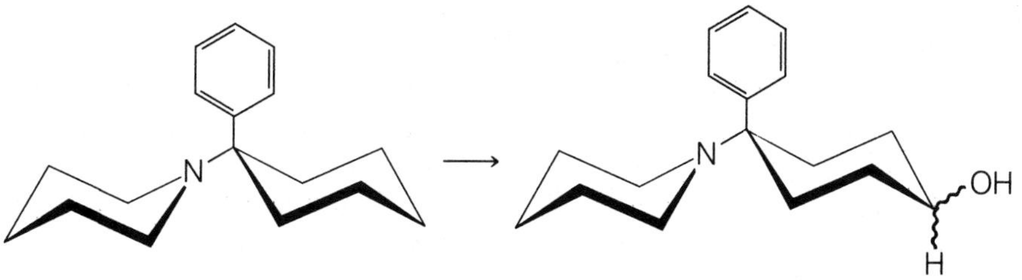

Phencyclidine

4-Hydroxycyclohexyl Metabolite

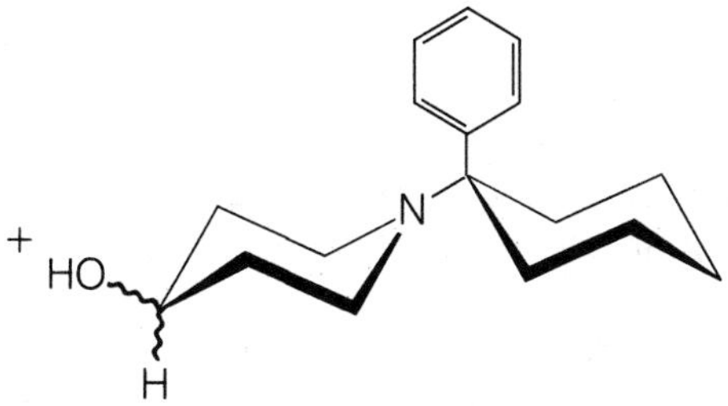

4-Hydroxypiperidyl Metabolite

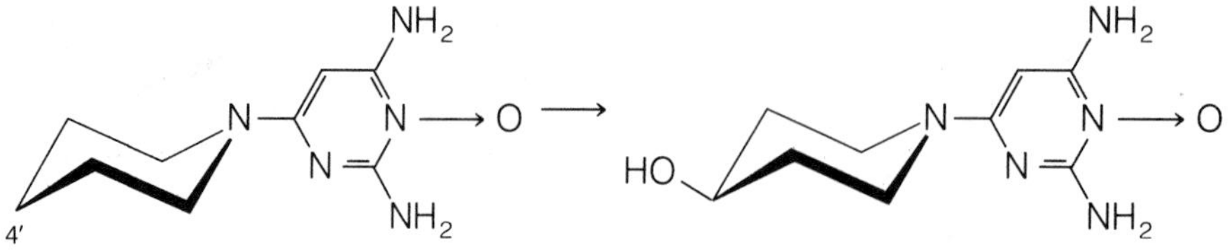

Minoxidil → 4′-Hydroxyminoxidil

Two human urinary metabolites of phencyclidine (PCP) have been identified as the 4-hydroxypiperidyl and 4-hydroxycyclohexyl derivatives of the parent compound.[125a] Thus, from these results, it appears that "alicyclic" hydroxylation of the six-membered piperidyl moiety may closely parallel the hydroxylation pattern of the cyclohexyl moiety. The stereochemistry of the hydroxylated centers in the two metabolites has not been clearly established. Biotransformation of the antihypertensive agent minoxidil (Loniten) yields the 4′-hydroxypiperidyl metabolite. In the dog, this product is a major urinary metabolite (29% to 47%), whereas in humans it is detected in small amounts (approximately 3%).[125b]

Several structural features frequently determine which pathway will predominate, especially in carbon–nitrogen systems. Metabolism of some nitrogen-containing compounds is complicated by the fact that carbon- or nitrogen-hydroxylated products may undergo secondary reactions to form other, more complex, metabolic products (e.g., oxime, nitrone, nitroso, imino). Other oxidative processes that do not fall under the foregoing two basic categories will be discussed individually in the appropriate carbon–heteroatom section. The metabolism of carbon–nitrogen systems will be discussed first, followed by the metabolism of carbon–oxygen and carbon–sulfur systems.

OXIDATION INVOLVING CARBON – HETEROATOM SYSTEMS

Nitrogen and oxygen functionalities are commonly found in most drugs and foreign compounds, whereas sulfur functionalities occur only occasionally. Metabolic oxidation of carbon–nitrogen, carbon–oxygen, and carbon–sulfur systems principally involves two basic types of biotransformation processes:

1. Hydroxylation of the α-carbon atom attached directly to the heteroatom (N, O, S). The resulting intermediate is often unstable and decomposes with the cleavage of the carbon–heteroatom bond:

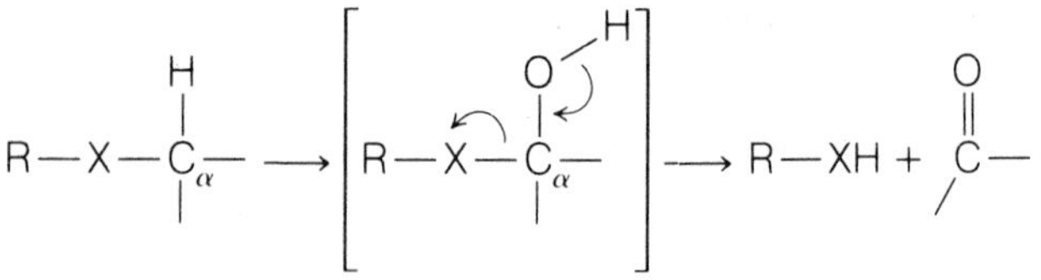

$$R{-}X{-}\overset{H}{\underset{|}{C_\alpha}}{-} \longrightarrow \left[R{-}X{-}\overset{OH}{\underset{|}{C_\alpha}}{-}\right] \longrightarrow R{-}XH + \overset{O}{\overset{\|}{C}}{-}$$

Where X = N,O,S   Usually Unstable

   Oxidative N-, O-, and S-dealkylation, as well as oxidative deamination reactions, fall under this mechanistic pathway.

2. Hydroxylation or oxidation of the heteroatom (N, S only) e.g., N-hydroxylation, N-oxide formation, sulfoxide, and sulfone formation).

OXIDATION INVOLVING CARBON – NITROGEN SYSTEMS

Metabolism of nitrogen functionalities (e.g., amines, amides) is of importance, because such functional groups are found in many natural products (e.g., morphine, cocaine, nicotine) and in numerous important drugs (e.g., phenothiazines, antihistamines, tricyclic antidepressants, β-adrenergic agents, sympathomimetic phenylethylamines, barbiturates, benzodiazepines).[126] The discussion to follow divides nitrogen-containing compounds into three basic classes:

1. Aliphatic (tertiary, secondary, and primary) and alicyclic (tertiary, secondary) amines
2. Aromatic and heterocyclic nitrogen compounds
3. Amides

The susceptibility of each class of these nitrogen compounds to either α-carbon hydroxylation or N-oxidation and the metabolic products that are formed will be discussed.

The hepatic enzymes responsible for carrying out α-carbon hydroxylation reactions are the cytochrome P-450 mixed function oxidases. However, the N-hydroxylation or N-oxidation reactions appear to be catalyzed not only by cytochrome P-450 but also by a second class of hepatic mixed function oxidases called amine oxidases (sometimes called

N-oxidases).[127] These enzymes are NADPH-dependent flavoproteins and do not contain cytochrome P-450.[128] They require NADPH and molecular oxygen to carry out N-oxidation.

Tertiary aliphatic and alicyclic amines

The oxidative removal of alkyl groups (particularly methyl groups) from tertiary aliphatic and alicylic amines is carried out by hepatic cytochrome P-450 mixed function oxidase enzymes. This reaction is commonly referred to as oxidative N-dealkylation.[129] The initial step involves α-carbon hydroxylation to form a carbinolamine intermediate, which is unstable and undergoes spontaneous heterolytic cleavage of the C — N bond to give a secondary amine and a carbonyl moiety (aldehyde or ketone).[130] In general, small alkyl groups, such as methyl, ethyl, and isopropyl, are rapidly removed.[129] N-Dealkylation of the t-butyl group is not possible by the carbinolamine pathway because α-carbon hydroxylation cannot occur. Removal of the first alkyl group from a tertiary amine occurs more rapidly than the removal of the second alkyl group. In some instances, bisdealkylation of the tertiary aliphatic amine to the corresponding primary aliphatic amine occurs very slowly.[129] For example, the tertiary amine imipramine (Tofranil) is monodemethylated to desmethylimipramine (desipramine).[98, 131] This major plasma metabolite is pharmacologically active in humans and contributes substantially to the antidepressant activity of the parent drug.[132] Very little of the bisdemethylated metabolite of imipramine is detected. In contrast, the local anesthetic and antiarrhythmic agent, lidocaine, is extensively metabolized by N-deethylation to both monoethylglycylxylidine and glycyl-2,6-xylidine in humans.[133]

Numerous other tertiary aliphatic amine drugs are metabolized principally by oxidative N-dealkylation. Some of these include the antiarrhythmic disopyramide (Norpace),[134] the antiestrogenic agent tamoxifen (Nolvadex),[135] diphenhydramine (Benadryl),[136] chlorpromazine (Thorazine),[137] and

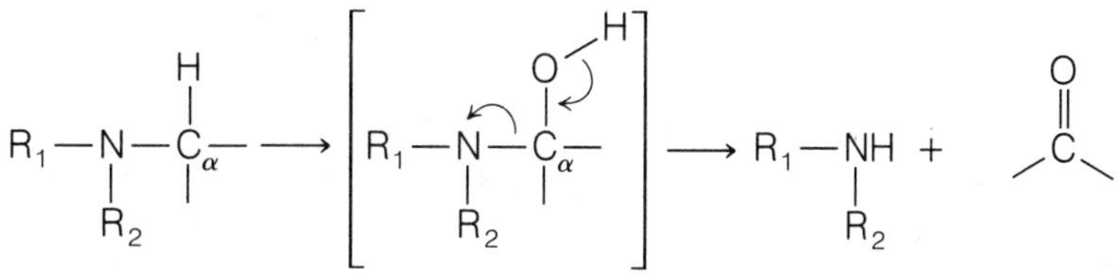

Tertiary Amine — Carbinolamine — Secondary Amine — Carbonyl Moiety (aldehyde or ketone)

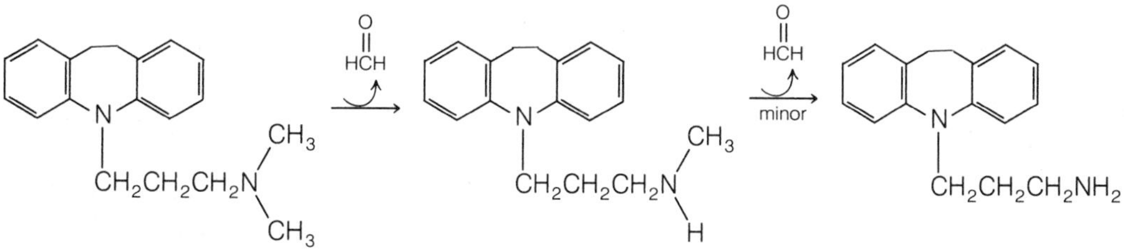

Imipramine — Desmethylimipramine (desipramine) — Bisdesmethylimipramine

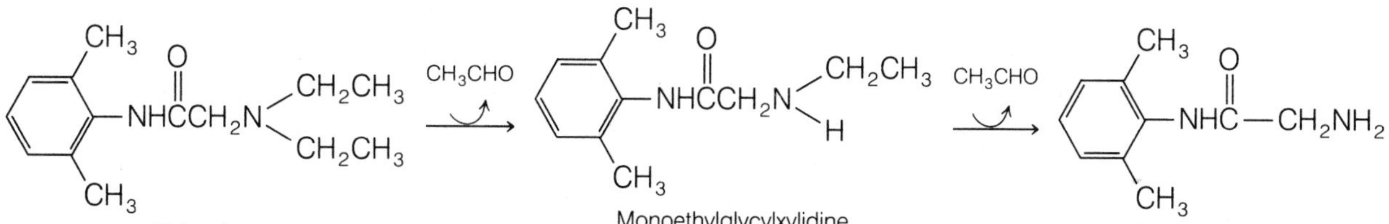

Lidocaine — Monoethylglycylxylidine (MEGX) — Glycyl-2,6-xylidine

(+)-α-propoxyphene (Darvon).[138] When the tertiary amine contains several different substituents capable of undergoing dealkylation, the smaller alkyl group is preferentially and more rapidly removed. For example, in benzphetamine (Didrex), the methyl group is removed much more rapidly than the benzyl moiety.[139]

An interesting cyclization reaction occurs with methadone upon *N*-demethylation. The demethylated metabolite, normethadone, undergoes a spontaneous cyclization reaction to form the enamine metabolite, 2-ethylidene-1,5-dimethyl-3,3-diphenylpyrrolidine (EDDP).[140] Subsequent *N*-demethylation of EDDP and isomerization of the double bond leads to 2-ethyl-5-methyl-3,3-diphenyl-1-pyrroline (EMDP).

Many times, bisdealkylation of a tertiary amine leads to the corresponding primary aliphatic amine metabolite, which is susceptible to further oxidation. For example, the bisdesmethyl metabolite of the H_1-histamine antagonist, brompheniramine (Dimetane), undergoes oxidative deamination and further oxidation to the corresponding propionic acid metabolite.[141] Oxidative deamination will be discussed in greater detail when we examine the metabolic reactions of secondary and primary amines.

Similar to their aliphatic counterparts, alicyclic tertiary amines are susceptible to oxidative *N*-dealkylation reactions. For example, the analgesic meperidine (Demerol) is metabolized principally by this pathway to yield normeperidine as a major

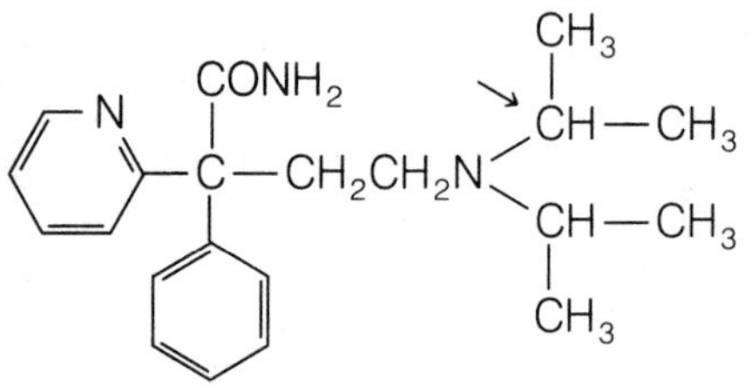

Disopyramide

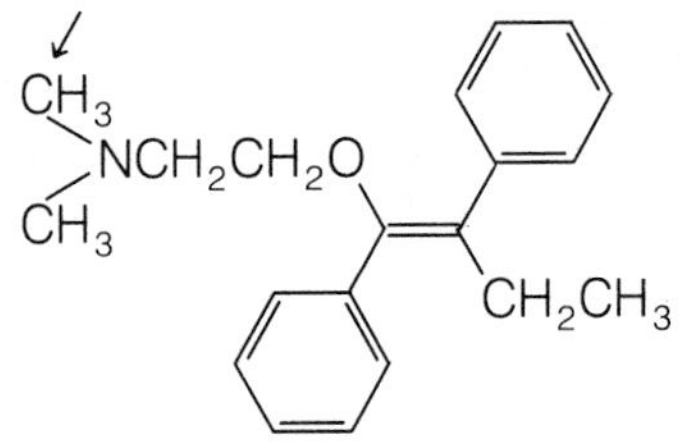

Tamoxifen

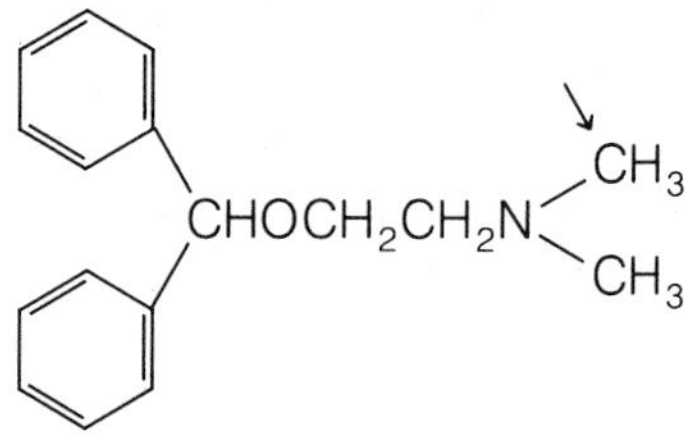

Diphenhydramine

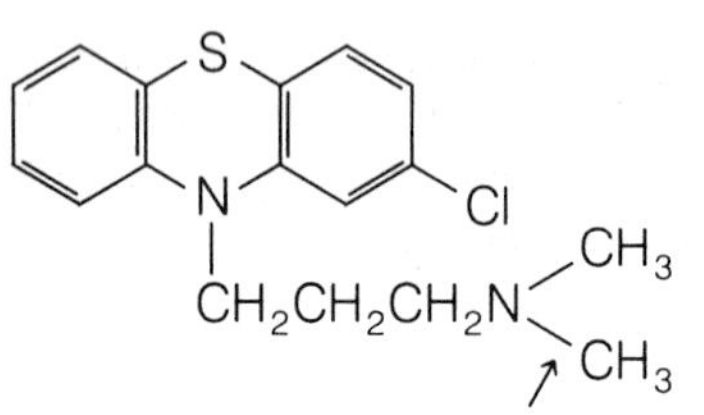

Chlorpromazine

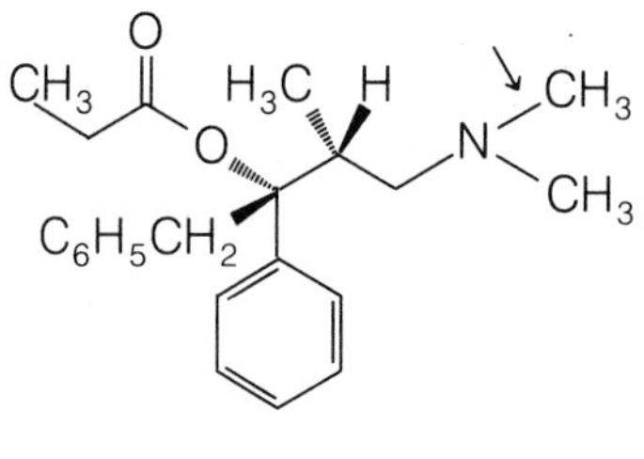

(+)-α-Propoxyphene

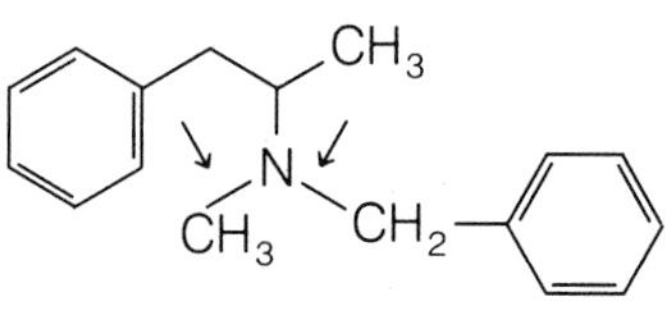

Benzphetamine (*N*-demethylation and *N*-debenzylation)

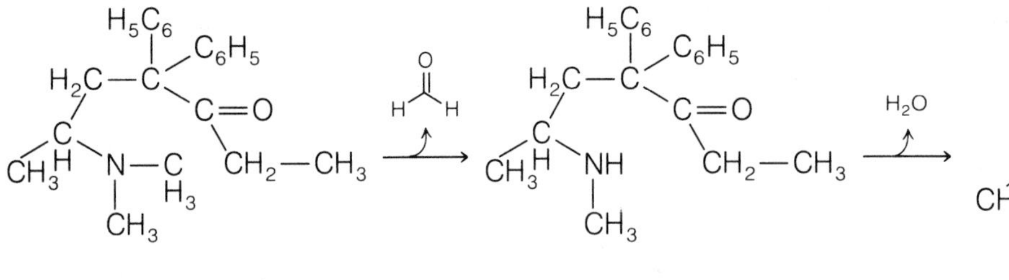

Methadone

Normethadone

2-Ethylidene-1,5-dimethyl-3,3-diphenylpyrrolidine (EDDP)

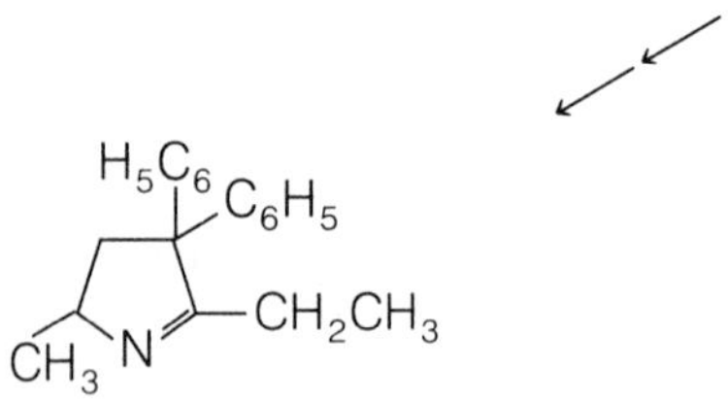

2-Ethyl-5-methyl-3,3-diphenyl-1-pyrroline (EMDP)

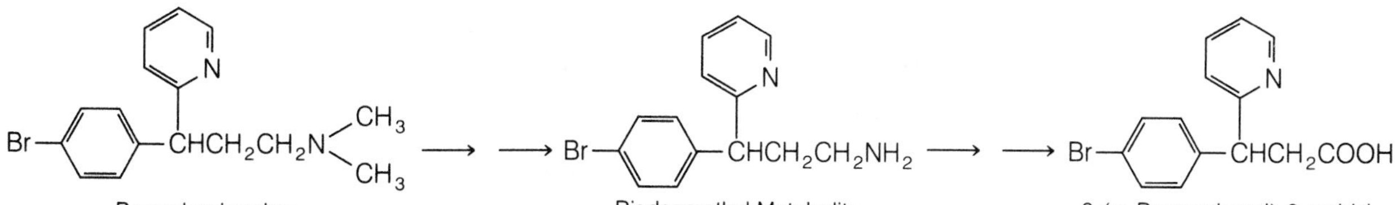

Brompheniramine → → Bisdesmethyl Metabolite → → 3-(p-Bromophenyl)-3-pyridylpropionic acid

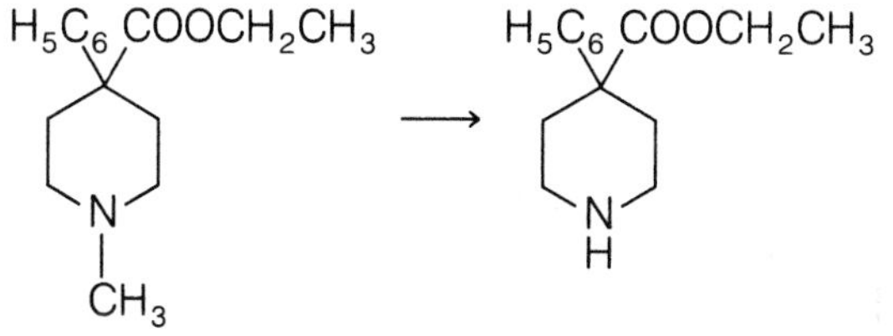

Meperidine → Normeperidine

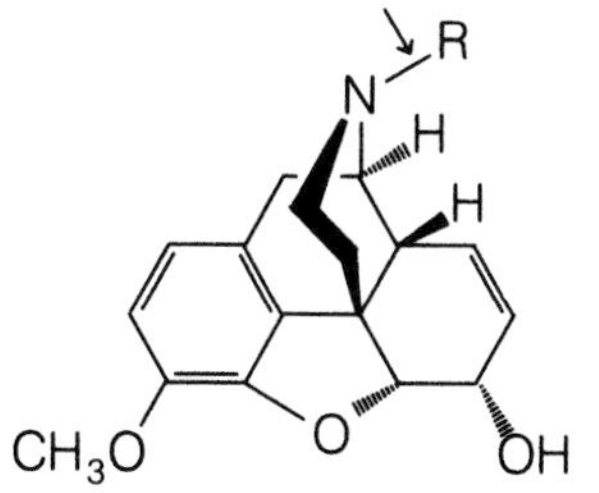

Morphine R = CH_3
N-Ethylnormorphine R = CH_2CH_3

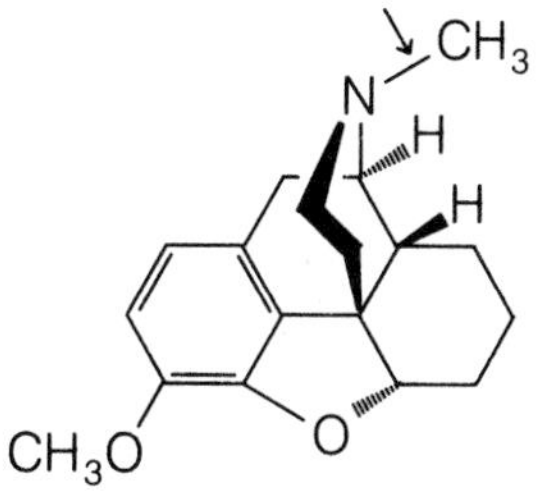

Dextromethorphan

plasma metabolite in humans.[142] Morphine, N-ethylnormorphine, and dextromethorphan also undergo N-dealkylation to some extent.[143]

Direct N-dealkylation of t-butyl groups, as discussed earlier, is not possible by the α-carbon hydroxylation pathway. However, in vitro studies indicate that N-t-butylnorchlorocyclizine is indeed metabolized to significant amounts of norchlorocyclizine, whereby the t-butyl group is lost.[144] Careful studies showed that the t-butyl group is removed by initial hydroxylation of one of the methyl groups of the t-butyl moiety to the carbinol or alcohol product.[145] Further oxidation generates the corresponding carboxylic acid, which upon decarboxylation forms the N-isopropyl derivative. The N-isopropyl intermediate is dealkylated by the normal α-carbon hydroxylation (i.e., carbinolamine) pathway to give norchlorocyclizine and acetone. Whether this is a general method for the loss of t-butyl groups from amines is still unclear. It appears that indirect N-dealkylation of t-butyl groups is not significantly observed. The N-t-butyl group, present in many β-adrenergic antagonists such as terbutaline and salbutamol, remains intact and does not appear to undergo any significant metabolism.[146]

Alicyclic tertiary amines often generate lactam metabolites by α-carbon hydroxylation reactions. For example, the tobacco alkaloid, nicotine, is initially hydroxylated at the ring carbon atom α to the nitrogen to yield a carbinolamine intermediate. Furthermore, enzymatic oxidation of this cyclic carbinolamine generates the lactam metabolite, cotinine.[147]

Formation of lactam metabolites has also been reported to occur to a minor extent for the antihistamine cyproheptadine (Periactin)[148] and the antiemetic diphenidol (Vontrol).[149]

N-Oxidation of tertiary amines occurs with several drugs.[150] The true extent of N-oxide formation is often complicated by the susceptibility of N-oxides to undergo in vivo reduction back to the parent tertiary amine. Tertiary amines such as H_1-histamine antagonists (e.g., orphenadrine, tripelenamine), phenothiazines (e.g., chlorpromazine), tricyclic an-

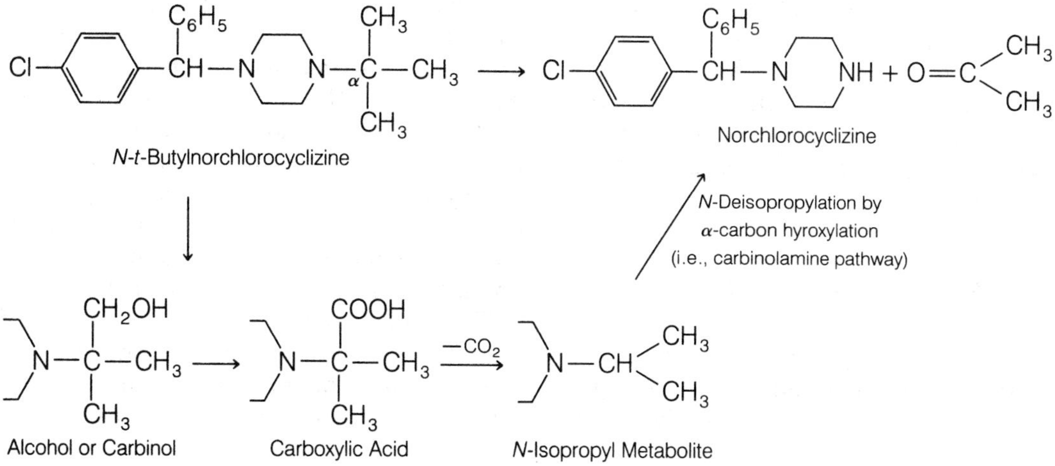

N-t-Butylnorchlorocyclizine → Norchlorocyclizine + $O{=}C(CH_3)_2$

N-Deisopropylation by α-carbon hyroxylation (i.e., carbinolamine pathway)

Alcohol or Carbinol → Carboxylic Acid $\xrightarrow{-CO_2}$ N-Isopropyl Metabolite

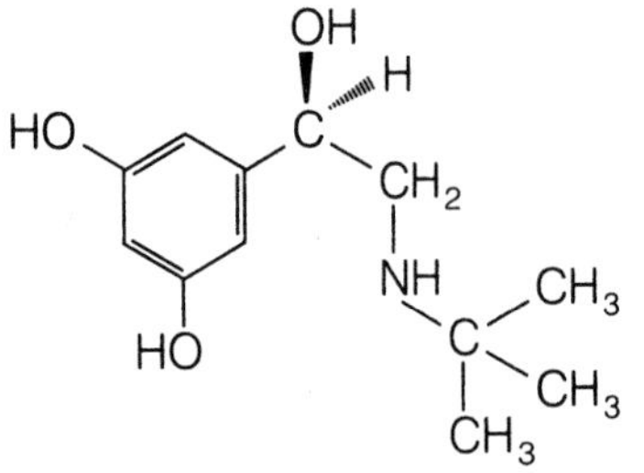

Terbutaline 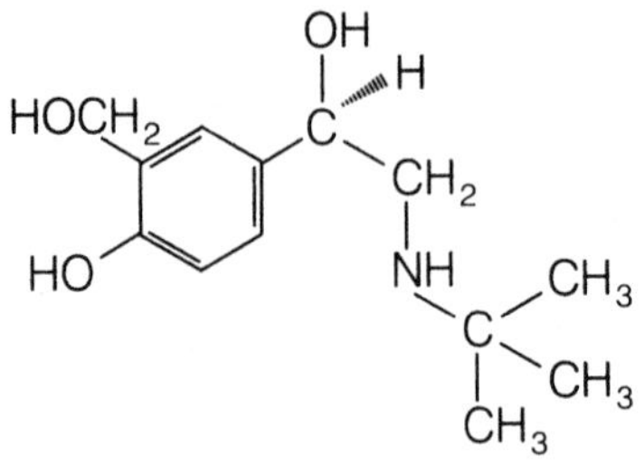   Salbutamol

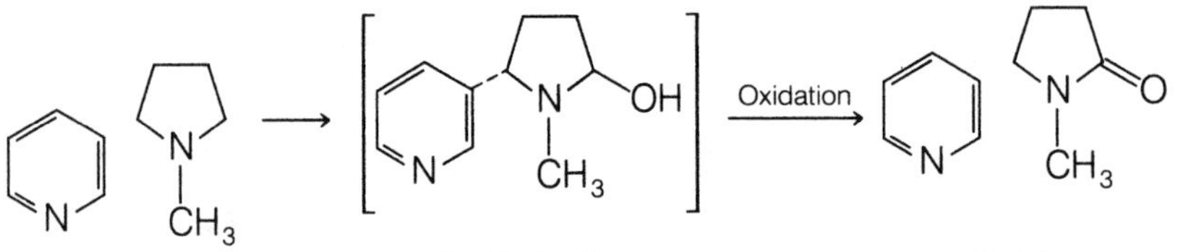

Nicotine → [Carbinolamine] —Oxidation→ Cotinine

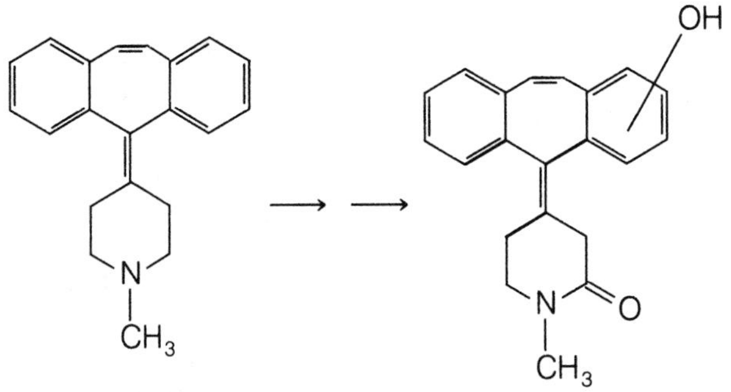

Cyproheptadine → → Lactam Metabolite

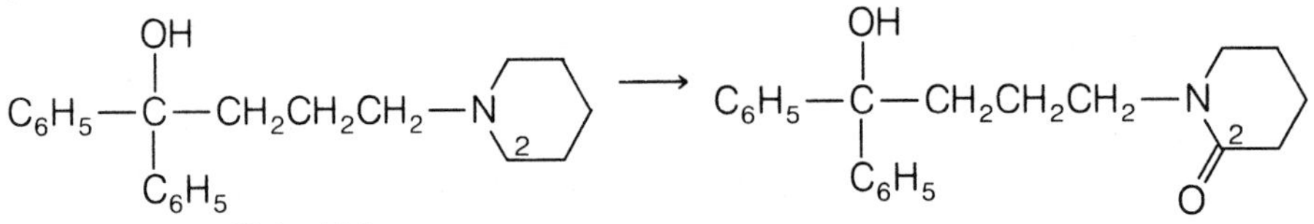

Diphenidol → 2-Oxodiphenidol

tidepressants (e.g., imipramine), and narcotic analgesics (e.g., morphine, codeine, and meperidine) have been reported to form *N*-oxides products. In some instances, *N*-oxides possess pharmacologic activity.[151] For example, comparison of imipramine *N*-oxide with imipramine indicates that the *N*-oxide itself possesses antidepressant and cardiovascular activity similar to the parent drug.[152]

SECONDARY AND PRIMARY AMINES

Secondary amines (either parent compounds or metabolites) are susceptible to oxidative *N*-dealkylation, oxidative deamination, and *N*-oxidation reactions.[129,153] As in tertiary amines, *N*-dealkylation of secondary amines proceeds by the carbinolamine pathway. Dealkylation of secondary amines gives rise to the corresponding primary amine metabolite. For example, the β-adrenergic blockers propranolol[39] and oxprenolol[154] undergo *N*-deisopropylation to the corresponding primary amines. *N*-Dealkylation appears to be a significant biotransformation pathway for the secondary amine drugs, methamphetamine[155] and ketamine,[156] yielding amphetamine and norketamine, respectively.

The primary amine metabolites formed from oxidative dealkylation are susceptible to *oxidative deamination*. This process is similar to *N*-dealkylation, in that it involves an initial α-carbon hydroxyl-

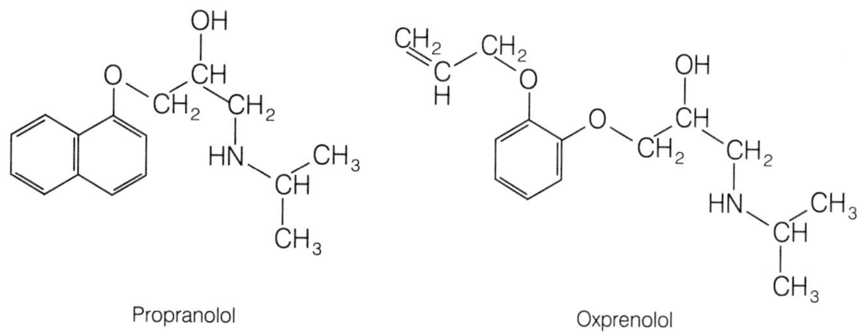

Propranolol   Oxprenolol

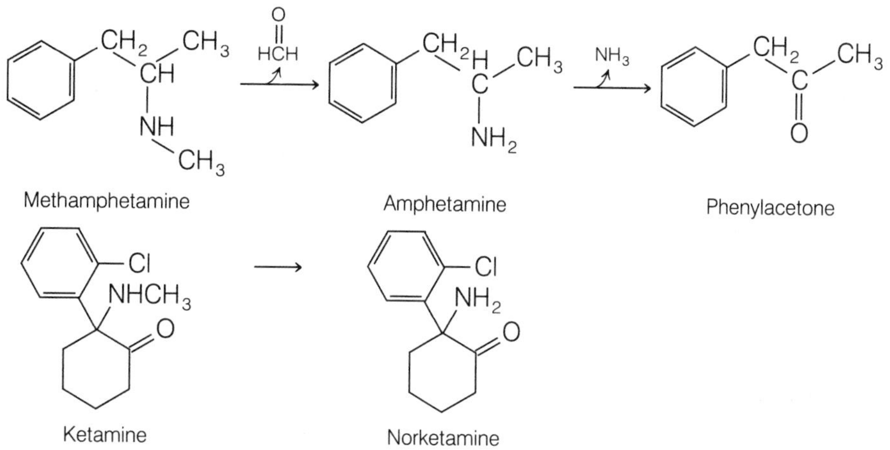

Methamphetamine   Amphetamine   Phenylacetone

Ketamine   Norketamine

ation reaction to form a carbinolamine intermediate, which then undergoes subsequent carbon–nitrogen cleavage to the carbonyl metabolite and ammonia. If α-carbon hydroxylation cannot occur, then oxidative deamination is not possible. For example, deamination does not occur for norketamine because α-carbon hydroxylation cannot take place.[156] For methamphetamine, oxidative deamination of primary amine metabolite amphetamine produces phenylacetone (see preceding).[155]

In general, dealkylation of secondary amines is believed to take place before oxidative deamination occurs. However, there is some evidence that this may not always be true. Direct deamination of the secondary amine also has occurred. For example, in addition to undergoing deamination through its desisopropyl primary amine metabolite, propranolol can undergo a direct oxidative deamination reaction (also by α-carbon hydroxylation) to yield the aldehyde metabolite and isopropylamine (Fig. 3-9).[157] How much direct oxidative deamination contributes to the metabolism of secondary amines remains unclear.

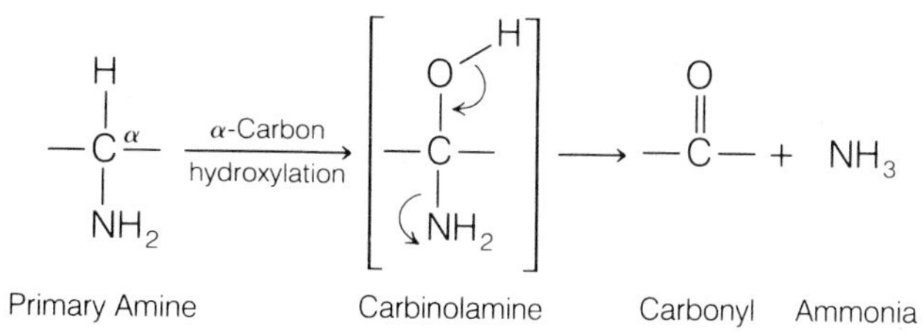

Some secondary alicyclic amines, similarly to their tertiary amine analogues, are metabolized to their corresponding lactam derivatives. For example, the anorectic agent phenmetrazine (Preludin), is principally metabolized to the lactam product 3-oxophenmetrazine.[158] In humans, this lactam metabolite is a major urinary product. Methylphenidate (Ritalin) has also been reported to yield a lactam metabolite, 6-oxoritalinic acid, by oxidation of its hydrolyzed metabolite, ritalinic acid, in humans.[159]

Metabolic N-oxidation of secondary aliphatic and alicyclic amines leads to several N-oxygenated products.[153] N-Hydroxylation of secondary amines generates the corresponding N-hydroxylamine metabolites. Often these hydroxylamine products are susceptible to further oxidation (either spontaneous or enzymatic) to the corresponding nitrone derivatives. For example, N-benzylamphetamine has been observed to undergo metabolism to both the corresponding N-hydroxylamine and the nitrone metabolites.[160] In humans, the nitrone metabolite of phenmetrazine (Preludin) found in the urine is believed to be formed by further oxidation of the N-hydroxylamine intermediate, N-hydroxyphen-

α-Carbon hydroxylation

Primary Amine   Carbinolamine   Carbonyl   Ammonia

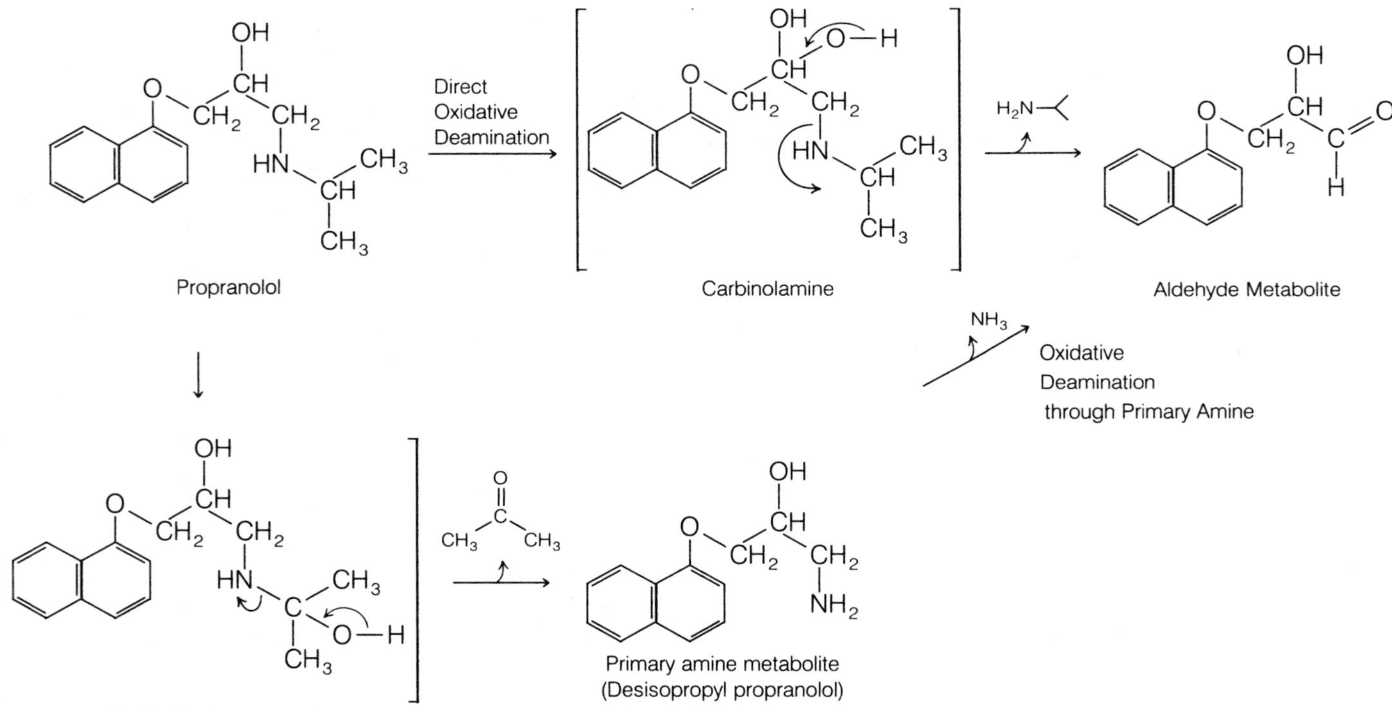

FIG. 3-9. *The metabolism of propranolol to its aldehyde metabolite by direct deamination of the parent compound and by deamination of its primary amine metabolite, desisopropyl propranolol.*

Phenmetrazine → [Carbinolamine Intermediate] → 3-Oxophenmetrazine

Methylphenidate —Hydrolysis→ Ritalinic acid → → 6-Oxoritalinic Acid

Secondary amine → Hydorxylamine → Nitrone

metrazine.[158] Importantly, in comparison with oxidative dealkylation and deamination, *N*-oxidation occurs to a much lesser extent for secondary amines.

Primary aliphatic amines (whether parent drugs or metabolites) are biotransformed by oxidative deamination (through carbinolamine pathway) or by *N*-oxidation. In general, oxidative deamination of most exogeneous primary amines is carried out by the mixed function oxidases discussed earlier. However, endogenous primary amines, such as dopamine, norepinephrine, tryptamine, and serotonin, are metabolized through oxidative deamination by a specialized family of enzymes called monoamine oxidases (MAO).[161] These enzymes are involved primarily in inactivating the foregoing neurotransmitter amines. Monoamine oxidase apparently plays no significant role in the metabolism of xenobiotic primary amines.

Structural features, especially the α-substituents of the primary amine, often determine whether carbon or nitrogen oxidation will occur. For example, compare amphetamine with its α-methyl homologue, phentermine. In amphetamine, α-carbon hydroxylation can occur to form the carbinolamine intermediate, which is converted to the oxidatively deaminated product, phenylacetone.[46] With phentermine, α-carbon hydroxylation is not possible and precludes oxidative deamination for this drug. Consequently, phentermine would be expected to undergo *N*-oxidation readily. In humans, *p*-hydroxylation and *N*-oxidation are the main pathways for biotransformation of phentermine.[162]

N-Benzylamphetamine → Hydroxylamine Metabolite → Nitrone Metabolite

Phenmetrazine → *N*-Hydroxyphenmetrazine → Nitrone Metabolite

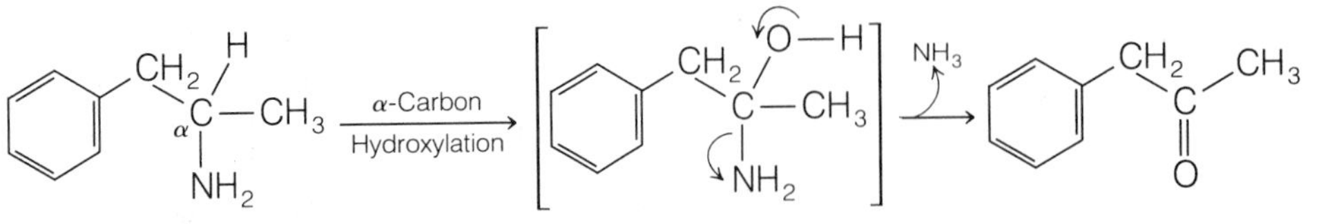

Amphetamine — Carbinolamine — Phenylacetone

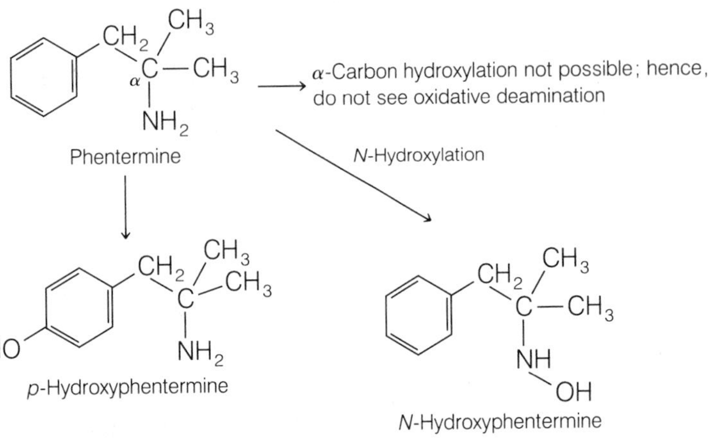

Phentermine

p-Hydroxyphentermine — *N*-Hydroxyphentermine

Indeed, *N*-hydroxyphentermine is an important (5%) urinary metabolite in humans.[162] As shall be discussed shortly, *N*-hydroxylamine metabolites are susceptible to further oxidation to yield other *N*-oxygenated products.

Xenobiotics, such as the hallucinogenic agents mescaline[163] and 1-(2,5-dimethoxy-4-methylphenyl)-2-aminopropane (DOM or "STP"),[97] are oxidatively deaminated. Primary amine metabolites arising from *N*-dealkylated or decarboxylation reactions also undergo deamination. The example of the bisdesmethyl primary amine metabolite derived from brompheniramine was discussed earlier (see section on tertiary aliphatic and alicyclic amines).[141] In addition, many tertiary aliphatic amines (e.g., antihistamines) and secondary aliphatic amines (e.g., propranolol) are dealkylated to their corresponding primary amine metabolites, which are amenable to oxidative deamination. (*S*)(+)-α-Methyldopamine resulting from decarboxylation of the antihypertensive agent (*S*)(−)-α-methyldopa (Aldomet) is deaminated to the corresponding ketone metabolite, 3,4-dihydroxyphenylacetone.[164] In humans, this ketone is a major urinary metabolite.

The *N*-hydroxylation reaction is not restricted to α-substituted primary amines such as phentermine. Amphetamine has been observed to undergo some *N*-hydroxylation in vitro to *N*-hydroxyamphetamine.[165] However, *N*-hydroxyamphetamine is susceptible to further conversion to the imine or oxidation to the oxime intermediate. Note, that the oxime intermediate arising from this *N*-oxidation pathway can undergo hydrolytic cleavage to yield phenylacetone, the same product obtained by the α-carbon hydroxylation (carbinolamine) pathway.[166] Thus, it is apparent that conversion of amphetamine to phenylacetone may arise through either the α-carbon hydroxylation or *N*-oxidation pathway. The debate concerning the relative importance of the two pathways remains ongoing.[168, 167] The general consensus, however, is that both metabolic pathways

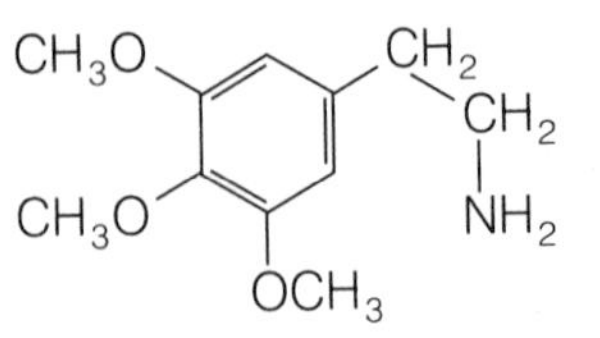

Mescaline 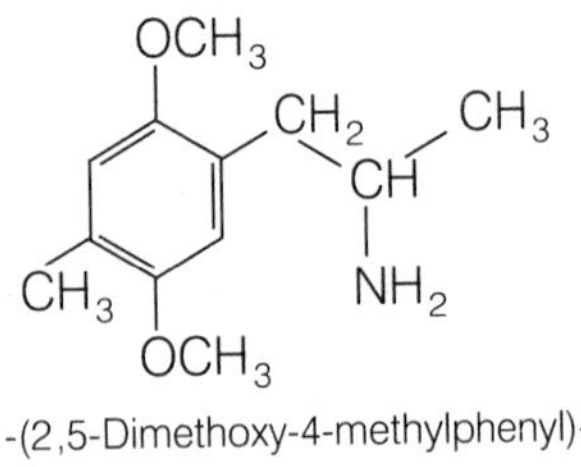 — 1-(2,5-Dimethoxy-4-methylphenyl)-2-aminopropane DOM or "STP"

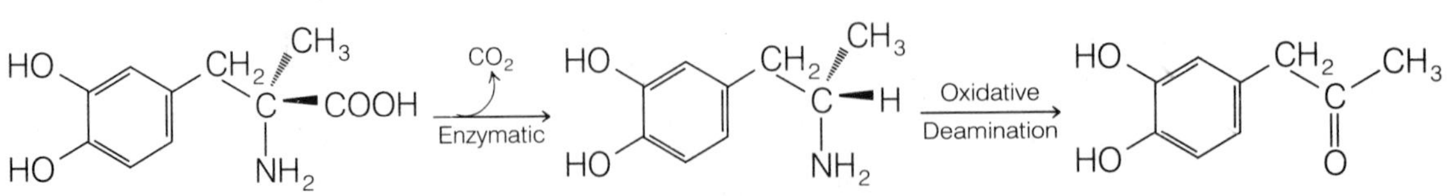

S(−)-α-Methyldopa — *S*(+)-α-Methyldopamine — 3,4-Dihydroxyphenylacetone

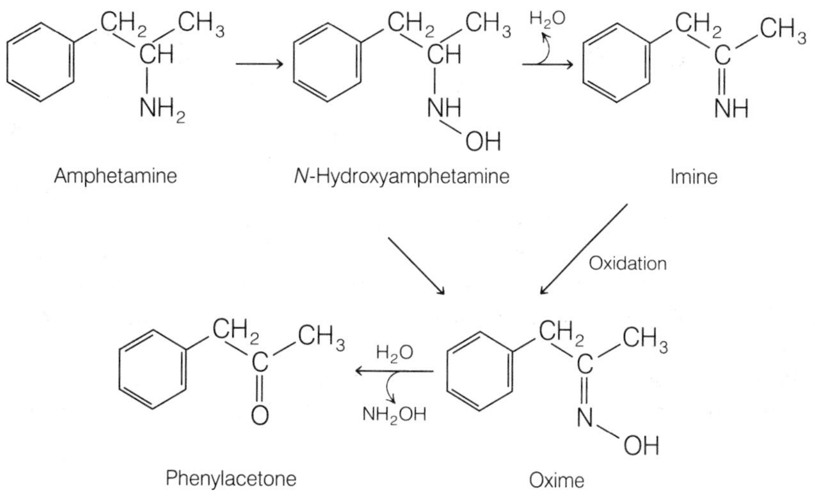

Amphetamine → N-Hydroxyamphetamine → Imine; Oxidation → Oxime → Phenylacetone (H_2O, NH_2OH)

(carbon and nitrogen oxidation) are probably operative. Whether α-carbon or nitrogen oxidation predominates in the metabolism of amphetamine appears to be species-dependent.

In primary aliphatic amines, such as phentermine,[162] chlorophentermine (*p*-chlorophentermine),[168] and amantadine,[169] *N*-oxidation appears to be the major biotransformation pathway because α-carbon hydroxylation cannot occur. In humans, chlorphentermine is extensively *N*-hydroxylated. About 30 percent of a dose of chlorphentermine is found in the urine (48 hours) as *N*-hydroxychlorphentermine (free and conjugated) and an additional 18 percent as other products of *N*-oxidation (presumably the nitroso and nitro metabolites).[168] In general, *N*-hydroxylamines are chemically unstable and susceptible to spontaneous or enzymatic oxidation to the nitroso and nitro derivatives. For example, the *N*-hydroxylamine metabolite of phentermine undergoes further oxidation to the nitroso and nitro products.[162] The antiviral and antiparkinsonian agent, amantadine (Symmetre), has been reported to undergo *N*-oxidation to yield the corresponding *N*-hydroxy and nitroso metabolites in vitro.[169]

Aromatic amines and heterocyclic nitrogen compounds

The biotransformation of aromatic amines parallels the carbon and nitrogen oxidation reactions seen for aliphatic amines.[170,171] For tertiary aromatic amines, such as *N*,*N*-dimethylaniline, oxidative *N*-dealkylation as well as *N*-oxide formation takes place.[172] Secondary aromatic amines may undergo *N*-dealkylation or *N*-hydroxylation to give the corresponding *N*-hydroxylamines. Further oxidation of the *N*-hydroxylamine leads to nitrone products, which in turn may be hydrolyzed to primary hydroxylamines.[173] Tertiary and secondary aromatic amines are rarely encountered in medicinal agents. In contrast, primary aromatic amines are found in many drugs and often are generated from enzymatic reduction of aromatic nitro compounds, reductive cleavage of azo compounds, and hydrolysis of aromatic amides.

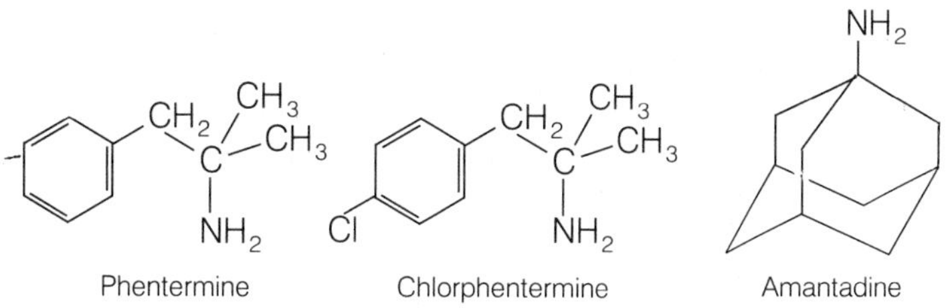

Phentermine   Chlorphentermine   Amantadine

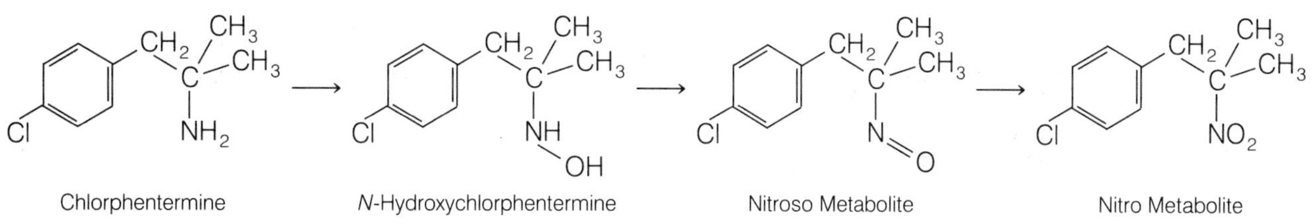

Chlorphentermine → N-Hydroxychlorphentermine → Nitroso Metabolite → Nitro Metabolite

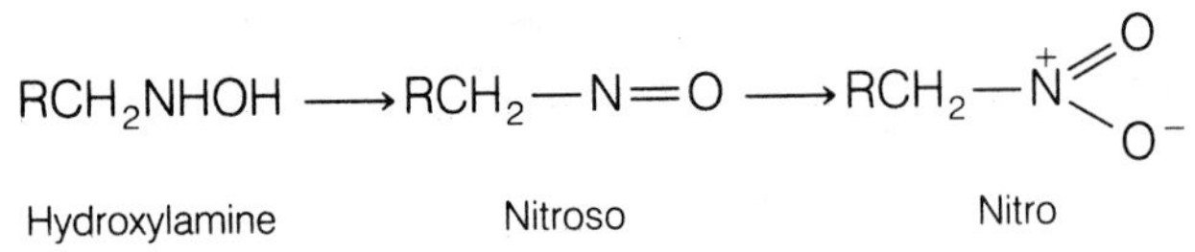

$$RCH_2NHOH \longrightarrow RCH_2{-}N{=}O \longrightarrow RCH_2{-}\overset{+}{N}(=O)O^-$$

Hydroxylamine Nitroso Nitro

N-Oxidation of primary aromatic amines generates the *N*-hydroxylamine metabolite. For example, aniline is metabolized to the corresponding *N*-hydroxy product.[171] Oxidation of the hydroxylamine derivative to the nitroso derivative can also occur. When considering primary aromatic amine drugs or metabolites, *N*-oxidation constitutes only a minor pathway in comparison with other biotransformation pathways, such as *N*-acetylation and aromatic hydroxylation, in humans. However, some *N*-oxygenated metabolites have been reported. For example, the antileprotic agent, dapsone, and its *N*-acetylated metabolite are significantly metabolized to their corresponding *N*-hydroxylamine derivatives.[174] The *N*-hydroxy metabolites are further conjugated with glucuronic acid.

Methemoglobinemia toxicity caused by several aromatic amines, including aniline and dapsone, is a result of the bioconversion of the aromatic amine to its *N*-hydroxy derivative. Apparently the *N*-hydrox-

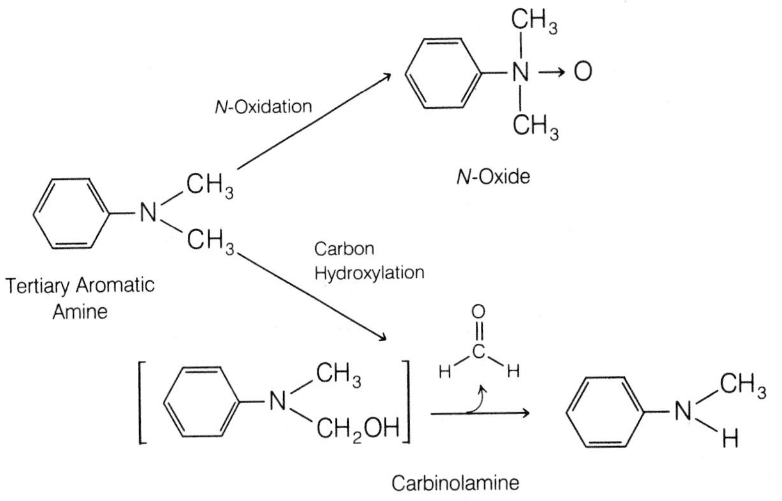

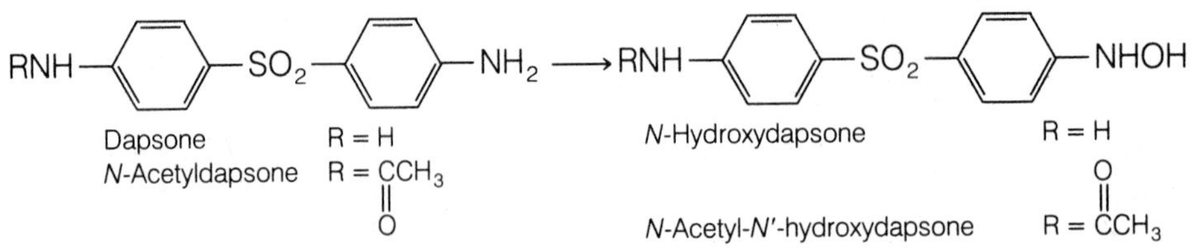

Tertiary Aromatic Amine

N-Oxidation

N-Oxide

Carbon Hydroxylation

Carbinolamine

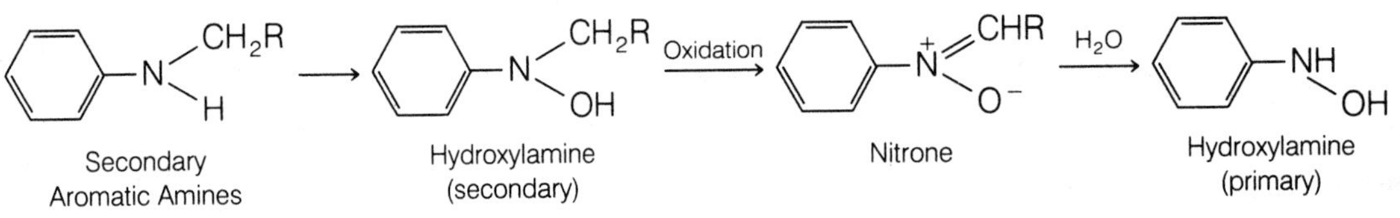

Secondary Aromatic Amines

Hydroxylamine (secondary)

Oxidation

Nitrone

H_2O

Hydroxylamine (primary)

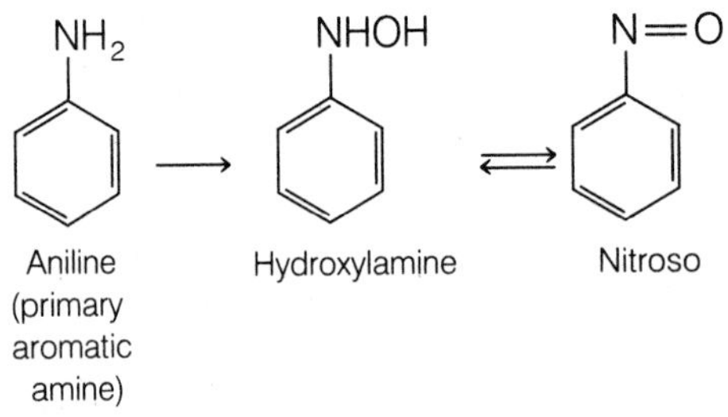

Aniline (primary aromatic amine)

Hydroxylamine

Nitroso

Dapsone R = H

N-Acetyldapsone R = $\overset{O}{\|}CCH_3$

N-Hydroxydapsone R = H

N-Acetyl-*N'*-hydroxydapsone R = $\overset{O}{\|}CCH_3$

ylamine oxidizes the Fe^{2+} form of hemoglobin to its Fe^{3+} form. This oxidized (Fe^{3+}) state of hemoglobin (called methemoglobin or ferrihemoglobin) is no longer capable of transporting oxygen and leads to serious hypoxia or anemia.[175]

Divers aromatic amines (especially azoamino dyes) are known to be carcinogenic. *N*-Oxidation plays an important role in bioactivating these aromatic amines to potentially reactive electrophilic species that covalently bind to cellular protein, DNA, or RNA. A well-studied example is the carcinogenic agent *N*-methyl-4-aminoazobenzene.[176] *N*-Oxidation of this compound leads to the corresponding hydroxylamine, which undergoes sulfate conjugation. Owing to the good leaving group ability of the sulfate (SO_4^{2-}) anion, this sulfate conjugate can spontaneously ionize to form a highly reactive, resonance-stabilized nitrenium species. Covalent adducts between this species and DNA, RNA, and protein have been characterized.[177] The sulfate ester is believed to be the ultimate carcinogenic species. Thus, the example indicates that certain aromatic amines can be bioactivated to reactive intermediates by *N*-hydroxylation and *O*-sulfate conjugation. Whether primary hydroxylamines can be similarly bioactivated is unclear. In addition, it is not known if this biotoxification pathway plays any substantial role in the toxicity of aromatic amine drugs.

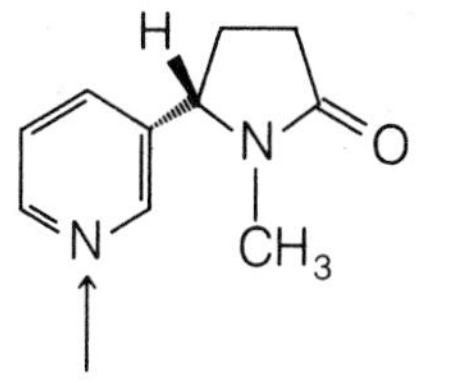

Cotinine

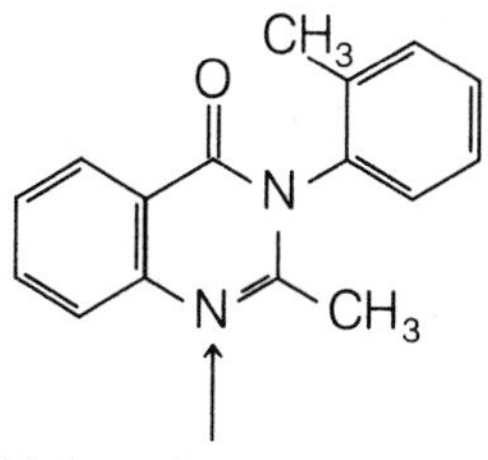

Methaqualone

N-Oxidation of the nitrogen atoms present in aromatic heterocyclic moieties of many drugs occurs to a minor extent. For example, in humans, *N*-oxidation of the folic acid antagonist trimethoprim (Proloprim, Trimpex) has yielded approximately equal amounts of the isomeric 1-*N*-oxide and 3-*N*-oxide as minor metabolites.[178] The pyridinyl nitrogen atom present in cotinine (the major metabolite of nicotine) undergoes oxidation to yield the corresponding *N*-oxide metabolite.[179] Formation of

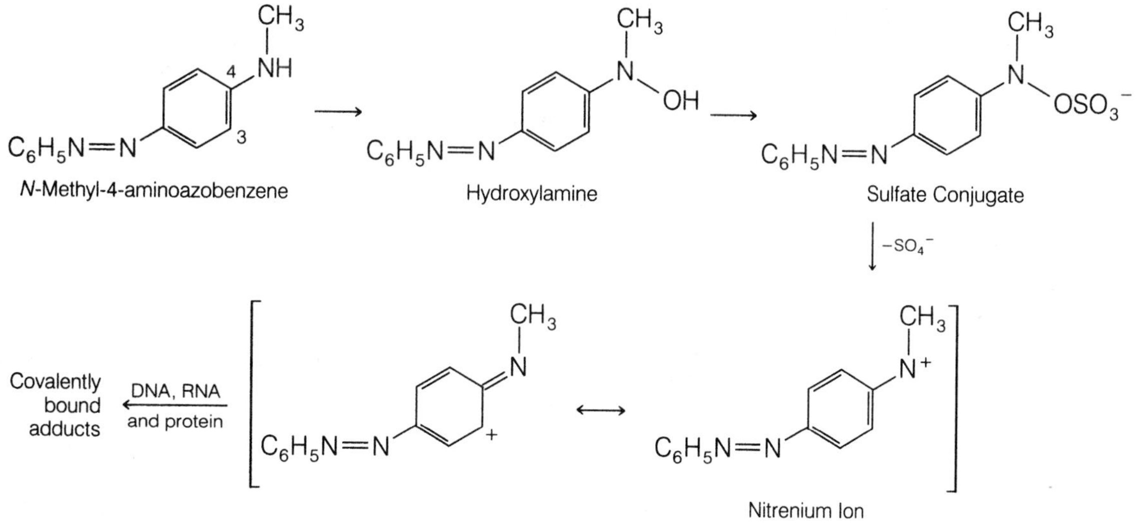

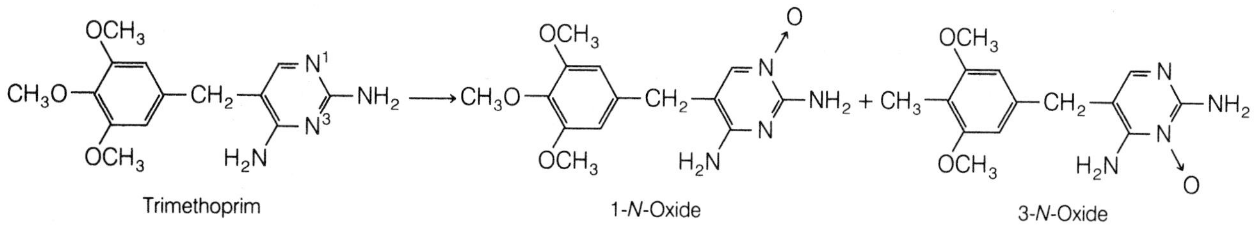

N-Methyl-4-aminoazobenzene → Hydroxylamine → Sulfate Conjugate

$-SO_4^-$

Covalently bound adducts ← DNA, RNA and protein ← [resonance forms] Nitrenium Ion

Trimethoprim → 1-*N*-Oxide + 3-*N*-Oxide

an *N*-oxide metabolite of methaqualone (Quaalude, Parest) has also been observed in humans.[180]

Amides

Amide functionalities are susceptible to oxidative carbon–nitrogen bond cleavage (by way of α-carbon hydroxylation) and *N*-hydroxylation reactions. Oxidative dealkylation of many *N*-substituted amide drugs and xenobiotics has been reported. Mechanistically, oxidative dealkylation proceeds by way of an initially formed carbinolamide, which is unstable and fragments to form the *N*-dealkylated product. For example, diazepam undergoes extensive *N*-demethylation to the pharmacologically active metabolite, desmethyldiazepam.[181]

Various other *N*-alkyl substituents present in benzodiazepines (e.g., flurazepam)[111] and in barbiturates (e.g., hexobarbital and mephobarbital)[105] are similarly oxidatively *N*-dealkylated. Alkyl groups attached to the amide moiety of some sulfonylureas, such as the oral hypoglycemic chlorpropamide,[182] also are subject to dealkylation to a minor extent.

In the cyclic amides or lactams, hydroxylation of the alicyclic carbon α to the nitrogen atom also leads to carbinolamides. An example of this pathway is the conversion of cotinine to 5-hydroxycotinine. Interestingly, the latter carbinolamide in – termediate is in tautomeric equilibrium with the ring-opened metabolite, γ-(3-pyridyl)-γ-oxo-*N*-methylbutyramide.[183]

Metabolism of the important cancer chemotherapeutic agent cyclophosphamide (Cytoxan) follows a hydroxylation pathway similar to that just described for cyclic amides. This drug is a cyclic phosphoramide derivative and, for the most part, is the phosphorous counterpart of a cyclic amide. Because cyclophosphamide itself is pharmacologically inactive,[184] metabolic bioactivation is required for the drug to mediate its antitumorigenic or cytotoxic effects. The key biotransformation pathway leading to the active metabolite involves an initial carbon hydroxylation reaction at C-4 to form the carbinolamide, 4-hydroxycyclophosphamide.[185] 4-Hydroxycyclophosphamide is in equilibrium with the ring-opened dealkylated metabolite, aldophosphamide.

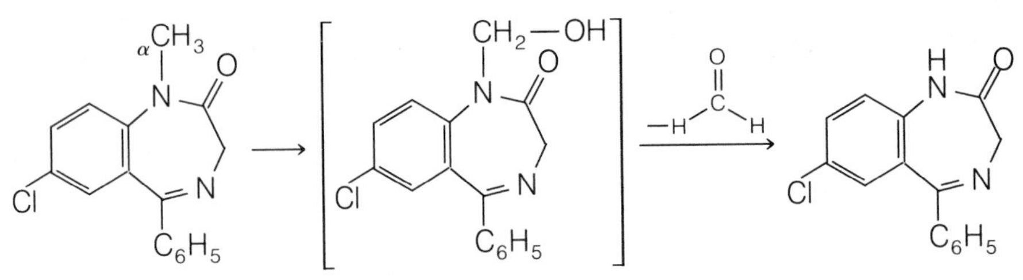

Diazepam → Carbinolamide → ($-HCHO$) Desmethyldiazepam

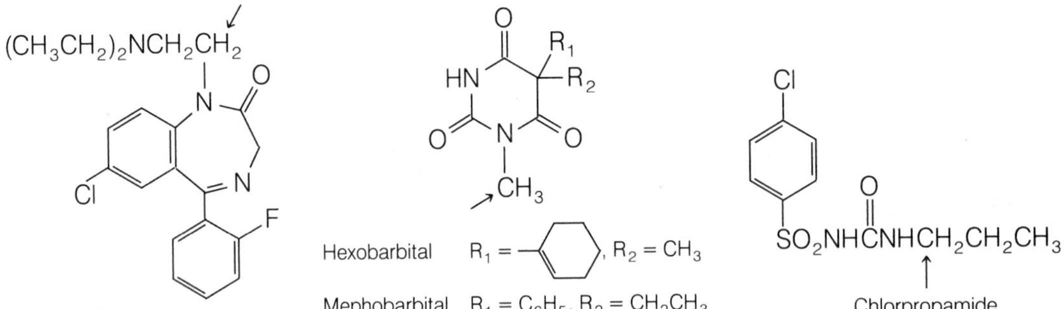

Flurazepam

Hexobarbital R_1 = cyclohexenyl, $R_2 = CH_3$

Mephobarbital $R_1 = C_6H_5$, $R_2 = CH_2CH_3$

Chlorpropamide

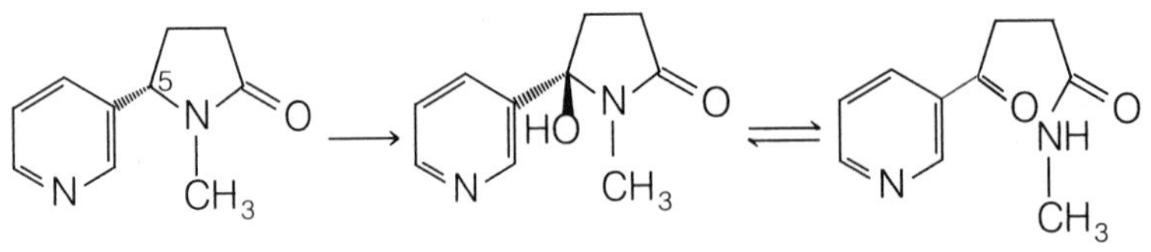

Cotinine → 5-Hydroxycotinine ⇌ γ-(3-Pyridyl)-γ-oxo-*N*-methylbutyramide

Cyclophosphamide → 4-Hydroxycyclophosphamide —Oxidation→ 4-Ketocyclophosphamide

Phosphoramide Mustard N,N-bis(2-Chloroethyl)-phosphorodiamidic Acid + Acrolein ← Aldophosphamide —Oxidation→ Carboxyphosphamide

Although it has potent cytotoxic properties, aldophosphamide undergoes a further elimination reaction (reverse Michael reaction) to generate acrolein and the phosphoramide mustard [*N,N*-bis(2-chloroethyl)phosphorodiamidic acid]. The latter metabolite is the principal species responsible for cyclophosphamide's antitumorigenic properties and chemotherapeutic effect. Enzymatic oxidation of 4-hydroxycyclophosphamide and aldophosphamide leads to the relatively nontoxic metabolites 4-ketocyclophosphamide and carboxycyclophosphamide, respectively.

N-Hydroxylation of aromatic amides, which occurs to a minor extent, is of some toxicologic interests, since this biotransformation pathway may lead to the formation of chemically reactive intermediates. Several examples of cytotoxicity or carcinogenicity that have been clearly associated with metabolic *N*-hydroxylation of the parent aromatic amide have been reported. For example, the well-known hepatocarcinogenic 2-acetylaminofluorene (AAF) undergoes an *N*-hydroxylation reaction catalyzed by cytochrome P-450 to form the corresponding *N*-hydroxy metabolite (also called a hydroxamic acid).[186] Further conjugation of this hydroxamic acid produces the corresponding *O*-sulfate ester, which ionizes to generate the electrophilic nitrenium species. The covalent binding of this reactive intermediate to DNA is known to occur and is likely to be the initial event that ultimately leads to malignant

2-Acetylaminoflourene (AAF) —N-Hydroxylation→ N-Hydroxy AAF —Sulfate Conjugation→ O-Sulfate ester of N-hydroxy AAF

$-SO_4^{2-}$

Nitrenium Species —Nu→

Nu = Nucleophile e.g., DNA

tumor formation.[187] Sulfate conjugation plays an important role in this biotoxification pathway (see section on sulfate conjugation for further discussion).

Acetaminaphen is a relatively safe and nontoxic analgesic agent, if used at therapeutic doses. When large doses of this drug are ingested, extensive liver necrosis is produced in humans and animals.[188] Considerable evidence argues that this hepatotoxicity is dependent upon the formation of a metabolically generated reactive intermediate.[189] Until recently,[190] the accepted bioactivation pathway was believed to involve an initial *N*-hydroxylation reaction to form *N*-hydroxyacetaminophen.[191] Spontaneous dehydration of this *N*-hydroxyamide produces *N*-acetylimidoquinone, the proposed reactive metabolite. Usually, the glutathione present in the liver combines with this reactive metabolite to form the corresponding glutathione conjugate. If glutathione levels are sufficiently depleted by large doses of acetaminophen, covalent binding of the reactive intermediate occurs with macromolecules present in the liver, thereby leading to cellular necrosis. However, studies indicate that the reactive *N*-acetylimidoquinone intermediate is not formed from *N*-hydroxyacetaminophen.[189, 190] It probably arises through some other oxidative process. Therefore, the mechanistic formation of the reactive metabolite of acetaminophen remains unclear.

OXIDATION INVOLVING CARBON–OXYGEN SYSTEMS

Oxidative *O*-dealkylation of carbon–oxygen systems (principally ethers) is catalyzed by microsomal mixed function oxidases.[129] Mechanistically, the biotransformation involves an initial α-carbon hydroxylation to form either a hemiacetal or hemiketal, which undergoes spontaneous carbon–oxygen bond cleavage to yield the dealkylated oxygen species (phenol or alcohol) and a carbon moiety (aldehyde or ketone). Small alkyl groups (e.g., methyl or ethyl) attached to oxygen are rapidly *O*-dealkylated. For example, morphine is the metabolic product resulting from *O*-demethylation of codeine.[192] The antipyretic and analgesic activities of phenacetin in humans appear to be a consequence of *O*-deethylation to the active metabolite acetaminophen.[193] Several other drugs containing ether groups, such as indomethacin (Indocin),[194] prazosin (Minipress),[195] and metoprolol (Lopressor),[96] have been reported to undergo significant *O*-demethylation to their corresponding phenolic or alcoholic metabolites, which are further conjugated. In many drugs that have several nonequivalent methoxy groups, one particular methoxy group often appears to be selectively or preferentially *O*-demethylated. For example, the 3,4,5-trimethoxyphenyl moiety in both mescaline[196] and trimethoprim[178] undergoes *O*-demethylation to

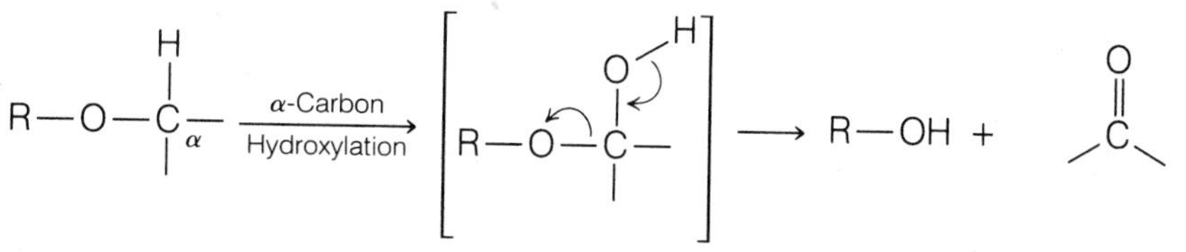

Ether — α-Carbon Hydroxylation → [Hemiacetal or Hemiketal] → Phenol or Alcohol + Carbonyl Moiety (aldehyde or ketone)

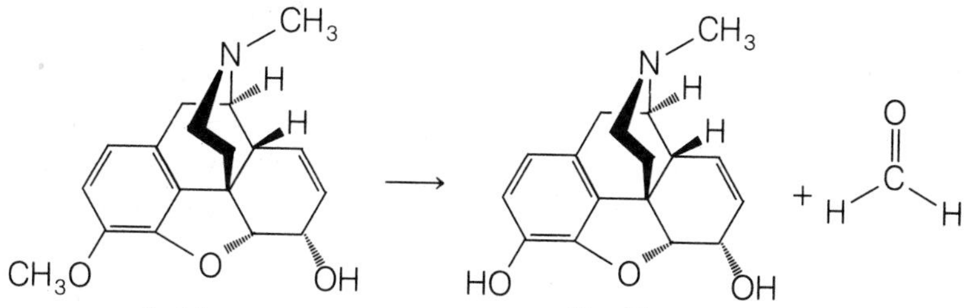

Codeine → Morphine + $H_2C{=}O$

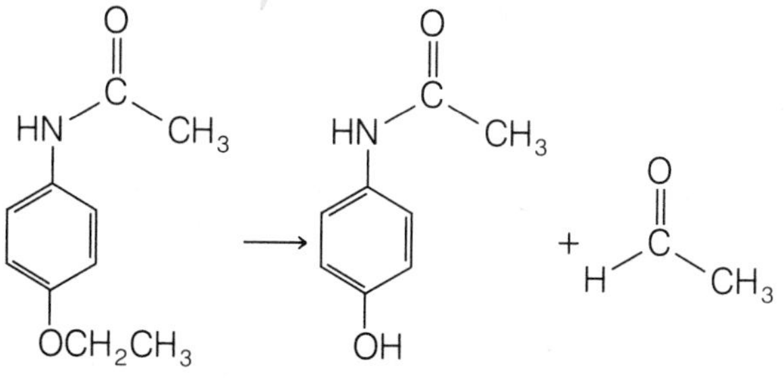

Phenacetin → Acetaminophen + CH_3CHO

yield predominantly the corresponding 3-*O*-demethylated metabolites. 4-*O*-Demethylation also occurs to a minor extent for both drugs. The phenolic and alcoholic metabolites formed from oxidative *O*-demethylation are susceptible to conjugation, particularly glucuronidation.

OXIDATION INVOLVING CARBON-SULFUR SYSTEMS

Carbon–sulfur functional groups are susceptible to metabolic *S*-dealkylation, desulfuration, and *S*-oxidation reactions. The first two processes involve oxidative carbon–sulfur bond cleavage. *S*-Dealkylation is analogous to *O*- and *N*-dealkylation mechanistically (i.e., involves α-carbon hydroxylation) and has been observed for various sulfur xenobiotics.[129, 197] For example, 6-(methylthio)purine is oxidatively demethylated in rats to 6-mercaptopurine.[198] *S*-Demethylation of methitural[199] and *S*-debenzylation of 2-benzylthio-5-trifluoromethylbenzoic acid have also been reported. In contrast to *O*- and *N*-dealkylation, examples of drugs undergoing *S*-dealkylation in humans are limited, owing to the small number of sulfur-containing medicinals and to competing metabolic *S*-oxidation processes (see diagram.)

Oxidative conversion of carbon–sulfur double bonds (C=S) (thiono) to the corresponding carbon–oxygen double bond (C=O) is called desulfuration. A well-known drug example of this metabolic process is the biotransformation of thiopental to its corresponding oxygen analogue pentobarbital.[200] An analogous desulfuration reaction also occurs with the P=S moiety present in a number of organophosphate insecticides, such as parathion.[201]

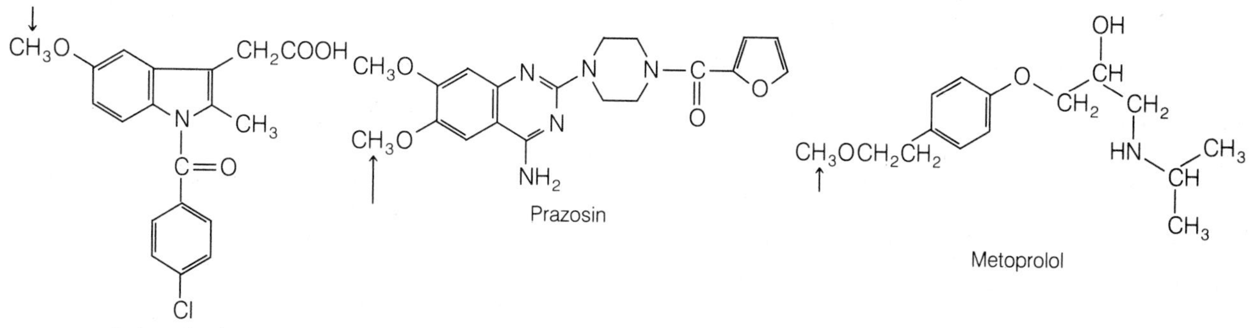

Indomethacin

Prazosin

Metoprolol

Trimethoprim

Mescaline

Desulfuration of parathion leads to the formation of paraoxon, which is the active metabolite responsible for the anticholinesterase activity of the parent drug. The mechanistic details of desulfuration are poorly understood, but it appears to involve microsomal oxidation of the C=S or P=S double bond.[202]

Organosulfur xenobiotics commonly undergo *S*-oxidation to yield sulfoxide derivatives. Several phenothiazine derivatives are metabolized by this pathway. For example, both sulfur atoms present in thioridazine (Mellaril)[203] are susceptible to *S*-oxidation. Oxidation of the 2-methylthio group yields the active sulfoxide metabolite, mesoridazine. Interestingly, mesoridazine is twice as potent an antipsychotic agent as thioridazine in humans and has been introduced into clinical use as Serentil.[204]

6-(Methylthio)-purine

6-Mercaptopurine

Methitural

2-Benzylthio-5-trifluoromethylbenzoic Acid

S-Oxidation constitutes an important pathway in the metabolism of the H_2-histamine antagonists cimetidine (Tagamet)[205] and metiamide.[206] The corresponding sulfoxide derivatives are the major urinary metabolites found in humans.

Sulfoxide drugs and metabolites may be further oxidized to sulfones ($—SO_2—$). The sulfoxide group present in the immunosuppressive agent oxisuran is metabolized to a sulfone moiety.[207] In humans,

Thiopental

Pentobarbital

Parathion

Paraoxon

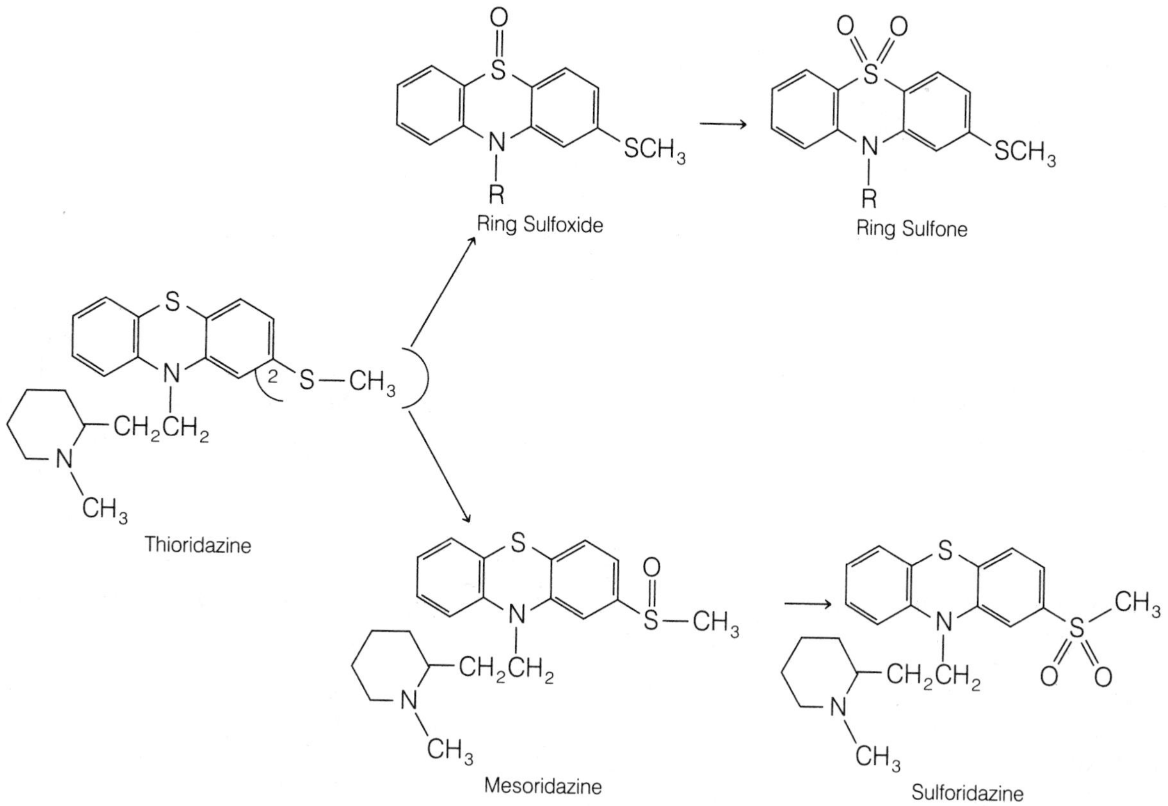

Ring Sulfoxide

Ring Sulfone

Thioridazine

Mesoridazine

Sulforidazine

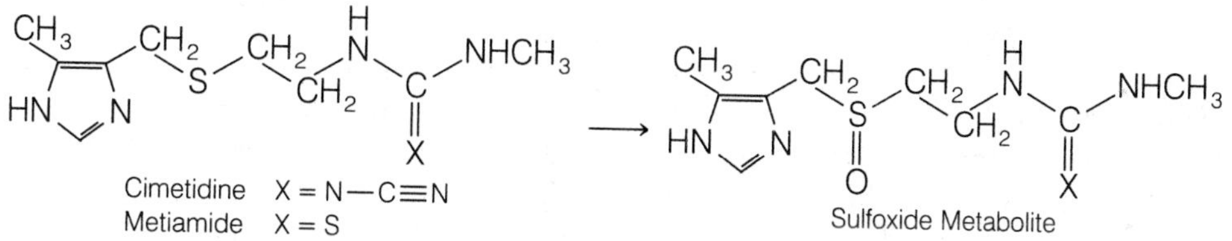

Cimetidine X = N—C≡N
Metiamide X = S

Sulfoxide Metabolite

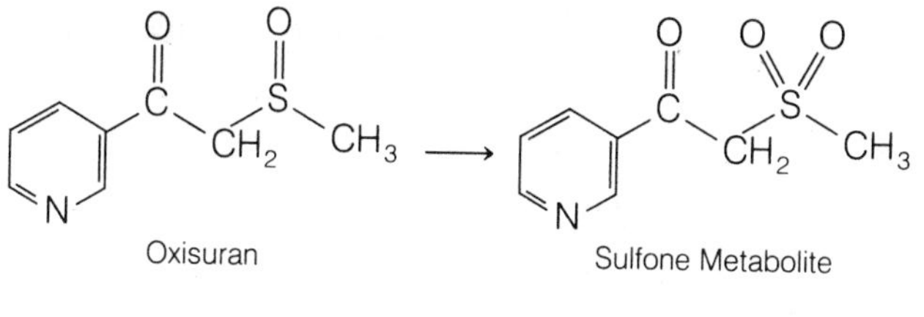

Oxisuran

Sulfone Metabolite

dimethylsulfoxide is primarily found in the urine as the oxidized product dimethylsulfone. Sulfoxide metabolites such as those of thioridazine have been reported to undergo further oxidation to their sulfone derivatives (see earlier diagram).[203]

OXIDATION OF ALCOHOLS AND ALDEHYDES

Many oxidative processes (e.g., benzylic, allylic, alicyclic, or aliphatic hydroxylation) generate alcohol or carbinol metabolites as intermediate products. If not conjugated, these alcohol products are further oxidized to aldehydes (if primary alcohols) or to ketones (if secondary alcohols). Aldehyde metabolites resulting from oxidation of primary alcohols or from oxidative deamination of primary aliphatic amines often undergo facile oxidation to generate polar carboxylic acid derivatives.[91] As a general rule, primary alcoholic groups and aldehyde functionalities are quite vulnerable to oxidation. Several drug examples in which primary alcohol metabolites and

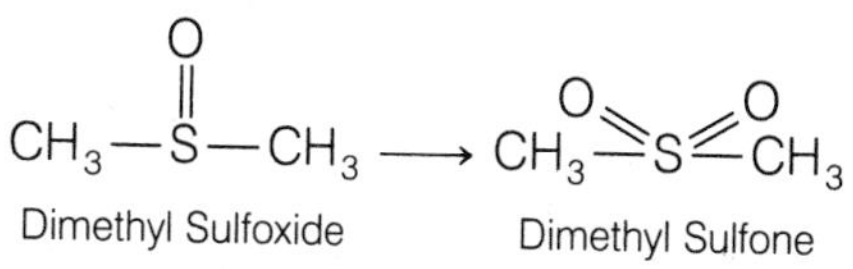

$$CH_3-S(=O)-CH_3 \longrightarrow CH_3-S(=O)_2-CH_3$$

Dimethyl Sulfoxide   Dimethyl Sulfone

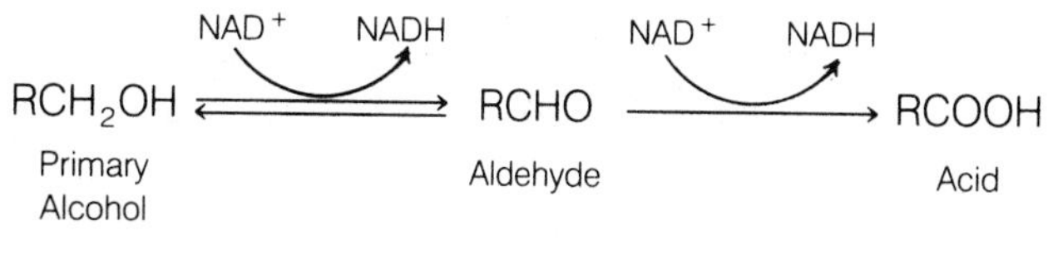

$$RCH_2OH \underset{}{\overset{NAD^+ \;\; NADH}{\rightleftharpoons}} RCHO \xrightarrow{NAD^+ \;\; NADH} RCOOH$$

Primary Alcohol   Aldehyde   Acid

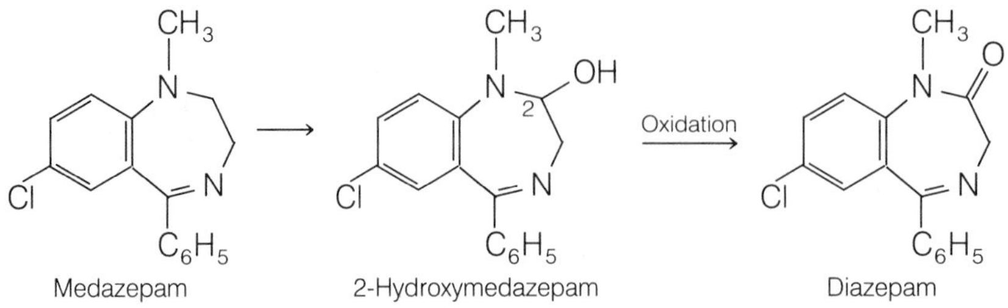

Medazepam → 2-Hydroxymedazepam → (Oxidation) Diazepam

aldehyde metabolites are oxidized to carboxylic acid products were cited in earlier sections.

Although secondary alcohols are susceptible to oxidation, this reaction is not often important because the reverse reaction, namely, reduction of the ketone back to the secondary alcohol, occurs quite readily. In addition, the secondary alcohol group being polar and functionalized is more likely to be conjugated than the ketone moiety.

The bioconversion of alcohols to aldehydes and ketones is catalyzed by soluble alcohol dehydrogenases present in the liver and other tissues. NAD^+ is required as a coenzyme, although $NADP^+$ may also serve as a coenzyme. The reaction catalyzed by alcohol dehydrogenase is reversible, but often proceeds to the right because the aldehyde formed is further oxidized to the acid. Several aldehyde dehydrogenases, including aldehyde oxidase and xanthine oxidase, carry out the oxidation of aldehydes to their corresponding acids.[91,209]

The metabolism of cyclic amines to their lactam metabolites has been observed for various drugs (e.g., nicotine, phenmetrazine, and methylphenidate). It appears that soluble or microsomal dehydrogenase and oxidases are involved in oxidizing the carbinol group of the intermediate carbinolamine to a carbonyl moiety.[209] For example, in the metabolism of medazepam to diazepam, the intermediate carbinolamine (2-hydroxymedazepam) undergoes oxidation of its 2-hydroxy group to a carbonyl moiety. A microsomal dehydrogenase carries out this oxidation.[210]

OTHER OXIDATIVE BIOTRANSFORMATION PATHWAYS

In addition to the many oxidative biotransformations discussed already, oxidative aromatization or dehydrogenation and oxidative dehalogenation reactions also occur. Metabolic aromatization has been reported for norgestrel. Aromatization or dehydrogenation of the A ring present in this steroid leads to the corresponding phenolic product, 17α-ethinyl-18-homoestradiol as a minor metabolite in women.[211] In mice, the terpene ring of Δ^1-THC or $\Delta^{1,6}$-THC undergoes aromatization to give cannabinol.[212]

Many halogen-containing drugs and xenobiotics are metabolized by oxidative dehalogenation. For example, the volatile anesthetic agent halothane is metabolized principally to trifluoroacetic acid in humans.[213] It has been postulated that this metabolite

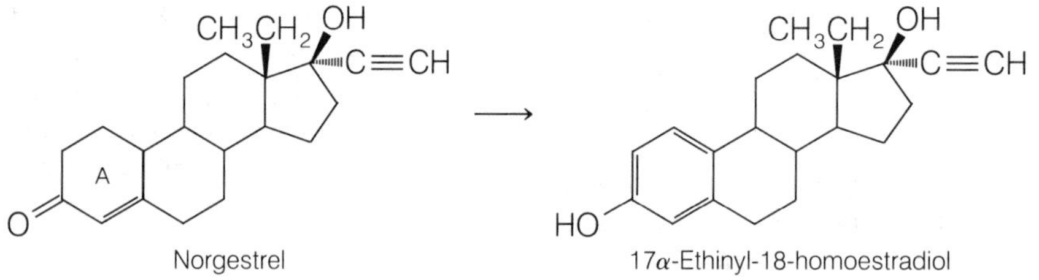

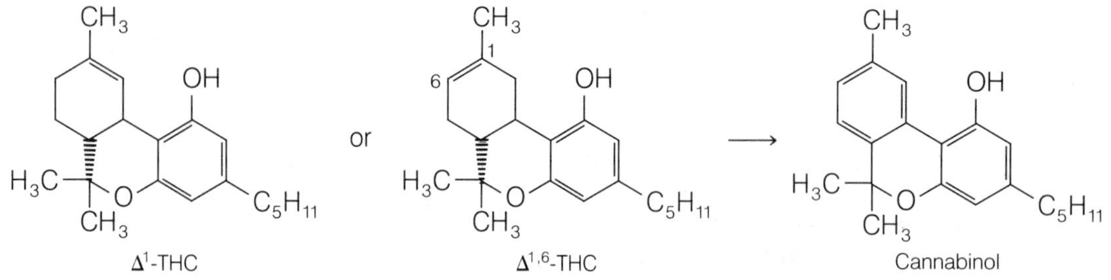

Norgestrel → 17α-Ethinyl-18-homoestradiol

Δ^1-THC or $\Delta^{1,6}$-THC → Cannabinol

$$F_3C{-}CHBrCl \longrightarrow [F_3C{-}C(OH)BrCl] \xrightarrow{-HBr} F_3C{-}C(=O){-}Cl \xrightarrow{H_2O} F_3C{-}C(=O){-}OH + HCl$$

Halothane — Carbinol Intermediate — Trifluoroacetyl Chloride — Trifluoroacetic Acid

arises from cytochrome P-450-mediated hydroxylation of halothane to form an initial carbinol intermediate that spontaneously eliminates hydrogen bromide (dehalogenation) to yield trifluoroacetyl chloride. The latter acyl chloride is chemically reactive and reacts rapidly with water to form trifluoroacetic acid. Alternatively, it can acylate tissue nucleophiles. Indeed, in vitro studies indicate that halothane is metabolized to a reactive intermediate (presumably trifluoroacetyl chloride) that covalently binds to liver microsomal proteins. Chloroform also appears to be oxidatively metabolized by a similar dehalogenation pathway to yield the chemically reactive species, phosgene. It has been postulated that phosgene may be responsible for the hepato- and nephrotoxicity associated with chloroform.[215]

A final example of oxidative dehalogenation concerns the antibiotic chloramphenicol. In vitro studies have demonstrated that the dichloroacetamide portion of the molecule undergoes oxidative dechlorination to yield a chemically reactive oxamyl chloride intermediate that is capable of reacting with water to form the corresponding oxamic acid metabolite or capable of acylating microsomal proteins.[216] Thus, it appears that in several instances oxidative dehalogenation can lead to the formation of toxic and reactive acyl halide intermediates.

REDUCTIVE REACTIONS

Reductive processes play an important role in the metabolism of many compounds containing carbonyl, nitro, and azo groups. Bioreduction of carbonyl compounds generates alcohol derivatives,[91,217] whereas nitro and azo reduction lead to amino derivatives.[218] The hydroxyl and amino moieties of the metabolites are much more susceptible to conjugation than the functional groups of the parent compounds. Hence, reductive processes, as such, facilitate drug elimination.

Reductive pathways that are less frequently encountered in drug drug metabolism include reduction of *N*-oxides to their corresponding tertiary amines and reduction of sulfoxides to sulfides. Reductive cleavage of disulfide linkages and reduction of carbon–carbon double bonds also occur but constitute only minor pathways in drug metabolism.

$$Cl{-}CHCl_2 \longrightarrow [Cl{-}C(OH)Cl_2] \xrightarrow{-HCl} Cl{-}C(=O){-}Cl \xrightarrow{H_2O} H_2CO_3 + HCl$$

Phosgene $\xrightarrow{\text{Tissue Nucleophiles}}$ Covalent Binding

Chloroform — Phosgene

Dichloroacetamide portion

$$O_2N{-}C_6H_4{-}CH(OH){-}CH(CH_2OH){-}NH{-}C(=O){-}CHCl{-}Cl \longrightarrow [R{-}NHC(=O){-}C(OH)Cl{-}Cl] \xrightarrow{-HCl} R{-}NH{-}C(=O){-}C(=O){-}Cl$$

Chloramphenicol — Oxamyl Chloride Derivative

$$R{-}NH{-}C(=O){-}C(=O){-}Cl \xrightarrow{H_2O} R{-}NH{-}C(=O){-}C(=O){-}OH$$

Oxamyl Chloride Derivative $\xrightarrow{\text{Tissue Nucleophiles}}$ Covalent Binding (toxicity?)

Oxamic Acid Derivative

REDUCTION OF ALDEHYDE AND KETONE CARBONYLS

The carbonyl moiety, particularly the ketone group, is frequently encountered in many drugs. In addition, metabolites containing ketone and aldehyde functionalities often arise from oxidative deamination of xenobiotics (e.g., propranolol, chlorpheniramine, amphetamine). Owing to their ease of oxidation, aldehydes are mainly metabolized to carboxylic acids. Occasionally, aldehydes are reduced to primary alcohols. Ketones, on the other hand, are generally resistant to oxidation and are primarily reduced to secondary alcohols. Alcohol metabolites arising from reduction of carbonyl compounds generally undergo further conjugation (e.g., glucuronidation).

$$\text{R-C(=O)-H (Aldehyde)} \longrightarrow \text{R-CH(OH)-H (Primary Alcohols)}$$

$$\text{R-C(=O)-R' (Ketone)} \longrightarrow \text{R-C(OH)(H)-R' (Secondary Alcohols)}$$

Aldehyde — Primary Alcohols

Ketone — Secondary Alcohols (stereoisomeric products possible)

Divers soluble enzymes called aldo–keto reductases carry out bioreduction of aldehydes and ketones.[91,219] They are found in the liver and other tissues (e.g., kidney). As a general class these soluble enzymes have similar physiochemical properties and broad substrate specificities and require NADPH as a cofactor. Oxidoreductase enzymes that carry out both oxidation and reduction reactions also are capable of reducing aldehydes and ketones.[219] For example, the important liver alcohol dehydrogenase is an NAD^+-dependent oxidoreductase that oxidizes ethanol and other aliphatic alcohols to aldehydes and ketones. However, in the presence of NADH or NADPH, the same enzyme system is capable of reducing carbonyl derivatives to their corresponding alcohols.[91]

Few aldehydes undergo bioreduction because of the relative ease of oxidation of aldehydes to carboxylic acids. However, one frequently cited example of a parent aldehyde drug undergoing extensive enzymatic reduction is the sedative–hypnotic chloral hydrate. Bioreduction of this hydrated aldehyde yields trichloroethanol as the major metabolite in humans.[220] Interestingly, this alcohol metabolite is pharmacologically active. Further glucuronidation of

$$Cl_3C\text{-}CH(OH)\text{-}OH \underset{-H_2O}{\overset{H_2O}{\rightleftharpoons}} Cl_3C\text{-}CH{=}O \longrightarrow Cl_3C\text{-}CH_2OH$$

Chloral Hydrate — Chloral — Trichloroethanol

the alcohol leads to an inactive conjugated product that is readily excreted in the urine.

Aldehyde metabolites resulting from oxidative deamination of drugs also have been observed to undergo reduction to a minor extent. For example, in humans the β-adrenergic blocker, propranolol, is converted to an intermediate aldehyde by *N*-dealkylation and oxidative deamination. Although the aldehyde is primarily oxidized to the corresponding carboxylic acid (naphthoxylactic acid), a small fraction is also reduced to the alcohol derivative (propranolol glycol).[221]

Two major polar urinary metabolites of the histamine H_1 antagonist chlorpheniramine have recently been identified in dogs, as the alcohol and carboxylic acid products (conjugated), derived respectively from reduction and oxidation of an aldehyde metabolite. The aldehyde precursor arises from bis-*N*-demethylation and oxidative deamination of chlorpheniramine.[222]

Bioreduction of ketones often leads to the creation of an asymmetric center and, thereby, two possible stereoisomeric alcohols.[91,223] For example, reduction of acetophenone by a soluble rabbit kidney reductase leads to the enantiomeric alcohols, (*S*)(−)- and (*R*)(+)-methylphenylcarbinol with the (*S*)(−)-isomer predominating (3 : 1 ratio).[224] The preferential formation of one stereoisomer over the other is termed *product stereoselectivity* in drug metabolism.[223] Mechanistically, ketone reduction involves a "hydride" transfer from the reduced nicotinamide moiety of the cofactor NADPH or NADH to the carbonyl carbon atom of the ketone. It is generally agreed that this step proceeds with considerable *stereoselectivity*.[91,223] Consequently, it is not surprising to find many reports of xenobiotic ketones that are preferentially reduced to a predominant stereoisomer. Oftentimes, ketone reduction yields alcohol metabolites that are pharmacologically active.

Although many ketone-containing drugs undergo significant reduction, only a few selected examples will be presented in detail here. Those xenobiotics that are not discussed in the text have been structurally tabulated in Figure 3-10. The keto group undergoing reduction has been designated with an arrow.

Ketones lacking asymmetric centers in their molecules, such as acetophenone or the oral hypo-

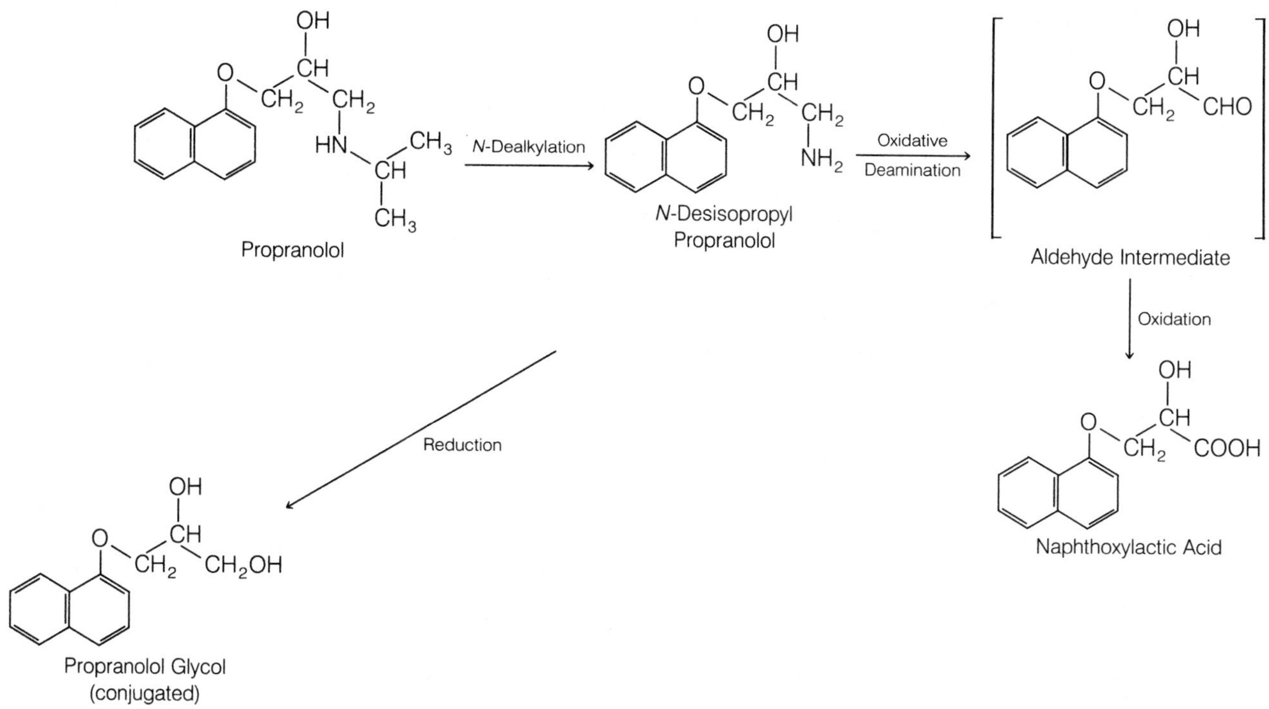

Propranolol → N-Dealkylation → N-Desisopropyl Propranolol → Oxidative Deamination → [Aldehyde Intermediate] → Oxidation → Naphthoxylactic Acid

Reduction → Propranolol Glycol (conjugated)

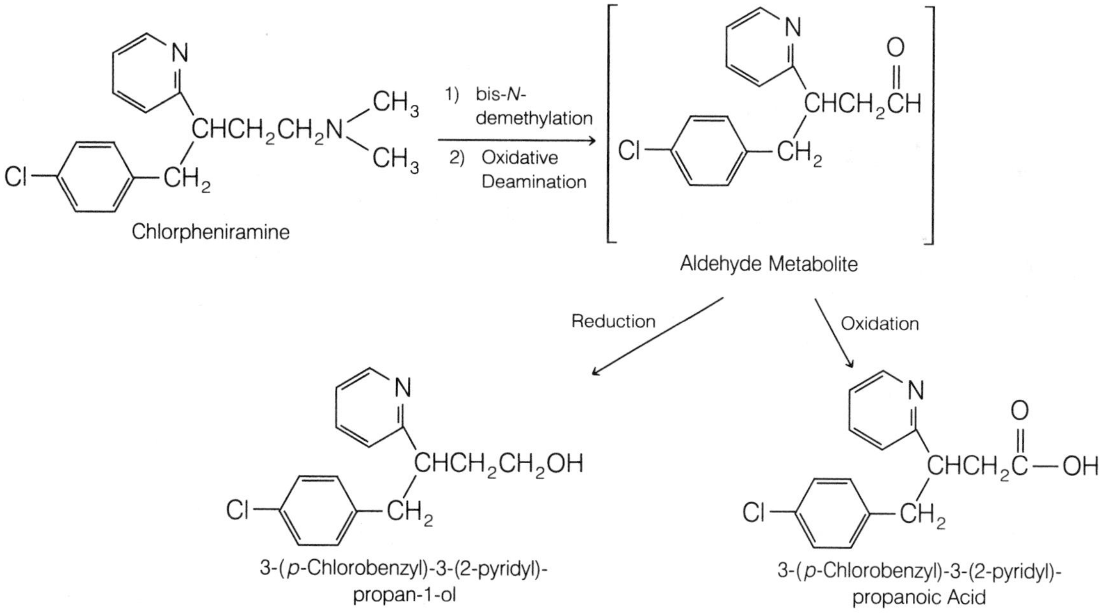

Chlorpheniramine → 1) bis-N-demethylation 2) Oxidative Deamination → [Aldehyde Metabolite]

Reduction → 3-(p-Chlorobenzyl)-3-(2-pyridyl)-propan-1-ol

Oxidation → 3-(p-Chlorobenzyl)-3-(2-pyridyl)-propanoic Acid

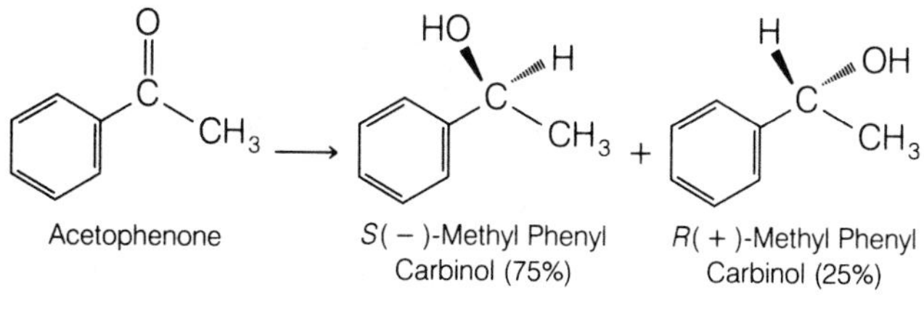

Acetophenone → S(−)-Methyl Phenyl Carbinol (75%) + R(+)-Methyl Phenyl Carbinol (25%)

glycemic acetohexamide, usually give rise to predominantly one enantiomer upon reduction. In humans, acetohexamide is rapidly metabolized in the liver to give principally (S)(−)-hydroxyhexamide.[225] This metabolite is as active a hypoglycemic agent as its parent compound and is further eliminated through the kidney.[226] Acetohexamide usually is not recommended in diabetic patients with renal failure, owing to the possible accumulation of its active metabolite, hydroxyhexamide.

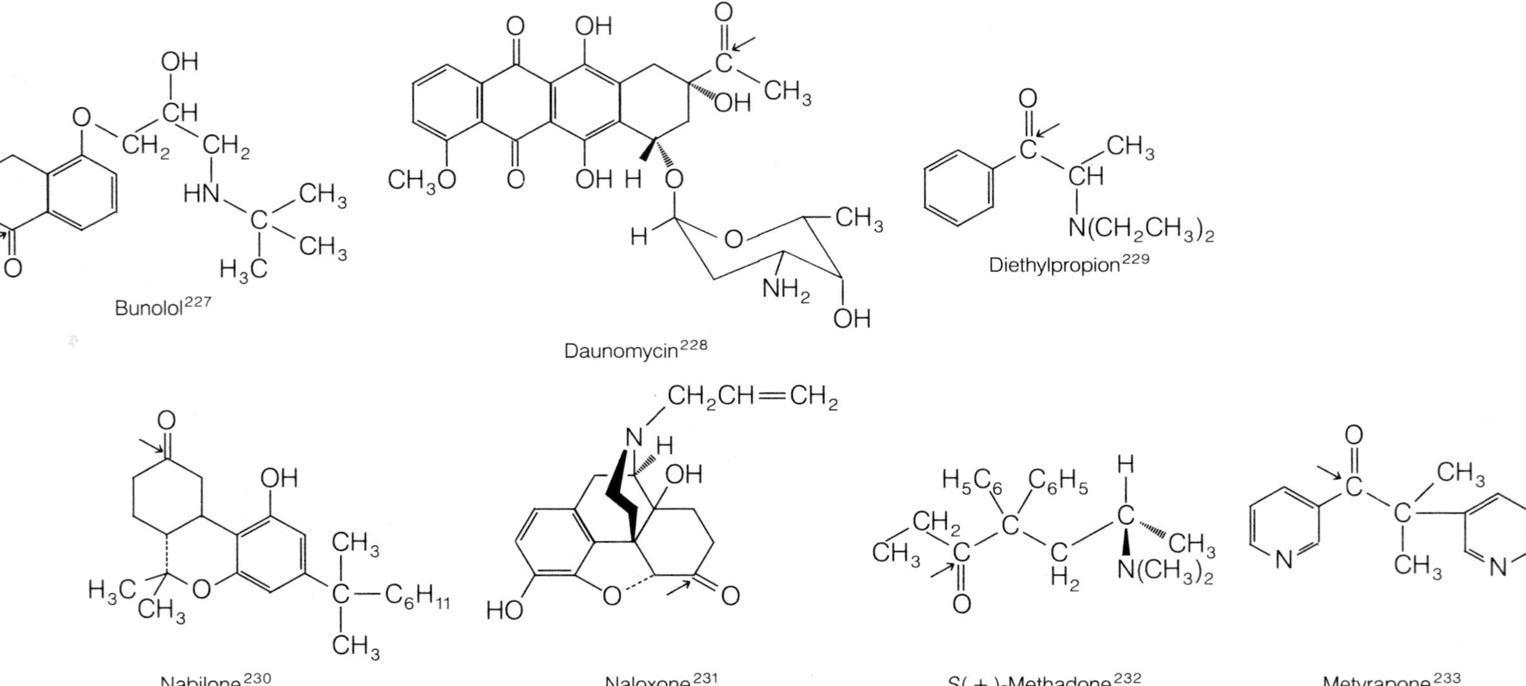

FIG. 3-10. *Additional examples of xenobiotics that undergo extensive ketone reduction, which were not covered in the text. Arrow indicates the keto group undergoing reduction.*

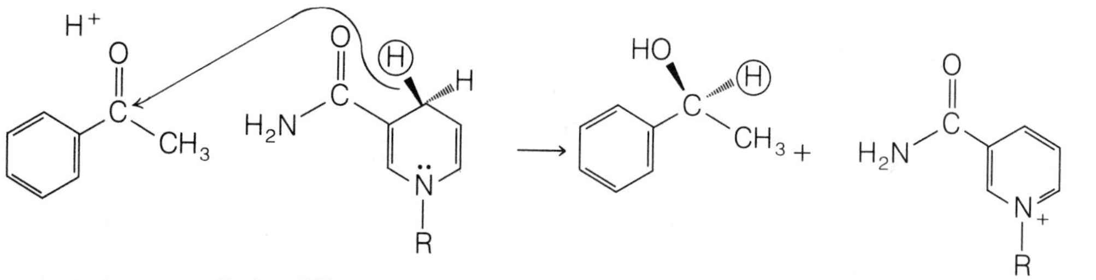

Acetophenone — Reduced Nicotamide Moiety of NADPH or NADH — $S(-)$-Methyl Phenyl Carbinol — Oxidized Nicotamide Moiety of $NADP^+$ or NAD^+

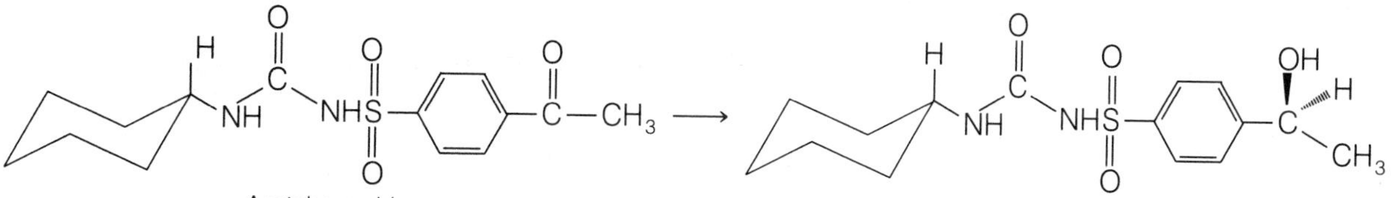

Acetohexamide — $S(-)$-Hydroxyhexamide

When chiral ketones are reduced, they yield two possible diastereomeric or epimeric alcohols. For example, the $(R)(+)$-enantiomer of the oral anticoagulant warfarin undergoes extensive reduction of its side chain keto group to generate the $(R,S)(+)$-alcchol as the major plasma metabolite in humans.[45,234] Small amounts of the $(R,R)(+)$-diastereomer are also formed. In contrast, the $(S)(-)$-enantiomer undergoes little ketone reduction and is primarily 7-hydroxylated (i.e., aromatic hydroxylation) in humans.

Reduction of the 6-keto functionality in the narcotic antagonist, naltrexone, can lead to either the epimeric 6α- or 6β-hydroxy metabolites depending on the animal species.[231] In human and rabbit, bioreduction of naltrexone is highly stereoselective and generates only 6β-naltrexol, whereas in chicken, reduction occurs to yield only 6α-naltrexol.[231,235] However, in monkey and guinea pig, both epimeric alcohols are formed (predominantly 6β-naltrexol).[236] It appears that in the latter two species, reduction of naltrexone to the epimeric 6α- and 6β-alcohols is carried out by two distinctly different reductases found in the liver.[235,236]

Reduction of oxisuran appears not be an important pathway by which the parent drug mediates its immunosuppressive effects. Studies indicate that oxisuran has its greatest immunosuppressive effects in

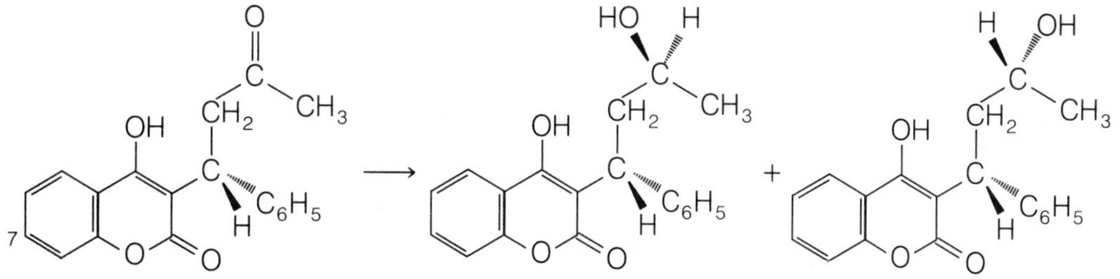

$R(+)$-Warfarin — R,S-$(+)$-Alcohol Major Diastereomer — R,R-$(+)$-Alcohol Minor Diastereomer

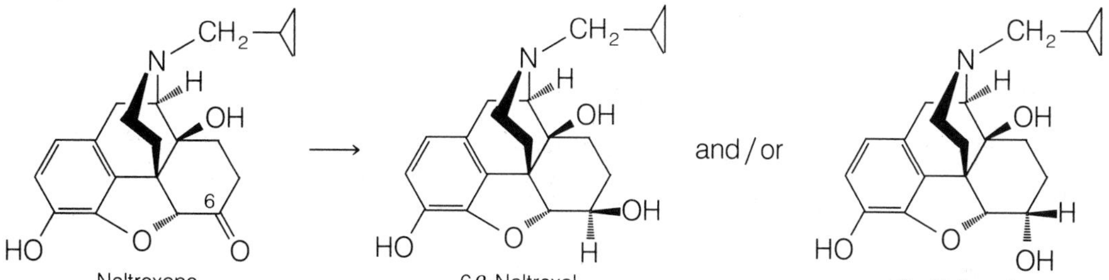

Naltrexone — 6β-Naltrexol — and/or — 6α-Naltrexol

those species that form alcohols as their major metabolic products (e.g., human, rat).[237,238] In species in which reduction is a minor pathway (e.g., dog), oxisuran shows little immunosuppressive activity.[238] These findings tend to indicate that the oxisuran alcohols (oxisuranols) are pharmacologically active and contribute substantially to the overall immunosuppressive effect of the parent drug. The sulfoxide group in oxisuran is chiral, by virtue of the lone pair of electrons on sulfur. Therefore, reduction of oxisuran leads to diastereomeric alcohols.

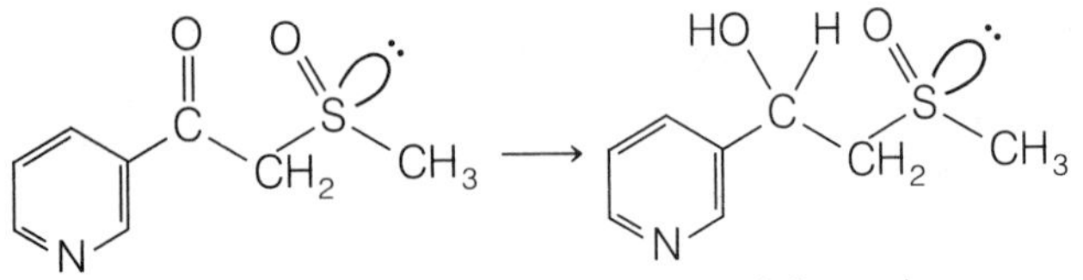

Oxisuran → Oxisuranols (diastereomeric mixture)

Reduction of α,β-unsaturated ketones results not only in reduction of the ketone group, but in reduction of the carbon–carbon double bond as well. Steroidal drugs often fall into this class, including norethindrone, a synthetic progestin found in many oral contraceptive drug combinations. In women, the major plasma and urinary metabolite of norethindrone is the $3\beta,5\beta$-tetrahydro derivative.[239]

Ketones resulting from metabolic oxidative deamination processes are also susceptible to reduction. For example, rabbit liver microsomal preparations metabolize amphetamine to phenylacetone, which is subsequently reduced to 1-phenyl-2-propanol.[240] In humans, a minor urinary metabolite of (−)-ephedrine has been identified as the diol derivative formed from keto reduction of the oxidatively deaminated product, 1-hydroxy-1-phenylpropan-2-one.[241]

REDUCTION OF NITRO AND AZO COMPOUNDS

The reduction of aromatic nitro and azo xenobiotics leads to aromatic primary amine metabolites.[218] Aromatic nitro compounds are initially reduced to the nitroso and hydroxylamine intermediates as shown in the metabolic sequence below:

$$Ar—\overset{+}{N}O_2 \longrightarrow Ar—N{=}O \longrightarrow Ar—NHOH \longrightarrow Ar—NH_2$$

Nitro → Nitroso → Hydroxylamine → Amine

Azo reduction on the other hand is believed to proceed via a hydrazo intermediate (—NH—NH—) that is subsequently reductively cleaved to yield the corresponding aromatic amines:

$$AR—N{=}N—Ar' \longrightarrow Ar—NH—NH—Ar' \longrightarrow Ar—NH_2 + H_2N—Ar'$$

Azo → Hydrazo → Amines

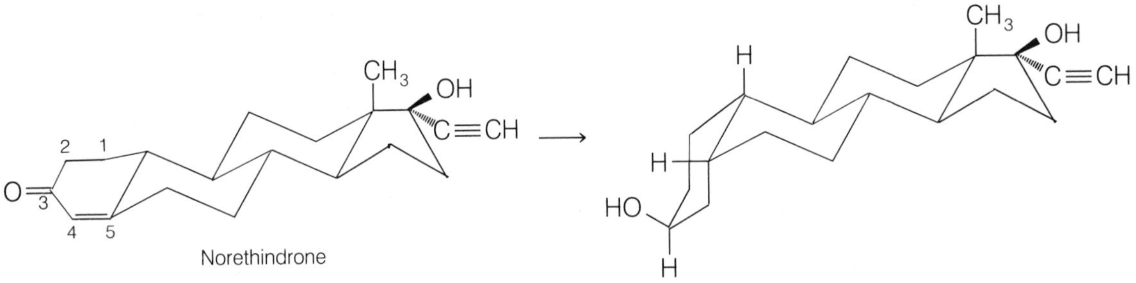

Norethindrone → $3\beta,5\beta$-Tetrahydronorethindrone

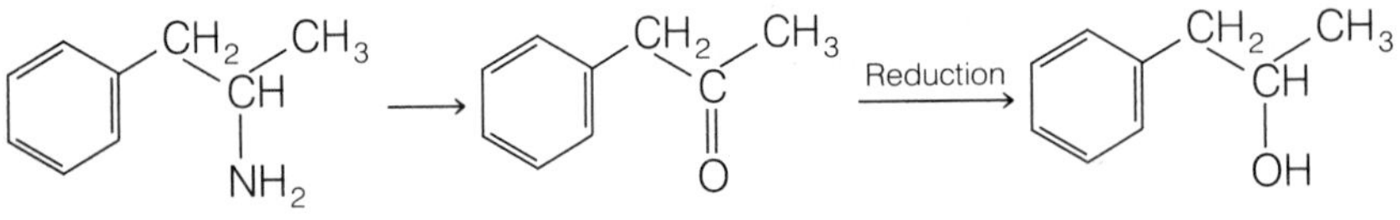

Amphetamine → Phenylacetone —Reduction→ 1-Phenyl-2-propanol

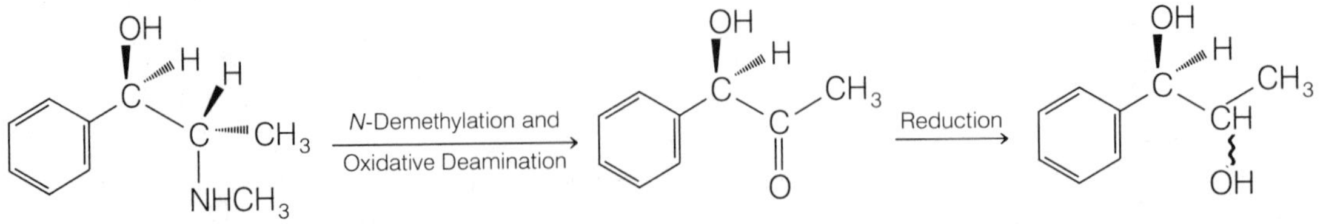

(−)-Ephedrine —*N*-Demethylation and Oxidative Deamination→ 1-Hydroxy-1-phenylpropan-2-one —Reduction→ 1-Phenyl-1,2-propanediol (as glucuronide conjugate)

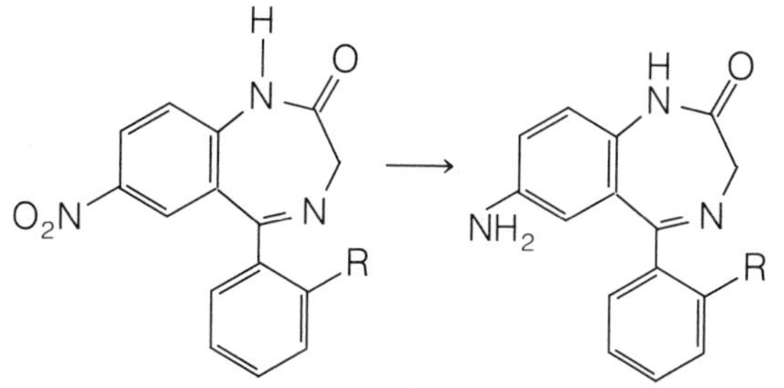

Clonazepam, R = Cl
Nitrazepam, R = H

7-Amino Metabolite

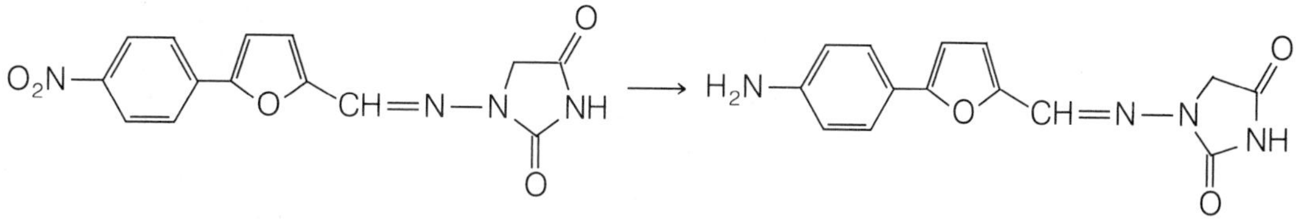

Dantrolene

Aminodantrolene

Bioreduction of nitro compounds is carried out by NADPH-dependent microsomal and soluble nitro reductases present in the liver. A multicomponent hepatic microsomal reductase system requiring NADPH appears to be responsible for azo reduction. In addition, bacterial reductases present in the intestine are also capable of reducing nitro and azo compounds, especially those that are poorly absorbed or that are excreted mainly in the bile.[244]

Various aromatic nitro drugs undergo enzymatic reduction to the corresponding aromatic amines. For example, the 7-nitro benzodiazepine derivatives, clonazepam and nitrazepam, are extensively metabolized to their respective 7-amino metabolites in humans.[245, 246] The skeletal muscle relaxant dantrolene (Dantrium) also has been reported to undergo reduction to aminodantrolene in humans.[247]

For some nitro xenobiotics, bioreduction appears to be a minor metabolic pathway in vivo, owing to competing oxidative and conjugative reactions. However, under artificial anaerobic in vitro incubation conditions, these same nitro xenobiotics are rapidly enzymatically reduced. For example, most of the urinary metabolites of metronidazole found in humans are either oxidation or conjugation products. Reduced metabolites of metronidazole have not been detected.[248] However, when incubated anaerobically with guinea pig liver preparations, metronidazole undergoes considerable nitro reduction.[249]

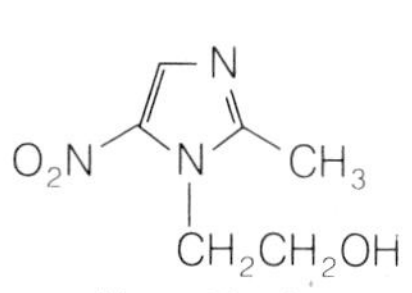

Bacterial reductase present in the intestine also tends to complicate in vivo interpretations of nitro reduction. For example, in rats the antibiotic chloramphenicol is not reduced in vivo by the liver but is excreted in the bile and, subsequently, reduced by intestinal flora to form the amino metabolite.[250]

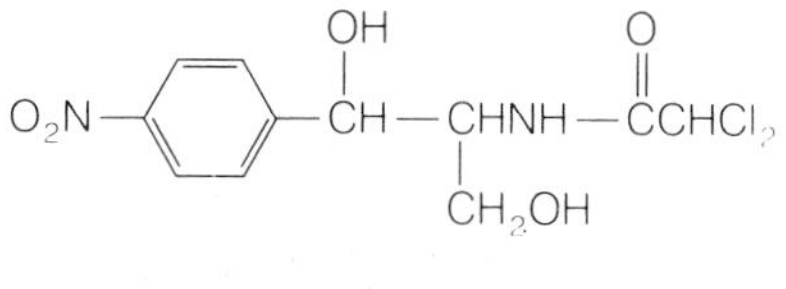

The enzymatic reduction of azo compounds is best exemplified by the conversion of sulfamidochrysoidine (Prontosil) to the active sulfanilamide metabolite in the liver.[251] This reaction has historical significance, for it led to the discovery of sulfanilamide as an antibiotic and eventually to the development of many of the therapeutic sulfonamide drugs. Bacterial reductases present in the intestine play a significant role in reducing azo xenobiotics, particularly those that are poorly absorbed.[244] For example, the two azo dyes tartrazine[252] and amaranth[253] have poor oral absorption as a result of the many polar and ionized sulfonic acid groups present in their structures. Therefore, these two azo compounds are primarily metabolized by bacterial reductases present in the intestine. The importance of intestinal reduction is further revealed in the metabolism of sulfasalazine (formerly salicylazosulfapyridine) (Azulfidine), a drug used in the treatment of ulcerative colitis. The drug is poorly absorbed and undergoes reductive cleavage of the azo linkage to yield sulfapyridine and 5-aminosalicylic acid.[254] The reaction occurs primarily in the colon and is carried out principally by intestinal bacteria. Studies in germ-free rats, lacking intestinal flora, have demonstrated that sulfasalazine is not reduced to any appreciable extent.[255]

Metronidazole

Chloramphenicol

Sulfamidochrysoidine (Prontosil) → Sulfanilamide + 1,2,4-Triaminobenzene

Tartrazine

Amaranth

Sulfasalazine → Sulfapyridine + 5-Aminosalicylic Acid

Imipramine *N*-Oxide → Imipramine

MSCELLANEOUS REDUCTIONS

Several minor reductive reactions also occur. Reduction of *N*-oxides to the corresponding tertiary amine occurs to some extent. This reductive pathway is of interest, because several tertiary amines are oxidized to form polar and water-soluble *N*-oxide metabolites. If reduction of *N*-oxide metabolites occurs to a significant extent, drug elimination of the parent tertiary amine would be impeded. *N*-Oxide reduction is often assessed by administering the pure synthetic *N*-oxide in vitro or in vivo and then attempting to detect the formation of the tertiary amine. For example, imipramine *N*-oxide undergoes reduction in rat liver preparations.[256]

Reduction of sulfur-containing functional groups, such as the disulfide and sulfoxide moieties, also constitutes minor reductive pathways. Reductive cleavage of the disulfide bond in disulfiram (Antabuse) yields *N,N*-diethyldithiocarbamic acid (free or glucronidated) as a major metabolite in humans.[257] Although sulfoxide functionalities are mainly oxidized to sulfones ($-SO_2-$), they sometimes undergo reduction to sulfides. A recent example demonstrating the importance of this reductive pathway is seen in the metabolism of the anti-inflammatory agent sulindac (Clinoril). Studies in humans show that sulindac undergoes reduction to an active sulfide that is responsible for the overall anti-inflammatory effect of the parent drug.[258] Sulindac or its sulfone metabolite exhibits little anti-inflammatory activity. Another example of sulfide formation involves the reduction of dimethyl

Disulfiram → *N,N*-Diethylthiocarbamic Acid

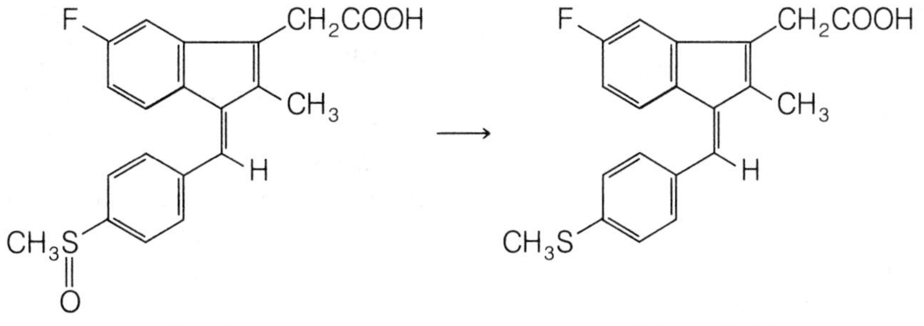

Sulindac → Sulindac Sulfide Metabolite

sulfoxide (DMSO) to dimethyl sulfide. In humans, DMSO is metabolized to a minor extent by this pathway. The characteristic unpleasant odor of dimethyl sulfide is evident on the breath of patients who use this agent.[259]

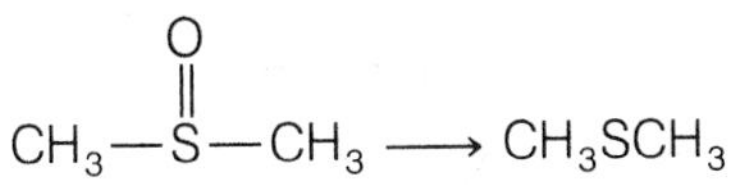

$$CH_3-\overset{O}{\overset{\|}{S}}-CH_3 \longrightarrow CH_3SCH_3$$

Dimethyl Sulfoxide → Dimethyl Sulfide

HYDROLYTIC REACTIONS

HYDROLYSIS OF ESTERS AND AMIDES

The metabolism of ester and amide linkages in many drugs is catalyzed by hydrolytic enzymes present in various tissues and in plasma. The metabolic products formed, namely carboxylic acids, alcohols, phenols, and amines, generally are polar and functionally more susceptible to conjugation and excretion than the parent ester or amide drugs. The enzymes carrying out ester hydrolysis include several nonspecific esterases found in the liver, kidney, and intestine as well as the pseudocholinesterases present in plasma.[260,261] Amide hydrolysis appears to be mediated by liver microsomal amidases, esterases, and deacylases.[261]

Hydroylsis is a major biotransformation pathway for drugs containing the ester functionality. This is because of the relative ease of hydrolyzing the ester linkage. A classic example of ester hydrolysis is the metabolic conversion of aspirin (acetylsalicylic acid) to salicylic acid.[262] Of the two ester moieties present in cocaine, it appears that, in general, the methyl group is preferentially hydrolyzed to yield benzoylecgonine as the major urinary metabolite in humans.[263] However, the hydrolysis of cocaine to methylecgonine has also been demonstrated to occur in plasma and blood to a minor extent.[264] Methylphenidate (Ritalin) is rapidly biotransformed by hydrolysis to yield ritalinic acid as the major urinary metabolite in humans.[265] Often, ester hydrolysis of the parent drug leads to pharmacologically active metabolites. For example, hydrolysis of diphenoxylate in humans leads to diphenoxylic acid (difenoxin), which is apparently five times more potent an antidiarrheal agent than the parent ester.[266] The rapid metabolism of clofibrate (Atromid-S) yields *p*-chlorophenoxyisobutyric acid (CPIB) as the major plasma metabolite in humans.[267] Studies in rats indicate that the free acid CPIB is responsible for clofibrate's hypolipidemic effect.[268]

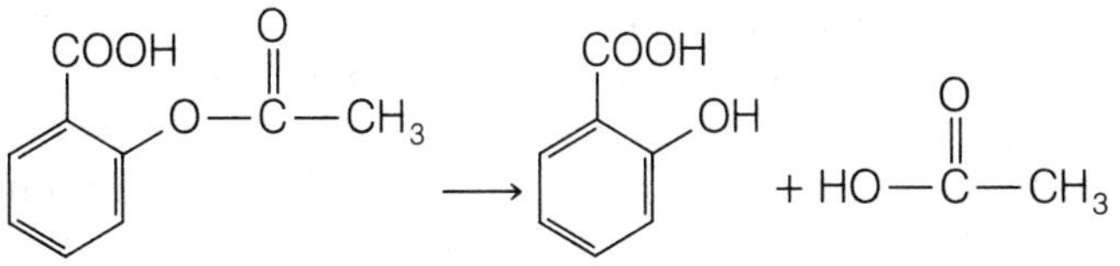

Aspirin (Acetylsalicylic acid) → Salicylic Acid + $HO-\overset{O}{\overset{\|}{C}}-CH_3$ Acetic Acid

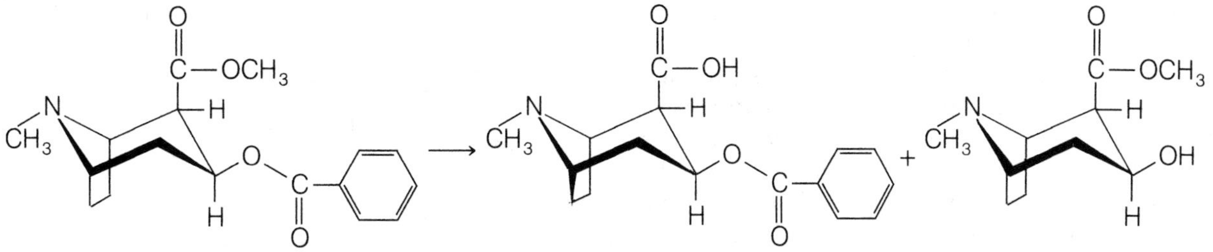

Cocaine → Benzoylecgonine + Methylecgonine

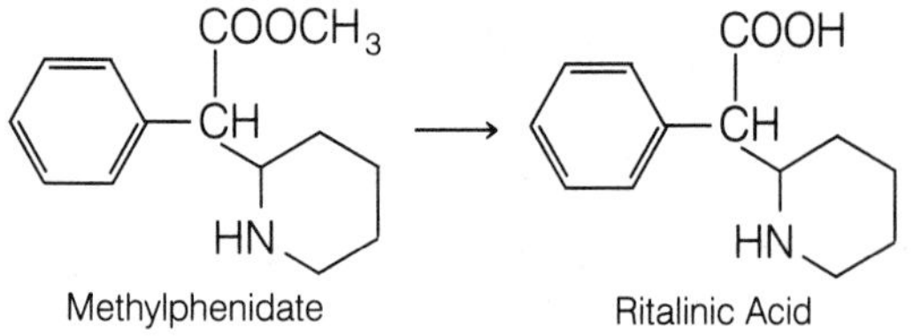

Methylphenidate → Ritalinic Acid

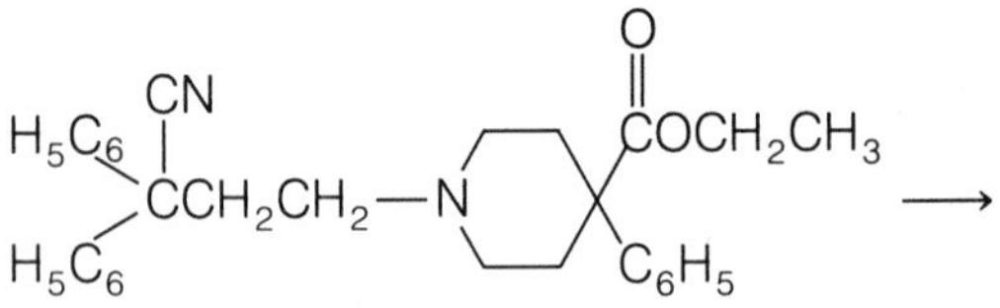

Diphenoxylate 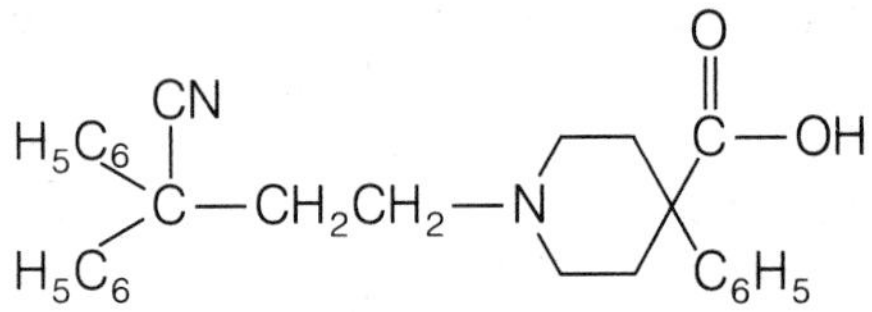 → Diphenoxylic Acid (Difenoxin)

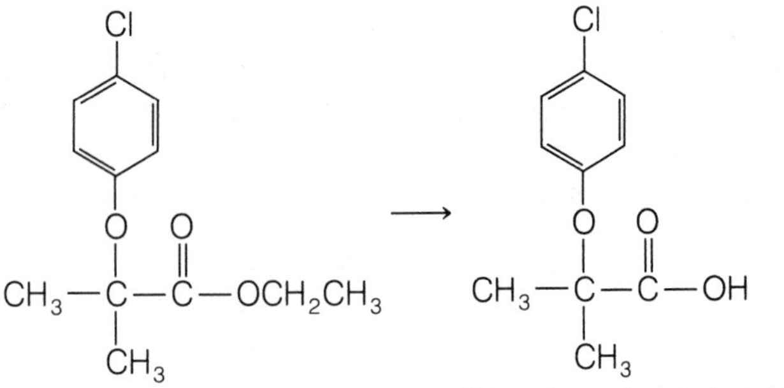

Clofibrate → *p*-Chlorophenoxyisobutyric Acid

In recent years, many parent drugs have been chemically modified or derivatized to generate so-called *prodrugs* to overcome some undesirable property (e.g., bitter taste, poor absorption, poor solubility, irritation at site of injection). The rationale behind the prodrug concept was to develop an agent that once inside the biologic system will be biotransformed to the active parent drug.[17] The presence of esterases in many tissues and in plasma makes ester derivatives logical prodrug candidates because hydrolysis would cause the ester prodrug to revert to the parent compound. For example, antibiotics, such as chloramphenicol and clindamycin, have been derivatized as their palmitate esters to minimize their bitter taste and to improve their palatability in pediatric liquid suspensions.[269] After oral administration, intestinal esterases and lipases hydrolyze the palmitate esters to the free antibiotics. To improve the poor oral absorption of carbenicillin, a lipophilic indanyl ester has been formulated (Geocillin).[270] Once orally absorbed, the ester is rapidly hydrolyzed to the parent drug. A final example involves derivatization of prednisolone to its C-21 hemisuccinate sodium salt. This water-soluble derivative is extremely useful for parenteral administration and is metabolized to the parent steroid drug by plasma and tissue esterases.[271]

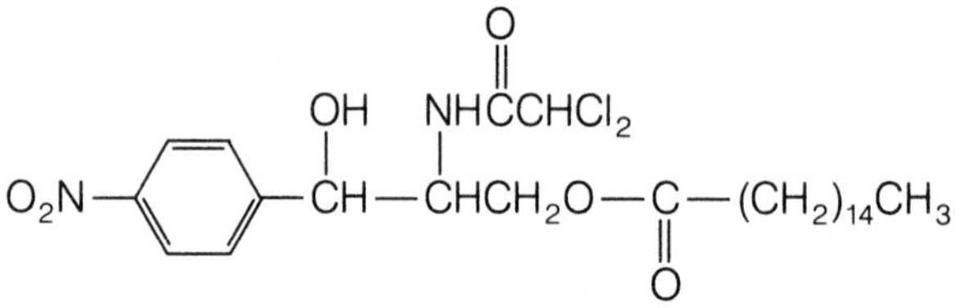

Chloramphenicol Palmitate

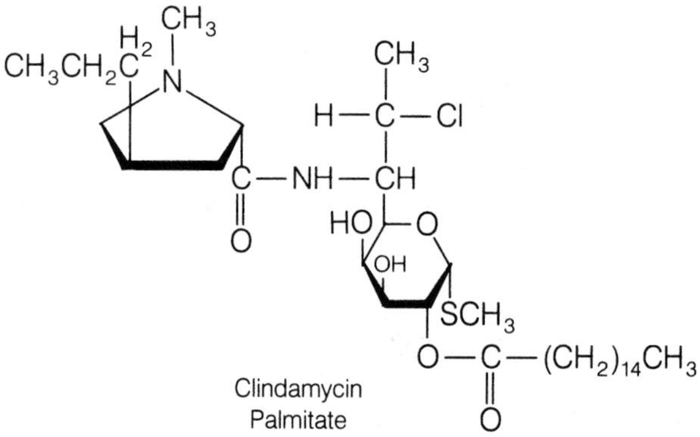

Clindamycin Palmitate

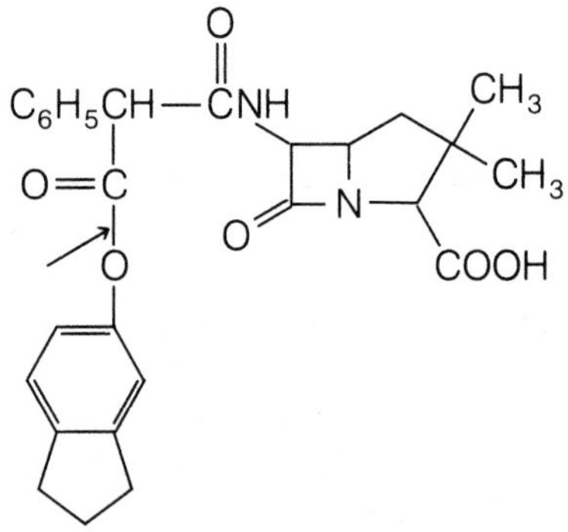

Carbenicillin Indanyl Ester

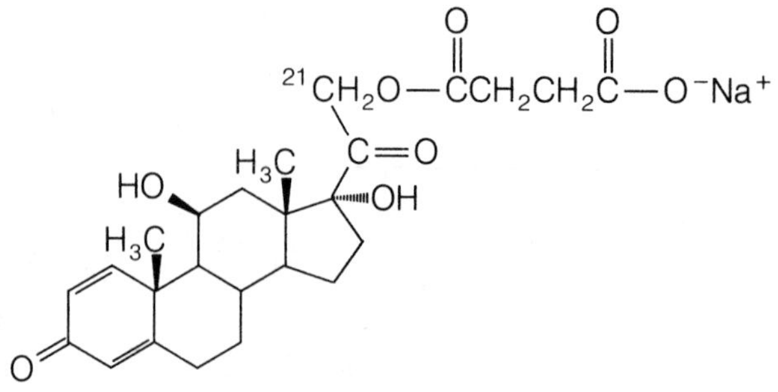

Prednisolone Hemisuccinate Sodium Salt

Amides are slowly hydrolyzed in comparison to esters.[261] For example, hydrolysis of the amide bond of procainamide is relatively slow compared with the hydrolysis of the ester linkage in procaine.[260,272] Drugs in which amide cleavage has been reported to occur to some extent include lidocaine,[273] carbamazepine,[72] indomethacin,[194] and prazosin (Minipress).[195] Amide linkages present in barbiturates (e.g., hexobarbital)[274] as well as in hydantoins (e.g., 5-phenylhydantoin)[275] and succinimides (phensuximide)[275] are also susceptible to hydrolysis.

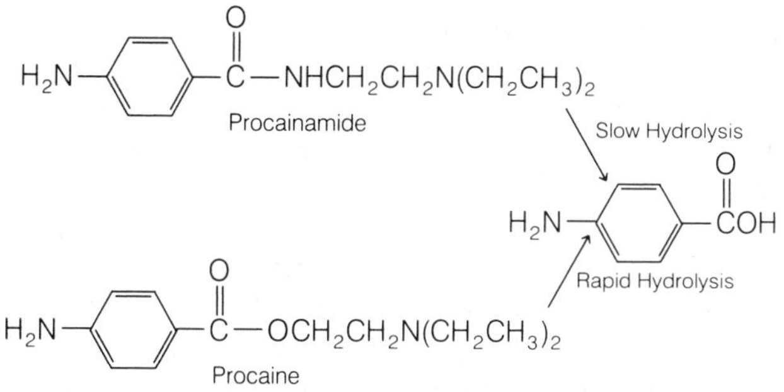

$H_2N-C_6H_4-C(=O)-NHCH_2CH_2N(CH_2CH_3)_2$

Procainamide

Slow Hydrolysis

$H_2N-C_6H_4-COOH$

Rapid Hydrolysis

$H_2N-C_6H_4-C(=O)-OCH_2CH_2N(CH_2CH_3)_2$

Procaine

MSCELLANEOUS HYDROLYTIC REACTIONS

In addition to hydrolysis of amides and esters, hydrolytic cleavage of other moieties also occur to a minor extent in drug metabolism.[8] These include the hydrolysis of phosphate esters (e.g., diethylstilbestrol diphosphate), sulfonylureas, cardiac glycosides, carbamate esters, and organophosphate compounds. Glucuronide and sulfate conjugates are also capable of undergoing hydrolytic cleavage by β-glucuronidase and sulfatase enzymes. These hydrolytic reactions will be discussed later under conjugation reactions. Finally, the hydration or hydrolytic cleavage of epoxides and arene oxides by epoxide hydrase is sometimes considered under hydrolysis reactions.

PHASE II OR CONJUGATION REACTIONS

Phase I or functionalization reactions do not always produce hydrophilic or pharmacologically inactive metabolites. However, various phase II or conjuga-

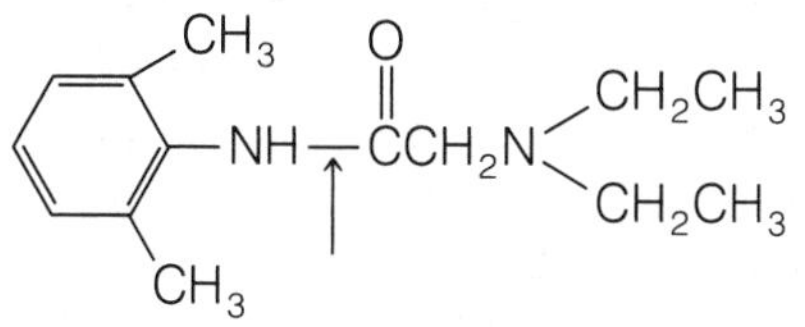

Lidocaine

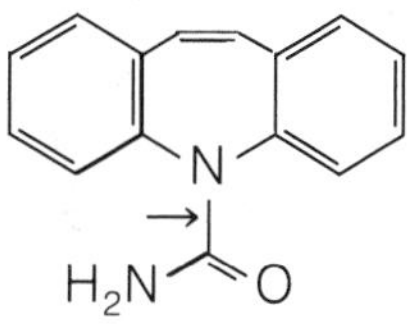

Carbamazepine

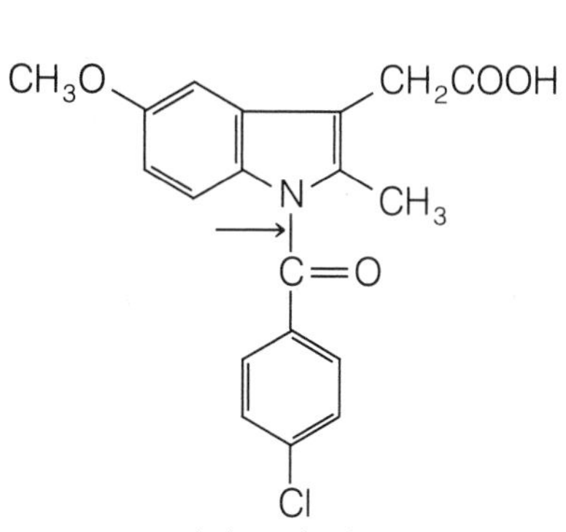

Indomethacin

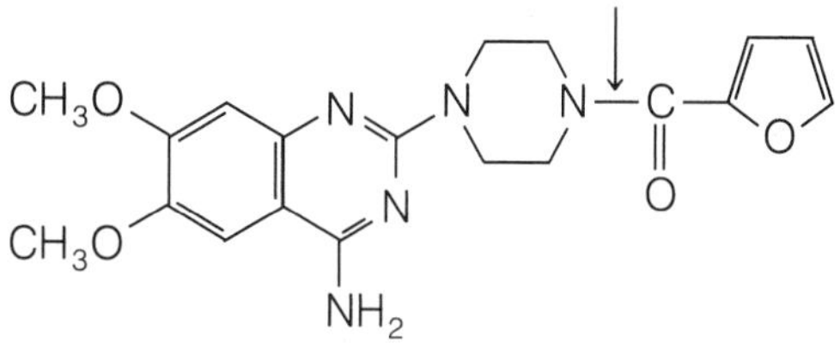

Prazosin

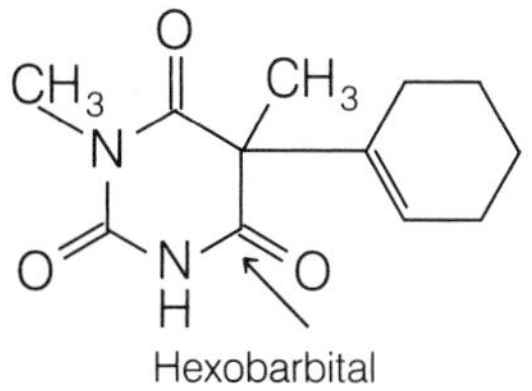

Hexobarbital

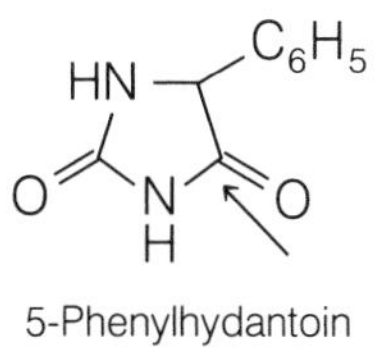

5-Phenylhydantoin

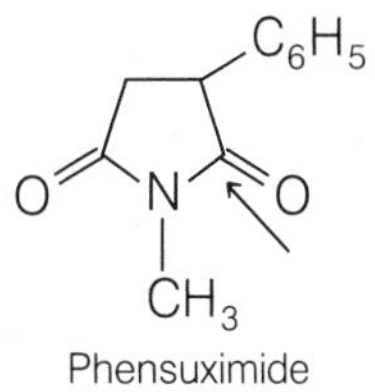

Phensuximide

tion reactions are capable of converting these metabolites to more polar and water-soluble products. Many conjugative enzymes accomplish this objective by attaching small, polar, and ionizable endogenous molecules, such as glucuronic acid, sulfate, glycine, and glutamine, to the phase I metabolite or parent xenobiotic. The resulting conjugated products are relatively water-soluble and readily excretable. In addition, they generally are biologically inactive and nontoxic. Other phase II reactions, such as methylation and acetylation, do not generally increase water solubility, but mainly serve to terminate or attenuate pharmacologic activity. The role of glutathione is to combine with chemically reactive compounds to prevent damage to important biomacromolecules such as DNA, RNA, and proteins. Thus, phase II conjugation reactions can be regarded as truly detoxifying pathways in drug metabolism, with a few exceptions.

A distinguishing feature of most phase II reactions is that the conjugating group (glucuronic acid, sulfate, methyl, and acetyl) is initially activated in the form of a coenzyme before transfer or attachment of the group is made to the accepting substrate by the appropriate transferase enzyme. In other cases, such as glycine and glutamine conjugation, the substrate is initially activated. Many endogenous compounds, such as bilirubin, steroids, catecholamines, and histamine, also undergo conjugation reactions and utilize the same coenzymes, although they appear to be mediated by more specific transferase enzymes. The phase II conjugative pathways to be discussed include those listed earlier in this chapter. Although other conjugative pathways (e.g., conjugation with glycosides, phosphate, other amino acids, conversion of cyanide to thiocyanate) exist, they are of only minor importance in drug metabolism and will not be covered in this chapter.

GLUCURONIC ACID CONJUGATION

Glucuronidation is the most common conjugative pathway in drug metabolism for several reasons: (1) a readily available supply of D-glucuronic acid (derived from D-glucose); (2) numerous functional groups that can combine enzymatically with glucuronic acid; and (3) the glucuronyl moiety (with its ionized carboxylate [pK_a 3.2] and polar hydroxyl groups) when attached to xenibiotic substrates, greatly increases the water solubility of the conjugated product.[92,276,277] Formation of β-glucuronides involves two steps, synthesis of an activated coenzyme, uridine-5′-diphospho-α-D-glucuronic acid (UDPGA) and subsequent transfer of the glucuronyl group from UDPGA to an appropriate substrate.[92,277] The transfer step is catalyzed by microsomal enzymes called UDP-glucuronyltransferases. They are found primarily in the liver, but also occur in many other tissues, including kidney, intestine, skin, lung, and brain.[277]

The sequence of events involved in glucuronidation is summarized in Figure 3-11.[92,277] The synthesis of the coenzyme UDPGA utilizes α-D-glucose-1-phosphate as its initial precursor. Note, that all glucuronide conjugates have the β-configuration or β-linkage at C-1 (hence, the term *β-glucuronides*). In contrast, the coenzyme UDPGA has an α-linkage. In the enzymatic transfer step, it appears that nucleophilic displacement of the α-linked UDP moiety from UDPGA by the substrate RXH proceeds with complete inversion of configuration at C-1 to

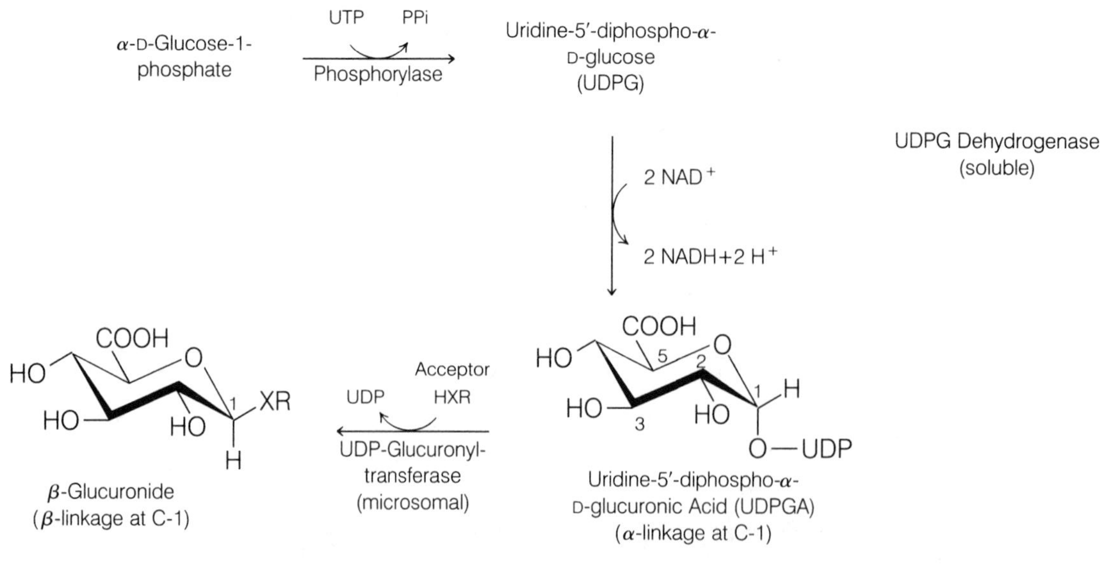

FIG. 3-11. *Formation of UDPGA and β-glucuronide conjugates.*

BOX 3-2. TYPES OF COMPOUNDS FORMING OXYGEN, NITROGEN, SULFUR, AND CARBON GLUCURONIDES

OXYGEN GLUCURONIDES

Hydroxyl compounds

Penols: morphine, acetaminophen, *p*-hydroxyphenytoin
Alcohols: trichloroethanol, chloramphenicol, propranolol
Enols: 4-hydroxycoumarin
N-Hydroxylamines: *N*-hydroxydapsone
N-Hydroxylamides: *N*-hydroxy-2-acetylaminofluorene

Carboxyl compounds

Aryl acids: benzoic acid, salicylic acid
Arylalkyl acids: naproxen, fenoprofen

NITROGEN GLUCURONIDES

Arylamines: 7-amino-5-nitroindazole
Alkylamines: desipramine
Amides: meprobamate
Sulfonamides: sulfisoxazole
Tertiary amines: cyproheptadine, tripelennamine

SULFUR GLUCURONIDES

Sulfhydryl groups: methimazole, propylthiouracil, diethylthiocarbamic acid

CARBON GLUCURONIDES

3,5-Pyrazolidinedione: phenylbutazone, sulfinpyrazone
(For structures and site of β-glucuronide attachment, see Fig. 3-12.)

give the β-glucuronide. Glucuronidation of one functional group is usually sufficient to effect excretion of the conjugated metabolite; diglucuronide conjugates usually do not occur.

The diversity of functional groups undergoing glucuronidation is illustrated by the examples given in Box 3-2. Metabolic products are classified as oxygen-, nitrogen-, sulfur-, or carbon-glucuronide, according to the heteroatom attached to the C-1 atom of the glucuronyl group. Two important functionalities, the hydroxy and carboxy, form *O*-glucuronides. Phenolic and alcoholic hydroxyls are the most common functional groups undergoing glucuronidation in drug metabolism. As we have seen, phenolic and alcoholic hydroxyl groups are present in many parent compounds and arise through various phase I metabolic pathways. Morphine,[278] acetaminophen,[279] and *p*-hydroxyphenytoin (the major metabolite of phenytoin)[41] represent a few examples of phenolic compounds that undergo considerable glucuronidation. Alcoholic hydroxyls, such as those present in trichloroethanol (major metabolite of chloral hydrate),[220] Chloramphenicol,[280] and propranolol,[281] are also commonly glucuronidated. Occurring less frequently is glucuronidation of other hydroxyl groups such as enols ($-\overset{|}{C}=\overset{|}{C}-OH$),[282] *N*-hydroxylamines (RNHOH),[174] and *N*-hydroxylamides ($R\underset{\underset{O}{\|}}{C}NHOH$).[186] For examples refer to the list of glucuronides (see Box 3-2).

The carboxy group is also subject to conjugation with glucuronic acid. For example, arylaliphatic acids, such as the anti-inflammatory agents naproxen[283] and fenoprofen,[284] are primarily excreted as their *O*-glucuronide derivatives in humans. Carboxylic acid metabolites such as those arising from chlorpheniramine[222] and propranolol[221] (see section on reduction of aldehyde and ketone carbonyls) have been observed to form *O*-glucuronide conjugates. Aryl acids (e.g., benzoic acid,[285] salicylic acid[286]) also undergo conjugation with glucuronic acid, but a more important pathway for these compounds appears to be conjugation with glycine.

The formation of *N*-glucuronides with aromatic amines, aliphatic amines, amides, and sulfonamides occurs occasionally. Representative examples are found in the list of glucuronides given in Box 3-2. Glucuronidation of aromatic and aliphatic amines is generally a minor pathway in comparison with *N*-acetylation or oxidative processes (e.g., oxidative deamination). More recently, tertiary amines such as the antihistaminic agents cyproheptadine (Periactin)[291] and tripelennamine[292] have been observed to form interesting quaternary ammonium glucuronide metabolites.

Because the thiol group (SH) does not commonly occur in xenobiotics, *S*-glucuronide products have been reported for only a few drugs. For instance, the thiol groups present in methimazole (Tapazole),[293] propylthiouracil,[294] and *N,N*-diethyldithiocarbamic acid (major reduced metabolite of disulfiram, Antabuse)[295] have been demonstrated to undergo conjugation with glucuronic acid.

The formation of glucuronides attached directly to a carbon atom is relatively novel in drug metabolism. Recent studies in humans have shown that conjugation of phenylbutazone (Butazolidin)[296] and sulfinpyrazone (Anturane)[297] yield the corresponding *C*-glucuronide metabolites:

C_6H_5 N-N C_6H_5; O, O; H R → HOOC, HO, HO, HO, O; C_6H_5 N-N C_6H_5; O, O; R; H

Phenylbutazone: $R = CH_2CH_2CH_2CH_3$
Sulfinpyrazone: $R = CH_2CH_2\underset{\underset{O}{\|}}{S}C_6H_5$

C-Glucuronide Metabolite

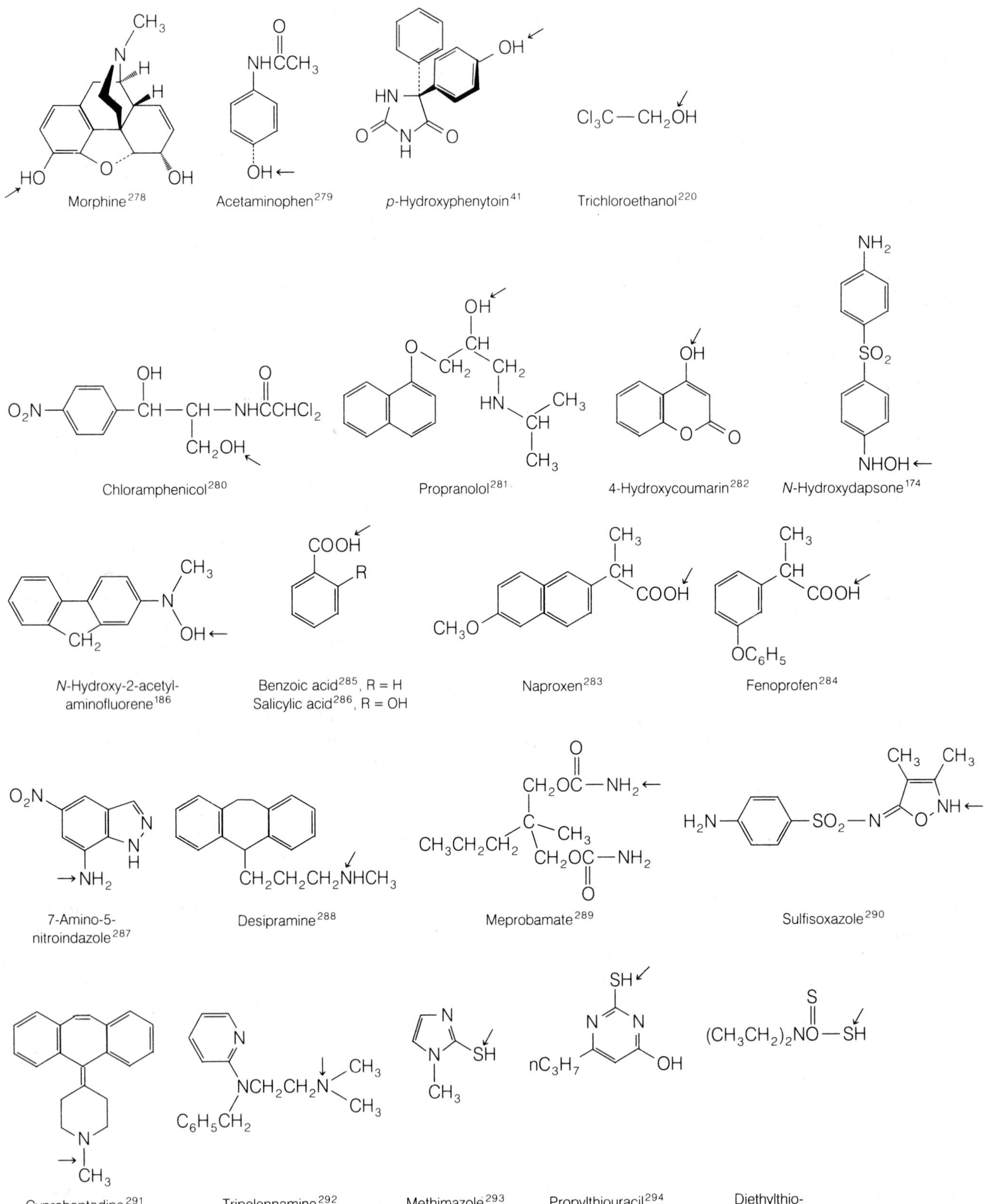

FIG. 3-12. *Structure of compounds that undergo glucuronidation (arrow indicates the site of β-glucuronide attachment).*

Besides xenobiotics, a number of endogenous substrates, notably bilirubin[298] and steroids,[299] are eliminated as glucuronide conjugates. Glucuronide conjugates are primarily excreted in the urine. However as the relative molecular mass of the conjugate exceeds 300 Da, biliary excretion may become an important route of elimination.[300] Glucuronides that are excreted in the bile are susceptible to hydrolysis by β-glucuronidase enzymes present in the intestine. The hydrolyzed product may be reabsorbed in the intestine, thus leading to enterohepatic recycling.[19] β-Glucuronidases are also present in many other tissues, including the liver, the endocrine system, and the reproductive organs. Although the function of these hydrolytic enzymes in drug metabolism is unclear, it appears that, in terms of hormonal and endocrine regulation, β-glucuronidases may be functioning to liberate active hormones (e.g., steroids) from their inactive glucuronide conjugates.[19]

In neonates and children, glucuronidating processes often are not fully developed. In such subjects, drugs and endogenous compounds (e.g., bilirubin) that are normally metabolized by glucuronidation may accumulate and cause serious toxicity. For example, neonatal hyperbilirubinemia may be attributable to the inability of newborns to conjugate bilirubin with glucuronic acid.[301] Similarly, the inability of infants to glucuronidate chloramphenicol has been suggested to be responsible for the "gray baby syndrome," which results from accumulation of toxic levels of the free antibiotic.[302]

SULFATE CONJUGATION

Conjugation of xenobiotics with sulfate occurs primarily with phenols and, occasionally, with alcohols, aromatic amines, and *N*-hydroxy compounds.[303–305] In contrast to glucuronic acid, the amount of available sulfate is rather limited. A significant portion of the sulfate pool is utilized by the body to conjugate numerous endogenous compounds, such as steroids, heparin, chondroitin, catecholamines, and thyroxine. The sulfate conjugation process involves activation of inorganic sulfate to the coenzyme 3′-phosphoadenosine-5′-phosphosulfate (PAPS). Subsequent transfer of the sulfate group from PAPS to the accepting substrate is catalyzed by various soluble sulfotransferases present in the liver and other tissues (e.g., kidney, intestine).[305] The sequence of events involved in sulfoconjugation is depicted in Figure 3-13. Sulfate conjugation generally leads to water-soluble and inactive metabolites. However, it appears that the *O*-sulfate conjugates of some *N*-hydroxy compounds give rise to chemically reactive intermediates that are toxic.[186]

Phenols compose the main group of substrates undergoing sulfate conjugation. Thus, drugs containing phenolic moieties are often susceptible to sulfate formation. For example, the antihypertensive agent α-methyldopa (Aldomet) is extensively metabolized to its 3-*O*-sulfate ester in humans.[306] The β-adrenergic bronchodilators salbutamol (albuterol)[307] and terbutaline (Brethine, Bricanyl)[308] also undergo sulfate conjugation as their principal

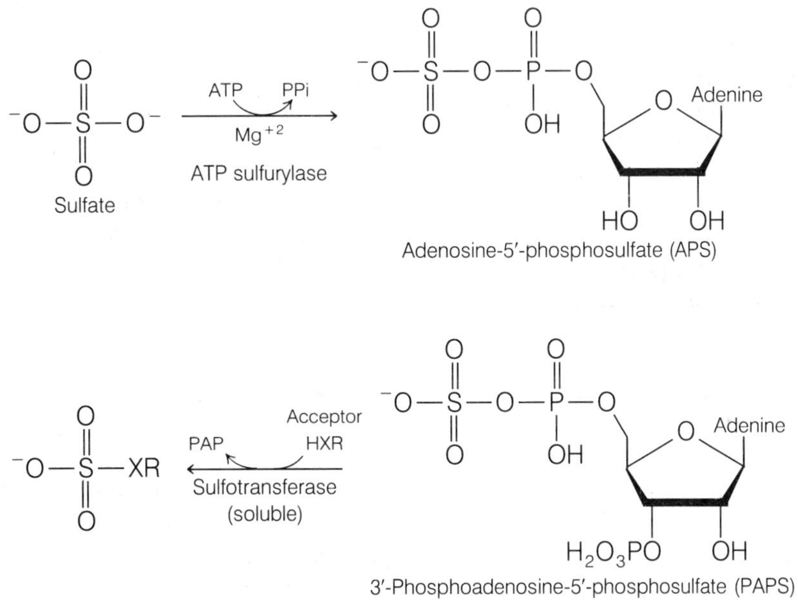

Sulfate

ATP PPi Mg^{+2} ATP sulfurylase

Adenosine-5′-phosphosulfate (APS)

Acceptor HXR PAP Sulfotransferase (soluble)

3′-Phosphoadenosine-5′-phosphosulfate (PAPS)

FIG. 3-13. *Formation of PAPS and sulfate conjugates.*

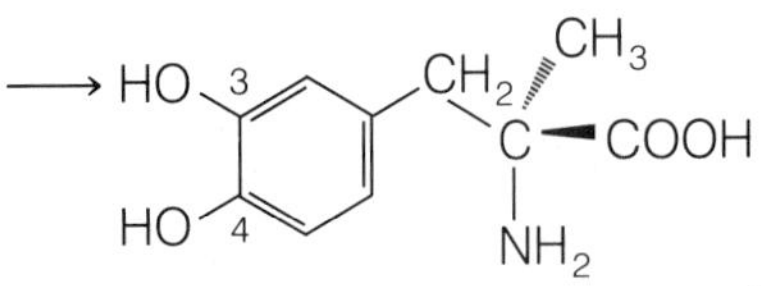

α-Methyldopa

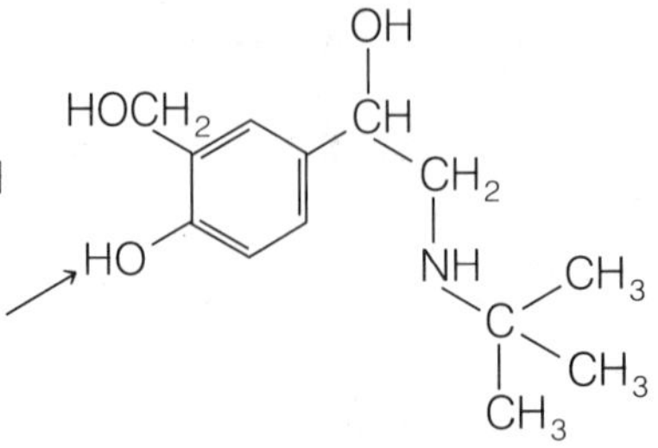

Salbutamol (Albuterol)

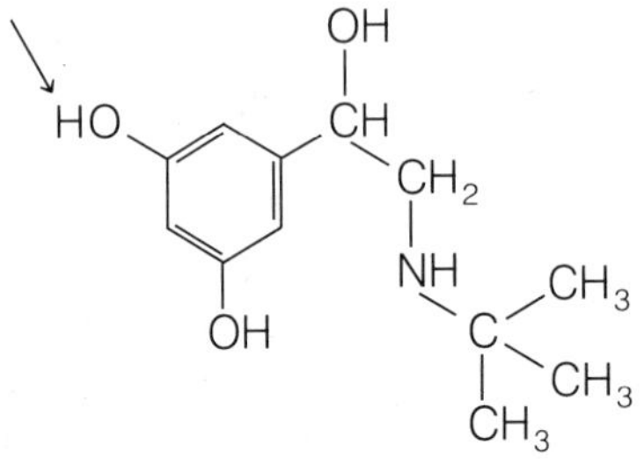

Terbutaline

route of metabolism in humans. However, for many phenols, sulfoconjugation may represent only a minor pathway. Glucuronidation of phenols is frequently a competing reaction and may predominate as the conjugative route for some phenolic drugs. In adults, the major urinary metabolite of the analgesic acetaminophen is the *O*-glucuronide conjugate, with the *O*-sulfate conjugate being formed in small amounts.[279] Interestingly, in infants and young children (ages three to nine years) a different urinary excretion pattern is observed: the *O*-sulfate conjugate is the main urinary product.[309] The explanation for this reversal stems from the fact that neonates and young children have a decreased glucuronidating capacity owing to undeveloped glucuronyltransferases or low levels of these enzymes. Sulfate conjugation, however, is well developed and becomes the main route by which acetaminophen is conjugated in this pediatric-aged group.

Other functionalities such as alcohols (e.g., aliphatic C_1 to C_5 alcohols, diethylene glycol)[310] and aromatic amines (e.g., aniline, 2-naphthylamine)[311]

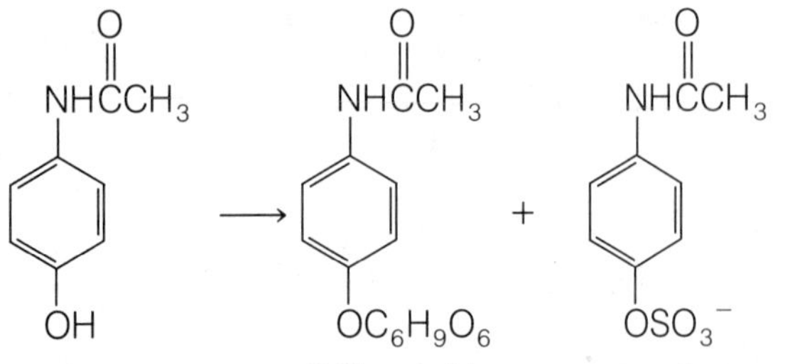

Acetaminophen → *O*-Glucuronide Conjugate ($OC_6H_9O_6$) + *O*-Sulfate Conjugate (OSO_3^-)

are also capable of forming sulfate conjugates. These reactions, however, have only minor importance in drug metabolism. The sulfate conjugation of *N*-hydroxylamines and *N*-hydroxylamides takes place occasionally as well. *O*-Sulfate ester conjugates of *N*-hydroxy compounds are of considerable toxicologic concern because they can lead to reactive intermediates that are responsible for cellular toxicity. The carcinogenic agents *N*-methyl-4-aminoazobenzene and 2-acetylaminofluorene are believed to mediate their toxicity through *N*-hydroxylation to the corresponding *N*-hydroxy compounds (see earlier section on *N*-hydroxylation of amines and amides). Sulfoconjugation of the *N*-hydroxy metabolites yields *O*-sulfate esters, which presumably are the ultimate carcinogenic species. Loss of SO_4^{2-} from the foregoing sulfate conjugates generates electrophilic nitrenium species, which may react with nucleophilic groups (e.g., NH_2, OH, SH), present in proteins, DNA, and RNA, to form covalent linkages that lead to structural and functional alteration of these crucial biomacromolecules. The consequences of this are cellular toxicity (tissue necrosis) or alteration of genetic code material leading eventually to cancer. Some evidence supporting the role of sulfate conjugation in metabolic activation of *N*-hydroxy compounds to reactive intermediates comes from the observation that the degree of hepatotoxicity and hepatocarcinogenicity of *N*-hydroxy-2-acetylaminofluorene is markedly dependent on the level of sulfotransferase activity in the liver.[313]

The analgesic phenacetin is metabolized to *N*-hydroxyphenacetin and subsequently conjugated with sulfate.[314] The *O*-sulfate conjugate of *N*-hydroxyphenacetin has been demonstrated to covalently bind to microsomal proteins.[315] It has been suggested that this pathway may represent one possible route leading to reactive intermediates that are responsible for the hepatotoxicity and nephrotoxicity associated with phenacetin. Other pathways (e.g., arene oxides) leading to reactive electrophilic intermediates are also possible.[6]

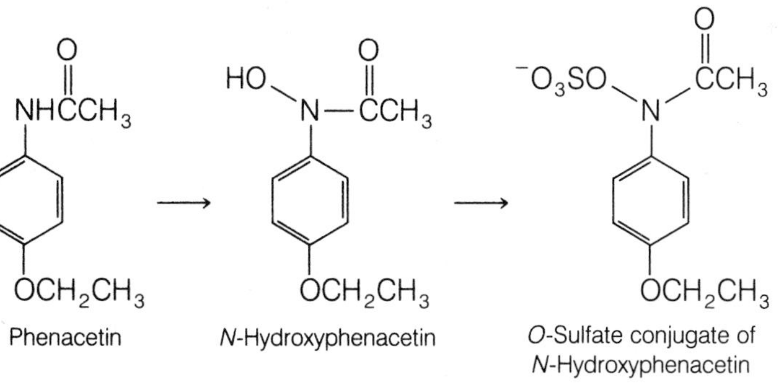

Phenacetin N-Hydroxyphenacetin O-Sulfate conjugate of N-Hydroxyphenacetin

CONJUGATION WITH GLYCINE, GLUTAMINE, AND OTHER AMINO ACIDS

The amino acids glycine and glutamine are utilized by mammalian systems to conjugate carboxylic acids, particularly aromatic acids and arylalkyl acids.[316, 317] Glycine conjugation is common to most mammals, whereas glutamine conjugation appears mainly confined to humans and other primates. The quantity of amino acid conjugates formed from xenobiotics is minute because of the limited availability of amino acids in the body and competition with glucuronidation for carboxylic acid substrates. In contrast with glucuronic acid and sulfate, glycine, and glutamine are not converted to activated coenzymes. Instead, the carboxylic acid substrate is activated with ATP and coenzyme A (CoA) to form an acyl–Co A complex. The latter intermediate, in turn, acylates glycine or glutamine under the influence of specific glycine or glutamine *N*-acyltransferase enzymes. The activation and acylation steps take place in the mitochondria of liver and kidney cells. The sequence of metabolic events associated with glycine and glutamine conjugation of phenylacetic acid is summarized in Figure 3-14. Amino acid conjugates, being polar and water-soluble, are mainly excreted renally and sometimes in the bile.

Aromatic acids and arylalkyl acids are the major substrates undergoing glycine conjugation. The conversion of benzoic acid to its glycine conjugate, hippuric acid, is a well-known metabolic reaction in many mammalian systems.[319] The extensive metabolism of salicylic acid (75% of dose) to salicyluric acid in humans is another illustrative example.[320] Carboxylic acid metabolites resulting from oxidation or hydrolysis of many drugs are also susceptible to glycine conjugation. For example, the H_1-histamine antagonist, brompheniramine, is oxidized to a propionic acid metabolite that is conjugated with glycine in both human and dog.[141] Similarly, *p*-fluorophenylacetic acid, derived from the metabolism of the antipsychotic agent haloperidol (Haldol), is found as

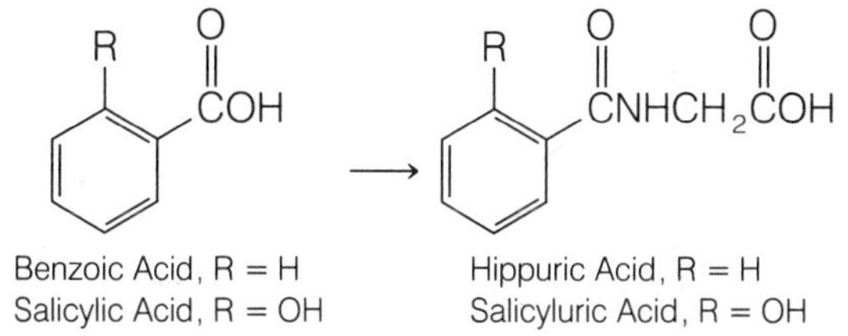

Benzoic Acid, R = H
Salicylic Acid, R = OH

Hippuric Acid, R = H
Salicyluric Acid, R = OH

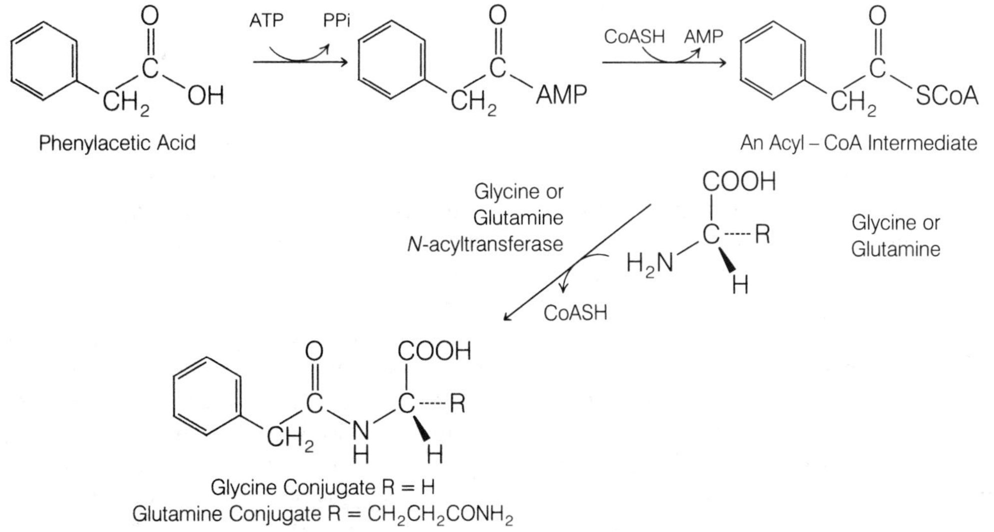

Phenylacetic Acid → An Acyl – CoA Intermediate

Glycine or Glutamine *N*-acyltransferase

Glycine or Glutamine

Glycine Conjugate R = H
Glutamine Conjugate R = $CH_2CH_2CONH_2$

FIG. 3-14. *Formation of glycine and glutamine conjugates of phenylacetic acid.*

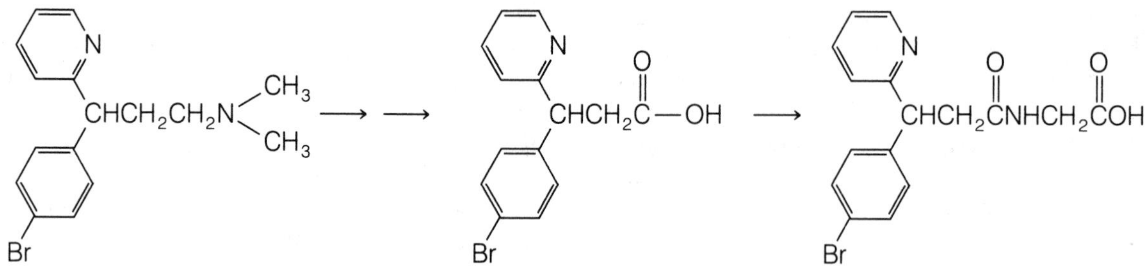

Brompheniramine → → 3-(*p*-Bromophenyl)-3-(2-pyridyl)-propionic Acid → Glycine Conjugate

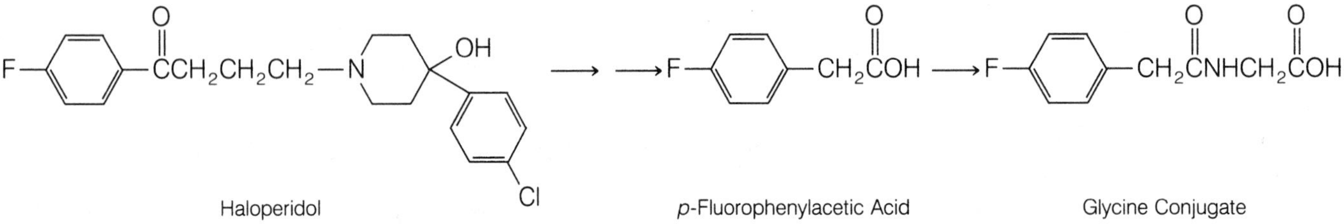

Haloperidol → → *p*-Fluorophenylacetic Acid → Glycine Conjugate

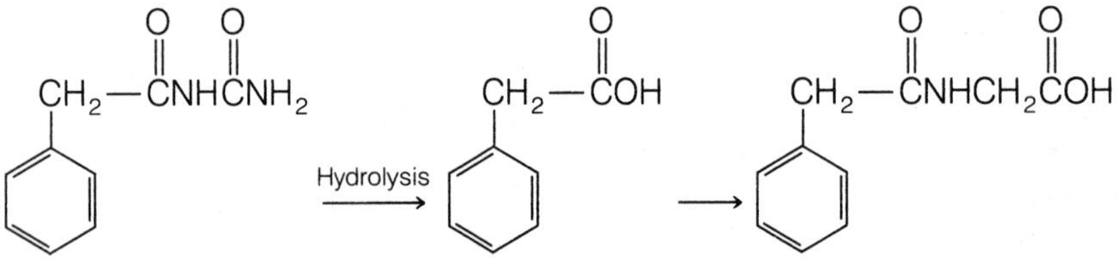

Phenacemide —Hydrolysis→ Phenylacetic acid → Glycine Conjugate

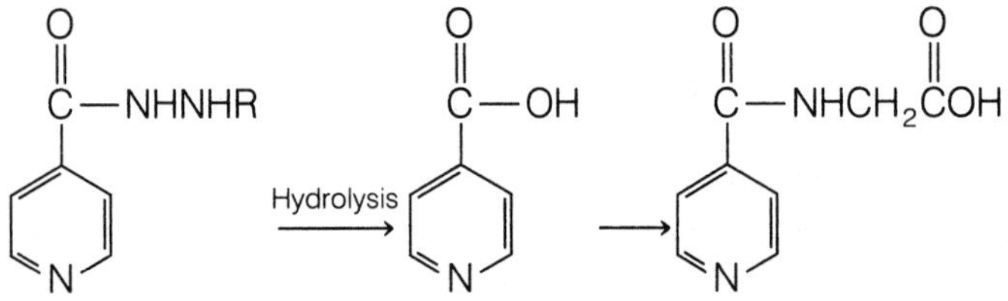

Isoniazid (R = H) or *N*-Acetylisoniazid (R = COH_3) —Hydrolysis→ Isonicotinic Acid → Glycine Conjugate

the glycine conjugate in the urine of rats.[321] Phenylacetic acid and isonicotinic acid resulting, respectively, from the hydrolysis of the anticonvulsant phenacemide (Phenurone)[322] and the antituberculosis agent isoniazid[323] are also conjugated with glycine to some extent.

Glutamine conjugation occurs mainly with arylacetic acids, including endogenous phenylacetic[318] and indolylacetic acid.[324] A few glutamine conjugates of drug metabolites have been reported. For example, in humans the 3,4-dihydroxy-5-methoxyphenylacetic acid metabolite of mescalineis found as a conjugate of glutamine.[325] Diphenylmethoxyacetic acid, a metabolite of the antihistamine diphenhydramine (Benadryl), is biotransformed further to the corresponding glutamine derivative in the rhesus monkey.[326]

Several other amino acids are known to be involved in the conjugation of carboxylic acids, but these reactions occur only occasionally and appear to be highly substrate- and species-dependent.[317,327] Ornithine (in birds), aspartic acid and serine (in rats), alanine (in mouse and hamster), taurine ($H_2NCH_2CH_2SO_3H$) (in mammals and pigeons) and histidine (in African bats) are among these amino acids.[327]

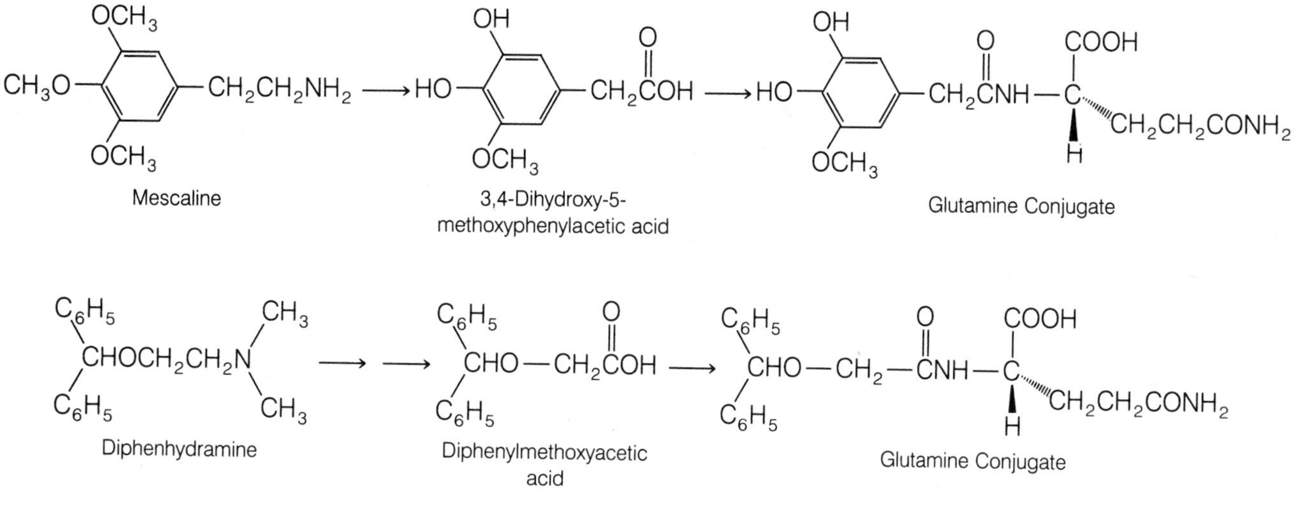

Mescaline → 3,4-Dihydroxy-5-methoxyphenylacetic acid → Glutamine Conjugate

Diphenhydramine → → Diphenylmethoxyacetic acid → Glutamine Conjugate

GLUTATHIONE OR MERCAPTURIC ACID CONJUGATES

Glutathione conjugation is an important pathway by which chemically reactive electrophilic compounds are detoxified.[328,329] It is now generally accepted that reactive electrophilic species manifest their toxicity (e.g., tissue necrosis, carcinogenicity, mutagenicity, teratogenicity) by combining covalently with nucleophilic groups present in vital cellular proteins and nucleic acids.[4,330] Many serious drug toxicities may be explainable also in terms of cova-

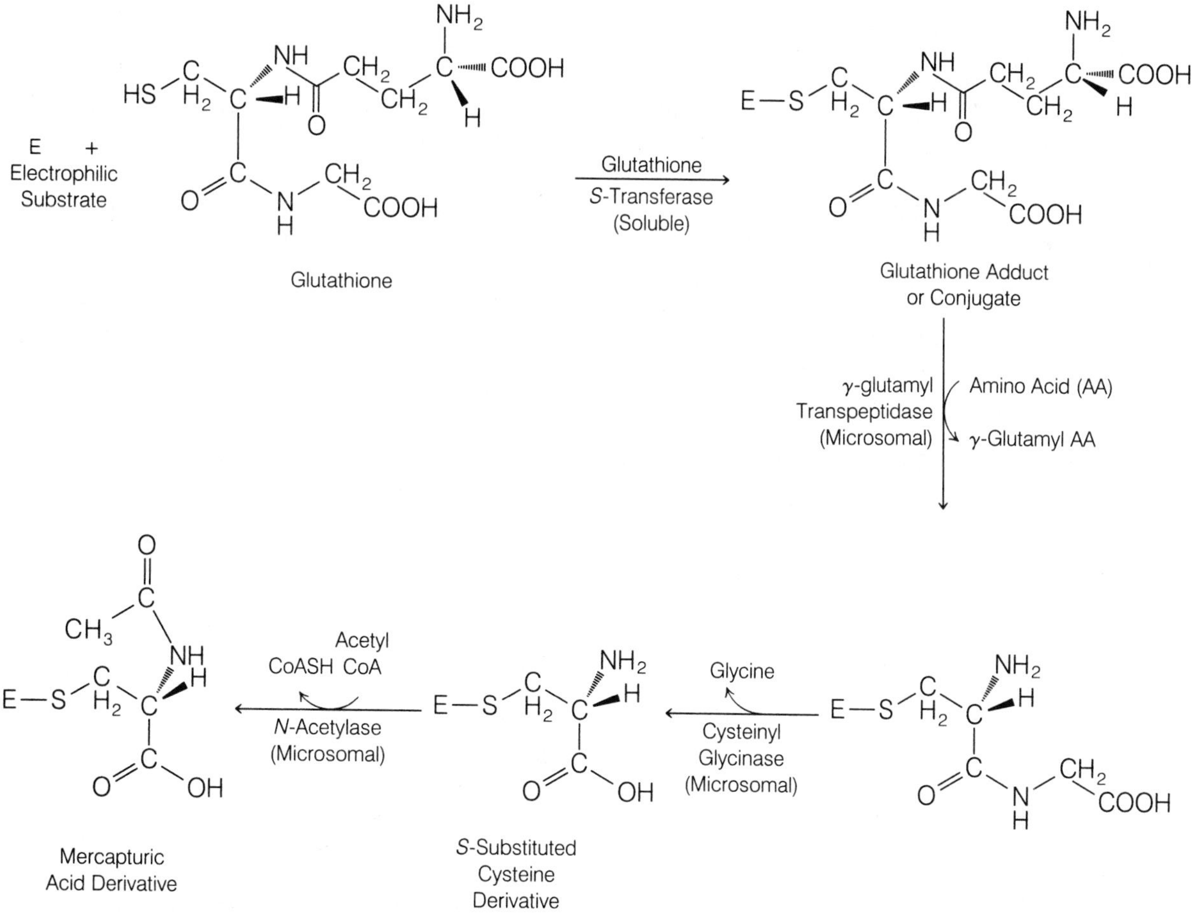

E + Electrophilic Substrate; Glutathione → (Glutathione S-Transferase (Soluble)) → Glutathione Adduct or Conjugate → (γ-glutamyl Transpeptidase (Microsomal); Amino Acid (AA) → γ-Glutamyl AA) → (Cysteinyl Glycinase (Microsomal); Glycine) → S-Substituted Cysteine Derivative → (N-Acetylase (Microsomal); Acetyl CoASH CoA) → Mercapturic Acid Derivative

FIG. 3-15. *Formation of glutathione conjugates of electrophilic xenobiotics or metabolites (E) and their conversion to mercapturic acids.*

lent interaction of metabolically generated electrophilic intermediates with cellular nucleophiles.[5,6] Glutathione protects vital cellular constituents against chemically reactive species by virtue of its nucleophilic sulfhydryl group. It is the sulfhydryl (SH) group that reacts with electron-deficient compounds to form *S*-substituted glutathione adducts (Fig. 3-15).[328,329]

Glutathione (GSH) is a tripeptide (γ-glutamylcysteinylglycine) found in most tissues. Xenobiotics conjugated with GSH usually are not excreted as such but undergo further biotransformation to give *S*-substituted *N*-acetylcysteine products called mercapturic acids.[63,71,329] This process involves enzymatic cleavage of two amino acids (namely, glutamic acid and glycine) from the initially formed glutathione adduct and subsequent *N*-acetylation of the remaining *S*-substituted cysteine residue. The formation of glutathione conjugates and their conversion to mercapturic acid derivatives are outlined in Figure 3-15.

Conjugation of a wide spectrum of substrates with GSH is catalyzed by a family of cytoplasmic enzymes known as glutathione *S*-transferases.[62] These enzymes are found in most tissues, particularly the liver and kidney. Degradation of GSH conjugates to mercapturic acids is carried out principally by renal and hepatic microsomal enzymes (see Fig. 3-15).[63] Unlike other conjugative phase II reactions, GSH conjugation does not require the initial formation of an activated coenzyme or substrate. The inherent reactivity of the nucleophilic GSH toward an electrophilic substrate usually provides sufficient driving force. The substrates susceptible to GSH conjugation are quite varied and encompass many chemically different classes of compounds. A major prerequisite is that the substrate be sufficiently electrophilic. Compounds that react with gutathione do so by two general mechanisms: (1) nucleophilic displacement at an electron-deficient carbon or heteroatom, or (2) nucleophilic addition to an electron-deficient double bond.[328]

Many aliphatic and arylalkyl halides (Cl, Br, I), sulfates (OSO_3^-), sulfonates (OSO_2R), nitrates (NO_2), and organophosphates (O-P[OR]$_2$) possess electron-deficient carbon atoms that react with GSH (by aliphatic nucleophilic displacement) to form glutathione conjugates, as shown:

$$\text{GSH} + \overset{+\delta}{\text{CH}_2}(\text{R})-\overset{-\delta}{\text{X}} \longrightarrow \text{GS}-\text{CH}_2(\text{R}) + \text{HX}$$

R = Alkyl, Aryl, Benzylic, Allylic
X = Br, Cl, I, OSO_3^-, OSO_2R, $OPO(OR)_2$

The carbon center is rendered electrophilic as a result of the electron-withdrawing group (e.g., halide, sulfate, phosphate) attached to it. Nucleophilic displacement is often facilitated when the carbon atom is benzylic or allylic or when X is a good leaving group (e.g., halide, sulfate). Many industrial chemicals such as benzyl chloride ($C_6H_5CH_2Cl$), allyl chloride ($CH_2{=}CHCH_2Cl$), and methyl iodide are known to be toxic and carcinogenic. The reactivity of these three halides toward GSH conjugation in mammalian systems is demonstrated by the formation of the corresponding mercapturic acid derivatives.[329] Organophosphate insecticides, such as methyl parathion, are detoxified by two different glutathione pathways.[331] Pathway *a* involves aliphatic nucleophilic substitution and yields *S*-methylglutathione. Pathway *b*, on the other hand, involves aromatic nucleophilic substitution and produces *S*-*p*-nitrophenylglutathione. Aromatic or heteroaromatic nucleophilic substitution reactions with GSH occur only when the ring is rendered sufficiently electron-deficient by the presence of one or more strongly electron-withdrawing substituents (e.g., NO_2, Cl). For example, 2,4-dichloronitrobenzene is susceptible to nucleophilic substitution by glutathione, whereas chlorobenzene is not.[332]

Methyl Parathion: $CH_3O-P(=S)(OCH_3)-O-C_6H_4-NO_2$ (GSH attack: a, b)

Pathway a → $GSCH_3 + {}^-O-P(=S)(OCH_3)-O-C_6H_4-NO_2$
S-Methylglutathione

Pathway b → $GS-C_6H_4-NO_2 + CH_3O-P(=S)(OCH_3)-OH$
S-*p*-Nitrophenylglutathione

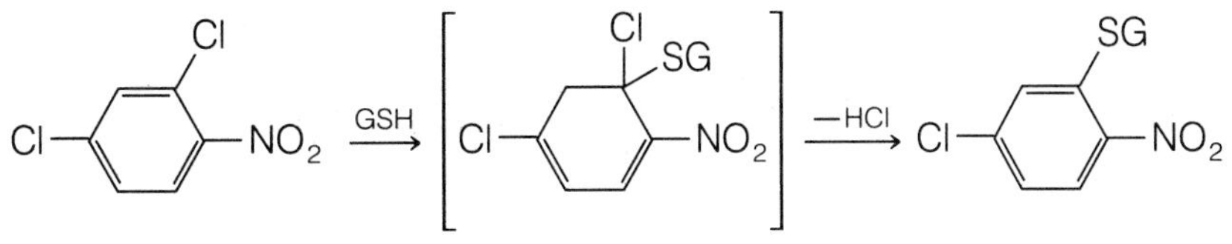

2,4-Dichloronitrobenzene

The metabolism of the immunosuppressive drug azathioprine (Imuran) to 1-methyl-4-nitro-5-(*S*-glutathionyl)imidazole and 6-mercaptopurine is an example of heteroaromatic nucleophilic substitution reaction involving glutathione.[333] Interestingly, 6-mercaptopurine formed in this reaction appears to be responsible for azathioprine's immunosupressive activity.[334]

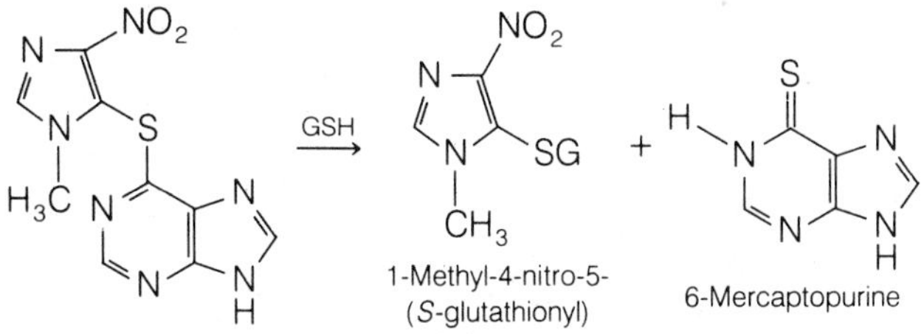

Azathioprine

1-Methyl-4-nitro-5-(*S*-glutathionyl) imidazole

6-Mercaptopurine

Arene oxides and aliphatic epoxides (or oxiranes) represent a very important class of substrates that are conjugated and detoxified by glutathione.[335] The three-membered oxygen-containing ring in these compounds is highly strained and, therefore, is reactive toward ring cleavage by nucleophiles (e.g., GSH, H_2O, or nucleophilic groups present on cellular macromolecules). As discussed previously, arene oxides and epoxides are intermediary products formed from cytochrome P-450 oxidation of aromatic compounds (arenes) and olefins, respectively. If reactive arene oxides (e.g., benzo[*a*]pyrene-4,5-oxide, 4-bromobenzene oxide) and aliphatic epoxides (e.g., styrene oxide) are not "neutralized" or detoxified by glutathione *S*-transferase, epoxide hydrase, or other pathways, they ultimately covalently bind to cellular macromolecules to cause serious cytotoxicity and carcinogenicity. The isolation of glutathione or mercapturic acid adducts from benzo[*a*]pyrene, bromobenzene, and styrene clearly demonstrates the importance of GSH in reacting with the reactive epoxide metabolites generated from these compounds.

Glutathione conjugation involving substitution at heteroatoms, such as oxygen, is often seen with organic nitrates. For example, nitroglycerin (Nitrostat) and isosorbide dinitrate (Isordil) are metabolized by a pathway involving an initial glutathione conjugation reaction. However, the GSH conjugate products are not metabolized to mercapturic acids, but instead, are converted enzymatically to the corresponding alcohol derivatives and glutathione disulfide (GSSG).[336]

The nucleophilic addition of glutathione to electron-deficient carbon–carbon double bonds occurs mainly in compounds with α,β-unsaturated double bonds. In most instances, the double bond is rendered electron deficient by resonance or conjugation with a carbonyl group (ketone or aldehyde), ester, nitrile, or other. Such α,β-unsaturated systems undergo so-called Michael addition reactions with GSH to yield the corresponding glutathione adduct.[328, 329] For example, in rats and dogs the diuretic agent ethacrynic acid (Edecrin) reacts with GSH to form the corresponding glutathione or mercapturic acid derivatives.[337] The compound diethyl maleate is

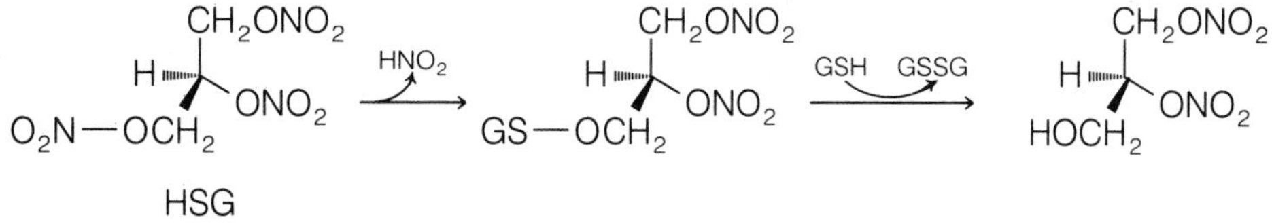

HSG

Nitroglycerin

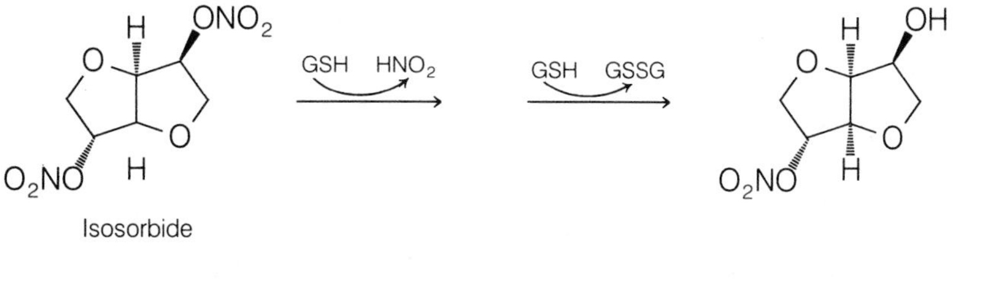

Isosorbide

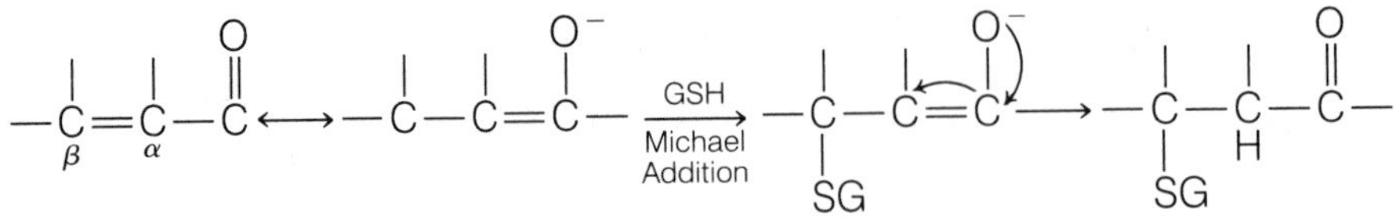

α-β-Unsaturated System

GSH Michael Addition

Glutathione Adduct

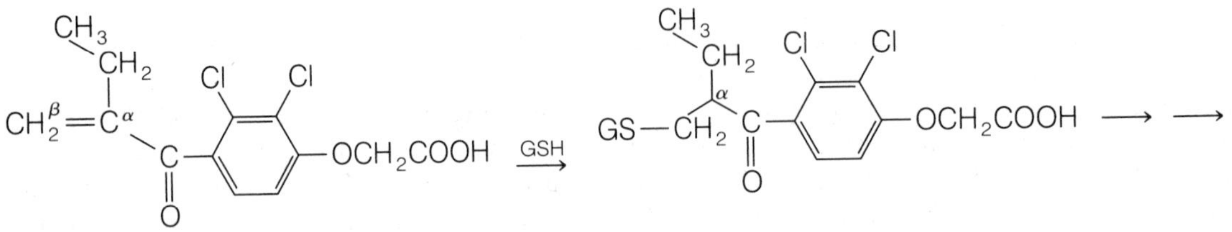

Ethacrynic Acid (note α,β-unsaturated e tone moiety)

GSH

Glutathione adduct of Ethacrynic Acid

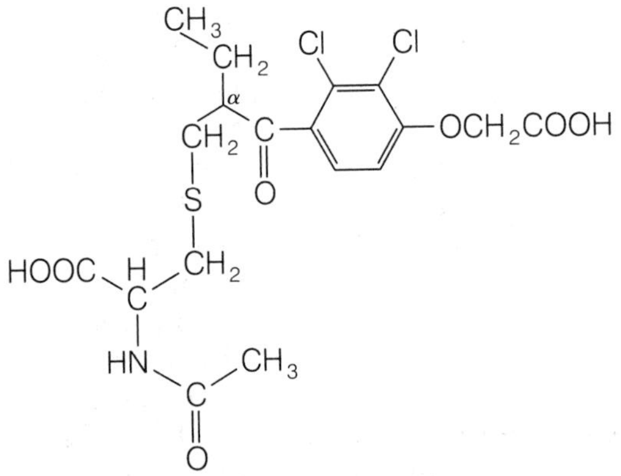

Mercapturic Acid Derivative

readily conjugated with glutathione and has been used experimentally to deplete hepatic GSH stores in laboratory animals.[338] Several other α,β-unsaturated compounds, such as acrolein, crotonaldehyde, o-chlorobenzylidenemalononitrile, and arecoline, form mercapturic acid or glutathione derivatives.[329] It should be emphasized that not all α,β-unsaturated compounds are conjugated with glutathione. Many steroidal agents possessing α,β-unsaturated carbonyl moieties, such as prednisone and digitoxigenin, have not been observed to undergo any significant conjugation with glutathione.

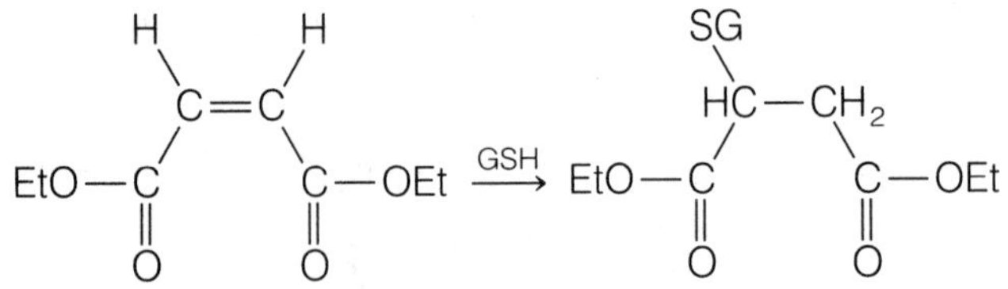

GSH

Diethyl Maleate

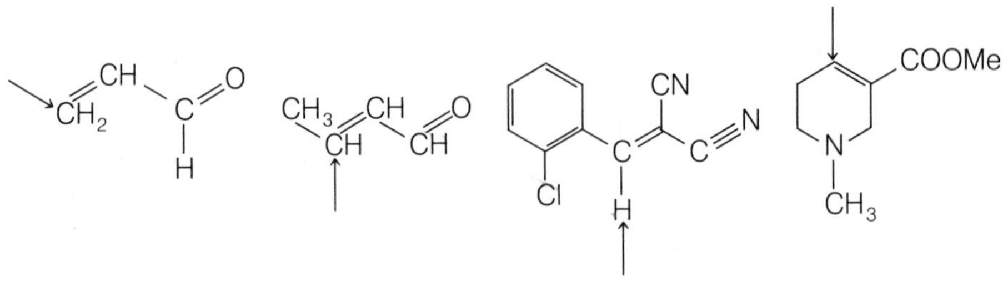

Acrolein

Crotonaldehyde

o-Chlorobenzylidene-malononitrile

Arecoline

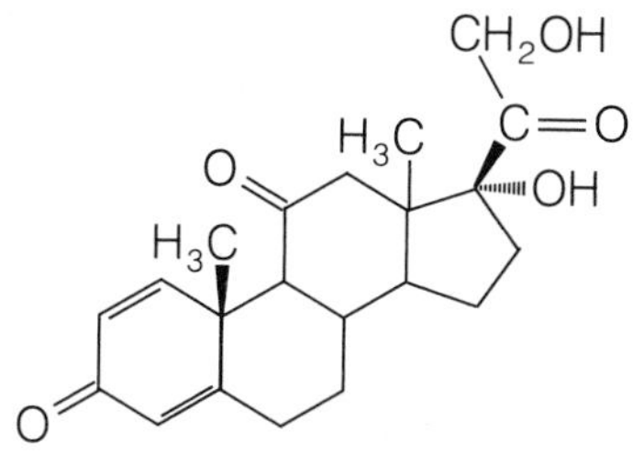

Prednisone

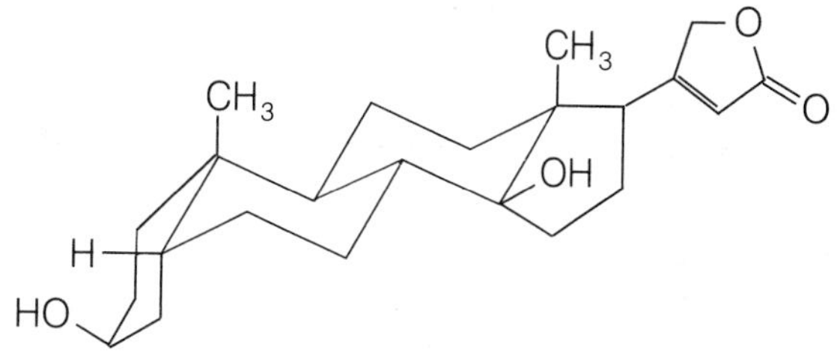

Digitoxigenin

Steric factors, decreased reactivity of the double bond as well as other factors (e.g., susceptibility to metabolic reduction of the ketone or the C=C double bond) may account for these observations.

Occasionally, metabolic oxidative biotransformation reactions may generate chemically reactive α,β-unsaturated systems that react with glutathione. For example, metabolic oxidation of acetaminophen presumably generates the chemically reactive intermediate, *N*-acetylimidoquinone. Michael addition of GSH to the imidoquinone leads to the corresponding mercapturic acid derivative in both animals and humans.[189,191] 2-Hydroxyestrogens, such as 2-hydroxy-17β-estradiol, undergo conjugation with glutathione to yield the two isomeric mercapturic acid or glutathione derivatives. Although the exact mechanism is unclear, it appears that the 2-hydroxyestrogen is oxidized to a chemically reactive orthoquinone or semiquinone intermediate that reacts with GSH at either the electrophilic C-1 or C-4 position.[339]

In most instances, glutathione conjugation is regarded as a detoxifying pathway that functions to protect cellular macromolecules, such as protein and DNA, against harmful electrophiles. In a few cases, GSH conjugation has been implicated in causing toxicity. Often this is because the GSH conjugates are themselves electrophilic (e.g. vicinal dihaloethanes) or give rise to metabolic intermediates (e.g., cysteine metabolites of haloalkenes) that are electrophilic.[329] 1,2-Dichloroethane, for example, reacts with GSH to produce *S*-(2-chloroethyl)glutathione; the nucleophilic sulfur group in this conjugate can internally displace the chlorine group to give rise to a electrophilic three-membered ring episulfonium ion. The covalent interaction of the episulfonium intermediate with the guanosine moiety of DNA may contribute to the mutagenic and carcino-

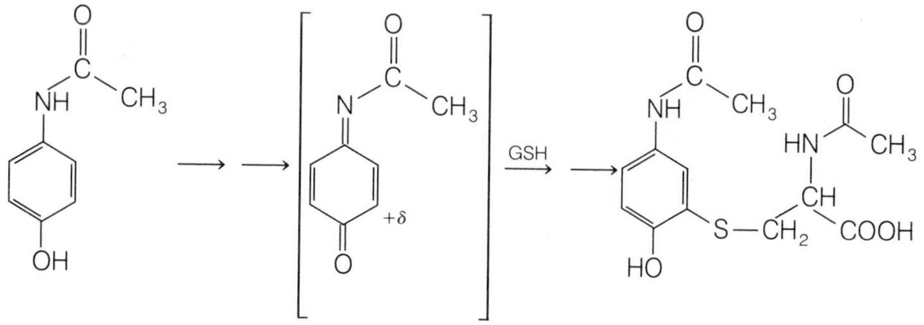

Acetaminophen

N-Acetylimidoquinone

Mercapturic Acid Derivative

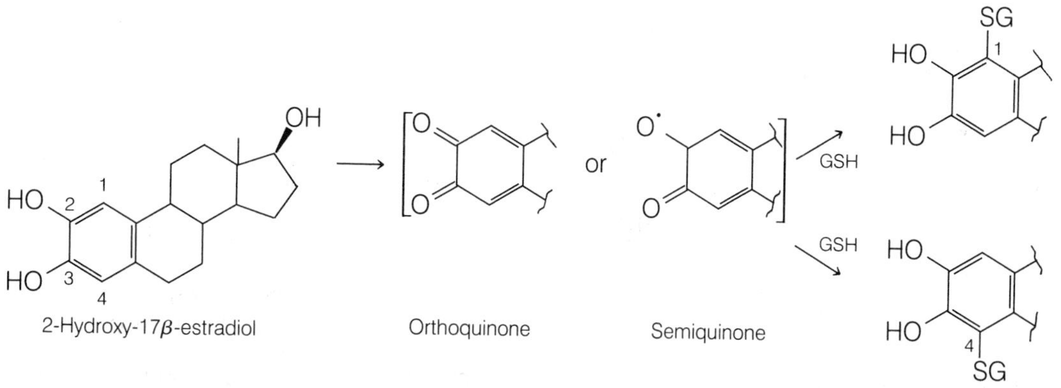

2-Hydroxy-17β-estradiol

Orthoquinone

Semiquinone

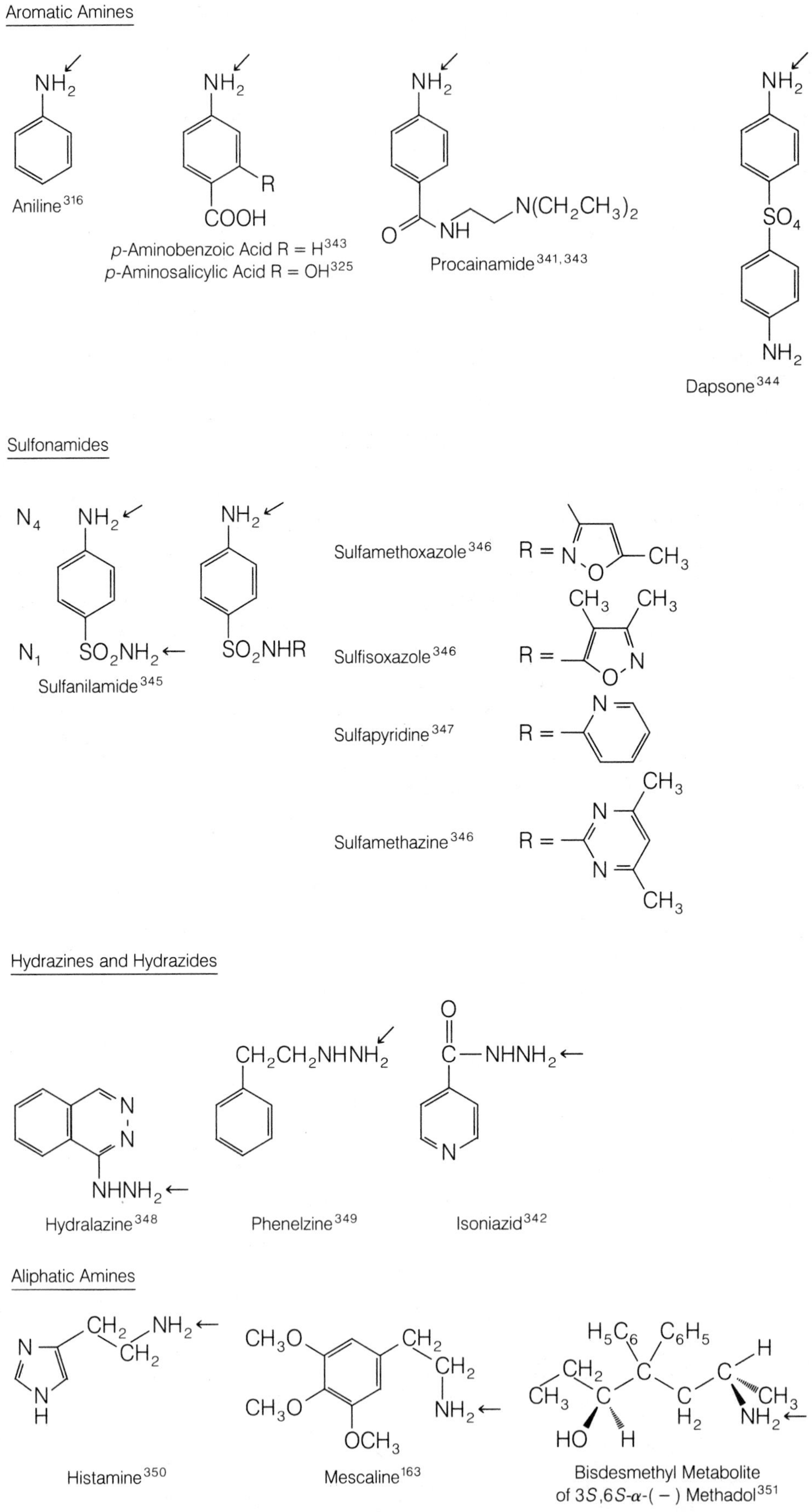

Aromatic Amines

NH_2 NH_2 NH_2 NH_2

R

Aniline[316]

COOH

N(CH_2CH_3)$_2$

O NH

SO_4

p-Aminobenzoic Acid R = H[343]
p-Aminosalicylic Acid R = OH[325]

Procainamide[341, 343]

NH_2

Dapsone[344]

Sulfonamides

N_4 NH_2 NH_2

Sulfamethoxazole[346] R = N O CH_3

CH_3 CH_3

N_1 SO_2NH_2← SO_2NHR Sulfisoxazole[346] R = O N

Sulfanilamide[345]

Sulfapyridine[347] R = N

CH_3

Sulfamethazine[346] R = N N

CH_3

Hydrazines and Hydrazides

O

$CH_2CH_2NHNH_2$ C—$NHNH_2$←

N N N

$NHNH_2$←

Hydralazine[348] Phenelzine[349] Isoniazid[342]

Aliphatic Amines

CH_2 NH_2← CH_3O CH_2 H_5C_6 C_6H_5 H

N CH_2 CH_2 CH_2 C C

N CH_3O NH_2← CH_3 C C CH_3

H OCH_3 HO H H_2 NH_2←

Histamine[350] Mescaline[163] Bisdesmethyl Metabolite of 3*S*,6*S*-α-(−) Methadol[351]

FIG. 3-16. *Examples of different types of compounds undergoing N-acetylation (arrow indicates the site of N-acetylation).*

genic effects observed for 1,2-dichloroethane.[329b] The metabolic conversion of GSH conjugates to reactive cysteine metabolites has been shown to be responsible for the nephrotoxicity associated with some halogenated alkanes and alkenes.[329c] The activation pathway appears to involve γ-glutamyl transpeptidase and cysteine conjugate β-lyase, two enzymes that apparently target the conjugates to the kidney.

ACETYLATION

Acetylation constitutes an important metabolic route for drugs containing primary amino groups.[316,340] This encompasses primary aromatic amines ($ArNH_2$), sulfonamides ($H_2NC_6H_4SO_2NHR$), hydrazines ($-NHNH_2$), hydrazides ($-CONHNH_2$), and primary aliphatic amines. The amide derivatives formed from acetylation of these amino functionalities are generally inactive and nontoxic. Because water solubility is not greatly enhanced by *N*-acetylation, it appears that the primary function of acetylation is one of termination of pharmacologic activity and detoxification. However, a few reports indicate that acetylated metabolites may be as active (e.g., *N*-acetylprocainamide)[341] or more toxic (e.g., *N*-acetylisoniazid)[342] than their corresponding parent compounds.

The acetyl group utilized in *N*-acetylation of xenobiotics is supplied by acetyl-CoA.[316] Transfer of the acetyl group from this cofactor to the accepting amino substrate is carried out by soluble *N*-acetyltransferases present in hepatic reticuloendothelial cells. Other extrahepatic tissues, such as the lung, spleen, gastric mucosa, red blood cells, and lymphocytes, also show acetylation capability. *N*-Acetyltransferase enzymes display broad substrate specificity and catalyze the acetylation of several drugs and xenobiotics (Fig. 3-16).[340] Aromatic compounds possessing a primary amino group, such as aniline,[316] *p*-aminobenzoic acid,[343] *p*-aminosalicylic acid,[325] procainamide (Pronestyl),[341,343] and dapsone (Avlosulfon),[344] are especially susceptible to *N*-acetylation. Aromatic amine metabolites resulting from the reduction of aryl nitro compounds are also *N*-acetylated. For example, the anticonvulsant clonazepam (Clonopin) undergoes nitro reduction to its 7-amino metabolite, which in turn is *N*-acetylated.[245] Another related benzodiazepam analogue, nitrazepam, follows a similar pathway.[246]

The metabolism of a number of sulfonamides such as sulfanilamide,[345] sulfamethoxazole (Gantanol),[346] sulfisoxazole (Gantrisin),[346] sulfapyridine[347] [major metabolite from azo reduction of sulfasalazine (Azulfidine)] and sulfamethazine[316] occurs mainly by acetylation at the N-4 position. With sulfanilamide, acetylation also takes place at the sulfamido N-1 position.[345] *N*-Acetylated metabolites of sulfonamides tend to be less water-soluble than their parent compounds and have the potential of crystallizing out in renal tubules (crystalluria), thereby causing kidney damage. The frequency of crystalluria and renal toxicity is especially prominent with older sulfonamide derivatives, such as sulfathiazole.[1,327] However, newer sulfonamides, such as sulfisoxazole and sulfamethoxazole, are metabolized to relatively water-soluble acetylated derivatives, which are less likely to precipitate out.

The biotransformation of hydrazine and hydrazide derivatives also proceeds by acetylation. The antihypertensive hydralazine (Apresoline)[348] and the monoamine oxidase (MAO) inhibitor phenelzine (Nardil)[349] are two representative hydrazine compounds that are metabolized by this pathway. It should be noted that the initially formed *N*-acetyl derivative of hydralazine is unstable and cyclizes intramolecularly to form 3-methyl-*s*-triazolo[3,4-*a*]phthalazine as the major isolable hydralazine metabolite in humans.[348] The antituberculosis drug isoniazid or isonicotinic acid hydrazide (INH) is extensively metabolized to *N*-acetylisoniazid.[342]

The acetylation of some primary aliphatic amines such as histamine,[350] mescaline,[163] and the bis-*N*-demethylated metabolite of $\alpha(-)$-methadol[351] also has been reported. In comparison with oxidative deamination processes, *N*-acetylation is only a minor pathway in the metabolism of this class of compounds.

Clonazepam, R = Cl
Nitrazepam, R = H

7-Amino Metabolite

7-Acetamido Metabolite
or
N-Acetylated Metabolite

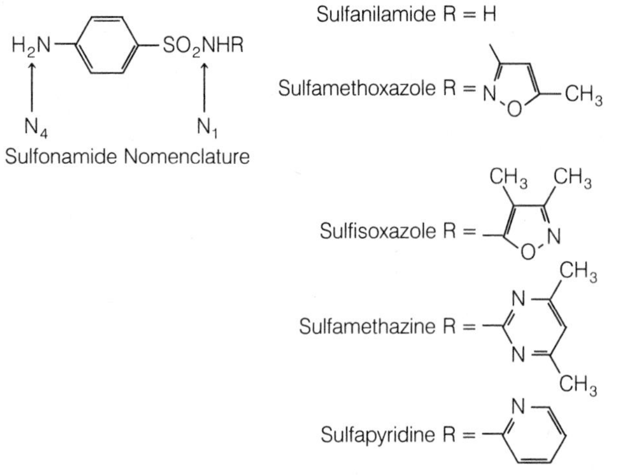

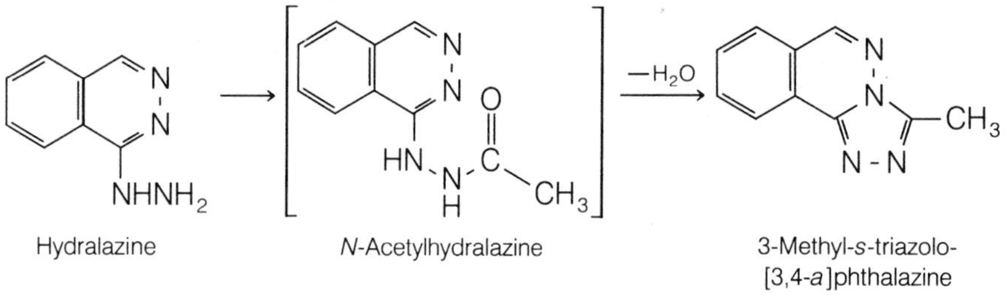

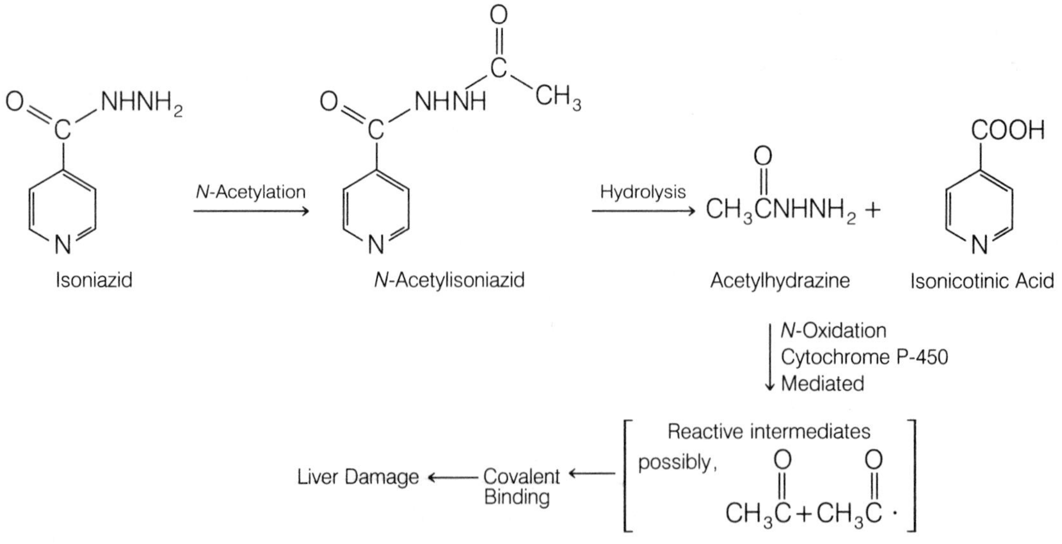

The acetylation pattern of several drugs (e.g., isoniazid, hydralazine, procainamide) in the human population displays a bimodal character in which the drug is either rapidly or slowly conjugated with acetyl-CoA.[352,353] This phenomenon is termed acetylation polymorphism. Individuals are classified as being either slow or rapid acetylator phenotypes. This variation in acetylating ability is genetic and is mainly caused by differences in *N*-acetyltransferase activity. The proportion of rapid and slow acetylators varies widely among different ethnic groups throughout the world. For example, a high proportion of Eskimos and Orientals are rapid acetylators, whereas Egyptians and some Western European groups are mainly slow acetylators.[353] Other populations are intermediate between these two extremes. Because of the bimodal distribution of the human population into rapid and slow acetylators, there appears to be significant individual variations in therapeutic and toxicologic responses to drugs displaying acetylation polymorphism.[316,352,353] It seems that slow acetylators are more likely to develop adverse reactions, whereas rapid acetylators seem more likely to show an inadequate therapeutic response to standard doses of the drug. The antituberculosis drug isoniazid illustrates many of these points. The plasma half-life of isoniazid in rapid acetylators ranges from 45 to 80 minutes, whereas in slow acetylators the half-life is about 140 to 200 minutes.[354] Thus, for a given fixed-dosing regimen, slow acetylators tend to accumulate higher plasma concentrations of isoniazid than do rapid acetylators. Higher concentrations of isoniazid may explain the greater therapeutic response (i.e., higher cure rate) among slow acetylators, but they probably also account for the greater incidence of adverse effects (e.g., peripheral neuritis and drug-induced systemic lupus erythematosus syndrome) observed among slow acetylators.[353] Slow acetylators of isoniazid apparently are also more susceptible to certain drug interactions involving drug metabolism. For example, phenytoin toxicity associated with concomitant use with isoniazid appears to be more prevalent in slow acetylators than in rapid acetylators.[355] Isoniazid inhibits the metabolism of phenytoin, thereby leading to an accumulation of high and toxic plasma levels of phenytoin.

Interestingly, patients who are rapid acetylators appear to be more likely to develop isoniazid-associated hepatitis.[342] This liver toxicity presumably arises from initial hydrolysis of the *N*-acetylated metabolite *N*-acetylisoniazid to acetylhydrazine. The latter metabolite is further converted (by cytchrome P-450 enzyme systems) to chemically reactive acylating intermediates that convalently bind to hepatic tissue, causing necrosis. Pathologic and biochemical studies in experimental animals appear to support this hypothesis. Therefore, rapid acetylators run a greater risk of incurring liver injury by virtue of producing more acetylhydrazine.

It appears that the tendency of drugs such as hydralazine and procainamide to cause lupus erythematosus syndrome and to elicit formation of antinuclear antibodies (ANA) is related to acetylator phenotype, with greater prevalence in slow acetylators.[356] Rapid acetylation may be preventing the immunologic triggering of ANA formation and the lupus syndrome. Interestingly, the *N*-acetylated metabolite of procainamide has been shown to be as active an antiarrhythmic agent as the parent drug[341] and has a half-life twice as long in humans.[357] These findings indicate that *N*-acetylprocainamide may be a promising alternative to procainamide as an antiarrhythmic agent with decreased lupus-inducing potential.

METHYLATION

Methylation reactions play an important role in the biosynthesis of many endogeneous compounds (e.g., epinephrine and melatonin) and in the inactivation of numerous physiologically active biogenic amines (e.g., norepinephrine, dopamine, serotonin, and histamine).[358] However, methylation constitutes only a minor pathway for conjugating drugs and xenobiotics. Methylation generally does not lead to polar or water-soluble metabolites, except when it creates a quaternary ammonium derivative. Most methylated products tend to be pharmacologically inactive, although there are a few exceptions.

The coenzyme involved in methylation reactions is *S*-adenosylmethionine (SAM). The transfer of the activated methyl group from this coenzyme to the acceptor substrate is catalyzed by various cytoplasmic and microsomal methyltransferases (Fig. 3-17).[358,359] Methyltransferases having particular importance in the metabolism of foreign compounds include catechol-*O*-methyltransferase (COMT), phenol-*O*-methyltransferase, and nonspecific *N*-methyltransferases and *S*-methyltransferases.[358] One of these enzymes, COMT, should be familiar, because it carries out *O*-methylation of such important neurotransmitters as norepinephrine and dopamine, the consequence of which is termination of their activity. Besides being present in the CNS and peripheral nerves, COMT is widely distributed in other mammalian tissue, particularly the liver and kidney. The other methyltransferases mentioned are primarily located in the liver, kidney, or lungs. Transferases that specifically methylate histamine, serotonin, and epinephrine usually are not involved in the metabolism of xenobiotics.[358]

Foreign compounds undergoing methylation include catechols, phenols, amines, *N*-heterocyclic, and

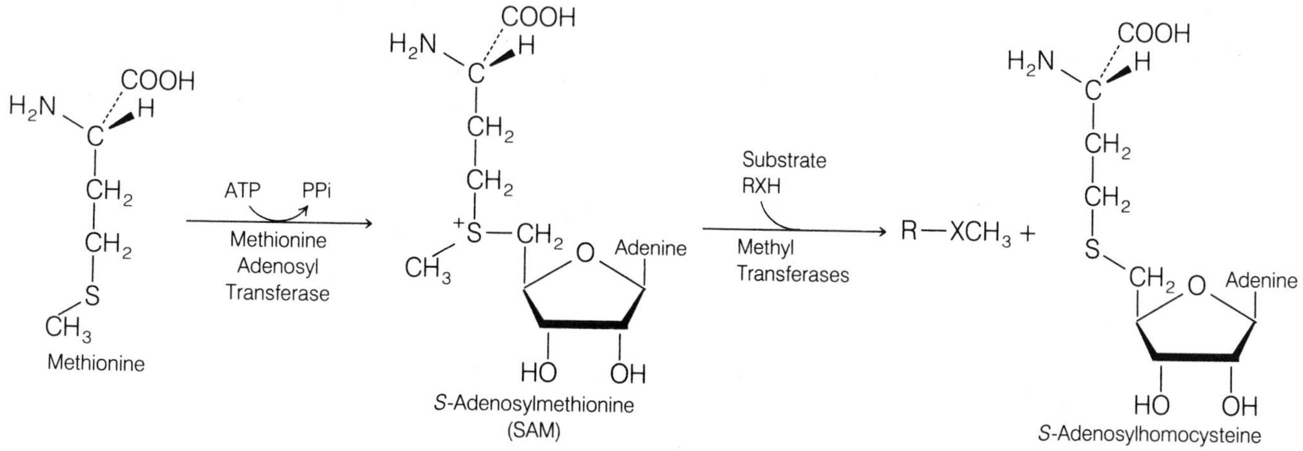

FIG. 3-17. *Conjugation of exogenous and endogenous substrates (RXH) by methylation.*

Norepinephrine, R = OH
Dopamine, R = H

COMT

Normetanephrine, R = OH
3-Methoxytyramine, R = H

S(−)-α-Methyldopa

S(−)-Dopa

Isoproterenol

Dobutamine

thiol compounds. Catechol and catecholaminelike drugs are metabolized by COMT to inactive monomethylated catechol products. Examples of drugs undergoing significant *O*-methylation by COMT in man include the antihypertensive (*S*)(−)α-methyldopa (Aldomet),[360] the antiparkinsonism agent (*S*)(−)-dopa (levodopa),[361] isoproterenol (Isuprel),[362] and dobutamine (Dobutrex).[363] The student should note the marked structural similarities between these drugs and the endogenous catecholamines such as norepinephrine and dopamine. In the foregoing four drugs, COMT selectively *O*-methylates only the phenolic OH at C-3. Bismethylation does not occur. Catechol metabolites arising from aromatic hydroxylation of phenols (e.g., 2-hydroxylation of 17α-ethinylestradiol)[44] and from the arene oxide dihydrodiol–catechol pathway (see earlier section on oxidation of aromatic moieties, e.g., the catechol mebolite of phenytoin)[364] also undergo *O*-methylation. Substrates undergoing *O*-methylation by COMT are required to contain an aromatic 1,2-dihydroxy group (i.e., catechol group). Resorcinol (1,3-dihydroxybenzene) or *p*-hydroquinone (1,4-dihydroxybenzene) derivatives are not substrates for COMT. This would explain why isoproterenol undergoes extensive *O*-methylation,[362] but terbutaline (which contains a resorcinol moiety) does not.[308]

Occasionally, phenols have also been reported to undergo *O*-methylation, but only to a minor extent.[358] One interesting example involves the conversion of morphine to its *O*-methylated derivative, codeine, in humans. This metabolite is formed in significant amounts in tolerant subjects and may account for up to 10% of the morphine dose.[365]

Although *N*-methylation of endogenous amines (e.g., histamine, norepinephrine) occurs commonly, biotransformation of nitrogen-containing xenobiotics to *N*-methylated metabolites occurs to only a limited extent. Some examples reported include the *N*-methylation of the antiviral and antiparkinsonism agent amantadine (Symmetrel) in dogs[366] and the in vitro *N*-methylation of norephedrine in rabbit lung preparations.[358] *N*-Methylation of nitrogen atoms present in heterocyclic compounds (e.g., pyridine derivatives) also takes place. For example, the pyridinyl nitrogens of nicotine[147] and nicotinic acid[367] are *N*-methylated to yield quaternary ammonium products.

Thiol-containing drugs, such as propylthiouracil,[368] 2,3-dimercapto-1-propanol (BAL),[369] and 6-

2-Hydroxy-17α-ethinylestradiol

Catechol Metabolite
of Phenytoin

Terbutaline
(not a substrate for COMT)

Propylthiouracil 2,3-Dimercapto-1-propanol (BAL) 6-Mercaptopurine

Morphine

O-Methylation

Codeine

Amantadine

Norephedrine

Nicotine

Nicotinic Acid

Trigonelline

mercaptopurine,[370] have also been reported to undergo *S*-methylation.

FACTORS AFFECTING DRUG METABOLISM

Drugs and xenobiotics often are metabolized by several different phase I and phase II pathways to give a number of metabolites. The relative amount of any particular metabolite is determined by the concentration and activity of the enzyme(s) responsible for the biotransformation. The rate of metabolism of a drug is particularly important for its pharmacologic action as well as its toxicity. For example, if the rate of metabolism of a drug is decreased, this generally leads to an increase in the intensity and duration of the drug. In addition, decreased metabolic elimination may also lead to accumulation of toxic levels of the drug. Conversely, an increase in the rate of metabolism leads to decreases in intensity and duration of action, as well as to decreased efficacy. Many factors may affect drug metabolism, and they will be discussed in the following sections. These include age, species and strain, genetic or hereditary factors, sex, enzyme induction, and enzyme inhibition.[29,371]

AGE DIFFERENCES

Age-related differences in drug metabolism are generally quite apparent in the newborn.[372,373] In most fetal and newborn animals, undeveloped or deficient oxidative and conjugative enzymes are chiefly responsible for the reduced metabolic capability seen. In general, the ability to carry out metabolic reactions increases rapidly after birth and approaches adult levels in about one to two months. An illustration of the influence of age on drug metabolism is seen in the duration of action (sleep time) of hexobarbital in newborn and adult mice.[374] When given a dose of 10 mg/kg body weight, the newborn mouse sleeps more than six hours. In contrast, the adult mouse sleeps for fewer than five minutes when given the same dose.

In humans, oxidative and conjugative (e.g., glucuronidation) capabilities of newborns are also low compared with adults. For example, the oxidative (cytochrome P-450) metabolism of tolbutamide appears to be markedly decreased in newborns.[375] In comparison with the half-life of eight hours in adults, the plasma half-life of tolbutamide in infants is greater than 40 hours. As discussed earlier, infants possess poor glucuronidating ability owing to a deficiency in glucuronyltransferase activity. The inability of infants to conjugate chloramphenicol with glucuronic acid appears to be responsible for the accumulation of toxic levels of this antibiotic, resulting in the so-called gray baby syndrome.[302] Similarly, neonatal hyperbilirubinemia (or kernicterus) results from the inability of newborn babies to glucuronidate bilirubin.[301]

The effect of old age on drug metabolism has not been as well studied as in the young. There is some evidence in animals and in humans that suggests that drug metabolism diminishes with old age.[376] However, much of the evidence is based on prolonged plasma half-life of drugs that are totally or mainly metabolized by hepatic microsomal enzymes (e.g., antipyrine, phenobarbital, acetaminophen). The quantitative importance of old age on drug metabolism is not now known.

SPECIES AND STRAIN DIFFERENCES

The metabolism of many drugs and foreign compounds is often species-dependent. Different animal species may biotransform a particular xenobiotic by similar or markedly different metabolic pathways. Even within the same species, there may be individual variations (strain differences) that may result in significant differences in a specific metabolic pathway.[377, 378]

Species variation has been observed in many oxidative biotransformation reactions. For example, metabolism of amphetamine occurs by two main pathways; oxidative deamination or aromatic hydroxylation. In human, rabbit, and guinea pig, oxidative deamination appears to be the predominant pathway, whereas, in the rat, aromatic hydroxylation appears to be the more important route.[379] Phenytoin is another drug example showing marked species difference in metabolism. In humans, phenytoin undergoes aromatic oxidation to yield primarily (*S*)(−)-*p*-hydroxyphenytoin, whereas in dogs, oxidation occurs mainly to give (*R*)(+)-*m*-hydroxyphenytoin.[380] There not only is a dramatic difference in the position (i.e., *meta* or *para*) of aromatic hydroxylation, but also a pronounced difference in which of the two phenyl rings (at C-5 of phenytoin) undergoes aromatic oxidation.

Species differences in many conjugation reactions have also been observed. Often these differences are

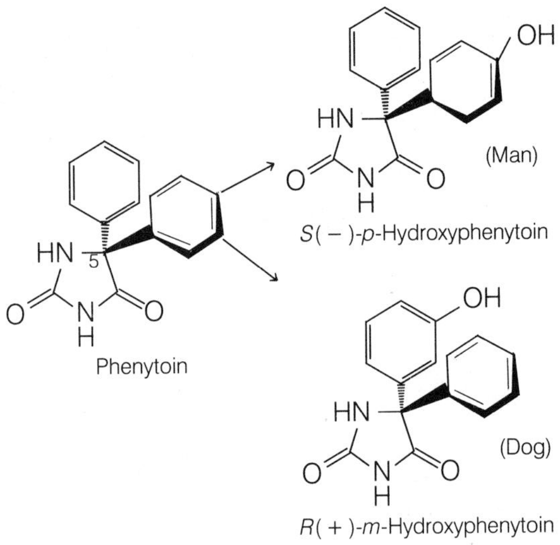

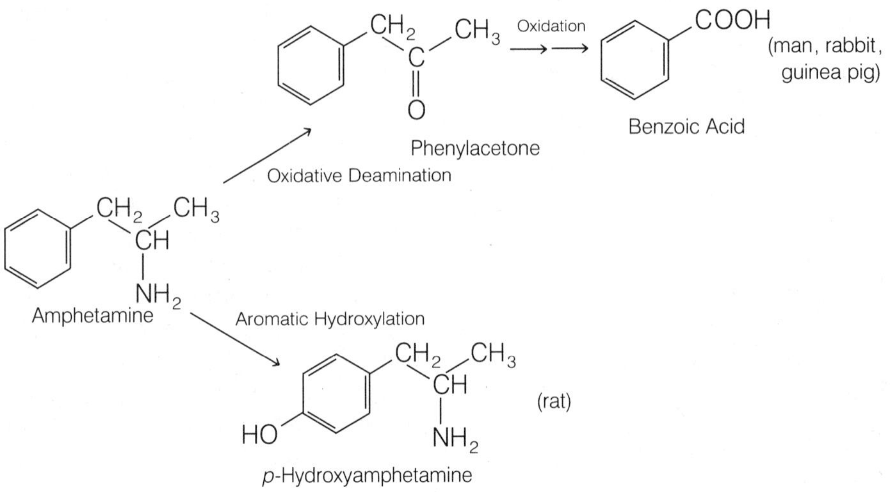

Phenytoin → *S*(−)-*p*-Hydroxyphenytoin (Man); *R*(+)-*m*-Hydroxyphenytoin (Dog)

Amphetamine → Oxidative Deamination → Phenylacetone → Oxidation → Benzoic Acid (man, rabbit, guinea pig)

Amphetamine → Aromatic Hydroxylation → *p*-Hydroxyamphetamine (rat)

caused by the presence or absence of transferase enzymes involved in the conjugative process. For example, cats lack glucuronyltransferase enzymes and, therefore, tend to conjugate phenolic xenobiotics by sulfation instead.[381] In pigs, the situation is reversed. Pigs are not able to conjugate phenols with sulfate (owing to lack of sulfotransferase enzymes), but appear to have good glucuronidation capability.[381] The conjugation of aromatic acids with amino acids (e.g., glycine, glutamine) has been noted to be dependent on the animal species, as well as on the substrate. For example, glycine conjugation is a common conjugation pathway for benzoic acid in many animals. However, in certain birds (e.g., duck, goose, turkey), glycine is replaced by the amino acid, ornithine.[382] Phenylacetic acid is a substrate for both glycine and glutamine conjugation in humans and other primates. However, in nonprimates, such as the rabbit and rat, phenylacetic acid is excreted only as the glycine conjugate.[383]

Strain differences in drug metabolism, particularly in inbred mice and rabbits, have been noted. These differences are apparently caused by genetic variations in the amount of metabolizing enzyme present among the different strains. For example, in vitro studies indicate that cottontail rabbit liver microsomes metabolize hexobarbital about ten times more rapidly than New Zealand rabbit liver microsomes.[384] In humans, interindividual differences in drug metabolism will be considered under hereditary or genetic factors.

HEREDITARY OR GENETIC FACTORS

Marked individual differences in the metabolism of several drugs exist in humans.[354] Apparently genetic or hereditary factors are mainly responsible for the large differences seen in the rate of metabolism of these drugs. One frequently cited example that dramatically illustrates the influence of genetic factors on drug metabolism concerns the biotransformation of the antituberculosis agent, isoniazid or isonicotinic acid hydrazide (INH). Metabolism of isoniazid occurs primarily by *N*-acetylation.[342] Studies indicate that individuals differ markedly in their ability to acetylate the drug, either slowly or rapidly. Rapid acetylators appear to have more hepatic *N*-acetyltransferase enzymes than do slow acetylators. In addition, the level of *N*-acetyltransferase is genetically determined and is transmitted as an autosomal recessive trait in humans. As discussed earlier in the section on acetylation, the proportion of rapid and slow acetylators varies widely among different ethnic groups. For instance, a high proportion (90%) of Eskimos and Orientals are rapid acetylators, whereas Egyptians and Mediterranean Jews are mainly slow acetylators.[353]

The rate at which isoniazid is acetylated is clinically important in terms of therapeutic response and toxicity. In general, it seems that rapid acetylators are more likely to show an inadequate therapeutic response (lower cure rate against tuberculosis), whereas slow acetylators are more likely to develop a greater incidence of adverse effects (e.g., peripheral neuritis, lupus erythematosus syndrome).[353] Other drugs, such as hydralazine, procainamide, and dapsone, also show similar bimodal distribution in the rate at which they are acetylated.[353]

Genetic factors also appear to influence the rate of oxidation of drugs like phenytoin, phenylbutazone, dicumarol, and nortriptyline.[385, 386] The rate of oxidation of these drugs varies widely among different individuals; however, these differences do not appear to be distributed bimodally, as in acetylation. In general, individuals who tend to oxidize one drug rapidly are also likely to oxidize other drugs rapidly. Numerous studies in twins (identical and fraternal) and in families tend to indicate that oxidation of these drugs is under genetic control.[386]

SEX DIFFERENCES

The rate of metabolism of xenobiotics also varies according to sex in some animal species. For example, marked difference is observed in female and male rats. Adult male rats metabolize several foreign compounds at a much faster rate than female rats (e.g., *N*-demethylation of aminopyrine, hexobarbital oxidation, glucuronidation of *o*-aminophenol). Apparently, this sex difference is also dependent on substrate, because some xenobiotics are metabolized at the same rate in both female and male rats. Differences in microsomal oxidation have been shown to be under the control of sex hormones, particularly androgens. The anabolic action of androgens seems to increase metabolism.[387]

Sex differences in drug metabolism appear to be species-dependent. Rabbits and mice, for example, do not show significant sex difference in drug metabolism.[387] In humans, there have been a few reports of sex difference in metabolism. For instance, nicotine and aspirin seem to be metabolized differently in women and men.[388]

ENZYME INDUCTION

The activity of hepatic microsomal enzymes, such as the cytochrome P-450 mixed function oxidase system, can be markedly increased upon exposure to

TABLE 3-1

DRUGS THAT INDUCE METABOLISM IN MAN

Inducing Agent	*Enhances Metabolism of*
Phenobarbital and other barbiturates	Coumarin anticoagulants, phenytoin, cortisol, testosterone, bilirubin, vitamin D, acetaminophen, oral contraceptives
Glutethimide	Glutethimide, warfarin
Phenylbutazone	Aminopyrine, cortisol
Meprobamate	Meprobamate
Ethanol	Pentobarbital, tolbutamide
Phenytoin	Cortisol, nortriptyline oral contraceptives
Rifampin	Rifampin, hexobarbital, tolbutamide, coumarin anticoagulants, oral contraceptives, methadone, digitoxin, cortisol
Griseofulvin	Warfarin
Carbamazepine	Carbamazepine, warfarin, phenytoin

(From Nelson, S. D.: In, Wolff, M. E. (ed.). Burger's Medicinal Chemistry, 4th ed., Part I, p. 227, New York, Wiley-Interscience, 1980. Reprinted with permission.)

divers drugs, pesticides, polycyclic aromatic hydrocarbons, and environmental xenobiotics. The process by which the activity of these drug-metabolizing enzymes is increased is referred to as enzyme induction.[389-391] The increase in activity is apparently caused by an increase in the amount of newly synthesized enzyme. Enzyme induction often leads to an increase in the rate of drug metabolism and to a decrease in the duration of drug action.

Inducing agents may increase the rate of their own metabolism as well as those of other unrelated drugs or foreign compounds (Table 3-1).[29] Concomitant administration of two or more drugs may often lead to serious drug interactions as a result of enzyme induction. For instance, a clinically critical drug interaction occurs with phenobarbital and warfarin.[392] Induction of microsomal enzymes by phenobarbital causes an increase in the metabolism of warfarin and, consequently, a marked decrease in the anticoagulant effect. Therefore, if a patient is receiving warfarin anticoagulant therapy and begins taking phenobarbital, careful attention must be paid to readjustment of the warfarin dose. Dosage readjustment must also be made if a patient receiving both warfarin and phenobarbital therapy suddenly stops taking the barbiturate. The ineffectiveness of oral contraceptives in women on concurrent phenobarbital or rifampin therapy has been attributed to the enhanced metabolism of estrogens (e.g., 17α-ethinylestradiol) caused by phenobarbital[393] and rifampin[394] induction.

Inducers of microsomal enzymes may also enhance the metabolism of endogenous compounds such as steroidal hormones and bilirubin. For example, phenobarbital has been observed to increase the metabolism of cortisol, testosterone, vitamin D, and bilirubin in humans.[389] The enhanced metabolism of vitamin D_3 induced by phenobarbital and phenytoin appears to be responsible for the osteomalacia seen in patients on long-term use of these two anticonvulsant drugs.[395] Interestingly, phenobarbital causes the induction of glucuronyltransferase enzymes, thereby enhancing the conjugation of bilirubin with glucuronic acid. Phenobarbital has been used occasionally to treat hyperbilirubinemia in neonates.[396]

In addition to drugs, other chemicals such as polycyclic aromatic hydrocarbons (e.g., benzo[*a*]-pyrene, 3-methylcholanthrene) and environmental pollutants (e.g., pesticides, polychlorinated biphenyls, TCDD) also may induce certain oxidative pathways and, thereby, alter drug response.[389,391] Cigarette smoke contains minute amounts of polycyclic aromatic hydrocarbons, such as benzo[*a*]-pyrene, which are potent inducers of microsomal cytochrome P-450 enzymes. This induction has been noted to increase the oxidation of some drugs in smokers. For example, theophylline is metabolized more rapidly in smokers than in nonsmokers. This difference is reflected in the marked difference in the plasma half-life of theophylline between smokers ($T_{1/2}$ 4.1 hours) and nonsmokers ($T_{1/2}$ 7.2 hours).[397] Other drugs, such as phenacetin, pentazocine, and propoxyphene, also have been reported to undergo more rapid metabolism in smokers than in nonsmokers.[398]

Occupational and accidental exposures to chlorinated pesticides and insecticides have also stimulated drug metabolism. For instance the half-life of antipyrine in workmen occupationally exposed to the insecticides lindane and DDT has been reported to be significantly shorter (7.7 hours versus 11.7 hours) in contrast with control subjects.[399] A case has also been reported in which a worker exposed to chlorinated insecticides showed a lack of response (i.e., decreased anticoagulant effect) to a therapeutic dose of warfarin.[400]

The existence of multiple forms of cytochrome P-450 have been demonstrated.[28] Many chemicals have been shown to selectively induce one or more distinct forms of cytochrome P-450.[28] These inducers fall into two categories: "phenobarbitallike" inducers or "polycyclic aromatic hydrocarbonlike" inducers (e.g., benzo-[*a*]pyrene, 3-methylcholanthrene). Phenobarbitallike compounds generally induce one or more forms of cytochrome P-450, in which the spectral maximum of the reduced cytochrome P-450 carbon monoxide complex occurs at 450 nm. In contrast, polycyclic aromatic hydrocarbonlike chemicals induce a different form of cytochrome P-450 in which the reduced cytochrome

P-450–carbon monoxide complex occurs at 448 nm.[25,28] This distinct enzyme form is often referred to as cytochrome P-448. Xenobiotics, such as benzo[*a*]pyrene, 3-methylcholanthrene, and TCDD, induce cytochrome P-448.[389] Cytochrome P-448 is particularly interesting in that it shows a greater selectivity for the oxidation of polycyclic aromatic hydrocarbons.

Enzyme induction may also affect toxicity of some drugs by enhancing the metabolic formation of chemically reactive metabolites. Particularly important is the induction of cytochrome P-450 enzymes involved in the oxidation of drugs to reactive intermediates. For example, the oxidation of acetaminophen to a reactive imidoquinone metabolite appears to be carried out by a phenobarbital-inducible form of cytochrome P-450 in rats and mice. Numerous studies in these two animals indicate that phenobarbital pretreatment leads to an increase in vivo hepatotoxicity and covalent binding, as well as to an increase in the formation of reactive metabolite in microsomal incubation mixtures.[188,189,191] Induction of cytochrome P-448 is of toxicologic concern, because it is now well established that this particular enzyme is involved in the metabolism of polycyclic aromatic hydrocarbons to reactive and carcinogenic intermediates.[67,401] For example, the metabolic bioactivation of benzo[*a*]-pyrene to its ultimate carcinogenic diol epoxide intermediate is carried out by cytochrome P-448 (see earlier section on aromatic oxidation for the bioactivation pathway of benzo[*a*]pyrene to its diol epoxide).[401] Thus, it is becoming increasingly apparent that many times enzyme induction may enhance the toxicity of some xenobiotics by increasing the rate of formation of reactive metabolites.

ENZYME INHIBITION

Several drugs and xenobiotics are capable of inhibiting drug metabolism.[29,371] With metabolism decreased, drug accumulation often occurs, thereby leading to prolonged drug action and serious adverse effects. Enzyme inhibition can occur by divers mechanisms, including substrate competition, interference with protein synthesis, inactivation of drug metabolizing enzymes, hepatotoxicity leading to impairment of enzyme activity, and others. Some drug interactions resulting from enzyme inhibition have been reported in humans.[402] For example, phenylbutazone has been noted to stereoselectively inhibit the metabolism of the more potent (*S*)(−)-enantiomer of warfarin. This inhibition may explain the excessive hypoprothrombinemia (increased anticoagulant effect) and many instances of hemorrhaging seen in patients on both warfarin and phenylbutazone therapy.[45] The metabolism of phenytoin is inhibited by drugs such as chloramphenicol, disulfiram, and isoniazid.[392] Interestingly, phenytoin toxicity as a result of enzyme inhibition by isoniazid appears to occur primarily in slow acetylators.[355] Several drugs, such as dicumarol, chloramphenicol, and phenylbutazone,[392] have been observed to inhibit the biotransformation of tolbutamide, which may lead to a possible hypoglycemic response.

Other compounds, such as SK & F-525A (proadifen hydrochloride), metyrapone, piperonyl butoxide, and cobaltous chloride, have been used experimentally in animals as general inhibitors of microsomal enzymes.[29,403]

MISCELLANEOUS FACTORS AFFECTING DRUG METABOLISM[29,371]

Other factors also may influence drug metabolism. Dietary factors, such as the protein/carbohydrate ratio in the diet affect the metabolism of a few drugs. Indoles present in vegetables, such as brussel sprouts, cabbage, and cauliflower, and polycyclic aromatic hydrocarbons present in charcoal-broiled beef cause enzyme induction and stimulate the metabolism of some drugs. Vitamins, minerals, starvation, and malnutrition also apparently have an influence on drug metabolism. Finally, physiologic factors, such as the pathologic state of the liver (e.g., hepatic cancer, cirrhosis, hepatitis), pregnancy, hormonal disturbances (e.g., thyroxine, steroids), and circadian rhythm, may also markedly affect drug metabolism.

STEREOCHEMICAL ASPECTS OF DRUG METABOLISM

Many drugs (e.g., warfarin, propranolol, hexobarbital, glutethimide, cyclophosphamide, ketamine, and ibuprofen) are often administered as racemic mixtures in humans. The two enantiomers present in a racemic mixture may differ from one another in pharmacologic activity. Usually, one enantiomer tends to be much more active than the other. For example, the (*S*)(−)-enantiomer of warfarin is five times more potent as an oral anticoagulant than is the (*R*)(+)-enantiomer.[404] In some instances, the two enantiomers may have totally different pharmacologic activities. For example, (+)-α-propoxyphene (Darvon) is an analgesic, whereas (−)-α-propoxyphene (Novrad) is an antitussive.[405] Such differences in activity between stereoisomers should not be surprising, since in Chapter 2 we learned that stereochemical factors generally have a dramatic influence on how the drug molecule interacts with the target receptors to elicit its pharmacologic response. By the same token, the preferential interaction of one stereoisomer with drug metabolizing enzymes may lead one to anticipate differences in metabolism for the two enantiomers of a racemic mixture. Indeed, individual enantiomers of a racemic drug are often metabolized at different rates. For example, studies in humans indicate that the less active (+)-enantiomer of propranolol undergoes more rapid metabolism than the corresponding (−)-enantiomer.[406] Allylic hydroxylation of hexobarbital has been observed to occur more rapidly with the $R(-)$-enantiomer in humans.[407] The term *substrate stereoselectivity* is frequently used to denote the preference of one stereoisomer as a substrate for a metabolizing enzyme or metabolic process.[223]

Individual enantiomers of a racemic mixture may also be metabolized by different pathways. For instance, in dogs the (+)-enantiomer of the sedative hypnotic glutethimide (Doriden) is primarily hydroxylated α to the carbonyl to yield 4-hydroxyglutethimide, whereas the (−)-enantiomer undergoes aliphatic $\omega - 1$ hydroxylation of its C-2 ethyl group.[113] Dramatic differences in the metabolic profile of two enantiomers of warfarin have also been noted. In humans, the more active $(S)(-)$-isomer is 7-hydroxylated (aromatic hydroxylation), whereas the $(R)(+)$-isomer undergoes keto reduction to yield primarily the (RS) warfarin alcohol as the major plasma metabolite.[45, 234] Although numerous other examples of substrate stereoselectivity or enantioselectivity in drug metabolism exist, the examples presented will suffice to emphasize the point.[223, 408]

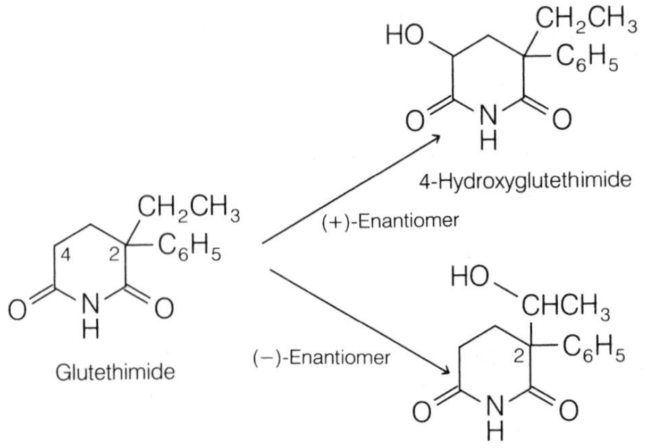

Glutethimide → (+)-Enantiomer → 4-Hydroxyglutethimide; (−)-Enantiomer → ($\omega-1$ hydroxylation of C-2 ethyl)

Drug biotransformation processes often lead to the creation of a new asymmetric center in the metabolite (i.e., stereoisomeric or enantiomeric products). The preferential metabolic formation of a stereoisomeric product is called *product stereoselectivity*.[223] For example, bioreduction of ketone xenobiotics, as a general rule, produces predominantly one stereoisomeric alcohol (see section on reduction of keto carbonyls).[91, 223] The preferential formation of $(S)(-)$-hydroxyhexamide from the hypoglycemic agent acetohexamide[225] and the exclusive generation of 6β-naltrexol from naltrexone[231] (see section on reduction of keto carbonyls for structure) are two examples of highly stereoselective bioreduction processes in humans.

Oxidative biotransformations have been noted to display product stereoselectivity, too. For example, phenytoin contains two phenyl rings in its structure, both of which a priori should be susceptible to aromatic hydroxylation. However, in humans, *p*-hy-

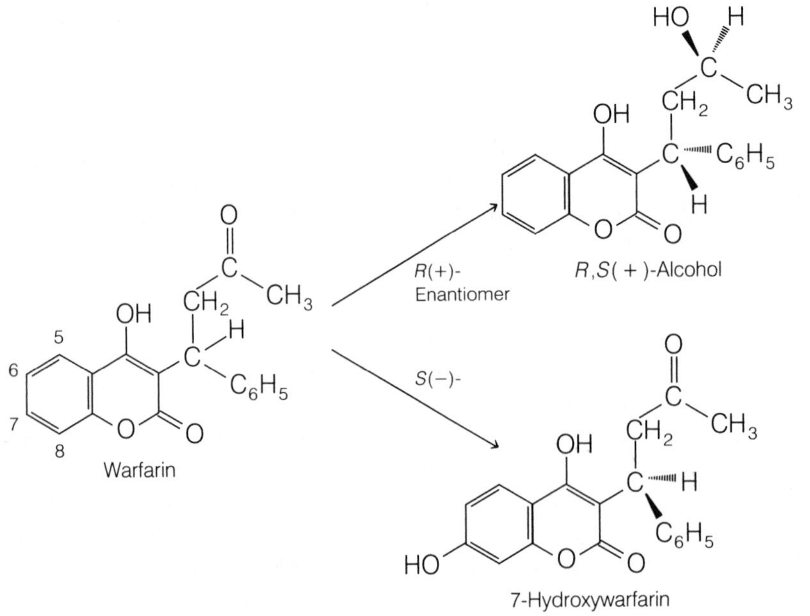

Warfarin → $R(+)$-Enantiomer → $R,S(+)$-Alcohol; $S(-)$- → 7-Hydroxywarfarin

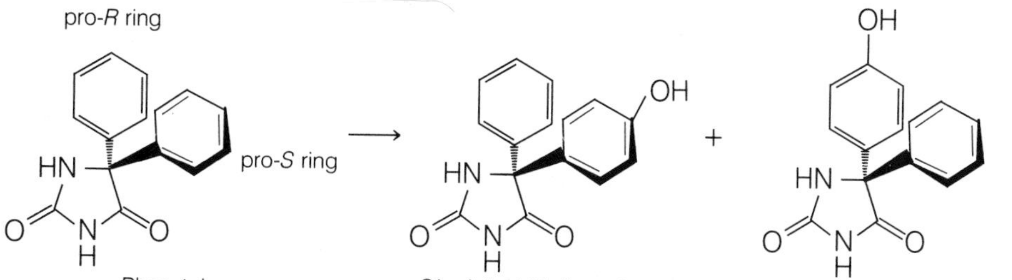

Phenytoin → $S(-)$-5-(4-Hydroxyphenyl)-5-phenylhydantoin + $R(+)$-5-(4-Hydroxyphenyl)-5-phenylhydantoin

droxylation occurs preferentially (approximately 90%) at the pro- (S) phenyl ring to give primarily (S)(−)-5-(4-hydroxyphenyl)-5-phenylhydantoin. Although the other phenyl ring is also p-hydroxylated, it occurs to only to a minor extent (10%).[380] Microsomal hydroxylation of the C-3 carbon of diazepam and desmethyldiazepam (using mouse liver preparations) has been reported to proceed with remarkable stereoselectivity to yield optically active metabolites having the 3(S) absolute configuration.[112] Interestingly, these two metabolites are also pharmacologically active and one of them, oxazepam, is marketed as a drug (Serax). The allylic hydroxylation of the N-butenyl side group of the analgesic pentazocine (Talwin) leads to two possible alcohols (*cis* and *trans* alcohols). In human, mouse, and monkey, pentazocine is metabolized predominantly to the *trans* alcohol metabolite, whereas the rat primarily tends to form the *cis* alcohol.[106] The product stereoselectivity observed in this biotransformation involves *cis* and *trans* geometric stereoisomers.

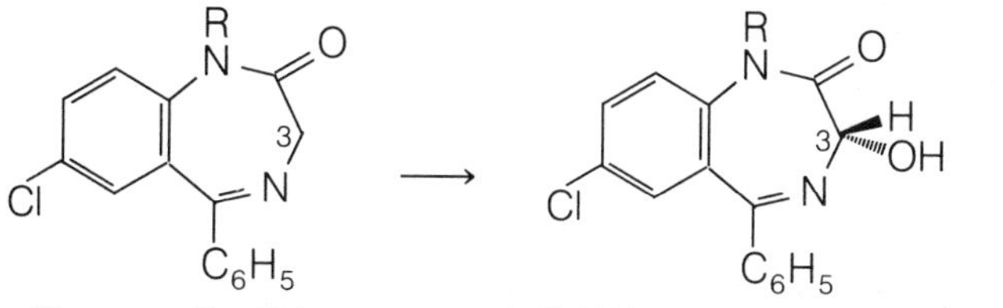

Diazepam, R = CH_3
Desmethyldiazepam, R = H

(3S) N-Methyloxazepam, R = CH_3
$S(+)$-Oxazepam, R = H

More recently, the term *regioselectivity*[409] has been introduced in drug metabolism to denote the selective metabolism of two or more similar functional groups (e.g., OCH_3, OH, NO_2) or two or more similar atoms that are positioned in different regions of a molecule. The examples now described should make this concept clear. For example, of the four methoxy groups present in papaverine, the 4′-OCH_3 group is regioselectively O-demethylated in several species (e.g., rat, guinea pig, rabbit, and dog).[410] Trimethoprim (Trimpex, Proloprim) has two heterocyclic sp^2 nitrogen atoms (N^1 and N^3) in its structure. In dogs, it appears that oxidation occurs regioselectively at N^3 to give the corresponding 3-N-oxide.[178] Nitroreduction of the 7-nitro group in 5,7-dinitroindazole to yield the 7-amino derivative in the mouse and rat occurs with a high degree of regioselectivity.[287] Substrates amenable to O-methylation by COMT appear to proceed with remarkable regioselectivity, as typified by the cardiotonic agent dobutamine (Dobutrex). O-Methylation occurs exclusively with the phenolic hydroxy group at C-3.[363]

PHARMACOLOGICALLY ACTIVE METABOLITES

The traditional notion that drug metabolites are inactive and are insignificant in drug therapy has dramatically changed in recent years. There is now increasing evidence to indicate that many drugs are

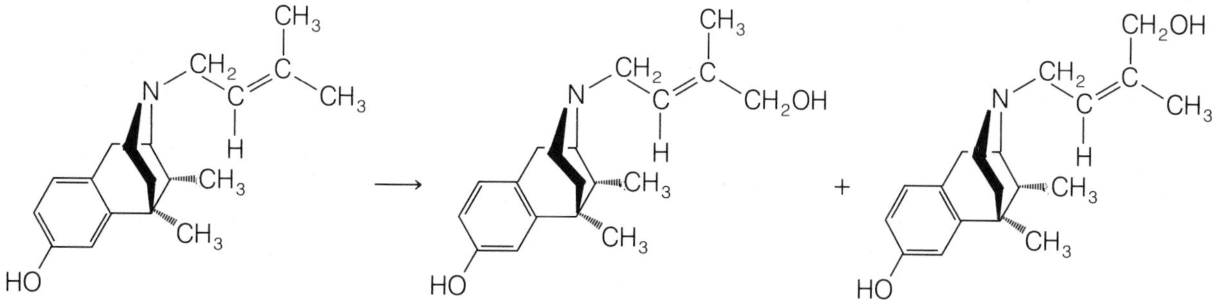

Pentazocine → *trans*-Alcohol + *cis*-Alcohol

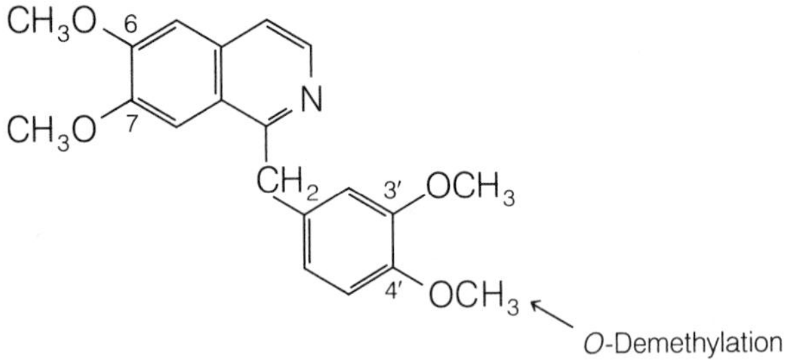

Papaverine

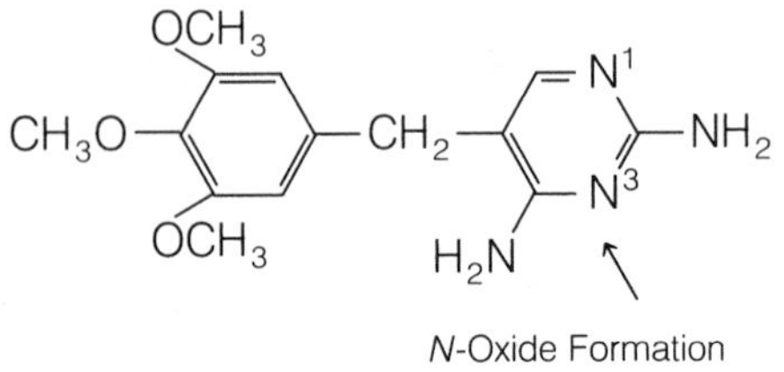

Trimethoprim

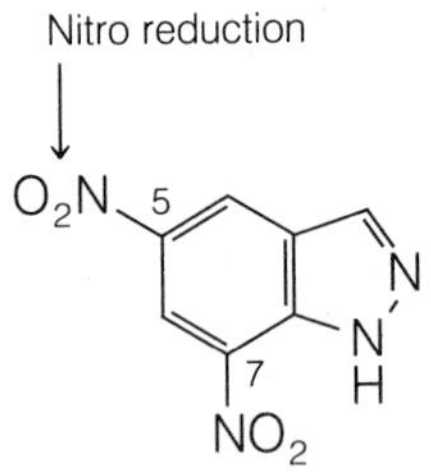

5,7-Dinitroindazole

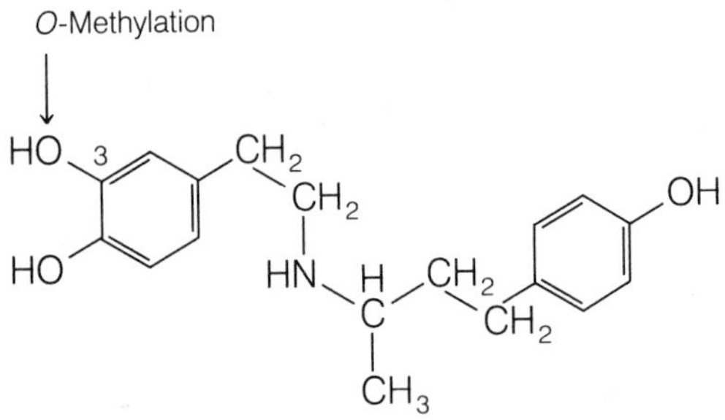

Dobutamine

TABLE 3-2

PHARMACOLOGICALLY ACTIVE METABOLITES IN HUMANS

Parent Drug	Metabolite	Biotransformation Process
Acetohexamide	Hydroxyhexamide	Ketone reduction
Acetylmethadol	Noracetylmethadol	N-Demethylation
Amitriptyline	Nortriptyline	N-Demethylation
Azathioprine	6-Mercaptopurine	Glutathione conjugation
Carbamazepine	Carbamazepine-9, 10-epoxide	Epoxidation
Chloral hydrate	Trichloroethanol	Aldehyde reduction
Clofibrate	Chlorophenoxyisobutyric acid	Ester hydrolysis
Chlorpromazine	7-Hydroxychlorpromazine	Aromatic hydroxylation
Cortisone	Hydrocortisone	Ketone reduction
Diazepam	Desmethyldiazepam and oxazepam	N-Demethylation and 3-hydroxylation
Digitoxin	Digoxin	Alicylic hydroxylation
Diphenoxylate	Diphenoxylic acid	Ester hydrolysis
Imipramine	Desipramine	N-Demethylation
Mephobarbital	Phenobarbital	N-Demethylation
Metoprolol	α-Hydroxymethyl-metoprolol	Benzylic hydroxylation
Phenacetin	Acetaminophen	O-Deethylation
Phenylbutazone	Oxybutazone	Aromatic hydroxylation
Prednisone	Prednisolone	Ketone reduction
Primidone	Phenobarbital	Hydroxylation and oxidation to ketone
Procainamide	N-Acetylprocainamide	N-Acetylation
Propranolol	4-Hydroxypropranolol	Aromatic hydroxylation
Quinidine	3-Hydroxyquinidine	Allylic hyroxylation
Sulindac	Sulfide metabolite of sulindac	Sulfoxide reduction
Thioridazine	Mesoridazine	S-oxidation
Warfarin	Warfarin alcohols	Ketone reduction

biotransformed to pharmacologically active metabolites that contribute to the therapeutic as well as toxic effects of the parent compound. Metabolites that have been shown to have significant therapeutic activity in humans are listed in Table 3-2.[2,411] The parent drug from which the metabolite is derived and the biotransformation process involved are also given.

How significantly an active metabolite contributes to the therapeutic or toxic effects ascribed to the parent drug will be dependent on its relative activity and its quantitative importance (e.g., plasma concentration). In addition, whether the metabolite accumulates after repeated administration (e.g., desmethyldiazepam in geriatric patients) or in patients with renal failure are also determinants.

From a clinical standpoint, active metabolites are especially important in patients with decreased renal function. If renal excretion is the major pathway for elimination of the active metabolite, then accumulation is likely to occur in patients with renal failure. Especially with drugs, such as procainamide, clofibrate, and digitoxin, caution should be exercised in treating patients with renal failure.[2,411] Many of the toxic effects seen for these drugs have been attributed to high plasma levels of their active metabolites. For example, the severe muscle weakness and tenderness (myopathy) seen with clofibrate in renal failure patients is believed to be caused by high levels of the active metabolite, chlorophenoxyisobutyric acid.[412] Cardiovascular toxicity owing to digitoxin and procainamide in anephric subjects has been attributed to high plasma levels of digoxin and *N*-acetylprocainamide, respectively. In such situations, appropriate reduction in dosage and careful monitoring of plasma levels of the parent drug and its active metabolite are often recommended.

The pharmacologic activity of some metabolites has led many manufacturers to synthesize these metabolites and to market them as separate drug entities. For example, oxyphenbutazone (Tandearil, Oxalid) is the *p*-hydroxylated metabolite of the anti-inflammatory agent phenylbutazone (Butazolidin, Azolid); notriptyline (Aventyl) is the *N*-demethylated metabolite of the tricyclic antidepressant amitriptyline (Elavil); oxazepam (Serax) is the *N*-demethylated and 3-hydroxylated metabolite of diazepam (Valium); mesoridazine (Serentil) is the sulfoxide metabolite of the antipsychotic agent thioridazine (Mellaril).

REFERENCES

1. Williams, R. T.: Detoxication Mechanisms. 2nd ed. New York, Wiley & Sons, 1959.
2. Drayer, D. E.: Clin. Pharmacokinet. 1:426, 1976.
3. Drayer, D. E.: Drugs 24:519, 1982.
4. Jollow, D. J., Kocsis, J., Snyder, R., Vainio, H.: Biological Reactive Intermediates. New York, Plenum Press, 1977.
5. Gillette, J. R., et al.: Annu. Rev. Pharmacol. 14:271, 1974.
6. Nelson, S. D., et al.: In Jerina, D. M. (ed.). Drug Metabolism Concepts, p. 155. Washington, DC, American Chemical Society, 1977.
7. Testa, B., Jenner, P.: Drug Metab. Rev. 7:325, 1978.
8. Low, L. K., Castagnoli, N., Jr.: In Wolff, M. E. (ed.). Burger's Medicinal Chemistry, 4th ed. Part 1, p. 107. New York, Wiley-Interscience, 1980.
9. Gill, E. W., et al.: Biochem. Pharmacol. 22:175, 1973.
10. Wall, M. E., et al.: J. Am. Chem. Soc. 94:8579, 1972; Lemberger, L.: Drug Metab. Dispos. 1:641, 1973.
11. Green, D. E., et al.: In Vinson, J. A. (ed.) Cannabinoid Analysis in Physiological Fluids, p. 93. Washington, DC, American Chemical Society, 1979.
12. Williams, R. T.: In Brodie, B. B., Gillette, J. R. (eds.). Concepts in Biochemical Pharmacology, Part 2, p. 226. Berlin, Springer-Verlag, 1971.
13. Rowland, M.: In Melmon, K. L., Morelli, H. F. (eds.). Clinical Pharmacology: Basic Principles in Therapeutics, 2nd ed. p. 25. New York, Macmillan, 1978.
14. Benowitz, N. L., Meister, W.: Clin. Pharmacokinet. 3:177, 1978.
15. Connolly, M. E., et al.: Br. J. Pharmacol. 46:458, 1972.
16. Gibaldi, M., Perrier, D.: Drug Metab. Rev. 3:185, 1974.
17. Sinkula, A. A., Yalkowsky, S. H.: J. Pharm. Sci. 64:181, 1975.
18. Scheline, R. R.: Pharmacol. Rev. 25:451, 1973; Peppercorn, M. A., Goldman, P.: J. Pharmacol. Exp. Ther. 181:555, 1972.
19. Levy, G. A., Conchie, J.: In Dutton, G. J. (ed.). Glucuronic Acid, Free and Combined, p. 301. New York, Academic Press, 1966.
20. Testa, B., Jenner, P.: Drug Metabolism: Chemical and Biochemical Aspects, p. 419. New York, Marcel Dekker, 1976.
21. Powis, G., Jansson, I.: Pharmacol. Ther. 7:297, 1979.
22. Mason, H. S.: Annu. Rev. Biochem. 34:595, 1965.
23. Hayaishi, O.: In Hayaishi, O. (ed.). Oxygenases, p. 1. New York, Academic Press, 1962.
24. Mannering, G. J.: In LaDu, B. N., et al. (eds.). Fundamentals of Drug Metabolism and Disposition, p. 206. Baltimore, Williams & Wilkins, 1971.
25. Sato, R., Omura, T. (eds.): Cytochrome P-450. Tokyo, Kodansha, and New York, Academic Press, 1978.
26. Omura, T., Sato, R.: J. Biol. Chem. 239:2370, 1964.
27. Gillette, J. R.: Adv. Pharmacol. 4:219, 1966.
28. Johnson, E. F.: Rev. Biochem. Toxicol. 1:1. 1979; Guengerich, F. P.: Pharmacol. Ther. 6:99, 1979.
29. Nelson, S. D.: In Wolff, M. E. (ed.).: Burger's Medicinal Chemistry, 4th ed., Part 1, p. 227. New York, Wiley-Interscience, 1980.

29a. For recent review on nomenclature for cytochrome P-450 isozymes, see Waxman, D. J.: In Ortiz de Montellano, P. R. (ed.). Cytochrome P-450, p. 525. New York, Plenum Press, 1986.

30. Claude, A.: In Gillette, J. R., et al. (eds.). Microsomes and Drug Oxidations, p. 3. New York, Academic Press, 1969.
31. Hansch, D.: Drug Metab. Rev. 1:1, 1972.
32. Ortiz de Montellano, P. R.: Cytochrome P-450, p. 217. New York, Plenum Press, 1986.
33. Estabrook, R. W., Werringloer, J.: In Jerina, D. M. (ed.). Drug Metabolism Concepts, p. 1. Washington, DC, American Chemical Society, 1977; Trager, W. F.: In Jenner, P., Testa, B. (eds.). Concepts in Drug Metabolism, Part A, p. 177. New York, Marcel Dekker, 1980.
34. Ullrich, V.: Top. Curr. Chem. 83:68, 1979.
35. White, R. E., Coon, M. J.: Annu. Rev. Biochem. 49:315, 1980.
36. Guengerich, F. P. (ed.): Mammalian Cytochromes P-450, Vol. 1 and 2. Boca Raton, FL, CRC Press, 1987; Schenkman, J. B., Kupfer, D. (eds.): Hepatic Cytochrome P-450 Monooxygenase System. New York, Pergamon Press, 1982.
37. Daly, J. W., et al.: Experientia 28:1129, 1972.
38. Jerina, D. M., Daly, J. W.: Science 185:573, 1974; Kaminsky, L. S.: In Anders, M. W. (ed.). Bioactivation of Foreign Compounds, p. 157. New York, Academic Press, 1985.
39. Walle, T., Gaffney, T. E.: J. Pharmacol. Exp. Ther. 182:83, 1972; Bond, P.: Nature 213:721, 1967.
40. Whyte, M. P., Dekaban, A. S.: Drug Metab. Dispos. 5:63, 1977.
41. Witkin, K. M., et al.: Ther. Drug Monit. 1:11, 1979; Richens, A.: Clin. Pharmacokinet. 4:153, 1979.
42. Burns, J. J., et al.: J. Pharmacol. Exp. Ther. 113:481, 1955; Yü, T. F., et al.: J. Pharmacol. Exp. Ther. 123:63, 1958.
43. Murphy, P. J., Wick, A. N.: J. Pharm. Sci. 57:1125, 1968; Beckmann, R.: Ann. N.Y. Acad. Sci. 148:820, 1968.
44. Williams, M. C., et al.: Steroids 25:229, 1975; Ranney, R. E.: J. Toxicol. Environ. Health 3:139, 1977.
45. Lewis, R. J., et al.: J. Clin. Invest. 53:1697, 1974.
46. Beckett, A. H., Rowland, M.: J. Pharm. Pharmacol. 17:628, 1965; Dring, L. G., et al.: Biochem. J. 116:425, 1970.
47. Daly, J.: In Brodie, B. B., Gillette, J. R. (eds.). Concepts in Biochemical Pharmacology, Part 2, p. 285. Berlin, Springer-Verlag, 1971.
48. Lowenthal, D. T.: J. Cardiovasc. Pharmacol. 2(Suppl. 1):S29, 1980; Davies, D. S., et al.: Adv. Pharmacol. Ther. 7:215, 1979.
49. Dayton, P. G., et al.: Drug Metab. Dispos. 1:742, 1973.
50. Schreiber, E. E.: Annu. Rev. Pharmacol. 10:77, 1970.
51. Hollister, L. E., et al.: Res. Commun. Chem. Pathol. Pharmacol. 2:330, 1971.
52. Safe, S., et al.: J. Agric. Food Chem. 23:851, 1975.
53. Allen, J. R., et al.: Food Cosmet. Toxicol. 13:501, 1975; Vinopal, J. H., et al.: Arch. Environ. Contamin. Toxicol. 1:122, 1973.
54. Hathway, D. E. (Sr. Reporter): Foreign Compound Metabolism in Mammals, Vol. 4, p. 234. London, Chemical Society, 1977.

55. Guroff, G., et al.: Science, 157:1524, 1967.
56. Daly, J., et al.: Arch. Biochem. Biophys. 128:517, 1968.
57. Oesch, F.: Progr. Drug Metab. 3:253, 1978; Lu, A. A. H., Miwa, G. T.: Annu. Rev. Pharmacol. Toxicol. 20:513, 1980.
58. Ames, B. N., et al.: Science 176:47, 1972.
59. Maguire, J. H., et al.: Ther. Drug Monit. 1:359, 1979.
60. Harvey, D. G., et al.: Res. Commun. Chem. Pathol. Pharmacol. 3:557, 1972.
61. Stillwell, W. G.: Res. Commun. Chem. Pathol. Pharmacol. 12:25, 1975.
62. Jakoby, W. B., et al.: In Arias, I. M., Jakoby, W. B. (eds.). Glutathione, Metabolism and Function, p. 189. New York, Raven Press, 1976.
63. Boyland, E.: In Brodie, B. B., Gillette, J. R. (eds.). Concepts in Biochemical Pharmacology, Part 2, p. 584. Berlin, Springer-Verlag, 1971.
64. Brodie, B. B., et al.: Proc. Natl. Acad. Sci. USA 68:160, 1971.
65. Jollow, D. J., et al.: Pharmacology 11:151, 1974.
66. Gelboin, H. V., et al.: In Jollow, D. J., et al. (eds.). Biological Reactive Intermediates, p. 98. New York, Plenum Press, 1977.
67. Thakker, D. R., et al.: Chem. Biol. Interact. 16:281, 1977.
68. Weinstein, I. B., et al.: Science 193:592, 1976; Jeffrey A. M., et al.: J. Am. Chem. Soc. 98:5714, 1976; Koreeda, M., et al.: J. Am. Chem. Soc. 98:6720, 1976.
69. Straub, K. M., et al.: Proc. Natl. Acad. Sci. USA 74:5285, 1977.
70. Kasai, H., et al.: J. Am. Chem. Soc. 99:8500, 1977.
71. Chausseaud, L. F.: Drug Metab. Rev. 2:185, 1973.
72. Pynnönen, S.: Ther. Drug Monit. 1:409, 1979.
73. Eadie, M. J., Tyrer, J. H.: Anticonvulsant Therapy, 2nd ed., p. 142. Edinburgh, London, Churchill-Livingston, 1980.
74. Hucker, H. B., et al.: Drug Metab. Dispos. 3:80, 1975.
75. Hintze, K. L., et al.: Drug Metab. Dispos. 3:1, 1975.
76. Slack, J. A., Ford-Hutchinson, A. W.: Drug Metab. Dispos. 8:84, 1980.
77. Waddell, W. J.: J. Pharmacol. Exp. Ther. 149:23, 1965.
78. Seutter-Berlage, F., et al.: Xenobiotica 8:413, 1978.
79. Marniemi, J., et al.: In Ullrich, V., et al. (eds.). Microsomes and Drug Oxidations, p. 698. Oxford, Pergamon Press, 1977.
80. Garner, R. C.: Prog. Drug Metab. 1:77, 1976.
81. Swenson, D. H., et al.: Biochem. Biophys. Res. Commun. 60:1036, 1974; 53:1260, 1973.
82. Essigman, J. M., et al.: Proc. Natl. Acad. Sci. USA 74:1870, 1977; Croy, R. G., et al.: Proc. Natl. Acad. Sci. USA 75:1745, 1978.
83. Henschler, D., Bonser, G.: Adv. Pharmacol. Ther. 9:123, 1979.
84. Watabe, T., Akamatsu, K.: Biochem. Pharmacol. 24:442, 1975.
85. Metzler, M.: J. Toxicol. Environ. Health 1(Suppl.):21, 1976; Neuman, H. G., Metzler, M.: Adv. Pharmacol. Ther. 9:113, 1979.
86. Ortiz de Montellano, P. R., Correia, M. A.: Annu. Rev. Pharmacol. Toxicol. 23:481, 1983; Ortiz de Montellano, P. R.: In Anders, M. W. (ed.). Bioactivation of Foreign Compounds, p. 121. New York, Academic Press, 1985.
87. DeMatteis, F.: Biochem. J. 124:767, 1971; Levin, W., et al.: Arch. Biochem. Biophys. 148:262, 1972.
88. Levin, W., et al.: Science 176:1341, 1972; Levin, W., et al.: Drug Metab. Dispos. 1:275, 1973.
89. Ivanetich, K. M., et al.: In Ullrich, V., et al. (eds.). Microsomes and Drug Oxidations, p. 76. Oxford, Pergamon Press, 1977.
90. Ortiz de Montellano, P. R., et al.: Biochem. Biophys. Res. Commun. 83:132, 1978; Ortiz de Montellano, P. R., et al.: Arch. Biochem. Biophys. 197:524, 1979; Ortiz de Montellano, P. R., et al.: Biochemistry 21:1331, 1982.
91. McMahon, R. E.: In Brodie, B. B., Gillette, J. R. (eds.). Concepts in Biochemical Pharmacology, Part 2, p. 500. Berlin, Springer-Verlag, 1971.
92. Dutton, G.: In Brodie, B. B., Gillette, J. R. (eds.). Concepts in Biochemical Pharmacology, Part 2, p. 378. Berlin, Springer-Verlag, 1971.
93. Thomas, R. C., Ikeda, G. J.: J. Med. Chem. 9:507, 1966.
94. Selley, M. L., et al.: Clin. Pharmacol. Ther. 17:599, 1975; Sumner, D. D., et al.: Drug Metab. Dispos. 3:283, 1975.
95. Permisohn, R. C., et al.: J. Forensic Sci. 21:98, 1976.
96. Borg, K. O., et al.: Acta Pharmacol. Toxicol. 36(Suppl. 5):125, 1975; Hoffmann, K. J.: Clin. Pharmacokinet. 5:181, 1980.
97. Ho, B. T., et al.: J. Med. Chem. 14:158, 1971; Matin, S., et al.: J. Med. Chem. 17:877, 1974.
98. Crammer, J. L., Scott, B.: Psychopharmacologia 8:461, 1966.
99. Hucker, H. B.: Pharmacologist 4:171, 1962; Kraak, J. C., Bijster, P.: J. Chromatogr. 143:499, 1977.
100. Harvey, D. J., et al.: In Frigerio, A., Ghisalberti, E. L. (eds.). Mass Spectrometry in Drug Metabolism, p. 403. New York, Plenum Press, 1977.
101. Allen, J. G., et al.: Drug Metab. Dispos. 3:332, 1975.
102. Sims, P.: Biochem. Pharmacol. 19:795, 1970; Takahashi, G., Yasuhria, K.: Cancer Res. 32:710, 1972.
103. Lemberger, L., et al.: Science 173:72, 1971; 177:62, 1972.
104. Carroll, F. I., et al.: J. Med. Chem. 17:985, 1974; Drayer, D. E., et al.: Clin. Pharmacol. Ther. 27:72, 1980.
105. Drayer, D. E., et al.: Clin. Pharmacol. Ther. 24:31, 1978.
106. Bush, M. T., Weller, W. L.: Drug Metab. Rev. 1:249, 1972; Thompson, R. M., et al.: Drug Metab. Dispos. 1:489, 1973.
107. Breimer, D. D., Van Rossum, J. M.: J. Pharm. Pharmacol. 25:762, 1973.
108. Pittman, K. A., et al.: Biochem. Pharmacol. 18:1673, 1969; 19:1833, 1970.
109. Miller, J. A., Miller, E. C.: In Jollow, D. J., et al. (eds.). Biological Reactive Intermediates, p. 14. New York, Plenum Press, 1977.
110. Wislocki, P. G., et al.: Cancer Res. 36:1686, 1976.
111. Garattini, S., et al.: In Usdin, E., Forrest, I. (eds.). Psychotherapeutic Drugs, Part 2, p. 1039. New York, Marcel Dekker, 1977; Greenblatt, D. J., et al.: Clin. Pharmacol. Ther. 17:1, 1975; Yanagi, Y., et al.: Xenobiotica 5:245, 1975.
112. Corbella, A., et al.: J. Chem. Soc. Chem. Commun. 721, 1973.
113. Keberle, H., et al.: Arch. Int. Pharmacodyn. 142:117, 1963; Keberle, H., et al.: Experientia 18:105, 1962.
114. Ferrandes, B., Eymark, P.: Epilepsia, 18:169, 1977; Kuhara, T., Matsumoto, J.: Biomed. Mass Spectrom. 1:291, 1974.
115. Maynert, E. W.: J. Pharmacol. Exp. Ther. 150:117, 1965.
116. Palmer, K. H., et al.: J. Pharmacol. Exp. Ther. 175:38, 1970; Holtzmann, J. L., Thompson, J. A.: Drug Metab. Dispos. 3:113, 1975.
117. Carroll, F. J., et al.: Drug Metab. Dispos. 5:343, 1977.
118. Thomas, R. C., Judy, R. W.: J. Med. Chem. 15:964, 1972.
119. Adams, S. S., Buckler, J. W.: Clin. Rheum. Dis. 5:359, 1979.
120. Ludwig, B. J., et al.: J. Med. Pharm. Chem. 3:53, 1961.
121. Horning, M. G., et al.: Drug Metab. Dispos. 1:569, 1973.
122. Dieterle, W., et al.: Arzneim. Forsch. 26:572, 1976.
123. McMahon, R. E., et al.: J. Pharmacol. Exp. Ther. 149:272, 1965.
124. Fuccella, L. M., et al.: J. Clin. Pharmacol. 13:68, 1973.
125a. Lin, D. C. K., et al.: Biomed. Mass Spectrom. 2:206, 1975; Wong, L. K., Biemann, K.: Clin. Toxicol. 9:583, 1976.
125b. Thomas, R. C., et al.: J. Pharm. Sci. 64:1360, 1366, 1975; Gottlieb, T. B., et al.: Clin. Pharmacol. Ther. 13:436, 1972.
126. Gorrod, J. W. (ed.): Biological Oxidation of Nitrogen. Amsterdam, Elsevier-North Holland, 1978.
127. Gorrod, J. W.: Chem. Biol. Interact. 7:289, 1973.
128. Ziegler, D. M., et al.: Drug Metab. Dispos. 1:314, 1973; Arch. Biochem. Biophys. 150:116, 1972.
129. Gram, T. E.: In Brodie, B. B., Gillette, J. R. (eds.). Concepts in Biochemical Pharmacology, Part 2, p. 334. Berlin, Springer-Verlag, 1971.
130. Brodie, B. B., et al.: Annu. Rev. Biochem. 27:427, 1958; McMahon, R. E.: J. Pharm. Sci. 55:457, 1966.

131. Nagy, A., Johansson, R.: Arch. Pharm. 290:145, 1975.
132. Gram, L. F., et al.: Psychopharmacologia 54:255, 1977.
133. Collinsworth, K. A., et al.: Circulation 50:1217, 1974; Narang, P. K., et al.: Clin. Pharmacol. Ther. 24:654, 1978.
134. Hutsell, T. C., Kraychy, S. J.: J. Chromatogr. 106:151, 1975; Heel, R. C., et al.: Drugs 15:331, 1978.
135. Adam, H. K., et al.: Biochem. Pharmacol. 27:145, 1979.
136. Chang, T. K., et al.: Res. Commun. Chem. Pathol. Pharmacol. 9:391, 1974; Glazko, A. J., et al.: Clin. Pharmacol. Ther. 16:1066, 1974.
137. Hammar, C. G., et al.: Anal. Biochem. 25:532, 1968; Beckett, A. H., et al.: J. Pharm. Pharmacol. 25:188, 1973.
138. Due, S. L., et al.: Biomed. Mass Spectrom. 3:217, 1976.
139. Beckett, A. H., et al.: J. Pharm. Pharmacol. 23:812, 1971.
140. Pohland, A., et al.: J. Med. Chem. 14:194, 1971.
141. Bruce, R. B., et al.: J. Med. Chem. 11:1031, 1968.
142. Szeto, H. H., Inturri, C. E.: J. Chromatogr. 125:503, 1976.
143. Misra, A. L.: In Adler, M. L., et al. (eds.). Factors Affecting the Action of Narcotics, p. 297. New York, Raven Press, 1978.
144. Kamm, J. J., et al.: J. Pharmacol. Exp. Ther. 182:507, 1972.
145. Kamm, J. J., et al.: J. Pharmacol. Exp. Ther. 184:729, 1973.
146. Goldberg, M. E., (ed.): Pharmacological and Biochemical Properties of Drug Substances, Vol. 1, p. 257, 311. 1977.
147. Gorrod, J. W., Jenner, P.: Essays Toxicol. 6:35, 1975; Beckett, A. H., Triggs, E. J.: Nature 211:1415, 1966.
148. Hucker, H. B., et al.: Drug Metab. Dispos. 2:406, 1974; Wold, J. S., Fischer, L. J.: J. Pharmacol. Exp. Ther. 183:188, 1972.
149. Kaiser, C., et al.: J. Med. Chem. 15:1146, 1972.
150. Bickel, M. H.: Pharmacol. Rev. 21:325, 1969.
151. Jenner, P.: In Gorrod, J. W. (ed.). Biological Oxidation of Nitrogen, p. 383. Amsterdam, Elsevier-North Holland, 1978.
152. Faurbye, A., et al.: Am. J. Psychiatr. 120:277, 1963; Theobald, W., et al.: Med. Pharmacol. Exp. 15:187, 1966.
153. Coutts, R., Beckett, A. H.: Drug Metab. Rev. 6:51, 1977.
154. Leinweber, F.-J., et al.: J. Pharm. Sci. 66:1570, 1977.
155. Beckett, A. H., Rowland, M.: J. Pharm. Pharmacol. 17:109S, 1965; Caldwell, J., et al.: Biochem. J. 129:11, 1972.
156. Wieber, J., et al.: Anaesthology 24:260, 1975; Chang, T., Glazko, A. J.: Anaesthology 36:401, 1972.
157. Tindell, G. L., et al.: Life Sci. 11:1029, 1972.
158. Franklin, R. B., et al.: Drug Metab. Dispos. 5:223, 1977.
159. Barlett, M. F., Egger, H. P.: Fed. Proc. 31:537, 1972.
160. Beckett, A. H., Gibson, G. G.: Xenobiotica 8:73, 1978.
161. Tipton, K. F., et al.: In Wolsterholme, G. E. W., Knight, J. (eds.). Monoamine Oxidase and Its Inhibition, p. 5. Amsterdam, Elsevier-North Holland, 1976.
162. Beckett, A. H., Brookes, L. G.: J. Pharm. Pharmacol. 23:288, 1971.
163. Charalampous, K. D., et al.: J. Pharmacol. Exp. Ther. 145:242, 1964; Psychopharmacologia 9:48, 1966.
164. Au, W. Y. W., et al.: Biochem. J. 129:110, 1972.
165. Parli, C. J., et al.: Biochem. Biophys. Res. Comm. 43:1204, 1971; Lindeke, B., et al.: Acta Pharm. Suecica 10:493, 1973.
166. Hucker, H. B.: Drug Metab. Dispos. 1:332, 1973; Parli, C. H., et al.: Drug Metab. Dispos. 3:337, 1973.
167. Wright, J., et al.: Xenobiotica 7:257, 1977; Gal, J., et al.: Res. Commun. Chem. Pathol. Pharmacol. 15:525, 1976.
168. Beckett, A. H., Bélanger, P. M.: J. Pharm. Pharmacol. 26:205, 1974.
169. Bélanger, P. M., Grech-Bélanger, O.: Can. J. Pharm. Sci. 12:99, 1977.
170. Weisburger, J. H., Weisburger, E. K.: Pharmacol. Rev. 25:1, 1973; Miller, E. C., Miller, J. A.: Pharmacol. Rev. 18:805, 1966.
171. Weisburger, J. H., Weisburger, E. K.: In Brodie, B. B., Gillette, J. R. (eds.). Concepts in Biochemical Pharmacology, Part 2, p. 312. Berlin, Springer-Verlag, 1971.
172. Uehleke, H.: Xenobiotica 1:327, 1971.
173. Beckett, A. H., Bélanger, P. M.: Biochem. Pharmacol. 25:211, 1976.
174. Israili, Z. H., et al.: J. Pharmacol. Exp. Ther. 187:138, 1973.
175. Kiese, M.: Pharmacol. Rev. 18:1091, 1966.
176. Lin, J.-K., et al.: Cancer Res. 35:844, 1975; Poirer, L. A., et al.: Cancer Res. 27:1600, 1967.
177. Lin, J.-K., et al.: Biochemistry 7:1889, 1968; Biochemistry 8:1573, 1969.
178. Schwartz, D. E., et al.: Arzneim. Forsch. 20:1867, 1970.
179. Dagne, E., Castagnoli, N., Jr.: J. Med. Chem. 15:840, 1972.
180. Ericsson, O., Danielsson, B.: Drug Metab. Dispos. 5:497, 1977.
181. Garrattini, S., et al.: Drug Metab. Rev. 1:291, 1972.
182. Brotherton, P. M., et al.: Clin. Pharmacol. Ther. 10:505, 1969.
183. Langone, J. J., et al.: Biochemistry 12:5025, 1973.
184. Grochow, L. B., Colvin, M.: Clin. Pharmacokinet. 4:380, 1979.
185. Colvin, M., et al.: Cancer Res. 33:915, 1973; Connors, T. A., et al.: Biochem. Pharmacol. 23:115, 1974.
186. Irving, C. C.: In Fishman, W. H. (ed.). Metabolic Conjugation and Metabolic Hydrolysis, Vol. 1, p. 53. New York, Academic Press, 1970.
187. Miller, J., Miller, E. C.: In Jollow, D. J., et al. (eds.). Biological Reactive Intermediates, p. 6. New York, Plenum Press, 1970.
188. Jollow, D. J., et al.: J. Pharmacol. Exp. Ther. 187:195, 1973; Prescott, L. F., et al.: Lancet 1:519, 1971.
189. Hinson, J. A.: Rev. Biochem. Toxicol. 2:103, 1980.
190. Hinson, J. A., et al.: Life Sci. 24:2133, 1979; Nelson, S. D., et al.: Biochem. Pharmacol. 29:1617, 1980.
191. Potter, W. Z., et al.: J. Pharmacol. Exp. Ther. 187:203, 1973.
192. Adler, T. K., et al.: J. Pharmacol. Exp. Ther. 114:251, 1955.
193. Brodie, B. B., Axelrod, J.: J. Pharmacol. Exp. Ther. 97:58, 1949.
194. Duggan, D. E., et al.: J. Pharmacol. Exp. Ther. 181:563, 1972; Kwan, K. C., et al.: J. Pharmacokinet. Biopharm. 4:255, 1976.
195. Brogden, R. N., et al.: Drugs 14:163, 1977; Taylor, J. A., et al.: Xenobiotica 7:357, 1977.
196. Daly, J., et al.: Ann. N.Y. Acad. Sci. 96:37, 1962.
197. Mazel, P., et al.: J. Pharmacol. Exp. Ther. 143:1, 1964.
198. Sarcione, E. J., Stutzman, L.: Cancer Res. 20:387, 1960; Elion, G. B., et al.: Proc. Am. Assoc. Cancer Res. 3:316, 1962.
199. Taylor, J. A.: Xenobiotica 3:151, 1973.
200. Brodie, B. B., et al.: J. Pharmacol. Exp. Ther. 98:85, 1950; Spector, E., Shideman, F. E.: Biochem. Pharmacol. 2:182, 1959.
201. Neal, R. A.: Arch. Intern. Med. 128:118, 1971; Neal, R. A.: Biochem. J. 103:183, 1967.
202. Neal, R. A.: Rev. Biochem. Toxicol. 2:131, 1980.
203. Gruenke, L., et al.: Res. Commun. Chem. Pathol. Pharmacol. 10:221, 1975; Zehnder, K., et al.: Biochem. Pharmacol. 11:535, 1962.
204. Aguilar, S. J.: Dis. Nerv. Syst. 36:484, 1975.
205. Taylor, D. C., et al.: Drug Metab. Dispos. 6:21, 1978.
206. Taylor, D. C.: In Wood, C. J., Simkins, M. A. (eds.). International Symposium on Histamine H_2-Receptor Antagonists, p. 45. Welwyn Garden City, U.K., Smith, Kline & French, 1973.
207. Crew, M. C., et al.: Xenobiotica 2:431, 1972.
208. Hucker, H. B., et al.: J. Pharmacol. Exp. Ther. 154:176, 1966; 155:309, 1967.
209. Hathway, D. E., (Sr. Reporter): Foreign Compound Metabolism in Mammals, Vol. 3, p. 512. London, Chemical Society, 1975.
210. Schwartz, M. A., Kolis, S. J.: Drug Metab. Dispos. 1:322, 1973.
211. Sisenwine, S. F., et al.: Drug Metab. Dispos. 3:180, 1975.
212. McCallum, N. K.: Experientia 31:957, 1975; 31:520, 1975.
213. Cohen, E. N., Van Dyke, R. A.: Metabolism of Volatile

Anesthetics. Reading, MA, Addison-Wesley, 1977; Cohen, E. N., et al.: Anesthesiology 43:392, 1975.
214. Van Dyke, R. A., et al.: Drug Metab. Dispos. 3:51, 1975; 4:40, 1976.
215. Pohl, L.: Rev. Biochem. Toxicol. 1:79, 1979.
216. Pohl, L., et al.: Biochem. Pharmacol. 27:335, 1978; 27:491, 1978.
217. Parke, D. V.: The Biochemistry of Foreign Compounds, p. 218. Oxford, Pergamon Press, 1968.
218. Gillette, J. R.: In Brodie, B. B., Gillette, J. R. (eds.). Concepts in Biochemical Pharmacology, Part 2, p. 349. Berlin, Springer-Verlag, 1971.
219. Bachur, N. R.: Science 193:595, 1976.
220. Sellers, E. M., et al.: Clin. Pharmacol. Ther. 13:37, 1972.
221. Pritchard, J. F., et al.: J. Chromatogr. 162:47, 1979.
222. Osterloh, J. D., et al.: Drug Metab. Dispos. 8:12, 1980.
223. Jenner, P., Testa, B.: Drug Metab. Rev. 2:117, 1973.
224. Culp, H. W., McMahon, R. E.: J. Biol. Chem. 243:848, 1968.
225. McMahon, R. E., et al.: J. Pharmacol. Exp. Ther. 149:272, 1965; Galloway, J. A., et al.: Diabetes 16:118, 1967.
226. Yü, T. F., et al.: Metabolism 17:309, 1968.
227. DiCarlo, F. J., et al.: Clin. Pharmacol. Ther. 22:858, 1977.
228. Bachur, N. R., Gee, M.: J. Pharmacol. Exp. Ther. 177:567, 1971; Huffmann, D. H., et al.: Clin. Pharmacol. Ther. 13:895, 1972.
229. Testa, B., Beckett, A. H.: J. Pharm. Pharmacol. 25:119, 1973; Testa, B.: Acta Pharm. Suecica 10:441, 1973.
230. Billing, R. E., et al.: Xenobiotica 10:33, 1980.
231. Pollock, S. H., Blum, K.: In Blum, K., et al. (eds.). Alcohol and Opiates, p. 359. New York, Academic Press, 1977; Chatterjie, N., et al.: Drug Metab. Dispos. 2:401, 1974.
232. Sullivan, H. R., Due, S. L.: J. Med. Chem. 16:909, 1973.
233. Hollands, T. R., Johnson, W. J.: Biochem. Med. 7:288, 1973.
234. Chan, K. K., et al.: J. Med. Chem. 15:1265, 1972.
235. Dayton, H., Inturrisi, C. E.: Drug Metab. Dispos. 4:474, 1974.
236. Roerig, S., et al.: Drug Metab. Dispos. 4:53, 1976; Malsperis, L., et al.: Res. Commun. Chem. Pathol. Pharmacol. 14:393, 1976.
237. Bachur, N. R., Felsted, R. L.: Drug Metab. Dispos. 4:239, 1976; Crew, M. C., et al.: Clin. Pharmacol. Ther. 14:1013, 1973.
238. DiCarlo, F. J., et al.: Xenobiotica 2:159, 1972; Crew, M. C., et al.: Xenobiotica 2:431, 1972; DiCarlo, F. J., et al.: J. Reticuloendothel. Soc. 14:387, 1973.
239. Gerhards, E., et al.: Acta Endocrinol. 68:219, 1971.
240. Wright, J., et al.: Xenobiotica 7:257, 1977.
241. Kawai, K., Baba, S.: Chem. Pharm. Bull. (Jpn.) 24:2728, 1976.
242. Gillette, J. R., et al.: Mol. Pharmacol. 4:541, 1968; Fouts, J. R., Brodie, B. B.: J. Pharmacol. Exp. Ther. 119:197, 1957.
243. Hernandez, P. H., et al.: Biochem. Pharmacol. 16:1877, 1967.
244. Scheline, R. R.: Pharmacol. Rev. 25:451, 1973; Walker, R.: Food Cosmet. Toxicol. 8:659, 1970.
245. Min, B. H., Garland, W. A.: J. Chromatogr. 139:121, 1977.
246. Rieder, J., Wendt, G.: In Garattini, S., et al. (eds.). The Benzodiazepines, p. 99. New York, Raven Press, 1973.
247. Conklin, J. D., et al.: J. Pharm. Sci. 62:1024, 1973; Cox, P. L., et al.: J. Pharm. Sci. 58:987, 1969.
248. Stambaugh, J. E., et al.: J. Pharmacol. Exp. Ther. 161:373, 1968.
249. Mitchard, M.: Xenobiotica 1:469, 1971.
250. Glazko, A. J., et al.: J. Pharmacol. Exp. Ther. 96:445, 1949; Smith, G. N., Worrel, G. S.: Arch. Biochem. Biophys. 24:216, 1949.
251. Tréfouël, J., et al.: C. R. Seances Soc. Biol., Paris 190:756, 1935.
252. Jones, R., et al.: Food Cosmet. Toxicol. 2:447, 1966; 4:419, 1966.
253. Ikeda, M., Uesugi, T.: Biochem. Pharmacol. 22:2743, 1973.
254. Peppercorn, M. A., Goldman, P.: J. Pharmacol. Exp. Ther. 181:555, 1972; Das, E. M.: Scand. J. Gastroenterol. 9:137, 1974.
255. Schröder, H., Gustafsson, B. E.: Xenobiotica 3:225, 1973.
256. Bickel, M. H., Gigon, P. L.: Xenobiotica 1:631, 1971.
257. Eldjarn, L.: Scand. J. Clin. Lab. Invest. 2:202, 1950; Staub, H.: Helv. Physiol. Acta 13:141, 1955.
258. Duggan, D. E., et al.: Clin. Pharmacol. Ther. 21:326, 1977; Duggan, D. E., et al.: J. Pharmacol. Exp. Ther. 20:8, 1977.
259. Kolb, H. K., et al.: Arzneim. Forsch. 15:1292, 1965.
260. LaDu, B. N., Snady, H.: In Brodie, B. B., Gillette, J. R. (eds.). Concepts in Biochemical Pharmacology, Part 2, p. 477. Berlin, Springer-Verlag, 1971.
261. Junge, W., Krisch, K.: CRC Crit. Rev. Toxicol. 3:371, 1975.
262. Davison, C.: Ann. N.Y. Acad. Sci. 179:249, 1971.
263. Kogan, M. J., et al.: Anal. Chem. 49:1965, 1977.
264. Inaba, T., et al.: Clin. Pharmacol. Ther. 23:547, 1978; Stewart, D. J., et al.: Life Sci. 1557, 1977.
265. Wells, R., et al.: Clin. Chem. 20:440, 1974.
266. Rubens, R., et al.: Arzneim. Forsch. 22:256, 1972.
267. Gugler, L., Jensen, C.: J. Chromatogr. 117:175, 1976.
268. Houin, G., et al.: Eur. J. Clin. Pharmacol. 8:433, 1975.
269. Sinkula, A. A., et al.: J. Pharm. Sci. 62:1106, 1973; Martin, A. R.: In Wilson, C. O., et al. (eds.). Textbook of Organic Medicinal and Pharmaceutical Chemistry, 7th ed., p. 304. Philadelphia, J. B. Lippincott, 1977.
270. Knirsch, A. K., et al.: J. Infect. Dis. 127:S105, 1973.
271. Stella, V.: In Higuchi, T., Stella, V. (eds.): Prodrugs as Novel Drug Delivery Systems, p. 1. Washington, DC, American Chemical Society, 1975.
272. Mark, L. C., et al.: J. Pharmacol. Exp. Ther. 102:5, 1951.
273. Nelson, S. D., et al.: J. Pharm Sci. 66:1180, 1977.
274. Tsukamoto, H., et al.: Pharm. Bull. (Tokyo) 3:459, 1955; 3:397, 1955.
275. Dudley, K. H., et al.: Drug Metab. Dispos. 6:133, 1978; 2:103, 1974.
276. Dutton, G. J., et al.: In Parke, D. V., Smith, R. L. (eds.). Drug Metabolism: From Microbe to Man, p. 71. London, Taylor & Francis, 1977.
277. Dutton, G. J.: In Dutton, G. J. (ed.). Glucuronic Acid, Free and Combined, p. 186. New York, Academic Press, 1966; Dutton, G. D., et al.: Progr. Drug Metab. 1:2, 1977.
278. Brunk, S. F., Delle, M.: Clin. Pharmacol. Exp. Ther. 16:51, 1974; Berkowitz, B. A., et al.: Clin. Pharmacol. Exp. Ther. 17:629, 1975.
279. Andrews, R. S., et al.: J. Int. Med. Res. 4(Suppl. 4):34, 1976.
280. Thies, R. L., Fischer, L. J.: Clin. Chem. 24:778, 1978.
281. Walle, T., et al.: Fed. Proc. 35:665, 1976; Walle, T., et al.: Clin. Pharmacol. Ther. 26:167, 1979.
282. Roseman, S., et al.: J. Am. Chem. Soc. 76:1650, 1954.
283. Segre, E. J.: J. Clin. Pharmacol. 15:316, 1975.
284. Rubin, A., et al.: J. Pharm. Sci. 61:739, 1972; Rubin, A., et al.: J. Pharmacol. Exp. Ther. 183:449, 1972.
285. Bridges, J. W., et al.: Biochem. J. 118:47, 1970.
286. Gibson, T., et al.: Br. J. Clin. Pharmacol. 2:233, 1975; Tsuichiya, T., Levy, G.: J. Pharm. Sci. 61:800, 1972.
287. Woolhouse, N. M., et al.: Xenobiotica 3:511, 1973.
288. Bickel, M. H., et al.: Experientia, 29:960, 1973.
289. Tskamoto, H., et al.: Chem. Pharm. Bull. (Tokyo) 11:421, 1963.
290. Uno, T., Kono, M.: J. Pharm. Soc. (Jpn.) 82:1660, 1962.
291. Porter, C. C., et al.: Drug Metab. Dispos. 3:189, 1975.
292. Chaundhuri, N. K., et al.: Drug Metab. Dispos. 4:372, 1976.
293. Sitar, D. S., Thornhill, D. P.: J. Pharmacol. Exp. Ther. 184:432, 1973.
294. Lindsay, R. H., et al.: Pharmacologist 18:113, 1976; Sitar, D. S., Thornhill, D. P.: J. Pharmacol. Exp. Ther. 183:440, 1972.
295. Dutton, G. J., Illing, H. P. A.: Biochem. J. 129:539, 1972.
296. Dieterle, W., et al.: Arzneim. Forsch. 26:572, 1976; Richter, J. W., et al.: Helv. Chim. Acta 58:2512, 1975.
297. Dieterle, W., et al.: Eur. J. Clin. Pharmacol. 9:135, 1975.
298. Schmid, R., Lester, R.: In Dutton, G. J. (ed.). Glucuronic

Acid, Free and Combined, p. 493. New York, Academic Press, 1966.
299. Hadd, H. E., Blickenstaff, R. T.: Conjugates of Steroid Hormones. New York, Academic Press, 1969.
300. Smith, R. L.: The Excretory Function of Bile: The Elimination of Drugs and Toxic Substance in Bile. London, Chapman & Hall, 1973.
301. Stern, L., et al.: Am. J. Dis. Child. 120:26, 1970.
302. Weiss, C. F., et al.: N. Eng. J. Med. 262:787, 1960.
303. Dodgson, K. S.: In Parke, D. V., Smith, R. L. (eds.). Drug Metabolism: From Microbe to Man, p. 91. London, Taylor & Francis, 1977; Roy, A. B.: In Brodie, B. B., Gillette, J. R. (eds.). Concepts in Biochemical Pharmacology, Part 2, p. 536. Berlin, Springer-Verlag, 1971.
304. Williams, R. T.: In Bernfeld, P. (ed.). Biogenesis of Natural Products, 2nd ed., p. 611. Oxford, Pergamon Press, 1967.
305. Roy, A. B.: Adv. Enzymol. 22:205, 1960.
306. Kwan, K. C., et al.: J. Pharmacol. Exp. Ther. 198:264, 1976; Stenback, O., et al.: Eur. J. Clin. Pharmacol. 12:117, 1977.
307. Lin, C., et al.: Drug Metab. Dispos. 5:234, 1977.
308. Nilsson, H. T., et al.: Xenobiotica 2:363, 1972.
309. Miller, R. P., et al.: Clin. Pharmacol. Ther. 19:284, 1976; Levy, G., et al.: Pediatrics 55:818, 1975.
310. James, S. P., Waring, R. H.: Xenobiotica 1:572, 1971; Bostrum, H., Vestermark, A.: Acta Physiol. Scand. 48:88, 1960.
311. Boyland, E., et al.: Biochem. J. 65:417, 1957; Roy, A. B.: Biochem. J. 74:49, 1960.
312. Irving, C. C.: In Gorrod, J. W. (ed.). Biological Oxidation of Nitrogen, p. 325. Amsterdam, Elsevier-North Holland, 1978.
313. Irving, C. C.: Cancer Res. 35:2959, 1975; Jackson, C. D., Irving, C. C.: Cancer Res. 32:1590, 1972.
314. Hinson, J. A., Mitchell, J. R.: Drug Metab. Dispos. 4:430, 1975.
315. Mulder, G. J., et al.: Biochem. Pharmacol. 26:189, 1977.
316. Weber, W.: In Brodie, B. B., Gillette, J. R. (eds.). Concepts in Biochemical Pharmacology, Part 2, p. 564. Berlin, Springer-Verlag, 1971.
317. Williams, R. T., Millburn, P.: In Blaschko, H. K. F. (ed.). MTP International Review of Science, Biochemistry Series One; Vol. 12, Physiological and Pharmacological Biochemistry, p. 211. Baltimore, University Park Press, 1975.
318. James, M. O., et al.: Proc. R. Soc., Ser. B 182:25, 1972.
319. Bridges, J. W., et al.: Biochem. J. 118:47, 1970.
320. Wan, S. H., Riegelman, S.: J. Pharm. Sci. 61:1284, 1972; VonLehmann, B., et al.: J. Pharm. Sci. 62:1483, 1973.
321. Braun, G. A., et al.: Eur. J. Pharmacol. 1:58, 1967.
322. Tatsumi, K., et al.: Biochem. Pharmacol. 16:1941, 1967.
323. Weber, W. W., Hein, D. W.: Clin. Pharmacokinet. 4:401, 1979.
324. Smith, R. L., Caldwell, J.: In Parke, D. V., Smith, R. L. (eds.). Drug Metabolism: From Microbe to Man, p. 331. London, Taylor & Francis, 1977.
325. Williams, R. T.: Clin. Pharmacol. Ther. 4:234, 1963.
326. Drach, J. C., et al.: Proc. Soc. Exp. Biol. Med. 135:849, 1970.
327. Caldwell, J.: In Jenner, P., Testa, B. (eds.). Concepts in Drug Metabolism, Part A, p. 211. New York, Marcel Dekker, 1980.
328a. Jerina, D. M., Bend, J. R.: In Jollow, D. J., et al. (eds.). Biological Reactive Intermediates, p. 207. New York, Plenum Press, 1977.
328b. Chasseaud, L. F.: In Arias, I. M., Jakoby, W. B. (eds.). Glutathione: Metabolism and Function, p. 77. New York, Raven Press, 1976.
328c. Ketterer, B.: Mutat. Res. 202:343, 1988.
329a. Mantle, T. J., Pickett, C. B., Hayes, J. D. (eds.): Glutathione S-Transferases and Carcinogenesis. London, Taylor & Francis, 1987.
329b. Foureman, G. L., Reed, D. J.: Biochemistry 26:2028, 1987; Rannung, U., et al.: Chem. Biol. Interact. 20:1, 1978; Olson, W. A., et al.: J. Natl. Cancer Instit. 51:1993, 1973.
329c. Monks, T. J., Lau, S. S.: Toxicology 52:1, 1988.
330. Weisburger, E. K.: Annu. Rev. Pharmacol. Toxicol. 18:395, 1978.
331. Hollingworth, R. M., et al.: Life Sci. 13:191, 1973; Benke, G. M., Murphy, S. D.: Toxicol. Appl. Pharmacol. 31:254, 1975.
332. Bray, H. G., et al.: Biochem. J. 67:607, 1957.
333. Chalmers, A. H.: Biochem. Pharmacol. 23:1891, 1974; de Miranda, P., et al.: J. Pharmacol. Exp. Ther. 187:588, 1973; 195:50, 1975.
334. Elion, G. B.: Fed. Proc. 26:898, 1967.
335. Jerina, D. M.: In Arias, I. M., Jakoby, W. B. (eds.). Glutathione: Metabolism and Function, p. 267. New York, Raven Press, 1976.
236. Needleman, P.: In Needleman, P. (ed.). Organic Nitrates, p. 57. Berlin, Springer-Verlag, 1975.
337. Klaasen, C. D., Fitzgerald, T. J.: J. Pharmacol. Exp. Ther. 191:548, 1974.
338. Boyland, E., Chasseaud, L. F.: Biochem. J. 109:651, 1968.
339. Kuss, E., et al.: Hoppe-Seylers Z. Physiol. Chem. 352:817, 1971; Nelson, S. D., et al.: Biochem. Biophys. Res. Commun. 70:1157, 1976.
340. Weber, W.: In Fishman, W. H. (ed.). Metabolic Conjugation and Metabolic Hydrolysis, Vol. 3, p. 250. New York, Academic Press, 1973; Williams, R. T.: Fed. Proc. 26:1029, 1967.
341. Elson, J., et al.: Clin. Pharmacol. Ther. 17:134, 1975; Drayer, D. E., et al.: Proc. Soc. Exp. Biol. Med. 146:358, 1974.
342. Mitchell, J. R., et al.: Ann. Intern. Med. 84:181, 1976; Nelson, S. D., et al.: Science 193:193, 1976.
343. Giardina, E. G., et al.: Clin. Pharmacol. Ther. 19:339, 1976; 17:722, 1975.
344. Peters, J. H., Levy, L.: Ann. N.Y. Acad. Sci. 179:660, 1971.
345. Reimerdes, E., Thumim, J. H.: Arzneim. Forsch. 20:1171, 1970.
346. Vree, T. B., et al.: In Merkus, F. W. H. M. (ed.). The Serum Concentration of Drugs, p. 205. Amsterdam, Excerpta Medica, 1980.
347. Garrett, E. R.: Int. J. Clin. Pharmacol. 16:155, 1978.
348. Reidenberg, M. W., et al.: Clin. Pharmacol. Ther. 14:970, 1973; Israili, Z. H., et al.: Drug Metab. Rev. 6:283, 1977.
349. Evans, D. A. P., et al.: Clin. Pharmacol. Ther. 6:430, 1965.
350. Tabor, H., et al.: J. Biol. Chem. 204:127, 1953.
351. Sullivan, H. R., Due, S. L.: J. Med. Chem. 16:909, 1973; Sullivan, H. R., et al.: J. Am. Chem. Soc. 94:4050, 1972; Life Sci. 11:1093, 1972.
352. Drayer, D. E., Reidenberg, M. M.: Clin. Pharmacol. Ther. 22:251, 1977.
353. Lunde, P. K. M., et al.: Clin. Pharmacokinet. 2:182, 1977.
354. Kalow, W.: Pharmacogenetics: Heredity and the Response to Drugs. Philadelphia, W. B. Saunders, 1962.
355. Kutt, H., et al.: Am. Rev. Resp. Dis. 101:377, 1970.
356. Alarcon-Segovia,D.: Drugs 12:69, 1976; Reidenberg, M. M., Martin, J. H.: Drug Metab. Dispos. 2:71, 1974.
357. Strong, J. M., et al.: J. Pharmacokinet. Biopharm. 3:233, 1975.
358. Axelrod, J.: In Brodie, B. B., Gillette, J. R. (eds.). Concepts in Biochemical Pharmacology, Part 2, p. 609. Berlin, Springer-Verlag, 1971.
359. Mudd, S. H.: In Fishman, W. H. (ed.). Metabolic Conjugation and Metabolic Hydrolysis, Vol. 3, p. 297. New York, Academic Press.
360. Young, J. A., Edwards, K. D. G.: Med. Res. 1:53, 1962; J. Pharmacol. Exp. Ther. 145:102, 1964.
361. Shindo, H., et al.: Chem. Pharm. Bull. (Tokyo) 21:826, 1973.
362. Morgan, C. D., et al.: Biochem. J. 114:8P, 1969.
363. Weber, R., Tuttle, R. R.: In Goldberg, M. E. (ed.). Pharmacological and Biochemical Properties of Drug Substances, Vol. 1, p. 109. Washington, DC, American Pharmaceutical Association, 1977.
364. Glazko, A. J.: Drug Metab. Dispos. 1:711, 1973.
365. Börner, U., Abbott, S.: Experientia 29:180, 1973.

366. Bleidner, W. E., et al.: J. Pharmacol. Exp. Ther. 150:484, 1965.
367. Komori, Y., Sendju, Y.: J. Biochem. 6:163, 1926.
368. Lindsay, R. H., et al.: Biochem. Pharmacol. 24:463, 1975.
369. Bremer, J., Greenberg, D. M.: Biochim. Biophys. Acta 46:217, 1961.
370. Allan, P. W., et al.: Biochim. Biophys. Acta 114:647, 1966; Elion, G. B.: Fed. Proc. 26:898, 1967.
371. Testa, B., Jenner, P.: Drug Metabolism: Chemical and Biochemical Aspects, p. 329–418. New York, Marcel Dekker, 1976.
372. Ward, R. M., et al.: In Avery, G. S. (ed.). Drug Treatment, 2nd ed., p. 76. Sydney, ADIS Press, 1980.
373. Morselli, P. L.: Drug Disposition During Development. New York, Spectrum, 1977.
374. Jondorf, W. R., et al.: Biochem. Pharmacol. 1:352, 1958.
375. Nitowsky, H. M., et al.: J. Pediatr. 69:1139, 1966.
376. Crooks, J., et al.: Clin. Pharmacokinet. 1:280, 1976; Crooks, J., Stevenson, I. H. (eds.): Drugs and the Elderly. London, Macmillan, 1979.
377. Williams, R. T.: Ann. N.Y. Acad. Sci. 179:141, 1971.
378. Williams, R. T.: In LaDu, B. N., et al. (eds.). Fundamentals of Drug Metabolism and Disposition, p. 187. Baltimore, Williams & Wilkins, 1971.
379. Williams, R. T., et al.: In Snyder, S. H., Usdin, E. (eds.). Frontiers in Catecholamine Research, p. 927. New York, Pergamon Press, 1973.
380. Butler, T. C., et al.: J. Pharmacol. Exp. Ther. 199:82, 1976.
381. Williams, R. T.: Biochem. Soc. Trans. 2:359, 1974.
382. Bridges, J. W., et al.: Biochem. J. 118:47, 1970.
383. Williams, R. T.: Fed. Proc. 26:1029, 1967.
384. Cram, R. L., et al.: Proc. Soc. Exp. Biol. Med. 118:872, 1965.
385. Kutt, H., et al.: Neurology 14:542, 1962.
386. Vesell, E. S.: Progr. Med. Genet. 9:291, 1973.
387. Kato, R.: Drug Metab. Rev. 3:1, 1974.
388. Beckett, A. H., et al.: J. Pharm. Pharmacol. 23:62S, 1971; Menguy, R., et al.: Nature 239:102, 1972.
389. Conney, A. H.: Pharmacol. Rev. 19:317, 1967; Snyder, R., Remmer, H.: Pharmacol. Ther. 7:203, 1979.
390. Parke, D. V.: In Parke, D. V. (ed.). Enzyme Induction, p. 207. London, Plenum Press, 1975.
391. Estabrook, R. W., Lindenlaub, E. (eds.): The Induction of Drug Metabolism. Stuttgart, Schattauer Verlag, 1979.
392. Hansten, P. D.: Drug Interactions, 4th ed., p. 38. Philadelphia, Lea & Febiger, 1979.
393. Laenger, H., Detering, K.: Lancet 600, 1974.
394. Skolnick, J. L., et al.: JAMA 236:1382, 1976.
395. Dent, C. E., et al.: Br. Med. J. 4:69, 1970.
396. Yeung, C. Y., Field, C. E.: Lancet 135, 1969.
397. Jenne, J., et al.: Life Sci. 17:195, 1975.
398. Pantuck, E. J., et al.: Science 175:1248, 1972; Clin. Pharmacol. Ther. 14:259, 1973; Vaughan, D. P., et al.: Br. J. Clin. Pharmacol. 3:279, 1976.
399. Kolmodin, B., et al.: Clin. Pharmacol. Ther. 10:638, 1969.
400. Jeffrey, W. H., et al.: JAMA 236:2881, 1976.
401. Gelboin, H. V., TS'o P. O. P. (eds.): Polycyclic Hydrocarbons and Cancer: Environment, Chemistry, Molecular and Cell Biology. New York, Academic Press, 1978.
402. Vesell, E. S., Passananti, G. T.: Drug Metab. Dispos. 1:402, 1973; Anders, M. W.: Annu. Rev. Pharmacol. 11:37, 1971.
403. Mannering, G. J.: In Brodie, B. B., Gillette, J. R. (eds.). Concepts in Biochemical Pharmacology, Part 2, p. 452. Berlin, Springer-Verlag, 1971.
404. Hewick, D., McEwen, J.: J. Pharm. Pharmacol. 25:458, 1973.
405. Casy, A. F.: In Burger, A. (ed.). Medicinal Chemistry, 3rd ed., Part I, p. 81. New York, Wiley-Interscience, 1970.
406. George, C. F., et al.: Eur. J. Clin. Pharmacol. 4:74, 1972.
407. Breimer, D. D., Van Rossum, J. M.: J. Pharm. Pharmacol. 25:762, 1973.
408. Low, L. K., Castagnoli, N., Jr.: Annu. Rep. Med. Chem. 13:304, 1978.
409. Testa, B., Jenner, P.: J. Pharm. Pharmacol. 28:731, 1976.
410. Belpaire, F. M., et al.: Xenobiotica 5:413, 1975.
411. Drayer, D. E.: US Pharmacist (Hosp. Ed.) 5:H15, 1980.
412. Pierides, A. M., et al.: Lancet 2:1279, 1975; Gabriel, R., Pearce, J. M. S.: Lancet 2:906, 1976.

SELECTED READINGS

Aitio, A. (ed.): Conjugation Reactions in Drug Biotransformation. Amsterdam, Elsevier, 1978.

Anders, M. W. (ed.): Bioactivation of Foreign Compounds. New York, Academic Press, 1985.

Benford, D. J., Bridges, J. W., Gibson, G. G. (eds.): Drug Metabolism—From Molecules to Man. London, Taylor & Francis, 1987.

Brodie, B. B., Gillette, J. R. (eds.): Concepts in Biochemical Pharmacology, Part 2. Berlin, Springer-Verlag, 1971.

Caldwell, J.: Conjugation reactions in foreign compound metabolism. Drug Metab. Rev. 13:745, 1982.

Caldwell, J., Jakoby, W. B. (eds.): Biological Basis of Detoxification. New York, Academic Press, 1983.

Caldwell, J., Paulson, G. D. (eds.): Foreign Compound Metabolism. London, Taylor & Francis, 1984.

Creasey, W. A.: Drug Disposition in Humans. New York, Oxford University Press, 1979.

DeMatteis, F., Lock, E. A. (eds.): Selectivity and Molecular Mechanisms of Toxicity. London, Macmillan, 1987.

Dipple, A., Micheijda, C. J., Weisburger, E. K.: Metabolism of chemical carcinogens. Pharmacol. Ther. 27:265, 1985.

Drayer, D. E.: Pharmacologically active metabolites of drugs and other foreign compounds. Drugs 24:519, 1982.

Dutton, G. J.: Glucuronidation of Drugs and Other Compounds. Boca Raton, FL, CRC Press, 1980.

Estabrook, R. W., Lindenlaub, E. (eds.): Induction of Drug Metabolism. Stuttgart, Schattauer Verlag, 1979.

Gibson, G. G., Skett, P.: Introduction to Drug Metabolism. London, Chapman & Hall, 1986.

Gorrod, J. W. (ed.): Drug Toxicity. London, Taylor & Francis, 1979.

Gorrod, J. W., Damani, L. A. (eds.): Biological Oxidation of Nitrogen in Organic Molecules. Chichester, Engl., Ellis Horwood, 1985.

Gorrod, J. W., Oelschlager, H., Caldwell, J. (eds.): Metabolism of Xenobiotics. London, Taylor & Francis, 1988.

Gram, T. E. (ed.): Extrahepatic Metabolism of Drugs and Other Foreign Compounds. New York, SP Medical and Scientific, 1980.

Guengerich, F. P. (ed.): Mammalian Cytochromes P-450, Vols. 1 and 2. Boca Raton, FL, CRC Press, 1987.

Hathway, D. E.: Mechanisms of Chemical Carcinogenesis. London, Butterworths, 1986.

Hodgson, E., Levi, P. E. (eds.): A Textbook of Modern Toxicology. New York, Elsevier, 1987.

Jakoby, W. B. (ed.): Enzymatic Basis of Detoxication, Vols. 1 and 2. New York, Academic Press, 1980.

Jakoby, W. B. (ed.): Detoxification and drug metabolism. Methods Enzymol., Vol. 77. New York, Academic Press, 1981.

Jakoby, W. B., Bend, J. R., Caldwell, J. (eds.): Metabolic Basis of Detoxification: Metabolism of Functional Groups. New York, Academic Press, 1982.

Jenner, P., Testa, B.: The influence of stereochemical factors on drug disposition. Drug Metab. Rev. 2:117, 1973.

Jenner, P., Testa, B. (eds.): Concepts in Drug Metabolism, Parts A and B. New York, Marcel Dekker, 1980, 1981.

Jerina, D. M. (ed.): Drug Metabolism Concepts. Washington, DC, American Chemical Society, 1977.

Jollow, D. J., Kocsis, J. J., Snyder, R., Vainio, H. (eds.): Biological Reactive Intermediates: Formation, Toxicity and Inactivation. New York, Plenum Press, 1977.

Klaassen, C. D., Amdur, M. O., Doull, J. (eds.): Casarett and Doull's Toxicology: The Basic Science of Poisons, 3rd ed. New York, Macmillan, 1986.

La Du, B. N., Mandel, H. G., Way, E. L. (eds.): Fundamentals of

Drug Metabolism and Drug Disposition. Baltimore, Williams & Wilkins, 1971.

Low, L. K., Castagnoli, N.: Drug biotransformations. In Wolff, M. E. (ed.). Burger's Medicinal Chemistry, 4th ed., Part 1, p. 107. New York, Interscience-Wiley, 1980.

Mitchell, J. R., Horning, M. G. (eds.): Drug Metabolism and Drug Toxicity. New York, Raven Press, 1984.

Monks, T. J., Lau, S. S.: Reactive intermediates and their toxicological significance. Toxicology 52:1, 1988.

Mulder, G. J. (ed.): Sulfate Metabolism and Sulfate Conjugation. London, Taylor & Francis, 1985.

Nelson, S. D.: Chemical and biological factors influencing drug biotransformation. In Wolff, M. E. (ed.). Burger's Medicinal Chemistry, 4th ed., Part 1, p. 227. New York, Interscience-Wiley, 1980.

Nelson, S. D.: Metabolic activation and drug toxicity. J. Med. Chem. 25:753, 1982.

Ortiz de Montellano, P. R. (ed.): Cytochrome P-450: Structure, Mechanism, and Biochemistry. New York, Plenum Press, 1986.

Parke, D. V.: The Biochemistry of Foreign Compounds. New York, Pergamon Press, 1968.

Parke, D. V., Smith, R. L. (eds.): Drug Metabolism: From Microbe to Man. London, Taylor & Francis, 1977.

Paulson, G. D., Caldwell, J., Hutson, D. H., Mann, J. J. (eds.): Xenobiotic Conjugation Chemistry, Washington, DC, American Chemical Society, 1986.

Reid, E., Leppard, J. P. (eds.): Drug Metabolite Isolation and Detection. New York, Plenum Press, 1983.

Sato, R., Omura, T. (eds.): Cytochrome P-450. New York, Academic Press; Tokyo, Kadansha, 1978.

Schenkman, J. B., Kupfer, D. (eds.): Hepatic Cytochrome P-450 Monooxygenase System. New York, Pergamon Press, 1982.

Siest, G. (ed.): Drug Metabolism: Molecular Approaches and Pharmacological Implications. New York, Academic Press, 1985.

Singer, B., Grunberger, D.: Molecular Biology of Mutagens and Carcinogens. New York, Plenum Press, 1983.

Snyder, R., Parke, D. V., Kocsis, J. J., et al. (eds.): Biological Reactive Intermediates II, Parts A and B. New York, Plenum Press, 1982.

Snyder, R., Jollow, D. J., Kocsis, J. J., et al. (eds.): Biological Reactive Intermediates III: Animal Models and Human Disease, Part A and B. New York, Plenum Press, 1986.

Tagashira, Y., Omura, T. (eds.): P-450 and Chemical Carcinogenesis. New York, Plenum Press; Tokyo, Japan Scientific Societies Press, 1985.

Testa, B., Jenner, P.: Drug Metabolism: Chemical and Biochemical Aspects. New York, Marcel Dekker, 1976.

Timbrell, J. A.: Principles of Biochemical Toxicology. London, Taylor & Francis, 1982.

Williams, R. T.: Detoxication Mechanisms, 2nd ed. New York, Wiley, 1959.

CHAPTER 4

Anti-Infective Agents

Arnold R. Martin

Selective toxicity, the property of certain chemicals to destroy one form of life without harming another, is the cornerstone of modern antimicrobial chemotherapy. This concept is largely credited to Paul Ehrlich, who discovered the selective-staining properties of certain antibacterial dyes and the antiparasitic activity of organic arsenicals, shortly after the turn of this century. Although the compounds discovered by Ehrlich have largely been replaced by safer and more effective agents, his ideas paved the way for the advent of the sulfonamides and penicillin and the elucidation of the mechanisms for their selective toxicity. The local antimicrobial properties of phenol and iodine were discovered before the beginning of the century, but the only selective systemically useful chemotherapeutic agents known before Ehrlich's time were herbal remedies, such as cinchona for malaria and ipecac for amebic dysentery.

During the first quarter of the 20th century, the development of useful chemotherapeutic agents was largely confined to organic compounds containing heavy metals, such as mercury, arsenic, and antimony; dyes, such as gentian violet and methylene blue; and a few modifications of the quinine molecule. Although some of these discoveries represented significant advances, there were many drawbacks to them as well. The second quarter of this century ushered in the period of greatest progress in antimicrobial chemotherapy. The sulfonamides and sulfones (see Chap. 5), improved phenolic compounds such as hexachlorophene, synthetic antimalarial agents (see Chap. 6), the surfactants and, of greatest importance, numerous antibiotics (see Chap. 7) were introduced into medicine during this time. The first indication that certain antineoplastic agents might exhibit selective toxicity to tumor cells also began to appear toward the end of this period.

Chemotherapeutic agents may be classified according to their chemical type, their biologic properties, or their therapeutic indications. A combination of these classification systems is used to organize the chapters covering chemotherapeutic agents in this book. Where a variety of chemically divergent compounds are indicated for a specific disease or group of diseases, the medical classification is used, and the drugs are subclassified according to chemical type. On the other hand, when the information is best unified and presented in a chemical or biologic classification system, as for the sulfonamides or antibacterial antibiotics, then one of these systems is used.

This chapter covers a broad range of anti-infective drugs including local anti-infective agents (alcohols, phenols, oxidizing agents, halogen-containing compounds, cationic surfactants, dyes, mercury compounds, and nitrofurans), preservatives, antifungal agents, urinary antibacterial agents, antitubercular agents, antiviral agents, antiprotozoal agents, and anthelmintics. Those groups of antibiotics employed for the specific treatment of tuberculosis (cycloserine, viomycin, capreomycin, and rifampin) and fungal infections (griseofulvin and the polyenes) are also covered in this chapter. A separate chapter is devoted to antibacterial antibiotics (Chap. 7), whereas antineoplastic antibiotics are presented in Chapter 8.

LOCAL ANTI-INFECTIVE AGENTS

Local anti-infectives, or germicides, may be classified as antiseptics and disinfectants and constitute an important, if underappreciated, group of drugs. Antiseptics are compounds that kill (*-cidal*) or prevent the growth of (*-static*) microorganisms when applied to living tissue. An ideal antiseptic would exert a rapid and sustained lethal action against microorganisms (the spectrum may be narrow or broad depending on the use); have a low surface tension; retain activity in the presence of body fluids, including pus; be nonirritating to tissues; be nonallergenic; lack systemic toxicity when applied to skin or mucous membranes; and not interfere with healing. It is doubtful that any antiseptic available today fully meets all of these criteria.

A few antibiotics, generally ones that are poorly absorbed through the skin and mucous membranes, have been used topically for the treatment of localized infections for which they are uniquely effective. In general, the topical use of antibiotics has been limited because of the possibility that allergic reactions may result, or that resistant strains of microorganisms may emerge, thus reducing their usefulness for the treatment of more serious, systemic infections.

A *disinfectant* is an agent that prevents infection by the destruction of pathogenic microorganisms when applied to inanimate objects. The ideal disinfectant exerts a rapidly lethal action against all potentially pathogenic microorganisms and spores, has good penetrating power into organic matter, is compatable with organic compounds (especially soaps), is not inactivated by living tissue, is noncorrosive, and is esthetically desirable (nonstaining, odorless, and such).

Local anti-infective drugs continue to be widely used by the lay public and by members of the medical profession despite that the effectiveness of most such agents has not been firmly established. In certain situations, the use of a disinfectant or antiseptic may actually be harmful. Because of the availability of over-the-counter (OTC) germicides, the pharmacist is in a unique position to advise the public concerning the rational use of disinfectants and antiseptics. Although the effectiveness for most germicides is either lacking or yet to be established, there are a few compounds that have been shown to be effective in controlled studies. A major problem is that adequate standardized methods for evaluating antiseptics, in particular, have been established only very recently.

Numerous classes of chemically divergent compounds possess local anti-infective properties.

ALCOHOLS AND RELATED COMPOUNDS

Various alcohols and aldehydes have been used as antiseptics and disinfectants. Ethyl and isopropyl alcohols are still widely used for these purposes.

Antimicrobial Action and Chemical Structure

The antibacterial potencies of primary alcohols (against *Staphylococcus aureus*) increase with molecular weight up to C_8, where the "cutoff" is reached. Beyond this point, water solubility is less than the minimum effective concentration, and the apparent potency decreases with molecular weight. Branching decreases antibacterial potency; hence, the isomeric alcohols follow the order primary > secondary > tertiary. Nonetheless, isopropyl alcohol is used commercially instead of normal propyl alcohol, because it is cheaper. Isopropyl alcohol is slightly more active than ethyl alcohol against vegetative bacterial growth, but both alcohols are largely ineffective against spores.

Alcohol, USP, ethanol, *spiritus vini rectificatus*, wine spirit. Ethanol is a clear, colorless, volatile liquid having a burning taste and a characteristic pleasant odor. It is flammable and miscible with water and most organic solvents. The commercial product contains about 95% ethanol by volume because this concentration forms an azeotrope that distills at 78.2°F. Alcohol has been known for centuries as a fermentation product from grain and other carbohydrate sources. It can also be prepared synthetically by the sulfuric acid-catalyzed hydration of ethylene.

Control of the use of alcohol in the United States is exercised by the Treasury Department which has provided the following definition:

> The term 'alcohol' means that substance known as ethyl alcohol, hydrated oxide of ethyl, or spirit of wine, from whatever source or whatever process produced, having a proof of 160 or more, and not including the substances commonly known as whiskey, brandy, rum or gin.

Denatured alcohol is ethanol that has been rendered unfit for use in intoxicating beverages by the addition of other substances. Completely denatured alcohol contains added wood alcohol (methanol) and benzene and is unsuitable for either internal or external use. Specially denatured alcohol is ethanol treated with one or more substances so that its use may be permitted for a specialized purpose. Examples are: iodine in alcohol for tincture of iodine, methanol and other substances in mouthwashes and after-shave lotions, and methanol in alcohol for preparing plant extracts.

The primary medicinal uses of alcohol are external: as an antiseptic, preservative, mild counterirritant or solvent. Rubbing alcohol USP is employed as an astringent, rubefacient, refrigerant, and mild local anesthetic. Ethanol has been injected near nerves and ganglia to alleviate pain of neuralgias. It has low narcotic potency and has been used internally in diluted forms as a mild sedative, as a weak vasodilator, as a carminative, and as a source of energy.

Alcohol is, by far, the most widely abused of all recreational drugs. It is metabolized in the body by a series of oxidations, first to acetaldehyde, then to acetic acid (or active acetate), and finally to carbon dioxide and water. A widely employed form of aversion therapy for the prevention of alcohol abuse by alcoholics utilizes the aldehyde dehydrogenase inhibitor disulfiram, whose effectiveness results from the accumulation of acetaldehyde and its attendant toxicity.

Alcohol is widely employed in the practice of pharmacy for the preparation of a variety of pharmaceutical preparations, including spirits, tinctures, and fluid extracts. Spirits are liquid pharmaceutical preparations containing alcohol as the sole solvent, whereas tinctures are hydroalcoholic mixtures. Many fluid extracts also contain some alcohol.

The widely accepted optimal bactericidal concentration of 70% alcohol is not supported by a study that found that the kill rates of microorganisms suspended in alcohol concentrations between 60% and 95% were not significantly different.[1] Concentrations lower than 60% are also effective, but longer contact times are necessary. Concentrations above 70% can be used safely for preoperative sterilization of the skin.[2]

Dehydrated Alcohol, USP. Dehydrated ethanol, absolute alcohol contains not less than 99% by weight of C_2H_5OH. It is prepared commercially by azeotropic distillation of an ethanol and benzene mixture. It has a very high affinity for water and must be stored in tightly sealed containers. Absolute alcohol is used primarily as a chemical agent, but it has also been injected for the local relief of pain in carcinomas and neuralgias.

Isopropyl Alcohol, USP. 2-Propanol is a clear, colorless, volatile liquid having a characteristic odor and a slightly bitter taste. It is considered to be a suitable substitute for ethyl alcohol for most external uses, but it must not be taken internally.

$$CH_3CH{=}CH_2 \xrightarrow[H_2SO_4]{H_2O} CH_3CHOHCH_3$$

Isopropyl Alcohol

Isopropyl alcohol is prepared commercially by the sulfuric acid-catalyzed hydration of propylene. It forms a constant-boiling mixture with water containing 91% by volume of 2-propanol. It is primarily used to disinfect the skin and surgical instruments. Isopropyl alcohol is rapidly bactericidal in the concentration range of 50% to 95%. A 40% concentration is considered to be equal in antiseptic power to a 60% ethanol concentration. Azeotropic isopropyl alcohol, USP is used by diabetics for the sterilization of hypodermic needles and syringes. Isopropyl alcohol also finds use in pharmaceuticals and toiletries as a solvent and preservative.

Ethylene Oxide (C_2H_4O) is a colorless flammable gas that liquifies at 12°C. It has been used to sterilize temperature-sensitive medical equipment and certain pharmaceuticals that cannot be autoclaved. Ethylene oxide readily diffuses through porous material and effectively destroys all forms of microorganisms at ambient temperatures.[3]

Ethylene oxide forms explosive mixtures in air in concentrations ranging from 3% to 80% by volume. The hazard is eliminated when the gas is mixed with sufficient concentrations of carbon dioxide. Carboxide is a product consisting of 10% ethylene oxide and 90% carbon dioxide, by volume, that can be released in air without danger of explosion.

The mechanism of germicidal action of ethylene oxide involves alkylation of functional groups in nucleic acids and proteins by nucleophilic ring opening. Owing to its nonselective action, ethylene oxide is potentially very toxic and possibly carcinogenic. Personnel using the gas mixture should be cautioned to use a gas mask and avoid exposure to skin and mucous membranes.

Formaldehyde Solution, USP (formalin; formol) is a colorless aqueous solution containing not less than 37% of formaldehyde (CH_2O) with methanol added to retard polymerization. It is miscible in water and alcohol and is characteristically pungent. Formaldehyde readily undergoes oxidation and polymerization, leading to formic acid and paraformaldehyde, respectively. It should be preserved in tightly closed containers and stored at temperatures above 15°C to prevent the cloudiness that occurs at lower temperatures.

Formaldehyde solution exerts a slow, but powerful germicidal action. The mechanism of this effect is believed to involve the direct, nonspecific alkylation of nucleophilic functional groups (amino, hydroxyl, and sulfhydryl) in proteins to form carbinol derivatives. The action of formaldehyde is not confined to microorganisms. The compound is irritating to mucous membranes and causes hardening of the skin. Oral ingestion of the solution leads to severe gastric distress.

Formalin, diluted in water, had been employed to harden the skin, to prevent excessive perspiration, and to disinfect the hands before surgery. The gas has been employed to disinfect rooms, clothing, and

TABLE 4-1
ALCOHOL PRODUCTS

Product	Approximate Alcohol Content, by Volume (%)	Category	Application
Alcohol USP	95	Topical anti-infective, pharmaceutic aid (solvent)	Topically to the skin, as a 70% solution
Rubbing alcohol USP	70	Rubefacient	
Diluted alcohol USP	50	Pharmaceutic aid (solvent)	
Isopropyl alcohol USP	100	Local anti-infective, pharmaceutic aid (solvent)	
Azeotropic isopropyl alcohol USP	91	Local anti-infective	
Isopropyl rubbing alcohol USP	70	Rubefacient	

surgical instruments. A high frequency of allergic reactions associated with formaldehyde and formaldehyde-based products and the designation of formaldehyde as a suspect carcinogen suggest that it should not be used for the indicated foregoing purposes.

Glutarol Disinfectant Solution, NF (glutaraldehyde; Cidex). This dialdehyde is used as a dilute sterilizing solution for equipment and instruments that cannot be autoclaved. The commercial product (Cidex) is a stabilized alkaline glutaraldehyde solution, which actually consists of two components that are mixed together immediately before use. The activated solution thus prepared contains 2% glutaraldehyde buffered to *p*H 7.5 to 8.0. Stabilized solutions retain 86% of their original activity 30 days after preparation,[4] whereas the nonstabilized alkaline solutions lose 44% of their activity after 15 days. At high *p*H (greater than 8.5) glutaraldehyde rapidly polymerizes. Nonbuffered solutions of the compound are acidic, possibly owing to the existence of the cyclic hydrated (hemiacetal) form. Such acidic solutions are very stable, but lack sporicidal activity.

$$H-\overset{O}{\overset{\|}{C}}-CH_2CH_2CH_2-\overset{O}{\overset{\|}{C}}-H$$

Glutaraldehyde

PHENOLS AND THEIR DERIVATIVES

The standard by which most germicidal substances are compared is the activity of phenol USP. The *phenol coefficient* is defined as the ratio of a disinfectant to the dilution of phenol required to kill a given strain of the bacterium *Salmonella typhi* (*Eberthella typhosa*), under carefully controlled conditions over a given time period. If, for example, the dilution of a test compound is ten times as great as the dilution of phenol, then the phenol coefficient (PC) is 10. The PC of phenol is, of course, unity. There are many shortcomings of this testing method. Different microorganisms differ in their sensitivities to phenols, as compared with other germicides, and, therefore, different phenol coefficients would be expected. Also, the conditions used to conduct the test are difficult to reproduce exactly; hence, a high degree of variability between different laboratories is common.

Several phenols are more bactericidal than phenol itself. Substitution with alkyl, aryl, and halogen (especially *para*) groups increases bactericidal activity. Straight-chain alkyl groups are more effective than branched ones. Alkylated phenols and resorcinols are actually less toxic than the parent compounds. Phenols are believed to precipitate bacterial proteins at low concentrations. Lysis of bacterial cell membranes occurs at higher concentrations.

Phenol, USP (carbolic acid). Phenol occurs as a colorless to pale pink crystalline solid with a characteristic "medicinal" odor. It is soluble 1 : 15 in water, very soluble in alcohol, and soluble in most organic solvents. Phenol forms eutectic mixtures with many substances including thymol, menthol, and solol.

HO—C₆H₅ structure

Phenol

Phenol was introduced as a surgical antiseptic by Sir Joseph Lister in 1867. In addition to its germicidal activity, phenol has caustic and local anesthetic actions. It is a general protoplasmic poison that is toxic to all cells. Phenol is corrosive to the skin and must be diluted to avoid tissue destruction and dermatitis.

Phenol is still occasionally used as an antipruritic in phenolated calamine lotion in 0.1% to 1% concentrations. A 4% glycerin solution has been used to cauterize small wounds. Small amounts of pure phenol have also been used for this purpose. The use of phenol as either an antiseptic or disinfectant is largely obsolete.

Liquified Phenol, USP is phenol containing 10% water. The liquid form provides a convenient method for using phenol in a variety of pharmaceutical preparations. However, the water content precludes its use in fixed oils or liquid petrolatum.

p-Chlorophenol, USP. 4-Chlorophenol is used in combination with camphor in liquid petrolatum. It has a phenol coefficient of about 4.

***p*-Chloro-*m*-xylenol.** *p*-Chloro-*m*-xylenol (PC-MX, Metasep) is a relatively nonirritating antiseptic agent with broad-spectrum antibacterial and antifungal properties. It is available in 2% concentration as a medicated shampoo. It has also been used in topical preparations for the treatment of tinea infections, such as athlete's foot and jock itch.

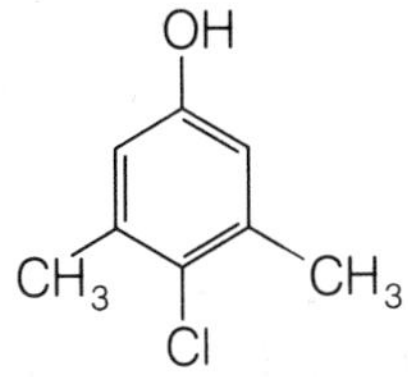

Parachlorometaxylenol

Hexachlorophene, USP. 2,2′-Methylenebis(3,4,6-trichlorophenol); 2,2′dihydroxy-3,5,6,3′,5′,6′-hexachlorodiphenylmethane (Gamophen, Surgicon, pHisoHex, Hex-O-San, Germa-Medica). It occurs as a white to light tan, crystalline powder that is insoluble in water, but soluble in alcohol and most other organic solvents.

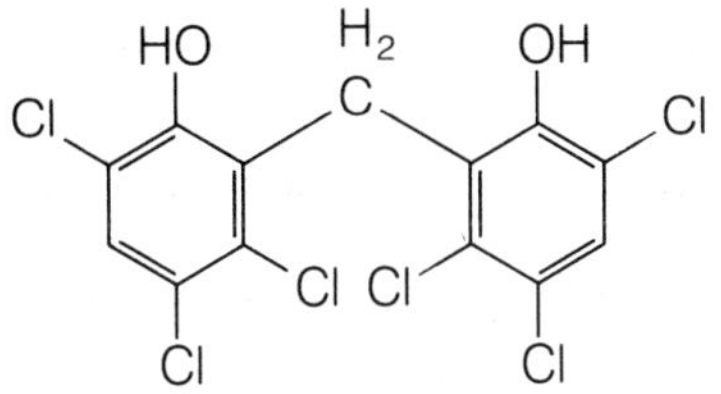

Hexachlorophene

Biphenolic compounds, such as hexachlorophene, are generally more potent than the corresponding monophenolic counterparts. Moreover, the chlorine content of hexachlorophene further increases its potency. The physical properties of this substance are such that it is readily deposited on the skin and in sebaceous glands. Therefore, it creates a prolonged antiseptic effect in low concentrations when applied topically. Hexachlorophene is employed in concentrations of 2% to 3% in soaps, detergent creams, lotions, and other forms for a variety of antiseptic uses. It is generally very effective against gram-positive bacteria, but many gram-negative organisms are resistant to its action.

Although the systemic toxicity of hexachlorophene in animals following oral and parenteral administration had been known for some time, it was not until the late 1960s and the early 1970s that reports of neurologic toxicity in infants and burn patients prompted the Food and Drug Administration (FDA) to ban its use in OTC antiseptic and cosmetic preparations.[5]

Cresol, NF is actually a mixture of the three isomeric cresols. It occurs as a yellowish to brownish-yellow liquid that has a characteristic creosote odor. Cresol is obtained from coal tar or petrolatum by alkaline extraction, acidification, and fractional distillation. Cresol is an inexpensive antiseptic and disinfectant with a phenol coefficient of 2.5. It is sparingly soluble in water.

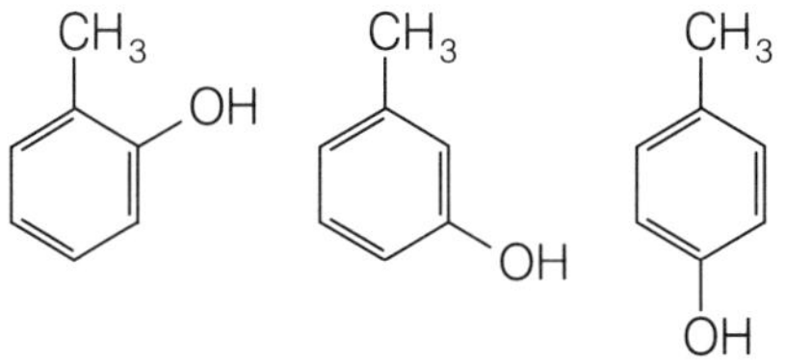

o-Cresol *m*-Cresol *p*-Cresol

Thymol, NF. Isopropyl *m*-cresol is obtained from oil of thyme by alkaline extraction followed by acidification. Thymol occurs as large colorless crystals, with an aromatic thymelike odor. It is sparingly soluble in water, but soluble in alcohol and organic solvents. Thymol has fungicidal properties and is used in alcoholic solutions and in dusting powders for the treatment of tinea infections.

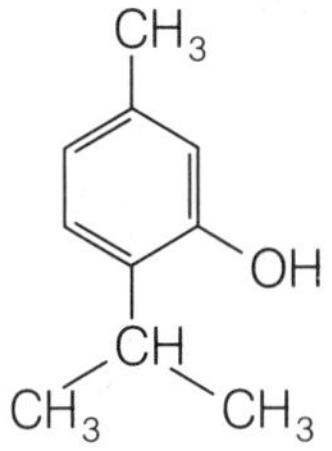

Thymol

Eugenol, USP. 4-Allyl-2-methoxyphenol is obtained from clove oil and other volatile oils. It occurs as a pale yellow liquid having an aromatic odor of cloves and a pungent taste. It is slightly soluble in water and miscible in alcohol and other organic solvents.

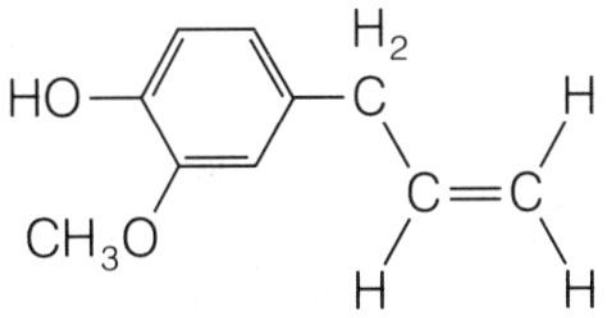

Eugenol

Eugenol has local anesthetic, as well as antiseptic, activity and is used to relieve toothaches. It also finds use in mouthwashes because of these properties. The phenol coefficient of eugenol is 14.4.

Resorcinol, USP; *m*-dihydroxybenzene; resorcin. Resorcinol is synthetically prepared. It occurs as white needle-shaped crystals or powder that is soluble in water and alcohol. Resorcinol is light-sensitive and readily oxidizes. It should be stored in light-resistant containers. It is much less stable in solution, especially at alkaline *p*H.

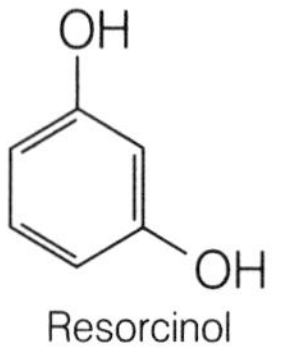

OH

OH

Resorcinol

Despite its weak antiseptic properties (phenol coefficient, 0.4), resorcinol is used in 1% to 3% solutions and in ointments and pastes, in concentrations of 10% to 20% for the treatment of skin conditions such as ringworm, eczema, psoriasis, and seborrheic dermatitis; resorcinol possesses keratolytic properties to augment its antiseptic action.

Resorcinol Monoacetate, USP; euresol, is prepared by monoacetylation of resorcinol. It is a viscous, pale yellow to amber liquid, with a faint phenolic odor, that is sparingly soluble in water. Similar to other esters of phenols, resorcinol monoacetate is readily hydrolyzed in aqueous alkaline solution. Although monoacetylation confers increased chemical stability, the compound remains light-sensitive and must be stored in light-resistant containers.

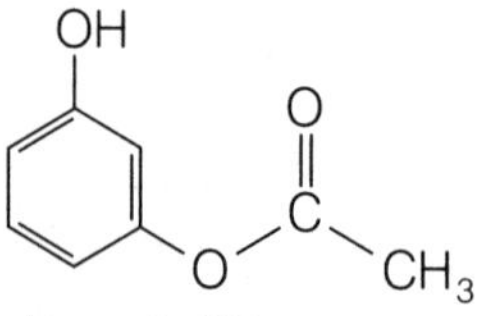

OH O C O CH_3

Resorcinol Monoacetate

Acetylation of resorcinol increases its germicidal potency. The ester is employed topically for skin conditions, such as seborrhea, alopacia, acne, sycosis, and chilblains for its keratolytic and antiseptic actions. It also finds use in the treatment of superficial fungal infections. Resorcinol monoacetate is available in ointments and pastes in concentrations of 5% to 20%. It is also found in lotions in concentrations of 3% to 5% to be applied to the scalp.

Hexylresorcinol, USP; 4-hexylresorcinol (Crystoids). Hexylresorcinol is found as a white, needle-like crystalline solid with a faint phenolic odor and an astringent taste. When placed on the tongue, it produces a sensation of numbness. It is freely soluble in alcohol, but sparingly soluble in water (1 : 20,000).

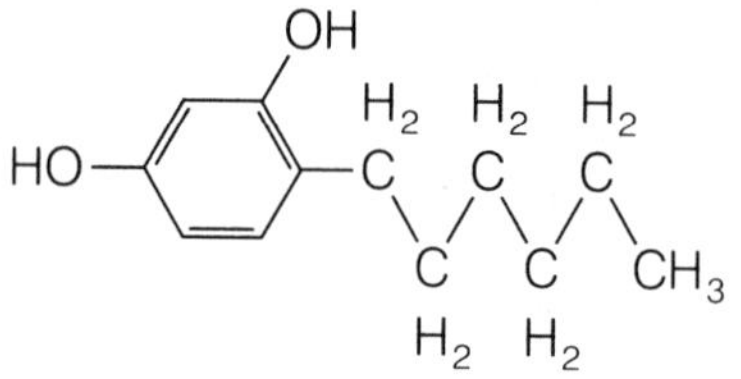

OH

HO H_2 H_2 H_2 C C C C C CH_3 H_2 H_2

Hexylresorcinol

Hexylresorcinol is an effective antiseptic, having bactericidal and fungicidal properties. It has a phenol coefficient of 98 against *Staphylococcus aureus*. As with other alkylated phenols and resorcinols, hexylresorcinol has surface-active properties. It also has local anesthetic activity. These properties may contribute to its effectiveness as an anthelmintic for the treatment of ascaris and hookworm infestations. However, more effective and better-tolerated anthelmintics are currently available for these purposes. Hexylresorcinol is also found in throat lozenges because of its local anesthetic and antiseptic properties. Such preparations are of dubious value and may be harmful because the local concentration of hexylresorcinol is probably not bactericidal, and the larynx may be anesthetized causing temporary laryngitis.

Anthralin, USP. 1,8,9-Anthracenetriol occurs as a yellowish-brown crystalline powder that is insoluble in water, slightly soluble in alcohol, and soluble in most nonpolar organic solvents. It finds application in the treatment of psoriasis and other chronic skin conditions because of its antiseptic, irritant, and keratolytic properties.

OH O OH

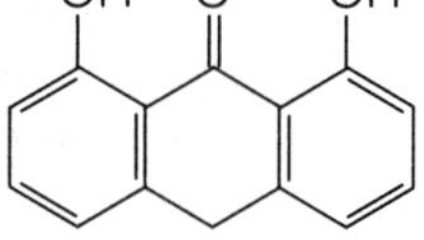

Anthralin

OXIDIZING AGENTS

Most oxidizing agents that are of value as germicides depend upon the liberation of oxygen in the tissues. Many are inorganic compounds and include hydrogen peroxide, various metal peroxides, and sodium perborate. Other oxidizing agents, such as potassium permanganate, denature proteins through

TABLE 4-2

PHENOL PRODUCTS

Name Proprietary Name	***Preparations***	***Category***	***Application****
Phenol USP		Pharmaceutic aid (preservative)	
Liquefied phenol USP		Topical antipruritic	Topically to the skin, as a 0.5–2% lotion or ointment
p-Chlorophenol USP	Camphorated *p*-chlorophenol USP	Anti-infective (dental)	Topically to root canals and the periapical region
Hexachlorophene USP	Hexachlorophene cleansing emulsion USP	Topical anti-infective; detergent	Topically to the skin, as the sole detergent, followed by thorough rinsing
pHisoHex, Presulin Cleanser	Hexachlorophene liquid soap USP		
Thymol NF		Pharmaceutic aid (stabilizer)	
Anthralin USP	Anthralin ointment USP	Topical antipsoriatic	Topical, to the skin, as a 0.1–1% ointment (qd or bid)
Eugenol USP	Zinc–eugenol cement USP	Dental protective	Topical to dental cavities
Resorcinol USP		Keratolytic	Topically to the skin, as a 2–20% ointment
	Resorcinol ointment USP	Local antifungal; keratolytic	Topical, to the skin as a 2–20% ointment or lotion
Resorcinol monoacetate USP *Euresol*	Resorcinol lotion USP	Antiseborrheic; keratolytic	Topical, for application to the scalp
Hexylresorcinol USP *Crystoids*	Hexylresorcinol pills USP	Anthelmintic (intestinal roundworms and trematodes)	1 g, may be repeated at weekly intervals if necessary

*See USP DI for complete dosage information.

a direct oxidant action. Oxidizing agents are particularly active against anaerobic bacteria and find use in the cleansing of contaminated wounds. The effectiveness of these agents is somewhat limited by their poor penetrability into tissues and organic matter and their transient action.

Carbamide Peroxide Solution, USP. Carbamide peroxide is a stable complex of urea and hydrogen peroxide, having the formula $H_2NCONH_2 \cdot H_2O_2$. The official solution contains 12.6% of the complex in anhydrous glycerin. It releases hydrogen peroxide when the solution is mixed with water. It is used as an antiseptic and disinfectant.

Hydrous Benzoyl Peroxide, USP. (Benoxyl, Oxy-5, Oxy-10, Persadox, Vanoxide) is a white, granular powder with a characteristic odor. It contains about 30% water to make it safer to handle.

Benzoyl Peroxide

Benzoyl peroxide is employed in concentrations of 5% to 10% as a keratolytic and keratogenic agent for the control of acne. Similar to other peroxides, it is chemically unstable and can explode when heated. Nonstabilized aqueous solutions slowly decompose to hydrogen peroxide and benzoic acid. Lotions containing benzoyl peroxide are stabilized with the addition of two parts of dicalcium phosphate.

The value of benzoyl peroxide in acne treatment is believed to derive from its irritant properties.[6] It induces proliferation of epithelial cells, leading to sloughing and repair.

HALOGEN-CONTAINING COMPOUNDS

Iodophors

Iodine (I_2) is one of the oldest known germicides in use today. It was listed in 1830 in *USP II* as a tincture and as a linament. Iodine tincture (a 2% solution of iodine in 50% alcohol with sodium iodide), strong iodine solution (Lugol's solution, 5% iodine in water with potassium iodide), and iodine solution (2% iodine in water with sodium iodide) are currently official in the *USP*. Inorganic iodide salts are present to solubilize the iodine and reduce its volatility. Iodine remains one of the most effective and useful germicides available today. It is believed to inactivate proteins by iodination (phenylalanyl and tyrosyl residues) and oxidation (sulfhydryl groups).

Various nonionic and cationic surfactants have been found to act as solubilizers for iodine by forming complexes that retain the germicidal properties of iodine. These complexes also reduce the volatility of iodine and essentially remove its irritant properties. It is estimated that approximately 80% of the dissolved iodine remains as available bacteriologically active iodine in the more effective, nonionic surfactant complexes.[7] Such complexes, called iodophors, are bactericidal and fungicidal.

Povidone-Iodine, USP (Betadine, Isodine) is a complex with the nonionic surfactant polymer, polyvinylpyrrolidone. This water-soluble complex releases iodine slowly. It provides a nontoxic, nonirritating, nonvolatile, and nonstaining form of iodine. Approximately 10% of available iodine is present in the complex.

Povidone-iodine is used as an aqueous solution as an antiseptic for application to the skin before surgery and injections, for the treatment of infected wounds and lacerations, and for local bacterial and fungal infections. Several other topical preparations containing providone-iodine are available, including aerosols, ointments, surgical scrubs, antiseptic gauze pads, sponges, whirlpool concentrates, and mouthwashes.

Chlorine-Containing Compounds

Chlorine and its derivatives have been used to disinfect water for more than a century. The discovery that hypochlorous acid (HClO) was the active germicidal species formed when chlorine was dissolved in water led to the use of first, inorganic hypochlorite salts such as NaOCl and $Ca(OCl)_2$ and later, *N*-chloro organic compounds as disinfectants. These compounds release hypochlorous acid when dissolved in water, especially in the presence of acid. Two equally plausible mechanisms have been proposed for the germicidal action of hypochlorous acid:

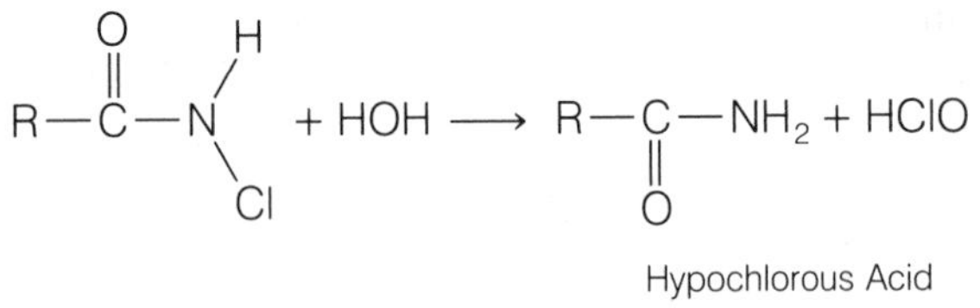

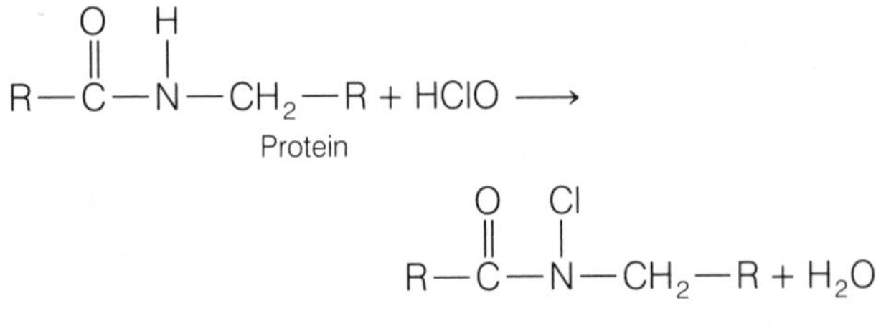

$$R{-}C(=O){-}N(H)(Cl) + HOH \longrightarrow R{-}C(=O){-}NH_2 + HClO$$

Hypochlorous Acid

$$R{-}C(=O){-}NH{-}CH_2{-}R + HClO \longrightarrow R{-}C(=O){-}N(Cl){-}CH_2{-}R + H_2O$$

Protein

the chlorination of amide nitrogen atoms in proteins (see diagram) and the oxidation of sulfhydryl groups in proteins.

Organic compounds that form reasonably stable *N*-chloro derivatives include amides, imides, and amidines. *N*-Chloro derivatives slowly release HOCl in water forming the parent amide in the process. The antiseptic power of these compounds is optimal at *p*H 7.

Halazone, USP. *p*-Dichlorosulfamoylbenzoic acid is a white crystalline light-sensitive compound with a faint chlorinelike odor. It is slightly soluble in water, but very soluble in alkaline solution. The sodium salt is used to disinfect drinking water.

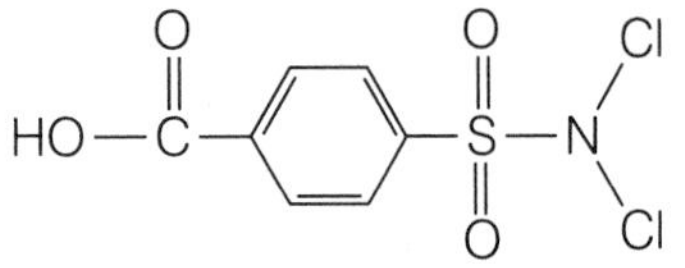

$$HO{-}C(=O){-}C_6H_4{-}S(=O)_2{-}NCl_2$$

Halazone

Chloroazodin. *N,N*-Dichlorodicarbonamidine (Azochloramid) is a bright yellow crystalline solid with a faint odor of chlorine. It is not very soluble in water or in most organic solvents and is unstable to light or heat. Chloroazodin is reported to explode when heated above 155°C.

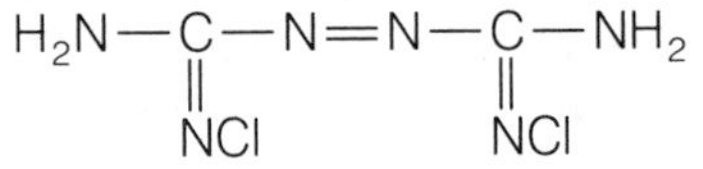

$$H_2N{-}C(=NCl){-}N{=}N{-}C(=NCl){-}NH_2$$

Chloroazodin

Dilute solutions are used to disinfect wounds, as a packing for cavities, and for lavage and irrigation. A glyceryltriacetate solution is used as a dressing. The antiseptic action of chloroazodin is prolonged because of its relatively slow reaction with water.

Oxychlorosene Sodium (Chlorpactin) is a complex of the sodium salt of dodecylbenzene sulfonic acid and hypochlorous acid. The complex slowly releases hypochlorous acid in solution. It is available as an amorphous white powder with a faint chlorinelike smell.

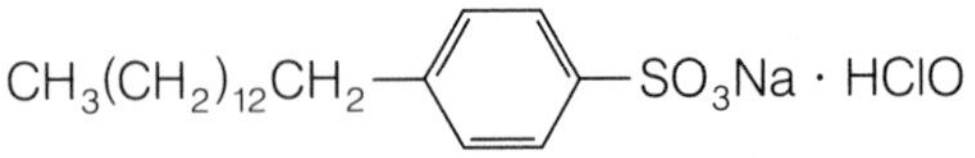

$$CH_3(CH_2)_{12}CH_2{-}C_6H_4{-}SO_3Na \cdot HClO$$

Oxychlorosene

Oxychlorosene combines the germicidal properties of hypochlorous acid with the emulsifying, wetting, and kerotolytic properties of an anionic detergent. The preparation has a marked and rapid cidal action against most microorganisms, including both gram-positive and gram-negative bacteria, molds, yeasts, viruses, and spores. It is used to treat localized infections (particularly when resistant organisms are present), to remove necrotic tissue in massive infections or from radiation necrosis, to counteract odorous discharges, as an irritant, and for disinfecting fistulas, empyemas, and wounds.

Oxychlorosene is supplied as a powder for solution. Typical applications use 0.1% to 0.5% concentrations in water. Dilutions of 0.1% to 0.2% are used in urology and ophthamology.

CATIONIC SURFACTANTS

The cationic surfactants are quaternary ammonium compounds that ionize in water and exhibit surface-active properties. The surface activity of these compounds, exemplified by lauryl triethylammonium sulfate, results from two structural features: (1) the cationic head, which has a high affinity for water, and (2) a long hydrocarbon tail, which has a high affinity for lipids and nonpolar solvents. These disparate solvent affinities cause molecules of lauryl trimethylammonium chloride to concentrate at the interface between immiscible solvents, such as water and oil, with the cationic group in the aqueous phase and the hydrocarbon group in the oil phase.

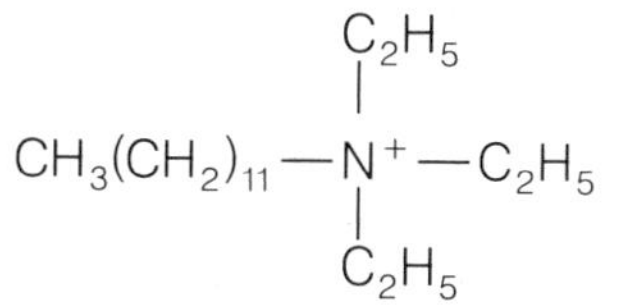

$$CH_3(CH_2)_{11}-N^+(C_2H_5)_3$$

Lauryl Trimethylammonium Chloride

The synthesis and antimicrobial actions of members of this class of compounds were first reported in 1908, but it was not until after the pioneering work of Gerhard Domagk in 1935[8] that attention was directed to their usefulness as antiseptics, disinfectants, and preservatives.

The cationic surfactants exert a bactericidal action against a broad spectrum of gram-positive and gram-negative bacteria. They are also active against several pathogenic species of fungi and protozoa. Spore-forming microorganisms, on the other hand, are resistant.

Mechanisms by which the cationic surfactants exert their bactericidal actions center around their surface-active properties. One suggestion is that they are adsorbed onto the surface of the bacterial cell, at which they cause lysis by interfering with enzymes in the cell wall and cell membrane.

The cationic surfactants possess several other desirable properties, in addition to their broad-spectrum antimicrobial activity, that are advantageous for germicidal use. They are highly water-soluble, relatively nontoxic, stable in solution, nonstaining, and noncorrosive. Their surface-active properties provide a keratolytic action and relatively good tissue penetration.

Despite all of the foregoing advantages, the cationic surfactants have numerous serious disadvantages. They are inactivated by soaps and other anionic detergents. All traces of soap must be removed from skin and other surfaces before they are applied. Tissue constituents, blood, serum, and pus tend to reduce the effectiveness of these substances. Cationic surfactants are also adsorbed on glass, talc, and kaolin to reduce or prevent their action. The bactericidal action of cationic surfactants is slower than that of iodine, and these compounds are not active against spores. Solutions of cationic surfactants intended for disinfecting surgical instruments, gloves, and the like, should never be reused because reused solutions have been reported to be a source of infection (especially by *Pseudomonas* and *Enterobacter* species).

Benzalkonium Chloride, NF. Alkylbenzyldimethylammonium chloride (Zephiran, Germicin, Benza) is a mixture of alkylbenzyldimethylammonium chlorides having the general formula $[C_6H_5CH_2N(CH_3)_2R]^+Cl^-$, where R represents a mixture of alkyl members beginning with C_8H_{17} and extending to higher homologues, with $C_{12}H_{25}$, $C_{14}H_{29}$, and $C_{16}H_{33}$ composing the major portion. Although variations in physical and antimicrobial properties between individual members of the mixture exist, they are of little importance for the usefulness of the mixture.

Benzalkonium chloride occurs as a white gel that is soluble in water, alcohol, and most organic solvents. Aqueous solutions are colorless, slightly alkaline to litmus paper, and highly foamy.

Benzalkonium chloride possesses detergent, emulsifying, and wetting actions. It is employed as an antiseptic for skin and mucosa in concentrations of 1:750 to 1:20,000. For irrigation, 1:20,000 to 1:40,000 concentrations are employed. For storage of surgical instruments, 1:750 to 1:5000 concentrations are used with 0.5% sodium nitrate added as a preservative.

Benzethonium Chloride, USP. Benzyldimethyl[2-[2-[*p*-(1,1,3,3,-tetramethylbutyl)phenoxy]-ethoxy]ethyl]ammonium chloride (Phemerol Chloride) is a colorless crystalline powder that is soluble in water, alcohol, and most organic solvents. The

TABLE 4-3

ANALOGUES OF DIMETHYLBENZYLAMMONIUM CHLORIDE

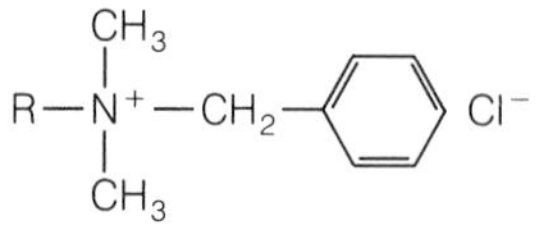

$$R-N^{+}(CH_3)_2-CH_2-C_6H_5 \quad Cl^-$$

Compound	R
Benzalkonium Chloride	n—C_8H_{17} to $C_{16}H_{33}$
Benzethonium Chloride	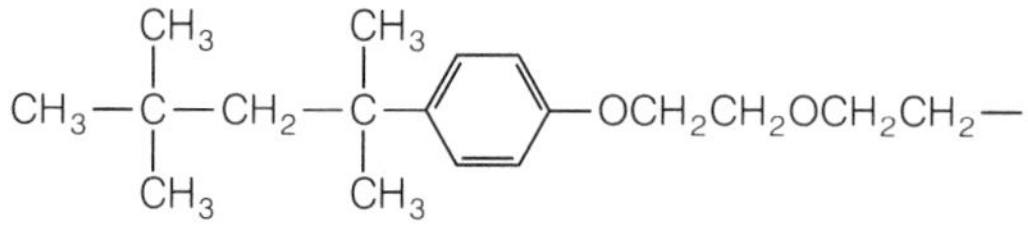 $CH_3-C(CH_3)_2-CH_2-C(CH_3)_2-C_6H_4-OCH_2CH_2OCH_2CH_2-$
Methylbenzethonium Chloride	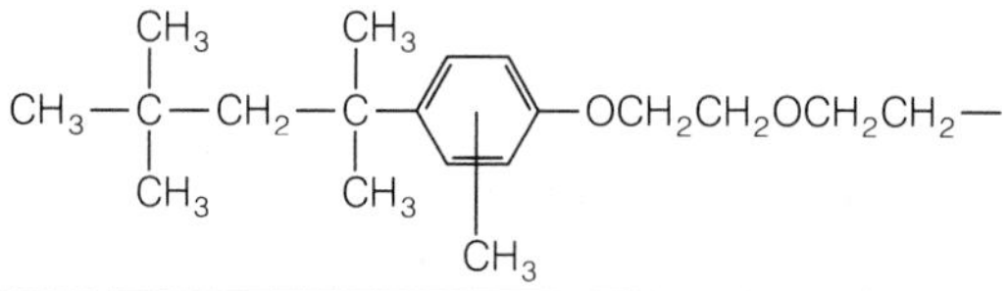 $CH_3-C(CH_3)_2-CH_2-C(CH_3)_2-C_6H_3(CH_3)-OCH_2CH_2OCH_2CH_2-$

structure of this agent and its relationship to other analogues of the class are shown in Table 4-3.

The actions and uses of this agent are similar to those of benzalkonium chloride. It is used as a 1:750 concentration for skin antisepsis. For the irrigation of mucous membranes, a 1:5000 solution is employed. A 1:500 alcoholic tincture is also available.

Methylbenzethonium Chloride, USP. Benzyldimethyl[2-[2-[[4-(1,1,3,3-tetramethylbutyl)tolyl]-oxy]ethoxy]ethyl]ammonium chloride (Diaparene). This mixture of methylated derivatives of methylbenzethonium chloride is used for the specific control of diaper rash in infants caused by the intestinal bacterium *Bacterium ammoniagenes*, which causes the liberation of ammonia in decomposed urine. It is also employed as a general antiseptic. Its properties are virtually identical with those of Phemerol chloride.

Cetylpyridinium Chloride, USP. 1-Hexadecylpyridinium chloride (Ceepryn) occurs as a white powder that is very soluble in water and alcohol. In this quaternary ammonium compound, the nitrogen is part of a pyridine ring.

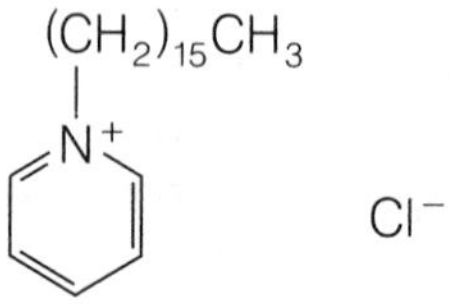

$$C_5H_5N^{+}-(CH_2)_{15}CH_3 \quad Cl^-$$

Cetylpyridinium Chloride

The cetyl derivative is the most active of a series of alkylpyridinium derivatives. It is used as general antiseptic solutions in concentrations of 1:100 to 1:1000 for intact skin, 1:1000 for minor lacerations, and 1:2000 to 1:10,000 for irrigation of mucous membranes. Ceepryn is also available in the form of throat lozenges and a mouthwash in 1:20,000 concentration.

Chlorhexidine Gluconate. 1,6-Di(4′-chlorophenyldiguanidino)hexane gluconate (Hibiclens) is the most effective of a series of antibacterial biguanides originally developed in Great Britain.[9] The antimicrobial properties of the biguanides were discovered as a result of earlier investigations of these substances as potential antimalarial agents (see Chap. 6). Although the biguanides are technically not bisquaternary ammonium compounds and, therefore, should perhaps be separately classified, they share many of the same physical, chemical, and antimicrobial properties with the cationic surfactants. The biguanides are strongly basic compounds that exist as di-cations at physiologic *p*H. Similar to cationic surfactants, they are inactivated by anionic detergents and by complex anions such as phosphate, carbonate, and silicate.

Chlorhexidine has broad-spectrum antibacterial activity, but it is not active against acid-fast bacteria, spores, or viruses. It has been employed for such topical antiseptic uses as preoperative skin disinfection, wound irrigation, bladder irrigation, mouthwashes, and general sanitation. Chlorhexidine is not absorbed through skin or mucous membranes and does not cause systemic toxicity.

TABLE 4-4

CATIONIC SURFACTANTS

Name Proprietary Name	***Preparations***	***Category***	***Application***	***Usual Dose***
Benzalkonium chloride NF *Zephiran Chloride, Benasept, Germicin, Pheneen*	Benzalkonium chloride solution NF	Pharmaceutic aid (antimicrobial preservative)		
Benzethonium chloride USP *Phemerol*	Benzethonium chloride solution USP Benzethonium chloride tincture USP	Local anti-infective; pharmaceutic aid (preservative)	Topical, 1 : 750 solution or 0.2% tincture to the skin or 1 : 5000 solution nasally	
Methylbenzethonium chloride USP *Diaparene*	Methylbenzethonium chloride lotion USP Methylbenzethonium chloride ointment USP Methylbenzethonium chloride powder USP	Local anti-infective	Topical, 0.067% lotion, 0.1% ointment, or 0.055% powder	
Cetylpyridinium chloride USP *Ceepryn*	Cetylpyridinium chloride lozenges USP Cetylpyridinium chloride solution USP	Local anti-infective; pharmaceutic aid (preservative)	Topical, 1 : 100–1 : 1,000 solution to intact skin, 1 : 1000 solution for minor lacerations, and 1 : 2000–1 : 10,000 solution to mucous membranes	Sublingual, 1 : 1500 lozenge

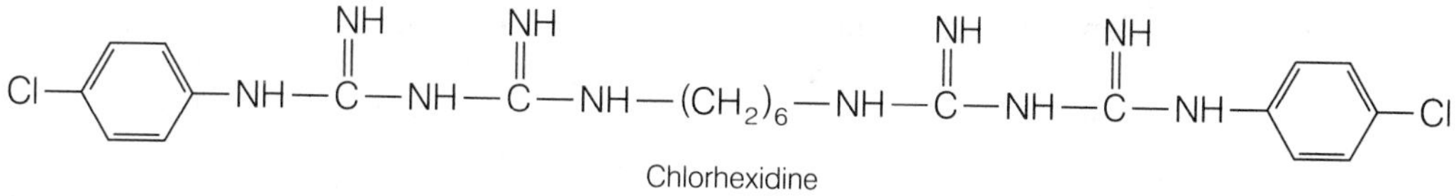

Chlorhexidine

DYES

Before the advent of the sulfonamides and the antibiotics, the organic dyes were used far more extensively than they are today. Only a handful of cationic dyes still find limited use as anti-infective agents (Table 4-5). They include the triphenylmethane dyes—gentian violet and basic fuchsin—and the thiazine dye, methylene blue. These dyes form colorless leucobase forms under alkaline conditions.

Cationic dyes are active against gram-positive bacteria and many fungi; gram-negative bacteria are generally resistant. The difference in susceptibility is thought to be related to the cellular characteristics that underlie the differential Gram stain.

Gentian Violet, USP. Hexamethyl-*p*-rosaniline chloride; crystal violet; methyl violet; methylrosaniline chloride (Genepax). Gentian violet occurs as a green powder or green flakes with a metallic luster. It is soluble in water (1 : 35) and alcohol (1 : 10), but insoluble in nonpolar organic solvents.

Gentian violet is available as vaginal suppositories for the treatment of yeast infections. It is also used as a 1% to 3% solution for the treatment of

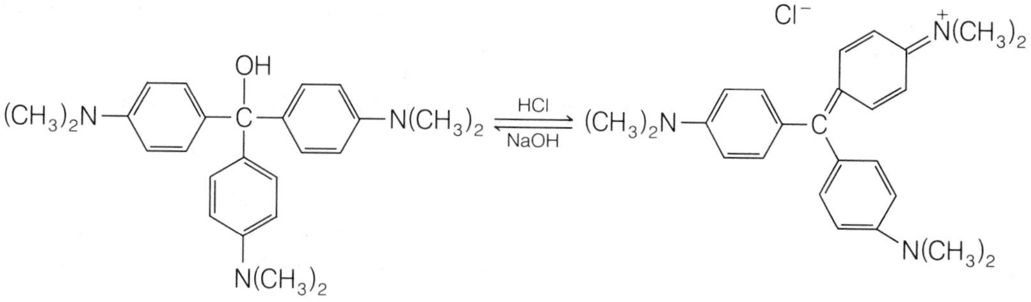

Leucobase

Hexamethyl Pararosaniline Chloride

TABLE 4-5

PHARMACEUTIC DYES

Name	Preparations	Category	Application*	Usual Dose*	Usual Pediatric Dose*
Gentian violet USP	Gentian violet vaginal suppositories USP	Topical anti-infective	Topically to the vagina, as a 1.35% cream once every 2 days		
	Gentian violet solution USP		Topically to the skin and mucous membranes, as a 0.5–2% solution bid or tid for 3 days		
Methylene blue USP	Methylene blue injection USP	Antidote to cyanide poisoning; antidote to methemoglobinemia		IV, 1–2 mg/kg of body weight	1–2 mg/kg of body weight or 25–50 mg/m^2 of body surface

*See USP DI for complete dosage information.

tinea and yeast infections. Gentian violet has also been used orally as an anthelmintic for strongyloidiasis (threadworm) and oxyuriasis.

Basic Fuchsin, USP is a mixture of the chlorides of rosaniline and *p*-rosaniline. It exists as a metallic green crystalline powder that is soluble in water and in alcohol, but insoluble in ether.

Basic fuchsin is an ingredient of carbol–fuchsin solution (Castellani's paint), which is used topically in the treatment of fungal infections, such as ringworm and athlete's foot.

Methylene Blue, USP. 3,7-Bis(dimethylamino)-phenazathionium chloride, occurs as a lustrous dark green crystalline powder that is soluble in water (1 : 25) and in alcohol (1 : 65).

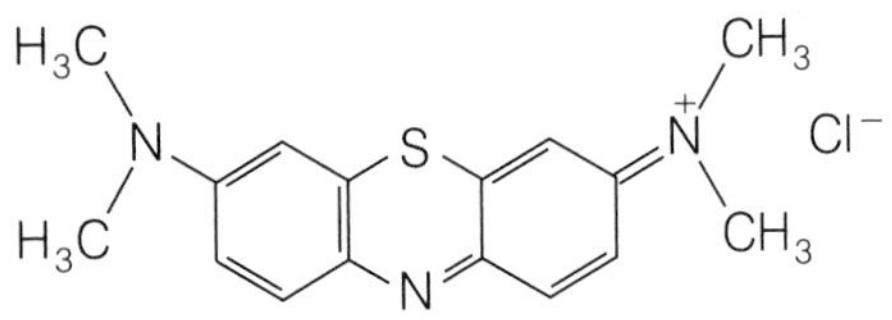

Methylene Blue

The redox properties of methylene blue provide the basis of its use as an antidote in cyanide poisoning. In high concentrations it promotes the conversion of hemoglobin to methemoglobin, which because of its high affinity for cyanide ion diverts it from inactivating cytochrome C. In low concentrations, methylene blue has the opposite effect and has been used to treat drug-induced methemoglobinemia.

Methylene blue has weak antiseptic properties that make it useful for the treatment of cystitis and urethritis. Its action is considered to be bacteriostatic. It colors the urine and sometimes the stool blue green. Methemoglobinemia and anemia may occur with prolonged use.

MERCURY COMPOUNDS

Mercury and its deviatives have been used in medicine for centuries. Elemental mercury incorporated in ointment bases was used topically for the treatment of local infections and syphilis. Several inorganic salts of mercury, such as mercuric chloride ($HgCl_2$) and mercurous chloride (Calomel, Hg_2Cl_2), were at one time widely used as antiseptics. Ammoniated mercury [$Hg(NH_2)Cl$] is still occasionally used for skin infections such as impetigo, psoriasis, and ringworm. Mercuric oxide is sometimes used to treat inflammatory infections of the eye. Although the potential interaction of mercuric ion with the tissues is greatly reduced by the low water solubility of these agents, they can be irritating to the tissues and can cause hypersensitivity reactions; therefore, their use is not recommended.

The comparatively few organic mercurials still in use are employed as antiseptics, preservatives, or diuretics (see Chap. 13). Organic mercurials are of two general types: (1) compounds with at least one carbon-mercury bond, which does not generally ionize readily; and (2) compounds with mercury bonded to heteroatoms, such as oxygen, nitrogen, or sulfur, that ionize partially or completely. In addition to its effect on ionization, the organic moiety may increase the lipid solubility of an organomercurial compound, thereby facilitating its penetration into microorganisms and host tissues.

The antibacterial action of mercury compounds is believed to result from their reaction with sulfhydyl (SH) groups in enzymes and other proteins to form covalent compounds of the type $R — S — Hg — R'$. This action is reversible by thiol-containing compounds, such as cysteine and dimercaprol (BAL); thus, organomercurials are considered to be largely bacteriostatic. The antibacterial activity of mercurial antiseptics is greatly reduced in the presence of serum because of the presence of proteins that inactivate mercury compounds. Organomercury antiseptics are not particularly effective against spores.

The disadvantages of mercurials for antiseptic and disinfectant use far outweigh any possible advantages they might have. Hence, other more effective and less potentially toxic agents are now preferred.

Nitromersol, USP. 3-(Hydroxymercuri)-4-nitro-*o*-cresol inner salt (Metaphen) occurs as a yellow powder that is practically insoluble in water and is sparingly soluble in alcohol and in most organic solvents. The somewhat improbable formula for the neutral form of nitromersol shown below is given in the *USP*. The sodium salt presumably has the "inner salt" structure.

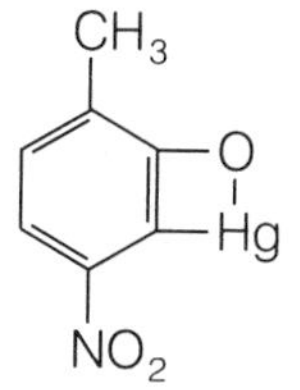

CH_3 O Hg NO_2

Nitromersol

Nitromersol is nonirritating to mucous membranes and is nonstaining. Therefore, at one time, it was a very popular topical antiseptic for skin and ocular infections. However, it has largely been replaced by more effective agents.

Thimerosal, USP. Sodium[(*o*-carboxyphenyl)-thio]ethylmercury (Merthiolate) occurs as a cream-colored, water-soluble powder. It is nonstaining and nonirritating to tissues. Thimerosal is a weakly bacteriostatic antiseptic that is applied topically as in aqueous solutions or as an ointment.

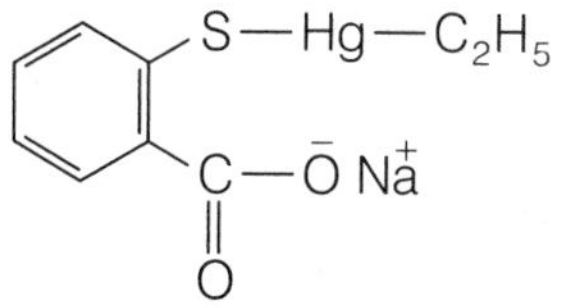

$S—Hg—C_2H_5$ $C—\bar{O}\ \overset{+}{Na}$ O

Thimerosal

Merbromin (Mercurochrome) occurs as iridescent green scales or granules of the sodium salt trihydrate. It is freely soluble in water to give a carmine-red solution, but is practically insoluble in alcohol and most organic solvents.

A 2% solution is available as an antiseptic, but the weak bacteriostatic action of merbromin renders it virtually useless for this purpose. Its property of staining the skin is primarily of psychologic benefit.

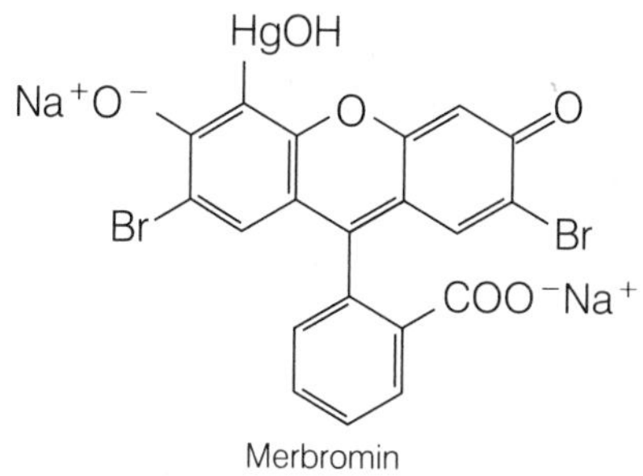

HgOH Na^+O^- O O Br Br COO^-Na^+

Merbromin

NITROFURAN DERIVATIVES

The first nitroheterocyclic compounds to be introduced into chemotherapy were the nitrofurans. Three of these compounds: nitrofurazone, furazolidone, and nitrofurantoin have been used for the treatment of bacterial infections of various kinds for more than 40 years. A fourth nitrofuran, nifurtimox, is used as an antiprotozoal agent to treat trypanosomiasis and leishmaniasis. Other nitroheterocycles of value as medicinal agents include metronidazole, employed as an amebicide, a trichomonicide, and for the treatment of systemic infections caused by anaerobic bacteria, and niridazole, an anthelmintic used in schistosomiasis treatment.

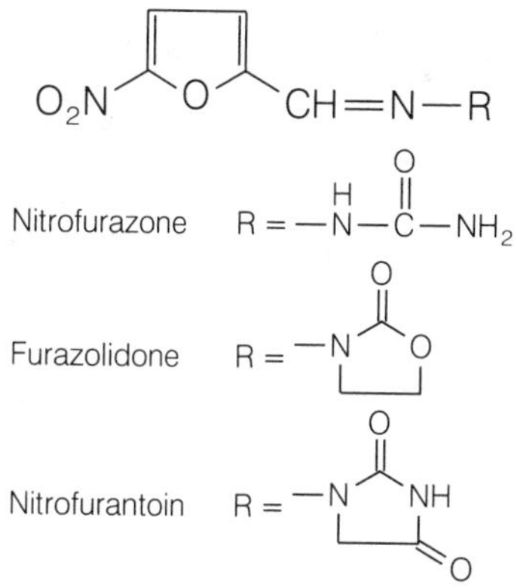

O_2N O $CH{=}N{—}R$

Nitrofurazone $R = —NH—C(=O)—NH_2$

Furazolidone R = —N (oxazolidinone ring)

Nitrofurantoin R = —N (imidazolidinedione ring, NH, O)

The nitrofurans are derivatives of 5-nitro-2-furaldehyde formed on reaction with the appropriate hydrazine or amine derivative. Antimicrobial activity is present only when the nitro group is in the 5-position.

The mechanism of antimicrobial action of the nitrofurans has been extensively studied, but it still is not fully understood. In addition to their antimicrobial actions, they are known to be mutagenic and carcinogenic under certain conditions. It is thought that DNA damage caused by metabolic reduction products may be involved in these cellular effects.

Nitrofurazone, USP. 5-Nitro-2-furaldehyde semicarbazone (Furacin) occurs as a lemon-yellow

TABLE 4-6

ANTISEPTICS AND DISINFECTANTS

Name Proprietary Name	***Preparations***	***Category***	***Application***
Alcohol USP		Topical anti-infective	Topically to the skin, as a 70% solution
Isopropyl alcohol USP		Local anti-infective	
Ethylene oxide		Disinfectant	
Formaldehyde solution USP		Disinfectant	Full strength or as a 10% solution to inanimate objects
Glutaral concentrate USP *Cidex*		Disinfectant	
Hexachlorophene USP *pHisoHex, Presulin Cleanser*	Hexachlorophene detergent lotion USP Hexachlorophene liquid soap USP	Topical anti-infective; detergent	Topically to the skin, as the sole detergent, followed by thorough rinsing
Carbamide peroxide solution USP		Local anti-infective (dental)	Several drops onto affected area; expectorate after 2–3 min
Povidone–iodine USP *Betadine*	Povidone–iodine solution USP	Topical anti-infective	Topically to the skin and mucous membranes, as the equivalent of a 0.75–1% solution of iodine
Halazone USP	Halazone tablets for solution USP	Disinfectant	2–5 ppm in drinking water
Benzethonium chloride USP *Phemerol*	Benzethonium chloride solution USP Benzethonium chloride tincture USP	Local anti-infective; pharmaceutic aid (preservative)	Topical, 1 : 750 solution or 0.2% tincture to the skin or 1 : 5000 solution nasally
Methylbenzethonium chloride USP *Diaparene*	Methylbenzethonium chloride lotion USP Methylbenzethonium chloride ointment USP Methylbenzethonium chloride powder USP	Local anti-infective	Topical, 0.067% lotion, 0.1% ointment, or 0.055% powder
Cetylpyridinium chloride USP *Ceepryn*	Cetylpyridinium chloride lozenges USP Cetylpyridinium chloride solution USP	Local anti-infective; pharmaceutic aid (preservative)	Topical, 1 : 100–1 : 1000 solution to intact skin, 1 : 1000 solution for minor lacerations, and 1 : 2000–1 : 10,000 solution to mucous membranes
Nitromersol USP *Metaphen*	Nitromersol solution USP Nitromersol tincture USP	Local anti-infective	Topical, solution or tincture
Thimerosal USP *Merthiolate*	Thimerosal aerosol USP Thimerosal solution USP Thimerosal tincture USP	Local anti-infective	Topical, a 0.1% aerosol, solution or tincture

crystalline solid that is sparingly soluble in water and practically insoluble in organic solvents. Nitrofurazone is chemically stable, but moderately light-sensitive.

It is employed topically in the treatment of burns, especially when bacterial resistance to other agents may be a problem. It may also be used to prevent bacterial infection associated with skin grafts. Nitrofurazone has a broad spectrum of activity against gram-positive and gram-negative bacteria, but it is not active against fungi. It is bactericidal against most bacteria commonly causing surface infections, including *Staphylococcus aureus*, *Streptococcus* sp., *Escherichia coli*, *Clostridium perfringens*, *Enterobacter* (*Aerobacter*) *aerogenes*, and *Proteus* sp; however, *Pseudomonas aeruginosa* strains are resistant.

Nitrofurazone is available in solutions, ointments, and suppositories in the usual concentration of 0.2%.

Furazolidone. 3-[(5-Nitrofurylidene)amino]-2-oxazolidinone (Furoxone) occurs as a yellow crystalline powder with a bitter aftertaste. It is insoluble in water or alcohol.

Furazolidone has bactericidal activity against a relatively broad range of intestinal pathogens including *S. aureus*, *E. coli*, *Salmonella*, *Shigella*, *Proteus*, *Enterobacter*, and *Vibrio cholerae*. It is also active against *Giardia lamblia*. It is recommended for the oral treatment of bacterial or protozoal diarrhea caused by susceptible organisms. The usual adult dosage is 100 mg four times daily.

Only a small fraction of an orally administered dose of furazolidone is absorbed. Approximately 5%

of the oral dose is detectable in the urine in the form of several metabolites. Some gastrointestinal distress has been reported with its use. Alcohol should be avoided when furazolidone is being used because the drug can inhibit aldehyde dehydrogenase.

PRESERVATIVES

Preservatives are added to various liquid dosage forms and cosmetic preparations to prevent microbial contamination. In parenteral and ophthalmic preparations, preservatives are used to maintain sterility in the event of accidental contamination during use.

The ideal preservative would be effective at low concentrations against all possible microorganisms, nontoxic, compatible with other constituents used in the preparation, and stable for the shelf life of the preparation. The ideal preservative does not exist, but there is extensive experience with some of them. Sometimes, combinations of preservative agents are employed.

p-HYDROXYBENZOIC ACID DERIVATIVES

Esters of *p*-hydroxybenzoic acid (parabens) have antifungal properties. Their toxicity is generally low owing to rapid hydrolysis in vivo to *p*-hydroxybenzoic acid, which is rapidly conjugated and excreted. These properties make parabens useful as preservatives for liquid dosage forms. The preservative effect tends to increase with molecular weight, but the methyl ester is more effective against molds, whereas the propyl ester is more effective against yeasts. The more oil-soluble propyl ester is the preferred preservative for oils and fats.

Methylparaben, NF. Methyl *p*-hydroxybenzoate, methylben, occurs as a white crystalline powder. It is soluble in water and alcohol, but only slightly soluble in nonpolar organic solvents. It is used as a preservative primarily to protect against molds.

Propylparaben, NF. Propyl *p*-hydroxybenzoate, propylben, occurs as a white crystalline powder that is only slightly soluble in water but soluble in most organic solvents. It is used as a preservative, primarily against yeasts.

HO–C_6H_4–C(=O)–O–CH_3

Methylparaben

Ethylparaben, NF. Ethyl-*p*-hydroxybenzoate is a white crystalline powder that is slightly soluble in water, but soluble in alcohol and most organic solvents.

Butylparaben, NF. Butyl-*p*-hydroxybenzoate occurs as a white crystalline powder that is sparingly soluble in water but very soluble in alcohols and in nonpolar organic solvents.

OTHER PRESERVATIVES

Chlorobutanol, NF. 1,1,1-Trichloro-2-methyl-2-propanol (Chloretone) is a white crystalline solid with a camphorlike odor. It occurs in an anhydrous form and a hemihydrate form, both of which sublime at room temperature and pressure. Chlorobutanol is slightly soluble in water and soluble in alcohol and in organic solvents.

CH_3–C(CH_3)(CCl_3)–OH

Chlorobutanol

Chlorobutanol is employed as a bacteriostatic agent in pharmaceuticals for injection, ophthalmic use, and intranasal administration. It is unstable when heated in aqueous solutions, especially at *p*Hs above 7. It undergoes a haloform reaction. Solutions of *p*H 5 or lower are reasonably stable at 25°C. It is stable in oils and organic solvents.

TABLE 4-7
PARABENS

Name	Preparations	Category	Solubility in Water
Methylparaben NF	Hydrophilic ointment USP	Pharmaceutic aid (antifungal preservative)	1 : 400
Ethylparaben NF		Pharmaceutic aid (antifungal preservative)	1 : 600
Propylparaben NF	Hydrophilic ointment USP	Pharmaceutic aid (antifungal preservative)	1 : 2,500
Butylparaben NF		Pharmaceutic aid (antifungal preservative)	1 : 5,000

Benzyl Alcohol, NF. Phenylcarbinol; phenylmethanol. Benzyl alcohol is found unesterified in oil of Jasmine and as esters of acetic, cinnamic, and benzoic acids in gum benzoin, storax resin, Peru balsam, tolu balsam, and in some volatile oils. It is soluble in water and alcohol. It is a clear liquid with an aromatic odor.

CH_2OH

Benzyl Alcohol

Benzyl alcohol is commonly used as a preservative in vials of injectable drugs in concentrations of 1% to 4% in water or saline solution. It has the added advantage of having a local anesthetic action. Benzyl alcohol is also used in ointments and lotions as an antiseptic in the treatment of various pruritic skin conditions.

Phenylethyl Alcohol, USP. 2-Phenylethanol; orange oil; rose oil; $C_6H_5CH_2CH_2OH$ is a clear liquid that is sparingly soluble in water (2%). It occurs naturally in rose oil, pine needle oil, and Neroli. It is used primarily in perfumery.

Benzoic Acid, USP. Benzoic acid and its esters occur naturally in gum benzoin and in Peru and tolu balsams. It is found as a white crystalline solid that slowly sublimes at room temperature and is steam distillable. It is slightly soluble in water (0.3%), but more soluble in alcohol and other polar organic solvents. It has a pK_a of 4.2.

COOH

Benzoic Acid

Benzoic acid is employed externally as an antiseptic in lotions, ointments, and mouthwashes. It is more effective as a preservative in foods and pharmaceutic products at low pH (below the pK_a). When used as a preservative in emulsions, its effectiveness depends upon both pH and distribution into the two phases.[10]

Sodium Benzoate, NF is a white crystalline solid that is soluble in water and alcohol. It is used as a preservative in acidic liquid preparations in which benzoic acid is released.

Sodium Propionate, NF occurs as transparent colorless crystals that are soluble in water and in alcohol. It is an effective antifungal agent that is used as a preservative. Sodium propionate is most effective at low pH.

Sorbic Acid, NF. 2,4-Hexadienoic acid is an effective antifungal preservative. It is sparingly soluble in water and has a pK_a of 4.8. Sorbic acid is used to preserve syrups, elixers, ointments, and lotions containing components, such as sugars, that support mold growth.

$CH_3CH{=}CHCH{=}CHCOOH$

Sorbic Acid

Potassium Sorbate, NF occurs as a white crystalline powder that is soluble in water and in alcohol. It is used in the same way as sorbic acid when greater water solubility is required.

Phenylmercuric Nitrate, NF is a mixture of phenylmercuric nitrate and phenylmercuric hydroxide. It occurs as a white crystalline powder that is sparingly soluble in water and slightly soluble in alcohol. It is used in concentrations of 1 : 10,000 to 1 : 50,000 to preserve injectables against bacterial contamination. Organomercurials have the disadvantage of having their bacteriostatic effectiveness reduced in the presence of serum.

$-HgNO_3$ $-HgOH$

Phenylmercuric Nitrate — Phenylmercuric Hydroxide

Phenylmercuric Acetate, NF. Acetoxyphenylmercury occurs as white prisms that are soluble in alcohol but only slightly soluble in water. It is used as a preservative.

TABLE 4-8

PRESERVATIVES

Name	Use
Chlorobutanol NF	Antimicrobial
Benzyl alcohol NF	Bacteriostatic (injections)
Phenylethyl alcohol USP	Bacteriostatic
Sodium benzoate NF	Antifungal
Sodium propionate NF	Antifungal
Sorbic acid NF	Antimicrobial
Potassium sorbate NF	Antimicrobial
Phenylmercuric nitrate NF	Bacteriostatic
Phenylmercuric acetate NF	Bacteriostatic

ANTIFUNGAL AGENTS

Most fungal infections (mycoses) involve superficial invasion of the skin or the mucous membranes of body orifices. These diseases, which can usually be controlled by local application of an antifungal agent, are conveniently divided into two etiologic groups: (1) the dermatophytoses (tinea infections), which are contagious superficial epidermal infections caused by various *Epidermophyton*, *Microsporum*, and *Trichophyton* species; and (2) mycoses caused by pathogenic saphrophytic yeasts, which are contagious and usually superficial infections involving the skin and mucous membranes. Some species of saphrophytic yeasts (*Aspergillus*, *Blastomyces*, *Candida*, *Coccidioides*, *Cryptococcus*, *Histoplasma*) under certain conditions are capable of invading deeper body cavities and causing systemic mycoses. Such infections may become serious and occasionally life-threatening, and they are frequently difficult to treat.

Fatty acids in perspiration have been found to be fungistatic, and this discovery has led to the introduction of fatty acids in therapy. The use of copper and zinc salts provides the added antifungal activity of the metal ion. Aromatic acids, especially salicylic acid, which also has a useful keratolytic action, and its derivatives, are employed for their topical fungistatic effect. A variety of alkylated or halogenated phenols and their derivatives are useful for the treatment of local fungal infections. The antifungal activity of the aforementioned compounds is largely confined to local dermatophytic infections. Deep dermatophytic infections, resistant to topical therapy, may be treated systemically with the antibiotic griseofulvin. Several years ago, it was discovered that local and gastrointestinal yeast infections, which became prevalent as superinfections that resulted from the misuse of broad-spectrum antibiotics, such as the tetracyclines, could be combated effectively with the polyene antibiotic nystatin. More recently, two additional polyene antibiotics, first candicidin and later pimaricin, have been introduced for the topical treatment of yeast infections. Research in Europe led to the discovery that certain highly substituted imidazole derivatives possessed broad-spectrum antifungal activity. Five substituted imidazoles are currently marketed in the United States.

Very few compounds have yet been found that combine the properties required for the treatment of systemic yeast infections, namely, effectiveness against the causative organisms and a reasonable margin of safety. For many years, amphoteracin B (a polyene antibiotic) and flucytosine were the only two agents available for systemic fungal infections. More recently, two substituted imidazoles, miconazole and ketoconazole, have been approved for systemic administration.

SYNTHETIC ANTIFUNGAL AGENTS

Fatty Acids

All fatty acids have fungicidal properties. The higher-molecular-weight members have the advantage of having lower volatility. The salts of fatty acids are also fungicidal and provide nonvolatile forms for topical administration. Because of their availability, the fatty acids and salts thereof that are the most widely used are propionic, caprylic, and undecylenic.

Propionic Acid is a readily available fungicide that is nonirritating and nontoxic. It is present in sweat in low concentrations (around 0.01%). Various salts, such as those of sodium, potassium, calcium, ammonium, and zinc are also fungicidal.

Propionic acid is a clear, corrosive liquid, with a characteristic odor. It is soluble in water and in alcohol. The salts are generally used, instead of the free acid, because they are nonvolatile and odorless.

Zinc Propionate occurs as both an anhydrous form and as a monohydrate. It is very soluble in water, but only sparingly soluble in alcohol. The salt is unstable to moisture, forming zinc hydroxide and propionic acid. It is employed as a fungicide, particularly on adhesive tape.

Sodium Caprylate is prepared from caprylic acid, which is found in coconut and palm oils. The salt occurs as cream-colored granules that are soluble in water and sparingly soluble in alcohol.

$$CH_3(CH_2)_5CH_2COO^-Na^+$$

Sodium Caprylate

Sodium caprylate is used topically to treat superficial fungal infections caused by *Candida albicans* and *Trichophyton*, *Microsporum*, and *Epidermophyton* species. It is found in solution, powder, and ointment forms.

Zinc Caprylate is a fine white powder that is virtually insoluble in water or alcohol. It is used as a topical fungicide. The copper salt is also found in various proprietary antifungal preparations. These salts must be kept in tightly closed containers to protect them from moisture.

Undecylenic Acid, USP. 10-Undecenoic acid (Desenex, Cruex, Decylenes) has the formula $CH_2{=}CH(CH_2)_8CO_2H$. It is obtained from the destructive distillation of castor oil. Undecylenic acid occurs as viscous yellow liquid having a characteristic odor. The acid is nearly insoluble in water, but soluble in alcohol and most organic solvents.

Undecylenic acid is one of the better fatty acids available as a fungicide. It may be used in concentrations up to 10% in solutions, ointments, emulsions, and powders for topical administration. It is considered too irritating, however, to be applied to mucous membranes. Several salts including those of sodium, potassium, zinc, and copper are also used, some in combination with the acid.

Triacetin, USP. Glyceryl triacetate (Enzactin, Fungacetin) is a colorless oily liquid, with a slight odor and a bitter taste. It is soluble in water and miscible with alcohol and most organic solvents.

The fungicidal properties of triacetin are due to acetic acid formed by enzymatic hydrolysis by esterases of the skin. The rate of acetic acid formation is self-limiting because the esterases are inactivated below a *p*H of 4.

Salicylic Acid, USP. *o*-Hydroxybenzoic acid is a strong acid (pK_a = 2.5) with antiseptic and keratolytic properties. It occurs as white, needlelike crystals or as a fluffy, crystalline powder. It is only slightly soluble in water, but soluble in most organic solvents. The greater acidity of salicylic acid and its lower solubility in water, as contrasted with *p*-hydroxybenzoic acid, are the consequence of intramolecular hydrogen bonding.

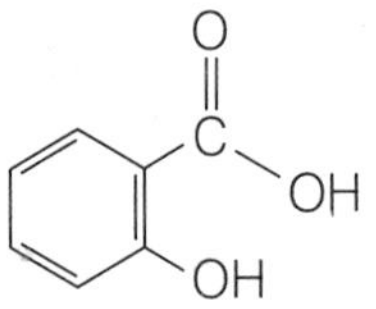

Salicylic Acid

Salicylic acid is used externally in ointments and solutions for its antiseptic and escharotic properties. Many foot remedies for corns and athlete's foot contain salicylic acid.

Tolnaftate, USP. *O*-2-Naphthyl *m*,*N*-dimethylthiocarbanilate (Tinactin, Aftate, Footwork, Fungatin, NP-27) occurs as a white crystalline solid that is insoluble in water, sparingly soluble in alcohol and soluble in most organic solvents.

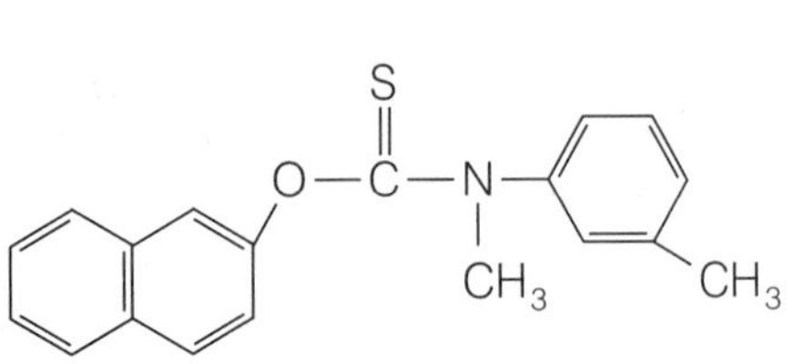

Tolnaftate

This compound, which is a thioester of β-naphthol, is fungicidal against dermatophytes, such as *Trichophyton*, *Microsporum*, and *Epidermophyton* species that cause superficial tinea infections. It is available in a concentration of 1% in creams, powders, aerosols, gels, and solutions for the treatment of ringworm, jock itch, and athlete's foot.

Acrisorcin, USP. 9-Aminoacridinium 4-hexylresorcinolate (Akrinol) is the salt of 9-aminoacridine and hexylresorcinol. It occurs as yellow crystals that are slightly soluble in water and soluble in alcohol. Both components of the salt possess antibacterial and antifungal properties.

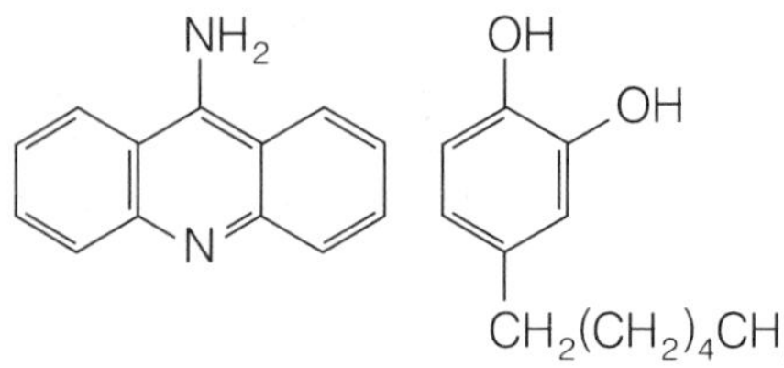

Acrisorcin

Acrisorcin is used to treat tinea versicolor caused by *Pityrosporon orbiculare* (*Malassezia furfur*). A six-week treatment period is recommended.

Haloprogin, USP. 3-Iodo-2-propynyl 2,4,5-trichlorophenyl ether (Halotex) occurs as white to pale yellow crystals that are sparingly soluble in water and easily soluble in ethanol. It is used as a 1% cream or solution for the treatment of superficial tinea infections. Formulations of haloprogin should be protected from the light owing to the sensitivity of the compound. It is available as a solution and as a cream, both containing 1% of the fungicide.

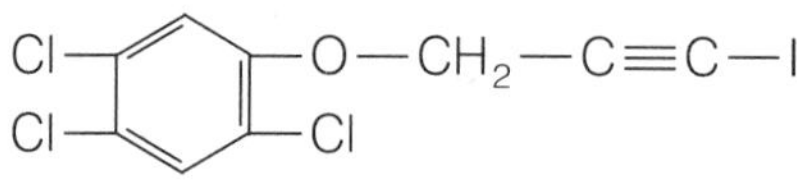

Haloprogin

Ciclopirox Olamine, USP. 6-Cyclohexyl-1-hydroxy-4-methyl-2-(1*H*)-pyridinone ethanolamine salt (Loprox) is a broad-spectrum antifungal agent intended for topical use. It is active against dermatophytes, as well as pathogenic yeasts, such as *C. albicans*, that cause superficial fungal infections. It is supplied as a cream containing 1% of the salt.

Ciclopirox is believed to act on the cell membranes of susceptible fungi at low concentrations to block the transport of amino acids into the cell. At higher concentrations it disrupts membrane integrity causing loss of cellular constituents.[11]

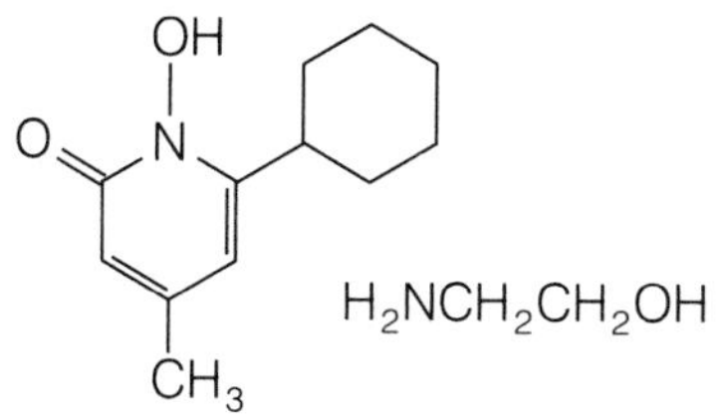

Ciclopirox Olamine

Substituted Imidazoles

The substituted imidazoles represent a relatively new class of versatile antifungal agents with an apparently unique mechanism of action. The properties of early members of the class, such as clotrimazole[12] and miconazole,[13] were reported independently in 1969. In general, the substituted imidazoles are effective against most fungi that cause superficial infections of the skin and mucous membranes, including the dermatophytes, such as *Trichophyton*, *Epidermophyton*, and *Microsporum* spp., and yeasts, such as *C. albicans*. They also exhibit activity against yeasts that cause systemic infections, including *Coccidioides immitis*, *Cryptococcus neoformans*, *Paracoccioides brasiliensis*, *Petriellidium boydii*, *Blastomyces dermatitidis*, and *Histoplasma capsulatum*.

The actions of the substituted imidazoles on mycotic biochemistry and physiology have been studied extensively. Nonetheless, the mechanisms by which they exert their antifungal effects remain to be fully elucidated.[14] At high concentrations (micromolar) the substituted imidazoles are fungicidal; at low concentrations (nanomolar) they are fungistatic. The fungicidal effect is associated with damage to the cell membrane, with the loss of essential cellular constituents, such as potassium ion and amino acids. Attempts have been made to correlate the inhibition of membrane-bound enzymes by low concentrations of the substituted imidazoles to their fungistatic effects. Some candidate enzymes are cytochrome P-450 oxidases that are required for the oxidative demethylation of 14α-methyl steroids (essential for cell membrane integrity); and mitochondrial cytochrome C oxidase and ATPase required for cellular respiration.

The primary structural requirement for members of this class is a weakly basic imidazole or thiazine ring bonded by nitrogen–carbon linkage to the rest of the molecule. The more potent compounds possess two or three aromatic rings, at least one of which is halogen substituted.

Clotrimazole, USP. 1-(*o*-Chloro-α,α-diphenylbenzyl)imidazole (Lotrimin, Mycelex) is a broad-spectrum antifungal agent used topically for the treatment of tinea infections and candidiasis. It occurs as a white crystalline solid that is sparingly soluble in water, but soluble in alcohol and most organic solvents. It is a weak base that is solubilized by dilute mineral acids.

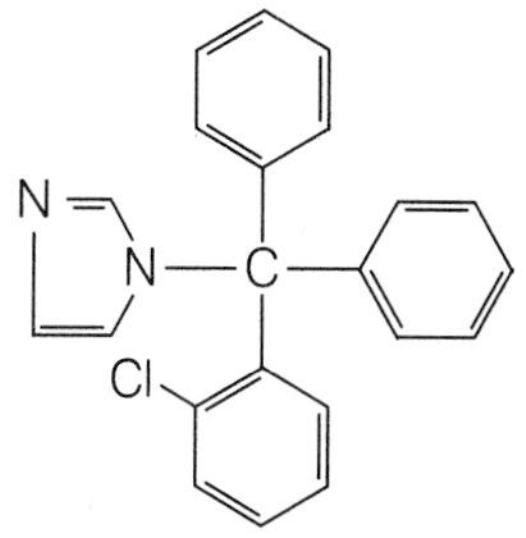

Clotrimazole

Clotrimazole is supplied as a solution in polyethylene glycol 400, a lotion, and a cream in a concentration of 1% for the treatment of tinea pedis, tinea cruis, tinea capitis, tinea versicolor, or candidiasis. A 1% vaginal cream and tablets of 100 and 500 mg are available for vaginal candidiasis. The compound is very stable, having a shelf life of more than five years.

Although clotrimazole is effective against a variety of pathogenic yeasts and is reasonably well absorbed orally, it causes severe gastrointestinal disturbances and is thus not considered suitable for the treatment of systemic infections.

Econazole Nitrate. 1-[2-[(4-Chlorophenyl)methoxy]-2-(2,4-dichlorophenyl)ethyl]-1*H*-imidazole (Spectrazole) is a white crystalline nitric acid salt of econazole. It is only slightly soluble in water and most common organic solvents.

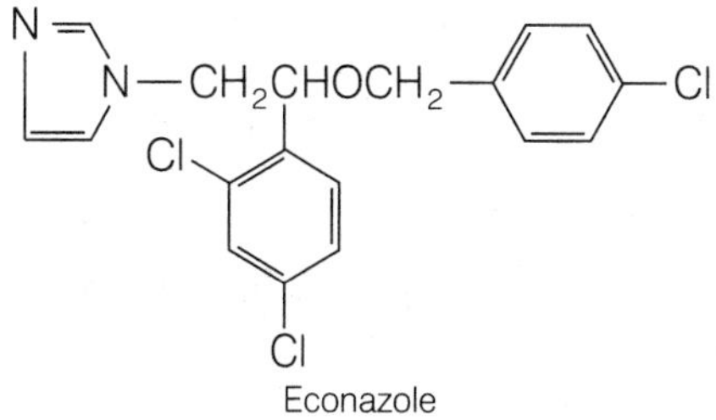

Econazole

Econazole is used as a 1% cream for the topical treatment of local tinea infections and cutaneous candidiasis.

Butoconazole Nitrate. 1-[4-(4-Chlorophenyl)-2-[(2,6-dichlorophenyl)-thio]butyl]-1*H*-imidazole (Femstat) is a broad-spectrum antifungal agent that is particularly effective against *C. albicans*. It is supplied as a vaginal cream containing 2% of the salt

salt and is intended for the treatment of vaginal candidiasis.

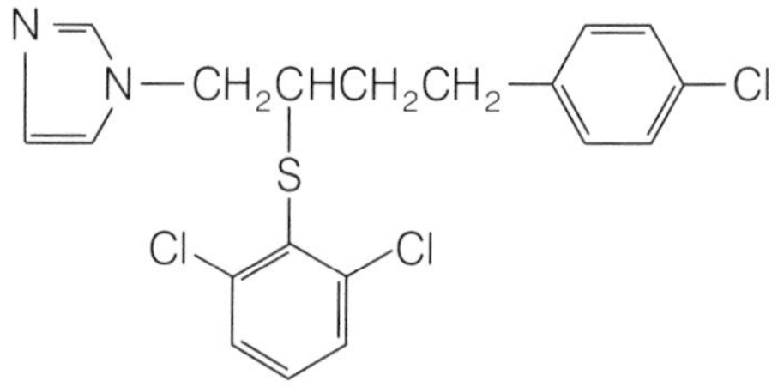

Butoconazole

Miconazole Nitrate, USP. 1-[2-(2,4-Dichlorophenyl)-2-[2, 4-dichlorophenyl]methoxy]ethyl]-1*H*-imidazole mononitrate (Monistat, Micatin) is a weak base with a pK_a of 6.65. The nitric acid salt occurs as a white crystalline salt that is sparingly soluble in water and most organic solvents.

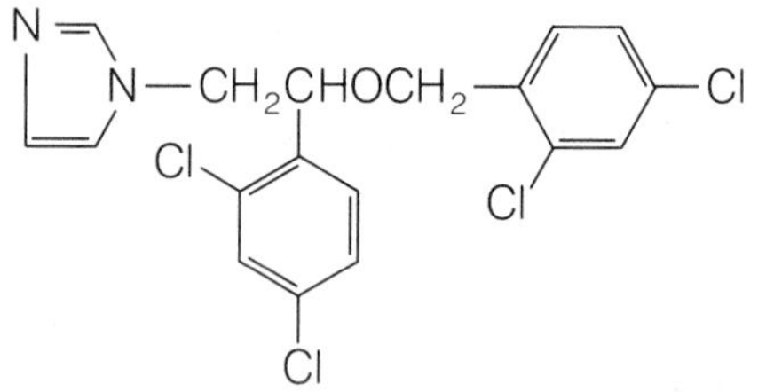

Miconazole

The free base is available in an injectable form, solubilized with polyethylene glycol and castor oil, intended for the treatment of serious systemic fungal infections, such as candidiasis, coccidioidomycosis, cryptococcosis, petriellidiosis, and paracoccidioidomycosis. It may also be used for the treatment of chronic mucocutaneous candidiasis. Although serious toxic effects from the systemic administration of miconazole are comparatively rare, thrombophlebitis, pruritus, fever, and gastrointestinal upset are relatively common.

Miconazole nitrate is supplied in a variety of dosage forms (a cream, a lotion, a powder, and a spray) for the treatment of tinea infections and cutaneous candidiasis. Vaginal creams and suppositories are also available for the treatment of vaginal candidiasis. A concentration of 2% of the salt is used in most topical preparations.

Ketoconazole, USP. 1-Acetyl-4-[4-[[2-(2,4-dichlorophenyl)-2-(1*H*-imidazole-1-ylmethyl)-1,3-dioxalan-4-yl]methoxy]phenyl]piperazine (Nizoral) is a broad-spectrum antifungal agent that is administered orally for the treatment of systemic fungal infections. It is a weakly basic compound that occurs as a white crystalline solid that is very slightly soluble in water.

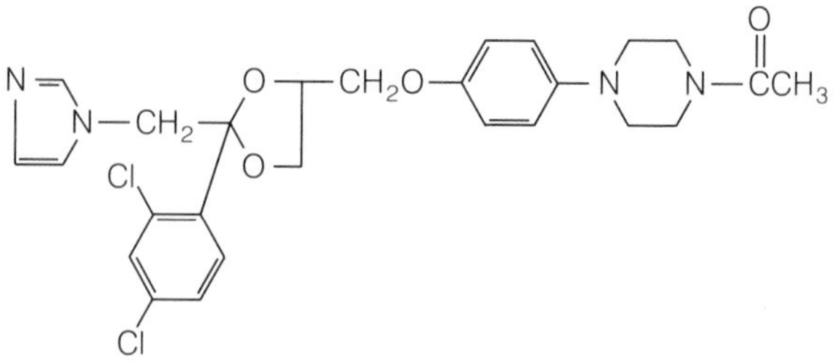

Ketoconazole

The oral bioavailability of ketoconazole depends upon an acidic *p*H for dissolution and absorption. Antacids and drugs, such as H_2-histamine antagonists and anticholinergics that inhibit gastric secretion, interfere with its oral absorption. Ketoconazole is extensively metabolized to inactive metabolites, and the primary route of excretion is enterohepatic. It is estimated to be 95% to 99% bound to protein in the plasma. Hepatotoxicity, primarily of the hepatocellular type, is the most serious adverse effect of ketoconazole. High doses have also been reported to lower testosterone and corticosterone levels.

Ketoconazole is recommended for the treatment of the following systemic fungal infections: candidiasis (including oral thrush and the chronic mucocutaneous form), coccidioidomycosis, blastomycosis, histoplasmosis, chromomycosis, and paracoccidioidomycosis. It is also used to treat severe refractory cutaneous dermatophytic infections not responsive to topical therapy or oral griseofulvin. The antifungal actions of ketoconazole and the polyene antibiotic amphoteracin B are reported to antagonize each other.

Flucytosine, USP. 5-Fluorocytosine; 5-FC; 4-amino-5-fluoro-2(1*H*)-pyrimidinone; 2-hydroxy-4-amino-5-fluoropyrimidine (Ancobon) is an orally active antifungal agent having a narrow spectrum of activity. It is indicated for the treatment of serious systemic infections caused by susceptible strains of *Candida* and *Cryptococcus*.

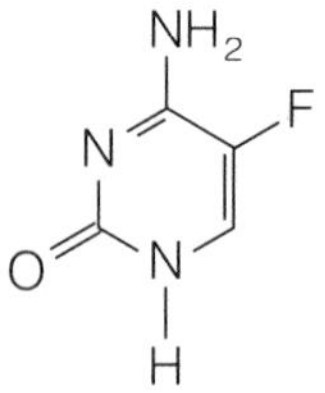

Flucytosine

TABLE 4-9

SYNTHETIC ANTIFUNGAL AGENTS

Name Proprietary Name	***Preparations***	***Application****	***Usual Adult Dose****	***Usual Pediatric Dose****
Undecylenic acid USP Zinc undecylenate USP *Desenex, Cruex*	Compound undecylenic acid Ointment USP Cream Aerosol powder	Topically to the skin, qd, hs as needed		
Triacetin USP *Enzactin, Fungacetin*	Ointment Cream Solution	Topically to the skin bid in a 25% ointment or cream		
Salicylic Acid USP	Salicylic acid ointment USP	Keratolytic topically to the skin, as a 3–10% ointment		
Clotrimazole USP *Lotromin* *Mycelex*	Clotrimazole cream	Topical, as a 1% cream bid for tinea		
	Clotrimazole topical solution USP Clotrimazole lotion	Topical, as a 1% solution or lotion bid for tinea		
	Clotrimazole vaginal cream USP	Intravaginal, 5 g of a 1% cream at bedtime for 7–14 consecutive days		
	Clotrimazole vaginal tablets	Intravaginal, 100 mg qd, preferably at bedtime, for 7 consecutive days		
Miconazole nitrate USP *Monistat* *Micatin*	Miconazole nitrate cream USP lotion	Topical, as a 2% cream or lotion to the affected areas morning and evening		
	Miconazole nitrate powder spray	Spray or sprinkle a 2% spray or powder liberally over affected area in the morning and in the evening		
	Miconazole nitrate vaginal cream USP	Intravaginal, one applicatorful of a 2% cream qd, hs, for 7 days		
	Miconazole nitrate vaginal suppositories	Intravaginal, one 200 mg suppository qd, hs, for 3 days		
Miconazole injection USP *Monistat IV*	Miconazole injection USP		Slow IV infusion of the equivalent of miconazole of 200 mg–3.6 g/day. Each 200 mg of miconazole injected must be diluted with a minimum of 200 mL of fluid (0.9 sodium chloride or 5% dextrose)	
Econazole nitrate *Spectazole*	Econazole nitrate cream	Topically to the skin as a 1% cream bid		
Butoconazole nitrate *Femstat*	Butaconazole nitrate vaginal cream	Intravaginal, 15 g applicatorful of a 2% cream qd, hs, for 6 days.		
Ketoconazole USP *Nizoral*	Ketoconazole USP Tablets Oral suspension		Oral: 200 mg qd initially, may be increased to 400 mg qd	Children under 2 years: 3.3–6.6 mg/kg daily
Tolnaftate USP *Tinactin* *Aftate*	Tolnaftate powder USP Tolnaftate aerosol powder USP Tolnaftate cream USP Tolnaftate topical solution USP	Topical, as a 1% powder, aerosol, cream or solution bid for tinea		

*See USP DI for complete dosage information.

TABLE 4-9 *Continued*

SYNTHETIC ANTIFUNGAL AGENTS

Name Proprietary Name	*Preparations*	*Application**	*Usual Adult Dose**	*Usual Pediatric Dose**
Acrisorcin USP *Akrinol*	Acrisorcin cream USP	Topical, as a 0.2% cream to the affected area bid for tinea		
Haloprogin USP *Halotex*	Haloprogin cream USP Haloprogin topical solution USP	Topical, as a 1% cream of solution to the affected area bid for tinea		
Clioquinol USP *Vioform*	Clioquinol cream USP Clioquinol ointment USP	Topical, as a 3% cream or ointment to the affected area for tinea		
Ciclopirox olamine USP *Loprox*	Ciclopirox olamine cream USP	Topical, as a 1% cream to the affected area bid for tinea		
Flucytosine USP	Flucytosine capsules USP		Orally, 50–150 mg/kg daily at 6-hr intervals	

*See USP DI for complete dosage information.

The mode of action of 5-flucytosine in susceptible fungi and mechanisms of resistance to the drug in nonsusceptible strains have been studied in some detail.[15] Incorporation of fluorinated pyrimidine into RNA following selective deamination to 5-fluorouracil (5-FU) in the fungus appears to be required for antifungal activity. Resistant strains appear to either be deficient in enzymes required for the bioactivation of 5-FC (e.g., uridinemonophosphate pyrophosphorylase or cytosine deaminase) or have a surplus of de novo pyrimidine-synthesizing capacity. The comparative lack of toxicity of 5-FC in humans is apparently because it is not deaminated to the toxic antimetabolite 5-FU in cells of the host after oral administration. The half-life of flucytosine, which is excreted largely unchanged, is three to six hours.

ANTIFUNGAL ANTIBIOTICS

The Polyenes

A number of structurally complex antifungal antibiotics isolated from soil bacteria were discovered to contain a conjugated system of double bonds in large lactone rings. They differ from the antibacterial macrocyclic lactones (macrolides; see Chap. 7) of the erythromycin type in the size of the lactone ring and the presence of the conjugated ene system and, hence, are referred to as the polyenes. These antibiotics fall into two groupings, based on the size of the macrolide ring: the 26-membered ring polyenes, such as natamycin (pimaricin) and the 38-membered ring polyenes such as nystatin, amphotericin B, and candicidin. A glycosidically linked deoxyaminohexose sugar, mycosamine, is common to the currently available polyenes. They also differ in the number of double bonds present in the lactone ring: natamycin is a tetraene; nystatin a hexaene; and amphotericin B and candicidin are heptaenes.

The polyenes are broad-spectrum antifungal agents with potent activity against pathogenic yeasts, molds, and dermatophytes. Important pathogenic fungi inhibited by low concentrations of these agents in vitro include: *Candida* sp., *Coccidioides immitis*, *Cryptococcus neoformans*, *Histoplasma capsulatum*, *Blastomyces dermatitidis*, *Rhodotorula* sp., *Sporothrix schenckii*, *Mucor mucedo*, *Aspergillis fumigatus*, *Cephalosporium* sp., and *Fusarium* sp. Polyenes are without effect against bacteria, rickettsia, or viruses. They do, however, possess activity against certain protozoa, such as *Leishmania* sp. Less expensive compounds are used to treat tinea infections. The usefulness of the polyenes for the treatment of systemic infections is limited by their toxicities, low water solubilities, and poor chemical stabilities. Thus, amphotericin B, the only polyene available for the treatment of serious systemic infections, must be solubilized with the aid of an emulsifying agent. The use of other members of the class is confined to the topical treatment of superficial fungal infections.

The mode of action of the polyene antibiotics has been extensively investigated.[15] They appear to bind with sterols in the cell membrane to cause disorganization and loss of cell constituents, especially potassium ion. The action may be fungistatic or fungicidal, suggesting an initial effect on membrane-bound enzymes (such as ATPase) at low concentrations and a generalized membrane disruption at higher concentrations.

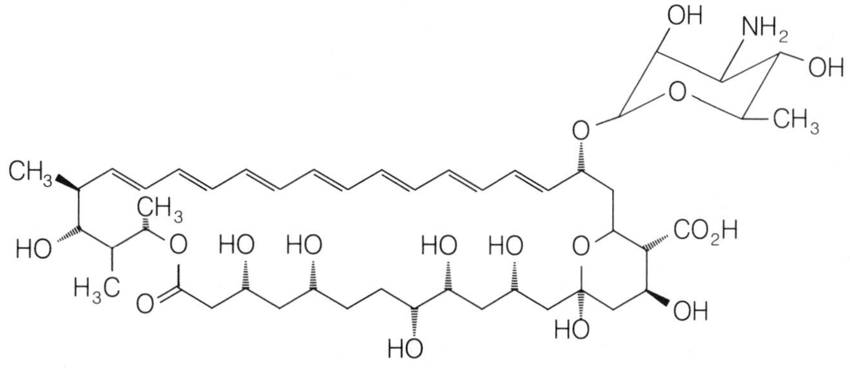

Amphotericin B

Amphotericin B, USP (Fungizone). A potent antifungal substance with a polyene structure was isolated in 1956 by Gold et al.[16] from an actinomycete, *Streptomyces nodosus*, found in a soil sample from Venezuela. The antibiotic material was a mixture of two closely related compounds designated amphotericins A and B. The B compound was more active and is consequently the one employed therapeutically. Its structure and absolute stereochemistry have been determined[17–19] as shown.

As the name implies, amphotericin B is an amphoteric substance containing a primary amino group in the sugar moiety, mycosamine, and a carboxyl group attached to the macrolide ring. It occurs as a deep yellow crystalline solid that is sparingly soluble in polar organic solvents, but insoluble in water. Although it forms salts with both acids and alkalis, these salts are only slightly soluble in water (about 0.1 mg/mL) and thus are not suitable for systemic administration. The parenteral dosage form is an aqueous colloidal dispersion stabilized with sodium deoxycholate. The compound is light- and heat-sensitive.

Parenterally, amphotericin B is indicated for the treatment of serious, potentially life-threatening fungal infections, including disseminated forms of coccidioidomycosis and histoplasmosis, sporotrichosis, North American blastomycosis, cryptococcosis (torulosis), mucormycosis, and aspergillosis. It may be beneficial in the treatment of American mucocutaneous leishmaniasis, but it is not a drug of choice.

A high prevalence of adverse reactions limits the usefulness of amphotericin B. Some form of nephrotoxicity occurs in nearly 80% of the patients. Generalized toxic reactions, such as fever, headache, anorexia, gastrointestinal distress, muscle and joint pain, and malaise, are also common. Pain at the injection site and thrombophlebitis frequently occur when the drug is injected intravenously. It must never be administered intramuscularly. The hemolytic action of amphotericin B may be the result of its affinity for membrane lipids and cholesterol.

Amphotericin B for injection is supplied as a sterile lyophylized cake or powder containing 50 mg of active ingredient with 41 mg of sodium deoxycholate to be dispersed in 10 mL of water. The infusion solution providing 0.1 mg/mL is then obtained by further dilution (1 : 50) with 5% dextrose injection. Normal saline should not be used because it will destroy the suspension. The suspension should be freshly prepared and used within 24 hours. The drug powder should be refrigerated and protected from light.

Amphotericin B is also used topically to treat cutaneous and mucocutaneous mycotic infections caused by *C. albicans*. It is supplied in a variety of topical forms including a cream, a lotion, and an ointment. The concentration of antibiotic in these preparations is 3%.

Nystatin, USP (Mycostatin, Nilstat, Mykinac, Nystex) is a polyene antibiotic first isolated in 1951 from a strain of *S. noursei* by Hazen and Brown.[20] It occurs as a yellow to light tan powder with a cereallike odor. Nystatin is only very slightly soluble in water and sparingly soluble in organic solvents. It is unstable to moisture, heat, and light.

Nystatinolide, the aglycon portion of nystatin, consists of a 38-membered lactone ring with single tetracene and diene chromophores isolated from each other by a methylene group, one carboxyl, one keto, and eight hydroxyl groups. It is glycosidically linked to the amino sugar mycosamine. The complete structure of nystatin has been determined by x-ray crystallographic and chemical degradation procedures.[21,22] A minor revision of the originally proposed structure[19] shows a hemiacetal formed between the keto group at C-13 and the hydroxyl group at C-17.

Nystatin has been a valuable agent for the treatment of local and gastrointestinal monilial infections caused by *C. albicans* and other *Candida* species for nearly three decades. For the management of

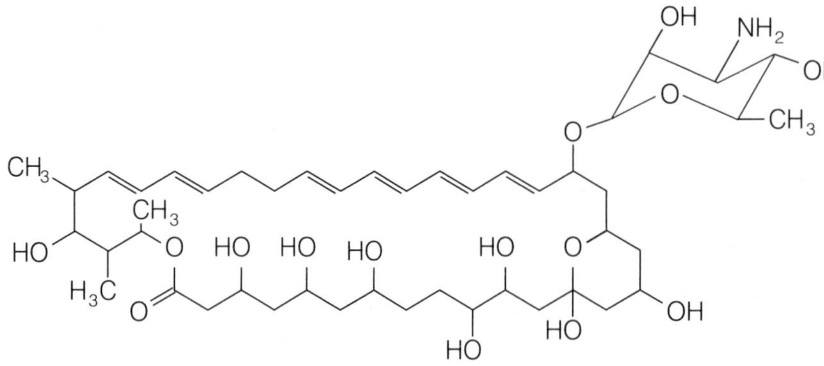

Nystatin

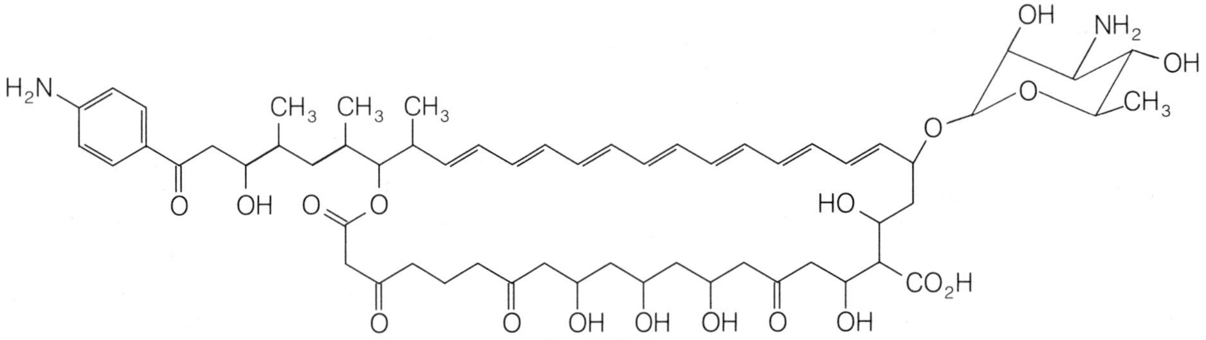

Candicidin D

cutaneous and mucocutaneous candidiasis it is supplied as a cream, an ointment, and a powder. Vaginal tablets are available for the control of vaginal candidiasis. Oral tablets and troches are used in the treatment of gastrointestinal and oral candidiasis. The systemic absorption of nystatin following oral administration is practically nil. Combinations of nystatin with tetracyclines have been long employed to prevent monilial overgrowth caused by destruction of the bacterial microflora of the intestine. Most informed medical opinion now appears to favor treatment of intestinal candidiasis only after it occurs secondary to tetracycline therapy.

Despite that nystatin is a pure compound of known structure, its dosage is still expressed in terms of units. One milligram of nystatin contains not less than 2000 USP units.

Candicidin, USP (Candeptin). The macrolide polyene antibiotic candicidin was isolated in 1953 by Lechevalier et al.[23] from a strain of *S. griseus*. Although its potent antifungal properties had been known for some time, it was not until 1964 that candicidin became available for medical use in the United States. It is recommended for use in the treatment of vaginal candidiasis.

The commercial product is actually a mixture of four closely related heptaene macrolides, candicidins A, B, C, and D, similar to amphotericin B. The structure of the principal component, candicidin D, was determined in 1979 by Zielinski et al.[24]

Candicidin is available as a 3-mg vaginal tablet and a vaginal ointment containing 3 mg of candicidin in 5 g of ointment.

Natamycin, USP (pimaricin; Natacyn) is a polyene macrolide antibiotic obtained from *S. natalensis*. It was first isolated by Struyk et al.,[25] and its structure was elucidated eight years later.[26] A minor revision of the originally proposed structure has been subsequently reported.[19] The natamycin structure consists of a 26-membered lactone ring containing a tetraene chromophore, a double bond conjugated with the lactone carbonyl group, three hydroxyl groups, one keto group forming a hemiacetal with one of the hydroxyl groups,[19] an epoxide,

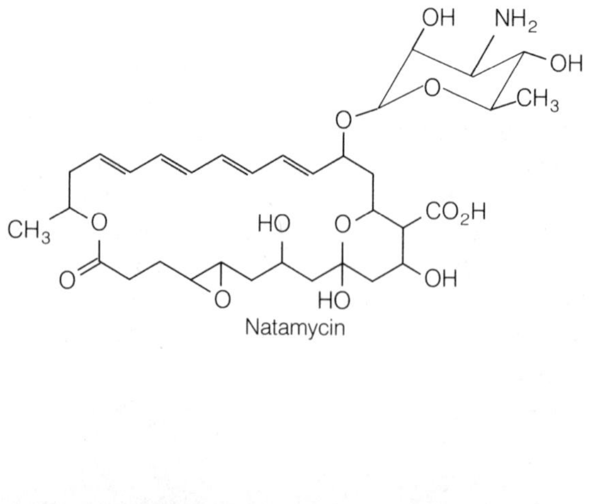

Natamycin

and a carboxyl group. The sugar mycosamine, common to all of the polyenes, is present. Of course, natamycin is an amphoteric substance.

Studies on the mechanism of action of the polyene antibiotics indicate differences in their effects on fungal membranes between smaller-ring macrolides, such as natamycin, and the larger-ring macrolides, such as amphotericin B and nystatin.[27] The 26-membered ring polyenes cause both potassium ion leakage and cell lysis at the same concentration, whereas the 38-membered ring polyenes cause potassium ion leakage at low, fungistatic concentra-

TABLE 4-10

ANTIFUNGAL ANTIBIOTICS

Name Proprietary Name	***Preparations***	***Application****	***Usual Adult Dose****	***Usual Pediatric Dose****
Nystatin	Nystatin vaginal tablets USP		100,000 units qid or bid for 2 wk	
	Nystatin topical powder USP Nystatin cream USP Nystatin lotion USP	Topically to the skin, as a 100,000 units/g powder, cream, or lotion, bid to tid		
Nystatin USP *Mycostatin, Nilstat*	Nystatin ointment USP	Topically to the skin, as 100,000 units/g ointment bid or tid		
	Nystatin oral suspension USP		400,000–600,000 units qid	Premature and low-birth-weight infants: 100,000 units 4 qid; older infants: 200,000 units qid; children: see usual adult dose
	Nystatin tablets USP		500,000–1 million units tid	
Amphotericin B USP *Fungizone*	Amphotericin B for injection USP		IV infusion, 250 μg/kg of body weight in 500 mL of 5% dextrose injection, over 6 hr	
	Amphotericin B cream USP Amphotericin B lotion USP Amphotericin B ointment USP	Topically to the skin, as a 3% cream, lotion, or ointment		
Candicidin USP *Candeptin, Vanobid*	Candicidin ointment USP	Vaginal, 0.06% ointment bid for 14 days		
	Candicidin vaginal tablets USP	Vaginal, 3-mg suppository bid for 14 days		
Natamycin USP *Natacyn*	Natamycin ophthalmic suspension USP	Topically to the conjunctival sac, one drop of a 5% suspension every 1 to 2 hr		
Griseofulvin USP *Fulvicin U/F, Grifulvin V, Grisactin*	Griseofulvin tablets USP Griseofulvin capsules USP Griseofulvin oral suspension USP		Tinea corporis, tinea cruris, or tinea capitis: oral, 500 mg/day, as a single dose or in divided doses; tinea pedis or tinea unguium: 1 g/day in divided doses	Oral, 10 mg/kg of body weight or 300 mg/m^2 of body surface daily, as a single dose or in divided doses, or for children 14–23 kg: oral, 125–250 mg/day as a single dose or in divided doses; children 23 kg and over: 250–500 mg/day
	Ultramicrosize griseofulvin USP		Tinea corporis, tinea cruris, or tinea capitis: oral, 330–375 mg/day as a single dose or in divided doses; tinea pedis or tinea unguium: 660–750 mg/day	7.3 mg/kg daily

*See USP DI for complete dosage information.

tions and cell lysis at high, fungicidal concentrations. The smaller-ring polyenes are fungistatic and fungicidal within the same concentration range.

Natamycin possesses in vitro activity against a variety of yeasts and filamentous fungi including *Candida*, *Aspergillis*, *Cephalosporium*, *Penicillium*, and *Fusarium* species. It is supplied as a 5% ophthalmic suspension intended for the treatment of fungal conjunctivitis, blepharitis, and keratitis.

Griseofulvin, USP (Grisactin, Fulvicin, Grifulvin, Gris-PEG). Griseofulvin was first reported in 1939 by Oxford et al.[28] as an antibiotic obtained from the mold *Penicillium griseofulvum* Dierckx. Although it had been used for its antifungal action in plants and animals for many years, it was not until 1959 that griseofulvin was introduced into human medicine in the United States for the systemic treatment of tinea infections.

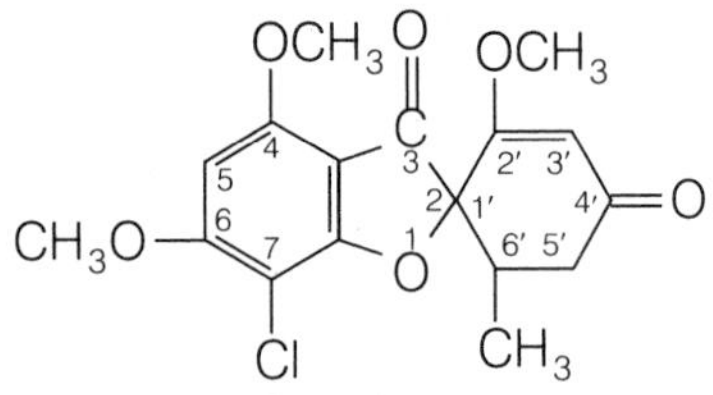

Griseofulvin

The structure of griseofulvin was determined by Grove et al.[29] to be 7-chloro-2′,4,6-trimethoxy-6′,β-methylspiro[benzofuran-2(3*H*),-1′-[2]cyclohexene]-3,4′-dione. It occurs as a white, bitter, heat-stable powder or crystalline solid that is sparingly soluble in water, but relatively soluble in alcohol and in nonpolar organic solvents. It is very stable in the dry state.

Griseofulvin is recommended for the systemic treatment of refractory ringworm infections of the body, nails, hair, and feet (tinea corporus, tinea unguium, tinea capitis, and tinea pedis) caused by various species of dermatophytic fungi, including *Trichophyton*, *Microsporum*, and *Epidermophyton*. Following oral absorption, some griseofulvin is carried by the systemic circulation to the skin, hair, and fingernails where it concentrates in keratin precursor cells, which are gradually exfoliated and replaced by new tissue. As the new tissue develops, the fungistatic action of the griseofulvin present prevents further infection within it. Because the old tissue may continue to support fungal growth, the treatment must be continued until all of the infected tissue has been exfoliated. Consequently, therapy of infections in slow-growing tissues, such as the nails, must be continued for several months.

Relatively few adverse effects have been reported for griseofulvin. The most common are mild allergic reactions, such as skin rashes and urticaria. Occasional gastrointestinal upset, headache, dizziness, or insomnia may also occur. Griseofulvin has no antibacterial activity, nor is it effective against pathogenic yeasts, including *Pityosporum obiculare* (*Malassezia furfur*), the organism that causes tinea versicolor.

Studies on the mechanism of action of griseofulvin have concentrated on its ability to arrest cell division in metaphase in vitro. The drug causes a rapid, reversible dissolution of the mitotic spindle apparatus, apparently by binding with the tubulin dimer required for microtubule assembly.[30] The basis for its selective toxicity is not known, but may be related to the tendency of the drug to concentrate in tissues rich in keratin.

The oral bioavailability of griseofulvin is notoriously bad. It is a very lipophilic compound with very low water solubility. The most successful attempts to improve oral absorption have concentrated on improving the dissolution of the antibiotic in the gastrointestinal tract by reducing its particle size. Griseofulvin is supplied in "microsize" and "ultramicrosize" forms. The efficiency of gastrointestinal absorption of the ultramicrosize form is 1.5 times that of the microsize form, permitting a dosage reduction of one-third. The bioavailability of griseofulvin may also be increased by administering the drug with a fatty meal. Although several structural analogues have been synthesized, none of them has shown activity superior to that of griseofulvin.

URINARY TRACT ANTI-INFECTIVES

Certain antibacterial agents that, by virtue of their solubility properties, concentrate in the urine are effective in the treatment of infections of the urinary tract. Generally speaking, the more typical first-time infections, such as acute cystitis, can be eradicated with use of either a short-acting sulfonamide (see Chap. 5) or the antibiotic tetracycline (see Chap. 7). Acute renal tissue infections such as glomerulonephritis require use of an antibiotic that achieves high concentrations in the kidneys and to which the causative organism is highly susceptible. A β-lactam (penicillin or cephalosporin) or an aminoglycoside (see Chap. 7) may be used, depending on the causative organism. Chronic, recurring infections tend to be less responsive to therapy and often require long-term treatment or the use of several drugs. Enteric bacterial strains resistant to particular chemotherapeutic agents are frequently the cause. Cotrimoxazole, a combination of the dihydrofolate reductase inhibitor, trimethoprim, and the sulfonamide, sulfamethiazole (see Chap. 5), is a treatment of choice for chronic urinary tract infections. β-Lactam and aminoglycoside antibiotics are also frequently employed. In addition, there is a group of antibacterial agents described in this chapter that, because they achieve significantly higher concentrations in the urine and in the kidneys compared with other body fluids and tissues, are well suited for the treatment of chronic urinary tract infections. They are usually employed when antibiotics or sulfonamides are either ineffective, owing to the emergence of bacterial resistance, or contraindicated because the patient is allergic to them.

QUINOLONES

The quinolones comprise a series of synthetic antibacterial agents patterned after nalidixic acid, a naphthyridine derivative introduced for the treatment of urinary tract infections in 1963. Isosteric heterocyclic groupings in this class include the quinolines (e.g., oxolinic acid, norfloxacin, ciprofloxacin, perfloxacin, and amifloxacin), the naphthyridines (e.g., nalidixic acid and enoxacin), and the cinnolines (e.g., cinoxacin). Up to the present time, the clinical usefulness of the quinolones has been largely confined to the treatment of urinary tract infections. For this, good oral absorption, activity against common gram-negative urinary pathogens, and comparatively higher urinary (compared with plasma and tissue) concentrations are useful properties. However, as a result of extensive structure–activity investigations leading to compounds with enhanced potency, extended spectrum of activity, and improved distribution properties, the class has evolved to the point that certain newer members have potential utility for the treatment of a variety of serious systemic infections.

Structure–activity studies have shown that the 1,4-dihydro-4-oxo-3-pyridinecarboxylic acid moiety is essential for antibacterial activity. The pyridone system must annulate with an aromatic ring. Isosteric replacements of nitrogen for carbon atoms at positions 2 (cinnolines), 5 (1,5-naphthyridines), 6 (1,6-naphthyridines), and 8 (1,8-naphthyridines) are consistent with retained antibacterial activity. Although the introduction of substituents at position 2 greatly reduces or abolishes activity, positions 5, 6, 7 (especially), and 8 of the annulated ring may be substituted to good effect. For example, piperazinyl substitution at position 7 has been identified with the enhanced activity of members of the quinolone class against *Pseudomonas aeruginosa*. Fluorine atom substitution at position 6 is also associated with enhanced antibacterial activity. Alkyl substitution at the 1-position is essential for activity with lower alkyl (methyl, ethyl, cyclopropyl) compounds having the greater potency. Ring condensations at the 1,8-, 5,6-, 6,7-, and 7,8-positions also lead to active compounds.

The effective antibacterial spectrum of nalidixic acid and early members of the quinolone class (e.g., oxolinic acid and cinoxacin) is largely confined to gram-negative bacteria, including common urinary pathogens, such as *Escherichia coli*, *Klebsiella*, *Enterobacter*, *Citrobacter*, and *Proteus* species. *Shigella*, *Salmonella*, and *Providencia* are also susceptible. Strains of *P. aeruginosa*, *Neisseria gonorrhoeae*, and *Haemophilus influenzae* are resistant, as are the gram-positive cocci and anaerobes. Newer members of this class possessing 6-fluoro and 7-piperazinyl substituents exhibit an extended spectrum of activity that includes effectiveness against additional gram-negative pathogens (such as *P. aeruginosa*, *H. influenzae*, and *N. gonorrhoeae*), gram-positive cocci (such as *Staphylococcus aureus*), and anaerobes (such as *Bacteriodes fragilis*).

The bactericidal action of nalidixic acid and its congeners is known to result from the inhibition of DNA synthesis. This effect is believed to be due to the inhibition of bacterial DNA gyrase (topoisomerase II), an enzyme responsible for introducing negative supercoils into circular duplex DNA.[31] Negative supercoiling relieves the torsional stress of unwinding helical DNA and, thereby, allows transcription and replication to occur. Although nalidixic acid inhibits gyrase activity, it binds only to single-stranded DNA and not to either the enzyme or double-helical DNA.[32] The binding of nalidixic acid to single-stranded DNA apparently does not require metal ions as had previously been supposed.

Nalidixic Acid, USP. 1-Ethyl-1,4-dihydro-7-methyl-4-oxo-1,8-naphthyridine-3-carboxylic acid (NegGram) occurs as a pale buff crystalline powder that is sparingly soluble in water and ether, but soluble in most polar organic solvents. It is a very strong acid (pK_a, 1) and, therefore, is readily solubilized by weak alkalis.

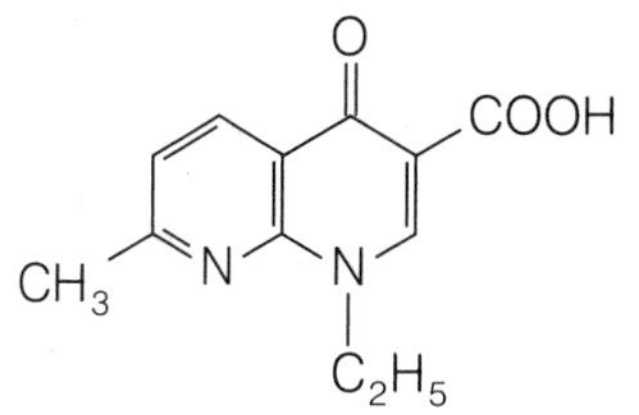

O
COOH
CH_3 N N
C_2H_5

Nalidixic Acid

Nalidixic acid is useful in the treatment of infections of the urinary tract in which gram-negative bacteria are predominant. The activity against indole-positive *Proteus* species is particularly noteworthy, and nalidixic acid and its congeners represent important alternatives for the treatment of urinary tract infections caused by strains of these bacteria resistant to other agents.

Nalidixic acid is rapidly absorbed, extensively metabolized, and rapidly excreted after oral administration. The 7-hydroxymethyl metabolite is significantly more active than the parent compound. Further metabolism of the active metabolite to inactive glucuronide and 7-carboxylic acid metabolites also occurs.

Cinoxacin, USP. 1-Ethyl-1,4-dihydro-4-oxo [1,3]dioxolo[4,5*g*]cinnoline-3-carboxylic acid (Cinobac). This close congener (isostere) of oxolinic acid has antibacterial properties similar to those of nalidixic and oxolinic acids and is recommended for the treatment of urinary tract infections caused by strains of gram-negative bacteria susceptible to these agents. Early clinical studies indicate that the drug possesses pharmacokinetic properties superior to those of either of its predecessors. Thus, higher urinary concentrations of cinoxacin are achieved following oral administration, compared with nalidixic acid or oxolinic acid. Cinoxacin appears to be more completely absorbed and less protein bound than is nalidixic acid and significantly less metabolized to inactive metabolites than is oxolinic acid (which has been subsequently withdrawn from the market in the United States).

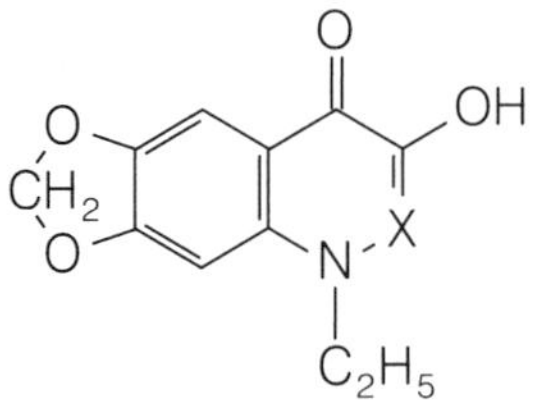

X = CH Oxolinic Acid
X = N Cinoxacin

Norfloxacin. 1-Ethyl-6-fluoro-1,4-dihydro-4-oxo-7-(1-piperazinyl)-3-quinoline carboxylic acid (Noroxin). This quinoline has broad-spectrum activity against gram-negative and gram-positive aerobic bacteria. The fluorine atom provides increased potency against gram-positive organisms, whereas the piperazine moiety improves antipseudomonal activity. Norfloxacin is indicated for the treatment of urinary tract infections caused by *E. coli*, *K. pneumoniae*, *Enterobacter cloacae*, *Proteus mirabilis*, indole-positive Proteus species, including *P. vulgaris*, *Providencia rettgeri*, *Morganella morganii*, *Pseudomonas aeruginosa*, *Staphylococcus aureus*, *S. epidermidis*, and group D streptococci. It is generally not effective against obligate anaerobic bacteria.

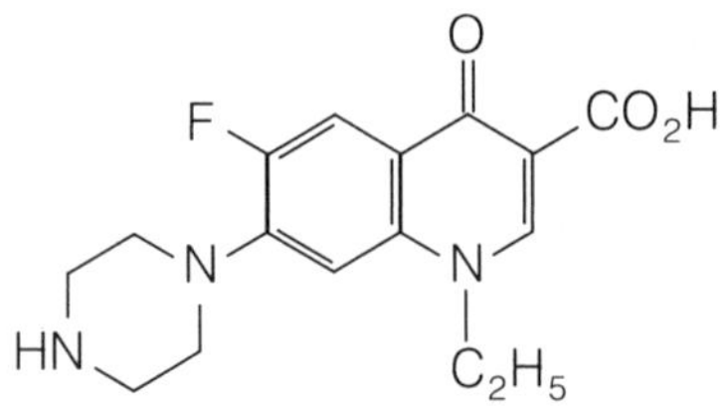

Norfloxacin

The oral absorption of norfloxacin is rapid and reasonably efficient. Approximately 30% of an oral dose is excreted in the urine in 24 hours, along with 5% to 8% consisting of less active metabolites. There is significant biliary excretion with approximately 30% of the original drug appearing in the feces.

Norfloxacin also exhibits good activity against *Neisseria* and *Legionella* species, but has not been approved for use in the treatment of either gonorrhea or Legionnaires' disease.

Investigational Quinolones. Several compounds of the quinolone class are undergoing clinical trials in the United States. Many of them are marketed in Europe and Japan. Enoxacin is a 1,8-naphthyridine congener (isostere) of norfloxacin. It is slightly less potent in vitro, but experiences better oral absorption than norfloxacin. Ciprofloxacin is generally the most potent member of the group of norfloxacin analogues. This compound also exhibits better oral absorption than norfloxacin and is converted to active metabolites. Another norfloxacin congener, perfloxacin, has better activity against gram-positive bacteria such as *S. aureus*. It is also active against the obligate anaerobe *Bacillus fragilis*. Perfloxacin has a longer half-life (approximately 11 hours) than other quinolones and is metabolized

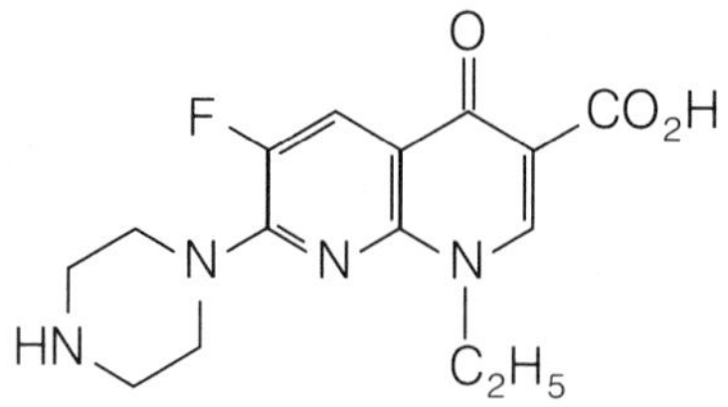

Enoxacin

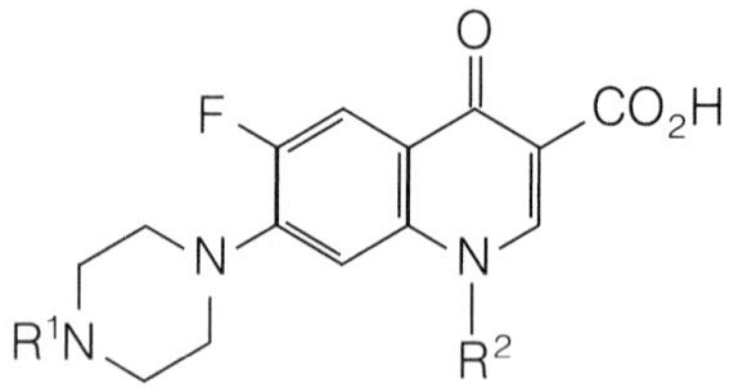

Ciprofloxacin $R^1 = H$, $R^2 =$ cyclopropyl
Perfloxacin $R^1 = CH_3$, $R^2 = C_2H_5$
Amifloxacin $R^1 = CH_3$, $R^2 = NHCH_3$

to norfloxacin. Amifloxacin is similar to norfloxacin in potency and antibacterial spectrum, but possibly has better oral bioavailability. It is converted to an active metabolite. Ofloxacin is a quinolone derivative that has an oxazine ring bridged between the 1- and 8-positions. Ofloxacin has better oral bioavailability than the other quinolones, but it is generally less active against the *Enterobacteriaceae* and *P. aeruginosa*.

Methenamine, USP. Hexamethylenetetramine (Urotropin, Uritone) depends upon the liberation of formaldehyde for its activity. It is manufactured by evaporating a solution of formaldehyde to dryness with strong ammonia water.

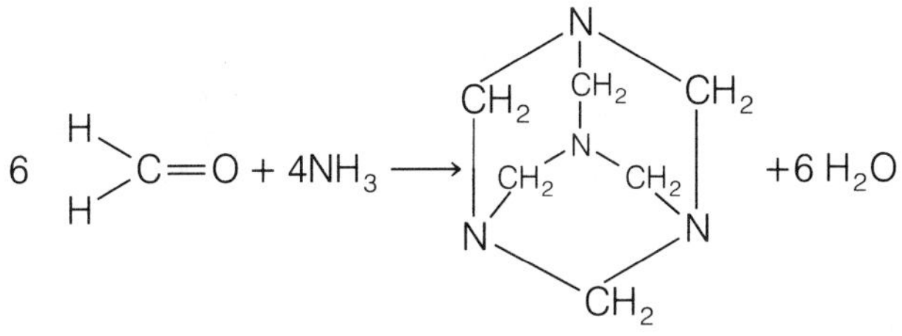

$$6\ H_2C{=}O + 4NH_3 \longrightarrow (CH_2)_6N_4 + 6\ H_2O$$

Methenamine

The free base exists as an odorless white crystalline powder that sublimes at about 260°C. It dissolves in water to form an alkaline solution and liberates formaldehyde when warmed with mineral acids.

Methenamine is employed internally as a urinary antiseptic for the treatment of chronic urinary tract infections. The free base has practically no bacteriostatic power; acidification to release formaldehyde in the comparatively lower *p*H of the kidney is required. To optimize the antibacterial effect, the administration of methenamine is generally accompanied by an acidifying agent such as sodium biphosphate or ammonium chloride.

Certain bacterial strains are resistant to the action of methenamine because they elaborate urease, an enzyme that hydrolyzes urea to form ammonia. The resultant high urinary *p*H prevents the activation of methenamine, rendering it ineffective. This problem can be overcome by the coadministration of the urease inhibitor acetohydroxamic acid (Lithostat).

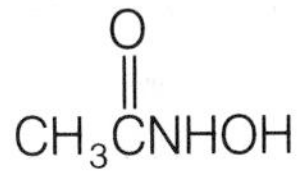

$CH_3C(=O)NHOH$

Acetohydroxamic Acid

Methenamine Mandelate, USP. Hexamethylenetetramine mandelate (Mandelamine) is a white crystalline powder with a sour taste and practically no odor. It is very soluble in water and has the advantage of furnishing its own acidity, although in its use the custom is to carry out a preliminary acidification of the urine for 24 to 36 hours before administration. It is effective with smaller amounts of mandelic acid and, thus, avoids the gastric disturbances attributed to the acid when used alone.

Methenamine Hippurate, USP (Hiprex) is the hippuric acid salt of methenamine. It is readily absorbed after oral administration and is concentrated in the urinary bladder, where it exerts its antibacterial activity. Its activity is increased in acid urine.

Nitrofuratoin, USP. 1-[(5-Nitrofurfurylidene)-amino]hydantoin (Furadantin, Macrodantin) is a nitrofuran derivative that is suitable for oral use. It is recommended for the treatment of urinary tract infections caused by susceptible strains of *E. coli*, enterococci, *S. aureus*, Klebsiella, *Enterobacter*, and *Proteus* species. The most common side effects are gastrointestinal (anorexia, nausea, and vomiting); however, hypersensitivity reactions (pneumonitis, skin rashes, hepatitis, and hemolytic anemia) have occasionally been observed. A large crystalline form (Macrodantin) is claimed to improve gastrointestinal tolerance without interfering with oral absorption.

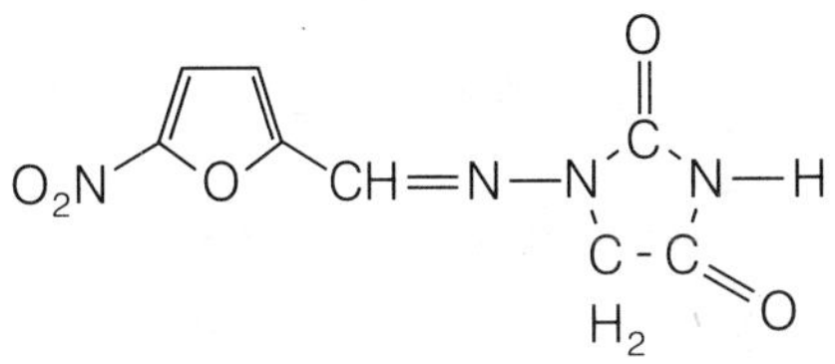

Nitrofurantoin

Phenazopyridine Hydrochloride, USP. 2,6-Diamino-3-(phenylazo)pyridine monohydrochloride (Pyridium) is a brick-red fine crystalline powder. It is slightly soluble in alcohol, in chloroform, and in water.

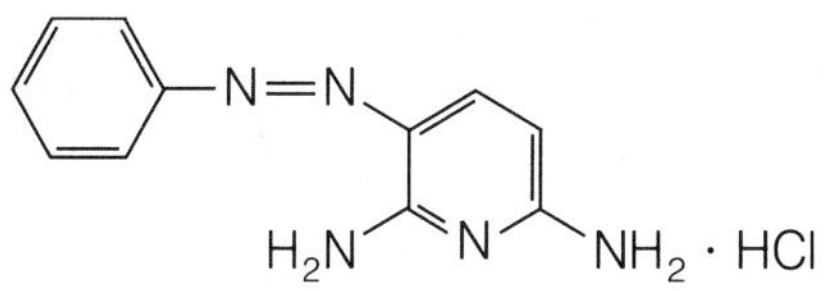

Phenazopyridine Hydrochloride

Phenazopyridine hydrochloride was formerly used as a urinary antiseptic. Although it is active in vitro against staphylococci, streptococci, gonococci, and *E. coli*, it has no useful antibacterial activity in the urine. Thus, its present utility lies in its local analgesic effect on the mucosa of the urinary tract.

TABLE 4-11

URINARY TRACT ANTIBACTERIAL AGENTS

Name Proprietary Name	***Preparations***	***Usual Adult Dose****	***Usual Dose Range****	***Usual Pediatric Dose****
Nalidixic acid USP *NegGram*	Nalidixic acid tablets USP	1 g qid for 1–2 wk; thereafter, for prolonged treatment, the dose may be reduced to 500 mg qid		
Cinoxacin USP *Cinobac*	Cinobac pulvules USP	1 g/day in 2 or 4 divided doses for 7–14 days		
Norfloxacin *Noroxin*	Norfloxacin tablets	400 mg bid for 7–10 days		
Methenamine USP	Methenamine elixir USP Methenamine tablets USP	1 g every 6 hr	Up to 12 g/day	
Methenamine Mandelate USP	Methenamine mandelate oral suspension USP	1 g every 6 hr		Under 6 yr of age: 18.3 mg/kg of body weight every 6 hr; 6–12 yr: 500 mg every 6 hr
Mandelamine	Methenamine mandelate tablets USP			
	Hiprex, Urex Methenamine hippurate tablets USP	1 g every 12 hr	Up to 4 g/day	Children 6–12 yr: 500 mg–1 g every 12 hr
Nitrofurantoin USP *Furadantin, N-Toin*	Nitrofurantoin oral suspension USP Nitrofurantoin tablets USP Nitrofurantoin capsules USP	50–100 mg every 6 hr; or 1.25–1.75 mg/kg of body weight every 6 hr	Up to 600 mg/day; or up to 10 mg/kg of body weight daily	Use for infants below 1 mo of age is not recommended; older infants and children: 1.25–1.75 mg/kg of body weight every 6 hr

*See USP DI for complete dosage information.

Usually, it is now given in combination with urinary antiseptics. For example, it is available as Azo-Gantrisin, a fixed-dose combination with the sulfonamide antibacterial sulfisoxazole (see Chap. 5). The drug is rapidly excreted in the urine, to which it gives an orange-red color. Stains in fabrics may be removed by soaking in a 0.25% solution of sodium dithionite.

Category: analgesic (urinary tract)
Usual dose: 100 mg three or four times daily
Occurrence: Phenazopyridine Hydrochloride Tablets, USP

ANTITUBERCULAR AGENTS

Ever since Koch identified the tubercle bacillus, *Mycobacterium tuberculosis*, there has been keen interest in the development of antitubercular drugs. The first breakthrough in antitubercular chemotherapy occurred in 1938 with the observation that sulfanilamide had weak bacteriostatic properties. Later, the sulfone derivative dapsone (4,4′-diaminodiphenylsulfone) was investigated clinically. Unfortunately, this drug, which is still considered one of the most effective drugs for the treatment of leprosy and also has useful antimalarial properties (see Chap. 6), was considered to be too toxic because of the high dosages used. The discovery of the antitubercular activity of the antibiotic streptomycin by Waksman and his associates in 1944 ushered in the modern era of tuberculosis treatment. This development was quickly followed by discoveries of the antitubercular properties of first, *p*-aminosalicylic acid and then, in 1952, of isoniazid. Later, the usefulness of the synthetic drug ethambutol, and, eventually, of the semisynthetic antibiotic rifampin was discovered.

Combination therapy, with the use of two or more antitubercular drugs, has been well documented to reduce the emergence of strains of *M. tuberculosis* resistant to individual agents and has become standard medical practice. The choice of antitubercular combination is dependent on a variety of factors, including the location of the disease (pulmonary, urogenital, gastrointestinal, or neural); the results of susceptibility tests and the pattern of resistance in the locality; the physical condition and age of the patient; and the toxicities of the individual agents. For some time, a combination of isoniazid and ethambutol, with or without streptomycin, had been the preferred choice of treatment among clinicians in this country. However, one or more of a relatively large group of compounds may be substituted for ethambutol or streptomycin. Antibiotics in this group include rifampin, cycloserine, kanamycin, viomycin, and capreomycin. Therapy with the antibiotic streptomycin or a suitable substitute is usually discontinued when the sputum becomes negative so that toxic effects of the drug may be minimized.

A major advance in the treatment of tuberculosis was signaled by the introduction into therapy of the antibiotic rifampin. Clinical studies indicated that when rifampin is included in the regimen, particularly in combination with isoniazid and ethambutol, a significant shortening of the period required for successful therapy is possible. Previous treatment schedules without rifampin required maintenance therapy for at least two years, whereas those based on the isoniazid–rifampin combination achieve equal or better results in six to nine months.

SYNTHETIC ANTITUBERCULAR AGENTS

Aminosalicylic Acid, USP. 4-Aminosalicylic acid (PAS, Parasal, Pamisyl). The acid occurs as a white to yellowish white crystalline solid that darkens on exposure to light or air. It is slightly soluble in water, but more soluble in alcohol. Alkali metal salts and the nitric acid salt are soluble in water, but the salts of hydrochloric acid and sulfuric acid are not. The acid undergoes decarboxylation when heated. An aqueous solution has a *p*H of about 3.2.

Aminosalicylic acid is administered orally, usually in tablet or capsule form. Symptoms of gastrointestinal irritation are common with both the acid and the sodium salt. A variety of enteric-coated dosage forms have been used in an attempt to overcome this disadvantage. Other forms that are claimed to improve gastrointestinal tolerance include the calcium salt, the phenyl ester, and a combination with an anion exchange resin (Rezi-PAS). An antacid, such as aluminum hydroxide, is frequently prescribed.

The oral absorption of PAS is rapid and nearly complete. It is widely distributed into most of the body fluids and tissues, with the exception of the cerebrospinal fluid in which levels are significantly lower. It is excreted primarily in the urine, both as unchanged drug and as metabolites. The *N*-acetyl derivative is the principal metabolite, with significant amounts of the glycine conjugate also being formed. When administered with isoniazid (which also undergoes *N*-acetylation), PAS increases the level of free isoniazid. The biologic half-life of PAS is about two hours.

The mechanism of antibacterial action of PAS is similar to that of the sulfonamides (see Chap. 5). Thus, it is believed to prevent the incorporation of *p*-aminobenzoic acid (PABA) into the dihydrofolic acid molecule catalyzed by the enzyme dihydrofolate synthetase. Structure–activity studies have shown that the amino and carboxyl groups must be *para* to each other and free; thus, esters and amides must readily undergo hydrolysis in vivo to be effective. The hydroxyl group may be *ortho* or *meta* to the carboxyl group, but optimal activity is seen in the former.

For many years, PAS was considered a first-line drug for the chemotherapy of tuberculosis and was generally included in combination regimens with isoniazid and streptomycin. However, the introduction of the more effective and generally better-tolerated agents, ethambutol and rifampin, have relegated it to an alternative drug status. The acid is taken orally, usually in tablet form. Often, severe gastrointestinal irritation accompanies the use of PAS or its sodium salt. To overcome this disadvantage, coated tablets, capsules, and granules are used. Often, an antacid, such as aluminum hydroxide, is prescribed concurrently.

Aminosalicylate Sodium, USP. Sodium 4-aminosalicylate (Parasal Sodium; Pasara Sodium; Pasem Sodium; Parapas Sodium; Paraminose). This salt occurs in the dihydrate form as a yellow-white powder or crystalline solid. It is very soluble in water in the *p*H range of 7.0 to 7.5, at which it is the most stable. Aqueous solutions decompose readily and darken. Two *p*H-dependant types of reactions occur: decarboxylation (more rapid at low *p*H) and oxidation (more rapid at high *p*H); therefore, solutions should be prepared within 24 hours of administration.

Aminosalicylate Potassium, USP. Potassium 4-aminosalicylate (Paskalium, Paskate). This salt has properties similar to those of the sodium salt. Its use is indicated in patients whose sodium intake must be controlled.

Aminosalicylate Calcium, USP. Calcium 4-aminosalicylate (Parasa Calcium) is available in a variety of solid dosage forms for oral administration. Gastrointestinal irritation is reported to be less with this salt form relative to the acid or sodium salt.

Isoniazid, USP. Isonicotinic acid hydrazide; isonicotinyl hydrazide; INH (Nydrazid) occurs as a nearly colorless crystalline solid that is very soluble in water. It is prepared by reacting the methyl ester of isonicotinic acid with hydrazine.

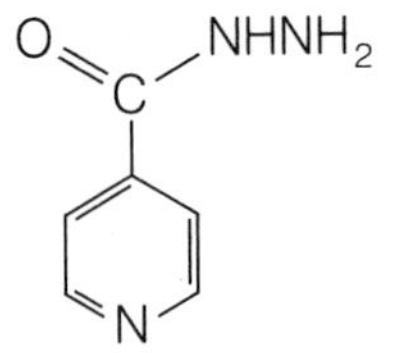

Isoniazid

Isoniazid is a remarkably effective agent and continues to be one of the primary drugs (along with rifampin and ethambutol) for the treatment of tuberculosis. It is not, however, uniformly effective against all forms of the disease. The frequent emergence of strains of the tubercle bacillus resistant to isoniazid during therapy was seen as the major shortcoming of the drug. This problem has been largely overcome with the use of combinations.

The activity of isoniazid is manifested on the growing tubercle bacilli and not on resting forms. Its action, which is considered bactericidal, is to cause the bacilli to lose lipid content by a mechanism that has not been fully elucidated. One theory suggests that the principal effect of isoniazid is to inhibit the synthesis of mycolic acids, branched β-hydroxy fatty acids that constitute important components of the cell walls of mycobacteria.

Despite that treatment regimens generally require long-term administration of isoniazid, the incidence of toxic effects is remarkably low. The principal toxic reactions are peripheral neuritis, gastrointestinal disturbances (such as constipation, loss of appetite), and hepatotoxicity. Coadministration of pyridoxine is reported to prevent the symptoms of peripheral neuritis, suggesting that this adverse effect may be the result of antagonism of a coenzyme action of pyridoxal phosphate. Pyridoxine does not appear to interfere with the antitubercular effect of isoniazid. Severe hepatotoxicity rarely occurs with isoniazid alone. However, the incidence is much higher when it is used in combination with rifampin.

Isoniazid is rapidly and almost completely absorbed following oral administration. It is widely distributed to all tissues and fluids within the body, including the cerebrospinal fluid. Approximately 60% of an oral dose is excreted in the urine within 24 hours in the form of numerous metabolites as well as the unchanged drug. Although the metabolism of isoniazid is very complex, the principal path of inactivation involves acetylation of the primary hydrazine nitrogen. In addition to acetylisoniazid, the isonicotinyl hydrazones of pyruvic and α-ketoglutaric acids—isonicotinic acid and isonicotinuric acid—have been isolated as metabolites in humans.[33] The capacity to inactivate isoniazid by acetylation is an inherited characteristic in humans. Approximately half the population are fast acetylators (plasma half-life, 45 to 80 minutes) and the remainder slow acetylators (plasma half-life, 140 to 200 minutes).

Ethambutol, USP. (+)-2,2′-(Ethylenediimino)-di-1-butanol dihydrochloride; EBM (Myambutol) is a white crystalline powder freely soluble in water and slightly soluble in alcohol.

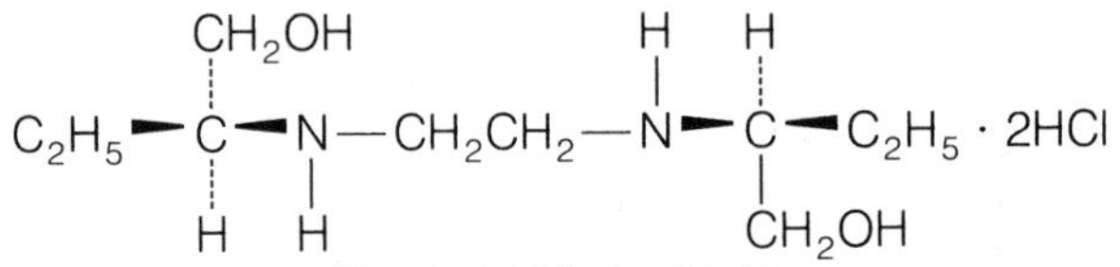

Ethambutol Dihydrochloride

Ethambutol is active only against dividing mycobacteria. It has no effect on encapsulated or other nonproliferating forms. The in vitro effect may be bacteriostatic or bactericidal, dependent on the conditions. Its selective toxicity toward mycobacteria appears to be related to the inhibition of the incorporation of mycolic acids into the cell walls of these organisms by some unknown mechanism.

This compound is remarkably stereospecific. Tests have shown that, although the toxicities of the *dextro*, *levo*, and *meso* isomers are about equal, their activities vary considerably. The *dextro* isomer is 16 times as active as the *meso* isomer. In addition, the length of the alkylene chain, the nature of the branching of the alkyl substituents on the nitrogens, and the extent of *N*-alkylation all have a pronounced effect on the activity.

Ethambutol is rapidly absorbed after oral administration, and peak serum levels occur in about two hours. It is rapidly excreted, mainly in the urine. Up to 80% is excreted unchanged, with the balance being metabolized and excreted as 2,2′-(ethylenediimino)dibutyric acid and as the corresponding dialdehyde.

It is not recommended for use alone, but in conjunction with other antitubercular drugs in the chemotherapy of pulmonary tuberculosis.

Pyrazinamide, USP. Pyrazinecarboxamide (Aldinamide) occurs as a white crystalline powder that is sparingly soluble in water and slightly soluble in polar organic solvents. Its antitubercular properties were discovered as a result of an investigation of heterocyclic analogues of nicotinic acid, with which it is isosteric.

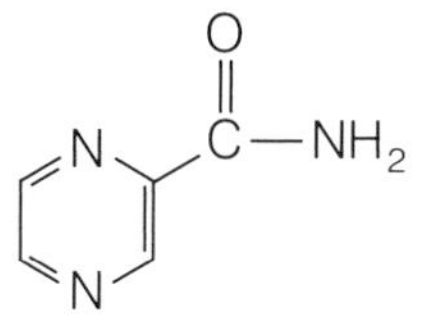

Pyrazinamide

Although pyrazinamide is an effective antitubercular drug, its potential to cause hepatotoxicity limits its use to the treatment of hospitalized patients whose tuberculosis is resistant to isoniazid. Pyrazinamide is used only in combination with other agents because resistance to its action rapidly develops.

Pyrazinamide is well absorbed orally and widely distributed throughout the body. It is excreted rapidly as the unchanged drug and as the carboxylic acid (pyrazinoic acid) and 5-hydroxylated metabolites. Pyrazinamide (or a metabolite) is reported to interfere with uric acid excretion and, therefore, should be used with caution in patients with gout.

Ethionamide, USP. 2-Ethylthioisonicotinamide (Trecator SC) occurs as a yellow crystalline material that is sparingly soluble in water. This nicotinamide has weak bacteriostatic activity in vitro but, because of its lipid solubility, is effective in vivo. In contrast to the isoniazid series, 2-substitution enhances activity in the thioisonicotinamide series.

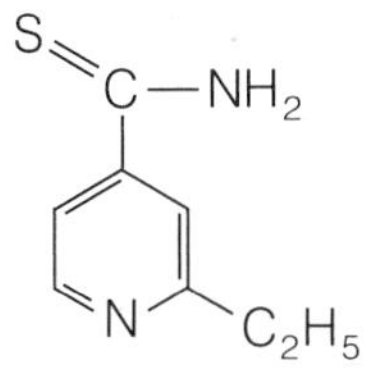

Ethionamide

Ethionamide is rapidly and completely absorbed following oral administration. It is widely distributed throughout the body and extensively metabolized to predominately inactive forms that are excreted in the urine. Less than 1% of the parent drug appears in the urine.

Ethionamide is considered a secondary drug for the treatment of tuberculosis. It is used in the treatment of isoniazid-resistant tuberculosis or when the patient is intolerant to isoniazid and other drugs. Because of its low potency, the highest tolerated dose is usually recommended. Gastrointestinal intolerance is the most common side effect associated with its use. Visual disturbances and hepatotoxicity have also been reported.

Clofazimine (Lamprene) is a basic red dye that exerts a slow, bactericidal effect on *M. leprae*, the bacterium that causes leprosy. It occurs as a dark red crystalline solid that is insoluble in water.

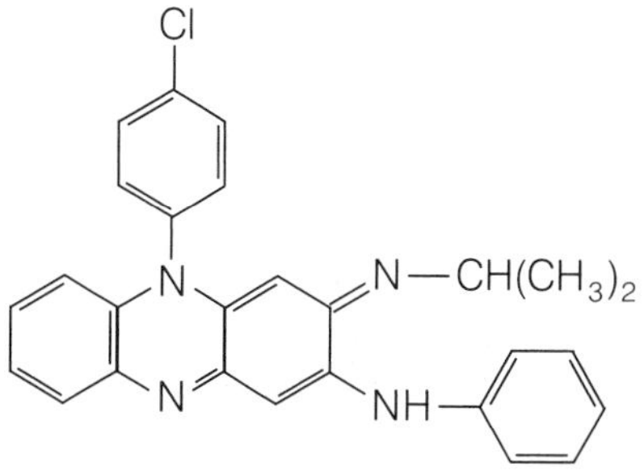

Clofazimine

Clofazimine is used in the treatment of lepromatous leprosy, including dapsone-resistant forms of the disease. In addition to its antibacterial action, the drug appears to possess anti-inflammatory and immune-modulating effects that are of value in controlling neuritic complications and in suppressing erythema nodosum leprosum reactions associated with lepromatous leprosy. It is frequently used in combination with other drugs, such as dapsone (see Chap. 5) or rifampin.

The mechanisms of antibacterial and anti-inflammatory actions of clofazimine are not known. The drug is known to bind to nucleic acids and concentrate in reticuloendothelial tissue. It can also act as an electron acceptor and may interfere with electron transport processes.

The oral absorption of clofazimine is estimated to be about 50%. It is a highly lipid-soluble drug that is distributed into lipoidal tissue and the reticuloendothelial system. Urinary excretion of unchanged drug and metabolites is negligible. Its half-life after repeated dosage is estimated to be about 70 days. Severe gastrointestinal intolerance to clofazimine is relatively common. Skin pigmentation, ichthyosis and dryness, rash, and pruritus also occur frequently.

Clofazamine has also been used to treat skin lesions caused by *M. ulcerans*.

ANTITUBERCULAR ANTIBIOTICS

Cycloserine, USP. D-(+)-4-Amino-3-isoxazolidinone (Seromycin) is an antibiotic that has been isolated from three different *Streptomyces* species: *S. orchidaceus*, *S. garyphalus*, and *S. lavendulus*. It occurs as a white to pale yellow crystalline material that is very soluble in water. It is stable in alkaline, but unstable in acidic solutions. The com-

TABLE 4-12

TUBERCULOSTATIC AGENTS

Name Proprietary Name	*Preparations*	*Usual Adult Dose**	*Usual Dose Range**	*Usual Pediatric Dose**
Aminosalicylic acid USP *Pamisyl, Parasal*	Aminosalicylic acid tablets USP	3 g qid	10–20 g/day	
Aminosalicylate sodium USP *Rezipas, NatriPas, Pamisyl Sodium, Pasara Sodium, Pasna*	Aminosalicylate sodium tablets USP	4–5 g tid	8–15 g/day	100 mg/kg of body weight or 2.7 g/m^2 of body surface tid
Aminosalicylate potassium USP *Parasal Potassium, Paskalium*	Aminosalicylate potassium tablets USP	3 g qid	10–20 g/day	
Aminosalicylate calcium USP *Parasal Calcium*	Aminosalicylate calcium capsules USP Aminosalicylate calcium tablets USP	4 g qid	10–25 g/day in 4 divided doses	
Benzoylpas calcium USP *Benzapas*	Benzoylpas calcium tablets USP		10–15 g/day in 2 or 3 divided doses	
Isoniazid USP *Hyzyd, Niconyl, Nydrazid*	Isoniazid injection USP	IM, 5 mg/kg of body weight qd up to 300 mg/day		
	Isoniazid syrup USP Isoniazid tablets USP	Prophylaxis: oral, 300 mg qd; treatment, in combination with other tuberculostatics: 5 mg/kg of body weight up to 300 mg qd	Up to 20 mg/kg of body weight, not to exceed 600 mg/day	Prophylaxis: oral, 10 mg/kg of body weight, up to 300 mg qd; treatment, in combination with other tuberculostatics: 10–20 mg/kg of body weight, up to 500 mg qd.
Pyrazinamide USP *Aldinamide*	Pyrazinamide tablets USP	5–8.75 mg/kg of body weight qid	1 to a maximum of 3 g/day	
Ethionamide USP *Trecator SC*	Ethionamide tablets USP	250 mg bid to qid	500 mg–1 g/day	4–5 mg/kg of body weight, up to a maximum of 250 mg tid
	Ethambutol hydrochloride tablets USP	In combination with other tuberculostatics: initial treatment, oral, 15 mg/kg of body weight qd; retreatment: 25 mg/kg of body weight qd	Initial treatment: 500 mg–1.5 g/day; retreatment: 900 mg–2.5 g/day	Children under 13 yr: use is not recommended; children 13 yr and over: see Usual Adult Dose

*See USP DI for complete dosage information.

pound slowly dimerizes to 2,5-bis(aminoxymethyl)-3,6-diketopiperazine in solution or standing.

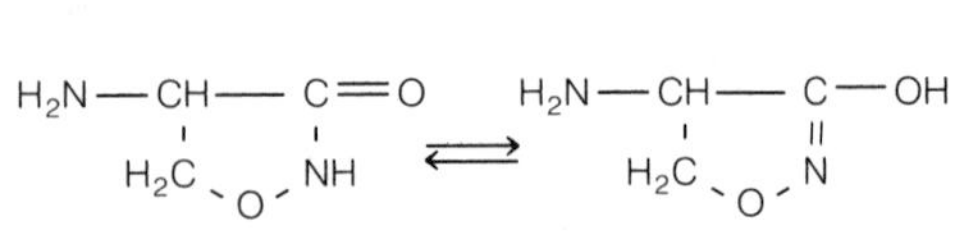

$$\begin{matrix} H_2N-CH-C{=}O \\ H_2C\diagdown_{O}\diagup NH \end{matrix} \rightleftharpoons \begin{matrix} H_2N-CH-C-OH \\ H_2C\diagdown_{O}\diagup N \end{matrix}$$

Cycloserine

The structure of cycloserine was reported simultaneously by Kuehl et al.[34] and Hidy et al.[35] to be D-4-amino-3-isoxazolidindione. It has been synthesized by Stammer et al.[36] and by Smrt et al.[37]. Cycloserine is stereochemically related to D-serine. However, the L-form has similar antibiotic activity.

Cycloserine is presumed to exert its antibacterial action by preventing the synthesis of cross-linking peptide in the formation of bacterial cell walls.[38] Rando[39] has recently suggested that it is an antimetabolite for alanine, which acts as a suicide substrate for the pyridoxal phosphate-requiring enzyme alanine racemase. Irreversible inactivation of the enzyme thereby deprives the cell of D-alanine required for the synthesis of the cross-linking peptide.

Although cycloserine enhibits antibiotic activity in vitro against a wide spectrum of both gram-negative and gram-positive organisms, its relatively weak potency and frequent toxic reactions limit its use to the treatment of tuberculosis. It is recommended for cases who fail to respond to other tuberculostatic drugs or are known to be infected with organisms resistant to other agents. It is usually administered orally in combination with other drugs, commonly isoniazid.

Sterile Viomycin Sulfate, USP (Viocin Sulfate). Viomycin is a cyclic peptide isolated from a number of *Streptomyces* species. Its use is confined to the treatment of tuberculosis, for which it is a second-line agent, occasionally substituted for streptomycin in infections resistant to that antibiotic. Viomycin exerts a bacteriostatic action against the tubercle bacillus by a mechanism that has not been determined. It is significantly less potent than streptomycin, and its toxicity is greater. Toxic effects of viomycin are primarily associated with damage to the eighth cranial nerve and to the kidney.

Viomycins are strongly basic peptides. At least two components have been obtained from *S. vinaceus* and have been named vinactins A and B. A closely related substance, identified as vinactin C, has also been found to be present. Vinactin A appears to be the major component of viomycin. Some disagreement remains concerning details of the chemical structure of the viomycins. Early work by Haskell et al.[40] and Mayer et al.[41] showed that vinactin A had no free α-amino groups and, on vigorous acid hydrolysis, yielded carbon dioxide, ammonia, urea, L-serine, α,β-diaminopropionic acid, β-lysine, and a guanidino compound. On the basis of additional chemical and spectroscopic evidence, at least three different structures have been proposed for vinactin A.[42–44] Doubt about the peptide sequence of the antibiotic, raised as a result of x-ray crystallographic studies on a closely related antibiotic, tuberactinomycin N,[45] appears to have been resolved by the chemical studies of Noda et al. who have suggested the structure of vinactin A shown.[44] It is perhaps noteworthy that the more recently proposed structures lack the fused hetero aromatic ring system of structures suggested earlier[42] to explain the ultraviolet spectrum of the antibiotic. A possible explanation for this could be the existence of the antibiotic in a different chemical form in solution, compared with that in its solid state.

Viomycin sulfate is an odorless powder that varies in color from white to slightly yellow. It is freely soluble in water, forming solutions ranging in pH from 4.5 to 7.0. It is insoluble in alcohol and other organic solvents. Because it is slightly hygroscopic, it should be stored in closed containers. It is administered in aqueous solutions intramuscularly.

Sterile Capreomycin Sulfate, USP (Capastat Sulfate). Capreomycin is a strongly basic cyclic peptide isolated from *S. capreolus* in 1960 by Herr et al.[46] It was released in the United States in 1971 exclusively as a tuberculostatic drug. Capreomycin, which resembles viomycin chemically and pharmacologically, is a second-line agent employed in combination with other antitubercular drugs. In particular, it may be used in place of streptomycin when either the patient is sensitive to, or the strain of *M. tuberculosis* is resistant to, streptomycin. Similar to viomycin, capreomycin is a potentially toxic drug. Damage to the eighth cranial nerve and renal damage, as with viomycin, are the more serious toxic effects associated with capreomycin therapy. There is, as yet, insufficient clinical data on which to reliably compare the relative toxic potential of capreomycin with either viomycin or streptomycin. Cross-resistance among strains of tubercle bacilli is probable between capreomycin and viomycin, but rare between either of these antibiotics and streptomycin.

Four capreomycins, designated 1A, 1B, IIA, and IIB, have been isolated from *S. capreolus*. The clinical agent contains primarily 1A and 1B. The close

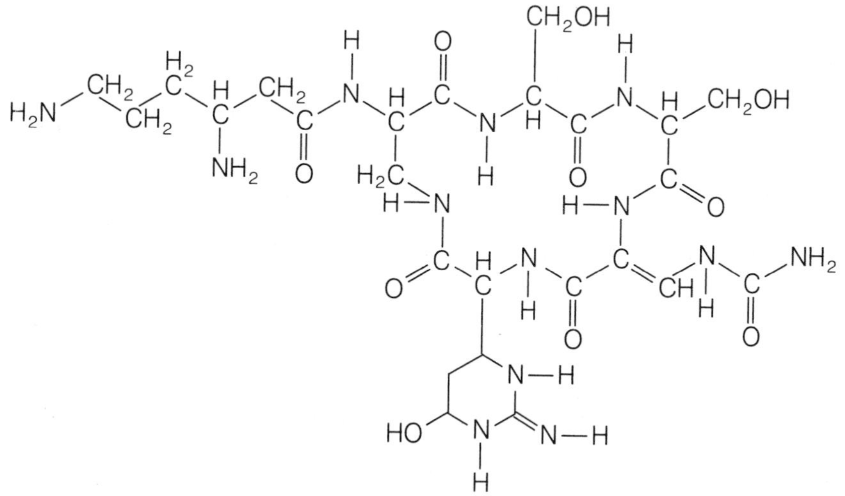

Vinactin A

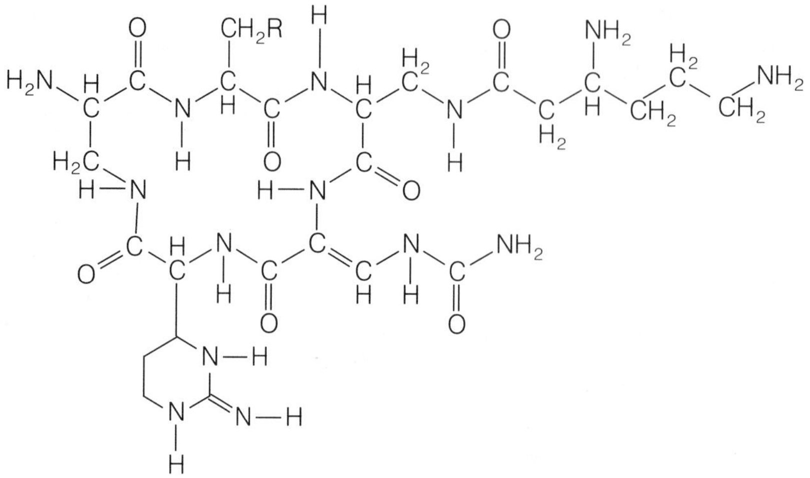

Capreomycin 1A R = OH
1B R = H

chemical relationship between capreomycins 1A and 1B and viomycin was established,[47] and the total synthesis and proof of structure of the capreomycins later accomplished.[48] The structures of capreomycins IIA and IIB correspond to those of IA and IB, but lack the β-lysyl residue. The sulfate salts are freely soluble in water.

Rifampin, USP (Rifadin; Rimactane; rifampicin). The rifamycins are a group of chemically related antibiotics obtained from *S. mediterranei*. They belong to a new class of antibiotics that contain a macrocyclic ring bridged across two nonadjacent (*ansa*) positions of an aromatic nucleus and called ansamycins. The rifamycins and many of their semisynthetic derivatives have a broad spectrum of antimicrobial activity. They are most notably active against gram-positive bacteria and *M. tuberculosis*. However, they are also active against some gram-negative bacteria and many viruses. Rifampin, a semisynthetic derivative of rifamycin B, was released as an antitubercular agent in the United States in 1971. Its structure is shown below.

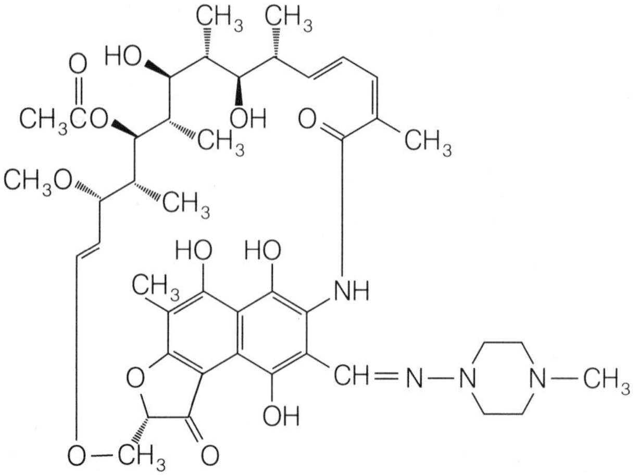

Rifampin

The chemistry of rifamycins and other ansamycins has been reviewed.[49] All of the rifamycins (A, B, C, D, and E) are biologically active. Some of the semisynthetic derivatives of rifamycin B are the most potent known inhibitors of DNA-directed RNA-polymerase in bacteria[50] and their action is bactericidal. They have no activity against the mammalian enzyme. The mechanism of action of rifamycins as inhibitors of viral replication appears to be different from that for their bactericidal action. Their net effect is to inhibit the formation of the virus particle, apparently by the prevention of a specific polypeptide conversion.[51] Rifamycin B (which lacks a substituent at C-4 and has a glycolic acid attached by an ether linkage at C-3), rifamycin SV (which lacks a C-4 substituent and the glycolic acid linked at C-3), and rifamide (the amide of rifamycin B) have antibacterial activity. However, only rifampin is well absorbed orally and finds clinical use in the United States. Rifamide is available in Europe for the treatment of hepatobiliary infections. It is 80% excreted in the bile after parenteral administration (IM). Some derivatives of 4-formylrifamycin SV are active against RNA-dependent DNA-polymerase in several RNA tumor viruses.[52] *N*-Demethylrifampin, *N*-demethyl-*N*-benzylrifampin, and 2,6-dimethyl-*N*-demethyl-*N*-benzylrifampin were very active in this system, whereas rifampin was ineffective. The clinical usefulness of these agents as antitumor agents has not been established.

Rifampin occurs as an orange to reddish-brown crystalline powder that is soluble in alcohol, but only sparingly soluble in water. It is unstable to moisture and a dessicant (silica gel) should be in-

TABLE 4-13

ANTITUBERCULAR ANTIBIOTICS

Name Proprietary Name	**Preparations**	**Usual Adult Dose***	**Usual Dose Range***	**Usual Pediatric Dose***
Cycloserine USP *Seromycin*	Cycloserine capsules USP	250 mg bid to qid	250 mg–1 g/day	5 mg/kg of body weight or 150 mg/m^2 of body surface, bid initially, then titrate the dose to yield a blood level of 20–30 μg/mL
Viomycin sulfate USP *Viocin*	Sterile viomycin sulfate USP	IM, the equivalent of 1 g of viomycin bid, twice weekly	4 to a maximum of 14 g/wk	Use in children is not recommended unless crucial to therapy. The equivalent of 20 mg/kg of body weight or 600 mg/m^2 of viomycin bid, twice weekly
Capreomycin sulfate USP *Capastat*	Sterile capreomycin sulfate USP	IM, the equivalent of 1 g of capreomycin qd for 2–4 mo, then 1 g, 2 or 3 times weekly		
Rifampin USP *Rifadin, Rimactane*	Rifampin capsules USP	In combination with other tuberculostatics: oral—600 mg qd		Children 5 yr and over: in combination with other tuberculostatics — oral, 10–20 mg/kg of body weight qd

*See USP DI for complete dosage information.

cluded with rifampin capsule containers. The expiration date for capsules thus stored is two years. Rifampin is stable in the solid state, but undergoes a variety of chemical changes in solution, the rates and nature of which are *p*H- and temperature-dependent.[53] In alkaline *p*H, it oxidizes to the quinone in the presence of oxygen; in acidic solutions, it hydrolyzes to 3-formylrifamycin SV. Slow hydrolysis of ester functions also occurs, even at neutral *p*H. Rifampin is well absorbed after oral administration to provide effective blood levels for eight hours or more. However, food markedly reduces its oral absorption, and rifampin should be administered on an empty stomach. It is distributed in effective concentrations to all body fluids and tissues except the brain, despite that it is 70% to 80% protein bound in the plasma. The principal excretory route is through the bile and feces, and high concentrations of rifampin and its primary metabolite, deacetylrifampin, are found in the liver and biliary system. Deacetylrifampin is also microbiologically active. Equally high concentrations of rifampin are found in the kidney, and although substantial amounts of the drug are passively reabsorbed in the renal tubules, its urinary excretion is sizable.

Rifampin is the most active agent in clinical use for the treatment of tuberculosis. As little as 5 μg/mL are effective against sensitive strains of *M. tuberculosis*. However, resistance to it develops rapidly in most species of bacteria, including the tubercle bacillus. Consequently, rifampin is used only in combination with other antitubercular drugs, and it is ordinarily not recommended for the treatment of other bacterial infections when other antibacterial agents are available. Toxic effects associated with rifampin are relatively infrequent. It may, however, interfere with liver function in some patients and should not be combined with other potentially hepatotoxic drugs, nor employed in patients with impaired hepatic function (e.g., chronic alcoholics). The incidence of hepatotoxicity was significantly higher when rifampin was combined with isoniazid than when either agent was combined with ethambutol. Allergic and sensitivity reactions to rifampin have been reported, but they are infrequent and usually not serious.

Rifampin is also employed to eradicate the carrier state in asymptomatic carriers of *Neisseria meningitidis* to prevent outbreaks of meningitis in high-risk areas such as military camps. Serotyping and sensitivity tests should be performed before its use because resistance develops rapidly. However, a daily dosage of 600 mg of rifampin for four days is sufficient to eradicate sensitive strains of *N. meningitidis*. Rifampin has also been very effective against *M. leprae* in experimental animals and in humans. Its usefulness in the treatment of human leprosy remains to be fully established.

ANTIVIRAL AGENTS

PROPERTIES OF VIRUSES

Viruses are obligate cellular parasites composed of a nucleic acid core surrounded by a proteinaceous outer shell. As cellular parasites they are dependent on the host cell for energy and the biochemical substrates required for replication and protein syn-

TABLE 4-14

CLASSIFICATION OF VIRUSES CAUSING DISEASE IN MAN

Group Agent	*Disease*	*Vaccine (Effectiveness)*	*Chemotherapy*
	RNA VIRUSES		
Picornavirus			
Enterovirus	Polio; three serotypes cause paralysis	Live and killed vaccines (very effective)	None
Rhinovirus	"Common cold"; over 100 serotypes	None	None
Coxsackie	Variety of symptoms	None	None
Togavirus			
Alphavirus	Equine encephalitis, mosquito borne	Inactivated virus	None
Flavivirus	Yellow fever, mosquito and tick borne	Attenuated virus (very effective)	None
Rubivirus	Rubella	Attenuated virus (95% effective)	None
Rhabdovirus			
Rabies virus	Rabies	Inactivated virus (effective)	None
Paramyxovirus			
Parainfluenza virus 1	Acute respiratory symptoms	None	None
Morbillivirus	Measles	Attenuated virus (90% effective)	None
Pneumovirus (respiratory syncytial virus, RSV)	Acute respiratory symptoms (children)	Attenuated virus (effectiveness uncertain)	Ribavirin
Orthomyxovirus			
Influenza viruses	Acute respiratory symptoms; A, B, and D serotypes	Attenuated viruses (70% effective)	Amantadine
Reovirus			
Human reovirus	Mild respiratory and gastro-intestinal symptoms	None	None
Orbivirus	Colorado tick fever, tick borne	Inactivated virus (effectiveness unknown)	None
Retrovirus			
Oncorna virus	Human T-cell leukemia	None	None
HIV(HTLV-III/LAV) virus	Acquired immune deficiency syndrome (AIDS) and AIDS-related complex (ARC)	None	AZT
Unclassified			
Hepatitis A (human entero-virus, HVA)	Acute hepatitis, usually mild and rarely chronic	None	None
Delta agent	Acute hepatitis, requires HBV for replication	None	None
	DNA VIRUSES		
Herpesvirus			
Herpes simplex type 1 (HSV1)	Fever blisters	Inactivated viruses (efficacy uncertain)	IUdR, Ara-A
Herpes simplex type 2 (HSV2)	Genital herpes	Immunity and antigenic structure is poorly understood	Acyclovir
Herpes zoster (varicella)	Chickenpox (children), shingles (adults)		
Epstein-Barr virus (EBV)	Infectious mononucleosis, Burkitt's lymphoma		
Cytomega-lovirus	Fetal damage		
Paporavirus			
Papillomavirus	Warts	None	None
Polyomavirus			

TABLE 4-14 *Continued*

CLASSIFICATION OF VIRUSES CAUSING DISEASE IN MAN

Group Agent	Disease	Vaccine (Effectiveness)	Chemotherapy
	DNA VIRUSES *Continued*		
Adenovirus			
Human adenovirus	Upper respiratory symptoms, conjunctivitis	None	None
Poxvirus			
Variola	Smallpox	Vaccinia (very effective)	Methisazone
Unclassified			
Hepatitis B virus (HBV)	Acute hepatitis, may become chronic	Inactivated subunit (effective)	None

thesis. They also utilize the biochemical apparatus of the cell to synthesize the virus-specific proteins required for production of the mature virus particle. Adult viruses possess only one type of nucleic acid [the genome may be either deoxyribonucleic acid (DNA) or ribonucleic acid (RNA)] and their organized structure is lost during replication of the genome in the host cell. These features differentiate viruses from other cellular parasites, such as the chlamydiae (which possess both DNA and RNA and replicate within the cell) and the more complex rickettsiae (which, in addition, have autonomous energy-generating and protein-synthesizing systems).

VIRAL CLASSIFICATION

Viral classification is based on various characteristics, such as nucleic acid content (RNA viruses, DNA viruses), morphology, site of replication in the cell (cytoplasm, nucleus), shell composition (nonenveloped, enveloped), and serologic typing. The RNA viruses include picornaviruses, togoviruses, rhabdoviruses, myxoviruses, reoviruses, and retroviruses. They cause such diseases as polio, colds, influenza, encephalitis, yellow fever, rubella, measles, mumps, rabies, and acquired immune deficiency syndrome (AIDS). The DNA-containing viruses include the herpesviruses, papoviruses, adenoviruses, and poxviruses. Diseases caused by DNA viruses include cold sores, conjunctivitis, genital herpes, encephalitis, shingles, chickenpox, mononucleosis, smallpox, and upper respiratory infections. It has been estimated that more than 60% of the infectious diseases that occur in developing countries are caused by viruses. Bacterial infections account for only 15%. A list of the more important viruses causing diseases in man is provided in Table 4-14.

PREVENTION OF VIRAL INFECTIONS: IMMUNIZATION

Prevention of viral infection by active immunization continues to be the mainstay of therapy of most viral diseases. Effective vaccines are available for prophylaxis of such diseases as polio, rubella, measles, mumps, influenza, yellow fever, rabies, smallpox (now considered eradicated worldwide), and hepatitis B. On the other hand, vaccines developed against herpes and Epstein-Barr viruses have thus far proved to be unreliable, and no vaccines that are effective against viruses causing AIDS, hepatitis A, or the common cold have yet been found. The development of a new vaccine that is effective against a disease-causing virus, such as the AIDS (HIV) virus, is not a simple task. The new vaccine must be sufficiently antigenic to induce an effective antibody response. At the same time, it must not cause disease, either the one that it is designed to prevent or some other toxic manifestation. Some viruses undergo rapid mutation, leading to numerous serotypes, thereby confounding the production of a broadly effective vaccine.

ANTIVIRAL CHEMOTHERAPY: BIOCHEMICAL TARGETS

The discovery of useful antiviral agents has historically lagged behind progress in antibacterial chemotherapy. There are a variety of reasons for this state of affairs. Unlike bacteria, viruses will not grow on synthetic culture media. The development of specialized cell culture techniques required for the screening of potential antiviral drugs has been accomplished only relatively recently. The comparative biochemical simplicity of viruses and their utilization of the biochemical processes of the host cell provide fewer targets for potential attack by chemotherapeutic agents. The spectacular success of immunization procedures for the prevention of certain viral diseases may have also contributed to the lack of interest in chemotherapy. Another problem that occurs in mild viral infections, such as the common cold, is that clinical symptoms do not appear until the infection is well established and the immune processes of the host have begun to mount a successful challenge. Thus, for many common viral infections, chemotherapy is not an appropriate choice of treatment. Chemotherapeutic agents are needed, however, against viruses that cause severe or chronic infections such as encephalitis, AIDS, slow viral disease, and herpes, particularly in patients with depressed immune systems.

Despite their simplicity relative to bacteria, viruses nonetheless possess a variety of biochemical targets for potential attack by chemotherapeutic agents. An understanding of the specific biochemical events that occur during viral infection of the host cell should provide the basis for the future discovery of site-specific antiviral agents. The process of viral infection can be conveniently divided into five stages: (1) attachment of the virus to the surface of the host cell and its penetration into the cell; (2) release of viral nucleic acid from the protein coat; (3) replication of the genome and synthesis of viral proteins; (4) assembly of the virus particle; and (5) release of the mature virus from the cell.

The initial attachment of virus particles to cells is believed to involve electrostatic interactions between components of the viral capsid or outer envelope and receptors on the cell surface. The high degree of specificity observed for such interactions appears to be responsible for the tissue tropism displayed by many viruses. There is good evidence that the cellular receptor for influenza viruses is *N*-acetylmuramic acid, which binds a protein molecule, hemagglutin, projecting from the viral surface.[54] Viruses are believed to enter cells by endocytosis, a process that involves fusion of the viral envelope with the cell membrane.

Before a virus can replicate within the cell, it must first shed its outer envelope and capsid to release its nucleic acid genome. For complex DNA viruses such as vaccinia, the uncoating process occurs in two stages. First, host cell enzymes partially degrade the envelope and capsid to reveal a portion of viral DNA, which serves as a template for mRNA synthesis. The mRNAs direct the synthesis of viral enzymes that then complete the degradation of the protein coat. Proteins of the viral envelope and capsule are the primary targets of the antibodies synthesized in response to immunization techniques. Protein synthesis inhibitors, such as cycloheximide and puromycin, inhibit the viral-uncoating process. However, they are not sufficiently selective to be useful as antiviral agents.

In the critical third state of infection, the virus commandeers the energy-producing and synthetic functions of the cell to replicate its genome and synthesize viral enzymes and structural proteins. Simple RNA viruses conduct both replication and protein synthesis in the cytoplasm of the cell. They contain specific RNA polymerases (RNA replicases) responsible for the replication of the genome. Some single-stranded RNA viruses, such as poliovirus, have a (+)-RNA genome that serves the dual function of the messenger for protein synthesis and the template for the synthesis of a complementary strand of (−)-RNA from which the (+)-RNA is replicated. In the poliovirus, the message is translated as a single large protein that is enzymatically cleaved to specific viral enzymes and structural proteins. Other RNA viruses, such as influenza viruses, contain (−)-RNA, which serves as the template for the synthesis of a complementary strand of (+)-RNA. The (+)-RNA strand directs viral protein synthesis and provides the template for the replication of the (−)-RNA genome. Certain antibiotics, such as the rifamycins, inhibit viral RNA polymerases in vitro, but none have yet proved useful clinically. Bioactivated forms of the nucleoside analogue ribavirin variously inhibit ribonucleotide synthesis, RNA synthesis, or RNA capping in RNA viruses. Ribavirin has recently been approved for aerosol treatment of severe lower respiratory infections caused by respiratory syncytial virus (RSV).

Retroviruses are a special class of RNA viruses that possess a RNA-dependent DNA polymerase (reverse transcriptase) required for viral replication. In these viruses, a single strand of DNA is synthesized on the RNA genome (reverse transcription), duplicated, and circularized to a double-stranded proviral DNA. The proviral DNA is then integrated into the host cell chromosomal DNA to form the template (termed *aprovirus* or *virogene*) required for the synthesis of mRNAs and replication of the viral RNA genome. Oncogenic (cancer-causing) viruses, such as the human T-cell leukemia viruses (HTLV) and the related human immune deficiency (HIV; AIDS) virus are retroviruses. Retroviral reverse transcriptase is believed to be inhibited by the triphosphates of certain deoxynucleosides, such as 2′,3′-deoxy-3′-azidothymidine (AZT), an antiviral agent that has exhibited some effectiveness in AIDS patients.[55]

The DNA viruses constitute a heterogeneous group of viruses that use DNA as the genome and replicate in the nucleus of the host cell. Some DNA viruses are simple structures consisting of a single DNA strand and a few enzymes surrounded by a capsule (e.g., parvovirus) or a lipoprotein envelope (e.g., hepatitis B virus). Others, such as the herpesviruses and poxviruses, are large complex structures with a double-stranded DNA genome and several enzymes encased in a capsule and surrounded by an envelope consisting of several membranes. DNA viruses contain DNA-dependent RNA polymerases (transcriptases), DNA polymerases, and various other enzymes (depending on the complexity of the virus) that may provide targets for antiviral drugs. The more successful chemotherapeutic agents discovered thus far are directed against replication of herpesviruses. The nucleotide analogues idoxuridine, trifluridine, and vidarabine appear to block replication in herpesviruses by three general mechanisms: first, as the monophosphates, they inhibit the formation of precursor nucleotides required for DNA synthesis; second, as triphosphates, they inhibit DNA polymerase; and third, the triphosphates are incorporated into an abnormal DNA that does not function normally.

Late in the infection process, the viral components are assembled into the mature virus particle (virion). For simple nonenveloped viruses (e.g., the poliovirus), the genome and a few enzymes are encased by capsid proteins to complete the virion. Other, more complex viruses are enveloped by one or more membranes containing carbohydrate and lipoprotein components derived from the host cell membrane.

Once the mature virion has been assembled, it is ready for release from the cell. The release of certain viruses (e.g., the poliovirus) is accompanied by lysis of the cell membrane and cell death. On the other hand, some enveloped viruses are released by exocytosis, a process involving fusion between the viral envelope and the cell membrane. The cell membrane remains intact under these conditions and the cell may survive.

Amantadine Hydrochloride, USP. 1-Adamantanamine hydrochloride (Symmetrel) is a white, bitter, crystalline powder that is soluble in water and insoluble in alcohol. It is useful in the prevention but not the treatment of influenza caused by influenza A viruses. It is not effective against influenza B virus or other viruses.

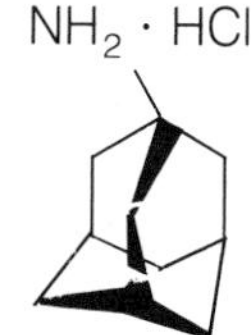

Amantadine Hydrochloride

The basis for the specific prophylactic action of amantadine against influenza A is not completely understood. Its mode of action appears to involve the inhibition of uncoating of the RNA virus to thereby block the transfer of viral RNA into the host cell. It may also prevent the penetration of the intact virus into the host cell. Amantadine has no chemotherapeutic effect once viral infection of the host cell has occurred.

Amantadine is estimated to be about 80% effective in the prevention of illness caused by strains of influenza A virus. When administered 24 to 48 hours after the onset of illness, it also reduces the duration of fever and other symptoms by preventing infection of additional cells. It is generally used in high-risk patients who have underlying debilitating disease or in close household or hospital contacts of patients with severe influenza A illness.

Amantadine is well absorbed following oral administration. It is distributed to all body fluids and tissues and is excreted largely unchanged in the urine. It is supplied as capsules and in a syrup for oral administration. The drug also finds occasional use as an adjunct for the treatment of Parkinson's disease in patients who do not tolerate full therapeutic doses of levodopa.

Side effects associated with amantadine therapy are primarily related to its ability to promote the release of dopamine from dopaminergic neurones (the basis for its use in Parkinson's disease). The most common side effects are depression, dizziness, urinary retention, hallucinations, anxiety, and gastrointestinal upset.

Idoxuridine, USP. 2′-Deoxy-5-iodouridine; IUdR (Stoxil, Herplex) was introduced in 1963 for the treatment of herpes simplex keratitis. At that time it was the only suitable chemotherapy for this infection.

Idoxuridine is active only against DNA viruses such as herpes and vaccinia. It is phosphorylated by

Idoxuridine

viral thymidylate kinase to the monophosphate, which is further bioactivated to the triphosphate. The triphosphate is believed to be both an inhibitor and a substrate of viral DNA polymerase, causing the inhibition of viral DNA synthesis and producing the synthesis of DNA that contains the iodinated pyrimidine. This bogus DNA is more susceptible to strand breakage and to miscoded errors in RNA and protein synthesis. The ability of idoxuridylic acid to substitute for deoxythymidylic acid in the synthesis of DNA may be due to the similar van der Waals radii of iodine (2.15 Å) and the methyl group (2.00 Å).

Idoxuridine occurs as a pale yellow crystalline solid that is soluble in water and alcohol, but poorly soluble in most organic solvents. It is a weak acid with a pK_a of 8.25. Aqueous solutions are slightly acidic, having a *p*H of about 6, and are stable for up to one year if refrigerated. Idoxuridine is light- and heat-sensitive. Its solutions may not be autoclaved. It is supplied as a 0.1% ophthalmic solution and a 0.5% ophthalmic ointment.

Trifluridine. *α,α,α*-Trifluorothymidine; 2-deoxy-5-(trifluoromethyl)uridine (Viroptic). This compound is similar to idoxuridine, having a trifluoromethyl group in place of the iodine atom at the 5-position. Trifluridine is employed as a 1% sterile ophthalmic solution in the treatment of keratoconjunctivitis caused by herpes simplex viruses types 1 and 2. It has been reported to be effective in cases resistant to idoxuridine.

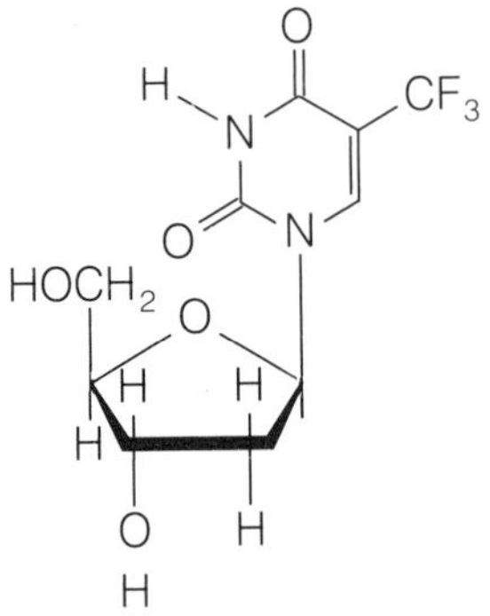

Trifluridine

The spectrum of antiviral activity and mechanism of action of trifluridine are similar to that of idoxuridine. However, the van der Waals radius of the trifluoromethyl group is 2.44 Å, which is somewhat larger than that of the iodine atom.

Trifluridine solutions are heat-sensitive, requiring refrigeration before and after dispensing. The product will lose 10% of its potency in one month when kept at 25°C. Less than 10% of its potency is lost in three years, when the solution is refrigerated at 4° to 8°C.

Vidarabine. 9-*β*-D-Arabinofuranosyladenine; adenine arabinoside; ara-A (Vira-A). Originally synthesized in 1960 as a potential anticancer agent, ara-A was later found to have broad-spectrum activity against DNA viruses.[56] First marketed in 1977 as an alternative to idoxuridine for the treatment of herpes simplex keratitis, it received FDA approval one year later for the treatment of herpes simplex encephalitis. Commercial vidarabine is now obtained as an antibiotic from cultures of *S. antibioticus*.

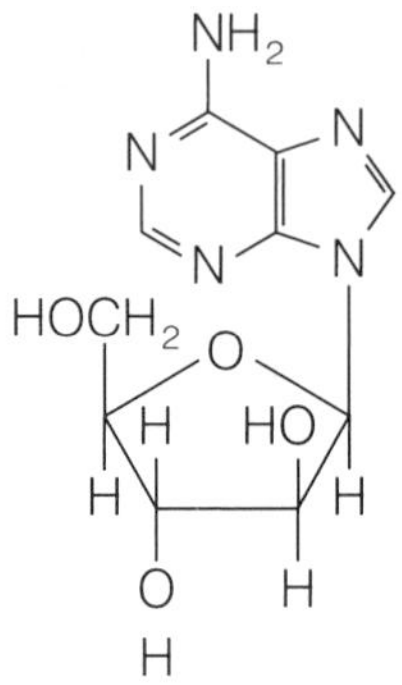

Vidarabine

In cases of viral encephalitis, vidarabine must be administered by continuous intravenous infusion because of its poor water solubility and relatively rapid metabolic conversion in vivo. The drug undergoes deamination, catalyzed by adenosine deaminase, to form the considerably less active hypoxanthine derivative, hypoxanthine arabinoside (ara-H).

The antiviral action of vidarabine is confined to DNA viruses. It is converted by viral enzymes to the triphosphate, which is a potent inhibitor of ribonucleotide reductases and DNA polymerases. The triphosphate may also be incorporated into viral DNA.

Vidarabine occurs as a white crystalline monohydrate that is sparingly soluble in water (0.45 mg/mL at 25°C). The injection intended for slow intravenous infusion is a 200-mg/mL suspension in water. It is recommended that the intravenous infusion fluid be preheated to 35° to 40°C before the addition of the suspension to facilitate dissolution of the drug. A 3% opthalmic ointment is also available for the treatment of herpes simplex keratitis.

Acyclovir. 2-Amino-1,9-dihydro-9-[(2-hydroxyethoxy)methyl]-6*H*-purin-6-one; acycloguanosine; ACV; (Zovirax). Acyclovir is the most effective member of a series of acyclic nucleosides with antiviral activity. In contrast with true nucleosides, which have a ribose or a deoxyribose sugar attached to a purine or pyrimidine base, acyclic nucleosides have only a portion of the sugar present. The relationship between acylovir (acycloguanosine) and 2′-deoxy guanosine is shown in the structural diagrams.

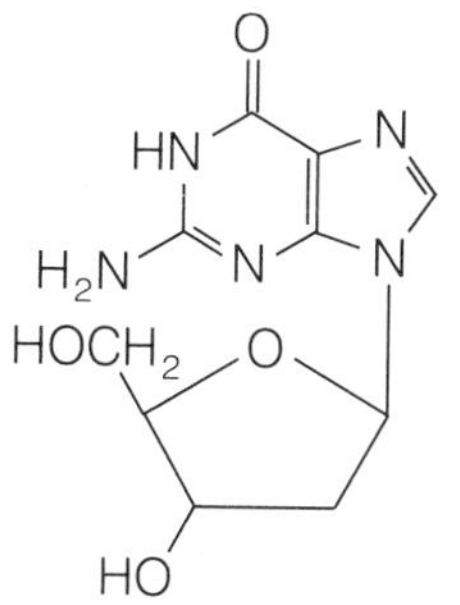

Deoxyguanosine 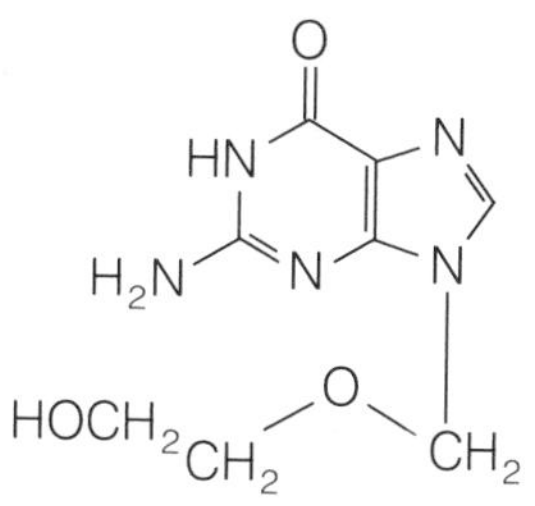 Acycloguanosine

Acyclovir has potent activity against several DNA viruses, including herpes simplex type 1 and type 2 viruses, varicella–zoster virus, and Epstein-Barr virus. It has become the drug of choice for the treatment of genital herpes. The mode of action of acyclovir in DNA viruses has been studied extensively.[57] The acyclic nucleoside is transported into infected cells where it is selectively phosphorylated by viral thymidylate kinase to the monophosphate. Further phosphorylation by cellular enzymes forms the triphosphate, which has an affinity 100 times greater for viral DNA polymerase than for human DNA polymerase. Viral DNA with acyclovir incorporated at its 3′-terminus is a potent inhibitor of viral DNA polymerase. Chain termination occurs because the 3′,5-phosphodiester bond cannot form. The selectivity of acyclovir for herpesvirus results from the properties of viral DNA polymerase and the virus-specified thymidylate kinase.

Two systemic dosage forms of acyclovir are available, oral and parenteral. Oral acyclovir is employed in the initial treatment of genital herpes and to control mild recurrent episodes. Intravenous administration is indicated for initial and recurrent infections in immunocompromised patients and for the prevention and treatment of severe recurrent episodes. The drug is slowly and incompletely absorbed from the gastrointestinal tract. Its oral absorption is reduced by food. Acyclovir is widely distributed to tissues and biologic fluids. Most of it is excreted in the urine unchanged, but a small fraction (approximately 10%) is excreted as the carboxy metabolites.

Acyclovir occurs as a chemically stable, white crystalline solid that is slightly soluble in water. Because of its amphoteric properties (pK_a values 2.27 and 9.25), the solubility of acyclovir is increased by both strong acids and strong bases. The injectable form is actually the sodium salt, which is supplied as a lyophylized powder, equivalent to 50 mg of active ingredient dissolved in 10 mL of sterile water for injection. Because the resulting solution is strongly alkaline, with a pH of approximately 11, it must be administered by slow intravenous infusion to avoid irritation and thrombophlebitis at the injection site.

Adverse reactions associated with acylovir are surprisingly few. Occasional gastrointestinal upset, dizziness, headache, lethargy, and joint pain are experienced by some patients. Thrombophlebitis following intravenous administration has been reported. An ointment containing 5% acyclovir in a polyethylene glycol base is available for the treatment of initial, mild episodes of herpes genitalis. It is not effective in preventing recurrent episodes.

Ribavirin. 1β-D-Ribofuranosyl-1H-1,2,4-triazole-3-carboxamide (Virazole). This synthetic nucleoside exhibits in vitro activity against a wide variety of DNA and RNA viruses[58] including adenovirus, herpesvirus, vaccinia virus, myxoma virus, influenza virus, parainfluenza virus, respiratory syncytial virus, measles virus, and rhinovirus. Despite the broad spectrum of activity of ribavirin, the drug has been approved for only one therapeutic indication: namely, the treatment of severe lower respiratory infections caused by respiratory syncytial virus (RSV) in carefully selected hospitalized infants and young children.

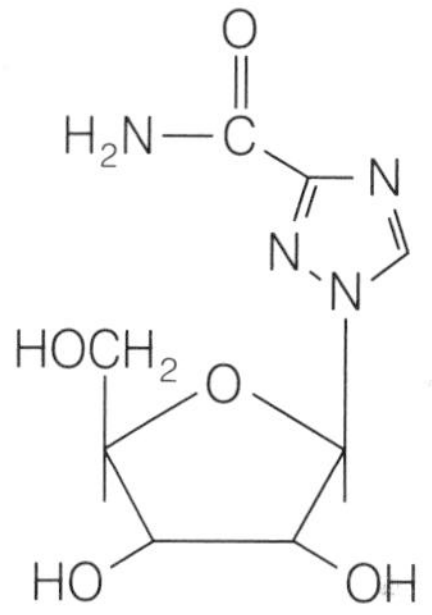

Ribavirin

The broad antiviral spectrum of ribavirin suggests multiple modes of action.[59] The nucleoside is known to be bioactivated by viral and cellular kinases and cellular phosphorylating enzymes to the monophosphate (RMP) and the triphosphate. RMP inhibits inosine monophosphate dehydrogenase, thereby preventing the conversion of inosine monophosphate (IMP) to xanthine monophosphate (XMP). XMP is required for guanosine triphosphate (GTP) synthesis. RTP inhibits viral RNA polymerases. It also prevents the capping of viral messenger RNA by inhibiting guanyl N^7-methyltransferase.

Ribavirin occurs as a white crystalline polymorphic solid that is soluble in water and chemically stable. It is supplied as a powder to be reconstituted in an aqueous aerosol containing 20 mg/mL of sterile water. The aerosol is administered with a special small-particle aerosol (SPAG) generator. Deterioration of respiratory function, bacterial pneumonia, pneumothorax, and apnea have been reported in severely ill infants and children with RSV infection. The role of ribavirin in these events has not been determined. Anemia, headache, abdominal pain, and lethargy have been reported in patients receiving oral ribavirin. It is teratogenic in some animal species and embryocidal in others.

Unlabeled uses of ribavirin include aerosol treatment of influenza A and B, and oral treatment of hepatitis, genital herpes, measles, and Lassa fever. It does not protect cells against the cytotoxic effects of the AIDS virus.

Zidovudine. 3′-Azido-3′,2′-deoxythymidine; azidothymidine; AZT (Retrovir). This nucleoside was synthesized in 1978 by Lin and Prusoff[60] as an intermediate in the preparation of amino acid analogues of thymidine. A screening program directed toward the identification of agents potentially effective for the treatment of AIDS patients led to the discovery of its unique antiviral properties some seven years later.[61] The following year, the clinical effectiveness of AZT in patients with AIDS and AIDS-related complex was demonstrated.[55]

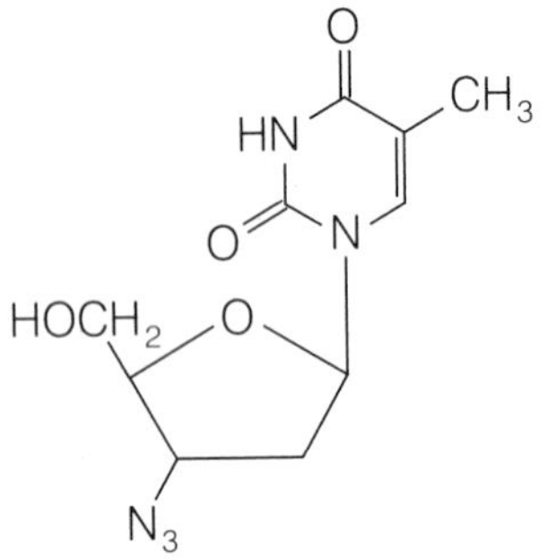

Zidovudine

Zidovudine is active against retroviruses, a group of RNA viruses responsible for AIDS and certain types of leukemia. Retroviruses possess a reverse transcriptase, or RNA-directed DNA polymerase, that directs the synthesis of a DNA copy (proviral DNA) of the viral RNA genome that is duplicated, circularized, and incorporated into the DNA of the infected cell. The AZT is converted to the monophosphate by thymidylate kinase and eventually to the triphosphate, which is utilized by reverse transcriptase for incorporation into an incomplete proviral DNA.[62] The DNA chain terminates at the site of AZT incorporation because a 3′,5′-phosphodiester bond cannot form with another nucleoside triphosphate, thereby inhibiting the reverse transcriptase. Cellular α-DNA polymerase is also inhibited by AZT triphosphate, but only at concentrations 100 times greater than those required for the viral DNA polymerase.

Zidovudine is recommended for the management of adult patients with symptomatic human immune deficiency virus (HIV) infection (AIDS or ARC) who have a history of confirmed *Pneumocystis carinii* pneumonia or an absolute CD4 (T4-helper/inducer) lymphocyte count of fewer than 200/mm^3 before therapy. The hematologic toxicity of the drug precludes its use in asymptomatic patients. Anemia and granulocytopenia are the most common toxic effects associated with zidovudine.

The drug is supplied as 100-mg capsules for oral administration. It is rapidly absorbed from the gastrointestinal tract and is well distributed into most body fluids and tissues, including the cerebrospinal fluid. It is rapidly metabolized to the inactive glucuronide metabolite in the liver. Only about 15% is excreted unchanged. Because zidovudine is an aliphatic azide, it is heat- and light-sensitive. It should be protected from light and stored at 15° to 25°C.

Inosiplex (Isoprinosine). A 1 : 3 complex of inosine and the 1-(dimethylamino)-2-propanol salt of 4-acetamidobenzoic acid. This investigational agent exhibits activity against a wide variety of both DNA and RNA viruses. Its broad-spectrum activity is thought to result from a two-pronged mode of action, namely, the simultaneous stimulation of host T-cell-mediated immunity and the direct inhibition of viral replication.

Clinical studies indicated potentially useful activity for inosiplex against herpes-, rhino-, and influenza viruses. More recent studies in pre-AIDS

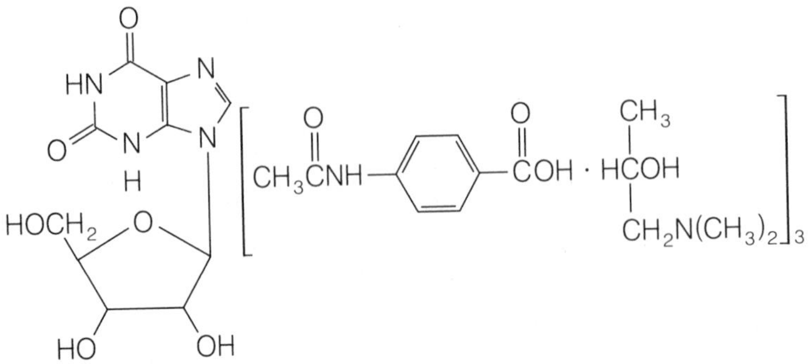

Inosiplex

TABLE 4-15

ANTIVIRAL AGENTS

Name Proprietary Name	**Preparations**	**Application***	**Usual Adult Dose***	**Usual Pediatric Dose***
Amantadine hydrochloride USP *Symmetrel*	Amantadine hydrochloride capsules USP Amantadine hydrochloride syrup USP		Orally, 200 mg/day in a single dose or in two divided doses	Children 9–12 yr: 100 mg bid Children 1–9 yr: 4.4–8.8 mg/kg daily qd or divided bid, not to exceed 150 mg/day
Idoxuridine USP *Herplex, Stoxil*	Idoxuridine ophthalmic solution USP Idoxuridine ophthalmic ointment USP	Topically to the conjunctiva, as 0.1 mL of a 0.1% solution 10–20 times daily or as an 0.5% ointment 5 times daily		
Vidarabine USP *Vira-A*	Sterile vidarabine USP		Slow IV infusion of 15 mg/kg daily over 12–24 hr	
	Vidarabine ointment USP	Topically to the conjunctiva as 2.8% equivalent of vidarabine ointment 5 times daily at 3-hr intervals		
Trifluridine *Viroptic*	Trifluridine ophthalmic solution	Topically, 1 drop onto the cornea of the affected eye every 2 hr while awake for a maximum of 9 days		

*See USP DI for complete dosage information.

patients suggest that inosiplex may delay the progression of pre-AIDS to AIDS. The drug, therefore, could be of value in long-term treatment because it is active orally and is apparently free of serious adverse effects.

ANTIPROTOZOAL AGENTS

In the United States and other countries of the temperate zone, protozoal diseases are of minor importance, whereas bacterial and viral diseases are widespread and the cause of considerable concern. On the other hand, protozoal diseases are highly prevalent in tropical Third-World countries, where they inflict both human and animal populations, causing suffering, death, and enormous economic hardship. The most common protozoal diseases in the United States are malaria, amebiasis, trichomoniasis, and toxoplasmosis. Antimalarial agents are covered in Chapter 6.

Although amebiasis is generally thought of as a tropical disease, it actually has a worldwide distribution. In some areas with temperate climates, in which sanitation is poor, the prevalence of amebiasis has been estimated to be as high as 20% of the population. The causative organism, *Entamoeba histolytica*, can invade the wall of the colon, or other parts of the body (such as liver, lungs, or skin). An ideal chemotherapeutic agent would be effective against both the intestinal and extraintestinal forms of the parasite.

Amebicides that are effective against both intestinal and extraintestinal forms of the disease are limited to the alkaloids emetine and dehydroemetine; the nitroimidazole derivative metronidazole; and the antimalarial agent chloroquine (see Chap. 6). A second group of amebicides that are effective only against intestinal forms of the disease includes the aminoglycoside antibiotic paramomycin (see Chap. 7); the 8-hydroxyquinoline derivative iodoquinolol; the arsenical compound carbarsone; and diloxanide furoate.

Trichomoniasis, a venereal disease caused by the flagellated protozoan *Trichomonas vaginalis*, is common in the United States and throughout the world. Although it is not generally considered to be serious, this affliction can cause serious physical discomfort and sometimes has a chilling effect on sexual relations. Oral metronidazole provides effective treatment against all forms of the disease. It is also employed to eradicate the organism from asymptomatic male carriers.

Various forms of trypanosomiasis, chronic tropical diseases caused by pathogenic members of the family Trypanosomidae, occur both in humans and in livestock. The principal disease in humans, sleeping sickness, can be broadly classified into two main geographic and etiologic groups: African sleeping sickness caused by *Trypanosoma gambiense* (West African), *T. rhodesiense* (East African) or *T. congolense*, and South American sleeping sickness (Chagas' disease) caused by *T. cruzi*. Of the various forms of trypanosomiasis, Chagas' disease is the most serious and generally the most resistant to chemotherapy. Leishmaniasis is a chronic tropical disease caused by various flagellate protozoa of the genus *Leishmania*. The more common visceral form caused by *L. donovani*, called kala-azar, is similar to Chagas' disease. Fortunately, although these diseases are widespread in tropical areas of Africa and South and Central America, they are of minor importance in the United States, Europe, and Asia.

The successful chemotherapy of trypanosomiasis and leishmaniasis remains somewhat primitive and often less than effective. In fact, it is doubtful that these diseases can be controlled by chemotherapeutic measures alone, without successful control of the intermediate hosts and vectors that transmit them. Heavy metal compounds, such as the arsenicals and antimonials, are sometimes effective, but frequently toxic. The old standby suramin, the newer bisamides (hydroxystilbamidine, USP, and pentamidine) appear to be of some value in long- and short-term prophylaxis. The nitrofuran derivative nifurtimox may be a major breakthrough in the control of these diseases. However, its potential toxicity remains to be fully assessed.

Emetine Hydrochloride, USP. The alkaloid emetine is obtained by isolation from ipecac or synthetically by methylation of the phenolic alkaloid cephaeline. It occurs as a levorotatory, light-sensitive, white powder that is insoluble in water. The alkaloid readily forms water-soluble salts. Solutions of the hydrochloride salt intended for intramuscular injection should be adjusted to *p*H 3.5 and stored in light-resistant containers.

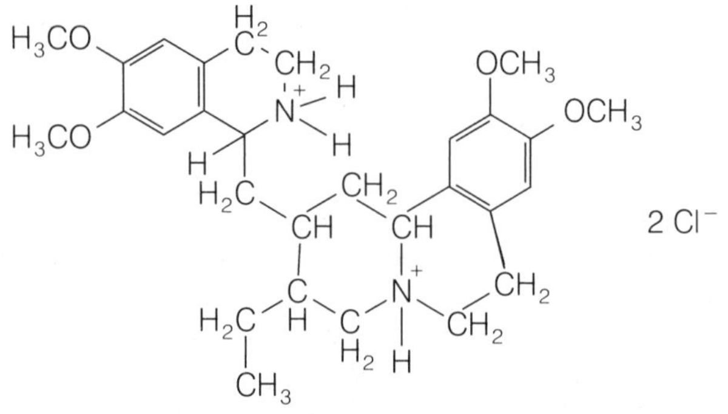

Emetine Hydrochloride

Emetine exerts a direct amebicidal action on various forms of *E. histolytica*. It is a protoplasmic poison that inhibits protein synthesis in protozoal and mammalian cells by preventing protein elongation. Because the effect of emetine in intestinal amebiasis is solely symptomatic and the cure rate is only 10% to 15%, it should be used exclusively in combination with other agents. The high concentrations of emetine achieved in the liver and other tissues after intramuscular injection provide the basis for its high degree of effectiveness against hepatic abscesses and other extraintestinal forms of the disease. Toxic effects limit the usefulness of emetine. It causes a high frequency of gastrointestinal distress (especially nausea and diarrhea), cardiovascular effects (hypotension and arrhythmias), and neuromuscular effects (pain and weakness). A lower incidence of cardiotoxicity has been associated with the use of the closely related alkaloid dehydroemetine (Mebadin), available from the Centers for Disease Control (CDC), which is also amebicidal.

Emetine hydrochloride has also been used to treat balantidial dysentery and fluke infestations such as fascioliasis and paragonimiasis.

Metronidazole, USP. 2-Methyl-5-nitroimidazole-1-ethanol (Flagyl) is the most useful of a multitude of antiprotozoal nitroimidazole derivatives that have been synthesized in various laboratories throughout the world. Metronidazole was first marketed for the topical treatment of *T. vaginalis* vaginitis. It has since been shown to be effective orally against both the acute and carrier states of the disease. The drug also possesses useful amebicidal activity and is, in fact, effective against both intestinal and hepatic amebiasis. It has also found use in the treatment of such other protozoal diseases as giardiasis and balantidiasis.

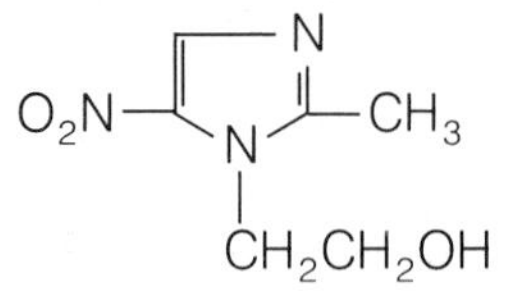

Metronidazole

More recently, metronidazole has been found to be effective against obligate anaerobic bacteria, but ineffective against facultative anaerobes or obligate aerobes. It is particularly active against gram-negative anaerobes, such as *Bacteriodes* and *Fusobacterium* species. It is also effective against gram-positive anaerobic bacilli (such as *Clostridium* sp.) and cocci (such as *Peptococcus* and *Peptidostreptococcus* sp.). Because of its bactericidal action, metronidazole has become an important agent for the treatment of serious infections (such as septicemia, peritonitis, abscesses, meningitis, and others) caused by anaerobic bacteria.

The common characteristic of microorganisms (bacteria and protozoa) sensitive to metronidazole is that they are anaerobic. It has been speculated that a reactive intermediate formed in the microbial reduction of the 5-nitro group of metronidazole covalently binds to the DNA of the microorganism triggering the lethal effect.[63] Potential reactive intermediates include the nitroxide, nitroso, hydroxylamine, and amine. The ability of metronidazole to act as a radiosensitizing agent is also related to its reduction potential.

Metronidazole is a pale yellow crystalline substance that is sparingly soluble in water. It is stable in air, but is light-sensitive. Despite its low water solubility, metronidazole is well absorbed following oral administration. It has a large apparent volume of distribution and achieves effective concentrations in all body fluids and tissues. Approximately 20% of an oral dose is metabolized to oxidized or conjugated forms. The 2-hydroxy metabolite is active; other metabolites are inactive.

Diloxanide Furoate (Furamide) is the 2-furoate ester of 2,2-dichloro-4-hydroxy-*N*-methylacetanilide. It was developed as a result of the discovery that various α,α-dichloroacetamides possessed amebicidal activity in vitro. Diloxanide itself and many of its esters are also active, and drug metabolism studies indicate that hydrolysis of the ester is required for the amebicidal effect. Nonpolar esters of diloxanide are more potent than polar ones. Diloxanide furoate has been used in the treatment of asymptomatic carriers of *E. histolytica*. Its effectiveness against acute intestinal amebiasis or heptaic abscesses, however, has not been established. Diloxanide furoate is a white crystalline powder. It is administered orally only as 500-mg tablets and may be obtained in the United States from the Centers for Disease Control in Atlanta, Georgia.

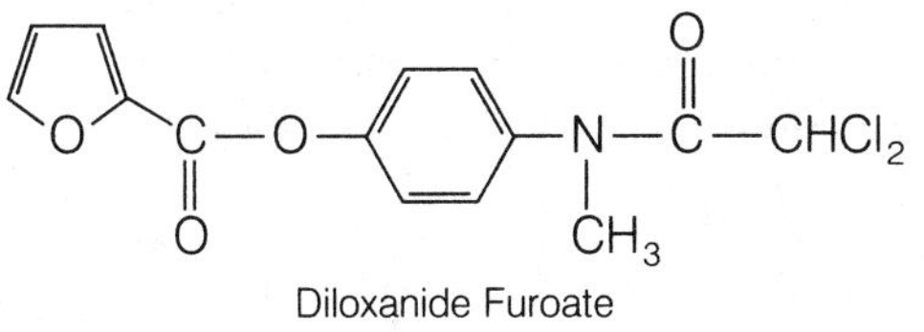

Diloxanide Furoate

8-Hydroxyquinoline. Oxine; quinophenol; oxyquinoline is the parent compound from which the antiprotozoal oxyquinolines have been derived. The antibacterial and antifungal properties of oxine and its derivatives, which are believed to result from the ability to chelate metal ions (see Chap. 2), are well known. Aqueous solutions of acid salts of oxine, particularly the sulfate (Chinosol, Quinosol), in concentrations of 1 : 3000 to 1 : 1000, have been used as topical antiseptics. The substitution of an iodine atom at the 7-position of 8-hydroxy-quinolines produces compounds with broad-spectrum antimicrobial properties.

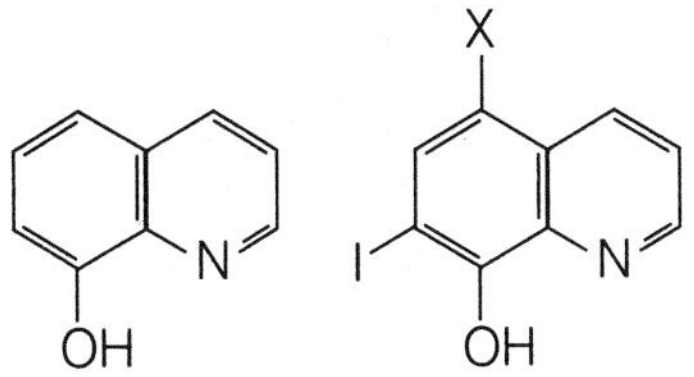

Oxine

Clioquinol X = Cl
Iodoquinol X = I

Clioquinol, USP. 5-Chloro-7-iodo-8-quinolinol; 5-chloro-8-hydroxy-7-iodoquinoline; iodochlorhydroxyquin (Vioform) is found as a spongy, voluminous, light-sensitive, yellowish-white powder that is virtually insoluble in water.

This compound was originally introduced as an odorless substitute for iodoform in the belief that it liberated iodine in the tissues. It has been used as powder for a variety of skin conditions, such as atopic dermatitis, eczema, psoriasis, and impetigo. It has also been applied vaginally as a 2% to 3% ointment or paste and in suppository form for the treatment of *T. vaginalis* vaginitis. However, its principal current use is in the local treatment of fungal infections, such as athlete's foot.

Category: Topical antifungal and antitrichomonal agent
Preparation
Clioquinol ointment, USP, 3%
Clioquinol cream, USP, 3%

Iodoquinol. 5,7,-Diiodo-8-quinolinol; 5,7-diiodo-8-hydroxyquinoline; diiodohydroxyquin (Yodoxin) is a yellowish to tan microcrystalline, light-sensitive substance that is insoluble in water. It is recommended for acute and chronic intestinal amebiasis, but is not effective in extraintestinal disease. Because a relatively high incidence of topic neuropathy has occurred with its use, iodoquinol should no longer be routinely used for traveler's diarrhea.

Hydroxystilbamidine Isethionate, USP. 2-Hydroxy-4,4′-stilbenedicarboxamidine diisethionate; 2-hydroxy-4,4′-diamidinostilbene occurs as a yellow, light-sensitive crystalline substance that is soluble in water. Parenteral solutions should be freshly prepared and stored in light-resistant containers.

Hydroxystilbamidine is considered a drug of choice for the prophylaxis and treatment of African trypanosomiasis. The drug does not penetrate the central nervous system and, therefore, is of doubtful value for the treatment of late stages of the disease. Hydroxystilbamidine has also been used as an anti-

Hydroxystilbamidine Isethionate

fungal agent in the treatment of systemic and pulmonary North American blastomycosis.

Pentamidine Isoethionate. 4,4′-(Pentamethylenedioxy)dibenzamidine diisethionate (Lomidine, Pentam 300) is a water-soluble, crystalline salt that is stable to light and air. The principal use of pentamidine is for the treatment of pneumonia caused by the opportunistic pathogenic protozoan *Pneumocystis carinii*, a frequent secondary invader associated with AIDS.

Pentamidine Isoethionate

Pentamidine has been used for the prophylaxis and treatment of African trypanosomiasis. It is also of some value for treating visceral leishmaniasis. Pentamidine rapidly disappears from the plasma after intravenous injection and is distributed to the tissues, where it is stored for a long period. This property probably contributes to the usefulness of the drug as a prophylactic agent.

Nifurtimox. 4-[(5-Nitrofurfurylidene)amino]-3-methylthiomorpholine 1,1-dioxide; Bayer 2502; (Lampit). The observation that various derivatives of 5-nitrofuraldehyde possessed, in addition to their antibacterial and antifungal properties, significant and potentially useful antiprotozoal activity eventually led to discovery of particular nitrofurans with antitrypanosomal activity. The most important of such compounds is nifurtimox because of its demonstrated effectiveness against *T. cruzi*, the parasite responsible for South American trypanosomiasis. In fact, use of this drug represents the only clinically proven treatment for both acute and chronic forms of the disease. Nifurtimox is available in the United States from the Centers for Disease Control.

Nifurtimox

Suramin Sodium is a high-molecular-weight bisurea derivative containing six sulfonic acid groups as their sodium salts. It was developed in Germany shortly after World War I as a by-product of research efforts directed toward the development of potential antiparasitic agents from dyestuffs. The drug has been used for more than half a century for the treatment of early cases of trypanosomiasis. It was not until several decades later, however, that suramin was discovered to be a long-term prophylactic agent the effectiveness of which, after a single intravenous injection, is maintained for periods of up to three months. The drug is tightly bound to plasma proteins, causing its excretion in the urine to be almost negligible.

Tissue penetration of the drug does not occur, apparently because of its high molecular weight and

Suramin Sodium

highly ionic character. Thus, an injected dose remains in the plasma for a very long period. Newer, more effective drugs are now available for short-term treatment and prophylaxis of African sleeping sickness. Suramin is also used for prophylaxis of onchocerciasis. It is available from the Centers for Disease Control.

Carbarsone, USP. *N*-Carbamoylarsanilic acid; *p*-ureidobenzenearsonic acid is a pentavalent organic arsenical that occurs as a white crystalline powder. It is slightly soluble in water, but is readily solubilized by dilute alkali.

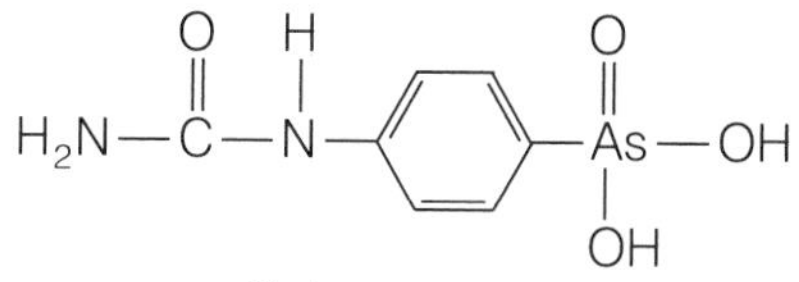

Carbarsone

Carbarsone is used orally for the treatment of intestinal amebiasis. Its lethal action on the trophozoite forms of *E. histolytica* is presumed to result from reaction of the trivalent form of the arsenical with sulfhydryl-containing enzymes in the parasite. When administered orally, substantial amounts of carbarsone are absorbed and, because the drug is slowly excreted in the urine, toxic levels can accumulate. Consequently, patients who are using carbarsone should be monitored for signs of arsenic toxicity.

Glycobiarsol, USP. (Hydrogen *N*-glycoloylarsanilato)oxobismuth; bismuthyl *N*-glycoloylarsanilate (Milibis) exists as a yellow to pink powder that is unstable to heat and is sparingly soluble in water. The compound is an acid that dissolves in dilute alkalis.

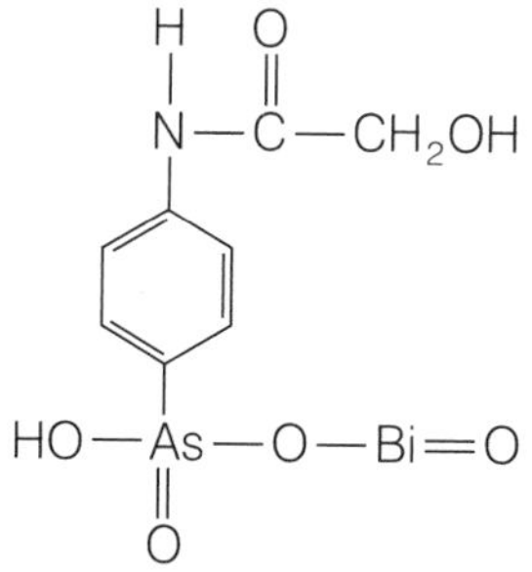

Bismuth Glycoloylarsanilate

Glycobiarsol is available in the form of vaginal suppositories for the treatment of trichomonal and monilial vaginitis. It has also been used orally for the treatment of intestinal amebiasis. It imparts a black color to the feces, a result of the formation of bismuth sulfide.

Melarsoprol. 2-*p*-(4,6-Diamino-*s*-triazin-2-ylamino)phenyl-4-hydroxymethyl-1,3,2-dithiarsoline (Mel B, Arsobal) is prepared by reduction of the corresponding pentavalent arsenilate to the trivalent arsenoxide followed by reaction of the latter with 2,3-dimercaptopropanol (BAL). It has become the drug of choice for the treatment of the latter stages of both forms of African trypanosomiasis. Melarsoprol has the advantage of excellent penetration into the central nervous system and, therefore, is effective against mengioencephalitic forms of *T. gambiense* and *T. rhodesiense*. Trivalent arsenicals tend to be more toxic to the host (as well as the parasites) than the corresponding pentavalent compounds. The bonding of arsenic with sulfur atoms tends to reduce host toxicity, increase chemical stability (to oxidation), and improve distribution of the compound to the arsenoxide. However, melarsoprol shares the toxic properties of other arsenicals and its use must be monitored for signs of arsenic toxicity.

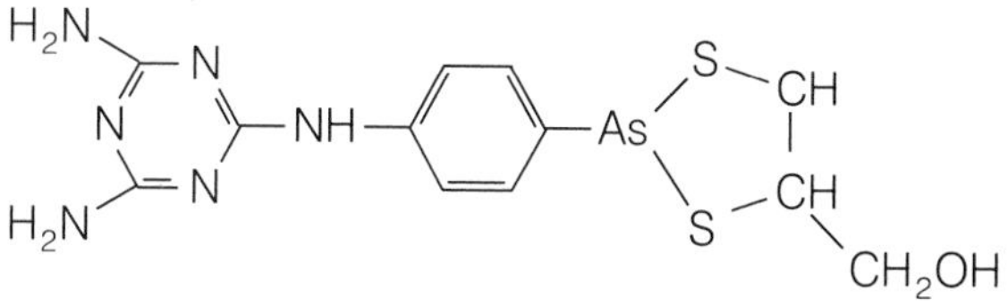

Melarsoprol

Sodium Stibogluconate. Sodium antimony gluconate (Pentostam) is a pentavalent antimonial compound intended primarily for the treatment of various forms of leishmaniasis. It is available from the Centers for Disease Control as the disodium salt, which is chemically stable and freely soluble in water. The 10% aqueous solution used for either intramuscular or intravenous injection has a *p*H of about 5.5. Similar to all antimonial drugs, this drug has a low therapeutic index, and patients undergoing therapy with it should be monitored carefully for signs of heavy metal poisoning. Other organic antimonial compounds are employed primarily for the treatment of schistosomiasis and other flukes.

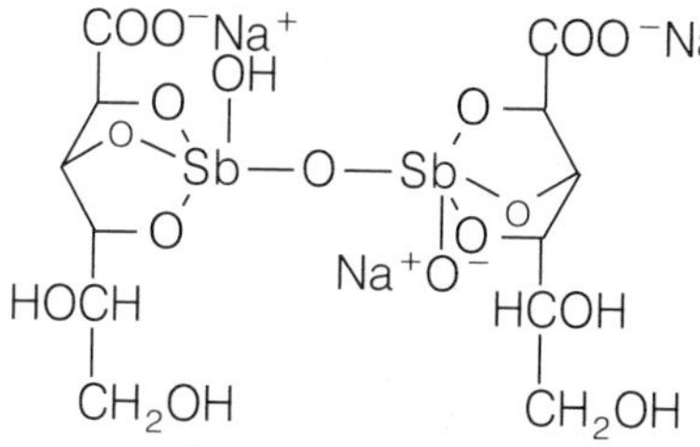

Sodium Stibogluconate

TABLE 4-16

ANTIPROTOZOAL AGENTS

Name Proprietary Name	***Preparations***	***Category***	***Usual Adult Dose****	***Usual Dose Range****	***Usual Pediatric Dose****
Emetine hydrochloride USP	Emetine hydrochloride injection USP	Amebicide	IM or SC, 1 mg/kg of body weight, but not exceeding a total of 65 mg qd for 3–10 days	Not exceeding 65 mg/day or a total dose of 650 mg in 10 days	SC 500 μg/kg of body weight or 15 mg/m^2 of body surface, bid for 4–6 days, but not exceeding a total of 65 mg/day
Metronidazole USP *Flagyl*	Metronidazole tablets	Amebicide, trichomonacide	Antiamebic: 500–750 mg tid for 5–10 days		35–50 mg/kg per 24 hr, in divided doses, for 10 days
			Antitrichomonal: 250 mg tid		
		Anaerobic bacteriocide	Antibacterial: slow IV infusion: loading dose of 15 mg/kg over 1 hr; maintenance dose of 7.5 mg/kg over 1 hr every 6 hr	4 g/24 hr	
Iodoquinol USP *Diodoquin, Yodoxin*	Diiodohydroxyquin tablets USP	Amebicide	650 mg tid for 20 days		13.3 mg/kg of body weight tid, not to exceed 1.95 g in 24 hr
Diloxanide furoate		Amebicide	500 mg orally tid for 10 days	500 mg orally tid for 10 days	
Carbarsone USP	Carbarsone capsules USP	Amebicide	250 mg bid to tid for 10 days	100–250 mg	
Glycobiarsol USP *Milibis*	Glycobiarsol tablets USP	Amebicide	500 mg tid for 7–10 days		
Hydroxystilbamidine isethionate USP	Sterile hydroxystilbamidine	Leishmanicide, trypanocide, fungicide	IM, 225 mg every 24 hr, administered in 10 mL of 5% dextrose injection or 0.9% sodium chloride injection; IV infusion, 225 mg every 24 hr, administered in 200 mL of 5% dextrose injection or 0.9% sodium chloride injection over a period of 2–3 hr		IV infusion, 3–4.5 mg/kg of body weight every 24 hr, administered in an appropriate amount of 5% dextrose injection or 0.9% sodium chloride injection over a period of 2–3 hr

Pentamidine isethionate USP *Pentam 300*	Sterile pentamidine USP	Antibacterial, trypanocide, leishmanicide	IV or deep IM 4 mg/kg qd. for 14 days		Same as adults
Nifurtimox		Trypanocide	Complicated dosage schedule available from the CDC		
Suramin		Trypanocide	100–200 mg, IV test dose weekly, then 1 g IV on days 1, 3, 7, 14, and 21	20 mg/kg IV on days 1, 3, 7, 14, and 21	
Melarsoprol		Trypanocide	2–3.6 mg/kg IV per day in 3 doses; after 1 wk 3.6 mg/kg per day for 3 doses. Repeat after 10–21 days	18–25 mg/kg IV over 1 mo. Initial dose of 0.36 mg/kg IV; intervals of 1–5 days depending on reactions for a total of 9–10 doses	
Stibogluconate sodium		Leishmanicide	600 mg IM or IV 1–6 times daily	10 mg/kg IM or IV 1–6 times daily	
Dimercaprol USP	Dimercaprol injection USP, 10%	Antidote to arsenic, gold, and mercury poisoning; metal-complexing agent	IM, 2.5 mg/kg of body weight 4–6 times daily on the first 2 days, then bid for the next 8 days, if necessary	2.5–5 mg/kg	IM, 2–3 mg/kg of body weight 6 times daily on the 1st day, qid on the 2nd day bid on the 3rd day, then qd for the next 10 days, if necessary

*See USP DI for complete dosage information.

The antileishmanial action of sodium stibogluconate requires reduction to the trivalent form, which is believed to inhibit phosphofructokinase in the parasite.

Dimercaprol, USP. 2,3-Dimercapto-1-propanol; BAL (British anti-Lewisite); dithioglycerol is a foul-smelling, colorless liquid. It is soluble in water (1 : 20) and alcohol. It was developed by the British during World War II as an antidote for "Lewisite," hence the name British anti-Lewisite or BAL. Dimercaprol is effective topically and systemically as an antidote for poisoning caused by arsenic, antimony, mercury, gold, and lead.

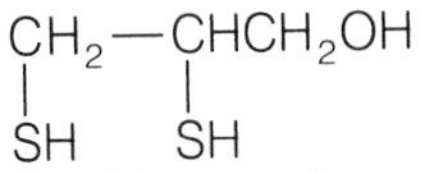

$$\underset{\displaystyle SH}{CH_2}-\underset{\displaystyle SH}{CH}CH_2OH$$

Dimercaprol

The antidotal properties of BAL are associated with the property of heavy metals to react with sulfhydryl (SH) groups in proteins (such as the enzyme pyruvate oxidase) and interfere with their normal function. 1,2-Dithiol compounds, such as BAL, compete effectively with such proteins for the metal by reversibly forming metal ring compounds of the type:

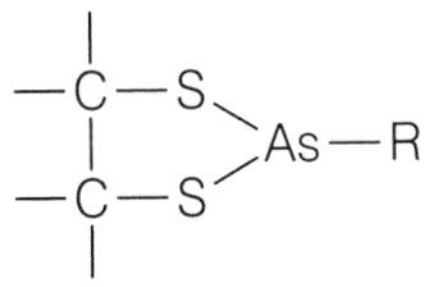

$$\begin{matrix} -\overset{|}{C}-S \\ | \\ -\underset{|}{C}-S \end{matrix} \!\!>As-R$$

which are relatively nontoxic, metabolically conjugated (as glucuronides), and rapidly excreted.

BAL may be applied topically as an ointment or injected intramuscularly as a 5% or 10% solution in peanut oil.

Eflornithine. DL-α-Difluoromethylornithine. This amino acid derivative is an enzyme-activated inhibitor of ornithine decarboxylase, a pyridoxal phosphate-requiring enzyme responsible for catalyzing the rate-limiting step in the biosynthesis of the diamine putrescene and the polyamines spermine and spermidine. Polyamines are essential for the regulation of DNA synthesis and cell proliferation in animal tissues and microorganisms.

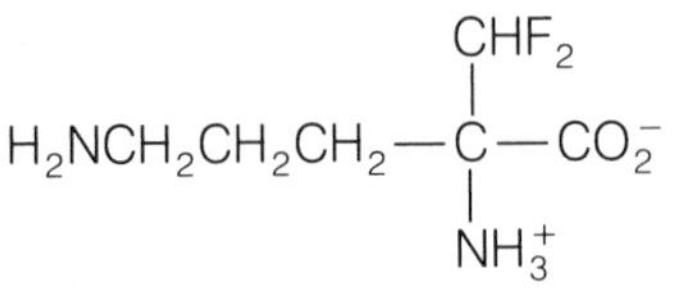

$$H_2NCH_2CH_2CH_2-\overset{\displaystyle CHF_2}{\underset{\displaystyle NH_3^+}{C}}-CO_2^-$$

Eflornithine

Eflornithine is currently being supplied as an orphan drug for the treatment of pneumonia in AIDS patients caused by *P. carinii*. It is reported to also be very effective against *T. gambiensie*, the organism that causes African sleeping sickness.

The irreversible inactivation of ornithine decarboxylase by eflornithine is accompanied by decarboxylation and release of fluoride ion from the inhibitor,[64] suggesting enzyme-catalyzed activation of the inhibitor. Only the (−)-isomer, stereochemically related to L-ornithine, is active. The drug is provided as the hydrochloride salt.

ANTHELMINTICS

Anthelmintics are drugs that have the capability of ridding the body of parasitic worms or helminths. The prevalence of human helminthic infestations is widespread throughout the globe and represents a major world health problem, particularly in Third-World countries. Helminths parasitic to humans and other animals are derived from two phyla: Platyhelminthes and Nemathelminthes. Cestodes (tapeworms) and trematodes (flukes) belong to the former, and the nematodes or true roundworms to the latter. The helminth infestations of major concern on the North American continent are caused by roundworms (i.e., hookworm, pinworm, and *Ascaris*). Human tapeworm and fluke infestations are rarely seen in the United States.

Several classes of chemicals are used as anthelmintics and include (1) chlorinated hydrocarbons, (2) phenols and derivatives, (3) dyes, (4) piperazine and related compounds, (5) antimalarial compounds (see Chap. 6), (6) various heterocyclic compounds, (7) alkaloids and other natural products, and (8) antimonial compounds.

Tetrachloroethylene, USP. Perchloroethylene; tetrachloroethene; $Cl_2C{=}CCl_2$ is a colorless liquid with an ethereal odor. It is soluble in most organic solvents, but insoluble in water. It is unstable in air, moisture, and light, forming phosgene and hydrochloric acid. The *USP* permits up to 1% ethanol as a stabilizing agent.

The effectiveness of tetrachloroethylene as an anthelmintic is due to its local irritant properties. Its specific use in medicine is in the treatment of hookworm infestations. It is preferred over carbon tetrachloride because it is less toxic. However, all halogenated hydrocarbons cause liver and kidney degeneration. Tetrachloroethylene is usually administered after the gastrointestinal tract is cleared. It is usually followed by a saline cathartic.

Piperazine, USP. Hexahydropyrazine; diethylenediamine (Arthriticine, Dispermin) occurs as colorless, volatile crystals of the hexahydrate that are freely soluble in water. After the discovery of the anthelmintic properties of a derivative diethylcarbamazine, the activity of piperazine itself was established. Piperazine is still employed as an anthelmintic for the treatment of pinworm [*Enterobius* (*Oxyuris*) *vermicularis*] and roundworm (*Ascaris lumbricoides*) infestations. It is available in a variety of salt forms including the citrate (official in the *USP*) in syrup and tablet forms.

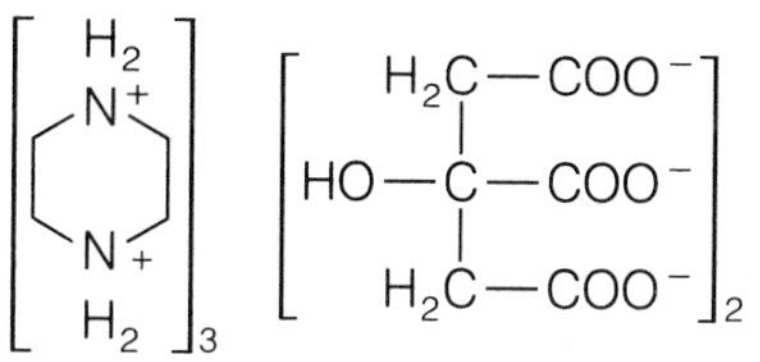

Piperazine Citrate

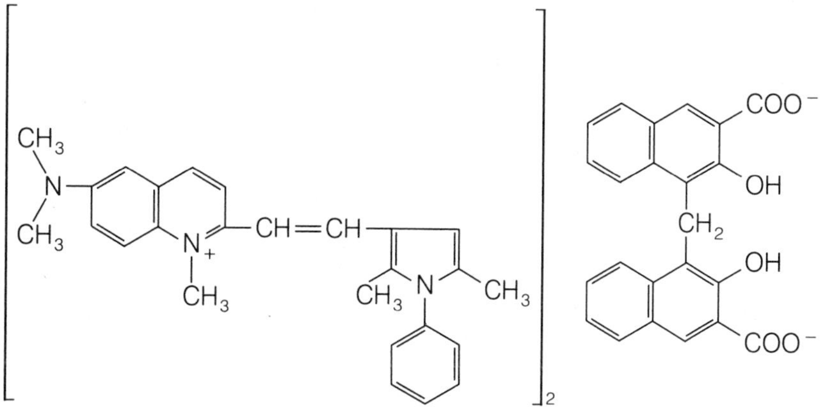

Pyrvinium Pamoate

Piperazine blocks the response of the ascaris muscle to acetylcholine, causing a flaccid paralysis in the worm, which is dislodged from the intestinal wall and expelled in the feces.

Diethylcarbamazine Citrate, USP. *N,N*-Diethyl-4-methyl-1-piperazinecarboxamide citrate; 1-diethylcarbamyl-4-methylpiperazine dihydrogen citrate (Hetrazan) is a highly water-soluble crystalline compound that has selective anthelmintic activity. It is effective against various forms of filariasis, including Bancroft's, onchocerciasis, and laviasis. It is also active against ascariasis. Relatively few adverse reactions have been associated with diethylcarbamazine.

Pyrvinium Pamoate, USP. 6-(Dimethylamino)-2-[2-(2,5-dimethyl-1-phenylpyrrol-3-yl)vinyl]-1-methylquinolinium with 4,4′-methylenebis[3-hydroxy-2-naphthoate] (Povan) is a red cyanine dye. It is used in the chemotherapy of pinworm infestations. The drug is believed to exert an anticholinergic effect on the worms, causing paralysis and expulsion by peristalsis. Pyrvinium is poorly soluble and not absorbed from the gastrointestinal tract. Its local irritant action may cause nausea and epigastric pain. It causes the feces to be stained a reddish brown.

Pyrantel Pamoate, USP. *trans*-1,4,5,6,-Tetrahydro-1-methyl-2-[2-(2-thienyl)vinyl]pyrimidine pamoate (Antiminth) is a depolarizing neuromuscular-blocking agent that causes spastic paralysis in susceptible helminths. It is employed in the treatment of infestations caused by pinworms and roundworms (ascariasis). Because its action opposes that of piperazine, the two anthelmintics should not be

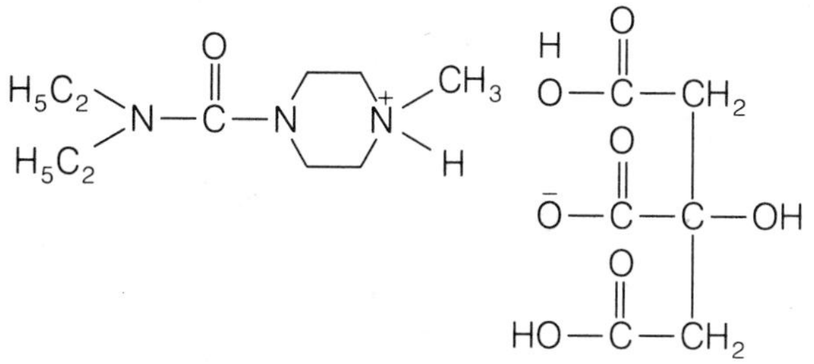

Diethylcarbamazine Citrate

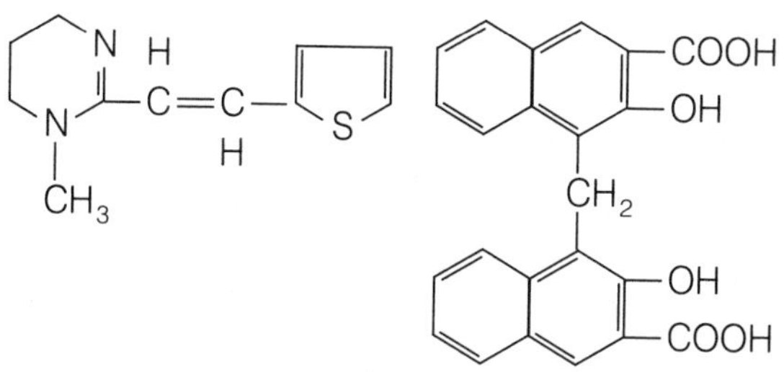

Pyrantel Pamoate

used together. Pyrantel is poorly absorbed from the gastrointestinal tract; over 50% of an oral dose is excreted unchanged. Adverse effects associated with its use are primarily gastrointestinal.

Thiabendazole, USP. 2-(4-Thiazolyl)benzimidazole (Mintezol) occurs as a white crystalline substance that is only slightly soluble in water, but soluble in strong mineral acids. Thiabendazole is a basic compound with a pK_a of 4.7 that forms complexes with metal ions.

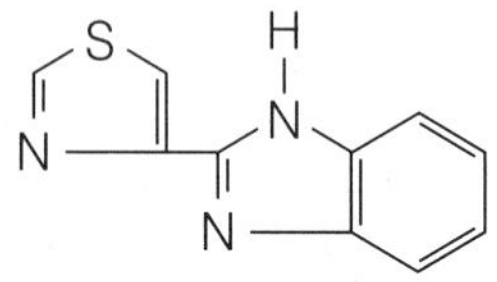

Thiabendazole

Thiabendazole inhibits the helminth-specific enzyme fumarate reductase.[65] It is not known whether metal ions are involved, or if the inhibition of the enzyme is related to thiabendazole's anthelmintic effect. Benzimidazole anthelmintic drugs such as thiabendazole and mebendazole also arrest nemotode cell division in metaphase by interfering with microtubule assembly.[66] They exhibit a high affinity for tubulin, the precursor protein for microtubule synthesis.

Thiabendazole has broad-spectrum anthelmintic activity. It is used to treat enterobiasis, strongyloidiasis (threadworm infection), ascariasis, uncinariasis (hookworm infection), and trichuriasis (whipworm infection). It has also been used to relieve symptoms associated with cutaneous larva migrans (creeping eruption) and the invasive phase of trichinosis.

In addition to its use in human medicine, thiabendazole is widely employed in veterinary practice to control intestinal helminths in livestock.

Mebendazole, USP. Methyl 5-benzoyl-2-benzimidazolecarbamate (Vermox) is a broad-spectrum anthelmintic that is effective against a variety of nematode infestations including whipworm, pinworm, roundworm, and hookworm. Mebendazole irreversibly blocks glucose uptake in susceptible helminths, thereby depleting glycogen stored in the parasite. It apparently does not affect glucose metabolism in the host. It also inhibits cell division in nematodes.[66]

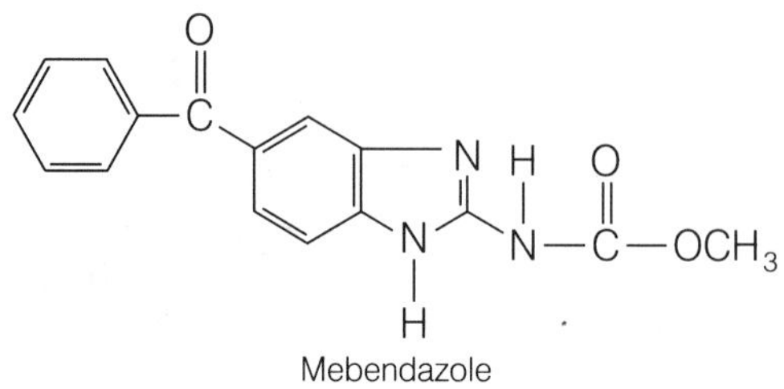

Mebendazole

Mebendazole is poorly absorbed by the oral route. Adverse reactions are uncommon and usually consist of abdominal discomfort. It is teratogenic in laboratory animals and, therefore, should not be given during pregnancy.

Niridazole. 1-(5-Nitro-2-thiazolyl)-2-imidazolidinone; 1-(5-nitro-2-thiazolyl)-2-oxotetrahydroimidazole (Ambilhar) is an antischistosomal drug that was synthesized as part of a systematic investigation of heterocyclic nitro compounds as potential antiparasitic agents. Its principal therapeutic application is in the treatment of infections caused by *Schistosoma haematobium* (urinary schistosomiasis), but it is also moderately effective against intestinal schistosomiasis (*S. mansoni*) and guinea worm (*Dracunculus medinensis*). Niridazole also has antiprotozoal activity and has been used effectively for intestinal amebiasis. The drug is generally tolerated and appears to be relatively nontoxic. However, occasional neuropsychiatric reactions (i.e., mental disorientation, mania, and convulsions) and electroencephalographic changes have been reported after oral administration in humans, and animal studies indicate that it may be both mutagenic and carcinogenic. Its use in the mass chemotherapy of schistosomiasis, therefore, requires further evaluation.

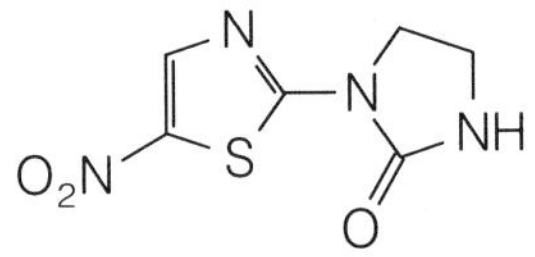

Niridazole

Niridazole occurs as a yellow crystalline powder that is sparingly soluble in water. It is supplied as 500-mg tablets and can be obtained from the Centers for Disease Control.

Bithionol. 2,2′-Thiobis(4,6-dichlorophenol); bis-(2-hydroxy-3,5-dichlorophenyl)sulfide (Actamer, Bitin). This chlorinated bisphenol was formerly used in soaps and cosmetics for its antimicrobial properties, but was removed from the market for topical use because of reports of contact photodermatitis. Bithionol has useful anthelmintic properties and has been employed as a fasciolicide and taeniacide. It is

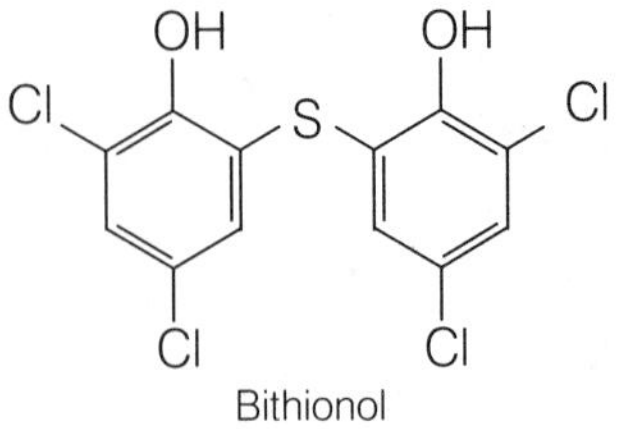

Bithionol

still considered the agent of choice for the treatment of infestations caused by the liver fluke *Fasciola hepatica* and the lung fluke *Paragonimus westermani*. Niclosamide is believed to be superior to it for the treatment of tapeworm infestations.

Niclosamide. 5-Cloro-*N*-(2-chloro-4-nitrophenyl)-2-hydroxybenzamide; 2,5′-dichloro-4′-nitrosalicylanilide (Cestocide, Mansonil, Yomesan) occurs as a yellowish white, water-insoluble powder. It is a potent taeniacide that causes rapid disintegration of worm segments and the scolex. Penetration of the drug into various cestodes appears to be facilitated by the digestive juices of this host because very little of the drug is absorbed by the worms in vitro. Niclosamide is well tolerated following oral administration, and little or no systemic absorption of it occurs. A saline purge one to two hours after the ingestion of the taeniacide is recommended to remove the damaged scolex and worm segments. This procedure is mandatory in the treatment of pork tapeworm infestations to prevent possible cysticercosis resulting from release of live ova from worm segments damaged by the drug.

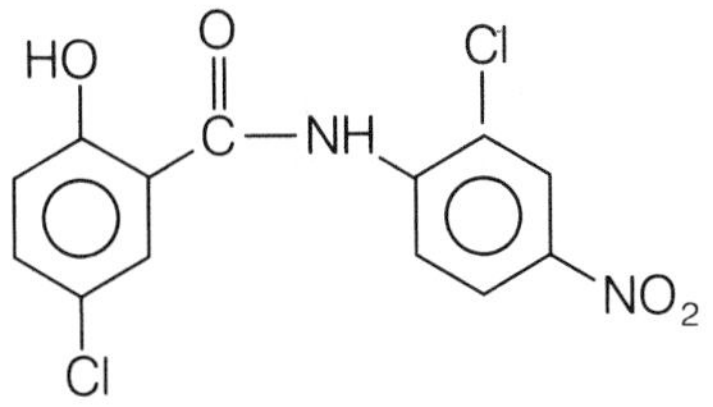

Niclosamide

Antimony Potassium Tartrate, USP. Antimonyl potassium tartrate, tartar emetic, occurs in two forms: as the colorless transparent crystalline hydrate or as the anhydrous powder. The hydrate effloresces on exposure to air. It is soluble in water (1 : 12), but insoluble in alcohol.

The precise structure of antimony potassium tartrate has been the subject of considerable controversy through the years. X-ray crystallographic[67] and chemical[68] techniques have both been applied to this problem. The monomeric structure shown has been suggested for tartar emetic in the solid state, based on x-ray crystallography.[67] This representation suffers from two major shortcomings: first, the bicyclic ring system is highly strained, and second, the position of the water molecule is not specified. Chemical studies indicate that tartar emetic readily dimerizes in solution.[68] The dimeric structure recently proposed by Steck[69] is nonstrained, defines the position of the water molecule of crystallization, and satisfies the bonding properties of antimony.

The compound has been used orally as an expectorant and an emetic, but has largely been replaced by other agents for these purposes. It still finds limited use in the treatment of leishmaniasis and schistosomiasis. It is particularly effective against *S. japonicum* (Asiatic intestinal schistosomiasis).

Sodium Stibocapate. Sodium antimony dimercaptosuccinate (Astiban) is a trivalent antimonial derivative of 2,3-dimercaptosuccinic acid in the form of the hexasodium salt. It provides a water-soluble derivative that is effective against all three forms of schistosomiasis following intramuscular administration. Because of the toxicity commonly associated with antimonial compounds, this drug is not suitable for mass chemotherapy of schistosomiasis, and its use must be carefully monitored. The drug is available as a 3.6% sterile solution in propylene glycol for injection from the Centers for Disease Control. It is only sparingly soluble in water.

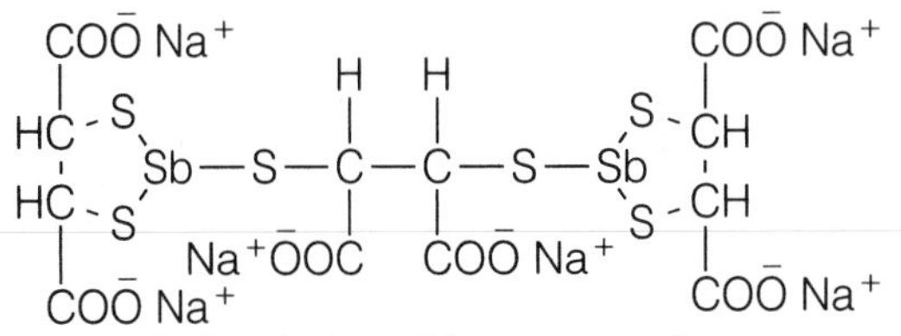

Sodium Antimony Dimercaptosuccinate

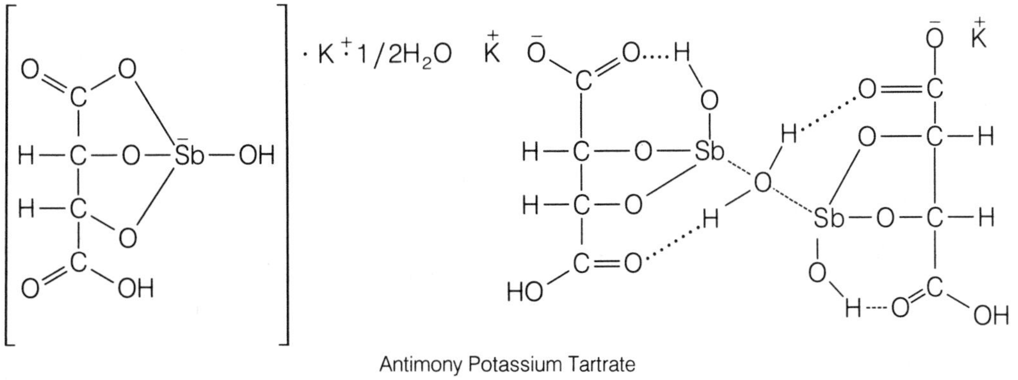

Antimony Potassium Tartrate

Oxamniquine, USP. 1,2,3,4-Tetrahydro-2-[(isopropylamino)methyl]-7-nitro-6-quinolinemethanol (Vansil) is an antischistosomal agent that is indicated for the treatment of *S. mansoni* (intestinal schistosomiasis) infection. It has been shown to inhibit DNA, RNA, and protein synthesis in schistosomes.[70] The 6-hydroxymethyl group is critical for activity; metabolic activation of precursor 6-methyl derivatives is critical. The oral bioavailability of oxamniquine is good; effective plasma levels are achieved in 1 to 1.5 hours. The plasma half-life is 1 to 2.5 hours. The drug is extensively metabolized to inactive metabolites, the principal of which is the 6-carboxy derivative.

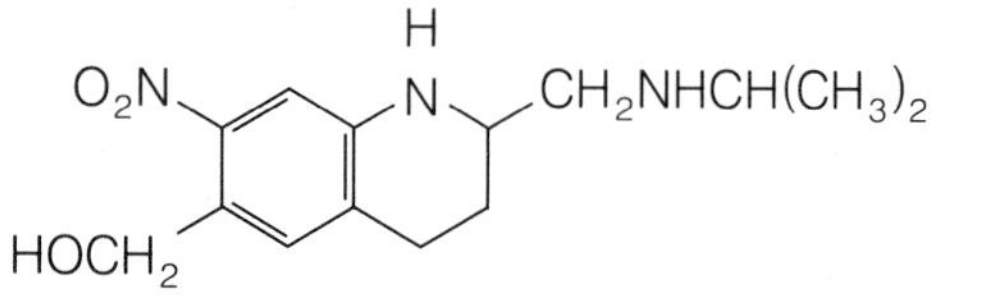

H
O_2N N $CH_2NHCH(CH_3)_2$
$HOCH_2$

Oxamniquine

The free base occurs as a yellow crystalline solid that is slightly soluble in water, but soluble in dilute aqueous mineral acids and soluble in most organic solvents. It is available in capsules containing 250 mg of the drug. Oxamniquine is generally well tolerated. Dizziness and drowsiness are common, but transitory with its use. Serious reactions, such as epileptiform convulsions, are rare.

Praziquantel. 2-(Cyclohexylcarbonyl)-1,2,3,6,7,11b-hexahydro-4*H*-pyrazino[2,1-*a*]isoquinolin-4-one (Biltricide) is a broad-spectrum agent that is effective against a variety of trematodes (flukes). It has become the agent of choice for the treatment of infections caused by schistosomes (blood flukes). The drug also provides effective treatment for fasciolopsiasis (intestinal fluke), clonorchiasis (Chinese liver fluke), fascioliasis (sheep liver fluke), opisthorchosis (liver fluke), and paragonimiasis (lung fluke). Praziquantel increases cell membrane permeability of susceptible worms, resulting in the loss of extracellular calcium. Massive contractions and ultimate paralysis of the fluke musculature occurs, followed by phagocytosis of the parasite.

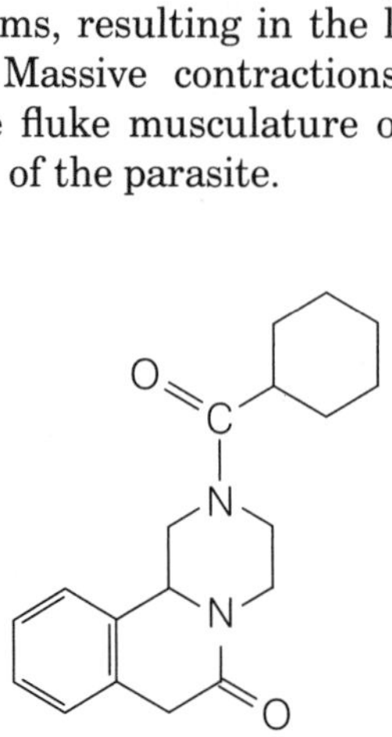

O
C
N
N
O

Praziquantel

Following oral administration, approximately 80% of the dose is absorbed. Maximal plasma concentrations are achieved in one to three hours. The drug is rapidly metabolized in the liver in the first pass. It is likely that some of the metabolites are also active. Praziquantel occurs as a white crystalline solid that is insoluble in water. It is available as 600-mg film-coated tablets. The drug is generally well tolerated.

Metrifonate (Bilarcil) is an organophosphate compound used as an alternate drug for the treatment of urinary schistosomiasis (bilharziasis). This acetylcholinasterase inhibitor causes a persistent depolarization of the musculature of the parasite, causing it to become paralyzed and susceptible to phagocytosis. Side effects characteristic of parasympathetic nerve stimulation are common with this agent. It is available from the Centers for Disease Control.

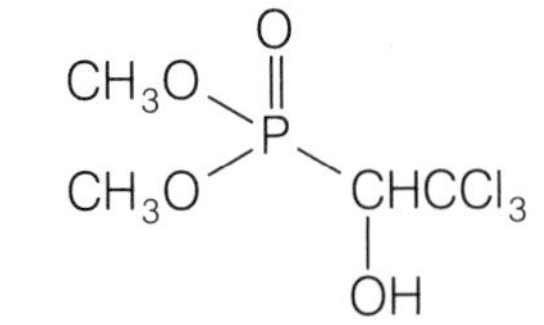

O
CH_3O P
CH_3O $CHCCl_3$
OH

Metrifonate

Ivermectin (Cardomec, Eqvalan, Ivomec) is a mixture of 22,23-dihydro derivatives of avermectins B_{1a} and B_{1b} prepared by catalytic reduction. Avermectins are members of a family of structurally complex antibiotics produced by a strain of *S. avermitilis*. Their discovery is the result of an intensive search for anthelmintic agents from natural sources.[71] Ivermectin is active in low dosage against a wide variety of nematodes and arthropods that parasitize animals.[72]

The structures of the avermectins were established, by a combination of spectroscopic[73] and x-ray crystallographic[74] techniques, to contain pentacyclic 16-membered ring aglycones glycosidically linked at the 3-position to a dissacharide that comprises two oleandrose sugar residues. The side chain at the 25-position of the aglycone is *sec*-butyl in avermectin B_{1a}, while in avermectin B_{1b} it is isopropyl. Ivermectin contains at least 80% of 22,23-dihydroavermectin B_{1a} and no more than 20% of 22,23-dihydroavermectin B_{1b}.

Ivermectin has achieved widespread use in veterinary practice in the United States and many countries throughout the world for the control of endoparasites and ectoparasites in domestic animals.[72] It has been found to be effective for the treatment of onchoceriasis ("river blindness") in humans,[75] an important disease caused by the roundworm *Oncocerca volvulus*, prevalent in West and Central Africa, the Middle East, and South and Central America. Ivermectin destroys the microfilariae, immature

TABLE 4-17

ANTHELMINTICS

Name Proprietary Name	***Preparations***	***Effective Against***	***Usual Dose****	***Usual Pediatric Dose****
Tetrachloroethylene USP	Tetrachloroethylene capsules USP	Hookworms and some trematodes	0.12 mL/kg of body weight as a single dose, up to a maximum of 5 mL	0.1 mL/kg of body weight or 3 mL/m^2 of body surface, as a single dose, up to a maximum of 5 mL
Piperazine citrate USP *Antepar Citrate, Multifuge Citrate, Ta-Verm, Vermidole*	Piperazine citrate syrup USP	Intestinal pinworms and roundworms	Against *Enterobius*: the equivalent of 2 g of piperazine hexahydrate qd for 7 days Against *Ascaris*: 3.5 g qd for 2 days	Against *Enterobius*: the following amounts, or 1 g/m^2 of body surface, are usually given qd for 7 days: up to 7 kg of body weight, 250 mg; 7–14 kg, 500 mg; 14–27 kg, 1 g; over 27 kg, 2 g Against *Ascaris*: the following amounts, or 2 g/m^2 of body surface, are usually given qd for 2 days: up to 14 kg of body weight, 1 g; 14–23 kg, 2 g; 23–45 kg, 3 g; over 45 kg, 3.5 g
	Piperazine citrate tablets USP	Intestinal pinworms and roundworms	Against *Enterobius*: the equivalent of 2 g of piperazine hexahydrate qd for 7 days Against *Ascaris*: 3.5 g qd for 2 days	Against *Enterobius*: the following amounts, or 1 g/m^2 of body surface, are usually given qd for 7 days: up to 7 kg of body weight, 250 mg; 7–14 kg, 500 mg; 14–27 kg, 1 g; over 27 kg, 2 g Against *Ascaris*: the following amounts, or 2 g/m^2 of body surface, are usually given qd for 2 days: up to 14 kg of body weight, 1 g; 14–23 kg, 2 g; 23–45 kg, 3 g; over 45 kg, 3.5 g
Piperazine phosphate USP *Antepar, Vermizine*	Piperazine phosphate tablets USP	Intestinal roundworms and trematodes	Antienterobiasis, an amount of piperazine phosphate equivalent to 2 g of piperazine hexahydrate daily for 7 days; antiascariasis, an amount of piperazine phosphate equivalent to 3.5 g of piperazine hexahydrate daily for 2 days	
Pyrvinium Pamoate USP *Povan*	Pyrvinium pamoate oral suspension USP Pyrvinium pamoate tablets USP	Intestinal pinworms	The equivalent of 5 mg/kg of body weight, of pyrvinium, as a single dose	See under Usual Dose, or the equivalent of 150 mg of pyrvinium per square meter of body surface, as a single dose

TABLE 4-17 *Continued*

ANTHELMINTICS

Name Proprietary Name	***Preparations***	***Effective Against***	***Usual Dose****	***Usual Pediatric Dose****
Pyrantel Pamoate USP *Antiminth*	Pyrantel pamoate oral suspension USP	Intestinal pinworms and roundworms	11 mg/kg of body weight	Children: 11 mg/kg of body weight
Thiabendazole USP *Mintezol*	Thiabendazole oral suspension USP	Pinworms, threadworms, whipworms, roundworms, hookworms, and in cutaneous larva migrans	Adults under 68 kg: 25 mg/kg of body weight bid for 1–4 days; adults 68 kg and over: 1.5 g bid for 1–4 days; up to a maximum of 3 g/day for 1–4 days	22 mg/kg of body weight, or 650 mg/m^2 of body surface bid for 1–4 days
Mebendazole USP *Vermox*	Mebendazole tablets USP		100 mg morning and evening for 3 consecutive days	
Bephenium hydroxynaphthoate USP *Alcopara*	Bephenium hydroxynaphthoate for oral suspension USP	Hookworms	Against *Ancylostoma duodenale*: the equivalent of 2.5 g of bephenium bid for 1 day; against *Necator americanus*: 2.5 g bid for 3 days	Under 23 kg of body weight: 500 mg–1.25 g bid for one day; over 23 kg: see Usual Dose
Niclosamide	Tablets	Tapeworms (fish, beef, pork, dwarf)	A single oral dose of 2 g (4 tablets) chewed thoroughly	A single oral dose of 1 g (2 tablets) for 11–34 kg
Bithionol		Flukes (lung, liver)	30–50 mg/kg orally on alternate days for 10–15 doses	30–50 mg/kg orally on alternate days for 10–15 doses
Niridazole	Tablets	Antischistosomal	25 mg/kg orally per day	25 mg/kg orally per day
Antimony potassium tartrate USP		Schistosomicide	IV, as a 0.5% solution given once every-other-day, the first dose 40 mg, each succeeding dose increased by 20 mg until 140 mg is reached, then 140 mg every-other-day to a total of 2 g	

*See USP DI for complete dosage information.

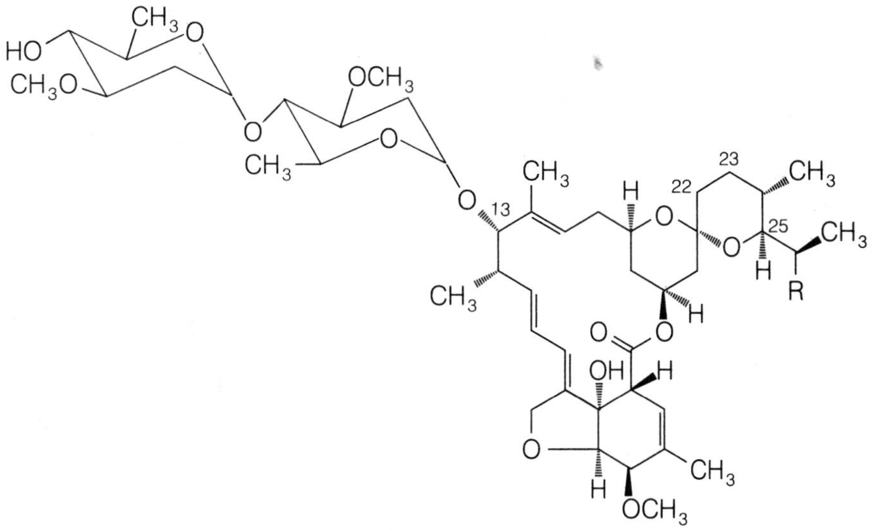

R = CH_3 22,23-Dihydroavermectin B_1
R = C_2H_5 22,23-Dihydroavermectin B_{1a}

forms of the nematode that create the skin and tissue nodules characteristic of the infestation and can lead to blindness. It also inhibits the release of microfilariae by the adult worms living in the host. Studies on the mechanism of action of ivermectin indicate that it blocks interneuron–motorneuron transmission in nematodes by stimulating the release of the inhibitory neurotransmitter γ-aminobutyric acid (GABA).[72] The drug has been made available by the manufacturer on a humanitarian basis to qualified treatment programs through the World Health Organization.

ANTISCABIOUS AND ANTIPEDICULAR AGENTS

Scabicides (antiscabious agents) are compounds used to control the mite *Sarcoptes scabei*, an organism that thrives under conditions of poor personal hygiene. The incidence of scabies is believed to be increasing in the United States and worldwide and has, in fact, reached pandemic proportions.[76] Pediculocides (antipedicular agents) are employed to eliminate head, body, and crab lice. Ideal scabicides and pediculocides must kill both the adult parasites and destroy their eggs.

Benzyl Benzoate, USP is a naturally occurring ester obtained from Peru balsam and other resins. It is also prepared synthetically from benzyl alcohol and benzoyl chloride. The ester is a clear colorless liquid with a faint aromatic odor. It is insoluble in water, but soluble in organic solvents.

Benzyl benzoate is an effective scabicide when topically applied. Immediate relief from itching probably results from a local anesthetic effect; however, a complete cure is frequently achieved with a single application of a 25% emulsion of benzyl benzoate in oleic acid, stabilized with triethanolamine. This preparation has the additional advantages of being essentially odorless, nonstaining, and nonirritating to the skin. It is applied topically as a lotion over the entire dampened body, except the face.

Lindane, USP. 1,2,3,4,5,6-Hexachlorocyclohexane; γ-benzene hexachloride; benzene hexachloride (Kwell, Scabene, Kwildane, G-Well). This halogenated hydrocarbon is prepared by the chlorination of benzene. A mixture of isomers are obtained in this process, five of which have been isolated: α, β, γ, δ, and ε. The γ-isomer, present to the extent of 10% to 13% in the mixture, is responsible for the insecticidal activity. The γ-isomer may be separated by a variety of extraction and chromatographic techniques.

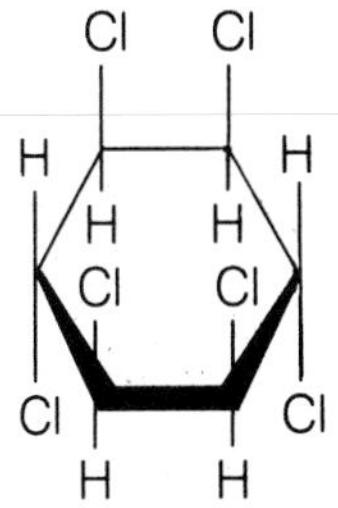

Lindane

Lindane occurs as light buff to tan powder, with a persistent musty odor and is bitter. It is insoluble in water, but soluble in most organic solvents. It is stable under acidic or neutral conditions, but undergoes elimination reactions under alkaline conditions.

The action of lindane against insects is threefold: it is a direct contact poison; it has a fumigant effect; and it acts as a stomach poison. The effect of lindane on insects is similar to that of DDT. Its toxicity in humans is somewhat lower than that of DDT. However, because of its lipid solubility properties, lindane tends to accumulate in the body when ingested.

Lindane is employed locally as a cream, lotion, or shampoo for the treatment of scabies and pediculosis.

Crotamiton, USP. *N*-Ethyl-*N*-(2-methylphenyl)-2-butenamide; *N*-ethyl-*o*-crotonotoluidide (Eurax) is a colorless, odorless oily liquid. It is virtually insoluble in water, but soluble in most organic solvents.

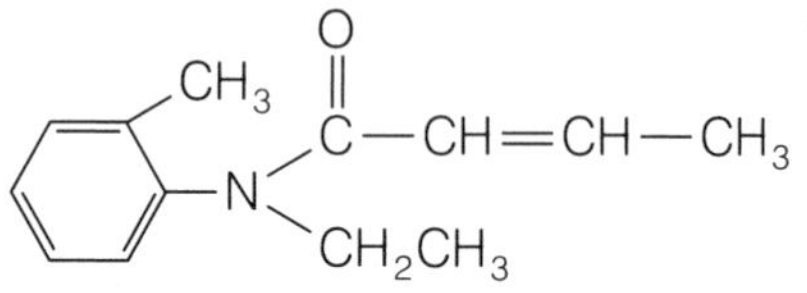

Crotamiton

Crotamiton is available in 10% concentration in a lotion and a cream intended for the topical treatment of scabies. Its antipruritic effect is probably due to a local anesthetic action.

Permethrin. 3-(2,2-Dichloroethenyl)-2,2-dimethylcyclopropanecarboxylic acid (3-phenoxyphenyl)-methyl ester; 3-(phenoxyphenyl)methyl (±)-*cis*, *trans*-3-(2, 2-dichloroethenyl)-2, 2-dimethylcyclopropane carboxylate (Nix). This synthetic pyrethrenoid compound is more stable chemically than most natural pyrethrins and is at least as active as an insecticide. Of the four isomers present, the *1R*, *trans*- and *1R*, *cis*-isomers are primarily responsible for the insecticidal activity. The commercial product is a mixture consisting of 60% *trans* and 40% *cis* racemic isomers. It occurs as colorless to pale yellow low-melting crystals or as a pale yellow liquid and is insoluble in water, but soluble in most organic solvents.

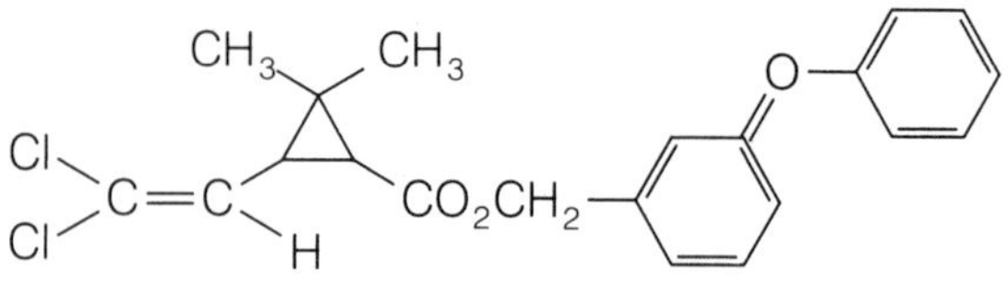

Permethrin

Permethrin exerts a lethal action against lice, ticks, mites, and fleas. It acts on the nerve cell membranes of the parasites to disrupt sodium channel conductance. It is employed as a pediculicide for the treatment of head lice. A single application of a 1% solution is known to effect cures in more than 99% of the cases. The most frequent side effect is pruritus, which occurred in about 6% of the patients tested.

Diethyltoluamide, USP. *N,N*-Diethyl-3-methylbenzamide; *N,N*-diethyl-*m*-toluamide is an insect repellent. It is a colorless liquid with a faint, pleasant odor that is virtually insoluble in water, but miscible with most organic solvents. Of the three possible regioisomers, only the *meta*-isomer has insect-repellent properties.

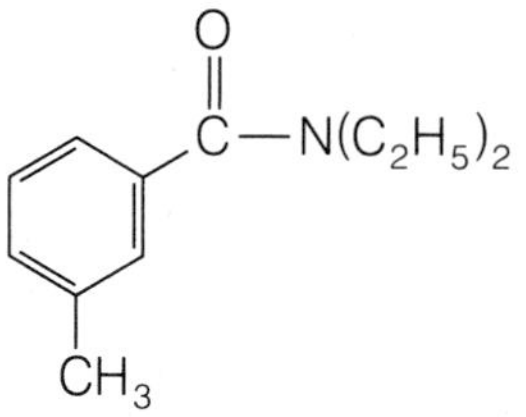

Diethyltoluamide

TABLE 4-18

SCABICIDES AND PEDICULICIDES

Name Proprietary Name	***Preparations***	***Category***	***Application***
Benzyl benzoate USP	Benzyl benzoate lotion USP	Scabicide	Topical, as lotion over previously dampened skin of entire body, except face
Lindane USP *Kwell*	Lindane cream USP	Pediculicide; scabicide	Topically to the skin, as a 1% cream once or twice weekly
	Lindane lotion USP	Pediculicide; scabicide	Topically to the skin, as a 1% lotion once or twice weekly
	Lindane shampoo USP		Topical to the scalp, as a 1% shampoo for one application, repeated after 7 days if necessary
Crotamiton	Crotamiton cream USP	Scabicide	Topical, massaged thoroughly into skin from the chin down, repeat 24 hr later
	Crotamiton lotion USP	Scabicide	See above
Permethrin *Nix*	Permethrin liquid	Pediculicide	Topical to the scalp after shampoo
Diethyltoluamide		Arthropod repellent	Topical to skin and clothing, 15% ointment

REFERENCES

1. DuMez, A. G.: J. Am. Pharm. A. 28:416, 1939.
2. Leech, P. N.: JAMA 109:1531, 1937.
3. Gilbert, G. L., et al.: Appl. Microbiol. 12:496, 1964.
4. Miner, N. A., et al.: Am. J. Pharm. 34:376, 1977.
5. U. S. Food and Drug Administration: "Hexachlorophene and Newborns," bulletin, December, 1971.
6. Vasarenish, A.: Arch. Dermatol. 98:183, 1968.
7. Gershenfeld, L.: Milk Food Technol. 18:223, 1955.
8. Domagk, G.: Dtsch. Med. Wochenschr. 61:250, 1935.
9. Rose, F. L., Swain, G. J.: J. Chem. Soc. 442, 1956.
10. Garrett, E. R., Woods, O. R.: J. Am. Pharm. A. (Sci. Ed.) 42:736, 1953.
11. Rieth, H.: Arzneim. Forsch. 31:1309, 1981.
12. Plempel, M., et al.: Dtsch. Med. Wochenschr. 94:1356, 1969.
13. Godefroi, E. F., et al.: J. Med. Chem. 12:784, 1969.
14. Thomas, A. H.: J. Antimicrob. Chemother. 17:269, 1986.
15. Polak, A., Scholer, H. J.: Chemotherapy 21:113, 1975.
16. Gold, W., et al.: Antibiotics Annual 1955–1956, p. 579. New York, Medical Encyclopedia, 1956.
17. Mechlinski, W., et al.: Tetrahedron Lett. 3873, 1970.
18. Borowski, E., et al.: Tetrahedron Lett. 3909, 1970.
19. Pandey, R. C., Rinehart, K. L.: J. Antibiot. 29:1035, 1976.
20. Hazen, E. L., Brown, R.: Proc. Soc. Exp. Biol. Med. 76:93, 1951.
21. Chong, C. N., Richards, R. W.: Tetrahedron Lett. 5145, 1970.
22. Borowski, E., et al.: Tetrahedron Lett. 685, 1971.
23. Lechevalier, H. A., et al.: Mycologia 45:155, 1953.
24. Zielinski, et al.: Tetrahedron Lett. 1791, 1979.
25. Struyk, et al.: Antibiotics Annual 1957–1958, p. 857. New York, Medical Encyclopedia, 1958.
26. Golding, B. T., et al.: Tetrahedron Lett. 3551, 1966.
27. Kotler-Brajtburg, J., et al.: Antimicrob. Agents Chemother. 15:716, 1979.
28. Oxford, A. E., et al.: Biochem. J. 33:240, 1939.
29. Grove, J. F., et al.: J. Chem. Soc. 3977, 1952.
30. Sloboda, R. D., et al.: Biochem. Biophys. Res. Commun. 105:882, 1982.
31. Gellert, M.: Annu. Rev. Biochem. 50:879, 1981.
32. Shen, L. L., Pernet, A. G.: Proc. Natl. Acad. Sci. USA 82:307, 1985.
33. Boxenbaum, H. G., Riegelman, S.: J. Pharm. Sci. 63:1191, 1974.
34. Kuehl, F. A., Jr., et al.: J. Am. Chem. Soc. 77:2344, 1955.
35. Hildy, P. H., et al.: J. Am. Chem. Soc. 77:2345, 1955.
36. Stammer, C. H., et al.: J. Am. Chem. Soc. 77:2346, 1955.
37. Smrt, J.: Experientia 13:291, 1957.
38. Neuhaus, F. C., Lynch, J. L.: Biochemistry 3:471, 1964.
39. Rando, R. R.: Biochem. Pharmacol. 24:1153, 1975.
40. Haskell, T. H., et al.: J. Am. Chem. Soc. 74:599, 1952.
41. Mayer, R. L., et al.: Experientia 10:335, 1954.
42. Bowie, J. H., et al.: Tetrahedron Lett. 3305, 1964.
43. Bancroft, B. W., et al.: Experientia 27:501, 1971.
44. Noda, T., et al.: J. Antibiot. 25:427, 1971.
45. Yoshioka, H., et al.: Tetrahedron Lett. 2043, 1971.
46. Herr, E. B., et al.: Indiana Acad. Sci. 69:134, 1960.
47. Bancroft, B. W., et al: Nature 231:301, 1971.
48. Nomoto, S., et al.: J. Antibiot. 30:955, 1977.
49. Rinehart, K. L.: Acc. Chem. Res. 5:57, 1972.
50. Hartmann, G., et al.: Angew. Chem. Int. Ed. Engl. 24:1009, 1985.
51. Katz, E., Moss, B.: Proc. Natl. Acad. Sci. USA 66:677, 1970.
52. Gurgo, C., et al.: Nature 229:111, 1971.
53. Gallo, G. G., Radaelli, P.: In Florey, K. (ed.). Analytical Profiles of Drug Substances, Vol. 5, p. 491. 1976.
54. Lamb, R. A., Choppin, R. W.: Annu. Rev. Biochem. 52:467, 1983.
55. Yarchoan, et al.: Lancet 575, 1986.
56. Pavan-Langston, D., et al.: Adenosine Arabinoside: An Antiviral Agent. New York, Raven Press, 1975.
57. Schaeffer, H. J., et al.: Nature 272:583, 1978.
58. Sidwell, R. W., et al.: Science 177:705, 1972.
59. Robins, R. K.: Chem. Eng. News, Jan. 27, 1986, pp. 28–40.
60. Lin, T. S., Prusoff, W. H.: J. Med. Chem. 21:109, 1978.
61. Mitsuya, H., et al.: Proc. Natl. Acad. Sci. USA 82:7096, 1985.
62. Grollman, A. P.: J. Biol. Chem. 243:4089, 1968.
63. Knight, R. C., et al.: Biochem. Pharmacol. 27:2089, 1978.
64. Metcalf, B. W., et al.: J. Am. Chem. Soc. 100:2551, 1978.
65. Prichard, R. K.: Nature 228:684, 1970.
66. Friedman, P. A., Platzer, E. G.: Biochim. Biophys. Acta 544:605, 1978.
67. Grdenic, D., Kamenar, B.: Acta Crystallogr. 19:192, 1965.
68. Banerjee, A. K., Chari, K. V. R.: J. Inorg. Nucl. Chem. 31:2958, 1969.
69. Steck, E. A.: Progr. Drug Res. 18:304, 1974.
70. Pica-Mattoccia, L., Cioli, D.: Am. J. Trop. Med. Hyg. 34:112, 1985.
71. Burg, R. W., et al.: Antimicrob. Agents Chemother. 15:361, 1979.
72. Campbell, W. C.: Science 221:823, 1983.
73. Albers-Schonberg, G., et al.: J. Am. Chem. Soc. 103:4216, 1981.
74. Springer, J. P., et al.: J. Am. Chem. Soc. 103:4221, 1981.
75. Aziz, M. A., et al.: Lancet 2:171, 1982.
76. Orkin, M., Maibach, H. I.: N. Engl. J. Med. 298:496, 1978.

SELECTED READINGS

Bambury, R. E.: Synthetic antibacterial agents. In Wolff, M. E. (ed.). Burger's Medicinal Chemistry, Part II, 4th ed., p. 41. New York, Wiley-Interscience, 1979.

Campbell, W. C., Rew, R. S. (eds.): Chemotherapy of Parasitic Diseases. New York, Plenum Press, 1986.

De Clercq, E.: Chemotherapeutic approaches to the treatment of acquired immune deficiency syndrome (AIDS). J. Med. Chem. 29:1561, 1986.

De Clercq, E., Walker, R. T. (eds.): Targets for the Design of Antiviral Agents. New York, Plenum Press, 1983.

Dixon, R. E., et al.: Aqueous quaternary ammonium antiseptics and disinfectants. Use and misuse. JAMA 236:2415, 1976.

Islip, P. J.: Anthelmintic agents. In Wolff, M. E. (ed.). Burger's Medicinal Chemistry, Part II, 4th ed. p. 481. New York, Wiley-Interscience, 1979.

Knight, R.: The chemotherapy of amoebiasis. J. Antimicrob. Chemother. 6:577, 1980.

Kunin, C. M.: Detection, Prevention and Management of Urinary Tract Infections, 3rd ed. Philadelphia, Lea & Febiger, 1979.

Lawrence, C. A., Block, S. S. (eds.): Disinfection, Sterilization and Preservation, 2nd ed. Philadelphia, Lea & Febiger, 1977.

Mandell, G. L., Douglas, R. G., Bennett, J. E. (eds.): Principles and Practice of Infectious Diseases, 2nd ed. New York, John Wiley & Sons, 1985.

Oxford, J. S., Oberg, B.: Conquest of Viral Diseases. Amsterdam, Elsevier, 1985.

Pratt, W. B., Fekety, R.: The Antimicrobial Drugs. New York, Oxford University Press, 1986.

Ross, W. J.: Antiamebic agents. In Wolff, M. E. (ed.). Burger's Medicinal Chemistry, Part II, 4th ed. p. 415. New York, Wiley-Interscience, 1979.

Ross, W. J.: Chemotherapy of trypanosomiasis and other protozoan diseases. In Wolff, M. E. (ed.). Burger's Medicinal Chemistry, Part II, 4th ed., p. 439. New York, Wiley-Interscience, 1979.

Sensi, P., Gialdroni-Grassi, G.: Antimycobacterial agents. In Wolff, M. E. (ed.). Burger's Medicinal Chemistry, Part II, 4th ed., p. 289. New York, Wiley-Interscience, 1979.

Sidwell, R. W., Wilkowski, J. T.: Antiviral agents. In Wolff, M. E. (ed.). Burger's Medicinal Chemistry, Part II, 4th ed., p. 543. New York, Wiley-Interscience, 1979.

Verderame, M. (ed.): Handbook of Chemotherapeutic Agents, Vols. 1 and 2. CRC Press, Boca Raton, FL, 1986.

CHAPTER 5

Sulfonamides, Sulfones, and Folate Reductase Inhibitors with Antibacterial Action

Dwight S. Fullerton

SULFONAMIDES AND FOLATE REDUCTASE INHIBITORS

PRONTOSIL AND GERHARD DOMAGK (1895–1964)

The founding of chemotherapy, drug design, and medicinal chemistry by Paul Ehrlich (1854–1915) and the antisyphilitic drug Salvarsan in 1908 will be discussed in Chapter 6. Ehrlich's discovery led to intensive investigations of dyes as antimicrobial agents, especially in Germany. Although Salvarsan and related drugs were revolutionary in treating some protozoal infections and syphilis, they were not useful in treating a major killer of the times—streptococcal and staphylococcal infections.

Fritz Mietzsch and Joseph Klarer of the I.G. Farbenindustrie (Bayer) laboratories began a systematic synthesis of azo dyes as possible antimicrobials. Sulfonamide azo dyes were included because they were relatively easy to synthesize and had improved staining properties. The Bayer pathologist–bacteriologist who evaluated the new Mietzsch-Klarer dyes was Gerhard Domagk, who, like Ehrlich, was a physician by training.[1–3] In 1932, Domagk began a study of a bright red dye, later to be named Prontosil, and found that it caused remarkable cures of streptococcal infections of mice.[1] However, Prontosil was inactive on bacterial cultures. Domagk's studies on Prontosil continued, and in 1933, the first of many human cures of severe staphyloccal septicemias was reported.[4] Domagk even saved the life of his own daughter from a severe streptococcal infection. For his pioneering efforts in chemotherapy, Gerhard Domagk was awarded the Nobel prize for medicine and physiology in 1939. The Gestapo prevented him from actually accepting the award, but he received it in Stockholm in 1947.

Prontosil's inactivity in vitro, but excellent activity in vivo, attracted much attention. In 1935, Trefouel, Trefouel, Nitti, and Bovet[5] reported their conclusion from a structure–activity study of sulfonamide azo dyes that the azo linkage was metabolically broken to release the active ingredient, sulfanilamide. Their reported finding was confirmed in 1937 when Fuller[6] isolated sulfanilamide from the blood and urine of patients being treated with Prontosil. Modern chemotherapy and the concept of the prodrug (see Chap. 2) were firmly established.

THE MODERN ERA

Following Prontosil's dramatic successes, a cascade of sulfanilamide derivatives began to be synthesized and tested—over 4500 by 1948 alone.[7] From these only about two dozen have actually been used in clinical practice. In the late 1940s, penicillins began to replace the sulfanilamides in chemotherapy. This was largely because of the sulfanilamides' toxicity for some patients and because sulfanilamide-resistant bacterial strains were becoming an increasing

$$H_2N-C_6H_3(NH_2)-N{=}N-C_6H_4-SO_2-NH_2 \xrightarrow{\text{in vivo}} H_2N-C_6H_4-SO_2-NH_2$$

Prontosil   Sulfanilamide

problem—the result of indiscriminant use worldwide.

Today, a few sulfonamides and, especially, sulfonamide–trimethoprim combinations are extensively used for urinary tract infections or for burn therapy.[8-12] They are also the drugs of choice or alternates for a few other types of infections (Table 5-1), but their overall use is otherwise quite limited in modern antimicrobial chemotherapy,[8-12] having been largely replaced by antibiotics.

TABLE 5-1

CURRENT THERAPY WITH SULFONAMIDE ANTIBACTERIALS[10,11]

Disease/Infection	*Sulfonamides Commonly Used*
WIDE USE	
First attack of urinary tract infection	Sulfamethoxazole and trimethoprim Sulfisoxazole
Burn therapy: prevention and treatment of bacterial infection	Silver sulfadiazine Mafenide
Conjunctivitis and related superficial ocular infections	Sodium sulfacetamide
Chloroquine-resistant malaria (Chapter 6)	Combinations with quinine, others Sulfadoxine Sulfalene
LESS COMMON INFECTIONS/DISEASES: DRUGS OF CHOICE OR ALTERNATES	
Nocardiosis	Trimethoprim–sulfamethoxazole
Toxoplasmosis	Pyrimethamine and trisulfapyrimidines
Severe travelers' diarrhea[13]	Trimethoprim–sulfamethoxazole
Meningococcal infections	Sulfonamides, only if proved to be sulfonamide-sensitive; otherwise penicillin G, ampicillin, or (for penicillin allergic patients) chloramphenicol should be used
GENERALLY NOT USEFUL	
Streptococcal infections	Most are resistant to sulfonamides
Prophylaxis of rheumatic fever recurrences	Most are resistant to sulfonamides
Other bacterial infections	Penicillin's low cost and bacterial resistance to sulfonamides have decreased sulfonamide use worldwide, but still used in a few countries.
Vaginal infections	FDA Bulletin[12] and USP DI find no evidence of effectiveness
Reduction of bowel flora	Effectiveness not established[9]
Ulcerative colitis	Corticosteroid therapy often preferred Relapses common with sulfanilamides Phthalylsulfathiazole Salicylazosulfapyridine Side effects of the sulfanilamides sometimes seem like the ulcerative colitis[14]

CHEMISTRY AND NOMENCLATURE

The term *sulfonamide* is commonly used to refer to antibacterials that are (1) aniline-substituted sulfonamides, the "sulfanilamides" (Fig. 5-1); (2) prodrugs that produce sulfanilamides (e.g., sulfasalazine); and (3) nonaniline sulfonamides (e.g., mafenide). However, several other widely used drugs are also sulfonamides or sulfanilamides. Included among these nonantibacterial sulfonamides are tolbutamide (an oral diabetic drug, see Chap. 14), furosemide (a potent diuretic, see Chap. 13), and chlorthalidone (also a diuretic, see Chap. 13).

As reviewed in Chapter 2, pK_bs are not used in pharmaceutical chemistry to compare compounds. If a pK_a of an amine is given, it refers to its salt acting as the conjugate acid, for example:

Aniline, pK_a 4.6 refers to:

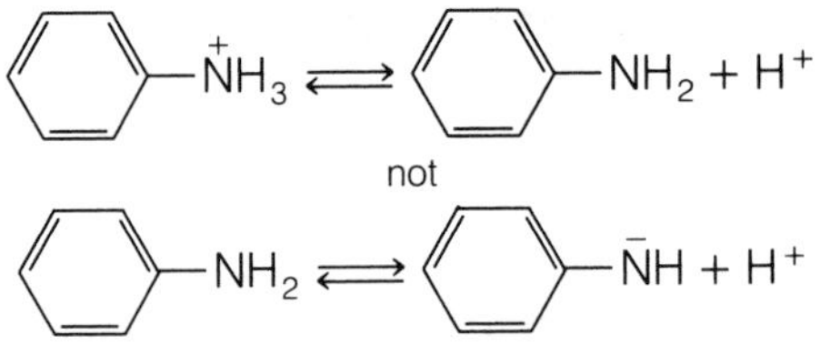

$$C_6H_5-\overset{+}{N}H_3 \rightleftharpoons C_6H_5-NH_2 + H^+$$

not

$$C_6H_5-NH_2 \rightleftharpoons C_6H_5-\bar{N}H + H^+$$

A minus charge on a nitrogen atom is normally not very stable, unless the minus charge can be greatly delocalized by resonance. This is exactly the case with the sulfanilamides. Thus, the single pK_a usually given with sulfanilamides refers to loss of an amide H^+, for example:

Sulfanilamide, pK_a 10.4 refers to:

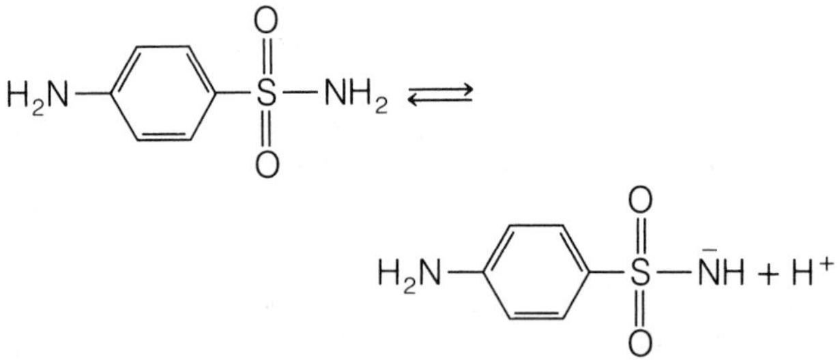

$$H_2N-C_6H_4-SO_2-NH_2 \rightleftharpoons H_2N-C_6H_4-SO_2-\bar{N}H + H^+$$

Salt

and Sulfisoxazole, pK_a 5.0 refers to:

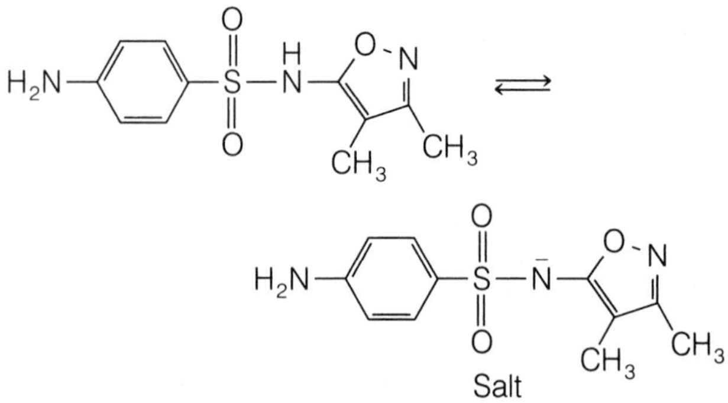

$$H_2N-C_6H_4-SO_2-NH-(\text{3,4-dimethylisoxazol-5-yl}) \rightleftharpoons H_2N-C_6H_4-SO_2-\bar{N}-(\text{3,4-dimethylisoxazol-5-yl})$$

Salt

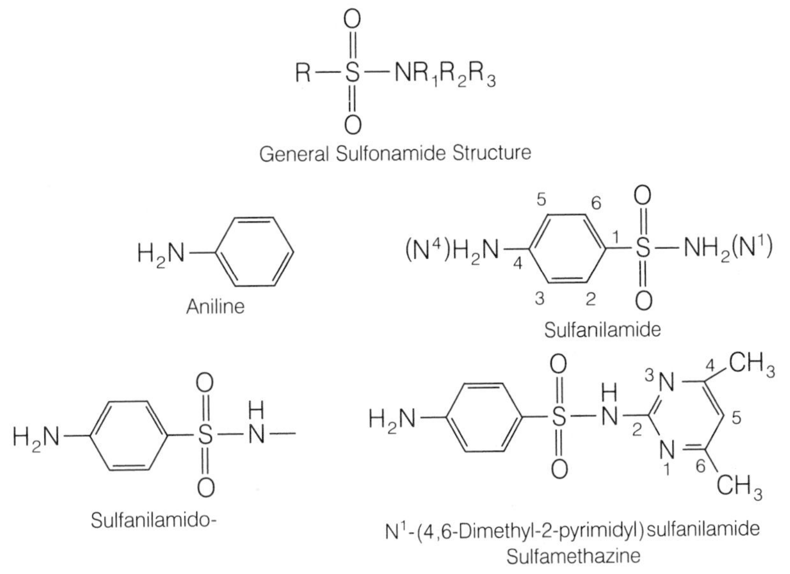

General Sulfonamide Structure

Aniline

Sulfanilamide

Sulfanilamido-

N^1-(4,6-Dimethyl-2-pyrimidyl)sulfanilamide
Sulfamethazine

FIG. 5-1. *Nomenclature and numbering.*

Thus, sulfisoxazole (pK_a 5.0) is a slightly weaker acid than acetic acid (pK_a 4.8).

REDUCING CRYSTALLURIA BY LOWERING pK_a

Sulfanilamide, although revolutionary in the early 1930s, often caused severe kidney damage from crystals of sulfanilamide forming in the kidneys. Sulfanilamides and their metabolites (usually acetylated at N^4) are excreted almost entirely in the urine. Unfortunately, sulfanilamide is not very water-soluble. Unless the *p*H is above the pK_a (that is, above *p*H 10.4), little of its water-soluble salt is present. Because urine *p*H is typically about 6, and often slightly lower during bacterial infections, essentially all sulfanilamide is in the relatively insoluble nonionized form in the kidneys.

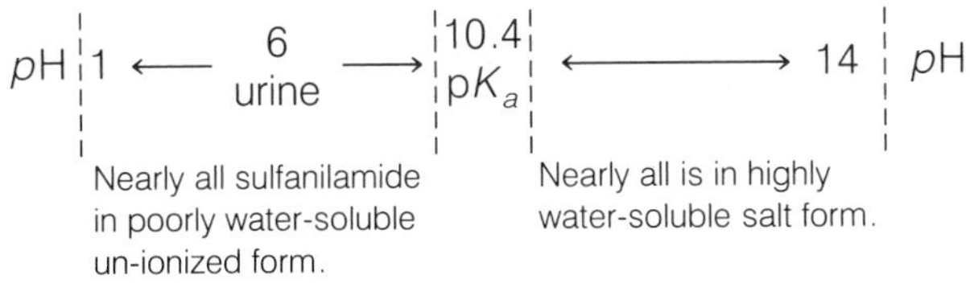

*p*H 1 ⟵ 6 urine ⟶ 10.4 pK_a ⟵⟶ 14 *p*H

Nearly all sulfanilamide in poorly water-soluble un-ionized form.

Nearly all is in highly water-soluble salt form.

At *p*H = pK_a, that is, at *p*H 10.4 for sulfanilamide, there will be a 1 : 1 mixture of nonionized and salt forms.

How can a sulfanilamide be made "more soluble" in the urine? There are several options:

1. Greatly increase urine flow. Thus, during the early days of sulfanilamide and sulfanilamide derivative use, patients were warned to "force fluids."
2. Raise the *p*H of the urine. The closer that the *p*H of the urine gets to 10.4 (for sulfanilamide itself), the more of the highly water-soluble salt form will be present. Thus, sometimes oral sodium bicarbonate was, and occasionally still is, given to raise urine *p*H.
3. Make derivatives of sulfanilamide that have lower pK_as, closer to the *p*H of urine. This has been the approach taken with virtually all sulfonamides clinically used today, for example:

Sulfanilamide	pK_a
Sulfadiazine	6.5
Sulfamerazine	7.1
Sulfamethazine	7.4
Sulfisoxazole	5.0
Sulfamethoxazole	6.1

4. Mix sulfonamides to reach the total dose. Because solubilities of sulfanilamides are independent, more of a *mixture* of sulfanilamides can stay in water solution at a particular *p*H than a single sulfonamide. Thus, trisulfapyrimidines, USP ("triple sulfas") contains a mixture of sulfadiazine, sulfamerazine, and sulfamethazine. However, such mixtures are little used today because the individual agents have sufficiently low pK_as to be adequately urine-soluble *providing at least normal urine flow is maintained*. Patients must still be cautioned to maintain a normal fluid intake, even if they do not feel like drinking during the illness. Forcing fluids, however, is no longer necessary.

It would be reasonable to ask, "Why do the modern sulfonamides have such low pK_as?" The answer is that the heterocyclic rings attached to N^1 are electron withdrawing—providing additional stability for the salt form. Therefore, the nonionized forms can more easily give up a H^+, so the pK_as are

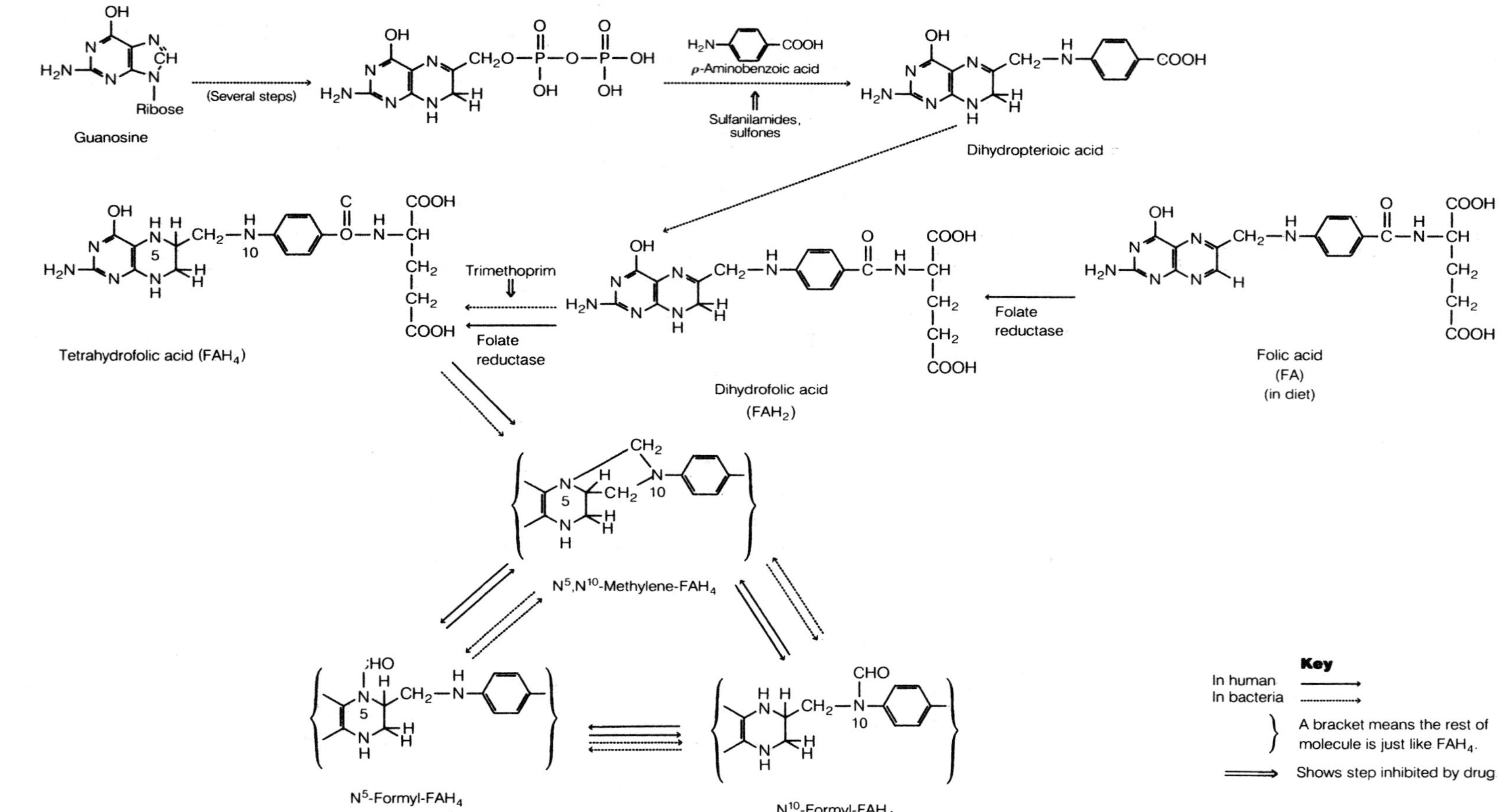

FIG. 5-2. *Sites of action of sulfanilamides and trimethoprim.*

lower. Why are not simpler electron-withdrawing groups used, such as *p*-nitrophenyl—the usual example studied in introductory organic chemistry classes? Such compounds were, in fact, extensively investigated, as were thousands of others. However, the other sulfonamides generally were too toxic, not sufficiently active, or both.

MECHANISM OF ACTION

Folinic acid (N^5-formyltetrahydrofolic acid, Fig. 5-2), N^5,N^{10}-methylenetetrahydrofolic acid, and N^{10}-formyltetrahydrofolic acid are indispensable for several biosynthetic pathways in humans, bacteria, animals, and plants (Fig. 5-3). Without these folate coenzymes, for example, thymidine monophosphate will not be available to produce nucleic acids needed for cell division. Other reactions requiring the folate coenzymes are shown in Figure 5-3. The result of any drug blocking the biosynthesis of folate coenzymes in bacteria, for example, is that growth and cell division are stopped. Such drugs—including the sulfonamides and trimethoprim—are thus *bacteriostatic*.

As shown in the biosynthetic pathway in Figure 5-2, folate coenzymes are biosynthesized from dietary folic acid in humans (and other animals). However, bacteria (and protozoa, see Chap. 6) must make them from *p*-aminobenzoic acid (PABA). The microbes cannot use dietary folic acid from the host, for reasons not yet completely understood.[3] It may be that folic acid cannot penetrate the cell wall.

The sulfonamide and sulfone antibacterials act as competitive inhibitors for the incorporation of PABA to form dihydropteroic acid (see Fig. 5-2). Trimethoprim is an inhibitor of folate reductase, needed to convert dihydrofolic acid (FAH_2) into tetrahydrofolic acid (FAH_4) in bacteria. Evidence for these inhibitions has been summarized in detail by Anand.[3]

Although trimethoprim does not have a very high affinity for malaria protozoa's folate reductase (see Chap. 6), it does have high affinity for the bacterial folate reductase. The reverse situation exists for the antimalarial drug pyrimethamine.[15] Trimethoprim still does have some affinity for human folate reductase—the cause of some toxic effects discussed later in this chapter and in Chapter 6.

Drugs with additional or modified mechanisms of action (e.g., silver sulfadiazine) will be discussed with individual descriptions, which follow later in the chapter.

SYNERGISM OF SULFONAMIDES AND FOLATE REDUCTASE INHIBITORS

Blocking the biosynthesis of folate coenzymes at more than one point in the biosynthetic pathway of bacteria (or protozoa; see Chap. 6) will result in a synergistic antimicrobial effect. An additional benefit is that the microbe will not be able to develop resistance as quickly as with a single pathway blocker. This synergistic approach is widely used in antibacterial therapy with the combination of sul-

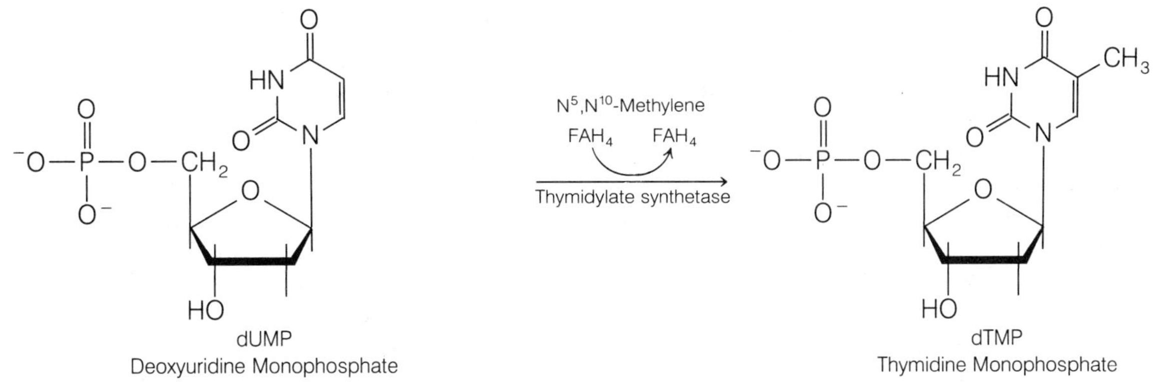

dUMP
Deoxyuridine Monophosphate

N^5,N^{10}-Methylene FAH_4 → FAH_4
Thymidylate synthetase

dTMP
Thymidine Monophosphate

Other examples

Coenzyme	**Reaction**
,N^{10}-Formyl FAH_4	Met-tRNA → Formyl Met-tRNA
N^5,N^{10}-Methylene FAH_4	Glycine → Serine
N^5-Formyl FAH_2	Homocysteine → Methionine

FIG. 5-3. *Reactions requiring the folate coenzymes.*

famethoxazole and trimethoprim[16,17] and in antimalarial therapy with pyrimethamine plus a sulfanilamide or quinine (see Table 5-1). Additional explanations for the synergistic antimicrobial actions of trimethoprim and sulfamethoxazole have been at times vigorously debated.[3,16,17,18]

Other combinations of trimethoprim have also been investigated (e.g., with rifampin).[19,20]

MECHANISMS OF RESISTANCE

As noted earlier in this chapter, wide and unselective use of sulfonamides led to the emergence of many drug-resistant strains of bacteria. The cause of the resistance is probably increased production of PABA by the resistant bacteria,[21] although other mechanisms may account for resistance in some cases.[3,17] If a microbe is resistant to one sulfonamide, it is generally resistant to all. A special concern is the finding that sulfonamide resistance can be transferred from a resistant bacterial species to a previously sulfanilamide-sensitive species. The transferring substances have been called "R-factors."[22]

Several explanations have been presented to account for bacterial resistance to trimethoprim,[17] including natural (intrinsic) resistance, development of an ability by the bacteria to use the host's dTMP (see Fig. 5-3), and R-factor transmission.

TOXICITY AND SIDE EFFECTS

A variety of serious toxic and hypersensitivity problems have been reported with the sulfonamide and sulfonamide–trimethoprim combinations. Weinstein[9] notes that these occur in about 5% of all patients. Hypersensitivity reactions include drug fever, Stevens–Johnson syndrome, skin eruptions, allergic myocarditis, photosensitization, and related conditions. Hematologic side effects also sometimes occur, especially hemolytic anemia in individuals with a deficiency of glucose-6-phosphate (discussed further in Chap. 6). Other hematologic side effects that have been reported include agranulocytosis, aplastic anemias, and others. Crystalluria may still occur, even with modern sulfonamides, when the patient does not maintain a normal fluid intake. Nausea and related gastrointestinal side effects are sometimes noted.

Detailed summaries of incidences of side effects with trimethoprim–sulfamethoxazole have been published by Wormser and co-workers[16] and by Gleckman and associates.[17]

METABOLISM, PROTEIN BINDING, AND DISTRIBUTION

With the exception of the poorly absorbed sulfonamides used for ulcerative colitis and reduction of bowel flora, and the topical burn preparations (e.g., mafenide), sulfonamides and trimethoprim tend to be quickly absorbed and well distributed. As Weinstein noted, sulfonamides can be found in the urine "within 30 minutes after oral ingestion."[9]

The sulfonamides vary widely in plasma protein binding—e.g., sulfisoxazole 76%, sulfamethoxazole 60%, sulfamethoxypyridazine 77%, and sulfadiazine 38%. (An excellent table comparing the percentage of protein bound, lipid solubility, plasma half-life, and percentage N^4-metabolites has been published by Anand.[3]) The fraction that is protein bound is not active as an antibacterial, but because the binding is reversible, free, and therefore active, sulfonamide eventually becomes available. Generally, the more lipid-soluble a sulfonamide is, at physiologic pH, the more it will be protein bound. Fujita and Hansch[23] have found that among sulfonamides with similar pK_a, the lipophilicity of the N^1 group has the largest effect on protein binding.[23] N^4-Acetate metabolites of the sulfonamides are more lipid-soluble and, therefore, better protein bound than the starting drugs themselves (which have a free 4-NH_2 group that decreases lipid solubility). Surprisingly, the N^4-acetylated metabolites, although more strongly protein bound, are more rapidly excreted than are the starting drugs.

Currently, the relationship between plasma protein binding and biologic half-life is not clear. Many competing factors are involved, as reflected in sulfadiazine with a serum half-life of 17 hours being much less protein bound than sulfamethoxazole with a serum half-life of 11 hours.[3]

Sulfonamides are excreted primarily as mixtures of unmetabolized drugs, N^4-acetates, and glucuronides.[24] The N^4-acetates and glucuronides are inactive. Sulfisoxazole, for example, is excreted about 80% unchanged, and sulfamethoxazole, 20% unchanged. Sulfadimethoxine is about 80% excreted as the glucuronide. The correlation between structure and route of metabolism has not yet been delineated, although progress has been made by Fujita.[25] Vree and co-workers, however, have described the excretion kinetics and pK_as of N^1 and N^4 acetylsulfamethoxazole and other sulfonamides.[26]

Trimethoprim and sulfamethoxazole are both partially plasma protein bound—about 45% of trimethoprim, and about 66% of sulfamethoxazole. Whereas about 80% of excreted trimethoprim and its metabolites are active as antibacterials, only

20% of sulfamethoxazole's are active, mostly unmetabolized sulfamethoxazole. Six metabolites of trimethoprim are known.[27] It is likely, therefore, that sulfonamide–trimethoprim combinations using a sulfonamide with a higher active urine concentration will be developed in the future for urinary tract infections. Sulfamethoxazole and trimethoprim have a similar half-life, about 10 to 12 hours, but the half-life of the active fraction of sulfamethoxazole is less—about nine hours.[27] (Ranges of half-lives have been summarized by Gleckman,[17] and a detailed summary of pharmacokinetics has been made by Hansen.[27]) In patients with impaired renal function, sulfamethoxazole and its metabolites may greatly increase in the plasma. Fixed combination of sulfamethoxazole and trimethoprim should not be used with patients with low creatinine clearances.

STRUCTURE–ACTIVITY RELATIONSHIPS

As noted earlier in this chapter, several thousand sulfonamides have been investigated as antibacterials (and many as antimalarials, see Chap. 6). From these efforts, several structure–activity relationships have been proposed, elegantly summarized by Anand.[3] The aniline (N^4) amino group is very important for activity because any modification of it other than to make prodrugs results in a loss of activity. As noted earlier, for example, the N^4-acetylated metabolites of sulfonamides are all inactive.

A variety of studies have shown that the active form of sulfonamides is the N^1-ionized salt form. Thus, although many modern sulfonamides are much more active than unsubstituted sulfanilamide, they are only two to six times more active when comparing amounts of N^1-ionized forms.[28] Maximal activity seems to be exhibited by sulfonamides between pK_as 6.6 and 7.4.[28–31] This reflects, in part, the need for enough nonionized (i.e., more lipid-soluble) drug to be present at physiologic pH to be able to pass through bacterial cell walls.[32] Fujita and Hansch[23] also related pK_a, partition coefficients, and electronic (Hammett) parameters with sulfonamide activity.

WELL-ABSORBED, SHORT-, AND INTERMEDIATE-ACTING SULFONAMIDES

Common preparations and doses are given in Table 5-2.

Sulfamethizole, USP. 4-Amino-*N*-(5-methyl-1,3,4-thiadiazole-2yl)benzenesulfonamide; N^1-(5-methyl-1,3,4-thiadiazol-2-yl)sulfanilamide; 5-methyl-2-sulfanilamido-1,3,4-thiadiazole. Its plasma half-life is 2.5 hours. This compound is a white crystalline powder soluble 1 : 2000 in water.

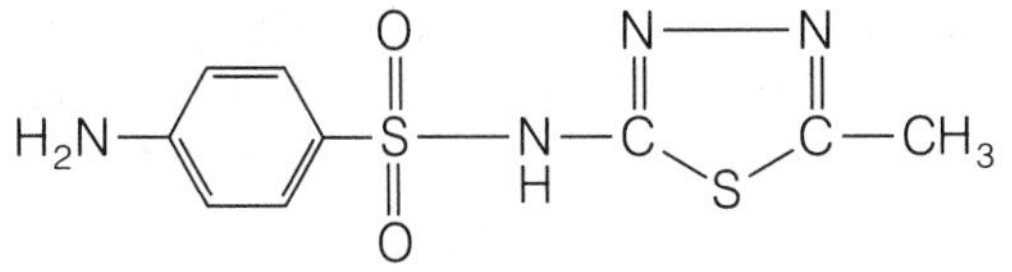

Sulfamethizole

Sulfisoxazole, USP. 4-Amino-*N*-(3,4-dimethyl-5-isoxazolyl)benzenesulfonamide; N^1-(3,4-dimethyl-5-isoxazolyl)sulfanilamide; 5-sulfanilamido-3,4-dimethylisoxazole. Its plasma half-life is six hours. This compound is a white, odorless, slightly bitter, crystalline power. Its pK_a is 5.0. At pH 6 this sulfonamide has a water solubility of 350 mg in 100 mL, and its acetyl derivative has a solubility of 110 mg in 100 mL of water.

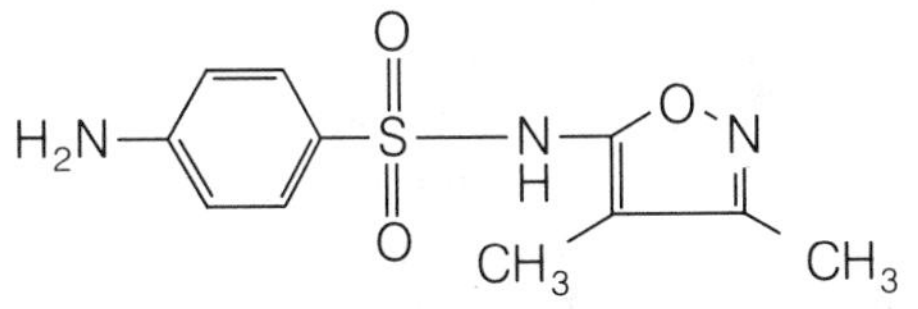

Sulfisoxazole

Sulfisoxazole possesses the action and the uses of other sulfonamides and is used for infections involving sulfonamide-sensitive bacteria. It is claimed to be effective in treatment of gram-negative urinary infections.

Sulfisoxazole Acetyl, USP. *N*-[(4-Aminophenyl)sulfonyl]-*N*-(3,4-dimethyl-5-isoxazolyl)acetamide; *N*-(3, 4-dimethyl-5-isoxazolyl)-*N*-sulfanilylacetamide; N^1-acetyl-N^1-(3, 4-dimethyl-5-isoxazolyl)sulfanilamide, shares the actions and the uses of the parent compound, sulfisoxazole. The acetyl derivative is tasteless and, therefore, suitable for oral administration, especially in liquid preparations of the drug. The acetyl compound is split in the intestinal tract and absorbed as sulfisoxazole; that is, it is a *prodrug* for sulfisoxazole.

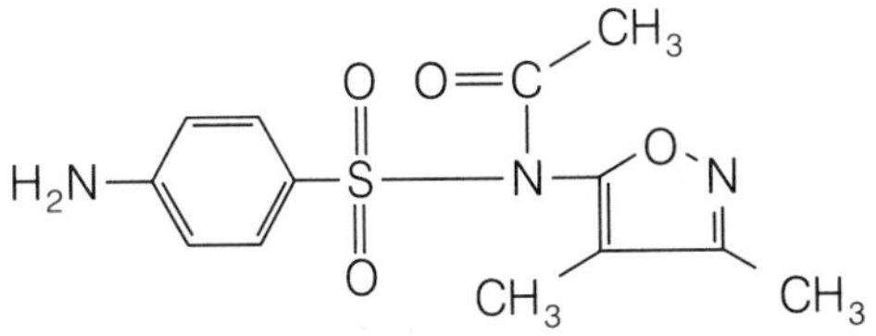

Sulfisoxazole Acetyl

Sulfisoxazole Diolamine, USP. 4-Amino-*N*-(3,5-dimethyl-5-isoxazolyl)benzenesulfonamide compound with 2,2′-iminobis[ethanol] (1 : 1); 2,2′-iminodi-ethanol salt of N^1-(3,4-dimethyl-5-isoxazolyl)sulfanilamide. This salt is prepared by adding enough diethanolamine to a solution of sulfisoxazole to bring the *p*H to about 7.5. It is used as a salt to make the drug more soluble at physiologic *p*H range of 6.0 to 7.5 and is used in solution for systemic administration of the drug by slow intravenous, intramuscular, or subcutaneous injection when sufficient blood levels cannot be maintained by oral administration alone. It also is used for instillation of drops or ointment in the eye for the local treatment of susceptible infections.

Sulfamethazine, USP. 4-Amino-*N*-(4,6-dimethy-2-pyrimidinyl)benzenesulfonamide; N^1-(4,6-dimethyl-2-pyrimidinyl)sulfanilamide; 2-sulfanilamido-4,6-dimethylpyrimidine. Its plasma half-life is seven hours. This compound is similar in chemical properties to sulfamerazine and sulfadiazine, but does have greater water solubility than either of them. Its pK_a is 7.2. Because it is more soluble in acid urine than is sulfamerazine, the possibility of kidney damage from use of the drug is decreased. The human body appears to handle the drug unpredictably; hence, there is some disfavor to its use in this country except in combination sulfa therapy (in trisulfapyrimidines, USP) and in veterinary medicine. (The structure of sulfamethazine in Fig. 5-1.)

Sulfacetamide. *N*-[(4-Aminophenyl)sulfonyl]-acetamide; *N*-sulfanilylacetamide; N^1-acetylsulfanilamide. Its plasma half-life is seven hours. This compound is a white crystalline powder, soluble in water (1 : 62.5 at 37°) and in alcohol. It is very soluble in hot water, and its water solution is acidic. It has a pK_a of 5.4.

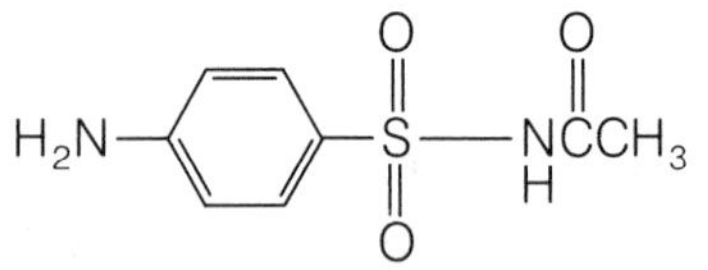

Sulfacetamide

Sulfachloropyridazine. N^1-(6-Chloro-3-pyridazinyl)sulfanilamide. Its plasma half-life is eight hours.

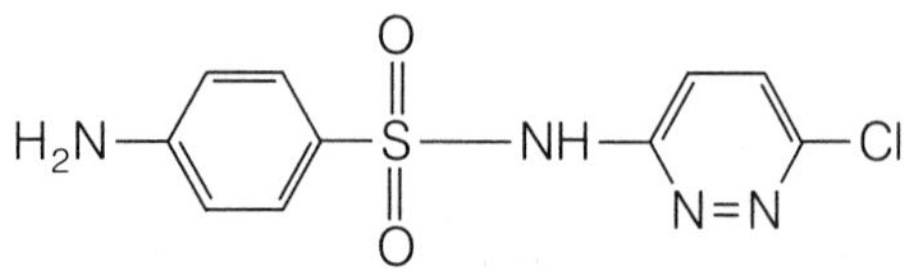

Sulfachloropyridazine

Sulfapyridine, USP. 4-Amino-*N*-2-pyridinylbenzenesulfonamide; N^1-2-pyridylsulfanilamide. Its plasma half-life is nine hours. This compound is a white, crystalline, odorless, and tasteless substance. It is stable in air but slowly darkens on exposure to light. It is soluble in water (1 : 3500), in alcohol (1 : 440), and in acetone (1 : 65) at 25°C. It is freely soluble in dilute mineral acids and aqueous solutions of sodium and potassium hydroxide. The pK_a is 8.4. Its outstanding effect in curing pneumonia was first recognized by Whitby; however, because of its relatively high toxicity it has been supplanted largely by sulfadiazine and sulfamerazine. Several cases of kidney damage have resulted from acetylsulfapyridine crystals deposited in the kidneys. It also causes severe nausea in most patients. Because of its toxicity it is used only for dermatitis herpetiformis.

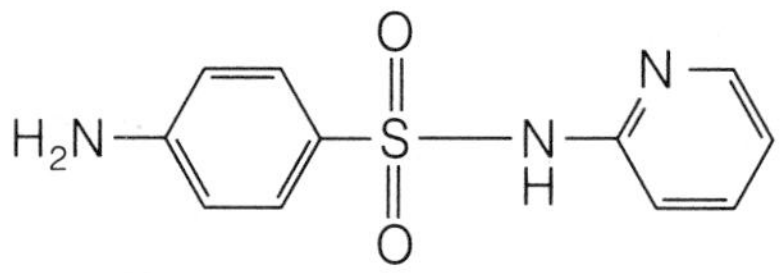

Sulfapyridine

Sulfapyridine was the first drug to have an outstanding curative action on pneumonia. It gave impetus to the study of the whole class of N^1-heterocyclically substituted derivatives of sulfanilamide.

Sulfamethoxazole, USP. 4-Amino-*N*-(5-methyl-3-isoxazolyl)benzenesulfonamide; N^1-(5-methyl-3-isoxazolyl)sulfanilamide (Gantanol). Its plasma half-life is 11 hours.

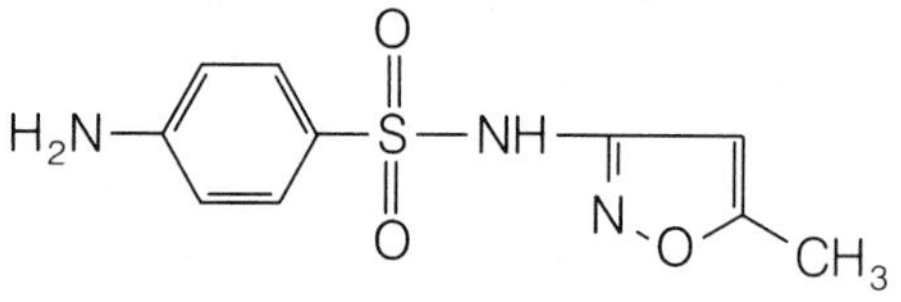

Sulfamethoxazole

Sulfamethoxazole is a sulfonamide drug closely related to sulfisoxazole in chemical structure and antimicrobial activity. It occurs as a tasteless, odorless, almost white crystalline powder. The solubility of sulfamethoxazole at the *p*H range of 5.5 to 7.4 is slightly less than that of sulfisoxazole but greater than that of sulfadiazine, sulfamerazine, or sulfamethazine.

Following oral administration, sulfamethoxazole is not as completely or as rapidly absorbed as sulfisoxazole and its peak blood level is only about 50% as high.

Sulfadiazine, USP. 4-Amino-*N*-2-pyrimidinylbenzenesulfonamide; N^1-2-pyrimidinylsulfanilamide; 2-sulfanilamidopyrimidine. Its plasma half-life is 17 hours. Sulfadiazine is a white, odorless, crystalline powder soluble in water to the extent of 1 : 8100 at 37°C, 1 : 13,000 at 25°C, in human serum to the extent of 1 : 620 at 37°C, and sparingly soluble in alcohol and acetone. It is readily soluble in dilute mineral acids and bases. Its pK_a is 6.3.

Sulfadiazine

Sulfadiazine Sodium, USP. Soluble sulfadiazine. This compound is an anhydrous, white, colorless, crystalline powder soluble in water (1 : 2) and slightly soluble in alcohol. Its water solutions are alkaline (*p*H 9 to 10) and absorb carbon dioxide

TABLE 5-2

SHORT- AND INTERMEDIATE-ACTING SULFONAMIDES

Name Proprietary Name	***Preparations***	***Usual Adult Dose****	***Usual Dose Range****	***Usual Pediatric Dose***†
Sulfamethizole USP *Thiosulfil, Utrasul*	Sulfamethizole oral suspension USP Sulfamethizole tablets USP	0.5–1 g 3–4 tid or qid		In children older than 2 mo: 30–45 mg/kg of body weight in 4 equally divided doses
Sulfisoxazole USP *Gantrisin, Sk-Soxazole*	Sulfisoxazole tablets USP	2–4 g initially, then 750 mg–1.5 g every 4 hr; or 1–2 g every 6 hr	Up to 12 g/day	In children older than 1 mo: 75 mg/kg of body weight initially, then 25 mg/kg every 4 hr, not exceeding 6 g/day
Sulfisoxazole acetyl USP *Gantrisin Acetyl, Lipo Gantrisin*	Sulfisoxazole acetyl oral suspension USP extended-release oral suspension 4–5 g every 12 hr	See Sulfisoxazole tablets USP		Use in infants under 1 mo of age is not recommended. In children older than 1 mo: 60–75 mg/kg of body weight initially, followed by 60–70 mg/kg every 12 hr
Sulfisoxazole diolamine USP *Gantrisin Diethanolamine*	Sulfisoxazole diolamine injection USP	IM or SC: 50 mg/kg of body weight initially, then 33.3 mg/kg of body weight every 8 hr. IV infusion: 50 mg/kg of body weight initially, then 25 mg/kg of body weight every 6 hr		See Usual Adult Dose
Sulfamethazine USP	See Mixed Sulfonamides			
Sulfapyridine	Sulfapyridine tablets	500 mg qid until improvement is noted; then reduce		
Sulfamethoxazole USP *Gantanol Urobak*	Sulfamethoxazole oral suspension USP Sulfamethoxazole tablets USP	Initial, 2 g, then 1 g every 8–12 hr		In children older than 1 mo: 50–60 mg/kg of body weight (max. 2 g) initially, then 25–30 mg/kg every 12 hr
Sulfadiazine USP	See Mixed Sulfonamides			
MIXED SULFONAMIDES				
Sulfamerazine USP Sulfamethazine USP Sulfadiazine USP *Terfonyl, Sulfalose, Trisulfazine, Trionamide, Neotrizine, Sulfose, Truozine*	Trisulfapyrimidines oral suspension Trisulfapyrimidines tablets	2–4 g initially, then 2–4 g/day in 3–6 divided doses	2–7 g/day	For children over 2 mo: 75 mg/kg of body weight initially, then 150 mg/kg per day in 4–6 divided doses

*See USP DI for complete dosage information.

†Sulfonamides are generally contraindicated in infants under 1 mo in age; these drugs may cause kernicterus.

from the air with precipitation of sulfadiazine. It is administered as a 5% solution in sterile water intravenously for patients requiring an immediate high blood level of the sulfonamide.

MIXED SULFONAMIDES

The danger of crystal formation in the kidneys on administration of sulfonamides has been reduced greatly through the use of the more soluble sulfonamides, such as sulfisoxazole. This danger may be diminished still further by administering mixtures of sulfonamides. When several sulfonamides are administered together, the antibacterial action of the mixture is the summation of the activity of the total sulfonamide concentration present, but the solubilities are independent of the presence of similar compounds. Thus, by giving a mixture of sulfadiazine, sulfamerazine, and sulfacetamide, the same therapeutic level can be maintained with much less danger of crystalluria, because only one-third the amount of any one compound is present. Some of the mixtures employed are the following:

Trisulfapyrimidines Oral Suspension. This mixture contains equal weights of sulfadiazine, USP, sulfamerazine, USP, and sulfamethazine, USP, either with or without an agent to increase the *p*H of the urine.

Trisulfapyrimidines Tablets. These tablets contain essentially equal quantities of sulfadiazine, sulfamerazine, and sulfamethazine.

SULFONAMIDES FOR OPHTHALMIC INFECTIONS

Sulfacetamide Sodium, USP. *N*-Sulfanilylacetamide monosodium salt (Sodium Sulamyd) is obtained as the monohydrate and is a white, odorless, bitter, crystalline powder that is very soluble (1 : 2.5) in water. Because the sodium salt is highly soluble at the physiologic *p*H of 7.4, it is especially suited, as a solution, for repeated topical applications in the local management of ophthalmic infections susceptible to sulfonamide therapy.

Sulfacetamide Sodium

Sulfisoxazole Diolamine, USP. Also used in intravenous and intramuscular preparations, this salt of sulfisoxazole was described along with the short- and intermediate-acting sulfonamides.

TABLE 5-3

SULFONAMIDES FOR OPHTHALMIC INFECTIONS

Name Proprietary Name	*Preparations*	*Application**
Sulfacetamide sodium USP *Blefcon, Bleph-10 Liquifilm, Bufopto Sulfacel-15, Cetamide, Isopto Cetamide, Sodium Sulamyd, Sulf-30*	Sulfacetamide sodium ophthalmic ointment USP	Topically to the conjunctiva, a thin strip of 10% ointment every 6 hr and hs
	Sulfacetamide sodium ophthalmic solution USP	Topically to the conjunctiva, 1 drop every 1–3 hr during the day and less frequently at night; 10%–30% strengths available.
Sulfisoxazole diolamine USP *Gantrisin Diethanolamine*	Sulfisoxazole diolamine ophthalmic ointment USP	Topically to the conjunctiva, a thin strip of 4% ointment every 8–24 hr and hs
	Sulfisoxasole diolamine ophthalmic solution USP	Topically to the conjunctiva, 1 drop every 8 hr or more frequently; 4% strength available

*See USP DI for complete dosage information.

SULFONAMIDES FOR BURN THERAPY

Mafenide Acetate. 4-(Aminomethyl)benzenesulfonamide acetate (Sulfamylon) is a homologue of the sulfanilamide molecule. It is not a true sulfanilamide-type compound, as it is not inhibited by *p*-aminobenzoic acid. Its antibacterial action involves a mechanism that is different from that of true sulfanilamide-type compounds. This compound is particularly effective against *Clostridium welchii* in topical application and was used during World War II by the German army for prophylaxis of wounds. It is not effective by mouth. It is employed currently alone or with antibiotics in the treatment of slow-healing infected wounds.

Mafenide

Some patients treated for burns with large quantities of this drug developed metabolic acidosis. To overcome this side effect, a series of new organic salts was prepared.[16] The acetate in an ointment base proved to be the most efficacious.

Silver Sulfadiazine (Silvadene). The silver salt of sulfadiazine applied in a water-miscible cream base has proved to be an effective topical antimicrobial agent, especially against *Pseudomonas* species. This is of particular significance in burn therapy because pseudomonad are often responsible for failures in therapy. The salt is only very slightly soluble and does not penetrate the cell wall but acts on the external cell structure. Studies using radioactive silver have shown essentially no absorption into body fluids. Sulfadiazine levels in the serum were about 0.5 to 2 mg/100 mL.

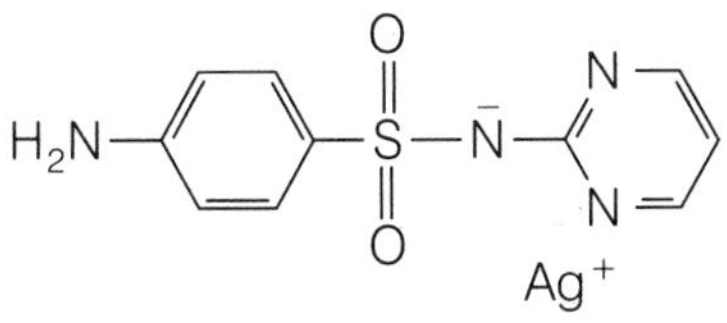

Silver Sulfadiazine

This preparation is reported to be simpler and easier to use than other standard burn treatments, such as application of freshly prepared dilute silver nitrate solutions or mafenide ointment.

TABLE 5-4

SULFONAMIDES FOR BURN THERAPY

Name Proprietary Name	***Preparations***	***Application****
Mafenide acetate *Sulfamylon*	Mafenide acetate cream	Topically to the skin, with sterile gloved hand to a thickness of about 1.6 mm; burned area should be covered with cream at all times
Silver Sulfadiazine *Silvadene*	Silver sulfadiazine cream	Topical (see Mafenide acetate)

*See USP DI for complete dosage information.

SULFONAMIDES FOR INTESTINAL INFECTIONS, ULCERATIVE COLITIS, OR REDUCTION OF BOWEL FLORA

Each of the sulfonamides in this group is a prodrug, which is designed to be poorly absorbable, although usually, in practice, a little is absorbed. Therefore, usual precautions with sulfonamide therapy should be observed. In the large intestine the N^4-protecting groups are cleaved, releasing the free sulfonamide antibacterial agent. Today only one example is used clinically—sulfasalazine.

Sulfasalazine, USP. 2-Hydroxy-5[[4-[(2-pyridinylamino)sulfonyl]phenyl]azo]benzoic acid; 5-[[*p*-(2-pyridylsulfamoyl)phenyl]azo]salicylic acid. This compound is a brownish-yellow, odorless powder, slightly soluble in alcohol, but practically insoluble in water, ether, and benzene.

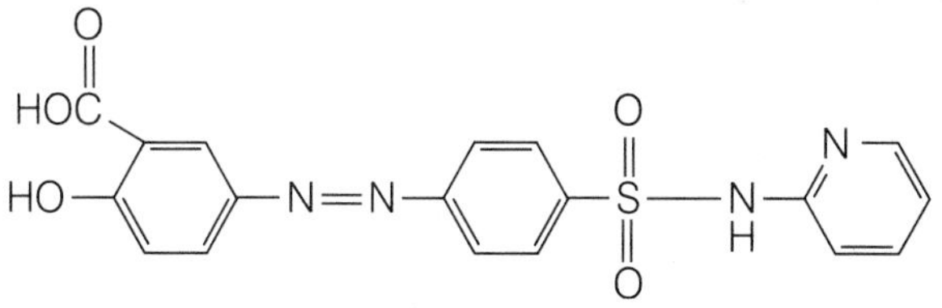

Sulfasalazine

It is broken down in the body to *m*-aminosalicylic acid and sulfapyridine. The drug is excreted through the kidneys and is colorimetrically detectable in the urine, producing an orange yellow color when the urine is alkaline and no color when the urine is acid.

FOLATE REDUCTASE INHIBITORS

Trimethoprim, USP. 5-[(3,4,5-Trimethoxyphenyl)methyl]-2,4-pyrimidinediamine; 2,4-diamino-5-(3,4,5-trimethoxybenzyl)pyrimidine. Trimethoprim is closely related to several antimalarials, but it does not have good antimalarial activity by itself;

TABLE 5-5

SULFONAMIDES FOR INTESTINAL INFECTIONS, ULCERATIVE COLITIS, OR REDUCTION OF BOWEL FLORA

Name Proprietary Name	***Preparations***	***Usual Adult Dose****	***Usual Dose Range****
Sulfasalazine *Azulfidine* *Salazopyrin*	Sulfasalazine tablets	Initial, 1–2 g every 6–8 hr; maintenance, 500 mg qid	Up to 12 g/day; or up to 500 mg/hr
	Sulfasalazine oral suspension	See dose for tablets	

*See USP DI for complete dosage information.

TABLE 5-6

FOLATE REDUCTASE INHIBITORS

Name Proprietary Name	**Preparations**	**Usual Adult Dose***	**Usual Pediatric Dose*†**
Trimethoprim USP *Proloprim, Trimpex*	Trimethoprim tablets USP	100 mg every 12 hr for 10 days for uncomplicated infections	
Trimethoprim and sulfamethoxazole USP *Bactrina, Septra, Bactrim-D5, Septra-D5*	Sulfamethoxazole and trimethoprim tablets	Antibacterial: 60 mg trimethoprim and 800 mg sulfamethoxazole every 12 hr	Antibacterial: For children up to 40 kg of body weight: 4 mg/kg trimethoprim and 20 mg/kg sulfamethoxazole every 12 hr
	Sulfamethoxazole and trimethoprim oral suspension USP	Antiprotozoal (*Pneumocystitis carinii* pneumonia): 5 mg trimethoprim and 25 mg sulfamethoxazole per kg of body weight every 6 hr	Antiprotozoal: For children up to 32 kg of body weight: 5 mg trimethoprim and 25 mg sulfamethoxazole per kg every 12 hr
	Sulfamethoxazole and trimethoprim concentrate for injection USP	Antibacterial: IV infusion, 10–12.5 mg of sulfamethoxazole and 2–2.5 mg of trimethoprim per kg of body weight every 6 hr	See Usual Adult Dose
		Antiprotozoal (*Pneumocystitis carinii* pneumonia): IV infusion, 25 mg sulfamethoxazole and 5 mg trimethoprim per kg of body weight every 6 hr	See Usual Adult Dose

*See USP DI for complete dosage information.
†Use with infants under 1 month of age is contraindicated.

however, it is a potent antibacterial. Originally introduced in combination with sulfamethoxazole, it is now available as a single agent. Approved by the FDA in 1980, trimethoprim as a single agent is used only for the treatment of uncomplicated urinary tract infections. The argument for trimethoprim to be a single agent was summarized in 1979 by Wormser and Deutsch.[16] They point out, for example, that several studies comparing trimethoprim with trimethoprim–sulfamethoxazole for treatment of chronic urinary tract infections found no statistically relevant difference between the two treatments. Furthermore, some patients cannot take sulfonamide products for the reasons discussed previously in this chapter, especially hypersensitivity. In contrast with these and similar arguments, the concern is that, when used as a single agent, bacteria now susceptible to trimethoprim will rapidly develop resistance. However, in combination with a sulfonamide, the bacteria will be less likely to do so. That is, they will not survive long enough to easily develop resistance to both drugs.

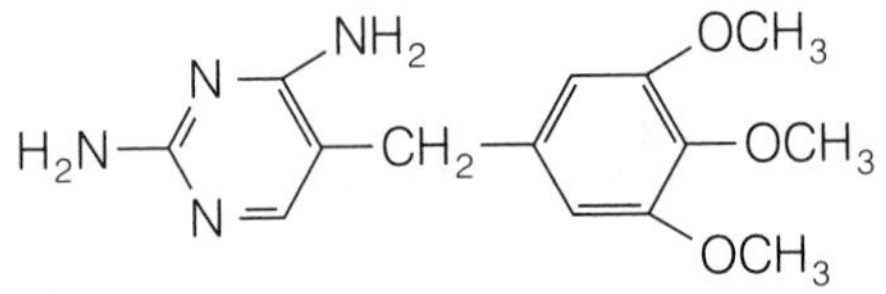

NH_2 OCH_3 H_2N CH_2 OCH_3 OCH_3

Trimethoprim

Sulfamethoxazole and Trimethoprim. The synergistic action of the combination of these two drugs has been discussed previously in this chapter.

SULFONES

The sulfones are primarily of interest as antibacterial agents, although there are some reports of their use in the treatment of malarial and rickettsial infections. They are less effective agents than are the sulfonamides. *p*-Aminobenzoic acid partially antagonizes the action of many of the sulfones, suggesting that the mechanism of action is similar to that of the sulfonamides. It also has been observed that infections that arise in patients being treated with sulfones are cross-resistant to sulfonamides. Several sulfones have found use in the treatment of leprosy, but among them only dapsone is generally used today.[33]

It has been estimated that there are about 11 million cases of leprosy in the world, of which about 60% are in Asia (with 3.5 million in India alone).[33] The first reports of dapsone resistance have prompted the use of multidrug therapy, with dapsone, rifampin, and clofazimine combinations, in some geographic areas.[33]

The search for antileprotic drugs has been hampered by the inability to cultivate *Mycobacterium leprae* on artificial media and by the lack of experimental animals susceptible to human leprosy. A method of isolating and growing *M. leprae* in the foot pads of mice and in armadillos has been reported and has permitted a much wider range of research. Sulfones were introduced into the treatment of leprosy after it was found that sodium glucosulfone was effective in experimental tuberculosis in guinea pigs.

The parent sulfone, dapsone (4,4′-sulfonyldianiline), is the prototype for a variety of analogues that have been widely studied. Four types of variations on this structure have given active compounds:

1. Substitution on both the 4- and 4′-amino functions
2. Monosubstitution on only one of the amino functions
3. Nuclear substitution on one of the benzenoid rings
4. Replacement of one of the phenyl rings with a heterocyclic ring

The antibacterial activity and the toxicity of the disubstituted sulfones are thought to be due to chiefly the formation in vivo of dapsone. Hydrolysis of disubstituted derivatives to the parent sulfone apparently occurs readily in the acid medium of the stomach, but only to a very limited extent following parenteral administration. Monosubstituted and nuclear substituted derivatives are believed to act as entire molecules.

Products

Dapsone, USP. 4,4′-Sulfonylbisbenzenamine; 4,4′-sulfonyldianiline; *p,p*′-diaminodiphenylsulfone; DDS (Avlosulfone) occurs as an odorless, white crystalline powder that is very slightly soluble in water and sparingly soluble in alcohol. The pure compound is light-stable, but the presence of traces of impurities, including water, makes it photosensitive and thus susceptible to discoloration in light. Although no

TABLE 5-7

LEPROSTATIC SULFONES

Name Proprietary Name	*Preparations*	*Usual Adult Dose**	*Usual Dose Range**	*Usual Pediatric Dose**
Dapsone USP *Avlosulfon*	Dapsone tablets USP	Leprosy: 100 mg, qd, or 1.4 mg/kg of body weight qd Dermatitis herpetiformis suppressant: initially 50 mg/day, increased up to 300 mg/day if symptoms are not completely controlled; then reduced to lowest effective maintenance dose as soon as possible		Leprosy: 1.4 mg/kg of body weight qd

*See USP DI for complete dosage information.

chemical change is detectable following discoloration, the drug should be protected from light.

$H_2N-C_6H_4-SO_2-C_6H_4-NH_2$

Dapsone

Dapsone is used in the treatment of both lepromatous and tuberculoid types of leprosy. Dapsone is widely used for all forms of leprosy, often in combination with clofazimine and rifampin. Initial treatment often includes rifampin with dapsone, followed by dapsone alone. It is also used to prevent the occurrence of multibacillary leprosy when given prophylactically. For a more detailed discussion of drug treatment, the review of Shepard is recommended.[33] The National Hansen's Disease Center can provide current information (telephone: (504) 642-8325).

Dapsone is also the drug of choice for dermatitis herpetiformis, with pyrimethamine for treatment of malaria (see Chap. 6), with trimethoprim for *Pneumocystis carnii* pneumonia, and has been used for rheumatoid arthritis.

Serious side effects can include hemolytic anemia, methemoglobinemia, and toxic hepatic effects. Hemolytic effects can be pronounced in patients with glucose-6-phosphate dehydrogenase deficiency. With all patients, blood counts during therapy are important.

REFERENCES

1. Domagk, G.: Dtsch. Med. Wochenschr. 61:250, 1935.
2. Baumler, E.: In Search of the Magic Bullet. London, Thames and Hudson, 1965.
3. Anand, N.: Sulfonamides and sulfones. In Wolff, M. E. (ed.). Burger's Medicinal Chemistry, Part II, 4th ed., Chap. 13. New York, Wiley-Interscience, 1979.
4. Forester, J.: Z. Haut. Geschlechtskr. 45:459, 1933.
5. Trefouel, J., et al.: C. R. Seances Soc. Biol. 120:756, 1935.
6. Fuller, A. T.: Lancet 1:194, 1937.
7. Northey, E. H.: The Sulfonamides and Allied Compounds. ACS Monogr. Ser. American Chemical Society, Washington, DC, 1948.
8. Drug Information, American Society of Hospital Pharmacists, 1990.
9. Weinstein, L.: In Gilman, A. G., et al. (eds.). The Pharmacological Basis of Therapeutics, 7th ed., New York, Macmillan, 1985.
10. Krupp, M. A., Chatton, M. J.: Current Medical Diagnosis and Treatment. Los Altos, CA, Lange Medical Publications, 1990.
11. Med. Lett. 30:33, 1988.
12. FDA Drug Bulletin, U.S. Department of Health, Education and Welfare, Food and Drug Administration, Feb., 1980.
13. Med. Lett. 29:53, 1987.
14. Werlin, S. L., Grand, R. J.: J. Pediatr. 92:450, 1978.
15. Bushby, S. R., Hitchings, G. H.: Br. J. Pharmacol. Chemotherap. 33:72, 1968.
16. Wormser, G. P., Deutsch, G. T.: Ann. Intern. Med. 91:420, 1979.
17. Gleckman, R., et al.: Am. J. Hosp. Pharm. 36:893, 1979.
18. Letters of Burchall, J. J., Then, R., Poe, M.: Science 197:1300–1301, 1977.
19. Palminteri, R., Sassella, D.: Chemotherapy 25:181, 1979.
20. Harvey, R. J.: J. Antimicrob. Chemotherap. 4:315, 1978.
21. White, P. J., Woods, D. D.: J. Gen. Microbiol. 40:243, 1965.
22. Watanabe, T.: Bacteriol. Rev. 27:87, 1963.
23. Fujita, T., Hansch, C.: J. Med. Chem. 10:991, 1967.
24. Zbinden, G.: In Gould, R. F. (ed.). Molecular Modification in Drug Design. Washington, DC, American Chemical Society, 1964.
25. Fujita, T.: In Gould, R. F. (ed.). Molecular Modification in Drug Design. Washington, DC, American Chemical Society, 1964.
26. Vree, T. B., et al.: Clin. Pharmacokinet. 4:310, 1979.
27. Hansen, I.: Antibiot. Chemother. 25:217, 1978.
28. Fox, C. L., Ross, H. M.: Proc. Soc. Exp. Biol. Med. 50:142, 1942.
29. Yamazaki, M., et al.: Chem. Pharm. Bull. 18:702, 1970.
30. Bell, P. H., Roblin, R. O.: J. Am. Chem. Soc. 64:2905, 1942.
31. Cowles, P. B.: Yale J. Biol. Med. 14:599, 1942.
32. Brueckner, A. H.: J. Biol. Med. 15:813, 1943.
33. Shepard, C. C.: N. Engl. J. Med. 307:1640, 1982.

CHAPTER **6**

Antimalarials

Dwight S. Fullerton

> Few people realize that there are far more kinds of parasitic than nonparasitic organisms in the world. Even if we exclude viruses, rickettsias . . . and the many kinds of parasitic bacteria and fungi, the parasites are still in the majority. The parasitic way of life . . . is a highly successful one. Humans are hosts to over 100 kinds of parasites, again not counting viruses, bacteria and fungi . . . . At least 45,000 species of protozoa have been described to date, many of which are parasitic. Parasitic protozoa still kill, mutilate, and debilitate more people in the world than any other group of disease organisms . . . .
>
> G. D. Schmidt and L. S. Roberts
> *Foundations of Parasitology*[1]

Malaria, African sleeping sickness, leishmaniasis, Chagas' disease, and other protozoal disease (Table 6-1) of humans and their livestock continue to have a devastating impact worldwide.[1–16] Over 12 million South Americans suffer from Chagas' disease alone, and millions in Africa and the Mediterranean areas from African sleeping sickness or the disfigurement of leishmaniasis.

However, none of these protozoal diseases has had the enormous effect upon civilization, either historically or in modern times, as has malaria. One million African children alone still die each year from the disease. Loss of productivity from the debilitating and cyclic clinical stages of malaria is enormous. It has been noted by Schmidt and Roberts[1] that a single day of malaria fever requires the caloric equivalent of two days of hard labor (and thus of food). With malaria protozoa in many areas becoming resistant to commonly used antimalarial drugs,[1–10] and with insecticide-related problems increasing (resistance by the mosquito vector, human and environmental harm), the adverse impact of malaria upon the world is likely to continue. About 1000 cases are reported in the United States each year.[9]

Although some protozoal infections of farm animals (e.g., coccidiosis in chickens) have been at times a serious problem in the United States, and trichomonal vaginitis in humans is common, malaria and other life-threatening protozoal infections are not. However, the ease of international travel has caused increased awareness of the prevention and treatment of protozoal infections by American physicians and pharmacists. During the Vietnam War several thousand cases of malaria were reported in the United States—largely attributed to returning servicemen.

The epidemiology, diagnosis, microbiology, medicinal chemistry, and chemotherapy of malaria and other parasitic disease have recently been reviewed.[1–8] Current approaches to prevention and treatment of parasitic infections including malaria have been summarized and drugs of choice and the doses given in *The Medical Letter*.[3]

ETIOLOGY

Malaria in humans is caused by four species of *Plasmodium* protozoa, which, as shown in Figure 6-1, spend half their life cycle in female *Anopheles* mosquitos. (Male *Anopheles* mosquitos do not feed on vertebrate blood.) Several hundred *Anopheles* species are known, several of which are commonly found in the United States. Resistance to DDT, dieldrin, and other insecticides has been reported for an increasing number of *Anopheles* species—making malaria control more difficult. Once infected, the mosquito carries sporozoites for life.

As can be seen in the simplified life cycle illustrated in Figure 6-1, the malaria protozoa undergo several morphologic changes in the human host. (Detailed descriptions of these changes are con-

TABLE 6-1

MALARIA AND OTHER COMMON PROTOZOAL INFECTIONS IN HUMANS AND FARM ANIMALS

Disease	*Protozoa*	*Insect Vector*	*Primary Occurrence*	*Clinical Notes*
Malaria	*Plasmodium vivax*, others	Mosquitoes	Tropical	High fever and chills, cyclic
African sleeping sickness	*Trypanosoma*[17] *rhodesiense, T. gambiense*, others	Tsetse flies	Tropical Africa	*T. rhodesiense* infection usually causes death before CNS depression, but this is commonly seen with *T. gambiense*
Chagas' disease	*T. cruzi*	Common "bedbug"	South America	Bug usually bites victim close to mouth, so is called the "kissing bug." Protozoa invade many tissues including the heart. Disfiguring edema
Leishmaniasis (kala-azar)	*Leishmania donovani*, others	Sand fly	Middle East, tropical Africa, tropical South America	Progressive wasting and anemia, severely enlarged spleen and liver
Amebiasis (amebic dysentery)	*Entamoeba histolytica, E. vaginalis*	None — transmitted by human and animal wastes	Tropical regions	
Trichomonal vaginitis	*Trichomonas vaginalis*	None — usually transmitted sexually	Worldwide	Can be serious in women and men (see discussion by Kreier)[17]
Coccidiosis in farm animals	*Eimeria* sp.	None — usually by animal wastes	Worldwide	Great economic losses even in United States
Toxoplasmosis	*Toxoplasma* sp.	None — usually by contact with infected cats	Worldwide	
Babesiasis in cattle	*Babesia* sp.	Ticks	Worldwide	

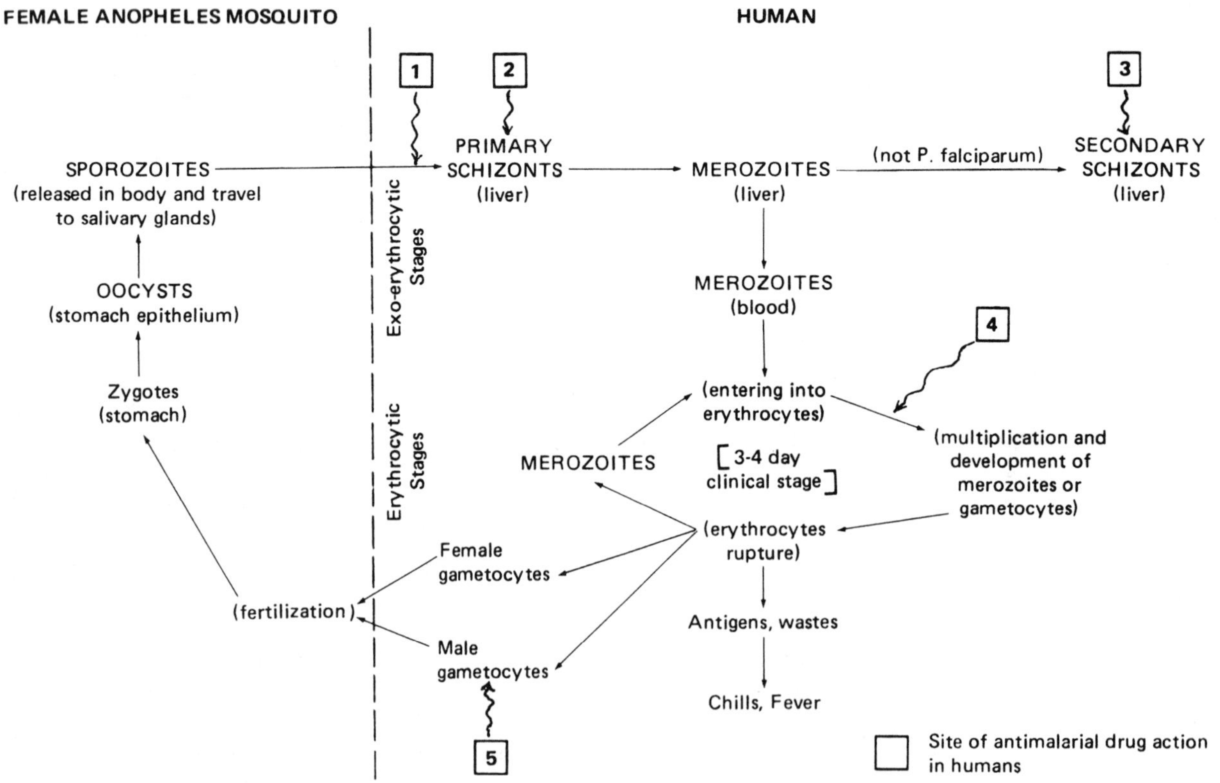

FIG. 6-1. *Life cycle of Plasmodium: (1) sporozoitocide — no drugs effective; (2) primary (exoerythrocytic) schizonticide — primaquine, pyrimethamine, chloroguanide, cycloguanil pamoate; (3) secondary (exoerythrocytic) schizonticide — primaquine; (4) erythrocytic schizonticide (fast-acting — chloroquine, quinine, amodiaquine; slow-acting — chloroguanide, pyrimethamine, sulfonamides, cycloguanil pamoate); (5) gametocytocide — primaquine.*

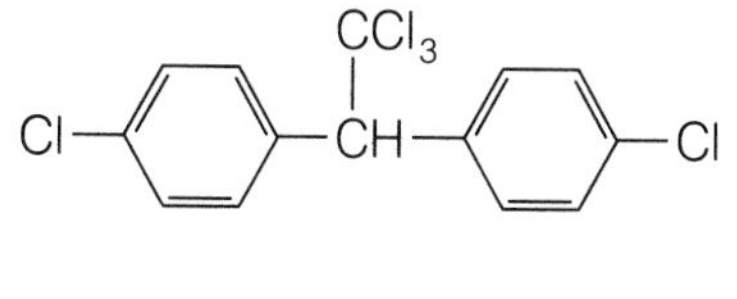

DDT

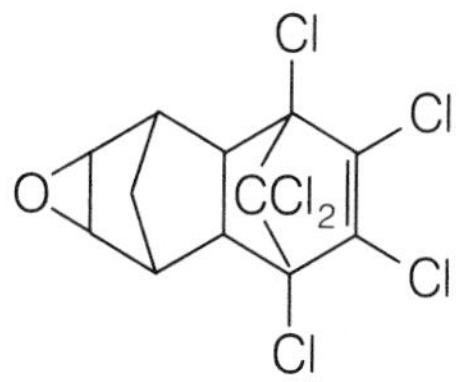

Dieldrin

tained in several texts.[1,2]) The rates at which these changes take place vary among the *Plasmodium* species. The patient generally has no adverse symptoms until erythrocytes rupture (generally 1.5 to 2 weeks after the initial bite), releasing antigenic cell residues and protozoal wastes—leading to (1) a recurring (every three to four days) attack of nausea, vomiting, severe chills, delirium, and high fevers (38–40°C); (2) severe anemia (and thrombocytopenia) from hemolysis of erythrocytes as more and more merozoites are produced; and (3) jaundice from excess bilirubin (a metabolite of hemoglobin) production.

The four *Plasmodium* species are as follows:

***P. falciparum* ("malignant tertian, subtertian").** Of the four, only *P. falciparum* does not have a secondary schizont (secondary exoerythrocytic) stage. However, it is the most lethal. Enormously high concentrations of the protozoa in the hosts' blood are often found—with over 65%[1] of the erythrocytes affected in some cases. Unlike malaria from the other three species, the patient may feel quite ill between acute attacks. Because all the merozoites are released from the primary schizont stage at the same time, reinfection from a secondary schizont stage is not a problem. About half of all malaria is caused by this species.

***P. vivax* ("benign tertian malaria").** This form of malaria is called tertian because clinical symptoms usually recur every 48 hours (i.e., if the patient is sick on day 1, sickness will occur again on day 3, the *tertiary* day). Not as many merozoites are produced as with *P. falciparum*, but many reenter new liver cells to form secondary schizonts, which can cause relapses for several years. Drugs specific for this stage (site 3) must be given for the patient to be truly cured. Only young erythrocytes are attacked, thus limiting the total erythrocytic involvement. About 40% of all malaria is caused by *P. vivax*.

***P. malariae* ("quartan malaria").** This species has a life cycle similar to that of *P. vivax*. Relapses may occur for decades.

***P. ovale* ("mild tertian malaria").** *P. ovale* is the least common of the four types of malaria, and, similarly to *P. vivax* and *P. malariae*, has a long-lasting secondary schizont stage.

BIOCHEMICAL DEPENDENCE ON HOST ERYTHROCYTES

Malaria protozoa are very species-specific, some infecting only birds and others only humans. Only *Anopheles* mosquitos harbor the protozoa in half their life cycles—none of the several thousand other mosquitos of other genera do so. This specificity reflects specialized biochemical dependence upon vertebrate host and vector host. The biochemistry of this specialized metabolic and biochemical dependence has been intensively studied, especially with the goal of discovering biochemical (drug-attackable) vulnerability of the malaria protozoa. Nevertheless, many questions remain.[1,2]

Most research has focused on the biochemical dependence of malaria protozoa on erythrocytes. The reasons are primarily that parasite–erythrocyte biochemistry can be studied with fewer technical problems than that of other tissues, and drug treatment and study at this stage are direct (it is easy to get drugs into blood, to obtain blood samples, and such).

Malaria protozoa need host erythrocytes to make (replicate) their own DNA and RNA, needed for the rapid production of numerous merozoites. The malaria parasites can, and must, synthesize their own pyrimidines (cytosine, uracil, thymine) because they can not use pyrimidines of the host. However, the protozoa cannot synthesize purines (adenine and guanine) and must obtain them from erythrocytes of the host. Phosphate is also obtained from the host.

Host hemoglobin and plasma are digested by proteolytic enzymes of the malaria protozoa and used as sources of several amino acids. Up to 75% of hemoglobin of infected erythrocytes is digested, with *haemozoin* (an insoluble pigment) formed as a byproduct. The protozoa also can synthesize some amino acids of their own.

It also appears that the malaria protozoa may be dependent upon the host as a source of pentoses (for DNA and RNA synthesis). There remains disagreement[1] on whether or not these protozoa have an operable pentose phosphate pathway.

Malaria protozoa also cannot synthesize their own cholesterol and fatty acids, needed for cell membrane and glyceride biosynthesis. In addition, they require most of the same vitamins and coenzymes needed by their hosts. Similar to many bacteria, they must also synthesize their own folic acid. Consequently, drugs, such as sulfonamides, which block folic acid biosynthesis, can also block malaria protozoal growth.

Large amounts of glucose are used by these protozoa, particularly during the erythrocytic stage. Hypoglycemia can result in some patients, requiring treatment with parenteral glucose in critical cases.

The roles of sulfonamides and related drugs in blocking folic acid biosynthesis are discussed in Chap. 5. The 1990 review, "Dihydrofolate reductase as a therapeutic target," by Bertino and co-workers[4] is especially recommended for further reading.

HISTORY

EARLY CINCHONA USE

The general antifebrile properties of the bark of the cinchona undoubtedly were known to the Incas before the arrival of the Spaniards early in the 16th century. However, it was probably the observations of early Jesuit missionaries that led to the discovery that infusions of cinchona bark were effective for the treatment of the tertian "ague," which was common in tropical Central and South America even then, and to the introduction of the crude drug into Western Europe. The first recorded use in South America was about 1630 and in Europe 1639. Thus began the first era in the chemotherapy of malaria. Cinchona and the purified alkaloids obtained from it were to remain the only drugs of significance in the treatment of malaria for three centuries.

PAUL EHRLICH AND THE FIRST SYNTHETIC ANTIMALARIALS

Much has been written about Paul Ehrlich (1854–1915), who can be considered the founder of modern medicinal chemistry, chemotherapy, and molecular pharmacology.[18–21] (Films about his life are also available for classroom use.[21]) His Nobel laureate for medicine in 1908 [with Elie Metchnikoff (Ilya Mechnikov)], however, was for his research on immunology and the development of diphtheria and other antitoxins.

In the late 1800s, the German dye industry was producing hundreds of new dyes. Ehrlich's first work

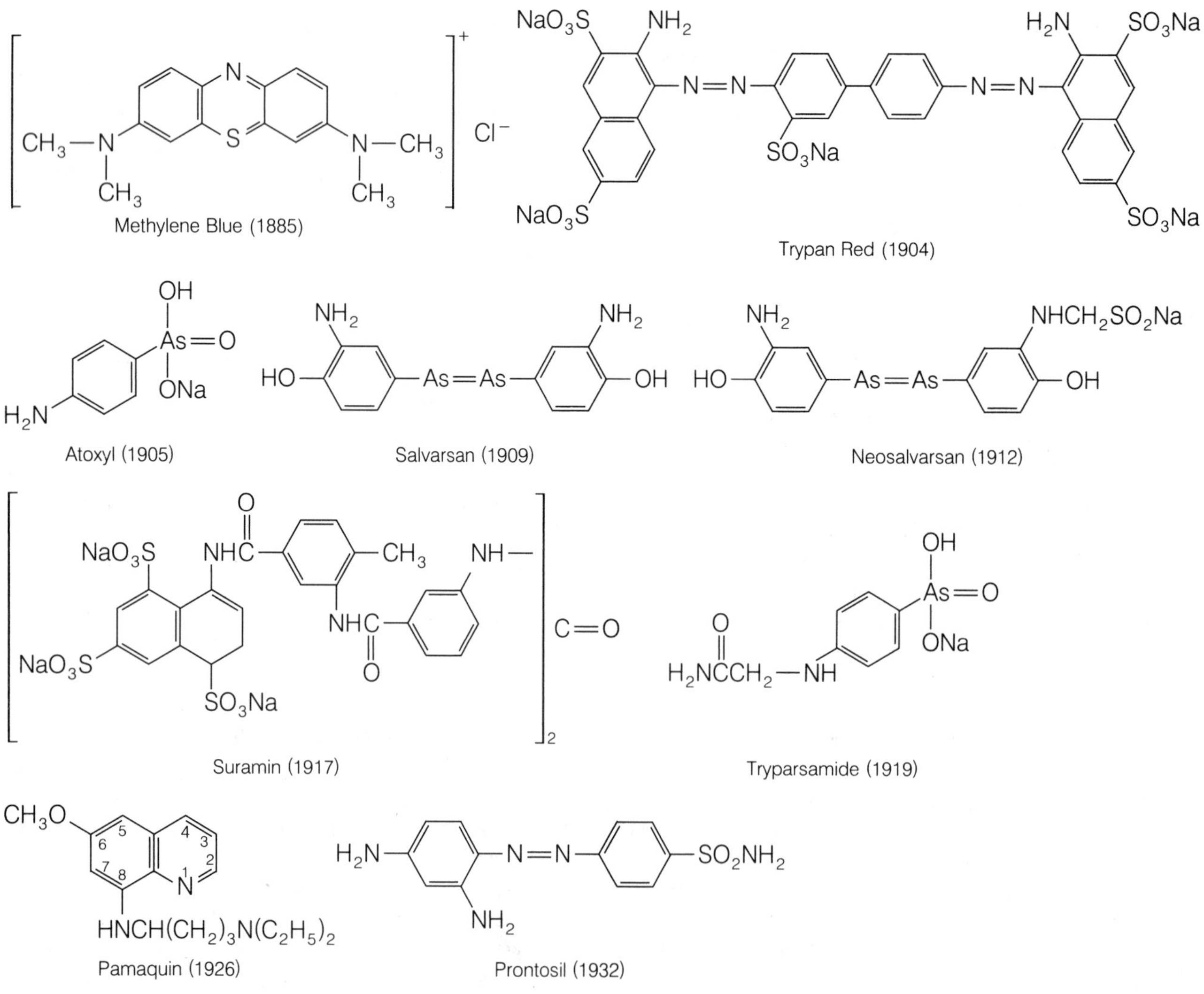

Methylene Blue (1885)

Trypan Red (1904)

Atoxyl (1905)

Salvarsan (1909)

Neosalvarsan (1912)

Suramin (1917)

Tryparsamide (1919)

Pamaquin (1926)

Prontosil (1932)

FIG. 6-2. *Antimalarial and related drugs of Paul Ehrlich and other early medicinal chemists.*

on the biologic properties of these dyes was on blood cells, demonstrating the specific staining of parts of certain leukocytes—the basis of modern hematology. Could pathogenic microorganisms also be specifically stained and perhaps killed without harm to the host ("magic bullet" concept)? Ehrlich's laboratory set about finding out, focusing first on malaria protozoa using methylene blue and related dyes. (Methylene blue, synthesized by Hoechst in 1885, was known to stain nerve tissues selectively without toxic effect.) For five years, Ehrlich and his associates studied hundreds of dyes—many of which they synthesized—leading to the first synthetic antimalarial, trypan red (Fig. 6-2) in 1904. However, trypan red did not have the antimalarial potency needed for an effective human cure. Perhaps a potent poison, such as arsenic, could be combined with an effective protozoal stain to carry the poison selectively to the malaria protozoa: one part of the drug for good affinity, one part of the drug for good intrinsic activity. Ehrlich had simultaneously conceived the basis of modern drug design and chemotherapy of the live receptor (see the discussion by Albert[20]).

TABLE 6-2

DRUGS OF CHOICE FOR HUMAN MALARIA*

Malaria Species	*Therapeutic Goal*	*Drug of Choice*	*Alternative*
AREAS NOT KNOWN TO HAVE CHLOROQUINONE-RESISTANT *P. FALCIPARUM* *			
All	Suppression while visiting endemic area (begin 1 wk before visit, continue for 4 wk after leaving)†	Chloroquine phosphate once a week	Hydroxychloroquine sulfate
P. vivax, *P. ovale*, *P. malariae*‡	Prevention of relapses	Primaquine phosphate§	
P. vivax, *P. ovale*, *P. malariae*	Elimination of secondary schizonts after leaving endemic area (take for 2 wk — the last 2 wk of chloroquine therapy above)‖	Primaquine phosphate§	
All	Treatment of uncomplicated attack (clinical stage)	Chloroquine phosphate	Hydroxychloroquine sulfate
All	Parenteral treatment of severe attack when patient cannot take an oral dose	Quinine dihydrochloride	Chloroquine hydrochloride
AREAS WITH CHLOROQUINONE-RESISTANT *P. FALCIPARUM*			
P. falciparum	Suppression while visiting endemic area†	Mefloquine, or chloroquine plus pyrimethamine–sulfadoxine,#, ** or Chloroquine plus chloroguanide (proguanil)**, ††	
P. falciparum	Treatment of uncomplicated attack (clinical stage)	Quinine sulfate and pyrimethamine or sulfadiazine	Mefloquine
P. falciparum	Parenteral treatment of severe illness, when patient cannot take an oral dose	Quinine dihydrochloride	

*Modified after *Med. Lett.* 32:23, 1990, and 32:36, 1990.

†Avoiding mosquito bites is the most important first line of prevention. Insect repellents containing at least 30% diethyltoluamide should be used and mosquito netting used at night.

‡*P. falciparum* does not have secondary schizonts; hence, unless the patient is reinfected by a mosquito bite, relapses do not occur.

§Primaquine phosphate may cause hemolytic anemias in persons who are glucose-6-phosphate dehydrogenase (G6PD)-deficient. Patients should be tested for this deficiency before primaquine is prescribed.

‖Because of hemolylic anemia in patients who are G6PD-deficient, some physicians do not recommend primaquine. Rather, they monitor their patients for signs of relapse and then treat accordingly.

#The pyrimethamine–sulfadoxine is taken as a single dose for self-treatment until medical care can be obtained.

**See *Med. Lett.* 32:36, 1990 for a description of which geographic areas have a risk for malaria.

††The chloroguanide (proguanil) is taken during exposure and 4 weeks afterward.

‡‡Use of doxycycline is investigational and, as with all tetracyclines, this drug should not be taken by young children or pregnant women.

Atoxyl was Ehrlich's first antimalarial success using the concept, but it was still too neurotoxic to the host. In 1906, he began focusing on spirochetes, specifically on *Treponema pallidum* (*S. pallida*), identified in 1905 as the cause of syphilis. His organoarsenicals seemed to be as toxic to these spirochetes as to malaria protozoa, and syphilis was a far greater cause of death in Europe than was malaria. In 1909, his #606 (marketed in 1910 as Salvarsan) was the world's first successful antisyphilitic drug. Ehrlich's Neosalvarsan, with greater water solubility (permitting larger intravenous doses) and less toxicity followed in 1912. In the subsequent five years, the incidence of syphilis dropped by 50% to 80% from its pre-1910 levels in Europe.

Erhlich's early success with drug design and chemotherapy with organic dyes and organoarsenicals prompted an explosion of research in drug design and chemotherapy: suramin in 1917 (still a drug of choice[4] for African sleeping sickness), the antimalarial pamaquin in 1926 (see Table 6-2), and the first sulfonamide antimalarial–antibacterial, Prontosil (sulfamidochrysoidine), in 1932. The development of other clinically useful organoarsenic and antimony antiprotozoal agents is also a direct outcome of Ehrlich's drug design success with arsphenamine (Salvarsan).

STIMULATION OF ANTIMALARIAL RESEARCH BY WAR

Following the pioneering work of Ehrlich and his immediate successors in antimalarial drug design, the necessities of war provided the greatest stimulus to the development of new antimalarials. During 1941 to 1946, for example, over 15,000 substances were synthesized and screened as possible antimalarial agents by the United States, Australia, and Great Britain. Activity increased again during the Vietnam War, especially because of the increasing problem of resistance to commonly used antimalarials. During the decade 1968–1978, over 250,000 compounds were investigated as part of a U.S. Army search program.[22]

MODERN MALARIA CHEMOTHERAPY: AN OVERVIEW

Most drugs used in modern malaria chemotherapy (see Table 6-2)—chloroquine, amodiaquine, pyrimethamine, quinine, and sulfonamides—act primarily at the erythrocytic stage in the malaria life cycle (see Fig. 6-1). Because the severe and life-threatening clinical symptoms of malaria occur at this stage, these drugs are very useful in (1) treating all four human malarias and (2) in preventing clinical symptoms of all four human malarias. However, cures from these drugs can result only with *P. falciparum*. The other three species, *P. vivax*, *P. malariae*, and *P. ovale*, have a "secondary exoerythrocytic" (secondary schizont) stage that can periodically release new merozoites for years or decades. An additional drug that is effective at this stage—usually primaquine—is prescribed by many physicians, when the patient leaves the endemic area, to result finally in a cure for these three species (see Table 6-2). Unfortunately, primaquine can cause hemolytic anemia in patients who are glucose-6-phosphate dehydrogenate (G6PD) deficient.

It obviously would be desirable to have drugs available that would protect humans from initial infection by the mosquito sporozoites (see Fig. 6-1). Unfortunately, no drugs are yet available that are effective at this stage in the life cycle.

Primaquine is quite active at the exoerythrocytic primary schizont stage, so it could be used as a prophylactic against all forms of human malaria. However, its toxicity generally precludes its long-term prophylactic use. Pyrimethamine and chloroguanide are not as effective. Primaquine is also effective as a gametocytocide.

A complicating factor in modern malarial chemotherapy is that drug-resistant strains of *Plasmodium* (especially *P. falciparum*) have been reported in many geographic areas, especially chloroquine-resistant *P. falciparum*. Chloroquine-resistant *P. vivax* has been reported in some areas; quinine-resistant *P. falciparum* in others. Similarly, resistance of certain *Anopheles* species to insecticides is a growing problem.

DEVELOPMENT OF MALARIA VACCINES: PROMISING OR A FAILED PROMISE?

In 1988, results were published on human clinical trials of the first vaccine for protection against the schizont–merozoite stages of malaria.[23] Since the early 1980s, there has been a tremendous effort worldwide to design malaria vaccines, largely based on cell surface proteins of sporozoites, merozoites, or schizonts.[23–36] The use of recombinant DNA techniques to determine the structure of these proteins, and then to direct their syntheses, have been the foundation of most of these studies.

Three primary lines of research include

- Development of sporozoite–merozoite vaccines to block clinical stages of the disease

- Development of sporozoite vaccines to stop infection and spread of the disease
- Development of vaccines that inactivate or block specific metabolic steps in the parasite after infecting humans

Since 1984, the advantages of using vaccines to control diseases of the Third World have received significant attention, although there are major obstacles as highlighted by Robbins and Freeman in 1988.[37] In view of the difficulties of improving sanitation on a wide scale, or often in controlling insect vectors, vaccines offer major advantages. As Hoffman and co-workers noted in their 1986 study on the feasibility of developing a sporozoite vaccine, "There is now little hope that malaria, which affects 100 to 300 million persons . . . per year, can be adequately controlled without vaccines."[33] The enormous costs involved in developing a vaccine, particularly with recombinant DNA techniques, remains the most serious problem. Stability of vaccine products in areas without refrigeration is another concern. The World Health Organization has also expressed concern about the possibility of clinical trials that might not be well planned or developed. Guidelines for clinical trials were therefore drafted.[35]

Most developmental vaccine work with malaria has centered on *P. falciparum*, because it is the primary cause of malaria mortality worldwide. In 1984, the first gene coding the major surface proteins for *P. falciparum* sporozoite was cloned,[36] opening the possibility for a vaccine that would stop the spread of malaria. The general approach was described by Godson in a well-illustrated *Scientific American* article the following year.[11] Oral vaccines developed from radiation-attenuated sporozoites also have been investigated and may possibly be feasible.[29]

More recently, Certa and co-workers[25] have reported that the immunity developed as a result of injection of purified proteins of irradiated merozoites and schizonts may result from antibodies to the malaria parasite's aldolase. (Aldolase catalyzes the reaction of fructose-1,6-diphosphate to glyceraldehyde-3-phosphate and dihydroxyacetone phosphate.) As highlighted by Cox,

> The importance of this enzyme is that the blood stages of the malaria parasite lack a citric acid cycle, and so use vast amounts of glucose . . . . The inhibition of aldolase activity would therefore totally inhibit the maturation of the parasite and subsequent invasion of fresh blood cells . . . . There is [therefore] a real possibility that aldolase could be a basis for a vaccine against many, if not all types of malaria.[24]

Research on malaria vaccines continues, but as summarized in *Science* in early 1990 ("Malaria Vaccines: The Failed Promise"[27]), the challenges are still considerable: "The parasite has shown a surprising immunologic variability, and vaccine strategies that once seemed straightforward have proven frustratingly ineffective in recent years . . . . The sporozoite has acquired the capacity to change its CS (coating) protein in a myriad of ways . . . . The merozoite is [also] extremely variable immunologically . . . . A vaccine remains the likeliest way of preventing entire populations from having a severe attack of malaria. When might such a boom arrive? After the frustrated hopes of the mid-1980's, no one is willing to hazard a guess."

QUINOLINES AND ANALOGUES

All the antimalarial agents in Table 6-3 have one common structural feature—a quinoline ring, or a "quinoline with an additional benzene added" (an *acridine ring*). Some, but not all, have quinine's CH_3O group on the quinoline ring. None except the cinchona alkaloids has a quinuclidine ring.

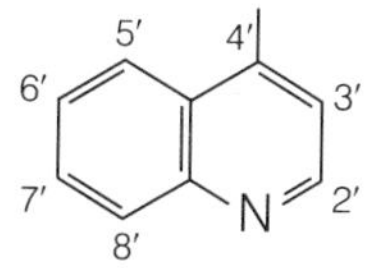

Quinoline 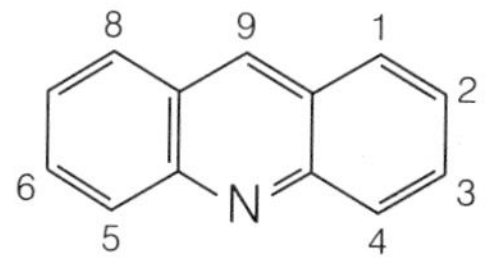 | Acridine 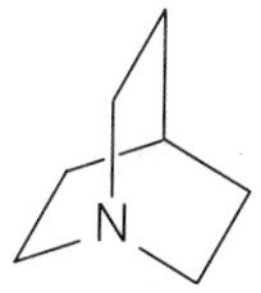 | Quinuclidine

Quinine's structure (see Table 6-3) has been known ever since 1908[38] (and proved by total synthesis by Woodward and Doering in 1945).[39] It was the structural model for all the other quinoline antimalarials shown in Table 6-3, along with the methylene blue–trypan red structures. Changing methylene blue's two amino groups from

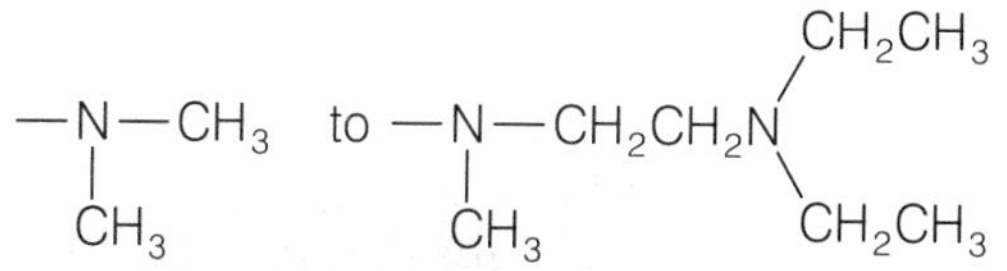

$$-N(CH_3)-CH_3 \quad \text{to} \quad -N(CH_3)-CH_2CH_2N(CH_2CH_3)_2$$

improved the antimalarial activity.[14] The approach used by Schulemann and co-workers in developing pamaquine in 1926[40] (see Fig. 6-2 and Table 6-3) was to synthesize combinations of the 6-methoxy quinoline moiety of quinine with variations of the "improved methylene blue side group" shown above. Later followed the development of all the other *8-aminoquinoline* antimalarials in Table 6-3.

The same general approach with the acridine structure (since methylene blue has three rings and acridine is a quinoline analogue) led to the less toxic quinacrine (see Table 6-3) in 1932.[41,42] Including a

TABLE 6-3

QUINOLINES AND ANALOGUES

CINCHONA ALKALOIDS AND OTHER 4-QUINOLINEMETHANOLS

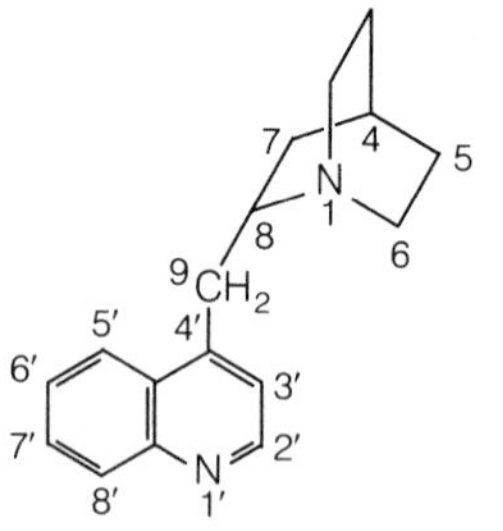

Rubane

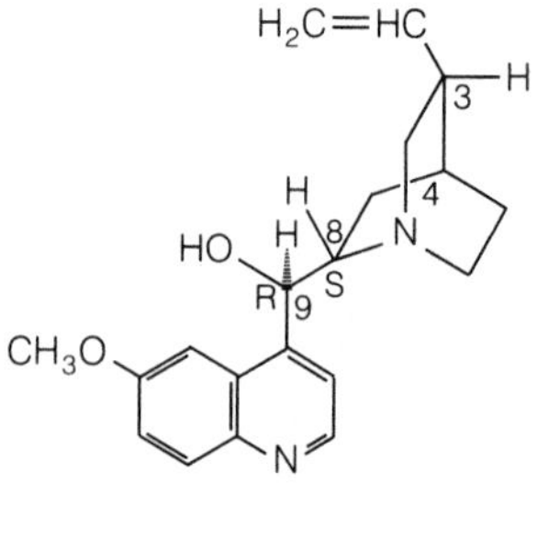

Quinine

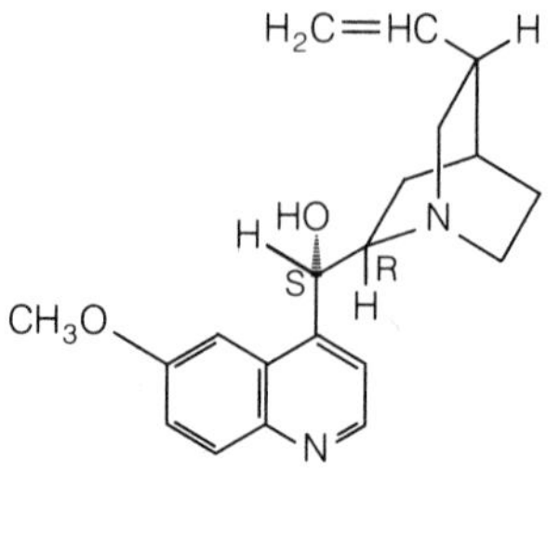

Quinidine

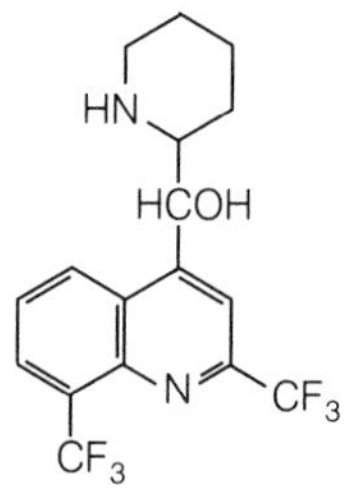

Mefloquine

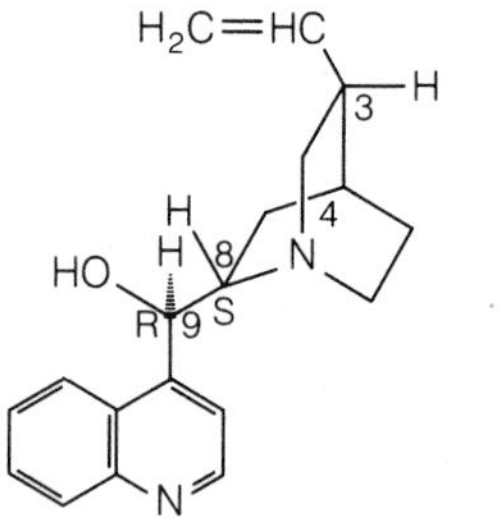

Cinchonidine

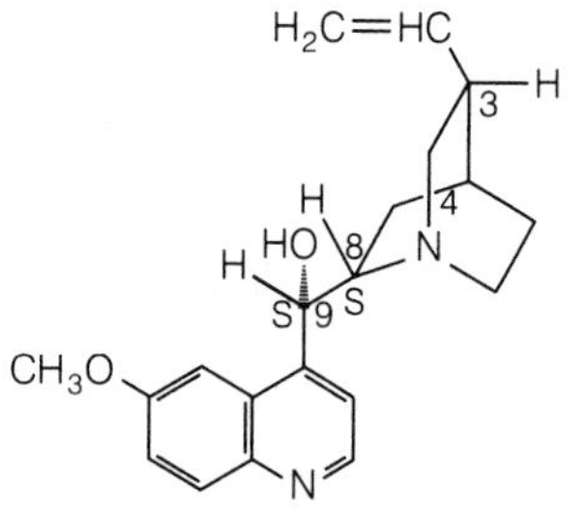

Epiquinine

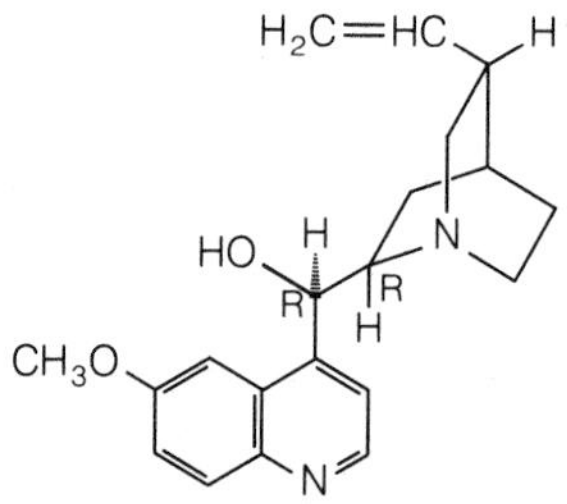

Epiquinidine

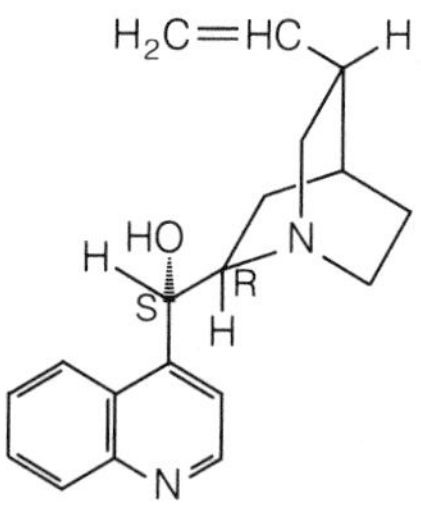

Cinchonine

8-AMINOQUINOLINES

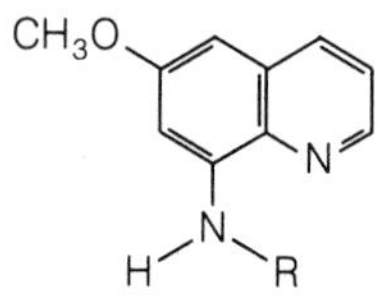

Compound	*R*
Pamaquine	$-CH(CH_3)-CH_2CH_2CH_2-N(C_2H_5)_2$
Primaquine	$-CH(CH_3)-CH_2CH_2CH_2-NH_2$
Pentaquine	$-CH_2CH_2CH_2CH_2CH_2-NH-CH(CH_3)_2$
Isopentaquine	$-CH(CH_3)CH_2CH_2CH_2-NH-CH(CH_3)_2$

TABLE 6-3 *Continued*

QUINOLINES AND ANALOGUES

9-AMINOACRIDINES

Quinacrine Hydrochloride	[structure: 9-aminoacridine bearing CH_3O and Cl, ring $\overset{+}{N}$–H, Cl^-; side chain $H{-}N{-}C(H)(H_3C){-}CH_2CH_2CH_2{-}N^+(H)(C_2H_5)_2$ Cl^-]

7-CHLORO-4-AMINOQUINOLINES

[General structure: 7-chloroquinoline with 4-NH(R_1) and 3-R_2]

Compound	*R_1*	*R_2*
Chloroquine	$-C(CH_3)-CH_2CH_2CH_2-N(C_2H_5)_2$	H
Hydroxychloroquine	$-C(CH_3)-CH_2CH_2CH_2-N(C_2H_4OH)(C_2H_5)$	H
Sontoquine	$-CH(CH_3)-CH_2CH_2CH_2-N(C_2H_5)_2$	CH_3
Amodiaquine	phenyl with –OH and $-CH_2-N(C_2H_5)_2$	H

chlorine atom in the structure was entirely fortuitous, but quinacrine's good activity stimulated the synthesis of many chloroquinolines in the decades that followed. Many other 9-aminoacridines were synthesized in later years, but none was superior to quinacrine.

The next logical step in drug design taken by German chemists was to "divide" the quinacrine structure into its two *4-aminoquinoline* "parts," one of which is chloroquine itself.

Although chloroquine was active, it was considered too toxic, and many other *7-chloro-4-*

[Scheme: Quinacrine (CH_3O, Cl acridine with $NHCH(CH_3)(CH_2)_3N(C_2H_5)_2$) ---→ 6-methoxy-4-aminoquinoline with $NHCH(CH_3)(CH_2)_3N(C_2H_5)_2$ + 7-chloro-4-aminoquinoline with $NHCH(CH_3)(CH_2)_3N(C_2H_5)_2$]

Quinacrine

(2)
Chloroquine

aminoquinolines (see Table 6-3), such as chloroquine, were investigated by the Germans in the late 1930s. Sontoquine (see Table 6-3) was actually used by German soldiers during World War II. After samples of sontoquine were captured by the Allied forces,[43] over 200 7-chloro-4-aminoquinolines were then synthesized in the United States. Hundreds of 9-aminoacridines analogues were also synthesized.

Another drug design approach, begun in the late 1930s, was to make derivatives of *4-quinolinemethanol*—the central part of the quinine structure:

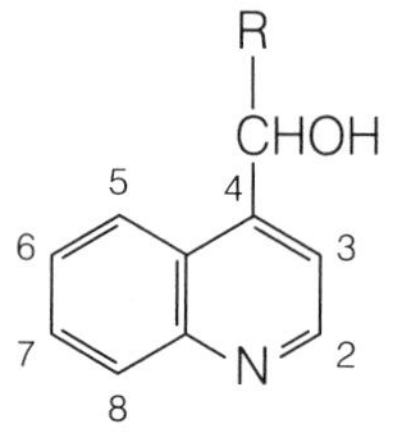

R
CHOH
5 4
6 3
7 N 2
8

Many antimalarials of this structure have been synthesized, with the most promising being mefloquine (Lariam) approved by the FDA in 1989.[46,47] Mefloquine is discussed later in this chapter.

STRUCTURE–ACTIVITY RELATIONSHIPS

Quinine has four asymmetric centers—C-8, C-9, C-3, and C-4. Quinidine is the diastereomer formed by "inverting" C-8 and C-9. (Inversion is used only in the discussion sense. The biosyntheses of the cinchona alkaloids have been elegantly elucidated by Leete.[48]) Cinchonidine is the desmethoxy derivative of quinine, and cinchonine of quinidine. All four of these naturally occurring cinchona alkaloids are active antimalarials (activity varying depending upon the malarial species used in biologic testing). Thus the 6-CH_3O group is not essential for activity, as illustrated also by mefloquine[44–47] and the 7-chloro-4-aminoquinolines (see Table 6-3).

Quinine is found in highest concentration in cinchona bark (about 5%; quinidine, 0.1%; conchonine, 0.3%; cinchonidine, 0.4%), and is the commercially used antimalarial of the group. Quinidine's cardiac effects (see Chap. 14) also preclude its use as an antimalarial. Inversion of only C-9 gives epiquinine and epiquinidine, both of which are inactive.

The good antimalarial activity of the 8-aminoquinolines, quinacrine, and the 7-chloro-4-aminoquinolines shows that the CH_2OH (aminoalcohol) of the cinchona alkaloids is also not essential for activity. Similarly, a chlorine atom on the quinoline ring is not a necessity, but it substantially increases activity for some analogues.

What are the exact structural requirements for good antimalarial activity of the quinoline–quinoline analogue group? Although thousands of quinoline analogues have been synthesized in the more than 50 years since the development of pamaquine in 1926, the question cannot yet be satisfactorily answered. Major progress in development of a quantitative structural model was reported by Kim, Hansch, and co-workers in 1979.[49] The difficulty is compounded because a variety of different malaria protozoan species—many species that do not attack humans—have been used for evaluation of new antimalarials. Furthermore, as discussed in the next section, some quinoline analogues' antimalarial action may not involve a specific receptor.

It is also noteworthy that other aromatic ring systems, besides quinoline and acridine, have been used to make active antimalarials, for example:

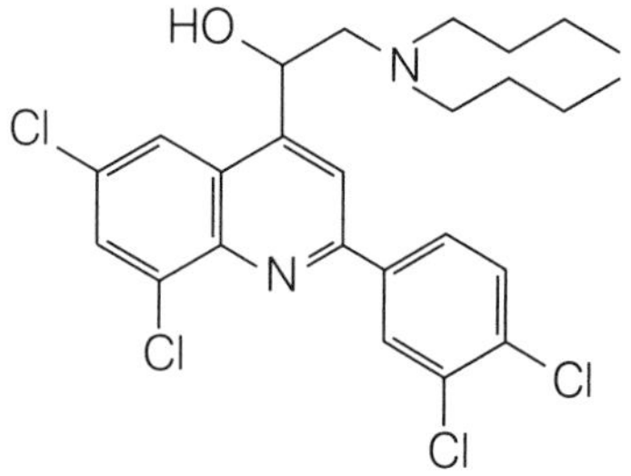

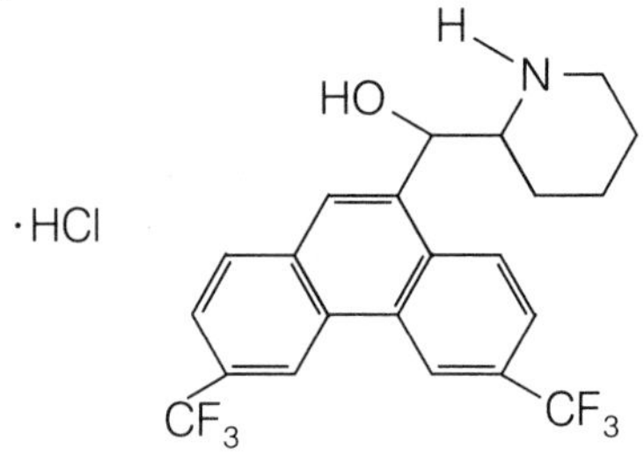

HO N Cl Cl N Cl Cl

H N HO ·HCl CF_3 CF_3

The intensive investigations of quinolines and acridines have revealed useful structure–activity guidelines. The quinoline ring system is more active for a given set of substituents than for the acridine ring system. The dialkylaminoalkylamino $—\underset{R}{N}(CH_2)_nNR_2$ side groups seen throughout compounds in Table 6-3 have been found to provide maximal activity, particularly when n = 2 to 5 or 6. However, as illustrated with the active analogues below, considerable structural variation is possible:

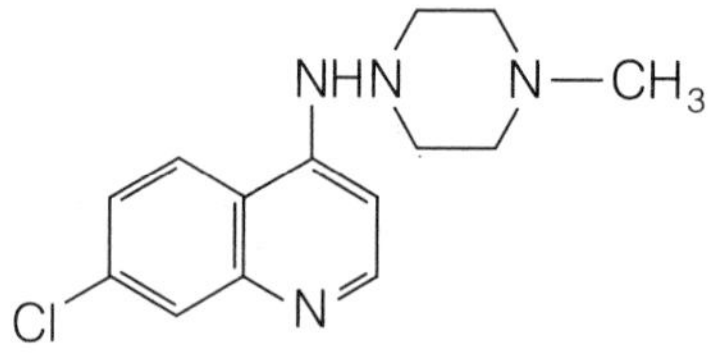

NHN N—CH_3 Cl N

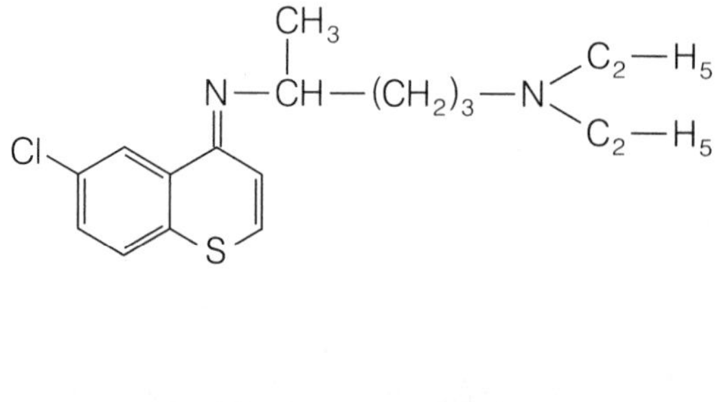

CH_3
N—CH—$(CH_2)_3$—N C_2—H_5 C_2—H_5
Cl S

A chlorine group at C-7 generally provides maximal activity (and toxicity) with chloro-4-aminoquinolines, but some C-6-chloro analogues are equally active. Other relationships have also been found (e.g., with 8-aminoquinolines), 6-CH_3O analogues are more active (and toxic) than 2- or 4-CH_3O analogues.

MECHANISMS OF ACTION

Interest in delineating the mechanism of action of the quinoline antimalarials has intensified as the problem of chloroquine resistance has increased worldwide.[1-10] The recent reviews by Ward[5] and by Peters[2] are recommended for further reading. A wide variety of possible mechanisms has been explored, and many discounted—from intercalation of the aminoquinoline into plasmodial DNA to direct inhibition of protein synthesis.

Two models that have recently attracted considerable attention include

- **Binding of chloroquine to ferriprotoporphyrin IX (FPIX):** FPIX can cause erythrocytes and malaria protozoal cells to lyse, and it can bind to the aminoquinolines with good affinity. However, even after binding to aminoquinolines in vitro, it can still lyse cells. This finding and related studies suggest that this probably, by itself, is not the operative mechanism of action.
- **Chloroquine (pK_a 8.1, 9.9) gets trapped in the malaria lysosome (*p*H 4.8 to 5.2), raises *p*H and, thereby, inactivates hemoglobin-digesting enzymes:** At the acidic *p*H of the lysosome, chloroquine would exist as the salt and, as such, might not easily pass through the lysosomal membrane. The *p*H of the lysozyme would increase, and *p*H-sensitive proteolytic enzymes would not be able to function efficiently. Additionally, it has been suggested that the trapped chloroquine may inactivate the lysosomal proton pump. In fact, about 1000 times more chloroquine accumulates in lysozymes than could be accounted by *p*H or pK_a considerations alone; therefore, other chloroquine-concentrating mechanisms must exist. FPIX binding may be an additional component.

Other investigators have recently proposed that a cytoplasmic proton carrier or permease is inactivated by chloroquine.

MECHANISMS OF RESISTANCE

Resistant strains of malaria protozoa accumulate less chloroquine than do chloroquine-sensitive strains, even though the rate of chloroquine uptake is similar.[1,5,6] Recent work by Van Dyke and Ye[50-52] has shown that drug-resistant malaria protozoa contain increased amounts of a membrane protein that appears to pump drugs out of the protozoa. It seems that the protein may be very similar to the P-glycoprotein that causes multidrug resistance in cancer cells.[53] Verapamil, in addition to being a calcium channel blocker, can reverse drug resistance in malaria, apparently by blocking the P-glycoprotein pump. However, the large doses of verapamil required to reverse the drug resistance of malaria also result in cardiac arrhythmias. A search for selective blockers for the proposed drug efflux pump has led to the intensive study of tetrandrine, a natural product from *Stephania tetrandra*—a plant used for centuries in treating a variety of illnesses in Chinese herbal medicine.

It has also been proposed that resistant strains may metabolize chloroquine at a faster rate, probably related to increased activity of cytochrome P-450.

CINCHONA ALKALOIDS

ABSORPTION, DISTRIBUTION, AND EXCRETION

After oral administration, the cinchona alkaloids are absorbed rapidly and nearly completely, with peak blood level concentrations occurring in one to four hours. About 70% is protein bound. Blood levels fall off very quickly after administration is stopped. A single dose of quinine is disposed of in about 24 hours. Various tissues contain enzymes capable of metabolizing the cinchona alkaloids, but the principal action apparently takes place in the liver, in which an oxidative process results in the addition of a hydroxyl group to the 2′-position of the quinoline ring. The resulting degradation products, called carbostyrils, are much less toxic, are eliminated more rapidly, and possess lower antimalarial activity than the parent compounds. The carbostyrils may be further oxidized to dihydroxy compounds. Excretion is mainly in the urine.

TOXICITY

The toxic reactions to the cinchona alkaloids have been studied extensively. These drugs are not as effective as chloroquine and are more toxic; therefore, they are not generally used, except as noted in Table 6-2 (and in some poor countries). Acute poisoning with quinine is not common. In one case, a death was reported after administration of 18 g; in another case, it was reported that the patient recovered after administration of 19.8 g of quinine. A fatality resulted after the intravenous administration of 1 g of quinine. The toxic manifestations that are most common are due to hypersensitivity to the alkaloids and are referred to collectively as cinchonism. Frequent reactions are allergic skin reactions, tinnitus, slight deafness, vertigo, and slight mental depression. The most serious is amblyopia, which may follow the administration of very large doses of quinine, but is not common; usual therapeutic regimens do not produce this effect.

OTHER ROUTES OF ADMINISTRATION AND DOSAGE FORMS

In addition to antimalarial action, cinchona alkaloids are antipyretic. The action of quinine on the central temperature-regulating mechanism causes peripheral vasodilation. This effect accounts for the traditional use of quinine in cold remedies and fever treatments. Quinine has been used as a diagnostic agent for myasthenia gravis (by accentuating the symptoms). Also, it has been used for the treatment of night cramps or "restless legs." The antifibrillating effect of quinidine in the treatment of cardiac arrhythmias is discussed in Chapter 14.

The antimalarial action of cinchona alkaloids may be obtained by oral, intravenous, or intramuscular administration. Administration by injection, particularly intravenous injection, is not without hazard and should be used cautiously. For intramuscular injection, quinine dihydrochloride is usually used.

Crude extract preparations containing the alkaloids of cinchona have been used widely as economical antimalarials for oral administration. During World War II, a mixture known as quinetum, containing a large amount of quinine, was used. As the interest in pure quinine increased, another crude mixture ("cinchona fibrifuge"), composed of the alkaloids remaining after quinine removal, was introduced to replace quinetum. Subsequently, the Malaria Commission of the League of Nations introduced "totaquina." The *National Formulary* (*NF*) *X* defined it as containing 7% to 12% of anhydrous crystallizable cinchona alkaloids. Totaquine now is the most widely used of inexpensive antimalarial drugs. The usual dose is 600 mg.

Quinine. Quinine is obtained from quinine sulfate prepared by extraction from the crude drug. To obtain it from solutions of quinine sulfate, a solution of the sulfate is alkalinized with ammonia or sodium hydroxide. Another method is to pour an aqueous solution of quinine bisulfate into excess ammonia water, with stirring. In either procedure the precipitated base is washed and recrystallized. The pure alkaloid crystallizes with three molecules of water of crystallization. It is efflorescent, losing one molecule of water at 20°C under normal conditions and losing two molecules in a dry atmosphere. All water is removed at 100°C.

It occurs as a levorotatory, odorless, white crystalline powder, possessing an intensely bitter taste. It is only slightly soluble in water (1 : 1500), but it is quite soluble in alcohol (1 : 1), chloroform (1 : 1), or ether.

It behaves as a diacidic base and forms salts readily. These may be of two types, the *acid* or *bisalts* and the *neutral* salts. The neutral salts are formed by involvement of only the tertiary nitrogen in the quinuclidine nucleus, and the acid salts are the result of involvement of both basic nitrogens. Inasmuch as the quinoline nitrogen is very much less basic than the quinuclidine nitrogen, involvement of both nitrogens results in a definitely acidic compound.

PRODUCTS

Quinine Sulfate, USP. 6′-Methoxycinchonan-9-ol sulfate dihydrate; quininium sulfate. Quinine sulfate is the usual salt of quinine and is ordinarily the "quinine" asked for by the layman.

It is prepared in one of two ways: from the crude bark or from the free base. When prepared from the crude bark, the powdered cinchona is alkalinized and then extracted with a hot, high-boiling petroleum fraction to remove the alkaloids. By carefully adding diluted sulfuric acid to the extract, the alkaloids are converted to sulfates, the sulfate of quinine crystallizing out first. The crude alkaloidal sulfate is decolorized and recrystallized to obtain the article of commerce. Commercial quinine sulfate is not pure, but contains from 2% to 3% of impurities, which consist mainly of hydroquinine and cinchonidine.

To obtain quinine sulfate from the free base, it is neutralized with dilute sulfuric acid. The resulting sulfate, when recrystallized from hot water, forms masses of crystals with the approximate formula $(C_{20}H_{24}O_2N_2)_2 \cdot H_2SO_4 \cdot 8H_2O$. This compound readily effloresces in dry air to the official dihydrate, which occurs as fine, white needles of a somewhat bulky nature.

Quinine sulfate often is prescribed in liquid mixtures. From a taste standpoint, it is better to suspend the salt rather than to dissolve it. However, in the event that a solution is desired, it may be accomplished by the use of alcohol or, more commonly, by addition of a small amount of sulfuric acid to convert it to the more soluble bisulfate. The capsule form of administration is the most satisfactory for masking the taste of quinine when it is to be administered orally.

The sulfate salts of cinchonidine and cinchonine may be used as antimalarials. The dextrorotatory cinchonine salt is of value in the treatment of patients who display a sensitivity to the levorotatory cinchona alkaloids.

7-CHLORO-4-AMINOQUINOLINES

ABSORPTION, DISTRIBUTION, AND EXCRETION

Chloroquine is absorbed readily from the gastrointestinal tract, but amodiaquin gives lower plasma levels than others in the group. Peak plasma concentrations are reached in one to three hours, with blood levels falling off rather rapidly after administration is stopped. About half the drug in the plasma is protein bound. These drugs tend to concentrate in the liver, the spleen, the heart, the kidney, and the brain. Half-life of chloroquine is about three days from a single dose and a week or more following daily dosage for two weeks. Small amounts of 4-aminoquinolines have been found in the skin, but probably not in sufficient quantity to account for their suppressant action on polymorphous light dermatoses. These compounds are excreted rapidly, with most of the unmetabolized drug being accounted for in the urine.

TOXICITY

The toxicity of 4-aminoquinolines is quite low in the usual antimalarial regimen. Side effects can include nausea, vomiting, anorexia, abdominal cramps, diarrhea, headache, dizziness, pruritus, and urticaria. Long-term administration in high doses (for other uses than malaria) may have serious effects on the eyes, and ophthalmologic examinations should be carefully carried out. Also, periodic blood examinations should be made. Patients with liver diseases particularly should be watched when 4-aminoquinolines are used.

OTHER USES, ROUTES OF ADMINISTRATION, AND DOSAGE FORMS

The 4-aminoquinolines, particularly chloroquine and hydroxychloroquine, have been used in the treatment of extraintestinal amebiasis. They are of value in the treatment of chronic discoid lupus erythematosus, but are of questionable value in the treatment of the systemic form of the disease. Symptomatic relief has been secured through the use of 4-aminoquinolines in the treatment of rheumatoid arthritis. A 1979 paper by Marks and Power[54] suggests that the use of chloroquine for this purpose is *not* obsolete. Although the mechanism for their effect in collagen diseases has not been established, these drugs appear to suppress the formation of antigens that may be responsible for hypersensitivity reactions that cause the symptoms to develop. Long-term therapy of at least four to five weeks is usually required before beneficial results are obtained in the treatment of collagen diseases.

For the treatment of malaria, these drugs usually are given orally as salts of the amines in tablet form. If nausea or vomiting occurs after oral administration, intramuscular injection may be used. For prophylactic treatment, the drugs may be incorporated into table salt. To protect the drugs from the high humidity of tropical climates, coating of the granules with a combination of cetyl and stearyl alcohols has been employed. These drugs are sometimes combined with other drugs, such as chloroguanide or pyrimethamine to obtain a broader spectrum of activity (see Table 6-2).

PRODUCTS

Chloroquine, USP. N^4-(7-Chloro-4-quinolinyl)-N^1,N^1-diethyl-1,4-pentanediamine; 7-chloro-4-[[4-(diethylamino)-1-methylbutyl]amino]-quinoline; CQ. Chloroquine occurs as a white or slightly yellow crystalline powder that is odorless and has a bitter taste. It is usually partly hydrated, very slightly soluble in water, and soluble in dilute acids, chloroform, and ether.

Chloroquine Phosphate, USP. N^4-(7-Chloro-4-quinolinyl)-N^1,N^1-diethyl-1,4-pentanediamine phosphate (1 : 2); 7-chloro-4-[[4-(diethylamino)-1-methylbutyl]-amino]quinoline phosphate (Aralen, Resochin). Chloroquine phosphate occurs as a white crystalline powder that is odorless, bitter, and slowly discolors on exposure to light. It is freely soluble in water, and aqueous solutions have a *p*H of about 4.5. It is almost insoluble in alcohol, ether, and chloroform. It exists in two polymorphic forms, either of which (or a mixture of both) may be used medicinally.

Hydroxychloroquine Sulfate, USP. 2-[[4-[(7-Chloro-4-quinolyl)amino]pentyl]ethylamino]ethanol sulfate (1 : 1) (Plaquenil Sulfate). Hydroxychloroquine sulfate occurs as a white, or nearly white, crystalline powder that is odorless, but bitter. It is freely soluble in water, producing solutions with a

pH of about 4.5. It is practically insoluble in alcohol, ether, and chloroform.

Although successful as an antimalarial, hydroxychloroquine has achieved greater use than chloroquine in the control and the treatment of collagen diseases, because it is somewhat less toxic.

Amodiaquine Hydrochloride, USP. 4-[(7-Chloro-4-quinolynyl)amino]-2-[(diethylamino)methyl]phenol dihydrochloride dihydrate; 4-[(7-chloro-4-quinolyl)amino]-α-(diethylamino)-*o*-cresol dihydrochloride dihydrate (Camoquin Hydrochloride). Amodiaquine dihydrochloride occurs as a yellow, odorless, and bitter crystalline powder. It is soluble in water, sparingly soluble in alcohol, and very slightly soluble in ether, chloroform, and benzene. The pH of a 1% solution is between 4 and 4.8. The synthesis of amodiaquine is more expensive than that of chloroquine.

This compound is an economically important antimalarial. Amodiaquine is highly suppressive in *P. vivax* and *P. falciparum* infections, being three to four times as active as quinine. However, it has no curative activity, except against *P. falciparum*. Amodiaquine is altered rapidly in vivo to yield products that appear to be excreted slowly and have a prolonged suppressive activity.

8-AMINOQUINOLINES

ABSORPTION, DISTRIBUTION, AND EXCRETION

The 8-aminoquinolines are absorbed rapidly from the gastrointestinal tract, to the extent of 85% to 95% within two hours after oral administration. Peak plasma concentration is reached within two hours after ingestion, after which the drug rapidly disappears from the blood. The drugs are localized mainly in liver, lung, brain, heart, and muscle tissue. Metabolic changes in the drug are produced very rapidly and, on excretion, metabolic products account for nearly all of the drug. Only about 1% of the drug is eliminated unchanged through the urine. It may be that the antiplasmodial and the toxic properties of these drugs are produced by metabolic transformation products. To maintain therapeutic blood level concentrations, frequent administration of 8-aminoquinolines may be necessary.

TOXICITY

The toxic effects of the 8-aminoquinolines are principally in the central nervous system and the hematopoietic system. Occasionally, anorexia, abdominal pain, vomiting, and cyanosis may be produced. The toxic effects related to the blood system are more common; hemolytic anemia (particularly in dark-skinned people), leukopenia, and methemoglobinemia are the usual findings. A genetic deficiency of glucose-6-phosphate dehydrogenase—found in up to 100 million people,[55] but rarely in Caucasians—weakens erythrocytes and makes them more easily damaged by drugs such as the 8-aminoquinolines. Patients are often tested for G6PD deficiency before primaquine is prescribed. Since primaquine is the drug of choice in preventing relapse of *P. vivax*, *P. ovale*, and *P. malariae* (see Table 6-2), some physicians do not recommend primaquine unless signs of relapse actually occur. For most Caucasians, primaquine is quite nontoxic. Toxicity is increased by quinacrine; therefore, the simultaneous use of quinacrine and 8-aminoquinolines must be avoided.

USES, ROUTES OF ADMINISTRATION, AND DOSAGE FORMS

Primaquine is used mainly to prevent relapses caused by the exoerythrocytic forms of the parasites (see Table 6-1). Primaquine is usually administered orally, in tablet form, as salts such as hydrochlorides or phosphates. Pamaquine is used as the methylene-bis-β-hydroxynaphthoate (naphthoate or pamoate), because this salt is of low solubility and is absorbed slowly and, thus, blood levels are maintained for longer periods and are more uniform.

PRODUCTS

Primaquine Phosphate, USP. N^4-(6-Methoxy-8-quinolinyl)pentanediamine phospate; primaquinium phosphate; 8-[(4-amino-1-methylbutyl)-amino]-6-methoxyquinoline phosphate. Primaquine phosphate is an orange red crystalline substance with a bitter taste. It is soluble in water and insoluble in chloroform and ether. Its aqueous solutions are acid to litmus. It may be noted that it is the primary amine homologue of pamaquine.

Primaquine is the most effective and the best tolerated of the 8-aminoquinolines. Against *P. vivax*, it is four to six times as active an exoerythrocytic schizontocide as pamaquine and about one-half as toxic. When 15 mg of the base are administered daily for 14 days, radical cure is achieved in most *P. vivax* infections. Success has been achieved against some very resistant strains of *P. vivax* by administering 45 mg of the base once a week for eight weeks, with simultaneous administration of

300 mg of chloroquine base. This regimen also tends to lessen the toxic hemolytic effects produced in primaquine-sensitive individuals.

9-AMINOACRIDINES

Quinacrine, the most useful compound in this class, is now rarely used for malaria. A brief discussion on its use has been consolidated in the following discussion.

PRODUCTS

Quinacrine Hydrochloride, USP. N^4-(6-Chloro-2-methoxy-9-acridinyl)-N^1, N^1-diethylpentanediamine dihydrochloride, dihydrate; 6-chloro-9-[[4-(diethylamino)-1-methylbutyl]amino]-2-methoxyacridine dihydrochloride; mepacrine hydrochloride (Atabrine, Atabrin).

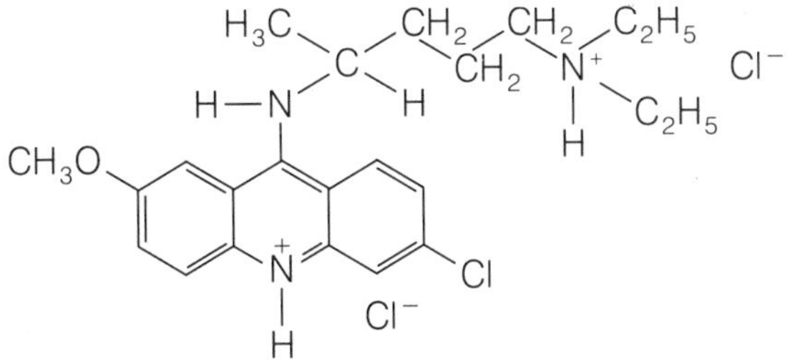

Quinacrine Hydrochloride

The wide use of this compound during the early 1940s resulted in many synonyms for quinacrine in various countries throughout the world.

The dihydrochloride salt is a bitter, yellow crystalline powder. It is sparingly soluble (1 : 35) in water and soluble in alcohol. A 1 : 100 aqueous solution has a pH of about 4.5 and shows a fluorescence. Solutions of the dihydrochloride are not stable and should not be stored. A dimethanesulfonate salt produces somewhat more stable solutions, but they too should not be kept for any length of time.

The yellow color that quinacrine imparts to the urine and the skin is temporary and should not be mistaken for jaundice. Quinacrine may produce toxic effects in the central nervous system, such as headaches, epileptiform convulsions, and transient psychoses that may be accompanied by nausea and vomiting. Hematopoietic disturbances, such as aplastic anemia, may occur. Skin reactions and hepatitis are other symptoms of toxicity. Deaths have occurred from exfoliative dermatitis caused by quinacrine.

As an antimalarial, quinacrine acts as an erythrocytic schizontocide in all kinds of human malaria. It has some effectiveness as a gametocytocide in *P. vivax* and *P. malariae* infections. It may be employed in the treatment of blackwater fever when the use of quinine is contraindicated. It is also an effective curative agent for the treatment of giardiasis caused by *Giardia lamblia*, eliminating the parasite from the intestinal tract. It is an important drug for use in the elimination of intestinal cestodes such as *Taenia saginata* (beef tapeworm), *T. solium* (pork tapeworm), and *Hymenolepis nana* (dwarf tapeworm). Like the 4-aminoquinolines, quinacrine may also be used to treat light-sensitive dermatoses such as chronic discoid lupus erythematosus.

MEFLOQUINE

Mefloquine (Lariam)[44–47] is the most promising of a group of new antimalarial drugs. Since its FDA approval in 1989, mefloquine has become a drug of choice for malaria supression, particularly in areas of known chloroquine-resistant strains. It has also been very effective in curing multidrug-resistant *P. falciparum* malaria, apparently at the erythrocytic stage in the life cycle. Single doses have cured 90 to 100% of patients. However, resistance to mefloquine has already been reported.[47] Primaquine is often recommended after leaving areas with *P. vivax* (see Table 6-2). (As noted earlier in this chapter, primaquine phosphate may cause hemolytic anemia in patients who are G6PD deficient. Patients should usually be tested for this deficiency before primaquine is prescribed.)

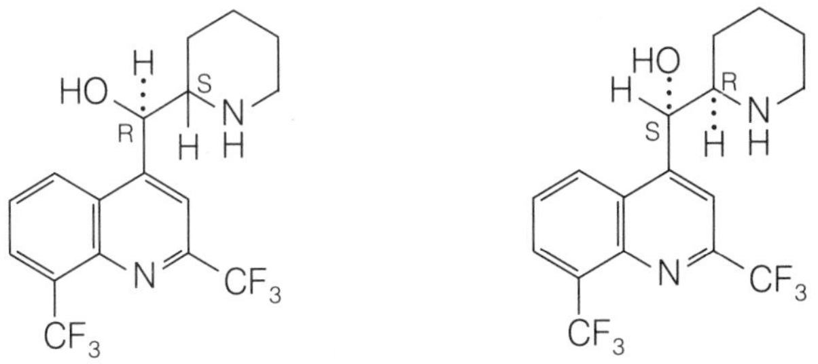

Mefloquine (racemic mixture)

TETRAHYDROFOLATE SYNTHESIS INHIBITORS

USES AND MECHANISMS OF ACTION

The biosyntheses and roles of p-aminobenzoic acid (PABA), folic acid (FA), dihydrofolic acid (FAH_2), tetrahydrofolic acid (FAH_4), and FAH_4 cofactors (such as folinic acid, FAH_4-CHO) have been recently reviewed by Schweitzer et al.[4] and are described in

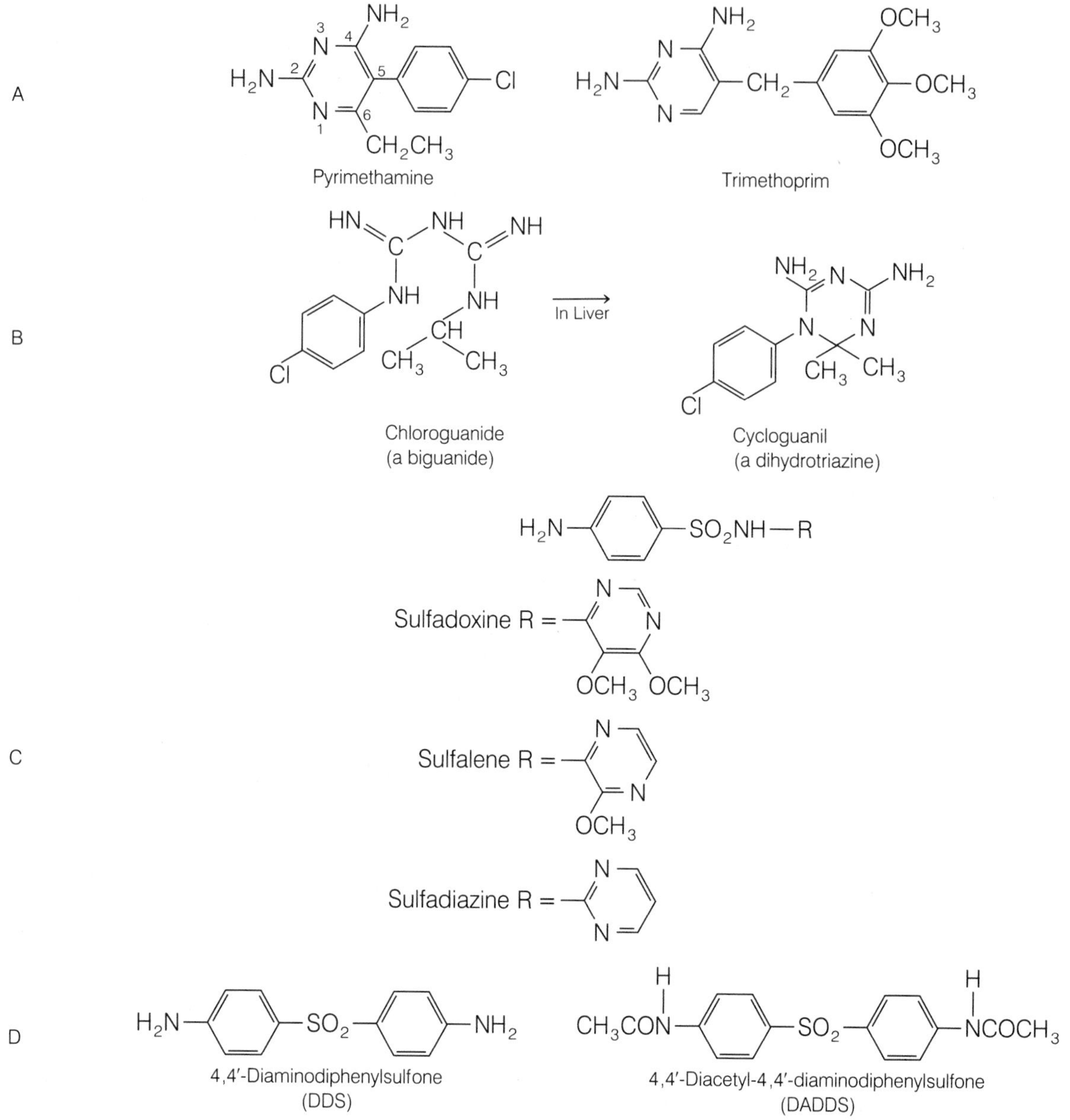

A

Pyrimethamine

Trimethoprim

B

Chloroguanide (a biguanide) —In Liver→ Cycloguanil (a dihydrotriazine)

C

$H_2N-C_6H_4-SO_2NH-R$

Sulfadoxine R =

Sulfalene R =

Sulfadiazine R =

D

4,4′-Diaminodiphenylsulfone (DDS)

4,4′-Diacetyl-4,4′-diaminodiphenylsulfone (DADDS)

FIG. 6-3. *Tetrahydrofolate synthesis inhibitors. (A) Diaminopyrimidines; (B) biguanides and dihydrotriazines; (C) sulfonamides most commonly used for malaria; (D) sulfones.*

Chapter 5. The FAH_4 cofactors are essential as one-carbon donors on several biosynthetic pathways, especially the conversion of uridine to thymidine nucleic acids needed for DNA synthesis. Malaria protozoa cannot convert FA to FAH_4, but can convert FAH_2 to FAH_4.[56]

Ferone and co-workers[57,58] have found that malarial dihydrofolate reductase is structurally different from mammalian dihydrofolate reductase—and up to 2000 times more sensitive to the antimalarial drugs discussed in this section (much less so with trimethoprim). As discussed earlier in this

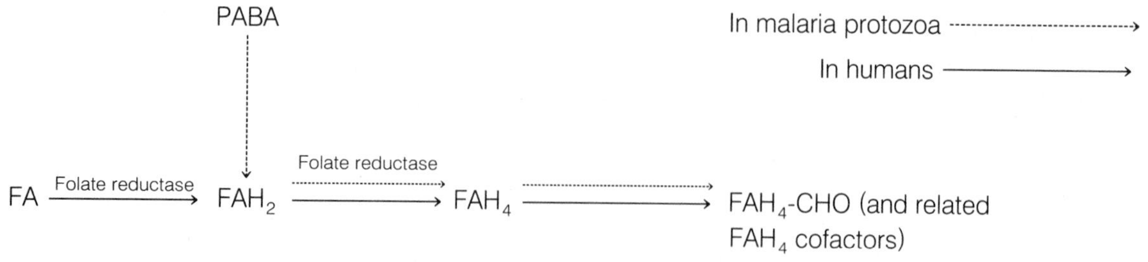

PABA

In malaria protozoa ┈┈┈→

In humans ——→

FA —Folate reductase→ FAH_2 —Folate reductase→ FAH_4 ——→ FAH_4-CHO (and related FAH_4 cofactors)

chapter, the malaria protozoa are also unable to use the host's pyrimidine nucleosides (but can use purine nucleosides) and must synthesize their own. Synthesizing thymidine nucleotides requires FAH_4-CHO. Thus, any drug that can inhibit the malarial protozoa's biosynthesis of FAH_2 or can selectively inhibit the protozoa's dihydrofolate reductase can inhibit the growth of and kill the protozoa.

A variety of structurally different small molecules have been very effective in inhibiting competitively one of the steps involved in the formation of FAH_4 in malaria protozoa (Fig. 6-3). An examination of the drugs of choice (see Table 6-2), however, shows that these drugs are usually reserved for treatment of malaria strains resistant to one or more of the quinoline-type antimalarials. Because the FAH_4 synthesis inhibitors tend to be slow acting, they are often given in combination with a quinoline antimalarial for treatment of acute clinical attacks. Combinations of FAH_4 synthesis inhibitors are often used, but it should be noted that indiscriminate use may lead to resistant strains (as with any antimalarial) of malaria and resistant bacterial strains as well.

The biguanides, diaminopyrimidines, and dihydrotriazines are selective inhibitors of malarial protozoa dihydrofolate reductase. The sulfonamides and sulfones apparently competitively block the incorporation of PABA into the malaria protozoans' dihydropteroic acid (FAH_2)—the same mechanism as with bacteria.

As shown in the malaria life cycle (see Fig. 6-1), pyrimethamine and chloroguanide are effective against both the primary (exoerythrocytic) schizont (site 2) and erythrocytic (site 4) stages. The sulfonamides and sulfones are effective only against the erythrocytic (site 4) stage. Thus, all can cure *P. falciparum* infections, but not *P. vivax*, *P. malarie*, or *P. ovale*. They can suppress clinical symptoms of all four species. Pyrimethamine and chloroguanide both could theoretically be effective as prophylactic agents (because they act at the primary schizont stage), but chloroguanide's slow onset of action and short half-life limit its use for this purpose. Pyrimethamine, however, is an effective prophylactic and excellent clinical suppressive.

The number of FAH_4 synthesis inhibitors in actual clinical use is small, but extensive structure–activity studies have been completed. Interested readers are encouraged to consult a review[14] for detailed information.

DIAMINOPYRIMIDINES

After the observations made in the late 1940s that some 2,4-diaminopyrimidines were capable of interfering with the utilization of folic acid by *Lactobacillus casei*, a property also shown by chloroguanide, these compounds received intensive study as potential antimalarials. It was noted that certain 2,4-diamino-5-phenoxypyrimidines possessed a structural resemblance to chloroguanide, and a series of such compounds was synthesized and found to possess good antimalarial action. Subsequently, a large series of 2,4-diamino-5-phenylpyrimidines was prepared and tested for activity. Maximum activity was obtained when an electron-attracting group was present in the 6-position of the pyrimidine ring and when a chlorine atom was present in the *para* position of the phenyl ring. If the two rings were separated by either an oxygen atom or a carbon atom, antimalarial action decreased. The best in the series of compounds was the one that became known as pyrimethamine.

PRODUCTS

Pyrimethamine, USP. 5-(4-Chlorophenyl)-6-ethyl-2,4-pyrimidinediamine; 2,4-diamino-5-(*p*-chlorophenyl)-6-ethylpyrimidine (Daraprim). Pyrimethamine is an effective erythrocytic schizontocide against all human malarias. It also acts as a primary exoerythrocytic schizontocide in most infections.

Pyrimethamine is slowly, but completely, absorbed from the gastrointestinal tract. It is localized in the liver, the lungs, the kidney, and the spleen and is slowly excreted through the urine, chiefly in metabolized form. A single weekly dose of 25 mg is sufficient for suppression. It is relatively nontoxic, but overdoses may lead to depression of cell growth by inhibition of folic acid activity.

It is administered in the form of the free base, a relatively tasteless powder.

Trimethoprim, USP. 5-[(3,4,5-Trimethoxyphenyl)methyl]-2,4-pyrimidinediamine; 2,4-diamino-5-(3,4,5-trimethoxybenzyl)pyrimidine. Trimethoprim is marketed by itself and in combination with sulfonamides (e.g., in the combination products Bactrim, Septra, and others) primarily as antibacterial products (see Chap. 5). It was developed as an antibacterial agent, but subsequent tests as an antimalarial showed it to be active. Studies by Ferone and co-workers[56–58] have shown trimethoprim to be not as selective for protozoal dihydrofolate reductase as is pyrimethamine. As a result, when used as an antimalarial, it is usually combined with another drug. Its antimalarial effectiveness for human malarias can best be described as mixed. (For a review of the early studies, see the discussion of Peters.[2]) An additional limiting factor in its antimalarial use is trimethoprim's much shorter half-life (about 24 hours) than pyrimethamine.

BIGUANIDES AND DIHYDROTRIAZINES

Although in several malaria species the biguanides have some antimalarial activity, they are largely prodrugs for their active metabolites, the dihydrotriazines. For example, as shown in Figure 6-3, chloroguanide is rapidly metabolized to the potent antimalarial dihydrotriazine, cycloguanil. The dihydrotriazines such as cycloguanil are so rapidly metabolized that, with one exception (cycloguanil pamoate), they are not used for human infections. Cycloguanil pamoate is an intramuscularly injectable depot preparation that can provide antimalarial protection for several months from a single dose.

Numerous biguanide and dihydrotriazine antimalarial agents have been synthesized. Useful structure–activity relationships have been found; for example, substitution of a halogen on the *para* position of the phenyl ring significantly increases activity. Chlorine is used in chloroguanide, but the bromine analogue also is very active. Later, it was observed that a second chlorine added to the 3 position of the phenyl ring of chloroguanide further enhanced activity. However, the dichloro compound, chlorproguanil, is more toxic than chloroguanide itself.

The biguanides are absorbed from the gastrointestinal tract very quickly, but not as rapidly as quinine or chloroquine. They concentrate in the liver, the lungs, the spleen, and the kidney, but appear not to cross the blood–brain barrier. Of the amount in plasma, about 75% is protein bound. They are metabolized mostly in the body and are eliminated very rapidly, principally in the urine. As a result, frequent administration of these drugs is necessary.

The toxic manifestations of biguanides are very mild in humans. Some gastrointestinal disturbances may occur if the drugs are taken on an empty stomach, but not if they are taken after meals. With excessive doses (1 g of chloroguanide), some renal disorders such as hematuria and albuminuria may develop.

PRODUCTS

Chloroguanide Hydrochloride. 1-(*p*-Chlorophenyl)-5-isopropylbiguanide hydrochloride; proguanil hydrochloride (Paludrine) occurs as a white crystalline powder or as colorless crystals that are soluble in water (1 : 75) and alcohol (1 : 30). It is odorless, bitter, and stable in air, but slowly darkens on exposure to light.

Cycloguanil Pamoate. 4,6-Diamino-1-(*p*-chlorophenyl)-1,2-dihydro-2,2-dimethyl-*s*-triazine (2 : 1) with 4,4′-methylenebis[3-hydroxy-2-naphthoic acid]; CI-501; cycloguanil embonate (Camolar). A single intramuscular injection of cycloguanil pamoate can provide protection against all four human malarial species for several months. As with all antimalarials,

TABLE 6-4

ANTIMALARIALS

Name Proprietary Name	***Preparations***	***Usual Adult Dose****	***Usual Dose Range****	***Usual Pediatric Dose****
Quinine sulfate USP	Quinine sulfate capsules USP Quinine sulfate tablets USP	Therapeutic: 200 mg–1 g tid for 6–12 days For chloroquine-resistant *P. falciparum*: 650 mg every 8 hr for at least 3 days with concurrent or consecutive administration of tetracycline or pyrimethamine and sulfadiazine		The following dose is to be divided into 2 or 3 portions and continued for 7–10 days: up to 1 yr: 100–200 mg; 1–3 yr: 200–300 mg; 4–6 yr: 300–500 mg; 7–11 yr: 500 mg–1 g; 12–15 yr: 1–2 g
Chloroquine USP	Chloroquine hydrochloride injection USP	Antiamebic: IM, 200–250 mg qd for 10–12 days	160–800 mg/day	Extraintestinal antiamebic: IM, 15 mg/kg of body weight daily for 2 days, followed by 7.5 mg/kg for at least 2–3 weeks
		Antimalarial: IM, initially 200–250 mg, repeated in 6 hr if necessary; not to exceed 1 g in the first 24 hr		Antimalarial: 6.25 mg/kg of body weight, repeated in 6 hr if necessary. In no instance should a single dose exceed 6.25 mg/kg or the total daily dose exceed 12.5 mg/kg

*Refer to the USP DI for complete dosage information, along with "Antimalarial Drugs of Choice" with doses from *The Medical Letter*.[3]
†Also used as an antiamebic and as a lupus erythematosus suppressant.
‡Also used as a lupus erythematosus suppressant.
§Also used as an anthelmintic (intestinal tapeworms) and as an antiprotozoal (giardiasis).

TABLE 6-4 *Continued*

ANTIMALARIALS

Name Proprietary Name	**Preparations**	**Usual Adult Dose***	**Usual Dose Range***	**Usual Pediatric Dose***
Chloroquine phosphate USP *Aralen, Resochin*	Chloroquine phosphate tablets USP†	Antiamebic: 250 mg qid for 2 days, followed by 250 mg bid for at least 2–3 wk Antimalarial, suppressive: 500 mg once weekly; therapeutic: 1 g initially followed by 500 mg qd on the 2nd and 3rd days Lupus erythematosus suppressant: Dosage not established		Extraintestinal antiamebic: initial, the equivalent of 6 mg/kg of body weight twice daily for 2 days; maintenance, the equivalent of 6 mg/kg of body weight qd Antimalarial, suppressive: 8.3 mg/kg of body weight up to 500 mg once weekly; therapeutic: 41.7 mg/kg of body weight administered over 3 days as follows: 16.7 mg/kg not to exceed a single dose of 1 g, then 8.3 mg/kg not to exceed a single dose of 500 mg 6, 24, and 48 hr later
Hydroxychloroquine sulfate USP *Plaquenil Sulfate*	Hydroxychloroquine sulfate tablets USP‡	Antimalarial, suppressive: 400 mg once every 7 days; therapeutic: initially 800 mg followed by 400 mg in 6–8 hr, and 400 mg qd on the 2nd and 3rd days Lupus erythematosus suppressant: 400 mg qd or bid		Antimalarial, suppressive: 6.4 mg/kg, not to exceed the adult dose, once every 7 days; therapeutic: 32 mg/kg of body weight is administered over a 3-day period as follows: 12.9 mg/kg not to exceed a single dose of 800 mg, then 6.4 mg/kg not to exceed a single dose of 400 mg 6, 24, and 48 hr later
Amodiaquine hydrochloride USP *Camoquin Hydrochloride*	Amodiaquine hydrochloride tablets USP	Suppressive: the equivalent of amodiaquine, 300–600 mg as a single dose once every 7 days; therapeutic: the equivalent of 600 mg of amodiaquine initially, then the equivalent of 300 mg 6, 24 and 48 hr later		
Primaquine phosphate USP	Primaquine phosphate tablets USP	26.3 mg qd for 14 days		
Quinacrine hydrochloride USP *Atabrine Hydrochloride*	Quinacrine hydrochloride tablets USP§	Antiprotozoal (giardiasis): 100 mg tid for 5–7 days Anthelmintic (tapeworms): 200 mg of quinacrine hydrochloride with 650 mg of sodium bicarbonate every 10 min for 4 doses Antimalarial, suppressive: 100 mg qd; therapeutic: 200 mg of quinacrine hydrochloride with 1 g of sodium bicarbonate every 6 hr for 5 doses, then 100 mg tid, up to a maximum total dose of 2.8 g in 7 days	300–900 mg/day	Antiprotozoal (giardiasis): 2.7 mg/kg of body weight or 83.3 mg/m^2 of body surface, tid, up to a maximum of 300 mg/day Anthelmintic (tapeworms): 7.5 mg/kg of body weight or 250 mg/m^2 of body surface every hour for 2 doses, up to a maximum of 800 mg Antimalarial, suppressive: 50 mg qd; therapeutic: children 1–4 yr: 100 mg tid on the 1st day, then 100 mg qd for 6 days; children 4–8 yr: 200 mg tid on the 1st day, then 100 mg bid for 6 days
Chloroguanide hydrochloride *Paludrine*	Tablets	100 mg/day		Children under 2 yr: 25–50 mg/day; 2–6 yr: 50–75 mg/day; 6–10 yr: 100 mg/day
Pyrimethamine USP *Daraprim*	Pyrimethamine tablets USP	Suppressive: 25 mg once weekly; therapeutic: 25–50 mg qd for 2 days	25 mg/wk to 75 mg/day	Suppressive: infants and children under 4 yr: 6.25 mg once weekly; 4–10 yr: 12.5 mg once weekly; over 10 yr: see Usual Dose. Therapeutic: 4–10 yr: 25 mg qd for 2 days

*Refer to the USP DI for complete dosage information, along with "Antimalarial Drugs of Choice" with doses from *The Medical Letter*.[3]
†Also used as an antiamebic and as a lupus erythematosus suppressant.
‡Also used as a lupus erythematosus suppressant.
§Also used as an anthelmintic (intestinal tapeworms) and as an antiprotozoal (giardiasis).

effectiveness is dependent upon the malaria strain not being resistant to the drug or its structural analogues (e.g., chloroguanide). Unfortunately, resistance develops quickly to cycloguanil pamoate, and injection sites can be painful. The problems do not seem to be significantly improved by combinations with other antimalarial drugs.[59]

SULFONAMIDES

As can be seen in Table 6-2, sulfonamides, such as sulfadoxine, are used in antimalarial therapy against drug-resistant malarial strains. They are effective against erythrocytic stages of the malaria protozoa (see Fig. 6-1, site 4). The azo dye Prontosil (see discussion in Chap. 5) was found to have antimalarial activity in vivo against both *P. falciparum* and *P. vivax* in the late 1930s. Later, when it was discovered that Prontosil was a prodrug for sulfanilamide, many other related sulfonamides were investigated as antibacterials and antimalarials.

Medium- or long-acting sulfonamides have been used clinically as antimalarials, particularly sulfadiazine, sulfadoxine, and sulfalene (see Chap. 5). However, each is much more effective when given in combination with pyrimethamine. Trimethoprim combinations have also been investigated.

SULFONES

It has been known for some time that 4,4′-diaminodiphenylsulfone, dapsone USP (DDS), was active against several of the *Plasmodium* species causing malaria.[60] However, it was considered to be an inferior antimalarial drug until it was discovered that it served effectively as a chemoprophylactic agent against chloroquine-resistant *P. falciparum* infections in southeast Asia.

The effectiveness of DDS has prompted the development of programs seeking the synthesis of sulfone compounds of superior activity and with longer duration of action. Among the compounds tested, *N,N*-diacetyl-4,4′-diaminodiphenylsulfone (DADDS) has been promising. Its more prolonged activity and lower toxicity, compared with DDS, are probably related to its slow conversion to either the monoacetyl derivative or DDS itself, both of which act as the antimalarial agents. Long-acting depot combinations of DADDS and cycloguanil pamoate are also under clinical investigation.

OTHER ANTIMALARIALS

The emergence of drug-resistant strains of malaria has prompted a reinvestigation of antibiotics and an intensive investigation into new types of antimalarials.[1–6, 10, 12] Tetracyclines in combination with other antimalarials have been effective against chloroquine-resistant strains of *P. falciparum* (see Table 6-2). As noted earlier, tricyclic antidepressants and Ca^{2+} channel blockers also may offer new approaches to the treatment of chloroquine-resistant strains.

REFERENCES

1. Schmidt, G. D., Roberts, L. S.: Foundation of Parasitology. St. Louis, C. V. Mosby, 1988.
2. Peters, W.: Chemotherapy and Drug Resistance in Malaria, Vols. 1 and 2, 2nd ed. New York, Academic Press, 1987.
3. Med. Lett. 32:23, 1990 and 32:52, 1990.
4. Schweitzer, B. I., Dicker, A. P., Bertino, J. R.: The FASEB Journal 4:2441, 1990.
5. Ward, S. A.: Trends Pharmacol. Sci. 9:241, 1988.
6. Bitonti, A. J., et al.: Science 242:1301, 1988.
7. Gilman, A. G., Rall, T. W., Nies, A. S., Taylor, P. (eds.): The Pharmacological Basis of Therapeutics, 8th ed. New York, Macmillan, 1990.
8. Foye, W. O.: Principles of Medicinal Chemistry. Philadelphia, Lea & Febiger, 1989.
9. Bruce-Chwatt, L. J.: Drugs Exp. Clin. Res. II:899, 1985.
10. Ginger, C. D.: Antiparasitic agents. In Hess, H. J. (ed.). Annual Reports of Medicinal Chemistry, Vol. 1. New York, Academic Press, 1982.
11. Godson, G. N.: Sci. Am. 252:52, May 1985.
12. Banks, B. J.: Antiparasitic agents. In Bailey, D. M. (ed.). Annual Reports in Medicinal Chemistry, Vol. 19. New York, Academic Press, 1984.
13. Donelson, J. E., Turner, M. J.: Sci. Am. 252:44, Feb. 1985.
14. Sweeney, T. R., Strube, R. E.: Antimalarials. In Wolff, M. E. (ed.). Burger's Medicinal Chemistry, Part II, 4th ed. New York, Wiley-Interscience, 1979.
15. Ross, W. J.: Antiamebic agents. In Wolff, M. E. (ed.). Burger's Medicinal Chemistry, Part II, 4th ed. New York, Wiley-Interscience, 1979.
16. Ross, W. J.: Chemotherapy of trypanosomiasis and other protozoan disease. In Wolff, M. E. (ed.). Burger's Medicinal Chemistry, Part II, 4th ed. New York, Wiley-Interscience, 1979.
17. Kreier, J. P. (ed.): Parasitic Protozoa, Vols. 1–4. New York, Academic Press, 1977.
18. Browning, C. H.: Nature 1975:570, 616, 1955.
19. Baumler, E.: In Search of the Magic Bullet, pp. 15–39. London, Thames & Hudson, 1965.
20. Albert, A.: Selective Toxicity, pp. 130–137. London, Chapman & Hall, 1973.
21. "Dr. Erlich's Magic Bullet," 103-min Warner Brothers film, 1940. Available from: United Artists, Los Angeles, CA 90048. A 33-min version, "Magic Bullets," is available from Utah State University, Audio Visual Services, Logan, UT, 84321.
22. Van den Bossche, H.: Nature 273:626, 1978.
23. Patarroyo, M. E., et al.: Nature 332:158, 1988.
24. Cox, F. E. G.: Nature 333:702, 1988.
25. Certa, U., et al.: Science 240:1036, 1988.
26. Sinigaglia, F., et al.: Nature 336:778, 1988.
27. Cherfas, J.: Science 247:402, 1990.
28. Young, J. F., et al.: Microb. Pathogen. 2:237, 1987.
29. Sadoff, J. C., et al.: Science 240:336, 1988.
30. Crisanti, A., et al.: Science 240:1324, 1988.
31. Kaslow, D. C., et al.: Nature 333:74, 1988.
32. Bruce-Chwatt, L. J.: Lancet 1:371, 1987.
33. Hoffman, S. L., et al.: N. Engl. J. Med. 315:601, 1986.
34. Brown, G. V.: Med. J. Aust. 144:703, 1986.
35. McGregor, I.: Parasitol. Today 1:31, 1985.
36. Marx, J. L.: Science 225:607, 1984.
37. Robbins, A., Freeman, P.: Sci. Am. p. 126, Nov 1988.
38. Rabe, P.: Chem. Ber. 41:62, 1908.

39. Woodward, R. B., Von E. Doering, W. J.: Am. Chem. Soc. 67:860, 1945.
40. Schulemann, W.: Proc. R. Soc. Med. 25:897, 1932.
41. Kikuth, W.: Dtsch. Med. Wochenschr. 58:530, 1932.
42. Mauss, H., Mietzsch, F.: Klin. Wochenschr. 12:1276, 1933.
43. Kikuth, W., in Sweeney, T. R., and Strube, R. E.: Antimalarials, p. 151. In Wolff, M. E. (ed.): Burger's Medicinal Chemistry, Part 2, 4th ed. New York, Wiley-Interscience, 1979.
44. Sweeney, T. R.: Med. Res. Rev. 1:281, 1981.
45. Roxman, R. S., Canfield, C. J.: Adv. Pharmacol. 16:1, 1979.
46. FDA Drug Bulletin 19:17, July 1989.
47. Med. Lett. 31:13, 1990.
48. Leete, E.: Account. Chem. Res. 1:59, 1969.
49. Kim, K. H., et al.: J. Med. Chem. 22:366, 1979.
50. Swyers, J. P.: Research Resources Reporter, National Institutes of Health: XIV (2):1–4 (February, 1990).
51. Ye, Z., Van Dyke, K., Spearman, T., Safa, A. T.: Biochem. Biophys. Research Commun. 162:809, 1989.
52. Ye, Z., Van Dyke, K.: Biophys. Research Commun. 159:242, 1989.
53. Kartner, N., Ling, V.: Scientific American 260:44, 1989.
54. Marks, J. S., Power, B. J.: Lancet X:371, 1979.
55. Marks, P. A., Banks, J.: Ann. N.Y. Acad. Sci. 123:198, 1965.
56. Ferone, R., Hitchings, G. H.: J. Protozool. 13:504, 1966.
57. Ferone, R., et al.: Mol. Pharmacol. 5:49, 1969.
58. Ferone, R.: J. Biol. Chem. 245:850, 1970.
59. Clyde, D. F.: J. Trop. Med. Hyg. 72:81, 1969.
60. Powell, R. D., et al.: Int. J. Lepr. 35:590, 1967.

CHAPTER 7

Antibiotics

Arnold R. Martin

HISTORICAL BACKGROUND

Sir Alexander Fleming's accidental discovery of the antibacterial properties of penicillin in 1929[1] is largely credited for initiating the modern antibiotic era. However, it was not until 1938, when Florey and Chain introduced penicillin into therapy, that practical medical exploitation of this important discovery began to be realized. Centuries earlier, man had learned to use crude preparations empirically for the topical treatment of infections, which we now assume to be effective because of the antibiotic substances present. As early as 500 to 600 B.C. molded curd of soybean was used in Chinese folk medicine to treat boils and carbuncles. Moldy cheese had also been employed for centuries by Chinese and Ukrainian peasants to treat infected wounds. The discovery by Pasteur and Joubert, in 1877, that anthrax bacilli were killed when grown in culture in the presence of certain bacteria, along with similar observations by other microbiologists, led Vullemin[2] to define *antibiosis* (literally against life) as the biologic concept of survival of the fittest, in which one organism destroys another to preserve itself. It is from this root that the word *antibiotic* has been derived. The use of the term by the lay public, as well as the medical and scientific community, has become so widespread that its original meaning has become obscured.

In 1942, Waksman[3] proposed the widely cited definition that "an antibiotic or antibiotic substance is a substance produced by microorganisms, which has the capacity of inhibiting the growth and even of destroying other microorganisms." Later proposals[4,5] have sought to both expand and restrict the definition to include any substance produced by a living organism that is capable of inhibiting the growth or survival of one or more species of microorganisms in low concentrations. With the advances made by medicinal chemists to modify naturally occurring antibiotics and to prepare synthetic analogues, it has become necessary to permit the inclusion of semisynthetic and synthetic derivatives in the definition. Therefore, a substance is classified as an antibiotic if

1. It is a product of metabolism (although it may be duplicated or even have been anticipated by chemical synthesis)
2. It is a synthetic product produced as a structural analogue of a naturally occurring antibiotic
3. It antagonizes the growth or the survival of one or more species of microorganisms
4. It is effective in low concentrations.

The isolation of the antibacterial antibiotic tyrocidin from the soil bacterium *Bacillus brevis* by Dubois suggested the probable existence of many antibiotic substances in nature and provided the impetus for the search for them. An organized search of the order *Actinomycetales* led Waksman and his associates to isolate streptomycin from *Streptomyces griseus*. The discovery that this antibiotic possessed in vivo activity against *Mycobacterium tuberculosis*, in addition to numerous species of gram-negative bacilli, was electrifying. It was now evident that soil microorganisms would provide a rich source of antibiotics. Broad screening programs were instituted to find antibiotics that might be effective in the treatment of infections hitherto resistant to existing chemotherapeutic agents, as well as to provide safer and more effective chemotherapy. The discovery of the broad-spectrum antibacterial antibiotics, such as chloramphenicol and the tetracyclines; the antifungal antibiotics, such as nystatin and griseofulvin (see Chap. 4); and the ever-increasing number of antibiotics that may be used to treat infections that have developed resistance to some of the older antibiotics attest to the spectacular success of this approach as it has been applied in research programs throughout the world.

CURRENT STATUS

Commercial and scientific interest in the antibiotic field has led to the isolation and identification of antibiotic substances that may be numbered in the thousands. Numerous semisynthetic and synthetic derivatives have been added to the total. Very few such compounds have found application in general medical practice, however, because in addition to an ability to combat infections or neoplastic disease, an antibiotic must possess other attributes as well. First, it must exhibit sufficient selective toxicity to be decisively effective against pathogenic microorganisms or neoplastic tissue, on the one hand, without causing significant toxic effects, on the other. Second, an antibiotic should possess sufficient chemical stability that it can be isolated, processed, and stored for a reasonable length of time without deterioration of potency. The amenability of an antibiotic for oral or parenteral administration to be converted into suitable dosage forms that will provide active drug in vivo is also important. Finally, the rates of biotransformation and elimination of the antibiotic should be sufficiently slow to allow a convenient dosing schedule, yet be sufficiently rapid and complete to facilitate removal of drug and metabolites from the body soon after administration has been discontinued. Some groups of antibiotics, because of certain unique properties, have been designated for specialized uses, such as the treatment of tuberculosis or fungal infections. Others are employed for cancer chemotherapy. These antibiotics are described along with other drugs of the same therapeutic class. Thus, antifungal and antitubercular antibiotics are discussed in Chapter 4 and antineoplastic antibiotics appear in Chapter 8.

The spectacular success of antibiotics in the treatment of human diseases has prompted the expansion of their use into a number of related fields. Extensive use of their antimicrobial power is made in veterinary medicine. The discovery that low-level administration of antibiotics to meat-producing animals resulted in faster growth, lower mortality, and better quality has led to use of these products as feed supplements. Several antibiotics are being used to control bacterial and fungal diseases of plants. Their use in food preservation is being studied carefully. Indeed, such uses of antibiotics have made necessary careful studies of their long-term effects on humans and their effect on various commercial processes. For example, foods that contain low-level amounts of antibiotics may be capable of producing allergic reactions in hypersensitive persons, or the presence of antibiotics in milk may interfere in the manufacture of cheese.

The success of antibiotics in therapy and related fields has made them one of the most important products of the drug industry today. The quantity of antibiotics produced in the United States each year may now be measured in several millions of pounds and valued at billions of dollars. With research activity stimulated to find new substances to treat viral infections that are now combated with only limited success and with the promising discovery that some antibiotics are active against cancers that may be viral in origin, the future development of more antibiotics and increase in the amounts produced seems to be assured.

COMMERCIAL PRODUCTION

The commercial production of antibiotics for medicinal use follows a general pattern, differing in detail for each antibiotic. The general scheme may be divided into six steps: (1) preparation of a pure culture of the desired organism for use in inoculation of the fermentation medium; (2) fermentation during which the antibiotic is formed; (3) isolation of the antibiotic from the culture medium; (4) purification; (5) assay for potency, tests for sterility, absence of pyrogens, other necessary data; (6) formulation into acceptable and stable dosage forms.

SPECTRUM OF ACTIVITY

The ability of some antibiotics, such as chloramphenicol and the tetracyclines, to antagonize the growth of numerous pathogens has resulted in their being designated as *broad-spectrum* antibiotics. Designations of spectrum of activity are of somewhat limited use to the physician, unless they are based on clinical effectiveness of the antibiotic against specific microorganisms. Many of the broad-spectrum antibiotics are active only in relatively high concentrations against some of the species of microorganisms often included in the "spectrum."

MECHANISMS OF ACTION

The manner in which antibiotics exert their actions against susceptible organisms is varied. The mechanisms of action of some of the more common antibiotics are summarized in Table 7-1. In many instances, the mechanism of action is not fully known; for a few (e.g., penicillins), the site of action is known, but precise details of the mechanism are still under investigation. The biochemical processes of microorganisms are lively subjects for research, for an understanding of those mechanisms that are peculiar to the metabolic systems of infectious organisms is the basis for the future development of modern chemotherapeutic agents. Antibiotics that interfere with those metabolic systems found in microorganisms and not in mammalian cells are the most successful anti-infective agents. For example, those antibiotics that interfere with the synthesis of bacterial cell walls have a high potential for selective toxicity. The fact that some antibiotics structurally resemble some essential metabolites of microorganisms has suggested that competitive antagonism may be the mechanism by which they exert their effects. Thus, cycloserine is believed to be an antimetabolite for D-alanine, a constituent of bacterial cell walls. Many antibiotics selectively interfere with microbial protein synthesis (e.g., the aminoglycosides, the tetracyclines, the macrolides, chloramphenicol, and lincomycin) or nucleic acid synthesis (e.g., rifampin). Others, such as the polymyxins and the polyenes, are believed to interfere with the integrity and function of cell membranes of microorganisms. The mechanism of action of an antibiotic determines, in general, whether the agent exerts a *cidal* or a *static* action. The distinction may be important for the treatment of serious, life-threatening infections, particularly if the natural defense mechanisms of the host are either deficient or overwhelmed by the infection. In such situations, a cidal agent is obviously indicated. Much work remains to be done in this area and, as mechanisms of actions are revealed, the development of improved structural analogues of effective antibiotics will probably continue to increase.

TABLE 7-1

MECHANISMS OF ANTIBIOTIC ACTION

Site of Action	Antibiotic	Process Interrupted	Type of Activity
Cell wall	Bacitracin	Mucopeptide synthesis	Bactericidal
	Cephalosporins	Cell wall cross-linking	Bactericidal
	Cycloserine	Synthesis of cell wall peptides	Bactericidal
	Penicillin	Cell wall cross-linking	Bactericidal
	Vancomycin	Mucopeptide synthesis	Bactericidal
Cell membrane	Amphotericin B	Membrane function	Fungicidal
	Nystatin	Membrane function	Fungicidal
	Polymyxins	Membrane integrity	Bactericidal
Ribosomes	Chloramphenicol	Protein synthesis	Bacteriostatic
50S subunit	Erythromycin	Protein synthesis	Bacteriostatic
	Lincomycins	Protein synthesis	Bacteriostatic
30S subunit	Aminoglycosides	Protein synthesis and fidelity	Bactericidal
	Tetracyclines	Protein synthesis	Bacteriostatic
Nucleic acids	Actinomycin	DNA and mRNA synthesis	Pancidal
	Griseofulvin	Cell division; microtuble assembly	Fungistatic
DNA and/ or RNA	Mitomycin C	DNA synthesis	Pancidal
	Rifampin	mRNA synthesis	Bactericidal

CHEMICAL CLASSIFICATION

The chemistry of antibiotics is so varied that a chemical classification is of limited value. However, it is noteworthy that some similarities can be found, indicating, perhaps, that some antibiotics are the products of similar mechanisms in different organisms and that these structurally similar products may exert their activities in a similar manner. For example, several important antibiotics have in common a macrolide structure, that is, a large lactone ring. In this group are erythromycin and oleandomycin. The tetracycline family presents a group of compounds very closely related chemically. Several compounds contain closely related amino sugar moieties such as those found in streptomycins, kanamycins, neomycins, paromomycins, and gentamicins. The antifungal antibiotics nystatin and the amphotericins (see Chap. 4) are examples of a group of conjugated polyene compounds. The bacitracins, tyrothricin, and polymyxin are among a large group of polypeptides that exhibit antibiotic action. The penicillins and cephalosporins are β-lactam ring-containing antibiotics derived from amino acids.

MICROBIAL RESISTANCE

The normal biologic processes of microbial pathogens are varied and complex. Thus, it seems reasonable to assume that there are many ways in which they may be inhibited and that different microorganisms that elaborate antibiotics antagonistic to a common "foe" produce compounds that are chemically dissimilar and that act on different processes. In fact, nature has produced many chemically different antibiotics that are capable of attacking the same microorganism by different pathways. The diversity of structure in antibiotics has proved to be of real clinical value. As the pathogenic cell is called on to combat the effect of one antibiotic and, thus, develops drug resistance, another antibiotic, attacking another metabolic process of the resisting cell, will deal it a crippling blow. The development of new and different antibiotics has been a very important step in providing the means for treating resistant strains of organisms that previously had been susceptible to an older antibiotic. More recently, the elucidation of biochemical mechanisms of microbial resistance to antibiotics, such as the inactivation of penicillins and cephalosporins by β-lactamase-producing bacteria, has stimulated research in the development of semisynthetic analogues that resist microbial biotransformation. The evolution of nosocomial (hospital-acquired) strains of staphylococci, resistant to penicillin, and of gram-negative bacilli (e.g., *Pseudomonas* and *Klebsiella* sp., *Escherichia coli*, and others), often resistant to several antibiotics, has become a serious medical problem. No doubt, the promiscuous and improper use of antibiotics has contributed to the emergence of resistant bacterial strains. The successful control of diseases caused by resistant strains of bacteria will require not only the development of new and improved antibiotics, but also the rational use of the currently available agents.

β-LACTAM ANTIBIOTICS

Antibiotics that contain the β-lactam (a four-membered cyclic amide) ring structure constitute the dominant class of agents currently employed for the chemotherapy of bacterial infections. The first antibiotic to be used in therapy, penicillin (penicillin G or benzyl penicillin) and a close biosynthetic relative, phenoxymethyl penicillin (penicillin V), remain the agents of choice for the treatment of infections caused by most species of gram-positive bacteria. The discovery of a second major group of β-lactam antibiotics, the cephalosporins, and chemical modifications of naturally occurring penicillins and cephalosporins, have provided semisynthetic derivatives that are variously effective against bacterial species known to be resistant to penicillin, in particular, penicillinase-producing staphylococci and gram-negative bacilli. Thus, apart from a few strains that have either inherent or acquired resistance, almost all bacterial species are sensitive to one or more of the β-lactam antibiotics currently available.

MECHANISM OF ACTION

In addition to a broad spectrum of antibacterial action, two additional properties contribute to the unequaled importance of β-lactam antibiotics in chemotherapy; namely, a potent and rapid cidal action against bacteria in the growth phase and a very low frequency of toxic and other adverse reactions in the host. The uniquely lethal antibacterial action of these agents has been attributed to a selective inhibition of bacterial cell wall synthesis.[6] Specifically, inhibition of the biosynthesis of the dipeptidoglycan that is needed to provide strength and rigidity to the cell wall is the basic mechanism involved. Penicillins and cephalosporins acylate a specific bacterial D-alanine transpeptidase,[7] thereby rendering it inactive for its role in forming peptide cross-links of two linear peptidoglycan strands by transpeptidation and loss of D-alanine. Bacterial D-alanine carboxypeptidases are also inhibited by β-lactam antibiotics.

Binding studies with tritiated benzyl penicillin have shown that the mechanisms of action of various β-lactam antibiotics are much more complex than previously had been assumed. Studies in *E. coli* revealed the existence of as many as seven different functional proteins, each having an important role in cell wall biosynthesis.[8] These penicillin-binding proteins (PBPs) have the following functional properties:

- PBPs 1_a and 1_b are transpeptidases involved in peptidoglycan synthesis associated with cell elongation. Inhibition results in spheroblast formation and rapid cell lysis[8,9] caused by autolysins (bacterial enzymes that function to create nicks in the cell wall for attachment of new peptidoglycan units or for separation of daughter cells during cell division[10]).
- PBP 2 is a transpeptidase involved in maintaining the rod shape of bacilli.[11] Inhibition results in ovoid or round forms that undergo delayed lysis.
- PBP 3 is a transpeptidase required for septum formation during cell division.[9] Inhibition results in the formation of filamentous forms containing rod-shaped units that cannot separate. It is not yet clear if inhibition of PBP 3 is lethal to the bacterium.
- PBPs 4 through 6 are carboxypeptidases responsible for the hydrolysis of D-alanine–D-alanine terminal peptide bonds of the cross-linking peptides. Inhibition of these enzymes is apparently not lethal to the bacterium,[12] despite the fact that cleavage of the terminal D-alanine bond is required before peptide cross-linkage.

The various β-lactam antibiotics differ in their affinities for PBPs. Penicillin G binds preferentially to PBP 3, whereas the first-generation cephalosporins bind with higher affinity to PBP 1_a. In contrast to other penicillins and to cephalosporins, which can bind to PBPs 1, 2, and 3, amdinocillin binds only to PBP 2.

THE PENICILLINS

Commercial Production and Unitage

Until 1944, it was assumed that the active principle in penicillin was a single substance and that variation in activity of different products was due to the amount of inert materials in the samples. Now it is known that during the biologic elaboration of the antibiotic several closely related compounds may be produced. These compounds differ chemically in the acid moiety of the amide side chain. Variations in this moiety produce differences in antibiotic effect and in physicochemical properties, including stability. Thus, it has become proper to speak of penicillins, referring to a group of compounds, and to identify each of the penicillins specifically. As each of the different penicillins was first isolated, letter designations were used in America; the British used Roman numerals.

Over 30 penicillins have been isolated from fermentation mixtures. Some of these occur naturally; others have been biosynthesized by altering the culture medium to provide certain precursors that may be incorporated as acyl groups. Commercial production of biosynthetic penicillins today depends chiefly on various strains of *Penicillium notatum* and *P. chrysogenum*. In recent years, many more penicillins have been prepared semisynthetically and, undoubtedly, many more will be added to the list in attempts to find superior products.

Because penicillin, when it was first used in chemotherapy, was not a pure compound and exhibited varying activity among samples, it was necessary to evaluate it by microbiologic assay. The procedure for assay was developed at Oxford, England, and the value became known as the Oxford unit: 1 *Oxford unit* is defined as the smallest amount of penicillin that will inhibit, in vitro, the growth of a strain of *Staphylococcus* in 50 mL of culture medium under specified conditions. Now that pure crystalline penicillin is available, the *United States Pharmacopoeia* (*USP*) defines *unit* as the antibiotic activity of 0.6 μg of USP Penicillin G Sodium Reference Standard. The weight–unit relationship of the penicillins will vary with the nature of the acyl substituent and with the salt formed of the free acid: 1 mg of penicillin G sodium is equivalent to 1667 units; 1 mg of penicillin G procaine is equivalent to 1009 units; 1 mg of penicillin G potassium is equivalent to 1530 units.

The commercial production of penicillin has increased markedly since its introduction. As production increased, the cost of penicillin dropped correspondingly. When penicillin was first available, 100,000 units sold for 20 dollars. Currently, the same quantity costs less than a penny. Fluctuations in the production of penicillins through the years have reflected changes in the popularity of broad-spectrum antibiotics compared with penicillins, the development of penicillin-resistant strains of several pathogens, the more recent introduction of semisynthetic penicillins, the use of penicillins in animal feeds and for veterinary purposes, and the increase in marketing problems in a highly competitive sales area.

Table 7-2 shows the general structure of the penicillins and relates the structures of the more familiar ones to their various designations.

TABLE 7-2
STRUCTURE OF PENICILLINS

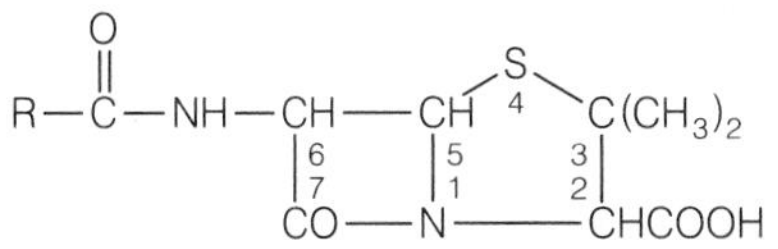

```
      O
      ||             S
R—C—NH—CH——CH  4  C(CH3)2
          |6     |5    3|
          |7     |1    2|
          CO——N———CHCOOH
```

Generic Name	Chemical Name	R Group
Penicillin G	Benzylpenicillin	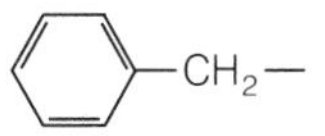 —CH_2—
Penicillin V	Phenoxymethylpenicillin	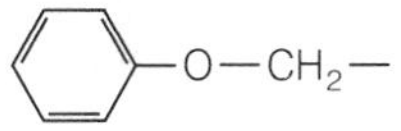 —O—CH_2—
Methicillin	2,6-Dimethoxyphenylpenicillin	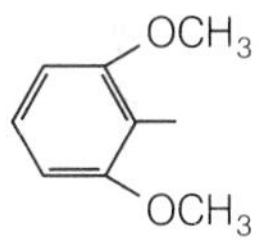 OCH_3, OCH_3
Nafcillin	2-Ethoxy-l-naphthylpenicillin	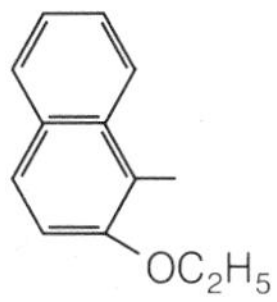 OC_2H_5
Oxacillin	5-Methyl-3-phenyl-4-isoxazolylpenicillin	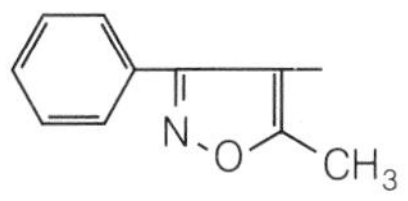 N, O, CH_3
Cloxacillin	5-Methyl-3-(2-chlorophenyl)-4-isoxazolylpenicillin	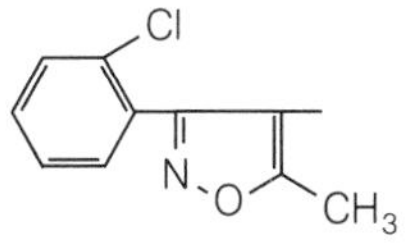 Cl, N, O, CH_3
Dicloxacillin	5-Methyl-3-(2,6-dichlorophenyl)-4-isoxazolylpenicillin	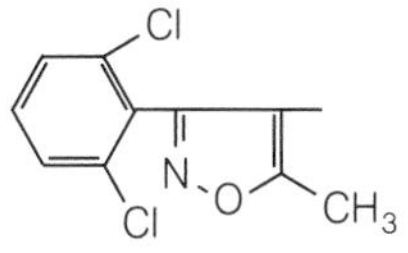 Cl, Cl, N, O, CH_3

TABLE 7-2 *Continued*

STRUCTURE OF PENICILLINS

R—C(=O)—NH—CH(6)—CH(5)(—S—4)—C(3)(CH_3)$_2$; CO(7)—N(1)—CHCOOH(2)

Generic Name	*Chemical Name*	*R Group*
Ampicillin	D-α-Aminobenzylpenicillin	phenyl—CH(NH_2)—
Amoxicillin	D-α-Amino-*p*-hydroxybenzylpenicillin	HO—phenyl—CH(NH_2)—
Cyclacillin	1-Aminocyclohexylpenicillin	cyclohexyl(NH_2)
Carbenicillin	α-Carboxybenzylpenicillin	phenyl—CH(CO_2H)—
Ticarcillin	α-Carboxy-3-thienylpenicillin	3-thienyl—CH(CO_2H)—
Piperacillin	α-(4-Ethyl-2,3-dioxo-1-piperazinylcarbonylamino)benzylpenicillin	phenyl—CH(NH—C=O—N(2,3-dioxopiperazine)—N—CH_2CH_3)—
Azlocillin	α-(2-oxoimidazolidinocarbonylamino)benzylpenicillin	phenyl—CH(NH—C=O—N(2-oxoimidazolidine)NH)—
Mezlocillin	α-(1-methanesulfonyl-2-oxoimidazolidinocarbonylamino)benzylpenicillin	phenyl—CH(NH—C=O—N(2-oxoimidazolidine)N—SO_2CH_3)—

Nomenclature

The systematic nomenclature of penicillins is somewhat complex and very cumbersome. Two numbering systems for the fused bicyclic heterocyclic system are in existence. The *Chemical Abstracts* system initiates the numbering with the sulfur atom and assigns the ring nitrogen as the four position. Thus, penicillins are named as 4-thia-1-azabicyclo[3.2.0]heptanes, according to this system. The numbering system adopted by the *USP* is the reverse of the *Chemical Abstracts* procedure, assigning the nitrogen atom as atom number 1 and the sulfur atom as number 4. Three simplified forms of penicillin nomenclature have been adopted for general use. One utilizes the name *penam* for the unsubstituted bicyclic system, including the amide carbonyl group, with one of the foregoing numbering systems as just described. Thus, penicillins are generally designated according to the *Chemical Abstracts* system as 5-acylamino-2,2-dimethylpenam-3-carboxylic acids. The second, seen more frequently in the medical literature, uses the name *penicillanic acid* to describe the ring system with substituents that are generally present (i.e., 2,2-dimethyl and 3-carboxyl). A third form followed in this chapter uses trivial nomenclature to name the entire 6-carbonylamino-penicillanic acid portion of the molecule as penicillin and then distinguishes compounds on the basis of the R group of the acyl portion of the molecule. Penicillin G is thus named benzylpenicillin, penicillin V is phenoxymethylpenicillin, methicillin is 2,6-dimethoxyphenylpenicillin and so on. For the most part, the latter two systems serve well for naming and comparing closely similar penicillin structures, but they are too restrictive to be applied to compounds with unusual substituents or to ring-modified derivatives.

Stereochemistry

The penicillin molecule contains three asymmetric carbon atoms (i.e., C-3, C-5, and C-6). All naturally occurring and microbiologically active synthetic and semisynthetic penicillins have the same absolute configuration about these three centers. The carbon atom bearing the acylamino group (C-6) has the L-configuration, whereas the carbon to which the carboxyl group is attached has the D-configuration. Thus, the acylamino and carboxyl groups are *trans* to each other, with the former in the α- and the latter in the β-orientation relative to the penam ring system. The atoms comprising the 6-aminopenicillanic acid portion of the structure are biosynthetically derived from two amino acids, L-cysteine (S-1, C-5, C-6, C-7, and 6-amino) and L-valine (2,2-

dimethyl, C-2, C-3, and 3-carboxyl). The absolute stereochemistry of the penicillins is designated as 3S : 5R : 6R as shown below.

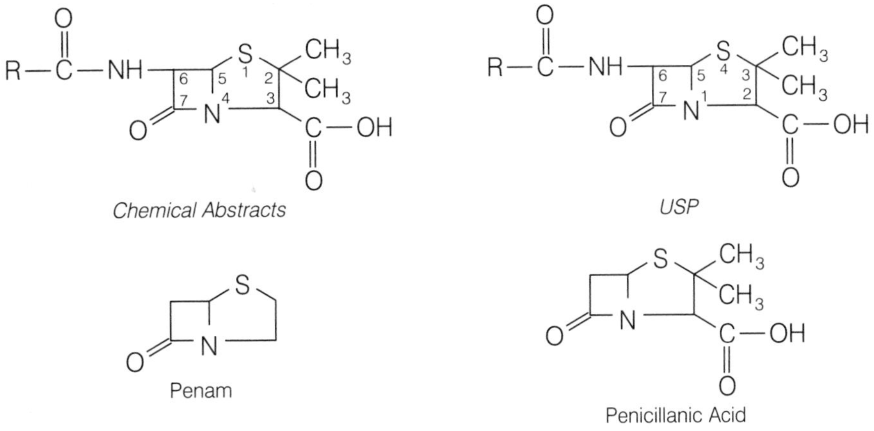

Chemical Abstracts

USP

Penam

Penicillanic Acid

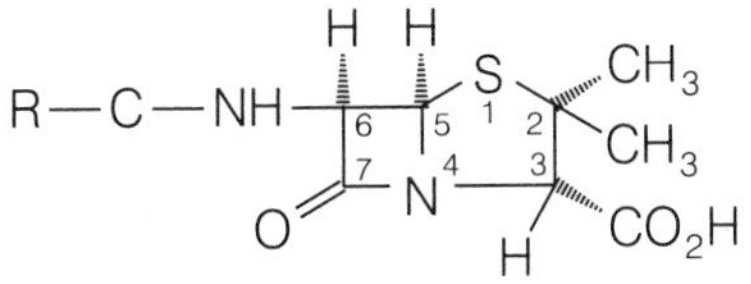

Synthesis

Examination of the structure of the penicillin molecule shows it to contain a fused ring system of unusual design, the β-lactam thiazolidine structure. The nature of the β-lactam ring delayed the elucidation of the structure of penicillin, but its determination was reached as a result of a collaborative research program involving research groups in Great Britain and the United States during the years 1943 to 1945.[13] Attempts to synthesize these compounds resulted, at best, in only trace amounts until Sheehan and Henery-Logan[14] adapted techniques developed in peptide syntheses to the synthesis of penicillin V. This procedure is not likely to replace the established fermentation processes because the last step in the reaction series develops only 10% to 12% of penicillin. It is of advantage in research because it provides a means of obtaining many new amide chains, hitherto not possible to achieve by biosynthetic procedures.

Two other developments have provided additional means for making new penicillins. A group of British scientists, Batchelor et al.,[15] have reported the isolation of 6-aminopenicillanic acid from a culture of *P. chrysogenum*. This compound can be converted to penicillins by acylation of the 6-amino group. Sheehan and Ferris[16] provided another route to synthetic penicillins by converting a natural penicillin, such as penicillin G potassium, to an intermediate (Fig. 7-1) from which the acyl side chain has been removed, which then can be treated to form biologically active penicillins with a variety of new side chains. By these procedures, new penicillins,

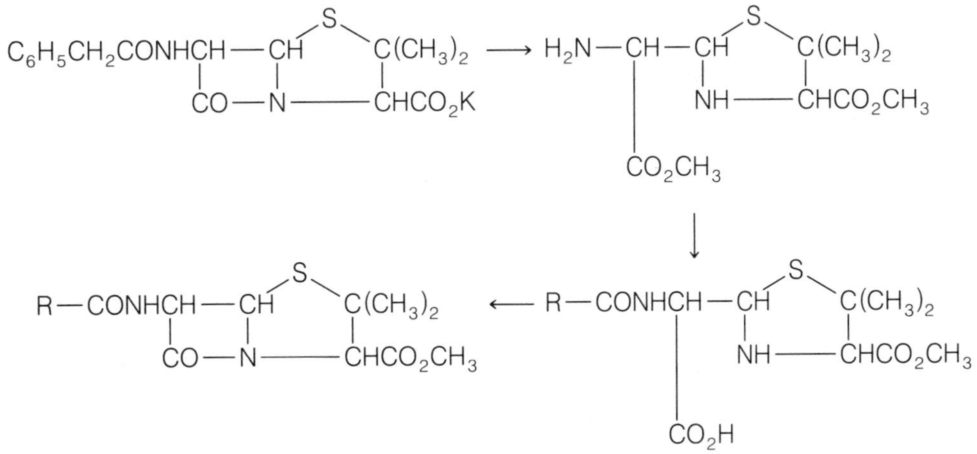

FIG. 7-1. *Conversion of natural penicillin to synthetic penicillin.*

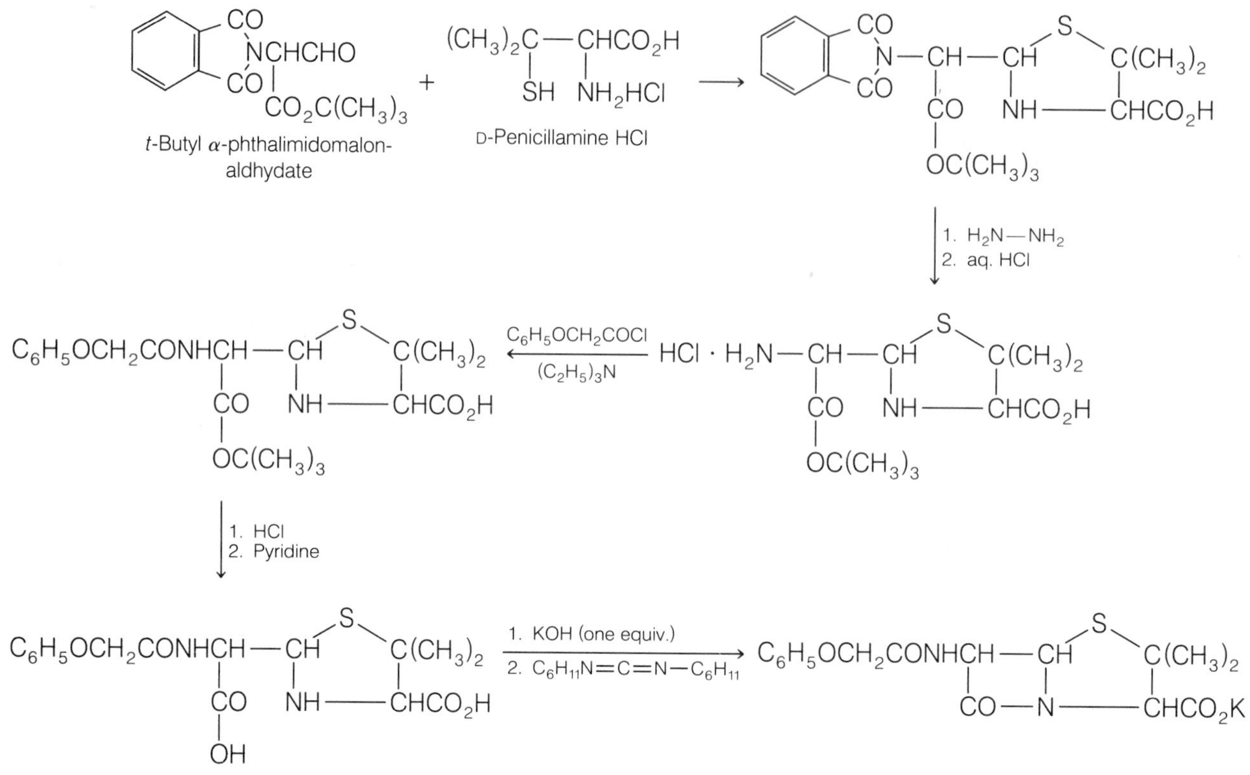

t-Butyl α-phthalimidomalonaldhydate

D-Penicillamine HCl

FIG. 7-2. *Synthesis of phenoxymethylpenicillin.*

superior in activity and stability to those formerly in wide use, have been found and, no doubt, others will be produced. The first commercial products of these research activities were phenoxyethylpenicillin (phenethicillin) (Fig. 7-2) and dimethoxyphenylpenicillin (methicillin).

Chemical Degradation

The early commercial penicillin was a yellow-to-brown amorphous powder that was so unstable that refrigeration was required to maintain a reasonable level of activity for a short time. Improved procedures for purification provide a white crystalline material in use today. The crystalline penicillin must be protected from moisture, but, when kept dry, the salts will remain stable for years without refrigeration. Many penicillins have an unpleasant taste, which must be overcome in the formation of pediatric dosage forms. All of the natural penicillins are strongly dextrorotatory. The solubility and other physicochemical properties of the penicillins are affected by the nature of the acyl side chain and by the cations used to make salts of the acid. Most penicillins are acids with pK_as in the range of 2.5 to 3.0, but some are amphoteric. The free acids are not suitable for oral or parenteral administration. However, the sodium and potassium salts of most penicillins are soluble in water and are readily absorbed orally or parenterally. Salts of penicillins with organic bases, such as benzathine, procaine, and hydrabamine, have limited water solubility and are, therefore, useful as depot forms to provide effective blood levels over a long period in the treatment of chronic infections. Some of the crystalline salts of the penicillins are hygroscopic, making it necessary to store them in sealed containers.

The main cause of deterioration of penicillin is the reactivity of the strained lactam ring, particularly to hydrolysis. The course of the hydrolysis and the nature of the degradation products are influenced by the pH of the solution.[17,18] Thus, the β-lactam carbonyl group of penicillin readily undergoes nucleophilic attack by water or (especially) hydroxide ion to form the inactive penicilloic acid, which is reasonably stable in neutral to alkaline solutions, but readily undergoes decarboxylation and further hydrolysis reactions in acidic solutions. Other nucleophiles, such as hydroxylamine, alkylamines, and alcohols react to open the β-lactam ring to form the corresponding hydroxyamic acids, amides, and esters. It has been speculated[19] that one of the causes of penicillin allergy may be the formation of antigenic penicilloyl proteins formed in vivo by the reaction of nucleophilic groups (e.g., ε-amino) on specific body proteins with the β-lactam carbonyl group. In strongly acidic solutions (pH < 3) penicillin undergoes a complex series of reactions leading to a variety of inactive degradation products

(Fig. 7-3).[18] The first step appears to involve rearrangement to the penicillenic acid. This process is initiated by protonation of the β-lactam nitrogen followed by nucleophilic attack of the acyl oxygen atom on the β-lactam carbonyl carbon. The subsequent opening of the β-lactam ring destabilizes the thiazoline ring, which then also suffers acid-catalyzed ring opening to form the penicillenic acid. The latter is very unstable and experiences two major degradation pathways. The most easily understood path involves the hydrolysis of the oxazolone ring to form the unstable penamaldic acid. Because it is an enamine, penamaldic acid easily hydrolyzes to penicillamine (a major degradation product) and the penaldic acid. The second path involves a complex rearrangement of the penicillenic acid to a penillic acid through a series of intramolecular processes that remain to be completely elucidated. Penillic acid (an imidazoline-2-carboxylic acid) readily decarboxylates and also suffers hydrolytic ring opening under acidic conditions to form a second major end product of acid-catalyzed penicillin degradation, the penilloic acid. The penicilloic acid, the major product formed under weakly acidic to alkaline (as well as enzymatic) hydrolytic conditions, cannot be detected as an intermediate under strongly acidic conditions. However, it is known to exist in equilibrium with the penamaldic acid and to undergo decarboxylation to form the penilloic acid in acid. The third major final product of the degradation is penicilloaldehyde, formed by decarboxylation of penaldic acid (a derivative of malonaldehyde).

By controlling the pH of aqueous solutions within a range of 6.0 to 6.8, and by refrigeration of the solutions, aqueous preparation of the soluble penicillins may be stored for periods up to several weeks. The relationship of these properties to the pharmaceutics of penicillins has been reviewed by Schwartz and Buckwalter.[20] It has been noted that some buffer systems, particularly phosphates and citrates, exert

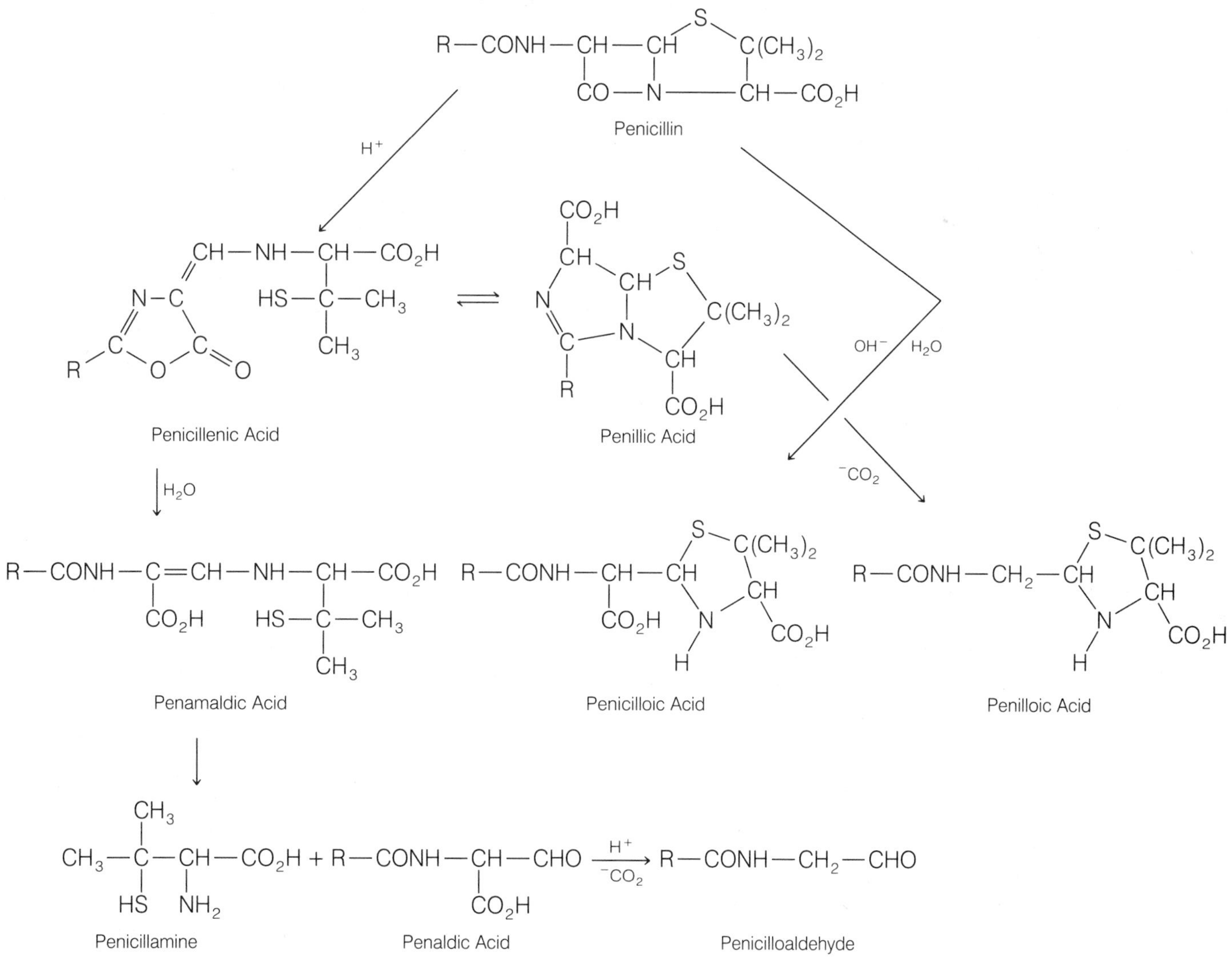

FIG. 7-3. *Degradation of penicillin.*

a favorable effect on penicillin stability, independently of the pH effect. However, Finholt et al.[21] have shown that these buffers may catalyze penicillin degradation if the pH is adjusted to obtain the requisite ions. Hydroalcoholic solutions of penicillin G potassium show about the same degree of instability as do aqueous solutions.[22] Because penicillins are inactivated by metal ions, such as zinc and copper, it has been suggested that the phosphates and the citrates combine with these metals to prevent their existing as ions in solution.

Oxidizing agents also inactivate penicillins, but reducing agents have little effect on them. Temperature affects the rate of deterioration; although the dry salts are stable at room temperature and do not require refrigeration, prolonged heating will inactivate the penicillins.

Acid-catalyzed degradation in the stomach contributes in a major way to the poor oral absorption of penicillin. Thus, efforts to obtain penicillins with improved pharmacokinetic and microbiologic properties have sought to find acyl functionalities that would minimize sensitivity of the β-lactam ring to acid hydrolysis and, at the same time, maintain antibacterial activity. Substitution of an electron-withdrawing group in the α-position of benzylpenicillin has been shown to markedly stabilize the penicillin to acid-catalyzed hydrolysis.[23] Thus, the phenoxymethyl-, α-aminobenzyl-, and α-halobenzylpenicillins are significantly more stable than benzylpenicillin in acid solutions. The increased stability imparted by such electron-withdrawing groups has been attributed to a decrease in reactivity (nucleophilicity) of the side chain amide carbonyl oxygen atom toward participation in β-lactam ring opening to form the penicillenic acid. Obviously, α-aminobenzylpenicillin (ampicillin) exists as the protonated form in acidic (as well as neutral) solutions, and the ammonium group is known to be powerfully electron withdrawing.

Bacterial Resistance

Some bacteria, in particular most species of gram-negative bacilli, are naturally resistant to the action of penicillins. Other normally sensitive species are capable of developing penicillin resistance (either through natural selection of resistant individuals or through mutation). The best understood, and probably the most important, biochemical mechanism of penicillin resistance is the bacterial elaboration of enzymes that inactivate penicillins. Such enzymes, which have been given the nonspecific name penicillinases, are of two general types: β-lactamases and acylases. By far the most important of these are the β-lactamases, enzymes that catalyze the hydrolytic opening of the β-lactam ring of penicillins to produce inactive penicilloic acids. Synthesis of bacterial β-lactamases may be under chromosomal or plasmid R-factor control and may be either constitutive or inducible (stimulated by the presence of the substrate), depending upon the bacterial species. The well-known resistance among strains of *Staphylococcus aureus* is apparently entirely due to the production of an inducible β-lactamase. Resistance among gram-negative bacilli, on the other hand, may be a result of other, poorly characterized "resistance factors" or of constitutive β-lactamase elaboration. β-Lactamases produced by gram-negative bacilli appear to be cytoplasmic enzymes that remain in the bacterial cell, whereas those elaborated by *Staph. aureus* are synthesized in the cell wall and released extracellularly. β-Lactamases from different bacterial species may be distinguished by their substrate and inhibitor specificities, their physical properties (pH optimum, isoelectric point, molecular weight, and such), and their immunologic properties.[24]

Specific acylases, enzymes that are capable of hydrolyzing the acylamino side chain of penicillins, have been obtained from several species of gram-negative bacteria, but their possible role in bacterial resistance has not been well defined. These enzymes find some commercial use in the preparation of 6-aminopenicillanic acid (6-APA) for the preparation of semisynthetic penicillins. The 6-APA is less active and more rapidly hydrolyzed (enzymatically and nonenzymatically) than is penicillin.

Another important resistance mechanism, especially in gram-negative bacteria, is decreased permeability to penicillins. The cell envelope in most gram-negative bacteria[25] is more complex than it is in gram-positive bacteria. It contains an outer membrane (linked by lipoprotein bridges to the peptidoglycan cell wall), not present in gram-positive bacteria, which creates a physical barrier to the penetration of antibiotics, especially those that are hydrophobic.[26] Small hydrophilic molecules, on the other hand, can traverse the outer membrane through pores formed by proteins called porins.[25] Alteration of the number or nature of porins in the cell envelope[27] could also be an important mechanism of antibiotic resistance. Bacterial resistance can also result from changes in the affinity of PBPs for penicillins.[28] Altered PBP binding has been demonstrated in non–β-lactamase-producing strains of penicillin-resistant *Neisseria gonorrhoeae*[29] and methicillin-resistant *Staph. aureus*.[30]

Certain strains of bacteria are resistant to the lytic properties of penicillins, but remain susceptible to their growth-inhibiting effects. Thus, the action of the antibiotic has been converted from bactericidal to bacteriostatic. This mechanism of resistance has been termed *tolerance* and apparently results from impaired autolysin activity in the bacterium.

Penicillinase-Resistant Penicillins

The availability of 6-APA on a commercial scale made possible the synthesis of numerous semisynthetic penicillins modified at the acyl amino side chain. Much of the early work done in the 1960s was directed toward the preparation of derivatives that would resist destruction by β-lactamases, particularly those produced by penicillin-resistant strains of *Staph. aureus*, which constituted a very serious health problem at that time. In general, it was found that increasing the steric hindrance at the α-carbon of the acyl group increased resistance to staphylococcal β-lactamase, with maximal resistance being observed with quaternary substitution.[31] More fruitful from the standpoint of antibacterial potency, however, was the observation that the α-acyl carbon could be part of an aromatic (e.g., phenyl or naphythyl) or heteroaromatic (e.g., 4-isoxazoyl) system.[32] Substitutions at the *ortho* positions of a phenyl ring [e.g., 2,6-dimethoxyl (methicillin)] or the 2-position of a 1-naphthyl system [e.g., 2-ethoxyl (nafcillin)] increase the steric hindrance of the acyl group and confer increased β-lactamase resistance over the unsubstituted compounds or those substituted at positions more distant from the α-carbon. Bulkier substituents are required to confer effective β-lactamase resistance among five-membered ring heterocyclic derivatives.[33] Thus, members of the 4-isoxazoyl penicillin family (e.g., oxacillin, cloxacillin, and dicloxacillin) require both the 3-aryl and 5-methyl (3-methyl and 5-aryl) substituents for effectiveness against β-lactamase-producing *Staph. aureus*.

Increasing the bulkiness of the acyl group is not without its price, however, because all of the clinically available penicillinase-resistant penicillins are significantly less active than either penicillin G or penicillin V against most non–β-lactamase-producing bacteria normally sensitive to the penicillins. The isoxazoyl penicillins, particularly those with an electronegative substituent in the 3-phenyl group (cloxacillin, dicloxacillin, and fluoxicillin), are also resistant to an acid-catalyzed hydrolysis of the β-lactam, for the reasons described earlier. However, steric factors that confer β-lactamase resistance do not necessarily also confer stability to acid. Accordingly, methicillin, which has electron-donating groups (by resonance) *ortho* to the carbonyl carbon, is even more labile to acid-catalyzed hydrolysis than is penicillin G, because of the more rapid formation of the penicillenic acid derivative.

Extended-Spectrum Penicillins

Another highly significant advance arising from the preparation of semisynthetic penicillins has been the discovery that the introduction of an ionized or polar group into the α-position of the side chain benzyl carbon atom of penicillin G confers activity against gram-negative bacilli. Hence, derivatives with an ionized α-amino group, such as ampicillin and amoxicillin, are generally effective against such gram-negative genera as *Escherichia*, *Klebsiella*, *Haemophilis*, *Salmonella*, *Shigella*, and non–indole-producing *Proteus*. Activity against penicillin G-sensitive, gram-positive species is, furthermore, largely retained. The introduction of an α-amino group in ampicillin (or amoxicillin) creates an additional asymmetric center. It is noteworthy that the extension of antibacterial spectrum brought about by the substituent applies only to the D-isomer, which is two to eight times more active than either the L-isomer or benzylpenicillin (which are equiactive) against various species of the aforementioned genera of gram-negative bacilli.

The basis for the expanded spectrum of activity associated with the ampicillin group is not related to β-lactamase inhibition, as ampicillin (and amoxicillin) are even more labile than is penicillin G to the action of β-lactamases elaborated by both *Staph. aureus* and by various species of gram-negative bacilli, including strains among the "ampicillin-sensitive" group. Hydrophilic penicillins, such as ampicillin, penetrate gram-negative bacteria with greater facility than do penicillin G, penicillin V, or methicillin. This selective penetration is believed to take place through the porin channels of the cell membrane.[34]

α-Hydroxy substitution also yields "expanded-spectrum" penicillins with similar activity and stereoselectivity to that of the ampicillin group. However, the α-hydroxybenzylpenicillins are about two to five times less active than their corresponding α-aminobenzyl counterparts and, unlike the latter, are not very stable under acidic conditions.

The incorporation of an acidic substituent at the α-benzyl carbon atom of penicillin G also imparts clinical effectiveness against gram-negative bacilli and, furthermore, extends the spectrum of activity to include organisms that are resistant to ampicillin. Thus, α-carboxybenzylpenicillin (carbenicillin) is active against ampicillin-sensitive, gram-negative species and additional gram-negative bacilli of the genera *Pseudomonas*, *Klebsiella*, *Enterobacter*, indole-producing *Proteus*, *Serratia*, and *Providencia*. The potency of carbenicillin against most species of penicillin G-sensitive, gram-positive bacteria is several orders of magnitude lower than that of either penicillin G or ampicillin, presumably because of poorer penetration of a more highly ionized molecule into these bacteria. (It will be noted that α-aminobenzylpenicillins exist as zwitterions over a broad *p*H range and, as such, are considerably less polar than is carbenicillin.) This increased polarity is apparently an advantage for the penetration of carbenicillin through the cell envelope of gram-negative bacteria by porin channels.[34]

Carbenicillin is active against both β-lactamase-producing and non–β-lactamase-producing strains of gram-negative bacteria. It is known to be somewhat resistant to a few of the β-lactamases produced by gram-negative bacteria, especially members of the *Enterobacteriaceae* family.[35] Resistance to β-lactamases elaborated by gram-negative bacteria, therefore, may be an important component of carbenicillin's activity against some ampicillin-resistant organisms. However, β-lactamases produced by *Pseudomonas* species readily hydrolyze carbenicillin. Although carbenicillin is also somewhat resistant to staphylococcal β-lactamase, it is considerably less so than methicillin or the isoxazoyl penicillins, and its inherent antistaphylococcal activity is also less impressive, compared with the β-lactamase-resistant penicillins.

When compared with the aminoglycoside antibiotics, the potency of carbenicillin against such gram-negative bacilli as *Ps. aeruginosa*, *Pr. vulgaris*, and *K. pneumoniae* is much less impressive. Large parenteral doses are thus required to achieve bactericidal concentrations in the plasma and in the tissues. However, the low toxicity of carbenicillin (and the penicillins in general) usually permits (in the absence of allergy) the use of such high doses without untoward effects. Furthermore, carbenicillin (and other penicillins), when combined with aminoglycosides, exerts a synergistic cidal action against bacterial species sensitive to both agents, frequently allowing the use of a lower dose of the more toxic aminoglycoside than normally required for treatment of a life-threatening infection. The chemical incompatability of penicillins and aminoglycosides requires that the two antibiotics be administered separately; otherwise both are inactivated. Iyengar et al.[36] have shown that acylation of amino groups in the aminoglycoside by the β-lactam of the penicillin occurs.

Unlike the situation with ampicillin, the introduction of asymmetry at the α-benzyl carbon in carbenicillin imparts little or no stereoselectivity of antibacterial action; the individual enantiomers are nearly equally active and, furthermore, are readily epimerized to the racemate in aqueous solution. Because it is a derivative of phenylmalonic acid, carbenicillin readily decarboxylates to benzylpenicillin in the presence of acid; therefore, it is not active (as carbenicillin) orally and must be administered parenterally. Esterification of the α-carboxyl group (e.g., as the 5-indanyl ester) partially protects the compound from acid-catalyzed destruction and provides an orally active derivative that is hydrolyzed to carbenicillin in the plasma. However, the plasma levels of free carbenicillin achievable with oral administration of such esters may not be sufficiently high to effectively treat serious infections caused by some species of gram-negative bacilli, such as *Ps. aeruginosa*.

A series of α-acylureido-substituted penicillins, exemplified by azlocillin, mezlocillin, and piperacillin, exhibit enhanced activity against certain gram-negative bacilli, compared with carbenicillin. Although the acylureidopenicillins are acylated derivatives of ampicillin, the antibacterial spectrum of activity of the group is more like that of carbenicillin. The acylureidopenicillins are, however, superior to carbenicillin against *Klebsiella*, *Enterobacter*, and *Ps. aeruginosa*. This enhanced activity is apparently not due to β-lactamase resistance because both inducible and plasmid-mediated β-lactamases hydrolyze these penicillins. More facile penetration through the cell envelope of these particular bacterial species is the most likely explanation for the greater potency. The acylureidopenicillins, unlike ampicillin, are unstable under acidic conditions; therefore, they are not available for oral administration.

Protein Binding

The nature of the acylamino side chain also determines the extent to which penicillins are plasma protein bound. Quantitative structure activity (QSAR) studies of the binding of penicillins to human serum[37,38] indicate that hydrophobic groups (positive π-dependence) in the side chain appear to be largely responsible for increased binding to serum proteins. Penicillins with polar or ionized substituents in the side chain exhibit low to intermediate fractions of protein binding. Accordingly, ampicillin, amoxicillin, and cyclacillin experience 25% to 30% protein binding; carbenicillin and ticarcillin show 45% to 55%. Those with nonpolar, lipophilic substituents (nafcillin and isoxazoylpenicillins) are highly protein bound, with fraction exceeding 90%. The penicillins with less complex acyl groups (benzylpenicillin, phenoxymethylpenicillin, and methicillin) fall in the range of 35% to 80%. Protein binding is thought to restrict the tissue availability of drugs if the fraction of binding is sufficiently high; thus, the tissue distribution of the penicillins in the highly bound group may be inferior to that of other penicillins. The similarity of biologic half-lives for various penicillins, however, indicates that plasma protein binding has little effect on their durations of action. All of the commercially available penicillins are actively secreted by the renal active transport system for anions. The reversible nature of protein binding does not compete very effectively with the active tubular secretion process.

Allergy to Penicillins

Allergic reactions to various penicillins, ranging in severity from a variety of skin and mucous membrane rashes to drug fever and anaphylaxis, constitute the major problem associated with the use of this class of antibiotics. Estimates place the prevalence of hypersensitivity to penicillin G throughout the world at between 1% and 10% of the population. In the United States and other industrialized countries, it is nearer to the higher figure, ranking penicillin as the most common cause of drug-induced allergy. The penicillins most frequently implicated as causes of allergic reactions are penicillin G and ampicillin. However, virtually all commercially available penicillins have been reported to cause such reactions and, in fact, cross-sensitivity among most chemical classes of 6-acylaminopenicillanic acid derivatives has been demonstrated.[39]

The chemical mechanisms by which penicillin preparations become antigenic have been studied extensively.[19] Evidence suggests that penicillins, or their rearrangement products formed in vivo (e.g., penicillenic acids),[40] react with lysine-ε-amino groups of proteins to form penicilloyl proteins, which are major antigenic determinants.[41,42] Early clinical observations with the biosynthetic penicillins, penicillin G and penicillin V, indicated a higher incidence of allergic reactions with unpurified amorphous preparations, compared with highly purified crystalline forms, giving rise to the suggestion that small amounts of highly antigenic penicilloyl proteins present in unpurified samples were a cause. Polymeric impurities in ampicillin dosage forms have been implicated as possible antigenic determinants and as a possible explanation for the high frequency of allergic reactions with this particular semisynthetic penicillin. Ampicillin is known to undergo *p*H-dependent polymerization reactions (especially in concentrated solutions) that involve nucleophilic attack of the side chain amino group of one molecule on the β-lactam carbonyl carbon atom of a second molecule, and so on.[43] The high frequency of antigenicity shown by ampicillin polymers together with their isolation and characterization in some ampicillin preparations supports the theory that they contribute to ampicillin-induced allergy.[44]

Classification

A variety of designations have been used for classifying penicillins, based on their sources, chemistry, pharmacokinetic properties, resistance to enzymatic inactivation, antibacterial spectrum of activity, and clinical uses (Table 7-3). Thus, penicillins may be biosynthetic, semisynthetic, or (potentially) synthetic; acid resistant or not; orally or (only) parenterally active; and resistant to β-lactamases (penicillinases) or not. They may have a narrow, intermediate, or broad spectrum of antibacterial activity and may be intended for multipurpose or limited clinical use. In the latter two connections, it is important to emphasize that designations of the activity spectrum as narrow, intermediate, and broad are relative and, furthermore, do not necessarily imply the breadth of therapeutic applications of a particular antibiotic. Indeed, the classification of penicillin G as a "narrow-spectrum" antibiotic has meaning only relative to other penicillins. Although the β-lactamase-resistant penicillins have a spectrum of activity similar to that of penicillin G, they are generally reserved for the treatment of infections caused by penicillin G-resistant, β-lactamase-producing *Staph. aureus*, because their activity against most penicillin G-sensitive bacteria is significantly inferior. Similarly, carbenicillin and ticarcillin are usually reserved for the treatment of infections caused by ampicillin-resistant, gram-negative bacilli, because they offer no advantage (and have some disadvantages) to ampicillin or penicillin G in infections sensitive to them.

Products

Penicillin G; benzylpenicillin. For years, the most popular penicillin has been benzylpenicillin. In fact, with the exception of patients allergic to it, penicillin G remains the agent of choice for the treatment of more different kinds of bacterial infections than any other antibiotic. It was first made available in the form of the water-soluble salts of potassium, sodium, and calcium. These salts of penicillin are inactivated by the gastric juice, and are not effective when administered orally unless antacids, such as calcium carbonate, aluminum hydroxide, and magnesium

TABLE 7-3

CLASSIFICATION AND PROPERTIES OF PENICILLINS

Penicillin	Source	Acid Resistance	Oral Absorption	Plasma Protein Binding (%)	β-Lactamase Resistant (S. aureus)	Spectrum of Activity	Clinical Use
Benzylpenicillin	Biosynthetic	Poor	Poor	50–60	No	Intermediate	Multipurpose
Phenoxymethyl-penicillin	Biosynthetic	Good	Good	55–80	No	Intermediate	Multipurpose
Methicillin	Semisynthetic	Poor	Poor	30–40	Yes	Narrow	Limited use
Nafcillin	Semisynthetic	Fair	Variable	90	Yes	Narrow	Limited use
Oxacillin	Semisynthetic	Good	Good	85–94	Yes	Narrow	Limited use
Cloxacillin	Semisynthetic	Good	Good	88–96	Yes	Narrow	Limited use
Dicloxacillin	Semisynthetic	Good	Good	95–98	Yes	Narrow	Limited use
Ampicillin	Semisynthetic	Good	Fair	20–25	No	Intermediate	Multipurpose
Amoxicillin	Semisynthetic	Good	Good	20–25	No	Intermediate	Multipurpose
Cyclacillin	Semisynthetic	Good	Good	20–25	No	Intermediate	Multipurpose
Carbenicillin	Semisynthetic	Poor	Poor	50–60	No	Broad	Limited use
Ticarcillin	Semisynthetic	Poor	Poor	45	No	Broad	Limited use
Azlocillin	Semisynthetic	Poor	Poor	20	No	Broad	Limited use
Mezlocillin	Semisynthetic	Poor	Poor	50	No	Broad	Limited use
Piperacillin	Semisynthetic	Poor	Poor	50	No	Broad	Limited use
Amdinocillin	Semisynthetic	Poor	Poor	5–10	No	Intermediate	Limited use

trisilicate, or a strong buffer, such as sodium citrate, are added. Also, because penicillin is poorly absorbed from the intestinal tract, oral doses must be very large—about five times the amount necessary with parenteral administration. Only after the production of penicillin had increased sufficiently that low-priced penicillin was available did the oral dosage forms become popular. The water-soluble potassium and sodium salts are used orally and parenterally to achieve rapid high plasma concentrations of penicillin G. The more water-soluble potassium salt is usually preferred when large doses are required. However, situations in which hyperkalemia is a danger, as in renal failure, require use of the sodium salt; the potassium salt is preferred for patients on "salt-free" diets or with congestive heart conditions.

The rapid elimination of penicillin from the bloodstream through the kidneys by active tubular secretion and the need for maintaining an effective blood level concentration have led to the development of "repository" forms of this drug. Suspensions of penicillin in peanut oil or sesame oil with white beeswax added were first employed for prolonging the duration of injected forms of penicillin. This dosage form was replaced by a suspension in vegetable oil to which aluminum monostearate or aluminum distearate was added. Today, most repository forms are suspensions of high-molecular-weight amine salts of penicillin in a similar base.

Penicillin G Procaine, USP (Crysticillin, Duracillin, Wycillin). The first widely used amine salt of penicillin G was made with procaine. It can be made readily from penicillin G sodium by treatment with procaine hydrochloride. This salt is considerably less soluble in water than are the alkaline metal salts, requiring about 250 mL to dissolve 1 g. The free penicillin is released only as the compound dissolves and dissociates. It has an activity of 1009 units/mg. A large number of preparations for injection of penicillin G procaine are commercially available. Most of these are either suspensions in water to which a suitable dispersing or suspending agent, a buffer, and a preservative have been added, or suspensions in peanut oil or sesame oil that have been gelled by the addition of 2% aluminum monostearate. Some of the commercial products are mixtures of penicillin G potassium or sodium with penicillin G procaine to provide a rapid development of a

$$C_6H_5\text{—}CH_2CONHCH\text{—}CH(S)C(CH_3)_2 \;/\; CO\text{—}N\text{—}CHCOO^- \quad H\text{—}N^+(C_2H_5)_2\text{—}CH_2CH_2\text{—}O\text{—}C(=O)\text{—}C_6H_4\text{—}NH_2$$

Penicillin G Procaine

high plasma concentration of penicillin through the use of the water-soluble salt plus the prolonged duration of effect obtained from the insoluble salt. In addition to the injectable forms, penicillin G procaine is available in oral dosage forms. It is claimed that the rate of absorption of this salt is as rapid as that of other forms usually administered orally.

Penicillin G Benzathine, USP. *N,N′*-Dibenzylethylenediamine dipenicillin G (Bicillin, Permapen). Since it is the salt of a diamine, two moles of penicillin are available from each molecule of the salt. It is very insoluble in water, requiring about 3000 mL to dissolve 1 g. This property gives the compound great stability and prolonged duration of effect. At the *p*H of gastric juice it is quite stable, and food intake does not interfere with its absorption. It is available in tablet form and in a number of parenteral preparations. The activity of penicillin G benzathine is equivalent to 1211 units/mg.

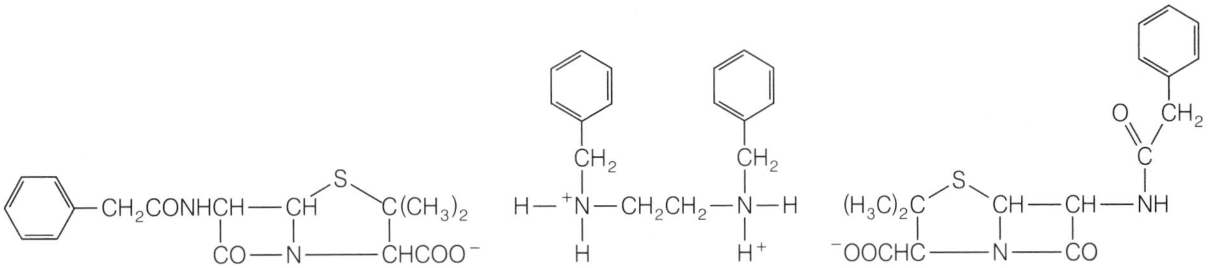

Penicillin G Benzathine

Several other amines have been used to make penicillin salts, and research is continuing on this subject. Other amines that have been used include 2-chloroprocaine; L-*N*-methyl-1,2-diphenyl-2-hydroxyethylamine (L-ephenamine); dibenzylamine; tripelennamine (Pyribenzamine); and *N,N′*-bis-(dehydroabietyl)ethylenediamine (hydrabamine).

Penicillin V, USP. Phenoxymethylpenicillin (Pen Vee, V-Cillin) was reported by Behrens et al.[45] in 1948 as a biosynthetic product. However, it was not until 1953 that its clinical value was recognized by some European scientists. Since then it has enjoyed wide use because of its resistance to hydrolysis by gastric juice and its ability to produce uniform concentrations in blood (when administered orally). The free acid requires about 1200 mL of water to dissolve 1 g, and it has an activity of 1695 units/mg. For parenteral solutions, the potassium salt usually is employed. This salt is very soluble in water. Solutions of it are made from the dry salt at the time of administration. Oral dosage forms of the potassium salt are also available, providing rapid effective plasma concentrations of this penicillin. The salt of phenoxymethylpenicillin with *N,N′*-bis(dehydroabietyl)ethylenediamine (hydrabamine; Compocillin-V) provides a very long-acting form of this compound. Its high degree of water insolubility makes it a desirable compound for aqueous suspensions used as liquid oral dosage forms.

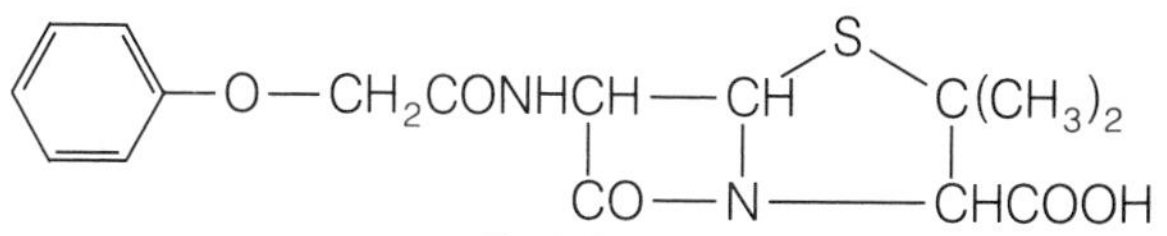

Penicillin V

Methicillin Sodium, USP. 2,6-Dimethoxyphenylpenicillin sodium (Staphcillin). During 1960, the second penicillin produced as a result of the research that developed synthetic analogues was introduced for medicinal use. By reacting 2,6-dimethoxybenzoylchloride with 6-aminopenicillanic acid, 6-(2,6-dimethoxybenzamido)penicillanic acid forms. The sodium salt is a white crystalline solid that is extremely soluble in water, forming clear neutral solutions. As with other penicillins, it is very sensitive to moisture, losing about half of its activity in five days at room temperature. Refrigeration at 5°C reduces the loss in activity to about 20% in the same period. Solutions prepared for parenteral use may be kept as long as 24 hours if refrigerated. It is extremely sensitive to acid, a *p*H of 2 causing a 50% loss in activity in 20 minutes; thus it cannot be used orally.

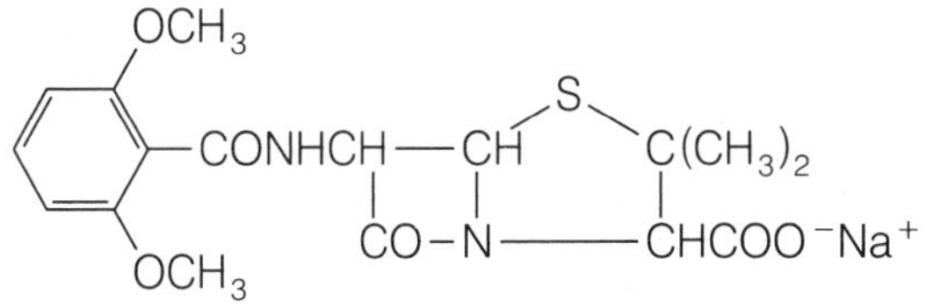

Methicillin Sodium

Methicillin sodium is particularly resistant to inactivation by penicillinase found in staphylococcal organisms and somewhat more resistant than penicillin G to penicillinase from *Bacillus cereus*. Methicillin and many other penicillinase-resistant penicillins are inducers of penicillinase, an observation that has implications against the use of these agents in the treatment of penicillin-G-sensitive infections. Clearly, the use of a penicillinase-resistant penicillin should not be followed by penicillin G.

It may be assumed that the absence of the benzylmethylene group of penicillin G and the steric protection afforded by the 2- and 6-methoxy groups makes this compound particularly resistant to enzyme hydrolysis.

Methicillin sodium has been introduced for use in the treatment of staphylococcal infections caused by strains found resistant to other penicillins. It is recommended that it not be used in general therapy to avoid the possible widespread development of organisms resistant to it.

The incidence of interstitial nephritis is reportedly higher with methicillin than with other penicillins.

Oxacillin Sodium, USP. (5-Methyl-3-phenyl-4-isoxazolyl)penicillin sodium monohydrate (Prostaphlin). Oxacillin sodium is the salt of a semisynthetic penicillin that is highly resistant to inactivation by penicillinase. Apparently, the steric effects of the 3-phenyl and 5-methyl groups of the isoxazolyl ring prevent the binding of this penicillin to the β-lactamase active site and, thereby, protect the lactam ring from degradation in much the same way as has been suggested for methicillin. It is also relatively resistant to acid hydrolysis and, therefore, may be administered orally with good effect.

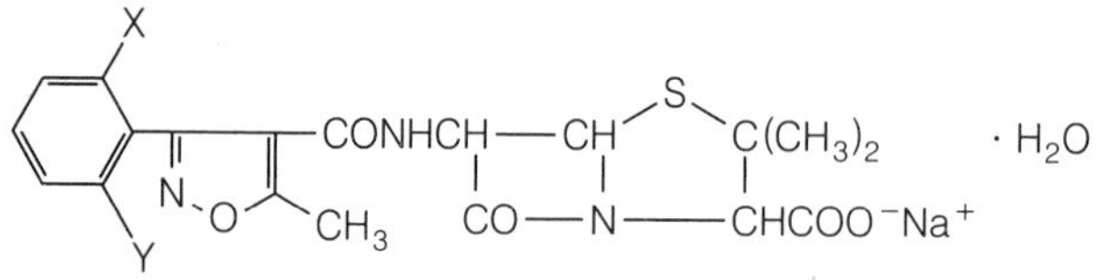

Oxacillins

X, Y = H: Sodium Oxacillin
X = Cl; Y = H: Sodium Cloxacillin
X, Y = Cl: Sodium Dicloxacillin
X = Cl; Y = F: Sodium Floxicillin

Oxacillin sodium, which is available in capsule form, is well absorbed from the gastrointestinal tract, particularly in fasting patients. Effective plasma levels of oxacillin are obtained in about one hour, but despite extensive plasma protein binding, it is rapidly excreted through the kidneys.

The use of oxacillin and other isoxazolylpenicillins should be restricted to the treatment of infections caused by staphylococci resistant to penicillin G. Although their spectrum of activity is similar to that of penicillin G, the isoxazolylpenicillins are, in general, inferior to it and the phenoxymethylpenicillins for the treatment of infections caused by penicillin G-sensitive bacteria. Because they cause allergic reactions similar to those produced by other penicillins, the isoxazolylpenicillins should be used with great caution in patients that are penicillin-sensitive.

Cloxacillin Sodium, USP. [3-(*o*-Chlorophenyl)-5-methyl-4-isoxazolyl]penicillin sodium monohydrate (Tegopen). The chlorine atom *ortho* to the position of attachment of the phenyl ring to the isoxazole ring enhances the activity of this compound over that of oxacillin, not by an increase in intrinsic activity of absorption, but by achieving higher plasma levels. In almost all other respects it resembles oxacillin.

Dicloxacillin Sodium, USP. [3-(2,6-Dichlorophenyl)-5-methyl-4-isoxazolyl]penicillin sodium monohydrate (Dynapen, Pathocil, Veracillin). The substitution of chlorine atoms on both carbons *ortho* to the position of attachment of the phenyl ring to the isoxazole ring is presumed to further enhance the stability of this oxacillin congener and to produce high plasma concentrations of dicloxacillin. Its medicinal properties and use are similar to those of cloxacillin sodium. However, progressive halogen substitution also increases the fraction of protein binding in the plasma, potentially reducing the concentration of free antibiotic in the plasma and in the tissues. Its medicinal properties and use are the same as those of cloxacillin sodium.

Nafcillin Sodium, USP. 6-(2-Ethoxyl-1-naphthyl)penicillin sodium (Unipen). Nafcillin sodium is another semisynthetic penicillin produced as a result of the search for penicillinase-resistant compounds. Similarly to methicillin, nafcillin has substituents in positions *ortho* to the point of attachment of the aromatic ring to the carboxamide group of penicillin. No doubt, the ethoxy group and the second ring of the naphthalene group play steric roles in stabilizing nafcillin against penicillinase. Very similar structures have been reported to produce similar results in some substituted 2-biphenylpenicillins.[32]

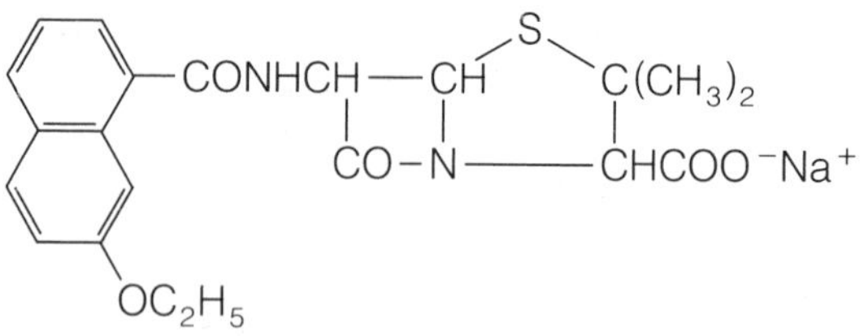

Nafcillin Sodium

Unlike methicillin, nafcillin is sufficiently stable to acid to permit its use by oral administration. When given orally, its absorption is somewhat slow and incomplete, but satisfactory plasma levels may be achieved in about one hour. Relatively small amounts are excreted through the kidneys, with the major portion excreted in the bile. Even though some cyclic reabsorption from the gut may thus occur, nafcillin should be readministered every four to six hours when given orally. This salt is readily soluble in water and may be administered intramuscularly or intravenously to obtain high plasma concentrations quickly for the treatment of serious infections.

Nafcillin sodium may be used in infections caused solely by penicillin G-resistant staphylococci or when streptococci are present also. Although it is recommended that it be used exclusively for such resistant infections, it is effective also against pneumococci and group A β-hemolytic streptococci. Because, as with other penicillins, it may cause allergic side effects, it should be administered with care.

Ampicillin, USP. 6-[D-α-Aminophenylacetamido]penicillanic acid; D-α-aminobenzylpenicillin (Penbritin, Polycillin, Omnipen, Amcil, Principen). With ampicillin, another goal in the research on semisynthetic penicillins—an antibacterial spectrum broader than that of penicillin G—has been attained. This product is active against the same gram-positive organisms that are susceptible to other penicillins, and it is more active against some gram-negative bacteria and enterococcal infections. Obviously, the α-amino group plays an important role in the broader activity, but the mechanism for its action is unknown. It has been suggested that the amino group confers an ability to cross cell wall barriers that are impenetrable to other penicillins. It is noteworthy that D-(−)-ampicillin, prepared from D-(−)-α-aminophenylacetic acid is significantly more active than L-(+)-ampicillin.

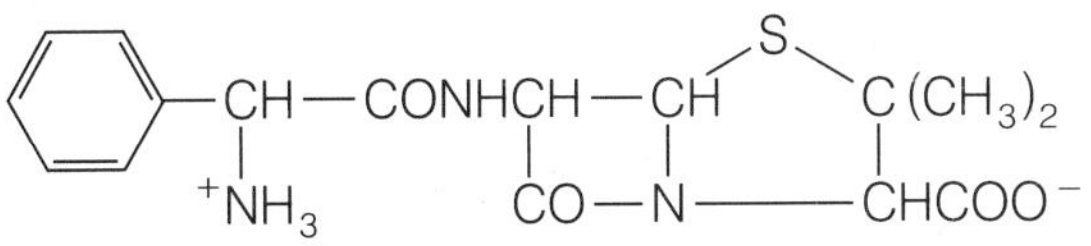

$$C_6H_5-\underset{\underset{^{+}NH_3}{|}}{CH}-CONHCH-CH\overset{S}{\frown}C(CH_3)_2 \quad | \; CO-N-CHCOO^-$$

Ampicillin

Ampicillin is not resistant to penicillinase, and it produces the allergic reactions and other untoward effects that are found in penicillin-sensitive patients. However, because such reactions are relatively few, it may be used in some infections caused by gram-negative bacilli for which a broad-spectrum antibiotic, such as a tetracycline or chloramphenicol, may be indicated but not preferred because of undesirable reactions or lack of bactericidal effect. However, ampicillin is not so widely active that it should be used as a broad-spectrum antibiotic in the same manner as the tetracyclines. It is particularly useful for the treatment of acute urinary tract infections caused by *E. coli* or *Pr. mirabilis* and is the agent of choice against *Haemophilus influenzae* infections. Ampicillin, together with probenecid to inhibit its active tubular excretion, has also become a treatment of choice for gonorrhea in recent years. However, β-lactamase-producing strains of gram-negative bacteria that are highly resistant to ampicillin appear to be increasing in the world population. The threat from such resistant strains is particularly great with *H. influenzae* and *Neisseria gonorrhoeae* because few alternative therapies for infections caused by these organisms are available. Incomplete absorption together with excretion of effective concentrations in the bile may contribute to the effectiveness of ampicillin in the treatment of salmonellosis and shigellosis.

Ampicillin is water-soluble and stable to acid. The protonated α-amino group of ampicillin has a pK_a of 7.3,[45] and it is thus extensively protonated in acidic media, which explains ampicillin's stability toward acid hydrolysis and instability toward alkaline hydrolysis. It is administered orally and is absorbed from the intestinal tract to produce peak plasma concentrations in about two hours. Oral doses must be repeated about every six hours, because it is rapidly excreted unchanged through the kidneys. It is available as a white, crystalline, anhydrous powder that is sparingly soluble in water, or as the colorless or slightly buff-colored crystalline trihydrate that is soluble in water. Either form may be used for oral administration either in capsules or as a suspension. Earlier claims of higher plasma levels for the anhydrous form, compared with the trihydrate, following oral administration have been disputed.[46,47] The white crystalline sodium salt is very soluble in water, and solutions for injections should be administered within one hour after being made.

Hetacillin, USP (Versapen) is prepared by the reaction of ampicillin with acetone. In aqueous solution it is rapidly converted back to ampicillin and acetone. The spectrum of antibacterial action is identical with that of ampicillin and probably is due to the hydrolysis product, ampicillin. Although hetacillin is more slowly excreted than ampicillin, initial plasma levels are lower after equivalent oral doses. Thus, it appears that hetacillin represents only another form in which to administer ampicillin and offers no advantages over it. Hetacillin occurs as a fine, off-white powder that is freely soluble in water and in alcohol. It is available for intramuscular administration in a preparation together with lidocaine for patients unable to take it orally. The water-soluble potassium salt is used for intravenous administration.

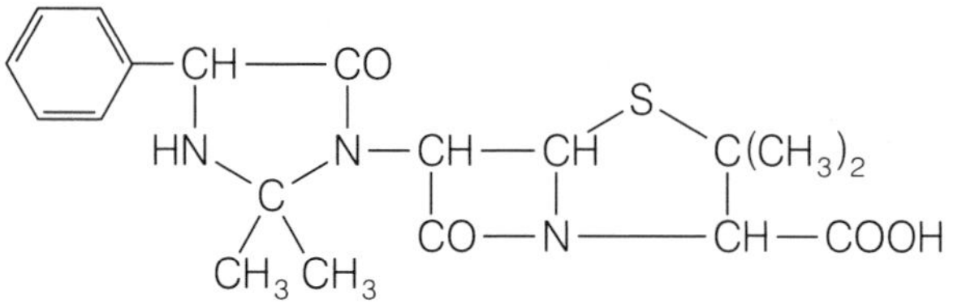

Hetacillin

Bacampicillin Hydrochloride, USP (Spectrobid) is the hydrochloride salt of the 1-ethoxycarboxyloxyethyl ester of ampicillin. It is a prodrug of ampicillin with no antibacterial activity. After oral absorption, bacampicillin is rapidly hydrolyzed by esterases in the plasma to form ampicillin. The oral absorption of bacampicillin is more rapid and more complete than that of ampicillin and less affected by food. Plasma levels of ampicillin from oral bacampicillin exceed those of oral ampicillin or amoxicillin for the first 2.5 hours, but, thereafter, are the same as for ampicillin and amoxicillin.[48] Effective plasma levels are sustained for 12 hours, allowing twice-a-day dosing.

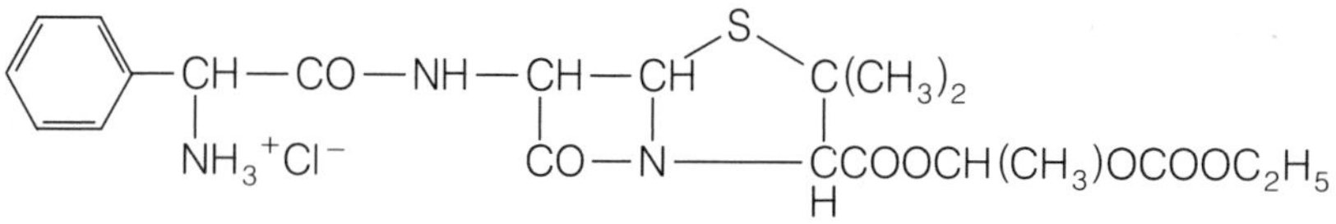

Bacampicillin Hydrochloride

Amoxicillin, USP. 6-[D-(−)-α-Amino-*p*-hydroxyphenylacetamido]penicillanic acid (Amoxil, Larotid, Polymox). Amoxicillin, a semisynthetic penicillin introduced in 1974, is simply the *p*-hydroxy analogue of ampicillin prepared by the acylation of 6-APA with *p*-hydroxyphenylglycine. Its antibacterial spectrum is nearly identical with that of ampicillin and, as is ampicillin, it is also resistant to acid, susceptible to alkaline and β-lactamase hydrolysis, and weakly protein bound. Early clinical reports[49] indicated that orally administered amoxicillin possesses significant advantages over ampicillin, including more complete gastrointestinal absorption to give higher plasma and urine levels, less diarrhea, and little or no effect of food on absorption. Thus, it appears that amoxicillin may replace ampicillin for the treatment of certain systemic and urinary tract infections wherein oral administration is desirable, particularly if relative costs become more competitive. Amoxicillin is reported to be less effective than ampicillin in the treatment of bacillary dysentery, presumably because of its greater gastrointestinal absorption.

Amoxicillin is a fine, white to off-white crystalline powder that is sparingly soluble in water. It is available in a variety of oral dosage forms. Aqueous suspensions are stable for one week at room temperature.

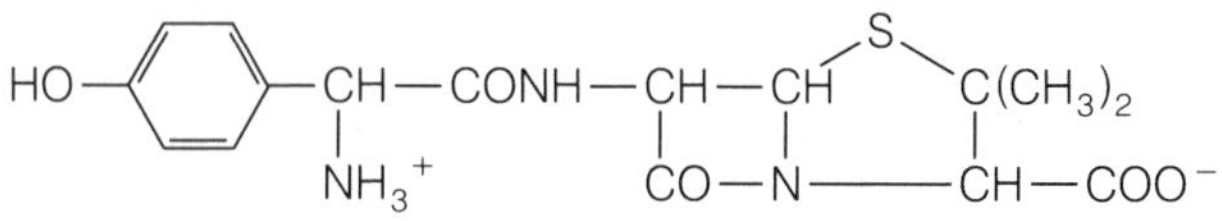

Amoxicillin

Cyclacillin, USP. 1-Aminocyclohexylpenicillin (Cyclapen) was approved for introduction into the American market in 1979. This agent has properties that are very similar to those of ampicillin. Its spectrum of antibacterial activity is virtually identical. Although it is somewhat resistant to most β-lactamases, cyclacillin is not particularly active against β-lactamase-producing strains of *Staph. aureus* or gram-negative bacilli. Furthermore, its potency against most species of ampicillin-sensitive bacteria is 25% to 50% lower than ampicillin's. The major advantages of cyclacillin appear to derive from faster and more complete oral absorption and less tendency to form potentially antigenic polymers. Indeed, higher plasma levels after oral administration and significantly lower frequency of skin rashes from cyclacillin, compared with ampicillin, have been reported in early clinical studies.

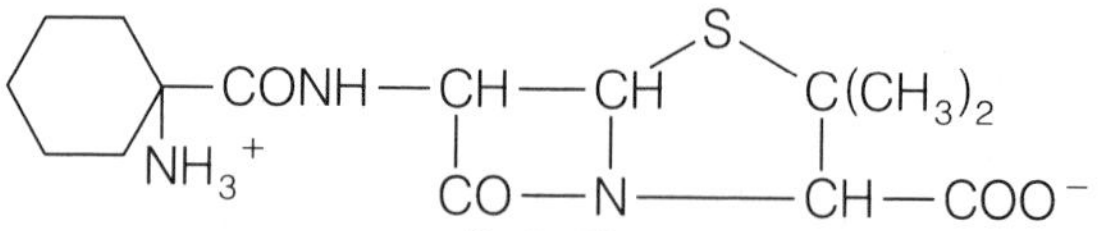

Cyclacillin

Carbenicillin Disodium, Sterile, USP. Disodium α-carboxybenzylpenicillin (Geopen, Pyopen) is a semisynthetic penicillin released in the United States in 1970, which was introduced in England and first reported by Acred et al.[50] in 1967. Examination of its structure shows that it differs from ampicillin by having an ionizable carboxyl group substituted on the α-carbon atom of the benzyl side chain rather than an amino group. Carbenicillin has a broad range of antimicrobial activity, broader than any other known penicillins, a property attributed to the unique carboxyl group. It has been proposed that the carboxyl group confers improved penetration of the molecule through cell wall barriers of gram-negative bacilli, compared with other penicillins.

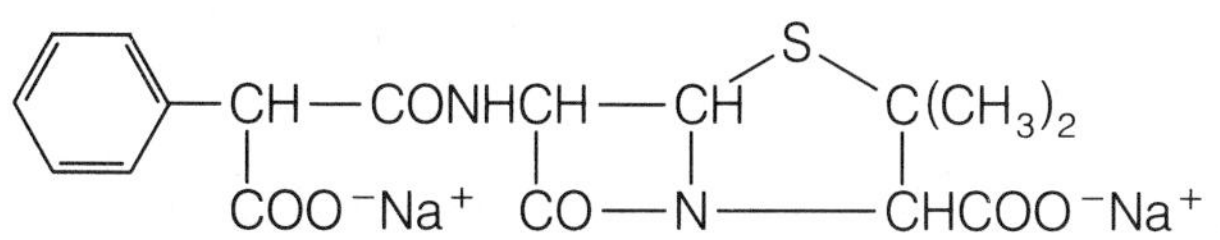

Carbenicillin Disodium

Carbenicillin is not stable in acids and is inactivated by penicillinase. It is a malonic acid derivative and, as such, decarboxylates readily to penicillin G, which is acid-labile. Solutions of the disodium salt should be freshly prepared, but may be kept for two weeks when refrigerated. It must be administered by injection and is usually given intravenously.

Carbenicillin has been effective in the treatment of systemic and urinary tract infections caused by *P. aeruginosa*, indole-producing *Proteus*, and Providencia species, all of which are resistant to ampicillin. The low toxicity of carbenicillin, with the exception of allergic sensitivity, permits the use of large dosages in serious infections. Most clinicians prefer to use a combination of carbenicillin and gentamicin for serious pseudomonal and mixed coliform infections. However, the two antibiotics are chemically incompatible and should never be combined in the same intravenous solution.

Carbenicillin Indanyl Sodium, USP. 6-[2-Phenyl-2-(5-indanyloxycarbonyl)acetamido]penicillanic acid (Geocillin). Efforts to obtain orally active forms of carbenicillin led to the eventual release of the 5-indanylester in 1972. Approximately 40% of the usual oral dose of indanyl carbenicillin is absorbed. After absorption, the ester is rapidly hydrolyzed by plasma and tissue esterases to yield carbenicillin. Thus, despite that the highly lipophilic and highly protein-bound ester has in vitro activity comparable with carbenicillin, its activity in vivo is due to carbenicillin. Indanyl carbenicillin thus provides an orally active alternative for the treatment of carbenicillin-sensitive systemic and urinary tract infections caused by *Pseudomonas*, indole-positive *Proteus* species, and selected species of gram-negative bacilli.

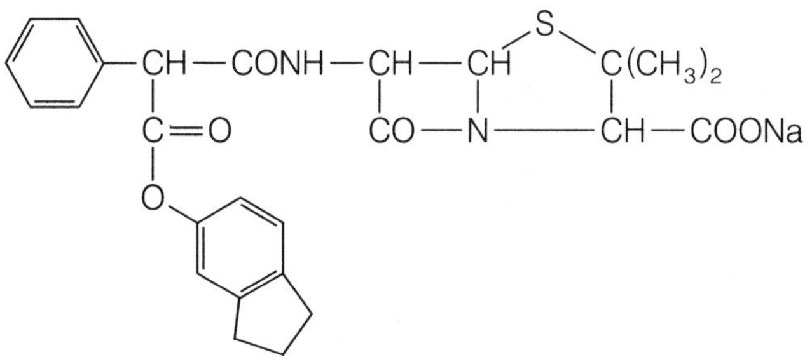

Carbenicillin Indanyl Sodium

In clinical trials with indanyl carbenicillin, a relatively high frequency of gastrointestinal symptoms (nausea, occasional vomiting, and diarrhea) was reported. It seems doubtful the high doses required for the treatment of serious systemic infections could be tolerated by most patients. Indanyl carbenicillin occurs as the sodium salt, an off-white, bitter powder that is freely soluble in water. It is stable to acid. It should be protected from moisture to prevent hydrolysis of the ester.

Ticarcillin Disodium, Sterile, USP. α-Carboxy-3-thienylpenicillin (Ticar) is an isostere of carbenicillin, wherein the phenyl group is replaced by a thienyl group. This semisynthetic penicillin derivative, as with carbenicillin, is unstable in acid and, therefore, must be administered parenterally. It is similar to carbenicillin in antibacterial spectrum and pharmacokinetic properties. Two advantages for ticarcillin are claimed: (1) slightly better pharmacokinetic properties, including higher serum levels and a longer duration of action; and (2) greater in vitro potencies against several species of gram-negative bacilli, most notably *Ps. aeruginosa* and *Bacteroides fragilis*. These advantages can be crucial in the treatment of serious infections requiring high-dose therapy.

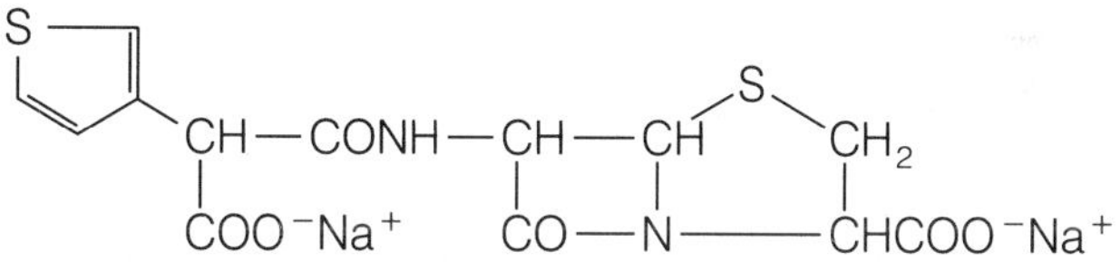

Ticarcillin Disodium

Azlocillin Sodium, Sterile, USP (Azlin) is one of a series of semisynthetic derivatives of ampicillin developed in Japan. It has been discovered that α-acylureido- and α-guanylbenzyl substituents (e.g., 2-oxoimidazolidinylcarbonylamino, for azlocillin) in ampicillin create penicillins that exhibit broad-spectrum activity in vitro against gram-positive and gram-negative bacteria.[51] Thus, the antibacterial properties of azlocillin and other ureidopenicillins (such as mezlocillin and piperacillin) more closely resemble the spectrum of carbenicillin than that of ampicillin.

The primary indication for azlocillin is in the treatment of serious infections caused by susceptible strains of *Ps. aeruginosa*. It is destroyed by β-lactamases produced by both gram-positive and gram-negative bacteria. Improved penetration through the cell envelope is believed to be responsible for its 8 to 16 times greater potency against *Ps. aeruginosa*, compared with carbenicillin. Azlocillin

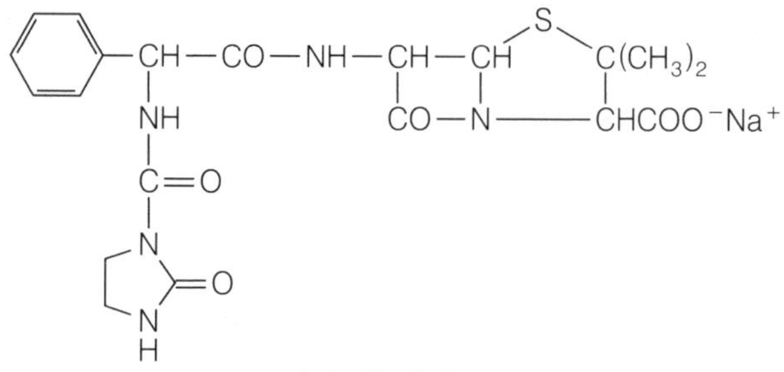

Azlocillin Sodium

is also very active against most non–β-lactamase-producing aerobic and anaerobic bacterial species, but generally less so than mezlocillin or piperacillin. It is not very stable under acidic conditions; therefore, it is not used orally.

Azlocillin sodium is supplied as a crystalline, water-soluble powder for injection. It is stable in the dry state, but solutions for injection should be used within 24 hours or refrigerated.

Mezlocillin Sodium, Sterile, USP (Mezlin) is an acylureido penicillin with an antibacterial spectrum similar to that of carbenicillin and ticarcillin; however, there are some major differences. It is much more active against most *Klebsiella* species, *Ps. aeruginosa*, anaerobic bacteria (such as *Streptococcus faecalis* and *B. fragilis*), and *H. influenzae*. It is recommended for the treatment of serious infections caused by these organisms.

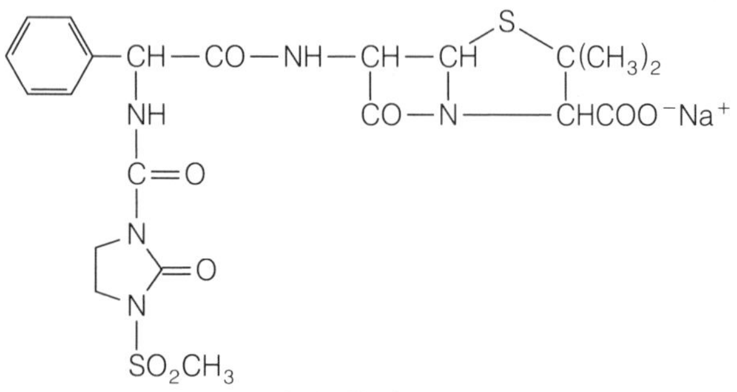

Mezlocillin Sodium

Mezlocillin is not generally effective against β-lactamase-producing bacteria, nor is it active orally. It is available as a white crystalline, water-soluble sodium salt for injection. Solutions should be freshly prepared and refrigerated if not used within 24 hours. Mezlocillin and other acylureidopenicillins, unlike carbenicillin, exhibit nonlinear pharmacokinetics. Peak plasma levels, half-life, and area under the time curve increase with increased dosage. The effect of mezlocillin on bleeding time is less than carbenicillin's, and it is less likely to cause hypokalemia.

Piperacillin Sodium, Sterile, USP (Pipracil) is the more generally useful of the extended-spectrum acylureidopenicillins. It is more active than mezlocillin against susceptible strains of gram-negative aerobic bacilli such as *Serratia marcescens*, *Proteus*, *Enterobacter*, *Citrobacter*, and *Ps. aeruginosa*. However, mezlocillin appears to be more active against *Providencia* species and *K. pneumoniae*. Piperacillin is also active against anaerobic bacteria, especially *B. fragilis* and *Strep. faecalis* (enterococcus). β-Lactamase-producing strains of these organisms are, however, resistant to piperacillin, and it is also hydrolyzed by *Staph. aureus* β-lactamase. β-Lactamase susceptibility of piperacillin is not absolute, because β-lactamase-producing ampicillin-resistant strains of *N. gonorrhoeae* and *H. influenzae* are susceptible to piperacillin.

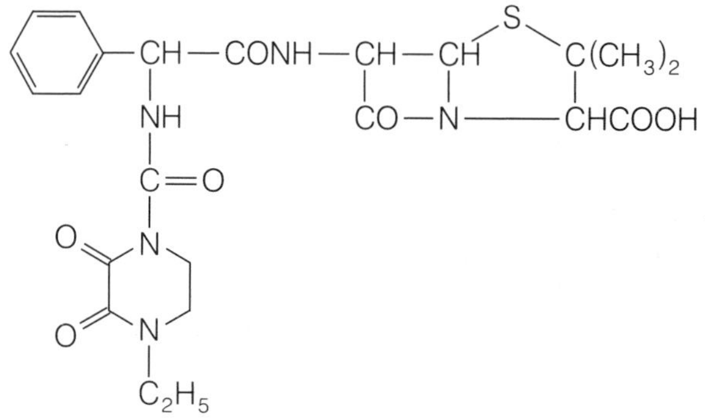

Piperacillin

Piperacillin is rapidly destroyed by stomach acid; therefore, it is active only by intramuscular or intravenous administration. The injectable form is provided as the white crystalline, water-soluble sodium salt. Its pharmacokinetic properties are very similar to those of the other acylureido penicillins.

Amdinocillin, USP. 6β-(Hexahydro-1*H*-azepin-1-yl)methyleneaminopenicillanic acid; mecillinam (Coactin) and its congeners differ structurally from other penicillins in that they are not acylamino derivatives but rather alkylideneamino (or amidino) derivatives of 6-APA. This structural difference confers unique biochemical and microbiologic properties to the 6β-amidinopenicillins that are of some significance. In contrast with most other penicillins, amdinocillin has significantly greater gram-negative, compared with gram-positive, antibacterial activity.[52] It is particularly active against members of the *Enterobacteriaceae*, including some ampicillin-resistant strains. Unfortunately, the activity of amdinocillin against *Haemophilus* and *Neisseria* species is low, and *Pseudomonas* species are resistant to it.

Amdinocillin is unique in that it interacts only with a single penicillin-binding protein of *E. coli*,

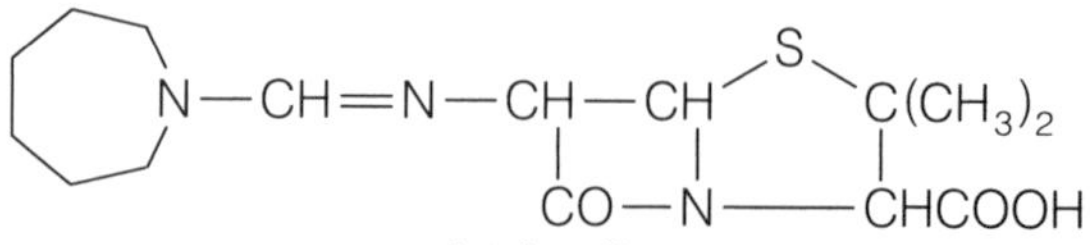

Amdinocillin

namely PBP 2.[53] It does not bind to the PBPs of gram-positive bacteria. Synergy occurs when amdinocillin and penicillins or cephalosporins are combined, but not when amdinocillin is combined with an aminoglycoside. The synergism between amdinocillin and other β-lactam antibiotics may be of therapeutic significance in the treatment of serious infections.

Amdinocillin, itself, is poorly absorbed from the gastrointestinal tract. The water-soluble crystalline sodium salt is available for parenteral administration. The pivaloyloxymethyl ester is marketed in Europe for oral administration. It is reasonably well absorbed from the gastrointestinal tract and undergoes esterase-catalyzed hydrolysis in the plasma. Amdinocillin is hydrolyzed by many β-lactamases produced by strains of gram-negative bacteria and, therefore, is not active against these strains. Although it is somewhat resistant to *Staph. aureus* β-lactamase, this is of no clinical importance.

Temocillin. 6α-Methoxy-6β-[2-carboxy-2-(3-thienyl)acetamido]penicillanic acid is the 6α-methoxy derivative of ticarcillin. This semisynthetic derivative is unique among penicillins in being resistant to β-lactamases, including those elaborated by gram-negative bacilli. Evidently, the 6α-methoxy group confers β-lactamase resistance in penicillins in a manner analogous to the effect of 7α-methoxyl substitution in the cephalosporins (see the discussion on cephamycins later in this chapter). Thus, temocillin possesses excellent activity against members of the *Enterobacteriaceae* family, *H. influenzae*, pathogenic *Neisseria* species, and *Branhamella catarrhalis*, including β-lactamase-producing strains resistant to ampicillin and carbenicillin. Unfortunately, its potency against gram-positive cocci and *Bacteriodes fragilis* is low, and it is inactive against *Ps. aeruginosa*. Low affinity for PBPs of gram-positive cocci and poor penetration of *Ps. aeruginosa* are possible causes for the lack of activity of temocillin against these organisms.

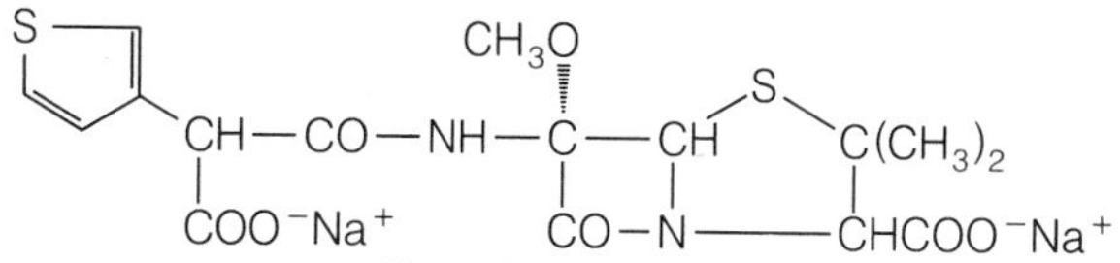

Temocillin Disodium

Another property of temocillin that distinguishes it from most other penicillins is its comparatively long duration of action. The plasma half-life of temocillin is 4.5 to 5 hours, which allows it to be administered on either a once- or twice-a-day dosing schedule. The long half-life of temocillin has been attributed to a combination of a high fraction of plasma protein binding (estimated to be 87%) and excretion by glomerular filtration, but not active secretion. Temocillin is unstable under acidic conditions and, therefore, is not active by the oral route. It must be administered parenterally.

TABLE 7-4

PENICILLINS

Name Proprietary Name	***Preparations***	***Usual Adult Dose****	***Usual Dose Range****	***Usual Pediatric Dose****
Penicillin G USP	Penicillin G potassium for injection USP Sterile penicillin G potassium USP Sterile penicillin G sodium USP	IM or IV, 1 million–5 million units every 4–6 hr	Up to 160 million units/day	Premature and full-term newborn infants IM or IV: 30,000 units/kg of body weight every 12 hr; older infants and children, IM or IV: 4167–16,667 units/kg of body weight every 4 hr; 6250–25,000 units/kg of body weight every 6 hr
	Penicillin G sodium for injection USP Penicillin G potassium tablets USP Penicillin G potassium tablets for oral	IM, 400,000 units qid; IV, 10 million units/day Oral, 200,000–500,000 units every 6–8 hr	Up to 2 million units/day	Infants and children under 12 yr of age: 4167–15,000 units/kg of body weight every 6 hr, or 8333–30,000 units/kg of body weight every 8 hr; children 12 yr and older: see Usual Adult Dose
Crysticillin, Duracillin, Wycillin	Sterile penicillin G procaine suspension USP	IM, 600,000–1.2 million units/day	Up to 4.8 million units/day	Infants and children up to 32 kg of body weight, I.M.: 10,000 units/kg of body weight daily for 10 days
Bicillin, Permapen	Sterile penicillin G benzathine suspension USP	IM, 1.2 million–2.4 million units as a single dose; 1.2 million units once a month or 600,000 units every 2 wk	Up to 2.4 million units/day	Up to 2 yr of age: 50,000 units/kg of body weight as a single dose

*See USP DI for complete dosage information.

TABLE 7-4 *Continued*

PENICILLINS

Name Proprietary Name	***Preparations***	***Usual Adult Dose****	***Usual Dose Range****	***Usual Pediatric Dose****
	Penicillin G benzathine tablets USP Penicillin G benzathine oral suspension USP	400,000–600,000 units every 4–6 hr	Up to 12.6 million units/day	Infants and children up to 12 yr: 4167–15,000 units/kg of body weight every 4 hr; 6230–28,500 units/kg of body weight every 6 hr; or 6333–30,000 units/kg of body weight every 8 hr
Penicillin V USP *Pen-Vee, V-Cillin*	Penicillin V for oral suspension USP	The equivalent of 125–500 mg (200,000–800,000 units) of penicillin every 6–8 hr	Up to 7.2 g/day	Infants and children up to 12 yr: 2.5–4 mg (4167–16,667 units)/kg of body weight every 4 hr; 3.75–15.6 mg (6250–25,000 units)/kg of body weight every 6 hr; or 5–20.8 mg (8333–33,333 units)/kg of body weight every 8 hr; children 12 yr and older: see Usual Adult Dose
Penicillin V potassium USP *Pen-Vee-K, V-Cillin-K*	Penicillin V potassium for oral solution USP Penicillin V potassium tablets USP			
Penicillin V benzathine USP *Pen-Vee*	Penicillin V benzathine oral suspension USP	125–250 mg every 6–8 hr	125–375 mg every 6–8 hr	
Methicillin sodium USP *Staphcillin*	Methicillin sodium for injection USP	IM, 1 g every 4–6 hr; IV, 1 g every 6 hr	Up to 24 g/day	IM, 25 mg/kg of body weight every 6 hr; IV, 16.7–33.3 mg/kg of body weight every 4 hr; or 25–50 mg/kg of body weight every 6 hr
Oxacillin sodium USP *Prostaphlin*	Oxacillin sodium capsules USP Oxacillin sodium for oral solution USP	Oral, the equivalent of 500 mg–1 g of oxacillin every 4–6 hr	Up to 6 g/day	Children up to 40 kg of body weight: 12.5–25 mg/kg of body weight every 6 hr daily; children 40 kg and over: see Usual Adult Dose
	Oxacillin sodium for injection USP	IM or IV, the equivalent of 250 mg–1 g of oxacillin every 4–6 hr	Up to 20 g/day	IM or IV, up to 40 kg of body weight: the equivalent of 12.5–25 mg of oxacillin/kg of body weight every 4 hr; 40 kg of body weight and over: see Usual Adult Dose
Cloxacillin sodium USP *Tegopen*	Cloxacillin sodium capsules USP Cloxacillin sodium for oral solution USP	The equivalent of 250–500 mg every 6 hr	Up to 6 g/day	Up to 20 kg of body weight: 12.5–25 mg/kg of body weight every 6 hr; 20 kg of body weight and over: see Usual Adult Dose
Dicloxacillin sodium USP *Dynapen, Pathocil, Veracillin*	Dicloxacillin sodium capsules USP Dicloxacillin sodium for oral suspension USP Sterile dicloxacillin sodium USP	The equivalent of 125–250 mg of dicloxacillin every 6 hr	Up to 6 g/day	Up to 40 kg of body weight: 3.125–6.25 mg/kg of body weight every 6 hr; 40 kg and over: see Usual Adult Dose
Nafcillin sodium USP *Unipen*	Nafcillin sodium capsules USP Nafcillin sodium for oral solution USP Nafcillin sodium tablets USP	The equivalent of 250 mg–1 g of nafcillin every 4–6 hr	Up to 6 g/day	Newborn infants: the equivalent of nafcillin 10 mg/kg of body weight every 4–6 hr; older infants and children: the equivalent of nafcillin 6.25–12.5 mg/kg of body weight every 6 hr
	Nafcillin sodium for injection USP	IM, the equivalent of 500 mg of nafcillin every 4–6 hr; IV, 500 mg–1 g every 4 hr	IM, up to 12 g/day; IV, up to 20 g/day	Newborn infants, IM: the equivalent of nafcillin 10 mg/kg of body weight bid; older infants and children, IM: the equivalent of nafcillin 25 mg/kg of body weight or 750 mg/m^2 of body surface, bid

*See USP DI for complete dosage information.

TABLE 7-4 *Continued*

PENICILLINS

Name Proprietary Name	***Preparations***	***Usual Adult Dose****	***Usual Dose Range****	***Usual Pediatric Dose****
Ampicillin USP *Penbritin, Polycillin, Omnipen, Amcill, Principen*	Ampicillin capsules USP Ampicillin chewable tablets USP Ampicillin for oral suspension USP Sterile ampicillin for suspension USP	250–500 mg every 6 hr	Up to 6 g/day	Infants and children up to 20 kg of body weight: 12.5–25 mg/kg of body weight every 6 hr, or 16.7–33.3 mg/kg of body weight every 8 hr; children 20 kg of body weight and over: see Usual Adult Dose
Penbritin-S, Polycillin-N, Omnipen-N	Sterile ampicillin sodium USP	IM or IV, the equivalent of ampicillin, 250–500 mg every 6 hr	Up to 16 g/day; or up to 300 mg/kg of body weight/day	IM or IV, the equivalent of ampicillin, infants up to 20 kg: 6.25–25 mg/kg of body weight every 6 hr; infants and children 20 kg and over: see Usual Adult Dose
Hetacillin USP *Versapen*	Hetacillin for oral suspension USP Hetacillin tablets USP	Oral, the equivalent of ampicillin, 225–450 mg of every 6 hr	Up to 6 g/day	Oral, the equivalent of ampicillin, infants and children up to 40 kg: 5.625–11.25 mg/kg of body weight every 6 hr; children 40 kg and over: see Usual Adult Dose
Hetacillin potassium USP *Versapen-K*	Hetacillin potassium capsules USP	See dose for oral suspension	See dose for oral suspension	See dose for oral suspension
Bacampicillin hydrochloride USP Spectrobid	Bacampicillin hydrochloride tablets USP Bacampicillin hydrochloride oral suspension USP	Oral, 400–800 every 12 hr (equivalent to 280–560 mg of ampicillin)		25–50 mg/kg daily every 12 hr
Amoxicillin USP *Amoxil, Larotid*	Amoxicillin capsules USP Amoxicillin for oral suspension USP	The equivalent of anhydrous amoxicillin, 250–500 mg every 8 hr	The equivalent of anhydrous amoxicillin, up to 4.5 g/day	Oral, the equivalent of anhydrous amoxicillin, infants weighing up to 6 kg: 25–50 mg every 8 hr; infants 6–8 kg: 50–100 mg every 8 hr; infants and children 8–20 kg: 6.7–13.3 mg/kg of body weight every 8 hr; children 20 kg and over: see Usual Adult Dose
Cyclacillin *Cyclapen*	Tablets, oral suspension	Oral, 250 mg qid	1–2 g/day	50–100 mg/kg per day
Sterile carbenicillin disodium USP *Geopen, Pyopen*	Sterile carbenicillin disodium USP	Septicemia, meningitis, respiratory tract or soft-tissue infections, IM or IV: the equivalent of carbenicillin, 50–83.3 mg/kg of body weight every 4 hr; urinary tract infections, IM or IV: the equivalent of carbenicillin, 1–2 g every 6 hr or up to 50 mg/kg of body weight every 6 hr	The equivalent of carbenicillin up to 42 g/day	Neonates weighing up to 2 kg, septicemia, meningitis, respiratory tract or soft-tissue infections, IM or IV: the equivalent of carbenicillin, 100 mg/kg of body weight initially, then 75 mg/kg of body weight every 8 hr during the first week of life; 100 mg/kg of body weight every 6 hr thereafter; neonates weighing 2 kg and over, septicemia, meningitis, respiratory tract or soft-tissue infections, IM or IV: the equivalent of carbenicillin, 100 mg/kg of body weight initially, then 75 mg/kg of body weight every 6 hr during the first 3 days of life; 100 mg/kg of body weight every 6 hr thereafter; older infants and children, septicemia, meningitis, respiratory tract or soft-tissue infections: see Usual Adult Dose; urinary tract infections, IM or IV: the equivalent of carbenicillin, 12.5–50 mg/kg of body weight every 6 hr, or 8.3–33.3 mg/kg of body weight every 4 hr

*See USP DI for complete dosage information.

TABLE 7-4 *Continued*

PENICILLINS

Name Proprietary Name	***Preparations***	***Usual Adult Dose****	***Usual Dose Range****	***Usual Pediatric Dose****
Carbenicillin indanyl sodium USP *Geocillin*	Carbenicillin indanyl sodium tablets USP	Oral, the equivalent of carbenicillin, 382–764 mg every 6 hr		
Sterile ticarcillin disodium USP *Ticar*	Sterile ticarcillin disodium USP	Septicemia, respiratory tract, skin, and soft-tissue infections, IV infusion: the equivalent of ticarcillin, 3 g every 3–6 hr; 25–37.5 mg/kg of body weight every 3 hr; 33.5–50 mg/kg of body weight every 4 hr; or 50–75 mg/kg of body weight every 6 hr	The equivalent of ticarcillin, up to 500 mg/kg of body weight/day	
Sterile azlocillin sodium USP *Azlin*	Sterile azlocillin sodium USP	Lower respiratory tract infections, bone and joint infections, septicemia: the equivalent of azlocillin of 16–18 g/day (225–300 mg/kg daily); 3 g every 4 hr or 4 g every 8 hr IV. Uncomplicated urinary tract infections: 8 g/day (100–125 mg/kg daily); 2 g every 6 hr IM or IV. Complicated urinary tract infections: 12 g/day (150–200 mg/kg daily); 3 g every 6 hr	24 g/day	Cystic fibrosis: 75 mg/kg every 4 hr (450 mg/kg daily) IV over 30 min. Do not exceed 24 g/day
Sterile mezlocillin sodium USP *Mezlin*	Sterile mezlocillin sodium USP	Life-threatening infections: the equivalent of mezlocillin of up to 350 mg/kg daily; 4 g every 4 hr IV. Lower respiratory tract, intra-abdominal, gynecologic, skin and skin structure infections and septicemia: 225–300 mg/kg daily (16–18 g/day); 4 g every 6 hr or 3 g every 4 hr IV. Uncomplicated urinary tract infections: 100–125 mg/kg daily (6–8 g/day); 1.5–2 g every 6 hr IM or IV. Complicated urinary tract infections: 150–200 mg/kg daily; 3 g every 6 hr IV	24 g/day	
Sterile piperacillin sodium USP *Pipracil*	Sterile piperacillin sodium USP	Serious infections (septicemia, intra-abdominal infections, pneumonia, gynecologic infections, skin and soft-tissue infections): The equivalent of 12–18 g/day IV (200–300 mg/kg daily) in divided doses every 4–6 hr. Uncomplicated urinary tract infections: 6–8 g/day IM or IV (100–125 mg/kg daily) in divided doses every 6–12 hr. Complicated	24 g/day	

*See USP DI for complete dosage information.

TABLE 7-4 *Continued*

PENICILLINS

Name Proprietary Name	***Preparations***	***Usual Adult Dose****	***Usual Dose Range****	***Usual Pediatric Dose****
		urinary tract infections: 8–16 g/day IV (125–200 mg/kg daily) in divided doses every 6–8 hr. Prophylaxis: Intra-abdominal surgery: 2 g IV just before surgery; 2 g during surgery; 2 g every 6 hr after surgery for no more than 24 hr		
Amdinocillin USP *Coactin*	Sterile amdinocillin USP	Serious infections: 60 mg/kg daily in divided doses (10 mg/kg every 4 hr). Combination therapy with another β-lactam antibiotic: reduce to 40 mg/kg daily in divided doses (10 mg/kg every 6 hr)		

*See USP DI for complete dosage information.

β-LACTAMASE INHIBITORS

The strategy of using a β-lactamase inhibitor in combination with a β-lactamase-sensitive penicillin in the therapy of infections caused by β-lactamase-producing bacterial strains has, until relatively recently, failed to live up to its obvious promise. Early attempts to obtain synergy against such resistant strains, utilizing combinations consisting of a β-lactamase-resistant penicillin (such as methicillin or oxacillin) as a competitive inhibitor and a β-lactamase-sensitive penicillin (such as ampicillin or carbenicillin) to kill the organisms, met with limited success. Factors that may contribute to the failure of such combinations to achieve synergy include (1) failure of most lipophilic penicillinase-resistant penicillins to penetrate the cell envelope of gram-negative bacilli in effective concentrations; (2) the reversible binding of penicillinase-resistant penicillins to β-lactamase requires high concentrations to prevent substrate binding and hydrolysis; and (3) the induction of β-lactamases by some penicillinase-resistant penicillins.

The discovery of the naturally occurring, mechanism-based inhibitor clavulanic acid, which causes a potent and progressive inhibition of β-lactamases (Fig. 7-4), has created renewed interest in β-lactam combination therapy. This interest has led to the design and synthesis of additional mechanism-based β-lactamase inhibitors, such as sulbactam, and the isolation of naturally occurring β-lactams, such as the thienamycins, that both inhibit β-lactamases and interact with penicillin-binding proteins.

The chemical events leading to the inactivation of β-lactamases by mechanism-based inhibitors are very complex. In a review of the chemistry of β-lactamase inhibition, Knowles[54] has described two classes of β-lactamase inhibitors: class I inhibitors, such as clavulanic acid and sulbactam, that have a hetero atom leaving group at position 1; and class II inhibitors, such as the carbapenems, that do not. Unlike competitive inhibitors, which bind reversibly with the enzyme they inhibit, mechanism-based inhibitors react with the enzyme in much the same way that the substrate does. With the β-lactamases, an acylenzyme intermediate is formed by reaction of the β-lactam with an active site serine hydroxyl group of the enzyme. For normal substrates the acylenzyme intermediate readily undergoes hydrolysis, destroying the substrate and freeing the enzyme to attack more substrate. The acylenzyme intermediate formed when a mechanism-based inhibitor is attacked by the enzyme is diverted by tautomerism to a more stable imine form that hydrolyzes more slowly to eventually free the enzyme (transient inhibition) or, for a class I inhibitor, may attack a second group on the enzyme to inactivate it. Because these inhibitors are also substrates for the enzymes that they inactivate, they are sometimes referred to as "suicide substrates."

Because they cause permanent inactivation of β-lactamases, class I inhibitors are particularly useful in combination with extended-spectrum, β-lactamase penicillins to treat infections caused by β-lactamase-producing bacteria. Two such inhibitors, clavulanic acid and sulbactam, are currently mar-

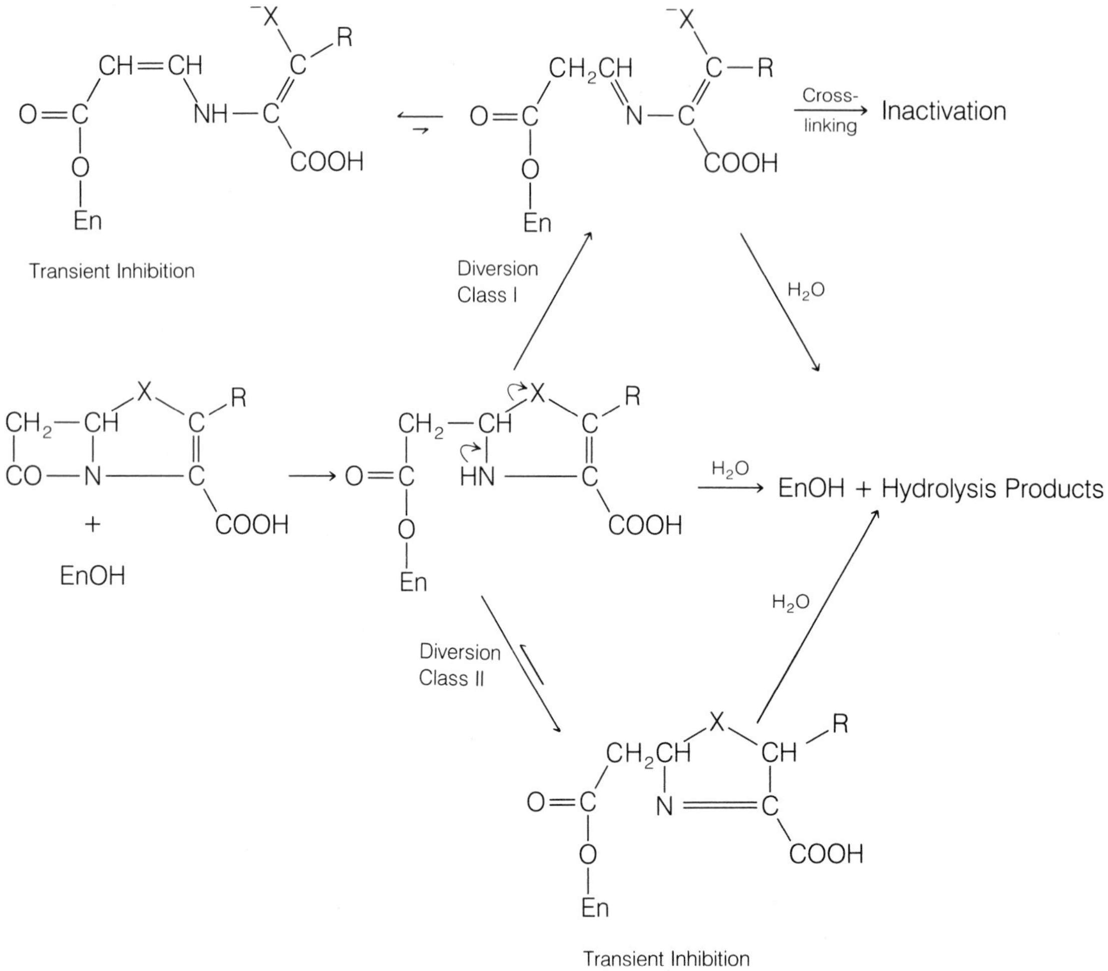

FIG. 7-4. Mechanism-based inhibition of β-lactamases.

keted in the United States for this purpose. A class II inhibitor, the carbapenem derivative imipenem, has potent antibacterial activity, in addition to its ability to cause transient inhibition of some β-lactamases. Certain antibacterial cephalosporins, with a leaving group at the C-3 position, can cause transient inhibition of β-lactamases by forming stabilized acylenzyme intermediates. These will be discussed more fully later in this chapter.

Clavulanate Potassium, USP. Clavulanic acid is an antibiotic isolated from *Streptomyces clavuligeris*. Structurally it is a 1-oxopenam lacking the 6-acylaminoside chain of penicillins, but possessing a 2-hydroxyethylidene moiety at C-2. Clavulanic acid exhibits very weak antibacterial activity, comparable with 6-APA and, therefore, is not useful as an antibiotic. It is, however, a potent inhibitor of *Staph. aureus* β-lactamase and plasmid-mediated β-lactamases elaborated by gram-negative bacilli.

Combinations of amoxicillin and the potassium salt of clavulanic acid are available (Augmentin) in a variety of fixed-dose, oral dosage forms intended for

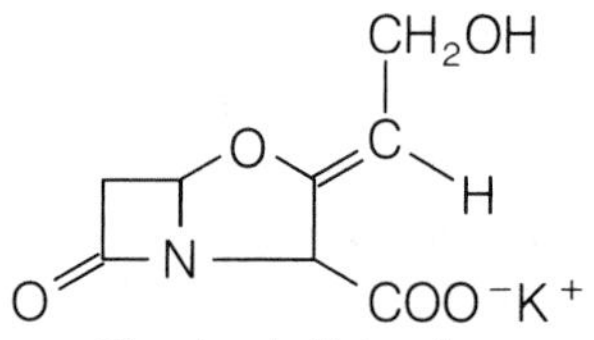

Clavulanate Potassium

the treatment of skin, respiratory, ear, and urinary tract infections caused by β-lactamase-producing bacterial strains. These combinations are effective against β-lactamase-producing strains of *Staph. aureus*, *E. coli*, *K. pneumoniae*, *Enterobacter*, *H. influenzae*, *Br. catarrhalis*, and *H. ducreyi* that are resistant to amoxicillin alone. The oral bioavailability of amoxicillin and potassium clavulinate are similar. It should be noted that clavulanic acid is acid-stable. It cannot undergo penicillanic acid formation because it lacks an amide side chain.

Potassium clavulanate and the extended-spectrum penicillin ticarcillin have been combined in a fixed-dose injectable form for the control of serious infections caused by β-lactamase-producing bacterial strains. This combination has been recommended for septicemia, lower respiratory tract infections, and urinary tract infections caused by β-lactamase-producing *Klebsiella*, *E. coli*, *Ps. aeruginosa* and other *Pseudomonas* species, *Citrobacter*, *Enterobacter*, *Serratia marcescens*, and *Staph. aureus*. It is also used in bone and joint infections caused by these organisms. The combination contains 3 g of ticarcillin disodium and 100 mg of potassium clavulinate in a sterile powder for injection (Timentin).

Sulbactam is penicillanic acid sulfone or 1,1-dioxopenicillanic acid. This synthetic penicillin derivative is a potent inhibitor of *Staph. aureus* β-lactamase, as well as many β-lactamases elaborated by gram-negative bacilli. Sulbactam has weak intrinsic antibacterial activity, but potentiates the activity of ampicillin and carbenicillin against β-lactamase-producing *Staph. aureus* and members of the *Enterobacteriaceae* family. However, it does not synergize with either carbenicillin or ticarcillin against *Ps. aeruginosa* strains resistant to these agents. Failure of sulbactam, a rather polar β-lactam, to penetrate the cell envelope is a possible explanation for the lack of synergy.

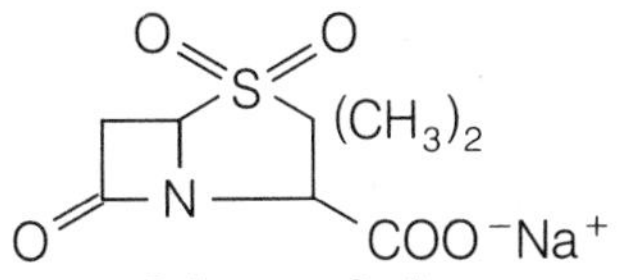

Sulbactam Sodium

Fixed-dose combinations of ampicillin sodium and sulbactam sodium, marketed under the trade name of Unasyn as sterile powders for injection, have been approved for use in the United States. These combinations are recommended for the treatment of skin, tissue, intra-abdominal, and gynecologic infections caused by β-lactamase-producing strains of *Staph. aureus*, *E. coli*, *Klebsiella*, *Pr. mirabilis*, *B. fragilis*, *Enterobacter*, and *Acinetobacter*.

Carbapenems

Thienamycin is a novel β-lactam antibiotic first isolated and identified by researchers at Merck[55] from fermentation of cultures of *Streptomyces cattleya*. Its structure and absolute configuration were established by both spectroscopic and total synthesis procedures.[56,57] Two structural features of thienamycin are shared with the penicillins and cephalosporins: a fused bicyclic ring system containing a β-lactam and an equivalently attached 3-carboxyl group. In other respects, the thienamycins represent a significant departure from the established β-lactam antibiotics. The bicyclic system consists of a carbapenam containing a double bond between C-2 and C-3 [i.e., it is a 2-carbapenem (or Δ^2-carbapenem) system]. The presence of the double bond in the bicyclic structure creates considerable ring strain and increases the reactivity of the β-lactam to ring-opening reactions. The side chain is unique in two respects: it is a simple-1-hydroxyethyl group, instead of the familiar acylamino side chains; and it is oriented α to the bicyclic ring system rather than having the usual β-orientation of the penicillins and cephalosporins. The remaining feature is a 2-aminoethylthioether function at C-2. The absolute stereochemistry of thienamycin has been determined to be 5R : 6S : 8S. Several additional structurally related antibiotics have been isolated from various *Streptomyces* species, including the four epithienamycins, which are isomeric to thienamycin at C-5, C-6, or C-8, and derivatives wherein the 2-aminoethylthio side chain is modified.

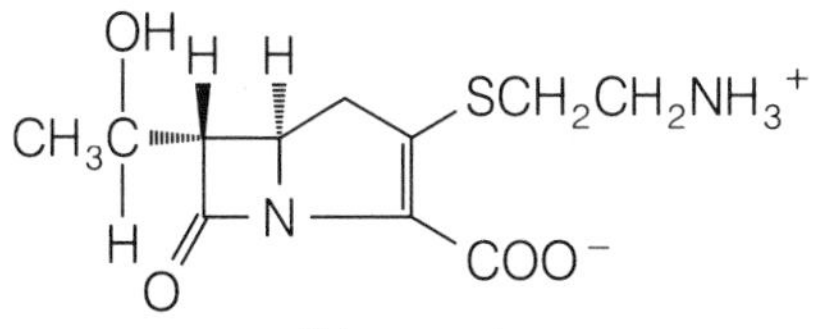

Thienamycin

Thienamycin displays outstanding broad-spectrum antibacterial properties in vitro.[58] It is highly active against most aerobic and anaerobic gram-positive and gram-negative bacteria, including *Staph. aureus*, *Ps. aeruginosa*, and *B. fragilis*. Furthermore, thienamycin is resistant to inactivation by most β-lactamases elaborated by gram-negative and gram-positive bacteria and, therefore, is effective against many strains resistant to penicillins and cephalosporins. Resistance to lactamases appears to be a function of the α-1-hydroxyethyl side chain, because this property is lost in the 6-nor derivative and epithienamycins having β-stereochemistry show variable resistance to the different β-lactamases.

An unfortunate property of thienamycin is its chemical instability in solution. It is more susceptible to hydrolysis in both acidic and alkaline solution than most β-lactam antibiotics owing to the strained nature of its fused ring system containing an endocyclic double bond. Furthermore, at its optimally stable *p*H of between 6 and 7, thienamycin undergoes a concentration-dependent inactivation. This inactivation is believed to result from intermolecular aminolysis of the β-lactam by the cysteamine side chain of a second molecule. Another shortcoming of thienamycin is its susceptibility to hydrolytic inactivation by renal dehydropeptidase-1 (DHP-1),[59] which causes it to have an unacceptably short half-life in vivo.

Imipenem–Cilastatin (Primaxin). Imipenem is *N*-formimidoylthienamycin, the most successful of a series of chemically stable derivatives of thienamycin in which the primary amino group is converted to a nonnucleophilic basic function.[60] Cilastatin is an inhibitor of DHP-1. The combination provides a chemically stable form of thienamycin that has clinically useful pharmacokinetic properties.

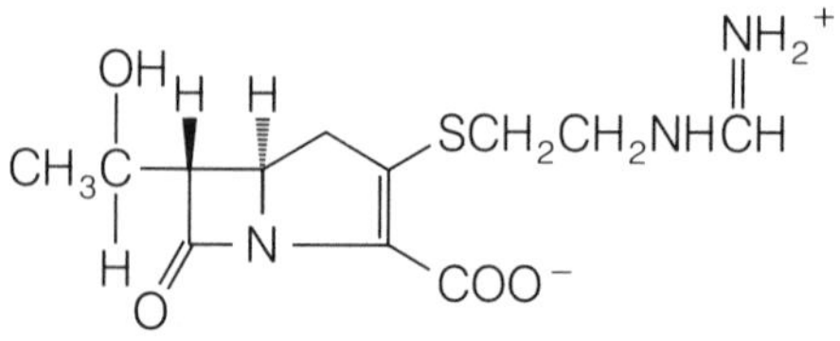

Imipenem

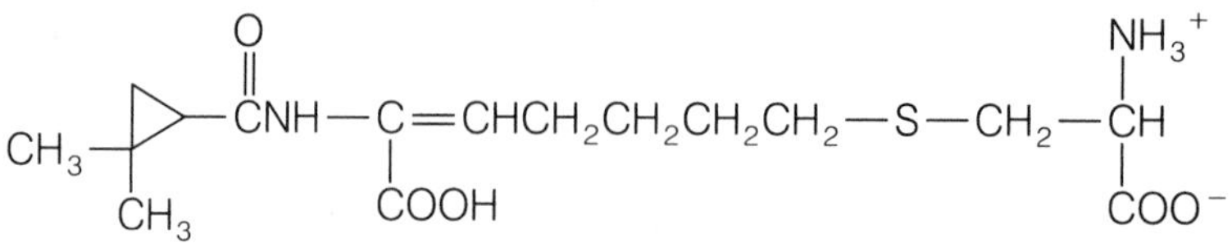

Cilastatin

Imipenem retains the extraordinary broad-spectrum antibacterial properties of thienamycin. Its bactericidal activity results from the inhibition of cell wall synthesis associated with bonding to PBPs 1_b and 2. Imipenem is very stable to most β-lactamases. It is an inhibitor of β-lactamases from certain gram-negative bacteria resistant to other β-lactam antibiotics, for example, *Ps. aeruginosa*, *Ser. marcescens*, and *Enterobacter*.

Imipenem is indicated for the treatment of a wide variety of bacterial infections of the skin and tissues, lower respiratory tract, bones and joints, and genitourinary tract, septicemia, and endocarditis caused by β-lactamase-producing strains of susceptible bac-

TABLE 7-5

β-LACTAMASE INHIBITORS

Name Proprietary Name	*Preparations*	*Usual Adult Dose**	*Usual Dose Range**	*Usual Pediatric Dose**
Ampicillin sodium and sulbactam sodium *Unasyn*	Sterile ampicillin sodium and sulbactam sodium	1.5 g (1 g ampicillin plus 0.5 g sulbactam) to 3 g every 6 hr IV or IM	4 g/day sulbactam, 6 g/day ampicillin	
Amoxicillin and potassium clavulanate USP *Augmentin*	Amoxicillin and potassium clavulanate tablets USP Amoxicillin and potassium clavulanate oral suspension USP	Usual dose: Orally one 250-mg tablet (contains 250 mg amoxicillin and 125 mg potassium clavulanate) every 8 hr. Severe infections and respiratory tract infections: One 500-mg tablet (contains 500 mg amoxicillin and 125 mg potassium clavulanate) every 8 hr. Chancroid: One 500-mg tablet tid		Usual dose: children less than 40 kg: 20 mg/kg daily in divided oral doses every 8 hr. Severe infections, respiratory tract infections, otitis media, sinusitis, and lower respiratory infections: 40 mg/kg daily in divided oral doses every 8 hr
Ticarcillin and clavulanate potassium *Timentin*	Sterile ticarcillin and clavulanate potassium	Systemic and urinary tract infections: 3.1 g (3 g ticarcillin and 100 mg clavulanate potassium) every 4–6 hr IV for adults ≥ 60 kg. For adults < 60 kg: 200–300 mg/kg daily (based on ticarcillin content) in divided doses every 4–6 hr		
Imipenem–cilastatin *Primaxin*	Imipenem–cilastatin	The equivalent of 250–500 mg of imipenem every 6 hr IM or IV for mild infections and uncomplicated urinary tract infections; 500 mg–1 g every 6 hr IM or IV for moderately severe infections and complicated urinary tract infections; 500 mg–1 g every 6 hr IV for life-threatening infections	4 g/day	

*See USP DI for complete dosage information.

teria. These include aerobic gram-positive organisms such as *Staph. aureus*, *Staph. epidermidis*, enterococci, and *Streptococcus viridans*; aerobic gram-negative bacteria such as *E. coli*, *Klebsiella*, *Serratia*, *Providencia*, *Haemophilus*, *Citrobacter*, indole-positive *Proteus*, *Morganella morgani*, *Acinetobacter*, *Enterobacter*, and *Ps. aeroginosa*; and anaerobes such as *B. fragilis*, *Clostridium*, *Peptococcus*, *Peptidostreptococcus*, *Eubacterium*, and *Fusobacterium*. Some *Pseudomonas* species are resistant, such as *Ps. maltophilia* and *Ps. cepacia*, as are some methicillin-resistant staphylococci. Of course, imipenem is effective against non-β-lactamase-producing strains of these and additional bacterial species, but other less expensive and equally effective antibiotics are preferred for the treatment of infections caused by these organisms.

The imipenem–cilastatin combination is marketed as a sterile powder intended for the preparation of solutions for intravenous infusion. Such solutions are stable for four hours at 25°C and up to 24 hours when refrigerated. The concomitant administration of imipenem with an aminoglycoside antibiotic results in synergistic antibacterial activity in vivo. The two types of antibiotics are, however, chemically incompatible and should never be combined in the same intravenous bottle.

CEPHALOSPORINS

Historical Background

The cephalosporins are β-lactam antibiotics isolated from *Cephalosporium* species or prepared semisynthetically. Most of the antibiotics introduced since 1965 have been semisynthetic cephalosporins. Interest in *Cephalosporium* fungi began in 1945 with Giuseppe Brotzu's discovery that cultures of *C. acremonium* inhibited the growth of a wide variety of gram-positive and gram-negative bacteria. Abraham and his colleagues in Oxford, having been supplied cultures of the fungus in 1948, isolated three principal antibiotic components: cephalosporin P1, a steroid with minimal antibacterial activity; cephalosporin N, later discovered to be identical with synnematin N (a penicillin derivative now called penicillin N that had earlier been isolated from *C. salmosynnematum*); and cephalosporin C.

The structure of penicillin N was discovered to be D-(4-amino-4-carboxybutyl)penicillanic acid. The amino acid side chain confers increased activity against gram-negative bacteria, particularly *Salmonella* species, but reduced activity against gram-positive organisms compared with penicillin G. It has been used successfully in clinical trials for the treatment of typhoid fever, but was never released as an approved drug.

Cephalosporin C turned out to be a close congener of penicillin N, containing a dihydrothiazine ring instead of the thiazolidine ring of the penicillins. Despite the observation that cephalosporin C was resistant to *Staph. aureus* β-lactamase, early interest in it was not great because its antibacterial potency was inferior to that of penicillin N and other penicillins. However, the discovery that the α-aminoadipoyl side chain could be removed to efficiently produce 7-aminocephalosporanic acid (7-ACA)[61,62] prompted investigations that led to semisynthetic cephalosporins of medicinal value. The relationship of 7-ACA and its acyl derivatives to 6-APA and the semisynthetic penicillins is obvious. Woodward et al.[63] have prepared both cephalosporin C and the clinically useful cephalothin by an elegant synthetic procedure, but the commercially available drugs are obtained as semisynthetic products from 7-ACA.

Nomenclature

The systematic chemical nomenclature of the cephalosporins is slightly more complex than even that of the penicillins, because of the presence of a double bond in the dihydrothiazine ring. The fused ring system is designated by *Chemical Abstracts* as 5-thia-1-azabicyclo[4.2.0]oct-2-ene. With this system, cephalothin is 3-(acetoxymethyl)-7-(2-thienyl)-8-oxo-5-thia-1-azabicyclo[4.2.0]oct-2-ene-2-carboxylic acid. A simplification that retains some of the systematic nature of the *Chemical Abstracts*' procedure is to name the saturated bicyclic ring system with the lactam carbonyl oxygen as cepham (*cf.*, penam for penicillins). According to this system all of the commercially available cephalosporins and cephamycins are named as 3-cephems (or Δ^3-cephems) to designate the position of the double bond. (Interestingly, all known 2-cephems are inactive, presumably be-

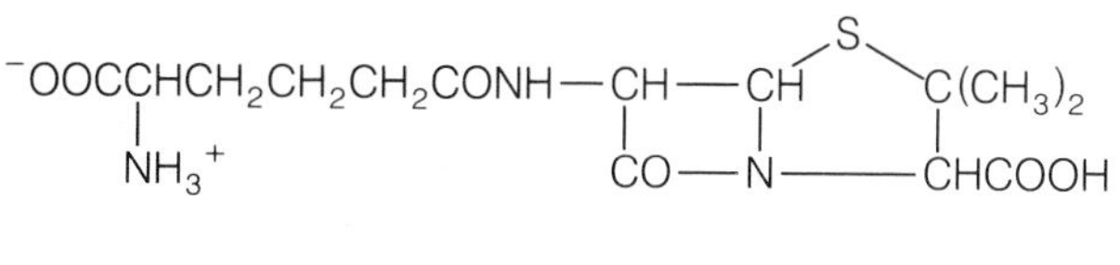

$^-OOCCH(NH_3^+)CH_2CH_2CH_2CONH$—CH—CH—S—$C(CH_3)_2$—CHCOOH—N—CO

Penicillin N
(Cephalosporin N, Synnematin B)

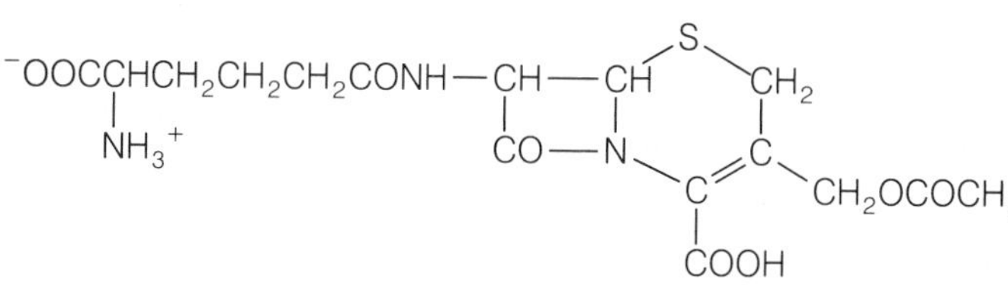

$^-OOCCH(NH_3^+)CH_2CH_2CH_2CONH$—CH—CH—S—$CH_2$—C($CH_2OCOCH_3$)=C(COOH)—N—CO

Cephalosporin C

cause the β-lactam lacks the necessary ring strain to be sufficiently reactive.) The trivialized forms of nomenclature of the type that have been applied to the penicillins are not consistently applicable to the naming of cephalosporins because of variations in the substituent at the 3-position. Thus, although some cephalosporins have been named as derivatives of cephalosporanic acids, this practice applies only to the derivatives that have a 3-acetoxymethyl group.

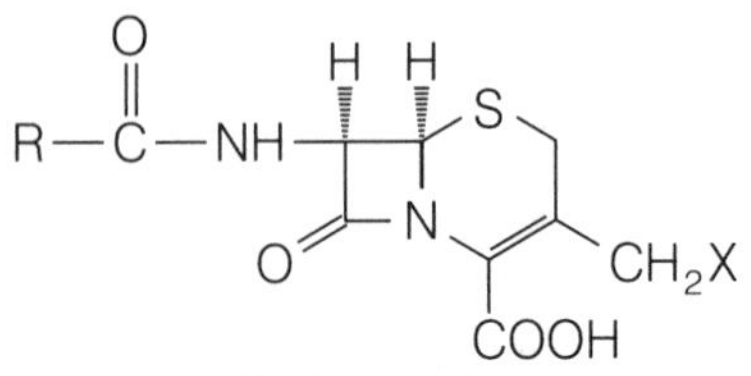

Cephalosporins

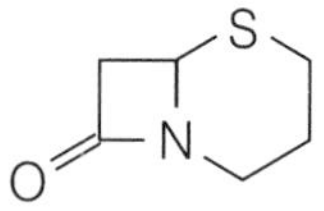

Cepham

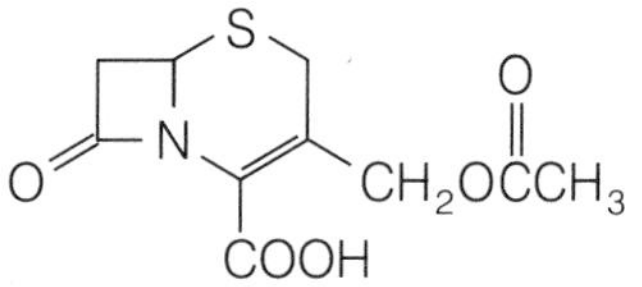

Cephalosporanic Acid

Semisynthetic Derivatives

To date, the more useful semisynthetic modifications of the basic 7-ACA nucleus have resulted from acylations of the 7-amino group with different acids or nucleophilic substitution or reduction of the acetoxyl group. Structure–activity relationships among the cephalosporins appear to parallel those among the penicillins insofar as the acyl group is concerned. However, the presence of an allylic acetoxyl function in the 3-position provides a reactive site at which various 7-acylamino cephalsporanic acid structures can easily be varied by nucleophilic displacement reactions. Reduction of the 3-acetoxymethyl to a 3-methyl substituent to prepare 7-aminodesacetylcephalosporanic acid (7-ACDA) derivatives can be accomplished by catalytic hydrogenation, but the process currently employed for the commercial synthesis of 7-ACDA derivatives involves the rearrangement of the corresponding penicillin sulfoxide.[64] Perhaps the most noteworthy development thus far is the discovery that 7-phenylglycyl derivatives of 7-ACA and especially 7-ACDA are active orally.

In the preparation of semisynthetic cephalosporins, the following improvements are sought: (1) increased acid stability; (2) improved pharmacokinetic properties, particularly better oral absorption; (3) broadened antimicrobial spectrum; (4) increased activity against resistant microorganisms (as a result of resistance to enzymatic destruction, improved penetration, increased receptor affinity, and

TABLE 7-6

STRUCTURE OF CEPHALOSPORINS

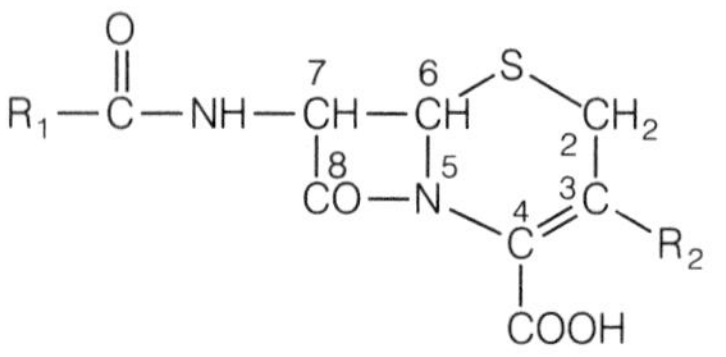

Generic Name	R_1	R_2
ORAL CEPHALOSPORINS		
Cephalexin	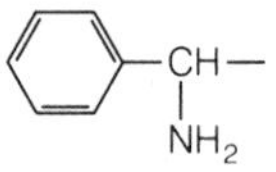	$-CH_3$
Cephradine	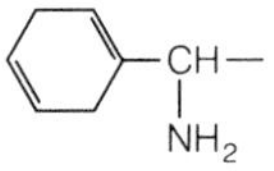	$-CH_3$
Cefadroxil	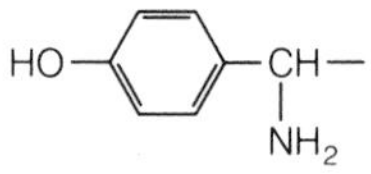	$-CH_3$
Cefaclor	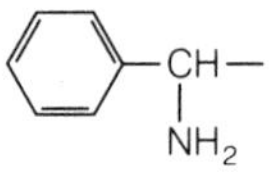	$-Cl$
PARENTERAL CEPHALOSPORINS		
Cephalothin	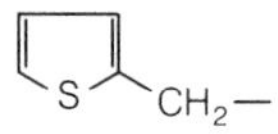	$-CH_2OCOCH_3$
Cephapirin	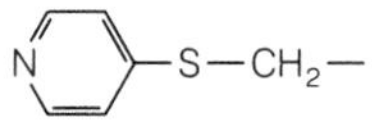	$-CH_2OCOCH_3$
Cefazolin	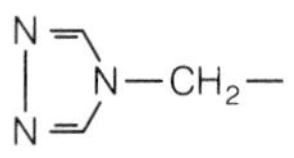	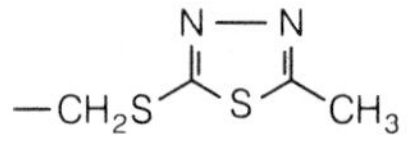
Cefamandole	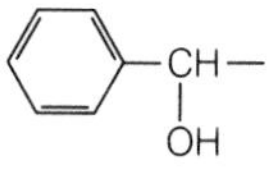	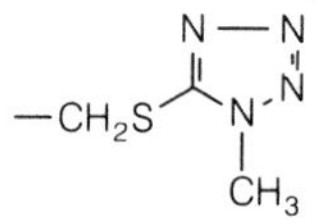
Cefonacid	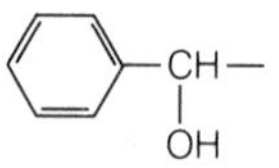	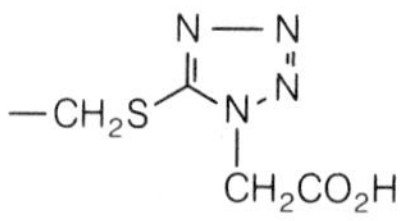
Ceforanide	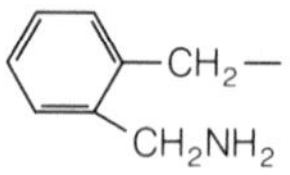	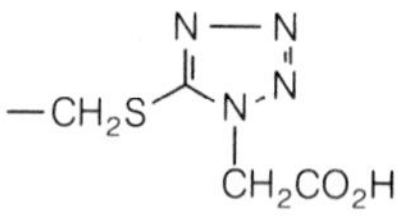
Cefuroxime	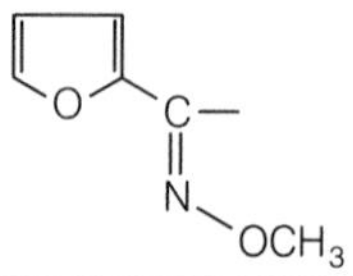	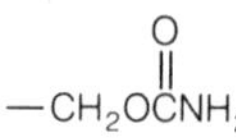 $-CH_2OCNH_2$

TABLE 7-6 *Continued*

STRUCTURE OF CEPHALOSPORINS

PARENTERAL CEPHALOSPORINS

	R_1	R_2
Cefotaxime	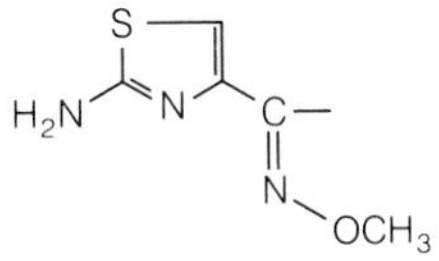 H_2N–thiazolyl–$C(=N–OCH_3)$–	$-CH_2OC(O)NH_2$
Ceftizoxime	H_2N–thiazolyl–$C(=N–OCH_3)$–	$-H$
Ceftriaxone	H_2N–thiazolyl–$C(=N–OCH_3)$–	$-CH_2S$–[ring structure with CH_3, N, N, OH, O]
Ceftazidime	H_2N–thiazolyl–$C(=N–O–C(CH_3)_2–COOH)$–	$-CH_2N^+$[pyridinium]
Cefoperazone	HO–C_6H_4–CH(NH–C(=O)–N[ring structure with O, O, N–CH_2CH_3])–	$-CH_2S$–[tetrazole ring, N–CH_3]

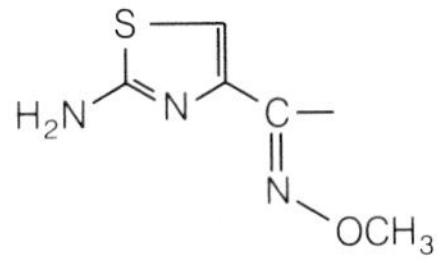

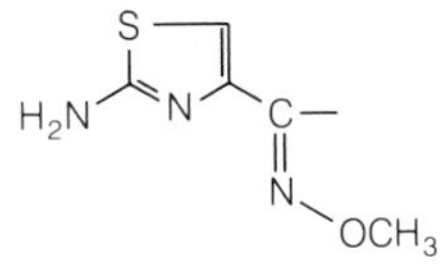

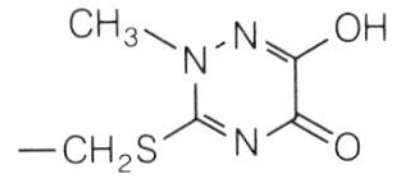

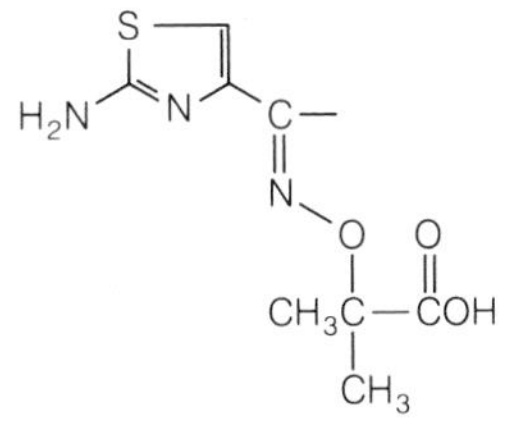

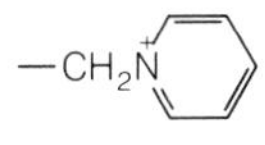

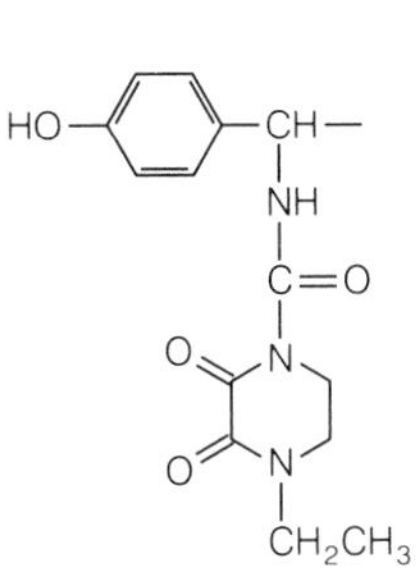

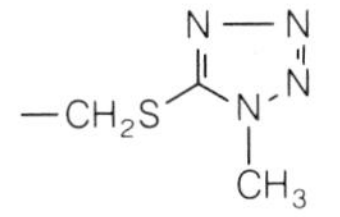

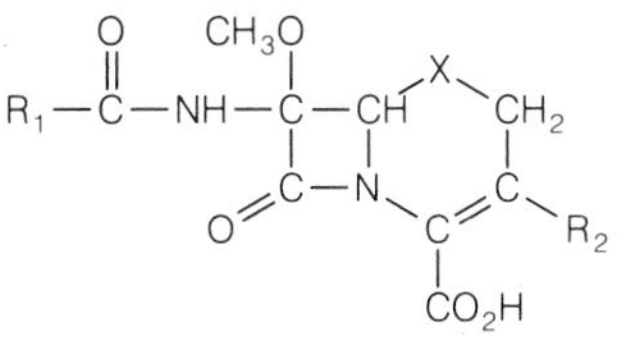

PARENTERAL CEPHAMYCINS

$R_1-C(O)-NH-C(OCH_3)$–[bicyclic ring structure: CH, X, CH_2, C–R_2, C–CO_2H, N, C=O]

	R_1	R_2	X
Cefoxitin	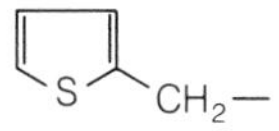 thienyl–CH_2–	$-CH_2OC(O)NH_2$	S
Cefotetan	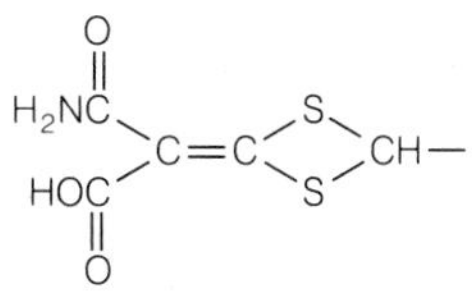 $H_2NC(O)$ / $HOC(O)$ > C=C<[ring structure with S, S]CH–	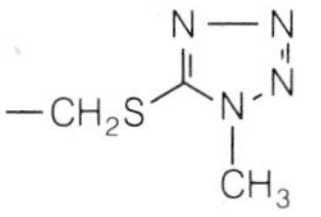 $-CH_2S$–[tetrazole ring, N–CH_3]	S

TABLE 7-6 *Continued*

STRUCTURE OF CEPHALOSPORINS

PARENTERAL CEPHAMYCINS

	R_1	R_2	X
Moxalactam	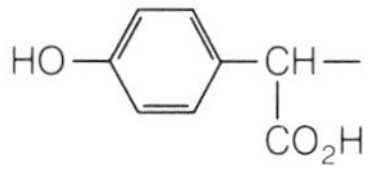 HO–C_6H_4–CH(CO_2H)–	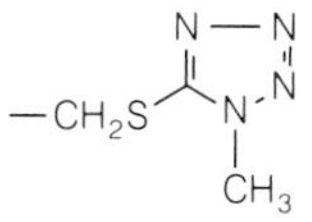 $-CH_2S$–[tetrazole ring, N–CH_3]	O

such); (5) decreased allergenicity; and (6) increased tolerance after parenteral administration.

Structures of cephalosporins currently marketed in the United States are represented in Table 7-6.

Chemical Degradation

Cephalosporins experience a variety of hydrolytic degradation reactions, the specific nature of which depends on the individual structure (see Table 7-6).[65] Among 7-acylaminocephalosporanic acid derivatives, the 3-acetoxylmethyl group is the most reactive site. In addition to its reactivity to nucleophilic displacement reactions, the acetoxyl function of this group of cephalosporins also readily undergoes solvolysis in strongly acidic solutions to form the desacetylcephalosporin derivatives. The latter lactonize to form the desacetylcephalosporin lactones, which are virtually inactive. The 7-acylamino group of some cephalosporins can also be hydrolyzed under enzymatic (acylases) and possibly nonenzymatic conditions to give 7-ACA (or 7-ADCA) derivatives. 7-Aminocephalosporanic acid, following hydrolysis or solvolysis of the 3-acetoxylmethyl group, also lactonizes under acidic conditions (Fig. 7-5).

The reactive functionality common to all cephalosporins is the β-lactam. Hydrolysis of the β-lactam of cephalosporins is believed to give initially cephalosporoic acids (in which the R′ group is stable, e.g., R′ = H or S heterocycle) or possibly anhydrodesacetylcephalosporoic acids (for the 7-acylaminocephalosporanic acids). It has not been possible to isolate either of these initial hydrolysis products in aqueous systems, however. Apparently, both types of cephalosporanic acids undergo fragmentation reactions that have not been fully characterized. However, studies of the in vivo metabolism[66] of orally administered cephalosporins have demonstrated the formation of arylacetylglycines and arylacetamidoethanols, which are believed to be formed from the corresponding arylacetylaminoacetaldehydes by metabolic oxidation and reduction, respectively. The aldehydes, no doubt, arise from nonenzymatic hydrolysis of the corresponding cephalosporoic

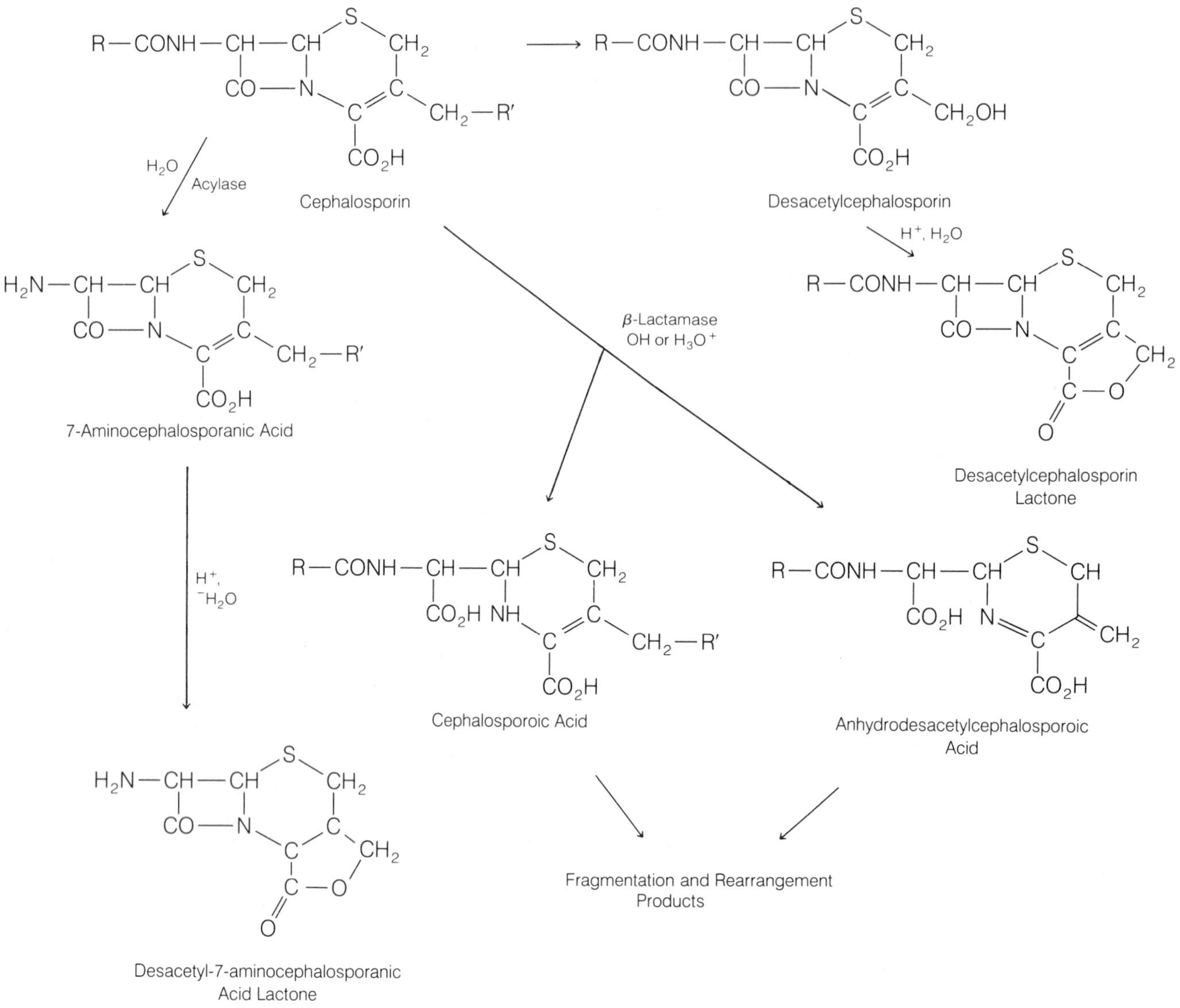

Cephalosporin

Desacetylcephalosporin

H_2O Acylase

H^+, H_2O

β-Lactamase OH or H_3O^+

7-Aminocephalosporanic Acid

Desacetylcephalosporin Lactone

H^+, $-H_2O$

Cephalosporoic Acid

Anhydrodesacetylcephalosporoic Acid

Fragmentation and Rearrangement Products

Desacetyl-7-aminocephalosporanic Acid Lactone

FIG. 7-5. *Degradation of cephalosporins.*

acids. No evidence for the intramolecular opening of the β-lactam ring by the 7-acylamino oxygen to form oxazolones of the penicillanic acid type has been found in the cephalosporins. However, at neutral to alkaline pH intramolecular aminolysis of the β-lactam ring by the α-amino group in the 7-ADCA derivatives, cephaloglycin, cephradine, and cefadroxil, occurs, forming diketopiperazine derivatives.[67,68] The formation of dimers, and possibly also polymers, from 7-ADCA derivatives containing an α-amino group in the acylamino side chain may also occur, especially in concentrated solutions and at alkaline pHs.

Oral Cephalosporins

The oral activity conferred by the phenylglycyl substituent is attributed to increased acid stability of the lactam ring resulting from the presence of a protonated amino group on the 7-acylamino portion of the molecule. The situation, then, is analogous to that of the α-aminobenzylpenicillins (e.g., ampicillin). Also important for high acid stability (and, therefore, good oral activity) of the cephalosporins is the absence of the leaving group at the 3-position. Thus, despite the presence of the phenylglycyl side chain in its structure, the cephalosporanic acid derivative, cephaloglycin, is poorly absorbed orally, presumably because of solvolysis of the 3-acetoxyl group in the low pH of the stomach. The resulting 3-hydroxyl derivative is known to undergo lactonization under acidic conditions. The 3-hydroxyl derivatives, and especially the corresponding lactones, are considerably less active in vitro than the parent cephalosporins. Generally, acyl derivatives of 7-ACDA show lower in vitro antibacterial potencies than the corresponding 7-ACA analogues.

Parenteral Cephalosporins

Hydrolysis of the ester function, catalyzed by hepatic and renal esterases, is responsible for some in vivo inactivation of parenteral cephalosporins containing a 3-acetoxymethyl substituent (e.g., cephalothin, cephapirin, and cefotaxime). The extent of such inactivation (20% to 35%) is not sufficiently great to compromise seriously the in vivo effectiveness of acetoxyl cephalosporins. Parenteral cephalosporins lacking a hydrolyzable group at the 3-position are not subject to hydrolysis by esterases. Cephradine is the only cephalosporin that is used both orally and parenterally.

Spectrum of Activity

The cephalosporins are considered broad-spectrum antibiotics with patterns of antibacterial effectiveness comparable with ampicillin's. Several significant differences exist, however. Cephalosporins are much more resistant to inactivation by β-lactamases, particularly those produced by gram-positive bacteria, than is ampicillin. However, ampicillin is generally more active against non–β-lactamase-producing strains of gram-positive and gram-negative bacteria sensitive to both it and the cephalosporins. Cephalosporins, among β-lactam antibiotics, exhibit uniquely potent activity against most species of *Klebsiella*. Differential potencies of cephalosporins, compared with penicillins, against different species of bacteria have been attributed to several variable characteristics of individual bacterial species and strains, the most important of which probably are (1) resistance to inactivation by β-lactamases, (2) permeability of bacterial cells, and (3) intrinsic activity against bacterial enzymes involved in cell wall synthesis and cross-linking.

β-Lactamase Resistance

The susceptibility of cephalosporins to various β-lactamases varies considerably with the source and properties of these enzymes. Cephalosporins are significantly less sensitive than all but the β-lactamase-resistant penicillins to hydrolysis by the enzymes from *Staph. aureus* and *Bacillus subtilis*. The "penicillinase" resistance of cephalosporins appears to be a property of the bicyclic cephem ring system, rather than of the acyl group. Despite natural resistance to staphylococcal β-lactamase, the different cephalosporins exhibit considerable variation in rates of hydrolysis by the enzyme.[69] Thus, cephalothin and cefoxitin are the most resistant, and cephaloridine and cefazolin are the least resistant of several cephalosporins tested in vitro. The same acyl functionalities that impart β-lactamase resistance in the penicillins unfortunately render cephalosporins virtually inactive against *Staph. aureus* and other gram-positive bacteria.

β-Lactamases elaborated by gram-negative bacteria present an exceedingly complex picture. Nearly a dozen distinct enzymes from various species of gram-negative bacilli have been identified and characterized,[35] differing widely in specificity for various β-lactam antibiotics. Most of these enzymes hydrolyze penicillin G and ampicillin at faster rates than they do the cephalosporins. However, some ("cephalosporinases") actually hydrolyze cephalosporins more rapidly, and inactivation by β-lactamases is an important factor in determining resistance to cephalosporins in many strains of gram-negative bacilli.

The introduction of polar substituents in the aminoacyl moiety of cephalosporins appears to confer stability to some β-lactamases.[70] Thus, cefamandole and cefonacid, which contain an α-hydroxyphenylacetyl (or mandoyl) group, and ceforanide, which has an *o*-aminophenyl acetyl group, are resistant to a few β-lactamases. Steric factors may also be important because cefoperazone, an acylureidocephalosporin that contains the same 4-ethyl-2,3-dioxo-1-piperazinylcarbonyl group present in piperacillin, is resistant to many β-lactamases. Oddly enough, piperacillin is hydrolyzed by most of these enzymes.

Two structural features confer broadly based resistance to β-lactamases among the cephalosporins:[70] (1) an alkoximino function in the aminoacyl group, and (2) a methoxyl substituent at the 7-position of the cephem nucleus having α-stereochemistry. The structures of several β-lactamase-resistant cephalosporins feature a methoximino acyl group, including cefuroxime, cefotaxime, ceftizoxime, and ceftriaxone. β-Lactamase resistance is modestly enhanced if the oximino substituent also features a polar function, as in ceftazidime, which has a 2-methylpropionic acid substituent on the oximino group. Both steric and electronic properties of the alkoximino group may contribute to the β-lactamase resistance conferred by this functionality, since *syn*-isomers are more potent than *anti*-isomers.[70] β-Lactamase-resistant 7α-methoxylcephalosporins, also called cephamycins because they are derived from cephamycin C (an antibiotic isolated from *Streptomyces*), are represented by cefoxitin and cefotetan and the 1-oxocephalosporin, moxalactam, which is prepared by total synthesis.

Base- or β-lactamase-catalyzed hydrolysis of cephalosporins containing a good leaving group at the 3′-position is known to be accompanied by elimination of the leaving group. The enzymatic process

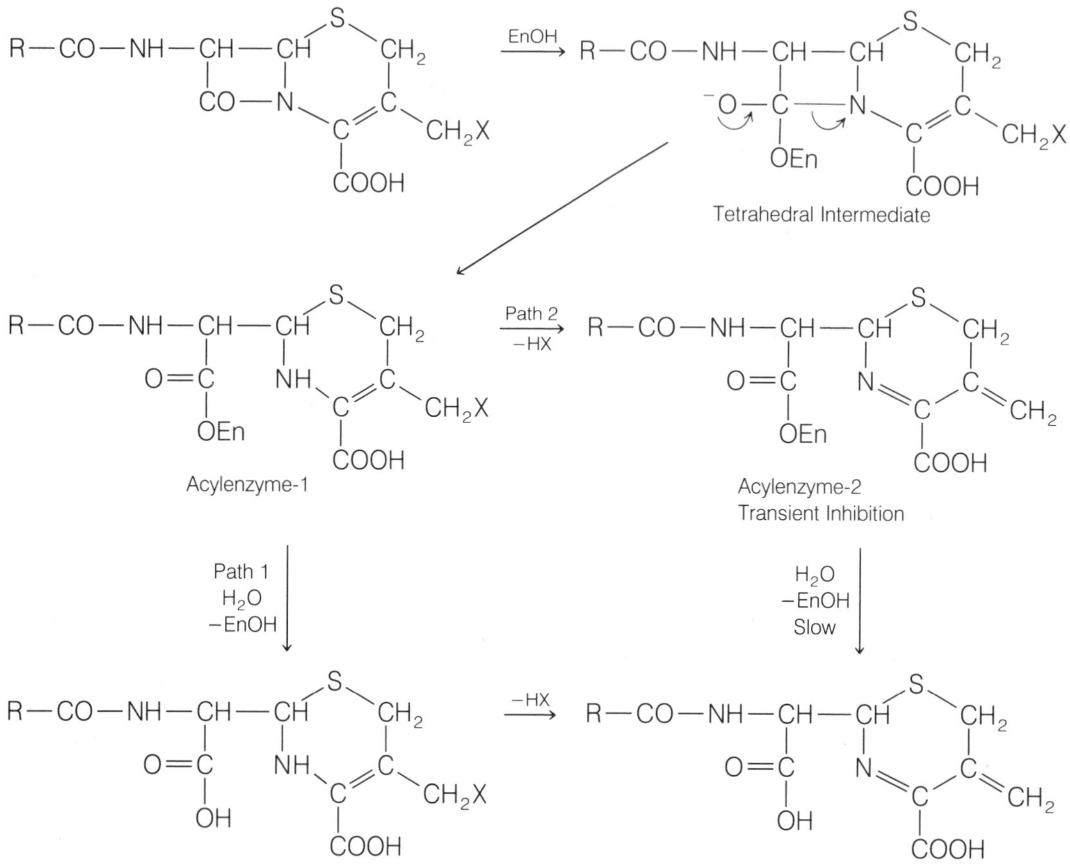

FIG. 7-6. Inhibition of β-lactamases by cephalosporins.

occurs in a stepwise fashion beginning with the formation of a tetrahedral transition state, which quickly collapses into an acylenzyme intermediate (Fig. 7-6). This intermediate can then either undergo hydrolysis to free the enzyme (path 1), or suffer elimination of the leaving group to form a relatively stable acylenzyme having a conjugated imine structure (path 2). Because of the stability of the acylenzyme intermediate, path 2 leads to transient inhibition of the enzyme. Faraci and Pratt[71] have shown that cephalothin and cefoxitin inhibit β-lactamases by this mechanism, whereas analogues lacking a 3′-leaving group do not.

Antipseudomonal Cephalosporins

Species of *Pseudomonas*, especially *Ps. aeruginosa*, represent a special public health problem because of their ubiquity in the environment and their propensity to develop resistance to antibiotics, including the β-lactams. The primary mechanisms of β-lactam resistance appear to involve destruction of the antibiotics by β-lactamases and interference with their penetration through the cell envelope. Apparently not all β-lactamase-resistant cephalosporins penetrate the cell envelope of *Ps. aeruginosa*, as only cefoperazone, moxalactam, cefotaxime, ceftizoxime, ceftriaxone, and ceftazidime have useful antipseudomonal activity. Two cephalosporins, namely moxalactam and cefoperazone, contain the same polar functionalities, such as carboxy and *N*-acylureido, that facilitate penetration into *Pseudomonas* species in the penicillins (see carbenicillin, ticarcillin, and piperacillin). Unfortunately, strains of *Ps. aeruginosa* resistant to moxalactam and cefotaxime have been found in clinical isolates. The war between man and microbe continues!

Classification

Cephalosporins are divided into first-, second-, and third-generation agents, based roughly on their time of discovery and their antimicrobial properties (Table 7-7). In general, progression from first to third generation is associated with a broadening of the gram-negative antibacterial spectrum, reduction in activity against gram-positive organisms, and enhanced resistance to β-lactamases. Individual

TABLE 7-7

CLASSIFICATION AND PROPERTIES OF CEPHALOSPORINS

Cephalosporin	Generation	Route of Administration	Acid Resistant	Plasma Protein Binding (%)	β-Lactamase Resistance	Spectrum of Activity	Antipseudomonal Activity
Cephalexin	First	Oral	Yes	5–15	Poor	Intermediate	No
Cepharadine	First	Oral	Yes	8–17	Poor	Intermediate	No
Cefadroxil	First	Oral, parenteral	Yes	20	Poor	Intermediate	No
Cefachlor	First	Oral	Yes	22–25	Poor	Intermediate	No
Cephalothin	First	Parenteral	No	65–80	Poor	Intermediate	No
Cephapirin	First	Parenteral	No	40–54	Poor	Intermediate	No
Cefazolin	First	Parenteral	No	70–86	Poor	Intermediate	No
Cefamandole	Second	Parenteral	No	56–78	Poor to average	Broad	No
Cefonacid	Second	Parenteral	No	98	Poor to average	Broad	No
Ceforanide	Second	Parenteral	No	80	Average	Broad	No
Cefoxitin	Second	Parenteral	No	13–22	Good	Broad	No
Cefuroxime	Second	Parenteral	No	33–50	Good	Broad	No
Cefoperazone	Third	Parenteral	No	82–93	Average to good	Broad	Yes
Moxalactam	Third	Parenteral	No	40–57	Good	Broad	Yes
Cefotaxime	Third	Parenteral	No	30–51	Good	Broad	Yes
Ceftizoxime	Third	Parenteral	No	30	Good	Broad	Yes
Ceftrixone	Third	Parenteral	No	85–95	Good	Broad	Yes
Cefotetan	Third	Parenteral	No	78–91	Good	Broad	No
Ceftazidime	Third	Parenteral	No	80–90	Good	Broad	Yes

cephalosporins differ in their pharmacokinetic properties, especially plasma protein binding and half-life, but the structural bases for these differences are not obvious.

Products

Cephalexin, USP. 7α-(D-Amino-α-phenylacetamido)-3-methylcephemcarboxylic acid (Keflex, Keforal). Cephalexin was purposely designed as an orally active semisynthetic cephalosporin. The oral inactivation of cephalosporins has been attributed to two causes: instability of the β-lactam ring to acid hydrolysis (cephalothin and cephaloridine) and solvolysis or microbial transformation of the 3-methylacetoxy group (cephalothin, cephaloglycin). The α-amino group of cephalexin renders it acid-stable, and reduction of the 3-acetoxymethyl to a methyl group circumvents reaction at that site.

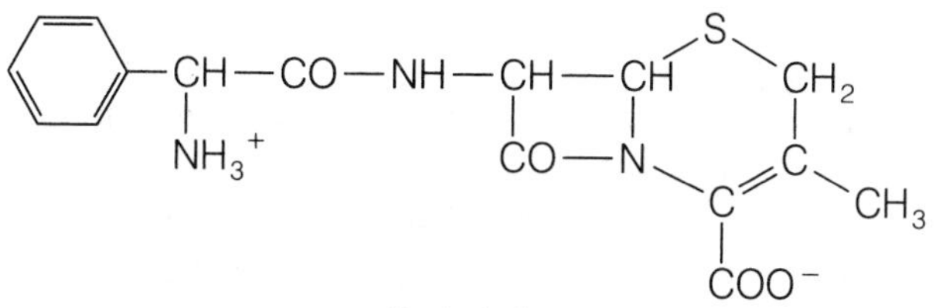

C_6H_5—CH(NH_3^+)—CO—NH—CH—CH(S)—CH_2—C(CH_3)=C(COO^-)—N—CO (β-lactam ring)

Cephalexin

Cephalexin occurs as the white crystalline monohydrate. It is freely soluble in water, resistant to acid, and well absorbed orally. Food does not interfere with its absorption. Because of minimal protein binding and nearly exclusive renal excretion, cephalexin is particularly recommended for the treatment of urinary tract infections. It is also sometimes employed for upper respiratory tract infections. Its spectrum of activity is very similar to those of cephalothin and cephaloridine. Cephalexin is somewhat less potent than these two agents after parenteral administration and, therefore, is inferior to them for the treatment of serious systemic infections.

Cephradine, USP (Anspor, Velosef). Cephradine is the only cephalosporin derivative that is available in both oral and parenteral dosage forms. It closely resembles cephalexin chemically (it may be regarded as a partially hydrogenated derivative of cephalexin) and has very similar antibacterial and pharmacokinetic properties. It occurs as the crystalline hydrate, which is readily soluble in water. Cephradine is stable to acid and almost completely absorbed after oral administration. It is minimally protein bound and is excreted almost exclusively through the kidney. It is recommended for the treatment of uncomplicated urinary tract infections and upper respiratory tract infections caused by susceptible organisms.

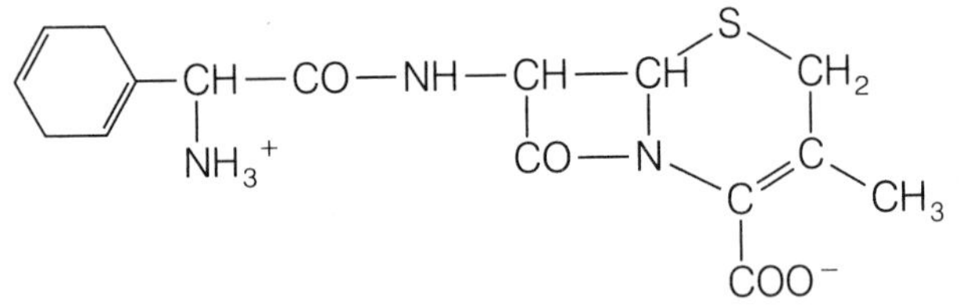

Cephradine

Cephradine is available in both oral and parenteral dosage forms.

Cefadroxil, USP (Duricef) is an orally active semisynthetic derivative of 7-ACDA wherein the 7-acyl group is the D-hydroxylphenylglycyl moiety. This compound is well absorbed after oral administration to give plasma levels that reach 75% to 80% of those of an equal dose of its close structural analogue cephalexin. The main advantage claimed for cephadroxil results from its somewhat prolonged duration of action that permits once-a-day dosing. The prolonged duration of action of this compound is related to relatively slow urinary excretion of the drug, compared with other cephalosporins, but the basis for the latter effect remains to be completely explained. The antibacterial spectrum of action and therapeutic indications of cefadroxil are very similar to those of cephalexin and cephradine. The D-*p*-hydroxyphenylglycyl isomer is much more active than the L isomer.

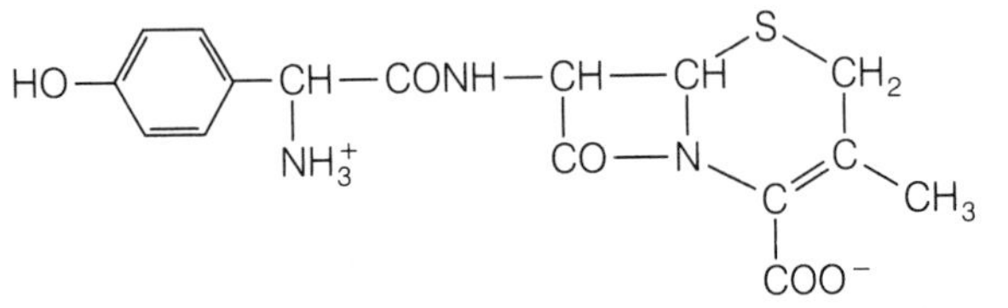

Cefadroxil

Cefaclor, USP (Ceclor) is an orally active semisynthetic cephalosporin that was introduced in the American market in 1979. It differs structurally from cephalexin in that the 3-methyl group has been replaced by a chlorine atom. It is synthesized from the corresponding 3-methylenecepham sulfoxide ester by ozonolysis, followed by halogenation of the resulting β-ketoester.[72] The 3-methylenecepham sulfoxide esters are prepared, in turn, by rearrangement of the corresponding 6-acylaminopenicillanic acid derivative. Cefaclor is moderately stable in acid and achieves sufficient oral absorption to provide effective plasma levels (equal to about two-thirds of those obtained with cephalexin). The compound is apparently unstable in solution, because about 50% of its antimicrobial activity is lost in two hours in serum at 37°C.[73] The antibacterial spectrum of activity is similar to cephalexin's, but it is claimed to be more potent against some species sensitive to both agents. The drug is currently being recommended for the treatment of non–life-threatening infections caused by *H. influenzae*, particularly strains resistant to ampicillin.

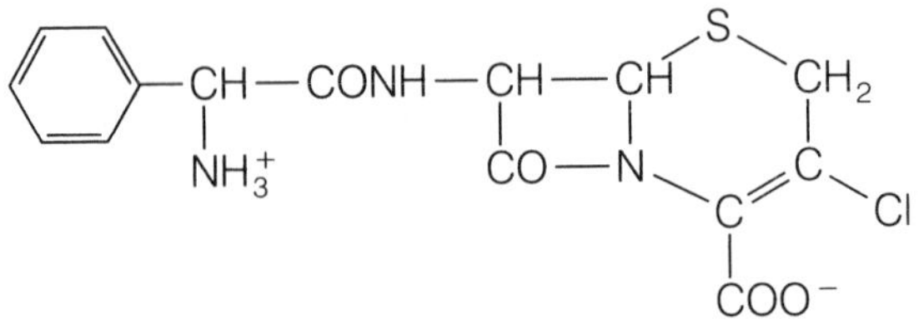

Cefaclor

Cephalothin Sodium, USP (Keflin). Cephalothin sodium occurs as a white to off-white crystalline powder that is practically odorless. It is freely soluble in water and insoluble in most organic solvents. Although it has been described as a broad-spectrum antibacterial compound, it is not in the same class as the tetracyclines. Its spectrum of activity is broader than that of penicillin G, and more similar to that of ampicillin. Unlike ampicillin, cephalothin is resistant to penicillinase produced by *Staph. aureus* and provides an alternative to the use of penicillinase-resistant penicillins for the treatment of infections caused by such strains.

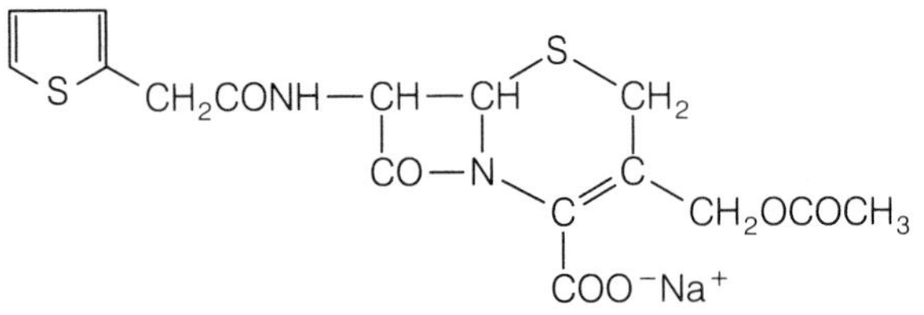

Cephalothin Sodium

Cephalothin is poorly absorbed from the gastrointestinal tract and must be administered parenterally for systemic infections. It is relatively nontoxic and is acid-stable. It is excreted rapidly through the kidneys, about 60% being lost within six hours of administration. Pain at the site of intramuscular injection and thrombophlebitis following intravenous injection of cephalothin have been reported. Hypersensitivity reactions from cephalothin have been observed, and there is some evidence of cross-sensitivity in patients noted previously to be penicillin-sensitive.

Sterile Cefazolin Sodium, USP (Ancef, Kefzol). Cefazolin is one of a series of semisynthetic cephalosporins in which the C-3 acetoxy function has been replaced by a thiol-containing heterocycle, here, 5-methyl-2-thio-1,3,4-thiadiazole. It also con-

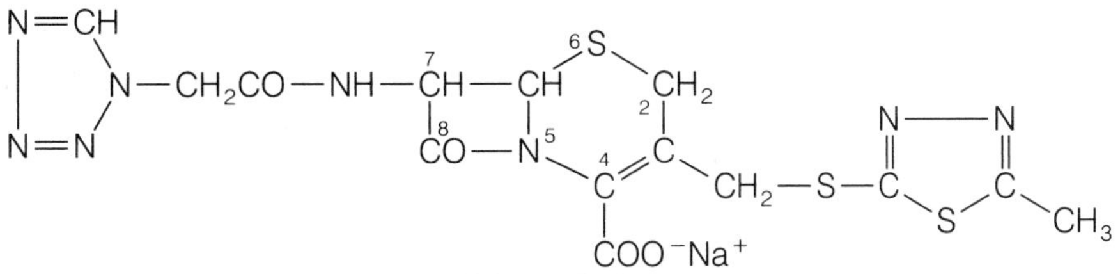

Cefazolin Sodium

tains the somewhat unusual tetrazolylacetyl acylating group. Cefazolin was released in 1973 as the water-soluble sodium salt. It is active only by parenteral administration.

In comparison with other first-generation cephalosporins, cefazolin provides higher serum levels, slower renal clearance, and a longer half-life. It is approximately 75% protein bound in the plasma, a value that is higher than for most other cephalosporins. Early in vitro and clinical studies suggest that cefazolin is more active against gram-negative bacilli, but less active against gram-positive cocci than either cephalothin or cephaloridine. Thrombophlebitis following intravenous injection and pain at the site of intramuscular injection of cefazolin appear to be the lowest of the parenteral cephalosporins.

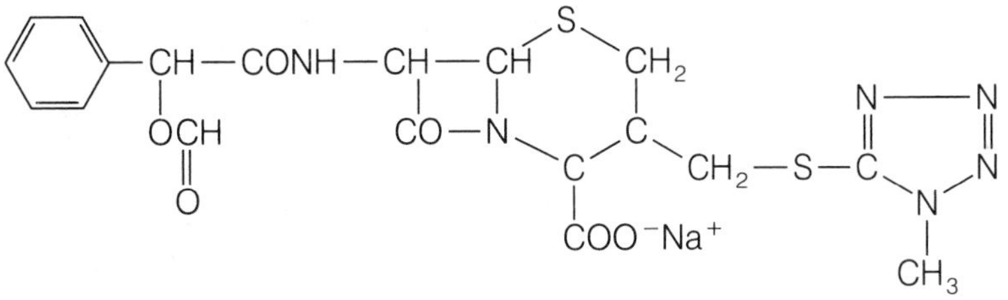

Sterile Cephapirin Sodium, USP (Cefadyl). Cephapirin is a semisynthetic 7-ACA derivative released in the United States in 1974. It closely resembles cephalothin in chemical and pharmacokinetic properties. Like cephalothin, cephapirin is unstable to acid and must be administered parenterally in the form of an aqueous solution of the sodium salt. It is moderately protein bound (45% to 50%) in the plasma and is rapidly cleared by the kidney. Cephapirin and cephalothin are very similar in antimicrobial spectrum and potency. Conflicting reports concerning the relative occurrence of pain at the site of injection and thrombophlebitis after intravenous injection of cephapirin and cephalothin are difficult to assess on the basis of available clinical data.

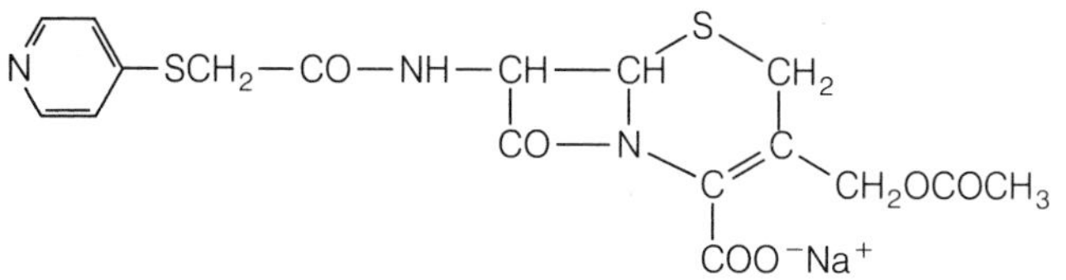

Cephapirin Sodium

Cefamandole Nafate, USP (Mandol) is the formate ester of cefamandole, a semisynthetic cephalosporin that incorporates D-mandelic acid as the acyl portion and a thiol-containing heterocycle (5-thio-1,2,3,4-tetrazole) in place of the acetoxyl function on the C-3 methylene carbon atom. Esterification of the α-hydroxyl group of the D-mandeloyl function overcomes the instability of cefamandole in solid-state dosage forms[74] and provides satisfactory concentrations of the parent antibiotic in vivo through spontaneous hydrolysis of the ester in neutral to alkaline pH. Cefamandole is the first second-generation cephalosporin to be marketed in the United States.

Cefamandole Nafate

The D-mandeloyl moiety of cefamandole appears to confer resistance to a few β-lactamases, since some β-lactamase-producing, gram-negative bacteria (particularly *Enterobacteriaceae*) that show resistance to cefazolin and other first-generation cephalosporins are sensitive to cefamandole. Additionally, it is active against some ampicillin-resistant strains of *Neisseria* and *Haemophilus*. Although resistance to β-lactamases may be a factor in determining sensitivity of individual bacterial strains to cefamandole, an early study[75] indicated that other factors, such as permeability and intrinsic activity, are frequently more important. It should be noted that the L-mandeloyl isomer is significantly less active than the D-isomer.

Cefamandole nafate is very unstable in solution, hydrolyzing rapidly to release cefamandole and formate. However, there is no loss of potency when such solutions are stored for 24 hours at room temperature or for up to 96 hours when refrigerated. Air oxidation of the released formate to carbon dioxide can cause pressure to build up in the injection vial.

Sterile Cefonacid Sodium, USP (Monicid) is a second-generation cephalosporin that is structurally similar to cefamandole, except that it contains a methane sulfonic acid group attached to N-1 position of the tetrazole ring. The antimicrobial spectrum and limited β-lactamase stability of cefonacid are essentially identical to those of cefamandole.

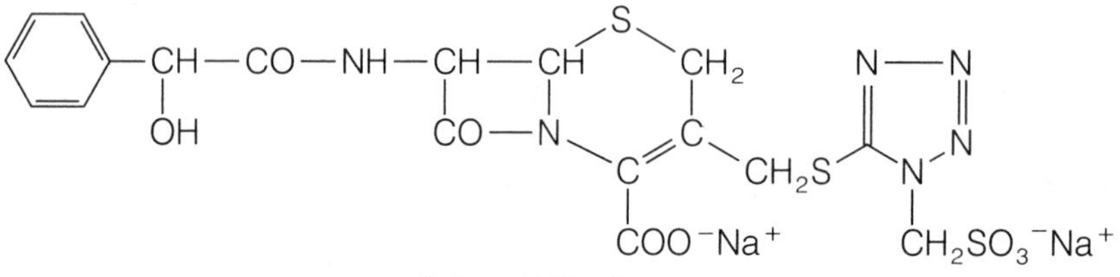

Cefonacid Disodium

Cefonacid is unique among the second-generation cephalosporins in that it has an unusually long serum half-life of approximately 4.5 hours. A high fraction of plasma protein binding, coupled with slow renal tubular secretion, are apparently responsible for the long duration of action. Despite the high fraction of drug bound in the plasma, cefonacid is widely distributed throughout body fluids and tissues, with the exception of the cerebrospinal fluid.

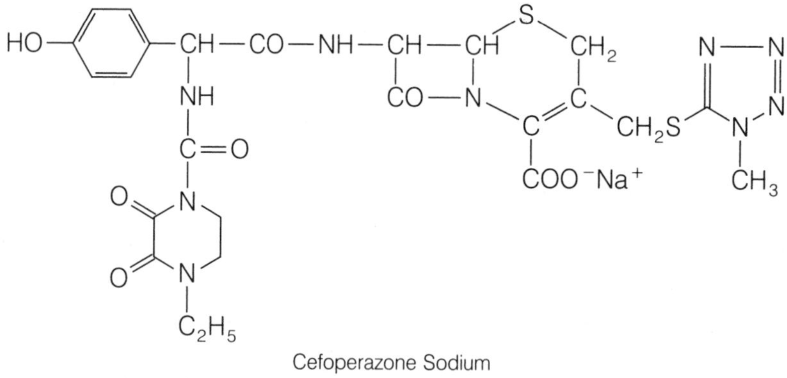

Cefoperazone Sodium

Cefonacid is supplied as the highly water-soluble disodium salt, in the form of a sterile powder to be reconstituted for injection. Solutions are stable for 24 hours at 25°C and for 72 hours when refrigerated.

Sterile Ceforanide, USP (Precef) was approved for clinical use in the United States in 1984. Despite this recent introduction, however, ceforanide is classified as a second-generation cephalosporin because its antimicrobial properties are similar to those of cefamandole. It exhibits excellent potency against most members of the *Enterobacteriaceae*, especially *K. pneumoniae*, *E. coli*, *Pr. mirabilis*, and *Enterobacter cloacae*. However, ceforanide is less active than cephamandole against *H. influenzae*.

The duration of action of ceforanide lies between that of cefamandole and cefonacid. It has a serum half-life of around three hours, permitting twice-a-day dosing for most indications. Ceforanide is supplied as the sterile, crystalline disodium salt. Parenteral solutions are stable for four hours at 25°C and for up to five days when refrigerated.

Sterile Cefoperazone Sodium, USP (Cefobid) is a third-generation antipseudomonal cephalosporin that resembles piperacillin chemically and microbiologically. Cefoperazone is active against many strains of *Ps. aeruginosa*, indole-positive *Proteus*, *Enterobacter*, and *Ser. marcescens* that are resistant to cefamandole. It is less active than cephalothin against gram-positive bacteria and less active than cefamandole against most of the *Enterobacteriaceae*. Similarly to piperacillin, cefoperazone is hydrolyzed by many of the β-lactamases that hydrolyze peni-

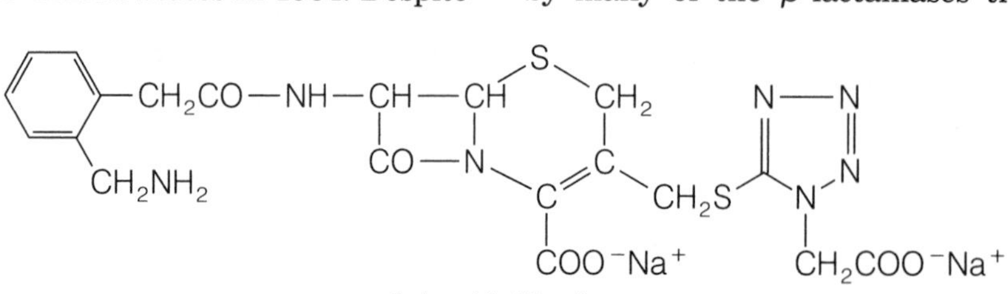

Ceforanide Disodium

cillins. Unlike piperacillin, however, it is resistant to many (but not all) of the β-lactamases that hydrolyze cephalosporins.

Cefoperazone is excreted primarily in the bile. Hepatic dysfunction can affect its clearance from the body. Although only 25% of the free antibiotic is recovered in the urine, urinary concentrations are high enough to be effective in the management of urinary tract infections caused by susceptible organisms. The relatively long half-life (two hours) of cefoperazone allows for dosing twice a day. Solutions prepared from the crystalline sodium salt are stable up to four hours at room temperature. If refrigerated they will last five days without appreciable loss of potency.

Sterile Cefoxitin Sodium, USP (Mefoxin). Cefoxitin is a semisynthetic derivative obtained by modification of cephamycin C, a 7α-methoxy-substituted cephalosporin isolated independently from various *Streptomyces* by research groups in Japan[76] and the United States.[77] Although it is less potent than cephalothin against gram-positive bacteria and less potent than cefamandole against most of the *Enterobacteriaceae*, cefoxitin is effective against certain strains of gram-negative bacilli (for example, *E. coli*, *K. pneumoniae*, *Providencia*, *Ser. marcescens*, indole-positive *Proteus*, and *Bacteriodes*) that are resistant to these cephalosporins. It is also effective against penicillin-resistant *Staph. aureus* and *N. gonorrhoeae*.

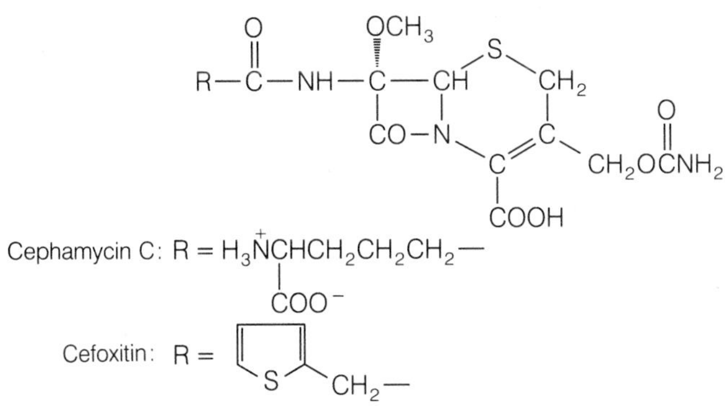

Cephamycin C: $R = H_3\overset{+}{N}CH(COO^-)CH_2CH_2CH_2—$

Cefoxitin: R = 2-thienyl–$CH_2—$

The activity of cefoxitin and cephamycins, in general, against resistant bacterial strains is due to their resistance to hydrolysis by β-lactamases conferred by the 7α-methoxyl substituent.[70] Cefoxitin is a potent competitive inhibitor of many β-lactamases. It is also a potent inducer of chromosomally mediated β-lactamases. The temptation to exploit the β-lactamase-inhibiting properties of cefoxitin by combining it with β-lactamase-labile β-lactam antibiotics should be tempered by the possibility that antagonism may occur. In fact, cefoxitin antagonizes the action of cefamandole against *Enterobacter cloacae* and that of carbenicillin against *Ps. aeruginosa*.[78] Cefoxitin, alone, is essentially ineffective against these organisms.

The pharmacokinetic properties of cefoxitin resemble those of cefamandole. Because its half-life is relatively short, cefoxitin must be administered three or four times daily. Solutions of the sodium salt intended for parenteral administration are stable for 24 hours at room temperature and for one week if refrigerated. 7α-Methoxyl substitution stabilizes the β-lactam to alkaline hydrolysis to some extent.

The principal role of cefoxitin in therapy seems to be for the treatment of certain anaerobic and mixed aerobic–anaerobic infections. It is also used to treat gonorrhea caused by β-lactamase-producing strains. It is classified as a second-generation agent because of its spectrum of activity.

Cefotetan Disodium (Cefotan) is a third-generation cephalosporin that is structurally similar to cefoxitin. Like cefoxitin, cefotetan is resistant to destruction by β-lactamases. It is also a competitive inhibitor of many β-lactamases and causes transient inactivation of some of these enzymes. Cefotetan is reported to synergize with β-lactamase-sensitive β-lactams, but unlike cefoxitin does not appear to cause antagonism.[79]

The antibacterial spectrum of cefotetan closely resembles cefoxitin's. It is, however, generally more active against *Staph. aureus* and members of the *Enterobacteriaceae* sensitive to both agents. It also exhibits excellent potency against *H. influenzae* and *N. gonorrhoeae*, including β-lactamase-producing strains. Cefotetan is slightly less active than cefoxitin against *B. fragilis* and other anaerobes. *Enterobacter* species are generally resistant to cefotetan, and it is without effect against *Pseudomonas*.

Cefotetan has a relatively long half-life of around 3.5 hours. It is administered on a twice-daily-dosing schedule. It is largely excreted unchanged in the

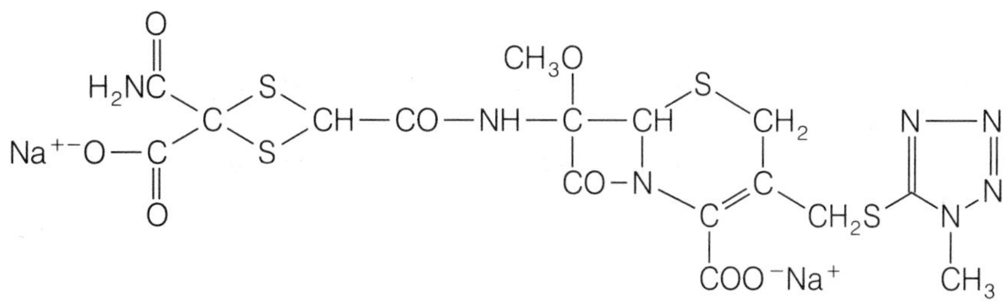

Cefotetan Disodium

urine. Aqueous solutions for parenteral administration maintain potency for 24 hours at 25°C. Refrigerated solutions are stable for four days.

Moxalactam Disodium, USP (Moxam) is a 1-oxacephalosporin prepared by semisynthetic procedures. It is considered to be a third-generation compound because of its antimicrobial properties. Moxalactam was approved for use in the United States in 1982.

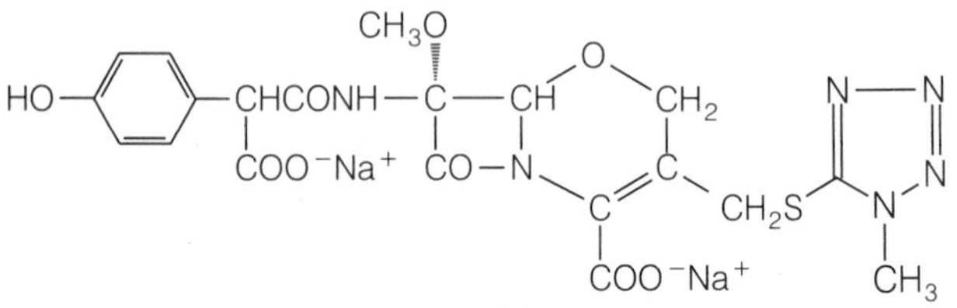

Moxalactam Disodium

As a consequence of the application of new synthetic methods for the modification of penam and cepham ring systems and for total syntheses inspired by Woodward's synthesis of cephalosporins,[63] several classes of isosteres of the bicyclic ring systems of penicillins and cephalosporins are now available. One of the more promising groups of such isosteric compounds is the 1-oxacephalosporin class. These compounds have been prepared by semisynthetic procedures involving the thiazolidine ring opening of penicillin-*S*-oxides[80] and also by total synthesis methods.[81] Thus, oxygen analogues of 7-ACA, 7-ADCA, and the cephamycins have been prepared and evaluated for antimicrobial properties.

In general, the 1-oxacephalosporins parallel the corresponding cephalosporin and cephamycin derivatives in antimicrobial spectrum of activity and in pharmacokinetic properties. The antibacterial potencies of the 1-oxa derivatives, however, tend to be consistently higher than those of their sulfur isosteres. The "1-oxacephalosporin" that has received the most attention to date is moxalactam, a parenterally active compound the structure of which combines three useful features present in other β-lactam antibiotics: (1) a 7α-methoxyl group (note the analogy to the cephamycins); (2) an α-carboxy-*p*-hydroxyphenylacyl group (note the analogy to carbenicillin); and (3) the same 5-thio heterocyclic moiety on the C-3 methylene carbon atom that is present in cefamandole.

Moxalactam exhibits significantly greater activity against most species of *Enterobacteriaceae* and anaerobic gram-negative bacilli than the cephalosporins and cefoxitin. Furthermore, it is four times more active than carbenicillin against *Ps. aeruginosa*, a property possessed by very few cephalosporins. The expanded spectrum of activity of moxalactam compared with conventional cephalosporins appears to be related to greater β-lactamase stability, resulting from the 7α-methoxyl group. However other factors, such as permeability through porin channels in gram-negative bacilli facilitated by the ionized carboxyl group, probably also contribute. Moxalactam is active against many strains of *Ps. aeruginosa*. However, some strains have developed resistance, either by elaborating β-lactamases that hydrolyze it or by excluding the antibiotic.

Moxalactam is widely distributed throughout the body fluids, including the cerebrospinal fluid. It achieves sufficiently high concentrations in the cerebrospinal fluid to be useful in the treatment of gram-negative bacillary meningitis. The chemical stability of moxalactam is similar to that of the cephamycins.

Cefuroxime Sodium, USP (Zinacef) is the first of a series of α-methoximinoacyl-substituted cephalosporins that constitute most of the third-generation agents currently available for clinical use. A *syn*-alkoximino substituent is associated with β-lactamase stability in these cephalosporins.[70] Cefuroxime is actually classified as a second-generation cephalosporin because its spectrum of antibacterial activity more closely resembles that of cefamandole. It is, however, active against β-lactamase-producing strains that are resistant to cefamandole, such as *E. coli*, *K. pneumoniae*, *N. gonorrhoeae*, and *H. influenzae*. Other important gram-negative pathogens such as *Serratia*, indole-positive *Proteus*, *Ps. aeruginosa*, and *B. fragilis* are resistant.

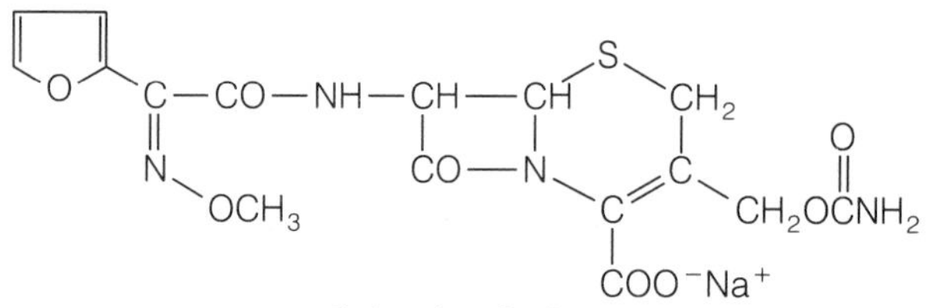

Cefuroxime Sodium

Cefuroxime is widely distributed throughout the body. It penetrates inflamed meninges in sufficiently high concentrations to be effective in meningitis caused by susceptible organisms. Thrice-daily dosing of cefuroxime is required to maintain effective plasma levels for most sensitive organisms.

Sterile Cefotaxime Sodium, USP (Claforan) is the first of the third-generation cephalosporins to be introduced. It possesses excellent broad-spectrum activity against gram-positive and gram-negative aerobic and anaerobic bacteria. It is more active than moxalactam against gram-positive organisms. Many β-lactamase-producing bacterial strains are sensitive to cefotaxime, including *N. gonorrhoeae*,

Klebsiella, *H. influenzae*, *Staph. aureus*, and *Enterobacter cloacae*. Some, but not all, *Pseudomonas* strains are sensitive. Enterococci and *Listeria monocytogenes* are resistant.

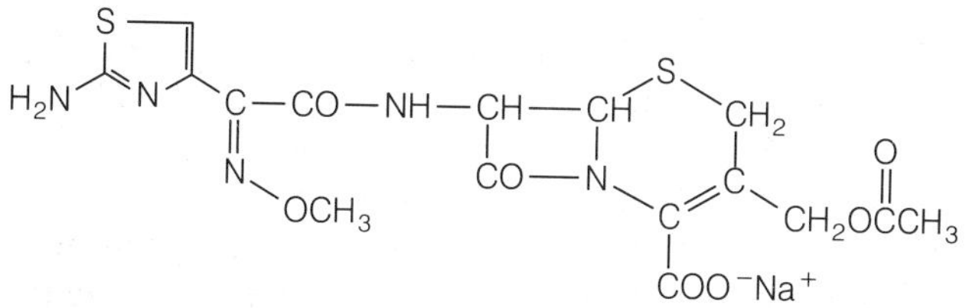

Cefotaxime Sodium

The *syn* isomer of cefotaxime is significantly more active than the *anti* isomer against β-lactamase-producing bacteria. This potency difference, in part, is due to greater resistance of the *syn* isomer to the action of β-lactamases.[70] However, higher affinity of the *syn* isomer for penicillin-binding proteins may also be a factor.[82]

Cefotaxime is metabolized, in part, to the less-active desacetyl metabolite. Approximately 20% of the metabolite and 25% of the parent drug are excreted in the urine. The drug reaches the cerebrospinal fluid in sufficient concentrations to be effective in the treatment of meningitis.

Solutions of cefotaxime sodium should be used within 24 hours. If they are stored they should be refrigerated. Refrigerated solutions maintain potency for up to ten days.

Sterile Ceftizoxime Sodium, USP (Ceftizox) is a third-generation cephalosporin that was introduced in 1984. This β-lactamase-resistant agent exhibits excellent activity against the *Enterobacteriaceae*, especially *E. coli*, *K. pneumoniae*, *Enterobacter cloacae*, *E. aerogenes*, indole-positive and indole-negative *Proteus*, and *Ser. marcescens*. Ceftizoxime is claimed to be more active than cefoxitin against *B. fragilis*. It is also very active against gram-positive bacteria. The activity of ceftizoxime against *Ps. aeruginosa* is somewhat variable and less than that of either cefotaxime or cefaperazone.

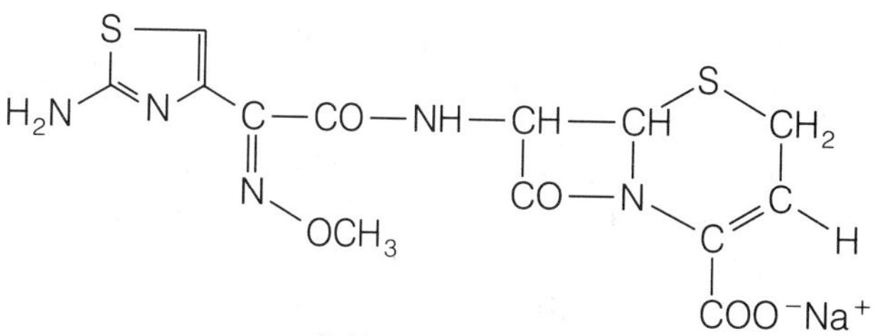

Ceftizoxime Sodium

Ceftizoxime is not metabolized in vivo. It is largely excreted in the urine unchanged. Adequate levels of the drug are achieved in the cerebrospinal fluid for the treatment of gram-negative or gram-positive bacterial meningitis. Ceftizoxime must be administered on a three-times-daily dosing schedule because of its relatively short half-life. Ceftizoxime sodium is very stable in the dry state. Solutions maintain potency for up to 24 hours at room temperature and for ten days when refrigerated.

Sterile Ceftriaxone Disodium, USP (Rocephin) is a β-lactamase-resistant cephalosporin with an extremely long serum half-life. Once-daily dosing of ceftriaxone suffices for most of its indications. Two factors contribute to the prolonged duration of action of ceftriaxone: a high fraction of protein binding in the plasma and a slow urinary excretion. Ceftriaxone is excreted both in the bile and in the urine. Its urinary excretion is not affected by probenecid. Despite the comparatively low volume of distribution of ceftriaxone, it reaches the cerebrospinal fluid in concentrations that are effective in meningitis. Nonlinear pharmacokinetics are observed for ceftriaxone.

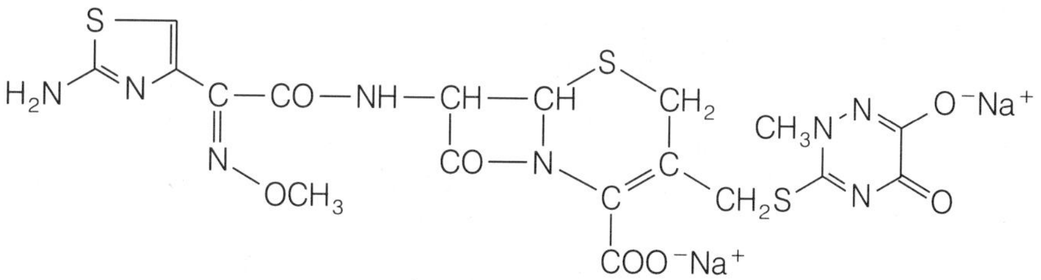

Ceftriaxone Disodium

Ceftriaxone contains a highly acidic heterocyclic system on the 3-thiomethyl group. This unusual ring system is believed to confer the unique pharmacokinetic properties to this agent.

Ceftriaxone exhibits excellent broad-spectrum antibacterial activity against both gram-positive and gram-negative organisms. It is highly resistant to most chromosomally and plasmid-mediated β-lactamases. The activity of ceftriaxone against *Enterobacter*, *Citrobacter*, *Serratia*, indole-positive

Proteus, and *Pseudomonas* is particularly impressive. It is also effective in the treatment of ampicillin-resistant gonorrhea and *H. influenzae* infections. Ceftriaxone is generally less active than cefotaxime against gram-positive bacteria and *B. fragilis*.

Solutions of ceftriaxone sodium should be used within 24 hours. They may be stored up to ten days if refrigerated.

Sterile Ceftazidime Sodium, USP (Fortax, Taxidime) is a new β-lactamase-resistant third-generation cephalosporin that is noted for its antipseudomonal activity. It is active against some strains of *Ps. aeruginosa* that are resistant to cefoperazone and ceftriaxone. Ceftazidime is also highly effective against β-lactamase-producing strains of the *Enterobacteriaceae* family. It is generally less active than cefotaxime against gram-positive bacteria and *B. fragilis*.

Ceftazidime has been used effectively for the treatment of meningitis caused by *H. influenzae* and *N. meningitidis*.

Cefixime is an orally active third-generation cephalosporin that is currently undergoing clinical trials. It differs from the orally active first-generation agents by being resistant to β-lactamases. Consequently, cefixime is effective against strains of *E. coli*, *K. pneumoniae*, *Enterobacter*, *Citrobacter*, *Salmonella*, *Shigella*, *Proteus*, and *Providencia* that are resistant to first- and second-generation cephalosporins. It is also active against ampicillin-resistant *H. influenzae* and *N. gonorrhoeae*. The activity of cefixime against *Ps. aeruginosa*, *Ser. marcescens*, and anaerobic bacteria is much less impressive.

The oral bioavailability of cefixime is reasonably good. It is estimated that 27% to 55% of an oral dose is absorbed, depending on the conditions. The place of cefixime in therapy awaits the results of clinical trials.

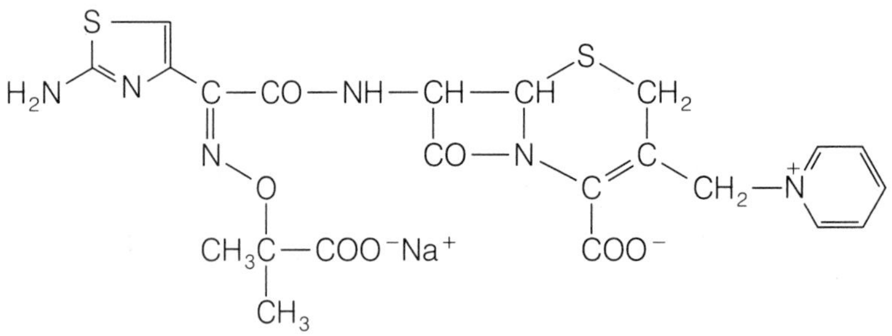

Ceftazidime Sodium

The structure of ceftazidime contains two noteworthy features: (1) a 2-methylpropionicoxaminoacyl group that confers β-lactamase resistance and possibly increased permeability through the porin channels of the cell envelope and (2) a pyridinium group at the 3′-position that confers zwitterionic properties to the molecule.

Ceftazidime is administered parenterally two or three times daily, depending on the severity of the infection. Its serum half-life is about 1.8 hours.

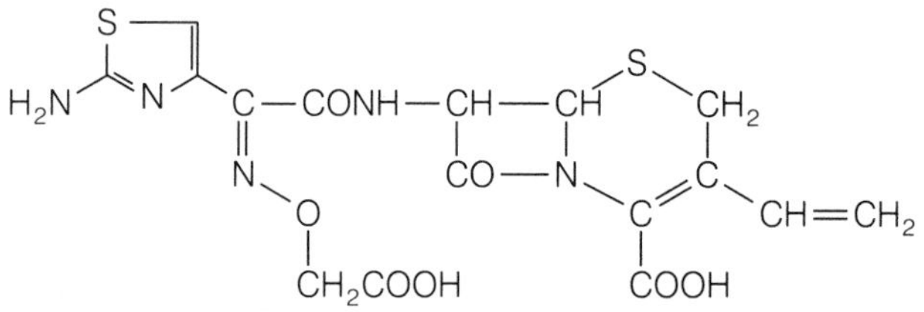

Cefixime

TABLE 7-8

CEPHALOSPORINS

Name Proprietary Name	*Preparations*	*Usual Adult Dose**	*Usual Dose Range**	*Usual Pediatric Dose**
Cephalexin USP *Keflex, Keforal*	Cephalexin capsules USP	The equivalent of 250 mg of cephalexin qid	1–4 g/day	6–12 mg/kg of body weight qid
	Cephalexin capsules USP Cephalexin for oral suspension USP Cephalexin tablets USP	Oral, the equivalent of anhydrous cephalexin, 250–500 mg every 6 hr	Up to 4 g or more daily	Oral, the equivalent of anhydrous cephalexin, 6.25–25 mg/kg of body weight every 6 hr

*See USP DI for complete dosage information.

TABLE 7-8 *Continued*

CEPHALOSPORINS

Name Proprietary Name	***Preparations***	***Usual Adult Dose****	***Usual Dose Range****	***Usual Pediatric Dose****
Cephradine USP *Anspor, Velosef*	Sterile cephradine USP Cephradine capsules USP Cephradine for oral suspension USP	IM or IV, 2 equally divided doses qid	2–4 g in equally divided doses qid	50–100 mg/kg per day up to 300 mg/kg per day
	Cephradine tablets USP	Oral, 250–500 mg every 6 hr or 500 mg every 12 hr	Up to 6 g/day	
	Cephradine for injection USP	IM or IV, 500 mg–1 g every 6 hr	Up to 8 g/day	Infants and children 1 yr of age and over, IM or IV: 12.5–25 mg/kg of body weight every 6 hr
Cefadroxil USP *Duricef*	Cefadroxil capsules USP	Oral, the equivalent of cefadroxil, skin and skin-structure infections: 500 mg every 12 hr, or 1 g qd; urinary tract infections: 1 g every 12 hr	Up to 6 g/day	
Cefaclor USP *Ceclor*	Cefaclor capsules USP			
	Cefaclor for oral suspension USP	Oral, the equivalent of anhydrous cefaclor, 250–500 mg every 8 hr	Up to 4 g/day	Infants 1 mo of age and over: oral, the equivalent of anhydrous cefaclor, 6.7–13.4 mg/kg of body weight every 8 hr
Cephalothin sodium USP *Keflin*	Cephalothin sodium for injection USP	IM or IV, the equivalent of 500 mg–1 g of cephalothin every 4–6 hr	Up to 12 g/day	IM or IV, the equivalent of cephalothin: 13.3–26.6 mg/kg of body weight every 4 hr, or 20–40 mg/kg of body weight every 6 hr
Sterile cefazolin sodium USP *Amcef, Kelzol*		IM or IV, the equivalent of cefazolin, 250 mg–1 g every 6–8 hr	The equivalent of cephazolin, up to 12 g/day	Infants and children 1 mo of age and over, IM or IV: the equivalent of cefazolin, 6.25–25 mg/kg of body weight every 6 hr, or 8.3–33.3 mg/kg of body weight every 8 hr
Sterile cephapirin sodium USP *Cefadyl*		IM or IV, the equivalent of cephapirin, 500 mg–1 g every 4–6 hr	The equivalent of cephapirin, up to 12 g/day	Infants and children 3 mo of age and over, IM or IV: the equivalent of cephapirin, 10–20 mg/kg of body weight every 6 hr
Cefamandole Naftate USP *Mandol*	Cefamandole naftate USP	The equivalent of 500 mg–1 g of cefamandole every 4–8 hr IM or IV	12 g/day	50–100 mg/kg/day IM or IV in equally divided doses every 4–8 hr up to 150 mg/kg/day for severe infections
Sterile cefonicid sodium USP *Monocid*	Sterile cefonicid sodium USP	The equivalent of 1 g of cefonicid every 24 hr IV or deep IM. Up to 2 g (24 hr for severe infections	2 g/day	
Sterile ceforanide USP	Sterile ceforanide USP	The equivalent of 0.5–1 g of ceforanide every 12 hr IM or IV		Children: 20–40 mg/kg/day in equally divided doses every 12 hr
Sterile cefoxitin sodium USP *Mefoxin*	Sterile cefoxitin sodium USP	Pneumonia, urinary tract infections, cutaneous infections: The equivalent of 3–4 g of cefoxitin per day; 1 g every 6– 8 hr IV or IM. Moderately severe to severe infections: 6–8 g/day; 1 g every 4 hr or 2 g every 6–8 hr IV. Life-threatening infections: 12 g/day; 2 g every 4 hr or 3 g every 6 hr IV	12 g/day	

*See USP DI for complete dosage information.

TABLE 7-8 *Continued*

CEPHALOSPORINS

Name Proprietary Name	***Preparations***	***Usual Adult Dose****	***Usual Dose Range****	***Usual Pediatric Dose****
Cefuroxime sodium USP *Zinacef*	Cefuroxime sodium USP	The equivalent of 750 mg–1.5 g of cefuroxime every 8 hr IM or IV	9 g/day	Infants and children over 3 mo: 50–100 mg/kg/day in equally divided doses every 6–8 hrs IV or IM Bacterial meningitis: 200–400 mg/kg/day IV in divided doses every 6–8 hr
Sterile cefoperazone sodium USP *Cefobid*	Sterile cefoperazone sodium USP	The equivalent of 2–4 g of cefoperazone daily in equally divided doses every 12 hr IM or IV	16 g/day	
Moxalactam disodium USP *Moxam*	Moxalactam disodium USP	The equivalent of 2–4 g of moxalactam daily in divided doses every 8 hr IM or IV	12 g/day	Neonates (up to 1 wk): 50 mg/kg every 12 hr. Neonates (1–4 wk): 50 mg/kg every 8 hr. Infants: 50 mg/kg every 6 hr. Children: 50 mg/kg every 6–8 hr IM or IV
Sterile cefotaxime sodium USP *Claforan*	Sterile cefotaxime sodium USP	Uncomplicated infections: The equivalent of cefotaxime of 2 g/day, 1 g every 12 hr IM or IV. Moderate to severe infections: 3–6 g/day, 2 g every 6–8 hr IV. Septicemia: 6– 8 g/day; 2 g every 6–8 hr IV. Life-threatening infections: Up to 12 g/day; 2 g every 4 hr IV	12 g/day	Neonates (up to 1 wk): 50 mg/kg every 12 hr Neonates (1–4 wk): 50 mg/kg every 8 hr. Infants and children (1 mo–12 yr): 50–180 mg/kg/day in 4–6 divided doses IV or IM
Sterile ceftizoxime sodium USP *Cefizox*	Sterile ceftizoxime sodium USP	The equivalent of 1–2 g of ceftizoxime every 8–12 hr IM or IV	9–12 g/day	Children over 6 mo: 50 mg/kg every 6–8 hr IM or IV up to 200 mg/kg/day.
Sterile ceftriaxone sodium USP *Rocephin*	Sterile ceftriaxone sodium USP	The equivalent of 1–2 g of ceftriaxone daily IM or IV	4 g/day	Children: 50–75 mg/kg/day IV for serious infections other than meningitis. Not to exceed 2 g/day. Meningitis: 100 mg/kg/day IV in divided doses every 12 hr. Not to exceed 4 g/day
Sterile cefotetan disodium USP *Cefotan*	Sterile cefotetan disodium USP	The equivalent of 1–2 g of cefotetan IV or IM every 12 hr	6 g/day	
Sterile ceftazidime sodium USP *Fortaz, Tazidime*	Sterile ceftazidime sodium	The usual dose is equivalent to 1 g of ceftazidime IV or IM every 8–12 hr. Uncomplicated urinary tract infections: 250 mg every 12 hr IM or IV. Complicated UTI require 500 mg every 8–12 hr IM or IV. Pneumonia and mild skin infections: 500 mg–1 g IM or IV every 8 hr. Bone and joint infections: 2 g IV every 12 hr. Serious gynecologic and intra-abdominal infections and meningitis: 2 g IV every 8 hr	6 g/day	Neonates (0–4 wk): 30 mg/kg IV every 12 hr. Infants and children (1 mo–12 yr) 30–50 mg/kg IV every 8 hr to a maximum of 6 g/day

*See USP DI for complete dosage information.

MONOBACTAMS

The development of useful monobactam antibiotics began with the independent isolation of sulfazecin (SQ 26,445) and other monocyclic β-lactam antibiotics from saprophytic soil bacteria in Japan[83] and the United States.[84] Sulfazecin was found to be weakly active as an antibacterial agent, but highly resistant to β-lactamases. Extensive structure–activity studies[85] eventually led to the development of aztreonam, which has useful properties as an antibacterial agent. Early work established that the 3-methoxy group, which was in part responsible for β-lactamase stability in the series, contributed to the low antibacterial potency and poor chemical stability of these antibiotics. A 4-methyl group, on the other hand, increases stability to β-lactamases while increasing activity against gram-negative bacteria at the same time. Unfortunately, potency against gram-positive bacteria decreases. 4,4-Gem-dimethyl substitution confers oral activity with a slight decrease in antibacterial potency.

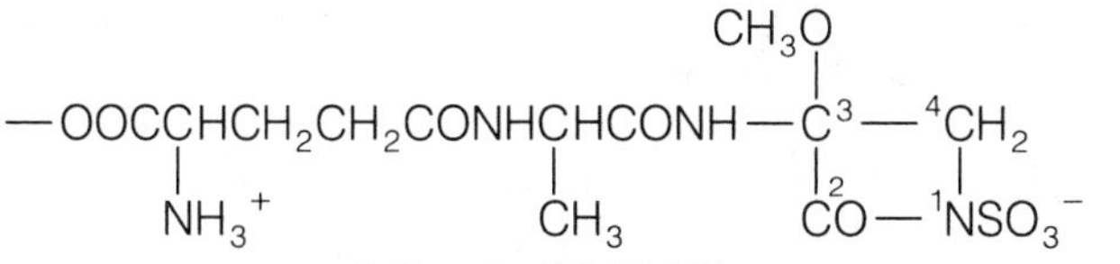

Sulfazecin (SQ 26,445)

Aztreonam Disodium (Azactam) is a monobactam prepared by total synthesis. It binds with high affinity to PBP 3 in gram-negative bacteria only. Aztreonam is inactive against gram-positive bacteria and anaerobes. The β-lactamase resistance of aztreonam is similar to that of ceftazidime, which has the same isobutyric acid oximinoacyl group. Aztreonam is not an inducer of chromosomally mediated β-lactamases.

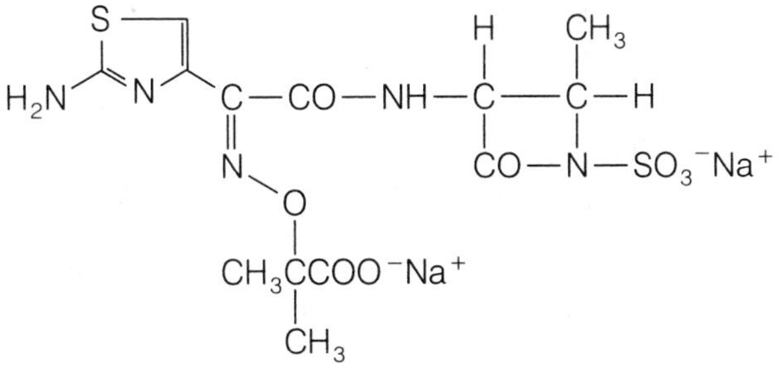

Aztreonam Disodium

Aztreonam is particularly active against aerobic gram-negative bacilli, including *E. coli*, *K. pneumoniae*, *K. oxytoca*, *Pr. mirabilis*, *Ser. marcescens*, *Citrobacter*, and *Ps. aeruginosa*. It is used to treat infections of the urinary tract, lower respiratory tract, intra-abdominal, and gynecologic, as well as septicemias caused by these organisms. Aztreonam is also effective against, but is not currently used to treat, infections caused by *Haemophilus*, *Neisseria*, *Salmonella*, indole-positive *Proteus*, and *Yersinia*. It is not active against gram-positive bacteria, anaerobic bacteria or other species of *Pseudomonas*.

The urinary excretion of aztreonam is about 70% of the administered dose. Some is excreted through the bile. The serum half-life is 1.7 hours, thereby allowing aztreonam to be administered two or three times daily, depending on the severity of the infection. Less than 1% of an orally administered dose of aztreonam is absorbed, prompting the suggestions that this β-lactam could be used to treat intestinal infections.

The disodium salt of aztreonam is very soluble in water. Solutions for parenteral administration containing 2% or less are stable for 48 hours at room temperature. Refrigerated solutions retain full potency for one week.

Tigemonam is an investigational monobactam that is orally active.[86] It is highly resistant to β-lactamases. The antibacterial spectrum of activity of tigemonam resembles that of aztreonam. It is very active against the *Enterobacteriaceae*, including *E. coli*, *Klebsiella*, *Proteus*, *Citrobacter*, *Serratia*, and *Enterobacter* species. It also exhibits good potency against *H. influenzae* and *N. gonorrhoeae*. Tigemonam is not particularly active against gram-positive or anaerobic bacteria and is inactive against *Ps. aeruginosa*.

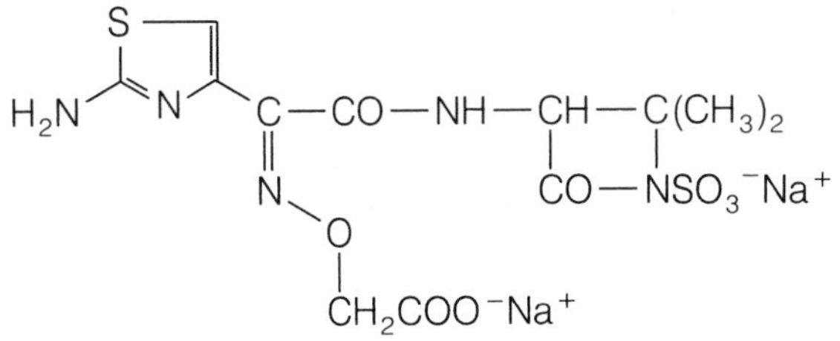

Tigemonam Disodium

In contrast with the poor oral bioavailability of aztreonam, the oral absorption of tigemonam in animals is excellent. It could become a valuable agent for the oral treatment of urinary tract infections and other non–life-threatening infections caused by β-lactamase-producing gram-negative bacteria.

THE AMINOGLYCOSIDES

The discovery of streptomycin, the first aminoglycoside antibiotic to be used in chemotherapy, was the result of a planned and deliberate search begun in

1939 and brought to fruition in 1944 by Waksman and his associates.[87] This success stimulated worldwide searches for antibiotics from the actinomycetes and, particularly, from the genus *Streptomyces*. Among the many antibiotics isolated from that genus, several are compounds closely related in structure to streptomycin. Six of them, kanamycin, neomycin, paromomycin, gentamicin, tobramycin, and netilmicin are currently marketed in the United States. Amikacin, a semisynthetic derivative of kanamycin A, has been added, and it is possible that additional aminoglycosides will be introduced in the future.

All aminoglycoside antibiotics are very poorly absorbed (less than 1% under normal circumstances) following oral administration and some of them (kanamycin, neomycin, and paromomycin) are administered by that route for the treatment of gastrointestinal infections. Because of the potent broad-spectrum nature of their antimicrobial activity, they are also used for the treatment of systemic infections. However, their undesirable side effects, particularly oto- and nephrotoxicity, have led to restrictions in their systemic use to serious infections or infections caused by bacterial strains resistant to other agents. When administered for systemic infections, aminoglycosides must be given parenterally, usually by intramuscular injection. An additional antibiotic obtained from *Streptomyces*, spectinomycin, is also an aminoglycoside, but differs chemically and microbiologically from other members of the group. It is employed exclusively for the treatment of uncomplicated gonorrhea.

CHEMISTRY

Aminoglycosides are so named because their structures consist of amino sugars linked glycosidically. All have at least one aminohexose and some have a pentose lacking an amino group (e.g., streptomycin, neomycin, and paromomycin). Additionally, each of the clinically useful aminoglycosides contains a highly substituted 1,3-diaminocyclohexane central ring: in kanamycin, neomycin, gentamicin, and tobramycin, it is deoxystreptamine; and in streptomycin it is streptadine. The aminoglycosides are thus strongly basic compounds that exist as polycations at physiologic *p*H. Their inorganic acid salts are very soluble in water. All are available as sulfates. Solutions of the aminoglycoside salts are stable to autoclaving. The high water solubility of the aminoglycosides no doubt contributes to their pharmacokinetic properties. They distribute well into most body fluids, but not into the central nervous system, bone, or fatty or connective tissues. They tend to concentrate in the kidney and are excreted by glomerular filtration. Metabolism of aminoglycosides in vivo apparently does not occur.

SPECTRUM OF ACTIVITY

Although the aminoglycosides are classified as broad-spectrum antibiotics, their greatest usefulness lies in the treatment of serious systemic infections caused by aerobic gram-negative bacilli. The choice of agent is generally between kanamycin, gentamicin, tobramycin, netilmicin, and amikacin. Aerobic gram-negative and gram-positive cocci (with the exception of staphylococci) tend to be less sensitive and, thus, the β-lactam and other antibiotics tend to be preferred for the treatment of infections caused by these organisms. Anaerobic bacteria are invariably resistant to the aminoglycosides. Streptomycin is the most effective of the group for the chemotherapy of tuberculosis, brucellosis, tularemia, and yersina infections. Paromomycin is used primarily in the chemotherapy of amebic dysentery. Under certain circumstances, aminoglycoside and β-lactam antibiotics are known to exert a synergistic action in vivo against certain bacterial strains when the two are administered jointly. Thus, carbenicillin and gentamicin are synergistic against gentamicin-sensitive strains of *Ps. aeruginosa* and several other species of gram-negative bacilli; penicillin G and streptomycin (or gentamicin or kanamycin) tend to be more effective than either agent alone in the treatment of enterococcal endocarditis, and so on. The two antibiotic types should not be combined in the same solution because they are chemically incompatible. Damage to the cell wall caused by the β-lactam is believed to increase the penetration of the aminoglycoside into the bacterial cell.

MECHANISM OF ACTION

Most of the studies concerning the mechanism of antibacterial action of the aminoglycosides have been carried out with streptomycin. However, the specific actions of other aminoglycosides are thought to be qualitatively similar. The aminoglycosides act directly on the bacterial ribosome to inhibit the initiation of protein synthesis and to interfere with the fidelity of translation of the genetic message. They bind to the 30S ribosomal subunit to form a complex that is unable to initiate proper amino acid polymerization.[88] The binding of streptomycin and other aminoglycosides to ribosomes also causes misreading mutations of the genetic code, apparently resulting from failure of specific aminoacyl-RNAs to recognize the proper codons on mRNA and the incorporation of improper amino acids into the peptide chain.[89] Evidence suggests[90] that the deoxystreptamine-containing aminoglycosides differ quantitatively from streptomycin in causing misreading at lower concentrations than are required to prevent initiation of protein synthesis, whereas streptomycin inhibits initiation and causes misreading equally effectively. Spectinomycin, on the other hand, prevents the initiation of protein synthesis, but apparently does not cause misreading. All of the commercially available aminoglycoside antibiotics are bactericidal, except spectinomycin. The mechanistic basis for the bactericidal action of the aminoglycosides, however, has not been elucidated.

MICROBIAL RESISTANCE

The development of strains of *Enterobacteriaceae* resistant to antibiotics has become well recognized as a serious medical problem. Nosocomial (hospital-acquired) infections caused by these organisms are often resistant to antibiotic therapy. Research has clearly established that multiple resistance among gram-negative bacilli to a variety of antibiotics occurs and can be transmitted to previously nonresistant strains of the same species and, indeed, to different species of bacteria. The mechanism of transfer of resistance from one bacterium to another has been directly attributed to extrachromosomal R-factors (DNA) which are self-replicative and transferable by conjugation (direct contact). The aminoglycoside antibiotics, because of their potent bactericidal action against gram-negative bacilli, are now preferred for the treatment of many serious infections caused by coliform bacteria. However, a pattern of bacterial resistance has developed to each of the aminoglycoside antibiotics as the clinical use of them has become more widespread. Consequently, there are bacterial strains resistant to streptomycin, kanamycin, and gentamicin. Strains carrying R-factors for resistance to these antibiotics synthesize enzymes capable of acetylating, phosphorylating, or adenylating key amino or hydroxyl groups of the aminoglycosides. Much of the recent effort in aminoglycoside research is directed toward identification of new, or modification of existing, antibiotics that are resistant to inactivation by bacterial enzymes.

Resistance of individual aminoglycosides to specific inactivating enzymes can, in large measure, be understood on the basis of chemical principles. As a first principle, it can be assumed that if the target functional group is absent in a position of the structure normally attacked by an inactivating enzyme, then the antibiotic will be resistant to the enzyme. Second, steric factors may confer resistance to attack at functionalities otherwise susceptible to enzymatic attack. For example, the conversion of a primary amino group to a secondary amine has been shown to inhibit *N*-acetylation by certain aminoglycoside acetyl transferases. At least nine different aminoglycoside-inactivating enzymes have been identified and partially characterized.[91] The sites of attack of these enzymes and the biochemistry of the inactivation reactions will be briefly described, using the kanamycin B structure (which holds the dubious distinction of being a substrate for all of the enzymes described) for illustrative purposes. Aminoglycoside-inactivating enzymes include (1) aminoacetyltransferases that acetylate the 6′-NH_2 of ring I, the 3-NH_2 of ring II or the 2′-NH_2 of ring I; (2) phosphotransferases that phosphorylate the 3′-OH of ring I; and nucleotidyltransferases that adenylate the 2″-OH of ring III, the 4′-OH of ring I, or the 4″-OH of ring III.

The gentamicins and tobramycin lack a 3′-hydroxyl group in ring I (see the section on the individual products for structures) and consequently are not inactivated by the phosphotransferase enzymes that phosphorylate that group in the kanamycins. Gentamicin C_1, but not gentamicins C_{1a} or C_2 or tobramycin, is resistant to the acetyltransferase that acetylates the 6′-amino group in ring I of kanamycin B. All gentamicins are resistant to the nucleotidyltransferase enzyme that adenylates the secondary equatorial 4″-hydroxyl group of kanamycin B, because the 4″-hydroxyl group in the gentamicins is tertiary and has an *axial* orientation. Removal of functional groups susceptible to attacking an aminoglycoside can, occasionally, lead to derivatives that resist enzymatic inactivation and retain activity. For example, the 3′-deoxy-, 4′-deoxy-, and 3,4′-dideoxykanamycins are more similar to the gentamicins and tobramycin in their patterns of activity against clinical isolates that resist one or more of the aminoglycosides.

The most significant breakthrough yet achieved in the search for aminoglycosides resistant to bacterial enzymes has been amikacin, the 1-*N*-L(−)-amino-α-butyryl (L-AHBA) derivative of kanamycin A. This remarkable compound retains most of the intrinsic potency of kanamycin A and is resistant to all aminoglycoside-inactivating enzymes known, except the aminoacetyltransferase that acetylates the 6′-amino group of ring I.[92] The cause of resistance of amikacin to enzymatic inactivation is not known, but it has been suggested that the introduction of the L-AHBA group into kanamycin A markedly decreases its affinity for the inactivating enzymes. The importance of amikacin's resistance to enzymatic inactivation is reflected in the results of an investigation of the comparative effectiveness of amikacin and other aminoglycosides against clinical isolates of bacterial strains known to be resistant to one or more of the aminoglycosides.[93] In this study, amikacin was effective against 91% of the isolates (with a range of 87% to 100% depending on the species). The strains found susceptible to other systemically useful aminoglycosides were kanamycin, 18%; gentamicin, 36%; and tobramycin, 41%.

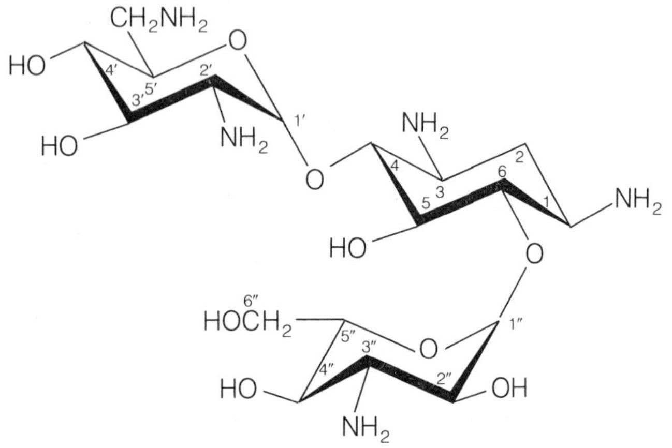

Kanamycin B

Low-level resistance associated with diminished aminoglycoside uptake has been observed in certain strains of *Ps. aeruginosa* isolated from nosocomial infections.[94] Bacterial susceptibility to aminoglycosides requires uptake of the drug by an energy-dependent active process.[95] Uptake is initiated by the binding of the cationic aminoglycoside to anionic phospholipids of the cell membrane. Electron transport-linked transfer of the aminoglycoside through the cell membrane then occurs. Divalent cations, such as Ca^{2+} and Mg^{2+}, antagonize the transport of aminoglycosides into bacterial cells by interfering with their binding to cell membrane phospholipids. The resistance of anaerobic bacteria to the lethal action of the aminoglycosides is due to the absence of the respiration-driven active transport process for transporting the antibiotics.

STRUCTURE–ACTIVITY RELATIONSHIPS

Despite the complexity inherent in various aminoglycoside structures, some conclusions on structure–activity relationships (SAR) in this antibiotic class have been made.[96] Such conclusions have been formulated on the basis of comparisons of naturally occurring aminoglycoside structures, the results of selective semisynthetic modifications, and the elucidation of sites of inactivation by bacterial enzymes. It is convenient to discuss sequentially aminoglycoside SAR in terms of substituents in rings I, II, and III.

Ring I is crucially important for characteristic broad-spectrum antibacterial activity, and it is the primary target for bacterial inactivating enzymes. Aminofunctions at 6′ and 2′ are particularly important, as kanamycin B (6′-amino, 2′-amino) is more active than kanamycin A (6′-amino, 2′-hydroxyl) which, in turn, is more active than kanamycin C (6′-hydroxyl, 2′-amino). Methylation at either the 6′-carbon or the 6′-amino positions does not appreciably lower antibacterial activity and confers resistance to enzymatic acetylation of the 6′-amino group. Removal of the 3′-hydroxyl or the 4′-hydroxyl group, or both, in the kanamycins (e.g., 3′,4′-dideoxykanamycin B or dibekacin) does not reduce antibacterial potency. The gentamicins also lack oxygen functions at these positions, as do sisomicin and netilmicin, which also have a 4′,5′-double bond. None of these derivatives is inactivated by phosphotransferase enzymes that phosphorylate the 3′-hydroxyl group. Evidently the 3′-phosphorylated derivatives have very low affinity for aminoglycoside-binding sites in bacterial ribosomes.

Few modifications of ring II (deoxystreptamine) functional groups are possible without appreciable loss of activity in most of the aminoglycosides. However, the 1-amino group of kanamycin A can be acylated (e.g., amikacin), and activity is largely retained. Netilmicin (1-*N*-ethylsisomicin) retains the antibacterial potency of sisomicin and is resistant to several additional bacterial-inactivating enzymes. 2″-Hydroxysisomicin is claimed to be resistant to bacterial strains that adenylate the 2″-hydroxyl group of ring III, whereas 3-deoxysisomicin exhibits good activity against bacterial strains that elaborate 3-acetylating enzymes.

Ring III functional groups appear to be somewhat less sensitive than those of either ring I or ring II to structural changes. Although the 2″-deoxygentamicins are significantly less active than their 2″-hydroxyl counterparts, the 2″-amino derivatives (seldomycins) are highly active. The 3″-amino group of gentamicins may be primary or secondary with high antibacterial potency. Furthermore, the 4″-hydroxyl group may be *axial* or *equatorial* with little change in potency.

Despite improvements in antibacterial potency and spectrum among newer naturally occurring and semisynthetic aminoglycoside antibiotics, efforts to find agents with improved margins of safety over the earlier-discovered antibiotics have clearly been disappointing. The potential for toxicity of these important chemotherapeutic agents continues to restrict their use largely to the hospital environment.

The discovery of agents with higher potency/toxicity ratios remains an important goal of aminoglycoside research. In a relatively recent review, however, Price[97] expresses doubt that many significant clinical breakthroughs in aminoglycoside research will occur in the future.

PRODUCTS

Streptomycin Sulfate, Sterile, USP. Streptomycin sulfate is a white, odorless powder that is hygroscopic but stable toward light and air. It is freely soluble in water, forming solutions that are slightly acidic or nearly neutral. It is very slightly soluble in alcohol and is insoluble in most other organic solvents. Acid hydrolysis of streptomycin yields streptidine and streptobiosamine, the compound that is a combination of L-streptose and *N*-methyl-L-glucosamine.

Streptomycin acts as a triacidic base through the effect of its two strongly basic guanidino groups and the more weakly basic methylamino group. Aqueous solutions may be stored at room temperature for one week without any loss of potency, but they are most stable if the *p*H is adjusted between 4.5 and 7.0. The solutions decompose if sterilized by heating, so sterile solutions are prepared by adding sterile distilled water to the sterile powder. The early salts of streptomycin contained impurities that were difficult to remove and caused a histaminelike reaction. By forming a complex with calcium chloride, it was possible to free the streptomycin from these impurities and to obtain a product that was generally well tolerated.

The organism that produces streptomycin, *S. griseus*, also produces several other antibiotic compounds, hydroxystreptomycin, mannisidostreptomycin, and cycloheximide (*q.v.*). None of these has achieved importance as a medicinally useful substance. The term *streptomycin A* has been used to refer to what is commonly called streptomycin, and mannisidostreptomycin has been called streptomycin B. Hydroxystreptomycin differs from streptomycin in having a hydroxyl group in place of one of the hydrogens of the streptose methyl group. Mannisidostreptomycin has a mannose residue attached by glycosidic linkage through the hydroxyl group at C-4 of the *N*-methyl-L-glucosamine moiety. The work of Dyer[98,99] to establish the complete stereostructure of streptomycin has been completed with the total synthesis of streptomycin and dihydrostreptomycin[100] by Japanese scientists.

A clinical problem that sometimes develops with the use of streptomycin is the early development of resistant strains of bacteria, making necessary a change in therapy of the disease. Another factor that limits its therapeutic efficacy is its chronic toxicity. Certain neurotoxic reactions have been observed after the use of streptomycin. They are characterized by vertigo, disturbance of equilibrium, and diminished auditory acuity. Minor toxic effects include skin rashes, mild malaise, muscular pains, and drug fever.

As a chemotherapeutic agent, the drug is active against numerous gram-negative and gram-positive bacteria. One of the greatest virtues of streptomycin is its effectiveness against the tubercle bacillus. In itself, it is not a cure, but it is a valuable adjunct to the standard treatment of tuberculosis. The greatest drawback to the use of this antibiotic is the rather rapid development of resistant strains of microorganisms. In infections that may be due to both streptomycin- and penicillin-sensitive bacteria, the combined administration of the two antibiotics has been advocated. The possible development of damage to the otic nerve by the continued use of streptomycin-containing preparations has led to the discouragement of the use of such products. There is an increasing tendency to reserve the use of streptomycin products for the treatment of tuberculosis. However, it remains one of the agents of choice for the treatment of certain "occupational" bacterial infections, such as brucellosis, tularemia, bubonic plague, and glanders. Because streptomycin is not absorbed when given orally and is not significantly

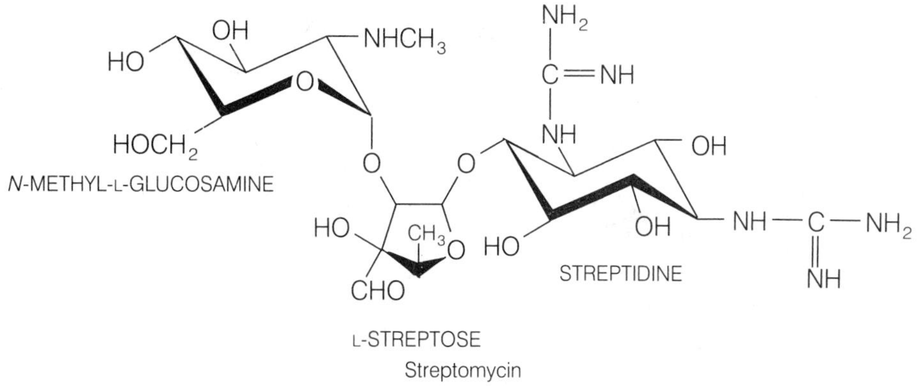

Streptomycin

destroyed in the gastrointestinal tract is why that, at one time, it was rather widely used in the treatment of infections of the intestinal tract. For systemic action, streptomycin usually is given by intramuscular injection.

Neomycin Sulfate, USP (Mycifradin, Neobiotic). In a search for less toxic antibiotics than streptomycin, Waksman and Lechevalier[101] obtained neomycin in 1949 from *S. fradiae*. Since then, neomycin has increased steadily in importance and, today, it is considered to be one of the most useful antibiotics in the treatment of gastrointestinal infections, dermatologic infections, and acute bacterial peritonitis. Also, it is employed in abdominal surgery to reduce or avoid complications caused by infections from bacterial flora of the bowel. It has a broad-spectrum activity against a variety of organisms. It shows a low incidence of toxic and hypersensitivity reactions. It is very slightly absorbed from the digestive tract, so its oral use does not ordinarily produce any systemic effect. Neomycin-resistant strains of pathogens have seldom been reported to develop from those organisms against which neomycin is effective.

Neomycin as the sulfate salt is a white to slightly yellow crystalline powder that is very soluble in water. It is hygroscopic and photosensitive (but stable over a wide *p*H range and to autoclaving). Neomycin sulfate contains the equivalent of 60% of the free base.

Neomycin, as produced by *S. fradiae*, is a mixture of closely related substances. Included in the "neomycin complex" is neamine (originally designated neomycin A) and neomycins B and C. *Streptomyces fradiae* also elaborates another antibiotic called fradicin that has some antifungal properties, but no antibacterial activity. This substance is not present in "pure" neomycin.

The structures of neamine and neomycin B and C are known, and the absolute configurational structures of neamine and neomycin have been reported by Hichens and Rinehart.[102] Neamine may be obtained by methanolysis of neomycins B and C, during which the glycosidic link between the deoxystreptamine and D-ribose is broken. Therefore, neamine is a combination of deoxystreptamine and neosamine C, linked glycosidically (α) at the 4-position of deoxystreptamine. According to Hichens and Rinehart, neomycin B differs from neomycin C by the nature of the sugar attached terminally to D-ribose. That sugar, called neosamine B, differs from neosamine C in its stereochemistry. It has been suggested by Rinehart et al.[103] that, in neosamine B, the configuration is that of 2,6-diamino-2,6-dideoxy-L-idose in which the orientation of the 6-aminomethyl group is inverted to that of the 6-amino-6-deoxy-D-glucosamine in neosamine C. In both instances, the glycosidic links are assumed to be α. However, Huettenrauch[104] later suggested that both of the diamino sugars in neomycin C have the L-idose configuration and that the glycosidic link is β in the one attached to D-ribose. The latter stereochemistry has been confirmed by the total synthesis of neomycin C by the Umezawa group.[105]

Paromomycin Sulfate, USP (Humatin). The isolation of paromomycin was reported in 1956 as an antibiotic obtained from a *Streptomyces* species (P-D 04998) that is said to resemble closely *S. rimosus*. The parent organism had been obtained from soil samples collected in Colombia. However, paromomycin more closely resembles neomycin and streptomycin, in antibiotic activity, than it does oxytetracycline, the antibiotic obtained from *S. rimosus*.

The general structure of paromomycin was first reported by Haskell et al.[106] as one compound. Subsequently, chromatographic determinations have

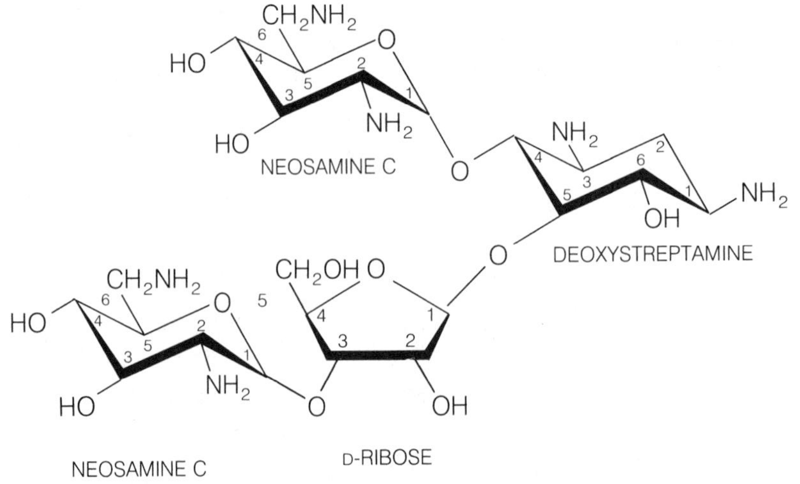

Neomycin C

shown paromomycin to consist of two fractions that have been named paromomycin I and paromomycin II. The absolute configurational structures for the paromomycins were suggested by Hichens and Rinehart[102] as shown in the structural formula and have been confirmed by DeJongh et al.[107] by mass spectrometric studies. It may be noted that the structure of paromomycin is the same as that of neomycin B, except that paromomycin contains D-glucosamine instead of the 6-amino-6-deoxy-D-glucosamine found in neomycin B. The same relationship in structures is found between paromomycin II and neomycin C. The combination of D-glucosamine with deoxystreptamine is obtained by partial hydrolysis of both paromomycins and is called paromamine [4-(2-amino-2-deoxy-α-4-glucosyl)deoxystreptamine].

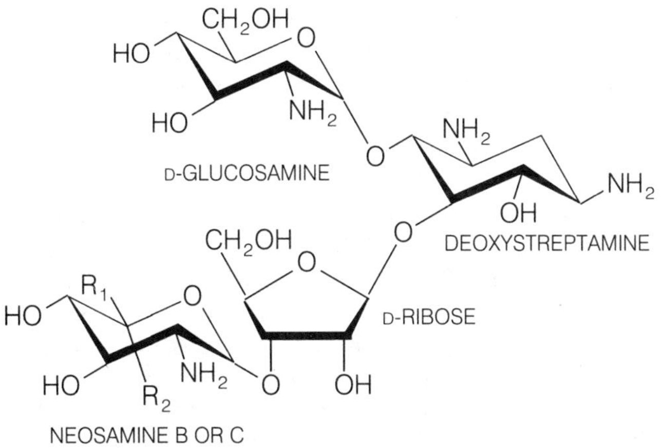

D-GLUCOSAMINE

DEOXYSTREPTAMINE

D-RIBOSE

NEOSAMINE B OR C

Paromomycin I: R_1 = H; R_2 = CH_2NH_2
Paromomycin II: R_1 = CH_2NH_2; R_2 = H

Paromomycin has broad-spectrum antibacterial activity and has been employed for the treatment of gastrointestinal infections caused by *Salmonella*, *Shigella*, and enteropathogenic *E. coli*. However, currently, its use is largely restricted to the treatment of intestinal amebiasis. Paromomycin is soluble in water and stable to heat over a wide pH range.

Kanamycin Sulfate, USP (Kantrex). Kanamycin was isolated in 1957 in Japan by Umezawa and co-workers[108] from *S. kanamyceticus*. Its activity against mycobacteria and many intestinal bacteria, as well as a number of pathogens that show resistance to other antibiotics, brought a great deal of attention to this antibiotic. As a result, kanamycin was tested and released for medical use in a very short time.

Research activity has been focused intensively on the determination of the structures of the kanamycins. It has been determined by chromatography that *S. kanamyceticus* elaborates three closely related structures that have been designated kanamycins A, B, and C. Commercially available kanamycin is almost pure kanamycin A, the least toxic of the three forms. The kanamycins differ by only the nature of the sugar moieties attached to the glycosidic oxygen on the 4-position of the central deoxystreptamine. The absolute configuration of the deoxystreptamine in kanamycins has been reported by Tatsuoka et al.,[109] as represented in the following diagram. The chemical relationships among the kanamycins, the neomycins, and the paromomycins have been reported by Hichens and Rinehart.[102] It may be noted that the kanamycins do not have the D-ribose molecule that is present in neomycins and paromomycins. Perhaps this structural difference is significant in the lower toxicity observed with kanamycins. The kanosamine fragment linked glycosidically to the 6-position of deoxystreptamine is 3-amino-3-deoxy-D-glucose (3-D-glucosamine) in all three kanamycins. The structures of the kanamycins have been proved by total synthesis.[110,111] It may be seen that they differ in the nature of the substituted D-glucoses attached glycosidically to the 4-position of the deoxystreptamine ring. Kanamycin A contains 6-amino-6-deoxy-D-glucose; kanamycin B contains 2,6-diamino-2,6-dideoxy-D-glucose; and kanamycin C contains 2-amino-2-deoxy-D-glucose (see diagram).

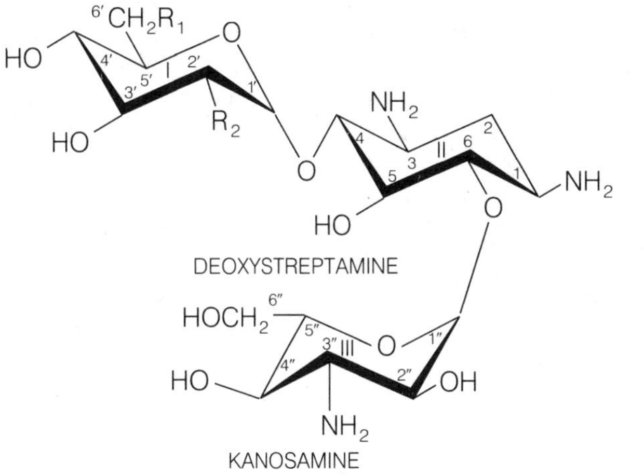

DEOXYSTREPTAMINE

KANOSAMINE

Kanamycin A: R_1 = NH_2; R_2 = OH
Kanamycin B: R_1 = NH_2; R_2 = NH_2
Kanamycin C: R_1 = OH; R_2 = NH_2

Kanamycin is basic and forms salts of acids through its amino groups. It is water soluble as the free base, but it is used in therapy as the sulfate salt, which is very soluble. It is stable to both heat and chemicals. Solutions resist both acids and alkali within the pH range of 2.0 to 11.0. Because of possible inactivation of either agent, kanamycin and penicillin salts should not be combined in the same solution.

The use of kanamycin in the United States is usually restricted to infections of the intestinal tract (such as bacillary dysentery) and to systemic infections arising from gram-negative bacilli (e.g., *Klebsiella*, *Proteus*, *Enterobacter*, and *Serratia*) that have developed resistance to other antibiotics. It has been recommended also for antisepsis of the bowel preoperatively. It is poorly absorbed from the intestinal tract; consequently, systemic infections must be treated by intramuscular or, for serious infections, intravenous injections. Injections of it are rather painful, and the concomitant use of a local anesthetic is indicated. The use of kanamycin in treatment of tuberculosis has not been widely advocated since the discovery that mycobacteria develop resistance to it very rapidly. In fact, clinical experience as well as experimental work of Morikubo[112] indicate that kanamycin does develop cross-resistance in the tubercle bacilli with dihydrostreptomycin, viomycin, and other antitubercular drugs. Similarly to streptomycin, kanamycin may cause a decrease in, or complete loss of, hearing. Upon development of such symptoms, its use should be stopped immediately.

Amikacin, USP. 1-*N*-Amino-α-hydroxybutyrylkanamycin A (Amikin) is a semisynthetic aminoglycoside first prepared in Japan. The synthesis formally involves simple acylation of the 1-amino group of the deoxystreptamine ring of kanamycin A with L-amino-α-hydroxybutyric acid (L-AHBA). This particular acyl derivative retains about 50% of the original activity of kanamycin A against sensitive strains of gram-negative bacilli. The L-AHBA derivative is much more active than the D-isomer.[113] The remarkable feature of amikacin is that it resists attack by most bacterial-inactivating enzymes and, therefore, is effective against strains of bacteria that are resistant to other aminoglycosides,[93] including gentamicin and tobramycin. In fact, it is resistant to all known aminoglycoside-inactivating enzymes, except the aminotransferase that acetylates the 6′-amino group of the aminoglycosides.[92]

Preliminary studies indicate that amikacin may be less ototoxic than either kanamycin or gentamicin.[114] It is noteworthy, however, that higher dosages of amikacin are generally required for the treatment of most gram-negative bacillary infections. For this reason, and to discourage the proliferation of bacterial strains resistant to it, amikacin is currently recommended for the treatment of serious infections caused by bacterial strains resistant to other aminoglycosides.

Gentamicin Sulfate, USP (Garamycin). Gentamicin was isolated in 1958 and reported in 1963 by Weinstein et al.[115] to belong to the streptomycinoid (aminocyclitol) group of antibiotics. It is obtained commercially from *Micromonospora purpurea*. Similar to the other members of its group, it has a broad spectrum of activity against many common pathogens of both gram-positive and gram-negative types. Of particular interest is its high degree of activity against *Ps. aeruginosa* and other gram-negative enteric bacilli.

Gentamicin is effective in the treatment of a variety of skin infections for which a topical cream or ointment may be used. However, because it offers no real advantage over topical neomycin in the treatment of all but pseudomonal infections, it is recommended that topical gentamicin be reserved for use in such infections and in the treatment of burns complicated by pseudomonemia. An injectable solution containing 40 mg of gentamicin sulfate per milliliter may be used for serious systemic and genitourinary tract infections caused by gram-negative bacteria, particularly *Pseudomonas*, *Enterobacter*, and *Serratia* species. Because of the development of strains of these bacterial species resistant to previously effective broad-spectrum antibiotics, gentamicin has been employed for the treatment of hospital-acquired infections caused by such organisms. However, resistant bacterial strains that inactivate

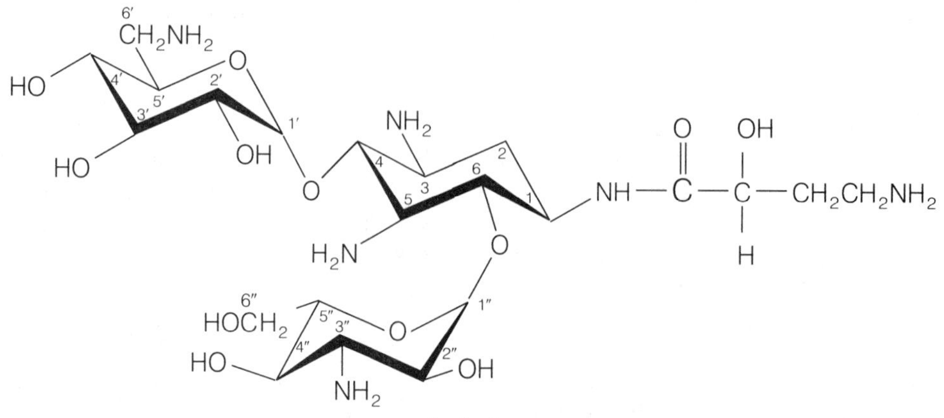

Amikacin

gentamicin by adenylation and acetylation appear to be emerging with increasing frequency.

Gentamicin sulfate is a mixture of the salts of compounds identified as gentamicins C_1, C_2, and C_{1a}. These gentamicins have been reported by Cooper et al.[116] to have the structures shown in the following diagram. Furthermore, the absolute stereochemistries of the sugar components and the geometries of the glycosidic linkages have been established.[117]

Coproduced, but not a part of the commercial product, are gentamicins A and B. Their structures have been reported by Maehr and Schaffner[118] and are closely related to the gentamicins C. Although the gentamicin molecules are similar in many ways to other aminocyclitols, such as streptomycins, they are sufficiently different that their medical effectiveness is significantly greater. Gentamicin sulfate is a white to buff substance that is soluble in water and insoluble in alcohol, acetone, and benzene. Its solutions are stable over a wide pH range and may be autoclaved. It is chemically incompatible with carbenicillin and the two should not be combined in the same intravenous solution.

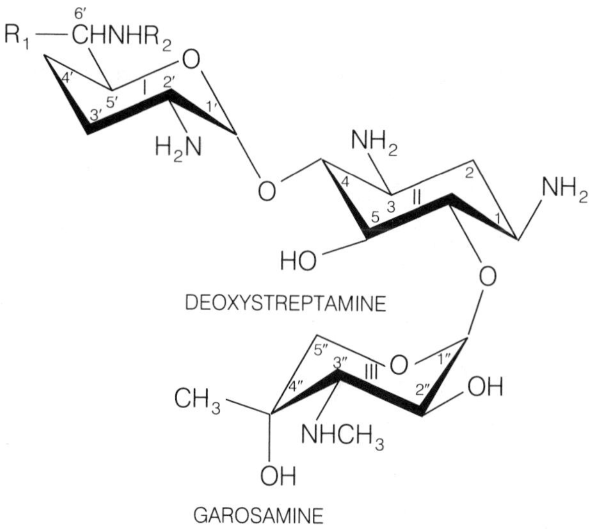

DEOXYSTREPTAMINE

GAROSAMINE

Gentamicin C_1: $R_1 = R_2 = CH_3$
Gentamicin C_2: $R_1 = CH_3$; $R_2 = H$
Gentamicin C_{1a}: $R_1 = R_2 = H$

Tobramycin Sulfate, USP (Nebcin), introduced in 1976, is the most active of chemically related aminoglycosides called nebramycins, obtained from a strain of *S. tenebrarius*. Five members of the nebramycin complex have been identified chemically.[119] Factors 4 and 4′ are 6″-*O*-carbamoylkanamycin B and kanamycin B, respectively; factors 5′ and 6 are 6″-*O*-carbamoyltobramycin and tobramycin; and factor 2 is apramycin, a tetracyclic aminoglycoside with an unusual bicyclic central ring structure. Kanamycin B and tobramycin probably do not occur in fermentation broths per se but are formed by hydrolysis of the 6-*O*″-carbamoyl derivatives in the isolation procedure.

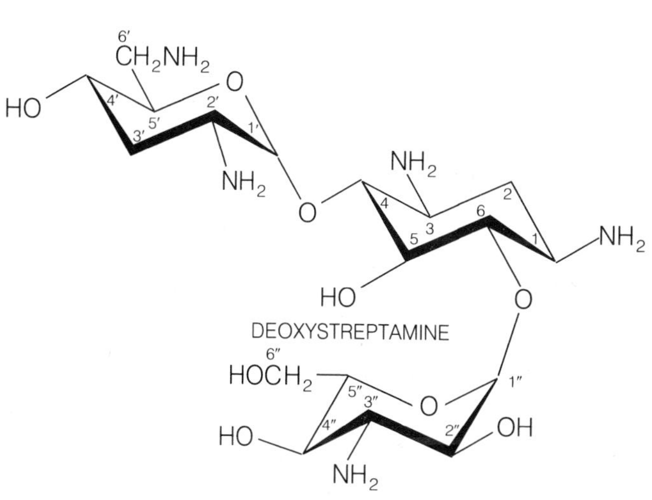

DEOXYSTREPTAMINE

Tobramycin

The most important property of tobramycin is an activity against most strains of *Ps. aeruginosa*, exceeding that of gentamicin by two- to fourfold. Some gentamicin-resistant strains of this troublesome organism are sensitive to tobramycin, but others are resistant to both antibiotics.[120] Other gram-negative bacilli and staphylococci are generally more sensitive to gentamicin. Tobramycin more closely resembles kanamycin B in structure (it is 3′-deoxykanamycin B).

Netilmicin Sulfate, USP. 1-*N*-Ethylsisomicin (Netromycin) is a semisynthetic derivative prepared by reductive ethylation[121] of sisomicin, an aminoglycoside antibiotic obtained from *Micromonosporum inyoensis*.[122] Structurally, sisomicin and netilmicin resemble gentamicin C_{1a}, a component of the gentamicin complex.

Against most strains of *Enterobacteriaceae*, *Ps. aeruginosa*, and *Staph. aureus*, sisomicin and netilmicin are comparable with gentamicin in potency.[123] However, netilmicin is active against many gentamicin-resistant strains, in particular among *E. coli*, *Enterobacter*, *Klebsiella*, and *Citrobacter*. A few strains of gentamicin-resistant *Ps. aeruginosa*, *Ser. marcescens*, and indole-positive *Proteus* are also sensitive to netilmicin. On the other hand, very few gentamicin-resistant bacterial strains are sensitive to sisomicin. The potency of netilmicin against certain gentamicin-resistant bacteria is attributed to its resistance to inactivation by bacterial enzymes that adenylylate or phosphorylate gentamicin and sisomicin. Evidently, the introduction of a 1-ethyl group in sisomicin markedly decreases the affinity of these enzymes for the molecule in a manner similar to that observed in the 1-*N*-γ-amino-α-hydroxybutyryl amide of kanamycin A (amikacin). Netilmicin, how-

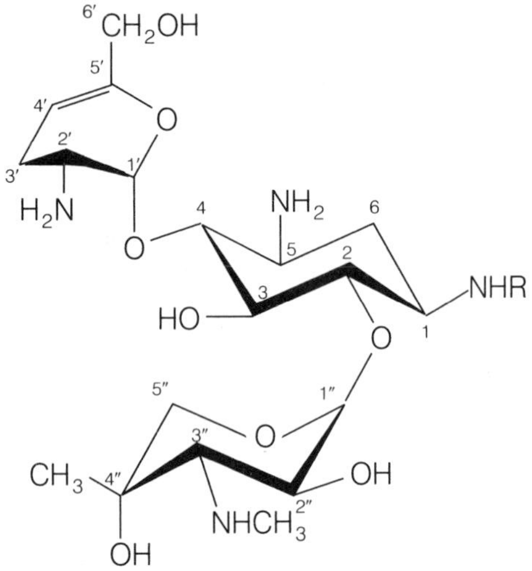

Sisomicin: R = H
Netilmicin: R = C_2H_5

ever, is inactivated by most of the bacterial enzymes that acetylateaminoglycosides, whereas amikacin is resistant to most of these enzymes.

The pharmacokinetic and toxicologic properties of netilmicin and gentamicin appear to be similar clinically, although animal studies indicated greater nephrotoxicity for gentamicin.

Sisomicin Sulfate, USP. Although it has been approved for human use in the United States, sisomicin has not been marketed in this country. Its antibacterial potency and effectiveness against aminoglycoside-inactivating enzymes resemble those of gentamicin. Sisomicin also exhibits pharmacokinetics and pharmacologic properties similar to those of gentamicin.

Spectinomycin Hydrochloride, Sterile, USP (Trobicin). This aminocyclitol antibiotic, isolated from *S. spectabilis*, and once called actinospectocin, was first described by Lewis and Clapp.[124] Its structure and absolute stereochemistry have recently been

TABLE 7-9

AMINOGLYCOSIDE ANTIBIOTICS

Name Proprietary Name	***Preparations***	***Usual Adult Dose****	***Usual Dose Range****	***Usual Pediatric Dose****
Streptomycin sulfate USP	Streptomycin sulfate injection USP Sterile streptomycin sulfate USP	IM, the equivalent of streptomycin, in combination with other antibacterials, 250 mg– 1 g every 6 hr, or 500 mg–2 g every 12 hr	Tuberculosis: 1 g twice weekly to 2 g/day; other infections: up to 4 g/day	IM, the equivalent of streptomycin, in combination with other antibacterials, 5–10 mg/kg of body weight every 6 hr; or 10–20 mg/kg of body weight every 12 hr
Neomycin sulfate USP *Mycifradin, Neobiotic*	Neomycin sulfate ointment USP Neomycin sulfate cream USP	For external use, topically to the skin, as the equivalent of a 0.35% ointment or cream of neomycin qd to tid		
	Neomycin sulfate ophthalmic ointment USP	For external use, topically to the conjunctiva, as the equivalent of a 0.35% ointment of neomycin every 8–24 hr		
	Neomycin sulfate oral solution USP Neomycin sulfate tablets USP	Hepatic coma: the equivalent of 700 mg–2.1 g of neomycin every 6 hr for 5–6 days; infectious diarrhea: 8.75 mg/kg of body weight every 6 hr for 2–3 days; preoperative preparation: 700 mg/hr for 4 doses, then 700 mg every 4 hr for the balance of 24 hr; or 10.3 mg/kg of body weight every 4 hr for 2–3 days	Antibacterial, the equivalent of neomycin, up to 8.4 g/day for 24–48 hr in preoperative bowel preparation	Adjunct in hepatic coma: 437.5 mg–1.225 g/m^2 of body surface every 6 hr for 5–6 days
	Sterile neomycin sulfate USP	IM, the equivalent of neomycin, 1.3–2.6 g/kg of body weight every 6 hr	Up to a maximum of 10.5 mg/kg of body weight daily, but not to exceed 700 mg/day for more than 10 days	

*See USP DI for complete dosage information.

TABLE 7-9 *Continued*

AMINOGLYCOSIDE ANTIBIOTICS

Name Proprietary Name	***Preparations***	***Usual Adult Dose****	***Usual Dose Range****	***Usual Pediatric Dose****
Paromomycin sulfate USP *Humatin*	Paromomycin sulfate capsules USP Paromomycin sulfate syrup USP	The equivalent of 500 mg of paromomycin every 6 hr taken with meals	500 mg–1 g of paromomycin	
Kanamycin sulfate USP *Kantrex, Klebcil*	Kanamycin sulfate capsules USP	Intestinal infections: the equivalent of 1 g of kanamycin every 6–8 hr for 5–7 days; preoperative preparation: the equivalent of 1 g of kanamycin every hr for 4 hr, then 1 g every 6 hr for 36–72 hr	3–12 g/day	Intestinal infections: 12.5 mg/kg of body weight or 375 mg/m^2 of body surface, qid
	Kanamycin sulfate injection USP	IM, the equivalent of kanamycin: 3.75 mg/kg of body weight every 6 hr; 5 mg/kg of body weight every 6 hr; 5 mg/kg of body weight every 8 hr; or 7.5 mg/kg of body weight every 12 hr	Up to 15 mg/kg of body weight daily, but not to exceed 1.5 g/day	IM or IV, the equivalent of kanamycin, premature and full-term neonates up up to 1 yr: 7.5 mg/kg of body weight every 12 hr; older infants and children: 3–7.5 mg/kg of body weight every 12 hr
Gentamicin sulfate USP *Garamycin*	Gentamicin sulfate cream USP Gentamicin sulfate ointment USP	For external use, topically to the skin, the equivalent of 0.1% gentamicin as a cream or ointment tid or qid		
	Gentamicin sulfate injection USP	IM or IV, the equivalent of gentamicin, 1–1.7 mg/kg of body weight every 8 hr; or 750 μg–1.25 mg/kg of body weight every 6 hr	The equivalent of gentamicin up to 8 mg/kg of body weight daily in severe, life-threatening infections	IM or IV, the equivalent of gentamicin, premature or full-term neonates 1 wk of age or less: 2.5–3 mg/kg of body weight every 12 hr; older neonates and infants: 2–2.5 mg/kg of body weight every 8 hr; children: 1–2.5 mg/kg of body weight every 8 hr
	Gentamicin sulfate ophthalmic ointment USP	Topical, to the conjunctiva, the equivalent of gentamicin, a thin strip (approximately 1 cm) of a 0.3% ointment every 6–12 hr		
	Gentamicin sulfate ophthalmic solution USP	Topical, to the conjunctiva, the equivalent of gentamicin, 1 drop of a 0.3% solution every 4–8 hr; topical to the ear canal, 3 or 4 drops of a 0.3% solution every 4–8 hr		
Amikacin USP *Amakin*	Amikacin sulfate injection USP	IM or IV, the equivalent of amikacin, 5 mg/kg of body weight every 8 hr, or 7.5 mg/kg of body weight every 12 hr	The equivalent of amikacin, up to 15 mg/kg of body weight daily, but not to exceed 1.5 g/day for more than 10 days	IM or IV, the equivalent of amikacin, neonates: initially, 10 mg/kg of body weight, then 7.5 mg/kg of body weight every 12 hr
Tobramycin USP *Nebcin*	Tobramycin sulfate injection USP	IM or IV, the equivalent of tobramycin, 750 μg–1.25 mg/kg of body weight every 6 hr, or 1–1.7 mg/kg of body weight every 8 hr	Up to 8 mg/kg of body weight daily in severe, life-threatening infections	IM or IV, the equivalent of tobramyacin, neonates 1 wk of age or less: up to 2 mg/kg of body weight every 12 hr
Trobicin	Sterile spectinomycin hydrochloride USP	IM, the equivalent of 2–4 g of spectinomycin		Dosage in infants and children not established

*See USP DI for complete dosage information.

confirmed by x-ray crystallography.[125] It occurs as the white crystalline dihydrochloride pentahydrate, which is stable in the dry form and very soluble in water. Solutions of spectinomycin, a hemiacetal, slowly hydrolyze on standing and should be freshly prepared and used within 24 hours. It is administered by deep intramuscular injection.

Spectinomycin is a broad-spectrum antibiotic with moderate activity against many gram-positive and gram-negative bacteria. It differs from streptomycin and the streptamine-containing aminoglycosides in chemical and antibacterial properties. Similarly to streptomycin, spectinomycin interferes with the binding of tRNA to the ribosomes and, thereby, interferes with the initiation of protein synthesis. Unlike streptomycin or the streptamine-containing antibiotics, however, it does not cause misreading of the messenger. Spectinomycin exerts a bacteriostatic action and is inferior to other aminoglycosides for most systemic infections. Currently, it is recommended as an alternative to penicillin G salts for the treatment of uncomplicated gonorrhea. A cure rate of greater than 90% has been observed in clinical studies for this indication. Many physicians prefer to use a tetracycline or erythromycin for prevention or treatment of suspected gonorrhea in penicillin-sensitive patients because, unlike these agents, spectinomycin is ineffective against syphilis. Furthermore, it is considerably more expensive than erythromycin and most of the tetracyclines.

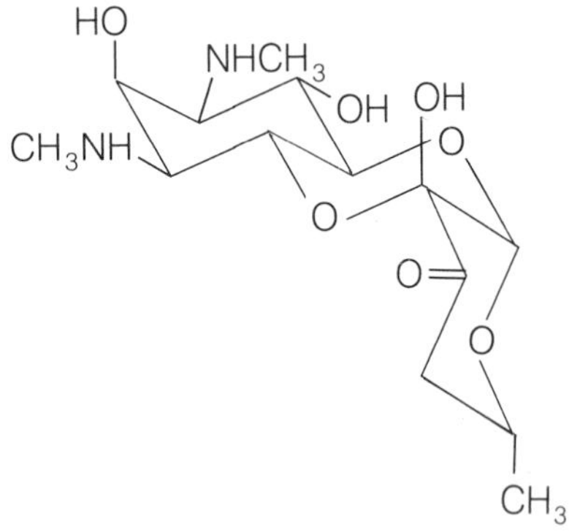

Spectinomycin

THE TETRACYCLINES

CHEMISTRY

Among the most important broad-spectrum antibiotics are members of the tetracycline family. Nine such compounds—tetracycline, rolitetracycline, oxytetracycline, chlortetracycline, demeclocycline, meclocycline, methacycline, doxycycline, and minocycline—have been introduced into medical use. Several others possess antibiotic activity. The tetracyclines are obtained by fermentation procedures from *Streptomyces* species or by chemical transformations of the natural products. Their chemical identities have been established by degradation studies and confirmed by the synthesis of three members of the group, oxytetracycline,[126,127] 6-demethyl-6-deoxytetracycline,[128] and anhydrochlortetracycline[129] in their (±) forms. The important members of the group are derivatives of an octahydronaphthacene, a hydrocarbon system that comprises four annelated six-membered rings. It is from this tetracyclic system that the group name is derived. The antibiotic spectra and chemical properties of these compounds are very similar, but not identical.

The stereochemistry of the tetracyclines is very complex. Carbon atoms 4, 4a, 5, 5a, 6, and 12a are potentially asymmetric depending on substitution. Oxytetracycline and doxycycline, each with a 6α-hydroxyl substituent, have six asymmetric centers, whereas the others, lacking asymmetry at C-6, have only five. Determination of the complete absolute stereochemistry of the tetracyclines was a difficult problem. Detailed x-ray diffraction analysis[130–132] established that the stereochemical formula shown in Table 7-10 represents the orientations found in the natural and semisynthetic tetracyclines. These studies also confirmed that conjugated systems exist in the structure from C-10 through C-12 and from C-1 through C-3 and that the formula represents only one of several canonical forms existing in those portions of the molecule.

The tetracyclines are amphoteric compounds, forming salts with either acids or bases. In neutral solutions these substances exist mainly as zwitterions. The acid salts, which are formed through protonation of the enol group on C-2, exist as crys-

TABLE 7-10

STRUCTURE OF TETRACYCLINES

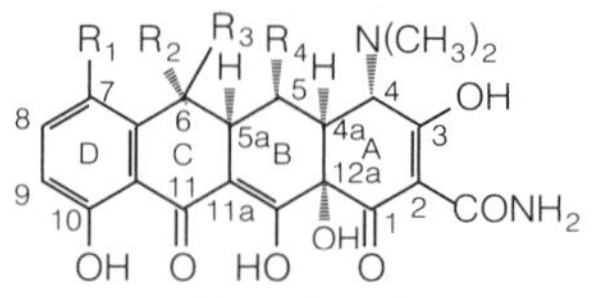

	R_1	R_2	R_3	R_4
Tetracycline	H	CH_3	OH	H
Chlortetracycline	Cl	CH_3	OH	H
Oxytetracycline	H	CH_3	OH	OH
Demeclocycline	Cl	H	OH	H
Methacycline	H	CH_2		OH
Doxycycline	H	H	CH_3	OH
Minocycline	$N(CH_3)_2$	H	H	H

talline compounds that are very soluble in water. However, these amphoteric antibiotics will crystallize out of aqueous solutions of their salts unless stabilized by an excess of acid. The hydrochloride salts are used most commonly for oral administration and are usually encapsulated because they are bitter. Water-soluble salts may be obtained also from bases, such as sodium or potassium hydroxides, but they are not stable in aqueous solutions. Water-insoluble salts are formed with divalent and polyvalent metals.

The unusual structural groupings in the tetracyclines produce three acidity constants in aqueous solutions of the acid salts (Table 7-11). The particular functional groups responsible for each of the thermodynamic pK_a values have been determined by Leeson et al.[133] to be as shown in the following formula diagram. These groupings have been identified by Stephens et al.[134] previously as the sites for protonation, but their earlier assignments for which produced the values responsible for pK_{a2} and pK_{a3} were opposite those of Leeson et al. This latter assignment has been substantiated by Rigler et al.[135]

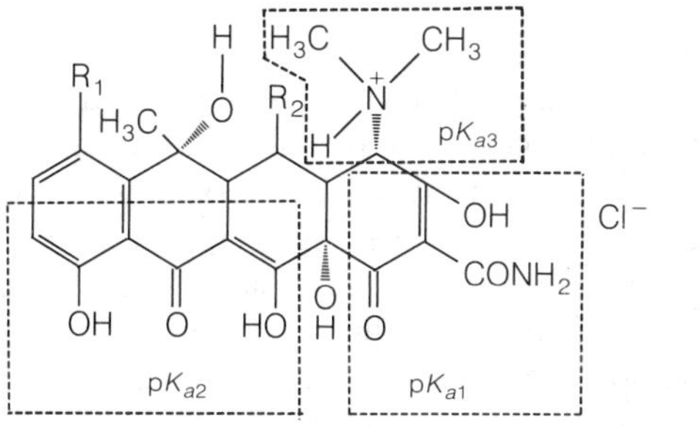

The approximate pK_a values for each of these groups in the six tetracycline salts in common use are shown (see Table 7-11). The values are taken from Stephens et al.,[134] from Benet and Goyan,[136] and from Barringer et al.[137] The pK_a of the 7-dimethylamino group of minocycline (not listed) is 5.0.

An interesting property of the tetracyclines is their ability to undergo epimerization at C-4 in solutions of intermediate pH range. These isomers are called *epi*tetracyclines. Under the influence of the acidic conditions, an equilibrium is established in about one day and consists of approximately equal amounts of the isomers. The partial structures below indicate the two forms of the epimeric pair. The 4-*epi* tetracyclines have been isolated and characterized. They exhibit much less activity than the "natural" isomers, thus accounting for a decrease in therapeutic value of aged solutions.

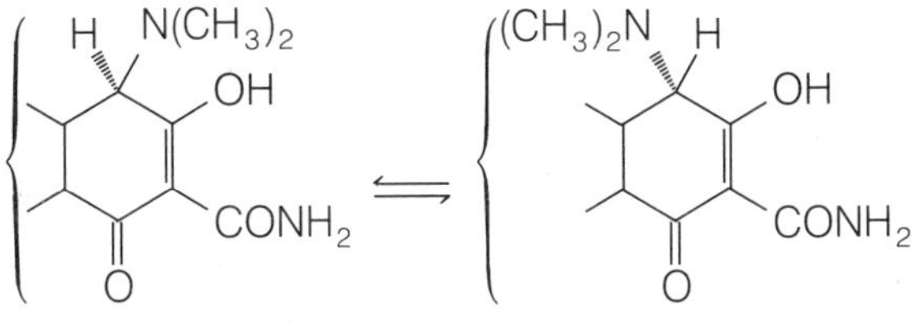

Strong acids and strong bases attack the tetracyclines having a hydroxyl group on C-6, causing a loss in activity through modification of the C ring. Strong acids produce a dehydration through a reaction involving the 6-hydroxyl group and the 5a-hydrogen. The double bond thus formed between positions 5a and 6 induces a shift in the position of the double bond between C-11a and C-12 to a position between C-11 and C-11a, forming the more energetically favored resonant system of the naphthalene group found in the inactive anhydrotetracyclines. Bases promote a reaction between the 6-hydroxyl group and the ketone group at the 11-position, causing the bond between the 11 and 11a atoms to cleave and to form the lactone ring found in the inactive isotetracycline. These two unfavorable reactions stimulated the research that has led to the development of the more stable and longer-acting compounds—6-deoxytetracycline, methacycline, doxycycline, and minocycline.

Stable chelate complexes are formed by the tetracyclines with many metals, including calcium, magnesium, and iron. Such chelates are usually very insoluble in water, accounting for the impairment in absorption of most (if not all) tetracyclines in the presence of milk, calcium-, magnesium-, and aluminum-containing antacids, and iron salts. Soluble alkalinizers, such as sodium bicarbonate, also decrease the gastrointestinal absorption of the tetracycline.[138] Deprotonation of tetracyclines to more ionic species, and their observed instability in alkaline solutions, may account for this observation. The affinity of tetracyclines for calcium causes them to be laid down in newly formed bones and teeth as tetracycline–calcium orthophosphate complexes. Deposits of these antibiotics in teeth cause a yellow discoloration that darkens (a photochemical reaction) over time. Tetracyclines are distributed into the milk of lactating mothers and also cross the

TABLE 7-11

pK_a VALUES (OF HYDROCHLORIDES) IN AQUEOUS SOLUTION AT 25%

	pK_{a_1}	pK_{a_2}	pK_{a_3}
Tetracycline	3.3	7.7	9.5
Chlortetracycline	3.3	7.4	9.3
Demeclocycline	3.3	7.2	9.3
Oxytetracycline	3.3	7.3	9.1
Doxacycline	3.4	7.7	9.7
Minocycline	2.8	7.8	9.3

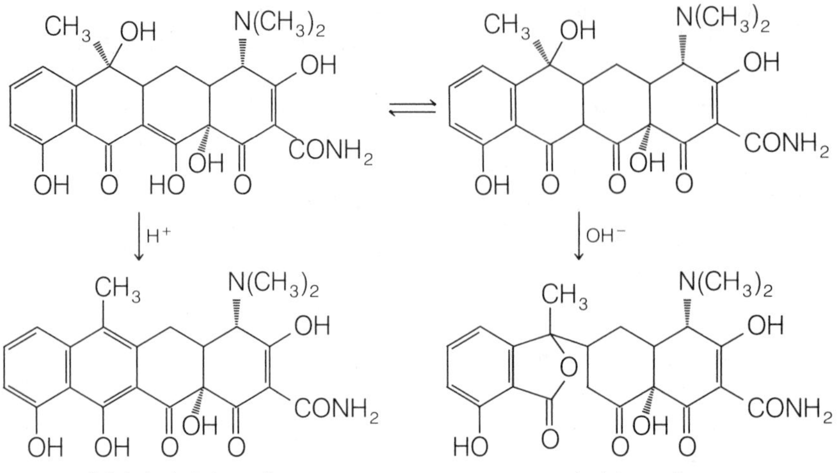

5,6-Anhydrotetracycline

Isotetracycline

placental barrier into the fetus. The possible effects of these agents on bones and teeth of the child should be taken into consideration before their use in pregnancy or in children under eight years of age is instituted.

MECHANISM OF ACTION AND RESISTANCE

The strong binding properties of the tetracyclines with metal ions caused Albert[139] to suggest that their antibacterial properties may be due to an ability to remove essential metal ions as chelated compounds. Elucidation of details of the mechanism of action of the tetracyclines,[140] however, has more clearly defined more specific roles for magnesium ion in molecular processes affected by these antibiotics in bacteria. Tetracyclines are specific inhibitors of bacterial protein synthesis. They bind to the 30S ribosomal subunit and, thereby, prevent the binding of aminoacyl tRNA to the mRNA–ribosome complex. The binding of aminoacyl tRNA and the binding of tetracyclines at the ribosomal-binding site both require magnesium ion.[141] Tetracyclines also bind to mammalian ribosomes, but with lower affinities, and they apparently do not achieve sufficient intracellular concentrations to interfere with protein synthesis. The selective toxicity of the tetracyclines against bacteria is strongly dependent upon the self-destructive capacity of bacterial cells to concentrate these agents in the cell. The active uptake of tetracyclines by bacterial cells is an energy-dependent process that requires ATP and magnesium ions.[142] Bacterial resistance to the action of the tetracyclines appears to result primarily, if not exclusively, from inability of the compounds to penetrate the bacterial cell wall. Loss of the capacity to transport tetracyclines actively[143] and the presence of tetracycline-binding proteins at the cell surface[144] have both been implicated as mechanisms preventing cell wall penetration in resistant bacterial strains. Inactivation of tetracyclines by bacterial enzymes has not been observed.

SPECTRUM OF ACTIVITY

The tetracyclines are truly broad-spectrum antibiotics with the broadest spectrum of any known antibacterial agents. They are active against a wide range of gram-positive and gram-negative bacteria, spirochetes, mycoplasmas, rickettsiae, and clamydiae. Their potential indications are, therefore, numerous. However, their bacteriostatic action is a disadvantage in the treatment of life-threatening infections, such as septicemia, endocarditis, and meningitis, wherein the aminoglycosides are usually preferred for gram-negative, and the penicillins for gram-positive, infections. Because of incomplete absorption and effectiveness against the natural bacterial flora of the intestine, tetracyclines may induce superinfections caused by the pathogenic yeast *Candida albicans*. Resistance to tetracyclines among both gram-positive and gram-negative bacteria is relatively common. Superinfections caused by resistant *Staph. aureus* and *Ps. aeruginosa* have resulted from the use of these agents over time. Parenteral tetracyclines may cause severe liver damage, especially when given in excessive dosage to pregnant women or to patients with impaired renal function.

STRUCTURE–ACTIVITY RELATIONSHIPS

As a result of the large amount of research carried out to prepare semisynthetic modifications of the tetracyclines and to obtain individual compounds by total synthesis, several interesting structure–activity relationships have emerged. Reviews are available that discuss structure–activity relationships among the tetracyclines in detail,[145–147] as well as their synthesis and chemical properties.[146–149] Only a brief review of the salient structure–activity features will be presented here. All derivatives containing fewer than four rings are inactive or nearly inactive. The simplest tetracycline derivative that retains the characteristic broad-spectrum activity associated with this antibiotic class is 6-demethyl-6-deoxytetracycline. It has become evident that many of the precise structural features present in this molecule must remain unmodified for derivatives to retain activity. Consequently, the integrity of substituents at carbon atoms 1, 2, 3, 4, 10, 11, 11a, and 12 cannot be drastically violated without deleterious effects on the antimicrobial properties of the resulting derivatives.

Only very slight modifications of A-ring substituents can be made without dramatic loss of antibacterial potency. The enolized tricarbonylmethane system at C-1 to C-3 must be intact for good activity. Replacement of the amide at C-2 with other functions, such as aldehyde, nitrile, or such, reduces or abolishes activity. Monoalkylation of the amide nitrogen reduces activity proportionately to the size of the alkyl group. Aminoalkylation of the amide nitrogen, accomplished by the Mannich reaction, yields derivatives that are substantially more water-soluble than the parent tetracycline and are hydrolyzed to it in vivo (e.g., rolitetracycline). The dimethylamino group at the 4-position must have the α-orientation: 4-epitetracyclines are very much less active than the natural isomers. Removal of the 4-dimethylamino group reduces activity even further. Activity is largely retained in the primary and *N*-methyl secondary amines, but rapidly diminishes in the higher alkylamines. A *cis* A/B ring fusion with a β-hydroxyl group at C-12a is apparently also essential. Esters of the C-12a hydroxyl group are all inactive, with the exception of the formyl ester, which readily hydrolyzes in aqueous solutions. Alkylation at C-11a also leads to inactive compounds, demonstrating the importance of an enolizabel β-diketone functionality at C-11 and C-12. The importance of the shape of the tetracyclic ring system is further illustrated by a substantial loss in antibacterial potency resulting from epimerization at C-5a. Dehydrogenation to form a double bond between C-5a and C-11a markedly decreases activity, as does aromatization of ring C to form anhydrotetracyclines.

In contrast, substituents at positions 5, 5a, 6, 7, 8, and 9 can be modified with varying degrees of impunity, resulting in retention and, sometimes, improvement of antibiotic activity. A 5-hydroxyl group, as in oxytetracycline and doxycycline, may influence pharmacokinetic properties, but does not change antimicrobial activity, when compared with the 5-deoxy compounds. 5a-Epitetracyclines (prepared by total synthesis), although highly active in vitro, are unfortunately much less impressive in vivo. Acid-stable 6-deoxytetracyclines and 6-demethyl-6-deoxytetracyclines have been used to prepare a variety of mono- and disubstituted derivatives by electrophilic substitution reactions at C-7 and C-9 of the D ring. The more useful results have been achieved with the introduction of substituents at C-7. Oddly, both strongly electron-withdrawing groups [e.g., chloro (chlortetracycline) and nitro] and strongly electron-donating groups [e.g., dimethylamino (minocycline)] enhance activity. This unusual circumstance is reflected in quantitative structure–activity (QSAR) studies of 7- and 9-substituted tetracyclines,[146,150] which indicated a squared dependence on σ, Hammet's electronic substituent constant, and in vitro inhibition of an *E. coli* strain. The effect of introducing substituents at C-8 has not been studied because this position cannot be directly substituted by classic electrophilic aromatic substitution reactions and, thus, 8-substituted derivatives are available only through total synthesis.[151]

The most fruitful site for semisynthetic modification of the tetracyclines has been the 6-position. Neither the 6α-methyl nor the 6β-hydroxyl group is essential for antibacterial activity. In fact, doxycycline and methacycline are more active in vitro against most bacterial strains than their parent oxytetracycline. The conversion of oxytetracycline to doxycycline, which can be accomplished by reduction of methacycline,[152] gives a 1 : 1 mixture of doxycycline and epidoxycycline (which has a β-oriented methyl group), whereas if the C-11a α-fluoro derivative of methacycline is employed, the β-methyl epimer is exclusively formed.[153] 6-Epidoxycycline is much less active than doxycycline. 6-Demethyl-6-deoxytetracycline, synthesized commercially by catalytic hydrogenolysis of the 7-chloro and 6-hydroxyl groups of 7-chloro-6-demethyltetracycline, obtained by fermentation of a mutant strain of *S. aureofaciens*,[154] is slightly more potent than tetracycline. More successful from a clinical standpoint, however, is 6-demethyl-6-deoxy-7-dimethylaminotetracycline (minocycline)[155] because of its activity against tetracycline-resistant bacterial strains.

6-Deoxytetracyclines also possess important chemical and pharmacokinetic advantages over their 6-oxy counterparts. Unlike the latter, they are incapable of forming anhydrotetracyclines under acidic conditions because they cannot dehydrate at C-5a and C-6. They are also more stable in base because they do not readily undergo β-ketone cleavage, followed by lactonization, to form isotetracyclines. Despite the fact that it lacks a 6-hydroxyl group, methacycline shares the instability of the 6-oxytetracyclines. It suffers prototropic rearrangement to the anhydrotetracycline in acid and β-ketone cleavage followed by lactonization to the isotetracycline in base. Reduction of the 6-hydroxyl group also brings about a dramatic change in the solubility properties of tetracyclines. This effect is reflected in the significantly higher oil/water partition coefficients of the 6-deoxytetracyclines, compared with the tetracyclines (Table 7-12).[156,157] The greater lipid solubility of the 6-deoxy compounds, in turn, has important pharmacokinetic consequences.[146] Hence, doxycycline and minocycline are more completely absorbed following oral administration, exhibit higher fractions of plasma protein binding, and have higher volumes of distribution and lower renal clearances than the corresponding 6-oxytetracyclines.

Polar substituents (i.e., hydroxyl groups) at C-5 and C-6 contribute decreased lipid versus water solubility to the tetracyclines. However, the 6-position is considerably more sensitive than is the 5-position to this effect. Ergo, doxycycline (6-deoxy-5-oxytetracycline) has a much higher partition coefficient than either tetracycline or oxytetracycline. Nonpolar substituents (those with positive values; see Chap. 2), for example, 7-dimethylamino, 7-chloro, and 6-methyl, have the opposite effect. Accordingly, the partition coefficient of chlortetracycline is substantially greater than tetracycline's and slightly greater than demeclocycline's. Interestingly, minocycline (5-demethyl-6-deoxy-7-dimethylaminotetracycline) has the highest partition coefficient of the commonly used tetracyclines.

The poorer oral absorption of the more water-soluble compounds, tetracycline and oxytetracycline, can be attributed to several factors. In addition to their comparative difficulty in penetrating lipid membranes, the polar tetracyclines probably experience more complexation with metal ions in the gut and also undergo some acid-catalyzed destruction in the stomach. Poorer oral absorption, coupled with biliary excretion of some tetracyclines, is also thought to cause a higher incidence of superinfections caused by resistant microbial strains. On the other hand, the more polar tetracyclines are excreted in higher concentrations in the urine (e.g., 70% for tetracycline and 60% for oxytetracycline) than the more lipid-soluble compounds (e.g., 33% for doxycycline and only 11% for minocycline). Significant passive renal tubular reabsorption, coupled with higher fractions of protein binding, contribute to the lower renal clearance and more prolonged durations of action of doxycycline and minocycline compared with the other tetracyclines, especially tetracycline and oxytetracycline. Although all tetracyclines are widely distributed into tissues, the more polar ones have larger volumes of distribution than do the nonpolar compounds. The more lipid-soluble tetracyclines, however, distribute better to poorly vascu-

TABLE 7-12

PHARMACOKINETIC PROPERTIES* OF TETRACYCLINES

Tetracycline	Substituents C-5α	C-6α	C-6β	C-7	$K_{pc}^{\dagger}$ Octanol/Water pH 5.6	% Absorbed Orally	% Excreted Feces	% Excreted Urine	% Protein Bound	Volume of Distribution (% body weight)	Renal Clearance (ml/min/1.73 m²)	Half-Life (hr)
Tetracycline	H	CH_3	OH	H	0.056	58	20–50	60	24–65	156–306	50–80	10
Oxytetracycline	OH	CH_3	OH	H	0.075	77–80	~ 50	70	20–35	189–305	99–102	9
Chlortetracycline	H	CH_3	OH	Cl	0.41	25–30	> 50	18	42–54	149	32	7
Demeclocycline	H	H	OH	Cl	0.25	66	23–72	42	68–77	179	35	15
Doxycycline	OH	CH_3	H	H	0.95	93	20–40	27–39	60–91	63	18–28	15
Minocycline	H	H	H	$N(CH_3)_2$	1.10	~ 100	40	5–11	55–76	74	5–15	19

*Values taken from Brown and Ireland[146] and references cited therein.
†Values taken from Colazzi and Klink.[156]

larized tissue. It is also claimed that the distribution of doxycycline and minocycline into bone is less than with other tetracyclines.[158]

Nearly two decades have passed since the last tetracycline derivative, minocycline, was introduced into medicine for systemic administration, and it might appear that the possibilities for discovery of improved compounds prepared by structural modification or total synthesis have largely been exhausted. As mentioned earlier, however, the structure–activity relationships of 8-substituted compounds have not been explored. Interestingly, "6-thiatetracycline" (actually the 6-thia isostere of 6-deoxy-6-demethyltetracycline) has been claimed to be superior to all known tetracyclines and, furthermore, to exert a bactericidal action against gram-negative bacteria.[159] The total synthesis of this compound has also been reported.[160] Because the standard tetracyclines exert a bacteriostatic action and are far from ideal from a pharmakokinetic point of view, it is possible that this new lead will provide compounds with improved clinical properties.

PRODUCTS

Tetracycline, USP (Achromycin, Cyclopar, Panmycin, Tetracyn). During the chemical studies on chlortetracycline, it was discovered that controlled catalytic hydrogenolysis would selectively remove the 7-chloro atom and thereby produce tetracycline. This process was patented by Conover[161] in 1955. Later, tetracycline was obtained from fermentations of *Streptomyces* species, but the commercial supply is still chiefly dependent upon the hydrogenolysis of chlortetracycline.

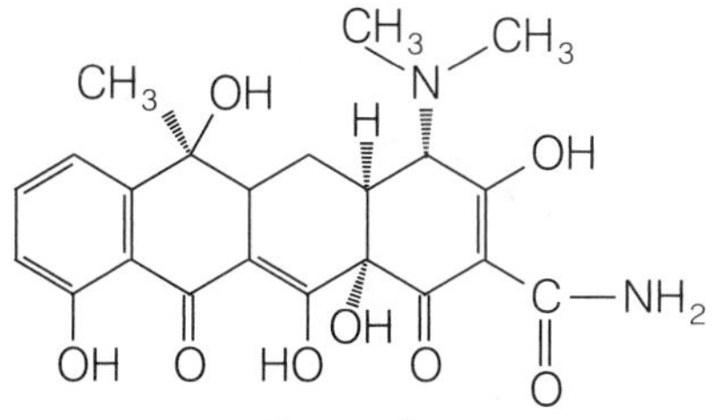

CH_3 OH H CH_3 N CH_3 OH OH O HO OH O C—NH_2 O

Tetracycline

Tetracycline is 4-dimethylamino-1,4,4a,5,5a,6,11,12a-octahydro-3,6,10,12,12a-pentahydroxy-6-methyl-1,11-dioxo-2-naphthacenecarboxamide. It is a bright yellow crystalline salt that is stable in air, but darkens in color upon exposure to strong sunlight. Tetracycline is stable in acid solutions having a pH higher than 2. It is somewhat more stable in alkaline solutions than chlortetracycline, but, like those of the other tetracyclines, such solutions rapidly lose their potencies. One gram of the base requires 2500 mL of water and 50 mL of alcohol to dissolve it. The hydrochloride salt is most commonly used in medicine, although the free base is absorbed from the gastrointestinal tract about equally well. One gram of the hydrochloride salt dissolves in about 10 mL of water and in 100 mL of alcohol. Tetracycline has become the most popular antibiotic of its group, largely because its plasma concentration appears to be higher and more enduring than that of either oxytetracycline or chlortetracycline. Also, it is found in higher concentration in the spinal fluid than are the other two compounds.

A number of combinations of tetracycline with agents that increase the rate and the height of plasma concentrations are on the market. One such adjuvant is magnesium chloride hexahydrate (Panmycin). Also, an insoluble tetracycline phosphate complex (Tetrex) is made by mixing a solution of tetracycline, usually as the hydrochloride, with a solution of sodium metaphosphate. A variety of claims concerning the efficacy of these adjuvants has been made. The mechanisms of their actions are not clear, but it has been reported[162,163] that these agents enhance plasma concentrations over those obtained when tetracycline hydrochloride alone is administered orally. Remmers et al.[164,165] have reported on the effects that selected aluminum–calcium gluconates complexed with some tetracyclines have on the plasma concentrations when administered orally, intramuscularly, or intravenously. Such complexes enhanced plasma levels in dogs when injected, but not when given orally. They also observed enhanced plasma levels in experimental animals when complexes of tetracyclines with aluminum metaphosphate, with aluminum pyrophosphate, and aluminum–calcium phosphinicodilactates were administered orally. As has been noted previously, the tetracyclines are capable of forming stable chelate complexes with metal ions, such as calcium and magnesium, that would retard absorption from the gastrointestinal tract. The complexity of the systems involved has not permitted unequivocal substantiation of the idea that these adjuvants act by competing with the tetracyclines for substances in the alimentary tract that would otherwise be free to complex with these antibiotics and thereby retard their absorption. Certainly, there is no evidence that the metal ions act by any virtue they possess as buffers, an idea alluded to sometimes in the literature.

Tetracycline hydrochloride is also available in ointments for topical and ophthalmic administration. A topical solution is used for the management of acne vulgaris.

Rolitetracycline, USP. *N*-(Pyrrolidinomethyl)-tetracycline (Syntetrin) was introduced for use by intramuscular and intravenous injection. This derivative is made by condensing tetracycline with pyrrolidine and formaldehyde in the presence of *t*-butyl alcohol. It is very soluble in water, 1 g dissolving in about 1 mL, and provides a means of injecting the antibiotic in a small volume of solution. It has been recommended in cases for which the oral dosage forms are not suitable, but it is no longer widely used.

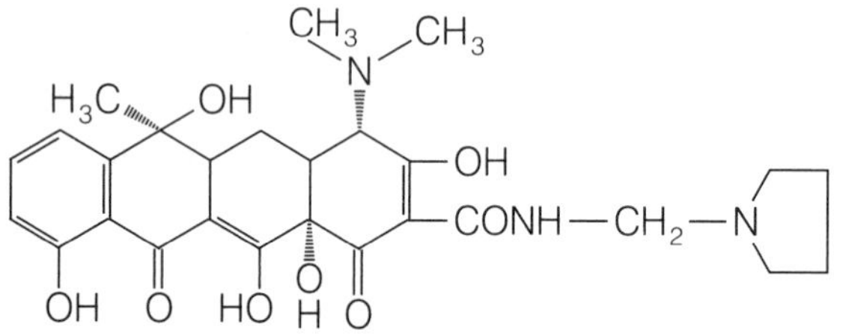

N-(pyrrolidinomethyl)tetracycline

Chlortetracycline Hydrochloride, USP (Aureomycin Hydrochloride). Chlortetracycline was isolated by Duggar[166] in 1948 from *S. aureofaciens*. This compound, which was produced in an extensive search for new antibiotics, was the first of the group of highly successful tetracyclines. It soon became established as a valuable antibiotic with broad-spectrum activities. It is used in medicine chiefly as the acid salt of the compound whose systematic chemical designation is 7-chloro-4-(dimethylamino)-1,4,4a,5a,6,11,12a-octahydro-3,6,10,12,12a-pentahydroxy-6-methyl-1,11-dioxo-2-naphthacenecarboxamide. The hydrochloride salt is a crystalline powder having bright yellow color that suggested its brand name Aureomycin. It is stable in air, but is slightly photosensitive and should be protected from light. It is odorless and bitter. One gram of the hydrochloride salt will dissolve in about 75 mL of water, producing a *p*H of about 3. It is only slightly soluble in alcohol and practically insoluble in other organic solvents.

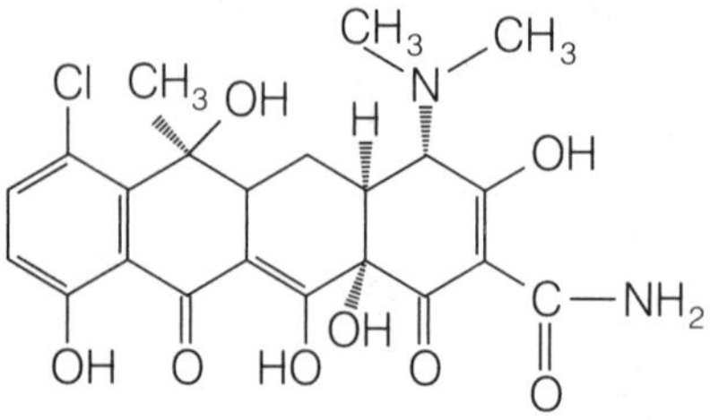

Chlortetracycline

Oral and parenteral forms of chlortetracycline are no longer used because of the poor bioavailability and inferior pharmacokinetic properties of the drug. It is still marketed in ointment forms for topical and ophthalmic use.

Oxytetracycline Hydrochloride, USP (Terramycin). Early in 1950, Finlay et al.[167] reported the isolation of oxytetracycline from *S. rimosus*. It was soon established that this compound was a chemical analogue of chlortetracycline and showed similar antibiotic properties. The structure of oxytetracycline was elucidated by Hochstein et al.,[168] and this work provided the basis for the confirmation of the structure of the other tetracyclines.

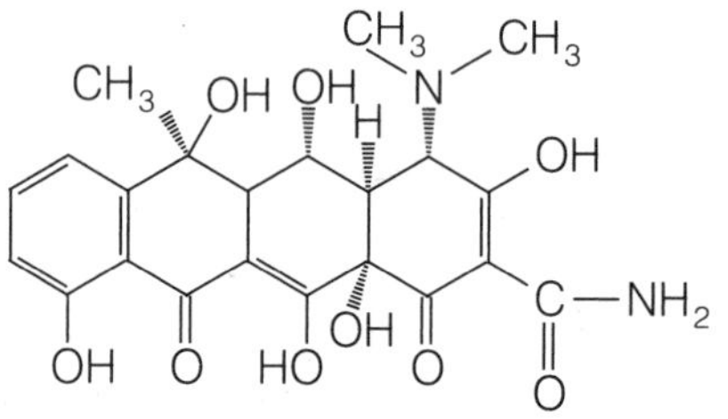

Oxytetracycline

Oxytetracycline hydrochloride is a pale yellow, bitter crystalline compound. The amphoteric base is only slightly soluble in water and slightly soluble in alcohol. It is odorless and stable in air, but darkens upon exposure to strong sunlight. The hydrochloride salt is a stable yellow powder that is more bitter than the free base. It is much more soluble in water, 1 g dissolving in 2 mL, and also it is more soluble in alcohol. Both compounds are inactivated rapidly by alkali hydroxides and by acid solutions below *p*H 2. Both forms of oxytetracycline are absorbed from the digestive tract rapidly and equally well, so that the only real advantage the free base offers over the hydrochloride salt is that it is less bitter. Oxytetracycline hydrochloride also is used for parenteral administration (IV and IM).

Methacycline Hydrochloride, USP. 6-Deoxy-6-demethyl-6-methylene-5-oxytetracycline hydrochloride (Rondomycin). The synthesis of methacycline, reported by Blackwood et al.[169] in 1961, was accomplished by chemical modification of oxytetracycline. It has an antibiotic spectrum similar to that of the other tetracyclines, but has a greater potency; about 600 mg of methacycline is equivalent to 1 g of tetracycline. Its particular value lies in its longer serum half-life; doses of 300 mg produce continuous serum antibacterial activity for 12 hours. Its toxic manifestations and contraindications are similar to those of the other tetracyclines.

The greater stability of methacycline, both in vivo and in vitro, is a result of the modification at C-6. The removal of the 6-hydroxy group markedly increases the stability of ring C to both acids and bases, preventing the formation of isotetracyclines

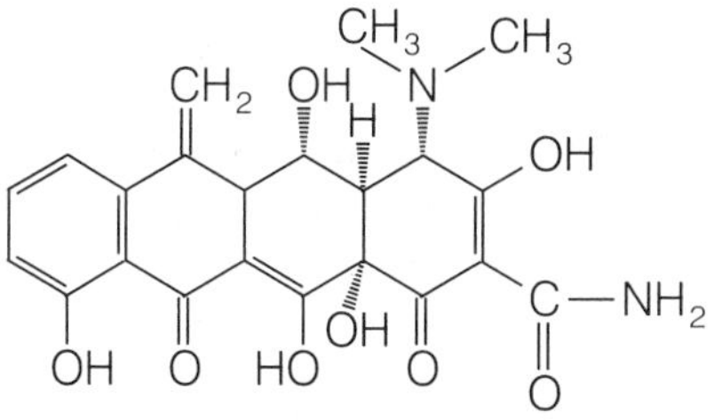

Methacycline

by bases. However, anhydrotetracyclines can still form by acid-catalyzed isomerization. Methacycline hydrochloride is a yellow to dark yellow crystalline powder that is slightly soluble in water and insoluble in nonpolar solvents. It should be stored in tight, light-resistant containers in a cool place.

Demeclocycline, USP. 7-Chloro-6-demethyltetracycline (Declomycin) was isolated in 1957 by McCormick et al.[154] from a mutant strain of *S. aureofaciens*. Chemically, it is 7-chloro-4-(dimethylamino)-1,4,4a,5,5a,6,11,12a-octahydro-3,6,10,12,12a-pentahydroxy-1,11-dioxo-2-naphthacenecarboxamide. Thus, it differs from chlortetracycline only by the absence of the methyl group on C-6.

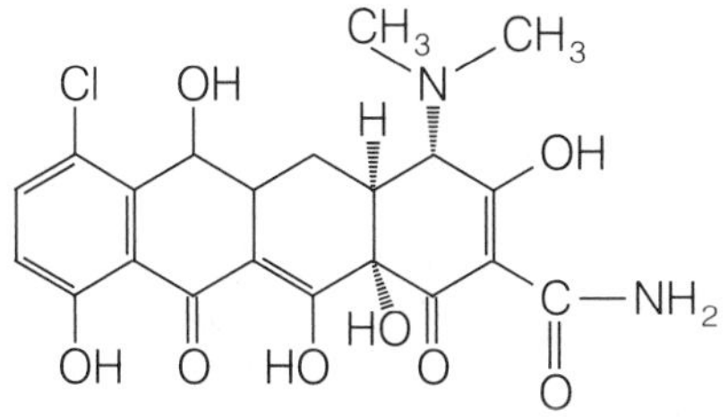

Demeclocycline

Demeclocycline is a yellow crystalline powder that is odorless and bitter. It is sparingly soluble in water. A 1% solution has a *p*H of about 4.8. It has an antibiotic spectrum similar to that of other tetracyclines, but it is slightly more active than the others against most of the microorganisms for which they are used. This, together with its slower rate of elimination through the kidneys, gives demeclocycline an effectiveness comparable with that of the other tetracyclines, at about three-fifths of the dose. Similar to the other tetracyclines, it may cause infrequent photosensitivity reactions that produce erythema after exposure to sunlight. It appears that demeclocyline may produce the reaction somewhat more frequently than the other tetracyclines. The incidence of discoloration and mottling of the teeth in youths from demeclocyline appears to be as low as with the other tetracyclines.

Meclocylcine Sulfosalicylate, USP. 7-Chloro-6-deoxy-6-demethyl-6-methylene-5-oxytetracycline sulfosalicylate (Meclan) is a semisynthetic derivative prepared from oxytetracycline.[169] Although meclocycline has been used in Europe for many years, it was only relatively recently marketed in the United States for a single therapeutic indication, namely the treatment of acne. It is available as the sulfosalicylate salt in a 1% cream.

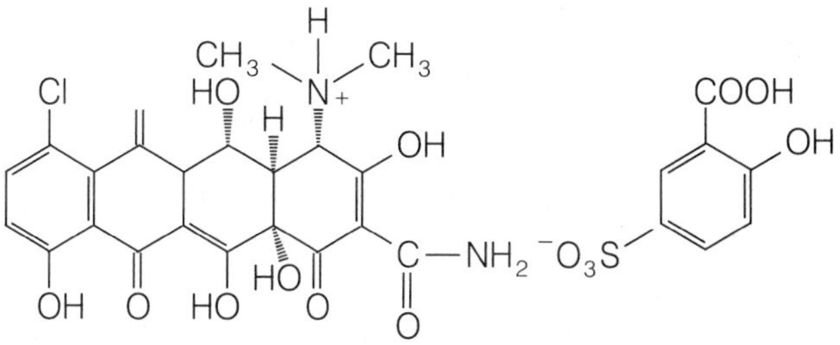

Meclocycline Sulfosalicylate

Meclocycline sulfosalicylate is a bright yellow crystalline powder that is slightly soluble in water and insoluble in organic solvents. It is light-sensitive and should be stored in light-resistant containers.

Doxycycline, USP. α-6-Deoxy-5-oxytetracycline (Vibramycin). A more recent addition to the tetracycline group of antibiotics available for antibacterial therapy is doxycycline, first reported by Stephens et al.[170] in 1958. It was first obtained in small yields by a chemical transformation of oxytetracycline, but it is now produced by catalytic hydrogenation of methacycline or by reduction of a benzylmercaptan derivative of methacycline with Raney nickel. In the latter process, a nearly pure form of the 6α-methyl epimer is produced. It is noteworthy that the 6-α-methyl epimer is more than three times as active as its β-epimer.[152] Apparently, the difference in orientation of the methyl groups, slightly affecting the shapes of the molecules, causes a substantial difference in biologic effect. Also, as in methacycline, the absence of the 6-hydroxyl group produces a compound that is very stable to acids and bases and that has a long biologic half-life. In addition, it is very well absorbed from the gastrointestinal tract, thus allowing a smaller dose to be administered. High tissue levels are obtained with it and, unlike other tetracyclines, doxycycline apparently does not accumulate in patients with impaired renal function. Therefore it is preferred for uremic patients with infections outside the urinary tract. However, its low renal clearance may limit its effectiveness in urinary tract infections.

Doxycycline is available as the hyclate salt, a hydrochloride salt solvated as the hemiethanolate hemihydrate, and as the monohydrate. The hyclate form is sparingly soluble in water and is used in the capsule dosage form; the monohydrate is water-insoluble and is used for aqueous suspensions, which

Doxycycline

are stable for periods up to two weeks when kept in a cool place.

Minocycline Hydrochloride, USP. 7-Dimethylamino-6-demethyl-6-deoxytetracycline (Minocin, Vectrin). Minocycline, the most potent tetracycline currently employed in therapy, is obtained by reductive methylation of 7-nitro-6-demethyl-6-deoxytetracycline.[155] It was released for use in the United States in 1971. Because minocycline, as does doxycycline, lacks the 6-hydroxyl group it is stable to acids and does not dehydrate or rearrange to anhydro or lactone forms. Minocycline is well absorbed orally to give high plasma and tissue levels. It has a very long serum half-life resulting from slow urinary excretion and moderate protein binding. Doxycycline and minocycline, along with oxytetracycline, show the least in vitro calcium binding of the clinically available tetracyclines. The improved distribution properties of the 6-deoxytetracyclines have been attributed to a greater degree of lipid-solubility.

Minocycline

Perhaps the most outstanding property of minocycline is its activity toward gram-positive bacteria, especially staphylococci and streptococci. In fact, minocycline has been effective against staphylococcal strains that are resistant to methacillin and all other tetracyclines, including doxycycline.[171] Although it is doubtful the minocycline will replace bactericidal agents for the treatment of life-threat-

TABLE 7-13

TETRACYCLINES

Name Proprietary Name	***Preparations***	***Usual Adult Dose****	***Usual Dose Range****	***Usual Pediatric Dose****
Tetracycline USP *Achromycin, Cyclopar, Panmycin, Steclin, Tetracyn, Robitet, Bristacycline*	Tetracycline oral suspension USP Tetracycline for oral suspension USP	The equivalent of 250–500 mg of tetracycline hydrochloride every 6 hr or 500 mg–1 g every 12 hr	Up to 4 g/day	Oral, the equivalent of tetracycline hydrochloride, children 8 yr and over: 6.25–12.5 mg/kg of body weight every 6 hr; or 12.5–25 mg/kg of body weight every 12 hr
Tetracycline hydrochloride USP *Achromycin, Bristacycline, Panmycin, Steclin, Sumycin, Tetracyn*	Tetracycline hydrochloride capsules USP	Oral, the equivalent of tetracycline hydrochloride, 250–500 mg every 6 hr; or 500 mg–1 g every 12 hr	Up to 4 g/day	Oral, the equivalent of tetracycline hydrochloride, children 8 yr and over: 6.25–12.5 mg/kg of body weight every 6 hr; or 12.5–25 mg/kg of body weight every 12 hr
	Tetracycline hydrochloride tablets USP			
	Tetracycline hydrochloride ointment USP	Topical, to the skin, as a 3% ointment qd or bid		
	Tetracycline hydrochloride for topical solution USP	Topical, to the skin, as a 0.22% solution bid, morning and evening		
	Tetracycline hydrochloride ophthalmic ointment USP	Topical, to the conjunctiva, a thin strip (approximately 1 cm) of a 1% ointment every 2–4 hr or more frequently		

*See USP DI for complete dosage information.

TABLE 7-13 *Continued*

TETRACYCLINES

Name Proprietary Name	***Preparations***	***Usual Adult Dose****	***Usual Dose Range****	***Usual Pediatric Dose****
	Tetracycline hydrochloride ophthalmic suspension USP	Topical, to the conjunctiva, 1 drop of a 1% suspension every 6–12 hr or more frequently		
	Tetracycline hydrochloride for intramuscular injection USP	IM, 100 mg every 8 hr; 150 mg every 12 hr; or 250 mg qd	Up to 1 g/day	Children 8 yr and over: IM, 5–8.3 mg/kg of body weight every 8 hr; or 7.5–12.5 mg/kg of body weight every 12 hr; maximal dose should not exceed 250 mg
	Tetracycline hydrochloride for intravenous injection USP	250–500 mg every 12 hr	Up to 2 g/day	Children 8 yr and over: IV, 5–10 mg/kg of body weight every 12 hr
	Tetracycline phosphate complex USP			
	Tetracycline phosphate complex capsules USP	See Tetracycline oral suspension USP	See Tetracycline oral suspension USP	See Tetracycline oral suspension USP
Rolitetracycline USP *Syntetrin*	Rolitetracycline for injection USP		IM, 150–350 mg every 12 hr; IV infusion, 350–700 mg every 12 hr	Children 8 yr and over: oral, 6.25–12.5 mg/kg of body weight every 6 hr
Chlortetracycline hydrochloride USP *Aureomycin*	Chlortetracycline hydrochloride capsules USP	Oral, 250–500 mg every 6 hr	Up to 4 g/day	
	Chlortetracycline hydrochloride ophthalmic ointment USP	Topical, to the conjunctiva, a thin strip (approximately 1 cm) of a 1% ointment every 2–4 hr or more often		See Usual Adult Dose
	Chlortetracycline hydrochloride ointment USP	Topical, to the skin, as a 3% ointment qd or bid		See Usual Adult Dose
Oxytetracycline USP *Terramycin*	Oxytetracycline tablets USP	Oral, the equivalent of oxytetracycline, 250–500 mg every 6 hr	Up to 4 g/day	Children 8 yr and over: oral, the equivalent of oxytetracycline, 6.25–12.5 mg/kg of body weight every 6 hr
	Oxytetracycline injection USP	IM, 100 mg every 8 hr; 150 mg every 12 hr; or 250 mg qd	Up to 500 mg/day	Children 8 yr and over: IM, 5–8.3 mg/kg of body weight every 8 hr; or 7.5–12.5 mg/kg of body weight every 12 hr; maximal daily dose should not exceed 250 mg for single injection
Oxytetracycline calcium USP *Terramycin*	Oxytetracycline calcium oral suspension USP	See Oxytetracycline tablets USP	See Oxytetracycline tablets USP	See Oxytetracycline tablets USP
Oxytetracycline hydrochloride USP *Terramycin*	Oxytetracycline hydrochloride capsules USP	See Oxytetracycline tablets USP	See Oxytetracycline tablets USP	See Oxytetracycline tablets USP
	Oxytetracycline hydrochloride for injection USP	IV, the equivalent of oxytetracycline, 250–500 mg every 12 hr	Up to 2 g/day	Children 8 yr and over: IV, the equivalent of oxytetracycline, 5–10 mg/kg of body weight every 12 hr

*See USP DI for complete dosage information.

TABLE 7-13 *Continued*

TETRACYCLINES

Name Proprietary Name	***Preparations***	***Usual Adult Dose****	***Usual Dose Range****	***Usual Pediatric Dose****
Methacycline hydrochloride USP *Rondomycin*	Methacycline hydrochloride capsules USP	Oral, 150 mg every 6 hr; or 300 mg every 12 hr	Up to 2.4 g/day	Children 8 yr and over: oral, 1.65–3.3 mg/kg of body weight every 6 hr; or 3.3–6.6 mg/kg of body weight every 12 hr
	Methacycline hydrochloride oral suspension USP	See Methacycline hydrochloride capsules USP	See Methacycline hydrochloride capsules USP	See Methacycline hydrochloride capsules USP
Demeclocycline USP *Declomycin*	Demeclocycline oral suspension USP	Oral, the equivalent of demeclocycline hydrochloride, 150 mg every 6 hr; or 300 mg every 12 hr	Up to 2.4 g/day	Children 8 yr and over: oral, the equivalent of demeclocycline hydrochloride, 1.65–3.3 mg/kg of body weight every 6 hr; or 3.3–6.6 mg/kg of body weight every 12 hr
Demeclocycline hydrochloride USP *Declomycin*	Demeclocycline hydrochloride capsules USP	See Demeclocycline oral suspension USP	See Demeclocycline oral suspension USP	See Demeclocycline oral suspension USP
Meclocycline sulfosalicylate USP *Meclan*	Meclocycline sulfosalicylate USP	Topical, to the skin, as a 1% cream qd or bid		
Doxycycline USP *Vibramycin*	Doxycycline for oral suspension USP	Oral, the equivalent of anhydrous doxycycline, 100 mg every 12 hr the first day, then 100–200 mg qd; or 50–100 mg every 12 hr	Up to 300 mg/day or up to 600 mg/day for 1 day in acute gonococcal infections	Oral, the equivalent of anhydrous doxycycline, children 45 kg and under: 2.2 mg/kg of body weight every 12 hr the first day, then 2.2–4.4 mg/kg of body weight qd; or 1.1–2.2 mg/kg of body weight every 12 hr
Doxycycline Hyclate USP *Vibramycin*	Doxycycline hyclate for injection USP Doxycycline hyclate tablets USP	IV infusion, the equivalent of doxycycline, 200 mg qd or 100 mg every 12 hr the first day, then 100–200 mg qd; or 50–100 mg every 12 hr	Up to 300 mg/day	IV infusion, the equivalent of doxycycline, children 45 kg and under: 4.4 mg/kg of body weight qd or 2.2 mg/kg of body weight every 12 hr the first day; then 2.2–4.4 mg/kg of body weight qd or 1.1–2.2 mg/kg of body weight every 12 hr
	Doxycycline calcium oral suspension USP	See Doxycycline for oral suspension USP	See Doxycycline for oral suspension USP	See Doxycycline for oral suspension USP
Minocycline hydrochloride USP *Minocin, Vectrin*	Minocycline hydrochloride capsules USP	Oral, the equivalent of minocycline, 200 mg initially, then 100 mg every 12 hr; or 100–200 mg initially, then 50 mg every 6 hr	Up to 150 mg the first day; then up to 200 mg/day	Oral, the equivalent of minocycline, children 8 yr and over: 4 mg/kg of body weight initially, then 2 mg/kg of body weight every 12 hr
	Minocycline hydrochloride tablets USP	See Minocycline hydrochloride capsules USP	See Minocycline hydrochloride capsules USP	See Minocycline hydrochloride capsules USP

*See USP DI for complete dosage information.

ening staphylococcal infections, it may become a useful alternative for the treatment of less serious tissue infections. Minocycline has been recommended for the treatment of chronic bronchitis and other upper respiratory tract infections. Despite its relatively low renal clearance, partially compensated for by high serum and tissue levels, it has also been recommended for the treatment of urinary tract infections. It has been effective in the eradication of *N. meningitidis* in asymptomatic carriers.

THE MACROLIDES

Among the many antibiotics isolated from the actinomycetes is the group of chemically related compounds called the macrolides. It was in 1950 that picromycin, the first of this group to be identified as a macrolide compound, was first reported. In 1952, erythromycin and carbomycin were reported as new antibiotics, and these were followed in subsequent years by other macrolides. Currently, more than three dozen such compounds are known, and new ones are likely to appear in the future. Of all of these, only two, erythromycin and oleandomycin, have been consistently available for medical use in the United States. One other, carbomycin, has been available, but, because of its poor and irregular absorption from the gastrointestinal tract and its inferior antibacterial activity when compared with erythromycin, it never enjoyed wide use and was withdrawn. Spiramycin is used in Europe and other parts of the world, but its activity in vitro is inferior to that of erythromycin, and it is difficult to account for its reputed therapeutic success. Various members of the leucomycin group have been used clinically in various parts of the world. Josamycin (leucomycin A_3), isolated from *S. narbonensis* var. *josamyceticus*, has been marketed in Japan and in Europe primarily for the treatment of respiratory and genitourinary infections caused by gram-positive bacteria. It has recently been subjected to clinical evaluations in the United States and compares favorably with erythromycin for many indications. Rosamicin, a macrolide antibiotic obtained from *Micromonospora rosaria*, has shown some promise for the treatment of genitourinary infections in in vitro tests. Both josamycin and rosamicin remain on investigational status in the United States. Tylosin is a macrolide that is of considerable commercial importance as a feed supplement to enhance weight and growth of meat animals.

CHEMISTRY

The macrolide antibiotics have three common chemical characteristics: (1) a large lactone ring (which prompted the name *macrolide*), (2) a ketone group, and (3) a glycosidically linked amino sugar. Usually the lactone ring has 12, 14, or 16 atoms in it and is often partially unsaturated, with an olefinic group conjugated with the ketone function. (The polyene macrocyclic lactones, such as natamycin and amphoteracin B, the ansamycins, such as rifampin, and the polypeptide lactones generally are not included among the macrolide antibiotics.) They may have, in addition to the amino sugar, a neutral sugar that is glycosidically linked to the lactone ring (see erythromycin). Because of the presence of the dimethylamino group on the sugar moiety, the macrolides are bases that form salts with pK_a values between 6.0 and 9.0. This feature has been employed to make clinically useful salts. The free bases are only slightly soluble in water, but dissolve in somewhat polar organic solvents. They are stable in aqueous solutions at or below room temperature, but are inactivated by acids, bases, and heat.

The chemistry of macrolide antibiotics has been the subject of several reviews.[172–175]

SPECTRUM OF ACTIVITY

The spectrum of antibacterial activity of the more potent macrolides, such as erythromycin, resembles that of penicillin. They are frequently active against bacterial strains that are resistant to the penicillins. The macrolides are generally effective against most species of gram-positive bacteria, both cocci and bacilli, and also exhibit useful effectiveness against gram-negative cocci, especially *Neisseria* species. Many of the macrolides are also effective against *Treponema pallidum*. In contrast to penicillin, macrolides are also effective against *Mycoplasma*, *Chlamydia*, *Campylobacter*, and *Legionella* species. Their activity against most species of gram-negative bacilli is generally low and often unpredictable, although some strains of *Haemophilus influenzae* and *Brucella* species are sensitive.

PRODUCTS

Erythromycin, USP (E-Mycin, Erythrocin, Ilotycin). Early in 1952, McGuire et al.[176] reported the isolation of erythromycin from *S. erythreus*. It achieved a rapid early acceptance as a well-tolerated antibiotic of value for the treatment of a variety of upper respiratory and soft-tissue infections caused by gram-positive bacteria. It is also effective against many venereal diseases, including gonorrhea and syphilis, and provides a useful alternative to penicillin for the treatment of many infections in patients allergic to penicillins. More recently,[177] erythromycin has been shown to be effective therapy for Eaton agent pneumonia (*Mycoplasma pneumoniae*), venereal diseases caused by chlamydiae, bacterial enteritis caused by *Campylobacter jejuni*, and Legionnaires' disease.

The commercial product is actually erythromycin A, which differs from its biosynthetic precursor erythromycin B in having a hydroxyl group at the 12-position of the aglycone. The chemical structure of erythromycin A was reported by Wiley et al.[178] in 1957 and its stereochemistry by Celmer[179] in 1965. An elegant synthesis of erythronolide A, the aglycone present in erythromycin A, has been described by Corey and his associates.[180]

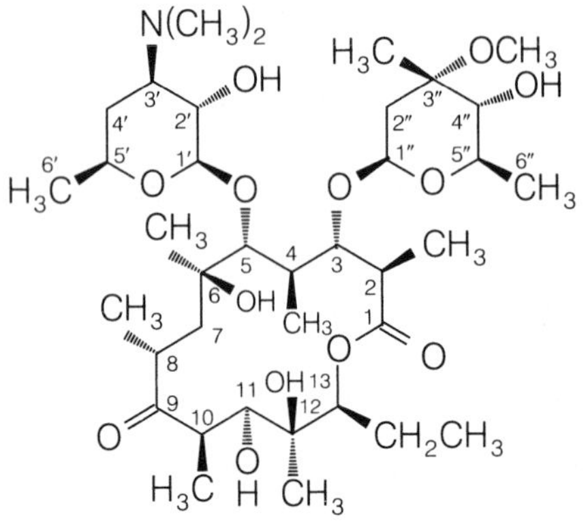

Erythromycin A

The amino sugar attached through a glycosidic link to C-5 is desosamine, a structure found in a number of other macrolide antibiotics. The tertiary amine of desosamine (3,4,6-trideoxy-3-dimethylamino-D-*xylo*-hexose) confers a basic character to erythromycin and provides the means by which acid salts may be prepared. The other carbohydrate structure linked as a glycoside to C-3 is called cladinose (2,3,6-trideoxy-3-methoxy-3-*C*-methyl-L-*ribo*-hexose) and is unique to the erythromycin molecule.

As is common with other macrolide antibiotics, compounds closely related to erythromycin have been obtained from culture filtrates of *S. erythreus*. Two such analogues have been found and are designated as erythromycins B and C. Erythromycin B differs from erythromycin A only at C-12, at which a hydrogen has replaced the hydroxyl group. The B analogue is more acid-stable but has only about 80% of the activity of erythromycin. The C analogue differs from erythromycin by the replacement of the methoxyl group on the cladinose moiety by a hydrogen atom. It appears to be as active as erythromycin, but is present in very small amounts in fermentation liquors.

Erythromycin is a very bitter, white or yellowish white crystalline powder. It is soluble in alcohol and in the other common organic solvents, but only slightly soluble in water. Saturated aqueous solutions develop an alkaline *p*H in the range of 8.0 to 10.5. It is extremely unstable at the *p*H of 4 or lower. The optimum *p*H for stability of erythromycin is at or near neutrality.

Erythromycin may be used as the free base in oral dosage forms and for topical administration. In attempts to overcome its bitterness and irregular oral absorption (resulting from acid destruction and adsorption onto food), a variety of enteric-coated and delayed-release dose forms of erythromycin base have been developed. These forms have been fully successful in overcoming the bitterness, but have only marginally solved problems of oral absorption of the antibiotic. Chemical modifications of erythromycin have been made with primarily two different goals in mind: (1) to increase either its water or lipid solubility for parenteral dosage forms; and (2) to increase its acid stability (and possibly increase its lipid solubility) for improved oral absorption. Modified derivatives of the antibiotic are of two types: acid salts of the dimethylamino group of the desosamine moiety, such as the glucoheptonate, the lactobionate, and the stearate; and esters of the 2′-hydroxyl group of the desosamine, such as the ethylsuccinate and the propionate (available as the laurylsulfate salt and known as the estolate).

The stearate salt and the ethylsuccinate and propionate esters are used in oral dose forms intended to improve absorption of the antibiotic. The stearate releases erythromycin base in the intestinal tract, which is then absorbed. The ethylsuccinate and the estolate are absorbed largely intact and are partially hydrolyzed by plasma and tissue esterases to give free erythromycin. The question of bioavailability of the antibiotic from its various oral dosage and chemical forms has been the subject of considerable concern and dispute over the past two decades.[181–185] It is generally believed that the 2′-esters, of themselves, have little or no intrinsic antibacterial activity[186] and, therefore, must be hydrolyzed to the parent antibiotic in vivo. Although the ethylsuccinate is hydrolyzed more efficiently than the estolate in vivo and, in fact, provides higher levels of erythromycin following intramuscular administration, an equal dose of the estolate gives higher levels of the free antibiotic following oral administration.[182] Superior oral absorption of the estolate is attributed to both its greater acid stability and higher intrinsic absorption compared with the ethylsuccinate. Also the oral absorption of the estolate, unlike that of both the stearate and the ethylsuccinate, is not affected by food or fluid volume content of the gut. Superior bioavailability of active antibiotic from oral administration of the estolate over the ethylsuccinate, stearate, or erythromycin base cannot necessarily be assumed, however, because the estolate is more extensively protein bound than erythromycin itself.[187,188] Measured fractions of plasma protein binding for erythromycin-2′-propionate and erythromycin base range from 0.94 to 0.98 for the former and 0.73 to 0.90 for the latter, indicating a much higher level of free erythromycin in the plasma. Bioavailability studies comparing equivalent doses of the enteric-coated base, the stearate salt, the ethylsuccinate ester, and the estolate ester in human volunteers[184,185] showed delayed but slightly higher bioavailability for the free base than for the stearate, ethylsuccinate, or estolate.

Furthermore, studies comparing the clinical effectiveness of recommended doses of the stearate, estolate, ethylsuccinate, or free base in the treatment of respiratory tract infections have failed to demonstrate substantial differences among them.[189]

The water-soluble ethylsuccinate ester is also available as a suspension for intramuscular injection. The glucoheptonate and lactobionate salts, on the other hand, are highly water-soluble derivatives that provide high plasma levels of the active antibiotic immediately after intravenous injection. Aqueous solutions of these salts may also be administered by intramuscular injection, but this is not a common practice.

Some details of the mechanism of antibacterial action of erythromycin are known. It selectively binds to a specific site on the 50S ribosomal subunit to prevent the translocation step of bacterial protein synthesis.[190] Erythromycin does not bind to mammalian ribosomes. Broadly based, nonspecific resistance to the antibacterial action of erythromycin among many species of gram-negative bacilli appears to be related, in large part, to the inability of the antibiotic to penetrate effectively the cell walls of these organisms.[191] In fact, the sensitivities of members of the *Enterobacteriaceae* family are *p*H-dependent, with minimum inhibitory concentrations (MICs) decreasing as a function of increasing *p*H. Furthermore, protoplasts from gram-negative bacilli, lacking cell walls, are sensitive to erythromycin. A highly specific resistance mechanism to the macrolide antibiotics occurs in erythromycin-resistant strains of *Staph. aureus*.[192] Such strains produce an enzyme that methylates a specific adenine residue at the erythromycin-binding site of the bacterial 50S ribosomal subunit. The thus methylated ribosomal RNA remains active in protein synthesis, but no longer binds erythromycin. Bacterial resistance to the lincomycins apparently also occurs by this mechanism.

Erythromycin Stearate, USP (Ethril, Wyamycin S, Erypar) is the stearic acid salt of erythromycin. Like erythromycin base, the stearate is acid-labile. It is film coated to protect it from acid degradation in the stomach. In the alkaline *p*H of the duodenum, the free base is liberated from the stearate and is absorbed. Erythromycin stearate occurs as a crystalline powder that is practically insoluble in water, but soluble in alcohol and ether.

Erythromycin Ethylsuccinate, USP (E.E.S., Pediamycin, Eryped) is the ethylsuccinate mixed ester of erythromycin in which the 2′-hydroxyl group of the desosamine is esterified. Erythromycin ethylsuccinate is absorbed as the ester and slowly hydrolyzed in the body to form erythromycin. It is somewhat acid-labile and its absorption is enhanced by the presence of food. The ester is insoluble in water, but soluble in alcohol and ether.

Erythromycin Estolate, USP (Ilosone). Erythromycin propionate laurylsulfate is the laurylsulfate salt of the 2′-propionate ester of erythromycin. Erythromycin estolate is acid-stable and is absorbed as the propionate ester. The ester undergoes slow hydrolysis in vivo. Only the free base binds to bacterial ribosomes. However, there is some evidence to suggest that the ester is taken up by bacterial cells more rapidly than the free base and undergoes hydrolysis by bacterial esterases within the cells. The incidence of cholestatic hepatitis is reportedly higher with the estolate than with other erythromycin preparations.

Erythromycin estolate occurs as long needles that are sparingly soluble in water, but soluble in organic solvents.

Erythromycin Gluceptate Sterile, USP. Erythromycin glucoheptonate (Ilotycin Gluceptate) is the glucoheptonic acid salt of erythromycin. It is a crystalline substance that is freely soluble in water and practically insoluble in organic solvents. Erythromycin gluceptate is intended for intravenous administration for the treatment of serious infections, such as Legionnaires' disease, or when oral administration is not possible. The solutions are stable for one week when refrigerated.

Erythromycin Lactobionate, USP is a water-soluble salt prepared by reacting erythromycin base with lactobiono-δ-lactone. It occurs as an amorphous powder that is freely soluble in water and alcohol and slightly soluble in ether. Erythromycin lactobionate is intended, after reconstitution in sterile water, for intravenous administration to achieve high plasma levels in the treatment of serious infections.

Troleandomycin. Triacetyloleandomycin (TAO). Oleandomycin, as its triacetyl derivative trolean-

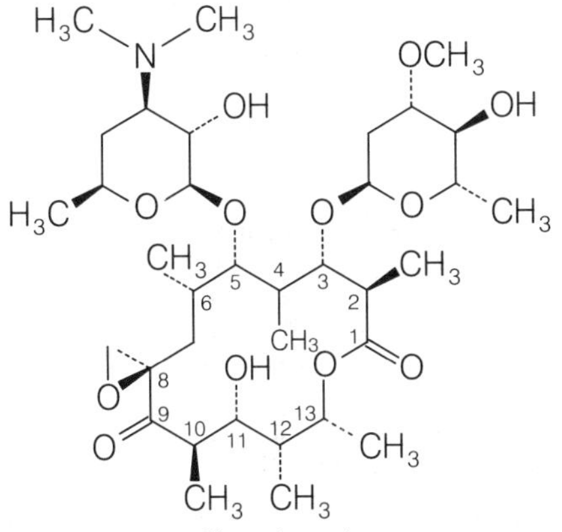

Oleandomycin

domycin, remains available as an alternative to erythromycin for limited indications permitting use of an oral dosage form. Oleandomycin was originally isolated by Sobin and associates.[193] The structure of oleandomycin was first proposed by Hochstein et al.[194] and its absolute stereochemistry elucidated by Celmer.[195] The oleandomycin structure (diagram) consists of two sugars and a 14-member lactone ring designated *oleandolide*. One of the sugars is desosamine, also present in erythromycin; the other is L-oleandrose. The sugars are glycosidically linked to the 5- and 3-positions, respectively, of oleandolide.

TABLE 7-14

MACROLIDE ANTIBIOTICS

Name Proprietary Name	***Preparations***	***Usual Adult Dose****	***Usual Dose Range****	***Usual Pediatric Dose****
Erythromycin USP *Erythrocin, Ilotycin, E-Mycin*	Erythromycin ointment USP	For external use, topically to the skin, as a 1% ointment tid or qid		
	Erythromycin ophthalmic ointment USP	For external use, topically to the conjunctiva, as a 0.5% ointment one or more times daily		
	Erythromycin tablets USP	Antibacterial, oral, 250 mg every 6 hr or 500 mg every 12 hr	Antibacterial, up to 4 g/day or more	Antibacterial, oral, 7.5–25 mg/kg of body weight every 6 hr; or 15–50 mg/kg of body weight every 12 hr
Erythromycin ethylsuccinate USP *Erythrocin Ethylsuccinate, Pediamycin*	Erythromycin ethylsuccinate oral suspension USP	Antibacterial, oral, the equivalent of erythromycin, 400 mg every 6 hr, or 800 mg every 12 hr	Antibacterial, the equivalent of erythromycin, up to 4 g/day	Antibacterial, oral, the equivalent of erythromycin, 7.5–25 mg/kg of body weight every 6 hr; or 15–50 mg/kg of body weight every 12 hr
	Erythromycin ethylsuccinate for oral suspension USP	See Erythromycin ethylsuccinate for oral suspension USP	See Erythromycin ethylsuccinate for oral suspension USP	See Erythromycin ethylsuccinate for oral suspension USP
	Erythromycin ethylsuccinate tablets USP			
	Erythromycin ethylsuccinate chewable tablets USP			
Ilotycin Gluceptate	Sterile erythromycin gluceptate USP	IV infusion, the equivalent of 250–500 mg of erythromycin every 6 hr, or 3.75–5 mg/kg of body weight every 6 hr	Up to 4 g/day	IV infusion, the equivalent of 3.75–5 mg/kg of body weight every 6 hr
Erythrocin Lactobionate	Erythromycin lactobionate for injection USP	See Sterile erythromycin gluceptate USP	See Sterile erythromycin gluceptate USP	See Sterile erythromycin gluceptate USP
Erythromycin stearate USP *Erythrocin Stearate, Bristamycin, Ethril*	Erythromycin stearate tablets USP Erythromycin stearate for oral suspension USP	See Erythromycin tablets USP	See Erythromycin tablets USP	See Erythromycin tablets USP
Erythromycin estolate USP *Ilosone*	Erythromycin estolate capsules Erythromycin estolate tablets USP Erythromycin estolate for oral suspension USP Erythromycin estolate oral suspension USP Erythromycin estolate chewable tablets USP	See Erythromycin tablets USP		
Troleandomycin *TAO*	Troleandomycin capsules Troleandomycin oral suspension	Antibacterial, oral, 250–500 mg qid		Antibacterial, oral, 125–250 mg qid

*See USP DI for complete dosage information.

Oleandomycin contains three hydroxyl groups that are subject to acylation, one in each of the sugars and one in the oleandolide. The triacetyl derivative retains the in vivo antibacterial activity of the parent antibiotic, but possesses superior pharmacokinetic properties. It is hydrolyzed in vivo to oleandomycin. Troleandomycin achieves more rapid and higher plasma concentrations following oral administration than oleandomycin phosphate, and it has the additional advantage of being practically tasteless. Troleandomycin occurs as a white crystalline solid that is nearly insoluble in water. It is relatively stable in the solid state, but undergoes chemical degradation in either aqueous acidic or alkaline conditions.

Approved medical indications for troleandomycin are currently limited to the treatment of upper respiratory infections caused by such organisms as *Streptococcus pyogenes* and *Strep. pneumoniae*. It may be considered as an alternative drug to oral forms of erythromycin. It is available in capsules and as a suspension.

THE LINCOMYCINS

The lincomycins are sulfur-containing antibiotics isolated from *S. lincolnensis*. Lincomycin is the most active and medically useful of the compounds obtained from fermentation. Extensive efforts to modify the lincomycin structure to improve its antibacterial and pharmacologic properties resulted in the preparation of the 7-chloro-7-deoxy derivative clindamycin. Of the two antibiotics, clindamycin appears to have the greater antibacterial potency and better pharmacokinetic properties as well. Lincomycins resemble the macrolides in antibacterial spectrum and biochemical mechanisms of action. They are primarily active against gram-positive bacteria, particularly the cocci, but are also effective against non–spore-forming anaerobic bacteria, actinomycetes, mycoplasma, and some species of *Plasmodium*. Lincomycin binds to the 50S ribosomal subunit to inhibit protein synthesis. Its action may be bacteriostatic or bactericidal depending on a variety of factors, which include the concentration of the antibiotic. A pattern of bacterial resistance and cross-resistance to lincomycins similar to that observed with the macrolides has been emerging.

PRODUCTS

Lincomycin Hydrochloride, USP (Lincocin). This antibiotic, which differs chemically from other major antibiotic classes, was first isolated by Mason et al.[196] Its chemistry has been described by Hoeksema and his co-workers[197] who assigned the structure, later confirmed by Slomp and MacKellar,[198] given in the diagram. Total syntheses of the antibiotic were independently accomplished in 1970 through research efforts in England and the United States.[199,200] The structure contains a basic function, the pyrrolidine nitrogen, by which water-soluble salts having an apparent pK_a of 7.6 may be formed. When subjected to hydrazinolysis, lincomycin is cleaved at its amide bond into *trans*-L-4-*n*-propylhygric acid (the pyrrolidine moiety) and methyl α-thiolincosamide (the sugar moiety). Lincomycin-related antibiotics have been reported by Argoudelis[201] to be produced by *S. lincolnensis*. These antibiotics differ in structure at one or more of three positions of the lincomycin structure: (1) the *N*-methyl of the hygric acid moiety is substituted by a hydrogen; (2) the *n*-propyl group of the hygric acid moiety is substituted by an ethyl group; and (3) the thiomethyl ether of the α-thiolincosamide moiety is substituted by a thioethyl ether.

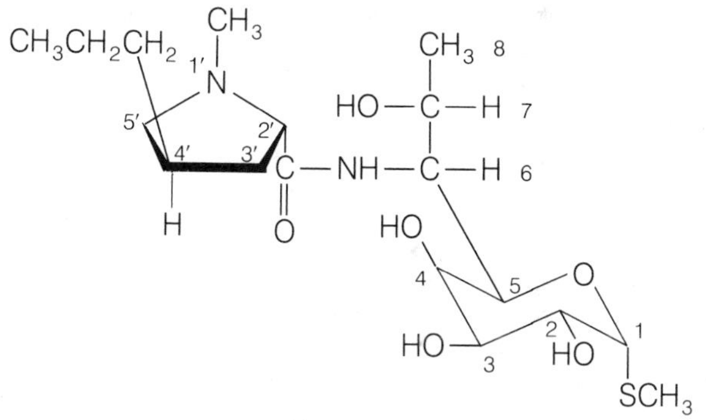

Lincomycin

Lincomycin is employed for the treatment of infections caused by gram-positive organisms, notably staphylococci, β-hemolytic streptococci, and pneumococci. It is moderately well absorbed orally and is widely distributed in the tissues. Effective concentrations are achieved in bone for the treatment of staphylococcal osteomyelitis, but not in the CSF for the treatment of meningitis. At one time, lincomycin was thought to be a very nontoxic compound, with a low incidence of allergy (skin rashes) and occasional gastrointestinal complaints (nausea, vomiting, and diarrhea) as the only adverse effects. However, recent reports of severe diarrhea and the development of pseudomembranous colitis in patients treated with lincomycin (or clindamycin) have brought about the need for reappraisal of the position these antibiotics should have in therapy. In any event, clindamycin is superior to lincomycin for the treatment of most infections for which these antibiotics are indicated.

Lincomycin hydrochloride occurs as the monohydrate, a white crystalline solid that is stable in the dry state. It is readily soluble in water and alcohol, and its aqueous solutions are stable at room temperature. It is slowly degraded in acid solutions but is well absorbed from the gastrointestinal tract. Lincomycin diffuses well into peritoneal and pleural fluids and into bone. It is excreted in the urine and the bile. It is available in capsule form for oral administration and in ampules and vials for parenteral administration.

Clindamycin Hydrochloride, USP. 7(*S*)-Chloro-7-deoxylincomycin (Cleocin). In 1967, Magerlein et al.[202] reported that replacement of the 7(*R*)-hydroxy group of lincomycin by chlorine with inversion of configuration resulted in a compound with enhanced antibacterial activity in vitro. Clinical experience with this semisynthetic derivative, called clindamycin and released in 1970, has established that its superiority over lincomycin is even greater in vivo. Improved absorption and higher tissue levels of clindamycin, and its greater penetration into bacteria, have been attributed to its higher partition coefficient, compared with that of lincomycin. Structural modifications at C-7, for example 7(*S*)-chloro and 7(*R*)-OCH_3, and of the C-4 alkyl groups of the hygric acid moiety,[203] appear to influence activity of congeners more through an effect on the partition coefficient of the molecule than through a stereospecific binding role. On the other hand, changes in the α-thiolincosamide portion of the molecule appear to markedly decrease activity, as is evidenced by the marginal activity of 2-deoxylincomycin, its anomer) and 2-*O*-methyllincomycin.[203,204] Exceptions to this are fatty acid and phosphate esters of the 2-hydroxyl group of lincomycin and clindamycin, which are rapidly hydrolyzed in vivo to the parent antibiotics.

Clindamycin is recommended by the manufacturer for the treatment of a wide variety of upper respiratory, skin, and tissue infections caused by susceptible bacteria. Certainly, its activity against streptococci, staphylococci, and pneumococci is undisputably high; and it is one of the most potent agents available against some non–spore-forming anaerobic bacteria, the *Bacteriodes* species in particular. However, an ever-increasing number of reports of clindamycin-associated gastrointestinal toxicity, which range in severity from diarrhea to an occasionally serious pseudomembranous colitis, have caused some clinical experts to call for a reappraisal of the appropriate position of this antibiotic in therapy. Clindamycin- (or lincomycin)-associated colitis may be particularly dangerous in elderly or debilitated patients and has caused deaths in such individuals. This condition, which is usually reversible when the drug is withdrawn, is now believed to result from an overgrowth of a clindamycin-resistant strain of the anaerobic intestinal bacterium *Clostridium difficile*.[205] Damage to the intestinal lining is caused by a glycoprotein endotoxin released by lysis of this organism. Vancomycin has been effective in the treatment of clindamycin-induced pseudomembranous colitis and in the control of the experimentally induced bacterial condition in animals. Clindamycin should be reserved for staphylococcal tissue infections, such as cellulitis and osteomyelitis in penicillin-allergic patients, and for severe anaerobic infections outside the central nervous system. It should not ordinarily be used to treat upper respiratory tract infections caused by bacteria sensitive to other, safer antibiotics or in prophylaxis.

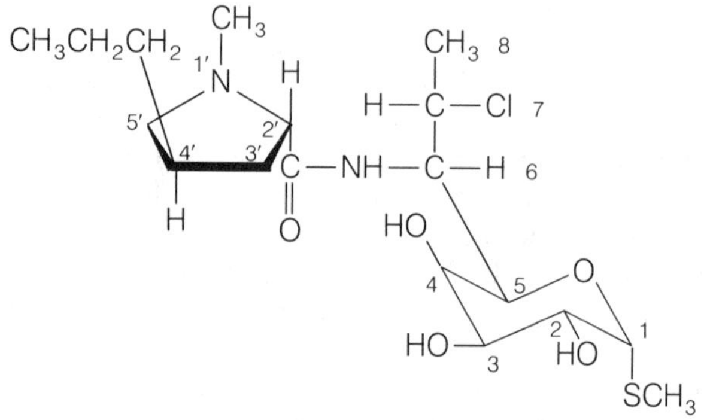

Clindamycin

Clindamycin is rapidly absorbed from the gastrointestinal tract, even in the presence of food. It is available as the crystalline, water-soluble hydrochloride hydrate (hyclate) and the 2-palmitate ester hydrochloride salts in oral dosage forms, and as the 2-phosphate ester in solutions for intramuscular and intravenous injection. All forms are chemically very stable in solution and in the dry state.

Clindamycin Palmitate Hydrochloride, USP (Cleocin Pediatric) is the hydrochloride salt of the palmitic acid ester of cleomycin. The ester bond is to the 2-hydroxyl group of the lincosamine sugar. The ester serves as a tasteless prodrug form of the antibiotic which hydrolyzes to clindamycin in the plasma. The salt form confers water solubility to the ester which is available as granules for reconstitution into an oral solution for pediatric use. Although the absorption of the palmitate is slower than that of the free base, there is little difference in overall bioavailability of the two preparations. Reconstituted solutions of the palmitate hydrochloride are stable for two weeks at room temperature. Such solutions should not be refrigerated because thickening occurs, making the preparation difficult to pour.

Clindamycin Phosphate, USP (Cleocin Phosphate) is the 2-phosphate ester of clindamycin. It

TABLE 7-15

LINCOMYCINS

Name Proprietary Name	***Preparations***	***Usual Adult Dose****	***Usual Dose Range****	***Usual Pediatric Dose****
Lincomycin hydrochloride USP *Lincocin*	Lincomycin hydrochloride injection USP	IM, the equivalent of 600 mg of lincomycin qd or bid; IV infusion, the equivalent of 600 mg–1 g of lincomycin over not less than 1 hr, bid or tid	600 mg–8 g/day	Dosage is not established in children under 1 mo of age. Over 1 mo: IM, 10 mg/kg of body weight or 300 mg/m^2 of body surface, qd or bid; IV infusion, 5–10 mg/kg or 150–300 mg/m^2 over not less than 1 hr, bid
	Lincomycin hydrochloride capsules USP Lincomycin hydrochloride syrup	The equivalent of 500 mg of lincomycin every 6–8 hr		Infants 1 mo of age and over: oral, the equivalent of lincomycin, 7.5–15 mg/kg of body weight every 6 hr; or 10–20 mg/kg of body weight every 8 hr
Clindamycin hydrochloride USP *Cleocin*	Clindamycin hydrochloride capsules USP	The equivalent of 150–450 mg of clindamycin every 6 hr		Infants under 1 mo: use with caution; infants 1 mo and over: oral, the equivalent of clindamycin, 2–6.3 mg/kg of body weight every 6 hr; or 2.7–8.3 mg/kg of body weight every 8 hr
Clindamycin palmitate hydrochloride USP *Cleocin Palmitate*	Clindamycin palmitate hydrochloride for oral solution USP	Equivalent of clindamycin, 12 mg/kg, as clindamycin palmitate hydrochloride, tid or qid	Clindamycin, present as clindamycin palmitate hydrochloride, 8–25 mg/kg of body weight, divided into 3 or 4 equal doses. In children weighing 10 kg or less, 37.5 mg of clindamycin tid is the minimum recommended dose	
Clindamycin phosphate USP *Cleocin Phosphate*	Clindamycin phosphate injection USP	IM or IV, 300 mg of clindamycin, as the phosphate, bid to qid	600 mg–2.7 g of clindamycin, as the phosphate, daily, divided into 2, 3, or 4 equal doses; in children over 1 mo of age, of clindamycin, 10–40 mg/kg of body weight daily, divided into 3 or 4 equal doses	

*See USP DI for complete dosage information.

exists as a zwitterionic structure that is very soluble in water. Clindamycin phosphate is intended for parenteral (IV or IM) administration for the treatment of serious infections and when oral administration is not feasible. Solutions of clindamycin phosphate are stable at room temperature for 16 days and for up to 32 days when refrigerated.

THE POLYPEPTIDES

Among the most powerful bactericidal antibiotics are those possessing a polypeptide structure. Many of them have been isolated, but, unfortunately, their clinical use has been limited by their undesirable side reactions, particularly renal toxicity. The chief source of the medicinally important members of this class has been various species of the genus *Bacillus*. A few have been isolated from other bacteria, but have not gained a place in medical practice. Three medicinally useful polypeptide antibiotics have been isolated from a *Streptomyces* species.

Polypeptide antibiotics are of three main types: neutral, acidic, and basic. It had been presumed that the neutral compounds, such as the gramicidins, possessed cyclopeptide structures and, thus, had no free amino or carboxyl groups. It has been shown that the neutrality is due to the formylation of a terminal group and that the neutral gramicidins are linear rather than cyclic. The acidic compounds have free carboxyl groups,[206] indicating that at least part of the structure is noncyclic. The basic compounds have free amino groups and, similarly, are noncyclic, at least in part. Some, such as the gramicidins, are active against only gram-positive organisms; others, such as the polymyxins, are active against gram-negative organisms and, hence, have achieved a special place in antibacterial therapy. Significant comments about the biosynthesis and structure–activity relationships of peptide antibiotics have been published by Bodanszky and Perlman.[207]

Gramicidin, USP. Gramicidin is obtained from tyrothricin, a mixture of polypeptides usually obtained by extraction of cultures of *Bacillus brevis*. Tyrothricin was isolated in 1939 by Dubos[208] in a planned search to find an organism growing in soil that would have antibiotic activity against human pathogens. Having only limited use in therapy now, it is of historical interest as the first in the series of modern antibiotics. Tyrothricin is a white to slightly gray or brownish white powder, with little or no odor or taste. It is practically insoluble in water and is soluble in alcohol and in dilute acids. Suspensions for clinical use can be prepared by adding an alcoholic solution to calculated amounts of distilled water or isotonic saline solutions.

Tyrothricin is a mixture of two groups of antibiotic compounds, the gramicidins and the tyrocidines. Gramicidins are the more active components of tyrothricin, and this fraction, occurring in 10% to 20% quantities in the mixture, may be separated and used in topical preparations for the antibiotic effect. Five gramicidins, A_3, A_2, B_1, B_2, and C have been identified. Their structures have been proposed and confirmed through synthesis by Sarges and Witkop.[206] It may be noted that the gramicidins A differ from the gramicidins B by having a tryptophan moiety substituted by an L-phenylalanine moiety. In gramicidin C, a tyrosine moiety substitutes for a tryptophan moiety. In both of the gramicidin A and B pairs, the only difference is the amino acid located at the end of the chain having the neutral formyl group on it. If that amino acid is valine, the compound is either valine-gramicidin A or valine-gramicidin B. If that amino acid is isoleucine, the compound is isoleucine-gramicidin, either A or B.

Tyrocidine is a mixture of tyrocidines A, B, C, and D whose structures have been determined by Craig and co-workers.[209] The synthesis of tyrocidine A has been reported by Ohno et al.[210]

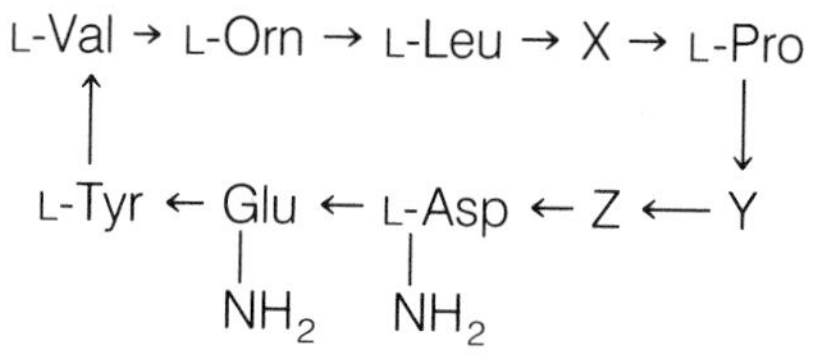

L-Val → L-Orn → L-Leu → X → L-Pro
↑ ↓
L-Tyr ← Glu ← L-Asp ← Z ← Y
(Glu–NH_2, L-Asp–NH_2)

	X	Y	Z
Tyrocidine A:	D-Phe	D-Phe	D-Phe
Tyrocidine B:	D-Phe	L-Tyr	D-Phe
Tyrocidine C:	D-Tyr	L-Tyr	D-Phe
Tyrocidine D:	D-Tyr	L-Tyr	D-Tyr

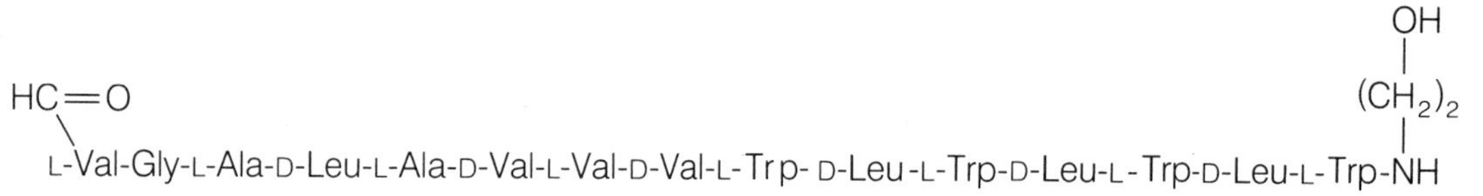

HC=O–L-Val-Gly-L-Ala-D-Leu-L-Ala-D-Val-L-Val-D-Val-L-Trp-D-Leu-L-Trp-D-Leu-L-Trp-D-Leu-L-Trp-NH–$(CH_2)_2$–OH

Valine-gramicidin A

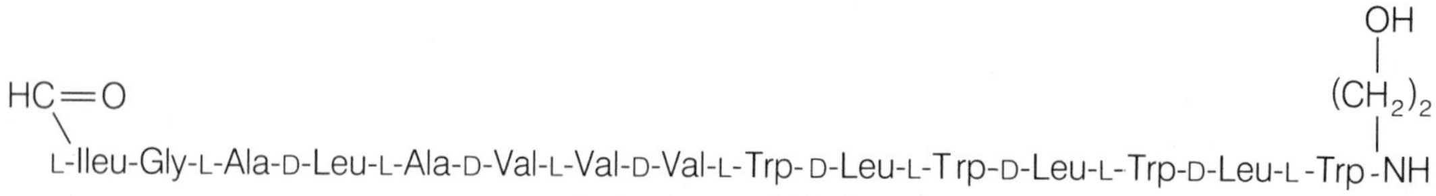

HC=O–L-Ileu-Gly-L-Ala-D-Leu-L-Ala-D-Val-L-Val-D-Val-L-Trp-D-Leu-L-Trp-D-Leu-L-Trp-D-Leu-L-Trp-NH–$(CH_2)_2$–OH

Isoleucine-gramicidin A

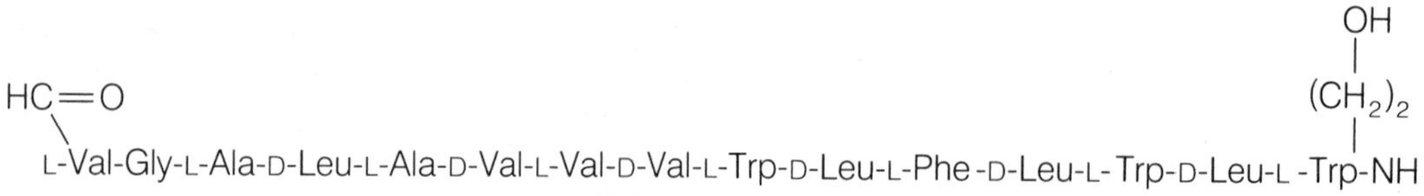

HC=O–L-Val-Gly-L-Ala-D-Leu-L-Ala-D-Val-L-Val-D-Val-L-Trp-D-Leu-L-Phe-D-Leu-L-Trp-D-Leu-L-Trp-NH–$(CH_2)_2$–OH

Valine-gramicidin B

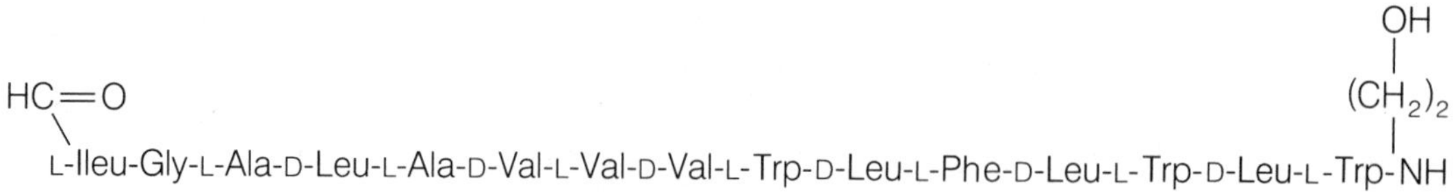

HC=O–L-Ileu-Gly-L-Ala-D-Leu-L-Ala-D-Val-L-Val-D-Val-L-Trp-D-Leu-L-Phe-D-Leu-L-Trp-D-Leu-L-Trp-NH–$(CH_2)_2$–OH

Isoleucine-gramicidin B

The mechanism of action of gramicidin is to act as a ionophore in bacterial cell membranes to cause the loss of potassium ion from the cell.[211] It exerts a bactericidal effect.

Tyrothricin and gramicidin are effective primarily against gram-positive organisms. Their use is restricted to local applications. The ability of tyrothricin to cause lysis of erythrocytes makes it unsuitable for the treatment of systemic infections. Its applications should avoid direct contact with the bloodstream through open wounds or abrasions. It is ordinarily safe to use tyrothricin in troches for throat infections, as it is not absorbed from the gastrointestinal tract. Gramicidin is available in a variety of topical preparations containing other antibiotics such as bacitracin and neomycin.

Bacitracin, USP. The organism from which Johnson, Anker, and Meleney[212] produced bacitracin in 1945 is a strain of *B. subtilis*. The organism had been isolated from debrided tissue from a compound fracture in 7-year-old Margaret Tracy, hence the name bacitracin. Production of bacitracin is now accomplished from the licheniformis group (sp. *B. subtilis*). Like tyrothricin, the first useful antibiotic obtained from bacterial cultures, bacitracin is a complex mixture of polypeptides. So far, at least 10 polypeptides have been isolated by countercurrent distribution techniques: A, A′, B, C, D, E, F_1, F_2, F_3, and G. It appears that the commercial product known as bacitracin is a mixture principally of A, with smaller amounts of B, D, E, and F.

The official product is a white to pale buff powder that is odorless or nearly so. In the dry state, bacitracin is stable, but it rapidly deteriorates in aqueous solutions at room temperature. Because it is hygroscopic, it must be stored in tight containers, preferably under refrigeration. The stability of aqueous solutions of bacitracin is affected by *p*H and temperature. Slightly acidic or neutral solutions are stable for as long as one year if kept at a temperature of 0° to 5°C. If the *p*H rises above 9, inactivation occurs very rapidly. For greatest stability, the *p*H of a bacitracin solution is best adjusted at 4 to 5 by the simple addition of acid. The salts of heavy metals precipitate bacitracin from its solutions, with resulting inactivation. However, EDTA also inactivates bacitracin, leading to the discovery that a divalent ion (i.e., Zn^{2+}) is required for activity. In

Bacitracin A

addition to being soluble in water, bacitracin is soluble in low-molecular-weight alcohols, but is insoluble in many other organic solvents, including acetone, chloroform, and ether.

The principal work on the chemistry of the bacitracins has been directed toward bacitracin A, the component in which most of the antibacterial activity of crude bacitracin resides. The structure shown in the diagram is that proposed by Stoffel and Craig[212] and subsequently confirmed by Ressler and Kashelikar.[214]

The activity of bacitracin is measured in units. The potency per milligram is not less than 40 USP units except for material prepared for parenteral use, which has a potency of not less than 50 units/mg. It is a bactericidal antibiotic that is active against a wide variety of gram-positive organisms, very few gram-negative organisms, and some others. It is believed to exert its bactericidal effect through an inhibition of mucopeptide cell wall synthesis. Its action is enhanced by zinc. Although bacitracin has found its widest use in topical preparations for local infections, it is quite effective in a number of systemic and local infections when administered parenterally. It is not absorbed from the gastrointestinal tract; accordingly, oral administration is without effect, except for the treatment of amebic infections within the alimentary canal.

Polymyxin B Sulfate, USP (Aerosporin). Polymyxin was discovered in 1947 almost simultaneously in three separate laboratories in America and Great Britain.[215–217] As often happens when similar discoveries are made in widely separated laboratories, differences in nomenclature referring both to the antibiotic-producing organism and the antibiotic itself appeared in references to the polymyxins. Since it now has been shown that the organisms first designated as *Bacillus polymyxa* and *B. aerosporus* Greer are identical species, the one name, *B. polymyxa*, is used to refer to all of the strains that produce the closely related polypeptides called polymyxins. Other organisms (see colistin, for example) also produce polymyxins. Identified so far are polymyxins A, B_1, B_2, C, D_1, D_2, M, colistin A (polymyxin E_1), colistin B (polymyxin E_2), circulins A and B, and polypeptin. The known structures of this group and their properties have been reviewed by Vogler and Studer.[218] Of these, polymyxin B as the sulfate is usually used in medicine because, when used systemically, it causes less kidney damage than the others.

Polymyxin B sulfate is a nearly odorless, white to buff powder. It is freely soluble in water and slightly soluble in alcohol. Its aqueous solutions are slightly acidic or nearly neutral (*p*H 5 to 7.5) and, when refrigerated, are stable for at least six months. Alkaline solutions are unstable. Polymyxin B has been shown by Hausmann and Craig,[219] who used countercurrent distribution techniques, to contain two fractions that differ in structure only by one fatty acid component. Polymyxin B_1 contains (+)-6-methyloctan-1-oic acid (isopelargonic acid), a fatty acid isolated from all of the other polymyxins. The B_2 component contains an isooctanoic acid, $C_8H_{16}O_2$, of undetermined structure. The structural formula for polymyxin B has been proved by the synthesis accomplished by Vogler et al.[220]

Polymyxin B sulfate is useful against many gram-negative organisms. Its main use in medicine has been in topical applications for local infections

```
                    NH—CO
                    |   |
       C6H5CH2—CH   CH—CH2CH2NH2
                    |   |
                    CO  NH
                    |   |
                    NH  CO          CH2CH2NH2                 CH2CH2NH2                       CH3
                    |   |           |                         |                               |
    (H3C)2CHCH2—CH   CH—NHCO—C—NH—CO—CH—NH—CO—CH—NH—CO—(CH2)4—CHCH2CH3
                    |   |           |            |
                    CO  CH2         H            CHOHCH3
                    |   |
                    NH  CH2
                    |   |
      H2NCH2CH2—CH   NH
                    |   |
                    CO  CO
                    |   |
                    NH  CH—CHOHCH3
                    |   |
      H2NCH2CH2—CH   NH
                     \ /
                     CO
```

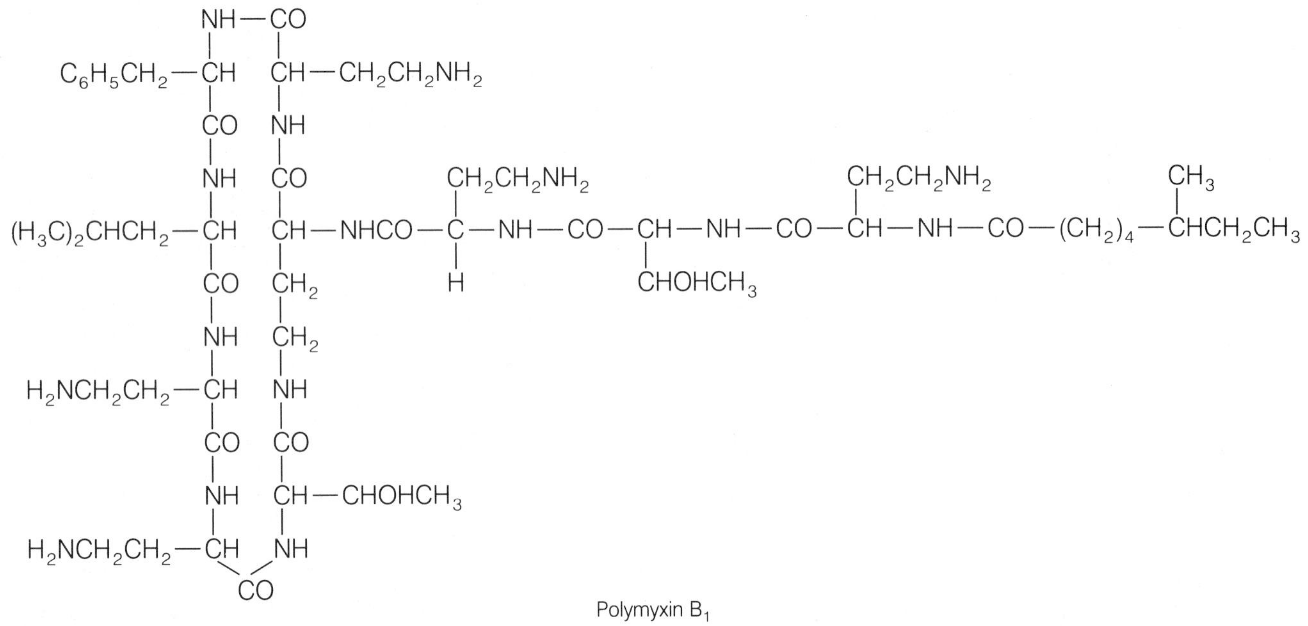

Polymyxin B_1

in wounds and burns. For such use it is frequently combined with bacitracin, which is effective against gram-positive organisms. Polymyxin B sulfate is poorly absorbed from the gastrointestinal tract; therefore, oral administration of it is of value only in the treatment of intestinal infections such as pseudomonal enteritis or those due to *Shigella*. It may be given parenterally by intramuscular or intrathecal injection for systemic infections. The dosage of polymyxin is measured in USP units. One milligram contains not less than 6000 USP units.

Colistin Sulfate, USP (Coly-Mycin S). In 1950, Koyama and co-workers[221] isolated an antibiotic from *Aerobacillus colistinus* (*B. polymyxa* var. *colistinus*) that has been given the name colistin. It had been used in Japan and in some European countries for several years before it was made available for medicinal use in the United States. It is especially recommended for the treatment of refractory urinary tract infections caused by gram-negative organisms such as *Aerobacter*, *Bordetella*, *Escherichia*, *Klebsiella*, *Pseudomonas*, *Salmonella*, and *Shigella*.

Chemically, colistin is a polypeptide that has been reported by Suzuki et al.[222] to be heterogeneous, with the major component being colistin A. They proposed the structure shown here for colistin A, which may be noted to differ from polymyxin B_1 only by the substitution of D-leucine for D-phenylalanine as one of the amino acid fragments in the cyclic portion of the structure. Wilkinson and Lowe[223] have corroborated the structure and have shown colistin A to be identical with polymyxin E_1.

Some additional confusion on nomenclature for this antibiotic exists, as Koyama et al. originally named the product colimycin, and that name is still used. Particularly, it has been the basis for variants used as brand names as Coly-Mycin, Colomycin, Colimycine, and Colimicina.

Two forms of colistin have been made, the sulfate and methanesulfonate, and both forms are available for use in the United States. The sulfate is used to make an oral pediatric suspension; the methanesulfonate is used to make an intramuscular injection. In the dry state, the salts are stable, and their aqueous solutions are relatively stable at acid *p*H from 2 to 6. Above *p*H 6, solutions of the salts are much less stable.

Sterile Colistimethate Sodium, USP. Pentasodium colistinmethanesulfonate; sodium colistimethanesulfonate (Coly-Mycin M). In colistin, five of the terminal amino groups of the α_1-aminobutyric acid fragment may be readily alkylated. In colistimethate sodium, the methanesulfonate radical is the attached alkyl group and, through each of them, a sodium salt may be made. This provides a highly water-soluble compound that is very suitable for injection. In the injectable form, it is given intramuscularly and is surprisingly free from toxic reactions when compared with polymyxin B. Colistimethate sodium does not readily induce the development of resistant strains of microorganisms, and no evidence of cross-resistance with the common broad-spectrum antibiotics has been shown. It is used for the same conditions as those mentioned for colistin.

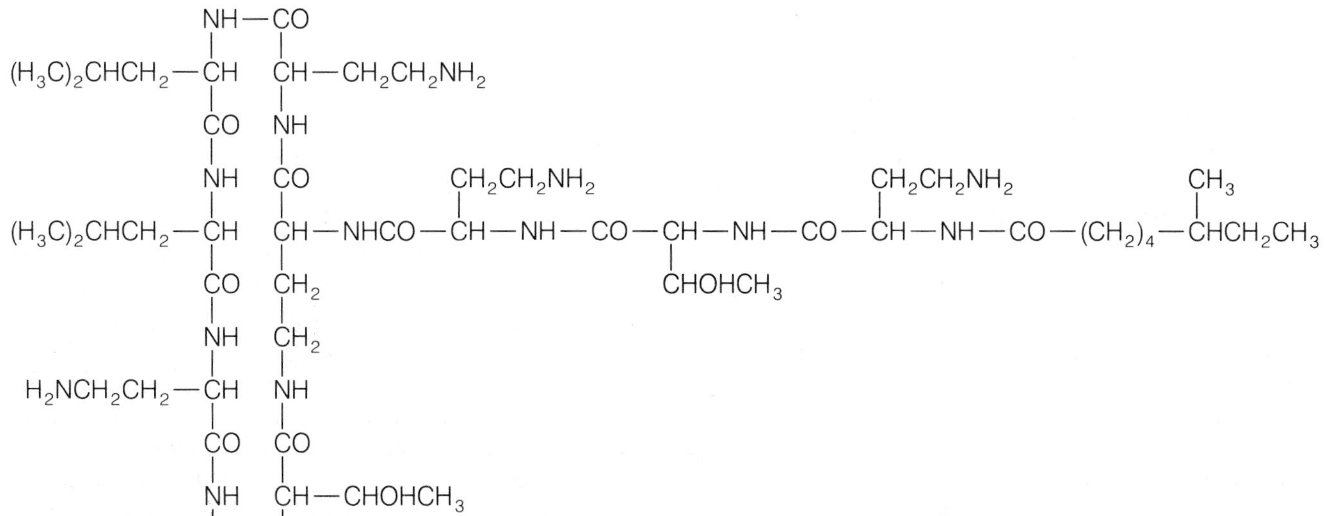

```
                 NH—CO
                 |   |
(H3C)2CHCH2—CH   CH—CH2CH2NH2
             |   |
             CO  NH
             |   |
             NH  CO         CH2CH2NH2                     CH2CH2NH2                   CH3
             |   |          |                             |                           |
(H3C)2CHCH2—CH   CH—NHCO—CH—NH—CO—CH—NH—CO—CH—NH—CO—(CH2)4—CHCH2CH3
             |   |                  |
             CO  CH2                CHOHCH3
             |   |
             NH  CH2
             |   |
 H2NCH2CH2—CH    NH
             |   |
             CO  CO
             |   |
             NH  CH—CHOHCH3
             |   |
 H2NCH2CH2—CH    NH
             \  /
              CO
```

Colistin A (Polymyxin E_1)

TABLE 7-16

POLYPEPTIDE ANTIBIOTICS

Name Proprietary Name	***Preparations***	***Application***	***Usual Adult Dose****	***Usual Dose Range****	***Usual Pediatric Dose****
Gramicidin USP		Topical, 0.05% solution			
Bacitracin USP *Baciguent*	Bacitracin ointment USP	Topically to the skin, bid or tid			
	Bacitracin ophthalmic ointment USP	Topically to the conjunctiva, bid or tid			
	Sterile bacitracin USP		IM, 10,000–20,000 units bid or tid	30,000–100,000 units/day	Premature infants: 300 units/kg of body weight tid; full-term newborn infants to 1 yr: 330 units/kg tid; older infants and children: 500 units/kg or 15,000 units/m^2 of body surface, qid
Bacitracin zinc USP	Bacitracin ointment USP	Topically to the skin, bid or tid			
Polymyxin B sulfate USP *Aerosporin*	Sterile polymyxin B sulfate USP		IM, 6250–7500 units/kg of body weight qid; intrathecal, 50,000 units qd for 3 or 4 days, then 50,000 units once every 2 days; IV infusion, 7500–12,500 units/kg of body weight in 300–500 mL of 5% dextrose injection as a continuous infusion, bid. The total daily dose must not exceed 25,000 units/kg daily		IM, see Usual Dose. Intrathecal, children under 2 yr: 20,000 units qd for 3 or 4 days or 25,000 units once every 2 days; children over 2 yr: see Usual Dose. IV infusion, 7500–12,500 units/kg of body weight in 300–500 mL of 5% dextrose injection over a period of 60–90 min, bid. The total daily dose must not exceed 25,000 units/kg daily
	Polymixin B sulfate otic solution USP				
Colistin sulfate USP *Coly-Mycin S*	Colistin sulfate for oral suspension USP			3–15 mg/kg daily	The equivalent of colistin 2–5 mg/kg of body weight tid
Coly-Mycin M	Sterile colistimethate sodium USP		IM or IV, the equivalent of colistin, 1.25 mg/kg of body weight bid to qid	1.5–5 mg/kg daily	See Usual Dose

*See USP DI for complete dosage information.

UNCLASSIFIED ANTIBIOTICS

Among the many hundreds of antibiotics that have been evaluated for activity are several that have gained significant clinical attention, but that do not fall into any of the previously considered groups. Some of these have quite specific activities against a narrow spectrum of microorganisms. Some have found a useful place in therapy as substitutes for other antibiotics to which resistance has developed.

Chloramphenicol, USP (Chloromycetin, Amphicol). The first of the widely used broad-spectrum antibiotics, chloramphenicol, was isolated by Ehrlich et al.[224] in 1947. They obtained it from *S. venezuelae*, an organism that was found in a sample of soil collected in Venezuela. Since that time, chloramphenicol has been isolated as a product of several organisms found in soil samples from widely separated places. More importantly, its chemical structure was soon established and in 1949, Controulis, Rebstock, and Crooks[225] reported its synthesis. This opened the way for the commercial production of chloramphenicol by a totally synthetic route. It was the first and still is the only therapeutically important antibiotic to be so produced in competition with microbiologic processes. Diverse synthetic proce-

dures have been developed for chloramphenicol. The commercial process most generally used has started with p-nitroacetophenone.[226]

Chloramphenicol is a white crystalline compound that is very stable. It is very soluble in alcohol and other polar organic solvents but is only slightly soluble in water. It has no odor, but is very bitter.

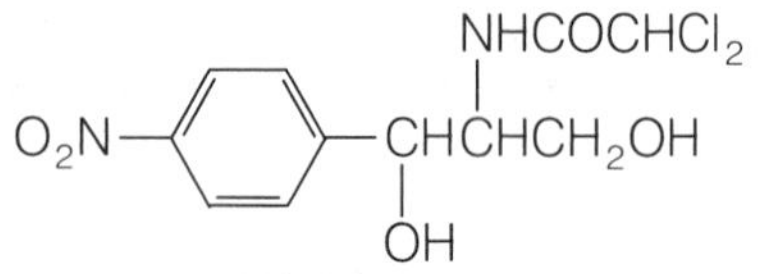

$NHCOCHCl_2$

O_2N—C₆H₄—$CHCHCH_2OH$

OH

Chloramphenicol

Note that chloramphenicol possesses two asymmetric carbon atoms in the acylamidopropanediol chain. Biologic activity resides almost exclusively in the D-*threo* isomer; the L-*threo* and the D- and L-*erythro* isomers are virtually inactive.

Chloramphenicol is very stable in the bulk state and in solid dosage forms. In solution, however, it slowly undergoes various hydrolytic and light-induced reactions.[227] The rates of these reactions are dependent on pH, heat, and light. Hydrolytic reactions include general acid–base-catalyzed hydrolysis of the amide to give 1-(p-nitrophenyl)-2-aminopropan-1,3-diol and dichloracetic acid, and alkaline hydrolysis (above pH 7) of the α-chloro groups to form the corresponding α,α-dihydroxy derivative.

The metabolism of chloramphenicol has been investigated thoroughly.[228] The main path involves formation of the 3-*O*-glucuronide. Minor reactions include reduction of the p-nitro group to the aromatic amine, hydrolysis of the amide, and hydrolysis of the α-chloracetamido group, followed by reduction to give the corresponding α-hydroxyacetyl derivative.

Strains of certain bacterial species are resistant to chloramphenicol by virtue of the ability to produce chloramphenicol acetyltransferase, an enzyme that acetylates the hydroxy groups at the 1- and 3-positions. Both the 3-acetoxy and the 1,3-diacetoxy metabolites are devoid of antibacterial activity.

Numerous structural analogues of chloramphenicol have been synthesized to provide a basis for correlation of structure to antibiotic action. It appears that the p-nitrophenyl group may be replaced by other aryl structures without appreciable loss in activity. Substitution on the phenyl ring with several different types of groups for the nitro group, a very unusual structure in biologic products, does not cause a great decrease in activity. However, all such compounds yet tested are less active than chloramphenicol. As part of a quantitative SAR study, Hansch et al.[229] reported that the 2-$NHCOCF_3$ derivative is 1.7 times as active as chloramphenicol against *E. coli*. Modifications of the side chain show it to possess a high degree of specificity in structure for antibiotic action. A conversion of the alcohol group on C-1 of the side chain to a keto group causes an appreciable loss in activity. The relationship of the structure of chloramphenicol to its antibiotic activity will not be clearly seen until the mode of action of this compound is known. The review article by Brock[230] reports on the large amount of research that has been devoted to this problem. It has been established that chloramphenicol exerts its bacteriostatic action by a strong inhibition of protein synthesis. The details of such inhibition are as yet undetermined, and the precise point of action is unknown. Some process lying between the attachment of amino acids to sRNA and the final formation of protein appears to be involved.

The broad-spectrum activity of chloramphenicol and its singular effectiveness in the treatment of some infections not amenable to treatment by other drugs made it an extremely popular antibiotic. Unfortunately, instances of serious blood dyscrasias and other toxic reactions have resulted from the promiscuous and widespread use of chloramphenicol in the past. Because of these reactions, it is now recommended that it not be used in the treatment of infections for which other antibiotics are as effective and not as hazardous. When properly used, with careful observation for untoward reactions, chloramphenicol provides some of the very best therapy for the treatment of serious infections.[231]

Chloramphenicol is specifically recommended for the treatment of serious infections caused by strains of gram-positive and gram-negative bacteria that have developed resistance to penicillin G and ampicillin, such as *H. influenzae*, *Salmonella typhi*, *Strep. pneumoniae*, *B. fragilis*, and *N. meningitidis*. Because of its penetration into the central nervous system, chloramphenicol represents a particularly important alternative therapy for meningitis. It is not recommended for the treatment of urinary tract infections because 5% to 10% of the unconjugated form is excreted in the urine. Chloramphenicol is also employed for the treatment of rickettsial infections, such as Rocky Mountain spotted fever.

Because it is bitter, this antibiotic is administered orally either in capsules or as the palmitate ester. Chloramphenicol palmitate USP is insoluble in water and may be suspended in aqueous vehicles for liquid dosage forms. The ester forms by reaction with the hydroxyl group on C-3. In the alimentary tract it is slowly hydrolyzed to the active antibiotic. Parenteral administration of chloramphenicol is

made by use of an aqueous suspension of very fine crystals or by use of a solution of the sodium salt of the succinate ester of chloramphenicol. Sterile chloramphenicol sodium succinate has been used to prepare aqueous solutions for intravenous injections.

Chloramphenicol Palmitate, USP is the palmitic acid ester of chloramphenicol. It is a tasteless prodrug of chloramphenicol intended for pediatric use. The ester must hydrolyze in vivo following oral absorption to provide the active form. Erratic serum levels have been associated with early formulations of the palmitate, but it is claimed by the manufacturer that the bioavailability of the current preparation is comparable with that of chloramphenicol itself.

Chloramphenicol Sodium Succinate, USP is the water-soluble sodium salt of the hemisuccinate ester of chloramphenicol. Because of the low solubility of chloramphenicol, the sodium succinate is preferred for intravenous administration. The availability of chloramphenicol from the ester following intravenous administration is estimated to be 70% to 75%; the remainder is excreted unchanged.[231,232] Poor availability of the active form from the ester following intramuscular injection precludes the attainment of effective plasma levels of the antibiotic by this route. Orally administered chloramphenicol or its palmitate ester actually gives higher plasma levels of the active antibiotic than does intravenously administered chloramphenicol sodium succinate.[232,233] Nonetheless, effective concentrations are achieved by either route.

Vancomycin Hydrochloride, USP (Vancocin, Vancoled). The isolation of vancomycin from *S. orientalis* was described in 1956 by McCormick et al.[234] The organism was originally obtained from cultures of an Indonesian soil sample and subsequently has been obtained from Indian soil. It was introduced in 1958 as an antibiotic active against gram-positive cocci, particularly streptococci, staphylococci, and pneumococci. It is not active against gram-negative bacteria. It is recommended for use when infections have not responded to treatment with the more common antibiotics or when the infection is known to be caused by a resistant organism. It is particularly effective for the treatment of endocarditis caused by gram-positive bacteria.

Vancomycin hydrochloride is a free-flowing, tan to brown powder that is relatively stable in the dry state. It is very soluble in water and insoluble in organic solvents. The salt is quite stable in acidic solutions. The free base is an amphoteric substance, the structure of which was determined on the basis of a combination of chemical degradation and nuclear magnetic resonance (NMR) studies and an x-ray crystallographic analysis of a close analogue.[235] Slight stereochemical and conformational revisions in the originally proposed structure have subsequently been made.[236,237] Vancomycin is a glycopeptide containing two glycosidically linked sugars, glucose and vancosamine, and a complex cyclic peptide aglycon containing aromatic residues linked together in a unique resorcinol ether system.

Vancomycin inhibits cell wall synthesis by preventing the synthesis of cell wall mucopeptide polymer. It does so by binding with the D-alanine-D-alanine terminus of the uridine diphosphate-*N*-acetylmuramyl peptides required for mucopeptide polymerization.[238] Details of the binding have been elucidated by the elegant NMR studies of Williamson

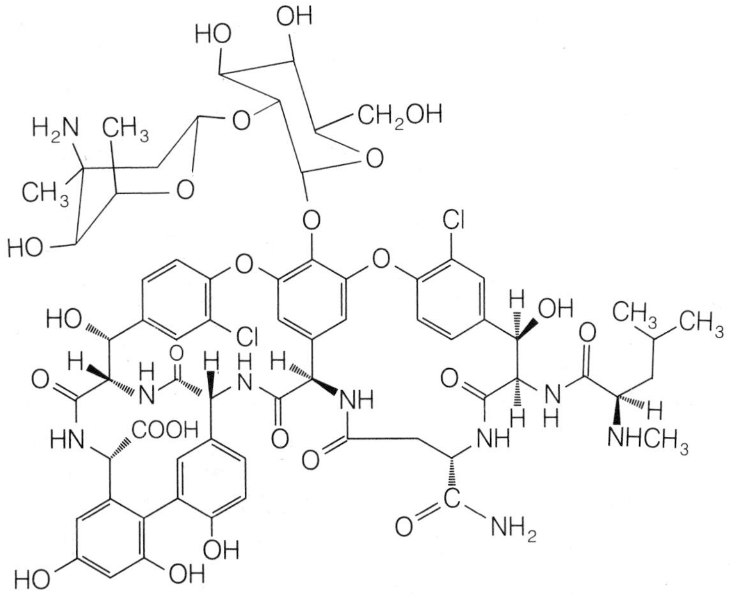

Vancomycin

et al.[239] The action of vancomycin leads to lysis of the bacterial cell. The antibiotic does not exhibit cross-resistance to β-lactams, bacitracin, or cycloserine from which it differs in mechanism.

Vancomycin hydrochloride is always administered intravenously (never intramuscularly), either by slow injection or by continuous infusion, for the treatment of systemic infections. In short-term therapy, the toxic side reactions are usually slight, but continued use may lead to impairment of auditory acuity, renal damage, and to phlebitis and skin rashes. Because it is not absorbed, vancomycin may be administered orally for the treatment of staphylococcal enterocolitis and for pseudomembranous colitis associated with clindamycin therapy. It is likely that some conversion to aglucovancomycin occurs in the low *p*H of the stomach. The latter retains about three-fourths of the activity of vancomycin.

Novobiocin Sodium, USP. Streptonivicin (Albamycin). In the search for new antibiotics, three different research groups independently isolated novobiocin from *Streptomyces* species. It was first reported in 1955 as a product from *S. spheroides* and from *S. niveus*. It is currently produced from cultures of both species. Until the common identity of the products obtained by the different research groups was ascertained, confusion in the naming of this compound existed. Its chemical identity has been established as 7-[4-(carbamoyloxy)tetrahydro-3-hydroxy-5-methoxy-6,6-dimethylpyran-2-yloxy]-4-hydroxy-3-[4-hydroxy-3(3-methyl-2-butenyl)benzamido]-8-methylcoumarin by Shunk et al.[240] and Hoeksema, Caron, and Hinman[241] and confirmed by Spencer et al.[242, 243]

Chemically novobiocin has a unique structure among antibiotics, although, as do several others, it possesses a glycosidic sugar moiety. The sugar in novobiocin, devoid of its carbamate ester, has been named noviose and is an aldose having the configuration of L-lyxose. The aglycon moiety has been termed novobiocic acid.

Novobiocin is a pale yellow, somewhat photosensitive compound that crystallizes in two chemically identical forms with different melting points. It is soluble in methanol, ethanol, and acetone, but is quite insoluble in less polar solvents. Its solubility in water is affected by *p*H. It is readily soluble in basic solutions, in which it deteriorates, and is precipitated from acidic solutions. It behaves as a diacid, forming two series of salts. The enolic hydroxyl group on the coumarin moiety behaves as a rather strong acid and is the group by which the commercially available sodium and calcium salts are formed. The phenolic —OH group on the benzamido moiety also behaves as an acid but is weaker than the former. Disodium salts of novobiocin have been prepared. The sodium salt is stable in dry air but decreases in activity in the presence of moisture. The calcium salt is quite water-insoluble and is used to make aqueous oral suspensions. Because of its acidic characteristics, novobiocin combines to form salt complexes with basic antibiotics. Some of these salts have been investigated for their combined antibiotic effect, but none has been placed on the market, as no advantage is offered by them.

The action of novobiocin is largely bacteriostatic. Its mode of action is not known with certainty, although it does inhibit bacterial protein and nucleic acid synthesis. Studies indicate that novobiocin and related coumarin-containing antibiotics bind to the subunit of DNA gyrase to possibly interfere with DNA supercoiling[244] and energy transduction in bacteria.[245] The effectiveness of novobiocin is largely confined to gram-positive bacteria and a few strains of *Pr. vulgaris*. Its low activity against gram-negative bacteria is apparently due to poor cellular penetration.

Although cross-resistance to other antibiotics is reported not to develop to novobiocin, resistant *Staph. aureus* strains are known. Consequently, the medical use of novobiocin is reserved for the treatment of staphylococcal infections resistant to other antibiotics and sulfas and for patients allergic to these drugs. Another shortcoming that limits the usefulness of novobiocin is the relatively high frequence of adverse reactions, such as urticaria, aller-

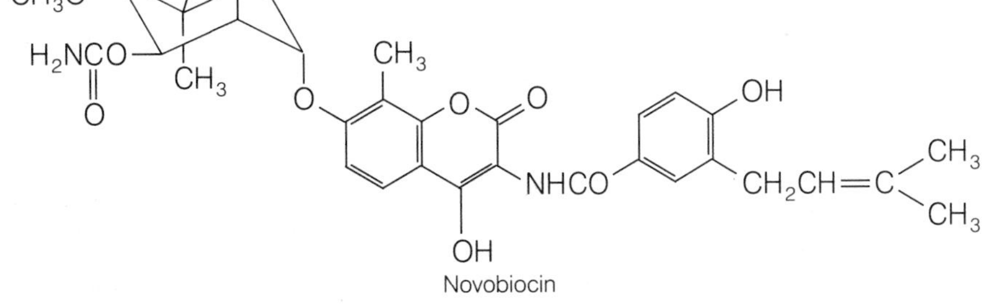

Novobiocin

TABLE 7-17

UNCLASSIFIED ANTIBIOTICS

Name Proprietary Name	***Preparations***	***Usual Adult Dose****	***Usual Dose Range****	***Usual Pediatric Dose****
Chloramphenicol USP *Chloromycetin*	Chloramphenicol capsules USP	Oral, 50 mg/kg every 6 hr for typhoid fever and rickettsial infections	Up to 100 mg/kg daily for serious infections such as meningitis or brain abscesses	
	Chloramphenicol ophthalmic USP Chloramphenicol for ophthalmic solution USP	Topical, to the conjunctiva, 1–2 drops of a 0.16–0.5% solution every 15–30 min, initially; bid to qid		
	Chloramphenicol ophthalmic ointment USP	Topical, to the conjunctiva, (1 cm) of a 1% ointment tid to qid		
	Chloramphenicol otic USP	In the ear, 2 or 3 drops of a 0.5% solution tid;		
	Chloramphenicol injection USP	IV, 12.5 mg/kg every 6 hr	Up to 100 mg/kg daily	
Chloramphenicol palmitate USP	Chloramphenicol palmitate oral suspension USP	Oral, the chloramphenicol equivalent of 50 mg/ every 6 hr	Up to 100 mg/kg daily	Children: 50–75 mg/kg daily in divided doses every 6 hr. For meningtis: 50–100 mg/kg daily in divided doses every 6 hr
Sterile chloramphenicol sodium succinate	Sterile chloramphenicol sodium succinate USP	IV, the chloramphenicol equivalent of 50 mg/kg every 6 hr	Up to 100 mg/kg daily	Neonates less than 2 kg: 25 mg/kg qd Neonates from birth to 7 days: 50 mg/kg qd Neonates over 7 days and greater than 2 kg: 50 mg/kg daily in divided doses every 12 hr
Vancomycin hydrochloride USP *Vancocin, Vancoled*	Vancomycin hydrochloride for oral solution USP	Oral, the equivalent of vancomycin 500 mg every 6 hr or 1 g every 12 hr	1–2 g	Children: 40 mg/kg daily in 4 divided doses; do not exceed 2 g/day Neonates: 10 mg/kg daily in divided doses
	Sterile vancomycin hydrochloride USP	IV infusion, over at least 60 min of 500 mg the equivalent of vancomycin every 6 hr or 1 g every 12 hr	1–2 g	Children: 40 mg/kg daily in divided doses by slow IV infusion Neonates and infants: initial dose of 15 mg/kg followed by 10 mg/kg every 12 hr in the first week of life up to the age of 1 mo and every 8 hr thereafter
Novobiocin sodium, USP *Albamycin*	Novobiocin sodium capsules USP	Oral, the equivalent of novobiocin of 250 mg every 6 hr or 500 mg every 12 hr	Up to 0.5 g every 6 hr or 1 g every 12 hr	15 mg/kg daily up to 30–45 mg/kg daily in divided doses every 6–12 hr
Novobiocin calcium USP	Novobiocin calcium suspension			

*See USP DI for complete dosage information.

gic skin rashes, hepatotoxicity, and blood dyscrasias associated with its use.

REFERENCES

1. Fleming, A.: Br. J. Exp. Pathol. 10:226, 1929.
2. Vuillemin, P.: Assoc. Franc Avance Sc. Part 2:525–543, 1889.
3. Waksman, S. A.: Science 110:27, 1949.
4. Benedict, R. G., Langlykke, A. F.: Annu. Rev. Microbiol. 1:193, 1947.
5. Baron, A. L.: Handbook of Antibiotics, p. 5. New York, Reinhold, 1950.
6. Yocum, R. R., et al.: J. Biol. Chem. 255:3977, 1980.
7. Waxman, D. J., Strominger, J. L.: Annu. Rev. Pharmacol. 52:825, 1983.
8. Spratt, B. G.: Proc. Natl. Acad. Sci. USA 72:2999, 1975.
9. Spratt, B. G.: Eur. J. Biochem. 72:341, 1977.
10. Tomasz, A.: Annu. Rev. Microbiol. 33:113, 1979.

11. Spratt, B. G.: Nature 254:516, 1975.
12. Suzuki, H., et al.: Proc. Natl. Acad. Sci. USA 75:664, 1978.
13. Clarke, H. T., et al.: The Chemistry of Penicillin, p. 454. Princeton, N.J., Princeton University Press, 1949.
14. Sheehan, J. C., Henery-Logan, K. R.: J. Am. Chem. Soc. 81:3089, 1959.
15. Batchelor, F. R., et al.: Nature 183:257, 1959.
16. Sheehan, J. C., Ferris, J. P.: J. Am. Chem. Soc. 81:2912, 1959.
17. Hou, J. P., Poole, J. W.: J. Pharm. Sci. 60:503, 1971.
18. Blaha, J. M., et al.: J. Pharm. Sci. 65:1165, 1976.
19. Schwartz, M.: J. Pharm. Sci. 58:643, 1969.
20. Schwartz, M. A., Buckwalter, F. H.: J. Pharm. Sci. 51:1119, 1962.
21. Finholt, P., Jurgensen, G., Kristiansen, H.: J. Pharm. Sci. 54:387, 1965.
22. Segelman, A. B., Farnsworth, N. R.: J. Pharm. Sci. 59:725, 1970.
23. Doyle, F. P., et al.: Nature 191:1091, 1961.
24. Sykes, R. B., Matthew, M.: J. Antimicrob. Chemother. 2:115, 1976.
25. Nikaido, H., Nakae, T.: Adv. Microb. Physiol. 20:163, 1979.
26. Zimmerman, W., Rosselet, A.: Antimicrob. Agents Chemother. 12:368, 1977.
27. Yoshimura, F., Nikaido, H.: Antimicrob. Agents Chemother. 27:84, 1985.
28. Malouin, F., Bryan, L. E.: Antimicrob. Agents Chemother. 30:1, 1986.
29. Dougherty, et al.: Antimicrob. Agents Chemother. 18:730, 1980.
30. Hartman, B., Tomasz, A.: Antimicrob. Agents Chemother. 19:726, 1981.
31. Brain, E. G., et al.: J. Chem. Soc. 1445, 1962.
32. Stedmen, R. J., et al.: J. Med. Chem. 7:251, 1964.
33. Nayler, J. H. C.: Adv. Drug Res. 7:52, 1973.
34. Nikaido, H., Rosenberg, E. Y., Foulds, J.: J. Bacteriol. 153:232, 1983.
35. Matthew, M.: J. Antimicrob. Chemother. 5:349, 1979.
36. Iyengar, B. S., et al.: J. Med. Chem. 29:611, 1986.
37. Hansch, C., Deutsch, E. W.: J. Med. Chem. 8:705, 1965.
38. Bird, A. E., Marshall, A. C.: Biochem. Pharmacol. 16:2275, 1967.
39. Stewart, G. W.: The Penicillin Group of Drugs. Amsterdam, Elsevier, 1965.
40. Corran, P. H., Waley, S. G.: Biochem. J. 149:357, 1975.
41. Batchelor, F. R., et al.: Nature 206:362, 1965.
42. DeWeck, A. L.: Int. Arch. Allergy 21:20, 1962.
43. Smith, H., Marshall, A. C.: Nature 232:45, 1974.
44. Monroe, A. C., et al.: Int. Arch. Appl. Immunol. 50:192, 1976.
45. Behrens, O. K., et al.: J. Biol. Chem. 175:793, 1948.
46. Mayersohn, M., Endrenyi, L.: Can. Med. Assoc. J. 109:989, 1973.
47. Hill, S. A., et al.: J. Pharm. Pharmacol. 27:594, 1975.
48. Neu, H. C.: Rev. Infect. Dis. 3:110, 1981.
49. Neu, H. C.: J. Infect. Dis. 12S:1, 1974.
50. Ancred, P., et al.: Nature 215:25, 1967.
51. Fu, K. P., Neu, H. C.: Antimicrob. Agents Chemother. 13:358, 1978.
52. Gedes, A. M., et al.: J. Antimicrob. Chemother. 3 (Suppl. B):1, 1977.
53. Spratt, B. G.: J. Antimicrob. Chemother. 3:13, 1977.
54. Knowles, J. R.: Acc. Chem. Res. 18:97, 1985.
55. Merck & Co., Inc.: U.S. Patent 3,950,357 (April 12, 1976).
56. Johnston, D. B. R., et al.: J. Am. Chem. Soc. 100:313, 1978.
57. Albers-Schonberg, G., et al.: J. Am. Chem. Soc. 100:6491, 1978.
58. Kahan, J. S., et al.: Abstr. 227, 16th Conference on Antimicrobial Agents and Chemotherapy, Chicago, 1976.
59. Kropp, H., et al.: Antimicrob. Agents Chemother. 22:62, 1982.
60. Leanza, W. J., et al.: J. Med. Chem. 22:1435, 1979.
61. Morin, R. B., et al.: J. Am. Chem. Soc. 84:3400, 1962.
62. Fechtig, B., et al.: Helv. Chim. Acta 51:1108, 1968.
63. Woodward, R. B., et al.: J. Am. Chem. Soc. 88:852, 1966.
64. Morin, R. B., et al.: J. Am. Chem. Soc. 85:1896, 1963.
65. Yamana, T., Tsuji, A.: J. Pharm. Sci. 65:1563, 1976.
66. Sullivan, H. R., McMahon, R. E.: Biochem. J. 102:976, 1967.
67. Indelicato, J. M., et al.: J. Med. Chem. 17:523, 1974.
68. Tsuji, A., et al.: J. Pharm. Sci. 70:1120, 1981.
69. Fong, I., et al.: Antimicrob. Agents Chemother. 9:939, 1976.
70. Cimarusti, C. M.: J. Med. Chem. 27:247, 1984.
71. Faraci, W. S., Pratt, R. F.: Biochemistry 24:903, 1985; 25:2934, 1986.
72. Kukolja, S.: In Elks, J. (ed.). Recent Advances in the Chemistry of Beta-Lactam Antibiotics, p. 181. Chichester, Engl., The Chemical Society (London), Burlington House, 1977.
73. Gillett, A. P., et al.: Postgrad. Med. 55(Suppl. 4):9, 1979.
74. Indelicato, J. M., et al.: J. Pharm. Sci. 65:1175, 1976.
75. Ott, J. L., et al.: Antimicrob. Agents Chemother. 15:14, 1979.
76. Nagarajan, R., et al.: J. Am. Chem. Soc. 93:2308, 1971.
77. Stapley, E. O., et al.: Antimicrob. Agents Chemother. 2:122, 1972.
78. Goering, R. V., et al.: Antimicrob. Agents Chemother. 21:963, 1982.
79. Grassi, G. G., et al.: J. Antimicrob. Chemother. 11 (Suppl. A):45, 1983.
80. Uyeo, S., et al.: J. Am. Chem. Soc. 101:4403, 1979.
81. Cama, L. D., Christensen, B. G.: J. Am. Chem. Soc. 96:7583, 1974.
82. Labia, R., et al.: Drugs Exp. Clin. Res. 10:27, 1984.
83. Imada, A., et al.: Nature 289:590, 1981.
84. Sykes, R. B., et al.: Nature 291:489, 1981.
85. Bonner, D. B., Sykes, R. B.: J. Antimicrob. Chemother. 14:313, 1984.
86. Tanaka, S. N., et al.: Antimicrob. Agents Chemother. 31:219, 1987.
87. Schatz, A., et al.: Proc. Soc. Exp. Biol. Med. 55:66, 1944.
88. Weisblum, B., Davies, J.: Bacteriol. Rev. 32:493, 1968.
89. Davies, J., Davis, B. D.: J. Biol. Chem. 243:3312, 1968.
90. Lando, D., et al.: Biochemistry 12:4528, 1973.
91. Reynolds, A. V., Smith, J. T.: Recent Adv. Infect. 1:165, 1979.
92. Chevereau, P. J. L., et al.: Biochemistry 13:598, 1974.
93. Price, K. E., et al.: Antimicrob. Agents Chemother. 5:143, 1974.
94. Bryan, L. E., et al.: J. Antibiot. 29:743, 1976.
95. Hancock, R. E. W.: J. Antimicrob. Chemother. 8:249, 1981.
96. Cox, D. A., et al.: The aminoglycosides. In Sammes, P. G. (ed.). Topics in Antibiotic Chemistry, Vol. 1, p. 44. Chichester, Engl., Ellis Harwood, 1977.
97. Price, K. E.: Antimicrob. Agents Chemother. 29:543, 1986.
98. Dyer, J. R., Todd, A. W.: J. Am. Chem. Soc. 85:3896, 1963.
99. Dyer, J. R., et al.: J. Am. Chem. Soc. 87:654, 1965.
100. Umezawa, S., et al.: J. Antibiot. 27:997, 1974.
101. Waksman, S. A., Lechevalier, H. A.: Science 109:305, 1949.
102. Hichens, M., Rinehart, K. L., Jr.: J. Am. Chem. Soc. 85:1547, 1963.
103. Rinehart, K. L., Jr., et al.: J. Am. Chem. Soc. 84:3218, 1962.
104. Huettenrauch, R.: Pharmazie 19:697, 1964.
105. Umezawa, S., Nishimura, Y.: J. Antibiot. 30:189, 1977.
106. Haskell, T. H., et al.: J. Am. Chem. Soc. 81:3482, 1959.
107. DeJongh, D. C., et al.: J. Am. Chem. Soc. 89:3364, 1967.
108. Umezawa, H., et al.: J. Antibiot. [A]10:181, 1957.
109. Tatsuoka, S., et al.: J. Antibiot. [A]17:88, 1964.
110. Nakajima, M.: Tetrahedron Lett. 623, 1968.
111. Umezawa, S., et al.: J. Antibiot. 21:162, 367, 424, 1968.
112. Morikubo, Y.: J. Antibiot. [A]12:90, 1959.
113. Kawaguchi, H., et al.: J. Antibiot. 25:695, 1972.
114. Paradelis, A. G., et al.: Antimicrob. Agents Chemother. 14:514, 1978.
115. Weinstein, M. J., et al.: J. Med. Chem. 6:463, 1963.
116. Cooper, D. J., et al.: J. Infect. Dis. 119:342, 1969.
117. Cooper, D. J., et al.: J. Chem. Soc. C. 3126, 1971.
118. Maehr, H. and Schaffner, C. P.: J. Am. Chem. Soc. 89:6788, 1968.
119. Koch, K. F., et al.: J. Antibiot. 26:745, 1963.
120. Lockwood, W., et al.: Antimicrob. Agents Chemother. 4:281, 1973.
121. Wright, J. J.: J. Chem. Soc. Chem. Commun. 206, 1976.
122. Wagman, G. M., et al.: J. Antibiot. 23:555, 1970.

123. Braveny, I., et al.: Arzneim. Forsch. 30:491, 1980.
124. Lewis, C., Clapp, H.: Antibiot. Chemother. 11:127, 1961.
125. Cochran, T. G., Abraham, D. J.: J. Chem. Soc. Chem. Commun. 494, 1972.
126. Muxfeldt, H., et al.: J. Am. Chem. Soc. 90:6534, 1968.
127. Muxfeldt, H., et al.: J. Am. Chem. Soc. 101:689, 1979.
128. Korst, J. J., et al.: J. Am. Chem. Soc. 90:439, 1968.
129. Muxfeldt, H., et al.: Angew. Chem. Int. Ed. 12:497, 1973.
130. Hirokawa, S., et al.: Z. Krist. 112:439, 1959.
131. Takeuchi, Y. and Buerger, M. J.: Proc. Natl. Acad. Sci. USA 46:1366, 1960.
132. Cid-Dresdner, H.: Z. Krist. 121:170, 1965.
133. Leeson, L. J., Krueger, J. E., Nash, R. A.: Tetrahedron Lett. 1155, 1963.
134. Stephens, C. R., et al.: J. Am. Chem. Soc. 78:4155, 1956.
135. Rigler, N. E., et al.: Anal. Chem. 37:872, 1965.
136. Benet, L. Z., Goyan, J. E.: J. Pharm. Sci. 55:983, 1965.
137. Barringer, W., et al.: Am. J. Pharm. 146:179, 1974.
138. Barr, W. H., et al.: Clin. Pharmacol. Ther. 12:779, 1971.
139. Albert, A.: Nature 172:201, 1953.
140. Jackson, F. L.: Mode of action of tetracyclines. In Schnitzer, R. J., Hawking, F. (eds.). Experimental Chemotherapy, Vol. 3, p. 103. New York, Academic Press, 1964.
141. Bodley, J. W., Zieve, P. J.: Biochem. Biophys. Res. Commun. 36:463, 1969.
142. Dockter, M. E., Magnuson, A.: Biochem. Biophys. Res. Commun. 42:471, 1973.
143. Izaka, K., Arima, K.: Nature 200:384, 1963.
144. Sompolinsky, D., Krausz, J.: Antimicrob. Agents Chemother. 4:237, 1973.
145. Durckheimer, W.: Angew. Chem. Int. Ed. 14:721, 1975.
146. Brown, J. R., Ireland, D. S.: Adv. Pharmacol. Chemother. 15:161, 1978.
147. Mitscher, L. A.: The Chemistry of Tetracycline Antibiotics. New York, Marcel Dekker, 1978.
148. Cline, D. L. J.: Chemistry of tetracyclines. Q. Rev. 22:435, 1968.
149. Hlavka, J. J., Boothe, J. H. (eds.): The Tetracyclines. New York, Springer-Verlag, 1985.
150. Cammarata, A., Yau, S. J.: J. Med. Chem. 13:93, 1970.
151. Glatz, B., et al.: J. Am. Chem. Soc. 101:2171, 1979.
152. Schach von Wittenau, M., et al.: J. Am. Chem. Soc. 84:2645, 1962.
153. Stephens, C. R., et al.: J. Am. Chem. Soc. 85:2643, 1963.
154. McCormick, J. R. D., et al.: J. Am. Chem. Soc. 79:4561, 1957.
155. Martell, M. J., Jr., Booth, J. H.: J. Med. Chem. 10:44, 1967.
156. Colazzi, J. L., Klink, P. R.: J. Pharm. Sci. 58:158, 1969.
157. Schumacher, G. E., Linn, E. E.: J. Pharm. Sci. 67:1717, 1978.
158. Schach von Wittenau, M.: Chemotherapy 13S:41, 1968.
159. Teare, E. L., et al.: Drugs Exp. Clin. Res. 7:307, 1981.
160. Kirchlechner, R., Rogalalaski, W.: Tetrahedron Lett. 247, 251, 1979.
161. Conover, L. H.: U.S. Patent 2,699,054, Jan. 11, 1955.
162. Bunn, P. A., Cronk, G. A.: Antibiot. Med. 5:379, 1958.
163. Gittinger, W. C., Weinger, H.: Antibiot. Med. 7:22, 1960.
164. Remmers, E. G., et al.: J. Pharm. Sci. 53:1452, 1534, 1964.
165. Remmers, E. G., et al.: J. Pharm. Sci. 54:49, 1965.
166. Duggar, B. B.: Ann. N.Y. Acad. Sci. 51:177, 1948.
167. Finlay, A. C., et al.: Science 111:85, 1950.
168. Hochstein, F. A., et al.: J. Am. Chem. Soc. 75:5455, 1953.
169. Blackwood, R. K., et al.: J. Am. Chem. Soc. 83:2773, 1961.
170. Stephens, C. R., et al.: J. Am. Chem. Soc. 80:5324, 1958.
171. Minuth, J. N.: Antimicrob. Agents Chemother. 6:411, 1964.
172. Wiley, P. F.: Res. Today (Eli Lilly & Co.) 16:3, 1960.
173. Miller, M. W.: The Pfizer Handbook of Microbial Metabolites. New York, McGraw–Hill, 1961.
174. Morin, R., Gorman, M.: Kirk-Othmer Encyl. Chem. Technol., ed. 2, 12:637, 1967.
175. Mitscher, L. A., et al.: The Chemistry of Macrolide Antibiotics. New York, Marcel-Dekker, 1982.
176. McGuire, J. M., et al.: Antibiot. Chemother. 2:821, 1952.
177. Malmborg, A.-S.: J. Antibiot. Chemother. 18:293, 1986.
178. Wiley, P. F.: J. Am. Chem. Soc. 79:6062, 1957.
179. Celmer, W. D.: J. Am. Chem. Soc. 87:1801, 1965.
180. Corey, E. J., et al.: J. Am. Chem. Soc. 101:7131, 1979.
181. Stephens, C. V., et al.: J. Antibiot. 22:551, 1969.
182. Bechtol, L. D., et al.: Curr. Ther. Res. 20:610, 1976.
183. Welling, P. G., et al.: J. Pharm. Sci. 68:150, 1979.
184. Yakatan, G. J., et al.: J. Clin. Pharmacol. 20:625, 1980.
185. Tjandramaga, T. B., et al.: Pharmacology 29:305, 1984.
186. Tardew, P. L., et al.: Appl. Microbiol. 18:159, 1969.
187. Wiegand, R. G., Chun, A. H.: J. Pharm. Sci. 61:425, 1972.
188. Janicki, R. S., et al.: Clin. Pediatr. 14:1098, 1975.
189. Nicholas, P.: N.Y. State J. Med. 77:2088, 1977.
190. Wilhelm, J. M., et al.: Antimicrob. Agents Chemother. 236, 1967.
191. Gutman, L. T., et al.: Lancet 1:464, 1967.
192. Lai, C. J., Weisblum, B.: Proc. Soc. Natl. Acad. Sci. USA 68:856, 1971.
193. Sobin, B. A., et al.: Antibiotics Annual 1954–1955, p. 827. New York, Medical Encyclopedia, 1955.
194. Hochstein, F. A., et al.: J. Am. Chem. Soc. 82:3227, 1960.
195. Celmer, W. D.: J. Am. Chem. Soc. 87:1797, 1965.
196. Mason, D. J., et al.: Antimicrob. Agents Chemother. 544, 1962.
197. Hoeksema, H., et al.: J. Am. Chem. Soc. 86:4223, 1964.
198. Slomp, G., MacKellar, F. A.: J. Am. Chem. Soc. 89:2454, 1967.
199. Howarth, G. B., et al.: J. Chem. Soc. C 2218, 1970.
200. Magerlein, B. J.: Tetrahedron Lett. 685, 1970.
201. Argoudelis, A. D., et al.: J. Am. Chem. Soc. 86:5044, 1964.
202. Magerlein, B. J., et al.: J. Med. Chem. 10:355, 1967.
203. Bannister, B.: J. Chem. Soc. Perkin Trans. 1:1676, 1973.
204. Bannister, B.: J. Chem. Soc. Perkin Trans. 1:3025, 1972.
205. Bartlett, J. G.: Rev. Infect. Dis. 1:370, 1979.
206. Sarges, R., Witkop, B.: J. Am. Chem. Soc. 86:1861, 1964.
207. Bodanszky, M., Perlman, D.: Science 163:352, 1969.
208. Dubos, R. J.: J. Exp. Med. 70:1, 1939.
209. Paladini, A., Craig, L. C.: J. Am. Chem. Soc. 76:688, 1954; King, T. P., Craig, L. C.: J. Am. Chem. Soc. 77:6627, 1955.
210. Ohno, M., et al.: Bull. Soc. Chem. Jpn. 39:1738, 1966.
211. Finkelstein, A., Anderson, O. S.: J. Membr. Biol. 59:155, 1981.
212. Johnson, B. A., et al.: Science 102:376, 1945.
213. Stoffel, W., Craig, L. C.: J. Am. Chem. Soc. 83:145, 1961.
214. Ressler, C., Kashelikar, D. V.: J. Am. Chem. Soc. 88:2025, 1966.
215. Benedict, R. G., Langlykke, A. F.: J. Bacteriol. 54:24, 1947.
216. Stansly, P. J., et al.: Bull. Johns Hopkins Hosp. 81:43, 1947.
217. Ainsworth, G. C., et al.: Nature 160:263, 1947.
218. Vogler, K., Studer, R. O.: Experientia 22:345, 1966.
219. Hausmann, W., Craig, L. C.: J. Am. Chem. Soc. 76:4892, 1952.
220. Volger, K., et al.: Experientia 20:365, 1964.
221. Koyama, Y., et al.: J. Antibiot. [A]3:457, 1950.
222. Suzuki, T., et al.: J. Biochem. 54:414, 1963.
223. Wilkinson, S., Lowe, L. A.: J. Chem. Soc. 4107, 1964.
224. Ehrlich, J., et al.: Science 106:417, 1947.
225. Controulis, J., et al.: J. Am. Chem. Soc. 71:2463, 1949.
226. Long, L. M., Troutman, H. D.: J. Am. Chem. Soc. 71:2473, 1949.
227. Szulcewski, D., Eng, F.: Anal. Profiles Drug Subst. 4:47, 1972.
228. Glazko, A.: Antimicrob. Agents Chemother. 655, 1966.
229. Hansch, C., et al.: J. Med. Chem. 16:917, 1973.
230. Brock, T. D.: Chloramphenicol. In Schnitzer, R. J., Hawking, F. (eds.). Experimental Chemotherapy, Vol. 3, p. 119. New York, Academic Press, 1964.
231. Shalit, I., Marks, M. I.: Drugs 28:281, 1984.
232. Kauffman, R. E., et al.: J. Pediatr. 99:963, 1981.
233. Kramer, W. G., et al.: J. Clin. Pharmacol. 24:181, 1984.
234. McCormick, M. H., et al.: Antibiotics Annual 1955–1956, p. 606, New York, Medical Encyclopedia, 1956.
235. Sheldrick, G. M., et al.: Nature 271:233, 1978.
236. Williamson, M. P., Williams, D. H.: J. Am. Chem. Soc. 103:6580, 1981.
237. Harris, C. M., et al.: J. Am. Chem. Soc. 105:6915, 1983.
238. Perkins, H. R., Nieto, M.: Ann. N.Y. Acad. Sci. 235:348, 1974.
239. Williamson, M. P., et al.: Tetrahedron Lett. 40:569, 1984.

240. Shunk, C. H., et al.: J. Am. Chem. Soc. 78:1770, 1956.
241. Hoeksma, H., et al.: J. Am. Chem. Soc. 78:2019, 1956.
242. Spencer, C. H., et al.: J. Am. Chem. Soc. 78:2655, 1956.
243. Spencer, C. H., et al.: J. Am. Chem. Soc. 80:140, 1958.
244. Gellert, M., et al.: Proc. Natl. Acad. Sci. USA 73:4474, 1976.
245. Sugino, A., et al.: Proc. Natl. Acad. Sci. USA 75:4842, 1978.

SELECTED READINGS

Bryan, J. E. (ed.): Antimicrobial Drug Resistance. New York, Academic Press, 1984.

Demain, A. L., Solomon, N. A. (eds.): Antibiotics Containing the β-Lactam Structure, Vols. 1 and 2. New York, Springer-Verlag, 1983.

Drusano, G. L., et al.: Extended spectrum penicillins. Rev. Infect. Dis. 6:13, 1984.

Gale, E. F., et al.: Molecular Basis of Antibiotic Action, 2nd ed. New York, John Wiley & Sons, 1981.

Gordon, E. M., Sykes, R. B. (eds.): Chemistry and Biology of β-Lactam Antibiotics. New York, Academic Press, 1982.

Hlavka, J. J., Boothe, J. H. (eds.): The Tetracyclines. New York, Springer-Verlag, 1985.

Hollstein, U.: Nonlactam antibiotics. In Wolff, M. E. (ed.). Burger's Medicinal Chemistry, Part II, 4th ed., p. 173. New York, Wiley-Interscience, 1979.

Hoovey, J. R. E., Dunn, G. L.: The β-Lactam antibiotics. In Wolff, M. E. (ed.). Burger's Medicinal Chemistry, Part II, 4th ed., p. 83. New York, Wiley-Interscience, 1979.

Mandell, G. L., Douglas, R. G., Bennett, J. E. (eds.): Principles and Practice of Infection Diseases, 2nd ed., New York, John Wiley & Sons, 1985.

Pratt, W. B., Fekety, R.: The Antimicrobial Drugs. New York, Oxford University Press, 1986.

Umezawa, H., Hooper, I. R. (eds.): Aminoglycoside Antibiotics. New York, Springer-Verlag, 1982.

Verderame, M. (ed.): Handbook of Chemotherapeutic Agents, Vol. 1. Boca Raton, FL, CRC Press, 1986.

CHAPTER **8**

Antineoplastic Agents

William A. Remers

The chemotherapy of neoplastic disease has become increasingly important in recent years. An indication of this importance is the establishment of a medical specialty in oncology, wherein the physician practices various protocols of adjuvant therapy. Most cancer patients now receive some form of chemotherapy, even though often it is merely palliative.

Cancer chemotherapy has received no spectacular breakthrough of the kind that the discovery of penicillin provided for antibacterial chemotherapy. However, there has been substantial progress in many aspects of cancer research. In particular, an increased understanding of tumor biology has led to elucidation of the mechanisms of action for antineoplastic agents. It also has provided a basis for the more rational design of new agents. Recent advances in clinical techniques, including large cooperative studies, are allowing more rapid and reliable evaluation of new drugs. The combination of these advantages with improved preliminary screening systems is enhancing the emergence of newer and more potent compounds.

At least ten different neoplasms can now be "cured" by chemotherapy in most patients. *Cure* is defined here as an expectation of normal longevity. These neoplasms are acute leukemia in children, Burkitt's lymphoma, choriocarcinoma in women, Ewing's sarcoma, Hodgkin's disease, lymphosarcoma, mycosis fungoides, rhabdomyosarcoma, retinoblastoma in children, and testicular carcinoma.[1] Unfortunately, only these relatively rare neoplasms are readily curable. Considerable progress is being made in the treatment of breast cancer by combination drug therapy; however, for carcinoma of the pancreas, colon, liver, or lung (except small-cell carcinoma) the outlook is bleak. Short-term remissions are the best that can be expected for most patients with these diseases.

There are cogent reasons why cancer is more difficult to cure than bacterial infections. One is that there are qualitative differences between human and bacterial cells. For example, bacterial cells have distinctive cell walls and their ribosomes are different from those of human cells. In contrast, the differences between normal and neoplastic human cells are merely quantitative. Another difference is that immune mechanisms and other host defenses are very important in killing bacteria and other foreign cells, whereas they play a negligible role in killing cancer cells. By their very nature, the cancer cells have eluded or overcome the immune surveillance system of the body. Thus, it is necessary for chemotherapeutic agents to kill every single clonogenic malignant cell, because even one can reestablish the tumor. This kind of kill is extremely difficult to effect because antineoplastic agents kill cells by first-order kinetics. That is, they kill a constant fraction of cells. Suppose that a patient had a trillion leukemia cells. This amount would cause a serious debilitation. A potent anticancer drug might reduce this population 10,000-fold, in which case the symptoms would be alleviated and the patient would be in a state of remission. However, the remaining 100 million leukemia cells could readily increase to the original number after cessation of therapy. Furthermore, a higher proportion of resistant cells would be present, which would mean that retreatment with the same agent would achieve a smaller response than before. Hence, multidrug regimens are used to reduce drastically the number of neoplastic cells. Typical protocols for leukemia contain four different anticancer drugs, usually with different modes of action. The addition of immunostimulants to the therapeutic regimen helps the body's natural defense mechanisms identify and eliminate the remaining few cancer cells.

A further complication to chemotherapy is the relative unresponsiveness of slow-growing solid tumors. Current antineoplastic agents are most effective against cells with a high-growth fraction. They act to block the biosynthesis or transcription of nucleic acids or to prevent cell division by interfering with mitotic spindles. Cells in the phases of synthesis or mitosis are highly susceptible to these agents. In contrast, cells in the resting state are resistant to many agents. Slow-growing tumors characteristically have many cells in the resting state.[2]

Most antineoplastic drugs are highly toxic to the patient and must be administered with extreme caution. Some of them require a clinical setting in which supportive care is available. The toxicity usually involves rapidly proliferating tissues such as bone marrow and the intestinal epithelium. However, individual drugs produce distinctive toxic effects on the heart, lungs, kidneys, and other organs. Chemotherapy is seldom the initial treatment used against cancer. If the cancer is well-defined and accessible, surgery is the preferred method. Skin cancers and certain localized tumors are treated by radiotherapy. Even some widely disseminated tumors such as Hodgkin's disease are treated by radiation, although chemotherapy might be equally effective. Generally, chemotherapy is important when the tumor is inoperable or when metastasis has occurred. Chemotherapy is finding increasing use after surgery to ensure that no cells remain to regenerate the parent tumor.

The era of chemotherapy of malignant disease was born in 1941, when Huggins demonstrated that the administration of estrogens produced regressions of metastatic prostate cancer.[3] In the following year, Gilman and others began clinical studies on the nitrogen mustards and discovered that mechlorethamine was effective against Hodgkin's disease and lymphosarcoma.[4] These same two diseases were treated with cortisone acetate in 1949, and dramatic, although temporary, remissions were observed.[5] The next decade was marked by the design and discovery of antimetabolites: methotrexate in 1949, 6-mercaptopurine in 1952, and 5-fluorouracil in 1957. Additional alkylating agents, such as melphalan and cyclophosphamide, were developed during this period, and the activity of natural products such as actinomycin, mitomycin C, and the vinca alkaloids was discovered. During the 1960s progress continued in all of these areas with the discovery of cytarabine (cytosine arabinoside), bleomycin, doxorubicin, and carmustine. Novel structures such as procarbazine, dacarbazine, and cisplatin complexes were found to be highly active. In 1965, Kennedy reported that remissions occurred in 30% of postmenopausal women with metastatic breast cancer upon treatment with high doses of estrogen.[6]

Much of the leadership and financial support for the development of antineoplastic drugs derives from the National Cancer Institute. In 1955, this organization established the Cancer Chemotherapy National Service Center (now the Division of Cancer Treatment) to coordinate a national voluntary cooperative cancer chemotherapy program. By 1958, this effort had evolved into a targeted drug development program. A massive screening system was established to discover new lead compounds, and thousands of samples have been submitted. Currently, the primary screen is P388 lymphocytic leukemia in mice. Compounds active in this screen are tested further against lymphoid leukemia, melanoma, and lung carcinoma in mice. A panel of human tumor xenographs in mice is available for additional preliminary screening.[7] In the future, it is anticipated that the primary assay protocol will involve a panel of cloned human cancer cells. This kind of assay can be automated and, in principle, it should be more predictive of clinical anticancer activity. Compounds of significant interest are subjected to preclinical pharmacologic and toxicologic evaluation in mice and dogs. Clinical trials are generally underwritten by the National Cancer Institute (NCI). They involve three discrete phases. Phase I is the clinical pharmacology stage. The dosage schedule is developed, and toxicity parameters are established in it. Phase II involves the determination of activity against a "signal" tumor panel, which includes both solid and hematologic types.[8] A broad-based multicenter study is usually undertaken in Phase III. It features randomization schemes designed to validate the efficacy of the new drug statistically in comparison with alternative modalities of therapy. As might be anticipated, the design of clinical trials for antineoplastic agents is complicated, especially in the matter of controls. Ethical considerations do not permit patients to be left untreated if any reasonable therapy is possible.

Several pharmaceutical industry laboratories and foreign institutions have made significant contributions to the development of anticancer drugs. Frequently, their research is in collaboration with the NCI Division of Cancer Treatment.

ALKYLATING AGENTS

Toxic effects of sulfur mustard and ethylenimine on animals were described in the 19th century.[9] The powerful vesicant action of sulfur mustard led to its use in World War I, and medical examination of the victims revealed that tissues were damaged at sites distant from the area of contact.[10] Systemic effects included leukopenia, bone marrow aplasia, lymphoid tissue suppression, and ulceration of the gastrointestinal tract. Sulfur mustard was shown to be active against animal tumors, but it was too nonspecific for clinical use. A variety of nitrogen mustards were synthesized between the two world wars. Some of these compounds, for example, mechlorethamine, showed selective toxicity, especially to lymphoid tissue. This observation led to the crucial suggestion that nitrogen mustards be tested against tumors of the lymphoid system in animals. Success in this area was followed by cautious human trials that showed methchlorethamine to be useful against Hodgkin's disease and certain lymphomas. This work was classified during World War II, but was finally published in a classic paper by Gilman and Philips in 1946.[4] In this paper, the chemical transformation of nitrogen and sulfur mustards to cyclic "onium" cations was described, and the locus of their interaction with cancer cells was established to be the nucleus. The now familiar pattern of toxicity to rapidly proliferating cells in bone marrow and the gastrointestinal tract was established.

Alkylation is defined as the replacement of hydrogen on an atom by an alkyl group. The alkylation of nucleic acids or proteins involves a substitution reaction in which a nucleophilic atom (nu) of the biopolymer displaces a leaving group from the alkylating agent.

$$\text{nu-H} + \text{alkyl-Y} \longrightarrow \text{alkyl-nu} + H^+ + Y^-$$

The reaction rate depends on the nucleophilicity of the atom (S, N, O), which is greatly enhanced if the nucleophile is ionized. A hypothetical order of reactivity at physiologic *p*H would be ionized thiol, amine, ionized phosphate, and ionized carboxylic acid.[11] Rate differences among various amines would depend on the degree to which they are protonated and their conjugation with other functional groups. The N-7 position of guanine in DNA (see Scheme 5) is strongly nucleophilic. Reaction orders depend on the structure of the alkylating agent. Methanesulfonates, epoxides, and aziridines give second-order reactions that depend on concentrations of the alkylating agent and nucleophile. The situation is more complex with β-haloalkylamines (nitrogen mustards) and β-haloalkylsulfides (sulfur mustards) because these molecules undergo neighboring-group reactions in which the nitrogen or sulfur atom displaces the halide to give strained, 3-membered onium intermediates. These onium ions react with nucleophiles in second-order processes. However, the overall reaction kinetics depend on the relative rates of the two steps. For mechlorethamine, the aziridinium ion is rapidly formed in water, but reaction with biologic nucleophiles is slower. Thus, the kinetics will be second order.[12]

$$CH_3N(CH_2CH_2Cl)_2 \xrightarrow{\text{very fast}} CH_3\overset{+}{N}(\triangle)CH_2CH_2Cl \xrightarrow[\text{moderate}]{\text{nu-H}} CH_3N(CH_2CH_2Cl)(CH_2CH_2nu)$$

In contrast, sulfur mustard forms the less stable episulfonium ion more slowly than this ion reacts with biologic nucleophiles. Thus, the neighboring-group reaction is rate-limiting and the kinetics are first order.[13]

$$S(CH_2CH_2Cl)_2 \xrightarrow{\text{slow}} \overset{+}{S}(\triangle){-}CH_2CH_2Cl \xrightarrow[\text{moderate}]{\text{nu-H}} S(CH_2CH_2Cl)(CH_2CH_2nu)$$

Aryl-substituted nitrogen mustards, such as chlorambucil, are relatively stable toward aziridinium ion formation because the aromatic ring decreases the nucleophilicity of the nitrogen atom. These mustards react according to first-order kinetics.[13] The stability of chlorambucil allows it to be taken orally, whereas mechlorethamine is given by intravenous administration of freshly prepared solutions. The requirement for freshly prepared solutions is based on the gradual decomposition of the aziridinium ion by interaction with water.

$$HO_2C(CH_2)_3{-}C_6H_4{-}N(CH_2CH_2Cl)_2$$

$$CH_3\overset{+}{N}(\triangle)CH_2CH_2Cl \xrightarrow{H_2O} CH_3N(CH_2CH_2Cl)(CH_2CH_2OH) \rightarrow \rightarrow CH_3N(CH_2CH_2OH)_2$$

Ethylenimines and epoxides are strained ring systems, but they do not react as readily as aziridinium or episulfonium ions with nucleophiles. Their reactions are second order and are enhanced by the presence of acid.[11]

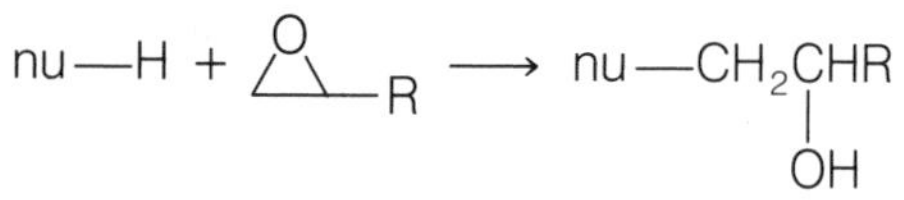

$$\text{nu—H} + \text{(epoxide-R)} \longrightarrow \text{nu—CH}_2\text{CH(OH)R}$$

Examples of antitumor agents containing ethylenimine groups are triethylene melamine and thiotepa.

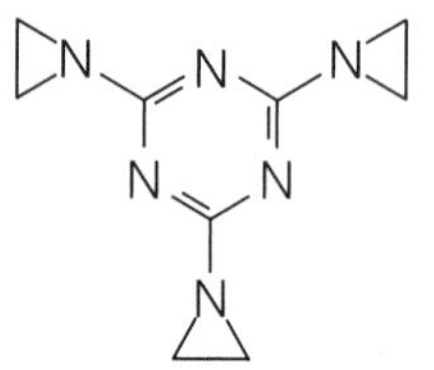

Triethylene Melamine 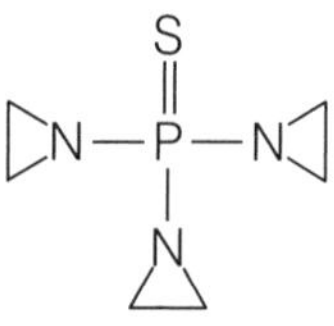   Thiotepa

The use of epoxides as cross-linking agents in textile chemistry suggested that they be tried in cancer chemotherapy. Simple diepoxides such as erythritol anhydride (1,2 : 3,4-diepoxybutane) showed clinical activity against Hodgkin's disease,[14] but none of these compounds became an established drug. Mitobromitol (dibromomannitol) is presently under clinical study. It gives the corresponding

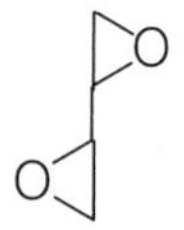

Erythritol Anhydride

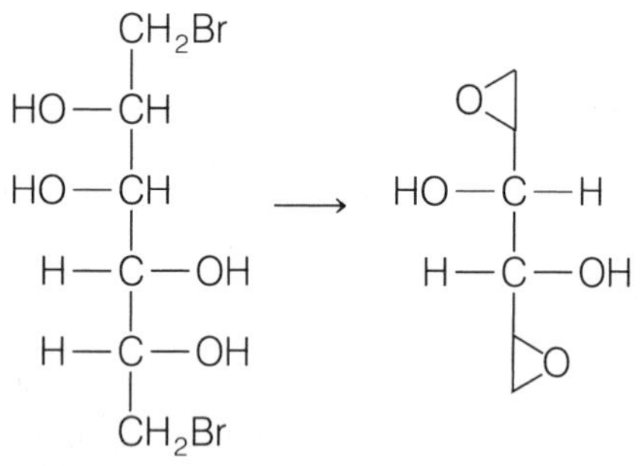

Mitobromitol   Dianhydro-D-mannitol

diepoxide upon continuous titration at *p*H 8. This diepoxide (1,2 : 5,6-dianhydro-D-mannitol) shows potent alkylating activity against experimental tumors.[15] Thus, it is supposed that mitobromitol and related compounds such as mitolactol (dibromodulcitol) act by way of the diepoxides. However, this reaction sequence has not yet been verified in vivo.

A somewhat different type of alkylating agent is the *N*-alkyl-*N*-nitrosourea. Compounds of this class are unstable in aqueous solution under physiologic conditions. They produce carbonium ions (also called carbenium ions) that can alkylate, and isocyanates that can carbamoylate. For example, methylnitrosourea decomposes initially to form isocyanic acid and methyldiazohydroxide. The latter species decomposes further to methyldiazonium ion and finally to methyl carbonium ion, the ultimate alkylating species.[16]

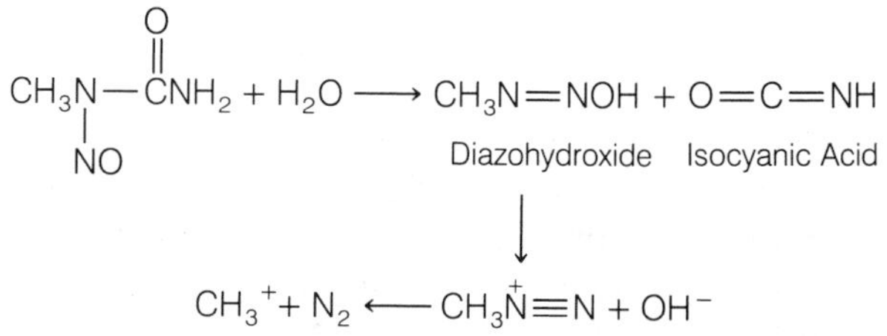

$$CH_3N(NO)—C(=O)NH_2 + H_2O \longrightarrow CH_3N{=}NOH + O{=}C{=}NH$$

Diazohydroxide   Isocyanic Acid

$$CH_3^+ + N_2 \longleftarrow CH_3\overset{+}{N}{\equiv}N + OH^-$$

Substituents on the nitrogen atoms of the nitrosourea influence the mechanism of decomposition in water, which determines the species generated and controls the biologic effects. Carmustine (BCNU) undergoes an abnormal decomposition in which the urea oxygen displaces a chlorine to give a cyclic intermediate (Scheme 1). This intermediate decomposes to vinyl diazohydroxide, the precursor to vinylcarbonium ion, and 2-chloroethylisocyanate. The latter species gives 2-chloroethylamine, an additional alkylating agent.[16]

Some clinically important alkylating agents are not active until they have been transformed by metabolic processes. The leading example of this

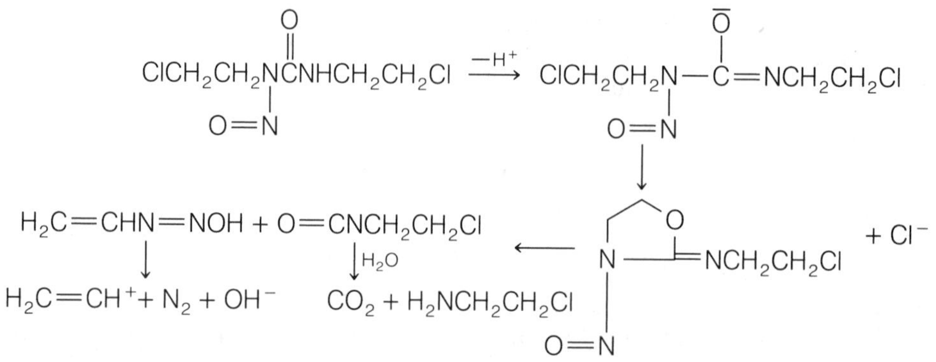

$$ClCH_2CH_2N(N{=}O)C(=O)NHCH_2CH_2Cl \xrightarrow{-H^+} ClCH_2CH_2N(N{=}O)—C(\bar{O}^-){=}NCH_2CH_2Cl$$

$$\longrightarrow \text{cyclic intermediate (N-nitroso oxazolidine}{=}NCH_2CH_2Cl) + Cl^-$$

$$\longrightarrow H_2C{=}CHN{=}NOH + O{=}CNCH_2CH_2Cl$$

$$H_2C{=}CHN{=}NOH \longrightarrow H_2C{=}CH^+ + N_2 + OH^-$$

$$O{=}CNCH_2CH_2Cl \xrightarrow{H_2O} CO_2 + H_2NCH_2CH_2Cl$$

Scheme 1. *Decomposition of carmustine.*

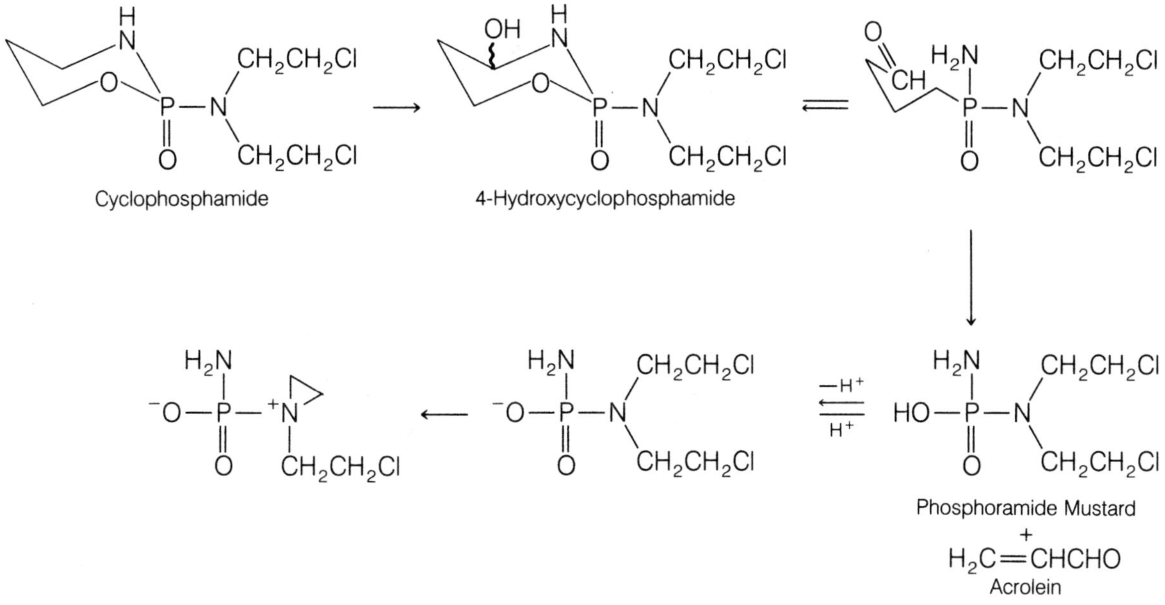

Cyclophosphamide → 4-Hydroxycyclophosphamide ⇌ Phosphoramide Mustard + $H_2C{=}CHCHO$ Acrolein

Scheme 2. *Activation of cyclophosphamide.*

group is cyclophosphamide, which is converted by hepatic cytochrome P-450 into the corresponding 4-hydroxy derivative by way of the 4-hydroperoxy intermediate (Scheme 2). The 4-hydroxy derivative is a carbinolamine in equilibrium with the open-chain aminoaldehyde form. Nonenzymatic decomposition of the latter form generates phosphoramide mustard and acrolein. Recent studies[17] that are based on nuclear magnetic resonance with ^{31}P (^{31}P-NMR) have shown that the conjugate base of phosphoramide mustard cyclizes to the aziridinium ion,[18] which is the principal cross-linking alkylator formed from cyclophosphamide. The maximal rate of cyclization occurs at *p*H 7.4. It was suggested that selective toxicity toward certain neoplastic cells might be based on their abnormally low *p*H. This would afford a slower formation of aziridinium ions, and they would persist longer because of decreased inactivation by hydroxide ions.[17]

Cyclophosphamide has been resolved and the enantiomers have been tested against tumors. The levorotatory form has twice the therapeutic index of the dextrorotatory form.[19]

Ifosfamide (iphosphamide), an isomer of cyclophosphamide in which one of the 2-chloroethyl substituents is on the ring nitrogen, also has potent antitumor activity. It must be activated by hepatic enzymes, but its metabolism is slower than that of cyclophosphamide.[20]

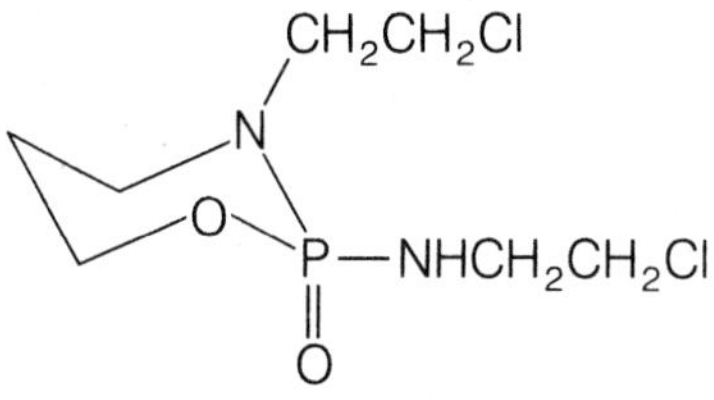

Ifosfamide

Other examples of alkylating species are afforded by carbinolamines, as found in maytansine, and vinylogous carbinolamines, as found in certain pyrrolizine diesters.[21]

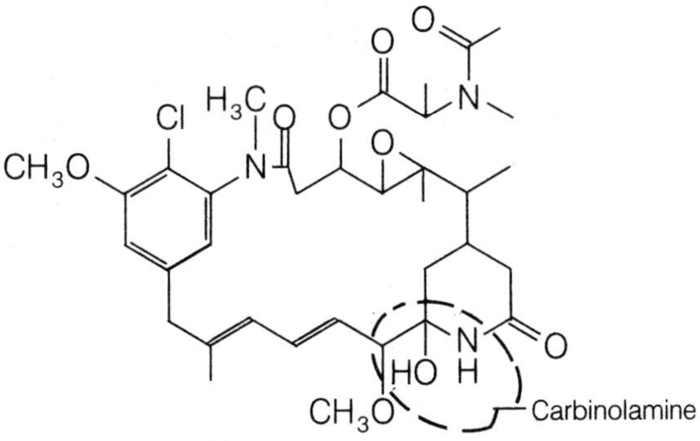

Maytansine (Carbinolamine)

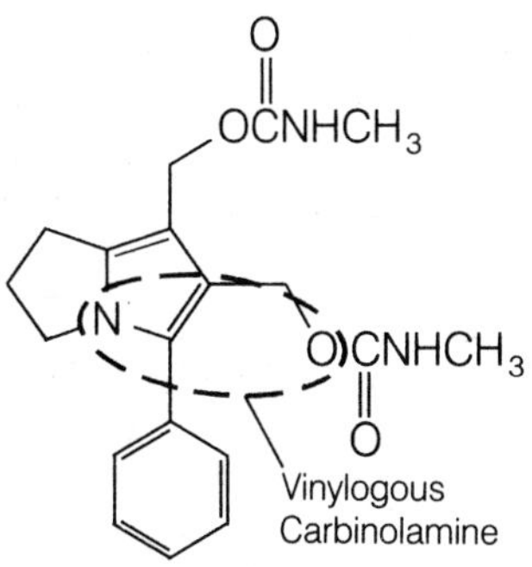

Pyrrolizine Diester (Vinylogous Carbinolamine)

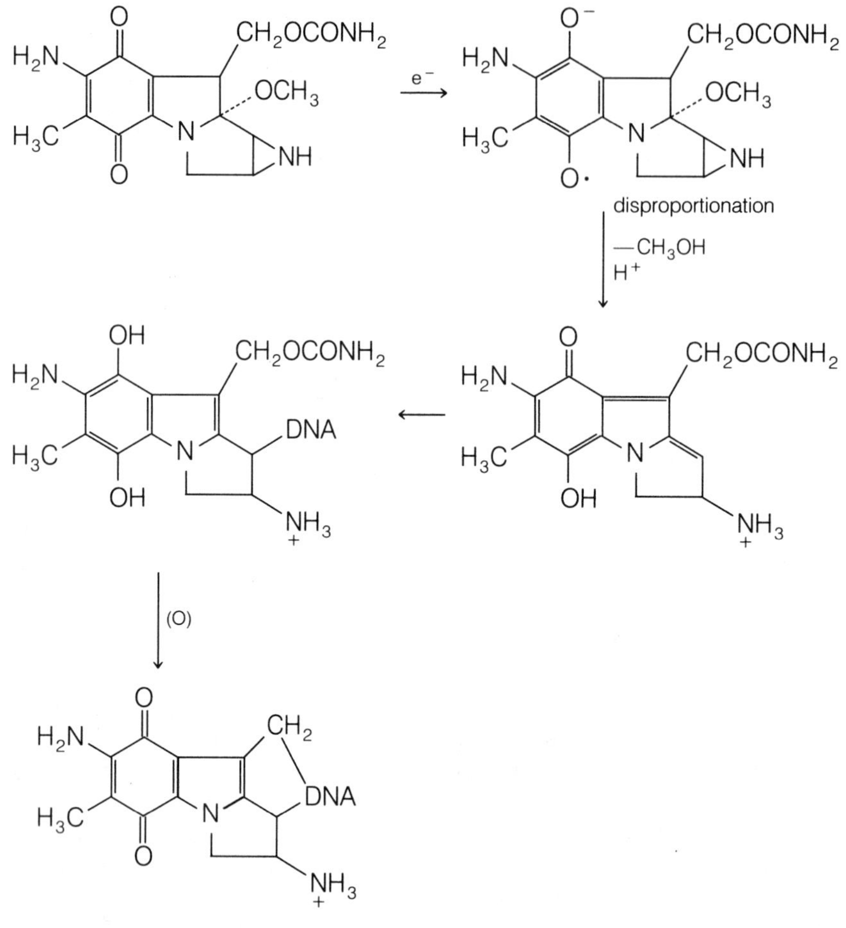

Scheme 3. Mitomycin C activation and DNA alkylation.

When mitomycin (mitomycin C) is reduced enzymatically to its semiquinone radical, which disproportionates to the quinone and hydroquinone in water, the spontaneous elimination of methanol affords the vinylogous carbinolamine system. Loss of the carbamoyloxy group from this system gives a stabilized carbonium ion that is capable of alkylating DNA (Scheme 3). The aziridine ring of mitomycin provides a second alkylating group that allows mitomycin to cross-link double helical DNA.[22] Molecules such as mitomycin are said to act by "bioreductive alkylation."[23]

Another type of alkyating species occurs in α,β-unsaturated carbonyl compounds. These compounds can alkylate nucleophiles by conjugate addition. Although there are no established clinical agents of this type, many natural products active against experimental tumors contain α-methylene lactone or α,β-unsaturated ketone functionalities. For example, the sesquiterpene helenalin has both of these systems.[24]

$$\text{nu—H} + H_2C{=}CH\overset{O}{\overset{\|}{C}}R \rightarrow \text{nu}CH_2CH_2\overset{O}{\overset{\|}{C}}R$$

Alkylation can also occur by free radical reactions. The methylhydrazines are a chemical class prone to decomposition in this manner. These compounds were tested as antitumor agents in 1963, and one of them, procarbazine, had a pronounced, but rather specific, effect on Hodgkin's disease.[25] Procarbazine is relatively stable at pH 7, but air oxidation to azoprocarbazine occurs readily in the presence of metalloproteins. Isomerization of this azo compound to the corresponding hydrazone, followed by hydrolysis, gives methylhydrazine and p-formyl-N-isopropylbenzamide. The formation of methylhydrazine from procarbazine has been demonstrated in living organisms.[26] Methylhydrazine is known to be oxidized to methyldiazine, which can decompose to nitrogen, methyl radical, and hydrogen radical.[27]

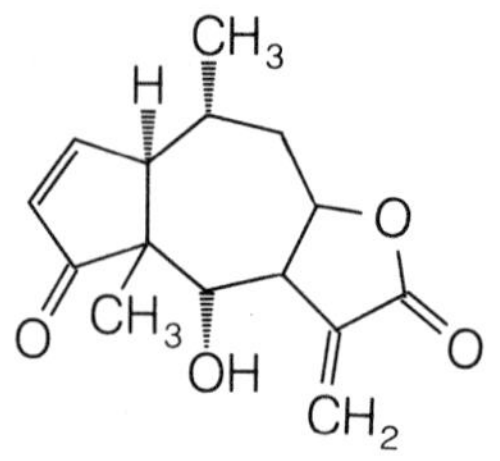

Helenalin

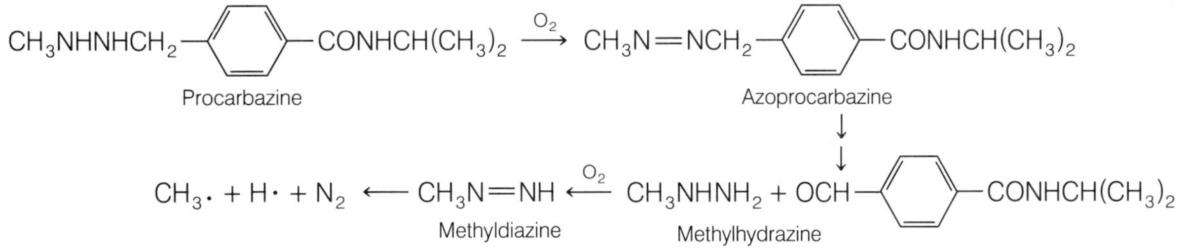

$$CH_3NHNHCH_2-C_6H_4-CONHCH(CH_3)_2 \xrightarrow{O_2} CH_3N{=}NCH_2-C_6H_4-CONHCH(CH_3)_2$$

Procarbazine → Azoprocarbazine

$$CH_3\cdot + H\cdot + N_2 \longleftarrow CH_3N{=}NH \xleftarrow{O_2} CH_3NHNH_2 + OCH-C_6H_4-CONHCH(CH_3)_2$$

Methyldiazine ← Methylhydrazine

The methyl group of procarbazine is incorporated intact into cytoplasmic RNA.[28] However, the methylating species has not been conclusively established. Formation of the methyl radical seems certain because methane is generated. The metabolism of procarbazine is a complex process involving more than one pathway. In humans, the conversion to azoprocarbazine is very rapid, with procarbazine having a half-life of seven to ten minutes. The major metabolite is *N*-isopropylterephathalamic acid and the *N*-methyl group appears as both carbon dioxide and methane.[29]

Dacarbazine was originally thought to be an antimetabolite because of its close resemblance to 5-aminoimidazole-4-carboxamide, an intermediate in purine biosynthesis (Scheme 4). However, it now appears to be an alkylating agent.[30] The isolation of an *N*-demethyl metabolite suggested that there might be a sequence in which this metabolite was hydrolyzed to methyldiazohydroxide, a precursor to methylcarbonium ion,[31] but this metabolite was less active than starting material against the Lewis lung tumor. An alternative mode of action was proposed in which dacarbazine undergoes acid-catalyzed hydrolysis to a diazonium ion, which can react in this form or decompose to the corresponding carbonium ion. Support for the latter mechanism was afforded by a correlation between the hydrolysis rates of phenyl-substituted dimethyltriazines and their antitumor activities.[32]

The interaction of alkylating agents with biopolymers has been studied extensively. However, no mode of action for the lethality to cancer cells has been conclusively established. A good working model has been developed for the alkylation of bacteria and viruses, but there are uncertainties in extrapolating it to mammalian cells. The present working hypothesis is that most alkylating agents produce cytotoxic, mutagenic, and carcinogenic effects by reacting with cellular DNA. They also react with RNA and proteins, but these effects are thought to be less significant.[33] The most active clinical alkylating agents are bifunctional compounds capable of cross-linking DNA. Agents such as methylnitrosourea that give simple alkylation are highly mutagenic relative to their cytotoxicity. The cross-linking process can be either interstrand or intrastrand. Interstrand links can be verified by a test that is based on the thermal denaturation and renaturation of DNA. When double-helical DNA is heated in water, it unwinds, and the strands separates. Renaturation, in which the strands recombine in the double helix, is slow and difficult. In contrast, if the two strands are cross-linked they cannot separate. Hence, they renaturate rapidly on cooling. Interstrand cross-linking occurs with mechlorethamine and other "two-armed" mustards, but busulfan appears to give intrastrand links, according to this test.[34]

In DNA, the 7-position (nitrogen) of guanine is especially susceptible to alkylation by mechloretha-

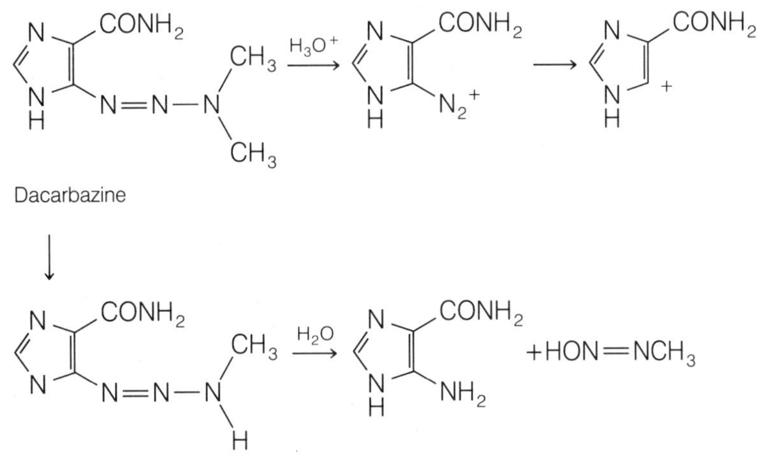

Dacarbazine (imidazole-4-carboxamide, 5-$N{=}N{-}N(CH_3)_2$) $\xrightarrow{H_3O^+}$ imidazole-4-carboxamide 5-diazonium (N_2^+) $\longrightarrow$ imidazole-4-carboxamide 5-carbonium ion (+)

Dacarbazine ↓ 5-$N{=}N{-}NH(CH_3)$ metabolite $\xrightarrow{H_2O}$ 5-aminoimidazole-4-carboxamide (NH_2) $+ HON{=}NCH_3$

Scheme 4. *Activation of dacarbazine.*

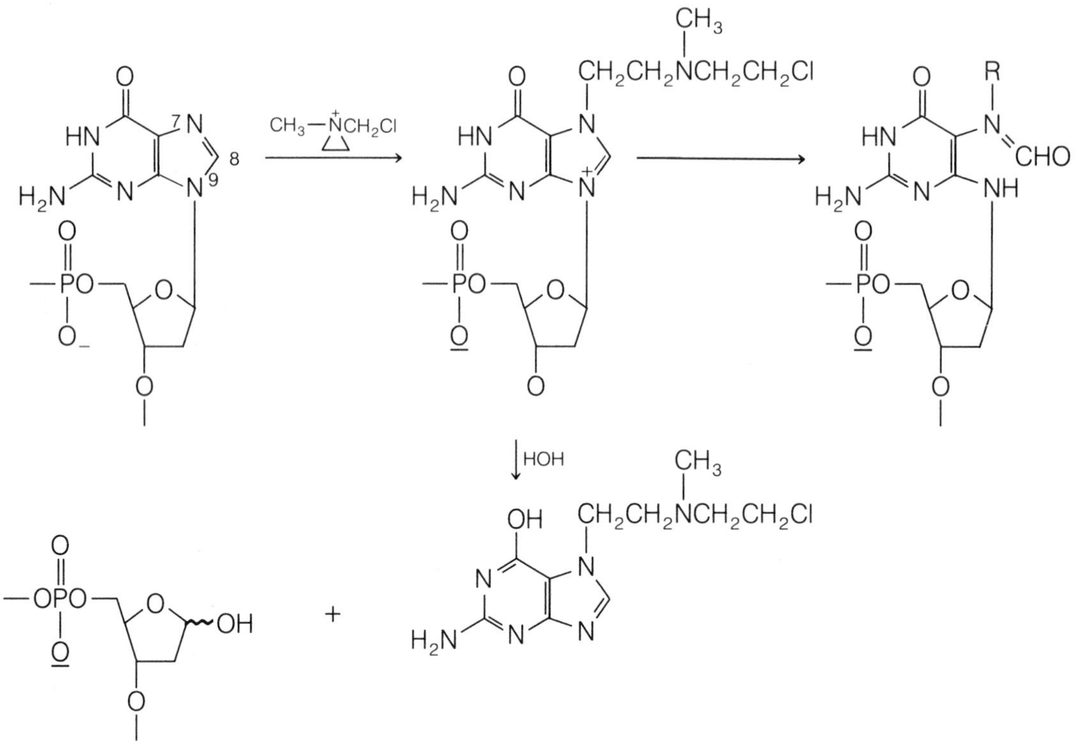

Scheme 5. Alkylation of guanine in DNA.

mine and other nitrogen mustards (Scheme 5).[35] The alkylated structure has a positive charge in its imidazole ring, which renders the guanine–ribose linkage susceptible to cleavage. This cleavage results in the deletion of guanine, and the resulting "apurinic acid" ribose–phosphate link is readily hydrolyzable. Alkylation of the imidazole ring also activates it to cleavage of the 8,9-bond.[11]

Other consequences of the positively charged purine structure are facile exchange of the 8-hydrogen, which can be used as a probe for 7-alkylation,[36] and a shift to the enolized pyrimidine ring as the preferred tautomer. The latter effect has been cited as a possible basis for abnormal base pairing in DNA replication, but this has not been substantiated. One example in which alkylation of guanine does lead to abnormal base pairing is the O-6-ethylation produced by ethyl methanesulfonate. This ethyl derivative pairs with thymine, whereas guanine normally pairs with cytosine.[37]

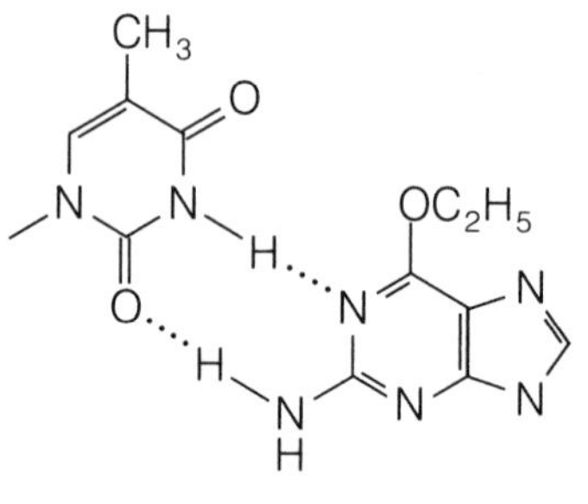

Other base positions of DNA attacked by alkylating agents are N-2 and N-3 of guanine, N-3, N-1, and N-7 of adenine, O-6 of thymine, and N-3 of cytosine. The importance of these minor alkylation reactions is difficult to assess. The phosphate oxygens of DNA are alkylated to an appreciable extent, but the significance of this feature is unknown.[38]

Guanine is also implicated in the cross-linking of double-helical DNA. Di(guanin-7-yl) derivatives have been identified among the products of reaction with mechlorethamine.[39] Busulfan alkylation has given 1′,4′-di(guanin-7-yl)-butane, but this product is considered to have resulted from intrastrand linking.[34] Enzymatic hydrolysis of DNA cross-linked by mitomycin has given fragments in which the antibiotic is covalently bound to the 2-amino groups of two guanosine residues, presumably from opposite strands of the double helix.[36]

Alkylating agents also interact with enzymes and other proteins. Thus, the repair enzyme, DNA nucleotidyltransferase of L1210 leukemia cells, was inhibited strongly by carmustine (BCNU), lomustine (CCNU), and 2-chloroethyl isocyanate. Because 1-(2-chloroethyl)-1-nitrosourea was a poor inhibitor of this enzyme, it was concluded that the main interaction with the enzyme was carbamoylation by the

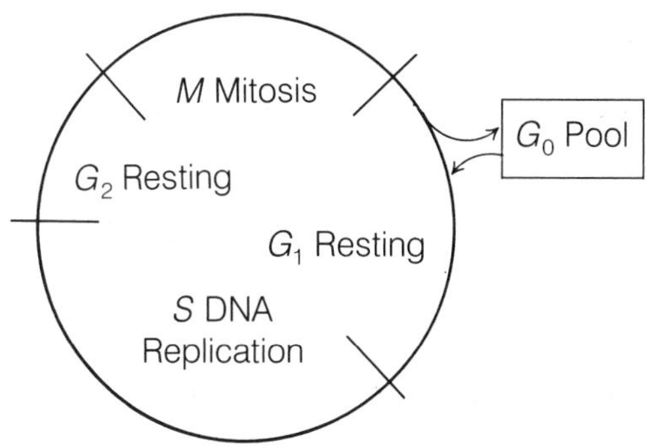

FIG. 8-1. The cell life cycle.

alkyl isocyanates generated in the decomposition of carmustine and lomustine.[40]

Alkylating agents can damage tissues with low mitotic indices, but they are most cytotoxic to rapidly proliferating tissues that have large proportions of cells in cycle. Nucleic acids are especially susceptible to alkylation when their structures are changed or unpaired in the process of replication. Thus, alkylating agents are most effective in the late G_1 or S phases. Alkylation may occur to some degree at any stage in the cell cycle, but the resulting toxicity is usually expressed when cells enter the S phase (Fig. 8-1). Progression through the cycle is blocked at G_2, the premitotic phase, and cell division fails.[41,42]

If cells can repair damage to their DNA before the next cell division, the effects of alkylation will not be lethal. Cells have developed a complex mechanism to accomplish this repair. First, a recognition enzyme discovers an abnormal region in the DNA. This recognition brings about the operation of an endonuclease, which makes a single-strand breakage in the DNA. An exonuclease then removes a small segment of DNA containing the damaged bases. Finally, the DNA is restored to its original structure by replacing the bases and rejoining the strand.[43,44] It is evident that tumor cells with efficient repair mechanisms will be relatively resistant to alkylating agents. Tumor cells outside the cell cycle in the resting phase (G_0) will have a rather long time to repair their DNA. Thus, slow-growing tumors should not respond well to alkylating agents. This limitation is observed clinically.

PRODUCTS

Mechlorethamine Hydrochloride, USP. 2,2-Dichloro-*N*-methyldiethylamine hydrochloride; nitrogen mustard; HN_2; NSC-762 (Mustargen). This compound is prepared by treating 2,2′-(methylimino)diethanol with thionyl chloride.[45] It occurs as hygroscopic leaflets that are very soluble in water. The dry crystals are stable at temperatures up to 40°C. They are very irritating to mucous membranes and harmful to eyes. The compound is supplied in rubber-stoppered vials containing a mixture of 10 mg mechlorethamine hydrochloride and 90 mg of sodium chloride. It is diluted with 10 mL of sterile water immediately before injection into a rapidly flowing intravenous infusion.

Mechlorethamine is effective in Hodgkin's disease. Current practice is to give it in combination with other agents. The combination with vincristine (Oncovin), procarbazine, and prednisone (the MOPP regimen) is considered the treatment of choice. Other lymphomas and mycosis fungoides can be treated with mechlorethamine. The most serious toxic reaction is bone marrow depression, which results in leukopenia and thrombocytopenia. Emesis is prevalent and lasts about eight hours. Nausea and anorexia persist longer. These gastrointestinal effects may be prevented by administering a short-acting barbiturate and prochlorperazine. Extravasation produces intense local reactions at the site of injection. If it occurs, the immediate application of sodium thiosulfate solution can protect the tissues, because thiosulfate ion reacts very rapidly with the aziridinium ion formed from mechlorethamine.

Cyclophosphamide, USP. *N,N*-Bis(2-chloroethyl)tetrahydro-2*H*-1,3,2-oxazaphosphorin-2-amine-2-oxide; NSC-26271 (Cytoxan). It is prepared by treating bis(2-chloroethyl)phosphoramide dichloride with propanolamine.[46] The monohydrate is a low-melting solid that is very soluble in water. It is supplied as 25- and 50-mg white tablets, 50-mg unit-dose cartons, and as a powder (100, 200, or 500 mg) in sterile vials.

Cyclophosphamide has advantages over other alkylating agents in that it is active orally and parenterally and can be given in fractionated doses over prolonged periods. It is active against multiple myeloma, chronic lymphocytic leukemia, and acute leukemia of children. In combination with other chemotherapeutic agents, it has given complete remissions and even cures in Burkitt's lymphoma and acute lymphoblastic leukemia in children.[47] The most frequently encountered toxic effects are alopecia, nausea, and vomiting. Leukopenia occurs, but thrombocytopenia is less frequent than with other alkylating agents. Sterile hemorrhagic cystitis may result and even be fatal. Gonadal suppression has been reported in several patients.

Melphalan, USP. 4-[Bis(2-chloroethyl)amino]-L-phenylalanine; L-sarcolysin; L-mustard; NSC-8808 (Alkeran). This compound is prepared by treating L-*N*-phthalimido-*p*-aminophenylalanine ethyl ester with ethylene oxide, followed by phosphorus oxychloride, and finally hydrolysis with hydrochloric acid.[48] It is soluble in alcohols, but poorly soluble in water. Oral absorption is good, and the drug is equally effective whether given orally or intravenously. Scored 2-mg tablets are available.

Melphalan is active against multiple myeloma. Recent clinical studies have shown that it is highly effective in preventing the recurrence of cancer in premenopausal women who undergo radical mastectomy.[49] The clinical toxicity is mainly hematologic, which means that the blood count must be carefully followed. Nausea and vomiting are infrequent, and alopecia does not occur.

Chlorambucil, USP. 4-[*p*-[Bis(2-chloroethyl)amino]phenyl]butyric acid; chloraminophene; NSC-3088 (Leukeran). It is prepared by treating *p*-aminophenyl butyric acid with ethylene oxide, followed by thionyl chloride.[50] Chlorambucil is soluble in ether and aqueous alkali. Its oral absorption is efficient and reliable. Sugar-coated 2-mg tablets are supplied.

Chlorambucil is the slowest acting and least toxic of any nitrogen mustard derivative in use. It is indicated especially in treatment of chronic lymphocytic leukemia and primary macroglobulinemia. Other indications are lymphosarcoma and Hodgkin's disease.[51] Many patients develop progressive, but reversible, lymphopenia during treatment. Most patients also develop a dose-related and rapidly reversible neutropenia. Hence, weekly blood counts are done to determine the total and differential leukocyte levels. The hemoglobin levels are also determined.

Busulfan, USP. 1,4-Butanediol dimethanesulfonate; 1,4-di(methanesulfonyloxy)butane; NSC-750 (Myleran). This compound is synthesized by treating 1,4-butanediol with methanesulfonyl chloride in the presence of pyridine.[52] It is obtained as crystals that are soluble in acetone and alcohol. Although practically insoluble in water, it dissolves slowly upon hydrolysis. However, it is stable in dry form. It is supplied as scored 2-mg tablets.

Busulfan is well absorbed orally and metabolized rapidly. Much of the drug undergoes a process known as "sulfur stripping," in which its interaction with thiol compounds such as glutathione or cysteine results in loss of two equivalents of methanesulfonic acid and formation of a cyclic sulfonium intermediate involving the sulfur atom of the thiol.[53] Such sulfonium intermediates are stable in vitro, but in vivo they are readily converted into the metabolite 3-hydroxythiolane-1,1-dioxide.[54] That the sulfur atom of this thiolane does not come from a methanesulfonyl group is shown by the nearly quantitative isolation of labeled methanesulfonic acid in the urine when [^{35}S]busulfan is administered to animals.[55]

$$CH_3SO_2O{-}CH_2CH_2CH_2CH_2{-}OSO_2CH_3 + HSCH_2CH(NHR')COR \longrightarrow \text{(cyclic sulfonium, } \overset{+}{S}CH_2CH(NHR')COR) + 2CH_3SO_3^- + H^+$$

$$\downarrow$$

$$\text{HO–CH(ring)–CH}_2\text{–SO}_2\text{–CH}_2\text{–CH}_2 \text{ (3-hydroxythiolane-1,1-dioxide)}$$

The main therapeutic use of busulfan is in chronic granulocytic leukemia. Remissions in 85% to 90% of patients are observed after the first course of therapy. However, it is not curative. Toxic effects are mostly limited to myelosuppression, in which the depletion of thrombocytes may lead to hemorrhage. Blood counts should be taken, not fewer than weekly. The rapid destruction of granulocytes can cause hyperuricemia that might result in kidney damage. This complication is prevented by using allopurinol, a xanthine oxidase inhibitor.[56]

Carmustine. 1,3-Bis(2-chloroethyl)-1-nitrosourea; BCNU; NSC-409,962 (BiCNU). This compound was synthesized at the Southern Research Institute by treating 1,3-bis(2-chloroethyl)urea with sodium nitrite and formic acid.[57] It is a low-melting white powder that changes to an oily liquid at 27°C. This change is considered a sign of decomposition, and such samples should be discarded. Carmustine is most stable in petroleum ether or water at *p*H 4. It is administered intravenously because metabolism is very rapid. However, some of the degradation products have prolonged half-lives in plasma. Carmustine is supplied as a lyophilized powder. When it is diluted with 3 mL of the supplied sterile diluent, ethanol, and further diluted with 27 mL of sterile water, a 10% ethanolic solution containing 3.3 mg/mL is obtained.

Because of its ability to cross the blood–brain barrier, carmustine is used against brain tumors and other tumors, such as leukemias, that have metastasized to the brain.[58] It is used also as secondary therapy in combination with other agents for Hodgkin's disease and other lymphomas. Multiple myeloma responds to a combination of carmustine and prednisone. Delayed myelosuppression is the most frequent and serious toxicity. This condition usually occurs four to six weeks after treatment. Thrombocytopenia is the most pronounced effect, followed by leukopenia. Nausea and vomiting frequently occur about two hours after treatment.

Carmustine is given as a single dose by intravenous injection at 100 to 200 mg/m^2. A repeat course is not given until the blood elements return to normal levels. This process requires about six weeks.

Lomustine. 1-(2-Chloroethyl)-3-cyclohexyl-1-nitrosourea; CCNU; NSC-79037 (CeeNU). This compound also was synthesized at the Southern Research Institute. The procedure involved treating ethyl 5-(2-chloroethyl)-3-nitrosohydantoate with cyclohexylamine, followed by renitrosation of the resulting intermediate, 1-(2-chloroethyl)-3-cyclohexylurea.[59] Lomustine is a yellow powder that is soluble in absolute or 10% ethanol, but is relatively insoluble in water. It is sufficiently metabolically stable to be administered orally. The high lipid solubility of lomustine allows it to cross the blood–brain barrier rapidly. Levels in the cerebrospinal fluid (CSF) are 50% higher than those in plasma. Lomustine is supplied in dose packs that contain two each of color-coded 100-, 40-, and 10-mg capsules. The total dose prescribed is obtained by appropriate combination of these capsules.

Lomustine is used against both primary and metastatic brain tumors and as secondary therapy in Hodgkin's disease. The most common adverse reactions are nausea and vomiting, thrombocytopenia, and leukopenia. As with carmustine, the myelosuppression caused by lomustine is delayed.[60]

The recommended dosage of lomustine is 130-mg/m^2 orally every six weeks. A reduced dose is given to patients with compromised bone marrow function.

Thiotepa, USP. Tris(1-aziridinyl)phosphine sulfide; *N,N′,N″*-triethylenethiophosphoramide; TSPA; NSC-6396. This compound is prepared by treating trichlorophosphine sulfide with aziridine[61] and is obtained as a white powder that is soluble in water. It is supplied in vials containing 15 mg of thiotepa, 80 mg of sodium chloride, and 50 mg of sodium bicarbonate. Sterile water is added to make an isotonic solution. Both the vials and solutions must be stored at 2° to 8°C. The solutions may be stored five days without loss of potency.

Thiotepa has been tried against a wide variety of tumors, and it has given palliation in many types, although with varying frequency. The most consistent results have been obtained in breast, ovarian, and bronchogenic carcinomas and malignant lymphomas. It also is used to control intracavity effusions resulting from neoplasms. Thiotepa is highly toxic to bone marrow, and blood counts must be done during therapy.

Procarbazine Hydrochloride, USP. *N*-Isopropyl-α-(2-methylhydrazino)-*p*-toluamide; MIH; NSC-77213 (Matulane). It is prepared from *N*-isopropyl-*p*-toluamide in a process involving condensation with diethyl azodicarboxylate, methylation with methyl iodide and base, and acid hydrolysis.[62] Although soluble in water, it is unstable in solution. Capsules containing the equivalent of 50 mg of procarbazine as its hydrochloride are supplied.

Procarbazine has demonstrated activity against Hodgkin's disease. For this condition it is used in combination with other agents, such as mechlorethamine, vincristine, and prednisone (MOPP program). Toxic effects such as leukopenia, thrombocytopenia, nausea, and vomiting occur in most patients. Neurologic and dermatologic effects also occur. Concurrent intake of alcohol or certain amines are contraindicated. The weak monoamine oxidase-inhibiting properties of procarbazine may potentiate amines to produce hypertension.

Dacarbazine, USP. 5-(3,3-Dimethyl-1-triazenyl)-1*H*-imidazole-4-carboxamide; DIC; DTIC; NSC-45388 (DTIC-Dome). This compound is prepared by treating the diazonium salt, prepared from 5-aminoimidazole-4-carboxamide, with dimethylamine in methanol.[63] It is obtained as a colorless to ivory solid that is very sensitive to light. It does not melt, but decomposes explosively when heated above 250°C. Water solubility is good, but solutions must be protected from light. Dacarbazine is supplied in vials containing either 100 or 200 mg. When reconstituted with 9.9 and 19.7 mL, respectively, of sterile water, these samples give solutions containing 10 mg/mL at *p*H 3.0 to 4.0. Such solutions may be stored at 4°C for 72 hours.

Dacarbazine is indicated for the treatment of metastic malignant melanoma. Combination with other antineoplastic drugs is superior to its use as a single agent. Anorexia, nausea, and vomiting are the most frequent toxic reactions. However, leukopenia and thrombocytopenia are the most serious effects. Blood counts should be made and therapy temporarily suspended if the counts are too low.

The recommended daily dosage is 2 to 4.5 mg/kg for ten days with repetition at four-week intervals. Extravasation of the drug during injection may result in tissue damage and severe pain.

Uracil Mustard, USP. 5-[Bis(2-chloroethyl)amino]uracil; 2,6-dihydroxy-5-bis(2-chloroethyl)-aminopyrimidine; NSC-34462 (Uramustine) is prepared by treating 5-aminouracil with ethylene oxide, followed by thionyl chloride.[64] This crystalline solid is sparingly soluble in water. It is supplied as capsules containing 1 mg of uracil mustard.

Uracil mustard is an analogue of nitrogen mustard in which the uracil moiety was designed to serve as a carrier group. However, it shows little selectivity in this respect. It is used in the palliative treatment of chronic lymphocytic leukemia, lymphosarcoma, giant follicular lymphoma, and Hodgkin's disease. Nausea, vomiting, and diarrhea are the most common untoward effects. They are dose-related. Complete blood counts should be done once or twice weekly during the first month of therapy. Cumulative, irreversible bone marrow damage may occur as the total dose reaches 1 mg/kg of body weight.

ANTIMETABOLITES

Antimetabolites are compounds that prevent the biosynthesis or utilization of normal cellular metabolites. They usually are closely related in structure to the metabolite that is antagonized. Many antimetabolites are enzyme inhibitors. In this capacity they may combine with the active site, as if they are the substrate or cofactor. Alternatively, they may bind to an allosteric regulatory site, especially when they resemble the end product of a biosynthetic pathway under feedback control.[11] Sometimes, the antimetabolite must be transformed biosynthetically (anabolized) into the active inhibitor. For example, 6-mercaptopurine is converted into the corresponding ribonucleotide, which is a potent inhibitor of the conversion of 5-phosphoribosylpyrophosphate into 5-phosphoribosylamine, a rate-controlling step in the de novo synthesis of purines[65] (see Scheme 5). An antimetabolite and its transformation products may inhibit several

TABLE 8-1

ALKYLATING AGENTS

Name Proprietary Name	*Preparations*	*Usual Adult Dose**	*Usual Dose Range**	*Usual Pediatric Dose**
Mechlorethamine hydrochloride USP *Mustargen*	Mechlorethamine hydrochloride for injection USP	IV, 400 μg/kg of body weight as a single dose; or divided into 2 or 4 daily doses		See Usual Adult Dose
Cyclophosphamide USP *Cytoxan, Endoxan*	Cyclophosphamide for injection USP Cyclophosphamide tablets USP	Initial, IV, 40–50 mg/kg of body weight in divided doses given over 2–5 days; maintenance, 10–15 mg/kg every 7–10 days or 3–5 mg/kg twice weekly		
Melphalan USP *Alkeran*	Melphalan tablets USP	150 μg/kg of body weight for 7 days, followed by a rest period of at least 3 wks		
Chlorambucil USP *Leukeran*	Chlorambucil tablets USP	Initial, 100–200 μg/kg of body weight qd		100–200 μg/kg of body weight or 4.5 mg/m^2 of body surface daily
Uracil Mustard USP	Uracil mustard capsules USP	1–2 mg/day until bone marrow depression or clinical improvement occurs		
Busulfan USP *Myleran*	Subsulfan tablets USP	4–6 mg qd	4–12 mg/day	1.8–4.6 mg/m^2 of body surface daily
Thiotepa USP	Thiotepa for injection USP	IV, 30 mg at 1- to 4-wk intervals, or 800 μg/kg of body weight every 4 wk, or 200–400 μg/kg of body weight every 2 wk	IV, 60 mg/dose	
Procarbazine hydrochloride USP *Matulane*	Procarbazine hydrochloride capsules USP	Initial, the equivalent of procarbazine, 2–4 mg/kg of body weight, followed by 4–6 mg/kg		Initial, the equivalent of procarbazine, 50 mg/day, followed by 100 mg/m^2 of body surface

*See USP DI for complete dosage information.

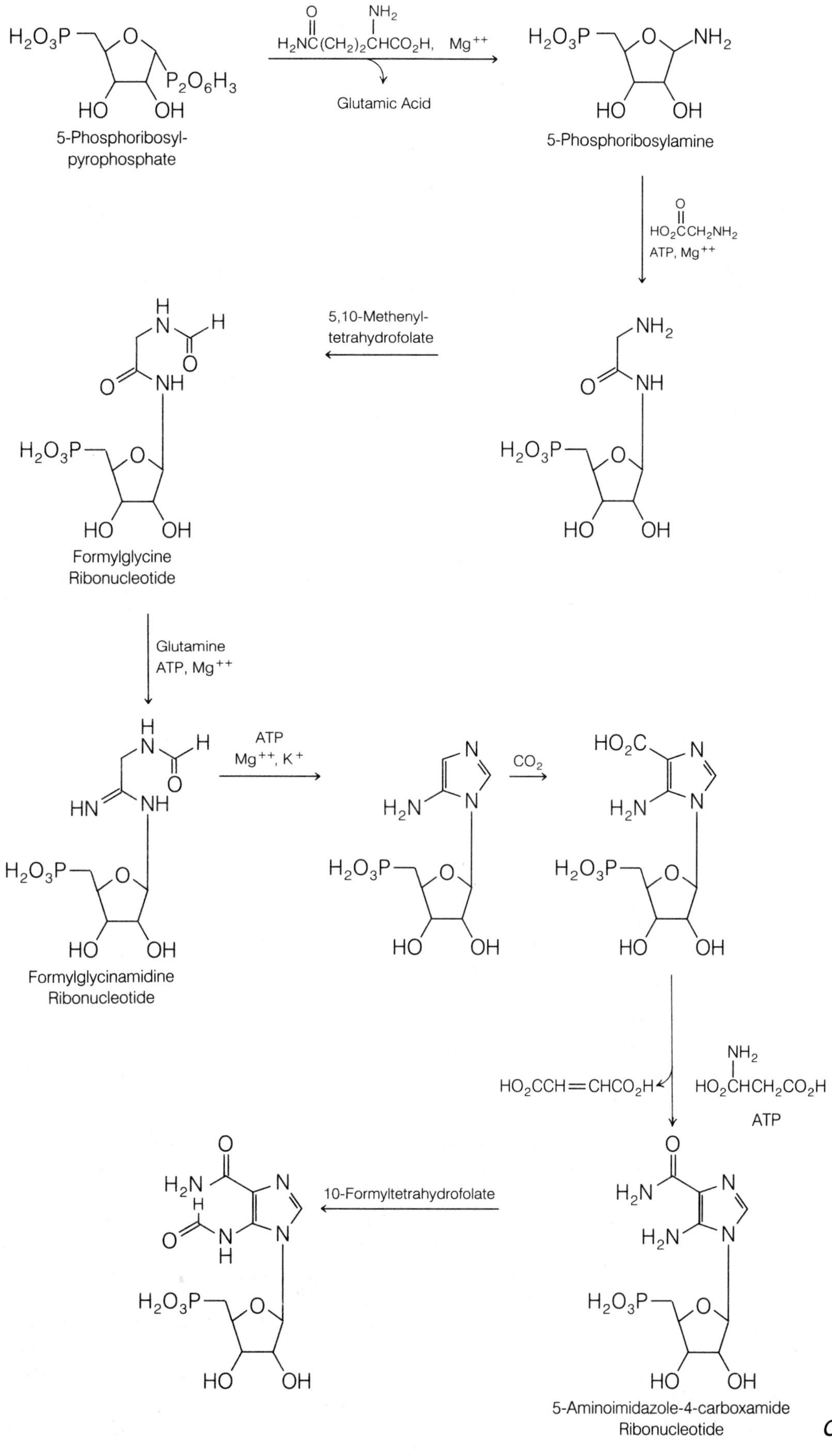

5-Phosphoribosyl-pyrophosphate

$H_2NC(O)(CH_2)_2CH(NH_2)CO_2H$, Mg^{++}

Glutamic Acid

5-Phosphoribosylamine

$HO_2CCH_2NH_2$

ATP, Mg^{++}

5,10-Methenyl-tetrahydrofolate

Formylglycine Ribonucleotide

Glutamine

ATP, Mg^{++}

ATP

Mg^{++}, K^{+}

CO_2

$HO_2CCH{=}CHCO_2H$

$HO_2CCH(NH_2)CH_2CO_2H$

ATP

Formylglycinamidine Ribonucleotide

10-Formyltetrahydrofolate

5-Aminoimidazole-4-carboxamide Ribonucleotide

Continued

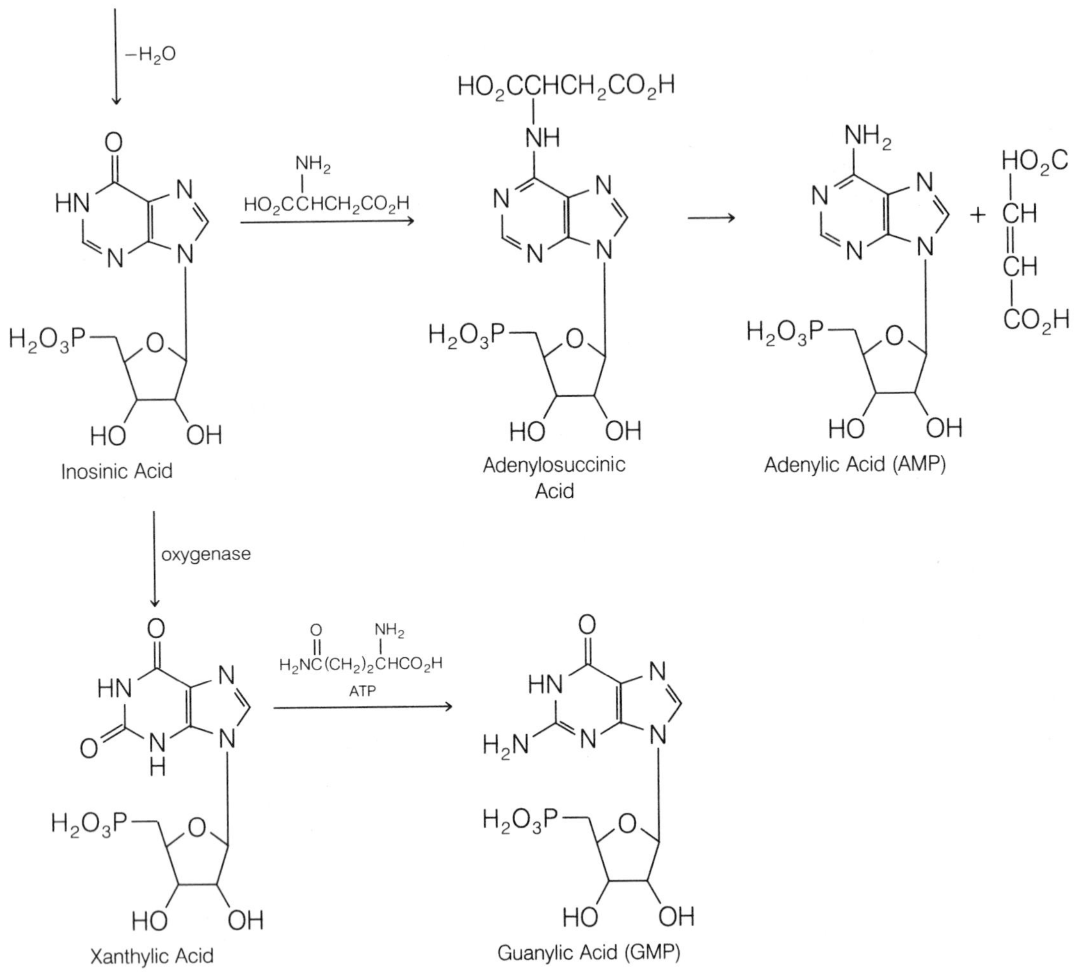

$-H_2O$

Inosinic Acid → ($HO_2CCHCH_2CO_2H$ with NH_2) → Adenylosuccinic Acid → Adenylic Acid (AMP) + fumaric acid ($HO_2C-CH=CH-CO_2H$)

Inosinic Acid → (oxygenase) → Xanthylic Acid → ($H_2NC(O)(CH_2)_2CH(NH_2)CO_2H$, ATP) → Guanylic Acid (GMP)

different enzymes. Thus, 6-mercaptopurine and its anabolites interact with more than 20 enzymes. This multiplicity of effects makes it difficult to decide which ones are crucial to the antitumor activity. The anabolites of purines and pyrimidine antagonists may be incorporated into nucleic acids. In this event, part of their antitumor effect might result from malfunction of the nucleic acids.[66] Although antimetabolites of every type have been tested against neoplasms, nearly all of the clinically useful agents are related to metabolites and cofactors in the biosynthesis of nucleic acids.

Following the formulation of the antimetabolite theory by Woods and Fildes in 1940,[67] antimetabolites that were based on a variety of known nutrients were prepared. The first purine analogue to show antitumor activity in mice, 8-azaguanine, was synthesized by Roblin in 1945.[68] This compound was introduced into clinical trials, but was abandoned in favor of newer and more effective agents such as 6-mercaptopurine and 6-thioguanine developed by Elion and Hitchings.[69] 6-Mercaptopurine was synthesized in 1952[70] and it was shown to be active against human leukemia in the following year.

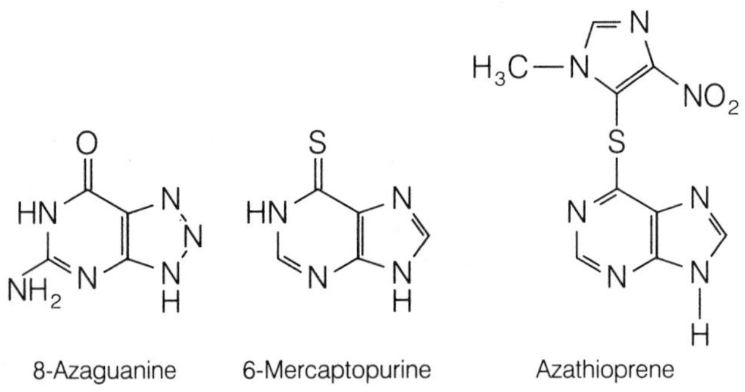

8-Azaguanine 6-Mercaptopurine Azathioprene

To be active against neoplasms, 6-mercaptopurine must be converted into its ribonucleotide, 6-thioinosinate, by the enzyme hypoxanthine phosphoribosyltransferase. Neoplasms that lack this enzyme are resistant to the drug.[71] 6-Thioinosinate is a potent inhibitor of the conversion of 5-phosphoribosylpyrophosphate into 5-phosphoribosylamine, as mentioned earlier. It also inhibits the conversion of inosinic acid to adenylic acid at two stages: (1) the reaction of inosinic acid with aspartate to give adenylosuccinic acid, and (2) the loss of fumaric acid from adenylosuccinic acid to give adenylic acid.[72]

Furthermore, it inhibits the oxidation of inosinic acid to xanthylic acid.[73] The mode of action of 6-mercaptopurine is further complicated because its ribose diphosphate and triphosphate anabolites are also active enzyme inhibitors.[72] Still more complex is the ability of 6-thioinosinate to act as a substrate for a methyl transferase, requiring *S*-adenosylmethionine, that converts it into 6-methylthioinosinate. The latter compound is responsible for certain of the antimetabolite activities of 6-mercaptopurine.[74]

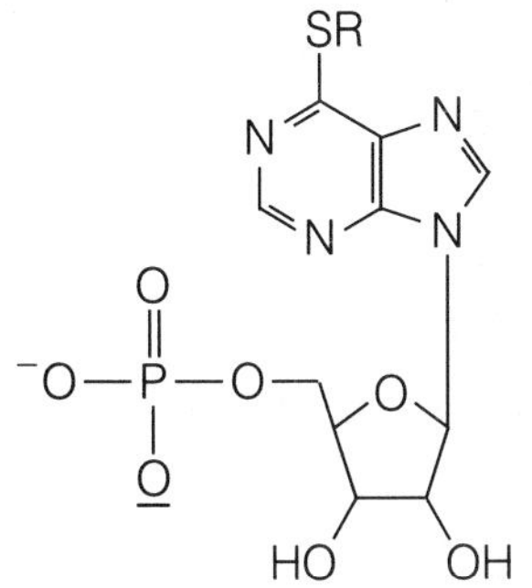

6-Thioinosinate (R = H)
6-Methylthioinosinate (R = CH_3)

Metabolic degradation (catabolism) of 6-mercaptopurine by guanase gives 6-thioxanthine, which is oxidized by xanthine oxidase to afford 6-thiouric acid.[75] Allopurinol, an inhibitor of xanthine oxidase, increases both the potency and the toxicity of 6-mercaptopurine. However, its main importance is as an adjuvant to chemotherapy because it prevents the uric acid toxicity caused by the release of purines from destroyed cancer cells. Heterocyclic derivatives of 6-mercaptopurine such as azathioprene (Imuran) were designed to protect it from catabolic reactions.[76] Although azathioprene has antitumor activity, it is not significantly better than 6-mercaptopurine. However, it has found an important role as an immunosuppressive agent in organ transplants.[69]

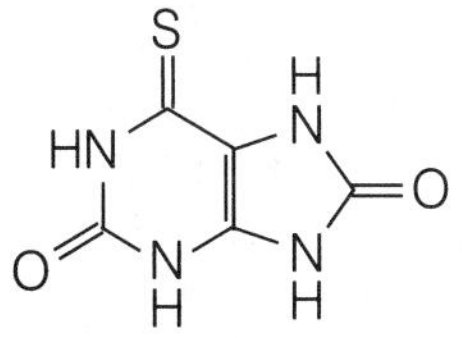

6-Thiouric Acid

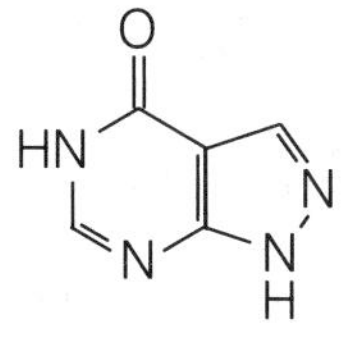

Allopurinol

Thioguanine is converted into its ribonucleotide by the same enzyme that acts on 6-mercaptopurine. It is converted further into the di- and triphosphates.[77] These species inhibit most of the same enzymes that are inhibited by 6-mercaptopurine.

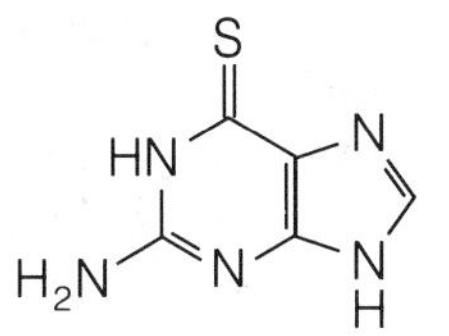

6-Thioguanine

Thioguanine is incorporated into RNA and its 2′-deoxy metabolite is incorporated into DNA. The importance of these "fraudulent" nucleic acids in lethality to neoplasms is uncertain.[78]

Vidarabine (adenine arabinoside) was first prepared by chemical synthesis[79] and later isolated from cultures of *Streptomyces antibioticus*.[80] It has a sugar, D-arabinose, that is epimeric with D-ribose at the 2′-position. This structural change makes it a competitive inhibitor of DNA polymerase.[81] In addition to its antineoplastic activity, vidarabine has potent antiviral action. The 5′-phosphate of vidarabine is included in the NCI listing of compounds in development. Vidarabine and its derivatives are limited in their antitumor effect by susceptibility to adenosine deaminase. This enzyme converts them into hypoxanthine arabinoside derivatives. The resistance of certain tumors correlates with their levels of adenosine deaminase.[82] Among the purines currently undergoing clinical study is the 2-fluoro derivative of vidarabine phosphate (fludarabine).[83]

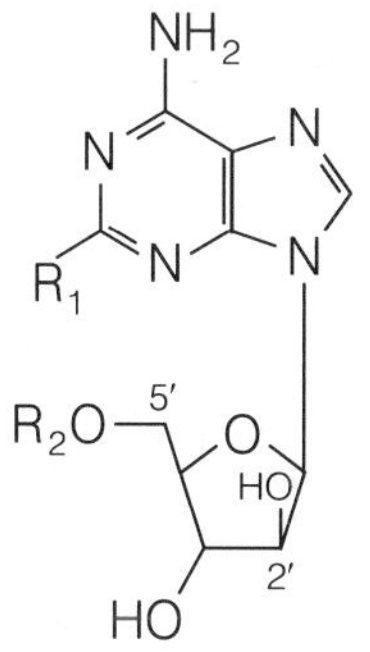

$R_1 = R_2 = H$; Vidarabine
$R_1 = F$, $R_2 = HOPO_2$; Fludarabine

The invention of 5-fluorouracil as an antimetabolite of uracil by Heidelberger in 1957 provided one of our foremost examples of rational drug design.[84] Starting with the observation that in certain tumors uracil was used more than orotic acid—the major precursor for nucleic acid pyrimidine biosynthesis in normal tissue—he decided to synthesize an antimetabolite of uracil with only one modification in the structure. The 5-position was chosen for a substituent to block the conversion of uridylate to thymidylate (Scheme 6), thereby diminishing DNA biosynthesis. Fluorine was chosen as the substituent because the increased acidity caused by its inductive effect was expected to cause the molecule to bind strongly to enzymes. These choices were well founded, as 5-fluorouracil soon became one of the most widely used antineoplastic agents.

5-Fluorouracil is activated by anabolism to 5-fluoro-2′-deoxyuridylic acid. This conversion may proceed by two routes. In one route, 5-fluorouracil re-

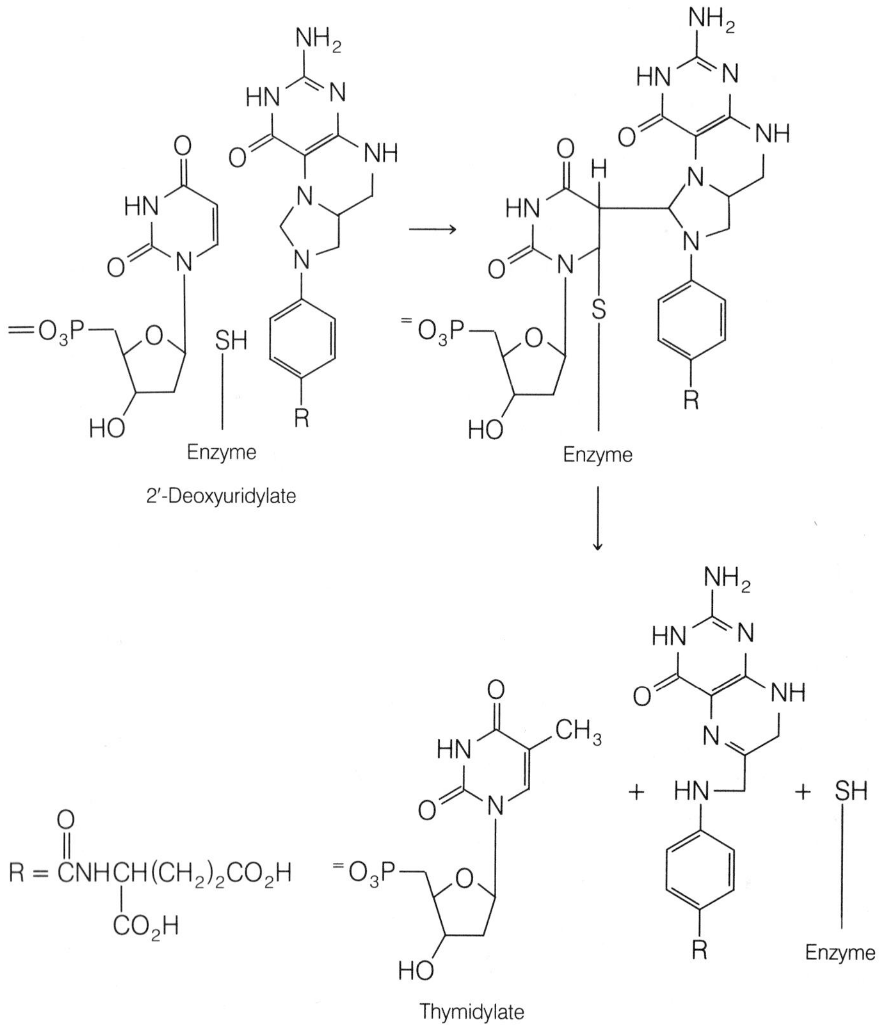

2′-Deoxyuridylate

Thymidylate

R = $\overset{O}{\overset{\|}{C}}NHCH(CH_2)_2CO_2H$ with CO_2H on the CH

Scheme 6. *Conversion of uridylate into thymidylate.*

acts with ribose-1-phosphate to give its riboside, which is phosphorylated by uridine kinase.[85] The resulting compound, 5-fluorouridylic acid, is converted into its 2′-doxy derivative by ribonucleotide reductase. 5-Fluorouracil also may be transformed directly into 5-fluorouridylic acid by a phosphoribosyltransferase, which is present in certain tumors.[86] An alternative pharmaceutical, based on 5-fluorouracil, is its 2-deoxyriboside.[84] This compound is phosphorylated by 2′-deoxyuridine kinase. The resulting 5-fluoro-2′-deoxyuridylic acid is a powerful competitive inhibitor of thymidylate synthetase, the enzyme that converts 2′-deoxyuridylic acid to thymidylic acid. This blockade is probably the main lethal effect of 5-fluorouracil and its metabolites.[87] In the inhibiting reaction, the sulfhydryl group of a cysteine residue in the enzyme adds to the 6-position of the fluorouracil moiety. The 5-position then binds to the methylene group of 5,10-methylenetetrahydrofolate. Ordinarily, this step would be followed by the transfer of the 5-hydrogen of uracil to the methylene group, resulting in the formation of thymidylate and dihydrofolate. However, the 5-fluorine is stable to transfer, and a terminal product results involving the enzyme, cofactor, and substrate, all covalently bonded. Thus, 5-fluoro-2′-deoxyuridylic acid would be classified as a K_{cat} inhibitor.[88]

The tetrahydrofuranyl derivative of 5-fluorouracil, known as tegafer (Ftorafur), was prepared in the Soviet Union.[89] It is active in clinical cancer and less myelosuppressive than 5-fluorouracil. However, it has gastrointestinal and CNS toxicity. Tegafer is slowly metabolized to 5-fluorouracil; thus, it may be considered a prodrug.[90]

Trifluridine (trifluorothymidine) was designed by Heidelberger as an antimetabolite of thymine.[84] The riboside is essential because mammalian cells are

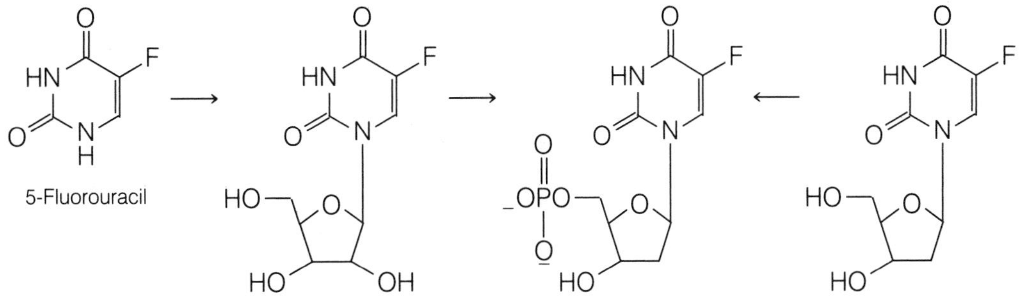

5-Fluorouracil

5-Fluorouracil Riboside

5-Fluorodeoxyuridylic Acid

5-Fluorouracil 2-Deoxyriboside

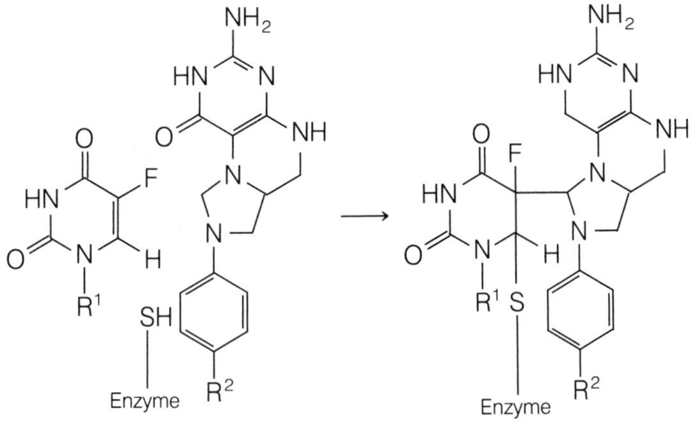

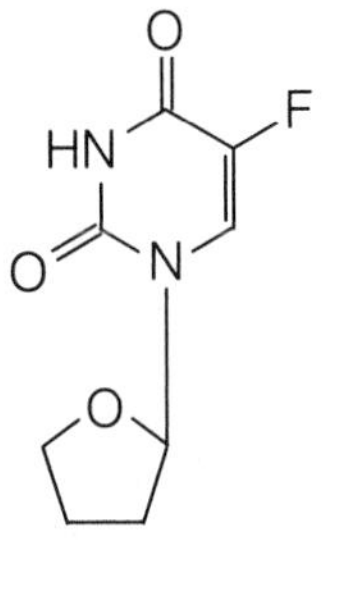

Enzyme

Enzyme

Tegafer (Ftorafur)

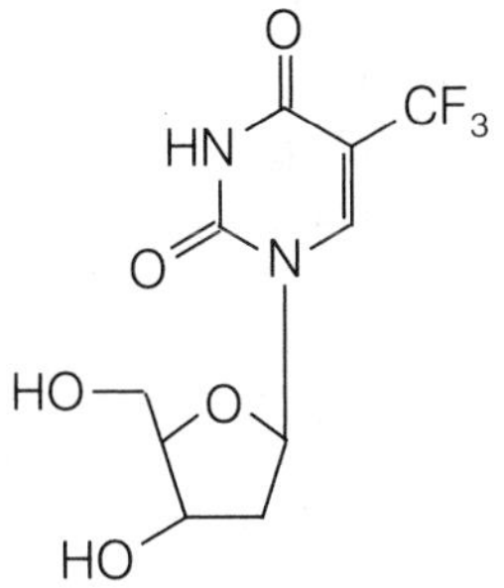

Trifluridine (Trifluorothymidine)

unable to convert thymine and certain analogues into thymidine and its analogues. Thymidine kinase converts trifluridine into trifluorothymidylic acid, which is a potent inhibitor of thymidylate synthetase.[84] In contrast with the stability of most trifluoromethyl groups, the one of trifluorothymidylic acid is extraordinarily labile. It reacts with glycine to give an amide at neutral *p*H.[91] Kinetic studies have shown that this reaction involves initial nucleophilic attack at position 6, followed by loss of HF to give the highly reactive difluoromethylene group.[92] Glycine then adds to this group and hydrolysis of the remaining two fluorine atoms follows (Scheme 7). The interaction of trifluorothymidylic acid with thymidylate synthetase apparently follows a similar course. Thus, after preincubation it becomes irreversibly bound to the enzyme, and the kinetics are noncompetitive.[84]

Cytarabine (cytosine arabinoside) was synthesized in 1959[93] and later found as a fermentation product.[94] It is noteworthy structurally in that its arabinose moiety is epimeric at the 2′-position with ribose. This modification, after anabolism to the triphosphate, causes it to inhibit the conversion of cytidylic acid to 2′-deoxycytidylic acid.[95] For several years this inhibition was believed to be the main mode of action of cytarabine triphosphate; however, it was shown recently that various deoxyribonucleosides were just as effective as cytarabine in reducing cellular levels of 2′-deoxycytidylic acid.[96] Other modes of action include the inhibition of DNA-dependent DNA polymerase[97] and miscoding following incorporation into DNA and RNA.[98] Cytarabine is readily transported into cells and phosphorylated by deoxycytidine kinase. It acts predominantly in the S phase of the cell cycle. Tumor cell resistance is based on low levels of deoxycytidine kinase and the elaboration of deaminases that convert cytarabine into uridine arabinoside.[99] Cytidine deaminase has been partially purified and found to be inhibited by tetra-

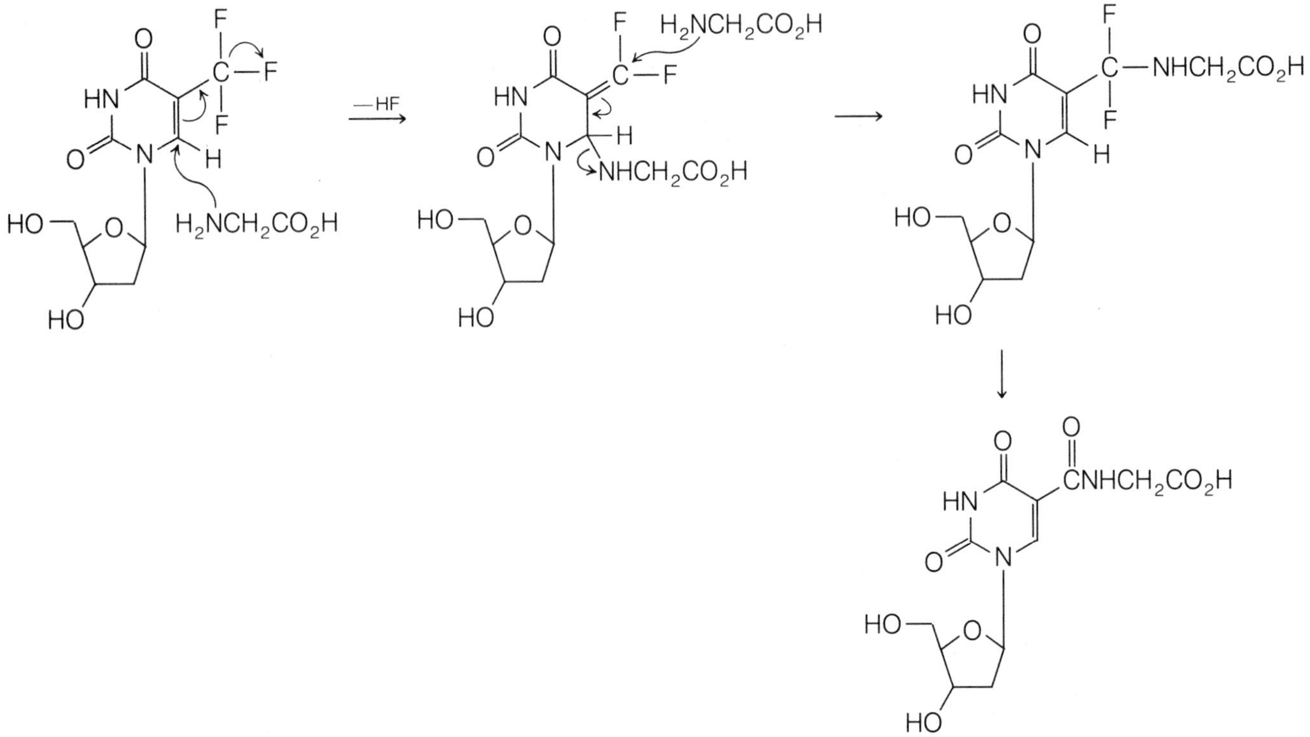

Scheme 7. Reaction of trifluorothymidine with glycine.

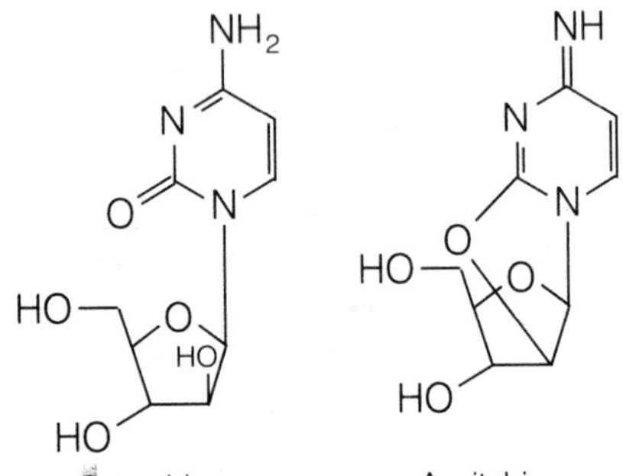

Cytarabine (Cytosine arabinoside)

Ancitabine (Cyclocytidine)

hydrouridine.[100] A combination of cytarabine and tetrahydrouridine is presently in clinical trial.

An analogue of cytarabine is ancitabine (cyclocytidine). This analogue apparently is a prodrug that is slowly converted into cytarabine. It is reported to be resistant to deamination and to have a therapeutic index superior to that of the parent compound.[101]

Several pyrimidine nucleoside analogues have one more or one less nitrogen in the heterocyclic ring. They are known as azapyrimidine or deazapyrimidine nucleosides. One of these antimetabolites is azacitidine (5-azacytidine). It was synthesized in 1964 by Sorm in Czechoslovakia[102] and later was isolated as an antibiotic by Hanka.[103] The mode of action of this compound is complex, involving anabolism to phosphate derivatives and deamination to 5-azauridine. In certain tumor systems it is incorporated into nucleic acids that possibly cause misreading.[104] One of its main effects is the inhibition of orotidylate decarboxylase (see Scheme 8), which prevents the new synthesis of pyrimidine nucleotides.[105] Tumor resistance is based on decreased phosphorylation of the nucleoside, decreased incorporation into nucleic acids, and increase in RNA and DNA polymerase activity.[106]

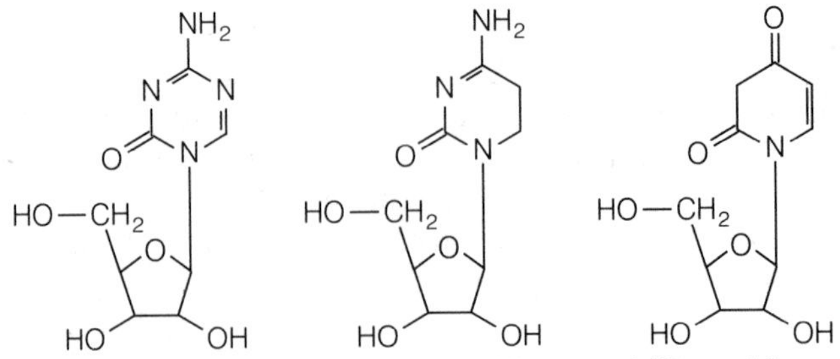

Azacitidine (5-Azacytidine)

Dihydro-5-azacytidine

3-Deazauridine

Pyrimidine nucleoside antagonists active against experimental tumors and presently in clinical study include dihydro-5-azacytidine and 3-deazauridine.[83]

The resistance to purine and pyrimidine antimetabolites such as vidarabine and cytarabine by

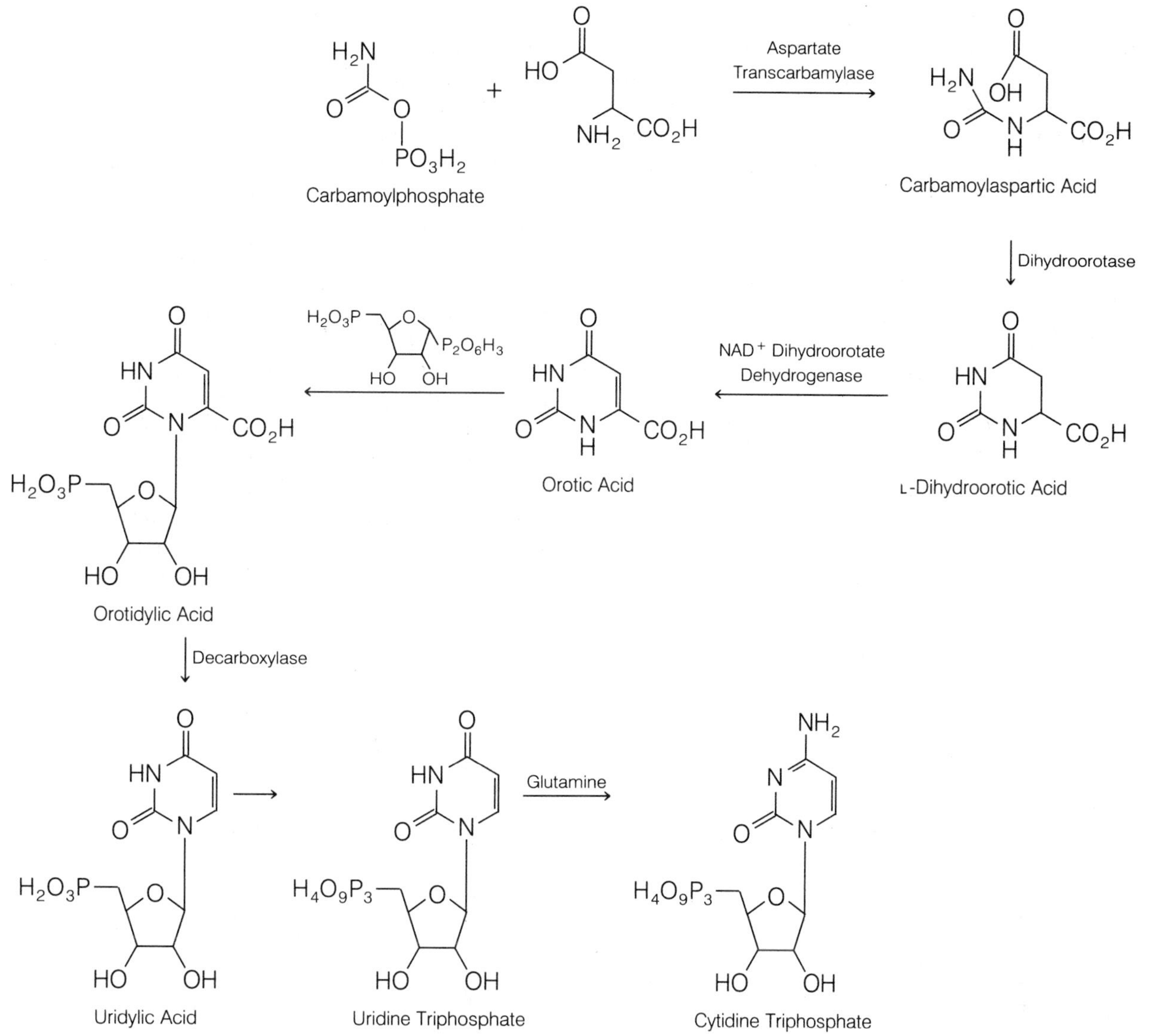

Scheme 8. *De novo synthesis of pyrimidine nucleotides (simplified).*

neoplastic cells that produce deaminases has stimulated a search for compounds that might inhibit these deaminases. In principle, a potent deaminase inhibitor would produce a synergistic effect on the antitumor activity of the antimetabolite, even though it might not be active itself. Two types of deaminase inhibitors have emerged. One type is the purine analogue in which the pyrimidine ring has been expanded to a seven-membered ring. The first example of this type was coformycin, an unusual nucleoside produced in the same cultures as the antibiotic formycin.[107] It strongly synergized the action of formycin against organisms that produce deaminases. In clinical trials it showed a synergistic effect on the activities of vidarabine and cytarabine. Currently, 2′-deoxycoformycin is undergoing clinical evaluation.[83] A second type of adenosine deaminase inhibitor has the adenine portion unchanged, but is modified in the ribose moiety. Such modifications have been designed to probe the active site of the enzyme and take advantage of strong binding to adjacent lipophilic regions.[108] The compound erythro-9-(2-hydroxy-3-nonyl)adenine (EHNA) is an example of a rationally designed inhibitor.

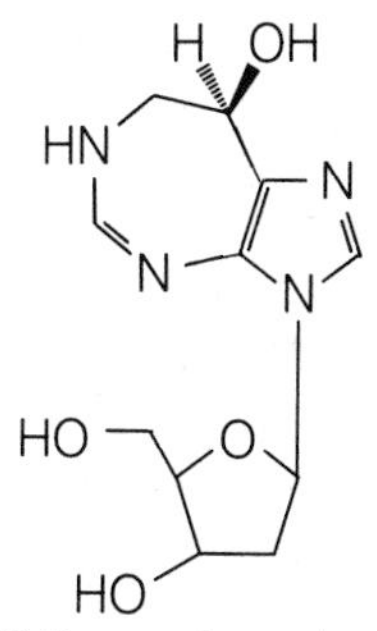

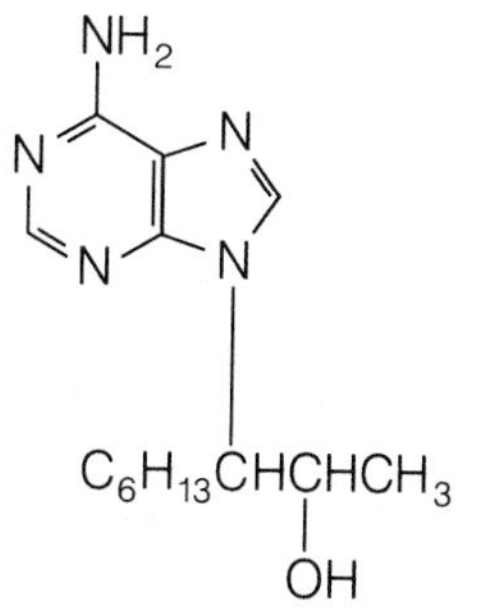

Following the discovery of folic acid, several analogues based on its structure were synthesized and tested as antimetabolites. The N^{10}-methyl derivative of folic acid was an antagonist, but it had no antitumor activity. Antitumor activity was finally found for the 4-amino-4-deoxy derivative, aminopterin and its N^{10}-methyl homologue, methotrexate (amethopterin).[109]

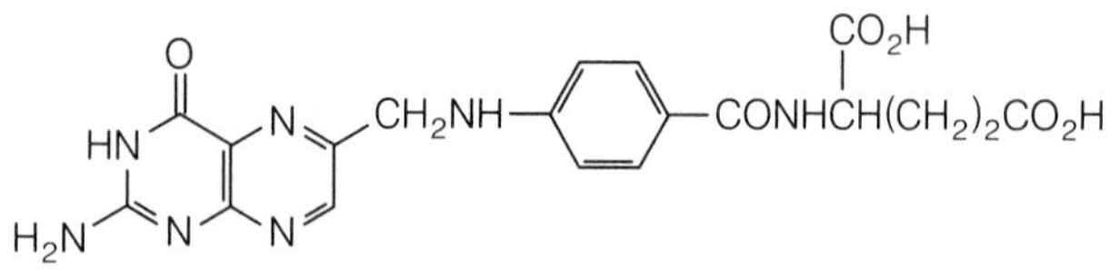

Folic Acid

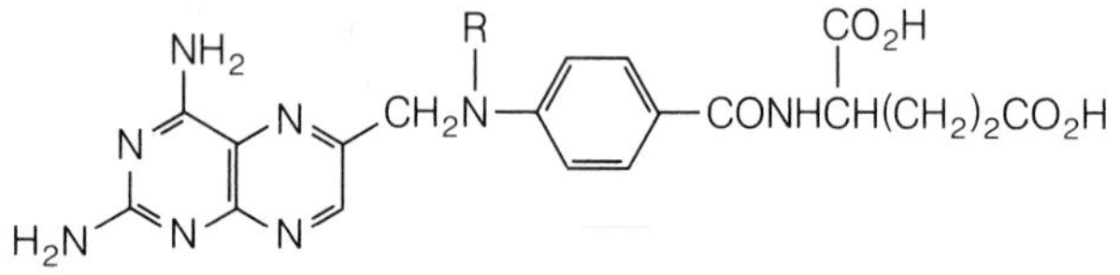

Aminopterin, R = H
Methotrexate, R = CH_3

Methotrexate and related compounds inhibit the enzyme dihydrofolate reductase. They bind so tightly to it that their inhibition has been termed "pseudo-irreversible." The basis of this binding strength is in the diaminopyrimidine ring, which is protonated at physiological *p*H. At *p*H 6, methotrexate binds stoichiometrically with dihydrofolate reductase (K_i 10^{-10} M), but at higher *p*H the binding is weaker and competitive with the substrate.[110]

Folate antagonists kill cells by inhibiting DNA synthesis in the S phase of the cell cycle. Thus, they are most effective in the log growth phase.[111] Their effect on DNA synthesis results partially from the inhibition of dihydrofolate reductase, which depletes the pool of tetrahydrofolic acid. Folic acid is reduced stepwise to dihydrofolic acid and tetrahydrofolic acid, with dihydrofolate reductase thought to catalyze both steps.[112] As shown in Scheme 9, tetrahydrofolic

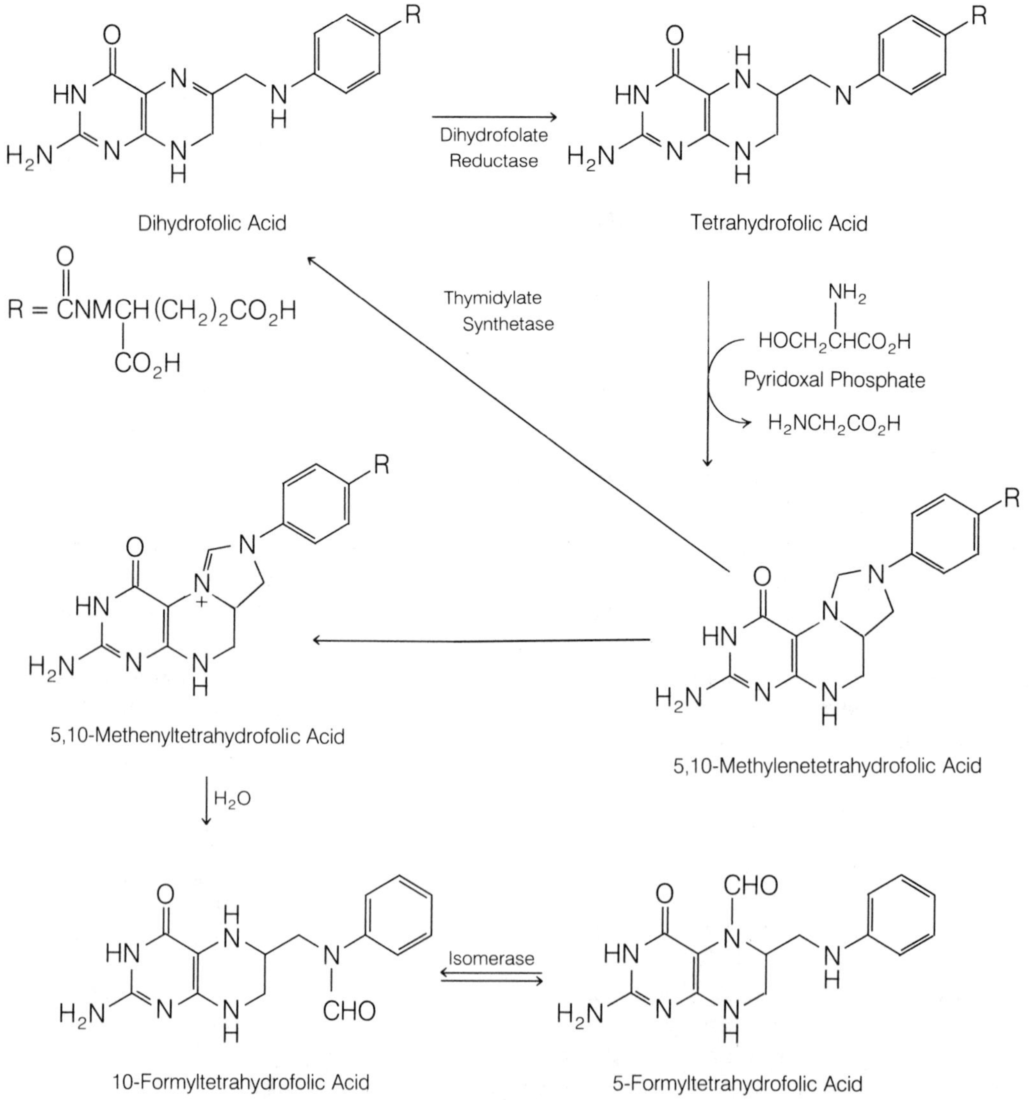

Dihydrofolic Acid → (Dihydrofolate Reductase) → Tetrahydrofolic Acid

R = $\overset{O}{\overset{\|}{C}}NMCH(CH_2)_2CO_2H$ with CO_2H

Tetrahydrofolic Acid → ($HOCH_2CH(NH_2)CO_2H$, Pyridoxal Phosphate, → $H_2NCH_2CO_2H$) → 5,10-Methylenetetrahydrofolic Acid

5,10-Methylenetetrahydrofolic Acid → (Thymidylate Synthetase) → Dihydrofolic Acid

5,10-Methylenetetrahydrofolic Acid → 5,10-Methenyltetrahydrofolic Acid

5,10-Methenyltetrahydrofolic Acid → (H_2O) → 10-Formyltetrahydrofolic Acid

10-Formyltetrahydrofolic Acid ⇌ (Isomerase) ⇌ 5-Formyltetrahydrofolic Acid

Scheme 9. *Interconversions of folic acid derivatives.*

acid accepts the β-carbon atom of serine, in a reaction requiring pyridoxal phosphate, to give N^5,N^{10}-methylenetetrahydrofolic acid. The last compound transfers a methyl group to 2′-deoxyuridylate to give thymidylate in a reaction catalyzed by thymidylate synthetase. Dihydrofolic acid is generated in this reaction, and it must be reduced back to tetrahydrofolic acid before another molecule of thymidylate can be synthesized. It is partly by their effect in limiting thymidylate synthesis that folic acid analogues prevent DNA synthesis and kill cells. This effect has been termed the "thymineless death."[113]

The inhibition of dihydrofolate reductase produces other limitations on nucleic acid biosynthesis. Thus, N^5,N^{10}-methylenetetrahydrofolic acid is oxidized to the corresponding methenyl derivative, which gives N^{10}-formyltetrahydrofolic acid on hydrolysis (see Scheme 9). The latter compound is a formyl donor to 5-aminoimidazole-4-carboxamide ribonucleotide in the biosynthesis of purines.[114] N^5-Formyltetrahydrofolic acid, also known as leucovorin and citrovorum factor, is interconvertible with the N^{10}-formyl analogue by way of an isomerase-catalyzed reaction. It carries the formimino group for the biosynthesis of formiminoglycine, a precursor of purines (see Scheme 5). Leucovorin is utilized in "rescue therapy" with methotrexate. It prevents the lethal effects of methotrexate on normal cells by overcoming the blockade of tetrahydrofolic acid production. In addition, it inhibits the active transport of methotrexate into cells and stimulates its efflux.[115]

It was shown that giving tymidine with methotrexate to mice bearing L1210 leukemia increased their survival time. This finding contradicts the idea that thymine deficiency is the most lethal effect of methotrexate on tumors. It suggests that the blockade of purine biosynthesis might have greater effects on tumor cells than on normal cells.[116] Consequently, the administration of thymidine might protect the normal cells relative to the tumor cells. This reasoning has been applied to the clinical situation with the result that "thymidine rescue" has shown promise in humans receiving methotrexate.[117]

Numerous analogues closely related to methotrexate have been prepared and tested against neoplasms[118, 119]. Most structural variations, such as alkylation of the amino groups, partial reduction, and removal or relocation of heterocyclic nitrogens lead to decreased activity. The analogue with homologation in the side chain, homofolic acid, is a potent inhibitor of thymidylate synthetase. It is presently in clinical studies.[83] 10-Ethyl-10-deazaaminopterin has received clinical study against lung cancer.[120]

Although the active sites of dihydrofolate reductases from normal and neoplastic cells are identical, Baker proposed that regions adjacent to the active sites of these enzymes might be different. He designed inhibitors to take advantage of these differences, thus affording species specificity. One of these inhibitors, known as "Baker's antifol," shows activity against experimental tumors that are resistant to methotrexate.[121]

Glutamine and glutamate are the donors of the 3- and 9-nitrogen atoms of purines and the 2-amino group of guanine.[122] They also contribute the 3-nitrogen atom and the amino group of cytosine[123]

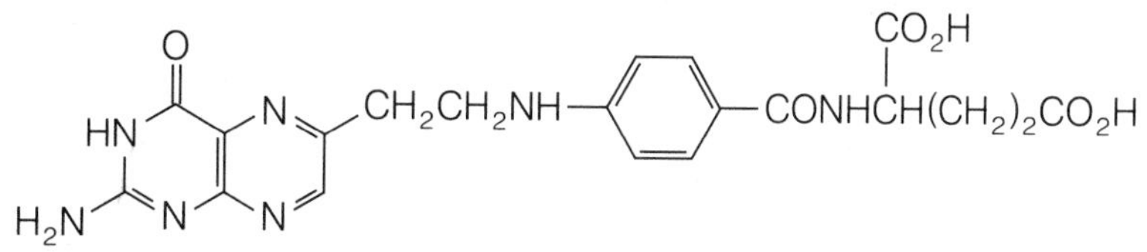

Homofolic Acid

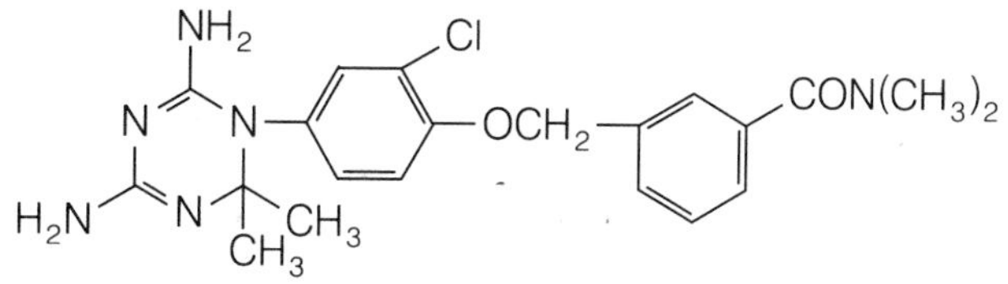

Baker's Antifol

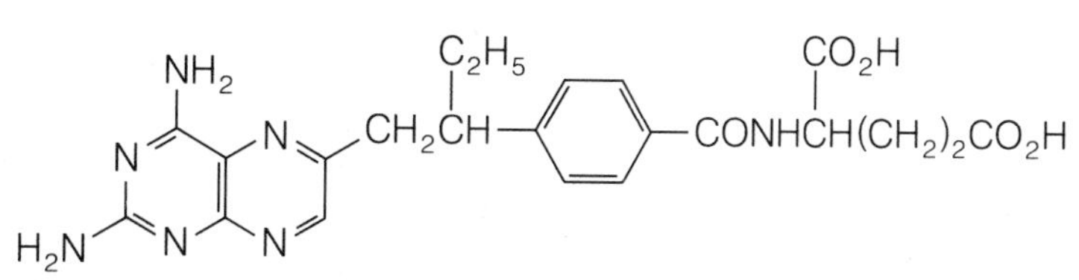

10-Ethyl-10-deazaaminopterin

(see Schemes 5 and 8). Thus, they are involved at five different sites of nucleic acid biosynthesis. Although glutamine is not an essential nutrient for normal cells, many tumors are dependent upon exogenous sources for it. This provides a rationale for the selective action of agents that interfere with the uptake, biosynthesis, or functions of glutamine.

In 1954, azaserine was isolated from a *Streptomyces* species.[124] It was found to antagonize many of the metabolic processes involving glutamine, with the most important effect being the conversion of formylglycine ribonucleotide into formyglycinamidine ribonucleotide (see Scheme 5).[125] A related compound, 6-diazo-5-oxo-L-norleucine (DON) was isolated in 1956 and found to possess similar antagonism.[126] A study involving incubation with [^{14}C]azaserine followed by digestion with proteolytic enzymes and acid hydrolysis produced *S*-[^{14}C]carboxymethylcysteine, which showed that azaserine had reacted covalently with a sulfhydryl group of cystein on the enzyme.[127]

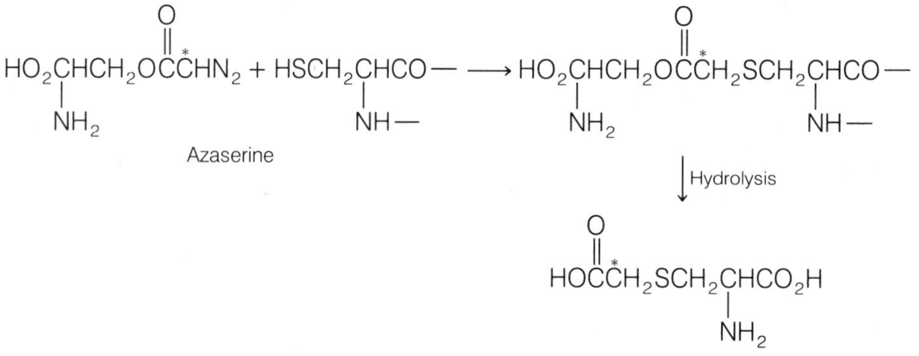

$$HO_2C\underset{NH_2}{CH}CH_2O\overset{O}{\overset{\|}{C}}\overset{*}{C}HN_2 + HSCH_2\underset{NH-}{CH}CO- \longrightarrow HO_2C\underset{NH_2}{CH}CH_2O\overset{O}{\overset{\|}{C}}\overset{*}{C}H_2SCH_2\underset{NH-}{CH}CO-$$

Azaserine

$$\downarrow \text{Hydrolysis}$$

$$HO\overset{O}{\overset{\|}{C}}\overset{*}{C}H_2SCH_2\underset{NH_2}{CH}CO_2H$$

DON is a more potent inhibitor of this enzyme than is azaserine and of the enzyme that converts uridine nucleosides into cytidine nucleosides.[128,129] Although both compounds show good antitumor activity in animal models, they have been generally disappointing in clinical trials. DON is classified as a compound in development by the NCI.[83]

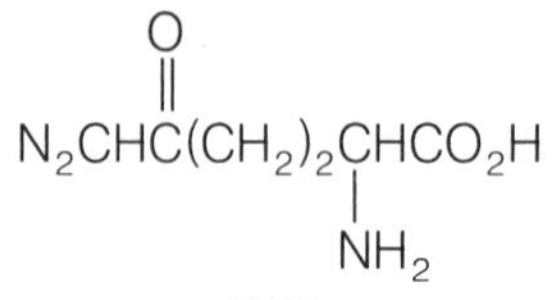

$$N_2CH\overset{O}{\overset{\|}{C}}(CH_2)_2\underset{NH_2}{CH}CO_2H$$

DON

PRODUCTS

Mercaptopurine, USP. 6-Mercaptopurine; 6 MP, NSC-755, 6-purinethiol (Purinethol; Leukerin, Mercaleukin). This compound is prepared by treating hypoxanthine with phosphorus pentasulfide[130] and is obtained as yellow crystals of the monohydrate. Solubility in water is poor. It dissolves in dilute alkali, but undergoes slow decomposition. Scored 50-mg tablets are supplied.

Mercaptopurine is used primarily for treating acute leukemia. A higher proportion of children than adults respond.[131] The chief toxic effect is leukopenia. Thrombocytopenia and bleeding occur in high doses. Because the leukopenia is delayed, it is important to discontinue the drug temporarily at the first sign of an abnormally large drop in the white cell count.

The tolerated dose varies with the individual patient. Allopurinol potentiates the effect of mercaptopurine by inhibiting its metabolism. However, it also increases its toxicity. If allopurinol is given for potentiation or reduction of hyperuricemia resulting from the death of leukemic cells, the doses of mercaptopurine must be decreased.[132]

Thioguanine, USP. 2-Aminopurine-6(1*H*)thione; 6-thioguanine; TG; NSC-752 (Tabloid). The preparation of this compound is by treating guanine with phosphorus pentasulfide in pyridine.[133] Scored 40-mg tablets are supplied.

Thioguanine is used in treating acute leukemia, especially in combination with cytarabine.[134] Cross-resistance exists between thioguanine and mercaptopurine. The chief toxic effect is delayed bone marrow depression, resulting in leukopenia and, eventually, thrombocytopenia and bleeding.

The usual initial dose is 2 mg/kg daily by the oral route. If there is no clinical improvement or leukopenia after four weeks, the dosage is increased to 3 mg/kg daily. In contrast with mercaptopurine, thioguanine may be continued in the usual dose when allopurinol is used to inhibit uric acid formation.

Fluorouracil, USP. 5-Fluoro-2,4(1*H*,3*H*)-pyrimidinedione; 2,4-dioxo-5-fluoropyrimidine; 5-FU (Fluorouracil Ampuls; Fluoroplex; Efudex). This compound is prepared by condensing *S*-ethylisothiouronium bromide with the potassium salt (enolate) of ethyl 2-fluoro-2-formylacetate.[135] The preparation of fluorouracil by direct fluorination of uracil also has been demonstrated.[136] Fluorouracil is obtained as white crystals. It is supplied in 10-mL ampuls containing 500 mg of fluorouracil in a colorless to faint-yellow water solution at *p*H 9. These ampuls should be stored at room temperature and protected from light. Topical formulations of fluorouracil are Efudex Solution, which contains 2% or 5% fluorouracil compounded with propylene glycol, tris(hydroxymethyl)aminomethane, hydroxypropyl cellulose, methyl and propyl parabens and disodium edetate, and Efudex Cream, which contains 5% fluorouracil in a vanishing cream base consisting of white petrolatum, stearyl alcohol, propylene glycol, polysorbate 60, and methyl and propyl parabens.[137]

Fluorouracil is effective in the palliative management of carcinoma of the breast, colon, pancreas, rectum, and stomach in patients who cannot be cured by surgery or other means.[138] The topical formulations are used with favorable results for the treatment of premalignant keratoses of the skin and superficial basal cell carcinomas.[139] Parenteral administration almost invariably produces toxic effects. Leukopenia usually follows every course of therapy, with the lowest white blood cell counts occurring between days 9 and 14 after the first course. Gastrointestinal hemorrhage may occur and may even be fatal. Stomatitis, esophagopharyngitis, diarrhea, nausea, and vomiting are commonly seen; alopecia and dermatitis also occur. Therapy must be discontinued if leukopenia or gastrointestinal toxicity becomes too severe. Topical administration is contraindicated in patients who develop hypersensitivity. Prolonged exposure to ultraviolet radiation may increase the intensity of topical inflammatory reactions.

Floxuridine, USP. FUDR; fluorodeoxyuridine; NSC-27640; 2′-deoxy-5-fluorouridine; 1-(2-deoxy-β-D-ribofuranosyl)-5-fluorouracil. This compound is prepared by condensing monomercuri-5-fluorouracil with 3,5-di-*O*-*p*-toluyl-2-deoxyribosyl-1-chloride, followed by alkaline hydrolysis.[140] It is supplied in 5-mL vials containing 500 mg of floxuridine as sterile powder. Reconstitution is by the addition of 5 mL of sterile water. The resulting solutions should be stored under refrigeration for not more than two weeks.

Floxuridine is used for palliation of gastrointestinal adenocarcinoma metastatic to the liver in patients who are considered incurable by surgery or other means.[141] It is administered by continuous regional intra-arterial infusion. Because floxuridine is rapidly catabolized to fluorouracil, it gives the same toxic reactions as flourouracil.

Cytarabine, USP. Ara-c; cytosine arabinoside; NSC-63878; 1-β-D-aribinofuranosylcytosine (Cytosar-U). It is synthesized from uracil arabinoside in a route involving acetylation, treatment with phosphorus pentasulfide, and heating with ammonia.[142] It is supplied as the freeze-dried solid in vials containing 100 mg or 500 mg. The 100-mg sample is reconstituted with 5 mL of sterile water containing 0.9% benzyl alcohol to give 20 mg/mL of cytarabine, whereas the 500-mg sample is reconstituted with 10 mL of sterile water containing 0.9% benzyl alcohol to give 50 mg/mL cytarabine. These solutions may be stored at room temperature for 48 hours.

Cytarabine is indicated primarily for inducing the remission of acute granulocytic leukemia of adults. It also is used for other acute leukemias of adults and children.[143] Remissions have been brief unless followed by maintenance therapy or given in combination with other antineoplastic agents.[144] Side effects include severe leukopenia, thrombocytopenia, and anemia. Gastrointestinal disturbances also are relatively frequent.

Methotrexate, USP. *N*-[4-[[(2,4-Diamino-6-pteridinyl)methyl]methylamino]benzoyl]-L-glutamic acid; 4-amino-N^{10}-methylpteroylglutamic acid; amethopterin; methylaminopterin; NSC-740. This compound is prepared by combining 2,4,5,6-tetrahydropyrimidine, 2,3-dibromopropionaldehyde, disodium *p*-(methylamino)-benzoylglutamate, iodine, and potassium iodide, followed by heating with lime water.[112] It is isolated as the monohydrate, a yellow solid. Studies have indicated that the commercial preparation contains several impurities including 4-amino-N^{10}-methylpteroic acid and N^{10}-methylfolic acid.[145] Methotrexate is soluble in alkaline solutions, but decomposes in them. It is supplied as 25-mg tablets and in vials containing either 5 mg or 50 mg of methotrexate sodium in 2 mL of solution. The 5-mg sample contains 0.90% of benzyl alcohol as preservative, 0.63% of sodium chloride, and sodium hydroxide to give *p*H 8.5. The 50-mg sample contains 0.90% of benzyl alcohol, 0.26% of sodium chloride, and sodium hydroxide to give *p*H 8.5.

Methotrexate was the first drug to produce substantial (although temporary) remissions in leukemia.[146] It is still used for this purpose against acute lymphocytic leukemia and acute lymphoblastic leukemia. Because it has some ability to enter the CNS, it is used in the treatment and prophylaxis of meningeal leukemia. The discovery that methotrexate afforded a high percentage of apparently permanent remissions in choriocarcinoma in women justified the use of the term "cure" in cancer chemotherapy.[147] Methotrexate is used in combination chemotherapy for palliative management of breast cancer, epidermoid cancers of the head and neck, and lung cancer. It also is used against severe, disabling psoriasis. The most common toxic reactions are ulcerative stomatitis, leukopenia, and abdominal distress. A high dose of methotrexate combined with leucovorin "rescue" produces renal failure in some patients. This condition is thought to result from crystallization of the drug or its metabolites in acidic urine; it is countered by hydration and alkalinization.[148]

Azathioprine, USP. 6-[(1-Methyl-4-nitroimidazole-5-yl)thio]purine (Imuran). This compound is prepared from 6-mercaptopurine and 5-chloro-1-methyl-4-nitroimidazole.[149] It is supplied as 50-mg scored tablets. The injectable sodium salt is available in 20-mL vials containing 100 mg of azathioprine.

Azathioprine is well absorbed when taken orally. It is converted extensively to 6-mercaptopurine. The main indication for azathioprine is as an adjunct to prevent the rejection of renal homotransplants. It is contraindicated in patients who show hypersensitivity to it. The chief toxic effects are hematologic, expressed as leukopenia, anemia, and thrombocytopenia. Complete blood counts should be performed at least weekly, and the drug should be discontinued if there is a rapid fall or persistent decrease in leukocytes. Patients with impaired renal function

TABLE 8-2

ANTIMETABOLITES

Name Proprietary Name	***Preparation***	***Usual Adult Dose****	***Usual Dose Range****	***Usual Pediatric Dose****
Mercaptopurine USP *Purinethol*	Mercaptopurine tablets USP	2.5 mg/kg of body weight daily in single or divided doses		Children 5 yr and older: 2.5 mg/kg of body weight or 50 mg/m^2 of body surface daily in single or divided doses
Azathioprine USP *Imuran*	Azathioprine tablets USP	Initial, 3–5 mg/kg of body weight qd; maintenance, 1–4 mg/kg qd		
	Azathioprine sodium for injection USP	IV, the equivalent of azathioprine, 3–5 mg/kg of body weight daily		
Thioguanine USP	Thioguanine tablets USP	Initial, 2 mg/kg of body weight qd, increase to 3 mg/kg if no toxicity or improvement after 4 wk		See Usual Adult Dose
Fluorouracil USP *Efudex*	Fluorouracil injection USP	Initial, IV, 12 mg/kg of body weight qd for 4 days; if no toxicity occurs, 7–10 mg/kg is given every 3–4 days for a total course of 4 wk	Up to 800 mg daily	
	Fluorouracil topical solution USP Fluorouracil cream USP	Superficial basal-cell carcinomas: topical, to the skin, as a 5% cream or solution bid in a sufficient amount to cover the lesions, for at least 2–6 wk		
Floxuridine USP	Sterile floxuridine USP	Intra-arterial, 100–600 μg/kg of body weight continuously over 24 hr, continued until toxicity or a response occurs		
Cytarabine USP *Cytosar-U*	Sterile cytarabine USP	IV infusion, 0.5–1 mg/kg of body weight daily over 1–24 hr, for 10 days, increasing to 2 mg/kg daily thereafter, if neither response nor toxicity occurs		See Usual Adult Dose
Methotrexate USP	Methotrexate tablets USP	Antineoplastic: oral, 15–50 mg/m^2 of body surface once or twice a week, depending on condition and other agents being used concurrently		Antineoplastic: oral, 20–30 mg/m^2 of body surface once a week
	Methotrexate sodium injection USP	Antineoplastic: IM or IV, 15–50 mg/m^2 of body surface once or twice a week		Antineoplastic: IM, the equivalent of methotrexate, 20–30 mg/m^2 of body surface once a week

*See USP DI for complete dosage information.

might have slower elimination of the drug, which requires appropriate reduction of the dose.

ANTIBIOTICS

Seven different antibiotics now are established clinical anticancer agents, and several other antibiotics are undergoing clinical development. Most of these agents have been approved within the last few years. Thus, the recognition of antibiotics as an important class of antineoplastic drugs is very recent. However, some of the compounds in this class have been known for a long time. For example, dactinomycin (actinomycin D) was first isolated in 1940 by Waksman and Woodruff,[150] although its activity against neoplasms was not described until 1958. Furthermore, plicamycin, originally discovered as aureolic acid in 1953, had to be rediscovered twice before its antitumor activity was established in 1962.[151] These compounds were originally rejected as antibacterial agents because of their cytotoxicity. Only later was it found that this toxicity could be turned to an advantage in the chemotherapy of cancer. The discovery of antitumor activity is much simpler today, and some laboratories routinely screen extracts of microorganism cultures for cytotoxicity and lysogenic phage induction in bacteria (a predictor of potential antitumor activity). Such assays can be performed on small quantities of the extracts.[152]

The production of antitumor agents from microbial fermentations has some special advantages and disadvantages over chemical synthesis. Occasionally, the biosynthesis can be controlled to afford novel analogues. This has been true for actinomycins[153] and bleomycins.[154] Strain selection and fermentation conditions can optimize the formation of a particular component of an antibiotic mixture. Thus, *Streptomyces parvullus* produces dactinomycin almost exclusively, in contrast with other species that form complex mixtures of actinomycins.[152] The fermentation in *S. caespitosus* has been developed similarly, to produce almost all mitomycin. In some cases, such as with doxorubicin, improvement of the antibiotic yield has been difficult. This results in an expensive product and intensive research on chemical synthesis.

The actinomycins comprise numerous closely related structures. All of them contain the same chromophore, a substituted 3-phenoxazone-1,9-dicarboxylic acid known as actinocin. Each of the carboxyl groups is bonded to a pentapeptide lactone by way of the amino group of an L-threonine unit of this pentapeptide. The hydroxyl group of the L-threonine forms part of the lactone, along with L-methylvaline, the fifth amino acid from the chromophore. D-Valine or D-alloisoleucine is the second amino acid, and the fourth amino acid usually is sarcosine. The third amino acid is more variable, consisting of L-proline, L-hydroxyproline, L-oxoproline, or others, produced by controlled biosynthesis. Actinomycins that have two identical pentapeptide lactones are called isoactinomycins, whereas those with different pentapeptide lactones are called anisoactinomycins. The individual pentapeptide lactones are designated α and β, depending on their attachment to the 9- or 1-carboxylic acids, respectively.[155] Dactinomycin (actinomycin D, actinomycin C_1) is an isoactinomycin with an amino acid sequence of L-threonine, D-valine, L-proline, sarcosine, and L-*N*-methylvaline. Actinomycin C_3, which is used in Germany, differs from actinomycin D by a D-alloisoleucine unit instead of D-valine in both the α and β chains.[155]

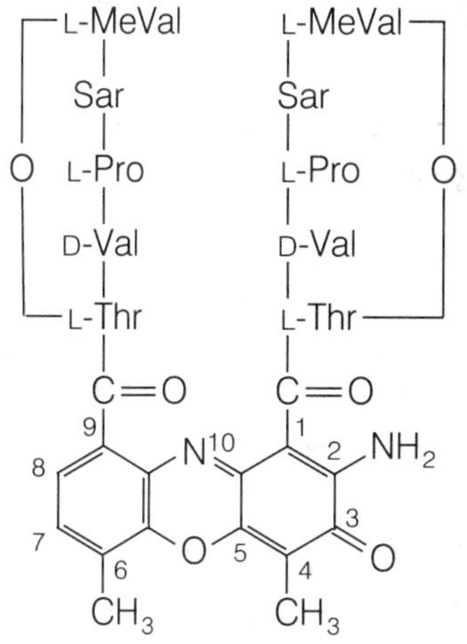

Dactinomycin

The mode of action of actinomycins has been studied extensively, and it now is generally accepted that they intercalate into double-helical DNA. In the intercalation process, the helix unwinds partially to permit the flat phenoxazone chromophone to fit in between successive base pairs. Adjacent G-C pairs are especially suitable because the 2-amino groups of the guanines can hydrogen bond with the carbonyl groups of threonines in the actinomycin. This bonding reinforces the π-bonding between the heterocyclic chromophores. Additional stability is conferred by the interaction between the pentapeptide lactone chains and DNA. These chains lie in the minor groove of the double helix, running in opposite directions, and they make numerous van der Waals interactions with the DNA.[156]

Intercalation into DNA changes its physical properties in characteristic ways. Thus, the length, viscosity, and melting temperature increase, whereas the sedimentation coefficient decreases.[157,158] Changes in substituents on the actinomycins influence their binding to DNA, usually by making it less effective. Opening of a lactone ring or changing the stereochemistry of an amino acid abolishes activity, and replacement of the 4- and 6-methyl groups by other substituents reduces it. Replacement of the 2-amino group also reduces activity.[152]

The main biochemical consequence of the intercalation of actinomycins into DNA is the inhibition of DNA and RNA synthesis. This inhibition eventually results in depletion of RNA and proteins and leads to cell death.[159]

Anthracyclines are another large and complex family of antibiotics. Many members of this family were isolated before a useful antitumor agent, daunorubicin, was found. This significant discovery was made independently in France and Italy in 1963.[160,161] Daunorubicin proved to be active against acute leukemias and it became an established clinical agent. However, it was pushed into the background by the discovery of doxorubicin (Adriamycin) in 1969.[162] Doxorubicin is active against a broad spectrum of tumors including both solid and hematologic types. It is presently the most widely used antineoplastic agent.[163]

Many anthracyclines, including all of those with antitumor activity, occur as glycosides of the anthracyclinones. The glycosidic linkage usually involves the 7-hydroxyl group of the anthracyclinone and the β-anomer of a sugar with L-configuration. *Anthracyclinone* refers to an aglycone containing the anthraquinone chromophore within a linear hydrocarbon skeleton, related to that of the tetracyclines.[164] The anthracyclinones differ from each other in the number and location of phenolic hydroxyl groups, the degree of oxidation of the two-carbon side chain at position 9, and the presence of a carboxylic acid ester at position 10. Thus, daunorubicin is a glycoside formed between daunomycinone and L-daunosamine, whereas doxorubicin is its 14-hydroxy analogue.[165] In contrast, aclarubicin (aclacinomycin A), a compound currently under development, has aklavinone in combination with a trisaccharide chain.[166]

Daunorubicin and doxorubicin exhibit biologic effects similar to those of actinomycin, and they are thought to intercalate into double-helical DNA.[167] However, the relationship of this intercalation to their antineoplastic activity is uncertain. Thus, the *N*-trifluoracetyl derivative of doxorubicin 14-valerate does not bind to DNA or even penetrate the nuclei of cells, yet it has antitumor activity.[168] It has been suggested that it either reacts by a different mechanism or is metabolized to an active form. Reduction of doxorubicin, followed by intercalation, causes DNA-strand scission. This scission is thought to result from the attack of hydroxyl radicals generated from redox cycles involving doxorubicin.[169] In contrast with daunorubicin, aclarubicin and related compounds do not induce lysogenic phage in bacteria. They are believed to interfere with RNA synthe-

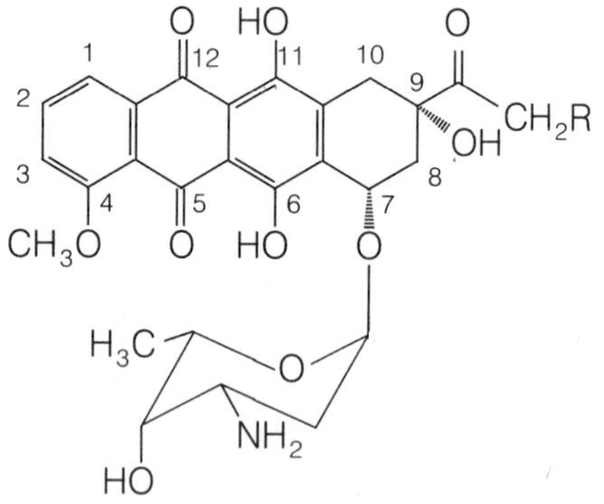

Daunorubicin, R = H
Doxorubicin, R = OH

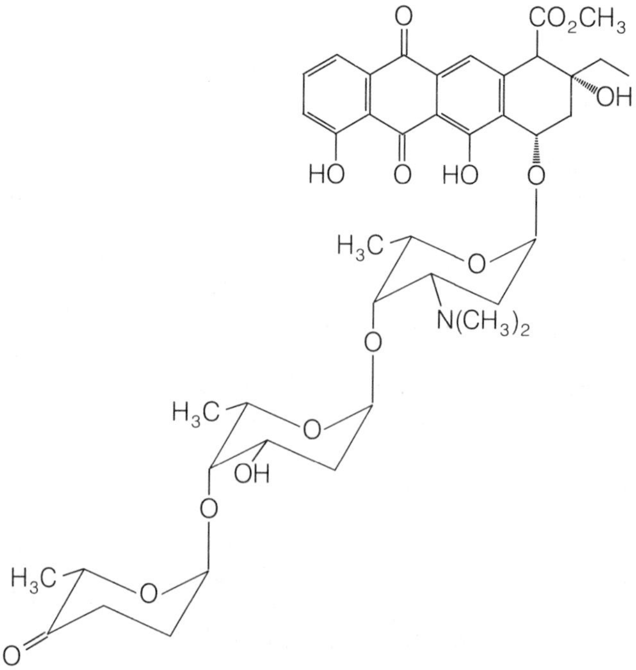

Aclarubicin
(Aclacinomycin A)

ses more than with DNA synthesis. Aclarubicin also lacks the cardiotoxicity shown by daunorubicin and doxorubicin.[166]

In contrast with the actinomycins, anthracyclines are metabolized in the liver. Daunorubicin is readily converted into its 13-hydroxy analogue, daunorubicinol, which is further cleaved to the aglycone.[170] The 14-hydroxyl group of doxorubicin makes it less susceptible to reduction of the 13-carbonyl group. However, the 13-hydroxy derivative, adriamycinol, is found among the metabolites, along with the 4-demethyl-4-sulfate. Both daunomycinol and adriamycinol are active against neoplastic cells, but their rates of uptake are low.[171]

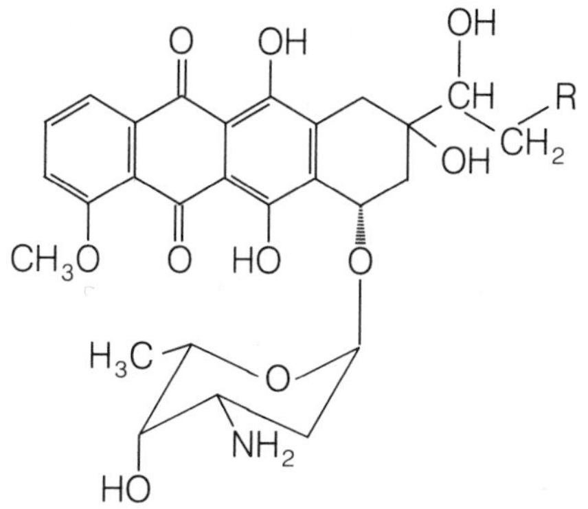

Many derivatives and analogues of daunorubicin and doxorubicin have been prepared. They include the 4-deoxy analogue of doxorubicin, which is orally active, and the benzoylhydrazone of daunorubicin (rubidazone). Transformations in the sugar moiety, including the 4′-deoxy and 4′-epi analogues and the *N*-trifluoroacetyl derivative of the 14-valerate ester, also give compounds with interesting activity.[172] The 4-hydroxy analogue of daunorubicin, carminomycin, isolated from *Actinomadura carminata*, has been evaluated in the Soviet Union.[173]

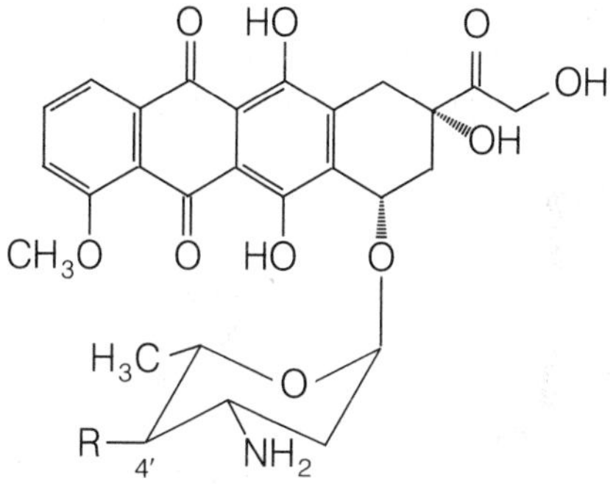

Another anthracycline with significant antitumor activity is nogalamycin, which is obtained from *S. nogalater*.[174] It differs from other anthracyclines in that the amino sugar is joined to the nucleus by a carbon–carbon bond and a cyclic acetal linkage. However, there is a nonamino sugar, nogalose, at the usual 7-position. Although nogalamycin itself is not an established antineoplastic drug, a semisynthetic analogue, menogaril, has received Phase II clinical trials and is under consideration for approval. Menogaril has the nogalose moiety replaced by a methoxy group and reversed chirality at the 7-position. It also differs from nogalamycin by the absence of the 10-carbomethoxy group.[175] These structural changes result in a change in the mode of action from intercalation into DNA, as found for nogalamycin, to some other site and type of cytotoxic process. Thus, menogaril localizes in the cytoplasm, rather than the nucleus, and it has very little effect on DNA and RNA synthesis at cytotoxic doses.[176] Menogaril is not more effective than doxorubicin against tumors, but its much lower cardiotoxicity and potential oral activity might offer clinical advantages.

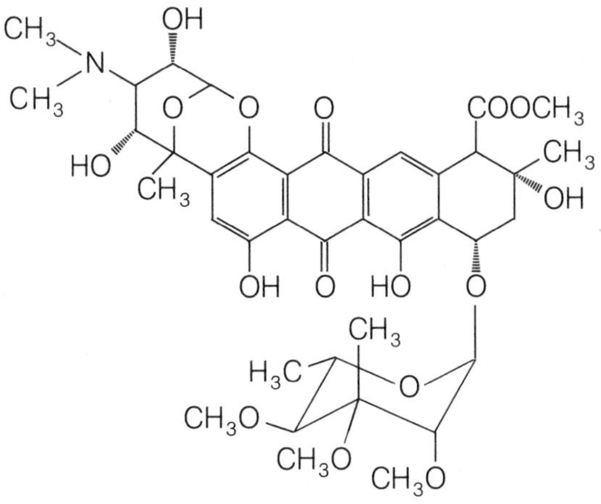

Daunomycinol, R = H
Adriamycinol, R = OH

4′-Deoxydoxorubicin, R = H
4′-Epidoxorubicin, R = OH

Nogalamycin

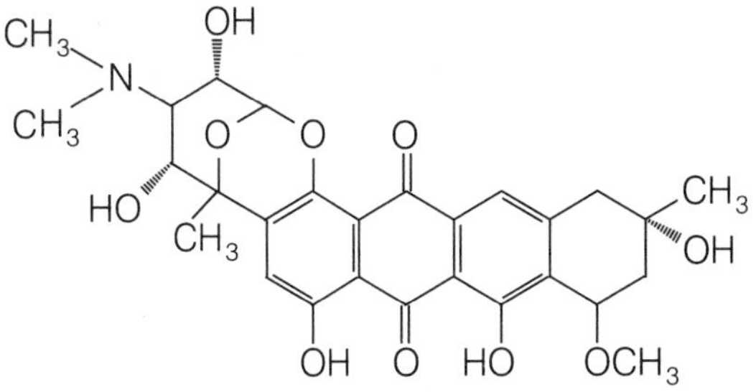

Menogaril

The aureolic acid group of antitumor antibiotics includes aureolic acid (plicamycin, mithramycin), the olivomycins, the chromomycins, variamycin, and related compounds. Plicamycin is the only member approved for clinical use in the United States. It is restricted to testicular carcinoma and hypercalcemia that is resistant to other drugs. However, chromomycin A_3 is used in Japan and olivomycin A is used in the Soviet Union.[152] Aureolic acid group compounds have complex structures consisting of an aglycone and two carbohydrate chains. The aglycones are tetrahydroanthracene derivatives with phenolic hydroxyl groups at positions 6, 8, and 9, and a pentanyl side chain that is highly oxygenated. The carbohydrate chains contain either two or three 2,6-dideoxy sugars of novel structures.[177]

Plicamycin and related compounds are weakly acidic owing to the phenolic groups (pK_a 5). They readily form sodium salts that show brilliant yellow fluorescence.[178] The chromophore is responsible for complex formation with divalent metals such as magnesium and calcium. Such complex formation is required before aureolic acids can bind with DNA.[179] The nature of this DNA binding is currently uncertain. Intercalation has been suggested, but the evidence for this process is incomplete.[180] Whatever the exact nature of the binding, plicamycin and other aureolic acids inhibit DNA-dependent RNA polymerase, and this effect leads to cell death.[181]

The discovery of bleomycin in 1966 resulted from the establishment by H. Umezawa of a program for screening microbial culture filtrates against experimental tumors.[182] Bleomycin is a mixture of closely related compounds that is partly resolved before formulation for clinical use.[183] The presently used commercial product, Blenoxane, contains bleomycins A_2 and B_2. A variety of other antibiotics have structures similar to those of the bleomycins. They include the phleomycins (which differ from bleomycins in having one thiazole ring partly reduced), zorbamycin and the zorbonamycins, antibiotic YA-56, victomycin, the tallysomycins, and the platomycins.[152] New bleomycin analogues also have been prepared by controlled biosynthesis from bleomycinic acid.

Bleomycins and their analogues occur naturally as blue copper chelates. Removal of the copper by chemical reduction or complexing agents affords the antibiotics as white solids.[184,185] Copper-free bleomycin is preferred for chemotherapy because of its decreased toxicity. The complexation of bleomycin with metal ions occurs readily and is a key factor in its mode of action. Inside the cell, bleomycin forms a chelate with Fe(II) that has square pyramidal geometry.[186] Nitrogen atoms from bleomycin occupy five of the positions in this structure. The sixth position may be occupied by the carbonyl group of the carbamate function, but this group is readily displaced by

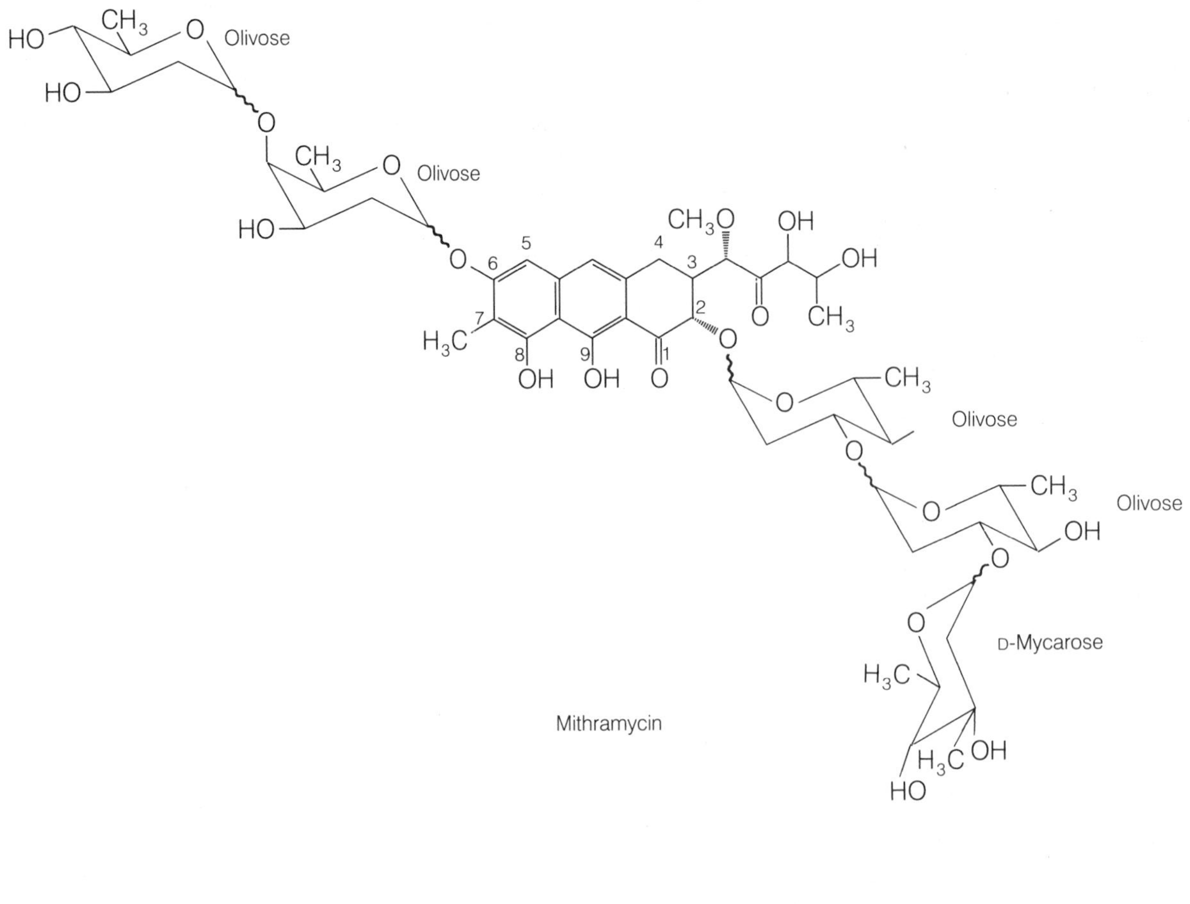

Mithramycin

Bleomycinic Acid, R = OH
Bleomycin A_2, R = $NH(CH_2)_3\overset{+}{S}(CH_3)_2$
Bleomycin B_2, R = $NH(CH_2)_4NHC(=NH)NH_2$

molecular oxygen. The resulting complex may give rise to hydroxyl radicals and superoxide radicals. Because the bithiazole portion of bleomycin can intercalate partially into DNA, these highly reactive radicals are generated close to the double helix, and they cause cleavage of the phosphodiester bonds. This degradation of DNA strands is thought to be the lethal event in cells.[187]

Bleomycin is inactivated by an intracellular enzyme named bleomycin hydrolase, an aminopeptidase that hydrolyzes the carboxamide group of the β-aminoalanine carboxamide residue to the corresponding carboxylate. This structural change increases the pK_a of the α-amino group from 7.3 to 9.4, which results in poorer binding to DNA.[185] Chelation with Fe(II) still occurs, but the production of hydroxyl radicals is drastically reduced.[187] Bleomycin hydrolase levels in tumor cells help to determine their resistance to bleomycin. Thus, squamous cell carinoma is characterized by ready uptake of bleomycin and low levels of the hydrolase. It is especially sensitive to bleomycin.[188]

Bleomycins undergo two different inactivating reactions under mildly alkaline conditions. One is migration of the carbamoyl group to an adjacent hydroxyl group of the mannose residue; the resulting product is called an isobleomycin.[189] Copper-chelated bleomycins do not undergo this reaction. However, they are slowly transformed into epibleomycins, which are racemized at the carbon atom substituted at the 2-position of the pyrimidine ring.[190] Epibleomycins retain about 25% of the antitumor activity of the parent bleomycins.

Bleomycinic acid is obtained by chemical degradation of bleomycin A_2 or enzymatic degradation of bleomycin B_2. It can be readily transformed into semisynthetic bleomycins such as PEP-bleomycin, which possesses reduced pulmonary toxicity.[183]

The mitomycins were discovered in Japan in the late 1950s, and one of them, mitomycin C, was rapidly developed as an anticancer drug.[191] However, the initial clinical experience with this compound in the United States was disappointing. It was not approved until 1974, following extensive studies and the establishment of satisfactory dosage schedules. Porfiromycin, the *N*-methyl homologue of mitomycin C, was discovered at the Upjohn Company.[192] It has received clinical study, but it is not yet an approved agent.

Structures of the mitomycins were elucidated at Lederle Laboratories. These compounds have an unusual combination of three different carcinostatic functions: quinone, carbamate, and aziridine.[193] They are arranged in such a way that the molecule is relatively unreactive in its natural state. However, chemical or enzymatic reduction to the correspond-

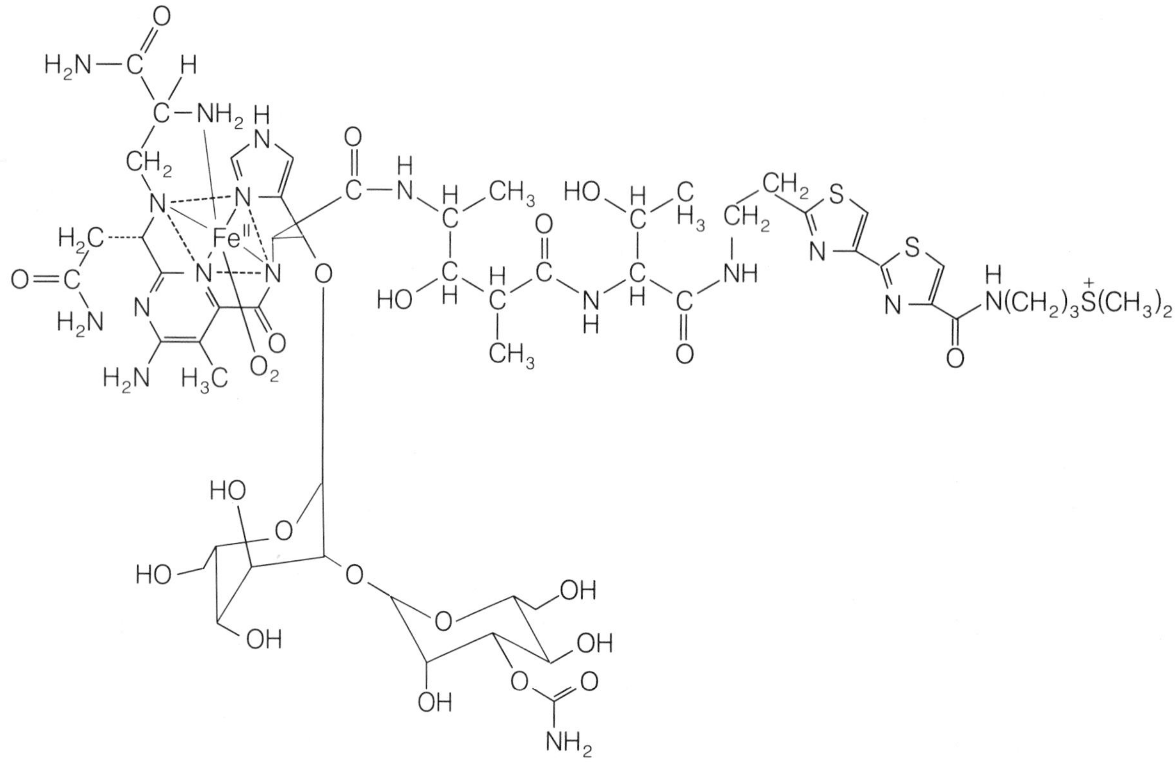

Bleomycin A_2 Fe(II) Chelate

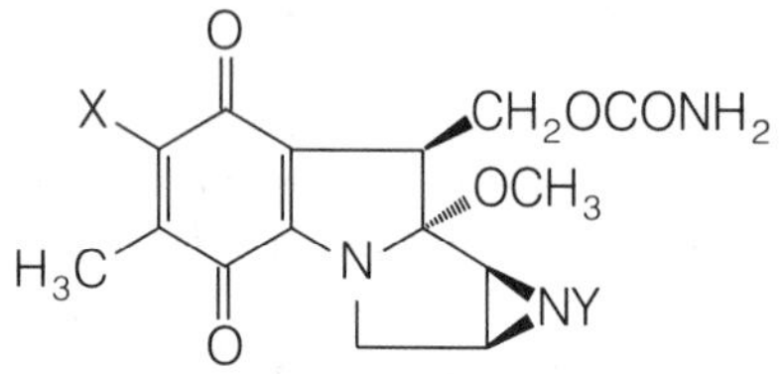

Mitomycin A, $X = CH_3O$, $Y = H$
Mitomycin C, $X = H_2N$, $Y = H$
Porfiromycin, $X = H_2N$, $Y = CH_3$
BMY-25067, $X = O_2NC_6H_4SS(CH_2)_2NH$

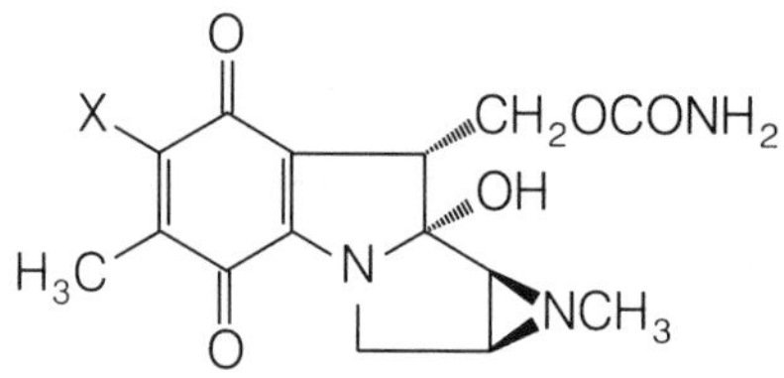

Mitomycin B, $X = CH_3O$
Mitomycin D, $X = H_2N$

ing hydroquinone is followed by the loss of methanol (water from mitomycin B), and the resulting indolohydroquinone becomes a bifunctional alkylating agent capable of cross-linking double-helical DNA (see Scheme 3).[194] Mitomycins bound to DNA may undergo successive redox cycles, each of which results in the generation of hydrogen peroxide. This potent oxidizing agent can cause single-strand cleavage of the DNA.[195]

Mitomycins are unstable in both acids and bases. Mild acid hydrolysis results in opening of the aziridine ring and loss of methanol or water to give mitosenes such as 2,7-diamino-1-hydroxymitosene.[196] Catalytic hydrogenation followed by reoxidation gives aziridinomitosenes, which retain a significant amount of antitumor activity in animals.[197] Many mitomycin analogues have been prepared by partial synthesis, and two of them have received clinical trials.[198, 199] However, unexpected toxicity has led to their withdrawal. The present clinical candidate, BMY-25067 contains a disulfide substituent on the 7-amino group. Control of the quinone reduction potential is especially stressed in analogue studies, because reduction is the key step in bioactivation of these molecules.[200]

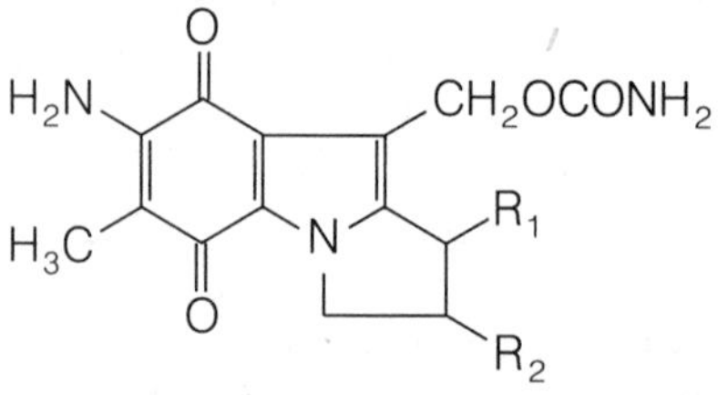

2,7-Diamino-1-hydroxymitosene, $R^1 = OH$, $R^2 = NH_2$

1,2-Aziridino-7-aminomitosene, $R^1, R^2 = \rangle NH$

Streptozocin was isolated from *S. achromogenes* in 1960.[201] It is the nitrosomethylurea derivative of a 2-deoxyglucose.[202] The simplicity of its structure and the cost of preparing it by fermentation have led to the development of practical syntheses from 2-amino-2-deoxyglucose.[203] Streptozocin is an alkylating agent similar in reactivity to other nitrosomethylureas, except that its glucose moiety causes it to be especially taken up in the pancreas. This effect is detrimental in that it produces diabetes, but it makes the molecule especially effective against malignant insulinomas.[204] It is an approved clinical agent for this specific use. The chloroethyl analogue of streptozocin, called chlorozotocin, shows good antitumor activity in animals and is not diabetogenic.[205]

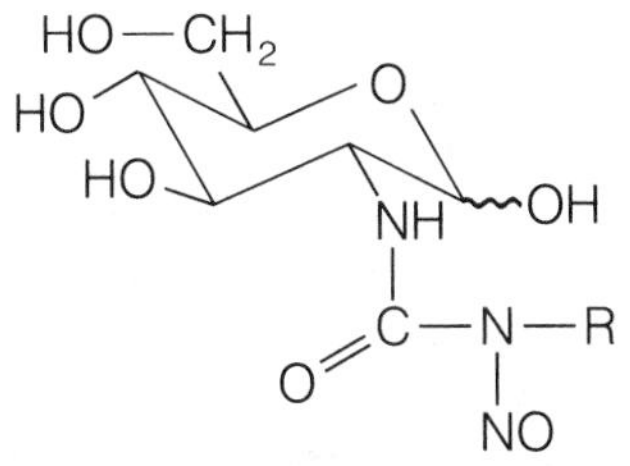

Streptozocin, R = CH_3
Chlorozotocin, R = CH_2CH_2Cl

Acivicin is another antibiotic that is currently receiving clinical study. It is obtained from *S. sivicens*, and it functions as an inhibitor of the amidotransferases involved in purine and pyrimidine biosynthesis.[206] The structure of acivicin shows a chlorine atom that can be replaced readily because it is located on an imine group. A cysteine residue at the active site of an amidotransferase replaces this chlorine, affording alkylation and irreversible inhibition of the enzyme. Phase I clinical studies revealed CNS toxicity for acivicin. It is thought that conversion of the antibiotic to ibotenic acid, a known CNS toxin found in mushrooms, by exchange of the chlorine for a hydroxyl group, might be responsible for this toxicity.

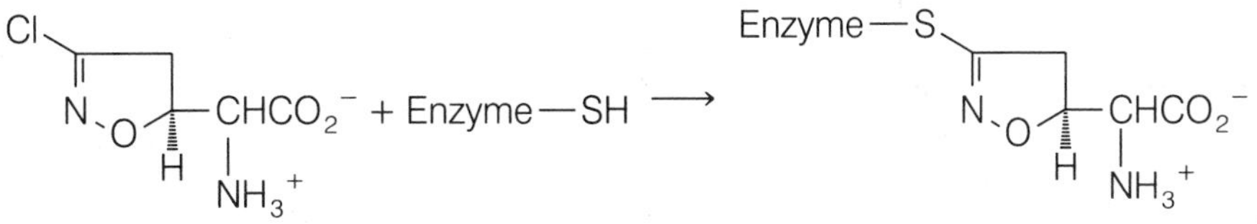

Acivicin

PRODUCTS

Dactinomycin, USP. Actinomycin D; actinomycin C_1; actinomycin IV; NSC-3053 (Cosmegan). This compound is obtained from the fermentation of selected strains of *S. parvullus*. It is soluble in alcohols and alcohol–water mixtures; however, these solutions are very sensitive to light. Vials containing 0.5 mg of lyophilized powder are supplied.

Dactinomycin is used against rhabdomyosarcoma and Wilms' tumor in children.[207] It can be lifesaving for women with choriocarcinoma resistant to methotrexate. In combination with vincristine and cyclophosphamide, it has received some use for solid tumors in children. Toxic reactions include anorexia, nausea, and vomiting. Bone marrow depression, resulting in pancytopenia may occur within a week after therapy. Alopecia, erythema, and tissue injury at the site of injection may occur.

Daunorubicin Hydrochloride, USP. Daunomycin; rubidomycin; NSC-82151 (Cerubidine). It is obtained from the fermentation of *S. peucetius*.[161] The hydrochloride salt is a red crystalline compound that is soluble in water and alcohols. The pink solution turns blue at alkaline *p*H. Daunorubicin hydrochloride is available as lyophilized powder in 20-mg vials. In this form it is stable at room temperature, but after reconstitution with 5 to 10 mL of sterile water, it should be used within six hours.

Daunorubicin is used in the treatment of acute lymphocytic and granulocytic leukemias.[208] Toxic effects include bone marrow depression, stomatitis, alopecia, and gastrointestinal disturbances. At higher doses cardiac toxicity may develop. Severe and progressive congestive heart failure may follow initial tachycardia and arrythmias.

The usual dose of daunorubicin is 30 to 60 mg/m^2 daily for three days. It is administered intravenously, taking care to prevent extravasation.

Doxorubicin Hydrochloride, USP. NSC-123127; 14-hydroxydaunomycin (Adriamycin). This compound is obtained from cultures of *S. peucetius* var. *caesius*.[209] The orange-red needles are soluble in water and alcohols. Above *p*H 9 the orange solutions turn blue-violet. Doxorubicin hydrochloride is supplied as a freeze-dried powder in two different sizes: 10 mg plus 50 mg of lactose, USP, and 50 mg plus 250 mg of lactose, USP. These amounts are reconstituted with 5 mL and 25 mL, respectively, of sodium chloride injection, USP.

Doxorubicin is one of the most effective antitumor agents. It has been used successfully to produce regressions in acute leukemias, Hodgkin's disease and other lymphomas, Wilms' tumor, neuroblastoma, soft-tissue and bone sarcomas, breast carcinoma, ovarian carcinoma, transitional-cell bladder carcinoma, thyroid carcinoma, and small-cell bronchogenic carcinoma.[210] Combination chemotherapy with various other agents is being developed for specific tumors. The dose-limiting toxicities are myelosuppression and cardiotoxicity. There is a high incidence of bone marrow depression, primarily of leukocytes, that usually reaches its nadir at 10 to 14 days. Red blood cells and platelets also may be depressed. Consequently, careful blood counts are essential. Acute left ventricular failure has occurred, particularly in patients receiving a total dose exceeding the currently recommended 550 mg/m^2. Cardiomyopathy and congestive heart failure may be encountered several weeks after discontinuing doxorubicin. Toxicity is augmented by impaired liver function because this is the site of metabolism. Thus, evaluation of liver function by conventional laboratory tests is recommended before individual dosing.

The recommended dosage schedule is 60 to 75 mg/m^2, intravenously at 21-day intervals. This dose is decreased if liver function or bone marrow reserves are inadequate. Care must be taken to avoid extravasation.

Bleomycin Sulfate, Sterile, USP. NSC-125066 (Blenoxane). This product is a mixture of cytotoxic glycopeptides isolated from a strain of *S. verticillus*.[211] The main component is bleomycin A_2, and bleomycin B_2 also is present. Bleomycin is a whitish powder that is readily soluble in water. It occurs naturally as a blue copper complex, but the copper is removed from the pharmaceutical form. It is supplied in ampuls containing 15 units of sterile bleomycin sulfate.

Bleomycin is used for the palliative treatment of squamous cell carcinomas of the head and neck, esophagus, skin, and genitourinary tract including penis, cervix, and vulva.[212] It also is used against testicular carcinoma, especially in combination with cisplatin and vinblastine.[213] The principal toxicities of bleomycin are for skin and lungs. Other tissues contain an aminopeptidase that rapidly inactivates it. Bleomycin has very little bone marrow toxicity; hence, it may be used in combination with myelosuppressive agents. Pulmonary toxicity is induced in about 10% of treated patients, with pulmonary fibrosis and death occurring in about 1%. Skin or mucous membrane toxicity occurs in about half of the patients. Anaphylactoid reactions are possible in lymphoma patients.

The recommended dosage is 0.25 to 0.50 units/kg (10 to 20 units/m^2) given intravenously, intramuscularly, or subcutaneously once or twice weekly. For maintenance of Hodgkin's disease patients in remission, a dose of 1 unit daily or 5 units weekly is given. Blenoxane is stable for 24 hours at room temperature in sodium chloride or 5% dextrose solutions for injection.

Mitomycin, USP. Mitomycin C; NSC-26980 (Mutamycin). This compound is obtained from cultures of *S. caespitosus* as blue-violet crystals.[214] It is soluble in water and polar organic solvents. Vials containing either 5 mg of mitomycin and 10 mg of mannitol or 20 mg of mitomycin and 40 mg of mannitol are supplied. The unreconstituted product is stable at room temperature for at least two years.

Mitomycin is useful in treating disseminated breast, gastric, pancreatic, or colorectal adenocarcinomas in combination with fluorouracil and doxorubicin (Adriamycin) (FAM program). It is used in combination with cyclophosphamide and doxorubicin for lung cancer. Complete remissions of superficial transitional-cell carcinomas of the bladder have been obtained in 60% of patients.[215] The dose-limiting toxicity is myelosuppression, characterized by delayed, cumulative pancytopenia. Fever, anorexia, nausea, and vomiting also occur.

Mitomycin at 10 to 20 mg/m^2 is given as a single dose by intravenous catheter. No repeat dose should be given until the leukocyte and platelet counts have recovered (about eight weeks).

Plicamycin, USP. Aureolic acid; mithramycin; NSC-24559 (Mithracin). This antibiotic is obtained from *S. plicatus*[178] or *S. argillaceus* as a yellow solid that melts at 180° to 183°C. It is soluble in polar organic solvents and aqueous alkali; however, it is susceptible to air oxidation in alkali. Plicamycin readily forms complexes with magnesium and other divalent metal ions, and these complexes have drastically altered optical rotations. Vials containing 2.5 mg of plicamycin as a freeze-dried powder are supplied.

Plicamycin is used largely in the treatment of advanced embryonal tumors of the testes.[204] However, it has been largely superseded by newer agents such as bleomycin and cisplatin. The main present use of plicamycin is in Paget's disease, in which it gives reduction of alkaline phosphatase activity and relief of bone pain.[216] It also is useful in treating patients with severe hypercalcemia or hypercalciuria resulting from advanced metastatic cancer involving bones. Plicamycin may produce severe hemorrhaging. Bone marrow, liver, and kidney toxicity also occur. The lower total dose used for hypercalcemia results in less toxicity.

Streptozocin. 2-Deoxy-2-(3-methyl-3-nitrosoureido)-D-glucopyranose; NSC-85998 (Zanosar) is obtained from cultures of *S. achromogenus* subspecies *streptozoticus* or synthesized from D-glucosamine. It is readily soluble in water or saline. Vials containing 1.0 g of lyophilized powder are supplied. They should be refrigerated at 35° to 46°F and protected from light.

Streptozocin is indicated only for metastatic islet cell carcinoma of the pancreas.[217] Therapy is limited to patients with symptomatic or progressive disease because of inherent renal toxicity of the drug. Up to two-thirds of patients treated with it experience renal toxicity. Nausea and vomiting occur in over 90% of patients, which occasionally requires discontinuation of drug therapy. Liver dysfunction also occurs. It is mutagenic, carcinogenic, and teratogenic in animals. Carcinogenesis following topical exposure is a possible hazard. After rapid injection, unchanged drug is rapidly cleared from the plasma. The half-life is 35 minutes.

TABLE 8-3

ANTIBIOTICS

Name Proprietary Name	***Preparation***	***Usual Adult Dose****
Daunorubicin hydrochloride *Cerubidine*	Powder for injection	IV, 60 mg/m² of body surface given on days 1, 2, and 3 every 3–4 wk
Doxorubicin hydrochloride USP *Adriamycin*	Powder for injection	IV, 60–75 mg/m² of body surface as a single injection at 21-day intervals, or 30 mg/m² on each of 3 successive days, repeated every 4 wk
Mitomycin USP *Mutamycin*	Lyophilized powder for injection	IV, 20 mg/m² of body surface as a single dose or 2 mg/m² daily for 5 days
Dactinomycin USP *Cosmegen*	Dactinomycin for injection USP	IV, 10–15 μg/kg of body weight daily for a minimum of 5 days every 4–6 wk
Plicamycin USP *Mithracin*	Powder for injection	IV, 25–30 μg/kg of body weight daily for 8–10 days
Streptozocin *Zanosar*	Powder for injection	IV, 500 mg/m² of body surface daily for 5 consecutive days every 6 wk

*See the USP DI for complete dosage information.

PLANT PRODUCTS

The use of higher plants in treating neoplastic disease dates to antiquity. Dioscorides described the use of colchicine for this purpose in the first century. In more recent years, scientists have attempted to select and systematically screen plants reputed to have antitumor activity. If the presence of activity is established for one member of a plant family, other members of this family are selected and tested. A major impetus to this research was given by Hartwell at the NCI, who established an extensive system of plant collection, screening, and isolation.[218] Approximately 100,000 plants have been screened under this program.

Resin of the mayapple, *Podophyllum peltatum*, has long been used as a remedy for warts. One of its constituents, podophyllotoxin, has antineoplastic activity, but it is highly toxic.[219] This lignan inhibits mitosis by destroying the structural organization of the mitotic apparatus.[220] Early derivatives of podophyllotoxin showed poor clinical activity, but newer analogues, such as the epipodophyllotoxin derivatives etiposide (VP-16,213; also designated VP-16) and teniposide (VM-26), are much better. The former compound has been approved for use against small-cell lung carcinoma. Both of these analogues differ from podophyllotoxin in that they are inhibitors of topoisomerase II, rather than microtubule assembly.[221]

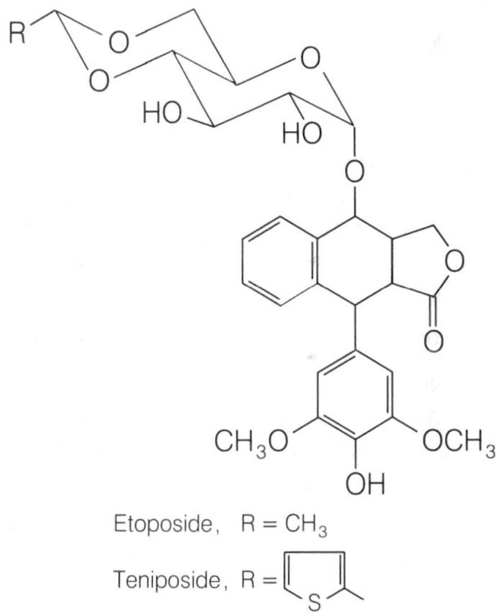

Etoposide, R = CH_3

Teniposide, R = (2-thienyl)

The most important antitumor agents from plants are the vinca alkaloids (Table 8-4). These compounds were isolated from the periwinkle *Catharanthus rosea* at the Eli Lilly Company.[222] They have complex structures composed of an indole-containing moiety named catharanthine and an indoline-containing moiety named vindoline.[223] Four closely related compounds have antitumor activity: vincristine, vinblastine, vinrosidine, and vinleurosine. Among this group, vincristine and vinblastine are proved clinical agents. These two compounds are used against different types of tumors, despite the similarity of their structures. Several semisynthetic compounds have been prepared. Among them, vinglycinate and 6,7-dihydrovinblastine have clinical

TABLE 8-4

VINCA ALKALOIDS AND THEIR ANALOGUES

	R	R_1	R_2	R_3	R_4
Vincristine	CH_3CO	CHO	H	OH	OCH_3
Vinblastine	CH_3CO	CH_3	H	OH	OCH_3
Vinrosidine	CH_3CO	CH_3	OH	H	OCH_3
Vinleurosine	CH_3CO	CH_3	Oxide		OCH_3
Vinglycinate	$(CH_3)_2NCH_2CO$	CH_3	H	OH	OCH_3
Vindesine	H	CH_3	H	OH	NH_2

potential.[224] Vindesine has undergone Phase II clinical studies. It is considered to resemble vincristine, but to be less neurotoxic.[225]

Vinca alkaloids cause mitotic arrest by promoting the dissolution of microtubules in cells. Microtubule crystals containing the alkaloids are formed in the cytoplasm.[226] Vinblastine is the most active compound, whereas vincristine is the only compound to cause irreversible inhibition of mitosis.[227] Cells can resume mitosis after brief exposure to other vinca alkaloids after these compounds are withdrawn.[228]

As noted in the foregoing, colchicine, obtained from the crocus *Colchicum autumnale*, has long been known for its antitumor activity. However, it is not now used clinically for this purpose. Its main use is in terminating acute attacks of gout. Among colchicine derivatives, demecolcine (Colcemid) is active against myelocytic leukemia, but only at near-toxic doses. Colchicines have an unusual tricyclic structure containing a tropolone ring. They inhibit

Colchicine, R = $COCH_3$
Colcemid, R = CH_3

mitosis at metaphase by disorienting the organization of the spindle and asters.[229]

Many other plant constituents show significant antitumor activity in animals and some of them have been given clinical evaluation. The more important compounds are homoharringtonine, anguidine, bouvardin, bruceantin, camptothecin, indicene-*N*-oxide lapachol, taxol, and thalicarpine.[83] Taxol is currently in Phase II clinical study.

PRODUCTS

Etoposide. VP-16,213; NSC-141540 (Ve Pesid) is a semisynthetic derivative of podophyllotoxin. It is supplied in 5-mL ampules containing 20 mg/mL of the drug plus 30 mg benzyl alcohol, 80 mg polysorbate 80, 650 mg of polyethyleneglycol 300, and 30.5% alcohol. This mixture is diluted with either 5% dextrose or 0.9% saline to give a final concentration of 0.2 or 0.4 mg/mL. Etoposide also is supplied as 50-mg capsules, also containing sorbitol. They must be stored at 36° to 46°F.

Etoposide in combination with other chemotherapeutic agents is the first-choice treatment for small-cell lung cancer. It also is effective in combination with other agents for refractory testicular tumors, and it has been used alone or in combination against acute nonlymphocytic leukemias, Hodgkin's disease, non-Hodgkin's lymphomas, and Kaposi's sarcoma. It is contraindicated in patients who develop hypersensitivity. Dose-limiting bone marrow suppression is the most critical toxicity, and reversible alopecia occurs frequently. Nausea and vomiting are usually controlled with standard therapy. On intravenous administration the disposition of etoposide is biphasic, with a distribution half-life of about 1.5 hours and an elimination half-life of 4 to 11 hours.

Vinblastine Sulfate, USP. Vincaleukoblastine; VLB; NSC-49842 (Velban). This antitumor alkaloid is isolated from *Vinca rosea* Linnaeus, the periwinkle plant.[222] It is soluble in water and alcohol. Vials containing 10 mg of vinblastine sulfate as a lyophilized plug are supplied. It is reconstituted by the addition of sodium chloride solution for injection preserved with phenol or benzyl alcohol.

Vinblastine has been used for the palliation of a variety of neoplastic diseases. It is one of the most effective single agents against Hodgkin's disease, and it may be used in combination chemotherapy for patients who have relapses after treatment by the MOPP program. Advanced testicular germinal cell tumors respond to vinblastine alone or in combination.[208] Beneficial effects are also obtained against lymphocytic lymphoma, histiocytic lymphoma, mycosis fungoides, Kaposi's sarcoma, Letterer-Siwe disease, resistant choriocarcinoma, and carcinoma of the breast. The limiting toxicity is leukopenia, which reaches its nadir in five to ten days after the last dose. Gastrointestinal and neurologic symptoms occur and are dose-dependent. Extravasation during injection can lead to cellulitis and phlebitis.

Vincristine Sulfate, USP. Leurocristine; VCR; LCR; NSC-67574 (Oncovin). This alkaloid is isolated from *V. rosea* Linnaeus.[222] The sulfate is a crystalline solid that is soluble in water. It is supplied in vials containing either 1 mg of vincristine sulfate and 10 mg of lactose or 5 mg of vincristine and 50 mg of lactose. Each size has an accompanying vial of 10 mL of bacteriostatic sodium chloride solution containing 90 mg of sodium chloride and 0.9% benzyl alcohol. The reconstituted pharmaceutical may be stored 14 days in a refrigerator.

Vincristine is effective against acute leukemia. In combination with prednisone it produces complete remission in 90% of children with acute lymphoblastic leukemia.[143] It is used in the MOPP program of combination chemotherapy for Hodgkin's disease.[230] Other tumors that respond to vincristine in combination with other antineoplastic agents include lym-

TABLE 8-5

OTHER ANTINEOPLASTIC AGENTS

Name Proprietary Name	***Preparation***	***Usual Adult Dose****	***Usual Dose Range****	***Usual Pediatric Dose****
Vinblastine sulfate USP *Velban*	Sterile vinblastine sulfate USP	IV, 3.7 mg/m^2 of body surface initially, followed at weekly intervals by doses increased gradually to 18.5 mg/m^2 of body surface		Initial, 2.5 mg/m^2 of body surface gradually increasing until a maximum of 7.5 mg/m^2
Vincristine sulfate USP *Oncovin*	Vincristine sulfate for injection USP	IV, 1.4 mg/m^2 of body surface given weekly as a single dose	2 mg/dose	1.5–2 mg/m^2 of body surface, given weekly as a single dose
Hydroxyurea USP *Hydrea*	Hydroxyurea capsules USP	For solid tumors, 80 mg/kg of body weight as a single dose every third day, or 20–30 mg/kg of body weight daily as a single dose		
Pipobroman USP *Vercyte*	Pipobroman tablets USP	Initial, 1–2.5 mg/kg of body weight: maintenance, 100–200 μg/kg or 7–175 mg		
Etoposide *Ve Pesid*	Ampule for injection	IV, 35–50 mg/m^2 of body surface daily for 4 consecutive days every 3–4 wk for small-cell lung cancer; 500–100 mg/m^2 on days 1–5 for testicular cancer		
		Oral, approximately two times the parenteral dose		
Asparaginase *Elspar*	Powder for injection	Single agent therapy: 200 IU/kg of body weight IV for 28 days.		
		Combination therapy: 1000 IU/kg of body weight IV for 10 successive days beginning on day 22, or 6000 IU/kg every 3 days beginning on day 4		
Testolactone USP *Teslac*	Testolactone tablets USP	250 mg qid		
	Sterile testolactone suspension USP	IM, 100 mg 3 times per week		
Dromostanolone propionate USP *Drolban*	Dromostanolone propionate injection USP	Parenteral, 100 mg 3 times weekly		
Leuprolide acetate *Lupron*	Solution for injection	Subcutaneous, 1 mg/day		
Interferon alfa-2a *Roferon-A*	Powder for injection	Subcutaneous or IM, 3 million IU/day for 16–24 wk for induction; 3 million IU 3 times per week for maintenance		
Interferon alfa-2b *Intron*	Powder for injection, plus sterile water	Subcutaneous or IM, 3 million IU 3 times per week		

*See USP DI for complete dosage information.

phosarcoma, reticulum cell sarcoma, rhabdomyosarcoma, neuroblastoma, and Wilms' tumor. Although the tumor spectra of vinblastine and vincristine are similar, there is a lack of cross-resistance between the two. Because vincristine is less myelosuppressive than vinblastine, it is preferred in combination with myelotoxic agents. The most serious clinical toxicity of vincristine is neurologic, with paresthesias, loss of deep-tendon reflexes, pain, and muscle weakness occurring. These symptoms can usually be reversed by lowering the dose or suspending therapy. Constipation and alopecia also occur. The rapid action of vincristine in destroying cancer cells may result in hyperuricemia. This complication can be prevented by administering allopurinol.

MISCELLANEOUS COMPOUNDS

In 1965 Rosenberg investigated the effects of electrical fields on bacteria and found that *Escherichia coli* formed long filaments instead of dividing.[231] He subsequently discovered that this effect was caused not by the electrical current, but by a complex, $[Pt(Cl_4)(NH_3)_2]^0$, formed from the platinum electrode in the presence of ammonium and chloride ions.[232] This discovery was followed by testing a variety of platinum-neutral complexes against tumors, with the result that *cis*-diamminedichloroplatinum II (cisplatin) eventually became established as a clinical agent.[233]

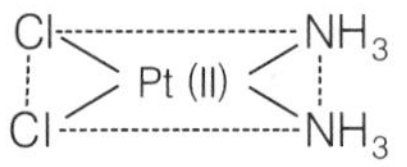

Cisplatin

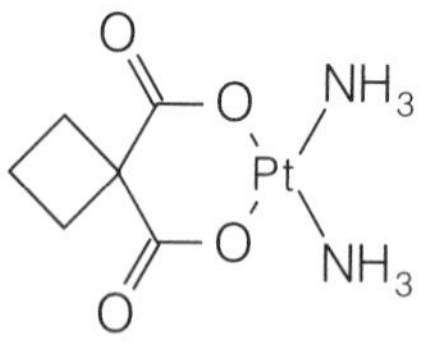

Carboplatin

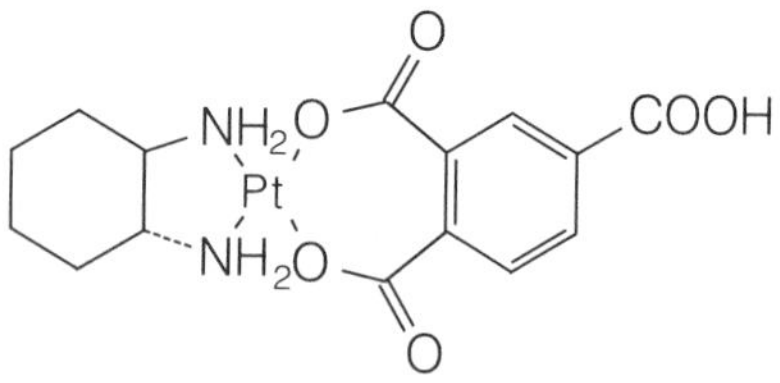

DACH

This platinum complex is a potent inhibitor of DNA polymerase. Its activity and toxicity resemble those of the alkylating agents. Considerable evidence has been obtained for DNA cross-linking by the platinum complex, wherein the two chlorides are displaced by nitrogen or oxygen atoms of purines. This evidence includes facilitated renaturation, increased sedimentation coefficient, hyperchromicity of the DNA ultraviolet spectrum, and selective reaction of the complex with guanine over other bases.[234]

Many other platinum complexes have been found active against tumors. Generally, they fall into the classification of *cis*-isomers in which one pair of ligands are monodentate anions of intermediate-leaving ability (such as chloride) or bidentate anions (such as malonate), and the other pair are mono- or bidentate amines.[235] Among the more significant analogues are carboplatin, which has received clinical studies in lung, ovarian, genitourinary, and head and neck cancer,[236] and DACH.

Hydroxyurea has been known for more than 100 years, but its antitumor activity was not discovered until 1963.[237] It is active against rapidly proliferating cells in the synthesis phase, during which it prevents the formation of deoxyribonucleotides from ribonucleotides. The mode of its action is inhibition of ribonucleotide diphosphate reductase, an enzyme consisting of two protein subunits.[238] It does this by interfering with the iron-containing portion of one of these subunits.[239]

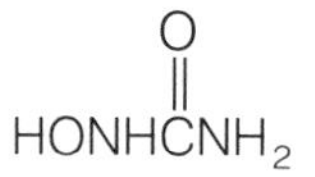

Hydroxyurea

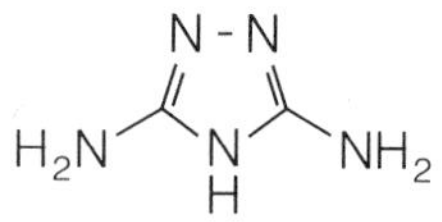

Guanazole

Another very old compound recently found active against tumors is guanazole.[240] This diaminotriazole resembles hydroxyurea in its ability to limit DNA synthesis by inhibiting the reduction of ribonucleotides. It is clinically active in inducing remissions of acute adult leukemia.[241]

In 1953, Kidd found that injections of guinea pig serum caused regressions of certain transplanted tumors in mice and rats.[242] Subsequent investigation revealed that these tumors required L-asparagine as a nutrient, but the presence of the enzyme L-asparaginase in the guinea pig serum created a deficiency in this amino acid.[243] The practical preparation of L-asparaginase for clinical trials follows the discovery that *Escherichia coli* produces a form of it that has antineoplastic activity.[244] Thus, mass cultures are harvested and treated with ammonium sulfate to rupture the cells, and the liberated enzyme is isolated by solvent extraction and chromatography. Very pure material is obtained by gel filtration or affinity chromatography, followed by crystallization. The *E. coli* enzyme has a molecular mass of 120,000 to 141,000 Da, an isoelectric point of 4.9 to 5.2, and a K_m of 1.2×10^{-5}.[245]

Earlier preparations of L-asparaginase contained endotoxins from *E. coli*, but these are absent in the purer new preparations. Clearance of the enzyme from plasma is due to an immunologic reaction in which it combines with protein. This reaction may lead to sensitization in some patients. Patients who cannot tolerate L-asparaginase from *E. coli* might be treated by the preparation from *Erwinia carotovora*.[246] Tumor resistance is based on the development of asparagine synthetase by the tumor cells.[247]

A variety of thiosemicarbazones and guanylhydrazones have antitumor activity, although currently none is an established clinical agent. 5-Hydroxypyridine-2-carboxaldehyde thiosemicarbazone and related heterocyclic compounds are powerful chelating agents for transition metals, including iron.[248] There is a direct correlation between chelating ability and antitumor activity. It has been suggested that these compounds inhibit ribonucleoside diphosphate reductase by coordinating with the iron that it contains.[249]

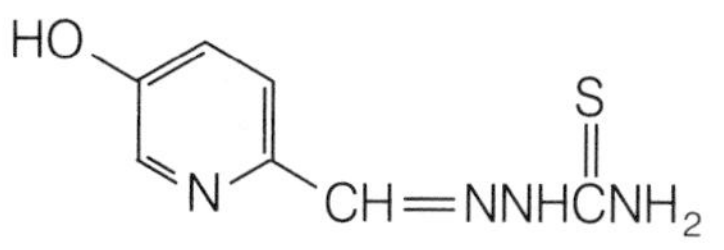

5-Hydroxypyridine-2-carboxaldehyde Thiosemicarbazone

Bis(thiosemicarbazones) of α-ketoaldehydes and α-diketones also have antineoplastic activity. Among the most thoroughly studied compounds of this type are the derivatives of methylglyoxal and 3-ethoxy-2-oxobutyraldehyde.[250] These compounds form strong chelates with copper and zinc, but the function of these species in causing cytotoxicity is unknown.[251]

Mitoguazone [Methylglyoxal bis(guanylhydrazone)] has antitumor activity in humans. It interferes with nuclear and mitochondrial metabolism.[252] Many of its actions are related to the functions of spermidine, which it resembles in structure. Thus, it

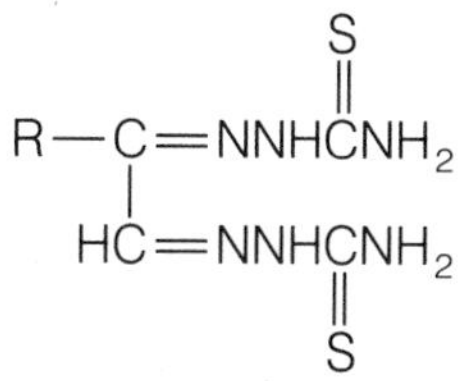

[Methylglyoxal Bis(thiosemicarbazone)], R = CH_3
3-Ethoxy-2-oxobutyraldehyde Bis(thiosemicarbazone), R = $CH_3CHOC_2H_5$

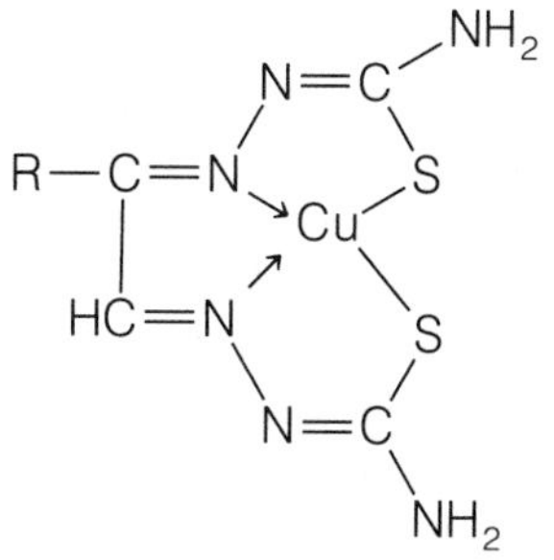

Copper Chelate

competes with spermidine for the transport carrier and intracellular binding site. It also inhibits spermidine biosynthesis. Its antiproliferative effects on cells can be prevented by administering spermidine.[253] Many other bis(guanylhydrazones) have been prepared, but none has proved superior to the methylglyoxal derivative.

Phthalanilides, such as NSC-60,339, showed antitumor activity, but they were highly toxic in clinical trials. They inhibit oxidative phosphorylation in mitochondria and are thought to influence the phospholipid–histone–DNA equilibrium in the nucleus and mitochondria.[254]

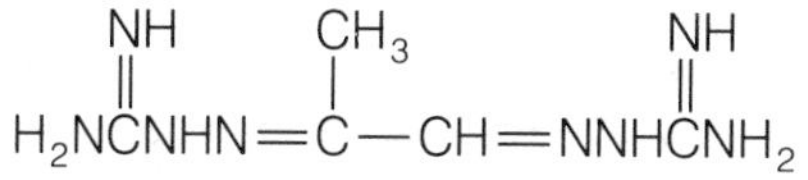

$H_2NC(=NH)NHN{=}C(CH_3){-}CH{=}NNHC(=NH)NH_2$

Mitoguazone
[Methylglyoxal Bis(guanylhydrazone)]

$H_2NCH_2CH_2CH_2NHCH_2CH_2CH_2CH_2NH_2$

Spermidine

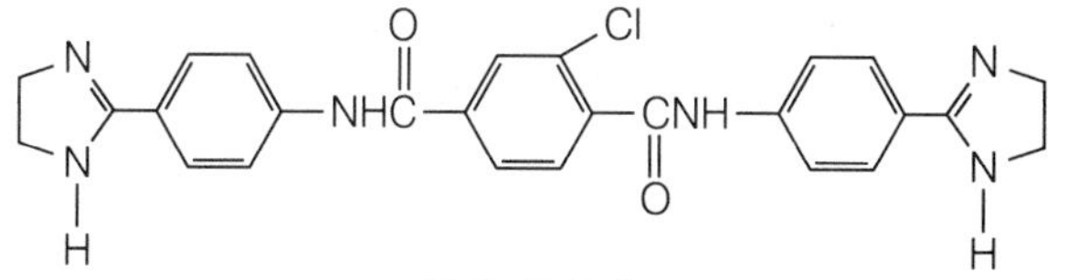

NSC-60, 339

The antischistosomal drug hycanthone has antitumor activity in animals, and it is relatively nontoxic.[255] It is an intercalating agent that inhibits DNA and RNA synthesis.[256]

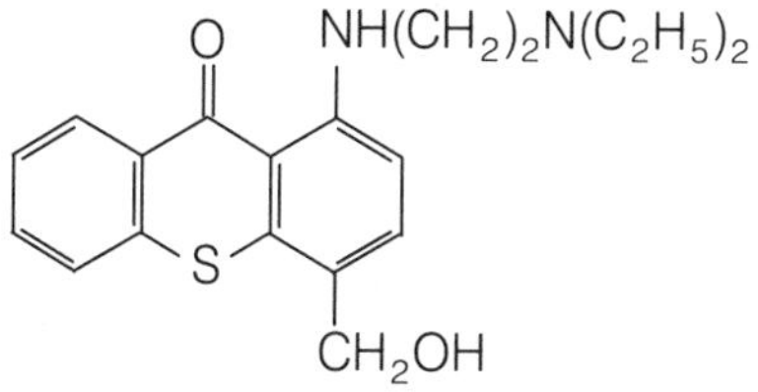

Hycanthone

Unusual antitumor properties are shown by the compound known as razoxane (ICRF 159). It inhibits the metastases of Lewis lung tumor in mice without affecting the primary tumor.[257] This activity is thought to be caused by normalization of the developing blood vessels at the invading margins of the primary tumor. Another antineoplastic action of razoxane is potent inhibition of DNA synthesis.[258] Phase II clinical trials indicated activity against lymphoma, Kaposi's sarcoma, osteosarcoma, and advanced head and neck cancer. Toxic effects included leukopenia, nausea and vomiting, and alopecia.[259] A

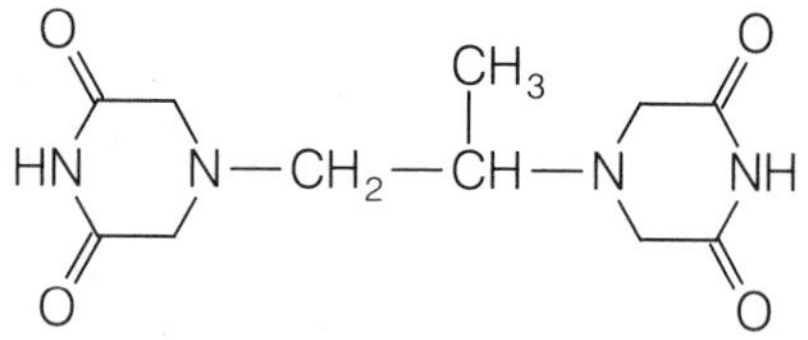

Razoxane
(ICRF 159)

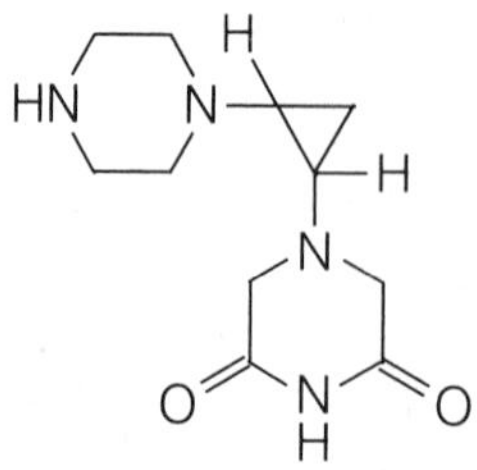

Cis-isomer

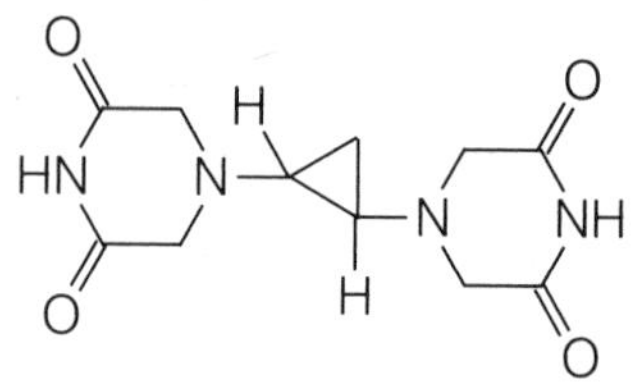

Trans-isomer

recent study on cyclopropyl analogues of razoxane revealed that, in one tumor model, controlling the drug geometry resulted in a separation of antitumor activity (*cis*-isomer) from tumor-potentiating activity (*trans*-isomer).[260]

Pipobroman is used against hematologic cancers, especially polycythemia vera and chronic granulocytic leukemia.[261] Its mode of action is unknown, although it has sometimes been included among the alkylating agents. It has not been possible to demonstrate alkylating properties in vitro under physiologic conditions.

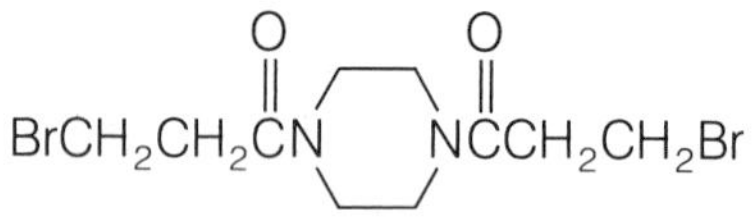

Pipobroman

Hexamethylmelamine is clinically active against melanoma. It is metabolized to the water-soluble compound pentamethylmelamine and eight other compounds. Pentamethylmelamine also is active against melanoma, but it is unclear if its formation represents a necessary conversion to activate hexamethylmelamine.[262]

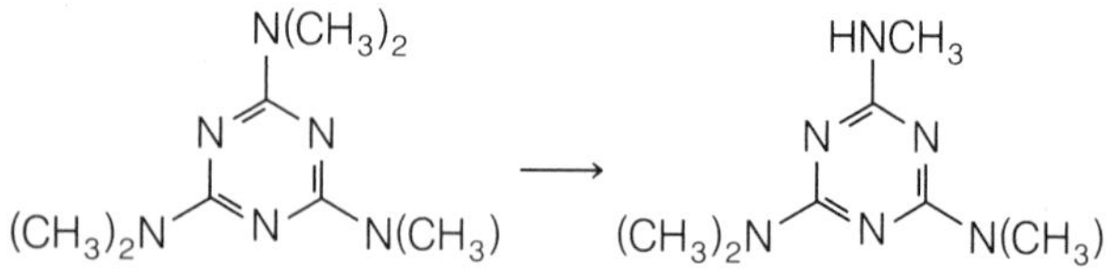

Hexamethylmelamine　　Pentamethylmelamine

Among the newer antineoplastic drugs, amsacrine (4-[(9-acridinyl)amino]methanesulfon-*m*-anisidide; *m*-AMSA) appears to be valuable because it showed a wide spectrum of activity in early clinical trials. Phase II trials are underway in a panel of signal tumors, with some remissions already shown in refractory cases of breast cancer, malignant melanoma, and acute myelocytic leukemia. Leukopenia is the limiting toxicity.[259]

Amsacrine is an acridine derivative that is thought to bind to DNA through intercalation. However, it does not affect DNA synthesis.[263] This compound was rationally designed as one member of a group of acridinylaminomethanesulfonamides.[264] Previously, several acridine derivatives had shown antitumor activity.

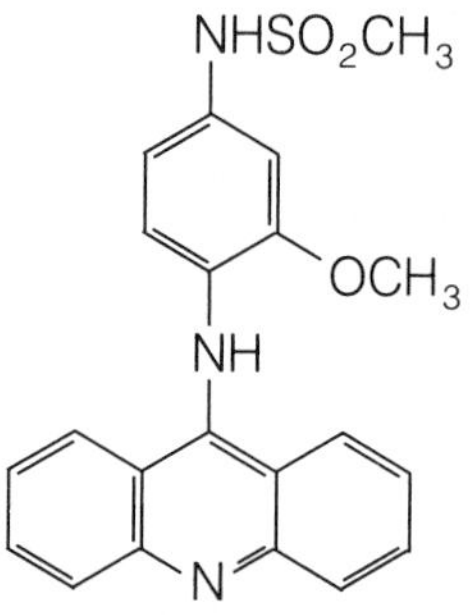

Amsacrine
(m-AMSA)

PRODUCTS

Cisplatin, USP. NSC-119875; CDDP (Platinol). This compound is prepared by treating potassium chloroplatinate with ammonia.[233] It is a water-soluble (1 mg/1 mL) white solid. Amber vials containing 10 mg of cisplatin as a lyophilized powder are supplied. For reconstitution, 10 mL of sterile water is added and the resulting solution is diluted in 2 L of 5% dextrose in 0.5 or 0.33 *N* saline containing 37.5 g of mannitol.[265]

Cisplatin is used in combination with bleomycin and vinblastine for metastatic testicular tumors. This combination represents a significant improvement over previous treatments.[266] As a single agent or in combination with doxorubicin, cisplatin is used for the remission of metastatic ovarian tumors. Other tumors that have shown sensitivity to cisplatin include penile cancer, bladder cancer, cervical cancer, head and neck cancer, and small-cell cancer of the lung. The major dose-limiting toxicity is cumulative renal insufficiency associated with renal tubular damage. Hydrating patients with intravenous fluids before and during cisplatin treatment significantly reduces the incidence of renal toxicity.[267] Myelosuppression, nausea and vomiting, and ototoxicity also occur frequently.

The usual dosage for metastatic testicular tumors is 20 mg/m^2 intravenously daily for five days, once every three weeks for three courses. Metastatic ovarian tumors are treated with 50 mg/m^2 intravenously once every three weeks. Pretreatment hydration is recommended for both regimens.[265]

Hydroxyurea, USP. Hydroxycarbamide; NSC-32065 (Hydrea). This compound is prepared from hydroxylamine hydrochloride and potassium cyanide.[268] It is a crystalline solid with good solubility in water. Capsules containing 500 mg of hydroxyurea are supplied.

Hydroxyurea is active against melanoma, chronic myelocytic leukemia, and metastatic ovarian carcinoma. It is used in combination with radiotherapy for head and neck cancer. The main toxicity is bone marrow depression expressed as leukopenia, anemia, and, occasionally, thrombocytopenia. Gastrointestinal toxicity and dermatologic reactions also occur.

Pipobroman, USP. 1,4-Bis(3-bromopropionyl)piperazine; NSC-25154 (Vercyte). This compound is prepared from piperazine and 3-bromopropionyl bromide.[269] It is supplied as 10-mg or 25-mg grooved tablets.

Pipobroman is used primarily for treating polycythemia vera. It also is used in patients with chronic granulocytic leukemia refractory to busulfan. Nausea, vomiting, abdominal cramping, diarrhea, and skin rash occur.

Asparaginase (Elspar) is a preparation from *Escherichia coli* containing the enzyme L-asparagine amidohydrolase [EC 3.5.1.1]. It is supplied in 10-mL vials containing 10,000 IU of the drug, and it is soluble in sterile water or saline used for reconstitution. For infusion, saline or 5% dextrose solution may be used only if clear. Filtration through a 5.0-μm filter will remove gelatinous particles that sometimes form.

Asparaginase is used in patients with acute lymphocytic and other leukemias. It can induce remissions as a single agent or in combination with prednisone and vincristine sulfate. Hypersensitive reactions occur frequently, and they may include anaphylaxis. Consequently, it must be used only in a hospital setting under the direction of an oncologist. Hepatotoxicity and renal toxicity occur frequently. Asparaginase serum levels after intramuscular administration are approximately one-half of those obtained on intravenous administration. Elimination is biphasic with an initial half-life of four to nine hours and a terminal half-life of 1.4 to 1.8 days.

HORMONES

Steroid hormones, including estrogens, androgens, progestins, and glucocorticoids, act on the appropriate target tissues at the level of transcription. Generally, the effect is derepression of genetic template operation, which stimulates the cellular process. However, glucocorticoids act in lymphatic tissues to impair glucose uptake and protein synthesis. Target cells contain specific protein receptors in their cytoplasm with very high affinities for the hormones. Binding of the hormone to the receptor causes a transformation in the receptor structure, which is followed by migration of the resulting complex into the nucleus. In the nucleus, the complex interacts with an acceptor site to influence transcription.[270]

Normal and well-differentiated neoplastic target cells have several hormone receptors, and they are dependent upon the hormones for stimulation.[271–274] Less-differentiated neoplastic cells become independent of hormonal control and lose their specific receptors. Thus, some neoplasms are hormone-dependent and responsive to hormone-based therapy, whereas others are independent and unresponsive. Assays of the number of hormone receptors present in the neoplastic cells should be valuable in predicting the probability of a favorable response.

Hormonal effects in breast cancer are complex and not completely understood. The hormone dependency of breast cancer has been known since 1889,[275] and removal of the ovaries of premenopausal women, which results in decreased estrogens, is an established treatment. Some patients who do not respond to this procedure do respond to adrenalectomy, which suggests that the hormone dependence is not simply related to estrogens.[276] Remission after adrenalectomy occurs more often in patients with estrogen receptors than in those lacking receptors. Administration of estrogens to postmenopausal women with metastatic breast cancer resulted in objective remissions in about 30% of the cases.[277] This response appears paradoxical, but the estrogen levels resulting from drug treatment are much higher than physiologic levels. A recent suggestion is that high estrogen levels interfere with the peripheral action of prolactin, a pituitary hormone that also stimulates breast tissue.[278] Ethinyl estradiol is given orally in the treatment of breast cancer in postmenopausal women, and estradiol dipropionate or benzoate are used parenterally. Tamoxifen is an antiestrogen that has been used suc-

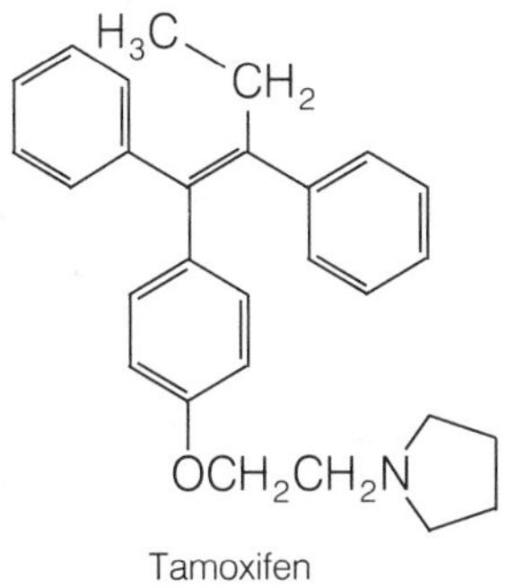

H_3C CH_2 OCH_2CH_2N

Tamoxifen

cessfully in the treatment of premenopausal women. It has very low toxicity.[279]

Androgens are active against metastatic breast cancer in about 20% of postmenopausal women. Their mode of action is not completely understood. Inhibition of the release of pituitary gonadotropins has been suggested, but the situation must be more complicated than this because certain androgens are active in hypophysectomized patients.[280] Other useful effects of androgens in advanced breast cancer are stimulation of the hematopoietic system and reversal of bone demineralization. Testosterone propionate is the androgen most frequently used against breast cancer. Other compounds are 2α-methyltestosterone, fluoxymesterone, and 19-nor-17α-methyltestosterone. Testolactone is preferred in some patients because it has no androgenic side effects.

Estrogens can be used to induce remissions of disseminated prostatic cancer. It is not certain whether their effect is due to direct interference with peripheral androgens, inhibition of pituitary gonadotropin, or both.[281] Diethylstilbestrol is the compound most widely used for advanced prostatic cancer, and it benefits over 60% of patients. Chlorotrianisine also is used.

Progesterone and its analogues are active against certain neoplasms that are stimulated by estrogens. They appear to exert antiestrogenic effects of uncertain mechanism. The neoplasms treated by progestins are metastatic endometrial carcinoma and advanced renal cell carcinoma.[282] Progesterone suspension in oil, megesterol acetate, and medroxyprogesterone acetate are used against endometrial cancer. They provide regressions of several months to three years in about 30% of women.[282] Medroxyprogesterone acetate causes regression of renal cell carcinoma in about 20% of men and 8% of women.

Glucocorticoids cause pronounced immediate changes in lymphoid tissues. Lymphocytes in the thymus and lymph nodes are dissolved and lymphopenia occurs in peripheral bloods.[283] This property is used to advantage in the treatment of leukemia and Hodgkin's disease, wherein profound temporary regressions are observed following the administration of cortisone derivatives or corticotropin (ACTH).[284] Prednisone is usually the corticoid chosen for this purpose and it is almost always used in combination with other chemotherapeutic agents such as mechlorethamine, vincristine, and procarbazine. Such combinations are effective in maintaining the remissions in many cases. Glucocorticoids also are useful in treating metastatic prostate cancer of patients who have relapsed after castration. The rationale for this use is that they inhibit release of ACTH from the pituitary, which leads to adrenal atrophy and decreased adrenal production of androgens.[285] Prednisone and cortisone acetate are used in the treatment of metastatic breast cancer. Their value in this condition derives not from an antineoplastic effect, but in alleviating specific complications such as hypercalcemia and anemia.[286]

Mitotane is unique among antitumor agents in its highly selective effect on one gland, the adrenal cortex. It has a direct cytotoxic action on adrenal cortical cells, in which it extensively damages the mitochondria.[287] This leads to cell death and atrophy of the gland. Mitotane is used specifically against adrenocortical carcinoma.[288]

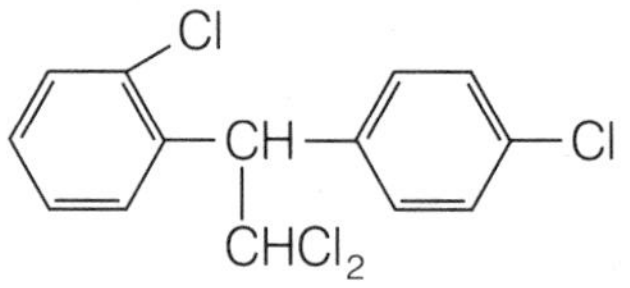

Cl
CH
$CHCl_2$
Cl

Mitotane

Another way in which the proliferation of hormone-dependent tumors can be limited is to inhibit the release of gonadotropins from the anterior pituitary gland. This release is controlled by gonadotropin-releasing hormone (LH-RH), a nonapeptide, and it can be blocked effectively by continuous administration of certain analogues of this hormone. Leuprolide is a nonapeptide that is identical in structure with LH-RH, except that it has D-leucine replacing the natural L-leucine as the sixth amino acid. It is used for palliative treatment of advanced prostatic cancer.

A thorough discussion of the structures, nomenclature, properties, and dose forms of the steroid hormones is presented in Chapter 18. Only the products not included in that chapter are described in the following section.

PRODUCTS

Mitotane, USP. 1,1-Dichloro-2-(*o*-chlorophenyl)-2-(*p*-chlorophenyl)ethane; *o,p*-DDD; CB133 (Lysodren). This compound is obtained as a constituent of commercial DDD, which is prepared from 2,2-dichloro-1-(*o*-chlorophenyl)ethanol, chlorobenzene, and sulfuric acid.[289] Isolation from commercial DDD gives mitotane as crystals that are soluble in alcohol and other organic solvents.[290] Scored 500-mg tablets are supplied.

Mitotane is indicated only for treating inoperable adrenal cortical carcinoma. Frequently occurring side effects include gastrointestinal disturbances, CNS depression, and skin toxicity.

The usual regimen is 8 to 10 g daily, divided into three or four doses.

Dromostanolone Propionate, USP. 17β-Hydroxy-2α-methyl-5α-androstan-3-one propionate; 2α-methyldihydrotestosterone propionate (Drolban). This semisynthetic androgen is prepared from dihydrotestosterone in a route involving condensation with ethyl formate followed by hydrogenation to give the 2α-methyl derivative, and then reaction with propionic anhydride.[291] The compound is supplied in rubber-stoppered vials containing 500 mg of dromostanolone propionate in 10 mL of sesame oil, with 0.5% phenol as a preservative.

Dromostanolone propionate is used in the palliative treatment of metastatic breast carcinoma in postmenopausal women. It is contraindicated in premenopausal women and in carcinoma of the male breast. The most usual side effect is virilism, although this is less intense than that afforded by testosterone propionate. Edema occurs occasionally.

Testolactone, USP. D-Homo-17α-oxaandrosta-1,4-diene-3,17-dione; 1-dehydrotestolactone (Teslac) is prepared by microbial transformation of progesterone.[292] It is soluble in alcohol and slightly soluble in water. The compound is supplied as a sterile aqueous suspension providing 100 mg of testolactone per milliliter in multidose vials of 5 mL. Tablets containing 50 mg or 250 mg of testolactone also are supplied.

Testolactone is used in the palliative treatment of advanced or disseminated breast cancer in postmenopausal women. It is contraindicated in breast cancer in men. Testolactone is devoid of androgenic activity in the commonly used doses.

Megestrol Acetate. 17-(Acetoxy)-6-methylpregna-4,6-diene-3,20-dione (Megace). This compound is prepared by a multistep synthesis from 17α-hydroxypregnadienolone.[293] It is supplied as light blue scored tablets containing 20 or 40 mg of megestrol acetate.

Megestrol acetate is indicated for the palliative treatment of advanced breast or endometrial carcinoma when other methods of treatment are inappropriate. No serious side effects or adverse reactions have been reported. However, there is an increased risk of birth defects in children whose mothers take the drug during the first four months of pregnancy. The usual doses are 160 mg/day in four equal doses for breast cancer and 40 to 320 mg/day in divided doses for endometrial cancer.

Tamoxifen Citrate. (*Z*)-2-[*p*-(1,2-Diphenyl-1-butenyl)phenoxy]-*N*,*N*-dimethylethylamine citrate (Nolvadex) is prepared by treating 2-ethyldeoxybenzoin with 4-[(2-*N*,*N*-dimethylamino)ethoxy]-phenylmagnesium bromide,[294] followed by dehydration and separation of the *E* and *Z* isomers.[295] The citrate salt of the *Z* isomer is soluble in water. Tablets containing 15.2 mg of tamoxifen citrate, which is equivalent to 10 mg of tamoxifen, are supplied. They should be protected from heat and light.

Tamoxifen is a nonsteroidal agent that has shown potent antiestrogenic properties in animals. In the rat model, it appears to exert its antitumor effects by binding to estrogen receptors.[296] Tamoxifen is useful in the palliative treatment of advanced breast cancer in postmenopausal women. There are no known contraindications. The most frequent side effects are hot flashes, nausea, and vomiting. They are rarely severe enough to require dose reduction. The usual dose is one or two 10-mg tablets twice daily.

Leuprolide Acetate (Lupron) is a synthetic nonapeptide analogue of naturally occurring gonadotropin-releasing hormone (LH-RH). It is supplied in 2.8-mL multidose vials containing 5 mg/mL of the drug and benzyl alcohol. These vials should be refrigerated until dispensed.

Leuprolide is used for palliative treatment of advanced prostatic cancer. There are no known contraindications. Worsening of symptoms may occur during the first few weeks of treatment, with increase in bone pain as the usual manifestation. Hot flashes and irritation at the injection site also occur.

IMMUNOTHERAPY

It is now generally accepted that cells of neoplastic potential are continually produced in the human body and that our immune surveillance system destroys them. The development of tumors implies that this system is not functioning properly. Evidence for this factor in carcinogenesis includes a high rate of cancer in organ-transplant patients whose immune systems are suppressed by drugs, such as azathioprene, and a high correlation between cancer and immunodeficiency diseases, such as bacterial and viral infections.[297] Stimulation of the body's immune system should provide a valuable method of cancer treatment, because it is capable of eradicating the neoplastic cells completely. Research in this area is expanding rapidly, and some promising leads are emerging.

The first attempt at immunotherapy was made in the 1890s by Coley, who injected bacterial toxins in cancer patients. His results were generally unaccepted because of rather extravagant claims. However, his techniques have been revived in recent years. Most oncologists now use a live bacteria tuberculosis vaccine, bacille Calmette-Guerin (BCG).[297] This vaccine is given to certain patients who show a functioning immune system, as determined by sensitivity toward dinitrochlorobenzene.[298] Remissions have been obtained in malignant melanoma, breast cancer, and leukemia. Unfortunately, BCG causes several undesirable effects including fever, hypersensitivity, and liver disorders. Other immunostimulants currently under investigation as anticancer agents are the methanol-extracted residue (MER) of BCG, *Corynebacterium parvulum*, *Bordetella pertussis* vaccine, and synthetic polynucleotides.[299] The activity of these bacterial products is thought to be mediated by a protein known as tumor necrosis factor (TNF). This factor produces hemorrhagic necrosis of sensitive transplanted tumor cells, and it is synergistic with interferons.[301] Unfortunately, TNF does not show activity against primary tumors when used alone. However, clinical trials based on expected synergism with interferons are in progress. Potentiation of TNF by agents such as mitomycin and vinblastine suggest that it might have a role in combination chemotherapy.

One approach to overcoming the difficulties of BCG therapy is to develop simpler chemical structures with immunostimulant properties. Two such compounds are presently under clinical investigation as potential anticancer drugs. One of them is levamisole, an anthelmintic agent found to be an immunostimulant by Renoux in 1972. It appears to be most effective in patients with small-tumor burdens, and it acts by stimulating the responsiveness of lymphocytes to tumor antigens. Advantages of levamisole include oral activity and few adverse reactions. Tilorone is one member of a large family of synthetic compounds that show immunostimulant activity. It stimulates the production of interferon, which affords antiviral activity. However, its effect is more general. The main cause of its antitumor action appears to be an effect on T cells originating in the thymus. Tilorone causes gastrointestinal disorders, dizziness, and headaches in many patients.[300]

called lymphokines, which include interferons, interleukins, and B-cell growth and differentiation factors. Lymphokines cause profound enhancing or suppressing effects on responding precursor cells of the immune system at nanomolar to picomolar concentrations.[302]

Interferons are secreted by cells in response to viral infections or other chemical or biologic inducers. Three major classes of interferons, α, β, and γ, have been identified. They bind to specific membranes on cell surfaces, which induces a sequence of intracellular events including the induction of enzymes. This process produces effects such as inhibition of virus replication, suppression of cell proliferation, enhancement of macrophage activity, and increase in the specific cytotoxic activity of lymphocytes for target cells.[302] Interferons alfa-2a and alfa-2b promote the immunologic response to neoplastic cells, which results in significant cytotoxicity in *some instances*. They are the drugs of choice for treating hairy cell leukemia.[303] They also are showing responses against renal cell cancer, multiple myeloma, melanoma, and Kaposi's sarcoma in clinical trials.

Another important lymphokine is interleukin-2 (IL-2). This glycoprotein interacts with specific receptors on T-effector cells to activate their cytotoxicity. It also stimulates the activation and proliferation of antigen-nonspecific natural killer cells, which are involved in immune functions associated with tumor surveillance. These effects are thought to be mediated through the induction of γ-interferon. Human IL-2 is now produced by recombinant gene technology, which has permitted extensive clinical trials against a variety of tumors. In some of these trials, the IL-2 is given in combination with lymphokine-activated killer cells. Although some encouraging results have been reported,[304] there is strong controversy about efficacy and toxic effects, including tachycardia and induced psychotic states.[305]

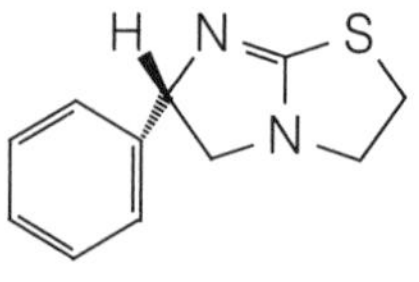

Levamisole

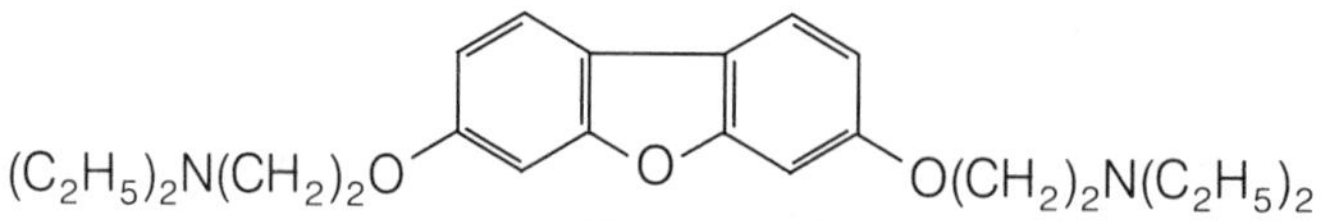

$(C_2H_5)_2N(CH_2)_2O$ — $O(CH_2)_2N(C_2H_5)_2$

Tilorone

The induction phase of the immune response of both B and T lymphocytes is regulated by interactions between macrophages and subpopulations of T lymphocytes known as helper T cells. This interaction induces the production of soluble glycoproteins

PRODUCTS

Interferon Alfa-2a. rIFN; IFLrA (Roferon-A) is a highly purified protein containing 165 amino acids. It is manufactured from a strain of *E. coli* bearing a genetically engineered plasmid containing an interferon alfa-2a gene from human leukocytes. Vials containing 3 million IU with 5 mg of human serum albumin and 3 mg of phenol are supplied. An 18 million IU preparation also is available. These preparations should be stored at 36° to 48°F.

Interferon alfa-2a is used in patients 18 years old or older for treatment of hairy cell leukemia. It is contraindicated in persons who develop hypersensitivity. Most patients develop flulike syndromes consisting of fever, fatigue, myalgias, headache, and chills. Gastrointestinal and CNS symptoms also occur. Caution must be used in administering the drug to patients with renal or hepatic disease, seizure disorders, or cardiac disease. Metabolism occurs by rapid proteolytic degradation during reabsorption in the kidney. The elimination half-life is 3.7 to 8.5 hours.

Interferon Alfa-2b. IFN-alfa 2; rIFN-α2, α-2-interferon (Interon) is a highly purified protein produced by *E. coli* containing a plasmid with an alfa-2b gene. This plasmid is obtained from recombinant DNA technology using human leukocytes. The drug is supplied in vials containing 3, 5, 10, or 25 IU. Sterile water diluent also is supplied. Reconstituted solutions are stable for one month at 36° to 48°F.

Hairy cell leukemia is the present indication for interferon alfa-2b. Hypersensitivity to this protein has not been observed. Patients develop a flulike syndrome, CNS effects, and cardiovascular effects including hypotension, arrythmia, or tachycardia.

REFERENCES

1. Zubrod, C. G.: Agents of choice in neoplastic disease. In Sartorelli, A. C., and Johns, D. J. (eds.). Handbook of Experimental Pharmacology, Vol. 38, part 1, p. 7. New York, Springer-Verlag, 1974.
2. Clarkson, B.: Clinical applications of cell cycle kinetics. In Sartorelli, A. C., and Johns, D. J. (eds.). Handbook of Experimental Pharmacology, Vol. 38, part 1, p. 156. New York, Springer-Verlag, 1974.
3. Huggins, C., and Hodges, C. V.: Cancer Res. 1:293, 1941.
4. Gilman, A., and Phillips, F. S.: Science 103:409, 1946.
5. Pearson, O. H.: Cancer 2:943, 1949.
6. Kennedy, B. J.: Cancer 18:1551, 1965.
7. Wood, H. B., Jr.: Cancer Chemother. Rep. 2:9, 1971.
8. Goldin, A., et al.: Evaluation of antineoplastic activity: Requirements of test systems. In Sartorelli, A. C., and Johns, D. J. (eds.). Handbook of Experimental Pharmacology, Vol. 38, part 1, p. 12. New York, Springer-Verlag, 1974.
9. Himmelweit, F. (ed.): The Collected Papers of Paul Ehrlich, vol. 1, pp. 596–618. London, Pergamon Press, 1956.
10. Lynch, V., et al.: J. Pharmacol. Exp. Ther. 12:265, 1918.
11. Montgomery, J. A., et al.: Drugs for neoplastic diseases. In Burger, A. (ed.). Medicinal Chemistry, 3rd ed., p. 680. New York, Wiley-Interscience, 1970.
12. Connors, T. A.: Mechanism of action of 2-chloroethylamine derivatives, sulfur mustards, epoxides and aziridines. In Sartorelli, A. C., and Johns, D. J. (eds.). Handbook of Experimental Pharmacology, Vol. 38, part 2, p. 19. New York, Springer-Verlag, 1975.
13. Price, C. C.: Chemistry of alkylation. In Sartorelli, A. C., and Johns, D. J. (eds.). Handbook of Experimental Pharmacology, Vol. 38, part 2, p. 4. New York, Springer-Verlag, 1975.
14. White, F. R.: Cancer Chemother. Rep. 4:55, 1959.
15. Jarmon, M., and Ross, W. C. J.: Chem. Ind. (Lond.) 1789, 1967.
16. Montgomery, J. A., et al.: J. Med. Chem. 10:668, 1967.
17. Engle, T. W., et al.: J. Med. Chem. 22:897, 1979.
18. Colvin, M., et al.: Cancer Res. 36:1121, 1976.
19. Karle, I. L., et al.: J. Am. Chem. Soc. 99:4803, 1977.
20. Ahmann, D. L., et al.: Cancer Chemother. Rep. 58:861, 1974.
21. Anderson, W. K., and Corey, P. F.: J. Med. Chem. 20:812, 1977.
22. Szybalski, W., and Iyer, V. N.: The mitomycins and porfiromysins. In Gottlieb, D., and Shaw, P. D. (eds.). Antibiotics, Vol. 1, p. 221. New York, Springer-Verlag, 1967.
23. Lin, A. J., et al.: Cancer Chemother. Rep. 4:23, 1974.
24. Buchi, G., and Rosenthal, D.: J. Am. Chem. Soc. 78:3860, 1956.
25. Bollag, W.: Cancer Chemother. Rep. 33:1, 1963.
26. Chabner, B. A., et al.: Proc. Soc. Exp. Biol. 132:1169, 1969.
27. Tsuji, T., and Kosower, E. M.: J. Am. Chem. Soc. 93:1992, 1971.
28. Kreis, W., and Yen, W.: Experientia 21:284, 1965.
29. Schwartz, D. E., et al.: J. Labelled Compd. 111:487, 1967.
30. Preussmann, R., and Von Hodenberg, A.: Biochem. Pharmacol. 19:1505, 1970.
31. Skibba, J. L., et al.: Cancer Res. 30:147, 1970.
32. Sava, G., et al.: Cancer Treat. Rep. 63:93, 1979.
33. Ludlum, D. B.: Molecular biology of alkylation: An overview. In Sartorelli, A. C., and Johns, D. J. (eds.). Handbook of Experimental Pharmacology, Vol. 38, part 2, p. 7. New York, Springer-Verlag, 1975.
34. Kohn, K. W., et al.: J. Mol. Biol. 19:226, 1966.
35. Ross, W. C. J.: Biological Alkylating Agents. London, Butterworths, 1962.
36. Thomasz, M., et al.: Proc. Natl. Acad. Sci. USA 83:6702, 1986.
37. Lawley, P. D., and Martin, C. M.: Biochem. J. 145:85, 1975.
38. Ludlum, D. B.: Biochim. Biophys. Acta 152:282, 1967.
39. Brooks, P., and Lawley, P. D.: Biochem. J. 80:496, 1961.
40. Wheeler, G. P.: Mechanism of action of nitrosoureas. In Sartorelli, A. C., and Johns, D. J. (eds.). Handbook of Experimental Pharmacology, Vol. 38, part 2, p. 75. New York, Springer-Verlag, 1975.
41. Levis, A. G., et al.: Nature 207:608, 1965.
42. Ludlum, D. B.: Biochim. Biophys. Acta 95:674, 1965.
43. Boyce, R. P., and Howard-Flanders, P.: Proc. Natl. Acad. Sci. USA 51:293, 1964.
44. Setlow, R. B., and Carrier, W. L.: Proc. Natl. Acad. Sci. USA 51:226, 1964.
45. Prelog, V., and Stepan, V.: Coll. Czech. Chem. Commun. 7:93, 1935.
46. Arnold, H., et al.: Nature 181:931, 1958.
47. Zubrod, C. G.: Cancer 21:553, 1968.
48. Bergel, F., and Stock, J. A.: J. Chem. Soc. 1954:2409.
49. Fisher, B., et al.: N. Engl. J. Med. 292:117, 1975.
50. Phillips, A. P., and Mentha, J. W.: U.S. Patent 3,046,301, Oct. 29, 1959.
51. Calabresi, P., and Welch, A. D.: Annu. Rev. Med. 13:147, 1962.
52. Timmis, G. M.: U.S. Patent 2,917,432, Dec. 15, 1959.
53. Parham, W. E., and Wilbur, J. M., Jr.: J. Org. Chem. 26:1569, 1961.
54. Roberts, J. J., and Warwick, G. P.: Biochem. Pharmacol. 6:217, 1961.
55. Warwick, G. P.: Cancer Res. 23:1315, 1963.
56. Physicians' Desk Reference, 33rd ed., p. 746. Oradell, NJ, Medical Economics, 1979.
57. Johnston, T. P., et al.: J. Med. Chem. 6:669, 1963.
58. Walker, M. D.: Cancer Chemother. Rep. 4:21, 1973.
59. Johnston, T. P., et al.: J. Med. Chem. 9:892, 1966.
60. Moertel, C. G.: Cancer Chemother. Rep. 4:27, 1973.

61. Kuh, E., and Seeger, D. R.: U.S. Patent 2,670,347, Feb. 23, 1954.
62. Hoffman-LaRoche & Co., A.-G.: Belg. Patent 618,638, Dec. 7, 1962.
63. Shealy, Y. F., et al.: J. Org. Chem. 27:2150, 1962.
64. Lyttle, D. A., and Petering, H. F.: J. Am. Chem. Soc. 80:6459, 1958.
65. Lukens, L. N., and Herrington, K. A.: Biochim. Biophys. Acta 24:432, 1957.
66. Patterson, A. R. P., and Tidd, D. M.: 6-Thiopurines. In Sartorelli, A. C., and Johns, D. J. (eds.). Handbook of Experimental Pharmacology, Vol. 38, part 2, p. 384. New York, Springer-Verlag, 1975.
67. Woods, D. D., and Fildes, P.: J. Chem. Soc. 59:133, 1960.
68. Roblin, R. O., et al.: J. Am. Chem. Soc. 67:290, 1945.
69. Hitchings, G. H., and Elion, G. B.: Acc. Chem. Res. 2:202, 1969.
70. Elion, G. B., et al.: J. Am. Chem. Soc. 74:411, 1952.
71. Brockman, R. W.: Adv. Cancer Res. 7:129, 1963.
72. Atkison, M. R., et al.: Biochem. J. 92:389, 1964.
73. Salser, J. S., et al.: J. Biol. Chem. 235:429, 1960.
74. Bennett, L. L., Jr., and Allan, P. W.: Cancer Res. 31:152, 1971.
75. Currie, R., et al.: Biochem. J. 104:634, 1967.
76. Elion, G. B.: Fed. Proc. 26:898, 1967.
77. Moor, E. C., and LePage, G. A.: Cancer Res. 18:1075, 1958.
78. LePage, G. A., et al.: Cancer Res. 24:835, 1964.
79. Lee, W. W., et al.: J. Am. Chem. Soc. 82:2648, 1960.
80. Parke-Davis and Co.: Belg. Patent 671,557, 1967.
81. Furth, J. J., and Cohen, S. S.: Cancer Res. 27:1528, 1967.
82. Brink, J. J., and LePage, G. A.: Cancer Res. 24:312, 1964.
83. Taylor, S., and Eyre, H. J.: Am. Soc. Clin. Oncol. Annu. Meet., Atlanta, GA, May 17–19, 1987, Abstr. 275.
84. Heidelberger, C.: Fluorinated pyrimidines and their nucleosides. In Sartorelli, A. C., and Johns, D. J. (eds.). Handbook of Experimental Pharmacology, Vol. 38, part 2, p. 193. New York, Springer-Verlag, 1975.
85. Skold, O.: Biochim. Biophys. Acta 29:651, 1958.
86. Reyes, P.: Biochemistry 8:2057, 1969.
87. Cohen, S. S., et al.: Proc. Natl. Acad. Sci. USA 44:1004, 1958.
88. Santi, D. V., and McHenry, C. S.: Proc. Natl. Acad. Sci. USA 69:1855, 1972.
89. Hiller, S. A., et al.: Dokl. Akad. Nauk. SSR 176–332, 1967.
90. Benvenuto, J., et al.: Cancer Res. 38:3867, 1978.
91. Heidelberger, C., et al.: J. Med. Chem. 7:1, 1964.
92. Santi, D. V., and Sakai, T. T.: Biochemistry 10:3598, 1971.
93. Cohen, S. S.: Prog. Nucleic Acid Res. 5:1, 1966.
94. Bergmann, W., and Feeney, R. J.: J. Org. Chem. 16:981, 1951.
95. Chu, M. Y., and Fischer, G. G.: Biochem. Pharmacol. 11:423, 1962.
96. Larsson, A., and Reichard, P. J.: Biol. Chem. 241:2540, 1966.
97. Creasey, W. A., et al.: Cancer Res. 28:1074, 1968.
98. Borun, T. W., et al.: Proc. Natl. Acad. Sci. USA 58:1977, 1967.
99. Creasey, W. A.: Arabinosylcytosine. In Sartorelli, A. C., and Johns, D. J. (eds.). Handbook of Experimental Pharmacology, Vol. 38, part 2, p. 245. New York, Springer-Verlag, 1975.
100. Stoller, R. G., et al.: Biochem. Pharmacol. 27:53, 1978.
101. Hoshi, A., et al.: Gann 62:145, 1971.
102. Sorm, F., et al.: Experientia 20:202, 1964.
103. Hanka, L. J., et al.: Antimicrob. Agents Chemother.: 619, 1966.
104. Paces, V., et al.: Biochim. Biophys. Acta 161:352, 1968.
105. Vesely, J., et al.: Biochem. Pharmacol. 17:519, 1968.
106. Vesely, J., et al.: Cancer Res. 30:2180, 1970.
107. Nakamura, H., et al.: J. Am. Chem. Soc. 96:4327, 1974.
108. Schaeffer, H. J., and Schwender, C. F.: J. Med. Chem. 17:6, 1974.
109. Seeger, D. R., et al.: J. Am. Chem. Soc. 71:1753, 1949.
110. Werkheiser, W.: J. Biol. Chem. 236:888, 1961.
111. Hryniuk, W. M., et al.: Mol. Pharmacol. 5:557, 1969.
112. Zakrzewski, S. F., et al.: Mol. Pharmacol. 2:423, 1969.
113. Cohen, S.: Ann. N.Y. Acad. Sci. 186:292, 1971.
114. Li, M. C., et al.: Proc. Soc. Exp. Biol. 97:29, 1958.
115. Bertino, J. R., et al.: Proc. 5th Int. Congr. Pharmacol. 3:376, 1973.
116. Semon, J. H., and Grindley, G. B.: Cancer Res. 38:2905, 1978.
117. Howell, S. B., et al.: Cancer Res. 38:325, 1978.
118. Goldin, A., et al.: J. Natl. Cancer Inst. 22:811, 1959.
119. Alt, F. W., et al.: J. Biol. Chem. 253:1357, 1978.
120. Shum, K. Y., et al.: Am. Soc. Clin. Oncol. Annu. Meet., Atlanta, GA, May 17–19, 1987, Abstr. 698.
121. Baker, B. R.: Acc. Chem. Res. 2:129, 1969.
122. Hartman, S. C.: Purines and pyrimidines. In Greenberg, D. M. (ed.). Metabolic Pathways, Vol. 4, p. 1. New York, Academic Press, 1970.
123. Eidinoff, M. L., et al.: Cancer Res. 18:105, 1958.
124. Fusari, S. A., et al.: J. Am. Chem. Soc. 76:2881, 1954.
125. DeWald, H. A., and Moore, A. M.: J. Am. Chem. Soc. 80:3941, 1958.
126. Dion, H. W., et al.: J. Am. Chem. Soc. 78:3075, 1956.
127. French, T. C., et al.: J. Biol. Chem. 238:2186, 1963.
128. Dawid, I. B., et al.: J. Biol. Chem. 238:2187, 1963.
129. Levenberg, B., et al.: J. Biol. Chem. 225:163, 1957.
130. Elion, G. B., et al.: J. Am. Chem. Soc. 74:441, 1952; Beaman, A. G., and Robbins, R. K.: J. Am. Chem. Soc. 83:4042, 1961.
131. Burchenal, J. H., et al.: Blood 8:965, 1953.
132. Physicians' Desk Reference, 33rd ed., p. 749. Oradell, NJ, Medical Economics, 1979.
133. Elion, G. B., and Hitchings, G. H.: J. Am. Chem. Soc. 77:1676, 1955.
134. Clarkson, B. D.: Cancer 5:227, 1970.
135. Duschinsky, R., et al.: J. Am. Chem. Soc. 79:4559, 1957.
136. Earl, R. A., and Townsend, L. B.: J. Heterocyclic Chem. 9:1141, 1972.
137. Physicians' Desk Reference, 33rd. ed., p. 749. Oradell, NJ, Medical Economics, 1979.
138. Moore, G. E., et al.: Cancer Chemother. Rep. 52:641, 1968.
139. Klein, E., et al.: Topical 5-fluorouracil chemotherapy for premalignant and malignant epidermal neoplasms. In Brodsky, I., and Kahn, S. B. (eds.). Cancer Chemotherapy II, p. 147. New York, Grune & Stratton, 1972.
140. Hoffer, M., et al.: J. Am. Chem. Soc. 81:4112, 1959.
141. Sullivan, R. D., and Miller, E.: Cancer Res. 25:1025, 1965.
142. Hunter, J. H.: U.S. Patent 3,116,282, Dec. 31, 1963.
143. Greenwald, E. S.: Cancer Chemotherapy. Flushing, NY, Medical Examination Publishing, 1973.
144. Clarkson, B. D.: Cancer 30:1572, 1972.
145. Hignite, C. E., et al.: Cancer Treat. Rep. 62:13, 1978.
146. Farber, S., et al.: N. Engl. J. Med. 238:787, 1948.
147. Hertz, R.: Ann. Intern. Med. 59:931, 1963.
148. Stoller, R. G., et al.: N. Engl. J. Med. 297:630, 1977.
149. Hitchings, G. H., and Elion, G. B.: U.S. Patent 3,056,785, Oct. 2, 1962.
150. Waksman, S. A., and Woodruff, H. B.: Proc. Soc. Exp. Biol. Med. 45:609, 1940.
151. Rao, K. V., et al.: Antibiot. Chemother. 12:182, 1962.
152. Remers, W. A.: The Chemistry of Antitumor Antibiotics, Vol. 1. New York, Wiley, 1979.
153. Schmidt-Kastner, G.: Naturwissenschaften 43:131, 1956.
154. Umezawa, H.: Bleomycin: Discovery, chemistry, and action. In Umezawa, H. (ed.). Bleomycin, Fundamental and Clinical Studies, p. 3. Tokyo, Gann Monogr. Cancer Res. 1976.
155. Brockman, H.: Fortschr. Chem. Org. Naturst. 18:1, 1960.
156. Sobell, H. M., and Jain, S. C.: J. Mol. Biol. 68:21, 1972.
157. Goldberg, I. H., and Friedman, P. A.: Pure Appl. Chem. 28:499, 1971.
158. Wells, R. D., and Larson, J. E.: J. Mol. Biol. 49:319, 1970.
159. Reich, E., et al.: Science 134:556, 1961.
160. DuBost, N., et al.: C. R. Acad. Sci. Paris 257:1813, 1963.
161. Grein, A., et al.: Giorn. Microbiol. 11:109, 1963.
162. DiMarco, A., et al.: Cancer Chemother. Rep. 53:33, 1969.

163. Rauscher, F. J., Jr.: Special Communication from the Director, National Cancer Program, June 20, 1975.
164. Brockman, H.: Fortschr. Chem. Org. Naturst. 21:1, 1963.
165. Arcamone, F., et al.: Tetrahedron Lett. 1968:3349.
166. Oki, T., et al.: J. Antibiot. 28:830, 1975.
167. DiMarco, A.: Daunomycin and related antibiotics. In Gottlieb, D., and Shaw, D. (eds.). Antibiotics, I, pp. 190–210. New York, Springer-Verlag, 1967.
168. Krishan, A., et al.: Cancer Res. 36:2108, 1976.
169. Lown, J. W., et al.: Biochem. Biophys. Res. Commun. 76:705, 1979.
170. Bachur, N. R., and Gee, M.: J. Pharmacol. Exp. Ther. 177:567, 1971.
171. Bachur, N. R., et al.: J. Med. Chem. 19:651, 1976.
172. Arcamone, F., et al.: J. Med. Chem. 19:1424, 1976.
173. Gause, G. F., et al.: Antibiotiki 18:675, 1973.
174. Bhuyan, B. K., and Dietz, A.: Antimicrob. Agents Chemother. 1965:836.
175. Eckle, E., et al.: Tetrahedron Lett. 21:507, 1980.
176. McGovren, J. P., et al.: Invest. New Drugs 2:359, 1984.
177. Berlin, Y. A.: Nature 218:193, 1968.
178. Rao, K. V., et al.: Antibiot. Chemother. 12:182, 1962.
179. Nayak, R., et al.: FEBS Lett. 30:157, 1973.
180. Gauze, G. F.: Olivomycin, chromomycin, and mithramycin. In Corcoran, J. W., and Hahn, F. E. (eds.). Antibiotics, III., p. 197–202. New York, Springer-Verlag, 1975.
181. Kersten, W.: Abh. Deut. Akad. Wiss. Berlin, Kl. Med. p. 593, 1968.
182. Umezawa, H., et al.: J. Antibiot. Ser. A. 19:260, 1966.
183. Umezawa, H., et al.: J. Antibiot. Ser. A. 19:210, 1966.
184. Ikekawa, T., et al.: J. Antibiot. Ser. A. 17:194, 1964.
185. Argoudelis, A. A., et al.: J. Antibiot. 24:543, 1971.
186. Takita, T., et al.: J. Antibiot. 31:1073, 1978.
187. Sugiura, Y., and Kikuchi, T.: J. Antibiot. 31:1310, 1978.
188. Umezawa, H., et al.: J. Antibiot. 25:409, 1972.
189. Nakayama, Y., et al.: J. Antibiot. 26:400, 1973.
190. Muraoka, Y., et al.: J. Antibiot. 29:853, 1976.
191. Wakaki, S., et al.: Antibiot. Chemother. 8:288, 1958.
192. DeBoer, C., et al.: Antimicrob. Agents Annu. 1960:17, 1961.
193. Webb, J. S., et al.: J. Am. Chem. Soc. 84:3185, 1962.
194. Iyer, V. N., and Szybalski, W.: Science 145:55, 1964.
195. Tomasz, M.: Chem. Biol. Interactions 13:89, 1976.
196. Taylor, W. G., and Remers, W. A.: J. Med. Chem. 18:307, 1975.
197. Patrick, J. B., et al.: J. Am. Chem. Soc. 86:1889, 1964.
198. Meguro, S., et al.: Invest. New Drugs 2:381, 1984.
199. Bradner, W. T., et al.: Cancer Res. 45:6475, 1985; Nakatsubo, F., et al.: J. Am. Chem. Soc. 99:8115, 1977.
200. Kinoshita, S., et al.: J. Med. Chem. 14:103, 1971.
201. Vavra, J. J., et al.: Antibiot. Annu. 1960:230, 1960.
202. Herr, R. R., et al.: J. Am. Chem. Soc. 89:4808, 1967.
203. Hessler, E. J., and Jahnke, H. K.: J. Org. Chem. 35:245, 1970.
204. Kennedy, B. J.: Cancer 26:755, 1970.
205. Johnston, T. P., et al.: J. Med. Chem. 18:104, 1975.
206. Jarjaram, H. N., et al.: Cancer Chemother. Rep. Part 1 59:481, 1975.
207. Farber, S.: JAMA 198:826, 1966.
208. Livingston, R. B., and Carter, S. K.: Single Agents in Cancer Chemotherapy. New York, Plenum Press, 1970.
209. Arcamone, F., et al.: Tetrahedron Lett. 1969:1007.
210. Chabner, B. A., et al.: N. Engl. J. Med. 292:1107, 1975.
211. Umezawa, H., et al.: J. Antibiot. Ser. A., 19:200, 1966.
212. Blum, R. H., et al.: Cancer 31:903, 1973.
213. Einhorn, L. H., and Donohue, J.: Ann. Intern. Med. 87:293, 1977.
214. Wakaki, S., et al.: Antibiot. Chemother. 8:288, 1958.
215. Baker, L. H.: The development of an acute intermittent schedule—mitomycin C. In Carter, S. K., and Crooke, S. T. (eds.). Mitomycin C: Current Status and New Developments, p. 77. New York, Academic Press, 1969.
216. Veldhius, J. D.: Lancet 1:1152, 1978.
217. Moertel, C. G., et al.: Cancer Chemother. Rep. 55:303, 1971.
218. Hartwell, L. J.: Lloydia 31:71, 1968.
219. Kelley, M. G., and Hartwell, J. L.: J. Natl. Cancer Inst. 14:967, 1953.
220. Sartorelli, A. C., and Creasey, W. A.: Annu. Rev. Pharmacol. 9:51, 1969.
221. Long, B. H., et al.: Cancer Res. 45:3106, 1985.
222. Svoboda, G.: Lloydia 24:173, 1961.
223. Neuss, N., et al.: J. Am. Chem. Soc. 86:1440, 1964.
224. Creasey, W. A.: Vinca alkaloids and colchicine. In Sartorelli, A. C., and Johns, D. J. (eds.). Handbook of Experimental Pharmacology, Vol. 38, part 2, pp. 670–694. New York, Springer-Verlag, 1975.
225. Barnett, C. J., et al.: J. Med. Chem. 21:88, 1978.
226. Bensch, K. G., and Malawista, S. E.: J. Cell. Biol. 40:95, 1969.
227. Journey, L. J., et al.: Cancer Chemother. Rep. 52:509, 1968.
228. Krishan, A.: J. Natl. Cancer Inst. 41:581, 1968.
229. Taylor, E. W.: J. Cell. Biol. 25:145, 1965.
230. DeVita, V. T., et al.: Cancer 30:1495, 1972.
231. Rosenberg, B., et al.: Nature 205:698, 1965.
232. Rosenberg, B., et al.: J. Bacteriol. 93:716, 1967.
233. Rosenberg, B., et al.: Nature 222:385, 1969.
234. Gale, G. R.: Platinum compounds. In Sartorelli, A. C., and Johns, D. J. (eds.). Handbook of Experimental Pharmacology, Vol. 38, part 2, p. 829–838. New York, Springer-Verlag, 1975.
235. Rozencweig, M., et al.: Cisplatin. In Pinedo, H. M. (ed.). Cancer Chemotherapy 1979, p. 107. Amsterdam, Excerpta Medica, 1979.
236. Kennedy, P., et al.: Am. Soc. Clin. Oncol. Annu. Meet., Atlanta, GA, May 17–19, 1986: Abstr. 533.
237. Stearns, B., et al.: J. Med. Chem. 6:201, 1963.
238. Krakoff, I. H., et al.: Cancer Res. 28:1559, 1968.
239. Brown, N. C., et al.: Biochem. Biophys. Res. Commun. 30:522, 1968.
240. Brockman, R. W., et al.: Cancer Res. 30:2358, 1970.
241. Hewlett, J. S.: Proc. Am. Assoc. Cancer Res. 13:119, 1972.
242. Kidd, J. G.: J. Exp. Med. 98:565, 1953.
243. McCoy, T. A., et al.: Cancer Res. 19:591, 1959.
244. Mashburn, L. T., and Wriston, J. C., Jr.: Arch. Biochem. Biophys. 105:450, 1964.
245. Wriston, J. C., Jr.: Enzymes 4:101, 1971.
246. Hrushesky, W. J., et al.: Med. Pediatr. Oncol. 2:441, 1976.
247. Broome, J. D., and Schwartz, J. H.: Biochim. Biophys. Acta 138:637, 1967.
248. Brockman, R. W., et al.: Proc. Soc. Exp. Biol. 133:609, 1970.
249. Michaud, R. L., and Sartorelli, A. C.: 155th Am. Chem. Soc. Natl. Meet., San Francisco, April 1968, Abstr. 54.
250. French, F. A., and Freedlander, B. L.: Cancer Res. 18:1298, 1958.
251. Booth, B. A., and Sartorelli, A. C.: Mol. Pharmacol. 3:290, 1967.
252. Pressman, B. C.: J. Biol. Chem. 238:401, 1963.
253. Mihich, E.: Pharmacologist 5:270, 1963.
254. Yesir, D. W., and Kensler, C. J.: The phthalanilides. In Sartorelli, A. C., and Johns, D. J. (eds.). Handbook of Experimental Pharmacology, Vol. 38, part 2, pp. 820–828. New York, Springer-Verlag, 1975.
255. Hirschberg, E., et al.: J. Natl. Cancer Inst. 20:567, 1959.
256. Waring, M. J.: Humangenetik 9:234, 1970.
257. Creighton, A. M., et al.: Nature 222:384, 1969.
258. Creighton, A. M., and Birnie, G. D.: Int. J. Cancer 5:47, 1970.
259. Von Hoff, D. D., et al.: New anticancer drugs. In Pinedo, H. M. (ed.). Cancer Chemotherapy 1979, pp. 126–166. Amsterdam, Excerpta Medica, 1979.
260. Witiak, D. T., et al.: J. Med. Chem. 21:1194, 1978.
261. Bond, J. V.: Proc. Am. Assoc. Cancer Res. 3:306, 1962.
262. Legha, S. S., et al.: Cancer 38:27, 1976.
263. Wilson, W. R.: Chem. N. Z. 37:148, 1973.
264. Atwell, G. J., et al.: J. Med. Chem. 15:611, 1972.
265. Bristol Laboratories: Platinol Product Monograph; Syracuse, N.Y.
266. Eihorn, L. H., and Donohue, J.: Ann. Intern. Med. 87:293, 1977.

267. Einhorn, L. H.: Combination chemotherapy with *cis*-diamminedichloroplatium, vinblastine, and bleomycin in disseminated testicular cancer. In Carter, S. K., et al. (eds.). Bleomycins: Current Status and New Developments. New York, Academic Press, 1978.
268. Hantzsch, A.: Ann. 299:99, 1898.
269. Horrom, B. W., and Carbon, J. A.: Ger. Pat. 1,138,781, Oct. 31, 1962.
270. Gorski, J., et al.: J. Cell. Comp. Physiol. 66:91, 1965.
271. Jensen, E. V., and Jacobson, H. I.: Recent Prog. Horm. Res. 18:387, 1962.
272. Bruchovsky, N., and Wilson, J. D.: J. Biol. Chem. 243:2012, 1968.
273. Sherman, R. R., et al.: J. Biol. Chem. 245:6085, 1970.
274. Wira, C., and Munck, A.: J. Biol. Chem. 245:3436, 1970.
275. Schinzinger, A.: 18th Kongr. Beilage Centralblatt Chir. 29:5, 1889.
276. Dao, T. L.: Some current thoughts on adrenalectomy. In Segaloff, A., et al. (eds.). Current Concepts of Breast Cancer, pp. 189–199. Baltimore, Williams & Wilkins, 1967.
277. Kennedy, B. J.: Cancer 18:1551, 1965.
278. Pearson, O. H., et al.: In Dao, T. L. (ed.). Estrogen Target Tissues and Neoplasia, pp. 287–305. Chicago, University of Chicago Press, 1972.
279. Heel, R. C., et al.: Drugs 16:1, 1978.
280. Beckett, V. L., and Brennan, M. J.: Surg. Gynecol. Obstet. 109:235, 1959.
281. Dao, T. L.: Pharmacology and clinical utility of hormones in hormone related neoplasma. In Sartorelli, A. C., and Johns, D. J. (eds.). Handbook of Experimental Pharmacology, Vol. 38, part 2, p. 172. New York, Springer-Verlag, 1975.
282. Bloom, H. J. G.: Br. J. Cancer 25:250, 1971.
283. Kelley, R., and Baker, W.: Clinical observations on the effect of progesterone in the treatment of metastatic endometrial carcinoma. In Pincus, G., and Vollmer, E. P. (eds.). Biological Activities of Steroids in Relation to Cancer, pp. 427–443. New York, Academic Press, 1960.
284. Dougherty, T. F., and White, A.: Am. J. Anat. 77:81, 1965.
285. Heilman, F. R., and Kendall, E. C.: Endocrinology 34:416, 1944.
286. Dao, T. L.: Third National Cancer Conference Proceedings, pp. 292–296. Philadelphia, J. B. Lippincott, 1957.
287. Hart, M. M., and Straw, J. A., Steroids 17:559, 1971.
288. Bergenstal, D. M., et al.: Ann. Intern. Med. 53:672, 1960.
289. Haller, H. L., et al.: J. Am. Chem. Soc. 67:1600, 1945.
290. Cueto, C., and Brown, J. H. V.: Endocrinology 62:326, 1958.
291. Ringold, H. E., Batres, E., Halpern, O., and Necoechea, E.: J. Am. Chem. Soc. 81:427, 1959.
292. Fried, J., Thoma, R. W., and Klingsberg, A.: J. Am. Chem. Soc. 75:5764, 1953.
293. Ringold, H. E., Ruelas, J. P., Batres, E., and Djerassi, C.: J. Am. Chem. Soc. 81:3712, 1959.
294. Imperial Chemical Industries: Belg. Patent 637,389, Mar. 13, 1964.
295. Bedford, G. R., and Richardson, D. N.: Nature 212:733, 1966.
296. Jordan, V. C., and Jaspan, T.: J. Endocrinology 68:453, 1976.
297. Morton, D.: Report to the American Association for the Advancement of Science, San Francisco, February, 1974.
298. O'Brien, P. H.: J. S. C. Med. Assoc. 68:466, 1972.
299. Goodnight, J. E., Jr., and Morton, D. L.: Annu. Rev. Med. 29:231, 1978.
300. Sanders, H. J.: Chem. Eng. News, p. 74, Dec. 23, 1974.
301. Old, L. J.: Science 230:630, 1985.
302. Farrar, J. J., et al.: Annu. Rep. Med. Chem. 19:191, 1984.
303. Facts and Comparisons. St. Louis, Facts and Comparisons, 1986, p. 683b.
304. Rosenberg, S.: N. Engl. J. Med. 313:1485, 1985.
305. Moertel, C. G.: JAMA 256:3117, 1986.

SELECTED READINGS

Baker, B. R.: Design of Active–Site-Directed Irreversible Enzyme Inhibitors. New York, John Wiley & Sons, 1967.
Baserga, R. (ed.): The Cell Cycle and Cancer. New York, Marcel Dekker, 1971.
Brodsky, I., and Kahn, S. B. (eds.): Cancer Chemotherapy II. New York, Grune & Stratton, 1972.
Calabresi, P., and Parks, R. E., Jr.: Chemotherapy of neoplastic diseases. In Gilman, A. G., Goodman, L. S., Rall, T. W., and Murad, F. (eds.). The Pharmacological Basis of Therapeutics, 7th ed., pp. 1240–1306. New York, Macmillan, 1985.
Cline, M. J., and Haskell, C. M.: Cancer Chemotherapy. Philadelphia, W. B. Saunders, 1980.
Colata, G.: Why do cancer cells resist drugs? Science 232:643, 1986.
De Vita, V. T., Jr., and Busch, H. (eds.): Methods in Cancer Research, Vol. 16. New York, Academic Press, 1979.
De Vita, V. T., and Kershner, L. M.: Cancer, the curable diseases. In Am. Pharm. N520:16, 1980.
Ferguson, L. N.: Cancer and chemicals. Chem. Rev. 75:289, 1975.
Frei, E. III.: Curative cancer chemotherapy. Cancer Res. 45:6523, 1985.
Greenspan, E. M. (ed.): Clinical Cancer Chemotherapy. New York, Raven Press, 1975.
Holland, J. F., and Frei, E. III (eds.): Cancer Medicine. Philadelphia, Lea & Febiger, 1973.
Livingston, R. B., and Carter, S. K.: Single Agents in Cancer Chemotherapy. New York, Plenum Press, 1970.
McKearns, K. W. (ed.): Hormones and Cancer. New York, Academic Press, 1974.
Montgomery, J. A., et al.: Drugs for neoplastic disease. In Wolff, M. E. (ed.). Burgers' Medicinal Chemistry, 4th ed., part II, pp. 595–670. New York, Wiley-Interscience, 1979.
Pinedo, H. M. (ed.): Cancer Chemotherapy 1979. Amsterdam-Oxford, Excerpta Medica, 1979.
Plants and cancer. In Proceedings of the 16th Annual Meeting of the Society for Economic Botany. Cancer Treat. Rep. 60:973, 1976.
Remers, W. A. (ed.): Antineoplastic Agents. New York, John Wiley & Sons, 1984.
Ross, W. C. J.: Biological Alkylating Agents. London, Butterworths, 1962.
Sartorelli, A. C., and Johns, D. J., (eds.): Antineoplastic and Immunosuppressive Agents: Handbook of Experimental Pharmacology, Vol. 38, parts 1 and 2. New York, Springer-Verlag, 1974, 1975.
Stoll, B. A., (ed.): Endocrine Therapy in Malignant Disease. Philadelphia, W. B. Saunders, 1972.
Suhadolnik, R. J.: Nucleoside Antibiotics. New York, John Wiley & Sons, 1970.

CHAPTER 9

Central Nervous System Depressants

Eugene I. Isaacson

The agents included in this chapter can all be characterized as possessing a depressant action on the central nervous system (CNS). The *general anesthetics*, most notably exemplified by the inhalation anesthetics and the intravenous barbiturates, produce a rather nonselective and general depression of the CNS. This, in turn, appears to be a consequence of generally nonspecific effects producing membrane stabilization at the molecular level. Classically, the area of *sedative–hypnotic agents* has overlapped the area of anesthetic agents (e.g., at low doses, sedation; at higher doses, hypnosis; and at still higher doses, general anesthesia), with correspondingly similar mechanisms of action at the molecular level. This is still true for many of the agents. However, the development of newer agents, especially the benzodiazepines, has introduced certain distinctions. Some of the latter appear to be rather selectively *anxiolytic* with markedly diminished sedative–hypnotic effects, and the agents overall have a decreased ability to produce general CNS-depressant effects. A greater specificity of action at the molecular level appears to correlate with these effects.

Anticonvulsants, or more accurately, *antiepileptics*, are depressants that have the ability to control epileptic seizures, which may or may not be convulsive. At the molecular level, some antiepileptic actions appear to be related to anxiolytic and sedative–hypnotic actions; the agents inhibit nerve transmission by a variety of mechanisms.

Antipsychotics possess CNS-depressant effects. For example, on the basis of their ability to produce decreased responsiveness to emotionally unpleasant stimuli, they are sometimes called *neuroleptics*. It should be noted, however, that they have little in common with most of the other drugs described in this chapter, including action at the molecular level. It is becoming increasingly accepted that these agents competitively antagonize dopamine (DA) at D-2 receptors (or, as do some experimental drugs, act as dopamine agonists at presynaptic DA receptors) in the mesolimbic system as the basis for their antipsychotic effects.

GENERAL ANESTHETICS

General anesthetics are agents that produce insensibility by a successive or progressive depression of CNS function. The successive levels of depression were first observed and described with diethyl ether as the anesthetic agent.

Diethyl ether is highly soluble in blood. Accordingly, the progression of effects on the CNS proceed slowly, which permits anesthesiologists to observe the successive phases of anesthesia. With modern anesthetics, and with the present use of one or more adjunctive agents, not all of the signs of the stages are now seen.[1] A review of the stages is still appropriate because it addresses the effects of gradual progressive general depression of the CNS.

- *Stage I* (*cortical stage*): This is the stage proceeding up to unconsciousness. The patient is sleepy, analgesia is produced, and some types of surgery that do not require muscle relaxation can be performed.
- *Stage II* (*delirium*): This is the stage between unconsciousness and surgical anesthesia. Depression of higher centers produces a variety of effects, including excitement, involuntary activity, and increased skeletal muscle tone.
- *Stage III* (*surgical anesthesia*): In this stage excitement is lost and skeletal muscle relaxation is produced. Most types of surgery are done in this

stage. There are four planes to this stage, based on eyeball movements, pupil size, presence of reflexes, and the nature of the respiration.

- *Stage IV (medullary depression)*: Respiratory and circulatory failure occur as depression of the vital centers of the medulla and brain stem occur.

For many years, only a few agents, such as opium and ethanol, were available to help patients withstand the trauma of surgery. They gave some relief, but could not safely produce total insensitivity to pain. The development of chemistry to the point at which pure chemical substances could be prepared, and the observation that some of these substances gave imperviousness to pain, resulted, in the 1840s, in the successful introduction of nitrous oxide, diethyl ether, and chloroform into use as general anesthetics in surgery.

Today, general anesthesia for surgery employs multiple drug regimens. Some drugs are intended to potentiate or otherwise augment the effect of the general anesthetic agent, for example, preanesthetic barbiturates, benzodiazepines, neuroleptics, and narcotic analgesics. Other drugs add an action, for example, the skeletal muscle relaxants, and finally, drugs to control side effects, such as anticholinergics, may be employed. For the ways in which these agents may be orchestrated into an effective harmonious whole, the interested reader should consult a pharmacology or medical textbook. The ensuing discussion will examine only those agents clearly used as general anesthetic agents.

These anesthetic agents can be broadly categorized as those useful by the inhalation route and those useful by the intravenous (IV) route.

The preference for these routes is readily apparent: they afford the flexibility in dosing and the short time between administration and effect that is required when administering agents with a low margin of safety (i.e., dosing can be rapidly changed in accordance with changing conditions). Given the inhalation route, the IV route, and the intent—a controlled general depression of the CNS—the drugs under consideration become the inhalation anesthetics and the water-soluble salts of the ultrashort-acting barbiturates given IV, with the IV barbiturates used almost exclusively for induction, not maintenance, of anesthesia, because their margin of safety is low.

The area of general anesthetics has been fruitful in the stimulation and development of useful theories of how chemical structure and properties relate to drug action. Theories of how general anesthetics act are given in Chapter 3. Suffice it to say that the various partitioning theories and their implied or stated effects (e.g., disordering of lipids or proteins in nerve cell membranes or ordering of synaptic water) are in accordance with both the observed phenomenon of general depression of the CNS and the chemical and physical properties of the inhalation agents.[2–4] Theories of how barbiturates may act are given in the section on sedative hypnotics.

INHALATION ANESTHETICS

The inhalation anesthetics in use today are halothane, enflurane, isoflurane, methoxyflurane, and nitrous oxide. (For a consideration of older agents, such as diethyl ether and cyclopropane, see the eighth edition of this textbook.) These agents became obsolete because of a fundamental chemical property. They are explosive or flammable and add an unacceptable level of danger to the production of anesthesia.

Halothane, USP. 2-Bromo-2-chloro-1,1,1-trifluroethane (Fluothane); $CH(Br)ClCF_3$. Halothane, a volatile halogenated hydrocarbon (bp, 50°C), is now taken as the standard inhalation anesthetic agent. It was introduced in 1956 and gained rapid acceptance. One major factor in its acceptance was its nonflammability. Additionally, the drug has high potency and a relatively low blood/gas partition coefficient. Accordingly, induction of, and recovery from, anesthesia are relatively rapid. In actual practice, intravenous sodium thiopental is usually used to induce anesthesia.

Most of halothane is eliminated intact in the inspired air. There is sufficient reactivity to oxidative processes, however, to allow up to 20% of the administered compound to undergo metabolism. The trifluoromethyl group is quite stable, in itself, because the high electronegativity of fluorine stabilizes the C—F bonds. The C—Cl, C—Br, and C—H bonds, however, are destabilized. The latter is the probable initial site of metabolic entry. Metabolites are chloride and bromide ions and trifluoroacetic acid. Additionally, loss of HF yields the olefin 1,1-difluro-2-chloro-2-bromoethylene, which reacts with the SH group of glutathione.

There has been a low incidence of hepatic necrosis associated with halothane. It is suggested that reactive metabolites formed, as outlined in the foregoing, might produce an immunoreactive response.[5, 6]

Hypotension is a common side effect of the drug and sometimes may be used to advantage.

Halothane does not have a wide margin of safety. Respiratory depression is notable, and mechanical ventilation and increased oxygen concentrations are often required. Opioids or nitrous oxide are often needed to obtain adequate surgical analgesia.

Methoxyflurane, USP. 2,2-Dichloro-1,1-difluoroethyl methyl ether; $CHCl_2CF_2—O—CH_3$ (Penthrane). Methoxyflurane is a volatile liquid (bp, 105°C). The agent does not have a high vapor pressure at room temperature; thus concentrations in the inspired air are low. This, together with a large blood/gas partition coefficient, produces a slow induction of anesthesia featuring an excitatory phase. Accordingly, induction is usually made with intravenous sodium thiopental. The compound is very soluble in lipids; consequently, recovery is slow. The agent produces excellent analgesia and good muscle relaxation.

Methoxyflurane is as much as 70% metabolized. Apparently all labile sites are attacked. Metabolites include dichloroacetic acid, difluoromethoxyacetic acid, oxalic acid, and fluoride ions. Fluoride ion especially, and oxalic acid also, are responsible for the renal damage the agent produces when used to produce deep anesthesia over prolonged periods. Accordingly, it has restricted application as an anesthetic.[7] It is said to be safe when used by intermittent inhalation as an analgesic during labor.

Enflurane, USP. 2-Chloro-1,1,2-trifluoroethyl difluoromethyl ether; HF_2COCF_2CHFCl (Ethrane). Enflurane is a volatile liquid with a vapor pressure at room temperature about three-fourths that of halothane and a blood/gas partition coefficient also about three-fourths that of halothane. Consequently, induction is relatively easy, although an ultrashort-acting barbiturate is usually employed for this purpose. The agent is said to be a relatively easy agent with which to work and to have a relatively low frequency of adverse cardiovascular effects. Respiration is depressed, and thus mechanical ventilation and oxygen supplementation are employed. At high doses in a small percentage of patients, tonic–clonic convulsive activity is seen. Accordingly, enflurane should not be used in patients with epileptic foci.

As much as 5% of administered drug is metabolized. Difluoromethoxydifluoroacetic acid and fluoride ion have been reported as metabolites. The fluoride concentration, however, is generally thought to lie within safe limits.[8]

Isoflurane, USP. 1-Chloro-2,2,2-trifluoroethyl difluoromethyl ether; $F_3CC(H)ClOCF_3$ (Forane). Isoflurane is a close structural relative of enflurane and shares many properties with it. Changes, however, in the electroencephalogram (EEG) and tonic–clonic activity are not reported. It does differ considerably in extent of metabolism; only about 0.2% is metabolized.[9] Metabolites are fluoride ion and trifluoroacetic acid. Neither kidney nor liver damage has been reported for the drug.

TABLE 9-1

GASEOUS AND LIQUID ANESTHETICS

Name Proprietary Name	***Physical State***	***Category***	***Application***
Enflurane USP *Ethrane*	Liquid (bp, 55.5–57.5°C)	General anesthetic (inhalation)	By inhalation, as required
Halothane USP *Fluothane*	Liquid (bp, 49–51°C)	General anesthetic (inhalation)	By inhalation, as required
Isoflurane USP *Forane*	Liquid (bp, 49°C)	General anesthetic (inhalation)	By inhalation, as required
Methoxyflurane USP *Penthrane*	Liquid (bp, 105°C)	General anesthetic (inhalation)	By inhalation, as required
Nitrous oxide USP	Gas	General anesthetic (inhalation)	By inhalation, as required

Nitrous Oxide, USP. Nitrogen monoxide; N_2O. Nitrous oxide is a gas at room temperature and is supplied as a liquid under pressure in metal cylinders. It is a good analgesic, but requires such high concentrations in the inspired mixture (up to 80%) to achieve anesthesia that there are attendant dangers of hypoxia. Accordingly, it is rarely used as the sole anesthetic agent. It is often useful in conjugation with other agents, permitting their use at decreased concentrations. Some studies have suggested that nitrous oxide analgesia is mediated through opioids.[10] Other studies have failed to substantiate this and suggest that analgesia may be a consequence of a general depressant effect on synaptic transmission of pain messages.[11]

ULTRASHORT-ACTING BARBITURATES

The sodium salts of the ultrashort-acting barbiturates may be administered intravenously in aqueous solutions for the induction of anesthesia. Thereafter, the maintaining anesthetic agent (halothane, enflurane, or isoflurane, with or without nitrous oxide) is employed. Respiratory depression is marked with the barbiturates at anesthetic doses; consequently, these agents are not used to maintain surgical anesthesia. Unconsciousness is produced within seconds of intravenous injection, and the duration of action is about 30 minutes. The rapid onset of action

TABLE 9-2

ULTRASHORT-ACTING BARBITURATES USED TO PRODUCE GENERAL ANESTHESIA

General Structure

(Barbiturate ring: positions 1–6; R_5 and R'_5 at C-5; $N-R_1$ at N-1; O at C-4 and C-6; $R_2^-Na^+$ at C-2)

Generic Name Proprietary Name	*Substituents* R_5	R'_5	R_1	R_2
Methohexital sodium *Brevital Sodium*	$CH_2{=}CH{-}CH_2{-}$	$CH_3CH_2C{\equiv}C{-}CH(CH_3){-}$	CH_3	O
Thiamylal sodium *Surital Sodium*	$CH_2{=}CH{-}CH_2{-}$	$CH_3CH_2CH_2CH(CH_3){-}$	H	S
Thiopental sodium *Pentothal Sodium*	$CH_3CH_2{-}$	$CH_3CH_2CH_2CH(CH_3){-}$	H	S

TABLE 9-3

ULTRASHORT-ACTING BARBITURATES

Name Proprietary Name	*Preparations*	*Category*	*Usual Adult Dose* *
Methohexital sodium *Brevital Sodium*	Methohexital sodium for injection USP	General anesthetic (intravenous)	IV, 5–12 mL of a 1% solution at the rate of 1 mL every 5 sec for induction, then 2–4 mL every 4–7 min as required
Thiamylal sodium *Surital Sodium*	Thiamylal sodium for injection USP	General anesthetic (systemic)	IV, induction: 3–6 mL of a 2.5% solution at the rate of 1 mL every 5 sec; maintenance: 500 μL to 1 mL as required
Thiopental sodium USP *Pentothal Sodium*	Thiopental sodium for injection USP	Anticonvulsant; general anesthetic (intravenous)	Anticonvulsant: IV, 3–10 mL of a 2.5% solution over a 10-min period. Anesthetic (induction): IV, 2–3 mL of a 2.5% solution at intervals of 30–60 sec as necessary

*See USP DI for complete dosage information.

is believed to be caused by the agents quickly partitioning from the blood across the blood–brain barrier into the sites of action in the brain. Thiopental, for example, has an exceptionally high lipid/water partition coefficient. The very short duration of action is attributed to partitioning from the brain into peripheral tissues—initially to well-perfused tissues and, subsequently, to body fat. Additionally, for methohexital, rapid metabolism may be involved as well. The structures of the compounds can be seen in Table 9-2.

Methohexital Sodium. Sodium-(±)-1-methyl-5-allyl-5-(1-methyl-2-pentynyl) barbiturate (Brevital Sodium). Methohexital is an *N*-methyl barbiturate and has a pK_a of 8.4 versus about 7.6 for the non–*N*-methylated compounds. This pK_a change increases the concentration of the lipid-soluble free acid form at physiologic hydrogen ion concentrations. The compound also has extensive hydrophobic character (total nine hydrocarbon carbons); consequently, the lipid/water partition coefficient of the free acid form is high. Finally, it has an accessible site of metabolism, the CH_2 α to the triple bond. Overall, the compound has the properties to rapidly penetrate the CNS after intravenous injection and, then, rapidly redistribute to other body sites and also undergo rapid metabolic inactivation.

Thiamylal Sodium. Sodium 5-allyl-5-(1-methylbutyl)-2-thiobarbiturate (Surital Sodium). Thiamylal is a highly hydrophobic thiobarbiturate and has structural features closely related to thiopental. It has biologic properties similar to thiopental. After intravenous administration, unconsciousness is produced within seconds, with recovery of consciousness within 30 minutes.

Thiopental Sodium, USP. Sodium 5-ethyl-5-(1-methylbutyl)-2-thiobarbiturate (Pentothal Sodium). Thiopental sodium is the most widely used ultrashort-acting anesthetic barbiturate. Additionally, the compound is the prototype for the ultrashort-acting barbiturates. Most discussions of how structure influences duration of action relate specifically to it. The compound's onset of action is about equal to the time required for it to travel to the brain from the site of administration. Consciousness is regained within 30 minutes.

DISSOCIATIVE ANESTHETICS

Ketamine Hydrochloride, USP. (±)-2-(*o*-Chlorophenyl)-2-methylaminocyclohexanone hydrochloride (Ketalar). Ketamine was designed as a structural relative of the medically discontinued agent phencyclidine (PCP) (see Chap. 10), with which it shares a number of biologic properties. The incidence of hallucinations, however, is much less with ketamine. Ketamine produces a sense of dissociation from events being experienced, followed by anesthesia, analgesia, and sometimes amnesia. The anesthetic state produced by ketamine has also been described as cataleptic anesthesia.[12] The prevalence of hallucinations and excitement is higher in adults than in children. The drug may be used as the sole agent (mainly for minor surgical procedures in children), it can be used to induce anesthesia that is then maintained by one of the potent inhalation agents, or it and nitrous oxide may be used together for general anesthesia.

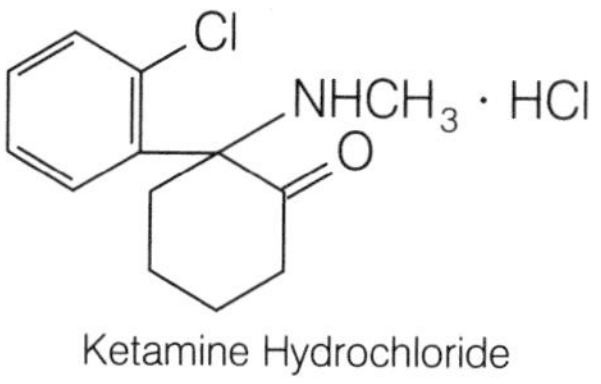

Cl
$NHCH_3 \cdot HCl$
O

Ketamine Hydrochloride

ANXIOLYTIC, SEDATIVE, AND HYPNOTIC AGENTS

Agents of the anxiolytic, sedative, and hypnotic group can encompass a wide variety of drugs. For example, most antipsychotic agents reduce anxiety and are sedative, possibly related to block of DA and norepinephrine (NE) receptors. They no longer are often used for minor tranquilizing effects (e.g., in the psychoneuroses) because of the danger of inducing tardive dyskinesia. Additionally, many H_1-antihistamines have antianxiety and sedative properties, possibly related to central NE receptor block and, at times, may be employed in these capacities. Many opioid analgesics produce sedation and reduction of anxiety, and these actions are important and often valuable components of their use in pain control. Also many antihypertensive drugs that act on noradrenergic systems have sedative side effects. And, finally, most anticonvulsant drugs also possess sedative or anxiolytic effects, or both.

The compounds considered here are used more-or-less exclusively for one or more of the following actions: anxiolytic, sedative, and hypnotic. Many agents in this group also have anticonvulsant properties. There are indications that, at least occasionally, the neuronal effects related to anxiolytic, sedative, and hypnotic effects also relate to the anticonvulsant effects. In particular, the two numerically largest groups, the benzodiazepines and the barbiturates, have received much study.

Benzodiazepines at relatively low concentrations appear to combine with a receptor, located postsynaptically, that in conjunction with γ-aminobutyric acid (GABA) and the GABA receptor, opens chloride conductance channels and exerts a postsynaptic inhibitory effect.[13–15] This action is generally taken to be the basis of antianxiety effects and, reportedly, may also be the basis of antiabsence seizure effects.[16] At higher concentrations, benzodiazepines limit sustained high-frequency repetitive firing by a nonsynaptic action. This may be related to antigeneralized tonic–clonic seizure effects.[16] Also, at these higher concentrations, they block the entry of Ca^{2+} into presynaptic neurons and alter neurotransmitter release. This effect may be related to sedation.[16]

Barbiturates act postsynaptically to promote GABA binding and prolong the mean open time of chloride channels by binding to a site other than the benzodiazepine-binding site or the GABA-binding site. Possibly a macromolecular complex involving these binding sites may be involved.[17] This action may also be involved in sedative and antianxiety effects. Also barbiturates appear to facilitate chloride conductance in the absence of GABA, presumably by binding to the GABA receptor.[18] Sedative–hypnotic and anesthetic barbiturates do this to a greater extent than do the anticonvulsant barbiturates.

Another action of barbiturates is to reduce calcium-dependent action potentials and thereby inhibit neurotransmitter release by blocking Ca^{2+} entry into neurons.[18] Again, sedative–hypnotic and anesthetic barbiturates tend to do this to a greater extent than the more selective anticonvulsant barbiturates. The anticonvulsant effect of barbiturates may, in part, be due to enhancing postsynaptic GABA responses and antagonizing postsynaptic excitatory responses.[16] Also, at high concentrations, such as those seen in status epilepticus, phenobarbital reduces repetitive firing.[16]

Antianxiety effects of barbiturates might be mediated by postsynaptic enhancement of the effect of GABA binding. The GABA effect in the absence of GABA and effects on Ca^{2+} may be involved in sedation, hypnosis, and anesthesia and account for the fact that the barbiturates are much more prone to produce these effects than are the benzodiazepines. They may also account for the much narrower therapeutic margin than that of the benzodiazepines.[18, 19]

In addition, other effects, such as dissolution in the lipids or proteins in nerve plasma membranes, have been proposed to account for some of the CNS-depressant actions for certain members of the group, which includes structurally simple alcohols, carbamates, cyclic ethers, and the like.

Finally, to illustrate further the various mechanisms involved, recent work has begun to focus on the role of agonists of serotonin at 5-HT_{1A} receptors in anxiety therapy.

BENZODIAZEPINES

Certain of the benzodiazepines are highly selective in producing anxiolytic effects, although, as a practical matter, some sedation is usually present. Other members are promoted as sedative–hypnotic agents. Overall, they are also relatively safe drugs. In overdose, unlike the barbiturates, respiratory depression is usually not marked.

The field of benzodiazepines was opened with the synthesis of chlordiazepoxide by Sternbach and the discovery of its unique pharmacologic properties by Randall.[20]

Chlordiazepoxide (refer to the discussion of individual compounds) is a 2-amino benzodiazepine, and other amino compounds have been synthesized. However, when it was discovered that chlordiazepoxide is rapidly metabolized to a series of benzodiazepin-2-ones (see the general scheme: metabolic relationships) that are active, emphasis shifted to the synthesis and testing of the later group. Empirical structure–activity relationships (SARs) for antianxiety activity have been tabulated for this group (analogous statements apply for the older 2-amino group).[20, 21] Reference can be made to the following general structure to visualize them.

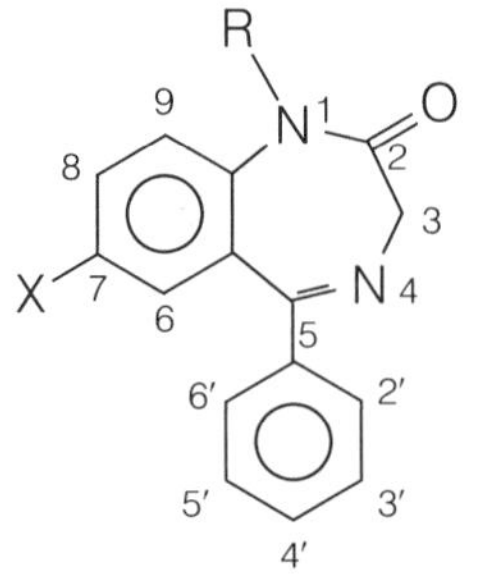

The presence of an electron-attracting substituent at position 7 is required for activity, and the more electron attracting it is, the higher is the activity. Positions 6, 8, and 9 should not be substituted. A phenyl at the 5-position promotes activity. If this group is *ortho* (2′) or di*ortho* (2′,6′) substituted with electron-attracting substituents, activity is increased. On the other hand, *para* substitution decreases activity greatly. Saturation of the 4,5 double bond or a shift to the 3–4-position decreases activity. Alkyl substitution at the 3-position decreases activity, whereas substitution of the 3-position with a hydroxy does not. The presence or absence of the 3-hydroxyl group is very important pharmacokinetically. Compounds without the 3-hydroxyl are very nonpolar and have long half-lives; those with the 3-hydroxyl are much more polar and are readily converted to the excreted glucuronide (see the overall metabolic relationship scheme). The 2-carbonyl function is optimal for activity, as is the nitrogen atom at position 1. The *N*-substituent should be small.

Additional research has yielded compounds with a fused triazolo ring represented by triazolam and alprazolam. These are metabolized mainly by hydroxylation of the methyl substituent on the triazolo ring. The resulting hydroxy compound is active, but is quickly conjugated. The compounds are also metabolized by 3-hydroxylation of the benzodiazepine ring. Interestingly, the presence of an electron-attracting group at position 7 is not required for activity among these compounds.

The metabolism of benzodiazepines has received much study.[22, 23] Some of the major metabolic relationships are shown in the scheme on page 365.

The benzodiazepines are well absorbed from the gastrointestinal tract, although the more polar com-

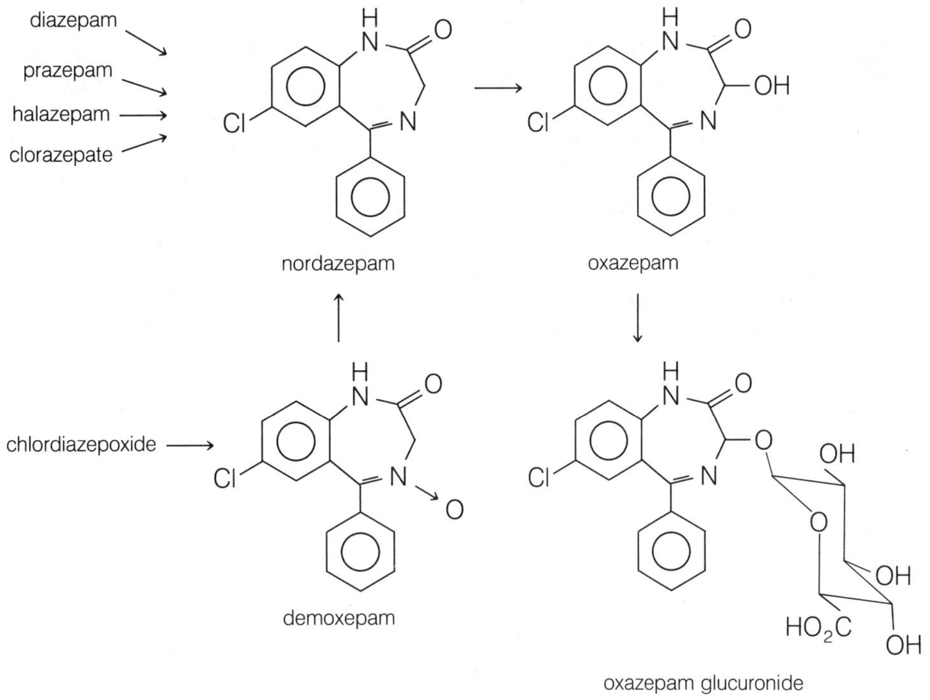

diazepam
prazepam
halazepam
clorazepate
nordazepam
oxazepam
chlordiazepoxide
demoxepam
oxazepam glucuronide

pounds (e.g., those with a hydroxyl at the 3-position) tend to be absorbed more slowly than the more nonpolar compounds.

The drugs tend to be highly bound to plasma proteins; in general, the more nonpolar the drug, the greater the extent of binding. They are also very effectively distributed to the brain. Again, generally, the more nonpolar the compound, the greater the extent of distribution to the brain.

Compounds without the 3-hydroxyl group usually have long half-lives and undergo conversion to the 3-hydroxy compounds by hepatic oxidation. Compounds with the 3-hydroxyl group have short half-lives because of rapid conjugation to the 3-glucuronide, which undergoes urinary excretion. In patients with impaired liver function, the 3-hydroxy compounds, which do not require hepatic oxidation, tend to present fewer hazards than the more nonpolar compounds.

The drugs have legitimate indications in several conditions that feature anxiety or insomnia. Critics of the agents feel that they are overused (e.g., there may be a tendency to continue using the agents when they are no longer needed). On the other hand, patients tend not to increase the dose or exhibit compulsive drug-seeking behavior after the drug is discontinued. Also, chronic drug abusers prefer ethanol or barbiturates.

In addition to lower abuse potential, and a much greater margin of safety than the barbiturates, the drugs have fewer drug interactions. Especially noteworthy is that they do not promote the metabolism of other drugs.

Chlordiazepoxide Hydrochloride, USP. 7-Chloro-2-(methylamino)-5-phenyl-3*H*-1,4-benzodiazepine 4-oxide monohydrochloride (Librium). Absorption from the gastrointestinal tract is good. Peak plasma levels are reached in two to four hours. *N*-Demethylation and hydrolysis of the condensed amidino group is rapid and extensive, producing demoxapam as a major metabolite. Demoxapam, in turn, is converted principally to nordazepam. Nordazepam, in turn, is converted principally to oxepam which undergoes conjugation to the excreted glucuronide. Of course, other routes of metabolism can occur, for example, opening of the seven-membered ring by hydrolysis of the lactam group.

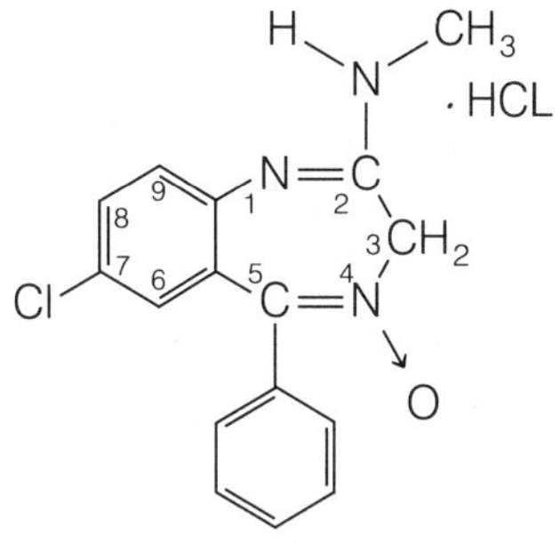

Chlordiazepoxide Hydrochloride

Diazepam, USP. 7-Chloro-1,3-dihydro-1-methyl-5-phenyl-2*H*-1,4-benzodiazepin-2-one (Valium). Diazepam was the first member of the benzodiazepin-2-one group to be introduced. It is very nonpolar and is rapidly absorbed. Diazepam has a long half-life and is metabolized by *N*-demethylation to nordazepam, which is then metabolized according to the general scheme. It is widely used for several anxiety states and has an additional wide range of uses (e.g., as an anticonvulsant, a premedication in anesthesiology, and in various spastic disorders).

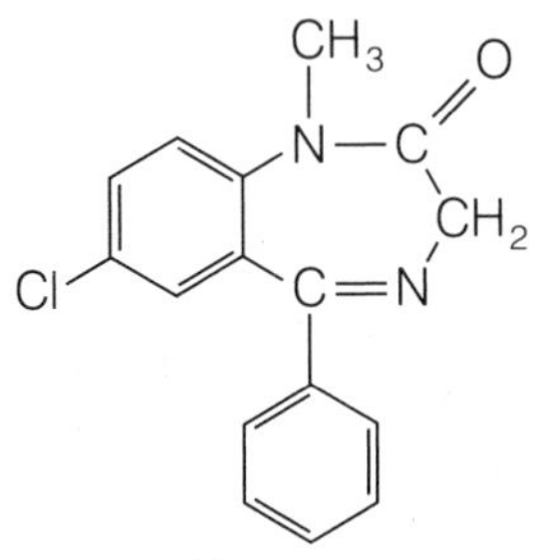

Diazepam

Oxazepam, USP. 7-Chloro-1,3-dihydro-3-hydroxy-5-phenyl-2*H*-1,4-benzodiazepin-2-one (Serax). Oxazepam can be considered a prototype for the 3-hydroxy compounds. It is much more polar than diazepam, for example. Metabolism is relatively uncomplicated, and the duration of action is short.

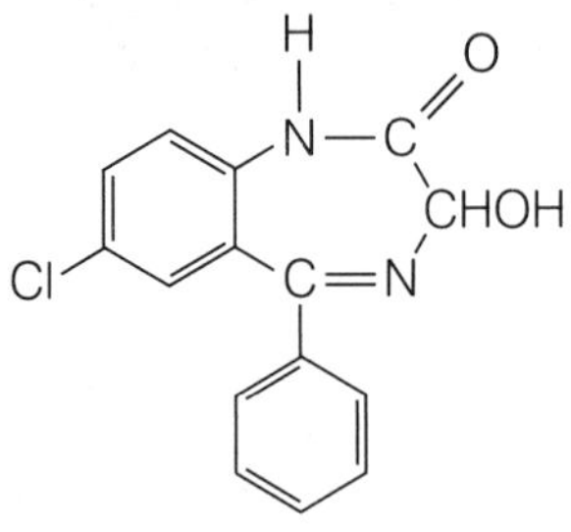

Oxazepam

Clorazepate Dipotassium. 7-Chloro-2,3-dihydro-2-oxo-5-phenyl-1*H*-1,4-benzodiazepine-3-carboxylic acid dipotassium salt monohydrate (Tranxene). Clorazepate can be considered to be a prodrug. Itself inactive, it undergoes rapid loss of water and decarboxylation to nordazepam, which has a long half-life and undergoes hepatic conversion to oxazepam. Despite the polar character of the drug as administered, because it is quickly converted to a nonpolar compound, it has an overall long half-life.

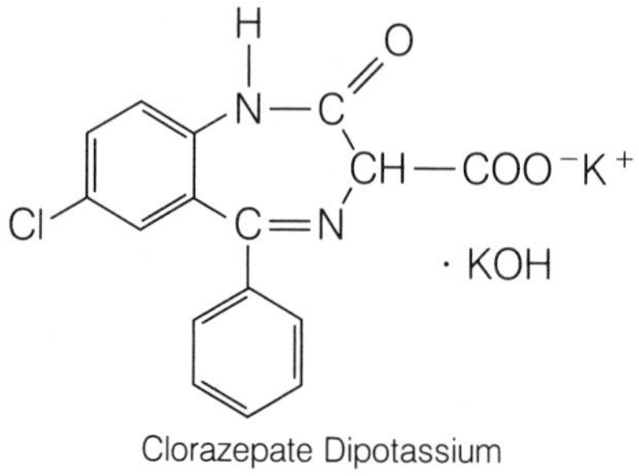

Clorazepate Dipotassium

Prazepam, USP. 7-Chloro-1-(cyclopropylmethyl)-1,3-dihydro-5-phenyl-2*H*-1,4-benzodiazepin-2-one (Verstran). The overall half-life is long. Extensive *N*-dealkylation occurs to yield nordazepam. 3-Hydroxylation of prazepam and of nordazepam occur.

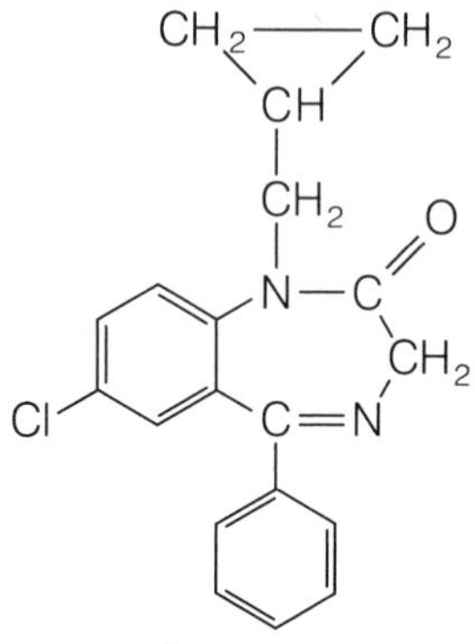

Prazepam

Lorazepam, USP. 7-Chloro-5-(2-chlorophenyl)-3-dihydro-3-hydroxy-2*H*-1,4-benzodiazepin-2-one (Ativan). Lorazepam can be recognized as the 2′-chloro substituted analogue of oxazepam. In keeping with overall SARs, the 2′-chloro substituent increases activity. Metabolism is relatively rapid and uncomplicated because of the presence of the 3-hydroxyl group.

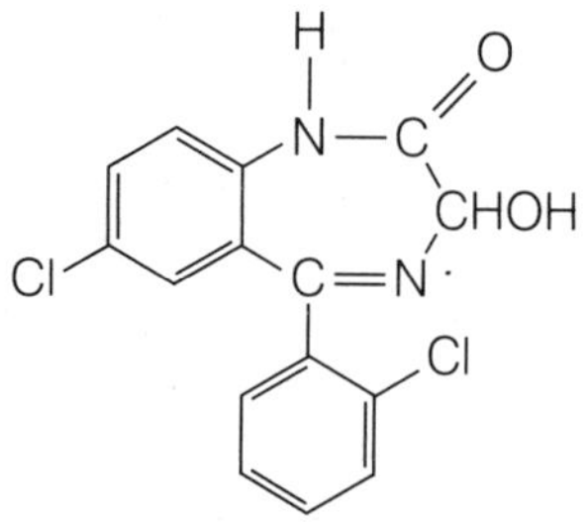

Lorazepam

Halazepam, USP. 7-Chloro-1,3-dihydro-5-phenyl-1-(2,2,2-trifluoroethyl)-2*H*-1,4-benzodiazpin-2-one (Paxipam). Halazepam is well absorbed. It is active and present in plasma, but much of its activity is due to the major metabolite (as well as oxazepam). The drug is marketed as an anxiolytic.

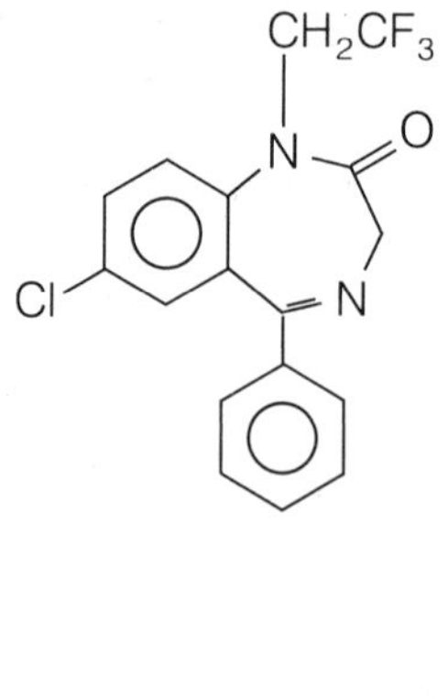

Temazepam. 7-Chloro-1,3-dihydro-3-hydroxy-1-methyl-5-phenyl-2*H*-1,4-benzodiazepin-2-one (Restoril). The compound is a minor metabolite of diazepam. It can also be visualized as *N*-methyl oxazepam. There is a small amount of *N*-demethylation. However, metabolism proceeds mainly through conjugation of the 3-hydroxyl group. The duration of action is short. It is marketed as a hypnotic with little or no residual effects.

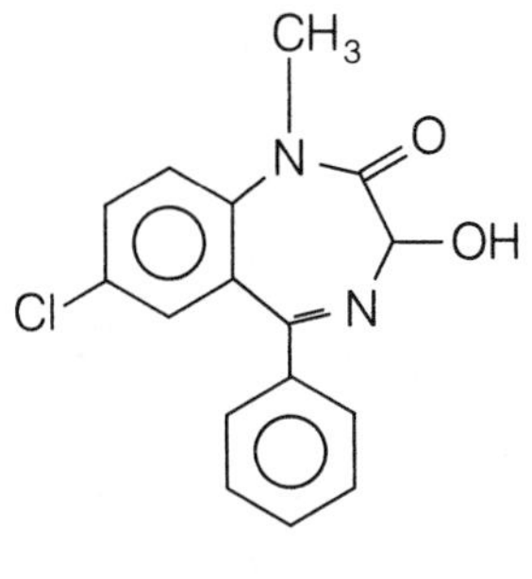

Flurazepam Hydrochloride, USP. 7-Chloro-1-[2-(diethylamino)ethyl]-5-(2-fluorophenyl)-1,3-dihydro-2*H*-1,4-benzodiazepin-2-one dihydrochloride (Dalmane). Flurazepam is notable as a benzodiazepine for being indicated almost exclusively in insomnia. Metabolism of the dialkyl aminoalkyl side chain is extensive. A major metabolite is N^1-dealkyl flurazepam, with a very long half-life, persisting for several days after the drug is administered.

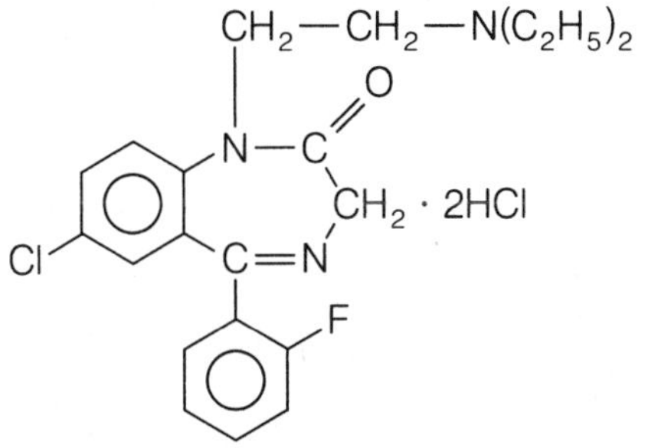

Flurazepam Hydrochloride

Alprazolam, USP. 8-Chloro-1-methyl-6-phenyl-4*H*-*s*-triazolo[4,3-*a*][1,4]benzodiazepine (Xanax) is rapidly absorbed. Protein binding is lower (~70%) than for most benzodiazepines. Oxidative metabolism of the methyl group to the methyl alcohol followed by conjugation is rapid and the duration of action is short. The drug is highly potent as an anxiolytic on a milligram basis.

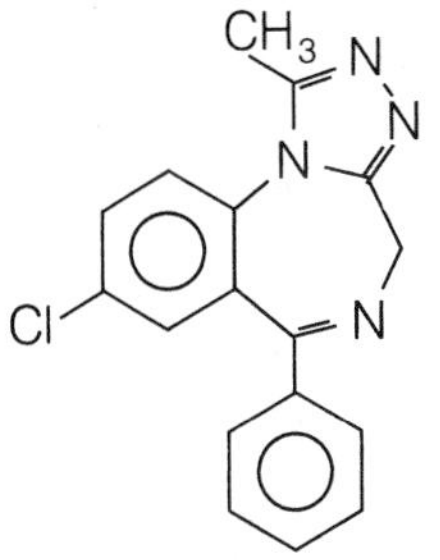

Triazolam, USP. 8-Chloro-6-(*o*-chlorophenyl)-1-methyl-4*H*-*s*-triazolo[4,3-*a*][1,4]benzodiazepine (Halcion). Triazolam has all of the characteristic benzodiazepine actions. It is marketed as a sedative–hypnotic drug said to produce little, if any, daytime impairment of function. It is rapidly metabolized to the 1-methyl alcohol, which is then excreted.

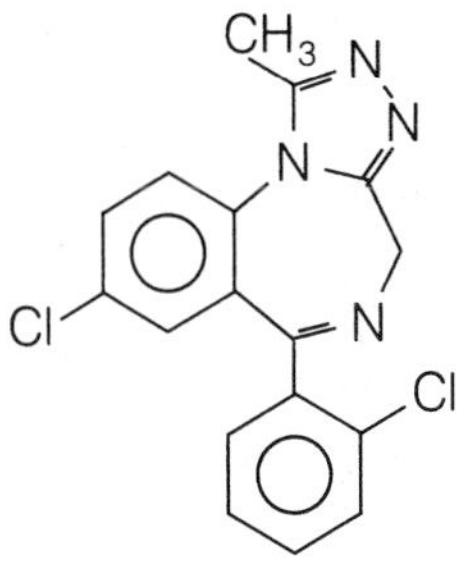

TABLE 9-4

BENZODIAZEPINE DERIVATIVES

Name Proprietary Name	***Preparations***	***Category***	***Usual Adult Dose***	***Usual Dose Range***	***Usual Pediatric Dose***
Alprazolam *Xanac*	Alprazolam tablets USP	Antianxiety agent	0.25–0.5 mg tid	Up to 4 mg/day	Dosage not established
Chlordiazepoxide USP *Libritabs*	Chlordiazepoxide tablets USP	Antianxiety agent	5–25 mg tid or qid		
Chlordiazepoxide hydrochloride USP *Librium*	Chlordiazepoxide hydrochloride capsules USP	Antianxiety agent	5–25 mg tid or qid		Children under 6 yr: dosage is not established; children 6 yr and over: 5–10 mg bid to qid
	Sterile chlordiazepoxide hydrochloride USP	Antianxiety; sedative	IM or IV, 50–100 mg initially, then 25–50 mg tid or qid	Up to 300 mg/day	Children 12 yr and over: 25–50 mg/dose

*See USP DI for complete dosage information.

TABLE 9-4 *Continued*

BENZODIAZEPINE DERIVATIVES

Name Proprietary Name	***Preparations***	***Category***	***Usual Adult Dose***	***Usual Dose Range***	***Usual Pediatric Dose***
Clorazepate dipotassium *Tranxene*	Capsules	Antianxiety agent	7.5–15 mg bid to qid	Up to 90 mg/day	
Diazepam USP *Valium*	Diazepam injection USP	Antianxiety agent	2–10 mg		
	Diazepam tablets USP		2–10 mg bid to qid	2–40 mg/day	Infants under 6 mo: use is not recommended; over 6 mo: 1–2.5 mg tid or qid
Flurazepam hydrochloride USP *Dalmane*	Flurazepam hydrochloride capsules USP	Hypnotic	15–30 mg hs		
Halazepam *Paxipam*	Halazepam tablets USP	Antianxiety agent	20–40 mg tid or bid		Dosage not established
Lorazepam *Ativan*	Lorazepam tablets	Antianxiety agent	1–3 mg bid or tid		Dosage not established
Oxazepam USP *Serax*	Oxazepam capsules USP Oxazepam tablets USP		10–30 mg tid or qid		
Prazepam USP *Verstran*	Prazepam tablets USP	Antianxiety agent	10 mg tid	20–60 mg/day	Dosage not established
Temazepam *Restoril*	Temazepam capsules	Hypnotic	15–30 mg		Dosage not established
Triazolam *Halcion*	Triazolam tablets USP	Hypnotic	0.25–0.5 mg		Dosage not established

*See USP DI for complete dosage information.

BARBITURATES

The first sedative–hypnotic barbiturate, 5,5-diethylbarbituric acid, was introduced in 1903. With time, many members were added, and the barbiturates dominated the sedative–hypnotic field until the advent of the benzodiazepines, which for reasons outlined earlier, but most notably a much greater margin of safety, displaced the barbiturates as the most broadly useful agents in sedative–hypnotic applications.

The barbiturates are 5,5-disubstituted barbituric acids. The following scheme shows how the 5,5-dialkyl compounds are synthesized. Substitution of thiourea for urea produces the 2-thiobarbiturates, useful as induction anesthetics.

$$H_2C(COOC_2H_5)_2 \xrightarrow[NaOC_2H_5]{R-X} RHC(COOC_2H_5)_2 \xrightarrow[NaOC_2H_5]{R'-X}$$

Diethylmalonate → Monoalkyl Diethylmalonate

$$RR'C(COOC_2H_5)_2 \xrightarrow[NaOC_2H_5]{H_2N-CO-NH_2}$$

Dialkyl Diethylmalonate

$\xrightarrow{H^+}$

Sodium 5,5-Dialkylbarbiturate → 5,5-Dialkylbarbituric Acid

A consideration of the structure of 5,5-disubstituted barbituric acids reveals their acidic character. Those without methyl substituents on the nitrogen have pK_as around 7.6, those with a methyl substituent have pK_as about 8.4. The free acids have poor water solubility and good lipid solubility (the latter largely a function of the two hydrocarbon substituents on the 5-position, although in the 2-thiobarbiturates the sulfur atom increases lipid solubility).

Sodium salts of the barbiturates are readily prepared and are water-soluble. Their solutions generate an alkaline *p*H. A classic incompatibility is the addition of an agent with an acidic *p*H in solution, which results in formation and precipitation of the free water-insoluble disubstituted barbituric acid.

Sodium salts of barbiturates in aqueous solution decompose at varying rates by base-catalyzed hydrolysis, generating ring-opened salts of carboxylic acids.

That the names of many barbiturates end in *al* (e.g., phenobarbital), appearing to denote an aldehydic compound, derives from the fact that chloral hydrate was widely known as a sedative–hypnotic when the first barbiturates were introduced. Accordingly, the suffix was an attempt to denote a therapeutic, not a chemical, class.

Structure–Activity Relationships

Extensive synthesis and testing of the barbiturates over a long time span has produced well-defined structure–activity relationships, which have been summarized.[24]

Both hydrogen atoms at the 5-position of barbituric acid must be replaced. If one hydrogen is available at position 5, tautomerization to a highly acidic trihydroxypyrimidine ($pK_a \sim 4$) can occur. Consequently the compound is largely in the anionic form at physiologic *p*Hs, with little nonionic lipid-soluble compound available to cross the blood–brain barrier.

Beginning with lower alkyls, there is an increase in onset and a decrease in duration of action with increasing hydrocarbon content up to about seven to nine total carbon atoms substituted on the 5-position. Lipophilicity and an ability to penetrate the brain in the first case and an ability to penetrate liver microsomes in the second may be involved. Also, for more hydrophobic compounds, partitioning out of the brain to other sites can be involved in the second instance. There is an inverse correlation between the total number of carbon atoms substituted on the 5-position and the duration of action, which is even better when the character of these substituents is taken into account, for example, the relatively polar character of a phenyl substituent (approximates a three- to four-carbon aliphatic chain), branching of alkyls, presence of an isolated double or triple bond, and so on. Additionally, these groups can influence the ease of oxidative metabolism by effects on bond strengths as well as by influencing partitioning.

Metabolism of the barbiturates is discussed in Chapter 3. Suffice it to say that increasing the lipid/water partition coefficient generally increases the rate of metabolism, except for compounds with an extremely high lipid/water partition coefficient (e.g., thiopental). Metabolism generally follows an ultimate (Ω) or penultimate (Ω-1) oxidation pattern. Ring-opening reactions are usually minor. *N*-methylation decreases duration of action, in large part, probably by increasing the concentration of the lipid-soluble free barbituric acid. 2-Thiobarbiturates have a very short duration of action because the lipid/water partition coefficient is extremely high, promoting depotization. Barbiturates find employment as sedatives, as hypnotics, for induction of anesthesia, and as anticonvulsants. Absorption from the gastrointestinal tract is good. Binding to blood proteins is substantial. Compounds with low lipid/water partition coefficients may be excreted intact in the urine. Those with higher lipid/water partition coefficients are excreted after metabolism to polar metabolites.

Some of the more frequently used barbiturates are described briefly in the following sections. For the structures, the usual dosages required to produce sedation and hypnosis, the times of onset, and the duration of action, see Table 9-5.

Barbiturates with a Long Duration of Action (Six Hours or More)

Barbital. 5,5-Diethylbarbituric acid. Barbital, although discontinued as a sedative–hypnotic, is interesting because of the biologic consequence of its low lipid/water partition coefficient. It is slowly eliminated, mostly in the intact form, by the kidney.

Metharbital. 5,5-Diethyl-1-methylbarbituric acid (Gemonil). This methyl derivative of barbital finds a limited employment in a variety of epilepsies.

Phenobarbital, USP. 5-Ethyl-5-phenylbarbituric acid (Luminal). The compound is a long-acting sedative and hypnotic. It is also a valuable anticonvulsant, especially in generalized tonic–clonic and partial seizures (see the discussion on anticonvulsants). Metabolism to the *p*-hydroxy compound followed by glucuronidation accounts for about 90% of a dose.

Barbiturates with an Intermediate Duration of Action (Three to Six Hours)

These compounds are used principally as sedative–hypnotics. They include **Amobarbital, USP**, 5-Ethyl-5-isopentylbarbituric acid (Amytal), and its water-soluble sodium salt, **Amobarbital Sodium, USP**, 5-Allyl-5-isopropylbarbituric acid (Aprobarbital, Alurate); **Butabarbital Sodium, USP**, the water-soluble sodium salt of 5-*sec*-butyl-5-ethylbarbituric acid (Butisol Sodium), and **Talbutal, USP**, 5-Allyl-5-*sec*-butylbarbituric acid (Lotusate).

Barbiturates with a Short Duration of Action (Less than Three Hours)

Included in this group, which have substituents in the 5-position promoting more rapid metabolism

TABLE 9-5

BARBITURATES USED AS SEDATIVES AND HYPNOTICS

General Structure

A. LONG DURATION OF ACTION (6 OR MORE HR)

Generic Name Proprietary Name	***Substituents*** R_5	R_5'	R_1	***Sedative Dose (mg)***	***Hypnotic Dose (mg)***	***Usual Onset of Action (min)***
Mephobarbital USP *Mebaral*	C_2H_5	C_6H_5	CH_3	30–100*	100	30–60
Metharbital USP *Gemonil*	C_2H_5	C_2H_5	CH_3	50–100*		30–60
Phenobarbital USP *Luminal*	C_2H_5	C_6H_5	H	15–30*	100	20–40
B. INTERMEDIATE DURATION OF ACTION (3–6 HR)						
Butalbital USP *Sandoptal*	$CH_2{=}CHCH_2-$	$(CH_3)_2CHCH_2-$	H		200–600	20–30
Amobarbital USP *Amytal*	CH_3CH_2-	$(CH_3)_2CHCH_2CH_2-$	H	20–40	100	20–30
Aprobarbital *Alurate*	$CH_2{=}CHCH_2-$	$(CH_3)_2CH-$	H	20–40	40–160	
Butabarbital Sodium USP *Butisol Sodium*	CH_3CH_2-	$CH_3CH_2CH(CH_3)-$	H	15–30	100	20–30
Talbutal USP *Lotusate*	$CH_2{=}CHCH_2-$	$CH_3CH_2CH(CH_3)-$	H	50	120	20–30
C. SHORT DURATION OF ACTION (LESS THAN 3 HR)						
Pentobarbital Sodium USP *Nembutal Sodium*	CH_3CH_2-	$CH_3CH_2CH_2CH(CH_3)-$	H	30	100	20–30
Secobarbital USP *Seconal*	$CH_2{=}CHCH_2-$	$CH_3CH_2CH_2CH(CH_3)-$	H	15–30	100	20–30

*Daytime sedative and anticonvulsant.

TABLE 9-6

SEDATIVE-HYPNOTIC BARBITURATES

Name Proprietary Name	***Preparations***	***Category***	***Usual Adult Dose****	***Usual Dose Range****	***Usual Pediatric Dose****
Mephobarbital USP *Mebaral*	Mephobarbital tablets USP	Anticonvulsant, sedative		Anticonvulsant, 400–600 mg/day; sedative, 32–100 mg tid or qid	
Metharbital NF *Gemonil*	Metharbital tablets NF	Anticonvulsant	Initial, 100 mg qd to tid	100–800 mg/day	

*See USP DI for complete dosage information.

TABLE 9-6 *Continued*

SEDATIVE-HYPNOTIC BARBITURATES

Name Proprietary Name	**Preparations**	**Category**	**Usual Adult Dose***	**Usual Dose Range***	**Usual Pediatric Dose***
Phenobarbital USP *Luminal*	Phenobarbital elixir USP Phenobarbital tablets USP	Anticonvulsant, hypnotic, sedative	Anticonvulsant, 50–100 mg bid or tid; hypnotic, 100–200 mg hs; sedative, 15–30 mg bid or tid	30–600 mg/day	Anticonvulsant, 15–50 mg bid or tid; sedative, 2 mg/kg of body weight or 60 mg/m^2 of body surface, tid
Phenobarbital sodium USP *Luminal Sodium, Phenalix*	Phenobarbital sodium injection USP Sterile phenobarbital sodium USP	Anticonvulsant, hypnotic, sedative	Anticonvulsant, IM or IV, 200–320 mg; may repeat in 6 hr as necessary; hypnotic, IM or IV, 130–200 mg; sedative, IM or IV, 100–130 mg; may repeat in 6 hr as necessary	30–600 mg/day	Anticonvulsant, IM, 3–5 mg/kg of body weight or 125 mg/m^2 of body surface; sedative, 2 mg/kg of body weight, or 60 mg/m^2 of body surface, tid
	Phenobarbital sodium tablets USP		Anticonvulsant, 50–100 mg bid or tid as necessary; hypnotic, 100–200 mg hs; sedative, 15–30 mg bid to tid	30–600 mg/day	Sedative, 2 mg/kg of body weight or 60 mg/m^2 of body surface, tid
Amobarbital USP *Amytal*	Amobarbital elixir USP Amobarbital tablets USP	Sedative	Sedative, 25 mg; hypnotic, 100 mg	25–200 mg	
Amobarbital sodium USP *Amytal Sodium*	Amobarbital sodium capsules USP Sterile amobarbital sodium USP	Hypnotic, sedative	Hypnotic, oral, IM, or IV, 65– 200 mg hs; sedative, oral, IM or IV, 30–50 mg bid or tid	Hypnotic, oral, IM or IV, 50–200 mg/day; sedative, oral, IM or IV, 15 mg–1 g/day	Sedative, oral, 2 mg/kg of body weight or 60 mg/m^2 of body surface, tid
Butabarbital sodium USP *Butisol Sodium*	Butabarbital sodium capsules USP Butabarbital sodium elixir USP Butabarbital sodium tablets USP	Sedative, hypnotic	Sedative, 15–30 mg tid or qid; hypnotic, 100 mg	Sedative, 7.5–60 mg; hypnotic, 100–200 mg	
Talbutal USP *Lotusate*	Talbutal tablets USP	Sedative, hypnotic	Hypnotic, 120 mg from 15–30 min before bedtime		
Pentobarbital sodium USP *Nembutal*	Pentobarbital sodium capsules USP	Hypnotic, sedative	Hypnotic, 100 mg hs; sedative, 30 mg tid or qid	50–200 mg/day	Sedative, 2 mg/kg of body weight or 60 mg/m^2 of body surface, tid
	Pentobarbital sodium elixir USP		Hypnotic, 100 mg hs; sedative, 20 mg tid or qid	50–200 mg/day	Same as for pentobarbital sodium capsules USP
	Pentobarbital sodium injection USP		Hypnotic, IM, 150–200 mg; IV, 100 mg repeated as necessary; sedative, IM, 30 mg tid or qid	50–500 mg/day	Same as for pentobarbital sodium capsules USP
Secobarbital USP *Seconal*	Secobarbital USP elixir	Hypnotic, sedative	Hypnotic, 100 mg hs; sedative, 30–50 mg tid or qid	90–300 mg/day	Sedative, 2 mg/kg of body weight or 60 mg/m^2 of body surface, tid
Secobarbital sodium USP *Seconal Sodium*	Secobarbital sodium capsules USP	Hypnotic, sedative	Hypnotic, 100 mg hs; sedative, 30–50 mg tid or qid	90–300 mg/day	Sedative, 2 mg/kg of body weight or 60 mg/m^2 of body surface, tid
	Secobarbital sodium injection USP Sterile secobarbital sodium USP		Hypnotic, IM or IV, 2.2 mg/kg of body weight; sedative, IM or IV, 1.1–1.65 mg/kg of body weight	1.1–4.4 mg/kg	Sedative, preoperative intramuscular, 4 to 5 mg/kg of body weight

*See USP DI for complete dosage information.

(e.g., by increasing the lipid/water partition coefficient) than the intermediate group, are **Pentobarbital Sodium, USP**, Sodium 5-ethyl-5-(1-methylbutyl)barbiturate (Nembutal); **Secobarbital, USP**, 5-Allyl-5-(1-methylbutyl)barbituric acid (Seconal); and the sodium salt, sodium secobarbital.

Barbiturates with an ultrashort duration of action are discussed under anesthetic agents.

MISCELLANEOUS SEDATIVE HYPNOTICS

A wide range of chemical structures (e.g., imides, amides, alcohols) can produce sedation and hypnosis resembling that produced by the barbiturates. Despite this apparent structural diversity, the compounds have generally similar structural characteristics and chemical properties: a hydrophobic portion and a polar or H-bonding portion. In most cases, modes of action are undetermined. As a working hypothesis, some of the agents could be envisioned to act by mechanisms similar to those proposed for general inhalation anesthetics (e.g., partitioning into and reversibly altering the structure and function of biologic macromolecules).

Amides and Imides

Glutethimide, USP. 2-Ethyl-2-phenylglutarimide (Doriden). This imide has many structural relationships with the barbiturates and resembles them in many respects biologically. It is an effective sedative–hypnotic. It is very hydrophobic and absorption from the gastrointestinal tract is somewhat erratic. Metabolism is extensive and the drug is an enzyme inducer. In the therapeutic dosage range, adverse effects tend to be infrequent. Toxic effects in overdose are as severe as, and possibly more troublesome than, those of the barbiturates.

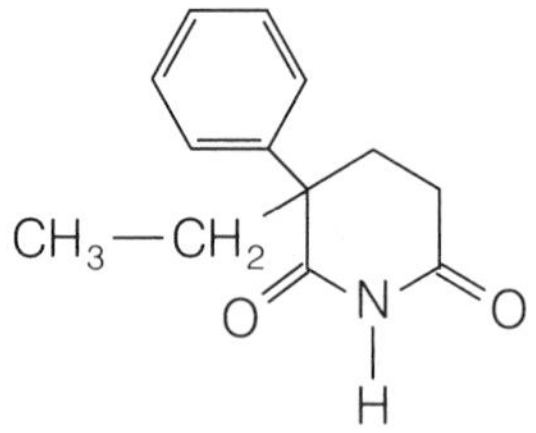

Glutethimide

Methyprylon, USP. 3,3-Diethyl-5-methyl-2,4-piperidinedione (Noludar). The drug is an effective sedative–hypnotic. Its effects resemble those of the barbiturates. Metabolism is extensive, and similar to the barbiturates and glutethimide, it is an enzyme inducer. Some of the effects of overdose and of withdrawal resemble those of the barbiturates.

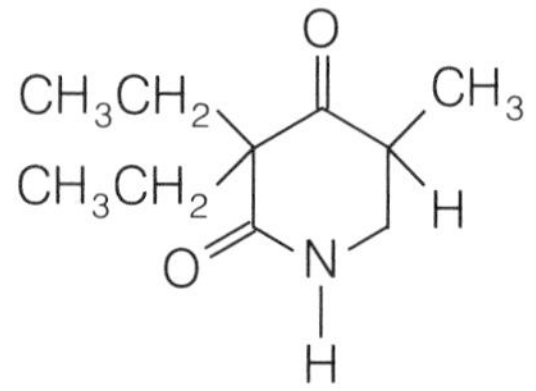

Methyprylon

Methaqualone Hydrochloride. 2-Methyl-3-*o*-tolyl-4(3*H*)-quinazolinone monohydrochloride. The drug is an effective sedative–hypnotic, but has been withdrawn because of abuse potential.

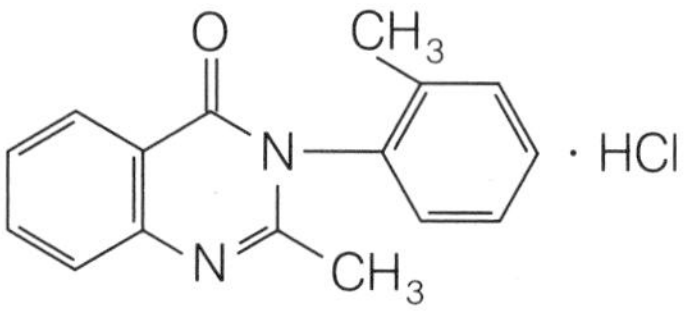

Methaqualone Hydrochloride

Metabolites include the *N*-oxide and oxidative metabolites of the tolyl methyl group. Some of the central effects have been reported to resemble those of the opioids. The effects of overdosage are complex with some of them resembling those of the barbiturates.

Alcohols and Their Carbamate Derivatives

A very simple alcohol, ethanol, has a long history of use as a sedative and hypnotic. Because its use in these capacities is associated with so many potential hazards (e.g., production of alcoholism and its ability to exert toxic effects on many organ systems), it is seldom a preferred agent medically, and other drugs have been developed.

As the homologous series of normal alcohols is ascended from ethanol, CNS depressant potency increases up to eight carbon atoms, with activity decreasing thereafter (the Meyer–Overton parabola, Chap. 2). Branching of the alkyl chain increases depressant activity and, in an isometric series, the order of potency is tertiary > secondary > primary. In part, this may be because tertiary and secondary alcohols are not metabolized by oxidation to corresponding carboxylic acids. Replacement of a hydrogen atom in the alkyl group by a halogen has the effect of increasing the alkyl portion and, accordingly, for the lower molecular weight compounds, increases potency. Carbamylation of alcohols generally increases depressant potency. Carbamate groups are generally much more resistant to metabolism than hydroxyl functions. Hydroxylic compounds and their carbamate derivatives used as sedative and hypnotics are described in the following:

Ethchlorvynol, USP. 1-Chloro-3-ethyl-1-penten-4-yn-3-ol (Placidyl). Ethchlorvynol is an effective sedative–hypnotic with a rapid onset and short duration of action. Metabolism, probably involving the hydroxyl group, accounts for about 90% of a dose. Acute overdose has several features in common with barbiturate overdose.

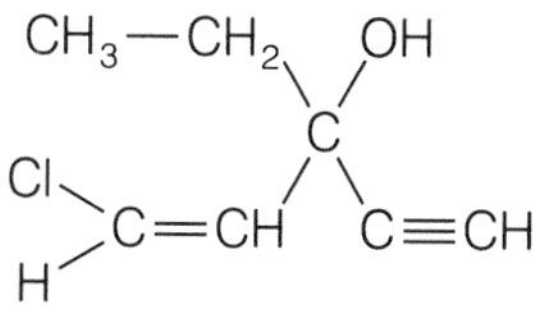

CH_3-CH_2, OH, C, Cl, H, C=CH, C≡CH

Ethchlorvynol

Ethinamate, USP. 1-Ethynyl-cyclohexanol carbamate (Valmid). The compound is a modestly active sedative and hypnotic agent. The onset of action is rapid and the duration of action is short. Metabolism involves hydroxylation of the cyclohexane ring.

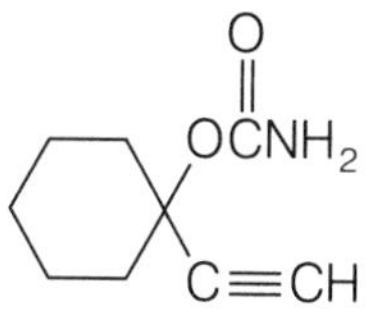

O, $OCNH_2$, C≡CH

Ethinamate

Meprobamate, USP. 2-Methyl-2-propyltrimethylene dicarbamate; 2-methyl-2-propyl-1,3-propanediol dicarbamate (Equanil, Miltown). Meprobamate is discussed here because its official indication is as an antianxiety agent, and it is also extensively used as a sedative–hypnotic. Whether or not its antianxiety properties are separate from sedative effects is uncertain. Another reason for considering the compound here is that it has many overall properties resembling benzodiazepines and barbiturates. The mechanism of action underlying anxiolytic effects is unknown but may involve effects on conductivity in specific brain areas.[25] It does not appear to act through effects on GABAergic systems.

The drug is effective against absence seizures and may worsen generalized tonic–clonic seizures.

Meprobamate could also be grouped with the centrally acting skeletal muscle relaxants, which are discussed later. Its relative, carisoprodol, is discussed there. This muscle relaxant action, in part, may be due to blockade of the polysynaptic reflexes in the spinal cord and appears to be associated with appropriately substituted glycols, other dihydric compounds, and their carbamate derivatives.[25] The major route of metabolism of mephenesin involves oxidative hydroxylation of the carbon next to the terminal atom of the *n*-propyl radical.

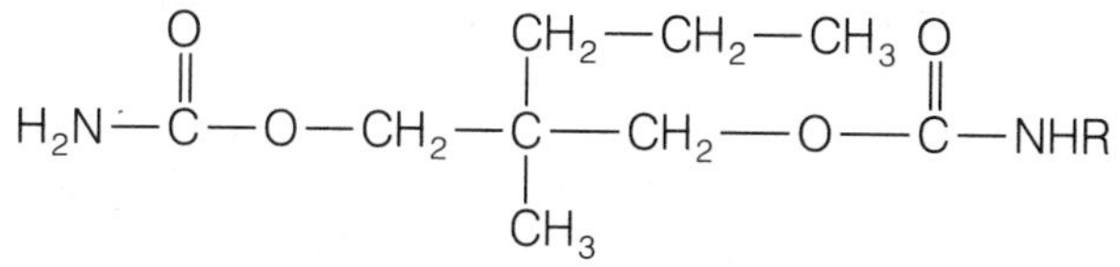

$H_2N-C(=O)-O-CH_2-C(CH_2-CH_2-CH_3)(CH_3)-CH_2-O-C(=O)-NHR$

Meprobamate: R = H
Carisoprodol: R = $-CH(CH_3)_2$

Adelhydes and Their Derivatives

This heading has the potential to mislead. Actually, few aldehydes are valuable hypnotic drugs. The one aldehyde in use, chloral (as the hydrate), is thought to act principally though a metabolite, trichloroethanol, and the derivative of acetaldehyde is a cyclic trimer, which could also be grouped as an ether.

Chloral Hydrate, USP. Trichloroacetaldehyde monohydrate; $CCl_3CH(OH)_2$ (Noctec). Chloral hydrate is an aldehyde hydrate sufficiently stable that it can be isolated. The relative stability of this *gem*-diol is largely due to an unfavorable dipole–dipole repulsion between the trichloromethyl carbon and the carbonyl carbon present in the parent carbonyl compound.[26]

The compound is unstable under alkaline conditions, undergoing the last step of the haloform reaction to yield chloroform and formate ion. In combination with ethanol, it forms the hemiacetal. Whether or not this compound is the basis for the notorious and potentially lethal effect of the combination of ethanol and chloral hydrate can be argued. Synergism between two different CNS depressants also could be involved. Additionally, ethanol, by increasing the concentration of NADH, enhances the reduction of chloral to trichloroethanol, and, finally, chloral can inhibit the metabolism of alcohol because it inhibits alcohol dehydrogenase. Although it is suggested that chloral hydrate, per se, may act as a hypnotic,[27] it is also recognized that chloral hydrate is very quickly converted to trichloroethanol and the latter compound is generally assumed to account for almost all of the hypnotic effect.

Triclofos Sodium. 2,2,2-Trichloroethanol dihydrogen phosphate monosodium salt (Triclos). Chloral hydrate is irritating to the gastrointestinal mucosa. Its active metabolite, trichloroethanol, also has unpleasant properties when given orally. Triclofos is the nonirritating sodium salt of the phosphate ester of trichloroethanol and is rapidly converted to trichloroethanol after administration. Accordingly, triclofos sodium has biologic properties similar to chloral hydrate.

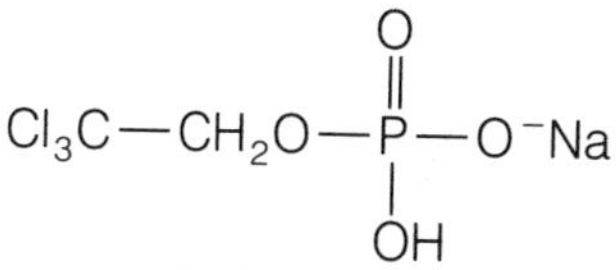

$Cl_3C-CH_2O-P(=O)(OH)-O^-Na^+$

Triclofos Sodium

TABLE 9-7

NONBARBITURATE SEDATIVES AND HYPNOTICS

Name Proprietary Name	***Preparations***	***Category***	***Usual Adult Dose****	***Usual Dose Range****	***Usual Pediatric Dose****
Glutethimide USP *Doriden*	Glutethimide capsules USP Glutethimide tablets USP	Sedative, hypnotic		Sedative, 125–250 mg qd to tid; hypnotic, 500 mg–1 g	
Methyprylon USP *Noludar*	Methyprylon capsules USP Methyprylon tablets USP	Hypnotic		50–400 mg hs	
Ethchlorvynol USP *Placidyl*	Ethchlorvynol capsules USP	Sedative, hypnotic	Sedative, 100 mg bid or tid; hypnotic, 500 mg	Sedative, 100–200 mg; hypnotic, 500 mg–1 g	
Ethinamate USP *Valmid*	Ethinamate capsules USP	Hypnotic	500 mg	500 mg–1 g	
Chloral Hydrate USP *Noctec,*	Chloral hydrate capsules USP	Hypnotic, sedative	Hypnotic, 500 mg–1 g hs; sedative, 250 mg tid	250 mg–2 g/day	
Felsules, Rectules, Aquachloral	Chloral hydrate syrup USP		Hypnotic, 500 mg–1 g hs; sedative, 250 mg tid	250 mg–2 g/day	Hypnotic, 50 mg/kg of body weight or 1.5 g/m^2 of body surface, up to 1 g/ dose, at bedtime; sedative, 8 mg/kg or 250 mg/m^2, up to 500 mg/dose, tid
Paraldehyde USP *Paral*		Hypnotic, sedative	Hypnotic, 10–30 mL; sedative, 5–10 mL	3–30 mL	Hypnotic, 0.3 mL/kg of body weight or 12 mL/m^2 of body surface, per dose; sedative, 0.15 mL/kg or 6 mL/m^2 of body surface, per dose
	Sterile paraldehyde USP	Hypnotic, sedative	Hypnotic, IM, 10 mL; IV infusion, diluted with several volumes of sodium chloride injection, 10 mL; sedative, IM, 5 mL; IV infusion, diluted with several volumes of sodium chloride injection, 5 mL	3–10 mL	Hypnotic, IM, 0.3 mL/kg of body weight or 12 mL/m^2 of body surface, per dose; sedative, IM, 0.15 mL/kg of body weight or 6 mL/m^2 of body surface, per dose

*See USP DI for complete dosage information.

Paraldehyde, USP. 2,4,6-Trimethyl-*s*-trioxane; paracetaldehyde. Paraldehyde is recognizable as the cyclic trimer of acetaldehyde. It has a strong characteristic odor detectable in the expired air and an unpleasant taste. These properties limit its use almost exclusively to an institutional setting (e.g., in the treatment of delirium tremens).

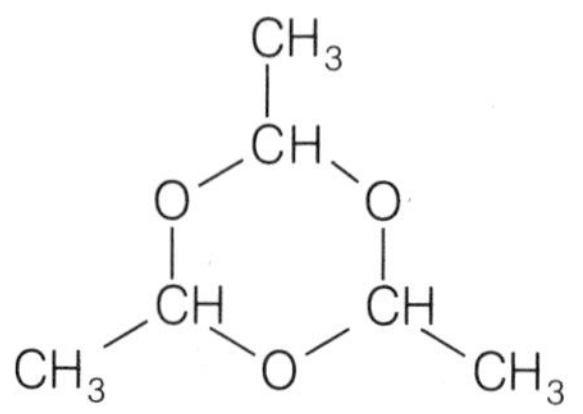

Paraldehyde

CENTRAL NERVOUS SYSTEM DEPRESSANTS WITH SKELETAL MUSCLE RELAXANT PROPERTIES

AGENTS USED IN ACUTE MUSCLE SPASMS

The agents in this group find use in a number of conditions, such as strains and sprains, that may produce acute muscle spasm. They have interneuronal-blocking properties at the level of the spinal cord, and these are said to be partly responsible for skeletal muscle relaxation.[25] Also, they have general CNS depressant properties that may contribute to, or be mainly responsible for, the skeletal muscle relaxant activity. Dihydric compounds and their carbamate (urethan) derivatives, as described earlier in the discussion of meprobamate, are prominent members of the group.

Chlorphenesin Carbamate. 3-(*p*-Chlorophenoxy)-1,2-propanediol 1-carbamate (Maolate). This is the *p*-chloro substituted and 1-carbamate derivative of the lead compound in the development of this group of agents, mephenesin.

Cl—C6H4—O—CH2—CH(OH)—CH2—O—C(=O)—NH2

Chlorphenesin Carbamate

Mephenesin is weakly active and short-lived largely because of facile metabolism of the primary hydroxyl group. Carbamylation of this group increases activity. *p*-Chlorination increases the lipid/water partition coefficient and seals off the *p*-position from hydroxylation.

Metabolism, still fairly rapid, involves glucuronidation of the secondary hydroxyl group. The biologic half-life in humans is 3.5 hours.

Methocarbamol, USP. 3-(*o*-Methoxyphenoxy)-1,2-propanediol 1-carbamate (Robaxin). Methocarbamol is said to be more sustained in effect than mephenesin. Likely sites for metabolic attack include the secondary hydroxyl group and the 2 ring positions opposite the ether functions. The dihydric parent compound, guaifenesin, is used as an expectorant.

OCH_3; O—CH_2—CH(OH)—CH_2OR

Guaifenesin: R = H

Methocarbamol: R = —C(=O)—NH_2

Carisoprodol, USP. *N*-Isopropyl-2-methyl-2-propyl-1,3-propanediol dicarbamate; 2-methyl-2-propyl-1,3-propanediol carbamate isopropylcarbamate (Soma). The compound is the mono-*N*-isopropyl substituted relative of meprobamate. The structure is given in the discussion of meprobamate. It is indicated in acute skelatomuscular conditions characterized by pain, stiffness, and spasm. As can be expected, a major side effect of the drug is drowsiness.

DRUGS USED IN SPASTICITY

Several drugs benefit the spasticity associated with diseases such as multiple sclerosis and cerebral palsy. Notable compounds are the centrally acting agents diazepam (discussed earlier) and baclofen and the peripherally (acting directly on skeletal muscle) active agent dantrolene.

Baclofen, USP. 3-(*p*-Chlorophenyl)-γ-aminobutyric acid (Lioresal). Baclofen, a substituted GABA analogue, is useful in spasticity involving diseases of the spinal cord. It depresses monosynaptic and polysynaptic transmission. The exact mechanisms of action are unclear. Some effects resemble those of GABA,[28] others do not.[29] This agent is included in Table 9-8.

H_2N—CH_2—CH(C6H4Cl)—CH_2—CO_2H

Dantrolene Sodium (Dantrium). Dantrolene decreases the release of calcium ion by the sarcoplasmic reticulum and thereby blocks contraction of skeletal muscle.[30] It is useful in cerebral palsy

TABLE 9-8

CENTRAL NERVOUS SYSTEM DEPRESSANTS WITH SKELETAL MUSCLE RELAXANT PROPERTIES

Name Proprietary Name	***Preparations***	***Category***	***Usual Adult Dose****	***Usual Dose Range****	***Usual Pediatric Dose****
Baclofen USP *Lioresal*	Baclofen tablets USP	Antispastic	5 mg tid, increased by 5 mg/dose every 3 days until response is reached	Up to 80 mg/day	Dosage not established
Carisoprodol USP *Soma*	Carisoprodol tablets USP	Skeletal muscle relaxant	350 mg qid		Dosage not established in children up to 5 yr; 5–12 yr: 6.25 mg/kg of body weight qid
Chlorphenesin carbamate *Maolate*	Chlorphenesin carbamate tablets	Skeletal muscle relaxant	800 mg tid; maintenance, 400 mg qid		

*See USP DI for complete dosage information.

TABLE 9-8 *Continued*

CENTRAL NERVOUS SYSTEM DEPRESSANTS WITH SKELETAL MUSCLE RELAXANT PROPERTIES

Name Proprietary Name	***Preparations***	***Category***	***Usual Adult Dose****	***Usual Dose Range****	***Usual Pediatric Dose****
Meprobamate USP *Equanil, Miltown, Bamate, Arcoban, Kesso-Bamate, Meprospan, Tranmep*	Meprobamate injection USP	Adjunct in tetanus (sedative)	IM, 400 mg 6–8 times daily	1.2–3.2 g/day	Infants: 125 mg qid; children: 200 mg 6–8 times daily
	Meprobamate tablets USP	Sedative	400 mg tid or qid	1.2–2.4 g/day	Dosage is not established in children under 6 yr; 6–12 yr: 100–200 mg bid or tid
	Meprobamate oral suspension USP	Tranquilizer	400 mg of meprobamate tid or qid		
Methocarbamol USP *Robaxin*	Methocarbamol injection USP Methocarbamol tablets USP	Skeletal muscle relaxant		Oral, 1.5–2 g qid for the first 2 or 3 days, then 1 g qid	

*See USP DI for complete dosage information.

and multiple sclerosis. Its effect on calcium ion release is also the basis for its usefulness in malignant hyperthermia. Peripherally acting, it is grouped here with centrally acting agents because of its muscle relaxant effects.

Dantrolene Sodium

The most common side effect is muscle weakness. Another serious problem that limits the use of the drug is hepatic toxicity.

MISCELLANEOUS

Buspirone. 8-[4-[4-(Pyrimidinyl)-1-piperazinyl]-butyl]azaspiro[4,5]decane-7,9-dione monohydrochloride (Buspar). Buspirone is the first clinically introduced anxiolytic representative of a group of compounds under study that affect serotoninergic systems. It is reported to act as an agonist of serotonin at 5-HT_{1A} receptors.[31,32] The compound also has antidopaminergic activity. There is concern that the latter activity could lead to extrapyramidal effects.

HCl

ANTIPSYCHOTICS

Antipsychotics are drugs that ameliorate mental aberrations that are characteristic of the psychoses. The psychoses differ from the milder behavioral disorders, the neuroses, in that thinking tends to be illogical, bizarre, and loosely organized. Importantly, patients have difficulty understanding reality and their own conditions. There are often hallucinations (usually auditory) and delusions.

Psychoses can be organic and related to a specific toxic chemical, as in delirium produced by some central anticholinergic agents, or to a definite disease process, such as dementia, or they can be idiopathic. Idiopathic psychoses may be acute or chronic. Idiopathic acute psychotic reactions have been reported to follow extremely severe acute stress. Schizophrenia is a group of chronic idiopathic psychotic disorders with the overall symptomology described earlier.

The term *antipsychotic* was slow in gaining acceptance. Now it is widely acknowledged that antipsychotics actually diminish the underlying thought disorder that is the chief characteristic of the schizophrenias.[33] The agents often have a calming effect in agitated psychotic patients; hence, they also have been referred to as *major tranquilizers*. Finally, because they induce a lessening of reactivity to emotional stimuli, with little effect on consciousness, they are referred to as *neuroleptics*.

The most frequent uses of the agents are in manic disorders and the schizophrenias. In the manic disorders, it could be conjectured that the agents block dopamine (DA) at D-2 receptors, reducing euphoria and hyperactivity. In the chronic idiopathic psychoses, the clinically active agents appear to act largely by blocking DA at D-2 receptors in the mesolimbic area of the brain.[34,35] The rauwolfia alkaloids, seldom used today as antipsychotics, are considered to act as antipsychotics by depleting neuronal DA, rather than acting by a postsynaptic D-2 receptor block. Also, selective presynaptic DA receptor agonists, the net effect of which is to reduce the release of DA into the synapse where it is available for interaction with postsynaptic D-2 receptors, are being studied as antipsychotic agents.

A useful paradigm for the schizophrenias is that there is excessive DA activity in the limbic system. Consequently, more information is supplied than can be interpreted, and an involuntary picking and choosing of the informational overload begins (i.e., the brain begins to receive in code, and perceptions of reality are altered). Finally, the person responds in code and, because we do not have the key to the code, we perceive bizarre and illogical behavior. It has ironically been observed that if the key were found and underlying pattern discerned, then there could be seen a certain logic or sanity to psychotic behavior.[33]

Extrapyramidal side effects (EPS) of the clinically useful antipsychotics include a parkinsonism that resembles the symptomology of Parkinson's disease; acute dystonic reactions, characterized by spasm of the tongue, face, and neck; and akathesia, characterized by a restlessness and a need to be in constant motion. The effects are thought to arise from a block of D-2 receptors in the striatum.

The tardive dyskinesia that is sometimes seen with use of these agents is probably related to prolonged or potent block of the D-2 striatal receptors. It appears to arise from biologic compensation for the D-2 striatal block of antipsychotic agents and is characterized by excessive striatal DA activity. Symptoms resemble those of Huntington's chorea and include involuntary movements of the lips, tongue, and mouth, and purposeless choreiform motions of the extremities. It is not clear that the condition is reversible. Since it has been recognized as a potential effect of therapy, there has been increased care in dosing; that is, give the smallest effective dose for the shortest possible time. Also, it has led to a reluctance to use the agents as tranquilizers or sedatives in the minor psychiatric disorders such as the neuroses.

Many, but not all, of the agents are strongly anticholinergic. This can produce typical peripheral anticholinergic effects. Centrally, this action can work to advantage. It may minimize drug-induced parkinsonism by a block of striatal muscarinic receptors that partly compensates for the DA striatal block.

The drugs increase prolactin levels by a block of DA receptors in the hypothalamus and have an antiemetic effect by a block of DA receptors in the chemoreceptor trigger zone.

Postural hypotension, possibly related to peripheral α_1-adrenoreceptor block, is a fairly common side effect. Skin reactions, including urticaria, contact dermatitis, and photosensitivity, are reported for the phenothiazine group of antipsychotics.

PHENOTHIAZINES

Many potentially useful phenothiazine derivatives have been synthesized and pharmacologically evaluated. Consequently, there is a large body of information permitting accurate statements about the structural features associated with activity. Many of the features were summarized and interpreted by Gordon et al.[36] The best position for substitution is the 2-position. Activity increases (with some exceptions) as electron-withdrawing ability of the substituent increases. Another possibly important structural feature in the more potent compounds is the presence of an unshared electron pair on an atom or atoms of the 2-substituent. Substitution at the 3-position can improve activity over nonsubstituted compounds, but not as significantly as substitution at the 2-position. Substitution at position 1 has a deleterious effect on antipsychotic activity, as to a lesser extent, does substitution at the 4-position.

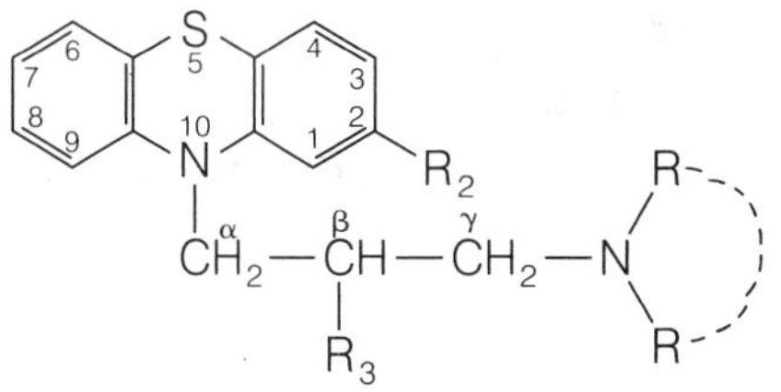

Phenothiazine Antipsychotic Agents–General Structure

The significance of these substituent effects could be that the hydrogen atom of the protonated amino group of the side chain H-bonds with an electron pair of an atom of the 2-substituent to develop a DA-like arrangement. Horn and Snyder, from x-ray crystallography, proposed that the chlorine-substituted ring of chlorpromazine base could be superimposed on the aromatic ring of dopamine base with the sulfur atom aligned with the *p*-hydroxyl of dopamine and the aliphatic amino groups of the two compounds also aligned.[37] The model used here is based on the interpretation of the SARs by Gordon et al.,[36] and on the Horn and Snyder proposal,[37] but involves the protonated species rather than the free base. The effect of the substituent at the 1-position might be to interfere with the side chain's ability to bring the protonated amino group into proximity with the 2-substituent. In the Horn and Snyder scheme,[37] the sulfur atom at position 5 is in a position analogous to the *p*-hydroxyl of dopamine, and it was also assigned a receptor-binding function by Gordon et al.[36] A substituent at position 4 might interfere with receptor binding by the sulfur atom.

The three-atom chain between position 10 and the amino nitrogen is required. Shortening or lengthening the chain at this position drastically decreases activity. The three-atom chain length may be necessary to bring the protonated amino nitrogen into proximity with the 2-substituent.

As expected, branching with large groups, such as phenyl, decreases activity, as does branching with polar groups. Methyl branching on the β-position has a variable effect on activity. More importantly, there is a high separation of antipsychotic potency between *levo-* (the more active) and *dextro*-isomers. This has long been taken to suggest that a precise fit, that is, receptor binding, is involved in the action of these compounds.

Decreases in size from a dimethylamino group, as in going to a monomethylamino, greatly decrease activity, as do effective size increases, such as the one that occurs with *N,N*-diethylamino. Once the fundamental requirement of an *effective* size of about that equivalent to a dimethylamino is maintained, as in fusing *N,N*-diethyl substituents to generate a pyrrolidino group, activity can be enhanced with increasing chain length, as in N_2-substituted piperizino compounds.

The criticalness of the size about the nitrogen suggests the importance of the amino group (here protonated) for receptor attachment. The effect of the added chain length, once the size requirement is met, could be to add receptor-binding forces. It appears to have been reasonably proved that the protonated species of the phenothiazines can bind to DA receptors.[38]

Metabolism of the phenothiazines is complex in detail, but simple overall. A major route is hydroxylation of the tricyclic system. The usual pattern, for which there are good chemical reasons, is hydroxylation para to the 10-nitrogen atom of the ring other than the ring bearing the electron-attracting substituent at the 2-position. Thus, the major initial metabolite is frequently the 7-hydroxy compound. This compound is further metabolized by conjugation with glucuronic acid, and the conjugate is excreted. Detailed reviews of the metabolites of phenothiazines (as well as SARs and pharmacokinetic factors) are available.[39]

Products

The structures of the phenothiazine derivatives described in the following section are given in Table 9-9.

Chlorpromazine Hydrochloride, USP. 2-Chloro-10-[3-(dimethylamino)propyl]phenothiazine monohydrochloride (Thorazine). Chlorpromazine is the earliest phenothiazine compound introduced into therapy. It is still useful as an antipsychotic. Other uses are in nausea and vomiting and hiccough. In activity comparisons, it is the reference compound,

TABLE 9-9

PHENOTHIAZINE DERIVATIVES (AMINOPROPYL SIDE CHAIN)

Generic Name Proprietary Name	R_{10}	R_2
PROPYL DIALKYLAMINO SIDE CHAIN		
Promazine hydrochloride USP *Sparine*	$-(CH_2)_3N(CH_3)_2 \cdot HCl$	H
Chlorpromazine hydrochloride USP *Thorazine*	$-(CH_2)_3N(CH_3)_2 \cdot HCl$	Cl
Triflupromazine hydrochloride USP *Vesprin*	$-(CH_2)_3N(CH_3)_2 \cdot HCl$	CF_3
Piperacetazine USP *Quide*	$-(CH_2)_3-N$(piperidine)$-CH_2-CH_2-OH$	$-C(=O)-CH_3$

TABLE 9-9 *Continued*

PHENOTHIAZINE DERIVATIVES (AMINOPROPYL SIDE CHAIN)

Generic Name Proprietary Name	R_{10}	R_2
ALKYL PIPERIDYL SIDE CHAIN		
Thioridazine hydrochloride USP *Mellaril*	$-(CH_2)_2-$(2-piperidyl, N-CH_3) · HCl	SCH_3
Mesoridazine besylate USP *Serentil*	$-(CH_2)_2-$(2-piperidyl, N-CH_3) · $C_6H_5SO_3H$	$O \leftarrow SCH_3$ (S→O)
PROPYL PIPERAZINE SIDE CHAIN		
Prochlorperazine maleate USP *Compazine*	$-(CH_2)_3-N$(piperazine)$N-CH_3 \cdot 2C_4H_4O_4$	Cl
Trifluoperazine hydrochloride USP *Stelazine*	$-(CH_2)_3-N$(piperazine)$N-CH_3 \cdot 2HCl$	CF_3
Thiethylperazine maleate USP *Torecan*	$-(CH_2)_3-N$(piperazine)$N-CH_3 \cdot 2C_4H_4O_4$	SCH_2CH_3
Perphenazine USP *Trilafon*	$-(CH_2)_3-N$(piperazine)$N-CH_2-CH_2-OH$	Cl
Fluphenazine hydrochloride USP *Permitil, Prolixin*	$-(CH_2)_3-N$(piperazine)$N-CH_2-CH_2-OH \cdot 2HCl$	CF_3
Acetophenazine maleate USP *Tindal*	$-(CH_2)_3-N$(piperazine)$N-CH_2-CH_2-OH$ $2C_4H_4O_4$	$-C(=O)-CH_3$
Carphenazine maleate USP *Proketazine*	$-(CH_2)_3-N$(piperazine)$N-CH_2CH_2OH \cdot 2C_4H_4O_4$	$-C(=O)-CH_2CH_3$

TABLE 9-10

PHENOTHIAZINE ANTIPSYCHOTICS

Name Proprietary Name	***Preparations***	***Category***	***Usual Adult Dose****	***Usual Dose Range****	***Usual Pediatric Dose****
Acetophenazine maleate USP *Tindal*	Acetophenazine maleate tablets USP	Antipsychotic	20 mg tid	Up to 120 mg/day	
Carphenazine maleate USP *Proketazine*	Carphenazine maleate solution USP Carphenazine tablets USP	Antipsychotic		12.5–50 mg tid; increased by 12.5–50 mg/day at intervals of from 4 days to 1 wk; the maximum daily dose recommended is 400 mg	

*See USP DI for complete dosage information.

TABLE 9-10 *Continued*

PHENOTHIAZINE ANTIPSYCHOTICS

Name Proprietary Name	***Preparations***	***Category***	***Usual Adult Dose****	***Usual Dose Range****	***Usual Pediatric Dose****
Chlorpromazine USP *Thorazine*	Chlorpromazine suppositories USP	Antiemetic, antipsychotic	Antiemetic, antipsychotic rectal, 50–100 mg tid or qid as necessary	50–400 mg/day	Antiemetic or antipsychotic; use not recommended in infants under 6 mo of age; children 6 mo and older: 1 mg/ kg of body weight tid or bid as necessary
Chlorpromazine hydrochloride USP *Thorazine* *Promapar*	Chlorpromazine hydrochloride injection USP	Antiemetic, antipsychotic	Antiemetic, IM, 25–50 mg repeated 6–8 times daily as necessary; antipsychotic, IM, 25–50 mg repeated in 1 hr, if necessary	Antipsychotic, 25 mg–1 g/day	Antiemetic, use not recommended in infants under 6 mo of age; children under 5 yr: 550 μg/kg of body weight tid or qid, as necessary, up to 40 mg/day; 5–12 yr: 550 μg/kg tid or qid, as necessary, up to 75 mg/day. Antipsychotic, use not recommended in infants under 6 mo of age; 550 μg/kg of body weight or 15 mg/m^2 of body surface, qid up to 40 mg/day for children 6 mo–5 yr and up to 75 mg/day for children 5–12 yr of age
	Chlorpromazine hydrochloride syrup USP Chlorpromazine hydrochloride tablets USP Chlorpromazine hydrochloride USP		30–300 mg qd to tid	Up to 1 g/day	Antiemetic, use not recommended in infants under 6 mo of age; over 6 mo: 550 μg/kg of body weight 4–6 times daily, as necessary; antipsychotic, use not recommended in infants under 6 mo of age; over 6 mo: 550 μg/ kg of body weight or 15 mg/m^2 qid
Fluphenazine decanoate *Prolixin*	Fluphenazine decanoate injection	Antipsychotic	IM or SC, 25 mg every 1–3 wk	Up to 100 mg/dose	
Fluphenazine enanthate USP *Prolixin*	Fluphenazin enanthate injection USP	Antipsychotic	IM or SC, 25 mg every 1–3 wk	Up to 100 mg every 1–3 wk	Dosage is not established in children under 12 yr of age; children over 12 yr: see Usual Adult Dose
Fluphenazine hydrochloride USP *Permitil,* *Prolixin*	Fluphenazine hydrochloride elixir USP Fluphenazine hydrochloride injection USP Fluphenazine hydrochloride oral solution USP	Antipsychotic	Oral, 0.5 mg–2.5 mg qd to qid; IM, 1.25–2.5 mg every 6–8 hr	Oral, up to 20 mg/day; IM, up to 10 mg/day	Oral, 0.25–0.75 mg qd to qid; IM, up to 12 yr of age: dosage not established; children over 12 yr: see Adult Dose

*See USP DI for complete dosage information.

TABLE 9-10 *Continued*

PHENOTHIAZINE ANTIPSYCHOTICS

Name Proprietary Name	***Preparation***	***Category***	***Usual Adult Dose****	***Usual Dose Range****	***Usual Pediatric Dose****
	Fluphenazine hydrochloride tablets USP				
Mesoridazine besylate USP *Serentil*	Mesoridazine besylate injection Mesoridazine besylate oral solution USP Mesoridazine besylate tablets USP	Antipsychotic	Oral, 10–50 mg tid; IM, 25 mg, repeated in 1/2–1 hr if needed	Oral, IM, 25–400 mg of mesoridazine, as the base daily	
Perphenazine USP *Trilafon*	Perphenazine injection USP Perphenazine oral solution USP Perphenazine tablets USP	Antipsychotic agent	Oral, 8–16 mg bid to qid		
Piperacetazine USP *Quide*	Piperacetazine tablets USP	Antipsychotic agent	Initial, 10 mg bid to qid, may be increased up to 160 mg/day within 3–5 days; maintenance, up to 160 mg/day in divided doses		
Prochlorperazine edisylate USP *Compazine*	Prochlorperazine edisylate injection USP	Antiemetic, antipsychotic	Antiemetic, IM, 5–10 mg (base) every 3 or 4 hr; antipsychotic, 10–20 mg every 4–6 hr	Antiemetic, 5 mg to not more than 40 mg daily; antipsychotic, 10–200 mg/day	Use in children under 9 kg of body weight or 2 yr of age is not recommended; antiemetic, IM, the equivalent of prochlorperazine 132 μg/kg of body weight; antipsychotic, IM, the equivalent of prochlorperazine 132 μg/kg
	Prochlorperazine edisylate syrup USP		5–10 mg (base) tid to qid	Up to 150 mg/day	Use in children under 9 kg of body weight or 2 yr of age is not recommended; antiemetic, 9–13 kg: the equivalent of 2.5 mg of prochlorperazine qd or bid, not exceeding 7.5 mg/day; 14–17 kg: the equivalent of 2.5 mg of prochlorperazine bid or tid, not exceeding 10 mg/day; 18–39 kg: the equivalent of 2.5 mg of prochlorperazine tid or 5 mg bid, not exceeding 15 mg/day; antipsychotic, the equivalent of 100 μg/kg of body weight of prochlorperazine or 2.5 mg/m^2 of body surface, qid

*See USP DI for complete dosage information.

TABLE 9-10 *Continued*

PHENOTHIAZINE ANTIPSYCHOTICS

Name Proprietary Name	***Preparations***	***Category***	***Usual Adult Dose****	***Usual Dose Range****	***Usual Pediatric Dose****
Prochlorperazine maleate USP *Compazine*	Prochlorperazine maleate tablets USP	Antiemetic, antipsychotic	Antiemetic, the equivalent of 5–10 mg of prochlorperazine tid or qid, as necessary; antipsychotic, the equivalent of 5–35 mg of prochlorperazine tid or qid	5 to 150 mg daily	Since tablets are not suitable for many pediatric patients, the oral syrup is usually preferred
Prochlorperazine USP *Compazine*	Prochlorperazine suppositories USP	Antiemetic, antianxiety	Rectal, 25 mg bid		Dosage in children under 9 kg of body weight or 2 years of age is not established; 9–13 kg: 2.5 mg qd or bid, not exceeding 7.5 mg/day; 14–17 kg: 2.5 mg bid or tid, not exceeding 10 mg/day; 18–39 kg: 2.5 mg tid or 5 mg bid, not exceeding 10 mg/day
Thioridazine hydrochloride USP *Mellaril*	Thioridazine hydrochloride oral solution USP Thioridazine hydrochloride tablets USP	Antipsychotic	Initial, 25–100 mg tid; maintenance, 10–200 mg bid to qid	20 to a maximum of 800 mg/day	Use in children under 2 yr of age is not recommended; 2 yr of age and over: 250 μg/kg of body weight or 7.5 mg/m^2 of body surface, qid
Trifluoperazine hydrochloride USP *Stelazine*	Trifluoperazine hydrochloride injection USP Trifluoperazine hydrochloride oral solution USP Trifluoperazine hydrochloride tablets USP	Antipsychotic		Oral, up to the equivalent of 40 mg (base) daily	
Triflupromazine hydrochloride *Vesprin*	Triflupromazine hydrochloride injection USP	Antipsychotic		IM, up to 150 mg/day	

*See USP DI for complete dosage information.

that is, the compound to which others are compared. The drug has significant sedative and hypotensive properties, possibly reflective of central and peripheral α_1-noradrenergic-blocking activity, respectively. Effects of peripheral anticholinergic activity are common. As with the other phenothiazines, the effects of other CNS depressant drugs, such as sedatives and anesthetics, can be potentiated.

Triflupromazine Hydrochloride, USP. 10-[3-(Dimethylamino)propyl]-2-(trifluoromethyl)phenothiazine monohydrochloride (Vesprin). Triflupromazine has decreased sedative and hypotensive effects relative to chlorpromazine and a greater milligram potency as an antipsychotic. Extrapyramidal symptoms (EPS) are higher. The 2-CF_3 versus the 2-Cl is associated with these changes. Overall, the drug has uses analogous to those of chlorpromazine.

Thioridazine Hydrochloride, USP. 10-[2-(1-Methyl-2-piperidyl)ethyl]-2-(methylthio)phenothiazine monohydrochloride (Mellaril). Thioridazine is a member of the piperidine subgroup of the phenothiazines. The drug has a relatively low tendency to produce EPS. The drug has high anticholinergic activity, and this activity in the striatum, counterbalancing a striatal DA block, has been said to be responsible for the low EPS. The drug has sedative and hypotensive activity in common with chlorpromazine. Antiemetic activity is decreased. At high doses pigmentary retinopathy has been observed. A metabolite of the drug is mesoridazine (discussed next).

Mesoridazine Besylate, USP. 10-[2-(Methyl-2-piperidyl)ethyl]-2-(methylsulfinyl)phenothiazine monobenzenesulfonate (Serentil). This drug shares many properties with thioridazine. Pigmentary retinopathy, however, has not been reported for the drug.

Piperacetazine, USP. 10-[3-[4-(2-Hydroxyethyl)piperidino]propyl]phenothiazin-2-yl methyl ketone; 2-acetyl-10-{-3-[4-(β-hydroxyethyl)piperidino]-propyl}phenothiazine (Quide). Piperacetazine has a piperidino moiety in common with thioridazine and mesoridazine. However, the molecular arrangement resembles the substituted piperazino compounds. Consequently, overall pharmacologic properties lie between the piperidine group and the piperazine group.

Prochlorperazine Maleate, USP. 2-Chloro-10-[3-(4-methyl-1-piperazinyl)propyl]phenothiazine maleate (Compazine). The piperazine subgroup of the phenothiazines is characterized by high milligram antipsychotic potency, a high prevalence of EPS, and low sedative and autonomic effects. Prochlorperazine is more potent on a milligram basis than its alkylamino counterpart, chlorpromazine. However, because of its high prevalence of EPS, it is used mainly for its antiemetic effect, rather than for its antipsychotic effect.

Trifluoperazine Hydrochloride, USP. (10-[3-(4-Methyl-1-piperazinyl)propyl]-2-trifluoromethyl)-phenothiazine dihydrochloride (Stelazine). Because it has both 2-trifluoromethyl and piperazine groups, trifluoperazine is a potent antipsychotic agent. Extrapyramidal symptoms are high, and sedative and hypotensive effects are low.

Other agents in the piperazine group that reportedly have overall pharmacologic profiles roughly similar to trifluoperazine are **Acetophenazine Maleate, USP,** 10-[3-[4-(2-Hydroxyethyl)-1-piperazinyl]propyl]phenothiazin-2-yl methyl ketone maleate (1:2) (Tindal); **Carphenazine Maleate, USP,** 1-[10-[3-[4-(2-Hydroxyethyl)-1-piperazinyl]propyl]-phenothiazin-2-yl]-1-propanone maleate (1 : 2) (Proketazine); and **Perphenazine, USP,** 4-[3-(2-Chlorophenothiazine-10-yl)propyl]-piperazineethanol; 2-chloro-10-[3-[4-(2-hydroxyethyl)piperazinyl]-propyl]phenothiazine (Trilafon).

Finally, the member of the piperazine subgroup with a trifluoromethyl group at the 2-position of the phenothiazine system and the most potent phenothiazine on a milligram basis is **Fluphenazine Hydrochloride, USP,** 4-[3-[2-(Trifluoromethyl)phenazin-10-yl]propyl]-1-piperazineethanol dihydrochloride; 10-[3-[4-(2-hydroxyethyl)piperazinyl]propyl]-2-trifluoromethylphenothiazine dihydrochloride (Permatil, Prolixin). It is available as two lipid-soluble esters for depot intramuscular injection, the enanthate (heptanoic acid ester) and the decanoate ester. These long-acting preparations can be crucial in treating psychotic patients who do not take their medication or who are subject to frequent relapse.

RING ANALOGUES OF PHENOTHIAZINES: THIOXANTHINES, DIBENZOXAZEPINES, AND DIBENZODIAZEPINES

This is a group of close structural relatives of the phenothiazine antipsychotics. They share many clinical properties with the phenothiazines. However, the dibenzodiazepine clozapine has some important differences, notably a low production of EPS.

Chlorprothixene, USP, 3-(2-Chloro-9*H*-thioxanthen-9-ylidene)-*N*,*N*-dimethyl-1-propanamine; 2-chloro-9-(3′-dimethylaminopropylidene)thioxanthene (Taractan) and **Thiothixene, USP,** *Z*-*N*,*N*-dimethyl-9-[3-(4-methyl-1-piperazinyl)propylidene]thioxanthene-2-sulfonamide (Navane). The thioxanthene system differs from the phenothiazine system by replacement of the N-H moiety with a carbon atom doubly bonded to the propylidene side chain. With the substituent in the 2-position, *Z*- and *E*-isomers are produced. In accordance with the concept that the presently useful antipsychotics can be superimposed on DA, the *Z*-isomers are the more active antipsychotic isomers. The compounds are very similar in pharmacologic properties to the corresponding phenothiazines. Thus, chlorprothixene displays properties similar to chlorpromazine, and thiothixene displays properties similar to the piperazine subgroup of the phenothiazines.

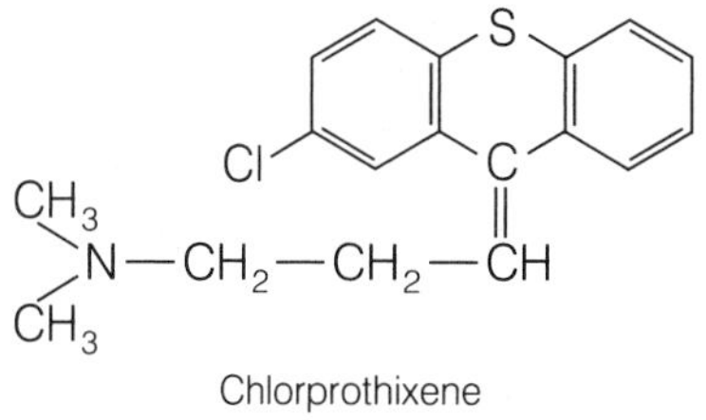

Chlorprothixene

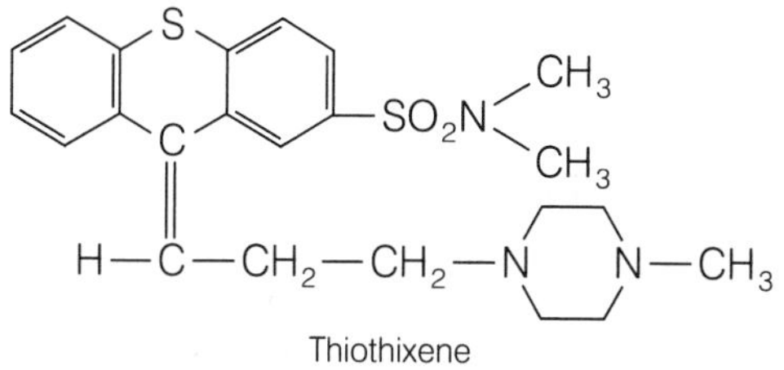

Thiothixene

A dibenzoxazepine derivative in use is **Loxapine Succinate**, 2-Chloro-11-(4-methyl-1-piperazinyl)dibenz[*b*,*f*][1,4]oxazepine succinate (Daxolin).

The structural relationship to the phenothiazine antipsychotics is apparent. It is an effective antipsychotic and has side effects similar to those reported for the phenothiazines.

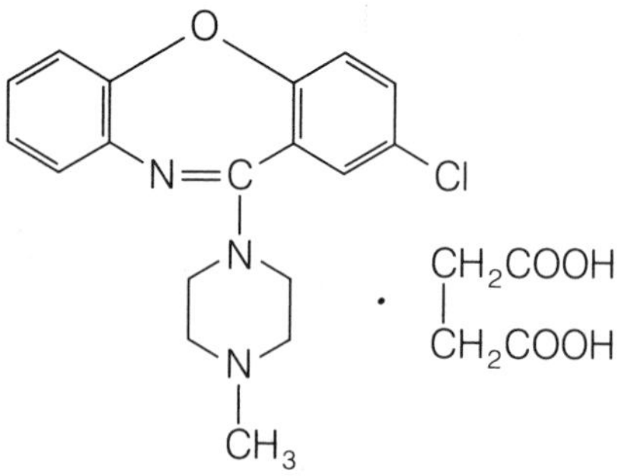

Loxapine Succinate

The dibenzodiazepine derivative is **clozapine**. It is not a potent antipsychotic on a milligram basis (note the "wrong" orientation of the *N*-methyl piperazino group relative to the chlorine atom), but it is effective and has received much attention because the production of EPS is low. It has been used extensively in Europe. It has been introduced (Clozaril) into the United States, but with restrictions, because of a high frequency of agranulocytosis associated with its use.

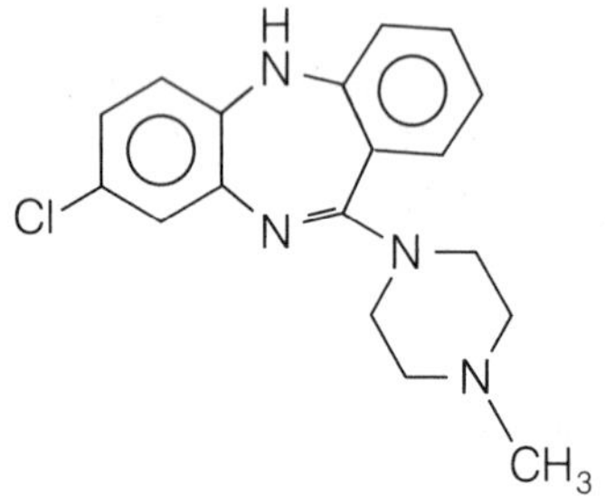

Clozapine

Several theories to account for the low EPS have been proposed. One is that it effectively blocks striatal cholinergic receptors, thus balancing out a striatal D-2 block.[40] It has also been proposed that it blocks limbic receptors more so than striatal D-2 receptors.[41,42] The compound has usefulness as a model for the development of other compounds that have low EPS potential.

FLUOROBUTYROPHENONES

The fluorobutyrophenones belong to a much studied class of compounds, with many compounds possessing high antipsychotic activity. Only two of these are used in the United States—haloperidol and droperidol, the latter usually for its neuroleptic properties as an adjunct to anesthetic agents. The structural requirements for antipsychotic activity in the group are well worked out.[43] General features are expressed in the following structure.

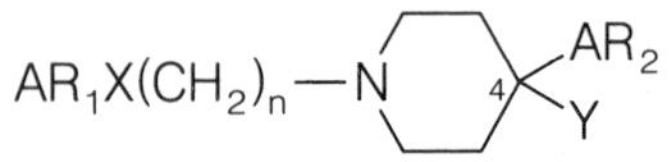

$$AR_1X(CH_2)_n\text{—N}\langle\ \rangle\text{4}\langle^{AR_2}_{Y}$$

Optimal activity is seen when AR_1 is an aromatic system. A *p*-fluoro substituent aids activity. When X = C=O, optimal activity is seen, although other groups, C(H)OH and C(H)aryl, also give good activity. When n = 3, activity is optimal: longer or shorter chains decrease activity. The aliphatic amino nitrogen is required, and highest activity is seen when it is incorporated into a cyclic form. AR_2 is an aromatic ring and is needed. It should be attached directly to the 4-position or occasionally separated from it by one intervening atom. The Y group can vary and assist activity. An example is the hydroxyl group of haloperidol.

The empirical SARs could be construed to suggest that the 4-aryl piperidino moiety is superimposable on the 2-phenylethylamino moiety of dopamine and, accordingly, could promote affinity for D-2 receptors. The long *N*-alkyl substituent could help promote affinity and produce antagonistic activity.

Some members of the class are extremely potent antipsychotic agents and D-2 receptor antagonists. The EPS are extremely marked in this class, which may, in part, be due to a potent DA block in the striatum and almost no compensatory striatal anticholinergic block. The compounds do not have the structural features associated with effective anticholinergic activity.

Haloperidol, USP. 4[4-(*p*-Chlorophenyl)-4-hydroxypiperidino]-4′-fluorobutyrophenone (Haldol). The compound is a potent antipsychotic useful in schizophrenia and in psychoses associated with brain damage. It is often chosen as the agent to terminate mania. Therapy for Gilles de la Tourette's syndrome often employs the drug.

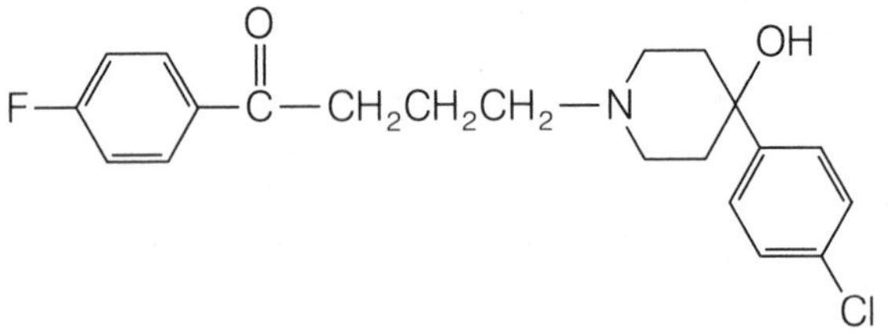

Haloperidol

Droperidol, USP. 1-{1-[3-(*p*-Fluorobenzoyl)propyl]-1,2,3,6-tetrahydro-4-pyridyl}-2-benzimidazolinone (Inapsine). The agent may be used alone as a preanesthetic neuroleptic or as an antiemetic. Its most frequent use is in combination (Innovar) with the narcotic agent fentanyl (Sublimaze) preanesthetically.

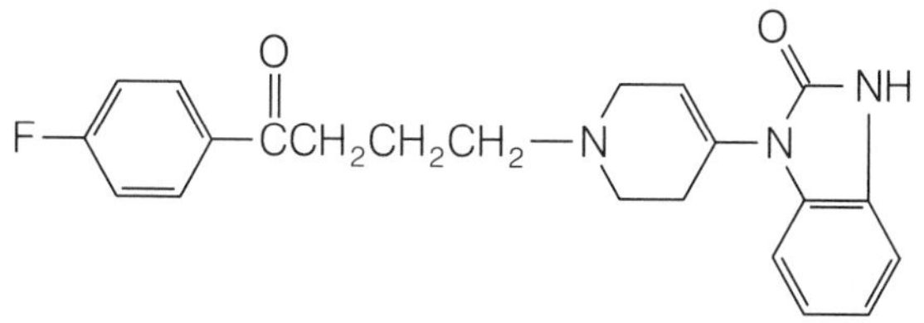

Droperidol

The diphenylbutylpiperidine class can be considered as a modification of the fluorobutyrophenone class. Because of their high hydrophobic character, the compounds are inherently long acting. Penfluridol has undergone clinical trials in the United States, and pimozide has been approved for antipsychotic use. Overall, side effects for the two compounds resemble those produced by the fluorobutyrophenones.

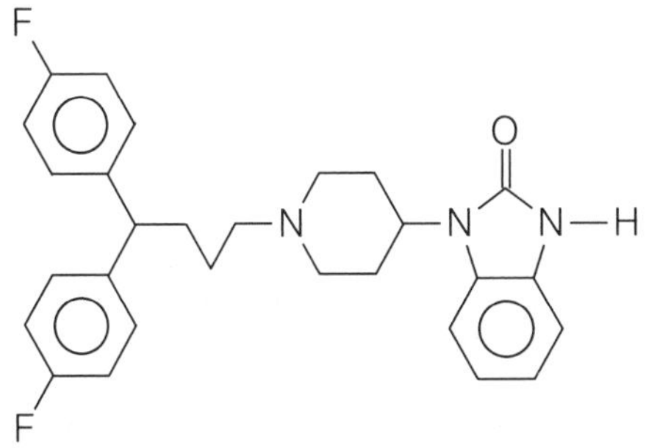

Pimozide

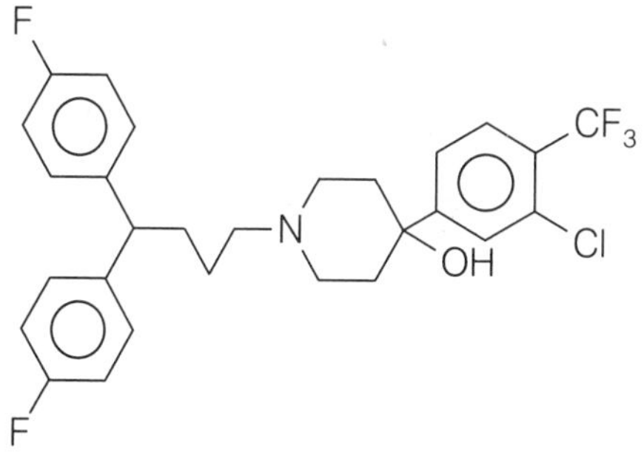

Penfluridol

β-AMINOKETONES

Several of these agents have been examined.[39] The overall structural features associated with activity can be seen in the structure of the one compound that is in clinical use, molindone. In addition to the β-aminoketone, there must be an aryl group positioned as in molindone. It might be conjectured that the proton on the protonated amino group in these compounds H-bonds with the electrons of the carbonyl oxygen atom. This would produce a cationic center, two-atom distance, and an aryl group that could be superimposed on the analogous features of protonated dopamine.

Molindone Hydrochloride. 3-Ethyl-6,7-dihydro-2-methyl-5-morpholinomethyl)indole-4(5*H*)-one monohydrochloride (Moban). The compound is about as potent an antipsychotic as trifluoperazine. Overall, side effects resemble those of the phenothiazines. Hypotension may not be a problem, and the compound is said to produce less weight gain than other antipsychotics.

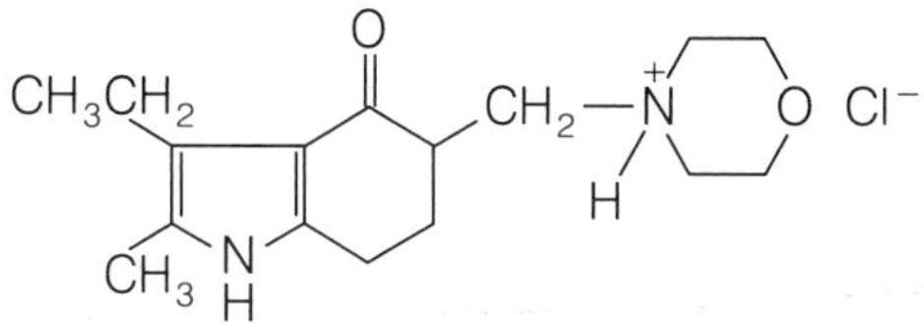

Molindone Hydrochloride

BENZAMIDES

The benzamides, evolved from observations that the gastroprokinetic and antiemetic agent metoclopramide (see Chap. 22), represented by sulpiride, have antipsychotic activity apparently related to D-2 receptor block. It is hoped that the group might yield compounds with diminished EPS liability, but this remains to be determined in the clinic. It is said that an H-bond between the amido H and the unshared electrons of the methoxy group, to generate a pseudo ring, is important for antipsychotic activity in these compounds. Presumably, when the protonated amine is superimposed on that of protonated dopamine, this pseudo ring would superimpose on dopamine's aromatic ring.[44]

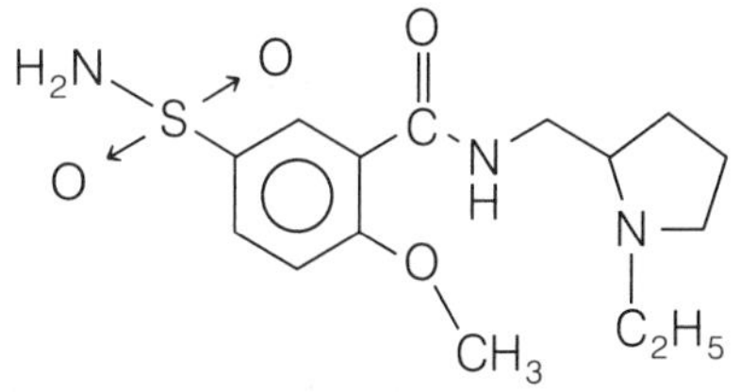

The general structural features associated with antipsychotic activity are well represented by sulpiride's structure, except that the sulfonamido group of sulpiride is not required for activity.

Apomorphine (see Chap. 17), at selected doses, can act principally as a presynaptic DA agonist rather than a postsynaptic agonist and has served to stimulate interest in presynaptic agonists as antipsychotic agents. Another approach to the DA system could be made through agonists of cholecystokinin.[45] Finally, although the block of mesolimbic D-2 receptors works well to explain the action of currently useful antipsychotic drugs, many of the agents have effects

TABLE 9-11

MISCELLANEOUS ANTIPSYCHOTIC AGENTS

Name Proprietary Name	***Preparations***	***Category***	***Usual Adult Dose****	***Usual Dose Range****	***Usual Pediatric Dose****
Chlorprothixene USP *Taractan*	Chlorprothixene injection USP Chlorprothixene oral suspension USP Chlorprothixene tablets USP	Antipsychotic		Oral, moderate anxiety, 10 mg tid or qid (up to 60 mg/day); severe neurotic and psychotic states, 25–50 mg tid or qid (up to 600 mg/day); IM, moderate anxiety, 12.5–25 mg; severe neurotic and psychotic states, 75–200 mg/day	
Droperidol *Inapsine*	Droperidol injection USP	Antipsychotic	IM or IV, 1.25–10 mg	1.25–10 mg	
Haloperidol USP *Haldol*	Haloperidol oral solution USP Haloperidol tablets USP	Antipsychotic	500 μg–5 mg bid or tid	1–100 mg day	Dosage is not estabished in infants and children
Loxapine succinate *Loxitane*	Loxapine succinate capsules	Antipsychotic	Initial, 10 mg bid; maintenance, 60–100 mg/day; total daily doses over 250 mg are not recommended		
Molindone *Moban*	Molindone hydrochloride oral solution	Antipsychotic	Initial, 50–75 mg/day	Usual limit is 225 mg/day	Dosage not established in children under 12 yr of age
	Molindone hydrochloride tablets	Antipsychotic	Initial, 50–75 mg/day	Usual limit is 225 mg/day	Dosage not established in children under 12 yr of age
Thiothixene USP *Navane*	Thiothixene capsules USP	Antipsychotic	20–30 mg/day in divided doses	6–60 mg/day in divided doses	
Thiothixene hydrochloride USP *Navane*	Thiothixene hydrochloride injection USP Thiothixene hydrochloride oral solution USP	Antipsychotic	Oral, 6–60 mg/day in divided doses; IM, 4 mg bid to qid	Oral, 20–60 mg/day in divided doses; IM, 16–30 mg/day in divided doses	

*See USP DI for complete dosage information.

on other systems, including noradrenergic and serotoninergic systems. These effects have also been reported to be involved in antipsychotic actions. Should an endogenous ligand for the PCP receptor be found, its antagonists could also be antipsychotic.

Additional information on the various antipsychotic agents just described are given in Table 9-11.

ANTIMANIC AGENT

Lithium Salts

The lithium salts used in the United States are the carbonate and the citrate. Lithium chloride is not employed because of its hygroscopic nature and because it is more irritating than the carbonate or citrate to the gastrointestinal tract.

The active species in these salts is the lithium ion. The classic explanation for its antimanic activity is that it resembles the sodium ion (as well as potassium, magnesium, and calcium ions) and can occupy the sodium pump. However, it cannot, unlike the sodium ion, maintain membrane potentials. Accordingly, it might prevent excessive release of neurotransmitters (e.g., dopamine) that characterize the manic state. Many of the actions of lithium ion have been reviewed.[46] The indications for lithium salts are acute mania (often with a potent neuroleptic agent to immediately control the mania because lithium is slow to take effect), and as a prophylactic to prevent occurrence of the mania of bipolar manic–depressive illness. Lithium salts are also being used in severe recurrent unipolar depressions. Present explanations for the dual nature (antimanic and antidepressant) of lithium revolve around its ability to act on G-proteins and the second messenger, inositol triphosphate.

Because of its water solubility, the lithium ion is extensively distributed in body water. It tends to become involved in the many physiologic processes involving sodium and potassium ions, hence, the side effects and potential drug interactions are many. The margin of safety is low; therefore lithium should be used only when plasma levels can be routinely determined. In the desired dose range, side effects can be adequately controlled.

Lithium Carbonate, USP. (Eskalith, Lithane); **Lithium Citrate** (Cibalith-S).

ANTICONVULSANT OR ANTIEPILEPTIC DRUGS

As is customary, these two terms, anticonvulsant and antiepileptic, will be used interchangeably in this discussion. Strictly speaking, however, the term *anticonvulsant* designates an agent that blocks experimentally produced seizures in laboratory animals, and an *antiepileptic* drug is a drug used medically to control the epilepsies, not all of which are convulsive, in humans.

A classification of the epilepsies has been widely accepted because its accuracy facilitates diagnosis, drug selection, and a precise discussion of the epilepsies.[47,48] The major classification types are (1) the *generalized seizures*, which essentially involve the entire brain and do not have an apparent local onset; (2) *unilateral seizures* (involve one entire side of the body); (3) *partial* (or *focal*) *seizures* that have a focus (i.e., begin locally); (4) *erratic seizures* of the newborn; and (5) *unclassified seizures* (severe seizures associated with a high mortality such that time does not permit a precise categorization).

Two major types of generalized seizures are the generalized tonic–clonic seizure (grand mal) and the nonconvulsive seizures or absence (petit mal) seizures. The typical generalized tonic–clonic seizure is often preceded by a series of bilateral muscular jerks; this is followed by loss of consciousness, which, in turn, is followed by a series of tonic and then clonic spasms. The typical absence seizure (classic petit mal) consists of a sudden brief loss of consciousness, sometimes with no motor activity, although there is often some minor clonic motor activity.

Major types of focal (partial) epilepsies are the simple focal and the complex focal seizures. A prototype simple partial seizure is jacksonian motor epilepsy in which the jacksonian march may be seen. As the abnormal discharge proceeds over the cortical site involved, the visible seizure progresses over the area of the body controlled by the cortical site. The complex partial seizure is represented by the psychomotor or temporal lobe seizure. There is an aura, then a confused or bizarre but purposefully appearing behavior lasting two to three minutes, often with no memory of the event. The seizure may be misdiagnosed as a psychotic episode.

For the purposes of broad considerations of how structure relates to antiepileptic activity, further condensation of the classification of the epilepsies is traditionally made (generalized tonic–clonic seizures, simple partial seizures, complex partial seizures, and absence seizures). The broad general pattern of structural features associated with antigeneralized tonic–clonic seizure activity is discernible. This SAR also applies to simple partial seizures. It applies with less certainty to complex partial seizures, which are relatively resistant to treatment. With fewer effective drug entities, overall structural conclusions are more tenuous. The other general seizure type for which a broad SAR pattern can be seen is the absence seizure. These features are cited under the heading: Mechanism of Anticonvulsant Action.

Likewise, animal models characteristically discern three types of activity: activity against electrically induced convulsions correlates with activity against generalized tonic–clonic and partial seizures, and activity against pentylenetetrazole (PTZ)-induced seizures correlates with antiabsence activity.

Each of the epilepsies is characterized by a typical abnormal pattern in the electroencephalogram (EEG). The EEG indicates that there is sudden and excessive electrical activity in the brain. The precise causes of these sudden and excessive electrical discharges may be many, and not all are understood. However, a working hypothesis is that there is a site or focus of damaged or abnormal, and consequently hyperexcitable, neurons in the brain. These can fire excessively and sometimes recruit adjacent neurons that, in turn, induce other neurons to fire. The location and extent of the abnormal firing determines the epilepsy. A recent refinement in this theory is the kindling model.[49] Experimentally, a brief and very localized electrical stimulus is applied to a site in the brain, with long intervals between application. As the process is repeated, neuronal afterdischarges grow both longer and more intense at the original site and at new sites far from the original site. It is thought that changes occur in neurons at the discharge site and these neurons, in turn, induce changes in neurons far from the site. Progressively more severe seizures can be induced, and these can arise from secondary foci that have been kindled far from the site of stimulation.

MECHANISM OF ANTICONVULSANT ACTION

The basic mechanism of anticonvulsants is to decrease electrical excitability at the site or at adjacent or recruited neurons. Some of the mechanisms by which anticonvulsants exert these effects are considered under the preceding discussions of benzodiazepines and barbiturates. As a broad summary, it has been proposed that blockade of sustained high frequency repetitive firing (SRF) may underlie the action of phenytoin, carbamazepine, phenobarbital, and valproic acid against generalized tonic–clonic seizures in humans and against maximal electroshock seizures in animals. Enhancement of GABAergic transmission may underlie the action of benzodiazepines and valproic acid against absence seizures in humans and against PTZ-induced seizures in experimental animals.[16]

Current research on the neurochemical roles of adenosine and excitatory amino acids is expected to yield additional information on eleptogenesis and mechanism of action of anticovulsant drugs.

Several major groups of drugs have the common structure shown below:

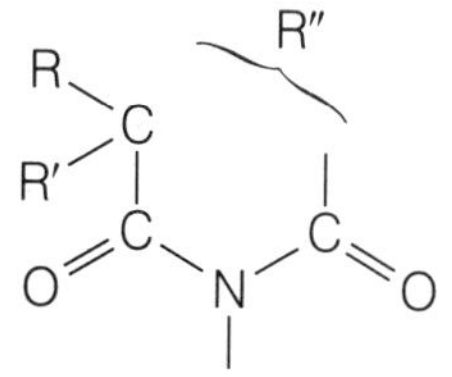

R, R′, R″, C, C, C, O, N, O

Structure common to anticonvulsant drugs.

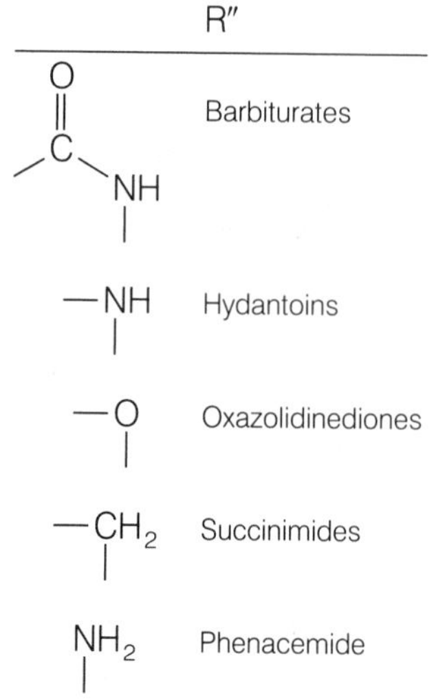

R″	
—C(=O)—NH—	Barbiturates
—NH—	Hydantoins
—O—	Oxazolidinediones
—CH_2—	Succinimides
NH_2	Phenacemide

An overall pattern in the foregoing is that R and R′ should both be hydrocarbon radicals. If both R and R′ are lower alkyls, the tendency is to be active against absence seizures (petit mal) and not active against generalized tonic–clonic (grand mal) or partial seizures. If one of the hydrocarbon substituents is an aryl group, activity tends to be directed toward generalized tonic–clonic and partial seizures, and not toward antiabsence activity.[50]

A conformational analysis of the aryl-containing antigeneralized tonic–clonic agents indicates that the conformational arrangement of the hydrophobic groups is important.[51]

TABLE 9-12

DRUGS USED IN THE TREATMENT OF EPILEPSY

Drug	*Types of Seizure*
BARBITURATES	
Phenobarbital	Generalized tonic–clonic
Mephobarbital	Generalized tonic–clonic
Metharbital	Generalized tonic–clonic
HYDANTOINS	
Phenytoin	Generalized tonic–clonic*
Mephenytoin	Generalized tonic–clonic*
Ethotoin	Generalized tonic–clonic*
OXAZOLIDINEDIONES	
Trimethadione	Absence
Paramethadione	Absence
SUCCINIMIDES	
Phensuximide	Absence
Methsuximide	Absence*
Ethosuximide	Absence
BENZODIAZEPINES	
Clonazepam	Absence
Diazepam IV	Status epilepticus
Clorazepate	Complex partial seizures (adjunctive)
MISCELLANEOUS	
Primidone	Generalized tonic–clonic*
Carbamazepine	Generalized tonic–clonic*
Phenacemide	Complex partial seizures
Valproic acid	Absence

*Some effectiveness against complex partial seizures.

BARBITURATES

Although sedative–hypnotic barbiturates commonly display anticonvulsant properties, only phenobarbital and mephobarbital (and, marginally, metharbital) display adequate anticonvulsant selectivity for use as antiepileptics.

For the structures of these agents, consult Table 9-5, and for discussion of chemical properties see the section on barbiturates under sedative–hypnotic–

anxiolytic agents. The metabolism of phenobarbital involves *p*-hydroxylation followed by conjugation.

Mephobarbital is extensively *N*-demethylated in vivo and is thought to owe most of its activity to the metabolite phenobarbital. In keeping with their structures, both agents are effective against generalized tonic–clonic and partial seizures. Metharbital has been said to act more as a sedative than as a specific antigeneralized tonic–clonic agent.

HYDANTOINS

The hydantoins are close structural relatives of the barbiturates, differing in lacking the 6-oxo moiety. They are cyclic monoacylureas rather than cyclic diacylureas. Consequently, they are weaker organic acids than the barbiturates (e.g., phenytoin $pK_a = 8.3$). Thus, aqueous solutions of sodium salts, such as of phenytoin sodium, generate strongly alkaline solutions.

The compounds have a trophism toward antigeneralized tonic–clonic rather than antiabsence activity. This is not an intrinsic activity of the hydantoin ring system. All of the clinically useful compounds (Table 9-13) possess an aryl substituent on the 5-position, corresponding to a branched atom of the general pharmacophore. Hydantoins with lower alkyl substituents have been reported to have antiabsence activity.

Phenytoin and Phenytoin Sodium, USP. 5,5-Diphenylhydantoin (Dilantin). Phenytoin is the first anticonvulsant in which it was clearly demonstrated that anticonvulsant activity could definitely be separated from sedative–hypnotic activity. The drug is useful against all seizure types except absence. It is sometimes noted that the drug is incompletely or erratically absorbed from sites of administration. This is due to the very low water solubility of the drug.

Metabolism proceeds by *p*-hydroxylation of an aromatic ring, followed by conjugation.

Mephenytoin, USP. 5-Ethyl-3-methyl-5-phenylhydantoin (Mesantoin). Mephenytoin is metabolically *N*-dealkylated to 5-ethyl-5-phenylhydantoin, believed to be the active agent. Interestingly, 5-ethyl-5-phenylhydantoin, the hydantoin counterpart of phenobarbital, was one of the first hydantoins introduced into therapy. It was introduced as a sedative–hypnotic and anticonvulsant under the name Nirvanol, but was withdrawn because of toxicity. Presumably, mephenytoin may be considered a prodrug, which ameliorates some of the toxicity—serious skin and blood disorders—of the delivered active drug.

Metabolic inactivation of mephenytoin and its demethyl metabolite is by, as expected, *p*-hydroxylation and conjugation of the hydroxyl group. The drug has a spectrum of activity similar to phenytoin. It may worsen absence seizures.

Ethotoin. 3-Ethyl-5-phenylhydantoin (Peganone). The compound is *N*-dealkylated and *p*-hydroxylated; the *N*-dealkyl metabolite, presumably the active compound, is likewise metabolized by *p*-hydroxylation. The hydroxyl group is then conjugated.

The compound is employed against generalized seizures, but usually on an adjunctive basis owing to its low potency. In general, agents that are not completely branched on the appropriate carbon are of lower potency than their more completely branched counterparts.

TABLE 9-13

THE ANTICONVULSANT HYDANTOIN DERIVATIVES

(Hydantoin ring: R_5, R'_5 at C-5; N—H at position 1; C-2 and C-4 carbonyls; R_3 on N-3)

Generic Name Proprietary Name	Substituents R_5	R'_5	R_3
Phenylethylhydantoin *Nirvanol*	C_6H_5—	$CH_3—CH_2—$	H
Phenytoin USP *Dilantin, Diphentoin*	C_6H_5—	C_6H_5—	H
Mephenytoin USP *Mesantoin*	C_6H_5—	$CH_3—CH_2—$	$CH_3—$
Ethotoin *Peganone*	C_6H_5—	H	$CH_3—CH_2—$

OXAZOLIDINEDIONES

Replacement of the N-H group at position 1 of the hydantoin system with an oxygen atom yields the oxazolidine-2,4-dione system.

The oxazolidinedione system is sometimes equated with antiabsence activity, but this trophism probably is more dictated by the fact that the branched atom of these compounds is substituted with lower alkyls. Aryl-substituted oxazolidine-2,4-diones have shown activity against generalized tonic–clonic seizures.

Trimethadione, USP. 3,5,5-Trimethyl-2,4-oxazolidinedione; 3,5,5-trimethadione (Tridione). Trimethadione was the first drug introduced specifically for treating absence seizures. It is important as a prototype structure for antiabsence compounds.

Dermatologic and hematologic toxicities limit its clinical use.

The drug is metabolized by *N*-demethylation to the putative active metabolite dimethadione.[52] Dimethadione is a water-soluble and lowly lipophilic compound and, thus, is excreted as such without further metabolism.

Paramethadione, USP. 5-Ethyl-3,5-dimethyl-2,4-oxazolidinedione (Paradione). Paramethadione is very closely related to trimethadione in structure and has similar actions, uses, and side effects, although it may be safer. The *N*-demethyl metabolite, which is excreted rather slowly, is thought to be the active drug.

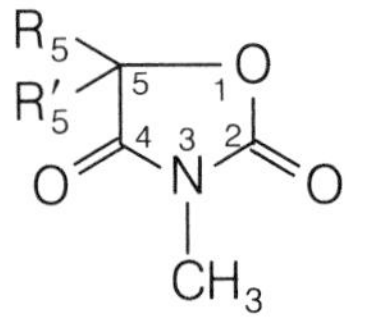

Trimethadione $R_5 = R'_5 = CH_3$
Paramethadione $R_5 = CH_3$; $R'_5 = C_2H_5$

SUCCINIMIDES

In view of the activity of antiepileptic agents such as the oxazolidine-2,4-diones, substituted succinimides (CH_2 replaces O) were a logical choice for synthesis and evaluation. Three are now in clinical use.

Phensuximide, USP. *N*-Methyl-2-phenylsuccinimide (Milontin). Some trophism toward antiabsence activity is produced by the succinimide system. The $-CH_2-$ could be viewed as an α-alkyl branch condensed into the ring. Phensuximide is used primarily against absence seizures, but it is of low potency and is relegated to a secondary status. The phenyl substituent confers some activity against generalized tonic–clonic and partial seizures. *N*-Demethylation occurs to yield the putative active metabolite. Both phensuximide and the *N*-demethyl metabolite are inactivated by *p*-hydroxylation and conjugation.

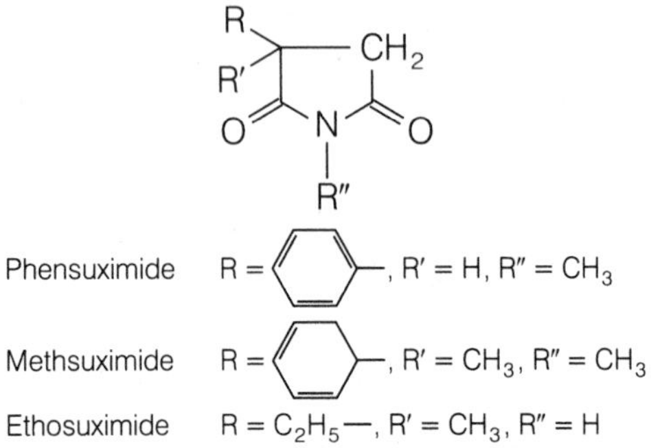

Phensuximide R = phenyl, R′ = H, R″ = CH_3
Methsuximide R = phenyl, R′ = CH_3, R″ = CH_3
Ethosuximide R = C_2H_5—, R′ = CH_3, R″ = H

Methsuximide. *N*,2-Dimethyl-2-phenylsuccinimide (Celontin). *N*-demethylation and *p*-hydroxylation of parent and metabolite occur. The drug has some use against absence and complex partial seizures.

Ethosuximide, USP. 2-Ethyl-2-methylsuccinimide (Zarontin). Ethosuximide conforms very well to the general structural pattern for antiabsence activity. The drug is more active and less toxic than trimethadione; consequently, it has emerged as the drug of choice for typical absence seizures. Toxicity primarily involves the skin and blood.

Some of the drug is excreted intact. The major metabolite is produced by oxidation of the ethyl group.

UREAS AND MONOACYLUREAS

The two chemical classes, ureas and monoacylureas, have a long history of producing compounds with anticonvulsant activity. However, the numerical yield of clinically useful compounds has not been great.

Phenacemide, USP. (Phenylacetyl)urea (Phenurone). Phenacemide, a rather broad-spectrum agent, finds some use in psychomotor epilepsy. Its principal problems are severe side effects, including personality changes and blood, renal, and skin disorders. Metabolism is by *p*-hydroxylation.

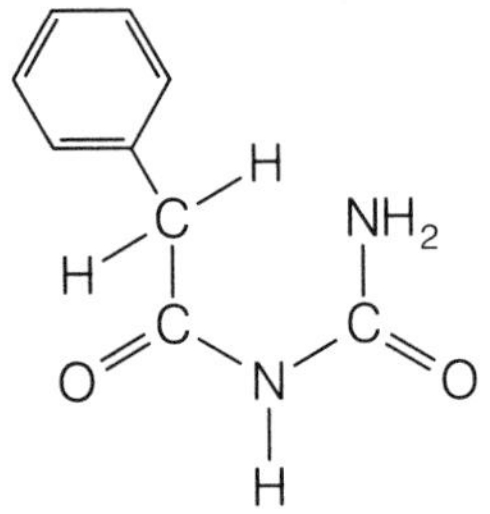

Phenacemide

Carbamazepine, USP. 5*H*-Dibenz[*b, f*]azepine-5-carboxamide (Tegretol). For SAR discussion purposes, carbamazepine can be viewed either as an ethylene-bridged 1,1-diphenylurea or an amido-substituted tricyclic system. Either view fits a very general activity pattern for anticonvulsants, namely, a hydrophobic moiety joined to a rather simple nonionic polar H-bonding group. The two phenyls on the nitrogen fit the pattern of antigeneralized

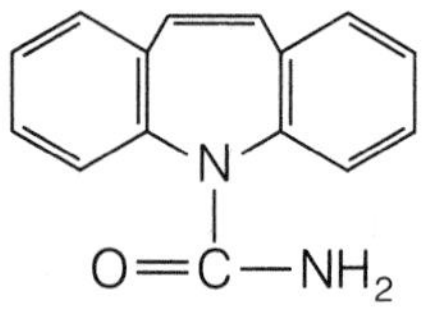

Carbamazepine

tonic–clonic activity. Carbamazepine is useful in generalized tonic–clonic and partial seizures.

The drug has the potential for serious hematologic toxicity, and it is used with caution.

Metabolism proceeds largely through the epoxide formed at the *cis*-stilbene double bond. In humans the epoxide reportedly is converted largely to the 10*S*,11*S* *trans*-diol.[53]

MISCELLANEOUS AGENTS

Primidone. 5-Ethyldihydro-5-phenyl-4,6-(1*H*,5*H*)-pyrimidinedione (Mysoline). Primidone is sometimes described as a 2-deoxybarbiturate. It appears to act as such, and through conversion to phenobarbital and to phenylethylmalonyldiamide (PEMA).[54] The efficacy is against all types of seizures except absence. The agent has good overall safety, but rare serious toxic effects do occur.

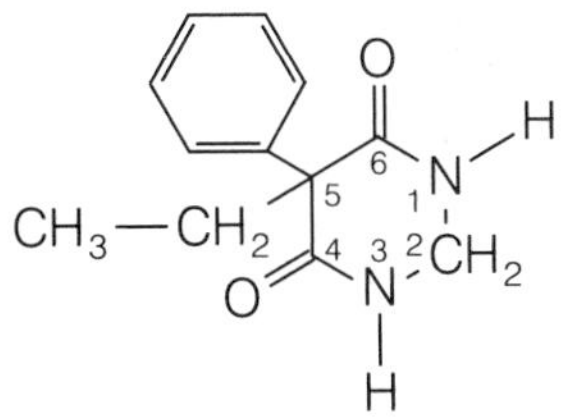

Primidone

Valproic Acid. 2-Propylpentanoic acid (Depakene). Many carboxylic acids have anticonvulsant activity, although often of a low order of potency, possibly in part because extensive dissociation at physiologic *p*H produces poor partitioning across the blood–brain barrier. Valproic acid has good potency and is used mainly against absence seizures, but also benefits several other seizure types. Metabolism is by conjugation of the carboxylic acid group and oxidation of one of the hydrocarbon chains. Many of the side effects are mild. However, a rare, but potentially fatal, fulminant hepatitis has caused concern.

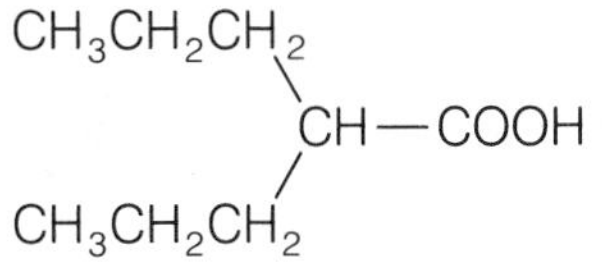

$$(CH_3CH_2CH_2)_2CH{-}COOH$$

Valproic Acid

BENZODIAZEPINES

For details of the chemistry and SARs for the benzodiazepines, see the discussion under anxiolytic–sedative–hypnotic drugs. The structural features associated with anticonvulsant activity are identical with those associated with anxiolytic–sedative–hypnotic activity.[20] Animal models predict benzodiazepines to be modestly effective against generalized tonic–clonic and partial seizures and very highly active against absence seizures. This difference in seizure control trophism is markedly different from the barbiturates, hydantoins, and most other chemical compounds when they are aryl- or diaryl-substituted. The difference in trophism is consistent with dissimilar mechanisms of action (see discussion of mechanisms of action). Despite the high effectiveness of benzodiazepines as a group in animal models, only three benzodiazepines have achieved established positions in anticonvulsant therapy.

Clonazepam, USP. 5-(2-Chlorophenyl)-1,3-dihydro-7-nitro-2*H*-1,4-benzodiazepin-2-one (Klonapin). Clonazepam is useful in absence seizures and in myoclonic seizures. Development of tolerance to the anticonvulsant effect often develops. This is a common problem with the benzodiazepines. Metabolism involves hydroxylation of the 3-position and nitro group reduction, followed by acetylation.

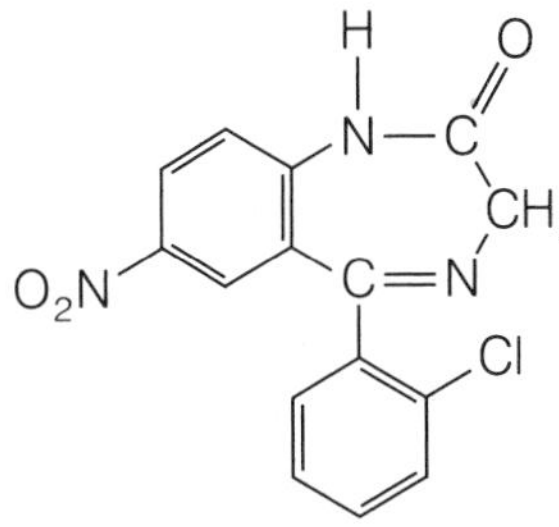

Clonazepam

TABLE 9-14
ANTICONVULSANT DRUGS

Name Proprietary Name	*Preparations*	*Category*	*Usual Adult Dose**	*Usual Dose Range**	*Usual Pediatric Dose**
Ethotoin *Peganone*	Ethotoin tablets	Anticonvulsant	Initially, 1 g/day in divided doses; maintenance, 2–3 g/day in 4–6 divided doses		Children, initially, 750 mg/day; maintenance, 500 mg–1 g/day in divided doses

*See USP DI for complete dosage information.

TABLE 9-14 *Continued*

ANTICONVULSANT DRUGS

Name Proprietary Name	***Preparations***	***Category***	***Usual Adult Dose****	***Usual Dose Range****	***Usual Pediatric Dose****
Mephenytoin USP *Mesantoin*	Mephenytoin tablets USP	Anticonvulsant	100 mg	200–600 mg/day	
Paramethadione USP *Paradione*	Paramethadione capsules USP Paramethadione oral solution USP	Anticonvulsant	Initial, 300 mg tid	900 mg–2.4 g/day	Under 2 yr of age: 100 mg tid; 2–6 yr: 200 mg tid; over 6 yr: see Usual Adult Dose
Phenytoin USP *Dilantin*	Phenytoin oral suspension USP Phenytoin tablets USP	Anticonvulsant, cardiac depressant (antiarrhythmic)	Anticonvulsant, initial, 100 mg tid; cardiac depressant, 100 mg bid to qid	200–600 mg/day	Anticonvulsant, 1.5–4 mg/kg of body weight or 125 mg/m^2 of body surface, bid, not to exceed 300 mg/day
Phenytoin sodium USP *Dilantin Sodium*	Phenytoin sodium capsules USP	Anticonvulsant, cardiac depressant (antiarrhythmic)	Anticonvulsant, initial, 100 mg tid; cardiac depressant, 100 mg bid to qid	200–600 mg/day	Anticonvulsant, 1.5–4 mg/kg of body weight or 125 mg/m^2 of body surface, bid, not to exceed 300 mg/day
	Sterile phenytoin sodium USP		Anticonvulsant, IV, 150–250 mg then 100–150 mg repeated in 30 min as necessary, at a rate not exceeding 50 mg/min; cardiac depressant, IV, 50–100 mg, repeated every 10–15 min as necessary, up to a maximum total dose of 10–15 mg/kg of body weight	50–800 mg/day	Anticonvulsant, IV, 1.5–4 mg/kg of body weight or 125 mg/m^2 of body surface, bid
Trimethadione USP *Tridione*	Anticonvulsant capsules USP	Anticonvulsant	Initial, 300 mg tid	900 mg–2.4 g/day	13 mg/kg of body weight or 335 mg/m^2 of body surface, or the following amounts, are usually given tid: infants, 100 mg; 2 yr, 200 mg; 6 yr, 300 mg; 13 yr, 400 mg
Carbamazepine USP *Tegretol*	Carbamazepine tablets USP	Anticonvulsant	Anticonvulsant, initial, 200 mg bid; trigeminal neuralgia, 100 mg bid	Anticonvulsant, 800–1200 mg/day; trigeminal neuralgia, 400–800 mg/day	Anticonvulsant, 6–12 yr, initial, 100 mg bid, increased not to exceed 1000 mg/day
Clonazepam USP *Klonapin*	Tablets	Anticonvulsant	Initial, 0.5 mg tid	Dose increased in increments of 0.5–1 mg every 3 days to a maximum of 20 mg/day	Initial, 0.01–0.03 mg/kg of body weight daily, increased not to exceed 0.1–0.2 mg/kg body weight daily
Clorazepate dipotassium *Tranxene*	Clorazepate dipotassium capsules and tablets	Anticonvulsant	Initial, up to 7.5 mg tid, increased by no more than 7.5 mg/wk not to exceed 90 mg/day		Children up to 9 yr, dosage is not established; children 9–12 yr, 7.5 mg bid, increased by no more than 7.5 mg/wk not to exceed 6 mg/day; children over 12 yr, see Usual Adult Dose

*See USP DI for complete dosage information.

TABLE 9-14 *Continued*

ANTICONVULSANT DRUGS

Name Proprietary Name	***Preparations***	***Category***	***Usual Adult Dose****	***Usual Dose Range****	***Usual Pediatric Dose****
Diazepam USP *Valium*	Diazepam injection USP	Anticonvulsant (status epilepticus)	Initial, 5–10 mg IV, if necessary repeated at 10–15 min intervals to a maximum of 30 mg		Infants, 30 days to 5 yr, 0.2–0.5 mg IV every 2–5 min to a maximum of 5 mg; children, 5 yr or older, 1 mg IV every 2–5 min to a maximum of 10 mg
Ethosuximide USP *Zarontin*	Ethosuximide capsules USP	Anticonvulsant	250 mg bid initially, increased as necessary every 4–7 days in increments of 250 mg	500 mg–1.5 g/day	Children under 6 yr, 250 mg qd; over 6 yr, 250 mg bid initially, increased as necessary every 4–7 days in increments of 250 mg
Methsuximide USP *Celontin*	Methsuximide capsules USP	Anticonvulsant	Initial, 300 mg/day; maintenance, 300 mg–1.2 g/day		
Phenacemide USP *Phenurone*	Phenacemide tablets USP	Anticonvulsant	Initial, 250–500 mg tid; maintenance, 250–500 mg 3–5 times daily	2–5 g/day in divided doses	
Phensuximide USP *Milontin*	Phensuximide capsules USP Phensuximide oral suspension USP	Anticonvulsant		500 mg–1 g bid or tid, irrespective of age	
Primidone USP *Mysoline*	Primidone oral suspension USP Primidone tablets USP	Anticonvulsant	Initial, week 1, 250 mg qd hs; week 2, 250 mg bid; week 3, 250 mg tid; week 4, 250 mg qid; maintenance, 250–500 mg tid	250 mg to not more than 2 g/day	Children under 8 yr, initial, week 1, 125 mg qd hs; week 2, 125 mg bid; week 3, 125 mg tid; week 4, 125 mg qid; maintenance, 250 mg bid or tid; children over 8 yr, see Usual Adult Dose
Valproic acid *Depakene*	Valproic acid capsules Valproate sodium syrup	Anticonvulsant	Initial, 15 mg/kg daily, increased weekly by 5–10 mg/kg daily to a maximum of 60 mg/kg daily		Children weighing over 10 kg, initial dose, 15 mg/kg daily, increased weekly by 5–10 mg/kg daily to a maximum of 60 mg/kg daily

*See USP DI for complete dosage information.

Diazepam (Valium). For details on the chemical entity, see its discussion under sedative–hypnotic–anxiolytics. The drug is mainly useful in treating status epilepticus, which is an ongoing and potentially fatal generalized tonic–clonic seizure.

Chlorazapate (Tranxene). See the detailed discussion of this agent in the sedative–hypnotic–anxiolytic section. Its principal anticonvulsant use is adjunctively in complex partial seizures.

REFERENCES

1. Smith, T. C., and Wollman, H.: History and principles of anesthesiology. In Gilman, A. F., Goodman, L. S., Rall, T. W., and Murad, F. (eds.). Goodman and Gilman's The Pharmacological Basis of Therapeutics, 7th ed., p. 339. New York, Macmillan, 1985.
2. Miller, K. W., Paton, W. D. M., Smith, E. B., and Smith, R. A.: Anesthesiology 36:339, 1972.
3. Johnson, F. H., Eyring, H., and Polissar, M. J.: The Kinetic Basis of Molecular Biology. New York, Wiley, 1954.
4. Pauling, L.: Science 134:15, 1961.
5. Cohen, E. N.: Br. J. Anaesth. 50:665, 1978.
6. Stock, J. G. L., and Strunin, L.: Anesthesiology 63:424, 1985.
7. Cousins, M. J., and Mazze, R. I.: JAMA 225:1611, 1973.
8. Hitt, B. A., et al.: J. Pharmacol. Exp. Ther. 203:193, 1977.
9. Holliday, D. A., et al.: Anesthesiology 43:325, 1975.
10. Yang, J. C., Clarke, W. C., and Ngai, S. H.: Anesthesiology 52:414, 1980.
11. Willer, J. C., Bergeret, S., Gaudy, J. H., and Dauthier, C.: Anesthesiology 63:467, 1985.
12. Winters, W. D., et al.: Neuropharmacology 11:303, 1972.
13. Squires, R.F., and Braestrup, C.: Nature 66:732, 1977.

14. Mohler, H., and Okada, T.: Science 198:849, 1977.
15. MacDonald, R. L., and Barker, J. L.: Neurology 29:432, 1979.
16. MacDonald, R. L., and McLean, M.J.: Anticonvulsant drugs, mechanisms of action. In Delgado-Escueta, A. V., Ward, A. H. Jr., Woodbury, D. M., and Porter, R. J. (eds.). Adv. Neurol. 44:713, 1986.
17. Olsen, R. W.: Annu. Rev. Pharmacol. Toxicol. 22:245, 1982.
18. MacDonald, R. L., and McLean, M. J.: Epilepsia 23(Suppl. 1):S7, 1982.
19. Harvey, S. C.: Hypnotics and Sedatives. In Gilman, A. F., Goodman, L. S., Rall, T. W., and Murad, F. (eds.). Goodman and Gilman's The Pharmacological Basis of Therapeutics, 7th ed., p. 339. New York, Macmillan, 1985.
20. Sternbach, L. H.: In Garattini, S., Mussini, E., and Randall, L. O. (eds.). The Benzodiazepines, p. 1. New York, Raven Press, 1972.
21. Childress, S. J.: Antianxiety agents. In Wolff, M. E. (ed.). Burger's Medicinal Chemistry, 4th ed., part III, p. 981. New York, John Wiley & Sons, 1981.
22. Greenblatt, D. J., and Shader, R. I.: Benzodiazepines in Clinical Practice, p. 17. New York, Raven Press, 1974 (and references therein).
23. Greenblatt, D. J., Shader, R. I., and Abernethy, D. R.: N. Engl. J. Med. 309:354; 410, 1983.
24. Daniels, T. C., and Jorgensen, E. C.: Central nervous system depressants. In Doerge, R. F. (ed.). Wilson and Gisvold's Textbook of Organic Medicinal and Pharmaceutical Chemistry, 8th ed., p. 335. Philadelphia, J. B. Lippincott, 1982.
25. Berger, F. M.: Meprobamate and other glycol derivatives. In Usdin, E., and Forrest, I. S. (eds.). Psychotherapeutic Drugs, part II, p. 1089. New York, Marcel Dekker, 1977.
26. Cram, D. J., and Hammond, G. S.: Organic Chemistry, 2nd ed., p. 295. New York, McGraw-Hill, 1964.
27. Mackay, F. J., and Cooper, J. R.: J. Pharmacol. Exp. Ther. 135:271, 1962.
28. DaPrada, M., and Keller, H. H.: Life Sci. 19:1253, 1976.
29. Nistri, A., and Constanti, A.: Experientia 31:64, 1975.
30. Van Winkle, W. B.: Science 193:1130, 1976.
31. New, J. S.: Buspirone and gepirone—$5HT_{1A}$ agonists with a novel mechanism of anxiolysis. Paper No. 96. Presented at Div. of Med. Chem. 194th ACS National Meeting, New Orleans, 1987.
32. Glennon, R. A.: J. Med. Chem. 30:1, 1987.
33. O'Brien, P.: The Disordered Mind. Englewood Cliffs, NJ, Prentice-Hall, 1978.
34. Seeman, P.: Pharmacol. Rev. 32:329, 1981.
35. Seeman, P., et al.: Biochem. Pharmacol. 34:151, 1985.
36. Gordon, M., Cook, L., Tedeschi, D. H., and Tedeshi, R. E.: Arzneim. Forsch. 13:318, 1963.
37. Horn, A. S., and Snyder, S. H.: Proc. Natl. Acad. Sci. USA 68:2325, 1971.
38. Miller, D. D., et al.: J. Med. Chem. 30:163, 1987.
39. Kaiser, C., and Setler, P.: Antipsychotic agents. In Wolff, M. E. (ed.). Burger's Medicinal Chemistry, 4th ed., part III, p. 859. New York, John Wiley & Sons, 1981.
40. Friedman, E., Gianutosos, G., and Kuster, J.: J. Pharmacol. Exp. Ther. 226:7, 1983.
41. Rupniak, N. M. J., et al.: Psychopharmacology 84:512, 1984.
42. White, F. J., and Wang, R. Y.: Science 221:1054, 1983.
43. Janssen, P. A. J., and Van Bever, W. F. M.: Butyrophenones and diphenylbutylamines. In Usdin, E., and Forrest, I. S. (eds.). Psychotherapeutic Drugs, part II, p. 869. New York, Marcel Dekker, 1977.
44. van de Waterbeemd, H., and Testa, B.: J. Med. Chem. 26:203, 1982.
45. Hökfelt, T., et al.: Nature 285:476, 1980.
46. Emrich, H. M., Aldenhoff, J. B., and Lux, H. D. (eds.): Basic Mechanisms in the Action of Lithium. Symposium Proceedings. Amsterdam, Excerpta Medica, 1981.
47. Gastaut, H., and Broughton, R.: In Anticonvulsant Drugs, Vol. 1, International Encyclopedia of Pharmacology and Therapeutics, p. 3. New York, Pergamon, 1973.
48. Commission on Classification and Terminology of the International League Against Epilepsy: Proposal for revised clinical and electroencephalographic classification of epileptic seizures. Epilepsia 22:489, 1981.
49. Wada, J. A. (ed.): Symposium: Kindling 2. New York, Raven Press, 1981.
50. Close, W. J., and Spielman, M. A.: In Hartung, W. H. (ed.). Medicinal Chemistry, Vol. 5, p. 1. New York, Wiley, 1961.
51. Wong, M. G., Defina, J. A., and Andrews, P. R.: J. Med. Chem. 29:562, 1986.
52. Frey, H. H., and Kretschmer, B. H.: Arch. Int. Pharmacodyn. Ther. 193:181, 1971.
53. Bellucci, G., Berti, G., Chiappe, C., et al.: J. Med. Chem. 30:768, 1987.
54. Spinks, A., and Waring, W. S.: In Ellis, G. P., and West, G. B. (eds.). Progress In Medicinal Chemistry, Vol. 3, p. 261. Washington, DC, Butterworth, 1963.

SELECTED READINGS

Glennon, R. A.: Central serotonin receptors as targets for drug research. J. Med. Chem. 30:1, 1987.

Arvidsson, L. E., Hacksell, U., and Glennon, R. A.: Recent advances in central 5-hydroxytryptyamine receptor agonists and antagonists. In Jucker, E. (ed.). Drug Research, Vol. 30, p. 365. Basel, Birkhauser Verlag, 1986.

Richards, L. E., and Burger, A.: Mechanism-based inhibitors of monoamine oxidase. In Jucker, E. (ed.). Drug Research, Vol. 30, p. 205. Basel, Birkhauser Verlag, 1986.

Janssen, P. A. J., and Van Bever, W. F. M.: Butyrophenones and diphenylbutylamines. In Usdin, E., and Forrest, I. S. (eds.). Psychotherapeutic Drugs, Vol. 2, part 2, p. 369. New York, Marcel Dekker, 1977.

Miller, K. W.: General anesthetics. In Wolff, M. E. (ed.). Burger's Medicinal Chemistry, 4th ed., part III, p. 623. New York, John Wiley & Sons, 1981.

Vida, J. A.: Sedative–hypnotics. In Wolff, M. E. (ed.). Burger's Medicinal Chemistry, 4th ed., part III, p. 182. New York, John Wiley & Sons, 1981.

Isaacson, E. I., and Delgado, J. N.: Anticonvulsants. In Wolff, M. E. (ed.). Burger's Medicinal Chemistry, 4th ed., part III, p. 829. New York, John Wiley & Sons, 1981.

Kaiser, C., and Setler, P. E.: Antipsychotic agents. In Wolff, M. E. (ed.). Burger's Medicinal Chemistry, 4th ed., part III, p. 859. New York, John Wiley & Sons, 1981.

Childress, S. J.: Antianxiety agents. In Wolff, M. E. (ed.). Burger's Medicinal Chemistry, 4th ed., part III, p. 951. New York, John Wiley & Sons, 1981.

Delgado-Escueta, A. V., Ward, A. A., Jr., Woodbury, D. M., and Porter, R. J. (eds.). Basic Mechanisms of the Epilepsies: Molecular and Cellular Approaches. Adv. Neurol. Ser., Vol. 44. New York, Raven Press, 1986.

Glaser, G. H., Penry, J. K., and Woodbury, D. M. (eds.). Antiepileptic Drugs: Mechanisms of Action. Adv. Neurol. Ser. Vol. 27. New York, Raven Press, 1980.

CHAPTER **10**

Central Nervous System Stimulants

Eugene I. Isaacson

This chapter discusses a broad range of agents that produce stimulation of the central nervous system (CNS). The *analeptics* are a group of agents with a very limited range of use (as respiratory stimulants) because of the general nature of their effects. The *methylxanthines* have interesting stimulatory properties, and caffeine is much used informally as a cortical stimulant. The *central sympathomimetic agents*, amphetamine and close relatives, have alerting and antidepressant properties, but are now more often used as anorexients. The *antidepressant drugs* are most frequently employed in serious depressive disorders and are broadly groupable into the monoamine oxidase inhibitors (MAOI) and the tricyclic (and mechanistically related) agents. The so-called *psychedelic drugs* have a broad range of CNS effects, and because one effect of several of these agents is CNS stimulation, they are discussed in this chapter.

ANALEPTICS

The analeptics are a group of potent and relatively nonselective CNS stimulants the convulsive dose of which lies near their analeptic dose. They once had some employment as respiratory stimulants in countering the effects of CNS depressant drugs; however, they are now obsolete for that use. Some members of the group retain a very small therapeutic niche in the treatment of chronic obstructive pulmonary disease (COPD). Certain of the compounds have usefulness as pharmacologic tools and have interesting mechanisms of action.

Picrotoxin is obtained from the seeds of *Anamirta cocculus*. The active ingredient is picrotoxinin, with the following structure:

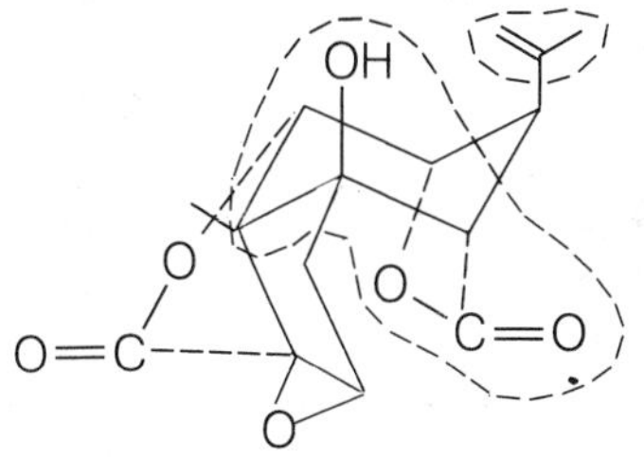

According to Jarboe,[1] the encircled hydroxylactonyl moiety is mandatory for activity with the encircled 2-propenyl group assisting. Picrotoxin exerts its effects by interfering with the inhibitory effects of γ-aminobutyric acid (GABA) at the level of the chloride channel (i.e., it is said to jam chloride channels). The drug is obsolete medically. Pharmacologically it is used as an aid in determining how certain sedative–hypnotics and anticonvulsants act at the molecular level.

Pentylenetetrazol. 6,7,8,9-Tetrahydro-5*H*-tetrazoloazepine; 1,5-pentamethylenetetrazole (Metrazol) has been used in conjunction with the electroencephalograph to help locate epileptic foci. It is routinely used as a laboratory tool in determining potencies of potential anticonvulsant drugs in experimental animals. The drug may act as a convulsant by interfering with chloride conductance.[2] Overall, it appears to share similar effects on chloride conductance with several other convulsive drugs, including picrotoxin.

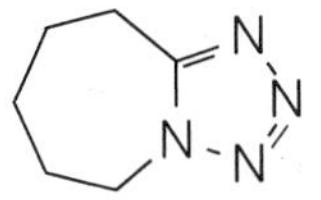

Pentylenetetrazol

Nikethamide. *N,N*-Diethylnicotinamide (Coramine) appears to act by facilitating excitatory processes rather than by depressing inhibitory ones. The overall effect resembles that of an amphetamine more than that of a drug such as picrotoxin.

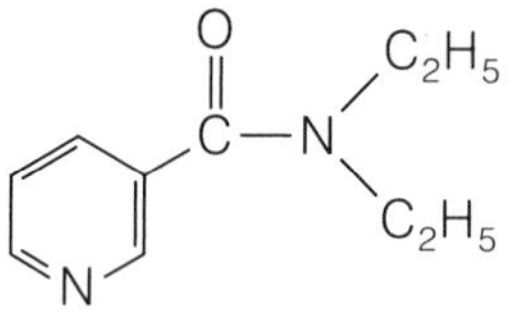

Nikethamide

It is possible to stimulate respiration with the drug without inducing generalized CNS stimulation. However, selectivity is still very low. The drug is obsolete in managing poisoning from sedative–hypnotic drugs. It may have a very limited place in treating acute respiratory insufficiency in COPD. It may also have value in correcting respiratory depression caused by oxygen therapy in COPD.

Doxapram Hydrochloride, USP. 1-Ethyl-4-(2-morpholinoethyl)-3,3-diphenyl-2-pyrrolidinone hydrochloride hydrate (Dopram). Doxapram has CNS stimulant properties resembling those of nikethamide more than those of picrotoxin or pentylenetetrazol. It appears to have greater selectivity as a respiratory stimulant than nikethamide, but symptoms of generalized CNS stimulation are still frequent. Uses of doxapram are as described for nikethamide.

Category: respiratory stimulant
Usual dose range: IV, 1 to 15 mg/kg of body weight
Occurrence: Doxapram Hydrochloride Injection, USP

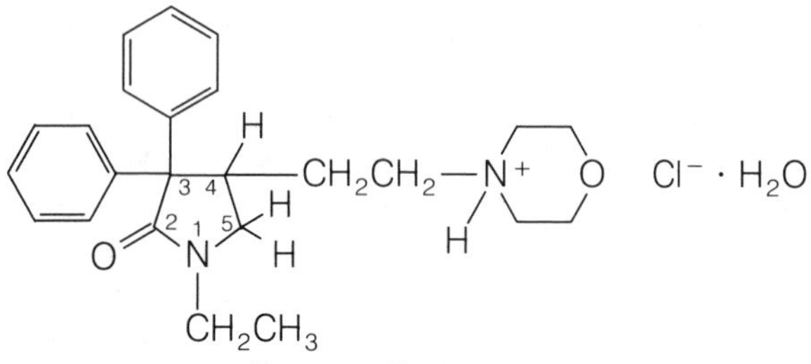

Doxapram Hydrochloride

METHYLXANTHINES

The naturally occurring methylxanthines are caffeine, theophylline, and theobromine. Refer to Table 10-1 for their structures and occurrence and to Table 10-2 for their relative potencies.

Caffeine enjoys wide use as a CNS stimulant. Theophylline has some use as a CNS stimulant (as will be discussed later); its CNS stimulant properties are more often encountered as side effects, sometimes severe and potentially life-threatening, of its use in bronchial asthma therapy. Theobromine has very little CNS activity and will not be discussed further as a CNS stimulant. Its 1-[5-oxohexyl] derivative pentoxifylline (Trental) is useful in intermittent claudication, presumably, in part, by improving red blood cell deformability. Its potential use in other occlusive disorders, such as in acute stroke, is under investigation.

Caffeine is often used as it occurs in brewed coffee (~ 85 mg/cup), brewed tea (~ 60 mg/cup), and cola beverages (~ 50 mg/12 fl oz). In most subjects 85 to 250 mg of caffeine acts as a cortical stimulant and facilitates clear thinking, wakefulness, promotes an ability to concentrate on the task at hand, and lessens fatigue. As the dose is increased, side effects indicative of excessive stimulation, such as restlessness, anxiety, nervousness, and tremulousness, become more marked. (They may be present in varying degrees at lower dose levels.) With further increases in dosage, convulsions can occur.

TABLE 10-1

XANTHINE ALKALOIDS

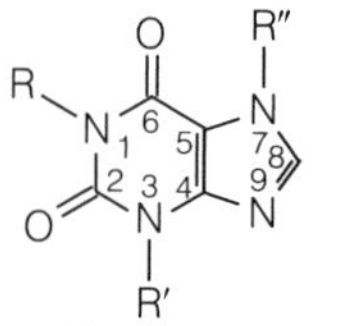

Xanthine
(R, R′ & R″ = H)

Compound	*R*	*R′*	*R″*	*Common Source*
Caffeine	CH_3	CH_3	CH_3	Coffee, Tea
Theophylline	CH_3	CH_3	H	Tea
Theobromine	H	CH_3	CH_3	Cocoa

TABLE 10-2

RELATIVE PHARMACOLOGIC POTENCIES OF THE XANTHINES

Xanthine	*CNS Stimulation*	*Respiratory Stimulation*	*Diuresis*	*Coronary Dilatation*	*Cardiac Stimulation*	*Skeletal Muscle Stimulation*
Caffeine	1*	1	3	3	3	1
Theophylline	2	2	1	1	1	2
Theobromine	3	3	2	2	2	3

*1 = most potent.

Theophylline's CNS effects at the lower dose levels have been little studied. At high doses, the tendency to produce convulsions is greater for theophylline than for caffeine.

In addition to being cortical stimulants, theophylline and caffeine are medullary stimulants, and both are used in treating sleep apnea in preterm infants. Caffeine (as caffeine and sodium benzoate) may be used rarely in treating poisoning from CNS depressant drugs, although it is not a preferred choice.

The important use of theophylline and its preparations in bronchial asthma is discussed elsewhere. Caffeine also is reported to have valuable bronchodilating properties. Finally, caffeine has value, presumably because of central vasoconstrictive effects, in treating migraine and tension headaches.

The basis for the CNS-stimulating effects of methylxanthines has often been attributed to their phosphodiesterase-inhibiting ability. For example, they retard the metabolism of cyclic adenine monophosphate (cAMP). This action is now considered probably irrelevant at therapeutic doses. The evidence now indicates that the CNS stimulant action is related more to the ability of these compounds to antagonize adenosine at A_1 receptors.[3-5]

Caffeine and theophylline are chemically interesting. Both are weak Bronsted bases. The reported pK_as are 0.8 and 0.6 for caffeine and 0.7 for theophylline. These pK_as represent the basicity of the imino nitrogen at position 9. As acids, caffeine has a pK_a over 14, and theophylline a pK_a of 8.8. In theophylline, a proton can be donated from position 7 (i.e., it can act as a Bronsted acid). Caffeine cannot donate a proton from position 7 and does not act as a Bronsted acid at pHs under 14. Caffeine does have electrophilic sites at positions 1, 3, and 7. In addition to its Bronsted acid site at 7, theophylline has electrophilic sites at 1 and 3. In condensed terms, both compounds are electron pair donors, but only theophylline is a proton donor in most pharmaceutic systems.

Although both compounds are quite soluble in hot water (e.g., caffeine 1 : 6 at 80°C), neither compound is very soluble in water at room temperature (caffeine about 1 : 40; theophylline about 1 : 120). Consequently, a variety of mixtures or complexes designed to increase solubility are available [e.g., citrated caffeine, caffeine and sodium benzoate, and theophylline ethylenediamine compound (aminophylline)].

Caffeine in blood is not highly protein bound, whereas theophylline is about 50% bound. Differences in the substituent at the 7-position may be involved. Additionally, caffeine is more lipophilic than theophylline and, reputedly, achieves higher brain concentrations. The half-life for caffeine is five to eight hours and for theophylline about 3.5 hours. About 1% of each compound is excreted unchanged. The compounds are metabolized in the liver. The major metabolite of caffeine is 1-methyluric acid and of theophylline, 1,3-dimethyluric acid.[6] Neither compound has been reported to be metabolized to uric acid, and they are not contraindicated in gout.

Occurrence	Percent Caffeine
Caffeine and sodium benzoate injection USP	45–52
Ergotamine tartrate and caffeine suppositories USP	
Ergotamine tartrate and caffeine tablets USP	

Caffeine and theophylline are both marketed in a variety of tablet strengths. Citrated caffeine is available in tablet form. Theophylline ethylenediamine is available in a variety of enteral and parenteral forms. Other complexes and derivatives of theophylline are available.

CENTRAL SYMPATHOMIMETIC AGENTS (PSYCHOMOTOR STIMULANTS)

Sympathomimetic agents whose effects are manifested mainly in the periphery are discussed in Chapter 11. A few simple structural changes in these peripheral agents produces compounds more resistant to metabolism and better able cross the blood–brain barrier. These effects increase the proportion of central to peripheral activity, and the agents are somewhat arbitrarily designated as central sympathomimetic agents.

In addition to CNS-stimulating effects, manifested as excitation and increased wakefulness, many central sympathomimetics exert an anorexient effect. Central sympathomimetic (noradrenergic) action is often the basis for these effects. However, other central effects, notably dopaminergic and serotoninergic, can be operative.[7] In some agents, the proportion of excitation and increased wakefulness to anorexient effects is decreased, and the agents are marketed as anorexients. Actually, in one drug, fenfluramine, acting as an anorexient by predominately serotoninergic mechanisms, sedation and drowsiness, rather than excitation, are typically seen. Representative structures of this group of compounds are given in Table 10-3. Additionally, the structure of the anorexients phenmetrazine, phendimetrazine, mazindol, and the alerting agents methylphenidate and pemoline, useful in attention-deficient disorders, are given in the text.

TABLE 10-3

SYMPATHOMIMETICS WITH SIGNIFICANT CENTRAL STIMULANT ACTIVITY

Generic Name	Base Structure		
Amphetamine	C_6H_5–C(H)(H)–	–C(H)(CH_3)–	–NH(H)
Methamphetamine	H	H	CH_3
Phentermine	H	CH_3	H
Benzphetamine	H	H	CH_3, $CH_2C_6H_5$
Diethylpropion	O*	H	C_2H_5, C_2H_5
Fenfluramine	3-CF_3-C_6H_4–C(H)(H)–	–C(H)(CH_3)–	–NH(CH_2CH_3)
Chlorphentermine	4-Cl-C_6H_4–C(H)(H)–	–C(CH_3)(CH_3)–	–NH(H)
Clortermine	2-Cl-C_6H_4–C(H)(H)–	–C(CH_3)(CH_3)–	–NH(H)

*Carbonyl.

Structural features for many of the agents can be easily visualized by considering that within their structure they contain a β-phenethylamine moiety, and this grouping can give some selectivity for pre- or postsynaptic noradrenergic systems.

β-Phenethylamine, given peripherally, is without central activity. Facile metabolic inactivation by monoamine oxidases is responsible. Branching with lower alkyl groups on the carbon atom adjacent (α) to the amino nitrogen increases CNS, rather than peripheral, activity (e.g., amphetamine, presumably by retarding metabolism). The α-branching generates a chiral center. The *dextro*(*S*)-isomer of amphetamine is up to 10 times as potent as the *levo*(*R*)-isomer for alerting activity and about twice as active as a psychotomimetic agent.[8] Hydroxylation of the ring or hydroxylation on the β-carbon (to the nitrogen) decreases activity, in large measure by decreasing ability to cross the blood–brain barrier. For example, phenylpropanolamine, with a β-OH, has about 0.001th the ability to cross the blood–brain barrier of its deoxy congener, amphetamine.

Halogenation (F, Cl, Br) of the aromatic ring decreases sympathomimetic activity. Other activities may increase. *p*-Chloroamphetamine has strong central serotoninergic activity (and is also a neurotoxin, destroying serotoninergic neurons in experimental animals).[9,10]

Methoxyl substitution on the ring tends to produce psychotomimetic agents, suggesting trophism for dopaminergic (D-2) receptors.

N-Methylation increases activity, as with methamphetamine compared with dextroamphetamine. Di-*N*-methylation decreases activity. Mono-*N*-substituents larger than methyl decrease excitatory properties, but many compounds retain anorexient properties. Consequently, some of these agents are useful anorexients, with decreased abuse potential relative to amphetamine.

There can be some departure from the basic β-phenethylamine structure when compounds act by indirect noradrenergic mechanisms such as block of norepinephrine (NE) uptake, as with cocaine, mazindol, and many tricyclic antidepressants. However, a concealed β-phenethylamine structure can be considered to also be present in these compounds.

The abuse potential of the more euphoriant and stimulatory of the amphetamines and amphetamine-like drugs, such as cocaine, is well documented. They produce an exceedingly destructive addiction. Apparently, both a euphoric "high" (possibly related to effects on dopaminergic systems) and a posteuphoric depression contribute to compulsive drug use in these agents.

Recognized medical indications for dextroamphetamine and some very close congeners include narcolepsy, Parkinson's disease, attention deficit disorders and, although not the preferred agents, as anorexients. In some conditions, such as in Parkinson's disease, for which its main use is to decrease rigidity, the antidepressant effects of dextroamphetamine can be beneficial. It has also been reported to be an effective antidepressant in terminal malignancies. However, in almost all cases of depression, and especially in major depressive disorders of the unipolar type, dextroamphetamine has been superseded by other agents, notably the monoamine oxidase inhibitors and the tricyclic and mechanistically related antidepressants.

The compounds and certain of their metabolites can have complex multiple actions. In a fundamental sense, however, the structural basis for actions is quite simple. The compounds and their metabolites resemble NE and can participate in the various neuronal and postsynapatic processes involving NE, such as synthesis, release, uptake, and pre- and postsynaptic receptor activation. Also, because dopamine (DA) and to a lesser extent serotonin

(5-HT) bear a structural resemblance to NE, processes in DA- and 5-HT-activated systems also can be affected.

PRODUCTS

Amphetamine Sulfate, USP. α-Methylbenzeneethanamine sulfate; ($\pm$)-1-phenyl-2-aminopropane (Benzedrine). The racemic mixture has a higher proportion of cardiovascular effects than the *dextro*-isomer. For most medical uses, the dextrorotatory isomer is preferred.

Dextroamphetamine Sulfate, USP and Dextroamphetamine Phosphate. (+)-(*S*)-Methylphenethylamine salts with sulfuric acid (Dexedrine) and with phosphoric acids, respectively. The phosphate salt is the more water-soluble salt and is preferred if parenteral administration is required. The dextrorotatory isomer has the (*S*) configuration and has less cardiovascular effects than the levorotatory (*R*) isomer. Additionally, it may be up to ten times as potent as the (*R*) isomer as an alerting agent and about twice as potent as a psychotomimetic agent. Although it is more potent as a psychotomimetic agent than the (*R*) isomer, it has a better ratio of alerting to psychotomimetic effects.

The alerting actions appear to relate to increased NE available to interact with postsynaptic receptors. The major mode of action of dextroamphetamine is release of NE from the nerve terminal.[7] Other mechanisms, such as inhibition of uptake, may make a small contribution to overall effects. The psychotomimetic effects are said to be linked to release of DA. Effects on 5-HT systems have also been linked to some behavioral effects of dextroamphetamine.

Dextroamphetamine is a strongly basic amine with a pK_a of 9.77 to 9.94. Absorption from the gastrointestinal tract occurs as the lipid-soluble amine. The drug is not extensively protein bound. Varying amounts of drugs are excreted intact under ordinary conditions. The amount is insignificant under conditions of alkaline urine. Under conditions producing systemic acidosis, 60% to 70% of the drug can be excreted unchanged. This fact can be used to advantage in treating drug overdose.

The α-methyl group retards, but does not terminate, metabolism by monoamine oxidases. Under most conditions, the bulk of a dose of dextroamphetamine is metabolized by *N*-dealkylation to phenylacetone and ammonia. Phenylacetone is further degraded to benzoic acid.

In experimental animals, about 5% of a dose accumulates in the brain, especially the cerebral cortex, thalamus, and the corpus callosum. It is first *p*-hydroxylated and then β-hydroxylated to produce *p*-hydroxynorephedrine, which may be the major active metabolite involved in NE and DA release.[11]

Methamphetamine Hydrochloride. (+)-*N*,α-Dimethylphenethylamine hydrochloride; deoxephedrine hydrochloride; (+)-1-phenyl-2-methylaminopropane hydrochloride (Desoxyn). Methamphetamine is the *N*-methyl analogue of dextroamphetamine. Methamphetamine has more marked central and decreased peripheral actions relative to dextroamphetamine. Methamphetamine has very high abuse potential, and by the intravenous route, its salts are known as "speed." Therapeutic uses of methamphetamine are analogous to those of dextroamphetamine.

Phentermine Ion-Exchange Resin (Ionamin); **Phentermine Hydrochloride, USP** (Wilpowr). The free base is α,α-dimethylphenethylamine; 1-phenyl-2-methylaminopropane. In the resin preparation, the base is bound with an ion-exchange resin to afford a slow-release product; the hydrochloride is a water-soluble salt.

Interestingly, because phentermine has a quaternary carbon atom with one methyl oriented analogously to the methyl of (*S*)-amphetamine and one methyl oriented analogously to the methyl of (*R*)-amphetamine, it is reported to have pharmacologic properties partaking of both the (*R*)- and (*S*)-isomers of amphetamine. The compound is used as an appetite suppressant and is a Schedule IV agent indicating less abuse potential than dextroamphetamine.

Chlorphentermine Hydrochloride. *p*-Chloro-α,α-dimethylphenethylamine hydrochloride; 1-(4-chlorophenyl)-2-methyl-2-aminopropane hydrochloride (Pre-Sate). Chlorphentermine is structurally interesting because of the *p*-chloro substituent on α,α-dimethyphenylethylamine. It is an effective anorexient with less abuse potential than dextroamphetamine.

Clortermine Hydrochloride. *o*-Chloro-α,α-dimethylphenethylamine hydrochloride; 1-(2-chlorophenyl)-2-methyl-2-aminopropane hydrochloride (Voranil). Clortermine is the *o*-chloro isomer of chlorphentermine. It is an effective appetite suppressant and, like its *p*-isomer, has less abuse potential than dextroamphetamine.

Benzphetamine Hydrochloride. (+)-*N*-Benzyl-*N*,α-dimethylphenethylamine hydrochloride; (+)-1-phenyl-2-(*N*-methyl-*N*-benzylamine)propane hydrochloride (Didrex). This compound is *N*-benzyl substituted methamphetamine. The large (benzyl) *N*-substituent decreases excitatory properties in keeping with the general SAR for the group. The compound has been observed to share mechanism-of-action characteristics with methylphenidate. The agent reduces appetite with fewer CNS excitatory effects than dextroamphetamine.

Diethylpropion Hydrochloride, USP. 2-(Diethylamino)propiophenone hydrochloride; 1-phenyl-2-diethylaminopropan-1-one hydrochloride (Tenuate, Tepanil). With two large (relative to H or methyl) *N*-alkyl substituents, diethylpropion has fewer cardiovascular and CNS stimulatory effects than amphetamine. It is said to be the anorexient agent best suited for the treatment of obesity in patients with hypertension and cardiovascular disease.

Fenfluramine Hydrochloride. *N*-Ethyl-α-methyl-*m*-(trifluoromethyl)-phenylethylamine HCl (Pondimin). Fenfluramine is unique in this group of drugs in that it tends to produce sedation rather than excitation. Effects are said to be mediated principally by central serotoninergic rather than central noradrenergic mechanisms. In large doses in experimental animals, the drug has been reported to be a serotonin neurotoxin.[12] The drug is favored in patients in whom CNS stimulation is to be avoided, and in weight reduction in non-insulin-dependent diabetes mellitis (NIDDM).

Phenmetrazine Hydrochloride, USP. 3-Methyl-2-phenylmorpholine hydrochloride (Preludin). A β-phenethylamine moiety is present in this Schedule II anorexient. The configuration was deduced by Clark.[13] Although less excitatory than dextroamphetamine, it still possesses abuse potential.

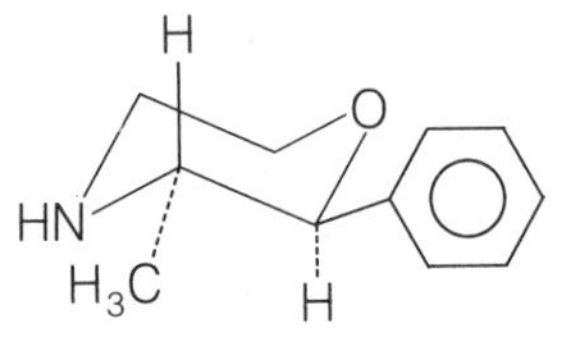

H HN O H_3C H

Phendimetrazine Tartrate, USP. (2*S*,3*S*)-3,4-Dimethyl-2-phenylmorpholine-L-(+)-tartrate (Plegine). This optically pure compound is considered an effective anorexient and is classed as a Schedule III compound. The stereochemistry of (+)phendimetrazine and (+)phenmetrazine are analogous.[14]

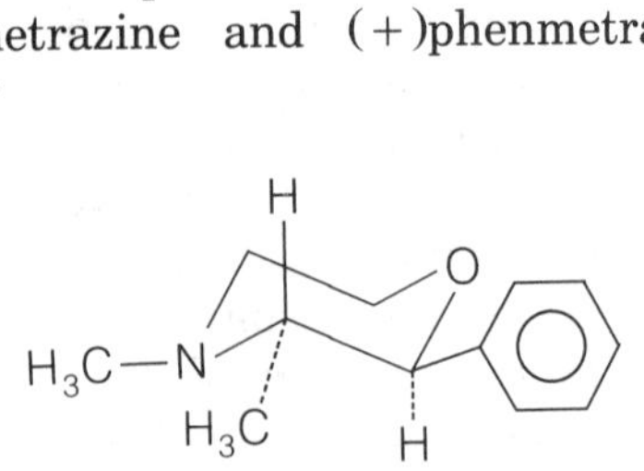

H H_3C—N O H_3C H

Mazindol, USP. 5-(*p*-Chlorophenyl)-2,5-dihydroxy-3*H*-imidazo[2,1-*a*]isoindole-5-ol (Sanorex). Mazindol can exist in two tautomeric forms. In acidic media, the imidazoisoindole structure is preferred. Seemingly, either structure is far removed from any resemblance to a protonated β-arylamine, but tak-

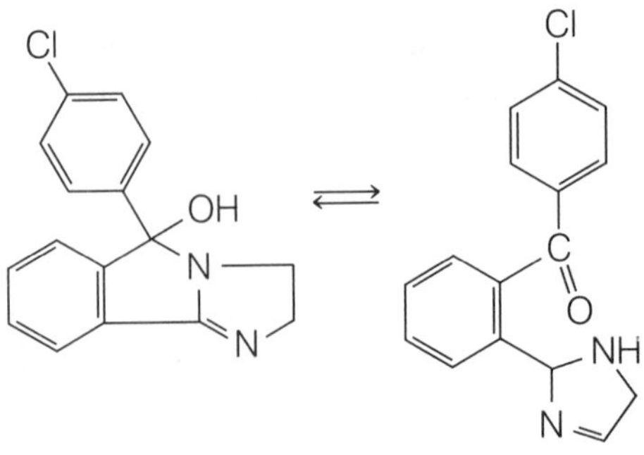

Cl OH N N ⇌ Cl C O NH N

Mazindol

ing the isoindole structure and protonating the cyclic amidino group produces a cationic center, about a two-atom distance and two aryl groups. This fits the structural features associated with block of neuronal NE and DA uptake. A major mode of action is block of NE uptake.[15] Mazindol is an effective anorexient.

Methylphenidate Hydrochloride, USP (Ritalin). There are two asymmetric centers in methylphenidate and four possible isomers. The *threo*-racemate is the marketed compound and is about 400 times as potent as the *erythro*-racemate.[16] The absolute configuration of each of the *threo*-methylphenidate isomers has been determined.[17] Considering that the structure is fairly complex, it is likely that one of the two components of the *threo*-racemate contains most of the activity. Evidence indicates that the *d*(2*R*,2′*R*) *threo*-isomer is principally involved in the behavioral and pressor effects of the racemate.[18] As is likely with many central psychomotor stimulants, the mode of action at the molecular level is multiple. Methylphenidate or its *p*-hydroxy metabolite, or both, blocks NE uptake; acts as a postsynaptic agonist; depletes the same NE pool as reserpine; and also has effects on DA systems such as a block of DA uptake.

Methylphenidate is an ester drug and has interesting pharmacokinetic properties arising from structure. The pK_a for the drug is reported as 8.5 and 8.8. The protonated form in the stomach reportedly resists ester hydrolysis. Absorption of the intact drug is very good. However, 80% to 90% of the drug is rapidly hydrolyzed to inactive ritalinic acid after absorption from the gastrointestinal tract.[19] (The extent of hydrolysis may be about fivefold that for *d* versus *l*.[20]) Another 2% to 5% of the racemate is oxidized by liver microsomes to the inactive cyclic amide. About 4% of a dose of the racemate reaches the brain in experimental animals and there is *p*-hydroxylated to yield the putative active metabolite.

Methylphenidate is a potent CNS stimulant. Indications include narcolepsy and the attention deficit disorder. The structure of the (2*R*,2′*R*)-isomer of the *threo*-racemic mixture is shown here.

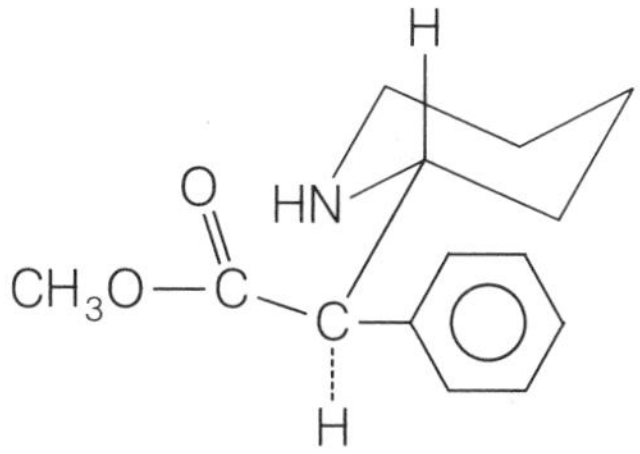

Pemoline. 2-Amino-5-phenyl-4(5*H*)-oxazolone (Cylert). The structure of this compound is unique as can be seen from its depiction. The compound is

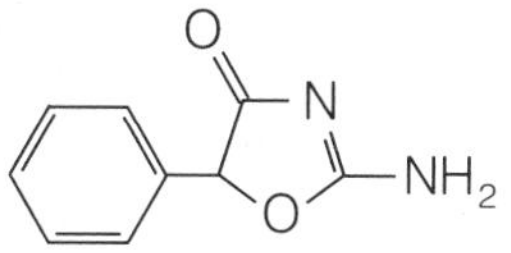

Pemoline

described as having an overall effect on the CNS similar to methylphenidate. However, the agent requires three to four weeks of administration to take effect. A partial explanation for the delayed effect may be that one of the actions of the agent, as observed in rats, is to increase the rate of synthesis of dopamine.

MONOAMINE OXIDASE INHIBITORS

Antidepressant therapy usually implies therapy directed against major depressive disorders of the unipolar type and is centered around two groups of chemical agents–the monoamine oxidase inhibitors (MAOI) and the tricyclic and mechanistically related antidepressants—and electroshock therapy. The highest cure or remission rate is achieved with electroshock therapy. In some patients, especially those who are suicidal, this is the treatment of choice. Monoamine oxidase inhibitors and tricyclic antidepressants have about the same response rate (about 60% to 70%). In the United States, the tricyclic antidepressants are considered the agents of choice in most antidepressant therapy.

Side effects of tricyclics include troublesome anticholinergic effects and effects of noradrenergic stimulation. In overdose, the combination of effects can be lethal. A severe problem associated with the MAOI, and one that is a major factor in relegating them to second-line drug status, originates in the fact that most presently available compounds inhibit liver monoamine oxidases (MAO) in addition to brain MAO, thereby allowing dietary pressor amines that would normally be inactivated to exert their effects. A number of severe hypertensive responses, sometimes fatal, have followed ingestion of foods high in pressor amines. It has been hoped that the development of agents, such as deprenyl, that presumably spare liver MAO might solve this problem. Another prominent side effect of the MAOI is orthostatic hypotension. It is said to arise from a bretyliumlike action. Actually, one MAOI, pargyline, is used clinically for its hypotensive action. Also, some of the first compounds produced serious hepatotoxicity. Compounds available today are much safer in this regard.

The history of the development of the MAOI illustrates the role of serendipity. Isoniazid is an effective antitubercular agent, but is a very polar compound. To gain better penetration into the *Mycobacterium tuberculosis* organism, a more hydrophobic compound, isoniazid substituted with an isopropyl group on the basic nitrogen (iproniazide), was designed and synthesized. It was introduced into clinical practice as an effective antitubercular agent. However, it was noted that CNS stimulation occurred, and the drug was withdrawn. Later it was determined in experimental animals, as well as in vitro experiments with purified MAO, that MAO inhibition, resulting in higher levels of NE and 5-HT, could account for the CNS effects. Thereafter, the compound was reintroduced into therapy as an antidepressant agent. It stimulated an intensive interest in hydrazines and hydrazides as antidepressants and inaugerated effective drug treatment of depression.[21] It continued in therapy several years, but eventually was withdrawn because of hepatotoxicity.

The currently available compounds can be considered to be mechanism-based inhibitors of MAO.[22] They are converted by MAO to agents that inhibit the enzyme. Inhibition by the agents is often irreversible. All appear to have the ability to form reactants that covalently bond with the enzyme or its cofactor. A consequence of irreversible inactivation is that the action of the agents may continue for up to two weeks after their administration is discontinued. Consequently, many drugs degraded by MAO cannot be administered during that time.

It is possible to have agents that act by competitive enzyme inhibition. This may be true for some of the presently clinically used agents before the production of irreversible noncompetitive inhibition. The harmala alkaloids, harmine and harmaline, are thought to act primarily as CNS stimulants by competitive inhibition of MAO. Reversible inhibitors selective for each of the two major MAO subtypes (A and B) are forthcoming. Reversible inhibitors of MAO-A reportedly are antidepressant without production of hypertensive crises. Presently, selective MAO-B inhibition has failed to correlate positively with antidepressant activity.

Most of the clinically useful MAOI antidepressants are nonselective between inhibiting metabol-

TABLE 10-4

MONOAMINE OXIDASE INHIBITORS

Generic Name Proprietary Name	*Structure*
Phenelzine sulfate USP *Nardil*	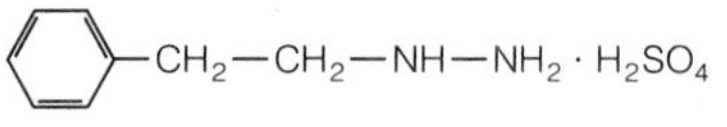 C_6H_5—CH_2—CH_2—NH—$NH_2 \cdot H_2SO_4$
Isocarboxazid USP *Marplan*	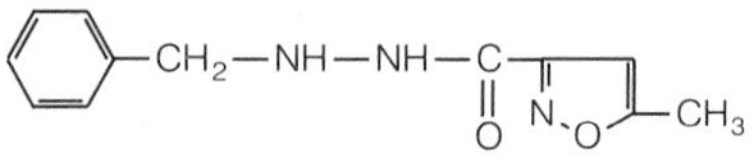 C_6H_5—CH_2—NH—NH—C(=O)—(5-methyl-3-isoxazolyl, CH_3)
Tranylcypromine sulfate USP *Parnate*	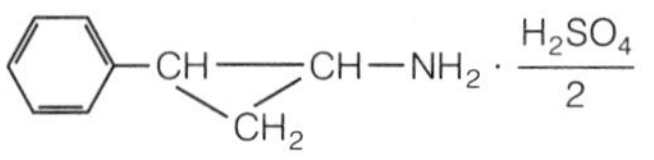 C_6H_5—CH—CH—$NH_2 \cdot \frac{H_2SO_4}{2}$ (ring closed by CH_2)
Pargyline hydrochloride USP *Eutonyl*	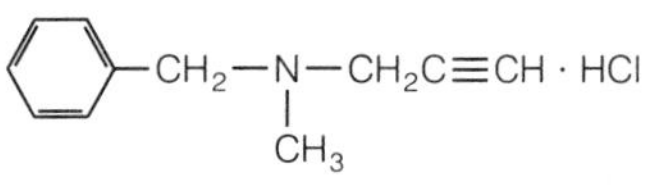 C_6H_5—CH_2—N(CH_3)—$CH_2C{\equiv}CH \cdot HCl$

ism of NE and 5-HT. Agents selective for an MAO that degrade 5-HT (e.g., clorgyline) have been under study for some time. The structures of the three clinically used antidepressant MAOI and of the MAOI used in hypertensive therapy are given in Table 10-4.

PRODUCTS

Phenelzine Sulfate, USP. 2-(Phenylethyl)hydrazine sulfate (Nardil). Phenelzine is an effective antidepressant agent. A mechanism-based inactivator, it irreversibly inactivates the enzyme or its cofactor, presumably after oxidation to the diazine, which can then break up into molecular nitrogen, a hydrogen atom, and a phenethyl free radical. The latter would be the active species in irreversible inhibition.[23]

Isocarboxazid, USP. 5-Methyl-3-isoxazolecarboxylic acid 2-benzylhydrazide; 1-benzyl-2-(5-methyl-3-isoxazolylcarbonyl)hydrazine (Marplan). It is generally held that hydrazides such as isocarboxazide are prodrugs: hydrolysis of the acyl group yields the active hydrazine. Thereafter, the action of the enzyme is thought to produce the active irreversibly inhibiting agent as described in the discussion of phenylzine. For isocarboxazid, the agent would be the corresponding benzyl radical.

Tranylcypromine Sulfate, USP. (±)-*trans*-2-Phenylcyclopropylamine sulfate (Parnate). Tranylcypromine was synthesized to be an amphetamine analogue (visualize the α-methyl of amphetamine condensed onto the β-carbon atom).[24] It does have some amphetaminelike properties that probably relate to the fact that the agent has more immediate CNS stimulant effects than do agents that act by MAO inhibition alone. For MAO inhibition, there may be two components to the action. One component is thought to act because the agent has structural features (the basic nitrogen and the *quasi-π* character of the α- and β-cyclopropane carbon atoms) that approximate the transition state in a route of metabolism of β-arylamines.[25,26] As α- and β-hydrogen atoms are removed from the substrate, quasi-π character develops over the α–β-carbon system. Duplication of the transition state permits extremely strong, but reversible, attachment to the enzyme. Additionally, tranylcypromine is a mechanism-based inactivator. It is metabolized by MAO with one electron of the nitrogen pair lost to flavin. This, in turn, produces a homolytic fission of a carbon–carbon bond of cyclopropane, one electron from the fission pairing with the remaining lone nitrogen electron to generate an imine (protonated), the other residing on a methylene carbon. Thus, a free radical ($C_6H_6\dot{C}HCH_2C{=}NH_2^+$) is formed that reacts covalently with either the enzyme or with reduced flavin, in either event inactivating the enzyme.[27]

Pargyline Hydrochloride, USP. *N*-Methyl-*N*-2-propynylbenzylamine hydrochloride (Eutonyl). Pargyline is said to be specific for an MAO that prefers phenylethylamine substrates rather than 5-HT substrates. Although it has CNS-stimulating properties, it is usually used for its hypotensive properties. Hypotensive properties for the MAOI are said, as stated earlier, to arise from a bretyliumlike effect and need not be a consequence of MAO inhibition. To exemplify, *N*-demethylpargyline is an active MAO inhibitor, but reportedly is not hypotensive.

Pargyline, as are other propargylamines, is a mechanism-based inactivator of MAO. Pargyline is converted by the enzyme to the Michael acceptor below which can react with the Michael donor, the number 5 nitrogen of flavin.[22]

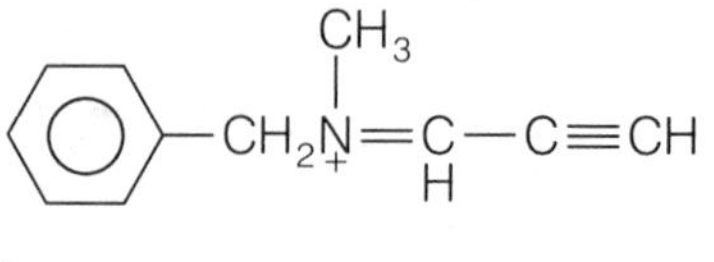

$$C_6H_5{-}CH_2\overset{+}{N}(CH_3){=}CH{-}C{\equiv}CH$$

L- or (*S*)-Deprenyl.

$$C_6H_5{-}CH_2{-}CH(CH_3){-}N(CH_3){-}CH_2{-}C{\equiv}CH$$

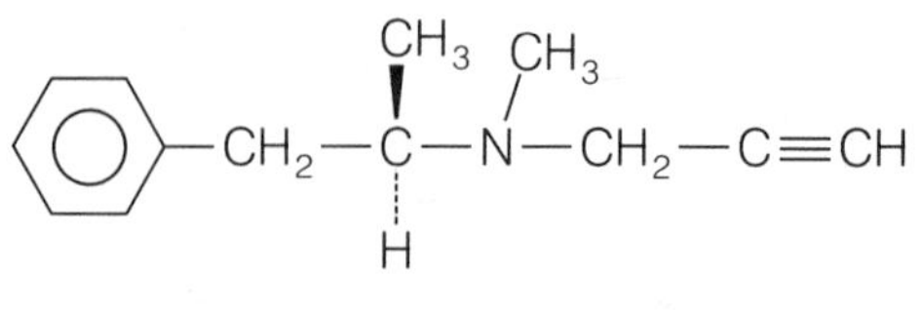

L-Deprenyl has received attention in recent years as an agent that might spare liver MAO and, thereby, avoid the "cheese effect" or hypertensive response to tyramine and other dietary pressor amines. There appears to be avoidance of the effect, but it has been suggested that an effect on tyramine other than by sparing of liver MAO may be involved. The agent has been employed in Parkinson's disease to decrease the severity of the "on–off" phenomenon, one of a variety of abrupt changes in motor status believed to be caused by fluctuations in DA levels. It has been introduced into therapy as selegiline (Eldepryl). It has also undergone trials in antidepressant therapy.

Clorgyline and Lilly 51641. In both of these compounds, a halogenated phenyl in conjunction with the amino nitrogen produces selectivity for 5-HT-metabolizing enzymes. The four-atom chain between the aromatic ring and nitrogen could fold to a two-atom chain when the nitrogen is protonated and the proton can H-bond with the ether oxygen. Finally, the nitrogen is substituted with a group that produces a mechanism-based inactivating effect, in one case a propargyl group and in the other a cyclopropyl group. The agents have undergone trials as potential antidepressant agents.

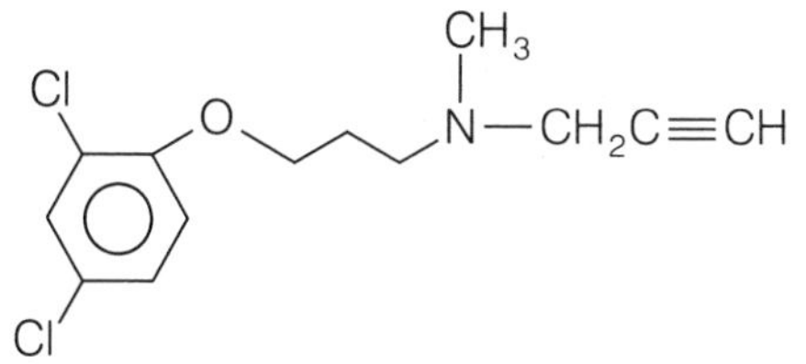

Clorgyline

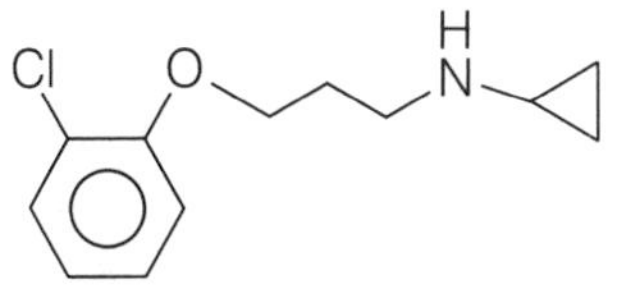

Lilly 51641

TRICYCLIC (AND MECHANISTICALLY RELATED) ANTIDEPRESSANT COMPOUNDS

This heading attempts to bring under one umbrella what was originally a small group of closely related agents (the "tricyclics" or "TCA"), but that, with time, has become a chemically diverse group of agents. Almost all of the agents block neuronal uptake of NE, 5-HT, and DA. Some agents appear not to block uptake of these transmitters, but still share with the uptake (reuptake) blockers the property of increasing synaptic availability of NE, 5-HT, and DA. As to which amine is mainly responsible for antidepressant activity, various opinions exist. Many references have implied, over recent years, that 5-HT would be shown to be primarily responsible.[28] Others have cited NE as the likely prime agent.[29] Selective serotonin reuptake inhibitors are proving to be effective antidepressants. Some feel a combination of the two is responsible. The involvement of DA in antidepressant effects is less certain. Some investigators would exclude it altogether. One basis for exclusion is that drugs that act largely by blocking uptake of DA, such as methylphenidate and cocaine, tend to produce euphoria or dysphoria and psychotic effects and do not have highly beneficial antidepressant properties.[29]

The structure–activity relationships (SARs) for the TCA are compiled in detail in the eighth edition of this text.[30] The interested reader is referred to this compilation. Overall, these SARs appear to be summarizable as a large bulky group encompassing two aromatic rings, preferably held in a skewed arrangement by a third central ring and a three- or sometimes two-atom chain to an aliphatic amino group that is monomethyl- or dimethyl-substituted. The features can be visualized by consulting the structures of imipramine and desipramine as examples. The overall arrangement has features within it approximating a fully extended *trans* conformation of the β-aryl amines. To relate these features to the mechanism of action,—block of reuptake—it can be visualized that the same basic arrangement is present as is found in the β-arylamines, plus extra bulky groups that block the uptake process. The overall concept of a β-arylaminelike system with added structural bulk appears to be applicable to many newer compounds that do not have a tricyclic grouping.

Still another way of rationalizing the SARs may be useful. Many of the antidepressant drugs are close structural relatives of postsynaptic DA (antipsychotic) and NE (sedative) blockers. Conceivably, as a consequence of small structural changes, an agent can begin to gain the ability to block a presynaptic event (uptake) and then lose ability to effect a postsynaptic block. In fact, in many tricyclics there is some retention of postsynaptic effects. Many antidepressants retain appreciable antipsychotic and sedative properties, which may largely be due to postsynaptic DA block and postsynaptic NE block, respectively.

The TCA and related agents are usually the preferred agents for treatment of major depressions. Some of the newer agents differ in side effects from the TCA and these side effects will be discussed under the headings for the individual agents. The TCA are structurally related to each other and, consequently, possess related biologic properties that can be summarized as characteristic of the group.

The dimethylamino compounds tend to be sedative, whereas the monomethyl relatives tend to be stimulatory. The dimethyl compounds tend toward higher ratios of 5-HT/NE uptake block; in the monomethyl, the proportion of NE uptake block tends to be higher. The compounds have anticholinergic properties, and these are usually higher in the dimethylamino compounds. When treatment is begun with a dimethyl compound, with time there is a significant accumulation of the monomethyl compound as *N*-demethylation proceeds.

As with the MAOI, there is a time lag before antidepressant effects are seen. It was once thought that the antidepressant effect depended on the buildup of the nor-metabolite. This is now known not to be true. Current views on the time lag focus on down-regulation or decrease in sensitivity of receptors (α_2, β-, 5-HT_2) as a sequela of increased levels of NE and 5-HT being responsible for actual antidepressant action.[31–35]

The TCA are extremely lipophilic and, accordingly, are very highly tissue bound outside the CNS. The TCA do have anticholinergic and noradrenergic effects, both central and peripheral, that are often unpleasant and sometimes dangerous. In overdose, the combination of anticholinergic and antiadrenergic effects, as well as a quinidinelike cardiac depressant effect, can be lethal. Overdose is complicated because the agents are so highly protein bound that dialysis is ineffective.

PRODUCTS

Imipramine Hydrochloride, USP. 5-[3-(Dimethylamino)propyl]-10,11-dihydro-5*H*-dibenz[*b,f*]azepine monohydrochloride (Tofranil). Imipramine may be considered the parent compound of the TCA. It is also a close relative of the antipsychotic phenothiazines (replace the 10–11 bridge with sulfur and the compound is the discontinued antipsychotic agent promazine). Relative to promazine, it has weak D-2 postsynaptic blocking activity and mainly presynaptic effects on amines (5-HT, NE, and DA) are seen. As is typical of dimethylamino compounds, anticholinergic and sedative effects tend to be marked. The compound has a tendency toward a high 5-HT/NE uptake block ratio. Metabolic inactivation proceeds mainly by oxidative hydroxylation in the 2-position, followed by conjugation with glucuronic acid of the conjugate. Urinary excretion predominates (about 75%), but some biliary excretion (up to 25%) can occur, probably because of the large nonpolar grouping. Oxidative hydroxylation is not as rapid or complete as that of the more nucleophilic ring phenothiazine antipsychotics; consequently, appreciable *N*-demethylation with a buildup of nor- (or des-) imipramine occurs.

The demethylated metabolite is less anticholinergic, less sedative, more stimulatory, and has a higher NE than 5-HT uptake-blocking capability. Consequently, a patient treated with imipramine has present at least two major bioactive metabolites that contribute to activity. The activity of des- or norimipramine is terminated by 2-hydroxylation followed by conjugation and excretion. Of course, a second *N*-demethylation also can occur that, in turn, is followed by 2-hydroxylation, conjugation, and excretion.

Desipramine Hydrochloride, USP. 10,11-Dihydro-*N*-methyl-5*H*-dibenz[*b,f*]azepine-5-propanamine monohydrochloride; 5-(3-methylaminopropyl)-10,11-dihydro-5*H*-dibenz[*b,f*]azepine hydrochloride (Norpramin; Pertofrane). The structure of this agent as well as its salient properties are discussed under imipramine. In choosing an antidepressant drug for a patient, desipramine would be considered when few anticholinergic effects or a low level of sedation are important.

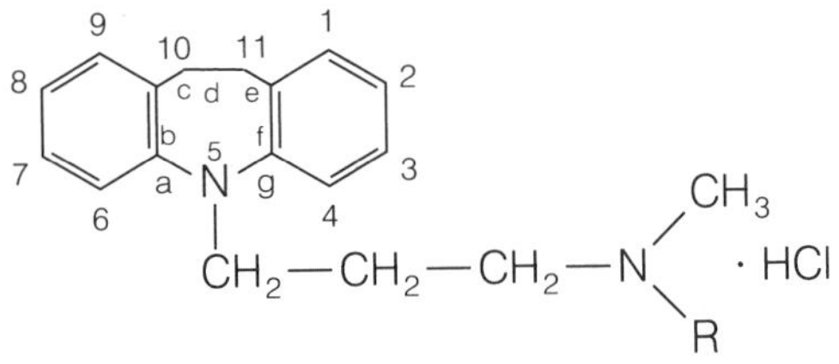

CH_2—CH_2—CH_2—N(CH_3)(R) · HCl

Imipramine: R = CH_3
Desipramine: R = H

Amitriptyline Hydrochloride, USP. 3-(10,11-Dihydro-5*H*-dibenzo[*a,d*]cyclohepten-5-ylidene)-*N*,*N*-dimethyl-1-propanamine hydrochloride; 5-(3-dimethylaminopropylidene)-10,11-dihydro-5*H*-dibenzo[*a,d*]cycloheptene hydrochloride (Elavil). Amitriptyline is one of the most anticholinergic and sedative of the TCA. Because it lacks the ring electron-enriching nitrogen atom of imipramine, metabolic inactivation mainly proceeds not at the analogous 2-position, but at the benzylic 10-position (i.e., toluenelike metabolism predominates). Because of the 5-exocyclic double bond, *E*- and *Z*-hydroxy isomers are produced by oxidation metabolism. Conjugation produces excretable metabolites. As is typical of the dimethyl compounds, *N*-demethylation occurs, and nortriptyline is produced, which has a less anticholinergic, less sedative, and more stimulant action than amitriptyline.

Nortriptyline Hydrochloride, USP. 3-(10,11-Dihydro-5*H*-dibenzo[*a,d*]cyclohepten-5-ylidene)-*N*-methyl-1-propanamine hydrochloride; 5-(3-methylaminopropylidene)-10,11-hydro-5*H*-dibenzo[*a,d*] cycloheptene hydrochloride (Aventyl; Pamelor). Pertinent biologic and chemical properties for this agent are given in the foregoing discussion for amitriptyline. Metabolic inactivation and elimination are analogous with those of amitriptyline.

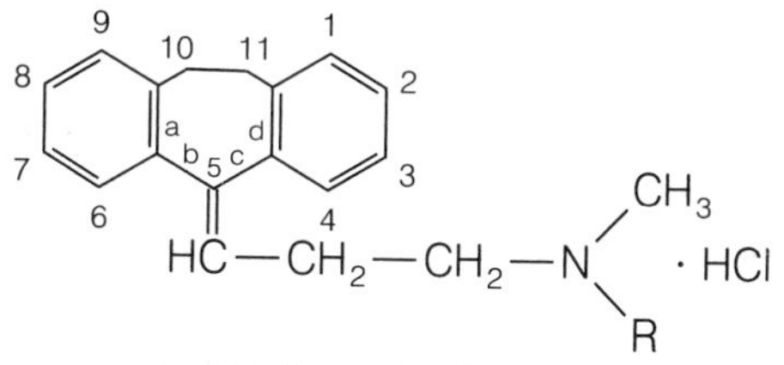

Amitriptyline: R = CH_3
Nortriptyline: R = H

Protriptyline Hydrochloride, USP. *N*-Methyl-5*H*-dibenzo[*a,d*]cycloheptene-5-propylamine hydrochloride; 5-(3-methylaminopropyl)-5*H*-dibenzo-[*a,d*]cycloheptene hydrochloride (Vivactil). As with the other compounds under consideration, protriptyline is an effective antidepressant. The basis for its chemical naming can be seen by consulting the naming and the structure for imipramine. It is a structural isomer of nortriptyline. Inactivation can be expected to involve the 10–11 double bond and the 2-position. Because it is a monomethyl compound, the sedative potential is low.

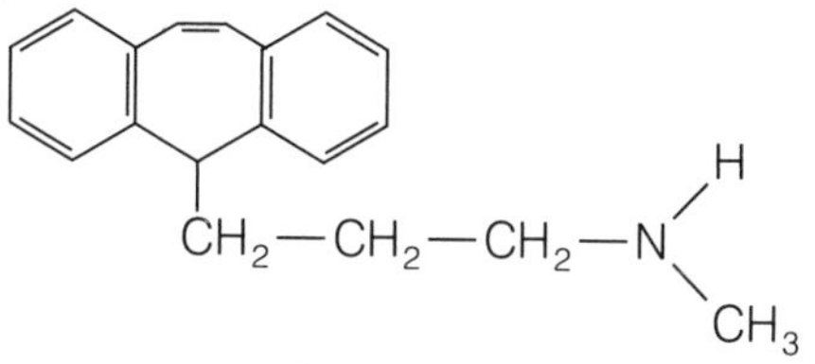

Protriptyline

Trimipramine Maleate (Surmontil). For details on chemical nomenclature, consult the description of imipramine. The replacement of hydrogen with an α-methyl substituent produces a chiral carbon, and the compound is employed as the racemic mixture. Biologic properties are said to resemble those of imipramine.

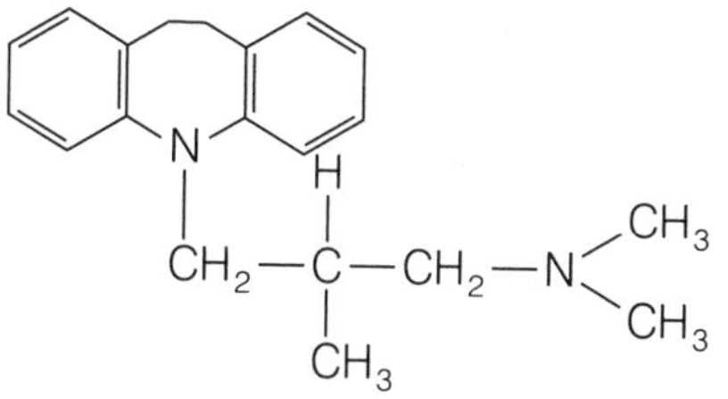

Trimipramine

Doxepin Hydrochloride, USP. 3-Dibenz[*b,e*]-oxepin-11(6*H*)ylidine-*N,N*-dimethyl-1-propanamine hydrochloride; *N,N*-dimethyl-3-(dibenz[*b,e*]oxepin-11(6*H*)-ylidene)propylamine (Sinequan; Adapin). Doxepin is an oxa congener of amitriptyline as can be seen from its structure.

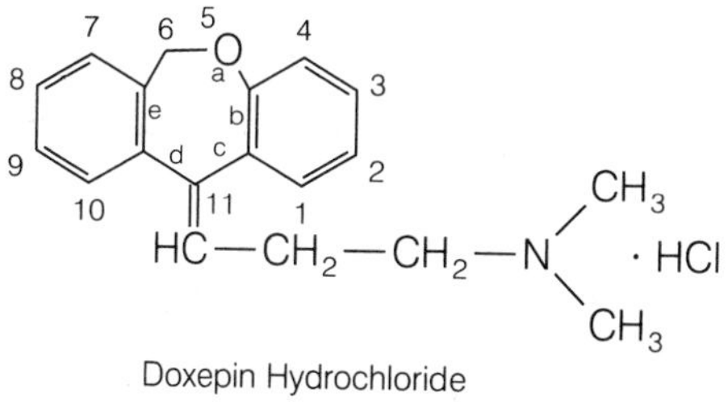

Doxepin Hydrochloride

The oxygen is interestingly placed and should influence oxidative metabolism as well as post- and presynaptic binding affinities. The *Z*-isomer is the more active isomer; however, the drug is marketed as the mixture of isomers. The drug is a NE- and 5-HT-uptake blocker and also has significant anticholinergic and sedative properties. It can be anticipated that the nor or des metabolite will contribute to the overall activity pattern.

Maprotiline Hydrochloride, USP. *N*-Methyl-9,10-ethanoanthracene-9(10*H*)-propanamine hydrochloride (Ludiomil). Maprotiline is sometimes described as a tetracyclic rather than a tricyclic antidepressant. The description is chemically accurate, but the compound, nonetheless, conforms to the overall TCA pharmacophore. It is a dibenzobicyclooctadiene and can be viewed as a TCA with an ethylene-bridged central ring. The compound is not strongly anticholinergic and has been noted to have stimulant properties. The drug can have effects on the cardiovascular system.

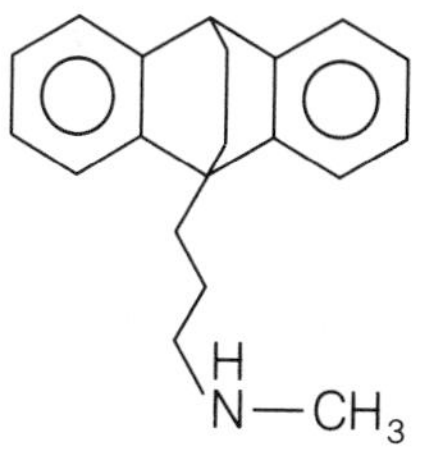

Amoxapine. 2-Chloro-11-(1-piperazinyl)dibenz-[*b,f*][1,4]oxazepine (Asendin). Consideration of the structure of amoxapine reinforces the observation that many TCA bear a very close resemblance to antipsychotics; indeed, some have significant effects at D-2 receptors. The *N*-methyl-substituted relative is the antipsychotic loxapine (Loxitane). It is reported that the 8-hydroxy metabolite of amoxapine is active as an antidepressant and as a D-2 receptor blocker, as is, for the latter, amoxapine.

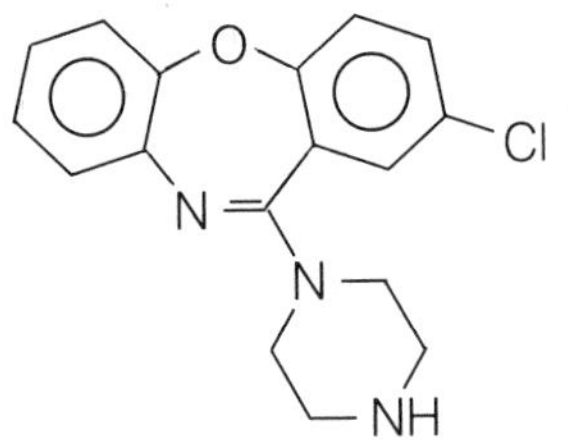

An agent, extensively employed in Europe (and recently introduced in the United States), is the very potent, on a milligram basis, 2-chloroimipramine (Anafranil). The close structural parallel with the antipsychotics (e.g., chlorpromazine) is evident. Anafranil is strongly sedative and a very strong 5-HT-uptake blocker. Its *N*-demethyl metabolite is reported to be both a 5-HT- and a NE-uptake blocker.

Trazodone Hydrochloride. 2-[3-[4-(3-Chlorophenyl)-1-piperazinyl]propyl]-1,2,4-triazolo[4,3-*a*]-pyridin-3(2*H*)-one (Desyrel). Consideration of the structure of trazodone reveals certain structural similarities with the fluorobutyrophenone antipsychotics, just as many tricyclic antidepressants relate to the phenothiazine and other tricyclic antipsychotics. Whereas the fluorobutyrophenone antipsychotics block DA postsynaptically, trazodone, as well as a major metabolite, *m*-chloro,4-phenylpiperazine, block presynaptic uptake of 5-HT. The agent is an effective antidepressant. As can be expected from the structure, unlike the TCA, anticholinergic effects are not usually a problem. The usual side effect is sedation.

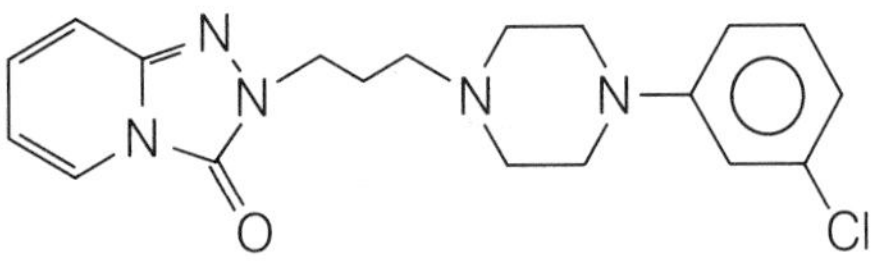

Finally, interesting examples of amine-uptake blockers that deviate from the fundamental TCA structure are **zimelidine** and **nomifensin**. Zimelidine was employed in Europe, but reportedly produced the Guillain-Barré syndrome and, accordingly, was withdrawn from use. Nomifensin was introduced in the United States, but was withdrawn because of reports of production of hemolytic anemia. The serotonin reuptake inhibitor, fluoxetine (Prozac) has received much favorable attention as an antidepressant. Additionally, it has been reported that α_2-adrenoreceptor antagonists (e.g., agents that antagonize clonidine) are potential antidepressants.[36]

Also, it appears that several β_2-NE stimulants (e.g., salbutamol and atenolol) are effective antidepressants.[37,38]

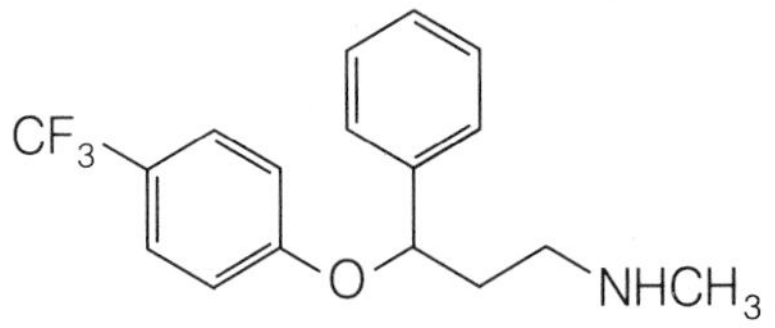

Fluoxetine

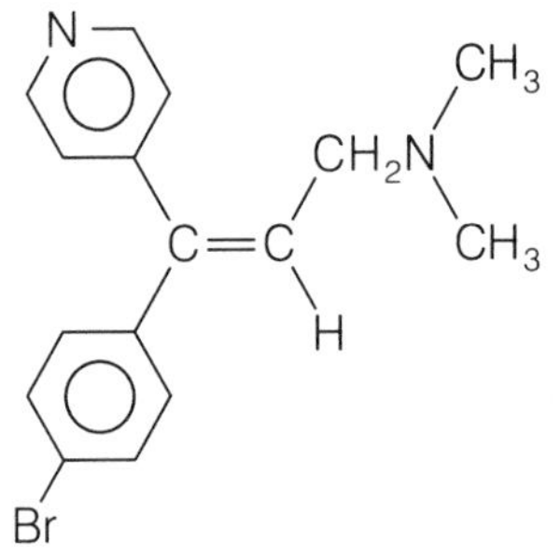

Zimelidine

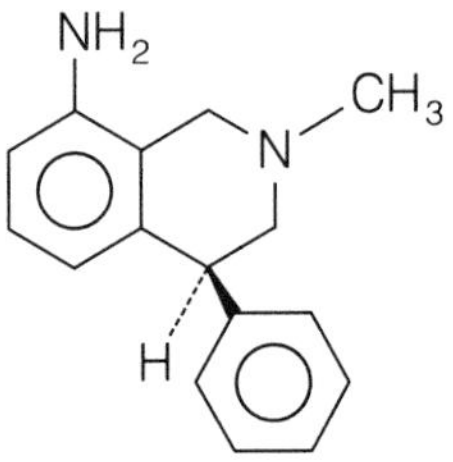

Nomifensin

Additional information on psychomotor stimulants, monoamine oxidase inhibitors, tricyclic (and mechanistically related) antidepressants is given in Table 10-5.

TABLE 10-5

PSYCHOMOTOR STIMULANTS, MONOAMINE OXIDASE INHIBITORS, AND TRICYCLIC (AND RELATED) ANTIDEPRESSANTS

Name Proprietary Name	***Preparations***	***Category***	***Usual Adult Dose****	***Usual Dose Range****	***Usual Pediatric Dose****
Amitriptyline hydrochloride USP *Elavil*	Amitriptyline hydrochloride injection	Antidepressant	IM, 20–30 mg qid	80–200 mg/day	Dosage is not established in children under 12 yr
	Amitriptyline hydrochloride tablets USP		25 mg bid to qid	30–300 mg/day	Dosage is not established in children under 12 yr
Amoxapine *Asendin*	Amoxapine tablets	Antidepressant	Initial, 150 mg/day	150–300 mg/day	Dosage is not established in children up to 16 yr

*See USP DI for complete dosage information.

TABLE 10-5 *Continued*

PSYCHOMOTOR STIMULANTS, MONOAMINE OXIDASE INHIBITORS, AND TRICYCLIC (AND RELATED) ANTIDEPRESSANTS

Name Proprietary Name	***Preparations***	***Category***	***Usual Adult Dose****	***Usual Dose Range****	***Usual Pediatric Dose****
Amphetamine phosphate, dextro sulfate USP *Dexedrine*	Amphetamine phosphate, dextro sulfate elixir USP Amphetamine phosphate, dextro sulfate tablets USP	Central stimulant	Narcolepsy: 5–20 mg qd to tid	2.5–60 mg/day	Hyperkinesia: children under 3 yr of age, use is not recommended; 3–5 yr, 2.5 mg qd, increased by 2.5 mg at weekly intervals; 6 yr and over, 5 mg qd or bid, increased by 5 mg at weekly intervals Narcolepsy: 6–12 yr, 2.5 mg bid, increased by 5 mg at weekly intervals; 12 yr and over, 5 mg bid, increased by 10 mg at weekly intervals
Amphetamine phosphate, dextro USP *Dextro-Profetamine*	Amphetamine phosphate, dextro tablets USP	Central stimulant	5 mg every 4–6 hr	5–10 mg	
Desipramine hydrochloride USP *Norpramin, Pertofrane*	Desipramine hydrochloride capsules USP Desipramine hydrochloride tablets USP	Antidepressant	150 mg/day in divided doses	50–200 mg/day	
Diethylpropion hydrochloride USP *Tenuate, Tepanil*	Diethylpropion hydrochloride tablets USP	Anorexic	25 mg tid		
Doxepin hydrochloride USP *Sinequan*	Doxepin hydrochloride capsules USP Doxepin hydrochloride oral solution USP	Antidepressant	Initial, 75 mg/day	75–150 mg/day	Not recommended for use in children under 12 yr
Imipramine hydrochloride USP *Tofranil, Presamine*	Imipramine hydrochloride injection USP	Antidepressant	IM, 25–50 mg tid or qid	50–300 mg/day	375 μg/kg of body weight or 11 mg/m^2 of body surface qid. Dose is not established in children under 12 yr
	Imipramine hydrochloride tablets USP		25–50 mg tid or qid	50–300 mg/day	375 μg/kg of body weight or 11 mg/m^2 of body surface qid. Dosage is not established in children under 12 yr
Isocarboxazid USP *Marplan*	Isocarboxazid tablets USP	Antidepressant	Initial, 30 mg/day as a single dose or in divided doses; maintenance, 10–20 mg/day		
Maprotiline hydrochloride USP *Ludiomil*	Maprotiline hydrochloride tablets USP	Antidepressant	Initial, 75 mg/day	75–150 mg/day	Dosage is not established in children up to 18 yr
Mazindol USP *Sanorex*	Mazindol tablets USP	Appetite suppressant	1 mg tid or 2 mg qd		Not recommended

*See USP DI for complete dosage information.

TABLE 10-5 *Continued*

PSYCHOMOTOR STIMULANTS, MONOAMINE OXIDASE INHIBITORS, AND TRICYCLIC (AND RELATED) ANTIDEPRESSANTS

Name Proprietary Name	**Preparations**	**Category**	**Usual Adult Dose***	**Usual Dose Range***	**Usual Pediatric Dose***
Methamphetamine hydrochloride *Desoxyn*	Methamphetamine hydrochloride tablets	Central stimulant	Narcolepsy: 5–60 mg/day in divided doses		
Methylphenidate hydrochloride USP *Ritalin*	Methylphenidate hydrochloride tablets USP	Central stimulant	Narcolepsy: 10 mg bid or tid	10–60 mg/day	Hyperkinesia: use in children under 6 yr is not recommended; over 6 yr, 5 mg bid, increased by 5–10 mg at weekly intervals
Nortriptyline hydrochloride USP *Aventyl*	Nortriptyline hydrochloride capsules USP Nortriptyline hydrochloride oral solution USP	Antidepressant		An amount of nortriptyline hydrochloride equivalent to 20–100 mg of nortriptyline daily in divided doses	
Phendimetrazine tartrate USP *Anorex, Adephen*	Phendimetrazine tablets and capsules	Appetite suppressant	17.5–70 mg bid or tid		Not recommended
Phenelzine sulfate USP *Nardil*	Phenelzine sulfate tablets USP	Antidepressant	The equivalent of 15 mg of phenelzine once daily or every other day	7.5–75 mg/day	
Phenmetrazine hydrochloride USP *Preludin*	Phenmetrazine hydrochloride tablets USP	Anorexic	25–75 mg/day in divided doses, 1 hr before meals		
Phentermine hydrochloride USP *Fastin, Adepex-P*	Phentermine hydrochloride tablets and capsules	Appetite suppressant	8 mg tid or 15–37.5 mg qd		Not recommended
Protriptyline hydrochloride USP *Vivactil*	Protriptyline hydrochloride tablets USP	Antidepressant	15–40 mg/day in 3 or 4 divided doses	15–60 mg/day in divided doses	
Tranylcypromine sulfate USP *Parnate*	Tranylcypromine sulfate tablets USP	Antidepressant		Initial, 10 mg in the morning and afternoon daily for 2 wk; if no response appears, increase dosage to 20 mg in the morning and 10 mg in the afternoon daily for another week; maintenance, 10–20 mg/day	
Trazodone hydrochloride *Desyrel*	Trazodone hydrochloride tablets	Antidepressant	Initial, 150 mg/day	150–400 mg/day	Dosage is not established in children up to 18 yr

*See USP DI for complete dosage information.

PSYCHEDELICS

The term *psychedelic* refers to agents that are subjectively described as producing an increased awareness and enhanced perception of sensory stimuli. Additionally, the stimuli may be perceived in unusual or novel ways (e.g., sound may be perceived as color). Also the user may both experience the sensation and also feel as though participating as an observer. These are the so-called mind-expanding drugs.

Additionally, the drugs can produce anxiety, fear, panic, hallucinations, and an overall symptomatology resembling a psychosis. Hence, they can be classed as hallucinogens and psychotomimetics.

Also included in this discussion are drugs that are mainly hallucinogenic (phencyclidine; PCP) and euphorient (cocaine) and mainly depressant or intoxicating (Δ^9- or Δ^1-tetrahydrocannabinol; THC).

Psychedelics are broadly groupable into those possessing an indolethylamine moiety and those with a phenylethylamine moiety. In the first group, there is a structural resemblance to the central neurotransmitter 5-HT and in the second to NE and DA. This resemblance is intriguing and, indeed, there may be some effects on the respective transmitter systems. However, with structures of the complexity found in many of these agents, there is also a likelihood that a given structure may affect not just the closest structurally related neurotransmitter, but other systems as well. Thus, a phenethylamine system can affect not only NE and DA systems, but also 5-HT systems, and an indolethylamine system could affect not only 5-HT, but also NE and DA systems.

INDOLETHYLAMINES

Dimethyltryptamine. The compound is a rather weak hallucinogen active only by inhalation or injection, with a short duration of action. It possesses pronounced sympathomimetic side effects.

Bufotenine. The compound is 5-hydroxydimethyltryptamine and is probably not a hallucinogen, as it was once thought to be. Rather, the agent acts as a cardiovascular stimulant by release of 5-HT.[39]

Psilocybin and Psilocyn. Psilocybin is the phosphoric acid ester of psilocyn and appears to be converted to psilocyn as the active species in vivo. It occurs in a mushroom, *Psilocybe mexicana*. Both drugs are orally active and of short duration of action.

Synthetic α-methyl-substituted relatives have a much longer duration of action and enhanced oral potency.[40]

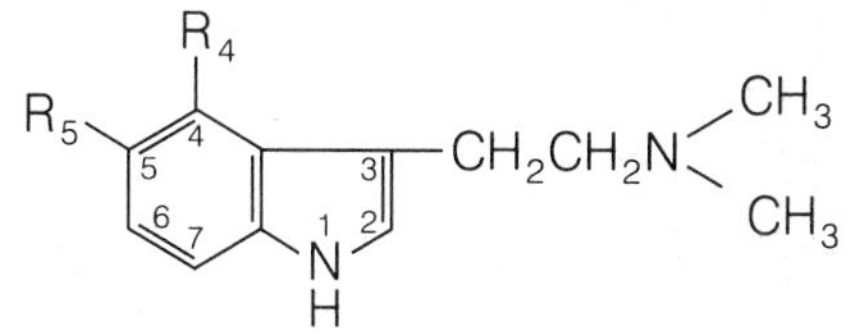

Dimethyltryptamine:	$R_4 = R_5 = H$
Bufotenine:	$R_4 = H$; $R_5 = OH$
Psilocybin:	$R_4 = OPO(OH)_2$; $R_5 = H$
Psilocyn:	$R_4 = OH$; $R_5 = H$

2-PHENYLETHYLAMINES

Mescaline. 3,4,5-Trimethoxyphenethylamine is a much-studied hallucinogen with many complex effects on the CNS. It occurs in the peyote cactus. The oral dose required for its hallucinogenic activities is very high, as much as 500 mg of the sulfate salt. The low oral potency is probably because of facile metabolism by MAO. α-Methylation increases activity. Synthetic α-methyl-substituted relatives, such as the two illicit drugs DOM and MDA, are more potent than mescaline.[40, 41]

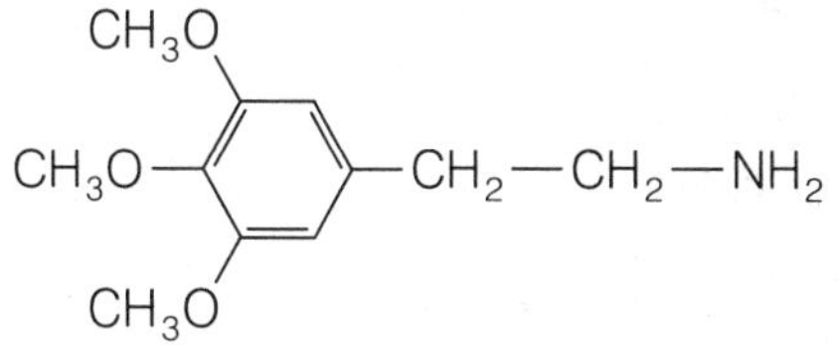

Mescaline

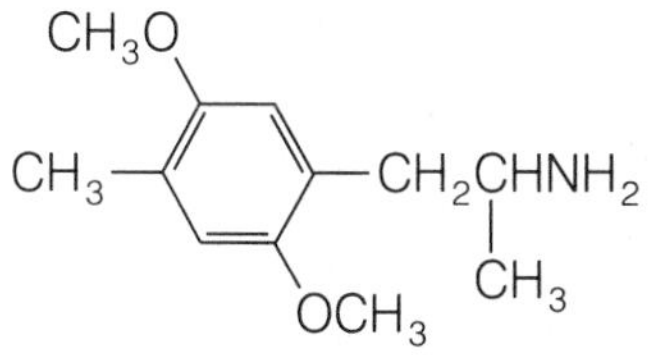

1-(2,5-Dimethoxy-4-methylphenyl)-2-aminopropane (DOM, STP)

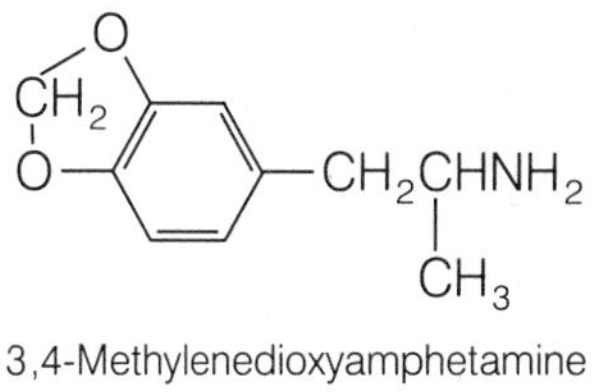

3,4-Methylenedioxyamphetamine (MDA)

AGENT POSSESSING BOTH AN INDOLETHYLAMINE AND A PHENYLETHYLAMINE MOIETY

(+)-Lysergic Acid Diethylamide (LSD). In the structure of the extraordinarily potent hallucinogen LSD can be seen both an indolethylamine group and a phenylethylamine group. The stereochemistry is exceedingly important. Chirality, as shown, must be maintained or activity is lost; likewise, the location of the double bond, as shown, is required.[43] Experimentally LSD has marked effects on serotoninergic neurons. However, the bases for all of its complex CNS actions are not completely understood.

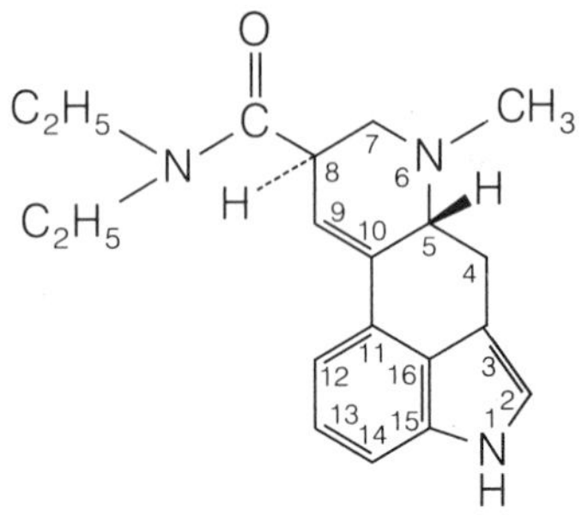

Lysergic Acid Diethylamide

DISSOCIATIVE AGENTS

Phencyclidine (PCP) was introduced as a dissociative anesthetic for animals. Its close structural relative ketamine is still so employed. The drug produces a sense of intoxication; there are hallucinogenic experiences, not unlike those produced by the anticholinergic hallucinogens, and often amnesia.

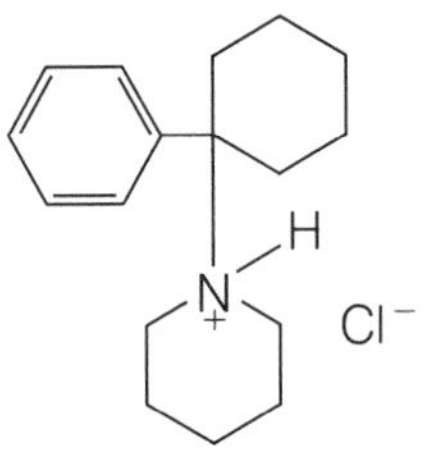

Phencyclidine Hydrochloride

The drug affects many systems including these of NE, DA, and 5-HT. It has been proposed that PCP (and certain other psychotomimetics) produce a unique pattern of activation of ventral tegumental area dopaminergic neurons.[44] It reportedly also blocks *N*-methyl aspartate-regulated ion channels by binding in the ion channel.[45] Additionally, there are sites in the brain, designated as μ-opioid receptors, to which the agent binds. Phencyclidine itself appears to be the main active agent (i.e., it acts mainly per se) producing its many CNS effects.

EUPHORIANT–STIMULANT

Cocaine as a euphoriant–stimulant, psychotomimetic, and drug of abuse could as well be discussed with amphetamine and methamphetamine with which it shares many biologic properties. At low doses, it produces feelings of well-being, decrease of fatigue, and increased alertness. The drug tends to produce compulsive drug-seeking behavior and a full-blown toxic psychosis may emerge. Many of these effects appear to be related to the effects of increased availability of DA for interaction with postsynaptic receptors. Cocaine is a potent DA-uptake blocker. (The amphetamines largely increase availability of DA by release.) As with many uptake blockers, the structural requirements do not appear to be highly stringent. A bulky phenethylamine moiety may suffice. If an interaction between a hydrogen atom on the nitrogen of the protonated form of cocaine and an oxygen of the ester group or, alternatively, between the unshared electron pair of the free-base nitrogen and the carbonyl of the ester group occurs, this grouping could be approximated.

For detailed discussions of the toxicology of cocaine including peripheral effects, which appear to be mediated by noradrenergic mechanisms, a review may be consulted.[46]

DEPRESSANT–INTOXICANT

Δ^1-Tetrahydrocannabinol (THC) or **Δ^9-THC.** There are two conventions for numbering THC. That convention arising from terpinoid chemistry produces Δ^1-THC, and that based on the dibenzopyran system results in a Δ^9-THC designation. The terpinoid convention is employed in the structure.

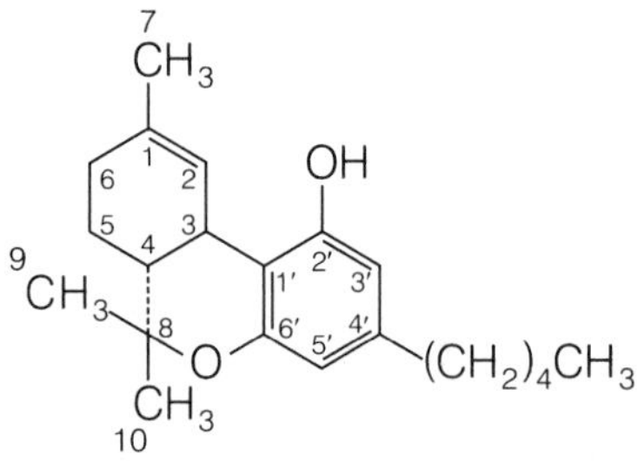

(−)-Δ^1-*trans*-Tetrahydrocannabinol

The drug is a depressant with stimulant sensations arising from depression of higher centers. Many effects, reputedly subjectively construed as pleasant, are evident at low doses. The interested reader may consult a pharmacology text for a detailed account of these effects. At higher doses, psychotomimetic actions, including dysphoria, hallucinations, and paranoia, can be marked. Structural features associated with activity have been reviewed.[47] Notably, the phenolic OH is required for activity. Some SARs (especially separation of potency between enantiomers) for cannabonoids suggest action at receptors.[48] The bulk of evidence suggests that actions are mediated through effects on cell membranes.[49] A receptor for THC has been discovered.[50]

Medically, in some areas of the United States, THC is available as an effective antiemetic in patients undergoing cancer chemotherapy. Also, the drug has served as a model for other potentially useful agents, such as anticonvulsants.

REFERENCES

1. Jarboe, C. H., Porter, L. A., and Buckler, R. T.: J. Med. Chem. 11:729, 1968.
2. Pellmar, T. C., and Wilson, W. A.: Science 197:912, 1977.
3. Daly, J. W.: J. Med. Chem. 25:197, 1982.
4. Williams, M., and Huff, J. R.: Adenosine as a neuromodulator. In McDermed, J. (ed.). The mammalian nervous system, Sect. I CNS Agents. Annu. Rep. Med. Chem. 18:1, 1983.
5. Snyder, S. H., Katims, J. J., Annau, Z., et al.: Proc. Natl. Acad. Sci. USA 78:3260, 1981.
6. Arnaud, M. J.: Products of metabolism of caffeine. In Dews, P. B. (ed.). Caffeine, Perspectives from Recent Research, p. 3. New York, Springer-Verlag, 1984.
7. Weiner, H.: Norepinephrine, epinephrine and the sympathomimetic amines. In Gilman, A. G., Goodman, L. S., Rall, T. W., and Morad, F. (eds.). Goodman and Gilman's The Pharmacological Basis of Therapeutics, 7th ed., p. 145. New York, Macmillan, 1985.
8. Snyder, S. H., Banerjee, S. P., Yamamura, H. I., and Greenberg, D.: Science 184:1243, 1974.
9. Fuller, R. W.: Ann. N.Y. Acad. Sci. 305:147, 1978.
10. Harvey, J. A.: Ann. N.Y. Acad. Sci. 305:289, 1978.
11. Groppetti, A., and Costa, E.: Life Sci. 8:635, 1969.
12. Clineschmidt, B. V., et al.: Ann. N.Y. Acad. Sci. 305:222, 1978.
13. Clarke, F. H.: J. Org. Chem. 27:3251, 1962.
14. Dvornik, D., and Schilling, G.: J. Med. Chem. 8:466, 1965.
15. Engstrom, R. G., Kelly, L. A., and Gogerty, J. H.: Arch. Int. Pharmacodyn. 214:308, 1975.
16. Weisz, I., and Dudas, A.: Monatsh. Chem. 91:840, 1960.
17. Shaffi'ee, A., and Hite, G.: J. Med. Chem. 12:266, 1969.
18. Patrick, K. S., et al.: J. Pharmacol. Exp. Ther. 241:152, 1987.
19. Perel, J. M., and Dayton, P. G.: Methylphenidate. In Usdin, E., and Forrest, I. S. (eds.). Psychotherapeutic Drugs, part II, p. 1287, New York, Marcel-Dekker, 1977.
20. Srinvas, N. R., et al.: J. Pharmacol. Exp. Ther. 241:300, 1987.
21. Whitelock, O. V. (ed.): Amine Oxidase Inhibitors. Ann. N.Y. Acad. Sci., vol. 80, 1959.
22. Richards, L. E., and Burger, A.: Mechanism-based Inhibitors of Monoamine Oxidase. In Jucker, E. (ed.). Progr. Drug Res. 30:205, 1986.
23. Green, A. L.: Biochem. Pharmacol. 13:249, 1964.
24. Burger, A.: J. Med. Pharm. Chem. 4:571, 1961.
25. Belleau, B., and Moran, J. F.: J. Am. Chem. Soc. 82:5752, 1960.
26. Belleau, B., and Moran, J. F.: J. Med. Pharm. Chem. 5:215, 1962.
27. Silverman, R. B.: J. Biol. Chem. 258:14766, 1983.
28. Lahti, R. A.: Antidepressants and antipsychotic agents. In Krapcho, J. (ed.). Sect. I, CNS Agents. Annu. Rep. Med. Chem. 12:1, 1977 (and references cited therein).
29. Baldessarini, R. J.: Drugs and the treatment of psychiatric disorders. In Gilman, A. G., Goodman, L. S., Rall, T. W., and Murad, F. (eds.). Goodman and Gilman's The Pharmacological Basis of Therapeutics, 7th ed., p. 416. New York, Macmillan, 1985.
30. Daniels, T. C., and Jorgensen, E. C.: Central nervous system stimulants. In Doerge, R. F. (ed.). Wilson and Gisvold's Textbook of Organic, Medicinal, and Pharmaceutical Chemistry, 8th ed., p. 383. Philadelphia, J. B. Lippincott, 1982.
31. Peroutka, S. J., and Snyder, S. H.: Science 210:88, 1980.
32. Crews, F. J., Scott, J. A., and Shorestein, N. H.: Neuropharmacology 22:1203, 1982.
33. Sugrue, M. F.: J. Neural Transm. 57:281, 1983.
34. Stanford, C., Nutt, D. J., and Cowen, P. J.: Neuroscience 8:161, 1983.
35. Yeh, H. H., and Woodward, D. J.: J. Pharmacol. Exp. Ther. 226:126, 1983.
36. Devoskin, L. P., and Sparber, S. P.: J. Pharmacol. Exp. Ther. 226:5, 1987.
37. Francis, H., Poncelet, M., Danti, S., et al.: Drug Dev. Res. 3:349, 1983.
38. Simon, P., Lecrubier, Y., Jouvent, R., et al.: In Usdin, E., Asberg, M., Bertilsson, L., and Sjoquist, F. (eds.). Frontiers in biochemical and pharmacological research in depression. Adv. Biochem. Psychopharmacol. 39:293, 1984.
39. Fischer, R.: Nature 220:411, 1968.
40. Murphree, H. B., Dippy, R. H., Jenney, E. H., and Pfeiffer, C. C.: Clin. Pharmacol. Ther. 2:722, 1961.
41. Shulgin, A. T.: Nature 201:120, 1964.
42. Hey, P.: Q. J. Pharmacol. 20:129, 1947.
43. Stoll, A., and Hofmann, A.: Helv. Chim. Acta 38:421, 1955.
44. Bowers, M. B., Bannon, M. J., and Hoffman, F. J., Jr.: Psychopharmacology 93:133, 1987.
45. Foster, A. C., and Fogg, G. E.: Nature 329:395, 1987.
46. Cregler, L. L., and Mark, H.: N. Engl. J. Med. 315:1495, 1986.
47. Edery, H., Grunfeld, Y., Zri, Z. B., and Mechoulam, R.: Ann. N.Y. Acad. Sci. 191:40, 1971.
48. Hollister, L. E., Gillespie, H. K., and Srebnik, M.: Psychopharmacology 92:505, 1987.
49. Bloom, J.: Pharmacol. Exp. Ther. 232:579, 1985.
50. Matsuda, L. A., Lolait, S. J., Brownstein, M. J., Young, A. C., and Bonner, T. I.: Nature 346:561, 1990.

SELECTED READINGS

Biel, J. H.: Some rationales for the development of antidepressant drugs. In Molecular Modification in Drug Design. Adv. Chem. Ser. 45. Am. Chem. Soc. Applied Pub., Washington, DC, 1964.

Nieforth, K. A., and Cohen, M. L.: Central nervous system stimulants. In Foye, W. O. (ed.). Principles of Medicinal Chemistry, 2nd ed., p. 303. Philadelphia, Lea & Febiger, 1981.

Dews, P. B., (ed.): Caffeine, Perspectives from Recent Research. Berlin, New York, Springer-Verlag, 1984.

Houlihan, W. J., and Babington, R. G.: Anorexigenics. In Wolff, M. E. (ed.). Burger's Medicinal Chemistry, part III, 4th ed., p. 1069, New York, Wiley-Interscience, 1981.

Kaiser, C., and Setler, P. E.: Antidepressant agents. In Wolff, M. E. (ed.). Burger's Medicinal Chemistry, part III, 4th ed., p. 997, New York, Wiley-Interscience, 1981.

Richards, L. E., and Burger, A.: Mechanism-based inhibitors of monoamine oxidase. In Jucker, E. (ed.). Progr. Drug Res. 30:205, 1986.

Shulgin, A. T.: Halucinogens. In Wolff, M. E. (ed.). Burger's Medicinal Chemistry, part III, 4th ed., p. 1109, New York, Wiley-Interscience, 1981.

Singer, A. J. (ed.): Marijuana: Chemistry, pharmacology and patterns of social use. Ann. N.Y. Acad. Sci., vol. 191, 1971.

Whitelock, O. V. (ed.): Amine oxidase inhibitors. Ann. N.Y. Acad. Sci., vol. 80, 1959.

CHAPTER 11

Adrenergic Agents

Patrick E. Hanna

Adrenergic drugs are those chemical agents that exert their principal pharmacologic and therapeutic effects by acting at peripheral sites to either enhance or reduce the activity of components of the sympathetic division of the autonomic nervous system. In general, those substances that produce effects similar to stimulation of sympathetic nervous activity are known as *sympathomimetics*, *adrenomimetics*, or *adrenergic stimulants*. Those that decrease sympathetic activity are referred to as *sympatholytics*, *antiadrenergics*, or *adrenergic-blocking agents*. In addition to their effects on sympathetic nerve activity, a number of adrenergic agents produce important effects on the central nervous system.

This chapter includes discussion of both adrenergic stimulants and adrenergic-blocking agents. We assume that the reader has a basic understanding of both the principal anatomic features and the principal functions of the autonomic nervous system, including the underlying concepts of neurochemical transmission. Useful reviews of these topics have published[1] and we refer the reader to them for supplemental background information.

ADRENERGIC NEUROTRANSMITTERS

FUNCTION

The adrenergic nerves in the autonomic nervous system are the postganglionic sympathetic fibers; preganglionic fibers and postganglionic parasympathetic fibers are cholinergic. Thus, the neurotransmitter that is released from preganglionic nerves and from postganglionic parasympathetic neurons is acetylcholine. The adrenergic neurotransmitter that is liberated from postganglionic sympathetic neurons, as a result of sympathetic nerve stimulation, is norepinephrine (NE).

Norepinephrine, after its release from the sympathetic nerve ending into the synaptic cleft, interacts with specific postsynaptic receptors on cells of the effector organ (Fig. 11-1). The effector organ is the organ or tissue (usually a gland, smooth muscle, or cardiac muscle) that is innervated by the postganglionic nerve. Interaction of norepinephrine with the adrenergic receptors of the effector cells ultimately results in the production of a physiologic response (muscle contraction or relaxation, glandular secretion, or other) that is characteristic of that organ or tissue.

The action of NE at adrenergic receptors is terminated by a combination of processes including uptake into the neuron and into extraneuronal tissues, diffusion away from the synapse, and metabolism. Usually the primary mechanism for termination of the action of NE appears to be reuptake (uptake$_1$) of the catecholamine into the nerve terminal. This is an energy-requiring process; it involves a membrane pump system that has a high affinity for NE. The uptake system will also transport certain amines other than NE into the nerve terminal. Some of the NE that reenters the sympathetic neuron is then transported by a second uptake process into the storage granules, where it is held in a stable complex with ATP and protein until sympathetic nerve activity or some other stimulus causes it to be released into the synaptic cleft.

In addition to the neuronal uptake of NE just discussed, there exists an extraneuronal uptake process that is commonly referred to as uptake$_2$. This uptake process, which was first discovered in cardiac muscle cells, is present in a variety of tissues. Its physiologic significance is not known, but it may

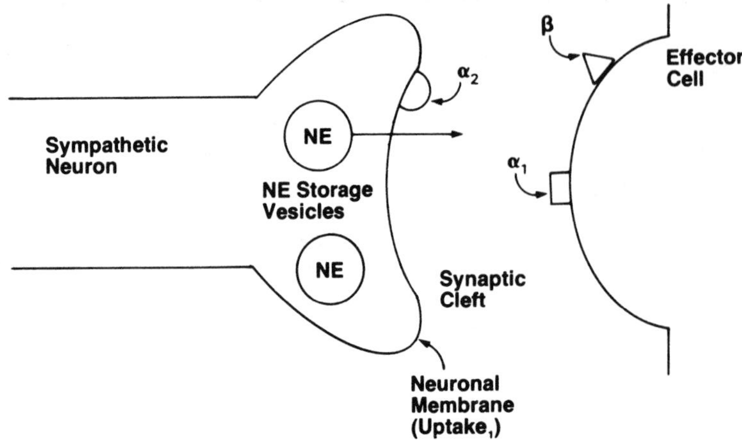

FIG. 11-1. Postganglionic sympathetic neuron, synaptic cleft, and effector cell.

serve to help regulate the concentration of NE and other catecholamines in the vicinity of adrenergic receptors.[2] Catecholamines that are taken up into extraneuronal tissues are rapidly metabolized.

Epinephrine is also an endogenous adrenergic neurotransmitter, but it is not released from peripheral sympathetic nerve endings in the fashion described for NE. Epinephrine is synthesized and stored in the adrenal medulla from which it is released into the circulation. Thus, epinephrine is often referred to as a neurohormone. Until the 1950s, it was generally believed, however, that epinephrine was not only a product of adrenal medullary cells, but also was the principal sympathetic neurotransmitter released at postganglionic adrenergic receptors. It is now known that epinephrine is distributed by the circulation to various organs and tissues at which it exerts its effects at adrenergic receptor sites. Epinephrine also is biosynthesized in certain neurons of the central nervous system, where both it and NE serve as neurotransmitters.

The termination of the action of epinephrine apparently is not highly dependent on the neuronal membrane pump ($uptake_1$) for which it has less affinity than NE. Current evidence indicates that the physiologic activity of epinephrine may be terminated, in large part, by metabolism in extraneuronal tissues after its removal from the vicinity of the adrenergic receptors by the $uptake_2$ mechanism.

STRUCTURE AND PHYSICOCHEMICAL PROPERTIES

The neurotransmitters, NE and epinephrine, belong to the chemical class of substances known as *catecholamines*. This name arises because the two hydroxyl groups present on the aromatic ring of each of these compounds are situated *ortho* to each other; the same arrangement of aromatic hydroxyl groups is found in catechol.

HO, HO–C_6H_3–CH(OH)–CH_2–NH_2
Norepinephrine

HO, HO–C_6H_3–CH(OH)–CH_2–$NHCH_3$
Epinephrine

HO, HO–C_6H_4
Catechol

Norepinephrine and epinephrine are relatively small, polar substances with a high degree of water solubility. These physicochemical characteristics are typical of several endogenous agonists such as acetylcholine, histamine, serotonin, and dopamine. The catecholamines contain both acidic (the aromatic hydroxyls) and basic (the aliphatic amine) functional groups. The magnitude of the dissociation constants of the phenolic and amino groups of the catecholamines has been the subject of considerable investigation and some controversy.[3] Ganellin has pointed out that the pK_a values of 9.9 and 8.7 (which are frequently attributed to a phenolic hydroxyl and the protonated amino group, respectively, of the epinephrine cation) have been shown by several research groups to be incorrectly assigned.[3] The lower pK_a value of the protonated form of epinephrine is due to ionization of a phenolic hydroxyl group. Ganellin has calculated the relative populations of the various ionized and nonionized species of NE and epinephrine at *p*H 7.4 and has found that the cation (shown in the diagram) is present to an extent slightly greater than 95% for both catecholamines. The zwitterionic form, in which both the aliphatic amine is protonated and one of the hydroxyl groups is ionized, is present to an extent of approximately 3%. Thus, at physiologic *p*H, less than 2% of either epinephrine or NE exists in the nonionized form. This largely accounts for the high degree of water solubility of these compounds as well as other catecholamines, such as isoproterenol and dopamine.

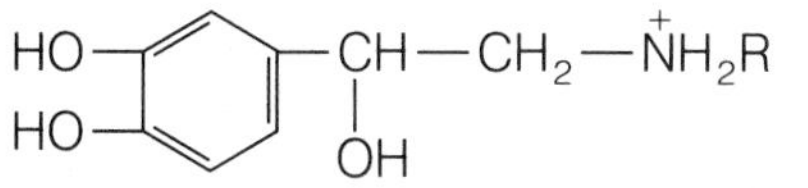

R = H or CH_3; Cationic Form of Norepinephrine and Epinephrine

Aromatic compounds that contain *ortho*-hydroxyl substituents are highly susceptible to oxidation. Therefore, catecholamines, such as epinephrine and NE, undergo oxidation in the presence of oxygen (air) or other oxidizing agents to produce a mixture of colored products. Hence, solutions of catecholamine drugs are often stabilized by the addition of an antioxidant (reducing agent) such as ascorbic acid or sodium bisulfite.

Tyrosine —Tyrosine Hydroxylase→ L-Dopa —L-Aromatic Amino Acid Decarboxylase→ Dopamine —Dopamine β-Hydroxylase→ Norepinephrine —Phenylethanolamine N-Methyltransferase→ Epinephrine

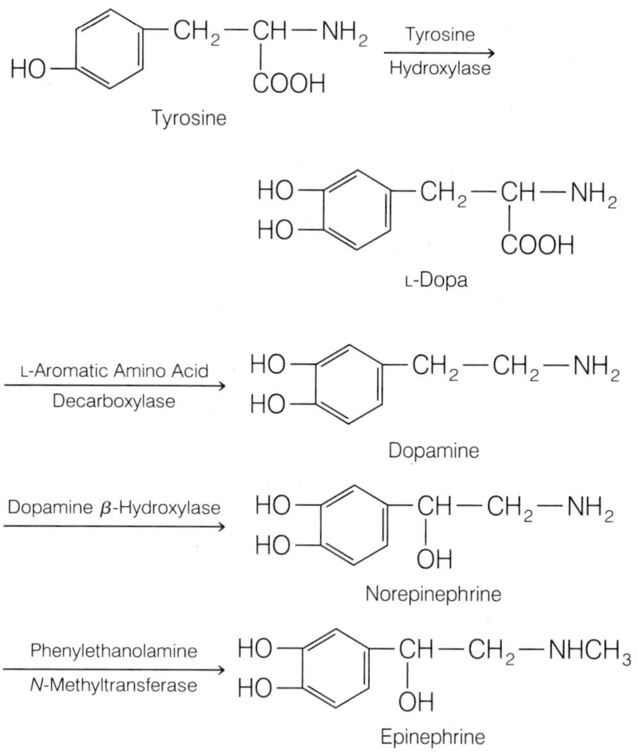

FIG. 11-2. *Catecholamine biosynthesis.*

BIOSYNTHESIS

Catecholamine biosynthesis takes place in adrenergic and dopaminergic neurons in the central nervous system, in sympathetic neurons in the autonomic nervous system, and in the adrenal medulla. Norepinephrine biosynthesis takes place through a three-step process, beginning with the conversion of tyrosine to L-dihydroxyphenylalanine (L-dopa) (Fig. 11-2). The tyrosine-3-monooxygenase (tyrosine hydroxylase)-catalyzed formation of dopa takes place in the cytoplasm of the neuron, and it is the rate-limiting process in catecholamine biosynthesis. End-product inhibition of the tyrosine hydroxylase reaction is believed to be a key mechanism for the regulation of catecholamine biosynthesis, because NE markedly reduces tyrosine hydroxylase activity.[1]

The second enzymatic process in NE biosynthesis is the decarboxylation of L-dopa to produce dopamine. This reaction is catalyzed by L-aromatic amino acid decarboxylase (dopa decarboxylase), an enzyme that also resides in the cell cytoplasm and that exhibits broad substrate specificity. Thus, a variety of aromatic amino acids that have the L-configuration can serve as substrates for this enzyme.

The dopamine that is formed in the cytoplasm of the neuron is transported by an uptake process into the storage vesicles, where it is stereospecifically hydroxylated by another enzyme of rather wide substrate specificity, dopamine β-monooxygenase (dopamine β-hydroxylase). The NE formed thereby is stored in the vesicles until it is released into the synaptic cleft to interact with the adrenergic receptors.

In the adrenal medulla, a fourth biosynthetic reaction takes place by which NE is converted to epinephrine. This reaction occurs in the cell cytoplasm and is catalyzed by phenylethanolamine-*N*-methyltransferase (PNMT). This enzyme is highly localized in the adrenal medulla, but small amounts are present in other tissues, such as heart and brain. The epinephrine is held in the storage granules of the chromaffin cells in the adrenal medulla.

There are two drugs in clinical use that were designed with the intent that they would produce their therapeutic effects by inhibition of NE and epinephrine biosynthesis. The first of these, methyldopa (α-methyldopa; L-α-methyl-3,4-dihydroxyphenylalanine; Aldomet) is an inhibitor of dopa decarboxylase and is used as an antihypertensive agent. Although methyldopa does inhibit this enzyme, apparently by serving as an alternative substrate, its mechanism of antihypertensive activity is rather complex. This drug is discussed in Chapter 14.

The second catecholamine biosynthesis inhibitor in clinical use is metyrosine (α-methyl-*p*-tyrosine; Demser). Metyrosine differs structurally from tyrosine only by the presence of the α-methyl group. This is the same structural feature that was incorporated into dopa to produce methyldopa. However, because methyltyrosine inhibits the first, and rate-limiting, step in catecholamine biosynthesis, it is a much more effective inhibitor of epinephrine and NE production than is methyldopa. Metyrosine has been approved for use in the United States in patients with pheochromocytoma. The latter condition involves the presence of chromaffin cell tumors that produce large amounts of NE and epinephrine. Although these tumors, which can occur in the adrenal medulla, are often benign, patients frequently suffer hypertensive episodes. Metyrosine reduces the frequency and severity of these episodes by lowering catecholamine production. Metyrosine, which is given orally in dosages as high as 2 to 3 g/day, is useful for preoperative management of pheochromocytoma.

contrast with the mechanism of termination of the cholinergic activity of acetylcholine, which involves metabolism of the neurotransmitter by acetylcholinesterase (see Chap. 12). However, both NE and epinephrine are extensively metabolized before excretion; little NE or epinephrine is excreted unchanged under ordinary circumstances.

The two principal enzymes involved in catecholamine metabolism are monoamine oxidase (MAO) and catechol *O*-methyltransferase (COMT). Both of these enzymes are widely distributed throughout the body. Monoamine oxidase is associated primarily with the outer membrane of mitochondria, and COMT is found in both the cellular cytoplasm and in membrane-bound forms. It appears that COMT is not present in sympathetic neurons, whereas the neuronal mitochondria do contain MAO. Accordingly, MAO has a role in the metabolism of intraneuronal catecholamines, whereas COMT acts primarily upon catecholamines that enter the circulation and the extraneuronal tissues after being released from nerves, from the adrenal medulla, or after being administered exogenously.

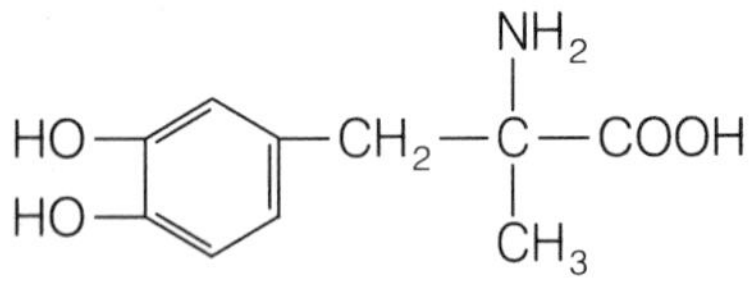

Methyldopa

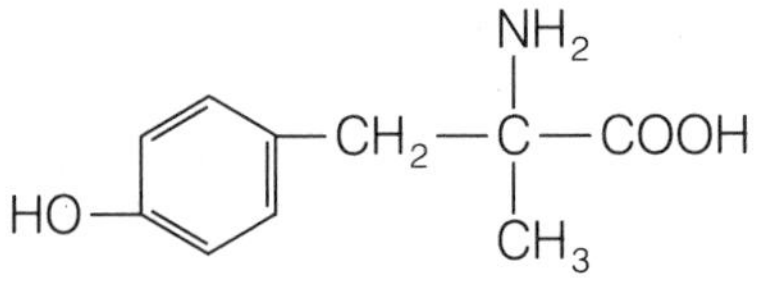

Metyrosine

The administration of D,L-*threo*-3,4-dihydroxyphenylserine to patients with dopamine β-hydroxylase deficiency has been shown to relieve the symptoms of severe orthostatic hypotension associated with this condition. The L-*threo*-isomer of 3,4-dihydroxyphenylserine is a good substrate for L-aromatic amino acid decarboxylase. Thus, its administration helps restore NE synthesis to normal because the production of NE is no longer dependent upon the conversion of dopamine to NE by dopamine β-hydroxylase.[4,5]

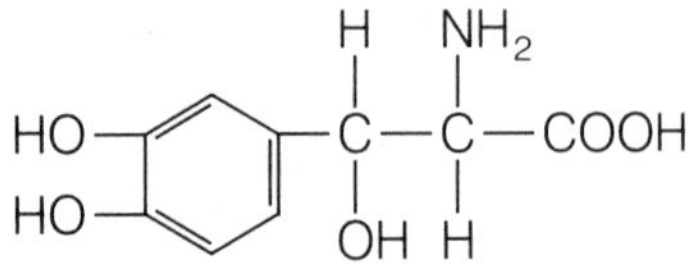

L-Dihydroxyphenylserine

METABOLISM

As discussed earlier, removal of NE from the synaptic cleft by uptake into the sympathetic neuron is believed to be the primary mode of termination of its action on the effector organ receptors. This is in

Neither COMT nor MAO exhibits a high degree of substrate specificity. Monoamine oxidase oxidatively deaminates a variety of compounds that contain an amino group attached to a terminal carbon. In addition, mammalian tissues contain more than one type of MAO, and the various types of MAO exhibit different substrate selectivities. Similarly, COMT catalyzes the methylation of a variety of catechol-containing molecules. The lack of substrate specificity of COMT and MAO is manifested in the metabolic disposition of NE and epinephrine (Fig. 11-3). Both MAO and COMT utilize not only NE and epinephrine as substrates, but they also each act upon the metabolites produced by the other.

The results of extensive research on catecholamine metabolism indicate that NE is oxidatively deaminated to form an aldehyde that is reduced to 3,4-dihydroxyphenylethylene glycol (see *B*, Fig. 11-3) in the neurons of human brain and peripheral tissues. It is primarily this glycol metabolite that is released into the circulation. The glycol *B* undergoes methylation by the COMT that it encounters in nonneuronal tissues. The product of methylation, 3-methoxy-4-hydroxyphenylethylene glycol (see *C*, Fig. 11-3), is oxidized by alcohol dehydrogenase and

FIG. 11-3. *Norepinephrine and epinephrine metabolism.* A. *R = H = norepinephrine, R = CH_3 = epinephrine;* B. *3,4-dihydroxyphenylethylene glycol;* C. *3-methoxy-4-hydroxyphenylethylene glycol;* D. *R = H = normetanephrine, R = CH_3 = metanephrine;* E. *3-methoxy-4-hydroxymandelic acid;* F. *3,4-dihydroxymandelic acid. MAO, monamine oxidase; COMT, catechol O-methyltransferase; Ald. Dehydr., aldehyde dehydrogenase; Ald. Red., aldehyde reductase; Alc. Dehydr., alcohol dehydrogenase.*

aldehyde dehydrogenase to yield 3-methoxy-4-hydroxymandelic acid (*E*). Metabolite *E* is commonly referred to as vanillylmandelic acid (VMA). Although metabolite *E* can be the end product of several pathways of metabolism of NE, the glycol *C* is its principal precursor.[6]

In vivo, COMT methylates almost exclusively the *meta*-hydroxyl group of catechols, regardless of whether the catechol is NE, epinephrine, or one of the metabolic products. Therefore, a converging pattern of metabolism of NE and epinephrine occurs in which 3-methoxy-4-hydroxymandelic acid (VMA) and 3-methoxy-4-hydroxyphenylethylene glycol (see *C*, Fig. 11-3) are common end products, regardless of whether the first metabolic step is oxidation by MAO or methylation by COMT.

Under normal circumstances, 3-methoxy-4-hydroxymandelic acid (see *E*, Fig. 11-3) is the principal urinary metabolite of NE; substantial amounts of the glycol *C* are excreted along with varying quantities of other metabolites (see Fig. 11-3), both in the free forms and as sulfate or glucuronide conjugates. Endogenous epinephrine is excreted primarily as metanephrine (*D*, R = CH_3) and *E*. 3,4-Dihydroxymandelic acid (*F*) is an intermediate in an alternative, minor pathway from *A* to *E*.

ADRENERGIC RECEPTORS

Much drug research activity has focused upon the development of selective adrenergic agonists and antagonists. This high level of effort and the progress that has stemmed from it have been, in large part, the result of the acceptance of Ahlquist's proposal that there exists more than one type of adrenergic receptor.[7] Ahlquist proposed that two general types of adrenergic receptors (adrenoceptors) exist in mammalian tissues, and he designated them α and β. The postsynaptic α-adrenergic receptors are involved primarily in contraction of smooth muscle, whereas β-receptors are associated with the relaxation of smooth muscle and the stimulation of cardiac muscle. There are also α- and β-receptors that mediate the effects of catecholamines on carbohydrate and lipid metabolism. Both α- and β-receptors mediate the relaxation of intestinal smooth muscle.

A key factor in the classification of adrenergic receptors is the grouping together of certain receptors on the basis of the order of potency of agonists that activate them. A second important factor that supports the concept of multiple types of adrenergic receptors is the existence of selective antagonists. Careful analysis of the potency order of agonists and the selectivity of antagonists has led to the further subclassification of both α- and β-receptors into groupings that have been designated α_1, α_2, β_1, and β_2.

α-ADRENERGIC RECEPTORS

For several years, the discussion and definition of α-adrenergic receptors was a rather simple matter. *α-Receptors* were designated as those postsynaptic receptors that exhibited the following order of agonist potency: epinephrine $\geq$ NE $\gg$ isoproterenol. Additionally, α-adrenergic receptors are blocked by antagonists such as phentolamine and phenoxybenzamine. The discovery that certain adrenergic agonists and antagonists exhibit various degrees of selectivity for presynaptic and postsynaptic adrenergic α-receptors led to suggestions that postsynaptic α-receptors be designated α_1 and that presynaptic α-receptors be referred to as α_2.[8,9] Further developments revealed, however, that neither a functional nor an anatomic subclassification of α-receptors was as generally useful as a subdivision based on the selectivity of agonists and antagonists.[10]

The prototypical α_2-receptor is the presynaptic α-receptor found on the terminus of the sympathetic neuron (see Fig. 11-1).[8,9] Interaction of this receptor with agonists such as NE and epinephrine results in inhibition of NE release from the neuron. Thus, α_2-receptors play a role in the regulation of NE release. Other α_2-receptors are postsynaptic (postjunctional), and some are associated with nonneural tissue. The postjunctional α_2-receptors in vascular smooth muscle mediate the production of contractile responses, an important function that is also served by α_1-receptors. α_1-Receptors usually predominate over α_2-receptors in mediating vasoconstriction. Although both prejunctional α_1- and prejunctional α_2-receptors are present in rat cardiac tissue, and both appear to be involved in feedback regulation of NE release, it is not known if prejunctional α_1-receptors are present in the human heart.[11] The presence of postjunctional α_1-receptors in the human heart has been established; however, in contrast with the importance of cardiac β-receptors in the mediation of adrenergic effects, the physiologic function of α_1-receptors in heart tissue is unknown.[12]

Among the numerous agents that have been evaluated for their interactions with α_1- and α_2-receptors, methoxamine and phenylephrine are two of the more important therapeutic agents that exhibit selective α_1-receptor agonist activity.[13] Epinephrine and NE are agonists at both types of α-adrenergic receptors. Clonidine and α-methylnorepinephrine are selective α_2-receptor agonists. Prazosin is a selective α_1-antagonist and yohimbine preferentially blocks agonist activity at α_2-receptors. Phentolamine and tolazoline, two α-receptor antagonists that are available for clinical use, are relatively nonselective in their antagonistic actions. The irreversible antagonist, phenoxybenzamine, exhibits moderate α_1-selectivity.[14]

Progress in the understanding of the biochemical mechanisms by which activation of α-adrenergic receptors produces the tissue responses with which they are associated has been less rapid than the development of insight into β-receptor mechanisms. It has been recognized for some time that the activation of α_1-receptors is related to an increase in cellular Ca^{2+} concentrations. The enhanced Ca^{2+} levels appear to result from both an influx of extracellular Ca^{2+} and from a mobilization of Ca^{2+} from intracellular storage sites. The increases in Ca^{2+} are now believed to be preceded by a sequence of events that includes agonist-induced activation of the α_1-receptor on the cell membrane, stimulation of phospholipase C, and hydrolysis of inositol phospholipids.[15,16] The products of the hydrolysis of inositol phospholipids are diacylglycerol and inositol phosphates, both of which are believed to function as intracellular messengers. For example, inositol 1,4,5-triphosphate has been proposed to cause the release of Ca^{2+} from intracellular stores, and diacylglycerol is believed to activate protein kinase C, an enzyme that phosphorylates proteins. Calcium ion also is believed to be involved in the stimulation of protein phosphorylation. The activation of α_2-adrenergic receptors results in a lowering of intracellular levels of cyclic 3′,5′-adenosine monophosphate (AMP).[16] Cyclic AMP, like diacylglycerol and inositol 1,4,5-triphosphate, is an intracellular second messenger the functions of which are discussed in the following paragraphs relative to its role in the mediation of β-adrenergic effects.

β-ADRENERGIC RECEPTORS

In 1967, almost 20 years after Ahlquist's landmark paper proposing the existence of α- and β-adrenergic receptors, Lands and co-workers suggested that β-receptors could be designated as β_1 and β_2 subtypes.[17] β_1-Receptors are those that exhibit the agonist potency order: isoproterenol > epinephrine = NE. The agonist potency order for β_2-receptors is: isoproterenol > epinephrine > NE. Norepinephrine is a somewhat more potent agonist at β_1-receptors than at β_2-receptors. Some researchers have proposed further subclassifications of β-receptors, but until more definitive evidence is developed, the distinction between β_1- and β_2-subtypes appears to be the most useful approach to classification. Further evidence for the existence of more than one type of β-receptor is derived from recent research on selective β_1- and β_2-receptor antagonists. This topic is discussed later in this chapter. Cardiac stimulation and lipolysis are typical physiologic effects associated with β_1-receptors; bronchodilatation and vasodilatation are primarily β_2 responses. The development of agonists and antagonists that selectively modulate these β-receptor-controlled responses has led to potentially important advances in drug therapy that will be discussed later.

An abundance of data indicates that the β-adrenergic effects of catecholamines and other β-receptor agonists are mediated by stimulation of the enzyme adenylate cyclase. This enzyme catalyzes the conversion of adenosine triphosphate (ATP) to cAMP. Cyclic AMP is released from the membrane-bound enzyme into the cell, in which it functions as a "second messenger." According to this concept, the β-adrenergic agonist acts as the "first messenger" in that it carries information to the cell by interacting with β-receptors on the cell membrane. The receptor–agonist interaction results in stimulation of adenylate cyclase and production of cAMP, which acts as a second messenger by carrying the information to intracellular sites. It is now known that a variety of hormones and drugs exert their effects by stimulating or depressing adenylate cyclase activity subsequent to their interaction with membrane receptors. Both of the β-adrenergic receptor subtypes are coupled to adenylate cyclase, and many of the details of the coupling process have been elucidated.[18,19] A key component in the process is a guanine nucleotide protein (referred to as the G_S or N_S protein). In the absence of agonist, guanosine diphosphate (GDP) is bound reversibly to the G_S protein. Interaction of the agonist with the receptor is believed to bring about a conformational change in the receptor protein. This conformational change appears to cause a reduction in the affinity of the G_S protein for GDP and a concomitant increase in affinity for guanosine triphosphate (GTP). The subunit of the G_S protein that contains the bound GTP then activates adenylate cyclase by interacting with it. The bound GTP then undergoes hydrolysis to GDP, and the receptor–G_S protein complex returns to the basal state.

The intracellular function of cAMP, the second messenger, appears to be the activation of a group of enzymes called protein kinases that phosphorylate specific proteins. Thus, the phosphorylated proteins mediate the actions of cAMP, which functions as the mediator of the action of the drug or neurotransmitter that originally interacted with the β-adrenergic receptor.[20] The action of cAMP is terminated by a class of enzymes known as phosphodiesterases; these enzymes catalyze the hydrolysis of cAMP to AMP.

A full understanding of the molecular characteristics of biologic receptors necessitates their purification in a way that allows them to retain their ability to bind in a highly specific fashion to agonists and antagonists. For membrane-bound macromolecules, such as the adrenergic receptor proteins, this is a formidable task. However, all four adrenergic receptor subtypes have been purified, and the genes for some of them have been cloned.[18,21] Additionally, it has been determined that the α-adrenergic receptors act through guanine nucleotide regulatory proteins (G proteins) in a fashion similar to that described in the foregoing for the β-receptors.[22] The rapid advances that are being made, both in understanding the interactions of drugs and endogenous ligands with adrenergic receptors and in elucidating the biochemical consequences of those interactions, can be expected to lead to the design of more specific and more effective therapeutic agents.

SYMPATHOMIMETIC AGENTS

MECHANISM OF ACTION AND STRUCTURE–ACTIVITY CONSIDERATIONS

In terms of mechanism of action, sympathomimetic agents may be classified as producing their effects by direct, indirect, or mixed mechanisms. Direct-acting agents elicit a sympathomimetic response by interacting directly with adrenergic receptors. Indirect-acting agents produce their effects primarily by causing the release of NE from adrenergic nerve terminals; the NE that is released by the indirect-acting agent then activates the receptors to produce the response. Those compounds with a mixed mechanism of action both interact directly with adrenergic receptors and cause the release of NE. Most sympathomimetics appear to have direct and indirect actions, although one mechanism or the other often is predominant. Unfortunately, complete mechanistic data are not available for many compounds.

Direct-Acting Sympathomimetics

The mechanism by which an agent produces a sympathomimetic effect is, usually, intimately related to its chemical structure. For example, the prototypical direct-acting compounds are NE, epinephrine, and isoproterenol. Each of these three substances is a catecholamine. More fundamentally, they are phenylethylamine derivatives that contain the appropriate substituents to impart direct receptor-activating capabilities. These "appropriate" substituents are the catechol hydroxyl groups in the *meta-* and *para*-positions of the aromatic ring and the β-hydroxyl group on the ethylamine portion of the molecule. It was previously believed that, as a general rule, a potent, direct-acting phenylethylamine sympathomimetic agent should contain at least the *meta-* and *para*-hydroxyl groups or, alternatively, the *meta*-hydroxyl and the β-hydroxyl. However, exceptions to the generalizations can readily be found. For example, β-receptor agonist activity is often retained in phenylethanolamines when certain groups are substituted for the *meta*-hydroxyl, as will be discussed later in this chapter.[23, 24]

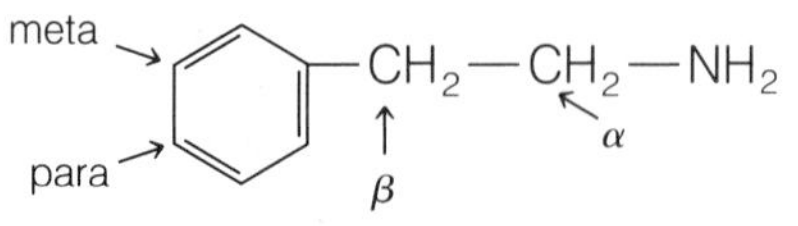

Phenylethylamine

The presence of the amino group in phenylethylamines is important for direct agonist activity. Both primary and secondary amines are found among the potent direct-acting agonists, but tertiary amines tend to be poor direct agonists. The amino group should be separated from the aromatic ring by two carbon atoms for optimal activity.

Direct receptor–agonist activity is enhanced by the presence of a hydroxyl group, of the correct stereochemical configuration, on the β-carbon, but it is reduced by the presence of a methyl group on the α-carbon. However, importantly, an α-methyl group increases the duration of action of the phenylethylamine agonist by making the compound resistant to metabolic deamination by MAO. Such compounds often exhibit enhanced oral effectiveness and greater central nervous system activity than their counterparts that do not contain an α-methyl group.

A highly critical factor in the interaction of adrenergic agonists with their receptors is that of stereoselectivity. Those direct-acting sympathomimetics that exhibit chirality by virtue of the presence of a β-hydroxyl group (phenylethanolamines) invariably exhibit a high degree of stereoselectivity in producing their agonist effects. That is, one enantiomeric form of the drug has greater affinity for the receptor than the other form. This is true for both α- and β-agonists. For epinephrine, NE, and related compounds, the more potent enantiomer has the (*R*)(−), or D(−), configuration. It appears that for all direct-acting agonists and antagonists that are structurally similar to NE, the more potent enantiomer is capable of assuming a conformation that results in the arrangement in space of the aromatic group, the amino group, and the β-hydroxyl group in a fashion resembling that of (*R*)(−) NE. This explanation of stereoselectivity is based upon the presumed interaction of these three critical pharmacophoric groups with three complementary binding areas on the receptor.

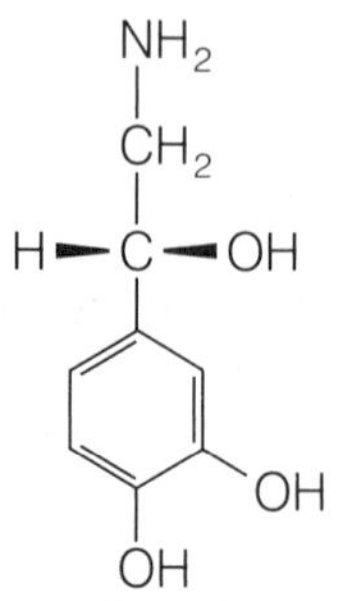

(*R*)(−) Norepinephrine

Indirect-Acting Sympathomimetics

Certain structural characteristics tend to impart indirect sympathomimetic activity to phenylethylamines. As with the direct-acting agents, the presence of the catechol hydroxyls enhances the potency of indirect-acting phenethylamines. In contrast with the direct-acting agents, the presence of a β-hydroxyl group decreases, and an α-methyl group increases, the effectiveness of indirect-acting agents. The presence of nitrogen substituents decreases indirect activity, with substituents larger than methyl rendering the compound virtually inactive. Phenylethylamines that contain a tertiary amino group are also ineffective as NE-releasing agents. Given the foregoing structure–activity considerations, it is easy to understand why amphetamine and *p*-tyramine are often cited as the prototype indirect-acting sympathomimetics. Although *p*-tyramine is not a clinically useful agent, its α-methylated derivative, hydroxyamphetamine, is an effective, indirect-acting sympathomimetic drug. Amphetamine-type drugs are discussed in more detail in the chapter on central nervous system stimulants (see Chap. 10).

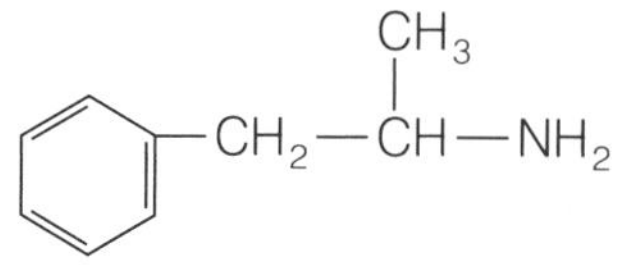

Amphetamine

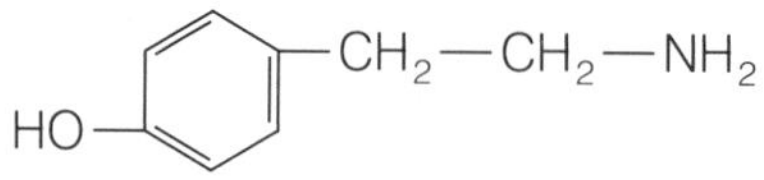

p-Tyramine

Sympathomimetic Agents with a Mixed Mechanism of Action

Those phenylethylamines that are considered to have a mixed mechanism of action usually have no hydroxyls on the aromatic ring, but usually do have a β-hydroxyl group. Thus, D(−)-ephedrine, which is discussed later in this chapter, is the classic example of a sympathomimetic with a mixed mechanism of action.

α- And β-Receptor Agonists

If a phenylethylamine has the structural prerequisites that cause it to act primarily by a direct-receptor activation mechanism, then it is the nature of the nitrogen substituent that determines whether it will act primarily at α- or β-receptors. In general, as the bulk of the nitrogen substituent increases, α-receptor agonist activity decreases, and β-receptor activity increases. Thus, isoproterenol is a potent β-receptor agonist, but has little affinity for α-receptors. However, the converse is not true; NE, which is an effective β_1-receptor agonist, is also a potent α-agonist. Epinephrine is a potent agonist at α-, β_1-, and β_2-receptors.

The fact that isoproterenol activates β-receptors, and not α-receptors, has resulted in its being widely used for the management of bronchial asthma. Isoproterenol is an effective bronchodilator, but it has several deficiencies as a therapeutic agent. It stimulates both β_1- and β_2-receptors. The β_1-component of its action imparts an undesirable cardiac stimulatory effect. After oral administration, its absorption is rather erratic and undependable, and it has a duration of action of only a few minutes, regardless of the route of administration. The principal reason for its poor absorption characteristics and short duration of action is its facile metabolic transformation by sulfate and glucuronide conjugation of the ring hydroxyls, and methylation by COMT. Unlike epinephrine and NE, isoproterenol does not appear to undergo oxidative deamination by MAO.

(HO)₂C₆H₃—CH(OH)—CH$_2$—NHCH(CH$_3$)$_2$

Isoproterenol

(HO)₂C₆H₃—CH(OH)—CH$_2$—NHC(CH$_3$)$_3$

N-tert-Butylnorepinephrine

The problems of lack of β-receptor selectivity and rapid metabolic inactivation associated with isoproterenol have at least partially been overcome by the design and development of selective β_2-adrenoceptor stimulants. In several instances it has been determined that an *N-tert*-butyl group enhances β_2-selectivity. For example, *N-tert*-butyl NE, which is not in clinical use, is nine to ten times as potent an agonist at tracheal β_2-receptors than at cardiac β_1-receptors.

Several modifications of the catechol portion of β-agonists have been useful in the development of selective drugs. The resorcinol structure has served as a successful replacement for catechol. Resorcinol is not a substrate for COMT. Therefore, β-agonists that contain this ring structure tend to have better absorption characteristics and a longer duration of action than their catechol-containing counterparts. Terbutaline is an example of a drug that contains both the *N-tert*-butyl substituent and the resorcinol structure. Unfortunately, terbutaline does not have as much β_2-adrenoceptor selectivity as is desired for optimal therapeutic use.

As mentioned earlier, a variety of functional groups can replace the *meta*-hydroxyl of the catechol structure in β_2-agonists. Advantage of this fact was taken in the design of albuterol, which is a saligenin derivative, is β_2-receptor selective, is orally effective, and has an acceptable duration of action. It is not metabolized by COMT.

HO—C₆H₄—OH (meta)

Resorcinol

(HO)₂C₆H₃—CH(OH)—CH$_2$—NHC(CH$_3$)$_3$

Terbutaline

HO(HOCH$_2$)C₆H₃—CH(OH)—CH$_2$—NHC(CH$_3$)$_3$

Albuterol

Another molecular modification that enhances β_2-agonist selectivity is the presence of an ethyl group on the carbon adjacent to the nitrogen. For example, the α-ethyl derivative of isoproterenol, which is known as isoetharine, is a β_2-selective agonist used in the treatment of bronchial asthma.

(HO)₂C₆H₃—CH(OH)—CH(CH$_2$CH$_3$)—NHCH(CH$_3$)$_2$

Isoetharine

Although numerous clinical studies have shown that β_2-agonists are useful in the management of bronchial asthma and that they produce less cardiac stimulation than nonselective agents, they should not be used indiscriminately or excessively. In sufficient doses, they can produce cardiovascular effects. Also, tachyphylaxis may develop in patients who use β_2-adrenergic agonists, resulting in diminished responsiveness to the drugs. Finally, an annoying, but apparently harmless, problem is the tremor that occurs as the result of the stimulation of β_2-receptors in skeletal muscle.

Extensive structure–activity investigations have shown that large *N*-substituents other than isopropyl and *tert*-butyl are useful and effective in the development of both β_1- and β_2-adrenoreceptor agonists. This fact, combined with the knowledge that dopamine has a direct stimulant effect on cardiac β-receptors, provides at least a partial explanation for the effectiveness of dobutamine as an adrenergic β_1-receptor agonist. Dobutamine, like dopamine, lacks a β-hydroxyl group on the side chain.

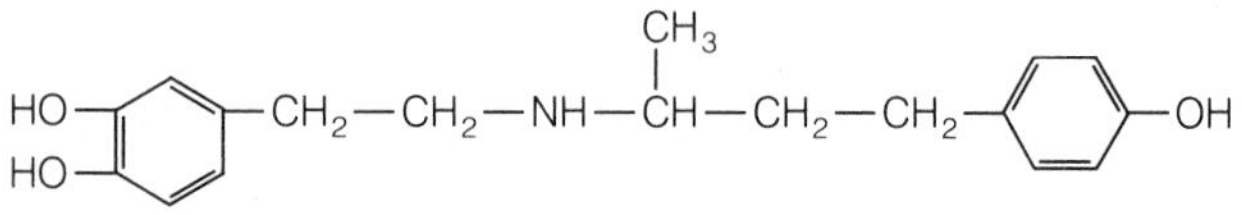

HO, HO — C6H3 —CH2—CH2—NH—CH(CH3)—CH2—CH2— C6H4 —OH

Dobutamine

PRODUCTS

Catecholamines, Phenylethanolamines, Phenylethylamines, and Related Agents

Epinephrine, USP. (−)-3,4-Dihydroxy-α-[(methylamino)methyl]benzyl alcohol (Adrenalin). This compound is a white, odorless crystalline substance that is light-sensitive. In the official product norepinephrine is present. Initially, epinephrine was isolated from the medulla of the adrenal glands of animals used for food. Although synthetic epinephrine became available soon after the structure of the hormone had been elucidated, the synthetic (±)-base has not been used widely in medicine because the natural levorotatory form is about 15 times as active as the racemic mixture.

Because of its catechol nucleus, it is oxidized easily and darkens slowly on exposure to air. Dilute solutions are partially stabilized by the addition of chlorobutanol and by reducing agents (e.g., sodium bisulfite or ascorbic acid). As the free amine, it is available in oil solution for intramuscular injection and in aqueous solution for inhalation. Like other amines, it forms salts with acids; for example, those now used include the hydrochloride, the borate, and the bitartrate. The bitartrate has the advantage of being less acid and, therefore, is used in the eye because its solutions have a *p*H close to that of lacrimal fluid. Epinephrine is destroyed readily in alkaline solutions, by aldehydes, weak oxidizing agents, and oxygen of the air.

Although an intravenous infusion of epinephrine has pronounced effects on the cardiovascular system, its use in the treatment of heart block or circulatory collapse is limited because of its tendency to induce cardiac arrhythmias. It increases systolic pressure by increasing cardiac output, and it lowers diastolic pressure by causing an overall decrease in peripheral resistance; the net result is little change in mean blood pressure.

It is sometimes useful in the treatment of glaucoma, because it apparently reduces the formation of aqueous humor, which results in a lowering of intraocular pressure.

Local application is limited, but is of value as a constrictor in hemorrhage or nasal congestion. One of its uses is to enhance the activity of local anesthetics. It is used by injection to relax the bronchial muscle in asthma and in anaphylactic reactions. Forms of administration are aqueous or oil solutions, ointment, suppositories, and inhalation.

Epinephrine has the following disadvantages: short duration of action; decomposition of its salts in solution; vasoconstrictive action frequently followed by vasodilatation; and inactivity on oral administration.

Epinephryl Borate, Ophthalmic Solution, USP. Epinephrine forms a soluble epinephryl borate complex at a neutral or slightly alkaline *p*H. The buffered solution has a *p*H of about 7.4 and the complex probably has the following structure:

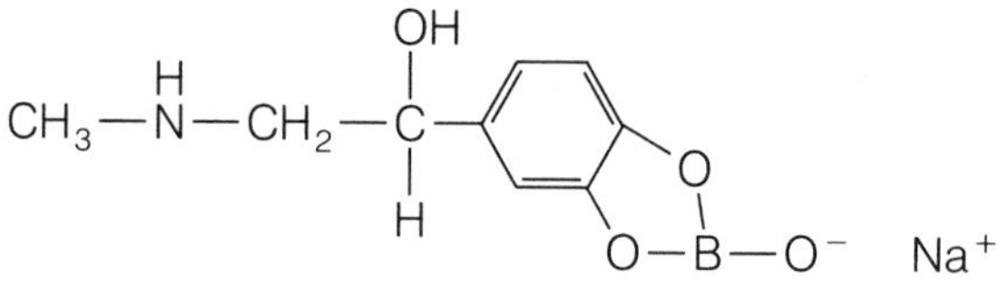

CH_3—NH—CH_2—C(OH)(H)— C6H3 (—O—B(—O⁻)—O—) Na^+

Epinephryl Borate

It is used like other epinephrine preparations by topical application in the treatment of primary open-angle glaucoma. It possesses the same limitations as the other preparations, but it is claimed to cause less stinging upon application. In the lacrimal fluid it immediately dissociates to yield free epinephrine.

Dipivefrin Hydrochloride (Propine). Dipivefrin is the pivalic acid ester prodrug of epinephrine. It is formed by the esterification of the catechol hydroxyl groups of epinephrine with pivalic acid. Dipivefrin is much more lipophilic than epinephrine, and it achieves much better penetration of the eye when administered topically as an aqueous solution for the treatment of primary open-angle glaucoma. It is converted to epinephrine by esterases in the cornea and anterior chamber. Dipivefrin offers the advantage of being less irritating to the eye than epinephrine, and it can be used in lower concentrations than epinephrine because of its more efficient transport into the eye.

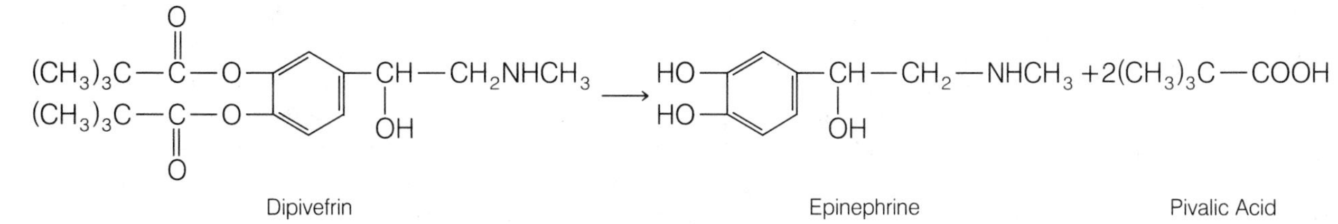

Dipivefrin → Epinephrine + Pivalic Acid

Norepinephrine Bitartrate, USP. (−)-α-(Aminomethyl)-3,4-dihydroxybenzyl alcohol bitartrate; (−)norepinephrine bitartrate (Levophed Bitartrate). Norepinephrine differs from epinephrine in that it is a primary amine rather than a secondary amine.

The bitartrate is a white crystalline powder that is soluble in water (1 : 2.5) and in alcohol (1 : 300). Solutions of the hydrochloride of norepinephrine are comparable with those of epinephrine hydrochloride in stability. The bitartrate salt is available as a more stable injectable solution. It has a pH of 3 to 4 and is preserved by using sodium bisulfite. It is used to maintain blood pressure in acute hypotensive states resulting from surgical or nonsurgical trauma, central vasomotor depression, and hemorrhage.

Isoproterenol Hydrochloride, USP. 3,4-Dihydroxy-α-[(isopropylamino)methyl]benzyl alcohol hydrochloride; isopropylarterenol hydrochloride; isoproterenolium chloride (Isuprel Hydrochloride; Aludrine Hydrochloride). This compound is a white, odorless, slightly bitter crystalline powder. It is soluble in water (1 : 3) and in alcohol (1 : 50). A 1% solution in water is slightly acidic (pH 4.5 to 5.5). It gradually darkens on exposure to air and light. Its aqueous solutions become pink on standing.

Isoproterenol is a potent β-adrenergic agonist that has virtually no effect on α-receptors. Because it is not selective for either β_1- or β_2-receptors, it causes an increase in cardiac output by stimulating cardiac β_1-receptors and brings about bronchodilatation by stimulating β_2-receptors in the respiratory tract. It also produces the metabolic effects expected of a potent β-agonist.

It is available for use by inhalation, or injection, in liquid form, and as sublingual tablets. Its principal clinical use is for the relief of bronchospasm associated with bronchial asthma. Cardiac stimulation is an undesirable, and occasionally dangerous, side effect. On the other hand, advantage is sometimes taken of isoproterenol's effect on the heart by using it for the treatment of heart block.

Isoproterenol Sulfate, USP. 3,4-Dihydroxy-α-[(isopropylamino)methyl]benzyl alcohol sulfate; 1-(3′,4′-dihydroxyphenyl)-2-isopropylaminoethanol sulfate (Medihaler-Iso; Norisodrine Sulfate). This compound is a white, odorless, slightly bitter crystalline powder. It is slightly soluble in alcohol and freely soluble in water and is hygroscopic. A 1% solution in water is acidic (pH 3.5 to 4.5). Its aqueous solutions become pink on standing. It is used for the same purpose as is the corresponding hydrochloride.

Phenylephrine Hydrochloride, USP. (−)-m-Hydroxy-α-[(methylamino)methyl]benzyl alcohol hydrochloride; phenylephrinium chloride (Neo-Synephrine Hydrochloride; Isophrin Hydrochloride). This compound is a white, odorless, crystalline, slightly bitter powder that is freely soluble in water and in alcohol. It is relatively stable in alkaline solution and is unharmed by boiling for sterilization.

The duration of action is about twice that of epinephrine. It is a vasoconstrictor and is active when given orally. It is relatively nontoxic and, when applied to mucous membrane, reduces congestion and swelling by constricting the blood vessels of the mucous membranes. It has little central nervous system stimulation and finds its main use in the relief of nasal congestion. It is also used as a mydriatic agent, as an agent to prolong the action of local anesthetics and to prevent a drop in blood pressure during spinal anesthesia. Phenylephrine is a direct-acting, α-receptor agonist.

Metaproterenol Sulfate, USP. 3,5-Dihydroxy-α-[(isopropylamino)methyl]benzyl alcohol sulfate (Alupent; Metaprel). Metaproterenol is the resorcinol counterpart of isoproterenol. It is used as the sulfate salt and, like isoproterenol, as the racemic mixture. It is more effective when given orally and has a longer duration of action than isoproterenol. Metaproterenol does exhibit significant β_2-adrenoreceptor selectivity; therefore, it does not produce cardiovascular effects similar to those caused by isoproterenol. It is not metabolized by COMT, its principal metabolite being the glucuronide conjugate. Metaproterenol is used in tablet, syrup, and inhalation forms for the management of bronchial asthma.

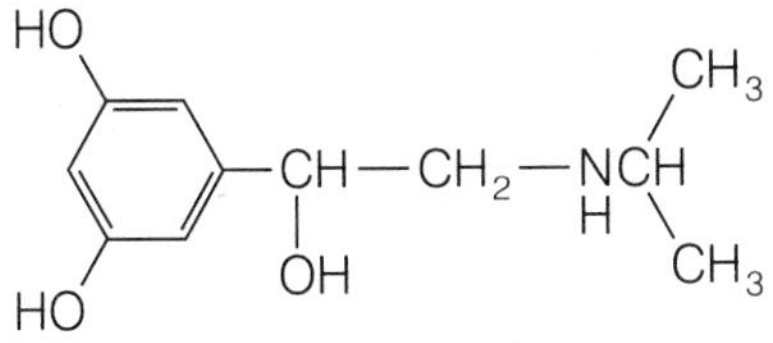

Metaproterenol

Albuterol (Proventil; Ventolin). Albuterol, the chemical structure of which was illustrated previously, is a bronchodilator that is known in most countries by the generic name salbutamol. It is a β_2-selective adrenergic agonist that produces significant bronchodilatation and less cardiac stimulation than isoproterenol when administered by inhalation. Albuterol has a longer duration of action than isoproterenol because it is not a substrate for COMT. Albuterol also is available as its sulfate salt in syrup and tablet form. It is rapidly absorbed from the gastrointestinal tract with peak plasma concentrations being achieved in about two hours. It is used primarily for the relief of bronchospasm associated with asthma.

Terbutaline Sulfate, USP. 5-[2-[(1,1-Dimethylethyl)amino]-1-hydroxyethyl]-1,3-benzenediol sulfate; 1-(3,5-dihydroxyphenyl)-2-*tert*-butylaminoethanol sulfate (Bricanyl; Brethine). This drug has been introduced with the implication that it acts preferentially at β_2-receptor sites, thus making it useful in the treatment of bronchial asthma and related conditions. However, the common cardiovascular effects that are associated with other adrenergic agents are also seen in the use of terbutaline sulfate. The drug is administered orally and is not metabolized by COMT. Evidence indicates that it is excreted primarily as a conjugate.

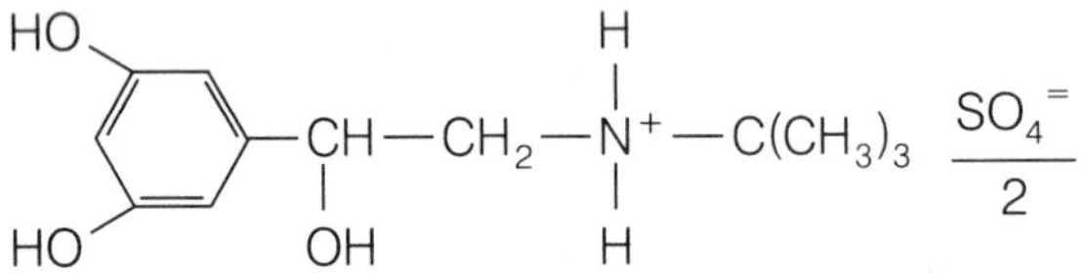

HO, HO — CH(OH)—CH_2—N^+H_2—$C(CH_3)_3$ $\frac{SO_4^{=}}{2}$

Terbutaline Sulfate

Bitolterol Mesylate (Tornalate). Bitolterol is a prodrug of the β_2-selective adrenergic agonist colterol which is the *N*-*t*-butyl analogue of norepinephrine. The presence of the two *p*-toluic acid esters in bitolterol makes it considerably more lipophilic than colterol. Bitolterol is administered by inhalation for bronchial asthma and for reversible bronchospasm. It is hydrolyzed by esterases in the lung and other tissues to produce the active agent, colterol. Bitolterol has a longer duration of action than isoproterenol and is metabolized, after hydrolysis of the esters, by COMT and by conjugation.

Dopamine Hydrochloride, USP (Intropin). Dopamine is the precursor in the biosynthesis of norepinephrine. It is used in the treatment of shock and, in contrast with the usual catecholamines, increases blood flow to the kidney in doses that have no chronotropic effect on the heart or cause no increase in blood pressure.

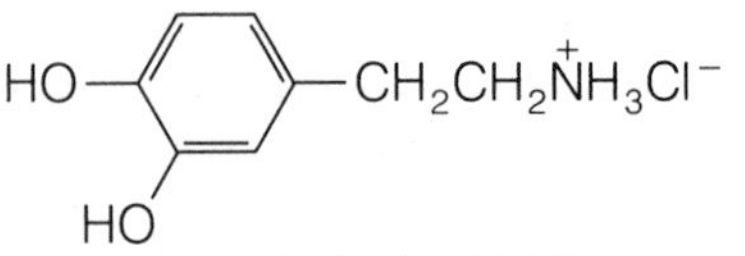

HO, HO — $CH_2CH_2\overset{+}{N}H_3Cl^-$

Dopamine Hydrochloride

The increased blood flow to the kidneys enhances the glomerular filtration rate, Na^+ excretion, and, in turn, urinary output. This is accomplished with doses of 1 to 2 $\mu g/kg$ per minute. The infusion is made using solutions that are neutral or slightly acidic in reaction.

In doses slightly higher than those required to increase renal blood flow, dopamine stimulates the β-receptors of the heart to increase cardiac output. Some of dopamine's effect on the heart is due to NE release. It is known that the dilatation of renal blood vessels produced by dopamine is the result of its agonist action on a specific dopaminergic receptor, rather than on β-receptors. Infusion at a rate greater than 10 $\mu g/kg$ per minute results in a stimulation of α-receptors leading to vasoconstriction and an increase in arterial blood pressure.

Ephedrine, USP. (−)-*erythro*-α-[(1-Methylamino)ethyl]benzyl alcohol. Ephedrine is an alkaloid that can be obtained from the stems of various species of *Ephedra*. The drug Ma Huang, containing ephedrine, was known to the Chinese in 2000 BC, but the active principle, ephedrine, was not isolated until 1885.

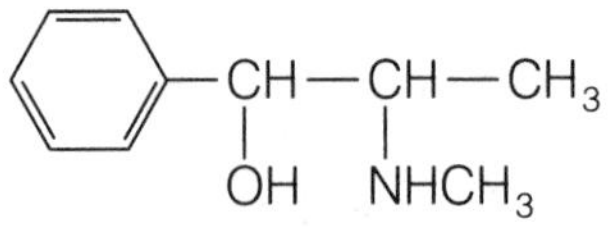

C_6H_5—CH(OH)—CH($NHCH_3$)—CH_3

Ephedrine

Ephedrine has two asymmetric carbon atoms; thus, there are four optically active forms. The *erythro* racemate is called ephedrine, and the *threo*

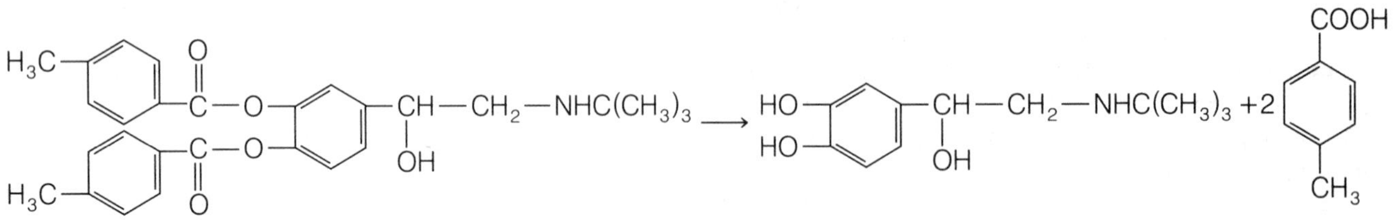

Bitolterol: (H_3C—C_6H_4—C(=O)—O)$_2$—C_6H_3—CH(OH)—CH_2—$NHC(CH_3)_3$ ⟶ Colterol: (HO)$_2C_6H_3$—CH(OH)—CH_2—$NHC(CH_3)_3$ + 2 *p*-Toluic Acid: COOH—C_6H_4—CH_3

Bitolterol   Colterol   *p*-Toluic Acid

TABLE 11-1

Isomer		Relative Pressor Activity
D	(−) Ephedrine	36
DL	(±) Ephedrine	26
L	(+) Ephedrine	11
L	(+) Pseudoephedrine	7
DL	(±) Pseudoephedrine	4
D	(−) Pseudoephedrine	1

racemate is known as *pseudo*ephedrine (ψ-ephedrine). Natural ephedrine is D(−) and is the most active of the four isomers as a pressor amine. Table 11-1 lists the relative pressor activity of isomers of ephedrine. Racemic ephedrine, racephedrine, is used for the same purpose as the optically active alkaloids.

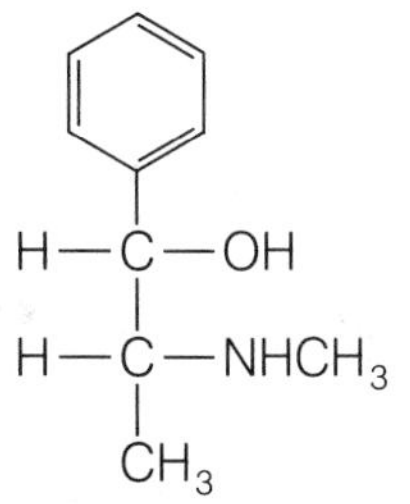

H—C—OH, H—C—$NHCH_3$, CH_3

Erythro form
Ephedrine

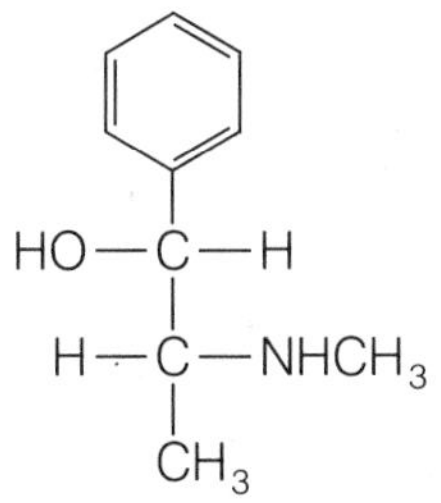

HO—C—H, H—C—$NHCH_3$, CH_3

Threo form
ψ-Ephedrine

The ephedrine alkaloid occurs as a waxy solid and as crystals or granules and has a characteristic pronounced odor. Because of its instability in light, it decomposes gradually and darkens. It may contain up to one-half molecule of water of hydration. It is soluble in alcohol, water (5%), some organic solvents, and liquid petrolatum. The free alkaloid is a strong base and an aqueous solution of the free alkaloid has a *p*H above 10. The salt form has a pK_a of 9.6.

Ephedrine simulates epinephrine in physiologic effects, but its pressor action and local vasoconstrictor action are of greater duration. It causes more pronounced stimulation of the central nervous system than does epinephrine, and it is effective when given orally or systemically.

Ephedrine and its salts are used orally, intravenously, intramuscularly, and topically in a variety of conditions such as allergic disorders, colds, hypotensive conditions, and narcolepsy. It is employed locally to constrict the nasal mucosa and cause decongestion, and to dilate the pupil or the bronchi. Systemically, it is effective for asthma, hay fever, and urticaria.

Pseudoephedrine Hydrochloride, USP. (+)-*threo*-α-[(1-Methylamino)ethyl]benzyl alcohol hydrochloride; isoephedrine hydrochloride (Sudafed). The hydrochloride salt is a white crystalline material, soluble in water, in alcohol, and in chloroform. Pseudoephedrine, like ephedrine, is a useful nasal decongestant, but is much less active in increasing blood pressure. However, it should be used with caution in hypertensive individuals.

Phenylpropanolamine Hydrochloride, USP. α-(1-Aminoethyl)benzenemethanol hydrochloride; (±)-1-phenyl-2-amino-1-propanol hydrochloride; (±)-norephedrine hydrochloride (Propadrine Hydrochloride). Phenylpropanolamine is the primary amine corresponding to ephedrine, and this modification gives an agent that has slightly higher vasopressor action and lower toxicity and central stimulation action than has ephedrine. It can be used in place of ephedrine for most purposes and is used widely as a nasal decongestion agent. For the latter purpose it is applied locally to shrink swollen mucous membranes; its action is more prolonged than that of ephedrine. It also is stable when given orally.

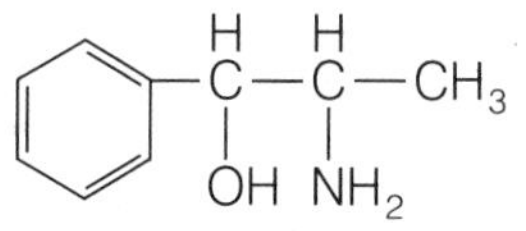

H H, —C—C—CH_3, OH NH_2

Phenylpropanolamine

Phenylpropanolamine appears to have a mixed mechanism of action. It is commonly used as the active component of over-the-counter appetite suppressants, but its efficacy as an anorectic agent is questionable. No sympathomimetic amines have been found to be effective for long-term use in weight-reduction programs.

Mephentermine Sulfate, USP. *N*,α,α-Trimethylbenzeneethanamine sulfate; *N*,α,α-trimethylphenethylamine sulfate (Wyamine Sulfate). The sulfate is a white crystalline powder with a faint fishy odor. It is soluble 1:20 in water and 1:50 in alcohol. A 1% solution in water is acidic, *p*H 5.5 to 6.2.

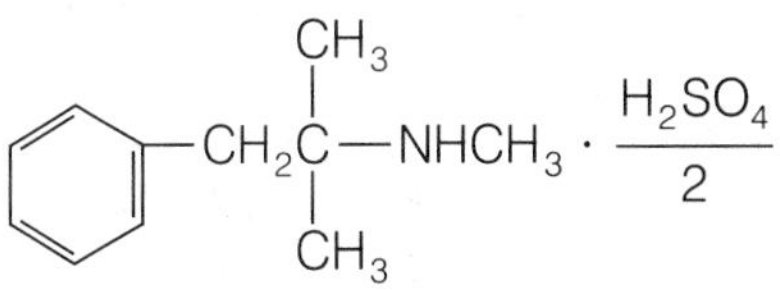

CH_3, —CH_2C—$NHCH_3$ · $\frac{H_2SO_4}{2}$, CH_3

Mephentermine Sulfate

It exhibits pressor amine properties and may be injected parenterally as a vasopressor agent in acute hypotensive states. Mephentermine is an indirect-acting agent with a prolonged duration of action.

Metaraminol Bitartrate, USP. (−)-α-(1-Aminoethyl)-*m*-hydroxybenzyl alcohol tartrate; (−)-*m*-hydroxynorephedrine bitartrate (Aramine). Metaraminol bitartrate is freely soluble in water and 1 : 100 in alcohol. This compound is a potent vasopressor with prolonged duration of action.

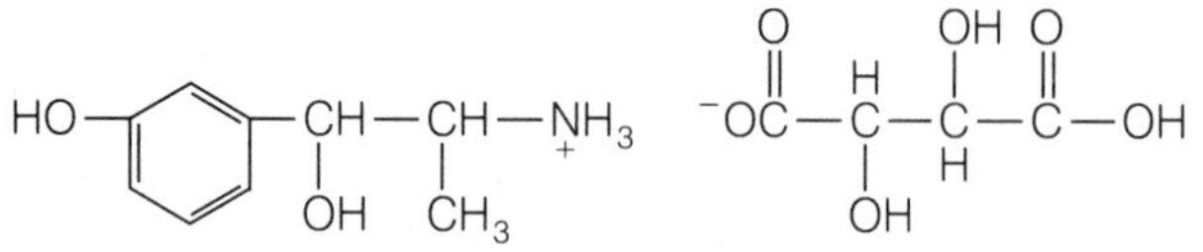

Metaraminol Bitartrate

Metaraminol bitartrate is useful for parenteral administration in hypotensive episodes during surgery, for sustaining blood pressure in patients under general or spinal anesthesia, and for the treatment of shock associated with trauma, septicemia, infectious diseases, and adverse reactions to medication. It does not produce central nervous system stimulation.

Hydroxyamphetamine Hydrobromide, USP. (±)-*p*-(2-Aminopropyl)phenol hydrobromide; 1-(*p*-hydroxyphenyl)-2-aminopropane hydrobromide (Paredrine). This compound is a white crystalline material that is very soluble in water (1 : 1) and in alcohol (1 : 2.5).

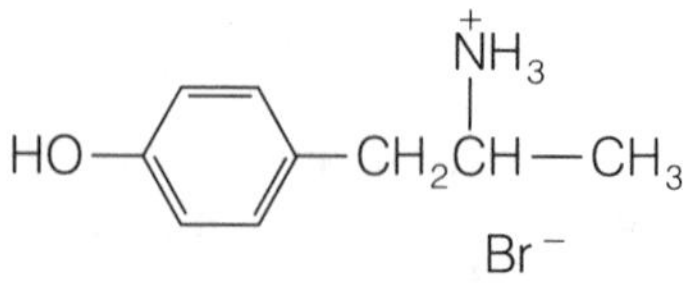

Hydroxyamphetamine Hydrobromide

It has been useful for its synergistic action with atropine in producing mydriasis. A more rapid onset, more complete dilatation, and more rapid recovery are observed with a mixture of atropine and hydroxyamphetamine hydrobromide than with atropine alone.

Hydroxyamphetamine has little or no ephedrine-like central nervous system-stimulating action, but retains the ability to shrink the nasal mucosa. Its actions as a bronchodilator or as an appetite-reducing agent are too weak to make it useful in these fields.

Levonordefrin, USP. (−)-α-(1-Aminoethyl)-3,4-dihydroxybenzyl alcohol (Neo-Cobefrin). This compound is a strong vasoconstrictor and is used with local anesthetics.

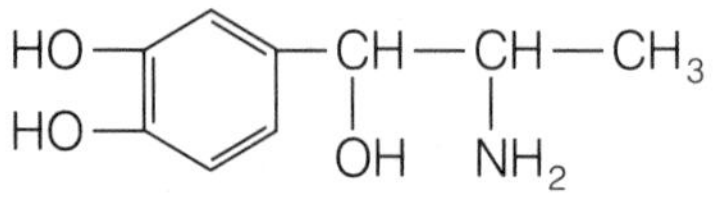

Levonordefrin

The structure has the catechol nucleus of epinephrine, but the side chain of norephedrine.

Isoetharine Hydrochloride, USP (Bronkosol). Isoetharine, which is also available as the mesylate salt, is a β_2-selective agent used as an inhalation aerosol for the treatment of bronchial asthma. It is metabolized by COMT.

Ethylnorepinephrine Hydrochloride, USP (Bronkephrine). This compound differs from NE by the presence of an ethyl group on the α-position of the side chain. There are two asymmetric centers, thus giving rise to four possible isomers. No information is available on the relative activities of the individual isomers. The predominant action of the drug is as a β-adrenergic stimulant. It is weaker than isoproterenol in this regard. Although not the drug of choice, it does have utility in the treatment of asthma. Bronkephrine sulfate is administered by subcutaneous or intramuscular injection.

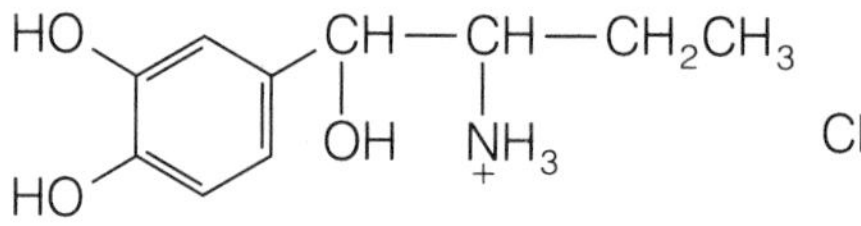

Ethylnorepinephrine Hydrochloride

Methoxyphenamine Hydrochloride, USP. 2-Methoxy-*N*,α-dimethylbenzeneethanamine hydrochloride; 2-(*o*-methoxyphenyl)isopropylmethylamine hydrochloride (Orthoxine Hydrochloride). Methoxyphenamine hydrochloride is a bitter, odorless, white crystalline powder that is freely soluble in alcohol and water. A 5% solution is slightly acidic (*p*H 5.3 to 5.7).

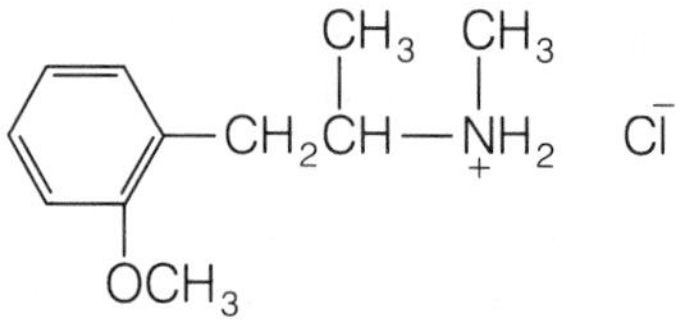

Methoxyphenamine Hydrochloride

It is a sympathomimetic compound, the predominant actions of which are bronchodilatation and inhibition of smooth muscle. Its effects on blood vessels are slight, its pressor effect being considerably less than that of ephedrine or epinephrine. Methoxyphenamine is used as a bronchodilator.

Administration of this drug produces no alterations in blood pressure and only slight cardiac stimulation. The actions on the central nervous system are minor.

Methoxamine Hydrochloride, USP. α-(1-Aminoethyl)-2,5-dimethoxybenzeneethanol hydrochloride; 2-amino-1-(2,5-dimethoxyphenyl)propanol hydrochloride (Vasoxyl). This compound is a white, platelike crystalline substance with a bitter taste. It is odorless or has only a slight odor. It is soluble in water 1:2.5 and in alcohol 1:12. A 2% solution in water is slightly acidic (pH 4.0 to 5.0), and it is affected by light.

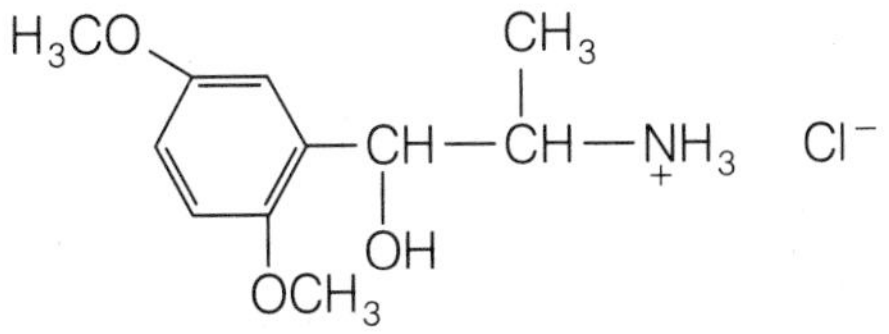

Methoxamine Hydrochloride

Methoxamine hydrochloride is a sympathomimetic amine that exhibits the vasopressor action characteristic of other agents of this class. The drug tends to slow the ventricular rate; it does not produce ventricular tachycardia, fibrillation, or an increased sinoatrial rate. It is free of cerebral-stimulating action.

It is used primarily during surgery to maintain adequate or to restore arterial blood pressure, especially in conjunction with spinal anesthesia. It is also used in myocardial shock and other hypotensive conditions associated with hemorrhage, trauma, and surgery.

Dobutamine Hydrochloride, USP (Dobutrex). Dobutamine is a catecholamine that exerts its principal agonist action on cardiac β_1-receptors. It is a considerably less potent agonist at β_2-receptor sites. Its cardiac stimulant effect does not involve an indirect NE-releasing action. In vivo, the inotropic action of dobutamine is predominant over its chronotropic effects. It does not act as an agonist at the dopaminergic receptors that mediate renal vasodilatation.

Dobutamine is used as an intravenous infusion in the treatment of heart failure.

Dobutamine has a plasma half-life of about two minutes. It is metabolized by COMT and by conjugation.

Nylidrin Hydrochloride, USP. p-Hydroxy-α-[1-[(1-methyl-3-phenylpropyl)amino]ethyl]benzyl alcohol hydrochloride (Arlidin). This compound is a white, odorless, practically tasteless crystalline powder. It is soluble in water 1:65 and in alcohol 1:40. A 1% solution in water is acidic (pH 4.5 to 6.5).

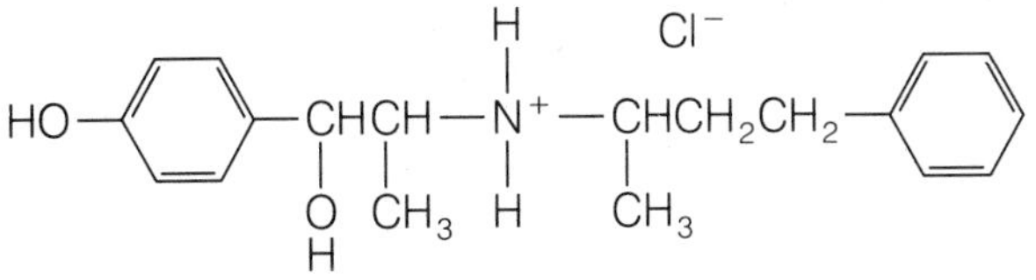

Nylidrin Hydrochloride

Nylidrin acts as a peripheral vasodilator. It is indicated in vascular disorders of the extremities that may be benefited as the result of increased blood flow. It is administered orally.

Isoxsuprine Hydrochloride, USP. p-Hydroxy-α-[1-[(1-methyl-2-phenoxyethyl)amino]ethyl]benzyl alcohol hydrochloride (Vasodilan). This compound is used as a vasodilator for symptomatic relief in peripheral vascular disease and cerebrovascular insufficiency. Isoxsuprine may activate the β_2-receptors in small blood vessels with a resulting dilatation of those vessels. However, some studies indicate that it is not a β-agonist.

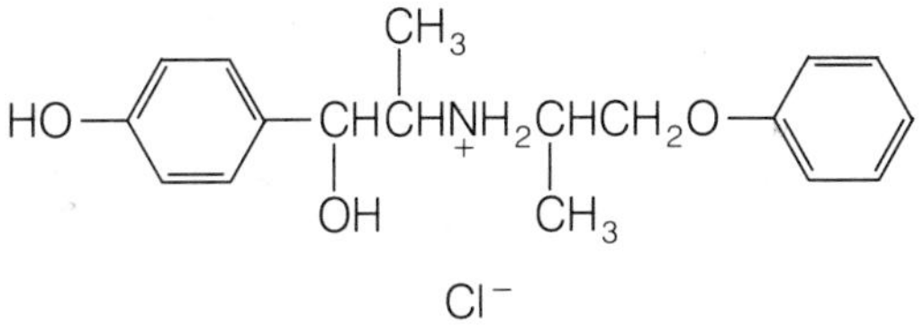

Isoxsuprine Hydrochloride

Ritodrine Hydrochloride, USP. 4-Hydroxy-α-[1-[[2-(4-hydroxyphenyl)ethyl]amino]ethyl]benzenemethanol hydrocloride; 1-(4-hydroxyphenyl)-2-[2-(4-hydroxyphenyl)ethylamino]propanol hydrochloride (Yutopar). Ritodrine is a selective β_2-receptor agonist that is used to control premature labor and to reverse fetal distress caused by excessive uterine activity. Its uterine inhibitory effects are more sustained than its effects on the cardiovascular system, which are minimal compared with those caused by nonselective β-agonists. The cardiovascular effects usually associated with its administration are mild tachycardia and slight diastolic pressure decrease. It is available as an injection and in tablet form for oral administration. It is usually administered initially by intravenous infusion to stop premature labor. Subsequently, it may be given orally.

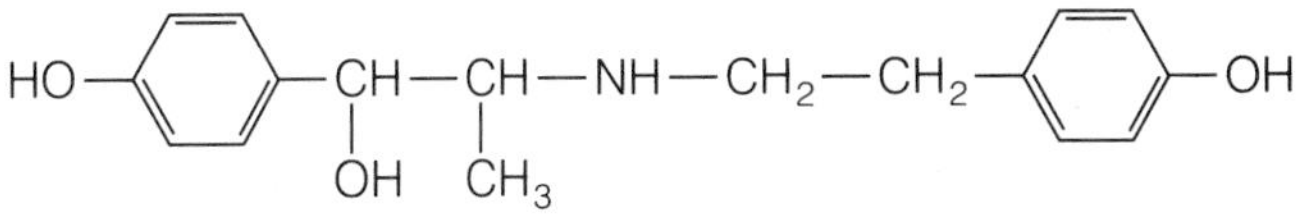

Ritodrine

Aliphatic Amines

In the classic work by Barger and Dale[25] in 1910, the pressor action of the aliphatic amines was described, but only since the early 1940s have these agents become of pharmaceutic importance. An investigation in 1944[26] determined the influence of the location of an amino group on an aliphatic carbon chain and the effect of branching of the carbon chain carrying an amino group on pressor action. Optimal conditions were found in compounds of seven to eight carbon atoms with a primary amino group in the 2-position. Branching of the chain increases pressor activity.

A series of secondary β-cyclohexylethyl- and β-cyclopentylethylamines have been shown to have sympathomimetic activity.[27]

Propylhexedrine, USP. *N*,α-Dimethylcyclohexaneethylamine (Benzedrex). This material is a clear, colorless liquid, with a characteristic fishy odor. Propylhexedrine is very soluble in alcohol and only slightly soluble in water. It volatilizes slowly at room temperature and absorbs carbon dioxide from air. Its uses and actions are similar to those of other volatile sympathomimetic amines. It produces vasoconstriction and a decongestant effect on the nasal membranes, but it has only about one-half the pressor effect of amphetamine and produces decidedly less effect on the nervous system. Therefore, its major use is for local shrinking effect on nasal mucosa in the symptomatic relief of nasal congestion caused by the common cold, allergic rhinitis, or sinusitis.

Propylhexedrine

Isometheptene Mucate. *N*,1,5-Trimethyl-4-hexenylamine mucate; 2-methylamino-6-methyl-5-heptene mucate. Isometheptene is a sympathomimetic amine that has limited use as a component of medications used to treat migraine headaches. It is claimed to constrict cranial and cerebral arterioles, which reduces the stimuli that lead to vascular headaches.

Isometheptene Mucate

Imidazoline Derivatives

Several important adrenergic agonists are derivatives of imidazoline. The imidazoline nucleus is associated primarily with α_2-receptor activity.

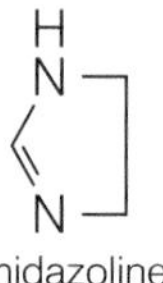

Imidazoline

Naphazoline Hydrochloride, USP. 2-(1-Naphthylmethyl)-2-imidazoline monohydrochloride (Privine Hydrochloride). This compound is a bitter, odorless, white crystalline powder that is a potent vasoconstrictor, similar to ephedrine in its action. It is freely soluble in water and in alcohol. When applied to nasal and ocular mucous membranes, it causes a prolonged reduction of local swelling and congestion. It is of value in the symptomatic relief of disorders of the upper respiratory tract.

Naphazoline Hydrochloride

Tetrahydrozoline Hydrochloride, USP. 2-(1,2,3,4-Tetrahydro-1-naphthyl)-2-imidazoline monohydrochloride (Tyzine; Visine). This compound is closely related to naphazoline hydrochloride in its pharmacologic action. When applied topically to the nasal mucosa, the drug causes vasoconstriction, which results in reduction of local swelling and congestion. It is also useful in a 0.05% solution (Visine) as an ocular decongestant. When used two or three times daily, there is no influence on pupil size. It does not appear to increase intraocular pressure; however, its use in the presence of glaucoma is not recommended.

Tetrahydrozoline Hydrochloride

Xylometazoline Hydrochloride, USP. 2-(4-*t*-Butyl-2,6-dimethylbenzyl)-2-imidazoline monohydrochloride (Ortrivin Hydrochloride). This compound is used as a nasal vasoconstrictor. Its duration of action is approximately four to six hours.

Xylometazoline Hydrochloride

Oxymetazoline Hydrochloride, USP. 6-*t*-Butyl-3-(2-imidazolin-2-ylmethyl)-2,4-dimethylphenol monohydrochloride; 2-(4-*t*-butyl-2,6-dimethyl-3-hydroxybenzyl)-2-imidazoline hydrochloride (Afrin). This compound, closely related to xylometazoline, is a long-acting vasoconstrictor. It is used as a topical aqueous nasal decongestant.

Clonidine Hydrochloride (Catapres) is an imidazoline derivative that is structurally related to vasoconstrictor agents such as naphazoline and xylometazoline. Although clonidine exhibits some vasoconstrictor properties, it is of principal interest because of its agonist effects on central α_2-receptors. The usefulness of clonidine as a centrally acting antihypertensive agent is discussed in the chapter on cardiovascular drugs. The imidazoline nucleus is not required for central α_2-agonist activity, as exemplified by the guanidine derivatives, **guanabenz** (Wytensin) and **guanfacine** (Tenex), both of which produce antihypertensive effects as a result of activation of α_2-receptors in the central nervous system.

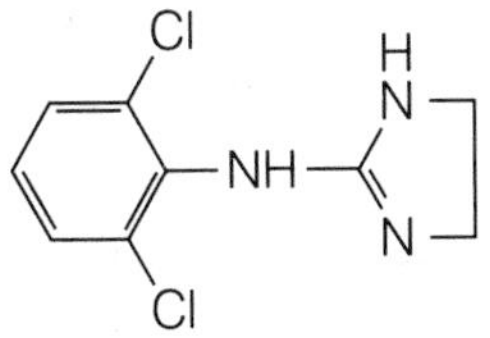

Clonidine

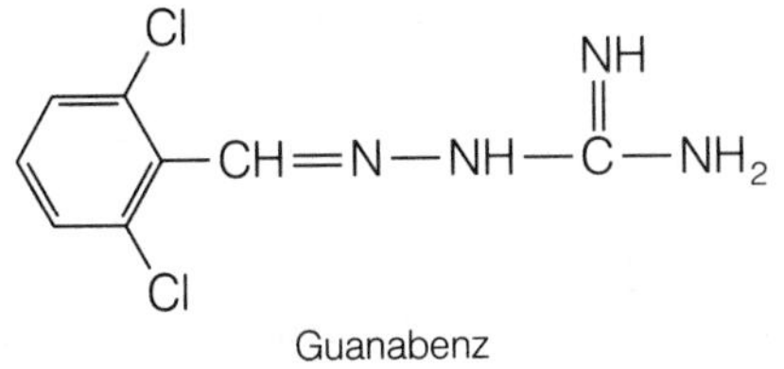

Guanabenz

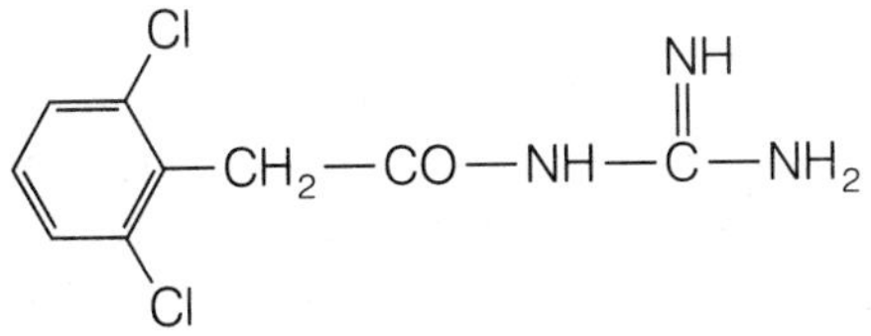

Guanfacine

ADRENERGIC-BLOCKING AGENTS

NEURONAL-BLOCKING AGENTS

Neuronal-blocking agents are drugs that produce their pharmacologic effects primarily by preventing the release of NE from sympathetic nerve terminals. The drugs of this type enter the adrenergic neuron by way of the uptake$_1$ process, and they cause the release of some of the stored NE. However, NE depletion does not appear to account for their antiadrenergic activity. Currently, it is believed that neuronal-blocking agents produce their effects by stabilization of the neuronal membrane or the membranes of the storage vesicles. This stabilization makes the membranes less responsive to nerve impulses, thereby inhibiting the release of NE into the synaptic cleft. Two neuronal blockers that are currently available for clinical use in the United States are bretylium tosylate, which is administered parenterally to control cardiac arrhythmias, and guanethidine sulfate, which is used as an antihypertensive agent. A third such agent is guanadrel (Hylorel), a guanidine derivative that is structurally related to guanethidine. These compounds are discussed in the chapter on cardiovascular drugs (see Chap. 14).

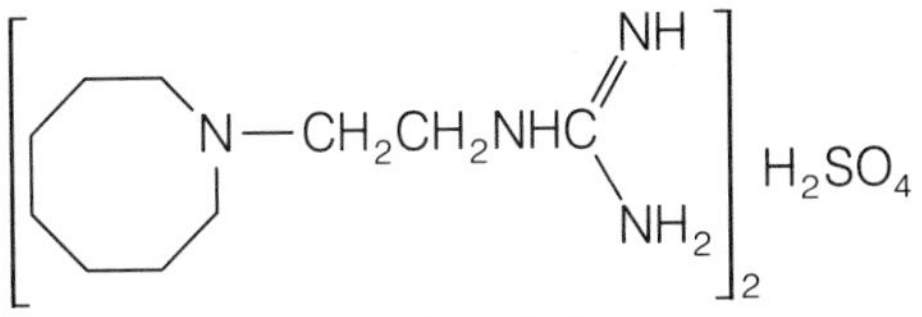

Guanethidine Sulfate

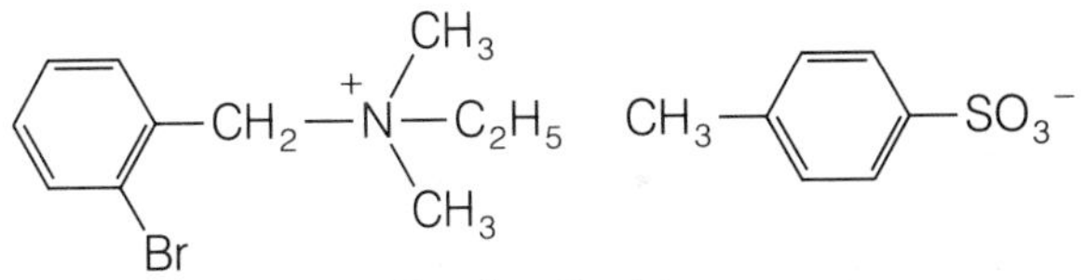

Bretylium Tosylate

α-ADRENERGIC-BLOCKING AGENTS

Structure–activity relationships among the various types of α-antagonists are difficult to perceive, if they indeed do exist. Unlike the β-antagonists, which bear clear structural similarities to the adrenergic agonists NE, epinephrine, and isoproterenol, the α-antagonists comprise a mixture of chemical classes that bears little obvious resemblance to the agonists.

Ergot Alkaloids

The oxytocic action of ergot was recognized as early as the 16th century, and it was used by midwives for years before its acceptance by the medical profession. Ergotoxine, isolated in 1906, and ergotamine, isolated in 1920, for many years were thought to be the principal alkaloids present. Since then, the former has been shown to be nonhomogeneous and composed of equal parts of three bases: ergocornine, ergocristine, and ergocryptine (Table 11-2). In 1933, sensibamine was reported as a new base, only to be

TABLE 11-2

ERGOT ALKALOIDS*

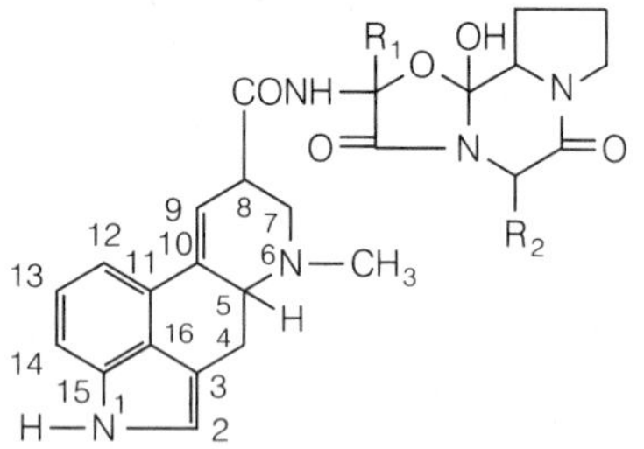

	R_1	R_2
Ergotamine group		
Ergotamine	$-CH_3$	$-CH_2-C_6H_5$
Ergosine	$-CH_3$	$-CH_2CH(CH_3)_2$
Ergotoxine group		
Ergocristine	$-CH(CH_3)_2$	$-CH_2-C_6H_5$
Ergocryptine	$-CH(CH_3)_2$	$-CH_2CH(CH_3)_2$
Ergocornine	$-CH(CH_3)_2$	$-CH(CH_3)_2$

*Each of the listed alkaloids has an inactive diastereoisomer derived from *iso*-lysergic acid which, in the above formulas, differs only in that the configuration of the hydrogen and the carboxyl groups at position 8 is interchanged. The nomenclature also differs, in that the suffix *in* is added to the name, e.g., ergotaminine instead of ergotamine. However, for ergonovine, the diastereoisomer is named "ergometrinine" because this derives from the name of ergonovine commonly used in England (i.e., ergometrine).

TABLE 11-3

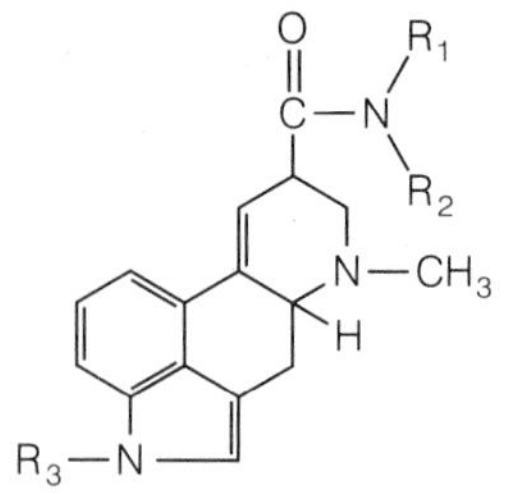

	R_1	R_2	R_3
A. Ergonovine	$-H$	$-CH(CH_3)CH_2OH$	$-H$
B. Methylergonovine	$-H$	$-CH(C_2H_5)CH_2OH$	$-H$
C. LSD	$-C_2H_5$	$-C_2H_5$	$-H$
D. Methysergide	$-H$	$-CH(C_2H_5)CH_2OH$	$-CH_3$

shown later to be a mixture of equal parts of ergotamine and ergotaminine. Similarly, ergoclavine (1934) has been shown to be a mixture of ergosine and ergosinine. In 1935, an active water-soluble alkaloid was reported simultaneously by four research groups and is the alkaloid now known as ergonovine (ergometrine in Great Britain).

The active alkaloids are all amides of lysergic acid, whereas the inactive diastereoisomeric counterparts are similarly derived from *iso*-lysergic acid. The only difference between the two acids is the configuration of the substituents at position 8 of the molecule. The structure of ergonovine is the simplest of these alkaloids, being the amide of lysergic acid derived from (+)-2-aminopropanol.

The isomers of ergonovine (*A*; Table 11-3) have been prepared for pharmacologic study. Only the propanolamides of (+)-lysergic acid were active. The optical configuration of the amino alcohol did not seem to be important to pharmacologic activity. Other partially synthetic derivatives of (+)-lysergic acid have been prepared, with two of them showing notable activity: methylergonovine, the amide formed from (+)-lysergyl chloride and 2-aminobutanol (*B*), and the *N*-diethyl amide of (+)-lysergic acid (*C*). The latter compound, also known as LSD, has an oxytocic action comparable with that of ergonovine and, in addition, is known to cause, in very small doses (100 to 400 μg), marked psychic changes combined with hallucinations and colored visions. The hydrogenation of the C-9 to C-10 double bond in the lysergic acid portion of the ergot alkaloids, other than ergonovine, enhances the adrenergic-blocking activity.

Pharmacologically, the naturally occurring ergot alkaloids may be placed in two classes: (1) the water-insoluble, polypeptidelike group comprising ergocryptine, ergocornine, ergocristine (ergotoxine group), ergosine, and ergotamine; and (2) the water-soluble alkaloid ergonovine. The members of the water-insoluble group are typical adrenergic-blocking agents in that they inhibit responses to the stimulation of adrenergic nerves and block the effects of circulating epinephrine. In addition, they cause a rise in blood pressure by constriction of the peripheral blood vessels owing to a direct action on the smooth muscle of the vessels. The most important action of these alkaloids, however, is their strongly stimulating action on the smooth muscle of the uterus, especially the gravid or puerperal uterus. This activity develops more slowly and lasts longer when the water-insoluble alkaloids are used than when ergonovine is administered. Toxic doses or the too frequent use of these alkaloids in small doses are responsible for the symptoms of ergotism. These alkaloids are rendered water-soluble by preparing salts of them with such organic acids as tartaric, maleic, ethylsulfonic, or methylsulfonic.

Ergonovine has little or no activity as an adrenergic-blocking agent and, indeed, has many of the pharmacologic properties (produces mydriasis in the rabbit's eye, relaxes isolated strips of gut, constricts blood vessels) of a sympathomimetic drug. It does not raise the blood pressure when injected intravenously into an anesthetized animal. It possesses a strong, prompt, oxytocic action. It is highly active orally and causes little nausea or vomiting.

Ergonovine Maleate, USP. (Ergotrate Maleate). This water-soluble alkaloid was isolated, as indicated, from ergot, in which it occurs to the extent of 200 $\mu g/g$ of ergot. The several research groups who almost simultaneously isolated the alkaloid named the alkaloid according to the dictates of each. Thus, the names ergometrine, ergotocin, ergostetrine, and ergobasine were assigned to this alkaloid. To clarify the confusion, the Council on Pharmacy and Chemistry of the American Medical Association adopted a new name, ergonovine, which is in general use today. Of course, commercial names differ from the Council-accepted name, the principal one in the United States being Ergotrate (ergonovine maleate).

The free base occurs as white crystals that are quite soluble in water or alcohol and levorotatory in solution. It readily forms crystalline, water-soluble salts, behaving as a monoacidic base. The nitrogen involved in salt formation obviously is not the one in the indole nucleus, for it is far less basic than the other nitrogen. The official salt is the maleate. It is said to be a convenient form in which to crystallize the alkaloid and is also quite stable.

Ergonovine maleate occurs in the form of a light-sensitive, white or nearly white, odorless crystalline powder. It is soluble in water (1 : 36) and in alcohol (1 : 20), but is insoluble in ether and in chloroform.

Ergonovine has a powerful stimulating action on the uterus and is used for this effect.

During the third stage of labor, ergonovine should be used after passage of the placenta. Ordinarily, 200 μg of ergonovine is injected at this stage to bring about prompt and sustained contraction of the uterus. The effect lasts about five hours and prevents excessive blood loss. It also lowers the incidence of uterine infection. A continued effect may be obtained by further administration of the alkaloid, either orally or parenterally.

Ergotamine Tartrate, USP. Ergotamine tartrate (2 : 1) (salt) (Gynergen). Ergotamine, one of the insoluble ergot alkaloids, is obtained from the crude drug by the usual isolation methods.

It occurs as colorless crystals or as a white to yellowish white crystalline powder. It is not especially soluble in water (1 : 500) or in alcohol (1 : 500), although the aqueous solubility is increased with a slight excess of tartaric acid. Ergotamine has both direct vasoconstrictor properties and an α-receptor antagonist action.

Ergotamine is usually administered orally or sublingually. Because it is not well absorbed after oral administration, it is also available for administration by inhalation. It is available in combination with caffeine, both in tablets and suppositories. Caffeine appears to enhance the absorption of ergotamine.

Previous to the discovery of ergonovine, ergotamine was the ergot drug of choice as a uterine stimulant, either orally or parenterally. Because it offered no advantage over ergonovine, except for a more sustained action and, in addition, was more toxic, it fell into disuse. However, it has been employed as a specific analgesic in the treatment of migraine headache, in which capacity it is reasonably effective. A combination of ergotamine tartrate and caffeine (Cafergot) is an available product. It is of no value in other types of headaches and sometimes fails to abort migraine headaches. It has no prophylactic value. Care should be exercised in its continued use to prevent signs of ergotism.

Dihydroergotamine Mesylate, USP. Dihydroergotamine monomethanesulfonate (DHE-45). This compound is produced by the hydrogenation of the easily reducible C9-C10 double bond in the lysergic acid portion of the ergotamine molecule. It occurs as a white, yellowish, or faintly red powder that is only slightly soluble in water and chloroform, but soluble in alcohol.

Dihydroergotamine, although very closely related to ergotamine, has less vasoconstrictor activity and less uterine stimulant activity than ergotamine. However, the adrenergic-blocking action is stronger. One of its principal uses has been in the relief of migraine headache in a manner similar to ergotamine. Because the drug is not particularly effective orally, it is administered intravenously or intramuscularly.

Methylergonovine Maleate, USP. *N*-[α-(Hydroxymethyl)propyl]-D-lysergamide (Methergine). This compound occurs as a white to pinkish-tan microcrystalline powder that is odorless and bitter. It is only slightly soluble in water and alcohol and very slightly soluble in chloroform and ether.

It is very similar to ergonovine in its pharmacologic actions, and it is used as a uterine stimulant after delivery of the placenta.

Methysergide Maleate, USP. (Sansert). This drug was introduced in 1962. It occurs as a white to yellowish white crystalline powder that is practically odorless. It is only slightly soluble in water and alcohol and very slightly soluble in chloroform and ether.

Although it is closely related in structure to methylergonovine, it does not possess the potent oxytocic action of the latter. It has been shown to be a potent serotonin antagonist and has found its principal use in the prevention of migraine headache, but the exact mechanism of prevention has not been elucidated. Methysergide is a weak α-antagonist and a weak vasoconstrictor.

Methysergide produces a variety of untoward side effects, although most of them are mild and will disappear with continued use. Some of the most common of these effects are nausea, epigastric pain, dizziness, restlessness, drowsiness, leg cramps, and psychic effects. However, it has become increasingly evident that this drug must be carefully administered under a physician's and pharmacist's watchful eye. This is because when administered on a long-term uninterrupted basis, it appears to be prone to induce retroperitoneal fibrosis, pleuropulmonary fibrosis, and fibrotic thickening of cardiac valves. As a consequence of these potential fibrotic manifestations, the drug has been reserved for "prophylaxis in patients whose vascular headaches are frequent and/or severe and uncontrollable and who are under close medical supervision." Because of its side effects, it should not be continuously administered for longer than a six-month period without a drug-free interval of three to four weeks between each six-month course of treatment. Furthermore, the dosage should be reduced gradually during the last two to three weeks of the six-month treatment period to avoid "headache rebound." The drug is not recommended for children.

Methysergide is an effective blocker of the effects of serotonin, which may be involved in the mechanism of vascular headaches. The complete mechanism and the involvement of the drug have not yet been completely clarified.

Ergoloid Mesylates (Hydergine, Alkergot, Deaphil-ST, Trigot). This preparation contains equal parts of the dihydrogenated alkaloids, dihydroergocornine, dihydroergocristine, and dihydroergocryptine. These alkaloids are not vasoconstrictors. The preparation is used in the treatment of senile dementias. The rationale for this use is based on the assumptions that the symptoms of senile dementia may be due to cerebral arteriosclerosis and that the condition may be alleviated by the cerebral vasodilator effect of the dihydrogenated ergot alkaloids. The benefits of the drug are modest, and the mechanism of action has not been proved to involve cerebrovascular dilatation. The use of ergot alkaloids for treatment of senile dementias is controversial. The ergoloid mesylates preparation is available in tablet form for oral administration.

β-Haloalkylamines

Although dibenamine (*N,N*-dibenzyl-β-chloroethylamine), the prototype of these compounds, was characterized in 1934 by Eisleb[28] incidental to a description of some other synthetic intermediates, it was the report of Nickerson and Goodman[29] in 1947 on the pharmacology of the compound that revealed the powerful adrenergic-blocking properties. The blockade produced by this group of compounds seems to be the most complete of the entire group of blocking agents. The differences in activity of the members of this group differ only quantitatively, being qualitatively the same. When given in adequate doses, they produce a slowly developing, prolonged adrenergic blockade that is not overcome by epinephrine. Much of the early work done with this group was confined to dibenamine, but this has been largely supplanted by the more orally useful and potent phenoxybenzamine.

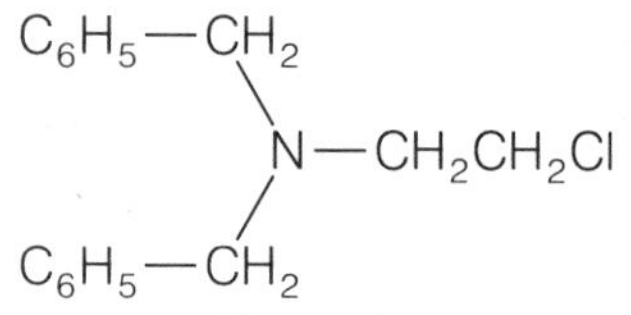

$$(C_6H_5{-}CH_2)_2N{-}CH_2CH_2Cl$$

Dibenamine

The mechanism whereby β-haloalkylamines produce a long-lasting, α-adrenoreceptor blockade involves the formation of an intermediate aziridinium ion (ethylene iminium ion, immonium ion). This positively charged electrophile reacts with a nucleophilic group on the receptor, resulting in the formation of a covalent bond between the drug and the receptor.

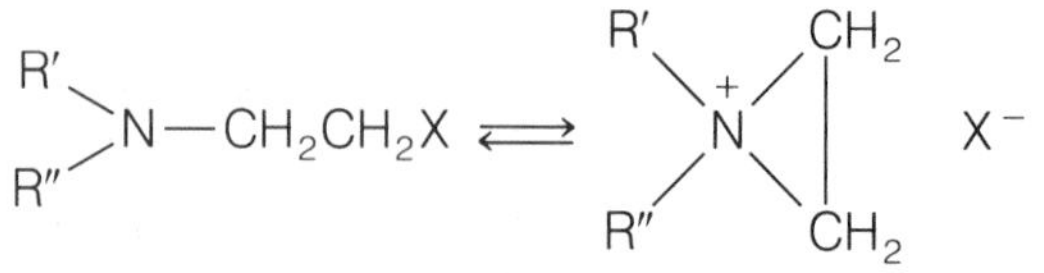

$$R'R''N{-}CH_2CH_2X \rightleftharpoons R'R''\overset{+}{N}(CH_2CH_2)\ X^-$$

This alkylation of the receptor through the formation of a covalent bond results in prolonged α-receptor blockade. Although the aziridinium ion intermediate has long been believed to be the active receptor alkylating species, only in recent years has it been demonstrated unequivocally that the aziridinium ions derived from dibenamine and phenoxybenzamine are capable of an α-receptor alkylation.[30]

Phenoxybenzamine Hydrochloride, USP. *N*-(2-Chloroethyl)-*N*-(1-methyl-2-phenoxyethyl)benzylamine hydrochloride (Dibenzyline). The compound exists in the form of colorless crystals that are soluble in water, freely soluble in alcohol and chloroform, and insoluble in ether. It slowly hydrolyzes in neutral and basic solutions, but is stable in acid solutions and suspensions.

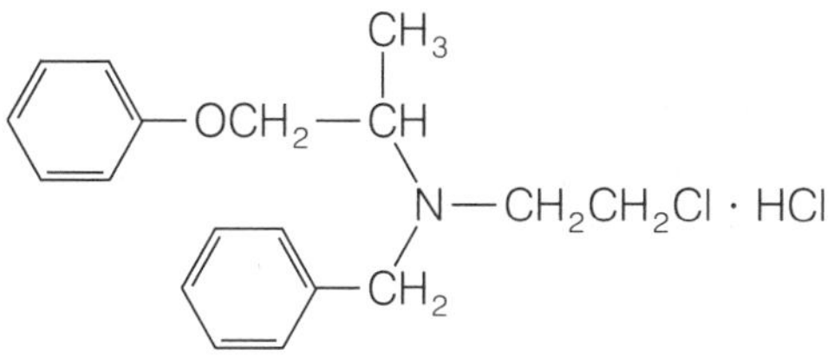

Phenoxybenzamine Hydrochloride

The action of phenoxybenzamine has been described as representing a "chemical sympathectomy" because of its selective blockade of the excitatory responses of smooth muscle and of the heart muscle.

Although phenoxybenzamine is capable of blocking acetylcholine, histamine, and serotonin receptors, its primary pharmacologic effects, especially vasodilatation, may be attributed to its α-adrenergic-blocking capability. As would be expected of a drug that produces such a profound α-blockade, the administration of phenoxybenzamine is frequently associated with reflex tachycardia, increased cardiac output, and postural hypotension. There is also evidence indicating that blockade of presynaptic α_2-receptors contributes to the increased heart rate produced by phenoxybenzamine.

The onset of action is slow, but the effects of a single dose of phenoxybenzamine may last three to four days. The principal effects following administration are an increase in peripheral blood flow, increase in skin temperature, and a lowering of blood pressure. It has no effect on the parasympathetic system and has little effect on the gastrointestinal tract. The most common side effects are miosis, tachycardia, nasal stuffiness, and postural hypotension, which are all related to the production of adrenergic blockade.

Oral phenoxybenzamine is used for the preoperative management of patients with pheochromocytoma and in the chronic management of patients whose tumors are not amenable to surgery. In some patients, the administration of a β-blocking agent after phenoxybenzamine blockade has been established as useful.

Phenoxybenzamine is used to treat peripheral vascular disease such as Raynaud's syndrome. It has also been used in shock and frostbite to improve blood flow to peripheral tissues. In low doses, it has been used in hypertensive patients who have developed vascular supersensitivity to sympathomimetics during treatment with adrenergic neuronal blocking agents.

Imidazolines

Similar to the ergot alkaloids, the imidazoline α-antagonists are competitive (reversible) blocking agents. They are structurally similar to the imidazoline α-agonists such as naphazoline, tetrahydrozoline, and xylometazoline. These agents have not proved to be useful for the treatment of hypertension, and their clinical applications have been limited largely to treatment of conditions involving peripheral vasospasm.

The two representatives of the imidazoline α-antagonists are tolazoline and phentolamine. Phentolamine is a more effective α-antagonist than tolazoline, but neither drug is useful as an antihypertensive agent. Theoretically, the vasodilator effects of an α-antagonist should be beneficial in the management of hypertension. However, tolazoline and phentolamine have both α_1- and α_2-antagonist activity and produce tachycardia. Presumably, the antagonist actions of these agents at presynaptic α_2-receptors contribute to their cardiac stimulant effects by enhancing the release of NE. Both agents have a direct vasodilator action on vascular smooth muscle that may be more prominent than their α-receptor antagonist effects.

Tolazoline Hydrochloride, USP. 2-Benzyl-2-imidazoline monohydrochloride (Priscoline Hydrochloride).

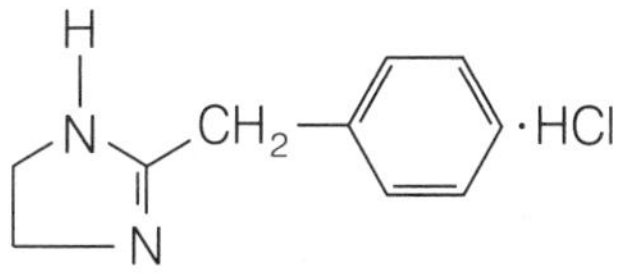

Tolazoline Hydrochloride

Tolazoline is well absorbed after oral administration, and it is largely excreted unmetabolized in the urine. Its α-antagonistic action is relatively weak, but its histamine-like and acetylcholine-like agonist actions probably contribute to its vasodilator activity. Its histamine-like effects include stimulation of gastric acid secretion, rendering it inappropriate for administration to patients who have gastric or peptic ulcers. It has been used to treat Raynaud's syndrome and other conditions involving peripheral vasospasm. Tolazoline is available in an injectable form, but the drug is rarely used in current clinical practice. It is indicated for use in persistent pulmonary hypertension of the newborn when supportive measures are not successful.

The drug occurs as a white or creamy white, bitter crystalline powder possessing a slight aromatic odor. It is freely soluble in water and alcohol. A 2.5% aqueous solution is slightly acidic (*p*H 4.9 to 5.3). It is only slightly soluble in ether and ethyl acetate, but is soluble in chloroform.

Phentolamine Mesylate, USP. *m*-[*N*-(2-Imidazolin-2-ylmethyl)-*p*-toluidino]phenol monomethanesulfonate (Regitine Mesylate). This compound occurs as a white, odorless, bitter powder that is freely soluble in alcohol and very soluble in water. Aqueous solutions are slightly acidic (*p*H 4.5 to 5.5) and deteriorate slowly. However, the chemical itself is stable when protected from moisture and light. The stability and the solubility of this salt of phentolamine are superior to those of the hydrochloride and account for the use of the methanesulfonate (mesylate) rather than the hydrochloride for parenteral injection.

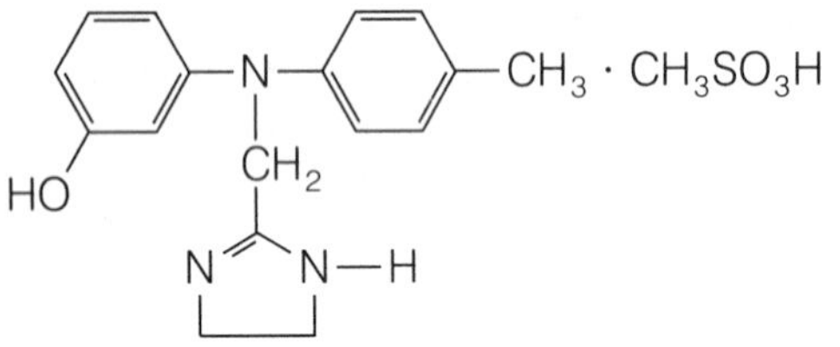

Phentolamine Mesylate

Phentolamine is used to prevent or control hypertensive episodes that occur in patients with pheochromocytoma. It can be used as an aid in the diagnosis of pheochromocytoma, but the measurement of catecholamine levels is a safer and more reliable method of diagnosis. Phentolamine also is used in combination with papaverine by intracavernosal injection in impotence therapy.

Prazosin Hydrochloride, USP. (Minipress) is a quinazoline derivative that has been found to be of value in the management of hypertension.

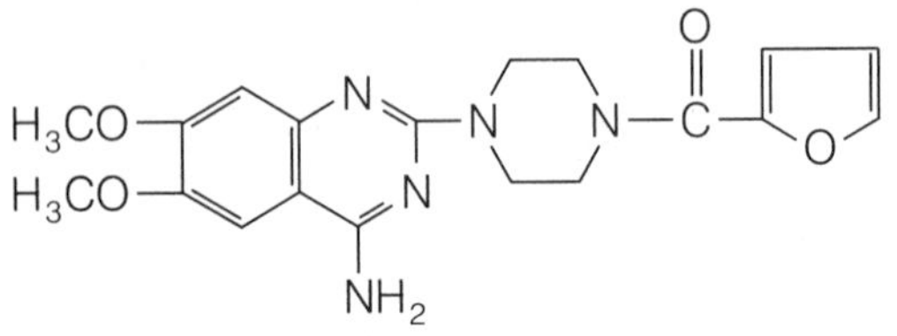

Prazosin

Prazosin is an α-adrenergic antagonist that offers distinct advantages over the other α-blockers. It causes peripheral vasodilatation without an increase in heart rate or cardiac output. This advantage, at least in part, is attributed to the fact that prazosin blocks postjunctional α_1-receptors selectively without blocking presynaptic α_2-receptors. Prazosin is effective after oral administration and has a plasma half-life of approximately four hours. It is extensively metabolized and is excreted mainly in the bile.

Although its side effects are usually minimal, the most frequent one, known as the *first-dose phenomenon*, is sometimes severe. This is a dose-dependent effect characterized by dizziness, palpitations, and syncope. The cause of the first-dose effect is unknown.

Terazosin Hydrochloride (Hytrin). Terazosin, like prazosin, is a selective α_1-adrenoceptor antagonist that is used for the treatment of hypertension. The only structural difference between the drugs is the replacement of the furan ring of prazosin with the tetrahydrofuran ring in terazosin. Terazosin is well absorbed after oral administration; its bioavailability is approximately 90%, whereas that of prazosin is approximately 55% to 60%. Terazosin has a longer duration of action than prazosin, its elimination half-life being approximately two to three times longer than that of prazosin. First-dose syncope can occur with terazosin, as well as with prazosin.

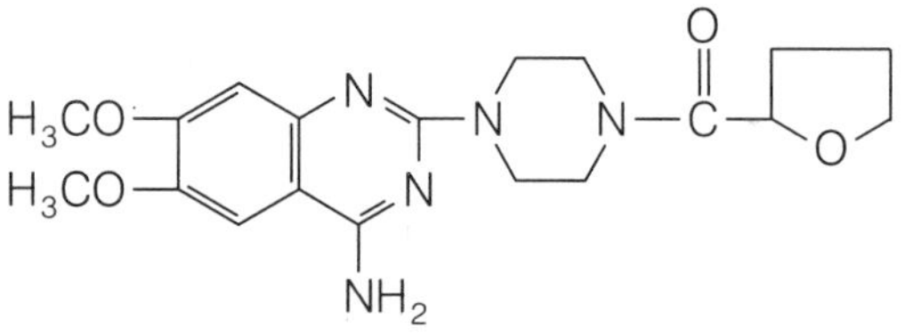

Terazosin

β-ADRENERGIC-BLOCKING AGENTS

Although the α-receptor antagonist action of agents such as the ergot alkaloids was discovered many years ago, the first β-blocker was not reported until 1958, when Powell and Slater described the activity of dichloroisoproterenol (DCI).[31] Shortly thereafter, Moran and Perkins reported that DCI blocked the effects of sympathomimetics on the heart.[32] The structure of DCI is identical with that of isoproterenol, with the exception that the catechol hydroxyl groups have been replaced by two chloro groups. This simple structural modification, involving the replacement of the aromatic hydroxyl groups, has provided the basis for nearly all of the approaches employed in subsequent efforts to design and synthesize therapeutically useful β-receptor antagonists. Unfortunately, DCI was not a pure antagonist, but rather, it was a partial agonist. The substantial, direct sympathomimetic action of DCI precluded its development as a clinically useful drug.

Cl, Cl — (3,4-dichlorophenyl)—CH(OH)—CH_2—$NHCH(CH_3)_2$

Dichloroisoproterenol

Pronethalol, whose structural similarities to isoproterenol and DCI are obvious, was the next important β-antagonist to be described.[33] Although pronethalol had much less intrinsic sympathomimetic activity than DCI, it was withdrawn from clinical testing because of reports that it caused thymic tumors in mice. However, within two years of the report on pronethalol, Black and co-workers described the β-blocking actions of propranolol, a close structural relative to pronethalol. Propranolol has become one of the most thoroughly studied and widely used drugs in the therapeutic armamentarium. It is the standard against which all other β-antagonists are compared.

(2-naphthyl)—CH(OH)—CH_2—$NHCH(CH_3)_2$

Pronethalol

(1-naphthyl)—O—CH_2—CH(OH)—CH_2—$NHCH(CH_3)_2$

Propranolol

Propranolol belongs to the group of β-blocking agents known as *aryloxypropanolamines*. This term reflects the fact that an $-OCH_2-$ group has been incorporated into the molecule between the aromatic ring and the ethylamino side chain. Because this structural feature is so frequently found in β-antagonists, the assumption is often made that the $-OCH_2-$ group is responsible for the antagonist properties of the molecules. However, this is not true; in fact, the $-OCH_2-$ group is present in several compounds that are potent β-agonists.[34] This latter fact again leads to the conclusion that it is the nature of the aromatic ring and its substituents that is the primary determinant of β-antagonist activity.

Propranolol, similar to the other β-receptor antagonists that will be discussed, is a competitive antagonist, the receptor-blocking actions of which can be reversed with sufficient concentrations of β-agonists. Currently, propranolol is approved for use in the United States for hypertension, cardiac arrythmias, angina pectoris caused by coronary atherosclerosis, hypertrophic subaortic stenosis, and prophylaxis of migraine headache. Propranolol is under investigation for the treatment of a variety of other conditions, including anxiety and schizophrenia.

Although some degree of controversy exists, it appears that nearly all of the pharmacologic effects of propranolol may be attributed to β-receptor blockade. By blocking the β-receptors of the heart, it slows the heart, reduces the force of contraction, and reduces cardiac output. Because of reflex sympathetic activity and blockade of vascular β_2-receptors, administration of propranolol may result in increased peripheral resistance. The antihypertensive action of propranolol, at least in part, may be attributed to its ability to reduce cardiac output, as well as to its suppression of renin release from the kidney.

Because propranolol exhibits no selectivity for β_1-receptors, it blocks β_2-receptors in the respiratory tract. Therefore, propranolol is contraindicated in the presence of conditions such as asthma or bronchitis.

The rationale for the use of propranolol or other β-antagonists in the prophylaxis of migraine largely resides in the assumption that the drugs cause a vasoconstriction that inhibits the vasodilating effects of chemical mediators. However, because little is known about the causal factors and processes in migraine, it is difficult to specify the precise reason for the effectiveness of β-blockers. Likewise, it is uncertain whether propranolol's effectiveness in anxiety states is due to an action in the central nervous system, a peripheral β-blocking effect, or a combination of these.

The use of propranolol in cardiovascular diseases is covered in the chapter on cardiovascular drugs (see Chap. 14).

A facet of the pharmacologic action of propranolol that has received a good deal of attention is its so-called *membrane-stabilizing activity*. This is a nonspecific effect (i.e., not mediated by a specific receptor), which is also referred to as a *local anesthetic* effect or a *quinidinelike* effect. Although various authors have implied that the membrane-stabilizing effect may be therapeutically important, it now seems clear that the concentrations required to produce this effect far exceed those obtained with normal therapeutic doses of propranolol and related β-blocking drugs. It is most unlikely that the nonspecific membrane-stabilizing activity plays any role in the clinical efficacy of β-blocking agents.

The β-blocking agents exhibit a high degree of stereoselectivity in the production of their β-blocking effects. As discussed earlier for sympathomimetic agents, the configuration of the hydroxyl-bearing carbon of the side chain plays a critical role in the interaction of β-antagonist drugs with β-receptors. The available data indicate that the pharmacologically more active enantiomer interacts with the receptor recognition site in a manner analogous to that of the agonists. However, the structural features of the aromatic portion of the antagonist appear to perturb the receptor, or to interact with it in a manner that inhibits activation. In spite of the fact that nearly all of the β-antagonist activity resides in one enantiomer, propranolol and most other β-blocking agents are available for clinical use as racemic mixtures. Both enantiomers of propranolol have membrane-stabilizing activity, but only the levorotatory isomer is a potent β-receptor antagonist, the dextrorotatory isomer being approximately 100 times less effective.

The metabolism of propranolol is a topic that has received intense study. Propranolol is well absorbed after oral administration, but it undergoes extensive "first-pass" metabolism before it reaches the systemic circulation. The term *first-pass metabolism* refers to the fact that the compound is efficiently extracted from the portal vein by the liver where it undergoes biotransformation. Lower doses of propranolol are more efficiently extracted than higher doses; this indicates that the extraction process may become "saturated" at higher doses. The outcome of this saturation of the extraction process is that a larger percentage of a high oral dose than of a lower dose of propranolol reaches the systemic circulation. The extensive first-pass metabolism of propranolol accounts for the fact that, when the drug is given intravenously, much smaller doses are required to achieve a therapeutic effect than when it is given orally.

Numerous metabolites of propranolol have been identified, but the major metabolite in people, after a single oral dose, is naphthoxylactic acid, which is formed by a series of metabolic reactions involving *N*-dealkylation, deamination, and oxidation of the resultant aldehyde.[35,36] A propranolol metabolite of particular interest is 4-hydroxypropranolol. This compound is a potent β-antagonist that has some intrinsic sympathomimetic activity.[37] Interestingly, 4-hydroxypropranolol has been detected in plasma following intravenous administration of propranolol, although the plasma levels of the metabolite are considerably higher after oral administration.[38] It is not known what contribution, if any, 4-hydroxypropranolol makes to the pharmacologic effects seen after administration of propranolol. It has been suggested that 4-hydroxypropranolol contributes to propranolol's pharmacologic effects after low single doses or long-term oral administration.[38] The metabolic oxidation of the side chain and the glucuronidation of propranolol are induced by cigarette smoking.[39]

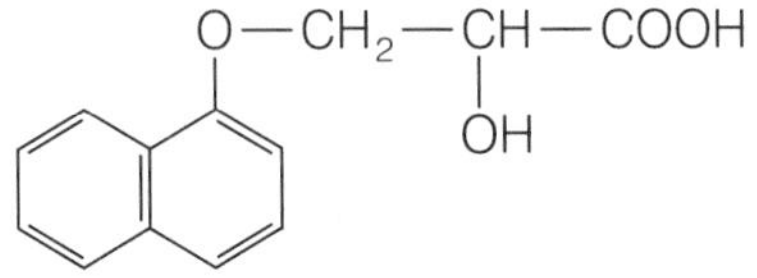

$O-CH_2-CH(OH)-COOH$ (on 1-naphthyl)

Naphthoxylactic Acid

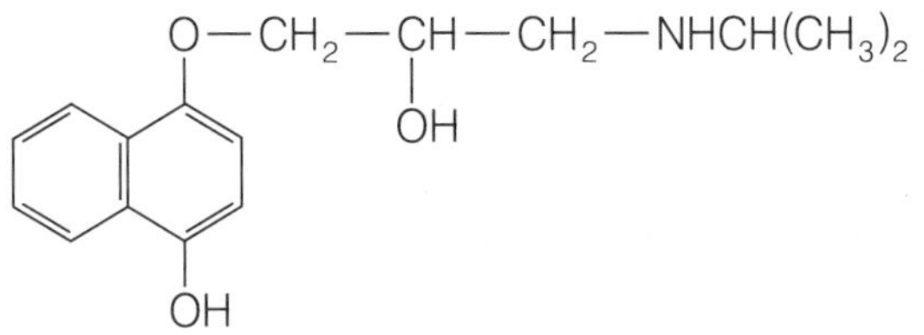

$O-CH_2-CH(OH)-CH_2-NHCH(CH_3)_2$ (on 1-naphthyl, 4-OH)

4-Hydroxypropranolol

The half-life of propranolol after a single oral dose is three to four hours, and it increases to four to six hours after long-term therapy. This relatively short half-life seems inconsistent with the observations that the pharmacologic effects of propranolol may persist for two to three days after discontinuation of the drug and that some patients can be treated effectively with one or two doses of propranolol per day. It has been proposed that propranolol glucuronide, which is the principal metabolite formed during long-term therapy, may serve as a storage pool for propranolol and, thereby, be at least partially responsible for the slow accumulation of propranolol.[40] According to this mechanism, propranolol glucuronide undergoes deconjugation in various tissues, as well as during enterohepatic circulation, thus providing for a slow release of propranolol from the stored glucuronide. Another mechanism that may contribute to the slow accumulation of propranolol is the storage of propranolol in sympathetic neurons from which it is released during sympathetic nerve stimulation.[41] Although several authors have commented on the interindividual variation of metabolism of propranolol, and upon the lack of correlation of propranolol plasma levels and interindividual therapeutic effects, there does not appear to be a high degree of patient variation in propranolol disposition during long-term therapy. This is somewhat surprising because, after oral administration, propranolol is almost completely metabolized. Very little is excreted unchanged.

The discovery that β-blocking agents are useful in the treatment of cardiovascular disease, such as hypertension, stimulated a search for cardioselective β-blockers. Cardioselective β-antagonists are drugs that have a greater affinity for the β_1-receptors of the heart than for β_2-receptors in other tissues. Such cardioselective agents should provide two important therapeutic advantages. The first advantage would be the lack of an antagonist effect on the β_2-receptors in the bronchi. Theoretically, this would make β_1-blockers safe for use in patients who have bronchitis or bronchial asthma. The second major therapeutic advantage of cardioselective agents would be the absence of blockade of the vascular β_2-receptors, which mediate vasodilatation. This would be expected to reduce or eliminate the increase in peripheral resistance that sometimes occurs after the administration of nonselective β-antagonists. Unfortunately, cardioselectivity is usually observed with β_1-antagonists at only relatively low doses. At normal therapeutic doses, much of the selectivity is lost.

The goal of complete cardioselectivity is one that may never be achieved, even if a drug is developed that retains its β_1-selectivity at therapeutic doses. One obstacle to the attainment of complete cardioselectivity may be the presence of both β_1- and β_2-receptors in cardiac and lung tissue. Although it is true that β_1-receptors are predominant in the heart and that the β-receptors of the lung are primarily β_2, evidence indicates that both types of receptors are present in both tissues.[42,43] Thus, theoretically, any β-blocker would block at least some of the receptors in either tissue.

The prototype β_1-antagonist is practolol. Although it was not released for use in the United States, it was the first cardioselective β_1-antagonist to be extensively used in humans. However, because it produced several toxic effects, it is no longer in general use in most countries. In Great Britain it is used only in hospitals; its principal application is in the management of cardiac arrythmias.

Although medicinal chemists have been successful in synthesizing several β_1-selective antagonists, it is not clear why some compounds exhibit selectivity, and others do not. One common structural feature of many cardioselective antagonists is the presence of a *para*-substituent on the aromatic ring along with the absence of *meta*-substituents. Practolol and metoprolol are examples of this structural type. However, there are a sufficient number of exceptions to show that this type of ring substitution is not the sole determinant of β_1-receptor selectivity.[44,45]

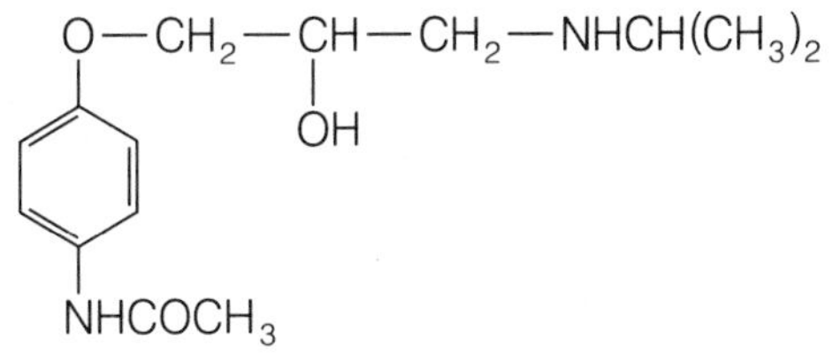

O—CH_2—CH(OH)—CH_2—$NHCH(CH_3)_2$; para-$NHCOCH_3$

Practolol

O—CH_2—CH(OH)—CH_2—$NHCH(CH_3)_2$; para-CH_2—CH_2—OCH_3

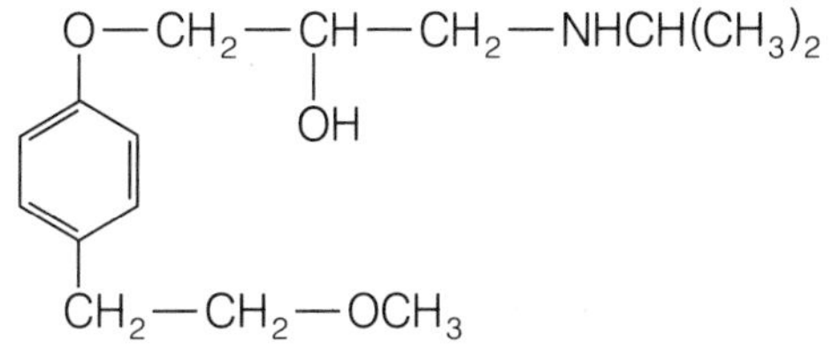

Metoprolol

Metoprolol (Lopressor) was released in 1978 for clinical use in the United States as an antihypertensive agent. Similarly to propranolol, metoprolol undergoes extensive first-pass metabolism after oral administration. Metoprolol has a bioavailability of approximately 50% of an oral dose compared with approximately 30% for propranolol. The elimination half-life of metoprolol is three to four hours, which is very similar to that of propranolol. But unlike propranolol, metoprolol does not appear to be converted to any pharmacologically important metabolites. Although metoprolol is considered to be a β_1-receptor-selective antagonist, it is important to be aware that most of its cardioselectivity is usually lost at full therapeutic dose. Therefore, in spite of its cardioselectivity at low doses, it is not normally indicated for patients with bronchial asthma.

Nadolol (Corgard) is a nonselective β-blocker approved late in 1979 for general use in the United States as an antihypertensive agent and in the management of angina pectoris. Unlike propranolol, nadolol has no membrane-stabilizing properties. It also has no intrinsic sympathomimetic activity, a characteristic it shares with both propranolol and metoprolol.

O—CH_2—CH(OH)—CH_2—$NHC(CH_3)_3$; HO, HO

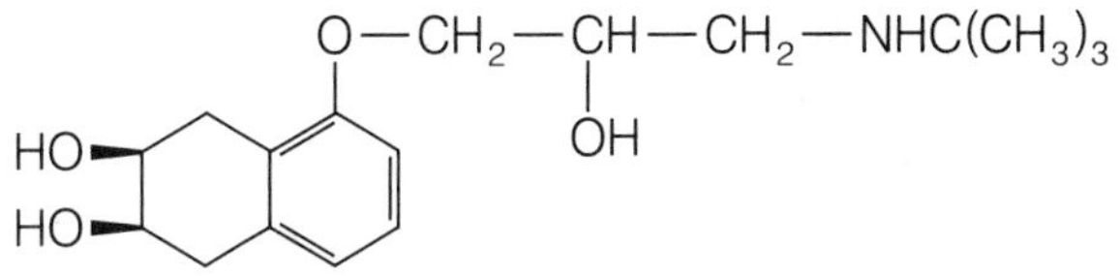

Nadolol

Nadolol exhibits profound differences from propranolol and metoprolol in its absorption and disposition. Whereas propranolol and metoprolol are almost completely absorbed after oral administration, only about 30% of an oral dose of nadolol is absorbed. In contrast with the extensive first-pass metabolism undergone by propranolol and metoprolol, nadolol appears to be excreted unchanged after either oral or intravenous administration. It is excreted primarily by the kidneys and, because of its lack of metabolism, has a long duration of action. The serum half-life of nadolol is 12 to 20 hours, a property that permits the drug to be administered in single daily doses.

Timolol is a nonselective β-blocker approved in 1978 for general use in the United States for the management of glaucoma. Timolol has no significant membrane-stabilizing (local anesthetic) action or intrinsic sympathomimetic activity. Although the chemical structure of the "aromatic ring" portion of timolol is substantially different from that found in other β-blockers, the side chain is structurally identical with the ones found in numerous other β-antagonists. It is supplied for clinical use as the more potent (S)-isomer.

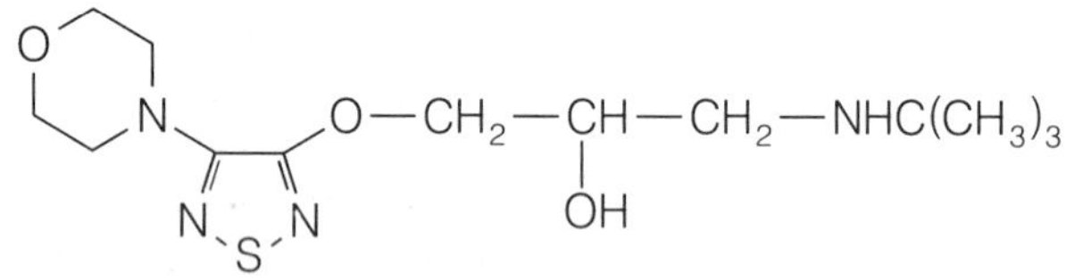

Timolol

Timolol is used topically to lower intraocular pressure in the treatment of glaucoma, and it is an effective antihypertensive agent. The mechanism whereby β-blockers lower intraocular pressure is not known with certainty, but it appears to be at least partially due to a β-receptor antagonism that results in a reduction in the formation and secretion of aqueous humor. It seems somewhat paradoxical that sympathomimetic agents, such as epinephrine, phenylephrine, norepinephrine, and isoproterenol, also are used to lower intraocular pressure because of their ability to inhibit aqueous humor production. β-Blockers offer an advantage over many other drugs used in the treatment of glaucoma: they do not have an effect on pupil size, and they do not cause spasm of the ciliary muscle. Timolol also offers an advantage over β-blockers, such as propranolol, in that it does not have a local anesthetic effect.

Two other drugs of interest that have β-receptor-blocking activities are labetalol and butoxamine.

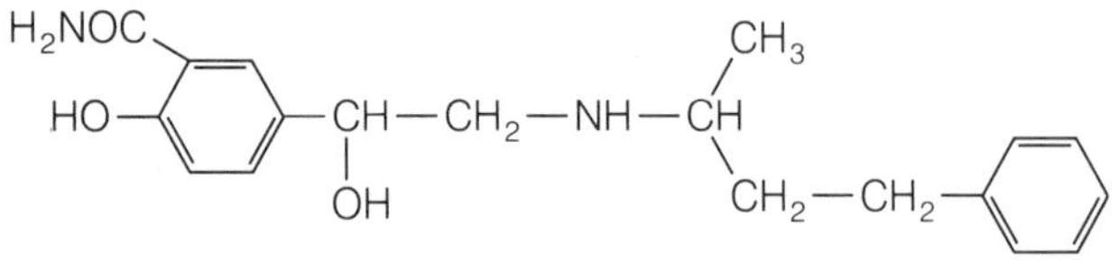

Labetalol

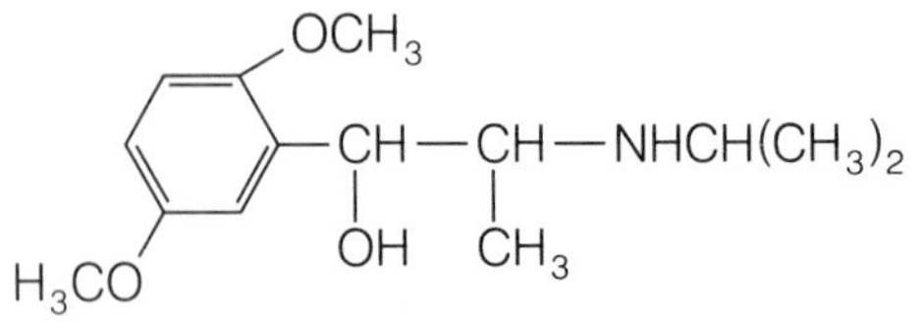

Butoxamine

Labetalol has both α-receptor and β-receptor antagonist properties. It is a more potent β-antagonist than α-antagonist, and it has membrane-stabilizing activity. Labetalol is a clinically useful antihypertensive agent. The rationale for its use in the management of hypertension is that its α-receptor-blocking effects produce vasodilatation and that its β-receptor-blocking effects prevent the reflex tachycardia that is usually associated with the vasodilatation.

Butoxamine is of interest because it is a selective β_2-receptor antagonist. It blocks the β_2-receptors in uterine smooth muscle, in bronchial smooth muscle, and in skeletal muscle. Because of its β_2-receptor selectivity, butoxamine is a useful research tool, but it now has no clinical use.

PRODUCTS

Propranolol Hydrochloride, USP. 1-(Isopropylamino)-3-(1-naphthyloxy)-2-propanol hydrochloride (Inderal). Propranolol hydrochloride is a white to off-white crystalline solid, soluble in water or ethanol and insoluble in nonpolar solvents. Its therapeutic indications were discussed earlier in this chapter. Propranolol is contraindicated in the following instances: bronchial asthma; allergic rhinitis during the pollen season; sinus bradycardia and greater than first-degree block; cardiogenic shock; right ventricular failure secondary to pulmonary hypertension; congestive heart failure, unless the failure is secondary to a tachyarrhythmia treatable with propranolol hydrochloride; patients who are taking adrenergic-augmenting psychotropic drugs. In patients with angina pectoris, the sudden withdrawal of propranolol or other β-blockers should be avoided to prevent exacerbation of the angina. This so-called propranolol-withdrawal rebound has been associated with fatal myocardial infarctions, and it is recommended that withdrawal from the drug be achieved by a gradual reduction of the dose.[46] Propranolol hydrochloride is supplied in both tablet and injectable form. For most therapeutic purposes, the dosage of propranolol must be individualized.

Metoprolol Tartrate, USP. 1-Isopropylamino-3-[p-(2-methoxyethyl)phenoxy]-2-propanol *dextro*-tartrate (Lopressor). Metoprolol tartrate is a water-soluble, white crystalline solid. It is a selective β_1-receptor antagonist that is indicated for use in the management of hypertension. Metoprolol tartrate is contraindictated in sinus bradycardia, heart block greater than first degree, cardiogenic shock, and overt cardiac failure. Abrupt withdrawal of the drug should be avoided, and it should not be used by patients who have bronchospastic diseases, unless the patients do not respond to, or cannot tolerate, other antihypertensive drugs. In such cases, the lowest dose possible of metoprolol tartrate should be used, and a β_2-agonist should be administered concomitantly.

Nadolol, USP. 2,3-*cis*-5-[3-(1,1-Dimethylethyl)amino-2-hydroxypropoxy]-1,2,3,4-tetrahydro-2,3-naphthalenediol (Corgard). Nadolol is a nonselective β-blocker that has been approved for the management of hypertension and angina pectoris. Similarly to propranolol and metoprolol, it is often used in combination with other antihypertensive agents, such as diuretics. It is contraindicated in the presence of bronchial asthma, sinus bradycardia and greater than first-degree conduction block, cardiogenic shock, and overt cardiac failure. Sudden cessation of nadolol therapy may exacerbate ischemic heart disease. Because it is excreted unchanged, it must be used with caution and in lower dose in patients with renal impairment. Nadolol is supplied in tablet form.

Timolol Maleate, USP. (*S*)-1(*tert*-Butyl-amino)-3-[(4-morpholino-1,2,5-thiadiazol-3-yl)oxy]-2-propanol maleate (Timoptic). Timolol maleate is a nonselective β-blocker used in the management of chronic open-angle glaucoma and ocular hypertension. It is available as an ophthalmic solution in 0.25% and 0.5% concentrations. The usual starting dose is one drop of 0.25% solution in each eye twice a day. The systemic effects of ophthalmically administered timolol are usually minor and may include slight bradycardia or acute bronchospasm in patients with bronchospastic disease.

Timolol maleate is also available in tablet form (Blocadren) for use as an antihypertensive and antianginal agent. Timolol is less lipid-soluble than propranolol, is well absorbed after oral administration, and undergoes first-pass metabolism. Its bioavailability is approximately 50%.

Atenolol (Tenormin). Atenolol is a cardioselective β-blocker that is used for the treatment of hypertension and for the management of angina pectoris. It has no intrinsic sympathomimetic (partial agonist) activity and no local anesthetic activity. Atenolol, as does nadolol, has a low lipid solubility and does not readily cross the blood–brain barrier. It is incompletely absorbed from the gastrointestinal tract, the oral bioavailability being approximately 50%. Little of the absorbed portion of the dose is metabolized; most of it is excreted unchanged in the urine. Atenolol has a long duration of action and may be administered in a single daily dose.

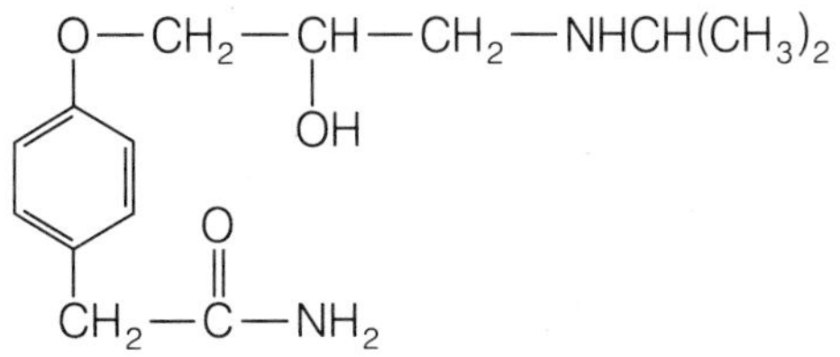

Atenolol

Acebutolol (Sectral). Acebutolol is a cardioselective β-antagonist with weak intrinsic sympathomimetic activity and weak membrane-stabilizing activity. It is a moderately lipophilic agent, similar to metoprolol. Acebutolol is well absorbed from the gastrointestinal tract, but it undergoes extensive first-pass metabolic conversion to diacetolol, a pharmacologically active metabolite. After oral administration, the plasma levels of diacetolol are higher than those of acebutolol. Diacetolol also is a selective β_1-receptor antagonist with partial agonist activity; it has little membrane-stabilizing activity. Diacetolol is formed by hydrolytic conversion of the amide group to the amine, followed by acetylation of the amine; it has a longer half-life than the parent drug, and it is excreted by the kidneys. Acebutolol is used clinically as an antihypertensive, antianginal, and antiarrhythmic agent.

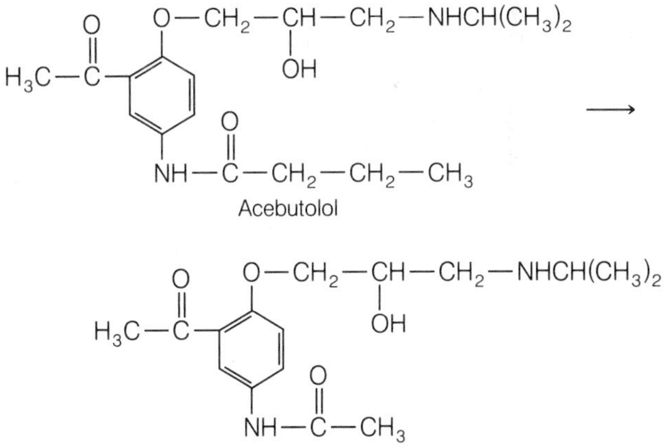

Acebutolol

Diacetolol

Pindolol, USP. (Visken). Pindolol is a moderately lipophilic, nonselective β-adrenergic antagonist that has modest membrane-stabilizing activity and significant intrinsic β-agonist activity. However, unlike the original β-antagonist, dichloroisoproterenol, the agonist activity of pindolol is not of sufficient magnitude to render it unsuitable for clinical use as an antihypertensive agent. β-Antagonists with partial agonist activity cause less slowing of the resting heart rate than do agents without this capability; the partial agonist activity may be beneficial in patients who are likely to exhibit severe bradycardia or who have little cardiac reserve. In general, pindolol appears to be as satisfactory as other β-antagonists for the treatment of hypertension, angina pectoris, and cardiac arrhythmias. Because of its lipophilicity, pindolol readily crosses the blood–brain barrier.

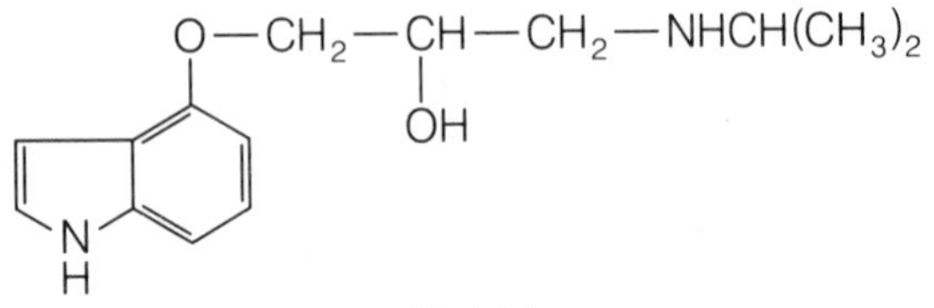

$O—CH_2—CH(OH)—CH_2—NHCH(CH_3)_2$

Pindolol

Pindolol is rapidly absorbed from the gastrointestinal tract and exhibits 90% to 100% bioavailability. Thus, it undergoes little first-pass metabolism. Approximately 40% to 50% of an oral dose is excreted unchanged in the urine and the remainder is excreted as metabolites.

Esmolol Hydrochloride (Brevibloc). Esmolol is a relatively cardioselective β-receptor antagonist that exhibits neither intrinsic sympathomimetic activity nor local anesthetic activity at clinically meaningful concentrations. The distinguishing characteristic of this drug is its very short duration of action; it has an elimination half-life of nine minutes. Esmolol is administered by continuous intravenous infusion for control of ventricular rate in patients with atrial flutter, atrial fibrillation, or sinus tachycardia. Its rapid onset and short duration of action render it useful during surgery, postoperatively, or during emergency situations for short-term control of heart rates. Its effects disappear within 20 to 30 minutes after the infusion is discontinued. The most common side effect of esmolol is hypotension; the hypotensive effects are usually reversed within 30 minutes of the termination of the infusion. Esmolol must be diluted with an injection solution before administration; it is incompatible with sodium bicarbonate.

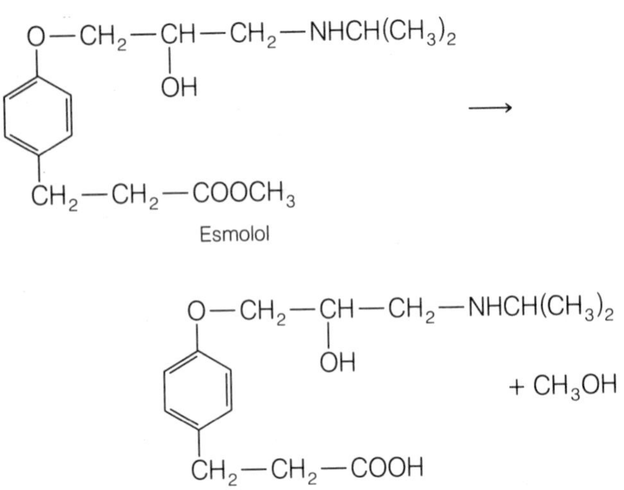

The short duration of action of esmolol is the result of rapid hydrolysis of its ester functionality by esterases that are present in erythrocytes. The resultant carboxylic acid is an extremely weak β-antagonist that does not appear to exhibit clinically significant effects. The acid metabolite has an elimination half-life of three to four hours and is excreted primarily by the kidneys.

Betaxolol Hydrochloride (Betoptic). This compound is a lipophilic, β_1-selective adrenergic antagonist that has no partial agonist activity and no significant membrane-stabilizing effects. Betaxolol is used for the treatment of chronic open-angle glaucoma and ocular hypertension. Although betaxolol and the other β-blockers that are used for the treatment of glaucoma are administered topically as ophthalmic solutions, systemic distribution and systemic effects can occur. Thus, side effects can result from the blockade of β-receptor in heart, lung, and other tissues. The β_1-selectivity of betaxolol may reduce the likelihood of respiratory reactions.

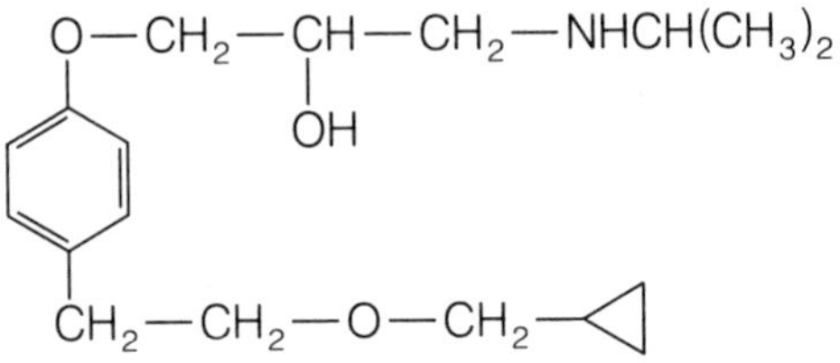

$O—CH_2—CH(OH)—CH_2—NHCH(CH_3)_2$; $CH_2—CH_2—O—CH_2$—cyclopropyl

Betaxolol

Levobunolol Hydrochloride (Betagan). This nonselective β-antagonist is used as an ophthalmic solution to lower intraocular pressure in patients with chronic open-angle glaucoma or ocular hypertension. It has a longer duration of action than either timolol or betaxolol. It has no partial agonist activity, no local anesthetic activity, and it is marketed as the levorotatory enantiomer.

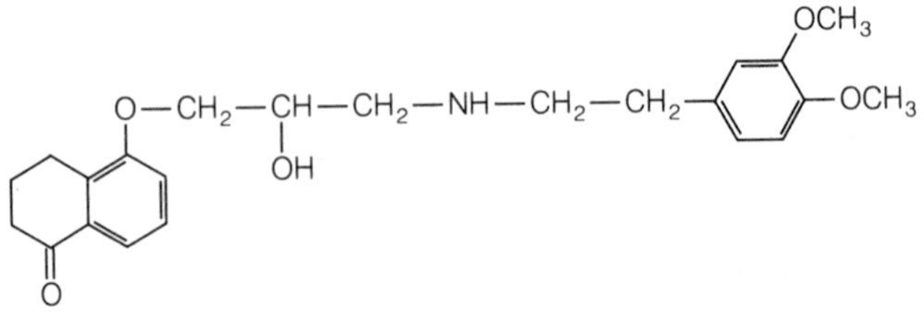

$O—CH_2—CH(OH)—CH_2—NHCH(CH_3)_2$; $CH_2—CH_2—COOCH_3$

Esmolol

$\longrightarrow$

$O—CH_2—CH(OH)—CH_2—NHCH(CH_3)_2$; $CH_2—CH_2—COOH$ $+ CH_3OH$

$O—CH_2—CH(OH)—CH_2—NH—CH_2—CH_2$—(3,4-$OCH_3$ phenyl)

Bunolol

Labetolol Hydrochloride (Trandate). The structure of labetolol was presented earlier in the discussion of β-antagonists. Labetolol is an antihypertensive agent that produces its beneficial effects through a complex of mechanisms that has not been completely elucidated. Although the drug is known to have both β- and α-antagonist activities, it also appears to exhibit β_2-antagonist effects on the vascular system. Most of the antihypertensive effect appears to be due to vasodilatation. Labetolol undergoes rather extensive first-pass metabolism, and is metabolized primarily by glucuronidation.

Labetolol contains two asymmetric centers, and the preparation that is used clinically is a mixture of the four possible stereoisomers. It has been demonstrated that the (−)-(*RR*) isomer is primarily responsible for the peripheral vasodilator activity and the β_1-antagonist activity. The (+)-(*SR*) isomer exhibits most of the α-blocking activity.[47, 48]

REFERENCES

1. Weiner, N., and Taylor, P.: In Gilman, A. G., Goodman, L. S., Rall, T. W., and Murad, F. (eds.). The Pharmacological Basis of Therapeutics, 7th ed., Chap 4. New York, Macmillan, 1985.
2. Trendelenburg, U.: Trends Pharmacol. Sci. 1:4, 1979.
3. Ganellin, C. R.: J. Med. Chem. 20:579, 1977.
4. Biaggioni, I., and Robertson, D.: Lancet 2:1170, 1987.
5. Manin'L Veld, A. J., et al.: Lancet 2:1172, 1987.
6. Kopin, I. J.: Pharmacol. Rev. 37:333, 1985.
7. Ahlquist, R. P.: Am. J. Physiol. 154:586, 1948.
8. Berthelsen, S., and Pettinger, W. A.: Life Sci. 21:595, 1977.
9. Langer, S. Z.: Biochem. Pharmacol. 23:1793, 1974.
10. McGrath, J. C.: Biochem. Pharmacol. 31:467, 1982.
11. Rand, M. J., et al.: J. Mol. Cell Cardiol. 18(Suppl. 5):17, 1986.
12. Davey, M. J.: J. Mol. Cell. Cardiol. 18(Suppl. 5):1, 1986.
13. Langer, S. Z., and Hicks, P. E.: J. Cardiovasc. Pharmacol. 6(Suppl.):S547, 1984.
14. Doxey, J. C., et al.: Br. J. Pharmacol. 60:91, 1977.
15. Hirasawa, K., and Nishizuka, Y.: Annu. Rev. Pharmacol. Toxicol. 25:147, 1985.
16. Exton, J. H.: Trends Pharmacol. Sci. 3:111, 1982.
17. Lands, A. M., et al.: Nature 214:597, 1967.
18. Lefkowitz, R. J., et al.: Trends Pharmacol. Sci. 7:444, 1986.
19. Brodde, O.-E.: ISI Atlas Sci. Pharmacol. 1:107, 1987.
20. Nester, E. J., et al.: Science 225:1357, 1984.
21. Kobilka, B. K., et al.: Science 238:650, 1987.
22. Lefkowitz, R. J., and Caron, M. G.: J. Biol. Chem. 263:4993, 1988.
23. Triggle, D. J., and Triggle, C. R.: Chemical Pharmacology of the Synapse, Chap. 3. New York, Academic Press, 1976.
24. Kaiser, C., et al.: J. Med. Chem. 18:674, 1975.
25. Barger, G., and Dale, H. H.: J. Physiol. 41:19, 1910.
26. Rohrmann, E., and Shonle, H.: J. Am. Chem. Soc. 66:1517, 1944.
27. Lands, A. M., et al.: J. Pharmacol. Exp. Ther. 89:271, 1947.
28. Eisleb, O.: U.S. Patent 1,949,247. Chem. Abstr. 28:2850, 1934.
29. Nickerson, M., and Goodman, L. S.: J. Pharmacol. Exp. Ther. 89:167, 1947.
30. Henkel, J. G., et al.: J. Med. Chem. 19:6, 1976.
31. Powell, C. E., and Slater, I. H.: J. Pharmacol. Exp. Ther. 122:480, 1958.
32. Moran N. C., and Perkins, M. E.: J. Pharmacol. Exp. Ther. 124:223, 1958.
33. Black, J. W., and Stephenson, J. S.: Lancet 2:311, 1962.
34. Kaiser, C., et al.: J. Med. Chem. 20:687, 1977.
35. Walle, T., et al.: Clin. Pharmacol. Ther. 26:548, 1979.
36. Walle, T., and Gaffney, T. E.: J. Pharmacol. Exp. Ther. 182:83, 1972.
37. Fitzgerald, J. D., and O'Donnell, S. R.: Br. J. Pharmacol. 43:222, 1971.
38. Walle, T., et al.: Clin. Pharmacol. Ther. 27:23, 1980.
39. Walle, T., et al.: J. Pharmacol. Exp. Ther. 241:928, 1987.
40. Walle, T., et al.: Clin. Pharmacol. Ther. 26:686, 1979.
41. Daniell, H. B., et al.: J. Pharmacol. Exp. Ther. 208:354, 1979.
42. Daly, M. J., and Levy, G. P.: The subclassification of β-adrenoceptors: Evidence in support of the dual β-adrenoceptor hypothesis. In Kalsner, S. (ed.). Trends in Autonomic Pharmacology, Vol. 3, p. 347. Baltimore, Urban and Schwarzenberg, 1979.
43. Barnett, D. B., et al.: Nature 273:167, 1978.
44. Clarkson, R., et al.: Annu. Rep. Med. Chem. 10:51, 1975.
45. Evans, D. B., et al.: Annu. Rep. Med. Chem. 14:81, 1979.
46. Miller, R. R., et al.: N. Engl. J. Med. 293:416, 1978.
47. Gold, E. H., et al.: J. Med. Chem. 25:1363, 1982.
48. Baum, T., and Sybertz, E. J.: Fed. Proc. 42:176, 1983.

SELECTED READINGS

Antihypertensive agents. In AMA Drug Evaluations, 6th ed., Chap. 28. Philadelphia, W. B. Saunders, 1986.

Gerbec, J. G., and Nies, A. S.: beta-Adrenergic blocking drugs. Annu. Rev. Med. 36:145, 1985.

Levitzki, A.: From epinephrine to cyclic AMP. Science 241:800, 1988.

Patil, P. N., Miller, D. D., and Trendelenburg, U.: Molecular geometry and adrenergic drug activity. Pharmacol. Rev. 26:323, 1975.

Riddell, J. G., et al.: Clinical pharmacokinetics of β-adrenoceptor antagonists: An update. Clin. Pharmacokin. 12:305, 1987.

Triggle, D. J.: Adrenergics: Catecholamines and related agents. In Wolff, M. E. (ed.). Burger's Medicinal Chemistry, 4th ed., part III, p. 225. New York, John Wiley & Sons, 1981.

van Zweiten, P. A.: Antihypertensive drugs interacting with α- and β-adrenoceptors: A review of basic pharmacology. Drugs 35(Suppl. 6):6, 1988.

Weiner, N.: In Gilman, A. G., Goodman, L. S., Rall, T. W., and Murad, F. (eds.). The Pharmacological Basis of Therapeutics, 7th ed., Chaps. 8 and 9. New York, Macmillan, 1985.

CHAPTER 12

Cholinergic Drugs and Related Agents

George H. Cocolas

Few systems, if any, have been studied as extensively as those innervated by neurons that release acetylcholine (ACh) at their endings. Since the classic studies of Dale,[1] who described the actions of esters and ethers of choline on isolated organs and their relationship to muscarine, pharmacologists, physiologists, chemists, and biochemists have applied their knowledge to understand the actions of the cholinergic nerve and its neurotransmitter. This chapter includes the drugs and chemicals that act on cholinergic nerves or the tissues that they innervate to either mimic or block the action of ACh. Drugs that mimic the action of ACh do so either by acting directly on the cholinergic receptors in the tissue or by inhibiting acetylcholinesterase (AChE), the enzyme that inactivates ACh at the nerve terminal. Cholinergic neurotransmission may be blocked by chemicals that bind or compete with ACh for binding to the cholinergic receptor.

$$(CH_3)_3-\overset{+}{N}-\overset{\alpha}{C}H_2-\overset{\beta}{C}H_2-O-\overset{\overset{O}{\|}}{C}-CH_3$$

Acetylcholine

Cholinergic nerves are found in the peripheral and the central nervous system of humans. Synaptic terminals in the cerebral cortex, corpus striatum, hippocampus, and several other regions in the central nervous system (CNS) are rich in ACh and in the enzymes that synthesize and hydrolyze this neurotransmitter. This serves as strong evidence that ACh is a neurotransmitter in the central nervous system. Although its function in the brain and brain stem is not clear, it has been implicated in memory and behavioral activity in humans.[2] The *peripheral nervous system* consists of those nerves outside the cerebrospinal axis and includes the somatic nerves and the autonomic nervous system. The *somatic nerves* are made up of a sensory (afferent) nerve and a motor (efferent) nerve. The *motor nerves* arise from the spinal cord and project uninterrupted throughout the body to all skeletal muscle. Acetylcholine mediates transmission of impulses from the motor nerve to skeletal muscle (i.e., neuromuscular junction).

The *autonomic nervous system* is composed of two divisions: the *sympathetic* and *parasympathetic*. Acetylcholine serves as a neurotransmitter at both sympathetic and parasympathetic preganglionic nerve endings, postganglionic nerve fibers in the parasympathetic, and some postganglionic fibers (e.g., salivary and sweat glands) in the sympathetic division of the autonomic nervous system. The autonomic system regulates the activities of smooth muscle and glandular secretions. These, as a rule, function below the level of consciousness (e.g., respiration, circulation, digestion, body temperature, metabolism). The two divisions have contrasting effects on the internal environment of the body. The sympathetic division frequently discharges as a unit, especially during conditions of rage or fright, and expends energy. The parasympathetic is organized for discrete and localized discharge and acts to store and conserve energy.

Drugs and chemicals that cause the parasympathetic system to react are termed *parasympathomimetic*, whereas those blocking the actions are called *parasympatholytic*. Agents that mimic the sympathetic system are *sympathomimetic*, and those that block the actions are *sympatholytic*. Another classification used to describe drugs and chemicals acting on the nervous system or the structures that the fibers innervate is based on the neurotransmitter released at the nerve ending. Drugs acting on the autonomic nervous system are divided into *adrener-*

gic for those postganglionic sympathetic fibers that release norepinephrine and epinephrine, and *cholinergic* for the remaining fibers in the autonomic nervous system and the motor fibers of the somatic nerves that release ACh.

CHOLINERGIC RECEPTORS

There are two distinct receptors for ACh that differ in composition, location, pharmacologic function, and have specific agonists and antagonists. Cholinergic receptors have been characterized as *nicotinic* and *muscarinic* on the basis of their ability to be bound selectively by the naturally occurring alkaloids nicotine and muscarine, respectively.

NICOTINIC ACETYLCHOLINE RECEPTOR

The nicotinic ACh receptor is the first neurotransmitter receptor that has been isolated and purified in an active form.[3,4] It is a glycoprotein embedded into the polysynaptic membrane that has been isolated from the electric organs of the marine ray *Torpedo californica* and the electric eel *Electrophorus electricus*. The receptor is pictured as a cylindric protein of about 250,000 Da and consists of five subunit polypeptide chains of which two appear to be identical.[5] The subunit stoichiometry of the polypeptide units from the *Torpedo* receptor is $\alpha_2, \beta, \gamma, \delta$.[6] The peptide chains of the receptor are arranged to form an opening in the center which is the ion channel. Each α-chain contains a negatively charged binding site for the quaternary ammonium group of ACh. The receptor is believed to exist as a dimer of the two five-subunit polypeptide chain monomers linked together through a disulfide bond between δ-chains (Fig. 12-1).

When the neurotransmitter, ACh, binds to the nicotinic receptor it causes a change in the permeability of the membrane to allow passage of small

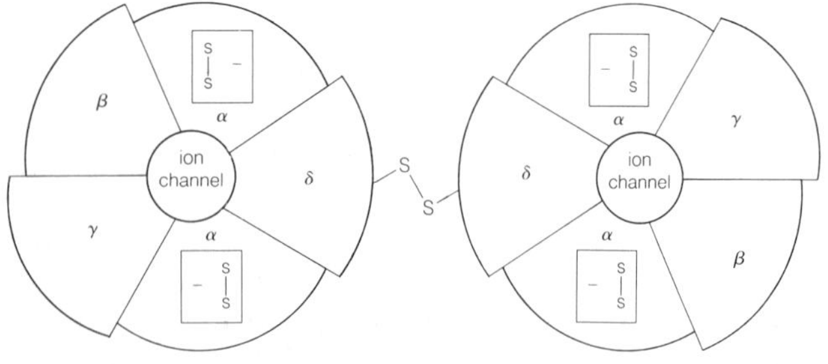

FIG. 12-1. Schematic representation of the polypeptide chains of the cholinergic nicotinic receptor (see Ref. 3).

cations, Ca^{2+}, Na^+, and K^+. The physiologic effect is to temporarily depolarize the end-plate. This depolarization results in muscular contraction at a neuromuscular junction, or in continuation of the nerve impulse, as occurs in an autonomic ganglion. Neuromuscular nicotinic ACh receptors are of interest as the target for autoimmune antibodies in myasthenia gravis and muscle relaxants used during the course of surgical procedures. Nicotinic receptors in autonomic ganglia, when blocked by drugs, play an important role in the control of hypertension.

MUSCARINIC ACETYLCHOLINE RECEPTOR

Muscarinic receptors play an essential role in regulating the functions of organs innervated by the autonomic nervous system to maintain homeostasis of the organism. The action of ACh on muscarinic receptors can be either stimulatory or inhibitory. Acetylcholine stimulates secretions from salivary and sweat glands, causes secretion and contraction of the gut, and constriction of the airways of the respiratory tract. Yet, it inhibits the contraction of the heart and relaxes smooth muscle of blood vessels. A number of cholinergically induced responses depend on K^+ channels, although not always in the same way. In the heart, these channels are opened by ACh, and the result is hyperpolarization and inhibition. In other neurons, the channels are closed by ACh and the result is depolarization (e.g., contraction of smooth muscle).

All muscarinic responses have a rather leisurely time course. This characteristic suggests that a second messenger must mediate each response.[7] It is proposed that one of two biochemical events may occur as a result of activation of muscarinic receptors. In some tissues, the primary response is an inhibition of adenylate cyclase activity, whereas in others, phosphoinositide turnover is enhanced. The biochemical response to activation of muscarinic cholinergic receptors is centered around a mechanism whereby ACh inhibits adenylate cyclase.[8] A second mechanism is associated with a GTP-dependent formation of an inositol phosphate (IP_2) and diacylglycerol[9] (Fig. 12-2).

Although nicotinic receptors are homogeneous, there is now strong evidence from both pharmacologic and biochemcial studies that there are at least two subtypes of muscarinic receptors, M_1 and M_2.[10,11] The M_1 muscarinic receptors are located in autonomic ganglia and select regions of the central nervous system. They can be distinguished by their high sensitivity to the agonist McN-A-343 and the antagonist pirenzepine. The M_2 muscarinic recep-

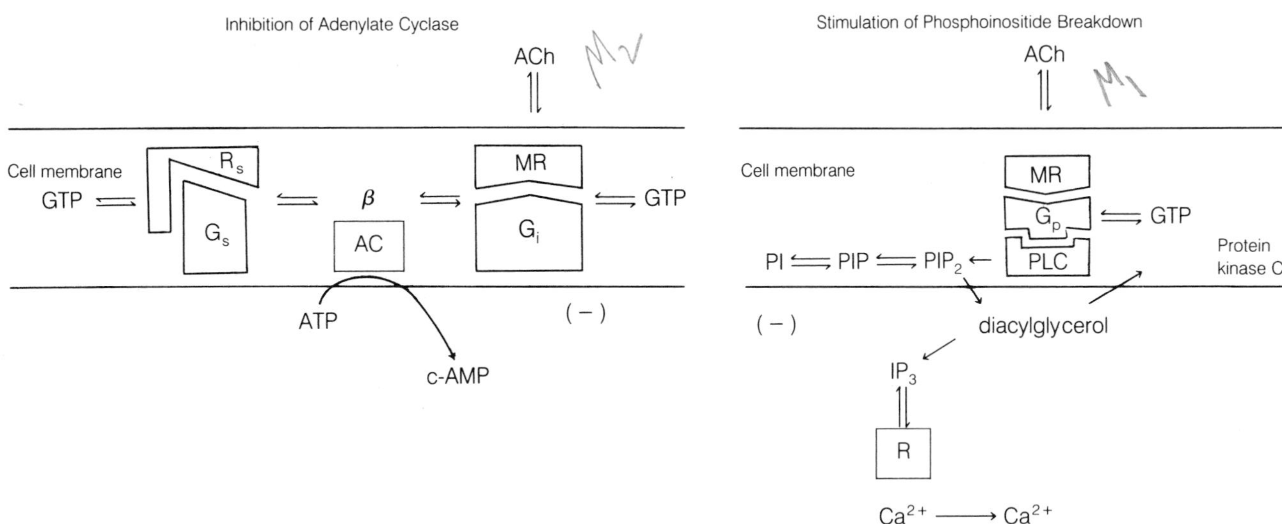

FIG. 12-2. *Proposed biochemical mechanisms of muscarinic cholinergic receptor action. Acetylcholine (ACh) acts on muscarinic receptor (MR) to activate the inhibitory guanine nucleotide regulatory protein (G_i); G_i deactivates the stimulatory guanine nucleotide regulatory protein (G_s) causing inhibition of the catalytic activity of adenylate cyclase (AC). A stimulatory hormone receptor (R) (e.g., β-adrenergic receptor) is included as part of the model. ACh may also stimulate MR to activate a guanine nucleotide regulatory protein (G_p). G_p activates a membrane phospholipase C (PLC) causing hydrolysis of phosphatidylinositol 4,5-biphosphate (PIP_2) to the second messengers, inositol 1,4,5-triphosphate (IP_3) and 1,2-diacylglycerol. IP_3 causes release of Ca^{2+} into the cytosol, whereas diacylglycerol activates phosphokinase C. (After Harden, T. K., et al.: Trends Pharmacol. Sci. 7(Suppl):14, 1986.)*

tors predominate in the periphery (e.g., heart, smooth muscle, exocrine glands) and have a much lower affinity than M_1 receptors for McN-A-343 and pirenzepine. It is suspected that the M_2 receptors

$(CH_3)_3\overset{+}{N}CH_2C{\equiv}CCH_2OC(=O){-}NH{-}C_6H_4Cl$

McN-A-343

Pirenzepine

may not be a homogeneous class of receptors and, as a result, M_2 receptors in various effector organs, such as the heart, may be selectively affected by agonists and antagonists. The M_2 receptors also are found in the CNS (Fig. 12-3).

Central Nervous System		**Peripheral Nervous System**	
Postsynaptic neurons	M_1, M_2	Heart	M_2
Presynaptic terminals	M_1, M_2	Smooth muscle	M_2
		Exocrine glands	M_2
		Autonomic ganglia	M_1, M_2

FIG. 12-3. *Muscarinic receptor subtypes.*

CHOLINERGIC NEUROCHEMISTRY

Cholinergic neurons synthesize, store, and release ACh (Fig. 12-4). The neurons also form choline acetyltransferase (ChAT) and acetylcholinesterase (AChE). These enzymes are synthesized in the soma of the neuron and transported down the axon through microtubules to the nerve terminal. Acetylcholine is prepared at the nerve ending by the transfer of an acetyl group from acetyl-CoA to choline. The reaction is catalyzed by choline acetyltransferase. Cell fractionation studies show that much of the ACh is contained in synaptic vesicles at the nerve ending, but some is also free in the cytosol. The major source of choline for ACh synthesis in vivo is from the hydrolysis of ACh in the synapse. The choline formed is recaptured by the presynaptic terminal by a high-affinity uptake system under the influence of sodium ions[12, 13] and used to synthesize the ACh released from the synaptic vesicles by the nerve action potential. Free ACh and phosphorylcholine found in the cytosol are prepared from

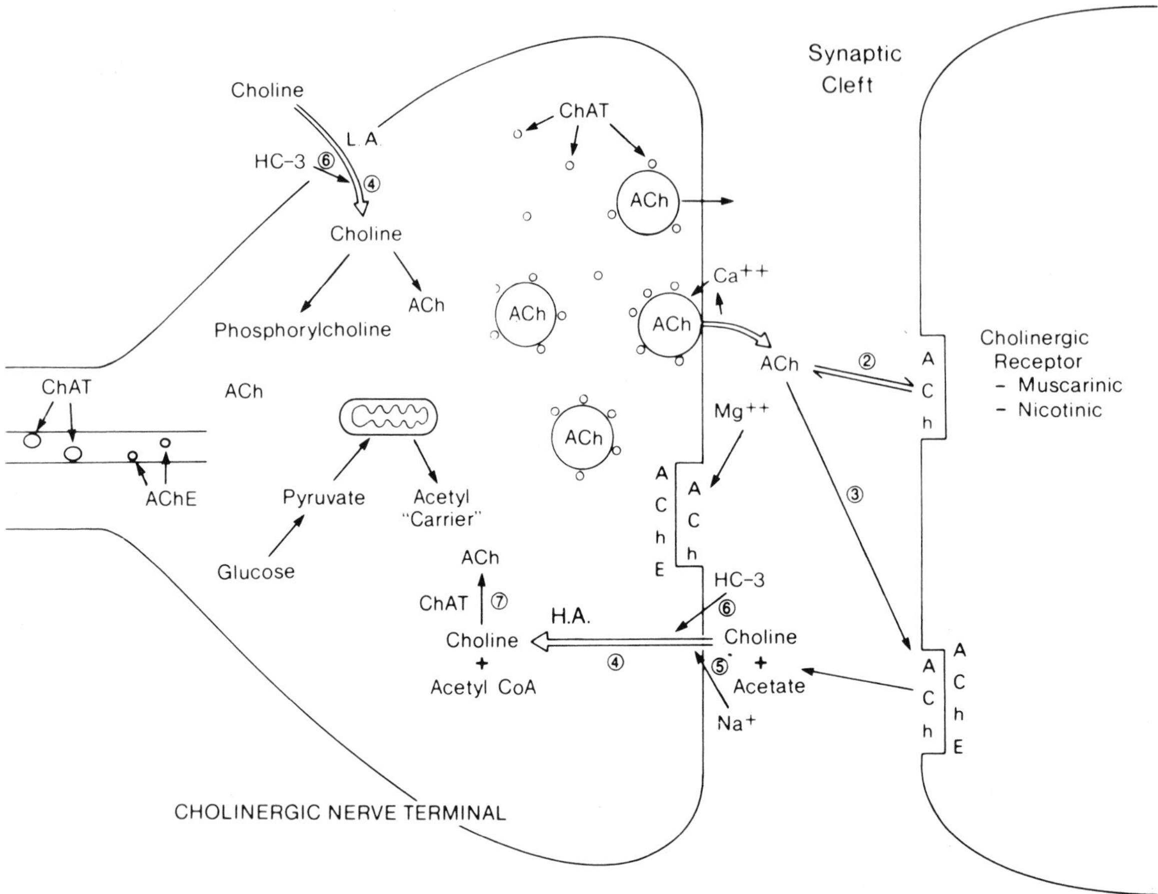

FIG. 12-4. *Hypothetical model for synthesis, storage, and release of ACh. 1. Newly formed ACh is preferentially released by diffusion through the membrane under influence of Ca^{2+} and inhibited by Mg^{2+}. 2. ACh crosses synaptic cleft and reacts reversibly with cholinergic receptors or is (3) hydrolyzed by AChE to choline and acetate. 4. Choline is taken up by the neuron by a high-affinity (HA) or low-affinity (LA) uptake system. 5. Na^+-dependent, high-affinity uptake system of choline stimulates ACh synthesis. 6. Hemicholinium (HC3) specifically blocks both high- and low-uptake systems. 7. ACh is synthesized by ChAT and stored in synaptic vesicles. 8. Acetyl-CoA is released from the mitochondria using acetate formed from glucose in the cytosol. 9. Choline from the low-affinity uptake system forms ACh (not released on nerve stimulation) and phosphorylcholine.*

choline brought into the neuron by the low-affinity uptake system, but are not active in the nerve transmission process.

Several quaternary ammonium bases act as competitive inhibitors of choline uptake. Hemicholinium (HC-3), a bis quaternary cyclic hemiacetal, and the triethyl analogue of choline, 2-hydroxyethyltriethylammonium, act at the presynaptic membrane to inhibit the high-affinity uptake system of choline into the neuron. These compounds cause a delayed paralysis at repetitively activated cholinergic synapses and can produce respiratory paralysis in test animals. The delayed block is due to the depletion of stored ACh that may be reversed by choline. The acetyl group used for the synthesis of ACh is obtained by conversion of glucose to pyruvate in the cytosol of the neuron and eventual formation of acetyl-CoA. Owing to the impermeability of the mitochondrial membrane to acetyl-CoA, this substrate

Hemicholinium (HC—3)

$HO—CH_2CH_2—\overset{+}{N}(C_2H_5)_3$

2-Hydroxyethyltriethylammonium

is brought into the cytosol by the aid of an acetyl "carrier."

The synthesis of ACh from choline and acetyl-CoA is catalyzed by ChAT. Transfer of the acetyl group from acetyl-CoA to choline may be by a random or by an ordered reaction of the Theorell–Chance type. In the ordered sequence, acetyl-CoA first binds to the enzyme, forming a complex (EA) that then binds

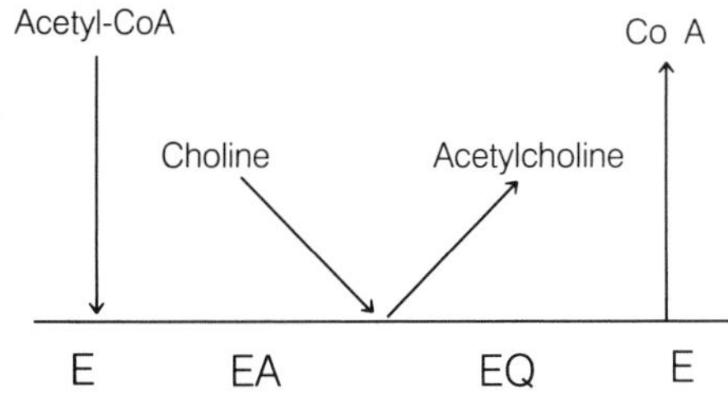

FIG. 12-5. *Ordered synthesis of ACh by choline acetyltransferase.*

to choline. The acetyl group is transferred and the ACh formed dissociates from the enzyme active site. The CoA then is released from the enzyme complex, EQ, to regenerate the free enzyme. The scheme is diagrammed in Figure 12-5. Choline acetyltransferase is inhibited in vitro by *trans-N*-methyl-4-(1-naphthylvinyl)pyridinium iodide;[14] however, its inhibitory activity in whole animals is unreliable.[15]

I^- $N^+—CH_3$

trans-N-Methyl-4(1-naphthylvinyl)pyridinium iodide

Newly formed ACh is released from the presynaptic membrane when a nerve action potential invades a presynaptic nerve terminal.[16] The release of ACh results from a depolarization of the nerve terminal by the action potential, which alters membrane permeability to Ca^{2+}. Calcium enters the nerve terminal and causes the release of the contents of several synaptic vesicles containing ACh into the synaptic cleft. This burst or quantal release of ACh causes depolarization of the postsynaptic membrane. The number of quanta of ACh released may be as high as several hundred at a neuromuscular junction, with each quantum containing between 12,000 to 60,000 molecules. Acetylcholine is also released spontaneously in small amounts from presynaptic membranes. This small amount of neurotransmitter serves to maintain muscle tone by acting on the cholinergic receptors on the postsynaptic membrane.

The released ACh may either diffuse out of the cleft and be hydrolyzed by AChE into choline and acetate, or it may bind to receptors in the postsynaptic membrane.

After ACh has been released into the synaptic cleft, its concentration decreases rapidly. It is generally accepted that there is enough AChE at nerve endings to hydrolyze into choline and acetate any ACh that has been liberated. The AChE activity from rat intercostal muscle is able to hydrolyze about 2.7×10^8 ACh molecules in one millisecond; this far exceeds the 3×10^6 molecules released there by one nerve impulse.[17] In skeletal muscle the greater part of AChE found at the end-plate region is located on the postsynaptic membrane. Acetylcholinesterase is present predominantly on presynaptic membranes in sympathetic ganglia.

The major sites of activity of ACh in the periphery are as follows: the postganglionic parasympathetic receptors, the autonomic ganglia, and the skeletal neuromuscular junction. The first is blocked by atopine, the second by hexamethonium, and the third by decamethonium or curare. The function of the receptors in each of these three different sites is to convert the binding of ACh into some change in cellular activity. To do this the receptor must modulate some effector system. It is suggested that this is accomplished by a conformational transition in the receptor. This assumes that the receptor can exist in at least two conformations, a ground and an excited state. Burgen states that the final conformations of the receptor to which agonists and antagonists bind are different, but the magnitude of the difference is now unknown.[18]

CHOLINERGIC AGONISTS

CHOLINERGIC STEREOCHEMISTRY

Three techniques have been used to study the conformational properties of ACh and other cholinergic chemicals: roentgenographic (x-ray) crystallography, nuclear magnetic resonance (NMR), and molecular orbital calculations. Each of these methods may report the spatial distribution of atoms in a molecule in terms of torsion angles. However, they provide only circumstantial evidence about the conformation of the chemical as it acts on the biologic receptor.

AcO τ_2 N^+ 4 5

AcO τ_2 N^+ H H H H

Newman projection

FIG. 12-6. *Gauche conformation of ACh.*

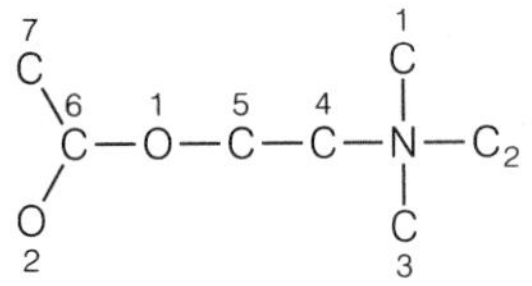

FIG. 12-7. *Torsion angles of ACh.*

A *torsion angle* is defined as the angle formed between two planes as, for example, by the O1-C5-C4-N atoms in ACh. The angle between the oxygen and nitrogen atoms is best depicted by means of Newman projections (Fig. 12-6).

A torsion angle has a positive sign when the bond to the front atom is rotated to the right to eclipse the bond of the rear atom. The spatial orientation of ACh is described by four torsion angles (Fig. 12-7).

The conformation of the choline moiety of ACh has drawn the most attention in attempting to relate structure and pharmacologic activity. The torsion angle (τ_2) determines the spatial orientation of the cationic head of ACh to the ester group. Roentgenographic diffraction studies have shown that the torsion angle (τ_2) on ACh has a value of +77°. Many compounds that are muscarinic receptor agonists containing a choline component [i.e., O-C-C-$N^+(CH_3)_3$] have a preferred synclinal (gauche) conformation with τ_2 values ranging from 68° to 89° (Table 12-1). Intermolecular-packing forces in the crystal, as well as electrostatic interactions between the charged nitrogen group and the ether oxygen of the ester group, are probably the two dominant factors that lead to a preference for the synclinal conformation in the crystal state. Some choline esters display an antiperiplanar (*trans*) conformation between the onium and ester groups. For example, carbamoylcholine chloride (τ_2 + 178°) is stabilized in this *trans* conformation by several hydrogen bonds,[19] acetylthiocholine iodide (τ_2 + 171°) is in this conformation because of the presence of the more bulky and less electronegative sulfur atom, and (+) *trans*-(1*S*,2*S*)-acetoxycyclopropyl trimethylammonium iodide (τ_2 + 137°) is fixed in this conformation by the rigidity of the cyclopropyl ring.

The NMR spectroscopy of cholinergic molecules in solution is more limited than crystallography in delineating the conformation of compounds and is restricted to determining the torsion angle O1-C5-C4-N. Most NMR data are in agreement with the results of roentgenographic diffraction studies. The NMR studies indicate that ACh and methacholine are apparently not in their most stable *trans* conformation, but exist in one of two gauche conformers[20] (Fig. 12-8). This is in agreement with the roentgenographic data and may be the result of strong intramolecular interactions that stabilize the conformation of these molecules in solution.[21]

Molecular orbital calculations based on the principles of quantum mechanics may be used to determine energy minima of rotating bonds and predict preferred conformations for the molecule. The measurements consider the molecules only in an isolated state and, therefore, do not take into account solvent interactions. Calculations[22] using the Hückel molecular orbital method have found that ACh has an energy minimum for the τ_2 torsion angle at about 80° and that the preferred conformation for the entire molecule of ACh corresponds closely to that found in the crystal state.

The exact conformation adopted by ACh to exert its cholinergic activity in vivo is unknown. The application of molecular modeling to produce a hypothetical picture of a cholinergic receptor drug interaction waits for computer-assisted analysis of the protein structure of the receptor and active agonists. In the meantime, we are left to use the more classic approaches of the stereochemical requirements of drug–receptor interactions. Studies of compounds

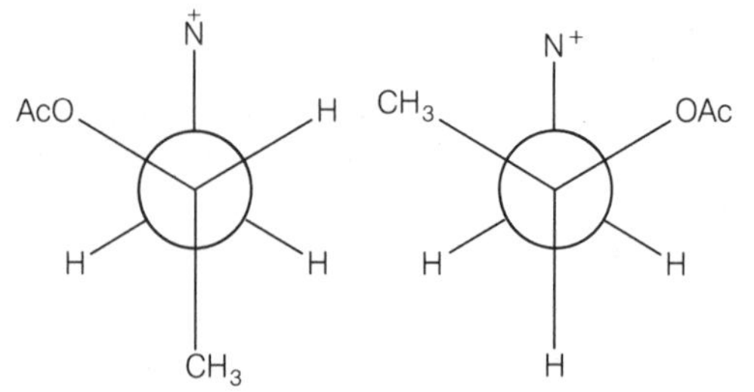

TABLE 12-1

CONFORMATIONAL PROPERTIES OF SOME CHOLINERGIC AGENTS

Compound	O1-C5-C4-N Torsion Angle
Acetylcholine bromide	+77
Acetylcholine chloride	+85
(+)-(2*S*,3*R*,5*S*)-muscarine iodide	+73
Methylfurmethide iodide	+83
(+) Acetyl(*S*)β-methylcholine iodide	+85
(−) Acetyl(*R*)α-methylcholine iodide	
Crystal form A	+89
Crystal form B	−150
(+) *cis*(2*S*)-methyl-(4*R*)-trimethylammonium-1,3-dioxolane iodide	+68
(+) *trans*(1*S*,2*S*)-acetoxycycloproplytrimethyl-ammonium iodide	+137
Carbamoylcholine bromide	+178
Acetylthiocholine bromide	+171
Acetyl(*R*α,*S*β)-dimethylcholine iodide (*erythro*)	+76

[From Shefter, E.: In Triggle, D. J., Moran, J. F., and Barnard, E. A., (eds.). Cholinergic Ligand Interactions. New York, Academic Press, 1971.]

FIG. 12-8. *Gauche conformers of acetyl β-methacholine.*

Fig. 12-9. Geometric isomers of muscarine.

that have significant muscarinic activity and contain semirigid or rigid structures have been used to suggest the active conformation of ACh. The muscarinic receptor is stereoselective in its interaction with cholinergic agents. From this information a hypothesis about the conformational properties of ACh at the muscarinic receptor has been developed.

The parasympathomimetic effects of muscarine were first reported in 1869,[23] but its structure was not elucidated until 1957.[24] Muscarine has four geometric isomers: muscarine, epimuscarine, allomuscarine, and epiallomuscarine (Fig. 12-9), none of which has a center or plane of symmetry. Each geometric isomer can exist as an enantiomeric pair.

The activity of muscarine is highly specific and resides primarily in the naturally occurring (+) muscarine. It is essentially free of nicotinic activity and apparently has the optimal stereochemistry to act on the muscarinic receptor. Synthetic molecules having a substituent on the carbon atom that corresponds to the β-carbon atom of ACh show great differences in muscarinic activity between their isomers. Acetyl (+) (*S*)-β-methylcholine, (+) *cis*-(2*S*)-methyl-(4*R*)-trimethylammonium-1,3-dioxolane, (+) *trans*-(1*S*,2*S*)-acetoxycyclopropyltrimethylammonium, and naturally occurring (+) (2*S*,3*R*,5*S*)-muscarine are more potent than their enantiomers and have very high ratios of activity between the (*S*) and (*R*) isomers (Table 12-2). Each of these more active isomers has the same configuration at the carbon atom adjacent to the ester group. Although ACh does not have an asymmetric center, its preferred conformation gives it the characteristics of an asymmetric molecule and perhaps acts in this orientation during its role as a neurotransmitter. A similar observation may be made of (+) acetyl(*S*)-β-methylcholine, (+) *cis*-(2*S*)-methyl-(4*R*)-trimethylammonium methyl-1,3-dioxolane, and (+) *trans*-(1*S*,2*S*)-acetoxycyclopropyltrimethylammonium, all of which have an (*S*) configuration at the carbon atom adjacent to the ester group. Each of these

TABLE 12-2

EQUIPOTENT MOLAR RATIOS OF ISOMERS ON GUINEA PIG ILEUM; RATIOS RELATIVE TO ACETYLCHOLINE

Compound	Guinea Pig Ileum	(S)/(R) Ratio
(+) Acetyl(*S*)β-methylcholine chloride	1.0*	24
(−) Acetyl(*R*)β-methylcholine chloride	24.0*	
(+) (2*S*,3*R*,5*S*)-Muscarine iodide	0.33†	394
(−) (2*R*,3*S*,5*R*)-Muscarine iodide	130†	
(+) *cis*(2*S*)-Methyl-(4*R*)-trimethylammonium-1,3-dioxolane Iodide	6.00‡	100
(−) *cis*(2*R*)-Methyl-(4*S*)-trimethylammonium-1,3-dioxolane Iodide	0.06‡	
(+) *trans*(1*S*,2*S*)-Acetoxycyclopropyltrimethylammonium Iodide	0.88§	517
(−) *trans*(1*R*,2*R*)-Acetoxycyclopropyltrimethylammonium Iodide	455§	

(*Beckett, A. H., Harper, N. J., Clitherow, J. W., and Lesser, E.: Nature 189:671, 1961; †Waser, P. I.: Pharmacol. Rev. 13:465, 1961; ‡Belleau, B., and Puranen, J.: J. Med. Chem. 6:235–328, 1963; §Armstrong, P. D., Cannon, J. G., and Long, J. P.: Nature 220:65–66, 1968.)

active muscarinic molecules may be deployed on the receptor, in the same manner as ACh and (+) muscarine. Their (*S*)/(*R*) ratios (see Table 12-2) show the great degree of stereoselectivity of the muscarinic receptor in guinea pig ileum for the configuration at the carbon adjacent to the ester group.

cis-2-Methyl-4-trimethylammonium-1,3-dioxolane

trans-2-Acetoxycyclopropyltrimethylammonium

The nicotinic receptor is not considered as highly stereoselective as its muscarinic counterpart.

STRUCTURE–ACTIVITY RELATIONSHIPS OF CHOLINERGIC AGONISTS

Acetylcholine is a relatively simple molecule. Both its chemistry and its ease of testing for activity have allowed numerous chemical derivatives to be made and studied. Alterations on the molecule may be divided into three categories: the onium group, the ester function, and the choline moiety.

The onium group is essential for the intrinsic activity and contributes to the affinity of the drug to the receptors, partially through the binding energy and partially because of its action as a detecting and directing group. The trimethylammonium group is the optimal functional group for activity, although some significant exceptions are known (e.g., pilocarpine, arecoline, nicotine, and oxytremorine). Phosphonium, sulfonium, arsenonium isosteres, or substituents larger than methyl on the nitrogen increase the size of the onium moiety, resulting not only in a diffusion of the charge and steric interference to proper drug–receptor interaction but also in a decrease in affinity (Table 12-3).

The ester group in ACh contributes to the binding of the compound to the parasympathetic terminal synapse, probably because of its hydrogen bond-forming capacity. A comparison of the cholinergic activity of a series of alkyl trimethylammonium compounds ($R — \overset{+}{N}(CH_3)_3$; $R = C_1–C_9$) shows that *n*-amyltrimethylammonium,[25] which may be considered to have a size and mass similar to ACh, is about one magnitude weaker as a muscarinic agonist. The presence of the acetyl group in ACh is not as critical as is the size of the molecule. In studying a series of *n*-alkyl trimethylammonium salts, it was noted[26] that for maximal muscarinic activity the quaternary ammonium group should be followed by a chain of five atoms; this was referred to as the *five-atom rule*. For the receptors for ACh at the ganglionic synapse, compounds with a chain length of four atoms are more active than similar compounds with a chain length of five atoms.

Beckett[27] observed that a trimethylammonium group and ether oxygen were present in most active muscarinic agonists. It was proposed that the muscarinic receptor contained an anionic-binding site (*site 1*, see Fig. 12-10) that accommodates the quaternary group of muscarine approximately 3.0 Å from a hydrogen-bonding site (*site 2*, see Fig. 12-10) for the ring oxygen of muscarine or the other oxygen of the choline moiety. A distance of 5.7 Å separates the quaternary nitrogen from the alcohol group of muscarine, which is the location of a subsidiary site (*site 3*, see Fig. 12-10). Site 3 can interact with the

TABLE 12-3

ACTIVITY OF ACETOXYETHYL ONIUM SALTS: EQUIPOTENT MOLAR RATIOS RELATIVE TO ACETYLCHOLINE

$CH_3COOCH_2CH_2^-$	Cat Blood Pressure	Intestine	Frog Heart
$\overset{+}{N}Me_3$	1	1 (Rabbit)	1
$\overset{+}{N}Me_2H$	50	40	50
$\overset{+}{N}MeH_2$	500	1,000	500
$\overset{+}{N}H_3$	2,000	20,000	40,000
$\overset{+}{N}Me_2Et$	3	2.5 (Guinea pig)	2
$\overset{+}{N}MeEt_2$	400	700	1,500
$\overset{+}{N}Et_3$	2,000	1,700	10,000*
$\overset{+}{P}Me_3$	13	12 (Rabbit)	12
$\overset{+}{As}Me_3$	66	90	83
$\overset{+}{S}Me_2$	50	30 (Guinea pig)	96

Size of quaternary atom:† $\overset{+}{X}$ bonded to two C atoms (X–C distance d; C↔C distance d′)

	d	d′
N	d = 1.47 Å	d′ = 2.4 Å
P	1.87	3.05
S	1.82	—
As	1.98	3.23

*Reduces effect of acetylcholine

†From Barlow, R. B.: Introduction to Chemical Pharmacology, London, Methuen and Co. 1964.)

(Welsh, A. D., and Roepke, M. H.; J. Pharmacol. Exp. Ther. 55:118, 1935; Stehle, K. L., Melville, K. J., and Oldham, F. K.: J. Pharmacol. Exp. Ther. 56:473, 1936; Holton, P., and Ing, H. R.: Br. J. Pharmacol. 4:190, 1949; Ing. H. R., Kordik, P., and Tudor Williams, D. P. H.: Br. J. Pharmacol. 7:103, 1952.)

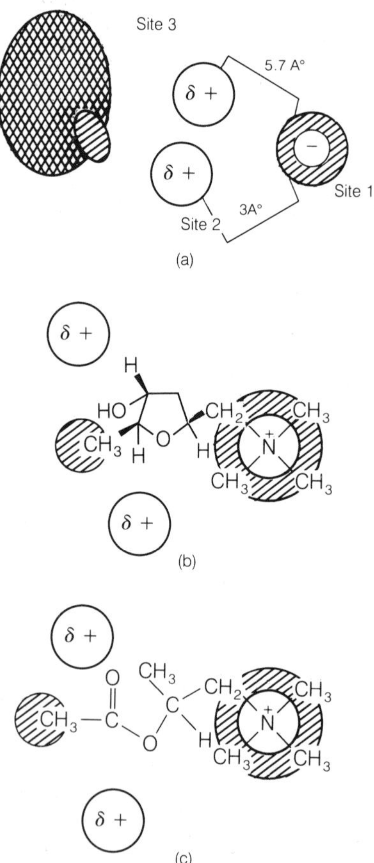

FIG. 12-10. *Hypothetical structure of the muscarinic receptor.* (a) *Diagram of distribution of charges and hydrophobic portions (shaded) and large flat area for van der Waals interaction (cross-hatched).* (b) *Diagram of interaction of muscarine with muscarinic receptor.* (c) *Diagram of interaction of acetyl β-methylcholine with muscarinic receptor. [Adapted from Michelson, M. J., and Zeimal, E. V. (eds.): Acetylcholine. Oxford, Pergamon Press, 1973.]*

carbonyl group of ACh, the ether oxygen of the dioxolane, and the double bond of furan.

The ether oxygen appears to be of primary importance for high muscarinic activity, because both choline ethyl ether and β-methylcholine ethyl ether have high muscarinic activity.[28,29] The attachment of the choline part of ACh to the receptor seems to be related to that of (+) muscarine, or vice versa.

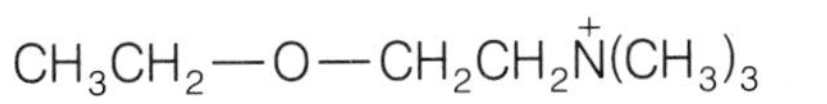

$$CH_3CH_2—O—CH_2CH_2\overset{+}{N}(CH_3)_3$$

Choline Ethyl Ether

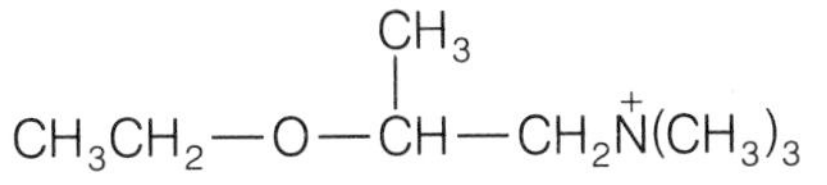

$$CH_3CH_2—O—CH(CH_3)—CH_2\overset{+}{N}(CH_3)_3$$

β-Methylcholine Ethyl Ether

Either shortening or lengthening the chain of atoms that separates the ester group from the onium moiety reduces muscarinic activity. α-Substitution on the choline moiety decreases both nicotinic and muscarinic activity, but the muscarinic activity is decreased to a greater extent. Nicotinic activity is decreased to a greater degree by substitution on the β-carbon. Therefore, acetyl α-methylcholine has nicotiniclike activity but little muscariniclike activity, and acetyl β-methylcholine (methacholine) has muscariniclike properties but little nicotinic activity. Hydrolysis by AChE is affected more by substitutions on the β- than on the α-carbon. The hydrolysis rate of racemic acetyl β-methylcholine is about 50% of that of ACh; racemic acetyl α-methylcholine is hydrolyzed about 90% as fast.

PRODUCTS

Acetylcholine Chloride. Acetylcholine exerts a powerful stimulation of the parasympathetic nervous system. Attempts have been made to use it as a cholinergic agent,[30,31] but its duration of action is too short for sustained effects, owing to rapid hydrolysis by esterases and lack of specificity when administered for systemic effects. Acetylcholine is a cardiac depressant and an effective vasodilator. The stimulation of the vagus and the parasympathetic nervous system produces a tonic action on smooth muscle and induces a flow from the salivary and lacrimal glands. Its cardiac-depressant effect results from (1) a negative chronotropic effect that causes a decrease in heart rate, and (2) a negative inotropic action on heart muscle that produces a decrease in the force of myocardial contractions. The vasodilator action of ACh is primarily on the arteries and the arterioles, with distinct effect on the peripheral vascular system. Bronchial constriction is a characteristic side effect when the drug is given systemically.

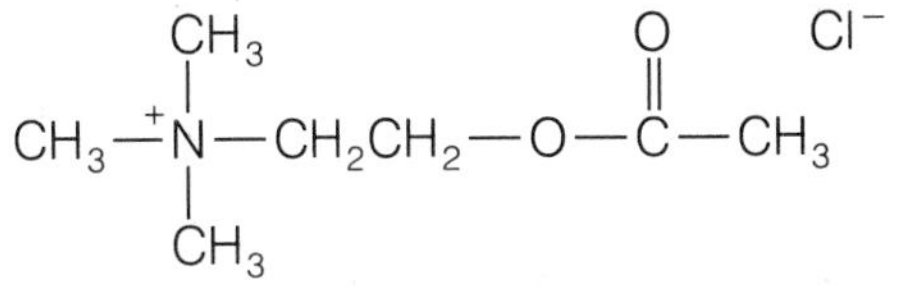

$$(CH_3)_3\overset{+}{N}—CH_2CH_2—O—\overset{O}{\overset{\|}{C}}—CH_3 \quad Cl^-$$

Acetylcholine Chloride

One of the most effective antagonists to the action of ACh is atropine. Atropine blocks out the depressant effect of ACh on cardiac muscle and its production of peripheral vasodilatation (i.e., muscarinic effects), but does not affect the skeletal muscle contraction (i.e., nicotinic effect) that is produced.

Acetylcholine chloride is a hygroscopic powder that is available in an admixture with mannitol to be dissolved in sterile water for injection shortly before use. It acts as a short-acting miotic when introduced into the anterior chamber of the eye and is especially useful after cataract surgery during the placement of sutures. When applied topically to the eye it has little therapeutic value because of poor corneal penetration and rapid hydrolysis by AChE.

Methacholine Chloride, USP. Acetyl β-methylcholine chloride; (2-hydroxypropyl)trimethylammonium chloride acetate. Methacholine is the acetyl ester of β-methylcholine. Unlike ACh, methacholine has sufficient stability in the body to give sustained parasympathetic stimulation. This action is accompanied by little (1/1000 that of ACh) or no "nicotinic effect." Methacholine chloride is rarely used clinically today.

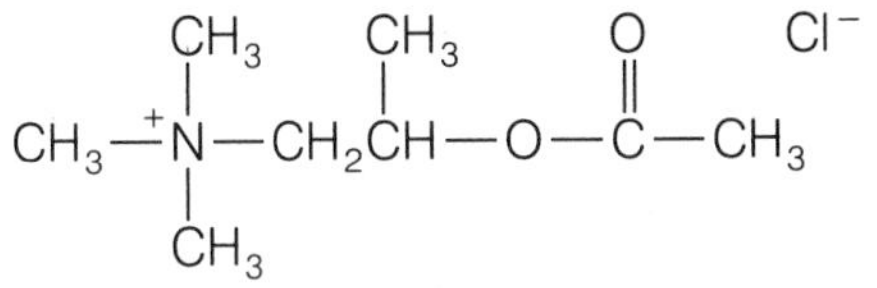

$$(CH_3)_3\overset{+}{N}—CH_2CH(CH_3)—O—\overset{O}{\overset{\|}{C}}—CH_3 \quad Cl^-$$

Methacholine Chloride

Methacholine can exist as (*S*) and (*R*) enantiomers. Although the chemical is used as the racemic mixture, its muscarinic activity resides principally in the (*S*)isomer. The (*S*)/(*R*) ratio of muscarinic potency for these enantiomers is 240 : 1.

(+) Acetyl (*S*)-β-methylcholine is hydrolyzed by AChE, whereas the (*R*)(−) isomer is not. (−) Acetyl (*R*)-β-methylcholine is a weak competitive inhibitor (K_i 4 × 10^{-4} M) of AChE obtained from the electric organ of the eel (*Electrophorus electricus*). The hydrolysis rate of the (*S*)(+) isomer is about 54% that of ACh. This rate probably compensates for any decreased association (affinity) owing to the β-methyl group with the muscarinic receptor site and may account for the fact that ACh and (+) acetyl β-methylcholine have equimolar muscarinic potencies in vivo. (−) Acetyl (*R*)-β-methylcholine weakly inhibits AChE and slightly reinforces the muscarinic activity of the (*S*)(+) isomer in the racemic mixture of acetyl β-methylcholine.

In the hydrolysis of the acetyl α- and β-methylcholines, the greatest stereochemical inhibitory effects occur when the choline is substituted in the β-position. This also appears to be true of organophosphorus inhibitors. The (*R*)(−) and (*S*)(+) isomers of acetyl α-methylcholine are hydrolyzed at 78% and 97% of the rate of ACh, respectively.

Methacholine chloride occurs as colorless or white crystals or as a white crystalline powder. It is odorless or has a slight odor and is very deliquescent. It is freely soluble in water, alcohol, or chloroform, and its aqueous solution is neutral to litmus and is bitter. It is rapidly hydrolyzed in alkaline solutions. Solutions are relatively stable to heat and will keep for at least two or three weeks when refrigerated to delay growth of molds.

Carbachol, USP. Choline chloride carbamate. The pharmacologic activity of carbachol is similar to that of ACh. It is an ester of choline and, thus, possesses both muscarinic and nicotinic properties. Carbachol stimulates the autonomic ganglia and causes contraction of skeletal muscle, but differs from a true muscarinic agent in that it does not have cardiovascular activity. Carbachol is a miotic and has been used to reduce the intraocular tension of glaucoma when a response cannot be obtained with pilocarpine or neostigmine. Penetration of the cornea by carbachol is poor, but can be enhanced by the use of a wetting agent in the ophthalmic solution. In addition to its topical use for glaucoma, carbachol is used during ocular surgery, when a more prolonged miosis is required than that which can be obtained with acetylcholine chloride.

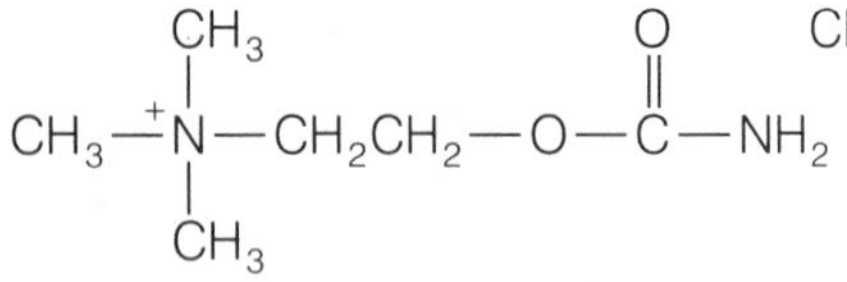

$$CH_3-\overset{+}{N}(CH_3)_2-CH_2CH_2-O-\overset{O}{\overset{\|}{C}}-NH_2 \quad Cl^-$$

Carbachol Chloride

Carbachol differs chemically from ACh in its stability to hydrolysis. The carbamyl group of carbachol decreases the electrophilicity of the carbonyl and, thus, can form resonance structures more easily than ACh. The result is that carbachol is less susceptible to hydrolysis and, therefore, more stable than ACh in aqueous solutions.

The pharmacologic properties of carbachol may be the result of its ability to cause release of ACh from cholinergic nerve endings. Carbachol also produces a cholinergic response by acting as an inhibitor of AChE. Carbachol acts as a semireversible inhibitor of AChE and serves to prolong the duration of ACh at the neuromuscular junction.

Bethanechol Chloride, USP. β-Methylcholine carbamate chloride; (2-hydroxypropyl)trimethylammonium chloride carbamate; carbamylmethylcholine chloride (Urecholine). Bethanechol has pharmacologic properties similar to methacholine. Both are esters of β-methylcholine and have feeble nicotinic activity. Bethanechol is more slowly inactivated than methacholine in vivo by AChE.

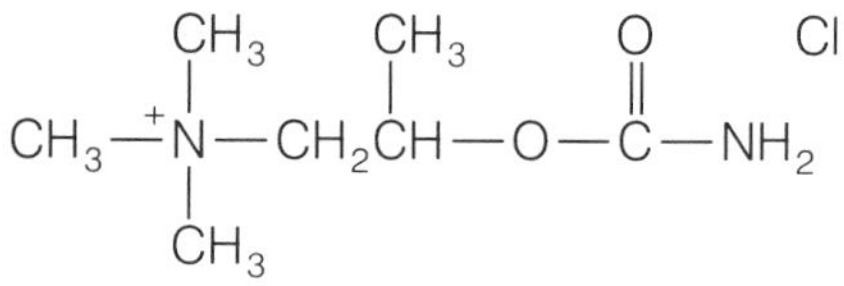

$$CH_3-\overset{+}{N}(CH_3)_2-CH_2\underset{}{\overset{CH_3}{\overset{|}{C}}}H-O-\overset{O}{\overset{\|}{C}}-NH_2 \quad Cl^-$$

Bethanechol Chloride

Bethanechol is a carbamyl ester and would be expected to have the same stability in aqueous solutions as carbachol.

The main use of bethanechol chloride is in the relief of urinary retention and abdominal distention after surgery. The drug is used orally and by subcutaneous injection. It must never be administered by intramuscular or intravenous injection. Administration of the drug is associated with low toxicity, and has no serious side effects. Bethanol chloride should be used with caution in asthmatic patients and, when used for glaucoma, does produce frontal headaches from the constriction of the sphincter muscle in the eye and from ciliary muscle spasms. Its duration of action is about one hour.

Bethanechol chloride occurs as a white crystalline solid with an aminelike odor. It is soluble in water 1:1 and 1:10 in alcohol, but is nearly insoluble in most organic solvents. An aqueous solution has a *p*H of 5.5 to 6.3.

Aceclidine Hydrochloride. 3-Quinuclidinol acetate hydrochloride is a parasympathomimetic drug under clinical investigation. It is amiotic and used in the treatment of glaucoma.

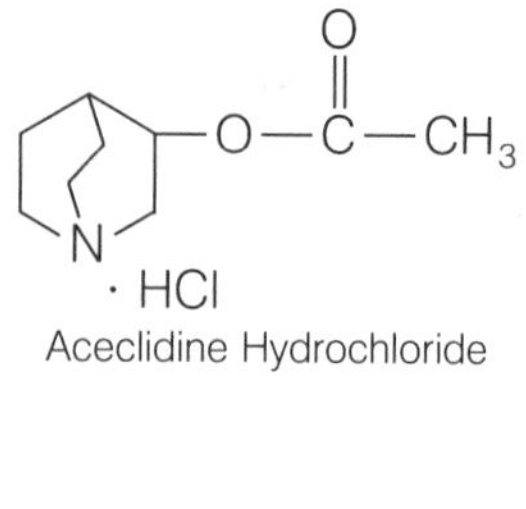

[quinuclidine ring]$-O-\overset{O}{\overset{\|}{C}}-CH_3$ · HCl

Aceclidine Hydrochloride

Pilocarpine Hydrochloride, USP. Pilocarpine monohydrochloride is the hydrochloride of an alkaloid obtained from the dried leaflets of *Pilocarpus jaborandi* or *P. microphyllus* in which it occurs to the extent of about 0.5% together with other alkaloids.

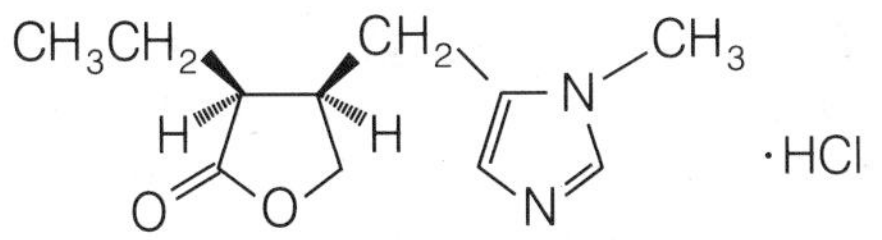

Pilocarpine Hydrochloride

It occurs as colorless, translucent, odorless, faintly bitter crystals that are soluble in water (1 : 0.3), alcohol (1 : 3), and chloroform (1 : 360). It is hygroscopic and affected by light; its solutions are acid to litmus and may be sterilized by autoclaving. Alkalies saponify the lactone group to give the pharmacologically inactive hydroxy acid (pilocarpic acid). Base-catalyzed epimerization at the ethyl group position occurs to an appreciable extent and is another major pathway of degradation.[32] Both routes result in loss of pharmacologic activity.

Pilocarpine has a physostigminelike action but appears to act by direct cell stimulation rather than by disturbance of the cholinesterase–ACh relationship, as with physostigmine.

Evidence supports the view that pilocarpine mimics the action of muscarine to stimulate ganglia through receptor site occupation similar to that of ACh.[33] Its overall molecular architecture and the interatomic distances of its functional groups in certain conformations are similar to those of muscarine (i.e., about 4 Å from the tertiary N—CH_3 nitrogen to the ether oxygen or carbonyl oxygen) and are compatible with this concept.

Outstanding pharmacologic effects are the production of copious sweating, salivation, and gastric secretion. Pilocarpine causes pupillary constriction (miosis) and a spasm of accommodation. These effects are valuable in the treatment of glaucoma. The pupil constriction and spasm of the ciliary muscle serve to reduce intraocular tension by establishing a better drainage of ocular fluid through the canal of Schlemm, located near the corner of the iris and cornea. Pilocarpine is used as a 0.5% to 6% solution (i.e., of the salts) in treating glaucoma. Secretion in the respiratory tract is noted following therapeutic doses and, therefore, the drug sometimes is used as an expectorant.

Pilocarpine Nitrate, USP. Pilocarpine mononitrate. This salt occurs as shining, white crystals that are not hygroscopic, but are light-sensitive. It is soluble in water (1 : 4), soluble in alcohol (1 : 75), and insoluble in chloroform and in ether. Aqueous solutions are slightly acid to litmus and may be sterilized in the autoclave. The alkaloid is incompatible with alkalies, iodides, silver nitrate, and reagents that precipitate alkaloids.

INDIRECT-ACTING CHOLINERGIC AGONISTS: CHOLINESTERASE INHIBITORS

Termination of the action of ACh at the synaptic junction is caused primarily by destruction of the neurotransmitter by AChE.

$$CH_3-\overset{O}{\overset{\|}{C}}-O-CH_2CH_2-\overset{+}{N}(CH_3)_3 + H_2O \xrightarrow{AChE} HO-CH_2CH_2-\overset{+}{N}(CH_3)_3 + CH_3COOH$$

Acetylcholinesterase catalyzes the hydrolysis of ACh to choline and acetic acid. Inhibition of AChE prolongs the life of the neurotransmitter in the junction and produces pharmacologic effects similar to those observed when ACh is administered. Anticholinesterases have been used in the treatment of myasthenia gravis, atony in the gastrointestinal tract, and glaucoma. They have also been employed as nerve gases and insecticides.

Butyrylcholinesterase (BuChE; pseudocholinesterase) is located in human plasma. Although its biologic function and purpose in humans are not clear, BuChE has catalytic properties similar to those of AChE. Acetylcholinesterase isolated from red blood cells has the same substrate properties as the junctional esterase and hydrolyzes only acetyl and propionyl esters. The substrate-specificity for BuChE is much more liberal (Table 12-4).

Three different chemical groupings—acetyl, carbamyl, and phosphoryl—may react with the esteratic site of AChE. Although the chemical reactions are similar, the kinetic parameters for each of these types of substrates are not the same and result in the differences between the toxicity and usefulness of these materials as drugs or chemicals.

The initial step in the hydrolysis of ACh by AChE is a reversible enzyme–substrate complex formation. The association rate (k_{+1}) and dissociation rate (k_{-1}) are relatively large. The enzyme–substrate complex, E · ACh, may also form an acetyl–enzyme intermediate at a rate (k_2) that is slower than either the association or dissociation rates. Choline is released from this complex with the formation of the acetyl enzyme intermediate, E–A. This intermediate is then hydrolyzed to regenerate the free enzyme and acetic acid. The acetylation rate, k_2, is the slowest step in this sequence and is rate-limiting (see later discussion).

TABLE 12-4

HYDROLYSIS OF VARIOUS SUBSTRATES BY AChE AND BuChE

	AChE		BuChE	
Enzyme Substrate	**Source**	**Relative Rate***	**Source**	**Relative Rate***
Acetylcholine	Human or bovine RBC	100	Human or horse plasma	100
Acetylthiocholine	Bovine RBC	149	Horse plasma	407
Acetyl β-methylcholine	Bovine RBC	18	Horse plasma	0
Propionylcholine	Human RBC	80	Horse plasma	170
Butyrylcholine	Human RBC	2.5	Horse plasma	250
Butyrylthiocholine	Bovine RBC	0	Horse plasma	590
Benzoylcholine	Bovine RBC	0	Horse plasma	67
Ethyl acetate	Human RBC	2	Human plasma	1
3,3-Dimethylbutyl acetate	Human RBC	60	Human plasma	35
2-Chloroethyl acetate	Human RBC	37	Human plasma	10
Isoamyl acetate	Human RBC	24	Horse plasma	7
Isoamyl propionate	Human RBC	10	Horse plasma	13
Isoamyl butyrate	Human RBC	1	Horse plasma	14

*Relative rates at approximately optimal substrate concentration; rate with acetylcholine = 100.

(Adapted from Heath, D. F.: Organophosphorus Poisons—Anticholinesterases and Related Compounds, 1st ed. New York, Pergamon Press, 1961.)

The active site of AChE consists of an anionic and an esteratic site. The anionic site is formed by the γ-carboxylate group (COO—) of a glutamic acid residue.[34] Also present in the active site of AChE are an acidic group, HA, pK_a 9.2, believed to be a tyrosine residue;[35] two imidazole groups from histidine residues Im_1 and Im_2, with pK_a values of 6.3 and 5.5, respectively; and a serine residue. Under physiologic conditions, free serine residues do not catalyze hydrolysis of esters. However, in conjunction with the imidazole group, Im_2, the serine hydroxyl in the esteratic site of AChE is activated to become a strong nucleophile.

It is proposed that ACh ester is hydrolyzed by a mechanism that involves imidazole groups of two histidine residues. One imidazole group Im_2 accepts a proton from a serine hydroxyl group located at the esteratic site, creating a strong nucleophile. The activated serine residue then initiates the hydrolysis mechanism by attack on the carbonyl group of ACh. Choline is released, leaving the acetylated serine residue on the enzyme, E–A. The enzyme undergoes a conformational change, which brings the acetylated serine in close proximity to a second imidazole residue, Im_1. Im_1, acting as a general base catalyst,

$$E + ACh \underset{k_{-1}}{\overset{k_{+1}}{\rightleftharpoons}} E \cdot ACh \xrightarrow[-\text{choline}]{k_2} E\text{–}A \xrightarrow[H_2O]{k_3} E + CH_3COOH$$

causes the hydrolysis of the acetylated enzyme (Fig. 12-11). The rate of this reaction is indicated by the k_3 rate constant (i.e., deacylation rate).

Carbamates such as in carbachol are also able to serve as substrates for AChE, which they carbamylate. The rate of carbamylation (k_2) is slower than the rate of acetylation. The hydrolysis (k_3, decarbamylation) of the carbamyl–enzyme intermediate is 10^7 times slower than its acetyl counterpart. Therefore, carbamyl esters acting as substrates for AChE form enzyme–carbamyl (E–C) intermediates that hydrolyze slowly and serve to limit the optimal functional capacity of AChE and may be considered to be semireversible inhibitors of AChE.

$$E + CX \underset{k_{-1}}{\overset{k_{+1}}{\rightleftharpoons}} E \cdot CX \xrightarrow[X]{k_2} E\text{–}C \xrightarrow{k_3} E + C$$

where CX = carbamylating substrate

In the mechanism above, k_3 is the rate-limiting step. The rate of k_2 depends not only on the nature of the alcohol moiety of the ester, but also on the type of carbamyl ester. Esters of carbamic acid, i.e.

$$R\text{—}O\text{—}\overset{\overset{\displaystyle O}{\|}}{C}\text{—}NH_2$$

are better carbamylating agents of AChE than the methylcarbamyl, i.e.,

$$R\text{—}O\text{—}\overset{\overset{\displaystyle O}{\|}}{C}\text{—}NHCH_3$$

and dimethylcarbamyl, i.e.,

$$R\text{—}O\text{—}\overset{\overset{\displaystyle O}{\|}}{C}\text{—}N(CH_3)_3$$

analogues. They also decarbamylate from the enzyme faster.[36]

Organophosphate esters of selected compounds are also able to esterify the serine residue in the active site of AChE (see Fig. 12-11). The hydrolysis rate (k_3) of the phosphorylated serine is extremely slow, and hydrolysis to the free enzyme and phosphoric acid derivative is so limited that the inhibition is considered irreversible. These organophosphorus compounds are used as agricultural insecticides, for the treatment of glaucoma, and have been employed as nerve gases in warfare.

$$E + PX \underset{k_{-1}}{\overset{k_{+1}}{\rightleftharpoons}} E \cdot PX \xrightarrow[X]{k_2} E\text{–}P \xrightarrow{k_3} E + P$$

where PX = phosphorylating substrate

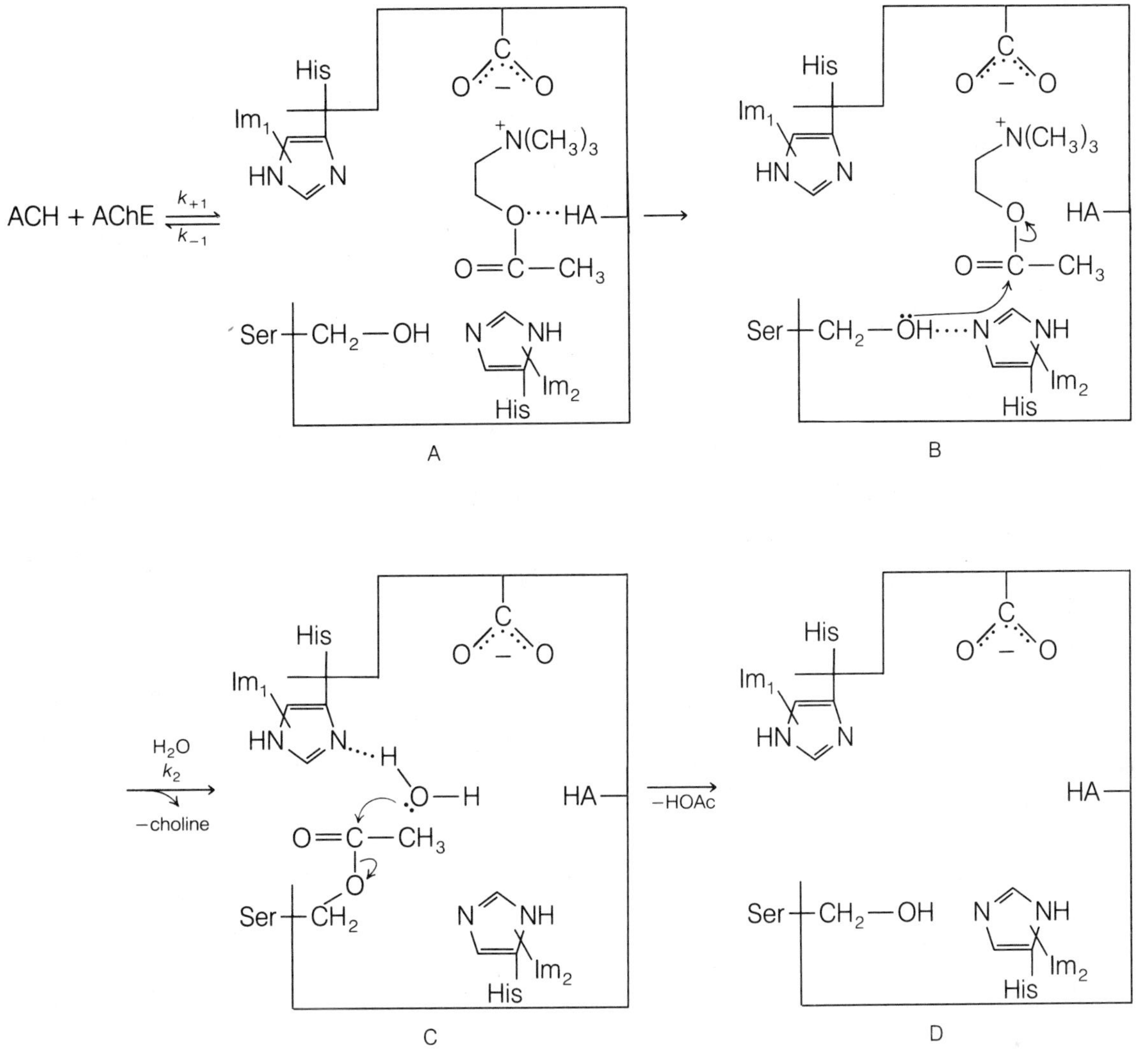

FIG. 12-11. *Mechanism of hydrolysis of ACh by AChE.* (A) *ACh-AChE reversible complex;* (B) *acetylation of esteratic site;* (C) *general base hydrolysis of acetylated enzyme;* (D) *free enzyme.*

It is possible for chemicals to inhibit AChE without interacting specifically with the esteratic site of the enzyme. Acetylcholinesterase as isolated from the electric eel is a macroanion. It possesses three distinct anionic sites[37] and perhaps other nonspecific anionic loci. Compounds containing an onium group within a large, flat, aromatic ring structure appear to be potentially good reversible inhibitors of the enzyme. The dissociation constant ($K_s = k_{-1}/k_{+1}$) is of the magnitude of 10^{-4} M. Reversible inhibitors with dissociation constant values (K_i) of 10^{-7} M have a 1000-fold greater affinity for the enzyme and, therefore, prevent ACh from forming the enzyme–substrate complex that precedes the hydrolysis of the compound. Table 12-5 shows the relative potencies of several types of AChE inhibitors.

TABLE 12-5

INHIBITION CONSTANTS FOR ANTICHOLINESTERASE POTENCY OF ACETYLCHOLINESTERASE INHIBITORS

Reversible and Semireversible Inhibitors	***K_I(M)***
Ambenonium	4.0×10^{-8}
Carbachol	1.0×10^{-4}
Demecarium	1.0×10^{-10}
Edrophonium	3.0×10^{-7}
Neostigmine	1.0×10^{-7}
Physostigmine	1.0×10^{-8}
Pyridostigmine	4.0×10^{-7}
Irreversible Inhibitors	***k_2(mol/min)***
Isoflurophate	1.9×10^{4}
Echothiophate	1.2×10^{5}
Paraoxon	1.1×10^{6}
Sarin	6.3×10^{7}
Tetraethylpyrophosphate	2.1×10^{8}

Products

Physostigmine, USP is an alkaloid usually obtained from the dried ripe seed of *Physostigma venenosum*. It occurs as a white, odorless, microcrystalline powder that is slightly soluble in water and freely soluble in alcohol, chloroform, and the fixed oils. This alkaloid as the free base is quite sensitive to heat, light, moisture, and bases, and readily undergoes decomposition. When used topically to the conjunctiva it is better tolerated than its salts. Its lipid solubility properties permit adequate absorption from ointment bases.

Physostigmine is a competitive inhibitor of AChE when ACh is simultaneously present. The mechanism proposed is one of a reversible competition for the active site on the enzyme. A noncompetitive inhibition is observed when the enzyme is preincubated with physostigmine.

Physostigmine — OH^- → Eseroline — (O) → Rubreserine → eserine blue ↓ eserine brown

Physostigmine Salicylate, USP. Eserine salicylate. The salicylate of physostigmine may be prepared by neutralizing an ethereal solution of the alkaloid with an ethereal solution of salicylic acid. Excess salicylic acid is removed from the precipitated product by washing it with ether. The salicylate is less deliquescent than the sulfate.

Physostigmine Salicylate ($C_7H_5O_3^-$)

Physostigmine salicylate occurs as a white, shining, odorless crystal, or white powder that is soluble in water (1 : 75), alcohol (1 : 16), or chloroform (1 : 6), but is much less soluble in ether (1 : 250). Upon prolonged exposure to air and light, the crystals turn red. The red may be removed by washing the crystals with alcohol, although this causes loss of the compound as well. Aqueous solutions are neutral or slightly acidic and take on a red coloration after a period. The coloration may be taken as an index of the loss of activity of physostigmine solutions.

Solutions of physostigmine salicylate are incompatible with the usual reagents that precipitate alkaloids, alkalies, and with iron salts. Incompatibilty also occurs with benzalkonium chloride and related wetting agents because of the salicylate ion. Physostigmine in solution is hydrolyzed to methylcarbamic acid and eseroline, neither of which inhibits AChE. Eseroline is oxidized readily to a red compound rubreserine[38] and is then converted to eserine blue and eserine brown. The addition of sulfite or ascorbic acid prevents the oxidation of the phenol, eseroline, to rubreserine. Hydrolysis does take place however and the physostigmine is inactivated. Solutions are most stable at *p*H 6 and should never be sterilized by heat, but rather by bacteriologic filtration.

Physostigmine is a relatively poor carbamylating agent of AChE and is often considered a reversible inhibitor of the enzyme. It has a K_i value (i.e., k_{-1}/k_{+1}) on the order of 10^{-8} M and is considered as a potent inhibitor of the enzyme. Its cholinesterase-inhibiting properties vary with *p*H (Fig. 12-12). The conjugate acid of physostigmine has a pK_a of about 8 and as the *p*H is lowered more is in the protonated form. The inhibitory action is enhanced at lower *p*Hs as shown in Figure 12-12; thus it is

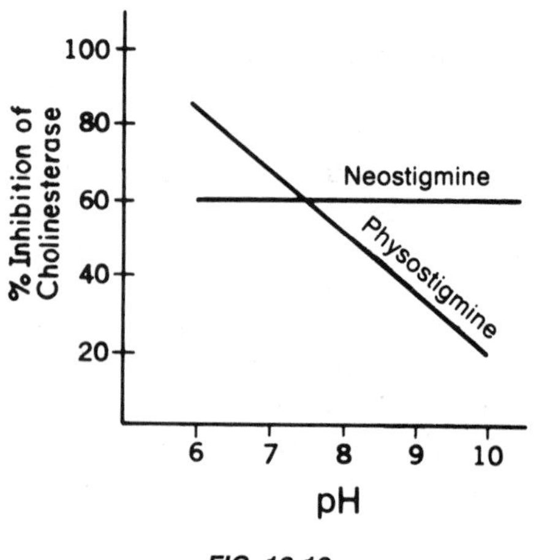

FIG. 12-12.

obvious that the protonated or salt form makes a marked contribution to its activity.

The ophthalmic effect (miotic) of physostigmine and related compounds is due to contraction of the ciliary body. This promotes drainage through the canal of Schlemm and decreases intraocular pressure. Hence, physostigmine is used in the treatment of glaucoma by direct instillation of a 0.1% to 1% solution in the eye. It is directly antagonistic to atropine in the eye, and it is sometimes used to help restore the pupil to normal size following atropine dilatation. Physostigmine also causes stimulation of the intestinal musculature and is used in conditions of depressed intestinal motility. In gaseous distention of the bowel, physostigmine often aids in the evacuation of gas as well as restoring normal bowel movement. It is administered by injection for this purpose. Much research has been done to find synthetic drugs with a physostigminelike action. This has resulted in compounds of the neostigmine type that, at least for intestinal stimulation, are superior to physostigmine.

Physostigmine Sulfate, USP occurs as a white, odorless, microcrystalline powder that is deliquescent in moist air. It is soluble in water 1 : 4, 1 : 0.4 in alcohol, and 1 : 1,200 in ether. It has the advantage over the salicylate salt in that it is compatible in solution with benzalkonium chloride and related compounds.

Neostigmine Bromide, USP. (*m*-Hydroxyphenyl)trimethylammonium bromide dimethylcarbamate; dimethylcarbamic ester of 3-hydroxyphenyltrimethylammonium bromide (Prostigmin Bromide). A method of preparation is from dimethylcarbamyl chloride and the postassium salt of 3-hydroxyphenyldimethylamine. Methyl bromide readily adds to the tertiary amine, forming the stable quaternary ammonium salt (see formula for neostigmine bromide). It occurs as a bitter, odorless, white crystalline powder. It is soluble in water (1 : 0.5) and in alcohol. The crystals are much less hygroscopic than are those of neostigmine methylsulfate and thus may be used in tablets. Solutions are stable and may be sterilized by boiling. Aqueous solutions are neutral to litmus.

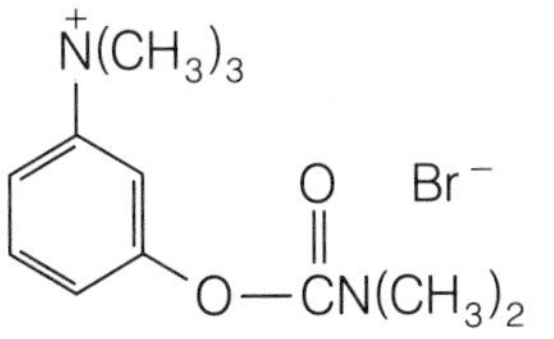

$\overset{+}{N}(CH_3)_3$ O Br^- O—CN$(CH_3)_2$

Neostigmine Bromide

Use of physostigmine as a prototype of an indirect-acting parasympathomimetic drug led to the development of stigmine in which a trimethylamine group was placed *para* to a dimethyl carbamate group in benzene. Better inhibition of cholinesterase was observed when these groups were placed *meta* to each other, giving the drug neostigmine, a more active and useful agent. Although physostigmine contains a methyl carbamate functional group, greater chemical stability toward hydrolysis was obtained with the dimethyl carbamate group in neostigmine.[39,40]

After oral or intravenous administration, neostigmine has a half-life of about 50 minutes. About 80% of a single intramuscular dose of the drug is excreted in urine within 24 hours, approximately 40% as unchanged and the remainder as metabolities. Of the neostigmine that reaches the liver, 98% is metabolized in ten minutes to 3-hydroxyphenyltrimethyl ammonium, which has activity similar to, but weaker than, neostigmine. Its transfer from plasma to liver cells and then to bile is probably passive. Because cellular membranes permit the passage of plasma proteins synthesized in the liver into the bloodstream through capillary walls or lymphatic vessels, they may not present a barrier to the diffusion of quaternary amines such as neostigmine. Possibly the rapid hepatic metabolism of neostigmine provides a downhill gradient for the continual diffusion of this compound.[41] A certain amount is hydrolyzed slowly by plasma cholinesterase.

Neostigmine has a mechanism of action quite similar to that of physostigmine. Neostigmine effectively inhibits cholinesterase at about 10^{-6} M concentration. Its activity does not vary with *p*H, and at all ranges it exhibits similar cationic properties (see Fig. 12-12). There may be a direct action of the drug on tissues innervated by cholinergic nerves, but this has not yet been confirmed.

The uses of neostigmine are similar to those of physostigmine, but they differ in that there is greater miotic activity, fewer and less unpleasant local and systemic manifestations, and greater chemical stability. The most frequent application of neostigmine is to prevent atony of the intestinal, skeletal, and bladder musculature. An important use is in the treatment of myasthenia gravis, a condition caused by an autoimmune mechanism that requires an increase in ACh in the neuromuscular junction to sustain normal muscular activity.

Neostigmine Methylsulfate, USP. (*m*-Hydroxphenyl)trimethylammonium methylsulfate dimethylcarbamate; dimethylcarbamic ester of 3-hydroxyphenyltrimethylammonium methylsulfate (Prostigmin Methylsulfate). Neostigmine is prepared as in the method previously described, and the quaternary ammonium salt is made with methyl sulfate. This compound is a bitter, odorless, white crystalline powder. It is very soluble in water and is soluble in alcohol. Solutions are stable and can be sterilized by boiling. The compound is too hygroscopic for use in a solid form and, thus, always is used in injection. Aqueous solutions are neutral to litmus.

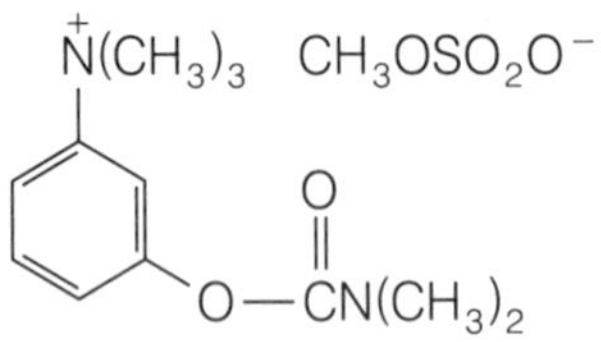

Neostigmine Methylsulfate

The methylsulfate salt is used postoperatively as a urinary stimulant, in the diagnosis and treatment of myasthenia gravis, and as an antiarrhythmic agent to terminate supraventricular tachycardia in patients who fail to respond to vagal stimulation.

Pyridostigmine Bromide, USP. 3-Hydroxy-1-methylpyridinium bromide dimethylcarbamate, pyridostigminium bromide (Mestinon). This compound occurs as a white, hygroscopic crystalline powder having an agreeable characteristic odor. It is freely soluble in water, alcohol, and chloroform.

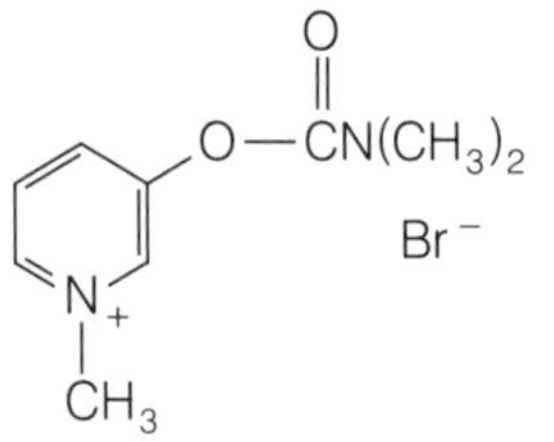

Pyridostigmine Bromide

Pyridostigmine bromide is about one-fifth as toxic as neostigmine. It appears to function in a manner similar to that of neostigmine and is the most widely used anticholinesterase agent employed to treat myasthenia gravis. The drug is metabolized by the liver enzymes and plasma cholinesterase. The principal metabolite is 3-hydroxy-*N*-methylpyridinium. Pyridostigmine has a half-life of 90 minutes and a duration of action between three and six hours when administered orally.

Ambenonium Chloride. [Oxalylbis(iminoethylene)]bis[(*o*-chlorobenzyl)diethylammonium] dichloride (Mytelase Chloride). This compound is a white, odorless powder, soluble in water and in alcohol, slightly soluble in chloroform, and practically insoluble in ether and in acetone. Ambenonium chloride is used for the treatment of myasthenia gravis in patients who do not respond satisfactorily to neostigmine or pyridostigmine.

This drug acts by suppressing the activity of AChE. It possesses a relatively prolonged duration of action and causes fewer side effects in the gastrointestinal tract than the other anticholinesterase agents. The dosage requirements vary considerably, and the dosage must be individualized according to the response and tolerance of the patient. Because of its quaternary ammonium structure, ambenonium chloride is poorly absorbed from the gastrointestinal tract. In moderate doses, the drug does not cross the blood–brain barrier. Ambenonium is not hydrolyzed by cholinesterases.

Demecarium Bromide, USP. (*m*-Hydroxyphenyl)trimethylammonium bromide decamethylenebis[methylcarbamate] (Humorsol) is the diester of (*m*-hydroxyphenyl)trimethylammonium bromide with decamethylene bis(methylcarbamic acid) and, thus, is comparable to a bis-prostigmine molecule.

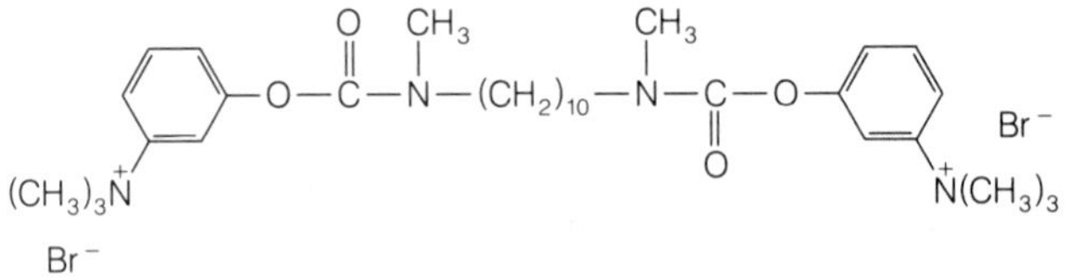

Demecarium Bromide

It occurs as a slightly hygroscopic powder that is freely soluble in water or alcohol. Ophthalmic solutions of the drug have a *p*H of 5 to 7.5. Aqueous solutions are stable and may be sterilized by heat. Its efficacy and toxicity are comparable with those of other potent anticholinesterase inhibitor drugs. It is a long-acting miotic used to treat wide-angle glaucoma and accommodative esotropia. Maximal effect occurs hours after administration, and the effect may persist for days.

Edrophonium Chloride, USP. Ethyl(*m*-hydroxphenyl)dimethylammonium chloride (Tensilon). Edrophonium chloride is a reversible anticholinesterase agent prepared by treating 3-dimethylaminophenol with ethyl iodide and converting the quaternary ammonium salt to the chloride with silver oxide and hydrochloric acid. The drug is bitter

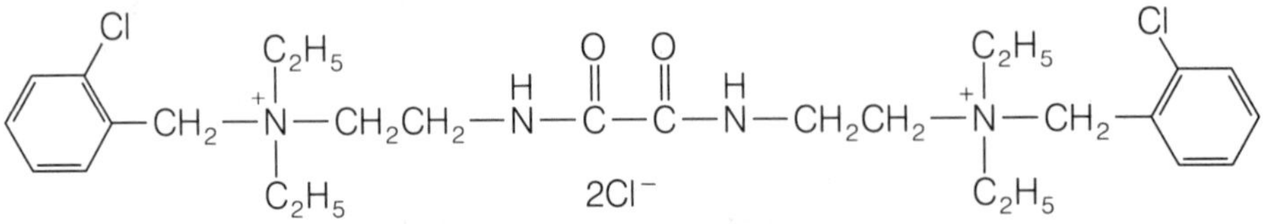

Ambenonium Chloride

and is very soluble in water and freely soluble in alcohol. Edrophonium chloride injection has a *p*H of 5.2 to 5.5. On parenteral administration, edrophonium has a more rapid onset and shorter duration of action than has neostigmine, pyridostigmine, or ambenonium. It is a specific anticurare agent and acts within one minute to alleviate overdose of *d*-tubocurarine, dimethyl *d*-tubocurarine, or gallamine triethiodide. The drug also is used to terminate the action of any one of these drugs when the physician so desires. However, it is of no value in terminating the action of the depolarizing (i.e., noncompetitive) blocking agents such as decamethonium and succinylcholine. In addition to inhibiting AChE, edrophonium has a direct cholinomimetic effect on skeletal muscle, which is greater than most other anticholinesterase drugs.

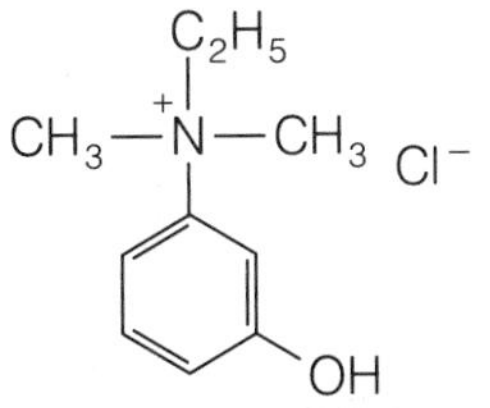

C_2H_5; $CH_3-\overset{+}{N}-CH_3$ Cl^-; OH

Edrophonium Chloride

Edrophonium chloride is structurally related to neostigmine methylsulfate and has been used as a potential diagnostic agent for myasthenia gravis. This is the only degenerative neuromuscular disease the condition of which can be temporarily improved by administration of an anticholinesterase agent. Edrophonium brings about a rapid increase in muscle strength without significant side effects. It is also used as an antiarrhythmic agent to terminate supraventricular arrhythmias that cannot be controlled by vagal stimulation alone.

IRREVERSIBLE INHIBITORS

Both AChE and BuChE can be inhibited irreversibly by a group of phosphate esters that are highly toxic (LD_{50} for humans is 0.1 to 0.001 mg/kg). These chemicals are nerve poisons and have been used in warfare and as agricultural insecticides. They permit ACh to accumulate at nerve endings and produce an exacerbation of ACh-like actions. The compounds belong to a class of organophosphorus esters. A general type formula of such compounds is as follows:

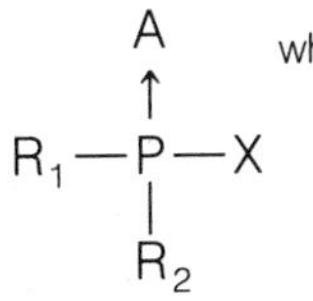

$$R_1-\overset{\overset{A}{\uparrow}}{P}-X \quad (\text{with } R_2 \text{ on P})$$

where R_1 = alkoxyl
R_2 = alkoxyl, alkyl, or tertiary amine
X = a good leaving group (e.g. F, CN, thiomalate, *p*-nitrophenoxy)

A is usually oxygen or sulfur, but may also be selenium. When *A* is other than oxygen, biologic activation is required before the compound becomes effective as an inhibitor of cholinesterases. Phosphorothionates ($R_1R_2P(S)X$) have much poorer electrophilic character than their oxygen analogues and are much weaker hydrogen bond-forming molecules because of the sulfur atom.[42] Their anticholinesterase activity is 10^5-fold weaker than their oxygen analogues. *X* is the "leaving group" when the molecule reacts with the enzyme. Typical leaving groups include fluoride, nitrile, and *p*-nitrophenoxy. The *R* groups may be alkyl, alkoxy, aryl, aryloxy, or amino. The *R* moiety imparts lipophilicity to the molecule and contributes to its absorption through the skin. Inhibition of AChE by organophosphorus compounds takes place in two steps, association of enzyme and inhibitor and the phosphorylation step, completely analogous to acylation by the substrate (Fig. 12-13). Stereospecificity is mainly due to interactions of enzyme and inhibitor at the esteratic site.

The serine residue in the esteratic site forms a stable phosphoryl ester with the organophosphorus inhibitors. This stability permits labeling studies[43] to be carried out on this and other enzymes (e.g., trypsin, chymotrypsin) that have the serine hydroxyl as part of their active site.

Although insecticides and nerve gases are irreversible inhibitors of cholinesterases by forming a phosphorylated serine in the esteratic site of the enzyme, it is possible to reactivate the enzyme if action is taken soon after the phosphorylation has occurred. Several compounds can provide a nucleophilic attack on the phosphorylated enzyme and cause regeneration of the free enzyme. Substances such as choline, hydroxylamine, and hydroxamic acid have led to the development of more effective cholinesterase reactivators such as nicotinic hydroxamic acid and pyridine-2-aldoxime methiodide (2-PAM). A proposed mode of action for the reactivation of cholinesterase (inactivated by isoflurophate) by 2-PAM is shown in Figure 12-13.

Cholinesterases that have been exposed to phosphorylating agents (e.g., sarin) become refractory to reactivation by cholinesterase reactivators. The process is called aging and occurs both in vivo and in vitro to AChE and BuChE. Aging occurs by partial hydrolysis of the phosphorylated moiety that is attached to the serine residue in the esteratic site of the enzyme (Fig. 12-14).

Phosphate esters used as insecticidal agents are toxic and must be handled with extreme caution. Symptoms of toxicity are nausea, vomiting, excessive sweating, salivation, miosis, bradycardia, low blood pressure, and respiratory difficulty, which is the usual cause of death.

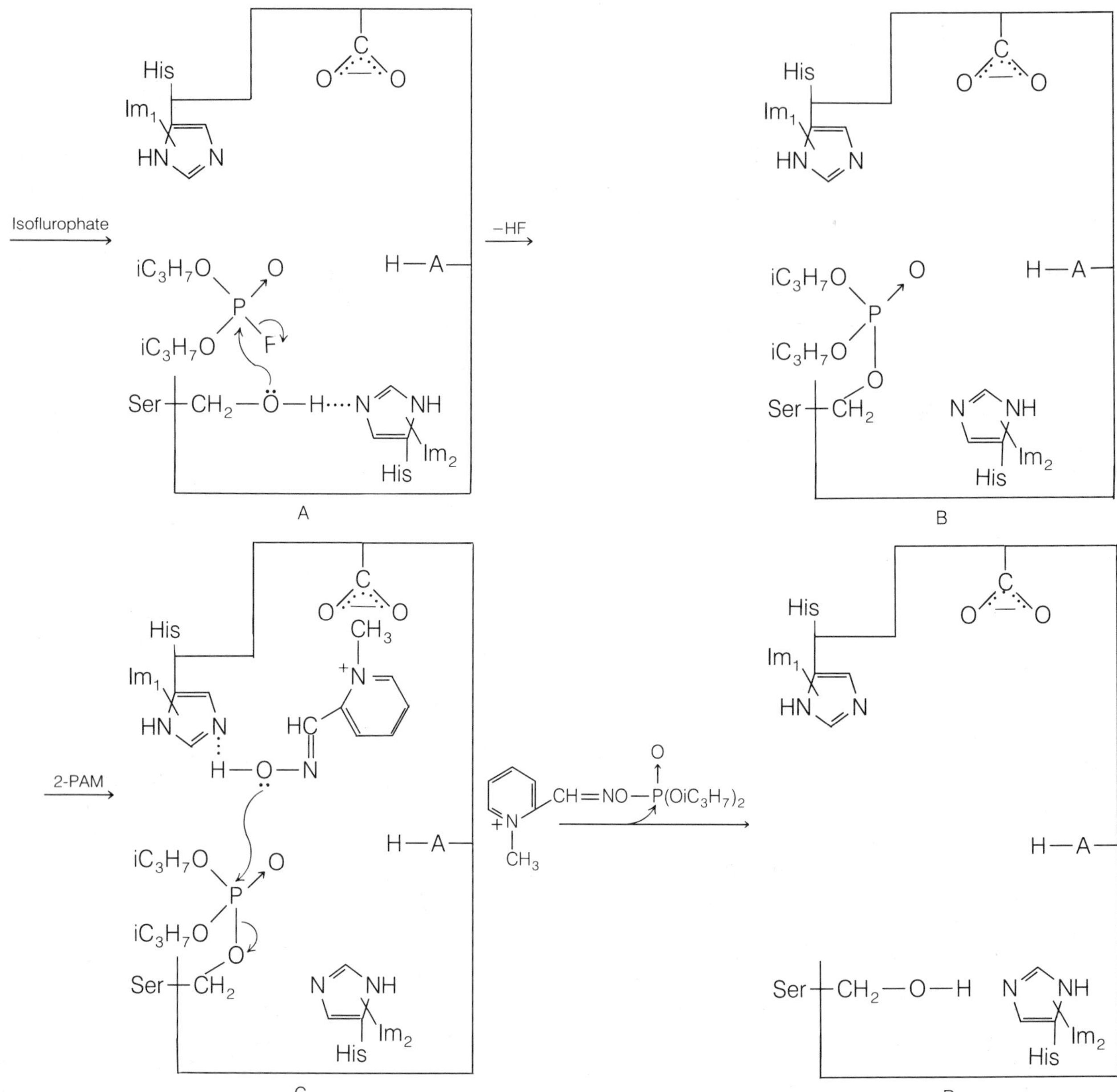

FIG. 12-13. Phosphorylation and reactivation of cholinesterase. (A) Phosphorylation of serine residue by isoflurophate; (B) phosphorylated serine residue in esteratic site; (C) nucleophilic attack on phosphorylated residue by 2-PAM; (D) removal of phosphorylated 2-PAM to generate free enzyme.

$$\text{Enz—P(}\rightarrow\text{O)(—CH}_3\text{)—O—CH(CH}_3\text{)}_2 \xrightarrow{H_2O} \text{Enz—P(}\rightarrow\text{O)(—CH}_3\text{)—OH} + (\text{CH}_3)_2\text{CH—OH}$$

FIG. 12-14. Aging of cholinesterase phosphorylated with sarin.

The organophosphate insecticides of low toxicity, such as malathion, generally cause poisoning only by ingestion of relatively large doses. On the other hand, parathion or methylparathion cause poisoning by inhalation or dermal absorption. Because these compounds are so long-acting, they are cumulative and serious toxic manifestations may result after several small exposures to them.

$$(\text{CH}_3)_2\text{CH—O—P(=O)(CH}_3\text{)—F}$$

Sarin

Products

Isoflurophate, USP. Diisopropyl phosphorofluoridate; (Floropryl). DFP is a colorless liquid, soluble in water to the extent of 1.54% at 25°C, and decomposes to give a pH of 2.5. It is soluble in alcohol and to some extent in peanut oil. Isoflurophate is stable in peanut oil for a period of one year, but decomposes in water in a few days. Solutions in peanut oil can be sterilized by autoclaving. The compound should be stored in hard glass containers. Continued contact with soft glass is said to hasten decomposition, as evidenced by a discoloration.

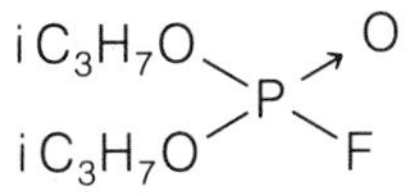

Isoflurophate

Isoflurophate must be handled with extreme caution. Avoid contact with eyes, nose, mouth, and even the skin, because it can be absorbed readily through intact epidermis and more so through mucous tissues.

Because isoflurophate irreversibly[44] inhibits cholinesterase, its activity lasts for days or even weeks. During this period new cholinesterase may be synthesized in plasma, erythrocytes, and other cells.

A combination of atropine sulfate and magnesium sulfate gives protection in rabbits against the toxic effects of isoflurophate. Atropine sulfate counteracts the muscarinic effect, and magnesium sulfate the nicotinic effect of the drug.[45] Isoflurophate has been used clinically in the treatment of glaucoma.

Echothiophate Iodide, USP. (2-Mercaptoethyl)trimethylammonium iodide, *S*-ester with *O,O*-diethyl phosphorothioate (Phospholine Iodide) occurs as a white, crystalline, hygroscopic solid that has a slight mercaptanlike odor. It is soluble in water 1 : 1, and 1 : 25 in dehydrated alcohol; aqueous solutions have a pH of about 4 and are stable at room temperature for about one month.

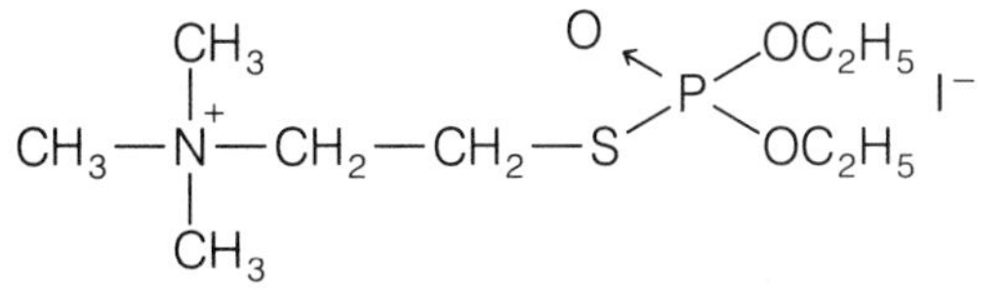

Echothiophate Iodide

Echothiophate iodide is a long-lasting cholinesterase inhibitor of the irreversible type such as isoflurophate. However, unlike the latter, it is a quaternary salt and, when applied locally, its distribution in tissues is limited, which can be very desirable. It is used as a long-acting anticholinesterase agent in the treatment of glaucoma.

Hexaethyltetraphosphate (HETP) and **Tetraethylpyrophosphate (TEPP).** These two substances are compounds that also show anticholinesterase activity. HETP was developed by the Germans during World War II and is used as an insecticide against aphids. When used as insecticides, these compounds have the advantage of being hydrolyzed rapidly to the relatively nontoxic water-soluble compounds phosphoric acid and ethyl alcohol. Fruit trees or vegetables sprayed with this type of compound retain no harmful residue after a period of a few days or weeks, depending on the weather conditions. The disadvantage of their use comes from their very high toxicity, which results from their anticholinesterase activity. Workers spraying with these agents should use extreme caution that none of the vapors are breathed and that none of the vapor or liquid comes in contact with the eyes or skin.

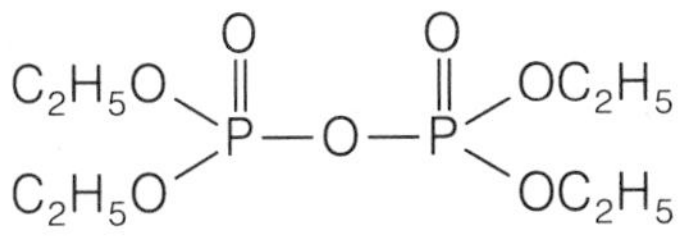

Tetraethylpyrophosphate

Malathion. 2[(Dimethoxyphosphinothioyl)thio]-butanedioic acid diethyl ester is a water-insoluble phosphodithioate ester that has been used as an agricultural insecticide. Malathion is a poor inhibitor of cholinesterases. Its effectiveness as a safe insecticide is due to the different rates at which humans and insects metabolize the chemical. Microsomal oxidation, which causes desulfuration, occurs slowly to form the phosphothioate (malaoxon), which is 10,000 times more active than the phosphodithioate (malathion) as a cholinesterase inhibitor. Insects detoxify the phosphothioate by a phosphatase-forming dimethylphosphorothioate that is inactive as an inhibitor. Humans, however, are able to rapidly hydrolyze malathion through a carboxyesterase enzyme yielding malathion acid, a yet poorer inhibitor of AChE. Phosphatases and carboxyesterases further metabolize malathion acid to dimethylphosphothioate. The metabolic reactions are shown in Figure 12-15.

Parathion. *O,O*-diethyl *O*-*p*-nitrophenyl phosphorothioate (Thiophos). This compound is a yellow liquid that is freely soluble in aromatic hydrocarbons, ethers, ketones, esters, and alcohols, but practically insoluble in water, petroleum ether, kerosene, and the usual spray oils. It is decomposed at a pH above 7.5. Parathion is used as an agricultural insecticide. It is highly toxic, the effects being cumulative. Special precautions are necessary to prevent skin contamination or inhalation. Parathion is a relatively weak inhibitor of cholinesterase. There are enzymes present in liver microsomes and insect tissues that convert parathion ($pI_{50} < 4$) to paraoxon (a more potent inhibitor of cholinesterase, $pI_{50} > 8$).[46] Parathion is also metabolized by liver

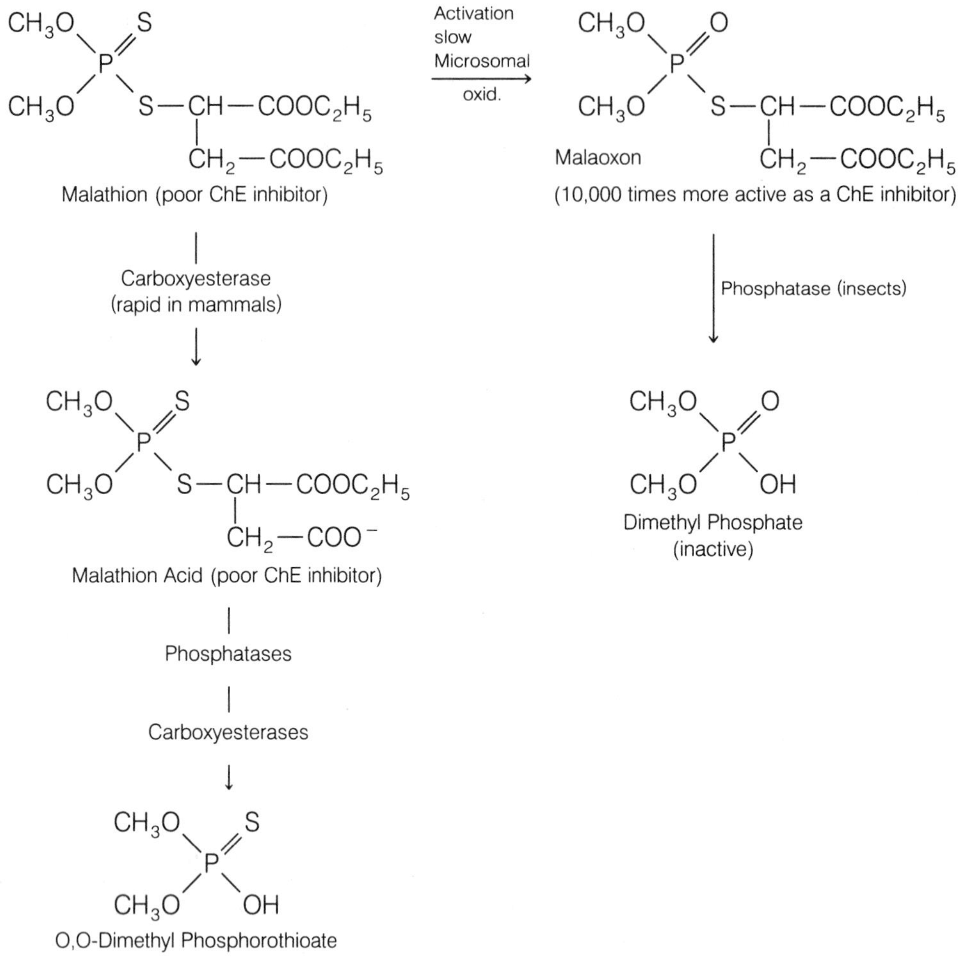

FIG. 12-15. *Metabolism of malathion in mammals and insects.*

microsomes to yield *p*-nitrophenol and diethylphosphate, which is inactive as a cholinesterase inhibitor.[47]

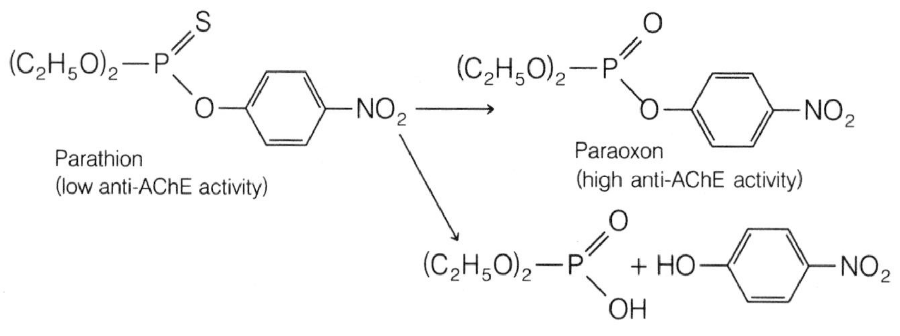

Schradan. Octamethylpyrophosphoramide; OMPA; *bis*[bisdimethylaminophosphonous] anhydride (Pestox III). This compound is a viscous liquid that is miscible with water and soluble in most organic solvents. It is not hydrolyzed by alkalies or water, but is hydrolyzed by acids. Schradan is used as a systemic insecticide for plants, being absorbed by the plants without appreciable injury. Insects feeding on the plant are incapacitated.

Schradan is a weak inhibitor of cholinesterases in vitro. In vivo it is metabolized to the very strong inhibitor hydroxymethyl OMPA.[48] Hydroxymethyl OMPA is not stable and is further metabolized to the N-methoxide, which is a weak inhibitor of cholinesterase.[49]

$$(CH_3)_2N-P(=O)(N(CH_3)_2)-O-P(=O)(N(CH_3)_2)_2 \xrightarrow[(O)]{\text{liver microsomes}} (CH_3)_2N-P(=O)(N(CH_3)_2)-O-P(=O)(N(CH_3)_2)-N(CH_3)CH_2OH \longrightarrow$$

OMPA
(weak cholinesterase inhibitor)

hydroxymethyl OMPA
(strong cholinesterase inhibitor)

$$(CH_3)_2N-P(=O)(N(CH_3)_2)-O-P(=O)(N(CH_3)_2)-N(CH_3)CH_2OCH_3$$

OMPA-*N*-methoxide
(weak cholinesterase inhibitor)

Pralidoxime Chloride, USP. 2-Formyl-1-methylpyridinium chloride oxime; 2-PAM chloride; 2-pyridine aldoxime methyl chloride (Protopam Chloride). It occurs as a white, nonhygroscopic crystalline powder that is soluble in water, 1 g in less than 1 mL.

Pralidoxime Chloride

Pralidoxime chloride is used as an antidote for poisoning by parathion and related pesticides. It may be effective against some phosphates that have a quaternary nitrogen. It also is an effective antagonist for some carbamates, such as neostigmine methylsulfate and pyridostigmine bromide.

The mode of action of pralidoxime is described in Figure 12-13.

The biologic half-life of 2-PAM chloride in humans is about two hours, and its effectiveness is a function of its concentration in plasma that reaches a maximum in two to three hours after oral administration. Concentrations of 4 and 8 $\mu g/mL$ of 2-PAM chloride in the blood plasma of rats significantly decrease the toxicity of sarin by factors of 2 and 2.5, respectively.[50]

Pralidoxime chloride, a quaternary ammonium compound, is most effective by intramuscular, subcutaneous, or intravenous administration. Treatment of poisoning by an anticholinesterase will be

TABLE 12-6

CHOLINERGIC AGENTS

Name Proprietary Name	***Preparations***	***Application***	***Usual Adult Dose****	***Usual Dose Range****	***Usual Pediatric Dose****
Acetylcholine chloride	Sterile powder		0.5–2 mL of a freshly prepared 1% solution instilled in the anterior chamber of the eye		
Carbachol USP *Isopto Carbachol, Carbacel*	Carbachol ophthalmic solution USP	Topically, to the conjunctiva	1 drop of a 1.5% solution bid to qid		
	Carbachol intraocular solution USP		Intraocular irrigation, 0.5 mL of a 0.01% solution instilled into the anterior chamber		
Bethanechol chloride USP *Urecholine Chloride*	Bethanechol chloride injection USP Bethanechol chloride tablets USP		Oral, 10–30 mg tid; SC, 5 mg tid	Oral, 20–200 mg/day; SC, 0.15–0.2 mg/kg of body weight tid	
Physostigmine USP	Physostigmine salicylate ophthalmic ointment USP	Topically, to the conjunctiva	1 cm of a 0.25% ointment qd to tid		

*See USP DI for complete dosage information.

TABLE 12-6 *Continued*

CHOLINERGIC AGENTS

Name Proprietary Name	***Preparations***	***Application***	***Usual Adult Dose****	***Usual Dose Range****	***Usual Pediatric Dose****
Physostigmine salicylate USP	Physostigmine salicylate ophthalmic solution USP	Topically, to the conjunctiva	1 drop of a 0.25 or 0.5% solution bid or tid		
Physostigmine sulfate USP *Isopto Eserine*	Physostigmine salicylate injection USP		0.5–2 mg		
	Physostigmine sulfate ophthalmic ointment USP	Topically, to the conjunctiva	1 cm of a 0.25% ointment qd to tid		
Neostigmine bromide USP *Prostigmin Bromide*	Neostigmine bromide tablets USP		15–45 mg 3–6 times daily	15–375 mg/d	330 μg/kg of body weight or 10 mg/m² of body surface, 6 times daily
Neostigmine methylsulfate USP *Prostigmin Methylsulfate*	Neostigmine methylsulfate injection USP		Antidote to curare principals: IV, 500 μg–2 mg repeated as necessary (may be administered in combination with 600 μg–1.2 mg of atropine sulfate injection); cholinergic: IM or SC, 250–500 μg 4–6 times daily as necessary	250 μg–5 mg/d	
Pyridostigmine bromide USP *Mestinon Bromide*	Pyridostigmine bromide syrup USP Pyridostigmine bromide tablets USP		60–180 mg 3–6 times daily	60 mg–1.5 g/d	1.2 mg/kg of body weight or 33 mg/m² of body surface, 6 times daily
Demecarium bromide *Humorsol*	Demecarium bromide ophthalmic solution USP	Topically, to the conjunctiva	0.125–0.25% solution qd or bid		
Ambenonium chloride *Mytelase*	Ambenonium Chloride Tablets USP		Initial, 5 mg, gradually increasing as required up to 5–25 mg tid or qid	5–50 mg	
Isoflurophate USP *Floropyrl*	Isoflurophate ophthalmic ointment USP	Topically, to the conjunctiva	0.025% ointment once every 3 days to 3 tid		
Echothiophate iodide USP *Phospholine Iodide*	Echothiophate iodide ophthalmic solution USP	Topically, to the conjunctiva	1 drop of a 0.03–0.25% solution qd or bid		
Edrophonium chloride USP *Tensilon Chloride*	Edrophonium chloride injection USP	Antidote to curare principals; diagnostic aid (myasthenia gravis)	Antidote: IV, 10 mg, repeated if necessary; diagnostic: IV, 2 mg followed by 8 mg if no response in 45 sec; IM, 10 mg	Antidote: 1–40 mg in one episode; diagnostic: IV, 2–10 mg/test; IM, 10–12 mg in one episode	Diagnostic: IM, children 34 kg of body weight and under, 2 mg; children over 34 kg of body weight, 5 mg; IV, infants, 500 μg; children 34 kg of body weight and under, 1 mg; if no response after 45 sec, 1 mg every 30–45 sec, up to 5 mg; children over 34 kg of body weight, 2 mg; if no response after 45 sec, 1 mg every 30–45 seconds, up to 10 mg

*See USP DI for complete dosage information.

TABLE 12-6 *Continued*

CHOLINERGIC AGENTS

Name Proprietary Name	***Preparations***	***Application***	***Usual Adult Dose****	***Usual Dose Range****	***Usual Pediatric Dose****
Pilocarpine hydrochloride USP *Almocarpine, Pilocel, Isopto Carpine, Pilocar, Pilomiotin*	Pilocarpine hydrochloride ophthalmic solution USP	Topically, to the conjunctiva	1 drop of a 10% solution 1–6 times daily		
Pilocarpine USP	Pilocarpine ocular system USP	Topically, to the conjunctiva	1 ocular system delivering 20 or 40 μg/hr, once every 7 days		
Pilocarpine nitrate USP *PV Carpine, Liquifilm*	Pilocarpine nitrate ophthalmic solution	Topically, to the conjunctiva	1 drop of a 0.5–6% solution qd to qid		

*See USP DI for complete dosage information.

most effective if given within a few hours. Little will be accomplished if the drug is used more than 36 hours after parathion poisoning has occurred.

CHOLINERGIC-BLOCKING AGENTS

Studies have shown that there is a wide variety of tissues the response of which is one of stimulation or inhibition to neuronally released ACh or exogenously administered chemicals that mimic the neurotransmitter. Peripheral cholinergic receptors that respond to ACh are located at parasympathetic postganglionic nerve endings in smooth muscle, sympathetic and parasympathetic ganglia, and neuromuscular junctions in skeletal muscle. Although these receptors are all activated exclusively by ACh, there are selective antagonists for each of these cholinergic receptors. Atropine is an effective blocking agent at parasympathetic postganglionic terminals. Like most classic blocking agents it acts on both M_1 and M_2 receptors. Hexamethonium blocks transmission at the nicotinic ACh receptor located in autonomic ganglia. *d*-Tubocurarine blocks the effect of ACh on skeletal muscle, which also is activated by an ACh nicotinic receptor. The cause of the selective action of these blocking agents is not clear, but it may be attributed to several factors that play a role in the accessibility of the blocking agent to the site. The receptors could be the same, but located in different environments (e.g., associated with other nonreceptor membrane proteins, different lipid environment). The receptor differences may be the result of dissimilar primary amino acid sequences or their linkage to subcellular components that control their conformational state and, hence, their affinity for different molecules.

Anticholinergic action by drugs and chemicals is apparently dependent upon their ability to reduce the number of free receptors that can interact with ACh. The theories of Clark and Stephenson[51] and Ariens[52] have explained the relationship between drug–receptor interactions and the observed biologic response (see Chap. 2). These theories indicate that the quantity of a drug–receptor complex formed at a given time depends upon the affinity of the drug for the receptor and that a drug that acts as an agonist must also possess another property called *efficacy* or *intrinsic activity*. Another explanation of drug–receptor interactions, the Paton rate theory,[53] defines a biologic stimulus as being proportional to the rate of drug–receptor interactions (see Chap. 2).

Both of these theories are compatible with the concept that a blocking agent that has a high affinity for the receptor may serve to decrease the number of available free receptors and decrease the efficiency of the endogenous neurotransmitter.

PARASYMPATHETIC POSTGANGLIONIC-BLOCKING AGENTS

These blocking agents are also known as *anticholinergics*, *parasympatholytics*, or *cholinolytics*. One might be more specific in stating that members typical of this group are antimuscarinic, because they block the action of muscarine.

Endogenous neurotransmitters, including ACh, are relatively small molecules. It was noted by Ariens[52] that competitive reversible antagonists generally are larger molecules capable of additional binding to the receptor surface. The most potent anticholinergic drugs are derived from muscarinic agonists that contain one, and sometimes two, large or bulky groups. Ariens suggested that molecules that act as competitive reversible antagonists were capable of binding to the active site of the receptor, but they also had an additional binding site that increased receptor affinity but did not contribute to the intrinsic activity (efficacy) of the drug. Consistent with this hypothesis, Bebbington and Brimblecombe[54] proposed that there is a relatively large area lying outside the agonist receptor-binding site where van der Waals interactions can take place (see Fig. 12-10) between the antagonist and the receptor area.

STRUCTURE–ACTIVITY RELATIONSHIPS

A wide variety of compounds possess anticholinergic activity. The development of such compounds has been largely empiric and based principally on atropine as the prototype. Nevertheless, structural permutations have resulted in compounds that do not have obvious relationships to the parent molecule. The following classification will serve to delineate the major chemical types that are encountered:

- Solanaceous alkaloids and synthetic analogues
- Synthetic aminoalcohol esters
- Aminoalcohol ethers
- Aminoalcohols
- Aminoamides
- Miscellaneous
- Papaveraceous alkaloids and their synthetic analogues*

The chemical classification of anticholinergics acting on parasympathetic postganglionic nerve endings is complicated somewhat because some agents, especially the quaternary ammonium derivatives, act on the ganglia that have a musarinic component to their stimulation pattern and, at high doses, act at the neuromuscular junction in skeletal muscle.

There are several ways in which the structure–activity relationship could be considered, but in this discussion we shall follow, in general, the considerations of Long et al.,[55] who based their postulations on the 1-hyoscyamine molecule as being one of the most active anticholinergics and, therefore, having an optimal arrangement of groups.

Anticholinergic compounds may be considered as chemicals that have some similarity to ACh, but contain additional substituents that enhance their binding to the cholinergic receptor.

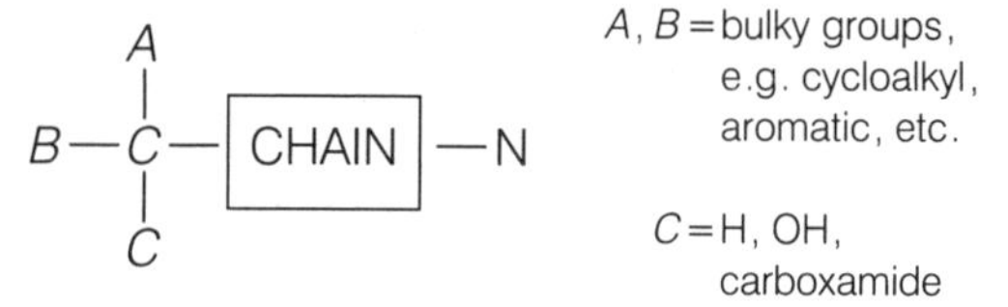

A
B—C—[CHAIN]—N
C

A, *B* = bulky groups, e.g. cycloalkyl, aromatic, etc.

C = H, OH, carboxamide

As depicted above, an anticholinergic may contain a quaternary ammonium function or possess a tertiary amine that is protonated in the biophase to form a cationic species. The nitrogen is separated from a pivotal carbon atom by a chain that may include an ester, ether, or hydrocarbon moiety. The substituent groups *A* and *B* contain at least one aromatic moiety capable of van der Waals interactions to the receptor surface and one cycloaliphatic or other hydrocarbon moiety for hydrophobic-bonding interactions. *C* may be hydroxyl or carboxamide to undergo hydrogen-bonding with the receptor.

The Cationic Head

It is generally considered that the anticholinergic molecules have a primary point of attachment to cholinergic sites through the *cationic head* (i.e., the positively charged nitrogen). For quaternary ammonium compounds, there is no question of what is implied, but for tertiary amines, one assumes, with good reason, that the cationic head is achieved by protonation of the amine at physiologic *p*H. The nature of the substituents on this cationic head is critical insofar as a parasympathomimetic response is concerned. Steric factors that cause a diffusion of the onium charge or produce less than an optimal drug–receptor interaction result in a decrease of parasympathomimetic properties and allow the drug to act as an antagonist because of other bonding interactions. It is undoubtedly true that a cationic head is better than none at all; it is possible to obtain a typical competitive block *without* such a head. Ariens and his co-workers[52] have shown that carbocholines (e.g., benzilylcarbocholine) show a typical competitive action with ACh, although they are less effective than the corresponding compounds possessing a cationic head.

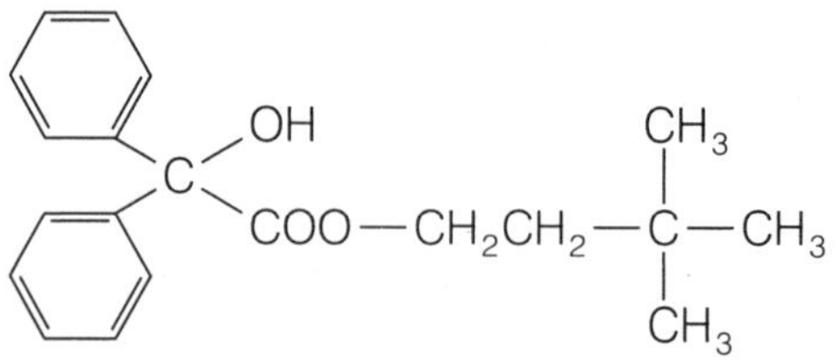

$(C_6H_5)_2C(OH)COO-CH_2CH_2-C(CH_3)_3$

Benzilylcarbocholine

* Although these are not anticholinergics acting at the postganglionic nerve endings of the parasympathetic system, they are included here as a matter of convenience because of their deployment as antispasmodics.

The Hydroxyl Group

Although not a requisite for activity, a suitably placed alcoholic hydroxyl group in an anticholinergic usually enhances the activity over a similar compound without the hydroxyl group. The position of the hydroxyl group relative to the nitrogen appears to be fairly critical, with the diameter of the receptive area being estimated at about 2 to 3 Å. It is assumed that the hydroxyl group contributes to the strength of binding, probably by hydrogen-bonding to an electron-rich portion of the receptor surface.

The Esteratic Group

Many of the highly potent anticholinergic compounds possess an ester grouping, and it may be a contributing feature for effective binding. This is reasonable because the agonist (i.e., ACh) possesses a similar function for binding to the same site. That an esteratic function is not necessary for activity is amply illustrated by the several types of compounds not possessing such a group (e.g., ethers, aminoalcohols).

Cyclic Substitution

It will be apparent from an examination of the active compounds discussed in the following sections that at least one cyclic substituent (phenyl, thienyl, or other) is a common feature in almost all anticholinergic molecules. Aromatic substitution is often used in connection with the acidic moiety of the ester function. However, it will be noted that virtually all acids employed are of the aryl-substituted acetic acid variety. Use of aromatic acids leads to low activity of these compounds as anticholinergics but with potential activity as local anesthetics. The question of the superiority of the cyclic species used (i.e., phenyl, thienyl, cyclohexyl) apparently has not been explored in depth, although phenyl rings seem to predominate. Substituents on the aromatic rings do not contribute to the activity of the molecule.

In connection with the apparent need for a cyclic group, it is instructive to consider the postulations of Ariens.[52] He points out that the "mimetic" molecules, richly endowed with polar groups, undoubtedly require a complementary polar receptor area for effective binding. As a consequence, it is implied that a relatively nonpolar area surrounds such sites. Thus, by increasing the binding of the molecule in this peripheral area by means of introducing flat nonpolar groups (e.g., aromatic rings), it should be possible to achieve compounds with excellent affinity, but not possessing intrinsic activity. This postulate is consistent with most anticholinergics, whether they possess an ester group or not.

Stereochemical Requirements

The stereochemistry of ACh and other cholinergic agonists has been discussed earlier in this chapter. The results are summarized in Table 12-2 and show the stereoselectivity of the muscarinic cholinergic receptor at the alcohol portion of ACh. Molecules with the (*S*) configuration are more potent agonists. In contrast, the benzilate esters of the isomeric β-methylcholines show only a small difference in competitive antagonistic activity [(*R*)/(*S*) ratio, 1.2], indicating that the stereochemical requirement for antagonism is small in the alcohol portion of the molecule. If the stereochemical difference is removed from the choline moiety and introduced into the acidic portion once again, a significant difference [(*R*)/(*S*) ratio, 100 : 1] in activity is noted in the (*R*) and (*S*) forms of cyclohexylphenylglycolate esters of choline. These observations are further reinforced by Ellenbroek's findings[56] on the comparative blocking activities of the four possible stereoisomers of the cyclohexylphenylglycolate esters of β-methylcholine as summarized in Table 12-7.

TABLE 12-7

EFFECT OF STEREOISOMERS ON ANTICHOLINERGIC ACTIVITY

Compound	pA_2*	(R)/(S) Ratio
Benzilylcholine	8.6	
(*R*)-Hyoscyamine[(*R*)-tropyltropeine]	6.9	
(*RS*)-Hyoscyamine (atropine)	8.7	115 (*S*)/(*R*)
(*S*)-Hyoscyamine[(*S*)-tropyltropeine]	9.0	
Benzilyl(*R*β)-methylcholine	8.1	
Benzilyl(*S*β)-methylcholine	8.0	
(*R*)-Cyclohexylphenylglycolylcholine	10.4	100
(*S*)-Cyclohexylphenylglycolylcholine	8.4	
(*R*)-Cyclohexylphenylglycolyl(*R*β)-methylcholine	8.9	100
(*S*)-Cyclohexylphenylglycolyl(*R*β)-methylcholine	6.9	
(*R*)-Cyclohexylphenylglycolyl(*S*β)-methylcholine	8.3	40
(*S*)-Cyclohexylphenylglycolyl(*S*β)-methylcholine	6.6	
(*R*)-Tropylcholine	6.5 (*S*)/(*R*)	15.2 (*S*)/(*R*)
(*S*)-Tropylcholine	7.7	

*pA_2 = the logarithm of the reciprocal molar concentration of antagonist that requires a doubling of the concentration of agonist to compensate for the action of the agonist.

(Data from Ellenbroek, B. W. J., Nivard, F. J. R., van Rossum, J. M., and Ariens, E. J.: J. Pharm. Pharmacol. 17:393, 1967; Rama Sastry, B. V., and Cheng, H. C.: J. Pharmacol. Exp. Ther. 202:105, 1977.)

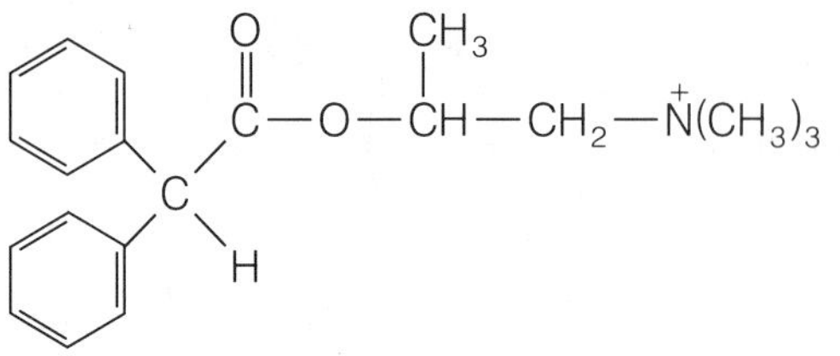

Benzilyl β-methylcholine

Similar relationships have been noted for tropic acid esters. Because of the Cahn rules for describing absolute configuration (S), tropic acid has the same spatial deployment of its aromatic, carboxyl, and hydroxy moieties as (R)-cyclohexylphenylglycolic acid. Hence, hyoscyamine has an (S)/(R) ratio of 115 and tropylcholine an (S)/(R) ratio of 15.2. One is thus drawn to the conclusion that, for muscarinic-blocking activity, the structural requirements are low for the aminoalcohol portion and high for the acidic portion of cholinergic parasympathetic postganglionic-blocking agents.

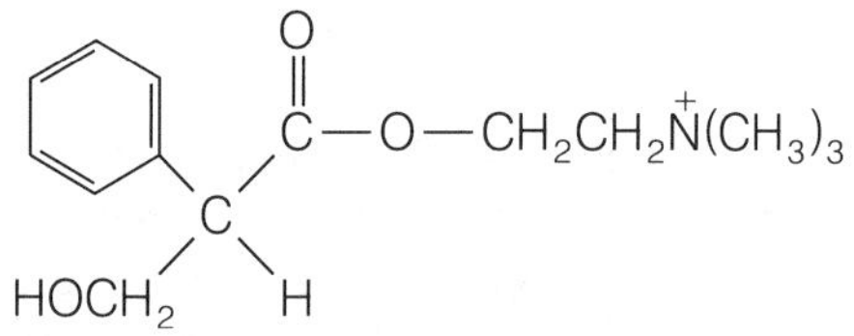

Tropylcholine

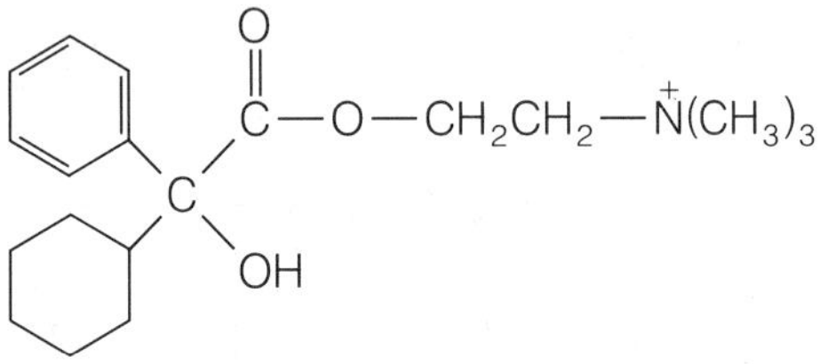

Cyclohexylphenylglycolyl choline

From another viewpoint there is a body of evidence[57] that suggests that muscarinic agonist and antagonist drugs do not interact with the same receptor. Competitive antagonism could result if the receptors were quite distinct, but the presence of an antagonist near the ACh receptor could modify the latter in such a way that the affinity of an agonist would be reduced.[58]

THERAPEUTIC ACTIONS

Because organs controlled by the autonomic nervous system are doubly innervated by both the sympathetic and the parasympathetic systems, it is believed that there is a continual state of dynamic balance between the two systems. Theoretically, one should achieve the same end result either by stimulation of one of the systems or blockade of the other. Unfortunately, there is usually a limitation to this type of generalization. However, there are three predictable and clinically useful results from blocking the muscarinic effects of ACh. These are

1. *Mydriatic effect*: dilatation of pupil of the eye; and *cycloplegia*, a paralysis of the ciliary structure of the eye, resulting in a paralysis of accommodation for near vision
2. *Antispasmodic effect*: lowered tone and motility of the gastrointestinal tract and the genitourinary tract
3. *Antisecretory effect*: reduced salivation (*antisialagogue*), reduced perspiration (*anhidrotic*), and reduced acid and gastric secretion

These three general effects of parasympatholytics can be expected in some degree from any of the known drugs, although occasionally it is necessary to administer rather heroic doses to demonstrate the effect. The mydriatic and cycloplegic effects, when produced by topical application, are not subject to any great undesirable side effects, because of limited systemic absorption. This is not true for the systemic antispasmodic effects obtained by oral or parenteral administration. It is generally understood that drugs having effective blocking action on the gastrointestinal tract are seldom free of undesirable side effects on the other organs. The same is probably true of drugs used for their antisecretory effects. Perhaps the most common side effects experienced from the oral use of these drugs, under ordinary conditions, are dryness of the mouth, mydriasis, and urinary retention.

Mydriatic and cycloplegic drugs are generally prescribed or used in the office by ophthalmologists. The principal purpose is for refraction studies in the process of fitting lenses. This permits the physician to examine the eye retina for possible discovery of abnormalities and diseases, as well as to provide controlled conditions for the proper fitting of glasses. Because of the inability of the iris to contract under the influence of these drugs, there is a definite danger to the patient's eyes during the period of drug activity unless they are protected from strong light by the use of dark glasses. These drugs also are used to treat inflammation of the cornea (keratitis), inflammation of the iris and the ciliary organs (iritis and iridocyclitis), and inflammation of the choroid (choroiditis). A dark-colored iris appears to be more difficult to dilate than a light-colored one and may require more concentrated solutions. Caution in the use of mydriatics is advisable because of their demonstrated effect in raising the intraocular pressure. The pressure rises because pupil dilation tends to cause the iris to restrict drainage of fluid through the canal of Schlemm by crowding the angular space, thereby leading to increased intraocular pressure. This is particularly true for patients having glaucomatous conditions.

It is noteworthy at this juncture that atropine is used widely as an antispasmodic because of its marked depressant effect on parasympathetically innervated smooth muscle. Atropine blocks both M_1 and M_2 muscarinic receptors. Indeed, atropine is the standard by which other similar drugs are measured. Also, the action of atropine is a blocking action on the transmission of the nerve impulse, rather than a depressant effect directly on the musculature. Therefore, its action is termed *neurotropic*, in contrast with the action of an antispasmodic such as papaverine, which appears to act by depression of the muscle cells and is termed *musculotropic*.

Papaverine is the standard for comparison of musculotropic antispasmodics and, although not strictly a parasympatholytic, will be treated together with its synthetic analogues later in this chapter. The synthetic antispasmodics appear to combine neurotropic and musculotropic effects in greater or lesser measure, together with a certain amount of ganglion-blocking activity for the quaternary derivatives.

Because of the widespread use of anticholinergics in the treatment of various gastrointestinal complaints, it is desirable to examine the pharmacologic basis on which this therapy rests. Smooth-muscle spasm, hypermotility, and hypersecretion, individually or in combination, are associated with many painful ailments of the gastrointestinal tract. Among these are peptic ulcer, pylorospasm, cardiospasm, and functional diarrhea. On the basis of Selye's original work on stress and Cannon's classic demonstration of the disruptive effects on normal digestive processes of anger, fear, and excitement, stress is considered as being the causative factor for the malfunction of the cholinergic apparatus. The excitatory (parasympathetic) nerve of the stomach and the gut is intimately associated with the hypothalamus (the so-called seat of feelings) as well as with the medullary and the sacral portions of the spinal cord. Emotions arising or passing through the hypothalamic area can transmit definite effects to the peripheral neural pathways, such as the vagus and other parasympathetic and sympathetic routes. The stomach appears to be influenced by emotions more readily and more extensively than any other organ, and it does not strain the imagination to establish a connection between emotional effects and malfunction of the gastrointestinal tract. Individuals under constant stress are thought to develop a condition of "autonomic imbalance" caused by repeated overstimulation of the parasympathetic pathways. The result is little rest and gross overwork on the part of the muscular and the secretory cells of the stomach and other viscera.

One of the earlier hypotheses advanced for the formation of ulcers proposed that strong emotional stimuli could lead to a spastic condition of the gut, with accompanying anoxia of the mucosa owing to prolonged vasoconstriction. The localized ischemic areas, combined with simultaneous secretion of hydrochloric acid and pepsin, could then provide the groundwork for gastric ulcer formation by repeated irritation of the involved mucosal areas. Lesions in the protective mucosal lining would, of course, then permit the normal digestive processes to attack the tissue of the organ. Hydrochloric acid is considered as the causative agent of duodenal ulcers because it is known that these ulcer patients secrete substantially higher quantities of the acid than do normal people. Duodenal ulcers also can be induced in dogs with normal stomachs if the gastric acidity level is raised to the level found in ulcer patients. Nervous influence is thought to be basic to the hypersecretion of acid resulting in duodenal ulcers, whereas humoral or hormonal influences are believed to be responsible for excessive secretion in gastric ulcers.[59]

The condition of overstimulation of the parasympathetic nervous supply (vagus) to the stomach is sometimes termed *parasympathotonia*. Reduction of this overstimulated condition can be achieved by surgery (surgical vagotomy) or by the use of anticholinergic drugs (chemical vagotomy), resulting in inhibition of both secretory and motor activity of the stomach. Although anticholinergic drugs can exert an antimotility effect, there is some question of whether they can correct disordered motility or counteract spontaneous "spasms" of the intestine. In addition, although these drugs can (in adequate dosage) diminish the basal secretion of acid, there is little effect on the acid secreted in response to food or to insulin hypoglycemia.

After the initial surge of chemical modification of the belladonna alkaloids, which began in the 1920s and ended in the early 1950s, few antispasmodics have been presented to the clinician to aid in the therapy of ulcers. It is suggested that neither the belladonna alkaloids nor the synthetic spasmolytics have achieved the degree of selectivity of cholinolytic action to have either class preferred for the management of gastrointestinal ailments.[60] For the present, the most rational therapy used by clinicians employing anticholinergic drugs seems to be a combination of a nonirritating diet to reduce acid secretion, antacid therapy, and reduction of emotional stress. Most of the anticholinergic drugs on the market are offered either as the chemical alone or in combination with a central nervous system depressant, such as phenobarbital, or with one of the tranquilizers to reduce the central nervous system contribution to parasympathetic hyperactivity.

There seems to be no advantage to the use of combination products of antispasmodics with phenobarbital or phenothiazine tranquilizers because of the difficulty in balancing the effects of the two different central nervous system drugs with the anticholinergic agent. If the two drugs must be given together, it is suggested that they be given in separated preparations to permit better control of central nervous system effects.[61] Some clinical findings tend to show that phenobarbital is preferable to the tranquilizers. Whereas combinations of anticholinergics with sedatives are considered rational, there is not complete agreement on combinations with antacids. This is because anticholinergic drugs primarily affect the fasting phases of gastrointestinal secretion and motility and are most efficient if administered at bedtime and well before mealtimes. Antacids neutralize acid largely present in the between-meal, digestive phases of gastrointestinal activity and are of more value if given after meals. H_2-receptor antagonists of histamine decrease hydrochloric acid secretion and are especially useful for those patients who suffer from the nocturnal secretory activity of the stomach (see Chap. 16).

In addition to the antisecretory effects of anticholinergics on hydrochloric acid and gastric secretion just described, there have been some efforts to employ them as *antisialagogues* (to suppress salivation) and *anhidrotics* (to suppress perspiration).

Paralysis agitans or parkinsonism, first described by the English physician James Parkinson in 1817, is another condition that is often treated with the anticholinergic drugs. It is characterized by tremor, "pill rolling," cog-wheel rigidity, festinating gait, sialorrhea, and masklike facies. Fundamentally, it represents a malfunction of the extrapyramidal system.[62] Skeletal muscle movement is controlled to a great degree by patterns of excitation and inhibition resulting from the feedback of information to the cortex and is mediated through the pyramidal and extrapyramidal pathways. The basal ganglia structures, such as the pallidum, corpus striatum, and substantia nigra, serve as data processors for the pyramidal pathways and also the structures through which the extrapyramidal pathways pass on their way from the spinal cord to the cortex. Lesions of the pyramidal pathways cause spasticity, weakness, and exaggerated tendon reflexes. Interruption of the extrapyramidal pathways leads to a persistent increase in muscle tone, resulting in an excess of spontaneous involuntary movements, along with changes in the reflexes. It is apparent, therefore, that the basal ganglia are functional in maintaining normal motor control. In parkinsonism there is a degeneration of the substantia nigra and corpus striatum, which involve controlled integration of muscle movement.

The neurochemical basis of parkinsonism has been characterized as (1) having decreased levels of dopamine and its metabolites, for example, homovanillic acid, in the basal ganglia; (2) a reduced concentration of dopamine-synthesizing enzymes tyrosine hydroxylase and dopa decarboxylase, the enzyme that coverts L-dopa to dopamine; and (3) a decrease in the number of specific presynpatic dopamine sites. Dopamine is believed to act principally as an inhibitory neurotransmitter and ACh as an excitatory neurotransmitter in the nigrostriatal pathway. Despite this information the clinical picture remains as it has for many years. The disease state is apparently not reversible and chemotherapy is, of necessity, palliative.

The usefulness of the belladonna group of alkaloids was an empiric discovery by Charcot. The several synthetic preparations were developed in an effort to retain the useful antitremor and antirigidity effects of the belladonna alkaloids while, at the same time, reducing the undesirable side effects. Incidentally, it was also discovered that antihistamine drugs (e.g., diphenhydramine) sometimes reduced tremor and rigidity. The antiparkinsonlike activity of the antihistamines has been attributed to their anticholinergic effects. The activity is confined to those drugs that can pass the blood–brain barrier (i.e., tertiary amines, not quaternary ammonium compounds).

Acetylcholine is widely found in the brain. Its concentration in the areas that control movement and behavior is higher than in the cortex. Large amounts of choline acetyltransferase and AChE are also found in the caudate nucleus. Therefore, all the components required for the synthesis and distribution of ACh acting as a neurotransmitter are present here. The present assumptions are that ACh acts as a neurotransmitter in the central nervous system and that anticholinergics can block its action as they do in the peripheral nervous system. One reason for this statement is data from the investigations on oxotremorine.[63]

Injections of tremorine (1,4-dipyrrolidino-2-butyne) or its active metabolite oxotremorine [1-(2-pyrrolidono)-4-pyrrolidino-2-butyne] increase the brain ACh level in rats up to 40%. This increase coincides roughly with the onset of tremors similar to those observed in parkinsonism. The mechanism of ACh increase in rats is uncertain, but has been shown not to be due to AChE inhibition or to activation of choline acetyltransferase. The tremors are stopped effectively by administration of the tertiary amine-type anticholinergic but not by the quaternaries.

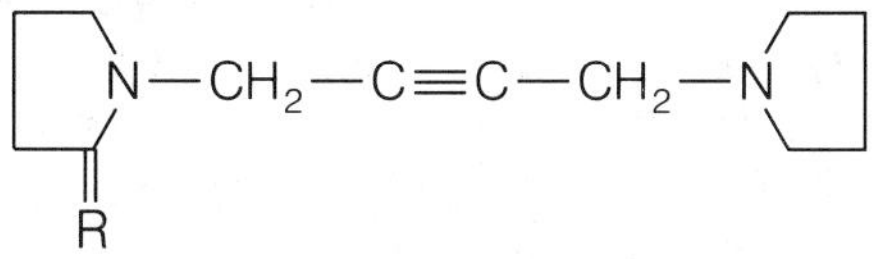

Tremorine: $R = H_2$
Oxotremorine: $R = O$

Although many compounds have been introduced to treat of parkinsonism, there is apparently a real need for compounds that will provide more potent action with fewer side effects and, also drugs that will not lose their efficacy in patients during long-term therapy for this disease.

The most significant advance in the treatment of parkinsonism stems from the discovery of the usefulness of L-dopa in managing the disease. This amino acid acts as a source of dopamine, known to be deficient in the patient afflicted with parkinsonism. Initially, L-dopa was given in rather large doses, with a concomitant increase in side effects. Combinations of L-dopa with a decarboxylase inhibitor (e.g., carbidopa) have, in many cases, allowed reduction of the dose to about one-fourth that required in the absence of the inhibitor. In particular, nausea and vomiting (caused by dopamine stimulation of the medullary vomiting center) have been sharply reduced, although the mechanism by which beneficial activity is produced is not clear.

Apomorphine has been examined as a possible parkinsonlytic, mainly because it may be considered to be a dopamine congener with the dopamine structure locked in a rigid conformation.

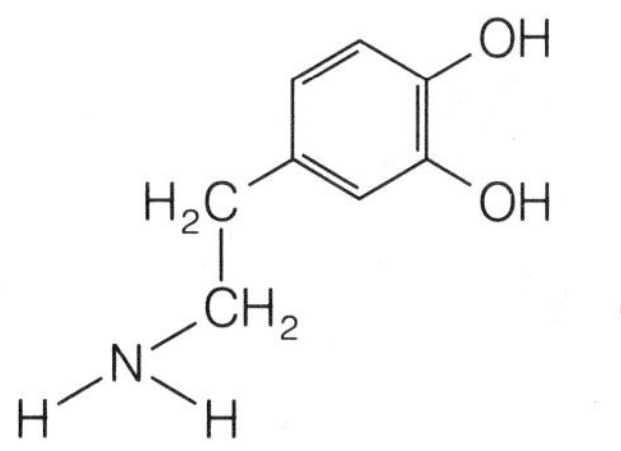

Dopamine

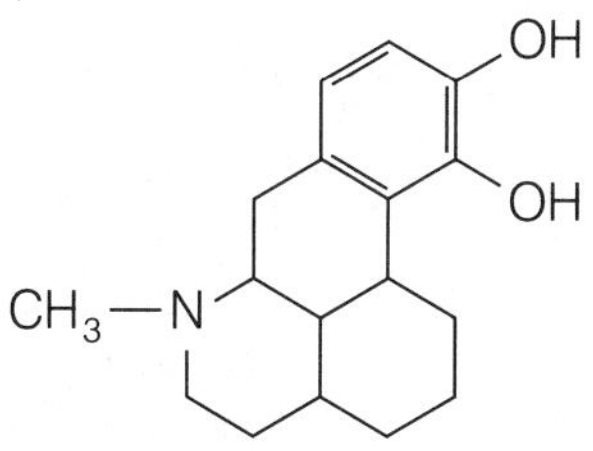

Apomorphine

A very interesting observation seems to indicate a link between the anticholinergic drugs and dopamine in the therapy of parkinsonism. It has been shown[64] that several antiparkinsonian drugs (e.g., benztropine, trihexyphenidyl, orphenadrine, diphenhydramine) were able to inhibit the uptake of catecholamines into synaptosomes from the corpus striatum. Subsequently, atropine, benztropine, and other antiparkinsonian drugs were shown to reduce the uptake of dopamine in rat striata.[65] The reduced uptake of dopamine presumably results in a potentiation of its effect and serves to ameliorate the symptoms of parkinsonism.

SOLANACEOUS ALKALOIDS AND SYNTHETIC ANALOGUES

The solanaceous alkaloids represented by (−)-hyoscyamine, atropine [(±)-hyoscyamine] and scopolamine (hyoscine) are the forerunners of the class of parasympatholytic drugs. These alkaloids are found principally in henbane (*Hyoscyamus niger*), deadly nightshade (*Atropa belladonna*), and jimson weed (*Datura stramonium*). There are other alkaloids that are members of the solanaceous group (e.g., apoatropine, noratropine, belladonnine, tigloidine, meteloidine), but are not of sufficient therapeutic value to be considered in this text.

The crude drugs containing these alkaloids have been used since early times for their marked medicinal properties, which depend largely on inhibition of the parasympathetic nervous system and stimulation of the higher nervous centers. Belladonna, probably as a consequence of the weak local anesthetic activity of atropine, has been used topically for its analgesic effect on hemorrhoids, certain skin infections, and various itching dermatoses. The application of sufficient amounts of belladonna or of its alkaloids results in mydriasis. Internally, the drug causes diminution of secretions, increases the heart rate (by depression of the vagus nerve), depresses the motility of the gastrointestinal tract, and acts as an antispasmodic on various smooth muscles (ureter, bladder, and biliary tract). In addition, it stimulates the respiratory center directly. The multiplicity of actions exerted by the drug causes it to be looked upon with some disfavor, because the physician seeking one type of response unavoidably also obtains the others. The action of scopolamine-containing drugs differs from those containing hyoscyamine and atropine in that there is no central nervous system stimulation, and a narcotic or sedative effect predominates. The use of this group of drugs is accompanied by a fairly high incidence of reactions because of individual idiosyncrasies; death from overdosage usually results from respiratory failure. A complete treatment of the pharmacology and the uses of these drugs is not within the scope of this text, and the reader is referred to the several excellent pharmacology texts that are available. However, the introductory pages of this chapter have reviewed briefly some of the more pertinent points in connection with the major activities of these drug types.

STRUCTURAL CONSIDERATIONS

All of the solanaceous alkaloids are esters of the bicyclic aminoalcohol, 3-hydroxytropane, or of related aminoalcohols.

The structural formulas that follow show the piperidine ring system in the commonly accepted chair conformation, because this form has the lowest energy requirement. However, the alternate boat form can exist under certain conditions, because the energy barrier is not great. Inspection of the 3-hydroxytropane formula also indicates that, even though there is no optical activity because of the plane of symmetry, two stereoisomeric forms (tropine and pseudotropine) can exist because of the rigidity imparted to the molecule through the ethylene chain across the 1,5-positions. In tropine, the axially oriented hydroxyl group, *trans* to the nitrogen bridge, is designated as α, and the alternate *cis* equatorially oriented hydroxyl group is β. The aminoalcohol derived from scopolamine, namely scopine, has the axial orientation of the 3-hydroxyl group but, in addition, has a β-oriented epoxy group bridged across the 6,7-positions as shown. Of the several different solanaceous alkaloids known, it has already been indicated that (−)-hyoscyamine, atropine, and scopolamine are the most important. Their structures are indicated, but it can be pointed out that antimuscarinic activity is associated with all of the solanaceous alkaloids that possess the tropinelike axial orientation of the esterified hydroxyl group. It will be noted in studying the formulas that tropic acid is, in each case, the esterifying acid. Tropic acid contains an easily racemized asymmetric carbon atom, the moiety accounting for optical activity in these compounds in the absence of racemization. The proper enantiomorph is necessary for high antimuscarinic activity, as illustrated by the potent (−)-hyoscyamine in comparison with the weakly active (+)-hyoscyamine. The racemate, atropine, has an intermediate activity. The marked difference in antimuscarinic potency of the optical enantiomorphs apparently does not extend to the action on the central nervous system, inasmuch as both seem to have the same degree of activity.[66]

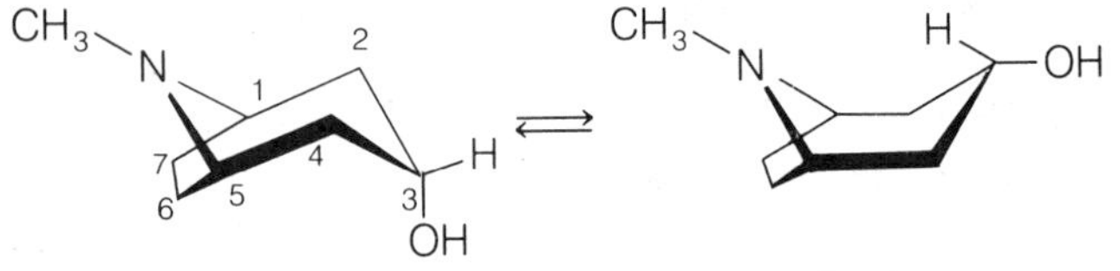

(chair) (boat)

TROPINE
(3α-Hydroxytropane or 3α-tropanol)

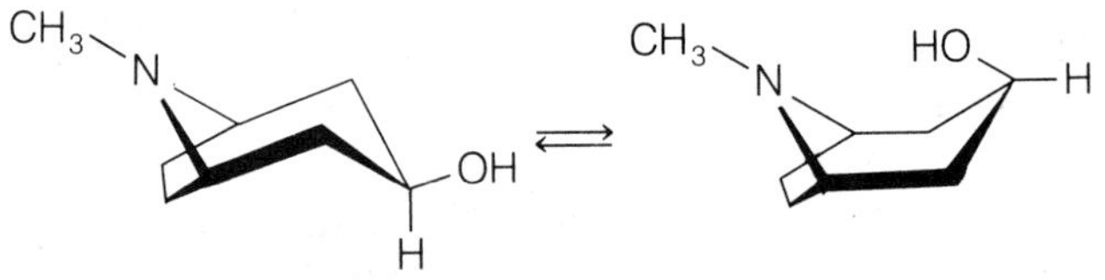

(chair) (boat)

PSEUDOTROPINE
(3β-Hydroxytropane or 3β-tropanol)

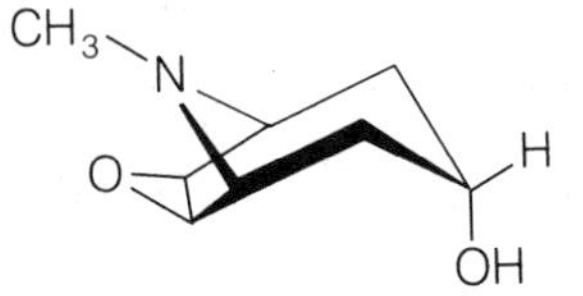

SCOPINE
(6 : 7 β-Epoxy-3α-hydroxytropane or 6 : 7 β-Epoxy-3α-tropanol)

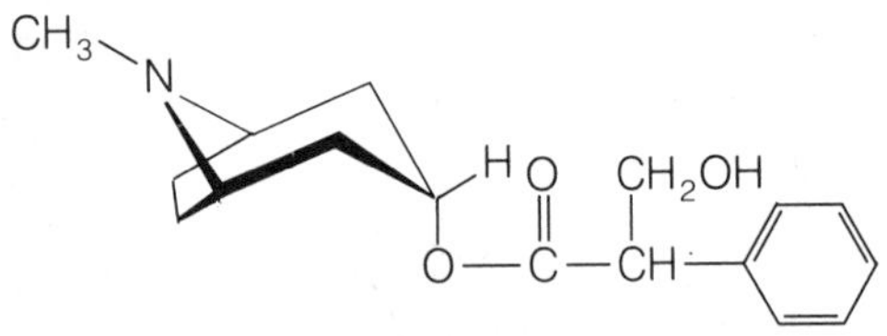

ATROPINE
(or Hyoscyamine)

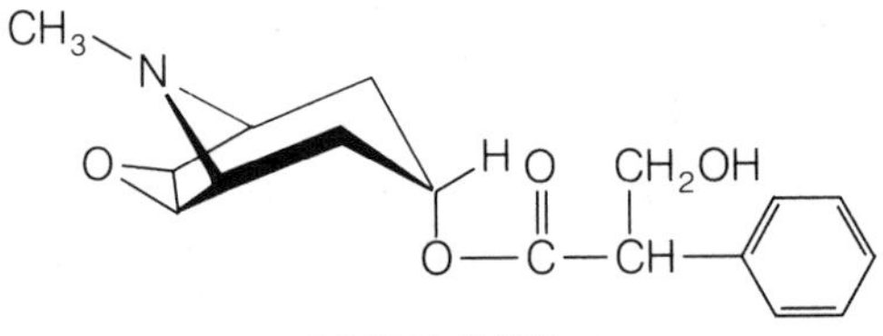

SCOPOLAMINE
(or Hyoscine)

The solanaceous alkaloids have been modified by preparing other esters of 3α-tropanol or making a quaternary of the nitrogen in tropanol or scopine with a methyl halide. These compounds were some of the initial attempts to separate the varied actions of atropine and scopolamine. It should be pointed out that few aminoalcohols have been found that impart the same degree of neurotropic activity as that exhibited by the ester formed by combination of tropine with tropic acid. Similarly, the tropic acid portion is highly specific for the anticholinergic action, and substitution by other acids results in decreased neurotropic potency, although the musculotropic action may increase. The earliest attempts to modify the atropine molecule retained the tropine portion and substituted various acids for tropic acid.

Besides changing the acid residue, other changes have been directed toward the quaternization of the nitrogen. Examples of this type of compound are methscopolamine bromide, homatropine methylbromide, and anisotropine methylbromide. Quaternization of the tertiary amine produces variable effects in terms of increasing potency. Decreases in activity are apparent in comparing atropine with methylatropine (no longer used) and scopolamine with methscopolamine. Ariens et al.[67] ascribe decreased activity, especially when the groups attached to nitrogen are larger than methyl, to a possible decrease in affinity for the anionic site on the cholinergic receptor. They attribute this decreased affinity to a combination of greater electron repulsion by such groups and greater steric interference to the approach of the cationic head to the anionic site. In general, however, the effect of quaternization is much greater in reduction of parasympathomimetic than parasympatholytic action. This may be partially due to the additional blocking at the parasympathetic ganglion induced by quaternization, which could serve to offset the decreased affinity at the postganglionic site. However, it also is to be noted that quaternization increases the curariform activity of these alkaloids and aminoesters, a usual consequence of quaternizing alkaloids. Another disadvantage in converting an alkaloidal base to the quaternary form is that the quaternized base is more poorly absorbed through the intestinal wall, with the consequence that the activity becomes erratic and, in some instances, unpredictable. The reader will find Brodie and Hogben's[68] comments on the absorption of drugs in the dissociated and the undissociated states of considerable interest. They point out that bases (such as alkaloids) are absorbed through the lipoidal gut wall only in the dissociated form, which can be expected to exist for a tertiary base, in the small intestine. On the other hand, quaternary nitrogen bases cannot revert to an undissociated form, even in basic media and, presumably, may have difficulty passing through the gut wall. That quaternary compounds can be absorbed indicates that other less efficient mechanisms for absorption probably prevail.

PRODUCTS

Atropine, USP. Atropine is the tropine ester of racemic tropic acid and is optically inactive. It possibly occurs naturally in various Solanaceae, although some claim, with justification, that whatever atropine is isolated from natural sources results from racemization of (−)-hyoscyamine during the isolation process. Conventional methods of alkaloid isolation are used to obtain a crude mixture of atropine and hyoscyamine from the plant material.[69] This crude mixture is racemized to atropine by refluxing in chloroform or by treatment with cold dilute alkali. Because atropine is made by the racemization process, an official limit is set on the hyoscyamine content by restricting atropine to a maximum levorotation under specified conditions.

Synthetic methods for preparing atropine take advantage of Robinson's synthesis, employing modifications to improve the yield of tropinone. Tropinone may be reduced, under proper conditions, to tropine, which is then used to esterify tropic acid. Other acids may be used in place of tropic acid to form analogues, and numerous compounds of this type have been prepared, which are known collectively as *tropeïnes*. The most important one, homatropine, is considered later in this section.

Atropine occurs in the form of optically inactive, white, odorless crystals possessing a bitter taste. It is not very soluble in water (1 : 460; 1 : 90 at 80°C) but is more soluble in alcohol (1 : 2; 1 : 1.2 at 60°C). It is soluble in glycerin (1 : 27), in chloroform (1 : 1) and in ether (1 : 25).* Saturated aqueous solutions are alkaline in reaction (approximate *p*H, 9.5). The free base is useful when nonaqueous solutions are to be made, such as in oily vehicles and ointment bases. Atropine has a plasma half-life of about two to three hours. It is metabolized in the liver to several products, including tropic acid and tropine.

Atropine Sulfate, USP (Atropisol) is prepared by neutralizing atropine in acetone or ether solution with an alcoholic solution of sulfuric acid, care being exercised to prevent hydrolysis.

The salt occurs as colorless crystals or as a white crystalline powder. It is efflorescent in dry air and should be protected from light to prevent decomposition.

Atropine sulfate is freely soluble in water (1 : 0.5), in alcohol (1 : 5; 1 : 2.5 at boiling point), and in glycerin (1 : 2.5). Aqueous solutions of atropine are not very stable, although it has been stated[70] that solutions may be sterilized at 120°C (15 lb pressure) in an autoclave if the *p*H is kept below 6. Sterilization probably is best effected by the use of aseptic technique and a bacteriologic filter. The foregoing reference suggests that no more than a 30-day supply of an aqueous solution should be made, and, for small

* In this chapter a solubility expressed as 1 : 460 indicates that 1 g is soluble in 460 mL of the solvent at 25°C. Solubilities at other temperatures will be so indicated.

quantities, the best procedure is to use hypodermic tablets and sterile distilled water. Kondritzer and his co-workers[71,72] have studied the kinetics of alkaline and proton-catalyzed hydrolyses of atropine in aqueous solution. The region of maximum stability lies between *p*H 3 and approximately 5. They also have proposed an equation to predict the half-life of atropine undergoing hydrolysis at constant *p*H and temperature.

The action of atropine or its salts is the same. It produces a mydriatic effect by paralyzing the iris and the ciliary muscles and, for this reason, is used by the oculist in iritis and corneal inflammations and lesions. Its use is rational in these conditions because one of the first rules in the treatment of inflammation is rest, which, of course, is accomplished by the paralysis of muscular motion. It use in the eye (0.5% to 1% solutions or gelatin disks) for fitting glasses is widespread. Atropine is administered in small doses before general anesthesia to lessen oral and air passage secretions and, when morphine is administered with it, it serves to lessen the respiratory depression induced by morphine.

Atropine causes restlessness, prolonged pupillary dilation, and loss of visual accommodation and, furthermore, gives rise to arrhythmias such as atrioventricular dissociation, ventricular extrasystoles, and even ventricular fibrillation. Even though there has been a gradual replacement of ether with halothane as a general anesthetic, thereby eliminating problems with respiratory secretions caused by ether and thus requiring atropine, surgeons and anesthesiologists today continue to use it as an anesthetic premedicant to "dry up secretions" and to prevent vagal reflexes.[73]

Its ability to dry secretions also has been utilized in the so-called rhinitis tablets for symptomatic relief in colds. In cathartic preparations, atropine or belladonna has been used as an antispasmodic to lessen the smooth-muscle spasm (griping) often associated with catharsis.

In a recent application of atropine it has been found that this drug may be used in treatment of some types of arrhythmias. Atropine increases the heart rate by blocking the effects of ACh on the vagus. In this context it is used to treat certain reversible bradyarrhythmias that may accompany acute myocardial infarction. It is also used as an adjunct to anesthesia to protect against bradycardia, hypotension, and even cardiac arrest induced by the skeletal muscle relaxant succinylcholine chloride.

Another use for atropine sulfate has emerged following the development of the organophosphates, which are potent inhibitors of AChE. Atropine is a specific antidote to prevent the "muscarinic" effects of ACh accumulation, such as vomiting, abdominal cramps, diarrhea, salivation, sweating, bronchoconstriction, and excessive bronchial secretions.[74] It is used intravenously but does not protect against respiratory failure caused by depression of the respiratory center and the muscles of respiration.

Hyoscyamine, USP, is a levorotatory alkaloid obtained from various solanaceous species. One of the commercial sources is Egyptian henbane (*Hyoscyamus muticus*), in which it occurs to the extent of about 0.5%. One method for extraction of the alkaloid utilizes *Duboisia* species.[75] Usually, it is prepared from the crude drug in a manner similar to that used for atropine and is purified as the oxalate. The free base is obtained easily from this salt.

It occurs as white needles that are sparingly soluble in water (1 : 281), more soluble in ether (1 : 69) or benzene (1 : 150), and very soluble in chloroform (1 : 1) or alcohol. It is official as the sulfate and hydrobromide. The principal reason for the popularity of the hydrobromide has been its nondeliquescent nature. The salts have the advantage over the free base in being quite water-soluble.

Hyoscyamine is the *levo*-form of the racemic mixture known as atropine. The *dextro*-form does not exist naturally, but has been synthesized. Comparison of the activities of (−)-hyoscyamine, (+)-hyoscyamine, and the racemate (atropine) was carried out by Cushny in 1903, wherein he found a greater peripheral potency for the (−)-isomer and twice the potency of the racemate. All later studies have essentially borne out these observations, namely, that the (+)-isomer is only weakly active and that the (−)-isomer is, in effect, the active portion of atropine. Inspection of the relative doses of Atropine Sulfate USP and Hyoscyamine Sulfate USP illustrates the differences very nicely. The principal criticism offered against the use of hyoscyamine sulfate exclusively is that it tends to racemize to atropine sulfate rather easily in solution so that atropine sulfate, then, becomes the more stable of the two. All of the isomers behave very much the same in the central nervous system. A preparation containing the levorotatory alkaloids of belladonna but consisting principally of (−)-hyoscyamine malate is on the market under the trade name of Bellafoline. It has been promoted extensively on the basis of less central activity and greater peripheral activity than atropine possesses.

Hyoscyamine is used to treat disorders of the urinary tract more so than any other antispasmodic, although there is no evidence that it has any advantages over the other belladonna preparations and the synthetic anticholinergics. It is used to treat spasms of the bladder and, in this manner, serves as a urinary stimulant. It is used together with a narcotic to counteract the spasm produced by the narcotic when the latter is used to relieve the pain of urethral colic. Hyoscyamine preparations are also used in therapy of peptic ulcers as antispasmodics.

Hyoscyamine Sulfate, USP (Levsin Sulfate). This salt is a white, odorless, crystalline compound of a deliquescent nature that is also affected by light. It is soluble in water (1 : 0.5) and alcohol (1 : 5), but almost insoluble in ether. Solutions of hyoscyamine sulfate are acidic to litmus.

This drug is used as an anticholinergic in the same manner and for the same uses as atropine and hyoscyamine, but possesses the disadvantage of being deliquescent.

Scopolamine. Hyoscine. This alkaloid is found in various members of the Solanaceae (e.g., *Hyoscyamus niger*, *Duboisia myoporoides*, *Scopolia* sp., and *Datura metel*). Scopolamine usually is isolated from the mother liquor remaining from the isolation of hyoscyamine.

The name *hyoscine* is the older name for this alkaloid, although *scopolamine* is the accepted name in the United States. Scopolamine is the *levo*-component of the racemic mixture, which is known as *atroscine*. The alkaloid is racemized readily in the presence of dilute alkali.

The alkaloid occurs in the form of a levorotatory, viscous liquid that is only slightly soluble in water, but very soluble in alcohol, chloroform, or ether. It forms crystalline salts with most acids, the hydrobromide being the most stable and the most popularly accepted. An aqueous solution of the hydrobromide, containing 10% of mannitol (Scopolamine Stable), is said to be less prone to decomposition than unprotected solutions. The commercially available transdermal system of scopolamine comprises an outer layer of polymer film and a drug reservoir containing scopolamine, polyisobutylene, and mineral oil that is interfaced with a microporous membrane to control diffusion of the drug.

Scopolamine Hydrobromide, USP. Hyoscine hydrobromide. This salt occurs as white or colorless crystals or as a white granular powder. It is odorless and tends to effloresce in dry air. It is freely soluble in water (1 : 1.5), soluble in alcohol (1 : 20), only slightly soluble in chloroform, and insoluble in ether.

Scopolamine is a competitive blocking agent of the parasympathetic nervous system as atropine but it differs markedly from atropine in its action on the higher nerve centers. Both drugs readily cross the blood–brain barrier and, even in therapeutic doses, cause confusion, particularly in the elderly.

Whereas atropine stimulates the central nervous system, causing restlessness and talkativeness, scopolamine usually acts as a CNS depressant. In this capacity, it has found use in the treatment of parkinsonism, although its value is depreciated by the fact that the effective dose is very close to the toxic dose. A sufficiently large dose of scopolamine will cause an individual to sink into a restful, dreamless sleep for a period about eight hours, followed by a period of approximately the same length in which the patient is in a semiconscious state. During this time, the patient does not remember events that take place. When scopolamine is administered with morphine, this temporary amnesia is termed *twilight sleep*.

Homatropine Hydrobromide, USP. $1\alpha H$, $5\alpha H$-Tropan-3α-ol mandelate (ester) hydrobromide (Homatrocel). Homatropine may be prepared by evaporating tropine (obtained from tropinone) in the presence of mandelic and hydrochloric acids. The hydrobromide is obtained readily from the free base by neutralizing with hydrobromic acid. The hydrochloride may be obtained in a similar manner.

The hydrobromide occurs as white crystals, or as a white crystalline powder that is affected by light. It is soluble in water (1 : 6) and in alcohol (1 : 40), less soluble in chloroform (1 : 420), and insoluble in ether.

Solutions are incompatible with alkaline substances, which precipitate the free base, and also with the common reagents that precipitate alkaloids. As in atropine, solutions are sterilized best by filtration through a bacteriologic filter, although it is claimed that autoclaving has no deleterious chemical effect.[76]

Homatropine hydrobromide is used topically to paralyze the ciliary structure of the eye (cycloplegia) and to effect mydriasis. It behaves very much like atropine but is weaker and less toxic. In the eye, it acts more rapidly, but less persistently, than atropine. The dilatation of the pupil takes place in about 15 to 20 minutes, and the action subsides in about 24 hours. By utilizing a miotic, such as physostigmine, it is possible to restore the pupil to normality in a few hours.

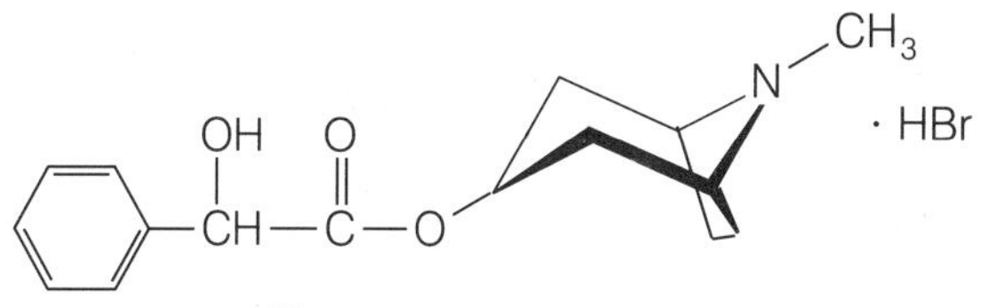

Homatropine Hydrobromide

Homatropine Methylbromide, USP. 3α-Hydroxy-8-methyl-$1\alpha H$,$5\alpha H$-tropanium bromide mandelate (Novatropine; Mesopin). This compound is the tropine methylbromide ester of mandelic acid. It may be prepared from homatropine by treating it with methyl bromide, to form the quaternary compound.

TABLE 12-8

ATROPINE AND RELATED COMPOUNDS

Name Proprietary Name	**Preparations**	**Category**	**Application**	**Usual Adult Dose***	**Usual Dose Range***	**Usual Pediatric Dose***
Atropine USP		Anticholinergic				
Atropine sulfate USP *Atropisol, Isopto Atropine*	Atropine sulfate injection USP	Anticholinergic; antidote to cholinesterase inhibitors		Anticholinergic: parenteral, 400–600 μg 4–6 times daily; antidote to cholinesterase inhibitors: IV, 2–4 mg initially, followed by IM, 2 mg repeated every 5–10 min until muscarinic symptoms disappear or signs of atropine toxicity appear	300 μg–5.0 mg/day	Anticholinergic: SC, 10 μg/kg of body weight or 300 μg/m² of body surface, up to 44 μg/dose, 4–6 times daily; antidote to cholinesterase inhibitors: IV or IM, 1 mg initially, followed by 500 μg–1 mg every 10–15 min until signs of atropine toxicity appear
	Atropine sulfate ophthalmic ointment USP	Anticholinergic (ophthalmic)	Topically, to the conjunctiva	0.3–0.5 cm of a 1% ointment qd to tid		
	Atropine sulfate ophthalmic solution USP	Anticholinergic (ophthalmic)	Topically, to the conjunctiva	1 drop of a 1% solution qd to tid		
	Atropine sulfate tablets USP	Anticholinergic		300–600 μg 3–6 times daily	300 μg–8 mg/day	
Hyoscyamine USP *Cystospaz*	Hyoscyamine tablets USP	Anticholinergic		250 μg qid		
Hyoscyamine sulfate USP *Levsin*	Hyoscyamine sulfate tablets USP	Anticholinergic			125–250 μg tid or qid	
Scopolamine hydrobromide USP *Isopto Hyoscine*	Scopolamine hydrobromide injection USP	Anticholinergic		Parenteral, 320 μg–650 μg as a single dose		SC, 6 μg/kg of body weight or 200 μg/m² of body surface, as a single dose
	Scopolamine hydrobromide ophthalmic solution USP	Anticholinergic (ophthalmic)	Topically, to the conjunctiva	1 drop of a 0.25% solution qd to tid		
	Scopolamine hydrobromide tablets USP	Anticholinergic		400–800 μg		6 μg/kg of body weight or 200 μg/m² of body surface, as a single dose
	Scopolamine hydrobromide ophthalmic ointment USP Scopolamine hydrobromide ophthalmic solution USP	Anticholinergic (ophthalmic)	Topically, to the conjunctiva	0.3–0.5 cm of a 0.2% ointment qd to tid		
Homatropine hydrobromide USP *Homatrocel, Isopto Homatropine*	Homatropine hydrobromide ophthalmic solution USP	Anticholinergic (ophthalmic)	Topically, to the conjunctiva	1 drop of a 2 or 5% solution bid or tid		
Homatropine methylbromide USP		Anticholinergic				
	Homatropine methylbromide tablets USP			2.5–5 mg qid		

*See USP DI for complete dosage information.

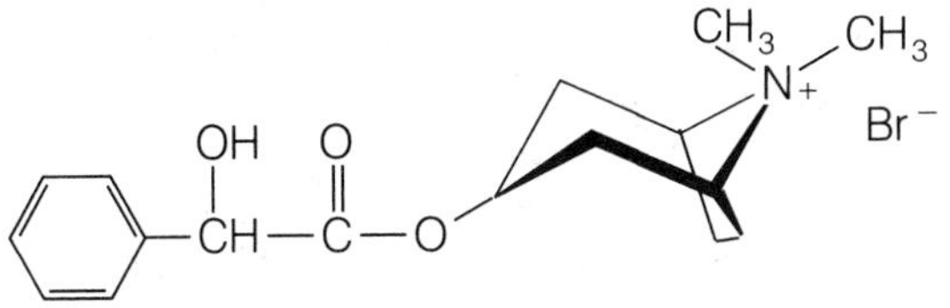

Homatropine Methylbromide

It occurs as a bitter, white, odorless powder and is affected by light. The compound is readily soluble in water and in alcohol, but is insoluble in ether. The *p*H of a 1% solution is 5.9 and of a 10% solution is 4.5. Although a solution of the compound yields a precipitate with alkaloidal reagents, such as mercuric–potassium–iodide test solution, the addition of alkali hydroxides or carbonates does not cause the precipitate that occurs with nonquaternary nitrogen salts (e.g., atropine, homatropine).

Homatropine methylbromide is said to be a poorer stimulant of the central nervous system than atropine, although retaining virtually all of its parasympathetic depressant action. It is used orally to reduce oversecretion and relieve gastrointestinal spasms.

Ipratropium Bromide. 3-(3-Hydroxy-1-oxo-2-phenylpropoxy)-8-methyl-8-(1-methylethyl)-8-aza-bicyclo[3.2.1]octane bromide (Atrovent). Ipratropium bromide is a quaternary ammonium derivative of atropine. It is freely soluble in water and ethanol, but insoluble in chloroform and ether. The salt is stable in neutral and acidic solutions, but rapidly hydrolyzed in alkaline solutions.

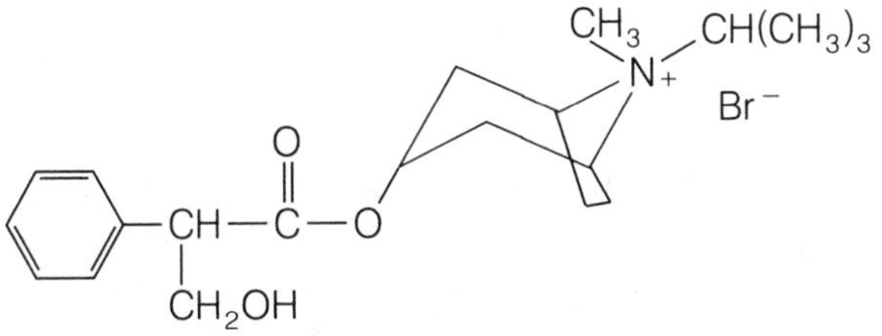

Ipratropium Bromide

Ipratropium bromide is used for inhalation therapy to produce dilatation of bronchial smooth muscle for acute asthmatic attacks. The drug produces the bronchodilatation by competitive inhibition of cholinergic receptors bound to smooth muscle of the bronchioles. Ipratropium may also act on the surface of mast cells to inhibit ACh-enhanced release of chemical mediators. The drug has a slow onset of action, within 5 to 15 minutes, after being administered by inhalation and should not be used alone for acute asthmatic attacks. The peak therapeutic effect from one dose is observed between one and two hours. The effects of the drug last for about six hours. It has a half-life of 3.5 hours.

SYNTHETIC CHOLINERGIC-BLOCKING AGENTS

AMINOALCOHOL ESTERS

It is generally agreed that the solanaceous alkaloids are potent parasympatholytics but that they have the undesirable property of producing a wide range of effects through their nonspecific blockade of autonomic functions. Efforts to use the antispasmodic effect of the alkaloids most often results in a side effect of dryness of the mouth. Therefore, synthesis of compounds possessing specific cholinolytic actions was a very desirable field of study. Few prototype drugs were as avidly dissected in the minds of researchers as was atropine in attempts to modify its structure to separate the numerous useful activities of the prototype (i.e., antispasmodic, antisecretory, mydriatic, and cycloplegic). Most early research was carried out in the pre- and post-World War II era.

Efforts at synthesis started with rather minor deviations from the atropine molecule, but a review of the commonly used drugs today indicates a marked departure from the rigid tropane aminoalcohol and tropic acid residue. An examination of the structures of antispasmodics shows that the acid portion has been designed to provide a large hydrophobic moiety rather than the stereospecific requirement of (*S*)-tropic acid in (−)-hyoscyamine that was once considered important. One of the major developments in the field of aminoalcohol esters was the successful introduction of the quaternary ammonium derivatives as contrasted with the tertiary amine-type esters synthesized originally. Although there are some effective tertiary amine esters in use today, the quaternaries, as a group, represent the most popular type and appear to be slightly more potent than their tertiary counterparts.

It has already been pointed out that the stereochemical arrangement in the rigid atropine molecule lends itself to high activity, presumably because of a good fit of its prosthetic groups with the muscarinic receptors. Early studies employing the empiric idea of structural dissection (so successful with local anesthetics) led to the conclusion that, even though atropine did seem to have a highly specific action, the tropine portion was nothing more than a highly complex aminoalcohol and was susceptible to simplification. The accompanying formula shows the portion of the atropine molecule (enclosed in the curved dotted line) believed to be responsible for its major activity. This group is sometimes called the "spasmophoric" group and compares with the "anesthesiophoric" group obtained by similar dissection of the cocaine molecule. The validity of this conclusion has been amply borne out by the many active compounds having only a simple diethylaminoethyl residue replacing the tropine portion.

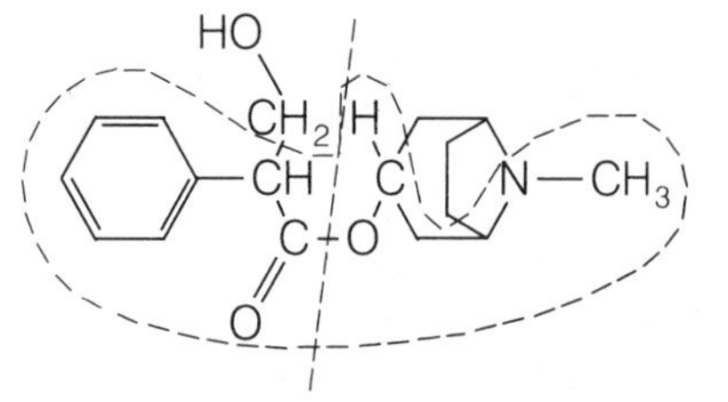

Tropic Acid Tropine

Eucatropine may be considered as a conservative approach to the simplification of the aminoalcohol portion of atropine. In eucatropine, the bicyclic tropine has been replaced by a monocyclic aminoalcohol and, in addition, mandelic acid replaces tropic acid.

One of the earliest compounds utilizing a simplified noncyclic aminoalcohol was amprotropine. This was prepared by Fromherz[77] in 1933 and, for many years, was widely used as a gastrointestinal antispasmodic, but it has been displaced by much more active compounds. In this particular compound, the tropic acid residue was retained, but the bulk of research on antispasmodics of this nature has been directed toward compounds in which both the acid and the aminoalcohol portions have been modified. Acids formally related to phenylacetic acid, particularly with a hydroxy function on the carbon adjacent to the carbonyl (e.g., mandelic and benzilic), were shown early to be among the most highly active acids to be employed. Table 12-9 depicts the relationship between substitution and spasmolytic potency of several compounds. It will be noted that, starting with a simple acetyl ester (which is spasmogenic), the activity increases with increasing aromatic substitution. An enhancing effect is apparent when hydroxylation of the acetyl carbon is employed, although the dangers of broad generalizations are noted in the decreased activity of I versus H. Likewise, two phenyl groups appear to be maximal, inasmuch as a sharp drop in activity is noted when three phenyls are employed. This is possibly caused by steric factors on binding or conformation of the molecule because of the triphenylacetyl moiety. Comparison of compounds E and H (see Table 12-9) would indicate also that enhancement of action results from bonding the phenyl groups together into the fluorene moiety, a compound which enjoyed some commercial success under the trade name of Pavatrine until it was withdrawn from the market some years ago because of more effective agents marketed by the same company.

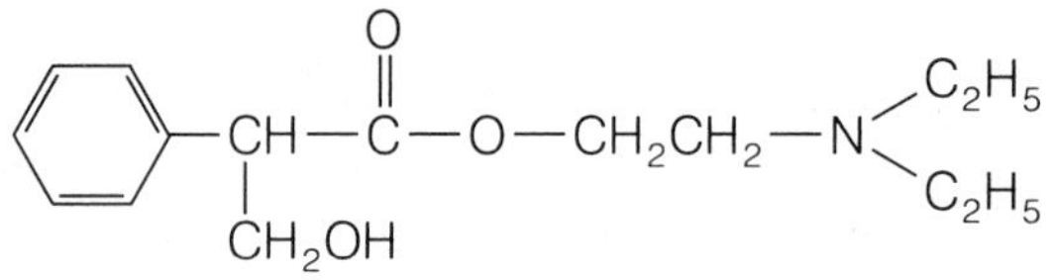

TABLE 12-9

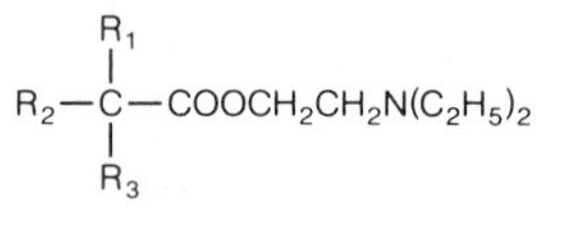

$R_2-C(R_1)(R_3)-COOCH_2CH_2N(C_2H_5)_2$

Compound	Structure: R_1	R_2	R_3	Spasmolytic Potency*: Acetylcholine $pA_2^{\dagger}$	Relative Potency (%)
A	H	H	H	Stimulates	
B	H	H	OH	4.0–4.3	1–2
C	Phenyl	H	H	5.0–5.3	10–20
D	Phenyl	H	OH	5.3–5.7	20–50
E‡	Phenyl	Phenyl	H	6.0	100
F§	Phenyl	Phenyl	OH	7.6	4000
G	Phenyl	Phenyl	Phenyl	5.0	10
H‖	Fluorene-9-carboxylic			6.8	600
I	Fluorene-9-hydroxy-9-carboxylic			6.7	500

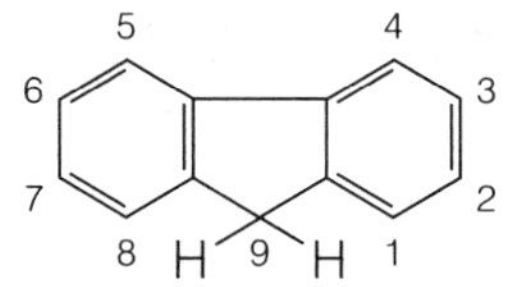

*All esters were tested as the hydrochlorides on rabbit small intestine (isolated segments).

†Logarithm of the reciprocal of the ED_{50}.

‡Trasentine.

§WIN 5606.

‖Pavatrine.

(Adapted from a table by Lands, A. M., et al.: J. Pharmacol. Exp. Ther. 100:19, 1950.)

$C_6H_5-CH(CH_2OH)-C(=O)-O-CH_2CH_2-N(C_2H_5)_2$

Amprotropine

Fluorene

It is evident that the acid portion (corresponding to tropic acid) should be somewhat bulky, especially when the aminoalcohol portion is simple. This is an indication for the need of at least one portion of the molecule to have the space-occupying, umbrellalike shape, which leads to firm binding at the receptor site area.

Although simplification of the aminoalcohol portion of the atropine prototype has been a guiding principle in most research, it is worth noting that many of the anticholinergics now used still include a cyclic aminoalcohol moiety. It should be noted that the aminoalcohol–ester anticholinergics are used primarily as antispasmodics or mydriatics, and that cholinolytic compounds classed as aminoalcohol or aminoalcohol ether analogues of atropine are, with few exceptions, employed as antiparkinsonian drugs.

Another important feature in many of the synthetic anticholinergics that are used as antispasmodics is that they contain a quaternary nitrogen, presumably to enhance activity. The initial synthetic quaternary compound methantheline bromide has served as a forerunner for many others. These compounds combine anticholinergic activity of the antimuscarinic type with some ganglionic blockade to reinforce the parasympathetic blockade. However, quaternization also introduces the possibility of blockade of voluntary synapses (curariform activity); this can become evident with sufficiently high doses.

The identification[78] of subclasses of muscarinic receptors revitalized research in selective antimuscarinic agents. Most earlier (classic) antimuscarinic drugs have not distinguished between M_1 and M_2 receptors. Atopine blocks both receptors. Recent studies, however, have uncovered new selective antagonists of M_1 and M_2 receptors. Pirenzepine has a much greater affinity for muscarinic receptors in autonomic ganglia than those in smooth muscle. It can inhibit gastric acid secretion without affecting gastrointestinal motility.[79,80] Secoverine, 1-cyclohexyl-4-[ethyl(*p*-methoxy-α-methylphenethyl)amino]-1-butanone, on the other hand, does not affect gastric or salivary secretions, but does inhibit gastrointestinal motility.[81,82] Additionally, 4-diphenylacetoxy-*N*-methylpiperidine methiodide (4-DAMP) has a greater affinity for ileal smooth-muscle muscarinic receptors than for muscarinic receptors on the heart or rat superior cervical ganglion.[81] Given these findings, the potential for developing selective antimuscarinic drugs is promising. These newer selective muscarinic antagonists are currently only investigational agents.

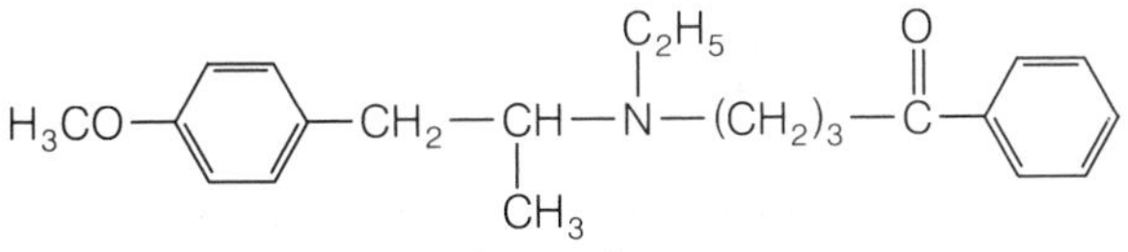

$H_3CO-C_6H_4-CH_2-CH(CH_3)-N(C_2H_5)-(CH_2)_3-C(=O)-C_6H_5$

Secoverine

$(C_6H_5)_2CH-C(=O)-O-C_5H_9N-CH_3$

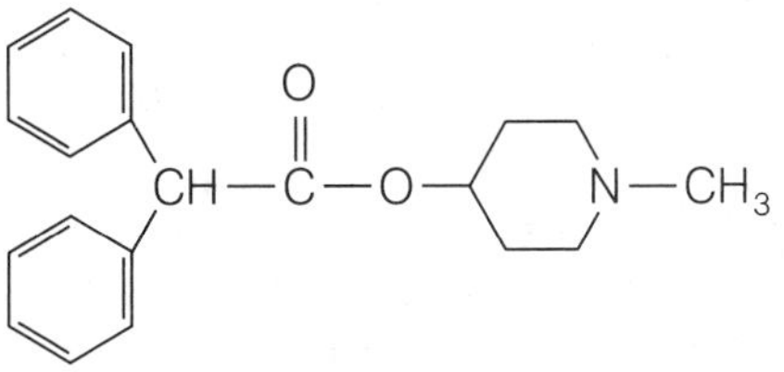

4-DAMP

Products

The antimuscarinic compounds now in use are described in the following monographs:

Clidinium Bromide, USP. 3-Hydroxy-1-methylquinuclidinium bromide benzilate (Quarzan). The preparation of this compound is described by Sternbach and Kaiser.[83,84] It occurs as a white or nearly white, almost odorless, crystalline powder that is optically inactive. It is soluble in water and in alcohol, but only very slightly soluble in ether and in benzene.

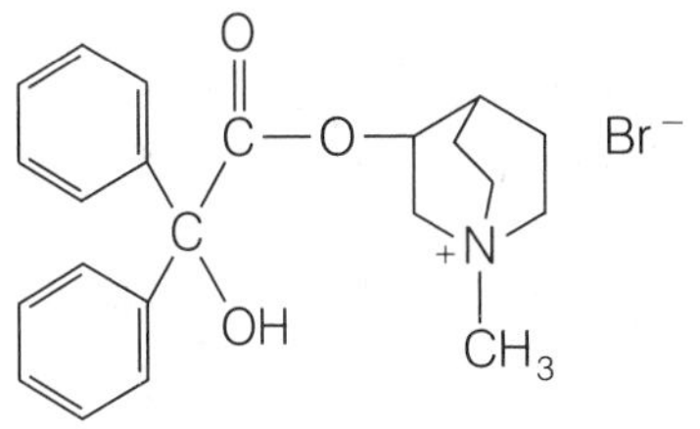

$(C_6H_5)_2C(OH)-C(=O)-O-$[quinuclidinium, N^+-CH_3] Br^-

Clidinium Bromide

This anticholinergic agent is marketed alone and in combination with the minor tranquilizer chlordiazepoxide (Librium), the resultant product being known as Librax. The rationale of the combination for the treatment of gastrointestinal complaints is the use of an anxiety-reducing agent together with an anticholinergic based on the recognized contribution of anxiety to the development of the diseased condition. It is suggested for peptic ulcer, hyperchlorhydria, ulcerative or spastic colon, anxiety states with gastrointestinal manifestations, nervous stomach, irritable or spastic colon, and others. The combination capsule contains 5 mg of chlordiazepoxide hydrochloride and 2.5 mg of clidinium bromide. Clidinium bromide is contraindicated in glaucoma and other conditions that may be aggravated by the parasympatholytic action, such as prostatic hypertrophy in the elderly man, which could lead to urinary retention. The usual recommended dose for adults is 2.5 mg or 5 mg four times a day before meals and at bedtime.

Cyclopentolate Hydrochloride, USP. 2-(Dimethylamino)ethyl 1-hydroxy-α-phenylcyclopentaneacetate hydrochloride (Cyclogyl). This chemical, together with a series of closely related compounds, was synthesized by Treves and Testa.[85] It is a crystalline, white, odorless solid that is very soluble in water, easily soluble in alcohol, and only slightly soluble in ether. A 1% solution has a *p*H of 5.0 to 5.4.

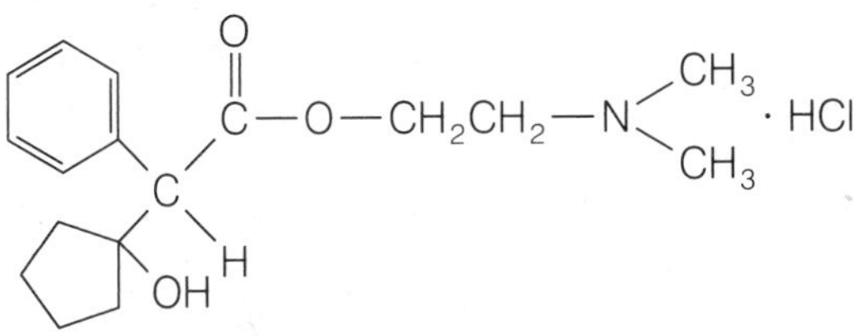

$C_6H_5-CH(C_5H_8OH)-C(=O)-O-CH_2CH_2-N(CH_3)_2 \cdot HCl$

Cyclopentolate Hydrochloride

It is used only for its effects on the eye, on which it acts as a parasympatholytic. It quickly produces cycloplegia and mydriasis when placed in the eye. Its primary field of usefulness is in refraction studies. However, cyclopentolate hydrochloride can be used as a mydriatic in the management of iritis, iridocyclitis, keratitis, and choroiditis. Although it does not seem to affect intraocular tension significantly, it is desirable to be very cautious with patients with high intraocular pressure and also with elderly patients with possible unrecognized glaucomatous changes.

Cyclopentolate hydrochloride has one-half the antispasmodic activity of atropine and has been shown to be nonirritating when instilled repeatedly into the eye. If not neutralized after the refraction studies, its effect dissipates within 24 hours. Neutralization with a few drops of pilocarpine nitrate solution, 1% to 2%, often results in complete recovery in six hours. It is supplied as a ready-made ophthalmic solution in concentrations of either 0.5% or 2%.

Dicyclomine Hydrochloride, USP. 2-(Diethylamino)ethyl bicyclohexyl-1-carboxylate hydrochloride (Bentyl). The synthesis of this drug is described by Tilford and his co-workers.[86] In common with similar salts, this drug is a white crystalline compound that is soluble in water.

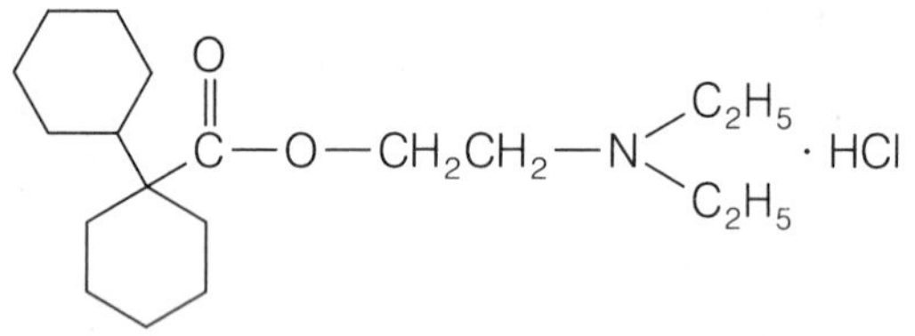

Dicyclomine Hydrochloride

Dicyclomine hydrochloride is reported to have one-eighth of the neurotropic activity of atropine and approximately twice the musculotropic activity of papaverine. Again, this preparation has minimized the undesirable side effects associated with the atropine-type compounds. It is used for its spasmolytic effect on various smooth-muscle spasms, particularly those associated with the gastrointestinal tract. It is also useful in dysmenorrhea, pylorospasm, and biliary dysfunction.

The drug, introduced in 1950, is marketed in the form of capsules, with or without 15 mg of phenobarbital, and also in the form of a syrup, with or without phenobarbital. For parenteral use (intramuscularly) it is supplied as a solution containing 10 mg/2 mL.

Eucatropine Hydrochloride, USP. Euphthalmine hydrochloride; 1,2,2,6-tetramethyl-4-piperidyl mandelate hydrochloride. This compound possesses the aminoalcohol moiety characteristic of one of the early local anesthetics (i.e., β-eucaine), but differs in the acidic portion of the ester by being a mandelate instead of a benzoate. The salt is an odorless, white, granular powder, providing solutions that are neutral to litmus. It is very soluble in water, freely soluble in alcohol and chloroform, but almost insoluble in ether.

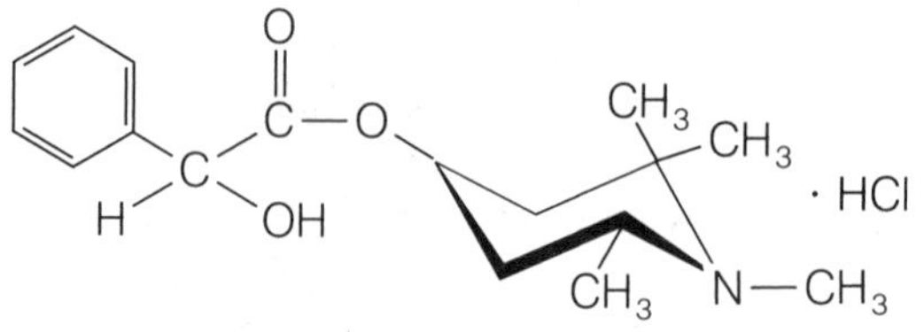

Eucatropine Hydrochloride

The action of eucatropine closely parallels that of atropine, although it is much less potent than the latter. It is used topically in a 0.1-mL dose as a mydriatic in 2% solution or in the form of small tablets. However, the use of concentrations of from 5% to 10% is not uncommon. Dilatation, with little impairment of accommodation, takes place in about 30 minutes, and the eye returns to normal in two to three hours.

Glycopyrrolate, USP. 3-Hydroxy-1,1-dimethylpyrrolidinium bromide α-cyclopentylmandelate (Robinul). The drug occurs as a white crystalline powder that is soluble in water or alcohol, but practically insoluble in chloroform or ether.

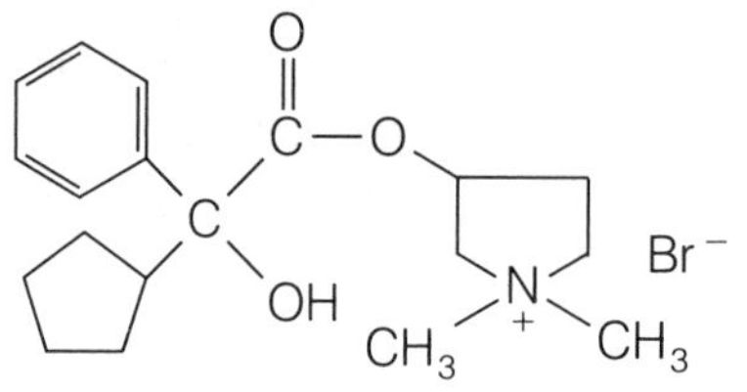

Glycopyrrolate

Glycopyrrolate is a typical anticholinergic and possesses, at adequate dosage levels, the atropinelike effects characteristic of this class of drugs. It has a spasmolytic effect on the musculature of the gastrointestinal tract as well as the genitourinary tract. It diminishes gastric and pancreatic secretions and diminishes the quantity of perspiration and saliva. Its side effects are typically atropinelike also (i.e., dryness of the mouth, urinary retention, blurred vision, constipation).[87] Because of its quaternary ammonium character, glycopyrrolate rarely causes central nervous system disturbances, although, in sufficiently high dosage, it can bring about ganglionic and myoneural junction block.

The drug is used as an adjunct in the management of peptic ulcer and other gastrointestinal ailments associated with hyperacidity, hypermotility, and spasm. In common with other anticholinergics its use does not preclude dietary restrictions or use of antacids and sedatives if these are indicated.

Mepenzolate Bromide. 3-Hydroxy-1,1-dimethylpiperidinium bromide benzilate (Cantil). This compound may be prepared by the method of Biel et al.[88] by the transesterification reaction with 1-methyl-3-hydroxypiperidine and methyl benzilate. The resulting base is quaternized with methyl bromide to give a white crystalline product that is water-soluble.

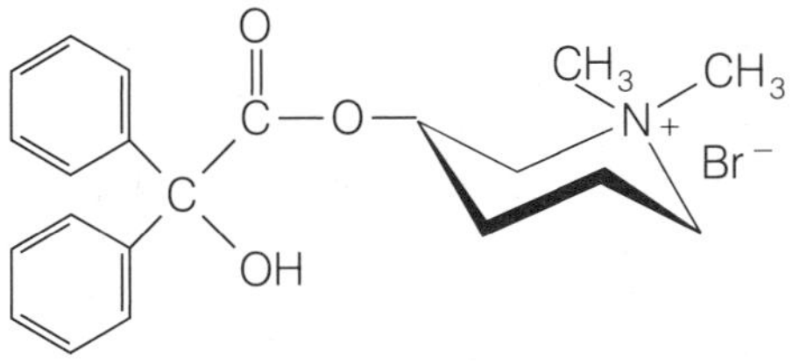

Mepenzolate Bromide

It has an activity of about one-half that of atropine in reducing ACh-induced spasms of the guinea pig ileum. The selective action on colonic hypermotility is said to relieve pain, cramps, and bloating and to help curb diarrhea.

Methantheline Bromide, USP. Diethyl(2-hydroxyethyl)methylammonium bromide xanthene-9-carboxylate (Banthine). Methantheline may be prepared according to the method outlined by Burtner and Cusic,[89] although this reference does not show the final formation of the quaternary salt. The compound from which the quaternary salt is prepared was in the series of esters from which aminocarbofluorene was selected as the best spasmolytic agent.

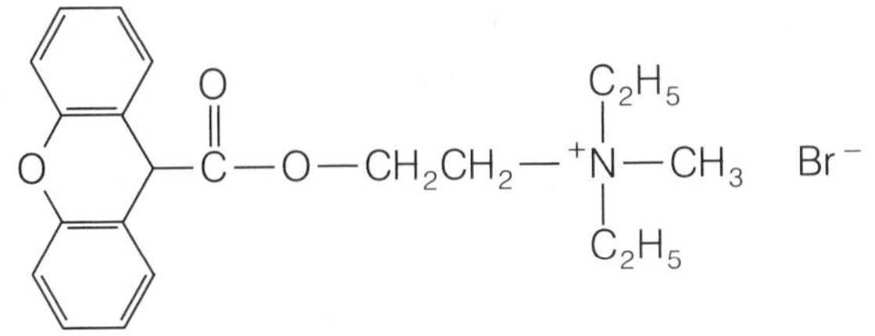

Methantheline Bromide

It is a white, slightly hygroscopic crystalline salt that is soluble in water to produce solutions with a pH of about 5. Aqueous solutions are not stable and hydrolyze in a few days. The bromide form is preferable to the very hygroscopic chloride.

This drug, introduced in 1950, is a potent anticholinergic agent and acts at the nicotinic cholinergic receptors of the sympathetic and the parasympathetic systems, as well as at the myoneural junction of the postganglionic cholinergic fibers. Similar to other quaternary ammonium drugs, methantheline bromide is incompletely absorbed from the gastrointestinal tract.

Among the conditions for which methantheline is indicated are gastritis, intestinal hypermotility, bladder irritability, cholinergic spasm, pancreatitis, hyperhidrosis, and peptic ulcer, all of which are manifestations of parasympathotonia.

Side reactions are atropinelike (mydriasis, cycloplegia, dryness of mouth). The drug is contraindicated in glaucoma. Toxic doses may bring about a curarelike action, a not too surprising fact when it is considered that ACh is the mediating factor for neural transmission at the somatic myoneural junction. This side effect can be counteracted with neostigmine methylsulfate.

Oxyphencyclimine Hydrochloride. (1,4,5,6-Tetrahydro-1-methyl-2-pyrimidinyl)methyl α-phenylcyclohexaneglycolate monohydrochloride (Daricon, Vistrax). The synthesis of this compound is described by Faust et al.[90] The product is a white crystalline compound that is sparingly soluble in water (1.2 g/100 mL at 25°C). It has a bitter taste.

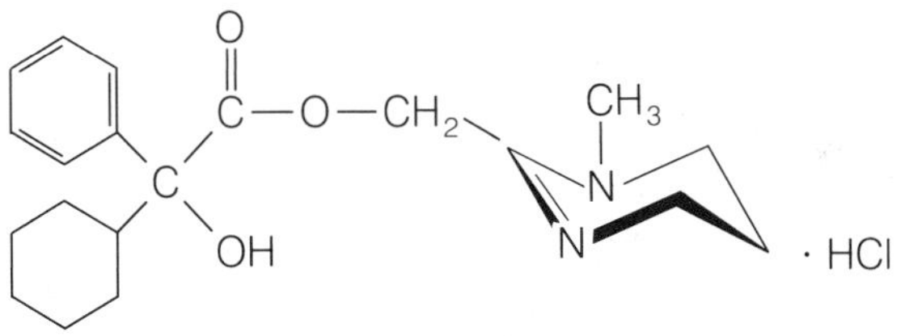

Oxyphencyclimine Hydrochloride

This compound, introduced in 1958, was promoted as a peripheral anticholinergic–antisecretory agent, with little or no curarelike activity and little or no ganglionic-blocking activity. That these activities are absent is probably because of the tertiary character of the molecule. This activity is in contrast with compounds that couple antimuscarinic action with ganglionic-blocking action. The tertiary character of the nitrogen promotes intestinal absorption of the molecule. Perhaps that most significant activity of this compound is its marked ability to reduce both the volume and the acid content of the gastric juices,[91] a desirable action in view of the more recent hypotheses pertaining to peptic ulcer therapy. Another important feature of this compound is its low toxicity in comparison with many of the other available anticholinergics. Oxyphencyclimine hydrochloride is hydrolyzed in the presence of excessive moisture and heat. It is absorbed from the gastrointestinal tract and has a duration of action of up to 12 hours.

Oxyphencyclimine is suggested for use in peptic ulcer, pylorospasm, and functional bowel syndrome. It is contraindicated, as are other anticholinergics, in patients with prostatic hypertrophy and glaucoma.

Propantheline Bromide, USP. (2-Hydroxyethyl)diisopropylmethylammonium bromide xanthene-9-carboxylate (Pro-Banthine). The method of preparation of this compound is exactly analogous to that used for methantheline bromide. It is a white, water-soluble crystalline substance, with properties quite similar to those of methantheline.

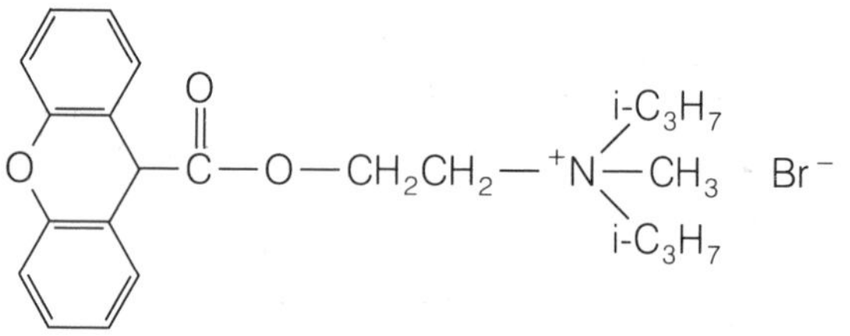

Propantheline Bromide

Its chief difference from methantheline is in its potency, which has been estimated variously as being from two to five times as great. This greater potency is reflected in its smaller dose. For example, instead of a 50-mg initial dose, a 7.5–15-mg initial dose is suggested for propantheline bromide. It is available in 7.5-mg and 15-mg sugar-coated tablets and in the form of a powder (30 mg) for preparing parenteral solutions.

AMINOALCOHOL ETHERS

The aminoalcohol ethers thus far introduced have been used as antiparkinsonian drugs rather than as conventional anticholinergics (i.e., as spasmolytics, mydriatics). In general, they may be considered as closely related to the antihistaminics and, indeed, do possess antihistaminic properties of a substantial order. In turn, the antihistamines possess anticholinergic activity and have been used as antiparkinsonian agents. Comparison of chlorphenoxamine and orphenadrine with the antihistaminic diphenhydramine illustrates the close similarity of structure. The use of diphenhydramine in parkinsonism has been cited earlier. Benztropine may also be considered as a structural relative of diphenhydramine, although the aminoalcohol portion is tropine and, therefore, more distantly related than chlorphenoxamine and orphenadrine. In the structure of benztropine, a three-carbon chain intervenes between the nitrogen and oxygen functions, whereas in the others, a two-carbon chain is evident. However, the rigid ring structure possibly orients the nitrogen and oxygen functions into more nearly the two-carbon chain interprosthetic distance than is apparent at first glance. This, combined with the flexibility of the alicyclic chain, would help to minimize the distance discrepancy.

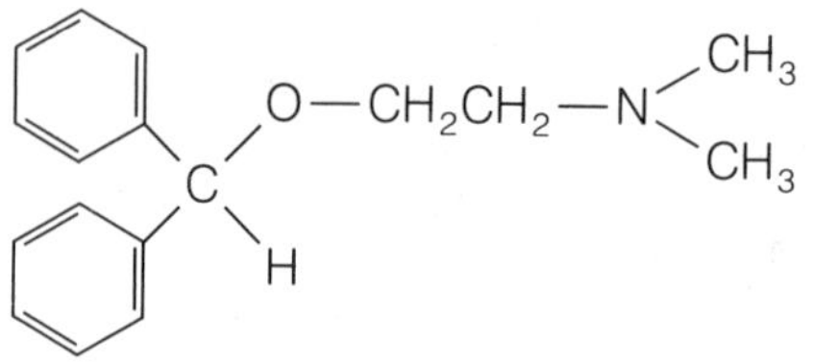

Diphenhydramine

Products

Benztropine Mesylate, USP. 3α-(Diphenylmethoxy)-1αH,5αH-tropane methanesulfonate (Cogentin). The compound occurs as a white, colorless, slightly hygroscopic crystalline powder. It is very soluble in water, freely soluble in alcohol, and very slightly soluble in ether. The pH of aqueous solutions is about 6. It is prepared according to the method of Phillips[92] by interaction of diphenyldiazomethane and tropine.

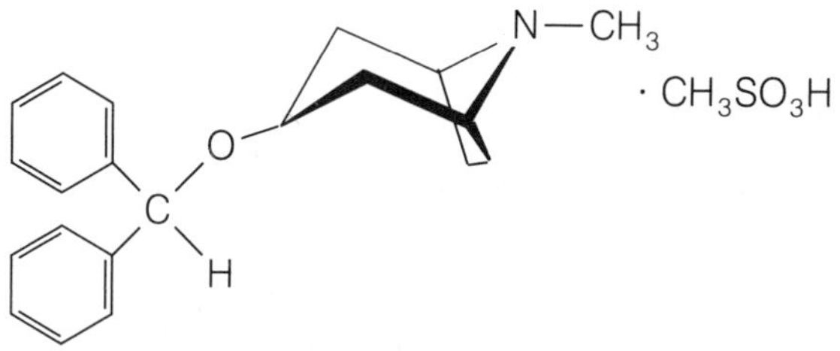

Benztropine Mesylate

Benztropine mesylate has anticholinergic, antihistaminic, and local anesthetic properties. Its anticholinergic effect makes it applicable in its use as an antiparkinsonian agent. It is about as potent as atropine as an anticholinergic and shares some of the side effects of this drug, such as mydriasis and dryness of mouth. Importantly, however, is that it does not produce central stimulation, but instead exerts the characteristic sedative effect of the antihistamines.

The tremor and rigidity characteristic of parkinsonism are relieved by benztropine mesylate, and it is of particular value for those patients who cannot tolerate central excitation (e.g., aged patients). It also may have a useful effect in minimizing drooling, sialorrhea, masklike facies, oculogyric crises, and muscular cramps.

The usual caution that is exercised with any anticholinergic in glaucoma and prostatic hypertrophy is observed with this drug.

Chlorphenoxamine Hydrochloride. 2-[(p-Chloro-α-methyl-α-phenylbenzyl)oxy]-N,N-dimethylethylamine hydrochloride (Phenoxene). It occurs in the form of colorless needles that are soluble in water. Aqueous solutions are stable.

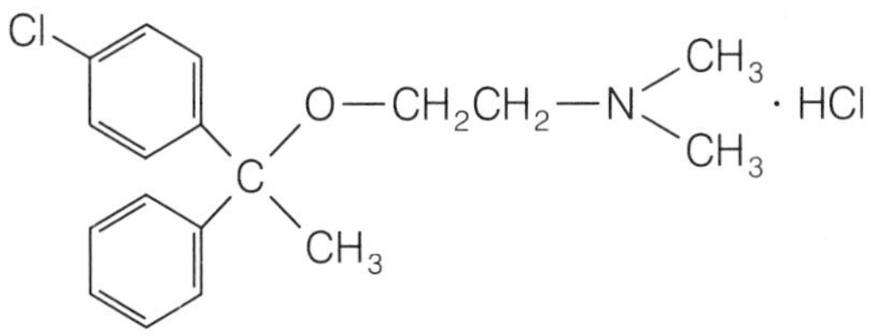

Chlorphenoxamine Hydrochloride

This drug was originally introduced in Germany as an antihistaminic. However, it is reported that this close relative of diphenhydramine (Benadryl) has its antihistaminic potency lowered by the *para*-Cl and the α-methyl group present in the molecule. At the same time, the anticholinergic action is increased. The drug has an oral LD_{50} of 410 mg/kg in mice, indicating a substantial margin of safety. It was introduced to medicine in the United States in 1959.

It is indicated for the symptomatic treatment of all types of Parkinson's disease and is said to be especially useful when rigidity and impairment of muscle contraction are evident. It is not as useful against tremor, and combined therapy with other agents may be necessary.

Orphenadrine Citrate, USP. *N,N*-Dimethyl-2-[(*o*-methyl-α-phenylbenzyl)oxyl]ethylamine citrate (1 : 1) (Norflex). This compound is synthesized according to the method in the patent literature.[93] It occurs as a white, bitter crystalline powder. It is sparingly soluble in water, slightly soluble in alcohol, and insoluble in chloroform, in benzene, and in ether. The hydrochloride salt is marketed as Disipal.

Although this compound, introduced in 1957, is closely related to diphenhydramine structurally, it has a much lower antihistaminic activity and a much higher anticholinergic action. Likewise, it lacks the sedative effects characteristic of diphenhydramine. Pharmacologic testing indicates that it is not primarily a peripherally acting anticholinergic because it has only weak effects on smooth muscle, on the eye, and on secretory glands. However, it does reduce voluntary muscle spasm by a central inhibitory action on cerebral motor areas, a central effect similar to that of atropine.

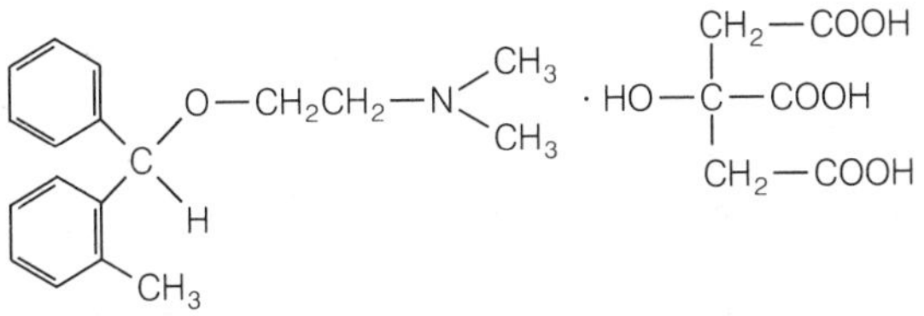

Orphenadrine Citrate

The drug is used for the symptomatic treatment of Parkinson's disease. It relieves rigidity better than it does tremor, and in certain cases it may accentuate the latter. The drug combats mental sluggishness, akinesia, adynamia, and lack of mobility, but this effect seems to be diminished rather rapidly on prolonged use. It is best used as an adjunct to the other agents, such as benztropine, procyclidine, cycrimine, and trihexyphenidyl in the treatment of paralysis agitans.

The drug has a low incidence of the common side effects, which are the usual ones for this group, namely, dryness of mouth, nausea, and mild excitation.

AMINOALCOHOLS

The development of aminoalcohols as parasympatholytics took place in the 1940s. It was soon established, however, that these antispasmodics were equally efficacious in parkinsonism.

Several of the drugs in this class of antimuscarinic agents have the structural characteristic of possessing bulky groups in the vicinity of hydroxyl and cyclic amino functional groups. These compounds are similar to the classic aminoester anticholinergic compounds derived from atropine. The presence of the alcohol group seems to adequately substitute as a prosthetic group for the carboxyl function in creating an effective parasympathetic blocking agent. It serves to emphasize that the aminoester group, per se, is not a necessary adjunct to cholinolytic activity, provided that other polar groupings, such as the hydroxyl, can substitute as a prosthetic group for the carboxyl function. Another structural feature common to all aminoalcohol anticholinergics is the γ-aminopropanol arrangement with three carbons intervening between the hydroxyl and amino functions. All of the aminoalcohols used for paralysis agitans are tertiary amines. Because the desired locus of action is central, quaternization of the nitrogen destroys the antiparkinsonian properties. However, quaternization of these aminoalcohols has been utilized to enhance the anticholinergic activity to produce an antispasmodic and antisecretory compound such as tridihexethyl chloride.

Products

Biperiden, USP. α-5-Norbornen-2-yl-α-phenyl-1-piperidinepropanol (Akineton). The drug consists of a white, practically odorless crystalline powder. It is practically insoluble in water and only sparingly soluble in alcohol, although it is freely soluble in chloroform. Its preparation is described by Haas and Klavehn.[94]

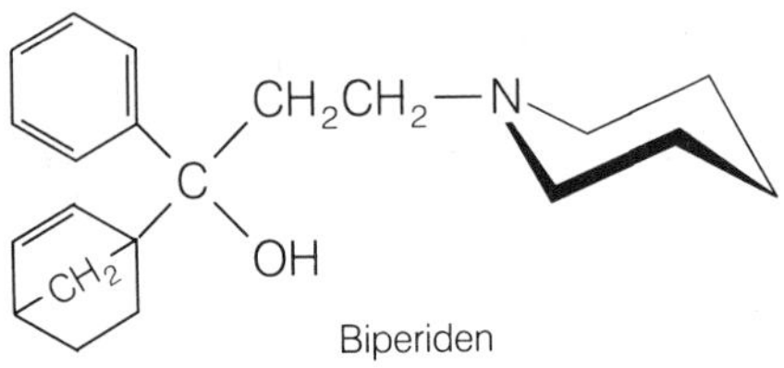

Biperiden

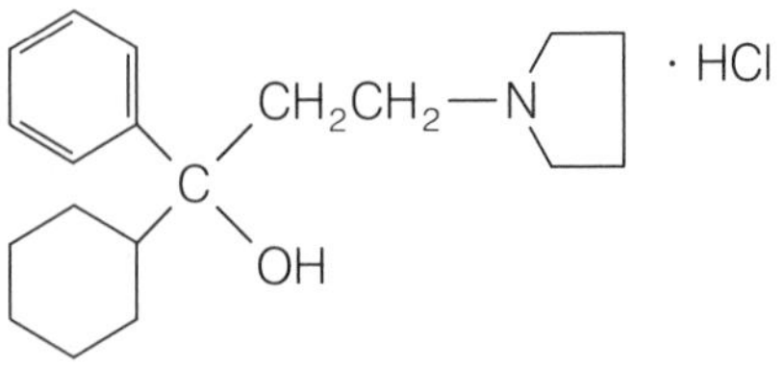

Procyclidine Hydrochloride

Biperiden, introduced in 1959, has a relatively weak visceral anticholinergic, but a strong nicotinolytic, action in terms of its ability to block nicotine-induced convulsions. Therefore, its neurotropic action is rather low on intestinal musculature and blood vessels. It has a relatively strong musculotropic action which is about equal to papaverine, in comparison with most synthetic anticholinergic drugs. Its action on the eye, although mydriatic, is much less than that of atropine. These weak anticholinergic effects serve to add to its usefulness in Parkinson's syndrome by minimizing side effects.

The drug is used in all types of Parkinson's disease (postencephalitic, idiopathic, arteriosclerotic) and helps to eliminate akinesia, rigidity, and tremor. It is also used in drug-induced extrapyramidal disorders by eliminating symptoms and permitting continued use of tranquilizers. Biperiden is also of value in spastic disorders not related to parkinsonism, such as multiple sclerosis, spinal cord injury, and cerebral palsy. It is contraindicated in all forms of epilepsy.

It is usually taken orally in tablet form, but the free base form is official to serve as a source for the preparation of Biperiden Lactate Injection, USP, a sterile solution of biperiden lactate in water for injection prepared from biperiden base with the aid of lactic acid. It usually contains 5 mg/ mL.

Biperiden Hydrochloride, USP. α-5-Norbornen-2-yl-α-phenyl-1-piperidinepropanol hydrochloride (Akineton Hydrochloride) is a white optically inactive, crystalline, odorless powder that is slightly soluble in water, ether, alcohol, and chloroform, and sparingly soluble in methanol.

Biperiden hydrochloride has all of the actions described for biperiden. The hydrochloride is used for tablets, because it is better suited to this dosage form than is the lactate salt. As with the free base and the lactate salt, xerostomia (dryness of the mouth) and blurred vision may occur.

Procyclidine Hydrochloride, USP. α-Cyclohexyl-α-phenyl-1-pyrrolidinepropanol hydrochloride (Kemadrin). This compound is prepared by the method of Adamson[95] or Bottorff[96] as described in the patent literature. It occurs as white crystals that are moderately soluble in water (3 : 100). It is more soluble in alcohol or chloroform and is almost insoluble in ether.

Although procyclidine, introduced in 1956, is an effective peripheral anticholinergic and, indeed, has been used for peripheral effects similarly to its methochloride (i.e., tricyclamol chloride), its clinical usefulness lies in its ability to relieve spasticity of voluntary muscle by its central action. Therefore, it has been employed with success in the treatment of Parkinson's syndrome.[97] It is said to be as effective as cycrimine and trihexyphenidyl and is used for reduction of muscle rigidity in the postencephalitic, the arteriosclerotic, and the idiopathic types of the disease. Its effect on tremor is not predictable and probably should be supplemented by combination with other similar drugs.

The toxicity of the drug is low, but side effects are noticeable when the dosage is high. At therapeutic dosage levels, dry mouth is the most common side effect. The same care should be exercised with this drug as with all other anticholinergics when administered to patients with glaucoma, tachycardia, or prostatic hypertrophy.

Tridihexethyl Chloride, USP. (3-Cyclohexyl-3-hydroxy-3-phenylpropyl)triethylammonium chloride (Pathilon). The preparation of this compound as the corresponding bromide is described by Denton and Lawson.[98] It occurs in the form of a white, bitter crystalline powder possessing a characteristic odor. The compound is freely soluble in water and alcohol, the aqueous solutions being nearly neutral in reaction.

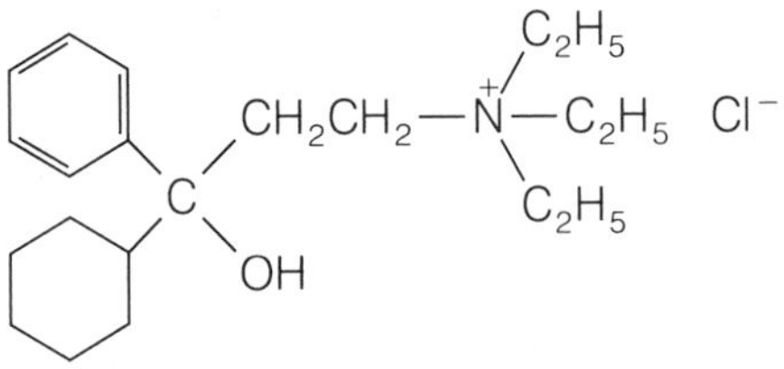

Tridihexethyl Chloride

Although this drug, introduced in 1958, has ganglion-blocking activity, it is said that its peripheral atropinelike activity predominates; therefore, its therapeutic application has been based on the latter activity. It possesses the antispasmodic and the antisecretory activities characteristic of this group but, because of its quaternary character, it is valueless in relieving the Parkinson syndrome.

The drug is useful for adjunctive therapy in a wide variety of gastrointestinal diseases, such as peptic ulcer, gastric hyperacidity, and hypermotility; and spastic conditions, such as spastic colon, functional diarrhea, pylorospasm, and other related conditions. Because its action is predominately antisecretory it is most effective in gastric hypersecretion rather than in hypermotility and spasm. It is best administered intravenously for the latter conditions.

The side effects usually found with effective anticholinergic therapy occur with the use of this drug. These are dryness of mouth, mydriasis, and such. As with other anticholinergics, care should be exercised when administering the drug in glaucomatous conditions, cardiac decompension, and coronary insufficiency. It is contraindicated in patients with obstruction at the bladder neck, prostatic hypertrophy, stenosing gastric and duodenal ulcers, or pyloric or duodenal obstruction.

The drug may be administered orally or parenterally. Oral therapy is preferable. The drug is supplied in 25-mg tablets and as powder for injection (10 mg/1 mL).

Trihexyphenidyl Hydrochloride, USP. α-Cyclohexyl-α-phenyl-1-piperidinepropanol hydrochloride (Artane; Tremin; Pipanol). This compound was synthesized by Denton and his co-workers.[99] It occurs as a white, odorless crystalline compound that is not very soluble in water (1:100). It is more soluble in alcohol (6:100) and chloroform (5:100), but only slightly soluble in ether and benzene. The pH of a 1% aqueous solution is about 5.5 to 6.0.

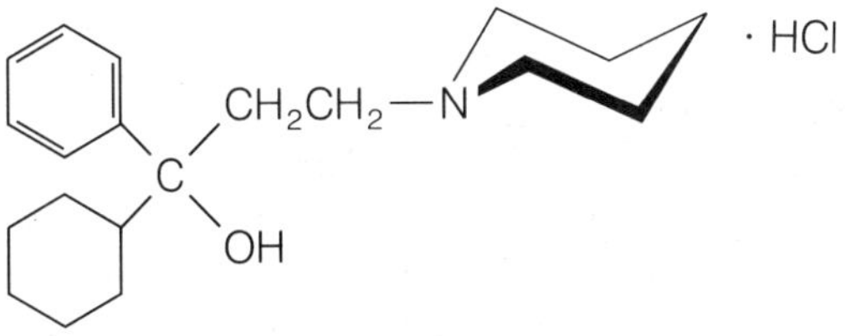

Trihexyphenidyl Hydrochloride

Introduced in 1949, it is approximately one-half as active as atropine as an antispasmodic, but is claimed to have milder side effects, such as mydriasis, drying of secretions, and cardioacceleration. It has a good margin of safety, although it is about as toxic as atropine. It has found a place in the treatment of parkinsonism and is claimed also to provide some measure of relief from the mental depression often associated with this condition. However, it does exhibit some of the side effects typical of the parasympatholytic-type preparation, although it is said that these often may be eliminated by adjusting the dose carefully.

AMINOAMIDES

From a structural standpoint, the aminoamide type of anticholinergic represents the same type of molecule as the aminoalcohol group, with the important exception that the polar amide group replaces the corresponding polar hydroxyl group. Aminoamides retain the same bulky structural features as are found at one end of the molecule or the other in all of the active anticholinergics. Isopropamide is the only drug of this class currently in use.

Another amide-type structure is that of tropicamide, formerly known as bistropamide, a compound having some of the atropine features.

Isopropamide Iodide, USP. (3-Carbamoyl-3,3-diphenylpropyl)diisopropylmethylammonium iodide (Darbid). This compound may be made according to the method of Janssen and his co-workers.[100] It occurs as a bitter, white to pale yellow crystalline powder and is only sparingly soluble in water, but is freely soluble in chloroform and alcohol.

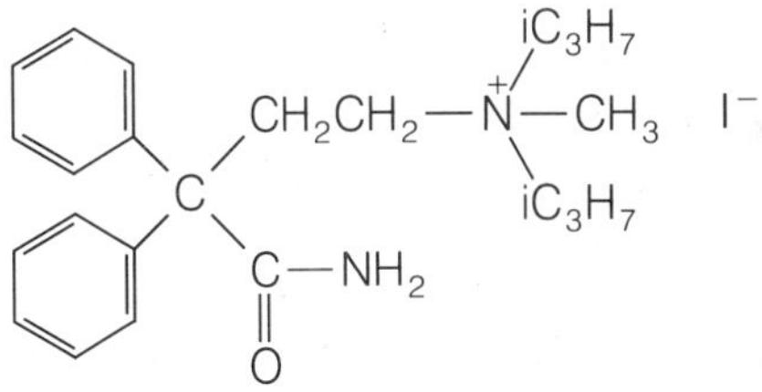

Isopropamide Iodide

This drug, introduced in 1957, is a potent anticholinergic producing atropinelike effects peripherally. Even with its quaternary nature, it does not cause sympathetic blockade at the ganglionic level except in high-level dosage. Its principal distinguishing feature is its long duration of action. It is said that a single dose can provide antispasmodic and antisecretory effects for as long as 12 hours.

It is used as adjunctive therapy in the treatment of peptic ulcer and other conditions of the gastrointestinal tract associated with hypermotility and hyperacidity. It has the usual side effects of anticholinergics (dryness of mouth, mydriasis, difficult urination) and is contraindicated in glaucoma, prostatic hypertrophy, etc.

Tropicamide, USP. N-Ethyl-2-phenyl-N-(4-pyridylmethyl)hydracrylamide (Mydriacyl). The preparation of this compound is described in the patent literature.[101] It occurs as a white or practically white crystalline powder that is almost odorless. It is only slightly soluble in water, but is freely soluble in chloroform and in solutions of strong acids. The pH of ophthalmic solutions ranges between 4.0 and 5.0, the acidity being achieved with nitric acid.

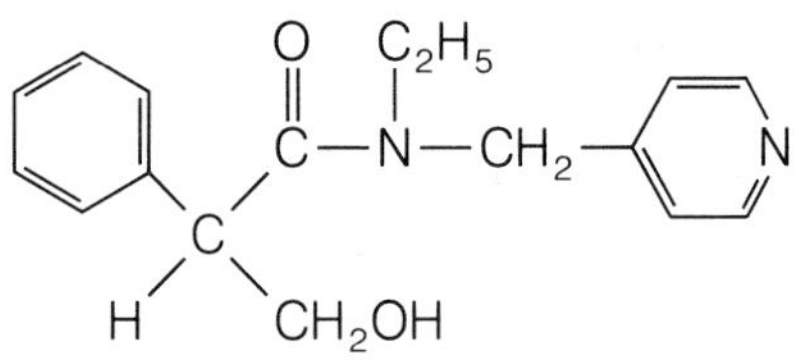

Tropicamide

Diphemanil Methylsulfate

This drug is an effective anticholinergic for ophthalmic use where mydriasis is produced by relaxation of the sphincter muscle of the iris, allowing the adrenergic innervation of the radial muscle to dilate the pupil. Its maximum effect is achieved in about 20 to 25 minutes and lasts for about 20 minutes, with complete recovery being noted in about six hours. Its action is more rapid in onset and wears off more rapidly than that of most other mydriatics. To achieve mydriasis either the 0.5% or 1.0% concentration may be used, although cycloplegia is achieved only with the stronger solution. Its uses are much the same as those described in general for mydriatics earlier, but opinions differ on whether or not the drug is as effective as homatropine, for example, in achieving cycloplegia. For mydriatic use, however, in examination of the fundus and treatment of acute iritis, iridocyclitis, and keratitis it is quite adequate, and because of its shorter duration of action, it is less prone to initiate a rise in intraocular pressure than the more potent longer-lasting drugs. However, as with other mydriatics, pupil dilatation can lead to increased intraocular pressure. In common with other mydriatics it is contraindicated in cases of glaucoma, either known or suspected, and should not be used in the presence of a shallow anterior chamber. Thus far, allergic reactions or ocular damage have not been observed with this drug.

MISCELLANEOUS

Further structural modification of classic antimuscarinic agents can be found in the drugs described in the following. Each of them has the typical bulky group that is characteristic of the usual anticholinergic molecule. One modification is represented by the diphenylmethylene moiety (e.g., diphemanil), the second, by a phenothiazine (e.g., ethopropazine), and the third, by a thioxanthene structure (e.g., methixene).

Diphemanil Methylsulfate, USP. 4-(Diphenylmethylene)-1,1-dimethylpiperidinium methylsulfate (Prantal). This compound may be prepared by two alternative syntheses as outlined by Sperber and coworkers.[102] It was introduced in 1951.

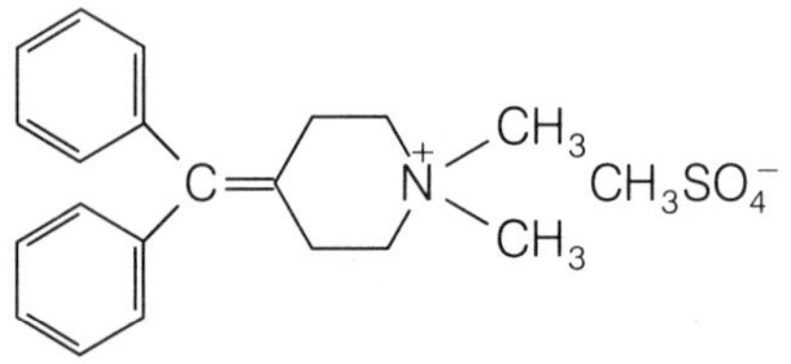

The drug is a white, crystalline, odorless compound that is sparingly soluble in water (50 mg/mL), alcohol and chloroform. The pH of a 1% aqueous solution is between 4.0 and 6.0.

The methylsulfate salt was chosen as the best because the chloride is hygroscopic and the bromide and iodide ions have exhibited toxic manifestations in clinical use.

Diphemanil methylsulfate is a potent cholinergic-blocking agent. In the usual dosage range it acts as an effective parasympatholytic by blocking nerve impulses at the parasympathetic ganglia, but it does not invoke a sympathetic ganglionic blockade. It is claimed to be highly specific in its action upon those innervations that activate gastric secretion and gastrointestinal motility. Although this drug is capable of producing atropinelike side effects, these rarely occur at recommended doses. The highly specific nature of its action on gastric functions makes the drug useful in the treatment of peptic ulcer, and its lack of atropinelike effects makes this use much less distressing than other antispasmodic drugs. In addition to its action in decreasing gastric hypermotility, diphemanil is valuable in hyperhidrosis in low doses (50 mg twice daily) or topically. The drug is not well absorbed from the gastrointestinal tract, particularly in the presence of food, and should be administered between meals.

Ethopropazine Hydrochloride, USP. 10-[2-(Diethylamino)propyl]phenothiazine monohydrochloride (Parsidol). The compound is prepared in several ways, among which is the patented method of Berg and Ashley.[103] It occurs as a white crystalline compound that has poor solubility in water at 20°C (1 : 400), but greatly increased solubility at 40°C (1 : 20). It is soluble in ethanol and chloroform, but almost insoluble in ether, benzene, and acetone. The pH of an aqueous solution is about 5.8.

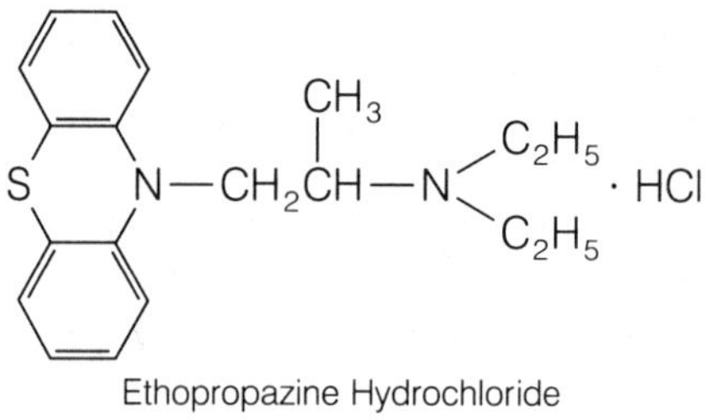

Ethopropazine Hydrochloride

TABLE 12-10

SYNTHETIC CHOLINERGIC BLOCKING AGENTS

Name Proprietary Name	***Preparations***	***Category***	***Application***	***Usual Adult Dose****	***Usual Dose Range****	***Usual Pediatric Dose****
Clidinium bromide USP	Clidinium bromide capsules USP	Anticholinergic		2.5–5.0 mg tid or qid	10–20 mg/day	
Cyclopentolate hydrochloride USP *Cyclogyl*	Cyclopentolate hydrochloride ophthalmic solution USP	Anticholinergic (ophthalmic)	Topically, to the conjunctiva	1 drop of a 1 or 2% solution repeated once in 5 min		
Dicyclomine hydrochloride USP *Bentyl*	Dicyclomine hydrochloride capsules USP	Anticholinergic		10–20 mg tid or qid	Up to 120 mg/day	
	Dicyclomine hydrochloride injection USP			IM, 20 mg every 4–6 hr		
	Dicyclomine hydrochloride syrup USP			10–20 mg tid or qid	Up to 120 mg/day	Infants: 5 mg tid or qid; children: 10 mg tid or qid
	Dicyclomine hydrochloride tablets USP			10–20 mg tid or qid	Up to 120 mg/day	
Eucatropine hydrochloride USP	Eucatropine hydrochloride ophthalmic solution USP	Anticholinergic (ophthalmic)				
Glycopyrrolate USP *Robinul*	Glycopyrrolate injection USP Glycopyrrolate tablets USP	Anticholinergic		Oral, 1 mg tid; IM, IV, or SC, 100–200 μg at 4-hr intervals tid or qid	1–2 mg	
Mepenzolate bromide *Cantil*	Mepenzolate bromide solution Mepenzolate bromide tablets	Anticholinergic		25 mg qid	25–50 mg	
Methantheline bromide USP *Banthine*	Sterile methantheline bromide USP Methantheline bromide tablets USP	Anticholinergic		Oral, 50 mg qid; IM or IV, 50 mg qid	50–100 mg	
Oxyphencyclimine hydrochloride *Daricon, Enarax, Vistrax*	Oxyphencyclimine hydrochloride tablets	Anticholinergic		10 mg bid	10–50 mg/day	
Propantheline bromide USP *Pro-Banthine Bromide*	Propantheline bromide tablets USP	Anticholinergic		15 mg tid and 30 mg hs	Up to 120 mg/day	375 μg/kg of body weight or 10 mg/m^2 of body surface, qid
	Propantheline bromide extended-release tablets USP			30 mg every 12 hr	Up to 120 mg/day	
Benztropine mesylate USP *Congentin Methanesulfate*	Benztropine mesylate injection USP	Antiparkinsonian		IM or IV, 1 or 2 mg qd or bid	Up to 6 mg/day	
	Benztropine mesylate tablets USP			1 or 2 mg qd or bid	Up to 6 mg/day	
Chlorphenoxamine hydrochloride *Phenoxene*	Chlorphenoxamine hydrochloride tablets	Skeletal muscle relaxant		50–100 mg tid or qid	150–400 mg/day	
Orphenadrine citrate USP *Norflex*	Orphenadrine citrate injection USP	Skeletal muscle relaxant (antihistaminic)		IM or IV, 60 mg every 12 hr as needed		
	Orphenadrine citrate tablets USP			Oral, 100 mg bid		

*See USP DI for complete dosage information.

TABLE 12-10 *Continued*

SYNTHETIC CHOLINERGIC BLOCKING AGENTS

Name Proprietary Name	***Preparations***	***Category***	***Application***	***Usual Adult Dose****	***Usual Dose Range****	***Usual Pediatric Dose****
Biperiden USP *Akineton*	Biperiden lactate injection USP	Anticholinergic		IM or IV, 2 mg of biperiden as the lactate which may be repeated every 1/2 hr until relief is obtained, but no more than 4 consecutive doses should be given in a 24-hr period		
Biperiden hydrochloride USP *Akineton Hydrochloride*	Biperiden hydrochloride tablets USP	Anticholinergic		Oral, 2 mg tid or qid		
Procyclidine hydrochloride USP *Kemadrin*	Procyclidine hydrochloride tablets USP	Skeletal muscle relaxant		Oral, 2 or 2.5 mg tid, the dosage being adjusted as needed and tolerated or until the total dose reaches 20–30 mg divided into 3 or 4 doses		
Tridihexethyl chloride USP *Pathilon*	Tridihexethyl chloride injection USP Tridihexethyl chloride tablets USP	Anticholinergic		Oral, 25 mg tid and 50 mg hs; parenteral, 10–20 mg every 6 hr	25–75 mg qd to qid	
Trihexyphenidyl hydrochloride USP *Artane, Pipanol, Tremin*	Trihexyphenidyl hydrochloride, elixir USP Trihexyphenidyl hydrochloride tablets USP	Antiparkinsonian		Initial, 1–2 mg the first day, with increases of 2 mg/day every 3–5 days until optimal effects are obtained; or 10–15 mg usually divided into 3 or 4 doses		
Isopropamide iodide USP *Darbid*	Isopropamide iodide tablets USP	Anticholinergic		5 mg bid	10–20 mg/day	
Tropicamide USP *Mydriacyl*	Tropicamide ophthalmic solution USP	Anticholinergic (ophthalmic)	Topically, to the conjunctiva, 1 drop of a 1% solution, repeated in 5 minutes			Topically, to the conjunctiva, 1 drop of a 0.5 or 1% solution, repeated once in 5 min
Ethopropazine hydrochloride USP *Parsidol*	Ethopropazine hydrochloride tablets USP	Antiparkinsonian		Initial, 50 mg qd or bid, the dose being gradually increased as necessary; maintenance, 100–150 mg qd to qid	50–600 mg daily	
Diphemanil methylsulfate USP *Prantal Methylsulfate*	Diphemanil methylsulfate tablets USP	Anticholinergic		100 mg every 4–6 hr	50–200 mg	

*See USP DI for complete dosage information.

This phenothiazine was introduced to therapy in 1954. It has antimuscarinic activity and is especially useful in the symptomatic treatment of parkinsonism. In this capacity, it has value in controlling rigidity, and it also has a favorable effect on tremor, sialorrhea, and oculogyric crises. It is often used in conjunction with other antiparkinsonian drugs for complementary activity.

Side effects are common with this drug, but not usually severe. Drowsiness and dizziness are the most common side effects at ordinary dosage levels, and as the dose increases, xerostomia, mydriasis, and others become evident. It is contraindicated in conditions such as glaucoma because of its mydriatic effect.

PAPAVERINE AND RELATED COMPOUNDS

Papaverine exerts an antispasmodic effect on smooth muscle, and consequently, it is customarily considered together with the solanaceous alkaloids. Papaverine does not interfere with the induction of the stimulus but rather with the response in the effector system. Because of its nonspecific action (i.e., to the ACh receptor) it is often called a nonspecific antagonist. This is sometimes referred to as a musculotropic type of spasmolysis, in contrast with the so-called neurotropic action of atropine and its congeners.

Papaverine interferes with the mechanism of muscle contraction by inhibiting the enzyme phosphodiesterase in smooth-muscle cells (Fig. 12-16). Cyclic AMP (cAMP) is formed by the action of the enzyme adenylate cyclase on the cellular nucleotide adenosine triphosphate (ATP). In turn the cAMP formed by this action is broken down by the cellular enzyme cyclic nucleotide phosphodiesterase (PDE). Papaverine is a potent inhibitor of vascular smooth-muscle PDE, and there is a significant elevation of cAMP following the administration of papaverine. The inhibition of PDE and elevation of cAMP were shown to precede and be associated with smooth-muscle relaxation. The inhibition of PDE and the subsequent increase of cAMP do not inactivate the contractile elements of the muscle because it is still possible to obtain a response after papaverine administration under certain conditions.

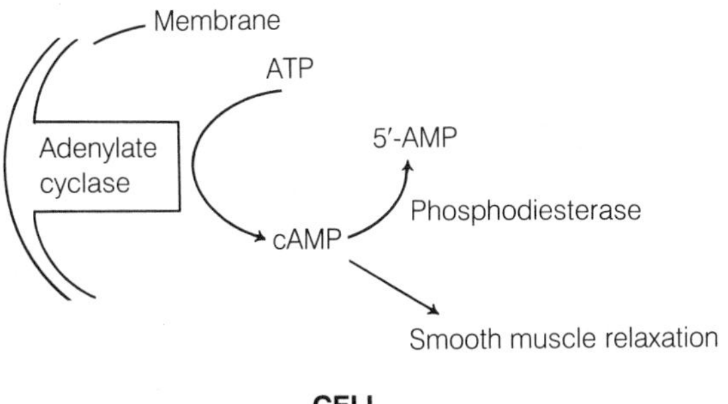

FIG. 12-16. Mechanism of antispasmodic activity of papaverine.

Regardless of the type of smooth muscle, papaverine acts as a spasmolytic, although its effectiveness is greater in some muscles than in others. It relaxes the smooth musculature of the larger blood vessels, especially coronary, systemic peripheral, and pulmonary arteries. Perhaps also, by its vasodilating action on cerebral blood vessels, papaverine increases cerebral blood flow and decreases cerebral vascular resistance. At the same time oxygen consumption is unaltered. These effects perhaps explain the benefits reported from the drug in cerebral vascular encephalopathy. Papaverine is devoid of the atropine-like effects on the central nervous system. The absence of such effects is a desirable characteristic of papaverine-type compounds, but, unfortunately, these compounds do not compare in potency to the atropine congeners.

Papaverine (see formula) is the principal naturally occurring member of this group that is of any therapeutic consequence as an antispasmodic.

Papaverine Hydrochloride, USP. 6,7-Dimethoxy-1-veratrylisoquinoline hydrochloride. This alkaloid was isolated first by Merck (1848) from opium, in which it occurs to the extent of about 1%.

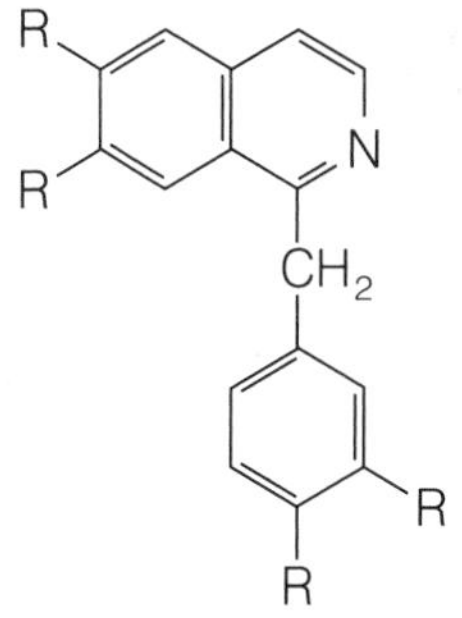

Papaverine
($R = OCH_3$)

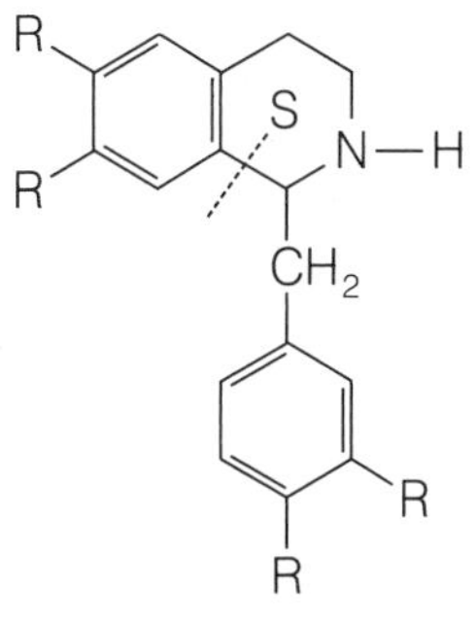

Tetrahydropapaverine
($R = OCH_3$)

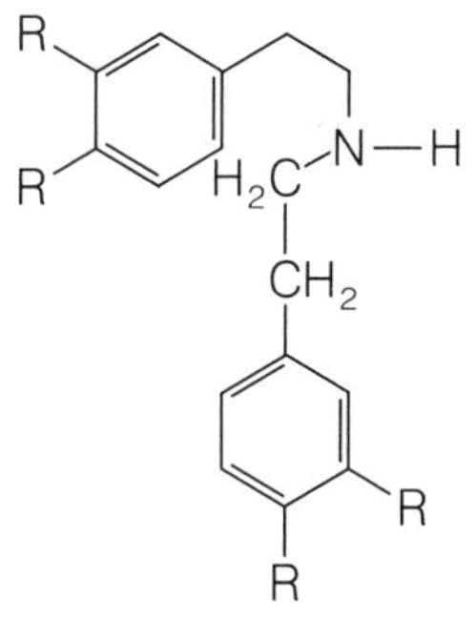

Bis-β-phenylethylamine
($R = H$)

Its structure was elucidated by the classic researches of Goldschmiedt, and its synthesis was effected first by Pictet and Gam in 1909.

Previous to World War II, papaverine had been obtained in sufficient quantities from natural sources. However, as a result of the war, the United States found itself, early in 1942, without a source of opium and, therefore, of papaverine. Consequently, the commercial synthesis of papaverine took on a new significance, and methods soon were developed to synthesize the alkaloid on a large scale.[104]

Papaverine itself occurs as an optically inactive white crystalline powder. It possesses one basic nitrogen and forms salts quite readily. The most important salt is the hydrochloride, which is official. The hydrochloride occurs as white crystals or as a crystalline white powder. It is odorless and slightly bitter. The compound is soluble in water (1 : 30), alcohol (1 : 120), or chloroform. It is not soluble in ether. Aqueous solutions are acid to litmus and may be sterilized by autoclaving. Unless properly handled and stored, extemporaneous solutions of papaverine salts deteriorate rapidly.

Because of the antispasmodic action of papaverine on blood vessels, it has become extremely valuable for relieving the arterial spasm associated with acute vascular occlusion. It is useful in the treatment of peripheral, coronary, and pulmonary arterial occlusions. Administration of an antispasmodic is predicated on the concept that the lodgment of an embolus causes an intense reflex vasospasm. This vasospasm affects not only the artery involved, but also the surrounding blood vessels. Relief of this neighboring vasospasm is imperative to prevent damage to these vessels and to limit the area of ischemia. Thus, it appears to increase collateral circulation in the affected area, rather than to act on the occluded vessel.

Other than its antispasmodic action on the vascular system, it is used for bronchial spasm and visceral spasm. In the latter type of spasm, it is not advisable to administer morphine simultaneously because it opposes the relaxing action of papaverine.

Because papaverine is a musculotropic drug, it has provided the starting point for synthetic analogues (Table 12-11) in which it has been hoped that a neurotropic activity could be combined with its musculotropic action. This combination of activities would be desirable, if possible, without the introduction of any atropinelike side effects. Comparing the

TABLE 12-11

PAPAVERINE ANALOGUES

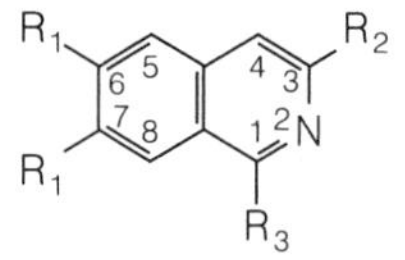

(Isoquinoline skeleton: R_1 at positions 6 and 7, R_2 at position 3, R_3 at position 1, N at position 2)

Compound	Structure R_1	R_2	R_3	Name
I	$-OC_2H_5$	$-H$	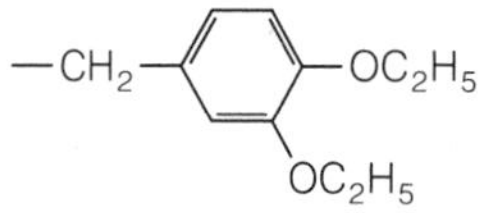 $-CH_2-C_6H_3(OC_2H_5)_2$ (3,4-diethoxybenzyl)	Ethaverine
II	$-OCH_3$	$-CH_3$	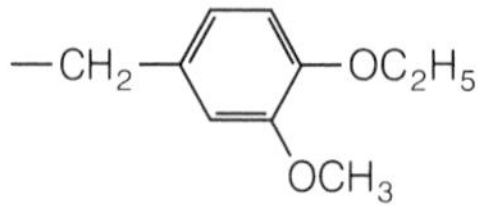 $-CH_2-C_6H_3(OC_2H_5)(OCH_3)$	Dioxyline
III	$-H$	$-CH_3$	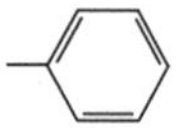 $-C_6H_5$	1-Phenyl-3-methylisoquinoline
IV	$-H$	$-CH_3$	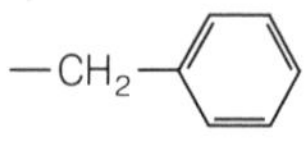 $-CH_2-C_6H_5$	1-Benzyl-3-methylisoquinoline
V	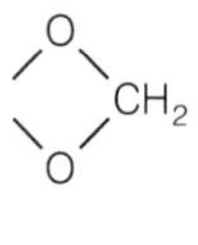 $-O-CH_2-O-$	$-CH_3$	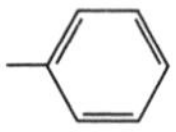 $-C_6H_5$	1-Phenyl-3-methyl-6,7-methylenedioxy-3,4-dihydroisoquinoline

results of this research with those of atropine analogues, it appears that the use of the latter has been more successful.

The four methoxyl groups in papaverine are easily altered functional groups. These have been changed to produce ethaverine, which has three times the activity of papaverine, but causes serious liver toxicity, and dioxyline, which has a combination of one ethoxyl group with three methoxyl groups. It is not entirely certain that the alkoxyl groups are necessary for activity, although they seem to be present in most of the accepted compounds. Activity is known to reside in both 1-phenyl-3-methyl-isoquinoline (III) and 1-benzyl-3-methyl-isoquinoline (IV) (see Table 12-11), but, on the other hand, spasmogenic properties are found in 1-phenyl-3-methyl-6,7-methylenedioxy-3,4-dihydroisoquinoline (V).

Ethaverine Hydrochloride. 1-[(3,4-Diethoxyphenyl)methyl]-6,7-diethoxyisoquinoline hydrochloride; 6,7-diethoxy-1-(3,4-diethoxybenzyl)isoquinoline hydrochloride (Isovex; Neopavrin). This well-known derivative of papaverine is synthesized in exactly the same way as papaverine, but intermediates that bear ethoxyl groups instead of methoxyl groups are utilized.[105]

The hydrochloride is soluble to the extent of 1 g/40 mL of water at room temperature. The aqueous solutions are acidic, with a 1% solution having a pH of 3.6 and 0.1% solution having a pH of 4.6.

The pharmacologic action of ethaverine is quite similar to that of papaverine, although the duration of its effect is said to be longer. It is used in peripheral and cerebral vascular insufficiency associated with arterial spasm in doses of 100–200 mg.

Dioxyline Phosphate. Dimoxyline phosphate; 1-(4-ethoxy-3-methoxybenzyl)-6,7-dimethoxy-3-methylisoquinoline phosphate (Paveril Phosphate). This may be prepared according to the usual Bischler-Napieralski isoquinoline synthesis followed by dehydrogenation.[106]

This compound is related quite closely to papaverine and gives the same type of antispasmodic action as papaverine, with less toxicity. By virtue of the lesser toxicity, it can be given in larger doses than papaverine if desired, although usually the same dosage regimen can be followed as with the natural alkaloid.

The drug is useful for mitigating the reflex vasospasm that already has been described for papaverine during peripheral, pulmonary, or coronary occlusion. The indications are the same as for papaverine.

GANGLIONIC-BLOCKING AGENTS

Autonomic ganglia have been the object of interest for many years for the study of the interactions occurring between drugs and nervous tissues. The first important account[107] was given by Langley and described the stimulating and blocking actions of nicotine on sympathetic ganglia. It was found that small amounts of nicotine stimulated ganglia and then produced a blockade of ganglionic transmission because of persistent depolarization. From these experiments Langley was able to outline the general pattern of innervation of organs by the autonomic nervous system. *Parasympathetic* ganglia are usually located near the organ they innervate and have preganglionic fibers that stem from the cervical and thoracic regions of the spinal cord. *Sympathetic* ganglia consist of 22 pairs that lie on either side of the vertebral column to form lateral chains. These ganglia are connected both to each other by nerve trunks and also to the lumbar or sacral regions of the spinal cord. The transmission of impulses in autonomic ganglia may be described as similar to the neurohumoral processes that occur at almost all nerve endings. The released ACh is normally not taken up by the nerve endings but hydrolyzed by AChE, and about half the choline formed is absorbed immedi-

TABLE 12-12

PAPAVERINE AND RELATED COMPOUNDS

Name Proprietary Name	***Preparations***	***Category***	***Usual Adult Dose****	***Usual Dose Range****
Papaverine hydrochloride USP *Cerespan, Pavabid, Vasospan*	Papaverine hydrochloride injection USP Papaverine hydrochloride tablets USP	Smooth-muscle relaxant	Oral, 150 mg; IM, 30 mg; IV, 120 mg	Oral, 100–300 mg; IM, 30–60 mg
Ethaverine hydrochloride *Ethaquin, Laverin, Isovex*	Ethaverine hydrochloride tablets Ethaverine hydrochloride injection Ethaverine hydrochloride elixir	Smooth-muscle relaxant	100 mg	100–200 mg tid
Dioxyline phosphate *Paveril Phosphate*	Dioxyline phosphate tablets	Smooth-muscle relaxant		100–400 mg tid or qid

*See USP DI for complete dosage information.

Nicotine

ately into the nerve by an active process that is blocked by hemicholinium-3.

By using the sympathetic cervical ganglion as a model, it has been found that transmission in the autonomic ganglion is more complex than formerly believed. Traditionally, stimulation of autonomic ganglia by ACh has been considered as the nicotinic action of the neurotransmitter. It is now understood that stimulation by ACh produces a triphasic response in sympathetic ganglia. Impulse transmission through the ganglion occurs when ACh is released from preganglionic fibers and activates the nicotinic receptors of the neuronal membrane. This triggers an increase in sodium and potassium conductances of a subsynaptic membrane resulting in an initial excitatory postsynaptic potential (EPSP) with a latency of one millisecond, followed by an inhibitory postsynaptic potential (IPSP) with a latency of 35 milliseconds and, finally, a slowly generating EPSP with a latency of several hundred milliseconds. The ACh released by preganglionic fibers also activates muscarinic receptors of the ganglion and probably of the small-intensity fluorescent (SIF) cell. This results in the appearance of a slow IPSP and slow EPSP in the neurons of the ganglion.[108] The initial EPSP is blocked by conventional competitive nondepolarizing ganglionic-blocking agents, such as hexamethonium, and is considered the primary pathway for ganglionic transmission.[109] The slowly generating or late EPSP is blocked by atropine but not by the traditional ganglionic-blocking agents. This receptor has muscarinic properties, because methacholine causes generation of the late EPSP without causing the initial spike characteristic of ACh. Atropine also blocks the late EPSP produced by methacholine. More recently, it has been observed that there may be more than one type of muscarinic receptor in sympathetic ganglia. Atropine blocks both high-affinity (M_1) and low-affinity (M_2) muscarinic receptors in the ganglion.[110] In addition to the cholinergic pathways, the cervical sympathetic ganglion was found to have a neuron that contains a catecholamine.[111] These neuronal cells, identified initially by fluorescence histochemical studies and shown to be smaller than the postganglionic neurons, are now referred to as small-intensity fluorescent cells or SIF cells. Dopamine has been identified as the fluorescent catecholamine in the SIF cells that are common to many other sympathetic ganglia. Dopamine apparently mediates an increase in cAMP, which causes hyperpolarization of postganglionic neurons (Fig. 12-17). The IPSP phase of the transmission of sympathetic ganglia following ACh administration can be blocked by both atropine and α-adrenergic-blocking agents.[112]

If a similar nontraditional type of ganglionic transmission occurs in the parasympathetic ganglia, it has not yet become evident.

With the anatomic and physiologic differences between sympathetic and parasympathetic ganglia, it should be no surprise that ganglionic agents may show some selectivity between the two types of ganglia. Although we do not have drug classifications such as "parasympathetic ganglionic blockers" and "sympathetic ganglionic blockers," we do find that

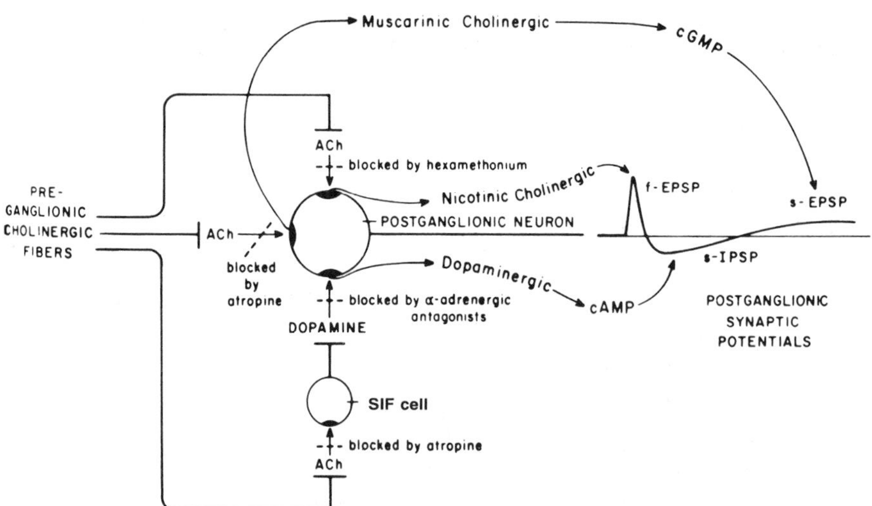

FIG. 12-17. Synaptic transmission in the mammalian superior cervical ganglion and the proposed role of the different receptors to the triphasic pattern of stimulation and selective blockade. (From Greengard, P., and Kebabian, J. W.: Fed. Proc. 33:1059, 1974.)

TABLE 12-13

RESULTS OF GANGLIONIC BLOCKERS ON ORGANS

Organ	Predominant System	Results of Ganglionic Blockade
Cardiovascular system		
Heart	Parasympathetic	Tachycardia
Arterioles	Sympathetic	Vasodilatation
Veins	Sympathetic	Dilatation
Eye		
Iris	Parasympathetic	Mydriasis
Ciliary muscle	Parasympathetic	Cycloplegia
GI tract	Parasympathetic	Relaxation
Urinary bladder	Parasympathetic	Urinary retention
Salivary glands	Parasympathetic	Dry mouth
Sweat glands	Sympathetic*	Anhidrosis

*Neurotransmitter is ACh.

(Adapted from Goth, A.: Medical Pharmacology, 9th ed. St. Louis, C. V. Mosby Co., 1978.)

certain ganglia have a predominant effect over certain organs and tissues and that a nondiscriminant blockade of autonomic ganglia results in a change in the effect of the autonomic nervous system on that organ (Table 12-13). Nevertheless there are drugs that have some selective action. Garrett[113] has shown that tetraethylammonium salts are nondiscriminating, whereas hexamethonium shows some selective-blocking action. None of the commonly known ganglionic blockers has yet been identified as having a selective blockade of parasympathetic ganglia.

Van Rossum[114, 115] has reviewed the mechanisms of ganglionic synaptic transmission, the mode of action of ganglionic stimulants, and the mode of action of ganglionic-blocking agents. He has conveniently classified the blocking agents in the following manner.

DEPOLARIZING GANGLIONIC-BLOCKING AGENTS

These blocking agents are actually ganglionic stimulants. Thus, for nicotine, it is well known that small doses give an action similar to that of the natural neuroeffector, ACh, an action known as the "nicotinic effect of ACh." However, larger amounts of nicotine bring about a ganglionic block, characterized initially by depolarization followed by a typical competitive antagonism. To conduct nerve impulses the cell must be able to carry out a polarization and depolarization process, and if the depolarized condition is maintained without repolarization, it is obvious that no conduction occurs. Acetylcholine itself, in high concentrations, will bring about an autoinhibition. Chemicals that cause this type of ganglionic block are not of therapeutic significance. The classes of ganglionic-blocking agents that will be described are therapeutically useful.

NONDEPOLARIZING COMPETITIVE GANGLIONIC-BLOCKING AGENTS

Compounds in this class possess the necessary affinity to attach to the nicotinic receptor sites that are specific for ACh, but lack the intrinsic activity necessary for impulse transmission (i.e., they are unable to effect depolarization of the cell). Under experimental conditions, in the presence of a fixed concentration of blocking agent of this type, a large enough concentration of ACh can offset the blocking action by competing successfully for the specific receptors. When such a concentration of ACh is administered to a ganglion preparation, it appears that the intrinsic activity of the ACh is as great as it was when no antagonist was present, the only difference being in the larger concentration of ACh required. It is evident, then, that such blocking agents are "competitive" with ACh for the specific receptors involved, and either the agonist or the antagonist can displace the other if present in sufficient concentration. Drugs falling into this class are tetraethylammonium salts, hexamethonium, and trimethaphan. Mecamylamine possesses a competitive component in its action but is also noncompetitive—a so-called dual antagonist.

NONDEPOLARIZING NONCOMPETITIVE GANGLIONIC-BLOCKING AGENTS

These blocking agents produce their effect, not at the specific ACh receptor site, but at some point further along the chain of events that is necessary for transmission of the nerve impulse. When the block has been imposed, increase of the concentration of ACh has no effect, and, thus, apparently ACh is not acting competitively with the blocking agent at the same receptors. Theoretically, a pure noncompetitive blocker should have a high specific affinity for the noncompetitive receptors in the ganglia, and it should have very low affinity for other cholinergic synapses, together with no intrinsic activity. Mecamylamine, as mentioned before, has a noncompetitive component, but is also a competitive blocking agent.

The first ganglionic-blocking agents employed in therapy were tetraethylammonium chloride and bromide (I; Table 12-14). Although one might assume that curariform activity would be a deterrent to their use, it has been shown that the curariform

TABLE 12-14
STRUCTURES OF GANGLIONIC-BLOCKING AGENTS

Compound	Structure	Name
I	$(C_2H_5)_4\overset{+}{N}\ X^-$	Tetraethylammonium chloride (X = Cl) Tetraethylammonium bromide (X = Br)
II	$(CH_3)_3\overset{+}{N}-(CH_2)_6-\overset{+}{N}(CH_3)_2\ \ 2X^-$	Hexamethonium chloride (X = Cl) Hexamethonium bromide (X = Br)
III	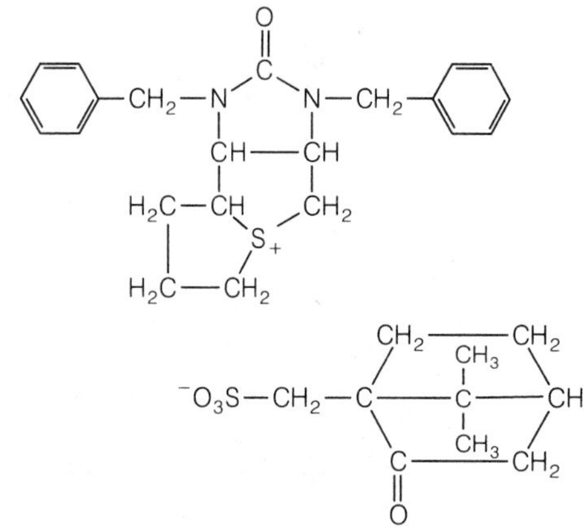 O ‖ C $C_6H_5-CH_2-N$ $N-CH_2-C_6H_5$ CH—CH H_2C-CH CH_2 S^+ H_2C-CH_2 CH_2 — CH_2 CH_3 $^-O_3S-CH_2-C$ — C — CH CH_3 C — CH_2 ‖ O	Trimethaphan camphorsulfonate
IV	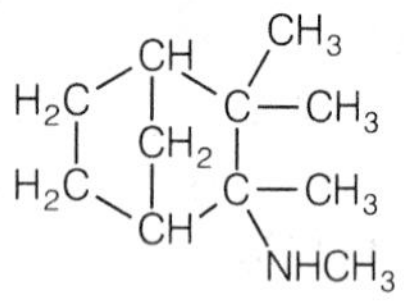 CH_3 CH H_2C $C-CH_3$ CH_2 H_2C $C-CH_3$ CH $NHCH_3$	Mecamylamine hydrochloride

activity of the tetraethyl compound is less than 1% of that of the corresponding tetramethylammonium compound. A few years after the introduction of the tetraethylammonium compounds, Paton and Zaimis[116] investigated the usefulness of the bistrimethylammonium polymethylene salts:

$\overset{+}{N}(CH_3)_3$ | $(CH_2)_n\ \ 2Br^-$ | $\underset{+}{N}(CH_3)_3$	n = 5 or 6, active as ganglionic blockers (feeble curariform activity) n = 9 to 12, weak ganglionic blockers (strong curariform activity)

As shown, their findings indicate that there is a critical distance of about five to six carbon atoms between the onium centers for good ganglionic-blocking action. Interestingly enough, the pentamethylene and the hexamethylene compounds are effective antidotes for counteracting the curare effect of the decamethylene compound. Hexamethonium (II; see Table 12-14), as the bromide and the chloride, emerged from this research as clinically useful products.

Trimethaphan camphorsulfonate (III; see Table 12-14), a monosulfonium compound, bears some degree of similarity to the quaternary ammonium types because it, too, is a completely ionic compound. Although it produces a prompt ganglion-blocking action on parenteral injection, its action is short, and it is used only for controlled hypotension during surgery. Almost simultaneously with the introduction of chlorisondamine (now long removed from the market), announcement was made of the powerful ganglionic-blocking action of mecamylamine (IV; see Table 12-14), a secondary amine *without* quaternary ammonium character. As expected, the latter compound showed uniform and predictable absorption from the gastrointestinal tract as well as a longer duration of action. The action was similar to that of hexamethonium.

Drugs of this class have a limited usefulness as diagnostic and therapeutic agents in the management of peripheral vascular diseases (e.g., thromboangiitis obliterans, Raynaud's disease, diabetic gangrene). However, the principal therapeutic application has been in the treatment of hypertension through blockade of the sympathetic pathways. Unfortunately, the action is nonspecific, and the parasympathetic ganglia, unavoidably, are blocked simultaneously to a greater or lesser extent, causing visual disturbances, dryness of the mouth, impotence, urinary retention, and constipation. Constipa-

TABLE 12-15

GANGLIONIC-BLOCKING AGENTS

Name Proprietary Name	*Preparations*	*Category*	*Usual Adult Dose**	*Usual Dose Range**
Trimethaphan camsylate USP *Arfonad*	Trimethaphan camsylate injection USP	Antihypertensive	IV infusion, 500 mg in 500 mL of 5% dextrose injection at a rate adjusted to maintain blood pressure at the desired level	200 μg–5 mg/min
Mecamylamine hydrochloride *Inversine*	Mecamylamine hydrochloride tablets	Antihypertensive	Initial, 2.5 mg bid, increased by 2.5-mg increments at intervals of not less than 2 days as required; maintenance, 7.5 mg tid	2.5–60 mg/day

*See USP DI for complete dosage information.

tion, in particular, probably caused by unabsorbed drug in the intestine (poor absorption), has been a drawback because the condition can proceed to a paralytic ileus if extreme care is not exercised. For this reason, cathartics or a parasympathomimetic (e.g., pilocarpine nitrate) are frequently administered simultaneously. Another serious side effect is the production of orthostatic (postural) hypotension (i.e., dizziness when the patient stands up in an erect position). Prolonged administration of the ganglionic-blocking agents results in their diminished effectiveness because of a buildup of tolerance, although some are more prone to this than others. Because of the many serious side effects, this group of drugs has been largely abandoned by researchers seeking effective hypotensive agents.

In addition to these side effects, there are several contraindications to the use of these drugs. For instance, they are all contraindicated in disorders characterized by severe reduction of blood flow to a vital organ (e.g., severe coronary insufficiency, recent myocardial infarction, retinal and cerebral thrombosis) as well as situations in which there have been large reductions in blood volume. In the latter, the contraindication exists because the drugs block the normal vasoconstrictor compensatory mechanisms necessary for homeostasis. A potentially serious complication, especially in older male patients with prostatic hypertrophy, is urinary retention. These drugs should be used with care or not at all in the presence of renal insufficiency, glaucoma, uremia, and organic pyloric stenosis.

Trimethaphan Camsylate, USP. (+)-1,3-Dibenzyldecahydro-2-oxoimidazo[4,5-*c*]thieno[1,2-*α*]-thiolium 2-oxo-10-bornanesulfonate (1 : 1) (Arfonad). The drug consists of white crystals or is a crystalline powder with a bitter taste and a slight odor. It is soluble in water and alcohol, but only slightly soluble in acetone and ether. The *p*H of a 1% aqueous solution is 5.0 to 6.0.

This ganglionic-blocking agent is short-acting and used only for certain neurosurgical procedures for which excessive bleeding obscures the operative field. Certain craniotomies are included among these operations. The action of the drug is a direct vasodilatation, and because of its transient action, it is subject to minute-by-minute control. On the other hand, this type of fleeting action makes it useless for hypertensive control. In addition, it is ineffective when given orally, and the usual route of administration is intravenous.

Mecamylamine Hydrochloride. *N*,2,3,3-Tetramethyl-2-norbornanamine hydrochloride (Inversine). The drug occurs as a bittersweet, white, odorless crystalline powder. It is freely soluble in water and chloroform, soluble in isopropyl alcohol, slightly soluble in benzene, and practically insoluble in ether. The *p*H of a 1% aqueous solution ranges from 6.0 to 7.5, and the solutions are stable to autoclaving.

This secondary amine has a powerful ganglionic-blocking effect that is almost identical with that of hexamethonium. It has an advantage over most of the ganglionic-blocking agents in that it is readily and smoothly absorbed from the gastrointestinal tract. It is rarely used, however, for the treatment of moderate to severe hypertension because of severe orthostatic hypotension that occurs when sympathetic ganglia are blocked by the drug.

NEUROMUSCULAR-BLOCKING AGENTS

Agents that block the transmission of ACh at the motor end-plate are called *neuromuscular blocking agents*. The therapeutic use of these compounds is primarily as adjuvants in surgical anesthesia to obtain relaxation of skeletal muscle. These drugs are also used in various orthopedic procedures such as alignment of fractures and correction of dislocations.

The therapeutically useful compounds in this group are sometimes referred to as possessing "curariform" or "curarimimetic" activity in reference to the original representatives of the class, which were obtained from curare. Since then, synthetic compounds have been prepared with a similar activity. Although all of the compounds falling into this category, natural and synthetic alike, bring about substantially the same end result (i.e., voluntary-muscle relaxation), there are some significant differences in the mechanisms whereby this is brought about.

The possible existence of a junction between muscle and nerve was raised as early as 1856 when Claude Bernard observed that the site of action of curare was neither the nerve nor the muscle. Since that time, it has been agreed that ACh mediates transmission at the neuromuscular junction by a sequence of events that has been described earlier in this chapter. The neuromuscular junction consists of the axon impinging onto a specialized area of the muscle known as the muscle end-plate. The axon is covered with a myelin sheath, containing the nodes of Ranvier, but is bare at the ending. The nerve terminal is separated from the end-plate by a gap of 200 Å. The subsynaptic membrane of the end-plate contains the cholinergic receptor, the ion-conducting channels (which are opened under the influence of ACh), and AChE.

One of the anatomic differences between the neuromuscular junction and other ACh-responsive sites is the absence in the former of a membrane barrier or sheath that envelopes the ganglia or constitutes the blood–brain barrier. This is of importance in the accessibility of the site of action to drugs, and particularly so for quaternary ammonium compounds, because they pass through living membranes with considerably greater difficulty and selectivity than do compounds that can exist in a nonionized species. The essentially bare nature (i.e., lack of lipophilic barriers) of the myoneural junction permits ready access by quaternary ammonium compounds. In addition, compounds with considerable molecular dimensions are also accessible to the receptors in the myoneural junction. As a result of this property, variations in the chemical structure of quaternaries have little influence on the potential ability of the molecule to reach the cholinergic receptor in the neuromuscular junction. Thus, the following types of neuromuscular junction blockers have been noted.

Nondepolarizing Blocking Agents. Traditionally, *nondepolarizing blocking agents* is a term applied to categorize drugs that compete with ACh for the recognition site on the nicotinic receptor by preventing the depolarization of the end-plate by the neurotransmitter. Thus, by decreasing the effective ACh–receptor combinations, the end-plate potential becomes too small to initiate the propagated action potential. This results in paralysis of neuromuscular transmission. The action of these drugs is quite analogous to that of atropine at the muscarinic receptor sites of ACh. Many experiments suggest that the agonist (ACh) and the antagonist compete on a one-to-one basis for the end-plate receptors. Drugs falling into this classification are tubocurarine, dimethyltubocurarine, pancuronium, and gallamine.

Depolarizing Blocking Agents. Drugs in this category are known to bring about a depolarization of the membrane of the muscle end-plate. This depolarizing is quite similar to that produced by ACh itself at ganglia and neuromuscular junctions (i.e., its so-called nicotinic effect), with the result that the drug, if in sufficient concentration, eventually will produce a block. It has been known for years that either smooth or voluntary muscle, when challenged repeatedly with a depolarizing agent, will eventually become insensitive. This phenomenon is known as *tachyphylaxis* or *desensitization* and is convincingly demonstrated under suitable experimental conditions with repeated applications of ACh itself, the results indicating that within a few minutes the end-plate becomes insensitive to ACh. The previous statements may imply that a blocking action of this type is quite clear-cut, but under experimental conditions it is not quite so clear and unambiguous because a block that initially begins with depolarization may regain the polarized state even before the block. Furthermore, a depolarization induced by increasing the potassium ion concentration does not prevent impulse transmission. For these and other reasons it is probably best to consider the blocking action as a desensitization until a clearer picture emerges. The drugs falling into this classification are decamethonium and succinylcholine.

Bovet[117] has classed antagonists at the neuromuscular junction according to their apparent structural characteristics. He identified a class of long, thin molecules as *leptocurares* and a group of more bulky molecules as *pachycurares*. The classification parallels considerably the mechanistic classification of these agents. Leptocurares are mechanistically depolarizing neuromuscular-blocking agents and are represented by molecules such as decamethonium and succinylcholine. Pachycurares are nondepolarizing (competitive) blocking agents and are represented by *d*-tubocurarine. The structural differences between leptocurares and pachycurares allow for a tentative explanation of mechanistic differences causing neuromuscular blockade, in that the former type of agent possesses a thin depolarizing group that is able to penetrate the muscle end-plate. For maximal depolarizing activity the onium head should bear substituents of minimal steric bulk. It is established that a change from a depolarizing to a competitive action of neuromuscular blockade is accompanied by a progressive increase in size of the substituents on the onium centers, although potency decreases concurrently.

CURARE AND CURARE ALKALOIDS

Originally *curare* was a term used to describe collectively the very potent arrow poisons used since early times by the South American Indians. The arrow poisons were prepared from numerous botanic sources and often were mixtures of several different plant extracts. Some were poisonous by virtue of a convulsant action and others by a paralyzant action. It is only the latter type that is of value in therapeutics and is ordinarily spoken of as "curare."

Chemical investigations of the curares were not especially successful because of the difficulties attendant on the obtaining of authentic samples of curare with definite botanic origin. It was not until 1935 that King was able to isolate a pure crystalline alkaloid, which he named *d*-tubocurarine chloride, from a curare of doubtful botanic origin[118] that possessed, in great measure, the paralyzing action of the original curare. Wintersteiner and Dutcher,[119] in 1943, also isolated the same alkaloid. However, they showed that the botanic source was *Chondodendron tomentosum* (Menispermaceae) and, thus, provided a known source of the drug.

Following the development of quantitative bioassay methods for determining the potency of curare extracts, a purified and standardized curare was developed and marketed under the trade name Intocostrin (Purified Chondodendron Tomentosum Extract), the solid content of which consisted of almost one-half (+)-tubocurarine solids. Following these essentially pioneering developments, (+)-tubocurarine chloride and dimethyltubocurarine iodide have appeared on the market as pure entities.

Tubocurarine Chloride, USP. (+)-Tubocurarine chloride hydrochloride pentahydrate. This alkaloid is prepared from crude curare by a process of purification and crystallization.

Tubocurarine chloride occurs as a white or yellowish white to grayish white, odorless crystalline powder, which is soluble in water. Aqueous solutions of it are stable to heat sterilization.

The structural formula for (+)-tubocurarine (see structure diagram) was long thought to be represented as in Ia. Through the work of Everette et al.,[120] the structure is now known to be that of Ib. The monoquaternary nature of Ib thus revealed has caused some reassessment of thinking concerning the theoretical basis for the blocking action because all previous assumptions had assumed a diquaternary structure (i.e., Ia). Nevertheless, this does not negate the earlier conclusions that a diquaternary nature of the molecule provides better blocking action than does a monoquaternary (e.g., compare the potency of Ib with dimethyl tubocurarine iodide and note the approximately fourfold difference). It may also be of interest that (+)-isotubocurarine chloride (Ic)[121] provides a compound with twice the activity of Ib in the particular test employed.

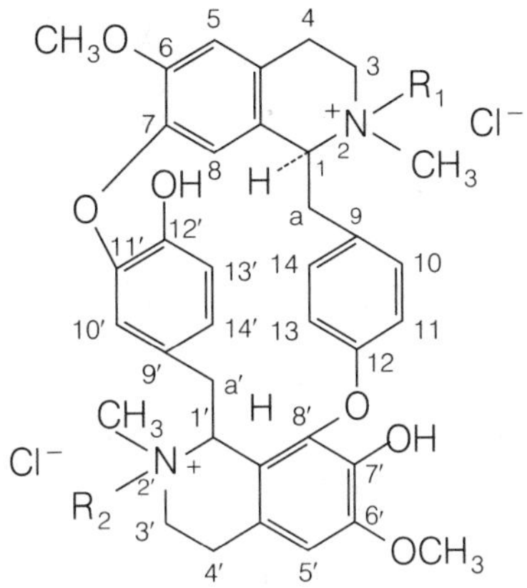

(Ia) $R_1 = R_2 = CH_3$
(Ib) $R_1 = H$; $R_2 = CH_3$
(Ic) $R_1 = CH_3$; $R_2 = H$

Tubocurarine is a nondepolarizing blocking agent used for its paralyzing action on voluntary muscles, the site of action being the neuromuscular junction. Its action is inhibited or reversed by the administration of AChE inhibitors, such as neostigmine, or by edrophonium chloride (Tensilon). Such inhibition of its action is necessitated in respiratory embarrassment caused by overdosage. Additionally, in somewhat higher concentrations, *d*-tubocurarine may also enter the open ion channel and add a noncompetitive blockade. This latter action is not easily or fully restored by cholinesterase inhibitors. It is often necessary to use artificial respiration as an adjunct until the maximum curare action has passed. The drug is inactive orally because of inadequate absorption through lipoidal membranes in the gastrointestinal tract and, when used therapeutically, is usually injected intravenously.

d-Tubocurarine binds for only one millisecond to the receptor, yet its pharmacologic effect of muscle paralysis, produced by administration of the drug intravenously during surgery, lasts for up to two hours. The basis of this action is the pharmacokinetics of the drug. *d*-Tubocurarine is given intravenously and, although 30% to 77% is bound to plasma proteins, the drug is rapidly distributed to central body compartments, including neuromuscular junctions. About 45% of *d*-tubocurarine is eliminated unchanged by the kidneys. Its half-life is 89 minutes.

Tubocurarine, in the form of a purified extract, was first used in 1943 as a muscle relaxant in shock therapy of mental disorders. By its use the incidence of bone and spine fractures and dislocations resulting from convulsions owing to shock was reduced markedly. Following this, it was employed as an adjunct in general anesthesia to obtain complete muscle relaxation, a use that persists to this day. Before its use began, satisfactory muscle relaxation in various surgical procedures (e.g., abdominal operations) was obtainable only with "deep" anesthesia with the ordinary general anesthetics. Tubocurarine permits a lighter plane of anesthesia, with no sacrifice in the muscle relaxation so important to the surgeon. A reduced dose of tubocurarine is administered with ether, because ether itself has a curarelike action.

Another recognized use of tubocurarine is in the diagnosis of myasthenia gravis because, in minute doses, it causes an exaggeration of symptoms by accentuating the already deficient ACh supply. It has been experimented with to a limited extent in the treatment of spastic, hypertonic, and athetoid conditions, but one of its principal drawbacks has been its relatively short duration of activity. When used intramuscularly its action lasts longer than when given by the intravenous route, although this characteristic has not made it useful in the foregoing conditions. Tubocurarine is frequently used with intravenous thiopental sodium anesthesia. Care should be used in the selection of the appropriate solution for this purpose because there are two available concentrations (i.e., 3 mg/mL and 15 mg/mL) of tubocurarine chloride injection. Most anesthesiologists employ the 15 mg/mL concentration injection, and although they note a transient cloudiness from precipitation of the free barbiturate, the condition clears up within minutes and, apparently, causes no problems.

Metocurine Iodide, USP. (+)-*O*,*O*′-Dimethylchondrocurarine diiodide (Metubine Iodide). This drug is prepared from natural crude curare by extracting the curare with methanolic potassium hydroxide. When the extract is treated with an excess of methyl iodide, the (+)-tubocurarine is converted to the diquaternary dimethyl ether and crystallizes out as the iodide (see tubocurarine chloride). Other ethers besides the dimethyl ether also have been made and tested. For example, the dibenzyl ether was one-third as active as tubocurarine chloride and the diisopropyl compound had only one-half the activity. This is compared with the dimethyl ether which has approximately four times the activity of tubocurarine chloride. It is only moderately soluble in cold water but more so in hot water. It is easily soluble in methanol, but insoluble in the water-immiscible solvents. Aqueous solutions have a *p*H of from 4 to 5, and the solutions are stable unless exposed to heat or sunlight for long periods.

The pharmacologic action of this compound is the same as that of tubocurarine chloride, namely, a nondepolarizing competitive blocking effect on the motor end-plate of skeletal muscles. However, it is considerably more potent than *d*-tubocurarine, and it has the added advantage of exerting much less effect on the respiration. The effect on respiration is not a significant factor in therapeutic doses. Accidental overdosage is counteracted best by forced respiration.

The drug is used for much the same purposes as tubocurarine chloride, but in a smaller dose. The dose ranges from 3 to 9 mg. The exact dosage is governed by the physician and depends largely on the depth of surgical relaxation. It is marketed in the form of a parenteral solution in ampules.

SYNTHETIC COMPOUNDS WITH CURARIFORM ACTIVITY

Over 100 years have passed since Crum-Brown and Fraser[122] described the curarimimetic properties induced in several tertiary alkaloids by quaternization with methyl sulfate. Their conclusion was that the quaternized forms of the tertiary alkaloids all had a more uniform pharmacologic activity (i.e., curarimimetic) than did the original tertiary forms, which in many cases (e.g., atropine, strychnine, morphine) had widely different and characteristic activities. Their findings are sometimes known as the *rule of Crum-Brown and Fraser*. Since that time, innumerable quaternary salts have been investigated in an effort to find potent, easily synthesized curarimimetics. The curarelike effect is a common property of all "onium" compounds. In the order of decreasing activity they are:

$$(CH_3)_4N^+ > (CH_3)_3S^+$$
$$(CH_3)_4P^+ > (CH_3)_4As^+ > (CH_3)_4Sb^+$$

Even ammonium, potassium, and sodium ions and other ions of alkali metals exhibit a certain amount of curare action. Thus far, however, it has been impossible to establish any quantitative relationships between the magnitude or the mobility of the cation and the intensity of action. One of the exceptions to the rule that "onium" compounds are necessary for curarelike activity has been the demonstrated activity of the *Erythrina* alkaloids, which are known to contain a tertiary nitrogen. Indeed, they seem to lose their potency when the nitrogen is quaternized. Whether or not two quaternary groups are necessary for maximum activity has led, through numerous studies, to the conclusion that the presence of two or more such quaternary groups permits higher activity by virtue of a more firm attachment at the site of action.[123, 124]

Curare, until relatively recent times, remained the only useful curarizing agent, and it, too, suffered from a lack of standardization. The original pronouncement in 1935 of the structure of (+)-tubocurarine chloride, unchallenged for 35 years, led other workers to hope for activity in synthetic substances of less complexity. The quaternary ammonium character of the curare alkaloids, coupled with the known activity of the various simple onium compounds, hardly seemed to be coincidental, and it was natural for research to follow along these lines.

One of the first approaches to the synthesis of this type of compound was based on the assumption that the highly potent effect of tubocurarine chloride was a function of some optimum spacing of the two quaternary nitrogens. Indeed, the bulk of the experimental work tended to suggest an optimum distance of 12 to 15 Å between quaternary nitrogen atoms in most of the bis-quaternaries for maximum curariform activity. However, other factors could modify this situation.[125, 126] Bovet and his coworkers[127–129] were the first to develop synthetic compounds of significant potency through a systematic structure–activity study based on (+)-tubocurarine as a model. One of their compounds, after consideration of potency/side effect ratios was marketed in 1951 as Flaxedil (gallamine triethiodide).

In 1948, another series of even simpler compounds was described independently by Barlow and Ing[130] and by Paton and Zaimis.[131] These were the bistrimethylammonium polymethylene salts (see formula, Fig. 12-18), and certain of them possessed a potency greater than that of (+)-tubocurarine chloride itself. Both groups concluded that the decamethylene compound was the best in the series and that the shorter chain lengths exhibited only feeble activity. Further investigations by Barlow and Zoller[132] revealed a second maximum of potency of 14 to 18 methylene groups in the bistriethylammonium polymethylene salt series. In particular, the compound with 16 methylene groups proved to be ten times more potent than decamethonium. Figure 12-18 shows the conclusions of Barlow and Ing and

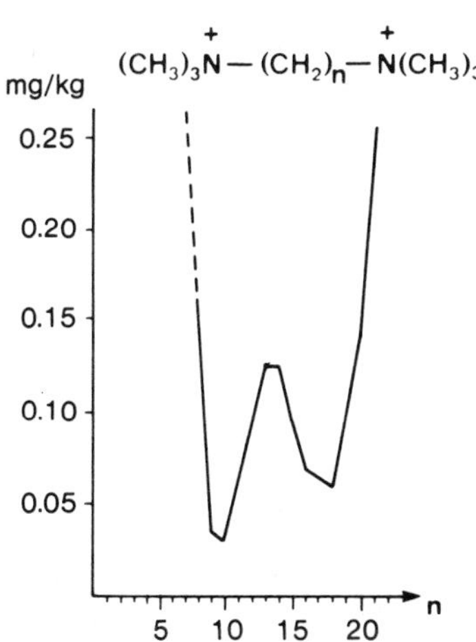

FIG. 12-18. *Curarizing dose in the series of polymethylene-bis-trimethylammonium compounds. Cat, m. tibialis.*[130–132]

Barlow and Zoller relative to these compounds. The commercially obtainable preparation known as decamethonium (represented by the formula in Fig. 12-18, where $n = 10$, salt may be I or Br) represents the decamethylene compound. An interesting finding is that the shorter-chain compounds, such as the pentamethylene and the hexamethylene compounds, are effective antidotes for counteracting the blocking effect of the decamethylene compound.

Cavallito et al.[124] introduced another type of quaternary ammonium compound with high curarelike activity. This type is represented by the following formula and may be designated as ammoniumalkylaminobenzoquinones. The distance between the onium centers is the same as in the other less active onium compounds without the quinone structure, and, therefore, it was speculated that the quinone itself may be involved in the activity. This appears to be reasonable because even the monoquaternary compounds and the corresponding nonquaternized amines show a significant curarelike activity. One compound (benzoquinonium chloride) was selected from this study and was marketed for several years as Mytolon Chloride ($n = 3$, $R_1 = C_2H_5$, $R_2 = CH_2C_6H_5$), but has since been withdrawn because of its anticholinesterase side effects.

$Cl^- \; R_2-\overset{+}{N}(R_1)_2-(CH_2)_n-NH-$[2,5-benzoquinone]$-NH-(CH_2)_n-\overset{+}{N}(R_1)_2-R_2 \; Cl^-$

R_1 = methyl, ethyl, etc.
R_2 = benzyl, methyl
n = 2,3,4,5

Ammonium Alkylaminobenzoquinones

One of the most interesting pathways that research on neuromuscular junction-blocking agents took was that which culminated in the widely used agent succinylcholine chloride, a dicholine ester of succinic acid. It is rather surprising to find that this compound had been examined pharmacologically as early as 1906 and that the muscle relaxant properties were not noticed until Bovet's pioneering study using (+)-tubocurarine as a model for interonium distances.[133] Others soon confirmed Bovet's observations and, in a commentary on the frequent outcomes of structure–activity studies, succinylcholine dichloride has withstood the test of time and is still the drug of choice as a depolarizing blocking agent. In retrospect, it may be looked upon as a "destabilized" decamethonium, as pointed out by Ariens et al.,[134] because it is metabolically disposed of by the action of cholinesterases, whereas decamethonium persists because it cannot be similarly metabolized.

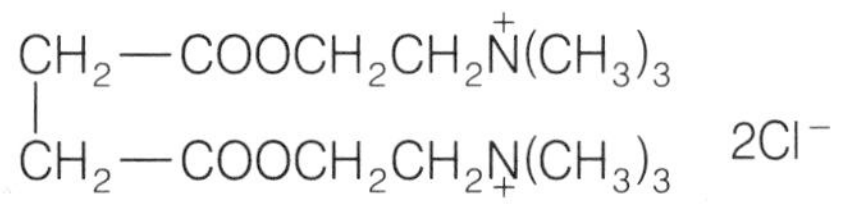

$$\begin{array}{l} CH_2-COOCH_2CH_2\overset{+}{N}(CH_3)_3 \\ | \\ CH_2-COOCH_2CH_2\underset{+}{N}(CH_3)_3 \end{array} \quad 2Cl^-$$

Succinylcholine Dichloride

Although new structural entities to replace succinylcholine are not envisioned, it is, nevertheless, interesting to consider the relative blocking activities of the dicholine esters of maleic and fumaric acid (*cis*- and *trans*-isomers, respectively). These were prepared by McCarthy et al.[135] with the objective of determining whether succinylcholine acts at the receptor in the "eclipsed" or "staggered" conformation. With the same objective in mind, Burger and Bedford[136] and McCarthy et al.[137] have prepared dicholine esters of the *cis*- and *trans*-cyclopropane dicarboxylic acids. In all compounds examined, it is apparent that the "staggered" conformation is the most effective, which tends to reinforce the concept that the binding points for blocking agents are spaced approximately 12 to 14 Å apart, as was suspected by the earliest workers.

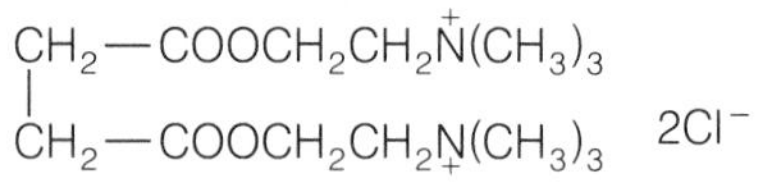

$$\begin{array}{l} CH_2-COOCH_2CH_2\overset{+}{N}(CH_3)_3 \\ | \\ CH_2-COOCH_2CH_2\underset{+}{N}(CH_3)_3 \end{array} \quad 2Cl^-$$

"Eclipsed"

$$\begin{array}{l} (CH_3)_3\overset{+}{N}CH_2CH_2OOC-CH_2 \\ \qquad\qquad\qquad\qquad\quad | \\ \qquad\qquad\qquad\qquad\quad CH_2-COOCH_2CH_2\underset{+}{N}(CH_3)_3 \end{array} \quad 2Cl^-$$

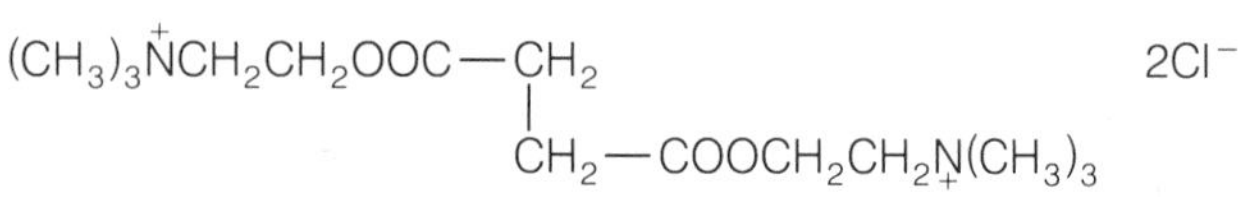

"Staggered"

Other structural studies that might have been anticipated as a consequence of the knowledge gained from the successes with succinylcholine would be the extension of the α,ω-dicholine ester concept to other diesters related to carbachol (i.e., choline carbamate). Although the direct analogue of succinylcholine (i.e., the dicarbaminoyl choline ester) did have blocking activity, it was best in the ester where six methylene groups had to be interspersed between the nitrogen atoms of two carbachol moieties. This led to a clinically useful drug known as imbretil, which seems to have an initial depolarizing action followed by a block that was typical of curarimimetics and was reversible with neostigmine. Although this blocking agent has been used abroad, it has not been marketed in the United States.

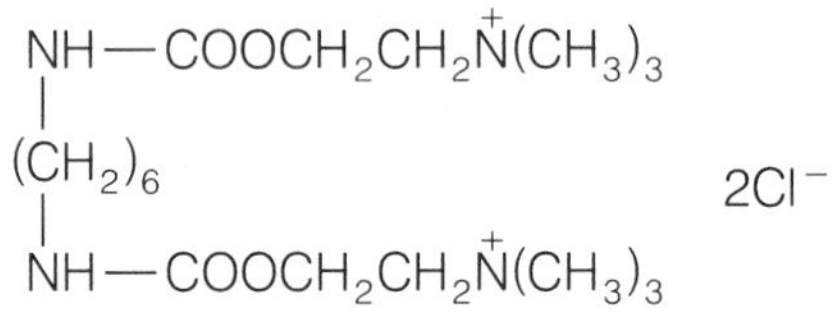

$$\begin{array}{l} NH-COOCH_2CH_2\overset{+}{N}(CH_3)_3 \\ | \\ (CH_2)_6 \\ | \\ NH-COOCH_2CH_2\overset{+}{N}(CH_3)_3 \end{array} \quad 2Cl^-$$

Imbretil

Another attempt to obtain a longer-lasting agent than succinylcholine was that of Phillips[138] in preparing the amide analogues to it, together with several closely related compounds. The blocking activity engendered was a disappointment, although it was noted that these compounds exerted a powerful action in prolonging the succinylcholine block when both were administered together.

Another neuromuscular-blocking agent to appear on the American scene has been pancuronium bromide, a bisquaternary derived from a steroidal framework. It is claimed to have about five times the potency of (+)-tubocurarine as a nondepolarizing agent in humans. Although it has only eight to nine atoms interspersed between the two onium heads, it appears to have unusually high potency, even though it does not conform to the usually expected interonium distances. However, it is probably safe to say that the steroidal skeleton makes little contribution to the activity, as contrasted with its role in estrogenic and androgenic activities.

Steric Factors

It is rather surprising that steric factors connected with neuromuscular junction (NMJ) blockade have not been examined more, in the face of the 20 to 60 times greater activity of the (+)-tubocurarine isomer over the (−)-enantiomer. Studies that bear on this problem have been forthcoming from Soine et al.,[139–141] as well as from the Stenlake group.[142–144] These findings indicate that when a monoquaternary species is under consideration, it appears almost inevitable that the (S)-configurational species is the most potent. On the other hand, it seems equally correct that bisquaternary species, even compounds derived from previously tested monoquaternary species, now show a decided (*R*)-configurational preference for blocking activity—directly opposite that of the monoquaternaries. In view of the evidence now accumulated, it appears that there is a definite (about 2 : 1) superiority of the (*S*)-configuration over the (*R*)-configuration in the monoquaternary forms, even extending to (+)-tubocurarine itself.[129] On the other hand, it seems equally evident that, in bisquaternary forms, those with an (*R*)-configuration for the carbon adjacent to the quaternary moiety are destined to be more active. The reasons for these differences are not immediately apparent.

Products

Atracurium Bresylate. 2-(2-Carboxyethyl)-1,2,3,4-tetrahydro-6,7-dimethoxy-2-methyl-1-veratrylisoquinolinium benzinesulfonate, pentamethylene ester (Tracrium). Atracurium bresylate is a nondepolarizing neuromuscular-blocking agent. Its duration of action (half-life, 0.33 hours) is much shorter than *d*-tubocurarine. Atracurium undergoes both spontaneous and enzymatic catalysis of the ester functions, which leads to a break in the chain between the quaternary nitrogen atoms in the molecule to form inactive metabolites. Hence, termination of its action is not dependent upon renal elimination of the drug, as with *d*-tubocurarine. Paralysis by atracurium is antagonized readily by neostigmine, edrophonium, or pyridostigmine.

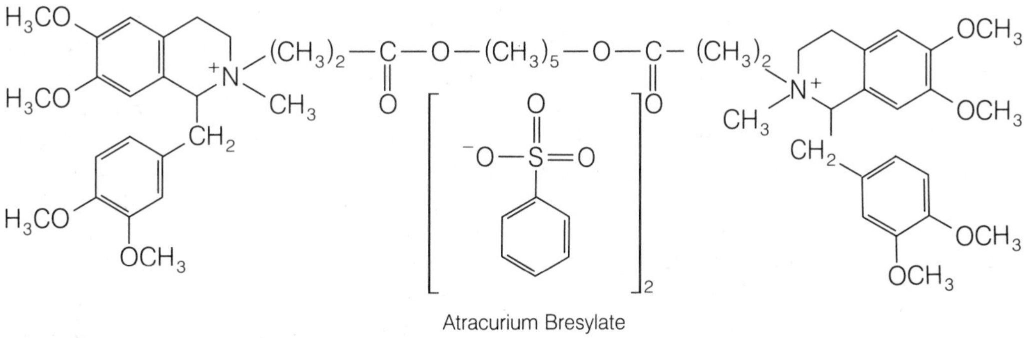

Atracurium Bresylate

Decamethonium Bromide, USP. Decamethylenebis[trimethylammonium]dibromide (Syncurine). This compound is prepared according to the method of Barlow and Ing.[130]

It is a colorless, odorless crystalline powder. It is soluble in water and alcohol, the solubility increasing with the temperature of the solvent. The compound is insoluble in chloroform and ether. It appears to be stable to boiling for at least 30 minutes in physiologic saline, either in the dark or in the sunlight. Twenty percent sodium hydroxide causes a white precipitate that is soluble on heating, but reappears on cooling. Solutions are compatible with procaine hydrochloride and sodium thiopental.

The drug has been used as a skeletal muscle relaxant, especially in combination with the anesthetic barbiturates. Decamethonium is about five times as potent as (+)-tubocurarine and is used in a dose of 500 μg to 3 mg. The antidote to overdosage is hexamethonium bromide or pentamethonium iodide.

Gallamine Triethiodide, USP. [*v*-Phenenyl-tris(oxyethylene)]tris[triethylammonium] triiodide (Flaxedil). This compound is prepared by the method of Bovet et al.[129] and was introduced in 1951. It is a slightly bitter, amorphous powder. It is very soluble in water, but is only sparingly soluble in alcohol. A 2% aqueous solution has a *p*H between 5.3 and 7.0.

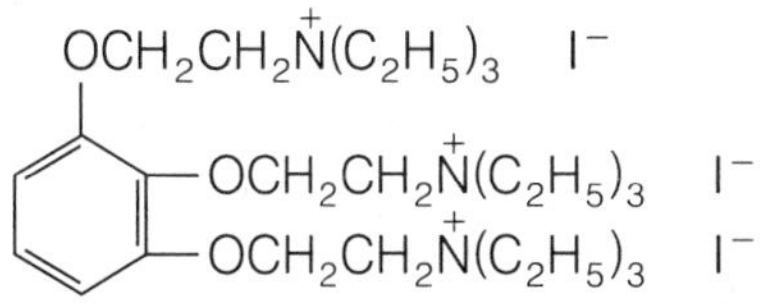

Gallamine Triethiodide

Pharmacologically, it is a relaxant of skeletal muscle by blocking neuromuscular transmission in a manner similar to *d*-tubocurarine (i.e., a nondepolarizing blocking agent). It does have some differences, however. It has a strong vagolytic effect and a persistent decrease in neuromuscular function after successive doses that cannot be overcome by cholinesterase inhibitors. Gallamine also has muscarinic antagonistic properties and binds with greater affinity to the M_2 receptors rather than the M_1 muscarinic receptor. This latter characteristic of gallamine may be the cause of its strong vagolytic action.[145]

The drug is contraindicated in patients with myasthenia gravis, and it should also be borne in mind that the drug action is cumulative, as with curare. The antidote for gallamine triethiodide is neostigmine.

Succinylcholine Chloride, USP. Choline chloride succinate (2 : 1) (Anectine; Sucostrin). This compound may be prepared by the method of Phillips.[146]

It is a white, odorless crystalline substance that is freely soluble in water to give solutions with a *p*H of about 4. It is stable in acidic solutions, but unstable in alkali. The aqueous solutions should be refrigerated to ensure stability.

Succinylcholine is characterized by a very short duration of action and a quick recovery because of its rapid hydrolysis after injection. It brings about the typical muscular paralysis caused by a blocking of nervous transmission at the myoneural junction. Large doses may cause a temporary respiratory depression, in common with other similar agents. Its action, in contrast with that of (+)-tubocurarine, is not antagonized by neostigmine, physostigmine, or edrophonium chloride. These anticholinesterase drugs actually prolong the action of succinylcholine and, on this basis, it is believed that the drug probably is hydrolyzed by cholinesterases. The brief duration of action of this curarelike agent is said to render an antidote unnecessary if the other proper supportive measures are available. However, succinylcholine has a disadvantage in that its action cannot be terminated promptly by the usual antidotes.

It is used as a muscle relaxant for the same indications as other curare agents. It may be used for either short or long periods of relaxation, depending on whether one or several injections are given. In addition, it is suitable for continuous intravenous drip administration.

Succinylcholine chloride should not be used with thiopental sodium because of the high alkalinity of the latter or, if used together, should be administered immediately after mixing. However, separate injection is preferable.

Hexafluorenium Bromide. Hexamethylenebis-[fluoren-9-yldimethylammonium]dibromide (Mylaxen). This compound occurs as a white, water-soluble crystalline material. Aqueous solutions are stable at room temperature but are incompatible with alkaline solutions. Hexafluorenium bromide is used clinically to modify the dose and extend the duration of action of succinylcholine chloride.

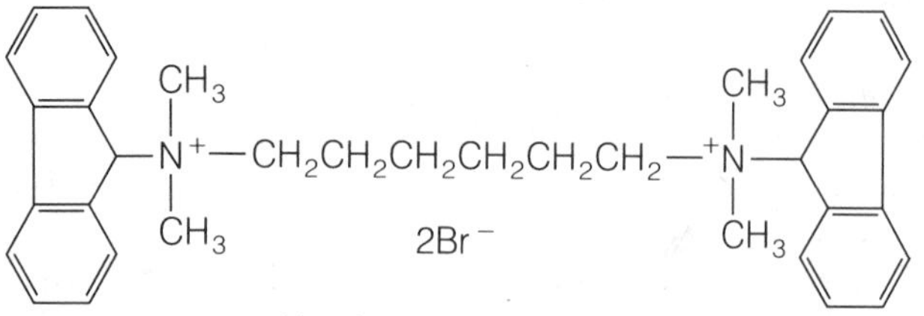

CH_3 / $N^+—CH_2CH_2CH_2CH_2CH_2CH_2—^+N$ / CH_3 $2Br^-$ CH_3 CH_3

Hexafluorenium Bromide

Generally, the dose of succinylcholine chloride can be lowered to about one-fifth the usual total dose when combined with hexafluorenium bromide. The combination produces profound relaxation and facilitates difficult surgical procedures. Its principal mode of action appears to be suppression of the enzymatic hydrolysis of succinylcholine chloride, thereby prolonging its duration of action.

Pancuronium Bromide. $2\beta,16\beta$-Dipiperidino-5α-androstane-$3\alpha,17\beta$-diol diacetate dimethobromide (Pavulon). Although pancuronium is a synthetic product, it is based upon the naturally occurring alkaloid malouetine, found in arrow poisons used by primitive Africans. Pancuronium acts on the nicotinic receptor and in the ion channel, inhibiting normal ion fluxes.

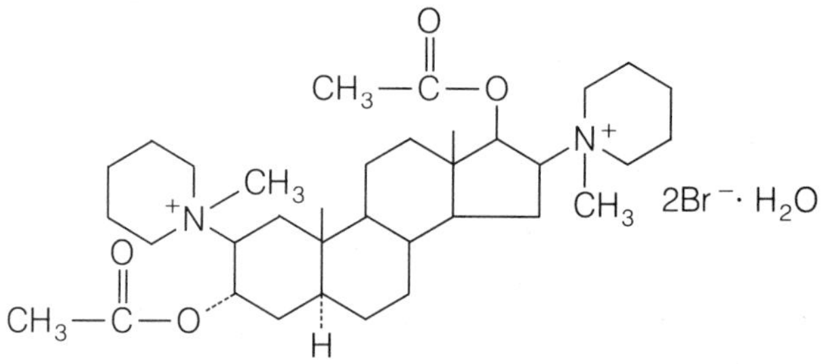

$CH_3—C(=O)—O$ N^+ CH_3 $2Br^- \cdot H_2O$ ^+N CH_3 $CH_3—C(=O)—O$ H

Pancuronium Bromide

This blocking agent is soluble in water and is marketed in concentrations of 1 mg/mL or 2 mg/mL for intravenous administration. It has been shown to be a typical nondepolarizing blocker, with a potency approximately five times that of (+)-tubocurarine chloride and a duration of action approximately equal to the latter. Studies indicate that it has little or no histamine-releasing potential or ganglion-blocking activity and that it has little effect on the circulatory system, except for causing a slight rise in the pulse rate. As one might expect, it is competitively antagonized by ACh, anticholinesterases, and potassium ion, whereas its action is increased by inhalation anesthetics such as ether, halothane, enflurane, and methoxyflurane. The latter enhancement in activity is especially important to the anesthetist because the drug is frequently administered as an adjunct to the anesthetic procedure to relax the skeletal muscle. Perhaps the most frequent adverse reaction to this agent is the occasional prolongation of the neuromuscular block beyond the usual time course, a situation that can usually be controlled with neostigmine or by manual or mechanical ventilation, since respiratory difficulty is a prominent manifestation of the prolonged blocking action.

TABLE 12-16

NEUROMUSCULAR JUNCTION-BLOCKING AGENTS

Name Proprietary Name	*Preparations*	*Category*	*Usual Adult Dose**	*Usual Dose Range**	*Usual Pediatric Dose**
Tubocurarine chloride USP	Tubocurarine chloride injection USP	Skeletal muscle relaxant	IM or IV, initial, 100–300 μg/kg of body weight, not exceeding 27 mg, then 25–100 μg/kg repeated as necessary	1–300 μg/kg	
Metocurine iodide USP	Metocurine iodide injection USP	Skeletal muscle relaxant		IV, initial, 1.5–8 mg given over a 60-sec period; maintenance, 500 μg–1 mg every 25–90 min	
Decamethonium bromide *Syncurine*	Decamethonium bromide injection	Skeletal muscle relaxant		IV, 40–60 μg/kg of body weight	Children: IV or IM, 50–80 μg/kg of body weight
Gallamine triethiodide USP *Flaxedil Triethiodide*	Gallamine triethiodide injection USP	Skeletal muscle relaxant	IV, 1 mg/kg of body weight, not exceeding 100 mg/dose, repeated at 30–40-min intervals if necessary	500 μg–1 mg/kg	Children less than 5 kg: use is not recommended; over 5 kg: see Usual Dose
Succinylcholine chloride *Anectine, Sucostrin*	Succinylcholine chloride injection	Skeletal muscle relaxant	IV, 20–80 mg IV infusion, 1 g in 500–1000 mL of 5% dextrose injection, sodium chloride injection, or sodium lactate injection at a rate of 500 μg–10 mg/min; IM, up to 2.5 mg/kg of body weight, not exceeding a total dose of 150 mg	IV, 10–80 mg	IV, 1–2 mg/kg of body weight; IM, see Usual Dose
	Sterile succinylcholine chloride USP		IV infusion, 1 g in 500–1000 mL of 5% dextrose injection, sodium chloride injection, or sodium lactate injection at a rate of 500 μg–10 mg/min		
Hexafluorenium bromide USP *Mylaxen*	Hexafluorenium bromide injection USP	Potentiator (succinylcholine chloride)	IV, initial, 400 μg/kg of body weight; maintenance, 100–200 μg/kg of body weight		

*See USP DI for complete dosage information.

As indicated, the principal use of pancuronium bromide is an adjunct to anesthesia to induce relaxation of skeletal muscle, but it is employed to facilitate the management of patients undergoing mechanical ventilation. It should be administered only by experienced clinicians equipped with facilities for applying artificial respiration, and the dosage should be carefully adjusted and controlled.

Vecuronium Bromide. 1-(3α,17β-Dihydroxy-2β-piperidino-5α-androstan-16β,5α-yl)-1-methylpiperidinium bromide, diacetate (Norcuron). Vecuronium bromide is the monoquaternary analogue of pancuronium. It belongs to the class of nondepolarizing neuromuscular-blocking agents producing similar effects to drugs in this class. Vecuronium is unstable in the presence of acids and undergoes gradual hydrolysis of its ester functions in aqueous solution. Aqueous solutions have a *p*H of about 4.0. This drug is used mainly to produce skeletal muscle relaxation during surgery and to assist in controlled respiration after general anesthesia has been induced.

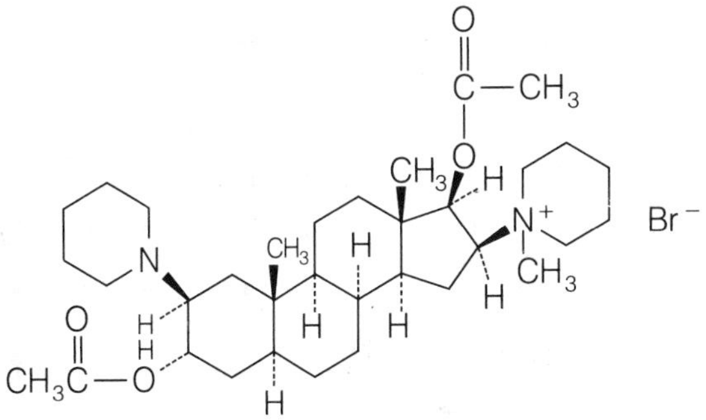

Vecuronium Bromide

REFERENCES

1. Dale, H. H.: J. Pharmacol. Exp. Ther. 6:147, 1914.
2. Karczmar, A. G., and Jenden, D. J. (eds.): In Cholinergic Mechanisms and Psychopharmacology. New York, Plenum Press, 1977.

3. Karlin, A.: Molecular properties of the nicotinic acetylcholine receptor. In Cotman, C. U., Poste, G., and Nicolson, G. L. (eds.). Cell Surface and Neuronal Function, pp. 191–260. Amsterdam, Elsevier Biomedical, 1980.
4. Changeaux, J. P., Devillers-Thiery, A., and Chermoulli, P.: Science 225:1335, 1984.
5. Anholt, R., et al.: In Martonosi, A. N. (ed.). The Enzymes of Biological Membranes, Vol. 3, pp. 335–401. New York, Plenum Press, 1985.
6. Raftery, M. A., et al.: Science 208:1454, 1980.
7. MacIntosh, F. C.: Trends Pharmacol. Sci. 5(Suppl.):100, 1984.
8. Lambrecht, G., and Mutschler, E.: In Hamprecht, B., and Neuhoff, V. (eds.). Neurochemistry, pp. 20–27. Berlin, Springer Verlag, 1985.
9. Abdel-Latif, A. Pharmacol. Rev. 38:227, 1986.
10. Birdsall, N. J. M., Hulme, E. C., and Stockton, J. M.: Trends Pharmacol. Sci. 5(Suppl.):4, 1984.
11. Ignarro, L. J., and Kadowitz, P. J.: Annu. Rev. Pharmacol. Toxicol. 25:171, 1985.
12. Haga, T., and Nada, H.: Biochim. Biophys. Acta 291:564, 1973.
13. Yamamura, H., and Snyder, S. H.: Neurochemistry 21:1355, 1973.
14. Cavallito, C. J., Yun, H. S., Crispin-Smith, T., and Foldes, F. F.: J. Med. Chem. 12:134, 1969.
15. Aquilonius, S. M., Frankenberg, L., Stensio, K. E., and Winbladh, B.: Acta Pharmacol. Toxicol. 30:129, 1971.
16. Whittaker, V. P.: Trends Pharmacol. Sci. 7:312, 1986.
17. Namba, T., and Grob, D.: J. Neurochem. 15:1445, 1968.
18. Burgen, A. S. V.: Trends Pharmacol. Sci. 7(Suppl.):1, 1986.
19. Barrans, Y., and Clastre, J.: C. R. Acad. Sci.: (C)270:306, 1970.
20. Partington, P., Feeney, J., and Burgen, A. S. V.: Mol. Pharmacol. 8:269, 1972.
21. Casey, A. F.: Prog. Med. Chem. 11:1, 1975.
22. Kier, L. B.: Mol. Pharmacol. 3:487, 1967.
23. Schmeideberg, O., and Koppe, R.: Das Muscarine, das Giftige Alka id des Fielgenpiltzes. Vogel, Leipzig, 1869.
24. Hardegger, E., and Lohse, F.: Helv. Chim. Acta 40:2383, 1957.
25. Ariens, E. J., and Simonis, A. M.: In deJong, H. (ed.). Quantitative Methods in Pharmacology, pp. 286–311. Amsterdam, North-Holland, 1961.
26. Ing, H. R.: Science 109:264, 1949.
27. Beckett, A. H., Harper, N. J., Clitherow, J. W., and Lesser, E.: Nature 189:671, 1961.
28. Beckett, A. H., et al.: J. Pharm. Pharmacol. 15:362, 1963.
29. Wilson, I. B.: Ann. N.Y. Acad. Sci. 135:177, 1968.
30. Welsh, H. H., and Taub, R.: Science, 112:47, 1950.
31. Schueler, F. W., and Keasling, H. H.: Am. Sci. 133:512, 1951.
32. Nunes, M. A., and Brochmann-Hanssen, E. J.: J. Pharm. Sci. 63:716, 1974.
33. Jones, A.: J. Pharmacol. Exp. Ther. 141:195, 1963.
34. Englehard, N., Prchal, K., and Nenner, M.: Angew. Chem. Int. Ed. 6:615, 1967.
35. Bergmann, F.: Adv. Cataly. 10:131, 1958.
36. Wilson, I. B., Harrison, M. A., and Ginsberg, S.: J. Biol. Chem. 236:1498, 1961.
37. Belleau, B., DiTullio, V., and Tsai, Y. H.: Mol. Pharmacol. 6:41, 1970.
38. Ellis, S.: J. Pharm. Exp. Ther. 79:309, 1943.
39. O'Brien, R. D., et al.: Mol. Pharmacol. 2:593, 1966; O'Brien, R. D., et al.: Mol. Pharmacol. 4:121, 1968.
40. Aeschlimann, J. A., and Reinert, M.: J. Pharmacol. Exp. Ther. 43:413, 1931.
41. Calvey, T. H.: Biochem. Pharmacol. 16:1989, 1967.
42. Heath, D. F.: Organophosphorus Poisons. Oxford, Pergamon Press, 1961.
43. Oosterban, R. A., and Cohen, J. A.: In Googwin, T. W., Harris, I. J., and Hartley, B. S. (eds.). Structure and Activity of Enzymes, 87. New York, Academic Press, 1964.
44. Tenn, J. G., and Toumarelli, R. C.: Am. J. Ophthalmol. 35:46, 1952.
45. McNammara, P., et al.: J. Pharmacol. Exp. Ther. 87:281, 1946.
46. Diggle, W. M., and Gage, J. C.: Biochem. J. 49:491, 1951; Metcalf, R. L., and March, R. B.: Ann. Entomol. Soc. Am. 46:63, 1953; Gage, J. C.: Biochem. J. 54:426, 1953.
47. Nakatsugawa, T., Tolman, N. M., and Dahm, P. A.: Biochem. Pharmacol. 17:1517, 1968.
48. O'Brien, R. D.: J. Agr. Food Chem. 11:163, 1963.
49. Mounter, L. A., and Cheatham, R. M.: Enzymology 25:215, 1963.
50. Zvirblis, P., and Kondritzer, A.: J. Pharmacol. Exp. Ther. 157:432, 1967.
51. Stephenson, R. P.: Br. J. Pharmacol. 11:378, 1956.
52. Ariens, E. J.: Adv. Drug Res. 3:235, 1966.
53. Paton, W. D. M.: Proc. R. Soc. B 154:21, 1961.
54. Bebbington, A., and Brimblecombe, R. W.: Adv. Drug Res. 2:143, 1965.
55. Long, J. P., et al.: J. Pharmacol. Exp. Ther. 117:29, 1956.
56. Ellenbroek, B. W.: J. Pharm. Pharmacol. 17:393, 1965.
57. Abramson, F. B., Barlow, R. B., Mustafa, M. G., and Stephenson, R. P.: Br. J. Pharmacol. 37:207, 1969; Brimblecombe, R. W., and Inch, T. D.: J. Pharm. Pharmacol. 22:881, 1970; Brimblecombe, R. W., Green, D. M., and Inch, T. D.: J. Pharm. Pharmacol. 22:951, 1970.
58. Goldstein, A., Aronow, L., and Kalman, S.: Principals of Drug Action. New York, Harper & Row, 1968.
59. Fordtran, J. S.: In Gastrointestinal Disease, p. 163. Philadelphia, W. B. Saunders, 1973.
60. Daniel, E. E.: In Bogoch, A. (ed.). Gastroenterology, p. 101. New York, McGraw-Hill, 1973.
61. AMA Drug Evaluations, 6th ed., p. 968. Philadelphia, PA, W. B. Saunders Co., 1986.
62. Pinder, R. M.: Prog. Med. Chem. 9:191, 1973.
63. Holmstedt, B., et al.: Biochem. Pharmacol. 14:189, 1965.
64. Coyle, J. T., and Snyder, S. J.: Science 166:899, 1969.
65. Farnebo, L. O., et al.: J. Pharm. Pharmacol. 22:733, 1970.
66. Gyermek, L., and Nador, K.: J. Pharm. Pharmacol. 9:209, 1957.
67. Ariens, E. J., Simonis, A. M., and Van Rossum, J. M.: In Ariens, E. J. (ed.). Molecular Pharmacology, p. 205. New York, Academic Press, 1964.
68. Brodie, B., and Hogben, C. A. M.: J. Pharm. Pharmacol. 9:345, 1957.
69. Chemnitius, F.: J. Prakt. Chem. 116:276, 1927; see also Hamerslag, F.: The Chemistry and Technology of Alkaloids, p. 264. New York, Van Nostrand, 1950.
70. J. Am. Pharm. Assoc. Pract. Ed. 8:37, 1947.
71. Zvirblis, P., et al.: J. Am. Pharm. Assoc. Sci. Ed. 45:450, 1956.
72. Kondritzer, A. A., and Zvirblis, P.: J. Am. Pharm. Assoc. Sci. Ed. 46:531, 1957.
73. Anesthesia 33:133, 1978.
74. Rodman, M. J.: Am. Prof. Pharm. 21:1049, 1955.
75. Ralph, C. S., and Willis, J. L.: Proc. R. Soc. N. S. Wales 77:99, 1944.
76. Pittenger, P. S., and Krantz, J. C.: J. Am. Pharm. Assoc. 17:1081, 1928.
77. Fromherz, K.: Arch. Exp. Pathol. Pharmakol. 173:86, 1933.
78. Barlow, R. B., Burnston, K. N., and Vis, A.: Br. J. Pharmacol. 68:141P, 1980.
79. Brown, D. A., Fong, J., and Marsh, J.: Br. J. Pharmacol. 71:362, 1980.
80. Zwangenmakers, J. M. A., and Classen, V.: Arzneim. Forsch. 30:1517, 1980.
81. Davison, J. S., Greenwood, B., Najafi-Farashah, A., and Read, N. W.: Br. J. Pharmacol. 79:525, 1983.
82. Hirschowitz, B. I., Fong, J., and Molina, E.: J. Pharmacol. Exp. Ther. 225:263, 1983.
83. Sternbach, L. H., and Kaiser, S.: J. Am. Chem. Soc. 74:2219, 1952.
84. U.S. Patent 2,648,667 (1953).
85. Treves, G. R., and Testa, F. C.: J. Am. Chem. Soc. 74:46, 1952.
86. Tilford, C. H., et al.: J. Am. Chem. Soc. 69:2902, 1947.
87. Med. Lett. 4:30, 1962.

88. Biel, J. H., et al.: J. Am. Chem. Soc. 77:2250, 1955; see also Long, J. P., and Keasling, H. K.: J. Am. Pharm. Assoc. Sci. Ed. 43:616, 1954.
89. Burtner, R. R., and Cusic, J. W.: J. Am. Chem. Soc. 65:1582, 1943.
90. Faust, J. A., et al.: J. Am. Chem. Soc. 81:2214, 1959.
91. Steigmann, F., et al.: Am. J. Gastroenterol. 33:109, 1960.
92. U.S. Patent 2,595,405 (1952).
93. U.S. Patent 2,567,351 (1951).
94. Haas, H., and Klavehn, W.: Arch. Exp. Pathol. Pharmakol. 226:18, 1955.
95. U.S. Patent 2,891,890 (1959); see also Adamson et al.: J. Chem. Soc. 52, 1951.
96. U.S. Patent 2,826,590 (1958).
97. Schwab, R. S., and Chafetz, M. E.: Neurology 5:273, 1955.
98. Denton, J. J., and Lawson, V. A.: J. Am. Chem. Soc. 72:3279, 1950; see also U.S. Patent 2,698,325 (1954).
99. Denton, J. J., et al.: J. Am. Chem. Soc. 71:2053, 1949.
100. Janssen, P., et al.: Arch. Intern. Pharmacodyn. 103:82, 1955.
101. U.S. Patent 2,726,245 (1955).
102. Sperber, N., et al.: J. Am. Chem. Soc. 73:5101, 1951.
103. U.S. Patent 2,607,773 (1952).
104. Caviezel, R., Eichenberger, E., Kunzle, F., and Schmutz, J.: Pharm. Acta Helv. 33:459, 1958.
105. Weijlard, J., et al.: J. Am. Chem. Soc. 71:1889, 1949.
106. U.S. Patent 2,728,769 (1955).
107. Karczmar, A. G., (ed.): International Encyclopedia of Pharmacology and Therapeutics. Sect. 12: Ganglionic Blocking and Stimulating Agents, Vol. 1. Oxford, Pergamon Press, 1966.
108. Skok, V. I.: In Kharkevich, D. A. (ed.). Pharmacology of Ganglionic Transmission, pp. 7–39. Berlin, Springer-Verlag, 1980.
109. Greenspan, P., and Kebabian, J. W.: Fed. Proc. 33:1059, 1974.
110. Hammer, R., and Giachetti, A.: Life Sci. 31:2291, 1982.
111. Volle, R. L., and Hancock, J. C.: Fed. Proc. 29:1913, 1970.
112. Greegard, P., and Kebabian, J. W.: Fed. Proc. 33:1059, 1974.
113. Garrett, J.: Arch. Int. Pharmacodyn. 144:381, 1963.
114. Van Rossum, J. M.: Int. J. Neuropharmacol. 1:97, 1962.
115. Van Rossum, J. M.: Int. J. Neuropharmacol. 1:403, 1962.
116. Paton, W. D. M., and Zaimis, E. J.: Br. J. Pharmacol. 4:381, 1949.
117. Bovet, D.: Ann. N.Y. Acad. Sci. 54:407, 1951.
118. King, H.: J. Chem. Soc. 1381, 1935; see also 265, 1948.
119. Wintersteiner, O., and Dutcher, J. D.: Science 97:467, 1943.
120. Everett, A. J., et al.: Chem. Commun. p. 1020, 1970.
121. Soine, T. O., and Naghaway, J.: J. Pharm. Sci. 63:1643, 1974.
122. Brown, A. C., and Fraser, T.: Trans. R. Soc. Edinb. 25:151, 693, 1868–1869.
123. Phillips, A. P., and Castillo, J. C.: J. Am. Chem. Soc. 73:3949, 1951.
124. Cavallito, C. J., et al.: J. Am. Chem. Soc. 72:2661, 1950.
125. Cavallito, C. J., et al.: J. Am. Chem. Soc. 76:1862, 1954.
126. Macri, F. J.: Proc. Soc. Exp. Biol. Med. 85:603, 1954.
127. Bovet, D., Courvoisier, S., Ducrot, R., and Horclois, R.: C.R. Acad. Sci. 223:597, 1946.
128. Bovet, D., Courvoisier, S., and Ducrot, R.: C.R. Acad. Sci. 224:1733, 1947.
129. Bovet, D., Depierre, F., and de Lestrange, Y.: C.R. Acad. Sci. 225:74, 1947.
130. Barlow, R. B., and Ing, H. R.: Nature 161:718, 1948.
131. Paton, W. D. M., and Zaimis, E. J.: Nature 161:718, 1948.
132. Barlow, R. B., and Zoller, A.: Br. J. Pharmacol. 23:131, 1964.
133. Bovet, D.: Ann. N.Y. Acad. Sci. 54:407, 1951.
134. Ariens, E. J.: Molecular pharmacology, a basis for drug design. In Jucker, E. (ed.). Progress in Drug Research, Vol. 10, p. 514. Basel, Birkhauser, 1966.
135. McCarthy, J. F., et al.: J. Pharm. Sci. 52:1168, 1963.
136. Burger, A., and Bedford, G. R.: J. Med. Chem. 6:402, 1963.
137. McCarthy, J. F., et al.: J. Med. Chem. 7:72, 1964.
138. Phillips, A. P.: J. Am. Chem. Soc. 74:4320, 1952.
139. Erhardt, P. W., and Soine, T. O.: J. Pharm. Sci. 64:53, 1975.
140. Genenah, A. A., Soine, T. O., and Shaath, N. A.: J. Pharm. Sci. 64:62, 1975.
141. Soine, T. O., Hanley, W. S., Shaath, N. A., and Genenah, A. A.: J. Pharm. Sci. 64:67, 1975.
142. Stenlake, J. B., Williams, W. D., Dhar, N. C., and Marshall, I. G.: Eur. J. Med. Chem. Chim. Ther. 9:233, 1974.
143. Stenlake, J. B., Williams, W. D., Dhar, N. C., and Marshall, I. G.: Eur. J. Med. Chem. Chim. Ther. 9:239, 1974.
144. Stenlake, J. B., Williams, W. D., Dhar, N. C., and Marshall, I. G.: Eur. J. Med. Chem. Chim. Ther. 9:243, 1974.
145. Burke, R. E.: Mol. Pharmacol. 30:58, 1986.
146. Phillips, A. P.: J. Am. Chem. Soc. 71:3264, 1949.

SELECTED READINGS

Ariens, E. J.: Receptor theory and structure action relationships. Adv. Drug Res. 3:235, 1966.

Barlow, R. B.: Introduction to Chemical Pharmacology, 2nd ed. London, Methuen & Co., 1964.

Barrett, E. F., and Magleby, K. L.: Physiology of cholinergic transmission. In Goldberg, A. M., and Hanin, I. (eds.). Biology of Cholinergic Function, p. 29. New York, Raven Press, 1976.

Bebbington, A., and Brimblecombe, R. W.: Muscarinic receptors in the peripheral and central nervous systems. Adv. Drug Res. 2:143, 1965.

Braun, A. P., and Sulakhe, P. V.: Muscarinic cholinergic receptors. In Boulton, A. A., Baker, G., and Hrdina, P. D., (eds.). Neuromethods: Receptor Binding 4:139, 1986.

Brimblecombe, R. W.: Review of tremorgenic agents. In Waser, P. G. (ed.). Cholinergic Mechanisms, p. 405. New York, Raven Press, 1975.

Brimblecombe, R. W.: Drug Actions on Cholinergic Systems. London, University Park Press, 1974.

Casy, A. F.: Stereochemical aspects of parasympathomimetics and their antagonists: Recent developments. Prog. Med. Chem. 11:1, 1975.

Enranko, O.: Small intensity fluorescent (SIF) cells and nervous transmission in sympathetic ganglia. Annu. Rev. Pharmacol. Toxicol. 18:417, 1978.

Goldberg, A. M., and Hanin, I. (eds.): Biology of Cholinergic Function. New York, Raven Press, 1976.

Hamphecht, B., and Neuhoff, V. (eds.): Neurobiochemistry. Berlin, Springer-Verlag, 1985.

Hirschowitz, B. I., et al.: Subtypes of muscarinic receptors. Trends Pharmacol. Sci. 5(Suppl.), 1984.

Levine, R. R., et al.: Subtypes of muscarinic receptors II. Trends Pharmacol. Sci. 7(Suppl.), 1986.

Nathanson, B. and Neuhoff, V. (eds.): Neurobiochemistry. Berlin, Springer-Verlag, 1985.

O'Brien, R. D.: Design of organophosphate and carbamate inhibitors of cholinesterases. In Ariens, E. J. (ed.). Drug Design 2:162, 1971.

Rama Sastry, B. V.: Stereoisomerism and drug action in the nervous system. Annu. Rev. Pharmacol. Toxicol. 13:253, 1973.

Siegel, G. J., et al.: Basic Neurochemistry, 2nd ed. Boston, Little Brown & Co., 1976.

Smith, C. U. M., Elements of Molecular Neurobiology. New York, Wiley, 1989.

Waser, P. (ed.): Cholinergic Mechanisms. New York, Raven Press, 1975.

Zaimis, E. (ed.): Neuromuscular Junction. Berlin, Springer-Verlag, 1976.

CHAPTER 13

Diuretics

Daniel A. Koechel

A *diuretic* is defined as a chemical that increases the rate of urine formation. It is convenient to refer to a diuretic as a *natriuretic*, *chloruretic*, *saluretic*, *kaliuretic*, *bicarbonaturetic*, or *calciuretic* agent, depending on whether it enhances the renal excretion of sodium, chloride, sodium chloride, potassium, bicarbonate, or calcium, respectively.

A relatively rigid set of structural features are generally required for a class of diuretics to carry out its primary action (i.e., the direct inhibition of sodium and water reabsorption at one or more of the four major anatomic sites along the nephron). Of additional importance are the secondary (or indirect) events that are triggered as a result of the diuretic's primary action. The secondary events are quite characteristic for each class of diuretics and are often highly predictable if the reader has an understanding of normal renal physiologic processes. Thus, the chemical structure of a substance will determine its diuretic efficacy and potency, as well as its anatomic site of action along the nephron, whereas the locus of action of the diuretic and the response of nephron sites "downstream" from the major site of diuretic action to an enhanced delivery of fluid, sodium, or other solutes will frequently determine the nature and magnitude of many of the observed secondary effects.

In this chapter, the normal function of the nephron will be presented, including the four major reabsorptive sites for sodium and other important solutes, and the renal physiologic events that occur when sodium and water reabsorption are altered by the patient's state of hydration, disease, or intake of diuretics. This will be followed by a discussion of each class of diuretics in current use. A knowledge of the important structural features and the site(s) of action of each class of diuretics should enable the reader to better understand the factors that dictate the nature and magnitude of the anticipated diuresis and the associated secondary effects.

ANATOMY AND PHYSIOLOGY OF THE NEPHRON

The functional unit of the kidney is the nephron with its accompanying glomerulus (Fig. 13-1, *upper right*). There are approximately one million nephrons in each kidney. The blood (or, more appropriately, the plasma), from which all urine is formed, is brought to each nephron within the glomerular capillary network (see Fig. 13-1, *upper left*). Most of the plasma components are filtered into Bowman's space. The resulting glomerular filtrate then flows through the proximal tubule (convoluted and straight portions), the descending limb of Henle's loop, the thin and thick ascending limb of Henle's loop, the area of the macula densa cells, the distal convoluted tubule, and the cortical and medullary collecting tubules during the process of urine formation. Each of these nephron segments consists of ultrastructurally and functionally unique cell types. The physiologic role of the glomerulus and each nephron segment will be discussed as it relates to the handling of important solutes and water in normally hydrated (normovolemic) and hypovolemic persons, and in patients inflicted with various edematous disorders (e.g., congestive heart failure, cirrhosis of the liver with ascites, and the nephrotic syndrome).

FUNCTION OF THE NEPHRON UNDER NORMOVOLEMIA (NORMAL PLASMA VOLUME)

As blood is delivered to each glomerulus, many (but not all) of its components are filtered into Bowman's space through the pores in the glomerular capillary barrier. Several physicochemical properties of each blood component dictate the extent to which it is removed from the blood by the glomerular filtration process. These include the component's (1) relative molecular mass (M_r), (2) overall charge (applies primarily to large molecules), and (3) degree and nature of binding to plasma proteins. For example, plasma proteins with an M_r in excess of 50,000 Da and red blood cells are not readily filtered, whereas low-M_r, non–protein-bound components (e.g., sodium, potassium, chloride, bicarbonate, glucose, and amino acids) are readily filtered.[1]

The *rate* of glomerular filtration of plasma components that possess an M_r of less than 50,000 Da and are not bound to plasma proteins is (1) directly dependent on the hydraulic (hydrostatic) pressure in the renal vasculature (created by the pumping heart), which tends to drive water and solutes out of the glomerular capillaries into Bowman's space; (2) inversely related to the plasma oncotic pressure (the osmotic pressure created by the plasma proteins within the vasculature), which tends to hold or prevent the filtration of water and solutes across the glomerular capillaries into Bowman's space;[1] and (3) governed by the intrarenal signals that allow each nephron to adjust the filtration rate through its own glomerular capillary network (i.e., tubuloglomerular feedback).[2] Clearly, the cardiovascular and renal functional status of an individual will also affect the rate of filtration of plasma components through the glomeruli. In addition, it is well known that neonates and the elderly usually have a reduced glomerular filtration rate—but for different reasons.[3,4]

The *fraction* of the total renal plasma flow that is collectively filtered by the glomeruli per unit time (i.e., the filtration fraction) is about one-fifth.[2] This means that only one-fifth (or 20%) of the plasma presented to the kidneys in a given period undergoes filtration at the glomeruli (i.e., about 650 mL of plasma flow through the kidneys each minute, and approximately 125 mL/min of this are filtered through the glomerular capillaries). The remaining four-fifths (or 80%) of the renal plasma flow is directed into the peritubular capillaries. Each minute only 1 mL of urine is formed from the 125 mL of glomerular filtrate.[5] Thus, approximately 99% of the glomerular filtrate is normally reabsorbed.

The *absolute quantity* of each filtrable plasma component that reaches Bowman's space—the filtered load of a substance—is directly dependent on the rate of glomerular filtration (GFR) and the plasma concentration of the filtrable substance, that is, filtered load of a substance = GFR (mL/min) × $[\text{filtrable substance}]_{\text{plasma}}$ (amount/mL).[5] The glomerular filtrate that houses the filtered load of a given solute will heretofore be referred to as the *luminal fluid*, since it enters the lumen of each nephron immediately after leaving Bowman's space. In the following discussion, attention will be focused on the percentage of the filtered load of sodium and other key solutes that are reabsorbed (i.e., transported from the luminal fluid into renal tubule cells with subsequent passage into the interstitium) at various nephron sites.

There are four major sites along the nephron that are responsible for the bulk of sodium reabsorption[6] (see Fig. 13-1, *upper right*): *Site 1*, the convoluted and straight portions of the proximal tubule; *site 2*, the thick ascending limb of Henle's loop; *site 3*, the distal convoluted tubule; and *site 4*, the terminal one-third to one-half of the distal convoluted tubule and the cortical collecting tubule. The actual transport processes involved with sodium reabsorption at each of these sites are highlighted in the drawings numbered 1 through 4 in the lower two-thirds of Figure 13-1.

The convoluted and straight portions of the proximal tubule (see *1*, Fig. 13-1) are responsible for the reabsorption of about (1) 65% to 70% of the filtered loads of sodium, chloride, water, and calcium;[6,7] (2) 80% to 90% of the filtered loads of bicarbonate,[6,8] phosphate,[7] and urate;[9] and (3) essentially 100% of the filtered loads of glucose, amino acids, and low-M_r proteins.[10] Thus, under normal circumstances, the proximal tubule has a tremendous capacity to reabsorb water and many different solutes. There are primarily two driving forces for this high degree of reabsorptive activity. First, because the plasma in the peritubular capillaries has a lower hydraulic pressure and a higher oncotic pressure than the luminal fluid or the plasma delivered to the glomerulus (owing to the removal of water, but not the protein, from plasma during the glomerular filtration process), there is a net movement of the luminal fluid contents in a reabsorptive direction.[1] Second, the Na^+,K^+–ATPase, strategically located on the antiluminal side (sometimes referred to as the basolateral, peritubular, or contraluminal side) of the proximal tubule cells, catalyzes the countertransport of intracellular sodium ions into the interstitium and extracellular potassium ions into the proximal tubule cells.[10] The stoichiometry for this countertransport is three sodium ions for two potassium ions. This activity creates a deficit of intracellular sodium, a surfeit of intracellular potassium, and a voltage-oriented negative inside proximal tubule cells[10] (see Fig. 13-1).

In response to the action of the Na^+,K^+–ATPase, sodium ions in the luminal fluid move down the concentration gradient into proximal tubule cells by at least three distinct processes. The *first* mechanism of sodium reabsorption at site 1 involves carbonic anhydrase, which is located in the cytoplasm and on the brush border of proximal tubule cells. Hydrogen ions generated as the result of the action of intracellular carbonic anhydrase are exchanged (i.e., countertransported) for the filtered sodium in the luminal fluid (see *1*, Fig. 13-1). The sodium

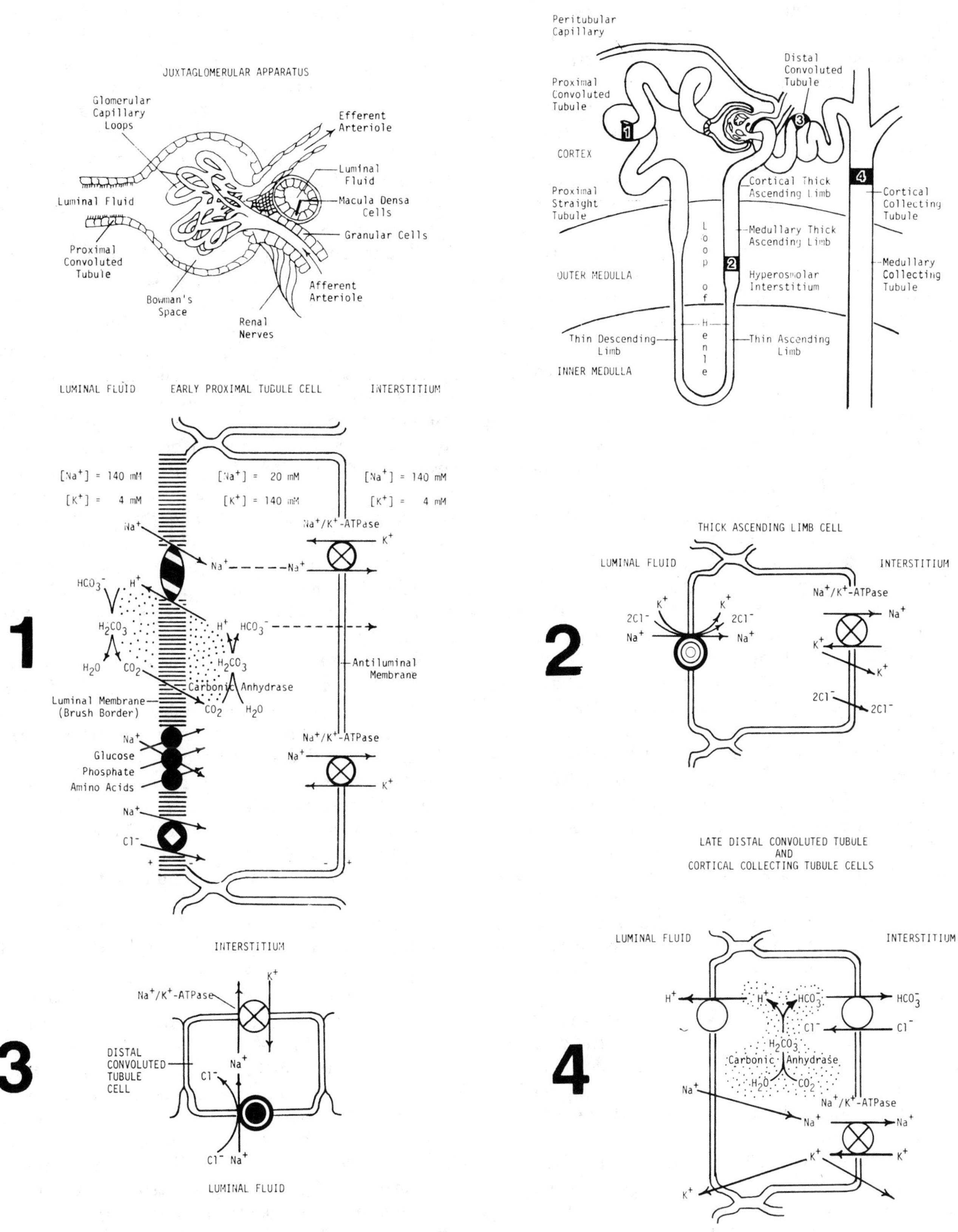

FIG. 13-1. *Upper left: The juxtaglomerular apparatus is of paramount importance for the operation of the tubuloglomerular feedback mechanism that allows a nephron to regulate the glomerular filtration rate of its own glomerulus. Upper right: The anatomy of the nephron with emphasis on the four major sites of sodium reabsorption. The figures labeled 1, 2, 3, and 4 on the lower two-thirds of the page represent the intricate transport processes that participate in the reabsorption of sodium at each of the four major sites of sodium handling.*

subsequently passes from the proximal tubule cells into the interstitium as a result of the action of the Na^+, K^+–ATPase in the antiluminal membrane. The hydrogen ions that are secreted (i.e., transported uphill or against their gradient) into the luminal fluid react therein with the bicarbonate ions to generate carbonic acid. The carbonic acid decomposes, either spontaneously or with the aid of the brush border-bound carbonic anhydrase, to carbon dioxide and water. The carbon dioxide diffuses into the proximal tubule cells and is converted back into bicarbonate, which subsequently passes from the proximal tubule cells into the interstitium. Carbonic anhydrase is very plentiful in the convoluted portion of the proximal tubule of the human, but is nonexistent in the straight portion.[8] Thus, the processes just described occur primarily in the convoluted portion of the proximal tubule and account for the reabsorption of about 20% to 25% of the filtered load of sodium (or about one-third of the filtered load of sodium that is reabsorbed at site 1) and about 80% to 90% of the filtered load of bicarbonate.[8,11]

The *second* mechanism by which sodium moves out of the luminal fluid at site 1 involves its cotransport into proximal tubule cells along with glucose, amino acids, or phosphate[10] (see *1*, Fig. 13-1). The latter three solutes enter proximal tubule cells against their concentration gradients. The amount of sodium that is reabsorbed by this type of cotransport is variable and dependent on the filtered loads of the three solutes. However, such cotransport is the mechanism by which 100% of the filtered loads of glucose and amino acids, and 80% to 90% of the filtered load of phosphate, are removed from the luminal fluid and subsequently reabsorbed.

Third, sodium is reabsorbed at site 1 along with chloride[10] (see Fig. 13-1). This latter process occurs throughout the entire length of the proximal tubule. Collectively, these three sodium-transporting processes remove 65% to 70% of the filtered load of sodium from the luminal fluid, and they do so in an isosmotic fashion (i.e., the osmolality of the luminal fluid entering the descending limb of Henle's loop is similar to that of the initial glomerular filtrate).

As the luminal fluid moves through the descending limb of Henle's loop, the high osmolality (or concentration of solutes) in the surrounding medullary interstitium draws water out of the luminal fluid by osmosis and allows sodium from the interstitium to be added to the luminal contents. In other words, the luminal fluid is concentrated as it flows through the descending limb of Henle's loop.[12]

When the luminal fluid enters the thick ascending limb of Henle's loop (see *2*, Fig. 13-1), it comes into contact with tubule cells that are not only impermeable to water, but possess a capacious luminal membrane-bound transport system for sodium. Here, sodium is cotransported from the luminal fluid into the cells of the thick ascending limb along with chloride and potassium in a ratio of $1Na^+/1K^+/2Cl^-$.[10] Again, (as occurred at site *1*, see Fig. 13-1), a major driving force for the transport of sodium at site 2 is the creation of an intracellular deficit of sodium by the antiluminal membrane-bound Na^+, K^+–ATPase. The combined activities of the Na^+, K^+–ATPase countertransport system on the antiluminal membrane and the $1Na^+/1K^+/2Cl^-$ cotransport system on the luminal membrane of the thick ascending limb cells normally account for the reabsorption of 20% to 25% of the filtered load of sodium;[6] the maintenance of the high osmolality of the medullary interstitium, which is absolutely critical for the normal functioning of the human nephron;[12] and the ability of this nephron segment to reabsorb more sodium than usual when proximal tubule sodium transport has been inhibited.[6,13] This latter compensatory phenomenon explains why diuretics that act primarily at site 1 are not particularly efficacious. By the time luminal fluid reaches the distal end of the thick ascending limb, most, if not all, of the filtered load of potassium has been reabsorbed.[6,11] In addition, the thick ascending limb is responsible for the reabsorption of 20% to 25% of the filtered load of calcium by, as yet, an undefined mechanism.[14]

The descending limb of Henle's loop is responsible for the concentration of luminal fluid (i.e., removal of water and addition of sodium), whereas the thick ascending limb is responsible for the dilution of luminal fluid (i.e., removal of solute from the luminal fluid without the concomitant removal of water). Hence, collectively, these two nephron segments produce a massive overall reduction of luminal fluid volume and solute content. Interestingly, the osmolality of the luminal fluid in the terminal portion of the thick ascending limb of Henle's loop is not much different than that which enters the descending portion of the loop (although drastic changes take place in between).

As the luminal fluid leaves the thick ascending limb of Henle's loop, it comes into contact with the *macula densa cells*—a specialized group of tubule cells that communicate with the granular cells of the afferent and efferent arterioles belonging to the same nephron[15] (see Fig. 13-1, *upper left*). Macula densa cells are capable of detecting increases in the flow rate of the luminal fluid that passes by them.[2] In response to such increases, the macula densa cells communicate with the granular cells and vascular components of the juxtaglomerular apparatus. Renin is then released, angiotensin II is formed, and the renal vasoconstriction that ensues produces a reduction in the glomerular filtration rate.[2] Because certain diuretics are capable of increasing the flow rate of luminal fluid past the macula densa cells, it is not surprising that some of these drugs reduce the glomerular filtration rate while enhancing the urinary loss of sodium. If the glomerular filtration rate is depressed too much, the magnitude of the diuresis will be blunted. Other stimuli that also trigger the release of renin into the vasculature may be received by the juxtaglomerular apparatus. These stimuli include increased sympathetic tone, reduced renal arterial pressure, and reduced vascular volume.[2]

Following its sojourn past the macula densa cells, the luminal fluid comes into contact with the third major site for the reabsorption of sodium—the relatively short distal convoluted tubule (see *3*, Fig. 13-1). Site 3 may be modified in the future to include the connecting tubule as well as the distal convoluted tubule. Again, a major driving force for sodium reabsorption from the luminal fluid at site 3 involves the deficit of intracellular sodium produced by the action of the antiluminal membrane-bound Na^+, K^+–ATPase. Sodium in the luminal fluid is cotransported into the distal convoluted tubule cells along with chloride. Approximately 5% to 8% of the filtered load of sodium is reabsorbed at site 3.[6,11]

The terminal one-third to one-half of the distal convoluted tubule and the cortical collecting tubule house the fourth and final site for the reabsorption of sodium from the luminal fluid[6] (see *4*, Fig. 13-1). At site 4 sodium enters the tubule cells in exchange for potassium and hydrogen ions, which are secreted into the luminal fluid. The exchange of sodium for potassium and hydrogen ions is not stoichiometric.[6] The handling of sodium in these nephron segments is unique in that there is no accompanying anion involved. Once again one of the driving forces for the entry of luminal sodium into the tubule cells at site 4 is the deficit of intracellular sodium created therein by the action of the antiluminal membrane-bound Na^+, K^+–ATPase. The exchange mechanism is partially controlled by the mineralocorticoids—the primary one being aldosterone. Although the exchange of luminal fluid sodium for intracellular hydrogen or potassium ions is associated with the reabsorption of only 2% to 3% of the filtered load of sodium,[6] the distal location of this exchange system along the nephron dictates the final acidity and potassium content of the urine.

The amount of sodium reabsorbed at site 4 and, therefore, the amount of hydrogen and potassium ions present in the final urine, is modulated by the (1) plasma and renal levels of aldosterone; (2) the luminal fluid flow rate and the percentage of the filtered load of sodium presented to the exchange sites—the greater the flow rate and the load of sodium, the greater the amount of exchange; (3) acid–base status of the individual—acidosis favors exchange of sodium and hydrogen ions, whereas alkalosis favors exchange of sodium and potassium ions.[6,16] The classes of diuretics that inhibit the reabsorption of sodium at sites 1, 2, or 3 (i.e., sites proximal to site 4) ultimately increase the luminal fluid flow rate and the percentage of the filtered load of sodium delivered to site 4. Thus, many diuretics sharply enhance the urinary loss of potassium and hydrogen ions and may be associated with the induction of hypokalemia and alkalosis.

FUNCTION OF THE NEPHRON DURING HYPOVOLEMIA (REDUCED PLASMA VOLUME)

An individual's plasma volume may be reduced by hemorrhage, diarrhea, vomiting, excessive sweating, or the overzealous use of diuretic agents. When this occurs, the renal processes previously discussed shift into a "conserve" mode to prevent further loss of vital body fluids and solutes.[2,17] These events include (1) the release of renin from the juxtaglomerular apparatus (in response to the reduced arterial pressure, reduced cardiac output, reduced renal perfusion, and increased sympathetic stimulation) and the subsequent generation of angiotensin II, which produces renovasoconstriction and reduction in renal blood flow and glomerular filtration rate; (2) an increase in the proximal tubule reabsorption of luminal fluid and solutes;[14] (3) angiotensin II-induced stimulation of aldosterone production,[17] which ultimately is associated with enhanced reabsorption (i.e., conservation) of sodium at site 4; and (4) release of antidiuretic hormone (ADH) into the bloodstream from the posterior pituitary gland in response to the reduced blood pressure and elevated plasma osmolality. The ADH increases the permeability of the cortical and medullary collecting ducts to water. In the presence of ADH, the high osmolality of the medullary interstitium (created in part by the collective actions of the luminal membrane-bound $1Na^+/1K^+/2Cl^-$ transport system and the antiluminal membrane-bound Na^+, K^+–ATPase of the thick ascending limb cells) draws the water out of the lumens of the collecting tubules by osmosis. Water is conserved, and the urine becomes extremely concentrated.

FUNCTION OF THE NEPHRON DURING DISEASE STATES ASSOCIATED WITH RETENTION OF BODY FLUIDS (EDEMATOUS STATES)

Frequently, the kidneys of individuals with congestive heart failure, cirrhosis of the liver with ascites, or the nephrotic syndrome receive messages that are interpreted to mean that they are being hypoperfused. This may occur whether or not there is an actual plasma volume reduction. The kidneys attempt to prevent urinary losses of body fluids and solutes by any one or a combination of the four processes just mentioned. Ultimately edema ensues.[18]

INTRODUCTION TO THE DIURETICS

It is important to address three issues before a discussion of the various classes of diuretics: (1) The difference between diuretic potency and efficacy; (2) the relative importance of the concentration of a diuretic in the interstitial fluid that bathes the antiluminal (peritubular) side of the tubule cell versus the concentration of the diuretic in renal luminal fluid; and (3) the shortcomings of many previous structure–activity relationship (SAR) studies involving the diuretics.

There is a clear distinction to be made between the use of the terms potency and efficacy.[19] The *potency* of a diuretic is related to the absolute amount of drug (e.g., milligrams; milligrams per kilogram) that is required to produce an effect. The *relative potency* is a convenient means of comparing two diuretics and is expressed as a ratio of equieffective doses. The potency of a diuretic is influenced by its absorption, distribution, biotransformation, excretion, and its inherent ability to combine with its receptor (i.e., its intrinsic activity). The potency of a diuretic is important for establishing its dosage, but is otherwise a relatively unimportant characteristic. On the other hand, the term *efficacy* relates to the maximal diuretic effect attainable (usually measured in terms of urine volume/time or urinary loss of sodium or sodium chloride/time). A diuretic's efficacy is closely related to its site of action along the nephron (i.e., site 1, 2, 3, or 4). Diuretics that inhibit the reabsorption of sodium at the same anatomic site are usually equiefficacious (i.e., evoke similar maximal responses), but may vary in potency (i.e., the amount of diuretic necessary to produce similar effects). Diuretics that act at site 1 are not very efficacious (even though they may inhibit the reabsorption of 20% to 25% of the filtered load of sodium) because the $1Na^+/1K^+/2Cl^-$ and Na^+,K^+–ATPase transport systems in the thick ascending limb of Henle's loop compensate by reabsorbing most of the extra sodium presented to them. Diuretics that inhibit the reabsorption of sodium at site 2 are the most efficacious of all diuretics because site 2 is normally responsible for the reabsorption of 20% to 25% of the filtered load of sodium, and there are no high-capacity sodium reabsorptive sites downstream that can markedly blunt such an action. Diuretics that act at site 2 are frequently referred to as "high-ceiling" or "loop" diuretics. Diuretics that act at site 3 or site 4 are less efficacious because these two sites are responsible for the reabsorption of only 5% to 8% and 2% to 3% of the filtered load of sodium, respectively.

The diuretic activity associated with some classes of diuretics may be related to their concentration in the interstitial fluid that bathes the antiluminal side of the renal cell (or perhaps to their concentration in the peritubular capillaries), whereas that of others is related to their concentration in renal luminal fluid. Because the major route of excretion of the diuretic agents is by the renal system, the kidneys are exposed to relatively high concentrations of these drugs. Whereas the concentration of a diuretic in the peritubular capillaries or in the interstitial fluid bathing the antiluminal side of the tubule cells is frequently a reflection of such factors as dose and bioavailability, the concentration of the diuretic in luminal fluid is determined by these same factors plus a combination of three distinct renal processes: glomerular filtration, active tubular secretion, and nonionic back diffusion.

First, all diuretics enter luminal fluid by the process of glomerular filtration, but to varying degrees. The amount of a diuretic that enters the luminal fluid by the filtration process is dependent on the glomerular filtration rate, the plasma concentration of the diuretic agent, and the extent that the diuretic is bound to the predominant unfiltrable plasma protein—albumin.

Second, all but a few of the diuretics attain relatively high concentrations in the luminal fluid of the proximal tubule by a two-step process commonly referred to as *active tubular secretion*[20] (Fig. 13-2). The antiluminal membrane of the proximal tubule houses a set of bidirectional active transport systems that participate in the first step of active tubular secretion of a diuretic. The organic anion transport system (OATS) transports endogenous and exogenous organic anions, whereas the organic cation transport system (OCTS) handles endogenous and exogenous organic cations. Because most diuretics are weak organic acids (e.g., carboxylic acids or sulfonamides) or weak organic bases (e.g., amines), they will exist as organic anions and cations, respectively, and are likely to be handled by the OATS or the OCTS. Although the OATS and OCTS are bidirectional, they transport diuretics primarily in a secretory direction (i.e., from the interstitium into proximal tubule cells). Even those diuretics that are extensively bound to plasma proteins may be avidly secreted. Importantly, neither the OATS nor the OCTS possesses rigid structural requirements for

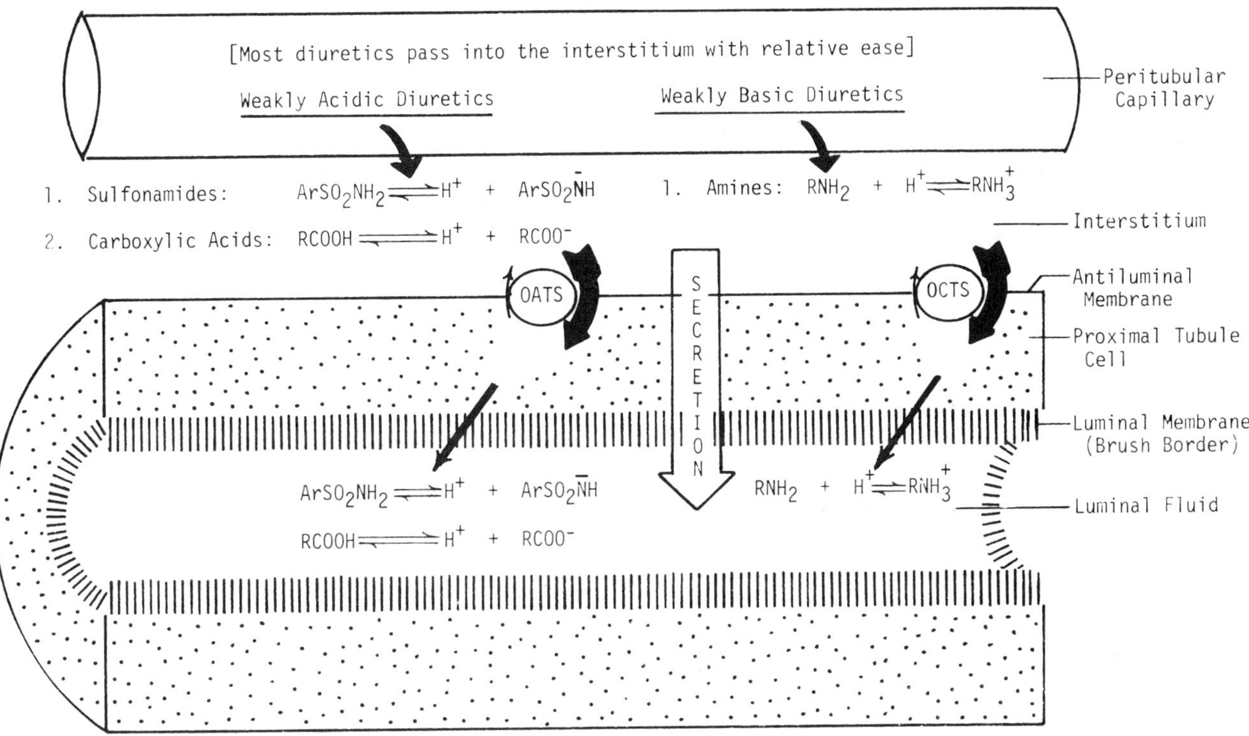

FIG. 13-2. *Active tubular secretion of the charged drug species that are generated during the ionization of weakly acidic and basic diuretics.*

the respective anion or cation being transported. The second step of active tubular secretion of a diuretic involves its passage from proximal tubule cells into the luminal fluid, probably by a combination of passive diffusion and active transport.

Third, the concentration of a diuretic in the luminal fluid of the more distal segments of the tubule is modulated by its lipid/water partition coefficient and pK, as well as by the urinary pH. Weakly acidic diuretics, the undissociated forms of which possess a favorable balance of lipid and water solubility, may undergo pH-dependent diffusion (referred to as *nonionic back diffusion*) from the distal tubular luminal fluid back into the bloodstream. This frequently decreases the renal excretion rate of the diuretic, but prolongs its plasma half-life. Diuretics that are weak bases will follow a similar course if the urinary pH is on the alkaline side, which favors the presence of the uncharged drug species. Weak organic acids or bases, the uncharged forms of which possess an unfavorable lipid/water partition coefficient, will not undergo nonionic back diffusion. These diuretics will be retained within the luminal fluid and, ultimately, will be excreted. Thus, diuretic agents may reach high concentrations in luminal fluid following glomerular filtration, active tubular secretion, and little, or no, subsequent nonionic back diffusion. It is extremely important for the diuretics that inhibit sodium reabsorptive processes on the luminal membrane (e.g., diuretics that act at site 2 and site 3, as well as some of those that act at site 4; see Fig. 13-1) to attain relatively high luminal fluid concentrations. For some diuretics a combination of adequate intracellular and luminal fluid levels is a prerequisite to the induction of a diuresis (e.g., the diuretics that act at site 1; see Fig. 13-1).

Many of the past SAR studies involving diuretics have been conducted in whole animals, and the results may be misinterpreted unless caution is exercised. Generally speaking, compounds of varied chemical structure are administered to animals and ranked according to their ability to produce changes in urine or sodium output over a prescribed period. Conclusions are then drawn about which functional groups are the most important for optimal diuretic activity. It is extremely important for the novice to remember that the results from such studies are not necessarily to be interpreted as a ranking of the intrinsic activity of the agents under study. Diuretic SAR studies conducted in whole animals yield results that are a composite of differences in the absorption, plasma protein binding, distribution, biotransformation, excretion, active tubular secretion, intrinsic activity, and secondary effects (e.g., changes in the glomerular filtration rate) of the various agents. Unfortunately, most, if not all, of these variables are neglected during initial diuretic screening procedures, and it may therefore be erroneously assumed that differences in diuretic activity are due to differences in intrinsic activity. If one is interested in the intrinsic activity of the members within a group of diuretics, then a closer approximation can be achieved by examining the agents on isolated nephron segments in which related or prototypic diuretics are known to act. Several such studies have been conducted.[21,22] It should not be a surprise when results from in vivo and in vitro SAR studies differ. This occurs because of the interplay between numerous parameters in the in vivo studies (i.e., absorption, distribution, and such) that can be eliminated in a properly designed in vitro study. Almost all structure–activity data cited in the upcoming portion of this chapter were obtained from whole animal and human investigations.

SITE 1 DIURETICS: CARBONIC ANHYDRASE INHIBITORS

Although the currently available carbonic anhydrase inhibitors are infrequently employed as diuretics, they played an important role in the development of

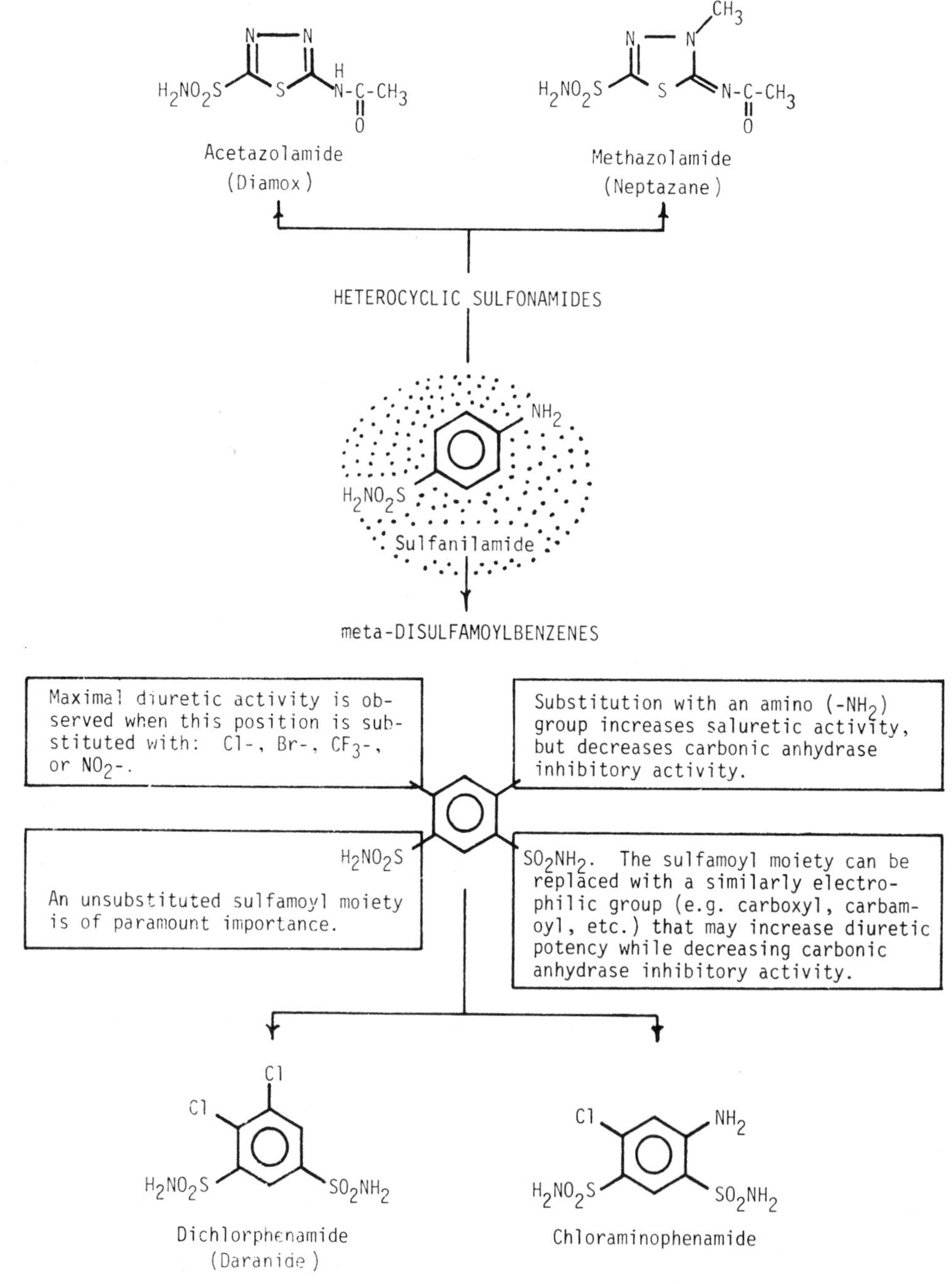

FIG. 13-3. Development of two classes of carbonic anhydrase inhibitors based on the actions of sulfanilamide.

other major classes of diuretics that are currently in widespread use, and they aided in our understanding of basic renal physiology.

Shortly after its introduction for the treatment of bacterial infections, sulfanilamide (Fig. 13-3) was observed to produce a mild diuresis characterized by the presence of urinary sodium and a substantial amount of bicarbonate.[23] It was subsequently shown that sulfanilamide induced this effect through the inhibition of renal carbonic anhydrase.[24,25] However, sulfanilamide was a relatively weak inhibitor of renal carbonic anhydrase, and the dose needed to exert an adequate diuresis was associated with severe adverse effects. In an attempt to improve upon the carbonic anhydrase inhibitory property of sulfanilamide, many sulfamoyl-containing compounds ($-SO_2NH_2$) were synthesized and screened for their diuretic activity in vivo and their ability to inhibit carbonic anhydrase in vitro. Two groups of carbonic anhydrase inhibitors emerged: simple heterocyclic sulfonamides and *meta*-disulfamoylbenzene derivatives.

STRUCTURE–ACTIVITY RELATIONSHIPS

The SAR studies involving the simple heterocyclic sulfonamides yielded the prototypic carbonic anhydrase inhibitor—acetazolamide[26–30] (see Fig. 13-3). The sulfamoyl group is absolutely essential for the in vitro carbonic anhydrase inhibitory activity and for the production of a diuresis in vivo. The sulfamoyl nitrogen atom must remain unsubstituted to retain both in vivo and in vitro activities. This feature explains why all of the antibacterial sulfonamides, except sulfanilamide, are incapable of inhibiting carbonic anhydrase or exerting a diuresis. In contrast, substitution of a methyl group on one of acetazolamide's ring nitrogens yields methazolamide (see Fig. 13-3), a product that retains carbonic anhydrase inhibitory activity. The moiety to which the sulfamoyl group is attached must possess aromatic character. In addition, within a given series of heterocyclic sulfonamides, the derivatives with the highest partition coefficients and the lowest p*K*s have the greatest carbonic anhydrase inhibitory activity and diuretic activity.

The SAR studies involving the *m*-disulfamoylbenzenes revealed that the parent 1,3-disulfamoylbenzene lacked diuretic activity, but key substitutions (which are summarized in Fig. 13-3) led to compounds with diuretic activity.[31] The first commercially available analogue, dichlorphenamide (see Fig. 13-3), is similar to acetazolamide in its carbonic anhydrase inhibitory activity, but it is also a chloruretic agent. Subsequently, chloraminophenamide (see Fig. 13-3) was shown to possess less carbonic anhydrase inhibitory activity, but greater chloruretic activity when given by the intravenous route. Poor diuretic activity following the oral administration of chloraminophenamide precluded its marketing.

PHARMACOKINETICS

The clinically available carbonic anhydrase inhibitors are well absorbed from the gastrointestinal tract, are distributed to the sites of major importance for carbonic anhydrase inhibition, undergo little, if any, biotransformation, and are excreted primarily by the kidneys. All carbonic anhydrase inhibitors attain relatively high concentrations in renal luminal fluid (by a combination of glomerular filtration and active tubular secretion) and in proximal tubule cells.

SITE AND MECHANISM OF ACTION

Carbonic anhydrase is located both intracellularly and in the luminal brush border membrane of proximal convoluted tubule cells (see *1*, see Fig. 13-1). Both of these site 1 locations are major targets of the carbonic anhydrase inhibitors.[8] This group of diuretics also inhibits intracellular carbonic anhydrase at site 4.

During the first four to seven days of continuous therapy with a carbonic anhydrase inhibitor, at least three noteworthy events occur that lead to an increase in sodium, bicarbonate, and potassium excretion: (1) Inhibition of the intracellular carbonic anhydrase in proximal tubule cells (i.e., site 1) results in a decrease in the available hydrogen ions that are normally exchanged for luminal fluid sodium ions; thus, a decrease in proximal tubule reabsorption of sodium occurs. (2) Inhibition of carbonic anhydrase on the luminal brush border membrane of proximal tubule cells (i.e., site 1) causes a decrease in the production of carbon dioxide within the luminal fluid and a decrease in the proximal tubule uptake of carbon dioxide. The net result is a decrease in the reabsorption of bicarbonate. (3) Inhibition of intracellular carbonic anhydrase at site 4 decreases the intracellular hydrogen ion concentration, which, in turn, favors the exchange of luminal fluid sodium for intracellular potassium (rather than for intracellular hydrogen ions).[32] It might be assumed that a massive diuresis would follow the inhibition of the portion of proximal tubule sodium reabsorption that is under the control of carbonic anhydrase (i.e., one-third of the 65% to 70% of the filtered load of sodium normally reabsorbed from the proximal luminal fluid or about 22% of the filtered load of sodium). However, sites downstream (especially site 2) can compensate for such an action by reabsorbing much of the additional sodium that is presented to them.[6,11,13] Some of the luminal fluid bicarbonate is also reabsorbed downstream by a noncarbonic anhydrase-mediated system.[32] Thus, the actions of the carbonic anhydrase inhibitors enhance the urinary loss of only 2% to 5% of the filtered load of sodium, up to 30% of the filtered load of bicarbonate, and a substantial amount of potassium.[11,33] The urinary loss of potassium increases because the proximal tubule actions of carbonic anhydrase inhibitors present a greater percentage of the filtered load of sodium to site 4, increase the flow rate of luminal fluid through the distal convoluted tubule and collecting tubule, and decrease the availability of intracellular hydrogen ions at site 4. All three changes result in an enhanced exchange of luminal fluid sodium for intracellular potassium at site 4. The urinary concentration of chloride actually decreases after the administration of carbonic anhydrase inhibitors.[32] Hence, carbonic anhydrase inhibitors are natriuretic, bicarbonaturetic, and kaliuretic agents.

Toward the end of the first week of continuous therapy with a carbonic anhydrase inhibitor, resistance develops to its diuretic effect.[34] This is primarily due to two factors. First, there is a marked reduction in the filtered load of bicarbonate. This occurs because the urinary loss of bicarbonate ultimately leads to a reduction in the plasma concentration of bicarbonate. In addition, the carbonic anhydrase inhibitors induce a 20% reduction in the glomerular filtration rate. When there is less bicarbonate present in the luminal fluid, there is less bicarbonate reabsorption to inhibit. Second, the metabolic acidosis created by these diuretics provides a sufficient amount of noncarbonic anhydrase-generated intracellular hydrogen ions to exchange for the luminal fluid sodium. The sodium reabsorption at site 1 progressively returns to a near-normal rate, and the diuresis wanes.

ADVERSE EFFECTS

Several highly predictable adverse effects are associated with the carbonic anhydrase inhibitors: (1) development of metabolic acidosis owing to the renal loss of bicarbonate, (2) hypokalemia owing to the renal loss of potassium, (3) up to a 20% reduction in the glomerular filtration rate that appears to be mediated by the juxtaglomerular apparatus because of the increased flow rate of luminal fluid past the macula densa cells (i.e., tubuloglomerular feedback),[33,35] and (4) typical sulfonamide-associated hypersensitivity reactions, such as urticaria, drug fever, blood dyscrasias, and interstitial nephritis.

Carbonic anhydrase inhibitors may also be associated with the production of paresthesias (tingling in the extremities), drowsiness, fatigue, anorexia, gastrointestinal disturbances, and urinary calculi. The latter occur because of a reduction in the urinary excretion rate of citrate—a normal urinary component that assists in maintaining urinary calcium salts in a solubilized form.[32,33]

Carbonic anhydrase inhibitors can exacerbate the symptoms associated with cirrhosis of the liver.[32,36] Consequently, their use should be avoided in patients with this disorder. The carbonic anhydrase-induced alkalinization of the urine decreases the normal luminal fluid trapping of ammonia (NH_3) in the form of ammonium ions (NH_4^+). This leads to a subsequent reduction in the urinary excretion of ammonium ions. Under these circumstances the highly diffusible ammonia is diverted from the luminal fluid into the systemic circulation where it may contribute to the development of hepatic encephalopathy.

USES

The major use of the carbonic anhydrase inhibitors is in the treatment of glaucoma.[36] Carbonic anhydrase is a functionally important enzyme in the eye, where it plays a key role in the formation of aqueous humor. Inhibition of this ocular enzyme reduces the rate of formation of the aqueous humor and, thereby, reduces the intraocular pressure associated with glaucoma. Interestingly, a reduction in the intraocular pressure usually persists at a time when resistance has developed to the renal effects of the carbonic anhydrase inhibitors.[33] Carbonic anhydrase inhibitors are infrequently employed as diuretics, prophylactically to counteract acute mountain sickness,[36] as adjuvants for the treatment of epilepsy, and to create an alkaline urine in an attempt to hasten the renal excretion of certain noxious weak acids or to maintain the urinary solubility of certain poorly water-soluble endogenous weak acids (e.g., uric acid).[36]

PRODUCTS

Acetazolamide, USP. *N*-[5-(Aminosulfonyl)-1,3,4-thiadiazol-2-yl]acetamide (Diamox) (see Fig. 13-3). Acetazolamide was introduced in 1953 as the first orally effective, nonmercurial diuretic available to the physician. It has a relatively restricted use today because of its limited efficacy and the refractoriness that develops to its diuretic action within the first week of continuous therapy. However, it remains the most important carbonic anhydrase inhibitor available and serves as the prototypic agent in its class. Acetazolamide is extremely well absorbed from the gastrointestinal tract, extensively bound to plasma proteins, and is not biotransformed. Peak plasma levels are attained within two to four hours. Its onset of action is about one hour, and its duration of action ranges from 6 to 12 hours. Acetazolamide is totally removed from the plasma by the kidneys within 24 hours. The renal handling of acetazolamide involves its filtration at the glomeruli, active tubular secretion in the proximal tubule, and a varying degree of *p*H-dependent nonionic back diffusion in the distal segments of the nephron. Acetazolamide is available as 500-mg extended-release capsules, 125- and 250-mg tablets, and as a sterile solution for parenteral use containing 500 mg of acetazolamide sodium.

Methazolamide, USP. *N*-[5-(Aminosulfonyl)-3-methyl-1,3,4-thiadiazol-2(3*H*)-ylidene]acetamide (Neptazane) (see Fig. 13-3). Although in vitro studies have shown methazolamide to be a more potent carbonic anhydrase inhibitor than the prototypic acetazolamide, it is seldom used as a diuretic (for the same reasons as those stated for acetazolamide). Methazolamide displays improved penetration into the eye,[28] a property that contributes to its usefulness in the treatment of glaucoma. Methazolamide is available as 50-mg tablets.

Dichlorphenamide, USP. 4,5-Dichloro-1,3-benzenedisulfonamide (Daranide) (see Fig. 13-3). Like the other carbonic anhydrase inhibitors, dichlorphenamide is seldom used as a diuretic. Little is known about its pharmacokinetics. It reduces intraocular pressure, as do other carbonic anhydrase inhibitors, and may be useful in the treatment of glaucoma. The importance of dichlorphenamide and chloraminophenamide is that they ultimately served as stepping stones away from the "pure" carbonic anhydrase-inhibiting diuretics and toward the development of the thiazide and thiazide-like diuretics, which are effective natriuretic and chloruretic agents with minimal carbonic anhydrase inhibitory activity.[28,30] Dichlorphenamide is available in 50-mg tablets.

SITE 3 DIURETICS: THE THIAZIDE AND THIAZIDE-LIKE DIURETICS

Chloraminophenamide became a logical key intermediate in the attempted development of diuretics that lacked the undesirable properties of the carbonic anhydrase inhibitors. When chloraminophenamide was treated with acylating reagents, cyclization occurred and 1,2,4-benzothiadiazine-1,1-dioxides were formed[28] (Fig. 13-4). The use of aldehydes or ketones in place of the acylating reagents yielded the corresponding dihydro-derivatives. The products of these reactions became known as thiazides and hydrothiazides, respectively. Hereafter, they will be referred to collectively as the *thiazide diuretics*. The thiazides were the first orally effective saluretic agents, the diuretic activity of which was not influenced by the acid–base status of the individual.

Cl, NH_2, H_2NO_2S, SO_2NH_2
Chloraminophenamide
Acylating Agents
Aldehydes (or Ketones)
Cl, N, R, H_2NO_2S, S, N, H, O_2
Thiazides
H, Cl, 5, 6, 7, 8, N, 4, 3, R, H, 1, 2, N, H_2NO_2S, S, H, O_2
Hydrothiazides

FIG. 13-4. *Development of thiazide and hydrothiazide diuretics from chloraminophenamide.*

STRUCTURE–ACTIVITY RELATIONSHIPS

Exhaustive SAR studies have been conducted with the thiazide diuretics.[28,30] Refer to Figure 13-4 for the numbering of the thiazide ring positions. Briefly, the 2-position can tolerate the presence of relatively small alkyl groups such as CH_3-. The 3-position is an extremely important site of molecular modification. Substitutents in the 3-position play a dominant role in determining the potency and duration of action of the thiazide diuretics. In addition, certain substituents in the 3-position have resulted in compounds that are relatively specific inhibitors of the diuretic action of the thiazides.[37–40] Loss of the carbon–carbon double bond between the 3- and 4-positions of the benzothiadiazine-1,1-dioxide nucleus increases the potency of this class of diuretics approximately three- to tenfold. Direct substitution of the 4-, 5-, or 8-position with an alkyl group usually results in diminished diuretic activity. Substitution of the 6-position with an "activating" group is essential for diuretic activity. The best substituents include Cl-, Br-, CF_3-, and NO_2- groups.

TABLE 13-1

CHLOROTHIAZIDE AND ANALOGUES

Generic Name	Proprietary Name	R	R_1	R_2
CHLOROTHIAZIDE AND ANALOGUES				
Chlorothiazide USP	Diuril, SK-Chlorothiazide	—Cl	—H	
Benthiazide USP	Exna, Aquatag, Hydrex	—Cl	$—CH_2—S—CH_2—C_6H_5$	

HYDROCHLOROTHIAZIDE AND ANALOGUES

Generic Name	Proprietary Name	R	R_1	R_2
Hydrochlorothiazide USP	HydroDIURIL, Esidrix, Oretic	—Cl	—H	—H
Hydroflumethiazide USP	Saluron, Diucardin	$—CF_3$	—H	—H
Bendroflumethiazide USP	Naturetin	$—CF_3$	$—CH_2—C_6H_5$	—H
Trichlormethiazide USP	Naqua, Triazide, Metahydrin	—Cl	$—CHCl_2$	—H
Methyclothiazide USP	Enduron, Aquatensen	—Cl	$—CH_2Cl$	$—CH_3$
Polythiazide USP	Renese	—Cl	$—CH_2—S—CH_2—CF_3$	$—CH_3$
Cyclothiazide USP	Anhydron	—Cl	(bicyclo[2.2.1]heptenyl)	—H

The sulfamoyl group in the 7-position is a prerequisite for diuretic activity. Table 13-1 depicts the commercially available diuretics that have evolved from the many alterations performed on the benzothiadiazine-1,1-dioxide nucleus.

When it was discovered that the sulfamoyl group *para* to the activating group in the *meta*-disulfamoylbenzenes could be replaced by several other electronegative groups with retention of diuretic activity (see Fig. 13-3), a host of diuretics emerged that have become known as *thiazide-like diuretics*. The diuretics shown in Figure 13-5 represent the most active member(s) of each series. Clearly, these diuretics are no longer benzothiadiazines, but their site of action, efficacy, electrolyte excretion pattern, and adverse effects resemble the thiazides. For these reasons, the thiazide and thiazide-like diuretics will be discussed as a group.

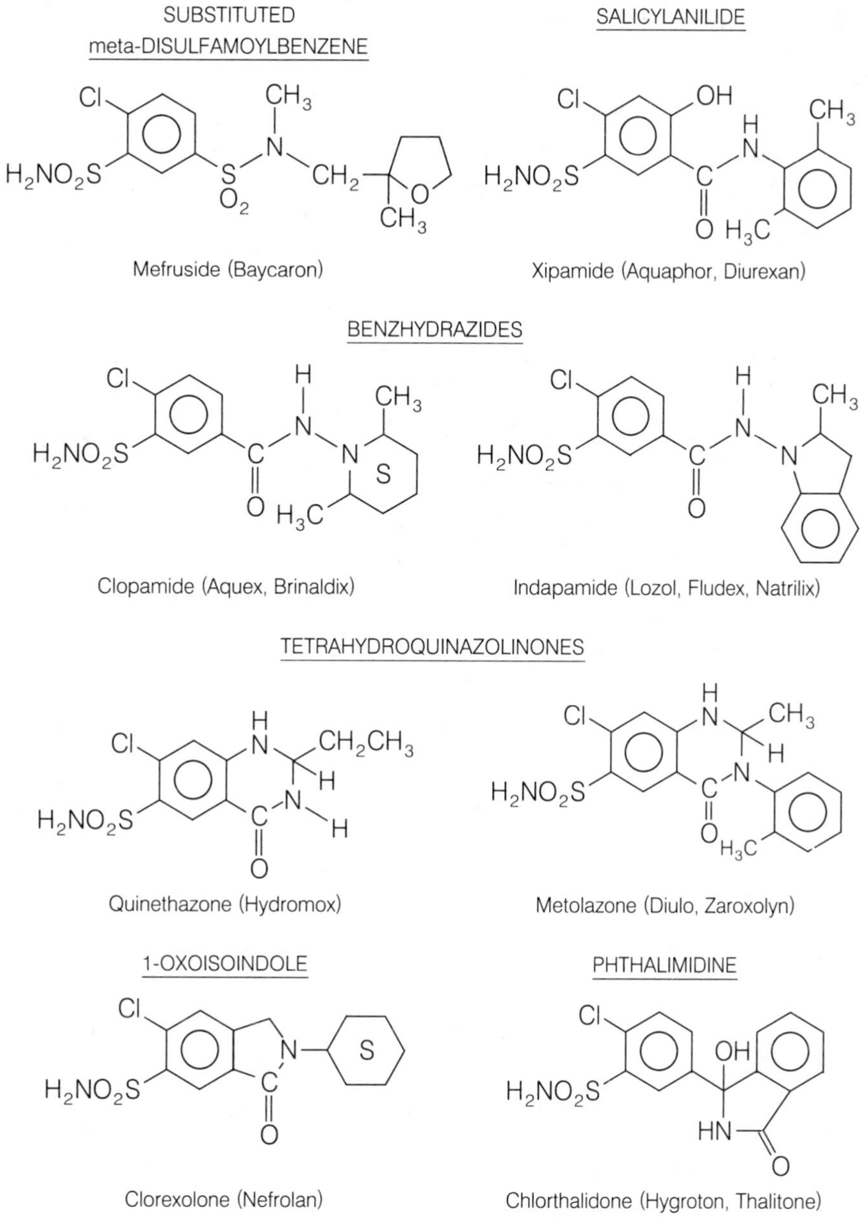

SUBSTITUTED meta-DISULFAMOYLBENZENE

Mefruside (Baycaron)

SALICYLANILIDE

Xipamide (Aquaphor, Diurexan)

BENZHYDRAZIDES

Clopamide (Aquex, Brinaldix)

Indapamide (Lozol, Fludex, Natrilix)

TETRAHYDROQUINAZOLINONES

Quinethazone (Hydromox)

Metolazone (Diulo, Zaroxolyn)

1-OXOISOINDOLE

Clorexolone (Nefrolan)

PHTHALIMIDINE

Chlorthalidone (Hygroton, Thalitone)

FIG. 13-5. *Representatives from six classes of thiazide-like diuretics. These diuretics were developed as an outgrowth of the thiazide research that involved molecular modification of aromatic sulfamoyl-containing compounds.*

PHARMACOKINETICS

Most of the thiazide and thiazide-like diuretics are well absorbed after oral administration, except chlorothiazide (which is only about 10% absorbed).[32] Their onset of action usually occurs within one to two hours, and their peak diuretic effect is expressed within three to six hours.[41] These diuretics do however differ drastically in their durations of action.[32,41–43] Most diuretics in this class are extensively bound to plasma proteins (or to red blood cell carbonic anhydrase for chlorthalidone and metolazone[32,44]), undergo little if any biotransformation (except mefruside and metolazone[44]), and are excreted primarily by the kidneys.[32,42,44] Relatively high luminal fluid concentrations of these diuretics are attained by a combination of glomerular filtration and active tubular secretion by the OATS in the proximal tubule (see Fig. 13-2). It appears that the luminal fluid concentration of these diuretics is more important for elicitation of a diuresis than their concentration in the interstitial fluid.[45]

The diuretics in this class differ primarily in their potencies and their durations of action. The differences in potency (which are reflected in their dosages[32,46]) are determined mainly by the chemical nature of the moiety attached to the 3-position of the benzothiadiazine nucleus, which modulates the overall lipophilicity of the diuretic.[28,30] The differences in their duration of action are dictated primarily by their degree of plasma protein binding (or red blood cell binding) and their lipid/water partition coefficients.[43,47] The latter values along with the pK of the drug and the pH of the luminal fluid determine the extent to which each member of the class undergoes reabsorption in the distal convoluted tubule by the process of nonionic back diffusion. Many of the diuretics in this class that have long half-lives undergo a significant degree of nonionic back diffusion. Pertinent pharmacologic data on the thiazide and thiazide-like diuretics are summarized in Tables 13-2 and 13-3, respectively.

TABLE 13-2

IMPORTANT PHARMACOLOGIC PARAMETERS OF THE THIAZIDE DIURETICS

Thiazide or Hydrothiazide (Trade Name)	Partition Coefficient (ether/H_2O)[47]	Plasma Protein Binding (%)[32,44]	Usual Daily Adult Oral Dosage Range (mg): Optimal Diuresis[32]	Usual Daily Adult Oral Dosage Range (mg): Hypertension[46]	Diuretic Effect: Onset (hr)[41]	Diuretic Effect: Peak (hr)[41]	Diuretic Effect: Duration (hr)[41–43]	% of the Parent Drug Excreted in the Urine[32,44]
Chlorothiazide USP *(Diuril, SK-Chlorothiazide)*	0.08	88–96	500–2000	250–1000	1–2	4	6–12	92
Benzthiazide USP *(Exna, Aquatag, Hydrex)*			50–200	25–200	2	4–6	12–18	
Hydrochlorothiazide USP *(HydroDIURIL, Esidrix, Oretic, Thiuretic)*	0.37	64	25–100	25–50	2	4	6–12	95(IV)
Hydroflumethiazide USP *(Diucardin, Saluron)*		95	25–200	25–200	1–2	3–4	18–24	65(oral) 85(IV)
Trichlormethiazide USP *(Metahydrin, Naqua, Triazide)*	1.53		1–4	2–4	2	6	up to 24	62–70
Bendroflumethiazide USP *(Naturetin)*		93	2.5–15	2.5–20	1–2	6–12	> 18	
Methyclothiazide USP *(Aquatensen, Enduron)*			2.5–10	2.5–5	2	6	24	
Polythiazide USP *(Renese)*		83.5	1–4	2–4	2	6	24–48	25(oral) 83(IV)
Cyclothiazide USP *(Anhydron, Fluidil)*			1–2	1–6	< 6	7–12	18–24	

TABLE 13-3

IMPORTANT PHARMACOLOGIC PARAMETERS OF THE THIAZIDE-LIKE DIURETICS

Thiazide-like Diuretic (Trade Name)	Partition Coefficient (Octanol/H_2O PO_4 Buffer)[30]	Plasma Protein Binding (%)[44]	Usual Daily Adult Oral Dosage Range (mg): Optimal Diuresis[32]	Usual Daily Adult Oral Dosage Range (mg): Hypertension[46]	Diuretic Effect: Onset (hr)[41]	Diuretic Effect: Peak (hr)[41]	Diuretic Effect: Duration (hr)[32,41–43]	% of the Parent Drug Excreted in the Urine[44]
Mefruside *(Baycaron)*								0.3–1.1
Xipamide *(Aquaphor, Diurexan)*		99						
Clopamide *(Aquex, Brinaldix)*	6.8	46						27
Indapamide *(Lozol, Fludex, Natrilix)*	31.7	71–79	2.5–5	2.5			24–36	5–7
Quinethazone USP *(Hydromox)*			50–200	50–200	2	6	18–24	
Metolazone *(Diulo, Zaroxolyn)*		50–70 (RBC)	2.5–20	2.5–5	1–2	2	12–24	80–95
Clorexolone *(Nefrolan)*								
Chlorthalidone USP *(Hygroton, Thalitone)*	5.0	94–99 (RBC)	25–200	25–50	2	2	24–72	44 (oral) 65 (IV)

SITE AND MECHANISM OF ACTION

The site of action of the thiazide and thiazide-like diuretics differs slightly from one species to another. However, it appears safe to conclude that all of these diuretics inhibit the reabsorption of sodium (and, thereby, the reabsorption of chloride) in the distal convoluted tubules (see *3*, Fig. 13-1) or in the connecting tubules (a short nephron segment that connects the distal convoluted tubule to the cortical collecting duct).[6,45,48,49] Thus, all diuretics in this class are responsible for the urinary loss of about 5% to 8% of the filtered load of sodium. Although they differ in their potencies (i.e., the amount of drug to produce a given diuretic response), it is important to note that they are all equally efficacious (i.e., they are all capable of exerting a similar maximal diuretic response).[28,30,50]

As a result of their site of action, the thiazide and thiazide-like diuretics alter the renal excretion rate of important ions other than sodium. Inhibition of sodium reabsorption at site 3 ultimately delivers more of the filtered load of sodium, at a faster rate, to site 4. At this latter site there is an enhanced exchange of the luminal fluid sodium for the intracellular potassium, and an increase in the urinary excretion rate of potassium follows. Most of the thiazide and thiazide-like diuretics possess a residual degree of carbonic anhydrase inhibitory activity that can be associated with a slight increase in the renal excretion rate of bicarbonate. Unlike the "pure" carbonic anhydrase inhibitors, resistance does not usually develop to the thiazide and thiazide-like diuretics as a result of drug-induced derangements in acid–base balance. Hence, diuretics in this class may be referred to as natriuretic, chloruretic, saluretic, kaliuretic, and extremely weak bicarbonaturetic agents. Importantly, short-term administration of thiazide and thiazide-like diuretics results in little, or no, change in the excretion of calcium; however, long-term use of these agents leads to a reduction in calcium excretion.[14]

ADVERSE EFFECTS

Some of the adverse effects associated with the thiazide and thiazide-like diuretics are highly predictable because of their chemical makeup or their site of action along the nephron. First, all of these diuretics possess a sulfamoyl moiety, which has been associated with hypersensitivity reactions such as urticaria, drug fever, blood dyscrasias, and interstitial nephritis. If an individual is hypersensitive to one of the agents in this class, then they will probably be hypersensitive to all of them. Cross-hypersensitivity may also occur between the thiazide and thiazide-like diuretics, carbonic anhydrase inhibitors, and the sulfamoyl-containing loop diuretics such as furosemide and bumetanide. Second, hypokalemia is a product of the diuretic-induced increase in the renal excretion of potassium. Third, initially, these diuretics produce a slight reduction in the cardiac output. A slight reduction in plasma volume and blood pressure occur upon continued use. These latter changes are frequently associated with an increase in the proximal tubule reabsorption of water and solutes, aldosterone secretion, and the renal release of renin. These changes usually assist in mitigating the diuretic effect, but the reduction of blood pressure persists.[51] Fourth, the primary site of uric acid and calcium reabsorption is the proximal tubule (i.e., site 1). When the recipient of a thiazide or a thiazide-like diuretic experiences a reduction in plasma volume, there is a concomitant increase in the proximal tubule reabsorption of luminal fluid and solutes. Because an increase in the proximal tubule reabsorption of calcium and uric acid occurs in this situation, it is not surprising to find an occasional case of hypercalcemia or hyperuricemia.[36] The seriousness of these two adverse effects is, in part, dependent on the duration and degree of the plasma volume reduction.

The precise mechanisms behind some of the adverse effects of the thiazide and thiazide-like diuretics are not well understood. These include a persistent reduction in the glomerular filtration rate (especially after intravenous administration[32]) and hyperglycemia. It is unlikely that the reduction in glomerular filtration rate is related to the tubuloglomerular feedback mechanism, because the major site of action of these diuretics is distal to the macula densa cells. Some investigators have suggested that the thiazide and thiazide-like diuretics act directly on the renal vasculature to depress the glomerular filtration rate.[32,42] Nonetheless, the reduction in the glomerular filtration rate involves all diuretics in this class, with the possible exception of metolazone and indapamide.[52] This is particularly important to those individuals who must commence diuretic therapy, but who have preexisting impairment of renal function. Thiazide and thiazide-like drugs are frequently ineffective in individuals who have a glomerular filtration rate of less than 15 to 25 mL/min. Metolazone[42,43,46,53,54] and indapamide[52] may be exceptions.

If lithium is administered to individuals on long-term thiazide or thiazide-like diuretic therapy, serious toxicity can occur. The proximal tubules handle lithium and sodium in a similar fashion. During long-term thiazide treatment, the resulting reduction in plasma volume triggers a compensatory increase in the proximal tubule reabsorption of fluid and solutes. Thus, more lithium is reabsorbed than would occur in normovolemic individuals. Elevated plasma levels of lithium may provoke serious adverse effects.[43]

USES

Thiazide and thiazide-like diuretics are extremely useful in the treatment of edema associated with mild to moderate congestive heart failure, cirrhosis of the liver, or the nephrotic syndrome. Because edema is a symptom of an underlying disease and not a disease itself, the underlying disease should be treated first if possible. If treatment of the underlying disease does not result in the removal of the edema fluid, then diuretic therapy may be indicated. Caution should always be exercised when thiazide or thiazide-like diuretics are coadministered with cardiac glycosides for the treatment of edema associated with congestive heart failure. These diuretics have a tendency to promote hypokalemia, a condition that enhances the general toxicity of the cardiac glycosides.[36,43] Combination diuretic therapy (i.e., a thiazide or thiazide-like diuretic plus a potassium-sparing diuretic) may avoid potassium loss under these circumstances. If combination diuretic therapy is instituted, the recipient should be advised not to take potassium supplements, to avoid serious hyperkalemia.[32]

Thiazide and thiazide-like diuretics are also useful in the treatment of certain nonedematous disorders. These include hypertension, diabetes insipidus (either the nephrogenic or the neurohypophyseal form), type II renal tubular acidosis, and hypercalciuria. These diuretics are primary agents in the control of hypertension—either alone or in combination with other drugs, depending on its severity. Thiazides generally lower blood pressure 10 to 15 mmHg within the first three to four days of continuous treatment.[55] After approximately a week of continuous treatment (about the time that there is a concomitant reduction in plasma volume), the kidney readjusts to the initial effects of the diuretic, and a waning of the diuretic effect is observed, whereas the reduction of blood pressure is maintained.[51] This readjustment occurs provided sodium intake is not increased.

Some individuals with hypercalciuria (i.e., an elevated urinary concentration of calcium) are prone to the formation of calcium-containing stones within the urinary tract. Because the chronic use of thiazide and thiazide-like diuretics decreases the urinary excretion rate of calcium, they may be helpful in preventing calcium-containing stone formation.[36,56]

PRODUCTS

Pertinent pharmacologic data on the thiazide and thiazide-like diuretics are cited in Tables 13-2 and 13-3.

Chlorothiazide, USP. 6-Chloro-2*H*-1,2,4-benzothiadiazine-7-sulfonamide 1,1-dioxide (Diuril, SK-Chlorothiazide). Chlorothiazide is usually supplied as 250- and 500-mg tablets, an oral suspension containing 50 mg/mL, and as the sodium salt for injection [500 mg (base)].

Benzthiazide, USP. 6-Chloro-3 [[(phenylmethyl)thio]methyl]-2*H*-1,2,4-benzothiadiazine-7-sulfonamide 1,1-dioxide (Aquatag; Exna; Hydrex). Benzthiazide is usually supplied as 25- and 50-mg tablets.

Hydrochlorothiazide, USP. 6-Chloro-3,4-dihydro-2*H*-1,2,4-benzothiadiazine-7-sulfonamide 1,1-dioxide (Esidrix; HydroDIURIL; Oretic; Thiuretic). Hydrochlorothiazide is usually supplied as 25-, 50-, and 100-mg tablets and as an oral solution (10 and 100 mg/mL). It is also available in fixed combinations with the potassium-sparing diuretics.

Hydroflumethiazide, USP. 3,4-Dihydro-6-(trifluoromethyl)-2*H*-1,2,4-benzothiadiazine-7-sulfonamide 1,1-dioxide (Diucardin; Saluron). Hydroflumethiazide is usually supplied as 50-mg tablets.

Trichlormethiazide, USP. 6-Chloro-3-(dichloromethyl)-3,4-dihydro-2*H*-1,2,4-benzothiadiazine-7-sulfonamide 1,1-dioxide (Metahydrin; Naqua; Triazide). Trichlormethiazide is usually supplied as 2- and 4-mg tablets.

Bendroflumethiazide, USP. 3-Benzyl-3,4-dihydro-6-(trifluoromethyl)-2*H*-1,2,4-benzothiadiazine-7-sulfonamide 1,1-dioxide (Naturetin). Bendroflumethiazide is usually supplied as 2.5-, 5-, and 10-mg tablets.

Methyclothiazide, USP. 6-Chloro-3-(chloromethyl)-3,4-dihydro-2-methyl-2*H*-1,2,4-benzothiadiazine-7-sulfonamide 1,1-dioxide (Aquatensen; Enduron). Methyclothiazide is usually supplied as 2.5- and 5-mg tablets.

Polythiazide, USP. 6-Chloro-3,4-dihydro-2-methyl-3-[[(2,2,2-trifluoroethyl)thio]methyl]-2*H*-1,2,4-benzothiadiazine-7-sulfonamide 1,1-dioxide (Renese). Polythiazide is usually supplied as 1-, 2-, and 4-mg tablets.

Cyclothiazide, USP. 3-Bicyclo[2.2.1]hept-5-en-2-yl-6-chloro-3,4-dihydro-2*H*-1,2,4-benzothiadiazine-7-sulfonamide 1,1-dioxide (Anhydron; Fluidil). Cyclothiazide is usually available as 2-mg tablets.

Mefruside. 4-Chloro-N^1-methyl-N^1-[(tetrahydro-2-methyl-2 furanyl)methyl]-1,3-benzenedisulfonamide (Baycaron). Not available in the United States.

Xipamide. 4-Chloro-5-sulfamoyl-2′,6′-salicyloxylidide (Aquaphor; Diurexan). Not available in the United States.

Clopamide. 3-(Aminosulfonyl)-4-chloro-*N*-(2,6-dimethyl-1-piperidinyl)benzamide (Aquex; Brinaldix). Not available in the United States.

Indapamide. 3-(Aminosulfonyl)-4-Chloro-*N*-(2,3-dihydro-2-methyl-1*H*-indol-1-yl)benzamide; 4-chloro-*N*-(2-methyl-1-indolinyl)-3-sulfamoylbenzamide (Fludex; Lozol; Natrilix). Indapamide is usually available as 2.5-mg tablets.

Quinethazone, USP. 7-Chloro-2-ethyl-1,2,3,4-tetrahydro-4-oxo-6-quinazolinesulfonamide (Hydromox). Quinethazone is usually supplied as 50-mg tablets.

Metolazone. 7-Chloro-1,2,3,4-tetrahydro-2-methyl-4-oxo-3-*o*-tolyl-6-quinazolinesulfonamide (Diulo; Zaroxolyn). Metolazone is usually supplied as 2.5-, 5-, and 10-mg tablets.

Clorexolone. 6-Chloro-2-cyclohexyl-2,3-dihydro-3-oxo-1*H*-isoindole-5-sulfonamide (Nefrolan). Not available in the United States.

Chlorthalidone, USP. 2-Chloro-5-(1-hydroxy-3-oxo-1-isoindolinyl)benzenesulfonamide (Hygroton; Thalitone). Chlorthalidone is usually supplied as 25-, 50-, and 100-mg tablets.

SITE 2 DIURETICS: HIGH-CEILING OR LOOP DIURETICS

The diuretics that belong to this class are of extremely diverse chemical structure.[57] Although brief mention will be made of the organomercurial diuretics, primary attention will be focused on the agents that currently enjoy widespread use. For example, furosemide (a 5-sulfamoyl-2-aminobenzoic acid or anthranilic acid derivative), bumetanide (a 5-sulfamoyl-3-aminobenzoic acid or metanilic acid derivative), and ethacrynic acid (a phenoxyacetic acid derivative).

ORGANOMERCURIALS

The organomercurials were the mainstay of diuretic therapy from 1920 to the early 1950s.[28] They elicit a diuresis by an inhibition of sodium reabsorption at site 2,[58] and they also block the subsequent exchange of sodium for potassium at site 4.[59] Thus, they are natriuretic and chloruretic without being kaliuretic. These properties are true attributes for any class of diuretics; however, the organomercurials have several serious limitations. First, they cannot be relied upon to elicit a diuresis when given orally because of poor and erratic absorption. Second, after their parenteral administration there is a one- to two-hour lag in the onset of the diuresis.[60] Third, their ability to trigger a diuretic response is dependent on the acid–base status of the individual (i.e., they are ineffective when the urine is alkaline).[61,62] Fourth, they are cardio- and nephrotoxic. The organomercurials became obsolete with the introduction of the thiazides, thiazide-like diuretics, furosemide, bumetanide, and ethacrynic acid. All of the latter agents are orally effective, equally effective in acidotic or alkalotic conditions, capable of inducing a relatively rapid diuresis when given parenterally, and relatively nontoxic. The reader is encouraged to consult the previous edition of this textbook for an in-depth discussion of the organomercurials.

5-SULFAMOYL-2- AND -3-AMINOBENZOIC ACID DERIVATIVES

Bumetanide. 3-(Butylamino)-4-phenoxy-5-sulfamoylbenzoic acid (Bumex) (Fig. 13-6).

Furosemide, USP. 4-Chloro-*N*-furfuryl-5-sulfamoylanthranilic acid (Lasix) (see Fig. 13-6).

Structure–Activity Relationships

The development of these diuretics is an outgrowth of the research involving the thiazide and thiazide-like diuretics.[30] There are important structural requirements that are common to the 5-sulfamoyl-2-aminobenzoic acid derivatives and the 5-sulfamoyl-3-aminobenzoic acid derivatives (see Fig. 13-6). First, the substitutent at the 1-position must be acidic. The carboxyl group provides optimal diuretic activity, but other groups, such as a tetrazole, may impart respectable diuretic activity. Second, a sulfamoyl group in the 5-position is a prerequisite for optimal high-ceiling diuretic activity. Third, the "activating" group (X-) in the 4-position can be Cl- or CF_3-, as was true for the thiazides and thiazide-like diuretics or, better yet, a phenoxy-, alkoxy-, anilino-, benzyl-, or benzoyl-group. Interestingly, when the latter five functional groups are substituted for the Cl- or CF_3-group in the thiazides or thiazide-like diuretics, their diuretic activity is decreased.

Major differences exist between these two series of 5-sulfamoyl-benzoic acids in the nature of the functional groups that can be substituted into the 2- and 3-positions with the retention of maximal diuretic activity (see Fig. 13-6). The substituents that can be tolerated on the 2-amino group of the 5-sulfamoyl-2-aminobenzoic acids series are extremely limited, and no deviations are allowed on the few moieties that are acceptable. For example, only furfuryl-, benzyl-, and thienylmethyl (in decreasing order) yield derivatives with maximal diuretic activity. On the other hand, the substituents allowable on

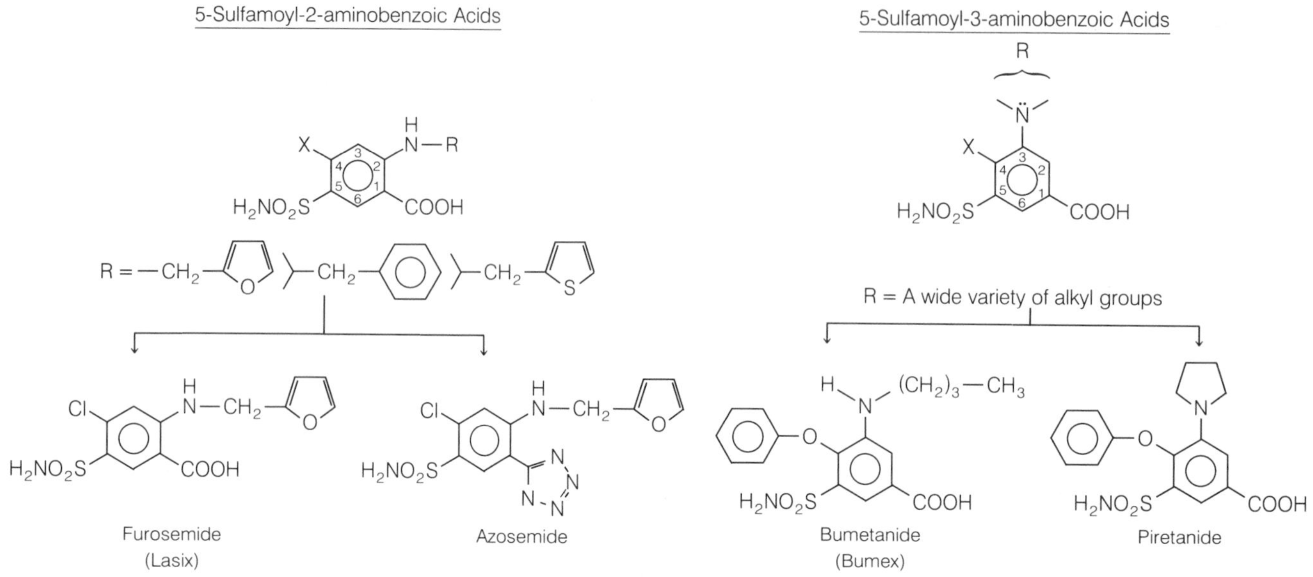

FIG. 13-6. *Results from structure–activity relationship studies that led to the development of furosemide and bumetanide.*

the 3-amino group of the 5-sulfamoyl-3-aminobenzoic acids can vary widely without jeopardizing optimal diuretic activity. High-ceiling diuretics that have emerged from the 5-sulfamoyl-2-aminobenzoic acid series include furosemide and azosemide, and from the 5-sulfamoyl-3-aminobenzoic acid series bumetanide and piretanide. Only furosemide and bumetanide are now commercially available in the United States.

Pharmacokinetics

Furosemide and bumetanide differ pharmacologically primarily in their potencies and bioavailabilities. When administered orally, the bioavailability of furosemide is about 60% to 69% in normal subjects, but is reduced to 43% to 46% in individuals with end-stage renal disease.[43] The bioavailability of bumetanide in normal individuals is 80% to 90%.[63] Bumetanide is also more potent than furosemide; it will produce an equieffective diuresis in about 1/40th the dose.[63,64]

Following parenteral administration, both furosemide and bumetanide have an extremely rapid onset of action (three to five minutes). Their durations of action following parenteral therapy are two hours for furosemide and 3.5 to 4 hours for bumetanide. Both diuretics have an onset of action of approximately 30 to 60 minutes after oral therapy, but furosemide has a slightly longer duration of action than bumetanide (six to eight hours compared with four to six hours).[41] Once these agents reach the bloodstream, they are extensively bound to plasma proteins (93% to 95%).[63] This degree of plasma protein binding severely limits the amount of each drug that can be removed from the plasma by glomerular filtration, but it does not prevent both drugs from attaining high renal luminal fluid concentrations by the process of active tubular secretion. Both diuretics are weak organic acids and are avidly secreted into the luminal fluid of the proximal tubule (see Fig. 13-2).[20,65] This is important for two reasons. First, it permits the relatively rapid renal excretion of both diuretics. Second, it provides for the delivery of substantial amounts of each diuretic to their luminal site of action.

The previously discussed factors that determine the luminal concentration of diuretics are critically important when employing these agents in individuals with uremia. Uremic individuals frequently have a depressed glomerular filtration rate and high circulating levels of endogenous weak acids, both of which lower the luminal fluid concentrations of the loop diuretics. The endogenous weak acids compete with the weakly acidic diuretics for active tubular secretion into proximal tubule luminal fluid. Often the effects of the endogenous weak acids can be overridden by increasing the dose of these diuretics. Caution must be exercised, however, because an increased frequency of adverse effects is likely to accompany increased doses.

A small percentage of furosemide is converted to the corresponding glucuronide, and 88% of the administered drug is excreted by the kidneys. Bumetanide undergoes more extensive biotransformation in the human, and 81% of it is excreted in the urine (45% as unchanged drug).[41]

Site and Mechanism of Action

The tremendous magnitude of the diuresis associated with furosemide and bumetanide is multifaceted. First, the high luminal fluid concentrations of furosemide and bumetanide allow these diuretics to inhibit the $1Na^+/1K^+/2Cl^-$ cotransport system located on the luminal membrane of cells of the thick ascending limb of Henle's loop[21,57] (i.e., *2*, Fig. 13-1). Approximately 20% to 25% of the filtered load of sodium that is normally reabsorbed in this nephron segment may be excreted. In addition, remember that it is the reabsorption of this 20% to 25% of the filtered load of sodium (and chloride) that is required to maintain the hypertonicity of the medullary interstitium.[12] The hypertonic medullary interstitium allows us to produce concentrated urine by drawing water out of the descending limb of the loop of Henle by osmosis and out of the collecting duct by osmosis when antidiuretic hormone is present. Thus, when these diuretics inhibit the reabsorption of 20% to 25% of the filtered load of sodium at site 2, within minutes they also destroy the hypertonicity of the medullary interstitium.[6] The net result is that when sodium and chloride fail to get reabsorbed at site 2, water is no longer removed from the luminal fluid in the descending limb of Henle's loop or from the collecting tubule by osmosis. Large amounts of water, sodium, and chloride are excreted. Second, although these diuretics increase the flow rate of luminal fluid past the macula densa cells, the expected reduction in glomerular filtration rate (which would normally mitigate the diuresis) does not occur. This is because these efficacious diuretics block the tubuloglomerular feedback mechanism.[66] Third, these diuretics also transiently increase total renal blood flow and induce a redistribution of intrarenal blood flow that is thought to participate in a positive way toward the magnitude of the diuresis.[57]

All diuretics that act at site 2 are equally efficacious and far more efficacious than diuretics that act at sites 1, 3, or 4. Because of their site of action and their efficacy, these agents are commonly referred to as *loop diuretics* and *high-ceiling diuretics*.[57]

The high-ceiling diuretics enhance the urinary loss of potassium by two mechanisms. First, they directly block the reabsorption of potassium at site 2 by inhibiting the $1Na^+/1K^+/2Cl^-$ cotransport complex. Second, inhibition of sodium reabsorption at site 2 ultimately delivers more of the filtered load of sodium, at a faster rate, to site 4. These events lead to an enhanced exchange of luminal fluid sodium for intracellular potassium and hydrogen ions at site 4.

When the loop diuretics are used in "submaximal" doses for the treatment of hypertension, it is intended that they create a diuresis that is similar in magnitude to that produced by the thiazide and thiazide-like diuretics. Under these circumstances the loop diuretics usually are associated with a lower frequency of hypokalemia than the thiazide and thiazide-like diuretics because their duration of action is not as prolonged and the kidneys have a greater portion of the day to readjust.[30,67] On the other hand, when the loop diuretics are used to treat acute edema, higher dosages are frequently employed, and the sodium and potassium losses exceed those accompanying thiazide therapy.[30]

For reasons that are not yet completely understood, the actions of furosemide and bumetanide in the thick ascending limb of Henle's loop are associated with an increase in magnesium and calcium excretion.[36] Approximately 20% to 25% of the filtered load of calcium will be excreted, provided the plasma volume is not allowed to decrease. If plasma volume decreases as a result of the diuresis, then there is an accompanying compensatory increase in the proximal tubule reabsorption of fluid and solutes. Because about 60% of the filtered load of calcium is normally reabsorbed in the proximal tubule, the percentage reabsorbed in a state of plasma volume reduction will increase and blunt the loop diuretic's calciuretic effect.

Adverse Effects

Highly predictable adverse effects are associated with furosemide and bumetanide.

1. Hypokalemic alkalosis occurs as a result of the enhanced exchange of luminal fluid sodium for intracellular potassium or hydrogen ions at site 4. Caution should be exercised when concurrent therapy with loop diuretics and cardiac glycosides is instituted because hypokalemia intensifies the toxicity of the cardiac glycosides.[41]
2. Initially, fluid and electrolyte losses may not be accompanied by changes in the glomerular filtration rate because of the effect of these agents on the tubuloglomerular feedback mechanism. However, long-term use of these diuretics may induce a reduction in plasma volume. If this condition is allowed to persist, then previously mentioned compensatory changes take place—one of which is a reduction in the glomerular filtration rate.
3. Because diuretic-induced plasma volume reduction leads to an increase in the reabsorption of solutes normally handled by the proximal tubule, it should not be surprising to find that some individuals develop hyperuricemia when placed on long-term loop diuretic therapy. For similar

reasons, coadministration of the loop diuretics and lithium may lead to severe lithium toxicity.[41]

4. Furosemide and bumetanide are similar to the carbonic anhydrase inhibitors, thiazides, and the thiazide-like diuretics in that they possess a sulfamoyl moiety. This functional group has been associated with hypersensitivity reactions such as urticaria, drug fever, blood dyscrasias, and interstitial nephritis.

Several adverse effects, which were unforeseen, are associated with these loop diuretics. For example, the loop diuretics are unique among diuretics in that they can produce ototoxicity. Usually, the hearing loss is temporary, but on occasion it may be permanent. Ototoxicity may be directly associated with rising plasma concentrations of the loop diuretics. Accordingly, individuals with impaired renal function appear to be at risk because they have a reduced ability to excrete these efficacious diuretics.[43] Although the milligram dose of bumetanide is 1/40th that of furosemide, both agents appear to be quite similar in their ototoxic potential. Caution must be exercised if loop diuretics must be administered to patients who are receiving any of the currently available aminoglycoside antibiotics. The ototoxicity of these two classes of drugs may be additive.[41] Other adverse effects of furosemide and bumetanide include hyperglycemia, nausea, vomiting, and myalgia.

Uses

High-ceiling diuretics are effective for the treatment of edema of congestive heart failure, cirrhosis of the liver, and nephrotic syndrome. A most important use of furosemide or bumetanide is in the treatment of pulmonary edema. No other group of diuretics is more effective than the loop diuretics in this situation. When loop diuretics are employed in the treatment of pulmonary edema that may accompany congestive heart failure, it is of paramount importance to avoid their overzealous use. Such use may lead to a sudden reduction in plasma volume, decreased venous return and cardiac output, and an exacerbation of the heart failure.[68]

Loop diuretics may be employed for the treatment of certain nonedematous disorders. Symptomatic hypercalcemia may be treated with a loop diuretic, provided that a reduction in plasma volume is not allowed to occur and the fluid used for replacement of the urinary losses is calcium-free.[69] In addition, furosemide has been used for the treatment of hypertension. However, some investigators believe that because of its relatively short duration of action, it may be less effective than the thiazide or thiazide-like diuretics. It has been suggested that furosemide be reserved for hypertensive patients with fluid retention refractory to thiazides or for patients with impaired renal function.[46]

In general, furosemide or bumetanide is preferred over ethacrynic acid (another site 2 diuretic) because they have a broader dose–response curve, less ototoxicity, and less gastrointestinal toxicity.[43]

Products

Bumetanide is available as 0.5- and 1.0-mg tablets and as a powder that is to be made into a solution for parenteral use (0.25 mg/mL).

Furosemide is available as an oral solution that is to be made up to contain 10 mg/mL; as 20-, 40-, and 80-mg tablets; and as a solution for injection that is to be made up to contain 10 mg/mL.

PHENOXYACETIC ACIDS

The phenoxyacetic acid group of high-ceiling diuretics was developed and introduced into clinical use about the same time as furosemide.

Ethacrynic Acid, USP. [2,3-Dichloro-4-(2-methylene-1-oxobutyl)phenoxy]acetic acid (Edecrin) (Fig. 13-7).

Structure–Activity Relationships

As mentioned previously, certain organomercurials are capable of eliciting a diuretic response, but because of their heavy metal content they are too toxic for widespread use. Consequently, a search was commenced for a nonmercury-containing compound that would react with sulfhydryl-containing receptors in renal tissue, like the organomercurials, but be devoid of heavy metal-type toxicity. Because one of the commercially available organomercurials (i.e., mersalyl; Salyrgan) possessed a phenoxyacetic acid moiety, the phenoxyacetic acids served as the chemical root for new nonmercury-containing diuretics. Hundreds of phenoxyacetic acids were examined.[70,71]

Within the phenoxyacetic acid series, optimal diuretic activity is achieved when (see Fig. 13-7) (1) an oxyacetic acid moiety is placed in the 1-position on the benzene ring, (2) a sulfhydryl-reactive acryloyl moiety is located *para* to the oxyacetic acid group, (3) activating groups (i.e., Cl- or CH_3-) occupy either the 3-position or the 2- and 3-positions, (4) alkyl substituents of two- to four-carbon atoms in length occupy the position α to the carbonyl on the acryloyl moiety, and (5) hydrogen atoms occupy the terminal

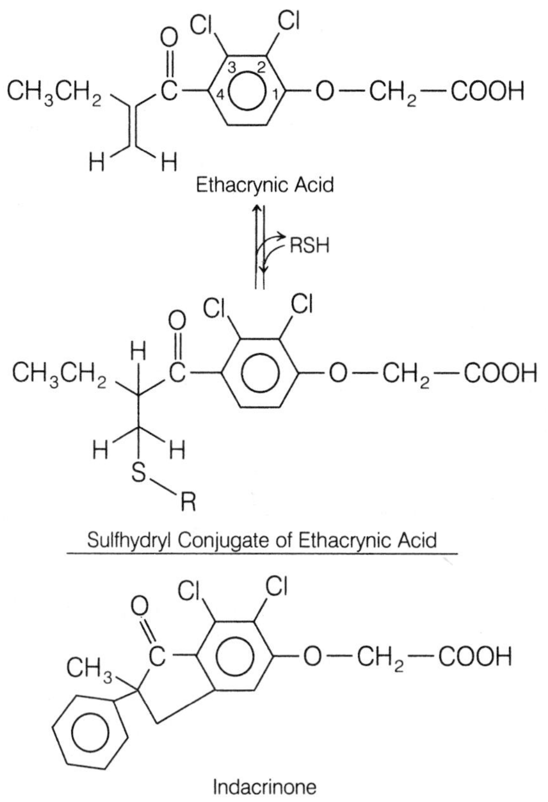

Ethacrynic Acid

Sulfhydryl Conjugate of Ethacrynic Acid

Indacrinone

FIG. 13-7. *The structure of ethacrynic acid and its reactivity with sulfhydryl-containing nucleophiles are shown in the upper portion of the figure. Indacrinone, a high-ceiling diuretic that lacks sulfhydryl reactivity, is shown in the lower portion of the figure.*

position of the carbon–carbon double bond of the acryloyl moiety. These structural features seemed to maximize both the diuretic activity and the in vitro rate of reactivity with various sulfhydryl-containing nucleophiles. The correlation between diuretic activity and chemical reactivity within this series of diuretics was further strengthened when it was found that reduction or epoxidation of the carbon–carbon double bond in the acryloyl moiety yielded compounds with little or no diuretic activity or chemical reactivity.[72,73] The design and synthesis of ethacrynic acid appeared to be the ultimate in terms of "the rational approach to drug design." However, the need for designing a diuretic with a high degree of sulfhydryl reactivity was foiled when indacrinone was found to be a highly efficacious diuretic, but incapable of reacting with sulfhydryl-containing nucleophiles (see Fig. 13-7).[74] Indacrinone has been withdrawn from clinical trials because of the appearance of abnormal liver function test results in certain individuals.

Pharmacokinetics

In spite of ethacrynic acid's unique chemical structure and its avid reactivity toward various nucleophiles, it has many pharmacologic features in common with the sulfamoyl-containing loop diuretics. First, after oral administration, its onset of action is about 30 minutes and its duration of action is six to eight hours. Second, after parenteral administration, ethacrynic acid's onset of action and duration of action are three to five minutes and two to three hours, respectively. Third, ethacrynic acid is highly bound to plasma proteins (i.e., $> 95\%$). Fourth, ethacrynic acid is handled and excreted predominately by the kidneys. Very little of the drug is removed from the plasma by the process of glomerular filtration because of its extensive binding to unfiltrable plasma proteins, such as albumin. However, the drug is avidly secreted into the luminal fluid of the proximal tubule with the assistance of the organic anion transport system (OATS; see Fig. 13-2).[65,75,76] High luminal fluid concentrations of ethacrynic acid are essential for its diuretic action and for its ultimate excretion.

Ethacrynic acid is biotransformed by a completely different pathway than furosemide or bumetanide. Ethacrynic acid alkylates the thiol group of glutathione in vivo (see Fig. 13-7; RSH = glutathione), and the resulting conjugate is subsequently converted to the ethacrynic acid–cysteine and ethacrynic acid–*N*-acetylcysteine (the mercapturic acid) conjugates. Ethacrynic acid–cysteine is quite unstable in vivo and in vitro; it readily releases cysteine and ethacrynic acid. Ethacrynic acid, ethacrynic acid–glutathione, and ethacrynic acid–cysteine are equiefficacious diuretics because of the aforementioned interconversions.[70] Approximately two-thirds of the ethacrynic acid appears in the urine in the various forms cited, whereas the remaining one-third is found in the bile.

Site and Mechanism of Action

Similar to furosemide and bumetanide, ethacrynic acid (1) inhibits the reabsorption of 20% to 25% of the filtered load of sodium at site 2[21] (i.e., the $1Na^+/1K^+/2Cl^-$ cotransport system located on the luminal membrane of cells in the thick ascending limb of Henle's loop); (2) blocks the tubuloglomerular feedback mechanism that normally would result in a sharp reduction of the glomerular filtration rate when the flow of luminal fluid is increased through the nephron segment possessing the macula densa cells;[77] and (3) transiently induces a redistribution of intrarenal blood flow, which contributes in a positive way toward the magnitude of the diuresis.[78] Because ethacrynic acid induces an acute increase in the renal excretion rate of sodium, chloride, potassium, and calcium, it is a natriuretic, chloruretic, saluretic, kaliuretic, and calciuretic agent.

Adverse Effects

Ethacrynic acid may produce all of the adverse effects noted with furosemide and bumetanide, except those related to the presence of a sulfamoyl group. The use of ethacrynic acid has been curtailed because it is more ototoxic than furosemide and bumetanide, and it produces more serious gastrointestinal effects (i.e., gastrointestinal hemorrhage) than observed with the sulfamoyl-containing loop diuretics.

Uses

Ethacrynic acid has the same indications as cited for furosemide and bumetanide. However, when a high-ceiling diuretic is indicated in the treatment of an individual who has a known hypersensitivity to sulfamoyl-containing drugs, ethacrynic acid may be an appropriate substitute.

Products

Ethacrynic acid is available as 25- and 50-mg tablets, and as ethacrynate sodium for injection which contains 50 mg (base).

SITE 4 DIURETICS: POTASSIUM-SPARING DIURETICS

A negative feature of all of the previously discussed classes of diuretics (except the organomercurials) is that they induce an increase in the renal excretion rate of potassium. Over the years, three chemically distinct diuretics have emerged that increase sodium and chloride excretion, without a concomitant increase in the urinary excretion rate of potassium. These agents are known as *potassium-sparing diuretics* or *antikaliuretic agents*. Although the potassium-sparing diuretics are derived from completely different chemical roots, they have a similar anatomic site of action in the nephron, efficacy, and electrolyte excretion pattern. They even share certain adverse effects. The potassium-sparing diuretics include the spirolactone, spironolactone; the 2,4,7-triamino-6-arylpteridine, triamterene; and the pyrazinoylguanidine, amiloride.

SPIROLACTONES, ALDOSTERONE ANTAGONISTS

Spironolactone, USP. 7α-(Acetylthio)-17β-hydroxy-3-oxopregn-4-ene-21-carboxylic acid γ-lactone (Aldactone) (Fig. 13-8).

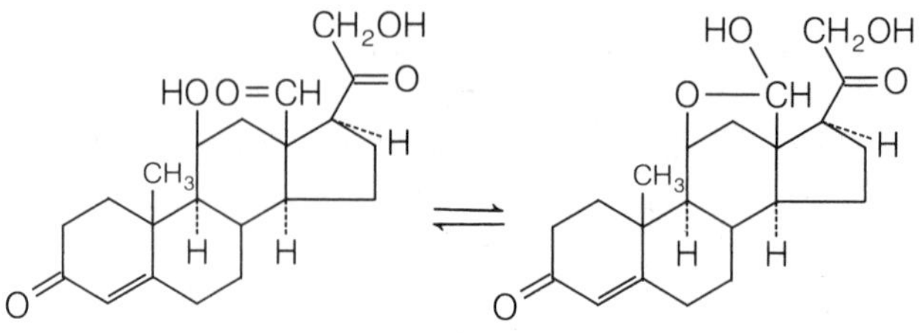

(Aldehyde Form) Aldosterone (Hemiacetal Form)

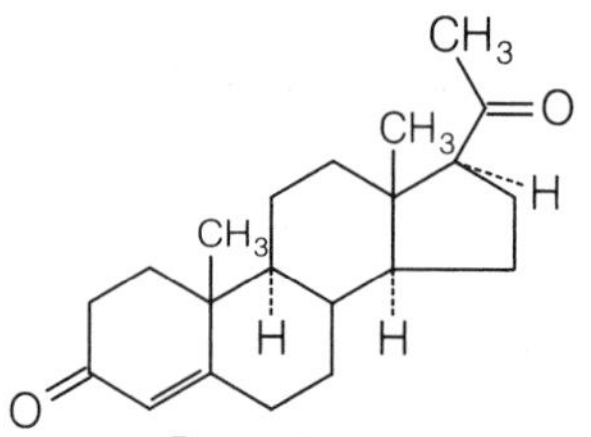

Progesterone

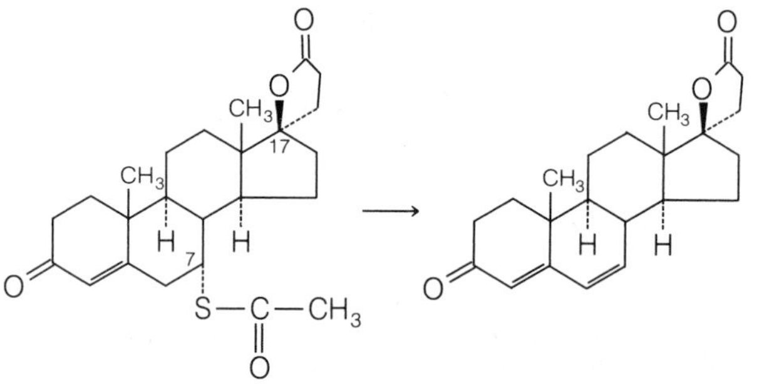

Spironolactone (Aldactone)

Canrenone (a major metabolite) (an active aldosterone antagonist)

FIG. 13-8. Aldosterone enhances the passage of sodium from the luminal fluid into tubular cells and the passage of intracellular potassium into the luminal fluid at site 4. Progesterone inhibits these actions of aldosterone, but has undesirable hormonal side effects. Spironolactone and canrenone also competitively inhibit the actions of aldosterone at site 4 and are associated with a much lower frequency of homonal side effects.

Structure–Activity Relationships

In the mid-1950s, it was observed that progesterone inhibited the antinatriuretic and kaliuretic effects of aldosterone—the primary mineralocorticoid in humans.[79,80] An intensive effort was launched to develop steroidal derivatives that possessed only the antimineralocorticoid activity of progesterone.[81–84] Spironolactone was selected from a host of derivatives for further examination.[83]

Pharmacokinetics

Spironolactone is well absorbed after oral administration (bioavailability is > 90%); rapidly and extensively biotransformed by the liver (about 80%) to canrenone—an active metabolite (see Fig. 13-8); extensively bound to plasma proteins (most likely as canrenone); and excreted primarily as metabolites in the urine. Some biliary excretion of metabolites also occurs. Its onset of action is slow (12 to 72 hours),[85] and its duration of action is quite long (two to three days).[41]

Site and Mechanism of Action

Spironolactone inhibits the reabsorption of 2% to 3% of the filtered load of sodium at site 4 by competitively inhibiting the actions of aldosterone.[6] Under normal circumstances, aldosterone enhances the synthesis of transport proteins, such as the Na^+, K^+–ATPase,[86,87] and the luminal membrane channels that are involved in the exchange of sodium for potassium or hydrogen ions at site 4. The permeability of the luminal membrane to sodium and, perhaps, to potassium and hydrogen ions increases. Passage of luminal fluid sodium into, and potassium and hydrogen ions out of, the late distal convoluted tubule and the early collecting tubule cells is enhanced. Increased intracellular levels of sodium stimulate the basolateral membrane-bound Na^+, K^+–ATPase.[11] Spironolactone acts by competitively inhibiting these actions of aldosterone.[87] In doing so, spironolactone enhances water, sodium, and chloride excretion and, therefore, is a natriuretic, chloruretic, saluretic, and an antikaliuretic agent. Unlike the other potassium-sparing diuretics, spironolactone requires the presence of endogenous aldosterone to exert its diuretic action. Because spironolactone inhibits the reabsorption of only 2% to 3% of the filtered load of sodium, it (and the other members of the site 4 potassium-sparing diuretics) has an extremely low efficacy.

Adverse Effects

It might be anticipated that inhibition of the exchange of luminal fluid sodium for intracellular potassium and hydrogen ions might lead to the retention of the latter two ions in certain individuals. The major adverse effects of spironolactone are hyperkalemia and mild metabolic acidosis—especially in individuals with poor renal function.[6,36] In addition, spironolactone may produce gynecomastia in men and breast tenderness and menstrual disturbances in women because of its residual hormonal activity.[88] Other adverse effects associated with spironolactone include minor gastrointestinal symptoms and skin rashes.[32,43]

Uses

Spironolactone may be used alone as an extremely mild diuretic to remove edema fluid in individuals with congestive heart failure, cirrhosis, or nephrotic syndrome, or as an antihypertensive agent. However, its primary use has been in combination with diuretics that act at site 2 or 3 in an attempt to reduce the urinary potassium loss associated with these latter groups of diuretics.

Products

Spironolactone is available in tablets of 25, 50, and 100 mg. It is also available as a fixed combination with hydrochlorothiazide (Aldactazide; generic). Tablets are available that contain either 25 mg of spironolactone and 25 mg of hydrochlorothiazide, or 50 mg of each.

2,4,7-TRIAMINO-6-ARYLPTERIDINES

Triamterene, USP. 2,4,7-Triamino-6-phenylpteridine (Dyrenium).

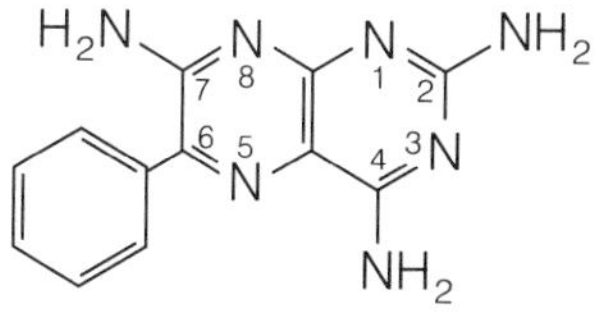

Triamterene

Structure–Activity Relationships

Triamterene is the primary compound selected from a host of pteridine analogues.[89] Although triamterene bears a structural resemblance to folic acid and certain dihydrofolate reductase inhibitors, it has little, if any, of their activities.[32]

Pharmacokinetics

Triamterene is rapidly but incompletely absorbed (30% to 70%) from the gastrointestinal tract,[41] bound to plasma proteins to the extent of about 60%, extensively biotransformed in the liver, and excreted primarily by the biliary route and secondarily by the renal route as unchanged drug (20%) and metabolites (80%). It enters the luminal fluid of the nephrons by glomerular filtration and active tubular secretion in the proximal tubule. Because triamterene is a weak organic base, it is assumed to be handled by the proximal tubule organic cation transport system[32] (OCTS; see Fig. 13-2).

Its onset of action after a single oral dose is two to four hours, and its duration of action is seven to nine hours.[41]

Site and Mechanism of Action

Triamterene acts from the luminal side of the cell membrane to selectively inhibit the electrogenic entry of 2% to 3% of the filtered load of sodium into the cells of the late distal convoluted tubule and early collecting tubule[6,11] (i.e., site 4). As the intracellular concentration of sodium decreases at site 4 as a result of triamterene's action, there is a decrease in the antiluminal membrane-bound Na^+, K^+–ATPase activity. This leads to a decrease in the cellular extrusion of sodium and in the cellular uptake of potassium. Because the secretion of potassium and hydrogen ions at site 4 is linked to sodium reabsorption, a concomitant reduction in the excretion rate of potassium and hydrogen ions occurs. The presence of aldosterone is not a prerequisite for its diuretic action. Triamterene, like the other potassium-sparing diuretics, has a low efficacy and may be referred to as a mild natriuretic, chloruretic, saluretic, and antikaliuretic agent.

Adverse Effects

As with the other potassium-sparing diuretics, the primary actions of which are elicited at site 4, triamterene's major adverse effect is hyperkalemia.[36] In addition, triamterene appears to be unique among the potassium-sparing diuretics in being associated with the formation of renal stones. Approximately 1 out of 1500 individuals taking a triamterene-containing diuretic experience nephrolithiasis.[90,91] The stones consist of triamterene (with or without its metabolite) or triamterene along with calcium oxalate or uric acid. Triamterene may also produce nausea, vomiting, leg cramps, and dizziness.[32]

Uses

Triamterene may be used alone in the treatment of mild edema associated with congestive heart failure or cirrhosis, but it should not be given to patients with impaired renal function.[92] Triamterene is not to be used alone in the treatment of hypertension.[46,92] Its primary use is in combination with hydrochlorothiazide (or other diuretics that act at site 2 or 3) to prevent hypokalemia that is associated with the latter diuretics.

Products

Triamterene is available as 50- and 100-mg capsules. Fixed combinations of triamterene and hydrochlorothiazide are also available: triamterene (50 mg)/hydrochlorothiazide (25 mg) (Dyazide); and triamterene (75 mg)/hydrochlorothiazide (50 mg) (Maxzide).

PYRAZINOYLGUANIDINES

Amiloride Hydrochloride, USP. 3,5-Diamino-*N*-(aminoiminomethyl)-6-chloropyrazinecarboxamide monohydrochloride dihydrate (Midamor).

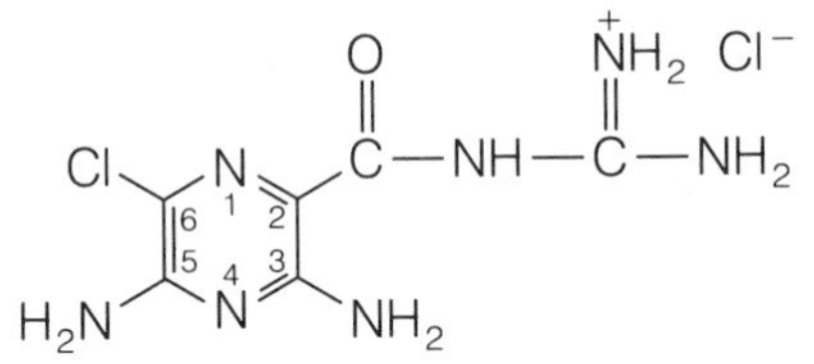

Amiloride Hydrochloride

Structure–Activity Relationships

An extensive screening procedure that examined over 25,000 agents was undertaken in an attempt to discover an antikaliuretic agent that did not have overlapping hormonal activity such as that of spironolactone.[93] Promising activity was noted with appropriately substituted pyrazinoylguanidines. Optimal diuretic activity in this series is observed when the 6-position is substituted with chlorine, the amino groups in the 3- and 5-positions are unsubstituted, and the guanidino nitrogens are not multiply substituted with alkyl groups. Amiloride emerged as the most active compound in the series.

Pharmacokinetics

Amiloride contains the strongly basic guanidine moiety, and possesses a pK of 8.7. Hence, it exists predominately as the charged guanidinium ion in the pH range of most body tissues and fluids. Therefore, it is not surprising that amiloride is incompletely and erratically absorbed (i.e., 15% to 20%) from the gastrointestinal tract—an event that occurs by passive diffusion of the uncharged form of most drugs. Amiloride is bound to plasma proteins to a moderate degree, is not biotransformed, and is excreted in the urine (20% to 50%) and in the feces (40%). The fecal amiloride content may represent unabsorbed drug. Amiloride reaches the luminal fluid by glomerular filtration and active tubular secretion. The proximal tubule organic cation transport system (OCTS; see Fig. 13-2) is involved in the latter process.[32] Amiloride's onset of action occurs within two hours after oral administration, and its duration of action may extend to 24 hours.[41]

Site and Mechanism of Action

Amiloride inhibits the electrogenic entry of 2% to 3% of the filtered load of sodium into the tubule cells at site 4. In turn, the driving force for potassium secretion is reduced or eliminated.[6,11] Similar to triamterene, amiloride does not require the presence of aldosterone to produce a diuresis. Amiloride induces the urinary loss of sodium, chloride, and water and, therefore, is a natriuretic, chloruretic, saluretic, and antikaliuretic agent—although with low efficacy.

Adverse Effects

The major adverse effect of amiloride is hyperkalemia, which may also be observed with the other potassium-sparing diuretics that act at site 4. Nausea, vomiting, diarrhea, and headache also have been noted after the use of amiloride.[32]

Uses

Amiloride may be used alone in the treatment of mild edema associated with congestive heart failure, cirrhosis, or nephrotic syndrome, or in the treatment of hypertension. Its most common use is in combination with diuretics that act at site 2 or 3, to circumvent the renal loss of potassium commonly associated with the latter agents.

Products

Amiloride hydrochloride is available for oral use as 5-mg tablets. In addition, a fixed combination of amiloride hydrochloride (5 mg) and hydrochlorothiazide (50 mg) (Moduretic) is available in tablet form.

MISCELLANEOUS DIURETICS

Mannitol, USP. The prototypic osmotic diuretic, D-mannitol is a water-soluble, but lipid-insoluble, hexahydroxy alcohol. Therefore, it does not diffuse across the gastrointestinal or renal tubule epithelium. Consequently, it is necessary to administer mannitol by the intravenous route. Mannitol enters renal luminal fluid only by glomerular filtration. Its high luminal fluid concentration creates an osmotic effect that may prevent the reabsorption of up to 28% of the filtered load of water.[85] Mannitol may be employed prophylactically in a hospital setting to avoid acute renal failure or for the reduction of cerebrospinal fluid volume and pressure. Because solutions of mannitol may expand the extracellular fluid volume, they should not be used in patients with severe renal disease or cardiac decompensation. Aqueous solutions of mannitol are available in a range of concentrations for intravenous use. The adult dosage range for the induction of a diuresis is from 50 to 200 g/24 hours.

Theophylline, the prototypic xanthine, is known to promote a weak diuresis by stimulation of cardiac function and by a direct action on the nephron. Although theophylline is infrequently used as a diuretic, a diuresis may be an observed side effect when it is used as a bronchodilator.

SUMMARY

The diuretic agent's chemical structure dictates its site of action within the nephron and its efficacy. The historical development of many diuretics has involved the molecular modification of the chemical structure of sulfamoyl-containing compounds. This has yielded carbonic anhydrase inhibitors, the thiazide and thiazide-like diuretics, and the high-ceiling diuretics that inhibit the reabsorption of sodium at site 1, site 3, and site 2, respectively. Increasing diuretic efficacy has occurred with the corresponding changes in the site of action of each of the three classes of diuretics. Predictable secondary effects that are dependent on a diuretic's site of action also have surfaced.

REFERENCES

1. Brenner, B. M., and Beeuwkes, R. III: Hosp. Pract. p. 35, July 1978.
2. Vander, A. J. (ed.): Control of renal hemodynamics. In Renal Physiology, 3rd ed., pp. 70–85. New York, McGraw-Hill, 1985.
3. Cafruny, E. J.: Am. J. Med. 62:490, 1977.
4. Holliday, M. A.: Hosp. Pract. p. 101, June 1978.
5. Valtin, H. (ed.): Sodium and water transport. Sodium balance. In Renal Function, Mechanisms Preserving Fluid and Solute Balance in Health, pp. 119–159. Boston, Little, Brown & Co., 1983.
6. Puschett, J. B.: Am. J. Cardiol. 57:6A, 1986.
7. Sullivan, L. P., and Grantham, J. J. (eds.): Physiology of the Kidney, 2nd ed., pp. 213–215. Philadelphia, Lea & Febiger, 1982.
8. DuBose, T. D., Jr.: Carbonic anhydrase-dependent bicarbonate transport in the kidney. In Tashian, R. E. and Hewett-Emmett, D. (eds.). Biology and Chemistry of the Carbonic Anhydrases. Ann. N.Y. Acad. Sci. 429:528, 1984.
9. Sullivan, L. P., and Grantham, J. J. (eds.): Physiology of the Kidney, 2nd ed., pp. 114–115. Philadelphia, Lea & Febiger, 1982.

10. Burg, M. B.: Renal handling of sodium, chloride, water, amino acids and glucose. In Brenner, B. M., and Rector, F. C., Jr. (eds.). The Kidney, 3rd ed., Vol. 1, pp. 145–175. Philadelphia, W. B. Saunders, 1986.
11. Steinmetz, P. R., and Koeppen, B. M.: Hosp. Pract. p. 125, Sept. 1984.
12. Sullivan, L. P., and Grantham, J. J. (eds.): Physiology of the Kidney, 2nd ed., pp. 135–137. Philadelphia, Lea & Febiger, 1982.
13. Buckalew, V. M., Jr., Walker, B. R., Puschett, J. B., and Goldberg, M.: J. Clin. Invest. 49:2336, 1970.
14. Dirks, J. H.: Hosp. Pract. p. 99, Sept. 1979.
15. Vander, A. J. (ed.): In Renal Physiology, 3rd ed., p. 17. New York, McGraw-Hill, 1985.
16. Eknoyan, G.: Drug Ther. (Hosp.) p. 87, Aug. 1981.
17. Vander, A. J. (ed.): In Renal Physiology, 3rd ed., pp. 111–142. New York, McGraw-Hill, 1985.
18. Levy, M.: Hosp. Pract. p. 95, Nov. 1978.
19. Ross, E. M., and Gilman, A. G.: Pharmacodynamics: mechanisms of drug action and the relationship between drug concentration and effect. In Gilman, A. G., Goodman, L. S., Rall, T. W., and Murad, F. (eds.). The Pharmacological Basis of Therapeutics, 7th ed., pp. 35–48. New York, Macmillan, 1985.
20. Moller, J. V., and Sheikh, M. I.: Pharmacol. Rev. 34:315, 1983.
21. Schlatter, E., Greger, R., and Weidtke, C.: Pflugers Arch 396:210, 1983.
22. Wittner, M., DiStefano, A., Wangemann, P., Delarge, J., Liegeois, J. F., and Greger, R.: Pflugers Arch. 408:54, 1987.
23. Strauss, M. B., and Southworth, H.: Bull. Johns Hopkins Hosp. 63:41, 1938.
24. Mann, T., and Keilin, K.: Nature 146:164, 1940.
25. Beckman, W. W., Rossmeisl, E. C., Pettengill, R. B., and Bauer, W.: J. Clin. Invest. 19:635, 1940.
26. Miller, W. H., Dessert, A. M., and Roblin, R. O.: J. Am. Chem. Soc. 72:4893, 1950.
27. Roblin, R. O., and Clapp, J. W.: J. Am. Chem. Soc. 72:4890, 1950.
28. Sprague, J. M.: Some results of molecular modifications of diuretics. In Gould, R. F. (ed.). Molecular Modification in Drug Design, Advances in Chemistry Series 45, pp. 87–101. Washington, DC, American Chemical Society, 1964.
29. Maren, T. H.: Physiol. Rev. 47:595, 1967.
30. Allen, R. C.: Sulfonamide diuretics. In Cragoe, E. J. (ed.). Diuretics—Chemistry, Pharmacology and Medicine, pp. 49–200. New York, John Wiley & Sons, 1983.
31. Beyer, K. H., and Baer, J. E.: Pharmacol. Rev. 13:517, 1961.
32. Weiner, I. M., and Mudge, G. H.: Diuretics and other agents employed in the mobilization of edema fluid. In Gilman, A. G., Goodman, L. S., Rall, T. W., and Murad, F. (eds.). The Pharmacological Basis of Therapeutics, 7th ed., pp. 887–907. New York, Macmillan, 1985.
33. Leaf, A., and Cotran, R. S. (eds.): Diuretics. In Renal Pathophysiology, 2nd ed., pp. 145–161. New York, Oxford University Press, 1980.
34. Whelton, A.: Am. J. Cardiol. 57:2A, 1986.
35. Erik, A., Persson, G., and Wright, F. S.: Acta Physiol. Scand. 114:1, 1982.
36. Berger, B. E., and Warnock, D. G.: Clinical uses and mechanisms of action of diuretic agents. In Brenner, B. M., and Rector, F. C., Jr. (eds.). The Kidney, Vol. 1, 3rd ed., pp. 433–455. Philadelphia, W. B. Saunders, 1986.
37. Ross, C. R., and Cafruny, E. J.: J. Pharmacol. Exp. Ther. 140:125, 1963.
38. Yeary, R. A., Brahm, C. A., and Miller, D. L.: Toxicol. Appl. Pharmacol. 7:598, 1965.
39. Terry, B., Hirsch, G., and Hook, J. B.: Eur. J. Pharmacol. 4:289, 1968.
40. Belair, E. J., Borrelli, A. R., and Yelnosky, J.: Proc. Soc. Exp. Biol. Med. 131:327, 1969.
41. United States Pharmacopeia Dispensing Information (USP DI), 7th ed., Vol. 1, Diuretics, pp. 796–816. Easton, PA, Mack Printing, 1987.
42. Am. J. Hosp. Pharm. 32:473, 1975.
43. AMA Drug Evaluations, 5th ed., pp. 741–765. Chicago, American Medical Association, 1983.
44. Ings, R. M. J., and Stevens, L. A.: Pharmacokinetics and metabolism of diuretics. Prog. Drug Metab., Vol. 7:57–171, 1983.
45. Shimizu, T., Yoshitomi, K., Nakamura, M., and Imai, M.: J. Clin. Invest. 82:721, 1988.
46. Med. Lett., 29 (Jan. 2), 1987.
47. Beyer, K. H., Jr., and Baer, J. E.: Med. Clin. North Am. 59:735, 1975.
48. Kunau, R. T., Weller, D. R., Jr., and Webb, H. L.: J. Clin. Invest. 56:401, 1975.
49. Hropot, M., Fowler, N., Karlmark, B., and Giebisch, G.: Kidney Int. 28:477, 1985.
50. Fuchs, M., Moyer, J. H., and Newman, B. E.: Ann. N.Y. Acad. Sci. 88:795, 1960.
51. Tobian, L.: Annu. Rev. Pharmacol. 7:399, 1967.
52. Greenberg, A.: Am. Fam. Physician 33:200, 1986.
53. Bennett, W. M., and Porter, G. A.: J. Clin. Pharmacol. 13:357, 1973.
54. Craswell, P. W., Ezzat, E., Kopstein, J., Varghese, Z., and Moorhead, J. F.: Nephron 12:63, 1973.
55. Med. Lett., 16 (Aug. 2), 1974.
56. Preminger, G. M.: Urol. Clin. North Am. 14:325, 1987.
57. Imbs, J. L., Schmidt, M., and Giessen-Crouse, E.: Pharmacology of loop diuretics: State of the art. In Grunfeld, J.-P., et al. (eds.). Advances in Nephrology, Vol. 16, pp. 137–158. Chicago, Year Book Medical, 1987.
58. Burg, M., and Green, N.: Kidney Int. 4:245, 1973.
59. Cafruny, E. J.: Geriatrics 22:107, 1967.
60. Cafruny, E. J., Cho, K. C., and Gussin, R. Z.: Ann. N.Y. Acad. Sci. 139:362, 1966.
61. Ethridge, C. B., Myers, D. W., and Fulton, M. N.: Arch. Intern. Med. 57:714, 1936.
62. Weiner, I. M., Levy, R. I., and Mudge, G. H.: J. Pharmacol. Exp. Ther. 138:96, 1962.
63. Ward, A., and Heel, R. C.: Drugs 28:426, 1984.
64. Feig, P. U.: Am. J. Cardiol. 57:14A, 1986.
65. Odlind, B.: J. Pharmacol. Exp. Ther. 211:238, 1979.
66. Gutsche, H.-U., Brunkhorst, R., Muller-Ott, K., et al.: Can. J. Physiol. Pharmacol. 62:412, 1984.
67. Finnerty, F. A., Jr., Maxwell, M. H., Lunn, J., and Moser, M.: Angiology 28:125, 1977.
68. Reineck, H. J., and Stein, J. H.: Mechanisms of action and clinical uses of diuretics. In Brenner, B. M., and Rector, F. C., Jr. (eds.). The Kidney, Vol. 1, 2nd ed., pp. 1097–1131. Philadelphia, W. B. Saunders, 1981.
69. Suki, W. N., Yium, J. J., Von Minden, M., et al.: N. Engl. J. Med. 283:836, 1970.
70. Koechel, D. A.: Annu. Rev. Pharmacol. Toxicol. 21:265, 1981.
71. Cragoe, E. J., Jr.: The (aryloxy)acetic acid family of diuretics. In Cragoe, E. J., Jr. (ed.). Diuretics—Chemistry, Pharmacology and Medicine, pp. 201–266. New York, John Wiley & Sons, 1983.
72. Koechel, D. A., Gisvold, O., and Cafruny, E. J.: J. Med. Chem. 14:628, 1971.
73. Koechel, D. A., Smith, S. A., and Cafruny, E. J.: J. Pharmacol. Exp. Ther. 203:272, 1977.
74. deSolm, S. J., Woltersdorf, O. W., Jr., Cragoe, E. J., Jr., et al.: J. Med. Chem. 21:437, 1978.
75. Beyer, K. H., Baer, J. E., Michaelson, J. K., and Russo, H. F.: J. Pharmacol. Exp. Ther. 147:1, 1964.
76. Gussin, R. Z., and Cafruny, E. J.: J. Pharmacol. Exp. Ther. 153:148, 1966.
77. Schnermann, J.: XII Symposium der Gesellschaft fur Nephrologie, Sept. 28–Oct. 1, 1977, Bonn, West Germany.
78. Birtch, A. G., Zakheim, R. M., Jones, L. G., and Barger, A.: Circ. Res. 24:869, 1967.
79. Landau, R. L., Bergenstahl, D. M., Lugibihe, K., and Kascht, M. E.: J. Clin. Endocrinol. 15:1194, 1955.
80. Rosemberg, E., and Engel, I.: Endocrinology 69:496, 1961.
81. Cella, J. A., and Kagawa, C. M.: J. Org. Chem. 79:4808, 1957.
82. Cella, J. A., Brown, E. A., and Burtner, R. R.: J. Org. Chem. 24:743, 1959.
83. Cella, J. A., and Tweit, R. C.: J. Org. Chem. 24:1109, 1959.
84. Brown, E. A., Muir, R. D., and Cella, J. A.: J. Org. Chem. 25:96, 1960.

85. Henry, D. A., and Coburn, J. W.: Diuretics. In Bevan, J. A., and Thompson, J. H. (eds.). Essentials of Pharmacology, 3rd ed., pp. 410–422. Philadelphia, Harper & Row, 1983.
86. Geering, K., Girardet, M., Bron, C., et al.: J. Biol. Chem. 257:10338, 1982.
87. Verrey, F., Schaerer, E., Zoerkler, P., et al.: J. Cell. Biol. 104:1231, 1987.
88. Smith, R. L.: Endogenous agents affecting kidney function: Their interrelationships, modulation, and control. In Cragoe, E. J., Jr., (ed.). Diuretics—Chemistry, Pharmacology and Medicine, pp. 571–651. New York, John Wiley & Sons, 1983.
89. Wiebelhaus, V. D., Weinstock, J., Maass, A. R., et al.: J. Pharmacol. Exp. Ther. 149:397, 1965.
90. Carey, R. A., Beg, M. M. A., McNally, C. F., and Tannenbaum, P.: Clin. Ther. 6:302, 1984.
91. Sorgel, F., Ettinger, B., and Benet, L. Z.: J. Pharm. Sci. 75:129, 1986.
92. Irish, J. M. III, and Stitzel, R. E.: Water, electrolyte metabolism, and diuretic agents. In Craig, C. R., and Stitzel, R. E. (eds.). Modern Pharmacology, 2nd ed., pp. 270–296. Boston, Little, Brown & Co., 1986.
93. Cragoe, E. J., Jr. (ed.): Pyrazine diuretics. In Diuretics—Chemistry, Pharmacology and Medicine, pp. 303–341. New York, John Wiley & Sons, 1983.

SELECTED READINGS

Am. J. Hosp. Pharm. 32:473–480, 1975.
Brenner, B. M., and Beeuwkes, R. III: Hosp. Pract. pp. 35–46, July 1978.
Dirks, J. H.: Hosp. Pract. pp. 99–110, Sept. 1979.
Greenberg, A.: Am. Fam. Physician 33:200–212, 1986.
Imbs. J.-L., Schmidt, M., and Giessen-Crouse, E.: Pharmacology of loop diuretics: State of the art. Grunfeld, J.-P., et al. (eds.). Adv. Nephrol. 16:137–158, 1987.
Puschett, J. B.: Am. J. Cardiol. 57:6A–13A, 1986.
Sprague, J. M.: Some results of molecular modifications of diuretics. In Gould, R. F. (ed.). Molecular Modification in Drug Design, Advances in Chemistry Series 45, pp. 87–101, Washington, DC, American Chemical Society, 1964.
Steinmetz, P. R., and Koeppen, B. M.: Hosp. Pract. pp. 125–134, Sept. 1984.
Med. Lett., 29 (Jan. 2), 1987.
Weiner, I. M., and Mudge, G. H.: Diuretics and other agents employed in the mobilization of edema fluid. In Gilman, A. G., Goodman, L. S., Rall, T. W., and Murad, F. (eds.). The Pharmacological Basis of Therapeutics, 7th ed., pp. 887–907. New York, Macmillan, 1985.

CHAPTER 14

Cardiovascular Agents

George H. Cocolas

The treatment and therapy of cardiovascular disease have undergone dramatic changes since the 1950s. Data show that since 1968 and continuing through the 1980s there has been a noticeable decline in mortality from cardiovascular disease. The basis for advances in the control of heart disease have been (1) a better understanding of the disease state, (2) the development of effective therapeutic agents, and (3) innovative medical intervention techniques to treat problems of the cardiovascular system.

The drugs discussed in this chapter are used for their action on the heart or other parts of the vascular system to modify the total output of the heart or the distribution of blood to the circulatory system. These drugs are employed in the treatment of (1) angina, (2) cardiac arrhythmias, (3) hypertension, (4) hyperlipidemias, and (5) disorders of blood coagulation. This chapter also includes a discussion of hypoglycemic agents, thyroid hormones, and antithyroid drugs.

ANTIANGINAL AGENTS AND VASODILATORS

Most coronary artery disease conditions are due to deposits of atheromas in the intima of large- and medium-sized arteries serving the heart. The process is characterized by an insidious onset of episodes of cardiac discomfort caused by ischemia from inadequate blood supply to the tissues. Angina pectoris (angina), the principal symptom of ischemic heart disease, is characterized by a severe constricting pain in the chest, often radiating from the precordium to the left shoulder and down the arm. The syndrome has been described since 1772, but it was not until 1867 that amyl nitrite was introduced for the symptomatic relief of angina pectoris.[1] It was believed at that time that anginal pain was precipitated by an increase in blood pressure and that the use of amyl nitrite reduced both blood pressure and, concomitantly, the work required of the heart. Later, it generally became accepted that nitrites relieved angina pectoris by dilating the coronary arteries and that changes in the work of the heart were of only secondary importance. However, it is now understood that the coronary blood vessels in the atherosclerotic heart are already dilated and that use of ordinary doses of dilator drugs does not significantly increase blood supply to the heart; instead, relief from anginal pain is by a reduction of cardiac consumption of oxygen.[2]

Although vasodilators are used in the treatment of angina, a more sophisticated understanding of the hemodynamic response to these agents has broadened their clinical usefulness to other cardiovascular conditions. Because of their ability to reduce peripheral vascular resistance, there is a rapidly growing interest in the use of organic nitrates to improve cardiac output in some patients with congestive heart failure.

The coronary circulation supplies blood to the myocardial tissues to maintain cardiac function. It is capable of reacting to the changing demands of the heart by dilatation of its blood vessels to provide sufficient oxygen and other nutrients and to remove metabolites. Myocardial metabolism is almost exclusively aerobic, which makes blood flow critical to the support of metabolic processes of the heart. This demand is met effectively by the normal heart, because it extracts a relatively large proportion of the oxygen delivered to it by the coronary circulation. The coronary blood flow is strongly dependent upon myocardial metabolism, which in turn is affected by work done by the heart and the efficiency of the heart. The coronary system normally has a reserve capacity that allows it to respond by vasodilatation to satisfy the needs of the heart during strenuous activity by the body.

Coronary atherosclerosis, one of the more prevalent cardiovascular diseases, develops with increasing age and may lead to a reduction of the reserve capacity of the coronary system. It most often results in multiple stenoses and makes it difficult for the coronary system to meet adequately the oxygen needs of the heart that occur during physical exercise or emotional duress. The insufficiency of the coronary blood flow (myocardial ischemia) in the face of increased oxygen demand produces angina pectoris.

The principal goal in the prevention and relief of angina is to limit the oxygen requirement of the heart so that the amount of blood supplied by the stenosed arteries is adequate. Nitrate esters such as nitroglycerin lower arterial blood pressure and, in turn, reduce the work of the left ventricle. This action is produced by the powerful vasodilating effect of the nitrates acting directly on the arterial system and, to an even greater extent, on the venous system. The result is a reduction of cardiac filling pressure and ventricular size. This reduces the work required of the ventricle and decreases the oxygen requirements, allowing the coronary system to satisfy the oxygen demands of myocardial tissue and relieve anginal pain.

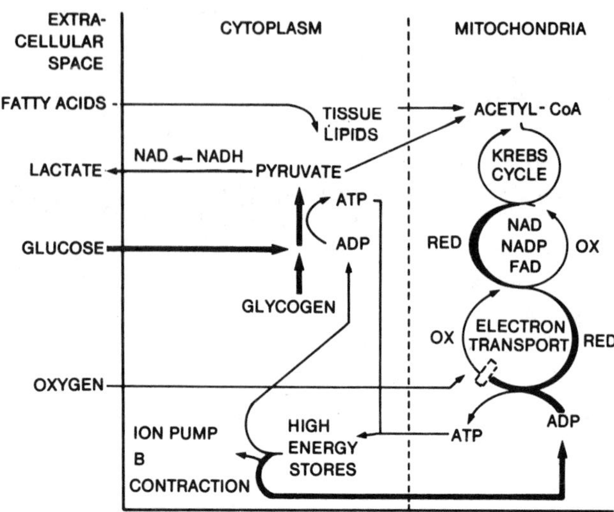

FIG. 14-1. Energy pathways in the hypoxic myocardium. With oxygen deprivation, there is relative blockade of the electron transport system; consequently, oxidative phosphorylation is inhibited, and high-energy phosphate stores may decline. Reduced flavin and adenine coenzymes accumulate so that citric acid cycle degradation of acetyl-CoA cannot proceed at a normal rate. Under these conditions, the utilization of glucose and glycogen increases, and pyruvate accumulates. The mitochondrial and cytoplasmic NAD / NADH ratios decline, and the lactate – pyruvate reaction becomes reversed, causing the formation of H^+. Fatty acids are deposited in tissue lipids. Increased glycolysis provides an inadequate compensatory mechanism for the formation of ATP. In myocardial hypoxia, coronary blood flow increases to assist in washing out lactate and H^+ from myocardial cells, preventing their accumulation in the heart. The size of labels and prominence of arrows in this figure are meant only to signify relative changes and do not have precise quantitative significance. (From Willerson, J. T., and Sanders, C. A.: Clinical Cardiology. New York, Grune & Strattan, 1979.)

INTERMEDIARY MYOCARDIAL METABOLISM

Normal myocardial metabolism is aerobic, and the rate of oxygen utilization parallels the amount of adenosine triphosphate (ATP) synthesized by the cells.[3] Free fatty acids are the principal fuel for myocardial tissue, but lactate, acetate, acetoacetate, and glucose are also oxidized to CO_2 and water. A large volume of the myocardial cell consists of mitochondria in which two-carbon fragments from free fatty acid breakdown are metabolized through the Krebs cycle. The reduced flavin and nicotinamide dinucleotides formed by this metabolism are reoxidized by the electron transport chain because of the presence of oxygen (Fig. 14-1). In the hypoxic or ischemic heart, the lack of oxygen inhibits the electron transport chain function and causes an accumulation of reduced flavin and nicotinamide coenzymes. As a result, fatty acids are converted to lipids rather than being oxidized. To compensate for this, glucose utilization and glycogenolysis increase, but the resulting pyruvate cannot be oxidized. A great loss of efficiency occurs as a result of the change of myocardial metabolism from aerobic to anaerobic pathways. Normally 36 moles of ATP are formed from the oxidation of 1 mole of glucose, but only 2 moles are formed from its glycolysis. This great loss of high-energy stores during hypoxia thus limits the functional capacity of the heart during stressful conditions and is reflected by the production of anginal pain.

NITROVASODILATORS

Mechanism of Smooth-Muscle Relaxation

Relaxation of smooth muscle may be the result of one or more biologic mechanisms. There appear to be receptors on smooth-muscle membranes that are activated by β-adrenergic agonists (e.g., isoproterenol); these, in turn, activate adenylate cyclase to increase the levels of cyclic adenosine monophosphate (cAMP) in the cell. The increased level of cAMP is associated with smooth-muscle relaxation. Drugs such as papaverine (see Chap. 12) and theophylline (see Chap. 13) also function to relax smooth muscle through a cAMP-mediated mechanism. These drugs inhibit phosphodiesterase and reduce the rate of conversion of cAMP to 5′-AMP in the cell.

Organic nitrates, nitrites, and nitroso compounds, and a variety of other nitrogen-containing substances such as sodium nitroprusside, cause relaxation of vascular smooth muscle by a mechanism that does not include cAMP.[4,5] These compounds have the common property of generating or releasing the unstable and lipophilic free radical nitric oxide (NO) in situ. The nitrogen oxide-containing vasodilators first react with thiols in the cell to form unstable *S*-nitrosothiols, which break down to yield nitric oxide.[6] The liberated nitric oxide activates guanylate cyclase and increases the cellular level of gaunosine 3′,5′-monophosphate (cGMP). Cyclic-GMP activates a cGMP-dependent protein kinase, which alters the phosphorylation state of several proteins. Included in this change is the dephosphorylation of the light chain of myosin. The altered state of the light chain of myosin now cannot play a normal role in the contractile process of smooth muscle and the result is relaxation.[7] The scheme in Figure 14-2 summarizes these events.

It has been observed that nitrogen oxide-containing vasodilators, such as the organic nitrates, lose their effectiveness on continual use. This tolerance is related to only nitrogen oxide-containing compounds, as other vasodilators, such as isoproterenol, papaverine, and the like, are not prevented from causing vasodilatation. It has been suggested that the vascular smooth-muscle receptor receptive to compounds such as nitroglycerin contains thiol groups. Tolerence to these types of drugs is believed to develop when the receptor is depleted of free thiol groups on its surface. Reversal of tolerance may be achieved by dithiothreitol, apparently by reduction of the oxidized thiol groups on the receptor membrane.

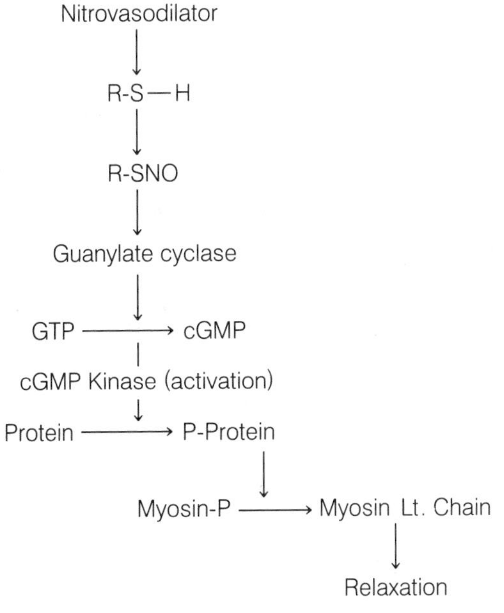

FIG. 14-2. Mechanism of nitrovasodilator relaxation of smooth muscle. (See Rapoport, R. M.: Circ. Res. 52:352, 1983.)

Esters of Nitrous and Nitric Acids

Inorganic acids, as do organic acids, will form esters with an alcohol. Pharmaceutically, the important ones are the sulfate, nitrite, and nitrate. Sulfuric acid forms organic sulfates, of which methyl sulfate and ethyl sulfate are examples.

Nitrous acid (HNO_2) esters may be formed readily from an alcohol and nitrous acid. The usual procedure is to mix sodium nitrite, sulfuric acid, and the alcohol. Organic nitrites are generally very volatile liquids that are only slightly soluble in water, but soluble in alcohol. Preparations containing water are very unstable because of hydrolysis.

The organic nitrates and nitrites and the inorganic nitrites have their primary utility in the prophylaxis and treatment of angina pectoris. They have a more limited application in treating asthma, gastrointestinal spasm, and certain cases of migraine headache. Nitroglycerin (glyceryl trinitrate) was one of the first members of this group to be introduced into medicine and still remains an important member of the group. By varying the chemical structure of the organic nitrates, differences in speed of onset, duration of action, and potency can be obtained (Table 14-1). Although the number of nitrate ester groups may vary from two to six or more, depending on the compound, there is no direct relationship between the number of nitrate groups and the level of activity.

It appears that the higher the oil/water partition coefficient of the drug, the greater the potency. The orientation of the groups within the molecule may also affect potency. Lipophilicity of the nitrogen oxide-containing compound produces a much longer response of vasodilator action. The highly lipophilic ester nitroglycerin permeates the cell membrane,

TABLE 14-1

RELATIONSHIP BETWEEN SPEED AND DURATION OF ACTION OF SODIUM NITRITE AND CERTAIN INORGANIC ESTERS

Compound	*Action Begins (min)*	*Maximum Effect (min)*	*Duration of Action (min)*
Amyl nitrite	$\frac{1}{4}$	$\frac{1}{2}$	1
Nitroglycerin	2	8	30
Isosobide dinitrate	3	15	60
Sodium nitrite	10	25	60
Erythrityl tetranitrate	15	32	180
Pentaerythritol tetranitrate	20	70	330

allowing continual formation of nitric oxide within the cell. The same effect appears to occur for sodium nitroprusside, nitroso compounds, and other organic nitrate esters and organic nitrite esters.[5]

Antianginal Action of Nitrovasodilators

The action of short-acting sublingual nitrates in the relief of angina pectoris is complex. Although the sublingual nitrates relax vascular smooth muscle and dilate the coronary arteries of normal humans, there is little improvement of coronary blood flow when these chemicals are administered to individuals with coronary artery disease. Nitroglycerin is an effective antianginal agent because it causes redistribution of coronary blood flow to the ischemic regions of the heart and also reduces myocardial oxygen demand. This latter effect is produced by a reduction of venous tone owing to the nitrate vasodilating effect and a pooling of blood in the peripheral veins that results in a reduction in ventricular volume, stroke volume, and cardiac output. It also causes reduction of peripheral resistance during myocardial contractions. The combined vasodilatatory effects cause a decrease in cardiac work and reduce oxygen demand.

Metabolism of Organic Nitrates

Organic nitrates are metabolized rapidly after oral administration. Metabolism of the organic nitrate occurs in the presence of reduced glutathione (GSH) and is catalyzed by hepatic glutathione–organic nitrate reductase. One molecule of nitroglycerin reacts with two GSH to release one inorganic nitrate ion to form either 1,2- or 1,3-glyceryl dinitrate. This product is further metabolized in the liver by the same enzyme, but at a slower rate to form a glyceryl mononitrate, the major urinary metabolite. The outcome of this metabolism is the conversion of a potent lipid-soluble vasodilator substance to a water-soluble metabolite of lower biologic potency that is readily excreted in the urine.

Buccal absorption reduces the immediate hepatic destruction of the organic nitrates because only 15% of the cardiac output is delivered to the liver; this allows a transient but effective circulating level of the intact organic nitrate before it is inactivated.[8]

Products

Amyl Nitrite, USP. Isopentyl nitrite. Amyl nitrite $[(CH_3)_2CHCH_2CH_2ONO]$ is a mixture of isomeric amyl nitrites, but is principally isoamyl nitrite. It may be prepared from amyl alcohol and nitrous acid by several procedures. Usually, amyl nitrite is dispensed in ampul form and used by inhalation, or orally in alcohol solution. Currently, it is recommended in treating cyanide poisoning; although not the best antidote, it does not require intravenous injections.

Amyl nitrite is a yellowish liquid having an ethereal odor and a pungent taste. It is volatile and inflammable at room temperature. Amyl nitrite vapor forms an explosive mixture in air or oxygen. Inhalation of the vapor may involve definite explosion hazards if a source of ignition is present, as both room and body temperatures are within the flammability range of amyl nitrite mixtures with either air or oxygen. It is nearly insoluble in water, but is miscible with organic solvents. The nitrite also will decompose into valeric acid and nitric acid.

Nitroglycerin. Glyceryl trinitrate is the trinitrate ester of glycerol and is official in tablet form in the *USP*. It is prepared by carefully adding glycerin to a mixture of nitric and fuming sulfuric acids. This reaction is exothermic and the reaction mixture must be cooled between 10° and 20°C.

The ester is a colorless oil, with a sweet, burning taste. It is only slightly soluble in water, but it is soluble in organic solvents.

$$\begin{array}{c} CH_2OH \\ | \\ CHOH \\ | \\ CH_2OH \end{array} \xrightarrow[H_2SO_4]{3HNO_3} \begin{array}{c} CH_2ONO_2 \\ | \\ CHONO_2 \\ | \\ CH_2ONO_2 \end{array} + 3H_2O$$

Nitroglycerin

Nitroglycerin is used extensively as an explosive in dynamite. A solution of the ester, if spilled or allowed to evaporate, will leave a residue of nitroglycerin. To prevent an explosion of the residue, the ester must be decomposed by the addition of alkali. It has a strong vasodilating action and, because it is absorbed through the skin, nitroglycerin is prone to cause headaches among workers associated with its manufacture. In medicine, it has the action typical of nitrites, but its action is developed more slowly and of longer duration. Of all the known coronary vasodilator drugs, nitroglycerin is the only one capable of stimulating the production of coronary collateral circulation and the only one able to prevent experimental myocardial infarction by coronary occlusion.

Previously, the nitrates were thought to be hydrolyzed and reduced in the body to nitrites, which then lowered the blood pressure. However, this is not true. The mechanism of vasodilatation of nitroglycerin through its formation of nitric oxide was described earlier.

Nitroglycerin tablet instability was reported in molded sublingual tablets.[9] The tablets, although uniform when manufactured, lost potency both because of volatilization of nitroglycerin into the surrounding materials in the container and intertablet migration of the active ingredient. Nitroglycerin may be stabilized in molded tablets by incorporating a "fixing" agent such as polyethylene glycol 400 or polyethylene glycol 4000.[10] In addition to sublingual nitroglycerin tablets, the drug has been formulated into an equally effective lingual aerosol for patients who have problems with dissolution of sublingual preparations because of dry mucous membranes. Transdermal nitroglycerin preparations appear to be less effective than other long-acting nitrates, as absorption from the skin is variable.

Diluted Erythrityl Tetranitrate, USP. Erythritol tetranitrate, 1,2,3,4-butanetetrol, tetranitrate, (R*,S*)-, (Cardilate), is the tetranitrate ester of erythritol and nitric acid, and it is prepared in a manner analogous to that used for nitroglycerin. The result is a solid crystalline material. This ester also is very explosive and is diluted with lactose or other suitable inert diluents to permit safe handling; it is slightly soluble in water and is soluble in organic solvents.

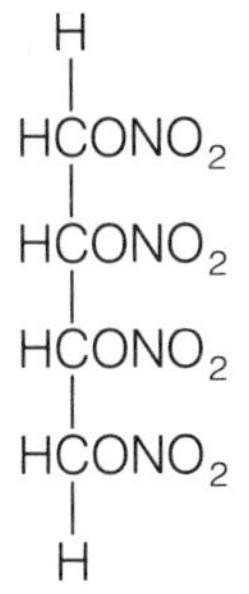

H
HCONO$_2$
HCONO$_2$
HCONO$_2$
HCONO$_2$
H

Erythrityl Tetranitrate

Erythrityl tetranitrate requires slightly more time than nitroglycerin to produce its effect, which is of longer duration. It is useful where mild, gradual, and prolonged vascular dilatation is warranted. The drug is used in the treatment of, and as a prophylaxis against, attacks of angina pectoris and to reduce blood pressure in arterial hypertonia.

Erythrityl tetranitrate produces a reduction of cardiac preload as a result of pooling of blood on the venous side of the circulatory system by its vasodilating action. This action results in a reduction of blood pressure on the arterial side during stressful situations and is an important factor in preventing the precipitation of anginal attacks.

Diluted Pentaerythritol Tetranitrate, USP. 2,2-Bis(hydroxymethyl)-1,3-propanediol tetranitrate (Peritrate; Pentritol). This compound is a white crystalline material with a melting point of 140°C. It is insoluble in water, slightly soluble in alcohol, and readily soluble in acetone. The drug is a nitric acid ester of the tetrahydric alcohol, pentaerythritol, and is a powerful explosive. Accordingly, it is diluted with lactose or mannitol or other suitable inert diluents to permit safe handling.

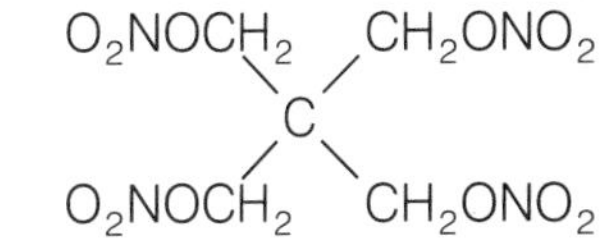

O_2NOCH_2 CH_2ONO_2
C
O_2NOCH_2 CH_2ONO_2

Pentaerythritol Tetranitrate

It relaxes smooth muscle of smaller vessels in the coronary vascular tree. Pentaerythritol tetranitrate is used prophylactically to reduce the severity and frequency of anginal attacks, and is usually administered in sustained-release preparations to increase its duration of action.

Diluted Isosorbide Dinitrate, USP. 1,4:3,6-Dianhydro-D-glucitol dinitrate (Isordil; Sorbitrate) occurs as a white crystalline powder. Its water-solubility is about 1 mg/mL.

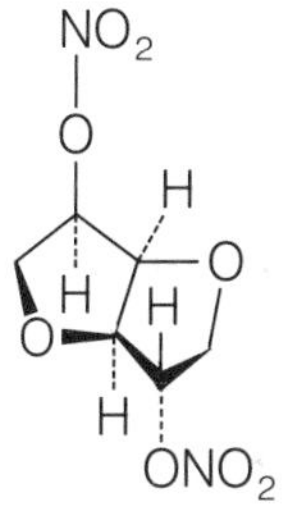

NO_2
O
H
O
H H
O
H
ONO_2

Isosorbide Dinitrate

Isosorbide dinitrate, as a sublingual or chewable tablet, is effective in treatment or prophylaxis of acute anginal attacks. When given sublingually, the effect begins in about two minutes with a shorter duration of action than when given orally. Oral tablets are not effective in acute anginal episodes; the onset of action ranges from 15 to 30 minutes.

CALCIUM ANTAGONISTS

Ion Channels and Calcium

Electrical excitation of axon and muscle membranes involves transient changes in permeability to Na^+, K^+, and Ca^{2+} ions. These voltage-dependent permeabilities can be altered by drugs and other chemical agents. It is commonly accepted that axon and muscle membranes contain at least three separate ionic pathways called Na, K, and Ca channels. A stimulus to an ionic channel by an applied electrical field,

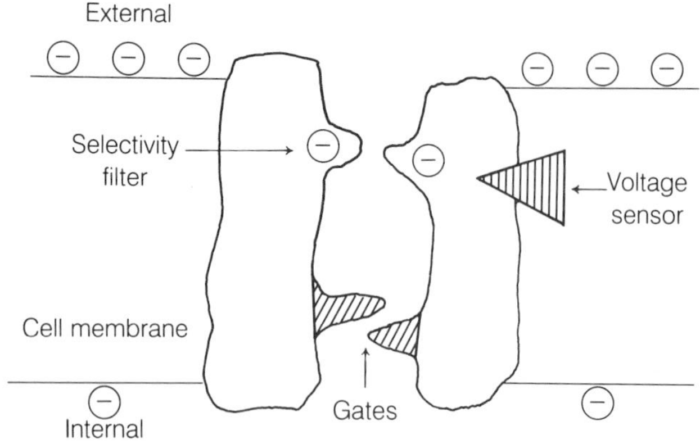

FIG. 14-3. Schematic diagram of a calcium channel.

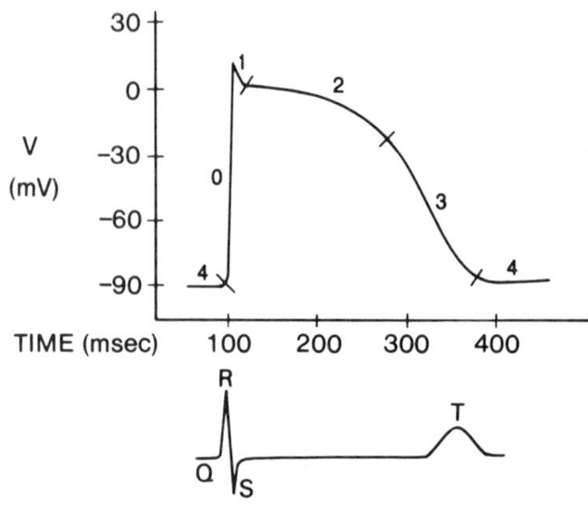

FIG. 14-4. Diagrammatic representation of the membrane action potential, as recorded from a Purkinje fiber, and an electrogram recorded from an isolated ventricular fiber. The membrane resting potential is 90 mv relative to the exterior of the fiber. At the point of depolarization, there is a rapid change (phase 0) to a more positive value. The phases of depolarization and repolarization are indicated by the numbers 0, 1, 2, 3, 4. Note that phases 0 and 3 of the membrane action potential correspond in time to the inscription of the QRS and T waves, respectively, of the local electrogram.

chemical transmitter, or other agent causes conformational changes in the ion channel called *gating*. This term is used because the conformational changes "gate" the movement of permanent ions by opening and closing the pore.[11] It has been determined that the actual gate that controls the flow of ions is near the inner end of the ion channel. Permeability of the ion occurs as the open pore provides an aqueous pathway for ions to pass through a sequence of energy barriers as it traverses the channel. The highest energy barrier serves as the rate-limiting "selectivity filter" for that ion (Fig. 14-3).

Calcium ions are vital to many biologic processes, such as bone metabolism, homeostasis, enzymatic reactions, and activation of excitable cells. In the specialized autonomic and conducting cells of the heart, Ca^{2+} is involved in the generation of the nerve action potential and also the link of excitation to muscle contraction in the myocardial contracting cells. Fortunately, the therapeutic effects of Ca^{2+} antagonists are found in concentrations that inhibit vascular smooth-muscle contraction or the atrioventricular (AV) node, allowing them to be used for their antianginal, antihypertensive, and antiarrhythmic action without exerting an undue negative inotropic effect on the heart.

Excitation–Contraction Coupling in Cardiac Muscle

Stimulation of the cardiac cell initiates the process of excitation, which has been related to ion fluxes through the cell membrane. Depolarization of the tissue in the atria of the heart is mediated by two inwardly directed ionic currents. When the cardiac cell potential reaches its threshold, ion channels in the membrane are opened and Na^+ enters the cell through ion channels. These channels give rise to the fast sodium current that is responsible for the rapidly rising phase, phase 0, of the ventricular action potential (Fig. 14-4). The second current is caused by the slow activation of an ion channel that allows the movement of Ca^{2+} into the cell. This "slow channel" contributes to the maintenance of the plateau phase (phase 2) of the cardiac action potential. It is also now understood that the Ca^{2+} that enters with the action potential initiates a second and larger release of Ca^{2+} from the sarcoplasmic reticulum in the cell. This secondary release of Ca^{2+} is sufficient to initiate the contractile process of cardiac muscle.

Contraction of cardiac and other muscle occurs from a reaction between actin and myosin. In cardiac muscle, a complex of proteins (troponin I, C, T and tropomyosin) attached to myosin modulate the interaction between actin and myosin. Free Ca^{2+} ions bind to troponin C, uncovering binding sites on

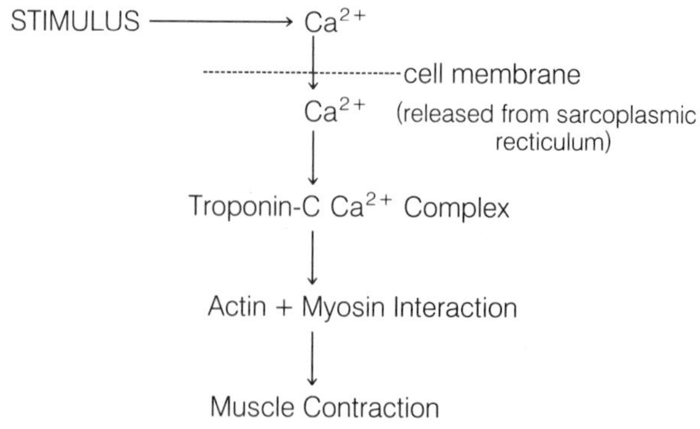

FIG. 14-5. Sequence of events showing excitation-contraction coupling in cardiac muscle.

the actin molecule and allowing interaction with myosin, causing contraction of the muscle. The schematic diagram in Figure 14-5 shows the sequence of events.[12]

Excitation–Contraction Coupling in Vascular Muscle

The contraction of vascular smooth, like cardiac, muscle is regulated by the concentration of cytoplasmic Ca^{2+} ions. However, the mechanism by which the contraction is effected differs in that it lacks a Ca^{2+}-sensitive troponin–tropomyosin complex. The activating effect is dependent on a different type of reaction. The elevated free cytosolic Ca^{2+} in vascular smooth muscle cells binds to a high-affinity binding protein, calmodulin. There are, in fact, two mechanisms responsible for increasing intracellular Ca^{2+} and, thereby, causing contraction of vascular smooth muscle: (1) *Electromechanical coupling*. Depolarization of the cell membrane causes voltage-sensitive Ca^{2+} channels in the membrane to open, allowing influx of Ca^{2+} into the cell to initiate the contractile process. (2) *Pharmacomechanical coupling* occurs without depolarization of the cell membrane, but requires an agonist–receptor interaction on the cell membrane to cause a release of intracellular Ca^{2+} from the sarcoplasmic reticulum. Regardless of the mechanism, the increased cellular Ca^{2+} levels result in an enhanced binding of Ca^{2+} to calmodulin. The Ca–calmodulin complex then activates a protein kinase (myosin light-chain kinase), resulting in the phosphorylation of the light chain of myosin. The phosphorylation activates the myosin–actin binding, leading to muscle contraction. Figure 14-6 describes the sequence leading to vascular smooth-muscle contraction.

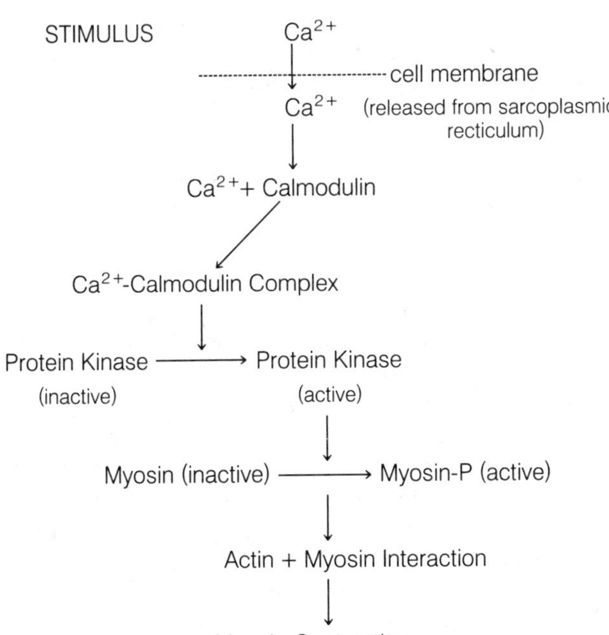

FIG. 14-6. *Sequence showing contraction of smooth muscle.*

β-Adrenergic Agonists and Calcium Ions

β-Adrenergic agonists have contrasting actions in myocardial cells and vascular smooth muscle through their effect on Ca^{2+} concentrations. In myocardial cells, β-agonists increase the levels of cAMP, which, in turn, increase the rate at which Ca^{2+} is available at contractile sites to stimulate contraction. β-Agonists also increase cAMP levels in vascular smooth muscle. Vasodilatation may be caused by two possible mechanisms: (1) cAMP-dependent protein kinase can lower intracellular Ca^{2+} levels by reducing transmembrane influx and enhancing Ca^{2+} efflux; and (2) cAMP inactivates myosin kinase, which reduces phosphorylation of the light chain of myosin and, in turn, causes relaxation of the vascular smooth muscle.

The Calcium Channel

The sarcolemma, the membrane surrounding each striated muscle, is an ion-impermeable lipid bilayer composed of phospholipid molecules. Present in the sarcolemma are large proteins that traverse the bilipid layer and selectively permit ions to move from one side of the membrane to the other. These pathways are known as *ion channels*. Although there is little known about the structure of the ion channels, they are highly specific for a given ion. Each aqueous "pore" within each channel has a selectivity filter that defines the ion that can pass through that channel (see Fig. 14-3). Channels also must have regions sensitive to electronegativity in the region (i.e., voltage sensors) to determine whether the channel is open or closed (i.e., gating). A propagated wave of depolarization that approaches the membrane containing the Ca^{2+} channel causes activation of the gate to open, allowing Ca^{2+} to pass into the cell. The gate closes when the interior of the cell again becomes electronegative. Because Ca^{2+} movement through these channels is controlled by electrical potentials, these are termed *voltage-dependent* channels.

The sarcolemma also contains receptors that respond to β_1-adrenergic agonists in cardiac muscle and α-adrenergic agonists on vascular smooth muscle. A drug–receptor interaction on these tissues causes Ca^{2+} influx by the slow inward current which, in turn, enhances contractility, frequency, and conduction velocity in the heart and alters the degree of contraction of arterioles. Activation of these receptors causes an increase in cellular levels of cAMP. The elevated cAMP levels facilitate activation of Ca^{2+} ion channels, increasing Ca^{2+} influx across the sarcolemma. These channels are termed *receptor-operated* channels.

Control of Calcium Ions in Myoplasm

Heart muscle contains much larger amounts of Ca^{2+} than are required to activate the contractile process. The Ca^{2+} is sequestered within many structures, including the sarcolemma (inner layer of the muscle membrane), mitochondria, and the sarcoplasmic reticulum. There are at least seven mechanisms that control the myoplasmic Ca^{2+} concentration and, thereby, muscle contraction.[13] They are as follows:

1. Voltage-dependent channel.
2. Receptor-operated channel.
3. Bidirectional Na^+–Ca^{2+} exchange across the sarcolemma. The energy required by this system to move Ca^{2+} across the sarcolemma and out of the cell is provided by the movement of Na^+ into the cell wall along its electrochemical gradient. The direction of the exchange depends upon the relative concentration of extracellular and intracellular Na^+ and Ca^{2+}. Inhibitors of Na^+, K^+–ATPase by cardiac glycosides (e.g., digitoxin) raise intracellular Na^+, resulting in Ca^{2+} entry into the cell because of Na^+–Ca^{2+} exchange, creating a positive inotropic effect.
4. Extrusion of Ca^{2+} from sarcolemma by a Ca^{2+}–ATPase in an energy-consuming process.
5. Calcium-stimulated magnesium–ATPase in the membrane of sarcoplasmic reticulum transports Ca^{2+} into the lumen of the sarcoplasmic reticulum. This process is a consequence of β-adrenergic stimulation and accounts for more rapid contraction and relaxation of cardiac muscle when exposed to catecholamines.
6. Calcium can be taken up by mitochondria and other intracellular structures using ATP generated by the mitochondria to effect the transfer into the organelle. As a result, excessive intracellular levels of Ca^{2+} may impede the function of mitochondria.
7. Ionophores, different from slow channels located on the membrane surface, are able to selectively pass Ca^{2+} through them along a concentration gradient.

Calcium Channel Blockers

It has been observed that several inorganic ions (e.g., manganese, cobalt, and lanthanum) can function as general Ca^{2+} antagonists to block a wide variety of biologic processes that depend upon Ca^{2+}. These inorganic ions apparently substitute for Ca^{2+} at a variety of Ca^{2+}-binding sites. The organic Ca^{2+} channel blockers produce a selective blockade of the slow inward channel and, thereby, lead to electromechanical uncoupling in the contraction process in heart muscle. Because organic Ca^{2+} channel blockers exert their actions at nanomolar concentrations and exhibit stereospecificity, it appears that they are recognized by specific structures in the Ca^{2+} channel. Nevertheless, the diversity of molecular structures of the different Ca^{2+} channel blockers is consistent with the understanding that there are different modes and sites of actions of these drugs (i.e., these drugs do not bind to a specific receptor).

Verapamil and diltiazem are ionized, water-soluble Ca^{2+}-entry blockers that may reach their binding sites by the hydrophilic pathway when the channel is open. Verapamil and diltiazem are *use-dependent* (i.e., their Ca^{2+}-blocking activity is a function of the frequency of the contractions). An increase in contraction frequency causes a reduction, rather than an augmentation, of contractions. Nifedipine, on the other hand, can "plug" the Ca^{2+} channel. It is a neutral compound at physiologic *p*H. This drug can block the channel in the open or closed state. In the closed state, nifedipine can traverse the phospholipid bilayer to reach its binding site because of its lipid solubility. In either state (i.e., open or closed), the Ca^{2+} channel blockers currently used prefer to interact with voltage-sensitive channels when the channels are in the inactive state.[14]

Cardiovascular Effects of Calcium Ion Channel Blockers

All Ca^{2+} antagonists yet developed are vasodilators. Vasodilatation is due to the uncoupling of the contractile mechanism of vascular smooth muscle which requires Ca^{2+}. Coronary artery muscle tone is reduced in healthy humans, but is particularly pronounced in a condition of coronary spasm. Peripheral arteriole resistance is reduced more than venous beds. The vasodilatation effect of these drugs is the basis for the use of these agents in control of angina and hypertension.[15]

The inhibition of Ca^{2+} influx into cardiac tissue by Ca^{2+} antagonists is also the basis of the use of these drugs as antiarrhythmic agents. The Ca^{2+}

channel blockers have a dampening effect on Ca^{2+}-dependent automaticity in the regular pacemaker cells in the sinoatrial (SA) node and also depress the origination of ectopic foci. Calcium antagonists can block reentry pathways in myocardial tissue, an integral component of arrhythmias. Numerous side effects on the heart, such as bradycardia, decrease in cardiac contractility, and reduction of AV conductance is traced to the Ca^{2+} channel-blocking activity of these drugs.

Products

Verapamil. 5-[3,4-Dimethoxyphenethyl)methylamino]-2-(3,4-dimethoxyphenyl)-2-isopropylvaleronitrile (Calan; Isoptin). Verapamil was originally introduced in 1962 as a coronary vasodilator and is the prototype of the Ca^{2+} antagonists used in cardiovascular diseases. It is employed in the treatment of angina pectoris, arrhythmias from ischemic myocardial syndromes, and supraventricular arrhythmias.

Verapamil's major effect is on the slow Ca^{2+} channel. The result is a slowing down of AV conduction and the sinus rate. This inhibition of the action potential inhibits one limb of the reentry circuit believed to underlie most paroxysmal supraventricular tachycardias that use the AV node as a reentry point. It is categorized as a class IV antiarrhythmic drug (see section Classes of Antiarrhythmic Drugs). Hemodynamically, verapamil causes a change in the preload, afterload, contractility, heart rate, and coronary blood flow. The drug reduces systemic vascular resistance and mean blood pressure, with minor effects on cardiac output.

Verapamil is a synthetic compound possessing slight structural similarity to papaverine. It can be separated into its optically active isomers, of which the levorotatory enantiomer is the most potent. Verapamil is rapidly absorbed after oral administration. The drug is quickly metabolized and, as a result, has low bioavailability. The liver is the main site of first-pass metabolism, forming several products. The preferential metabolic step involves *N*-dealkylation, followed by *O*-demethylation, and subsequent conjugation of the product before elimination. The metabolites of verapamil have no significant biologic activity. Verapamil has an elimination half-life of approximately five hours.

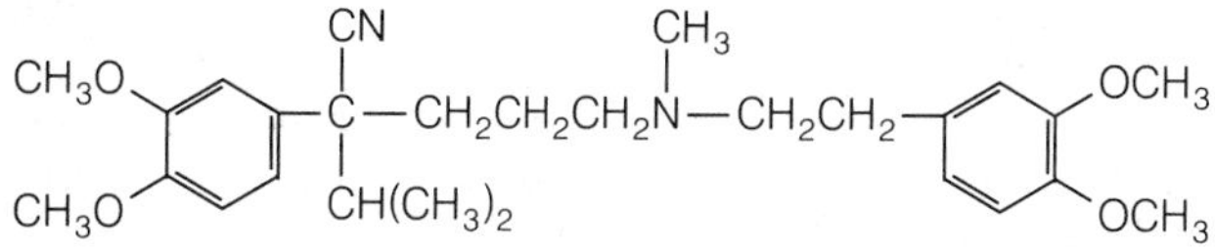

Verapamil

The route traveled by a Ca^{2+} channel blocker, such as verapamil, to its receptor site parallels that observed with many local anestheticlike antiarrhythmic agents. It is believed that verapamil, as do most of the Ca^{2+} channel blockers, crosses the cell membrane in an uncharged form to gain access to its site of action on the intracellular side of the membrane. Recent data show a much higher affinity of verapamil and other Ca^{2+} channel blockers to the inactivated channel.[16]

Diltiazem Hydrochloride. (+)-*cis*-3-(Acetoxy)-5-[2-(dimethylamino)ethyl]-2,3-dihydro-2-(4-methoxyphenyl)-1,5-benzothiazepin-4(5*H*)one hydrochloride (Cardizem). Diltiazem was originally developed and introduced in Japan as a cardiovascu-

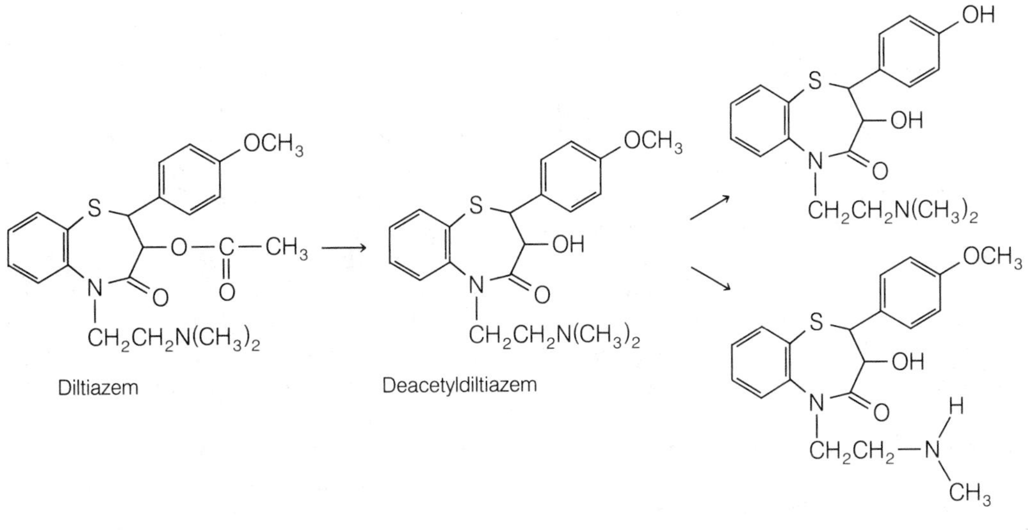

Diltiazem

Deacetyldiltiazem

FIG. 14-7. *Biotransformations of diltiazem.*

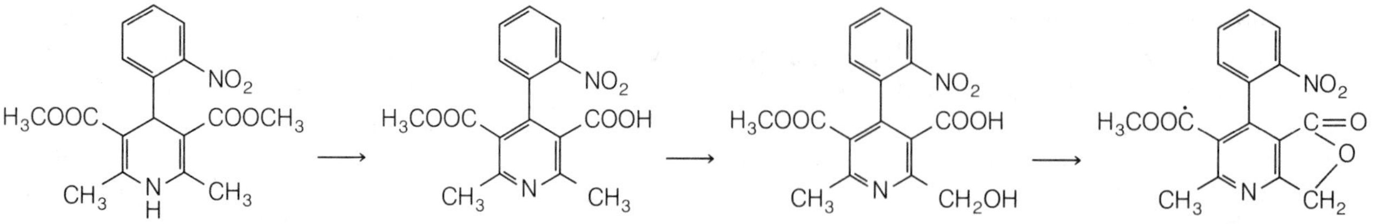

FIG. 14-8. Nifedipine metabolism.

lar agent to treat angina pectoris. It was observed to dilate peripheral arteries and arterioles. The drug increases myocardial oxygen supply by relieving coronary artery spasm and reduces myocardial oxygen demand by decreasing heart rate and reducing overload. Diltiazem is used in patients with variant angina. The drug has electrophysiologic properties similar to verapamil, being also employed in clinically similar treatment conditions as an antiarrhythmic agent, but it is less potent.

The drug is rapidly and almost completely absorbed from the digestive tract. It reaches peak plasma levels within one hour after administration of the drug in gelatin capsules. Oral formulations on the market are sustained-release preparations providing peak plasma levels three to four hours after administration.

Diltiazem is extensively metabolized after oral dosing by first-pass metabolism. As a result, the bioavailability is about 40% of the administered dose. The drug undergoes several biotransformations including deacetylation, oxidative *O*- and *N*-demethylations, and conjugation of the phenolic metabolites. Of the various metabolites (Fig. 14-7) only the primary metabolite, deacetyldiltiazem is pharmacologically active. Deacetyldiltiazem has about 40% to 50% of the potency of the parent compound.

Nifedipine. Dimethyl 1,4-dihydro-2,6-dimethyl-4-(2-nitrophenyl)-3,5-pyridinedicarboxylate (Adalat; Procardia). Nifedipine is a dihydropyridine derivative that bears no structural resemblance to the other calcium antagonists. It is not a nitrate, but its nitro group is essential for its antianginal effect.[17] The drug has potent peripheral vasodilator properties and little or no direct depressant effect on the SA or AV nodes. Nifedipine is more effective in patients whose anginal episodes are due to coronary vasospasm and is used in the treatment of vasospastic angina as well as classic angina pectoris. Because of its strong vasodilator properties, it is used in selected patients to treat hypertension.

Nifedipine is efficiently absorbed on oral or buccal administration. A substantial amount (90%) is protein bound. Systemic availability of an oral dose of the drug may be approximately 65%. Two inactive metabolites are the major product of nifedipine metabolism and are found in equilibrium with each other (Fig. 14-8). Only a trace of unchanged nifedipine is found in the urine.[18]

MISCELLANEOUS VASODILATORS

Nicotinyl Alcohol Tartrate. *β*-Pyridylcarbinol bitartrate; 3-pyridinemethanol tartrate (the alcohol corresponding to nicotinic acid) (Roniacol).

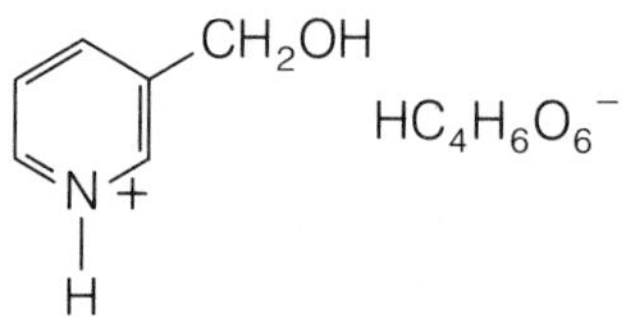

Nicotinyl Alcohol Tartrate

The free amine alcohol is a liquid having a boiling point of 145°C. It forms salts with acids. The bitartrate is crystalline and is soluble in water, alcohol, and ether. An aqueous solution has a sour taste, partly because of the bitartrate form of the salt.

In 1950, it was introduced as a vasodilator, following the lead that nicotinic acid dilated peripheral blood vessels. The drug, in fact, is converted by the body into nicotinic acid. The drug also has a direct relaxing effect on peripheral blood vessels, producing a flushing effect. It is given orally in tablets or as an elixir. Medicinal use includes the treatment of vascular spasm, Raynaud's disease, Buerger's disease, ulcerated varicose veins, chilblains (frostbite), migraine, Meniere's syndrome, and most conditions requiring a vasodilator effect. The usual dose is 150 to 300 mg.

Dipyridamole. 2,2′,2″,2‴-[(4,8-Di-1-piperidinopyrimido[5,4-*d*]pyrimidine-2,6-diyl)-dinitrilo]tetrakis ethanol (Persantine) is used for coronary and myocardial insufficiency. It is a bitter, yellow crystalline powder, soluble in dilute acids, methanol, or chloroform.

Dipyridamole is a long-acting vasodilator. Its vasodilating action is selective on the coronary system and is indicated for long-term therapy of chronic angina pectoris. The drug also inhibits adenosine deaminase in erythrocytes and interferes with the

Dipyridamole

uptake of the vasodilator adenosine by erythrocytes. These actions potentiate the effect of prostacyclin acting as an inhibitor to platelet aggregation (see p. 574).

The recommended oral dose is 50 mg two or three times daily before meals. Optimum response may not be apparent until the third or fourth week of therapy. Dipyridamole is available in 25-, 50-, and 75-mg sugar-coated tablets.

Cyclandelate. 3,5,5-Trimethylcyclohexyl mandelate (Cyclospasmol). This compound was introduced in 1956 for use especially in peripheral vascular disease in which there is vasospasm. It is a white to off-white crystalline powder, practically insoluble in water and readily soluble in alcohol and in other organic solvents. Its actions are similar to those of papaverine.

Cyclandelate

Cyclandelate produces peripheral vasodilatation by acting directly on vascular smooth muscle. When cyclandelate is effective, the improvement in peripheral circulation usually occurs gradually, and treatment often must be continued over long periods. At the maintenance dose of 100 mg four times daily, there is little incidence of serious toxicity. At higher doses, as high as 400 mg four times daily, which

TABLE 14-2

ANTIANGINAL AGENTS AND VASODILATORS

Name Proprietary Name	***Preparations***	***Category***	***Usual Adult Dose****	***Usual Dose Range****
Amyl nitrite USP	Amyl nitrite inhalant USP	Vasodilator	Inhalation, 300 μL, as required	Up to 10 mg/day
Diluted nitroglycerin USP	Nitroglycerin sublingual tablets USP	Antianginal	Sublingual, 150–600 μg as a sublingual tablet repeated at 5-min intervals as needed for relief of anginal attack	
	Nitroglycerin extended-release capsules USP		Oral, 2.5, 6.5, or 9.0 mg as an extended-release capsule every 12 hr, the dosage being increased every 8 hr if needed and tolerated	
	Nitroglycerin extended-release tablets USP		Oral, 1.3, 2.6, or 6.5 mg as an extended-release tablet every 12 hr, the dosage being increased to every 8 hr as needed and tolerated	
	Nitroglycerin ointment USP		Topically, to the skin, 2.5–5 cm (1–2 in.) of ointment as squeezed from the tube, every 3–4 hr as needed during the day and at bedtime	Up to 12.5 cm (5 in) of ointment as squeezed from the tube, per application
	Nitroglycerin transdermal systems	Topically, to the skin each day	2.5–15 mg/24 hr	
Diluted erythrityl tetranitrate USP *Cardilate*	Erythrityl tetranitrate tablets USP Erythrityl tetranitrate chewable tablets USP	Vasodilator	Oral, 10 mg as an oral (or chewable) tablet qid, the dosage being adjusted as needed or tolerated	Up to 100 mg/day
Diluted pentaerythritol tetranitrate USP *El-PETN, Pentritol*	Pentaerythritol tetranitrate tablets USP	Vasodilator	Oral, 10–20 mg qid, the dosage being adjusted as needed and tolerated	Up to 160 mg/day

*See USP DI for complete dosage information.

TABLE 14-2 *Continued*

ANTIANGINAL AGENTS AND VASODILATORS

Name Proprietary Name	***Preparations***	***Category***	***Usual Adult Dose****	***Usual Dose Range****
	Pentaerythritol tetranitrate extended-release capsules USP Pentaerythritol tetranitrate extended-release tablets USP		Oral, 30–80 mg as an extended-release capsule or tablet bid	Up to 160 mg/day
Diluted isosorbide dinitrate USP *Isordil, Sorbitrate*	Isosorbide dinitrate tablets USP	Antianginal	Oral, 10 mg qid, adjusting the dosage as needed and tolerated	
	Isosorbide dinitrate extended-release capsules USP		40 mg as an extended-release capsule every 12 hr, the dosage being increased up to 40 mg every 6 hr as needed and tolerated	
	Isosorbide dinitrate chewable tablets USP		Oral, 5–10 mg chewed well, every 2–3 hr, the dosage being adjusted as needed and tolerated	
	Isosorbide dinitrate extended-release tablets USP		Oral, 40 mg as an extended-release tablet every 12 hr, the dosage being increased up to 40 mg every 6 hr as needed and tolerated	
	Isosorbide dinitrate sublingual tablets USP		Sublingual, 5–10 mg as a sublingual tablet every 2–3 hr as needed	
Cyclandelate *Cyclospasmol*		Vasodilator	100–400 mg before meals and at bedtime	1.2–1.6 g/day
Dipyridamole *Persantine*		Vasodilator	50 mg tid	
Nicotinyl alcohol tartrate *Roniacol*		Vasodilator	50–100 mg tid	
Diltiazem hydrochloride *Cardizem*	Diltiazem hydrochloride tablets	Antianginal	30 mg tid or qid	
Nifedipine *Adalat, Procardia*	Nifedipine capsules	Antianginal	Up to 30 mg/day	
Verapamil hydrochloride *Calan, Isoptin*	Verapamil hydrochloride tablets	Antianginal	Up to 480 mg in divided doses	
	Verapamil hydrochloride injection	Antianginal	Initially 5–10 mg IV over a 2-min period	

*See USP DI for complete dosage information.

may be needed initially, there is a greater frequency of unpleasant side effects such as headache, dizziness, and flushing. It must be used with caution in patients with glaucoma. The oral dosage forms are 200-mg capsules and 100-mg tablets.

ANTIARRHYTHMIC DRUGS

Cardiac arrhythmias are caused by a disturbance in the conduction of the impulse through the myocardial tissue, by disorders of impulse formation, or by a combination of these factors. The antiarrhythmic agents used most commonly affect impulse conduction by altering conduction velocity and the duration of the refractory period. They also depress spontaneous diastolic depolarization, causing a reduction of automaticity by ectopic foci.

There are many pharmacologic agents currently available for the treatment of cardiac arrhythmias. Agents such as oxygen, potassium, and sodium bicarbonate relieve the underlying cause of some arrhythmias. Other agents, such as digitalis, propranolol, phenylephrine, edrophonium, and neostigmine, act on heart muscle or on the autonomic nerves to the heart and alter their influence on the cardiovascular system. Finally, there are drugs that alter the electrophysiologic mechanisms causing arrhythmias. The latter group of drugs is discussed in this chapter.

Within the last three decades, research on normal cardiac tissues and, in the clinical setting, on patients with disturbances of rhythm and conduction has brought to light information on the genesis of cardiac arrhythmias and the mode of action of antiarrhythmic agents. In addition, laboratory tests have been developed to measure blood levels of antiarrhythmic drugs, such as phenytoin, disopyramide, lidocaine, procainamide, and quinidine, to help evaluate the pharmacokinetics of these agents. As a result, it is possible to maintain steady-state plasma levels of these drugs that allow the clinician to use these and other agents more effectively and with greater safety. No other clinical intervention has been more effective in reducing mortality and morbidity in coronary care units.[19]

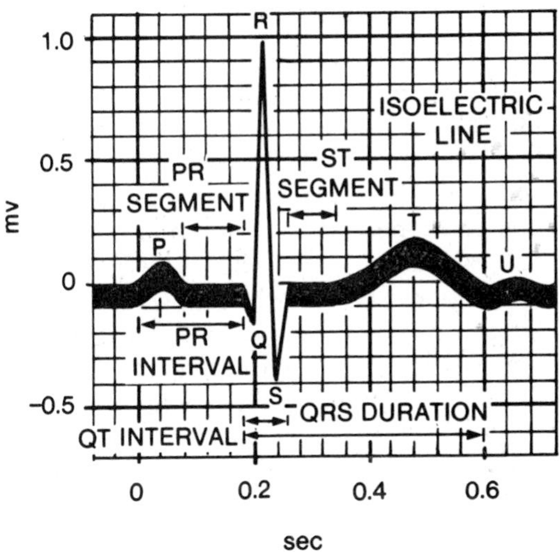

FIG. 14-9. Normal electrocardiogram. (From Ganong, W. F.: Review of Medical Physiology, 9th ed., San Francisco, Lange Medical Publications, 1985.)

CARDIAC ELECTROPHYSIOLOGY

The heart depends on the synchronous integration of electrical impulse transmission and myocardial tissue response to carry out its function as a pump. When the impulse is released from the sinoatrial (SA) node, excitation of the heart tissue takes place in an orderly manner by a spread of the impulse throughout the specialized automatic fibers in the atria, the AV node, and, finally, the Purkinje fiber network in the ventricles. This spreading of impulses produces a characteristic electrocardiographic pattern that can be equated to predictable myocardial cell membrane potentials and Na^+ and K^+ fluxes in and out of the cell.

A single fiber in the ventricle of an intact heart during the diastolic phase (see phase 4, Fig. 14-4) has a membrane potential (resting potential) of -90 mV. This potential is created by differential concentrations of K^+ and Na^+ in the intracellular and extracellular fluid. An active transport system (pump) on the membrane is responsible for concentrating the K^+ inside the cell and maintaining higher concentrations of Na^+ in the extracellular fluid. Diastolic depolarization is caused by a decreased K^+ ionic current into the extracellular tissue and a slow inward leakage of Na^+ until the threshold potential (-60 to -55 mV) is reached. At this time there is a sudden increase in the inward sodium current, and a self-propagated wave occurs to complete the membrane depolarization process. Pacemaker cells possess this property, which is termed *automaticity*. This maximal rate of depolarization (MRD) is represented by phase 0 or the spike action potential (see Fig. 14-4).

The form, duration, resting potential level, and amplitude of the action potential are characteristic for different types of myocardial cells. The rate of rise of the response (phase 0) is related to the level of the membrane potential at the time of stimulation and has been termed *membrane responsiveness*. Less negative potentials produce smaller slopes of phase 0 and are characterized by slower conduction times. The phase 0 spike of the SA node corresponds to the inscription of the P wave on the electrocardiogram (Fig. 14-9). Repolarization is divided into three phases. The greatest amount of repolarization is represented by phase 3, in which there is a passive flux of K^+ ions out of the cell. Phase 1 repolarization is caused by influx of chloride ions. During phase 2, a small inward movement of Ca^{2+} ions occurs through a slow channel mechanism that is believed to be important in the process of coupling excitation with contraction.[20] The process of repolarization determines the duration of the action potential and is represented by the *QT* interval. The action potential duration is directly related to the refractory period of cardiac muscle.

MECHANISMS OF ARRHYTHMIAS

The current understanding of the electrophysiologic mechanisms responsible for the origin and perpetuation of cardiac arrhythmias is that they are due to altered impulse formation, that is, change in automaticity; altered conduction; or both, acting simultaneously from different locations of the heart. The generation of cardiac impulses in the normal heart is usually confined to specialized tissues that spontaneously depolarize and initiate the action potential. These cells are located in the right atrium and are referred to as the *sinoatrial node* or the *pacemaker cells*. Although the spontaneous electrical depolarization of the SA pacemaker cells is independent of the nervous system, these cells are innervated by both sympathetic and parasympathetic fibers, which may cause an increase or decrease of the heart rate, respectively. Other special cells in the normal heart, which possess the property of automaticity, may influence cardiac rhythm when the normal pacemaker is suppressed or when pathologic changes occur in the myocardium to make these cells the dominant source of cardiac rhythm (i.e., ectopic pacemakers). Automaticity of subsidiary pacemakers may develop when myocardial cell damage occurs because of infarction or from digitalis toxicity, excessive vagal tone, excessive catecholamine release from sympathomimetic nerve fibers to the heart, or even high catecholamine plasma levels. The development of automaticity in specialized cells, such as that found in special atrial cells, certain AV node cells, bundle of His, and Purkinje fibers, may lead to cardiac arrhythmias. Because production of ectopic impulses is often due to a defect in the spontaneous phase 4 diastolic depolarization (i.e., "T wave"), drugs that are able to suppress this portion of the cardiac stimulation cycle are effective agents for these types of arrhythmias.

Arrhythmias are also caused by disorders in the conduction of impulses and changes in the refractory period of the myocardial tissue. Pharmacologic intervention is based on these two properties. The Purkinje fibers branch into a network of interlacing fibers, particularly at their most distant positions. This creates several pathways in which a unidirectional block in a localized area may establish circular (circus) micro- or macrocellular impulse movements that reenter the myocardial fibers and create an arrhythmia (Fig. 14-10). Unidirectional block results from localized myocardial disease (infarcts) or from a change in dependence of the tissue to Na^+ fluxes that causes a longer conduction time and allows the tissue to repolarize to propagate the retrograde impulse.

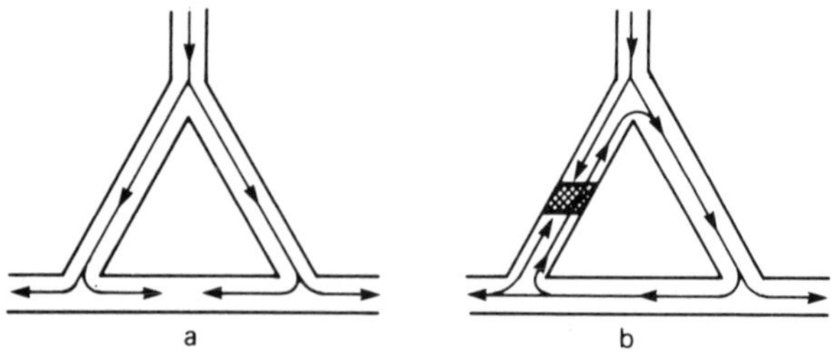

FIG. 14-10. *Reentry mechanism of Purkinje fibers.* (a) *Normal conduction of impulses through triangular arrangement of cardiac fibers.* (b) *Unidirectional block on left arm of triangular section allows impulse to reenter the regional conducting system and recycle.*

TABLE 14-3

CLASSES OF ANTIARRHYTHMIC DRUGS

Class	*Drugs*	*Mechanism of Action*
IA	Quinidine, procainamide, disopyramide	Lengthens refractory period
IB	Lidocaine, phenytoin, tocainide, mexiletine	Shortens duration of action potential
IC	Encainide, flecainide, lorcainide	Slows conduction
II	β-Adrenergic blockers (e.g., propranolol)	Slows AV conduction time; suppresses automaticity
III	Amiodarone, bretylium	Prolongs refractoriness
IV	Calcium channel blockers (e.g., verapamil, diltiazem)	Blocks slow inward Ca^{2+} channel

CLASSES OF ANTIARRHYTHMIC DRUGS

Antiarrhythmic drugs can be categorized into four separate classifications, based on their mechanism of action or the pattern of electrophysiologic effects they produce on heart tissue. Table 14-3 summarizes the four-part classification of antiarrhythmic drugs as first proposed by Vaughan Williams in 1970[21] and expanded in 1984.[22]

Class I. Membrane-Depressant Drugs

Class I antiarrhythmic agents are drugs that have membrane-stabilizing properties (i.e., shift membrane to more negative potentials). Drugs in this class act on the fast Na^+ channels and interfere with the process by which the depolarizing charge is transferred across the membrane. The prototype drugs in this class are quinidine and procainamide. During the 1970s several drugs were studied for their antiarrhythmic effects. Most of them were local anesthetics that affected Na^+ membrane channels and were grouped in a single class (class I) of antiarrhythmic drugs. More recent studies on the antiarrhythmic properties of these chemicals in this class have shown that there are sufficient differences to place them into separate subgroups.[22]

Class I antiarrhythmic drugs can be subdivided on the basis of the relative ease with which they dissociate from the Na^+ ion channel. Drugs, such as encainide and lorcainide, in class IC slowly dissociate from the Na^+ channel causing a slowing down of the conduction time of the impulse through the heart. Class IB drugs, which include lidocaine, tocainide, and mexiletine, have the property of rapidly dissociating from the Na^+ channels. These drugs shorten the action potential duration. Quinidine, procainamide, and disopyramide are drugs that have an intermediate rate of dissociation from Na^+ chan-

TABLE 14-4

DISSOCIATION CONSTANTS FOR CHANNEL STATES FOR VARIOUS SODIUM CHANNEL BLOCKING DRUGS

Drug	K_{dR}	K_{dA}	K_{dI}
Amiodarone	0.1	1,000	20
Lidocaine	2.5	30	40
Mexiletine	∞		4
Procainamide	1.85	37,000	370
Propafenone	0.3	1	0.5
Quinidine	∞	103	0

nels. These are categorized as class IA antiarrhythmic agents and lengthen the refractory period of cardiac tissue to cause cessation of arrhythmias.

Studies have shown that Na^+ channels on the membranes of Purkinje fiber cells normally exist in at least three states: R = rested, closed near the resting potential, but able to be opened by stimulation and depolarization; A = opened, allowing Na^+ ions to selectively pass through the membrane; and I = closed and unable to be opened (i.e., inactive).[23] The affinity of the antiarrhythmic drug for the receptor on the ion channel varies with the state of the channel or with the membrane potential. Because of this, a rested (R); activated (A); and inactive (I) ion channel can have different kinetics of interaction with antiarrhythmic drugs. A review of the recent literature shows that the antiarrhythmic drugs have a low affinity for rested (R) channels but a relatively high affinity for the activated (A) or inactivated (I) channels, or both. Table 14-4 summarizes the relative affinities of some of the class I antiarrhythmic drugs. Regardless of which channel state is blocked by class I antiarrhythmic drugs, the unblocking rate directly determines the amount of depression present at normal heart rates.

Class II. β-Adrenergic Blocking Agents

β-Adrenergic blocking drugs have the property of causing membrane-stabilizing or depressant effects on myocardial tissue. However, their antiarrhythmic properties are considered to be principally due to inhibition of adrenergic stimulation to the heart. The principal electrophysiologic property of these β-blocking agents is reduction of the phase 4 slope of potential sinus or ectopic pacemaker cells such that the heart rate decreases, and ectopic tachycardias are either slowed down or converted to sinus rhythm.

Class III. Repolarization Prolongators

Drugs in this class (e.g., amiodarone, bretylium) cause several different electrophysiologic changes on myocardial tissue, but share one common effect, that of prolonging the duration of the membrane action potential without altering the phase of depolarization or the resting membrane potential. Drugs in this class have been shown to (1) increase refractoriness without reducing conduction velocity in most tissues, and they have the ability to lessen heterogeneity in refractoriness and excitability in the heart; (2) have little or no negative inotropic activity; and (3) have a high efficacy in the treatment of "refractory" cardiac arrhythmias.

Class IV. Calcium Channel Blockers

Although not all Ca^{2+} channel blockers possess antiarrhythmic activity, some members of this class of antiarrhythmic drugs (verapamil, diltiazem) block the slow inward current of Ca^{2+} ions during phase 2 of the membrane action potential in cardiac cells. For example, the prototype drug in this group, verapamil, selectively blocks entry of Ca^{2+} into the myocardial cell and serves to depress AV conduction as well as block excitation–contraction coupling.

pH AND ACTIVITY

The action of class I local anesthetic-type antiarrhythmic drugs is pH-dependent and may vary with each drug.[24] Antiarrhythmic drugs are weak bases, with most having pK_a values ranging from 7.5 to 9.5. At physiologic pH of 7.40 these bases exist in an equilibrium mixture consisting of both the free base and the cationic form. Ionizable drugs such as lidocaine (pK_a 7.86) have stronger electrophysiologic effects in ischemic than normal myocardial cells. This potentiation has, in part, been attributed to the increase in H^+ concentration within the ischemic areas of the heart. Acidosis increases the proportion of Na^+ ion channels occupied by the protonated form of the antiarrhythmic agent. Nevertheless, the effect of pH on the antiarrhythmic activity of drugs can be complex, as both the free base and cationic species have been proposed as the active form of some drugs.

Small changes in pH can alter the drug's effectiveness by changing the charged/uncharged molecular ratio in the myocardial cells. Acidosis external to the myocardial cell promotes the cationic form. Because this species does not partition in the membrane as readily, onset of the drug's action would be delayed. Furthermore, concentration of the drug in the membrane would be reduced. Therefore, drugs that act on the channel only in the inactivated (closed) state would have a reduced effect from acidotic conditions. Acidosis also may cause prolongation of the effect of the drug. External acidosis facilitates protonation of receptor-bound drug. Because only neutral drug can dissociate from closed channels, recovery is prolonged by acidosis.

Alkalosis tends to cause hyperpolarization of the cell membrane and, thereby, reduces the effect of antiarrhythmic drugs. Because of this, alkalosis promotes the formation of more of the free-base antiarrhythmic agent, increasing the rate of recovery from the block. Alkalosis-inducing salts, such as sodium lactate, have been used to counteract toxicity caused by the antiarrhythmic quinidine.

PRODUCTS

Quinidine Sulfate, USP. Quinidinium sulfate is the sulfate of an alkaloid obtained from various species of *Cinchona* and their hybrids. It is a dextrorotary diastereoisomer of quinine. The salt crystallizes from water as the dihydrate in the form of fine, needlelike white crystals. Quinidine contains a hydroxymethyl group that serves as a link between a quinoline ring and a quinuclidine moiety. The structure contains two basic nitrogens, of which the quinuclidine nitrogen is the stronger base (pK_a 10). Quinidine sulfate is bitter and is light sensitive. Aqueous solutions are nearly neutral or slightly alkaline. It is soluble to the extent of 1% in water and is more highly soluble in alcohol or chloroform.

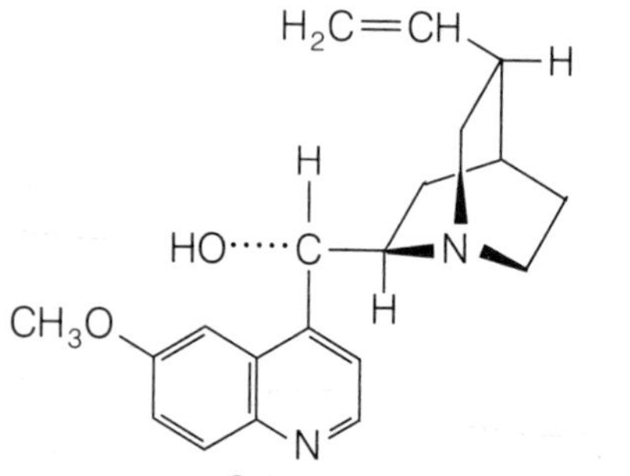

Quinidine

Quinidine is the prototype of antiarrhythmic drugs and a class IA antiarrhythmic agent according to the Vaughan Williams classification. It reduces Na^+ current by binding the open ion channels (i.e., state *A*). The decrease of Na^+ entry into the myocardial cell causes depression of phase 4 diastolic depolarization and shifts the intracellular threshold potential toward zero. These combined actions diminish spontaneous frequency of pacemaker tissues, depress the automaticity of ectoptic foci, and, to a lesser extent, reduce impulse formation in the SA node. This last action results in bradycardia. During the spike action potential, quinidine decreases transmembrane permeability to passive influx of Na^+, causing a slowing down of the process of phase 0 depolarization, which decreases conduction velocity. This is shown as a prolongation of the QRS complex of electrocardiograms. Quinidine also causes a prolongation of action potential duration, which results in a proportionate increase in the QT interval. It is used to treat supraventricular and ventricular ectopic arrhythmias, such as atrial and ventricular premature beats, atrial and ventricular tachycardia, atrial flutter, and atrial fibrillation.

Quinidine is used most frequently as an oral preparation and occasionally administered intramuscularly. Quinidine that has been absorbed from the gastrointestinal tract or from the site of intramuscular injection is bound 80% to serum albumin.[19] The drug is taken up quickly from the bloodstream by body tissues; consequently, a substantial concentration gradient is established within a few minutes. Onset of action begins within 30 minutes, the peak effect being attained in one to three hours. Quinidine is metabolized primarily by the liver by hydroxylation, and a small amount is excreted by the liver. One metabolic product has been identified, 2-hydroxyquinidine, which is equipotent to the activity of quinidine.[25] Because of serious side effects and the advent of more effective oral antiarrhythmic agents, quinidine is now used less, except in selected patients for long-term oral antiarrhythmic therapy.

Quinidine Gluconate, USP. Quinidinium gluconate (Duraquin; Quinaglute). This occurs as an odorless, very bitter white powder. In contrast with the sulfate salt, it is freely soluble in water. This is important because there are emergencies when the condition of the patient and the need for a rapid response may make the oral route of administration inappropriate. The high water solubility of the gluconate salt along with a low irritant potential makes it of value when an injectable form is needed in these emergencies. Quinidine gluconate forms a stable aqueous solution. When used for injection, it usually contains 80 mg/mL, equivalent to 50 mg of quinidine or 60 mg of quinidine sulfate.

Quinidine Polygalacturonate (Cardioquin). This is formed by reacting quinidine and polygalacturonic acid in a hydroalcoholic medium. It contains the equivalent of approximately 60% quinidine. This salt is only slightly ionized and slightly soluble in water, but studies have shown that although equivalent doses of quinidine sulfate give higher peak blood levels earlier, a more uniform and sustained blood level is achieved with the polygalacturonate salt.[26]

In many patients, the local irritant action of quinidine sulfate in the gastrointestinal tract causes pain, nausea, vomiting, and especially, diarrhea and often precludes oral use in adequate doses. It has been reported that in studies with the polygalacturonate salt, no evidence of gastrointestinal distress was encountered. It is available as 275-mg tablets. Each tablet is the equivalent of 200 mg of quinidine sulfate or 166 mg of free alkaloid.

Procainamide Hydrochloride, USP. *p*-Amino-*N*-[2-(diethylamino)ethyl]benzamide monohydrochloride; procainamidium chloride (Pronestyl). Procainamide has emerged as a major antiarrhythmic drug in the treatment of cardiac arrhythmias. It was developed in the course of research for compounds structurally similar to procaine, which had limited effect as an antiarrhythmic agent because of its central nervous system side effects and short-lived action resulting from the rapid hydrolysis of its ester linkage by plasma esterases. Procainamide is also more stable in water than procaine because of its amide structure. Aqueous solutions of procainamide hydrochloride have a *p*H of about 5.5. A kinetic study of the acid-catalyzed hydrolysis of procainamide has shown it to be unusually stable to hydrolysis in the *p*H range 2 to 7, even at elevated temperatures.[27]

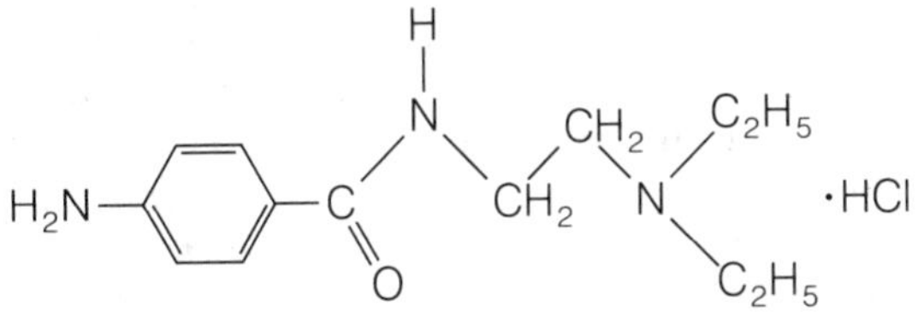

Procainamide Hydrochloride

Metabolism of procainamide occurs through the action of *N*-acetyltransferase. The product of enzymatic metabolism of procainamide is *N*-acetylprocainamide (NAPA), which possesses only 25% of the activity of the parent compound.[25] A study of the disposition of procainamide showed 50% of the drug to be excreted unchanged in the urine with 7% to 24% recovered as NAPA.[28,29] Unlike quinidine, procainamide is only minimally bound to plasma proteins. Between 75% and 95% of the drug is absorbed from the gastrointestinal tract. Plasma levels appear 20 to 30 minutes after administration and peak in about one hour.[30]

Procainamide appears to have all of the electrophysiologic effects of quinidine. It diminishes automaticity, decreases conduction velocity, increases action potential duration and, thereby, the refractory period of myocardial tissue. Clinicians have favored the use of procainamide for ventricular tachycardias and quinidine for atrial arrhythmias, even though the two drugs are effective in either type of disorder.

Disopyramide Phosphate, USP. α-[2(Diisopropylamino)ethyl]-α-phenyl-2-pyridineacetamide phosphate (Norpace) is an oral and intravenous class IA antiarrhythmic agent.[31] It is quite similar to quinidine and procainamide in its electrophysiologic properties, in that it decreases phase 4 diastolic depolarization, decreases conduction velocity, and also has vagolytic properties.[32] It is used clinically in the treatment of refractory, life-threatening ventricular tachyarrhythmias. Oral administration of the drug produces peak plasma levels within two hours. The drug is bound approximately 50% to plasma protein and has a half-life of 6.7 hours in humans. More than 50% of the drug is excreted unchanged in the urine. Therefore, patients with renal insufficiency should be carefully monitored for evidence of overdose. Disopyramide commonly exhibits side effects of dry mouth, constipation, urinary retention, and other cholinergic-blocking actions because of its structural similarity to anticholinergic drugs.

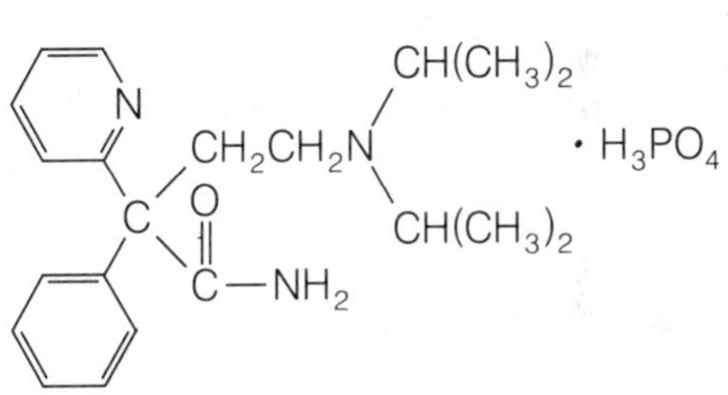

Disopyramide Phosphate

Lidocaine Hydrochloride, USP. 2-(Diethylamino)-2′,6′-acetoxylidide monohydrochloride (Xylocaine). This drug, which was initially conceived as a derivative of gramine (3-dimethylaminomethylindole) and introduced as a local anesthetic, is now being used intravenously as a standard parenteral agent for suppression of arrhythmias associated with acute myocardial infarction and with cardiac surgery.

Lidocaine is a class IB antiarrhythmic agent and has a different effect on the electrophysiologic properties of myocardial cells than do procainamide and quinidine. It binds with equal affinity to the active (*A*) and inactive (*I*) Na^+ ion channels. It depresses diastolic depolarization and automaticity in the Purkinje fiber network and increases the functional refractory period relative to action potential duration, as do procainamide and quinidine. However, it differs from the latter two drugs in that it does not decrease, but may even enhance, conduction velocity, and it increases membrane responsiveness to stimulation. There are fewer data available on the subcellular mechanisms responsible for the antiarrhythmic actions of lidocaine than on the more established drug quinidine. It has been proposed that lidocaine has little effect on membrane cation exchange of the atria. Sodium ion entrance into ventricular cells during excitation is not influenced by lidocaine because it does not alter conduction veloc-

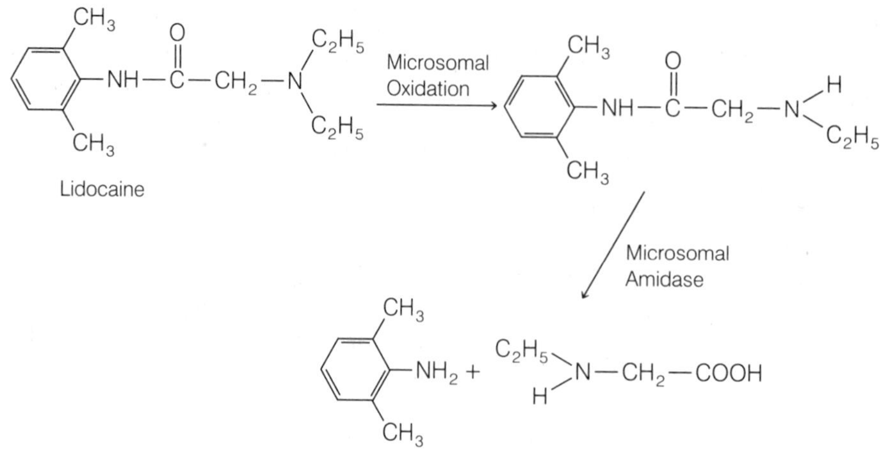

Lidocaine

Microsomal Oxidation

Microsomal Amidase

FIG. 14-11. Metabolism of lidocaine.

ity in this area. Lidocaine does depress Na^+ influx during diastole, as do all other antiarrhythmic drugs, to diminish automaticity in myocardial tissue. It also alters membrane responsiveness in Purkinje fibers, allowing an increase of conduction velocity and ample membrane potential at the time of excitation.[33]

Lidocaine administration is limited to the parenteral route and usually is given intravenously, although adequate plasma levels are achieved after intramuscular injections. Lidocaine is not bound to any extent to plasma proteins and is concentrated in the tissues. It is rapidly metabolized by the liver (Fig. 14-11). The first step is deethylation,[34] with the formation of monoethylglycinexylidide, followed by hydrolysis of the amide.[35] Metabolism is rapid, thus the half-life of a single injection ranges from 15 to 30 minutes. Lidocaine is a popular drug because of its rapid action and its relative freedom from toxic effects on the heart, especially in the absence of hepatic disease. Monoethylglycinexylidide, the initial metabolite of lidocaine, is an effective antiarrhythmic agent; however, its rapid hydrolysis by microsomal amidases prevents its use in humans.

Precautions must be taken that lidocaine hydrochloride solutions containing epinephrine salts are not used as cardiac depressants. Such solutions are intended only for local anesthesia and are not used intravenously. The aqueous solutions without epinephrine may be autoclaved several times, if necessary.

Phenytoin Sodium, USP. 5,5-Diphenyl-2,4-imidazolidinedione; 5,5-diphenylhydantoin; diphenylhydantoin sodium (Dilantin). This drug has been used for decades in the control of grand mal types of epileptic seizures. It is structurally analogous to the barbiturates, but does not possess the extensive sedative properties characteristic of the barbiturates. The compound is available as the sodium salt. Solutions for parenteral administration contain 40% propylene glycol and 10% alcohol to dissolve the sodium salt.

Phenytoin's cardiovascular effects were uncovered during observation of toxic manifestations of the drug in patients being treated for seizure disorders. Phenytoin was found to cause bradycardia, prolong the PR interval, and produce T-wave abnormalities on electrocardiograms. It is a class IB antiarrhythmic agent. Today phenytoin's greatest clinical use is in the treatment of digitalis-induced arrhythmias.[36] Its action is similar to that of lidocaine. It causes a depression of ventricular automaticity produced by digitalis, without adverse intraventricular conduction. Because it also reverses the prolongation of AV conduction by digitalis, phenytoin is useful in supraventricular tachycardias caused by digitalis intoxication.

Phenytoin is located in high amounts in the body tissues, especially fat and liver, leading to large gradients between the drug in tissues and the plasma concentrations. It is metabolized in the liver.

Mexiletine Hydrochloride. 1-Methyl-2-(2,6-xylyloxy)ethylamine hydrochloride (Mexitil). Mexiletine (pK_a 8.4) is a class IB antiarrhythmic agent. It resembles lidocaine in possessing a xylyl moiety, but otherwise is different chemically. Mexiletine is an ether and not subject to the hydrolysis common to the amides lidocaine and tocainide. Its mean half-life on oral administration is approximately ten hours. The drug is active orally and by intravenous admin-

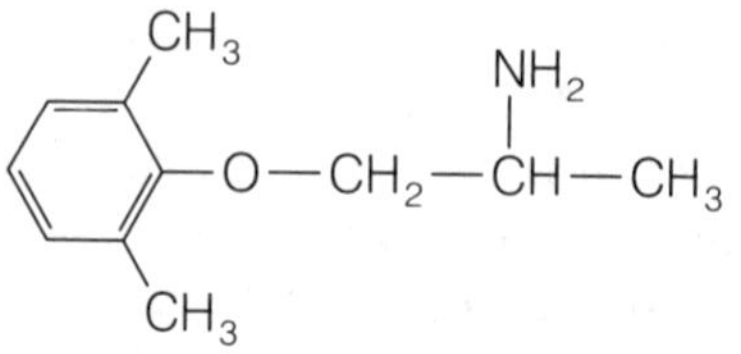

Mexiletine

istration. It has electrophysiologic properties similar to lidocaine.

Although not subject to hydrolysis, Mexiletine is metabolized by oxidative and reductive processes in the liver. Its metabolites, *p*-hydroxymexiletine and hydroxymethylmexiletine, are not pharmacologically active as antiarrhythmic agents.[37]

Mexiletine, similar to class I antiarrhythmic agents, blocks the fast Na^+ channel in cardiac cells. It is especially effective on the Purkinje fibers in the heart. The drug increases the threshold of excitability of myocardial cells by reducing the rate of rise and amplitude of the action potential and decreases automaticity.

Mexiletine is used for long-term oral prophylaxis of ventricular tachycardia. The drug is given in 200- to 400-mg doses every eight hours.

Tocainide Hydrochloride. 2-(Ethylamino-2′,6′-propionoxyxylidide hydrochloride (Tonocard). Tocainide (pK_a 7.7) is an analogue of lidocaine. It is orally active and has electrophysiologic properties similar to lidocaine.[38] Total body clearance of tocainide is only 166 mL/min, suggesting that hepatic clearance is not large. Because of low hepatic clearance, the hepatic extraction ratio must be small; therefore, tocainide is unlikely to be subject to a substantial first-pass effect. The drug differs from lidocaine in that it lacks two ethyl groups, which provides tocainide some protection from first-pass hepatic elimination after oral ingestion. Tocainide is hydrolyzed in a manner similar to lidocaine. None of its metabolites are active.

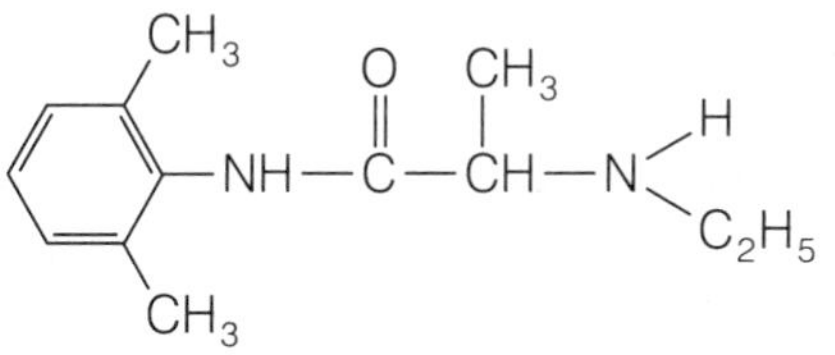

Tocainide

Tocainide is classed as a IB antiarrhythmic agent and used orally to prevent or treat ventricular ectopy and tachycardia. The drug is given in 400- to 600-mg doses every eight hours.

Encainide Hydrochloride. 4-Methoxy-*N*-[2-[2-(1-methyl-2-piperidinyl)ethyl]phenyl] benzamide, monohydrochloride (Enkaid). Encainide is one of the more recent antiarrhythmic agents and has been categorized as a class IC agent. The main effect of this compound on the heart, along with the other class IC drugs, is to depress the upstroke velocity of phase 0 of the action potential and lengthen the time-dependent refractoriness.

Encainide is a benzanilide derivative that has local anesthetic properties in addition to its class IC antiarrhythmic action. The drug is extensively metabolized, producing products that also have antiarrhythmic properties.[39] The metabolite, 3-methoxy-*O*-demethylencainide (MODE), is about equipotent to encainide. *O*-Demethylencainide (ODE) is considerably more potent than the parent drug. The half-life of encainide is two to four hours. The active metabolites have longer half-lives, estimated as being up to 12 hours, and may play an important part in use of this drug in long-term therapy. Encainide also undergoes *N*-demethylation to form *N*-demethylencainide (NDE). The probable metabolic pathway for encainide is shown in Figure 14-12.

Flecainide Acetate. *N*-(2-Piperidinylmethyl)-2,5-bis(2,2,2-trifluoroethoxy)benzamide monoacetate (Tambocor). Flecainide, like encainide, is a class IC antiarrhythmic drug with local anesthetic activity and is a chemical derivative of benzamide. The drug undergoes biotransformation forming a *meta-O*-dealkylated compound, the antiarrhythmic properties of which are one-half as potent as those of the parent drug, and a *meta-O*-dealkylated lactam of flecainide with little pharmacologic activity.[40] Flecainide is given orally to suppress chronic ventricu-

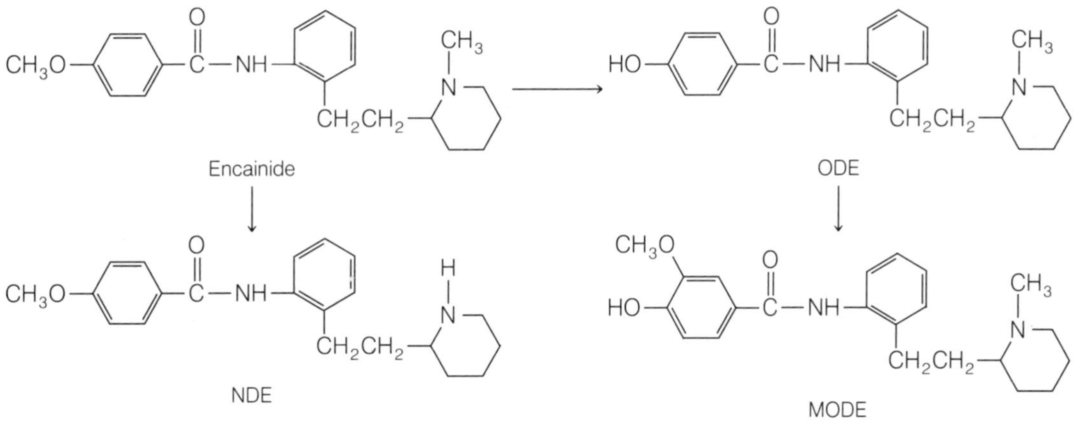

FIG. 14-12. Metabolism of encainide.

lar ectopy and ventricular tachycardia. The drug has some limitations because of central nervous system side effects.

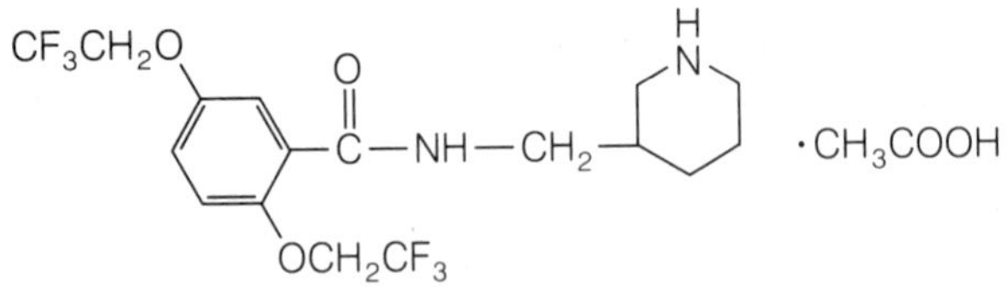

Flecainide Acetate

Lorcainide Hydrochloride. *N*-(4-Chlorophenyl)-*N*-[1-(1-methylethyl)-4-piperidinyl]benzeneacetamide monohydrochloride. Lorcainide is a local anesthetic-type antiarrhythmic agent. It is categorized as a class IC-type drug because it effectively causes a slowing down of conduction in the His–Purkinje fiber network and ventricles of the heart. It is used to suppress chronic ventricular ectopy. Lorcainide undergoes metabolic *N*-dealkylation, forming norlorcainide. The metabolism is the product of the first-pass clearance after oral administration of lorcainide. The basis for this observation is that norlorcainide is not produced in significant amounts in the body following intravenous administration. Norlorcainide is an important metabolite of the parent drug, as it is slowly cleared from the body and has a half-life that is approximately three times longer than lorcainide. Accumulation of norlorcainide is of considerable clinical importance because the metabolite is equipotent to the original drug.

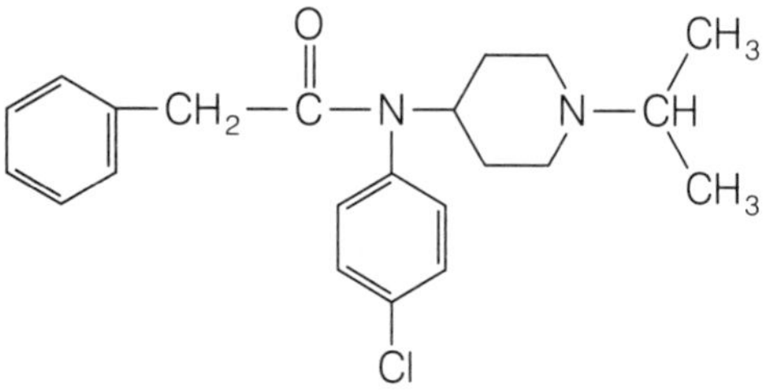

Lorcainide

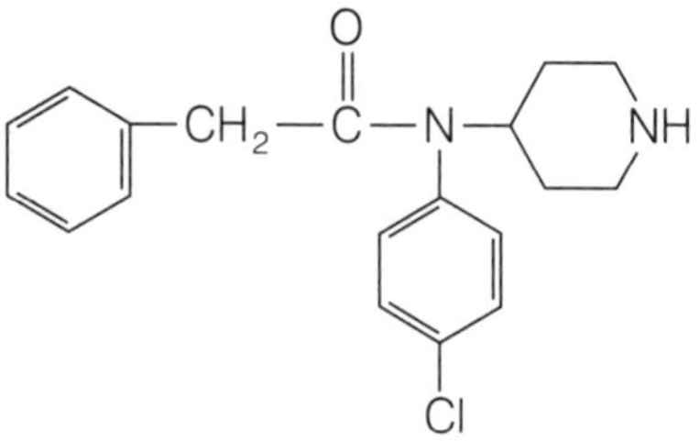

Norlorcainide

Amiodarone. 2-Butyl-3-benzofuranyl 4-[2-(diethylamino)ethoxy]-3,5-diiodophenyl ketone (Cordarone). Amiodarone was initially introduced as an antianginal agent. It has a very pronounced class III action and is especially effective in maintaining sinus rhythm in patients who have been treated by direct current shock for atrial fibrillation.[41] Similar to class III antiarrhythmic drugs, amiodarone lengthens the effective refractory period by prolonging the action potential duration in all myocardial tissues. Amiodarone is very slowly eliminated from the body, with a half-life of about 25 to 30 days after oral doses.[42] Although the drug has a broad spectrum of antiarrhythmic activity, its main limitation is a slow onset of action. The initiation of drug action may not be for several days, and the peak effect may not be obtained for several weeks.

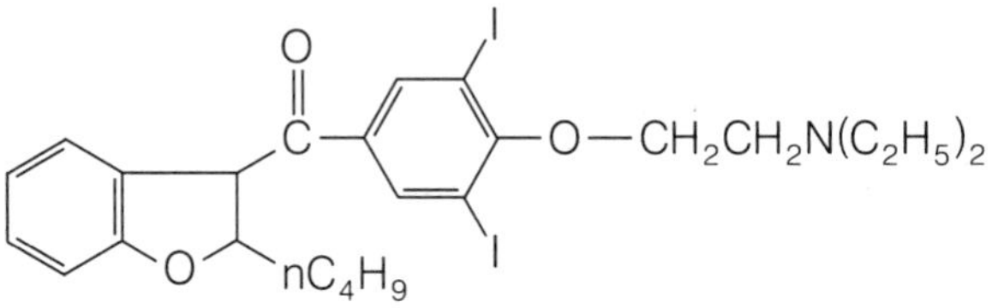

Amiodarone

Amiodarone contains iodine in its molecular structure and, as a result, has an effect on thyroid hormones. The principal effect is the inhibition of the peripheral conversion of T_4 to T_3. In most patients thyroid activity is not altered by amiodarone. However, serum reverse T_3 (rT_3) is increased as a function of the dose as well as the length of amiodarone therapy. As a result rT_3 levels have been used as a guide for judging adequacy of amiodarone therapy and predicting toxicity.[43]

Bretylium Tosylate. (*o*-Bromobenzyl)ethyl)dimethylammonium *p*-toluenesulfonate (Bretylol) is an extremely bitter, white crystalline powder. The chemical is freely soluble in water and alcohol. Bretylium tosylate is an adrenergic neuronal-blocking agent that accumulates selectively into the neurons and displaces norepinephrine. Because of this property, bretylium was used initially, under the trade name of Darenthin, as an antihypertensive agent. It caused postural decrease in arterial pressure.[44] This use was discontinued because of the rapid development of tolerance, erratic oral absorption of the quaternary ammonium compound, and persistent pain in the parotid gland on prolonged therapy. Currently, bretylium is reserved for use in ventricular arrhythmias that are resistant to other therapy. Bretylium does not suppress phase 4 depolarization, a common action of other antiarrhythmic agents. It prolongs the effective refractory period relative to the action potential duration but does not affect conduction time and is categorized as a class III antiarrhythmic agent. Because bretylium does not have properties similar to those of the other antiarrhythmic agents, it has been suggested that its action is due to its adrenergic neuronal-blocking properties; however, the antiarrhythmic properties of the drug are not affected by administration of reserpine. Bretylium is also a local anesthetic, but it has not been possible to demonstrate such an effect on atria of experimental animals, except at very high concentrations.[45] Therefore, the precise mechanism of the antiarrhythmic action of bretylium remains to be resolved.

Bretylium Tosylate

Verapamil and Diltiazem. Both of these drugs have the property of blocking the slow inward Ca^{2+} currents (voltage-sensitive channel) in cardiac fibers. This causes a slowing down of AV conduction and the sinus rate. These drugs are used in controlling atrial and paroxysmal tachycardias and are categorized as class IV antiarrhythmic agents according to the Vaughan Williams classification.[21,22] (A more detailed description of calcium channel blockers is given in a preceding section.)

TABLE 14-5

ANTIARRHYTHMIC AGENTS

Name Proprietary Name	***Preparations***	***Usual Adult Dose****	***Usual Dose Range****	***Usual Pediatric Dose****
Quinidine sulfate USP *Quinora, Quinidex*	Quinidine sulfate capsules USP Quindine sulfate tablets USP	Initial: oral, 200–800 mg every 2–3 hr up to 5 times daily as needed and tolerated	Up to 4 g/day	6 mg/kg of body weight or 180 mg/m^2 of body surface, 5 times daily
Quinidine gluconate USP	Quinidine gluconate injection USP	IM, 600 mg, then 400 mg repeated up to 12 times daily as necessary; IV infusion, 800 mg in 40 mL of 5% dextrose injection at the rate of 1 mL/min	Up to 5 g/day	
Quinidine polygalacturonate *Cardioquin*	Quinidine polygalacturonate tablets USP	Conversion of atrial and ventricular arrhythmias: oral, 275–825 mg initially, then 275–825 mg every 3–4 hr for 3 or 4 doses, with subsequent doses being increased by 137.5–275 mg every third or fourth dose until rhythm is restored or toxic effects occur; maintenance: oral, 275 mg bid or tid as needed and tolerated		
Procainamide hydrochloride USP *Pronestyl*	Procainamide hydrochloride capsules USP	Atrial arrhythmias: initial, 1.25 g followed in 1 hr by 750 mg if necessary, then 500 mg–1 g every 2–3 hr as necessary and tolerated; maintenace, 500 mg–1 g 4–6 times daily	Up to 6 g/day	12.5 mg/kg of body weight or 375 mg/m^2 of body surface, qid
	Procainamide hydrochloride tablets USP	Ventricular arrhythmias: 1 g initially, then 250–500 mg every 3 hr as needed and tolerated		
	Procainamide hydrochloride injection USP	IM, 500 mg–1 g qid; IV infusion, 500 mg–1 g at a rate of 25–50 mg/min followed by 2–6 mg/min to maintain therapeutic levels		
Disopyramide phosphate USP *Norpace*	Disopyramide phosphate capsules USP	100–200 mg qid	400–800 mg/day	
Phenytoin sodium USP *Dilantin*	Phenytoin sodium injection USP	Antiarrhythmic: IV, 50–100 mg every 10–15 min as necessary, but not to exceed a total dose of 15 mg/kg of body weight		
Bretylium tosylate *Bretylol*	Bretylium tosylate injection	5 mg/kg repeated in 1–2 hr	20–40 mg/kg daily	
Lidocaine hydrochloride USP *Xylocaine*	Lidocaine hydrochloride injection USP	Cardiac depressant (without epinephrine): IV, 50–100 mg, may be repeated in 5 min (up to 300 mg during a 1-hr period); IV infusion, 1–4 mg/min	Cardiac depressant (without epinephrine): 50–300 mg during a 1-hr period	

*See USP DI for complete dosage information.

TABLE 14-5 *Continued*

ANTIARRHYTHMIC AGENTS

Name Proprietary Name	***Preparations***	***Usual Adult Dose****	***Usual Dose Range****	***Usual Pediatric Dose****
Flecainide acetate *Tambocor*	Flecainide acetate tablets	100 mg every 10 hr	100–400 mg	
Mexiletine hydrochloride *Mexitil*	Mexiletine hydrochloride capsules	100–200 mg every 8 hr		
Tocainide hydrochloride *Tonocard*	Tocainide hydrochloride tablets	400 mg every 8 hr	1200–1800 mg/day	
Amiodarone hydrochloride *Cordarone*	Amiodarone hydrochloride tablets	200–400 mg	800–1600 mg/day	
Diltiazem hydrochloride *Cardizem*	Diltiazem hydrochloride tablets			
Verapamil hydrochloride *Calan, Isoptin*	Verapamil hydrochloride tablets	Oral, initially 80 mg tid to qid	Up to 720 mg/day	

*See USP DI for complete dosage information.

ANTIHYPERTENSIVE AGENTS

Hypertension is a consequence of many diseases. Hemodynamically, blood pressure is a function of the amount of blood pumped by the heart and the ease with which the blood flows through the peripheral vasculature (i.e., resistance to blood flow by peripheral blood vessels). Diseases of components of the central and peripheral nervous systems that regulate blood pressure, abnormalities of the hormonal system and of the kidney and peripheral vascular network that affect blood volume, all can create a hypertensive state in humans. Hypertension is generally defined as *mild* when the diastolic pressure is between 90 and 104 mmHg; *moderate* if the pressure is 105 to 114 mmHg; and *severe* when above 115 mmHg. It is estimated that about 15% of the adult population in the United States (about 40 million) are hypertensive.

Essential (primary) hypertension is the most common form of hypertension. Although advances have been made on the identification and control of essential hypertension, the etiology of this form of hypertension has not yet been resolved. *Renal hypertension* can be created by experimentally causing renal artery stenosis in animals. Renal artery stenosis may also occur in pathologic conditions of the kidney, such as nephritis, renal artery thrombosis, renal artery infarctions, or other conditions that have restricted blood flow through the renal artery. Hypertension may also originate from pathologic states in the central nervous system, such as malignancies. Tumors in the adrenal medulla that cause release of large amounts of catecholamines create a hypertensive condition known as *pheochromocytoma*. Excessive secretion of aldosterone by the adrenal cortex, often because of adenomas, also produces hypertensive disorders.

Arterial blood pressure is regulated by several physiologic factors such as heart rate, stroke volume, peripheral vascular network resistance, blood vessel elasticity, blood volume, and viscosity of blood. Endogenous chemicals also play an important part in the regulation of arterial blood pressure. The peripheral vascular system is greatly influenced by the sympathetic–parasympathetic balance of the autonomic nervous system, the control of which originates in the central nervous system. Enhanced adrenergic activity is recognized as a principal contributor to essential hypertension.

Therapy using antihypertensive agents evolved rapidly between 1950 and 1960. During that time a number of empiric discoveries were made that resulted in the marketing of drugs for the treatment and control of hypertensive disease. The first drugs of value were α-adrenergic blocking agents. These were intended to block the action of catecholamines, with the expectation that contraction of the smooth muscle of the vascular walls would be blocked. These drugs had their limitations, because the duration of action was far too short and side effects precluded long-term therapy. The clinical importance of α-adrenergic blocking agents lies in their value in the treatment of peripheral vascular disease and for diagnosis and short-term treatment of pheochromocytoma (see Chap. 11). Another type of chemical sympathectomy has been applied by the use of ganglionic blocking agents. Although ganglionic blockade is not selective, sympathetic control of the vascular smooth muscle tone results in a clinically useful lowering of blood pressure from the use of these blockers (see Chap. 12).

The antihypertensive drugs discussed in this section include (1) agents that affect the function of the peripheral sympathetic nervous system, (2) centrally acting adrenergic drugs, (3) vasodilators, and (4) angiotensin-converting enzyme inhibitors. Calcium channel blockers, in addition to their antianginal and antiarrhythmic uses, are also effective antihypertensive agents. These are discussed in an earlier section of this chapter.

AGENTS AFFECTING PERIPHERAL SYMPATHETIC NERVES

Several drugs are antihypertensive because they act on peripheral adrenergic nerves to alter their ability to function normally. These drugs act by either preventing the accumulation of norepinephrine into storage granules in adrenergic neurons or by displacing stores of norepinephrine from the granule with a less efficacious neurotransmitter.

Rauwolfia Alkaloids

Folk remedies prepared from the species of *Rauwolfia*, a plant genus belonging to the Apocynaceae family, have been reported as early as 1563. The root of the species *R. serpentina* has been used for centuries as an antidote to stings and bites of insects, to reduce fever, as a stimulant to uterine contractions, for insomnia, and particularly for the treatment of insanity. Its use in hypertension was recorded in the Indian literature in 1918, but it was not until 1949 that hypotensive properties of *Rauwolfia* species appeared in the Western literature.[46] Rauwolfia preparations were introduced in psychiatry in the treatment of schizophrenia in the early 1950s following confirmation of the folk remedy reports on their use in mentally deranged patients. By the end of the 1960s, however, the drug had been replaced by more efficacious neurotropic agents. Reserpine and its preparations still remain useful in control of mild essential hypertension.

Chemical investigations of the active components of *R. serpentina* roots have yielded several alkaloids (e.g., ajmaline, ajmalicine, ajmalinine, serpentine, serpentinine, and others). Reserpine, which is the

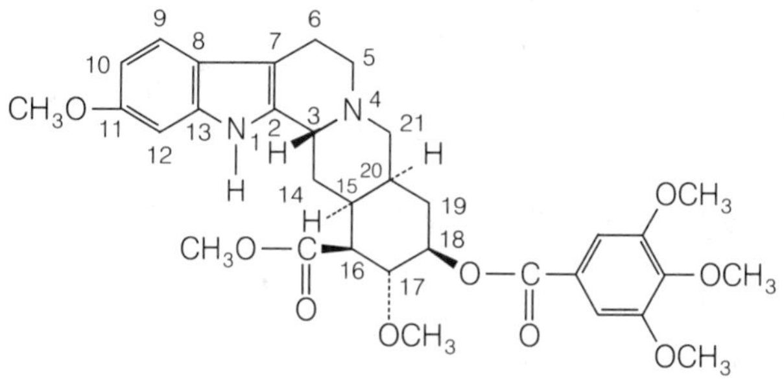

Reserpine

major active constituent of *Rauwolfia*, was isolated in 1952 by Muller et al.[47] and was a much weaker base than the alkaloids just mentioned. Reserpinoid alkaloids are yohimbinelike bases that have an additional functional group on C-18. Only three naturally occurring alkaloids possess reserpinelike activity strong enough for use in treating hypertension: reserpine, deserpidine, and rescinamine. Of these, only reserpine is official in the *USP*.

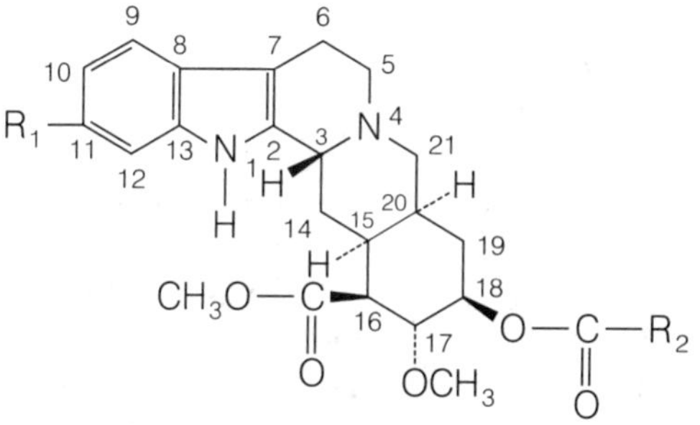

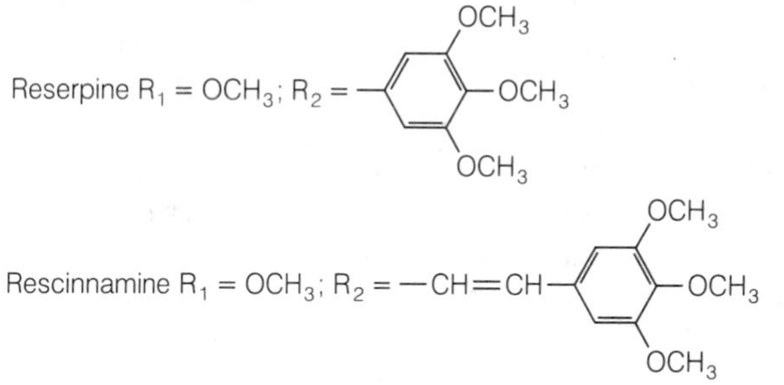

Reserpine $R_1 = OCH_3$; R_2 = 3,4,5-trimethoxyphenyl

Rescinnamine $R_1 = OCH_3$; R_2 = —CH=CH—(3,4,5-trimethoxyphenyl)

Reserpine is absorbed rapidly after oral administration. Fat tissue accumulates reserpine slowly, with a maximal level being reached between four and six hours. After 24 hours there are small amounts of reserpine in the liver and fat, but none in the brain or other tissues. Reserpine is metabolized by the liver and intestine to methyl reserpate and 3,4,5-trimethoxybenzoic acid (Fig. 14-13).

The effects of reserpine do not correlate well with the tissue levels of the drug. It was observed that the pharmacologic effects of reserpine were still manifest in animals at a time when reserpine could no longer be detected in the brain.[48] Subsequent to this observation it was found that reserpine causes depletion of catecholamines from postganglionic sympathetic nerves and the adrenal medulla. Both

FIG. 14-13. *Metabolism of reserpine. (From Rand, J. M., and Jurevics, H.: In Gross, F. (ed.). Antihypertensive Agents. Berlin, Springer-Verlag, 1977.)*

catecholamines and serotonin are depleted from the brain.[49,50] Even though reserpine has central nervous system activity, its hypertensive action is primarily due to depletion of catecholamines from the peripheral sympathetic nerves.

Reserpine depletes catecholamines at the nerve ending by inhibiting the ATP–Mg^{2+}-dependent uptake mechanism into the neuronal granules (Fig. 14-14). The catecholamines that are not taken up into the granule are metabolized by mitochondrial monoamine oxidase, causing the reduction of amine content (Fig. 14-14).

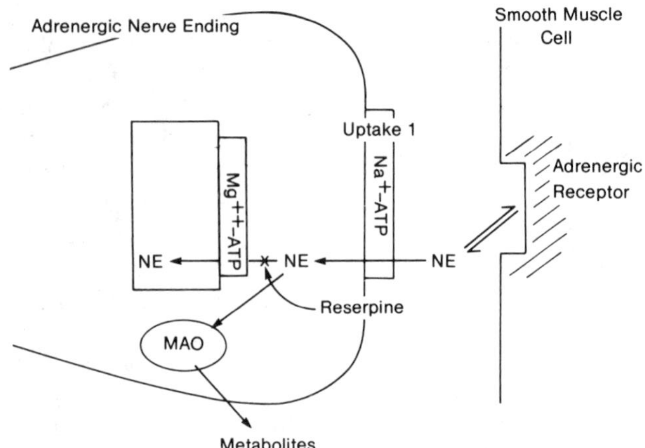

FIG. 14-14. *Action of reserpine at adrenergic nerve ending.*

PRODUCTS

Powdered Rauwolfia Serpentina, USP (Raudixin; Rauserpa; Rauval) is the powdered whole root of *R. serpentina* (Benth). It is a light tan to light brown powder, sparingly soluble in alcohol and only slightly soluble in water. It contains the total alkaloids, of which reserpine accounts for about 50% of the total activity. Orally, 200 to 300 mg is roughly equivalent to 500 μg of reserpine. It is used in the treatment of mild or moderate hypertension or in combination with other hypotensive agents in severe hypertension.

Reserpine, USP (Serpasil; Reserpoid; Rau-Sed; Sandril). This is a white to light yellow crystalline alkaloid, practically insoluble in water, obtained from various species of *Rauwolfia*. In common with other compounds with an indole nucleus, it is susceptible to decomposition by light and oxidation, especially when in solution. In the dry state discoloration occurs rapidly when exposed to light, but the loss in potency is usually small.[51] In solution, reserpine may break down when exposed to light, especially in clear glass containers, with no appreciable color change; thus, color change cannot be used as an index of the amount of decomposition.

There are several possible points of breakdown in the reserpine molecule. Hydrolysis may occur at C-16 and C-18.[52] Reserpine is stable to hydrolysis in acid media, but in alkaline media the ester group at

C-18 may be hydrolyzed to give methyl reserpate and trimethoxybenzoic acid (after acidification). If, in addition, the ester group at C-16 is hydrolyzed, reserpic acid (after acidification) and methyl alcohol are formed. Citric acid helps to maintain reserpine in solution and, in addition, stabilizes the alkaloid against hydrolysis.

Storage of solutions in daylight causes epimerization at C-3 to form 3-isoreserpine. In daylight, oxidation (dehydrogenation) also takes place, 3-dehydroreserpine being formed. It is green in solution, but, as the oxidative process progresses, the color disappears and, finally, a strongly orange color appears. Oxidation of solutions takes place in the dark at an increasing rate with increased amounts of oxygen and at an even faster rate when exposed to light. Sodium metabisulfite will stabilize the solutions if kept protected from light, but when exposed to light it actually oxidizes the reserpine so that the solutions are less stable than if the metabisulfite were absent. Nordihydroguaiaretic acid (NDGA) aids in stabilizing solutions when protected from light, but in daylight the degradation is retarded only slightly. Urethan in the solution stabilizes it in normally filled ampules, but affords no protection in daylight.

Reserpine is effective orally and parenterally in the treatment of hypertension. After a single intervenous dose, the onset of antihypertensive action usually begins in about one hour. After intramuscular injection, the maximum effect occurs within approximately four hours and lasts about ten hours. When given orally, the maximum effect occurs within about two weeks and may persist up to four weeks after the final dose. When used in conjunction with other hypotensive drugs in the treatment of severe hypertension, the daily dose varies from 100 to 250 μg.

Guanethidine Monosulfate, USP [2-(Hexahydro-1(2*H*)-azocinyl)ethyl]guanidine sulfate (Ismelin Sulfate) is a white crystalline material that is very soluble in water. It was one of a series of guanidine compounds prepared in the search for potent antitrypanosomal agents. There is an absence of central nervous system effects, such as depression, from guanethidine because the drug is highly polar and does not easily pass the blood–brain barrier. Guanethidine produces a gradual, prolonged fall in blood pressure. Usually two to seven days of therapy

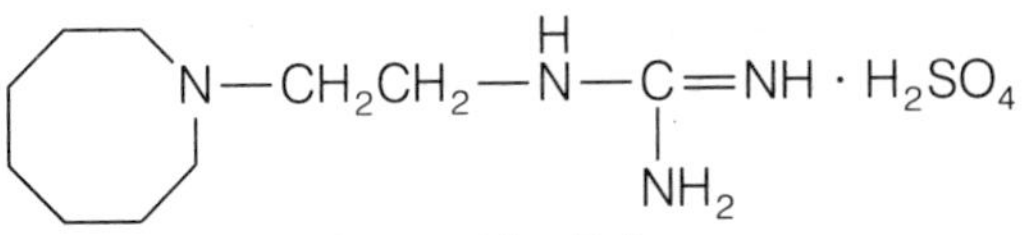

Guanethidine Sulfate

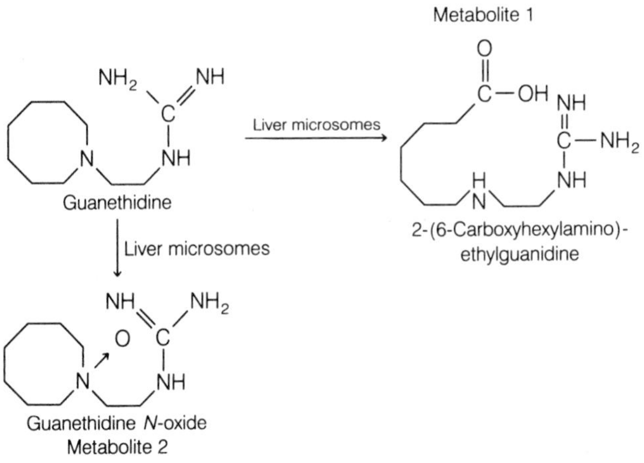

FIG. 14-15. *Metabolism of guanethidine.*

are required before the peak effect is reached, and usually, this peak effect is maintained for three or four days; then, if the drug is discontinued, the blood pressure returns to pretreatment levels over a period of one to three weeks. Because of this slow onset and prolonged duration of action, only a single daily dose is needed.

Guanethidine is metabolized by microsomal enzymes to 2-(6-carboxyhexylamino)ethylguanidine and guanethidine *N*-oxide (Fig. 14-15). Both metabolites have very weak antihypertensive properties. Guanethidine is taken up by the amine pump located on the neuronal membrane and retained in the nerve, displacing norepinephrine from its storage sites in the neuronal granules. The displaced norepinephrine is metabolized to homovanillic acid by mitochondrial monoamine oxidase, depleting the nerve ending of the neurotransmitter. The usefulness of guanethidine resides in the fact that although it is taken up by the nerve, it has a relatively low affinity for the amine pump and gives concentrations in the nerve endings that are too low to cause general depression of the neuronal membrane to stimulation[53] by the nerve action potential. Guanethidine stored in the granules is released by the nerve action potential, but has a very low intrinsic activity for the adrenergic receptors on the postjunctional membrane. Moderate doses for a prolonged period or large doses of guanethidine may produce undesirable side effects by causing neuromuscular blockade and adrenergic nerve conduction blockade.

Guanadrel Sulfate. (1,4-Dioxaspiro[4.5]dec-2-ylmethyl)guanadine sulfate (Hylorel). Guanadrel is similar to guanethidine in the manner in which it reduces elevated blood pressure. It acts as a postganglionic adrenergic-blocking agent by displacing norepinephrine in adrenergic neuron storage granules, thereby preventing release of the endogenous neurotransmitter on nerve stimulation. Guanadrel has a much shorter half-life (ten hours) than guanethidine, the half-life of which is measured in days. In the stepped-care approach to hypertension, guanadrel is usually a step 2 agent.

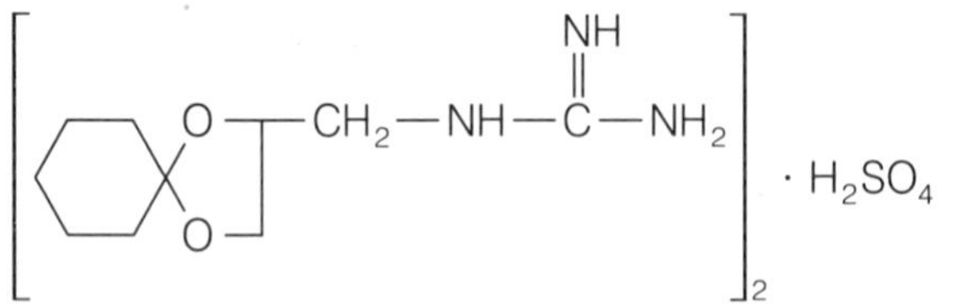

Guanadrel Sulfate

Prazosin Hydrochloride. 1-(4-Amino-6,7-dimethoxy-2-quinazolinyl)-4-(2-furoyl)piperazine monohydrochloride (Minipress). The antihypertensive effects of this drug are due to its peripheral vasodilatation as a result of its blockade of α_1-adrenergic receptors.

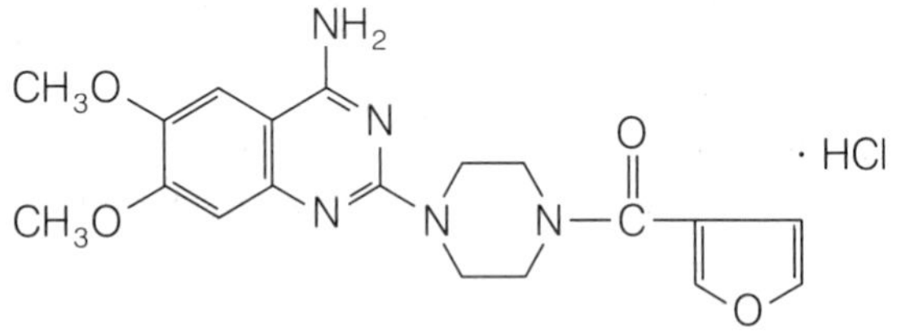

Prazosin Hydrochloride

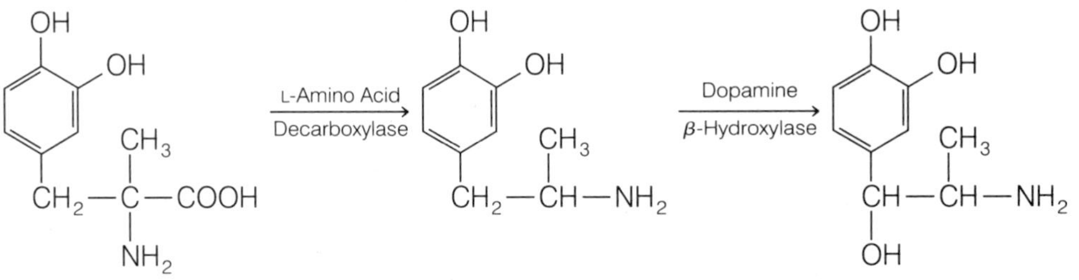

Prazosin is readily absorbed, and plasma concentrations reach a peak at about three hours after administration. Plasma half-life is between two and three hours. Prazosin is highly bound to plasma protein; however, it does not cause adverse drug reactions with drugs that might be displaced from their protein-binding sites (e.g., cardiac glycosides). Prazosin may cause severe orthostatic hypotension because of its α-adrenergic blocking action, which prevents the reflex venous constriction that is activated when an individual changes position. At least 10% of the patients reported dizziness as a side effect.

CENTRALLY ACTING ADRENERGIC DRUGS

The use of agents that directly affect the peripheral component of the sympathetic nervous system represents an important approach to the treatment of hypertension. A second approach to modifying sympathetic influence on the cardiovascular system is through inhibition or reduction of central nervous system control of blood pressure.[54]

Earlier studies have suggested that the hypotensive action of α-methyldopa was due to the peripheral properties of the drug as a decarboxylase inhibitor or as a false transmitter. The hypotensive action of this drug was initially attributed to its conversion to a α-methylnorepinephrine in adrenergic neurons in the central nervous system, where it was to act as a false transmitter on adrenergic neuron activation. α-Methylnorepinephrine, when released, was proposed to act on adrenergic receptors of the central blood pressure-regulating mechanism to exert a peripheral depressant effect because of its low efficacy (intrinsic activity).

The current hypothesis of hypotensive-producing activity of methyldopa involves the central nervous system as the site of action.[55] Methyldopa, on conversion to α-methylnorepinephrine, acts on α_2-adrenergic receptors to inhibit the release of norepinephrine, resulting in a decrease of sympathetic outflow from the central nervous system.

L-Amino Acid Decarboxylase → Dopamine β-Hydroxylase →

Methyldopa α-Methyldopamine α-Methylnorepinephrine

Methyldopa is used as a step 2 agent and recommended for patients with high blood pressure that is not responsive to diuretic therapy alone.

Methyldopate Hydrochloride, USP. L-3-(3,4-Dihydroxyphenyl)-2-methylalanine ethyl ester hydrochloride (Aldomet Ester Hydrochloride). Methyldopa, suitable for oral use, is a zwitterion and is not

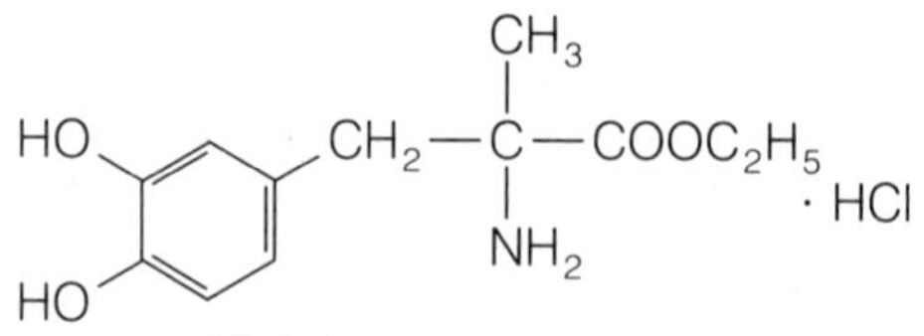

Methyldopate Hydrochloride

soluble enough for parenteral use. This problem was solved by making the ester, leaving the amine free to form the water-soluble hydrochloride salt. It is supplied as a stable, buffered solution, protected with antioxidants and chelating agents.

Clonidine Hydrochloride. 2-[(2,6-Dichlorophenyl)imino]imidazolidine monohydrochloride (Catapres) was the first antihypertensive known to act on the central nervous system. It was synthesized in 1962 as a derivative of the known α-sympathomimetic drugs naphazoline and tolazoline, potential nasal vasoconstrictors, but instead it has proved to be an effective drug in the treatment of mild to severe hypertension.

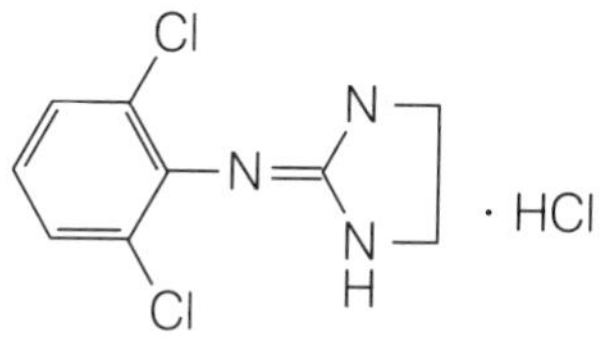

Clonidine Hydrochloride

Clonidine acts by both peripheral and central mechanisms in the body to affect blood pressure. It stimulates the peripheral α-adrenergic receptors to produce vasoconstriction, producing a brief period of hypertension. Clonidine acts centrally to inhibit the sympathetic tone and cause hypotension that is of much longer duration than the initial hypertensive effect. Administration of clonidine thus produces a biphasic change in blood pressure, beginning with a brief hypertensive effect and followed by a hypotensive effect that persists for about four hours. This biphasic response is altered by dose only, in that larger doses produce a greater hypertensive effect and delay the onset of the hypotensive properties of the drug. Clonidine acts on α_2-adrenoreceptors located in the hindbrain to produce its hypotensive action.[56] Clonidine also acts centrally to cause bradycardia and to reduce plasma levels of renin. Sensitization of baroreceptor pathways in the central nervous system appears to be responsible for the bradycardia transmitted by way of the vagus nerve. However, the central mechanism that results in plasma renin decrease is not known. The hypotensive properties of clonidine in animals can be blocked by applying α-adrenergic blocking agents directly to the brain.[57]

Clonidine has advantages over antihypertensive drugs such as guanethidine and prazosin in that it seldom produces orthostatic hypotensive side effects. However, clonidine does have some sedative properties that are undesirable; it also may cause constipation and dryness of the mouth.

Clonidine is distributed throughout the body, with the highest concentrations found in the organs of elimination: kidney, gut, and liver. Brain concentrations are low, but higher than plasma concentrations. The high concentration in the gut is due to an enterohepatic cycle wherein clonidine is secreted into the bile in rather high concentrations. The half-life in humans is about 20 hours. Clonidine is metabolized by the body to form two major metabolites, *p*-hydroxyclonidine and its glucuronide. *p*-Hydroxyclonidine does not cross the blood–brain barrier and has no hypotensive effect in people.

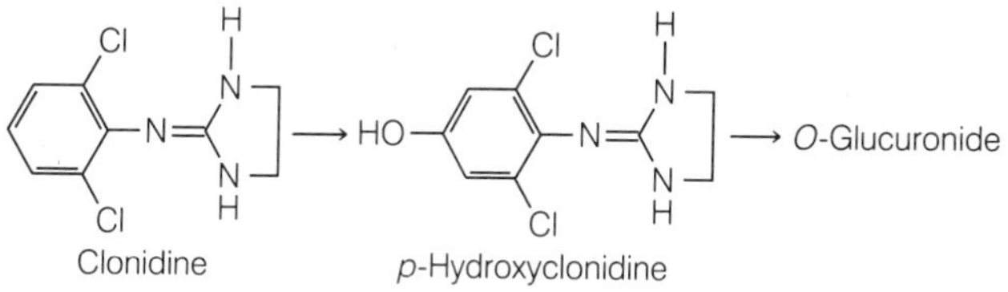

Clonidine → *p*-Hydroxyclonidine → *O*-Glucuronide

Metabolism of Clonidine

Guanabenz Acetate. [(2,6-Dichlorobenzylidene)amino]guanidine monoacetate (Wytensin). Guanabenz is a central α_2-adrenergic agonist that reduces the release of norepinephrine from the neuron when stimulated. The effect of the drug results in a decreased sympathetic tone in the heart, kidneys, and peripheral blood vessels. The drug does not produce orthostatic hypotension.

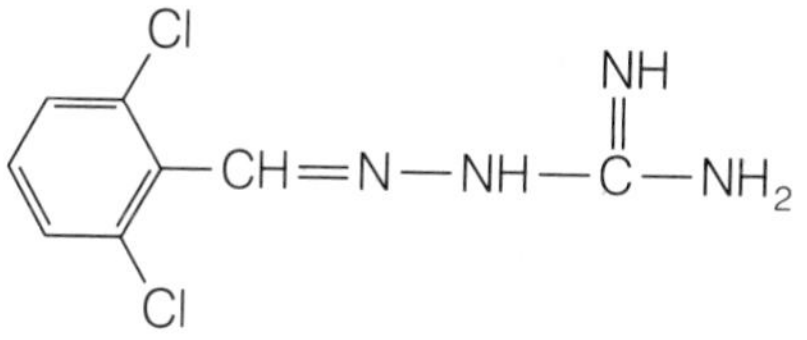

Guanabenz

DRUGS ACTING DIRECTLY ON SMOOTH MUSCLE (VASODILATORS)

Reduction of arterial smooth-muscle tone may occur by many mechanisms, such as reduction in sympathetic tone, stimulation of β-adrenergic receptors, or even direct action on the vasculature, without interference with the autonomic innervation. Drugs acting on the arteriolar smooth muscle also have the property of increasing sympathetic reflex activity, causing an increase in heart rate and cardiac output as well as stimulating renin release from the kidney, which increases sodium retention and expansion of plasma volume. As a result, it is common to coadminister saluretics and β-adrenergic blocking agents with these agents.

Diazoxide, USP (Hyperstat IV). This drug is used as the sodium salt of 7-chloro-3-methyl-2*H*-1,2,4-benzothiadiazine 1,1-dioxide.

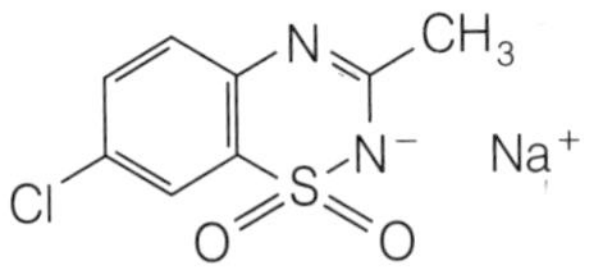

Sodium Diazoxide

Diazoxide is a direct vasodilating drug that lowers peripheral vascular resistance, increases cardiac output, and does not compromise renal blood flow.

This is a des-sulfamoyl analogue of the benzothiazine diuretics and has a close structural similarity to chlorothiazide. It was intentionally developed to increase the antihypertensive action of the thiazides and minimize the diuretic effect.

It is used by intravenous injection as a rapidly acting antihypertensive agent for the emergency reduction of blood pressure in hospitalized patients with accelerated or malignant hypertension. Over 90% is bound to serum protein, and caution should be exercised when it is used in conjunction with other protein-bound drugs that may be displaced by diazoxide. The injection is given rapidly by the intravenous route to ensure maximal effect. The initial dose is usually 1 mg per kg of body weight, with a second dose given if the first injection does not elicit a satisfactory lowering of blood pressure within 30 minutes. Further doses may be given at 4- to 24-hour intervals if needed. Oral antihypertensive therapy is begun as soon as possible.

The injection has a *p*H of about 11.5, which is necessary to convert the drug to its soluble sodium salt. There is no significant chemical decomposition after storage at room temperature for two years. When the solution is exposed to light, it darkens.

Most studies indicate that diazoxide acts directly on arteriolar smooth muscle to cause relaxation. The vasodilatation is independent of a direct stimulation of β-adrenergic receptors. The mechanism causing relaxation of smooth muscles is still under question. It has been suggested that diazoxide may act by depleting an intracellular pool of Ca^{2+}, inhibiting the release of Ca^{2+}, blocking α_1- or activating β-adrenergic receptors, or increasing cAMP levels.[58]

Hydralazine Hydrochloride, USP. 1-Hydrazinophthalazine monohydrochloride (Apresoline Hydrochloride) originated from the work of a chemist[59] attempting to produce some unusual chemical compounds and from the observation[60] that this compound had antihypertensive properties. It occurs as yellow crystals and is soluble in water to the extent of about 3%. A 2% aqueous solution has a *p*H of 3.5 to 4.5.

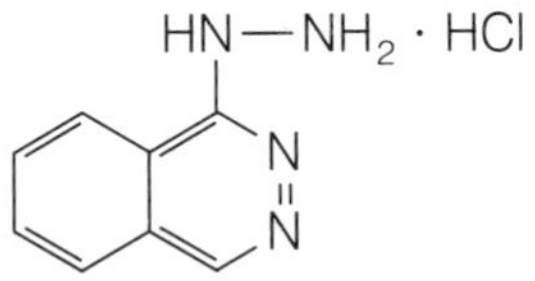

Hydralazine Hydrochloride

Hydralazine is useful in the treatment of moderate to severe hypertension. It is often used in conjunction with less potent antihypertensive agents, because when used alone in adequate doses there is a frequent occurrence of side effects. In combinations, it can be used in lower and safer doses. Its action appears to be centered on the smooth muscle of the vascular walls, with a decrease in peripheral resistance to blood flow. This results in an increased blood flow through the peripheral blood vessels. Also of importance is its unique property of increasing

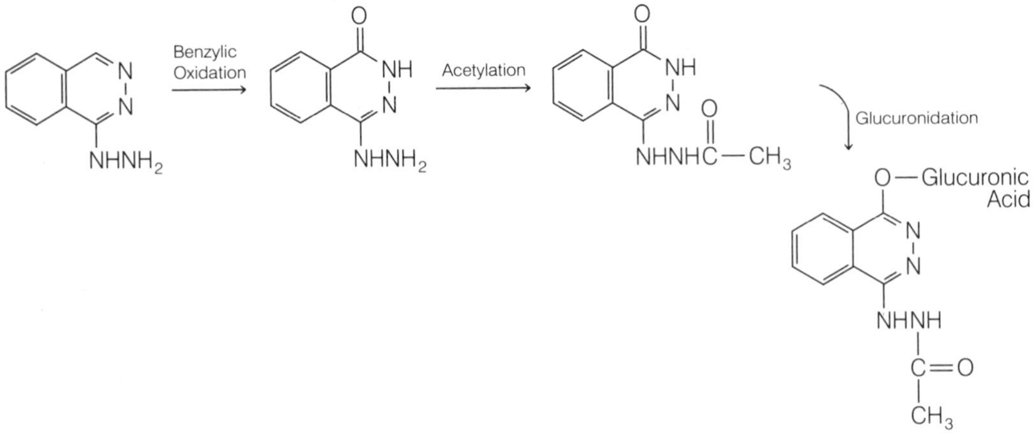

Benzylic Oxidation
Acetylation
Glucuronidation
O—Glucuronic Acid

FIG. 14-16. *Metabolism of hydralazine.*

renal blood flow, an important consideration in patients with renal insufficiency.

Hydralazine acts on vascular smooth muscle to cause relaxation. Its mechanism of action is unclear. It interferes with Ca^{2+} entry and Ca^{2+} release from intracellular stores and has also been reported to cause activation of guanylate cyclase, resulting in increased levels of cGMP. All of these biochemical events can cause vasodilatation.

Absorption of hydralazine hydrochloride taken orally is rapid and nearly complete. The maximal hypotensive effect is demonstrable within one hour. The drug is excreted rapidly by the kidney, and within 24 hours 75% of the total amount administered appears in the urine as metabolites or unchanged drug. Hydralazine undergoes benzylic oxidation, glucuronide formation, and *N*-acetylation by the microsomal enzymes in the tissues (Fig. 14-16). Acetylation appears to be a major determinant of the rate of hepatic removal of the drug from the blood and therefore of systemic availability.[61] Rapid acetylation results in a highly hepatic extraction ratio from blood and a greater first-pass elimination.[62]

Hydralazine is more effective clinically when coadministered with drugs that antagonize adrenergic transmission (e.g., β-adrenergic antagonists, reserpine, guanethidine, methyldopa, and clonidine). When given with diuretics, it is useful in the treatment of congestive heart failure.

Sodium Nitroprusside, USP. Sodium nitroferricyanide; disodium pentacyanonitrosylferrate(2) $Na_2[Fe(CN)_5NO]$ (Nipride) is one of the most potent blood pressure-lowering drugs. Its use is limited to hypertensive emergencies because of its short duration of action. The effectiveness of sodium nitroprusside as an antihypertensive has been known since 1928,[63] but it was not until 1955 that its efficacy as a drug was established.[64] This drug differs from other vasodilators in that vasodilatation occurs in both venous and arterial vascular beds. Sodium nitroprusside is a reddish brown, water-soluble powder that is decomposed by light when in solution. The hypotensive effect of the chemical is due to the formation of nitric oxide in situ (see earlier discussion, Nitrovasodilators), elevating cellular levels of cGMP. Sodium nitroprusside is metabolized by the liver, yielding thiocyanate. Because thiocyanate is excreted by the kidney, patients with impaired renal function may suffer from thiocyanate toxicity.

Minoxidil, USP. 2,4-Diamino-6-piperidinopyrimidine-3-oxide (Loniten) was developed as a result of isosteric replacement by triaminopyrimidine of a triaminotriazine moiety. The triaminotriazines were initially observed to be potent vasodilators in cats and dogs following their formation of *N*-oxides in these animals. The triazines were inactive in humans because of people's inability to form *N*-oxide metabolites of the triazines; this led to the discovery of minoxidil.

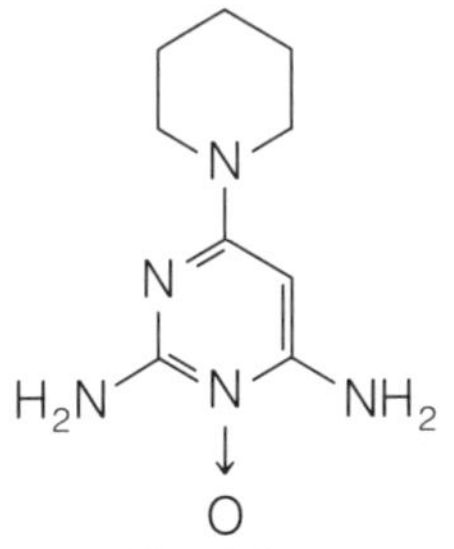

Minoxidil

The antihypertensive properties of minoxidil are similar to those of hydralazine in that minoxidil is able to decrease arteriolar vascular resistance. Minoxidil exerts its vasodilator action by a direct effect on arteriolar smooth muscle and appears to have no effect on the central nervous system or on the adrenergic nervous system in animals.[65] The serum half-life of the drug is 4.5 hours, and the antihypertensive effect may last up to 24 hours.

Minoxidil is used for severe hypertension that is difficult to control by other antihypertensive agents. The drug has some characteristic side effects of direct vasodilator drugs. It causes sodium and water retention and may require coadministration of a diuretic. Minoxidil also causes reflex tachycardia, which can be controlled by use of a β-adrenergic blocking agent.

ANGIOTENSIN-CONVERTING ENZYME INHIBITORS

Renal hypertension is controlled by the renin–angiotensin system. The *renin–angiotensin* system consists of peptidases and their substrates (found in blood and bound to tissue) that take part in the maintenance of blood pressure in low-volume or low-sodium states. Renal hypertension is created by the release of renin, an enzyme found primarily in the kidney, which forms a decapeptide, angiotensin I, from angiotensinogen, an α_2-globulin circulating in the blood. Angiotensin-converting enzyme (ACE) splits two amino acid residues from angiotensin I and forms the octapeptide, angiotensin II. Angiotensin II causes vasoconstriction and degrades bradykinin, a naturally occurring vasodilator. Angiotensin II also stimulates the synthesis of aldosterone in the adrenal cortex which, when released, enhances a retention of sodium ions that results in a rise in blood volume and an increase in blood pres-

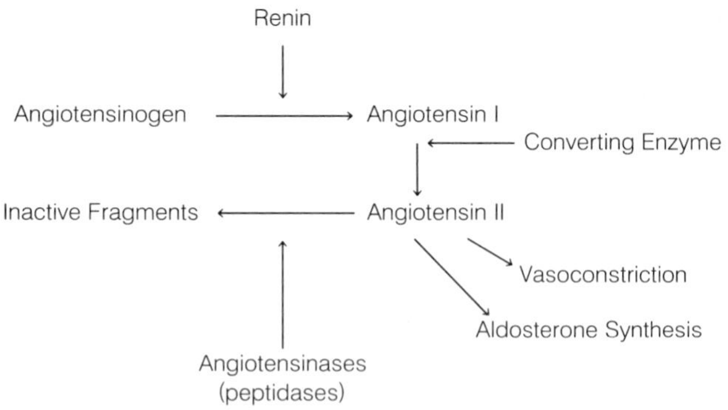

FIG. 14-17. Renin–angiotensin system.

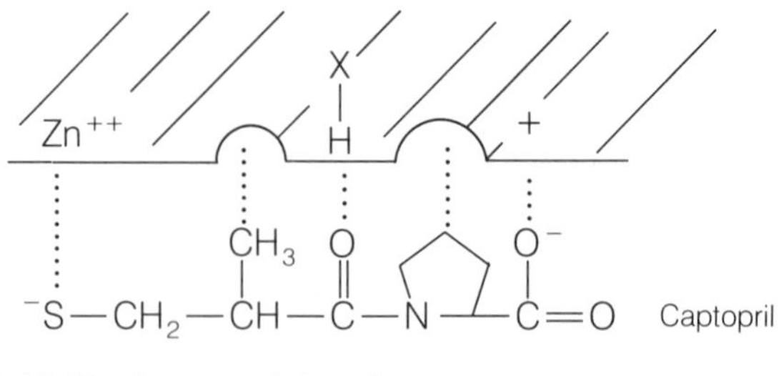

FIG. 14-18. Accommodation of captopril to active site of angiotensin-converting enzyme.

sure. The increase in blood pressure causes inhibition of renin release and limits the formation of angiotensin II.[66] Angiotensin II is inactivated by peptidases in plasma and tissues (Fig. 14-17).

Captopril. 1-[(2*S*)-3-Mercapto-2-methyl-1-oxopropionyl]-L-proline (Capoten). Captopril blocks the conversion of angiotensin I to angiotensin II by inhibiting the converting enzyme. Studies with substrates and inhibitors of angiotensin-converting enzyme suggest that the enzyme is a peptidase similar to pancreatic carboxypeptidase A.[67] The important binding points at the active site of ACE are thought to be a cationic site that can attract a carboxylate ion and a zinc ion that can polarize a carbonyl group of an amide function to make it more susceptible to hydrolysis, but with captopril, bind the thiol group. Hydrophobic pockets lie between these two polar areas on the enzyme, as does a functional group that can hydrogen-bond to an amide carbonyl. A hypothetical model of the active site of ACE in Figure 14-18 show the accommodation of captopril to the active site of ACE.[68]

Treatment with captopril reduces blood pressure in patients with renovascular disease and in patients with mild hypertension. The drug effect is evident in about 15 minutes after oral administration, with peak effect between 30 and 60 minutes. Its biologic half-life is about two hours.

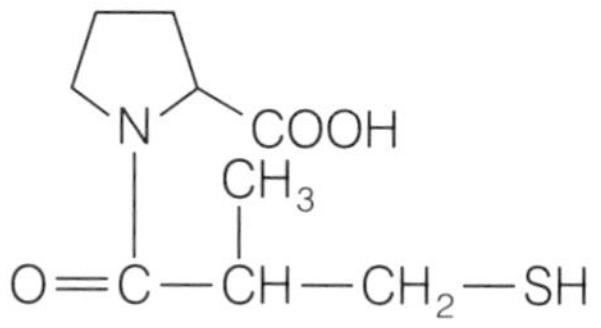

Captopril

Enalapril Maleate. 1-[*N*[(*S*)-1-Carboxy-3-phenylpropyl]-L-alanyl]-L-proline 1′-ethyl ester, maleate (Vasotec). Enalapril is a long-acting ACE inhibitor. It is, in fact, a prodrug, and requires activation by hydrolysis of its ethyl ester to form the diacid enalaprilat. Enalapril is devoid of the side effects of rashes and loss of taste seen with captopril. Investigators observed that these side effects were similar to the mercapto-containing drug penicillamine. It is suggested that the absence of the

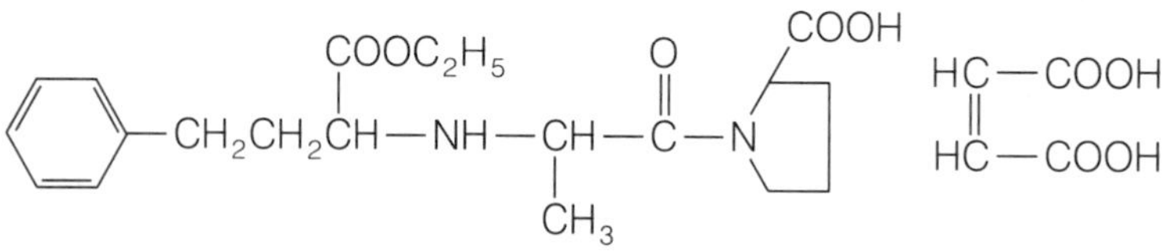

Enalapril Maleate

TABLE 14-6

ANTIHYPERTENSIVE AGENTS

Name Proprietary Name	***Preparations***	***Usual Adult Dose****	***Usual Dose Range****	***Usual Pediatric Dose****
Prazosin hydrochloride *Minipress*	Prazosin capsules	1 mg	1–3 mg/day	
Clonidine hydrochloride *Catapress*	Clonidine hydrochloride tablets	0.1 mg	0.1–0.3 mg/day	

*See USP DI for complete dosage information.

TABLE 14-6 *Continued*

ANTIHYPERTENSIVE AGENTS

Name Proprietary Name	***Preparations***	***Usual Adult Dose****	***Usual Dose Range****	***Usual Pediatric Dose****
Diazoxide USP *Hyperstat*	Diazoxide injection USP	300 mg	0.3–1.2 g/day	
Minoxidil USP *Loniten*	Minoxidil tablets	0.2 mg/kg	10–40 mg/day	
Powdered rauwolfia serpentina USP *Raudixin, Rauserpa, Rauval, Hyperloid, Rauja, Raulin, Veniber, Wolfina*	Rauwolfia serpentina tablets USP	Initial, 200 mg/day for 1–3 weeks; maintenance, 50–300 mg/day		
Reserpine USP *Serpasil, Reserpoid, Rau-Sed, Sandril, Lemiserp, Resercen, Rolserp, Sertina, Vio-Serpine*	Reserpine injection USP	IM, 500 μg–1 mg, followed by 2–4 mg 8 times daily as necessary	500 μg–32 mg/day	70 μg/kg of body weight or 2 mg/m^2 of body surface, qd or bid
	Reserpine tablets USP	Initial, 500 μg qd; maintenance, 100–250 μg qd		
	Reserpine elixir USP			
Guanethidine sulfate USP *Ismelin*	Guanethidine sulfate tablets USP	Ambulatory patients: initial, oral, 10 or 12.5 mg qd, the daily dosage being increased by 10–12.5 mg at 5- to 7-day intervals if necessary for control of blood pressure		Oral, 200 μg/kg of body weight or 6 mg/m^2 of body surface qd, the daily dosage being increased by 200 μg/kg of body weight or 6 mg/m^2 of body surface at 7- to 10-day intervals if necessary for control of blood pressure
Hydralazine hydrochloride USP *Apresoline*	Hydralazine hydrochloride injection USP	IM or IV, 20–40 mg repeated as necessary		IM or IV, 1.7–3.5 mg/kg of body weight or 50–100 mg/m^2 of body surface daily, divided into 4–6 doses
	Hydralazine hydrochloride tablets USP	Oral, 10 mg qid for the first 2–4 days; 25 mg qid for the balance of the first week, and 50 mg qid for the second and subsequent weeks, the dosage being adjusted to the lowest effective level	Up to 400 mg/day	Oral, 750 mg/kg of body weight or 25 mg/m^2 of body surface daily, divided into 4 doses, the dosage being increased gradually over 3–4 wk as needed
Methyldopa USP *Aldomet*	Methyldopa tablets USP	Initial: oral, 250 mg bid to tid for 2 days, the dosage then being adjusted, preferably at intervals of not less than 2 days until the desired response is obtained; maintenance: oral, 500 mg–2 g/day, divided into 2 to 4 doses		Oral, initially, 10 mg/kg of body weight or 300 mg/m^2 of body surface, divided into 2–4 doses, the dosage then being adjusted, preferably at intervals of not less than 2 days, until the desired response is obtained, but not exceeding 65 mg/kg of body weight or 3 g/day, whichever is less
Methyldopate hydrochloride USP *Aldomet Ester*	Methyldopate hydrochloride injection USP	IV infusion, 250–500 mg in 100 mL of 5% dextrose injection over a period of 30–60 min every 6 hr as necessary	Up to 1 g/day every 6 hours	IV infusion, 5–10 mg/kg of body weight in 5% dextrose injection, administered slowly over a 30–60 min period every 6 hr if necessary, but not exceeding 65 mg/kg of body weight or 3 g/day, whichever is less
Captopril *Capoten*	Captopril tablets	Oral, 12.5 mg bid or tid initially, 25 mg/day maintenance dose	25–100 mg bid or tid	
Enalapril maleate *Vasotec*	Enalapril maleate tablets	Oral, 5 mg qd initially, 10–40 mg/day maintenance dose	10–40 mg/day	

*See USP DI for complete dosage information.

thiol group in enalapril gives it its freedom from these side effects. The half-life of enalapril is 11 hours.

ANTIHYPERLIPIDEMIC AGENTS

The major cause of death in the Western world today is attributed to vascular disease, of which the most prevalent form is atherosclerotic heart disease. Although many causative factors of this disease are recognized (e.g., smoking, stress, diet), atherosclerotic disease can be treated through medication or surgery.

Hyperlipidemia is the most prevalent indicator for susceptibility to atherosclerotic heart disease; it is a term used to describe elevated plasma levels of lipids that are usually in the form of lipoproteins. Hyperlipidemia may be caused by an underlying disease involving the liver, kidney, pancreas, or thyroid; or it may not be attributable to any recognizable disease. Within recent years lipids have been indicated in the development of atherosclerosis in humans. *Atherosclerosis* may be defined as degenerative changes in the intima of medium and large arteries. The degeneration includes the accumulation of lipids, complex carbohydrates, blood, and blood products and is accompanied by the formation of fibrous tissue and calcium deposition on the intima of the blood vessels. These deposits or *plaques* decrease the lumen of the artery, reduce its elasticity, and may create foci for thrombi and subsequent occlusion of the blood vessel.

LIPOPROTEIN CLASSES

Lipoproteins are macromolecules consisting of lipid (cholesterol, triglycerides, and others) noncovalently bound with protein and carbohydrate. These combinations serve to solubilize the lipids and prevent them from forming insoluble aggregates in the plasma. The various lipoproteins found in plasma can be separated by ultracentrifugal techniques into chylomicrons, very low-density lipoprotein (VLDL), intermediate-density lipoprotein (IDL), low-density lipoprotein (LDL), and high-density lipoprotein (HDL). These correlate with the electrophoretic separations of the lipoproteins as follows: chylomicrons, pre-β-lipoprotein (VLDL), broad β-lipoprotein (IDL), β-lipoprotein (LDL), and α-lipoprotein (HDL).

Chylomicrons contain 90% triglycerides by weight and originate from exogenous fat from the diet. They are the least dense of the lipoproteins and migrate the least under the influence of an electric current. Chylomicrons are normally absent in plasma after 12 to 24 hours of fasting. The VLDL is composed of about 60% triglycerides, 12% cholesterol, and 18% phospholipids. The VLDL originates in the liver from free fatty acids. Although VLDL can be isolated from plasma, it is catabolized rapidly into IDL that is further degraded into LDL. Normally, IDL is also rapidly catabolized to LDL, but is not usually isolated from plasma. The LDL consists of 50% cholesterol and 10% triglycerides. This class is the major cholesterol-carrying protein. In normal persons this lipoprotein accounts for about 65% of the plasma cholesterol and is of major concern in hyperlipidemic-related disease states. The LDL is formed from the intravascular catabolism of VLDL. The HDL is composed of 25% cholesterol and 50% protein and accounts for about 17% of the total cholesterol in plasma.

LIPOPROTEIN METABOLISM

The rate at which cholesterol and triglycerides enter the circulation from the liver and small intestine depends upon the supply of the lipid and proteins necessary to form the lipoprotein complexes. Although the protein component must be synthesized, the lipids can be obtained either from a de novo biosynthesis in the tissues or from the diet. Reduction of plasma lipids by diet can delay the development of atherosclerosis. Furthermore, the use of drugs that decrease assimilation of lipids into the body plus a diet has been demonstrated to decrease mortality from cardiovascular disease.[69]

Lipid transport mechanisms that shuttle cholesterol and triglycerides among the liver, intestine, and other tissues exist. Normally, plasma lipids, including lipoprotein cholesterol, are cycled into and out of plasma and do not cause extensive accumulation of deposits in the walls of arteries. Genetic factors and changes in hormone levels affect lipid transport by altering enzyme concentrations and apoprotein content, as well as the number and activity of lipoprotein receptors. This complex relationship makes the treatment of all hyperlipoproteinemias by a singular approach difficult, if not impractical.

Lipids are transported by both *exogenous* and *endogenous* pathways. In the exogenous pathway, dietary fat (triglycerides and cholesterol) is incorporated into large lipoprotein particles (chylomicrons) that enter the lymphatic system and are then passed into the plasma. The chylomicrons are acted upon by lipoprotein lipase in the adipose tissue capillaries, forming triglycerides and monoglycerides. The free fatty acids cross the endothelial membrane of the capillary and are incorporated into triglycerides in

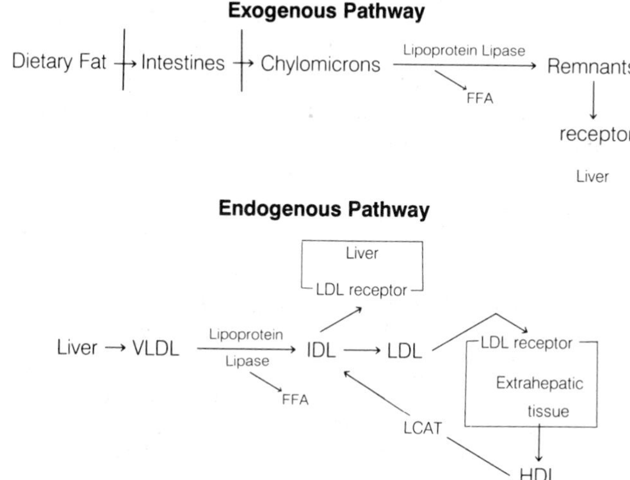

FIG. 14-19. *Exogenous and endogenous pathways of lipoprotein metabolism.*

the tissue for storage as fat or are used for energy by oxidative metabolism. The chylomicron remnant in the capillary reaches the liver and is cleared from the circulation by binding to a receptor that recognizes the apoprotein E and B-48 protein components of the chylomicron remnant.

In the endogenous pathway of lipid transport, lipids are secreted from the liver. These are triglycerides and cholesterol combined with apoprotein B-100 and apoprotein E to form VLDL. The VLDL is acted upon by lipoprotein lipase in the capillaries of adipose tissue to generate free fatty acids and an intermediate-density lipoprotein (IDL). Some IDL may bind to LDL receptors in the liver and is cleared from plasma by endocytosis. Approximately one-half of the circulating IDL is converted to LDL in the plasma by additional loss of triglycerides. This LDL has a half-life in plasma of about 1.5 days and represents 60% to 70% of the cholesterol in plasma. These LDL particles bind to LDL receptors in extrahepatic tissues and are removed from the plasma. Levels of LDL receptors vary according to the need by extrahepatic tissues to bind LDL for the purpose of utilizing their cholesterol. The extrahepatic tissue subsequently releases HDL. Free plasma cholesterol can be adsorbed onto HDL and the cholesterol esters formed by the enzyme lecithin-cholesterol acyltransferase (LCAT). These esters are transferred from HDL to VLDL or LDL in plasma to complete the cycle. The pathways for plasma lipoprotein metabolism by the exogenous and endogenous routes are shown in Figure 14-19.

HYPERLIPOPROTEINEMIAS

It is now recognized that lipid disorders are related to problems in lipoprotein metabolism[70] that create conditions of hyperlipoproteinemia. The hyperlipoproteinemias have been classified into six types, each of which is treated differently (Table 14-7).

The abnormal lipoprotein pattern characteristic of type I is caused by a decrease in the activity of lipoprotein lipase, an enzyme that normally hydrolyzes the triglycerides present in chylomicrons and clears the plasma of this lipoprotein fraction. Because triglycerides that are found in chylomicrons come primarily from exogenous sources, this type of hyperlipoproteinemia may be treated by decreasing the intake of dietary fat. There are no drugs presently available that can be used effectively to counteract type I hyperlipidemia.

TABLE 14-7

CHARACTERIZATION OF HYPERLIPOPROTEINEMIA TYPES

Hyperlipoproteinemia	*Abnormality*		*Appearance of Plasma**	*Triglycerides*	*Cholesterol (total)*
	Electrophoresis	Ultracentrifuge			
I	Massive chylomicronemia		Clear; on top creamy layer of chylomicronemia	Massively elevated	Slightly to moderately elevated
IIa	β-Lipoproteins elevated	LDL increased	Clear	Normal	Heavily elevated
IIb	Pre-β-lipoproteins elevated	LDL + VLDL increased	Slightly turbid	Slightly elevated	Heavily elevated
III	Broad β-lipoprotein band	VLDL/LDL of abnormal composition	Slightly turbid to turbid	Elevated	Elevated
IV	Pre-β-lipoproteins elevated	VLDL increased	Turbid	Moderately heavily elevated	Normal to elevated
V	Pre-β-lipoproteins elevated, chylomicronemia present	VLDL increased, chylomicronemia present	Turbid; on top chylomicronemia	Massively elevated	Slightly elevated

*After having been kept standing at 4°C for 25 hr.
(Adapted from Witte, E. C.: Prog. Med. Chem. 11:199, 1975.)

Type II hyperlipoproteinemia has been divided into type IIa and IIb. Type IIa is characterized by elevated levels of LDL (β-lipoproteins) and normal levels of triglycerides. This subtype disorder is very common and may be caused by disturbed catabolism of LDL. Type IIb differs from type IIa in that this hyperlipidemia has elevated VLDL levels in addition to LDL. Type II hyperlipoproteinemia is often clearly familial and frequently inherited as an autosomal dominant abnormality with complete penetrance and expression in infancy.[71] Patients have been treated by use of diet restrictions of cholesterol and also by a low intake of saturated fats. This type of hyperlipoproteinemia responds to some form of chemotherapy. The combined therapy may bring LDL levels back to normal.

Type III is a rare disorder that is characterized by a broad band of β-lipoprotein. Similar to the type II, it is also familial. Patients respond favorably to diet and drug therapy.

In type IV, hyperlipoproteinemia levels of VLDL are elevated. Because this type of lipoprotein is rich in triglycerides, plasma triglycerides are elevated. The metabolic defect that causes type IV is still unknown; however, this form of hyperlipidemia responds to diet and drug therapy.

Type V hyperlipoproteinemia has high levels of chylomicrons and VLDL, resulting in high levels of plasma triglycerides. The biochemical defect of type V hyperlipoproteinemia is not understood. Clearance of dietary fat is impaired, and a reduction of dietary fat is indicated along with drug therapy.

PRODUCTS

Clofibrate, USP. Ethyl 2-(*p*-chlorophenoxy)-2-methylpropionate (Atromid-S). Clofibrate is a stable, colorless to pale yellow liquid with a faint odor and a characteristic taste. It is soluble in organic solvents, but insoluble in water.

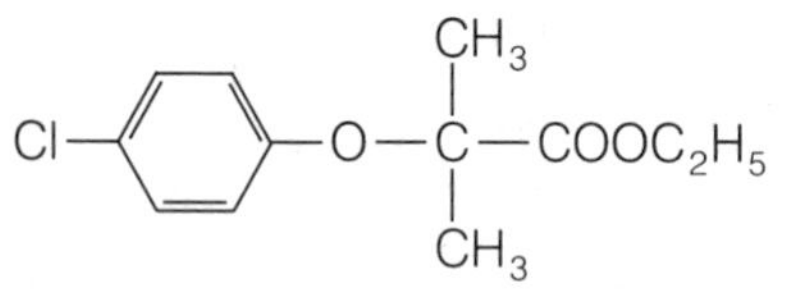

$$\mathrm{Cl{-}C_6H_4{-}O{-}C(CH_3)_2{-}COOC_2H_5}$$

Clofibrate

Clofibrate is prepared by a Williamson synthesis condensing *p*-chlorophenol with ethyl α-bromoisobutyrate[72] or by the interaction of a mixture of acetone, *p*-chlorophenol, and chloroform in the presence of excess potassium hydroxide. The acid obtained by either of these methods is esterified to give clofibrate. Both acid and ester are active; however, the latter is preferred for medicinal use. Clofibrate is hydrolyzed rapidly to 2-*p*-chlorophenoxy-2-methylpropionic acid by esterases in vivo and circulates in blood bound to serum albumin. The acid has been investigated as a hypolipidemic agent. It is absorbed more slowly and to a smaller extent than is the ester. The aluminum salt of the acid gave even lower blood levels than *p*-chlorophenoxy-2-methylpropionic acid.[73]

Clofibrate is the drug of choice in the treatment of type III hyperlipoproteinemias and may be useful also, to a lesser extent, in type IIb and IV hyperlipoproteinemias. The drug is not effective in types I and IIa.

Clofibrate can lower plasma concentrations of both triglycerides and cholesterol, but it has a more consistent clinical effect on triglycerides. It also affects lipoprotein plasma levels by enhancing removal of triglycerides from the circulation and causes reduction of VLDL by stimulating lipoprotein lipase to increase the catabolism of this lipoprotein to LDL.[74] Clofibrate lowers triglyceride levels in the serum much more so than those of cholesterol and also decreases free fatty acids and phospholipids. The lowering of cholesterol may be the result of more than one mechanism. Clofibrate inhibits the incorporation of acetate into the synthesis of cholesterol, between the acetate and mevalonate step, by inhibiting *sn*-glyceryl-3-phosphate acyltransferase. Clofibrate also regulates cholesterol synthesis in the liver by inhibiting microsomal reduction of 3-hydroxy-3-methylglutaryl-CoA (HMGCoA), catalyzed by HMGCoA reductase. Clofibrate may lower plasma lipids by other means than impairment of cholesterol biosynthesis, such as increasing excretion through the biliary tract.

Clofibrate is well tolerated by most patients, the most common side effects being nausea and, to a smaller extent, other gastrointestinal distress. The dosage of anticoagulants, if used in conjunction with this drug, should be reduced by one-third to one-half, depending on the individual response, such that the prothrombin time may be kept within the desired limits.

Usual adult dose: 1.5 to 2 g daily in two to four divided doses.

Usual pediatric dose: use in infants and children has not been established.

Occurrence: Clofibrate capsules USP

Gemfibrozil. 5-(2,5-Dimethylphenoxy)-2,2-dimethylpentanoic acid (Lopid). This is a congener of clofibrate that was first used in the treatment of hyperlipoproteinemia in the mid-1970s. Its mechanism of action and use are similar to those of clofibrate. Gemfibrozil reduces plasma levels of VLDL triglycerides and stimulates clearance of VLDL from plasma. The drug has little effect on cholesterol plasma levels, but does cause an increase of HDL.

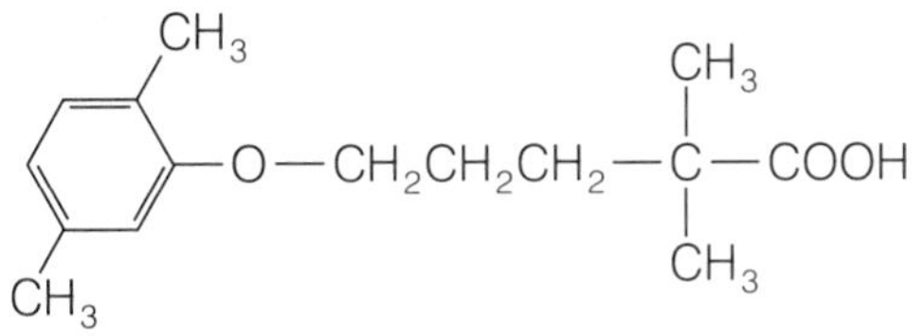

Gemfibrozil

Gemfibrozil is quickly absorbed from the gut and excreted unchanged in the urine. The drug has a plasma half-life of 1.5 hours, but the observed reduction of plasma VLDL concentration takes between two to five days to become evident. The peak effect of its hypolipidemic action may take up to four weeks to become manifest.

Lovastatin. 2-Methylbutanoic acid 1,2,3,7,8,8a-hexahydro-3,7-dimethyl-8-[2-(tetrahydro-4-hydroxy-6-oxo-2*H*-pyran-2-yl)ethyl]-1-naphthalenyl ester; mevinolin; MK-803 (Mevacor). Lovastatin (formerly mevinolin) is a potent inhibitor of 3-hydroxy-3-methylglutaryl-coA reductase (HMGCoAR) found in the liver. This enzyme is the rate-limiting catalyst for the irreversible conversion of HMGCoA to mevalonic acid in the synthesis of cholesterol. The activity of HMGCoAR is also under feedback regulation. When cholesterol is available in sufficient amounts for body needs, the enzyme activity of HMGCoAR is suppressed.

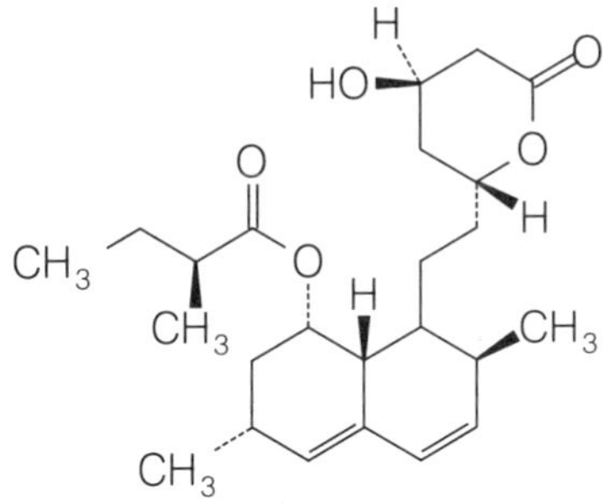

Lovastatin

Elevated plasma cholesterol levels have been correlated with an increase in cardiovascular disease. Of the plasma lipoproteins, the LDL fraction contains the most cholesterol. The source of cholesterol in humans is either from the diet or by de novo synthesis through HMGCoAR as the rate-limiting step. The ingested cholesterol as the free alcohol or ester is taken up after intestinal absorption and transported to the liver and other body organs through the exogenous pathway (see Fig. 14-19). The LDLs deliver cholesterol to peripheral cells. This process occurs after binding of LDL to specific LDL receptors located on the surface of cell membranes. After binding and endocytosis of the receptor and LDL, lysosomal degradation of this complex in the cell makes cholesterol available for use in cellular membrane synthesis. It is generally accepted that the lowering of total plasma cholesterol is most effectively accomplished by reducing LDL levels. Therefore, the population of LDL receptors is an important component of clearing the plasma of cholesterol. Lovastatin contributes to this by directly blocking the active site of HMGCoAR. This action has a twofold effect on cholesterol plasma levels: by causing a decrease of de novo cholesterol synthesis and by causing an increase in hepatic LDL receptors.[75] Lovastatin is an effective hypocholesteremic agent in patients with familial hypercholesteremia. The drug was originally obtained from the fermentation products of the fungi *Aspergillus terreus* and *Monascus ruber*. The lactone portion of the chemical structure of lovastatin bears a resemblance to HMGCoA.

Dextrothyroxine Sodium, USP. *O*-(4-Hydroxy-3,5-diiodophenyl)-3,5-diiodo-D-tyrosine monosodium salt hydrate; sodium D-3,3′,5,5′-tetraiodothyronine (Choloxin). This compound occurs as light yellow to buff powder. It is stable in dry air, but discolors on exposure to light; hence, it should be stored in light-resistant containers. It is very slightly soluble in water, slightly soluble in alcohol, and insoluble in acetone, in chloroform, and in ether.

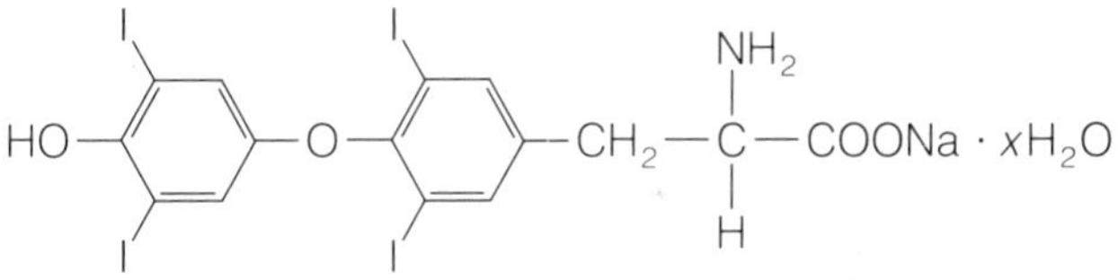

Dextrothyroxine Sodium

The hormones secreted by the thyroid gland have marked hypocholesterolemic activity along with their other well-known actions. With the finding that not all active thyroid principles possessed the same degree of physiologic actions, a search was made for congeners that would cause a decrease in serum cholesterol without other effects such as angina pectoris, palpitation, and congestive failure. D-Thyroxine has resulted from this search. However, at the dosage required, the L-thyroxine contamination must be minimal; otherwise it will exert its characteristic actions. One route to optically pure (at least 99% pure) D-thyroxine is the use of an L-amino acid oxidase from snake venom, which acts only on the L-isomer and makes separation possible.

The mechanism of action of D-thyroxine appears to be stimulation of oxidative catabolism of cholesterol in the liver through stimulation of 7α-cholesterol hydroxylase, the rate-limiting enzyme in the conversion of cholesterol to bile acids. The bile acids are conjugated with glycine or taurine and excreted by the biliary route into the feces. Although cholesterol biosynthesis is not inhibited by the drug, thyroxine increases the number of LDL receptors enhancing removal of LDL from plasma.

Use of thyroxine in the treatment of hyperlipidemias is not without adverse effects. The drug increases the frequency and severity of anginal attacks and may cause cardiac arrhythmias.

D-Thyroxine potentiates the action of anticoagulants, such as warfarin or dicumarol; thus, dosage of the anticoagulants should be reduced by one-third if used concurrently and then further modified, if necessary, to maintain the prothrombin time within the desired limits. Also, it may increase the dosage requirements of insulin or of oral hypoglycemic agents if used concurrently with them.

Usual dose: initial, 1 to 2 mg daily; maintenance, 4 to 8 mg daily
Usual dose range: 1 to 8 mg daily
Occurrence: Dextrothyroxine sodium tablets USP

Cholestyramine Resin, USP (Cuemid; Questran) is the chloride form of a strongly basic anion-exchange resin. It is a styrene copolymer with divinylbenzene with quaternary ammonium functional groups. After oral ingestion, cholestyramine resin remains in the gastrointestinal tract where it readily exchanges chloride ions for bile acids in the small intestine. This event increases the excretion of bile salts in the feces.[75] Cholestyramine resin is also useful in lowering plasma lipids. The reduction of the amounts of reabsorbed bile acids results in increased catabolism of cholesterol in bile acids in the liver. The decreased concentration of bile acids returning to the liver lowers the feedback inhibition by bile acids of 7α-hydroxylase, the rate-limiting enzyme in the conversion of cholesterol to bile acids, increasing the breakdown of hepatic cholesterol. Although the biosynthesis of cholesterol is increased, it appears that the rate of catabolism is greater, resulting in a net decrease in plasma cholesterol levels by affecting LDL clearance. The increase of LDL receptors in the liver that occurs when its content of cholesterol is lowered augments this biochemical event.[76]

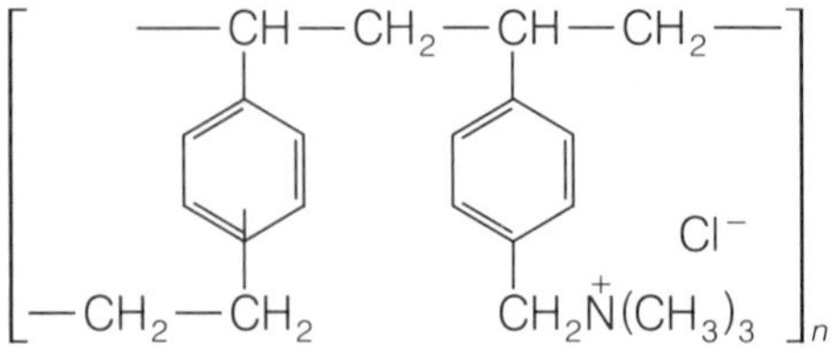

$$\left[\begin{array}{l} -CH-CH_2-CH-CH_2- \\ \quad (C_6H_4) \qquad\quad (C_6H_4) \qquad Cl^- \\ -CH_2-CH_2 \qquad CH_2\overset{+}{N}(CH_3)_3 \end{array}\right]_n$$

Cholestyramine

Cholestyramine resin does not bind with drugs that are neutral or with those that are amine salts; however, it is possible that acidic drugs (in the anion form) could be bound. For example, in animal tests absorption of aspirin given concurrently with the resin was only moderately depressed during the first 30 minutes.

Cholestyramine is the drug of choice for type IIa hyperlipoproteinemia. When used in conjunction with a controlled diet, it reduces β-lipoproteins. The drug is an insoluble polymer and, thus, probably one of the safest because it is not absorbed from the gastrointestinal tract to cause systemic toxic effects.

Category: ion-exchange resin (bile salts)
Usual dose: 4 g three times daily
Usual dose range: 10 to 16 g daily
Usual pediatric dose: children under 6 years of age, dosage is not established; over 6 years of age, 80 mg/kg of body weight or 2.35 g/m^2 of body surface, three times daily, or see Usual dose.

Colestipol Hydrochloride (Colestid) is a high-molecular-weight, insoluble, granular copolymer of tetraethylenepentamine and epichlorohydrin. It functions as an anion-exchange, resin-sequestering agent in a manner similar to that of cholestyramine. Colestipol reduces cholesterol levels without affecting triglycerides and seems to be especially effective in treatment of type II hyperlipoproteinemias.

Usual dose: 7.5 g two to four times daily
Usual dosage range: 15 to 30 g daily

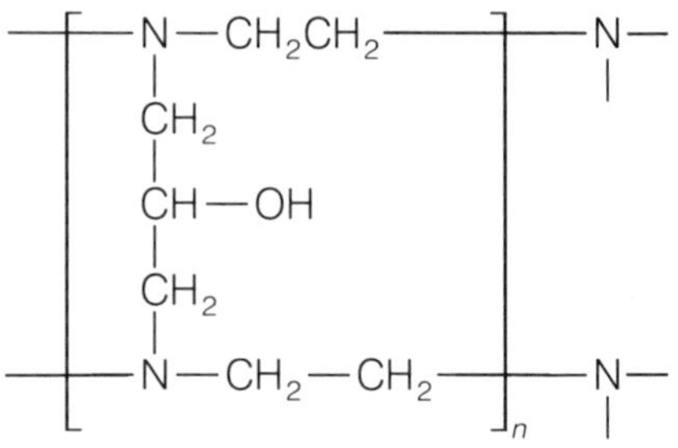

$$\left[\begin{array}{l} -N-CH_2CH_2- \\ \;\; | \\ \;\; CH_2 \\ \;\; | \\ \;\; CH-OH \\ \;\; | \\ \;\; CH_2 \\ \;\; | \\ -N-CH_2-CH_2- \end{array}\right]_n \begin{array}{l} -N- \\ \;\;| \\ \\ \\ \\ \\ \\ \\ -N- \\ \;\;| \end{array}$$

Colestipol

Niacin, USP. Nicotinic acid, 3-pyridinecarboxylic acid, is effective in the treatment of all types of hyperlipoproteinemias, with the exception of type I, at doses above those given as a vitamin supplement. The drug reduces VLDL synthesis and, subsequently, its plasma products, IDL and LDL. Plasma triglyceride levels are reduced because of the decreased VLDL production. Cholesterol levels are lowered, in turn, because of the decreased rate of LDL formation from VLDL. Although niacin is the drug of choice for type II hyperlipoproteinemias, its use is limited because of the vasodilating side effects. Flushing occurs in practically all patients, but generally subsides when the drug is discontinued.

The basis of the hypolipidemic effects of niacin may be due to its ability to inhibit lipolysis [i.e., prevent the release of free fatty acids (FFA) and glycerol from fatty tissues]. As a consequence there is a reduced reserve of FFA in the liver, and diminution of lipoprotein biosynthesis occurs, which results in a reduction of the production of VLDL. The decreased formation of lipoproteins leads to a pool of unused cholesterol normally incorporated in VLDL. This excess cholesterol is then excreted through the biliary tract.

Niacin (nicotinic acid) may be administered as aluminum nicotinate, Nicalex. This is a complex of aluminum hydroxy nicotinate and niacin. The aluminum salt is hydrolyzed to aluminum hydroxide and niacin in the stomach. There seems to be no advantage of the aluminum salt over the free acid. The frequency of hepatic reaction appears more prevalent than with niacin.

Nicotinic acid has been esterified with the purpose of prolonging its hypolipidemic effect. Pentaerythritol tetranicotinate has been more effective experimentally than niacin in reducing cholesterol in rabbits. Sorbitol and *myo*-inositol hexanicotinate polyesters have been employed in the treatment of patients with atherosclerosis obliterans.

The usual maintenance dose of niacin is 3 to 6 g/day given in three divided doses. The drug is usually given at mealtimes to reduce gastric irritation that often accompanies the large doses.

β-Sitosterol is a plant sterol, the structure of which is identical with that of cholesterol, except for the substituted ethyl group on C-24 of its side chain. Although the mechanism of its hypolipidemic effect is not clearly understood, it is suspected that the drug inhibits the absorption of dietary cholesterol from the gastrointestinal tract. Sitosterols are poorly absorbed from the mucosal lining and appear to compete with cholesterol for absorption sites in the intestine.

Usual dose: 3 g
Usual dosage range: 3 to 18 g daily

β-Sitosterol

Probucol, USP. 4,4′-[(1-Methylethylidene)-bis(thio)]bis[2,6-bis(1,1-dimethylethyl)]phenol; DH-581 (Lorelco) is a novel hypolipidemic agent. It causes reduction of both liver and serum cholesterol levels, but it does not alter plasma triglyceride levels. It reduces LDL and, to a lesser extent, HDL levels by a unique mechanism that is still not clearly identified. The reduction of HDL may be due to the ability of probucol to inhibit the synthesis of apoprotein A-1, a major protein component of HDL.[77] It is effective in reducing levels of LDL and is used in those hyperlipoproteinemias that are characterized by elevated LDL levels.

Usual dose: 500 mg twice daily
Usual pediatric dose: safety and effectiveness in children not established

Probucol

ANTICOAGULANTS

In 1905, Morawitz[78] introduced a theory of blood clotting based on the existence of four factors: thromboplastin (thrombokinase), prothrombin, fibrinogen, and ionized calcium. The clotting sequence proposed was that when tissue damage occurred, thromboplastin entered the blood from the platelets and reacted with prothrombin in the presence of calcium to form thrombin. Thrombin then reacted with fibrinogen to form insoluble fibrin that enmeshed red blood cells to create a clot. The concept remained unchallenged for almost 50 years, but

TABLE 14-8

THE ROMAN NUMERICAL NOMENCLATURE OF BLOOD-CLOTTING FACTORS AND SOME COMMON SYNONYMS

Factor	Synonyms
I	Fibrinogen
II	Prothrombin
III	Thromboplastin; tissue factor
IV	Calcium
V	Proaccelerin; accelerator globulin; labile factor
VI	(This number is not now used)
VII	Proconvertin; stable factor; autoprothrombin I, SPCA
VIII	Antihemophilic factor; antihemophilic globulin; platelet cofactor I; antihemophilic factor A
IX	Plasma thromboplastin component (PTC); Christmas factor; platelet cofactor II; autoprothrombin II; antihemophilic factor B
X	Stuart-Power factor; Stuart factor; autoprothrombin III
XI	Plasma thromboplastin antecedent (PTA); antihemophilic factor C
XII	Hageman factor
XIII	Fibrin stabilizing factor; fibrinase; Laki-Lorand factor

it has now been modified to accommodate the discovery of numerous additional factors that enter into the clotting mechanism (Table 14-8).

MECHANISM OF BLOOD COAGULATION

The fluid nature of blood can be attributed to the flat cells (endothelial) that maintain a nonthrombogenic environment in the blood vessels. This is a result of at least four phenomena: (1) the maintenance of a transmural negative electric charge that serves to prevent adhesion between platelets; (2) the release of a plasmalogen activator which activates the fibrinolytic pathway; (3) the release of thrombomodulin, a cofactor that activates protein C, a coagulation factor inhibitor; and (4) the release of prostacyclin (PGI_2), a potent inhibitor of platelet aggregation.

The process of blood coagulation (Fig. 14-20) involves a series of steps that occur in a cascade fashion and terminate in the formation of a fibrin clot. Blood coagulation occurs by activation of either an *intrinsic* pathway (a relatively slow process of clot formation) or an *extrinsic* pathway, which has a much faster rate of fibrin formation. Both pathways merge into a common pathway for the conversion of prothrombin to thrombin and subsequent transformation of fibrinogen to the insoluble strands of fibrin. Lysis of intravascular clots occurs through a plasminogen–plasmin system that consists of plasminogen, plasmin, urokinase, kallikrein, plasminogen activators, and also some undefined inhibitors of the plasminogen–plasmin system.

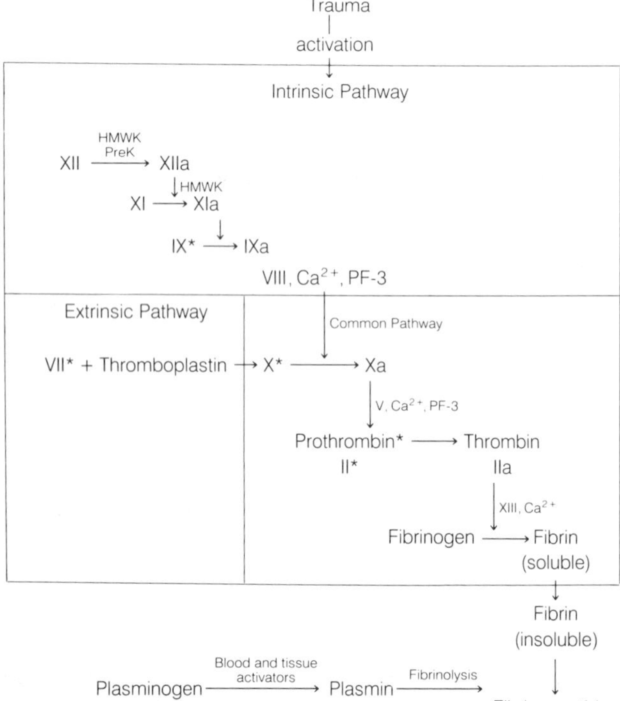

FIG. 14-20. *Scheme of blood coagulation and fibrinolysis. The asterisk denotes a vitamin K-dependent factor: HMWK, high-molecular-weight kininogen; PreK, prekallikrein; PF-3, platelet factor 3.*

The intrinsic pathway refers to the system for coagulation that occurs from the interaction of factors circulating in the blood. It is activated when blood comes into contact with a damaged vessel wall or a foreign substance. Each of the plasma coagulation factors (see Table 14-8), with the exception of factor III (tissue thromboplastin), circulates as an inactive proenzyme. Except for fibrinogen, which precipitates as fibrin, these factors are activated usually by enzymatic removal of a small peptide in the cascade of reactions that make up the clotting sequence (see Fig. 14-20). The extrinsic clotting system refers to the mechanism by which thrombin is generated in plasma after the addition of tissue extracts. When various tissues, such as brain or lung (containing thromboplastin), are added to blood, a complex between thromboplastin and factor VII in the presence of calcium ions activates factor X, bypassing the time-consuming steps of the intrinsic pathway that form factor X. The intrinsic and extrinsic pathways interact in vivo. Small amounts of thrombin formed early after stimulation of the extrinsic pathway accelerate clotting by the intrinsic pathway by activating factor VIII. Thrombin also speeds up the clotting rate by activation of factor V, located in the common pathway. Thrombin then converts the soluble protein, fibrinogen, into a soluble fibrin gel and also activates factor XIII, which stabilizes the fibrin gel in the presence of calcium by inducing the formation of cross-linking between the chains of the fibrin monomer to form an insoluble mass.

ANTICOAGULANT MECHANISMS

The biosynthesis of prothrombin is dependent upon an adequate supply of vitamin K. A deficiency of vitamin K results in the formation of a defective prothrombin molecule. The defective prothrombin is antigenically similar to normal prothrombin, but has reduced calcium-binding ability and no biologic activity. In the presence of calcium ions, normal prothrombin adheres to the surface of phospholipid vesicles and greatly increases its activity in the clotting mechanism. The defect in the abnormal prothrombin has been located in the NH_2-terminal portion, in which the second carboxyl residue has not been added to the γ-carbon atom of some glutamic acid residues on the prothrombin molecule to form

Precursor —(HCO_3^-, Carboxylase)→ Prothrombin

Glutamyl Residues: CH_2–CH_2–COOH → γ-Carboxyl Glutamyl Residues: CH_2–CH(COOH)(HOOC)

FIG. 14-21. *Vitamin K-dependent carboxylase system.*

γ-carboxylglutamic acid.[79] The administration of vitamin K antagonists results in the decreased synthesis of a biologically active prothrombin molecule and results in an increase in the clotting time of blood in humans. The reaction involved in forming an active prothrombin is a vitamin K-dependent carboxylase located in the microsomal fraction of liver cells. The enzyme uses bicarbonate as the source of the carboxyl group (Fig. 14-21). The role of vitamin K in the carboxylase reaction has not been resolved. It has been suggested that vitamin K acts to activate or carry bicarbonate or carbon dioxide in the carboxylation reaction. The increased clotting time is a consequence of the lack of prothrombin with γ-glutamyl residues to bind to phospholipid bilayer surfaces in the presence of calcium.[80]

PLATELET AGGREGATION AND INHIBITORS

Blood platelets play a pivotal role in hemostasis and thrombus formation. Actually, they have two roles in the cessation of bleeding, a hemostatic function in which platelets, through their mass, cause a physical occlusion of openings in blood vessels, and a thromboplastic function in which the chemical constituents of the platelets take part in the blood coagulation mechanism. The circulatory system is self-sealing because of the clotting properties of blood. On the other hand, the pathologic formation of clots within the circulatory system creates a potentially serious clinical situation that must be dealt with through the use of anticoagulants.

Platelets do not adhere to intact endothelial cells. They do become affixed to subendothelial tissues, which have been exposed by injury, to cause hemostasis. Platelets bind to collagen in the vessel wall and trigger other platelets to adhere to them. This adhesiveness is accompanied by a change in shape of the platelets and may be caused by mobilization of calcium bound to the platelet membrane. The growth of the platelet mass is dependent on ADP released by the first few adhering cells and enhances the aggregation process. The shape change and aggregation of platelets is termed *phase I aggregation* and is reversible. A secondary phase (phase II) immediately follows with additional platelet aggregation. In phase II, the platelets will undergo a secretory process during which enzymes, such as cathepsin and acid hydrolases along with fibrinogen, are released from α-granules in the platelets, and ADP, ATP, serotonin, and calcium are released from dense bodies in the platelets. The dense bodies are likened to the storage granules associated with adrenergic neurons. The selective process that releases the contents of dense bodies is called *release I* and the process that releases the contents of α-granules is referred to as *release II*. Increased levels of cAMP inhibit platelet aggregation. The cAMP activates specific dependent kinases, which form protein–phosphate complexes that chelate calcium ions. The reduced levels of calcium inhibit aggregation (Fig. 14-22). Inhibitors of platelet aggregation can increase cAMP levels by either stimulating adenylate cyclase or inhibiting phosphodiesterase.[81] Substances such as glucagon, adenosine, and isoproterenol increase cAMP levels and inhibit platelet aggregation. Drugs such as theophylline, aminophylline, dipyramidole, papaverine, and adenosine inhibit phosphodiesterase and also aggregation of platelets. Epinephrine, collagen, and serotonin inhibit adenylate cyclase and stimulate platelet aggregation.[82] The role of platelets in arterial thrombosis is similar to their role in hemostasis. The factors contributing to venous thrombosis are circulatory stasis, excessive generation of thrombin formation of fibrin, and, to a lesser extent than in the artery, platelet aggregation.

Aspirin, sulfinpyrazone, and indomethacin have an inhibitory effect on platelet aggregation. They inhibit cyclooxygenase, the enzyme that controls the formation of prostaglandin (PG) endoperoxides and increases the tendency for platelets to aggregate. Aspirin also inhibits the platelet-release reaction; it antagonizes phase II of ADP- or epinephrine-induced aggregation as well as collagen- and thrombin-induced aggregation. Dipyridamole inhibits adeno-

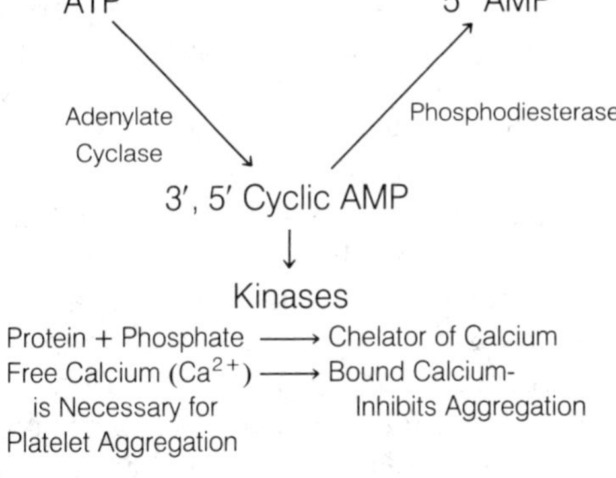

FIG. 14-22. *Role of adenosine 3′,5′-cyclic monophosphate (cAMP) in inhibition of platelet aggregation.*

sine deaminase and adenosine uptake by platelets. As a result, the increased plasma concentrations of adenosine inhibit ADP-induced aggregation of platelets.

Among the many pharmacologic actions of prostaglandins is the ability of some chemicals of this class to stimulate or inhibit the aggregation of platelets and alter the clotting time of blood. Prostaglandins are synthesized from 20-carbon polyunsaturated fatty acids containing from three to five double bonds. These fatty acids are present in the phospholipids of cell membranes of all mammalian tissues. The main precursor of prostaglandins is arachidonic acid. Arachidonic acid is released from membrane phospholipids by the enzyme phospholipase A_2. Once released, arachidonic acid is metabolized by cyclooxygenase synthetase to form unstable endoperoxides that are subsequently transformed into prostacyclin and thromboxane A_2 (TXA_2). The conversion to TXA_2 occurs with the aid of the enzyme thromboxane synthetase. The formation of prostacyclin can occur nonenzymatically. Blood platelets convert arachidonic acid to TXA_2 whereas prostacyclin is formed mainly by the vascular endothelium. Both prostacyclin and TXA_2 are unstable at physiologic pH and temperatures. Their half-lives are five minutes and 30 seconds, respectively.

Prostacyclin inhibits platelet aggregation by stimulating adenylate cyclase to increase cAMP levels in the platelets. Prostacyclin is also a vasodilator and, as a result, has potent hypotensive properties when given intravenously or by intra-arterial administration. Thomboxane A_2 induces aggregation of platelets. Together with prostacyclin, TXA_2 plays a role in the maintenance of vascular homeostasis. In addition to being a platelet aggregator, TXA_2 is also a potent vasoconstrictor.

Retardation of clotting is important in blood transfusions, to avoid thrombosis after surgery or from other causes, to prevent recurrent thrombosis in phlebitis and pulmonary embolism, and to lessen the propagation of clots in the coronary arteries. This retardation may be accomplished by agents that inactivate thrombin (heparin) or those substances that prevent the formation of prothrombin in the liver—the coumarin derivatives and the phenylindanedione derivatives.

Although heparin (see Chap. 19) is a useful anticoagulant, it has limited applications. Many of the anticoagulants in use today were developed following the discovery of dicumarol, an anticoagulant that is present in spoiled sweet clover. These compounds are orally effective, but there is a lag period of 18 to 36 hours before they significantly increase the clotting time. Heparin, in contrast, produces an immediate anticoagulant effect after intravenous injection. A major disadvantage of heparin is that the only effective therapeutic route is parenteral.

Dicumarol and related compounds are not vitamin K antagonists in the classic sense. They appear to act by interfering with the function of vitamin K in the liver cells, which are the sites of synthesis of the clotting factors, including prothrombin. This lengthens the clotting time by decreasing the amount of biologically active prothrombin in the blood.

The discovery of dicumarol and related compounds as potent reversible* competitors of vitamin K coagulant-promoting properties led to the development of antivitamin K compounds, such as phenindione, which, in part, was designed according to metabolite–antimetabolite concepts. The active compounds of the phenylindanedione series are

* At high levels dicumarol is not reversed by vitamin K.

TABLE 14-9

INDIVIDUAL CHARACTERISTICS OF COMMON ORAL ANTICOAGULANTS USED IN CLINICAL PRACTICE

Drug	Trade Name	Initial Dose (mg)*	Maintenance Dose (mg)	Time to Reach Therapeutic Level (hr)	Time to Return to Near-Normal "Pro Time" (days)
COUMARIN COMPOUNDS					
Bishydroxycoumarin	Dicoumarol	200–300	25–200	36–72	5–6
Warfarin	Coumadin, Panwarfin, Prothromadin	40–60	2–10	36–48	$3\frac{1}{2}$–$4\frac{1}{2}$
Ethyl biscoumacetate	Tromexan	900–1200	150–900	18–36	$1\frac{1}{2}$–$2\frac{1}{2}$
Phenprocoumon	Liquamar	18–24	1–6	30–48	7–10
INDANDIONE COMPOUNDS					
Phenindione	Danilone, Dindevan, Hedulin	200–300	25–100	36–48	3–4
Anisindione	Miradon	200–300	25–250	36–60	3–4

*When the traditional "loading dose" technique is used.
(From Deykin, D.: N. Engl. J. Med. 283:691, 1970.)

characterized by a phenyl, a substituted phenyl, or a diphenylacetyl group in the 2-position. Another requirement for activity is a keto group in the 1- and 3-positions, one of which may form the enol tautomer. A second substituent, other than hydrogen, at the 2-position prevents this keto–enol tautomerism and the resulting compounds are ineffective as anticoagulants.

PRODUCTS

Protamine Sulfate, USP, has an anticoagulant effect, but it counteracts the action of heparin if used in the proper amount and is used as an antidote for the latter in cases of overdosage. It is administered intravenously in a dose depending on the circumstances.

Usual adult dose: IV, 1 mg of protamine sulfate for every 90 USP units of beef lung heparin sodium or for every 115 USP units of porcine intestinal mucosa heparin sodium to be neutralized, administered slowly in one to three minutes, up to a maximum of 50 mg in any ten-minute period.

Usual pediatric dose: see Usual adult dose.

Occurrence: Protamine sulfate injection, USP; Protamine sulfate for injection, USP

Dicumarol, USP. 3,3′-Methylenebis[4-hydroxycoumarin] is a white or creamy white crystalline powder with a faint, pleasant odor and a slightly bitter taste. It is practically insoluble in water or alcohol, slightly soluble in chloroform and is dissolved readily by solutions of fixed alkalies. The effects after administration require 12 to 72 hours to develop and persist for 24 to 96 hours after discontinuance.

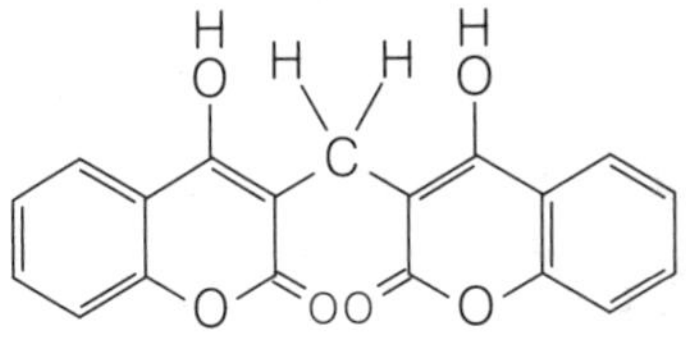

Dicumarol

Dicumarol is used alone or as an adjunct to heparin in the prophylaxis and treatment of intravascular clotting. It is employed in postoperative thrombophlebitis, pulmonary embolus, acute embolic and thrombotic occlusion of peripheral arteries, and recurrent idiopathic thrombophlebitis. It has no effect on an already formed embolus, but may prevent further intravascular clotting. Because the outcome of acute coronary thrombosis is largely dependent on extension of the clot and formation of mural thrombi in the heart chambers, with subsequent embolization, dicumarol has been used in this condition. It also has been administered to arrest impending gangrene after frostbite. The dose, after determination of the prothrombin clotting time, is 25 to 200 mg, depending on the size and the condition of the patient, the drug being given orally in the form of capsules or tablets. On the second day, and thereafter, it may be given in amounts sufficient to maintain the prothrombin clotting time at about 30 seconds. If hemorrhages should occur, 50 to 100 mg of menadione sodium bisulfite is injected, supplemented by a blood transfusion.

Warfarin Sodium, USP. 3-(α-Acetonylbenzyl)-4-hydroxycoumarin sodium salt (Coumadin; Panwarfin) is a white, odorless crystalline powder, having a slightly bitter taste; it is slightly soluble in chloroform, soluble in alcohol or water. A 1% solution has a pH of 7.2 to 8.5.

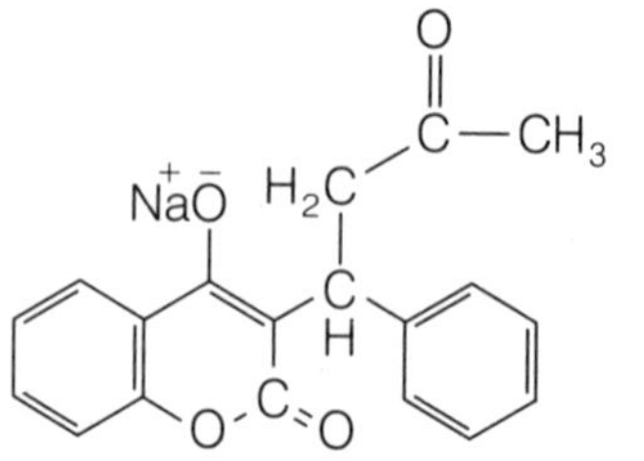

Warfarin Sodium

By virtue of its great potency, warfarin at first was considered unsafe for use in humans and was utilized very effectively as a rodenticide, especially against rats. However, when used in the proper dosage level, it can be used in humans, especially by the intravenous route.

Warfarin Potassium, USP. 3-(α-Acetonylbenzyl)-4-hydroxycoumarin potassium salt (Athrombin-K). Warfarin potassium is readily absorbed after oral administration, with a therapeutic hypoprothrombinemia being produced in 12 to 24 hours after administration of 40 to 60 mg. This salt is therapeutically interchangeable with warfarin sodium.

Phenprocoumon, USP. 3-(α-Ethylbenzyl)-4-hydroxycoumarin (Liquamar). This drug has been shown to possess marked and prolonged anticoagulant activity.

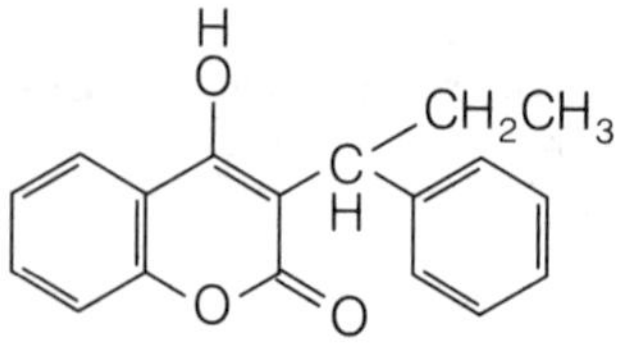

Phenprocoumon

Phenindione, USP. 2-Phenyl-1,3-indandione (Hedulin; Danilone) is an oral anticoagulant specifically designed to function as an antimetabolite for vitamin K. It is a pale yellow crystalline material that is slightly soluble in water but very soluble in alcohol. It is more prompt acting than dicumarol. The rapid elimination is presumed to make the drug safer than others of the class.

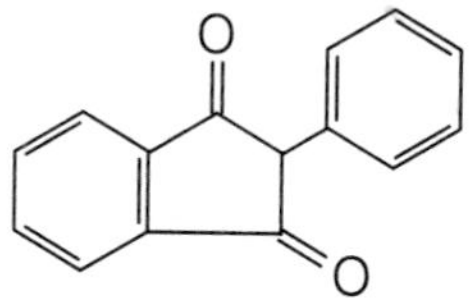

Phenindione

Anisindione, USP. 2-(p-Methoxyphenyl)-1,3-indandione; 2-(p-anisyl)-1,3-indandione (Miradon) is a p-methoxy congener of phenindione. It is a white crystalline powder, slightly soluble in water, tasteless, and well absorbed after oral administration.

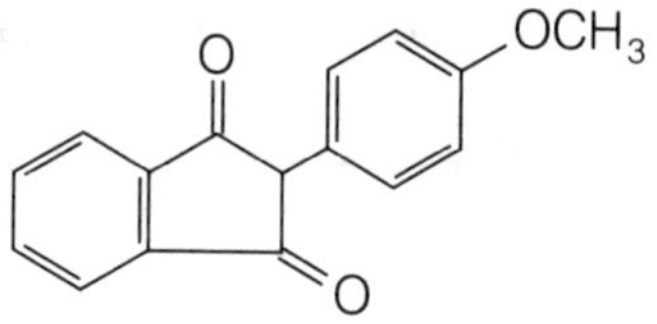

Anisindione

TABLE 14-10

ANTICOAGULANTS

Name Proprietary Name	***Preparations***	***Usual Adult Dose****	***Usual Dose Range****
Dicumarol USP	Dicumarol capsules USP Dicumarol tablets USP	25–200 mg qd, as indicated by prothrombin-time determinations	
Warfarin sodium USP *Coumadin, Panwarfin*	Warfarin sodium for injection USP Warfarin sodium tablets USP	Oral, IM, or IV, 10–15 mg/day for 2–3 days, then 2–10 mg/day, as indicated by prothrombin-time determinations	
Warfarin potassium USP *Athrombin-K*	Warfarin potassium tablets USP	Oral, 10–15 mg/day for 2 or 3 days, then 2–10 mg/day, as indicated by prothrombin-time determinations	
Phenprocoumon USP *Liquamar*	Phenprocoumon tablets USP		Initial, 21 mg the first day, 9 mg the second day, and 3 mg the third day; maintenance, 1–4 mg/day, according to prothrombin level
Phenindione USP *Hedulin*	Phenindione tablets USP	Oral, 750 μg–6 mg/day, as indicated by prothrombin-time determinations	
Anisindione USP *Miradon*	Anisindione tablets	Initial, 300 mg the first day, 200 mg the second day, 100 mg the third day; maintenance, 25–250 mg/day	

*See USP DI for complete dosage information.

In instances when the urine may be alkaline, an orange color may be detected. This is due to metabolic products of anisindione and is not hematuria.

SCLEROSING AGENTS

Several different kinds of irritating agents have been used for the obliteration of varicose veins. These are generally called sclerosing agents and include invert sugar solutions, dextrose, ethyl alcohol, iron salts, quinine and urea hydrochloride, fatty acid salts (soaps), and certain sulfate esters. Many of these preparations contain benzyl alcohol, which acts as a bacteriostatic agent and relieves pain after injection.

Morrhuate Sodium Injection, USP, is a sterile solution of the sodium salts of the fatty acids of cod-liver oil. The salt (a soap) was introduced first in 1918 as a treatment for tuberculosis and, in 1930, it was reported to be useful as a sclerosing agent. Morrhuate sodium is not a single entity, although morrhuic acid has been known for years. Morrhuate sodium is a mixture of the sodium salts of the saturated and unsaturated fatty acids from cod-liver oil.

The preparation of the free fatty acids of cod-liver oil is carried out by saponification with alkali and then acidulation of the resulting soap. The free acids are dried over anhydrous sodium sulfate before being dissolved in an equivalent amount of sodium hydroxide solution. Morrhuate sodium is obtained by careful evaporation of this solution. The result is a pale yellowish granular powder having a slight fishy odor.

Commercial preparations are usually 5% solutions, which vary in properties and in color from light yellow, to medium yellow, to light brown. They are all liquids at room temperature and have congealing points that range from $-11°$ to 7°C. A bacteriostatic agent, not to exceed 0.5%, and ethyl or benzyl alcohol to the extent of 3% may be added.

Usual dose: intravenous, by special injection, 1 ml to a localized area
Usual dose range: 500 μL to 5 ml

SYNTHETIC HYPOGLYCEMIC AGENTS

The discovery that certain organic compounds will lower the blood sugar level is not recent. In 1918, guanidine was shown to lower the blood sugar level. The discovery that certain trypanosomes need much glucose and will die in its absence was followed by the discovery that galegine lowered the blood sugar level and was weakly trypanocidal. This led to the development of several very active trypanocidal agents, such as the bisamidines, diisothioureas, bisguanidines, and others. Synthalin (trypanocidal at 1 : 250 million) and pentamidine are outstanding examples of very active trypanocidal agents. Synthalin lowers the blood sugar level in normal, depancreatized and completely alloxanized animals. This may be due to a reduction in the oxidative activity of mitochondria, resulting from inhibition of the mechanisms that simultaneously promote phosphorylation of ADP and stimulate oxidation by nicotinamide adenine dinucleotide (NAD) in the citric acid cycle. Hydroxystilbamidine Isethionate USP is used as an antiprotozoal agent.

$$(CH_3)_2C{=}CHCH_2NHC(=NH)NH_2$$

Galegine

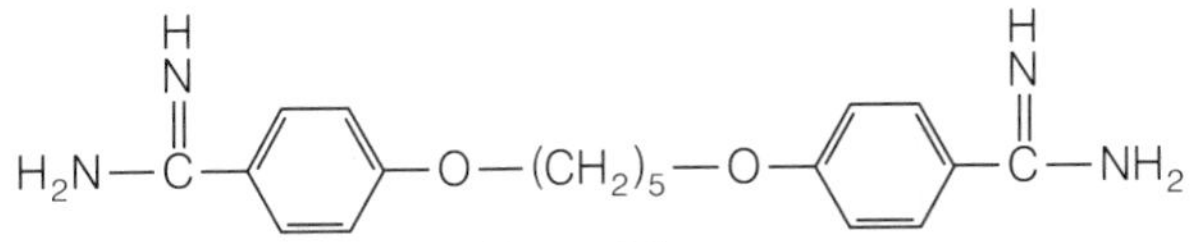

$$H_2N{-}C(=NH){-}C_6H_4{-}O{-}(CH_2)_5{-}O{-}C_6H_4{-}C(=NH){-}NH_2$$

Pentamidine

$$H_2N{-}C(=NH){-}NH{-}(CH_2)_{10}{-}NH{-}C(=NH){-}NH_2$$

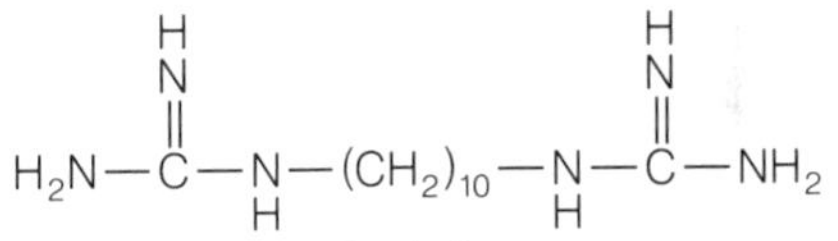

Synthalin

In 1942, *p*-aminobenzenesulfonamidoisopropylthiadiazole (an antibacterial sulfonamide) was found to produce hypoglycemia. These results stimulated the research for the development of synthetic hypoglycemic agents, several of which are in use today.

Sulfonylureas became widely available in 1955 for treatment of nonketosis-prone mild diabetes and are still the drugs of choice. A second class of compounds, the biguanides, in the form of a single drug, phenformin, was used since 1957. However, it has been withdrawn from the United States market because of its toxic effect. Phenformin causes lactic acidosis of which fatalities have been reported.

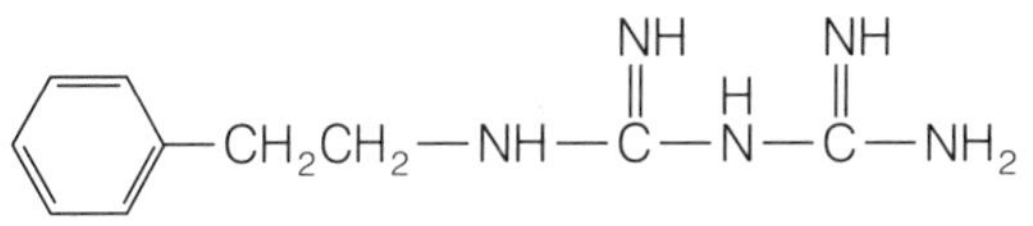

$$C_6H_5{-}CH_2CH_2{-}NH{-}C(=NH){-}NH{-}C(=NH){-}NH_2$$

Phenformin

SULFONYLUREAS

The sulfonylureas may be represented by the following general structure:

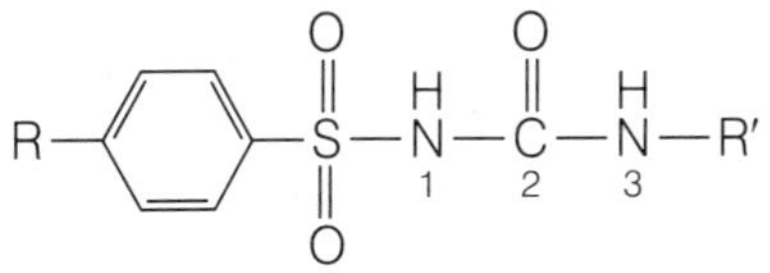

$$R{-}C_6H_4{-}SO_2{-}\underset{1}{NH}{-}\underset{2}{C}(=O){-}\underset{3}{NH}{-}R'$$

These are urea derivatives with an arylsulfonyl group in the 1-position and an aliphatic group at the 3-position. The aliphatic group, R′, confers lipophilic properties to the molecule. Maximal activity results when R′ consists of three to six carbon atoms as in chlorpropamide, tolbutamide, and acetohexamide. Aryl groups at R′ generally give toxic compounds. The R group on the aromatic ring primarily influences the duration of action of the compound. Tolbutamide disappears quite rapidly from the bloodstream through being metabolized to the inactive carboxy compound, which is rapidly excreted. On the other hand, chlorpropamide is metabolized more slowly and persists in the blood for a much longer time.

The mechanism of action of the sulfonylureas is to stimulate the release of insulin from the functioning β-cells of the intact pancreas. In the absence of the pancreas, they have no significant effect on blood glucose. The sulfonylureas may have other actions such as inhibition of secretion of glucagon and action at postreceptor intracellular sites to increase insulin activity.

For a time, tolbutamide, chlorpropamide, and acetohexamide were the only oral hypoglycemic agents. Subsequently, a second generation of these drugs became available. Although they did not present a new method of lowering blood glucose levels, they were more potent than the existing drugs. Glipizide and glyburide are the second-generation oral hypoglycemic agents.

Whether they are first- or second-generation oral hypoglycemic drugs, this group of agents remains as a valuable adjunct to therapy in the diabetic patient whose disease had its onset in adulthood. Accordingly, the group of sulfonylureas is not indicated in the juvenile-onset diabetic.

PRODUCTS

Tolbutamide, USP. 1-Butyl-3-(*p*-tolylsulfonyl)urea (Orinase) occurs as a white crystalline powder that is insoluble in water and soluble in alcohol or aqueous alkali. It is stable in air.

$CH_3-C_6H_4-SO_2-NH-C(=O)-NH-(CH_2)_3CH_3$

Tolbutamide

Tolbutamide is absorbed rapidly in responsive diabetic patients. The blood sugar level reaches a minimum after five to eight hours. It is oxidized rapidly in vivo to 1-butyl-3-(*p*-carboxyphenyl)sulfonylurea, which is inactive. The metabolite is freely soluble at urinary *p*H; however, if the urine is strongly acidified, as in the use of sulfosalicylic acid as a protein precipitant, a white precipitate of the free acid may be formed.

$CH_3-C_6H_4-SO_2NH-C(=O)-NH-(CH_2)_3CH_3$

in vivo

$HO(O=)C-C_6H_4-SO_2NH-C(=O)-NH-(CH_2)_3CH_3$

Tolbutamide should be used only when the diabetic patient is an adult or shows maturity-onset diabetes, and the patient should adhere to dietary restrictions.

Tolbutamide Sodium, USP. 1-Butyl-3-(*p*-tolylsulfonyl)urea monosodium salt (Orinase Diagnostic). Tolbutamide sodium is a white crystalline powder, freely soluble in water, soluble in alcohol and in chloroform, and very slightly soluble in ether.

$CH_3-C_6H_4-S(=O)_2-\underline{N}-C(=O)-NH-(CH_2)_3CH_3 \quad Na^+$

Tolbutamide Sodium

This water-soluble salt of tolbutamide is used intravenously for the diagnosis of mild diabetes mellitus and of functioning pancreatic islet cell adenomas. The sterile dry powder is dissolved in sterile water for injection to make a clear solution, which then should be administered within one hour. The main route of breakdown is to butylamine and sodium *p*-toluenesulfonamide.

Chlorpropamide, USP. 1-[(*p*-Chlorophenyl)-sulfonyl]-3-propylurea (Diabinese). Chlorpropamide is a white crystalline powder, practically insoluble in water, soluble in alcohol, and sparingly soluble in chloroform. It will form water-soluble salts in basic solutions. This drug is more resistant to conversion to inactive metabolites than is tolbutamide and, as a result, has a much longer duration of action. One study showed that about half of the drug is excreted

$Cl-C_6H_4-SO_2-NH-C(=O)-NH-C_3H_7$

Chlorpropamide

as metabolites, the principal one being hydroxylated in the 2-position of the propyl side chain.[83] After control of the blood sugar levels, the maintenance dose is usually on a once-a-day schedule.

Tolazamide, USP. 1-(Hexahydro-1*H*-azepin-1-yl)-3-(*p*-tolylsulfonyl)urea (Tolinase). This agent is an analogue of tolbutamide and is reported to be effective, in general, under the same circumstances for which tolbutamide is useful. However, tolazamide appears to be more potent than tolbutamide and is nearly equal in potency to chlorpropamide. In studies with radioactive tolazamide, investigators found that 85% of an oral dose appeared in the urine as metabolites that were more soluble than tolazamide itself.

Tolazamide

Acetohexamide, USP. 1-[(*p*-Acetylphenyl)sulfonyl]-3-cyclohexylurea (Dymelor). Acetohexamide is chemically and pharmacologically related to tolbutamide and chlorpropamide. Like the other sulfonylureas, acetohexamide lowers the blood sugar, primarily by stimulating the release of endogenous insulin.

Acetohexamide

Acetohexamide is metabolized in the liver to a reduced form—the α-hydroxyethyl derivative. This metabolite, the main one in humans, possesses hypoglycemic activity. Acetohexamide is intermediate between tolbutamide and chlorpropamide in potency and duration of effect on blood sugar levels.

Glipizide. 1-Cyclohexyl-3-[[*p*-[2-(methylpyrazinecarboxamido)ethyl]phenyl]sulfonyl]urea (Glucotrol). Structurally, glipizide is a cyclohexylsulfonylurea analogue similar to acetohexamide and glyburide. The drug is rapidly absorbed on oral administration. Its serum half-life is two to four hours, and it has a hypoglycemic effect that ranges from 12 to 24 hours.

Glipizide

Glyburide. 1-[*p*-[2-(5-Chloro-*o*-anisamido)ethyl]-phenyl]sulfonyl]-3-cyclohexylurea (DiaBeta; Euglu-

TABLE 14-11

SYNTHETIC HYPOGLYCEMIC AGENTS

Name Proprietary Name	***Preparations***	***Usual Adult Dose****
Tolbutamide USP *Orinase*	Tolbutamide tablets USP	Initial, 500 mg qd or bid, adjusted gradually according to patient response or until the total daily dose reaches 3 g
Tolbutamide sodium USP *Orinase Diagnostic*	Sterile tolbutamide sodium USP	IV, the equivalent of 1 g of tolbutamide over a 2- to 3-min period
Chlorpropamide USP *Diabinese*	Chlorpropamide tablets USP	Oral, 100–250 mg qd initially, the dosage being increased by 50–125 mg at 1-wk intervals until diabetic control is obtained or until the total daily dose reaches 750 mg
Tolazamide USP *Tolinase*	Tolazamide tablets USP	Oral, 100–250 mg qd initially, the dosage being adjusted gradually until diabetic control is obtained or until the total daily dose reaches 1 g
Acetohexamide USP *Dymelor*	Acetohexamide tablets USP	Oral, 250 mg qd initially, the dosage being adjusted gradually until diabetic control is obtained or until the total daily dose reaches 1.5 g
Glipizide *Glucotrol*	Glipizide tablets	Oral, initially 5 mg/day, the dosage being adjusted in increments of 2.5–5 mg at weekly intervals until daily dose reaches 40 mg
Glyburide *DiaBeta, Euglucon, Micronase*	Glyburide tablets	Oral, initially 2.5–5 mg/day, the dosage being adjusted in increments at weekly intervals until the daily dose reaches 20 mg

*See USP DI for complete dosage information.

con; Micronase). Similar to glipizide, this is a second-generation oral hypoglycemic agent. The drug has a half-life elimination of ten hours, but its hypoglycemic effect remains for up to 24 hours.

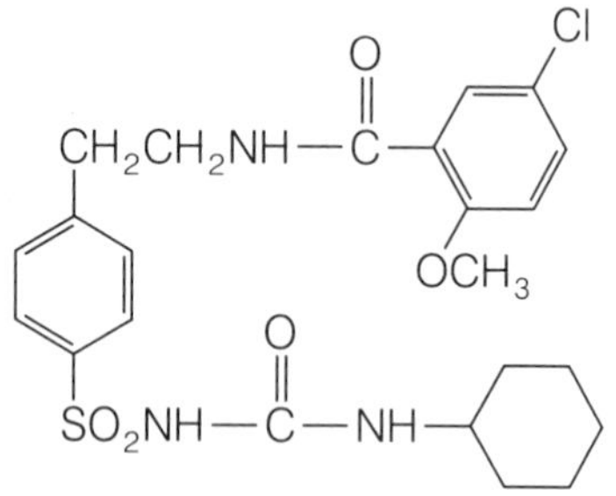

Glyburide

THYROID HORMONES

Desiccated, defatted thyroid substance has been used for many years as replacement therapy in thyroid gland deficiencies. The efficacy of the whole gland is now known to depend on its thyroglobulin content. This is an iodine-containing globulin. Thyroxine was obtained as a crystalline derivative by Kendall of the Mayo Clinic in 1916. It showed much the same action as the whole thyroid substance. Later thyroxine was synthesized by Harington and Barger in England. Later studies showed that an even more potent iodine-containing hormone existed, which is now known as triiodothyronine. There is now evidence that thyroxine may be the storage form of the hormone, whereas triiodothyronine is the circulating form. Another point of view is that, in the blood, thyroxine is more firmly bound to the globulin fraction than is triiodothyronine, which can then enter the tissue cells.

PRODUCTS

Levothyroxine Sodium, USP. *O*-[4-Hydroxy-3,5-diiodophenyl)-3,5-diiodo-2-tyrosine monosodium salt, hydrate (Synthroid; Letter; Levoroxine; Levoid). This compound is the sodium salt of the *levo*-isomer of thyroxine, which is an active physiologic principle obtained from the thyroid gland of domesticated animals used for food by humans. It is also prepared synthetically. The salt is a light yellow, tasteless, odorless powder. It is hygroscopic, but stable in dry air at room temperature. It is soluble in alkali hydroxides, 1 : 275 in alcohol and 1 : 500 in water to give a *p*H of about 8.9.

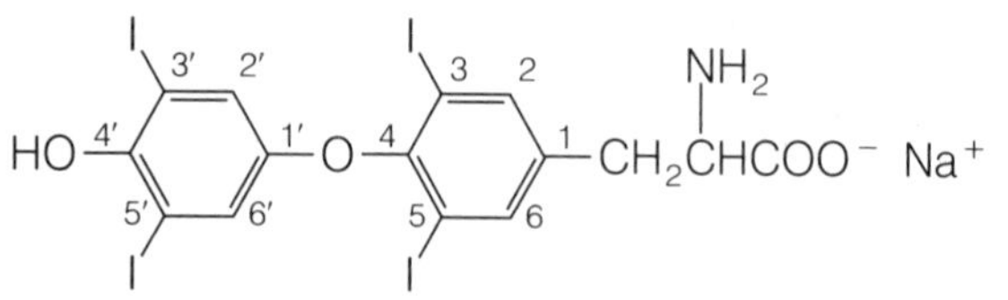

Levothyroxine Sodium

Levothyroxine sodium is used in replacement therapy of decreased thyroid function (hypothyroidism). In general, 100 μg of levothyroxine sodium is clinically equivalent to 30 to 60 mg of Thyroid USP.

Usual adult dose: initial, mild hypothyroidism, oral, 50 to 100 μg as a single daily dose, with increments of 50 to 100 μg at two-week intervals until the desired result is obtained; IV or IM, 100 to 300 μg as a single daily dose

Usual pediatric dose: children under 1 year old, oral, 25 to 50 μg/day or a single daily dose for the first year of life; children over 1 year, 3 to 5 μg/kg of body weight until the adult dose is reached; IV or IM, daily dose equal to 75% of the usual oral pediatric dose

Occurrence: Levothyroxine sodium tablets, USP

Liothyronine Sodium, USP. *O*-(4-Hydroxy-3-iodophenyl)-3,5-diiodo-L-thyroxine monosodium salt (Cytomel) is the sodium salt of L-3,3′,5-triiodothyronine. It occurs as a light tan, odorless crystalline powder, slightly soluble in water or alcohol and has a specific rotation of +18° to +22° in a mixture of diluted HCl and alcohol.

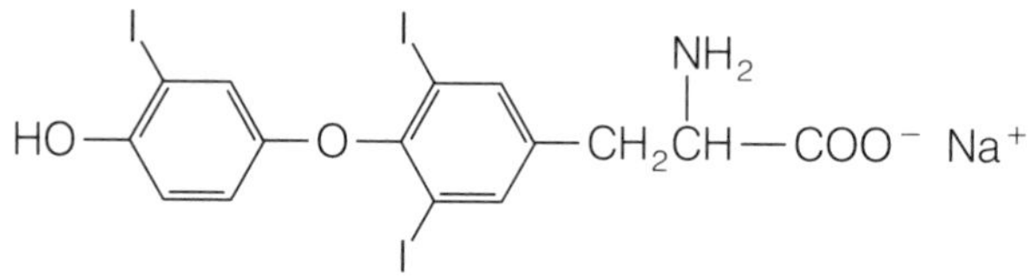

Liothyronine Sodium

Liothyronine occurs in vivo together with levothyroxine; it has the same qualitative activities as thyroxine but is more active. It is absorbed readily from the gastrointestinal tract, is cleared rapidly from the bloodstream, and is bound more loosely to plasma proteins than is thyroxine, probably because of the less acidic phenolic hydroxyl group.

Its uses are the same as those of levothyroxine, including treatment of metabolic insufficiency, male infertility, and certain gynecologic disorders.

Usual adult dose: mild hypothyroidism, initial: oral, 25 μg/day, with increments of 12.5 or 25 μg every one or two weeks until the desired result is obtained; maintenance: oral, 25 to 100 μg/day

Occurrence: Liothyronine sodium tablets, USP

ANTITHYROID DRUGS

When hyperthyroidism exists (excessive production of thyroid hormones), the condition usually requires surgery, but before surgery the patient must be prepared by preliminary abolition of the hyperthyroidism through the use of antithyroid drugs. Thiourea and related compounds show an antithyroid activity, but they are too toxic for clinical use. The more useful drugs are 2-thiouracil derivatives and a closely related 2-thioimidazole derivative. All of these appear to have a similar mechanism of action (i.e., prevention of the iodination of the precursors of thyroxine and triiodothyronine). The main difference in the compounds lies in their relative toxicities.[84]

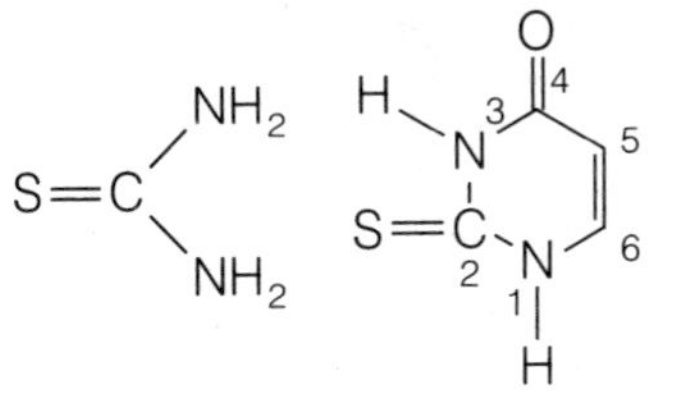

Thiourea 2-Thiouracil

These compounds are well absorbed after oral administration and are excreted in the urine.

The 2-thiouracils, 4-keto-2-thiopyrimidines, are undoubtedly tautomeric compounds and can be represented as follows:

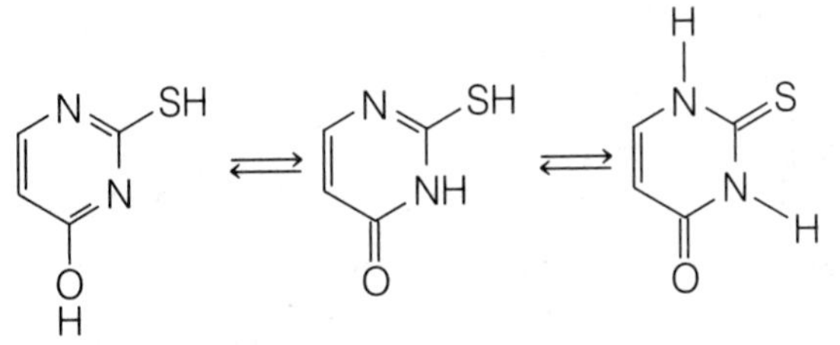

Some 300 related structures have been evaluated for antithyroid activity, but, of these, only the 6-alkyl-2-thiouracils and closely related structures possess useful clinical activity. The most serious side effect of thiouracil therapy is agranulocytosis.

Propylthiouracil, USP. 6-Propyl-2-thiouracil (Propacil). Propylthiouracil is a stable, white crystalline powder with a bitter taste. It is slightly soluble in water but is readily soluble in alkaline solutions (salt formation).

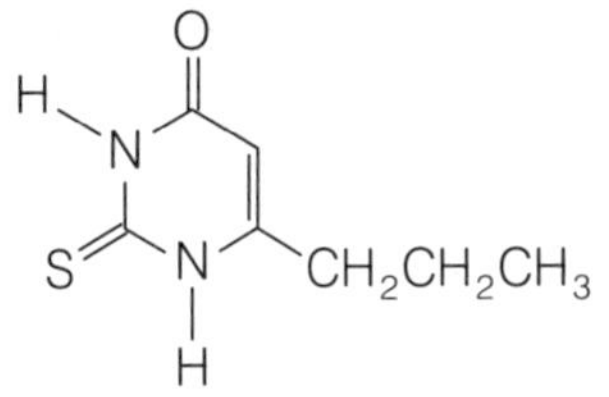

Propylthiouracil

This drug is useful in the treatment of hyperthyroidism. There is a delay in appearance of its effects, because propylthiouracil does not interfere with the activity of thyroid hormones already formed and stored in the thyroid gland. This lag period may vary from several days to weeks, depending on the condition of the patient. The need for three equally spaced doses during a 24-hour period is often stressed, but there is now evidence that a single daily dose is as effective as multiple daily doses in the treatment of most hyperthyroid patients.[85]

Methimazole, USP. 1-Methylimidazole-2-thiol (Tapazole) occurs as a white to off-white crystalline powder with a characteristic odor and is freely soluble in water. A 2% aqueous solution has a pH of 6.7 to 6.9. It should be packaged in well-closed, light-resistant containers.

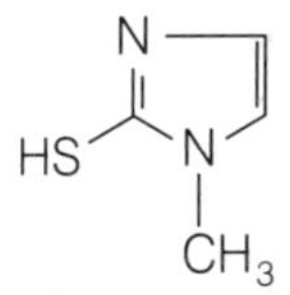

Methimazole

Methimazole is indicated in the treatment of hyperthyroidism. It is more potent than propylthiouracil. The side effects are similar to those of propylthiouracil. As with other antithyroid drugs,

TABLE 14-12

ANTITHYROID DRUGS

Name Proprietary Name	***Preparations***	***Usual Adult Dose****	***Usual Pediatric Dose****
Propylthiouracil USP *Propacil*	Propylthiouracil tablets USP	Initial: oral, 300–1,200 mg/day, divided into 3 doses at 8-hr intervals or 4 doses at 6-hr intervals, until patient becomes euthyroid; maintenance: oral, 50–800 mg/day in 2–4 divided doses	Initial: children 6–10 yr of age — oral, 50–200 mg/day in 2 or 3 divided doses; children 10 yr and over — 150–600 mg/day divided into 3 doses at 8-hr intervals; maintenance: determined by response
Methimazole USP *Tapazole*	Methimazole tablets USP	Initial: mild hyperthyroidism: oral, 15 mg/day, divided into 3 doses at 8-hr intervals; maintenance, 5–30 mg/day, in 2 or 3 divided doses	Initial: oral, 400 μg/kg of body weight daily, divided into 3 doses at 8-hr intervals

*See USP DI for complete dosage information.

patients using this drug should be under medical supervision. Also, similar to the other antithyroid drugs, methimazole is most effective if the total daily dose is subdivided and given at eight-hour intervals.

REFERENCES

1. Robinson, B. F.: Adv. Drug Res. 10:93, 1975.
2. Aronow, W. S.: Am. Heart J. 84:273, 1972.
3. Sonnenblick, E., et al.: Am. J. Cardiol. 22:238, 1968.
4. Ignarro, L. J., and Kadowitz, P. J.: Annu. Rev. Pharmacol. Toxicol. 25:171, 1985.
5. Ignarro, L. J., et al.: J. Pharmacol. Exp. Ther. 218:739, 1981.
6. Abshagen, U.: Clinical Pharmacology of Antianginal Drugs, p. 306. Berlin, Springer-Verlag, 1985.
7. Rapoport, R. M., and Murad, F.: J. Cyclic Nucleotide Protein Phosphor. Res. 9:281, 1983.
8. Needleman, P.: Annu. Rev. Pharmacol. Toxicol. 16:81, 1976.
9. Fusari, S. A.: J. Pharm. Sci. 62:123, 1973.
10. Fusari, S. A.: J. Pharm. Sci. 62:2012, 1973.
11. Hille, B.: Biophys. J. 22:283, 1978.
12. McCall, D., Walsh, R. A., Frohlich, E. D., and Rourke, R. A.: Curr. Prob. Cardiol. 10:1, 1985.
13. Braunwald, E.: N. Engl. J. Med. 307:1618, 1982.
14. McCall, D.: Curr. Res. Suppl., June 1987, p. v-3.
15. van Zwieten, P. A., van Meel, J. C., and Timmermans, P. B.: Prog. Pharmacol. 5:1, 1982.
16. M. C. Sanguinetti and Kass, R. S.: Circ. Res. 55:336, 1984.
17. Rosenkirchen, R. et al.: Nauyn-Schmiedebergs Arch. Pharmacol. 310:69, 1979.
18. Stone, P. H., and Antman, E. M.: Calcium Channel Blocking Agents in the Treatment of Cardiovascular Disorders, p. 179. Mount Kisco, NY, Futura, 1983.
19. Mason, D. T., et al.: Cardiovasc. Drugs 1:75, 1977.
20. Bealer, G. W., and Reuter, H.: Physiology (Lond.) 207:191, 1970.
21. Vaughan Williams, E. M.: Classification of antiarrhythmic drugs. In Sandoe, E., Flensted-Jansen, E., and Olesen, K. H. (eds.). Symposium on Cardiac Arrhythmias, pp. 449–472. Sodertalje, Sweden, B. Astra, 1970.
22. Vaughan Williams, E. M.: J. Clin. Pharmacol. 24:129, 1984.
23. Campbell, T. J.: Cardiovasc. Res. 17:344, 1983.
24. Hondeghem, L. M., and Katzung, B. G.: Annu. Rev. Pharmacol. Toxicol. 24:387, 1984.
25. Harrison, D. C.: In Morganroth, J., and Moore, E. N. (eds.). Cardiac Arrhythmias, p. 36. Boston, Martinus Nijhoff, 1985.
26. Sjoerdsma, A., et al.: Circulation 28:492, 1963.
27. Nies, A. S., and Shang, D. G.: Clin. Pharmacol. Exp. Ther. 14:823, 1973.
28. Koch-Wester, J.: Ann. N.Y. Acad. Sci. 179:370, 1971.
29. Giardinia, E. V., Dreyfuss, J., and Bigger, J. T.: Clin. Pharmacol. Ther. 19:339, 1976.
30. Elson, J., et al.: Clin. Pharmacol. Ther. 17:134, 1975.
31. Vismara, L. A., and Mason, D. T.: Clin. Pharm. Ther. 16:330, 1974.
32. Befeler, B., et al.: Am. J. Cardiol. 35:282, 1975.
33. Bigger, J. T., and Jaffe, C. C.: Am. J. Cardiol. 27:82, 1971.
34. Hollunger, G.: Acta Pharmacol. Toxicol. 17:356, 1960.
35. Hollunger, G.: Acta Pharmacol. Toxicol. 17:374, 1960.
36. Helfant, R. H., et al.: Am. Heart J. 77:315, 1969.
37. Beckett, A. H., and Chiodomere, E. C.: Postgrad. Med. J., 64(Suppl. 1):60, 1977.
38. Anderson, J. L.: Circulation 57:685, 1978.
39. Kates, R. E., Woosley, R. L., and Harrison, D. C.: Am. J. Cardiol. 53:248, 1983.
40. Guehler, J., et al.: Am. J. Cardiol. 55:807, 1985.
41. Olsson, S. B., Brorson, L., and Varnauskas, E.: Br. Heart J. 35:125, 1973.
42. Kannan, R., Nadermanee, K., Hendrickson, J. A., et al.: Clin. Pharmacol. 31:438, 1982.
43. Nadermanee, K., Singh, B. N., Hendrickson, J. A., et al.: Circulation 66:202, 1982.
44. Boura, A. L. A., et al.: Lancet 2:17, 1959.
45. Papp, J. G., and Vaughan Williams, E. M.: Br. J. Pharmacol. 37:380, 1969.
46. Vakil, R. J.: Br. Heart J. 11:350, 1949.
47. Muller, J. M., Schlittler, E., and Brin, H. J.: Experientia 8:338, 1952.
48. Hess, S. M., Shore, P. A., and Brodie, B. B.: J. Pharmacol. Exp. Ther. 118:54, 1956.
49. Kirshner, N.: J. Biol. Chem. 237:2311, 1962.
50. von Euler, U. S., and Lishajko, F.: Int. J. Neuropharmacol. 2:127, 1963.
51. Gold, H., et al.: JAMA 145:637, 1951; Brass, H.: J. Am. Pharm. Assoc. Pract. Ed. 4:310, 1943.
52. Halpern, A., et al.: Antibiot. Chemother. 9:97, 1959.
53. Elson, J., et al.: Clin. Pharmacol. Ther. 17:134, 1975.
54. Uvnas, B.: Physiol. Rev. 40(Suppl. 4):69, 1960.
55. Langer, S. Z., Cavero, I., and Massingham, R.: Hypertension 2:372, 1980.
56. Koblinger, W., and Klupp, H.: Recent Adv. Hypertension 1:53, 1975.
57. Starke, K., and Montel, H.: Neuropharmacology 12:1073, 1973.
58. Antonaccio, M. J.: Cardiovascular Pharmacology, p. 278. New York, Raven Press, 1983.
59. Druey, J., and Ringier, B. H.: Helv. Chim. Acta 34:195, 1951.
60. Gross, F., Druey, J., and Meier, R.: Experientia 6:19, 1950.
61. Zacest, R., and Koch-Weser, J.: Clin. Pharmacol. Ther. 13:420, 1972.
62. Gibaldi, M., Boyer, R. N., and Feldman, S. J.: Pharm. Sci. 60:1338, 1971.
63. Johnson, C. C.: Proc. Soc. Exp. Biol. Med. 26:102, 1928.
64. Corcoran, A. C., et al.: Circulation 2:188, 1955.
65. DuCharme, D. W., et al.: Pharmacol. Exp. Ther. 184:662, 1973.
66. Vallotton, M. B. Trends Pharmacol. Sci. 8:69, 1987.
67. Ondetti, M. A., Rubin, B., and Cushman, D. W.: Science 196:441, 1977.
68. Patchett, A., et al.: Nature 288:280, 1980.
69. Levy, R. I., et al.: Circulation 69:325, 1984.
70. Levy, R. L., and Rifkind, B. M.: Cardiovasc. Drugs 1:1, 1977.
71. Levy, R. L.: Annu. Rev. Pharmacol. Toxicol. 47:499, 1977.
72. Gilman, H., and Wilder, G. R.: J. Am. Chem. Soc. 82:1166, 1950.
73. Mannisto, P. T., et al.: Acta Pharmacol. Toxicol. 36:353, 1975.
74. Kesaniemi, Y. A., and Grundy, S. M.: JAMA 251:2241, 1984.
75. Lindenbaum, S., and Higuchi, T.: J. Pharm. Sci. 64:1887, 1975.
76. Goldstein, J. L., and Brown, M. S.: Med. Clin. North Am. 66:335, 1982.
77. Atmeh, R. F., et al.: J. Lipid Res. 24:585, 1983.
78. Morawitz, P.: Ergeb. Physiol. Biol. Chem. Exp. Pharmacol. 4:307, 1905.
79. Stenflo, J., et al.: Proc. Natl. Acad. Sci. USA 71:2730, 1974.
80. Jackson, C. M., and Suttie, J. W.: Prog. Hematol. 10:333, 1978.
81. Triplett, D. A. (ed.): Platelet Function. Chicago, Am. Soc. Clin. Pathol., 1978.
82. Hamberg, M., et al.: Proc. Natl. Acad. Sci. USA 11:345, 1974.
83. Thomas, R. C., and Rudy, R. W.: J. Med. Chem. 15:964, 1972.
84. McClintock, J. C., et al.: Surg. Gynecol. Obstet. 112:653, 1961.
85. Greer, M. A., et al.: N. Engl. J. Med. 272:888, 1965.

SELECTED READINGS

Antianginal Agents and Vasodilators

Baker, P. F.: Calcium in drug actions. Handbook of Experimental Pharmacology, Vol. 83. Berlin, Springer-Verlag, 1988.

Braunwald, E.: Mechanism of action of calcium channel blocking agents. N. Engl. J. Med. 307:1608, 1982.

Hugenholtz, P. G.: Calcium antagonists. In Abshagen, U. (ed.). Clinical Pharmacology of Antianginal Drugs, Vol. 76, p. 459. Berlin, Springer-Verlag, 1985.

Needham, P. (ed.): Organic Nitrates. Berlin, Springer Verlag, 1975.

Speralakis, N.: Calcium Antagonists. Boston, Martinus Nijhoff, 1984.

Stone, P. H., and Antman, E. M.: Calcium Channel Blocking Agents in the Treatment of Cardiovascular Disorders. Mount Kisco, NY, Futura, 1983.

Antiarrhythmic Agents

Hondeghem, L. M., and Katzung, B. G.: Antiarrhythmic agents: The modulated receptor mechanism of action of sodium and calcium channel blocking drugs. Annu. Rev. Pharmacol. Toxicol. 24:387, 1984.

Lucchesi, B. R., and Patterson, E. S.: Antiarrhythmic drugs. In Antonaccio, M. J. (ed.). Cardiovascular Pharmacology, p. 329. New York, Raven Press, 1984.

O'Rourke, R. A. (ed.): Calcium-entry blockade: Basic concepts and clinical implications. Circulation 75(Suppl.): June 1982.

Reiser, H. J., and Horowitz, L. M.: Mechanisms and Treatment of Cardiac Arrhythmias. Baltimore, Erban and Schwartzenberg, 1985.

Sasyniuk, B. I., and Ogilvie, R. I.: Antiarrhythmic drugs: Electrophysiological and pharmacokinetic considerations. Annu. Rev. Pharmacol. 15:131, 1977.

Steinberg, M. I., Lacefield, W. B., and Robertson, D. W.: Class I and III antiarrhythmic agents. Annu. Rep. Med. Chem. 21:97, 1986.

Szerkes, L.: Pharmacology of Antiarrhythmic Agents. Oxford, Pergamon Press, 1981.

Triggle, D. J.: Drugs acting on ion channels and membranes. In Emmett, J. C. (ed.). Comprehensvie Medicinal Chemistry, Vol. 3, p. 1047. Oxford, Pergamon Press, 1990.

Vaughan Williams, E. M.: Antiarrythmic drugs. Handbook of Experimental Pharmacology, Vol. 89. Berlin, Springer-Verlag, 1989.

Antihypertensive Agents

Cauvin, C., Loutzenhiser, R., and Van Breeman, C.: Mechanisms of calcium antagonist-induced vasodilatation. Annu. Rev. Pharmacol. Toxicol. 23:373, 1983.

Gross, F.: Antihypertensive agents. In Handbook of Experimental Pharmacology. Berlin, Springer-Verlag, 1977.

Ignarro, L. J., and Kadowitz, P. J.: The pharmacological and physiological role of cyclic GMP in vascular smooth muscle relaxation. Annu. Rev. Pharmacol. Toxicol. 25:717, 1985.

Janis, R. A., and Triggle, D. J.: New developments in Ca^{++} channel antagonists. J. Med. Chem. 26:775, 1983.

Sweet, C. S., and Blaine, E. H.: Angiotensin-converting enzyme and renin inhibitors. In Antonaccio, M. J. (ed.). Cardiovascular Pharmacology. New York, Raven Press, 1984.

Antihyperlipidemic Agents

Beneze, W. L.: Hypolipidemic agents. In Kritchevsky, D. (ed.). Hypolipidemic Agents, p. 349. Berlin, Springer-Verlag, 1975.

Cendella, R. J.: Cholesterol and hypolipidemic drugs. In Craig, C. R., and Stitzel, R. E. (eds.). Modern Pharmacology, 2nd ed., p. 329. Boston, Little Brown & Co., 1986.

Eisenberg, S., and Levy, R. I.: Lipoproteins and lipoprotein metabolism. In Krichevsky, D. (ed.). Hypolipidemic Agents, p. 191. Berlin, Springer-Verlag, 1975.

Goldstein, J. L., and Brown, M. S.: The low density lipoprotein pathway and its relation to atherosclerosis. Annu. Rev. Biochem. 46:897, 1977.

Krichevsky, D., et al.: Drugs, lipid metabolism and atherosclerosis. Adv. Exp. Med. Biol. 109:1978.

Anticoagulants

Fedan, J. S.: Anticoagulant, antiplatelet and fibrinolytic (thrombolytic) drugs. In Craig, C. R., and Stitzel, R. E. (eds.). Modern Pharmacology, 2nd ed., p. 415. Boston, Little Brown & Co., 1986.

Higgs, G. A., Higgs, E. A., and Moncada, S.: The arachidonic acid cascade. In Sammes, P. G. (ed.). Comprehensive Medicinal Chemistry, Vol. 2, p. 147. Oxford, Pergamon Press, 1990.

Salamon, J. A.: Inhibitors of prostaglandin, thromboxane and leukotriene biosynthesis. Adv. Drug Res. 15:11, 1986.

Seuter, F., and Sciabrine, A.: In Antonaccio, M. J. (ed.). Cardiovascular Pharmacology, p. 475. New York, Raven Press, 1984.

Taylor, M.: Enzyme cascades: Coagulation, fibrinolysis, and hemostasis. In Sammes, P. G. (ed.). Comprehensive Medicinal Chemistry, Vol. 2, p. 481. Oxford, Pergamon Press, 1990.

Whittle, B. J. R., and Moncada, S.: In Antonaccio, M. J. (ed.). Cardiovascular Pharmacology, 2nd ed., p. 519. New York, Raven Press, 1984.

CHAPTER 15

Local Anesthetic Agents

Charles M. Darling

Local anesthesia may be defined as the loss of sensation or the loss of motor function in a circumscribed area of the body. *Local anesthesics* are drugs that produce this condition by causing a block of nerve conduction. They must be applied locally to the nerve tissue in appropriate concentrations. To be useful clinically, the action should always be reversible.

The local anesthetic agents are useful chemical tools for the temporary relief of localized pain in dentistry and minor surgical procedures, as well as for producing a state of nonresistance (e.g., spinal anesthesia) without general anesthesia. Many over-the-counter agents are used topically for temporary relief of pain and itching caused by minor burns, insect bites, allergic responses, hemorrhoids, and other minor conditions.

An ideal local anesthetic has not yet been discovered, but several desirable properties may be stated:[1]

1. Nonirritating to tissue and not causing permanent damage. Most clinically available agents fulfill this requirement.
2. Low systemic toxicity because it is eventually absorbed from its site of application. Most local anesthetics are rapidly metabolized after absorption.
3. Effective whether injected into the tissue or applied locally to skin or mucous membranes. Intact skin is resistant generally to the action of most local anesthetics and requires high concentrations for prolonged periods. This is probably due to a slow rate of penetration and a rapid vascular diffusion following penetration. Thus, insufficient drug accumulates at the nerve endings in the dermis.
4. Rapid onset of anesthesia and a short duration of action. The action must last long enough to allow time for the contemplated surgery, yet not so long as to entail an extended period of recovery.

HISTORICAL DEVELOPMENT

Throughout history people have sought means to escape pain. History records the more popular methods[2] to be psychic influences, acupuncture, nerve compression, freezing, and drugs. Although the popularity of each method has been somewhat cyclic through the centuries, each persists today in one form of practice or another.

The modern-day local anesthetic drugs owe their origin to the isolation of cocaine in 1860. By nature, humans are experimenters in the environment, always seeking to identify natural phenomena observed through the senses. In earlier times people sought to explore the smell and taste of all plants endemic to their region. Sometimes the results were catastrophic (Jamestown weed[3]), sometimes valuable. As they explored other regions, they learned from the experiences of the natives of distant regions.

The natives of Peru learned to chew the leaves of the coca bush to stimulate the general feeling of well-being and to prevent hunger. Although no early literature citation is available to substantiate it, the suggestion is often made[2] that the Incas produced local anesthesia for surgical procedures by chewing a cud of the leaves and allowing the saliva to drop upon the site of incision.

The question of priority of discovery[2] is not pertinent to the present discussion; however, the use of a specific chemical substance to produce local anesthesia evolved slowly from the isolation of the crystalline alkaloid, cocaine, from coca leaves in 1860. A seemingly necessary historical component was the invention of the hypodermic syringe, which was introduced to the United States in 1856 and was used on a large scale for the first time during the Civil War.[4] Neither revolutionary concept can be assigned a historical precedence, because the fundamental observations were recorded over a span of centuries. Therefore, the modern use of local anesthetics evolved from the mid-19th century isolation of cocaine and the concomitant development of the hypodermic syringe.

In the late 19th century, cocaine was used as a topical anesthetic agent in ophthalmology, for the production of peripheral nerve blockade, and in spinal anesthesia.[5] From these early experiments, several adverse effects were observed. Cocaine has a high acute toxicity and addicting properties that served to stimulate efforts to find effective and less toxic local anesthetic agents.

Although the structure of cocaine was not identified until 1924, earlier work had shown that it contained a benzoic acid ester moiety. This led to the preparation of the ester, ethyl *p*-aminobenzoate (benzocaine), in 1890. The low aqueous solubility of benzocaine limited its usefulness as an injectable agent and it was neglected for many years. However, it is now recognized as an effective topical anesthetic agent for the production of surface anesthesia of mucous membranes.

During the next half-century, most compounds synthesized as local anesthetics were derivatives of benzoic acid, but, unlike benzocaine, the molecules contained a basic amino group from which water-soluble salts could be prepared. A few of these agents are still in wide clinical use as local anesthetic agents: procaine, 1905; dibucaine, 1929; and tetracaine, 1930. Also, these synthetic derivatives, in contrast to cocaine, caused no contraction of the blood vessels and were rapidly absorbed from the site of injection. However, in 1903, the practice was begun of adding a potent vasoconstrictor, epinephrine, to the solutions to prolong the duration of local anesthesia.

In the mid-1930s, the serendipitous discovery of the local anesthetic effect of an isomer of the alkaloid, gramine, gave new direction to the search for new agents. Isogramine, 2-(dimethylaminomethyl)indole, was observed to produce local anesthesia to the tongue when tasted. The discovery of activity in an indole-containing structure represented a significant departure from the benzoic acid series.[2]

During the following decade, numerous compounds were synthesized and investigated for useful activity. This work culminated in the development of lidocaine, which has structural characteristics of the indole, isogramine. For example, each contains a nitrogen atom attached to an sp^2 carbon atom. An aromatic group is attached to one side of this moiety and a tertiary aminoalkyl group is attached to the other side.

Essentially all of the clinically useful local anesthetic agents are of the cocaine (benzoic acid derivative) or isogramine (anilide or so-called reversed amide) lineage. The more recently introduced agents are of the latter type.

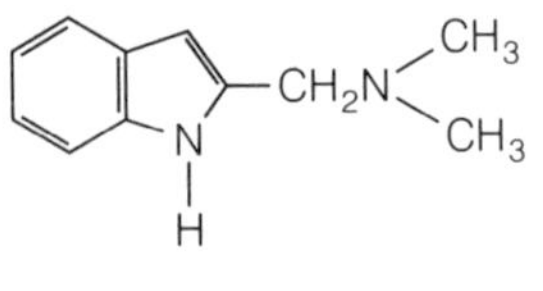

Isogramine

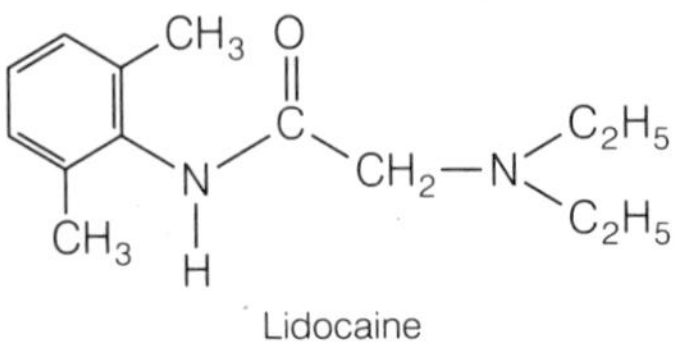

Lidocaine

NERVOUS TISSUE

The function of the nervous tissue is based on two fundamental properties. The first property is *irritability*, which refers to the ability to react to various stimulating agents. The second is *conductivity*, which is the ability to transmit the excitations. Both properties may be blocked by local anesthetic drugs.

The peripheral nerves are composed of fascicles (bundles) of nerve fibers of varying thickness, held together by connective tissue (Fig. 15-1). It is customary to classify nerve fibers according to their diameter. The speed of impulse transmission and the magnitude of the action potential are proportional to the size of the diameter. Fiber diameters cover a wide and continuous range from large myelinated (A fibers), through medium (B fibers), to small unmyelinated (C fibers).

The outer layer of connective tissue, *epineurium*, is made up of longitudinally arranged connective tissue cells and collagenous fibers interspersed with fat cells. Each of the smaller fascicles of a nerve is itself enclosed in a membrane of dense, concentric layers of connective tissue called *perineurium*. The individual nerve fibers (*axons*) are interspersed within longitudinally arranged strands of collagenous fibers called *endoneurium*. Blood vessels run throughout the various sheaths, which are essen-

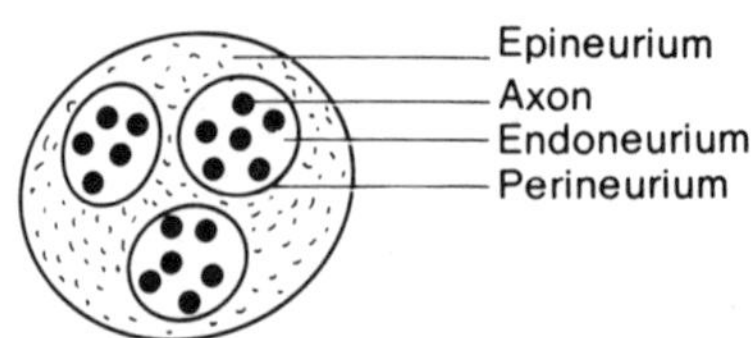

FIG. 15-1. Diagram showing the cross-sectional view of the parts of a peripheral nerve.

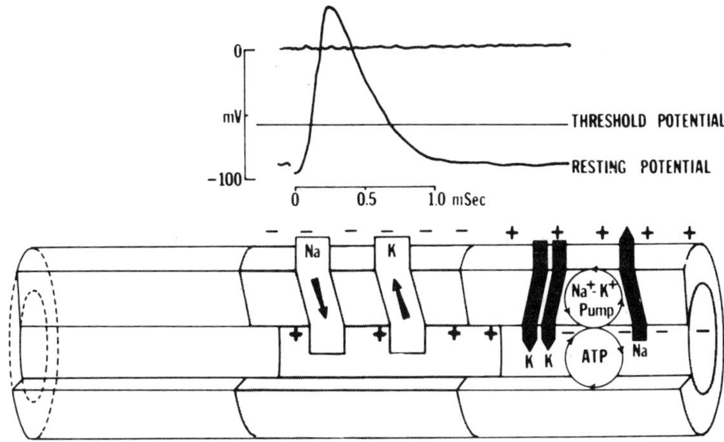

FIG. 15-2. Relationship between membrane action potential and ionic flux across the nerve membrane. (From Covino, B. G., and Vassallo, H. G.: Local Anesthetics, Mechanism of Action and Clinical Use, p. 1. New York, Grune & Stratton, 1976. By permission.)

tially lipid. The molecular structure of these sheaths is unknown.[5]

Each nerve fiber (axon) has its own cell membrane, the *axolemma*, which surrounds the cytoplasm (*axoplasm*) and serves as a barrier between the axoplasm and the endoneurium. The composition of the membrane is highly lipid with some protein.

During the period of nerve inactivity, the potential difference (resting potential) across the cell membrane is of the order of 50 to 100 mV inside negative (Fig. 15-2). Although this potential difference may seem rather small in absolute magnitude, it occurs across an interface of such extreme thinness that the resultant electric field approaches 100,000 V/cm.[6] It seems reasonable to suppose that such a high field strength is important to the orientation of molecules in the membrane structure and to the transport of ions across the lipid membrane.

Figure 15-2 illustrates the relationship between changes in membrane action potential and ionic flux across the nerve membrane when excitation occurs. An initial phase of depolarization is observed during which the electrical potential within the nerve cell becomes progressively less negative owing to the inflow of potassium ions. Although the membrane permits potassium ions to flow back and forth with relative freedom, it resists movement of the sodium ions from the extracellular fluid to the inside of the cell until depolarization reaches a critical level.

When the action potential reaches a critical level (the threshold potential or firing level), the resistance to sodium ions (*resistance* is the inverse expression of conductance) decreases, allowing sodium ions to flow into the cell along its concentration gradient. As sodium ions flow in, the membrane becomes more and more depolarized; this further decreases the resistance to movement of sodium ions. A result of this self-perpetuating process is an extremely rapid phase of depolarization. The process is self-limited and should be proportional to the original extracellular and intracellular concentrations of sodium ion. However, the explosive response to depolarization reduces the membrane potential not merely to zero, but overshoots zero, such that the inside of the membrane becomes positively charged relative to the outside. At the peak of the action potential, the interior has a positive electrical potential of approximately +40 mV relative to the exterior of the cell. The peak of the action potential would be even greater were it not for the increased flow of potassium ions along the concentration gradient from inside to outside the cell.

At the conclusion of the depolarization phase, repolarization begins (refractory phase) utilizing the so-called *sodium–potassium pump*, which essentially "pumps" the sodium ions from inside to outside the cell and the potassium ions from outside to inside. This metabolic pump requires energy and is thought to be catalyzed by a specific adenosinetriphosphatase.[7] The electrical potential within the cell again becomes progressively more negative to the outside until the resting potential of −60 to −90 mV is reestablished. Normally the entire process of depolarization and repolarization occurs within one millisecond.

MECHANISM OF ACTION

Several hypotheses on the mechanism of action of local anesthetic drugs have been examined.[1,8] It is generally accepted that bound calcium ions play a critical role in the generation of the nerve impulse. Possibly, the initial phase of depolarization (caused by a nerve impulse) results in the removal of calcium ions from sites in the nerve membrane leading to increased permeability to sodium ions, analogous to the opening of a pore through which sodium ions flow. Local anesthetics may inhibit the release of bound calcium ions or replace calcium ions at the site and thus stabilize the membrane to depolarization. The nerve impulse is not transmitted.

The association of local anesthetic molecules within the membrane may affect permeability by increasing the degree of disorder of the lipids that constitute the nerve membrane. Alternatively, the local anesthetic molecules may increase the surface pressure of the lipid layer and essentially close the pores through which ions move. The result of either alternative would restrict the openings to the sodium ion channels and cause a general decrease in permeability, the fundamental change necessary for the generation of the action potential and the propagation of an impulse.

Takman has proposed a classification of local anesthetic agents based on site of action.[9] In this classification, the four classes are agents acting (1) at receptor sites on the external surface of the nerve membrane, (2) at receptor sites on the internal surface of the nerve membrane, (3) by a receptor-independent physicochemical mechanism, and (4) by a combination of a receptor and receptor-independent mechanism. The first class (1) of agents includes two guanidine-type molecules possessing local anesthetic potency greater than that of other known compounds. They are thought to produce local anesthesia by blocking the opening of the sodium ion channel on the external surface of the nerve membrane.

Tetrodotoxin, obtained from the ovaries and other organs of the puffer fish (Tetraodontidae), and saxitoxin, produced by certain marine dinoflagellates (the rapid growth of which causes discoloration of the ocean, hence, the term *red tide*) that contaminate shellfish and cause paralytic shellfish poisoning in humans, are representatives of this group. Both highly hydrophilic substances are extremely toxic to humans and are not available for clinical use. However, tetrodotoxin is of experimental interest because it selectively blocks the increase in sodium conductance without affecting steady-state conductances.

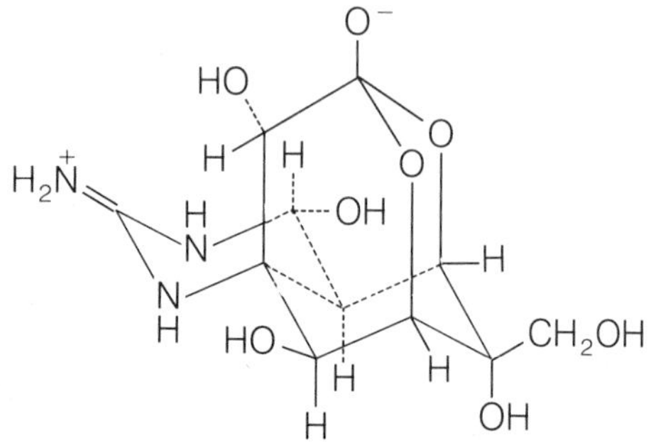

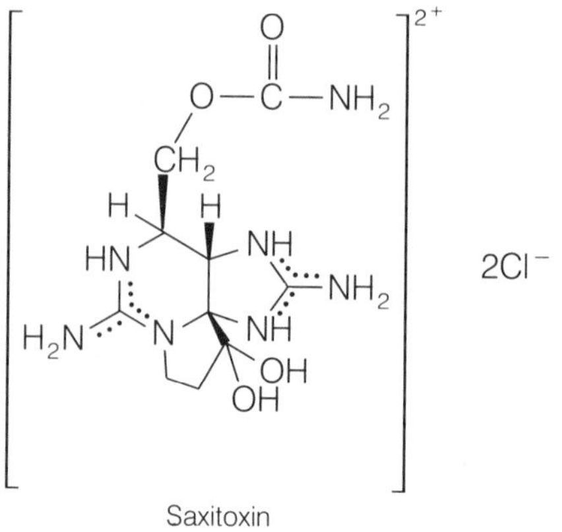

Tetrodotoxin

Saxitoxin

The second class (2) includes those agents acting at receptor sites near openings of the sodium ion channels on the internal axoplasmic surface of the membrane. Studies to determine the site of action and whether the cationic or neutral form is the active species have been reviewed.[5,8] Quaternary ammonium salts of some local anesthetics are representative of this class and were shown to block conduction when applied internally by perfusion. The same drugs applied externally did not gain access to the inside of the cell and were ineffective in blocking conduction. Interestingly, high concentrations of tetrodotoxin applied internally for long periods had no effect on sodium ion conductance. Because their probability of diffusion across lipid membranes is quite low, quaternary ammonium salts of local anesthetic drugs are not likely to have clinical value.

The compounds in the third class (3) are noncationic at physiologic *p*H; these include benzocaine, alcohols, and barbiturates. Members of the latter two groups are not useful clinically as local anesthetics, but they are of value as experimental tools in the study of the mechanisms of action. Similar to other local anesthetic drugs, alcohols (e.g., *n*-butyl alcohol and benzyl alcohol) in low concentrations (higher concentrations result in irreversible nerve damage), and barbiturates block conduction reversibly by decreasing sodium ion conductance.

Benzocaine, the only clinically useful member of this class, is known to exist in the uncharged form at physiologic *p*H, and its activity is not affected by changes in *p*H. Thus, its local anesthetic action is due to physicochemical properties other than ionic bonding. Takman suggests that compounds of this type exert their action by a receptor-independent physicochemical mechanism. They may restrict sodium ion conductance either by increasing the degree of disorder of the lipids of the nerve membranes or by increasing the surface pressure of the lipid layer.

The fourth class (4) of compounds includes by far the most clinically useful local anesthetic agents. These agents are believed to act by a combination of receptor and receptor-independent mechanisms. A mechanistic interpretation by Ritchie[8] of the results reported in the literature is not in disagreement with this classification.

Agents of this class are tertiary amines in which the neutral form exists in equilibrium with the cationic conjugate acid form at physiologic *p*H. The suggestion has been made that these agents act at the inside surface of the nerve membrane in the cationic form. The neutral form then more readily passes through the lipid membrane to gain access to the inside surface. If this is so, the *p*H of the axoplasm determines the active form of the drug in the region of the receptor. Because some reported results with mammalian C fibers do not support this view, Ritchie[8] suggested that the receptors with which local anesthetics interact may be within the membrane. The receptors may even be near the outer surface of the membrane, but in such a position that externally applied drugs cannot reach them.

In summary, drugs produce local anesthesia by inhibiting membrane conductance of sodium ions. The evidence suggests that the inhibition may be produced at more than one site. Tetrodotoxin and saxitoxin interact with receptors on the outer surface, whereas quaternary ammonium salts interact with receptors on the inner surface of the nerve membrane. The clinically useful local anesthetic agents (members of classes 3 and 4) interact with receptor site(s) on the inside surface of the membrane or those within the membrane itself. The precise mechanism of action at the molecular level is not known.

STRUCTURE–ACTIVITY RELATIONSHIPS

As indicated in the foregoing classification, a great variety of chemical structures may produce conduction blockade, and there appears to be no marked structural specificity. It would be difficult to define precisely the relationships between chemical structure and local anesthetic activity if all compounds were included without regard to site of action. Therefore, the following discussion of structure–activity relationships (SAR) is restricted to chemical classes of clinically useful local anesthetic agents (3 and 4), in which the salient features are quite similar.

The origin of the modern local anesthetic agents can be traced to the independent discoveries of two distinctly different alkaloids, cocaine and isogramine. Cocaine is an aminoalkyl ester of benzoic acid; isogramine is a 2-(aminoalkyl)indole. Insufficient data are available to make definitive statements about the relationship of the common structural features of the two molecules to local anesthetic activity.

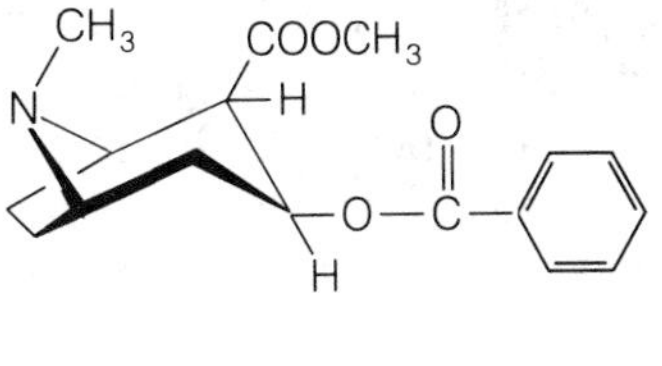

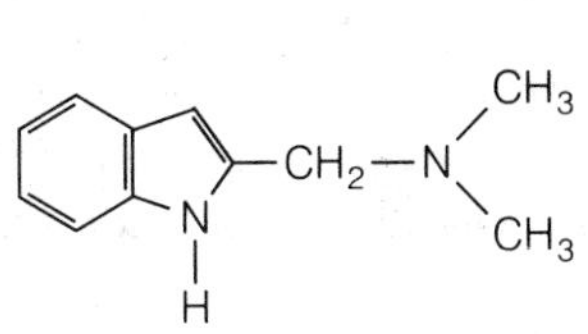

Cocaine

Isogramine

The structures of clinically useful agents contain a centrally located sp^2-hybridized carbon atom to which alkyl and aryl groups are attached either directly or through a heteroatom. When the aryl group is attached directly to the sp^2 carbon atom (cocaine type), activity appears to be relatively insensitive to the mode of attachment of the alkyl group. But when the alkyl group is attached directly to the sp^2 carbon atom (isogramine type), activity appears to require the attachment of an aryl group through a nitrogen bridge. The implications of these structural differences await further study. But apparently a positive mesomeric effect on the sp^2 carbon atom has a favorable effect on activity and was suggested[10] as leading to a greater affinity for the binding site at the receptor.

BENZOIC ACID DERIVATIVES

The benzoic acid derivatives are synthetic compounds derived from the structure of cocaine and may be represented as follows:

$$\text{Aryl}-\overset{\displaystyle O}{\overset{\|}{C}}-X-\text{Aminoalkyl}$$

Aryl Group

The clinically useful members of this series possess an aryl radical attached directly to the carbonyl group or attached through a vinyl group. Although alicyclic and aryl aliphatic carboxylic acid esters are active, conjugation of the aromatic group with the carbonyl enhances local anesthetic activity. Substitution of the aryl group with substituents that increase the electron density of the carbonyl oxygen enhance activity. Favorable substituents include alkoxy, amino, and alkylamino groups in the *para* or *ortho* positions. However, these substituents also alter the liposolubility of the molecule, the effect of which will be treated separately (see Coefficient of Distribution).

Aryl aliphatic radicals, which contain an alkylene group between the aryl radical and carbonyl group, result in compounds that have not found clinical use. In these compounds the mesomeric effect of the aryl radical does not extend to the carbonyl group.

Bridge, *X*

The bridge, *X*, may be carbon, oxygen, nitrogen, or sulfur. In an isosteric procaine series, conduction–anesthetic potency[10] decreased in the following order: sulfur, oxygen, carbon, nitrogen. These modifications largely determine the chemical class to which each derivative belongs; they also affect duration of action and relative toxicity. In general, amides (X = N) are more resistant to metabolic hydrolysis than esters (X = O). Thioesters (X = S) may cause dermatitis.

Aminoalkyl Group

The aminoalkyl group is not necessary for local anesthetic activity, but it is used to form water-soluble salts. For example, benzocaine (X = O; aminoalkyl = C_2H_5) is a local anesthetic, but the aqueous solutions of it must be highly acidic.

Because of its capacity for salt formation, the amino function is considered the hydrophilic part of the local anesthetic molecule. Tertiary amines result in more useful agents. The secondary amines appear to be longer-acting, but they are more irritating; primary amines are not very active and cause irritation.

The alkyl groups (including the intermediate chain linked to X) primarily influence relative lipid solubility (see Distribution Coefficient) as discussed later.

LIDOCAINE DERIVATIVES (ANILIDES)

Lidocaine derivatives are essentially anilide progenies of isogramine with the following general structural characteristics:

$$\text{Aryl—NH—}\overset{\overset{\displaystyle X}{\|}}{\text{C}}\text{—Aminoalkyl}$$

Aryl Group

The clinically useful local anesthetic agents of this type possess a phenyl group attached to the sp^2 carbon atom through a nitrogen bridge. Substitution of the phenyl with a methyl group in the 2- or 2- and 6-position enhances activity. The amide bond is more stable to hydrolysis than the ester bond. In addition, the methyl substituent(s) provide(s) steric hindrance to hydrolysis of the amide bond and increase(s) the coefficient of distribution.

Substituent *X*

In the foregoing general structure, X may be carbon (isogramine), oxygen (lidocaine), or nitrogen (phenacaine). Although the lidocaine series (X = O) has provided more useful products, insufficient data are available to state the relative contribution of X to activity.

Aminoalkyl Group

As found for the benzoic acid derivatives, the amino function has the capacity for salt formation and is considered the hydrophilic portion of the molecule. Tertiary amines are more useful clinically because the primary and secondary amines are more irritating to tissue.

Coefficient of Distribution

In general, nerve membranes consist primarily of lipids. Local anesthetic agents are injected near the site of action and must be capable of penetrating the nerve membrane. Increasing the lipid solubility of a series of compounds should result in facilitated penetration of nerve membranes. In in vitro experiments involving very simple systems such as isolated nerves, the potency of local anesthetic compounds is directly proportional to distribution coefficients.

The in vivo system is much more complicated, and often, within a congeneric series, increases in partition coefficients result in increasing potency to a maximum, after which the activity declines. Unfortunately, toxicity also increases concurrently.

Each structural component (aryl and aminoalkyl) contributes to the lipid solubility of the molecule and may be varied to produce derivatives of increasing distribution coefficient. Substitution of the aryl radical by alkyl, alkoxy, and alkylamino groups leads to homologous series in which partition coefficients increase with increasing number of methylene groups ($—CH_2—$). A review[10] of these series revealed that local anesthetic activity peaked with the C_4-, C_5-, or C_6- homologues, depending on the particular nature of the series being considered.

Similarly, variations in the aminoalkyl portion of the molecule lead to increases in activity and toxicity with increasing carbon number. The branching of *N*-alkyl groups is often accompanied by an intensification of activity. The aminoalkyl group may be part of an aliphatic heterocyclic ring.

The tertiary amino group may be diethylamino, piperidino, or pyrrolidino, leading to the products that exhibit essentially the same degree of activity. The more hydrophilic morpholino group usually leads to diminished potency. However, this weakening of the effect can be compensated fully by the introduction of additional lipophilic groups in other parts of the molecule.[11]

A given degree of lipid solubility does not assure local anesthetic activity, nor does a high degree of potency assure clinical usefulness. In the selection of a particular member of a series for clinical use, the relative potency must be weighed against the degree of toxicity.

Most of the clinically useful local anesthetic agents exhibit pK_a values in the range between 8.0 and 9.5. The implication of this is that compounds with higher pK_a values are ionized essentially 100% at physiologic pH and so have difficulty reaching the biophase. Substances having a lower pK_a value are not sufficiently ionized and are less effective in practice, even though they reach the biophase.[10]

PRODUCTS

The numerous local anesthetic products for which official monographs have been prepared are discussed under two major headings, Benzoic Acid Derivatives and Lidocaine Derivatives (anilides). If appropriate, the products are grouped by specific chemical class within each heading.

Except for cocaine, the clinically useful local anesthetics do not cause vasoconstriction. In clinical practice, therefore, a solution of a local anesthetic often contains epinephrine, norepinephrine, or a suitable synthetic congener, such as phenylephrine. The vasoconstrictor serves a dual purpose. By decreasing the rate of absorption, it not only localizes the anesthetic at the desired site, but also limits the rate at which the anesthetic is absorbed into the circulation. Thus, metabolism can reduce the systemic toxicity of the local anesthetic. In essence, then, the vasoconstrictor prolongs the action and lowers the systemic toxicity of local anesthetic agents.

BENZOIC ACID DERIVATIVES

Cocaine, USP. (−)3-(Benzoyloxy)-8-methyl-8-azabicyclo[3 · 2 · 1]octane-2-carboxylic acid methyl ester; [1(R),2(R),3(S)],methylbenzoylecgonine. Cocaine is an alkaloid obtained from the leaves of *Erythroxylon coca* Lamarck and other species of *Erythroxylon* (Erythroxylaceae). It occurs to the extent of approximately 1% in South American leaves and to the equivalent (as other derivatives) of about 2% in Java leaves. It was first isolated as an amorphous mixture by Gaedcke (1855) who thought it was an alkaloid related to caffeine, which he called erythroxyline. Later, Nieman (1860) obtained from coca leaves a crystalline alkaloid to which he gave the name cocaine.

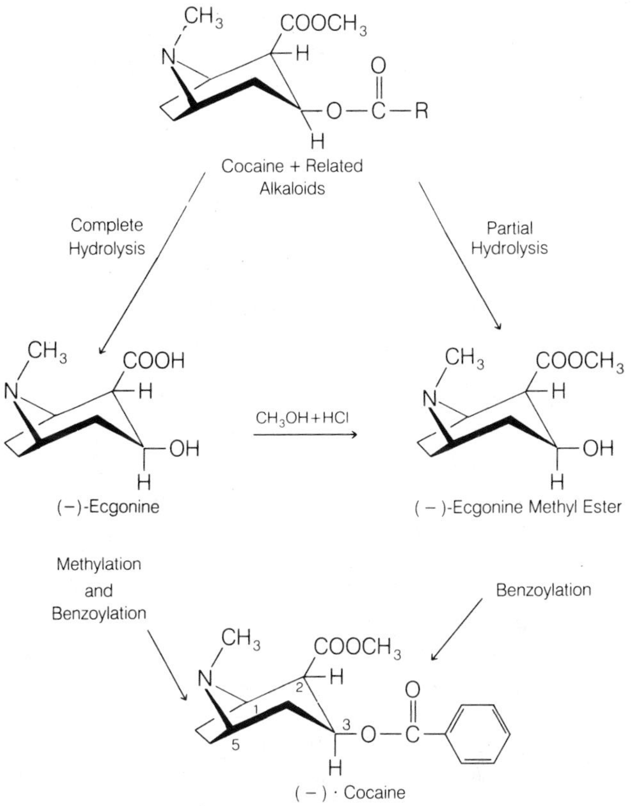

FIG. 15-3. *Commercial preparation of cocaine.*

Commercial production involves total extraction of bases followed by acid hydrolysis of the ester alkaloids to obtain the total content of (−)-ecgonine. After purification of the ecgonine, cocaine is synthesized by esterification with methanol and benzoic acid (Fig. 15-3).

The ecgonine portion of the cocaine molecule contains four asymmetric carbon atoms. Two of these (C-1 and C-5) are intramolecularly compensated; therefore, only eight optically active isomers (four racemates) exist. In cocaine, the benzoyloxy (C-3) and methoxycarbonyl (C-2) groups are *cis* to the nitrogen bridge. In (+)-pseudococaine, which is also active, the methoxycarbonyl is *trans*.[11]

Cocaine occurs as levorotatory, colorless crystals or as a white crystalline powder that numbs the lips and the tongue when applied topically. It is slightly soluble in water (1 : 600), more soluble in alcohol (1 : 7), and quite soluble in chloroform (1 : 1) and in ether (1 : 3.5). The crystals are fairly soluble in olive oil (1 : 12), but less so in mineral oil (1 : 80 to 1 : 100). Because of its solubility characteristics, it is used principally where oily solutions or ointments are indicated. Cocaine is basic and readily forms crystalline, water-soluble salts with organic and inorganic acids.

Cocaine Hydrochloride, USP. The salt occurs as colorless crystals or as a white crystalline powder that is soluble in water (1 : 0.5), alcohol (1 : 3.5), chloroform (1 : 15), or glycerin. It has a pK_a of 8.4.

Aqueous solutions of cocaine hydrochloride are stable if not subjected to elevated temperatures or stored for prolonged periods. Bacteriologic filtration is a better method of sterilization than autoclaving. At *p*H values near 7, solutions of the salt are very unstable and should be prepared freshly and not be autoclaved. Also, the solutions are incompatible with alkalis, the usual alkaloidal precipitants, silver nitrate, sodium borate, calomel, and mercuric oxide.

Koller, in 1884, reported[2] the first practical application of cocaine as a topical anesthetic in the eye. Its toxicity has prevented cocaine from being used for anything other than topical anesthesia, and even in this capacity, its use is limited for fear of causing systemic reactions and addiction.

Cocaine does not penetrate the intact skin, but is readily absorbed from mucous membranes, as is evidenced by the quick response obtained by addicts when cocaine ("snow") is snuffed into the nostrils. Cocaine blocks the uptake of catecholamines at adrenergic nerve endings and is thus a potent vasoconstrictor. This accounts for the ulceration of the nasal septum after cocaine has been "snorted" for long periods in high doses.[12]

As a local anesthetic, cocaine is applied topically to mucous membranes such as those of the eye, the nose, and the throat in solutions of 2% to 5%. The anesthesia produced by such concentrations lasts approximately half an hour. Its use in ophthalmology is marred by the fact that, not infrequently, cocaine causes corneal damage, with resultant opacity. For nose and throat work, concentrations higher than 10% are rarely used, 1% to 4% being a common concentration range. For external use, apply topically to mucous membranes, as a 2% to 20% solution.

Hexylcaine Hydrochloride, USP. 1-(Cyclohexylamino)-2-propanol benzoate (ester) hydrochloride (Cyclaine Hydrochloride). The hydrochloride salt is soluble in water (1 : 8) and is freely soluble in alcohol and in chloroform. Solutions are stable to boiling and autoclaving. A 1% solution has a *p*H of 4.4.

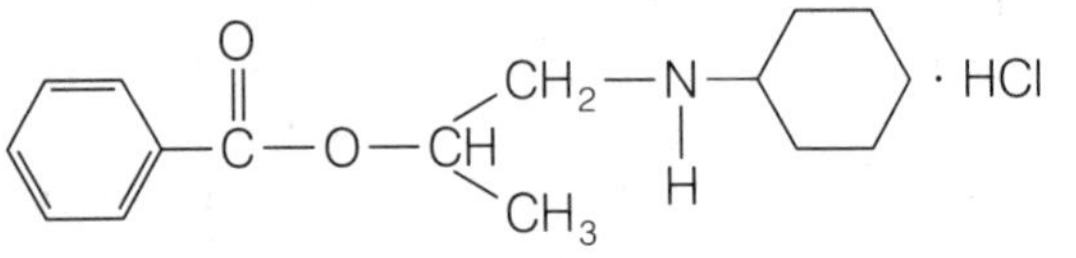

$$C_6H_5-\overset{O}{\overset{\|}{C}}-O-\underset{CH_3}{CH}-CH_2-\underset{H}{N}-C_6H_{11}\cdot HCl$$

Hexylcaine Hydrochloride

Hexylcaine has about the same toxicity as procaine; topical anesthesia (1% to 5%) is similar to cocaine and butacaine; for nerve-block anesthesia (1% to 2%), its toxicity is between that of butacaine and tetracaine. It has been employed for spinal anesthesia in a 1% to 2.5% solution containing 10% glucose.

Meprylcaine Hydrochloride, USP. 2-Methyl-2-(propylamino)-1-propanol benzoate (ester) hydrochloride (Oracaine Hydrochloride). The hydrochloride occurs as a white crystalline solid that is freely soluble in water, alcohol, or chloroform. A 2% solution has a *p*H of 5.7, and it has been reported that such a solution can be sterilized by autoclaving without decomposition.

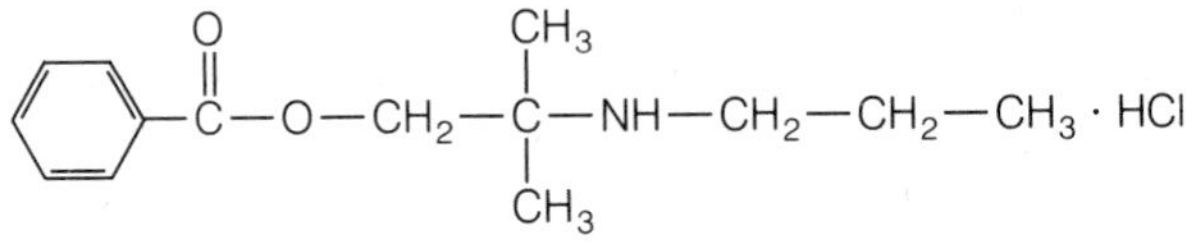

$$C_6H_5-\overset{O}{\overset{\|}{C}}-O-CH_2-\underset{CH_3}{\overset{CH_3}{C}}-NH-CH_2-CH_2-CH_3\cdot HCl$$

Meprylcaine Hydrochloride

Meprylcaine is used primarily in dentistry in a 2% solution containing epinephrine (1 : 50,000) as an infiltration and nerve-block anesthetic. It is substantially more potent and more rapidly hydrolyzed in human serum than procaine.

Isobucaine Hydrochloride, USP. 2-(Isobutylamino)-2-methyl-1-propanol benzoate (ester) hydrochloride. The salt is a white crystalline solid that is freely soluble in water, alcohol, or chloroform. The *p*H of a 2% solution is about 6.

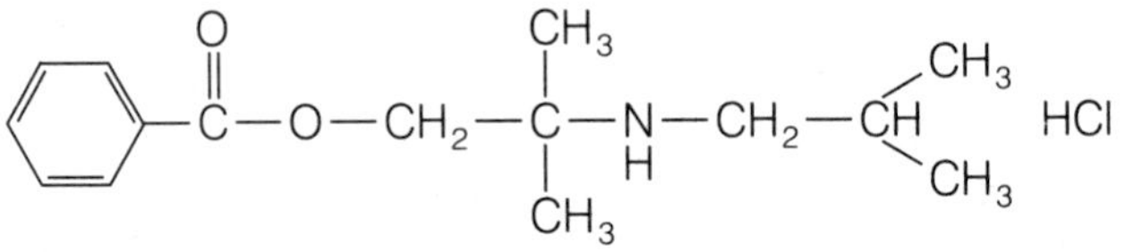

$$C_6H_5-\overset{O}{\overset{\|}{C}}-O-CH_2-\underset{CH_3}{\overset{CH_3}{C}}-\underset{H}{N}-CH_2-CH(CH_3)_2\quad HCl$$

Isobucaine Hydrochloride

Structurally, isobucaine differs from meprylcaine in having an *N*-isobutyl group instead of an *N*-propyl group. Similar to meprylcaine, it is more toxic and exhibits a shorter duration of action than does procaine. It is used primarily in dentistry in a 2% solution containing epinephrine (1 : 65,000) for infiltration and nerve block.

Cyclomethycaine Sulfate, USP. 3-(2-Methylpiperidino)propyl *p*-(cyclohexyloxy)benzoate sulfate

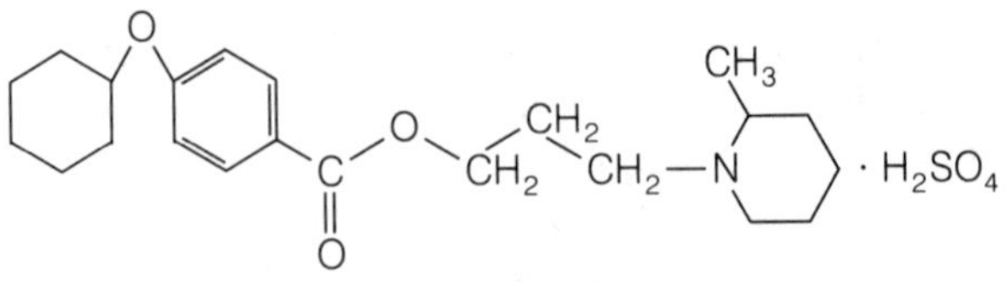

$$C_6H_{11}-O-C_6H_4-\overset{O}{\overset{\|}{C}}-O-CH_2-CH_2-CH_2-N(C_5H_9CH_3)\cdot H_2SO_4$$

Cyclomethycaine Sulfate

(1 : 1) (Surfacaine). The bisulfate occurs as a white crystalline powder and is sparingly soluble in water or alcohol and slightly soluble in chloroform.

Cyclomethycaine is an effective local anesthetic on damaged or diseased skin and on rectal mucous membrane. It is useful topically on burns (sunburn), abrasions, and mucosa of rectum and genitourinary tract, and is applied most commonly in 0.25% to 1% ointments, jellies, creams, or suppositories. It is not recommended for use on mucous membranes of the upper respiratory system or eye. As with all topical anesthetic agents, it carries a slight sensitizing potential.

Piperocaine Hydrochloride (Metycaine Hydrochloride). Piperocaine differs structurally from cyclomethycaine in that the 4-cyclohexyloxy group of cyclomethycaine is absent. It is soluble in water (1 : 1.5) and in alcohol (1 : 4.5). Aqueous solutions are faintly acid and are stable to sterilization by autoclaving. Piperocaine is recommended for application to the eye in 2% or 4% solutions; to the nose and throat in 2% to 10% solutions; for infiltration in 0.5% to 1.0% solutions; for nerve block in 0.5% to 2% solutions; and for spinal anesthesia in 1.5% solutions, with the maximal quantity of the drug limited to 1.65 mg/kg of body weight.

Structural relationships of the benzoic acid ester local anesthetics are shown in Table 15-1.

TABLE 15-1

BENZOIC ACID DERIVATIVES

$R_1-C_6H_4-C(=O)-O-R_2$

Generic Name	Proprietary Name	R_1	R_2
Cocaine		H	(tropane ring bearing $H_3CO-C(=O)-$ and $N-CH_3$)
Hexylcaine	Cyclaine	H	$-CH(CH_3)-CH_2-NH-C_6H_{11}$ (cyclohexyl)
Meprylcaine	Oracaine	H	$-CH_2C(CH_3)_2-NHCH_2CH_2CH_3$
Isobucaine		H	$-CH_2-C(CH_3)_2-NHCH_2CH(CH_3)_2$
Cyclomethycaine	Surfacaine	$C_6H_{11}-O-$ (cyclohexyloxy)	$-CH_2CH_2CH_2N$ (2-methylpiperidino, CH_3)
Piperocaine	Metycaine	H	$-CH_2CH_2CH_2N$ (2-methylpiperidino, CH_3)

AMINOBENZOIC ACID DERIVATIVES

Mixtures of sulfonamides and agents of this chemical class of local anesthetics should be avoided because of a potential pharmacologic incompatibility. A hydrolysis product of these agents is *p*-aminobenzoic acid (PABA). Sulfonamides are considered competitive inhibitors of the incorporation of PABA (see Chap. 5) in the biosynthesis of dihydrofolate.

Benzocaine, USP. Ethyl *p*-aminobenzoate (Americaine; Anesthesin). Benzocaine occurs as a white, odorless crystalline powder, stable in air. It is soluble in alcohol (1 : 5), ether (1 : 4), chloroform (1 : 2), glycerin, or propylene glycol, and is insoluble in water (1 : 2500).

$H_2N-C_6H_4-C(=O)-O-CH_2-CH_3$

Benzocaine

Unlike the other clinically employed local anesthetics, benzocaine and its homologues do not possess an aliphatic amino group that can be used for salt formation. The free electrons on the aromatic nitrogen are delocalized by the ring, and protonation at this position takes place less readily. Thus, the formation of water-soluble salts of benzocaine and its homologues is not possible, and it is unsuited for injection.

Benzocaine is sufficiently absorbed through abraded surfaces and mucous membranes to relieve the pain associated with ulcers, wounds, and inflamed mucous surfaces. It acts only as long as it is in contact with the skin or mucosal surface. It is used in ointment and cream preparations in concentrations up to 20% and is nonirritant and nontoxic.

Butamben, USP. Butyl *p*-aminobenzoate (Butesin). Butamben is the butyl ester analogue of benzocaine, but it is claimed to be more effective and more toxic. Butamben picrate, a complex thought to contain 2 moles of butamben per mole of picric acid, is a yellow amorphous powder. It is used in the eye as a saturated aqueous solution and for burns and denuded areas of the skin as a 1% ointment, **Butamben Picrate Ointment, USP.** A disadvantage is that the yellow stains the skin and clothing. The amyl ester is called Ultracain.

Procaine Hydrochloride, USP. 2-(Diethylamino)ethyl *p*-aminobenzoate monohydrochloride (Novocain). The odorless, white crystalline procaine hydrochloride is stable in air and soluble in water (1 : 1) and in alcohol (1 : 30). It was developed by Einhorn in 1906 and is one of the oldest and most used of the synthetic local anesthetics. It has a pK_a value of 8.9.

$$H_2N-C_6H_4-C(=O)-OCH_2CH_2N(C_2H_5)_2 \quad HCl$$

Procaine Hydrochloride

Procaine is most stable at *p*H 3.6 and becomes less stable as the *p*H is increased or decreased from this value.[14] Storage of buffered solutions at room temperature resulted in the following amounts of hydrolysis:

*p*H	Amount of Hydrolysis
3.7–3.8	0.5–1% in 1 yr
4.5–5.5	1.0–1.5% in 1 yr
7.5	1% in 1 day

Dosage forms are generally regarded as being satisfactory for use, as long as not more than 10% of the active ingredient has been lost, and there is no increase in toxicity. The following times for a 10% loss in potency at 20°C are based on kinetic studies of procaine solutions:

*p*H	Time in Days
3.6	2300
5.0	1200
7.0	7

It has also been calculated that the hydrolysis rate increases 3.1 times for each rise of 10° in the range of 20° to 70°C.

The following data show the effect on buffered 2% procaine hydrochloride solutions of autoclaving at 15 lb pressure for two hours.[15]

*p*H Before and After Autoclaving	Percent of Original Assay
2.4	97.5
2.6	97.9
2.8	98.1
3.0	98.4
3.2	98.5
3.4	98.5
3.6	98.3
3.8	98.2
4.0	97.8

Early studies suggested that neutral or slightly alkaline solutions of procaine had certain physiologic advantages over acid solutions in that there is less pain on injection, there is less tissue damage, and, most important, the rate of onset of anesthesia is more rapid and a smaller quantity of procaine is required.[16] However, neutral and alkaline solutions are very unstable and cannot be sterilized by autoclaving. The problem is even greater if epinephrine is to be used, because it is less stable than procaine in alkaline solutions.

The procaine molecule is a prototype of primary aromatic amines that are subject to oxidative decomposition. This decomposition can be controlled by nitrogen flushing of the solutions and by the addition of an antioxidant.

Another characteristic of the local anesthetics having a primary aromatic group is the reaction with glucose for the formation of procaine *N*-glucoside. There is no significant change in the clinical results, but there is the possibility of interference with the assay. Procaine also forms a soluble complex with sodium carboxymethyl cellulose.[17] However, sodium chloride displaces it. Again, there probably is no change in pharmacologic action, but there may be an analytical problem.

Procaine hydrochloride is not effective on intact skin or mucous membranes, but acts promptly when used by infiltration. The action may be prolonged by the concurrent administration of epinephrine or other vasoconstrictors that slow its release into the bloodstream, where procaine is rapidly inactivated by hydrolysis. Pseudocholinesterases are a group of plasma and tissue enzymes that catalyze the hydrolysis of esters, including procaine.

The toxicity of procaine is greatly decreased if absorption is restricted by vasoconstrictors, allowing the serum hydrolysis rate to exceed the rate of release from the tissues. Procaine induces a relatively high frequency of allergic reactions, and sensitive individuals usually react also to other aminobenzoates. For these patients, it is advisable to use an agent from a different chemical class. Unfortunately, tests for sensitivity are not reliable for predicting allergic responses.

Procaine is used to form an insoluble salt with penicillin G. The low solubility (7 : 1000) accounts for the prolonged action of this penicillin salt. Penicillin G is slowly released from the intramuscular depot.

Procainamide Hydrochloride, USP, the amide derivative of procaine, is a cardiac depressant (see Chap. 14).

Chloroprocaine Hydrochloride, USP. 2-(Diethylamino)ethyl 4-amino-2-chlorobenzoate monohydrochloride (Nesacaine). The salt, a white crys-

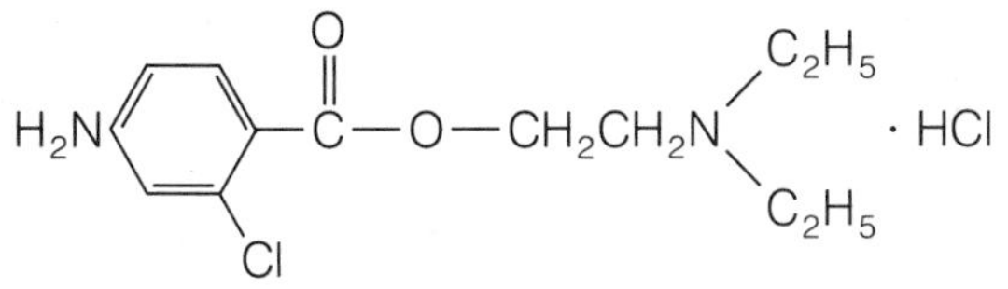

Chloroprocaine Hydrochloride

talline powder, is stable in air, and its solutions are acid to litmus. It is soluble in water (1 : 22) and slightly soluble in 95% alcohol (1 : 100). It has a pK_a value of 9.0.

Chloroprocaine differs structurally from procaine in having a chlorine substituent in the 2-position of the aromatic ring. The electron-withdrawing chlorine atom destabilizes the ester group to hydrolysis. Chloroprocaine is hydrolyzed by plasma more than four times faster than procaine. It is more rapid in onset of action and more potent than procaine.

Tetracaine, USP. 2-(Dimethylamino)ethyl *p*-(butylamino)benzoate (Pontocaine). Tetracaine, as the free base, occurs as a white to light yellow waxy solid that must be protected from light. It is very slightly soluble in water and soluble in alcohol and in lipid substances.

Of the ester derivatives of procaine, tetracaine is one of the most easily absorbed drugs. Absorption of tetracaine from mucous membranes is rapid and simulates intravenous injection closely,[13] whereas procaine is more slowly absorbed. The variation in absorption may be attributed to the difference in lipid solubility. The presence of a nonpolar *n*-butyl group on the aromatic nitrogen atom probably accounts for the greater lipid solubility of tetracaine. It is more slowly hydrolyzed in plasma than procaine. The free base is used to prepare ointments (0.5%) for topical and ophthalmic application.

Tetracaine Hydrochloride, USP, is a water-soluble salt (1 : 7) that occurs as a fine, white crystalline powder and is slightly bitter (pK_a 8.5). Its solutions are more stable to hydrolysis than are procaine hydrochloride solutions and they may be sterilized by boiling.

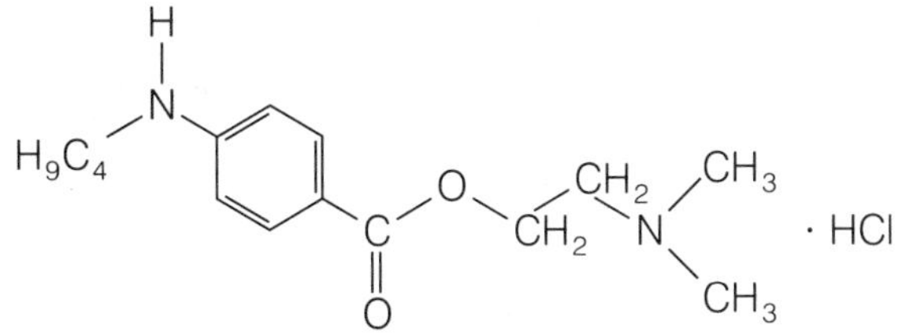

Tetracaine Hydrochloride

Butacaine, USP. 3-(Di-*n*-butylamino)-1-propanol *p*-aminobenzoate (Butyn) is a liquid. The sulfate salt occurs as a white crystalline powder that should be protected from light. It is freely soluble in water (1 : 1); the hydrochloride salt is less soluble. Structurally, butacaine differs from procaine in that the intermediate chain is increased to three carbons and the tertiary amino group has two *n*-butyl, instead of two ethyl, groups. Solutions of the salt may be sterilized by boiling and are effective on mucous membranes and in the eye. The rate of hydrolysis in plasma is comparable to that of tetracaine.

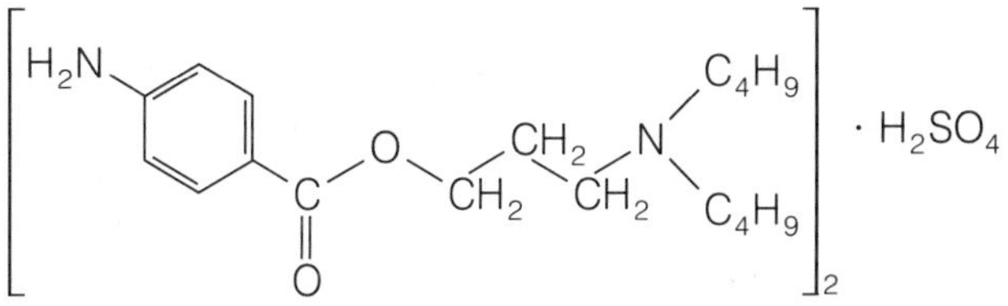

Butacaine Sulfate

The free base is used as a 4% ointment for the relief of pain associated with dental appliances.

Dosage form: Butacaine dental ointment, USP

Benoxinate Hydrochloride, USP. 2-(Diethylamino)ethyl 4-amino-3-*n*-butoxybenzoate hydrochloride (Dorsacaine). The salt occurs as a white crystalline powder and has a salty taste. It is very soluble in water and in chloroform, and it is soluble in alcohol. The powder is stable in air and not affected by light or heat. Aqueous solutions have a *p*H of 4.5 to 5.2.

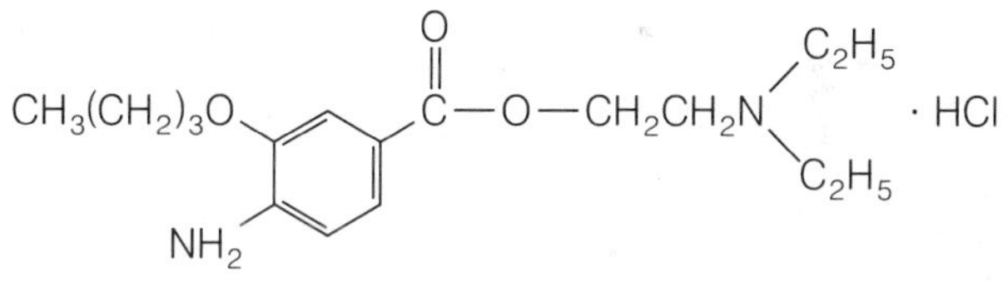

Benoxinate Hydrochloride

The chemical properties are similar to those of procaine, except that the 3-butoxy group appears to stabilize the molecule to hydrolysis. This is in marked contrast to the 2-chloroprocaine, which is less stable to hydrolysis than procaine. The rate of hydrolysis is *p*H-dependent, with the slowest rate at a *p*H of about 4.[18]

Benoxinate is employed primarily in ophthalmology as a 0.4% solution. Applications do not cause any significant irritation, constriction, or dilatation of the pupil, any noticeable light sensitivity, or symptoms indicating that absorption into the system has taken place.

Propoxycaine Hydrochloride, USP. 2-(Diethylamino)ethyl 4-amino-2-propoxybenzoate monohydrochloride (Blockain Hydrochloride). This salt is a white crystalline solid that is freely soluble in water and soluble in alcohol. The *p*H of a 2% solution is 5.4, which may be diluted and buffered at neutral *p*H without precipitation.

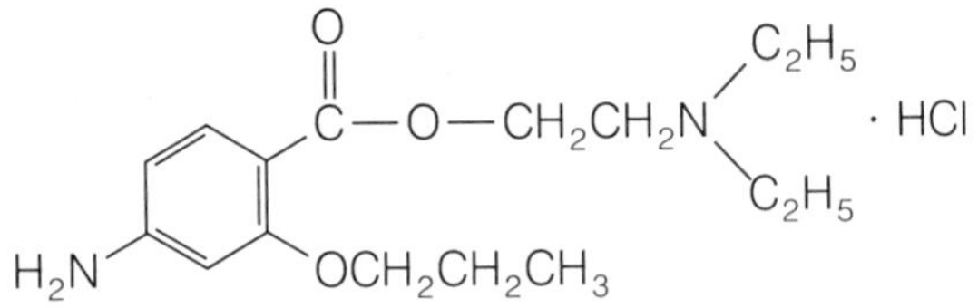

Propoxycaine Hydrochloride

Propoxycaine is not stable to autoclaving. The 2-propoxy group appears to labilize the ester group in much the same way as the 2-chloro group does in chloroprocaine. This is in contrast with the apparent stabilizing effect of a butoxy group placed in the 3-position (benoxinate). The inductive and steric effects of a 2-alkoxy substituent, which favor hydrolysis, apparently have a greater influence than the stabilizing positive mesomeric (resonance) effect. These effects are minimal when the alkoxy substituent is in the 3-position.

Propoxycaine has a quicker onset and longer duration of action and is considerably more potent than procaine. Its greater lipophilic character probably accounts for these advantages. It is administered by injection for nerve block and infiltration anesthesia.

Proparacaine Hydrochloride, USP. 2-(Diethyamino)ethyl 3-amino-4-propoxybenzoate monohydrochloride (Alcaine; Ophthaine; Ophthetic). The salt is a white to off-white crystalline powder that is soluble in water (1 : 30) and in warm alcohol (1 : 30). Solutions are neutral to litmus and discolor in the presence of air. Discolored solutions should not be used.

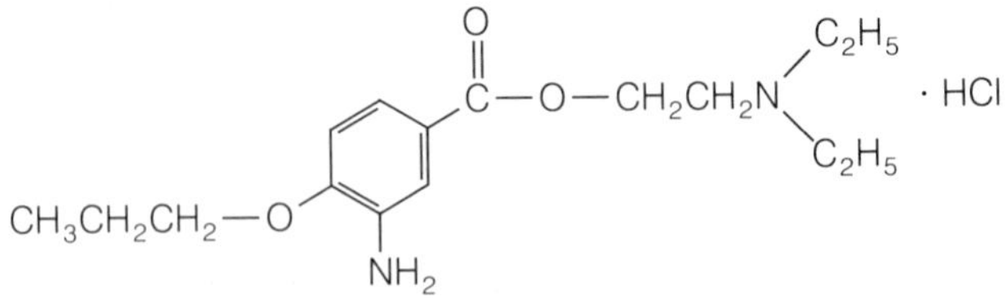

Proparacaine Hydrochloride

Proparacaine is a positional isomer of propoxycaine. Proparacaine is less soluble, slightly more potent, and more toxic than propoxycaine. Although it is too toxic for use as an injection anesthetic, proparacaine is suitable for use topically in ophthalmology.

Table 15-2 shows the structural relationships of the local anesthetics that are esters of *p*-aminobenzoic acid.

LIDOCAINE DERIVATIVES (ANILIDES)

As a chemical class of local anesthetics, the currently used anilides were derived from lidocaine. Agents of this class are more stable to hydrolysis, even under the conditions of prolonged autoclaving. They are more potent, have a lower frequency of side effects, and induce less local irritation than the procaine-type local anesthetic agents. Also, there seems to be no cross-sensitization between the anilides and the benzoic acid derivatives. Thus, the anilides are valuable alternatives for those individuals sensitive to the procaine-type local anesthetics.

TABLE 15-2

***p*-AMINOBENZOIC ACID DERIVATIVES**

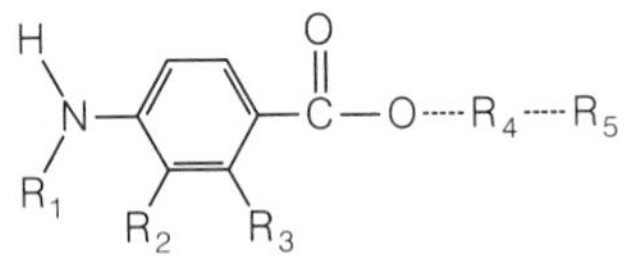

Generic Name	*Proprietary Name*	R_1	R_2	R_3	R_4	R_5
Benzocaine	Americaine	H	H	H	$-C_2H_5$	
Butamben	Butesin	H	H	H	$-CH_2CH_2CH_2CH_3$	
Procaine	Novocain	H	H	H	$-CH_2CH_2-$	$-N(C_2H_5)_2$
Chloroprocaine	Nesacaine	H	H	Cl	$-CH_2CH_2-$	$-N(C_2H_5)_2$
Tetracaine	Pontocaine	Bu	H	H	$-CH_2CH_2-$	$-N(CH_3)_2$
Butacaine	Butyn	H	H	H	$-CH_2CH_2CH_2-$	$-N(n\text{-}C_4H_9)_2$
Benoxinate	Dorsacaine	H	BuO —	H	$-CH_2CH_2-$	$-N(C_2H_5)_2$
Propoxycaine	Blockain	H	H	PrO —	$-CH_2CH_2-$	$-N(C_2H_5)_2$

TABLE 15-3

ANILIDES

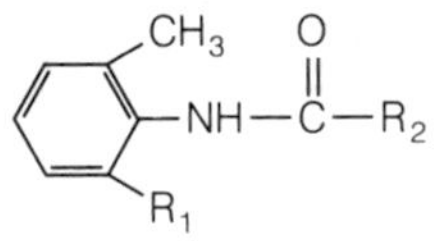

Generic Name	Proprietary Name	R_1	R_2
Lidocaine	Xylocaine	CH_3	$-CH_2-N(C_2H_5)_2$
Mepivacaine	Carbocaine	CH_3	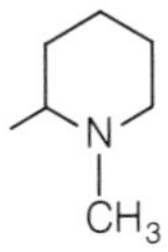 1-methyl-2-piperidyl ($N-CH_3$)
Prilocaine	Citanest	H	$-CH(CH_3)-NH-CH_2CH_2CH_3$
Bupivacaine	Marcaine	CH_3	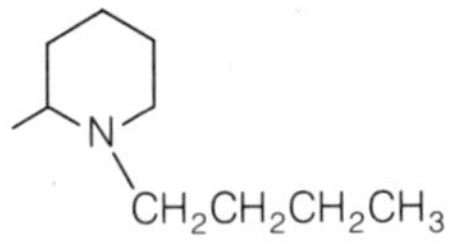 1-butyl-2-piperidyl ($N-CH_2CH_2CH_2CH_3$)
Etidocaine	Duranest	CH_3	$-CH(C_2H_5)-N(C_2H_5)CH_2CH_2CH_3$

Additionally, the anilides are effective with or without vasoconstrictors and are agents of choice in those instances when the dentist or patient may be sensitive to epinephrine and its congeners.

In contrast to the esters of benzoic acid derivatives, the anilides undergo enzymatic degradation, not in plasma, but primarily in the liver. The major route of metabolism of the anilides is enzymatic hydrolysis of the amide bond.

The difference in site of metabolism is of clinical relevance. Patients possessing a genetic deficiency in the enzyme pseudocholinesterase are unable to hydrolyze ester-type local anesthetic agents at a normal rate and show reduced tolerance to these drugs. For these patients, an anilide agent may be preferable.

On the other hand, the anilides are metabolized mainly by hepatic microsomal enzymes, and patients with liver disease may show reduced tolerance to this type of agent. In such cases, the benzoic acid derivatives provide suitable alternative agents.[5]

Structural relationships of the anilide-type local anesthetics are shown in Table 15-3.

Lidocaine, USP. 2-(Diethylamino)-*N*-(2,6-dimethylphenyl)acetamide; 2-(diethylamino)-2′,6′-acetoxylidide (Xylocaine). This compound may be considered the anilide of 2,6-dimethylaniline and *N*-diethylglycine. It occurs as a white to slightly yellow crystalline powder that has a characteristic odor and is stable in air. It is practically insoluble in water and very soluble in alcohol and lipid materials.

The hydrochloride salt (Lidocaine Hydrochloride, USP) occurs as a white, odorless crystalline powder that is very soluble in water and in alcohol. The salt has a pK_a of 7.86.

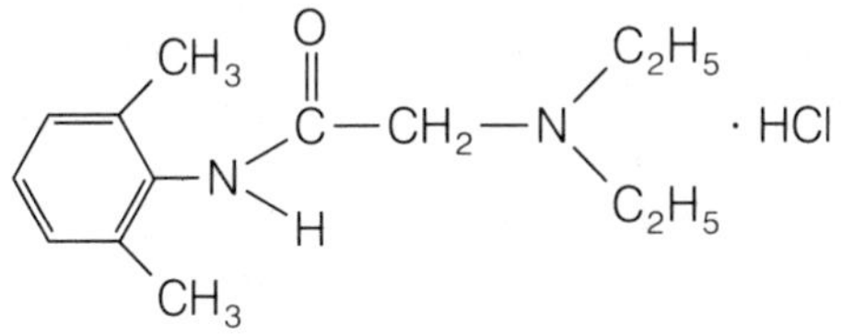

Lidocaine Hydrochloride

As opposed to ester-type local anesthetics, lidocaine is extremely resistant to hydrolysis. This is not unexpected because, in addition to the relative stability of the amide bond, the 2,6-dimethyl substituents provide steric hindrance to attack of the carbonyl. Solutions of the drug may be autoclaved and may be stored for months at room temperature without significant decomposition.

Lidocaine is about twice as potent as procaine and approximately 1.5 times as toxic. Systemic side reactions and local irritant effects are few. Lidocaine appears to be relatively free of sensitizing reactions characteristic of the *p*-aminobenzoates. Because there seems to be no cross-sensitization between the anilide and the benzoic acid types of local anesthetics, lidocaine is an agent of choice in individuals sensitive to procaine.

Also, lidocaine is effective with or without a vasoconstrictor. Hence, it is a valuable alternative for use in those individuals who are allergic to epinephrine and its congeners. The free base and its salts are effective topical local anesthetics. The hydrochloride is used for infiltration, peripheral nerve block, and epidural anesthesia.

Lidocaine is also an effective cardiac depressant and may be administered intravenously in cardiac surgery and life-threatening arrhythmias (see Chap. 14).

Mepivacaine Hydrochloride, USP. (±)-*N*-(2,6-Dimethylphenyl)-1-methyl-2-piperidinecarboxamide hydrochloride (Carbocaine). The salt occurs as a

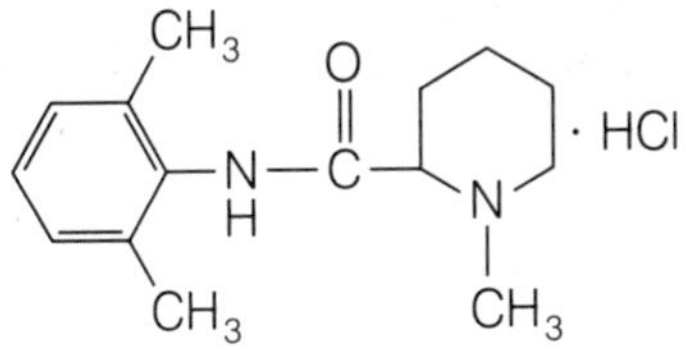

Mepivacaine Hydrochloride

white crystalline solid that is freely soluble in water. The pH of a 2% solution is about 4.5. It has a pK_a value of 7.69.

Similar to lidocaine and the other anilides, mepivacaine is highly resistant to hydrolysis, and its solutions can be autoclaved without appreciable decomposition.

Mepivacaine is used as the racemic mixture, because the two optical isomers were found[19] to have essentially the same toxicity and potency; the toxicity and potency of either are comparable with that of lidocaine. Its duration of action is considerably longer than that of lidocaine, even without a vasoconstrictor. Thus, mepivacaine hydrochloride is particularly useful when epinephrine and its congeners are contraindicated.

The effectiveness of mepivacine as a topical local anesthetic has not been established.

Prilocaine Hydrochloride, USP. *N*-(2-Methylphenyl)-2-(propylamino)propanamide hydrochloride (Citanest). The salt occurs as a bitter, white crystalline powder. It is freely soluble in water and in alcohol. It has a pK_a value of 7.9.

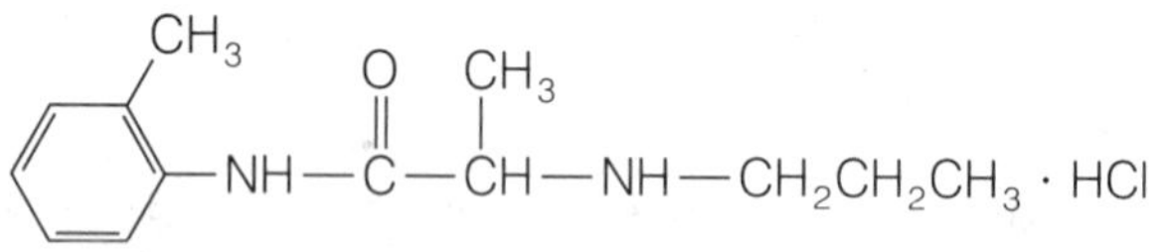

Prilocaine Hydrochloride

Prilocaine (originally named propitocaine) is similar in stability, potency, toxicity, and duration of action to the other anilides. Its duration of action is intermediate between that of lidocaine and mepivacaine. Except for methemoglobinemia, its side effects are similar to those observed with other anilides.

A primary metabolite of prilocaine is *o*-toluidine (2-methylaniline), which apparently is metabolized further to products that cause methemoglobinemia. Metabolic products of aniline (i.e., phenylhydroxylamine and nitrosobenzene) are known to produce methemoglobinemia.[20] Apparently, 2,6-dimethylaniline (2,6-xylidine), the hydrolysis product of lidocaine and certain other anilides, is resistant to *N*-oxidation, and use of these anilides does not result in methemoglobinemia. This side effect is not observed with other local anesthetic agents when used in normal dose ranges. However, when procaine was given intravenously in large doses to produce general anesthesia, methemoglobinemia was observed. It was suggested that aniline was formed by decarboxylation of PABA, a hydrolysis product of procaine.[21]

Prilocaine hydrochloride solutions can be used with or without the addition of vasoconstrictors; therefore, they are useful when such agents are contraindicated.

Bupivacaine Hydrochloride, USP (Marcaine). This local anesthetic differs from mepivacaine only in that the basic nitrogen contains a butyl instead of a methyl group. Its properties are very similar to those of mepivacaine. It has a pK_a value of 8.1.

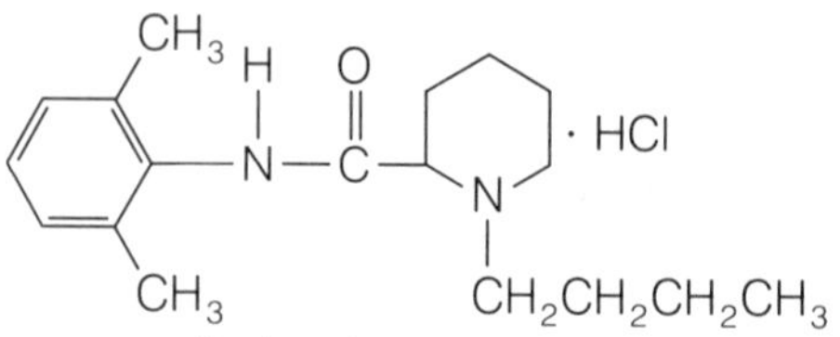

Bupivacaine Hydrochloride

The duration of action of bupivacaine is two to three times longer than that of lidocaine or mepivacaine and 20% to 30% longer than that of tetracaine. Its potency is comparable with that of tetracaine, but is about four times that of mepivacaine and lidocaine.

Etidocaine Hydrochloride, USP (Duranest) was clinically introduced in 1972. Structurally, it is closely related to lidocaine. Its physicochemical properties and pharmacologic actions are similar to those of lidocaine, but it possesses greater anesthetic potency and a longer duration of action.

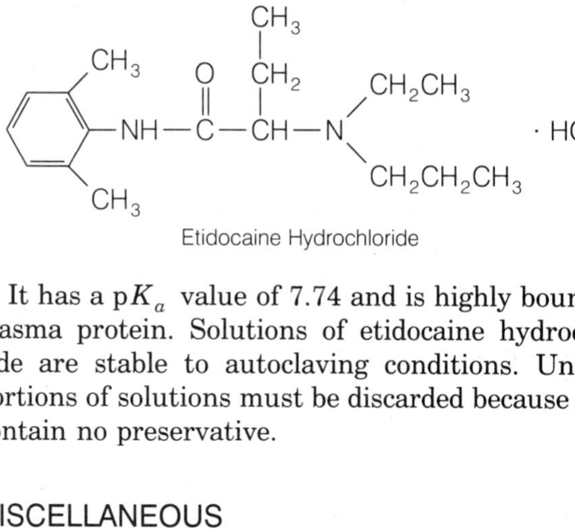

Etidocaine Hydrochloride

It has a pK_a value of 7.74 and is highly bound to plasma protein. Solutions of etidocaine hydrochloride are stable to autoclaving conditions. Unused portions of solutions must be discarded because they contain no preservative.

MISCELLANEOUS

Phenacaine Hydrochloride, USP. *N,N'*-Bis(4-ethoxyphenyl)acetamide monohydrochloride monohydrate (Holocaine Hydrochloride). The salt occurs as slightly bitter, white crystals that produce a numbness of the tongue. It is stable in air and

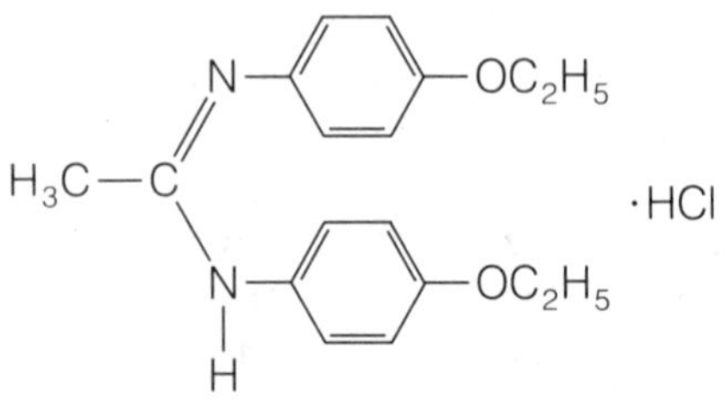

Phenacaine Hydrochloride

sparingly soluble in water (1 : 50), but freely soluble in alcohol. Aqueous solutions may be sterilized by boiling, but they are very sensitive to alkalis. The alkalinity of a soft glass container may be sufficient to cause precipitation of the free base from solution.

Phenacaine is structurally related to the anilides in that an aromatic ring is attached to an sp^2 carbon through a nitrogen bridge. Its local anesthetic properties were reported in 1897; this preceded the discovery of the action of isogramine. However, the appearance of this unique structural feature appears to be an isolated incident and did not exert a significant influence on the subsequent evolution of local anesthetic development until the discovery of isogramine.

Another isolated historical incident apparently had no effect on the evolution of the anilide-type local anesthetic agents. The anilide 5-(diethylaminoacetylamino)-2-hydroxybenzoic acid methyl ester (Nirvanine) was prepared as a water-soluble (salt) derivative of orthoform. It failed to stir the interest of clinicians because it was less active and more irritating than cocaine.

Phenacaine hydrochloride is slightly irritating, causing some discomfort to precede anesthesia. It is more toxic than cocaine and cannot be used for injection. However, it is fast-acting and very effective on mucous membranes. Thus, because of toxicity, it is used primarily in ophthalmology in 1% solutions or 1% to 2% ointments.

Diperodon, USP. 3-Piperidino-1,2-propanediol dicarbanilate monohydrate (Diothane). Diperodon occurs as a white to cream-colored powder having a characteristic odor. The free base is insoluble in water and relatively unstable to heat, but aqueous solutions of the hydrochloride salt at *p*H 4.5 to 4.7 are stable and may be autoclaved. Structurally, it is related to the anilides in that an aromatic ring is attached to an sp^2 carbon by a nitrogen bridge. Thus, a hydrolysis product is aniline, and methemoglobinemia is a potential toxic side effect.

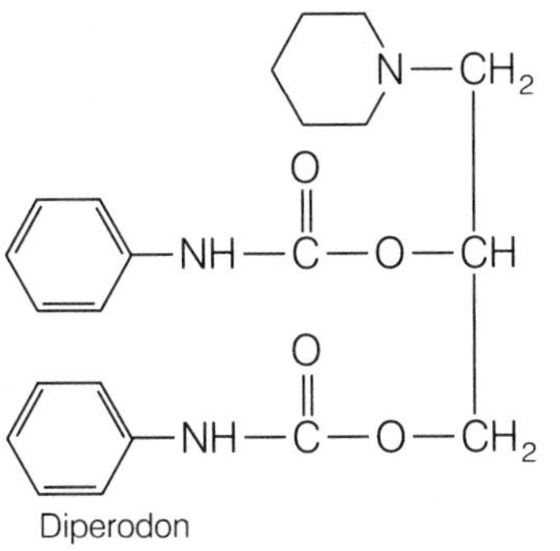

Diperodon

After intravenous injection (as the hydrochloride salt), diperodon possesses toxicity comparable with cocaine. It is recommended for the relief of pain and irritation in abrasions of the skin and mucous membranes, especially hemorrhoids.

Dimethisoquin Hydrochloride. 3-Butyl-1-[2-(dimethylamino)ethoxy]isoquinoline monohydrochloride (Quotane). The salt occurs as a white to off-white crystalline powder that is soluble in water (1 : 8), in alcohol (1 : 3), and in chloroform (1 : 2).

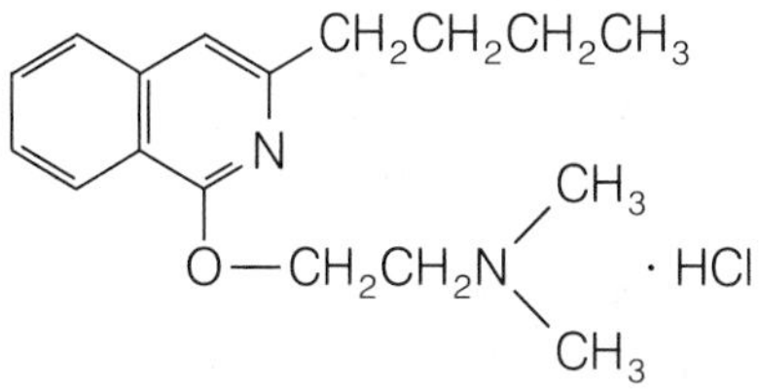

Dimethisoquin Hydrochloride

Structurally, dimethisoquin may be considered to be related to the benzoic acid derivatives in that the phenyl ring is attached directly to an sp^2 carbon atom and the basic side chain is attached by an oxygen bridge.

Dimethisoquin is a safe, effective compound for general application as a topical anesthetic. It is available as a water-soluble ointment (0.5%) for dry dermatologic conditions and in lotion form for moist skin surfaces.

Pramoxine Hydrochloride, USP. 4-[3-(4-Butoxyphenoxy)propyl]morpholine hydrochloride (Tronothane). The salt occurs as a white to off-white crystalline solid that may have a slight aromatic odor. It is freely soluble in water or alcohol. The *p*H of a 1% solution is about 4.5.

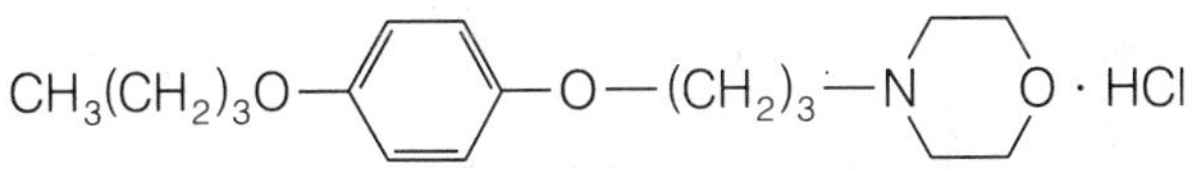

Pramoxine Hydrochloride

Pramoxine is too irritating for ophthalmic use, but is an effective topical local anesthetic of a low-sensitizing index with few toxic reactions. It is used for relief of pain and itching due to insect bites, minor wounds and lesions, and hemorrhoids.

Dyclonine Hydrochloride, USP. 4′-Butoxy-3-piperidinopropiophenone hydrochloride (Dyclone). The salt occurs as a white crystalline powder that may have a slight odor. It is soluble in water and in alcohol and stable in acidic solutions, if not autoclaved.

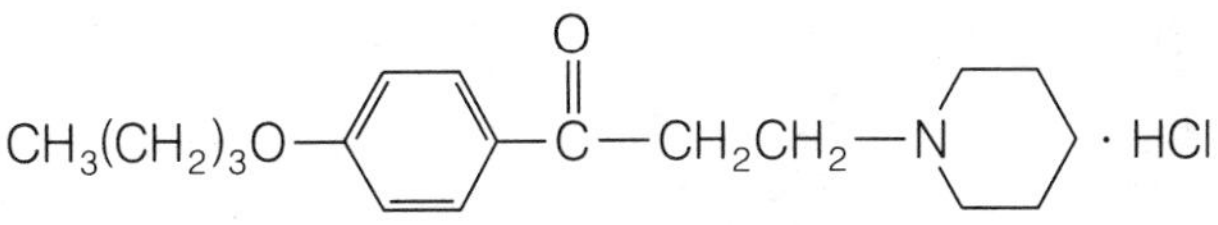

Dyclonine Hydrochloride

Dyclonine is a tissue irritant if injected, but it is an effective topical anesthetic for use on skin and mucous membranes. It is effective for anesthetizing mucous membranes before endoscopic and proctoscopic procedures. Anesthesia usually occurs in five to ten minutes after application and persists for 20 minutes to one hour. It has been approved for over-the-counter use in a rinse, mouthwash, gargle, spray, and lozenge.

Dibucaine, USP. 2-Butoxy-*N*-[2-(diethylamino)-ethyl]cinchoninamide (Nupercaine). The free base occurs as a white to off-white powder having a slight, characteristic odor. Chemically, the drug is an amide that darkens on exposure to light; it is slightly soluble in water and soluble in chloroform and in ether.

The less-soluble free base is used to prepare cream, ointment, and rectal suppository dosage forms for topical application. The more-soluble hydrochloride salt (1 : 2 in water) is suitable for the preparation of the injectable dosage form [Dibucaine Hydrochloride, USP (Nupercaine Hydrochloride)]. Aqueous solutions of the salt may be sterilized by autoclaving, but the pH should not be above 6.2. The pK_a value of the hydrochloride is 8.15.

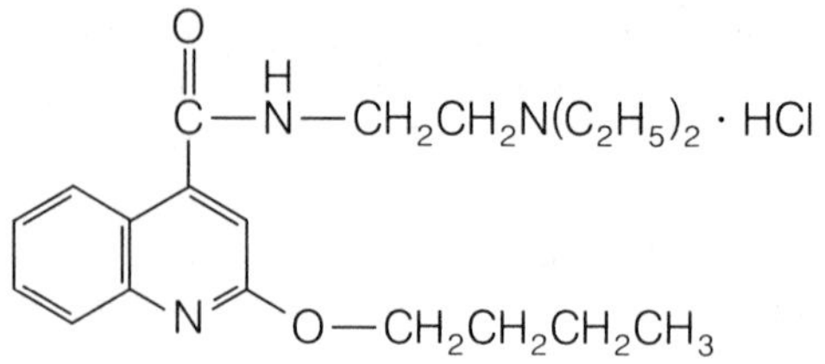

Dibucaine Hydrochloride

Because dibucaine is an amide, it is not hydrolyzed to any measurable extent in serum. It has been shown[13] to be metabolized rather slowly, which explains its high toxicity. It is the most potent, most toxic, and longest-acting of the commonly employed local anesthetics.

The official titles of the local anesthetics listed in the *United States Pharmacopeia* are designated simply as USP in Table 15-4; the proprietary name(s), official preparations, application, usual dose, and usual dose range are also given for some of the agents.

TABLE 15-4

LOCAL ANESTHETICS

Name Proprietary Name	***Preparations***	***Application***	***Usual Adult Dose****	***Usual Dose Range****
Cocaine hydrochloride USP		Topical, 2–20% solution to mucous membranes		
Hexylcaine hydrochloride USP *Cyclaine*	Hexylcaine hydrochloride topical solution USP		Topical	
Meprylcaine hydrochloride USP *Oracaine*	Meprylcaine hydrochloride and epinephrine injection USP	Dental		
Isobucaine hydrochloride USP	Isobucaine hydrochloride and epinephrine injection USP	Dental		
Cyclomethycaine sulfate USP *Surfacaine*	Cyclomethycaine sulfate cream USP Cyclomethycaine sulfate jelly USP Cyclomethycaine sulfate ointment USP	Topical, 0.5% cream, 0.75% jelly, or 1% ointment to the skin	Rectal, 10 mg	
Dibucaine USP *Nupercaine*	Dibucaine cream USP Dibucaine ointment USP	Topical, 0.5% cream or 1% ointment several times daily		
Dibucaine hydrochloride USP *Nupercaine Hydrochloride*	Dibucaine hydrochloride injection USP			
Benzocaine USP	Benzocaine cream USP Benzocaine ointment USP Benzocaine lotion USP	Topical, 1–20% aerosol, cream, or ointment to the skin		
Butamben USP *Butesin*	Butamben picrate ointment USP	Topical		
Procaine hydrochloride USP *Novocain*	Procaine hydrochloride injection USP			
Chloroprocaine hydrochloride USP *Nesacaine*	Chloroprocaine hydrochloride injection USP	Parenterally		

*See USP DI for complete dosage information.

TABLE 15-4 *Continued*

LOCAL ANESTHETICS

Name Proprietary Name	***Preparations***	***Application***	***Usual Adult Dose****	***Usual Dose Range****
Tetracaine USP *Pontocaine*	Tetracaine ointment USP Tetracaine ophthalmic ointment USP	Rectal, topical, 0.5% ointment to the conjunctiva to induce transient loss of corneal sensitivity		
Tetracaine hydrochloride USP *Pontocaine Hydrochloride*	Tetracaine hydrochloride injection USP Sterile tetracaine hydrochloride USP			
	Tetracaine hydrochloride ophthalmic solution USP	Topically to the conjunctiva, 0.05–0.1 mL of a 0.5 or 1% solution		
	Tetracaine hydrochloride topical solution USP	Topically to the nose and throat 1–2 mL of a 0.25–2% solution		
	Tetracaine hydrochloride cream USP	Topical, cream containing the equivalent of 1% tetracaine		
Butacaine USP *Butyn*	Butacaine dental ointment USP	Topical, 4% ointment		
Benoxinate hydrochloride USP *Dorsacaine*	Benoxinate hydrochloride ophthalmic solution USP	Topical, to the conjunctiva, 50–200 μL of a 0.4% solution		
Propoxycaine hydrochloride USP *Blockain*	Propoxycaine and procaine hydrochlorides and levonordefrin injection USP Propoxycaine and procaine hydrochlorides and norepinephrine bitartrate injection USP			
Proparacaine hydrochloride USP *Ophthaine*	Proparacaine hydrochloride ophthalmic solution USP	Topically to the conjunctiva		
Lidocaine USP *Xylocaine*	Lidocaine cream USP Lidocaine topical aerosol solution USP Lidocaine ointment USP Lidocaine topical solution USP	Topically to mucous membranes, as a 3–5% ointment, cream, solution, or spray		
Lidocaine hydrochloride USP *Xylocaine Hydrochloride*	Lidocaine hydrochloride injection USP Lidocaine hydrochloride and epinephrine injection USP			
	Lidocaine hydrochloride jelly USP Lidocaine hydrochloride cream USP	Topically to mucous membranes, as a 2 or 3% jelly cream, or solution		
Mepivacaine hydrochloride USP *Carbocaine*	Mepivacaine hydrochloride injection USP Mepivacaine hydrochloride and levonordefrin injection USP			
Prilocaine hydrochloride USP *Citanest*	Prilocaine hydrochloride injection USP			
Bupivacaine hydrochloride USP *Marcaine*	Bupivacaine hydrochloride injection USP Bupivacaine and epinephrine injection USP			

*See USP DI for complete dosage information.

TABLE 15-4 *Continued*

LOCAL ANESTHETICS

Name Proprietary Name	*Preparations*	*Application*	*Usual Adult Dose**	*Usual Dose Range**
Etidocaine hydrochloride USP *Duranest*	Etidocaine hydrochloride injection USP			
Diperodon USP *Diothane*	Diperodon ointment USP	Topical or intrarectal, a 1% ointment tid or qid		
Phenacaine hydrochloride USP		To the conjunctiva, 1 or 2% ointment or a 1% solution		
Dimethisoquin hydrochloride *Quotane*	Dimethisoquin hydrochloride lotion USP Dimethisoquin hydrochloride ointment USP	Topical, to the skin, 0.5% lotion or ointment bid to qid		
Pramoxine hydrochloride USP *Tronothane*	Pramoxine hydrochloride cream USP Pramoxine hydrochloride jelly USP	Topical, 1% cream or jelly every 3–4 hr		
Dyclonine hydrochloride USP *Dyclone*	Dyclonine hydrochloride topical solution USP	Topical, to mucous membranes, 0.5–1% solution		

*See USP DI for complete dosage information.

REFERENCES

1. Ritchie, J. M., and Greene, N. M.: Local anesthetics. In Gilman, A. G., Goodman, L. S., Rall, T. W., and Murad, F. (eds.). Goodman and Gilman's The Pharmacological Basis of Therapeutics, 7th ed., p. 302. New York, Macmillan, 1985.
2. Liljestrand, G.: The historical development of local anesthetics. In Lechat, P. (ed.). Local Anesthetics; International Encyclopedia of Pharmacology and Therapeutics, Sect. 8, Vol. 1, p. 1. New York, Pergamon Press, 1971.
3. Claus, E. P., Tyler, V. E., and Brady, L. R.: Pharmacognosy, 6th ed., p. 238. Philadelphia, Lea & Febiger, 1970.
4. Ray, O. S.: Drugs, Society, and Human Behavior, p. 18. Saint Louis, C. V. Mosby, 1972.
5. Covino, B. G., and Vassallo, H. G.: Local Anesthetics, Mechanisms of Action and Clinical Use, p. 1. New York, Grune & Stratton, 1976.
6. Brinley, F. J., and Mullins, L. J.: Ann. N.Y. Acad. Sci. 242:406, 1974.
7. Askari, A., (ed.): Properties and functions of ($Na^+ + K^+$)-activated adenosinetriphosphatase. Ann. N.Y. Acad. Sci. 242:1–741, 1974.
8. Ritchie, J. M.: The mechanism of action of local anesthetic agents. In Lechat, P. (ed.). Local Anesthetics; International Encyclopedia of Pharmacology and Therapeutics, Sect. 8, Vol. 1, p. 131. New York, Pergamon Press, 1971.
9. Takman, B.: Br. J. Anaesth. 47(Suppl.):183, 1975.
10. Büchi, J., and Perlia, X.: Structure–activity relations and physicochemical properties of local anesthetics. In Lechat, P. (ed.). Local Anesthetics; International Encyclopedia of Pharmacology and Therapeutics, Sect. 8, Vol. 1, p. 39. New York, Pergamon Press, 1971.
11. Büchi, J., and Perlia, X.: Design of local anesthetics. In Ariens, E. J. (ed.). Drug Design, Vol. 11-III, p. 243. New York, Academic Press, 1972.
12. Cohen, S.: JAMA 231:74, 1975.
13. Hansson, E.: Absorption, distribution, metabolism and excretion of local anesthetics. In Lechat, P. (ed.). Local Anesthetics; International Encyclopedia of Pharmacology and Therapeutics, Sect. 8, Vol. 1, p. 239. New York, Pergamon Press, 1971.
14. Terp, P.: Acta Pharmacol. Toxicol. 5:353 1949, through Chem. Abstr. 44:6576c, 1950.
15. Bullock, K., and Cannell, J. S.: Q. J. Pharm. Pharmacol. 14:241, 1941.
16. Bullock, K.: Q. J. Pharm. Pharmacol. 11:407, 1938.
17. Kennon, L., and Higuchi, T.: J. Am. Pharm. Assoc. Sci. Ed. 45:157, 1956.
18. Willi, A. V.: Pharm. Acta Helvet. 33:635, 1958.
19. Sadove, M., and Wessinger, G. D.: J. Int. Coll. Surg. 34:573, 1960.
20. Goldstein, A., Aronow, L., and Kalman, S. M.: Principles of Drug Action, p. 208. New York, Harper & Row, 1968.
21. Luduena, F. P.: Toxicity and irritancy of local anesthetics. In Lechat, P. (ed.). Local Anesthetics; International Encyclopedia of Pharmacology and Therapeutics, Sect. 8, Vol. 1, p. 319. New York, Pergamon Press, 1971.
22. Gutmann, G.: Deut. Med. Wochschr. 23:165, 1897.

SELECTED READINGS

Adriani, J., and Maraghi, M.: The pharmacologic principles of regional pain relief. Annu. Rev. Pharmacol. Toxicol. 17:223, 1977.

Büchi, J., and Perlia, X.: The design of local anesthetics. In Ariens, E. J. (ed.). Drug Design, Vol. 11-III, p. 243. New York, Academic Press, 1972.

Covino, B. G., and Vassallo, H. G.: Local Anesthetics, Mechanisms of Action and Clinical Use. New York, Grune & Stratton, 1976.

Lechat, P. (ed.): Local Anesthetics; International Encyclopedia of Pharmacology and Therapeutics. New York, Pergamon Press, 1971.

Lofgren, N.: Studies on Local Anesthetics. Stockholm, University of Stockholm, 1948.

Ritchie, J. M., and Cohen, P. J.: Local anesthetics. In Goodman, L. S., and Gilman, A. (eds.). The Pharmacological Basis of Therapeutics, 5th ed., p. 379. New York, Macmillan, 1975.

Takman, B. H., and Adams, H. J.: Local anesthetics. In Wolff, M. E. (ed.). Burger's Medicinal Chemistry, 4th ed., Part 3, p. 645. New York, John Wiley & Sons, 1981.

CHAPTER 16

Histamine and Antihistaminic Agents

Charles M. Darling

HISTAMINE

Histamine (β-imidazolylethylamine or 1*H*-imidazole-4-ethanamine) was synthesized[1] in 1907 before its presence in tissues was recognized. It is widespread in nature, being found in ergot and other plants and in all organs and tissues of the human body. Its physiologic importance is underscored by its classification as an autacoid, a word derived from the Greek *autos* (self) and *akos* (medicinal agent or remedy).

Noting species differences, Dale and Laidlaw[2–4] outlined the principal pharmacologic activities of histamine from 1910 to 1919. After more than half a century of painstaking study, several physiologic roles have been proposed for it.[5]

Histamine probably plays a basic role in the beginning of the inflammatory response of tissue to injury by dilating the capillaries and increasing their permeability. The usefulness of this effect, which usually leads to a "walling off" of the area of injury, may be counteracted by the effect of excess histamine released during the process. A conclusion was stated[6] that the inflammatory process would possibly develop faster and more favorably to the subject if an excess of histamine were not released.

The role of histamine in immediate hypersensitivity (antigen–antibody) reactions is complicated by release of other mediators[7] during such reactions. However, sufficient data[5] has been collected to suggest that histamine inhibits its own release and augments eosinophil chemotaxis.

Histamine exerts a variety of actions on the cardiovascular system.[8] Although the evidence is indirect, histamine continues to be implicated in the regulation of microcirculation. The available quantitative evidence indicates the presence of at least two distinct types of histamine receptors.

The increased synthesis of histamine in many tissues undergoing rapid growth or repair has been construed as evidence that "nascent" histamine may play a role in anabolic processes.[9] Histamine is distributed unevenly in the brain, as are other biogenic amines. Evidence on histamine as a central neurotransmitter has been reviewed.[6,10] The role of histamine in gastric secretion has been reviewed[11] and will be discussed in a later section. Other exocrine glands remain to be investigated for their quantitative responses to histamine.

In general, histamine appears to be necessary for many physiologic processes. Possibly, histamine has a homeostatic role and the histamine-forming capacity of various tissues is responsive to alterations in the concentration of the amine. However, certain conditions (e.g., shock) may lead to excessive production or release of the autacoid resulting in undesirable effects. Less drastic actions of histamine are implicated in other diseases, such as peptic ulceration and asthmatic conditions. Therefore, these and other actions of histamine have been studied for the purpose of developing drugs that will block selectively certain actions of histamine.

A review[12] of the reported evidence clearly indicates that there exists in the body and fluids of mammals one or more substances with antihistaminic activity. Natural antihistamine substance (NAS) appears to be highly potent, but neither the mechanism of action nor the chemical structure is known.

HISTAMINE RECEPTORS

The discovery of H_2-antagonists has confirmed that more than one type of histamine receptor exists in biologic systems. These antagonists have been shown to block gastric acid secretion effectively. The stage is set not only to revolutionize the treatment of peptic ulcers, but also for the birth of a new era of histamine antagonists that may elucidate the role of histamine in neurotransmission, microcirculatory control, inflammation, allergic disorders, and other phenomena. The setting of the stage began in the early years of this century.

Before the identification of histamine in the body, the testing of simple extracts from most organs and tissues of the body led to the use of such unspecific terms as *depressor substance*, *vasodilation*, *anaphylatoxin*, *H-substance*, and others. Although histamine had been identified chemically in tissue extracts, many were skeptical that histamine was a natural constituent of the body until Best and coworkers established its presence beyond doubt in 1927.[13]

Unlike anticholinergic or antiadrenergic agents, some of which are found in nature, there were no potent antihistaminic agents to be found in nature to serve as a model for developing synthetic antihistaminic drugs. This explains why the study of antihistaminic activity was delayed until the serendipitous discovery of synthetic antihistaminics.

Research on antihistaminic drugs was initiated in France in 1933. While screening chemicals for other activities, Fourneau and Bovet[14] observed that certain aryl ethers that contained a basic side chain protected guinea pigs against lethal doses of histamine. This initiated worldwide interest in the synthesis and study of related antihistaminic compounds during the next four decades. A review[15] of the chemistry of the more active antagonists reveals the surprising fact that these agents are structurally unrelated to histamine. Many of their physical and chemical properties differ greatly from those of histamine. With various degrees of potency, these classic antihistaminics inhibit the blood pressure-lowering, gut-stimulating, and undesirable side effects of histamine and its analogues. However, the gastric-stimulating effects of histamine are either unaffected or augmented by them.

Early on, it was recognized that not only did the antihistaminics fail to antagonize all of the pharmacologic actions of histamine, but they also exhibited other pharmacologic actions (e.g., anticholinergic and local anesthetic). This presented a challenge to establish the existence of receptors specific for histamine. A review[17] of the evidence for specific histamine receptors indicates that the postulated competitive antagonism of certain actions of histamine by the antihistaminics was supported by data collected in the late 1940s. The reader interested in the question of how to define an antihistaminic and its degree of activity is referred to an excellent review.[18]

Acceptance of the proposal in 1948[19] that there exist at least two adrenergic receptors gave impetus to the search for more than one type of histamine receptor. In 1962, Lin and co-workers[20] suggested that, because the antihistaminics inhibited some, but not all, of the actions of histamine, there must be more than one type of histaminergic receptor. The concept was formalized in 1966[21] by defining H_1*-receptors* as those that are specifically antagonized by low concentrations of antihistaminic drugs. The other actions of histamine then are unlikely to be mediated by H_1-receptors. The wide variation in relative agonist activities of several histamine analogues further supported the postulate that histamine receptors could be differentiated into at least two classes.

Of necessity, the definition of H_2-receptors awaited the discovery of a specific antagonist. In 1972, Black and associates[22] provided additional evidence for the postulate of more than one type of histaminergic receptor. They observed that 2-methylhistamine (Fig. 16-1) had significantly more activity on the tissues containing H_1-receptors (e.g., guinea pig ileum), whereas 4-methylhistamine had considerably more activity on guinea pig atrium (no H_1-receptors). They also reported that a new compound, burimamide, acted as a specific competitive antagonist of histamine on those tissues refractory to block by the antihistaminics. Therefore, they proposed that the receptors mediating these responses to histamine be termed H_2-receptors.

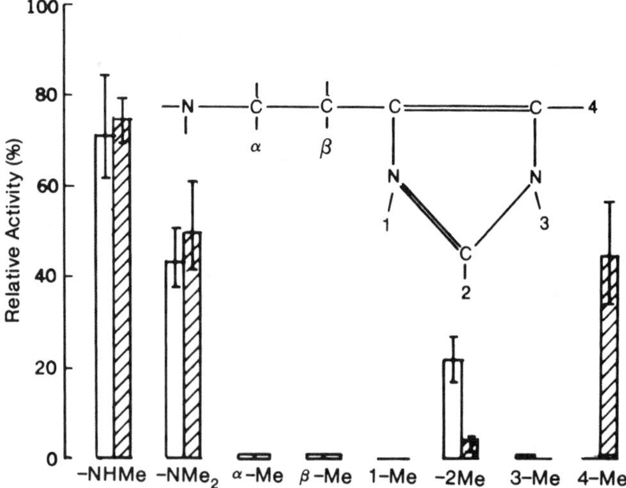

FIG. 16-1. Relative activity of monomethylhistamines. Bar graph showing activity of monomethyl derivatives of histamine relative to histamine (equal to 100), estimated on both ileum (white columns) and atrium (hatched columns). Each bar indicates mean relative activity (with 95% confidence limits). The action on ileum is considered to involve H_1-receptors, whereas that on atrium involves activation of H_2-receptors. (Reprinted by permission from Black, J. W., et al.: Nature 236:385, 1972, Macmillan Journals Limited.)

Briefly, the pharmacologic distinction of histamine receptors rests upon the actions of histamine and its antagonists. Histamine stimulates the contraction of smooth muscle from various organs, such as the gut and bronchi. Because low concentrations of antihistaminic drugs suppress this effect, the pharmacologic receptors that mediate this response are referred to as *H_1-receptors* and the drugs are said to be *H_1-antagonists*. Histamine also stimulates the secretion of acid by the stomach, increases the heart rate, and inhibits contractions in the rat uterus. Because the classic antihistaminics do not antagonize these effects, the receptors that mediate these actions are said to be *H_2-receptors*. Likewise, drugs that inhibit these responses to histamine are classified as *H_2-antagonists*.

Therapeutic areas that have evolved from research on antihistaminic drugs include tranquilizers, antipsychotics, antidepressants, antiparkinsonian agents, antiemetics, and others.[23] And now the events of the seventh decade of this century have set the stage for new discoveries to evolve from studies of the natural role of histamine in biologic systems.

BIOSYNTHESIS AND METABOLISM

Even though injected histamine can be taken up by various tissues, ingested histamine, or that formed by bacteria in the gastrointestinal tract, does not appear to contribute significantly to the endogenous pool. The major source of histamine in the body appears to be the decarboxylation of the naturally occurring amino acid, histidine. The reaction is catalyzed by histidine decarboxylase, an enzyme that requires the coenzyme, pyridoxal phosphate (pyridoxine). A general mechanism for amino acid decarboxylation catalyzed by pyridoxal phosphate has been proposed.[24] Part of the driving force for the decarboxylation of the chelated Schiff base may be the resonance stabilization of the imine chelate that is formed. The resulting structure is hydrolyzed, yielding the free amine and the coenzyme in its original form. The model illustrates the decarboxylation of histidine to form histamine. The histamine-forming capacity of some tissues is remarkably high.

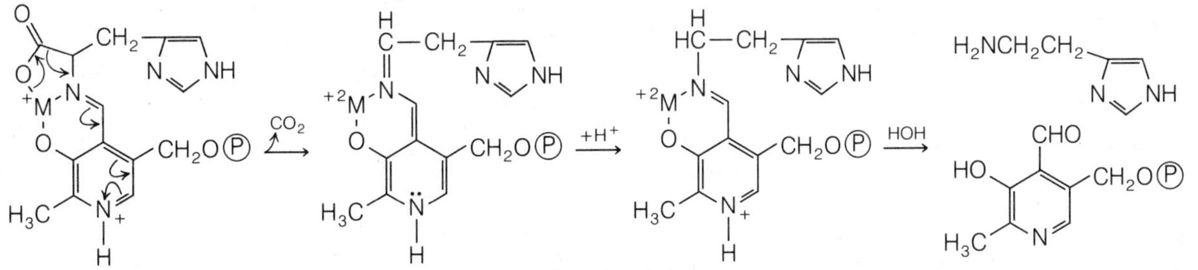

Histidine Decarboxylation

The chief sites of histamine storage are the mast* cells and the basophils, which are the blood-circulating counterpart of the fixed tissue mast cells. In nonmast cell sites, histamine usually is undergoing a rapid turnover and is released rather than stored. Very little of the released histamine is excreted unchanged. In humans, most of it is excreted as the polar metabolites shown in Figure 16-2. The numerical values represent the percentage recovery of histamine and metabolites in the urine in 12 hours after intradermal [^{14}C] histamine in human males.[25]

Recent reviews of the biogenesis[26] and metabolism and excretion[27] of histamine underscore the fact that nature has provided, on the one hand, some tissues with a high histamine-forming capacity and other tissues (mast cells and basophils) a great capacity to store histamine. On the other hand, the body has a strong defense against the unwanted effects of excessive histamine. The implication of this is that histamine is very important to many physiologic and pathologic conditions. But our present knowledge still lacks experimental evidence of the role of histamine in these processes.

Histamine Phosphate, USP. 1*H*-Imidazole-4-ethanamine phosphate (1 : 2); 4-(2-aminoethyl)imidazole bis(dihydrogen phosphate). The salt is stable in air, but affected by light, and it is freely soluble in water (1 : 4). The chief use of histamine clinically is to diagnose impairment of the acid-producing cells (parietal cells) of the stomach. It is the most powerful gastric secretory stimulant that is available. The absence of acid secretion after injection is considered

* *Mast* is the German word for *fattening* or *forced-feeding*. The cell was recognized as a well-fed connective tissue cell.

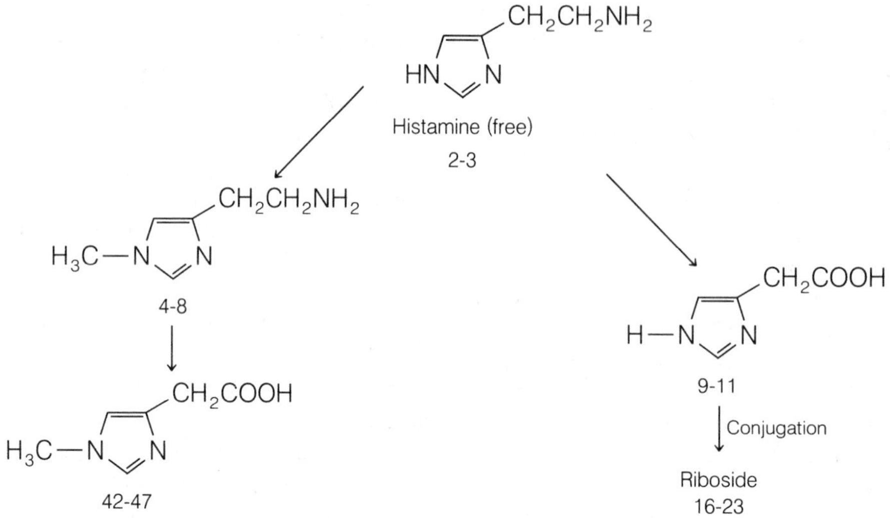

Histamine (free)
2-3

4-8

42-47

9-11

Conjugation

Riboside
16-23

FIG. 16-2. *Metabolism of histamine.*

proof that the gastric acid-secreting glands are nonfunctional, a condition (called *achlorhydria*) that is particularly symptomatic of pernicious anemia.

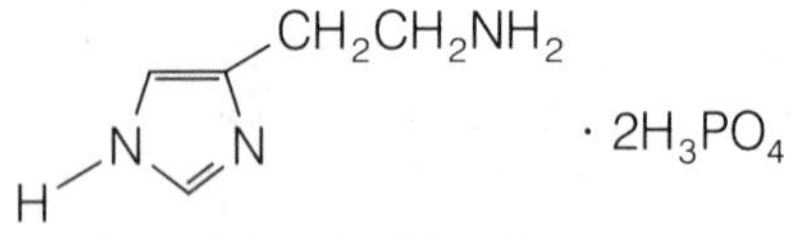

$\cdot 2H_3PO_4$

Histamine Phosphate

The wheal caused by intradermal injection of a 1:1000 solution has been suggested as a diagnostic test of local circulation. In normal individuals, the wheal appears in about 2.5 minutes, and any delay is considered a sign of vascular disease.

Usual dose: subcutaneous, 27.5 μg/kg (the equivalent of 10 μg of histamine base) of body weight
Usual dose range: 10 to 40 μg/kg
Dosage form: Histamine phosphate injection USP

Betazole Hydrochloride. 1*H*-Pyrazole-3-ethanamine dihydrochloride; 3-(2-aminoethyl)pyrazole dihydrochloride (Histalog). The salt of this analogue of histamine is a water-soluble, white, crystalline, nearly odorless powder. The *p*H of a 5% solution is about 1.5.

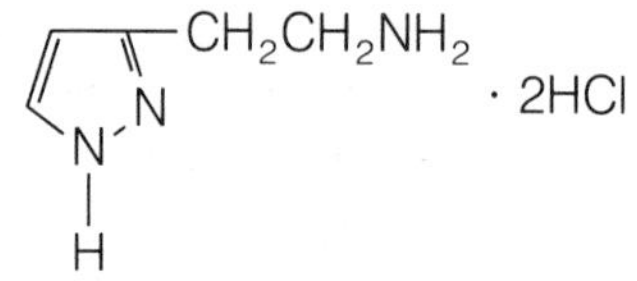

$\cdot 2HCl$

Betazole Hydrochloride

Betazole is less potent than histamine, but retains the ability to stimulate gastric secretion with much less tendency to produce the other effects that are usually observed after the use of histamine. The gastric secretory response to a dose of 5 mg of betazole is comparable with the response to 10 μg of histamine base.

Usual dose: subcutaneous or intramuscular, 50 mg
Usual dose range: 40 to 60 mg
Dosage form: Betazole hydrochloride injection

ANTIHISTAMINIC AGENTS

The discovery of the H_2-antagonist burimamide in the early 1970s opened a new era in the history of the attempt to explain histamine-related physiologic processes. Only history can record the importance of this discovery toward delineation of the role of histamine, as well as that of other autacoids, in homeostasis and disease. For a historical record of the progress thus far, we draw the reader's attention, in particular, to the literature surveys edited by Rocha e Silva[29] and Fordtran and Grossman.[30]

Even though the general term *antihistaminic* implies the inhibition of the action of histamine, the physicochemical properties, pharmacologic activities, and clinical uses of the H_1-antagonists (classic antihistamines) differ greatly from those of the H_2-antagonists.[31] Therefore, the H_1-antagonists and H_2-antagonists are discussed separately.

H_1-ANTAGONISTS

Mode of Action

H_1-antagonists may be defined as those drugs that, in low concentrations, competitively inhibit the ac-

tion of histamine on tissues containing H_1-receptors. H_1-antagonists are evaluated in terms of their ability to inhibit histamine-induced spasms in an isolated strip of guinea pig ileum. Also, antihistamines may be evaluated in vivo in terms of their ability to protect animals against the lethal effects of histamine aerosols.

To distinguish competitive antagonism of histamine from other modes of action, the index pA is applied to in vitro assays. The index pA_2, for example, is defined as the inverse of the logarithm of the molar concentration of the antagonist, which reduces the response of a double dose of the agonist to that of a single one. The more potent H_1-antagonists exhibit a pA_2 value significantly higher than 6. Although there are many pitfalls[32] to be avoided in the interpretation of structure–activity relationship (SAR) studies, the following example is illustrative in distinguishing competitive antagonism. It has been shown that pA_2 values for pyrilamine (mepyramine) antagonism ranged from 9.1 to 9.4 with human bronchii and guinea pig ileum.[33] By contrast, the pA_2 value in guinea pig atria (H_2-receptor) was 5.3. Thus, it may be concluded that pyrilamine is a weak, noncompetitive inhibitor of histamine at the atrial receptors and a competitive inhibitor at H_1-receptors.

For the most part, the therapeutic usefulness of the H_1-antagonists is in the symptomatic treatment of allergic disease. This action is clearly attributable to their antagonism of the action of histamine. In addition, the central properties of some of the agents are of clinical value, particularly in suppressing motion sickness.

H_1-antagonists are most useful in seasonal rhinitis (hay fever, pollinosis). The drugs best relieve the symptoms of these allergic diseases (sneezing, rhinorrhea, and itching of eyes, nose, and throat) at the beginning of the season when pollen counts are low.

Despite popular belief, antihistamines are of little value in treating the common cold.[31] Their weak anticholinergic effect may lessen rhinorrhea. Certain of the allergic dermatoses and the urticarial lesions of systemic allergies respond favorably to H_1-antagonists. However, the drugs are of little or no value in diseases such as systemic anaphylaxis and bronchial asthma, in which autacoids other than histamine are important.

Structure–Activity Relationships

From a study of the activity of the thousands of compounds synthesized and screened for antihistamine activity, the structural requirements for optimal activity may be suggested. A more detailed discussion is available.[15,33]

Most compounds that antagonize the action of histamine at H_1-receptor sites may be described by the general structure for which Ar is aryl (including phenyl, substituted phenyl, and heteroaryl groups such as 2-pyridyl), and Ar′ is a second aryl or arylmethyl group. Chemical classification of these agents is usually based on the unit X, which may be saturated carbon–oxygen (aminoalkyl ethers), nitrogen (ethylenediamines), or carbon (propylamines). Tricyclic derivatives, in which the two aromatic rings are bridged (e.g., phenothiazines), comprise additional chemical classes.

In general, the terminal nitrogen atom should be a tertiary amine for maximal activity. In contrast with many anticholinergics and local anesthetics, the dimethylamine derivatives appear to have a better therapeutic index. However, the terminal nitrogen atom may be part of a heterocyclic structure, as in antazoline and in chlorcyclizine, and may still result in a compound of high antihistaminic potency.

Extension or branching of the 2-aminoethyl side chain results in a less active compound. However, promethazine has a greater therapeutic index than its nonbranched counterpart.

Ar′ / Ar >X—C—C—N<

2-Aminoethyl side chain

Several properties of H_1-antagonists have been studied in an effort to relate them to activity.[15,33,34] Although physicochemical and steric parameters appear to be important to H_1-antagonist action (especially lipophilic characteristics), no direct correlation has been established between a property and the antihistaminic effect.

The relationships of structure to overlapping actions (H_1-antagonist, anticholinergic, and local anesthetic) have been analyzed.[18]

Aminoalkyl Ethers

Extensive studies of the qualitative and quantitative structure–activity relationships indicate that the critical length of the 2-aminoalkyl side chain shows more flexibility in the aminoalkylethers than in the ethylenediamines. However, the most active compounds have a chain length of two carbon atoms. Quaternization of the side chain nitrogen does not always result in less active compounds.

The drugs in this group possess significant anticholinergic activity, which enhances the H_1-blocking action on exocrine secretions. This action is not

TABLE 16-1

AMINOALKYL ETHERS

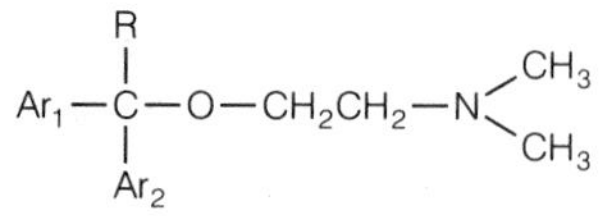

$Ar_1-C(R)(Ar_2)-O-CH_2CH_2-N(CH_3)_2$

Generic Name	Proprietary Name	Ar_1	Ar_2	R
Diphenhydramine	Benadryl	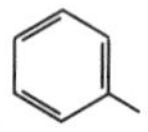 phenyl	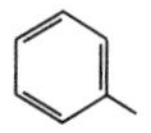 phenyl	H
Bromodiphenhydramine	Ambodryl	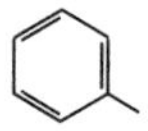 phenyl	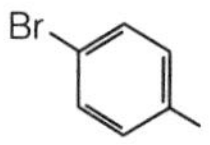 4-bromophenyl (Br)	H
Doxylamine	Decapryn	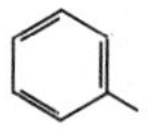 phenyl	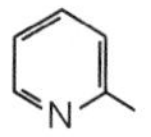 2-pyridyl (N)	CH_3
Carbinoxamine	Clistin	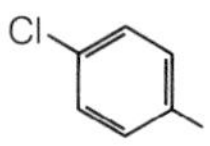 4-chlorophenyl (Cl)	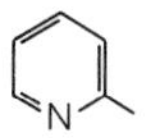 2-pyridyl (N)	H
Clemastine	Tavist	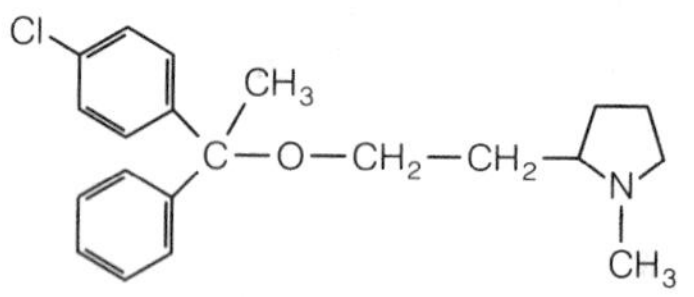 (Cl-phenyl)(phenyl)C(CH_3)$-O-CH_2-CH_2-$(N-CH_3 pyrrolidinyl)		

surprising in view of the structural similarity to the aminoalcohol ethers that exhibit cholinergic-blocking activity. Drowsiness is a side effect common to the tertiary aminoalkyl ethers. Although this side effect is exploited in over-the-counter (OTC) sleeping aids, it may interfere with the patient's performance of tasks requiring mental alertness. The frequency of gastrointestinal side effects is low. Structural relationships of the aminoalkyl ether derivatives are shown in Table 16-1, whereas the official preparations and dosages are given later (see Table 16-6).

Diphenhydramine Hydrochloride, USP. 2-(Diphenylmethoxy)-*N,N*-dimethylethanamine hydrochloride (Benadryl). The oily, lipid-soluble free base is available as the bitter-tasting hydrochloride salt, which is a stable, white crystalline powder, soluble in water (1:1), alcohol (1:2), and chloroform (1:2). The salt has a pK_a value of 9, and a 1% aqueous solution has a pH of about 5.

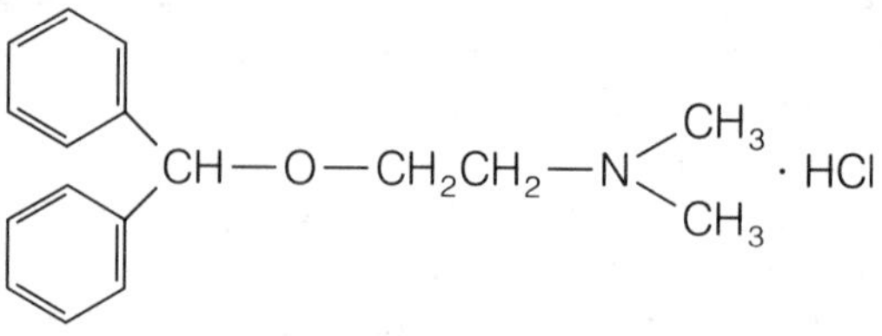

$(C_6H_5)_2CH-O-CH_2CH_2-N(CH_3)_2 \cdot HCl$

Diphenhydramine Hydrochloride

In addition to antihistaminic action, diphenhydramine exhibits antidyskinetic, antiemetic, antitussive, and sedative properties. It is used in OTC sleep-aid products.

In the usual dose range of 25 to 400 mg, diphenhydramine is not a highly active H_1-antagonist; it has anticholinergic and sedative properties. Conversion to a quaternary ammonium salt does not alter the antihistaminic action greatly, but does increase the anticholinergic action.

As an antihistaminic agent, diphenhydramine is recommended in various allergic conditions and, to a lesser extent, as an antispasmodic. It is administered either orally or parenterally in the treatment of urticaria, seasonal rhinitis (hay fever), and some dermatoses. The most common side effect is drowsiness, and the concurrent use of alcoholic beverages and other central nervous system depressants should be avoided.

Dimenhydrinate, USP. 8-Chlorotheophylline 2-(diphenylmethoxy)-*N,N*-dimethylethylamine compound (Dramamine). The 8-chlorotheophyllinate (theoclate) salt of diphenhydramine is a white crystalline, odorless powder that is slightly soluble in water and freely soluble in alcohol and chloroform.

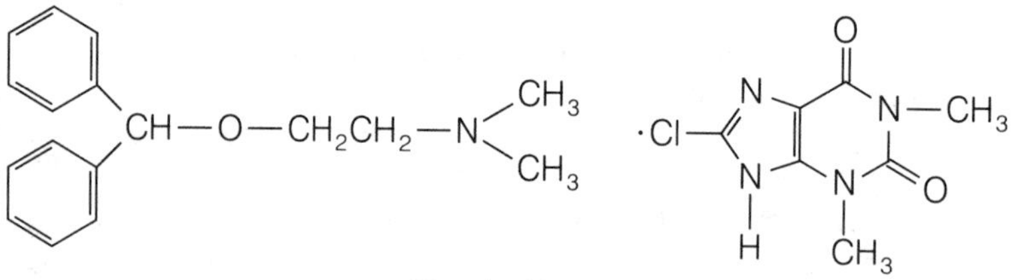

$(C_6H_5)_2CH-O-CH_2CH_2-N(CH_3)_2 \cdot$ 8-chlorotheophylline

Dimenhydrinate

Dimenhydrinate is recommended for the nausea of motion sickness and for hyperemesis gravidarum (nausea of pregnancy). For the prevention of motion sickness, the dose should be taken at least one-half hour before beginning the trip. The cautions listed for diphenhydramine should be observed.

Bromodiphenhydramine Hydrochloride, USP. 2-[(4-Bromophenyl)phenylmethoxy]-*N,N*-dimethylethanamine hydrochloride (Ambodryl Hydrochloride). The hydrochloride salt is a white to pale buff crystalline powder that is freely soluble in water and in alcohol.

Relative to diphenhydramine, bromodiphenhydramine is more lipid-soluble and was found to be twice as effective in protecting guinea pigs against the lethal effects of histamine aerosols.

Doxylamine Succinate, USP. 2-[α-[2-(Dimethylamino)ethoxy]-α-methylbenzyl]pyridine bisuccinate (Decapryn Succinate). The acid succinate salt (bisuccinate) is a white to creamy-white powder that has a characteristic odor and is soluble in water (1:1), alcohol (1:2), and chloroform (1:2). A 1% solution has a *p*H of about 5.

Doxylamine Succinate

Doxylamine succinate is comparable in potency to diphenhydramine. It is a good nighttime hypnotic when compared with secobarbital.[35] Concurrent use of alcohol and other central nervous system depressants should be avoided.

Carbinoxamine Maleate, USP. (*d,l*)-2-[*p*-Chloro-α-[2-(dimethylamino)ethoxy]benzyl]pyridine bimaleate (Clistin). The oily, lipid-soluble free base is available as the bitter bimaleate salt, a white crystalline powder that is very soluble in water and freely soluble in alcohol and in chloroform. The *p*H of a 1% solution is between 4.6 and 5.1.

Carbinoxamine Maleate

Carbinoxamine differs structurally from chlorpheniramine only in that an oxygen atom separates the asymmetric carbon atom from the aminoethyl side chain. The more active *levo* isomer of carbinoxamine has been shown to have the (*S*) absolute configuration[36] and to be superimposable upon the more active *dextro* isomer (*S* configuration[37]) of chlorpheniramine.

$Ar = p{-}ClC_6H_4$

(*S*) Carbinoxamine (*levo*)

(*S*) Chlorpheniramine (*dextro*)

Carbinoxamine is a potent antihistaminic and is available as the racemic mixture.

Clemastine Fumarate, USP. 2-[2-[1-(4-Chlorophenyl)-1-phenylethoxy]ethyl]-1-methylpyrrolidine hydrogen fumarate (1:1) (Tavist). Dextrorotatory clemastine has two chiral centers, each of which is of the (*R*) absolute configuration. A comparison of the activities of the antipodes indicates that the asymmetric center close to the side chain nitrogen is of lesser importance to antihistaminic activity.[33]

Clemastine Fumarate

This member of the ethanolamine series is characterized by a long duration of action, with an activity that reaches a maximum in five to seven hours and persists for 10 to 12 hours.

It is well absorbed when administered orally and is excreted primarily in the urine. The side effects are those usually encountered with this series of antihistamines. Clemastine is closely related to chlorphenoxamine, which is used for its central cholinergic-blocking activity. Therefore, it is not surprising that clemastine has significant antimuscarinic activity.

Ethylenediamines

As a chemical class, the ethylenediamines contain the oldest useful antihistamines. They are highly effective H_1-antagonists, with a relatively high frequency of central nervous system depressant and gastrointestinal side effects.[31]

The piperazine-, imidazoline-, and phenothiazine-type antihistamines contain the ethylenediamine moiety. These agents are discussed separately because they exhibit significantly different pharmacologic properties.

In most drug molecules, the presence of a nitrogen atom appears to be a necessary condition for the formation of a stable salt with mineral acids. The aliphatic amino group in the ethylenediamines is sufficiently basic for salt formation, but the nitrogen atom to which an aromatic ring is bonded is considerably less basic. The free electrons on the aryl nitrogen are delocalized by the aromatic ring. A general resonance structure depicting this electron delocalization is as follows.

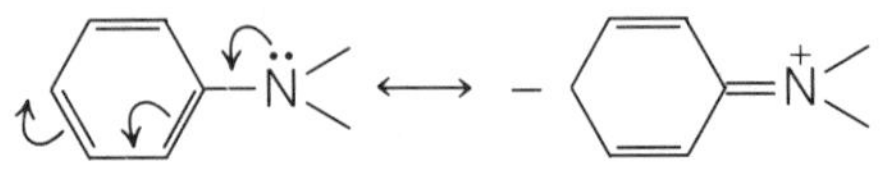

Because there is decreased electron density on nitrogen, it is less basic, and protonation at this position takes place less readily.

Table 16-2 shows structural relationships of the ethylenediamines (see Table 16-6 for the official preparations and dosages).

Tripelennamine Citrate, USP. 2-[Benzyl[2-(dimethylamino)ethyl]amino]pyridine citrate (1 : 1); PBZ (Pyribenzamine Citrate). The oily free base is available as the less-bitter monocitrate salt, which is a white crystalline powder freely soluble in water and in alcohol. A 1% solution has a *p*H of 4.25. For oral administration in liquid dose forms, the citrate salt is less bitter and, thus, more palatable than the hydrochloride. Because of the difference in molecular weights, the doses of the two salts must be equated: 30 mg of the citrate salt are equivalent to 20 mg of the hydrochloride salt.

TABLE 16-2

ETHYLENEDIAMINE DERIVATIVES

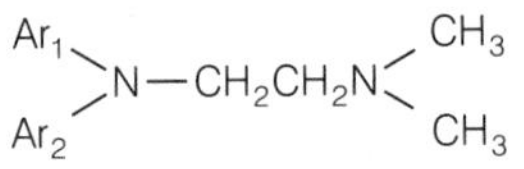

$Ar_1(Ar_2)N{-}CH_2CH_2N(CH_3)_2$

Generic Name	Proprietary Name	Ar_1	Ar_2
Pyrilamine		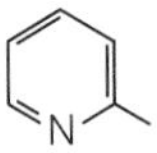 2-pyridyl	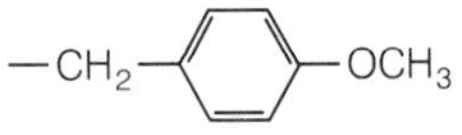 $-CH_2-C_6H_4-OCH_3$
Tripelennamine	Pyribenz-amine	2-pyridyl	$-CH_2-C_6H_5$
Methapyrilene	Histadyl	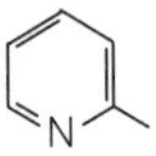 2-pyridyl	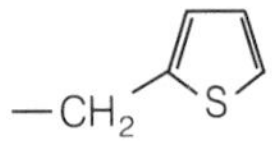 $-CH_2$-2-thienyl
Thonzylamine		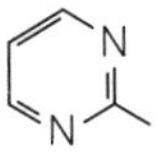 2-pyrimidinyl	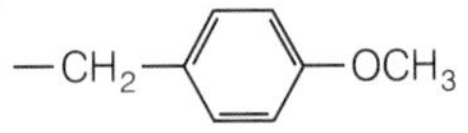 $-CH_2-C_6H_4-OCH_3$

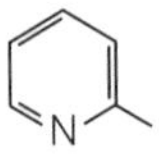

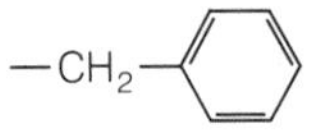

Tripelennamine Hydrochloride, USP is a white crystalline powder that darkens slowly on exposure to light. The salt is soluble in water (1 : 0.77) and in alcohol (1 : 6). It has a pK_a of about 9, and a 0.1% solution has a *p*H of about 5.5.

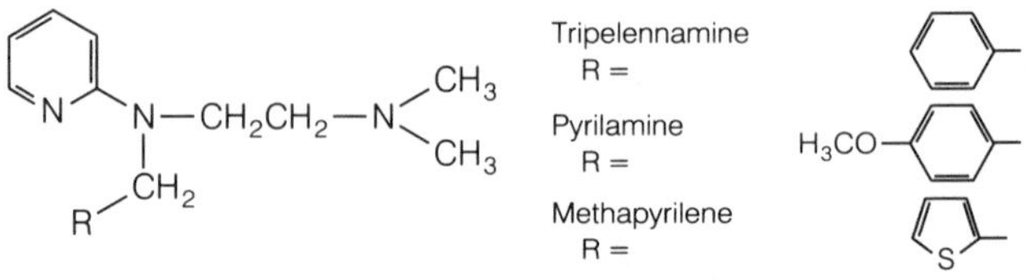

Tripelennamine R =
Pyrilamine R = H_3CO-
Methapyrilene R =

Tripelennamine, the first ethylenediamine developed in American laboratories, is well absorbed when given orally. On the basis of clinical experience, it appears to be as effective as diphenhydramine and may have the advantage of fewer and less severe side reactions. However, drowsiness may occur and may impair ability to perform tasks requiring alertness. The concurrent use of alcoholic beverages should be avoided.

Pyrilamine Maleate, USP. 2-[[2-(Dimethylamino)ethyl](*p*-methoxybenzyl)amino]pyridine maleate (1 : 1); mepyramine. The oily free base is available as the acid maleate salt, which is a white crystalline powder having a faint odor and a bitter, saline taste. The salt is soluble in water (1 : 0.4) and freely soluble in alcohol. A 10% solution has a *p*H of approximately 5. At a *p*H of 7.5 or above, the oily free base begins to precipitate.

Pyrilamine differs structurally from tripelennamine by having a methoxy group in the *para* position of the benzyl radical. It differs from its more toxic and less potent precursor phenbenzamine (Antergan) by having a 2-pyridyl group on the nitrogen atom in place of a phenyl group.

Clinically, pyrilamine and tripelennamine are considered to be among the less potent antihistaminics. They are highly potent, however, in antagonizing histamine-induced contractions of guinea pig ileum.[15] Because of the pronounced local anesthetic action, the drug should not be chewed, but taken with food.

Methapyrilene Hydrochloride. 2-[[2-(Dimethylamino)ethyl]-2-thenylamino]pyridine monohydrochloride (Histadyl). The oily free base is available as the bitter-tasting monohydrochloride salt, which is a white crystalline powder that is soluble in water (1 : 0.5), in alcohol (1 : 5), and in chloroform (1 : 3). Its solutions have a *p*H of about 5.5. It differs structurally from tripelennamine in having a 2-thenyl (thiophene-2-methylene) group in place of the benzyl group. The thiophene ring is considered isosteric with the benzene ring, and the isosteres exhibit similar activity.[38]

A study of the solid-state conformation of methapyrilene hydrochloride showed that the *trans* conformation is preferred for the two ethylenediamine nitrogen atoms. The Food and Drug Administration declared methapyrilene a potential carcinogen in 1979, and all products containing it have been recalled.

Thonzylamine Hydrochloride. 2-{[2-(Dimethylamino)ethyl](*p*-methoxybenzyl)amino}pyrimidine hydrochloride. The hydrochloride is a white crystalline powder, soluble in water (1 : 1), alcohol (1 : 6), and chloroform (1 : 4). A 2% aqueous solution has a *p*H of 5.5. It is similar in activity to tripelennamine but is claimed to be less toxic.

The usual dose is 50 mg up to four times daily. It is available in certain combination products.

TABLE 16-3

PROPYLAMINE DERIVATIVES

Generic Name	Proprietary Name	Ar_1	Ar_2
SATURATED			
	$Ar_1(Ar_2)CH{-}CH_2CH_2N(CH_3)_2$		
Pheniramine	Trimeton	2-pyridyl	phenyl
Chlorpheniramine, Dexchlorpheniramine	Chlortrimeton, Polaramine	2-pyridyl	4-Cl-phenyl
Brompheniramine, Dexbrompheniramine	Dimetane, Disomer	2-pyridyl	4-Br-phenyl
UNSATURATED			
	$Ar_1(Ar_2)C{=}C(H){-}CH_2{-}N$(pyrrolidine)		
Pyrrobutamine	Pyronil	phenyl	$-CH_2-$(4-Cl-phenyl)
Triprolidine	Actidil	2-pyridyl	4-CH_3-phenyl

Propylamine Derivatives

The saturated members of this group are sometimes referred to as the *pheniramines*. The pheniramines are chiral molecules, and the halogen-substituted derivatives have been resolved by crystallization of salts formed with *d*-tartaric acid.[15] The antihistamines in this group are among the most active H_1-antagonists. They are not so prone to produce drowsiness, but a significant proportion of patients do experience this effect.

In the unsaturated members, it has been suggested that a coplanar aromatic double-bond system ($ArC{=}CH{-}CH_2N$) is an important factor for antihistaminic activity. The pyrrolidino group is the side chain tertiary amine in the more active compounds.

The structural relationships of the propylamine derivatives are shown in Table 16-3. (For the official preparations and dosages, see Table 16-6.)

Pheniramine Maleate. 2-[α-[2-Dimethylaminoethyl]benzyl]pyridine bimaleate (Trimeton; Inhiston). This salt is a white crystalline powder, having a faint aminelike odor, that is soluble in water (1 : 5) and is very soluble in alcohol.

This drug is the least potent member of the series and is marketed as the racemate. The usual adult dose is 20 to 40 mg three times daily. It is available in certain combination products.

Chlorpheniramine Maleate, USP. (±)2-[*p*-Chloro-α-[2-(dimethylamino)ethyl]benzyl]pyridine bimaleate (Chlor-Trimeton). The bimaleate salt is a white crystalline powder that is soluble in water (1 : 3.4), in alcohol (1 : 10), and in chloroform (1 : 10). It has a pK_a of 9.2, and an aqueous solution has a *p*H between 4 and 5.

Chlorination of pheniramine in the *para* position of the phenyl ring gave a tenfold increase in potency with no appreciable change in toxicity. Most of the antihistaminic activity resides with the *dextro* enan-

$$(2\text{-pyridyl})(4\text{-R-phenyl})CHCH_2CH_2{-}N(CH_3)_2 \cdot \begin{matrix} CH{-}COOH \\ \| \\ CH{-}COOH \end{matrix}$$

Pheniramine R = H
Chlorpheniramine R = Cl
Brompheniramine R = Br

tiomorph (see dexchlorpheniramine below). The usual dose is 2 to 4 mg three or four times a day. It has a half-life of 12 to 15 hours.

Dexchlorpheniramine Maleate, USP (Polaramine). Dexchlorpheniramine is the dextrorotatory enantiomer of chlorpheniramine. In vitro and in vivo studies of the enantiomorphs of chlorpheniramine showed that the antihistaminic activity exists predominantly in the *dextro*-isomer.[15] As mentioned previously, the *dextro*-isomer has been shown[37] to have the (*S*) configuration, which is superimposable upon the (*S*) configuration of the more active levorotatory enantiomorph of carbinoxamine.

Brompheniramine Maleate, USP. (±)2-[*p*-Bromo-α-[2-(dimethylamino)ethyl]benzyl]pyridine bimaleate (Dimetane). This drug differs from chlorpheniramine by the substitution of a bromine atom for the chlorine atom. Its actions and uses are similar to those of chlorpheniramine. It has a half-life of 25 hours, which is almost twice that of chlorpheniramine.

Dexbrompheniramine Maleate, USP (Disomer). Like the chlorine congener, the antihistaminic activity exists predominantly in the *dextro*-isomer and is of comparable potency.

Pyrrobutamine Phosphate. (*E*)-1-[4-(4-Chlorophenyl)-3-phenyl-2-butenyl]pyrrolidine diphosphate (Pyronil).*

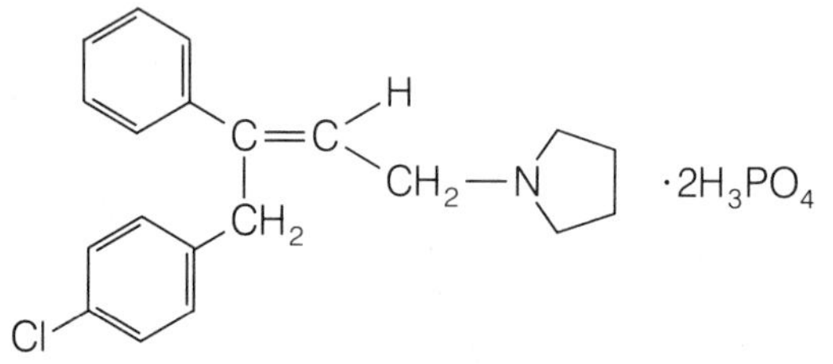

Pyrrobutamine Phosphate
*E** (*trans*) Isomer

The diphosphate occurs as a white crystalline powder that is soluble to the extent of 10% in warm water. Pyrrobutamine was investigated originally as the hydrochloride salt, but the diphosphate was absorbed more readily and completely. Clinical studies indicate that it is long acting with a comparatively slow onset of action.

The feeble antihistaminic properties of several analogues point to the importance of having a planar ArC=CH—CH_2N unit and a pyrrolidino group as the side chain tertiary amine.[15]

Triprolidine Hydrochloride, USP. (*E*)-2-[3-(1-Pyrrolidinyl)-1-*p*-tolylpropenyl]pyridine monohydrochloride monohydrate (Actidil). Triprolidine hydrochloride occurs as a white crystalline powder having no more than a slight, but unpleasant, odor. It is soluble in water and in alcohol, and its solutions are alkaline to litmus.

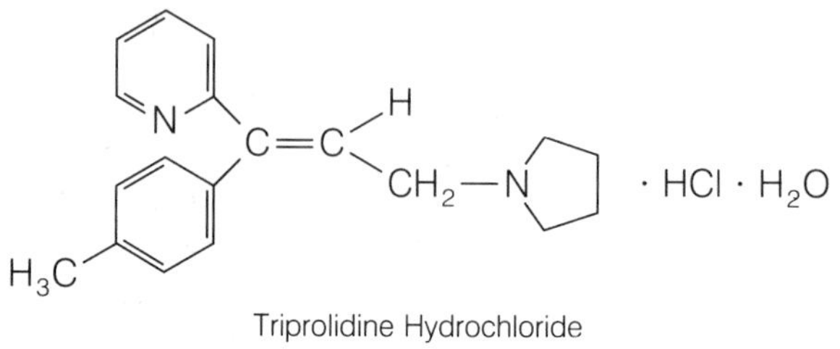

Triprolidine Hydrochloride

The activity is confined mainly to the geometric isomer in which the pyrrolidinomethyl group is *trans* to the 2-pyridyl group. Recent pharmacologic studies[40] confirm the high activity of triprolidine and the superiority of (*E*)- over corresponding (*Z*)-isomers as H_1-antagonists. At guinea pig ileum sites, the affinity of triprolidine for H_1-receptors was more than 1000 times the affinity of its (*Z*)-partner.

The relative potency of triprolidine is of the same order as that of dexchlorpheniramine. The peak effect occurs in about 3.5 hours after oral administration, and the duration of effect is about 12 hours.

Phenothiazine Derivatives

Beginning in the mid-1940s, several antihistaminic drugs have been discovered as a result of bridging the aryl units of agents related to the ethylenediamines. The search for effective antimalarials led to the investigation of phenothiazine derivatives in which the bridging entity is sulfur. In subsequent testing, the phenothiazine class of drugs was discovered to have not only antihistaminic activity, but also a pharmacologic profile of its own, considerably different from that of the ethylenediamines. Thus began the era of the useful psychotherapeutic agent.[41]

The structural relationships of the phenothiazines exhibiting antihistaminic action are shown in Table 16-4. (For the official preparations and dosages, see Table 16-6.)

Promethazine Hydrochloride, USP. (±)10-[2(Dimethylamino)propyl]phenothiazine monohydrochloride (Phenergan). The salt occurs as a white to faint yellow crystalline powder that is very soluble in water, in hot absolute alcohol, and in chloroform. Its aqueous solutions are slightly acid to litmus.

---

* Following IUPAC[39] stereochemical nomenclature, the choice of substituents used to denote configuration about an alkenic double bond is governed by the sequence rule. When the groups are *trans*, use the prefix (*E*) from the German word *entgegen* meaning "opposite"; when *cis*, use (*Z*) from the German word *zusammen* meaning "together."

TABLE 16-4
PHENOTHIAZINE DERIVATIVES

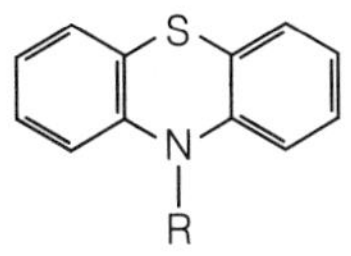

(Phenothiazine ring structure with R on N)

Generic Name	Proprietary Name	R
Promethazine	Phenergan	$-CH_2-CH(CH_3)-N(CH_3)_2$
Trimeprazine	Temaril	$-CH_2-CH(CH_3)-CH_2-N(CH_3)_2$
Methdilazine	Tacaryl	$-CH_2CH<(CH_2-CH_2)(CH_2-N-CH_3)$

$$\cdot HCl$$

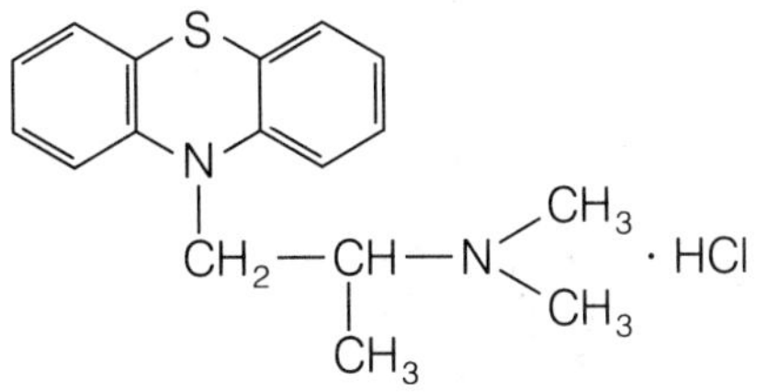

Promethazine Hydrochloride

Promethazine is moderately potent by present-day standards with prolonged action and pronounced sedative side effects. In addition to its antihistaminic action, it possesses an antiemetic effect, a tranquilizing action, and a potentiating action on analgesic and sedative drugs. In general, lengthening of the side chain and substitution of lipophilic groups in the 2-position of the aromatic ring results in compounds with decreased antihistaminic activity and increased psychotherapeutic properties.

Enantiomers of promethazine have been resolved and have similar antihistaminic and other pharmacologic properties.[42] This is in contrast with studies of the pheniramines and carbinoxamine compounds in which the chiral center is closer to the aromatic feature of the molecule. Asymmetry appears to be of less influence on antihistaminic activity when the chiral center lies near the positively charged side chain nitrogen.

The antihistaminic phenothiazines may cause drowsiness and so may impair the ability to perform tasks requiring alertness. Concurrent administration of alcoholic beverages and other central nervous system depressants should be avoided.

Trimeprazine Tartrate, USP. (±)-10-[3-(Dimethylamino)-2-methylpropyl]phenothiazine tartrate (Temaril). The salt occurs as a white to off-white crystalline powder that is freely soluble in water and soluble in alcohol. Its antihistaminic action is reported to be from 1.5 to 5 times that of promethazine. Clinical studies have shown it to have a pronounced antipruritic action. This action may be unrelated to its histamine-antagonizing properties.

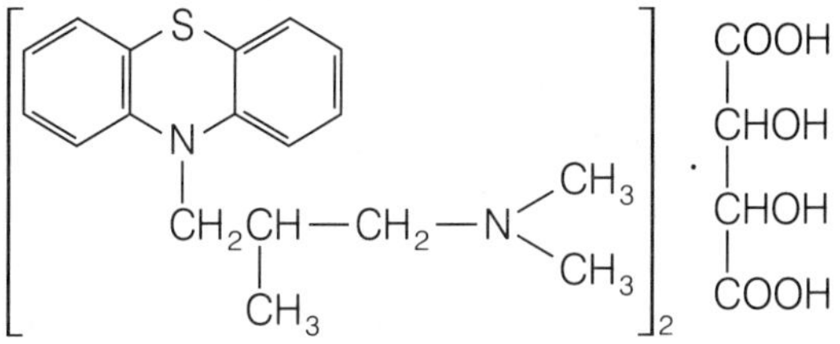

Trimeprazine Tartrate

Methdilazine, USP. 10-[(1-Methyl-3-pyrrolidinyl)methyl]phenothiazine (Tacaryl). This compound occurs as a light tan crystalline powder that has a characteristic odor and is practically insoluble in water. Methdilazine, as the free base, is used in chewable tablets because its low solubility in water contributes to its tastelessness. Some local anesthesia of the buccal mucosa may be experienced if the tablet is chewed and not swallowed promptly.

Methdilazine Hydrochloride, USP. 10-[(1-Methyl-3-pyrrolidinyl)methyl]phenothiazine monohydrochloride (Tacaryl Hydrochloride). The hydrochloride salt also occurs as a light tan crystalline powder having a slight characteristic odor. However, the salt is freely soluble in water and in alcohol.

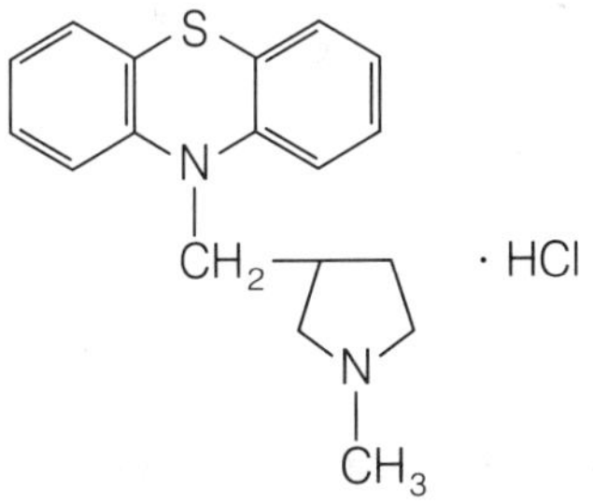

Methdilazine Hydrochloride

The activity is similar to that of methdilazine and is administered orally for its antipruritic effect.

Piperazine Derivatives

The activity of the piperazine-type antihistaminics (cyclic ethylenediamines) is characterized by a slow onset and long duration of action. These are moderately potent antihistaminics with a lower incidence of drowsiness. However, warning of the possibility of some dulling of mental alertness is advised.

TABLE 16-5

PIPERAZINE DERIVATIVES AND MISCELLANEOUS COMPOUNDS

Generic Name	Proprietary Name	Structure	Generic Name	Proprietary Name	Structure
PIPERAZINE DERIVATIVES					
Cyclizine	Marezine	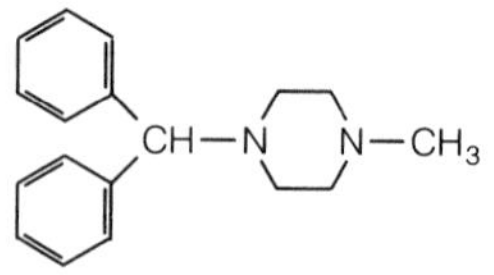	Meclizine	Bonine	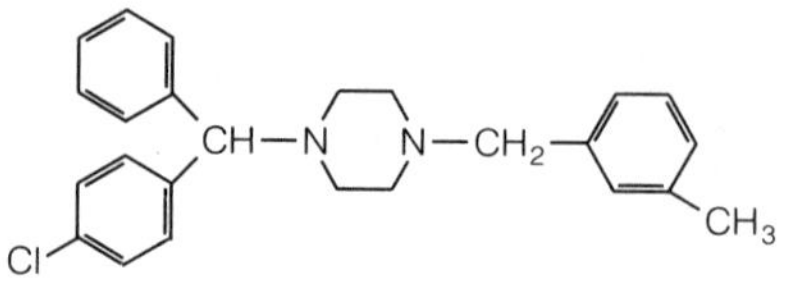
Chlorcyclizine		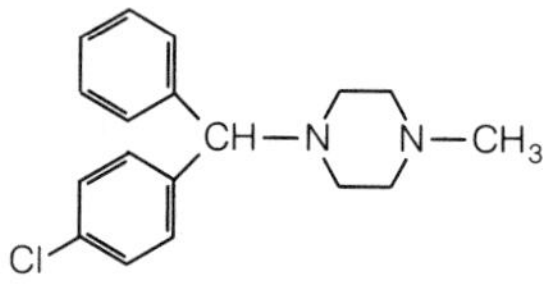	Buclizine	Bucladin-S	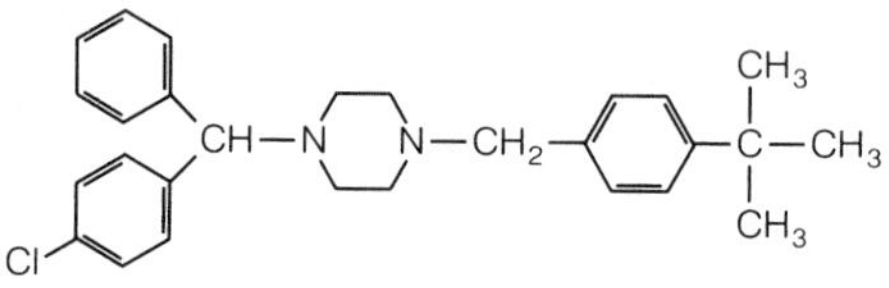
MISCELLANEOUS COMPOUNDS					
Diphenylpyraline	Hispril, Disfen	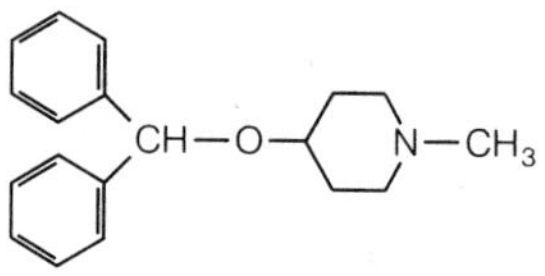	Antazoline		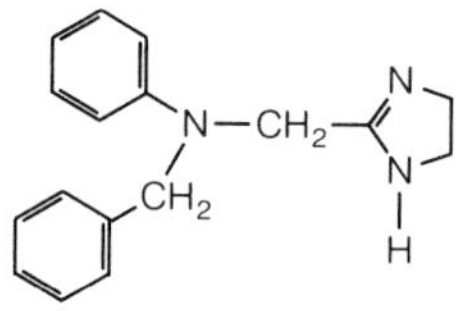
Phenindamine		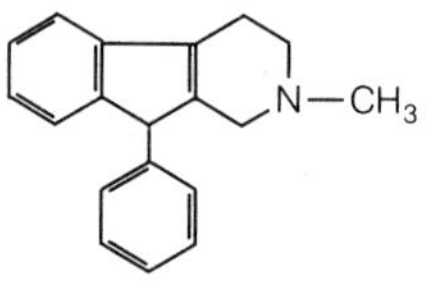	Cyproheptadine	Periactin	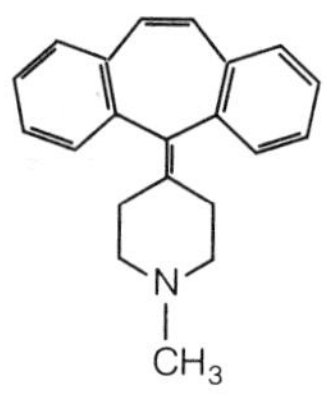
Dimethindene	Forhistal	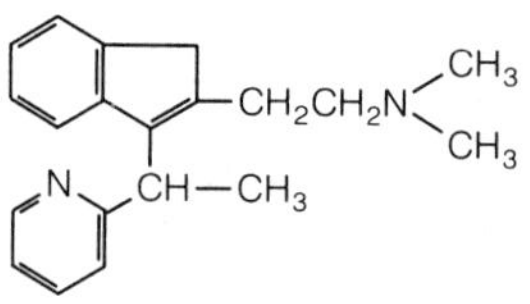	Azatadine	Optimine	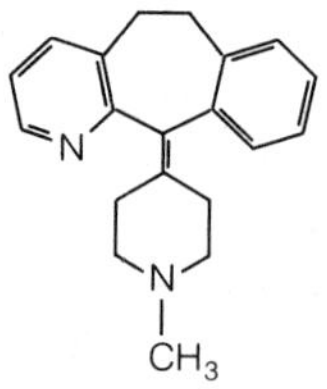

As a group, these agents are useful as antiemetics as well as antihistamines. They have exhibited a strong teratogenic potential, inducing a number of malformations in rats. Norchlorcyclizine, a metabolite of these piperazines, was proposed to be responsible for the teratogenic effects of the parent drugs.[43]

These agents exhibit peripheral and central antimuscarinic activity. The central action may be responsible for the antiemetic and antivertigo effects. The agents diminish vestibular stimulation and may act on the medullary chemoreceptor trigger zone.

Table 16-5 provides a structural comparison of the piperazine derivatives. (For official preparations and dosages, see Table 16-6.)

Cyclizine Hydrochloride, USP. 1-(Diphenylmethyl)-4-methylpiperazine monohydrochloride (Marezine). This drug occurs as a light-sensitive, white crystalline powder having a bitter taste. It is slightly soluble in water (1 : 115), alcohol (1 : 115), and chloroform (1 : 75). It is used primarily in the prophylaxis and treatment of motion sickness.

The lactate salt (**Cyclizine Lactate Injection, USP**) is used for intramuscular injection because of the limited water solubility of the hydrochloride. The injection should be stored in a cold place, because a slight yellow tint may develop if stored at room temperature for several months. This does not indicate a loss in biologic potency.

TABLE 16-6

ANTIHISTAMINIC AGENTS: H_1-Receptor Antagonists

Name Proprietary Name	***Preparations***	***Usual Adult Dose****	***Usual Dose Range****	***Usual Pediatric Dose****
Diphenhydramine hydrochloride USP *Benadryl*	Diphenhydramine hydrochloride capsules USP Diphenhydramine hydrochloride elixir USP Diphenhydramine hydrochloride syrup USP Diphenhydramine hydrochloride tablets USP	Oral, 25–50 mg tid or qid	Up to 400 mg/day	Use in premature and newborn infants is not recommended. 1.25 mg/kg of body weight or 37.5 mg/m^2 of body surface qid, not to exceed 300 mg/day
	Diphenhydramine hydrochloride injection USP	IM or IV, 10–50 mg	Up to a maximum of 400 mg/day	Use in premature and newborn infants is not recommended. IM or IV, 1.25 mg/kg of body weight or 37.5 mg/m^2 of body surface qid, not to exceed 300 mg/day
Dimenhydrinate USP *Dramamine*	Dimenhydrinate elixir USP Dimenhydrinate syrup USP Dimenhydrinate tablets USP	Oral, 50 mg–100 mg every 4 hr as needed		Use in premature and newborn infants is not recommended. 1.25 mg/kg of body weight or 37.5 mg/m^2 of body surface, every 6 hr as needed, not to exceed 300 mg/day
	Dimenhydrinate injection USP	IM, 50 mg up to every 4 hr as needed; IV, 50 mg in 10 mL of sodium chloride injection administered slowly over a period of at least 2 min, up to every 4 hr as needed		IM, 1.25 mg/kg of body weight or 37.5 mg/m^2 of body surface, every 6 hr as needed, not to exceed 300 mg/day; IV, 1.25 mg/kg of body weight or 37.5 mg/m^2 of body surface, in 10 mL of sodium chloride injection, administered slowly over a period of at least 2 min, every 6 hr as needed, not to exceed 300 mg/day
	Dimenhydrinate suppositories USP	Rectal, 100 mg qd or bid as needed		
Bromodiphenhydramine hydrochloride USP	Bromodiphenhydramine hydrochloride capsules USP Bromodiphenhydramine hydrochloride elixir USP	Oral, 25 mg every 4–6 hr as needed	Up to 150 mg/day	Oral, 6.25–25 mg every 4–6 hr as needed
Doxylamine succinate USP *Decapryn*	Doxylamine succinate syrup USP Doxylamine succinate tablets USP	Oral, 12.5–25 mg every 4–6 hr as needed		Use in children up to 6 yr in age is not recommended. Oral, 6.25–12.5 mg every 4–6 hr
Carbinoxamine maleate USP *Clistin*	Carbinoxamine maleate elixir USP Carbinoxamine maleate tablets USP	Oral, 4–8 mg tid or qid		Use in premature or full-term neonates is not recommended. Children 1–3 yr: oral, 2 mg tid or qid; age 4–6 yr: oral, 2–4 mg tid or qid; age 6 and over: oral, 4 mg tid or qid
Clemastine fumarate USP *Tavist*	Clemastine fumarate syrup USP Clemastine fumarate tablets USP	Oral, 1.34 mg bid or 2.68 mg qd to tid	Up to 8.04 mg/day	Use in premature and full-term neonates is not recommended. Children up to 12 yr: oral, 670 μg–1.34 mg every 8–12 hr
Pyrilamine maleate USP	Pyrilamine maleate tablets USP	Oral, 25–50 mg every 6–8 hr		Children 6 yr and over: oral, 12.5–25 mg every 6–8 hr as needed
Tripelennamine citrate USP *Pyribenzamine*	Tripelennamine citrate elixir USP	Oral, the equivalent of 25–50 mg of tripelennamine hydrochloride every 4–6 hr as needed	Up to the equivalent of 600 mg of tripelennamine hydrochloride daily	The equivalent of 1.25 mg/kg of body weight or 37.5 mg/m^2 of body surface of tripelennamine hydrochloride every 6 hr, not to exceed 300 mg/day; use in premature or full-term neonates is not recommended

*See USP DI for complete dosage information.

TABLE 16-6 *Continued*

ANTIHISTAMINIC AGENTS: H_1-Receptor Antagonists

Name Proprietary Name	***Preparations***	***Usual Adult Dose****	***Usual Dose Range****	***Usual Pediatric Dose****
Tripelennamine hydrochloride USP *Pyribenzamine*	Tripelennamine hydrochloride tablets USP	Oral, 25–50 mg every 4–6 hr as needed	Up to 600 mg/day	1.25 mg/kg of body weight or 37.5 mg/m^2 of body surface every 6 hr as needed, not to exceed 300 mg/day; use in premature or full-term neonates is not recommended
	Tripelennamine hydrochloride extended-release tablets USP	Oral, 100 mg every 8–12 hr as needed	Up to 600 mg/day	Not recommended for children
Chlorpheniramine maleate USP *Chlor-Trimeton*	Chlorpheniramine maleate extended-release capsules USP	Oral, 8 or 12 mg every 8–12 hr as needed		Not recommended for use in children up to 7 yr of age. Age 7 or over: oral, 8 mg every 12 hr as needed
	Chlorpheniramine maleate syrup USP Chlorpheniramine maleate tablets USP Chlorpheniramine maleate chewable tablets USP Chlorpheniramine maleate extended-release tablets USP	Oral, 4 mg every 4–6 hr	Up to 24 mg/day	Use in children up to 6 yr of age is not recommended. Oral, 87.5 μg/kg of body weight or 2.5 mg/m^2 of body surface every 6 hr; children to 12 yr of age: oral, 1–2 mg tid to qid, not to exceed 10 mg/day; not recommended for use in premature or full-term neonates. IM, IV, or SC, 87.5 μg/kg of body weight or 2.5 mg/m^2 of body surface every 6 hr
	Chlorpheniramine maleate injection USP	IM, IV, or SC, 5–40 mg as a single dose	Up to 40 mg/day	
Dexchlorpheniramine maleate USP *Polaramine*	Dexchlorpheniramine maleate syrup USP Dexchlorpheniramine maleate tablets USP	Oral, 2 mg tid or qid		Children up to 12 yr of age: oral, 500 μg–1 mg tid to qid as needed; not recommended for use in premature or full-term neonates; pediatric use is not recommended
	Dexchlorpheniramine maleate extended-release tablets USP	Oral, 4 or 6 mg every 8–12 hr as needed		
Brompheniramine maleate USP *Dimetane*	Brompheniramine maleate elixir USP Brompheniramine maleate injection USP Brompheniramine maleate tablets USP	Oral, 4 mg tid or qid; IM, IV, or SC, 10 mg every 8–12 hr		Not recommended for use in premature or full-term neonates. Children 2–6 yr: oral, 1 mg tid or qid as needed; children 6–12 yr: oral, 2 mg tid or qid as needed; children up to 12 yr: IM, IV, or SC, 125 μg/kg of body weight or 3.75 mg/m^2 of body surface tid or bid
	Brompheniramine maleate extended-release tablets USP	Oral, 8 or 12 mg every 8–12 hr		Not recommended for children under 6 yr of age. 6 yr and over: oral, 8 or 12 mg every 12 hr
Dexbrompheniramine maleate USP *Disomer*	Dexbrompheniramine maleate tablets			
Triprolidine hydrochloride USP *Actidil*	Triprolidine hydrochloride syrup USP Triprolidine hydrochloride tablets USP	Oral, 2.5 mg tid or qid	Up to 10 mg/day	Not recommended for use in premature or full-term neonates. Children up to 2 yr of age: oral, 625 μg bid or tid; 2 yr and over: oral, 1.25 mg bid or tid
Promethazine hydrochloride USP *Phenergan*	Promethazine hydrochloride syrup USP Promethazine hydrochloride tablets USP	12.5 mg every 4–6 hr as necessary, or 25 mg qd hs	Up to 150 mg/day	Oral, 125 μg/kg of body weight every 4–6 hr, or 500 μg/kg hs as needed
	Promethazine hydrochloride injection USP Promethazine hydrochloride suppositories USP	IM or IV, 12.5–25 mg every 4–6 hr as needed	Up to 150 mg/day	

*See USP DI for complete dosage information.

TABLE 16-6 *Continued*

ANTIHISTAMINIC AGENTS: H_1-Receptor Antagonists

Name Proprietary Name	***Preparations***	***Usual Adult Dose****	***Usual Dose Range****	***Usual Pediatric Dose****
Trimeprazine tartrate USP *Temaril*	Trimeprazine tartrate syrup USP; Trimeprazine tartrate tablets USP	Oral, 2.5 mg qid		6 mo–2 yr of age: the equivalent of 1.25 mg trimeprazine qd to qid as necessary, not exceeding 5 mg/day; 3–6 yr: 2.5 mg qd to qid as necessary, not exceeding 10 mg/day; 7–12 yr: 2.5–5 mg qd to tid as necessary, not exceeding 15 mg/day
Cyclizine USP *Marezine*	Cyclizine lactate injection USP	IM, 50 mg every 4–6 hr as needed		IM, 1 mg/kg of body weight or 33 mg/m^2 of body surface tid as needed
Cyclizine hydrochloride USP *Marezine*	Cyclizine hydrochloride tablets USP	50 mg every 4–6 hr as needed	Up to 200 mg/day	Oral, 1 mg/kg of body weight or 33 mg/m^2 of body surface tid; children 6–12 yr: oral, 25 mg every 4–6 hr as needed
Chlorcyclizine hydrochloride USP		50 mg qd to qid	25–100 mg	
Meclizine hydrochloride USP *Bonine, Antivert*	Meclizine hydrochloride tablets USP; Meclizine hydrochloride chewable tablets USP	Oral, 25–50 mg one hr before travel, repeated every 24 hr as needed		Pediatric dosage has not been established
Buclizine hydrochloride USP *Bucladin-S*	Buclizine hydrochloride tablets USP	Oral, 50 mg every 4–6 hr as needed	Up to 150 mg/day	Pediatric dosage has not been established
Diphenylpyraline hydrochloride USP *Hispril, Diafen*	Diphenylpyraline hydrochloride extended-release capsules USP	Oral, 5 mg every 12 hr		Not recommended for use in children under 6 yr of age. Children 6 yr and over: oral, 5 mg qd as needed
Cyproheptadine hydrochloride USP *Periactin*	Cyproheptadine hydrochloride syrup USP; Cyproheptadine hydrochloride tablets USP	Oral, 4 mg tid or qid	Up to 500 μg/kg of body weight daily	Use is not recommended in premature or full-term neonates. Oral, 125 μg/kg of body weight or 4 mg/m^2 of body surface bid as needed; children 2–6 yr: oral, 2 mg bid or tid, not to exceed 12 mg/day; children 7–14 yr: oral, 4 mg bid to tid, not to exceed 16 mg/day
Azatadine maleate USP *Optimine*	Azatadine maleate tablets USP	Oral, 1–2 mg bid as needed		Use is not recommended in children up to 12 yr of age. Children 12 yr: oral, 500 μg–1 mg bid as needed
Phenindamine tartrate *Nolahist*	Phenindamine tartrate tablets USP	Oral, 25 mg every 4–6 hr	Up to 150 mg/day	Children 6–12 yr of age: oral, 12.5 mg every 4–6 hr, not to exceed 75 mg/day; children 12 yr and older: see Usual Adult Dose
Terfenadine *Seldane*	Terfenadine tablets USP	Oral, 60 mg every 8–12 hr		Dosage has not been established
Astemizole USP *Hismanal*	Astemizole tablets USP	Oral, 10 mg daily		Children 6–12 yr of age: oral, 5 mg daily

*See USP DI for complete dosage information.

Chlorcyclizine Hydrochloride, USP. 1-(*p*-Chloro-α-phenylbenzyl)-4-methylpiperazine monohydrochloride. This salt, a light-sensitive, white crystalline powder, is soluble in water (1 : 2), in alcohol (1 : 11), and in chloroform (1 : 4). A 1% solution has a *p*H between 4.8 and 5.5.

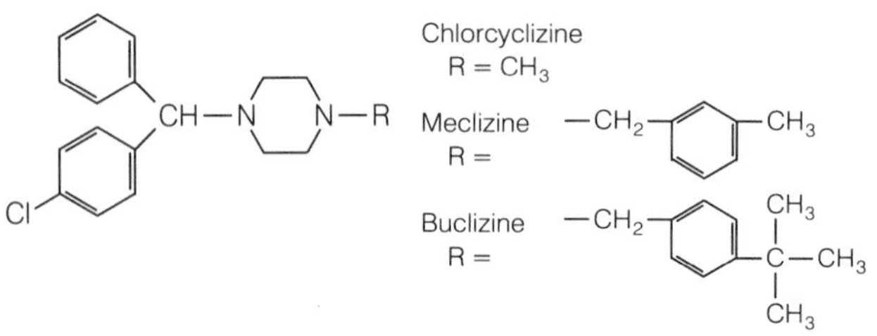

Chlorcyclizine R = CH_3
Meclizine R = $-CH_2-C_6H_4-CH_3$
Buclizine R = $-CH_2-C_6H_4-C(CH_3)_3$

Disubstitution or substitution of halogen in the 2- or 3-position of either of the benzhydryl rings results in a much less potent compound.

Chlorcyclizine is indicated in the symptomatic relief of urticaria, hay fever, and certain other allergic conditions.

Meclizine Hydrochloride, USP. 1-(*p*-Chloro-α-phenylbenzyl)-4-(*m*-methylbenzyl)piperazine dihydrochloride monohydrate (Bonine; Antivert). Meclizine hydrochloride is a tasteless, white or slightly yellowish crystalline powder that is practically insoluble in water (1 : 1000). It differs from chlorcyclizine by having an *N*-*m*-methylbenzyl group in place of the *N*-methyl group.

Although it is a moderately potent antihistaminic, meclizine is used primarily as an antinauseant in the prevention and treatment of motion sickness and in the treatment of nausea and vomiting associated with vertigo and radiation sickness.

Buclizine Hydrochloride, USP. 1-(*p-tert*-Butylbenzyl)-4-(*p*-chloro-α-phenylbenzyl)piperazine dihydrochloride (Bucladin-S). The salt occurs as a white to slightly yellow crystalline powder that is insoluble in water.

The highly lipid-soluble buclizine has central nervous system depressant, antiemetic, and antihistaminic properties. The salt is available in 50-mg tablets for oral administration. The usual dose is 50 mg 30 minutes before travel and repeated in four to six hours as needed.

MISCELLANEOUS COMPOUNDS

The miscellaneous group of compounds includes those agents that exhibit useful antihistaminic activity but do not fit conveniently into a chemical class (see Table 16-5 for structural comparison of these agents and Table 16-6 for the official preparations and dosages).

Diphenylpyraline Hydrochloride, USP. 4-(Diphenylmethoxy)-1-methylpiperidine hydrochloride (Hispril; Diafen). The salt occurs as a white or slightly off-white crystalline powder that is soluble in water or alcohol. Diphenylpyraline is structurally related to diphenhydramine with the aminoalkyl side chain incorporated in a piperidine ring.

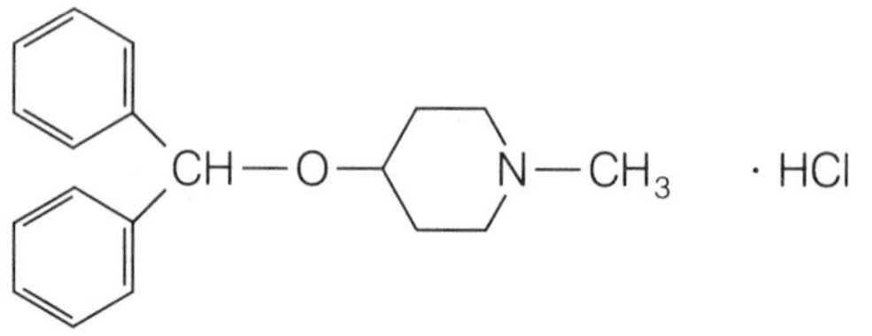

Diphenylpyraline Hydrochloride

Diphenylpyraline is a potent antihistaminic, and the usual dose is 2 mg three or four times daily. The hydrochloride is available as 5-mg sustained-release capsules.

Phenindamine Tartrate, USP. 2,3,4,9-Tetrahydro-2-methyl-9-phenyl-1*H*-indeno[2,1-*c*]pyridine bitartrate. The hydrogen tartrate occurs as a creamy white powder, usually having a faint odor, and sparingly soluble in water (1 : 40). A 2% aqueous solution has a *p*H of about 3.5. It is most stable in the *p*H range of 3.5 to 5.0 and is unstable in solutions of *p*H 7 or higher. Oxidizing substances or heat may cause isomerization to an inactive form.

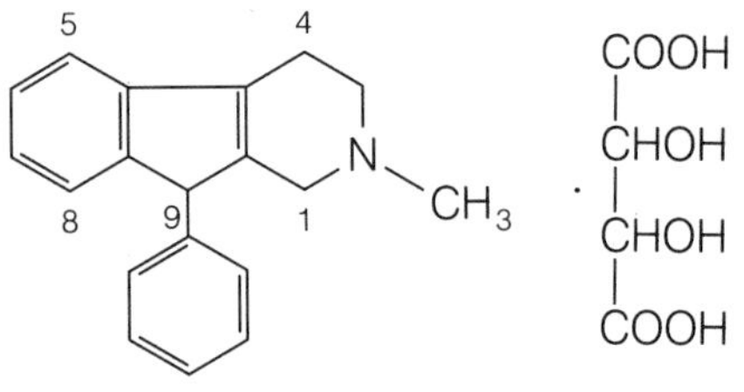

Phenindamine Tartrate

Structurally, phenindamine is related to the unsaturated propylamine derivatives in that the rigid ring system contains a distorted, *trans* $Ar-C=C-CH_2N$. Like the other commonly used antihistamines, it may produce drowsiness and sleepiness; but, also, it may cause a mildly stimulating action in some patients and insomnia when taken just before bedtime.[44]

Dimethindene Maleate. (±)2-[1-[2-[2-(Dimethylamino)ethyl]inden-3-yl]ethyl]pyridine bimaleate (1 : 1) (Forhistal Maleate). The salt occurs as a white to off-white crystalline powder that has a characteristic odor and is sparingly soluble in water. This potent antihistaminic agent may be considered as a derivative of the unsaturated propylamines. The principal side effect is some degree of sedation or drowsiness. The antihistaminic activity resides mainly in the levorotatory isomer.[15]

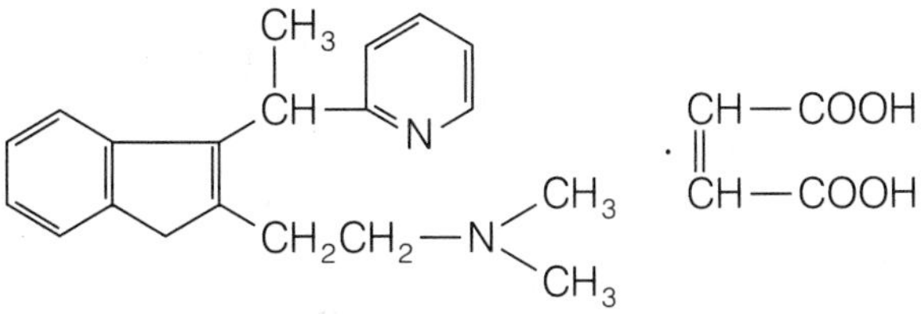

Dimethindene Maleate

Antazoline Phosphate. 2-[(*N*-Benzylanilino)-methyl]-2-imidazoline dihydrogen phosphate. The salt occurs as a bitter, white to off-white crystalline powder that is soluble in water. It has a pK_a of 10.0, and a 2% solution has a pH of about 4.5. Antazoline, similarly to the ethylenediamines, contains an *N*-benzylanilino group linked to a basic nitrogen through a two-carbon chain.

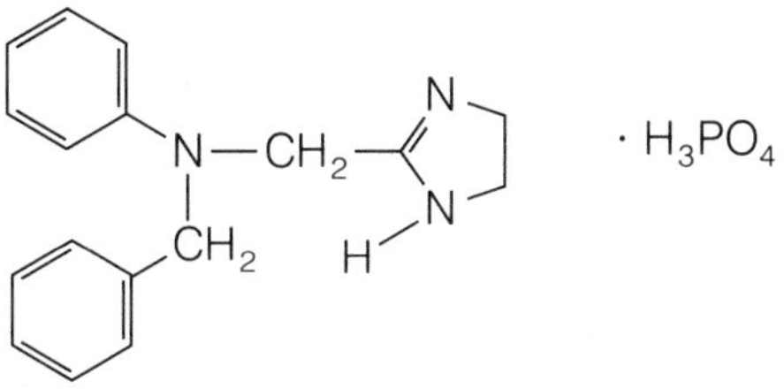

Antazoline Phosphate

Antazoline is less active than most of the other antihistaminic drugs, but it is characterized by the lack of local irritation. The more soluble phosphate salt is applied topically to the eye in a 0.5% solution. The less soluble hydrochloride is given orally. In addition to its use as an antihistamine, antazoline has over twice[45] the local anesthetic potency of procaine and also exhibits anticholinergic actions.

Cyproheptadine Hydrochloride, USP. 4-(5*H*-Dibenzo-[*a*, *d*]-cyclohepten-5-ylidene)-1-methylpiperidine hydrochloride sesquihydrate (Periactin). The salt is slightly soluble in water and sparingly soluble in alcohol.

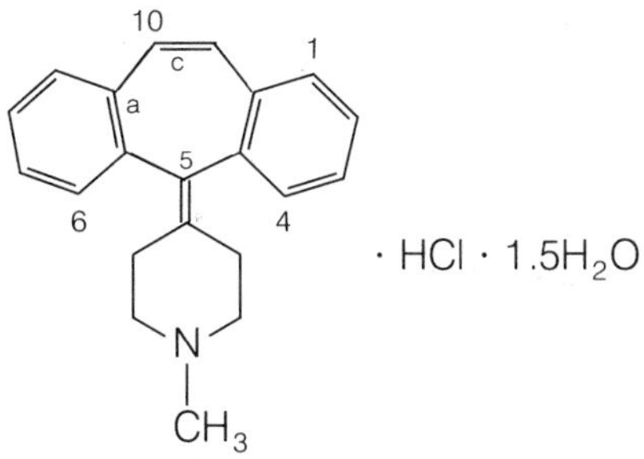

Cyproheptadine Hydrochloride

Cyproheptadine possesses both an antihistamine and an antiserotonin activity and is used as an antipruritic agent. Sedation is the most prominent side effect, and this is usually brief, disappearing after three or four days of treatment.

This dibenzocycloheptene may be regarded as a phenothiazine analogue in which the sulfur atom has been replaced by an isosteric vinyl group and the ring nitrogen replaced by an sp^2 carbon atom.

Azatadine Maleate, USP. 6,11-Dihydro-11-(1-methyl-4-piperidylidene)-5*H*-benzo[5,6]cyclohepta-[1,2-*b*]pyridine imaleate (1 : 2) (Optimine). Azatadine is an aza isostere of cyproheptadine in which the 10,11-double bond is reduced.

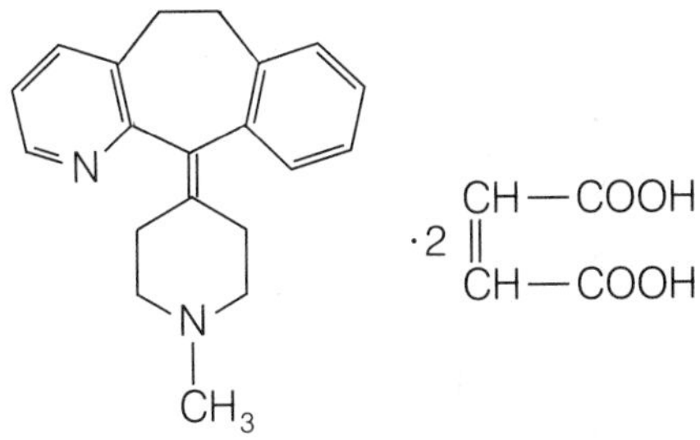

Azatadine Maleate

In early testing, azatadine exhibited more than three times the potency of chlorpheniramine in the isolated guinea pig ileum screen and more than seven times the oral potency of chlorpheniramine in protection of guinea pigs against a double lethal dose of intravenously administered histamine.[46]

It is a potent, long-acting antihistaminic with antiserotonin activity. The usual dosage is 1 to 2 mg twice daily. Azatadine is available in 1-mg tablets.

Terfenadine. Alpha-[4-(1,1-Dimethylethyl)-phenyl]-4-(hydroxydiphenylmethyl)-1-piperidinebutanol (Seldane) is a reduced butyrophenone derivative of an aminoalcohol-type anticholinergic agent. It occurs as a white to off-white crystalline powder that is soluble in alcohol and very slightly soluble in water.

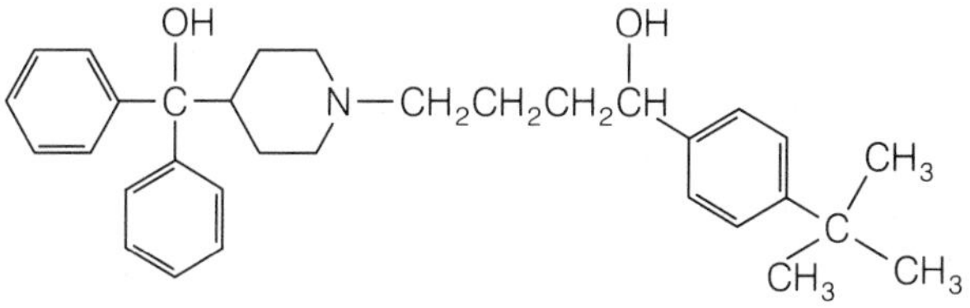

Terfenadine

Clinical studies[47,48] showed no significant anticholinergic or central nervous system side effects.

Terfenadine is 97% plasma protein bound with a half-life of about 20 hours. It has a slow onset of action (three to four hours to peak effect, orally) and a long duration of action (over 12 hours). It is primarily excreted in feces.

Usual adult dose: oral, 60 mg every 8 to 12 hours as needed. Pediatric dosage has not been established.
Dosage form: Terfenadine tablets, USP

Astemizole, USP. 1-(4-Fluorobenzyl)-2-((1-(4-methoxyphenyl)-4-piperidyl)amino)benzimidazole (Hismanal). Astemizole is often compared to terfenadine because both lack anticholinergic and sedative properties.

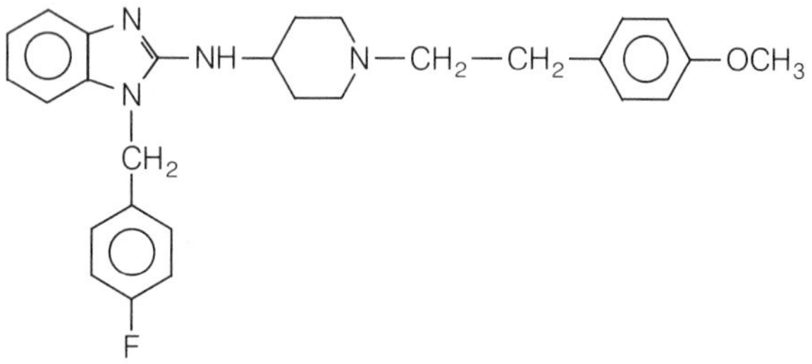

Astemizole

Astemizole is rapidly absorbed orally and should be administered 1 hour before meals. It is highly protein bound (96%) and has a plasma half-life of 1.6 days. The primary route of elimination is in the feces.

It is used for seasonal allergic rhinitis and chronic urticaria. It has a slow onset of action (2 to 3 days).

Usual adult dose: oral, 10 mg once daily. Usual pediatric dose: children 6 to 12 years of age, oral, 5 mg once daily.
Dosage form: Astemizole Tablets, USP, 10 mg.

INHIBITION OF HISTAMINE RELEASE

Interest has been generated in the suppression of release of autacoids as a therapeutic approach to the treatment of hypersensitivity.[31] The drug that has focused attention on this possibility is cromolyn.[49]

Cromolyn Sodium, USP. Disodium 1,3-bis(2-carboxychromon-5-yloxy)-2-hydroxypropane (Intal). The salt is a hygroscopic, white, hydrated crystalline powder that is soluble in water (1 : 10). It is tasteless at first, but leaves a very slightly bitter aftertaste. The pK_a of cromolyn is 2.0. Cromolyn belongs to a completely novel class of compounds and bears no structural relationship to other commonly used antiasthmatic compounds. Unlike its naturally occurring predecessor (khellin), cromolyn is not a smooth-muscle relaxant or a bronchodilator. It has no intrinsic bronchodilator, antihistaminic, or anti-inflammatory action.

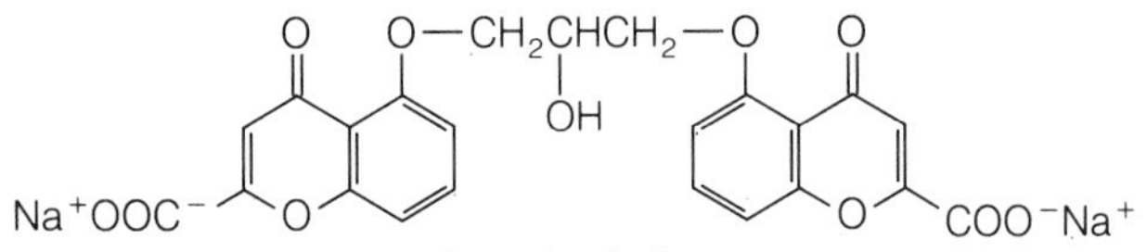

Cromolyn Sodium

Cromolyn inhibits release of histamine, leukotrienes, and other potent substances from mast cells during allergic responses. Apparently, its action is on the mast cell after the sensitization stage but before the antigen challenge. It does not seem to interfere with the antigen–antibody reaction, but it seems to suppress the responses to this reaction.

Although growing evidence[50] indicates that the mechanism of action is not all mast cell related, the benefits of the drug in asthma are exclusively prophylactic. It is of no value after an asthmatic attack has begun (status asthmaticus). Cromolyn is also indicated for the prevention and treatment of the symptoms of allergic rhinitis.

Usual adult dose: Intranasal, 5.2 mg (one metered spray) in each nostril three or four times daily at regular intervals. Dosage has not been established for use in children up to six years of age. For children six years of age and older, see the usual adult dose.
Dosage forms: Cromolyn sodium for nasal insufflation, USP; Cromolyn sodium nasal solution, USP

The antiallergic for ophthalmic use is relatively new.

Usual adult dose: Ophthalmic, one drop of a 2% to 4% solution four to six times daily. Dosage for children up to four years of age has not been established. For children over four years of age, administer the usual adult dose.
Dosage form: Cromolyn sodium ophthalmic solution, USP

H_2-ANTAGONISTS

Although centrally acting agents, such as the sedatives and antianxiety agents, remain important adjuncts in the therapy of peptic ulcer disease, they are not included in this section. Carbenoxolone sodium has been shown[51] to increase the rate of healing of ulcers by increasing the secretion of gastric mucoproteins. Prostanoids inhibit gastric acid secretion and enhance cytoprotection.

In a very short time, H_2-antagonists have become an important alternative in the therapy of peptic ulcers. To place this category of drugs in proper perspective, the discussion follows the general outline below.

Mechanisms of Inhibition of Pepsin Activity

- Chemical complexation
- *p*H control
  - —Antacid
  - —Antisecretory

The discussion emphasizes the advantages and disadvantages of the *mechanisms* involved and not the pros and cons of specific agents. When a side effect is recognized as being a consequence of the desired mechanism of action, the search for a more specifically acting agent among structural congeners is usually fruitless. Structural manipulation, here, can eliminate the side effect only if it results in an agent that acts by a different mechanism.

Aside from the degree of protection versus the degree of insult, there seems to be general agreement that the common denominator in the etiology of peptic ulceration is the presence of the active proteolytic enzyme, pepsin. Therefore, the mechanisms used to treat and prevent peptic ulcer disease are mechanisms of pepsin inhibition.

Chemical Complexation

The sulfate esters and sulfonate derivatives of polysaccharides and lignin form chemical complexes with the enzyme pepsin. These complexes have no proteolytic activity. Because polysulfates and polysulfonates are poorly absorbed from the gastrointestinal tract, specific chemical complexation appears to be a desirable mechanism of pepsin inhibition. Unfortunately, these polymers are also potent anticoagulants.

The properties of chemical complexation and anticoagulant action are separable by structural variation. In a comparison of selected sulfated saccharides of increasing number of monosaccharide units, from disaccharides through starch-derived polysaccharides of differing molecular size, three conclusions are supported by the data:[52] (1) the anticoagulant activity of sulfated saccharide is positively related to molecular size; (2) anticoagulant activity is absent in the disaccharides; and (3) the inhibition of pepsin activity and the protection against experimentally induced ulceration is dependent on the degree of sulfation and not on molecular size.

The readily available disaccharide, sucrose, has been used to develop a useful antiulcer agent, sucralfate.

Sucralfate. β-D-Fructofuranosyl-α-D-glucopyranoside, octakis (hydrogen sulfate), aluminum complex (Carafate) is the aluminum hydroxide complex of the octasulfate ester of sucrose. It is practically insoluble in water and soluble in strong acids and bases. It has a pK_a value between 0.43 and 1.19.

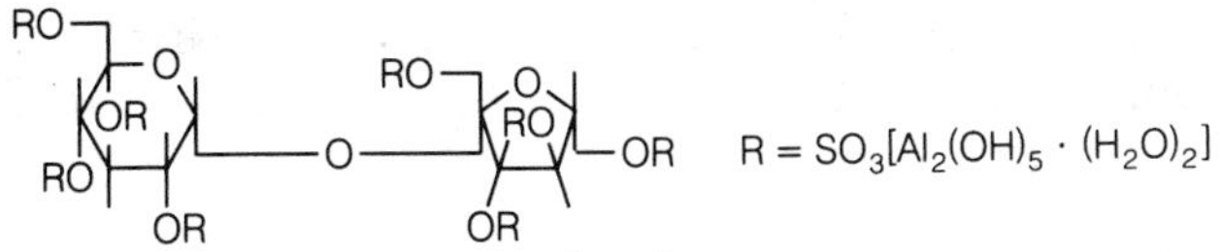

Sucralfate

Sucralfate is minimally absorbed from the gastrointestinal tract and thus exerts its antiulcer effect through local rather than systemic action. It has negligible acid-neutralizing or buffering capacity in therapeutic doses.

Its mechanism of action has not been established. The data[52] suggest that sucralfate binds preferentially to the ulcer site to form a protective barrier that prevents exposure of the lesion to acid and pepsin. In addition, it adsorbs pepsin and bile salts. Either would be very desirable modes of action.

The product labeling states that the simultaneous administration of sucralfate may reduce the bioavailability of certain agents (e.g., tetracycline, phenytoin, digoxin, or cimetidine). It further recommends restoration of bioavailability by separating administration of these agents from that of sucralfate by two hours. Presumably sucralfate binds these agents in the gastrointestinal tract.

The most frequently reported adverse reaction to sucralfate is constipation (2.2%). Antacids may be prescribed as needed, but should not be taken within one-half hour before or after sucralfate.

Usual adult dose: oral, 1 g four times a day on an empty stomach

Dosage form: 1-g sucralfate tablets

*p*H Control

Pepsin activity is *p*H dependent. In Figure 16-3, the solid line illustrates the *p*H dependency[53] of human pepsin activity incubated at 37°C. Maximal peptic activity is obtained at a *p*H of 1.5 to 2.5 (the actual optimal *p*H varies slightly with the method used). Seventy percent of maximal activity occurs in the *p*H range of 2.5 to 5, with almost no activity above *p*H 5.

The dotted line in Figure 16-3 illustrates the stability of the enzyme at various *p*H values. At *p*H levels up to 7.5, pepsin is stable and lowering of *p*H restores maximal pepsin activity. However, at *p*H levels above 7.5, the enzyme is irreversibly inactivated (alkaline denaturation).

The best value at which the *p*H of the gastric contents should be controlled in the treatment of peptic ulcer disease is not established. There is general agreement that antacids relieve pain of ulcers. The use of antacids in peptic ulcer treatment has been reviewed.[54] Two randomized, double-blind, multicenter trials[55,56] showed that intensive antacid therapy is superior to placebo in healing duodenal ulcer. In each trial, 210 mL of an Al-Mg antacid was administered daily in seven divided doses. The in vitro buffering capacity of each 30-mL dose was 123 mEq of hydrochloric acid. Possibly, peak *p*H

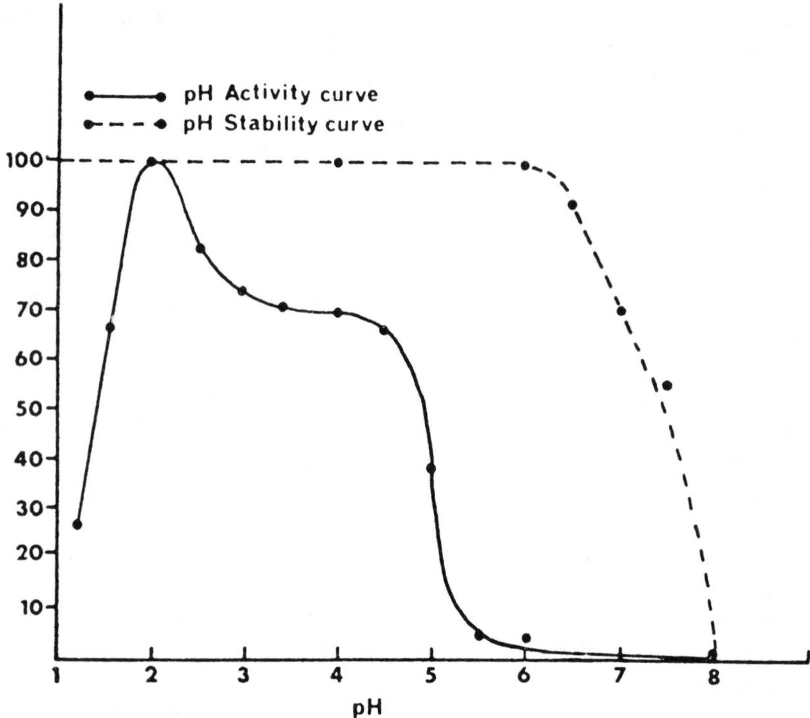

FIG. 16-3. The pH stability and pH activity of human pepsin. (From Piper, D. W., and Fenton, B. H.: Gut 6:506, 1965.)

values were sufficiently high to inactivate pepsin irreversibly.

Two different approaches to *p*H control of gastric contents are available to the practitioner: antacid administration and antisecretory agents.

Antacid Mechanism

Although the question of the best *p*H level to which the gastric contents should be buffered remains unanswered, the mechanism of antacid action is generally agreed to be neutralization of acid rather than chemical complexation with pepsin. Also, the antacid mechanism is desirable regardless of whether the attack factor in peptic ulceration is acid or acid–pepsin. However, larger doses of antacids would be required if the attack factor is acid–pepsin.[54]

Without regard to the side effects of specific agents, several disadvantages (theoretically at least) are associated with the antacid mechanism.

One complicating factor is the uncertainty of the dose interval. Except with intragastric drip, continuous buffering is difficult to achieve, and practical considerations force a compromise in which buffering is achieved discontinuously. On the one hand, the rate and quantity of acid secretion varies with the individual's thoughts of food (the cephalic phase, in which the secretions are richer in pepsin), eating habits, and rate of gastric emptying. On the other hand, antacids promote gastric emptying, which limits the duration of antacid action.

A second problem with the titration of gastric contents involves a reflex mechanism. The concept of acid rebound is of historical interest, but as a clinical problem, it is not universally accepted and apparently is difficult to document.[57] Theoretically, acid rebound is a potential problem because *p*H of the gastric contents influences the release of gastrin (a potent gastric stimulatory hormone). When the acidity of the gastric contents increases to a *p*H of 2.0, the gastrin mechanism for stimulating gastric secretion becomes totally blocked.[58] However, a rise in *p*H above 3 causes a release of gastrin. Therefore, the antacid mechanism indirectly stimulates acid secretion.

A third built-in difficulty associated with the antacid mechanism is related to the inhibition of secretion of intestinal factors. Excess acid in the duodenum stimulates the release of the hormone secretin, which stimulates the pancreas to release alkaline juices rich in digestive enzymes. The effect of antacid therapy on this complicated reflex mechanism has not been assessed. However, the potential does exist that antacids may interfere with release of intestinal hormones. Thus, they may indirectly prevent a slowing of gastric secretions, decrease the protection of the alkaline pancreatic juices to the duodenum, and adversely affect digestion of food products.

In addition, several problems, including systemic alkalosis, diarrhea, constipation, and a high sodium content, may be associated with the use of specific antacid products.

Antisecretory Mechanisms: Antimuscarinic Agents

The older, more familiar method of controlling gastric acid secretion involves blocking the vagus nerve. The neurogenic mechanism regulating gastric secretions is mediated through the parasympathetic fibers of the vagus nerves. Therefore, antimuscarinic agents have been used extensively as adjuncts to the treatment of peptic ulcer.

Except for its central nervous system effects, atropine is relatively specific for muscarinic sites. The synthetic quaternary ammonium drugs generally exhibit varying degrees of nicotinic-blocking activity. However, both types of agents are highly potent vagal blockers, the action of which is sometimes referred to as *chemical vagotomy*.

The disadvantages of the antimuscarinic mechanism of controlling gastric acid secretion are related to two important factors. For one, the antimuscarinic agents are not specific for vagal muscarinic receptors. These drugs block other muscarinic sites as well, and the resulting actions are considered to be side effects inherent in the mechanism of action. Additionally, other side effects may occur that are specifically related to the drug (e.g., central nervous system-stimulating actions of atropine).

Second, the antimuscarinic agents only partially inhibit gastric secretions. The greater volume of gastric secretion occurs during the gastric and intestinal phases, which are primarily under hormonal (gastrin) control. Furthermore, the drugs reduce gastric secretion, provided the dose is increased to the limits of tolerance of side effects. At doses sufficient to block vagally stimulated gastric secretions totally, intolerable side effects occur.

An additional theoretical factor is delayed gastric emptying caused by antimuscarinic action. The desirability of this action, which would prolong the neutralizing effect of antacids, might be counteracted by equally prolonged production of acid because of the continued presence of food in the stomach. (The antimuscarinic agents and their actions are discussed in Chap. 12.)

In spite of the fact that antacids provide only one of several therapeutic measures in the treatment of peptic ulcer, they are among the first agents employed. Antimuscarinic agents have been used extensively as adjuncts to dietary and antacid treatment. The value of these two mechanisms in the treatment and prevention of peptic ulcer disease is difficult to assess. However, the disadvantages of these mechanisms have demonstrated the need for the development of an alternative mechanism. The H_2-antagonists provide that alternative. It is too early to assess the value and impact of the H_2-antagonists on the therapy of this disease. The future will disclose the disadvantages, if any, of this alternative mechanism.

Antisecretory Mechanisms: H_2-Antagonists

Beginning with the discovery of the pharmacologic actions of histamine, many hypotheses have been proffered concerning the role of histamine in gastric secretion.[59] The most enduring hypothesis is the concept that histamine is the final common local mediator of the parietal cells, by which it is meant that all other stimulants of gastric secretion act through histamine. Although it has served very valuably to stimulate much highly productive research, the final common mediator hypothesis is no longer tenable. This certainly does not mean that histamine has no role in the normal physiology of gastric secretion.

The final common mediator hypothesis is not consistent with the patterns of action of H_2-antagonists and antimuscarinic agents. The hypothesis that histamine is the final common mediator cannot account for the observation that antimuscarinic agents inhibit the action of gastrin and histamine on gastric secretion. Likewise, the hypothesis that acetylcholine is the common mediator cannot account for the observation that H_2-antagonists inhibit the action of gastrin and cholinergic agents on gastric acid secretion.[60]

The discovery of H_2-antagonists opened another door to the secrets of gastric secretory mechanisms. Of the three most potent known gastric secretagogues (gastrin, histamine, acetylcholine), two have yielded to useful competitive antagonists. The tools are now available to study their normal physiologic roles and their potentiating interactions.

A *final common potentiator hypothesis* has been proposed[60] that is a modified version of the previous final common mediator hypothesis. The hypothesis states simply that H_2-antagonists inhibit the direct action of histamine on acid secretion and inhibit the potentiating action of histamine on acid secretion stimulated by gastrin or acetylcholine. According to the modified hypothesis, secretagogues have two efficacies: (1) an intrinsic efficacy, which refers to the maximal response it can produce in the absence of other agents; and (2) a potentiating efficacy, which refers to the magnitude of its response in the presence of a second agent that potentiates its action.

Histamine is assumed to have both intrinsic efficacy and potentiating efficacy, whereas gastrin and acetylcholine have only potentiating efficacy. This means that in the absence of histamine, neither gastrin nor cholinergic agents would increase acid secretion. Only histamine can increase acid secretion. Gastrin or acetylcholine increases acid secretion by potentiating the effect of histamine.

Thus, an H_2-antagonist inhibits local secretion and gastric secretion stimulated by gastrin and cholinergic agents by blocking the intrinsic and potentiating actions of histamine. An antimuscarinic agent (e.g., atropine) suppresses histamine-stimulated gastric secretion by blocking the potentiating action of acetylcholine.

Some in vitro studies,[61] using parietal cells isolated from canine fundic mucosa, partially support the hypothesis. From indices of response to stimulation other than secretion, the data suggest that the parietal cell contains receptors for each of the three secretagogues. Each secretagogue exhibited an intrinsic efficacy that could be blocked only by its specific antagonist. In other words, the intrinsic efficacy of histamine was blocked only by an H_2-antagonist, that of carbachol (cholinergic agent) was blocked only by atropine, and that of gastrin was not blocked by either an H_2-antagonist or atropine.

Also, each secretagogue markedly potentiated the action of each of the other two. Again, H_2-blockers inhibited the potentiating action only of histamine, and atropine inhibited only the potentiating effect of carbachol.

Structure–Activity Relationships

A review[62] of the characterization and development of cimetidine as a histamine H_2-receptor antagonist reveals a classic medicinal chemistry approach to problem-solving. Beginning with the study of the relative potencies of the methylhistamines (see Fig. 16-1), hundreds of compounds were synthesized and their actions studied, leading to the development of cimetidine. Since then, many more new compounds have been reported, many of which exhibit a degree of inhibition of gastric secretion. Some of the compounds inhibit gastric secretion by mechanisms other than competitive H_2-antagonism.

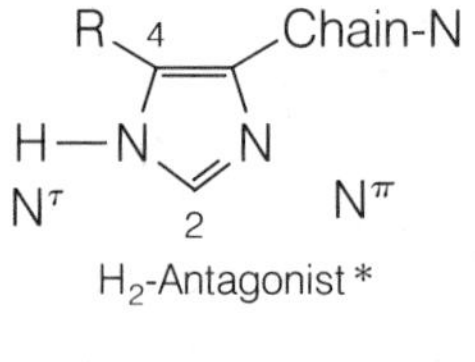

R 4 Chain-N

H—N N

N^{τ} 2 N^{π}

H_2-Antagonist*

A discussion of structure–activity relationships of the H_2-receptor antagonists is, at best, tentative. Structural variations and activity have been reviewed.[63]

Obviously, the imidazole ring of histamine is not required for competitive antagonism of histamine at H_2-receptors. Other heterocyclic rings may be used. However, if the imidazole ring is used, the N^{τ}-H tautomer should be the predominant species for maximal H_2-antagonist activity. The electronic effects of the ring substituents determine the tautomerism.

Separation of the ring and the nitrogen group with the equivalent of a four-carbon chain appears to be necessary for optimal antagonist activity. The isosteric thioether link is present in the four agents currently approved. Its purpose in cimetidine is to influence tautomerism of the imidazole ring.

The terminal nitrogen group should be a polar, nonbasic substituent for maximal antagonist activity. Groups that are positively charged at physiologic *p*H appear to confer agonist activity. In general, antagonist activity varies inversely with the hydrophilic character of the nitrogen group. However, the hydrophilic group, 1,1-diaminonitroethene, found in ranitidine and nizatidine is an exception[64] in that it is much more active than is predicted by partition.

Cimetidine, USP. *N″*-Cyano-*N*-methyl-*N′*-[2-[[(5-methylimidazol-4-yl)methyl]thio]ethyl]guanidine (Tagamet). The drug is a colorless crystalline solid that is slightly soluble in water (1.14% at 37°C). The solubility is greatly increased with the addition of dilute acid to protonate the imidazole ring (apparent pK_a of 7.09).[62] At *p*H 7, aqueous solutions are stable for at least seven days. Cimetidine is a relatively hydrophilic molecule having an octanol/water partition coefficient of 2.5.

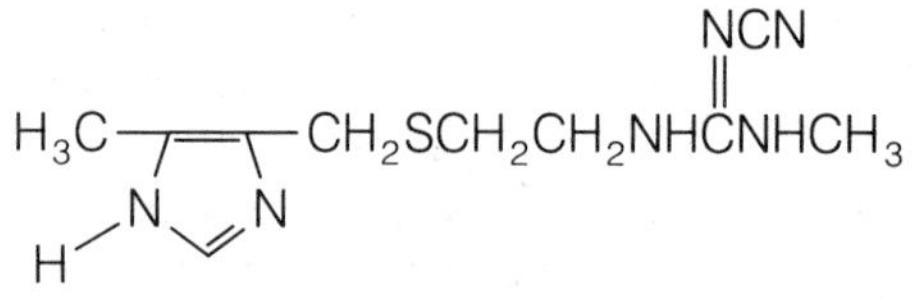

NCN

H_3C—(imidazole)—$CH_2SCH_2CH_2NHCNHCH_3$

H—N N

Cimetidine

Cimetidine inhibits hepatic cytochrome P-450 and P-448 mixed function oxidase systems. Concurrent administration of drugs such as benzodiazepines, warfarin, phenytoin, theophylline, and others, which are significantly metabolized by the hepatic microsomal enzymes, should be avoided or monitored closely. Also, it exhibits antiandrogenic action that sometimes leads to gynecomastia.

Antacids interfere with cimetidine absorption and should be administered at least one hour before or after a cimetidine dose.

* Numbering of the ring in the general structure of an H_2-antagonist is according to the system followed in the literature.

TABLE 16-7

ANTIHISTAMINIC AGENTS: H_2-Antagonists

Name Proprietary Name	*Preparations*	*Usual Adult Dose**	*Usual Dose Range**	*Usual Pediatric Dose**
Cimetidine USP *Tagamet*	Cimetidine tablets USP Cimetidine hydrochloride injection USP Cimetidine hydrochloride oral solution USP	Oral, 300 mg qid 300 mg (base) every 6 hr	Up to 2400 mg/day	20–40 mg/kg of body weight qid
Famotidine USP *Pepcid*	Famotidine tablets USP Famotidine injection USP	Oral, 20 mg bid or 40 mg hs		Children: dosage has not been established
Ranitidine USP *Zantac*	Ranitidine hydrochloride tablets USP Ranitidine hydrochloride injection USP	Oral, 150 mg (base) bid or 300 mg hs	Up to 400 mg/day	Children: dosage has not been established
Nizatidine *Axid*	Nizatidine capsules	Oral, 300 mg at bedtime or 150 mg twice daily		Children: dosage has not been established

*See USP DI for complete dosage information.

Cimetidine is eliminated primarily by renal excretion. About 48% of an oral dose is excreted unchanged in 24 hours.

Usual adult dose: Duodenal or gastric ulcer—oral, 300 mg four times a day, with meals and at bedtime; parenteral, the equivalent of 300 mg of cimetidine (as the hydrochloride salt) every six hours. The prescribing limit is 2400 mg daily by either route of administration. For prophylaxis of recurrent ulcer, the recommended oral dose is 400 mg at bedtime.

Pathologic hypersecretory conditions—oral, 300 mg four times a day, with meals and at bedtime, as long as clinically indicated.

Usual pediatric dose: oral, 20 to 40 mg (base) per kilogram of body weight four times a day, with meals and at bedtime.

Dosage forms: Cimetidine tablets, USP; Cimetidine hydrochloride injection, USP; and Cimetidine hydrochloride oral solution, USP

Famotidine, USP. *N'*-(Aminosulfonyl)-3-[[[2-[(diaminomethylene)amino]-4-thiazolyl]methyl]thio]-propanimidamide (Pepcid) is a white to pale yellow crystalline compound that is very slightly soluble in water and practically insoluble in ethanol.

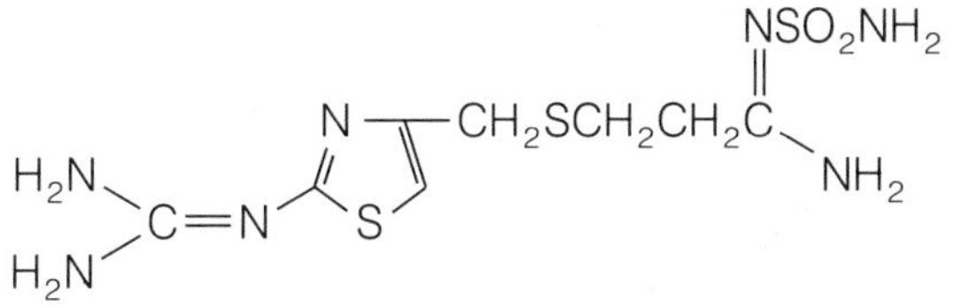

$$(H_2N)_2C{=}N\text{–(thiazolyl)–}CH_2SCH_2CH_2C({=}NSO_2NH_2)NH_2$$

Famotidine

Famotidine is a competitive inhibitor of histamine H_2-receptors and inhibits basal and nocturnal gastric secretion as well as secretion stimulated by food and pentagastrin. Its current labeling indications are for the short-term treatment of duodenal ulcer and treatment of pathologic hypersecretory conditions (e.g., Zollinger-Ellison syndrome).

No cases of gynecomastia, increased prolactin levels, or impotence have been reported, even at the higher dosage levels used in patients with pathologic hypersecretory conditions. Studies with famotidine in humans, in animal models, and in vitro have shown no significant interference with the disposition of compounds metabolized by the hepatic microsomal enzymes (e.g., cytochrome P-450 system).

Famotidine is incompletely absorbed. The bioavailability of oral doses is 40% to 45%. The effects of food or antacid on bioavailability are not clinically significant.

Usual adult dose: oral, 40 mg once daily at bedtime or 20 mg twice daily. Safety and effectiveness in children have not been established.

Dosage forms: Famotidine injection, USP; Famotidine tablets, USP

Ranitidine, USP. *N*-[2-[[[5-[Dimethylaminomethyl]-2-furanyl]methyl]thio]ethyl]-*N'*-methyl-2-nitro-1,1-ethenediamine (Zantac) is an aminoalkyl furan with pK_a values of 2.7 and 8.2. It is a white solid. The hydrochloride salt is highly soluble in water.

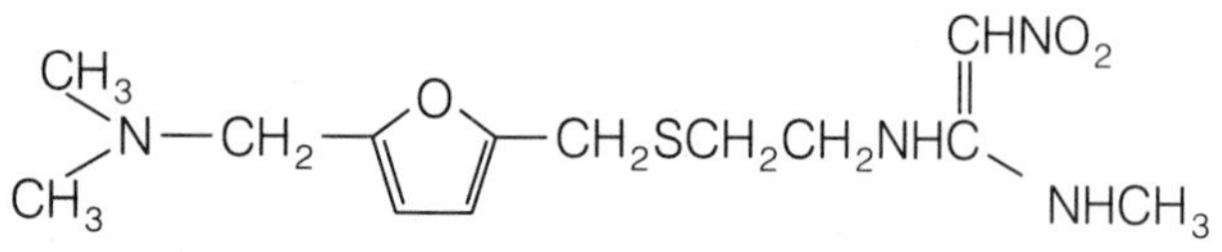

Ranitidine

Bioavailability of an oral dose of ranitidine is about 50% and is not significantly affected by the presence of food. Some antacids may decrease ranitidine absorption and should not be taken within one hour of its administration.

Its mechanism of inhibition of basal and nocturnal gastric acid secretion is competitive inhibition of the action of histamine at the H_2-receptors of the parietal cells. Ranitidine is a weak inhibitor of hepatic cytochrome P-450 mixed function oxidase system.

Usual adult dose: Oral, 150 mg (base) twice daily or 300 mg at bedtime. Pediatric dosage has not been established.

Dosage forms: Ranitidine hydrochloride tablets, USP; Ranitidine hydrochloride injection, USP

Nizatidine. *N*-[2-[[[2-[(dimethylamino)methyl]-4-thiazolyl]-methyl]thio]-ethyl]-*N*-methyl-2-nitro-1,1-ethenediamine, is an off-white to buff colored crystalline solid that is soluble in water and binds (35%) insignificantly to plasma proteins. It is a competitive inhibitor of

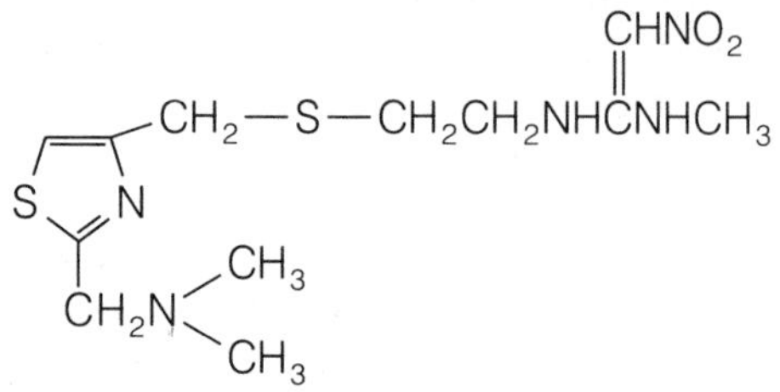

Nizatidine

histamine H_2-receptors and inhibits basal and nocturnal gastric secretion as well as secretion stimulated by betazole, pentagastrin, and food.

Bioavailability of nizatidine orally exceeds 90%. The effects of antacids or food on bioavailability are not clinically significant. The elimination half-life is 1–2 hours. It is excreted primarily (90%) in the urine and mostly (60%) as unchanged drug.

Nizatidine exhibited no demonstrable antiandrogenic action or inhibition of cytochrome P-450-linked drug-metabolizing enzyme system. Sweating and urticaria were reported significantly more frequently in nizatidine than in placebo patients.

Usual adult dose: Active Duodenal Ulcer—Oral, 300 mg once daily at bedtime or 150 mg twice daily. Maintenance: Oral, 150 mg once daily at bedtime. Pediatric dosage has not been established.

Antisecretory Mechanisms: Proton Pump Inhibitors

The final step in acid secretion in the parietal cell is the extrusion ("pumping") of protons. The membrane pump, H^+,K^+-ATPase, catalyzes the exchange of hydrogen ions for potassium ions.

Inhibition of this proton pump acts beyond the site of action of second messengers, e.g. calcium ion and cyclic AMP, and is independent of the action of secretogogues histamine, gastrin, and acetylcholine. Thus, acid pump inhibitors block basal and stimulated secretion.

An apparent disadvantage of this mechanism of gastric acid inhibition is the increase in circulating gastrin caused by the reflex reponse to a rise in gastric pH. If this hypergastrinemia is maintained, fundal mucosal hyperplasia will ensue. In rats this leads to carcinoid tumors.[65]

Also, total inhibition of gastric secretion is conducive to colonization of bacteria in the stomach. Intragastric bacteria may reduce dietary nitrate to nitrite and thus facilitate intragastric formation of N-nitroso compounds, some of which are carcinogenic.[65]

Substituted benzimidazoles are potent gastric acid pump inhibitors. One has been approved for use in humans.

Omeprazole. 5-Methoxy-2-(((4-methoxy-3,5-dimethyl-2-pyridinyl)methyl)sulfinyl)-1H-benzimidazole (Losec). Omeprazole is acid labile and therefore available in delayed-release capsules containing enteric coated granules. It is extensively metabolized with a plasma half-life of about 1 hour. Little, if any, unchanged drug is excreted in the urine.

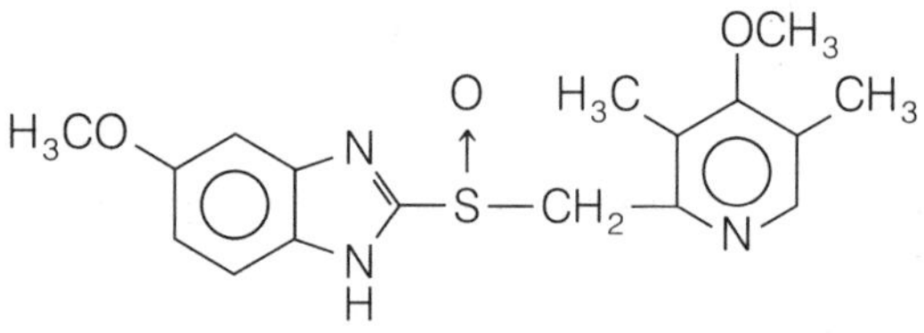

Omeprazole

Its action persists for 24 to 72 hours, long after the drug has disappeared from plasma. The suggested mechanism of action is irreversible inhibition of H^+,K^+-ATPase, involving oxidation of key sulfhydryl groups in the parietal cell membrane enzyme.

It is effective in the treatment of both gastric and duodenal ulcers as well as other hypersecretory diseases.

Usual adult dose: oral, 20 mg once daily.

Dosage form: delayed-release capsules containing 20 mg of omeprazole in enteric coated granules.

REFERENCES

1. Windaus, A., and Vogt, W.: Ber. Deutsch. Chem. Ges. 40:3691, 1907.
2. Dale, H. H., and Laidlaw, P. P.: J. Physiol. (Lond.) 41:318, 1910.
3. Dale, H. H., and Laidlaw, P. P.: J. Physiol. (Lond.) 43:182, 1911.
4. Dale, H. H., and Laidlaw, P. P.: J. Physiol. (Lond.) 52:355, 1919.
5. Beaven, M. A.: Monographs in Allergy. Histamine: Its Role in Physiological and Pathological Processes, Vol. 13, p. 22. New York, S. Karger, 1978.
6. Stern, P.: The relation of histamine to inflammation. In Rocha e Silva, M. (ed.). Handbook of Experimental Pharmacology, Vol. 18/1, p. 892. New York, Springer-Verlag, 1966.
7. Zweerink, H. J.: Early biochemical events leading to mast cell and basophil degranulation. In Hess, H. J. (ed.). Ann. Rep. Med. Chem. 18:247, 1983.
8. Altura, B. M., and Halevy, S.: Cardiovascular actions of histamine. In Rocha e Silva, M. (ed.). Handbook of Experimental Pharmacology, Vol. 18/2, p. 1. New York, Springer-Verlag, 1978.
9. Kahlson, G., and Rosengren, E.: Physiol. Rev. 48:155, 1968.
10. Schwartz, J. C.: Life Sci. 17:503, 1975.
11. Johnson, L. R.: Histamine and gastric secretion. In Rocha e Silva, M. (ed.). Handbook of Experimental Pharmacology, Vol. 18/2, p. 41. New York, Springer-Verlag, 1978.
12. Pelletier, G.: Naturally occurring antihistaminics in body tissues. In Rocha e Silva, M. (ed.). Handbook of Experimental Pharmacology, Vol. 18/2, p. 369. New York, Springer-Verlag, 1978.
13. Best, C. H., et al.: J. Physiol. (Lond.) 62:397, 1927.
14. Fourneau, E., and Bovet, D.: Arch. Int. Pharmacodyn. 46:178, 1933.
15. Casy, A. F.: Chemistry of anti-H_1 histamine antagonists. In Rocha e Silva, M. (ed.). Handbook of Experimental Pharmacology, Vol. 18/2, p. 175. New York, Springer-Verlag, 1978.
16. Witiak, D. T.: Antiallergenic agents. In Burger, A. (ed.). Medicinal Chemistry, 3rd ed., p. 1643. New York, Wiley-Interscience, 1970.
17. Paton, D. M.: Receptors for histamine. In Schachter, M. (ed.). Histamine and Antihistamines, p. 3. New York, Pergamon Press, 1973.
18. Rocha e Silva, M., and Antonio, A.: Bioassay of antihistaminic action. In Rocha e Silva, M. (ed.). Handbook of Experimental Pharmacology, Vol. 18/2, p. 381. New York, Springer-Verlag, 1978.
19. Ahlquist, R. P.: Am. J. Physiol. 153:586, 1948.
20. Lin, T. M., et al.: Ann. N.Y. Acad. Sci. 99:30, 1962.
21. Ash, A. S. F., and Schild, H. O.: Br. J. Pharmacol. Chemother. 27:427, 1966.
22. Black, J. W., et al.: Nature 236:385, 1972.
23. Biel, J. H., and Martin, Y. C.: Organic synthesis as a source of new drugs. In Gould, R. F. (ed.). Drug Discovery, Advances in Chemistry Series No. 108, p. 81. Washington, DC, American Chemical Society, 1971.
24. Metzler, D. E., Ikawa, M., and Snell, E. E.: J. Am. Chem. Soc. 76:648, 1954.
25. Schayer, R. W., and Cooper, J. A. D.: J. Appl. Physiol. 9:481, 1956.
26. Schayer, R. W.: Biogenesis of histamine. In Rocha e Silva, M. (ed.). Handbook of Experimental Pharmacology, Vol. 18/2, p. 109. New York, Springer-Verlag, 1978.
27. Wetterqvist, H.: Histamine metabolism and excretion. In Rocha e Silva, M. (ed.). Handbook of Experimental Pharmacology, Vol. 18/2, p. 131. New York, Springer-Verlag, 1978.
28. Ganellin, C. R., and Parsons, M. E. (eds.): Pharmacology of Histamine Receptors. Bristol, Wright, 1982.
29. Rocha e Silva, M. (ed.): Handbook of Experimental Pharmacology, Vol. 18/2. New York, Springer-Verlag, 1978.
30. Fordtran, J. S., and Grossman, M. I. (eds.): Third Symposium on Histamine H_2-Receptor Antagonists: Clinical Results with Cimetidine. Gastroenterology 74(2):339, 1978.
31. Douglas, W. W.: Histamine and 5-hydroxytryptamine (serotonin) and their antagonists. In Gilman, A. G., Goodman, L. S., Rall, T. W., and Murad, F. (eds.). Goodman and Gilman's The Pharmacological Basis of Therapeutics, 7th ed., p. 605. New York, Macmillan, 1985.
32. van den Brink, F. G., and Lien, E. J.: Competitive and noncompetitive antagonism. In Rocha e Silva, M. (ed.). Handbook of Experimental Pharmacology, Vol. 18/2, p. 333. New York, Springer-Verlag, 1978.
33. Nauta, W. T., and Rekker, R. F.: Structure-activity relationships of H_1-receptor antagonists. In Rocha e Silva, M. (ed.). Handbook of Experimental Pharmacology, Vol. 18/2, p. 215. New York, Springer-Verlag, 1978.
34. Ganellin, C. R.: Histamine receptors. In Hess, H. J. (ed.). Annu. Rep. Med. Chem. 14:91, 1979.
35. Sjoquist, F., and Lasagna, L.: Clin. Pharmacol. Ther. 8:48, 1967.
36. Barouh, V., et al.: J. Med. Chem. 14:834, 1971.
37. Shafi'ee, A., and Hite, G.: J. Med. Chem. 12:266, 1969.
38. Nobles, W. L., and Blanton, C. D.: J. Pharm. Sci. 53:115, 1964.
39. IUPAC. Tentative rules for the nomenclature of organic chemistry. Section E. Fundamental Stereochemistry. J. Org. Chem. 35:2849, 1970.
40. Ison, R. R., Franks, F. M., and Soh, K. S.: J. Pharm. Pharmacol. 25:887, 1973.
41. Zirkle, Charles L.: To tranquilizers and antidepressants from antimalarials and antihistamines. In Clarke, F. H. (ed.). How Modern Medicines are Discovered, p. 55. New York, Futura, 1973.
42. Toldy, L., et al.: Acta Chim. Acad. Sci. Hung. 19:273, 1959.
43. King, C. T. G., Weaver, S. R., and Narrod, S. A.: J. Pharmacol. Exp. Ther. 147:391, 1965.
44. Criep, L. H.: Lancet 68:55, 1948.
45. Landau, S. W., Nelson, W. A., and Gay. L. N.: J. Allergy 22:19, 1951.
46. Villani, et al.: J. Med. Chem. 15:750, 1972.
47. Nicholson, A. N.: Arzneim. Forsch. 32:1191, 1982.
48. Gastpar, H., and Dieterich, H. A.: Arzneim. Forsch. 32:1209 (1982).
49. Intal®, Cromolyn Sodium, A Monograph: Fisons Corporation, Bedford, Massachusetts, 1973.
50. Johnson, P. C., Gillespie, E., and Temple, D. L.: Pulmonary and antiallergy agents. In Hess, H. (ed.). Annu. Rep. Med. Chem. 17:55, 1982.
51. Baron, J. H., and Sullivan, F. M. (eds.): Symposium, Carbenoxolone Sodium. New York, Appleton-Century-Crofts, 1972.
52. Nagashima, R.: J. Clin. Gastroenterol. 3:103, 1981.
53. Piper, D. W., and Fenton, B. H.: Gut 6:506, 1965.
54. Harvey, S. C.: Gastric antacids, miscellaneous drugs for the treatment of peptic ulcers, digestants, and bile acids. In Gilman, A. G., Goodman, L. S., Rall, T. W., and Murad, F. (eds.). Goodman and Gilman's The Pharmacological Basis of Therapeutics, 7th ed., p. 980. New York, Macmillan, 1985.
55. Peterson, W. L., et al.: N. Engl. J. Med. 297:341, 1977.
56. Ippoliti, A. F., et al.: Gastroenterology 74:393, 1978.
57. Perevia-Lima, J., and Hollander, F.: Gastroenterology 37:145, 1959.
58. Guyton, A. C.: Textbook of Medical Physiology, 5th ed., p. 874. Philadelphia, W. B. Saunders, 1976.
59. Johnson, L. R.: Histamine and gastric secretion. In Rocha e Silva, M. (ed.). Handbook of Experimental Pharmacology, Vol. 18/2, p. 41. New York, Springer-Verlag, 1978.
60. Gardner, J. D., et al.: Gastroenterology 74:348, 1978.
61. Soll, A. H.: Gastroenterology 74:355, 1978.
62. Brimblecombe, R. W., et al.: Characterization and development of cimetidine as a histamine H_2-receptor antagonist. In Fordtran, J. S., and Grossman, M. I. (eds.). Third Symposium on Histamine H_2-Receptor Antagonists: Clinical Results with Cimetidine. Gastroenterology 74:339, 1978.
63. Daly, M. J., and Price, B. J.: Ranitidine and other H_2-receptor antagonists. In Ellis, G. P., and West, G. B. (eds.). Progress in Medicinal Chemistry, Vol. 20, p. 337. New York, Elsevier, 1983.

64. Ganellin, R.: J. Med. Chem. 24:913, 1981.
65. Kalant, H., and Roschlau, W. H. E.: Principles of Medical Pharmacology, 5th ed., p. 513. Philadelphia, Decker, 1989.

SELECTED READINGS

Bass, P.: Gastric antisecretory and antiulcer agents. In Harper, N. J., and Simmonds, A. B. (eds.). Adv. Drug Res. 8:206, 1974.

Beaven, M. A.: Histamine, its role in physiological and pathological processes. Monogr. Allergy, Vol. 13, New York, S. Karger, 1978.

Burland, W. L., and Simkins, M. A. (eds.): Cimetidine, Proceedings of the Second Int. Symposium on Histamine H_2-Receptor Antagonists. New York, Excerpta Medica/Elsevier, 1977.

Fordtran, J. S., and Grossman, M. I. (eds.): Third Symposium on Histamine H_2-Receptor Antagonists, Clinical Results with Cimetidine. Gastroenterology 74(2, Part 2):339, 1978.

Ganellin, C. R., and Parsons, M. E. (eds.): Pharmacology of Histamine Receptors. Bristol, Wright, 1982.

Ganellin, C. R., and Durant, G. J.: Histamine H_2 receptor agonists and antagonists. In Wolff, M. E. (ed.). Burger's Medicinal Chemistry, 4th ed., Part 3, p. 487. New York, John Wiley & Sons, 1981.

Rocha e Silva, M. (ed.): Histamine and antihistaminics. In Handbook of Experimental Pharmacology, Vol. 18/1, New York, Springer-Verlag, 1966.

Rocha e Silva, M. (ed.): Histamine II and antihistaminics. In Handbook of Experimental Pharmacology, Vol. 18/2, New York, Springer-Verlag, 1978.

Schachter, M. (ed.): Histamine and antihistamines. In International Encyclopedia of Pharmacology and Therapeutics, Section 74, Vol. 1. New York, Pergamon, 1973.

Sippy, B. W.: JAMA 250:2192, 1983.

Thompson, J. H.: Gastrointestinal disorders—peptic ulcer disease. In Rubin, A. A. (ed.). Search for New Drugs, Medicinal Res. Series, Vol. 6, p. 115. New York, Marcel Dekker, 1972.

Witiak, D. T., and Cavestri, R. C.: Inhibitors of the allergic response. In Wolff, M. E. (ed.). Burger's Medicinal Chemistry, 4th ed., Part 3, p. 553. New York, John Wiley & Sons, 1981.

CHAPTER 17

Analgesic Agents

Robert E. Willette

The struggle to relieve pain began with the origin of humanity. Ancient writings, both serious and fanciful, dealt with secret remedies, religious rituals, and other methods of pain relief. Slowly, there evolved the present, modern era of synthetic analgesics.*

Tainter[1] has divided the history of analgesic drugs into four major eras, namely:

1. The period of discovery and use of naturally occurring plant drugs
2. The isolation of pure plant principles (e.g., alkaloids) from the natural sources and their identification with analgesic action
3. The development of organic chemistry and the first synthetic analgesics
4. The development of modern pharmacologic techniques, making it possible to undertake a systematic testing of new analgesics

The discovery of morphine's analgesic activity by Sertürner, in 1806, ushered in the second era. It continues today only on a small scale. Wöhler introduced the third era indirectly with his synthesis of urea in 1828. He showed that chemical synthesis could be used to make and produce drugs. In the third era, the first synthetic analgesics used in medicine were the salicylates. These originally were found in nature (methyl salicylate, salicin) and then were synthesized by chemists. Other early, man-made drugs were acetanilid (1886), phenacetin (1887), and aspirin (1899).

These early discoveries were the principal contributions in this field until the modern methods of pharmacologic testing initiated the fourth era. The effects of small structural modifications on synthetic molecules now could be assessed accurately by pharmacologic means. This has permitted systematic study of the relationship of structure to activity during this era. The development of these pharmacologic testing procedures, coupled with the fortuitous discovery of meperidine by Eisleb and Schaumann,[2] has made possible the rapid strides in this field today.

The consideration of synthetic analgesics, as well as the naturally occurring ones, will be facilitated considerably by dividing them into two groups: morphine and related compounds and the antipyretic analgesics.

It should be called to the reader's attention that there are numerous drugs that, in addition to possessing distinctive pharmacologic activities in other areas, may also possess analgesic properties. The analgesic property exerted may be a direct effect or may be indirect, but is subsidiary to some other more pronounced effect. Some examples of these, which are discussed elsewhere in this text, are sedatives (e.g., barbiturates); muscle relaxants (e.g., mephenesin, methocarbamol); tranquilizers (e.g., meprobamate), and others. These types will not be considered in this chapter.

MORPHINE AND RELATED COMPOUNDS

HISTORICAL PERSPECTIVE

The discovery of morphine early in the 19th century and the demonstration of its potent analgesic properties led directly to the search for more of these potent principles from plant sources. In tribute to the remarkable potency and action of morphine, it has remained alone as an outstanding and indispensable analgesic from a plant source.

* An *analgesic* may be defined as a drug bringing about insensibility to pain without loss of consciousness. The etymologically correct term *analgetic* may be used in place of the incorrect but popular *analgesic*.

It is only since 1938 that synthetic compounds rivaling it in action have been found, although many earlier changes made on morphine itself gave more effective agents.

Modifications of the morphine molecule will be considered under the following headings:

1. Early changes on morphine before the work of Small, Eddy, and their co-workers.
2. Changes on morphine initiated in 1929 by Small, Eddy, and co-workers[3] under the auspices of the Committee on Drug Addiction of the National Research Council and extending to the present time.
3. The researches initiated by Eisleb and Schaumann[2] in 1938, with their discovery of the potent analgesic action of meperidine, a compound departing radically from the typical morphine molecule.
4. The researches initiated by Grewe, in 1946, leading to the successful synthesis of the morphinan group of analgesics.

Early Morphine Modifications

Morphine is obtained from **opium**, which is the partly dried latex from incised unripe capsules of *Papaver somniferum*. Opium contains numerous alkaloids (as meconates and sulfates), of which morphine, codeine, noscapine (narcotine), and papaverine are therapeutically the most important, and thebaine, which has convulsant properties, but is an important starting material for many other drugs. Other alkaloids, such as narceine, also have been tested medicinally, but are not of great importance. The action of opium is principally due to its morphine content. As an analgesic, opium is not as effective as morphine because of its slower absorption, but it has a greater constipating action and, thus, is better suited for antidiarrheal preparations (e.g., paregoric). Opium, as a constituent of Dover's powders and Brown mixture, also exerts a valuable expectorant action that is superior to that of morphine.

Two types of basic structures usually are recognized among the opium alkaloids [i.e., the *phenanthrene* (morphine) type and the *benzylisoquinoline* (papaverine) type] (see structures).

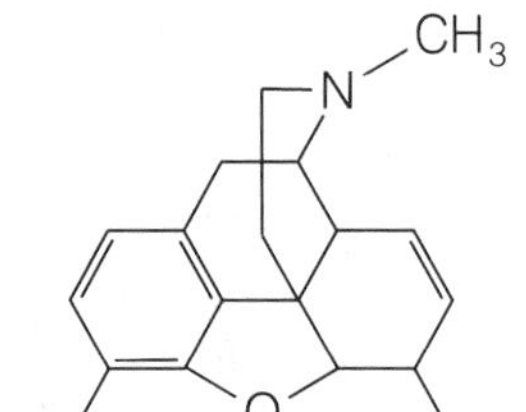

Phenanthrene Type
(Morphine, R & R′ = H)

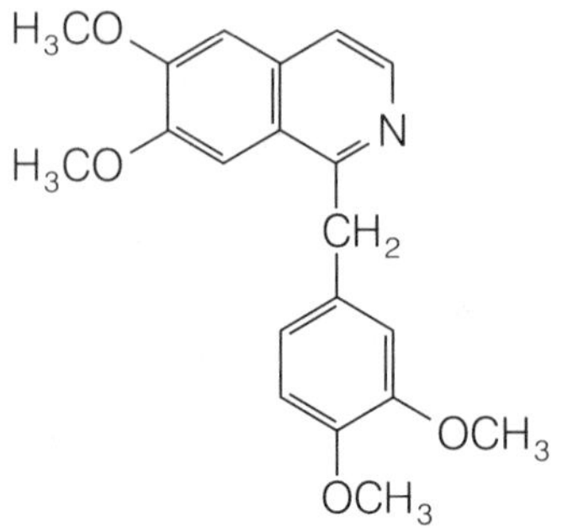

Benzylisoquinoline Type
(Papaverine)

The pharmacologic actions of the two types of alkaloids are dissimilar. The morphine group acts principally on the central nervous system as a depressant and stimulant, whereas the papaverine group has little effect on the nervous system, but has a marked antispasmodic action on smooth muscle.

Clinically, the depressant action of the morphine group is the most useful property, resulting in an increased tolerance to pain, a sleepy feeling, a lessened perception to external stimuli, and a feeling of well-being (euphoria). Respiratory depression, central in origin, is perhaps the most serious objection to this type of alkaloid, aside from its tendency to cause addiction. The stimulant action is well illustrated by the convulsions produced by certain members of this group (e.g., thebaine).

Before 1929, the derivatives of morphine that had been made were primarily the result of simple changes on the molecule, such as esterification of the phenolic or alcoholic hydroxyl group, etherification of the phenolic hydroxyl group, and similar minor changes. The net result was the discovery of some compounds with greater activity than morphine, but also with greater toxicities and addiction tendencies. No compounds were found that did not possess in some measure the addiction liabilities of morphine.*

Some of the compounds that were in common usage before 1929 are listed in Table 17-1, together with some other more recently introduced ones. All have the morphine skeleton in common.

Among the earlier compounds is codeine, the phenolic methyl ether of morphine, which also had been obtained from natural sources. It has survived as a good analgesic and cough depressant, together with the corresponding ethyl ether, which has found its principal application in ophthalmology. The diacetyl derivative of morphine, heroin, has been known for

* The term *addiction liability*, or the preferred term *dependence liability*, as used in this text, indicates the ability of a substance to develop true addictive tolerance and physical dependence and/or to suppress the morphine abstinence syndrome after withdrawal of morphine from addicts.

TABLE 17-1

SYNTHETIC DERIVATIVES OF MORPHINE

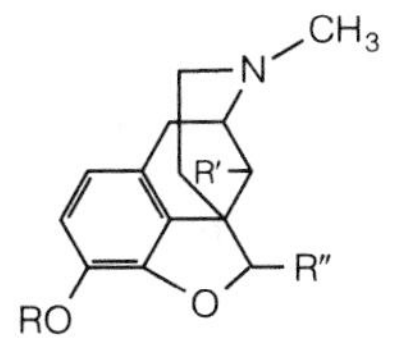

CH_3 N R' R'' RO O

Compound Proprietary Name	R	R'	R''	Principal Use
Morphine	H	H	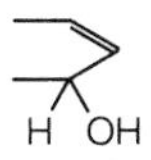 H OH	Analgesic
Codeine	CH_3	H	Same as above	Analgesic and to depress cough reflex
Ethylmorphine *Dionin*	C_2H_5	H	Same as above	Ophthalmology
Diacetylmorphine (heroin)	CH_3CO	H	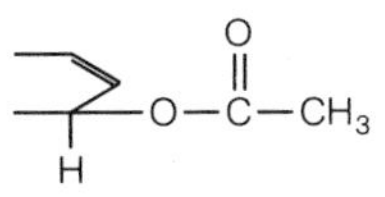 O ‖ —O—C—CH_3 H	Analgesic (prohibited in US)
Hydromorphone (dihydromorphinone) *Dilaudid*	H	H	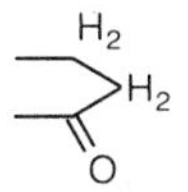 H_2 H_2 O	Analgesic
Hydrocodone (dihydrocodeinone) *Dicodid*	CH_3	H	Same as above	Analgesic and to depress cough reflex
Oxymorphone (dihydrohydroxymorphinone)	H	OH	Same as above	Analgesic
Oxycodone (dihydrohydroxycodeinone)	CH_3	OH	Same as above	Analgesic and to depress cough reflex
Dihydrocodeine *Paracodin*	CH_3	H	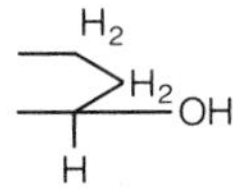 H_2 H_2 OH H	Depress cough reflex
Dihydromorphine	H	H	Same as above	Analgesic
Methyldihydromorphinone *Metopon*	H	H	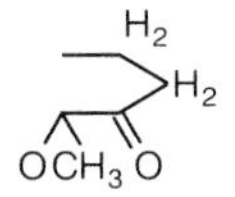 H_2 H_2 OCH_3 O	

a long time; it has been banished for years from the United States and is being used in decreasing amounts in other countries. It is the most widely used illicit drug among narcotic addicts. Among the reduced compounds were dihydromorphine and dihydrocodeine and their oxidized congeners, dihydromorphinone (hydromorphone) and dihydrocodeinone (hydrocodone). Derivatives of the last two compounds possessing a hydroxyl group in position 14 are dihydrohydroxymorphinone, or oxymorphone, and dihydrohydroxycodeinone, or oxycodone. These represent the principal compounds that either had been on the market or had been prepared before the studies of Small, Eddy, and co-workers.* It is well to note that no really systematic effort had been made to investigate the structure–activity relationships in the molecule, and only the easily changed peripheral groups had been modified.

Morphine Modifications Initiated by the Researches of Small and Eddy

The avowed purpose of Small, Eddy, and co-workers,[3] in 1929, was to approach the morphine problem from the standpoint that

1. It might be possible to separate chemically the addiction property of morphine from its other more salutary attributes. That this could be done with some addiction-producing compounds was shown by the development of the nonaddictive procaine from the addictive cocaine.
2. If it were not possible to separate the addictive tendencies from the morphine molecule, it might be possible to find other synthetic molecules without this undesirable property.

Proceeding on these assumptions, they first examined the morphine molecule in an exhaustive manner. As a starting point, it offered the advantages of ready availability, proven potency, and ease of alteration. In addition to its addictive tendency, it was hoped that other liabilities, such as respiratory depression, emetic properties, and gastrointestinal tract and circulatory disturbances, could be minimized or abolished as well. Because early modifications of morphine (e.g., acetylation or alkylation of hydroxyls, quaternization of the nitrogen, and so on) caused variations in the addictive potency, it was felt that the physiologic effects of morphine could be related, at least in part, to the peripheral groups.

It was not known if the actions of morphine were primarily a function of the peripheral groups or of the structural skeleton. This did not matter, however, because modification of the groups would alter activity in either case. These groups and the effects on activity by modifying them are listed in Table 17-2. The results of these and earlier studies[4] have

* The only exception is oxymorphone; this was introduced in the United States in 1959, but is mentioned here because it obviously is closely related to oxycodone.

TABLE 17-2

SOME STRUCTURAL RELATIONSHIPS IN THE MORPHINE MOLECULE

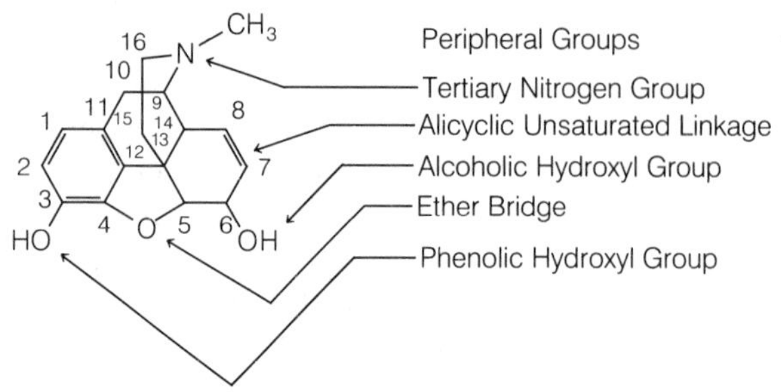

Peripheral Groups
Tertiary Nitrogen Group
Alicyclic Unsaturated Linkage
Alcoholic Hydroxyl Group
Ether Bridge
Phenolic Hydroxyl Group

Peripheral Groups of Morphine	Modification (On Morphine Unless Otherwise Indicated)	Effects on Analgesic Activity * (Morphine or Another Compound as Indicated = 100)	
Phenolic hydroxyl	$-OH \rightarrow -OCH_3$ (codeine)	15	
	$-OH \rightarrow -OC_2H_5$ (ethylmorphine)	10	
	$-OH \rightarrow -OCH_2CH_2-N$(morpholino)O (pholcodine)	1	
Alcoholic hydroxyl	$-OH \rightarrow -OCH_3$ (heterocodeine)	500	
	$-OH \rightarrow -OC_2H_5$	240	
	$-OH \rightarrow -OCOCH_3$	420	
	$-OH \rightarrow =O$ (morphinone)	37	
	†$-OH \rightarrow =O$ (dihydromorphine to dihydromorphinone)	600	(dihydromorphine vs dihydromorphinone)
	†$-OH \rightarrow =O$ (dihydrocodeine to dihydrocodeinone)	390	(dihydrocodeine vs dihydrocodeinone)
	†$-OH \rightarrow -H$ (dihydromorphine to dihydrodesoxymorphine-D)	1000	(dihydromorphine vs dihydrodesoxymorphine-D)
Ether bridge	‡$=C-O-CH- \rightarrow =C-OH\ HCH-$ (dihydrodesoxymorphine-D to tetrahydrodesoxymorphine)	13	(dihydrodesoxymorphine D vs tetrahydrodesoxymorphine)
Alicyclic unsaturated linkage	$-CH=CH- \rightarrow -CH_2CH_2-$ (dihydromorphine)	120	
	†$-CH=CH- \rightarrow -CH_2CH_2-$ (codeine to dihydrocodeine)	115	(codeine vs dihydrocodeine)
Tertiary nitrogen	$>N-CH_3 \rightarrow >N-H$ (normorphine)	5	
	$>N-CH_3 \rightarrow >N-CH_2CH_2-C_6H_5$	1400	
	§$>N-CH_3 \rightarrow >N-R$		Reversal of activity (morphine antagonism); R = propyl, isobutyl, allyl, methallyl
	$>N-CH_3 \rightarrow >\overset{+}{N}(CH_3)_2\ Cl-$	1	(strong curare action)
	Opening of nitrogen ring (morphimethine)		Marked decrease in action

*Percent ratio of the ED_{50} of morphine (or other compound indicated) to the ED_{50} of the compound as determined in mice. These conclusions have been adapted from data in Refs. 3 and 4. For a wealth of additional tabular material, the reader is urged to consult the original references.

†These represent cases in which, for various reasons, a direct comparison with morphine itself cannot be made. The alternative has been to compare the effect of modifying the group in a pair of compounds in which the changes can be made. It is felt that the direction of change in analgesic activity at least can be determined in this way.

‡See, however, discussion of *N*-methylmorphinan later in this chapter.

§Although many of these derivatives possess morphine antagonism, it has been shown that many of them also possess analgesic activity in their own right. Indeed, the ability to antagonize morphine in the rat is used as a screening method to assure low addiction potential in humans.[44]

#Not included in the studies of Small et al. (see Ref. 8).

TABLE 17-2 *Continued*

SOME STRUCTURAL RELATIONSHIPS IN THE MORPHINE MOLECULE

Peripheral Groups of Morphine	*Modification (On Morphine Unless Otherwise Indicated)*	*Effects on Analgesic Activity* (Morphine or Another Compound as Indicated = 100)*
Nuclear substitution	Substitution of:	
	—NH_2 (most likely at position 2)	Marked decrease in action
	—Cl or —Br (at position 1)	50
	—OH (at position 14 in dihydromorphinone)	250 (dihydromorphinone vs oxymorphone)
	—OH (at position 14 in dihydrocodeinone)	530 (dihydrocodeinone vs oxycodone)
	#—CH_3 (at position 6)	280
	#—CH_3 (at position 6 in dihydromorphine)	33 (dihydromorphine vs 6-methyldihydromorphine)
	#—CH_3 (at position 6 in dihydrodesoxy morphine D)	490 (dihydrodesoxymorphine D vs 6-methyldihydrodesoxymorphine)
	#=CH_2 (at position 6 in dihydrodesoxymorphine D)	600 (dihydrodesoxymorphine D vs-6-methylenedihydrodesoxymorphine)

*Percent ratio of the ED_{50} of morphine (or other compound indicated) to the ED_{50} of the compound as determined in mice. These conclusions have been adapted from data in Refs. 3 and 4. For a wealth of additional tabular material, the reader is urged to consult the original references.

†These represent cases in which, for various reasons, a direct comparison with morphine itself cannot be made. The alternative has been to compare the effect of modifying the group in a pair of compounds in which the changes can be made. It is felt that the direction of change in analgesic activity at least can be determined in this way.

‡See, however, discussion of *N*-methylmorphinan later in this chapter.

§Although many of these derivatives possess morphine antagonism, it has been shown that many of them also possess analgesic activity in their own right. Indeed, the ability to antagonize morphine in the rat is used as a screening method to assure low addiction potential in humans.[44]

#Not included in the studies of Small et al. (see Ref. 8).

not always shown quantitatively the effects of simple modifications on the analgesic action of morphine. However, they do indicate in which direction the activity is likely to go. The studies are far more comprehensive than Table 17-2 indicates, and the conclusions usually depend on more than one pair of compounds.

Unfortunately, these studies on morphine did not provide the answer to the elimination of addiction potentialities from these compounds. In fact, the studies suggested that any modification bringing about an increase in the analgesic activity caused a concomitant increase in addiction liability.

The second phase of the studies, engaged in principally by Mosettig and Eddy,[3] had to do with the attempted synthesis of substances with central narcotic and, especially, analgesic action. It is obvious that the morphine molecule contains in its makeup certain well-defined types of chemical structures. Among these are the phenanthrene nucleus, the dibenzofuran nucleus, and, as a variant of the latter, carbazole. These synthetic studies, although extensive and interesting, failed to provide significant findings and will not be discussed further in this text.

One of the more useful results of the investigations was the synthesis of 5-methyldihydromorphinone* (see Table 17-1). Although it possessed addiction liabilities, it was found to be a very potent analgesic with a minimum of the undesirable side effects of morphine, such as emetic action and mental dullness.

Later, the high degree of analgesic activity demonstrated by morphine congeners in which the alicyclic ring is either reduced or methylated (or both) and the alcoholic hydroxyl at position 6 is absent has prompted the synthesis of related compounds possessing these features. These include 6-methyldihydromorphine and its dehydrated analogue, 6-methyl-Δ^6-desoxymorphine or methyldesorphine,[6] both of which have shown high potency. Also of interest were compounds reported by Rapoport and his co-workers:[7] morphinone; 6-methylmorphine; 6-methyl-7-hydroxy-, 6-methyl-, and 6-methylenedihydrodesoxymorphine. In analgesic activity in mice, the last-named compound was 82 times more potent, milligram for milligram, than morphine. Its therapeutic index (TI_{50}) was 22 times as great as that of morphine.[8]

The structure–activity relationships of 14-hydroxymorphine derivatives have been reviewed,[3] and several new compounds have been synthesized.[9] Of these, the dihydrodesoxy compounds possessed the highest degree of analgesic activity. Also, esters of 14-hydroxycodeine derivatives have shown very high activity.[10] For example, in rats, 14-cinnamyloxycodeinone was 177 times more active than morphine.

In 1963, Bentley and Hardy[11] reported the synthesis of a novel series of potent analgesics derived

* The location of the methyl substituent was originally assigned to position 7.[5]

from the opium alkaloid thebaine. In rats the most active members of the series (I, R_1 = H, R_2 = CH_3, R_3 = isoamyl; and I, R_1 = $COCH_3$, R_2 = CH_3, R_3 = n-C_3H_7) were found to be several thousand times stronger than morphine.[12] These compounds exhibited marked differences in activity of optical isomers, as well as other interesting structural effects. It was postulated that the more rigid molecular structure might allow them to fit the receptor surface better. Extensive structural and pharmacologic studies have been reported.[13] Some of the N-cyclopropylmethyl compounds are the most potent antagonists yet discovered and have been studied very intensively.

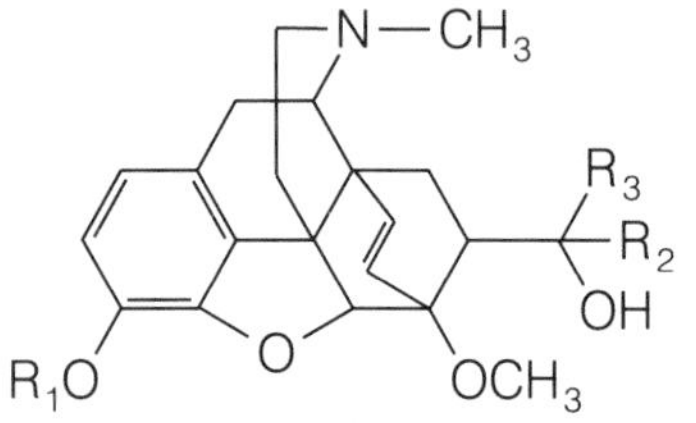

N—CH$_3$ R$_3$ R$_2$ OH R$_1$O O OCH$_3$ I

As indicated in Table 17-2, replacement of the N-methyl group in morphine by larger alkyl groups not only lowers analgesic activity, but confers morphine antagonistic properties on the molecule (discussed later). In direct contrast to this effect, the N-phenethyl derivative has 14 times the analgesic activity of morphine. This enhancement of activity by N-aralkyl groups has wide application, as will be shown later.

The morphine antagonists, such as nalorphine, are also strong analgesics.[14] The similarity of the ethylenic double bond and the cyclopropyl group has prompted the synthesis of N-cyclopropylmethyl derivatives of morphine and its derivatives.[15] This substituent usually confers strong narcotic antagonistic activity, with variable effects on analgesic potency. The dihydronormorphinone derivative had only moderate analgesic activity.

Morphine Modifications Initiated by the Eisleb and Schaumann Research

In 1938, Eisleb and Schaumann[2] reported the fortuitous discovery that a simple piperidine derivative, now known as meperidine, possessed analgesic activity. It was prepared as an antispasmodic, a property it shows as well. As the story is told, during the pharmacologic testing of meperidine in mice, it was observed to cause the peculiar erection of the tail known as the Straub reaction. Because the reaction is characteristic of morphine and its derivatives, the compound then was tested for analgesic properties and was about one-fifth as active as morphine. This discovery led not only to the finding of an active analgesic but, far more important, it served as a stimulus to research workers. The status of research in analgesic compounds with an activity comparable with that of morphine was at a low ebb in 1938. Many felt that potent compounds could not be prepared, unless they were very closely related structurally to morphine. However, the demonstration of high potency in a synthetic compound that was related only distantly to morphine spurred the efforts of various research groups.[16,17]

The first efforts, naturally, were made upon the meperidine-type molecule in an attempt to enhance its activity further. It was found that replacement of the 4-phenyl group by hydrogen, alkyl, other aryl, aralkyl, and heterocyclic groups reduced analgesic activity. Placement of the phenyl and ester groups at the 4-position of 1-methylpiperidine also gave optimum activity. Several modifications of this basic structure are listed in Table 17-3.

Among the simplest changes to increase activity is the insertion of a m-hydroxyl group on the phenyl ring. It is in the same relative position as in morphine. The effect is more pronounced on the keto compound (Table 17-3, A-4) than on meperidine (A-1). Ketobemidone is equivalent to morphine in activity and was once widely used.

More significantly, Jensen and co-workers[18] discovered that replacement of the carbethoxyl group in meperidine by acyloxyl groups gave better analgesic, as well as spasmolytic, activity. The "reversed" ester of meperidine, the propionoxy compound (A-6), was the most active, being five times as active as meperidine. These findings were validated and expanded upon by Lee et al.[19] In an extensive study of structural modifications of meperidine, Janssen and Eddy[20] concluded that the propionoxy compounds were always more active, usually about twofold, regardless of what group was attached to the nitrogen.

Lee[21] had postulated that the configuration of the propionoxy derivative (A-6) more closely resembled that of morphine, with the ester chain taking a position similar to that occupied by C-6 and C-7 in morphine. His speculations were based on space models and certainly did not reflect the actual conformation of the nonrigid meperidine. However, he did arrive at the correct assumption that introduction of a methyl group into position 3 of the piperidine ring in the propionoxy compound would yield two isomers, one with activity approximating that of desomorphine and the other with less activity. One of the two diastereoisomers (A-7), betaprodine, has an activity in mice of about nine times that of morphine and three times that of A-6. Beckett et al.[22] have established it to be the *cis* (methyl/phenyl)-form. The *trans*-form, alphaprodine, is twice

TABLE 17-3 *Continued*

COMPOUNDS RELATED TO MEPERIDINE

R_3, R_1, R_4-N, R_2, (R_5)

(R_5 = H except in trimeperidine, where it is CH_3)

Compound	Structure: R_1	R_2	R_3	R_4	Name (If Any)	Analgesic Activity* (Meperidine = 1)
A-1	$-C_6H_5$	$-COOC_2H_5$	$-CH_2CH_2-$	$-CH_3$	Meperidine	1.0
A-2	$-C_6H_4OH$	$-COOC_2H_5$	$-CH_2CH_2$	$-CH_3$	Bemidone	1.5
A-3	$-C_6H_5$	$-COOCH(CH_3)_2$	$-CH_2CH_2-$	$-CH_3$	Properidine	15
A-4	$-C_6H_5$	$-C(=O)-C_2H_5$	$-CH_2CH_2-$	$-CH_3$		0.5
A-5	$-C_6H_4OH$	$-C(=O)-C_2H_5$	$-CH_2CH_2-$	$-CH_3$	Ketobemidone	6.2
A-6	$-C_6H_5$	$-O-C(=O)-C_2H_5$	$-CH_2CH_2-$	$-CH_3$		5
A-7	$-C_6H_5$	$-O-C(=O)-C_2H_5$	$-CH_2CH(CH_3)-$	$-CH_3$	Alphaprodine Betaprodine	5 14
A-8	$-C_6H_5$	$-O-C(=O)-C_2H_5$	$-CH_2CH(CH_3)-$	$-CH_3(R_5 = CH_3)$	Trimeperidine	7.5
A-9	$-C_6H_5$	$-COOC_2H_5$	$-CH_2CH_2-$	$-CH_2CH_2C_6H_5$	Pheneridine	2.6
A-10	$-C_6H_5$	$-COOC_2H_5$	$-CH_2CH_2-$	$-CH_2CH_2-C_6H_4-NH_2$	Anileridine	3.5
A-11	$-C_6H_5$	$-COOC_2H_5$	$-CH_2CH_2-$	$-(CH_2)_3-NH-C_6H_5$	Piminodine	55†
A-12	$-C_6H_5$	$-O-C(=O)-C_2H_5$	$-CH_2CH_2-$	$-CH_2CH_2CH(C_6H_5)-O-C(=O)-C_2H_5$		1880†
A-13	$-C_6H_5$	$-COOC_2H_5$	$-CH_2CH_2-$	$-CH_2CH_2C(C_6H_5)_2CN$	Diphenoxylate	None

*Ratio of the ED_{50} of meperidine to the ED_{50} of the compound in mg/kg administered subcutaneously in mice. (Based on data in Refs. 4, 27, 33, 34.)

†In rats (see Ref. 21).

‡In rats (see Ref. 28).

TABLE 17-3 *Continued*

COMPOUNDS RELATED TO MEPERIDINE

Compound	Structure R1	R2	R3	R4	Name (If Any)	Analgesic Activity* (Meperidine = 1)
A-14	$-C_6H_4$-*p*-Cl	$-OH$	$-CH_2CH_2-$	$-CH_2CH_2C(C_6H_5)_2$ with $-C(=O)N(CH_3)_2$	Loperamide	None
A-15	$-C_6H_5$	$-COOC_2H_5$	$-CH_2CH_2CH_2-$	$-CH_3$	Ethoheptazine	1
A-16	$-C_6H_5$	$-O-C(=O)C_2H_5$	$-CH(CH_3)-$	$-CH_3$	Prodilidine	0.3
A-17	$-H$	$-N(C_6H_5)-C(=O)C_2H_5$	$-CH_2CH_2-$	$-CH_2CH_2C_6H_5$	Fentanyl	940
A-18	$-COOCH_3$	$-N(C_6H_5)-C(=O)C_2H_5$	$-CH_2CH(CH_3)-$	$-CH_2CH_2C_6H_5$	Lofentanil (R 34,995)	8400‡

*Ratio of the ED_{50} of meperidine to the ED_{50} of the compound in mg/kg administered subcutaneously in mice. (Based on data in Refs. 4, 27, 33, 34.)
†In rats (see Ref. 21).
‡In rats (see Ref. 28).

as active as morphine. Resolution of the racemates shows one enantiomer to have the predominant activity. In humans, however, the sharp differences in analgesic potency are not so marked. The *trans*-form is marketed as the racemate. The significance of the 3-methyl has been attributed to discrimination of the enantiotropic edges of these molecules by the receptor. This is even more dramatic in the 3-allyl and 3-propyl isomers, for which the α-*trans*-forms are considerably more potent than the β-isomers, indicating three-carbon substituents are not tolerated in the axial orientation. The 3-ethyl isomers are nearly equal in activity, further indicating that two or fewer carbons are more acceptable in the drug–receptor interaction.[23]

Until only the last few years it appeared that a small substituent, such as methyl, attached to the nitrogen was optimal for analgesic activity. This was believed to be true not only for the meperidine series of compounds, but also for all the other types. It is now well established that replacement of the methyl group by various aralkyl groups can increase activity markedly.[20] A few examples of this type of compound in the meperidine series are shown in Table 17-3. The phenethyl derivative (A-9) is seen to be about three times as active as meperidine (A-1). The *p*-amino congener, anileridine (A-10), is about four times more active. Piminodine, the phenylaminopropyl derivative (A-11), has 55 times the activity of meperidine in rats and, in clinical trials, is about five times as effective in humans as an analgesic.[24] The most active meperidine-type compounds to date are the propionoxy derivative (A-12), which is nearly 2000 times as active as meperidine, and the *N*-phenethyl analogue of betaprodine, which is over 2000 times as active as morphine.[22] Diphenoxylate (A-13), a structural hybrid of meperidine and methadone types, lacks analgesic activity although it reportedly suppresses the morphine abstinence syndrome in morphine addicts.[25,26] It is quite effective as an intestinal spasmolytic and is used for the treatment of diarrhea. Several other derivatives of it have been studied.[27] The related *p*-chloro analogue, loperamide (A-14), has been shown to bind to opiate receptors in the brain, but not to penetrate the blood–brain barrier sufficiently to produce analgesia.[28]

Another manner of modifying the structure of meperidine with favorable results has been the enlargement of the piperidine ring to the seven-membered hexahydroazepine (or hexamethylenimine) ring. As was true in the piperidine series, the most active compound was the one containing a methyl group on position 3 of the ring adjacent to the quaternary carbon atom in the propionoxy derivative, that is, 1,3-dimethyl-4-phenyl-4-propionoxy-hexahydroazepine, to which the name proheptazine has been given. In the study by Eddy and co-workers, previously cited, proheptazine was one of the more active analgesics included and had one of the highest addiction liabilities. The higher ring homologue of meperidine, ethoheptazine, has been marketed. Although originally thought to be inactive,[29] it is less active than codeine as an analgesic in humans, but has the advantages of being free of addiction liability and having a low incidence of side effects.[30] Because of its low potency it is not widely used.

Contraction of the piperidine ring to the five-membered pyrrolidine ring has also been successful. The lower ring homologue of alphaprodine, prodilidene (A-16), is an effective analgesic, 100 mg being equivalent to 30 mg of codeine, but because of its potential abuse liability, it has not been marketed.[31]

A more unusual modification of the meperidine structure may be found in fentanyl (A-17), in which the phenyl and the acyl groups are separated from the ring by a nitrogen. It is a powerful analgesic, 50 times stronger than morphine in humans, with minimal side effects.[32] Its short duration of action makes it well suited for use in anesthesia.[33] It is marketed for this purpose in combination with a neuroleptic, droperidol. The *cis*-(−)-3-methyl analogue with an ester group at the 4-position, like meperidine (A-18) was 8400 times more potent than morphine as an analgesic. In addition, it has shown the highest binding affinity to isolated opiate receptors of all other compounds tested.[28] Fentanyl and its 3-methyl and α-methyl analogues have found their way into the illicit drug market and are sold as substitutes for heroin. Because of their extreme potency, they have caused many deaths.

It should be recalled by the reader that when the nitrogen ring of morphine is opened, as in the formation of morphimethines, the analgesic activity virtually is abolished. On this basis, the prediction of whether a compound would or would not have activity without the nitrogen in a ring would be in favor of lack of activity or, at best, a low activity. The first report indicating that this might be a false assumption was based on the initial work of Bockmuehl and Ehrhart[34] wherein they claimed that the type of compound represented by B-1 in Table 17-4 possessed analgesic as well as spasmolytic properties. The Hoechst laboratories in Germany followed up this lead during World War II by preparing the ketones corresponding to these esters. Some of the compounds they prepared with high activity are represented by formulas B-2 through B-7. Compound B-2 is the well-known methadone. In the meperidine and bemidone types, the introduction of a *m*-hydroxyl group in the phenyl ring brought about slight to marked increase in activity, whereas the same operation with the methadone-type compound brought about a marked decrease in action. Phenadoxone (B-8), the morpholine analogue of methadone, has been marketed in England. The piperidine analogue, dipanone, was once under study in this country after successful results in England.

Methadone was first brought to the attention of American pharmacists, chemists, and allied workers by the Kleiderer report[35] and by the early reports of Scott and Chen.[36] Since then, much work has been done on this compound, its isomer known as isomethadone, and related compounds. The report by Eddy, Touchberry, and Lieberman[37] covers most of the points concerning the structure–activity relationships of methadone. It was demonstrated that the *levo*-isomer (B-3) of methadone (B-2) and the *levo*-isomer of isomethadone (B-4) were twice as effective as their racemic mixtures. It is also of interest that all structural derivatives of methadone demonstrated a greater activity than the corresponding structural derivatives of isomethadone. In other words, the superiority of methadone over isomethadone seems to hold, even through the derivatives. Conversely, the methadone series of compounds was always more toxic than the isomethadone group.

More extensive permutations, such as replacement of the propionyl group (R_3 in B-2) by hydrogen, hydroxyl, or acetoxyl, led to decreased activity. In a series of amide analogues of methadone, Janssen and Jageneau[38] synthesized racemoramide (B-12), which is more active than methadone. The (+)-isomer, dextromoramide (B-13), is the active isomer and has been marketed. A few of the other modifications that have been carried out, together with the effect on analgesic activity relative to methadone, are described in Table 17-4, which comprises most of the methadone congeners that are or were on the market. It can be assumed that much deviation in structure from these examples will result in varying degrees of activity loss.

Particular attention should be called to the two phenyl groups in methadone and the sharply decreased action resulting by removal of one of them. It is believed that the second phenyl residue helps to lock the $-COC_2H_5$ group of methadone in a position to simulate again the alicyclic ring of morphine, even though the propionyl group is not a particularly rigid group. However, in this connection it is interesting to note that the compound with a propionoxy group in place of the propionyl group (R_3 in B-2) is without significant analgesic action.[17] In direct contrast with this is (+)-propoxyphene (B-14), which is a propionoxy derivative with one of the

TABLE 17-4

COMPOUNDS RELATED TO METHADONE

$$R_1, R_2 > C < R_3, R_4$$

Compound	Structure: R_1	R_2	R_3	R_4	Name	Isomer, Salt	Analgesic Activity* (Methadone = 1)
B-1	$-C_6H_5$	$-C_6H_5$	$-COO-$Alkyl	$-CH_2CH_2N(CH_5)_2$		—	0.17
B-2	$-C_6H_5$	$-C_6H_5$	$-\underset{\underset{O}{\Vert}}{C}-C_2H_5$	$-CH_2\underset{\underset{CH_3}{\vert}}{CH}N(CH_3)_2$	Methadone	(±)-HCl	1.0
B-3	Same as in B-2				Levanone	(−)-bitartr.	1.9
B-4	$-C_6H_5$	$-C_6H_5$	$-\underset{\underset{O}{\Vert}}{C}-C_2H_5$	$-\underset{\underset{CH_3}{\vert}}{CH}CH_2N(CH_3)_2$	Isomethadone	(±)-HCl	0.65
B-5	$-C_6H_5$	$-C_6H_5$	$-\underset{\underset{O}{\Vert}}{C}-C_2H_5$	$-CH_2CH_2N(CH_3)_2$	Normethadone	HCl	0.44
B-6	$-C_6H_5$	$-C_6H_5$	$-\underset{\underset{O}{\Vert}}{C}-C_2H_5$	$-CH_2\underset{\underset{CH_3}{\vert}}{CH}N$(piperidine ring)	Dipanone	(±)-HCl	0.80
B-7	$-C_6H_5$	$-C_6H_5$	$-\underset{\underset{O}{\Vert}}{C}-C_2H_5$	$-CH_2CH_2N$(piperidine ring)	Hexalgon	HBr	0.50
B-8	$-C_6H_5$	$-C_6H_5$	$-\underset{\underset{O}{\Vert}}{C}-C_2H_5$	$-CH_2\underset{\underset{CH_3}{\vert}}{CH}N$(morpholine ring, O)	Phenadoxone	(±)-HCl	1.4
B-9	$-C_6H_5$	$-C_6H_5$	$-\underset{\underset{O-\underset{\underset{O}{\Vert}}{C}CH_3}{\vert}}{CH}C_2H_5$	$-CH_2\underset{\underset{CH_3}{\vert}}{CH}N(CH_3)_2$	Alphacetylmethadol	α,(±)-HCl	1.3
B-10	Same as in B-9				Betacetylmethadol	β,(±)-HCl	2.3
B-11	$-C_6H_5$	$-C_6H_5$	$-COOC_2H_5$	$-CH_2CH_2N$(morpholine ring, O)	Dioxaphetyl butyrate	HCl	0.25
B-12	$-C_6H_5$	$-C_6H_5$	$-\underset{\underset{O}{\Vert}}{C}-N$(pyrrolidine ring)	$-\underset{\underset{CH_3}{\vert}}{CH}CH_2N$(morpholine ring, O)	Racemoramide	(+)-base	3.6
B-13	Same as in B-12				Dextromoramide	(+)-base	13
B-14	$-C_6H_5$	$-CH_2C_6H_5$	$O-\underset{\underset{O}{\Vert}}{C}-C_2H_5$	$-\underset{\underset{CH_3}{\vert}}{CH}CH_2N(CH_3)_2$	Propoxyphene	(+)-HCl	0.21

(Adapted from Janssen, P. A. J.: Synthetic Analgesics, Part 1. New York, Pergamon Press, 1960.)

*Ratio of the ED_{50} of methadone to the ED_{50} of the compound in mg/kg administered subcutaneously to mice as determined by the hot-plate method.

phenyl groups replaced by a benzyl group. In addition, it is an analogue of isomethadone (B-4), making it an exception to the rule. This compound is lower than codeine in analgesic activity, possesses few side effects, and has a limited addiction liability.[39] Replacement of the dimethylamino group in (+)-propoxyphene with a pyrrolidyl group gives a compound that is nearly three-fourths as active as methadone and possesses morphinelike properties. The (−)-isomer of alphacetylmethadol (B-9), known as LAAM, is being investigated as a long-acting substitute for methadone in the treatment of addicts.[40]

Morphine Modifications Initiated by Grewe

Grewe, in 1946, approached the problem of synthetic analgesics from another direction when he synthesized the tetracyclic compound that he first named morphan and then revised to *N*-methylmorphinan. The relationship of this compound to morphine is obvious.

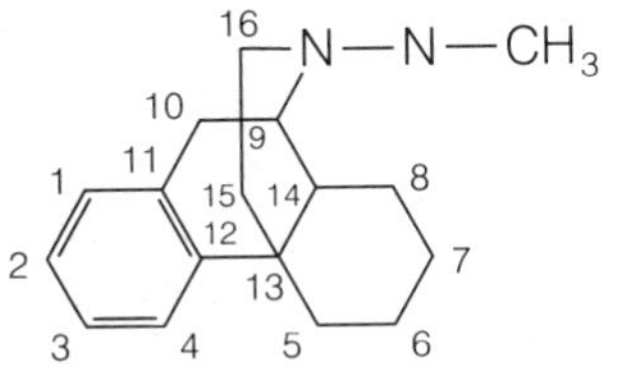

N-Methylmorphinan

N-Methylmorphinan differs from the morphine nucleus in the lack of the ether bridge between C-4 and C-5. Because this compound possesses a high degree of analgesic activity, it suggests the nonessential nature of the ether bridge. The 3-hydroxyl derivative of *N*-methylmorphinan (racemorphan) was on the market and had an intensity and duration of action that exceeded that of morphine. The original racemorphan was introduced as the hydrobromide and was the (±)-, or racemic, form as obtained by synthesis. Since then, realizing that the levorotatory form of racemorphan was the analgesically active portion of the racemate, the manufacturers have successfully resolved the (±)-form and have marketed the *levo* form as the tartrate salt (levorphanol). The *dextro* form has also found use as a cough depressant (see dextromethorphan). The ethers and acylated derivatives of the 3-hydroxyl form also exhibit considerable activity. The 2- and 4-hydroxyl isomers are, not unexpectedly, without value as analgesics. Likewise, the *N*-ethyl derivative is lacking in activity and the *N*-allyl compound, levallorphan, is a potent morphine antagonist.

Eddy and co-workers[41] have reported on an extensive series of *N*-aralkylmorphinan derivatives. The effect of the *N*-aralkyl substitution was more dramatic in this series than it was for morphine or meperidine. The *N*-phenethyl and *N*-*p*-aminophenethyl analogues of levorphanol are about 3 and 18 times, respectively, more active than the parent compound in analgesic activity in mice. The most potent member of the series was its *N*-β-furylethyl analogue, which was nearly 30 times as active as levorphanol or 160 times as active as morphine. The *N*-acetophenone analogue, levophenacylmorphan, was once under clinical investigation. In mice, it is about 30 times more active than morphine, and in humans a 2-mg dose is equivalent to 10 mg of morphine in its analgesic response.[42] It has a much lower physical-dependence liability than morphine.

The *N*-cyclopropylmethyl derivative of 3-hydroxymorphinan (cyclorphan) was reported to be a potent morphine antagonist, capable of precipitating morphine withdrawal symptoms in addicted monkeys, indicating that it is nonaddicting.[15] Clinical studies have indicated that it is about 20 times stronger than morphine as an analgesic, but has some undesirable side effects, primarily hallucinatory. However, the *N*-cyclobutyl derivative, butorphanol, possesses mixed agonist–antagonist properties and has been marketed as a potent analgesic.

Inasmuch as removal of the ether bridge and all the peripheral groups in the alicyclic ring in morphine did not destroy its analgesic action, May and co-workers[43] synthesized a series of compounds in which the alicyclic ring was replaced by one or two methyl groups. These are known as benzomorphan derivatives or, more correctly, as benzazocines. They may be represented by the following formula:

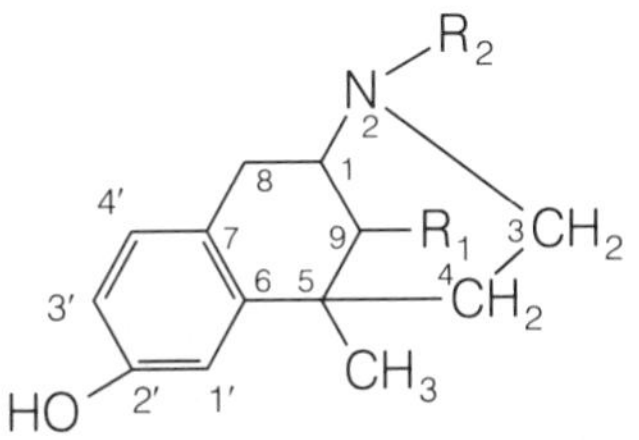

II

The trimethyl compound (II, $R_1 = R_2 = CH_3$) is about three times more potent than the dimethyl (II, $R_1 = H$, $R_2 = CH_3$). The *N*-phenethyl derivatives have almost 20 times the analgesic activity of the corresponding *N*-methyl compounds. Again, the more potent was the one containing the two ring methyls (II, $R_1 = CH_3$, $R_2 = CH_2CH_2C_6H_5$). Deracemization proved the *levo*-isomer of this compound to be more active, being about 20 times as potent as morphine in mice. The (±)-form, phenazocine, was on the market but was removed in favor of pentazocine.

May and his co-workers[44] have demonstrated an extremely significant difference between the two isomeric *N*-methyl benzomorphans in which the alkyl in the 5-position is *n*-propyl (R_1) and the alkyl in the 9-position is methyl (R_2). These have been termed the α-isomer and the β-isomer and have the groups oriented as indicated. The isomer with the alkyl *cis* to the phenyl has been shown to possess analgesic activity (in mice) equal to that of morphine, but has little or no capacity to suppress withdrawal symptoms in addicted monkeys. On the other hand, the *trans*-isomer has one of the highest analgesic potencies among the benzomorphans, but it is quite able to suppress morphine withdrawal symptoms. Further separation of properties is found between the enantiomers of the *cis*-isomer. The (+)-isomer has weak analgesic activity, but a high physical-dependence capacity. The (−)-isomer is a stronger analgesic, without the dependence capacity, and possesses antagonistic activity.[45] The same was true with the 5,9-diethyl and 9-ethyl-5-phenyl derivatives. The (−)-*trans*-5,9-diethyl isomer was similar, except it had no antagonistic properties. This demonstrates that it is possible to divorce analgesic activity comparable with morphine from addiction potential. That *N*-methyl compounds have shown antagonistic properties is of great interest as well. The most potent of these is the benzomorphan with an α-methyl and β-3-oxoheptyl group at position 9. The (−)-isomer shows greater antagonistic activity than naloxone and is still three times more potent than morphine as an analgesic.[46]

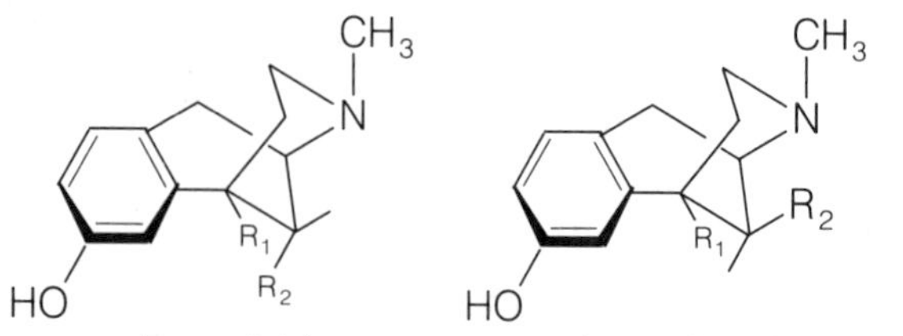

CH_3 N R_1 R_2 HO

α-isomer (*cis*)

CH_3 N R_2 R_1 HO

β-isomer (*trans*)

An extensive series of the antagonist-type analgesics in the benzomorphans has been reported.[47] Of these, pentazocine (II, $R_1 = CH_3$, $R_2 = CH_2CH = C(CH_3)_2$) and cyclazocine (II, $R_1 = CH_3$, $R_2 = CH_2$—cyclopropyl) have been the most interesting. Pentazocine has about half the analgesic activity of morphine, with a lower incidence of side effects.[48] Its addiction liability is much lower, approximating that of propoxyphene.[49] It is currently available in parenteral and tablet form. Cyclazocine is a strong morphine antagonist, showing about ten times the analgesic activity of morphine.[50] It was investigated as an analgesic and for the treatment of heroin addiction, but it was never marketed because of hallucinatory side effects.

It was mentioned previously that replacement of the *N*-methyl group in morphine by larger alkyl groups lowered analgesic activity. In addition, these compounds were found to counteract the effect of morphine and other morphinelike analgesics and are thus known as *narcotic antagonists*. The reversal of activity increases from ethyl, to propyl, to allyl, with the cyclopropylmethyl usually being maximal. This property was true not only for morphine, but with other analgesics as well. *N*-Allylnormorphine (nalorphine) was the first of these, but was taken off the market owing to side effects. Levallorphan, the corresponding allyl analogue of levorphanol, naloxone (*N*-allylnoroxymorphone), and naltrexone (*N*-cyclopropyl-noroxymorphone) are the three narcotic antagonists now on the market. Naloxone and naltrexone appear to be pure antagonists with no morphine- or nalorphinelike effects. They also block the effects of other antagonists. These drugs are used to prevent, diminish, or abolish many of the actions or the side effects encountered with the narcotic analgesics. Some of these are respiratory and circulatory depression, euphoria, nausea, drowsiness, analgesia, and hyperglycemia. They are thought to act by competing with the analgesic molecule for attachment at its, or a closely related, receptor site. The observation that some narcotic antagonists, which are devoid of addiction liability, are also strong analgesics has spurred considerable interest in them.[14] The *N*-cyclopropylmethyl compounds mentioned are the most potent antagonists, but appear to produce psychotomimetic effects and may not be useful as analgesics. However, one of these, buprenorphine, has shown an interesting profile and has been introduced in Europe and in the United States as a potent analgesic.[51] It is being studied as a possible treatment for narcotic addicts.[52]

Very intensive efforts were once under way to develop narcotic antagonists that can be used to treat narcotic addiction.[53] The continuous administration of an antagonist will block the euphoric effects of heroin, thereby aiding rehabilitation of an addict. The cyclopropylmethyl derivative of naloxone, naltrexone, has been marketed for this purpose. The oral dose of 100 to 150 mg three times a week is sufficient to block several usual doses of heroin.[54] Long-acting preparations were once also under study.[55]

Much research, other than that described in the foregoing discussion, has been carried out by the systematic dissection of morphine to give several interesting fragments. These approaches have not yet produced important analgesics; therefore, they are not discussed in this chapter. However, the interested reader may find a key to this literature from the excellent reviews of Eddy,[4] Bergel and Morrison,[17] and Lee.[21]

STRUCTURE–ACTIVITY RELATIONSHIPS

Several reviews on the relationship between chemical structure and analgesic action have been published.[4,25,56–64] Only the major conclusions will be considered here, and the reader is urged to consult these reviews for a more complete discussion of the subject.

From the time Small and co-workers started their studies on the morphine nucleus to the present, there has been much light shed on the structural features connected with morphinelike analgesic action. In a very thorough study made for the United Nations Commission on Narcotics in 1955, Braenden and co-workers[57] found that the features possessed by all known morphine-like analgesics were as follows:

1. A tertiary nitrogen, the group on the nitrogen being relatively small
2. A central carbon atom, of which none of the valences is connected with hydrogen
3. A phenyl group or a group isosteric with phenyl, which is connected to the central carbon atom
4. A two-carbon chain separating the central carbon atom from the nitrogen for maximal activity

From the foregoing discussion it is evident that several exceptions to these generalizations may be found in the structures of compounds that have been synthesized in the last several years. Eddy[25] has discussed the more significant exceptions.

Relative to the first feature mentioned, extensive studies of the action of normorphine have shown that it possesses analgesic activity that approximates that of morphine. In humans, it is about one-fourth as active as morphine when administered intramuscularly, but it was slightly superior to morphine when administered intracisternally. On the basis of the last-mentioned effect, Beckett and his co-workers[65] postulated that *N*-dealkylation was a step in the mechanism of analgesic action. This has been questioned.[66] Additional studies indicate that dealkylation does occur in the brain, although its exact role is not clear.[67] It is clear, from the previously discussed *N*-aralkyl derivatives, that a small group is not necessary.

Several exceptions to the second feature have been synthesized. In these series, the central carbon atom has been replaced by a tertiary nitrogen. They are related to methadone and have the following structures:

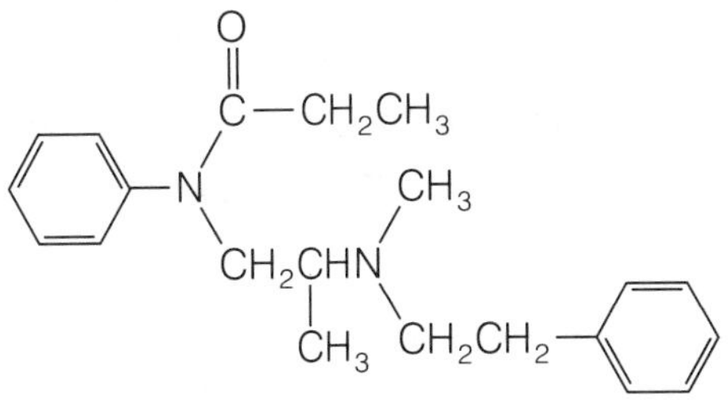

III

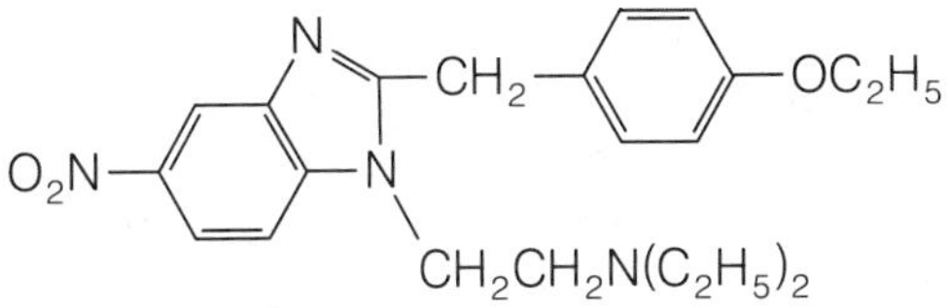

IV

Diampromide (III) and its related anilides have potencies that are comparable with those of morphine;[68] however, they have shown addiction liability and have not appeared on the market. The closely related cyclic derivative fentanyl (see A-17, Table 17-3), is used in surgery. The benzimidazoles, such as etonitazene (IV), are very potent analgesics, but show the highest addiction liabilities yet encountered.[69]

Possibly an exception to feature 3, and the only one that has been encountered, may be the cyclohexyl analogue of A-6 (see Table 17-3), which has significant activity.

Eddy[25] mentions two possible exceptions to feature 4 in addition to fentanyl.

As a consequence of the many studies on molecules of varying types that possess analgesic activity, it became increasingly apparent that activity was associated not only with certain structural features, but also with the size and the shape of the molecule. The hypothesis of Beckett and Casy[70] has dominated thinking for several years in the area of stereochemical specificity of these molecules. They initially noted that the more active enantiomers of the methadone- and thiambutene-type analgesics were related configurationally to (*R*)-alanine. This suggested to them that a stereoselective fit at a receptor could be involved in analgesic activity. To depict the dimensions of an analgesic receptor, they selected morphine (because of its semirigidity and high activity) to provide them with information on a complementary receptor. The features that were thought to be essential for proper receptor fit were

1. A basic center able to associate with an anionic site on the receptor surface

2. A flat aromatic structure, coplanar with the basic center, allowing for van der Waals bonding to a flat surface on the receptor site to reinforce the ionic bond
3. A suitably positioned projecting hydrocarbon moiety forming a three-dimensional geometric pattern with the basic center and the flat aromatic structure.

These features were selected, among other reasons, because they are present in *N*-methylmorphinan, which may be looked upon as a "stripped down" morphine [i.e., morphine without the characteristic peripheral groups (except for the basic center)]. Inasmuch as *N*-methylmorphinan possessed substantial activity of the morphine type, it was felt that these three features were the fundamental ones determining activity and that the peripheral groups of morphine acted essentially to modulate the activity.

In accord with the foregoing postulates, Beckett and Casy[70] proposed a complementary receptor site (Fig. 17-1) and suggested ways[71,72] in which the known active molecules could be adapted to it. Subsequent to their initial postulates, it was demonstrated that natural (−)-morphine was related configurationally to methadone and thiambutene, a finding that lent weight to the hypothesis. Fundamental to their proposal was that such a receptor was essentially inflexible and that a lock-and-key-type situation existed. More recently, the unnatural (+)-morphine was synthesized and shown to be inactive.[73]

Although the foregoing hypothesis appeared to fit the facts quite well and was a useful hypothesis for several years, it now appears that certain anomalies exist that cannot be accommodated by it. For example, the more active enantiomer of α-methadol is not

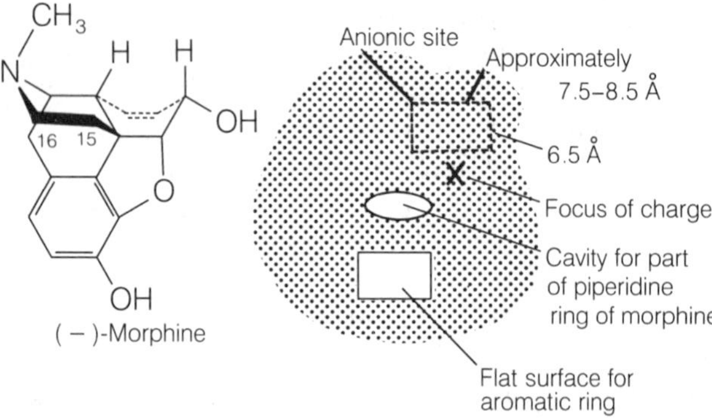

FIG. 17-1. *Diagram of the surface of the analgesic receptor site with the corresponding lower surface of the drug molecule. The three-dimensional features of the molecule are shown by the bonds: —, ---, and – –, which represent in front of, behind, and in the plane of the paper, respectively. (Gourley, D. R. H.: In Jucker, E. (ed.). Prog. Drug Res. 7:36, 1964.)*

related configurationally to (*R*)-alanine, in contrast with the methadone and thiambutene series. This is also true for the carbethoxy analogue of methadone (V) and for diampromide (III) and its analogues. Another factor that was implicit in considering a proper receptor fit for the morphine molecule and its congeners was that the phenyl ring at the 4-position of the piperidine moiety should be in the axial orientation for maximum activity. The fact that structure VI has only an equatorial phenyl group, yet possesses activity equal to that of morphine, would seem to cast doubt on the necessity for axial orientation as a receptor-fit requirement.

$(C_6H_5)_2C(COOC_2H_5)CH_2CH(CH_3)N(CH_3)_2$

V

VI

In view of the difficulty of accepting Beckett and Casy's hypothesis as a complete picture of analgesic–receptor interaction, Portoghese[74,76] has offered an alternative hypothesis. This hypothesis is based, in part, on the established ability of enzymes and other types of macromolecules to undergo conformational changes[77,78] on interaction with small molecules (substrates or drugs). The fact that configurationally unrelated analgesics can bind and exert activity is interpreted as meaning that more than one mode of binding may be possible at the same receptor. Such different modes of binding may be due to differences in positional or conformational interactions with the receptor. The manner in which the hypothesis can be adapted to the methadol anomaly is illustrated in Figure 17-2. Portoghese, after considering activity changes in various structural types (i.e., methadones, meperidines, prodines) as related to the identity of the *N*-substituent, noted that in certain series there was a parallelism in the direction of activity when identical changes in *N*-substituents were made. In others there appeared to be a nonparallelism. He has interpreted parallelism and nonparallelism, respectively, as being due to similar and to dissimilar modes of binding. As viewed by this hypothesis, although it is still a requirement that analgesic molecules be bound in a fairly precise manner, it nevertheless liberalizes the concept of

FIG. 17-2. An illustration of how different polar groups in analgesic molecules may cause inversion in the configurational selectivity of an analgesic receptor. A hydrogen-bonding moiety, denoted by X and Y, represents a site that is capable of being hydrogen bonded.

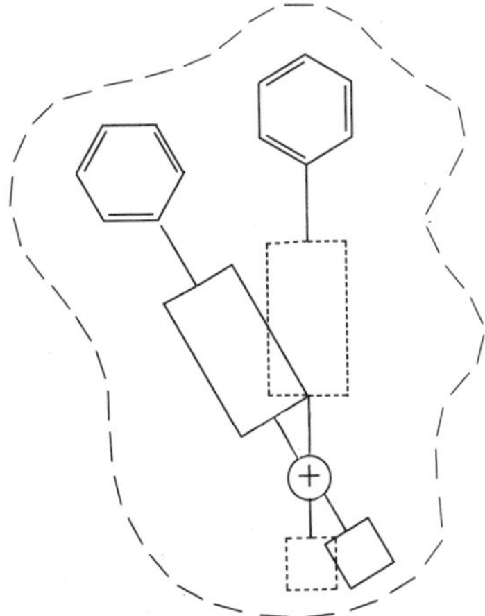

FIG. 17-3. A schematic illustration of two different molecular modes of binding to a receptor. The protonated nitrogen is represented by ⊕. The square denotes an N-substituent. The anionic sites lie directly beneath ⊕.

binding in that a response may be obtained by two different molecules binding stereoselectively in two different precise modes at the same receptor. A schematic representation of such different possible binding modes is shown in Figure 17-3. This representation will aid in visualizing the meaning of *similar* and *dissimilar* binding modes. If two different analgesiophores* bearing identical *N*-substituents are positioned on the receptor surface such that the *N*-substituent occupies essentially the same position, a similar pharmacologic response may be anticipated. Thus, as one proceeds from one *N*-substituent to another, the response should likewise change, resulting in a parallelism of effect. On the other hand, if two different analgesiophores are bound to the receptor such that the *N*-substituents are not arranged identically, one may anticipate nonidentical responses on changing the *N*-substituent (i.e., a nonparallel response). From the preceding statements, as well as the diagram, it is not to be inferred that the analgesiophore necessarily will be bound in the identical position within a series. They do, however, suggest that, in series with parallel activities, the pairs being compared will be bound identically to produce the parallel effect. Interestingly, when binding modes are similar, Portoghese has been able to demonstrate the existence of a linear free-energy relationship. There is also the possibility that more than one receptor is involved.

Considerable evidence is now available to demonstrate that multiple receptors exist. Martin has characterized and named these by responses from probe molecules [i.e., μ (mu) receptors for morphine-specific effects, σ (sigma) for cyclazocine, and κ (kappa) for ketocyclazocine].[79] Various combinations of these in different tissues could be responsible for the varying effects observed.[80, 81]

Although this hypothesis is new, it appears to embrace virtually all types of analgesic molecules presently known,* and it will be interesting to see whether it is of further general applicability as other molecules with activity are devised.

Another of the highly important developments in structure–activity correlations has been the development of highly active analgesics from the *N*-allyl-type derivatives that were once thought to be only morphine antagonists and devoid of analgesic properties. Serendipity played a major role in this discovery: Lasagna and Beecher,[82] in attempting to find some "ideal" ratio of antagonist (*N*-allylnormorphine, nalorphine) to analgesic (morphine) such as to maintain the desirable effects of morphine while minimizing the undesirable ones, discovered that nalorphine was, milligram for milligram, as potent an analgesic as morphine. Unfortunately, nalorphine has depersonalizing and psychotomimetic properties that preclude its use clinically as a pain reliever. However, the discovery led to the development of related derivatives such as pentazocine and cyclazocine. Pentazocine has achieved some success in providing an analgesic with low addiction potential, although it is not totally free of some of the

* The analgesic molecule less the *N*-substituent (i.e., the portion of the molecule giving the characteristic analgesic response).

* Two possible exceptions are 4-propionoxy-4-cyclohexyl-1-methylpiperidine and 1-tosyl-4-phenyl-4-ethylsulfone piperidine. (Helv. Chim. Acta 36:819, 1953.)

other side effects of morphine. The pattern of activity in these and other *N*-allyl and *N*-cyclopropylmethyl derivatives indicates that most potent antagonists possess psychotomimetic activity, whereas the weak antagonists do not. It is from this latter group that useful analgesics, such as pentazocine, butorphanol, and nalbuphine, have been found. The latter two possess *N*-cyclobutylmethyl groups.

What structural features are associated with antagonistlike activity has become uncertain. The *N*-allyl and dimethylallyl substituent does not always confer antagonist properties. This is true in the meperidine and thevinol series. Demonstration of antagonistlike properties by specific isomers of *N*-methyl benzomorphans has raised still further speculation. The exact mechanisms by which morphine and the narcotic antagonists act are not clearly defined, and a great amount of research is presently being carried on. Published reviews and symposia may be consulted for further discussions of these topics.[53,83–85]

A further problem also is demonstrated in the testing for analgesic activity. The analgesic activity of the antagonists was not apparent from animal testing, but was observed only in humans. Screening in animals can be used to assess the antagonistic action, which indirectly indicates possible analgesic properties in humans.[86]

It has been customary in the area of analgesic agents to attribute differences in their activities to structurally related differences in their receptor interactions. This rather universal practice continues, in spite of early warnings and recent findings. It now appears clear that much of the differences in relative analgesic potencies can be accounted for on the basis of pharmacokinetic or distribution properties.[84] For example, a definite correlation was found between the partition coefficients and the intravenous analgesic data for 17 agents of widely varying structures.[87] Usual test methods do not help define which structural features are related to receptor and which to distribution phenomena. Studies directed toward making this distinction are using the measurement of actual brain and plasma levels[88,89] or direct injection into the ventricular area,[87] the measurement of ionization potentials and partition coefficients,[90] and the application of molecular orbital theories and quantum mechanics.[85,91–93] These are providing valuable insight to the designing of new and more successful agents.

All of the foregoing work had strongly suggested for years the existence of specific binding sites or *receptors* in brain and other tissue. The demonstration of the high degree of steric and structural specificity in the action of the opiates and their antagonists led many investigators to search for such receptors.[94,95] Thus in 1971, Goldstein and co-workers demonstrated stereospecific binding in brain homogenates.[96] This was quickly followed by refinements and further discoveries by Simon, Terenius, and Snyder.[97–99] These receptor-binding studies have now become a routine assay for examining structure–activity relationships.

In addition to the binding studies, considerable attention continued on the use of in vitro models, in particular the isolated guinea pig ileum, rat jejunum, and mouse vas deferens.[100] While working with these preparations, Hughes was the first to discover the existence of an endogenous factor from pig brains that possessed opiatelike properties.[101] This factor, given the name *enkephalin*, consisted of two pentapeptides, called *methonine-*, or *met-enkephalin*, and *leucine-*, or *leu-enkephalin*. These two enkephalins have subsequently been shown to exist in all animals, including humans, and to possess all morphinelike properties. It was also observed that they exist as segments of a pituitary hormone, the 91-amino acid β-lipotropin, which is cleaved selectively to release specific segments that have now been found to have functions within the body. Thus, segment 61 to 65 is met-enkephalin, 61 to 76 is α-endorphin, 61 to 77 is γ-endorphin, and, probably the most important, 61 to 91 is β-endorphin.

```
                 5                  10                  15
H-Glu-Leu-Thr-Gly-Gln-Arg-Leu-Arg-Gln-Gly-Asp-Pro-Asn-Ala
                                                          |
30                25                 20                   |
Leu-Ser-His-Glu-Leu-Ala-Asn-Pro-Gly-Glu-Gly-Asp-Asn-Ala-Gly
|
|                   35                  40                45
Leu-Ala-Asp-Leu-Val-Ala-Ala-Glu-Lys-Lys-Asp-Glu-Gly-Pro-Tyr
                                                          |
                     55               50                  |
Arg-Lys-Asp-Lys-Pro-Pro-Ser-Gly-Trp-Asg-Phe-His-Glu-Met-Arg
             (  |                         )
             |  Tyr61                     |
             |  |                         |
Methionine  <   Gly-Gly                   |
Enkephalin   |       |                    |
             |  65   |                    |
             (  Met-Phe                   |
                |                         |
                |                 70      |
                Thr-Ser-Glu-Lys-Ser       |
                                  |        >   β-Endorphin
                Val-Leu-Pro-Thr-Gln       |
                |                         |
                |                 80      |
              76Thr-Leu-Phe-Lys-Asn       |
                                  |       |
                85                |       |
                Asn-Lys-Ile-Ile-Ala       |
                |                         |
                Ala-Tyr-Lys-Lys-Gly-Glu-OH
                                  91      )
```

β-Lipotropin, β-Endorphin, Methionine Enkephalin Relationships

The last endorphin (short for *end*ogenous mor*phine*) has 20 times the analgesic potency of morphine when injected into rat brain. It has also been shown that these substances can produce tolerance and dependence.

It is obvious that all of these techniques will lead to new concepts and understanding of the processes of analgesia, tolerance, and dependence. It is hoped that learning how these mechanisms operate will aid in the design and development of better analgesics.

PRODUCTS*

Morphine. This alkaloid was isolated first in 1803 by Derosne, but the credit for isolation generally goes to Serturner (1803), who first called attention to the basic properties of the substance. Morphine, incidentally, was the first plant base isolated and recognized as such. Although intensive research was carried out on the structure of morphine, it was only in 1925 that Gulland and Robinson[105] postulated the currently accepted formula. The total synthesis of morphine finally was effected by Gates and Tschudi[106] in 1952, thereby confirming the Gulland and Robinson formula.

Morphine is obtained only from the opium poppy, *Papaver somniferum*, either from opium, the resin obtained by lancing the unripe pod, or from poppy straw. The latter process is being favored as it helps to eliminate illicit opium from which heroin is readily produced. It occurs in opium in amounts varying from 5% to 20% (*USP* requires not less than 9.5%). It is isolated by various methods, of which the final step is usually the precipitation of morphine from an acid solution by using excess ammonia. The precipitated morphine then is recrystallized from boiling alcohol.

The free alkaloid occurs as levorotatory, odorless, white, needlelike crystals possessing a bitter taste. It is almost insoluble in water (1 : 5000,† 1 : 1100 at boiling point), ether (1 : 6250) or chloroform (1 : 1220). It is somewhat more soluble in ethyl alcohol (1 : 210, 1 : 98 at boiling point). Because of the phenolic hydroxyl group, it is readily soluble in solutions of alkali or alkaline earth metal hydroxides.

Morphine is a monoacidic base and readily forms water-soluble salts with most acids. Thus, because morphine itself is so poorly soluble in water, the salts are the preferred form for most uses. Numerous salts have been marketed, but the ones in use are principally the sulfate and, to a lesser extent, the hydrochloride. Morphine acetate, which is freely soluble in water (1 : 2.5), but is relatively unstable, has been used to a limited extent in liquid antitussive combinations.

Many writers have pointed out the "indispensable" nature of morphine, based on its potent analgesic properties toward all types of pain. It is properly termed a narcotic analgesic. However, because it causes addiction so readily, it should be used only in those cases for whom other pain-relieving drugs prove to be inadequate. It controls pain caused by serious injury, neoplasms, migraine, pleurisy, biliary and renal colic, and numerous other causes. It often is administered as a preoperative sedative, together with atropine to control secretions. With scopolamine, it is given to obtain the so-called twilight sleep. This effect is used in obstetrics, but care is exercised to prevent respiratory depression in the fetus. It is noteworthy that the toxic properties of morphine are much more evident in young and old people.

Morphine Hydrochloride. This salt may be prepared by neutralizing a hot aqueous suspension of morphine with diluted hydrochloric acid and then concentrating the resultant solution to crystallization. It is no longer commercially available.

It occurs as silky, white, glistening needles or cubical masses or as a crystalline, white powder. The hydrochloride is soluble in water (1 : 17.5, 1 : 0.5 at boiling point), alcohol (1 : 52, 1 : 46 at 60°), or glycerin, but it is practically insoluble in ether or chloroform. Solutions have a *p*H of approximately 4.7 and may be sterilized by boiling.

Its uses are the same as those of morphine.

The usual oral and subcutaneous dose is 15 mg every four hours as needed, with a suggested range of 8 to 20 mg.

Morphine Sulfate, USP. This morphine salt is prepared in the same manner as the hydrochloride (i.e., by neutralizing morphine with diluted sulfuric acid).

It occurs as feathery, silky, white crystals, as cubical masses of crystals, or as a crystalline, white powder. Although it is a fairly stable salt, it loses water of hydration and darkens on exposure to air and light. It is soluble in water (1 : 16, 1 : 1 at 80°), poorly soluble in alcohol (1 : 570, 1 : 240 at 60°), and insoluble in chloroform or ether. Aqueous solutions have a *p*H of approximately 4.8 and may be sterilized by heating in an autoclave.

This common morphine salt is used widely in England by oral administration for the management of pain in cancer patients. It has largely replaced

---

* In General Circular No. 253, March 10, 1960, the Treasury Department, Bureau of Narcotics, Washington, DC 20525 has published an extensive listing of narcotics of current interest in the drug trade. This listing will be much more extensive than the following monographic coverage of compounds primarily of interest to American pharmacists.

† In this chapter a solubility expressed as (1 : 5000) indicates that 1 g is soluble in 5000 mL of the solvent at 25°C. Solubilities at other temperatures will be so indicated.

Brompton's mixture, a combination of heroin and cocaine in chloroform water. In the United States this preparation has become mistakenly popular, substituting morphine sulfate for the heroin. Moreover, Twycross has advised that the stimulant cocaine is contraindicated because it interferes with sleep.[107]

Codeine, USP. Codeine is an alkaloid that occurs naturally in opium, but the amount present is usually too small to be of commercial importance. Consequently, most commercial codeine is prepared from morphine by methylating the phenolic hydroxyl group. The methylation methods usually are patented procedures and make use of reagents such as diazomethane, dimethyl sulfate, and methyl iodide. Newer methods are based on its synthesis from thebaine, which makes it possible to use *P. bracteatum* as a natural source (see beginning of chapter).

It occurs as levorotatory, colorless, efflorescent crystals or as a white crystalline powder. It is light-sensitive. Codeine is slightly soluble in water (1 : 120) and sparingly soluble in ether (1 : 50). It is freely soluble in alcohol (1 : 2) and very soluble in chloroform (1 : 0.5).

Codeine is a monoacidic base and readily forms salts with acids, the most important salts being the sulfate and the phosphate. The acetate and the methylbromide derivatives have been used to a limited extent in cough preparations. The free base is used little compared with the salts, its greatest use being in Terpin Hydrate and Codeine Elixir, USP.

The general pharmacologic action of codeine is similar to that of morphine but, as previously indicated, it does not possess the same degree of analgesic potency. Lasagna[108] once commented on the status of the drug as follows:

> Despite codeine's long use as an analgesic drug, it is amazing how little reliable information there is about its efficacy, particularly by the parenteral route.

There are studies that indicate that 30 to 120 mg of codeine are considerably less efficient parenterally than 10 mg of morphine, and the usual side effects of morphine—respiratory depression, constipation, nausea, and such—are apparent. Codeine is less effective orally than parenterally, and it has been stated by Houde and Wallenstein[109] that 32 mg of codeine is about as effective as 650 mg of aspirin in relieving terminal cancer pain. However, it also has been recognized that combinations of aspirin and codeine act additively as analgesics, thus giving some support to the common practice of combining the two drugs.

Codeine has a reputation as an antitussive, depressing the cough reflex, and is used in many cough preparations. It is one of the most widely used morphinelike analgesics. It is considerably less addicting than morphine and, in the usual doses, respiratory depression is negligible, although an oral dose of 60 mg will cause such depression in a normal person. It is probably true that much of codeine's reputation as an antitussive rests on subjective impressions rather than on objective studies. The average 5-mL dose of Terpin Hydrate and Codeine Elixir contains 10 mg of codeine. This preparation and many like it have been sold over-the-counter as exempt narcotic preparations. However, abuse or misuse of these preparations has led to their being placed on a prescription-only status in many states.

A combination of codeine and papaverine (Copavin) was advocated by Diehl[110] for the prophylaxis and treatment of common colds. When administered at the first signs of a cold, it was claimed to have aborted the cold in a significant percentage of the cases.

Codeine Phosphate, USP. This salt may be prepared by neutralizing codeine with phosphoric acid and precipitating the salt from aqueous solution with alcohol.

Codeine phosphate occurs as fine, needle-shaped, white crystals or as a white crystalline powder. It is efflorescent and is sensitive to light. It is freely soluble in water (1 : 2.5, 1 : 0.5 at 80°), but less soluble in alcohol (1 : 325, 1 : 125 at boiling point). Solutions may be sterilized by boiling.

Because of its high solubility in water, compared with the sulfate, this salt is used widely. It is often the only salt of codeine stocked by pharmacies and is dispensed, rightly or wrongly, on all prescriptions calling for either the sulfate or the phosphate.

Codeine Sulfate, USP. Codeine sulfate is prepared by neutralizing an aqueous suspension of codeine with diluted sulfuric acid and then effecting crystallization.

It occurs as white crystals, usually needlelike, or as a white crystalline powder. The salt is efflorescent and light-sensitive. It is soluble in water (1 : 30, 1 : 6.5 at 80°), much less soluble in alcohol (1 : 1280), and insoluble in ether or chloroform.

This salt of codeine is prescribed frequently, but is not as suitable as the phosphate for liquid preparations. Solutions of the sulfate and the phosphate are incompatible with alkaloidal reagents and alkaline substances.

Ethylmorphine Hydrochloride; dionin. This synthetic compound is analogous to codeine, but instead of being the methyl ether it is the ethyl ether. Ethylmorphine may be prepared by treating an alkaline alcoholic solution of morphine with diethyl sulfate. The hydrochloride is obtained from the free base by neutralizing it with diluted hydrochloric acid.

The salt occurs as a microcrystalline, white or faintly yellow odorless powder and is slightly bitter. It is soluble in water (1 : 10) and in alcohol (1 : 25), but only slightly soluble in ether and in chloroform.

The systemic action of this morphine derivative is intermediate between those of codeine and morphine. It has analgesic qualities and sometimes is used for the relief of pain. As a depressant of the cough reflex, it is as effective as codeine and, consequently, is found in some commercial cough syrups. However, the chief use of this compound is in ophthalmology. By an irritant dilating action on vessels, it stimulates the vascular and lymphatic circulation of the eye. This action is of value in chemosis (excessive edema of the ocular conjunctiva), and the drug is termed a *chemotic*.

Diacetylmorphine Hydrochloride; heroin hydrochloride; diamorphine hydrochloride. Although heroin is two to three times more potent than morphine as an analgesic, its sale and use are prohibited in the United States because of its intense addiction liability. It is available in some European countries, where it has a limited use as an antitussive and as an analgesic in terminal cancer patients. Because of its superior solubility over morphine sulfate, arguments have been raised for its availability. However, the other more potent analgesics described here have significant advantages in being more stable and longer acting. It remains as one of the most widely used narcotics for illicit purposes and still places major economic burdens on society.

Hydromorphone; dihydromorphinone. This synthetic derivative of morphine is prepared by the catalytic hydrogenation and dehydrogenation of morphine under acidic conditions, using a large excess of platinum or palladium.

The free base is similar in properties to those of morphine, being slightly soluble in water, freely soluble in alcohol, and very soluble in chloroform.

This compound, of German origin, was introduced in 1926. It is a substitute for morphine (five times as potent) but has approximately equal addicting properties and a shorter duration of action. It possesses the advantage over morphine of giving less daytime sedation or drowsiness. It is a potent antitussive and is often used for coughs that are difficult to control.

Hydromorphone Hydrochloride, USP. Dihydromorphinone hydrochloride (Dilaudid). Hydromorphone hydrochloride occurs as a light-sensitive, white crystalline powder that is freely soluble in water (1 : 3), sparingly soluble in alcohol, and practically insoluble in ether. It is used in about one-fifth the dose of morphine for any of the indications of morphine.

Hydrocodone Bitartrate, USP. Dihydrocodeinone bitartrate (Dicodid; Codone). This drug is prepared by the catalytic rearrangement of codeine or by hydrolyzing dihydrothebaine. It occurs as fine, white crystals or as a white crystalline powder. It is soluble in water (1 : 16), slightly soluble in alcohol, and insoluble in ether. It forms acidic solutions and is affected by light. The hydrochloride is also available.

Hydrocodone has a pharmacologic action midway between those of codeine and morphine, with 15 mg being equivalent to 10 mg of morphine in analgesic power. Although it possesses more addiction liability than codeine, it has been said to give no evidence of dependence or addiction with long-term use. Its principal advantage is in the lower frequency of side effects encountered with its use. It is more effective than codeine as an antitussive and is used primarily for this purpose. It is on the market in many cough preparations, as well as in tablet and parenteral forms. It has also been marketed in an ion-exchange resin complex form under the trade name of Tussionex. The complex releases the drug at a sustained rate and is said to produce effective cough suppression over a 10- to 12-hour period.

Hydrocodone is also marketed in combination with acetaminophen (e.g., Hydrocet, Vicodin, and Zydone) and with homatropine as Hycodan.

Although this drug found extensive use in antitussive formulations for many years, recently it has been placed under more stringent narcotic regulations, and it is being replaced gradually by codeine or dextromethorphan in most over-the-counter cough preparations.

Oxymorphone Hydrochloride, USP. (−)-14-Hydroxydihydromorphinone hydrochloride (Numorphan). Oxymorphone, introduced in 1959, is prepared by cleavage of the corresponding codeine derivative. It is used as the hydrochloride salt, which occurs as a white crystalline powder freely soluble in water and sparingly soluble in alcohol. In humans, oxymorphone is as effective as morphine in one-eighth to one-tenth the dosage, with good duration and a slightly lower frequency of side effects.[111] It has high addiction liability. It is used for the same purposes as morphine, such as control of postoperative pain, pain of advanced neoplastic diseases, as well as other types of pain that respond to morphine. Because of the risk of addiction, it should not be employed for relief of minor pains that can be controlled with codeine. It is also well to note that it has poor antitussive activity and is not used as a cough suppressant.

It may be administered orally, parenterally (intravenously, intramuscularly, or subcutaneously), or rectally, and for these purposes is supplied as a solution for injection (1.0 and 1.5 mg/mL), suppositories (2 and 5 mg), and in tablets (10 mg).

Nalbuphine Hydrochloride; *N*-Cyclobutylmethyl-14-hydroxy-*N*-nordihydromorphinone hydrochloride (Nubain). *N*-Cyclobutylmethylnoroxymorphone hydrochloride was introduced in 1979 as a potent analgesic of the agonist–antagonist type, with no to low abuse liability. It is somewhat less potent as an analgesic than its parent oxymorphone, but shares some of the antagonist properties of the closely related, but pure antagonists, naloxone and naltrexone. Nalbuphine hydrochloride occurs as a white to off-white crystalline powder that is soluble in water and sparingly soluble in alcohol. It is prepared from cyclobutylmethyl bromide and noroxycodone followed by cleavage of the *O*-methyl group.

This analgesic shows a very rapid onset with a duration of action of up to six hours. It has relatively low abuse liability, being judged to be less than that of codeine and propoxyphene. The injection was therefore introduced without narcotic controls, although caution is urged for long-term administration or use in emotionally disturbed patients. Abrupt discontinuation after prolonged use has given rise to withdrawal signs. Usual doses cause respiratory depression comparable with that of morphine, but no further decrease is seen with higher doses. It has fewer cardiac effects than pentazocine and butorphanol. The most frequent adverse effect is sedation, and, as with most other CNS depressants and analgesics, caution should be urged when administered to ambulatory patients who may need to drive a car or operate machinery.

Nalbuphine is marketed as an injectable (10 mg/mL). The usual dose is 10 mg administered subcutaneously, intramuscularly, or intravenously at three- to six-hour intervals, with a maximal daily dose of 160 mg.

Oxycodone Hydrocloride; dihydrohydroxycodeinone hydrochloride. This compound is prepared by the catalytic reduction of hydroxycodeinone, the latter compound being prepared by hydrogen peroxide (in acetic acid) oxidation of thebaine. This derivative of morphine occurs as a white crystalline powder that is soluble in water (1 : 10) or alcohol. Aqueous solutions may be sterilized by boiling. Although this drug is almost as likely to cause addiction as morphine, it is sold in the United States in Percodan and Tylox as a mixture of its hydrochloride and terephthalate salts in combination with aspirin. It is also marketed with hexobarbital as Percobarb.

It is used as a sedative, an analgesic, and a narcotic. Because it is believed to exert a physostigmine-like action, it is used externally in the eye in the treatment of glaucoma and related ocular conditions. To depress the cough reflex, it is used in 3- to 5-mg doses and as an analgesic in 5- to 10-mg doses. For severe pain, a dose of 20 mg is given subcutaneously.

Dihydrocodeine Bitartrate. Dihydrocodeine is obtained by the reduction of codeine. The bitartrate salt occurs as white crystals that are soluble in water (1 : 4.5) and only slightly soluble in alcohol. Subcutaneously, 30 mg of this drug is almost equivalent to 10 mg of morphine as an analgesic, giving more prompt onset and negligible side effects. It has addiction liability. It is available only as a cough preparation (Cophene-S). As an antitussive, the usual dose is 10 to 30 mg.

Normorphine. This drug may be prepared by *N*-demethylation of morphine.[112] In humans, by normal routes of administration, it is about one-fourth as active as morphine in producing analgesia, but has a much lower physical dependence capacity. Its analgesic effects are nearly equal by the intraventricular route. It does not show the sedative effects of morphine in single doses, but does so cumulatively. Normorphine suppresses the morphine abstinence syndrome in addicts, but after its withdrawal it gives a slow onset and a mild form of the abstinence syndrome.[113] It was once considered for possible use in the treatment of narcotic addiction.

Concentrated Opium Alkaloids (Pantopon) consist of a mixture of the total alkaloids of opium. It is free of nonalkaloidal material, and the alkaloids are said to be present in the same proportions as they occur naturally. The alkaloids are in the form of the hydrochlorides, and morphine constitutes 50% of the weight of the material.

This preparation is promoted as a substitute for morphine, the claim being that it is superior to the latter because of the synergistic action of the opium alkaloids. This synergism is said to result in less respiratory depression, less nausea and vomiting, and an antispasmodic action on smooth muscle. According to several authorities, however, the superiority to morphine is overrated, and the effects produced are comparable with the use of an equivalent amount of morphine. The commercial literature suggests a dose of 20 mg of Pantopon to obtain the same effect as is given by 15 mg of morphine.

Solutions prepared for parenteral use may be slightly colored, a situation that does not necessarily indicate decomposition.

Apomorphine Hydrochloride, USP. When morphine or morphine hydrochloride is heated at 140° under pressure with strong (35%) hydrochloric acid, it loses a molecule of water and yields a compound known as apomorphine.

Morphine $\xrightarrow[\text{pressure and heat}]{\text{HCl}}$ Apomorphine

The hydrochloride is odorless and occurs as minute, glistening, white or grayish white crystals or as a white powder. It is light-sensitive and turns green on exposure to air and light. It is sparingly soluble in water (1 : 50, 1 : 20 at 80°) and in alcohol (1 : 50) and is very slightly soluble in ether or chloroform. Solutions are neutral to litmus.

The change in structure from morphine to apomorphine causes a profound change in its physiologic action. The central depressant effects of morphine are much less pronounced, and the stimulant effects are enhanced greatly, thereby producing emesis by a purely central mechanism. It is administered subcutaneously to obtain emesis. It is ineffective orally. Apomorphine is one of the most effective, prompt (10 to 15 minutes), and safe emetics in use today. However, care should be exercised in its use because it may be depressant in already depressed patients.

Meperidine Hydrochloride, USP. Ethyl 1-methyl-4-phenylisonipecotate hydrochloride; ethyl 1-methyl-4-phenyl-4-piperidinecarboxylate hydrochloride (Demerol Hydrochloride). This is a fine, white, odorless crystalline powder that is very soluble in water, soluble in alcohol, and sparingly soluble in ether. It is stable in the air at ordinary temperature, and its aqueous solution is not decomposed by a short period of boiling. The free base may be made by heating benzyl cyanide with bis(β-chloroethyl)methylamine, hydrolyzing to the corresponding acid and esterifying the latter with ethyl alcohol.[2]

Meperidine first was synthesized to study its spasmolytic character, but it was found to have analgesic properties in far greater degree. The spasmolysis is primarily due to a direct papaverinelike depression of smooth muscle and, also, to some action on parasympathetic nerve endings. In therapeutic doses, it exerts an analgesic effect that lies between those of morphine and codeine, but it shows little tendency toward hypnosis. It is indicated for the relief of pain in most patients for whom morphine and other alkaloids of opium generally are employed, but it is especially of value where the pain is due to spastic conditions of intestine, uterus, bladder, bronchi, and so on. Its most important use seems to be in lessening the severity of labor pains in obstetrics and, with barbiturates or tranquilizers, to produce amnesia in labor. In labor, 100 mg is injected intramuscularly as soon as contractions occur regularly, and a second dose may be given after 30 minutes if labor is rapid or if the cervix is thin and dilated (2 to 3 cm or more). A third dose may be necessary an hour or two later, and at this stage a barbiturate may be administered in a small dose to ensure adequate amnesia for several hours. Meperidine possesses addiction liability. There is a development of psychic dependence in those individuals who experience a euphoria lasting for an hour or more. The development of tolerance has been observed, and it is significant that meperidine can be successfully substituted for morphine in addicts who are being treated by gradual withdrawal. Furthermore, mild withdrawal symptoms have been noted in certain persons who have become purposely addicted to meperidine. The possibility of dependence is great enough to put it under the federal narcotic laws. Nevertheless, it remains as one of the more widely used analgesics.

Alphaprodine Hydrochloride, USP. (±)-1,3-Dimethyl-4-phenyl-4-piperidinol propionate hydrochloride. This compound is prepared according to the method of Ziering and Lee.[114]

It occurs as a white crystalline powder that is freely soluble in water, alcohol, and chloroform, but insoluble in ether.

The compound is an effective analgesic, similar to meperidine, and is of special value in obstetric analgesia. It appears to be quite safe for use in this capacity, causing little or no depression of respiration in either mother or fetus.

Anileridine, USP. Ethyl 1-(*p*-aminophenethyl)-4-phenylisonipecotate (Leritine). It is prepared by the method of Weijlard et al.[115] It occurs as a white to yellowish white crystalline powder that is freely soluble in alcohol, but only very slightly soluble in water. It is oxidized on exposure to air and light. The injection is prepared by dissolving the free base in phosphoric acid solution.

Anileridine is more active than meperidine and has the same usefulness and limitations. Its dependence capacity is less, and it is considered a suitable substitute for meperidine.

Anileridine Hydrochloride, USP. Ethyl 1-(*p*-aminophenethyl)-4-phenylisonipecotate dihydrochloride (Leritine Hydrochloride). It is prepared as cited for anileridine, except that it is converted to the dihydrochloride by conventional procedures. It occurs as a white or nearly white, crystalline, odorless powder that is stable in air. It is freely soluble in water, sparingly soluble in alcohol, and practically insoluble in ether and chloroform.

This salt has the same activity as that cited for anileridine.

Diphenoxylate Hydrochloride, USP. Ethyl 1-(3-cyano-3,3-diphenylpropyl)-4-phenylisonipecotate monohydrochloride (Colonil; Diphenatol; Enoxa; Lofene; Loflo; Lomotil; Lonox; Lo-Trol; Nor-nil). It occurs as a white, odorless, slightly water-soluble powder with no distinguishing taste.

Although this drug has a strong structural relationship to the meperidine-type analgesics it has very little, if any, such activity itself. Its most pronounced activity is its ability to inhibit excessive gastrointestinal motility, an activity reminiscent of the constipating side effect of morphine itself. Investigators have demonstrated the possibility of addiction,[25,26] particularly with large doses, but virtually all studies using ordinary dosage levels show nonaddiction. Its safety is reflected in its classification as an exempt narcotic, with, however, the warning that it may be habit forming. To discourage possible abuse of the drug, the commercial product (Lomotil) once contained a subtherapeutic dose (25 μg) of atropine sulfate in each 2.5-mg tablet and in each 5 mL of the liquid, which contains a like amount of the drug. Atropine has now been removed because of unwarranted side effects.

It is indicated in the oral treatment of diarrheas resulting from a variety of causes. The usual initial adult dose is 5 mg, three or four times a day, with the maintenance dose usually being substantially lower and being individually determined. Appropriate dosage schedules for children are available in the manufacturer's literature.

The incidence of side effects is low, but the drug should be used with caution, if at all, in patients with impaired hepatic function. Similarly, patients taking barbiturates concurrently with the drug should be observed carefully, in view of reports of barbiturate toxicity under these circumstances.

Loperamide Hydrochloride, USP. 4-(4-Chlorophenyl)-4-hydroxy-*N*,*N*-dimethyl-α,α-diphenyl-1-piperidinebutanamide; 4-(4-*p*-chlorophenyl-4-hydroxypiperidino)-*N*, *N*-dimethyl-2, 2-diphenylbutyramide hydrochloride (Imodium). This hybrid of a methadonelike and meperidine molecule is closely related to diphenoxylate, being more specific, potent, and longer acting. It acts as an antidiarrheal by a direct effect on the circular and longitudinal intestinal muscles. After oral administration it reaches peak blood levels within four hours and has a very long plasma half-life (40 hours). Tolerance to its effects has not been observed.[116] Although it has shown minimal central nervous system effects, it has been controlled under schedule V.

Loperamide is available as 2-mg capsules (Loperamide hydrochloride capsules USP) for treatment of acute and chronic diarrhea. Dosage recommended is 4 mg initially, with 2 mg after each loose stool for a maximum of 16 mg/day.

Ethoheptazine Citrate. Ethyl hexahydro-1-methyl-4-phenyl-1*H*-azepine-4-carboxylate citrate; 1-methyl-4-carbethoxy-4-phenylhexamethylenimine citrate (Zactane Citrate). It is effective orally against moderate pain in doses of 50 to 100 mg, with minimal side effects. Parenteral administration is limited because of central stimulating effects. It appears to have no addiction liability, but toxic reactions have occurred with large doses. A double-blind study in humans rated 100 mg of the hydrochloride salt equivalent to 30 mg of codeine and found that the addition of 600 mg of aspirin increased analgesic effectiveness.[29] In another study, 150 mg was found to be equal to 65 mg of propoxyphene, both being better than placebo.[117] It is available as a 75-mg tablet and in combination with 600 mg of aspirin (Zactirin).

Fentanyl Citrate, USP. *N*-(1-Phenethyl-4-piperidyl)propionanilide citrate (Sublimaze). This compound occurs as a crystalline powder, soluble in water (1 : 40) and methanol, and sparingly soluble in chloroform.

This novel anilide derivative has demonstrated analgesic activity 50 times that of morphine in humans.[32] It has a very rapid onset (four minutes) and short duration of action. Side effects similar to those of other potent analgesics are common—in particular, respiratory depression and bradycardia. It is used primarily as an adjunct to anesthesia. For use as a neuroleptanalgesic in surgery, it is available in combination with the neuroleptic droperidol (Innovar). It has dependence liability.

Methadone Hydrochloride, USP. 6-(Dimethylamino)-4,4-diphenyl-3-heptanone hydrochloride (Dolophene Hydrochloride; Westadone). It occurs as a bitter, white crystalline powder. It is soluble in water, freely soluble in alcohol and chloroform, and insoluble in ether.

Methadone is synthesized in several ways. The method of Easton and co-workers[118] is noteworthy in that it avoids the formation of the troublesome isomeric intermediate aminonitriles. The analgesic effect and other morphinelike properties are exhibited chiefly by the (−)-form. Aqueous solutions are stable and may be sterilized by heat for intramuscular and intravenous use. Like all amine salts, it is incompatible with alkali and salts of heavy metals. It is somewhat irritating when injected subcutaneously.

The toxicity of methadone is three to ten times greater than that of morphine, but its analgesic effect is twice that of morphine and ten times that of meperidine. It has been placed under federal narcotic control because of its high addiction liability.

Methadone is a most effective analgesic, used to alleviate many types of pain. It can replace morphine for the relief of withdrawal symptoms. It produces less sedation and narcosis than does morphine and appears to have fewer side reactions in bedridden patients. In spasm of the urinary bladder and in the suppression of the cough reflex, methadone is especially valuable.

The *levo*-isomer, levanone, is said not to produce euphoria or other morphinelike sensations and has been advocated for the treatment of addicts.[119] Methadone itself is being used quite extensively in addict treatment, although not without some controversy.[120] It will suppress withdrawal effects and is widely used to maintain former heroin addicts during this rehabilitation. Large doses are often used to "block" the effects of heroin during treatment.

The use of methadone in treating addicts is subject to FDA regulations that require special registration of physicians and dispensers. Methadone is available, however, for use as an analgesic under the usual narcotic requirements.

***l*-α-Acetylmethadol.** (−)-α-6-(Dimethylamino)-4,4-diphenyl-3-heptyl acetate hydrochloride; methadyl acetate; LAAM. It occurs as a white crystalline powder that is soluble in water, but dissolves with some difficulty. It is prepared by hydride reduction of (+)-methadone followed by acetylation.

Of the four possible methadol isomers, the (3*S*,6*S*)-isomer LAAM has the unique characteristic of producing long-lasting narcotic effects. Extensive metabolism studies have shown that this is due to its *N*-demethylation to give (−)-α-acetylnormethadol, which is more potent than its parent, LAAM, and possesses a long half-life.[121] This is further accentuated by its demethylation to the dinor metabolite, which has similar properties.[121,122]

Because of the need to administer methadone daily, which leads to inconvenience to the maintenance patient and illicit diversion, the long-acting LAAM was actively investigated as an addict-maintenance drug to replace methadone. Generally, a 70-mg dose three times a week is sufficient for routine maintenance.[40,123] The drug is still undergoing clinical trials.

It is of interest to note that the racemate of the normetabolite, noracylmethadol, was once studied in the clinic as a potential analgesic.[124]

Propoxyphene Hydrochloride, USP. (2*S*,3*R*)-(+)-4(Dimethylamino)-3-methyl-1,2-diphenyl-2-butanol propionate hydrochloride (Darvon; Dolene; Doloxene; D-orafen; Progesic Compound-65; Proxagesic; Proxene; SK 65). This drug was introduced into therapy in 1957. It may be prepared by the method of Pohland and Sullivan.[125] It occurs as a bitter, white crystalline powder that is freely soluble in water, soluble in alcohol, chloroform, and acetone, but practically insoluble in benzene and ether. It is the α-(+)-isomer, the α-(−)-isomer and β-diastereoisomers being far less potent in analgesic activity. The α-(−)-isomer, *levo*-propoxyphene, is an effective antitussive (see below).

In analgesic potency, propoxyphene is approximately equal to codeine phosphate and has a lower frequency of side effects. It has no antidiarrheal, antitussive, or antipyretic effect, thus differing from most analgesic agents. It is able to suppress the morphine abstinence syndrome in addicts, but has shown a low level of abuse because of its toxicity. It is not very effective in deep pain and appears to be no more effective in minor pain than aspirin. Its widespread use in dental pain seems justified, since aspirin is reported to be relatively ineffective. It has been classified as a narcotic and controlled under federal law. It does give some euphoria in high doses and has been abused. It has been responsible for numerous overdosage deaths. Refilling of the drug should be avoided if misuse is suspected.

It is available in several combination products with aspirin (e.g., Darvon w/A.S.A.) or acetaminophen (e.g., Dolene A.P.; Wygesic).

Propoxyphene Napsylate, USP. (+)-α-4-Dimethylamino-3-methyl-1,2-diphenyl-2-butanol propionoate (ester) 2-naphthylenesulfonate (salt) (Darvon-N). It is very slightly soluble in water, but soluble in alcohol, chloroform, and acetone.

The napsylate salt of propoxyphene was introduced shortly before the patent on Darvon expired. As an insoluble salt form it is claimed to be less prone to abuse because it cannot be readily dissolved for injection, and upon oral administration gives a slower, less pronounced peak blood level.

Because of its mild narcoticlike properties it was once investigated as an addict-maintenance drug to be used in place of methadone. It was hoped that it would offer the advantage of providing an easier withdrawal and serve as an addict-detoxification drug. Unfortunately, toxicity at higher doses has limited this application.

It is available in combination with aspirin and acetaminophen, Darvocet-N.

Levorphanol Tartrate, USP. (−)-3-Hydroxy-*N*-methylmorphinan bitartrate (Levo-Dromoran). The basic studies in the synthesis of this type of compound were made by Grewe, as pointed out earlier. Schnider and Grüssner synthesized the hydroxymorphinans, including the 3-hydroxyl derivative, by similar methods. The racemic 3-hydroxy-*N*-methylmorphinan hydrobromide (racemorphan, (±)-Dromoran) was the original form in which this potent analgesic was introduced. This drug is prepared by resolution of racemorphan. It should be noted that the *levo* compound is available in Europe under the original name, Dromoran. As the tartrate, it occurs in the form of colorless crystals. The salt is sparingly soluble in water (1 : 60) and is insoluble in ether.

The drug is used for the relief of severe pain and is in many respects similar in its actions to morphine, except that it is from six to eight times as potent. The addiction liability of levorphanol is as great as that of morphine and, for that reason, caution should be observed in its use. It is claimed that the gastrointestinal effects of this compound are significantly less than those experienced with morphine. Naloxone is an effective antidote for overdosage. Levorphanol is useful for relieving severe pain originating from a multiplicity of causes (e.g., inoperable tumors, severe trauma, renal colic, biliary colic). In other words, it has the same range of usefulness as morphine and is considered an excellent substitute. It is supplied in ampules, in multidose vials, and in the form of oral tablets. The drug requires a narcotic form.

Butorphanol Tartrate, USP. 17-(Cyclobutylmethyl)morphinan-3,14-diol D-(−)-tartrate; (−)-*N*-cyclobutylmethyl-3,14-dihydroxymorphinan bitartrate (Stadol; Torbutrol). This potent analgesic occurs as a white crystalline powder soluble in water and sparingly soluble in alcohol. It is prepared from the dihydroxy-*N*-normorphinan obtained by a modification of the Grewe synthesis. It is the cyclobutyl analogue of levorphanol and levallorphan, being equally as potent as the former as an analgesic and somewhat less active as an antagonist than the latter.

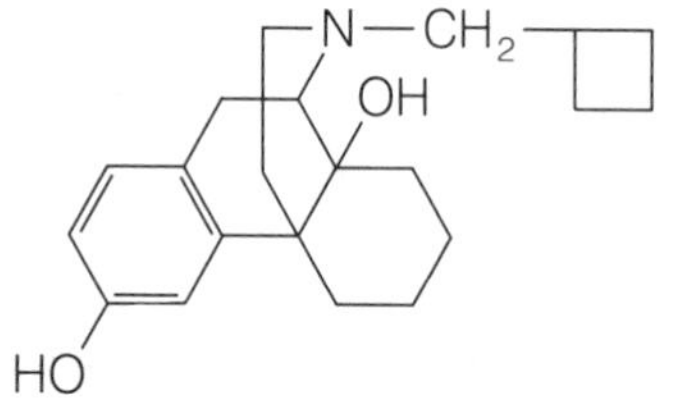

Butorphanol

The onset and duration of action of the drug is comparable with that of morphine, but it has the advantages of showing a maximal ceiling effect on respiratory depression and a greatly reduced abuse liability. The injectable form was marketed without narcotic controls; however, this product was considered for placement in schedule IV because of reported misuse and lack of recognition of its potential abuse liability. The drug has also been used illegally in the doping of race horses.

Butorphanol shares the adverse hemodynamic effects of pentazocine, causing increased pressure in specific arteries and on the heart work load. It should, therefore, be used with caution and only with patients hypersensitive to morphine for the treatment of myocardial infarction or other cardiac problems. Other adverse effects include a high incidence of sedation and, less frequently, nausea, headache, vertigo, and dizziness.[126]

It is available as a parenteral for intramuscular and intravenous administration in a dose of 1 or 2 mg every three to four hours, with a maximal single dose of 4 mg. A tablet form has been studied in clinical trials and is under review for marketing.

Pentazocine, USP. 1,2,3,4,5,6-Hexahydro-*cis*-6,11-dimethyl-3-(3-methyl-2-butenyl)-2,6-methano-3-benzazocin-8-ol; *cis*-2-dimethylallyl-5,9-dimethyl-2′-hydroxy-6,7-benzomorphan (Talwin). It occurs as a white crystalline powder that is insoluble in water and sparingly soluble in alcohol. It forms a poorly soluble hydrochloride salt, but is readily soluble as the lactate.

Pentazocine in a parenteral dose of 30 mg or an oral dose of 50 mg is about as effective as 10 mg of morphine in most patients. There is some evidence that the analgesic action resides principally in the (−)-isomer, with 25 mg being approximately equivalent to 10 mg of morphine sulfate.[127] Occasionally, doses of 40 to 60 mg may be required. Pentazocine's plasma half-life is about 3.5 hours.[128] At the lower dosage levels, it appears to be well tolerated, although some degree of sedation occurs in about one-third of those persons receiving it. The incidence of other morphinelike side effects is as high as with morphine and other narcotic analgesics. In patients who have been receiving other narcotic analgesics, large doses of pentazocine may precipitate withdrawal symptoms. It shows an equivalent or greater respiratory depressant activity. Pentazocine has given rise to a few cases of possible dependence liability. It has been placed under control, and its abuse potential should be recognized and close supervision of its use maintained. Levallorphan cannot reverse its effects, although naloxone can, and methylphenidate is recommended as an antidote for overdosage or excessive respiratory depression.

Pentazocine as the lactate is available in vials containing the base equivalent of 30 mg/mL, buffered to *p*H 4 to 5. It should not be mixed with barbiturates. Tablets of 50 mg (as the hydrochloride) are also available for oral administration.

Methotrimeprazine, USP. (−)-10-[3-(Dimethylamino)-2-methylpropyl]-2-methoxyphenothiazine (Levoprome). This phenothiazine derivative, closely related to chlorpromazine, possesses strong analgesic activity. An intramuscular dose of 15 to 20 mg is equal to 10 mg of morphine in humans. It has not shown any dependence liability and appears not to produce respiratory depression. The most frequent side effects are similar to those of phenothiazine tranquilizers, namely, sedation and orthostatic hypotension. These often result in dizziness and fainting, limiting the use of methotrimeprazine to nonambulatory patients. It is to be used with caution along with antihypertensives, atropine, and other sedatives. It shows some advantage in patients for whom addiction and respiratory depression are problems.[129]

Nefopam. 3,4,5,6-Tetrahydro-5-methyl-1-phenyl-1*H*-2,5-benzoxazocine; 5-methyl-1-phenyl-3,4,5,6-tetrahydro-[1*H*]-2,5-benzoxazocine (Acupan). This rather novel analgesic represents a departure from traditional structure–activity relationships, but shows activity comparable with that of codeine. It gives very rapid onset owing to rapid absorption, with 60 mg giving pain relief comparable with 600 mg of aspirin. Side effects were minimal.[130]

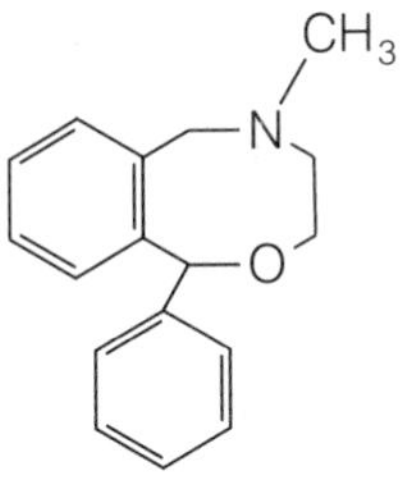

CH_3 N O

Nefopam

NARCOTIC ANTAGONISTS

Nalorphine Hydrochloride, USP. *N*-Allylnormorphine hydrochloride. This morphine derivative may be prepared according to the method of Weijlard and Erickson.[112] It occurs in the form of white or practically white crystals that slowly darken on exposure to air and light. It is freely soluble in water (1 : 8), but is sparingly soluble in alcohol (1 : 35), and is almost insoluble in chloroform and ether. The phenolic hydroxyl group confers water solubility in the presence of fixed alkali. Aqueous solutions of the salt are acid, having a *p*H of about 5.

Nalorphine has a direct antagonistic effect against morphine, meperidine, methadone, and levorphanol. However, it has little antagonistic effect toward barbiturate or general anesthetic depression.

Perhaps one of the most striking effects is on the respiratory depression accompanying morphine overdosage. The respiratory minute volume is quickly returned to normal by intravenous administration of the drug. However, it does have respiratory depressant activity itself, which may potentiate the existing depression. It affects circulatory disturbances in a similar way, reversing the effects of morphine. Other effects of morphine are affected similarly. It was observed that morphine addicts, when treated with the drug, exhibited certain of the withdrawal symptoms associated with abstinence from morphine. Thus, it was used as a diagnostic test agent to determine narcotic addiction. Administration of nalorphine along with morphine may occasionally prevent or minimize the development of dependence on morphine. It has strong analgesic properties, but it is not acceptable for such use owing to the high incidence of undesirable psychotic effects. Because of these properties and the availability of alternate antagonists, it was withdrawn from the market.

Levallorphan Tartrate, USP. 17-(2-Propenyl)-morphinan-3-ol tartrate; (−)-*N*-allyl-3-hydroxymorphinan bitartrate (Lorfan). This compound occurs as a white or practically white, odorless crystalline powder. It is soluble in water (1 : 20), sparingly soluble in alcohol (1 : 60), and practically insoluble in chloroform and ether. Levallorphan resembles nalorphine in its pharmacologic action, being about five times more effective as a narcotic antagonist. It has also been useful in combination with analgesics such as meperidine, alphaprodine, and levorphanol to prevent the respiratory depression usually associated with these drugs.

Naloxone Hydrochloride, USP. 4,5-Epoxy-3,14-dihydroxy-17-(2-propenyl)morphinan-6-one hydrochloride; *N*-allyl-14-hydroxynordihydromorphinone hydrochloride (Narcan). *N*-Allylnoroxymorphone hydrochloride is presently on the market as the agent of choice for treating narcotic overdosage. It lacks not only the analgesic activity shown by other antagonists, but also all of the other agonist effects. It is almost seven times more active than nalorphine in antagonizing the effects of morphine. It shows no withdrawal effects after long-term administration.

The duration of action is about four hours. It was briefly investigated for the treatment of heroin addiction. With adequate doses of naloxone, the addict does not receive any effect from heroin. It is given to an addict only after a detoxification period. Its long-term usefulness is currently limited because of its short duration of action, thereby requiring large oral doses. Long-acting and alternative antagonists are available.

Cyclazocine. 3-(Cyclopropylmethyl)-1,2,3,4,5,6-hexahydro-6,11-dimethyl-2,6-methano-3-benzazocin-8-ol; *cis*-2-cyclopropylmethyl-5,9-dimethyl-2′-hydroxy-6,7-benzomorphan is a potent narcotic antagonist that has shown analgesic activity in humans in 1-mg doses. It once was investigated as a clinical analgesic. It does possess hallucinogenic side effects at higher doses which has limited its usefulness as an analgesic. It was studied similarly to naloxone in the treatment of narcotic addiction. By voluntary treatment with cyclazocine, addicts are deprived of the euphorogenic effects of heroin. Its dependence liability is lower, and the effects of withdrawal develop more slowly and are milder. Tolerance develops to the side effects of cyclazocine, but not to its antagonist effects.[131] The effects are long-lasting and are not reversed by other antagonists such as nalorphine.

Naltrexone. 17-(Cyclopropylmethyl)-4,5α-epoxy-3,14-dihydroxymorphinan-6-one; *N*-cyclopropylmethyl-14-hydroxynordihydromorphinone; *N*-cyclopropylmethylnoroxymorphone; EN-1639 (Trexan). This naloxone analogue has been marketed as the preferred agent for treating former opiate addicts. Oral doses of 50 mg daily or 100 mg three times weekly are sufficient to "block" or protect a patient from the effects of heroin. Its metabolism,[132] pharmacokinetics,[133] and pharmacology[134] have been intensely studied because of the tremendous governmental interest in developing new agents for the treatment of addiction.[54]

Sustained-release or depot dosage forms of naltrexone were once investigated to avoid the recurrent decision on the part of the former addict of whether or not a protecting dose of antagonist is needed.[55, 135]

There are several other narcotic antagonists that have been investigated (e.g., diprenorphine[136] and oxilorphan)[137].

ANTITUSSIVE AGENTS

Cough is a protective, physiologic reflex that occurs in health as well as in disease. It is very widespread and commonly ignored as a mild symptom. However, in many conditions it is desirable to take measures to reduce excessive coughing. It should be stressed that many etiologic factors cause this reflex; and in a case where a cough has been present for an extended period or accompanies any unusual symptoms, the person should be referred to a physician. Cough preparations are widely advertised and often sold indiscriminately; hence, it is the obligation of the pharmacist to warn the public of the inherent dangers.

Among the agents used in the symptomatic control of cough are those that act by depressing the cough center located in the medulla. These have been termed *anodynes*, *cough suppressants*, and *centrally acting antitussives*. Until recently, the only effective drugs in this area were members of the narcotic analgesic agents. The more important and widely used ones are morphine, hydromorphone, codeine, hydrocodone, methadone, and levorphanol, which were discussed in the foregoing section.

In recent years, several compounds have been synthesized that possess antitussive activity without the addiction liabilities of the narcotic agents. Some of these act in a similar manner through a central effect. In a hypothesis for the initiation of the cough reflex, Salem and Aviado[138] proposed that bronchodilatation is an important mechanism for the relief of cough. Their hypothesis suggests that irri-

TABLE 17-5

NARCOTIC ANTAGONISTS

Name Proprietary Name	*Preparations*	*Usual Adult Dose**	*Usual Dose Range**	*Usual Pediatric Dose**
Levallorphan tartrate NF *Lorfan*	Levallorphan tartrate injection NF	IV, 1 mg, repeated twice at 10- to 15-min intervals, if necessary	500 μg–2 mg, repeated, if necessary	0.05–0.1 mg in neonates to decrease respiratory depression
Naloxone hydrochloride USP *Narcan*	Naloxone hydrochloride injection USP	Parenteral, 400 μg, repeated at 2- to 3-min intervals as necessary		0.01 mg as above

*See USP DI for complete dosage information.

tation of the mucosa initially causes bronchoconstriction, and this, in turn, excites the cough receptors.

Chappel and von Seemann[139] have pointed out that most antitussives of this type fall into two structural groups. The larger group has structures that bear a resemblance to methadone. The other group has large, bulky substituents on the acid portion of an ester, usually connected by means of a long, ether-containing chain to a *tertiary* amino group. The notable exceptions are benzonatate and sodium dibunate. Noscapine could be considered as belonging to the first group.

Many of the cough preparations sold contain various other ingredients in addition to the primary antitussive agent. The more important ones include antihistamines, useful when the cause of the cough is allergic, although some antihistaminic drugs (e.g., diphenhydramine) have a central antitussive action as well; sympathomimetics, which are quite effective owing to their bronchodilatory activity, the most useful being ephedrine, methamphetamine, phenylpropanolamine, homarylamine, isoproterenol, and isoöctylamine; parasympatholytics, which help to dry secretions in the upper respiratory passages; and expectorants. It is not known if these drugs potentiate the antitussive action, but they usually are considered as adjuvant therapy.

The more important drugs in this class will be discussed in the following section. For a more exhaustive coverage of the field, the reader is urged to consult the excellent review of Chappel and von Seemann.[139]

PRODUCTS

Some of the narcotic antitussive products have been discussed previously with the narcotic analgesics. Others are discussed below (see Table 17-6).

Noscapine, USP. (−)-Narcotine (Tusscapine). This opium alkaloid was isolated, in 1817, by Robiquet. It is isolated rather easily from the drug by ether extraction. It makes up 0.75% to 9% of opium.

Noscapine occurs as a fine, white or practically white crystalline powder that is odorless and stable in the presence of light and air. It is practically insoluble in water, freely soluble in chloroform, and soluble in acetone and benzene. It is only slightly soluble in alcohol and ether.

With the discovery of its unique antitussive properties, the name of this alkaloid was changed from narcotine to noscapine. It was realized that it would not meet with widespread acceptance as long as its name was associated with the narcotic opium alkaloids. The name "noscapine" was probably selected because a precedent existed in the name of (±)-narcotine, namely "gnoscopine."

Although noscapine had been used therapeutically as an antispasmodic (similar to papaverine), antineuralgic, and antiperiodic, it had fallen into disuse. It had also been used in malaria, migraine, and other conditions in the past in doses of 100 to 600 mg. Newer methods of testing for antitussive compounds were responsible for revealing the effectiveness of noscapine as an antitussive. In addition to its central action, it exerts bronchodilatation effects.

Noscapine is an orally effective antitussive, approximately equal to codeine in effectiveness. It is free of the side effects usually encountered with the narcotic antitussives and, because of its relatively low toxicity, may be given in larger doses to obtain a greater antitussive effect. Although it is an opium alkaloid, it is devoid of analgesic action and addiction liability. It is available in various cough preparations (e.g., Conar).

Dextromethorphan Hydrobromide, USP. (+)-3-Methoxy-17-methyl-9α,13α,14α-morphinan hydrobromide (Romilar). This drug is the *O*-methylated (+)-form of racemorphan left after the resolution necessary in the preparation of levorphanol. It occurs as practically white crystals, or as a crystalline powder, possessing a faint odor. It is sparingly soluble in water (1 : 65), freely soluble in alcohol and chloroform, and insoluble in ether.

It possesses the antitussive properties of codeine, without the analgesic, addictive, central depressant, and constipating features. Ten milligrams is suggested as being equivalent to a 15-mg dose of codeine in antitussive effect.

It affords an opportunity to note the specificity exhibited by very closely related molecules. Here, the (+)- and (−)-forms both must attach to receptors responsible for the suppression of cough reflex, but the (+)-form is apparently in a steric relationship, such that it is incapable of attaching to the receptors involved in analgesic, constipative, addictive, and other actions exhibited by the (−)-form. It has largely replaced many older antitussives, including codeine, in prescription and nonprescription cough preparations.

Levopropoxyphene Napsylate, USP. (−)-α-4-(Dimethylamino)-3-methyl-1,2-diphenyl-2-butanol propionate (ester) 2-naphthalenesulfonate (salt) (Novrad). This compound, the *levo*-isomer of propoxyphene, does not possess the analgesic properties of the (+)-form, but is equally effective as an antitussive, 50 mg being equivalent to 15 mg of codeine.[140] Side effects are infrequent. Levopropoxyphene napsylate is also available in suspension form and has the advantage of being virtually tasteless. Note that its tradename is Darvon spelled backward.

Benzonatate, USP. 2,5,8,11,14,17,20,23,26-Nonaoxaoctacosan-28-yl *p*-(butylamino)benzoate (Tessalon). This compound was introduced in 1956. It is a pale yellow, viscous liquid, insoluble in water and soluble in most organic solvents. It is chemically related to *p*-aminobenzoate local anesthetics, except that the aminoalcohol group has been replaced by a methylated polyethylene glycol group.

Benzonatate is said to possess both peripheral and central activity in producing its antitussive effect. It somehow blocks the stretch receptors thought to be responsible for cough. Clinically, it is not as effective as codeine, but produces far fewer side effects and has a very low toxicity. It is available in 50- and 100-mg capsules ("perles") and ampules (5 mg/mL).

Chlophedianol. 1-*o*-Chlorophenyl-1-phenyl-3-dimethylamino-1-propanol (ULO). This compound, which first was described as an antispasmodic, was found to be an effective antitussive agent.[141] It is useful in doses of 20 to 30 mg given three to five times daily, with a duration of effect for a single dose lasting up to five hours. It has a low incidence of side effects. It is available in several combinations (15 mg/mL) (Acutuss; Ulogesic).

Caramiphen Edisylate. 2-Diethylaminoethyl 1-phenylcyclopentane-1-carboxylate ethanedisulfonate. Caramiphen occurs in the form of water- and alcohol-soluble crystals. The antitussive activity of this compound is less than that of codeine. It has been shown to have both central and bronchodilator activity. The frequency of side effects is lower than with the narcotic antitussives. It is currently marketed as a combination under the tradenames of Tuss-Ornade, both in a liquid form and in a sustained-release form, and as De-Tuss, Tuss-Ade, Tuss-Liquid, and Tusscaps.

Carbetapentane Citrate. 2-[2-(Diethylamino)-ethoxy]ethyl 1-phenylcyclopentanecarboxylate citrate. This salt is a white, odorless, crystalline powder that is freely soluble in water (1 : 1), slightly soluble in alcohol, and insoluble in ether. It is similar to caramiphen chemically and is said to be equivalent to codeine as an antitussive. Introduced in 1956, it is well tolerated and has a low frequency of side effects. It is available as a syrup (7.5 mg/5 mL) in combination with codeine (Tussar-2) or alone (20 mg/5 mL; Cophene-X and -XP).

The tannate is also available (Rynatuss) and is said to give a more sustained action.

THE ANTI-INFLAMMATORY ANALGESICS

The growth of this group of analgesics was related closely to the early belief that the lowering or "curing" of fever was an end in itself. Drugs bringing about a drop in temperature in feverish conditions were considered to be quite valuable and were sought after eagerly. The decline of interest in these drugs coincided more or less with the realization that fever was merely an outward symptom of some other, more fundamental, ailment. However, during the use of the several antipyretics, it was noted that some were excellent analgesics for the relief of minor aches and pains. These drugs have survived to the present time on the basis of the analgesic, rather than the antipyretic, effect. Although these drugs are still widely utilized for the alleviation of minor aches and pains, they are also employed extensively in the symptomatic treatment of rheumatic fever, rheumatoid arthritis, and osteoarthritis. The dramatic effect of salicylates in reducing the inflammatory effects of rheumatic fever is time-honored, and, even with the development of the corticosteroids, these drugs are still of great value in this respect. It

TABLE 17-6

ANTITUSSIVE AGENTS

Name Proprietary Name	***Preparations***	***Usual Adult Dose****	***Usual Dose Range****
Dextromethorphan hydrobromide USP *Romilar*	Dextromethorphan hydrobromide syrup USP	15–30 mg qd to qid	
Levopropoxyphene napsylate USP *Novrad*	Levopropoxyphene napsylate capsules USP Levopropoxyphene napsylate oral suspension USP	50–100 mg of levopropoxyphene, as the napsylate, every 4 hr	
Benzonatate USP *Tessalon*	Benzonatate capsules USP	100 mg tid	100–200 mg

*See USP DI for complete dosage information.

has been reported that the steroids are no more effective than the salicylates in preventing the cardiac complications of rheumatic fever.[142]

The analgesic drugs that fall into this category have been disclaimed by some as not deserving the term "analgesic" because of the low order of activity in comparison with the morphine-type compounds. Indeed, Fourneau has suggested the name *antalgics* to designate this general category and, in this way, to make more emphatic the distinction from the narcotic or so-called true analgesics. Two of the principal features distinguishing these minor analgesics from the narcotic analgesics are the low activity for a given dose and that a higher dosage does not give any significant increase in effect.

Considerable research has continued in an effort to find new nonsteroidal anti-inflammatory agents (NSAIA). Long-term therapy with the corticosteroids is often accompanied by various side effects. Efforts to discover new agents have been limited, for the most part, to structural analogues of active compounds owing to a lack of knowledge about the causes and mechanisms of inflammatory diseases.[143] Although several new agents have been introduced for use in rheumatoid arthritis, aspirin appears to remain as one of the agents of choice.

Of considerable interest is the observation that prostaglandins appear to play a major role in the inflammatory processes.[144] Of particular significance are reports that drugs, such as aspirin and indomethacin, inhibit prostaglandin synthesis in several tissues.[145] Furthermore, almost all classes of nonsteroidal anti-inflammatory agents strongly inhibit the conversion of arachidonic acid into prostaglandin $E_2(PGE_2)$.[146,147] This has been shown to occur at the stage of conversion of arachidonic acid, released by the action of phospholipase A on damaged tissues, to the cyclic endoperoxides, PGG_2 and PGH_2, by prostaglandin synthetase. These are known to cause vasoconstriction and pain. They, in turn, are converted in part to PGE_2 and $PGF_{2\alpha}$, which can cause pain and vasodilatation. This effect of the nonsteroidal anti-inflammatory agents parallels their relative potency in various tests and is stereospecific.[146] The search for specific inhibitors of prostaglandin synthesis has opened a new area of research in this field.

Discussion of these drugs will be facilitated by considering them in their various chemical categories.

SALICYLIC ACID DERIVATIVES

Historically, the salicylates were among the first of this group to achieve recognition as analgesics. Leroux, in 1827, isolated salicin, and Piria, in 1838, prepared salicylic acid. Following these discoveries, Cahours (1844) obtained salicylic acid from oil of wintergreen (methyl salicylate); and Kolbe and Lautermann (1860) prepared it synthetically from phenol. Sodium salicylate was introduced in 1875 by Buss, followed by the introduction of phenyl salicylate by Nencki, in 1886. Aspirin, or acetylsalicylic acid, was first prepared in 1853 by Gerhardt, but remained obscure until Felix Hoffmann discovered its pharmacologic activities in 1899. It was tested and introduced into medicine by Dreser, who named it *aspirin* by taking the *a* from acetyl and adding it to *spirin*, an old name for salicylic or spiric acid, derived from its natural source of spirea plants.

The pharmacology of the salicylates and related compounds has been reviewed extensively by Smith.[148,149] Salicylates, in general, exert their antipyretic action in febrile patients by increasing heat elimination of the body through the mobilization of water and consequent dilution of the blood. This brings about perspiration, causing cutaneous dilatation. This does not occur with normal temperatures. The antipyretic and analgesic actions are believed to occur in the hypothalamic area of the brain. It is also thought by some that the salicylates exert their analgesia by their effect on water balance, reducing the edema usually associated with arthralgias. Aspirin has been shown to be particularly effective for this.

For an interesting account of the history of aspirin and a discussion of its mechanisms of action, the reader should consult an article on the subject by Collier,[150] as well as the reviews by Smith,[148,149] and Nickander et al.[147]

The possibility of hypoprothrombinemia and concomitant capillary bleeding in conjunction with salicylate administration accounts for the inclusion of menadione in some salicylate formulations. However, there is some doubt about the necessity for this measure. A more serious aspect of salicylate medication has been the possibility of inducing hemorrhage from direct irritative contact with the mucosa. Alvarez and Summerskill have pointed out a definite relationship between salicylate consumption and massive gastrointestinal hemorrhage from peptic ulcer.[151] Barager and Duthie,[152] on the other hand, in an extensive study found no danger of increase in anemia or in development of peptic ulcer. Levy[153] has demonstrated with the use of radiolabeled iron that bleeding does occur following administration of aspirin. The effects varied with the formulation. It is suggested by Davenport[154] that back diffusion of acid from the stomach is responsible for capillary damage.

Because of these characteristics of aspirin, it has been extensively studied as an antithrombotic agent in the treatment and prevention of clinical thrombosis.[155] It is thought to act by its selective action on the synthesis of the prostaglandin-related thromboxane and prostacyclin, which are the counterbalancing factors involved in platelet aggregation and are released when tissue is injured. Although aspirin has now been approved for the prevention of transient ischemic attacks, indicators of an impending stroke, it is not recommended for patients who have suffered heart attacks.[156]

The salicylates are readily absorbed from the stomach and the small intestine, being quite dependent on the pH of the media. Absorption is considerably slower as the pH rises (more alkaline) because of the acidic nature of these compounds and the necessity for the presence of undissociated molecules for absorption through the lipoidal membrane of the stomach and the intestines. Therefore, buffering agents administered at the same time in *excessive* amounts will decrease the rate of absorption. In small quantities, their principal effect may be to aid in the dispersion of the salicylate into fine particles. This would help to increase absorption and decrease the possibility of gastric irritation by the accumulation of large particles of the undissolved acid and their adhesion to the gastric mucosa. Levy and Haves[157] have shown that the absorption rate of aspirin and the incidence of gastric distress were a function of the dissolution rate of its particular dosage form. A more rapid dissolution rate of calcium and buffered aspirin was believed to account for faster absorption. They also established that significant variations exist in dissolution rates of different nationally distributed brands of plain aspirin tablets. This may account for some of the conflicting reports and opinions concerning the relative advantages of plain and buffered aspirin tablets. Lieberman and co-workers[158] have also shown that buffering is effective in raising the blood levels of aspirin. In a measure of the antianxiety effect of aspirin by means of electroencephalograms (EEG), differences between buffered, brand name, and generic aspirin preparations were found.[159]

Potentiation of salicylate activity by virtue of simultaneous administration of p-aminobenzoic acid or its salts has been the basis for the introduction of numerous products of this kind. Salassa and coworkers have shown this effect to be due to the inhibition both of salicylate metabolism and of excretion in the urine.[160] This effect has been proved amply, provided that the ratio of 24 g of p-aminobenzoic acid to 3 g of salicylate per day is observed. However, there is no strong evidence to substantiate any significant elevation of plasma salicylate levels when a smaller quantity of p-aminobenzoic acid is employed.

The derivatives of salicylic acid are of two types [I and II (a, b)]:

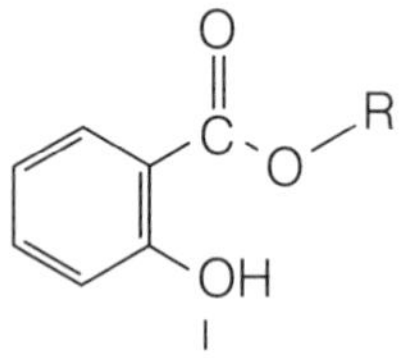

I

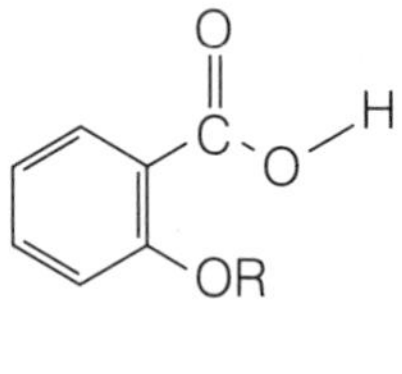

IIa

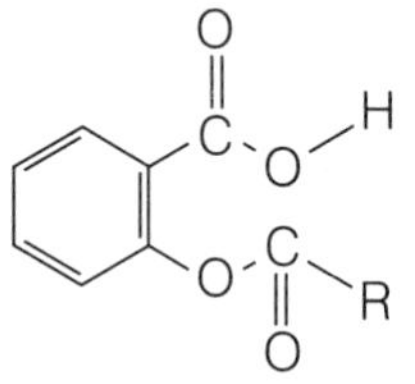

IIb

Type I represents those that are formed by modifying the carboxyl group (e.g., salts, esters, or amides). Type II (a and b) represents those that are derived by substitution on the hydroxyl group of salicylic acid. The derivatives of salicylic acid were introduced in an attempt to prevent the gastric symptoms and the undesirable taste inherent in the common salts of salicylic acid. Hydrolysis of type I takes place to a greater extent in the intestine, and most of the type II compounds are absorbed unchanged into the bloodstream (see aspirin).

Compounds of Type I

The alkyl and aryl esters of salicylic acid (type I) are used externally, primarily as counterirritants, where most of them are well absorbed through the skin. This type of compound is of little value as an analgesic.

A few inorganic salicylates are used internally when the effect of the salicylate ion is intended. These compounds vary in their irritation of the stomach. To prevent the development of pink or red coloration in the product, contact with iron should be avoided in their manufacture.

Sodium Salicylate, USP may be prepared by the reaction, in aqueous solution, between 1 mole each of salicylic acid and sodium bicarbonate; upon evaporating to dryness, the white salt is obtained.

Generally, the salt has a pinkish tinge or is a white microcrystalline powder. It is odorless or has a faint, characteristic odor, and it has a sweet, saline taste. It is affected by light. The compound is soluble in water (1 : 1), alcohol (1 : 10), and glycerin (1 : 4).

In solution, particularly in the presence of sodium bicarbonate, the salt will darken on standing (see salicyclic acid). This darkening may be lessened by the addition of sodium sulfite or sodium bisulfite. Also, a color change is lessened by using recently boiled distilled water and dispensing in amber-colored bottles. Sodium salicylate forms a eutectic mixture with antipyrine and produces a violet coloration with iron or its salts. Solutions of the compound must be neutral or slightly basic to prevent precipitation of free salicylic acid. However, the USP salt forms neutral or acid solutions.

This salt is the one of choice for salicylate medication and usually is administered with sodium bicarbonate to lessen gastric distress, or it is administered in enteric-coated tablets. The use of sodium bicarbonate[161] is ill-advised, because it decreases the plasma levels of salicylate and increases the excretion of free salicylate in the urine.

Sodium Thiosalicylate (Arthrolate; Nalate; Thiodyne; Thiolate; Thiosul; TH Sal) is the sulfur or thio analogue of sodium salicylate. It is more soluble and better absorbed, thereby requiring lower dosages. It is recommended for gout, rheumatic fever, and muscular pains in doses of 100 to 150 mg every three to six hours for two days, and then 100 mg once or twice daily.

Magnesium Salicylate, USP (Analate; Causalin; Lorisal; Mobidin; MSG-600; Triact; Magan) is a sodium-free salicylate preparation that may be used in conditions in which sodium intake is restricted. It is claimed to produce less gastrointestinal upset. The dosage and indications are the same as for sodium salicylate.

Choline Salicylate (Arthropan). This salt of salicylic acid is extremely soluble in water. It is claimed to be absorbed more rapidly than aspirin, giving faster peak blood levels. It is used in conditions for which salicylates are indicated in a recommended dose of 870 mg to 1.74 g four times daily.

Other salts of salicylic acid that have found use are those of ammonium, lithium, and strontium. They offer no distinct advantage over sodium salicylate.

Carbethyl Salicylate. Ethyl salicylate carbonate (Sal-Ethyl Carbonate) is an ester of ethyl salicylate and carbonic acid, and thus is a combination of a type I and type II compound.

It occurs as white crystals, insoluble in water and in diluted hydrochloric acid, slightly soluble in alcohol or ether and readily soluble in chloroform or acetone. The insolubility tends to prevent gastric irritation and makes it tasteless.

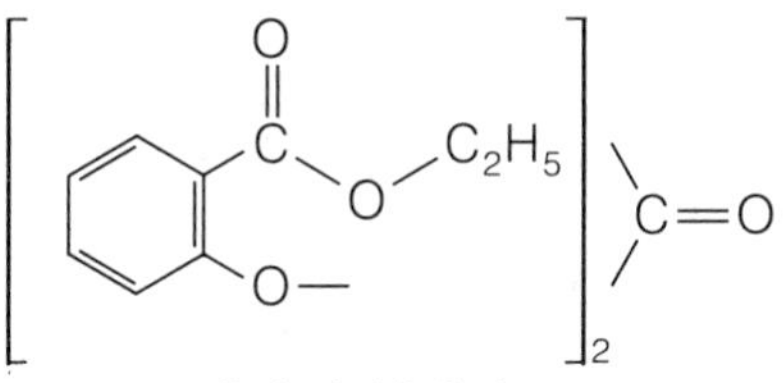

Carbethyl Salicylate

In action and uses it resembles aspirin and gives the antipyretic and analgesic effects of the salicylates. The pharmaceutic forms are powder, tablet, and a tablet containing aminopyrine.

The usual dose is 1.0 g.

The Salol Principle

Nencki introduced salol in 1886 and by so doing presented to the science of therapy the "Salol Principle." In salol, two toxic substances (phenol and salicylic acid) were combined into an ester that, when taken internally, will slowly hydrolyze in the intestine to give the antiseptic action of its components. This type of ester is referred to as a "full salol" or "true salol" when both components of the ester are active compounds. Examples are guaiacol benzoate, β-naphthol benzoate, and salol.

This Salol Principle can be applied to esters of which only the alcohol or the acid is the toxic, active or corrosive portion, and this type is called a "partial salol."

Examples of a partial salol containing an active acid are ethyl salicylate and methyl salicylate. Examples of a partial salol containing an active phenol are creosote carbonate, thymol carbonate, and guaiacol carbonate.

Although a host of the salol-type compounds have been prepared and used to some extent, none is presently very valuable in therapeutics, and all are surpassed by other agents.

Phenyl Salicylate; salol. Phenyl salicylate occurs as fine, white crystals or as a white crystalline powder with a characteristic taste and a faint, aromatic odor. It is insoluble in water (1 : 6700), slightly soluble in glycerin, soluble in alcohol (1 : 6), ether, chloroform, acetone, or fixed and volatile oils.

Damp or eutectic mixtures form readily with many organic materials, such as thymol, menthol, camphor, chloral hydrate, and phenol.

Salol is sold in combination with methenamine and atropine alkaloids as a urinary tract antiseptic and analgesic (e.g., Cystrea, Lanased, Renalgin, and Urised).

Salol is insoluble in the gastric juice but is slowly hydrolyzed in the intestine into phenol and salicylic acid. Because of this fact, coupled with its low melting point (41° to 43°C), it has been used in the past as an enteric coating for tablets and capsules. However, it is not efficient as an enteric-coating material, and its use has been superseded by more effective materials.

It also is used externally as a sun filter (10% ointment) for sunburn prevention (Rayderm).

Salicylamide; *o*-hydroxybenzamide. This is a derivative of salicylic acid that has been known for almost a century and has found renewed interest. It is readily prepared from salicyl chloride and ammonia. The compound occurs as a nearly odorless, white crystalline powder. It is fairly stable to heat, light, and moisture. It is slightly soluble in water (1 : 500), soluble in hot water, alcohol (1 : 15), and propylene glycol, and sparingly soluble in chloroform and ether. It is freely soluble in solutions of alkalies. In alkaline solution with sodium carbonate or triethanolamine, decomposition takes place, resulting in a yellow to red precipitate.

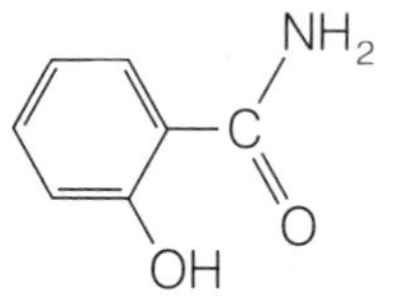

Salicylamide

Salicylamide is said to exert a moderately quicker and deeper analgesic effect than does aspirin. Long-term studies on rats revealed no untoward symptomatic or physiologic reactions. Its metabolism is different from that of other salicylic compounds, and it is not hydrolyzed to salicylic acid.[148] Its analgesic and antipyretic activity is probably no greater than that of aspirin, and possibly less. However, it can be used in place of salicylates and is particularly useful for those patients for whom there is a demonstrated sensitivity to salicylates. It is excreted much more rapidly than other salicylates, which probably accounts for its lower toxicity and, thus, does not permit high blood levels.

The dose for simple analgesic effect may vary from 300 mg to 1 g administered three times daily; but for rheumatic conditions the dose may be increased to 2 to 4 g, three times a day. However, gastric intolerance may limit the dosage. The usual period of this higher dosage should not extend beyond three to six days. It is available in several combination products.

Aspirin, USP. Acetylsalicylic acid (Aspro; Empirin). Aspirin was introduced into medicine by Dreser in 1899. It is prepared by treating salicylic acid, which was first prepared by Kolbe in 1874, with acetic anhydride.

The hydrogen atom of the hydroxyl group in salicylic acid has been replaced by the acetyl group; this also may be accomplished by using acetyl chloride with salicylic acid or ketene with salicylic acid.

Aspirin occurs as white crystals or as a white crystalline powder. It is slightly soluble in water (1 : 300) and soluble in alcohol (1 : 5), choloroform (1 : 17), and ether (1 : 15). Also, it dissolves easily in glycerin. Aqueous solubility may be increased by using acetates or citrates of alkali metals, although these are said to decompose it slowly.

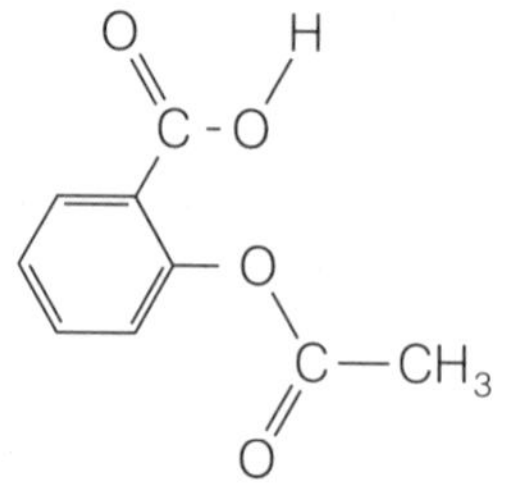

Aspirin

It is stable in dry air, but in the presence of moisture, it slowly hydrolyzes into acetic and salicylic acids. Salicylic acid will crystallize out when an aqueous solution of aspirin and sodium hydroxide is boiled and then acidified.

Aspirin itself is sufficiently acid to produce effervescence with carbonates and, in the presence of iodides, to cause the slow liberation of iodine. In the presence of alkaline hydroxides and carbonates, it decomposes, although it does form salts with alkaline metals and alkaline earth metals. The presence of salicyclic acid, formed upon hydrolysis, may be confirmed by the formation of a violet color upon the addition of ferric chloride solution.

Aspirin is not hydrolyzed appreciably on contact with weakly acid digestive fluids of the stomach, but on passage into the intestine is subjected to some hydrolysis. However, most of it is absorbed unchanged. The gastric mucosal irritation of aspirin has been ascribed by Garrett[162] to salicylic acid formation, the natural acidity of aspirin, or the adhesion of undissolved aspirin to the mucosa. He has also proposed the nonacidic anhydride of aspirin as a superior form for oral administration. Davenport[154] concludes that aspirin causes an alteration in mucosal cell permeability, allowing back diffusion of stomach acid which damages the capillaries. A number of proprietaries (e.g., Bufferin) employ compounds, such as sodium bicarbonate, aluminum glycinate, sodium citrate, aluminum hydroxide, or magnesium trisilicate, to counteract this acid property. One of the better antacids is dihydroxyaluminum aminoacetate USP. Aspirin has been shown to be unusually effective when prescribed with calcium glutamate. The more stable, nonirritant calcium acetylsalicylate is formed, and the glutamate portion (glutamic acid) maintains a *p*H of 3.5 to 5.

Preferably, dry dosage forms (i.e., tablets, capsules, or powders) should be used, since aspirin is somewhat unstable in aqueous media. In tablet preparations, the use of acid-washed talc has been shown to improve the stability of aspirin.[163] Also, it has been found to break down in the presence of phenylephrine hydrochloride.[164] Aspirin in aqueous media will hydrolyze almost completely in less than one week. However, solutions made with alcohol or glycerin do not decompose as quickly. Citrates retard hydrolysis only slightly. Some studies have indicated that sucrose tends to inhbit hydrolysis. A study of aqueous aspirin suspensions has indicated that sorbitol exerts a pronounced stabilizing effect.[165] Stable liquid preparations are available that use triacetin, propylene glycol, or a polyethylene glycol. Aspirin lends itself readily to combination with many other substances, but tends to soften and become damp with methenamine, aminopyrine, salol, antipyrine, phenol, or acetanilid.

Aspirin is one of the most widely used compounds in therapy and, for many years, was not associated with untoward effects. Allergic reactions to aspirin are now commonly observed. Asthma and urticaria are the most common manifestations and, when they occur, are extremely acute and difficult to relieve. Like sodium salicylate, it has been shown to cause congenital malformations when administered to mice.[166] Pretreatment with sodium pentobarbital or chlorpromazine resulted in a significant lowering of these effects.[167] Similar effects have been attributed to the consumption of aspirin in women, and its use during pregnancy should be avoided. However, other studies indicate that no untoward effects are seen. The reader is urged to consult the excellent review by Smith for an account of the pharmacologic aspects of aspirin.[148, 149]

Practically all salts of aspirin, except those of aluminum and calcium, are unstable for pharmaceutic use. These salts appear to have fewer undesirable side effects and to induce analgesia faster than aspirin.

A timed-release preparation (Measurin) of aspirin is available. It does not appear to offer any advantages over aspirin, except for bedtime dosage.

Aspirin is used as an antipyretic, analgesic, and antirheumatic, usually in powder, capsule, suppository, or tablet form. Its use in rheumatism has been reviewed, and it is said to be the drug of choice over all other salicylate derivatives.[168, 169] There is some anesthetic action when applied locally, especially in powder form in tonsilitis or pharyngitis, and in ointment form for skin itching and certain skin diseases. In the usual dose, 52% to 75% is excreted in the urine, in various forms, in a period of 15 to 30 hours. It is believed that analgesia is due to the unhydrolyzed acetylsalicylic acid molecule.[148–150]

Aluminum Aspirin. Hydroxybis(salicylato)aluminum diacetate. This salt of aspirin may be prepared by thoroughly mixing aluminum hydroxide gel, water, and acetylsalicylic acid, maintaining the temperature below 65°. Aluminum aspirin occurs as a white to off-white powder or granules and is odorless or has only a slight odor. It is insoluble in water and organic solvents, is decomposed in aqueous solutions of alkali hydroxides and carbonates, and is not stable above 65°C. It offers the advantages of being free of odor and taste and possesses added shelf-life stability. It is available in a flavored form for children (Dulcet).

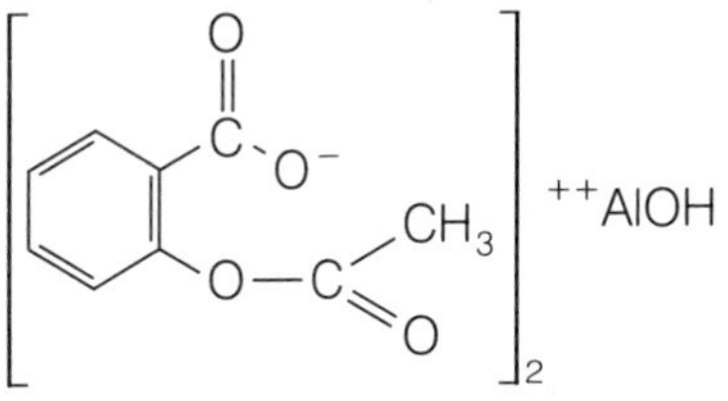

Aluminum Aspirin

Calcium Acetylsalicylate. Soluble aspirin; calcium aspirin. This compound is prepared by treating acetylsalicylic acid with calcium ethoxide or methoxide in alcohol or acetone solution. It is readily soluble in water (1 : 6), but only sparingly soluble in alcohol (1 : 80). It is more stable in solution than aspirin and is used for the same conditions.

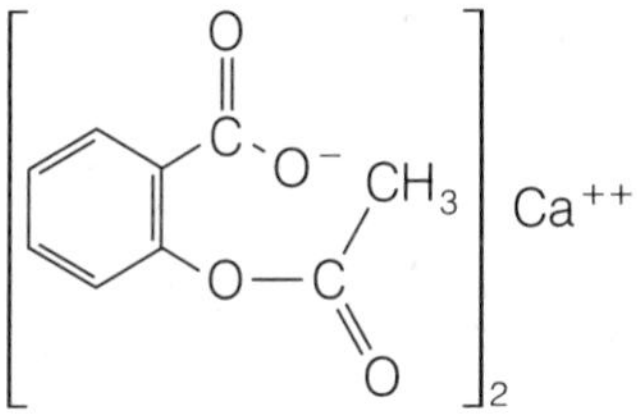

Calcium Acetylsalicylate

Calcium aspirin is marketed also as a complex salt with urea, calcium carbaspirin (Calurin), which is claimed to give more rapid salicylate blood levels and to be less irritating than aspirin, although no clear advantage has been shown.

The usual dose is 500 mg to 1.0 g.

Salsalate. Salicylsalicylic acid (Arcylate; Disalcid) is the ester formed between two salicylic acid molecules to which it is hydrolyzed following absorption. It is said to cause less gastric upset than aspirin because it is relatively insoluble in the stomach and is not absorbed until it reaches the small intestine. Limited clinical trails[170–172] suggest that it is as effective as aspirin and that it may have fewer side effects.[173] The recommended dose is 325 to 1000 mg two or three times a day.

Flufenisal. Acetyl-5-(4-fluorophenyl)salicylic acid; 5′-fluoro-4-hydroxy-3-biphenylcarboxylic acid acetate. Over the years several hundred analogues of aspirin have been made and tested to produce a compound that was more potent, longer acting, and with less gastric irritation. By the introduction of a hydrophobic group in the 5-position, flufenisal appears to meet these criteria. In animal tests it is at least four times more potent. In humans, it appears to be about twice as effective with twice the duration.[174] Like other aryl acids it is highly bound to plasma protein as its deacylated metabolite. It has not yet been marketed.

THE *N*-ARYLANTHRANILIC ACIDS

One of the early advances in the search for nonnarcotic analgesics was centered in the *N*-arylanthranilic acids. Their outstanding characteristic is that they are primarily nonsteroidal anti-inflammatory agents and, secondarily, that some possess analgesic properties.

Mefenamic Acid. *N*-2,3-Xylylanthranilic acid (Ponstel; Ponstan) occurs as an off-white crystalline powder that is insoluble in water and slightly soluble in alcohol. It appears to be the first genuine antiphlogistic analgesic discovered since aminopyrine. Because it is believed that aspirin and aminopyrine owe their general-purpose analgesic efficacy to a combination of peripheral and central effects,[175] a wide variety of arylanthranilic acids were screened for antinociceptive (analgesic) activity if they showed significant anti-inflammatory action. It has become evident that the combination of both effects is a rarity among these compounds. The mechanism of analgesic action is believed to be related to its ability to block prostaglandin synthetase. No relationship to lipid-plasma distribution, partition coefficient, or pK_a has been noted. The interested reader, however, will find additional information on antibradykinin and anti-UV erythema activities of these compounds, together with speculations on a receptor site, in the literature.[176]

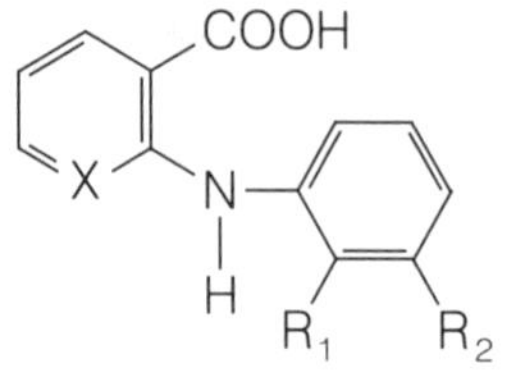

(a) $R_1 = CH_3$, X = CH, R_2 = H
(b) R_1 = H, $R_2 = CF_3$, X = CH

It has been shown[177] that mefenamic acid in a dose of 250 mg is superior to 600 mg of aspirin as an analgesic and that doubling the dose gives a sharp increase in efficacy. A study[178] examining this drug relative to gastrointestinal bleeding indicated that it has a lower incidence of this side effect than has aspirin. Diarrhea, drowsiness, and headache have accompanied its use. The possibility of blood disorders has prompted limitation of its administration to seven days. It is not recommended for children or during pregnancy. It has been approved for use in the management of primary dysmenorrhea, which is thought to be caused by excessive concentrations of prostaglandins and endoperoxides.

Meclofenamate Sodium (Meclomen) is sodium *N*-(2,6-dichloro-*m*-tolyl)anthranilate.

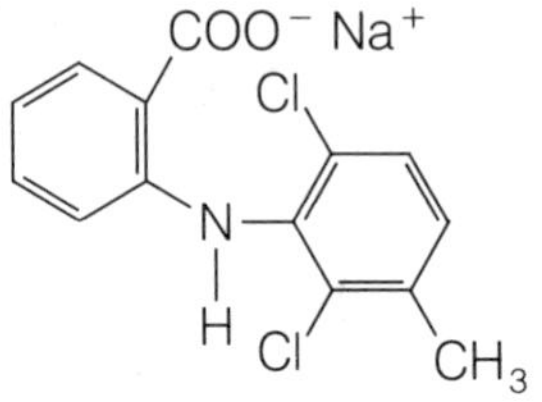

Meclofenamate Sodium

This drug is available in 50- and 100-mg capsules for use in the treatment of acute and chronic rheumatoid arthritis. The most significant side effects are gastrointestinal, including diarrhea.

ARYLACETIC ACID DERIVATIVES

This group of anti-inflammatory agents has received the most intensive attention for new clinical candidates. As a group they have the characteristic of showing high analgesic potency in addition to their anti-inflammatory activity.

Indomethacin, USP. 1-(*p*-Chlorobenzoyl)-5-methoxy-2-methylindole-3-acetic acid (Indocin) occurs as a pale yellow to yellow-tan crystalline powder that is soluble in ethanol and acetone and practically insoluble in water. It is unstable in alkaline solution and sunlight. It shows polymorphism, one form melting at about 155°C and the other at about 162°C. It may occur as a mixture of both forms with a melting range between these melting points.

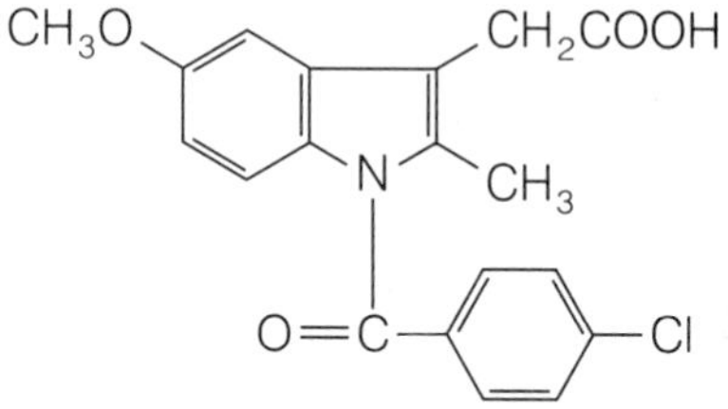

Indomethacin

Since its introduction in 1965, it has been widely used as an anti-inflammatory analgesic in rheumatoid arthritis, spondylitis, and osteoarthritis, and to a lesser extent in gout. Although both its analgesic and anti-inflammatory activities have been well established, it appears to be no more effective than aspirin.[179]

The most frequent side effects are gastric distress and headache. It has also been associated with peptic ulceration, blood disorders, and possible deaths. The side effects appear to be dose-related and sometimes can be minimized by reducing the dose. It is not recommended for use in children because of possible interference with resistance to infection. As do many other acidic compounds, it circulates bound to blood protein, requiring caution in the concurrent use of other protein-binding drugs.

Indomethacin is recommended only for those patients by whom aspirin cannot be tolerated, and in place of phenylbutazone in long-term therapy, for which it appears to be less hazardous than corticosteroids or phenylbutazone.

Sulindac, USP. (Z)-5-Fluoro-2-methyl-1-[[p-(methylsulfinyl)phenyl]methylene]-1H-indene-3-acetic acid (Clinoril) occurs as yellow crystals soluble in alkaline but insoluble in acidic solutions. The drug reaches peak blood levels within two to four hours and undergoes a complicated, reversible metabolism shown as follows:

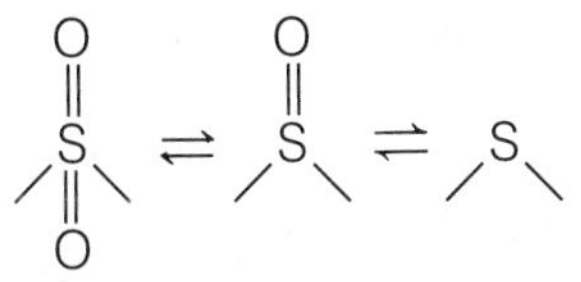

inactive parent active

The parent sulfinyl has a plasma half-life of eight hours, with that of the active sulfide metabolite being 16.4 hours. The more polar and inactive sulfoxide is virtually the sole form excreted. The long half-life is due to extensive enterohepatic recirculation.[180] In in vitro studies, only the sulfide species inhibits prostaglandin synthetase. Although these forms are highly protein bound, the drug does not appear to affect binding of anticoagulants or hypoglycemics. Coadministration of aspirin is contraindicated because it considerably reduces the sulfide blood levels.

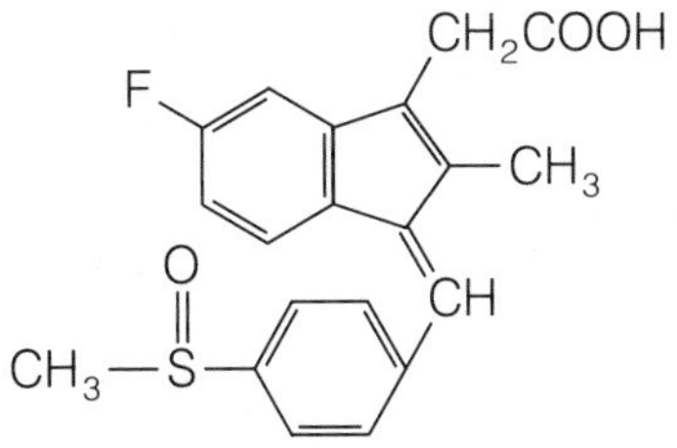

Sulindac

Careful monitoring of patients with a history of ulcers is recommended. Gastric bleeding, nausea, diarrhea, dizziness, and other adverse effects have been noted, but with a lower frequency than with aspirin. Sulindac is recommended for rheumatoid arthritis, osteoarthritis, and ankylosing spondolitis in 150- to 200-mg dose, twice daily.[181] It is available as tablets (150 and 200 mg).

Tolmetin Sodium, USP. 1-Methyl-5-(p-toluoyl)pyrrole-2-acetate dihydrate sodium; McN-2559 (Tolectin) is an arylacetic acid derivative with a pyrrole as the aryl group. This drug is rapidly absorbed with a relatively short plasma half-life (one hour). It is recommended for use in the management of acute and long-term rheumatoid arthritis. It shares similar, but less frequent, adverse effects with aspirin. It does not potentiate coumarinlike drugs nor alter the blood levels of sulfonylureas or insulin. As with other drugs in this class, it is known to inhibit prostaglandin synthetase and lower PGE blood levels.

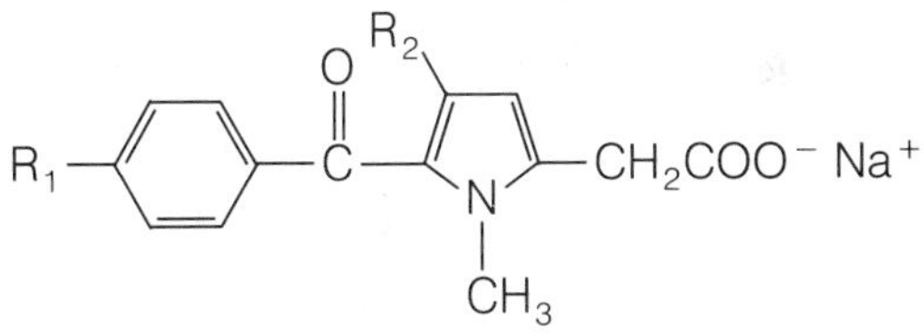

Tolmetin: $R_1 = CH_3$, $R_2 = H$
Zomepirac: $R_1 = Cl$, $R_2 = CH_3$

Available as tablets (200 mg), a dose of 400 mg three times daily, with a maximum of 2000 mg, is recommended. Clinical trials indicate a usual daily dose of 1200 mg is comparable in relief to 3.9 g of aspirin and 150 mg of indomethacin per day.[182]

Zomepirac Sodium, USP. 5-(4-Chlorobenzoyl)-1,4-dimethyl-1H-pyrrole-2-acetate dihydrate sodium; McN-2783 (Zomax) is the chloro analogue of tolmetin. It shows significantly longer plasma levels (seven hours),[183] thereby requiring less frequent dosing. In pain relief, 25 to 50 mg is reported to give relief equivalent to 650 mg of aspirin. In a study on cancer patients, oral doses of 100 to 200 mg were as effective as moderate parenteral doses of morphine.[184] This drug was marketed briefly in the United States in 100-mg tablets, but it was abruptly withdrawn by the manufacturer after reports of severe anaphylactoid reactions.

Ibuprofen, USP. 2-(4-Isobutylphenyl)propionic acid (Motrin; Advil; Nuprin). This arylacetic acid derivative was introduced into clinical practice following extensive clinical trials. It appears comparable to aspirin in the treatment of rheumatoid arthritis, with a lower incidence of side effects.[185] It has also been approved for use in primary dysmenorrhea.

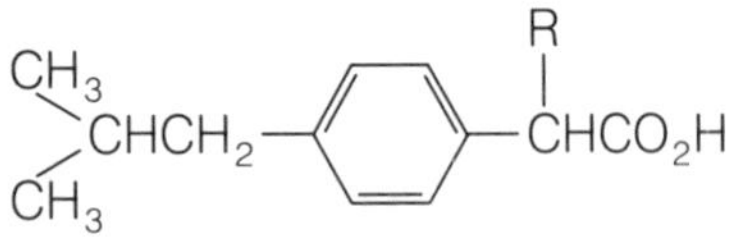

Ibufenac R = H
Ibuprofen R = CH_3

Of interest in this series of compounds is that it was noted that potency was enhanced by introduction of the α-methyl group on the acetic acid moiety. The precursor ibufenac (R = H), which was abandoned owing to hepatotoxicity, was less potent. Moreover, it was found that the activity resides in the (S)-(+)-isomer, not only in ibuprofen, but throughout the arylacetic acid series. Furthermore, it is these isomers that are the more potent inhibitors of prostaglandin synthetase.[146] The recommended dosage is 400 mg. Ibuprofen is also available over-the-counter as 200-mg tablets.

Namoxyrate. 2-(4-Biphenyl)butyric acid dimethylaminoethanol salt (Namol) is another phenylacetic acid derivative under investigation. Namoxyrate shows high analgesic activity, being about seven times that of aspirin and nearly as effective as codeine. It has high antipyretic activity, but appears to be devoid of anti-inflammatory activity. These effects are peripheral. The dimethylaminoethanol increases its activity by increasing intestinal absorption. The ester of these two components is less active.[186]

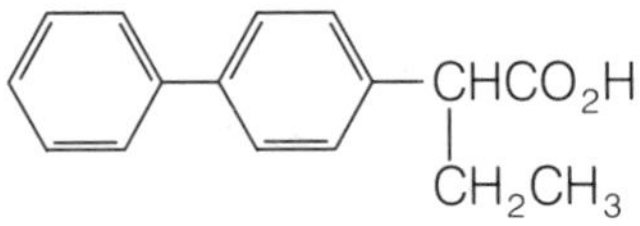

Namoxyrate

Naproxen, USP. (+)-6-Methoxy-α-methyl-2-naphthaleneacetic acid (Anaprox; Naprosyn) occurs as white to off-white crystals that are sparingly soluble in acidic solutions, freely soluble in alkaline solutions, and highly soluble in organic or lipidlike solutions. After oral administration, it is well absorbed, giving peak blood levels in two to four hours and a half-life of 13 hours. A steady-state blood level is usually achieved following four to five doses. This drug is very highly protein bound and displaces most protein-bound drugs. Dosages of these must be adjusted accordingly.

Naproxen is recommended for use in rheumatoid and gouty arthritis. It shows good analgesic activity,

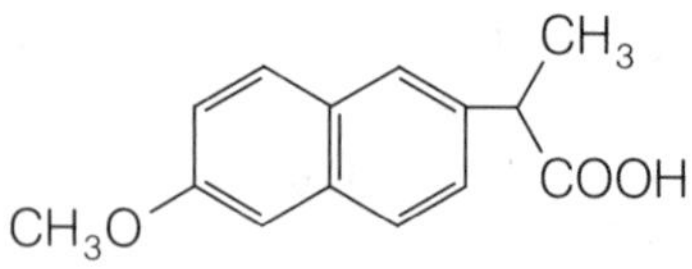

Naproxen

400 mg being comparable to 75 to 150 mg of oral meperidine and superior to 65 mg propoxyphene and 325 mg of aspirin plus 30 mg of codeine. A 220- to 330-mg dose was comparable to 600 mg of aspirin alone. It has been reported to produce dizziness, drowsiness, and nausea, with infrequent mentions of gastrointestinal tract irritation. Similarly to aspirin it inhibits prostaglandin synthetase and prolongs blood-clotting time. It is not recommended for pregnant or lactating women or in children under 16.[189]

Fenoprofen Calcium, USP. α-Methyl-3-phenoxybenzeneacetic acid dihydrate calcium (Nalfon) occurs as a white crystalline powder that is slightly soluble in water, soluble in alcohol, and insoluble in benzene. It is rapidly absorbed orally, giving peak blood levels within two hours and has a short plasma half-life (three hours). It is highly protein bound similar to the other acylacetic acids, and caution must be exercised when used concurrently with hydantoins, sulfonamides, and sulfonylureas. It shares many of the adverse effects common to this group of drugs, with gastrointestial bleeding, ulcers, dyspepsia, nausea, sleepiness, and dizziness reported at a lower incidence than with aspirin. It inhibits prostaglandin synthetase.[187]

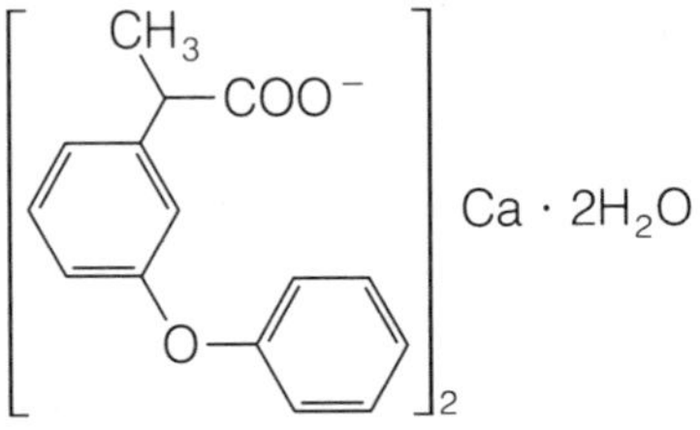

Fenoprofen

Available as capsules (200 and 300 mg) and tablets (600 mg), it is recommended for rheumatoid arthritis and osteoarthritis in divided doses four times a day for a maximum of 3200 mg/day. It should be taken at least 30 minutes before or two hours after meals. It is not yet recommended for the management of acute flare-ups. Doses of 2.4 g/day have been shown to be comparable to 3.9 g/day of aspirin in arthritis. For pain relief, 400 mg gave similar results to 650 mg of aspirin.[188]

Piroxicam, USP. 4-Hydroxy-2-methyl-N-2-pyridyl-2H-1,2-benzothiazine-3-carboximide 1,1-dioxide; CP-16,171 (Feldene) represents a class of acidic inhibitors of prostaglandin synthetase, although it does not antagonize PGE_2 directly.[190] This drug is very long acting with a plasma half-life of 38 hours, thereby requiring a dose of only 20 to 30 mg once daily. It is reported to give similar results to 25 mg of indomethacin or 400 mg of ibuprofen three times a day.[191,192]

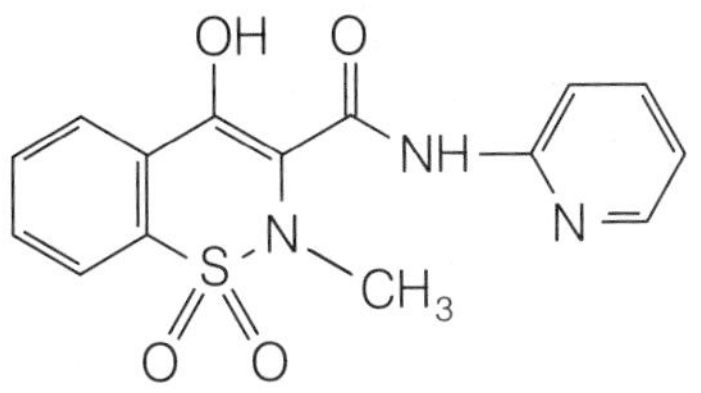

Piroxicam

Several other arylacetic acid derivatives have been under clinical evaluation. These include ketoprofen, alclofenac, fenclofenac, pirprofen, and prodolic and bucloxic acids. Although only early reports are available, many of these appear to show superiority over indomethacin and aspirin. Diclofenac Sodium, 2-[2,6-dichlorophenyl)amino]benzeneacetic acid monosodium salt (Volteran) has recently been introduced on the United States market as 75-mg tablets. The reader may consult the reviews of Evens and Scherrer and Whitehouse for further details.

ANILINE AND *p*-AMINOPHENOL DERIVATIVES

The introduction of aniline derivatives as analgesics is based on the discovery by Cahn and Hepp, in 1886, that aniline (C-1) and acetanilid (C-2) (Table 17-7) both have powerful antipyretic properties. The origin of this group from aniline has led to their being called "coal tar analgesics." Acetanilid was introduced by these workers because of the known toxicity of aniline itself. Aniline brings about the formation of methemoglobin, a form of hemoglobin that is incapable of functioning as an oxygen carrier. The acyl derivatives of aniline were thought to exert their analgesic and antipyretic effects by first being hydrolyzed to aniline and the corresponding acid, following which the aniline was oxidized to *p*-aminophenol (C-3). This is then excreted in combination with glucuronic or sulfuric acid.

The aniline derivatives do not appear to act upon the brain cortex; the pain impulse appears to be intercepted at the hypothalamus, wherein also lies the thermoregulatory center of the body. It is not clear if this is the site of their activity because most evidence suggests that they act at peripheral thermoceptors. They are effective in the return to normal temperature of feverish individuals. Normal body temperatures are not affected by the administration of these drugs.

It is notable that, of the antipyretic analgesic group, the aniline derivatives show little if any anti-inflammatory activity.

Table 17-7 shows some of the types of aniline derivatives that have been made and tested in the past. In general, any type of substitution on the amino group that reduces its basicity results also in a lowering of its physiologic activity. Acylation is one type of substitution that accomplishes this effect. Acetanilid (C-2) itself, although the best of the acylated derivatives, is toxic in large doses, but when administered in analgesic doses, it is probably without significant harm. Formanilid (C-4) is readily hydrolyzed and too irritant. The higher homologue of acetanilid are less soluble and, therefore, less active and less toxic. Those derived from aromatic acids (e.g., C-5) are virtually without analgesic and antipyretic effects. One of these, salicylanilide (C-6), is used as a fungicide and antimildew agent. Exalgin (C-7) is too toxic.

The hydroxylated anilines (*o*, *m*, *p*), better known as the aminophenols, are quite interesting from the standpoint of being considerably less toxic than aniline. The *para* compound (C-3) is of particular interest from two standpoints: namely, it is the metabolic product of aniline, and it is the least toxic of the three possible aminophenols. It also possesses a strong antipyretic and analgesic action. However, it is too toxic to serve as a drug and, for this reason, there were numerous modifications attempted. One of the first was the acetylation of the amine group to provide *N*-acetyl-*p*-aminophenol (acetaminophen) (C-8), a product that retained a good measure of the desired activities. Another approach to the detoxification of *p*-aminophenol was the etherification of the phenolic group. The best known of these are anisidine (C-9) and phenetidine (C-10), which are the methyl and ethyl ethers, respectively. However, it became apparent that a free amino group in these compounds, although promoting a strong antipyretic action, was also conducive to methemoglobin formation. The only exception to the preceding was for compounds in which a carboxyl group or sulfonic acid group had been substituted on the benzene nucleus. In these compounds, however, the antipyretic effect also had disappeared. The foregoing considerations led to the preparation of the alkyl ethers of *N*-acetyl-*p*-aminophenol of which the ethyl ether was the best and is known as phenacetin (C-11). The methyl and propyl homologues were undesirable from the standpoint of causing emesis, salivation, diuresis, and other reactions. Alkylation of the nitrogen with a methyl group has a potentiat-

TABLE 17-7

SOME ANALGESICS RELATED TO ANILINE

$R_1-C_6H_4-N(R_2)(R_3)$

Compound	Structure R_1	R_2	R_3	Name
C-1	$-H$	$-H$	$-H$	Aniline
C-2	$-H$	$-H$	$-C(=O)-CH_3$	Acetanilid
C-3	$-OH$	$-H$	$-H$	*p*-Aminophenol
C-4	$-H$	$-H$	$-C(=O)-H$	Formanilid
C-5	$-H$	$-H$	$-C(=O)-C_6H_5$	Benzanilid
C-6	$-H$	$-H$	$-C(=O)-C_6H_4(OH)$ (o-hydroxyphenyl)	Salicylanilide (not an analgesic, but is an antifungal agent)
C-7	$-H$	$-CH_3$	$-C(=O)-CH_3$	Exalgin
C-8	$-OH$	$-H$	$-C(=O)-CH_3$	Acetaminophen
C-9	$-OCH_3$	$-H$	$-H$	Anisidine
C-10	$-OC_2H_5$	$-H$	$-H$	Phenetidine
C-11	$-OC_2H_5$	$-H$	$-C(=O)-CH_3$	Phenacetin
C-12	$-OC_2H_5$	$-H$	$-C(=O)-CH(OH)CH_3$	Lactylphenetidin
C-13	$-OC_2H_5$	$-H$	$-C(=O)-CH_2NH_2$	Phenocoll
C-14	$-OC_2H_5$	$-H$	$-C(=O)-CH_2OCH_3$	Kryofine
C-15	$-OC(=O)-CH_3$	$-H$	$-C(=O)-CH_3$	*p*-Acetoxyacetanilid taceanilid
C-16	$-OC(=O)-C_6H_4(OH)$ (o-hydroxyphenyl)	$-H$	$-C(=O)-CH_3$	Phenetsal
C-17	$-OCH_2CH_2OH$	$-H$	$-C(=O)-CH_3$	Pertonal

ing effect on the analgesic action but, unfortunately, has a highly irritant action on mucous membranes.

The phenacetin molecule has been modified by changing the acyl group on the nitrogen with sometimes beneficial results. Among these are lactylphenetidin (C-12), phenocoll (C-13), and kryofine (C-14). None of these, however, is in current use.

Changing the ether group of phenacetin to an acyl type of derivative has not always been successful. *p*-Acetoxyacetanilid (C-15) has about the same activity and disadvantages as the free phenol. However, the salicyl ester (C-16) exhibits a diminished toxicity and an increased antipyretic activity. Pertonal (C-17) is a somewhat different type in which glycol has been used to etherify the phenolic hydroxyl group. It is very similar to phenacetin. None of these is currently on the market.

Relative to the fate in humans of the types of compounds just discussed, Brodie and Axelrod[193] point out that acetanilid and phenacetin are metabolized by two different routes. Acetanilid is metabolized primarily to *N*-acetyl-*p*-aminophenol, acetaminophen, and only a small amount to aniline, which they showed to be the precursor of phenylhydroxylamine, the compound responsible for methemoglobin formation. Phenacetin is mostly de-ethylated to acetaminophen, whereas a small amount is converted by deacetylation to *p*-phenetidine, also responsible for methemoglobin formation. With both acetanilid and phenacetin, the metabolite acetaminophen formed is believed to be responsible for the analgesic activity of the compounds.

Acetanilid (antifebrin; phenylacetamid) is the monoacetyl derivative of aniline, prepared by heating aniline and acetic acid for several hours.

It can be recrystallized from hot water and occurs as a stable, white crystalline compound. It is slightly soluble in water (1 : 190) and easily soluble in hot water, acetone, chloroform, glycerin (1 : 5), alcohol (1 : 4), or ether (1 : 17). It is available only in powdered form.

Acetanilid is a neutral compound and will not dissolve in either acids or alkalies.

It is prone to form eutectic mixtures with aspirin, antipyrine, chloral hydrate, menthol, phenol, pyrocatechin, resorcinol, salol, thymol, or urethan.

It is definitely toxic in that it causes formation of methemoglobin, affects the heart, and may cause skin reactions and a jaundiced condition. Nevertheless, in the doses used for analgesia, it is a relatively safe drug. However, it is recommended that it be administered in intermittent periods, no period exceeding a few days.[194]

The analgesic effect is selective for most simple headaches and for the pain associated with many muscles and joints. The usual dose is 200 mg.

Several compounds related to acetanilid have been synthesized in attempts to find a better analgesic, as previously indicated. They have not become particularly important in the practice of medicine, for they have little to offer over acetanilid. The physical and chemical properties are also much the same. Eutectic mixtures are formed with many of the same compounds.

Phenacetin, USP. Acetophenetidin; *p*-acetophenetidide may be synthesized in several steps from *p*-nitrophenol.

It occurs as stable, white, glistening crystals, usually in scales, or a fine, white crystalline powder. It is odorless and slightly bitter. It is very slightly soluble in water (1 : 1300), soluble in alcohol (1 : 15), and chloroform (1 : 15), but only slightly soluble in ether (1 : 130). It is sparingly soluble in boiling water (1 : 85).

In general properties and incompatibilities, such as decomposition by acids and alkalies, it is similar to acetanilid. Phenacetin forms eutectic mixtures with chloral hydrate, phenol, aminopyrine, pyrocatechin, or pyrogallol.

It was once used widely as an analgesic and antipyretic, having essentially the same actions as acetanilid. It should be used with the same cautions because the toxic effects are the same as those of acetaminophen, the active form to which it is converted in the body. Some feel there is little justification for its continued use,[193] and it is presently restricted to prescription use only and is available only in powdered form. In particular, a suspected nephrotoxic action[195] has been the basis for the present warning label requirements by the FDA (i.e., "This medication may damage the kidneys when used in large amounts or for a long period of time. Do not take more than the recommended dosage, nor take regularly for longer than 10 days without consulting your physician.") Some recent evidence suggests that phenacetin may not cause nephritis to any greater degree than aspirin, with which it has been most often combined.[196] However, it has been strongly indicated as being carcinogenic in rats and associated with tumors in abusers of phenocetin.[197,198] It has been removed from many combination products and replaced either with additional aspirin (e.g., Anacin) or with acetaminophen.

Acetaminophen, USP. *N*-Acetyl-*p*-aminophenol; 4′-hydroxyacetanilide (Datril; Tempra; Tylenol). This may be prepared by reduction of *p*-nitrophenol in glacial acetic acid, acetylation of *p*-aminophenol with acetic anhydride or ketene, or from *p*-hydroxyacetophenone hydrazone. It occurs as a white, odorless, slightly bitter crystalline powder. It is slightly soluble in water and ether, soluble in boiling water (1 : 20), alcohol (1 : 10), and sodium hydroxide T.S.

Acetaminophen has analgesic and antipyretic activities comparable with those of acetanilid and is used in the same conditions. Although it possesses the same toxic effects as acetanilid, they occur less frequently and with less severity; therefore, it is considered somewhat safer to use. However, the same cautions should be applied. The required FDA warning label reads: "Warning: Do not give to children under three years of age or use for more than ten days unless directed by a physician."[195]

It is available in several nonprescription forms and, also, is marketed in combination with aspirin and caffeine (Trigesic).

THE PYRAZOLONE AND PYRAZOLIDINEDIONE DERIVATIVES

The simple doubly unsaturated compound containing two nitrogen and three carbon atoms in the ring, and with the nitrogen atoms neighboring, is known as pyrazole. The reduction products, named as are other rings of five atoms, are pyrazoline and pyrazolidine. Several pyrazoline substitution products are used in medicine. Many of these are derivatives of 5-pyrazolone. Some can be related to 3,5-pyrazolidinedione.

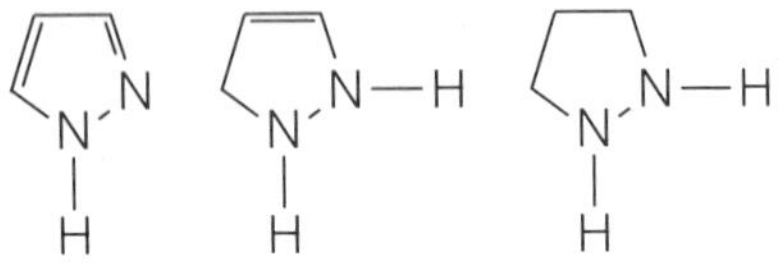

Pyrazole Pyrazoline Pyrazolidine

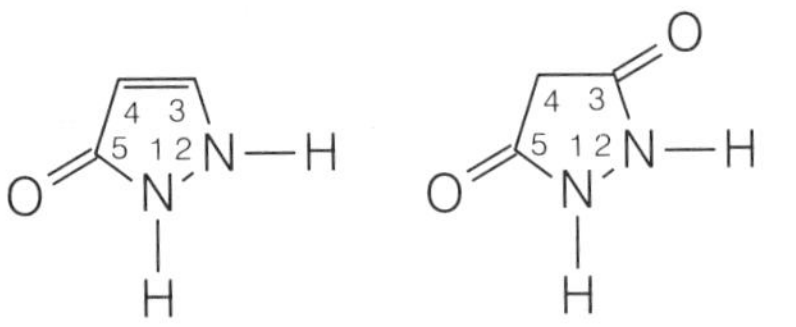

5-Pyrazolone 3,5-Pyrazolidinedione

Ludwig Knorr, a pupil of Emil Fischer, while searching for antipyretics of the quinoline type, in 1884, discovered the 5-pyrazolone now known as antipyrine. This discovery initiated the beginnings of the great German drug industry that dominated the field for approximately 40 years. Knorr, although at first mistakenly believing that he had a quinoline-type compound, soon recognized his error, and the compound was interpreted correctly as being a pyrazolone. Within two years, the analgesic properties of this compound became apparent when favorable reports began to appear in the literature, particularly with reference to its use in headaches and neuralgias. Since then, it has retained some of its popularity as an analgesic, although its use as an antipyretic has declined steadily. Since its introduction into medicine, there have been over 1000 compounds made in an effort to find others with a more potent analgesic action combined with less toxicity. Many modifications of the basic compound have been made. The few derivatives and modifications on the market are listed in Tables 17-8 and 17-9. Phenylbutazone, although analgesic itself, was originally developed as a solubilizer for the insoluble aminopyrine. It is now being used for the relief of many forms of arthritis, in which capacity it has more than an analgesic action in that it also reduces swelling and spasm by an anti-inflammatory action.

TABLE 17-8

DERIVATIVES OF 5-PYRAZOLONE

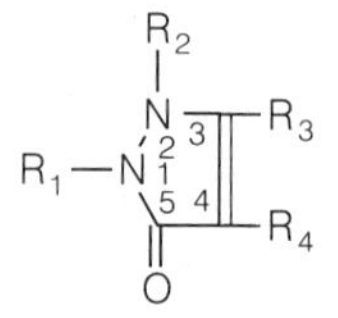

Compound Proprietary Name	R_1	R_2	R_3	R_4
Antipyrine *Phenazone*	$-C_6H_5$	$-CH_3$	$-CH_3$	$-H$
Aminopyrine *Amidopyrine*	$-C_6H_5$	$-CH_3$	$-CH_3$	$-N(CH_3)_2$
Dipyrone *Methampyrone*	$-C_6H_5$	$-CH_3$	$-CH_3$	$-N(CH_3)CH_2SO_3Na$

TABLE 17-9

DERIVATIVES OF 3,5-PYRAZOLIDINEDIONE

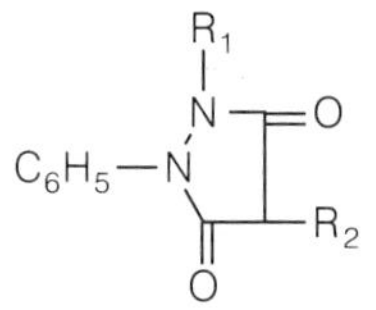

Compound Proprietary Name	R_1	R_2
Phenylbutazone *Azolid, Butazolidin*	$-C_6H_5$	$-C_4H_9$ (n)
Oxyphenbutazone *Oxalid, Tandearil*	$-C_6H_4(OH)$ (p)	$-C_4H_9$ (n)

Antipyrine, USP. 2,3-Dimethyl-1-phenyl-3-pyrazolin-5-one; phenazone (Felsol). This was one of the first important drugs to be made (1887) synthetically.

Antipyrine and many related compounds are prepared by the condensation of hydrazine derivatives with various esters. Antipyrine itself is prepared by the action of ethyl acetoacetate on phenylhydrazine and subsequent methylation.

It consists of colorless, odorless crystals or a white powder, with a slightly bitter taste. It is very soluble in water, alcohol, or chloroform, less so in ether, and its aqueous solution is neutral to litmus paper. However, it is basic in nature, which is due primarily to the nitrogen at position 2.

Locally, antipyrine exerts a paralytic action on the sensory and the motor nerves, resulting in some anesthesia and vasoconstriction, and it also exerts a feeble antiseptic effect. Systemically, it causes results that are very similar to those of acetanilid, although they are usually more rapid. It is readily absorbed after oral administration, circulates freely and is excreted chiefly by the kidneys without having been changed chemically. Any abnormal temperature is reduced rapidly by an unknown mechanism, usually attributed to an effect on the serotonin-mediated thermal regulatory center of the nervous system. It has a higher degree of anti-inflammatory activity than aspirin, phenylbutazone, and indomethacin. It also lessens perception to pain of certain types, without any alteration in central or motor functions, which differs from the effects of morphine. Very often it produces unpleasant and possibly alarming symptoms, even in small or moderate doses. These are giddiness, drowsiness, cyanosis, great reduction in temperature, coldness in the extremities, tremor, sweating, and morbilliform or erythematous eruptions; with very large doses there are asphyxia, epileptic convulsions, and collpase. Treatment for such untoward reactions must be symptomatic. It is probably less likely to produce collapse than acetanilid and is not known to cause the granulocytopenia that sometimes follows aminopyrine.

Antipyrine has been employed in medicine less often in recent years than formerly. It is administered orally to reduce pain and fever in neuralgia, the myalgias, migraine, other headaches, chronic rheumatism, and neuritis, but it is less effective than salicylates and more toxic. When used orally it is given as a 300-mg dose. It sometimes is employed in motor disturbances, such as the spasms of whooping cough or epilepsy. Occasionally, it is applied locally in 5% to 15% solution for its vasoconstrictive and anesthetic effects in rhinitis and laryngitis and sometimes as a styptic in nosebleed. Its use in otic preparations (e.g., Auralgan) has been questioned.[199]

The great success of antipyrine in its early years led to the introduction of a great many derivatives, especially salts with a variety of acids, but none of these has any advantage over the parent compound. Currently in use is the compound with chloral hydrate (Hypnal).

Aminopyrine. Amidopyrine; aminophenazone; 2,3-dimethyl-4-dimethylamino-1-phenyl-3-pyrazolin-5-one. It is prepared from nitrosoantipyrine by reduction to the 4-amino compound followed by methylation.

It consists of colorless, odorless crystals that dissolve in water and the usual organic solvents. It has about the same incompatibilities as antipyrine.

It has been employed as an antipyretic and analgesic, as is antipyrine, but is somewhat slower in action. However, it seems to be much more powerful, and its effects last longer. The usual dose is 300 mg for headaches, dysmenorrhea, neuralgia, migraine, and other like disorders, and it may be given several times daily in rheumatism and other conditions that involve continuous pain.

One of the chief disadvantages of therapy with aminopyrine is the possibility of producing agranulocytosis (granulocytopenia). It has been shown that this is caused by drug therapy with a variety of substances, including mainly aromatic compounds, but particularly with aminopyrine; indeed, several fatal cases have been traced definitely to this drug. The symptoms are a marked fall in leukocytes, absence of granulocytes in the blood, fever, sore throat, ulcerations on mucous surfaces, and prostration, with death in most cases of secondary complications. The treatment is merely symptomatic with penicillin to prevent any possible superimposed infection. The condition seems to be more or less an allergic reaction, because only a certain small percentage of those who use the drug are affected, but great caution must be observed to avoid susceptibility. Many countries have forbidden or greatly restricted its administration, and it has fallen more or less into disfavor.

Dipyrone, methampyrone, occurs as a white, odorless crystalline powder possessing a slightly bitter taste. It is freely soluble in water (1 : 1.5) and sparingly soluble in alcohol.

It is used as an analgesic, an antipyretic, and an antirheumatic. The recommended dose is 300 mg to 1 g orally and 500 mg to 1 g intramuscularly or subcutaneously. It is available as an injectable for veterinary use.

Phenylbutazone, USP. 4-Butyl-1,2-diphenyl-3,5-pyrazolidinedione (Azolid; Butazolidin; Phenylzone-A). This drug is a white to off-white, odorless, slightly bitter powder. It has a slightly aromatic odor and is freely soluble in ether, acetone, and ethyl acetate, very slightly soluble in water, and is soluble in alcohol (1 : 20).

According to the patents describing the synthesis of this type of compound, it can be prepared by condensing *n*-butyl malonic acid or its derivatives with hydrazobenzene to get 1,2-diphenyl-4-*n*-butyl-3,5-pyrazolidinedione. Alternatively, it can be prepared by treating 1,2-diphenyl-3,5-pyrazolidinedione, obtained by a procedure analogous to the foregoing condensation, with butyl bromide in 2 *N* sodium hydroxide at 70 °C or with *n*-butyraldehyde followed by reduction utilizing Raney nickel catalyst.

The principal usefulness of phenylbutazone lies in the treatment of the painful symptoms associated with gout, rheumatoid arthritis, psoriatic arthritis, rheumatoid spondylitis, and painful shoulder (peritendinitis, capsulitis, bursitis, and acute arthritis of the joint). Because of its many unwelcome side effects, this drug is not generally considered to be the drug of choice but should be reserved for trial in those patients who do not respond to less toxic drugs. It should be emphasized that, although the drug is an analgesic, it is not to be considered as one of the simple analgesics and is not to be used casually. The initial daily dosage in adults ranges from 300 to 600 mg divided into three or four doses. The manufacturer suggests that an average initial daily dosage of 600 mg/day administered for one week should determine whether the drug will give a favorable response. If no results are forthcoming in this time, it is recommended that the drug be discontinued to avoid side effects. In the event of favorable response, the dosage is reduced to a minimal effective daily dose, which usually ranges from 100 to 400 mg.

The drug is contraindicated in the presence of edema, cardiac decompensation, a history of peptic ulcer or drug allergy, blood dyscrasias, hypertension, and whenever renal, cardiac or hepatic damage is present. All patients, regardless of the history given, should be careful to note the occurrence of black or tarry stools, which might be indicative of reactivation of latent peptic ulcer and is a signal for discontinuance of the drug. The physician is well advised to read the manufacturer's literature and warnings thoroughly before attempting to administer the drug. Among the precautions the physician should take for the patient are to examine the patient periodically for toxic reactions, to check for increase in weight (due to water retention), and to make periodic blood counts to guard against agranulocytosis.

Oxyphenbutazone, USP. 4-Butyl-1-(*p*-hydroxyphenyl)-2-phenyl-3,5-pyrazolidinedione (Oxalid; Tandearil). This drug is a metabolite of phenylbutazone and has the same effectiveness, indications, side effects, and contraindications. Its only apparent advantage is that it causes acute gastric irritation less frequently.

The pharmacology of these and other analogues has been reviewed extensively.[200]

REFERENCES

1. Tainter, M. L.: Ann. N. Y. Acad. Sci. 51:3, 1948.
2. Eisleb, O., and Schaumann, O.: Dtsch. Med. Wochenschr. 65:967, 1938.
3. Small, L. F., Eddy, N. B., Mosettig, E., and Himmelsbach, C. K.: Studies on Drug Addiction, Supplement No. 138 to the Public Health Reports, Washington, DC, Supt. Doc., 1938.
4. Eddy, N. B., Halbach, H., and Braenden, O. J.: Bull. WHO 14:353, 1956.
5. Stork, G., and Bauer, L.: J. Am. Chem. Soc. 75:4373, 1953.
6. U.S. Patent 2,831,531; through Chem. Abstr. 52:13808, 1958.
7. Rapoport, H., Baker, D. R., and Reist, H. N.: J. Org. Chem. 22:1489, 1957; Chadha, M. S., and Rapoport, H.: J. Am. Chem. Soc. 79:5730, 1957.
8. Okun, R., and Elliott, H. W.: J. Pharmacol. Exp. Ther. 124:255, 1958.
9. Seki, I., Takagi, H., and Kobayashi, S.: J. Pharm. Soc. Jpn. 84:280, 1964.
10. Buckett, W. R., Farquharson, M. E., and Haining, C. G.: J. Pharm. Pharmacol. 16:174, 68T, 1964.
11. Bentley, K. W., and Hardy, D. G.: Proc. Chem. Soc. 220, 1963.
12. Lister, R. E.: J. Pharm. Pharmacol. 16:364, 1964.
13. Bentley, K. W., and Hardy, D. G.: J. Am. Chem. Soc. 89:3267, 1967.
14. Telford, J., Papadopoulos, C. N., and Keats, A. S.: J. Pharmacol. Exp. Ther. 133:106, 1961.
15. Gates, M., and Montzka, T. A.: J. Med. Chem. 7:127, 1964.
16. Schaumann, O.: Arch. Exp. Pathol. Pharmakol. 196:109, 1940.
17. Bergel, F., and Morrison, A. L.: Q. Rev. (Lond.) 2:349, 1948.
18. Jensen, K. A., Lindquist, F., Rekling, E., and Wolffbrandt, C. G.: Dan. Tidsskr. Farm. 17:173, 1943; through Chem. Abstr. 39:2506, 1945.
19. Lee, J., Ziering, A., Berger, L., and Heineman, S. D.: Jubilee Volume—Emil Barell, p. 267. Basel, Reinhardt, 1946; J. Org. Chem. 12:885, 894, 1947; Berger, L., Ziering, A., and Lee, J.: J. Org. Chem. 12:904, 1947; Ziering, A., and Lee, J.: J. Org. Chem. 12:911, 1947.
20. Janssen, P. A. J., and Eddy, N. B.: J. Med. Pharm. Chem. 2:31, 1960.
21. Lee, J.: Analgesics: B. Partial structures related to morphine. In American Chemical Society: Medicinal Chemistry, Vol. 1, pp. 438–466. New York, John Wiley & Sons, 1951.
22. Beckett, A. H., Casy, A. F., and Kirk, G.: J. Med. Pharm. Chem. 1:37, 1959.
23. Bell, K. H., and Portoghese, P. S.: J. Med. Chem. 16:203, 589, 1973; 17:129, 1974.
24. Groeber, W. R., et al.: Obstet. Gynecol. 14:743, 1959.
25. Eddy, N. B.: Chem. Ind. (Lond.), p. 1462, Nov. 21, 1959.
26. Fraser, H. F., and Isbell, H.: Bull. Narc. 13:29, 1961.
27. Janssen, P. A. J., et al.: J. Med. Pharm. Chem. 2:271, 1960.
28. Stahl, K. D., et al.: Eur. J. Pharmacol. 46:199, 1977.
29. Blicke, F. F., and Tsao, E.: J. Am. Chem. Soc. 75:3999, 1953.
30. Cass, L. J., et al.: JAMA 166:1829, 1958.
31. Batterham, R. C., Mouratoff, G. J., and Kaufman, J. E.: Am. J. Med. Sci. 247:62, 1964.

32. Finch, J. S., and DeKornfeld, T. J.: J. Clin. Pharmacol. 7:46, 1967.
33. Yelnosky, J., and Gardocki, J. F.: Toxicol. Appl. Pharmacol. 6:593, 1964.
34. Bockmuehl, M., and Ehrhart, G.: German Patent 711,069.
35. Kleiderer, E. C., Rice, J. B., and Conquest, V.: Pharmaceutical Activities at the I. G. Farbenindustrie Plant, Höchst-am-Main, Germany. Report 981, Office of the Publication Board, Dept. of Commerce, Washington, DC, 1945.
36. Scott, C. C., and Chen, K. K.: Fed. Proc. 5:201, 1946; J. Pharmacol. Exp. Ther. 87:63, 1946.
37. Eddy, N. B., Touchberry, C., and Lieberman, J.: J. Pharmacol. Exp. Ther. 98:121, 1950.
38. Janssen, P. A. J., and Jageneau, A. H.: J. Pharm. Pharmacol. 9:381, 1957; 10:14, 1958. See also Janssen, P. A. J.: J. Am. Chem. Soc. 78:3862, 1956.
39. Cass, L. J., and Frederik, W. S.: Antibiot. Med. 6:362, 1959, and references cited therein.
40. Blaine, J., and Renault, P. (eds): Rx 3 times/wk LAAM—Methadone Alternative. NIDA Res. Monogr. 8, DHEW, 1976.
41. Eddy, N. B., Besendorf, H., and Pellmont, B.: Bull. Narc., U.N. Dept. Social Affairs 10:23, 1958.
42. DeKornfeld, T. J.: Curr. Res. Anesth. 39:430, 1960.
43. Murphy, J. G., Ager, J. H., and May, E. L.: J. Org. Chem. 25:1386, 1960, and references cited therein.
44. Chignell, C. F., Ager, J. H., and May, E. L.: J. Med. Chem. 8:235, 1965.
45. May, E. L., and Eddy, N. B.: J. Med. Chem. 9:851, 1966.
46. Michne, W. F., et al.: J. Med. Chem. 22:1158, 1979.
47. Archer, S., et al.: J. Med. Chem. 7:123, 1964.
48. Cass, L. J., Frederik, W. S., and Teodoro, J. V.: JAMA 188:112, 1964.
49. Fraser, H. F., and Rosenberg, D. E.: J. Pharmacol. Exp. Ther. 143:149, 1964.
50. Lasagna, L., DeKornfeld, T. J., and Pearson, J. W.: J. Pharmacol. Exp. Ther. 144:12, 1964.
51. Houde, R. W.: B. J. Clin. Pharmacol. 7:297, 1979.
52. Mello, N. K., and Mendelson, J. H.: Science 207:657, 1980.
53. Martin, W. R.: Pharmacol. Rev. 19:463, 1967.
54. Julius, D., and Renault, P. (eds.): Narcotic Antagonists: Naltrexone, Progress Report. NIDA Res. Monogr. 9, DHEW, 1976.
55. Willette, R. E. (ed.): Narcotic Antagonists: The Search for Long-Acting Preparations. NIDA Res. Monogr. 4, DHEW, 1975.
56. deStevens, G. (ed.): Analgetics. New York, Academic Press, 1965.
57. Braenden, O. J., Eddy, N. B., and Halbach, H.: Bull. WHO 13:937, 1955.
58. Leutner, V.: Arzneim. Forsch. 10:505, 1960.
59. Janssen, P. A. J.: Br. J. Anaesth. 34:260, 1962.
60. Beckett, A. H., and Casy, A. F.: In Ellis, G. P., and West, G. B. (eds.). Prog. Med. Chem. 2:43–87, 1962.
61. Mellet, L. B., and Woods, L. A.: Prog. Drug Res. 5:156, 1963.
62. Casy, A. F.: In Ellis, G. P., and West, G. B. (eds.). Prog. Med. Chem. 7:229–284, 1970.
63. Lewis, J., Bentley, K. W., and Cowan, A.: Annu. Rev. Pharmacol. 11:241, 1970.
64. Eddy, N. B., and May, E. L.: Science 181:407, 1973.
65. Beckett, A. H., Casy, A. F., and Harper, N. J.: Pharm. Pharmacol. 8:874, 1956.
66. Lasagna, L., and DeKornfeld, T. J.: J. Pharmacol. Exp. Ther. 124:260, 1958.
67. Fishman, J., Hahn, E. F., and Norton, B. I.: Nature 261:64, 1976.
68. Wright, W. B., Jr., Brabander, H. J., and Hardy, R. A., Jr.: J. Am. Chem. Soc. 81:1518, 1959.
69. Gross, F., and Turrian, H.: Experientia 13:401, 1957; Fed. Proc. 19:22, 1960.
70. Beckett, A. H., and Casy, A. F.: J. Pharm. Pharmacol. 6:986, 1954.
71. Beckett, A. H.: J. Pharm. Pharmacol. 8:848, 860, 1958.
72. Beckett, A. H.: Pharm. J. p. 256, Oct. 24, 1959.
73. Jacquet, Y. F., et al.: Science 198:842, 1977.
74. Portoghese, P. S.: J. Med. Chem. 8:609, 1965.
75. Portoghese, P. S.: J. Pharm. Sci. 55:865, 1966.
76. Portoghese, P. S.: Acc. Chem. Res. 11:21, 1978.
77. Koshland, D. E., Jr.: Proc. First Int. Pharmacol. Meet. 7:161, 1963, and references cited therein.
78. Belleau, B.: J. Med. Chem. 7:776, 1964.
79. Martin, W. R., et al.: J. Pharmacol. Exp. Ther. 197:517, 1976.
80. Gilbert, P. E., and Martin, W. R.: J. Pharmacol. Exp. Ther. 198:66, 1976.
81. Beaumont, A., and Hughes, J.: Annu. Rev. Pharmacol. Toxicol. 19:245, 1979.
82. Lasagna, L., and Beecher, H. K.: J. Pharmacol. Exp. Ther. 112:356, 1954.
83. Soulairac, A., Cahn, J., and Charpentier, J. (eds.): Pain. New York, Academic Press, 1968.
84. Willette, R. E.: Am. J. Pharm. Ed. 34:662, 1970.
85. Barnett, G., Trsic, M., and Willette, R. E. (eds.): Quantitative Structure–Activity Relationships of Analgesics, Narcotic Antagonists, and Hallucinogens. NIDA Res. Monogr. 22, DHEW, 1978.
86. Archer, S., and Harris, L. S.: In Jucker, E. (ed.). Prog. Drug Res. 8:262, 1965.
87. Kutter, E., et al.: J. Med. Chem. 13:801, 1970.
88. Portoghese, P. S., et al.: J. Med. Chem. 14:144, 1971.
89. Portoghese, P. S., et al.: J. Med. Chem. 11:219, 1968.
90. Kaufman, J. J., Semo, N. M., and Koski, W. S.: J. Med. Chem. 18:647, 1975.
91. Kaufman, J. J., Kerman, E., and Koski, W. S.: Int. J. Quantum Chem. 289, 1974.
92. Loew, G. H., and Berkowitz, D. S.: J. Med. Chem. 21:101, 1978.
93. Loew, G. H., and Berkowitz, D. S.: J. Med. Chem. 22:603, 1979.
94. Simon, E. J., and Hiller, J. B.: Annu. Rev. Pharmacol. Toxicol. 18:371, 1978.
95. Goldstein, A.: Life Sci. 14:615, 1974.
96. Goldstein, A., Lowney, L. I., and nad Pal, B. K.: Proc. Natl. Acad. Sci. USA 68:1742, 1971.
97. Simon, E. J., Hiller, J. M., and Edelman, I.: Proc. Natl. Acad. Sci. USA 70:1947, 1973.
98. Terenius, L.: Acta Pharmacol. Toxicol. 32:317, 1973.
99. Pert, C. B., and Snyder, S. H.: Science 179:1011, 1973.
100. Kosterlitz, H. W., and Watt, A. J.: Br. J. Pharmacol. Chemother. 33:266, 1968.
101. Hughes, J.: Brain Res. 88:295, 1975; Neurosci. Res. Program Bull. 13:55, 1975.
102. Terenius, L.: Annu. Rev. Pharmacol. Toxicol. 18:189, 1978.
103. Goldstein, A.: Science 193:1081, 1976.
104. Kolanta, G. B.: Science 205:774, 1979.
105. Proc. Manchester Lit. Phil. Soc. 69–79, 1925.
106. Gates, M., and Tschudi, G.: J. Am. Chem. Soc. 74:1109, 1952; 78:1380, 1956.
107. Twycross, R. G.: Int. J. Clin. Pharmacol. 9:184, 1974.
108. Lasagna, L.: Pharmacol. Rev. 16:47, 1964.
109. Houde, R. W., and Wallenstein, S. L.: Minutes of the 11th Meeting, Committee on Drug Addiction and Narcotics, National Research Council, 1953, p. 417.
110. Diehl, H. S.: JAMA 101:2042, 1933.
111. Eddy, N. B., and Lee, L. E.: J. Pharmacol. Exp. Ther. 125:116, 1959.
112. Weijlard, J., and Erickson, A. E.: J. Am. Chem. Soc. 64:869, 1942.
113. Fraser, H. F., et al.: J. Pharmacol. Exp. Ther. 122:359, 1958; Cochin, J., and Axelrod, J.: J. Pharmacol. Exp. Ther. 125:105, 1959.
114. Ziering, A., and Lee, J.: J. Org. Chem. 12:911, 1947.
115. Weijlard, J., et al.: J. Am. Chem. Soc. 78:2342, 1956.
116. Med. Lett. 19:73, 1977.
117. Wang, R. I. H.: Eur. J. Clin. Pharmacol. 7:183, 1974.
118. Easton, N. R., Gardner, J. H., and Stevens, J. R.: J. Am. Chem. Soc. 69:2941, 1947. See also reference 34.
119. Freedman, A. M.: JAMA 197:878, 1966.

120. Med. Lett. 11:97, 1969.
121. Smits, S. E.: Res. Commun. Chem. Pathol. Pharmacol. 8:575, 1974.
122. Billings, R. E., Booher, R., Smits, S. E., et al.: J. Med. Chem. 16:305, 1973.
123. Jaffe, J. H., Senay, E. C., Schuster, C. R., et al.: JAMA 222:437, 1972.
124. Gruber, C. M., and Babtisti, A.: Clin. Pharmacol. Ther. 4:172, 1962.
125. Pohland, A., and Sullivan, H. R.: J. Am. Chem. Soc. 75:4458, 1953.
126. Med. Lett. 20:111, 1978.
127. Forrest, W. H., et al.: Clin. Pharmacol. Ther. 10:468, 1969.
128. Ehrnebo, M., Boreus, L. O., and Lönroth, V.: Clin. Pharmacol. Ther. 22:888, 1977.
129. Med. Lett. 9:49, 1967.
130. Klatz, A. L.: Curr. Ther. Res. 16:602, 1974; Workman, F. C., and Winter, L.: Curr. Ther. Res. 16:609, 1974.
131. Jasinski, D. R., Martin, W. R., and Sapira, J. D.: Clin. Pharmacol. Ther. 9:215, 1968.
132. Cone, E. J.: Tetrahedron Lett. 28:2607, 1973; Chatterjie, N., et al.: Drug Metab. Disp. 2:401, 1974.
133. Batra, V. K., Sams, R. A., Reuning, R. H., and Malspeis, L.: Acad. Pharm. Sci. 4:122, 1974.
134. Blumberg, H., and Dayton, H. B.: In Kosterlitz, H., and Villarreal, J. E. (eds.). Agonist and Antagonist Actions of Narcotic Analgesic Drugs, pp. 110–119. London, Macmillan, 1972.
135. Woodland, J. H. R., et al.: J. Med. Chem. 16:897, 1973.
136. Takemori, A. E. A., Hayashi, G., and Smits, S. E.: Eur. J. Pharmacol. 20:85, 1972.
137. Nutt, J. G., and Jasinsky, D. R.: Pharmacologist 15:240, 1973.
138. Salem, H., and Aviado, D. M.: Am. J. Med. Sci. 247:585, 1964.
139. Chappel, C. I., and von Seemann, C.: In Ellis, G. P., and West, G. B. (eds.). Prog. Med. Chem. 3:133–136, 1963.
140. Chernish, S. M.: Ann. Allergy 21:677, 1963.
141. Chen, J. Y. P., Biller, H. F., and Montgomery, E. G.: J. Pharmacol. Exp. Ther. 128:384, 1960.
142. Five Year Report. Br. Med. J. 2:1033, 1960.
143. Wong, S.: In Heinzelman, R. V. (ed.). Annu. Rep. Med. Chem. 10:172–181, 1975.
144. Collier, H. O. J.: Nature 232:17, 1971.
145. Vane, J. R.: Nature 231:232, 1971.
146. Shen, T. Y.: Angew. Chem. Int. Ed.: 11:460, 1972.
147. Nickander, R., McMahon, F. G., and Ridolfo, A. S.: Annu. Rev. Pharmacol. Toxicol. 19:469, 1979.
148. Smith, P. K.: Ann. N.Y. Acad. Sci. 86:38, 1960.
149. Smith, M. J. H., and Smith, P. K. (eds.): The Salicylates. A. Critical Bibliographic Review. New York, John Wiley & Sons, 1966.
150. Collier, H. O. J.: Sci. Am. 209:97, 1963.
151. Alvarez, A. S., and Summerskill, W. H. J.: Lancet 2:920, 1958.
152. Barager, F. D., and Duthie, J. J. R.: Br. Med. J. 1:1106, 1960.
153. Leonards, J. R., and Levy, G.: Abstr. of the 116th Meet. Am. Pharm. Assoc., p. 67, Montreal, May 17–22, 1969.
154. Davenport, H. W.: N. Engl. J. Med. 276:1307, 1967.
155. Weiss, H. J.: Schweiz. Med. Wochenschr. 104:114, 1974; Elwood, P. C., et al.: Br. Med. J. 1:436, 1974.
156. Aspirin Myocardial Infarction Study Research Group: JAMA 243:661, 1980.
157. Levy, G., and Hayes, B. A.: N. Engl. J. Med. 262:1053, 1960.
158. Lieberman, S. V., et al.: J. Pharm. Sci. 53:1486, 1492, 1964.
159. Pfeiffer, C. C.: Arch. Biol. Med. Exp. 4:10, 1967.
160. Salassa, R. M., Bollman, J. M., and Dry, T. J.: J. Lab. Clin. Med. 33:1393, 1948.
161. Smith, P. K., et al.: J. Pharmacol. Exp. Ther. 87:237, 1946.
162. Garrett, E. R.: J. Am. Pharm. Assoc. Sci. Ed. 48:676, 1959.
163. Gold, G., and Campbell, J. A.: J. Pharm. Sci. 53:52, 1964.
164. Troup, A. E., and Mitchner, H.: J. Pharm. Sci. 53:375, 1964.
165. Blaug, S. M., and Wesolowski, J. W.: J. Am. Pharm. Assoc. Sci. Ed. 48:691, 1959.
166. Obbink, H. J. K.: Lancet 1:565, 1964.
167. Goldman, A. S., and Yakovac, W. C.: Proc. Soc. Exp. Biol. Med. 115:693, 1964.
168. Anon.: Br. Med. J. 2:131T, 1963.
169. Med. Lett. 8:7, 1966.
170. Liyanage, S. P., and Tambar, P. K.: Curr. Med. Res. Opin. 5:450, 1978.
171. Deodhar, S. D., et al.: Curr. Med. Res. Opin. 5:185, 1978.
172. Regaldo, R. G.: Curr. Med. Res. Opin. 5:454, 1978.
173. Leonards, J. R.: J. Lab. Clin. Med. 74:911, 1969.
174. Bloomfield, S. S., Barden, T. P., and Hille, R.: Clin. Pharmacol. Ther. 11:747, 1970.
175. Winder, C. V.: Nature 184:494, 1959.
176. Scherrer, R. A.: In Scherrer, R. A., and Whitehouse, M. W. (eds.). Antiinflammatory Agents, p. 132. New York, Academic Press, 1974.
177. Cass, L. J., and Frederik, W. S.: J. Pharmacol. Exp. Ther. 139:172, 1963.
178. Lane, A. Z., Holmes, E. L., and Moyer, C. E.: J. New Drugs 4:333, 1964.
179. Med. Lett. 10:37, 1968.
180. Walker, R. W., et al.: Anal. Biochem. 95:579, 1979.
181. Brogden, R. N., et al.: Drugs 16:97, 1978. See also Med. Lett. 21:i, 1979.
182. Brogden, R. N., et al.: Drugs 15:429, 1978.
183. O'Neill, P. J., et al.: J. Pharmacol. Exp. Ther. 209:366, 1979.
184. Wallenstein, S. L.: Unpublished report.
185. Dornan, J., and Reynolds, W.: Can. Med. Assoc. J. 110:1370, 1974.
186. Emele, J. F., and Shanaman, J. E.: Arch. Int. Pharmacodyn. Ther. 170:99, 1967.
187. Brogden, R. N., et al.: Drugs 13:241, 1977.
188. Chernish, S. M., et al.: Arthritis Rheum. 22:376, 1979.
189. Brogden, R. N., et al.: Drugs 18:241, 1979.
190. Wiseman, E. H.: R. Soc. Med. Int. Congr. Ser. 1:11, 1978.
191. Weintraub, M., et al.: J. Rheumatol. 4:393, 1979.
192. Balogh, Z., et al.: Curr. Med. Res. Opin. 6:148, 1979.
193. Brodie, B. B., and Axelrod, J.: J. Pharmacol. Exp. Ther. 94:29, 1948; 97:58, 1949. See also Axelrod, J.: Postgrad. Med. 34:328, 1963.
194. Bonica, J. J., and Allen, G. D.: In Modell, W. (ed.). Drugs of Choice 1970–1971, p. 210. St. Louis, C. V. Mosby, 1970.
195. Med. Lett. 6:78, 1964.
196. Brown, D. M., and Hardy, T. L.: Br. J. Pharmacol. Chemotherap. 32:17, 1968.
197. Tomatis, L., et al.: Cancer Res. 38:877, 1978.
198. Bengsston, V., Johansson, S., and Angervall, L.: Kidney Int. 13:107, 1978. See also Science 240:129, 1979.
199. FDA Over-the-Counter Advisory Panel, Fed. Reg. 42:63562, 1977.
200. Burns, J. J., et al.: Ann. N.Y. Acad. Sci. 86:253, 1960; Domenjoz, R.: Ann. N.Y. Acad. Sci. 86:263, 1960.

SELECTED READINGS

American Chemical Society, First National Medicinal Chemistry Symposium, pp. 15–49, 1948.

Anon.: Codeine and Certain Other Analgesic and Antitussive Agents: A Review. Rahway, Merck & Co., 1970.

Archer, S., and Harris, L. S.: Narcotic antagonists. In Jucker, E. (ed.). Prog. Drug Res. 8:262, 1965.

Arrigoni-Martelli, E.: Inflammation and Antiinflammatories. New York, Spectrum, 1977.

Barlow, R. B.: Morphine-like analgesics. In Introduction to Chemical Pharmacology, pp. 39–56. New York, John Wiley & Sons, 1955.

Beckett, A. H., and Casy, A. F.: The testing and development of analgesic drugs. In Ellis, G. P., and West, G. B. (Eds.). Prog. Med. Chem. 2:43–87, 1963.

Bergel, F., and Morrison, A. L.: Synthetic analgesics. Q. Rev. (Lond.) 2:349, 1948.
Berger, F. M., et al.: Non-narcotic drugs for the relief of pain and their mechanism of action. Ann. N.Y. Acad. Sci. 86:310, 1960.
Braenden, O. J., Eddy, N. B., and Hallbach, H.: Relationship between chemical structure and analgesic action. Bull. WHO 13:937, 1955.
Braude, M. C., et al. (eds.): Narcotic Antagonists. New York, Raven Press, 1973.
Brümmer, T.: Die historische Entwicklung des Antipyrin und seiner Derivative. Fortschr. Ther. 12:24, 1936.
Casy, A. F.: Analgesics and their antagonists: Recent developments. In Ellis, G. P., and West, G. B. (eds.). Prog. Med. Chem. 7:229–284, 1970.
Chappel, C. I., and von Seemann, C.: Antitussive drugs. In Ellis, G. P., and West, G. B. (eds.). Prog. Med. Chem. 3:89–145, 1963.
Chen, K. K.: Physiological and pharmacological background, including methods of evaluation of analgesic agents. J. Am. Pharm. Assoc. Sci. Ed. 38:51, 1949.
Clouet, D. H.: Narcotic Drugs: Biochemical Pharamcology. New York, Plenum Press, 1971.
Collins, P. W.: Antitussives. In Burger, A. (ed.). Medicinal Chemistry, 3rd ed., pp. 1351–1364. New York, Wiley-Interscience, 1970.
Coyne, W. E.: Nonsteroidal antiinflammatory agents and antipyretics. In Burger, A. (ed.). Medicinal Chemistry, 3rd ed., pp. 953–975. New York, Wiley-Interscience, 1970.
deStevens, G. (ed.): Analgetics. New York, Academic Press, 1965.
Eddy, N. B.: Chemical structure and action of morphine-like analgesics and related substances. Chem. Ind. (Lond.), p. 1462, Nov. 21, 1959.
Eddy, N. B., Halbach, H., and Braenden, O. J.: Bull. WHO 14:353–402, 1956; 17:569–863, 1957.
Evens, R. P.: Drug therapy reviews: Antirheumatic agents. Am. J. Hosp. Pharm. 36:622, 1979.
Fellows, E. J., and Ullyot, G. E.: Analgesics: A. Aralkylamines. In American Chemical Society, Medicinal Chemistry, Vol. 1, pp. 390–437, New York, John Wiley & Sons, 1951.
Gold, H., and Cattell, M.: Control of pain. Am. J. Med. Sci. 246:590, 1963.
Greenberg, L.: Antipyrine: A Critical Bibliographic Review. New Haven, Hillhouse, 1950.
Gross, M.: Acetanilid: A Critical Bibliographic Review. New Haven, Hillhouse, 1946.
Hellerbach, J., Schnider, O., Besendorf, H., Dellmont, B., Eddy, N. B., and May, E. L.: Synthetic Analgesics: Part II. Morphinans and 6,7-Benzomorphans. New York, Pergamon Press, 1966.
Jacobson, A. E., May, E. L., and Sargent, L. J.: Analgetics. In Burger, A. (ed.). Medicinal Chemistry, 3rd ed., pp. 1327–1350. New York, Wiley-Interscience, 1970.
Janssen, P. A. J.: Synthetic Analgesics: Part I. Diphenylpropylamines. New York, Pergamon Press, 1960.
Janssen, P. A. J., and van der Eycken, C. A. M.: In Burger, A. (ed.). Drugs Affecting the Central Nervous System, pp. 25–85. New York, Marcel Dekker, 1968.
Lasagna, L.: The clinical evaluation of morphine and its substitutes as analgesics. Pharmacol. Rev. 16:47–83, 1964.
Lee, J.: Analgesics: B. Partial structures related to morphine. In American Chemical Society, Medicinal Chemistry, Vol. 1, pp. 438–466, New York, John Wiley & Sons, 1951.
Martin, W. R.: Opioid antagonists. Pharmacol. Rev. 19:463–521, 1967.
Mellet, L. B., and Woods, L. A.: Analgesia and addiction. Prog. Drug Res. 5:156–267, 1963.
Portoghese, P. S.: Stereochemical factors and receptor interactions associated with narcotic analgesics. J. Pharm. Sci. 55:865, 1966.
Reynolds, A. K., and Randall, L. O.: Morphine and Allied Drugs. Toronto, Univ. Toronto Press, 1957.
Salem, H., and Aviado, D. M.: Antitussive Agents, Vols. 1–3 (Sect. 27 of International Encyclopedia of Pharmacology and Therapeutics). Oxford, Pergamon Press, 1970.
Scherrer, R. A., and Whitehouse, M. W.: Antiinflammatory Agents. New York, Academic Press, 1974.
Shen, T. Y.: Perspectives in nonsteroidal anti-inflammatory agents. Angew. Chem. Int. Ed. 11:460, 1972.
Snyder, S. H.: Opiate receptors and internal opiates. Sci. Am. 237:236–244, 1977.
Winder, C. A.: Nonsteroid anti-inflammatory agents. In Jucker, E. (ed.). Prog. Drug Res. 10:139–203, 1966.

CHAPTER **18**

Steroids and Therapeutically Related Compounds

Dwight S. Fullerton

Steroids are widely distributed throughout the plant and animal kingdoms, and are formed by identical or nearly identical biosynthetic pathways in both plants and animals. Furthermore, because of their relatively rigid chemical structures, the steroids usually have easily predictable physical and chemical properties.

However, the similarity among the steroids ends with their fundamental chemical properties. The steroids have little in common therapeutically, except that as a group they are the most extensively used drugs in modern medicine. The major therapeutic classes of steroids are illustrated in Figure 18-1. The fact that minor changes in steroid structure can cause extensive changes in biologic activity has been a continual source of fascination for medicinal chemists and pharmacologists for some three decades.

No one has captured the excitement and fascination of steroid drugs better than Rupert Witzmann in his 1981 book *Steroids—Keys to Life*.[1] Witzmann vividly describes the drama of "the greatest [chemical and biologic research] attack on a single group of substances that the world has ever seen." His book is highly recommended for further reading.

In this chapter, we will consider the steroids used in modern medicine. Some nonsteroidal compounds that have similar therapeutic uses will also be discussed (e.g., nonsteroidal inotropic agents for treatment of heart failure, the diethylstilbestrol estrogens, the nonsteroidal chemical contraceptive agents, ovulation stimulants, and LH-RH analogues).

Many general reviews on steroid chemistry, synthesis and analysis, biochemistry and receptors, pharmacology, therapy, and metabolism have been published.[1-32] Additional reviews on particular classes of steroids will be cited in subsequent sections.

NEW INSIGHTS ON STEROID RECEPTORS

The number of steroid drugs has not substantially increased during the last decade, but the knowledge of their mechanisms of action has. New techniques in recombinant DNA studies, gene cloning, use of specific monoclonal antibodies against specific steroid receptors, receptor purification, and x-ray crystallography have all been used in unraveling the molecular basis of steroid actions. As King and Mainwaring[26] succinctly predicted in their 1974 review of early work on steroid receptors, "Many scientific discoveries appear delightfully simple at first but, as further experiments are performed, the simplicity disappears and a phase of maximum confusion occurs... this is [we hope] followed by an answer. ..." Many answers are now at hand, but much remains to be learned.

CELL SURFACE RECEPTORS

The discovery of the cholesteryl ester low-density lipoprotein (LDL) receptor in 1973, by Michael S. Brown and Joseph L. Goldstein, led to a virtual explosion of research worldwide. For their extraordinary contributions, Brown and Goldstein were awarded the Nobel Prize for Medicine in 1985. The research effort with cell surface receptors continues unabated to this day—particularly with the T4 cell surface receptor for the human immunodeficiency viruses (HIV), which cause the acquired immune deficiency syndrome (AIDS). Cell surface receptors, illustrated in Figure 18-2 for transport of cholesterol by LDL receptors,[33,34] are now known to be responsible for the transport of numerous large molecules into cells.[35] Included are a wide variety of

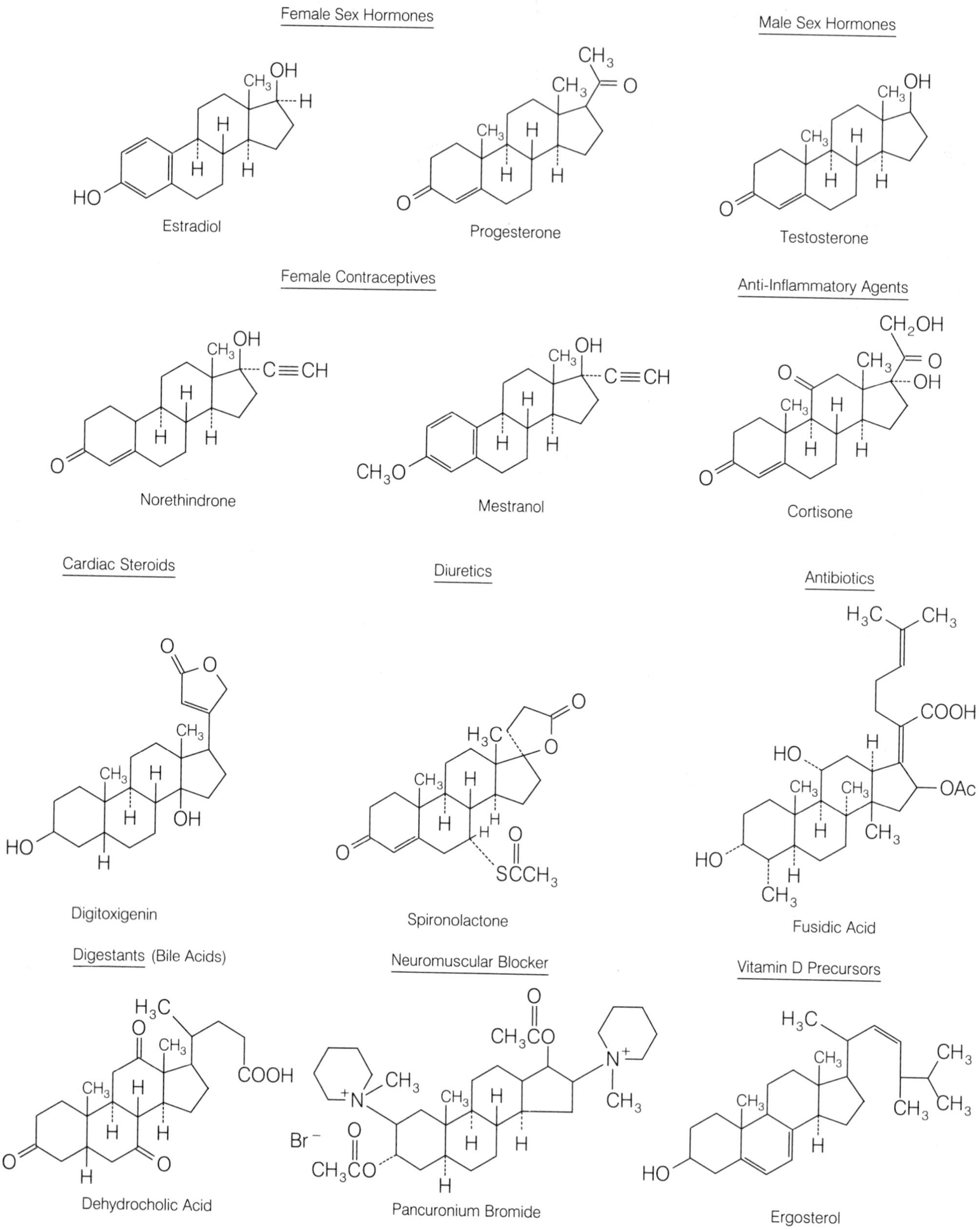

Female Sex Hormones

Estradiol

Progesterone

Male Sex Hormones

Testosterone

Female Contraceptives

Norethindrone

Mestranol

Anti-Inflammatory Agents

Cortisone

Cardiac Steroids

Digitoxigenin

Diuretics

Spironolactone

Antibiotics

Fusidic Acid

Digestants (Bile Acids)

Dehydrocholic Acid

Neuromuscular Blocker

Pancuronium Bromide

Vitamin D Precursors

Ergosterol

FIG. 18-1. *Representative examples of primary therapeutic classes of steroids.*

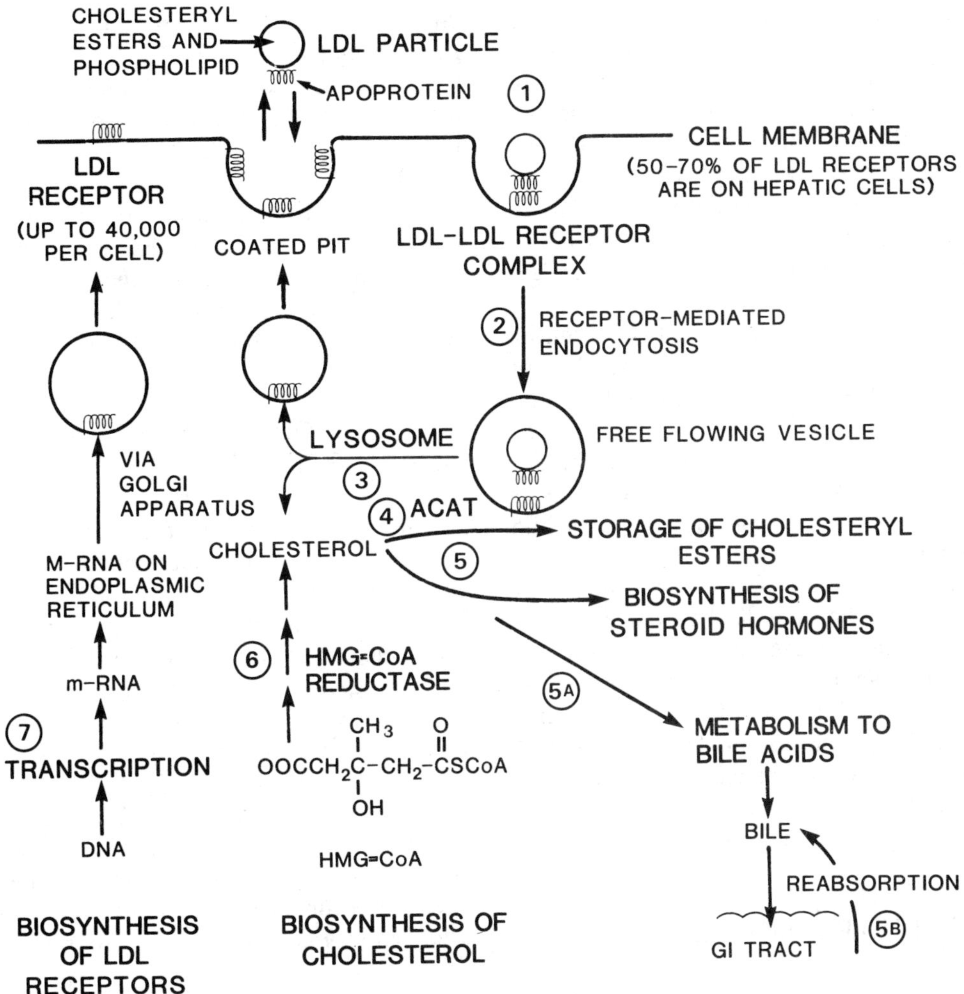

FIG. 18-2. *Cholesterol regulation pathways.*

hormones such as insulin, triiodothyronine (T_3), second messengers, and even nutrients.

Although it is generally believed that most steroid hormones enter cells by passive diffusion, a body of data is emerging that suggests that there may be cell surface receptors for these molecules as well.[36]

LOW-DENSITY LIPOPROTEIN RECEPTORS AND THEIR ROLES IN CHOLESTEROL REGULATION AND ATHEROSCLEROSIS

After discovering the LDL receptor in 1973, Brown and Goldstein unraveled the genetic basis for hypercholesterolemia and resulting heart disease.[33,34] As noted in the previous section, in 1985 they received the Nobel Prize in medicine for their research contributions to the health of mankind. As described in their vividly illustrated 1984 review in *Scientific American*,[34] Brown and Goldstein showed how LDL receptor biosynthesis, cholesterol biosynthesis, and cholesterol metabolism are intimately intertwined in the regulation of cholesterol (see Fig. 18-2).

The LDL particles bind to LDL receptors (step 1) contained in coated pits on the cell surface. The coated pits are composed primarily of a polypeptide called "clathrin" that can spontaneously form these pitlike structures. It has been estimated that at least 1500 to 3000 coated vesicles can form per minute per cell, making up about 2% of the cell surface.

The coated pits "pinch" off after the LDL–LDL receptor complex has formed, a process of receptor-mediated endocytosis (step 2) that leaves free-floating vesicles that transport the LDL inside the cell. Most large molecules circulating in the bloodstream are now known to enter cells by this process, including insulin, triiodothyronine (T_3), and other steroids.[33,35] The LDL is separated from the receptor and delivered to a lysosome filled with digestive enzymes (step 3). The LDL receptor is recycled and delivered to the cell surface. The LDL particle itself is degraded, releasing free cholesterol.

An excess of cholesterol in the cell results in a number of processes to reduce the amount. Acyl-CoA: cholesterol acyltransferase (ACAT) is activated (step 4) to attach fatty acids to the cholesterol, forming cholesterol esters that are stored as droplets in the cell. Hydroxymethylglutaryl (HMG)-CoA reductase, a key rate-limiting regulatory enzyme in cholesterol biosynthesis, is inhibited (step 6) to thereby reduce the cell's own biosynthesis of cholesterol.

In addition, the transcription of the LDL receptor gene is inhibited (step 7), reducing the number of LDL receptors available on the cell surface to transport cholesterol into the cell. There is also evidence that reabsorption of bile acids is reduced from the gastrointestinal tract, increasing the metabolism of cholesterol inside the cell to provide enough bile acids for digestion.

Understanding the role of LDL receptors in cholesterol regulation has resulted in a better understanding of the processes that result in atherosclerosis. In particular, familial hypercholesterolemia is a genetic form of atherosclerosis that affects about 1:500 people in the United States.[34] Individuals with this genetic disease have LDL levels about twice the normal level. The disease results from a defective gene for LDL receptors.

MEMBRANE-BOUND RECEPTORS

One important class of steroid drugs, the cardiac steroids, has a membrane-bound receptor. These steroids never enter the cell. The cardiac steroid receptor is the enzyme Na^+,K^+–ATPase.[37–45] As described in the section on cardiac steroids, these drugs inhibit the Na^+,K^+–ATPase, resulting in a small net increase in intracellular Ca^{2+} ions. Membrane-bound Na^+,K^+–ATPase has had its entire amino acid sequence determined through recombinant DNA techniques.[46] Deffo, Ahmed, Fullerton, and co-workers have recently located the receptor site for cardiac glycosides—the position on the Na^+,K^+–ATPase receptor at which they actually bind.[47]

STEROID HORMONES: REGULATORS OF GENE EXPRESSION

An enormous volume of research has shown that steroid hormones regulate gene expression.[48–59] The recent reviews of Evans,[59a] O'Malley,[59b] Miesfield,[59c] Beato,[59d] and Agarwal[59e] are recommended. Steroid hormones are quite tissue-specific at physiologic levels: estrogens increase uterine cell proliferation, but not that of prostate. Androgens do the reverse, but neither androgens nor estrogens affect stomach epithelium, for example.

Specific receptors (or DNA acceptor sites) found in some tissues, but rare in others, are the source of the tissue selectivity. Steroid hormones activate their receptors, thus enabling them to bind to DNA and regulate gene expression. As noted by O'Malley, "The precise mechanism by which steroid hormone receptors regulates gene expression is unknown, but the four main reactions appear to be: (1) ligand

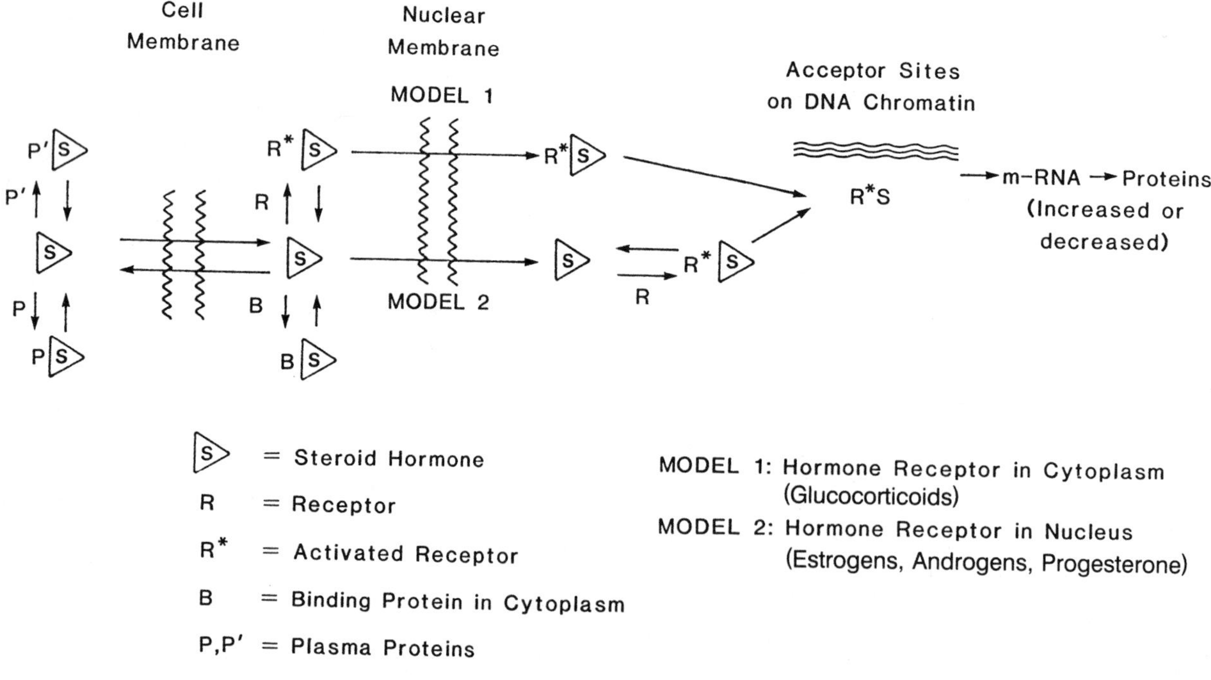

FIG. 18-3. Models for steroid hormone action.

[steroid] induced allosteric activation of the receptor, (2) specific binding [of the steroid-receptor complex] to "steroid response elements—SRE's" [located in genomic DNA adjacent to the target genes], (3) stable complex formation at these DNA enhancer sites, and (4) recruitment of transcription factors and RNA polymerase to initiate transcription of the target genes."[59b] The general process is illustrated in Fig. 18-3.

THE STEROID AND THYROID HORMONE RECEPTOR SUPERFAMILY

The cDNAs for all the primary steroid hormones have been cloned.[59a–f] Parts of the steroid hormone, thyroid hormone, vitamin D_3, and retinoic acid receptors have been found to be structurally very similar. In addition several human steroid hormone receptors contain a segment with a great structural similarity to the viral oncogene *erb*A. This was an extraordinary finding, but in 1986 Weinberger, Sap, and co-workers[60,61] made an even more remarkable discovery. They found that the *erb*A protooncogene product is, in fact, the thyroid hormone receptor. Subsequent research has shown a great commonality between the steroid hormone and thyroid hormone receptors. This steroid–thyroid hormone "superfamily" of receptors has been the focus of fascinating reviews.[59a–f,62]

The receptors have three primary sections:[59a–f] (1) **an N-terminal** portion with quite variable length and amino acid sequence. This section of the receptors appear to *enhance gene activation*, and may add to the selectivity of individual receptors for specific genes; (2) **a 66 amino acid sequence** in the center of the receptor that is responsible for *DNA binding*. It contains nine cysteines, with eight in two zinc "fingers," each with a single zinc atom. The amino acid sequence in the fingers is responsible for the receptor being able to recognize specific DNAs. Slight changes, for example, can convert a glucocorticoid receptor into an estrogen receptor; (3) **a C-terminal portion** that is responsible for *binding of the receptor to the steroid* (or thyroid) hormone. This portion contains two conserved amino acid sequences among the receptors in the superfamily, but as yet the function of these two sequences is not established.

STEROID HORMONE RECEPTORS — IN THE CYTOPLASM OR NUCLEUS?

For a number of years, there was an intense debate on whether steroid hormone receptors were located in the cytoplasm or in the nucleus. Today, a clearer picture has emerged. The glucocorticoid and mineralocorticoid receptors appear to be located primarily in the cytoplasm (Model 1, Fig. 18-3). After binding to a glucocorticoid molecule, the steroid–receptor complex passes through the nuclear membrane. The gonadal (estrogen, androgen, progestin) receptors appear to be located in the nucleus. Binding with the steroid hormone occurs after the steroid has entered the nucleus (Model 2, Fig. 18-3). In some cases, the steroid may actually activate the receptor, possibly by phosphorylation.

Some receptors are believed to act as dimers, with corresponding DNA-binding regions having a complementary (or "dyad") symmetry.[59] For example, the progesterone receptor is known to exist as a dimer of two dissimilar hormone-binding subunits of about 110 kDa and 80 kDa relative molecular mass (M_r). (The progesterone receptor dimer is graphically illustrated in the *Scientific American* review of steroid receptors by O'Malley and Schrader.[27])

Some steroid hormones are even known to control the biosynthesis of other hormones' receptors. For example, estrogens can dramatically stimulate production of progesterone receptor in some tissues.[50]

Sometimes steroid hormones act with other hormones to trigger the same receptor. For example, the glucocorticoid receptor–glucocorticoid steroid complex can activate the growth hormone gene, as can thyroid hormone (T_3). However, growth hormone gene expression is amplified and faster if thyroid hormone and the complex are present together.[58]

X-RAY CRYSTALLOGRAPHY AND STEROID FIT AT THE RECEPTOR

Medicinal chemists traditionally have assumed that there could be no relationship between the conformations of rigid molecules in crystals and their preferred conformations in solution with receptors. However, it is now clear from x-ray crystallography studies of steroids, prostaglandins, thyroid compounds, and many other drug classes that this technique can be a powerful tool in understanding drug action and in designing new drugs.[63–66] The ways in which x-ray crystallography and computer graphics revealed how cardiac steroids interact with their receptors are discussed later in this chapter. The relationship is simple: steroid drugs usually do not have a charge and, as a result, are held to their receptors by relatively weak forces of attraction (see Chap. 2). The same is true for steroid molecules as they "pack" into crystals. In both events, the binding energy is too small to hold any but low-energy conformations. In short, the steroid conformation observed in steroid crystals often is the same or very similar to that at the receptor.

MANY RECEPTORS PER CELL

There are many more than just a single receptor per cell in tissues sensitive to a particular steroid. This is probably not surprising to those who have used Avogadro's number to determine how many molecules are taken in a single dose of any drug. There are typically only 10^3 to 10^4 steroid hormone receptors per cell.[28, 59] However, for steroids that are not hormones, the number may be considerably larger.

For example, there are as many as 5.2 million digitalis receptor sites in a cell of a very digitalis steroid-sensitive tissue, such as cat ventricle, but there are many fewer in a less sensitive tissue (e.g., 100,000 per guinea pig atrium cell).[67]

Changes in steroid receptors and the number of receptors during development and puberty[68] and aging[69] have recently been reported, giving new insight into the roles of steroid hormones during life.

STEROID NOMENCLATURE, STEREOCHEMISTRY, AND NUMBERING

As shown in Figure 18-4, nearly all steroids are named as derivatives of cholestane, androstane, pregnane, or estrane. The standard system of numbering is illustrated with 5α-cholestane.

The absolute stereochemistry of the molecule and any substituents is shown with solid (β) and dashed (α) bonds. Most carbons have one β-bond and one

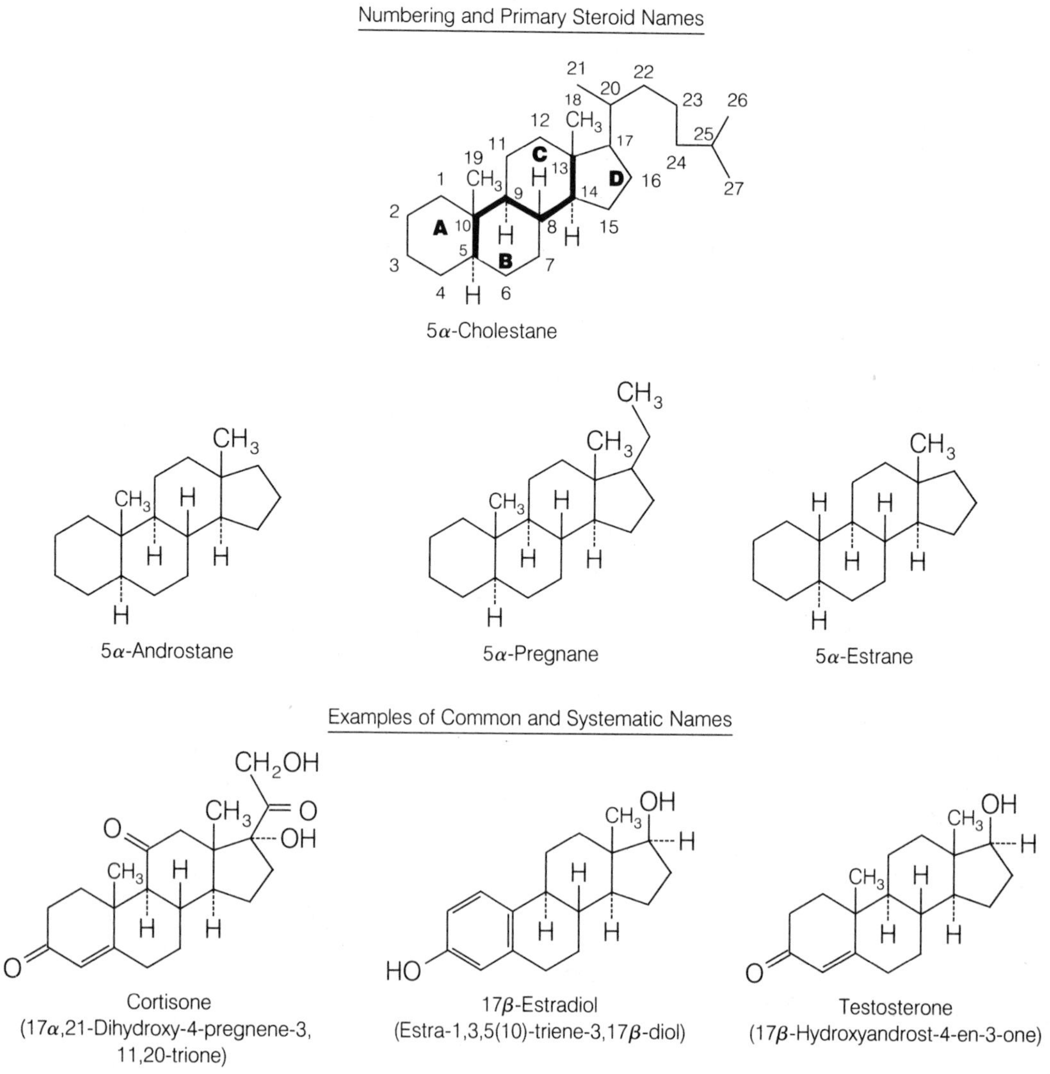

Numbering and Primary Steroid Names

5α-Cholestane

5α-Androstane

5α-Pregnane

5α-Estrane

Examples of Common and Systematic Names

Cortisone (17α,21-Dihydroxy-4-pregnene-3,11,20-trione)

17β-Estradiol (Estra-1,3,5(10)-triene-3,17β-diol)

Testosterone (17β-Hydroxyandrost-4-en-3-one)

FIG. 18-4.

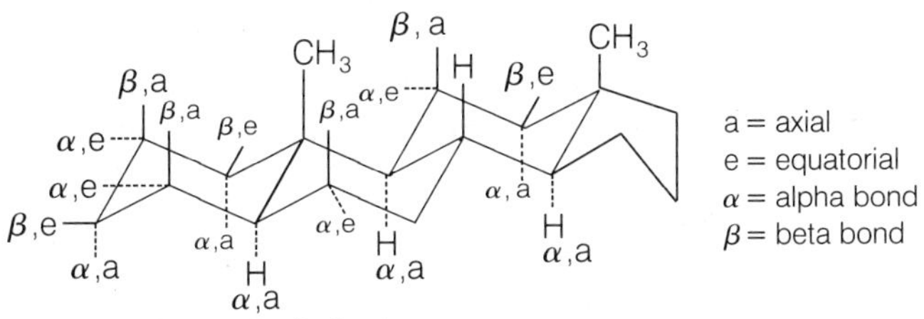

a = axial
e = equatorial
α = alpha bond
β = beta bond

5α-Androstane

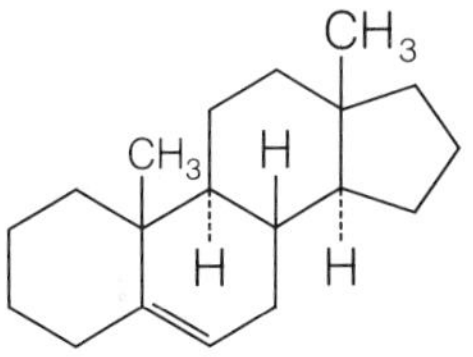

5-Androstene or
Δ^5-Androstene or
Androst-5-ene

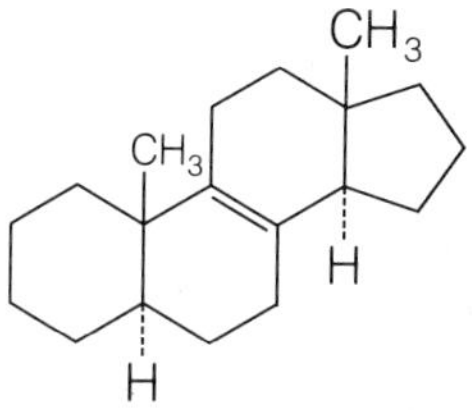

5α-Androst-8-ene or
5α-Δ^8-Androstene

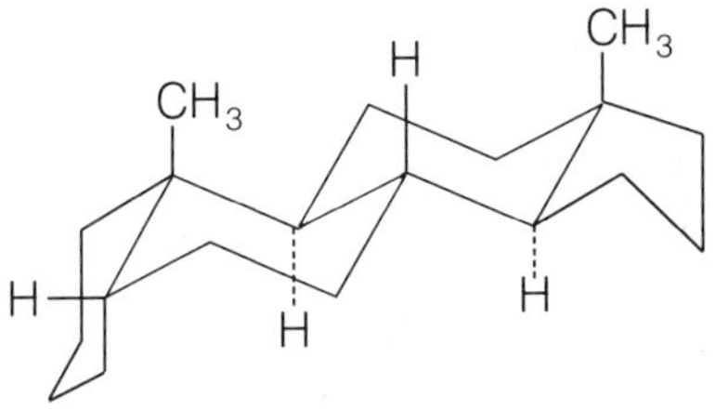

5β-Androstane

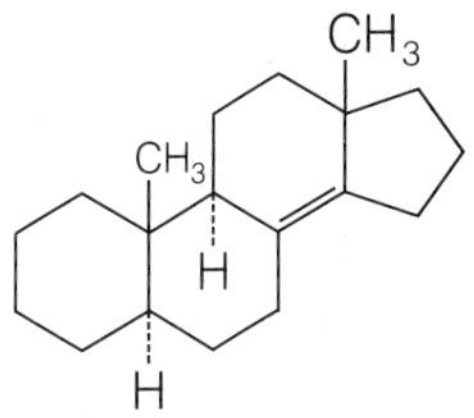

5α-Androst-8(14)-ene or
5α-$\Delta^{8(14)}$-Androstene

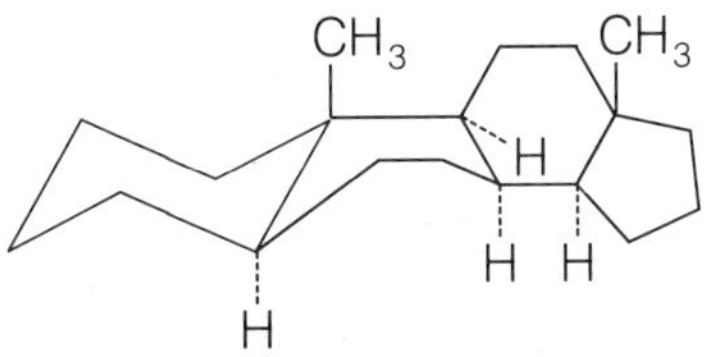

5α,8α-Androstane

α-bond, with the β-bond lying closer to the "top" or C-18 and C-19 methyl side of the molecule. Both α- and β-substituents may be axial or equatorial. This system of designating stereochemistry can best be illustrated using 5α-androstane.

The stereochemistry of the H at C-5 is always indicated in the name. Stereochemistry of the other H atoms is not indicated unless different from 5α-cholestane. Changing the stereochemistry of any of the ring juncture or backbone carbons (shown in Fig. 18-4 with a heavy line on 5α-cholestane) greatly changes the shape of the steroid.

Because of the immense effect that "backbone" stereochemistry has upon the shape of the molecule, IUPAC rules[70] require the stereochemistry at all backbone carbons to be clearly shown. That is, all *hydrogens* along the backbone must be drawn. When the stereochemistry is not known, a wavy line is used in the drawing, and the Greek letter xi (ξ) instead of α or β is used in the name. Methyls are always drawn as CH_3. Some authors also draw hydrogens at C-17.

The position of double bonds can be designated in any of the various ways shown below. Double bonds from C-8 may go toward C-9 or C-14, and those from C-20 may go toward C-21 or C-22. In such cases, both carbons are indicated in the name if the double bond is not between sequentially numbered carbons.

These principles of modern steroid nomenclature are applied to naming several common steroid drugs shown in Figure 18-4.

Such common names as "testosterone" and "cortisone" are obviously much easier to use than the long systematic names. However, substituents must always have their position and stereochemistry clearly indicated when common names are used (e.g., 17α-methyltestosterone, 9α-fluorocortisone).

The terms *cis* and *trans* are occasionally used in steroid nomenclature to indicate the backbone stereochemistry *between* rings. For example, 5α-steroids are A/B *trans*, and 5β-steroids are A/B *cis*. The terms *syn* and *anti* are used analogously to *trans* and *cis* for indicating stereochemistry in bonds *connecting* rings (e.g., the C-9:C-10 bond that connects rings A and C). The use of these terms is indicated below.

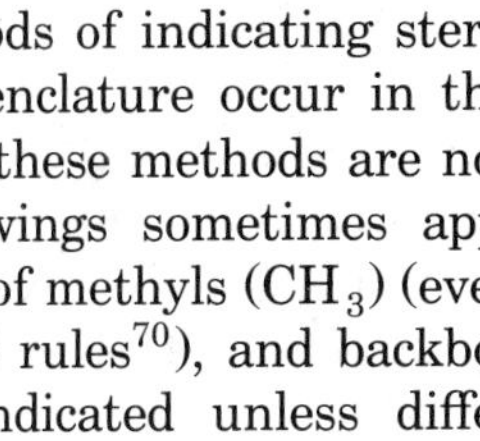

Other methods of indicating steroid stereochemistry and nomenclature occur in the early medical literature, but these methods are now seldom used.

Steroid drawings sometimes appear with lines drawn instead of methyls (CH_3) (even though incorrect by IUPAC rules[70]), and backbone stereochemistry is not indicated unless different from 5α-androstane, as follows.

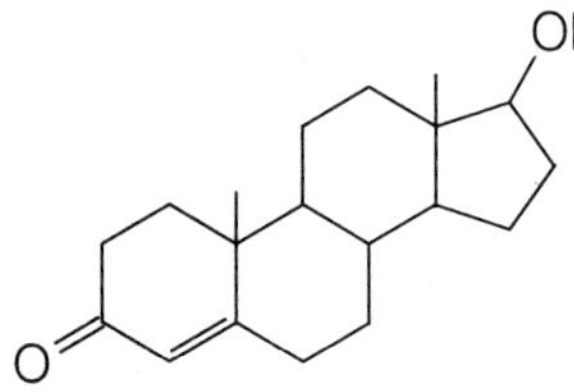

Testosterone

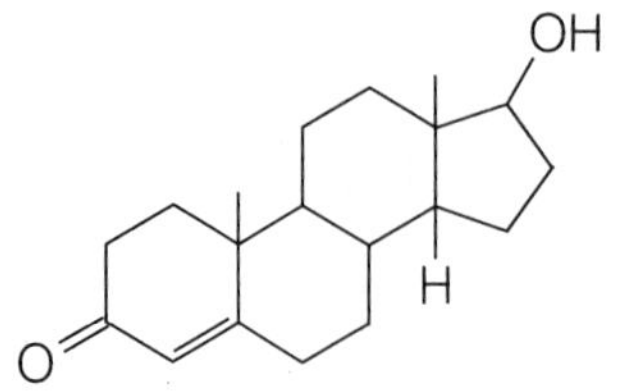

14β-Testosterone

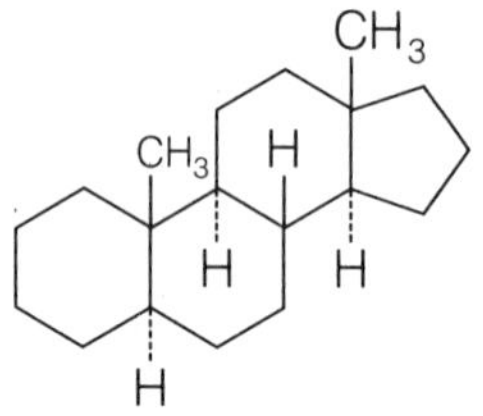

5α-Androstane

Finally, circles were sometimes used to indicate α-hydrogens and dark dots to indicate β-hydrogens.

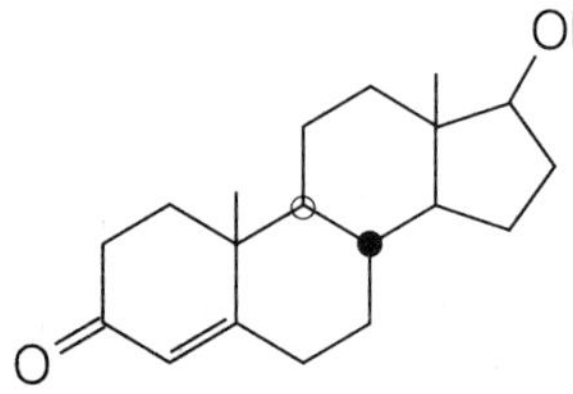

Testosterone

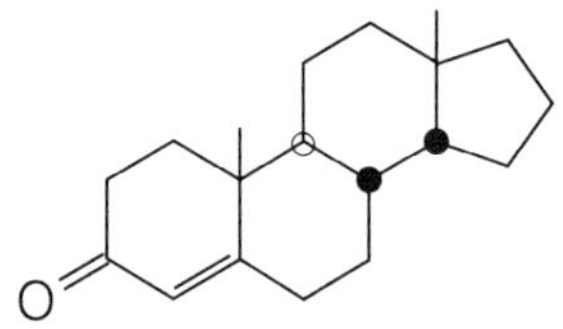

14β-Testosterone

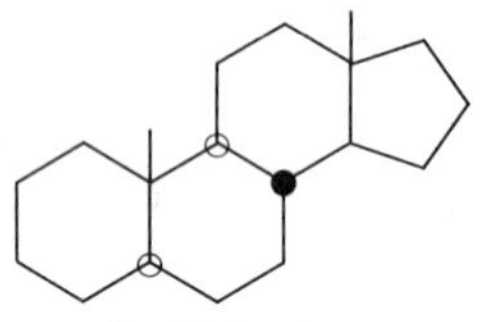

5α-Androstane

STEROID BIOSYNTHESIS

Steroid hormones in mammals are biosynthesized from cholesterol, which in turn is made in vivo from acetyl-coenzyme A (acetyl-CoA). About 1 g of cholesterol is biosynthesized per day in humans, and an additional 300 mg is provided in the diet. (The possible roles of cholesterol and diet in atherosclerosis will be discussed later.) A schematic outline of these biosynthetic pathways is shown in Figure 18-5.

HYDROXYMETHYLGLUTARYL-COA REDUCTASE INHIBITORS

HMG-CoA reductase converts 3-hydroxy-3-methylglutaryl-CoA (HMG-CoA) to mevalonate, a key step in the biosynthesis of cholesterol (see Fig. 18-5). For several decades, inhibition of this enzyme has been envisioned as a likely approach to reducing hypercholesterolemia. Ta-Jyh Lee has recently published a review of this work.[71] This approach gained momentum with the discoveries of Brown and Goldstein about the overall processes of cholesterol regulation in the cell. Mevastatin (Compactin), isolated from cultures of *Penicillium citrinum* and *P. brevicompactum* in the early 1970s, was found to be a very effective inhibitor of HMG-CoA reductase. A few years later, the methyl analogue (named lovastatin) was isolated from cultures of *Monascus ruber* and *Aspergillus terreus*. Numerous other analogues have been synthesized and are being investigated clinically (Fig. 18-6).

Lovastatin is the first HMG-CoA reductase inhibitor approved by the FDA,[72, 73] but many others are currently under study,[74, 74a] including MK-733, a particularly potent analogue (Fig. 18-6). Lovastatin reduces LDL cholesterol up to 41% in patients with genetic hypercholesterolemia.[71–73] Administration of bile acid sequestrants, such as colestipol (Colestid), with lovastatin reduces LDL levels even more than with lovastatin alone. Other drugs used to treat hypercholesterolemia have been recently reviewed[75, 75a, 75b] and are discussed in Chapter 14. The difficulties of accurately measuring plasma cholesterol have added to the difficulties of effective cholesterol management.[76]

Lovastin, USP. 2-Methylbutanoic acid, 1,2,3,7,8,8a-hexahydro-3,7-dimethyl-8-[2-(tetrahydro-4-hydroxy-6-oxo-2*H*-pyran-2-yl)-ethyl]-1-napthaleneyl ester (Mevacor) is the first HMG-CoA reductase inhibitor available clinically. Only about 30% is absorbed from the gastrointestinal tract, and much of this amount is metabolized in the first pass through the liver. The use of lovastin with bile acid sequestrants can lower LDL levels more than lovastin by itself. Reports of possible hepatic toxicity, myositis, and an uncertain relation to cataracts have led the USP DI and *Medical Letter*[72] to recommend liver function tests every four to six weeks during the first months of therapy and annual ophthalmic examinations.

INHIBITION OF OTHER STEPS IN CHOLESTEROL BIOSYNTHESIS

The success of lowering plasma cholesterol with HMG–CoA reductase inhibitors has led to the design of inhibitors of other steps in cholesterol biosynthesis. Inhibitors of cytoplasmic *S*-acetyl coenzyme A synthetase (which activates acetate, a step essential for the cholesterol biosynthetic pathway), mevalonate-5-pyrophosphate decarboxylase (one of the enzymes that convert mevalonate to 3,3-dimethyl pyrophosphate), and squalene synthetase are among those being studied.

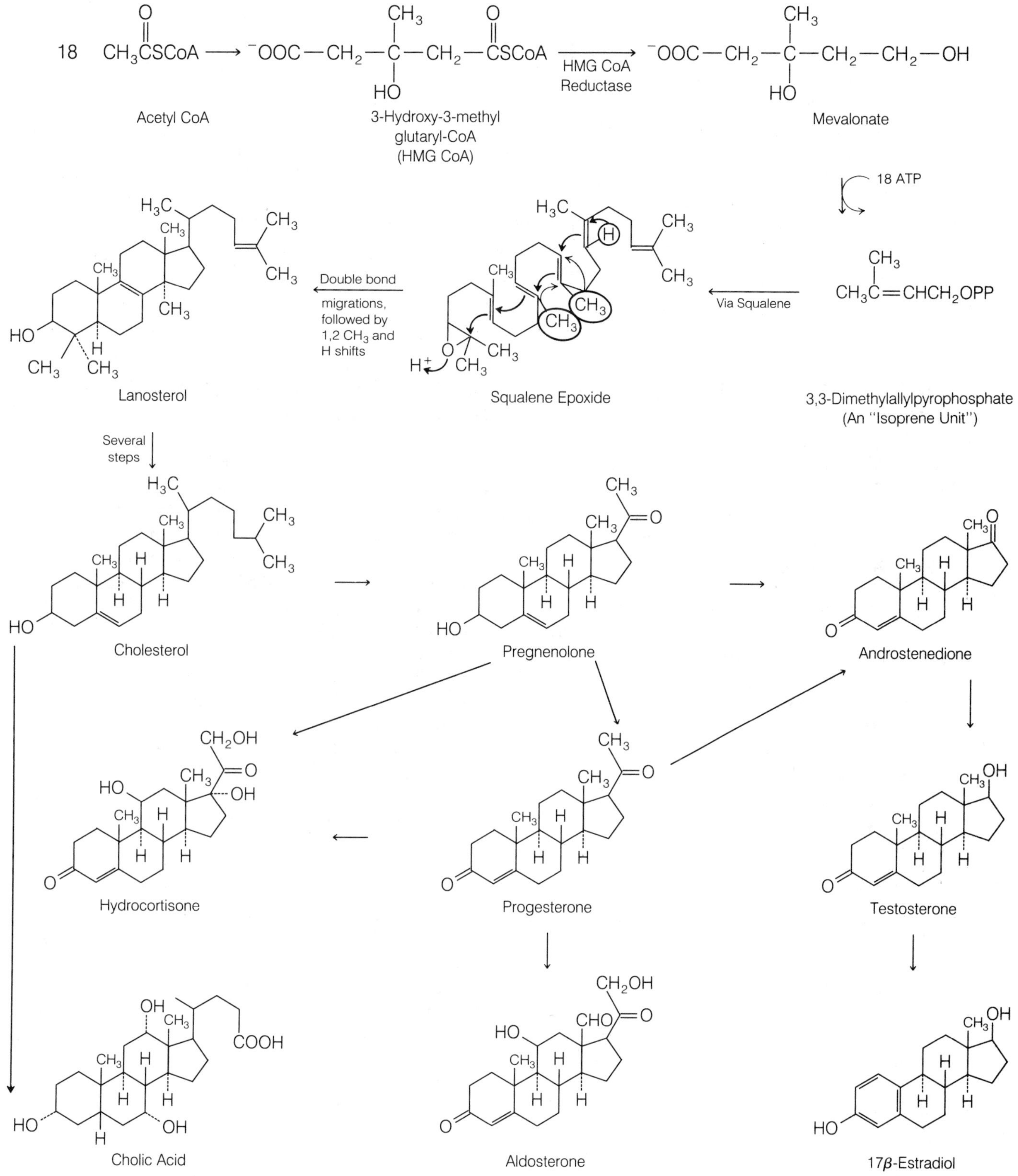

$$18\ CH_3\overset{O}{\overset{\|}{C}}SCoA \longrightarrow {}^-OOC-CH_2-\underset{HO}{\overset{CH_3}{C}}-CH_2-\overset{O}{\overset{\|}{C}}SCoA \xrightarrow[\text{Reductase}]{\text{HMG CoA}} {}^-OOC-CH_2-\underset{HO}{\overset{CH_3}{C}}-CH_2-CH_2-OH$$

Acetyl CoA

3-Hydroxy-3-methyl glutaryl-CoA (HMG CoA)

Mevalonate

18 ATP

$CH_3\overset{CH_3}{C}=CHCH_2OPP$

3,3-Dimethylallylpyrophosphate (An "Isoprene Unit")

Via Squalene

Squalene Epoxide

Double bond migrations, followed by 1,2 CH_3 and H shifts

Lanosterol

Several steps

Cholesterol

Pregnenolone

Androstenedione

Hydrocortisone

Progesterone

Testosterone

Cholic Acid

Aldosterone

17β-Estradiol

FIG. 18-5. *A schematic outline of the biosynthesis of steroids.*

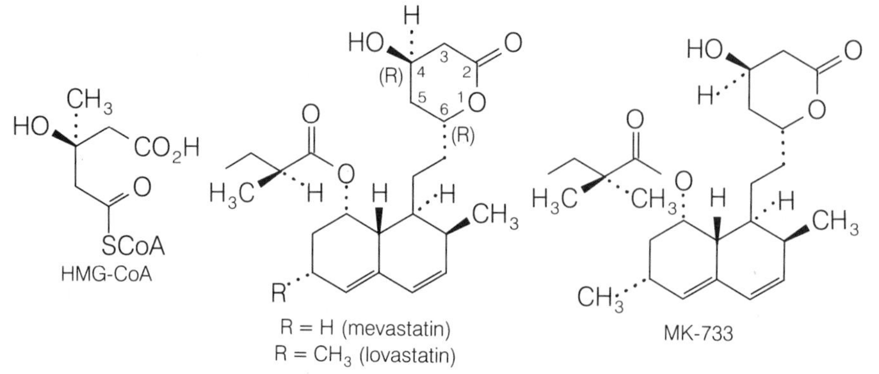

HMG-CoA

R = H (mevastatin)
R = CH_3 (lovastatin)

MK-733

FIG. 18-6. *HMG-CoA and HMG-CoA reductase inhibitors.*

CHEMICAL AND PHYSICAL PROPERTIES OF STEROIDS

With few exceptions, the steroids are white crystalline solids. They may be in the form of needles, leaflets, platelets, or amorphous particles depending upon the particular compound, solvent used in crystallization, and skill and luck of the chemist. As the steroids have 17 or more carbon atoms, it is not surprising that they tend to be water-insoluble. Addition of hydroxyl or other polar groups (or decreasing carbons) increases water solubility slightly as expected. Salts, of course, are the most water-soluble. Examples are shown in Table 18-1.

TABLE 18-1

SOLUBILITIES OF STEROIDS

	Solubility (g/100 mL)		
	$CHCl_3$	EtOH	H_2O
Cholesterol	22	1.2	Insoluble
Testosterone	50	15	Insoluble
Testosterone propionate	45	25	Insoluble
Dehydrocholic acid	90	0.33	0.02
Estradiol	1.0	10	Insoluble
Estradiol benzoate	0.8	8	Insoluble
Betamethasone	0.1	2	Insoluble
Betamethasone acetate	10	3	Insoluble
Betamethasone $NaPO_4$ salt	Insoluble	15	50
Hydrocortisone	0.5	2.5	0.01
Hydrocortisone acetate	1.0	0.4	Insoluble
Hydrocortisone $NaPO_4$ salt	Insoluble	1.0	75
Prednisolone	0.4	3	0.01
Prednisolone acetate	1.0	0.7	Insoluble
Prednisolone $NaPO_4$ salt	0.8	13	25

CHANGES TO MODIFY PHARMACOKINETIC PROPERTIES OF STEROIDS

As with many other compounds described in previous chapters, the steroids can be made more lipid-soluble or more water-soluble simply by making suitable ester derivatives of hydroxyl groups. Derivatives with increased lipid solubility are often made to decrease the rate of release of the drug from intramuscular injection sites (i.e., in depot preparations). More lipid-soluble derivatives also have improved skin absorption properties, and thus are preferred for dermatologic preparations. Derivatives with increased water solubility are needed for intravenous preparations. As hydrolyzing enzymes are found throughout mammalian cells, especially in the liver, converting hydroxyl groups to esters does not significantly modify the activity of most compounds.

Some steroids are particularly susceptible to rapid metabolism after absorption or rapid inactivation in the gastrointestinal tract before absorption. Often a simple chemical modification can be made to decrease these processes and, thereby, increase the drug's half-life—or make it possible to be taken orally.

Examples of common chemical modifications are illustrated in Figure 18-7. Drugs such as testosterone cyclopentylpropionate and methylprednisolone sodium succinate (which are converted in the body to more active drugs) are called *prodrugs*.

Counsell and co-workers have given particular attention to the tissue distribution of steroids and the implication of such in drug design. For example, it has long been known that cholesterol is found in the highest concentration in the adrenal gland; therefore, [^{131}I]19-iodocholesterol is now used therapeutically for the diagnosis of various adrenal cortical diseases.[77,78] Radioactive steroids, for many years, have been recognized as binding most selectively to tissues that respond to them; consequently, labeled

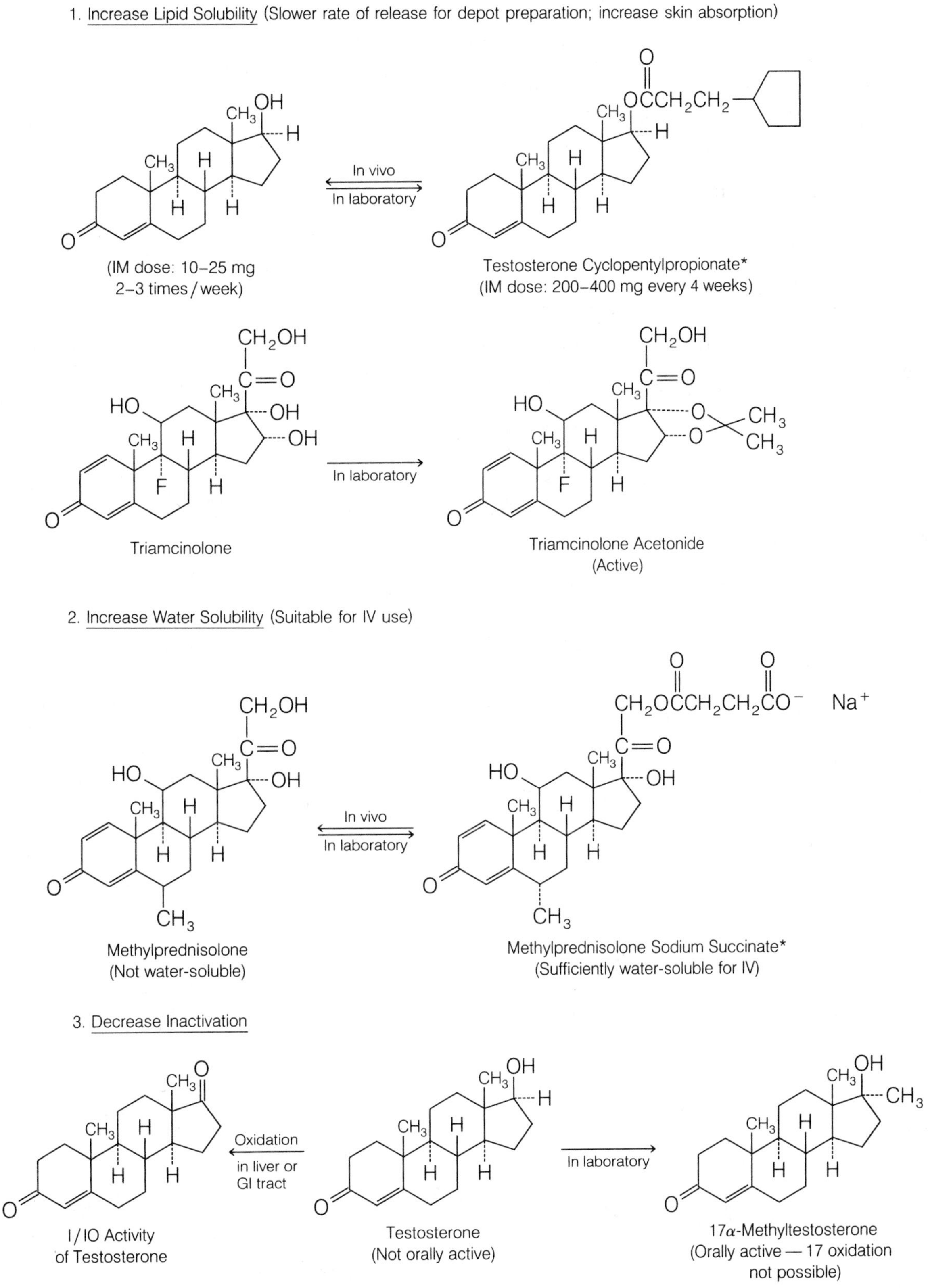

1. Increase Lipid Solubility (Slower rate of release for depot preparation; increase skin absorption)

2. Increase Water Solubility (Suitable for IV use)

3. Decrease Inactivation

FIG. 18-7. *Common steroid modifications to alter therapeutic utility: *, prodrug.*

steroids have been used for many receptor and tissue studies.

Drugs with high affinity for the adrenal glands or other hormone-synthesizing tissues also have been studied as potential blockers of biosynthetic pathways (e.g., to block the biosynthesis of cholesterol in hyperlipidemia and heart disease, or the biosynthesis of excessive hormones from cholesterol in adrenal gland diseases).

GONADOTROPINS

The gonadotropins are peptides that in vivo and in therapy are closely related to the steroids estrogens, progesterone, and testosterone. As shown in Figures 18-8 through 18-11, they control ovulation, spermatogenesis, development of sex organs, and maintain pregnancy. Included are the following:

- *Luteinizing-releasing hormone* (LH-RH or LRH), also called *gonadotropin-releasing hormone* (GnRH).
- *Leuteinizing hormone* (LH).
- *Follicle-stimulating hormone* (FSH).
- *Menotropins*, also called *human menopausal gonadotropin* (hMG), a purified preparation of FSH and LH obtained from the urine of postmenopausal women.
- *Chorionic gonadotropin* (CG; hCG is *human gonadotropin*), a glycopeptide produced by the placenta. Its pharmacologic actions are essentially the same as LH.

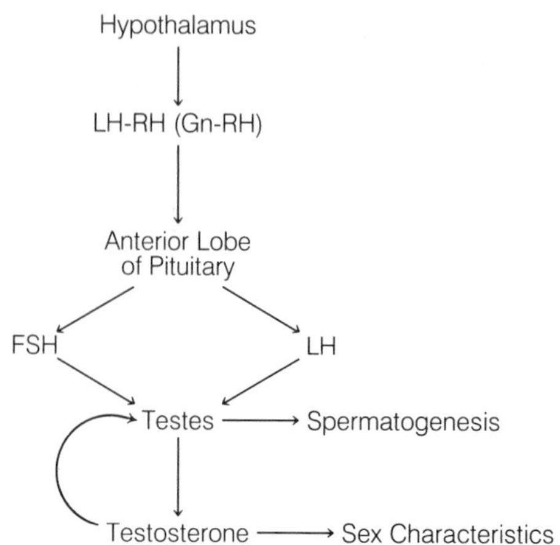

FIG. 18-9. Regulation of spermatogenesis in males by LH-RH (Gn-RH).

LUTEINIZING-RELEASING HORMONE ROLES IN MALES AND FEMALES

The hypothalamus releases LH-RH (GnRH), a decapeptide that stimulates the anterior pituitary to secrete LH and FSH in males and in females. This peptide controls and regulates both male and female reproduction (see Figs. 18-8 and 18-9). Its isolation, purification, and structure determination were achieved in the early 1970s by two groups headed by Guillemin and Schalley, an accomplishment that earned them a Nobel Prize in 1977. LH-RH is a decapeptide (10 amino acids):

$$\text{Glu-His-Trp-Ser-Tyr-Gly-Leu-Arg-Pro-Gly-NH}_2$$

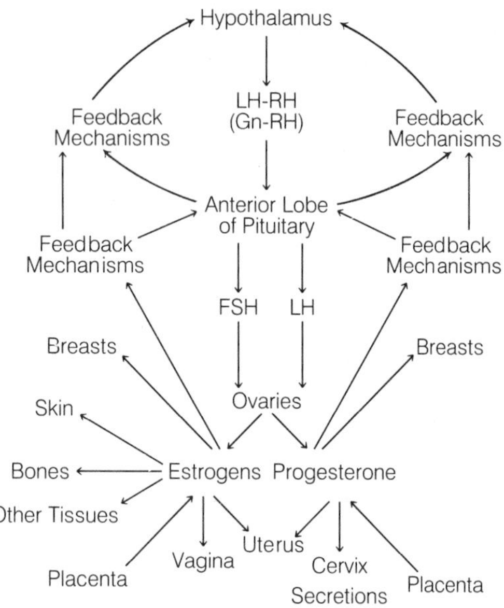

FIG. 18-8. Regulation of ovulation in females by LH-RH (GnRH).

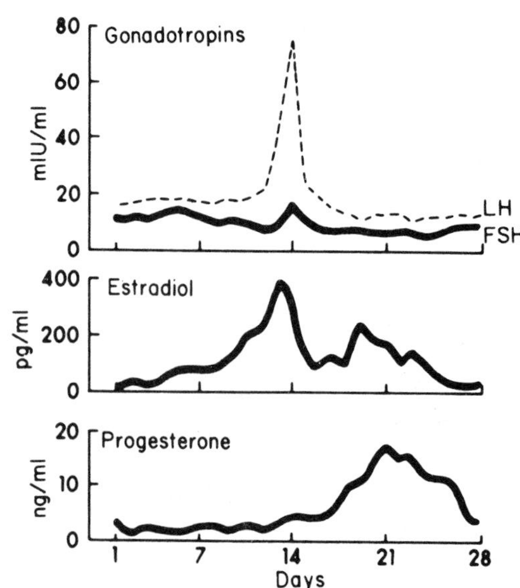

FIG. 18-10. Hormone changes in the normal menstrual cycle.

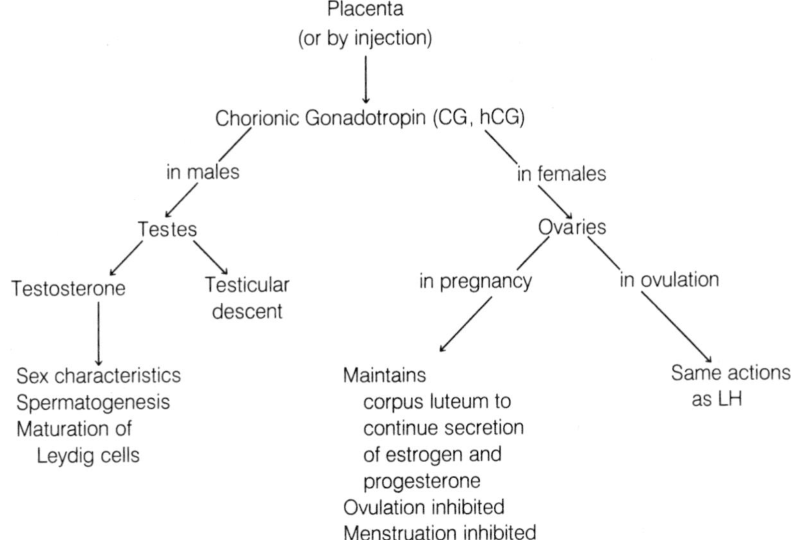

FIG. 18-11. *Natural and therapeutic roles of chorionic gonadotropin (CG) in females and males.*

Because of its simple structure and key roles in reproduction, there has been an enormous amount of interest in development of analogues as medicinal agents. Well over 1000 LH-RH analogues have now been synthesized.[79]

THE PITUITARY GONADOTROPINS — LH AND FSH

The pituitary gonadotropins LH and FSH, their structures, genes, receptors, biological roles, and their regulation (including by negative feedback actions of steroid hormones) have been intensively studied.[79a, b, c] FSH, LH, CG, and TSH (thyroid stimulating hormone) all are peptide dimers with similar α-subunits within a species, but different β-subunits. Each peptide unit bears one or more carbohydrate chains.

In females, LH and FSH regulate the menstrual cycle (see Figs. 18-8 and 18-10). At the start of the cycle (with day 1 being the first day of menstruation), plasma concentrations of estradiol and other estrogens (see Fig. 18-10) and progesterone are low. Follicle-stimulating hormone and LH stimulate several ovarian follicles to enlarge and begin developing more rapidly than others. After a few days, only one follicle continues the development process to the release of a mature ovum. The granulosa cells of the maturing follicles begin secreting estrogens, which then cause the uterine endometrium to thicken. Vaginal and cervical secretions increase. Gonadotropins and estrogen reach their maximum plasma concentrations at about the 14th day of the cycle. The release in LH causes the follicle to break open, releasing a mature ovum. Under the stimulation of LH, the follicle changes into the corpus luteum, which begins secreting progesterone as well as estrogen.

The increased concentrations of estrogens and progesterone regulate the hypothalamus and the anterior pituitary by a feedback inhibition process such that LH-RH, LH, and FSH production is diminished. The result is that further ovulation is inhibited. As described later in this chapter, this is the primary mechanism by which the steroid birth control products inhibit ovulation.

If fertilization does not occur by about day 25, the corpus luteum begins to degenerate, slowing down its production of hormones. The concentrations of estrogens and progesterone become too low to maintain the vascularization of the endometrium, and menstruation results.

In males (see Fig. 18-9), LH stimulates testosterone synthesis by the testes, and together testosterone and LH promote spermatogenesis (sperm production) and development of the testes. Testosterone is also essential for development of secondary sex characteristics in males. (Sharpe concludes that FSH is probably necessary for quantitative maintenance of spermatogenesis, but artificially high levels of testosterone may overcome the requirement.[79c])

HUMAN CHORIONIC GONADOTROPIN ROLES IN MALES AND FEMALES

Human gonadotropin (hCG) was originally commercially obtained from the urine of pregnant women, but more recently has been produced by recombinant DNA cloning techniques.[80] It is made up of two subunits (α and β), with the α-subunit structurally identical to the α-subunits of LH and FSH. The pharmacologic actions of hCG are essentially the same as LH. In females during pregnancy, the hCG secreted by the placenta maintains the corpus luteum to continue secretion of estrogen and progesterone, with the results that ovulation and menstruation are inhibited.

THERAPEUTIC APPLICATIONS

Pulsatile LH-RH (Gn-RH) Therapy to Induce Ovulation

Pulsatile subcutaneous administration of synthetic LH-RH (Gonadorelin) has been used in clinical trials to induce ovulation and pregnancy.[79,81,82] When the hypothalamus fails to produce enough LH-RH to stimulate the pituitary to produce sufficient LH and FSH, the result is often anovulatory infertility. LH-RH is released naturally by the hypothalamus in a pulsatile manner. When LH-RH is given continuously, for example, it *inhibits* pituitary production of LH and FSH. Mechanical pulsatile release of LH-RH, however, has been shown to be an effective means of treating infertility in women with ovarian dysfunction caused by insufficient LH and FSH production. Although a number of LH-RH analogues are more potent agonists than LH-RH itself, the increased potency has not been advantageous in inducing ovulation.

Clomiphene, a potent antiestrogen discussed later in this chapter, increases the LH-RH production in the hypothalamus—inducing ovulation in 80% of patients.[82a]

Pregnancy Tests Based on Urine Analysis for Human Chorionic Gonadotropin

Tests for Clinical Laboratory Professional Use

Bioassays for hCG were first developed in 1928. Since then, a wide variety of agglutination, antibody, and radioimmunoassay test kits have been developed for professional use.[83] They continue to be the mainstay of pregnancy diagnosis used in hospitals and for general clinical use. A recent multicenter trial with a variety of commercially available testing methods has shown that the tests are "capable of diagnosing pregnancy at the time of a missed period [equivalent to 14 days post conception]... However, routine clinical laboratories cannot diagnose pregnancy consistently before 14 days post conception. . . ."[84]

The great sensitivity of new radioimmunoassays (RIAs) for the β-subunit of hCG has made these β-hCG tests particularly useful for diagnosis of ectopic pregnancies. Ectopic pregnancies generally do not produce enough hCG to consistently result in positive tests by other analytical methods.[85,86]

Home Pregnancy Test Kits[83,87]

The nonradiological methods in professional use for hCG have been adopted for home use. Today, a wide range of products in available (marketed under trade names including Acu-Test, Advance, Answer, Daisy 2, e.p.t., e.p.t. Plus, Fact, Predictor). Several (including Advance and e.p.t. Plus) are based on assay of the α- or β-hCG subunit. One study[87] used three brands (Daisy 2, e.p.t., and Answer) and showed that the predictive value of positive test results (all positive test results minus false-positive results) were generally in the range of 78% to 84%, and the predictive value of negative test results (all negative test results minus pregnancies the tests did not detect) were in the range of 60%.

A recent review[83] has summarized the causes of false-positive results (including menstrual irregularities, menopause, and hCG-producing tumors, and a variety of drugs—phenothiazines, methadone, and penicillin among others) and of false-negative results (technical errors, reagent outdating, test taken too early after the missed period, and too dilute urine samples).

In short, the convenience and predictive value of positive tests has resulted in home test kits' wide popularity and use. However, because of the incidence of some false-positive and false-negative test results, it must be emphasized that *a physician should be consulted after positive results as well as after negative results with an overdue menstrual cycle*.[83]

Ovulation Tests Based on Analysis of Urinary Luteinizing Hormone

Numerous techniques have been claimed to predict ovulation, ranging from measurement of basal body temperature (BBT) to transvaginal ultrasound imaging, estradiol assays, and assays of gonadotropins in serum and urine.[88] These tests are used for a variety of diagnoses, including to assist in predicting ovulation initially evaluated as infertile. Radioimmunoassay (RIA) analyses have recently been developed for the LH plasma and urinary surge that takes place midcycle. More recently, home test kits have been used on nonradiologic enzymatic immunoassay (EIA) that uses the LH monoclonal antibody (marketed under trade names including First

Response, OvuStick, and QTEST for Ovulation). Preliminary studies[89] have shown an 88% to 94% accuracy in predicting ovulation when urine samples were taken between 11:00 AM and 3:00 PM, or 5:00 PM and 10:00 PM. The home test kits also differ in sensitivity, such that with one kit, the apparent LH surge seemed to occur up to 5 days before the actual day of ovulation.[88]

LH-RH for Assessing Hypothalamic–Pituitary–Gonadotropic Function

When given as a bolus subcutaneously (SC) or by intravenous (IV) injection, LH-RH will stimulate the anterior pituitary to release LH and FSH. Plasma LH levels are then measured by radioimmunoassay. Subnormal production of LH has been shown to correlate in about 70% of cases of documented disorders of the hypothalamic–pituitary–gonadal system in men and women. Women are given the test during the early follicular phase (days 1 to 7) of their menstrual cycles. The test must be given in the absence of other drugs that affect pituitary secretion of LH and RH, with birth control steroids the most obvious example, as well as glucocorticoids, phenothiazines, dopamine, and spironolactone. The test is contraindicated during pregnancy; hence caution must be taken with patients who are amenorrheic.

Chorionic Gonadotropin with Menotropins to Treat Infertility

Human chorionic gonadotropin is used with menotropins (hMG) to induce spermatogenesis in men with primary or secondary hypogonadotropic hygonadism. hCG is used in women to induce ovulation in patients for whom the anovulation is the result of pituitary insufficiency. Thyroid, adrenal, pituitary, or ovarian tumors may also be the cause of anovulation and should be ruled out before hCG therapy. The most common side effect in women is temporary ovarian enlargement that lasts up to two to three weeks.

Chorionic Gonadotropin or LH-RH to Cause Testicular Descent in Prepubertal Males

Chorionic gonadotropin treatment for young boys is used to treat prepubertal cryptorchidism not caused by anatomic obstruction. Cryptorchidism occurs with about 0.8% of boys. Internasal LH-RH therapy has also been very effective in treatment, with nearly an 80% success rate.[90]

LH-RH Agonists in Treatment of Prostate and Breast Cancers

As noted earlier, the decapeptide structure of LH-RH has made it a relatively easy target for synthetic modification. From among the thousands of analogues synthesized, a number of agonists have offered enough promise to enter clinical trials.[79]

Whereas low doses of some LH-RH analogues (e.g., leuprolide acetate), stimulate gonadal function if given by pulse therapy, higher continuous dosages greatly decrease gonadal function—including the medicinal equivalent of castration in men. The cause is receptor desensitization, also known as receptor down-regulation. The resultant drop in testosterone production has been shown to significantly aid the treatment of prostatic cancer, particularly when given with the antiandrogen flutamide.[79a–c] By a similar process in women, these agonist analogues cause a substantial drop in circulating estrogen levels, an important adjunct in treatment of breast cancer. They may also have a direct antitumor effect on breast cancers.[79,90]

LH-RH Agonists in the Treatment of Precocious Puberty in Boys and Girls

The reversible, gonadal suppressive actions of daily dosages of LH-RH agonist analogues has been used to treat precocious puberty in both sexes.[79] Leuprolide acetate is used by injection for this purpose.

LH-RH Agonists and Antagonists in Contraception

For a time, LH-RH analogues were viewed as certain future alternatives for steroid birth control products. Although an immense amount of research on these analogues has been completed, much more work needs to be done and major problems remain.[79] Development of orally effective formulations has been difficult, not unexpected for a drug that is a peptide. Although spermatogenesis has been suppressed by LH-RH agonists, libido and testosterone levels have also been unacceptably reduced. Whereas many agonists have been found with greater potency than LH-RH itself, fewer effective antagonists have been found. This has limited the progress, to date, for use of LH-RH antagonists in contraception.

SEX HORMONES

Although the estrogens and progesterone are usually called female sex hormones and testosterone is called a male sex hormone, it should be noted that all these steroids are biosynthesized in *both* males and females. For example, an examination of the biosynthetic pathway in Figure 18-5 will reveal that progesterone serves as a biosynthetic precursor to cortisone and aldosterone and, to a lesser extent, to testosterone and the estrogens. Testosterone is one of the precursors of the estrogens. However, the estrogens and progesterone are produced in much larger amounts in females, as is testosterone in males. These hormones play profound roles in reproduction, in the menstrual cycle, and in giving women and men their characteristic physical differences.

A larger number of synthetic or semisynthetic steroids having biologic activities similar to those of progesterone have been made, and these are commonly called progestins. Several nonsteroidal compounds also have estrogenic activity. Although the estrogens and progestins have had their most extensive use as chemical contraceptive agents for women, their wide spectrum of activity has given them many therapeutic uses in both women and men.

Testosterone has two primary kinds of activities —androgenic (or male physical characteristic-promoting) and anabolic (or muscle-building). Many synthetic and semisynthetic androgenic and anabolic steroids have been prepared. Much interest has focused on the preparation of anabolic agents (e.g., for use in aiding recovery from debilitating illness or surgery). However, the androgenic agents do have some therapeutic usefulness in women (e.g., in the palliation of certain sex organ cancers).

In summary, it can be said that although many sex hormone products have their greatest therapeutic uses in either women or in men, nearly all have some uses in both sexes. Nevertheless, the higher concentrations of estrogens and progesterone in women and of testosterone in men cause the development of the complementary reproductive systems and characteristic physical differences of women and men.

ESTROGENS

STRUCTURAL CLASSES

As shown in Figure 18-12, three structural classes of estrogens are used today.

Human Estrogens (Estradiol, Estrone, Estriol, and Derivatives)

These estrogens are found in humans and most other mammals. As will be discussed later in this chapter, estrone and estriol are metabolites of estradiol in vivo (Fig. 18-13). Estrone is about one-third as active as estradiol, and estriol about one-sixtieth as active. The addition of a 17α-alkyl group blocks metabolism to estrone. Ethinyl estradiol is, therefore, very effective orally, whereas estradiol, itself, is not. They and their derivatives, such as ethinyl

1. Human Estrogens and Derivatives

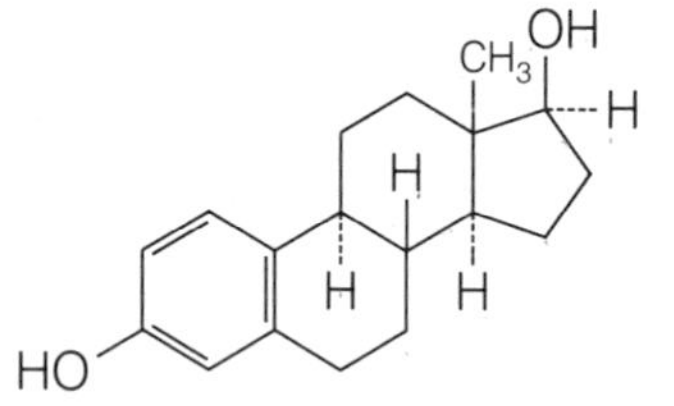

Estradiol
(IM and implantation pellets)

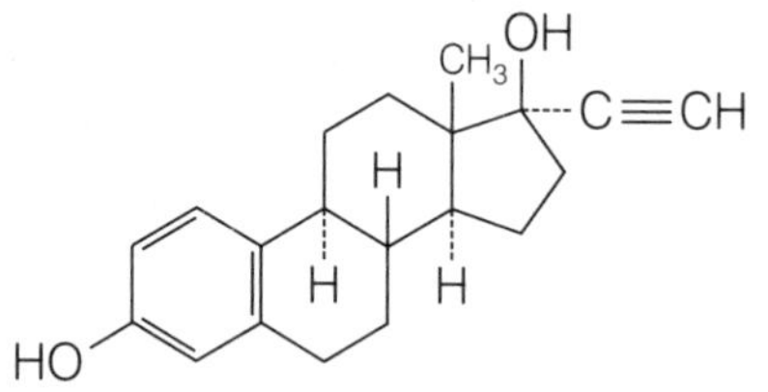

Ethinyl Estradiol
(Oral)

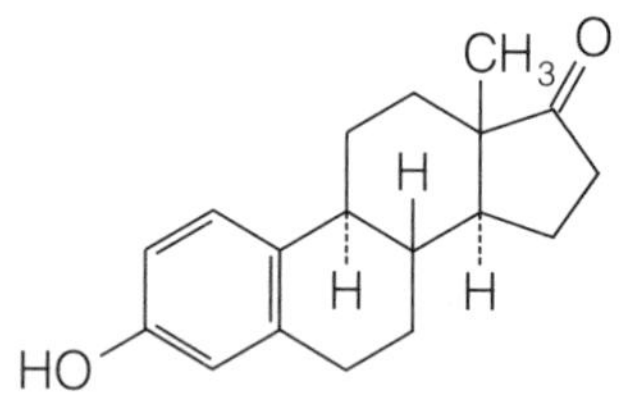

Estrone
(IM, vaginal, topical)

Esters for IM
Estradiol 3-benzoate
Estradiol 3,17β-dipropionate
Estradiol 17β-cyclopentylpropionate

Ethers for oral use
Ethinyl Estradiol 3-methylether (Mestranol)
Ethinyl Estradiol 3-cyclopentyl ether (Quinestrol)

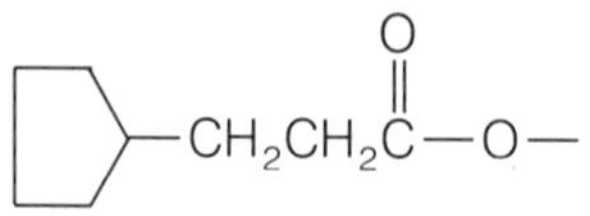

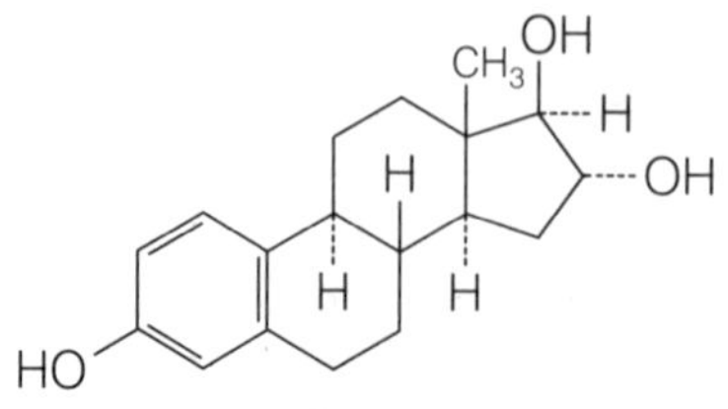

C_5H_9—$CH_2CH_2C(=O)$—O—

Estriol
(Oral)

2. Equine Estrogens (Oral, IM, topical, vaginal)

(a) Conjugated Estrogens: 50–65% Sodium Estrone Sulfate
20–35% Sodium Equilin Sulfate
plus nonestrogenic compounds

(b) Esterified Estrogens: 70–85% Sodium Estrone Sulfate
6.5–15% Sodium Equilin Sulfate
plus nonestrogenic compounds

Estrone Sodium Sulfate

Equilin Sodium Sulfate

Equilenin

Other salt available
Piperazine Estrone Sulfate

3. Synthetic Estrogens (Oral, IM, topical, vaginal)

Diethylstilbestrol

Dienestrol

Chlorotrianisene

Benzestrol

FIG. 18-12. *Natural and synthetic estrogens.*

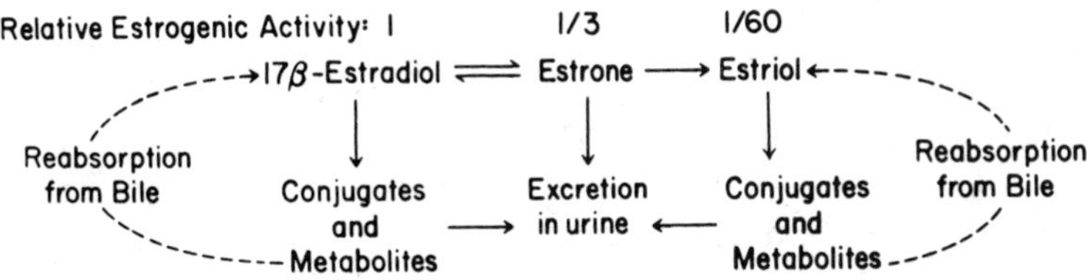

FIG. 18-13. *Interconversion and metabolism of natural estrogens.*

estradiol, all are produced semisynthetically from natural precursors such as diosgenin or cholesterol.

Equine Estrogens (Conjugated Estrogens, Esterified Estrogens)

Equine estrogens are estradiol metabolites originally obtained from the urine of horses, especially pregnant mares. Today, they are produced semisynthetically from estrogen intermediates synthesized from diosgenin and other natural precursors. Little or no equilin and equilenin are produced in humans. Equine estrogens are largely mixtures of estrone sodium sulfate and equilin sodium sulfate.

Diethylstilbestrol Derivatives

At first glance, it might be surprising that nonsteroidal molecules, such as diethylstilbestrol (DES), could have the same activity as estradiol or other estrogens. However, DES can be viewed as a form of estradiol with rings B and C open, and a six-carbon ring D. The activity of DES analogues was explained in 1946 by Schuler.[91] He proposed that the distance between the two DES phenol OH groups was the same as the 3-OH to 17-OH distance of estradiol; therefore, they could both fit the same receptor. Modern medicinal chemists have shown the OH-to-OH distance in DES is actually 12.1 Å, and 10.9 Å in estradiol. However, in water solution (or plasma), estradiol has two water molecules hydrogen bound to the 17-OH. If one of the two water molecules is included in the distance measurement, there is a perfect fit with the two OH groups of DES (Fig. 18-14). This suggests that water may have an important role for estradiol in its receptor site.

Thousands of DES analogues have been synthesized, and from them emerged many products including dienestrol, chlorotrianisene, and benzestrol. As long as the OH-to-OH distance relationship is maintained, significant estrogenic activity is usually found in the DES derivative. Without the central double bond and two ethyl or other alkyl groups, the molecule loses all its rigidity and shape, the OH-to-OH distance is not fixed, and activity is abolished (Fig. 18-15). Reduction of the double bond of DES results in two diastereomers of hexestrol. The *meso* form is active because the OH-to-OH distance is maintained. However in the *threo* isomer, there is steric repulsion of the two ethyl groups. The two phenol groups rotate to relieve the repulsion, the OH-to-OH distance is changed, and consequently, the *threo* isomer is inactive (see Fig. 18-15).

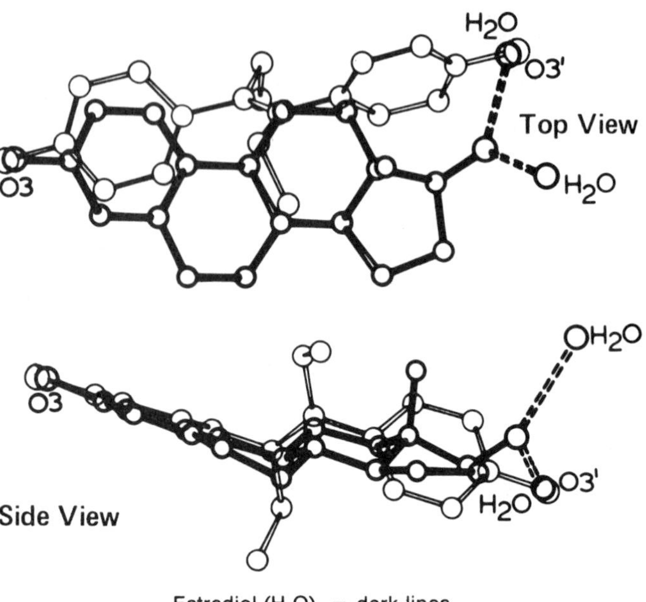

FIG. 18-14. *Computer graphics. Superposition of estradiol $(H_2O)_2$ (dark lines) with DES (light lines). (Drawing courtesy of Medical Foundation of Buffalo, Inc.)*

BIOSYNTHETIC SOURCES

The estrogens are normally produced in relatively large quantities in the ovaries and placenta, in lower amounts in the adrenal glands, and in trace quantities in the testes. About 50 to 350 μg/day of estradiol are produced by the ovaries (especially the corpus luteum) during the menstrual cycle. During the first months of pregnancy, the corpus luteum produces larger amounts of estradiol and other estrogens, whereas the placenta produces most of the circulating hormone in late pregnancy. During pregnancy, the estrogen blood levels are up to 1000 times higher than during the menstrual cycle.

BIOLOGIC ACTIVITIES OF ESTROGEN

In addition to having important roles in the menstrual cycle (described earlier), the estrogens and, to a lesser extent, progesterone are largely responsible for the development of secondary sex characteristics in women at puberty.

The estrogens cause a proliferation of the breast ductile system, and progesterone stimulates development of the alveolar system. The estrogens also stimulate the development of lipid and other tissues that contribute to breast shape and function. Pituitary hormones and other hormones are also involved. Fluid retention in the breasts during the

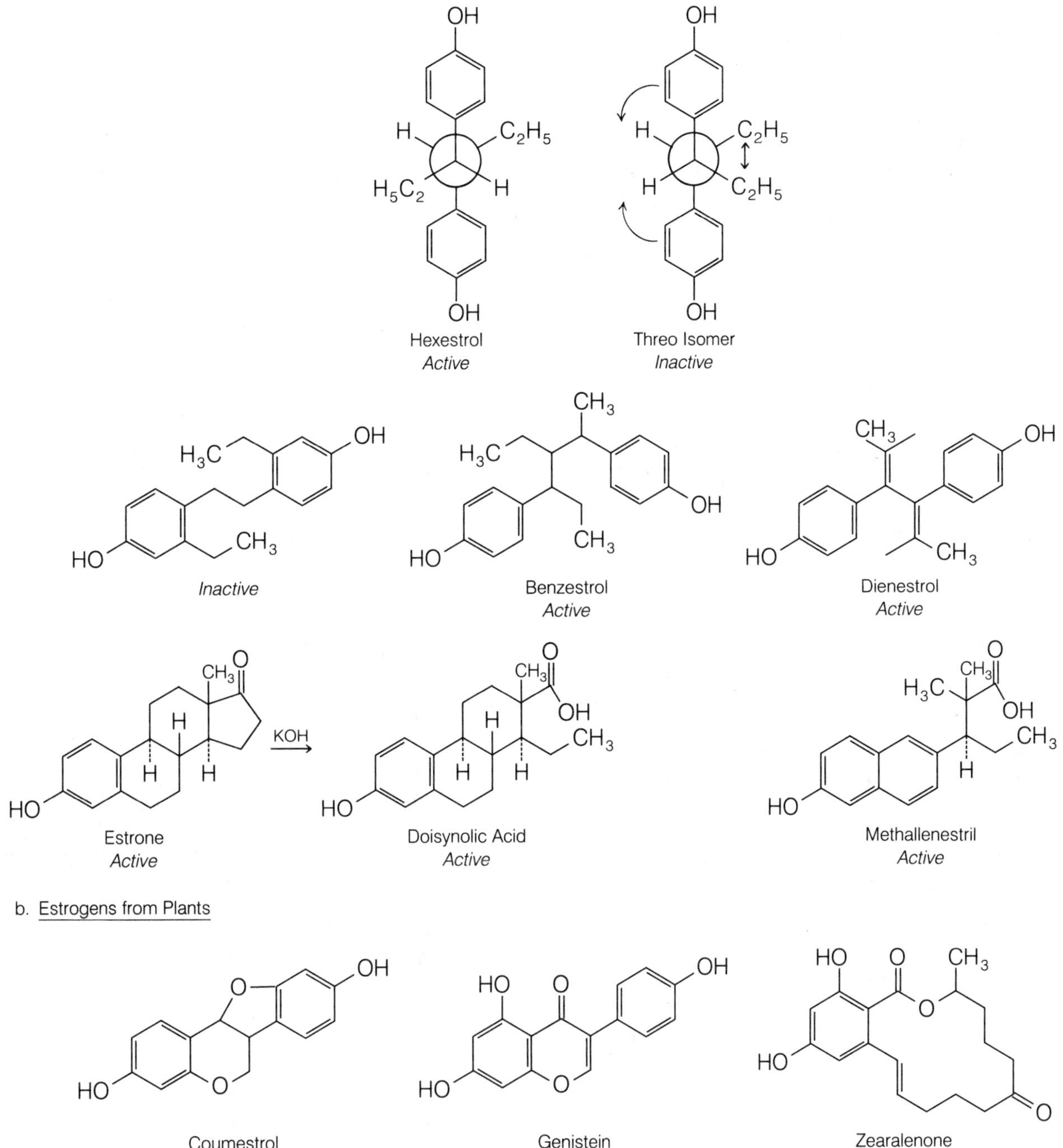

FIG. 18-15. *Examples of synthetic estrogens and estrogens from plants.*

later stages of the menstrual cycle is a common effect of the estrogens. Interestingly, the breast engorgement that occurs after childbirth (stimulated by prolactin, oxytocin, and other hormones) can be suppressed by administration of estrogen—probably owing to feedback inhibition of the secretion of pituitary hormones.

Murad and Kuret nicely summarized the important role of the estrogens in puberty in young women: "The estrogens ... go a long way toward accounting for the attributes called femininity."[12] The estrogens directly stimulate the growth and development of the vagina, uterus, and fallopian tubes and, in combination with other hormones, play a primary role in sexual arousal and in producing the body contours of the mature woman. Pigmentation of the nipples and genital tissues and growth stimulation of pubic and underarm hair (possibly with the help of small amounts of testosterone) are other results of estrogen action.

The physiologic changes at menopause emphasize the important roles of estrogens in the young woman. Breast and reproductive tissues atrophy, the skin loses some of its suppleness, coronary atherosclerosis and gout become potential health problems for the first time, and the bones begin to lose density because of decreased mineral content.

METABOLISM OF ESTROGENS

The metabolism of natural and synthetic (e.g., mestranol) estrogens has been reviewed in detail by Bolt in 1979[92] and Martin in 1985.[19]

The three primary estrogens in women are 17β-estradiol, estrone, and estriol. Although 17β-estradiol is produced in greatest amounts, it is quickly oxidized (see Fig. 18-13) to estrone, the estrogen found in highest concentration in the plasma. Estrone, in turn, is oxidized to estriol, the major estrogen found in human urine. During pregnancy, the placenta produces large amounts of estrone. However, in both pregnant and nonpregnant women, the three primary estrogens are also metabolized to small amounts of other derivatives (e.g., 2-hydroxyestrone, 2-methoxyestrone and 16β-hydroxy-17β-estradiol). Only about 50% of therapeutically administered estrogens (and their various metabolites) are excreted in the urine during the first 24 hours. The remainder is excreted into the bile and reabsorbed; consequently, several days are required for complete excretion of a given dose.

A low-fat diet has been shown to reduce the risk of breast cancer, whereas a high-fat diet increases the risk. Metabolism of estrone appears to be the primary determinant. Low-fat diets promote the conversion of estrone to 2-hydroxyestrone rather than 16-hydroxyestrone (and then to estriol), whereas high-fat diets promote the opposite.

Conjugation appears to be very important in estrogen transport and metabolism. Although the estrogens are unconjugated in the ovaries, in the plasma and other tissues significant amounts of conjugated estrogens may predominate. Most of the conjugation takes place in the liver.

The primary estrogen conjugates found in plasma and urine are the combination of estrogen with glucuronic acid and, to a lesser degree, with sulfate. The conjugates are called glucuronides and sulfates, respectively. As sodium salts, they are quite water-soluble. The sodium glucuronide of estriol and the sodium sulfate ester of estrone are shown below:

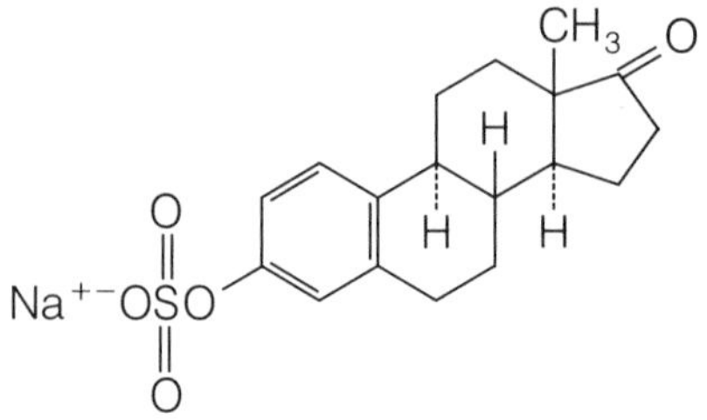

Sodium Sulfate Ester of Estrone

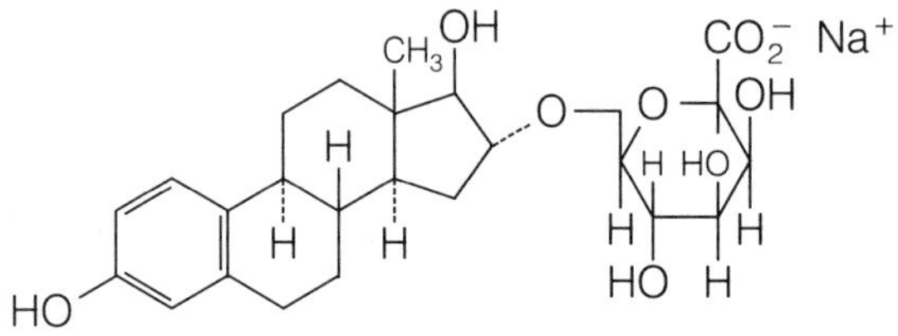

Sodium Glucuronide of Estriol

THERAPEUTIC USES

Birth Control

The greatest use of estrogens is for inhibition of ovulation, used in combination with progestins. Steroidal birth control agents containing estrogens will be discussed in the section on Chemical Contraception, later in this chapter.

Prevention and Treatment of Postmenopausal Osteoporosis

As recently summarized by Riggs and Melton in their excellent review,

> Osteoporosis is defined pathologically as an absolute decrease in the amount of bone, leading to fractures after minimal trauma. Osteoporosis is an enormous public health problem, responsible for at least 1.2 million fractures in the United States each year, [including] 538,000 fractures of the vertebrae and 227,000 fractures of the hip.... The direct and indirect costs of osteoporosis are estimated at $6.1 billion annually in the United States.[93, 94]

There is about a 0.3% to 0.5% loss in cortical bone each year after age 40, but immediately after menopause there is a 2% to 3% bone loss each year in women.[93, 94]

Several studies have shown that postmenopausal estrogen therapy reduces the incidence of osteoporosis and hip fracture by half or more.[95,96] Estrogen deficiency, not aging, has been shown to be the chief cause of osteoporosis.[97] Significantly, smoking has been shown to cause earlier menopause and to accelerate the development of osteoporosis by increasing the metabolic clearance of estradiol.[98]

The National Institutes of Health (NIH) Consensus Development Conference on Osteoporosis has recommended estrogen and calcium supplements to help prevent postmenopausal (or postoophorectomy) osteoporosis, along with a program of exercise.[95,99] Furthermore, it is generally believed that a high calcium intake and exercise throughout the premenopausal years may significantly prevent osteoporosis later in life. "Healthy women 35 to 50 years old (who often ingest less than 500 mg of dietary calcium per day) may require more than 1200 mg of calcium daily to maintain a positive calcium balance."[100] Nevertheless, several studies have shown that calcium supplementation alone does not necessarily prevent postmenopausal bone loss.[95,101] Bone weakness and generalized bone pain can result from vitamin D deficiency, but the incidence is much less than bone weakness from estrogen deficiency.

A 1990 study[101a] showed that intermittent cyclical therapy with etidronate (disodium ethane-1-hydroxy-1,1-bisphosphonate) can significantly increase spinal bone mass and decrease the incidence of vertebral fractures in women with postmenopausal osteoporosis. Calcitonin (Cibacalcin), a naturally occurring peptide hormone has been used to treat postmenopausal osteoporosis[100] by inhibition of bone resorption. However, it has not been shown to decrease postmenopausal bone fractures.[95,102] Calcitonin is given by subcutaneous (SC) or intramuscular (IM) injection for treatment of Paget's disease of bone.

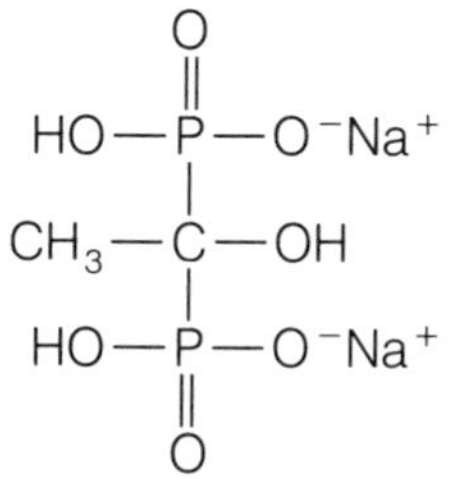

```
         O
         ||
   HO—P—O⁻Na⁺
         |
  CH₃—C—OH
         |
   HO—P—O⁻Na⁺
         ||
         O
```

Etidronate

A wide variety of other compounds are also being investigated to reduce or prevent osteoporosis. As reviewed in 1987 by Hayward and Caggiano,[103] they include inhibitors of carbonic anhydrase, inhibitors of K^+,H^+–ATPase, thiophene carboxylic acids, lysosomal enzyme-release inhibitors, and phosphodiesterase inhibitors.

Progestins Added to Postmenopausal Estrogens to Reduce the Risk of Endometrial Cancer

As will be discussed later in this section, the incidence of endometrial cancer has been shown to increase with postmenopausal women taking estrogens, but it is rarely fatal when managed appropriately.[99] Some studies suggest that the combination of a progestin plus an estrogen may significantly reduce this risk.[95,104] In contrast with most studies showing adverse effects of smoking on health, one study has shown that cigarette smoking actually reduces the incidence of endometrial cancer.[105]

Postmenopausal Estrogens in Lowering the Risk of Heart Disease

The potential benefits of postmenopausal estrogens in decreasing the risk of heart disease remain uncertain. When women who experience a natural menopause take estrogens, there appear to be no benefits,[106–108] although one prospective study suggested the opposite hypothesis.[109] However, for women who are taking estrogens after a bilateral oophorectomy, there actually may be a small increased risk of coronary heart disease. Because it is commonly believed that women are generally protected from heart disease by their natural production of estrogens before menopause, these findings remain puzzling.

Treatment of Vasomotor Symptoms of Menopause and Atrophic Vaginitis (Kraurosis Vulvae)

Estrogens have been very useful in treating the "hot flashes" associated with early menopause, as well as atrophic vaginitis and other vaginal symptoms of inadequate estrogen production. However, there is no evidence that they result in a more youthful appearance, nor that they help in the emotional symptoms, such as depression, sometimes associated with the onset of menopause.

Treatment of Estrogen Deficiency from Ovarian Failure or After Oophorectomy

Estrogen therapy, usually with a progestin, is common in cases of ovarian failure and after an oophorectomy.

Treatment of Advanced, Inoperable Breast Cancer in Men and Postmenopausal Women, and of Advanced, Inoperable Prostate Cancer in Men

Estrogens are used to treat inoperable breast cancer in men and in postmenopausal women, but estrogen therapy can actually stimulate existing breast cancers in premenopausal women. The antiestrogen tamoxifen is reported to have fewer side effects; hence, it is usually preferred. Estrogens also are often used to treat inoperable prostate cancer.

SIDE EFFECTS

As will be discussed in the section on Chemical Contraceptions, oral contraceptives containing moderate doses of estrogens have been associated with risk of thromboembolic and thrombotic disorders (e.g., thrombophlebitis, stroke, pulmonary embolism, and myocardial infarction). The low doses of estrogens and the cyclic therapy usually used in postmenopausal therapy explain the observation that thromboembolic disorders are not usually observed with postmenopausal patients. When increased doses of estrogen are used (e.g., for advanced, inoperable breast carcinoma of women or men), the risk can also increase.

Liver function may be altered during estrogen therapy, and endocrine function tests for thyroid function and glucose tolerance may appear abnormal. Liver tumors have been reported with patients taking estrogens, but very infrequently. Some patients may experience depression from estrogen therapy; therefore, those with psychiatric disorders should be carefully reviewed when considering such therapy.

Some contact lens wearers may experience lens discomfort when taking estrogens; such discomfort could indicate changes in corneal curvature. The patient's ophthalmologist or optometrist should be consulted for possible termination of lens use.

Estrogens may cause birth defects; consequently, they should be discontinued if pregnancy is suspected.

DES Babies

During the late 1930s through the early 1950s, it was believed that DES treatment could help those pregnant women who tended to miscarry to have full-term pregnancies. Not only was the belief incorrect, it was subsequently reported that daughters of women who had taken DES during pregnancy ("DES babies") had a high risk of vaginal or cervical cancer.[110–112] However, continuing studies of over 2000 women exposed to DES in utero revealed very few cases of cancer through early 1988, although a high percentage of the women had vaginal epithelial changes as of 1979.[110,113,114] There have been no increases in breast disease among DES-exposed women, when compared with women who have not been exposed.

PRODUCTS

Estrogens are commercially available in a wide variety of dosage forms: oral tablets, vaginal creams and foams, transdermal patches, and IM dosage preparations (Tables 18-2 and 18-3).

TABLE 18-2

ESTROGEN PRODUCTS (STEROIDAL ESTROGENS)

Name Proprietary Name	*Preparations*	*Application*	*Usual Adult Dose**	*Usual Dose Range**
Estrone USP *Theelin, Menformon, Urestrin*	Sterile estrone suspension USP Estrone injection USP	Available also in vaginal suppositories and transdermal patches	IM, 1 mg one or more times weekly as required; reduce to maintenance dose as soon as response is obtained	200 μg–5 mg/wk
Piperazine estrone sulfate USP *Ogen*	Piperazine estrone sulfate tablets USP Piperazine estrone sulfate vaginal cream USP		1.5 mg/day	750 μg–10 mg/day
Conjugated estrogens USP *Premarin, Menotabs*	Conjugated estrogens tablets USP	Available also as vaginal cream, and IV and IM dosage forms	1.25–2.5 mg qd to tid for 3 wk of every month	300 μg–30 mg/day

*See USP DI for complete dosage information.

TABLE 18-2 *Continued*

ESTROGEN PRODUCTS (NONSTEROIDAL ESTROGENS)

Name Proprietary Name	***Preparations***	***Application***	***Adult Usual Dose****	***Usual Dose Range****
Esterified estrogens USP *Amnestrogen, Menest, SK-Estrogens, Evex, Glyestrin*	Esterified estrogens tablets USP		1.25–2.5 mg qd to tid for 3 wk of every month	300 μg–30 mg/day
Estradiol USP *Estrace Estraderm*	Sterile estradiol suspension USP Estradiol pellets USP Estradiol tablets USP Estradiol Transdermal System			Implantation, 25 mg repeated when necessary; IM, 220 μg–1.5 mg, 2 or 3 times weekly. Oral, 200–500 μg qd to tid; transdermal patch: 0.05 or 0.1 mg/24 hr release rates
Estradiol benzoate USP *Progynon Benzoate*	Estradiol benzoate injection USP			IM, initial: 1.0–1.66 mg, 2 or 3 times weekly for 2 or 3 wk; maintenance: 330 μg–1.0 mg twice weekly
Estradiol valerate USP *DelEstrogen*	Estradiol valerate injection USP		IM, 5–30 mg every 2 wk	5–40 mg every 1–3 wk
Estradiol cypionate USP *Dep-Estradiol*	Estradiol cypionate injection USP		Initial: IM, 1–5 mg/wk for 2–3 wk; maintenance: 2–5 mg every 3 to 4 wk	
Ethinyl estradiol USP *Lynoral, Estinyl, Feminone*	Ethinyl estradiol tablets USP		50 μg qd to tid	20 μg–3 mg/day
Mestranol USP				See Table 18-6 for doses in oral contraceptives

*See USP DI for complete dosage information.

TABLE 18-3

ESTROGEN PRODUCTS (NONSTEROIDAL ESTROGENS)

Name Proprietary Name	***Preparations***	***Application***	***Adult Usual Dose****	***Usual Dose Range****
Diethylstilbestrol USP *Stilbetin*	Diethylstilbestrol tablets USP		Mammary carcinoma, 15 mg qd; carcinoma of prostate, 1–3 mg qd; estrogen, 200 μg–2 mg qd	Mammary carcinoma, 1–15 mg or more daily; carcinoma of prostate, 1–5 mg qd; estrogen, 100 μg–25 mg qd
	Diethylstilbestrol injection USP		Carcinoma of prostate, IM, 2–5 mg twice a week; estrogen, IM, 250 μg–1 mg, 2 or 3 times a week	Estrogen, 100 μg twice a week to 10 mg/day
	Diethylstilbestrol suppositories USP		Vaginal, 100 μg–1 mg qd	
Dienestrol USP *Synestrol*	Dienestrol cream USP	Vaginal, 5 g of a 0.01% cream qd or bid for 7–14 days, then once every 48 hr for 7–14 days	500 μg/day	100 μg–1.5 mg
Benzestrol USP *Chemestrogen*	Benzestrol tablets USP		1–2 mg/day	500 μg–5 mg
Chlorotrianisene USP *Tace*	Chlorotrianisene capsules USP		24 mg/day	12–144 mg/day, as determined by the practitioner for the condition being treated

*See USP DI for complete dosage information.

Dosages and Dosage Cycles

The general guideline for estrogen dosage is to use the lowest effective dose possible, and only for the duration in which it is truly needed. Estrogens are generally administered on a cyclic basis, typically one tablet daily for three weeks, followed by one week without any estrogen. Dosage cycles for contraceptive products containing estrogens will be discussed later in the chapter.

Estrone, USP. 3-Hydroxyestra-1,3,5(10)-trien-17-one is only one-third as active as its natural precursor, estradiol (see Fig. 18-13). As the salt of its 3-sulfate ester, estrone is the primary ingredient in Conjugated Estrogens USP and Esterified Estrogens USP. Although originally obtained from the urine of pregnant mares (about 10 mg/L), estrone is now prepared from the Mexican yam, discussed later in this chapter. Assay is usually by ultraviolet spectroscopy, using the maximum absorption 280 nm (EtOH). Radioimmunoassay procedures are also available for assay of estrone in plasma.

Piperazine Estrone Sulfate, USP. 3-Sulfoxy-estra-1,3,5(10)-trien-17-one piperazine salt. All the estrone 3-sulfate salts have the obvious pharmaceutic advantage of increased water solubility (as one would predict from Table 18-1) and better oral absorption. Acids will convert the salts to the free 3-sulfate esters, and, in addition, cause some hydrolysis of the ester. This does not seem to adversely affect absorption, but precipitation of the free sulfate esters in acidic pharmaceutic preparations should be avoided. The dibasic piperazine molecule acts as a buffer, giving it somewhat greater stability.

Conjugated Estrogens, USP and **Esterified Estrogens, USP.** *Conjugated estrogens* contain 50% to 65% of sodium estrone sulfate and 20% to 35% of sodium equilin sulfate (based on the total estrogen content of the product). *Esterified estrogens* have an increased amount of sodium estrone sulfate, 70% to 85%, often synthetically prepared from diosgenin and added to the urine extract. Although most commonly used to treat postmenopausal symptoms, the conjugated estrogens and esterified estrogens are used for the entire range of indications described previously.

The bioequivalency of the first conjugated estrogen to be marketed (Premarin) and generic products introduced in the late 1960s has been the subject of considerable discussion.[114a] In 1990, the FDA concluded that the "B" generics were not therapeutically equivalent to Premarin and withdrew them as candidates for substitution.[114b]

Estradiol, USP. Estra-1,3,5(10)-triene-3,17β-diol is the most active of the natural steroid estrogens (see Fig. 18-13). Although its 17β-OH group is vulnerable to bacterial and enzymatic oxidation to estrone (see Fig. 18-13), it can be temporarily protected as an ester or permanently protected by adding a 17α-alkyl group (giving 17α-ethinyl estradiol and the 3-methyl ether, mestranol, the most commonly used estrogen in oral contraceptives). 3-Esters increase the duration of activity. These derivatives illustrate the principles of steroid modification shown in Figure 18-7. The increased oil solubility of the 3- and 17β-esters (relative to estradiol) permits the esters to remain in oil at the injection site for extended periods. Transdermal estradiol products became available in 1986, with studies showing that they are as effective as oral estrogens for treating menopausal symptoms.[115] One study also showed that estradiol given transdermally may avoid the hepatic side effects sometimes noted with oral estrogens.[116] The commercially available estradiol esters are listed below.

Estradiol Benzoate, USP
Estradiol Valerate, USP
Estradiol Cypionate, USP

Ethinyl Estradiol, USP. 17α-Ethinyl estradiol has the great advantage over other estradiol products in that it is orally active. It is equal to estradiol in potency by injection, but 15 to 20 times more active orally. The 3-methyl ether of ethinyl estradiol is **mestranol, USP**, widely used in oral contraceptives. Mestranol is metabolized to ethinyl estradiol, with an oral dose of about 50 μg of mestranol providing an estrogenic action approximately equivalent to 35 μg of oral ethinyl estradiol. **Quinestrol**, the 3-cyclopentyl ether of estradiol, is a prodrug of ethinyl estradiol—inactive until dealkylated in vivo. Whereas about 60% of an oral dose of ethinyl estradiol is metabolized in the intestinal mucosa and during the first pass through the liver, most of an oral dose of quinestrol survives. The cyclopentyl group also enhances lipid solubility and, thereby, lipid storage. The overall result is that after initial loading in daily dosages for a week, quinestrol is typically taken just once weekly.

Estriol, USP, possesses estrogenic activity and is reported to be orally active.

Diethylstilbestrol, USP. α,α′-Diethyl-(*E*)-4,4′-stilbenediol, DES, is the most active of the nonsteroidal estrogens (see discussion of Estrogen Structural Classes), having about the same activity as estrone when given intramuscularly. The *cis*-isomer has only one-tenth the activity of the *trans*. The *trans*-isomer is also well absorbed orally and slowly metabolized; consequently, it has been a popular estrogen for many medical purposes (see

Therapeutic Uses). Side effects can be serious (see discussion of Side Effects for the Estrogens). The diphosphate ester, **Diethylstilbestrol Diphosphate, USP,** is used only for cancer of the prostate and is available for intravenous use. However, it has been reported that there may be an increased incidence of deaths from cardiovascular causes in men who received 5 mg of DES daily for prolonged periods. The diphosphate salt has great water solubility, as one would predict from Table 18-1. Diethylstilbestrol was extensively used in low doses as an aid to fatten cattle. Because DES has been implicated in cancer (albeit in higher doses), the United States Congress and Food and Drug Administration (FDA) began action in September 1975 to ban DES in animal feed.

Note: All stilbene derivatives, such as DES and dienestrol, are light-sensitive and must be kept in light-resistant containers.

Dienestrol, USP. 4,4′-(1,2-Diethylidene-1,2-ethanediyl)bisphenol has about the same activity as DES when taken orally. The cream is used to treat atrophic vaginitis.

Benzestrol, USP. 4,4′-(1,2-Diethyl-3-methyl-1,3-propanediyl)bisphenol, when drawn like DES in Figure 18-12, obviously resembles DES. Yet it has no double bonds such as have DES or dienestrol to keep the phenolic groups in a *trans* spatial arrangement. However, the adjacent ethyl groups do not prefer eclipsed conformation (much higher in energy than *trans*), thereby helping keep the phenolic groups *trans*. Benzestrol is used for all the usual indications for estrogens (see Therapeutic Uses).

Chlorotrianisene, USP. Chlorotris-(*p*-methoxyphenyl)ethylene is more active orally than by injection and is thought to be converted to a more active form hepatically. When given by injection, it is quite a weak estrogen. It has good lipid solubility and is slowly released from lipid tissues, thereby giving it a relatively longer duration of action. The fat storage can also delay its onset of action.

Estrogens from plants. Several natural plant substances that differ from DES in structure are also potent estrogens. These include genistein, from a species of clover;[117] coumestrol, found in certain legumes;[118] and zearalenone, from a *Fusarium* fungus.[119] These and others have antifertility activity, as reviewed by Briggs.[120]

ANTIESTROGENS AND RELATED DRUGS

Whereas estrogens have been very important in chemical contraception, estrogen antagonists (antiestrogens; Fig. 18-16) have been of great interest as ovulation stimulants. Although the term "antiestrogen" has been rather loosely applied to progestins and androgens, a few compounds, such as clomiphene, have a direct effect in increasing Gn-RH production by the hypothalamus. The mechanism is presumably a blocking of feedback inhibition of ovary-produced estrogens. The increased Gn-RH, in turn, induces ovulation as described before with pulsatile therapy.

In tests with experimental animals, clomiphene has no effect in the absence of a functioning pituitary gland. Its great structural similarity to chlorotrianisene can be seen easily.

Clomiphene causes several side effects, especially enlargement of the ovaries. Abdominal discomfort should immediately be discussed with the physician. Other side effects include nausea, visual disturbance, depression, breast soreness, and increased nervous tension. Multiple births occur in about 10% of patients.

Tamoxifen is another potent antiestrogen with a triphenylethylene structure. Although it has been

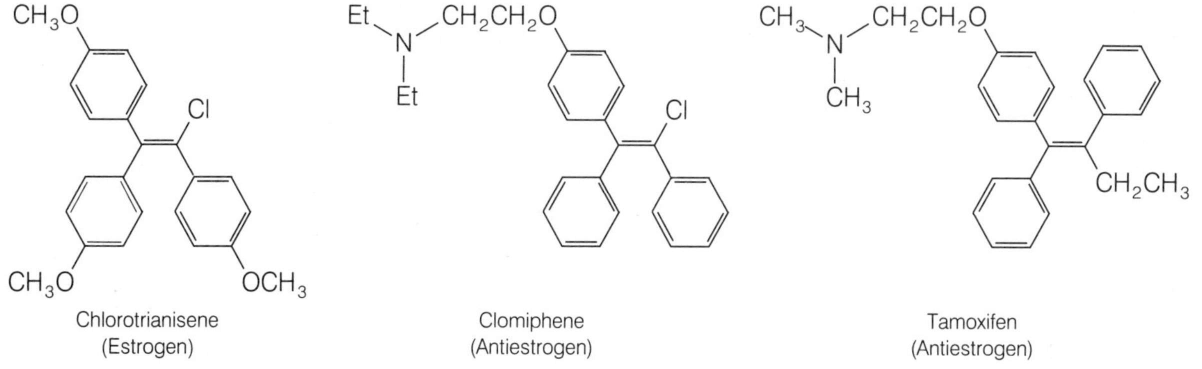

Chlorotrianisene (Estrogen) — Clomiphene (Antiestrogen) — Tamoxifen (Antiestrogen)

FIG. 18-16. Representative antiestrogens showing structural similarity with the estrogen chlorotrianisene.

investigated as an agent to induce ovulation, its primary use is for treatment of advanced breast carcinoma in postmenopausal women.

Why are some DES analogues estrogenic and others antiestrogenic? The answer is complicated by the fact that some progestins are also antiestrogens. As reviewed by Raynaud and Ojascoo,[121] the progestins norgestrel and promegestone (R 5020) are very potent antiestrogens in a variety of biologic test systems. The mechanism of antiestrogenic action of these progestins has not been delineated. One possibility is that the progestins inhibit estrogen receptor replenishment.

The DES analogues, such as clomiphene and tamoxifen, that act as antiestrogens can bind to the estrogen receptor. As shown by Duax and coworkers,[66] the phenyl A ring of antiestrogens like tamoxifen is in the same position and conformation as the A ring of estradiol. They believe therefore that the A ring is responsible for binding. The bulky dialkyl aminoethyl side groups may interfere with the receptor to prevent a hormonal response. The antiestrogens also bind to other receptors. Some inhibit prostaglandin synthase or glutamate dehydrogenase.

Clomiphene and other ovulation-inducing or antiabortifacient drugs have recently been reviewed.[122,123] The primary drugs used for induction of ovulation include clomiphene, hMG, pulsatile injections of LH-RH (Gn-RH), (discussed earlier in this chapter), and the ergot derivative bromocriptine:

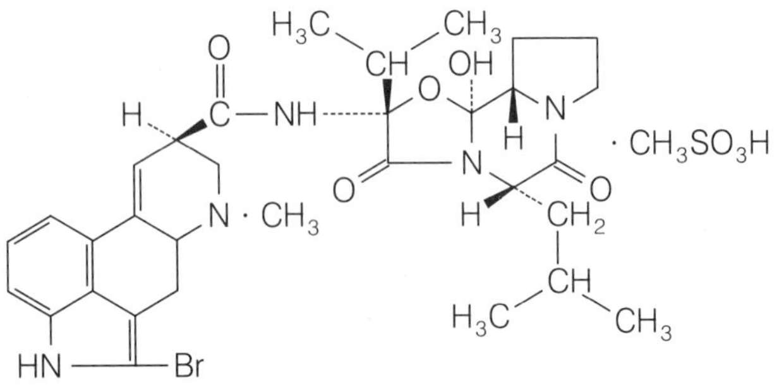

H_3C CH_3 CH OH O H C—NH N H N · CH_3SO_3H O N · CH_3 O H CH_2 CH H_3C CH_3 HN Br

Lactation has an antifertility effect of its own owing to prolactin inhibition of pituitary release of FSH and LH. Bromocriptine may have two actions to reduce serum prolactin levels: (1) a direct effect on the anterior pituitary to inhibit prolactin release; and (2) acting as a dopamine receptor agonist on dopamine receptors in the hypothalamus, causing the release of prolactin-release inhibitory hormone (PRO-RIH). Bromocriptine has also been used to treat males with prolactin-secreting adenomas. As discussed in Chapter 11, bromocriptine is also used for treatment of parkinsonism. The incidence of side effects with bromocriptine is relatively high and includes nausea, epigastric pain, headache, migraine, arrhythmias, and hypotension. Blood pressure monitoring is important.

hMG should be used with caution because ovarian enlargement is quite common. Multiple births occur in up to 20% of the cases, and pregnancies followed by spontaneous abortions occur in 20% to 30% of the cases.

In general, it is strongly recommended that product literature or detailed general references such as *Facts and Comparisons* or the *Hospital Formulary* be consulted before dispensing either clomiphene citrate or hMG.

Clomiphene citrate, USP. 2-[4-(2-Chloro-1,2-diphenylethenyl)phenoxy]-*N*, *N*-diethylethanamine (Clomid) is given to stimulate ovulation in the usual dosage of 50 mg daily for five days starting on the fifth day of the menstrual cycle. If ovulation does not occur, the dose is increased to 100 mg daily for five days in the next cycle. The patient should be warned to report any visual disturbances or abdominal pain to the physician. If menstruation does not occur at the end of the first full cycle following treatment, pregnancy tests should be conducted before additional clomiphene is taken. A careful physical examination before treatment is recommended, especially to determine the possible presence of ovarian cysts, because ovarian enlargement sometimes occurs.

Tamoxifen citrate, USP. 2-[4-(1,2-Diphenyl-1-butenyl)phenoxy]-*N*,*N*-dimethylethanamine (Novaldex) is an antiestrogen used to treat early and advanced breast carcinoma in postmenopausal women. It is as effective as androgen therapy without the androgenic side effects. "In trials of tamoxifen as a postsurgical adjuvant treatment of early breast cancer, disease-free survival is consistently prolonged."[123a] Tamoxifen is also being evaluated as a chemopreventative agent to prevent breast cancer.[123b] Antiestrogenic and estrogenic side effects can include hot flashes, nausea, vomiting, platelet reduction, and (in patients with bone metastases) hypercalcemia. As with all triphenylethylene derivatives, it should be protected from light.

Bromocriptine mesylate, USP. 2-Bromo-12′-hydroxy-2′-(1-methylethyl)-5′-(2-methylpropyl)ergotaman-3′-6′,18-trione (Parlodel) reduces serum prolactin concentrations, as a treatment of female infertility. It acts either directly on the anterior pituitary to inhibit prolactin release, or through its dopamine receptor agonist action to release prolactin-release inhibitory hormone. Bromocriptine is also used to treat parkinsonism as a dopamine agonist. In males, bromocriptine has been used to treat prolactin-secreting adenomas. The side effects can be serious, as discussed earlier in the chapter. It is light- and heat-sensitive and should be stored accordingly.

PROGESTINS

STRUCTURAL CLASSES

Progestins are compounds that have biologic activities similar to progesterone. They include two structural classes: *progesterone and derivatives*, and *testosterones* and *19-nortestosterones* (Fig. 18-17). Progesterone is not orally effective; with a plasma half-life of only about five minutes, it is almost completely metabolized in one passage through the liver.[19,124–127] Adding 17α-alkyl groups slows metabolism of the 20-one, whereas a 6-methyl or 6-chloro group enhances activity and reduces metabolism. Chlormadinone acetate, a progestin previously used in birth control products, is a particularly potent example (Table 18-4).

Duax and co-workers[65,66] have studied the structural requirements of the progesterone receptor in detail. They conclude that the progesterone 4-en-3-one ring A is a key to binding, but only when it is in a conformation quite different from that of testosterone or the glucocorticoids. Their reviews contain stereo drawings that show the requirement conformations in three dimensions.[65,66]

Although the 19-nortestosterones do have androgenic side effects, their primary activity, nevertheless, is progestational. Addition of 17α-alkyl groups to testosterone blocks oxidation at C-17. However, 17α-methyltestosterone has only half the androgenic activity of parenterally administered testosterone, and the 17α-ethyl analogue is nearly inactive. Adding the electron density of a triple bond, as in 17α-ethinyl, causes a marked increase in progestational activity and, simultaneously, blocks metabolic or bacterial oxidation to the corresponding 17-ones. Thus, by adding a 17α-ethinyl or propinyl group to testosterone, one can simultaneously decrease anabolic activity, promote good progestational activity, and have an orally active compound as well. Table 18-4 illustrates the relative progestational activity of a number of progestins.

The 19-nor derivatives also have marked ovulation-inhibiting activity, which does not necessarily parallel progestational activity. The endometrial proliferation (Clauberg-McPhail) test is most often used to evaluate progestational activity, whereas antiovulation activity is determined by examining treated female rabbits for ovulation rupture points in their ovaries. Other methods are discussed by Deghenghi and Givner.[4]

BIOSYNTHETIC SOURCES

Progesterone is produced in the ovaries, testes, and adrenal glands. Much of the progesterone that is synthesized is immediately converted to other hormonal intermediates and is not secreted. (Refer to the biosynthetic pathway, Fig. 18-5.) The corpus luteum secretes the most progesterone, 20 to 30 mg/day during the last or "luteal" stage of the menstrual cycle. Normal men secrete about 1 to 5 mg of progesterone daily.

METABOLISM OF PROGESTINS

As noted at the beginning of this section, progesterone has a half-life of about only five minutes when taken orally.

As shown in Figure 18-5, progesterone can be biotransformed to many other steroid hormones and, in that sense, it has numerous metabolic products. However, the principal excretory product of progesterone metabolism is 5β-pregnane-3α,20-diol and its conjugates.

The metabolism of progesterone is extremely rapid and, therefore, it is not effective orally. That fact has been a major stimulus in the development of the 19-nortestosterones with progesteronelike activity.

The great advantage of testosterone and 19-nortestosterone analogues is that, with a 17α-ethinyl group, these compounds are orally active. The 17-OH cannot be metabolized to the corresponding 17-ones that have little or no activity. Their 5-enes are metabolized to 5α-analogues; however, the metabolites are still quite active.

BIOLOGIC ACTIVITIES OF THE PROGESTINS

Crowley has recently captured the fascination and wonder of progesterone in a succinct overview in the *New England Journal of Medicine*.[128]

> Progesterone is a unique reproductive hormone. In women who are not pregnant, it is the chief secretory product . . . of the corpus luteum—itself a curious endocrine organ which is programmed for demise within a fortnight unless it is "rescued" from this fate by a fertilized ovum. When pregnancy occurs, hCG contributes to the persistence of the corpus luteum, which secretes the progesterone required for the maintenance of early pregnancy . . . . Progesterone's principal target organs are the uterus, breast, and brain. [Actions include] differentiation of the estrogen-primed, growing endometrium, and induction of protein secretion; [in breast tissue], differentiation of estrogen-prepared ductal tissue and

1. Progesterones and Derivatives

Progesterone
(IM, sublingual, vaginal)

17α-Hydroxyprogesterone
Caproate (IM)

Medroxyprogesterone
Acetate (Oral, IM)

Megestrol Acetate

Chlormadinone Acetate

2. Testosterones and 19-Nortestosterone Derivatives

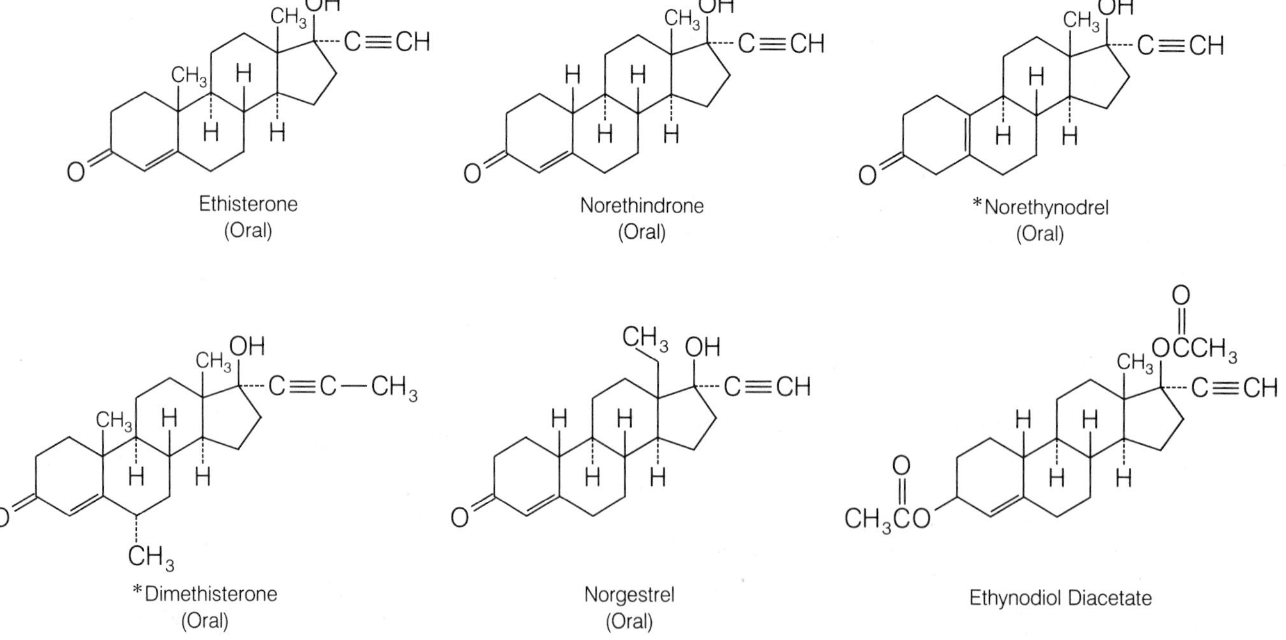

FIG. 18-17. *Natural and synthetic progestins. (*Available only in contraceptive products.)*

TABLE 18-4

COMPARATIVE PROGESTATIONAL ACTIVITY OF SELECTED PROGESTINS[125]

	Relative Oral Activity	*Activity SC*
Progesterone	(nil)	1
17α-Ethinyltestosterone (ethisterone)	1	0.1
17α-Ethinyl-19-nortestosterone (norethindrone)	5–10	0.5–1
Norethynodrel	0.5–1	0.05–1
17α-Hydroxyprogesterone caproate	2–10	4–10
Medroxyprogesterone acetate	12–25	50
19-Norprogesterone		5–10
Norgestrel		3
Dimethisterone	12	

(Data from Salhanick, H. A. et al.: Metabolic Effects of Gonadal Hormones and Contraceptive Steroids. New York, Plenum Press, 1969.[125])

> support [of] lactation. Progesterone's influence on the CNS is poorly understood, but appears to have diverse effects on the hypothalamic–pituitary axis, respiratory center, and perhaps cortical function . . . . Its thermogenic effect, also centrally mediated, leads to an increase in basal body temperature . . . [but] a considerable amount remains to be learned . . . [128]

THERAPEUTIC USES

Birth Control

The largest use of the progestins, as of the estrogens, is for inhibition of ovulation. Steroidal birth control agents are discussed in the section on Chemical Contraception.

To Reduce the Risk of Endometrial Cancer from Postmenopausal Estrogens

As discussed in the section on therapeutic uses of estrogens, several studies have suggested that the combination of a progestin with an estrogen may significantly reduce the risk of endometrial cancer in women taking postmenopausal estrogens.

Secondary Amenorrhea and Functional Uterine Bleeding Caused by Insufficient Progesterone Production or Estrogen–Progesterone Imbalance

Progestins have been used very effectively to treat secondary amenorrhea, functional uterine bleeding, and related menstrual disorders caused by hormonal deficiency or imbalance.

Treatment of Inoperable Endometrial Cancer

Surgery is usually the first treatment of endometrial or breast cancer, but in advanced cases for whom surgery would not be possible, progestins have provided palliative treatment.

Progestins for Premenstrual Syndrome

There have been claims that progesterone may reduce the effects of PMS. However, the *Medical Letter* has concluded that the claims are not supported by medical data.[129]

Progestins and Pregnancy — A Warning

Progestins have been used to prevent habitual abortions, but the FDA has strongly warned against the use of steroids during pregnancy.[15, 130] Large doses of progestins have also been given as a test for pregnancy, but the FDA warning would seem to discourage this practice as well. Adverse effects of sex hormones given during pregnancy were reviewed in 1988[15] and 1979.[131]

SIDE EFFECTS

Progestin therapy may cause menstrual irregularities, such as spotting or amenorrhea. Weight gain and acne have been associated with testosterone and 19-nortestosterone analogues, in part because of their slight androgenic effects. In combination with estrogens in oral contraceptives, the progestin components may contribute to the thromboembolic disorders primarily attributed to estrogens. As noted in the section on oral contraceptives, the frequency of these disorders in nonsmokers younger than 35 years is negligible.

PRODUCTS

The progestins are primarily used in oral contraceptive products for women, and they are also used to treat several gynecologic disorders: dysmenorrhea, endometriosis, amenorrhea, and dysfunctional uterine bleeding. Estrogens are given simultaneously in most of these situations.

The doses appropriate for the various foregoing indications can vary substantially, and detailed manufacturers' literature or general references should be consulted before advising physicians.

General USP DI doses are listed in Table 18-5.

TABLE 18-5

PROGESTIN PRODUCTS

Name Proprietary Name	***Preparations***	***Usual Adult Dose****	***Usual Dose Range****
Progesterone USP *Lipo-Lutin*	Progesterone injection USP Sterile progesterone suspension USP	IM, 5–25 mg qd beginning 8–10 days before menstruation	5–50 mg/day
Hydroxyprogesterone caproate USP *Delalutin, Corlutin, Hylutin*	Hydroxyprogesterone caproate injection USP	Menstrual disorders: IM, 375 mg once a month; uterine cancer: IM, 1 g or more, repeated 1 or more times per week	375 mg/mo to 7 g/wk
Medroxyprogesterone acetate USP *Provera*	Sterile medroxyprogesterone acetate suspension USP	Endometriosis: IM, 50 mg once a week; uterine cancer: IM, 400 mg–1 g once a week	50 mg–1 g/wk
	Medroxyprogesterone acetate tablets USP	Habitual and threatened abortion: 10–40 mg/day; menstrual disorders: 2.5–20 mg/day for 5–10 days, during the second half of the menstrual cycle	
Norethindrone USP *Norlutin*	Norethindrone tablets USP Norethindrone and ethinyl estradiol tablets USP Norethindrone and mestranol tablets USP Norethindrone acetate tablets USP	5–20 mg qd	5–40 mg/day
Norethindrone acetate USP *Norlutate*	Norethindrone acetate and ethinyl estradiol tablets USP†	1–2.5 mg of norethindrone acetate and 20–50 μg of ethinyl estradiol qd for 20 or 21 days, beginning on the 5th day of the menstrual cycle	
Norethynodrel USP	Norgestrel tablets USP	2.5–10 mg qd	2.5–30 mg/day
Norgestrel USP	Norgestrel and ethinyl estradiol tablets USP†	500 μg of norgestrel and 50 μg of ethinyl estradiol for 21 days, beginning on the 5th day of the menstrual cycle	
Ethynodiol diacetate USP	Ethynodiol diacetate and ethinyl estradiol tablets USP†	1 mg of ethynodiol diacetate and 50 μg of ethinyl estradiol qd for 21 days, beginning on the 5th day of the menstrual cycle	
Megestrol acetate	Megestrol acetate tablets USP	For palliative treatment of advanced breast carcinoma: 160 mg/day in four equally divided doses; for palliative treatment of advanced endometrial carcinoma: 40–320 mg/day in divided doses	10–80 mg qid

*See USP DI for complete dosage information.
†Oral contraceptive.

Progesterone, USP. Pregn-4-en-3,20-dione is so rapidly metabolized that it is not particularly effective orally, being only one-twelfth as active as intramuscularly. It can also be very irritating when given intramuscularly. Buccally it is only slightly more active than orally. As discussed later in this chapter (Commercial Production), progesterone was originally obtained from animal ovaries, but was prepared in ton quantities from diosgenin in the 1940s. This marked the start of the modern steroid industry, a fascinating history discussed later in this chapter. The discovery of 19-nortestosterones with progesterone activity made synthetically modified progestins of tremendous therapeutic importance.

Progesterone (and all other steroid 4-ene-3-ones) is light-sensitive and should be protected from light.

Hydroxyprogesterone Caproate, USP. 17-Hydroxypregn-4-ene-3,20-dione hexanoate is much more active and longer-acting than progesterone (see Table 18-4), probably because the 17α-ester function hinders reduction to the 20-ol. It is given only intramuscularly. The hexanoate ester greatly increases oil solubility, allowing it to be slowly released from depot preparations, as one would predict from Figure 18-7.

Medroxyprogesterone Acetate, USP. 17-Hydroxy-6α-methylpregn-4-ene-3,20-dione acetate adds a 6α-methyl group to the 17α-hydroxyprogesterone structure to greatly decrease the rate of reduction of the 4-ene-3-one system. The 17α-acetate group also decreases reduction of the 20-one, just as with the 17α-caproate. Medroxyprogesterone acetate is very active orally (see Table 18-4) and has such a long duration of action intramuscularly that it cannot be routinely used intramuscularly for treating many menstrual disorders.

Norethindrone, USP, and **Norethynodrel, USP.** 17α-Ethinyl-19-nortestosterone, and its $\Delta^{5(10)}$ isomer, respectively, might appear at first glance to be subtle copies of each other. One would predict that the $\Delta^{5(10)}$ double bond would isomerize in the stomach's acid to the Δ^3 position. In fact, however, the two drugs were simultaneously and independently developed; hence, neither can be considered a copy of the other. Furthermore, norethindrone is about ten times more active than norethynodrel (see Table 18-4), indicating that isomerization is not as facile in vivo as one might predict. Although they are less active than progesterone when given subcutaneously, they have the important advantage of being orally active. The discovery of the potent progestin activity of 17α-ethinyltestosterone (ethisterone) and 19-norprogesterone preceded the development of these potent progestins. All are orally active, with the 17α-ethinyl group blocking oxidation to the less active 17-one. The rich electron density of the ethinyl group and the absence of the 19-methyl group greatly enhance progestin activity. Both compounds have become of great importance as progestin ingredients of oral contraceptives, although norethindrone, USP, and norethindrone acetate, USP, are widely employed for all the usual indications of the progestins. Because these compounds retain the key feature of the testosterone structure—the 17β-OH—it is not surprising that they possess some androgenic side effects. The related compound, **norgestrel, USP**, has an ethyl group instead of the C-13-methyl, but has similar biologic properties. Norgestrel is used only in oral contraceptives. All these 19-nortestosterone derivatives will be discussed in the later section on chemical contraceptives.

Megestrol Acetate, USP. 17-Hydroxy-6-methylpregna-4,6-diene-3,20-dione acetate (Megace) is a progestin used primarily for the palliative management of recurrent, inoperable, or metastatic endometrial or breast carcinoma.

CHEMICAL CONTRACEPTIVE AGENTS

In a 1974 review of chemical contraception, Bennett notes that it has taken about 2 million years for the world's population to reach 3 billion. Only a few more years will be needed to increase the population to 6 billion at current growth rates. There is no doubt that the world's increasing population is a major concern, vastly exceeding fuel and food supplies in many parts of the world.[132]

Political, cultural, and research cost barriers have enormously complicated the development of contraceptive agents in modern times. The reviews by Djerassi,[133] former Research Director of Syntex and inventor of norethindrone, and by Lednicer[134] are important reading. (Their "insiders' viewpoint" of the research competition during the 1950s and 1960s to develop steroid products is especially interesting.)

The need for birth control was most dramatically and effectively expressed from 1910 to 1950 by Margaret Sanger.[135, 136] This remarkable American woman, founder of Planned Parenthood in 1916, made birth control information generally available in the United States. She made the medical profession better aware of the needs of women, and also raised the funds necessary for the early research on oral contraceptives.

During the 1940s and 1950s, great progress was made in the development of intravaginal spermicidal agents. However, the most notable achievement in chemical contraception came in the early 1960s with the development of oral contraceptive agents—"the pill." Since that time, a number of postcoital contraceptives and abortifacients have been developed. Hormone-releasing intrauterine devices are also being tested. However, progress has been much slower in the development of male contraceptive agents.

In the following pages, each of these approaches to chemical contraception will be discussed. Individual compounds have already been discussed with the estrogens and progestins.

Several excellent reviews on the mechanisms of action of birth control agents, and birth control in general, have recently been published. The Syntex guide for pharmacists, *Methods of Birth Control*,[137] is particularly recommended for the classroom. *Contraceptive Technology*, published annually by Irvington Publishers, provides a comprehensive overview of recent advances in contraceptive use and development.[138] Bell and co-workers[139] have summarized advances in the design of drugs for controlling fertility. The American Hospital Formulary Service *Drug Information*,[15] also published annually, contains a current review of information of importance to the clinician. A 1985 review, "Contraception for the 80's," appearing in *American Pharmacy*[140] provides a practical guide for the practicing pharmacist. Miller,[137] Stubblefield,[141] and the editors of *Drug and Therapeutics Bulletin*[142] and the *Medical Letter*[143] have published useful overviews of oral contraceptive side effects and optimal matching of products to particular patients. The overview by Youngkin and Miller provides current information on clinical use of the new triphasic oral contraceptive products.[144]

OVULATION INHIBITORS AND RELATED HORMONAL CONTRACEPTIVES

History[132–134]

In the 1930s, several research groups found that injections of progesterone inhibited ovulation in rats, rabbits, and guinea pigs.[145–148] Sturgis, Albright, and Kurzrok, in the early 1940s, are generally credited with the concept that estrogens, progesterone, or both could be used to prevent ovulation in women.[143, 149–151] In 1955, Pincus[152] reported that progesterone given from day 5 to day 25 of the menstrual cycle would inhibit ovulation in women. During this time, Djerassi and Rosenkranz[153] of Syntex, and Colton[154] of G. D. Searle and Co. reported the synthesis of norethindrone and norethynodrel. These progestins possessed very high progestational and ovulation-inhibiting activity. Most of the synthetic work was made possible by the development of the Birch reduction by Arthur J. Birch in 1950, and used by Birch to synthesize 19-nortestosterone itself.[155]

Extensive animal and clinical trials conducted by Pincus, Rock, and Garcia confirmed, in 1956, that Searle's norethynodrel and Syntex's norethindrone were effective ovulation inhibitors in women. In 1960 Searle marketed Enovid (a mixture of norethynodrel and mestranol), and in 1962 Ortho marketed Ortho Novum (a mixture of norethindrone and mestranol) under contract with Syntex. Norethynodrel and norethindrone have remained the most extensively used progestins in oral contraceptives, but several other useful agents have been developed. These will be discussed in the sections that follow.

Therapeutic Classes and Mechanism of Action

The ovulation inhibitors and modern hormonal contraceptives fall into several major categories (Table 18-6), each with its own mechanism of contraceptive action.[137–144] Individual compounds have been discussed with the estrogens and progestins in the previous section.

Combination Tablets: Mechanism of Action

Although, as noted earlier, Sturgis and Albright recognized in the early 1940s that either estrogens or progestins could inhibit ovulation, it was subsequently found that combinations were highly effective. Some problems, such as breakthrough (midcycle) bleeding, were also reduced by the use of a combination of progestin and estrogen.

Although all the details of the process are still not completely understood, it is now believed that the combination tablets suppress the production of LH or FSH, or both, by a feedback-inhibition process (see Fig. 18-8). Without FSH or LH, ovulation is prevented. The process is similar to the natural inhibition of ovulation during pregnancy, caused by the release of estrogens and progesterone from the placenta and ovaries. An additional effect comes from the progestin in causing the cervical mucus to become very thick, providing a barrier for the passage of sperm through the cervix. However, because pregnancy is impossible without ovulation, the contraceptive effects of thick cervical mucus or alterations in the lining of the uterus[156] (to decrease the probability of implantation of a fertilized ovum) would appear to be quite secondary. Nevertheless, some authors have reported that occasionally ovulation may occur,[157, 158] and thus the alterations of the cervical mucus and the endometrium may actually serve an important contraceptive function (especially, perhaps, when the patient forgets to take one of the tablets). During combination drug treatment, the endometrial lining develops sufficiently for withdrawal bleeding to occur about four or five days after taking the last active tablet of the series (see Table 18-6).

Monophasic (Fixed) Combinations. The monophasic combinations of a progestin and estrogen contain the same amount of drug in each active tablet (see Table 18-6). As discussed later in this chapter, the trend in prescribing has been toward lower doses of estrogen. However, as estrogen levels are reduced, breakthrough bleeding (or "spotting") becomes an annoying side effect for some patients at early to midcycle. Spotting after midcycle or amenorrhea appears to be related to too little progestin relative to the estrogen. The biphasic and triphasic combinations were developed to solve these breakthrough bleeding problems in some patients.[15, 17, 144]

Biphasic and Triphasic (Variable) Combinations. As illustrated in Figure 18-10, in the natural menstrual cycle, progesterone plasma concentrations peak late in the cycle. The higher estrogen/progesterone ratio early in the cycle is believed to assist in development of the endometrium. The higher progesterone concentration later contributes to proliferation of the endometrium and a resultant "normal" volume of menstrual flow. The biphasic and triphasic combinations attempt to mimic this variation in estrogen/progestin levels, and thereby to reduce the incidence of spotting associated with low-dose monophasic combinations.[15, 144] With

TABLE 18-6

COMPARISON OF STEROID CONTRACEPTIVE REGIMENS

1. COMBINATION — MONOPHASIC

Products are available in 20 (21)- or 28-day dispensers and refills. The first tablet is taken on the fifth day after menstruation has started, or on the first Sunday after or on which menstruation has started. The 28-day dispensers contain seven inert (or Fe^{2+}-containing) tablets of a different color, taken daily after the 20 or 21 days of active tablets. With 20 (21)-day regimens, if menstruation does not occur, the next cycle of 20 (21) tablets begins on the eighth day after the last active tablet was taken. Doses of active tablets are shown below.

Brand	Progestin	Estrogen
Loestrin 1/20	Norethindrone acetate, 1 mg	Ethinyl estradiol, 0.02 mg
Loestrin 1.5/30	Norethindrone acetate, 1.5 mg	Ethinyl estradiol, 0.03 mg
Lo/Ovral	Norgestrel, 0.3 mg	Ethinyl estradiol, 0.03 mg
Nordette	Norethindrone acetate, 1.5 mg	Ethinyl estradiol, 0.03 mg
Ovcon-35	Norethindrone, 0.4 mg	Ethinyl estradiol, 0.035 mg
Brevicon	Norethindrone, 0.5 mg	Ethinyl estradiol, 0.035 mg
Modicon	Norethindrone, 0.5 mg	Ethinyl estradiol, 0.035 mg
Demulen 1/35	Ethynodiol diacetate, 1 mg	Ethinyl estradiol, 0.035 mg
Ortho-Novum 1/35	Norethindrone, 1 mg	Ethinyl estradiol, 0.035 mg
Norinyl 1 + 35	Norethindrone, 1 mg	Ethinyl estradiol, 0.035 mg
Norlestrin 1/50	Norethindrone acetate, 1 mg	Ethinyl estradiol, 0.05 mg
Norlestrin 2.5/50	Norethindrone acetate, 2.5 mg	Ethinyl estradiol, 0.05 mg
Demulen	Ethynodiol diacetate, 1 mg	Ethinyl estradiol, 0.05 mg
Ovral	Norgestrel, 0.5 mg	Ethinyl estradiol, 0.05 mg
Norinyl 1 + 50	Norethindrone, 1 mg	Mestranol, 0.05 mg
Ortho-Novum 1/50	Norethindrone, 1 mg	Mestranol, 0.05 mg
Ovcon-50	Norethindrone, 1 mg	Ethinyl estradiol, 0.05 mg
Ortho-Novum 10 mg	Norethindrone, 10 mg	Mestranol, 0.06 mg
Enovid 5 mg	Norethynodrel, 5 mg	Mestranol, 0.075 mg
*Ortho-Novum 1/80	Norethindrone, 1 mg	Mestranol, 0.08 mg
*Norinyl 1 + 80	Norethindrone, 1 mg	Mestranol, 0.08 mg
*Ortho-Novum 2 mg	Norethindrone, 2 mg	Mestranol, 0.1 mg
*Norinyl 2 mg	Norethindrone, 2 mg	Mestranol, 0.1 mg
*Enovid-E	Norethynodrel, 2.5 mg	Mestranol, 0.1 mg
*Ovulen	Ethynodiol diacetate, 1 mg	Mestranol, 0.1 mg

2. COMBINATION — BIPHASIC

Products are available in 21- or 28-day dispensers and refills. They are taken on the same schedule of 21 days plus 7 days of no (or inert) tablets as with the foregoing monophasics. Doses of active tablets are shown below.

Brand	Progestin and Estrogen
Ortho-Novum 10/11	10 days: Norethindrone, 0.5 mg, and ethinyl estradiol, 0.035 mg then 11 days: Norethindrone, 1 mg, and ethinyl estradiol, 0.035 mg

3. COMBINATION — TRIPHASIC

Products are available in 21- or 28-day dispensers and refills. They are taken on the same schedule of 21 days plus 7 days of no (or inert) tablets as with the foregoing monophasics. Doses of active tablets are shown below.

Brand	Progestin and Estrogen	
Ortho 7/7/7	first 7 days:	Norethindrone, 0.5 mg, and ethinyl estradiol, 0.035 mg
	next 7 days:	Norethindrone, 0.75 mg, and ethinyl estradiol, 0.035 mg
	next 7 days:	Norethindrone, 1 mg, and ethinyl estradiol, 0.035 mg

*These high-dose estrogen products are being discontinued.

TABLE 18-6 *Continued*

COMPARISON OF STEROID CONTRACEPTIVE REGIMENS

3. COMBINATION — TRIPHASIC *(CONTINUED)*

Products are available in 21- or 28-day dispensers and refills. They are taken on the same schedule of 21 days plus 7 days of no (or inert) tablets as with the foregoing monophasics. Doses of active tablets are shown below.

Brand		Progestin and Estrogen
Tri-Norinyl	first 7 days:	Norethindrone, 0.5 mg, and ethinyl estradiol, 0.035 mg
	next 9 days:	Norethindrone, 1 mg, and ethinyl estradiol, 0.035 mg
	next 5 days:	Norethindrone, 0.5 mg, and ethinyl estradiol, 0.035 mg
Triphasil	first 6 days:	Levonorgestrel, 0.05 mg, and ethinyl estradiol, 0.03 mg
	next 5 days:	Levonorgestrel, 0.075 mg, and ethinyl estradiol, 0.04 mg
	next 10 days:	Levonorgestrel, 0.125 mg, and ethinyl estradiol, 0.03 mg

4. PROGESTIN ONLY

An active tablet is taken each day of the year.

Brand	Progestin	Dose	Dosage Cycle
Micronor	Norethindrone	0.35 mg	Continuous daily
Nor-Q.D.	Norethindrone	0.35 mg	Continuous daily
Ovrette	Norgestrel	0.075 mg	Continuous daily

5. INJECTABLE DEPOT HORMONAL CONTRACEPTIVES (NOT APPROVED FOR USE IN UNITED STATES)

Brand	Progestin	Dosage Cycle
Depo-Provera	Medroxyprogesterone acetate	400–1000 mg initially, with maintenance dose as low as 400 mg/mo

6. ONCE-A-MONTH ORAL COMBINATION CONTRACEPTIVE (NOT AVAILABLE IN UNITED STATES)

Progestin	Estrogen	Dosage Cycle
Norethindrone acetate 3-Cyclopentyl enol ether (quinestrol)	Ethinyl estradiol 3-Cyclopentyl ether (quingestanol)	1 tablet/mo

7. HORMONE-RELEASING IMPLANTS AND IUDS

Brand	Drug	Dosage Cycle
Progestasert	Progesterone-releasing IUD	38 mg dose in IUD lasts 1 yr
(in trials)	Subcutaneous Silastic implants containing estrogens and/or progestins implanted in forearms	Reported effective up to 1 yr per implant
(in trials)	Intravaginal Silastic rings containing progestins	Under study by Upjohn

8. PROGESTERONE ANTAGONISTS (IN CLINICAL TRIALS)

Brand	Drug	Dosage Cycle
Mifepristone	RU 486	50 mg once a month at midcycle

proper selection of patients, the goal has been achieved; but in other patients, the incidence of spotting has not appreciably decreased.[144]

The overall amount of progestin given in a month's cycle is also typically reduced with the biphasic and triphasic combination products. This has added to the marketing focus of these products, but the clinical relevance is uncertain.[141, 144] The reason is that some progestins may reduce high-density lipoprotein (HDL), but typically not at the progestin doses used in many low- to middose monophasic combinations.

How Safe?

The safety of the "pill" (typically meaning fixed-dose combinations) has been one of the most intensively discussed subjects in the press. Earlier studies, based largely on the earlier products that contained high doses of estrogen, showed an alarming incidence of thromboembolic disease. The results of these findings have been that (1) the sequential contraceptive products with their high doses of estrogen have been removed from American markets; (2) many combination contraceptives containing less than 0.050 mg estrogen per dose have recently been marketed (see Table 18-6); (3) progestin only or minipill products have appeared (see Table 18-6); and (4) a few groups of women have been identified who should definitely not take oral contraceptives (e.g., women older than 40 and women who are moderate to heavy smokers).

Several studies have shown that the risk of thromboembolism with preparations containing less than 0.05 mg estrogen per dose is less than with those containing 0.05 mg or more.[141, 143, 159]

The actual incidence of "pill-induced" cardiovascular death for nonsmoking young women appears to be quite small. An analysis by Tietze has shown that in the United States since adoption of the "pill" in the 1960s, deaths of reproductive-aged women from cardiovascular disease have declined much more steeply than those for men of the same-aged group.[160] Indeed, the risk of death from myocardial infarction for men (who obviously do not take the estrogen/progestin contraceptives) is significantly higher (Table 18-7). The risk of death by car accident is even higher for both sexes.

Jain has reported that the risk of death from myocardial infarction for "pill users" aged 30 to 39 is about 1.8 : 100,000 nonsmokers and 13.0 : 100,000 for heavy smokers (15 or more cigarettes per day).[161] The risks increase significantly for women 40 and older. For example, a 1979 study by Shapiro and co-workers[162] showed that the risk of myocardial infarction with oral contraceptive users of mean age 43 was four times that of nonusers. For heavy smokers of this age, the risk was 20 to 40 times greater. A recent study by Willette and co-workers has shown that smoking even one to four cigarettes a day will double a woman's risk of fatal coronary heart disease.[163]

TABLE 18-7

INCIDENCE OF DEATH FROM VARIOUS CAUSES, BY SEX AND AGE, 1986

		Death per 100,000 w/10-yr Range*	
Cause of Death	**Sex**	**Age Group (yr)**	**United States**
Myocardial infarction	M	15–24	0.2
	F	15–24	0.1
	M	25–34	2.7
	F	25–34	0.8
	M	35–44	22.4
	F	35–44	4.9
Malignant neoplasm genital organs	M	15–24	0.4
	F	15–24	0.3
	M	25–35	0.7
	F	25–35	2.1
	M	35–44	0.6
	F	35–44	7.2
Malignant neoplasm of breast	M	15–34†	—
	F	15–24	0.1
	F	25–34	3.1
	M	35–44	0.1
	F	35–44	18.3
Motor vehicle accident	M	15–24	58.2
	F	15–24	19.5
	M	25–34	37.7
	F	25–34	10.7
	M	34–44	24.8
	F	34–44	8.6

* Data supplied by Vital Statistics for US, 1986 Mortality, Vol. 2. Data are for all people in the sex and age group; therefore they do not reflect effects of other health factors, such as diet, exercise, smoking, and medicines.

† 20-year age group.

Nevertheless, the risk of cardiovascular disease from oral contraceptives appears to be *less* than the risks associated with pregnancy for women over 40. In April 1990, the FDA instructed manufacturers to revise labeling to reflect current scientific opinion that the benefits of oral contraception for healthy, nonsmoking women over 40 may outweigh the possible risks.

Oral contraceptives have been shown to decrease the risk of ovarian cancer[164] and not to increase the incidence of breast cancer.[165, 166] Oral contraceptives may alter serum proteins and liver function tests. Contact lens wearers taking oral contraceptives may experience more irritation or dryness than before taking them.

Which Combination?

Because the risk of thromboembolic disease with combination oral contraceptives is associated with estrogen dose, most physicians prescribe products with 0.05 mg or less estrogen. As shown in Table 18-9, lowering the estrogen content also slightly increases the incidence of pregnancy, although pregnancies are very rare. As noted in the previous sections, the frequency of spotting or breakthrough bleeding can also increase, particularly with monophasic products containing less than 0.035 mg estrogen per dose.

The FDA[166] and American Hospital Formulary Service[15] conclude:

> The combined oral contraceptive product prescribed should be one which contains the least amount of estrogen and progestin that is compatible with a low failure rate and needs of the individual patient.... The risk of serious morbidity or mortality is very small in healthy women without underlying risk factors (smoking, hypertension, hypercholesterolemia, obesity, diabetes...).[166]

> Preparations containing more than 0.05 mg of estrogen should generally be reserved for patients in whom lower dosage preparations are considered ineffective or are associated with bleeding irregularities.... Common adverse effects... disappear or diminish after 3 or 4 cycles.... If minor adverse effects persist after the fourth cycle, a different combination or a different dosage may be tried.[15]

In 1988, the FDA announced that the three manufacturers of high estrogen dose combination products were discontinuing the production and distribution of those products (see Table 18-6).[166a]

Oral Contraceptives and Pregnancy — Risks

Formerly, it was a common practice to prescribe progestins to prevent spontaneous abortions in some women, to administer high doses of progestins as a test for pregnancy, and to recommend "continue taking" oral contraceptives for an additional month if the patient missed a menstrual period. However, the FDA, USP, and independent investigators have strongly warned *against* the use of steroidal hormones for any purpose during early pregnancy[15,17,130] because of possible damage to the fetus. On the other hand, several studies have shown that there is no significant effect on progeny when women have taken "the pill" before becoming pregnant.[15,17,130,131,167–169]

The *USP DI*[17] and *AHFS Drug Information*[15] provide instructions for patients who have missed one or more doses of oral contraceptives.

Progestin Only (Minipill)

The estrogen component of sequential and combination oral contraceptive agents has been related to some side effects, with thromboembolism being a particular concern. One solution to this problem has been to develop new products with decreased estrogen content. In the "minipill," there is no estrogen at all.

Although higher doses of progestin are known to suppress ovulation, minipill doses of progestin are not sufficient to suppress ovulation in all women. Some studies have indicated that an increase in the viscosity of the cervical mucus (or sperm barrier) could account for much of the contraceptive effect, whereas other studies disagree.[132,137] Low doses of progestin have also been found to increase the rate of ovum transport and to disrupt implantation. There is a good probability that most, or all, of these factors contribute to the overall contraceptive effect of the minipill. The incidence of pregnancy with the minipill is higher than with combination products (see Table 18-9).

Injectable Depot Hormonal Contraceptives

In principle, there is no reason why long-acting depot preparation of a progestin or an estrogen–progestin combination could not be developed. Although there have been tests on a number of animals with various hormone preparations, only one drug (Depo-Provera) has been sufficiently tested to merit temporary FDA approval (later withdrawn). Other drugs still undergoing clinical or preclinical trials have been found to be effective contraceptive agents, but irregular menstrual cycles and menstrual "spotting" remain major problems.

Injectable medroxyprogesterone acetate was briefly approved by the FDA as a depot hormonal contraceptive for women who cannot use other methods of contraception (e.g., in mental institutions and low socioeconomic areas where patients probably would not follow the important dosage schedule of the oral contraceptive hormones).[170] However, the approval was stayed until its safety is more carefully evaluated.[114,171] Doses of 150 mg are injected once every three months, with other methods of contraception recommended for the first month after the initial injection. Prolonged infertility after stopping use of the drug is common.[127]

Once-a-Month and Once-a-Week Oral Contraceptives

The advantages of a once-a-month oral contraceptive are obvious, and good progress has been made in the development of such drugs. However, as reviewed by Diczfalusy, development of a long-acting, fertility-regulating agent is a difficult, costly, and time-consuming process.[171] Berman and co-workers have reported that a small oral dose of ethinyl estradiol 3-cyclopentyl ether (quinestrol) and norethindrone acetate 3-cyclopentyl enol ether (quingestanol) is effective in humans when given once a month.[172] However, full contraceptive protection is not achieved until the second month's dose has been taken. After that time, contraceptive efficiency is reported to be excellent.

Hormone-Releasing Implants and Intrauterine Devices[173, 174]

The low progestin doses of the minipill seem to have a direct effect on the uterus and associated reproductive tract. Therefore, it would seem possible to lower the progestin dose even more if the drug was released in the reproductive tract itself. Several devices employing these concepts are now being studied in clinical trials and are expected to be on the market in the future. These intravaginal and intracervical devices have been reviewed by the World Health Organization.[175]

In 1964, Folkman and Long[176] showed that chemicals can be released by diffusion through the walls of a silicone rubber capsule at a constant rate. A particularly attractive silicone rubber was found to be Silastic (Dow) which was nontoxic and apparently nonallergenic. During initial studies, capsules made of Silastic and containing estrogens or progestins, or both, were implanted subcutaneously in the forearms of women patients. These studies are still in progress, but it has been possible to obtain efficient contraception for one year with forearm transplants. Patient acceptance may be a significant problem.

Similar studies have been conducted with uterine-implanted Silastic capsules containing low doses of progestins. It was envisioned that progestin-containing intrauterine devices (IUDs) would have some particular advantages over other IUDs. First, the progestin should decrease uterine contractility (thus decreasing the number of IUDs ejected). Second, it should decrease the vaginal bleeding sometimes associated with IUDs. Additional studies are in progress to evaluate these predictions.

The Progestasert IUD (Progesterone Intrauterine Contraceptive System, USP) has 38 mg of microcrystalline progesterone dispersed in silicone oil. The dispersion is contained in a flexible polymer in the approximate shape of a *T*. The polymer acts as a membrane to permit 65 μg of progesterone to be slowly released into the uterus each day for one year. Contrary to prediction, the progesterone-containing IUD has had some of the therapeutic problems of other IUDs, including a relatively low patient continuation rate, some septic abortions, and some perforations of uterus and cervix. Clinical studies[177] have produced the following data on Progestasert, expressed as events per 100 women through 12 months of use.

	Parous	Nulliparous
Pregnancy	1.9	2.5
Expulsion	3.1	7.5
Medical removals	12.3	16.4
Continuation rate	79.1	70.9

Biodegradable Sustained-Release Systems

A very interesting and exciting approach to fertility control uses biodegradable polymers and microparticles to release the estrogen or progestin. As reviewed by Benagiano and Gabelnick[178] of the World Health Organization, the microparticles can be injected through regular needles. Release of the active drug occurs by erosion, diffusion, and cleavage of covalent bonds between the drug and polymer. Polymer matrices that have been investigated for these purposes include caprolactone, glutamic acid, lactic acid, and glycolic acid polymers.

OTHER METHODS OF CHEMICAL CONTRACEPTION

RU 486

A completely new drug design approach to oral contraceptives for women, progesterone antagonists, has received considerable attention. The first to be studied in clinical trials is RU 486 {mifepristone; 11β-[*p*-(dimethylamino)phenyl]-17β-hydroxy-17-(1-propynyl)estra-4,9-dien-3-one}.[179–182]

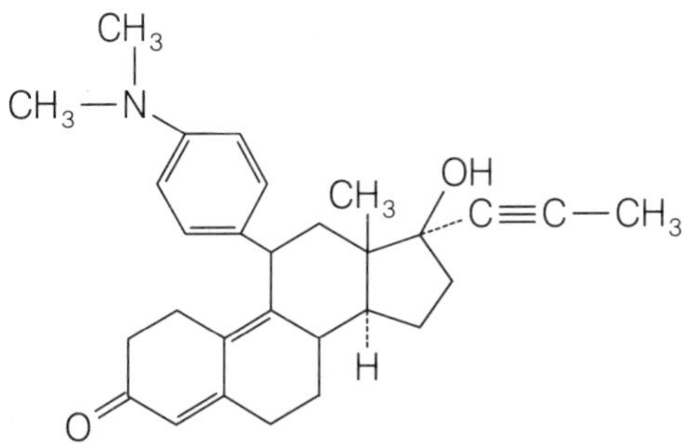

RU 486

As sumarized in *Science* in 1989,[182a] and *Scientific American* in 1990,[182b] "For a drug not yet a decade old, . . . RU 486 is causing quite a ruckus . . . . There is one thing no one argues about: RU 486 taken in conjunction with prostaglandins is an extremely effective method of terminating pregnancy within the first 9 weeks of gestation. And that could change the context of the debate over abortion and birth control . . . . It is not surprising, then, that RU

486 is viewed with alarm by antiabortion groups." France in mid-1990 was the only country in which RU 486 was widely available. Typically a PGE_1 or PGE_2 analogue has been used with the RU 486.

RU 486 binds strongly with both the glucocorticoid receptor and with the progesterone receptor. The story behind its design and development is fascinating reading.

When progesterone binds to its receptor, heat shock protein is released from the receptor and thereby opens the progesterone-receptor complex to DNA binding. However when RU 486 binds to the progesterone receptor, heat shock protein is not released; therefore no transcription of the DNA can occur. Alternatively, RU 486 may induce a conformational change in the progesterone receptor so that it does not fit its DNA site.

As shown in Figure 18-10, progesterone levels peak late in the menstrual cycle. Progesterone performs a variety of roles in maintaining secretory endometrium, and in inhibiting contractility of the uterus. Blocking these actions of progesterone results in breakdown of the endometrium, and detachment of the fertilized ovum (or embryo).

In addition to birth control applications, a variety of other therapeutic uses are also possible for RU 486, including treatment of progesterone sensitive cancers, and for Cushings syndrome (overproduction of glucocorticoids—discussed later in this chapter).

Postcoital Contraceptives

As reviewed by Bennett,[132] a variety of compounds have been found to be effective as postcoital contraceptives in animal studies. They range from steroid estrogens, such as DES, to compounds that bear little or no structural similarity to steroid hormones. Most act by an alteration of the mechanisms of fertilized ovum transport and implantation in the uterus.

Although the antiprogestin *RU 486* (see Table 18-6) may be approved for clinical use as a postcoital contraceptive, the most widely used agent today is the combination of *norgestrel and ethinyl estradiol* (Ovral).[138a] Two Ovral tablets are taken within 72 hours of unprotected intercourse, followed by another two tablets 12 hours later. Some patients experience nausea, which is usually mild. The treatment is successful with about 90% of patients in preventing pregnancy. It must be emphasized, however, that this treatment is intended only for use in short-term emergency situations.

Diethylstilbestrol has also been used as a postcoital contraceptive. However, the high estrogen doses (25 mg twice daily for five continuous days) are a significant concern—both for the patient and (if pregnancy results in spite of the treatment) for the unborn fetus (see section on Estrogen Side Effects). As with Ovral, treatment must begin no later than 72 hours after unprotected intercourse, and preferably within 24 hours.

Abortifacients

History records many different compounds that have been tried as abortifacients—everything from plant extracts to rusty nail water. Many chemicals have been very effective with animals, including metabolites, cytotoxic agents, 5-hydroxytryptamine, monoamine oxidase inhibitors, androgens, and others. Usually, these same compounds also have been found to be toxic or mutagenic or to cause severe hemorrhaging along with the abortion.

However, one compound, prostaglandin $F_{2\alpha}$ ($PGF_{2\alpha}$), has recently been approved by the FDA to induce second-trimester abortions. $PGF_{2\alpha}$ and PGE_2 concentrations significantly increase in amniotic fluid before normal labor and childbirth.

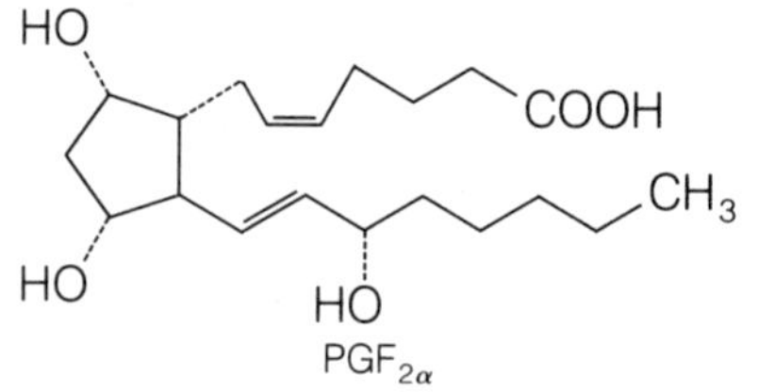

HO COOH CH$_3$ HO HO

$PGF_{2\alpha}$

Good surgical support is essential with $PGF_{2\alpha}$, because some clinicians report a high incidence of incomplete abortions that require dilatation and curettage. Furthermore, in those women in whom the placenta is retained, severe hemorrhage requiring transfusion may result. The drug is approved only for intra-amniotic injection in the second trimester. Suction is a common (and probably safer) method of clinical abortion during the first trimester, and saline-induced abortions are sometimes used during the second trimester. However, the saline method has been associated with disseminated intravascular coagulation in the patient, a problem not reported with $PGF_{2\alpha}$. (It should be noted that the β-adrenergic agonist *ritodrine* is used for the opposite effect—inhibiting preterm labor contractions.)

Spermicides

"As early as the 19th Century B.C., the Egyptians were mixing honey, natron [sodium carbonate], and crocodile dung to form a vaginal contraceptive paste. . . . During the middle ages, rock salt and alum were frequently used as vaginal contraceptives."[183] The history of spermicidal agents is indeed a long one. Modern spermicidal agents, or "vaginal contraceptives," fall into three categories: nonionic surfactants, bactericides, and acids. These agents have recently been reviewed by the FDA[184] and in the medical literature.[138,185] The new foaming tablet Encare Oval has also been studied.[186]

The FDA Advisory Panel on Nonprescription Products[184] found only three spermicides to be generally safe and effective: nonoxynol-9, octoxynol-9, and menfegol (Table 18-8). Other agents, including PEG 600 monolaurate and laureth-10S, were not classified because of insufficient data at the time the study was completed. Phenylmercuric acetate and phenylmercuric nitrate were found to be generally not safe and effective.

A 1987 study by Louik and co-workers[187] has confirmed several earlier reports showing that there is no increase in the overall frequency of birth defects in association with the use of spermicides. The study included use of spermicides during the first trimester of pregnancy.

TABLE 18-8

EXAMPLES OF COMMONLY USED SPERMICIDES

(In jellies, creams, suppositories, foaming tablets, aerosol foams, and soluble films — not all are available in the United States)

1. SAFE AND EFFECTIVE

Surface-active agents (also somewhat bactericidal)

a. 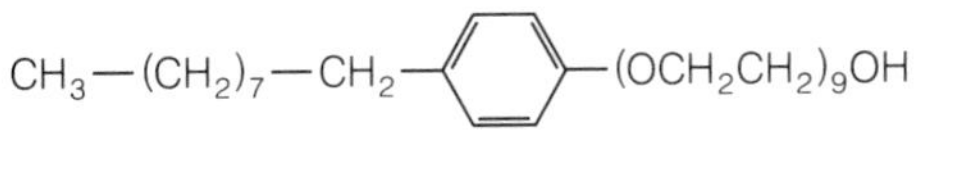 $CH_3-(CH_2)_7-CH_2-C_6H_4-(OCH_2CH_2)_9OH$

Nonoxynol-9, USP
(nonylphenoxypolyethoxyethanol)
(in Delfen, Immolin, Emko, Because, Encare Oval)

b. 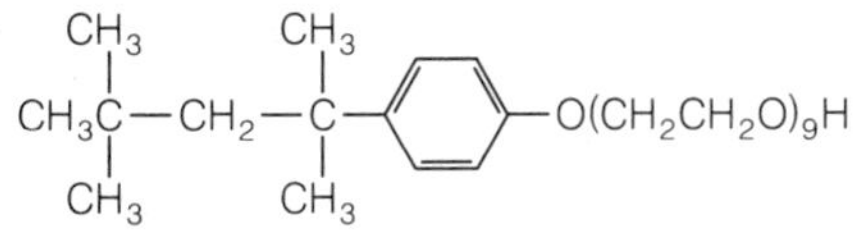 $CH_3C(CH_3)_2-CH_2-C(CH_3)_2-C_6H_4-O(CH_2CH_2O)_9H$

Octoxynol-9

2. NOT SAFE AND / OR EFFECTIVE

Bactericides

a. $C_6H_5HgOC(=O)CH_3$
Phenylmercuric acetate
(Lorophyn)

b. Others: benzethonium chloride, methylbenzethonium chloride, phenylmercuric borate

Acids

a. Boric acid

b. Others: tartaric acid, phenols, etc.

In addition to the inherent spermicidal properties of the active agent, the efficiency of spermicidal products depends upon many more factors. They must be inserted high into the vagina (usually with an applicator). They must (perhaps inconveniently) be used just before intercourse and reused if intercourse is to be repeated.

Furthermore, the formulation of these contraceptive products becomes almost as important as the active spermicidal agent itself. The product's formulation must permit diffusion into the cervix because some spermatozoa may be released directly into it. The product must also have a reasonable stability in the vagina so that enough active spermicide remains after intercourse. Finally, the ideal vaginal contraceptive must be nontoxic and nonirritating to both partners.

The primary action of the surface-active agents is to reduce the surface tension at the sperm cell surface and cause a lethal osmotic imbalance. They may also inhibit fructose metabolism. The bactericides also may alter the surface properties of the sperm cells, and after penetrating the cell membrane, they can disrupt metabolic processes. The acidic agents cause direct damage to the surface of the sperm cell membranes by denaturation of cell protein material. Examples of common spermicidal agents are shown in Table 18-8.

There are five primary types of vaginal contraceptive products containing spermicides: (1) creams, jellies, and pastes that are squeezed from a tube or applicator; (2) vaginal suppositories; (3) vaginal sponges; (4) foams (from aerosol pressurized containers or tablets); and (5) soluble films. Often, the vaginal contraceptives are used in combination with another contraceptive method (e.g., diaphragm, or "rhythm" method, or condom). The soluble films, primarily used in Europe, are transparent, water-soluble films that are impregnated with a spermicidal agent and then inserted into the vagina before intercourse.

BIOLOGICAL METHODS OF CONTRACEPTION

A Birth Control Vaccine in Clinical Trials

A birth control vaccine for women incorporating a synthetic peptide antigen representing the amino acid sequence 109–145 of the β-subunit of human chorionic gonadotropin (β-hCG) is now in Phase I clinical trials.[188] (An earlier approach using the entire β-subunit raised concern about antibodies that would also react with LH.)[189] As shown in Figure 18-11, hCG has some key roles in maintaining the corpus luteum after fertilization and in implantation. The stimulation of production of antibodies that would inactivate hCG could result in normal menstruation even after fertilization. Whatever the mechanism, earlier studies with primates showed, in principle, that this approach could be very effective. The contraceptive action is anticipated to last for several months.

LH-RH Agonists and Antagonists as Contraceptives

As discussed earlier in this chapter with gonadotropins, progress has been limited on the development of LH-RH analogues for use as contraceptive agents for men and women. An immense amount of research on these compounds has been conducted.[79]

MALE CONTRACEPTIVES

Future Prospects[190]

Much less research has been completed on the development of chemical contraceptives for men than for women. An excellent 1986 book, *Male Contraception: Advances and Future Prospects*, has summarized in detail progress made to date and the prospects for safe and effective agents in the future.[190] Earlier reviews were published in 1979[191] and in 1978.[192]

The lack of progress reflects various factors—an earlier concern that contraceptive products would not be used by men, the fact that spermatogenesis is not as cyclic as the menstrual cycle, and a lack of basic information about the biochemistry of the male reproductive system:

> We do not know how to interdict the production of sperm safely and reversibly. . . . We also have no existing products (as models). The result of this dilemma is that we do not know the adequacy of our experimental models to humans. We do not even know, for example, the exact requirements of androgens for maintenance of normal libido. The lack of existing male contraceptive drugs and knowledge about the male reproductive system should not cause pessimism. The same analogous situation existed with female oral contraceptives prior to the time of Pincus, Garcia, Chang, and Rock [in the early 1950s]. . . . [Bartke[193]]

> The male reproductive system seems less amenable to interference than does that of the female. . . . First, because the spermatogenic cycle is 74 days, months pass before a drug is effective. Second, because reproductive hormones are generally in a steady state in men, interruption of cyclicity is not an effective contraceptive approach. Third, because the testes are protected by a blood–testis barrier, many agents cannot reach the site of spermatogenesis. . . . [Alexander[194]]

> The chief reason . . . is the [earlier] presumption that men would not use a chemical contraceptive through fears of psychological or clinical effects upon libido and masculinity. . . . I have never agreed with this forecast. [Bennett[132]]

Ideally, one would like to have a drug that would inhibit spermatogenesis (without being mutagenic), would not decrease libido, and would not have any other effect on testicular function, e.g., hormone function. Alternatively, drugs that would affect only the spermatozoa *after* formation would be of great interest (e.g., drugs that could block the fertilizing ability of sperm stored in the epididymis).

Examples of drugs that have these properties in animals (and for a few in man) are shown in Figure 18-18.

Gossypol and Other Folklore Herbal Contraceptives

One compound that has received considerable attention as a male contraceptive is gossypol. A variety of other compounds have also been used as contraceptives in folk medicine.[195–198] Recent review about gossypol as a male contraceptive are recommended for further reading.[197, 197a]

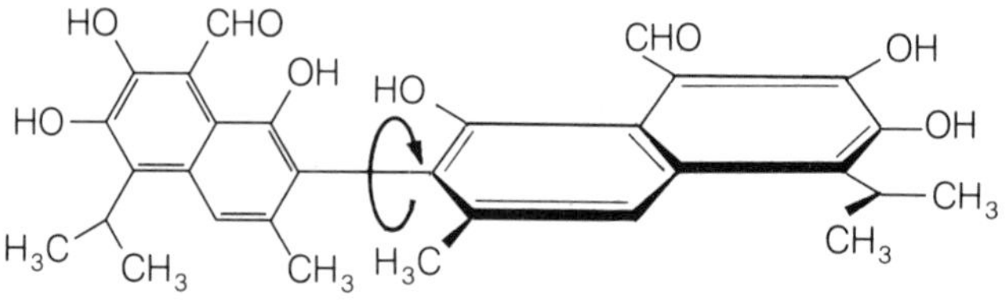

HO CHO OH HO CHO OH HO OH H_3C CH_3 CH_3 H_3C CH_3 CH_3

Gossypol, 1,1′,6.6′,7,7′-hexahydroxy-3,3′-dimethyl-5,5′-bis-(1-methylethyl)[2,2′-binaphthalene]-8,8′-dicarboxaldehyde, can exist in two different optical isomers. The optical isomers arise because of restricted rotation around the single bond connecting the two naphthalene ring systems. Most research has been done on the racemic mixture, but it appears that the (+) and (−) enantiomers may have significantly different activities. Gossypol is naturally found in cottonseed oil, used for cooking throughout the People's Republic of China. Today, gossypol is being evaluated in studies throughout the world, and it is claimed that over 12,000 men have been administered this drug.[196]

Gossypol has been found to be a very effective, but toxic, contraceptive agent. Claims of effectiveness have been as high as 99%,[195] with gossypol appearing to act on the spermatid stage of spermatogenesis. However, toxicity is a problem, particularly hypokalemic-induced paralysis. Dietary selenium may significantly enhance gossypol toxicity. Waller

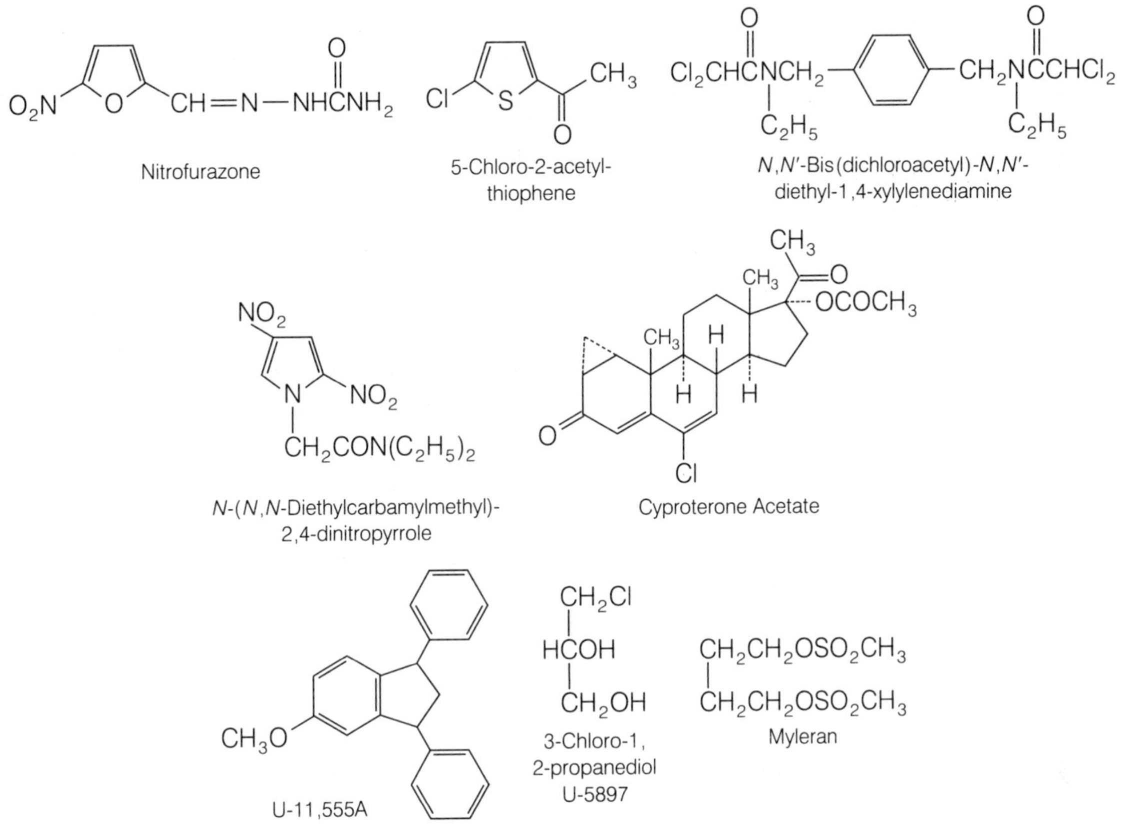

Nitrofurazone

5-Chloro-2-acetylthiophene

N,N'-Bis(dichloroacetyl)-*N,N'*-diethyl-1,4-xylylenediamine

N-(*N,N*-Diethylcarbamylmethyl)-2,4-dinitropyrrole

Cyproterone Acetate

U-11,555A

3-Chloro-1,2-propanediol U-5897

Myleran

FIG. 18-18. *Examples of chemical contraceptives for males. (Most have been tested only in animals. Some are quite toxic.) For a comprehensive review, Male Contraception: Advance and Future Prospects (Zatuchni, G. I., et al., eds.)*[190] *is recommended.*

and co-authors[197] conclude that:

> Gossypol in its present form is unsuitable and will not be approved by Western drug regulatory agencies as a male contraceptive agent. The therapeutic index ... is too small .... The poor distribution of gossypol to the testes, and high concentrations in organs such as the liver lead to significant exposures of tissues not involved [in the contraceptive action] ... [However,] gossypol may prove to be a very important compound even if it is never widely used as an antifertility agent.[197]

As a "lead compound" for studying the male reproductive system and for stimulating design of new and less toxic drugs, gossypol has been enormously important. Other herbal contraceptives[196] are also being investigated.

RELATIVE CONTRACEPTIVE EFFECTIVENESS OF VARIOUS METHODS

Some caution is required in interpreting data on the effectiveness of contraceptive methods. Even the "best" method can lead to pregnancy if not used consistently and correctly. Even the generally least effective method is better than no contraceptive at all. Table 18-9 presents some data on numbers of

TABLE 18-9

FAILURE RATE OF CONTRACEPTIVE METHODS*

(Data Based on Actual Users When Available)

Method	*Pregnancies/100 Woman-Years*
Abortion	0
Tubal ligation	0.04
Vasectomy	0.15
Combination oral contraceptive	
High estrogen dose (750 μg)	0.05–0.1
Mid estrogen dose (50 μg)	0.05–0.1
Low estrogen dose (30–35 μg)	0.03–0.7
Progestin only, oral (minipill)	.63–1.58
Progestasert IUD	1.9–2.5
IUD	1.6–5
Condom and spermicide	1–5
Spermicidal foam	3–29
Diaphragm	9–33
Condom	3–28
Coitus interruptus	18–22
Douche	33–60
Rhythm	1–47
No contraceptive method	60–80

*Table does not reflect relative safety, ease of use, and such.

pregnancies per method. The excellent review of Huff and Hernandez[185] on the over-the-counter contraceptives is suggested for a comparable study.

ANDROGENS AND ANABOLIC AGENTS

The commonly used androgenic and anabolic agents are shown in Figure 18-20. Several excellent reviews on androgens and anabolic agents have been published.[3,199] The text *Endocrine Physiology* by C. R. Martin[19] is highly recommended for a detailed review of androgen release, biosynthesis, and regulation.

STRUCTURAL CLASSES

Natural Hormones: Testosterone and Dihydrotestosterone

Although produced in small concentrations in females, testosterone and its potent metabolite dihydrotestosterone (Fig. 18-19) are produced in much greater amounts in males. Testosterone has two important activities: *androgenic activity* (or male sex characteristic-promoting) and *anabolic activity* (or muscle-building). Compounds that have these two activities are generally called androgens and anabolic agents. Because it would be very useful to have drugs that were anabolic (e.g., to aid the recovery of severely debilitated patients) but not androgenic, many compounds with increased anabolic activity have been synthesized. However, significant levels of androgenic activity have limited the therapeutic uses of all these compounds.

Semisynthetic Analogues

In his book *Androgens and Anabolic Agents*, Vida[199] has summarized the structures and biologic activity of over 500 different androgens and anabolic agents. Some, shown in Figure 18-19, have made it into clinical practice. The excellent discussion by Coun-

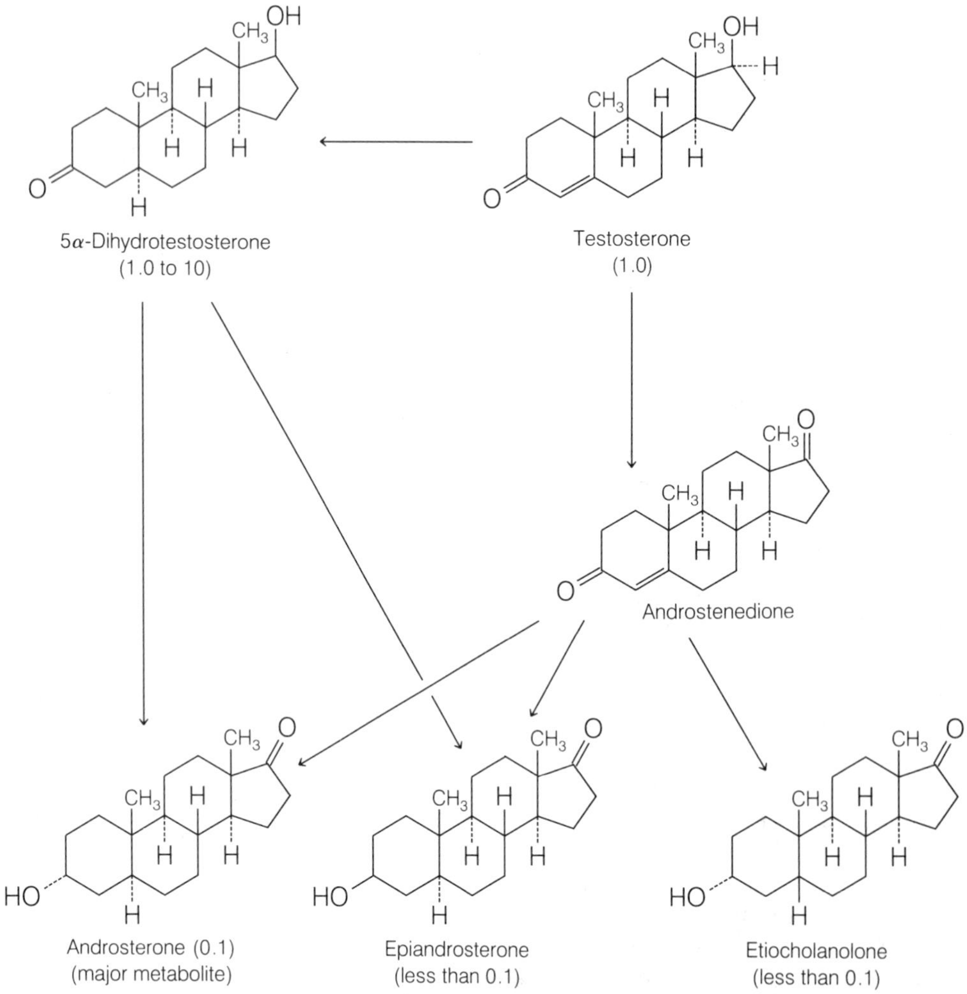

5α-Dihydrotestosterone (1.0 to 10)

Testosterone (1.0)

Androstenedione

Androsterone (0.1) (major metabolite)

Epiandrosterone (less than 0.1)

Etiocholanolone (less than 0.1)

FIG. 18-19. *Metabolism of testosterone and 5α-dihydrotestosterone (relative androgenic activity). Only primary metabolites are shown; for a full discussion, the review by Martin[19] is recommended.*

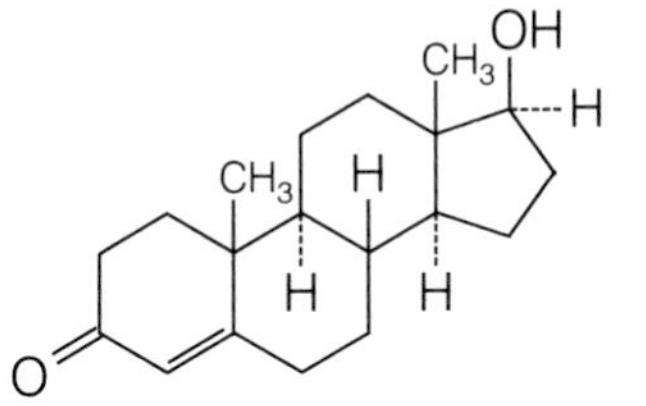

Testosterone
(1 : 1)

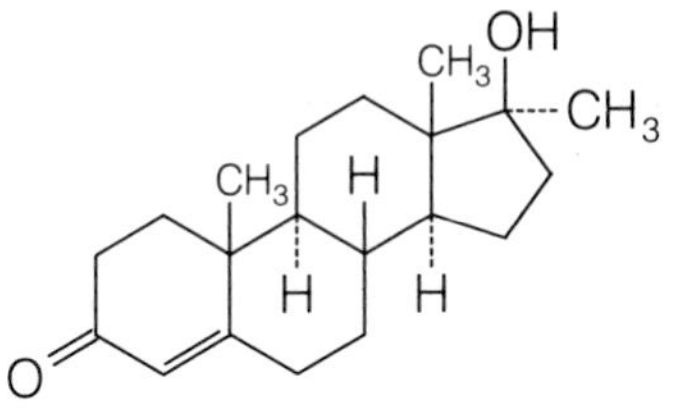

17α-Methyltestosterone
(1 : 1 but 1/2 as potent as testosterone)

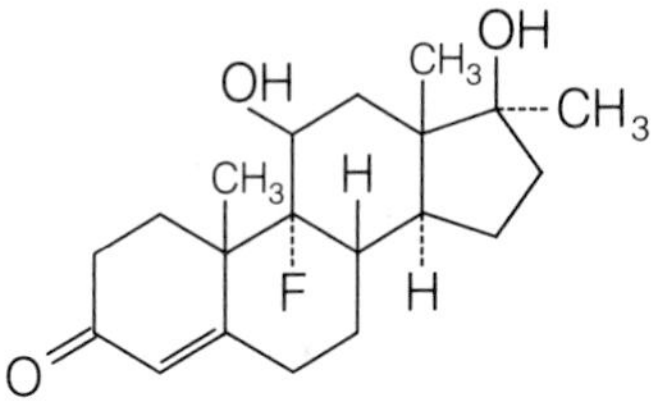

Fluoxymesterone
(1 : 1 to 2 : 1 and 5 to 10 times more potent than testosterone)

17β-Esters Commercially Available:

$-OC(=O)CH_2CH_3$ Testosterone Propionate

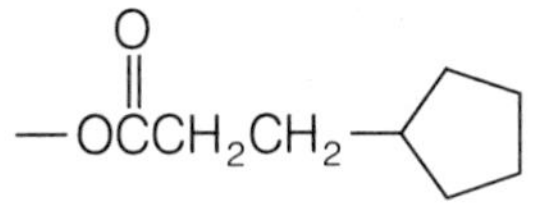

Testosterone Cyclopentylpropionate
(Cypionate)

$-OC(=O)CH_2CH_2CH_2CH_2CH_2CH_3$ Testosterone Enanthate

All are IM — some available as implantation pellets

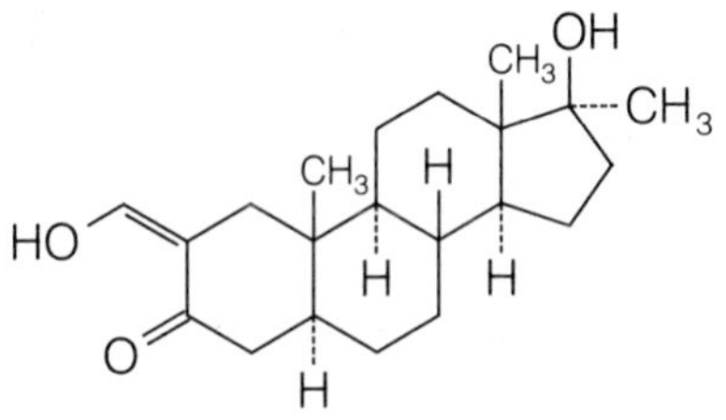

Oxymetholone
(2.5 : 1; 6 : 1 SC)

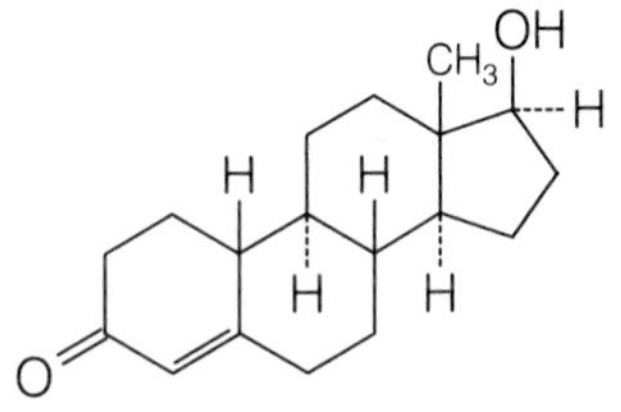

Nandrolone
(2.5 : 1 to 4 : 1)

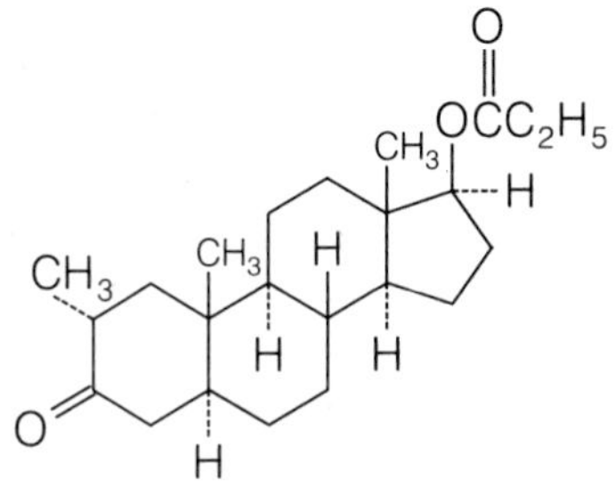

Dromostanolone
(Propionate, 3 : 1 to 4 : 1)

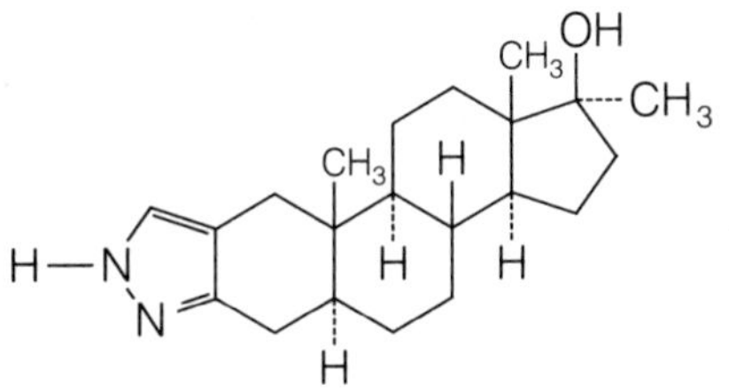

Stanozolol
(3 : 1 to 6 : 1)

17β-Esters Commercially Available:

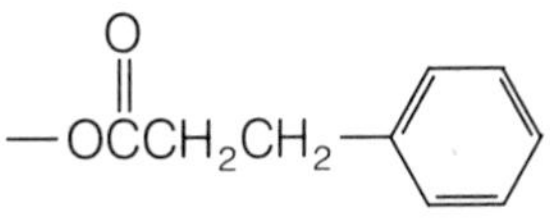

Nandrolone Phenpropionate

$-OC(=O)(CH_2)_8CH_3$ Nandrolone Decanoate

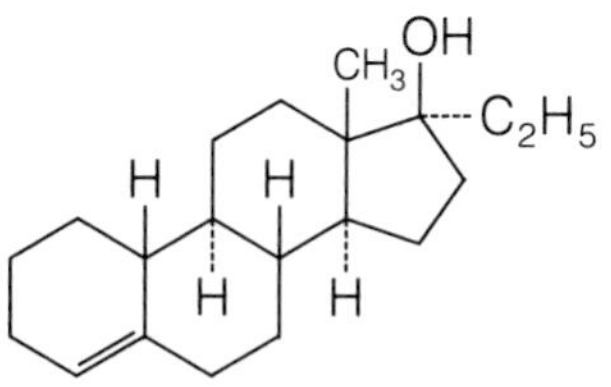

Ethylestrenol
(3 : 1)

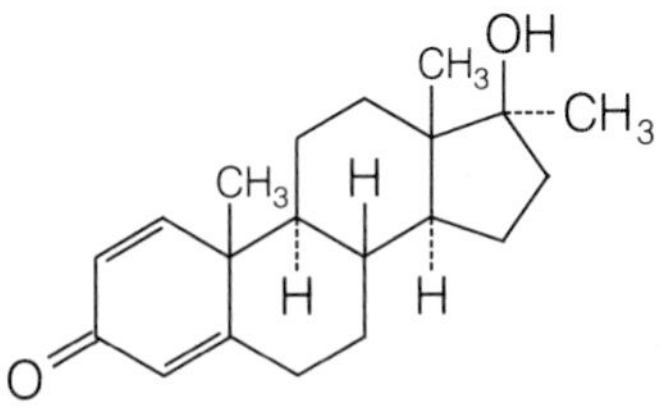

Methandrostenolone
(1 : 1)

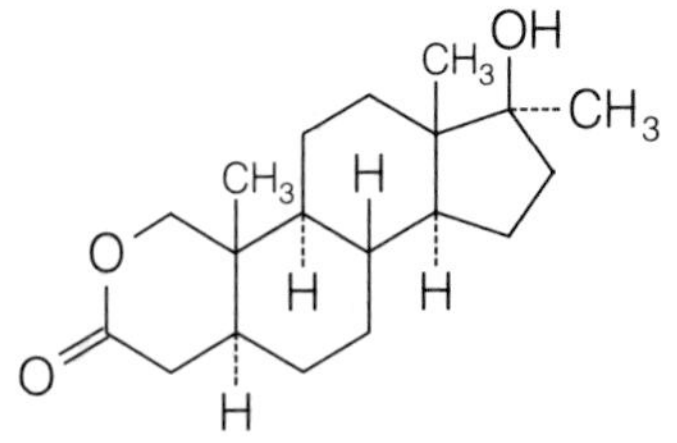

Oxandrolone

FIG. 18-20. *Androgens and anabolic agents (anabolic/androgenic ratio).*

sell and Brueggemier[3] also cites many compounds. One might suppose that the structure–activity relationships of these drugs have been well delineated. However, the structural requirements for selective anabolic activity are still unclear, and there is even uncertainty about the relationship of structure to androgenic activity.

Because bacterial and hepatic oxidation of the 17β-hydroxyl to the 17-one is the primary route of metabolic inactivation,[19] 17α-alkyl groups have been added. Even though 17α-methyltestosterone is only about half as active as testosterone, it can be taken orally. 17α-Ethyltestosterone has greatly reduced activity, as shown in Table 18-10. A disadvantage of the 17α-alkyl testosterones is that hepatic disturbances (and occasionally jaundice) may occur, particularly in the high doses typically used by athletes (see next section).

Table 18-10 illustrates other structure–activity effects of the androgen; for example, the greatly decreased activity of the 17β-ol isomer of testosterone.

Many hypotheses have been made in an attempt to summarize the structure–activity relationships of all the known androgens, including proposals of Vida,[199] Wolff,[200] and others. Counsell[3] and Klimstra[201] have published detailed discussions of the various hypotheses. They have also discussed the activity of the many substituted androstanes and testosterones studied to evaluate steric and electronic effects upon androgenic and anabolic activity.

Many drugs are available that have improved anabolic/androgenic activity ratios, but none is free of androgenic activity. This has greatly limited their therapeutic utility. Examples of drugs that have marked improvements in anabolic activity are illustrated in Figure 18-21, but these have not been used clinically because of hepatic toxicity or other side effects. For example, 19-nor steroids are quite anabolic, but their significant progestational activity has generally precluded their use.

TABLE 18-10

ANDROGENIC ACTIVITIES OF SOME ANDROGENS[153]

Compound	*Micrograms Equivalent to an International Unit*
Testosterone (17β-ol)	15
Epitestosterone (17α-ol)	400
17α-Methyltestosterone	25–30
17α-Ethyltestosterone	70–100
17α-Methylandrostane-3α,17β-diol	35
17α-Methylandrostane-3-one-17β-ol	15
Androsterone	100
Epiandrosterone	700
Androstane-3α,17β-diol	20–25
Androstane-3α,17α-diol	350
Androstane-3β,17β-diol	500
Androstane-17β-ol-3-one	20
Androstane-17α-ol-3-one	300
Δ^5-Androstene-3α,17β-diol	35
Δ^5-Androstene-3β,17β-diol	500
Androstanedione-3,17	120–130
Δ^4-Androstenedione	120

Generalizations about the structural changes that enhance anabolic activity are difficult to make, but Vida,[199] Counsell and Brueggemier,[3] and Klimstra[201] have presented detailed analyses. Albanese[117] has also made comparative studies of various anabolic agents in men and women. An examination of the compounds in Figure 18-19 shows that greater planarity and electron density in ring A seem to favor anabolic activity.

As with other compounds we have discussed, hydroxyl groups in the testosterones are often converted to the corresponding esters to prolong activity or to provide some protection from oxidation.

BIOSYNTHETIC SOURCES

As shown in Figure 18-5, testosterone can be synthesized through progesterone and androstenedione. About 7 mg/day is synthesized by young human

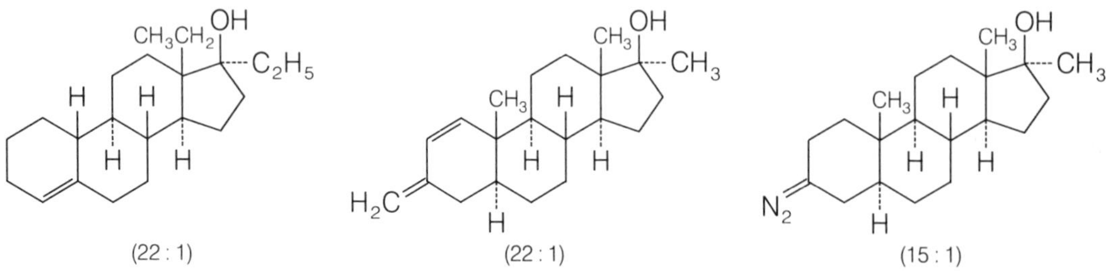

(22 : 1) (22 : 1) (15 : 1)

FIG. 18-21. Experimental compounds with improved anabolic activity (anabolic activity/androgenic activity ratio).

adult males.[19] Labeling experiments have also shown that it can be biosynthesized from androst-5-ene-3β,17β-diol.

Testosterone is primarily produced by the interstitial cells of the testes, synthesized largely from cholesterol made in sertoli cells.[19] The ovaries and adrenal cortex also synthesize androstenedione and 5-androsten-17-one-3β-ol (dehydroepiandrosterone), which can be rapidly converted to testosterone in many tissues.

Testosterone levels in the plasma of men are 5 to 100 times higher than the levels in the plasma of women.

Testosterone is produced in the testes in response to FSH and LH (interstitial cell-stimulating hormone, or ICSH) release by the anterior pituitary, as shown in Figure 18-9. Testosterone and dihydrotestosterone inhibit the production of LH and FSH by a feedback-inhibition process. This is quite similar to the feedback inhibition by estrogens and progestins in FSH and LH production.

METABOLISM

Testosterone is rapidly converted to 5α-dihydrotestosterone in many tissues, and 5α-dihydrotestosterone is also secreted by the testes. In fact, 5α-dihydrotestosterone is the active androgen in many tissues (e.g., in the prostate). The primary route for metabolic inactivation of testosterone and dihydrotestosterone is oxidation to the 17-one. The 3-one group is also reduced to the 3α- and 3β-ols. The metabolites are shown in Figure 18-19. Others have also been detected.[19]

BIOLOGIC ACTIVITIES

Testosterone and dihydrotestosterone cause pronounced masculinizing effects, even in the male fetus. They induce the development of the prostate, penis, and related sexual tissues.

At puberty, the secretion of testosterone by the testes increases greatly, leading to an increase in facial and body hair, a deepening of the voice, an increase in protein anabolic activity and muscle mass, a rapid growth of long bones, and a loss of some subcutaneous fat. Spermatogenesis begins, and the prostate and seminal vesicles increase in activity. Sexual organs increase in size. The skin becomes thicker and sebaceous glands increase in number, leading to acne in many young people. The androgens also play important roles in male psychology and behavior.

THERAPEUTIC USES

The primary use of androgens and anabolic agents is as androgen replacement therapy in men, either at maturity or in adolescence. The cause of testosterone deficiency may be either hypogonadism or hypopituitarism.

The use of the androgens and anabolic agents for their anabolic activity, or for uses other than androgen replacement, has been very limited because of their masculinizing actions. This has greatly limited their use in women and children. Although anabolic activity is often needed clinically, none of the products presently available has been found to be free of significant androgenic side effects.

The masculinizing (androgenic) side effects in females include hirsutism, acne, deepening of the voice, clitoral enlargement, and depression of the menstrual cycle. Furthermore, the androgens and anabolic agents generally alter serum lipid levels and increase the probability of atherosclerosis, characteristically a disease of men and postmenopausal women.

Androgens in low doses are sometimes used in the treatment of dysmenorrhea and postpartum breast enlargement. However, the masculinizing effects of the androgens and anabolic agents, even in small doses, preclude their use in most circumstances. Secondary treatment of advanced or metastatic breast carcinoma in selected patients is generally considered to be the *only* indication for large-dose, long-term androgen therapy in women.

Androgens and anabolic agents are also used to treat certain anemias, osteoporosis, and to stimulate growth in postpuberal boys. In all cases, use of these agents requires caution.

ANDROGENS AND SPORTS

The use of anabolic steroids by athletes began in the late 1940s[202] and is now widespread. One report estimated that up to 80% of competitive weightlifters and about 75% of professional football players use these drugs,[203] along with a variety of other athletes. One recent review puts the level of use at more than 90% of weightlifters and body builders.[204] During the 1972 Olympic games in Munich, 68% of interviewed track and field athletes (not including long-distance runners) had used anabolic steroids in preparation for the Olympics. Yet, international awareness did not stem the abuse of these drugs in subsequent Olympic Games. Canada's Ben Johnson was disqualified as the winner of the 1988 Olympic gold medal in the 100-yard dash for having traces of stanozolol in his urine. The shocking disqualification brought the international misuse of anabolic steroids to headlines worldwide: "The Doped-Up Games" (*Newsweek*, October 10, 1988), "The Drug of Champions" (*Science*, October 14, 1988). The subheadline in *Science* summarized the problem succinctly: "Athletes and body builders support a $100 million black market in steroids, while medical science has been slow to see why."

The understanding by athletes of the rational selection, risks, and risk/benefit ratios of the drugs they take appears to be quite limited.

Many studies have attempted to determine if taking anabolic steroids improves athletic performance.[205–217] However, some failed to use controls (athletes who trained in an identical manner, but who did not take anabolic steroids). Others failed to use placebos in at least a single-blind research design (neither the treated nor control groups knowing which they were taking).

Of the studies using at least a single-blind protocol, some have reported that anabolic steroids did increase athletic performance, whereas others found they did not. It would be fair to say, therefore, that the benefit of anabolic steroids to athletic performance is uncertain.

The risks of using these drugs appear to outweigh their uncertain benefits.

In late 1987, the FDA issued a special alert about the risks of anabolic steroid use and abuse by athletes,[218] along with a nationwide educational campaign. Other recent reviews on drug abuse and the athlete have shared a similar concern.[202–204, 218–221] The shocking book by Robert Goldman, *Death in the Locker Room*,[222] raised public perceptions about the risks of anabolic steroid abuse by athletes.

As summarized by the FDA[218] and recent reviews,[202–204, 218–221] the side effects include:

In both sexes

- Increased risk of coronary heart disease, stroke, or obstructed blood vessels
- Increased aggression and antisocial behavior (known as "steroid rage")
- Liver tumors, peliosis hepatis (blood-filled cysts), and jaundice

In men

- Testicular atrophy with consequent sterility or decreased sperm count and abnormal motility and morphology
- Impotence
- Enlarged prostate
- Breast enlargement

In women

- Clitoral enlargement
- Beard growth
- Baldness
- Deepened voice
- Breast diminution

Because of these risks, the International Olympic Committee has banned all anabolic drugs. The top winners in each Olympic event are tested for nontherapeutic drugs of all types.[223] The use of these drugs by athletes has also been criticized and discouraged by coaches,[224] physicians,[225] and other athletic governing bodies.[226]

PRODUCTS

Therapeutic uses of the androgens and anabolic agents have been previously discussed. 17β-Esters and 17α-alkyl products are available for a complete range of therapeutic uses (see Fig. 18-18). These drugs are contraindicated in men with prostatic cancer; in men or women with heart disease, kidney disease, or liver disease; and in pregnancy. Diabetics using the androgens and anabolic agents should be carefully monitored. Androgens potentiate the action of oral anticoagulants, causing bleeding in some patients, and they may also interfere with some laboratory tests.[15] Female patients may develop virilization side effects, and doctors should be warned that some of these effects may be irreversible (e.g., voice changes). Virtually all the anabolic agents now commercially available have significant androgenic activity; hence, virilization is a potential problem with all women patients. The 17α-alkyl products may cause cholestatic hepatitis in some patients.

Doses and dosage schedules for specific indications can vary markedly (see Therapeutic Uses for indications); accordingly, specialized dose-indication references such as *Drug Information* or *Facts and Comparisons*[16] should be consulted when advising physicians on doses. General USP DI doses are listed in Table 18-11.

All steroid 4-en-3-ones are light-sensitive and should be kept in light-resistant containers.

Testosterone, USP. 17β-Hydroxyandrost-4-en-3-one is a naturally occurring androgen in men. In women, it serves as a biosynthetic precursor to estradiol. However, it is rapidly metabolized to relatively inactive 17-ones (see Fig. 18-19), so it is not orally active. Testosterone 17β-esters are available in long-acting intramuscular depot preparations, illustrated in Figures 18-5 and 18-13, including the following:

Testosterone cypionate, USP. Testosterone 17β-cyclopentylpropionate

Testosterone enanthate, USP. Testosterone 17β-heptanoate

Testosterone propionate, USP. Testosterone 17β-propionate

TABLE 18-11

ANDROGENS AND ANABOLIC AGENTS

Name Proprietary Name	**Preparations**	**Usual Adult Dose***	**Usual Dose Range***
Testosterone, USP *Oreton, Neo-Hombreol (F)*	Testosterone pellets, USP Sterile testosterone suspension, USP	Implantation, 300 mg; IM, 25 mg twice weekly to once daily, depending on condition being treated	
Testosterone cypionate, USP *Depo-Testosterone, Malogen CYP, Durandro, T-Ionate-P.A.*	Testosterone cypionate injection, USP	IM, 200–400 mg once every 3–4 wk	100–400 mg
Testosterone enanthate, USP *Delatestryl, Malogen L.A., Testate, Testostroval-P.A.*	Testosterone enanthate injection, USP	IM, 200–400 mg once a month	100–400 mg
Testosterone propionate, USP *Neo-Hombreol, Oreton Propionate*	Testosterone propionate injection, USP Testosterone propionate tablets, USP	Replacement therapy: IM, 10–25 mg 2 or 3 times a week; inoperable mammary cancer: IM, 100 mg 3 times a week	20–300 mg/wk
Methyltestosterone, USP *Android, Metandren, Oreton Methyl, Testred, Neo-Hombreol (M)*	Methyltestosterone tablets, USP Methyltestosterone capsules, USP	Inoperable breast cancer: 100 mg bid; replacement therapy: 5–20 mg bid	5–200 mg/day
Fluoxymesterone, USP *Halotestin, Ora-Testryl, Ultandren*	Fluoxymesterone tablets, USP	Replacement therapy: 2–2.5 mg 1–4 times a day; inoperable mammary cancer: 5–10 mg tid	2–30 mg/day
Methandrostenolone, USP *Dianabol*	Methandrostenolone tablets, USP	2.5–5 mg/day	
Oxymetholone, USP *Adroyd, Anadrol, Anadrol-50*	Oxymetholone tablets, USP	5–10 mg/day	5–50 mg
Oxandrolone, USP *Anavar*	Oxandrolone tablets, USP		Initial: 5–10 mg/day; maintenance: 2.5–5 mg/day
Nandrolone decanoate, USP *Deca-Durabolin*	Nandrolone decanoate injection, USP	IM, 50–100 mg every 3–4 wk	
Nandrolone phenpropionate, USP *Durabolin, Durabolin-50*	Nandrolone phenpropionate injection, USP	IM, 25–50 mg/wk	
Stanozolol, USP *Winstrol*	Stanozolol tablets, USP	2 mg tid	

*See USP DI for complete dosage information.

Methyltestosterone, USP. 17β-Hydroxy-17-methylandrost-4-en-3-one is only about half as active as testosterone (when compared intramuscularly), but it has the great advantage of being orally active (see Fig. 18-5). (Methyltestosterone given by the buccal route is about twice as active as oral.) Both testosterone and methyltestosterone have high androgenic activity, limiting their usefulness where good anabolic activity–low androgenic activity is desired.

Fluoxymesterone, USP. 9-Fluoro-11,17-dihydroxy-17-methylandrost-4-en-3-one is a highly potent, orally active androgen, about five to ten times more potent than testosterone. It can be used for all the indications discussed previously, but its great androgenic activity has made it useful primarily for treatment of the androgen-deficient male.

Methandrostenolone, USP. 17β-Hydroxy-17-methylandrosta-1,4-dien-3-one is orally active and about equal in potency to testosterone.

Anabolic agents include the commercially available androgens with improved anabolic activity (see Fig. 18-19) and those that are still experimental (examples in Fig. 18-21). It should be emphasized that virtually all the commercial products have significant androgenic properties (ratios given in Fig. 18-19); consequently, virilization in women and children can be expected. Many of the anabolic agents are orally active, as one would predict by noting a 17α-alkyl group in many of them (see Fig. 18-19). Those without the 17α-alkyl (nandrolone and dromostalone) are active only intramuscularly. The commercially available anabolic agents include

Oxymetholone, USP. 17-Hydroxy-2-(hydroxymethylene)-17-methylandrostan-3-one

Oxandrolone, USP. 17β-Hydroxy-17-methyl-2-oxaandrostan-3-one

Stanozolol, USP. 17-Methyl-2′*H*-androst-2-eno[3,2,*c*]-pyrazol-17-ol

Nandrolone decanoate, USP and **nandrolone phenpropionate, USP.** 17β-Hydroxyestr-4-en-3-one 17-decanoate and 3′-phenylpropionate

ANTIANDROGENS

A variety of compounds (Fig. 18-22) have been intensively studied as androgen antagonists, or antiandrogens. As recently reviewed by Rasmussen,[227] a wide range of compounds has been evaluated.

Estrogens have been used as antiandrogens, but their feminizing side effects (e.g., loss of libido) have precluded their extensive use in men. Antiandrogens would be of therapeutic use in treating conditions of hyperandrogenism (e.g., hirsutism, acute acne, and premature baldness) or androgen-stimulated cancers (e.g., prostatic carcinoma). The ideal antiandrogen would be nontoxic, highly active, and devoid of any hormonal activity. Unfortunately, most of the compounds in Figure 18-22 have not met all these criteria completely,[228] although Sch 13521 (flutamide) and TSAA-291 (oxendolone) have had some notable successes.[229]

Flutamide used in combination with the LH-RH analogue leuprolide acetate has been found to be very effective in treating patients with androgen-dependent cancer of the prostate.[79a–c] As discussed earlier in the section on gonadotropins, high doses of LHRH agonists such as leuprolide greatly decrease gonadal function —the medical equivalent of castration in men. Thus, flutamide and a LHRH agonist taken together block testosterone and dihydrotestosterone receptors in the prostate. Prostatic carcinoma is known to be androgen sensitive.

Danazol and Other Drugs for Endometriosis

Endometriosis is characterized by endometrial tissue growing outside the uterus, especially in the pelvis.[230a, b] Constant pain starting two to seven days before menstruation, dyspareunia, and infertility of-

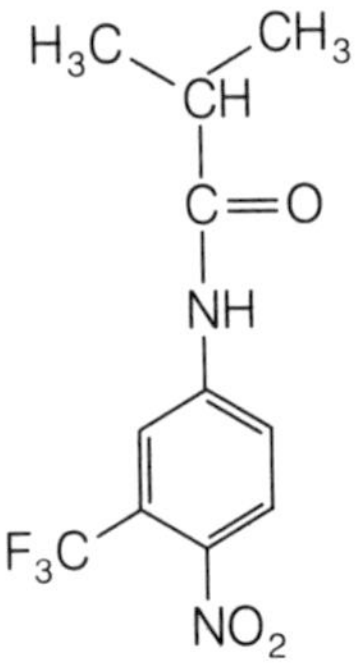

Sch 13521 (flutamide) (α,α,α-Trifluoro-2-methyl-4′-nitro-*m*-propionotoluidide)

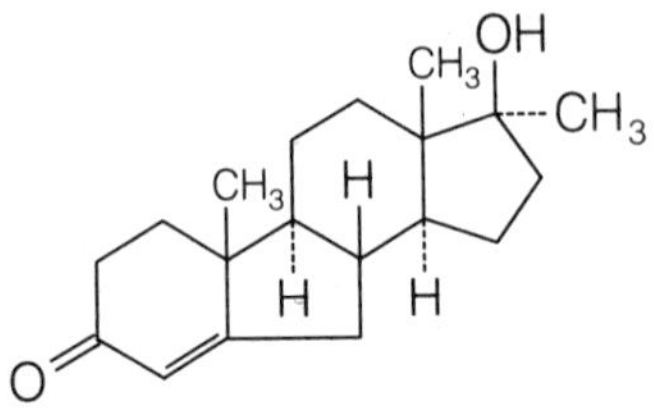

SK & F 7690 (benorterone) (17β-Hydroxy-17-*B* norandrost-4-en-3-one)

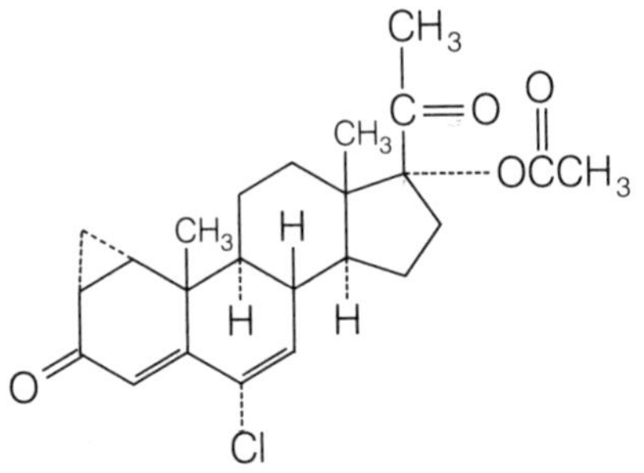

Cyproterone Acetate (6-Chloro-17-hydroxy-1α,2α-methylene-pregna-4,6-diene-3,20-dione17-acetate)

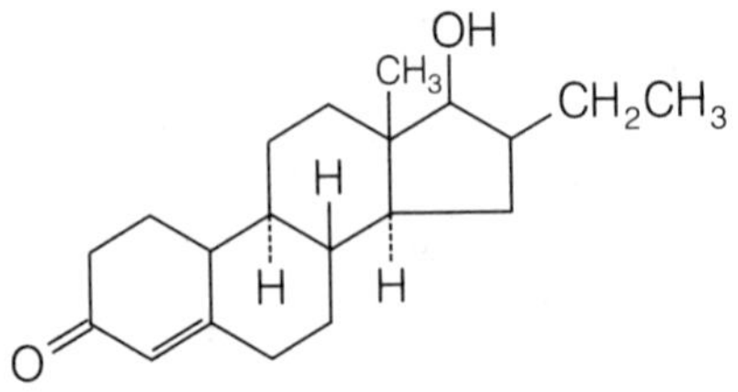

TSAA-291 (oxendolone) (16β-Ethyl-19-nortestosterone)

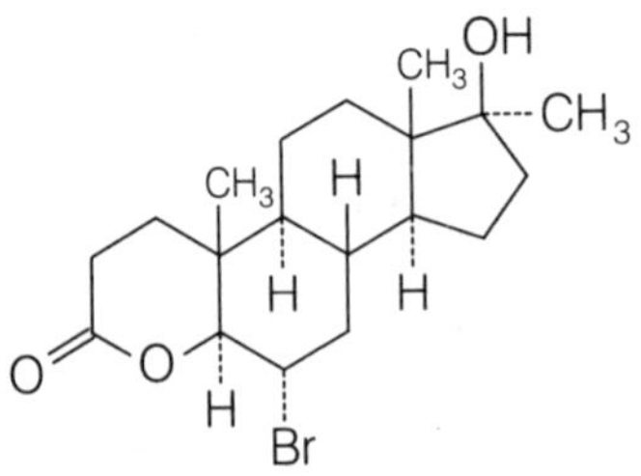

BOMT (6α-Bromo-17β-hydroxy-17α-methyl-4-oxa-5α-androstane-3-one)

FIG. 18-22. Antiandrogens.

ten result. The goal of drug treatment is to preserve the fertility of women wanting future pregnancies, ameliorate symptoms, and simplify future surgery.[230] Bilateral ovariectomy is often needed for severe endometriosis.

Danazol, 17α-pregna-2,4-dien-20-yno-[2,3-*d*]isoxazol-17-ol (Danocrine) is a weak androgen that, in spite of the 17α-ethinyl group, has no estrogenic or progestin activity. Danazol has been called "a synthetic steroid with diverse biologic effects."[230c] Danazol binds to sex hormone binding globulin (SHBG) and decreases the hepatic synthesis of this estradiol and testosterone carrier. Free testosterone thus increases.[230d] It also blocks a number of enzymes essential for estradiol, progesterone, testosterone, and glucocorticoid biosynthesis. Danazol inhibits FSH and LH production by the hypothalamus and pituitary. It binds to progesterone receptors, glucocorticoid receptors, testosterone receptors, and estrogen receptors. As a weak androgen, it suppresses endometrium development. As concluded by Barbieri, "The low-estrogen, high-androgen environment is hostile to the growth of endometrial tissue. In addition, the acyclic endocrine environment produced by danazol minimizes the chance of menstruation [and] the 'reseeding' of the peritoneum with new implants of endometrium."[230d]

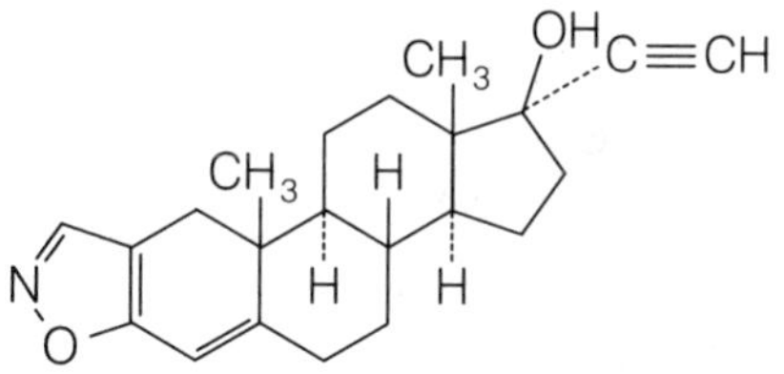

Danazol

Other drugs and drug combinations used to treat endometriosis include the following:

- Estrogen–progestin combination oral contraceptives for six to nine months
- Medroxyprogesterone acetate intramuscularly every two weeks
- Gn-RH analogues (under study)
- Analgesics

ADRENAL CORTEX HORMONES

The adrenal glands (which lie just above the kidneys) secrete over 50 different steroids, including precursors for other steroid hormones. However, the most important hormonal steroids produced by the adrenal cortex are aldosterone and hydrocortisone. Aldosterone is the primary *mineralocorticoid* in humans (i.e., it causes significant salt retention). Hydrocortisone is the primary *glucocorticoid* in humans (i.e., it has its primary effects on intermediary metabolism). The glucocorticoids have become very important in modern medicine, especially for their anti-inflammatory effects.

Medically important adrenal cortex hormones and synthetic mineralocorticoids and glucocorticoids are shown in Figure 18-23. Because salt-retention activity is usually undesirable, the drugs are classified by their salt-retention activities.

STRUCTURAL CLASSES

As illustrated in Figure 18-23, the adrenal cortex hormones are classified by their biologic activities into three major groups:

Mineralocorticoids

The mineralocorticoids are adrenal cortex steroids and analogues with high salt-retaining activity. They are used only for treatment of Addison's disease. The naturally occurring hormone **aldosterone** has an 11β-OH and an 18-CHO that naturally bridge to form a hemiacetal (as drawn in Fig. 18-23). Aldosterone is too expensive to produce commercially; therefore other semisynthetic analogues have taken its place for treatment of Addison's disease. The 11-deoxy analogue **deoxycorticosterone** has significant mineralocorticoid activity, even though it is less than a tenth of aldosterone's (Table 18-12), but still no glucocorticoid or anti-inflammatory activity. Adding a 9α-halogen to hydrocortisone (e.g., to produce **fludrocortisone**) greatly increases both salt retention and anti-inflammatory activity.

Table 18-13 summarizes the relative effects of various substituents on salt retention and glucocorticoid activity. The salt-retaining actions are approximately additive. For example, a 9α-fluoro group's + + + increase in salt retention can be eliminated by a 6α-methyl's − − −.

Glucocorticoids with Moderate to Low Salt Retention

The glucocorticoids with moderate to low salt retention include **cortisone, hydrocortisone**, and their 1-enes **prednisolone** and **prednisone**.

As shown in Table 18-12, an 11-OH maintains good topical anti-inflammatory activity, but 11-ones have little or none. The 1-ene of prednisolone and

1. Mineralocorticoids (High Salt Retention)

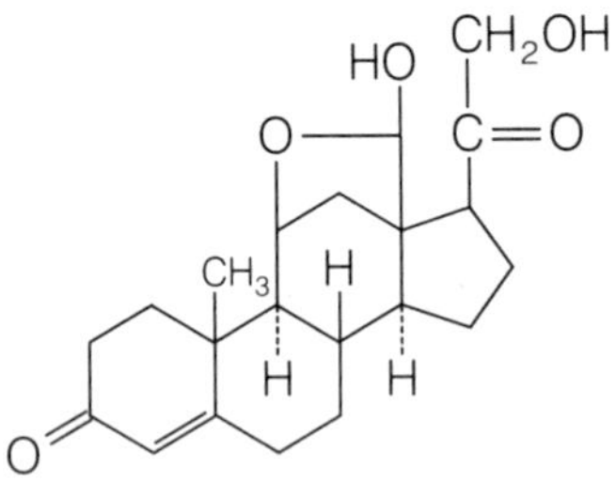

Aldosterone
(not commercially available)

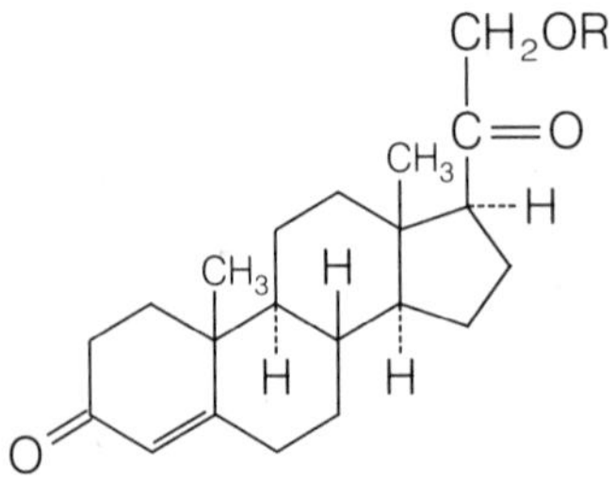

Deoxycorticosterone (R = H)

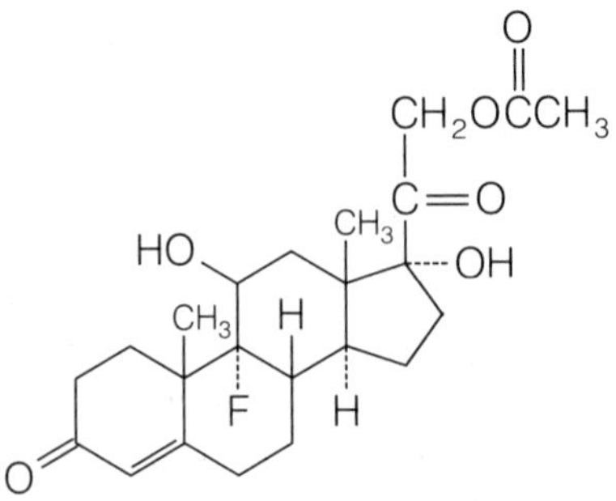

Fludrocortisone Acetate

Esters available:
Deoxycorticosterone Acetate: R = $COCH_3$
Deoxycorticosterone Pivalate: R = $COC(CH_3)_3$

2. Glucocorticoids with Moderate to Low Salt Retention

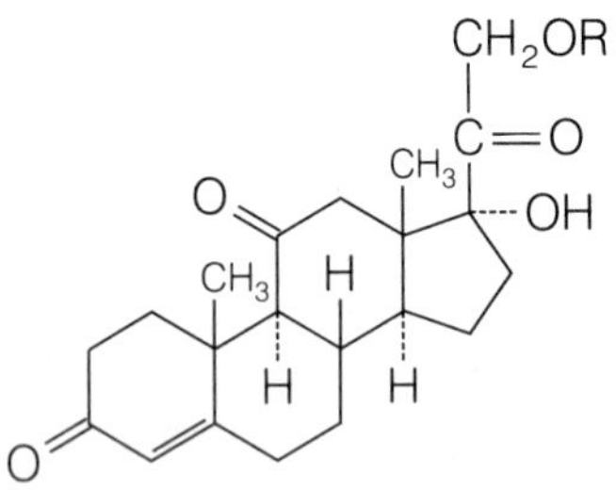

Cortisone (R = H)

Ester available:
Cortisone Acetate: R = $COCH_3$

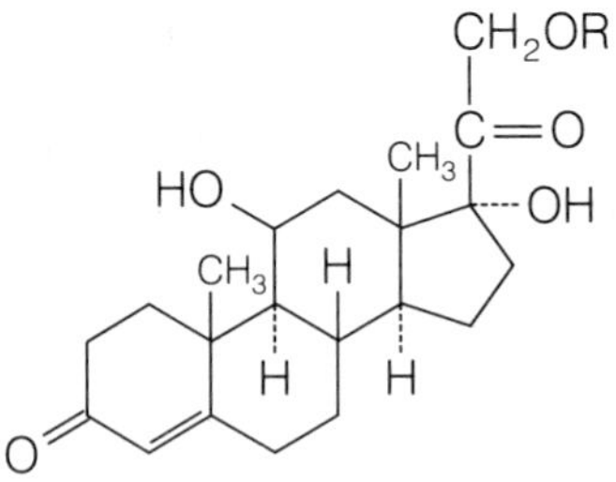

Hydrocortisone (R = H)
(or Cortisol)

Esters available:
Hydrocortisone Acetate: R = $COCH_3$
Hydrocortisone Cypionate: R = $COCH_2CH_2$–(cyclopentyl)

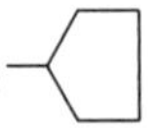

Salts available:
Hydrocortisone Sodium Phosphate: R = $PO_3^-(Na^+)_2$
Hydrocortisone Sodium Succinate:
R = $COCH_2CH_2COO^-Na^+$

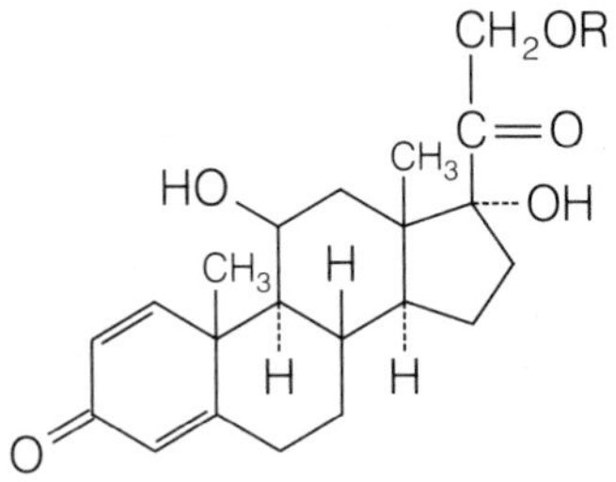

Prednisolone (R = H)

Salts available:
Prednisolone Sodium Phosphate: R = $PO_3^-(Na^+)_2$
Prednisolone Sodium Succinate: R = $COCH_2CH_2COO^-Na^+$

Esters available:
Prednisolone Acetate: R = Ac
Prednisolone Succinate: R = $COCH_2CH_2COOH$
Prednisolone Tebutate: R = $COCH_2C(CH_3)_3$

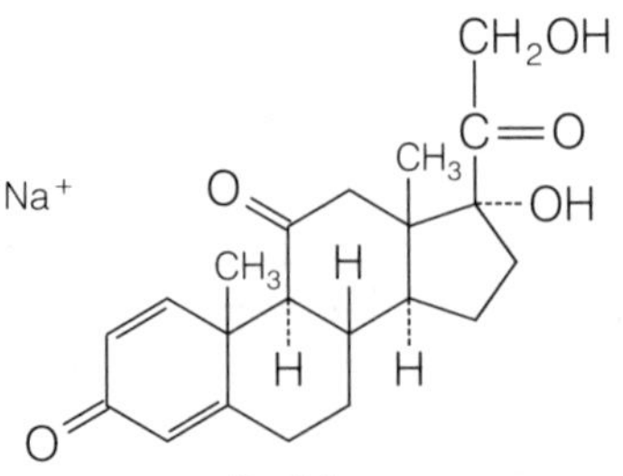

Prednisone

FIG. 18-23. *Natural and semisynthetic adrenal cortex hormones.*

3. Glucocorticoids with Low Salt Retention

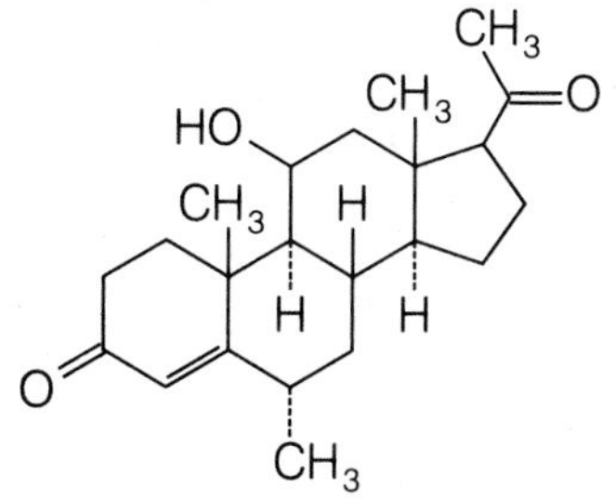

Medrysone

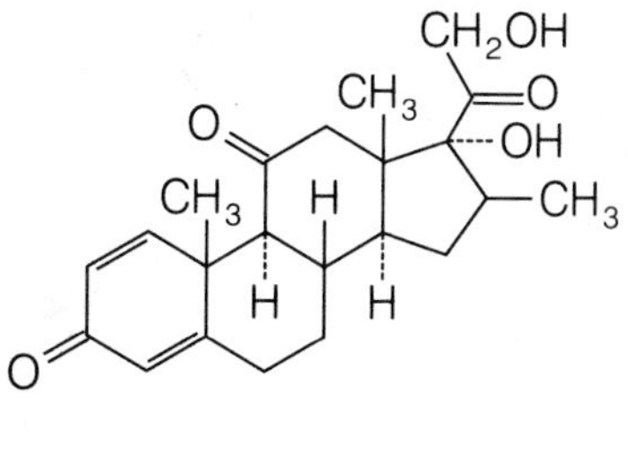

Meprednisone

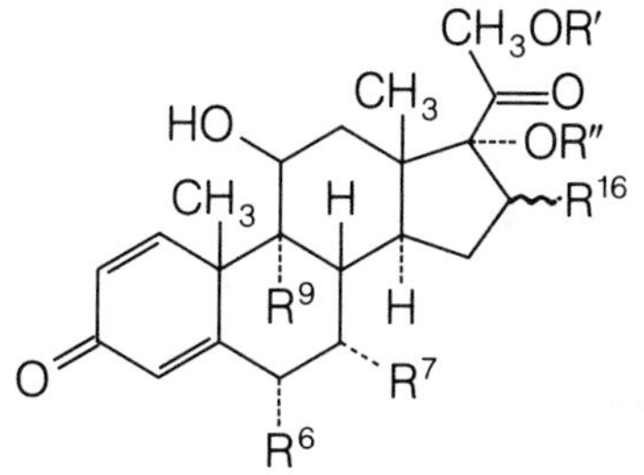

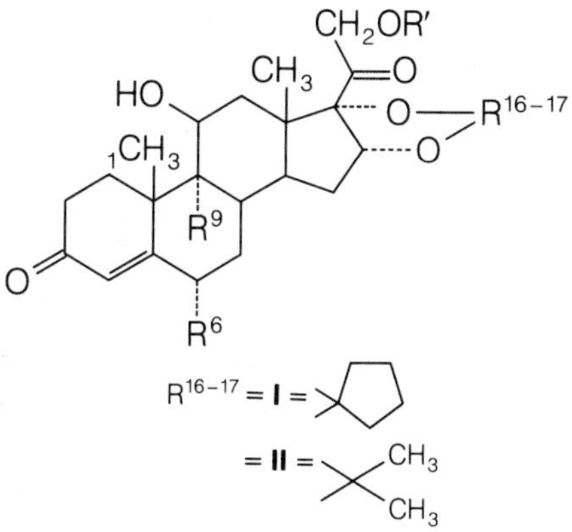

R^{16-17} = **I** = (cyclopentylidene)

= **II** = $C(CH_3)_2$

Alclometasone Dipropionate
$R^{16} = -CH_3$
$R^7 = Cl$
$R' = R'' = COCH_2CH_3$

Beclomethasone Dipropionate
$R^{16} = -CH_3$
$R^9 = Cl$
$R' = R'' = COCH_2CH_3$

Betamethasone
$R^{16} = -CH_3$
$R^9 = F$
$R' = R'' = COCH_2CH_3$

Dexamethasone
$R^{16} = -CH_3$
$R^9 = F$

Diflorasone
$R^{16} = -CH_3$
$R^6 = R^9 = F$
$R' = R'' = COCH_3$

Flumethasone
$R^{16} = -CH_3$
$R^6 = R^9 = F$

Methylprednisolone
$R^6 = CH_3$

Paramethasone
$R^{16} = -CH_3$
$R^6 = F$
$R' = COCH_3$

Amcinonide
R^{16-17} = **I**
1-ene
$R^9 = F$
$R' = COCCH_3$

Desonide
R^{16-17} = **II**
1-ene

Flucinolone Acetonide
R^{16-17} = **II**
1-ene
$R^6 = R^9 = F$
(The 16,17-diol is flucinonide)

Flunisolide
R^{16-17} = **II**
1-ene
$R^6 = F$

Flurandrenolide
R^{16-17} = **II**
$R^6 = F$

Triamcinolone Acetonide
R^{16-17} = **II**
1-ene
$R^9 = F$
(The 16,17-diol is triamcinolone)

FIG. 18-23. *Continued*

Clobetasol
$R^{16} = —CH_3$
1-ene
$R^9 = F$
$R^{17} = OCOCH_2CH_3$

Halcinonide
R^{16-17} = II (acetonide)
$R^9 = F$

Mometasone Furoate
$R^{16} = \cdots CH_3$
1-ene
R^{17} = OOC—(2-furyl)

Clocortolone Pivalate
$R^6 = F$
$R^9 = Cl$
$R^1 = CC(CH_3)_3$

Desoximetasone
$R^9 = F$

FIG. 18-23. Continued

prednisone increases anti-inflammatory activity by about a factor of 4 and somewhat decreases salt retention. Duax and co-workers have shown that the increase in activity may be due to a change in shape of ring A.[65, 66]

Specifically, it appears that analogues more active than hydrocortisone have their ring A bent underneath the molecule to a much greater extent than in hydrocortisone.

The 11β-OH of hydrocortisone is believed to be of major importance in binding to the receptors. Cortisone may be reduced in vivo to yield hydrocortisone as the active agent. The increase activity of 9α-halo derivatives may be due to the electron-withdrawing

TABLE 18-12
APPROXIMATE RELATIVE ACTIVITIES OF CORTICOSTEROIDS*

	Biological Half-Life (min)	Anti-Inflammatory Activity	Topical Activity	Salt-Retaining Activity	Equivalent Dose (mg)
MINERALOCORTICOIDS					
Aldosterone		0.2	0.2	800	
11-Deoxycorticosterone		0	0	40	
9α-Fluorohydrocortisone (fludrocortisone)		10	5–40	800	2
GLUCOCORTICOIDS					
Hydrocortisone	102	1	1	1	20
Cortisone		0.8	0	0.8	25
Prednisolone	200	4	4	0.6	5
Prednisone		3.5	0	0.6	5
6α-Methylprednisolone (methyprednisolone)		5	5	0	4
16β-Methylprednisone (meprednisone)	Used in eyes only; comparative data not available				
6α-Fluoroprednisolone (fluprednisolone)		15	7	0	1.5
Triamcinolone acetonide	300	5	5–100	0	4
Triamcinolone	100–200		1–5		
6α-Fluorotriamcinolone acetonide (fluocinolone acetonide)			over 40		
Flurandrenolide (flurandrenolone acetonide)			over 20		
Fluocinolone			over 40		
Fluocinolone 21-acetate (fluocinonide)			over 40–100		
Betamethasone		35	5–100	0	0.6
Dexamethasone (16α-isomer of betamethasone)	200	30	10–35	0	0.75

*The data in this table are only approximate. Blanks indicate that comparative data are not available to the author or that the product is used only for one use (e.g. topically). Data were taken from several sources, and there is an inherent risk in comparing such data. However, the table should serve as a guide of relative activities.

TABLE 18-13

EFFECTS OF SUBSTITUENTS ON GLUCOCORTICOID ACTIVITY

CLINICAL ANTIRHEUMATIC ENHANCEMENT FACTORS

Functional Group	Factor	Functional Group	Factor
1-Dehydro	2.8	16α-Methyl	1.6
6-Dehydro	0.9*	6β-Methyl	1.3*
6α-Methyl	0.9*	16α, 17α-Isopropyl-idenedioxy	0.6*
6α-Fluoro	1.9	17α-Acetoxy	0.3*
9α-Fluoro	4.9	21-Deoxy	0.2*
16α-Hydroxy	0.3	21-Methyl	0.3*

ENHANCEMENT FACTORS FOR VARIOUS FUNCTIONAL GROUPS OF CORTICOSTEROIDS

Functional Group	Glycogen Deposition	Anti-Inflam-matory Activity	Effects on Urinary Sodium†
9α-Fluoro	10	7–10	+++
9α-Chloro	3–5	3‡	++
9α-Bromo	0.4‡		+
12α-Fluoro	6–8§		++
12α-Chloro	4§		
1-Dehydro	3–4	3–4	–
6-Dehydro	0.5–0.7		+
2α-Methyl	3–6	1–4	++
6α-Methyl	2–3	1–2	– – –
16α-Hydroxy	0.4–0.5	0.1–0.2	– – – –
17α-Hydroxy	1–2	4	–
21-Hydroxy	4–7	25	++
21-Fluoro	2	2	– –

(From Rodig, O. R.: In Burger, A. (ed.). Medicinal Chemistry, Part 2, 3rd ed. New York, Wiley–Interscience, 1970. Used with permission.)

*Two observations or less.

† + = retention; – = excretion.

‡In 1-dehydrosteroids, this value is 4.

§In the presence of a 17α-hydroxyl group, this value is < 0.01.

inductive effect on the 11β-OH, making it more acidic and, therefore, better able to form noncovalent bonds with the receptor. A 9α-halo substituent also reduces oxidation of the 11β-OH to the less active 11-one.

Glucocorticoids with Very Little or No Salt Retention

Cortisone and hydrocortisone, and even prednisone and prednisolone, have too much salt-retaining activity in the doses needed for some therapeutic purposes. Over the last three decades, a number of substituents have been discovered that greatly decrease salt retention. They include 16α-hydroxy; 16α,17α-ketal; and 16α- and 16β-methyl. Other substituents have been found to significantly increase both glucocorticoid and mineralocorticoid activities: 1-ene; 2α-methyl; 9α-fluoro; 9α-chloro; and 21-hydroxy.

As a result of the great economic benefit in having a potent anti-inflammatory product on the market, pharmaceutical manufacturers have made every conceivable combination of these various substituents. In every case a 16-hydroxy or methyl (to eliminate salt retention) has been combined with another substituent to increase glucocorticoid or anti-inflammatory activity. The number of permutations and combinations has resulted in an redundant array of analogues with very low salt retention and high anti-inflammatory activity. Few have any significant advantage over any of the others, but prices vary enormously. This group of drugs represents a striking example of "me too" drug development in modern medicine.

A primary goal of these highly anti-inflammatory drugs has been to increase topical potency. As shown in Table 18-12, some are as much as 100 times more active topically than hydrocortisone. The uses, problems, costs, and effectiveness of topical corticosteroids have been recently reviewed.[15,231–234] *Drug Information*[15] and the *Medical Letter*[233,234] rank the topical steroids on the basis of potency of particular formulations. Examples include

High potency
Amcinonide, 0.1%
Betamethasone dipropionate, 0.05%
Desoximetasone, 0.25%
Diflorasone, 0.05%
Fluocinolone, 0.2%
Fluocinonide, 0.05%
Halcinonide, 0.1%
Triamcinolone acetonide, 0.5%

Intermediate potency
Betamethasone benzoate, 0.025%
Betamethasone valerate, 0.1%
Desoximetasone, 0.05%
Fluocinolone acetonide, 0.025%
Triamcinolone acetonide, 0.1%

Low potency
Alclometasone, 0.05%
Betamethasone valerate, 0.01%
Clocortolone, 0.1%
Fluocinolone acetonide, 0.01%
Triamcinolone acetonide, 0.025%

Lowest potency
Dexamethasone, 0.1%
Hydrocortisone, 1.0%
Methylprednisolone acetate, 0.25%

Although, as shown in Table 18-12, cortisone and prednisone are not active topically, most other glucocorticoids are active. Some compounds, such as triamcinolone acetonide, have striking activity topically. Skin absorption is favored by increased lipid solubility of the drug (see Fig. 18-7).

Absorption of topical glucocorticoids can also be greatly affected by the extent of skin damage, concentration of the glucocorticoid, cream or ointment base used, and similar factors. One must not, therefore, assume from a study of Table 18-12 that, for example, a 0.25% cream of prednisolone is necessarily exactly equivalent in anti-inflammatory potency to 1% hydrocortisone. Nevertheless, the table can serve as a preliminary guide. Furthermore, particular patients may seem to respond better to one topical anti-inflammatory glucocorticoid than to another, irrespective of the relative potencies shown in Table 18-12.

Risk of Systemic Absorption

Except for fludrocortisone, the topical corticosteroids do not cause absorption effects when used on small areas of intact skin. However, when these compounds are used on large areas of the body, systemic absorption may occur—especially if the skin is damaged or if occlusive dressings are used. Fludrocortisone is more readily absorbed than other topical corticosteroids; hence systemic problems can be expected more frequently with it. (Up to 20% to 40% of hydrocortisone given rectally may also be absorbed.)

BIOSYNTHESIS

As shown in a simplified scheme in Figure 18-24, aldosterone and hydrocortisone are biosynthesized from pregnenolone through a series of steps involving hydroxylations at C-17, C-11, and C-21 that convert cholesterol to hydrocortisone. Deficiencies in any of the enzymes are the cause of congenital *adrenal hyperplasias*. (The excellent discussions of Martin[19] and of White and colleagues[235,236] are recommended for further reading.)

These disorders are usually caused by an inability of the adrenal glands to carry out 11β-, 17α-, or 21-hydroxylations. The most common is a lack of 21-hydroxylase activity, which will result in decreased production of hydrocortisone and a compensatory increase in ACTH production. Furthermore, the resultant buildup of 17α-hydroxyprogesterone will lead to an increase of testosterone. When 11β-hydroxylase activity is low, large amounts of 11-deoxycorticosterone will be produced. Because 11-deoxycorticosterone is a potent mineralocorticoid, there will be symptoms of mineralocorticoid excess, including hypertension. When 17α-hydroxylase activity is low, there will be decreased production of testosterone and estrogens as well as hydrocortisone.

Although the details are not completely known, the polypeptide adrenocorticotropic hormone (corticotropin; ACTH) produced by the anterior pituitary enhances or is necessary for the conversion of cholesterol to pregnenolone. ACTH also stimulates the synthesis of hydrocortisone. (ACTH is discussed in Chap. 20.) Hydrocortisone then acts by feedback inhibition to suppress the formation of additional ACTH.

The release of the primary mineralocorticoid, aldosterone, is only slightly dependent upon ACTH. Aldosterone is an active part of the angiotensin–renin–blood pressure cycle that controls blood volume. A decrease in blood volume stimulates the juxtaglomerular cells of the kidneys to secrete the enzyme renin. Renin, in turn, converts angiotensinogen to angiotensin, then angiotensin stimulates the adrenal cortex to release aldosterone. Aldosterone then causes the kidneys to retain sodium, and blood volume increases. When the blood volume has increased sufficiently, there is a decreased production of renin, until blood volume drops again. These physiologic mechanisms are clearly illustrated in the text by Martin.[19]

Trilostane–An Inhibitor of Glucocorticoid and Mineralocorticoid Biosynthesis

Trilostane (4,5-epoxy-17-hydroxy-3-oxoandrostane-2-carbonitrile; Modrastane) completely inhibits the conversion of pregnenolone to progesterone, and of 17α-hydroxypregnenolone to 17α-hydroxyprogesterone blocks the hydroxylating enzymes (hydroxylases) involved in hydrocortisone and aldosterone biosynthesis (see Fig. 18-24). Thus, trilostane is used for treatment of Cushing's syndrome. However, this drug is teratogenic, and not all patients experience significant decreases in glucocorticoid and mineralocorticoid plasma levels.[237]

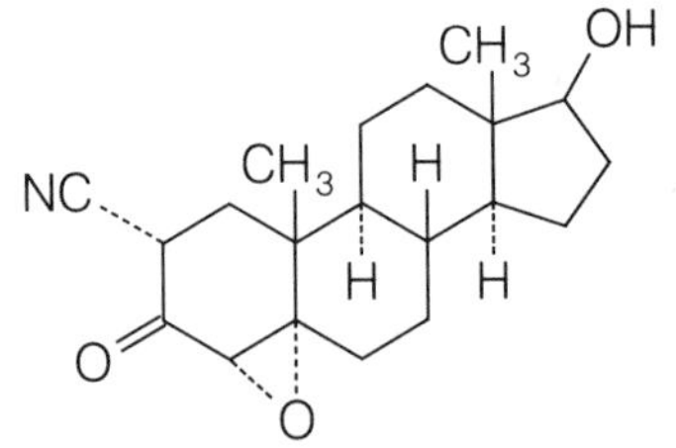

Trilostane

BIOCHEMICAL ACTIVITIES

The adrenocortical steroids permit the body to adjust to environmental changes, to stress, and to changes in the diet. As Sayers and Travis have succinctly stated in their fine 1975 review of the pharmacology of these steroids, "The adrenal cortex

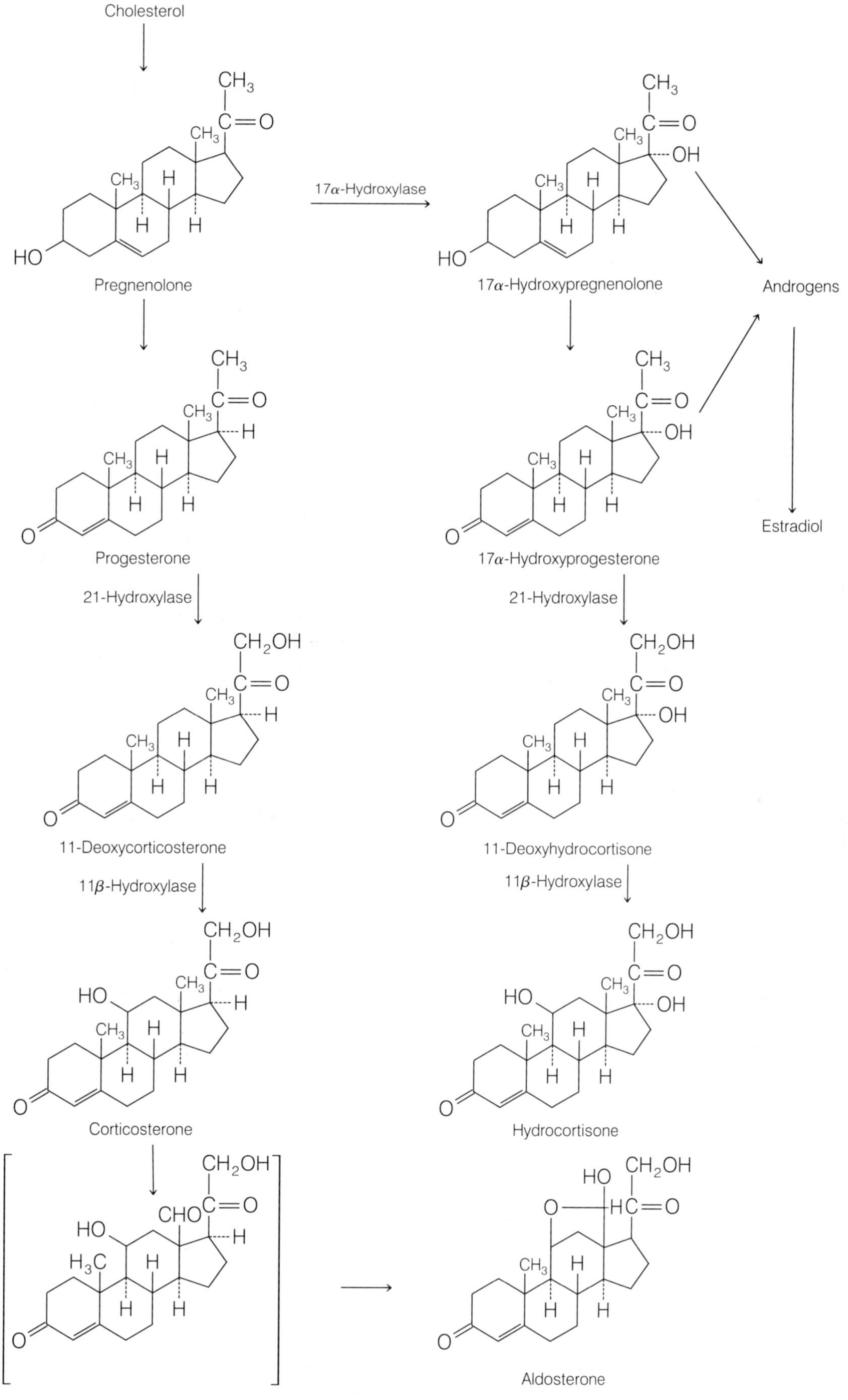

Cholesterol

Pregnenolone

17α-Hydroxylase

17α-Hydroxypregnenolone

Androgens

Progesterone

17α-Hydroxyprogesterone

Estradiol

21-Hydroxylase

11-Deoxycorticosterone

21-Hydroxylase

11-Deoxyhydrocortisone

11β-Hydroxylase

Corticosterone

11β-Hydroxylase

Hydrocortisone

Aldosterone

FIG. 18-24. *A simplified scheme of the biosynthesis of hydrocortisone and aldosterone. The biosynthetic pathways are more complex than presented here.*

is the organ, *par excellence*, of homeostasis."[238] Aldosterone and, to a lesser extent, other mineralocorticoids maintain a constant electrolyte balance and blood volume, and the glucocorticoids have key roles in controlling carbohydrate, protein, and lipid metabolism.

Aldosterone increases sodium reabsorption in the kidneys. An increase in plasma sodium concentration, in turn, will lead to increased blood volume, because blood volume and urinary excretion of water are directly related to the plasma sodium concentration. Simultaneously, aldosterone increases potassium ion excretion. 11-Deoxycorticosterone also is quite active as a mineralocorticoid. Similar actions are exhibited with hydrocortisone and corticosterone, but to a much smaller degree.

Aldosterone controls the movement of sodium ions in most epithelial structures involved in active sodium transport. Although aldosterone acts primarily on the distal convoluted tubules of the kidneys, it also acts on the proximal covoluted tubules and collecting ducts. Aldosterone also controls the transport of sodium in sweat glands, small intestine, salivary glands, and the colon. In all these tissues, aldosterone enhances the inward flow of sodium ions and promotes the outward flow of potassium ions.

Patients with Addison's disease exhibit muscle weakness and are easily fatigued. This primarily may be due to inadequate blood volume and aldosterone insufficiency, although changes in glucose availability may also be involved.

However, aldosterone (and other steroids with mineral activity) does not cause *immediate* changes in sodium excretion. There is a latent period after administration of any of the mineralocorticoids. This supports the view that aldosterone acts by stimulation of the synthesis of enzymes that, in turn, are actually responsible for active ion transport. This will be discussed in greater detail with the adrenocortical steroid receptors.

The glucocorticoids have many physiologic and pharmacologic actions. They control or influence carbohydrate, protein, lipid, and purine metabolism. They also affect the cardiovascular and nervous systems and skeletal muscle. They also regulate growth hormone gene expression.

Glucocorticoids stimulate glycogen storage synthesis by inducing the syntheses of glycogen synthase and stimulate gluconeogenesis in the liver. They have a catabolic effect on muscle tissue, stimulating the formation and transamination of amino acids into glucose precursors in the liver. As reviewed by Martin,[19] glucocorticoids induce tyrosine aminotransferase and tryptophan oxidase. The catabolic actions in Cushing's syndrome are demonstrated by a wasting of the tissues, osteoporosis, and reduced muscle mass. Lipid metabolism and synthesis are significantly increased in the presence of glucocorticoids, but the actions usually seem to be dependent on the presence of other hormones or cofactors. Glucocorticoids also protect the body from stress,[19] but the mechanism of this protective effect is unknown. High glucocorticoid production in response to stress can lead to a decrease in the size of the thymus gland by up to 95%. The roles of the thymus in protection against stress (by glucocorticoids stimulation) are, as yet, not fully delineated. Equally fascinating is the glucocorticoid's role in activating some parts of the immune system, but depressing others.[19] A lack of adrenal cortex steroids also causes depression, irritability, and even psychoses, reflecting significant effects on the nervous system. The glucocorticoids also play a role in maintaining bone, cartilage, and skin.

Glucocorticoids decrease lymphocyte production and are generally immunosuppressive. Of great importance therapeutically is that the glucocorticoids and ACTH decrease inflammation and the physiologic changes that occur to cause inflammation—edema, capillary dilatation, migration of phagocytes, capillary proliferation, deposition of collagen, and such.

METABOLISM

Cortisone and hydrocortisone are enzymatically interconvertible, and thus one finds metabolites from both. Most of the metabolic processes occur in the liver, with the metabolites excreted primarily in the urine. Although many metabolites have been isolated,[19] the primary routes of catabolism are (1) reduction of the C-4 double bond to yield 5β-pregnanes; (2) reduction of the 3-one to give 3α-ols; (3) reduction of the 20-one to the corresponding 20α-ol. The two primary metabolites are tetrahydrocortisol and tetrahydrocortisone (shown below) and their conjugates.

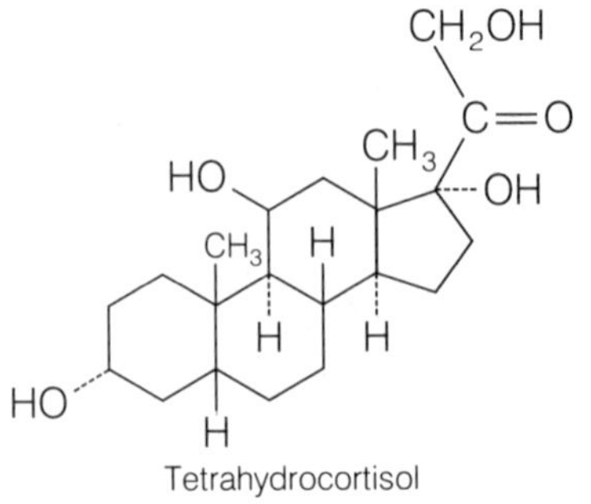

Tetrahydrocortisol

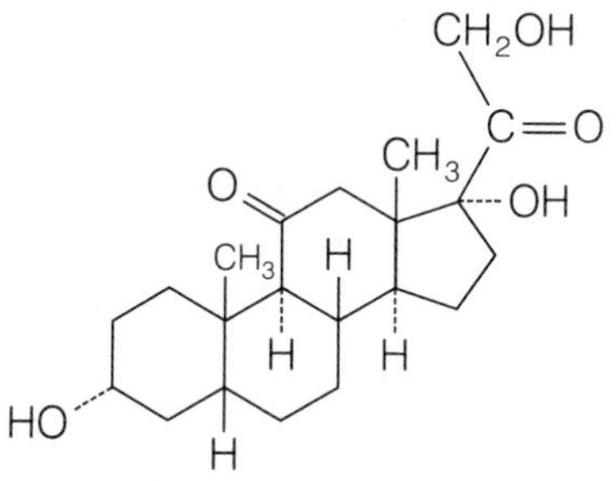

Tetrahydrocortisone

However, other metabolites include 20-ols and derivatives of side chain oxidation and cleavage, as shown below.

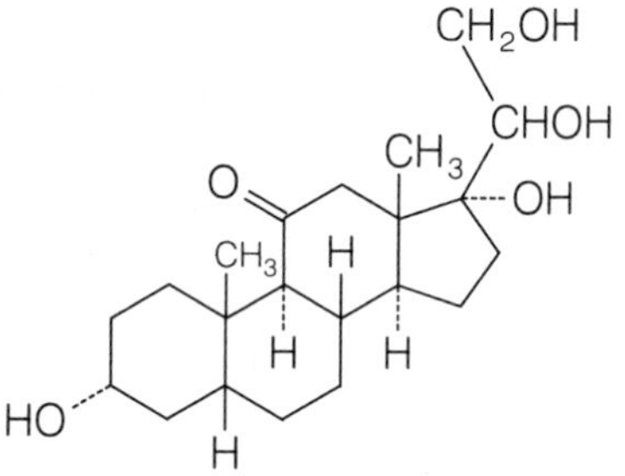

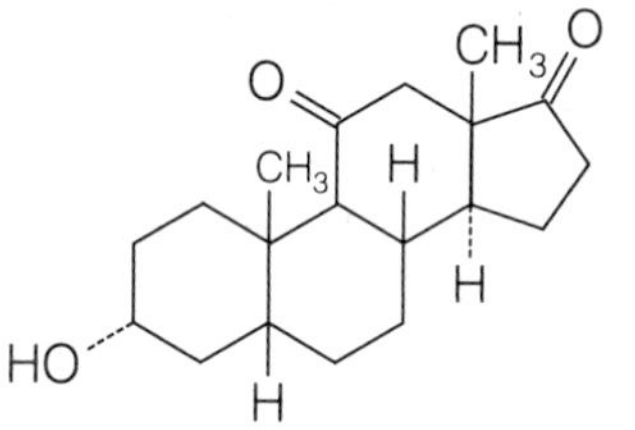

The C_{19} metabolites of the latter type are often androgenic.

GLUCOCORTICOID RECEPTORS

The glucocorticoids bind with great specificity to all tissues that elicit a "glucocorticoid response." They are readily metabolized, which has made definitive studies particularly difficult. Furthermore, plasma protein binding is quite high, leaving relatively little "plasma glucocorticoid" actually free to cause a physiologic response.

Glucocortcoid receptors have been studied intensively,[50, 56, 58, 239] and it has been shown that the glucocorticoids react with them essentially as illustrated in Figure 18-3.

The glucocorticoids stimulate the production or inhibition of a variety of enzymes, some of which are involved in the anabolic processes and others in catabolic processes of glucocorticoid activity. Tyrosine aminotransferase and other enzymes involved with transamination (necessary for amino acids to be converted to glucose and glycogen precursors) are induced by glucocorticoids.

The glucocorticoid receptor (90 to 95 kDa) has received considerable attention.[48, 50, 56, 58, 239] As noted by Ringold,[56] "Examples of glucocorticoid regulated genes abound in the literature . . . " Highly glucocorticoid-sensitive tissues can contain about 50,000 receptors per cell.[58] The remarkable similarity with the thyroid hormone receptor was discussed earlier in this chapter.

As described in a recent review,[239] "The [glucocorticoid] receptor is a hormone-activated 'switch,' made up of several independently functioning regulatory domains." As with other receptors in the steroid–thyroid hormone superfamily, the glucocorticoid receptor appears to have three distinct domains that can be cleaved by selective proteolysis. One fragment (40 kDa) contains both glucocorticoid and DNA-binding regions. Other fragments have major roles in binding to specific DNA sequences. The sequence T-G-T-T-C-T or related sequences[56] appears to be particularly involved with glucocorticoid receptor complex binding to DNA.

The glucocorticoids appear to inhibit phosphofructokinase (which converts fructose 6-phosphate to fructose 1,6-diphosphate in glycolysis), thereby decreasing glucose metabolism. They also may inhibit the conversion of pyruvate to acetyl-CoA, thus "forcing" pyruvate to be used in glyconeogenesis. Both inhibition processes are probable because of gene repression (i.e., repression of the formation of RNA templates needed to synthesize phosphofructokinase and other enzymes in glucose metabolism).

Glucocorticoid stimulation of lipid metabolism may be due to an increase in cAMP formation (by inducing adenylate cyclase or inhibiting phosphodiesterase).

The roles of the glucocorticoids in acting as anti-inflammatory agents are still not well-defined. Many different kinds of substances have been implicated in causing inflammation. These include histamine, serotonin, kinins, hyaluronic acid depolymerizers, acetylcholine, epinephrine, prostaglandins, antigen–antibody complexes, and others. It is known that the glucocorticoids' eosinopenic (reducing eosinophils) and hyperglycemic activities parallel their anti-inflammatory effectiveness. (Anti-inflammatory activity can be measured directly by using chemical irritants with experimental animals.) It is known that cortisone inhibits the release of inflammation-producing lysosomal enzymes; inhibits the ingestion of antigen–antibody complexes by white blood cells (which releases lysosomal enzymes); and reduces capillary permeability. How these effects are caused by glucocorticoids is still unknown.

MINERALOCORTICOID RECEPTORS

Aldosterone is secreted in very small amounts, much smaller than other steroid hormones. Human plasma concentrations of aldosterone are only about 8 ng/mL, with glucocorticoids up to 10^3 higher.

It now appears that aldosterone binds to cytoplasmic receptors of 8.5 S or 4.5 S size. This cytoplasmic receptor then undergoes a change into 3 S, which can be found in the nucleus. Finally, the 3 S receptor–steroid complex appears to act directly upon chromatin, specifically DNA. This overall process has been illustrated in Figure 18-3. The DNA would then be stimulated to produce RNA templates for the synthesis of enzymes necessary for active Na^+ transport. The latent period of aldosterone action is then easy to rationalize.

THERAPEUTIC USES

The adrenocortical steroids are used primarily for their glucocorticoid effects, including immunosuppression, anti-inflammatory activity, and antiallergic activity. The mineralocorticoids are used only for treatment of **Addison's disease**. Addison's disease is caused by chronic adrenocortical insufficiency and may be due to either adrenal or anterior pituitary failure. The anterior pituitary secretes ACTH, a polypeptide that stimulates the adrenal cortex to synthesize steroids.

The symptoms of Addison's disease illustrate the great importance of the adrenocortical steroids in the body and, especially, the importance of aldosterone. These symptoms include increased loss of body sodium, decreased loss of potassium, hypoglycemia, weight loss, hypotension, weakness, increased sensitivity to insulin, and decreased lipolysis.

Hydrocortisone is also used during postoperative recovery after surgery for **Cushing's syndrome**—excessive adrenal secretion of glucocorticoids. Cushing's syndrome can be caused by bilateral adrenal hyperplasia or adrenal tumors and is treated by surgical removal of the tumors or resection of hyperplastic adrenal gland(s).

The use of glucorcorticoids during recovery from surgery for Cushing's syndrome illustrates a most important principle of glucocorticoid therapy: *abrupt withdrawal of glucocorticoids may result in adrenal insufficiency*—showing clinical symptoms similar to Addison's disease. For that reason, patients who have been on long-term glucocorticoid therapy must have the dose *gradually* reduced. Furthermore, prolonged treatment with glucocorticoids can cause adrenal suppression, especially during times of stress. The symptoms are similar to those of Cushing's syndrome; for example, rounding of the face, hypertension, edema, hypokalemia, thinning of the skin, osteoporosis, diabetes, and even subcapsular cataracts. In doses of 45 mg/m^2 of body surface area or more daily, growth retardation occurs in children.

The glucocorticoids are used in the treatment of collagen vascular diseases, including rheumatoid arthritis, disseminated lupus erythematosus, and dermatomyositis. The uses of these drugs and nonsteroids in treatment of rheumatoid arthritis have recently been reviewed.[240]

Although there is usually prompt remission of redness, swelling, and tenderness by the glucocorticoids in rheumatoid arthritis, continued long-term use may lead to serious systemic forms of collagen disease. As a result, the glucocorticoids should be used infrequently in rheumatoid arthritis.

The glucocorticoids are used extensively topically, orally, and parenterally to treat inflammatory conditions. They also usually produce relief from the discomforting symptoms of many allergic conditions—intractable hay fever, exfoliative dermatitis, generalized eczema, and others. The glucocorticoids are also used to treat acute asthmatic symptoms unresponsive to bronchodilators.[241] They have been especially useful in aerosol preparations.[242]

The glucocorticoids' lymphocytopenic actions make them particularly useful for treatment of chronic lymphocytic leukemia in combination with other antineoplastic drugs.

The glucocorticoids are also used in the treatment of congenital adrenal hyperplasias.

The adrenocortical steroids are contraindicated or should be used with great caution in patients having (1) peptic ulcer (in which the steroids may cause hemorrhage), (2) heart disease, (3) infections (the glucocorticoids suppress the body's normal infection-fighting processes), (4) psychoses (since behavioral disturbances may occur during steroid therapy), (5) diabetes (the glucocorticoids increase glucose production, so more insulin may be needed), (6) glaucoma, (7) osteoporosis, and (8) herpes simplex involving the cornea.

When topically administered, the glucocorticoids present relatively infrequent therapeutic problems, but it should be remembered that their anti-inflammatory action can mask symptoms of infection. Many physicians prefer not giving a topical anti-inflammatory steroid until after an infection is controlled with topical antibiotics. The immunosuppressive activity of the topical glucocorticoids can also prevent natural processes from curing the infection. Topical steroids actually may also cause any of several dermatoses in some patients.

Finally, as discussed before with the oral contraceptives, steroid hormones should not be used during pregnancy. If absolutely necessary to use the glucocorticoids topically during pregnancy, they should be limited to small areas of intact skin and used for a limited time.

PRODUCTS

The adrenal corticosteroid products are shown in Figure 18-23. The structures illustrate the usual changes (see Fig. 18-7) made to modify solubility of the products—and, therefore, their therapeutic uses. In particular, the 21-hydroxyl can be converted to an ester to make it less water-soluble to modify absorption; or to a phosphate ester salt or hemisuccinate ester salt to make it more water-soluble and appropriate for intravenous use. The products also reflect the previously discussed structure–activity relationships changes to increase anti-inflammatory activity or potency, or to decrease salt retention.

Again, it must be emphasized that patients who have been on long-term glucocorticoid therapy must have the dose *gradually* reduced. This "critical rule" and indications have been previously discussed under Therapeutic Uses. Dosage schedules and gradual reduction of dose schedules can be quite complex and specific for each indication. For that reason, specialized indication and dose references such as *Facts and Comparisons* and *The Hospital Formulary* should be consulted before advising physicians on dosages.

General USP DI doses are shown in Tables 18-14–18-16.

Many of the glucocorticoids are available in topical dosage forms, including creams, ointments, aerosols, lotions, and solutions. They are usually applied three to four times a day to well-cleaned areas of affected skin. (The patient should be instructed to apply them with well-washed hands as well.) Ointments are usually prescribed for dry, scaly dermatoses. Lotions are well suited for weeping dermatoses. Creams are of general use for many other dermatoses. When applied to very large areas of skin or to damaged areas of skin, significant systemic absorption can occur. The use of an occlusive dressing can also greatly increase systemic absorption.

Deoxycorticosterone Acetate, USP. 21-Acetyloxypregn-4-ene-3,20-dione is a potent mineralocorticoid used only for the treatment of Addison's disease. It has essentially no anti-inflammatory (glucocorticoid) activity but has 100 times the salt-retention (mineralocorticoid) activity of hydrocortisone. (It has only 1/30 the activity of aldosterone.) Hydrocortisone or other glucocorticoids should be given simultaneously for patients with acute adrenal insufficiency. Its great salt-retaining activity can be expressed as edema and pulmonary congestion as toxic doses are reached. It is insoluble in water (as one would predict from Table 18-1). Because Addison's disease is essentially incurable, treatment continues for life. With a serum half-life of only 70 minutes, the drug is sometimes given in the form of subcutaneous pellets administered every 8 to 12 months. The duration of action of **deoxycorticosterone pivalate, USP,** when given intramuscularly in depot preparations is longer than the acetate, often being administered only once every four weeks.

TABLE 18-14

NATURAL AND SEMISYNTHETIC MINERALOCORTICOID HORMONES

Name Proprietary Name	***Preparations***	***Usual Adult Dose****	***Usual Dose Range****	***Usual Pediatric Dose****
Deoxycorticosterone acetate USP *Doca, Percorten Acetate*	Deoxycorticosterone acetate injection USP	IM, 1–6 mg qd	1–10 mg/day	IM, 1–5 mg qd or 1.5–2 mg/m^2 of body surface qd
	Deoxycorticosterone acetate pellets USP	Subcutaneous implantation: 125 mg (1 pellet)/500 μg of daily IM dose of deoxycorticosterone acetate		
Deoxycorticosterone pivalate USP *Percorten Pivalate*	Sterile deoxycorticosterone pivalate suspension NF			IM, 25 mg for each 1 mg of daily maintenance dose of deoxycorticosterone acetate injection, at 4-wk intervals
Fludrocortisone acetate USP *Florinef Acetate*	Fludrocortisone acetate tablets USP	100 μg qd	100–200 μg (0.1–0.2 mg)/day	

*See USP DI for complete dosage information.

TABLE 18-15

ANTI-INFLAMMATORY GLUCOCORTICOIDS WITH MODERATE TO LOW SALT RETENTION

Name Proprietary Name	***Preparations***	***Application***	***Usual Adult Dose****	***Usual Dose Range****	***Usual Pediatric Dose****
Cortisone acetate USP *Cortone*	Cortisone acetate tablets USP		25–300 mg/day as a single dose or in divided doses	10–400 mg/day	175 g–2.5 mg/kg of body weight or 5–75 mg/m^2 of body surface qid
	Sterile cortisone acetate suspension USP		20–30 mg/day	10–400 mg/day	200 µg–1.25 mg/kg of body weight or 7–37.5 mg/m^2 of body surface qd or bid
Hydrocortisone USP *Cortef, Hydrocortone*	Hydrocortisone enema USP		Rectal, 100 mg hs for 21 days, or until clinical and proctologic remission is obtained		
	Hydrocortisone ointment USP Hydrocortisone lotion USP Hydrocortisone cream USP	Topically to the skin, as a 0.25% ointment, 0.125–1% lotion, or 0.125–2.5% cream bid to qid			
	Hydrocortisone tablets USP Sterile Hydrocortisone suspension USP		5–60 mg qid	20–240 mg/day	140 µg–2 mg/kg of body weight or 4–60 mg/m^2 of body surface qid
Hydrocortisone acetate USP *Cortril Suspension, Hydrocortone Acetate Suspension*	Hydrocortisone acetate ointment USP Hydrocortisone acetate cream USP Hydrocortisone acetate lotion USP	Topically to the skin, as a 0.5–2.5% ointment qd to qid			
	Hydrocortisone acetate ophthalmic suspension USP	Topically to the conjunctiva, 0.05–0.1 ml of a 0.5–2.5% suspension 3–20 times daily			
	Hydrocortisone acetate opththalmic ointment USP	Topically to the conjunctiva, as a 0.5–1.5% ointment qd to qid			
	Sterile hydrocortisone acetate suspension USP		Intra-articular, intralesional, or soft-tissue injection, 5–75 mg, repeated at 2- to 3-wk intervals		
Hydrocortisone sodium succinate USP *Solu-Cortef, A-hydroCort*	Hydrocortisone sodium succinate for injection USP		IM or IV, the equivalent of 100–500 mg of hydrocortisone at 1- to 10-h intervals	100 mg–8 g/day	IM, 160 µg–1 mg/kg of body weight or 6–30 mg/m^2 of body surface qd or bid
Hydrocortisone cypionate USP *Cortef Fluid*	Hydrocortisone cypionate oral suspension USP		The equivalent of 15–30 mg of hydrocortisone tid or qid	The equivalent of 25–320 mg of hydrocortisone daily	
Hydrocortisone sodium phosphate USP *Hydrocortone Phosphate*	Hydrocortisone sodium phosphate injection USP		IM, IV, or SC, the equivalent of cortisol, 15–240 mg/day	25 mg–1 g/day	IM, 160 µg–1 mg/kg of body weight or 6–30 mg/m^2 of body surface qd or bid
Prednisolone USP *Delta-Cortef, Prednis, Sterane*	Prednisolone tablets USP Prednisolone cream USP		5–15 mg qd to qid	5–250 mg/day	35–500 µg/kg of body weight or 1–15 mg/m^2 of body surface qid
Prednisolone acetate USP *Meticortelone Acetate, Nisolone, Savacort*	Sterile prednisolone acetate suspension USP		Intra-articular, 12.5–25 mg at each site every 1–3 wk; IM, 2–30 mg bid		IM, 40–250 µg/kg of body weight or 1.5–7.5 mg/m^2 of body surface qd or bid

*See USP DI for complete dosage information.

TABLE 18-15 *Continued*

ANTI-INFLAMMATORY GLUCOCORTICOIDS WITH MODERATE TO LOW SALT RETENTION

Name Proprietary Name	**Preparations**	**Application**	**Usual Adult Dose***	**Usual Dose Range***	**Usual Pediatric Dose***
Prednisolone succinate USP	Available only as sodium salt (below)				
Meticortelone Soluble	Prednisolone sodium succinate for injection USP		IM or IV, the equivalent of 2–30 mg of prednisolone bid		IM or IV, the equivalent of 40–250 μg of prednisolone per kg of body weight or 1.5–7.5 mg/m^2 of body surface qd or bid
Prednisolone tebutate USP *Hydeltra-T.B.A.*	Sterile prednisolone tebutate suspension USP				
Prednisolone sodium phosphate USP *Sodasone*	Prednisolone sodium phosphate injection USP		IM or IV, the equivalent of 10–50 mg of prednisolone phosphate bid	10–400 mg/day	IM or IV, the equivalent of 40–250 μg of prednisolone phosphate per kg of body weight or 1.5–7.5 mg/m^2 of body surface qd or bid
Hydeltrasol	Prednisolone sodium phosphate ophthalmic solution USP	Topically to the conjunctiva, the equivalent of 0.05–0.1 ml of a 0.113–0.9% solution of prednisolone phosphate 2–20 times a day			
Prednisone USP *Delta-Dome, Orasone, Deltasone, Servisone, Meticorten, Paracort*	Prednisone tablets USP		5–60 mg/day as a single dose or in divided doses	5–250 mg/day	35–500 μg/kg of body weight or 1–15 mg/m^2 of body surface qid

*See USP DI for complete dosage information.

TABLE 18-16

ANTI-INFLAMMATORY GLUCOCORTICOIDS WITH LOW SALT RETENTION

Name Proprietary Name	**Preparations**	**Application**	**Usual Adult Dose***	**Usual Dose Range***	**Usual Pediatric Dose***
Aclometasone dipropionate *Aclovate*	Aclometasone dipropionate cream, 0.05% Aclometasone dipropionate ointment, 0.05%	Topically in thin film bid or tid			
Amcinonide USP *Cyclocort*	Amcinonide cream, 0.1% Amcinonide ointment, 0.1%	Topically in thin film to affected area bid or tid			
Beclomethasone dipropionate USP *Beconase, Vancenase, Vanceril*	Beclomethasone dipropionate aerosol Beclomethasone dipropionate nasal suspension	Nasal aerosol, 42 μg/spray Nasal suspension, 42 μg/dose	1 spray in each nostril bid to qid (168–336 μg total)		
Betamethasone USP *Celestone*	Betamethasone cream USP Betamethasone syrup USP Betamethasone tablets USP	Topical, 0.2% cream applied to the skin bid or tid	Oral, 600 μg (0.6 mg)–7.2 mg/day as a single dose or in divided doses	600 μg–7.2 mg/day	Oral, 17.5–250 μg (0.017–0.25 mg)/kg of body weight or 500 μg (0.5 mg)–7.5 mg/m^2 of body surface a day as a single dose or in divided doses

*See USP DI for complete dosage information.

TABLE 18-16 *Continued*

ANTI-INFLAMMATORY GLUCOCORTICOIDS WITH LOW SALT RETENTION

Name Proprietary Name	**Preparations**	**Application**	**Usual Adult Dose***	**Usual Dose Range***	**Usual Pediatric Dose***
Betamethasone acetate USP *Celestone Soluspan (with the sodium phosphate)*	Sterile betamethasone sodium phosphate and betamethasone acetate suspension USP		Intra-articular, 3–12 mg, repeated as needed Intrabursal, 6 mg, repeated as needed Intradermal or intralesional, 1.2 mg/cm^2 of affected skin up to a total amount of 6 mg, repeated at 1-wk intervals if necessary Intramuscular, 500 μg (0.5 mg)–9 mg/day		
Betamethasone benzoate USP *Unicort*	Betamethasone cream, 0.025% Betamethasone gel, 0.025% Betamethasone lotion, 0.025%	Topically in thin films qd to qid			
Betamethasone dipropionate USP *Diprosone (aerosol), Diprolene (cream), Diprosone (lotion)*	Betamethasone dipropionate topical aerosol, 0.1% USP Betamethasone dipropionate lotion, 0.5% USP Betamethasone dipropionate ointment, 0.05% USP	Topical aerosol, 60 μg/3 sec spray; cream, lotion, ointment in thin films qd to qid	Dosage of 0.05% preparation should not exceed 45 g/wk and not more than 14 days maximum		
Betamethasone valerate USP *Valisone*	Betamethasone valerate cream USP Betamethasone valerate lotion USP Betamethasone valerate ointment USP	Topical, qd to tid			
Betamethasone sodium phosphate USP *Celestone Soluspan (with the acetate)*	Sterile betamethasone sodium phosphate and betamethasone acetate suspension USP		See betamethasone acetate USP		
Clobetasol *Dermovate*	Clobetasol propionate cream, 0.05% Clobetasol propionate ointment, 0.05%	Topically in thin films bid			
Clocortolone *Cloderm*	Clocortolone pivalate cream, 0.1% USP	Topically in thin films bid or tid			
Desonide *DesOwen, Tridesilon*	Desonide cream, 0.05% Desonide ointment, 0.05%	Topically as a thin film bid to qid			
Desoximetasone *Topicort*	Desoximetasone cream USP, 0.05% and 0.25% Desoximetasone gel USP, 0.05% Desoximetasone ointment, 0.25%	Topically qd			

*See USP DI for complete dosage information.

TABLE 18-16 *Continued*

ANTI-INFLAMMATORY GLUCOCORTICOIDS WITH LOW SALT RETENTION

Name Proprietary Name	**Preparations**	**Application**	**Usual Adult Dose***	**Usual Dose Range***	**Usual Pediatric Dose***
Dexamethasone USP *Decadron, Deronil, Dexameth, Gammacorten*	Dexamethasone elixir USP Dexamethasone tablets USP		Oral, 750 μg (0.75 mg)–9 mg/day as a single dose or in divided doses		Oral, 23 μg (0.023 mg)–330 μg (0.33 mg)/kg of body weight or 670 μg (0.67 mg)–10 mg/m² of body surface a day as a single dose or in divided doses
	Dexamethasone topical aerosol USP, 0.1%, 0.04%	Topically to the skin tid or qid			
	Dexamethasone ophthalmic suspension USP, 0.1%	Opthalmic, 1 drop tid or qid			
	Dexamethasone gel USP, 0.1%				
Dexamethasone sodium phosphate USP *Decadron Phosphate, Hexadrol Phosphate*	Dexamethasone sodium phosphate cream USP, 0.1%	Topically to the skin, as the equivalent of a 0.1% cream tid or qid			
	Dexamethasone sodium phosphate injection USP		Intra-articular or soft-tissue injection, the equivalent of 200 μg–6 mg of dexamethasone phosphate, repeated at 3-day to 3-wk intervals, if necessary	Intra-articular or soft-tissue injection, 200 μg–6 mg at each site once every 3 days to 3 wk; IM or IV, 500 μg–80 mg/day	IM or IV, 6–40 μg/kg of body weight or 235 μg–1.25 mg/m² of body surface qd or bid
	Dexamethasone sodium phosphate inhalation aerosol USP	Oral inhalation, 300 μg, or 3 metered sprays tid or qid			
	Dexamethasone sodium phosphate ophthalmic ointment USP, 0.05%	Topically to the conjunctiva, a thin strip (about 1 cm) tid or qid			
	Dexamethasone sodium phosphate ophthalmic solution USP, 0.1%	Topically to the conjunctiva, 1 or 2 drops up to 6 times a day			
	Dexamethasone acetate USP	Sterile dexamethasone acetate suspension USP			
Diflorasone *Florone, Flutone*	Diflorasone diacetate cream USP, 0.05%	Topically qd			
Flumethasone pivalate USP *Locorten*	Flumethasone pivalate cream USP	Topical, 0.03% cream			
Flunisolide *Nasalide, Aerobid*	Flunisolide inhalation aerosol	Oral inhalation (500 μg or 2 metered inhalation sprays) bid			
	Flunisolide nasal solution USP	Nasal (500 μg or 2 metered nasal sprays) in each nostril bid			
Fluocinolone acetonide USP *Fluonid, Synalar*	Fluocinolone acetonide cream USP, 0.01, 0.025, 0.2% Fluocinolone acetonide ointment USP, 0.025% Fluocinolone acetonide topical solution USP, 0.01%	Topically bid to qid			

*See USP DI for complete dosage information.

TABLE 18-16 *Continued*

ANTI-INFLAMMATORY GLUCOCORTICOIDS WITH LOW SALT RETENTION

Name Proprietary Name	**Preparations**	**Application**	**Usual Adult Dose***	**Usual Dose Range***	**Usual Pediatric Dose***
Fluocinonide USP *Lidex*	Fluocinonide cream USP, 0.05% Fluocinonide ointment USP, 0.05% Fluocinonide gel USP, 0.05%	Topically bid to qid			
Fluprednisolone *Alphadrol*	Fluprednisolone tablets		Initial, 1.5–18 mg/day; maintenance, 1.5–12 mg/day		
Flurandrenolide USP *Cordran*	Flurandrenolide cream USP, 0.025, 0.05% Flurandrenolide ointment USP, 0.025, 0.05% Flurandrenolide lotion USP, 0.05%	Topically qd or bid			
	Flurandrenolide tape USP	Tape containing 4 μg/cm² replaced every 12–24 hr			
Halcinonide *Halog, Halciderm*	Halcinonide cream USP, 0.025, 0.1% Halcinonide ointment USP, 0.1% Halcinonide topical solution USP, 0.1%	Cream: topically bid or tid Ointment: topically qd Topical solution: topical to skin qd			
Medrysone USP *Medrocort*	Medrysone ophthalmic suspension USP, 1%	Topically to the conjunctiva, 1 drop up to every 4 hr			
Meprednisone *Betapar*	Meprednisone tablets		oral, 4 to 60 mg/day as a single dose or in divided dose		
Methylprednisolone USP *Medrol, Medrone(U.K.)*	Methylprednisolone tablets USP		4 mg qid	4–48 mg/day	117 μg (0.117 mg)/kg of body weight or 3.3 mg/m² of body surface a day in 3 divided doses
Methylprednisolone acetate USP *Depo-Medrol*	Methylprednisolone acetate ointment, 0.25, 1%	Topical qd to qid			
	Sterile methylprednisolone acetate suspension USP		Intra-articular, intralesional, or soft-tissue injection, 4–80 mg, repeated at 1–5-wk intervals, if necessary Intramuscular, 40–120 mg, repeated at 1-day to 2-wk intervals if necessary		Intramuscular, 117 μg/kg of body weight or 3.33 mg/m² of body surface a day every third day
	Methylprednisolone acetate for enema USP		Rectal, 40 mg, 3–7 times a week	120–280 mg/wk	
Methylprednisolone sodium succinate USP *Solu-Medrol*	Methylprednisolone sodium succinate for injection USP				
Mometasone furoate *Elocon*	Mometasone furoate cream, 0.1% Mometasone furoate ointment, 0.1%	Topically qd or bid in thin films			
Paramethasone acetate USP *Haldrone, Stemex*	Paramethasone acetate tablets USP		Oral, 2–24 mg/day as a single dose or in divided doses		Oral, 58–800 μg (0.058–0.8 mg)/kg of body weight or 1.67–25 mg/m² of body surface a day as a single dose or in divided doses

*See USP DI for complete dosage information.

TABLE 18-16 *Continued*

ANTI-INFLAMMATORY GLUCOCORTICOIDS WITH LOW SALT RETENTION

Name Proprietary Name	***Preparations***	***Application***	***Usual Adult Dose****	***Usual Dose Range****	***Usual Pediatric Dose****
Triamcinolone *Aristocort, Kenacort*	Triamcinolone tablets USP	Adrenocortical insufficiency: oral, 4–12 mg/day as a single dose or in divided doses, in conjunction with a mineralocorticoid			Adrenocortical insufficiency: 117 μg (0.117 mg)/kg of body weight or 3.3 mg/m^2 of body surface a day as a single dose or in divided doses in conjunction with a mineralocorticoid
Triamcinolone diacetate USP *Aristocort Diacetate*	Sterile triamcinolone diacetate suspension USP Triamcinolone diacetate syrup USP	See triamcinolone tablets	Oral, 4–30 mg/day		See triamcinolone tablets
Triamcinolone acetonide USP *Kenalog, Aristocort Acetonide, Aristoderm*	Triamcinolone acetonide cream USP Triamcinolone acetonide ointment USP Triamcinolone acetonide topical aerosol USP	Topically to the skin, as a 0.025–0.5% cream or ointment bid to qid			
	Triamcinolone acetonide dental paste USP	Topically to the oral mucous membranes, as a 0.1% paste bid to qid			
	Sterile triamcinolone acetonide suspension USP		Intra-articular, intrabursal, or tendon-sheath injection, 2.5–15 mg Intradermal or intralesional, up to 1 mg, repeated at weekly or less frequent intervals, if necessary Intramuscular, 40–80 mg, repeated at 4-wk intervals if necessary	Intra-articular or intrabursal, 2.5–80 mg; IM 20–100 mg	Children up to 6 yr of age–use is not recommended Children 6–12 yr of age–intra-articular, intrabursal, or tendon-sheath injection, 2.5–15 mg, repeated as needed Intramuscular, 40 mg, repeated at 4-wk intervals if necessary or 30–200 μg (0.03–0.2 mg)/kg of body weight or 1–6.25 mg/m^2 of body surface, repeated at 1- to 7-day intervals
Triamcinolone hexacetonide USP *Aristospan*	Triamcinolone hexacetonide suspension USP		Intra-articular, 2–20 mg at each site, once every 3–4 wk; intralesional or sublesional, up to 500 μg/m^2 of affected skin, repeated as needed		

*See USP DI for complete dosage information.

Fludrocortisone Acetate, USP. 9α-fluoro-11β,17,21-trihydroxypregn-4-ene-3,20-dione-21-acetate; 9α-fluorohydrocortisone is used only for the treatment of Addison's disease and for inhibition of endogenous adrenocortical secretions. As shown in Table 18-12, it has up to about 800 times the mineralocorticoid activity of hydrocortisone and about 11 times the glucocorticoid activity. Its potent activity stimulated the synthesis and study of the many fluorinated steroids shown in Figure 18-23. Although its great salt-retaining activity limits its use to Addison's disease, it has sufficient glucocorticoid activity that, in many cases of the disease, additional glucocorticoids need not be prescribed.

Cortisone Acetate, USP. 17α,21-Dihydroxy-4-pregnene-3,11,20-trione 21-acetate is a natural cortical steroid with good anti-inflammatory activity and low to moderate salt-retention activity. It is used for the entire spectrum of uses discussed previously under Therapeutic Uses—collagen diseases, especially rheumatoid arthritis; Addison's disease; severe shock; allergic conditions; chronic lymphatic leukemia; and many other indications. Cortisone acetate is relatively ineffective topically, in part because it must be reduced in vivo to hydrocortisone which is more active. Its plasma half-life is only about 30 minutes, compared to 90 minutes to three hours for hydrocortisone.

Hydrocortisone, USP. 11,17,21-Trihydroxypregn-4-ene-3,20-dione is the primary natural glucocorticoid in humans. Synthesis of 9α-fluorohydrocortisone during the synthesis of hydrocortisone has led to the array of semisynthetic glucocorticoids shown in Figure 18-23, many of which have greatly improved anti-inflammatory activity. Nevertheless, hydrocortisone, its esters, and its salts remain a mainstay of modern adrenocortical steroid therapy —and the standard for comparison of all other glucocorticoids and mineralocorticoids (see Table 18-12). It is used for all the indications previously mentioned. Its esters and salts illustrate the principles of chemical modification to modify pharmacokinetic utility shown in Figure 18-7. The commercially available salts and esters (see Fig. 18-23) include

Hydrocortisone acetate, USP
Hydrocortisone sodium succinate, USP
Hydrocortisone cypionate, USP
Hydrocortisone sodium phosphate, USP
Hydrocortisone valerate, USP

Prednisolone, USP. Δ^1-Hydrocortisone; 11, 17,21-trihydroxypregna-1,4-diene-3-20-dione has less salt-retention activity than hydrocortisone (see Table 18-12), but some patients have more frequently experienced complications such as gastric irritation and peptic ulcers. Because of low mineralocorticoid activity, it cannot be used alone for adrenal insufficiency. Prednisolone is available in a variety of salts and esters to maximize its therapeutic utility (see Fig. 18-7).

Prednisolone acetate, USP
Prednisolone succinate, USP
Prednisolone sodium succinate for injection, USP
Prednisolone sodium phosphate, USP
Prednisolone tebutate, USP

Prednisone, USP. Δ^1-Cortisone; 17,21-dihydroxypregna-1,4-diene-3,11,20-trione has systemic activity very similar to that of prednisolone, and because of its lower salt-retention activity, it is often preferred over cortisone or hydrocortisone.

GLUCOCORTICOIDS WITH LOW SALT RETENTION

Most of the key differences between the many glucocorticoids with low salt retention (see Fig. 18-23) have been summarized in Tables 18-12 and 18-13. The tremendous therapeutic and, therefore, commercial importance of these drugs has been a stimulus to the proliferation of new compounds and their products. Many compounds also are available as salts or esters to give the complete range of therapeutic flexibility illustrated in Figure 18-7. When additional pertinent information (other than that shown in Tables 18-12 and 18-14) is available, it will be given below.

Alclometasone, USP. 7-Chloro-11,17,21-trihydroxy-16-methylpregna-1,4-diene-3,20-dione.

Amcinonide, USP. 21-(Acetyloxy)-16,17-[cyclopentylidenebis(oxy)]-9-fluoro-11-hydroxypregna-1, 4-diene-3,20-dione.

Beclomethasone, USP. 9-Chloro-11,17,21-trihydroxy-16-methylpregna-1,4-diene-3,20-dione.

Beclomethasone Dipropionate, USP. 9-Chloro-11, 17, 21-trihydroxy-16-methylpregna-1, 4-diene-3,20-dione 17,21-diproprionate is available as a nasal inhaler and is used for seasonal rhinitis poorly responsive to conventional treatment.

Betamethasone, USP. 9-Fluoro-11,17,21-trihydroxy-16-methylpregna-1,4-diene-3,20-dione.

Betamethasone valerate, USP
Betamethasone acetate, USP
Betamethasone sodium phosphate, USP
Betamethasone benzoate, USP
Betamethasone dipropionate, USP

Clobetasol, USP. 21-Chloro-9-fluoro-11,17-dihydroxy-16-methylpregna-1,4-diene-3,20-dione.

Clocortolone, USP. 9-Chloro-6α-fluoro-11β, 21-dihydroxy-16α-methylpregna-1,4-diene-3,20-dione.

Desonide, USP. 16α-Hydroxprednisolone-16α,17-acetonide.

Desoximetasone, USP. 9-Fluoro-11,21-dihydroxy-16-methylpregna-1,4-diene-3,20-dione.

Dexamethasone, USP. 9α-Fluoro-11,17,21-trihydroxy-16-methylpregna-1,4-diene-3,20-dione is essentially the 16α-isomer of betamethasone.

Dexamethasone acetate, USP
Dexamethasone sodium phosphate, USP

Diflorasone, USP. 6,9-Difluoro-11,17,21-trihydroxy-16-methylpregna-1,4-diene-3,20-dione.

Flumethasone Pivalate, USP. 6,9-Difluoro-11, 17, 21-trihydroxy-16-methylpregna-1, 4-diene-3, 20-dione 21-pivalate.

Flunisolide, USP. 6α-Fluoro-11β,16α,17,21-tetrahydroxypregna-1,4-diene-3,20-dione 16,17-acetonide.

Fluocinolone Acetonide, USP. 6α-Fluorotriamcinolone acetonide; 6α,9α-difluoro-11β,16α,17,21-tetrahydroxypregna-1,4-diene-3,20-dione 16α,17α-acetone ketal, the 21-acetonide of **fluocinonide**. Fluocinonide is about five times more potent than the acetonide in the vasoconstrictor assay.

Flurandrenolide, USP. 6α-Fluoro-11β,21-dihydroxy-16α,17-[(1-methylethylidene)bis(oxy)]pregn-4-ene-3,20-dione has replaced the name **flurandrenolone**. Although a flurandrenolide tape product is available, it can stick to and remove damaged skin, so it should be avoided with vesicular or weeping dermatoses.

Halcinonide. 21-Chloro-9-fluoro-11β,16α,17-trihydroxypregn-4-ene-3,20-dione 16α,17α-ketal is the first *chloro*-glucocorticoid yet marketed. As with several other glucocorticoids (see Table 18-16), it is used only topically. In one double-blind study with betamethasone valerate cream, halcinonide was superior in the treatment of psoriasis. However, it can be used for the usual range of indications previously described.

Medrysone, USP. 11-Hydroxy-6-methylpregn-4-ene-3,20-dione is unique among the other corticosteroids shown in Figure 18-23 in that it does not have the usual 17α,21-diol system of the others. Currently, it is used only for treatment of inflammation of the eyes.

Meprednisone, USP. 17,21-Dihydroxy-16β-methylpregna-1,4-diene-3,11,21-trione; 16β-methylprednisolone.

Methylprednisolone, USP. 11,17,21-Trihydroxy-6 methyl-1,4-pregnadiene-3,20-dione

Methylprednisolone acetate, USP
Methylprednisolone hemisuccinate, USP
Methylprednisolone sodium succinate, USP

Mometasone Furoate, USP. 9α-21-Dichloro-16α-methylprednisolone 17α-furoate.

Paramethasone Acetate, USP. 6α-Fluoro-11β, 17, 21-trihydroxy-16α-methylpregna-1, 4-diene-3,20-dione 21-acetate.

Triamcinolone, USP. 9-Fluoro-11,16,17,21-tetrahydroxypregna-1,4-diene-3,20-dione.

Triamcinolone Acetonide, USP. Triamcinolone-16α,17α-acetone ketal; 16α,17α-[(1-methylethylidene)bis(oxy)]triamcinolone.

Triamcinolone Hexacetonide, USP. Triamcinolone acetonide 21-[3-(3,3-dimethyl)butyrate].

Triamcinolone Diacetate, USP. The hexacetonide is slowly converted to the acetonide in vivo and is given only by intra-articular injection. Only triamcinolone and the diacetate are given orally. When triamcinolone products are given intramuscularly, they are often given deeply into the gluteal region because local atrophy may occur with shallow injections. The acetonide and diacetate may be given by intra-articular or intrasynovial injection, and, additionally, the acetonide may be given by intrabursal or sometimes by intramuscular or subcutaneous injection. A single intramuscular dose of the diacetate or acetonide may last up to three or four weeks. Plasma levels with intramuscular doses of the acetonide are significantly higher than with triamcinolone itself.

Topically applied triamcinolone acetonide is a potent anti-inflammatory agent (see Table 18-12), about ten times more so than triamcinolone.

CARDIAC STEROIDS AND RELATED INOTROPIC DRUGS

A variety of drugs can increase the force of contraction of the heart.[243–258] This *inotropic* action can be particularly useful in treatment of congestive heart failure. A failing heart cannot pump sufficient blood to maintain body needs. With 2 million or more patients in the United States alone with congestive heart failure, and millions more throughout the world, inotropic drugs are extremely important. They prolong life, but even with drug treatment, the long-term outlook for these patients is poor.[244, 249]

The cardiac glycosides (Fig. 18-25) have been used for centuries, but their narrow therapeutic index has stimulated a wide search for alternatives. The first, amrinone, was reported in 1979.[250] The other nonsteroidal drugs in Figure 18-25 are in clinical or preclinical trials.

Primary inotropic drugs include the following:

- Cardiac steroids, such as digoxin (see Fig. 18-25), that act as Na^+,K^+–ATPase inhibitors. These steroids are also widely used to treat atrial fibrillation and flutter.
- Phosphodiesterase inhibitors, such as amrinone (see Fig. 18-25).
- Drugs, such as sulmazole, that increase the Ca^{2+} sensitivity of myocardial contractile proteins (see Fig. 18-25).
- Direct adenylate cyclase stimulants, such as forskolin (see Fig. 18-25).
- Adrenergic agonists (see Chapter 11), such as dopamine, levodopa, dobutamine.
- Indirect adenylate cyclase stimulants (see Chapter 21), such as glucagon and histamine.

This section of Chapter 18 will focus on the cardiac steroids, phosphodiesterase inhibitors, Ca^{2+} sensitivity stimulants, and the direct adenylate cyclase stimulant, forskolin. The use of the cardiac steroids in treating atrial fibrillation and flutter has been discussed in Chapter 14. Adrenergic agonists

a. The Cardenolides and Bufadienolides (Na^+,K^+– ATPase inhibitors)

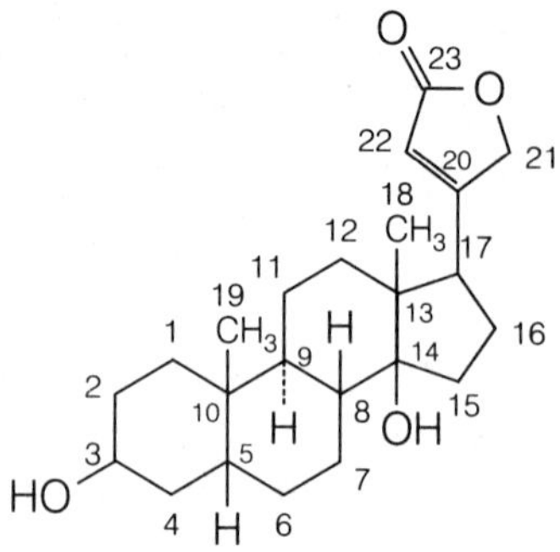

Digitoxigenin
(Cardenolide Prototype)
(Also in commercially available products)

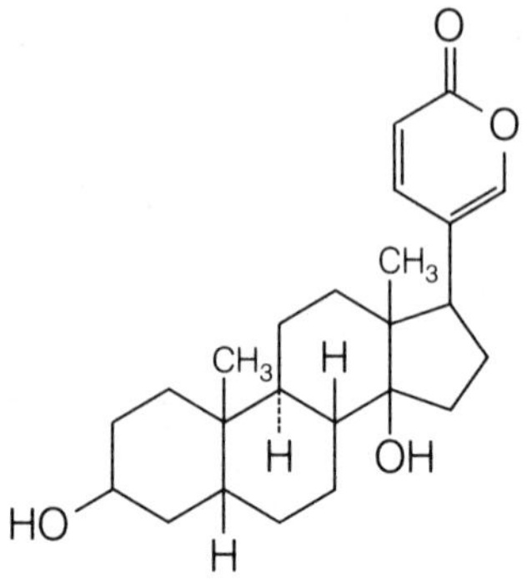

Bufalin
(Bufadienolide Prototype)

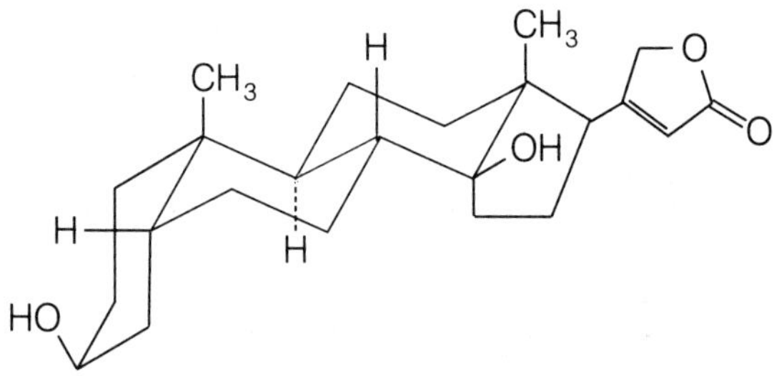

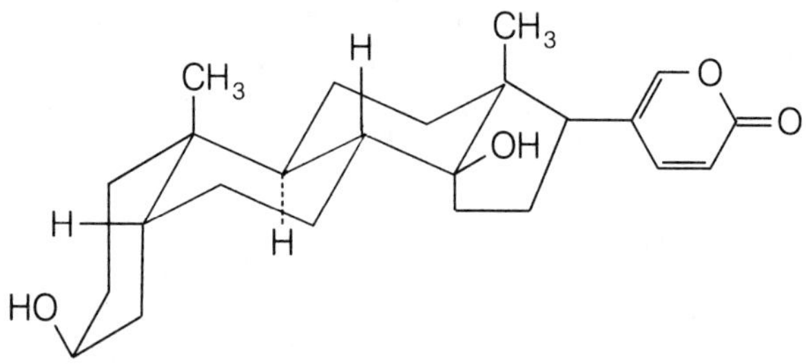

Cardenolide Aglycones
in Commercially Available Products

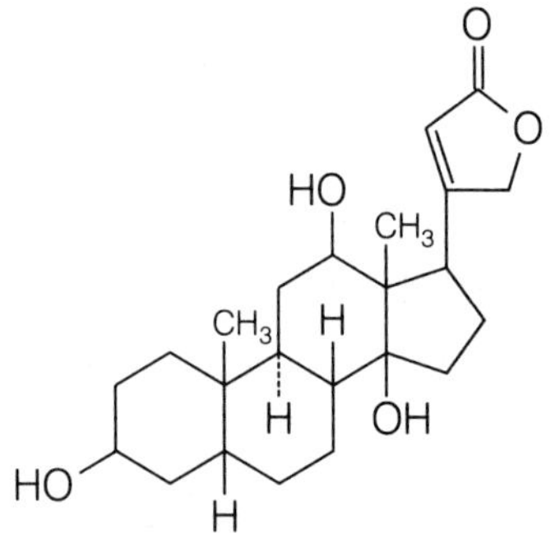

Digoxigenin

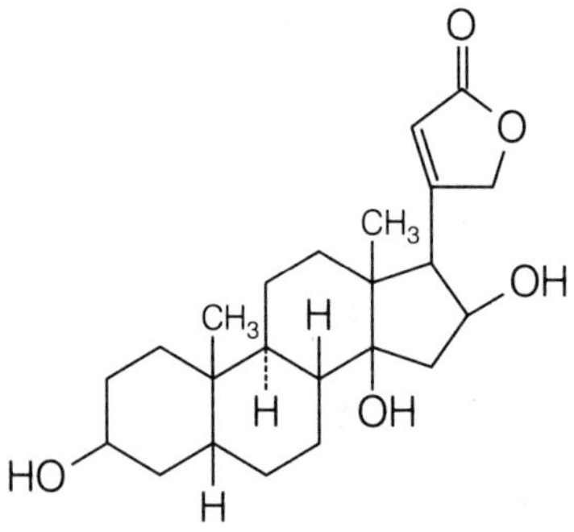

Gitoxigenin

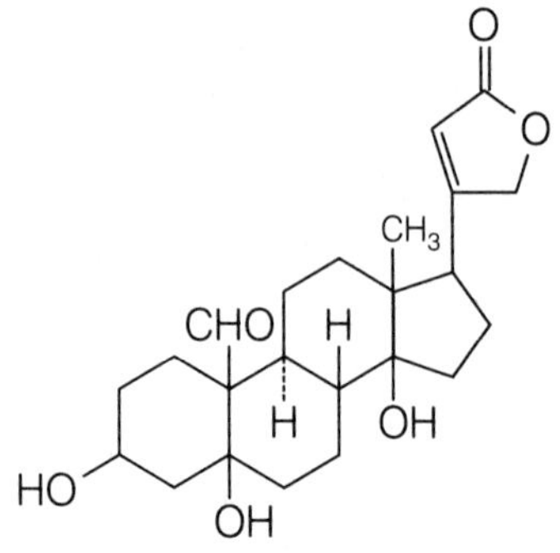

Strophanthidin

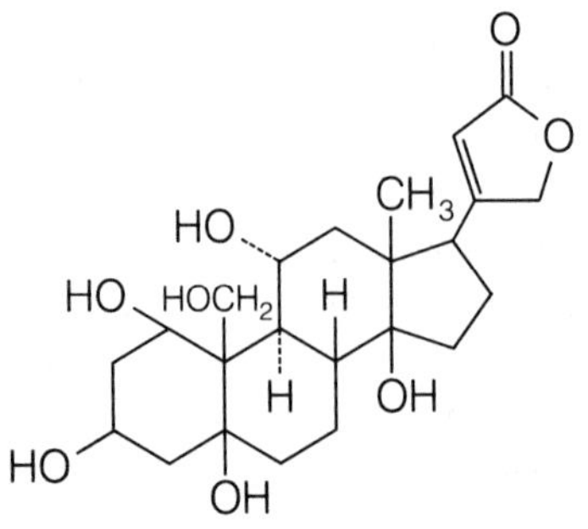

Ouabagenin

FIG. 18-25. Cardiac steroids and other inotropic drugs used for congestive heart failure. The cardiac steroids are also used to treat atrial fibrillation and flutter (discussed in Chap. 14).

b. Phosphodiesterase inhibitors

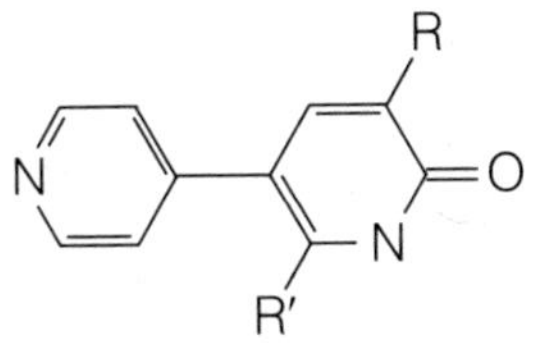

Amrinone $R = NH_2$, $R' = H$
Milrinone $R = CN$, $R' = CH_3$

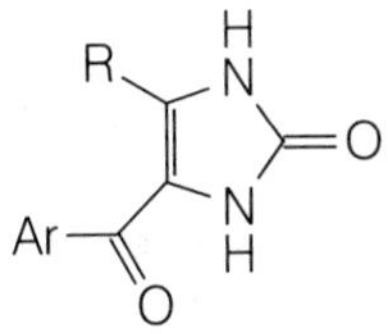

Enoximone $R = CH_3$
$Ar = H_3CS$–

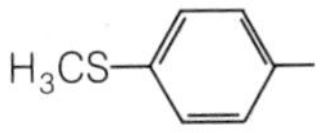

Piroximone $R = CH_2CH_3$

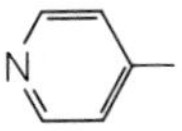

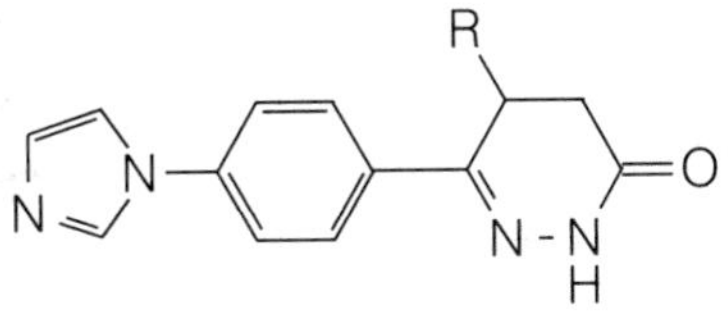

Imazodan $R = H$
CI-930 $R = CH_3$

c. Drugs that increase Ca^{2+} sensitivity of myocardial contractile proteins

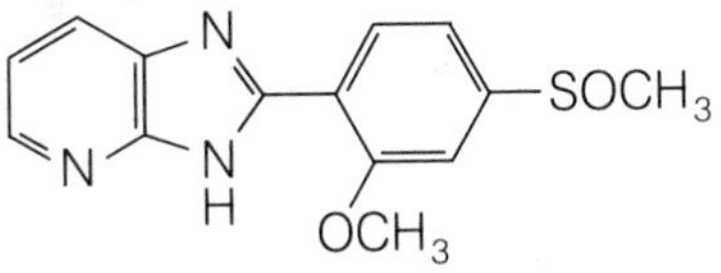

Sulmazole (AR-L 115)

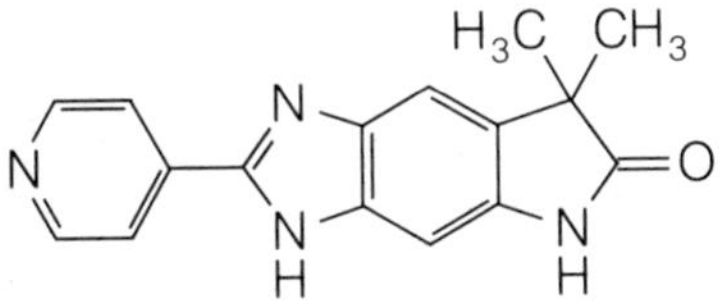

BM 14.478

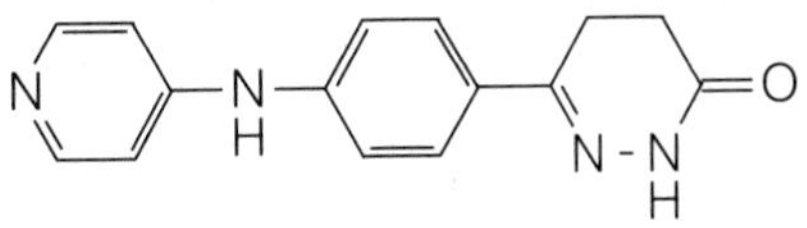

MCI-154

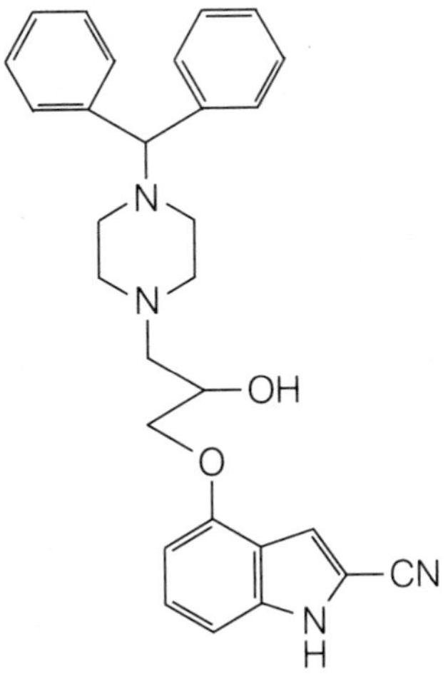

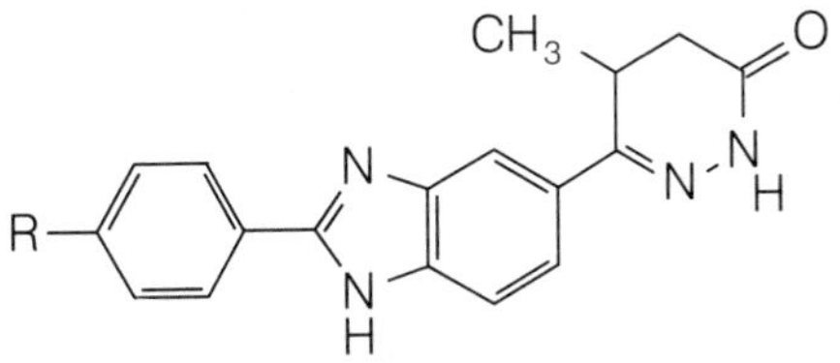

$R = OCH_3$ Pimobendan
$R = OH$ UD-CG 212

DPI 201-106

d. Adenylate cyclase stimulants

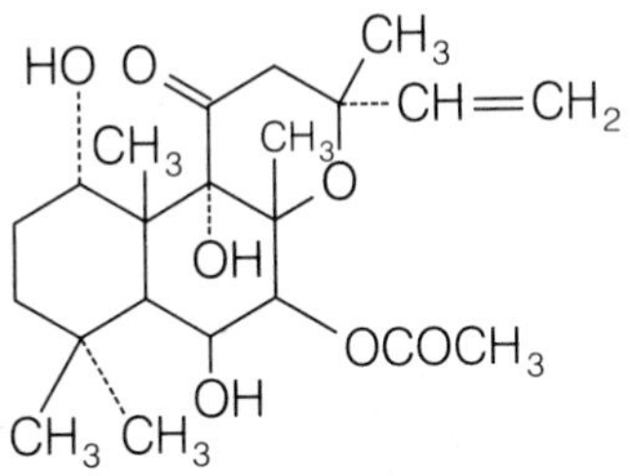

Forskolin

FIG. 18.25. *Continued*

have been discussed in Chapter 11. Although catecholamines have inotropic actions, their pharmacologic effects are too nonspecific and short in duration to be generally useful in treating congestive heart failure. However, levodopa has been reported to significantly improve the condition of patients with severe heart failure. Short-term therapy with dobutamine may result in long-term benefits with these patients as well. The indirect adenylate cyclase stimulants have varying problems that preclude their long-term use for congestive heart failure. The recent reviews of inotropic agents used in the treatment of congestive heart failure by Smith[40] and by Colucci and colleagues[243,244] discuss these clinical applications of drugs and their limitations in detail.

Poison and heart tonic—these are the "split personalities" of the cardiac steroids that have troubled and fascinated physicians and chemists for several centuries. Plants containing the cardiac steroids have been used as poisons and heart drugs at least since 1500 B.C., with squill appearing in the Ebers Papyrus of ancient Egypt. Throughout history these plants or their extracts have variously been used as arrow poisons, emetics, diuretics, and heart tonics. Toad skins containing cardiac steroids have been used as arrow poisons and even as aids for treating toothaches and as diuretics.

The poison–heart tonic dichotomy continues even today. Cardiac steroids are widely used in the modern treatment of congestive heart failure and for treatment of atrial fibrillation and flutter. Nevertheless, their toxicity remains a serious problem, although advances in biopharmaceutics have reduced the incidence of life-threatening overdoses.

ROLES IN MYOCARDIAL CALCIUM ION REGULATION

The inotropic drugs in Figure 18-25 all act by affecting the availability of intracellular Ca^{2+} for myocardial contraction or by increasing the sensitivity of myocardial contractile proteins. In normal heart function, Ca^{2+} is essential for heart contraction. During each myocardial action potential, Ca^{2+} enters through slow Ca^{2+} channels, and, in turn, there is a large release of stored Ca^{2+} from sequestration sites on the sarcoplasmic reticulum. The Ca^{2+} binds to troponin-C, and actin and myosin can interact to result in a contraction. (The mechanism involves other myofilament subunits as reviewed by Colucci et al.[243,244])

Cardiac Steroids

The cardiac steroids inhibit Na^+, K^+–ATPase (sodium- and potassium-dependent ATPase), the enzyme that catalyzes the "sodium pump" (Fig. 18-26). Its amino acid sequence was determined for the first time in 1985,[46] with the cardiac glycoside-binding site located in 1983.[47] This enzyme, by essentially *being* the sodium pump, maintains unequal distribution of Na^+ and K^+ ions across the cell membrane.

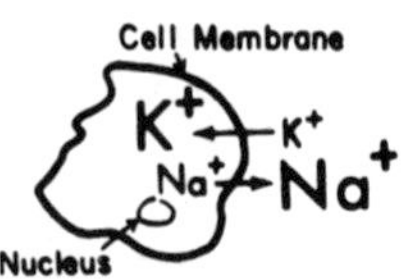

FIG. 18-26. *Movements of ions with the sodium pump. Sodium- and potassium-dependent ATPase catalyzes the pump. The hydrolysis of ATP to ADP provides the needed energy to move the ions against concentration gradients.*

Although the "pump" operates in all cells, it performs a critical function in heart contraction. During each contraction, there is an influx of Na^+ and an outflow of K^+. Before the next contraction, Na^+, K^+–ATPase must reestablish the concentration gradient, pumping Na^+ into the cell against a concentration gradient. This process requires energy, and the energy is obtained from hydrolysis of ATP to ADP by Na^+, K^+–ATPase.

The cardiac steroids inhibit Na^+, K^+–ATPase.[38–43] The review by Smith[40] is especially recommended for further reading about the cardiac glycosides' actions on Na^+, K^+–ATPase and their clinical uses. Inhibiting all molecules of Na^+, K^+–ATPase in the sarcolemmal membrane would be fatal, because there would be no way for the heart to reestablish the Na^+ and K^+ concentration gradients across the membrane. However, partial inhibition results in a small net increase in intracellular Na^+. Less than a usual amount of Na^+ is pumped out; consequently, a higher than normal amount of Na^+ remains.

As discussed in Chapter 14, several mechanisms regulate intracellular Ca^{2+}. The blocking of slow Ca^{2+} channels is the basis of the antiarrhythmic and antiangina actions of verapamil and related drugs (Chapter 14). For the cardiac steroids, the sodium–calcium exchanger is most important. When intracellular Na^+ increases in response to partially inhibited sarcolemmal sodium pumps, the sodium–calcium exchanger exchanges three Na^+ for each Ca^{2+}, with a net overall influx of Ca^{2+} across the sarcolemmal membrane. The increased Ca^{2+} triggers the contractile proteins, with a resulting inotropic action.

Inhibition of the sodium pump may result in additional effects that increase the availability of Ca^{2+} for heart contraction. As reviewed by Smith,[40] additional Ca^{2+} may enter through slow Ca^{2+} channels. The sodium–hydrogen exchange system may also facilitate additional Na^+ to be transported intracellularly, with increased Ca^{2+} resulting as well.

Phosphodiesterase Inhibitors

In normal heart function, cAMP performs important roles in regulating intracellular Ca^{2+}. Slow Ca^{2+} channels and storage sites for Ca^{2+} on the sarcoplasmic reticulum must be activated by cAMP-dependent protein kinases to result in (1) increased Ca^{2+} influx through the channels and (2) increased release and faster reaccumulation by the sarcoplasmic reticulum. If cAMP is increased by inhibiting its breakdown by phosphodiesterase to AMP, more Ca^{2+} becomes available for cardiac contraction. Drugs such as amrinone and milrinone (see Fig. 18-25b) exert their inotropic actions in this way.[243–250] Specifically, they inhibit phosphodiesterase F-III.

These drugs also cause significant reductions in cardiac-filling pressures and systemic vascular resistance, with a resulting drop in systemic blood pressure. The overall benefits of therapy, therefore, appear to result from a combination of their inotropic and vasodilator actions.[243,244] It is important to note that overall survival of patients with congestive heart failure does not appear to be significantly increased as a result of taking these agents. However, the quality of life may be substantially improved.

Drugs That Increase the Calcium Ion Sensitivity of Myocardial Contractile Proteins

Following the 1979 report[250] of **amrinone**'s inotropic action, there was an enormous effort made throughout the world to find other nonsteroidal inotropic agents. A number of agents (see Fig. 18-25c) were first thought to act pharmacologically as phosphodiesterase inhibitors, but a closer examination has revealed that the inotropic effect requires less drug than for phosphodiesterase inhibition. These drugs appear to increase the effect of *existing* Ca^{2+} levels, thereby resulting in an inotropic effect. DPI 201-106 has no phosphodiesterase-inhibiting activity at all. It may prolong the opening of Na^+ channels.[251]

Van Meel and co-workers[252] have found, for example, that the (+) and (−) stereoisomers of sulmazole have equivalent vasodilatation and phosphodiesterase-inhibiting potencies. However, the (+) isomer has good inotropic and Ca^{2+} sensitizing activity; the (−) isomer has weak activity for both.

Direct Adenylate Cyclase Stimulants: Forskolin

The diterpene **forskolin** (see Fig. 18-25d) directly stimulates adenylate cyclase or a closely related protein.[244,253,254] The cAMP levels increase in the myocardium, with resulting activation of protein kinases and increases in intracellular Ca^{2+}, as with the phosphodiesterase inhibitors. Vasodilatation also results. In addition, Hoshi and co-workers[255] have reported that forskolin has a direct effect on voltage-dependent K^+ channels, independent of cAMP activation. Wagoner and Pallotta[256] have also reported that some of forskolin's physiologic effects are not mediated by cAMP.[256]

STRUCTURAL CLASSES

As noted in the previous section, cardiac steroids are all Na^+,K^+–ATPase inhibitors. Thus, the classification and discussion that follow use the pharmacologic classification shown in Figure 18-25.

Cardiac Steroids: Na^+,K^+–ATPase Inhibitors

The cardiac steroids actually include two groups of compounds—the cardenolides and the bufadienolides. The cardenolides, illustrated by digitoxigenin in Figure 18-25a, have an unsaturated butyrolactone ring at C-17, whereas the bufadienolides have an α-pyrone ring. Both have essentially identical pharmacologic profiles and are found in a variety of plant species. The bufadienolides are commonly called "toad poisons" because several are found in the skin secretions of various toad species. Use of the toads as a source of cardiac steroid would quickly result in their extinction; therefore, by far the most historically and commercially important sources of cardiac steroids have been two species of *Digitalis*: *D. purpurea* and *D. lanata*. Whole-leaf digitalis preparations appeared in the *London Pharmacopeia* in the 1500s, but inconsistent results and common fatalities caused their removal. In 1785, William Withering published his classic, *An Account of the Foxglove and Its Medical Uses*, noting that digitalis could be used to treat cardiac insufficiency with its associated dropsy (edema). (The Withering story has been beautifully told by Witzman[1] and Skou.[259]) Nevertheless, it was not until the early 1900s that digitalis and the purified glycosides were commonly used for the treatment of heart disease.

The $5\beta,14\beta$-stereochemistry of the cardenolides and bufadienolides gives the molecules an interesting shape, caused by the resulting A/B *cis* and C/D *cis* ring junctures. This stereochemistry appears to be an important prerequisite for some, but possibly not all, cardiac steroid activity. This will be discussed later in this section.

The cardiac steroids are usually found in nature as the corresponding 3β-glycosides. One to four sugars are added to the steroid 3β-hydroxyl to form the

TABLE 18-17

CARDIAC GLYCOSIDES AND HYDROLYSIS PRODUCTS FROM COMMON SOURCES

Structure	Name
FROM *DIGITALIS PURPUREA* LEAF	
Glucose-digitoxose$_3$-digitoxigenin	Purpurea glycoside A
Digitoxose$_3$-digitoxigenin	Digitoxin
Glucose-digitoxose$_3$-ditoxigenin	Purpurea glycoside B
Digitoxose$_3$-gitoxigenin	Gitoxin
FROM *DIGITALIS LANATA* LEAF	
Glucose-3-acetyldigitoxose-digitoxose$_2$-digitoxigenin	Lanatoside A
Glucose-digitoxose$_3$-digitoxigenin	Desacetyl lanatoside A (same as purpurea glycoside A)
3-Acetyldigitoxose-digitoxose$_2$-digitoxigenin	Acetyldigitoxin
Digitoxose$_3$-digitoxigenin	Digitoxin
Glucose-3-acetyldigitoxose-digitoxose$_2$-gitoxigenin	Lanatoside B
Glucose-digitoxose$_3$-gitoxigenin	Desacetyl lanatoside B (same as purpurea glycoside B)
3-Acetyldigitoxose-digitoxose$_2$-gitoxigenin	Acetylgitoxin
Digitoxose$_3$-gitoxigenin	Gitoxin
Glucose-3-acetyldigitoxose-digitoxose$_2$-digoxigenin	Lanatoside C
Glucose-digitoxose$_3$-digoxigenin	Desacetyl lanatoside C
3-Acetyldigitoxose-digitoxose$_2$-digoxigenin	Acetyldigoxin
Digitoxose$_3$-digoxigenin	Digoxin
FROM *STROPHANTHUS GRATUS* SEED	
Rhamnose Ouabagenin	Ouabain

glycoside structure. Although hundreds of cardiac steroid glycosides have been found in nature, relatively few different cardenolide or bufadienolide aglycones have been found. For example, as shown in Table 18-17, only three make up the digitalis glycosides. (The substance that forms a glycoside with a sugar is called the aglycone. Thus, the cardenolides and bufadienolides are the *aglycones* or *genins* of the *cardiac glycosides*.) The structure of a representative cardenolide glycoside, lanatoside C, is shown in Figure 18-27.

The hydrolysis products of other naturally occurring cardiac glycosides are shown in Table 18-17. The two most commonly found glycosides are β-D-digitoxosides (including digoxin, digitoxin, gitoxin) and α-L-rhamnoside (ouabain).

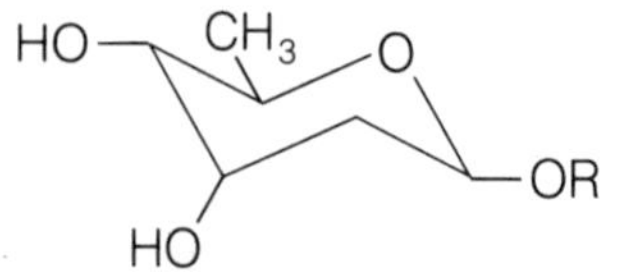

HO CH3 O OR HO

β-D-Digitoxoside

The sugars found as part of the cardiac glycosides are illustrated in Figure 18-28. As shown in Figure 18-27, the sugars may be removed by enzymatic-, acid-, or base-catalyzed hydrolysis. A wide variety of other sugars have been used to synthesize new cardiac glycosides (discussed later in this chapter).

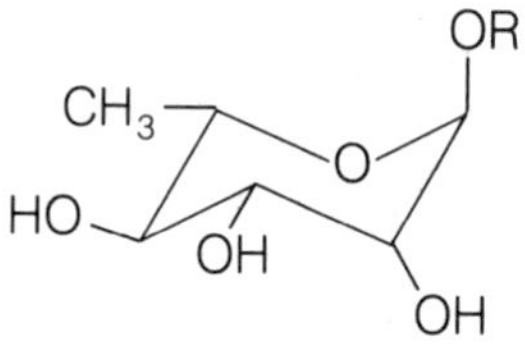

OR CH3 O HO OH OH

α-L-Rhamnoside
(R = Steroid Genin)

Phosphodiesterase Inhibitors

As shown in Figure 18-25c, a variety of small ring molecules have been found to be effective phosphodiesterase inhibitors. Hundreds of others have been synthesized,[249] but those in Figure 18-25c have been especially promising. Several are now in clinical trials. The compounds include 3,4′-bipyridin-6(1*H*)-ones, such as amrinone and milrinone; 4-aroyl-1,3-dihydro-2*H*-imidazol-2-ones, including fenoximone and piroximone; 4,5-dihydro-6-[4-(1*H*-imidazol-1-

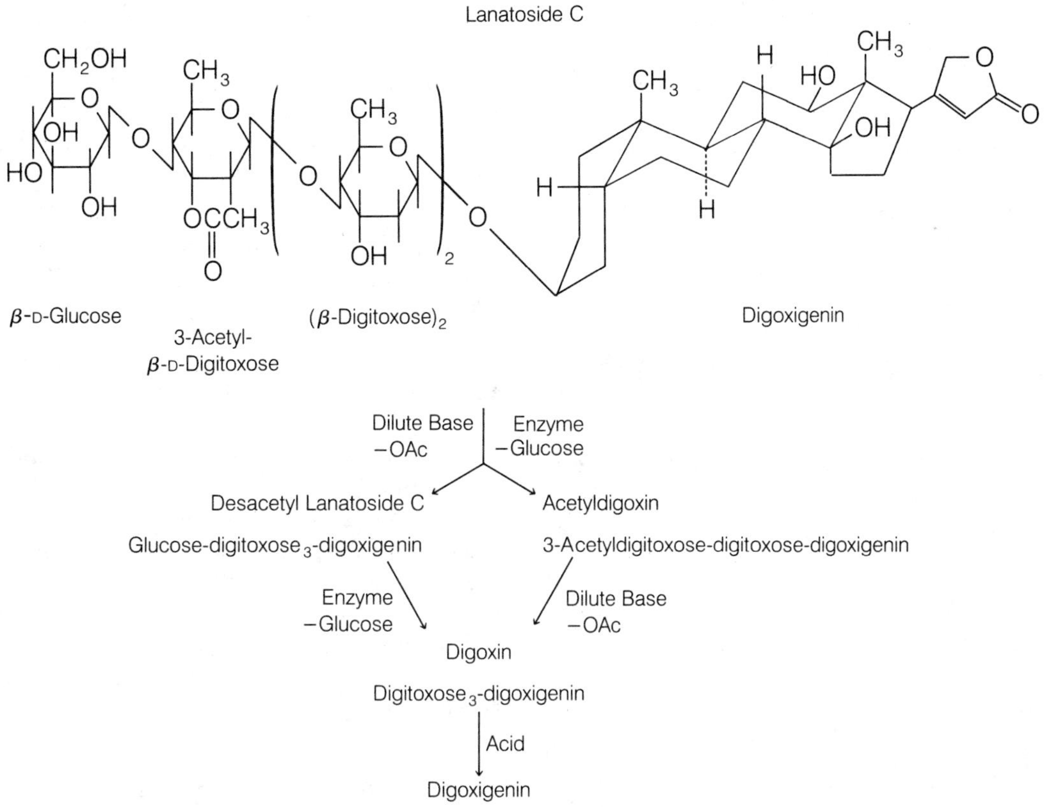

Lanatoside C

FIG. 18-27. *Selective hydrolysis of naturally occurring cardiac glycosides — a representative example. The sugars are exaggerated in size to clearly show their structure.*

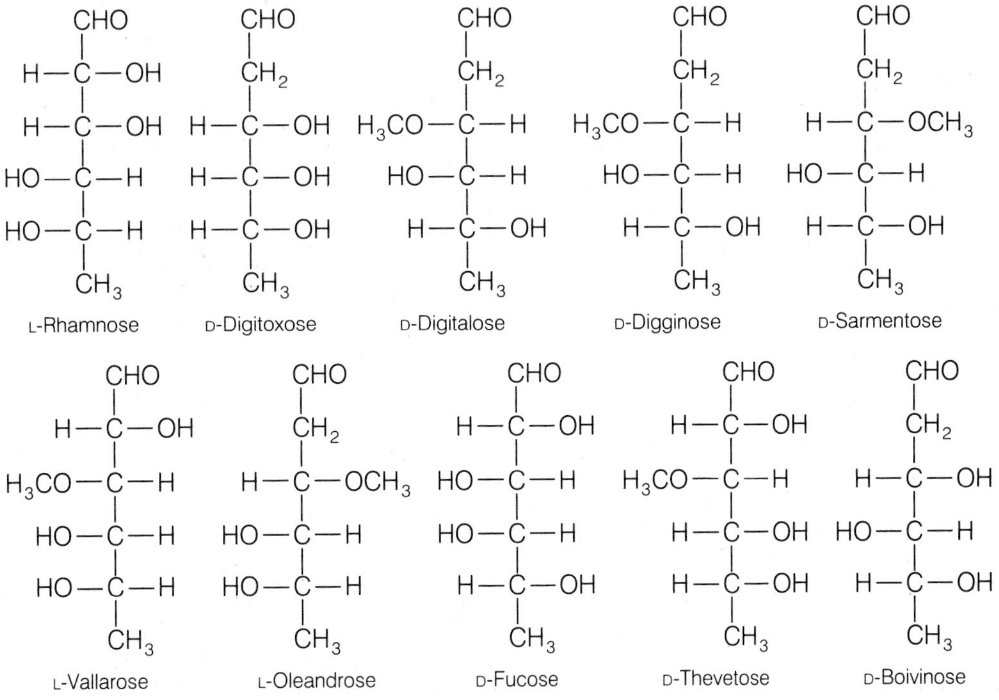

FIG. 18-28. *Sugars found in naturally occurring cardiac glycosides.*

yl)phenyl]-3(2*H*)-pyridazinones, including imazodan and CI-930; and a variety of other multiring heterocyclics, such as OPC-8212. Many of these compounds are much more active than amrinone. Milrinone, for example, is 10 to 20 times more inotropic than amrinone and causes fewer side effects.

More recently, it has been proposed that these compounds may cause their inotropic effects by acting as antagonists for A_1-adenosine receptors.[249a]

Drugs That Increase Sodium Ion Sensitivity of Myocardial Contractile Proteins

Included are imidazopyridines, including sulmazole and UN-LZ97- and 98; pyridazinones, such as MCI-154; and other multiring heterocyclics, including DPI201-106.

Direct Adenylate Cyclase Stimulant: Forskolin

Forskolin[253,254] is a naturally occurring diterpene from an Indian *Coleus* species. It causes a marked inotropic and vasodilatation response and has prompted considerable interest, including synthesis of new analogues.[253] Forskolin also has been shown[253a] to inhibit glucose transport, enhance nicotinic receptor desensitization, modulate voltage dependent K^+ channels, as well as a number of other cAMP-independent effects that may contribute to its inotropic action.

MODELING THE CARDIAC GLYCOSIDE RECEPTOR

The ultimate "structure–activity" questions to answer with any class of drugs are: (1) "How does the drug fit into the receptor?" (2) "What structures and conformations permit the best fit?" and (3) "Which parts of the drug molecule are responsible for binding (affinity) and for the pharmacologic response (intrinsic activity)?"

Unfortunately, as with most membrane-bound receptors, Na^+,K^+–ATPase has never been crystallized; nor has a cardiac steroid–Na^+,K^+–ATPase complex been cocrystallized. As a result, although the amino acid sequence of the subunits of Na^+,K^+–ATPase have been determined,[46,260,261] no one knows what Na^+,K^+–ATPase looks like three dimensionally. Although the receptor site has been identified by covalently linking a cardiac glycoside to the receptor,[47,262] no one knows exactly how the glycoside fits into the Na^+,K^+–ATPase. The review by Thomas et al.[6] summarizes progress in the investigations.

A combination of drug design, synthesis, conformational energy studies, and computer graphics have been used to indirectly learn about the structural requirements of the receptor site. This process is called *modeling the receptor* and is used with many other drugs besides the cardiac steroids. Recent reviews by Thomas[6] and by Fullerton and coworkers,[37,39,44] summarize progress made in modeling the Na^+,K^+–ATPase–cardiac glycoside receptor site.

Topology of Na^+,K^+–ATPase

Thomas et al.,[6] Ovchinnikov,[263,264] and Schwartz and Adams[265] have proposed model drawings of (1) Na^+,K^+–ATPase in the cell membrane and (2) the topology of Na^+,K^+–ATPase—how the subunits of the enzyme loop back and forth through the cell membrane. Na^+,K^+–ATPase is known to span the cell membrane, with the cardiac steroid-binding site on the extracellular surface and the ATP-binding site on the inside.[43,265]

Conformational Flexibility of Cardiac Glycosides

As shown with digoxin in Figure 18-29, cardiac glycosides are not rigid molecules. With these drugs, therefore, the concept of structure–activity study must include a consideration of *conformation*. If a newly synthesized analogue is active (or inactive), is the cause of its (in)activity its new structure, or conformations (im)possible at the receptor site?

Two regions of conformational flexibility are the bond connecting the C-17 side group to the steroid ring D, and the two bonds connecting C-3 to the sugar (see Fig. 18-29). The C-17 side group can rotate, but as shown in Figure 18-30, some conformations are very high in energy. Because cardiac glycosides are uncharged, and the forces that hold them to the receptor are therefore relatively weak, high-energy conformations are very unlikely at the receptor site. In short, low-energy conformations in Figure 18-30 are the most likely. A similar situation has been found for sugar conformations.[37,39,266,267] Very few combinations of the two bonds that connect the steroid to the glycoside sugar result in conformations that are likely to exist in appreciable quantity at the receptor site.

Roles of the Genin and C-17 Side Group

Many often-conflicting models have been proposed over the last two decades to describe structural and geometric features that govern the ability of a particular digitalis genin (cardenolide) or bufalin genin

FIG. 18-29. Digoxin: Enlargements show the regions of particular interest in modeling the cardiac glycoside receptor site. The two primary regions of flexibility important to biologic activity are rotation about the two glycosidic bonds (C1′—O3 and O3—C3) and rotation of the C-17 side group. (From Fullerton and co-workers.[37] Used with permission.)

(bufadienolide) to inhibit Na^+, K^+–ATPase (or cause an inotropic response). (The reviews of Thomas,[6] Dittrich et al.,[45] Schonfeld,[268, 269] Guntert and Linde,[270] Erdman,[271] and Flasch and Heinz[272] provide a variety of viewpoints.) Since hydrogenation of the side group carbon–carbon double bond was known to decrease activity significantly, an active role in Na^+, K^+–ATPase inhibition was generally envisioned for this double bond. (C-20 also changes from sp^2 to sp^3, greatly moving the side group carbonyl.)

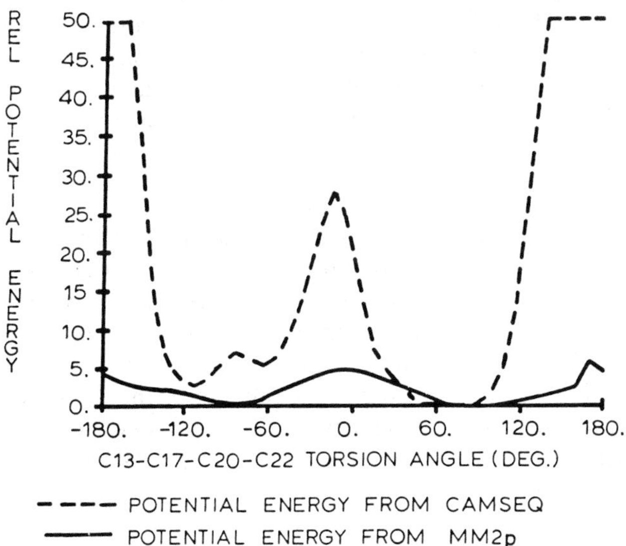

FIG. 18-30. A comparison of the energy calculated at 10° intervals for rotation about the digitoxigenin C17-C20 bond using two different computer programs: a rigid hard sphere model (CAMSEQ; dashed line) and a model in which bond distances and bond angles are allowed to relax at each interval (MM2p; solid line). (From Fullerton and co-workers.[37] Used with permission.)

Fullerton, Rohrer, Ahmed, From, and co-workers found several genins that they synthesized and tested did not fit these models.[37, 39, 273] This led them to a detailed study of cardenolide and bufadienolide genins with the use of the National Institutes of Health PROPHET computer system.

With PROPHET, the most stable conformations (from potential energy diagrams like those in Fig. 18-30) of each of a variety of genins were graphically superimposed with the cardenolide prototype, digitoxigenin. An example is shown in Figure 18-31. PROPHET was then used to measure distances between each superimposed analogue and the corresponding atoms in digitoxigenin. They found that the position of a particular genin's carbonyl oxygen (or nitrile nitrogen) relative to digitoxigenin's was a nearly perfect predictor of its activity (Fig. 18-32).

$$\log I_{50} = 0.44D - 6.36 \qquad r^2 = 0.98$$

where I_{50} = amount of genin required to inhibit 50% of a standard guinea pig brain Na^+, K^+–ATPase preparation in vitro. (A cardiac steroid with an I_{50} of 1×10^{-7} M is ten times more active than one with an I_{50} of 1×10^{-6} M.)

D equals distance between carbonyl oxygens of the analogue and digitoxigenin in angstroms. This

Digitoxigenin

Top View

4.08 Å

Analogue

A

Side View

B

FIG. 18-31. *(A) Digitoxigenin and its 14-ene, 22-methylene analogue. (B) PROPHET superposition of digitoxigenin and its analogue, both in their most stable C-17 side group conformations, showing distance between carbonyl oxygens; as shown in Fig. 18-33, as the distance between carbonyl oxygens increases, activity (NA$^+$, K$^+$–ATPase inhibition) decreases. With the superimposed pair above, it is 4.08 Å. (For additional drawings and discussion, see Fullerton, Rohrer, et al.*[27, 28a, 255]*)*

relationship means that for every 2.2 Å the carbonyl oxygen (or nitrile *N*, for one compound) moves, activity changes by one order of magnitude (10 or 0.1).

The r^2 is the statistical correlation. (Perfect would be 1.00.) As shown in Figure 18-32, a corresponding relationship was found for genins tested in a cat heart Na$^+$, K$^+$–ATPase system.

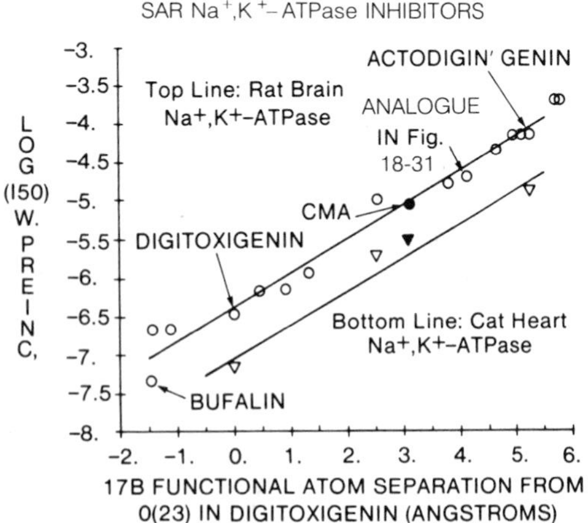

FIG. 18-32. *Relationship of cardiac–steroid–genin–carbonyl–oxygen position to inhibition of Na$^+$, K$^+$–ATPase. Circles: Genins studied with rat brain Na$^+$, K$^+$–ATPase. Triangles: Genins studied with cat heart Na$^+$, K$^+$–ATPase. Triangles directly under the circles represent the same genin. CMA, chlormadinone acetate. (Data from Fullerton, Ahmed, and co-workers.*[245, 256, 257]*)*

The genins graphed in Figure 18-32 included bufalin, cardenolide analogues, with and without a 14β-OH, with acyclic side groups such as esters and a nitrile, and also included the progesterone derivative chlormadinone acetate (CMA). From these studies the following conclusions may be made about digitalis genin–structure–activity relationships.

A-B-C-D Ring System

Both cardenolide and bufadienolide or pregnene ring systems appear to be able to fit the "digitalis" (ouabain) binding site on Na$^+$, K$^+$–ATPase. The activity-determining requirement appears to be primarily that the side group carbonyl oxygen (or nitrile *N*) be in the right spot relative to the A-B-C-D rings for maximal activity. It is not yet known what structural feature on Na$^+$, K$^+$–ATPase causes this requirement.

14β-Hydroxy

The 14β-OH is not necessary for activity. Some compounds that did not have a 14β-OH have been more active than otherwise identical genins that did. It appears that earlier studies with 14-dehydro analogues may have overlooked the stereochemical change in C-14 changing from sp^3 (with a 14β-OH) to sp^2 (with a double bond).

16β-Hydroxy

Gitaloxigenin (gitoxigenin 16β-formate) and its tridigitoxoside **gitaloxin** are naturally occurring cardenolide 16β-formates found in *D. purpurea*.[275] Gitaloxin has even been called the "forgotten" cardiac glycoside of *Digitalis* that may be responsible for most of the plant extract's therapeutic activity.[276] Fullerton, Griffin, and co-workers[277,278] synthesized 17 gitoxigenin 16β-formates, acetates, and methoxycarbonates to learn more about the role of the 16β-OH and 16β-OCHO (formate). A 16β-formate increased the activity 30 times, and a 16β-acetate 9 to 12 times. The increased activity was not the result of modified carbonyl oxygen position, which suggests the possibility of a separate binding site for C-16 esters.

Lactone Ring

The lactone ring is not necessary, as first shown by Thomas and co-workers[6] (see, for example, Fig. 18-33).

Side Group Carbon–Carbon Double Bond

It appears that this double bond may just be keeping the side group's carbonyl oxygen in the right place for most efficient Na^+, K^+–ATPase inhibition.

"Lead Structure"

Schonfeld and associates[268,269] and Dittrich and co-workers[45] have intensively studied binding energies for cardiac steroids at the Na^+, K^+–ATPase receptor site. They also have studied a variety of other compounds such as cassaine (see Fig. 18-33) that, to varying degrees, appear to fit in the same location on the Na^+, K^+–ATPase. From these studies, these investigators conclude that 5β,14β-androstane-3β,14β-diol, the steroid nucleus of the naturally occurring cardenolides and bufadienolides, is the minimum structure needed to inhibit Na^+, K^+–ATPase and cause a measurable response.

Roles of the Sugar

The roles of cardiac sugar structure have been intensively studied,[6,37,44,266,277,279–283] including the importance of sugar stereochemistry and conformation.[37,266,267,279,284,285] Yoda and Yoda were the first to show that there is a separate site for sugar binding on Na^+, K^+–ATPase.[282,283]

From these many studies, a model of Na^+, K^+–ATPase sugar site-binding requirements has begun to emerge.

1. *Sugar site binding is not necessary for good activity. Cardiac genins are usually very active.* The I_{50} for digitoxigenin, for example, is about 1.2×10^{-7} *M*—a very potent drug indeed. Adding a single β-D-digitoxose increases activity about 18 times ($I_{50} = 6.6 \times 10^{-9}$ *M*). Digitoxin, with three β-D-digitoxose sugars, is about 10 to 12 times more active than digitoxigenin. However, sugars certainly have a significant impact on the pharmacokinetic profiles of cardiac glycosides, affecting both their distribution and speed of elimination (Table 18-18).
2. *Stereochemistry of sugar OH groups is important.* The stereochemistry of the sugar 4'-OH is especially critical. As shown in Figure 18-34a, the β-D-glucoside of digitoxigenin is over five times more active than the β-D-galactoside.[286]
3. *Changing sugar stereochemistry changes activity.* For example, the four possible glycosides of glucose (see Fig. 18-34b) all have different activities. A similar dependence on stereochemistry has been seen with mannosides and other glycosides.[279]
4. *With β-L-sugars, an equatorial 4'-OH group is the only sugar OH needed for good activity.* As shown in Figure 18-34c, the α-L-dideoxyrhamnoside of digitoxigenin is as active as the α-L-rhamnoside (with two more sugars), and much more active than the α-L-glucoside and α-L-mannoside (with three more sugars).[284,285]
5. *With α-L- and β-D-sugars, an equatorial 4'-OH group is not enough; other OH groups help in sugar site binding.*[284,285] For example, the β-L-dideoxyrhamnoside of digitoxigenin is only three times more active than digitoxigenin; the β-L-rhamnoside is 25 times more active.[284,285]
6. *Binding is enhanced by 5'-CH_3 and reduced by 5'-CH_2OH.* As shown in Fig. 18-34c, adding an OH group to the 5'-CH_3 of rhamnose (thereby forming mannose) reduces activity significantly. It has been suggested that there is a hydrophobic-binding location for the CH_3 or, alternatively, that the additional OH introduces steric hindrance that diminishes binding.

STRUCTURE AND PARTITION COEFFICIENT

As shown in Table 18-18, commercially available cardiac steroids differ markedly in their degree of absorption, half-life, and time to maximal effect. Usually, this is due to polarity differences caused by

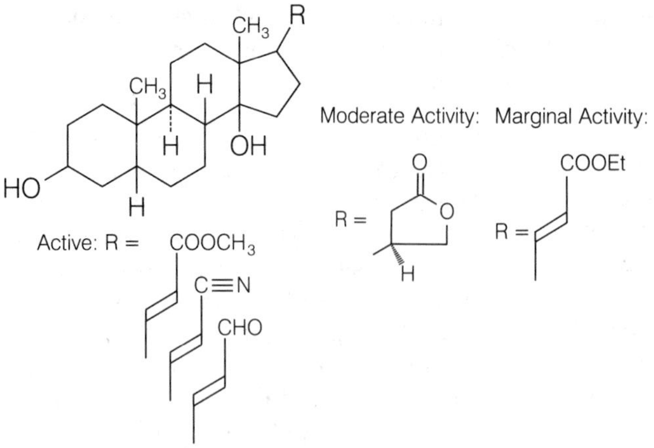

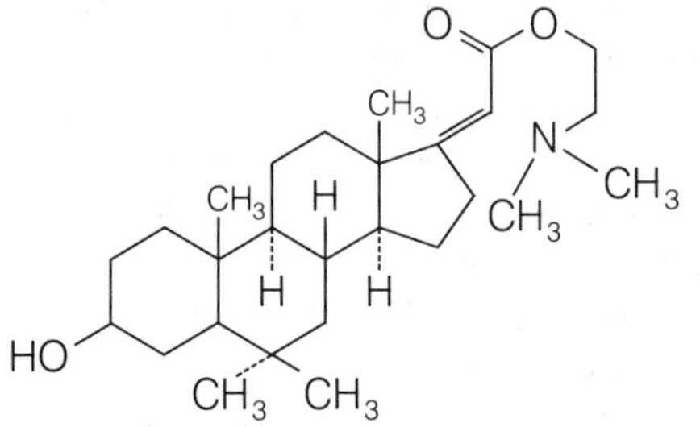
Cassaine

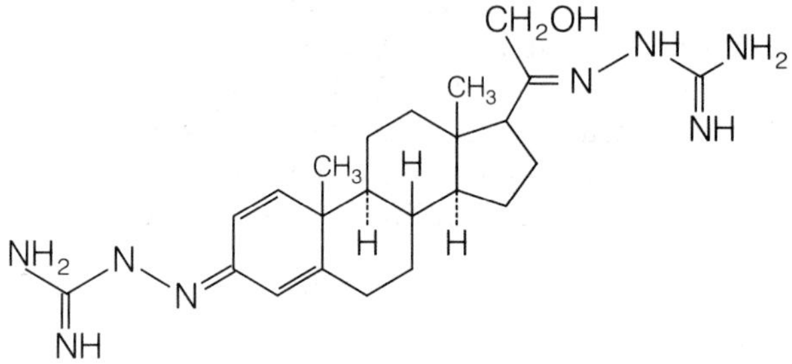
And Other Guanylhydrazones

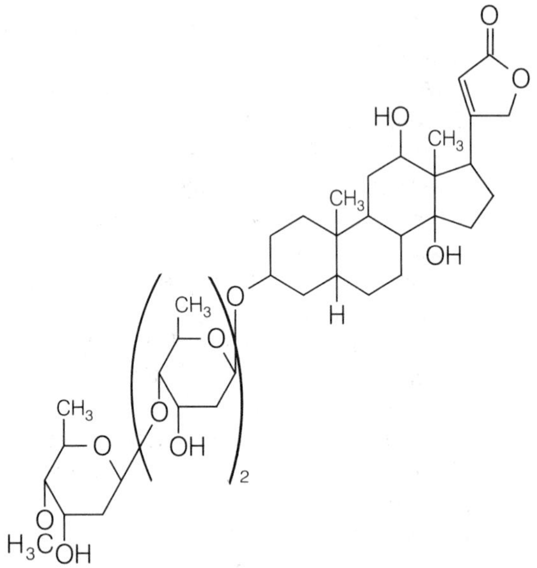
β-Methyldigoxin

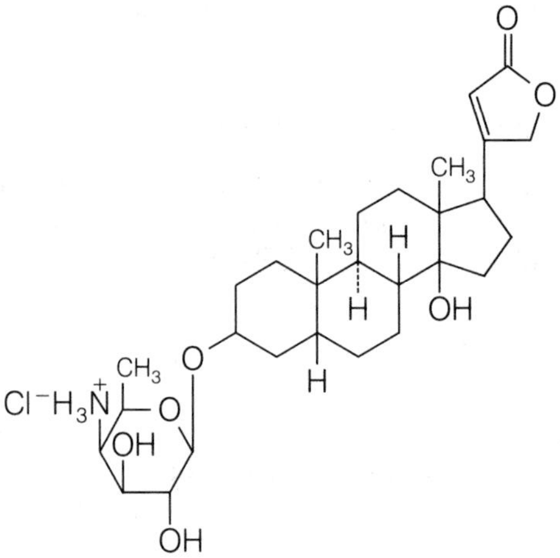
ASI-222

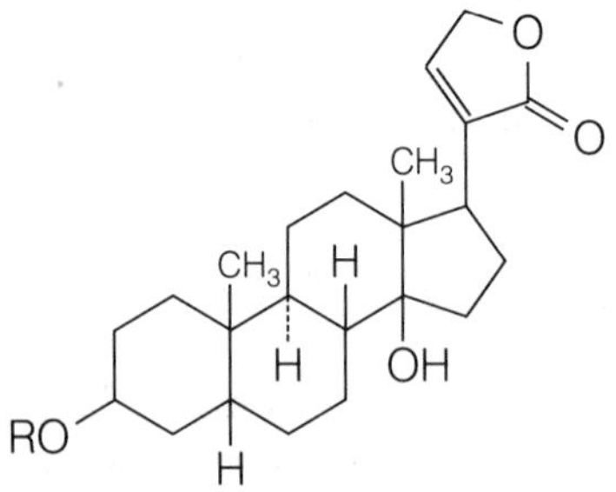
Actodigin (AY22,241)
R = Glucose

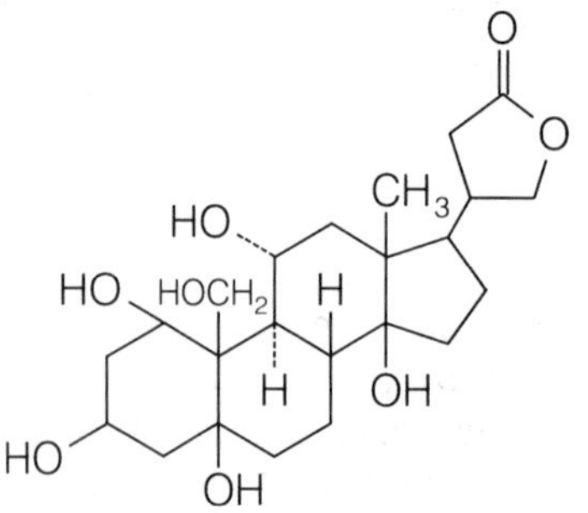
20,22-Dihydro-Ouabain
R = Rhamnose

FIG. 18-33. *Cardenolides and other steroids with modified structures that inhibit Na^+, K^+–ATPase.*

TABLE 18-18

CARDIAC GLYCOSIDE PREPARATIONS

Agent	Gastrointestinal Absorption	Onset of Action* (min)	Peak Effect (hr)	Average Half-Life†	Principal Metabolic Route (Excretory) Pathway	Average Digitalizing Dose		Usual Daily Oral Maintenance Dose‖
						Oral‡	IV§	
Ouabain	Unreliable	5–10	0.5–2	21 hr	Renal; some GI excretion		0.3–0.5 mg	
Deslanoside	Unreliable	10–30	1–2	33 hr	Renal		0.8 mg	
Digoxin	55–75%	15–30	1.5–5	36 hr	Renal; some GI excretion	1.25–1.5 mg	0.75–1.0 mg	0.25–0.5 mg
Digitoxin	90–100%	25–120	4–12	4–6 days	Hepatic; renal excretion of metabolites	0.7–1.2 mg	1.0 mg	0.1 mg
Digitalis leaf	About 40%			4–6 days	Similar to digitoxin	0.8–1.2 g		0.1 g

*For intravenous dose.

†For normal subjects (prolonged by renal impairment with digoxin, ouabain, and deslanoside and probably by severe hepatic disease with digitoxin and digitalis leaf).

‡Divided doses over 12–24 hr at intervals of 6–8 hr.

§Given in increments for initial subcomplete digitalization; supplement with additional increments p.r.n.

‖Average for adult patients without renal or hepatic involvement; varies widely among patients and requires close medical supervision.

(Table from Smith, T. W., and Haber, E. New Engl. J. Med. 289:1063, 1973. Used by permission.)

the number of sugars at C-3 and the presence of additional hydroxyls on the cardenolide.

Nevertheless, it has always been difficult to visualize how apparently minor structural variations can cause major differences in partition coefficient and absorption. For example, lanatoside C and digoxin differ only by the presence of an additional sugar in lanatoside C (see Table 18-17). One might expect that "one more sugar should not make much difference." However, the $CHCl_3$/16% aqueous MeOH partition coefficients for the compounds are very different indeed: 16.2 for lanatoside C, 81.5 for digoxin, 96.5 for digitoxin, and 10 for gitoxin.

The compounds with increased lipid solubility also are the slowest to be excreted. Additional hydroxyl groups in the more polar compounds provide additional sites for conjugate formation and other metabolic processes. In addition, it is commonly known that more lipid-soluble drugs tend to be excreted more slowly because of increased accumulation in lipid tissues.

METABOLISM

The cardiac glycosides are metabolized to a variety of products. However, with digoxin, a relatively small amount is actually metabolized. Reduction of the C-20(22) double bond results in 20,22-dihydro derivatives with about 1/100 the activity of the original glycosides (because of C-20 changing from sp^2 to sp^3 and thereby moving the carbonyl oxygen). Hydrolysis of digitoxose sugars also occurs, to result in a mixture of bisdigitoxosides and digitoxosides. These reactions can also be catalyzed by bacteria in the gastrointestinal tract, one of the primary reasons why absorption can be quite variable among patients. Whereas 60% to 85% of oral tablets are absorbed, liquid-filled gelatin capsules can be 90% or more absorbed.

CLINICAL USE

The cardiac glycosides are used to treat congestive heart failure, atrial fibrillation, and atrial flutter. (Their use with heart arrhythmias has been discussed in Chapter 14.) Not all patients with congestive heart failure, however, are appropriate candidates for cardiac glycoside ("digitalis") therapy. In his 1988 review of clinical uses of these drugs, Smith[40] notes that "the efficacy of digitalis in the management of heart failure in the presence of normal sinus rhythm has been called into question. . . . A consensus now exists that patients with dilated, failing hearts and impaired systolic function, often manifesting an S3 gallop, have subjective and objective improvement after receiving digitalis. . . ."

Drug Information concludes that "Cardiac glycosides are generally most effective in the management of low-output failure secondary to hypertension, coronary artery or atherosclerotic heart

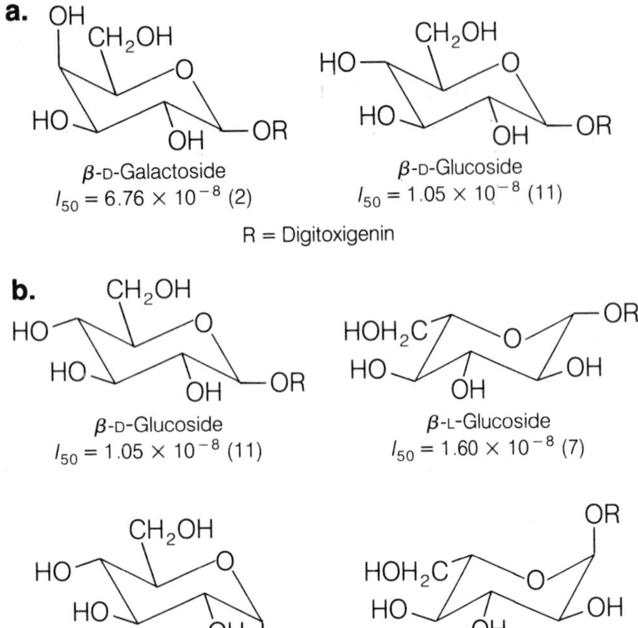

α-D-Glucoside
$I_{50} = 9.33 \times 10^{-8}$ (1.1)

α-L-Glucoside
$I_{50} = 6.03 \times 10^{-8}$ (2)

R = Digitoxigenin

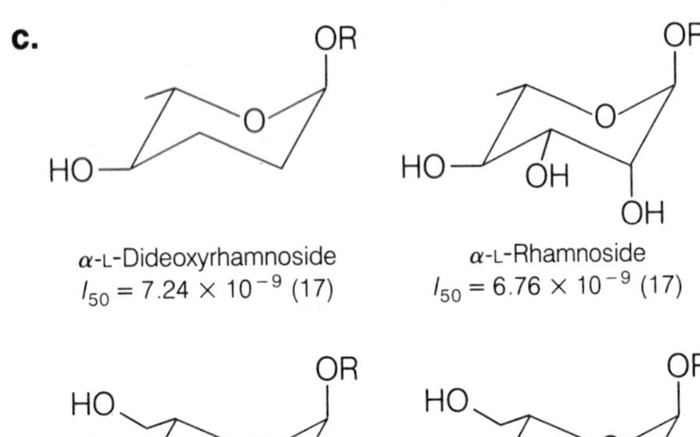

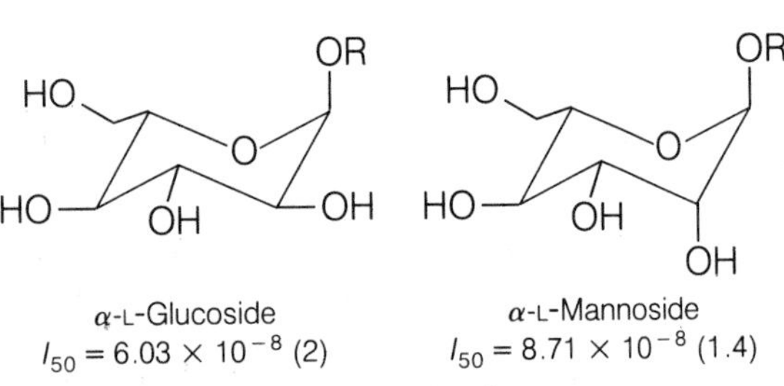

R = Digitoxigenin

FIG. 18-34. *Na^+,K^+-ATPase inhibitory activities of digitoxigenin glycosides with varying sugars, where R = digitoxigenin. Data in parentheses are activities relative to digitoxigenin. All data from Fullerton, Ahmed, and co-workers[284,285] using hog kidney Na^+,K^+-ATPase. Three sets of glycosides* (a – c) *illustrate structural relationships discussed in this section.*

disease, primary myocardial disease, nonobstructive cardiomyopathies, and valvular heart disease."[15] Their use in management of atrial fibrillation or flutter also depends on the particular clinical profile of the patient.

The safe and appropriate clinical use of cardiac glycosides requires a thorough knowledge of their clinical, pharmacologic, and pharmacokinetic profiles. Comprehensive discussions of Smith,[40] *Drug Information*,[15] and other leading clinical references must be thoroughly understood before recommending dosages or use.

TABLE 18-19

NONCARDIAC SYMPTOMS OF DIGITALIS TOXICITY

Symptoms	*Frequency*	*Manifestations*
Gastrointestinal	Most common	Anorexia, nausea, vomiting, diarrhea, abdominal pain, constipation
Neurologic	Common	Headache, fatigue, insomnia, confusion, vertigo
	Uncommon	Neuralgias (especially trigeminal), convulsions, paresthesias, delirium, psychosis
Visual	Common	Color vision, usually green or yellow; colored halos around objects
	Uncommon	Blurring, shimmering vision
	Rare	Scotomata, micropsia, macropsia, amblyopias (temporary or permanent)
Miscellaneous	Rare	Allergic (urticaria, eosinophilia), idiosyncrasy, thrombocytopenia, gastrointestinal hemorrhage, and necrosis

(From Gerbino,[287] Copyright 1973, American Society of Hospital Pharmacists, Inc. All rights reserved. Used with permission.)

It is important that the pharmacist recognize common symptoms of cardiac glycoside toxicity (Table 18-19) and be familiar with basic principles of cardiac glycoside dosing.

1. *A "loading" or "digitalizing" dose.* The potent activity of the digitalis steroids combined with their potential for toxicity makes selection of individual doses more complicated than for almost any other drug. Most importantly, one must carefully consider the renal function of the patient, because much of the dose is excreted in the urine. In patients with normal renal function, the average half-life of digoxin is much shorter than digitoxin (see Table 18-18).
2. *A maintenance dose.* As with the loading dose (if used at all), the maintenance dose must be carefully tailored to each individual patient. Average doses must not be used without accounting for individual patient variables—kidney function, age, potential drug interactions, presence of heart, thyroid, or hepatic disease.
3. *Avoiding or controlling drug interactions.* Hypokalemia, such as that brought about by diuretics, or hyperkalemia can cause cardiac arrhythmias—arrhythmias that can also be caused by digitalis. Calcium and digitalis glycosides are synergistic in their actions on the heart.

Many drugs can affect absorption of digitalis (e.g., cathartics and neomycin). Protein binding can be disturbed by coumarin anticoagulants, phenylbutazone, and some sulfonamides. However, these drug interactions are much less common than those involving potassium or calcium.

4. *Prompt treatment of digitalis toxicity*, if it occurs, by the physician. Digitalis therapy must be stopped until symptoms are under control.

Most important, **Digoxin Immune Fab** (Digibind, digoxin-specific antigen binding fragments) can rapidly reverse life-threatening digitalis glycoside intoxication. This antidote to digoxin toxicity is made from sheep that have been injected with digoxin and a carrier protein.

The antibody fragments' affinity for digoxin is higher than that of Na^+, K^+–ATPase. Since binding to Na^+, K^+–ATPase is reversible, the cardiac glycoside is removed from its binding site on the Na^+, K^+–ATPase and bound to the antibody. Digoxin immune Fab also has high affinity for digitoxin.

Digoxin immune Fab is given by intravenous infusion or by rapid intravenous injection for patients exhibiting severe digitalis-induced arrhythmias or hyperkalemia. The rapid withdrawal of digitalis from Na^+, K^+–ATPase, however, can result in low cardiac output, including congestive heart failure. Although allergic responses have generally not occurred, caution in using any ovine-based product with allergic patients is recommended. Skin testing for allergy to digoxin immune Fab should be considered with high-risk patients.

MODELING THE PHOSPHODIESTERASE INHIBITOR RECEPTOR

Moos and co-workers[287a] have synthesized and studied a variety of cAMP phosphodiesterase III inhibitors using x-ray crystallography, biologic studies, spectroscopy, molecular modeling, and electrostatic potential calculations. They have proposed that these inhibitors mimic the structural and electronic features of cAMP at the active site on phosphodiesterase III (PDE III). It appears that the pyridazone amide of the inhibitors mimics the 5′-phosphate of cAMP:

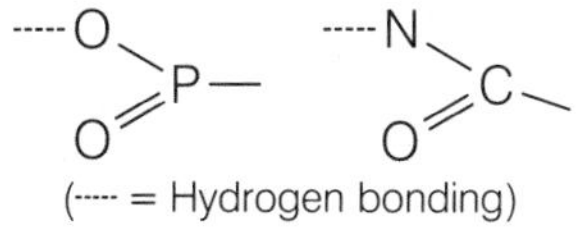

(----- = Hydrogen bonding)

and that, if the inhibitor molecule is long enough, one of its heterocyclic rings can overlap adenine's $6\text{-}NH_2$ (Fig. 18-35). Selective inhibition of PDE III depends upon the inhibitor having a relatively flat topography.

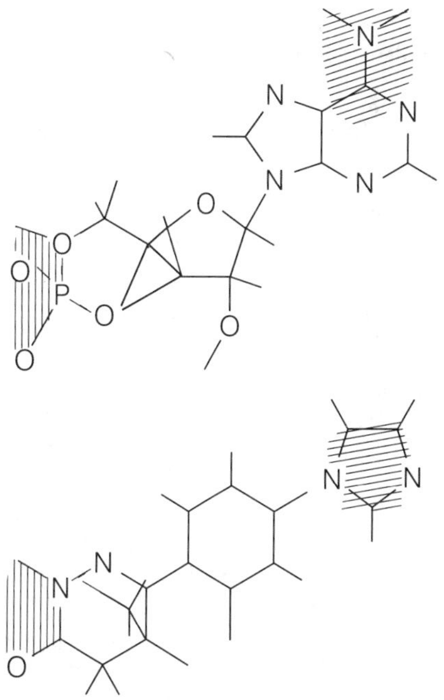

FIG. 18-35. Pharmacophore model of Moos and co-workers[287a] for active conformation of cAMP (top). CI-930 (bottom), as an example, is illustrated in its proposed matching conformation. Shading on left shows the mimicking of the pyridazone amide of all the inhibitors for the 5′-phosphate of cAMP. For extended inhibitors, such as imazodan, electron delocalized rings or atoms occupy the corresponding part of the adenine ring (shading on right).

PRODUCTS

Digitalis products are listed in Table 18-20, and the most important information on the digitalis steroids appears in Table 18-18. Structural differences are shown in Figure 18-25 and Table 18-17. Any additional pertinent information will be given in the following. The previous sections on digitalis toxicity and therapy should be carefully studied as well.

Powdered Digitalis, USP, is the dried, powdered leaf of *Digitalis purpurea*. When digitalis is prescribed, powdered digitalis is to be dispensed. One hundred milligrams is equivalent to one USP Digitalis Unit, used as a relative measure of activity in pigeon assays. Powdered digitalis contains digitoxin, gitoxin, and gitalin, of which digitoxin is usually in highest concentration. Because of the sizable presence of digitoxin, powdered digitalis has a slow onset of action and long half-life (see Table 18-18). The long half-life makes toxic symptoms more difficult to treat than with cardiotonic steroids with shorter half-lives.

Digoxin, USP, because of its moderately fast onset of action and relatively short half-life (see Table 18-18), has become the most frequently prescribed digitalis steroid. It is a *D. lanata* glycoside of digoxigenin (see Fig. 18-25), 3,12,14-trihydroxy-

TABLE 18-20

DIGITALIS PRODUCTS*

Name Proprietary Name	*Preparations*
Digitalis USP (*Note:* When digitalis is prescribed, powdered digitalis USP is to be dispensed)	The dried leaf of *Digitalis purpurea* — not used therapeutically until powered
Powdered digitalis USP *Digitora, Digifortis, Pil-Digis*	Digitalis capsules USP Digitalis tablets USP Digitalis tincture
Digitalis purpurea glycosides *Digiglusin, Gitaligin*	Tablets Injection (Each tablet or 1 ml of injection is equivalent to 1 USP unit, i.e., 100 mg of digitalis.)
Digoxin USP *Lanoxin*	Digoxin elixir USP Digoxin injection USP Digoxin tablets USP
Digitoxin USP *Crystodigin, Purodigin*	Digitoxin injection USP Digitoxin tablets USP
Acetyldigitoxin USP	Acetyldigitoxin tablets USP
Ouabain USP	Ouabain injection USP
Lanatoside C *Cedilanid*	Tablets
Deslanoside USP *Cedilanid-D*	Deslanoside injection USP

*For doses, see Table 18-18.

card-20(22)-enolide. Digoxin was first isolated by Smith,[288] in 1930. It may be given orally, intravenously, or intramuscularly (into deep muscle, followed by firm massage).

Digitoxin and digoxin are the most frequently prescribed digitalis steroids. Digoxin is more rapidly excreted and, therefore, also more rapidly accumulates in the presence of impaired renal function. Clinical changes from changing the maintenance dose are quickly observed. Digitoxin is excreted more slowly and accumulates more slowly. Because it is more slowly excreted, its kinetics are less affected by renal function. It has better absorption (bioavailability) than digoxin and, therefore, probably is more reproducible. The rapid excretion of digoxin is certainly useful when toxicity develops. However, if renal function should be impaired during long-term maintenance therapy, the risk of serious toxicity appears to be considerably greater for the patient receiving digoxin rather than digitoxin. Thus, Jelliffe suggests that digitoxin would be preferred with patients with potentially variable renal function.

Digitoxin, USP, is obtained from *D. purpurea* and *D. lanata*, as well as several other species of *Digitalis*. It was obtained in crystalline form in 1869 by Nativelle.[289] It is a glycoside of digitoxigenin (see Fig. 18-20), 3,14-dihydroxycard-20(22)-enolide. The properties of digitoxin have been compared with digoxin.

Acetyldigitoxin, USP, is obtained from the enzymatic hydrolysis of lanatoside A (see Table 18-17).

Ouabain, USP, also called G-strophanthin, is a glycoside obtained from the seeds of *Strophanthus gratus* or the wood of *Acokanthera schimperi*. It is too poorly and unreliably absorbed to be used orally, but its extremely fast onset of action (see Table 18-18) makes it useful for rapid digitalization in emergencies (e.g., nodal tachycardia, atrial flutter, or acute congestive heart failure). Its synonym G-strophanthin makes it easily confused with strophanthin (or K-strophanthin), a glycoside obtained from *S. kombe*. The aglycone of ouabain is ouabagenin, whereas the aglycone of strophanthin is strophanthidin.

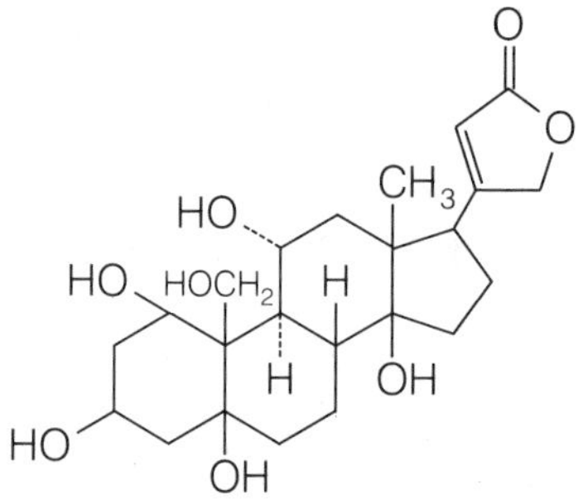

Ouabagenin

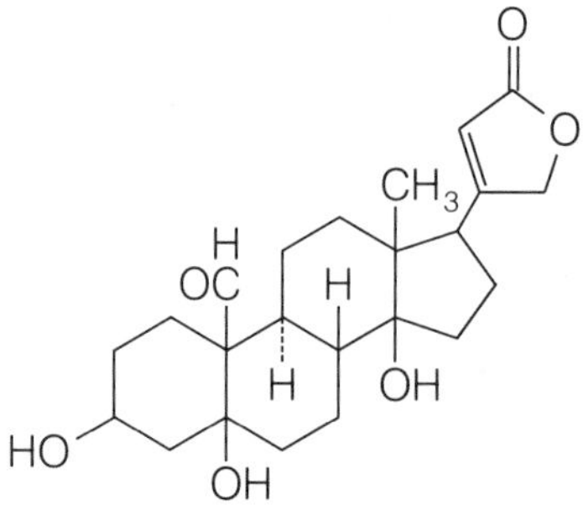

Strophanthidin

Lanatoside C is a digoxigenin glycoside obtained from the leaves of *D. lanata*. It is poorly and irregularly absorbed from the gastrointestinal tract and has a variable metabolic half-life.

Deslanoside, USP, is a digoxigenin glycoside obtained from lanatoside C by alkaline deacetylation (see Table 18-17). It is used only for rapid digitalization in emergency situations and may be given intravenously or intramuscularly.

STEROIDS WITH OTHER ACTIVITIES

As shown in Figure 18-1, there are a number of important steroids that do not fall into the previous classifications. (Vitamin D precursors are discussed in Chapter 22.) Because these compounds have di-

TABLE 18-21

STEROIDS WITH OTHER ACTIVITIES

Name Proprietary Name	*Preparations*	*Category*	*Usual Adult Dose**	*Usual Dose Range**	*Usual Pediatric Dose**
Cholesterol USP		Pharmaceutic aid (emulsifying agent)			
Spironolactone USP *Aldactone*	Spironolactone tablets USP	Diuretic	25 mg bid to qid	50–400 mg/day	20–60 mg/m^2 of body surface tid
Ox bile extract	Ox bile extract tablets	Digestant or choleretic	300 mg with water tid		
Dehydrocholic acid USP	Dehydrocholic acid tablets USP	Choleretic	500 mg tid	250–750 mg	
Fusidic acid *Fucidin*	Tablets, solution for infusion	Antibiotic (gram-positive only)	500 mg tid		20–40 mg/kg of body weight daily
Lanolin USP (Mixture of steroids, other fats and oils)		Water-in-oil emulsion ointment base			
Anhydrous lanolin USP (Mixture of steroids, other fats and oils)		Absorbent ointment base			

*See USP DI for complete dosage information.

verse activities and uses, they will be presented individually in the monographs that follow. Products are listed in Table 18-21.

The reader is also reminded of 19-[^{131}I]iodocholesterol compounds discussed in the section Changes to Modify Pharmacokinetic Properties.

PRODUCTS

Cholesterol, USP, is used as an emulsifying agent. Its biosynthesis and structure are shown in Figure 18-5, which also illustrates its essential role as a steroid hormone precursor. Cholesterol is the precursor of virtually all other steroid hormones.

It is important to note that significantly more cholesterol is biosynthesized in the body each day (about 1 to 2 g) than is contained in the usual Western diet (about 300 mg). Cholesterol has been implicated in coronary artery disease, but there is increasing evidence that genetic deficiencies in cholesterol metabolism (see section on LDL Receptors), high stress, low exercise, "junk" foods, and smoking are possibly primary causes of heart disease.

Cholesterol is found in most plants and animals. Brain and spinal cord tissues are rich in cholesterol. Gallstones are almost pure cholesterol. In fact, cholesterol was originally isolated from gallstones, by Paulleitier de Lasalle in about 1770. In 1815, Chevreul[289] showed that cholesterol was unsaponifiable and he called it cholesterin (*chole*, bile; *steros*, solid). In 1859, Berthelot[290] established its alcoholic nature, and since then it has been called cholesterol.

Cholesterol, lanosterol (structure shown in Fig. 18-5), fatty acids and their esters make up **Anhydrous Lanolin, USP** and **Lanolin, USP.** Lanolin (or hydrous wool fat) is the purified, fatlike substance from the wool of sheep, *Ovis aries*, and contains 25% to 30% water. Anhydrous lanolin (or wool fat) contains not more than 0.25% water.

Spironolactone, USP (see Fig. 18-1). 17-Hydroxy-7α-mercapto-3-oxo-17α-pregn-4-ene-3-one-21-carboxylic acid γ-lactone 7-acetate, is an aldosterone antagonist of great medical importance because of its diuretic activity. Spironolactone is discussed in Chapter 13.

Dehydrocholic Acid, USP. 3,7,12-Triketocholanic acid (see Fig. 18-1) is a product obtained by oxidizing bile acids. The bile acids serve as fat emulsifiers during digestion. About 90% of the cholesterol not used for biosynthesis of steroid hormones is degraded to bile acids. All are 5β-steroid-3α-ols, giving rise to the "normal" designation discussed previously in Nomenclature, Stereochemistry, and Numbering. As shown in Figure 18-36, cholesterol has part of its side chain oxidatively removed in the liver, and two or more hydroxyls are added.[290] The resulting bile acids are then converted to their glycine or taurine conjugate salts, which are secreted in the bile. After entering the large intestine, the conjugate salts are converted to cholic acid, deoxycholic acid, and several other bile acids. Many

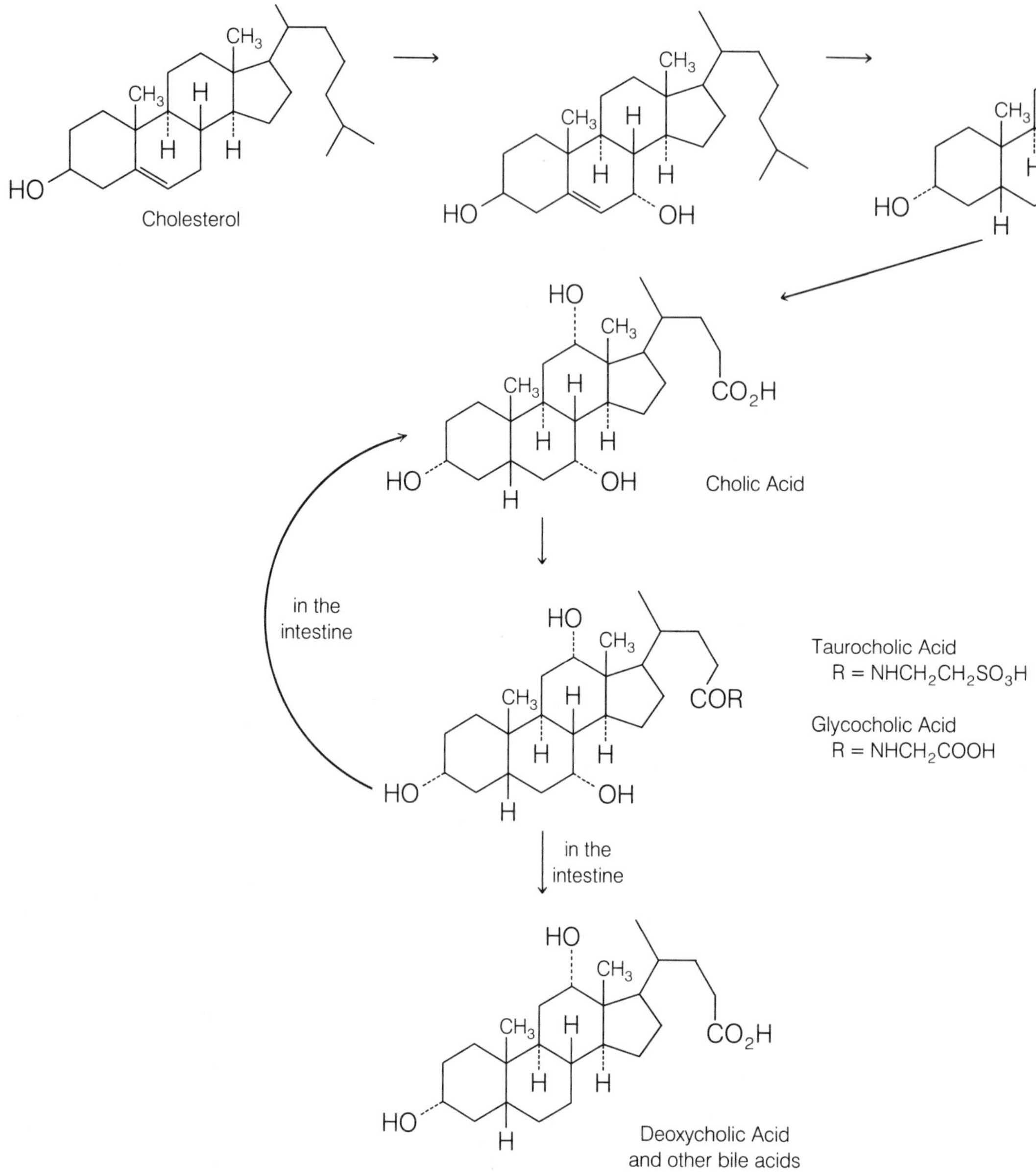

FIG. 18-36. *Metabolism of cholesterol to bile salts.*

of the bile acids are then reabsorbed, with cholic acid having a biologic half-life of about three days.[291]

The bile acids are anionic detergents that emulsify fats, fat-soluble vitamins, and other lipids so that they may be absorbed. Dehydrocholic acid also stimulates the production of bile (choleretic effect). It is used after surgery on the gallbladder or bile duct to promote drainage and for its lipid-solubilizing effects in certain manifestations of cirrhosis or steatorrhea. A related product, **ox bile extract**, contains not less than 45% cholic acid and is used for the same purposes as dehydrocholic acid.

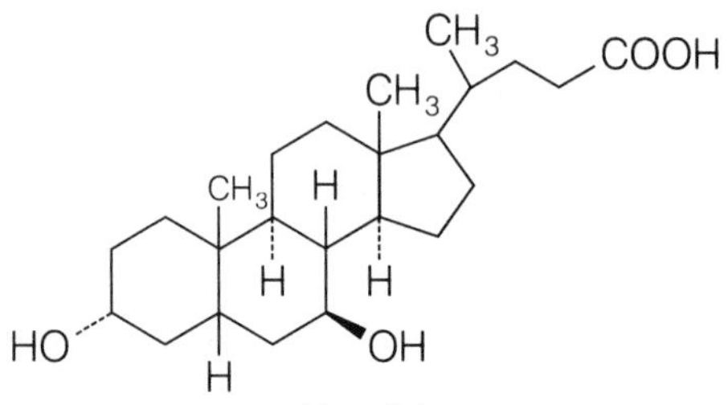

Ursodiol (ursodeoxycholic acid, Actigall), is a naturally occurring bile acid recently marketed to dissolve gallbladder stones.[291a] Ursodiol decreases cholesterol secretion into bile and may decrease absorption of dietary cholesterol. After absorption and subsequent conjugation by the liver, it is secreted into bile and then reabsorbed to thereby concentrate ursodiol in the circulating bile acids. In clinical trials, about 30% of patients have had their cholesterol gallstones completely dissolved. However, long treatment is required, and when therapy is stopped, the gallstones recur in up to 50% of patients.

Fusidic Acid (see Fig. 18-1) and its sodium salt are used in Europe as antibiotics for gram-positive bacterial infections, particularly with patients who are penicillin-sensitive. It acts by inhibition of G-factor during protein biosynthesis. It is also of interest because it appears that it is formed from an intermediate common to the biosynthesis of lanosterol during squalene epoxide cyclization. Structure–activity studies have been reported by Godtfredsen, in 1966, and it was found that just about any minor structural modification of the molecule will result in significantly decreased activity.[292] **Cephalosporin P_1** and **helvolic acid** are steroids with structures very similar to fusidic acid, and both are antibiotics useful in some gram-positive bacterial infections.

COMMERCIAL PRODUCTION OF STEROIDS

HISTORY

This chapter on steroids would not be complete without brief mention of the fascinating history of the steroid industry. In the 1930s, steroid hormones had to be obtained by extraction of cow, pig, and horse ovaries, adrenal glands, and urine. The extraction process was not only inefficient, it was expensive. Progesterone was valued at over 80 dollars per gram. However, by the late 1940s, progesterone was being sold for less than 50 cents a gram and was available in ton quantities. The man who made steroid hormones cheaply and plentifully available is Russell E. Marker, the "founding father" of the modern steroid industry.[1,293]

After leaving graduate school in 1925, Marker worked in a variety of areas in organic chemistry

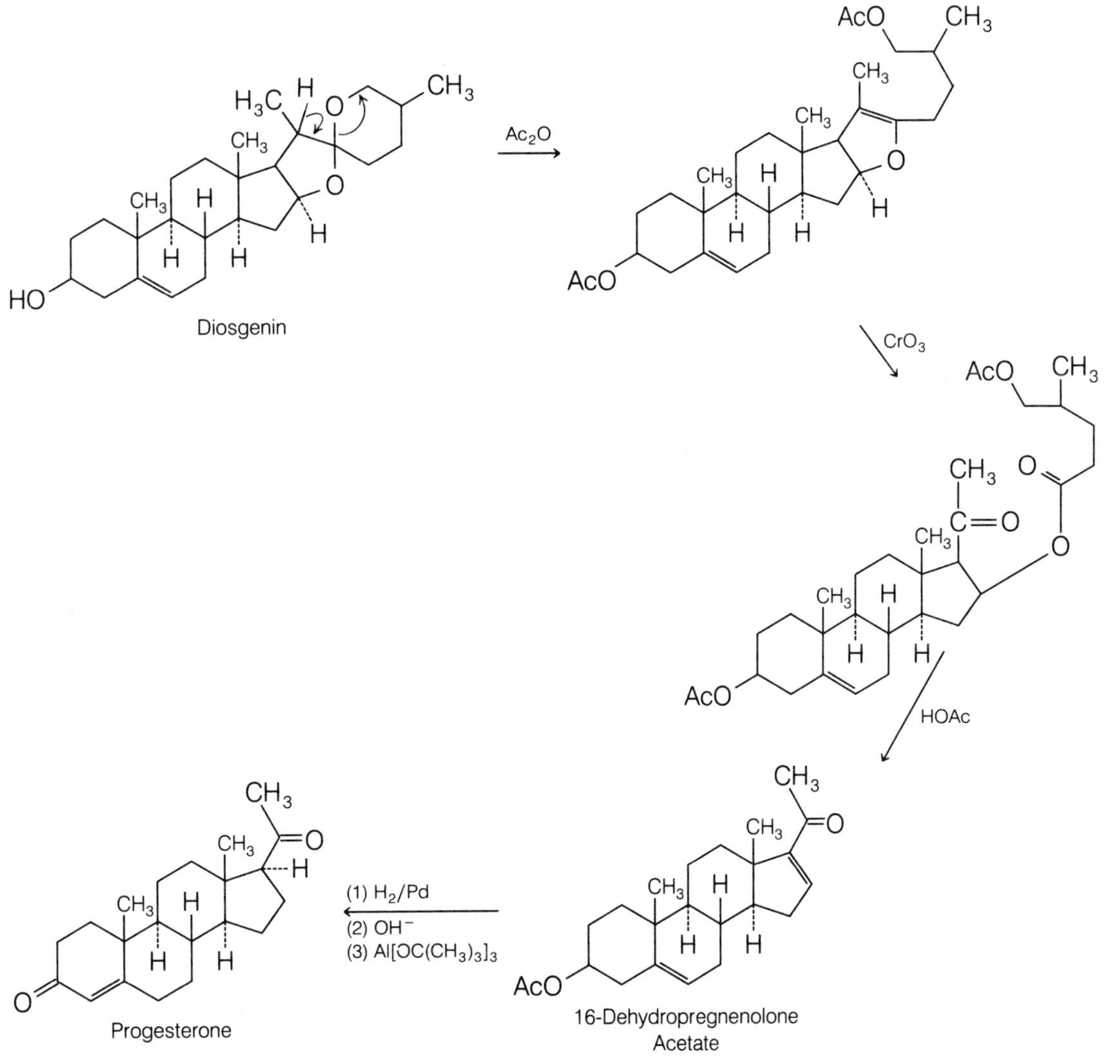

Diosgenin

Ac_2O

CrO_3

HOAc

(1) H_2/Pd
(2) OH^-
(3) $Al[OC(CH_3)_3]_3$

Progesterone

16-Dehydropregnenolone Acetate

FIG. 18-37. The Marker synthesis of progesterone from diosgenin.

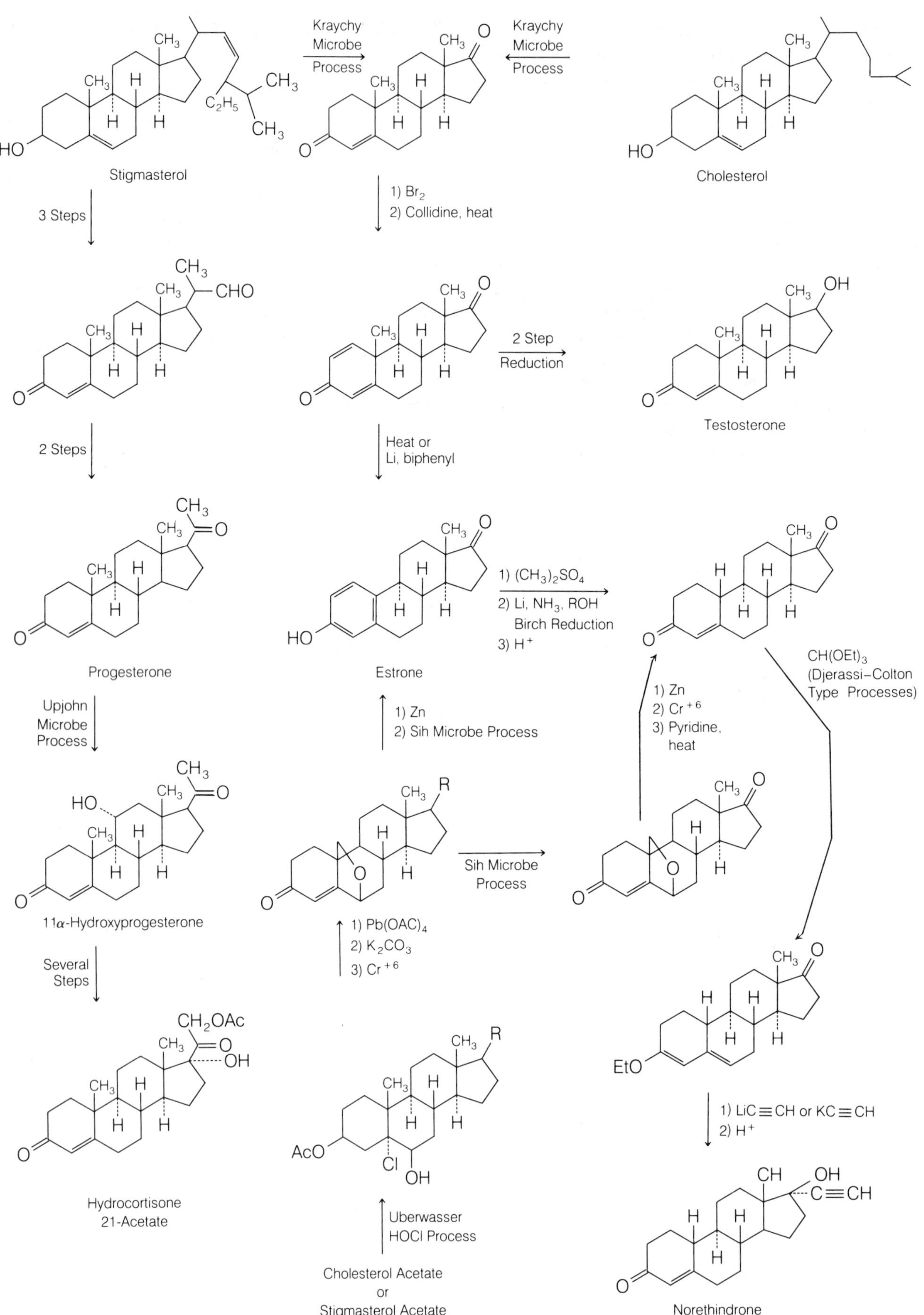

FIG. 18-38. Commercial production of steroid hormones from cholesterol and stigmasterol.

research. In 1935, he went to the Pennsylvania State University to begin studying steroids, turning his full attention to finding inexpensive starting materials for steroid hormone syntheses. In 1939, he correctly determined the structure of sarsasapogenin, a sapogenin (aglycone of a saponin, i.e., a glycoside that *foams* in water) whose structure had been incorrectly published by many other chemists a few years earlier.[294]

Marker quickly developed a procedure (Fig. 18-37) to degrade the side chain of sarsasapogenin to yield a pregnane. Soon thereafter, he degraded diosgenin (Δ^5-sarsasapogenin) to progesterone (see Fig. 18-36) in excellent yield.

The commercial potential of the process was obvious to Marker. He immediately launched a series of plant-collecting expeditions from 1939 to 1942 to find a high-yield source of diosgenin, isolated previously from a *Dioscorea* species in Japan.[295] Over 400 species were collected (over 40,000 kg of plant material) in Mexico and the American Southwest.

Two particularly high-yielding sources of diosgenin were found in Mexico—*Dioscorea composita* ("barbasco") and *D. macrostachya* ("cabeza de negro")—commonly called "the Mexican yams." Although barbasco had five times the diosgenin content of cabeza, it was in generally inaccessible areas, and so Marker concentrated on cabeza. He knew he had a high-yield, low-cost course of progesterone, but was unable to interest several American drug companies. In 1943, he returned to Mexico City and promptly made 3 kg of progesterone (valued at 240,000 dollars) from cabeza. On January 21, 1944, Marker, Lehmann, and Somlo incorporated Syntex Laboratories, and by 1951 Syntex was taking orders for 10-ton quantities of progesterone. The great demand for progesterone was compounded by it serving as a precursor for cortisone and hydrocortisone (Fig. 18-38) by a microbial process that was patented in 1951 by Upjohn.[11 pp. 672–678; 296, 297]

However, in 1945, Marker, Somlo, and Lehmann had a general "falling out," and Marker sold his 40% interest in Syntex to the other two partners. Syntex then brought in Rosenkranz, Djerassi, and other chemists to continue the synthesis of hormones from diosgenin. In 1951 and 1953, Frank Colton of G. D. Searle and Co.[298] and Djerassi and Rosenkranz of Syntex Laboratories[299] synthesized norethynodrel and norethindrone, respectively, thus beginning the era of oral contraceptives which continues to this day.

During the 1950s virtually all the steroid hormones had been made from diosgenin by chemists in North America and Europe. The Mexican yams have been "nationalized" by Mexico, thus blocking export. Attempts to grow high-yield barbasco or cabeza in other countries have been generally unsuccessful. Microorganisms have continued to play many key roles in the inexpensive commercial production of steroid drugs.

CURRENT METHODS

Today nearly all steroid hormones are made from diosgenin, stigmasterol (an inexpensive component of soybean oil), or cholesterol (available in ton quantities from wool fat). Microbiological side group-cleaving processes of Sih[300] and Kraychy[301] are the basis for the routes shown in Figure 18-38. The 19-methyl group is usually removed by the method of Uberwasser[302] (oxidation to form an oxygen bridge at C-6), but Birch reduction with Li/NH_3, an industrially more difficult process, is also used. The Upjohn microbial process[11, 296, 297] for converting progesterone to 11α-hydroxyprogesterone is used to make cortisone and hydrocortisone products. An overview of these processes has been published by Klimstra and Colton.[303] Some total synthetic routes are also used by some companies to make estrone.[304, 305]

ACKNOWLEDGMENT

I am indebted to my colleague Debra Jimmerson for her exceptional help in preparing the manuscript.

REFERENCES

1. Witzmann, R. F.: Steroids—Keys to Life. New York, Van Nostrand Reinhold, 1981.
2. Wolff, M. E. (ed.): Burger's Medicinal Chemistry, 4th ed., Vols. 1–3. New York, John Wiley & Sons, 1979–1980.
3. Counsell, R. W., and Brueggemier, R.: The male sex hormones. In: Wolff, M. E. (ed.). Burger's Medicinal Chemistry, Chap. 28, Vol. 2. New York, John Wiley & Sons, 1979–1980.
4. Deghenghi, R., and Givner, M. L.: The female sex hormones. In: Wolff, M. E. (ed.). Burger's Medicinal Chemistry, Chap. 29, Vol. 2. New York, John Wiley & Sons, 1979–1980.
5. Wolff, M. E.: The anti-inflammatory steroids. In: Wolff, M. E. (ed.). Burger's Medicinal Chemistry, Chap. 63, Vol. 3. New York, John Wiley & Sons, 1979–1980.
6. Thomas, R., Gray, P., and Andrews, J.: In: Testa, B. (ed.). Advances in Drug Research, Vol. 19. Academic Press, New York, 1990.
7. Foye, W. O. (ed.): Principles of Medicinal Chemistry, 3rd ed. London, Kimpton, 1989.
8. Hammer, R. J.: Cardiovascular drugs. In: Foye, W. O. (ed.). Principles of Medicinal Chemistry, 3rd ed., Chap. 17. London, Kimpton, 1989.
9. Witiak, D. T., and Miller, D. D.: Cholesterol, adrenocorticoids, and sex hormones. In: Foye, W. O. (ed.). Principles of Medicinal Chemistry, 3rd ed., Chap. 21. London, Kimpton, 1989.

10. Gilman, A. G., Rall, T. W., Nies, A. S., and Taylor, P. (eds.): The Pharmacological Basis of Therapeutics, 8th ed., New York, Macmillan, 1990.
11. Hoffman, B. F., and Bigger, J. T., Jr.: Digitalis and allied cardiac glycosides. In: Gilman, A. G., Rall, T. W., Nies, A. S., and Taylor, P. (eds.): The Pharmacological Basis of Therapeutics, 8th ed., Chap. 34, New York, Macmillan, 1990.
12. Murad, F., and Kuret, J.: Estrogens and progestins. In: Gilman, A. G., Rall, T. W., Nies, A. S., and Taylor, P. (eds.): The Pharmacological Basis of Therapeutics, 8th ed., Chap. 58, New York, Macmillan, 1990.
13. Wilson, J. D.: Androgens. In: Gilman, A. G., Rall, T. W., Nies, A. S., and Taylor, P. (eds.): The Pharmacological Basic of Therapeutics, 8th ed., Chap. 59, New York, Macmillan, 1990.
14. Haynes, R. C., Jr.: Adrenocorticotropic hormone; Adrenocortical steroids and their synthetic analogs; inhibitors of adrenocortical steroid biosynthesis. In: Gilman, A. G., Rall, T. W., Nies, A. S., and Taylor, P. (eds.): The Pharmacological Basis of Therapeutics, 8th ed., Chap. 60. New York, Macmillan, 1990.
15. American Society of Hospital Pharmacists, American Hospital Formulary Service, Drug Information 90. Bethesda, MD, ASHP, 1990.
16. Facts and Comparisons 1990 Edition, St. Louis: Facts and Comparisons Division, J. B. Lippincott, 1990.
17. United States Pharmacopoeial Convention, Inc., USP DI, 9th ed., Rockville, MD, 1990.
18. Greeff, K. (ed.): Cardiac Glycosides, Part I and Part II, Vol. 56, New York, Springer-Verlag, 1981.
19. Martin, C. R.: Endocrine Physiology. New York, Oxford University Press, 1985.
20. Dence, J. B.: Steroids and Peptides. New York, John Wiley & Sons, 1980.
21. Fieser, L. F., and Fieser, M.: Steroids. New York, Van Nostrand Reinhold, 1967.
22. Heftmann, E.: Steroid Biochemistry. New York, Academic Press, 1970.
23. Djerassi, C.: Steroid Reactions. San Francisco, Holden-Day, 1963.
24. Fried, J., and Edwards, J. A.: Organic Reactions in Steroid Chemistry, Vols. 1 and 2. New York, Van Nostrand Reinhold, 1972.
25. McKerns, K. W. (ed.): The Sex Steroids—Molecular Mechanisms. New York, Appleton-Century-Crofts, 1971.
26. King, R. J. G., and Mainwaring, W. I. P.: Steroid-Cell Interactions. Baltimore, University Park Press, 1974.
27. O'Malley, B. W., and Schrader, W. T.: Sci. Am. 234:32, 1976.
28. Baxter, J. D., and Funder, J. W.: N. Engl. J. Med. 301:1149, 1979.
29. Baxter, J. D., and MacLeod, K.: In Bondy, P. K. and Roxenbert, L. E. (eds.): Duncan's Diseases of Metabolism. Philadelphia, W. B. Saunders, 1979.
30. O'Malley, B., and Birnbaumer, L. (eds.): Receptors and Hormone Action, Vol. 2. New York, Academic Press, 1978.
31. Baxter, J. D., and Rousseau, G. G. (eds.): Glucocorticoid Hormone Action. New York, Springer-Verlag, 1979.
32. O'Malley, B. W., and Hardman, J. F.: Steroid Hormones, Methods in Enzymology, Vol. 36. New York, Academic Press, 1975.
33. Eastman, P., and Bello, M.: NIH Division of Research Resources, Res. Resources Rep. 10:1, 1986.
34. Brown, M. S., and Goldstein, J. L.: Sci. Am. (Nov), p. 58, 1984.
35. Raffa, R. D.: Trends Pharmacol. Sci. p. 133, 1985.
36. Sadler, S. E., and Maller, J. L.: Plasma membrane steroid hormone receptors. In: Conn, P. M. (ed.). The Receptors, Chap. 9, Vol. 1. New York, Academic Press, 1984.
37. Fullerton, D. S., Ahmed, K., From, A. H. L., et al.: Modelling the cardiac glycoside receptor. In: Burgen, A. S. V., Roberts, G. C. K., and Tute, M. S. (eds.). Topics in Molecular Pharmacology, Chap. 10. New York, Elsevier, 1986.
38. Abbott, A. C. (ed.): Two Hundred Years of Foxglove Therapy. Trends Pharmacol. Sci., Special Issue on Cardiac Glycosides 6 (July), 1985.
39. Fullerton, D. S., Rohrer, D. C., et al.: Trends Pharmacol. Sci. 6:279–282, 1985.
40. Smith, T. W.: N. Engl. J. Med. 318:358, 1988.
41. Carafoli, E., and Scarpa, A. (ed.): Transport ATPases, Vol. 42. New York, N.Y. Acad. Sci., 1982.
42. Mullins, L. J.: Ion Transport in Heart. New York, Raven Press, 1981.
43. Hoffman, J. F., and Forbush, B., III. (eds.): Structure, Function, and Mechanism of the Na/K Pump. Curr. Top. Membr. Transport, Vol. 19, New York, Academic Press, 1983.
44. Fullerton, D. S., Rohrer, D. C., et al.: Use of Prophet and MMS-X computer graphics in the study of the cardiac steroid receptor site of Na,K-ATPase. In: Hoffmann, J. F., and Forbush, B., III (eds.). Curr. Top. Membr. Transport 19:257–264, 1983.
45. Dittrich, F., Berlin, P., Kopke, K., and Repke, K. H. H.: Stereoelectronic interaction between cardiac steroids and Na,K-ATPase. In: Hoffman, J. F., and Forbush, B., III (eds.). Curr. Top. Membr. Transport 19:251–255, 1983.
46. Shull, G. E., Schwartz, A., and Lingrel, J. B.: Nature 316:691, 1985.
47. Deffo, T., Fullerton, D. S., Kihara, M., Schimerlik, M. I., et al.: Biochemistry 22:6303, 1983.
48. King, R. J. B.: Horm. Cell Regul. 139:55, 1986.
49. Walters, M. R.: Endocr. Rev. 6:512, 1985.
50. Scheidereit, C., Krauter, P., Von Der Ahe, D., et al.: J. Steroid Biochem. 24:19, 1986.
51. Gorski, J.: Experientia 42:744, 1986.
52. King, R. J. B., and Coffer, A. I.: Int. Breast Cancer Res. Conf., p. 267, 1985.
53. Baulieu, E.: Maturitas 8:133, 1986.
54. King, R. J. B.: J. Steroid Biochem. 25:451, 1986.
55. Bishop, J. M.: Nature 321:12, 1986.
56. Ringold, G. M.: Ann. Rev. Pharmacol. Toxicol. 25:529, 1985.
57. McEwen, B. S., De Kloet, E. R., and Rostene, W.: Physiol. Rev. 66:1121, 1986.
58. Slater, E. P., Anderson, T., Cattini, P., et al.: Adv. Exp. Med. Biol. 196:67, 1986.
59a. Evans, R. M.: Science 240:889, 1988.
59b. O'Malley, B.: Mol. Endocrinology 4:363, 1990.
59c. Miesfeld, R. L.: Clin. Rev. Biochem. Mol. Biol. 24:101, 1989.
59d. Beato, M.: Cell 56:335, 1989.
59e. Agarwal, M. K.: Naturwissenschaften 77:170, 1990.
59f. Berg, J. M.: Cell 57:1065, 1989.
60. Weinberger, C., et al.: Nature 324:641, 1986.
61. Sap, J., et al.: Nature 324:641, 1986.
62. Evans, R. M., and Hollenberg, S.: Cell 52:1, 1988.
63. Horn, A. S., and De Ranter, C. J. (ed.): X-ray Crystallography and Drug Action. Oxford, Clarendon Press, 1984.
64. Dix, C. J., and Jordan, V. C.: Endocrinology 107:2011, 1980.
65. Duax, W. L., Griffin, J. F., et al.: Biochemical Actions of Hormones, Vol. 11, New York, Academic Press, 1984.
66. Duax, W. L., Griffin, J. F., et al.: J. Steroid Biochem. 31:481, 1988.
67. Michael, L. H., et al.: Mol. Pharmacol. 16:135, 1979.
68. Greenstein, B. D., and Adcock, I. M.: Role of steroid hormone receptors in development. In Conn, P. M. (ed.). The Receptors, Vol. 2, Chap. 10. New York, Academic Press, 1985.
69. Chang, W. C., and Roth, G. S.: J. Steroid Biochem. 11:889, 1979.
70. Irving, H. M. N. H., Freiser, H., and West, T. S., (eds.): International Union of Pure and Applied Chemistry, Definitive Rules. New York, Pergamon Press, 1978.
71. Lee, T.: Trends Pharmacol. Sci. 8:442, 1987.
72. Med. Lett. 29:99, 1987.
73. USP DI Update: p. 6, 1988.
74. Hoffman, W. F., Alberts, A. W., Anderson, et al.: J. Med. Chem. 29:849, 1986.

74a. Roth, B. D., Sliskovic, D. R., and Trivedi, B. K.: Ann. Rept. Med. Chem. 24:147, 1989.
75. Med. Lett. 29:99, 1987.
75a. Shepard, J., and Packard, C. J.: Trends in Pharmacol. Sci. 9:326, 1989.
75b. Med. Lett. 30:81, 1988.
76. Roberts, L.: Science 238:482, 1987.
77. Counsell, R. E., et al.: J. Med. Chem. 16:945, 1973.
78. Counsell, R. E., et al.: J. Nucl. Med. 14:777, 1973.
79. Dutta, A. S., and Furr, J. A.: Annu. Rep. Med. Chem. 20:203, 1985.
79a. Gharib, S. D., Wierman, M. E., Shupnik, M. A., and Chin, W. W.: Endocrine Rev. 11:177, 1990.
79b. Combarnous, Y.: Reprod. Natr. Develop. 28 (2A):211, 1988.
79c. Sharpe, R. M.: J. Endocrin. 121:405, 1989.
80. Kolata, G.: Science 238:805, 1987.
81. Sopelak, V. M., Williams, R. F., Takeshi, A., Marut, E. L., and Hodgen, G. D.: JAMA 251:1477, 1984.
82. Vigayan, E.: J. Steroid Biochem. 23:827, 1985.
82a. Med. Lett. 30:91, 1988.
83. Sztraky, K., and Roberts, G. H.: Pharm. Tech. May/June:116, 1986.
83a. Meyer, S. P.: "In Home Diagnostic Products" in Handbook of Nonprescription Drugs, American Pharmaceutical Association, Washington, DC, 1990.
84. Jovanovic, L., et al.: Proc. Soc. Exp. Biol. Med. 184:201, 1987.
85. Cartwright, P. S., Victory, D. F., Wong, S. W., and Dao, A. H.: Am. J. Obstet. Gynecol. 153:730, 1985.
86. Nordenskyold, F., Ahlgren, M., Erreth, L., and Hultberg, B.: Fertil. Steril. 43:748, 1985.
87. Doshi, M. L.: Am. J. Public Health 76:512, 1986.
88. Vermesh, M., Kletzky, O. A., Davajan, V., and Israel, R.: Fertil. Steril. 47:259, 1987.
89. Corson, S. L.: J. Reprod. Med. 31:760, 1986.
90. Rajfer, J., et al.: N. Engl. J. Med. 314:466, 1986.
91. Schueler, F. S.: Science, 103:221, 1946.
92. Bolt, H. M.: Pharmacol. Ther. 4:155, 1979.
93. Riggs, B. L., and Melton, L. J. III: N. Engl. J. Med. 314:1676, 1986.
94. Holbrook, T. L., Grazier, K., Kelsey, J. L., and Stauffer, R. N.: American Academy of Orthopedic Surgeons, Chicago, 1984.
95. Med. Lett. 29:75, 1987.
96. Kiel, D. P., Felson, D. T., Anderson, J. J., Wilson, P. W. F., and Moskowitz, M. A.: N. Engl. J. Med. 317:1169, 1987.
97. Am. Pharm. NS25:18, 1985.
98. Jensen, J., Christiansen, C., and Rodbro, P.: N. Engl. J. Med. 313:973, 1985.
99. NIH Consensus Development Conference on Osteoporosis: Am. Pharm. NS24:25, 1984.
100. Med. Lett. Nov. 26, 1982.
101. Riis, B., Thomsen, K., and Christiansen, C.: N. Engl. J. Med. 316:173, 1987.
101a. Watts, N. B. et al: New Eng. J. Med. 323:73, 1990.
102. Med. Lett. 27:53, 1985.
103. Hayward, M. A., and Caggiano, T. J.: Medicinal chemistry opportunities in bone metabolism and osteoporosis, Chap. 17. In Annu. Rep. Med. Chem. New York, Academic Press, 1987.
104. Med. Lett. 27:91, 1985.
105. Lesko, S. M., et al.: N. Engl. J. Med. 313:593, 1985.
106. Wilson, P. W. F., Garrison, R. J., and Castelli, W. P.: N. Engl. J. Med. 313:1038, 1985.
107. Colditz, G. A., Willett, W. C., Stampfer, M. J., et al.: N. Engl. J. Med. 316:1106, 1987.
108. Kolata, G.: Science 220:1137, 1983.
109. Stampfer, M. J., Willett, W. C., Colditz, et al.: N. Engl. J. Med. 313:1044, 1985.
110. Mangan, C. E., et al.: Am. J. Obstet. Gynecol. 134:860, 1979.
111. Forsberg, J. G.: Arch. Toxicol. 2 (Suppl.):263, 1979.
112. FDA Drug Bull., May, 1973.
113. O'Brien, P. C., et al.: Obstet. Gynecol. 53:300, 1979.
114. Mayo Clinic, DESAD Project, DES Update, Jan. 1988.
114a. Ansbacher, R.: Amer. Pharm. NS30:27, 1990.
114b. Burlington, D. B.: ibid NS30:28, 1990.
115. Med. Lett. 28:119, 1986.
116. Chetkowski, R. J., et al.: N. Engl. J. Med. 314:1615, 1986.
117. Albanese, A. A.: N.Y. J. Med. 65:2116, 1965.
118. Heftmann, E.: J. Steroid Biochem. p. 141, 1970.
119. Mirocha, C. J., et al.: Microb. Toxins 7:107, 1971.
120. Briggs, M. H., and Christie, G. A.: (Estrogenic and nonestrogenic) antifertility substances in plants. Adv. Steroid Biochem. Pharmacol. 6:xi, 1977.
121. Raynaud, J., and Ojasoo, T.: J. Steroid Biochem. 25:811, 1986.
122. Med. Lett. 27:82, 1985.
123. Rahwan, R. G.: Am. J. Pharm. Educ. 49:86, 1985.
123a. Buckley, M. M., and Goa, K. L.: Drugs 37:451, 1989.
123b. Love, R. R.: Prev. Med. 18:661, 1989.
124. Dorfman, R. I., and Unger, F.: Metabolism of Steroid Hormones. New York, Academic Press, 1965.
125. Salhanick, H. A., et al.: Metabolic Effects of Gonadal Hormones and Contraceptive Steroids. New York, Plenum Press, 1969.
126. McKerns, K. W.: Steroid Hormones and Metabolism. New York, Appleton-Century-Crofts, 1969.
127. Aufrere, M. B., and Benson, H.: J. Pharm. Sci. 65:783, 1976.
128. Crowley, W. F.: N. Engl. J. Med. 315:1607, 1986.
129. Med. Lett. 26:101, 1984.
130. FDA Drug Bull. 5:1, 1975.
131. Briggs, M. H., and Briggs, M.: Sex hormone exposure during pregnancy and malformations. In Briggs, M. H., and Corbin, A. (eds.). Adv. Steroid Biochem. Pharmacol. 7:51, 1979.
132. Bennett, J. P.: Chemical Contraception. New York, Columbia University Press, 1974.
133. Djerassi, C.: The Politics of Contraception. New York, W. W. Norton, 1979.
134. Lednicer, D. (ed.): Contraception, The Chemical Control of Fertility. New York, Marcel Dekker, 1969.
135. Douglas, E. T.: Pioneer of the Future—Margaret Sanger. New York, Holt, Rinehart & Winston, 1975.
136. Lader, L.: The Margaret Sanger Story and the Fight for Birth Control. Garden City, Doubleday, 1955.
137. Miller, C., Reinders, T. P., Ruggiero, R. J., and Steadman, S. (ed.): Methods of Birth Control. Syntex, Inc., 1985.
138. Hatcher, R. A., Guest, F., Stewart, F., Stewart, G. K., Trussell, J., and Frank, E.: Contraceptive Technology 1990–1991. New York, Irvington Publishers, 1990.
138a. Med. Lett. 31:93, 1989.
139. Bell, M. R., Batzold, F. H., and Winneker, R. C.: Chemical Control of Fertility. Annu. Rep. Med. Chem. 21:169, 1986.
140. Ruggiero, R. J., Reinders, T. P., and Steadman, S.: Am. Pharm. NS25:31, 1985.
141. Stubblefield, P. G.: J. Reprod. Med. 31:922, 1986.
142. Drug Ther. Bull. 24:73, 1986.
143. Med. Lett. 30:105, 1988.
144. Youngkin, E. Q., and Miller, L. G.: Nurse Pract. 12:17, 1987.
145. Syntex, A Corporation and A Molecule. Palo Alto, CA: National Press, 1966.
146. Makepeace, A. W., et al.: Am. J. Physiol. 119:512, 1937.
147. Selye, H., et al.: Fertil. Steril. 22:735, 1971, and earlier references cited.
148. Dempsey, E. W.: Am. J. Physiol. 120:926, 1937.
149. Kurzrok, R.: J. Contracept. 2:27, 1937.
150. Albright, F.: In Musser, J. H. (ed.). Internal Medicine. Philadelphia, Lea & Febiger, 1945.
151. Sturgis, S. H., and Albright, F.: Endocrinology 26:68, 1940.
152. Pincus, G.: The Control of Fertility. New York, Academic Press, 1965, and cited references.
153. Djerassi, C., et al.: J. Chem. Soc. 76:4092, 1954.
154. Colton, F. B.: U.S. Patent 2,691,028, 1954 (applied May, 1953).
155. Birch, A. J.: Q. Rev. (Lond.) p. 12, 1958; pp. 4, 69, 1950.
156. Rudel, H. W., and Martinez-Manautou, J.: Oral Contraceptives. In Rabinowitz, J. L., and Myerson, R. M., (eds.).

Topics in Medicinal Chemistry. New York, Wiley-Interscience, 1967, p. 339.
157. Behrman, S. J.: In Austin, D. R., and Perry, J. S. (eds.): Agents Affecting Fertility. London, Churchill, 1965.
158. Goldzieher, J. W., et al.: JAMA 180:359, 1962.
159. Vessey, M., Mant, D., Smith, A., and Yeates, D.: Br. Med. J. 292:526, 1986.
160. Tietze, C.: Fam. Plann. Perspect. 11:80, 1979.
161. Jain, A. K.: Stud. Fam. Plann. 8:50, 1977.
162. Shapiro, S., et al.: Lancet 1:743, 1979.
163. Willett, W. C., et al.: N. Engl. J. Med. 317:1303, 1987.
164. Centers for Disease Control: N. Engl. J. Med. 316:650, 1987.
165. Centers for Disease Control: N. Engl. J. Med. 315:405, 1986.
166. FDA Drug Bull. 14:2, 1984.
166a. FDA Drug Bull. 18:19, 1988.
166b. FDA Drug Bull. 20:5, 1990.
167. Robinson, S. C.: Am. J. Obstet. Gynecol. 109:354, 1971.
168. Banks, A. L.: Int. J. Fertil. 13:346, 1968.
169. Poland, B. J., and Ash, K. A.: Am. J. Obstet. Gynecol. 116:1138, 1973.
170. Med. Lett. 16:37, 1974.
171. Diczfalusy, E.: J. Steroid Biochem. 11:443, 1979.
172. Berman, E.: J. Reprod. Med. 5:37, 1970.
173. Wheeler, R. G., et al.: Intrauterine Devices. New York, Academic Press, 1974.
174. Population Reports, Intrauterine Devices, Series B. Washington, DC, The George Washington University Medical Center, January 1975.
175. World Health Organization: J. Steroid Biochem. 11:461, 1979.
176. Folkman, J., and Long, D. M.: J. Surg. Res. 4:139, 1964.
177. Product insert for Progestasert, Alza,
178. Benagiano, G., and Gabelnick, H. L.: J. Steroid Biochem. 11:449, 1979.
179. Baulieu, E. E., and Ulmann, A.: Hum. Reprod. 1:107, 1986.
180. Nieman, L. K., et al.: N. Engl. J. Med. 316:187, 1987.
181. Couzinet, B., et al.: N. Engl. J. Med. 315:1565, 1986.
182. Collins, R. L., and Hodgen, G. D.: J. Clin. Endocrinol. Metab. 63:1270, 1986.
182a. Palca, J., Cherfas, J.: Science 245:1319, 1989.
182b. Ulmann, A., Teutsch, G., and Philibert, D.: Sci. Amer. 262:42, 1990.
183. Population Reports, Barrier Methods, Series H, No. 3, Department of Medical and Public Affairs, The George Washington University, January 1975 (see ref. 174).
184. FDA Advisory Panel on Nonprescription Contraceptive Products, Fed. Reg. 45:82014, Dec. 12, 1980.
185. Hernandez, L.: Contraceptive methods and products. In Handbook of Nonprescription Drugs, Chap. 18. Washington, DC, American Pharmaceutical Association, 1986.
186. Stone, S. C., and Cardinale, F.: Am. J. Obstet. Gynecol. 133:635, 1979.
187. Louik, C., et al.: N. Engl. J. Med., 317:474, 1987.
188. Jones, W. R., et al.: Lancet 1:1295, 1988.
189. Talwar, R. P., et al.: Proc. Natl. Acad. Sci. USA 73:218, 1976.
190. Zatuchni, G. I., Goldsmith, A., Spieler, J. M., and Sciarra, J. J. (eds.): Male Contraception: Advances and Future Prospects. Philadelphia, Harper & Row, 1986.
191. Jackson, H., and Morris, I. D.: Clin. Obstet. Gynecol. 6:129, 1979.
192. Barwin, B. N.: Can. Med. Assoc. J. 119:757, 1978.
193. Bartke, A., et al.: In Zatuchni, G., et al. (eds.). Male Contraception: Advances and Future Prospects. Philadelphia, Harper & Row, 1986.
194. Alexander, N. J.: In Zatuchni, G., et al.: (eds.). Male Contraception: Advances and Future Prospects. Philadelphia, Harper & Row, 1986.
195. Zhen-Gang, W., and Gan-Zhong, L.: Trends Pharmacol. Sci. 6:423, 1985.
196. Chaudhury, R. R.: Trends Pharmacol. Sci. 7:121, 1986.
197. Waller, D. P., Niu, X., and Kim, I.: In Zatuchni, G., et al. (eds.). Male Contraception: Advances and Future Prospects. Philadelphia, Harper & Row, 1986.
197a. Wu, D.: Drugs 38:333, 1989.
198. Matlin, S. A., Zhou, R., Bially, G., Blye, R. P., Maqvi, R. H., and Lindberg, M. C.: Contraception 31:141, 1985.
199. Vida, J. A.: Androgens and Anabolic Agents. New York, Academic Press, 1969.
200. Wolff, M. E., Ho, W., and Kwok, R.: J. Med. Chem. 1:577, 1964.
201. Klimstra, P. D.: In Chemistry and Biochemistry of Steroids, Vol. 3, Chap. 8. Los Altos, CA, Geron-X, 1969.
202. Hendershott, J.: Track Field News, Part 5 22:3, 1969.
203. Wade, N.: Science 176:1399, 1972.
204. Zurer, P. S.: Chem. Eng. News, April 30, 1984.
205. Casner, S., et al.: J. Sports Med. Phys. Fitness 11:98, 1971.
206. Fahey, T. D., and Brown, C. H.: Med. Sci. Sports 5:272, 1973.
207. Fowler, W. M., Jr., et al.: J. Appl. Physiol. 20:1038, 1965.
208. Freed, D. L., and Banks, A. J.: Br. J. Sports Med. 9:78, 1975.
209. Hervey, G. R., et al.: Lancet 2:699, 1976.
210. Johnson, L. C., et al.: Med. Sci. Sports 4:43, 1972.
211. Johnson, L. C., and O'Shea, J. P.: Science 64:957, 1969.
212. Johnson, L. C., et al.: Med. Sci. Sports 7:287, 1975.
213. O'Shea, J. P.: Nutr. Rep. Int. 4:363, 1971.
214. O'Shea, J. P., and Winkler, W.: Nutr. Rep. Int. 2:351, 1970.
215. Stamford, B. A., and Moffatt, R.: J. Sports Med. Phys. Fitness 14:191, 1974.
216. Stromme, S. B., et al.: Med. Sci. Sports 6:203, 1974.
217. Ward, P.: Med. Sci. Sports 5:227, 1973.
218. FDA Drug Bull. 17:27, 1987.
219. Voy, R.: Am. Pharm. NS26:39, 1986.
220. Brooks, R. V.: Anal. Proc. 24:70, 1984.
221. Med. Lett. 26:65, 1984.
222. Goldman, B.: Death in the Locker Room. Los Angeles, Price Stern Sloan, 1984.
223. Frischkorn, C. G., and Frichkorn, E. H.: J. Chromatogr. 151:33, 1978.
224. Wieder, B.: J. Sports Med. Phys. Fitness 13:131, 1973.
225. Novich, M. M.: N.Y. State J. Med. 73:2679, 1973.
226. American College of Sports Medicine: Position statement on the use and abuse of anabolic-androgenic steroids in sports, Med. Sci. Sports 9:xi, Winter, 1977.
227. Rasmussen, G. H.: Annu. Rep. Med. Chem. 21:179, 1986.
228. Cornforth, L. W.: Q. Rev. 23:125, 1969.
229. Mainwaring, W. I. P., et al.: Mol. Cell. Endocrinol. p. 133, 1974.
230. Schroeder, S. A., et al. (eds.): Current Medical Diagnosis and Treatment 1988, Norwalk, CT, Appleton & Lange, 1988.
230a. Rumore, M. M, and Rumore, J. S.: Amer. Pharm. NS29(9):49, 1989.
230b. Rumore, M. M., and Rumore, J. S.: ibid NS29(10):40, 1989.
230c. Dmowski, W. P.: J. Reprod. Med. 35(1 Supplement):69, 1990.
230d. Barbieri, R. L.: Curr. Concepts Endometriosis 241, 1990.
231. Blank, H., et al.: Round Table Discussion, Cutis, 24: Oct., Dec., 1979; Jan., Mar., Apr., 1980.
232. Med. Lett. 24:103, 1982.
233. Med. Lett. 28:57, 1986.
234. Med. Lett. 29:42, 1987.
235. White, P. C., New, M. I., and Dupont, B.: N. Engl. J. Med. 316:1519, 1987.
236. White, P. C., New, M. I., and Dupont, B.: N. Engl. J. Med 316:1580, 1987.
237. Med. Lett. 27:87, 1985.
238. Sayers, G., and Travis, R. H.: ACTH and adrenocortical steroids. In Goodman, L. S., and Gilman, A. (eds.). The Pharmacological Basis of Therapeutics, 5th ed., Chap. 72. New York, Macmillan, 1975.
239. Alexis, M. N.: Trends Pharmacol. Sci. 8:10, 1987.
240. Med. Lett. 29:21, 1987.
241. Med. Lett. 29:11, 1987.
242. Johnson, C. E.: Drug Intell. Clin. Pharm. 21:784, 1987.

243. Colucci, W. S., Wright, R. F., and Braunwald, E.: N. Engl. J. Med. 314:349, 1986.
244. Colucci, W. S., Wright, R. F., and Braunwald, E.: N. Engl. J. Med. 314:290, 1986.
245. Wetzel, B., and Hauel, N.: Trends Pharmacol. Sci. 9:166, 1988.
246. Erhardt, P. W.: J. Med. Chem. 30:231, 1987.
247. Weber, K. T., Santosh, K., Gill, J. S., et al.: Drugs 33:503, 1987.
248. LeJemtel, T. H., Keren, G., Reis, D., and Sonnenblick, E. H.: J. Cardiol. Pharmacol. 8:S47, 1986.
249. Taylor, M. D., Sircar, I., and Steffen, W. P.: Annu. Rep. Med. Chem. 22:85, 1987.
249a. Paton, D. M., and Manuel, J. M.: Trends in Pharmacol. Sci. 9:431, 1988.
250. Alousi, A. A., Farah, A. E., et al.: Circ. Res. 45:666, 1979.
251. Romey, G., Quast, U., Paroun, D., et al.: Proc. Natl. Acad. Sci. USA 84:896, 1987.
252. van Meel, J. C. A., Zimmerman, R., Diederen, W., et al.: Biochem. Pharmacol. 37:213, 1988.
253. Seamon, K. B., Daly, J. W., Metzger, H., et al.: J. Med. Chem. 26:436, 1983.
253a. Laurenza, A., Sutkowski, E. M., and Seamon, K. B., ibid 10:442, 1989.
254. Daly, J. W.: Adv. Cyclic Nucleotide Protein Phosphorylation Res. 17:81, 1984.
255. Hoshi, T., Garber, S. S., and Aldrich, R. W.: Science 240:1652, 1988.
256. Wagoner, P. K., and Pallotta, B. S.: Science 240:1655, 1988.
257. Bristol, J. A. (ed.): Cardiac Drugs, Chap. 6. New York, John Wiley & Sons, 1986.
258. Erdmann, E., Greeff, K., and Skou, J. C. (eds.): Cardiac Glycosides 1785–1985. New York, Springer-Verlag, 1986.
259. Skou, J. C.: William Withering—the man and his work. In Erdman, E., Greeff, K., and Skou, J. C. (eds.). Cardiac Glycosides 1785–1985. New York, Springer-Verlag, 1986, pp. 1–10.
260. Yu, A., Ovchinnikov, N. N., Modyanov, N. E., et al.: FEBS Newslett. 201:237, 1986.
261. Sverdlov, E. D., Monastyrskaya, G. S., and Broude, N. E.: FEBS Newslett. 217:275, 1987.
262. Ahmed, K., McParland, R., Becker, R., et al.: Fed. Proc. 45:1488, 1986.
263. Yu, A., Ovchinnikov, N. M., Arzamazova, E. A., et al.: FEBS Newslett. 217:269, 1987.
264. Yu, A., Ovchinnikov, N. M., Luneva, E. A., et al.: FEBS Newslett. 227:230, 1988.
265. Schwartz, A., and Adams, R. J.: Circ. Res. 46(Suppl. 1):154, 1980.
266. Rohrer, D. C., Kihara, J., Deffo, T., et al.: J. Am. Chem. Soc. 106:8269, 1984.
267. Chiu, F. C. K., and Watson, T. R.: J. Med. Chem. 28:509, 1985.
268. Schonfeld, W., Weiland, J., Lindig, C., et al.: Naunyn Schmiedebergs Arch. Pharmacol. 329:414, 1985.
269. Schonfeld, W., Menke, K., Schonfeld, R., and Repke, K. R. H.: J. Enzyme Inhibition 2:37, 1987.
270. Guntert, W., and Linde, H. H.: Arch. Int. Pharmacodyn. Ther. 233:53, 1978.
271. Erdmann, V. E.: Arzneim. Forsch. 35(11):12a, 1948, 1985.
272. Flasch, H. F., and Heinz, N.: Naunyn Schmiedeburgs Arch. Pharmacol. 304:37, 1978.
273. Fullerton, D. S., Yoshioka, K., Rohrer, D., et al.: Science 205:917, 1979.
274. Ahmed, K., Rohrer, D. C., Fullerton, D. S., et al.: J. Biochem. 258:8092, 1983.
275. Angliker, E., Barfuss, F., and Renz, J.: J. Chem. Ber. 41:479, 1958.
276. Woodcock, B. G., and Rietbrock, N.: Trends Pharmacol. Sci. p. 273, 1985.
277. Hashimoto, T., Ahmed, K., Fullerton, D. S., et al.: J. Med. Chem. 29:997, 1986.
278. Griffin, J. F., Rohrer, D. C., Fullerton, D. S., et al.: Mol. Pharmacol. 29:270, 1986.
279. Rathore, H., From, A. H. L., Ahmed, K., and Fullerton, D. S.: J. Med. Chem. 29:1945, 1986.
280. Brown, L., Thomas, R.: Arzneim. Forsch. 33:814, 1983.
281. Brown, L., Erdmann, E., and Thomas, R.: Biochem. Pharmacol. 32:2767, 1983.
282. Yoda, A.: Mol. Pharmacol. 12:399, 1976, and cited references.
283. Yoda, A.: Ann. N.Y. Acad. Sci. 242:598, 1974.
284. Fullerton, D. S., Rathore, H., Tjaharyanto, D., et al.: Am. Chem. Soc. Natl. Meet., New York, MEDI No. 69, 1986.
285. Ahmed, K., Fullerton, D. S., Rathore, H., and From, A. H. L.: Fed. Proc. 45:651, 1986.
286. Fullerton, D. S., Kihara, M., Deffo, T., et al.: J. Med. Chem. 27:256, 1984.
287. Gerbino, P. P.: Am. Hosp. Pharm. 30:499, 1973.
287a. Moos, W. H. et al.: J. Med. Chem 30:1963, 1987.
288. Smith, S.: J. Chem. Soc. p. 508, 1930; ibid., p. 23, 1931.
289. Chevreul, M.: Ann. Chim. Phys. (Ser. 1) 95:5, 1815.
290. Nair, P. P., and Kritchevsky, D. (eds.): The Bile Acids, Vols. 1 and 2. New York, Plenum Press, 1971–72, and cited references.
291. Lindstedt, S.: Acta Physiol. Scand. 40:1, 1957.
291a. Med. Lett. 30:81, 1988.
292. Godtfredsen, W. O., et al.: J. Med. Chem. 9:15, 1966.
293. Lehmann, P. A., et al.: J. Chem. Educ. 50:195, 1973.
294. Fieser, L. F., and Fieser, M.: Steroids. New York, Van Nostrand Reinhold, 1959, pp. 816–825.
295. Tsukamoto, T., et al.: J. Pharm. Soc. Jpn. 56:931, 1936.
296. Peterson, D. H., et al.: J. Am. Chem. Soc. 74:5933, 1952.
297. Peterson, D. H., et al.: J. Am. Chem. Soc. 75:408, 1953.
298. Colton, F. B.: U.S. Patents 2,655,518, 1952; 2,691,028, 1953; 2,725,378, 1953.
299. Djerassi, C., et al.: J. Am. Chem. Soc. 76:4092, 1954.
300. Sih, C., et al.: J. Am. Chem. Soc. 87:2765, 1965; Patent described in Chem. Abstr. 68:103880, 1968.
301. Kraychy, S., et al.: Appl. Microbiol. 23:72, 1972; Patent described in Chem. Abstr. 77:124777, 1972.
302. Uberwasser, H., et al.: Helv. Chim. Acta 46:344, 1963; 46:361, 1963.
303. Klimstra, P. D., and Colton, F. B.: Chemistry of the steroidal contraceptives. In Lednicer, D. (ed.). Contraception, The Chemical Control of Fertility, Chap. 3. New York, Marcel Dekker, 1969.
304. Velluz, L., et al.: Compt. Rend. 257:3086.
305. Ananchenko, S. M., et al.: Tetrahedron 18:1355, 1962.

CHAPTER 19

Carbohydrates

Jaime N. Delgado
Pedro L. Huerta, Jr.

Carbohydrates represent one of the four major classes of biomolecules that also include proteins, nucleic acids, and lipids. These compounds, usually called "sugars" (e.g., glucose, sucrose, starch, and glycogen), were thought to be correctly represented by the generalized formula, $C_x(H_2O)_y$, and thus the term *carbohydrate* became extensively used. However, many compounds now classified as carbohydrates (2-deoxyribose, digitoxose, glucuronic and gluconic acids, the amino sugars) possess structures that cannot be represented by such a formula. On a functional group basis, carbohydrates are characterized as polyhydroxy aldehydes or polyhydroxy ketones and their derivatives.

Carbohydrates are extensively distributed in both the plant and animal kingdoms. Chlorophyll-containing plant cells produce carbohydrates by photosynthesis which involves the fixation of CO_2 through reduction by H_2O and requires solar electromagnetic energy. Carbohydrates serve as a source of energy for plants and animals, and, in the form of cellulose and chitin, they function as the supporting structures of plants and insects or crustacea. In plants and microorganisms, carbohydrates are metabolized through various pathways leading to amino acids, purines, pyrimidines, fatty acids, vitamins, and such. Together with other dietary components (such as proteins, lipids, minerals, and vitamins) some carbohydrates are metabolically utilized by animals in many processes, degraded to acetyl coenzyme A (CoA) for the synthesis of lipids or oxidized to obtain adenosine triphosphate (ATP), and in plants they are used for the synthesis of other organic compounds. Most of the carbohydrate that is utilizable by the human consists of starch, glycogen, sucrose, maltose, or lactose, whereas cellulose, xylans, and pectins cannot be degraded by digestive processes because of the lack of the appropriate enzymes.

As the foregoing statements indicate, the biologic importance of carbohydrates is readily obvious. Various textbooks of biochemistry provide complete discussions of the chemistry and metabolism of carbohydrates.[1–6] Moreover, in medicinal chemistry it is recognized that many pharmaceutical products contain carbohydrates or modified carbohydrates as therapeutic agents or as pharmaceutical necessities. Certain antibiotics are carbohydrate derivatives. The streptomycins, neomycins, paromomycins, gentamicins, and kanamycins are basic carbohydrates that have significant antimicrobial properties.[7] The cardioactive glycosides represent another class of medicinal agents possessing carbohydrate moieties that contribute to their therapeutic efficacy (see Chapt. 18).[8] More recently, the neuromodulatory actions of adenosine agonists and antagonists, and the inhibition of gene expression by oligonucleotides have been reviewed to emphasize their use in rational drug design and biotechnology studies.[9,10]

Some knowledge of the interrelationships of carbohydrates with lipids and proteins in human metabolism is necessary for the study of the medicinal biochemistry of diabetes mellitus and the actions of antidiabetic agents. Accordingly, a brief discussion of these topics will be presented later in this chapter for purposes of emphasizing how some factors affecting carbohydrate metabolism also affect metabolic processes involving lipids and proteins.

CLASSIFICATION

A brief review of elementary characterizations of the more important carbohydrates is fundamental to the understanding of the structural and functional differences among the vast array of natural products that are classified as carbohydrates. The following summary is intended to delineate and exemplify the major classes and types of carbohydrates.

It is conventional to classify carbohydrates as *monosaccharides*, *oligosaccharides*, and *polysaccharides*, depending on the number of sugar residues present per molecule. Furthermore, monosaccharides containing three carbon atoms are called *trioses*, those containing four carbon atoms are *tetroses*, whereas *pentoses*, *hexoses*, and *heptoses* contain five, six, and seven carbon atoms, respectively. On a functional group basis, monosaccharides having a potential aldehyde group in addition to hydroxyl functions are known as *aldoses* and those bearing a ketone function are *ketoses*. For example, glyceraldehyde is an aldotriose and dihydroxyacetone is a ketotriose, whereas glucose is an aldohexose and fructose is a ketohexose.

Disaccharides, trisaccharides, and tetrasaccharides are oligosaccharides. Sucrose, lactose, maltose, cellobiose, gentiobiose, and melibiose are important disaccharides. Raffinose, melecotose, and gentianose are trisaccharides; stachyose is a tetrasaccharide.

Monosaccharides existing in the form of heterocycles are classified according to the size of the ring system; that is, the six-membered ring structures considered to be related to pyran are called *pyranoses* and the five-membered ring structures related to furan are called *furanoses*. This type of nomenclature can be applied to oligosaccharides and glycoside derivatives. Thus, maltose can be named 4-D-glucopyranosyl-α-D-glucopyranoside; lactose is 4-D-glucopyranosyl-β-D-galactopyranoside; and sucrose is a 1-α-D-glucopyranosyl-β-D-fructofuranoside. (Stereochemical classification of carbohydrates will be considered briefly as the basis for the aforementioned configurational designations.)

Most carbohydrate material in nature exists as high-molecular-weight polysaccharides that on hydrolysis yield monosaccharides or their derivatives. Glucose, mannose, galactose, arabinose, and glucuronic, galacturonic, and mannuronic acids, and some amino sugars occur as structural components of polysaccharides, glucose being the most common component.

Polysaccharides yielding only one variety of monosaccharide are called *homopolysaccharides* and those yielding a mixture of different monosaccharides are known as *heteropolysaccharides*. Homopolysaccharides of importance include the starches and glycogen, which are mobilizable stores of glucose, whereas cellulose is a structural polymer. All when hydrolyzed yield glucose. Heparin, hyaluronic acid, and the immunochemically specific polysaccharide of type III pneumococcus are representative examples of heteropolysaccharides. Heparin's polymeric structure is composed of α-D-glucuronic acid, α-L-iduronic acid, α-D-glucosamine, and *N*-acetyl-α-D-glucosamine (see the abbreviated structure for heparin in Fig. 19-1).[5] Hyaluronic acid contains glucuronic acid and *N*-acetyl glucosamine units, and the type III pneumococcus polysaccharide on hydrolysis yields glucose and glucuronic acid. These heteropolysaccharides contain two different sugars in each component monomer. Much more complex polysaccharides contain more than two monosaccharides; for example, gums and mucilages upon hydrolysis yield galactose, arabinose, xylose, and glucuronic and galacturonic acids.

Research in the field of structure–activity relationships among polysaccharides continues to increase the understanding of the relationship between their conformations in solution and biologic function. It has been noted that polysaccharides of the pyranose forms of glucose, galactose, mannose, xylose, and arabinose have conformations that are restricted by steric factors. Such polysaccharides have been characterized and classified on the basis of conformation properties: type A, extended and ribbonlike; type B, helical and flexible; type C, rigid and crumpled; and type D, very flexible and extended. Interestingly, most support materials are categorized in type A and most matrix materials belong to type B. Cellulose and chitin form rigid structures, and these polysaccharides are the most important structural polysaccharides in nature and possess β-1,4 linkages. Matrix materials form gels, and this property is fundamental to their biologic functions as ground material filling the extracellular spaces of tissue, synovial fluid, or as part of the vitreous humor.[5] It has been suggested that some matrix materials produce gels by forming double helices. Hyaluronic acid, which possesses β-1,3 and β-1,4 linkages (see Fig. 19-1), has been studied, and its gelling properties appear to be dependent on double-helix formation.[11] The foregoing summary and particularly the cited reference illustrate the significance of polysaccharides as the fibrous and matrix materials in support structures of plants and animal organisms. On the other hand, glycogen and starches, which serve as readily available energy sources for humans, possess mostly α-1,4-glycosidic bonds and some branching through α-1,6-glycosidic bonds. Amylose is the unbranched type of starch and possesses only α-1,4 linkages. Glycogen branching occurs at every eight to ten residues, whereas amylopectin branch points occur at about every 25 to 30 residues to affect their physicochemical properties.[5] The increased branching enhances the solubility of glycogen to increase its mobilizability (see Fig. 19-1).

Many different carbohydrates occur as components of glycoproteins. The term *glycoprotein* is

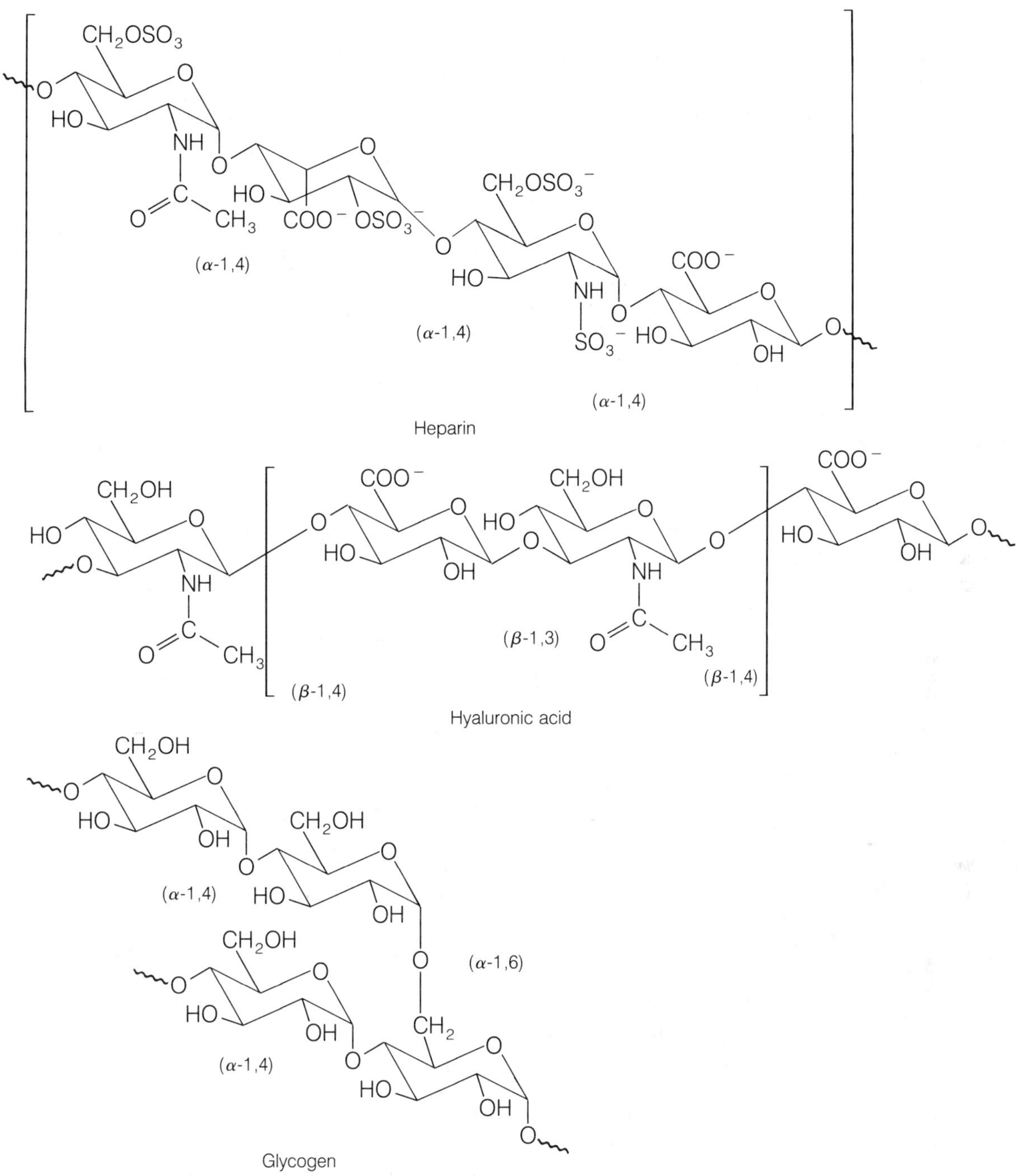

FIG. 19.1. *Structural relationships of three polysaccharides.*

used in a general sense and includes proteins that contain covalently bonded carbohydrates.* Glycoproteins are widely distributed in animal tissues, and some have been found in plants and microorganisms. All plasma proteins (except albumin), proteins of mucous secretions (antibodies, clotting factors), some hormones (e.g., thyroglobulin, chorionic gonadotropin), certain enzymes (e.g., serum cholinesterase, deoxyribonuclease), components of cellular and extracellular membranes (asialoglycoprotein and β-receptors), and constituents of connective tissue are classified as glycoproteins. The bonding of the carbohydrate moiety, oligosaccharide, to the peptide usually involves C-1 of the most internal sugar and a functional group of an amino acid within the peptide chain; for example, the linkage of N-acetylglucosamine through a β-glycosidic bond to the amide group of asparagine or by O-glycosidic

---

* Some textbooks have subcategorized glycoproteins and glycolipids as proteoglycans and peptidoglycans, which are massive aggregates and contain a much greater portion of carbohydrates than proteins or lipids. Hyaluronic acid and heparin have been included in this subcategory.[5]

linkage with the side-chain oxygen atom of serine or threonine residues.[6] A variety of oligosaccharides in different patterns have been identified on the outer surface of cell membranes, immunoglobulins, peptide hormones, antibodies, and clotting factors.[6, 12, 13] The significance of these findings is based on the intricacies of the inherent carbohydrate stereochemistry, different glycosidic linkages, and sequence variations of the basic units. The unique information supplied by each oligosaccharide pattern has initiated studies involving carbohydrates in molecular targeting and cell–cell recognition. One recent review of cell culture studies implicated the development of antisense oligonucleotide and analogues as potential antiviral or antitumor agents.[10] The synthesis and metabolism of glycoproteins have been studied by numerous investigators, and the major studies have been recently reviewed.[15–17]

Glycolipids are carbohydrates containing lipids, and some are derivatives of sphingosine. The carbohydrates containing derivatives of ceramides are called *glycosphingolipids*. Under normal circumstances there is a steady-state of balance between the synthesis and catabolism of glycosphingolipids in all cells. In the absence of any one of the hydrolases necessary for degradation, there is abnormal accumulation of intermediate metabolites, particularly in nervous tissue, which leads to various sphingolipodystrophies. There are three classes of glycosphingolipids: cerebrosides, gangliosides, and ceramide oligosaccharides.

Lipopolysaccharides of gram-negative bacteria have been studied with emphasis on structural elucidation. The peripheral portions of the lipopolysaccharides, called O-antigens, are composed of various carbohydrates arranged as oligosaccharide-repeating units forming high-molecular-weight polysaccharides. Structural details differ with the serotype of the organism. Some somatic O-antigens are highly toxic to animals. The lipopolysaccharide of the *Enterobacteriaceae* is one of the most complex of all polysaccharides, if not the most complex carbohydrate known. This polysaccharide has a gross structure: the carbohydrate moiety, which is the outermost portion, consists of abequose, mannose, rhamnose, galactose, and *N*-acetylglucosamine units; the lipid fraction includes glucosamine, phosphate, acetates, and β-hydroxymyristic acid.

BIOSYNTHESIS

Photosynthesis proceeds in the chlorophyll-containing cells of plants. The photosynthetic process involves the absorption of radiant energy by chlorophyll and the conversion of the absorbed light energy into chemical energy. This chemical energy is necessary for the reduction of CO_2 from the atmosphere to form glucose. The so-formed glucose may be metabolized by the plant cells, forming other carbohydrates, degraded to form precursors for the synthesis of other organic compounds, and oxidized as an energy source for the plant's physiology.

In higher plants, sucrose is synthesized through the activated form of glucose, uridine diphosphoglucose (UDPG), and fructose.

Polysaccharide biosynthesis also requires UDPG. For illustration of polysaccharide formation, consider glycogenesis in hepatic tissue: a glycogen synthetase enzyme catalyzes the polymerization of glucose units from UDPG. The latter is obtained from the reaction between glucose-1-phosphate and uridine triphosphate (UTP). It is noteworthy that UDPG performs an important role in the formation of the glycosidic linkage fundamental to the structure of oligosaccharides and polysaccharides. Analogous UDP compounds involving other monosaccharides are utilized in the biosynthesis of polysaccharides containing these sugars. The biosynthesis of cellulose is supposed to occur through the guanine-containing analogue of UDPG, guanosine diphosphoglucose (GDPG).

STEREOCHEMICAL CONSIDERATIONS

Basic organic chemistry textbooks cover the principles of stereoisomerism relevant to the study of carbohydrates.* The configurational and conformational aspects of carbohydrates have been reviewed by Bentley.[18] The stereochemistry of carbohydrates has presented many challenges to scientists, and there are several books[3, 19] that treat this subject comprehensively; hence, here only a brief resumé is presented. Stoddard[19] reviewed stereochemical studies, including nomenclature, on the basis of conformational analysis. In addition to configurational designations (e.g., β-D-glucopyranose), italic letters are used to specify conformation: *C*, chair; *B*, boat; *S*, twist boat; *H*, half-chair; and so on. As an illustration, consider the structure of β-D-glucose: accord-

* Symbols *d* and *l* or (+) and (−) are used to designate sign of rotation of plane-polarized light, and the configuration is designated by the symbols in small capitals D and L; monosaccharides are designated as D or L on the basis of the configuration of the highest-numbered asymmetric carbon, the carbonyl being at the top: D if the —OH is on the right and L if the —OH is on the left. (+)-Mannose, (−)-arabinose are assigned to the D-family because of their relation to D-(+)-glucose and D-(+)-glyceraldehyde. Thus, sugars configurationally related to D-glyceraldehyde are said to be members of the D-family, and those related to L-glyceraldehyde belong to the L-series.

ing to this system, conformation is defined by numerals indicating ring atoms lying above or below a defined reference plane; in structure 1 (see diagram) for β-D-(+)-glucose the reference plane contains C-2, C-3, C-5, and O; C-4 is above the plane and C-1 is below; hence, this conformation is designated as $^4C_1$ (compare with α-D-glucose). These symbols were proposed by the British Carbohydrate Nomenclature Committee, and they seem to be receiving general usage.

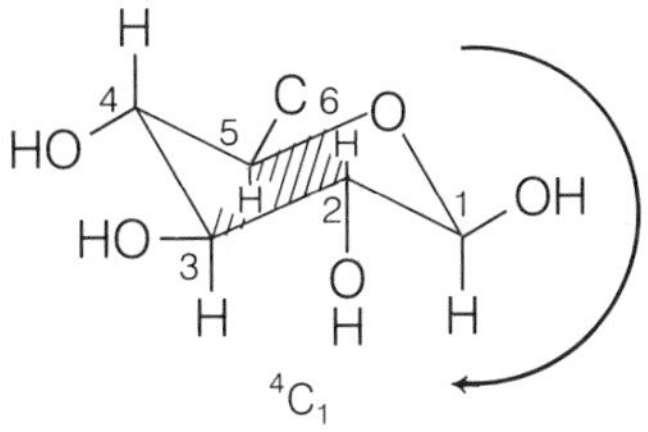

$^4C_1$

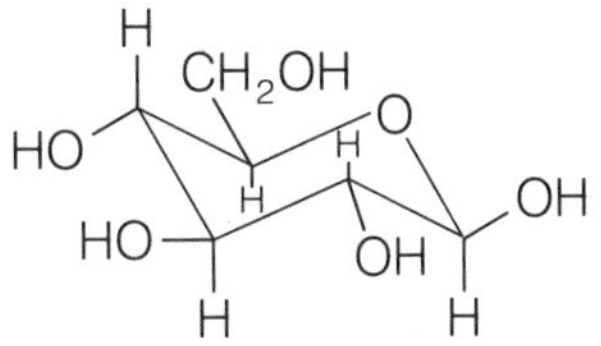

β-D-(+)-Glucose

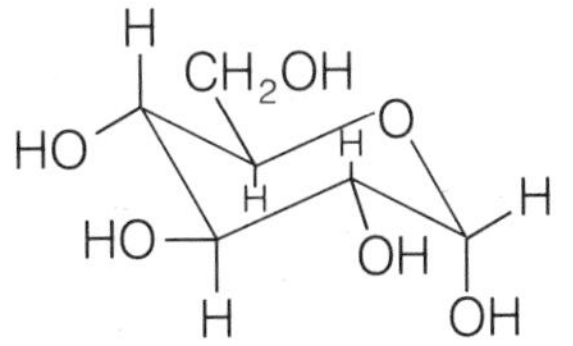

α-D-(+)-Glucose

Advances in studies of configurational and conformational features of carbohydrate structures have been facilitated by x-ray crystallography, nuclear magnetic resonance spectroscopy, and mass spectrometry, combined with gas chromatography.[20,21]

INTERRELATIONSHIPS WITH LIPIDS AND PROTEINS

Various interrelationships of carbohydrates with proteins and lipids have been noted in the foregoing as glycoproteins and glycolipids were characterized. Many other relationships exist among metabolic processes involving carbohydrates, lipids, and proteins. Some of these relationships exemplify how regulation of metabolism is maintained through numerous mechanisms (e.g., feedback regulation and hormonal regulation).

The Krebs citrate cycle requires acetyl-CoA, and this requirement is satisfied by glycolysis and pyruvic acid decarboxylation, by the β-oxidation of fatty acids, by oxidation of glycerol through the glycolytic pathway, and by pyruvic acid from alanine transamination. Even more correlations are found within the steps of the Krebs cycle. Oxaloacetate can be transformed into aspartate or into phosphoenolpyruvate and, subsequently, to carbohydrates. Such transformation of oxaloacetate into carbohydrates is the metabolic route of gluconeogenesis. Lactate from anaerobic glycolysis is the major starting material for gluconeogenesis; the lactate (from pyruvate and carboxylation of pyruvate) is transformed into the key intermediate phosphoenolpyruvate.

In anabolism, acetyl-CoA, which can originate from carbohydrates, proteins, and lipids, is utilized in the synthesis of important metabolites, such as steroids and fatty acids.

Appreciation of the foregoing correlations between so many important metabolic and anabolic processes facilitates the understanding of how and why factors affecting certain processes directly also affect other processes indirectly.

Glucose must be activated by formation of UDP-glucose (UDPG) before utilization in glycogenesis. The enzyme glycogen synthetase catalyzes the transformation of UDPG into glycogen. Glycogen catabolism to glucose proceeds through the action of phosphorylase a which catalyzes phosphorolysis of glycogen, providing glucose-1-phosphate; the latter then enters glycolysis. At this point, the dynamism of hormonal regulation can be illustrated by referring to the following phenomena. If and when the blood glucose concentration decreases below the normal level, epinephrine (from the adrenal medulla) binds with membrane receptors to activate adenylate cyclase by stimulation of G proteins. Adenylate cyclase then catalyzes the formation of cAMP (cyclic-3′,5′-adenosinemonophosphate). Cyclic AMP is a general intracellular mediator of many hormone actions, and herein stimulates the activation of phosphorylase b (inactive) providing phosphorylase a (active). Thus, the net effect of phosphorylase a action promotes glycogen breakdown, leading to glucose-1-phosphate and an increase in blood glucose concentration. This accounts for epinephrine's hyperglycemic action and also explains how epinephrine agonists can affect carbohydrate metabolism, whereas opposite effects can be expected from epinephrine antagonism. Newton and Hornbrook[22] investigated the metabolic effects exerted by adrenergic agonists and antagonists and concluded that the order of potency was isoproterenol > norepinephrine > salbutamol, for stimulation of rat liver adenylate cyclase; similar order of potency is reported for increased rat liver phosphorylase activity, but epinephrine produced a greater maximal response than isoproterenol. These authors also report that the β-adrenergic antagonist propranolol blocked the effects of isoproterenol or epinephrine on adenylate cyclase, whereas the α-adrenergic blockers ergotamine and phenoxybenzamine produced only partial inhibition. One explanation for this difference is based on the premise that α_1-receptors are associated with the phosphoinositide second-messenger system, whereas β_1- and β_2-receptors are involved in the adenylate cyclase cascade.[6] Therefore, it is noteworthy that the adrenergic metabolic receptor in the liver (rat) reacts to agonists and antagonists in parallel with the responses of those receptors in other tissues that have been designated β-adrenergic receptors. The rationalization behind this observation is based on the relationship between the β-receptor, a 64-kDa protein, and other receptors that are coupled to G proteins involving hormones that utilize cAMP as a second-messenger and possess seven transmembrane helices.[23-25]

Abnormally low blood glucose levels also stimulate pancreatic α-cells to release the hormone glucagon, another hyperglycemic hormone. Glucagon, much like epinephrine, activates adenylate cyclase, promoting cAMP formation, leading to enhancement of glycogen catabolism, but glucagon affects liver cells, and epinephrine affects both muscle and liver cells.

Adrenocortical hormones (i.e., the glucocorticoids) affect carbohydrate metabolism by promoting gluconeogenesis and glycogen formation. Because gluconeogenesis from amino acids is enhanced and because these hormones also inhibit protein synthesis in nonhepatic tissues, the precursor amino acids are made available for gluconeogenesis in the liver. Glucocorticoids stimulate synthesis of specific proteins in liver while inhibiting protein formation in muscle and other tissues. As protein catabolism continues in these tissues, the ultimate result is a net protein catabolic effect.

Sufficiently high blood glucose concentration stimulates pancreatic β-cells to secrete the hypoglycemic hormone insulin. Insulin exerts numerous biochemical actions,[26] affecting not only carbohydrate metabolism, but also lipid and protein metabolism; glycogenesis, lipogenesis, and protein synthesis are enhanced by insulin, whereas ketogenesis from fatty acids, glycogenolysis, and lipolysis are processes that are suppressed by insulin. (Insulin deficiency leads to the opposite effects on these processes.) It is clear that insulin modifies the reaction rates of many processes in its target cells, and highly specific insulin–receptor interactions have been implicated.[27] Insulin receptors on adipose and liver cells have been characterized, and they appear to have uniform characteristics. Experimental evidence limits insulin action to the plasma membrane of target cells. Insulin–receptor interactions can lead to modulation of other hormone actions through mechanisms involving cAMP phosphodiesterase activation and inhibition of adenylate cyclase activation. Cyclic AMP phosphodiesterase is responsible for catalyzing cAMP hydrolysis which inactivates cAMP; thus, insulin activation of this phosphodiesterase results in reversal of the metabolic effects of hormones that act through cAMP. On the other hand, many other compounds show capability of activating phosphodiesterase (e.g., cGMP and nicotinic acid), and even more compounds demonstrate inhibitory activity, (e.g., xanthine derivatives, papaverine and related isoquinoline compounds, and some adrenergic amines). The foregoing processes have been reviewed, but a more recent review demonstrated other pharmacologic properties for similar compounds. Adenosine receptor agonists and antagonists, through some overlapping biochemical processes, have produced important central nervous system (CNS) effects. Therefore, the adenosine agonists have been implicated as potential novel neuroleptic agents, whereas the antagonists may become potential CNS stimulants or antidepressant agents.[9,28]

The specific biochemical effects exerted by insulin and glucagon are delineated in more detail in Chapter 20. Suffice it to say here that medicinal agents that promote insulin availability exert actions through insulin and also a variety of other effects. Consider that the hypoglycemic sulfonylureas, which stimulate insulin secretion, also might act on phosphodiesterase, thereby inhibiting the inactivation of cAMP; moreover, cAMP has been implicated as a factor promoting insulin release. Reportedly some sulfonylureas also reduce glucagon secretion.

The biguanide phenformin was removed from the pharmaceutical market in 1977 because of the serious side effect called *lactic acidosis*.[29] The hypoglycemic action of phenformin involves various factors that promote glucose use. Although the exact molecular mechanism of action is unclear, it is known that phenformin promotes anaerobic glycolysis and exerts other effects. The action on anaerobic glycolysis is the effect that is responsible for excessive lactic acid formation (from pyruvic acid reduction) and the development of lactic acidosis.

Another effect of high blood glucose levels is the glycosylation of hemoglobin by small sugar units. Even though the process is not well understood, it has raised interesting questions concerning the diagnosis and treatment evaluation of diabetics. Diabetics appear to have elevated blood levels of glycosylated hemoglobin that return to accepted limits when the concentration of blood glucose is normalized.

The symptomatic significance of the elevated levels of glycohemoglobin is not clear, but studies have established that this type of hemoglobin has a lower affinity for oxygen.[5, 14]

Feedback regulation of enzyme-catalyzed reactions is another basic mechanism for the regulation of metabolism (i.e., allosteric inhibition of a key enzyme). Phosphofructokinase, the pacemaker enzyme of glycolysis, is inhibited allosterically by ATP, and through such modulation, ATP suppresses carbohydrate catabolism. Of course, there are other cases of feedback regulation. Atkinson's classic article[30] on phenomena associated with biologic feedback control at the molecular level should be consulted to compare negative and positive feedback regulation. In contrast to the effect of ATP, AMP can exert positive regulatory action on phosphofructokinase. The regulatory metabolite acting as modulator modifies the affinity of the enzyme for its substrates, and the terms positive and negative are used to indicate whether there is an increase or a decrease in affinity.

SUGAR ALCOHOLS

Sorbitol, glucitol, mannitol, galactitol, and dulcitol are natural products that are so closely related to the carbohydrates that it is traditional to classify them as carbohydrate derivatives (i.e., sugar alcohols). These compounds are reduction products of the corresponding aldohexoses—glucose, mannose, and galactose, respectively. Therefore, such sugar alcohols are characterized as hexahydroxy alcohols.

Sorbitol formation has recently been implicated as a factor contributing to complications of diabetes caused by high glucose concentration in nerve and eye lens cells. Cells that have excessive glucose tend to convert normally minor metabolic pathways to major processes (e.g., glucose reduction to form sorbitol). Sorbitol is not usually metabolized rapidly and cannot be effectively eliminated by the cell; because it accumulates in high concentrations in lens cells of diabetic rodents, osmotic swelling of the lens cells occurs. This osmotic swelling of lens cells has been associated with the development of cataracts as a complication of diabetes. In diabetic rodents, cataract formation can be prevented with agents that inhibit the aldose reductase enzyme that catalyzes glucose reduction to sorbitol.[31] Such aldose reductase inhibitors are, therefore, under investigation as potential medicinal agents.[32] In this connection it should be noted that some hydantoins have demonstrated aldose reductase inhibitory activity; for example, sorbinil is listed by the *USAN* (United States Adopted Names) *and USP Dictionary of Drug Names*, U.S. Pharmacopeia National Formulary, 1990.

Sorbitol, NF, is very water-soluble and produces sweet and viscous solutions. Hence, it is used in the formulation of some food products, cosmetics, and pharmaceuticals. Sorbitol Solution USP is a 70% w/w solution that contains at least 64% D-sorbitol, the balance being related sugar alcohols. Upon dehydration it forms tetrahydropyran and tetrahydrofuran derivatives, the fatty acid monoesters of which are the nonionic surface-active agents called Spans. Alternatively, these dehydration products react with ethylene oxide to form the Tweens, which are also useful surfactants.

Mannitol, USP, is a useful medicinal. It acts as an osmotic diuretic and is administered intravenously. After intravenous infusion (in the form of a sterile 25% solution), it is filtered by glomeruli and passes unchanged through the kidneys into the urine; however, while in the proximal tubules, the loops of Henle, the distal tubules, and the collecting ducts, mannitol increases the osmotic gradient against which these structures absorb water and solutes. Because of the foregoing osmotic effect, the urinary water, sodium, and chloride ions are increased. Mannitol is also indicated as an irrigating solution in transurethral prostatic resection.

Mannitol is also widely used as an excipient in chewable tablets. In contrast with sorbitol, it is nonhygroscopic. In addition, it has a sweet and cooling taste.

SUGARS

Dextrose, USP. D-(+)-Glucopyranose; grape sugar; D-glucose; glucose. Dextrose is a sugar usually obtained by the hydrolysis of starch. It can be either α-D-glucopyranose or β-D-glucopyranose or a mixture of the two. A large amount of the dextrose of commerce, whether crystalline or syrupy, usually is obtained by the acid hydrolysis of cornstarch, although other starches can be used.

Although some free glucose occurs in plants and animals, most of it occurs in starches, cellulose, glycogen, and sucrose. It also is found in other polysaccharides, oligosaccharides, and glycosides.

Dextrose occurs as colorless crystals or as a white crystalline or granular power. It is odorless and has a sweet taste. One gram of dextrose dissolves in about 1 mL of water and in about 100 mL of alcohol. It is more soluble in boiling water and in boiling alcohol.

Aqueous solutions of glucose can be sterilized by autoclaving.

Glucose can be used as a ready source of energy in various forms of starvation. It is the sugar found in the blood of animals and in the reserve polysaccharide glycogen that is present in the liver and muscle. It can be used in solution intravenously to supply fluid and to sustain the blood volume temporarily. It has been used in the management of the shock that may follow the administration of insulin used in the treatment of schizophrenia. This, because a "hypoglycemia" results from the use of insulin in this type of therapy, and the "hypoglycemic" state can be reversed by the use of dextrose intravenously. When dextrose is used intravenously, its solutions (5% to 50%) usually are made with physiologic salt solution or Ringer's solution. The dextrose used for intravenous injection must conform to the *USP* requirements for dextrose.

Liquid Glucose, NF, is a product obtained by the incomplete hydrolysis of starch. It consists chiefly of dextrose (D-glucose, $C_6H_{12}O_6$), with dextrins, maltose, and water. This glucose usually is prepared by the partial acid hydrolysis of cornstarch and, hence, the common name corn syrup and other trade names refer to a product similar to liquid glucose. The official product contains not more than 21% of water.

Liquid glucose is a colorless or yellowish, thick, syrupy liquid. It is odorless, or nearly so, and sweet. Liquid glucose is very soluble in water, but is sparingly soluble in alcohol.

Liquid glucose is used extensively as a food (sweetening agent) for both infants and adults. It is used in the massing of pills, in the preparation of pilular extracts, and for other similar uses. It is not to be used intravenously.

Calcium Gluconate, USP. The gluconic acid used in the preparation of calcium gluconate can be prepared by the electrolytic oxidation of glucose as follows:

$$\underset{\text{D-Glucose}}{\begin{array}{c} \text{H}\quad\text{O} \\ \diagdown\!\!\diagup\!\!\!\!/ \\ \text{C} \\ | \\ \text{HCOH} \\ | \\ \text{HOCH} \\ | \\ \text{HCOH} \\ | \\ \text{HCOH} \\ | \\ \text{CH}_2\text{OH} \end{array}} \xrightarrow[\text{Carbon Electrodes}]{\text{NaBr, CaCO}_3} \underset{\text{Calcium Gluconate}}{\left[\begin{array}{c} \text{COO}^- \\ | \\ \text{HCOH} \\ | \\ \text{HOCH} \\ | \\ \text{HCOH} \\ | \\ \text{HCOH} \\ | \\ \text{CH}_2\text{OH} \end{array}\right]_2 \text{Ca}^{2+}}$$

Gluconic acid is produced on a commercial scale by the action of a number of fungi, bacteria, and molds upon 25% to 40% solutions of glucose. The fermentation is best carried out in the presence of calcium carbonate and oxygen to give almost quantitative yields of gluconic acid. Several organisms can be used, for example, *Acetobacter oxydans*, *A. aceti*, *A. rancens*, *B. gluconicum*, *A. xylinum*, *A. roseus*, and *Penicillium chrysogenum*. The fermentation is complete in 8 to 18 days.

Calcium gluconate occurs as a white crystalline or granular powder, without odor or taste. It is stable in air. Its solutions are neutral to litmus paper. One gram of calcium gluconate dissolves slowly in about 30 mL of water and in 5 mL of boiling water. Each milliliter of 10% solution represents 9.3 mg of calcium. It is insoluble in alcohol and in many other organic solvents.

Calcium gluconate will be decomposed by the mineral acids and other acids that are stronger than gluconic acid. It is incompatible with soluble sulfates, carbonates, bicarbonates, citrates, tartrates, salicylates, and benzoates.

Calcium gluconate fills the need for a soluble, nontoxic well-tolerated form of calcium that can be employed orally, intramuscularly, or intravenously. Calcium therapy is indicated in conditions such as parathyroid deficiency (tetany), general calcium deficiency, and when calcium is the limiting factor in increased clotting time of the blood. It can be used both orally and intravenously.

Calcium Gluceptate, USP. Calcium glucoheptonate is a sterile, aqueous, approximately neutral solution of the calcium salt of glucoheptonic acid, a homologue of gluconic acid. Each milliliter of Calcium Gluceptate Injection USP represents 18 mg of calcium. Its uses and actions are the same as those of calcium gluconate.

Ferrous Gluconate, USP. Iron (2 +) gluconate (Fergon) occurs as a fine yellowish gray or pale greenish yellow powder with a slight odor like that of burnt sugar. One gram of this salt is soluble in 10 mL of water; however, it is nearly insoluble in alcohol. A 5% aqueous solution is acid to litmus.

Ferrous gluconate can be administered orally or by injection for the utilization of its iron content.

Glucuronic Acid occurs naturally as a component of many gums, mucilages, hemicelluloses, and in the mucopolysaccharide portion of a number of glycoproteins. It is used by animals and humans to detoxify such substances as camphor, menthol, phenol, salicylates, and chloral hydrate. None of the foregoing can be used to prepare glucuronic acid for commercial purposes. It is prepared by oxidizing the terminal primary alcohol group of glucose or a suitable derivative thereof, such as 1,2-isopropylidine-D-glucose. It is a white crystalline solid that is water-soluble and stable. It exhibits both aldehydic and acidic properties. It also may exist in a lactone form and, as such, is marketed under the name Glucurone, an abbreviation of glucuronolactone.

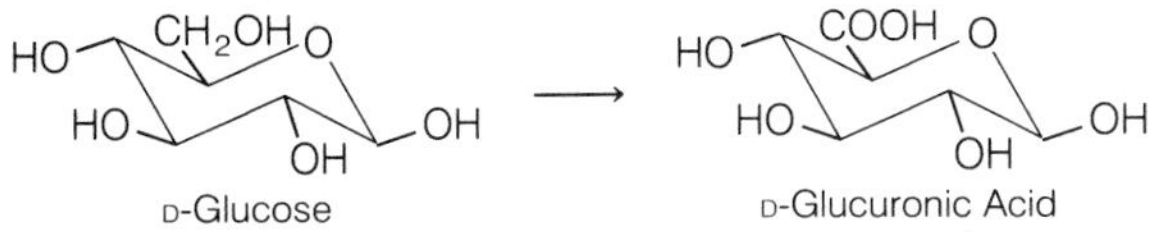

An average of 60% effectiveness was obtained in the relief of certain arthritic conditions by the use of glucuronic acid. A possible rationale for the effectiveness of glucuronic acid in the treatment of arthritic conditions is based upon the fact that it is an important component of cartilage, nerve sheath, joint capsule tendon, synovial fluid, and intercellular cement substances. The dose is 500 mg to 1.0 g orally four times a day or 3 to 5 mL of a 10% buffered solution given intramuscularly.

Fructose, USP. D-(−)-Fructose; levulose; β-D-(−)-fructopyranose is a sugar usually obtained by hydrolysis of aqueous solutions of sucrose and subsequent separation of fructose from glucose.* It occurs as colorless crystals or as a white or granular powder that is odorless and sweet. It is soluble 1 : 15 in alcohol and is freely soluble in water. Fructose is considerably more sensitive to heat and decomposition than is glucose and this is especially true in the presence of bases.

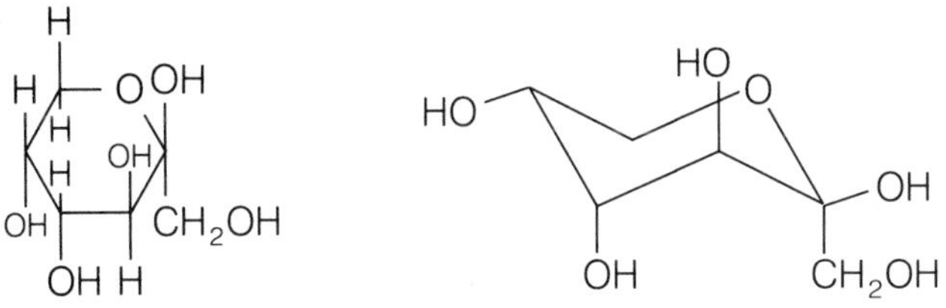

β-D-Fructopyranose

Fructose (a 2-ketohexose) can be utilized to a greater extent than glucose by diabetics and by patients who must be fed by the intravenous route.

Lactose, USP. Saccharum lactis; milk sugar. This is a sugar obtained from milk. Lactose is a by-product of whey, which is the portion of milk that is left after the fat and the casein have been removed for the production of butter and cheese. Cows' milk is 2.5% to 3% lactose, whereas that of other mammals contains 3% to 5%. Although common lactose is a mixture of the α- and β-forms, the pure β-form is sweeter than the slightly sweet-tasting mixture.

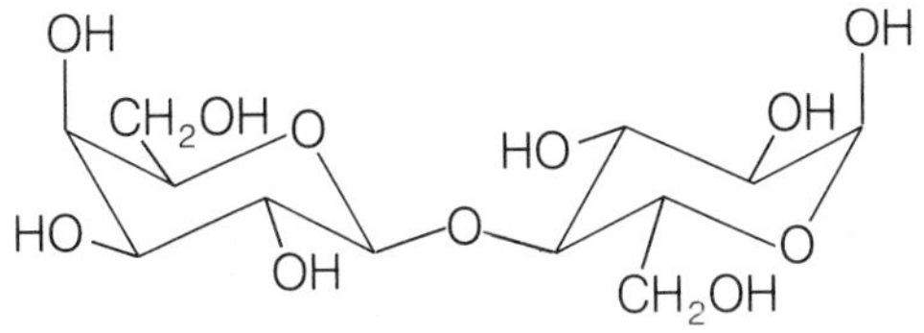

α-Lactose

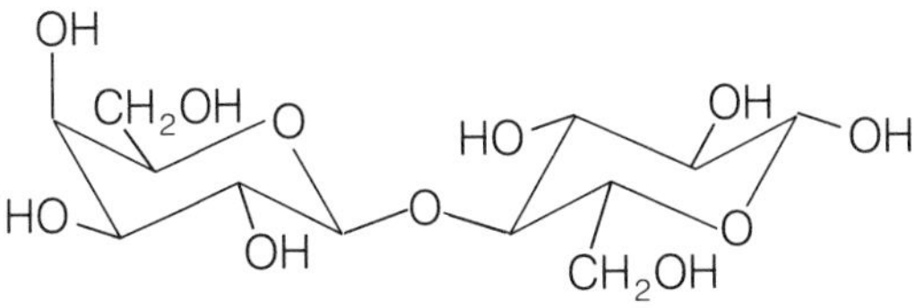

β-Lactose

Lactose occurs as white, hard, crystalline masses or as a white powder. It is odorless and faintly sweet. It is stable in air, but readily absorbs odors. Its solutions are neutral to litmus paper. One gram of lactose dissolves in 5 mL of water and in 2.6 mL of boiling water. Lactose is very slightly soluble in alcohol and is insoluble in chloroform and in ether.

Lactose is hydrolyzed readily in acid solutions to yield one molecule each of D-glucose and D-galactose. It reduces Fehling solution.

Lactose is used as a diluent in tablets and powders and as a nutrient for infants.

β-Lactose when applied locally to the vagina brings about a desirable lower *p*H. The lactose probably is fermented, with the production of lactic acid.

Maltose. Malt sugar; 4-D-glucopyranosyl-α-D-glucopyranoside is an end-product of the enzymatic hydrolysis of starch by the enzyme diastase. It is a reducing disaccharide that is fermentable and is hydrolyzed by acids or the enzyme maltose to yield two molecules of glucose.

Maltose is a constituent of malt extract and is used for its nutritional value for infants and adult invalids.

Malt Extract is a product obtained by extracting malt, the partially and artificially germinated grain of one or more varieties of *Hordeum vulgare* Linné (Gramineae). Malt extract contains maltose, dextrins, a small amount of glucose, and amylolytic enzymes.

Malt extract is used in the brewing industry because of its enzyme content which converts starches to fermentable sugars. It also is used in infant feeding for its nutritive value and laxative effect.

The usual dose is 15 g.

Dextrins are obtained by the enzymatic (diastase) degradation of starch. These degradation products vary in molecular weight in the following decreasing order: amylodextrin, erythodextrin, and achroodextrin. Lack of homogeneity precludes the

* The crystalline form of fructose is the β-anomer having a six-membered ring, but, when dissolved in water, it is converted not only to the α-form but the α- and β-forms of fructofuranose are formed also. The fructofuranose forms were called "gamma" sugars.

TABLE 19-1

SUGAR PRODUCTS

Name	Preparations	Category	Application	Usual Adult Dose*	Usual Dose Range*	Usual Pediatric Dose*
Dextrose USP	Dextrose injection USP	Fluid and nutrient replenisher		IV infusion, 1 L		
	Dextrose and sodium chloride injection USP	Fluid, nutrient, and electrolyte replenisher		IV infusion, 1 L		
	Anticoagulant citrate dextrose solution USP	Anticoagulant for storage of whole blood	For use in the proportion of 75 mL of solution A or 125 mL of solution B for each 500 mL of whole blood			
	Anticoagulant citrate phosphate dextrose solution USP	Anticoagulant for storage of whole blood	For use in the proportion of 70 mL of solution for each 500 mL of whole blood			
Calcium gluconate USP	Calcium gluconate injection USP	Calcium replenisher		IV, 10 mL of a 10% solution at a rate not exceeding 0.5 mL/min at intervals of 1–3 days	1 g/wk to 15 g/day	125 mg/kg of body weight or 3 g/m^2 of body surface, up to qid, diluted and given slowly
	Calcium gluconate tablets USP	Calcium replenisher		1 g 3 or more times daily	1–15 g/day	125 mg/kg of body weight or 3 g/m^2 of body surface, up to qid
Ferrous gluconate USP	Ferrous gluconate capsules USP Ferrous gluconate tablets USP	Iron supplement		300 mg tid	200–600 mg	
Fructose USP	Fructose injection USP	Fluid replenisher and nutrient		IV and SC, as required		
	Fructose and sodium chloride injection USP	Fluid replenisher, nutrient, and electrolyte replenisher		IV and SC, as required		
Lactose NF		Pharmaceutical aid (tablet and capsule diluent)				
Sucrose NF	Compressible sugar NF Confectioner's sugar NF	Pharmaceutical aid (sweetening agent; tablet excipient)				

*See USP DI for complete dosage information.

assignment of definite molecular weights. With the decrease in molecular weight, the color produced with iodine changes from blue to red to colorless.

Dextrin occurs as a white amorphous powder that is incompletely soluble in cold water, but freely soluble in hot water.

Dextrins are used extensively as a source of readily digestible carbohydrate for infants and adult invalids. They often are combined with maltose or other sugars.

Sucrose, NF. Saccharum; sugar; cane sugar; beet sugar. Sucrose is a sugar obtained from *Saccharum officinarum* Linné (Gramineae), *Beta vulgaris* Linné (Chenopodiaceae), and other sources. Sugar cane (15% to 20% sucrose) is expressed, and the juice is treated with lime to neutralize the plant acids. The water-soluble proteins are coagulated by heat and are removed by skimming. The resultant liquid is decolorized with charcoal and concentrated. Upon cooling, the sucrose crystallizes out. The mother liquor, upon concentration, yields more sucrose, brown sugar, and molasses.

Sucrose occurs as colorless or white crystalline masses or blocks, or as a white crystalline powder. It is odorless, sweet, and is stable in air. Its solutions are neutral to litmus. One gram of sucrose dissolves in 0.5 mL of water and in 170 mL of alcohol.

Sucrose does not respond to the tests for reducing sugars (i.e., reduction of Fehling solution and others). It is hydrolyzed readily, even in the cold, by acid solutions to give one molecule each of D-glucose and D-fructose. This hydrolysis also can be effected by the enzyme invertase. Sucrose caramelizes at about 210°C.

Sucrose is used in the preparation of syrups and as a diluent and sweetening agent in several pharmaceutic products (e.g., troches, lozenges, and powdered extracts). In a concentration of 800 mg/mL, sucrose is used as a sclerosing agent.

Invert Sugar (Travert) is a hydrolyzed product of sucrose (invert sugar) prepared for intravenous use.

Xylose, USP, is used as a diagnostic aid in testing for intestinal absorptive capacity in the diagnosis of celiac disease.

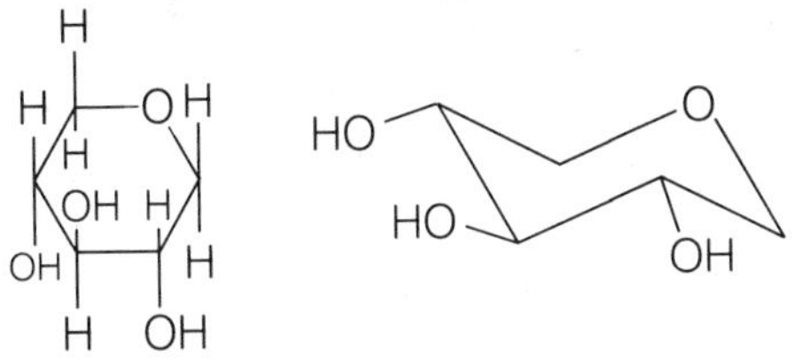

α-D-Xylose

STARCH AND DERIVATIVES

Starch, NF. Amylum; cornstarch consists of the granules separated from the grain of *Zea mays* Linné (Gramineae). Corn, which contains about 75% dry weight of starch, is first steeped with sulfurous acid and then milled to remove the germ and the seed coats. It then is milled with cold water, and the starch is collected and washed by screens and flotation. Starch is a high-molecular-weight carbohydrate composed of 10% to 20% of a hot water-soluble "amylose" and 80% to 90% of a hot water-insoluble "amylopectin." Amylose is hydrolyzed completely to maltose by the enzyme β-amylase, whereas amylopectin is hydrolyzed only incompletely (60%) to maltose. The glucose residues are in the form of branched chains in the amylopectin molecule. The chief linkages of the glucose units in starch are α-1,4, because β-amylase hydrolyzes only α-linkages and maltose is 4-D-glucopyranosyl-α-D-glucopyranoside.

Starch occurs as irregular, angular, white masses or as a fine powder, and consists chiefly of polygonal, rounded or spheroidal grains from 3 to 35 μm in diameter and usually with a circular or several-rayed central cleft. It is odorless and has a slight characteristic taste. Starch is insoluble in cold water and in alcohol.

Amylose gives a blue color on treatment with iodine, and amylopectin gives a violet to red-violet color.

Starch is used as an absorbent in starch pastes, as an emollient in the form of a glycerite, and in tablets and powders.

Pregelatinized Starch, NF. This is starch that has been modified to make it suitable for use as a tablet excipient. It has been processed in the presence of water to rupture most of the starch granules and then dried.

CELLULOSE AND DERIVATIVES

Cellulose is the name generally given to a group of very closely allied substances, rather than to a single entity. The celluloses are anhydrides of β-glucose, possibly existing as long chains that are not branched, consisting of 100 to 200 β-glucose residues. These chains may be cross-linked by residual valences (hydrogen bonds) to produce the supporting structures of the cell walls of plants. The cell walls found in cotton, pappi on certain fruits, and other sources are the purest forms of cellulose; however, because they are cell walls, they enclose varying amounts of substances that are proteinaceous, waxy, or fatty. These must be removed by proper treatment to obtain pure cellulose. Cellulose from almost all other sources is combined by ester linkages, glycoside linkages, and other combining forms with encrusting substances, such as lignin, hemicelluloses, pectins. These can be removed by steam under pressure, weak acid or alkali solutions, and sodium bisulfite and sulfurous acid. Plant celluloses, especially those found in wood, can be resolved into β-cellulose, which is soluble in 17.5% sodium hydroxide, and alkali-insoluble α-cellulose. The cellulose molecule can be depicted in part as shown in the diagram.

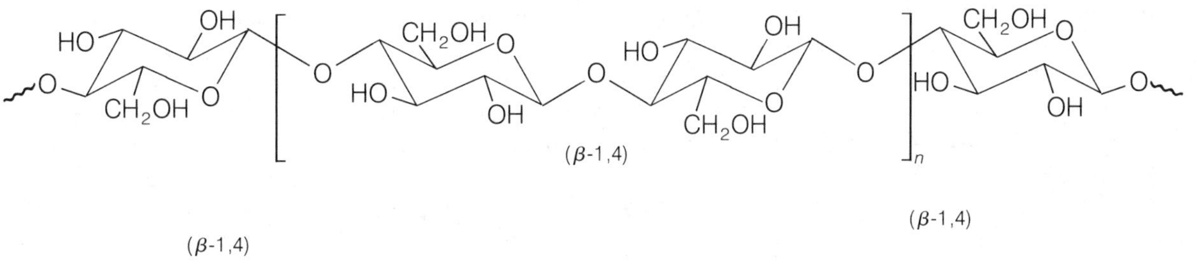

(β-1,4)

(β-1,4)

(β-1,4)

Cellulose

Purified Cotton, USP, is the hair of the seed of cultivated varieties of *Gossypium hirsutum* Linné, or of other species of Gossypium (Malvaceae), freed from adhering impurities, deprived of fatty matter, bleached, and sterilized.

Microcrystalline Cellulose, NF, is purified, partially depolymerized cellulose prepared by treating α-cellulose, obtained as a pulp from fibrous plant material, with mineral acids.

It occurs as a fine, white, odorless crystalline powder that is insoluble in water, in dilute alkalies, and in most organic solvents.

Methylcellulose, USP, (Syncelose, Cellothyl, Methocel) is a methyl ether of cellulose, the methoxyl content of which varies between 26% and 33%. A 2% solution has a centipoise (cp) range of not less than 80% and not more than 120% of the labeled amount when such is 100 cp or less, and not less than 75% nor more than 140% of the labeled amount for viscosity types higher than 100 cp.

Methyl- and ethylcellulose ethers (Ethocel) can be prepared by the action of methyl and ethyl chlorides or methyl and ethyl sulfates, respectively, on cellulose that has been previously treated with alkali. Purification is accomplished by washing the reaction product with hot water. The degree of methylation or ethylation can be controlled to yield products that vary in their viscosities when they are in solution. Seven viscosity types of methylcellulose are produced commercially and have the following centipoise values: 10, 15, 25, 100, 400, 1500, and 4000, respectively. Other intermediate viscosities can be obtained by the use of a blending chart. The ethylcelluloses have similar properties.

Methylated celluloses of a lower methoxy content are soluble in cold water, but, in contrast with the naturally occurring gums, they are insoluble in hot water and are precipitated out of solution at or near the boiling point. Solutions of powdered methylcellulose can be prepared most readily by first mixing the powder thoroughly with one-fifth to one-third of the required water as hot water (80° to 90°C) and allowing it to macerate for 20 to 30 minutes. The remaining water is then added as cold water. With the increase in methoxy content, the solubility in water decreases until complete water insolubility is reached.

Methylcellulose resembles cotton in appearance and is neutral, odorless, tasteless, and inert. It swells in water and produces a clear to opalescent, viscous, colloidal solution. Methylcellulose is insoluble in most of the common organic solvents. On the other hand, aqueous solutions of methylcellulose can be diluted with ethanol.

Methylcellulose solutions are stable over a wide range of *p*H (2 to 12) with no apparent change in viscosity. The solutions do not ferment and will carry large quantities of univalent ions, such as iodides, bromides, chlorides, and thiocyanates. However, smaller amounts of polyvalent ions, such as sulfates, phosphates, carbonates, and tannic acid or sodium formaldehyde sulfoxylate, will cause precipitation or coagulation.

The methylcelluloses are used as substitutes for the natural gums and mucilages, such as gum tragacanth, gum karaya, chrondrus or quince seed mucilage. They can be used as bulk laxatives and in nose drops, ophthalmic preparations, burn preparations, ointments, and like preparations. Although methylcellulose when used as a bulk laxative takes up water quite uniformly, tablets of methylcellulose have caused fecal impaction and intestinal obstruction. Commercial products include Hydrolose Syrup; Anatex; Cologel Liquid; Premocel Tablets; and Valocall. In general, methylcellulose of the 1500 or 4000 cp viscosity type is the most useful as a thickening agent when used in 2% to 4% concentrations. For example, a 2.5% concentration of a 4000 cp-type methylcellulose will produce a solution with a viscosity obtained by 1.25% to 1.75% of tragacanth.

Ethylcellulose, NF, is an ethyl ether of cellulose containing not less than 45% and not more than 50% of ethoxy groups and is prepared from ethyl chloride and cellulose. It occurs as a free-flowing, stable white powder that is insoluble in water, glycerin, and propylene glycol, but is freely soluble in alcohol, ethyl acetate, or chloroform. Aqueous suspensions are neutral to litmus. Films prepared from organic solvents are stable, clear, continuous, flammable, and tough.

Hydroxypropyl Methylcellulose 2208, USP. Propylene glycol ether of methyl cellulose contains a degree of substitution of no less than 19% nor more than 24% as methoxyl groups (OCH_3), and no less than 4% nor more than 12% as hydroxypropyl groups (OC_3H_6OH). It occurs as a white fibrous or granular powder that swells in water to produce a clear to opalescent, viscous, colloidal solution.

Oxidized Cellulose, USP, (Oxycel) when thoroughly dry contains no less than 16% nor more than 24% of carboxyl groups. Oxidized cellulose is cellulose in which some of the terminal primary alcohol groups of the glucose residues have been converted to carboxyl groups. Therefore, the product is possibly a synthetic polyanhydrocellobiuronide. Although the *USP* accepts carboxyl contents as high as 24%, it is reported that products that contain 25% carboxyl groups are too brittle (friable) and too readily soluble to be of use. Those products that have lower carboxyl contents are the most desirable. Oxidized cellulose is slightly off-white, is acid to the taste, and possesses a slight, charred odor. It is prepared by the action of nitrogen dioxide, or a mixture of nitrogen dioxide and nitrogen tetroxide, upon cellulose fabrics at ordinary temperatures. Because cellulose is a high-molecular-weight carbohydrate composed of glucose residues joined 1,4- to each other in their β-forms, the reaction must be as diagrammed on the cellulose molecule in part.

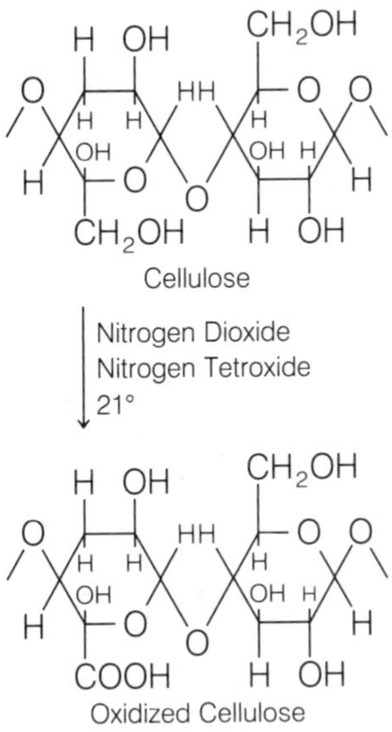

Cellulose

Nitrogen Dioxide
Nitrogen Tetroxide
21°

Oxidized Cellulose

The oxidized cellulose fabric, such as gauze or cotton, resembles the parent substance. It is insoluble in water and in acids, but is soluble in dilute alkalies. In weakly alkaline solutions, it swells and becomes translucent and gelatinous. When wet with blood, it becomes slightly sticky and swells, forming a dark brown gelatinous mass. Oxidized cellulose cannot be sterilized by autoclaving. Special methods are needed to render it sterile.

Oxidized cellulose has noteworthy hemostatic properties. However, when it is used in conjunction with thrombin, it should be neutralized previously with a solution of sodium bicarbonate. It is used in various surgical procedures in much the same way as gauze or cotton, by direct application to the oozing surface. Except when used for hemostasis, it is not recommended as a surface dressing for open wounds. Oxidized cellulose implants in connective tissue, muscle, bone, serous and synovial cavities, brain, thyroid, liver, kidney, and spleen were absorbed completely in varying lengths of time, depending on the amount of material introduced, the extent of operative trauma, and the amount of blood present.

Carboxymethylcellulose Sodium, USP. CMC, sodium cellulose glycolate, is the sodium salt of a polycarboxymethyl ether of cellulose, containing, when dried, 6.5 to 9.5% of sodium. It is prepared by treating alkali cellulose with sodium chloroacetate. This procedure permits a control of the number of $-OCH_2COO^- \, Na^+$ groups that are to be introduced. The number of $-OCH_2COO^- \, Na^+$ groups introduced is related to the viscosity of aqueous solutions of these products. CMC is available in various viscosities, i.e., 5 to 2000 centipoises in 1% solutions. Therefore, high-molecular-weight polysaccharides containing carboxyl groups have been prepared whose properties in part resemble those of the naturally occurring polysaccharides, whose carboxyl groups contribute to their pharmaceutic and medicinal usefulness.

Carboxymethylcellulose sodium occurs as a hygroscopic white powder or granules. Aqueous solutions may have a pH between 6.5 and 8. It is easily dispersed in cold or hot water to form colloidal solutions that are stable to metal salts and pH conditions from 2 to 10. It is insoluble in alcohol and organic solvents.

It can be used as an antacid, but is more adaptable for use as a nontoxic, nondigestible, unabsorbable, hydrophilic gel as an emollient-type bulk laxative. Its bulk-forming properties are not as great as those of methylcellulose; on the other hand, its lubricating properties are superior, with little tendency to produce intestinal blockage.

Pyroxylin, USP. Soluble guncotton is a product obtained by the action of nitric and sulfuric acids on cotton and consists chiefly of cellulose tetranitrate $[C_{12}H_{16}O_6(NO_3)_4]$. The glucose residues in the cellulose molecule contain three free hydroxyl groups that can be esterified. Two of these three hydroxyl groups are esterified to give the official pyroxylin, and, therefore, it is really a dinitrocellulose or cellulose dinitrate which conforms to the official nitrate content.

Pyroxylin occurs as a light yellow matted mass of filaments, resembling raw cotton in appearance, but harsh to the touch. It is exceedingly flammable and decomposes when exposed to light, with the evolution of nitrous vapors and a carbonaceous residue. Pyroxylin dissolves slowly but completely in 25 parts of a mixture of 3 volumes of ether and 1 volume of alcohol.

In the form of collodion and flexible collodion, it is used for coating purposes per se or in conjunction with certain medicinal agents.

Cellulose Acetate Phthalate, NF, is a partial acetate ester of cellulose that has been reacted with phthalic anhydride. One carboxyl of the phthalic acid is esterified with the cellulose acetate. The finished product contains about 20% acetyl groups and about 35% phthalyl groups. In the acid form, it is soluble in organic solvents and insoluble in water. The salt form is readily soluble in water. This combination of properties makes it useful in enteric coating of tablets because it is resistant to the acid

TABLE 19-2

PHARMACEUTICALLY IMPORTANT CELLULOSE PRODUCTS

Name Proprietary Name	***Preparations***	***Category***	***Usual Adult Dose****	***Usual Dose Range****
Purified cotton USP		Surgical aid		
Purified rayon USP		Surgical aid		
Microcrystalline cellulose NF		Pharmaceutical aid (tablet diluent)		
Powdered cellulose NF		Pharmaceutical aid (tablet diluent; adsorbant; suspending agent)		
Methylcellulose USP		Pharmaceutical aid (suspending agent; tablet excipient; viscosity-increasing agent)		
	Methylcellulose ophthalmic solution USP	Topical protectant (ophthalmic)		
	Methylcellulose tablets USP	Cathartic	1–1.5 g bid to qid	1–6 g/day
Ethylcellulose NF		Pharmaceutical aid (tablet binder)		
Hydroxypropyl methylcellulose USP		Pharmaceutical aid (suspending agent; tablet excipient; viscosity-increasing agent)	Topically to the conjunctiva, 0.05–0.1 mL of a 0.5–2.5% solution tid or qid, or as needed, as artificial tears or contact lens solution	
	Hydroxypropyl methylcellulose ophthalmic solution USP	Topical protectant (ophthalmic)		
Oxidized cellulose USP *Oxycel*		Local hemostatic	Topically as necessary to control hemorrhage	
Carboxymethylcellulose sodium USP		Pharmaceutical aid (suspending agent; tablet excipient; viscosity-increasing agent)		
	Carboxymethylcellulose sodium tablets USP	Cathartic	1.5 g tid	
Pyroxylin USP		Pharmaceutical necessity for Collodion USP		
Cellulose acetate phthalate NF		Pharmaceutical aid (tablet-coating agent)		

*See USP DI for complete dosage information.

condition of the stomach, but is soluble in the more alkaline environment of the intestinal tract.

HEPARIN

Heparin is a mucopolysaccharide composed of α-D-glucuronic acid, α-L-iduronic acid, α-D-glucosamine, and *N*-acetyl-α-D-glucosamine; these monosaccharide units are partially sulfated and are linked in the polymeric form through 1 → 4 linkages, as indicated by the structure shown in Figure 19-1. Heparin is present in animal tissue of practically all types, but mainly in lung and liver tissue.[33]

The chemistry and pharmacology of heparin have been reviewed by Ehrlich and Stivala.[34] This review comprehensively covers most topics pertinent to medicinal chemistry. Heparin is included among the *AMA Drug Evaluations* (1983) and *USPI* (1988) anticoagulants.[35,36] Its greatest use has been in the prevention and arrest of thrombosis (see Chap. 20 on the biochemical functions performed by thrombin, fibrinogen, and fibrin in normal blood coagulation).

The mechanism of anticoagulant action exerted by heparin has been investigated from various standpoints, and now it is recognized that the mechanism involves the plasma protein inhibitor of serine proteases, which is called *antithrombin III*. This naturally occurring inhibitor inactivates various critical clotting factors, which are enzymes designated as IXa, Xa, XIa, thrombin, and perhaps also XIIa, that have a serine residue within the reactive center. Antithrombin III interacts with and inhibits these factors irreversibly. Heparin interacts with antithrombin III and induces conformational changes that complement the interaction between antithrombin III and the aforementioned factors.[37]

Jaques[38] summarized studies that have shown that heparin is a biochemical representative of a class of compounds characterized as linear anionic polyelectrolytes. Such compounds demonstrate interesting specific reactions with biologically active proteins, forming stable complexes that change the bioactivity of these proteins. These complexations increase the negative charge of cell surfaces, including those of the blood vessel walls. The increase in the negative charge of the vessel wall is considered to be a factor that contributes to the prevention of thrombosis by heparin and similar compounds.

Heparin also affects fibrinolysis. It seems to reduce the inhibition of antifibrinolysin and, thereby, enhances fibrinolysis. Heparin's effects on platelets have been studied, and heparin was shown to prevent conversion of degenerated platelets in solution from forming a gel; heparin inhibits platelet adhesion to intercellular cement; it also prevents platelet disintegration and release of phospholipids. Another major effect of heparin is on blood lipids. Heparin stimulates the release of lipoprotein lipase, an enzyme that catalyzes the hydrolysis of triglycerides associated with chylomicrons and, through this action, promotes the clearing of lipemic plasma. Research on heparin has included the investigation of the possible effect of heparin on tumor growth and metastasis. Some studies show that heparin is a miotic inhibitor in Ehrlich's ascites tumor. Other investigations have produced negative data, and, hence, the question remains unanswered.[34]

Protamine has been characterized as a heparin antagonist, but it has the characteristic of prolonging clotting time on its own. Protamine (discussed also in Chap. 20) is basic enough to interact with heparin (which is acidic owing to its *O*-sulfate and *N*-sulfate groups). When protamine and heparin interact, they neutralize the action of each other.

It appears that the reticuloendothelial system may be involved in the disposition of heparin; that is, heparin may leave the plasma by uptake into the reticuloendothelial system. Recent data from kinetic studies of heparin removal from circulation of the minipig are consistent with this suggestion.[39]

Heparin is metabolized primarily in the liver by partial cleavage of the sulfate groups to form uroheparin and is excreted by the kidneys, primarily as a partially sulfated product. Up to 50% may be excreted unchanged when high doses are given. The partially desulfated product excreted in the urine has been shown to be one-half as active as heparin in anticoagulant properties.

Heparan sulfate is the polysaccharide found as a by-product in the preparation of heparin from lung and liver tissue. Heparan sulfate has a lower sulfate content, and its glucosamine residues are partially acetylated and *N*-sulfated. Heparan sulfate isolated from the aorta has negligible antithrombin activity.

Heparin Calcium, USP, and **Heparin Sodium, USP.** Heparin may be prepared commercially from lung and liver, employing the procedure of Kuizenga and Spaulding[30] combined with suitable methods for purifying the isolated heparin. More recently, the calcium salt has been prepared from porcine intestinal mucosa and the sodium salt from either porcine intestinal mucosa or bovine lung tissue. The sodium and calcium salts are white, amorphous, hygroscopic powders that are soluble (1 : 20) in water, but poorly soluble in alcohol. A 1% aqueous solution has an adjusted *p*H of 5 to 7.5. It is relatively stable to heat, and solutions may be sterilized by atuoclaving, but they should never be frozen.

For full-dose therapy, heparin is administered intravenously in two ways: (1) the intermittent injection method and (2) the continuous infusion method or by deep subcutaneous (intrafat) injection. The fixed-combination preparation containing dihydroergotamine mesylate and heparin sodium is administered *only* by deep subcutaneous injection. More importantly, heparin should not be administered intramuscularly because of the frequency of irritation, pain, and hematoma at the injection site. The continuous infusion method is to be preferred because it provides a more constant anticoagulating activity and lower incidence of bleeding complications. A constant-rate infusion pump is also recommended.[40]

The therapeutic use of subcutaneous heparin in low doses was extensively investigated, and some reports favorably evaluated this mode of administration. Therefore, a fixed low-dose therapy is also utilized that involves the administration of heparin calcium or heparin sodium by deep subcutaneous injection.[41–43]

A common side effect with heparin can be hemorrhage, but this can be minimized with the low-dose regimen.[44]

Category: anticoagulant

Usual full-dose: parenteral, the following amounts, as indicated by prothrombin-time determinations: IV, 10,000 USP heparin units initially, then 5000 to 10,000 U every four to six hours; infusion: 20,000–40,000 U/L at a rate of 1000 U/hour over a 24-hour period; subcutaneous: 10,000 to 20,000 U initially, then 8000 to 10,000 U every eight hours or 15,000 to 20,000 U every 12 hours.

Usual pediatric dose: IV injection: 50 U/kg of body weight initially, then 50–100 U/kg of body weight every four hours; infusion: 50 U/kg of body weight initially, followed by 100 U/kg, added and absorbed every four hours.

Occurrence: Heparin sodium injection USP; Heparin calcium injection USP

GLYCOSIDES

Because several plant constituents yielded glucose and an organic hydroxide upon hydrolysis, the term *glucoside* was introduced as a generic term for these substances. The fact that many plant constituents yielded sugars other than glucose led to the suggestion of the less specific general term *glycoside*. When the nature of the sugar residue is known, more specific terms can be used when desired, such as glucoside, fructoside, rhamnoside, and others, respectively. The nonsugar portion of the glycoside generally is referred to as the *aglycon* or *genin*.

Two general types of glycosides are known: namely, the nitrogen glycosides and the conventional type glycoside. The conventional type glycoside has an acetal structure and can be illustrated by the simplest type in which methyl alcohol is the aglycon or organic hydroxide. Two forms of this, as well as of all other glycosides, are possible: namely, α and β, because of the asymmetry centering about C-1 of the sugar residue that contains the acetal structure. It is thought that all naturally occurring glycosides are of the β-variety, because the enzyme emulsin, which cannot hydrolyze synthetic α-glycosides, hydrolyzes naturally occurring glycosides. Some of the β-glycosides also are hydrolyzed by amygdalase, cellobiase, gentiobiase, and the phenol glycosidases. The α-glycosides are hydrolyzed by maltase, mannosidase, and trehalase.

Glycosides usually are hydrolyzed by acids and are relatively stable toward alkalies. Some glycosides are much more resistant to hydrolysis than others. For example, those glycosides that contain a 2-desoxy sugar (see cardiac glycosides) are easily cleaved by weak acids, even at room temperature. On the other hand, most of the glycosides containing the normal-type sugars are quite resistant to hydrolysis, and of these, some may require rather drastic hydrolytic measures. The drastic treatment required for the hydrolysis of some glycosides causes chemical changes to take place in the aglycon portion of the molecule; these changes present problems in the elucidation of their structures. Conversely, those glycosides that are very easily hydrolyzed present problems in isolation and storage. Examples of the latter are the cardiac glycosides.

Although most glycosides are stable to hydrolysis by bases, the structure of the aglycon may determine its base sensitivity (e.g., picrocrocin has a half-life of three hours in 0.007 N KOH at 30°).

The sugar component of glycosides may be a mono-, di-, tri-, or tetrasaccharide. There is a wide variety of sugars found in the naturally occurring glycosides. Most of the unusual and rare sugars found in nature are components of glycosides.

The aglycons or nonsugar portions of glycosides are represented by a wide variety of organic compounds, as illustrated by the cardiac glycosides, the saponins, and others (see Chap. 18).

Because of the complexity of the structures of the naturally occurring glycosides, no generalizations are possible about their stabilities if the stabilities of the glycosidic linkages are excluded. It also follows that considerable deviations are met within their solubility properties. Many glycosides are soluble in water or hydroalcoholic solutions because the solubility properties of the sugar residues exert a considerable effect. Some glycosides, such as the cardiac glycosides, are slightly soluble or insoluble in water. For these, the steroid aglycon is markedly insoluble in water and offsets the solubility properties of the sugar residues. Most glycosides are insoluble in ether. Some glycosides are soluble in ethyl acetate, chloroform, or acetone.

Glycosides occur widely distributed in nature. They are found in varying amounts in seeds, fruits, roots, bark, and leaves. Occasionally, two or more glycosides are found in the same plant (e.g., cardiac glycosides and saponins). Glycosides often are accompanied by enzymes that are capable of synthesizing or hydrolyzing them. This phenomenon introduces problems in the isolation of glycosides because the disintegration of plant tissues, with no precautions to inhibit enzymatic activity, may lead to partial or complete hydrolysis of the glycosides.

Most glycosides are bitter, although there are many that are not. Glycosides per se or their hydrolytic products furnish a number of drugs, some of which are very valuable. Some plants that contain the cyanogenetic-type glycoside present an agricultural problem. Cattle have been poisoned by eating plants that are rich in the cyanogenetic-type glycosides.

REFERENCES

1. Harper, H. A.: Review of Physiological Chemistry, 17th ed. Los Altos, CA, Lange Medical, 1980.
2. Montgomery, R., et al.: Biochemistry, A Case-Oriented Approach. St. Louis, C. V. Mosby, 1983.
3. White, A., et al.: Principles of Biochemistry, 6th ed., pp. 423–567. New York, McGraw-Hill, 1978.
4. Lehninger, A. L.: Principles of Biochemistry, pp. 277–301. New York, Worth Publishing, 1982.
5. Bohinski, R. C.: Modern Concepts in Biochemistry, 5th ed., pp. 382–416. Newton, MA, Allyn & Bacon, 1987.
6. Stryer, L.: Biochemistry, 3rd ed., pp. 331–348, New York, W. H. Freeman, 1988.
7. Sensi, P., and Gialdroni-Grassi, G.: Antimycobacterial agents. In Wolff, M. E., (ed.). Burger's Medicinal Chemistry, 4th ed., pp. 311–321. New York, John Wiley & Sons, 1979.
8. Thomas, R., et al.: J. Pharm. Sci. 63:1649, 1974.
9. Bridges, A. J., et al.: Annu. Rep. Med. Chem. 23:39, 1988.

10. Miller, P. S., et al.: Annu. Rep. Med. Chem. 23:295, 1988.
11. Kirkwood, S.: Annu. Rev. Biochem. 43:401, 1974.
12. Steer, C. J., and Ashwell, G.: Prog. Liver Dis. 8:99, 1986.
13. Feizi, T., and Childs, R. A.: Trends Biochem. Sci. 10:24, 1985.
14. Bunn, H. F., et al.: Science 200:21, 1978.
15. Spiro, R. G.: Annu. Rev. Biochem. 39:599, 1970.
16. Ivatt, R. J. (ed.): The Biology of Glycoproteins. New York, Plenum Press, 1984.
17. Kornfeld, R., and Kornfeld, S.: Annu. Rev. Biochem. 54:631, 1985.
18. Bentley, R.: Annu. Rev. Biochem. 41:953, 1972.
19. Stoddard, J. F.: Stereochemistry of Carbohydrates. New York, Wiley-Interscience, 1971.
20. Sweeley, C. C., and Nunez, H.: Annu. Rev. Biochem. 54:765, 1985.
21. Barker, R., and Serianni, A. S.: Acc. Chem. Res. 19:307, 1986.
22. Newton, N. E., and Hornbrook, K.: J. Pharmacol. Exp. Ther. 181:479, 1972.
23. Kobilka, B. K., et al.: Science 240:1310, 1988.
24. Levitzki, A.: Science 241:800, 1988.
25. Ross, E. M.: J. Chem. Educ. 65:937, 1988.
26. Piles, S. J., and Parks, C. R.: Annu. Rev. Pharmacol. 14:365, 1974.
27. White, A., et al.: Principles of Biochemistry, 6th ed., pp. 1265–1279. New York, McGraw-Hill, 1978.
28. Amer, M. S., and Kreighbaum, W. E.: J. Pharm. Sci. 64:1, 1975.
29. Dept. of Health, Education and Welfare, Food and Drug Administration: FDA Drug Bull. 7:3, 1977; Science, 203:1094, 1979.
30. Atkinson, D. E.: Science 150:1, 1965.
31. Kolata, G. B.: Science 203:1098, 1979.
32. Blank, B.: In Wolff, M. E. (ed.). Burger's Medicinal Chemistry, 4th ed., Chap. 31. New York, John Wiley & Sons, 1979.
33. Kuizenga, M. H., and Spaulding, L. G.: J. Biol. Chem. 148:641, 1943.
34. Ehrlich, J., and Stivala, S. S.: J. Pharm. Sci. 52:517, 1973.
35. American Medical Association Department of Drugs: Drug Evaluations, 5th ed., p. 816. New York, John Wiley & Sons, 1983.
36. USP Convention, Inc.: USP DI, 8th ed., p. 1153. Rockville, MD, 1988.
37. von Kaulla, K. N.: In Wolff, M. E. (ed.). Burger's Medicinal Chemistry, 4th ed., Chap. 32, p. 1081. New York, John Wiley & Sons, 1979.
38. Jaques, L. B.: Science 206:528, 1979.
39. Harris, P. A., and Harris, K. L.: J. Pharm. Sci. 63:138, 1974.
40. ASHP: Drug Information, p. 734. AHFS, 1988.
41. Flemming, J. S., and MacNintch, J. E.: Annu. Rep. Med. Chem. 9:75, 1974.
42. Gallus, A. S., et al.: N. Engl. J. Med. 288:545, 1973.
43. Skillman, J. J.: Surgery 75:114, 1974.
44. Herrman, R. G., and Lacefield, W. B.: Annu. Rep. Med. Chem. 8:73, 1973.

CHAPTER 20

Amino Acids, Proteins, Enzymes, and Peptide Hormones

Jaime N. Delgado
Vilas A. Prabhu

Proteins are essential components of all living matter. As cellular components, proteins perform numerous functions. The chemical reactions fundamental to the life of the cell are catalyzed by proteins called enzymes. Other proteins are structural constituents of protoplasm and cell membranes. Some hormones are characterized as proteins or proteinlike compounds because of their polypeptide structural features.

Protein chemistry is essential not only to the study of molecular biology in understanding how cellular components participate in the physiologic processes of organisms, but also to medicinal chemistry. An understanding of the nature of proteins is necessary for the study of those medicinal agents that are proteins or proteinlike compounds and their physicochemical and biochemical properties relating to mechanisms of action. Also, in medicinal chemistry, drug–receptor interactions are implicated in the rationalization of structure–activity relationships and in the science of rational drug design. Drug receptors are considered to be macromolecules, some of which seem to be proteins or proteinlike.

Recombinant DNA technology[1] has had a dramatic impact on our ability to produce complex proteins and polypeptides, structurally identical with those found endogenously. Many of these endogenous proteins or polypeptides have exhibited neurotransmitter and hormonal properties that regulate a variety of physiologic processes. In the future, recombinant DNA technology should make complex, biologically active molecules routinely available for medicinal use.

This chapter reviews the medicinal chemistry of proteins and also includes some discussion of those amino acids that are products of protein hydrolysis. Some amino acids (e.g., dopa) are useful therapeutic agents, and their mode of action relates to amino acid metabolism. Some medicinals are amino acid antagonists, and their biochemical effects relate to their therapeutic uses; hence, brief mention of some representative cases of amino acid antagonism will be made in appropriate context. Moreover, the hormones with proteinlike structure are also discussed, with emphasis on their biochemical effects.

A study of medicinal chemistry cannot be made without including some enzymology, not only because many drugs affect enzyme systems, and vice versa, but also because fundamental lessons of enzymology have been applied to the study of drug–receptor interactions. Accordingly, this chapter includes a section on enzymes.

AMINO ACIDS

Proteins are biosynthesized from α-amino acids, and when proteins are hydrolyzed, amino acids are obtained. Some very complex (conjugated) proteins yield other hydrolysis products in addition to amino acids. α-Amino acids are commonly characterized with the generalized structure*:

$$R—C(H)(NH_2)—COOH$$

The most important amino acids are described in Table 20-1. Although the foregoing structure for

* All α-amino acids, except glycine, are optically active because the R for the generalized structure represents some moiety other than hydrogen: the amino acids of proteins have the same absolute configuration as L-alanine, which is related to L-glyceraldehyde. (The D and L designations refer to configuration rather than to optical rotation.)

TABLE 20-1

NATURALLY OCCURRING AMINO ACIDS

Name	Symbol	Formula
Glycine	Gly	H_2NCH_2COOH
Alanine	Ala	$CH_3CH(NH_2)COOH$
Valine	Val	$(CH_3)_2CHCH(NH_2)COOH$
Leucine	Leu	$(CH_3)_2CHCH_2CH(NH_2)COOH$
Isoleucine	Ile	$CH_3CH_2CH(CH_3)CH(NH_2)COOH$
Serine	Ser	$HOCH_2CH(NH_2)COOH$
Threonine	Thr	$CH_3CH(OH)CH(NH_2)COOH$
Cysteine	Cys	$HSCH_2CH(NH_2)COOH$
Cystine	Cys	$(—SCH_2CH(NH_2)COOH)_2$
Methionine	Met	$CH_3SCH_2CH_2CH(NH_2)COOH$
Proline	Pro	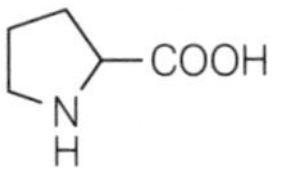 —COOH, N H
Hydroxyproline	Hyp	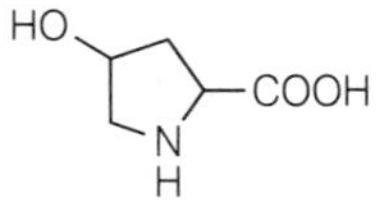 HO, —COOH, N H
Phenylalanine	Phe	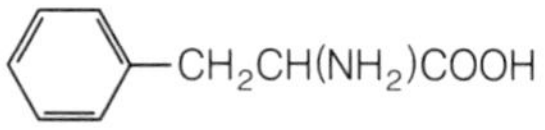 $—CH_2CH(NH_2)COOH$
Tyrosine	Tyr	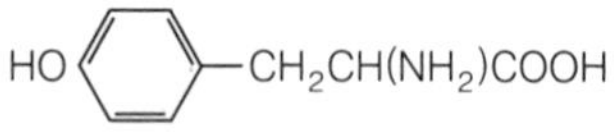 HO, $—CH_2CH(NH_2)COOH$
Tryptophan	Trp	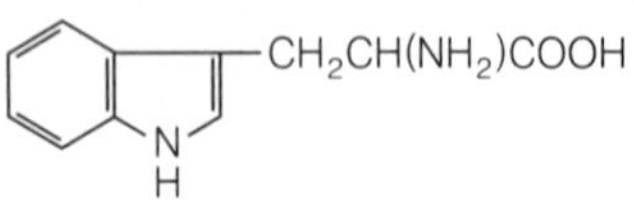 $—CH_2CH(NH_2)COOH$, N H
Aspartic acid	Asp	$HOOCCH_2CH(NH_2)COOH$
Glutamic acid	Glu	$HOOCCH_2CH_2CH(NH_2)COOH$
Lysine	Lys	$H_2NCH_2CH_2CH_2CH_2CH(NH_2)COOH$
Arginine	Arg	$H_2NC(=NH)NH_2CH_2CH_2CH_2$ $CH(NH_2)COOH$
Histidine	His	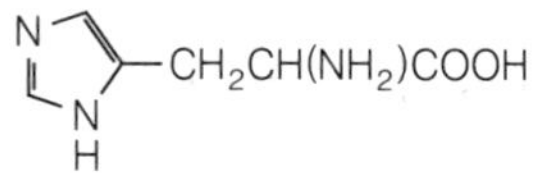 N, $—CH_2CH(NH_2)COOH$, N H

amino acids is widely used, physical, chemical, and some biochemical properties of these compounds are more consistent with a dipolar ion structure.

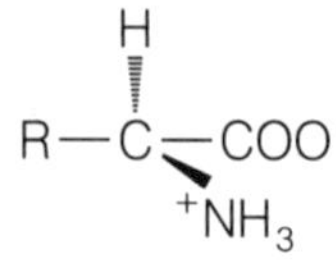

$$R—C(H)(^+NH_3)—COO^-$$

The relatively high melting point, solubility behavior, and acid–base properties characteristic of amino acids can be accounted for on the basis of the dipolar ion structure (commonly called *zwitterion*). Amino acids in the dry solid state are dipolar ions (inner salts).

Amino acids when dissolved in water can exist as dipolar ions and, in this form, would make no contribution to migration in an electric field. The concentration of the dipolar ion will vary depending on the pK_as of the amino acids and the hydronium ion concentration of the aqueous solution according to the following equilibrium:

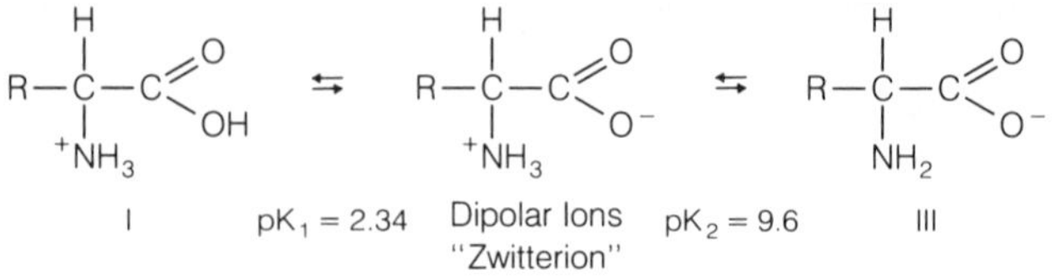

$$\underset{\text{I}}{R—C(H)(^+NH_3)—COOH} \underset{pK_1 = 2.34}{\rightleftharpoons} \underset{\text{Dipolar Ions "Zwitterion"}}{R—C(H)(^+NH_3)—COO^-} \underset{pK_2 = 9.6}{\rightleftharpoons} \underset{\text{III}}{R—C(H)(NH_2)—COO^-}$$

The hydronium ion concentration of the solution can be adjusted, and, if expressed in terms of pH, the pH at which the concentration of the dipolar form is maximal has been called the isoelectric point for the amino acid. (Because proteins are polymers of amino acids, they also have zwitterion character and isoelectric points.)

Glycine has pK_{a_1} = 2.34 for the carboxyl group and pK_{a_2} = 9.6 for the protonated amino group. The R groups of other amino acids change the pK_as slightly. The positive charge of I tends to repel a proton from the carboxyl group so that I is more strongly acidic than acetic acid (pK_a = 4.76). The pK_{a_2} value for III is less than methylamine because of the electron-withdrawing effect of the carboxyl group (see structure).

Table 20-1 demonstrates that most amino acids have complex side chains and that some amino acids have other functions (in addition to the α-carboxyl and α-amino groups) such as —OH, $—NH_2$, $—CO_2H$, —SH, phenolic —OH, guanidine, etc. These functions contribute to the physicochemical and biochemical properties of the respective amino acids or to their derivatives, including the proteins in which they are present. It has been customary to designate those amino acids that cannot be synthesized in the organism (animal), at a rate adequate to meet metabolic requisites, as *essential amino acids*. According to Lehninger,[2] nutritionally essential amino acids (for humans) are arginine, histidine, isoleucine, leucine, lysine, methionine, phenylalanine, threonine, tryptophan, and valine. At this point, it is important to note that some of these essential amino acids participate in the biosynthesis of other important metabolites; for example, histamine (from histidine); catecholamines and the thyroid hormones (from phenylalanine through tyrosine); serotonin from tryptophan; and others.

Amino acid antagonists have received the attention of many medicinal chemists. As antimetabolites these compounds interfere with certain metabolic processes and thereby exert, in some cases, therapeutically useful pharmacologic actions (e.g., α-

TABLE 20-2

SELECTED AMINO ACID ANTAGONISTS

Amino Acid Antagonist	Amino Acid Antagonized	Other Inhibitory Effects
D-Alanine	L-Alanine	Carboxypeptidase
D-Phenylalanine	L-Phenylalanine	D-Amino acid oxidase
α-Methyl-L-methionine	L-Methionine	D-Amino acid oxidase
α-Methyl-L-glutaric acid	L-Glutaric acid	Glutamic decarboxylase
Ethionine	Methionine	
α-Methyldopa	Dopa	Dopa decarboxylase
Allyl glycine	Methionine	Growth of *E. coli*
Propargylglycine	Methionine	Growth of *E. coli*
2-Amino-5-heptenoic acid	Methionine	Growth of *E. coli*
2-Thienylalanine	Phenylalanine	Growth of yeast
p-Fluorophenylalanine	Phenylalanine	Incorporation of phenylalanine into protein molecules
L-*O*-Methylthreonine	Isoleucine	Competitive incorporation of leucine into proteins
4-Oxalysine	Lysine	Growth of *E. coli, L. casei*, etc.
6-Methyltryptophan	Tryptophan	
5,5,5-Trifluoronorvaline	Leucine, methionine	Growth of *E. coli*, etc.
3-Cyclohexene-I-glycine	Isoleucine	Inhibits *E. coli*
O-Carbamyl-L serine	L-Glutamine	Inhibits *E. coli, S. lactis*

methyldopa as a dopa decarboxylase inhibitor). Table 20-2 lists some other amino acid antagonists. The study of such antimetabolites as potential chemotherapeutic agents continues. Research in cancer chemotherapy has involved experimentation with many antagonists of amino acids. Glutamine antagonists, azaserine and 6-diazo-5-oxonorleucine (DON), interfere with the metabolic processes that require glutamine and, thereby, disrupt nucleic acid synthesis (glutamine is required for nucleic acid formation; glutamine is derived from glutamic acid). The phenomenon *lethal synthesis* involves the incorporation of the antimetabolite into protein structure or into the structure of some other macromolecule, and this unnatural macromolecule alters metabolic processes dependent on it. *O*-Methylthreonine competes with isoleucine for incorporation into protein molecules, whereas *O*-ethylthreonine is incorporated into tRNA in *Escherichia coli*.

Although all of the naturally occurring amino acids have been synthesized, and several of them are available by the synthetic route, others are available more economically by isolation from hydrolyzed proteins. The latter are leucine, lysine, cystine, cysteine, glutamic acid, arginine, tyrosine, the prolines, and tryptophan.

PRODUCTS

Some pharmaceutically important amino acids are listed in Table 20-3.

Aminoacetic Acid, USP. Glycine (Glycocoll) contains not less than 98.5% and not more than 101.5% $C_2H_5NO_2$. It occurs as a white, odorless, crystalline powder, having a sweetish taste. It is insoluble in alcohol but soluble in water (1:4) to make a solution that is acid to litmus paper.

TABLE 20-3

PHARMACEUTICALLY IMPORTANT AMINO ACIDS

Name Proprietary Name	Preparations	Category	Application	Usual Adult Dose*	Usual Dose Range*
Aminoacetic acid USP	Aminoacetic acid irrigation USP	Irrigating solution	Topically to the body cavities, as a 1.5% solution		
Methionine USP *Amurex, Odor-Scrip, Oradash, Uranap*	Methionine capsules USP Methionine tablets USP	Acidifier (urinary)		400–600 mg/day	
Dihydroxyaluminum aminoacetate USP *Hyperacid*	Dihydroxyaluminum aminoacetate magma USP Dihydroxyaluminum aminoacetate tablets USP	Antacid		500 mg–1 g qid	500 mg–2 g
Aminocaproic acid USP *Amicar*	Aminocaproic acid injection USP Aminocaproic acid syrup USP Aminocaproic acid tablets USP	Hemostatic		Oral and IV, initial, 5 g followed by 1–1.25 g every hour to maintain a plasma level of 13 mg/100 mL. No more than 30 g/24-hr period is recommended	

*See USP DI for complete dosage information.

TABLE 20-3 *Continued*

PHARMACEUTICALLY IMPORTANT AMINO ACIDS

Name Proprietary Name	***Preparations***	***Category***	***Application***	***Usual Adult Dose****	***Usual Dose Range****
Acetylcysteine USP *Mucomyst*	Acetylcysteine solution USP	Mucolytic agent			By inhalation of nebulized solution, 3–5 mL of a 20% solution or 4–10 mL of a 10% solution tid or qid; by direct instillation, 1–2 mL of a 10 or 20% solution every 1–4 hr
Levodopa USP *Larodopa, Levopa, Dopar*	Levodopa capsules USP Levodopa tablets USP	Antiparkinsonian		Initial, 250 mg bid to qid, gradually increasing the total daily dose in increments of 100–750 mg every 3–7 days as tolerated	500 mg–8 g/day
Glutamic acid hydrochloride *Acidulin*	Glutamic acid hydrochloride capsules	Acidifier (gastric)			Oral, 340 mg–1 g tid before meals

*See USP DI for complete dosage information.

A 1.5% solution is preferred over the 2.1% isotonic solution for use as an irrigating solution during transurethral resection of the prostate gland. From 10 to 15 L of the solution may be used during the surgical operation.

Methionine, USP. DL-2-amino-4-(methylthio)-butyric acid (Amurex) occurs as white crystalline platelets or powder with a slight, characteristic odor; it is soluble in water (1 : 30), and a 1% solution has a *p*H of 5.6 to 6.1. It is insoluble in alcohol. In recent years, the racemic compound has been produced in ever-increasing quantities and at considerably reduced cost. The human body needs proteins that furnish methionine to prevent pathologic accumulation of fat in the liver, a condition that can be counteracted by administration of the acid or proteins that provide it. Methionine also has a function in the synthesis of choline, cystine, lecithin, and, probably, creatine. Deficiency not only limits growth in rats, but also inhibits progression of tumors.

In therapy, methionine has been employed in the treatment of liver injuries caused by poisons such as carbon tetrachloride, chloroform, arsenic, and trinitrotoluene. Although many physicians are enthusiastic about its value under such circumstances, this action has not been established satisfactorily.

Another use for methionine is as a urinary acidifier to help control the odor and dermatitis caused by ammoniacal urine in incontinent patients. It has been reported to be effective in both short- and long-term use. Treatment must be continued for three or four days before the ammoniacal odor is eliminated.

Dihydroxyaluminum Aminoacetate, USP. Basic aluminum glycinate (Hyperacid) may be represented by the formula $H_2NCH_2COOAl(OH)_2$. It is a white, odorless, water-insoluble powder that is faintly sweet and is employed as a gastric antacid in the same way as aluminum hydroxide gel. Over the latter, it is claimed to have the advantages of more prompt, greater, and more lasting buffering action. Also, it is said to have less astringent and constipative effects because of its smaller content of aluminum. However, all medical authorities are not yet satisfied that any of these claims are justified. The compound is furnished in powder, magma, or in tablets containing 500 mg.

Aminocaproic Acid, USP. 6-Aminohexanoic acid (Amicar) occurs as a fine, white crystalline powder that is freely soluble in water, slightly soluble in alcohol, and practically insoluble in chloroform.

Aminocaproic acid is a competitive inhibitor of plasminogen activators, such as streptokinase and urokinase. It is effective because it is an analogue of lysine, the position of which in proteins is attacked by plasmin. To a smaller degree, it also inhibits plasmin (fibrinolysin). Lowered plasmin levels lead to more favorable amounts of fibrinogen, fibrin, and other important clotting components.

Aminocaproic acid has been used in the control of hemorrhage in certain surgical procedures. It is of no value in controlling hemorrhage caused by thrombocytopenia or other coagulation defects or vascular disruption (e.g., bleeding ulcers, functional uterine bleeding, post-tonsillectomy bleeding). Because it inhibits the dissolution of clots, it may interfere with normal mechanisms for maintaining the patency of blood vessels.

Aminocaproic acid is well absorbed orally. Plasma peaks occur in about two hours. It is excreted rapidly, largely unchanged.

Acetylcysteine, USP (Mucomyst) is the *N*-acetyl derivative of L-cysteine. It is used primarily to reduce the viscosity of the abnormally viscid pulmonary secretions in patients with cystic fibrosis of the pancreas (mucoviscidosis) or various tracheobronchial and bronchopulmonary diseases.

Acetylcysteine is more active than cysteine, and its mode of action in reducing the viscosity of mucoprotein solutions, including sputum, may be by opening the disulfide bonds in the native protein.

Acetylcysteine is most effective in 10% to 20% solutions with a *p*H of 7 to 9. It is used by direct instillation or by aerosol nebulization. It is available as a 20% solution of the sodium salt in 10- and 30-mL containers. An opened vial of acetylcysteine must be covered, stored in a refrigerator, and used within 48 hours.

Glutamic Acid Hydrochloride (Acidulin) is essentially a pure compound that occurs as a white crystalline powder soluble 1 : 3 in water and insoluble in alcohol. It has been used in place of glycine in the treatment of muscular dystrophies, with rather unpromising results. It also is combined (8 to 20 g/day) with anticonvulsants for the petit mal attacks of epilepsy, a use that appears to depend on change in *p*H of the urine.

The hydrochloride, which releases the acid readily, has been recommended under a variety of names for furnishing acid to the stomach in the achlorhydria of pernicious anemia and other conditions. The usual dosage range is 600 mg to 1.8 g taken during meals.

Levodopa, USP (−)-3-(3,4-Dihydroxyphenyl)-L-alanine (Larodopa; Dopar; Levopa). It occurs as a colorless, crystalline material. It is slightly soluble in water and insoluble in alcohol. Levodopa is a precursor of dopamine and is of value in the treatment of Parkinson's disease. Dopamine does not cross the blood–brain barrier and, therefore, is ineffective. Levodopa does cross the blood–brain barrier and presumably is metabolically converted to dopamine in the basal ganglia. The dose must be carefully determined for each patient.

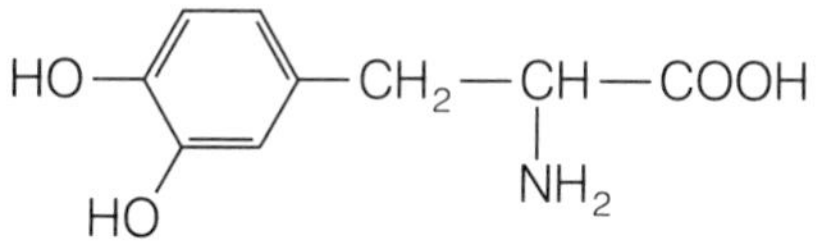

HO, HO, $CH_2—CH—COOH$, NH_2

Levodopa

Carbidopa, USP. The drug Sinemet is a combination of carbidopa and levodopa. The former is the hydrazine analogue of α-methyldopa, and it is an inhibitor of aromatic acid decarboxylation. Accordingly, when carbidopa and levodopa are administered in combination, carbidopa inhibits decarboxylation of peripheral levodopa, but carbidopa does not cross the blood–brain barrier and, hence, does not affect the metabolism of levodopa in the central nervous system. Because carbidopa's decarboxylase-inhibiting activity is limited to extracerebral tissues, it makes more levodopa available for transport to the brain. Thus, carbidopa reduces the amount of levodopa required by approximately 75%.

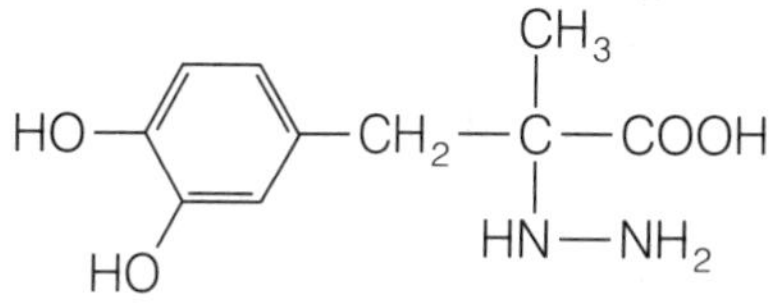

HO, HO, CH_3, $CH_2—C—COOH$, $HN—NH_2$

Carbidopa

Sinemet is supplied as tablets in two strengths: Sinemet-10/100, containing 10 mg of carbidopa and 100 mg of levodopa, and Sinemet-25/250, containing 25 mg of carbidopa and 250 mg of levodopa.

Management of acute overdosage with Sinemet is fundamentally the same as management of acute overdosage with levodopa; however, pyridoxine is not effective in reversing the actions of Sinemet.

PROTEIN HYDROLYSATES

In therapeutics, agents affecting volume and composition of body fluids include various classes of parenteral products. Idealistically, it would be desirable to have parenteral fluids available that would provide adequate calories and important proteins and lipids to mimic as closely as possible an appropriate diet. However, this is not so. Usually, sufficient carbohydrate is administered intravenously to prevent ketosis, and in some cases, it is necessary to give further sources of carbohydrate by vein to reduce the wasting of protein. Sources of protein are made available in the form of protein hydrolysates, and these can be administered to favorably influence the balance.

Protein deficiencies in human nutrition are sometimes treated with protein hydrolysates. The lack of adequate protein may result from several conditions, but the problem is not always easy to diagnose. The deficiency may be due to insufficient dietary intake; temporarily increased demands, as in pregnancy; impaired digestion or absorption; liver malfunction; increased catabolism; or loss of proteins and amino acids, as in fevers, leukemia, hemorrhage, after surgery, burns, fractures, or shock.

PRODUCTS

Protein Hydrolysate Injection, USP. Protein hydrolysates (intravenous) (Aminogen; Travamin). Protein hydrolysate injection is a sterile solution of amino acids and short-chain peptides that represent the approximate nutritive equivalent of the casein, lactalbumin, plasma, fibrin, or other suitable protein from which it is derived by acid, enzymatic, or other method of hydrolysis. It may be modified by partial removal and restoration or addition of one or more amino acids. It may contain dextrose or other carbohydrate suitable for intravenous infusion. Not less than 50% of the total nitrogen present is in the form of α-amino nitrogen. It is a yellowish to red-amber transparent liquid that has a *p*H of 4 to 7.

Parenteral preparations are employed for the maintenance of a positive nitrogen balance in patients for whom there is interference with ingestion, digestion, or absorption of food. In such patients, the material to be injected must be nonantigenic and must not contain pyrogens or peptides of high molecular weight. Injection may result in untoward effects, such as nausea, vomiting, fever, vasodilatation, abdominal pain, twitching and convulsions, edema at the site of injection, phlebitis, and thrombosis. Sometimes these reactions are due to inadequate care in cleanliness or too rapid administration.

Category: fluid and nutrient replenisher
Usual dose: IV infusion, 2 to 3 L of a 5% solution once daily at a rate of 1.5–2 mL/min initially, then increased gradually as tolerated to 3–6 mL/min
Usual dose range: 2 to 8 L/day
Usual pediatric dose: infants, IV infusion, 2 to 3 g of protein per kilogram of body weight in a 4% to 7% solution once daily at a rate not exceeding 0.2 mL/min initially, then increased gradually as tolerated to 0.2–0.6 mL/min. Children, IV infusion, 1 to 2 g of protein per kilogram in a 4% to 7% solution once daily at a rate not exceeding 0.2 mL/min initially, then increased gradually as tolerated to 1–3 mL/min.

AMINO ACID SOLUTIONS

These solutions contain a mixture of essential and nonessential crystalline amino acids with or without electrolytes (e.g., Aminosyn; Freeamine III; Procalamine; Trowasol; Novamine). Protein hydrolysates are being replaced by crystalline amino acid solutions for parenteral administration because the free amino acids are utilized more efficiently than the peptides produced by the enzymatic cleavage of protein hydrolysates.[4]

PROTEINS AND PROTEINLIKE COMPOUNDS

The chemistry of proteins is complex, and some of the most complex facets remain to be clearly understood. Protein structure is usually studied in basic organic chemistry and, to a greater extent, in biochemistry, but for the purposes of this chapter some of the more important topics will be summarized with emphasis on relationships to medicinal chemistry. Much progress has been made in the last 25 years in the understanding of the more sophisticated features of protein structure[3] and its correlation with physicochemical and biologic properties. With the total synthesis of ribonuclease in 1969, new approaches to the study of structure–activity relationships among proteins involve the synthesis of modified proteins.

Many types of compounds that are important in medicinal chemistry are structurally classified as proteins. Enzymes, antigens, and antibodies are pro-

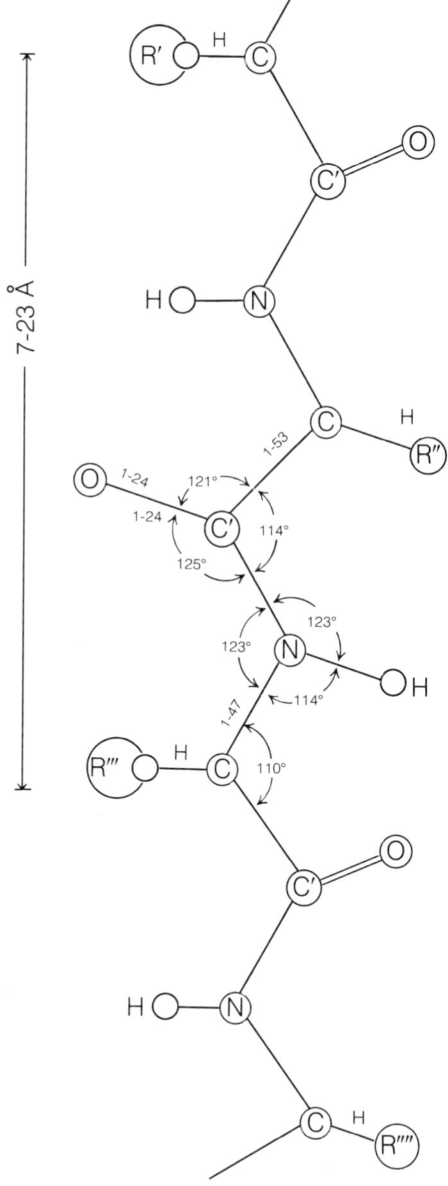

FIG. 20-1. A diagrammatic representation of a fully extended polypeptide chain with the bond lengths and the bond angles derived from crystal structures and other experimental evidence. (From Corey, R. B., and Pauling, L.: Proc. R. Soc. Lond. Ser. B 141:10, 1953.)

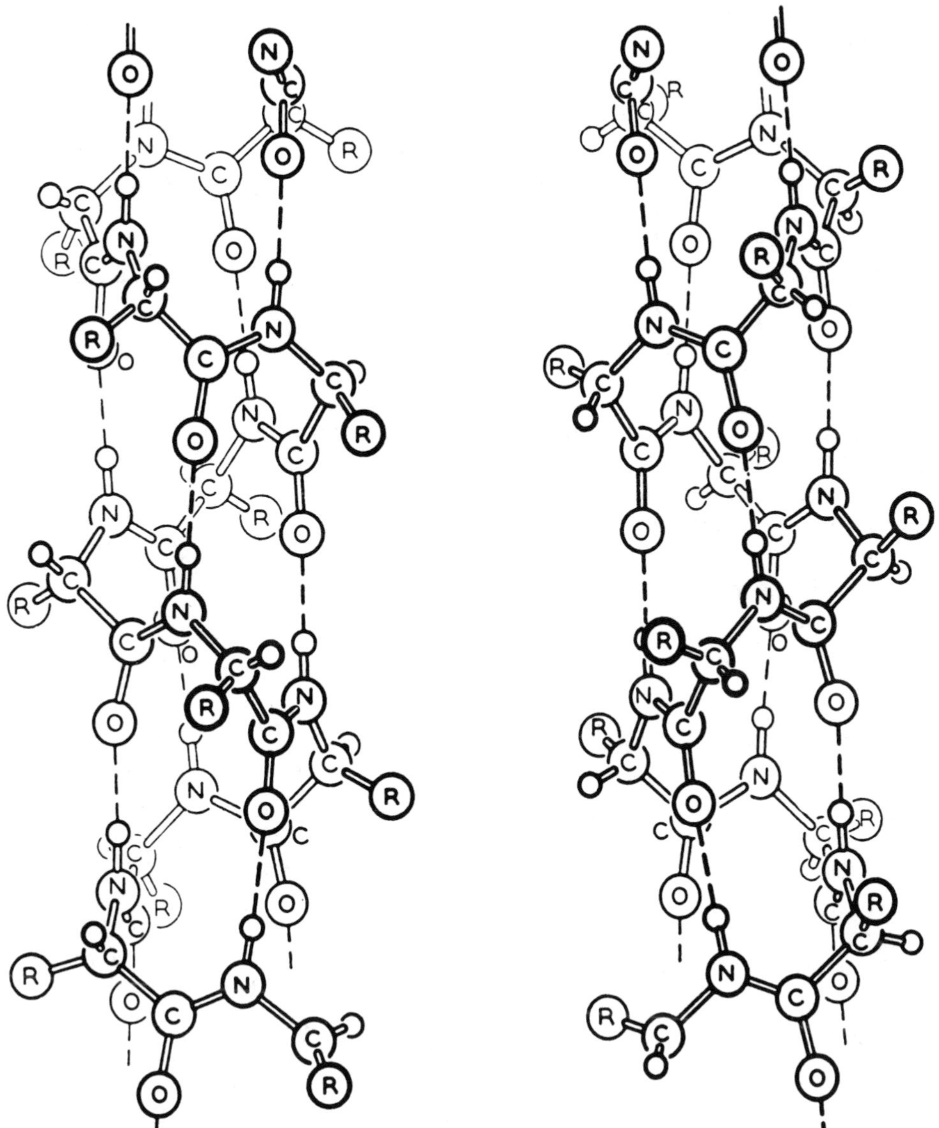

FIG. 20-2. Left-handed and right-handed α-helices. The R and H groups on the α-carbon atom are in the correct position corresponding to the known configuration of the L-amino acids in proteins. (L. Pauling and R. B. Corey, unpublished drawings.)

teins.* Numerous hormones are low relative molecular mass proteins; hence, relative to the foregoing they are called simple proteins. Fundamentally, all proteins are composed of one or more polypeptide chains; that is, the primary organizational level of protein structure is the polypeptide (polyamide) chain composed of naturally occurring amino acids bonded to one another by amide linkages. An extended polypeptide chain can be visualized with the aid of Figure 20-1. The specific physicochemical and biologic properties of proteins depend not only on the nature of the specific amino acids and on their sequence within the polypeptide chain, but also on conformational characteristics.

* The term *interferon* is generally applied to the antiviral proteins naturally produced by various cells. Since the characterization of interferon as an antiviral protein in 1957, much has been learned about the purification and characterization of human leukocyte interferon (e.g., purification by HPLC), antiviral properties of interferon from various species, cloning and expression of human interferons in bacteria, and monoclonal antibodies of human leukocyte interferon. The interested reader should refer to S. Pestka et al.: Annu. Rep. Med. Chem. 16:229, 1981.

CONFORMATIONAL FEATURES OF PROTEIN STRUCTURE

As just indicated, the polypeptide chain is considered to be the primary level of protein structure, and the folding of the polypeptide chains into a specific coiled structure is maintained through hydrogen-bonding interactions (intramolecular) (Fig. 20-2). The latter folding pattern is called the secondary level of protein structure. The intramolecular hydrogen bonds involve the partially negative oxygens of amide carbonyl groups and the partially positive hydrogens of the amide —NH (see later). Additional factors contribute to the stabilization of such folded structures (e.g., ionic bonding between positively charged and negatively charged groups, and disulfide bonds).

The arrangement and interfolding of the coiled chains into layers determine the tertiary and higher levels of protein structure. Such final conformational character is determined by various types of interactions, primarily hydrophobic forces and, to some extent, hydrogen bonding and ion pairing.[3] Hydrophobic forces are implicated in many biologic phenomena associated with protein structure and interactions.[6] The side chains (R groups) of various amino acids have hydrocarbon moieties that are hydrophobic, and they have minimal tendency to associate with water molecules, whereas water molecules are strongly associated through hydrogen bonding. Such hydrophobic R groups tend to get close to one another, with exclusion of water molecules, to form "bonds" between different segments of the chain or between different chains. These bonds are often termed *hydrophobic bonds*, *hydrophobic forces*, or *hydrophobic interactions*.

The study of protein structure has required several physicochemical methods of analysis.[3] Ultraviolet spectrophotometry has been applied to the assessment of conformational changes that proteins undergo. Conformational changes can be investigated by the direct plotting of the difference in absorption between the protein under various sets of conditions. X-ray analysis has been most useful in the elucidation of the structures of several proteins (e.g., myoglobulin and lysozyme). Absolute determinations of conformation and helical content can be made by x-ray diffraction analysis. Optical rotation of proteins also has been studied fruitfully. It is interesting that the specific rotations of proteins are always negative, but changes in *p*H (when the protein is in solution) and conditions that promote denaturation (urea solutions, increased temperatures) tend to augment the negative optical rotation. Accordingly, it is rationalized that the changes in rotation are due to conformational changes (i.e., changes in protein structure at the secondary and higher levels of organization). Optical rotatory dispersion (ORD) also has been experimented with in the study of conformation alterations and conformational differences among globular proteins. Additionally, circular dichroism methodology has been involved in structural studies. The shape and magnitude of rotatory dispersion curves and circular dichroism spectra are very sensitive to conformational alterations; thus, the effects of enzyme inhibitors on conformation can be analyzed. Structural studies have included the investigation of the tertiary structures of proteins in high-frequency nuclear magnetic resonance (NMR).[7] NMR spectroscopy has been of some use in the study of interactions between drug molecules and proteins such as enzymes, proteolipids, and others. NMR has been applied to the study of binding of atropine analogues to acetylcholinesterase[8] and of interactions involving cholinergic ligands to housefly brain and torpedo electroplax.[9] Kato[10,11] has also investigated the binding of inhibitors (e.g., physostigmine) to acetylcholinesterase utilizing NMR spectroscopy.*

FACTORS AFFECTING PROTEIN STRUCTURE

Conditions that promote the hydrolysis of amide linkages affect protein structure, as noted above under Protein Hydrolysates.

The highly ordered conformation of a protein can be disorganized (without hydrolysis of the amide linkages), and, in the process, the protein's biologic activity is obliterated. This process is customarily called *denaturation*, and it involves unfolding of the polypeptide chains, loss of the native conformation of the protein, and disorganization of the uniquely ordered structure, without the cleavage of covalent bonds. The rupture of native disulfide bonds is usually considered to be a more extensive and drastic change than denaturation. Criteria for the detection of denaturation involve detection of previously masked —SH, imidazole, and $—NH_2$ groups; decreased solubility; increased susceptibility to the action of proteolytic enzymes; decreased diffusion constant and increased viscosity of protein solution; loss of enzymatic activity if the protein is an enzyme; modification of antigenic properties.

For many years Eyring[12,13] has carried on studies of factors affecting protein structure and, therefore, biochemical processes. Eyring's studies,[13] involving interactions between general anesthetic molecules and proteins, are fundamental to medicinal chemistry and pharmacology. The interested reader should consult the references cited; however, herein brief mention must be made to exemplify the importance of hydrophobic phenomena in mechanisms of drug action involving proteins or other macromolecules.

---

* C. M. Deber, et al., have reviewed some modern approaches to the deduction of peptide conformation in solution: ^{13}CNMR, conformational energy calculations, and circular dichroism (Deber, C. M., et al.: Science 9:106, 1976.).

```
    \                      H   |
R—C—H                        \ C—R
    \C=O·······H—N/
H—N/        δ−     δ+      \ C=O
                                 |
```

Eyring proposes that anesthetics affect the action of proteins necessary for central nervous system function. It is emphasized that there are certain proteins needed for the maintenance of consciousness. To function normally, the protein must have a particular conformation. Anesthetic molecules are implicated as interacting with the hydrophobic regions of the protein, thus disrupting (unfolding) the conformation. These conformational changes in essential proteins affect their activities and function; hence, it is believed that these effects lead to blockade of synapses.[14]

PURIFICATION AND CLASSIFICATION

It may be said that it is old-fashioned to classify proteins according to the following system, as so much progress has been made in the understanding of protein structure. Nevertheless, an outline of this system of classification is given because the terms used are still found in the pharmaceutical and medical literature. Table 20-4 includes the classification and characterization of simple proteins. Before classification it must be ensured that the protein material is purified to the extent practically possible, and this is a very challenging task. Several criteria are used to determine homomolecularity; for example crystallinity; constant solubility at a given temperature; osmotic pressure in different solvents; diffusion rate; electrophoretic mobility; dielectric constant; chemical assay; spectrophotometry; quantification of antigenicity. The methodology of purification is complex; procedures can involve various techniques of chromatography (column), electrophoresis, ultracentrifugation, and others. In some cases high-performance liquid chromatography (HPLC) has been applied to the separation of peptides; for example, Folkers et al.[15] have reported the purification of some hypothalamic peptides by a combination of chromatographic methods including HPLC.

Conjugated proteins contain a nonprotein structural component in addition to the protein moiety, whereas *simple proteins* contain only the polypeptide chain of amino acid units. *Nucleoproteins* are conjugated proteins containing nucleic acids as structural components. *Glycoproteins* are carbohydrate-containing conjugated proteins (e.g., thyroglobulin). *Phosphoproteins* contain phosphate moieties (e.g., casein); *lipoproteins* are lipid-bearing; *metalloproteins* have some bound metal. *Chromoproteins*, such as hemoglobin or cytochrome, have some chromophoric moiety.

TABLE 20-4

SIMPLE (TRUE) PROTEINS

Class	Characteristics	Occurrence
Albumins	Soluble in water, coagulable by heat and reagents	Egg albumin, lactalbumin, serum albumin, leucosin of wheat, legumelin of legumes
Globulins	Insoluble in water, soluble in dilute salt solution, coagulable	Edestin of plants, vitelline of egg, serum globulin, lactoglobulin, amandin of almonds, myosin of muscles
Prolamines	Insoluble in water or alcohol, soluble in 60–80% alcohol, not coagulable	Found only in plants (e.g., gliadin of wheat, hordein of barley, zein of corn, and secalin of rye)
Glutelins	Soluble only in dilute acids or bases, coagulable	Found only in plants (e.g., glutenin of wheat and oryzenin of rice)
Protamines	Soluble in water or ammonia, strongly alkaline, not coagulable	Found only in the sperm of fish (e.g., salmine from salmon)
Histones	Soluble in water, but not in ammonia, predominantly basic, not coagulable	Globin of hemoglobin, nucleohistone from nucleoprotein
Albuminoids	Insoluble in all solvents	In keratin of hair, nails, and feathers; collagen of connective tissue; chondrin of cartilage; fibroin of silk; and spongin of sponges

PROPERTIES OF PROTEINS

The classification delineated in Table 20-4 is based on solubility properties. Fibrous proteins are water-insoluble and highly resistant to hydrolysis by proteolytic enzymes; the collagens, elastins, and keratins are in this class. On the other hand, globular proteins (albumins, globulins, histones, and protamines) are relatively water-soluble; they are also soluble in aqueous solutions containing salts, acids, bases, or ethanol. Enzymes, oxygen-carrying proteins, and protein hormones are globular proteins.

Another important characteristic of proteins is the amphoteric behavior. In solution, proteins migrate in an electric field, and the direction and rate of migration are a function of the net electrical charge of the protein molecule which, in turn, depends on the *p*H of the solution. The isoelectric point is the *p*H value at which a given protein does not migrate in an electric field, and it is a constant for any given protein and can be used as an index of characterization. Proteins differ in rate of migration and also in their isoelectric points. Electrophoretic analysis is used to determine purity and for quantitative estimation because proteins differ in electrophoretic mobility at any given pH.[3]

Because they are ionic in solution, proteins bind with cations and anions depending on the pH of the environment. Sometimes, complex salts are formed and precipitation takes place (e.g., trichloracetic acid is a precipitating agent for proteins and is used for deproteinizing solutions).

Proteins possess chemical properties characteristic of their component functional groups, but, in the native state, some of these groups are "buried" within the tertiary protein structure and may not readily react. Certain denaturation procedures can expose these functions and allow them to respond to the usual chemical reagents (e.g., an exposed $-NH_2$ group can be acetylated by ketene, $-CO_2H$ can be esterified with diazomethane).

COLOR TESTS, MISCELLANEOUS SEPARATION AND IDENTIFICATION METHODS

Proteins respond to the following color tests: (1) biuret, pink to purple with an excess of alkali and a small amount of copper sulfate; (2) ninhydrin, a blue color when boiled with ninhydrin (triketohydrindene hydrate) that is intensified by the presence of pyridine; (3) Millon's test for tyrosine, a brick-red color or precipitate when boiled with mercuric nitrate in an excess of nitric acid; (4) Hopkins–Cole test for tryptophan, a violet zone with a salt of glyoxylic acid and stratified over sulfuric acid; and (5) xanthoproteic test, a brilliant orange zone when a solution in concentrated nitric acid is stratified under ammonia.

Almost all so-called alkaloidal reagents will precipitate proteins in slightly acid solution.

The qualitative identification of the amino acids found in proteins and other substances has been simplified greatly by the application of paper chromatographic techniques to the proper hydrolysate of proteins and related substances. End-member degradation techniques for the detection of the sequential arrangements of the amino acid residues in polypeptides (proteins, hormones, enzymes, and such) have been developed to such a high degree with the aid of paper chromatography that very small samples of the polypeptides can be utilized. These techniques, together with statistical methods, have led to the elucidation of the sequential arrangements of the amino acid residues in oxytocin, vasopressin, insulin, hypertensin, glucagon, corticotropins, and others.

Ion-exchange chromatography has been applied to protein analysis and to the separation of amino acids. The principles of ion-exchange chromatography can be applied to the design of automatic amino acid analyzers with appropriate recording instrumentation.[3] One- or two-dimensional thin-layer chromatography also has been used to accomplish separations not possible with paper chromatography. Another method for separating amino acids and proteins involves a two-dimensional analytical procedure that uses electrophoresis in one dimension and partition chromatography in the other. The applicability of high-performance liquid chromatography was noted earlier.[15]

PRODUCTS

Gelatin, NF, is a protein obtained by the partial hydrolysis of collagen, an albuminoid found in bones, skins, tendons, cartilage, hoofs, and other animal tissues. The products seem to be of great variety, and from a technical standpoint, the raw material must be selected according to the purpose intended. This is because collagen usually is accompanied in nature by elastin and, especially, by mucoids, such as chondromucoid, which enter into the product in a small amount. The raw materials for official gelatin, and also that used generally for a food, are skins of calf or swine and bones. First, the bones are treated with hydrochloric acid to remove the calcium compounds and then are digested with lime for a prolonged period, which converts most other impurities to a soluble form. The fairly pure collagen is extracted with hot water at a pH of about 5.5, and the aqueous solution of gelatin is concentrated, filtered, and cooled to a stiff gel. Calf skins are treated in about the same way, but those from hogs are not given any lime treatment. The product derived from an acid-treated precursor is known as type A and exhibits an isoelectric point between pH 7 and 9, whereas that for which alkali is used is known as type B and exhibits an isoelectric point between pH 4.7 and 5. The minimum of gel strength officially is that a 1% solution, kept at 0° for six hours, must show no perceptible flow when the container is inverted.

Gelatin occurs in sheets, shreds, flakes, or coarse powder. It is white or yellowish, has a slight but characteristic odor and taste, and is stable in dry air, but subject to microbial decomposition if moist or in solution. It is insoluble in cold water, but swells and softens when immersed and gradually absorbs five to ten times its own weight of water. It dissolves in hot water to form a colloidal solution; it also dissolves in acetic acid and in hot dilute glycerin. Gelatin commonly is bleached with sulfur dioxide, but that used medicinally must have not over 40

TABLE 20-5

PHARMACEUTICALLY IMPORTANT PROTEIN PRODUCTS

Name Proprietary Name	***Category***
Gelatin, NF	Pharmaceutical acid (encapsulating agent; suspending agent; tablet binder and coating agent)
Gelatin film, absorbable, USP *Gelfilm*	Local hemostatic
Gelatin sponge, absorbable, USP *Gelfoam*	Local hemostatic

parts per million of sulfur dioxide. However, a proviso is made that for the manufacture of capsules or pills it may have certified colors added, may contain as much as 0.15% of sulfur dioxide, and may have a lower gel strength.

Gelatin is used in the preparation of capsules and the coating of tablets and, with glycerin, as a vehicle for suppositories. It also has been employed as a vehicle for other drugs when slow absorption is required. When dissolved in water, the solution becomes somewhat viscous, and such solutions are used to replace the loss in blood volume in cases of shock. This is accomplished more efficiently now with blood plasma, which is safer to use. In hemorrhagic conditions, it sometimes is administered intravenously to increase the clotting of blood or is applied locally for the treatment of wounds.

The most important value in therapy is as an easily digested and adjuvant food. It fails to provide any tryptophan at all and is lacking notably in adequate amounts of other essential amino acids; approximately 60% of the total amino acids consist of glycine and the prolines. Nevertheless, when supplemented, it is very useful in various forms of malnutrition, gastric hyperacidity or ulcer, convalescence, and in general diets of the sick. It is specially recommended in the preparation of modified milk formulas for feeding infants.

Gelatin Film, Absorbable, USP (Gelfilm) is a sterile, nonantigenic, absorbable, water-insoluble gelatin film. The gelatin films are prepared from a solution of specially prepared gelatin–formaldehyde combination, by spreading on plates and drying under controlled humidity and temperature. The film is available as light yellow, transparent, brittle sheets 0.076 mm to 0.228 mm thick. Although insoluble in water, they become rubbery after being in water for a few minutes.

Gelatin Sponge, Absorbable, USP (Gelfoam) is a sterile, absorbable, water-insoluble, gelatin-based sponge that is a light, nearly white, nonelastic, tough porous matrix. It is stable to dry heat at 150°C for four hours. It absorbs 50 times its own weight of water or 45 times oxalated whole blood.

It is absorbed in four to six weeks when it is used as a surgical sponge. When applied topically to control capillary bleeding, it should be moistened with sterile isotonic sodium chloride solution or thrombin solution.

Nonspecific Proteins. The intravenous injection of foreign protein is followed by fever, muscle and joint pain, sweating, and decrease and then increase in leukocytes; it even can result in serious collapse. The results have been used in the treatment of various infections, originally the chronic form. The method is presumed to be of value in acute and chronic arthritis, peptic ulcer, certain infections of the skin and eye, some vascular diseases, cerebrospinal syphilis, especially dementia paralytica, and other diseases. Because a fever is necessary in this system, the original program has developed into the use of natural fevers, such as malaria, of external heat, and of similar devices. However, the slightly purified proteins of milk still are recommended for some diseases; they are available commercially as Activin, Caside, Clarilac, Bu-Ma-Lac, Lactoprotein, Mangalac, Nat-i-lac, Neolacmanese, and Proteolac. Muscosol is a purified beef peptone, and Omniadin is a similar purified bacterial protein. Synodal contains nonspecific protein with lipoids, animal fats, and emetine hydrochloride and is designed for the treatment of peptic ulcer. One of the favorite agents of this class has been typhoid vaccine.

Venoms. Cobra (*Naja*) Venom Solution, from which the hemotoxic and proteolytic principles have been removed, has been credited with virtues owing to toxins and has been injected intramuscularly as a nonnarcotic analgesic in doses of 1 mL/day. Snake venom solution of the water moccasin is employed subcutaneously in doses of 0.4 to 1.0 mL as a hemostatic in recurrent epistaxis, thrombocytopenic purpura, and as a prophylactic before tooth extraction and minor surgical procedures. Stypven from the Russell viper is used topically as a hemostatic and as a thromboplastic agent in Quick's modified clotting-time test. Ven-Apis, the purified and standardized venom from bees, is furnished in graduated strengths of 32, 50, and 100 bee-sting units. It is administered topically in acute and chronic arthritis, myositis, and neuritis.

Nucleoproteins. The nucleoproteins previously mentioned are found in the nuclei of all cells and also in the cytoplasm. They can be deproteinized by several methods. Those compounds that occur in yeast usually are treated by grinding with a very dilute solution of potassium hydroxide, adding picric acid in excess, and precipitating the nucleic acids with hydrochloric acid, leaving the protein in solution. The nucleic acids are purified by dissolving in dilute potassium hydroxide, filtering, acidifying with acetic acid, and finally precipitating with a large excess of ethanol.

The nucleoproteins found in the nucleus of eukaryotic cells include a variety of enzymes, such as DNA and RNA polymerases (involved in nucleic acid synthesis), nucleases (involved in the hydrolytic cleavage of nucleotide bonds), isomerases, and others. The nucleus of eukaryotic cells also contains specialized proteins such as tubulin (involved in the formation of mitotic spindle before mitosis) and histones. Histones are proteins rich in the basic amino acids arginine and lysine, which together make up one-fourth of the amino acid residues. Histones combine with negatively charged double-helical DNA to form complexes that are held together by electrostatic interactions. Histones function to package and order the DNA into structural units called nucleosomes.

ENZYMES

Those proteins that have catalytic properties are called *enzymes* (i.e., enzymes are biologic catalysts of protein nature).* Some enzymes have full catalytic reactivity per se; these are considered to be simple proteins because they do not have a nonprotein moiety. On the other hand, other enzymes are conjugated proteins, and the nonprotein structural components are necessary for reactivity. Occasionally, enzymes require metallic ions. Because enzymes are proteins or conjugated proteins, the general review of protein structural studies presented earlier in this chapter (e.g., protein conformation and denaturation) is fundamental to the following topics. Conditions that effect denaturation of proteins usually have adverse effects on the activity of the enzyme.

General enzymology is discussed effectively in numerous standard treatises, and one of the most concise discussions appears in the classic work by Ferdinand.[16] Ferdinand includes reviews of enzyme structure and function, bioenergetics and kinetics, and appropriate illustrations with a total of 37 enzymes selected from the six major classes of enzymes. Accordingly, for additional basic studies of enzymology, the reader should refer to this classic monograph and to a recent comprehensive review of this topic.[17]

RELATION OF STRUCTURE AND FUNCTION

Koshland[18] has reviewed concepts concerning correlations of protein conformation and conformational flexibility of enzymes with enzyme catalysis. Enzymes do not exist initially in a conformation complementary to that of the substrate. The substrate induces the enzyme to assume a complementary conformation. This is the so-called induced fit theory. There is proof that proteins do possess conformational flexibility and undergo conformational changes under the influence of small molecules. It is emphasized that this does not mean that all proteins must be flexible; nor does it mean that conformationally flexible enzymes must undergo conformation changes when interacting with all compounds. Furthermore, a regulatory compound that is not directly involved in the reaction can exert control on the reactivity of the enzyme by inducing conformational changes (i.e., by inducing the enzyme to assume the specific conformation complementary to the substrate). (Conceivably, hormones as regulators function according to the foregoing mechanism of affecting protein structure.) So-called flexible enzymes can be distorted conformationally by molecules classically called inhibitors. Such inhibitors can induce the protein to undergo conformational changes disrupting the catalytic functions or the binding function of the enzyme. In this connection it is interesting to note how the work of Belleau and the molecular perturbation theory of drug action relate to Koshland's studies (see Chap. 2).

Evidence continues to support the explanation of enzyme catalysis on the basis of the *active site* (reactive center) of amino acid residues, which is considered to be that relatively small region of the enzyme's macromolecular surface involved in catalysis. Within this site, the enzyme has strategically positioned functional groups (from the side chains of amino acid units) that participate cooperatively in the catalytic action.[19]

Some enzymes have absolute specificity for a single substrate, but others catalyze a particular type of reaction that various compounds undergo. In the latter, the enzyme is said to have relative specificity. Nevertheless, when compared with other catalysts, enzymes are outstanding in their specificity for certain substrates.[20] The physical, chemical, conformational, and configurational properties of the substrate determine its complementarity to the enzyme's reactive center. These factors, therefore,

* Important factors limiting rates of enzyme-catalyzed reactions have been critically evaluated by W. W. Cleland (Acc. Chem. Res. 8:145, 1975).

determine whether a given compound satisfies the specificity of a particular enzyme. Enzyme specificity must be a function of the nature, including conformational and chemical reactivity, of the reactive center, but when the enzyme is a conjugated protein with a coenzyme moiety, the nature of the coenzyme also contributes to specificity characteristics.

It seems that in some instances the active center of the enzyme is complementary to the substrate molecule in a strained configuration, corresponding to the "activated" complex for the reaction catalyzed by the enzyme. The substrate molecule is attracted to the enzyme and is caused by the forces of attraction to assume the strained state, with conformational changes that favor the chemical reaction; that is, the activation energy requirement of the reaction is decreased by the enzyme to such an extent that the reaction is caused to proceed at an appreciably greater rate than it would in the absence of the enzyme. If the enzymes were always completely complementary in structure to the substrates, then no other molecule would be expected to compete successfully with the substrate in combination with the enzyme, which in this respect would be similar in behavior to antibodies. However, occasionally, an enzyme complementary to a strained substrate molecule might attract more strongly to itself a molecule resembling the strained substrate molecule itself; for example, the hydrolysis of benzoyl-L-tyrosylglycineamide was practically inhibited by an equal amount of benzoyl-D-tyrosylglycineamide. This example might also serve to illustrate a type of antimetabolite activity.

Several types of interactions contribute to the formation of enzyme–substrate complexes: attractions between charged (ionic) groups on the protein and the substrate; hydrogen bonding; hydrophobic forces (the tendency of hydrocarbon moieties of side chains of amino acid residues to associate with the nonpolar groups of the substrate in a water environment); and London forces (induced dipole interactions).

Many studies of enzyme specificity have been made on proteolytic enzymes (proteases). Configurational specificity can be exemplified by the aminopeptidase that cleaves L-leucylglycylglycine, but does not affect D-leucylglycylglycine. D-Alanylglycylglycine is slowly cleaved by this enzyme. These phenomena illustrate the significance of steric factors; at the active center of aminopeptidase, a critical factor is a matter of closeness of approach that affects the kinetics of the reaction.

One can easily imagine how difficult it is to study the reactivity of enzymes on a functional group basis, because the mechanism of enzyme action is so complex.[18] Nevertheless, it can be said that the

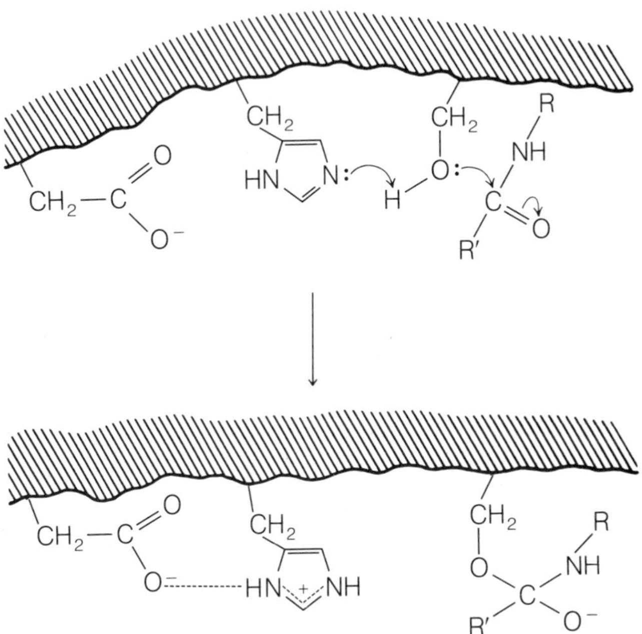

FIG. 20-3. *Generalized mechanism of protease catalysis. (Adapted from Chem. Eng. News, Apr. 16, 1979, p. 23.)*

—SH group probably is found in more enzymes as a functional group than are the other polar groups. It should be noted that in some enzymes (e.g., urease), the less readily available SH groups are necessary for biologic activity and cannot be detected by the nitroprusside test that is used to detect the freely reactive SH groups.

A free —OH group of the tyrosyl residue is necessary for the activity of pepsin. Both the —OH of serine and the imidazole portion of histidine appear to be necessary parts of the active center of certain hydrolytic enzymes, such as trypsin and chymotrypsin, and furnish the electrostatic forces involved in a proposed mechanism (Fig. 20-3), in which E denotes enzyme, the other symbols being self-evident.*

These two groups (i.e., —OH and =NH) could be located on separate peptide chains in the enzyme as long as the specific three-dimensional structure formed during activation of the zymogen brought them near enough to form a hydrogen bond. The polarization of the resulting structure would cause the serine oxygen to be the nucleophilic agent that attacks the carbonyl function of the substrate. The complex is stabilized by the simultaneous "ex-

* Alternative mechanisms have been proposed;[17] esterification and hydrolysis have been extensively studied by M. L. Bender (J. Am. Chem. Soc. 79:1258, 1957; 80:5338, 1958; 82:1900, 1960; 86:3704, 5330, 1964). More recently, D. M. Blow has reviewed studies concerning the structure and mechanism of chymotrypsin (Acc. Chem. Res. 9:145, 1976).

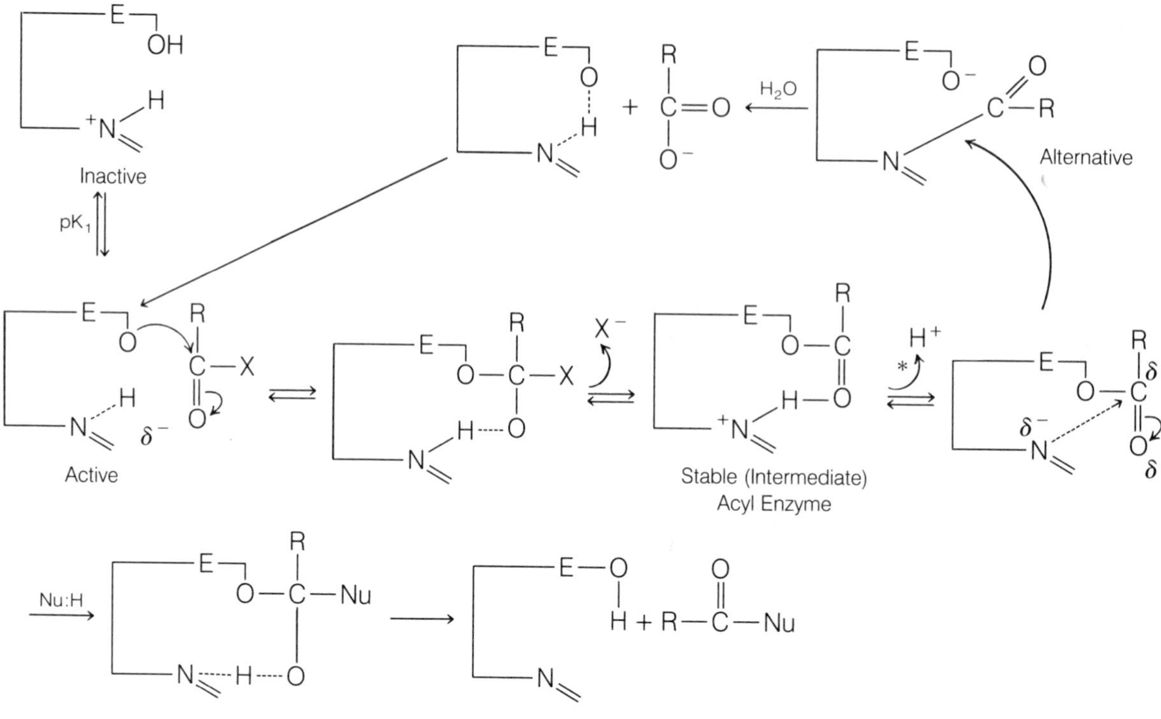

FIG. 20-4. Enzyme-catalyzed hydrolysis of $R-\overset{O}{\overset{\|}{C}}-X$: A proposed generalized mechanism.

change'' of the hydrogen bond from the serine oxygen to the carbonyl oxygen of the substrate.

The intermediate acylated enzyme is written with the proton on the imidazole nitrogen. The deacylation reaction involves the loss of this positive charge simultaneously with the attack of the nucleophilic reagent (abbreviated Nu:H).

Roberts[21] effectively used nitrogen-15 (^{15}N) NMR to study the mechanism of protease catalysis. A schematic summary of the generalized mechanism is represented in Figure 20-4. It is concluded that the tertiary N-1 nitrogen of the histidine unit within the reactive center of the enzyme deprotonates the hydroxyl of the neighboring serine unit and simultaneously the hydroxyl oxygen exerts nucleophilic attack on the carbonyl carbon of the amide substrate as depicted in the scheme. A tetrahedral intermediate is implicated, and the carboxylate group of the aspartate unit (the third functional group within the reactive center) stabilizes the developing imidazolium ion by a hydrogen bonding to the N-3 hydrogen. Finally, the decomposition of the anionic tetrahedral intermediate toward product formation (amine and acylated serine) is promoted by prior protonation of the amide nitrogen by the imidazolium group.

A possible alternative route to deacylation would involve the nucleophilic attack of the imidazole nitrogen on the newly formed ester linkage of the postulated acyl intermediate, leading to the formation of the acyl imidazole. The latter is unstable in water, hydrolyzing rapidly to give the product and regenerated active enzyme.

The reaction of an alkyl phosphate in such a scheme may be written in an entirely analogous fashion, except that the resulting phosphorylated

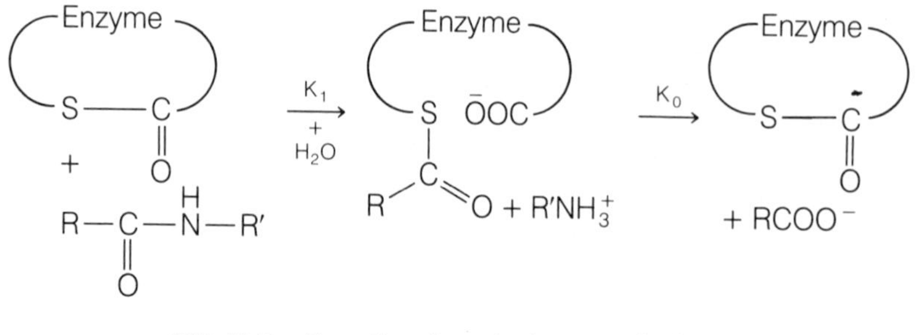

FIG. 20-5. The action of papain: A proposed scheme.

enzyme would be less susceptible to deacylation through nucleophilic attack. The following diagrammatic scheme (Fig. 20-5) has been proposed to explain the function of the active thiol ester site of papain. This ester site is formed and maintained by the folding energy of the enzyme (protein) molecule.

ZYMOGENS (PROENZYMES)

Zymogens, also called proenzymes, are enzyme precursors. These proenzymes are said to be activated when they are transformed to the enzyme. This activation usually involves catalytic action by some proteolytic enzyme. Occasionally, the activators merely effect a reorganization of the tertiary structure (conformation) of the protein so that the groups involved within the reactive center become functional (i.e., unmasked).

SYNTHESIS AND SECRETION OF ENZYMES

Exportable proteins (enzymes), such as amylase, ribonuclease, chymotrypsin(ogen), trypsin(ogen), insulin, and such, are synthesized on the ribosomes. They pass across the membrane of the endoplasmic reticulum into the cisternae and directly into a smooth vesicular structure which effects further transportation. They are finally stored in highly concentrated form within membrane-bound granules. These are called zymogen granules, the exportable protein content of which may reach a value of 40% of the total protein of the gland cell. In the foregoing enzyme sequences, the newly synthesized exportable protein (enzymes) is not free in the cell sap. The stored exportable proteins are released into the extracellular milieu for the digestive enzymes and into adjacent blood capillaries for hormones. The release of these proteins is initiated (triggered) by specific inducers: for example, cholinergic agents (but not epinephrine) and Ca^{2+} effect a discharge of amylase, lipase, or other, into the medium; increase in glucose levels stimulates the secretion of insulin and so on. This release of the reserve enzymes and hormones is completely independent of the synthetic process, as long as the stores in the granules are not completely depleted. Energy oxidative phosphorylation does not play an important role in these releases. Electron microscope studies indicate a fusion of the zymogen granule membrane with the cell membrane, such that a direct opening of the granule into the extracellular lumen of the gland is formed.

CLASSIFICATION

There are various systems for the classification of enzymes [e.g., the International Union of Biochemistry (IUB) system]. This system includes some of the terminology that is used in the literature of medicinal chemistry, and in many instances the terms are self-explanatory: for example, oxidoreductases; transferases (catalyze transfer of a group, such as methyltransferase); hydrolases (catalyze hydrolysis reactions, such as esterases and amidases); lyases (catalyze nonhydrolytic removal of groups leaving double bonds); isomerases; ligases. Other systems are sometimes used to classify and characterize enzymes, and the following terms are frequently encountered: lipases, peptidases, proteases, phosphatases, kinases, synthetases, dehydrogenases, oxidases, reductases.

PRODUCTS

Pharmaceutically important enzyme products are listed in Table 20-6.

Pancreatin, USP (Panteric) is a substance obtained from the fresh pancreas of the hog or of the ox and contains a mixture of enzymes, principally pancreatic amylase (amylopsin), protease, and pancreatic lipase (steapsin). It converts not less than 25 times its weight of USP Potato Starch Reference Standard into soluble carbohydrates, and not less

TABLE 20-6

PHARMACEUTICALLY IMPORTANT ENZYME PRODUCTS

Name Proprietary Name	***Preparations***	***Category***	***Application***	***Usual Adult Dose****	***Usual Dose Range****
Pancreatin USP *Panteric*	Pancreatin capsules USP Pancreatin tablets USP	Digestive aid		325 mg–1 g	
Trypsin crystallized USP	Trypsin crystallized for aerosol USP	Proteolytic enzyme		Aerosol, 125,000 USP units in 3 mL of saline daily	

*See USP DI for complete dosage information.

TABLE 20-6 *Continued*

PHARMACEUTICALLY IMPORTANT ENZYME PRODUCTS

Name Proprietary Name	***Preparations***	***Category***	***Application***	***Usual Adult Dose****	***Usual Dose Range****
Pancrelipase USP *Cotazym*	Pancrelipase capsules USP Pancrelipase tablets USP	Digestive aid			An amount of pancrelipase equivalent to 8,000–24,000 USP units of lipolytic activity before each meal or snack, or to be determined by the practitioner according to the needs of the patient
Chymotrypsin USP *Chymar*	Chymotrypsin for ophthalmic solution USP	Proteolytic enzyme (for zonule lysis)	1–2 mL by irrigation to the posterior chamber of the eye, under the iris, as a solution containing 75–150 U/mL		
Hyaluronidase for injection USP *Alidase, Wydase*	Hyaluronidase injection USP	Spreading agent		Hypodermoclysis, 150 USP hyaluronidase units	
Sutilains USP *Travase*	Sutilains ointment USP	Proteolytic enzyme	Topical, ointment, bid to qid		

*See USP DI for complete dosage information.

than 25 times its weight of casein into proteoses. Pancreatin of a higher digestive power may be brought to this standard by admixture with lactose, or with sucrose containing not more than 3.25% of starch, or with pancreatin of lower digestive power. Pancreatin is a cream-colored amorphous powder having a faint, characteristic, but not offensive, odor. It is slowly but incompletely soluble in water and insoluble in alcohol. It acts best in neutral or faintly alkaline media, and excessive acid or alkali renders it inert. Pancreatin can be prepared by extracting the fresh gland with 25% alcohol or with water and subsequently precipitating with alcohol. Besides the enzymes mentioned, it contains some trypsinogen, which can be activated by enterokinase of the intestines; chymotrypsinogen, which is converted by trypsin to chymotrypsin; and carboxypeptidase.

Pancreatin is used largely for the predigestion of food and for the preparation of hydrolysates. The value of its enzymes orally must be very small because they are digested by pepsin and acid in the stomach, although some of them may escape into the intestines without change. Even if they are protected by enteric coatings, it is doubtful if they could be of great assistance in digestion.

Trypsin Crystallized, USP, is a proteolytic enzyme crystallized from an extract of the pancreas gland of the ox, *Bos taurus*. It occurs as a white to yellowish white, odorless, crystalline or amorphous powder, and 500,000 USP trypsin units are soluble in 10 mL of water or saline TS.

Trypsin has been used for several conditions in which its proteolytic activities relieve certain inflammatory states, liquefy tenacious sputum, and such; however, the many side reactions encountered, particularly when it is used parenterally, mitigate against its use.

Pancrelipase, USP (Cotazym). This preparation has a greater lipolytic action than do other pancreatic enzyme preparations. Hence, it is used to help control steatorrhea and in other conditions in which pancreatic insufficiency impairs the digestion of fats in the diet.

Chymotrypsin, USP (Chymar). This enzyme is extracted from mammalian pancreas and is used in cataract surgery. A dilute solution is used to irrigate the posterior chamber of the eye to dissolve the fine filaments that hold the lens.

Hyaluronidase for Injection, USP (Alidase; Wydase) is a sterile, dry, soluble enzyme product prepared from mammalian testes and capable of hydrolyzing the mucopolysaccharide hyaluronic acid. It contains not more than 0.25 μg of tyrosine for each USP hyaluronidase unit. Hyaluronidase in solution must be stored in a refrigerator. Hyaluronic acid, an essential component of tissues, limits the spread of fluids and other extracellular material, and, because the enzyme destroys this acid, injected fluids and other substances tend to spread farther and faster than normal when administered with this enzyme. Hyaluronidase may be used to increase the spread and consequent absorption of hypodermoclysis solutions, to diffuse local anesthetics, especially in nerve blocking, and to increase diffusion and absorption of other injected materials, such as penicillin. It also enhances local anesthesia in surgery of the eye and is useful in glaucoma because it causes a temporary drop in intraocular pressure.

Hyaluronidase is practically nontoxic, but caution must be exercised in the presence of infection because the enzyme may cause a local infection to spread, through the same mechanism; it never should be injected in an infected area. Sensitivity to the drug is rare.

The activity of hyaluronidase is determined by measuring the reduction of turbidity that it produces on a substrate of native hyaluronidate and certain proteins, or by measuring the reduction in viscosity that it produces on a buffered solution of sodium or potassium hyaluronidate. Each manufacturer defines its product in turbidity or viscosity units, but they are not the same because they measure different properties of the enzyme.

Sutilains, USP (Travase) is a proteolytic enzyme obtained from cultures of *Bacillus subtilis* and is used to dissolve necrotic tissue occurring in second- and third-degree burns, as well as in bed sores and ulcerated wounds.

Many substances are contraindicated during the topical use of sutilains. These include detergents and anti-infectives that have a denaturing action on the enzyme preparation. The antibiotics penicillin, streptomycin, and neomycin do not inactivate sutilains. Mafenide acetate is also compatible with the enzyme.

Streptokinase (Kabikinase; Streptase) is a catabolic product secreted by group C β-hemolytic streptococci. It is a protein with no known enzymatic activity. Streptokinase activates plasminogen to plasmin, a proteolytic enzyme that hydrolyzes fibrin, and promotes the dissolution of thrombi.[22] Plasminogen is activated when streptokinase forms a 1 : 1 stoichiometric complex with it. Allergic reactions to streptokinase are a common occurrence because of antibody formation in individuals treated with it. Furthermore, the antibodies inactivate streptokinase and reduce its ability to prolong thrombin time. Streptokinase is indicated for acute myocardial infarction, for local perfusion of an occluded vessel, and before angiography, by intravenous, intra-arterial and intracoronary administration, respectively.

Urokinase (Abbokinase) is a glycosylated serine protease consisting of two polypeptide chains connected by a single disulfide bond. It is isolated from human urine or tissue culture of human kidneys. The only known substrate of urokinase is plasminogen, which is activated to plasmin, a fibrinolytic enzyme.[22] Unlike streptokinase, urokinase is a direct activator of plasminogen. Urokinase is nonantigenic because it is an endogenous enzyme and, therefore, may be used when streptokinase use is impossible because of antibody formation. It is administered intravenously or by the intracoronary route. Its indications are similar to those of streptokinase.

Altiplase (Activase) is a tissue plasminogen activator (t-PA) produced by recombinant DNA technology. It is a single-chain glycoprotein protease consisting of 527-amino acid residues. The native t-PA is isolated from a melanoma cell line. The single-chain molecule is susceptible to enzymatic digestion to a two-chain molecule, in which the two chains remain linked with a disulfide bond. Both forms of the native t-PA are equipotent in fibrinolytic (and plasminogen activating) properties.[22] It is an extrinsic plasminogen activator associated with vascular endothelial tissue, which preferentially activates plasminogen bound to fibrin. The fibrinolytic action of alteplase (t-PA) is confined to thrombi, with minimal systemic activation of plasminogen. It is commercially produced by recombinant DNA methods by inserting the alteplase gene (acquired from human melanoma cells) into ovarian cells of the Chinese hamster, serving as host cells. The melanoma-derived alteplase is immunologically and chemically identical with the uterine form.[22] Alteplase is indicated for the intravenous management of acute myocardial infarction.

Papain, USP (Papase), the dried and purified latex of the fruit of *Carica papava* L. (Caricaceae), has the power of digesting protein in either acid or alkaline media; it is best at a pH of from 4 to 7, and at 65° to 90°C. It occurs as light brownish gray to weakly reddish brown granules or as a yellowish gray to weakly yellow powder. It has a characteristic odor and taste and is incompletely soluble in water to form an opalescent solution. The commercial material is prepared by evaporating the juice, but the pure enzyme also has been prepared and crystallized. In medicine, it has been used locally in various conditions similar to those for which pepsin is employed. It has the advantage of activity over a wider range of conditions, but it is often much less reliable. Intraperitoneal instillation of a weak solution has been recommended to counteract a tendency to adhesions after abdominal surgery, and several enthusiastic reports have been made about its value under these conditions. Papain has been reported to cause allergies in persons who handle it, especially those who are exposed to inhalation of the powder.

Bromelains (Ananase) is a mixture of proteolytic enzymes obtained from the pineapple plant. It is proposed for use in the treatment of soft-tissue inflammation and edema associated with traumatic injury, localized inflammations, and postoperative tissue reactions. The swelling that accompanies inflammation may possibly be caused by occlusion of the tissue spaces with fibrin. If this be true, sufficient amounts of Ananase would have to be absorbed and reach the target area after oral administration to act selectively on the fibrin. This is yet to be firmly established, and its efficacy as an anti-inflammatory agent is inconclusive. On the other hand, an apparent inhibition of inflammation has been demonstrated with irritants such as turpentine and croton oil (granuloma pouch technique).

Ananase is available in 50,000-unit tablets for oral use.

Diastase (Taka-Diastase) is derived from the action of a fungus, *Aspergillus oryzae* Cohn (*Eurotium O.* Ahlburg), on rice hulls or wheat bran. It is a yellow, hygroscopic, almost tasteless powder that is freely soluble in water and can solubilize 300 times its weight of starch in ten minutes. It is employed in doses of 0.3 to 1.0 g in the same conditions as malt diastase. Taka-Diastase is combined with alkalies as an antacid in Takazyme, with vitamins in Taka-Combex, and in other preparations.

HORMONES

The hormones discussed in this chapter may be structurally classified as polypeptides, proteins, or glycoproteins. These hormones include metabolites elaborated by the hypothalamus, pituitary gland, pancreas, gastrointestinal tract, parathyroid gland, liver, and kidneys. A comprehensive review of the biochemistry of these polypeptides and other related hormones is beyond the scope of this chapter. For a detailed discussion, the reader should refer to a somewhat recent review[23] and to other literature cited throughout this chapter.

HORMONES FROM THE HYPOTHALAMUS

The physiologic and clinical aspects of hypothalamic-releasing hormones have been reviewed.[24] Through these hormones, the central nervous system regulates other essential endocrine systems, including the pituitary, which, in turn, controls other systems (e.g., the thyroid).

Thyroliberin (thyrotropin-releasing hormone; TRH) is the hypothalamic hormone that is responsible for the release of the pituitary's thyrotropin (TSH). Thyrotropin stimulates the production of thyroxine and liothyronine by the thyroid. The latter thyroid hormones, by feedback regulation, inhibit the action of TRH on the pituitary. Thyroliberin is a relatively simple tripeptide that has been characterized as pyroglutamyl-histidyl-prolinamide. TRH possesses interesting biologic properties. In addition to stimulating the release of thyrotropin, it promotes the release of prolactin. It also has some central nervous system effects that have been evaluated for antidepressant therapeutic potential, but, as yet, the results of clinical studies are not considered to be conclusive.[24,25]

Gonadoliberin, as the name implies, is the gonadotropin-releasing hormone (Gn-RH) and is also known as luteinizing hormone-releasing hormone (LH-RH). This hypothalamic decapeptide stimulates the release of luteinizing hormone (LH) and follicle-stimulating hormone (FSH) by the pituitary. LH-RH is considered to be of potential therapeutic importance in the treatment of hypogonadotropic infertility in both males and females.[26]

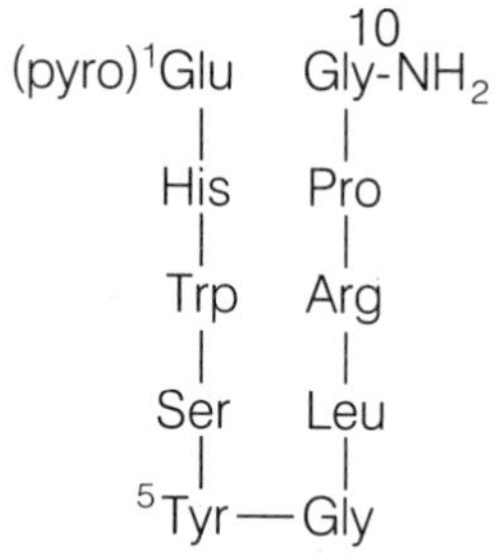

```
(pyro)¹Glu   Gly-NH₂ (10)
    |          |
   His        Pro
    |          |
   Trp        Arg
    |          |
   Ser        Leu
    |          |
  ⁵Tyr ——————Gly
```

Luteinizing Hormone-Releasing Hormone (LH-RH)

A hypothalamic growth-releasing factor (GRF), also called somatoliberin, continues to be under intensive investigation. Its identification and biologic characterization remain to be completed, but physiologic and clinical data support the existence of hypothalamic control of pituitary release of somatotropin.

Somatostatin is another very interesting hypothalamic hormone.[24] It is a tetradecapeptide possessing a disulfide bond linking two cysteine residues, 3-14, in the form of a 38-member ring. Somatostatin suppresses several endocrine systems. It inhibits the release of somatotropin and thyrotropin by the pituitary. It also inhibits the secretion of insulin and glucagon by the pancreas. Gastrin, pepsin, and secretin are intestinal hormones that are likewise affected by somatostatin. The therapeutic potential of somatostatin will be discussed later in relation to the role of glucagon in the pathology of human diabetes.

Other hypothalamic hormones include the luteinizing hormone release-inhibiting factor (LHRIF), prolactin-releasing factor (PRF), corticotropin-releasing factor (CRF), melanocyte-stimulating hormone-releasing factor (MRF), and melanocyte-stimulating hormone release-inhibiting factor (MIF).

As the foregoing discussion illustrates, the hypothalamic endocrine system performs many essential functions affecting other endocrine systems.[27] In turn, the thalamus and cortex exert control on the secretion of these (hypothalamic) factors. A complete review of this field is beyond the scope of this chapter; hence, the interested reader should refer to the literature cited.[23–27]

PITUITARY HORMONES

The pituitary gland, or the hypophysis, is located at the base of the skull and is attached to the hypothalamus by a stalk. The pituitary gland plays a major role[23] in regulating activity of the endocrine organs, including the adrenal cortex, the gonads, and the thyroid. The neurohypophysis (posterior pituitary), which originates from the brain, and the adenohypophysis (anterior pituitary), which is derived from epithelial tissue, are the two embryologically and functionally different parts of the pituitary gland. The adenohypophysis is under the control of hypothalamic regulatory hormones, and it secretes ACTH, GH, LH, FSH, prolactin, and others. The neurohypophysis is responsible for the storage and secretion of the hormones vasopressin and oxytocin, controlled by nerve impulses traveling from the hypothalamus.

Adrenocorticotropic Hormone

The adrenocorticotropic hormone (adrenocorticotropin; ACTH; corticotropin) is a medicinal agent that has been the center of much research. In the late 1950s its structure was elucidated, and the total synthesis was accomplished in the 1960s. Related peptides also have been synthesized, and some of these possess similar physiologic action. Human ACTH has 39-amino acid units within the polypeptide chain.

Structure–activity relationship studies of ACTH[28] showed that the COOH-terminal sequence is not particularly important for biologic activity. Removal of NH_2-terminal amino acid results in complete loss of steroidogenic activity. Full activity has been reported for synthetic peptides containing the first 20 amino acids. A peptide containing 24 amino acids has full steroidogenic activity, without allergenic reactions. This is of practical importance because natural ACTH preparations sometimes produce clinically dangerous allergic reactions.

Corticotropin exerts its major action on the adrenal cortex, promoting steroid synthesis by stimulating the formation of pregnenolone from cholesterol. An interaction between ACTH and specific receptors is implicated in the mechanism leading to stimulation of adenylate cyclase and acceleration of steroid production. The rate-limiting step in the biosynthesis of steroids from cholesterol is the oxidative cleavage of the side chain of cholesterol, which results in the formation of pregnenolone. This rate-limiting is step-regulated by cAMP.[29] Corticotropin, through cAMP, stimulates the biosynthesis of steroids from cholesterol by increasing the availability of free cholesterol. This involves activation of cholesterol esterase by phosphorylation. Corticotropin also stimulates the uptake of cholesterol from plasma lipoproteins. Other biochemical effects exerted by ACTH include stimulation of phosphorylase and hydroxylase activities. Glycolysis also is increased by this hormone. Enzyme systems that catalyze processes involving the production of NADPH are also stimulated. (It is noteworthy that NADPH is required by the steroid hydroxylations that take place in the overall transformation of cholesterol to hydrocortisone, the major glucocorticoid hormone.) Pharmaceutically important ACTH products are listed in Table 20-7.

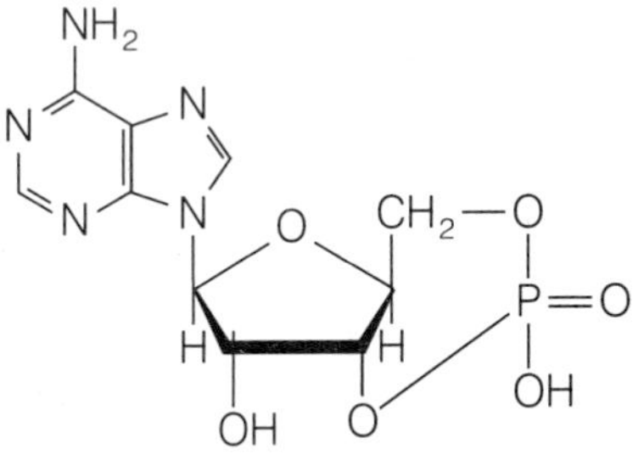

NH_2 N N N N O CH_2—O P=O OH H H OH O

cAMP

Corticotropin Injection, USP. ACTH injection; adrenocorticotropin injection is a sterile preparation of the principle or principles derived from the anterior lobe of the pituitary of mammals used for food by humans. It occurs as a colorless or light

[1]Ser	Val[20]—Lys		
Tyr	Pro	Val	Phe
Ser	Arg	Tyr	Glu
Met	Arg	Pro	Leu
[5]Glu	Lys	Asn[25]	Pro
His	Lys[15]	Gly	Phe[35]
Phe	Gly	Ala	Ala
Arg	Val	Glu	Glu
Tyr	Pro	Asp	Ala
[10]Gly—Lys		Gln[30]—Ser	

Corticotropin

straw-colored liquid, or soluble amorphous solid by drying such liquid from the frozen state. It exerts a tropic influence on the adrenal cortex. The solution has a *p*H range of 3.0 to 7.0 and is used for its adrenocorticotropic activity.

Repository Corticotropin Injection, USP. Corticotropin gel; purified corticotropin; ACTH purified is corticotropin in a solution of partially hydrolyzed gelatin to be used intramuscularly for a more uniform and prolonged maintenance of activity.

Sterile Corticotropin Zinc Hydroxide Suspension, USP, is a sterile suspension of corticotropin, adsorbed on zinc hydroxide and contains no less than 45 and no more than 55 μg of zinc for each 20 USP corticotropin units. Because of its prolonged activity owing to slow release of corticotropin, an initial dose of 40 USP units can be administered intramuscularly, followed by a maintenance dose of 20 units, two or three times a week.

Cosyntropin (Cortrosyn) is a synthetic peptide containing the first 24 amino acids of natural corti-

TABLE 20-7

PHARMACEUTICALLY IMPORTANT ACTH PRODUCTS

Preparation Proprietary Name	*Category*	*Usual Adult Dose**	*Usual Dose Range**	*Usual Pediatric Dose**
Corticotropin injection USP Corticotropin for injection USP *Acthar*	Adrenocorticotropic hormone; adrenocortical steroid (anti-inflammatory); diagnostic aid (adrenocortical insufficienccy)	Adrenocorticotropic hormone: parenteral, 20 USP units, qid. Adrenocortical steroid (anti-inflammatory): parenteral, 20 USP units qid. Diagnostic aid (adrenocortical insufficiency): rapid test — IM or IV, 25 USP units, with blood sampling in 1 hr; adrenocortical steroid output — IV infusion, 25 U in 500–1,000 mL of 5% dextrose injection over a period of 8 hr on each of 2 successive days, with 24-hr urine collection done on each day	Adrenocorticotropic hormone: 40–80 U/day; adrenocortical steroid (anti-inflammatory): 40–80 U/day	Parenteral, 0.4 U/kg of body weight or 12.5 U/m^2 of body surface, qid
Repository corticotropin injection USP *Acthar Gel,* *Cortrophin Gel*	Adrenocorticotropic hormone; adrenocortical steroid (anti-inflammatory); diagnostic aid (adrenocortical insufficiency)	Adrenocorticotropic hormone: IM or SC, 40–80 U every 24–72 hr; IV infusion, 40–80 U in 500 mL of 5% dextrose injection given over an 8-hr period, qd. Adrenocortical steroid (anti-inflammatory): IM or SC, 40–80 U every 24–72 hr; IV infusion, 40–80 U in 500 mL of 5% dextrose injection given over an 8-hr period, qd. Diagnostic aid (adrenocortical insufficiency): IM, 40 U bid on each of 2 successive days, with 24-hr urine collection done each day		Adrenocorticotropic hormone: parenteral, 0.8 U/kg of body weight or 25 U/m^2 of body surface per dose
Sterile corticotropin zinc hydroxide suspension USP *Cortrophin-Zinc*	Adrenocorticotropic hormone; adrenocortical steroid (anti-inflammatory); diagnostic aid (adrenocortical insufficiency)	Adrenocorticotropic hormone: IM, initial, 40–60 U/day, increasing interval to 48, then 72 hr; reduce dose per injection thereafter; maintenance, 20 U/day to twice weekly. Adrenocortical steroid (anti-inflammatory): IM, initial, 40–60 U/day, increasing interval to 48, then 72 hr; reduce dose per injection thereafter; maintenance, 20 U/day to twice weekly. Diagnostic aid (adrenocortical insufficiency): IM, 40 U on each of 2 successive 24-hr periods		
Cosyntropin *Cortrosyn*	Diagnostic aid (adrenocortical insufficiency)	IM or IV, 250 μg		Children 2 yr of age or less, 0.125 mg

*See USP DI for complete dosage information.

cotropin. Cosyntropin is used as a diagnostic agent to test for adrenal cortical deficiency. Plasma hydrocortisone concentration is determined before and 30 minutes after the administration of 250 μg of cosyntropin. Most normal responses result in an approximate doubling of the basal hydrocortisone concentration in 30 to 60 minutes. If the response is not normal, adrenal insufficiency is indicated. Such adrenal insufficiency could be due to either adrenal or pituitary malfunction, and further testing is required to distinguish between the two. Cosyntropin (250 μg infused within four to eight hours) or corticotropin (80 to 120 U/day for three to four days) is administered. Patients with functional adrenal tissue should respond to this dosage. Patients who respond accordingly are suspected of hypopituitarism and the diagnosis can be confirmed by other tests for pituitary function. On the other hand, little or no response is shown by patients who have Addison's disease.

MELANOTROPINS (MELANOCYTE-STIMULATING HORMONE)

Melanocyte-stimulating hormone (MSH) is elaborated by the intermediate lobe of the pituitary gland and regulates pigmentation of fish and amphibians' skin, and to a less extent in humans. Altered secretion of MSH has been implicated in causing changes in skin pigmentation during the menstrual cycle and pregnancy. The two major types of melanotropin, α-MSH and β-MSH, are derived from ACTH and β-lipotropin, respectively. α-MSH contains the same amino acid sequence as the first 13 amino acids of ACTH; β-MSH has 18-amino acid residues. A third melanotropin, γ-melanotropin, is derived from a larger peptide precursor, proopiomelanocortin (POMC). Some important endocrinologic correlations of interest include inhibitory actions of hydrocortisone on the secretion of MSH and the inhibitory effects of epinephrine and norepinephrine on MSH action.

LIPOTROPINS (ENKEPHALINS AND ENDORPHINS)

Opiates, such as opium and morphine, have been known to humans for centuries as substances that relieve pain and suffering. Neuropharmacologists have theorized that opiates interact with receptors in the brain that are affected by endogenous substances that function as regulators of pain perception. The important breakthrough came in 1975, with the isolation of two peptides with opiatelike activity[30] from pig brains. These related pentapeptides, called methionine-enkephalin and leucine-enkephalin, are abundant in certain nerve terminals and have been shown to occur in the pituitary gland.

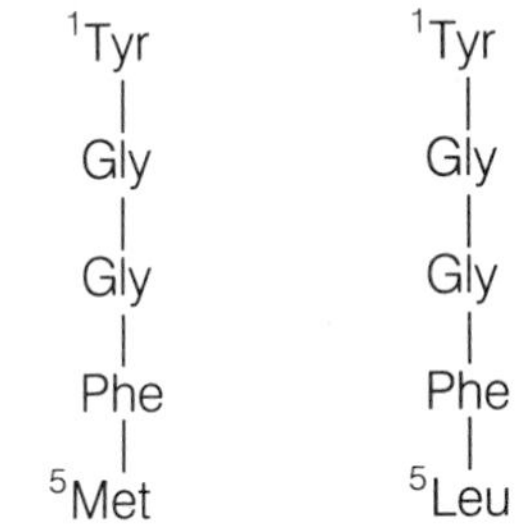

```
¹Tyr            ¹Tyr
 |               |
Gly             Gly
 |               |
Gly             Gly
 |               |
Phe             Phe
 |               |
⁵Met            ⁵Leu
[Met]Enkephalin  [Leu]Enkephalin
```

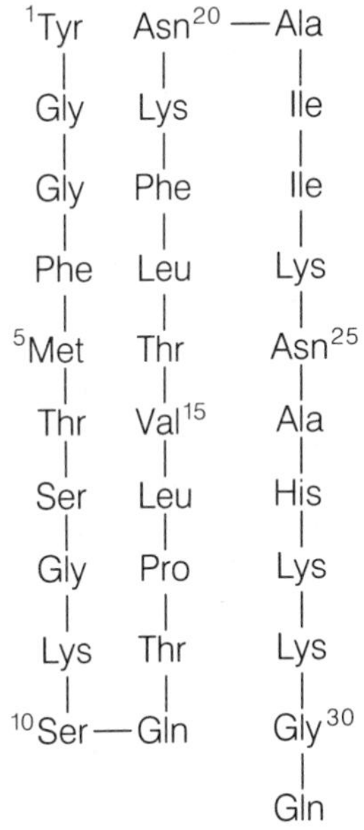

```
¹Tyr    Asn²⁰ — Ala
 |       |       |
Gly     Lys     Ile
 |       |       |
Gly     Phe     Ile
 |       |       |
Phe     Leu     Lys
 |       |       |
⁵Met    Thr     Asn²⁵
 |       |       |
Thr     Val¹⁵   Ala
 |       |       |
Ser     Leu     His
 |       |       |
Gly     Pro     Lys
 |       |       |
Lys     Thr     Lys
 |       |       |
¹⁰Ser — Gln     Gly³⁰
                 |
                Gln
```

β-Endorphin (sheep)

An examination of the structures of enkephalins revealed that the amino acid sequence of met-enkephalin was identical with the sequence of residues 61–65 of β-lipotropin (β-LPH), a larger peptide found in the pituitary gland. This discovery suggested that β-LPH may be a precursor for other larger peptides containing the met-enkephalin sequence. Soon after the structural relationship between β-LPH and met-enkephalin was established, longer peptides, called endorphins, were isolated from the intermediate lobe of the pituitary gland. The endorphins (α, β, and γ) contained the met-enkephalin amino acid sequence and possessed morphinelike activity.[31] The longest of these peptides, β-endorphin, a 31-residue peptide (residues 61–91 of β-LPH), is about 20 to 50 times more potent than morphine as an analgesic and has a considerably longer duration of action than that of enkephalins. Numerous enkephalin analogues and derivatives have been prepared and their biologic activity evaluated. Like morphine, β-endorphin and the enkephalins can induce tolerance and dependence.

In addition to the enkephalins and endorphins, several other opioid peptides have been extracted from pituitary, adrenal, and nervous tissue, including dynorphins and neoendorphins. It is now clear that β-LPH, ACTH, and melanocyte-stimulating hormone (γ-MSH) are all derived from the same precursor, proopiomelanocortin (POMC). Krieger and Liotta* have summarized and critically reviewed the available data concerning synthesis, distribution, regulation, and function of these hormones, and they conclude that pituitary hormones originating in the brain may be involved in central coordination of responses independently of those affected by peripheral secretion of such pituitary hormones. On the other hand, those hormones that appear to enter the brain by possible retrograde portal blood flow may participate in short-loop feedback regulation of anterior pituitary function. Another peptide, proenkephalin, is the primary precursor for met- and leu-enkephalins. The neoendorphins and dynorphins are derived from the same peptide precursor, prodynorphin.[31]

The endorphins and enkephalins have a wide range of biologic effects and most of their actions are in the central nervous system. Their actions include inhibition of release of dopamine in the brain tissue and inhibition of release of acetylcholine from neuromuscular junctions. The role of endorphins and enkephalins as inhibitory neurotransmitters agrees well with the observed biologic effects of these peptides in lowering response to pain and other stimuli. The role of endorphins and enkephalins as neurotransmitters and neuromodulators, with emphasis on receptor interactions, has been reviewed.[32]

GROWTH HORMONE (SOMATOTROPIN)

Growth hormone (GH) is a 191-residue polypeptide elaborated by the anterior pituitary. The amino acid sequence of GH has been determined, and comparison with growth hormones of different species have revealed a considerable amount of structural variation.[33]

The major biologic action of GH is to promote overall somatic growth. Deficiency in the secretion of this hormone can cause dwarfism, and an overproduction of this hormone can cause acromegaly and giantism. The secretion of this hormone is stimulated by growth hormone-releasing hormone (GH-RH), a 44-residue polypeptide secreted by the hypothalamus. Secretion of GH is inhibited by somatostatin (SRIF).

* See Krieger, D. T., and Liotta, A. S. Science 205:366, 1979.

Growth hormone stimulates protein synthesis, both in the skeletal muscles and in the liver. In the liver, GH stimulates uptake of amino acids and promotes the synthesis of all forms of RNA. It stimulates glucagon secretion by the pancreas, increases synthesis of glycogen in muscles, augments the release of fatty acids from adipose tissue, and increases osteogenesis. Growth hormone also causes acute hypoglycemia followed by elevated blood glucose concentration and, perhaps, glycosuria.

Growth hormone has been recognized as an effective replacement therapy for GH-deficient children. The supply of GH, however, was very limited because its source was the pituitary glands of human cadavers. There were also several reports of deaths in children with Crentzfeldt-Jacob disease (caused by viral contamination of GH), which halted the distribution of GH in 1977. Both these problems were solved recently with the application of recombinant DNA technology in the commercial production of somatrem and somatropin.

Somatrem (systemic) (Protropin) is a biosynthetic form of human GH that differs from the pituitary-derived GH and recombinant somatotropin by addition of an extra amino acid, methionine. Because of its structural difference from the natural GH, patients receiving somatrem may develop antibodies, which may result in a decreased response to it. Somatrem is administered intramuscularly or subcutaneously, and the therapy is continued as long as the patient is responsive, until the patient reaches a mature adult height, or until the epiphyses close. The dosage range is 0.05 to 0.1 IU.

Somatropin (rDNA origin) for injection (Humatrope) is natural-sequence human GH of recombinant DNA origin. Its composition and sequence of amino acids are identical with those of human GH of pituitary origin. It is administered intramuscularly or subcutaneously. The dosage range is from 0.05 to 0.1 IU.

PROLACTIN

Prolactin (PRL), a hormone secreted by the anterior pituitary, was discovered in 1928. It is 198-residue polypeptide with the general structural features similar to those of the growth hormone.[34]

Prolactin stimulates lactation of parturition.

GONADOTROPIC HORMONES

The two principal gonadotropins elaborated by the adenohypophysis are follicle-stimulating hormone (FSH) and luteinizing hormone (LH). LH is also known as interstitial cell-stimulating hormone

(ICSH). The gonadotropins, along with thyrotropin (TSH), form the glycoprotein group of hormones. FSH and LH may be produced by a single cell, the gonadotroph. The secretion of FSH and LH is controlled by the hypothalamus which produces LH-releasing hormone (LH-RH). The LH-RH stimulates the secretion of both FSH and LH, although its effects on the secretion of LH are more pronounced.

Follicle-Stimulating Hormone

Follicle-stimulating hormone (FSH) promotes the development of ovarian follicles to maturity and also promotes spermatogenesis in testicular tissue. It is a glycoprotein, the carbohydrate component of which is considered to be associated with its activity.

Luteinizing Hormone

Luteinizing hormone (LH) is another glycoprotein. It acts after the maturing action of FSH on ovarian follicles and stimulates production of estrogens and transforms the follicles into corpora lutea. LH also acts in the male of the species by stimulating the Leydig cells that produce testosterone.

Menotropins

Pituitary hormones, prepared from the urine of postmenopausal women whose ovarian tissue does not respond to gonadotropin, are available for medicinal use in the form of the product, menotropins (Pergonal). The latter has FSH and LH gonadotropin activity in a 1 : 1 ratio. Menotropins is useful in the treatment of anovular women whose ovaries are responsive to pituitary gonadotropins, but who have a gonadotropin deficiency caused by either pituitary or hypothalamus malfunction. Usually, menotropins is administered intramuscularly: initial dose of 75 IU of FSH and 75 IU of LH daily for 9 to 12 days, followed by 10,000 IU of chorionic gonadotropin one day after the last dose of menotropins.

THYROTROPIN

The thyrotropic hormone, also called thyrotropin (TSH), is a glycoprotein consisting of two polypeptide chains. This hormone promotes production of thyroid hormones by affecting the kinetics of the mechanism whereby the thyroid concentrates iodide ions from the bloodstream, thereby promoting incorporation of the halogen into the thyroid hormones and release of hormones by the thyroid.

Thyrotropin. Thyroid-stimulating hormone; TSH (Thytropar) appears to be a glycoprotein (M_r 26,000 to 30,000) containing glucosamine, galactosamine, mannose, and fucose, the homogeneity of which is yet to be established. It is produced by the basophil cells of the anterior lobe of the pituitary gland. TSH enters the circulation from the pituitary, presumably traversing cell membranes in the process. After exogenous administration it is widely distributed and disappears very rapidly from circulation. Some evidence suggests that the thyroid may directly inactivate some of the TSH by an oxidation mechanism that may involve iodine. TSH thus inactivated can be reactivated by certain reducing agents. TSH regulates the production by the thyroid gland of thyroxine which stimulates the metabolic rate. Thyroxine feedback mechanisms regulate the production of TSH by the pituitary gland.

The decreased secretion of TSH from the pituitary is a part of a generalized hypopituitarism that leads to hypothyroidism. This type of hypothyroidism can be distinguished from primary hypothyroidism by the administration of TSH in doses sufficient to increase the uptake of radioiodine or to elevate the blood or plasma protein-bound iodine (PBI) as a consequence of enhanced secretion of hormonal iodine (thyroxine). Interestingly, massive doses of vitamin A inhibit the secretion of TSH. Thyrotropin is used as a diagnostic agent to differentiate between primary and secondary hypothyroidism. Its use in hypothyroidism caused by pituitary deficiency has limited application; other forms of treatment are preferable.

Dose, intramuscular or subcutaneous, 10 IU.

SOMATOSTATIN

Somatostatin (SRIF) was first discovered in the hypothalamus. It is now established that it is elaborated by the δ-cells of the pancreas and elsewhere in the body. Somatostatin is an oligopeptide (14-amino acid residues) and is also referred to as somatotropin release-inhibiting factor (SRIF).[35]

Its primary action is inhibiting the release of GH from the pituitary gland. Somatostatin also suppresses the release of both insulin and glucagon. It causes a decrease in both cAMP levels and adenylate cyclase activity. It was also found to inhibit calcium ions influx into the pituitary cells and suppressed glucose-induced pancreatic insulin secretion by activating and deactivating potassium ion and calcium ion permeability, respectively. The chemistry, struc-

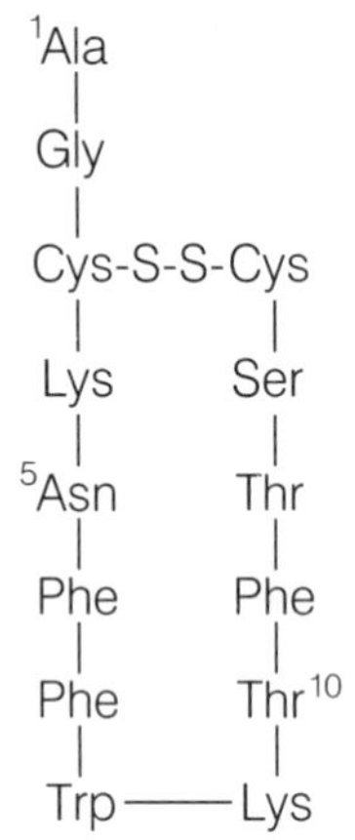

```
1Ala
 |
Gly
 |
Cys-S-S-Cys
 |       |
Lys     Ser
 |       |
5Asn    Thr
 |       |
Phe     Phe
 |       |
Phe     Thr10
 |       |
Trp-----Lys
```

Somatostatin

ture–activity relationships, and potential clinical applications have been reviewed.[24,36]

A powerful new synthetic peptide octreotide acetate (Sandostatin) that mimics the action of somatostatin has received approval from FDA for the treatment of certain rare forms of intestinal endocrine cancers, such as malignant carcinoid tumors and vasoactive intestinal peptide-secreting tumors (VIPomas).

PLACENTAL HORMONES

Human Chorionic Gonadotropin

Human chorionic gonadotropin (hCG) is a glycoprotein synthesized by the placenta. Estrogens stimulate the anterior pituitary to produce placentotropin, which in turn stimulates hCG synthesis and secretion. The hCG is produced primarily during the first trimester of pregnancy. It exerts effects that are similar to those of pituitary LH.

hCG is used therapeutically in the management of cryptorchidism in prepubertal boys. It is also used in women in conjunction with menotropins to induce ovulation when the endogenous availability of gonadotropin is not normal.

Human Placental Lactogen

Human placental lactogen (hPL) is also called human choriomammotropin and chorionic growth-hormone prolactin. This hormone exerts numerous actions. In addition to mammotropic and lactotropic effects, it also exerts somatotropic and luteotropic actions. It has been identified as a protein composed of 191-amino acid units in a single-peptide chain with two disulfide bridges.[26] hPL resembles human somatotropin.

NEUROHYPOPHYSEAL HORMONES (OXYTOCIN, VASOPRESSIN)

The posterior pituitary (neurohypophysis) is the source of vasopressin, oxytocin, α- and β-melanocyte-stimulating hormones, and coherin. The synthesis, transport, and release of these hormones have been reviewed by Brownstein et al.[37] It is herein noted that vasopressin and oxytocin are synthesized and released by neurons of the hypothalamic–neurohypophyseal system. These peptide hormones, and their respective neurophysin carrier proteins, are synthesized as structural components of separate precursor proteins, and these proteins appear to be partially degraded into smaller bioactive peptides in the course of transport along the axon.

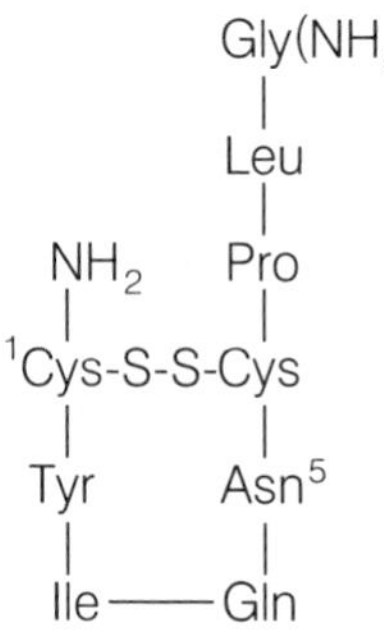

```
           Gly(NH2)                 Gly(NH2)
             |                        |
            Leu                      Arg
             |                        |
   NH2      Pro             NH2      Pro
    |        |               |        |
  1Cys-S-S-Cys             1Cys-S-S-Cys
    |        |               |        |
   Tyr      Asn5            Tyr      Asn5
    |        |               |        |
   Ile------Gln             Phe------Gln
```

Oxytocin 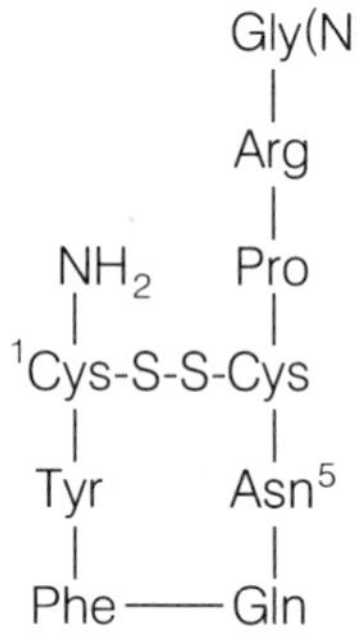 Vasopressin

The structures of vasopressin and oxytocin have been elucidated, and these peptides have been synthesized. Actually, three closely related nonapeptides have been isolated from mammalian posterior pituitary: oxytocin and argininevasopressin from most mammals, and lysinevasopressin from pigs. The vasopressins differ from one another in the nature of the eighth amino acid residues: arginine and lysine, respectively. Oxytocin has leucine at position 8 and its third amino acid is isoleucine instead of phenylalanine. Several analogues of vasopressin have been synthesized and their antidiuretic activity evaluated. Desmopressin, 1-desamino-8-arginine-vasopressin, is a synthetic derivative of vasopressin. It is a longer-acting and more potent antidiuretic than vasopressin, with much less pressor activity. Desmopressin is much more resistant to the actions of peptidases because of the deamination at position 1, which accounts for its longer duration of action. The substitution of D- for L-arginine in position 8 accounts for its sharply lower vasoconstrictor effects.[38]

Vasopressin is also known as the pituitary antidiuretic hormone (ADH). This hormone can effect graded changes in the permeability of the distal portion of the mammalian nephron to water, resulting in either conservation or excretion of water; thus, it modulates the renal tubular reabsorption of water. ADH has been shown to increase cAMP production in several tissues. Theophylline, which promotes cAMP by inhibiting the enzyme (phosphodiesterase) that catalyzes its hydrolysis, causes permeability changes similar to those caused by ADH. Cyclic-AMP also effects similar permeability changes; hence, it is suggested that cAMP is involved in the mechanism of action of ADH.

The nonrenal actions of vasopressin include its vasoconstrictor effects and neurotransmitter actions in the central nervous system, such as regulation of ACTH secretion, circulation, and body temperature.[39]

ADH is therapeutically useful in the treatment of diabetes insipidus of pituitary origin. It also has been used to relieve intestinal paresis and distention.

Oxytocin is appropriately named on the basis of its oxytocic action. Oxytocin exerts stimulant effects on the smooth muscle of the uterus and mammary gland. On the other hand, this hormone has a relaxing effect on vascular smooth muscle when administered in high doses. It is considered to be the drug of choice to induce labor and to stimulate labor in cases of intrapartum hypotonic inertia. Oxytocin also is used in inevitable or incomplete abortion after the 20th week of gestation. It also may be used to prevent or control hemorrhage and to correct uterine hypotonicity. In some cases, oxytocin is used to promote milk ejection; it acts by contracting the myoepithelium of the mammary glands. Oxytocin is usually administered parenterally by intravenous infusion, intravenous injection, or intramuscular injection. Oxytocin citrate buccal tablets are also available, but the rate of absorption is unpredictable and buccal administration is less precise. Topical administration (nasal spray) two or three minutes before nursing to promote milk ejection is sometimes recommended.[40] Refer to Table 20-8 for product listing.

Oxytocin Injection, USP, is a sterile solution in water for injection of oxytocic principle prepared by synthesis or obtained from the posterior lobe of the pituitary of healthy, domestic animals used for food by humans. The *p*H is 2.5 to 4.5; expiration date, three years.

Oxytocin preparations are widely used with or without amniotomy to induce and stimulate labor. Although injection is the usual route of administration, the sublingual route is extremely effective. Sublingual and intranasal spray (Oxytocin Nasal Solution USP) routes of administration also will stimulate milk let-down.

Vasopressin Injection, USP (Pitressin) is a sterile solution of the water-soluble pressor principle of the posterior lobe of the pituitary of healthy, domestic animals used for food by humans, or prepared by synthesis. Each milliliter possesses a pressor activity equal to 20 USP posterior pituitary units; expiration date, three years.

Vasopressin Tannate (Pitressin Tannate) is a water-insoluble tannate of vasopressin administered

TABLE 20-8

NEUROHYPOPHYSEAL HORMONES: PHARMACEUTICAL PRODUCTS

Preparation Proprietary Name	***Category***	***Usual Adult Dose****	***Usual Pediatric Dose****
Oxytocin injection USP *Pitocin, Syntocinon*	Oxytocic	IM, 3–10 U after delivery of placenta; IV, initially no more than 1–2 mU/min, increased every 15–30 min in increments of 1–2 mU	
Oxytocin nasal solution, USP *Syntocinon*	Oxytocic	1 spray or 3 drops in 1 or both nostrils 2–3 min before nursing or pumping of breasts	
Vasopressin injection USP *Pitressin*	Antidiuretic posterior pituitary hormone	IM or SC, 2.5–10 U tid or qid as necessary	IM or SC, 2.5–10 U tid or qid as necessary
Sterile vasopressin tannate oil suspension *Pitressin*	Antidiuretic posterior pituitary hormone	IM, 1.5–5 U every 1–3 days	IM, 1.25–2.5 U every 1–3 days
Desmopressin acetate nasal solution *DDAVP*	Antidiuretic posterior pituitary hormone	Maintenance: Intranasal, 2–4 μg/day, as a single dose or in 2–3 divided doses	Maintenance: Intranasal, 2–4 μg/kg of body weight per day or 5–30 mg/day or in 2–3 divided doses
Desmopressin acetate injection *DDAVP, Stimate*	Antidiuretic posterior pituitary hormone	IV or SC, 2–4 μg/day usually in 2 divided doses in the morning or evening	IV, 3 μg/kg of body weight diluted in 0.9% sodium chloride injection USP

*See USP DI for complete dosage information.

intramuscularly (1.5 to 5 pressor units daily) for its prolonged duration of action by the slow release of vasopressin. It is particularly useful for patients who have diabetes insipidus, but it never should be used intravenously.

Felypressin. 2-L-Phenylalanine-8-L-lysinevasopressin has relatively low antidiuretic activity and little oxytocic activity. It has considerable pressor (i.e., vasoconstrictor) activity which, however, differs from that of epinephrine (i.e., following capillary constriction in the intestine it lowers the pressure in the vena portae, whereas epinephrine raises the portal pressure). Felypressin also causes an increased renal blood flow in the cat, whereas epinephrine brings about a fall in renal blood flow. Felypressin is five times more effective as a vasopressor than is lysinevasopressin and is recommended in surgery to minimize blood flow, especially in obstetrics and gynecology.

Lypressin is synthetic 8-L-lysinevasopressin, a polypeptide similar to the antidiuretic hormone. The lysine analogue is considered to be more stable, and it is rapidly absorbed from the nasal mucosa. Lypressin (Diapid) is pharmaceutically available as a topical solution, spray, 50 pressor units (185 μg)/mL in 5-mL containers. Usual dosage, topical (intranasal), one or more sprays applied to one or both nostrils one or more times daily.[29]

Desmopressin Acetate (DDAVP; Stimate) is synthetic 1-desamino-8-D-argininevasopressin. Its efficacy, ease of administration (intranasal), long duration of action, and lack of side effects make it the drug of choice for the treatment of central diabetes insipidus. It may also be administered intramuscularly or intravenously. It is preferred to vasopressin injection and oral antidiuretics for use in children. It is also indicated in the management of temporary polydipsia and polyuria associated with trauma to, or surgery in, the pituitary region.

PANCREATIC HORMONES

Relationships between lipid and glucose levels in the blood and the general disorders of lipid metabolism found in diabetic subjects have received the attention of many chemists and clinicians. To understand diabetes mellitus, its complications, and its treatment, one has to begin at the level of the basic biochemistry of the pancreas and the ways carbohydrates are correlated with lipid and protein metabolism (see Chap. 19). The pancreas produces insulin, as well as glucagon; β-cells secrete insulin and the α-cells secrete glucagon. Insulin will be considered first.

Insulin

One of the major triumphs of this century occurred in 1922, when Banting and Best extracted insulin from dog pancreas.[41] Recent advances* in the biochemistry of insulin have been reviewed with emphasis on proinsulin biosynthesis, conversion of proinsulin to insulin, insulin secretion, insulin receptors, metabolism, effects by sulfonylureas, and so on.[42–44]

Insulin is synthesized by the islet β-cells from a single-chain, 86-amino acid polypeptide precursor, proinsulin.[45] Proinsulin itself is synthesized in the polyribosomes of the rough endoplasmic reticulum of the β-cells from an even larger polypeptide precursor, termed preproinsulin. The B-chain of preproinsulin is extended at the NH_2-terminus by at least 23 amino acids. Proinsulin then traverses the Golgi apparatus and enters the storage granules in which the conversion to insulin occurs.

The subsequent proteolytic conversion of proinsulin to insulin is accomplished by the removal of Arg-Arg residue at positions 31 and 32 and Arg-Lys residue at positions 64 and 65 by an endopeptidase that resembles trypsin in its specificity and a thiol-activated carboxypeptidase B-like enzyme.[46]

The actions of these proteolytic enzymes on proinsulin results in the formation of equimolar quantities of insulin and the connecting C-peptide. The resulting insulin molecule consists of chains A and B having 21- and 31-amino acid residues, respectively. The chains are connected by two disulfide linkages, with an additional disulfide linkage within chain A (Fig. 20.6).

The three-dimensional structure of insulin has been determined by x-ray analysis of single crystals. These studies have demonstrated that the high bioactivity of insulin depends on the integrity of the overall conformation. The biologically active form of the hormone is thought to be the monomer. The receptor-binding region, consisting of A-1 Gly, A-4 Glu, A-5 Gln, A-19 Tyr, A-21 Asn and B-12 Val, B-16 Tyr, B-24 Phe, B-26 Tyr, has been identified. The three-dimensional crystal structure appears to be conserved in solution and during its receptor interaction.[26]

The amino acid sequence of insulins from various animal species have been examined.[44] Details of these are shown in Table 20.9. It is apparent from the analysis that frequent changes in sequence occur within the interchain disulfide ring (positions 8, 9, and 10). The hormonal sequence for porcine insulin is the closest to that of humans and differs

* Galloway, J. A. et al. Diabetes Mellitus, 9th ed. Indianapolis, Eli Lily & Co., 1988.

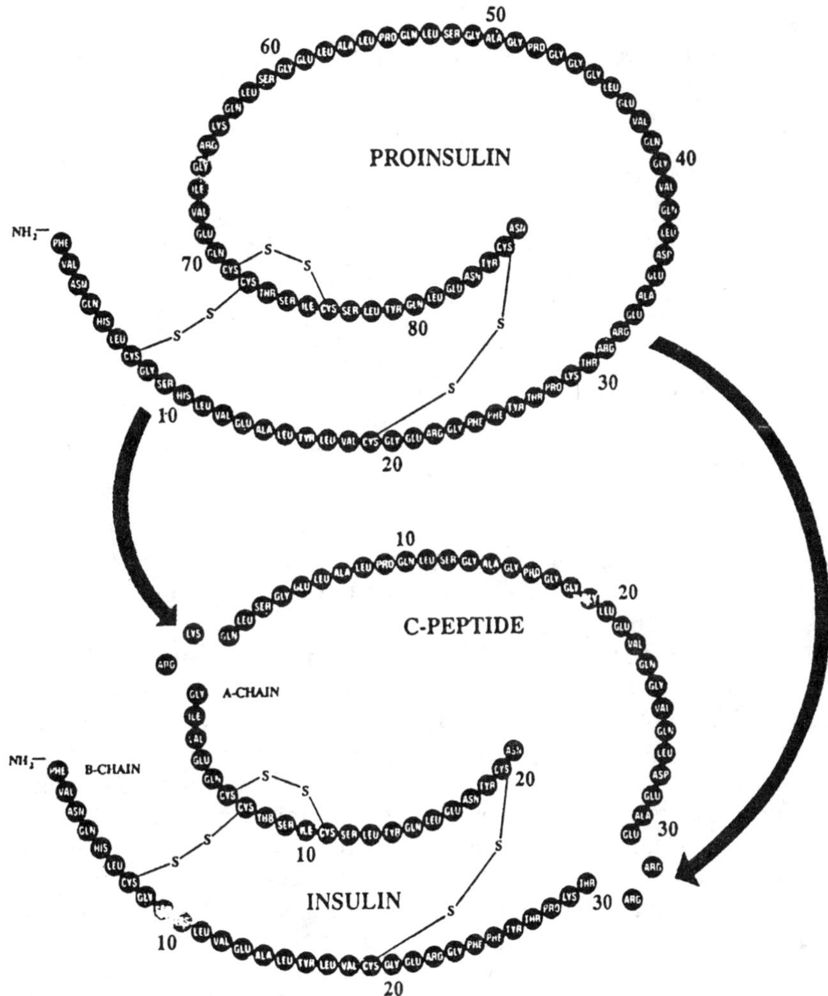

FIG. 20-6. Conversion of proinsulin to insulin.

only by the substitution of an alanine residue at the COOH-terminus of the B-chain. Porcine insulin, therefore, is a good starting material for the synthesis of human insulin.

Insulin composes 1% of pancreatic tissue, and secretory protein granules contain about 10% insulin. These granules fuse with the cell membrane with simultaneous liberation of equimolar amounts of insulin and the C-peptide. Insulin enters the portal vein and about 50% is removed in its first passage through the liver. The plasma half-life of insulin is approximately four minutes, compared with 30 minutes for the C-peptide.

Usually, exogenous insulin is weakly antigenic. Insulin antibodies have been observed to neutralize the hypoglycemic effect of injected insulin. The antibody-binding sites on insulin are quite different from those involved in binding of insulin with its receptors.[48]

Regulation of insulin secretion is affected by numerous factors, such as food, hormonal and neuronal stimuli, and ionic mechanisms.[49] In humans, the principal substrate that stimulates the release of insulin from the islet β-cells is glucose. In addition to glucose, other substrates (e.g., amino acids, free fatty acids, and ketone bodies) also can directly stimulate insulin secretion. Secretin and ACTH can directly stimulate the secretion of insulin. Glucagon and other related peptides can cause an increase in the secretion of insulin, whereas somatostatin inhibits its secretion.

Autonomic neuronal mechanisms also play an important role in regulating insulin release. In the sympathetic nervous system, α-adrenergic agonists inhibit insulin release, whereas β-adrenergic agonists stimulate the release of insulin. In the parasympathetic nervous system, cholinomimetic drugs stimulate the release of insulin.

"Clinical" insulin that has been crystallized five times and then subjected to countercurrent distribution (2-butanol: 1% dichloroacetic acid in water) yields about 90% insulin A, with varying amounts of

TABLE 20-9†

SOME SEQUENCE DIFFERENCES IN INSULINS OF VARIOUS SPECIES

	A-chain				B-chain			
Species	1	8	9	10	−1	1	29	30
Human	Gly.	Thr.	Ser.	Ile.		Phe.	Lys.	Thr.
Pork	Gly.	Thr.	Ser.	Ile.		Phe.	Lys.	Ala.
Beef	Gly.	Ala.	Ser.	Val.		Phe.	Lys.	Ala.
Sheep	Gly.	Ala.	Gly.	Val.		Phe.	Lys.	Ala.
Horse	Gly.	Thr.	Gly.	Ile.		Phe.	Lys.	Ala.
Rabbit	Gly.	Thr.	Ser.	Ile.		Phe.	Lys.	Ser.
Chicken	Gly.	His.	Asn.	Thr.		Ala.	Lys.	Ala.
Cod	Gly.	His.	Arg.	Pro.	Met.	Ala.	Lys.	—
Rat I*	Gly.	Thr.	Ser.	Ile.		Phe.	Lys.	Ser.
Rat II*	Gly.	Thr.	Ser.	Ile.		Phe.	Met.	Ser.

*Asp substitution for Glu at position 4 on A-chain.
†See Reference 44 for details.

insulin B, together with other minor components. A and B differ by an amide group and have the same activity. End-member analysis, sedimentation, and diffusion studies indicate an M_r of about 6000. The value of 12,000 M_r for insulin containing trace amounts of zinc (obtained by physical methods) is probably a bimolecular association product through the aid of zinc. Insulin was the first protein for which a complete amino acid sequence was determined. The extensive studies of Sanger[50] and others have elucidated the amino acid sequence and structure of insulin. Katsoyannis[51] and others followed with the synthesis of A- and B-chains of human, bovine, and sheep insulin. The A- and B-chains were combined to form insulin in 60% to 80% yields, with a specific activity comparable with that of the natural hormone.[35]

The total synthesis of human insulin has been reported by Rittel et al.* These workers were able to selectively synthesize the final molecule appropriately cross-linked by disulfide (—S—S—) groups in yields ranging between 40% and 50%, whereas earlier synthetic methods involved random combination of separately prepared A- and B-chains of the molecule.

An exciting new development, recombinant DNA technology, has been successfully applied in the production of human insulin on a commercial scale. This new technology involves cloning of human insulin gene in *Escherichia coli*.[52] Recently, Eli Lily and Co., in cooperation with Genentech Inc., began the commercial production of insulin by employing recombinant DNA technology. There are two available methods of applying recombinant DNA technology in the production of human insulin. The earlier method involved insertion of genes, for production of either the A- or the B-chain of the insulin molecule, into a special strain of *E. coli* (KI2) and subsequently combining the two chains chemically to produce an insulin that is structurally and chemically identical with pancreatic human insulin. The second, and most recent, method involves the insertion of genes for the entire proinsulin molecule into special *E. coli* bacteria, which are then grown in fermentation process. The connecting C-peptide is then enzymatically cleaved from proinsulin to produce human insulin.[53] Human insulin produced by recombinant DNA technology has been shown to be less antigenic than that from animal sources.

Although insulin is readily available from natural sources (e.g., porcine and bovine pancreatic tissue), partial syntheses and molecular modifications have been developed as the basis for structure–activity relationship (SAR) studies. Such studies have shown that amino acid units cannot be removed from the insulin peptide chain A without significant loss of hormonal activity. On the other hand, several amino acids of chain B are not considered to be essential for activity. Up to the first six and the last three-amino acid units can be removed without significant decrease in activity.[26]

Two insulin analogues, which differ from the parent hormone in that the NH_2-terminus of chain A (A^1) glycine has been replaced by L- and D-alanine, respectively, have been synthesized by Cosmatos et al.[54] for SAR studies. The relative potencies of the L- and D-analogues reveal interesting SARs. The L- and D-alanine analogues are 9.4% and 95%, respectively, as potent as insulin in glucose oxidation. The relative binding affinity to isolated fat cells is reported to be approximately 10% for the L- and 100% for the D-analogue. Apparently, substitution on the α-carbon of A^1 glycine of insulin with a methyl in a particular configuration interferes with the binding; hence, the resulting analogue (that of L-alanine) is

* Rittel, W., et al. Helv. Chim. Acta 67:2617, 1974.

much less active. Methyl substitution in the opposite configuration affects neither the binding nor the bioactivity.

It appears that molecular modifications of insulin on the amino groups lead to reduction of bioactivity, but modifications of the ε-amino group of lysine number 29 on chain B (B-29) may yield active analogues. Accordingly, May et al.[55] synthesized *N*-ε-(+)-biotinyl insulin, which was demonstrated to be equipotent with natural insulin. Complexes of this biotinyl–insulin derivative with avidin were also prepared and evaluated biologically; these complexes showed a potency decrease to 5% of that of insulin. Such complexes conjugated with ferritin are expected to be useful in the development of electron microscope stains of insulin receptors.

Alteration in the tertiary structure of insulin appears to drastically reduce biologic activity as well as loss of receptor binding. The three-dimensional structure provided by x-ray crystallography of insulin monomer has revealed an exposed hydrophobic face that is thought to be involved directly in interacting with the receptor.[56] Thus, loss of biologic activity in insulin derivatives, produced by chemical modification, can be interpreted in terms of adversely affecting this hydrophobic region. It is also observed that species variation in this hydrophobic region is very unusual.

Insulin is inactivated in vivo by (1) an immunochemical system in the blood of insulin-treated patients, (2) reduction of disulfide bonds (probably by glutathione), and (3) by insulinase (a proteolytic enzyme) that occurs in liver. Pepsin and chymotrypsin will hydrolyze some peptide bonds that lead to inactivation. It is inactivated by reducing agents such as sodium bisulfite, sulfurous acid, and hydrogen.

Recent advances in the area of insulin's molecular mechanisms have been reviewed* with emphasis on receptor interactions, effect on membrane structure and functions, effects on enzymes, and the role of second messengers.[36] The insulin receptor is believed to be a glycoprotein complex with a high relative molecular mass (M_r). The receptor is thought to consist of four subunits: two identical α-units with an M_r of about 130,000 Da and two identical β-units with an M_r of 95,000 Da, joined together by disulfide bonds. The α-subunits are primarily responsible for binding insulin to its receptor, and the β-subunits are thought to possess intrinsic protein kinase activity that is stimulated by insulin. The primary effect of insulin may be a kinase stimulation leading to phosphorylation of the receptor as well as other intracellular proteins.[57,60] Additionally, insulin binding to its receptors may result in the generation of a soluble intracellular second messenger (possibly a peptide) that may mediate some insulin activity relating to activation of enzymes such as pyruvate dehydrogenase and glycogen synthetase.[58] The insulin–receptor complex becomes internalized and may serve as a vehicle for translocating insulin to the lysosomes, in which it may be broken down and recycled back to the plasma membrane. The half-life of insulin is about ten hours.

The binding of insulin to its target tissue is determined by several factors. The number of receptors in the target tissue and their affinity for insulin are two important determinants. These factors vary substantially from tissue to tissue. Another important consideration is the concentration of insulin itself. Elevated levels of circulating insulin are known to decrease the number of insulin receptors on target cell surfaces and vice versa. Other factors that affect insulin binding to its receptors include *p*H, temperature, membrane lipid composition, and ionic strength.[59,60] It is conceivable, therefore, that conditions associated with insulin resistance, such as obesity, type I and type II diabetes mellitus, could be caused by altered receptor kinase activity or impaired generation of second messengers (low relative molecular mass peptides); increased degradation of the messenger; or decreased substrates (enzymes involved in metabolic activity) for the messenger or receptor kinase.[68]

Metabolic Effects of Insulin

Insulin has pronounced effects on the metabolism of carbohydrates, lipids, and proteins.[61] The major tissues affected by insulin are muscle (cardiac and skeletal), adipose tissue, and liver. The kidney is much less responsive, whereas others, such as brain tissue and red blood cells, do not respond at all. The actions of insulin are highly complex and diverse. Because many of the actions of insulin are mediated by second messengers, it is difficult to distinguish between its primary and secondary actions.

In muscle and adipose tissue, insulin promotes transport of glucose and other monosaccharides across cell membranes; it also facilitates transport of amino acids, potassium ions, nucleosides, and ionic phosphate. Insulin also activates certain enzymes—kinases and glycogen synthetase in muscle and adipose tissue. In adipose tissue, insulin decreases the release of fatty acids induced by epinephrine or glucagon. Cyclic-AMP promotes fatty acid release from adipose tissue; therefore, it is possible that insulin decreases fatty acid release by reducing tis-

* Insulin's molecular mechanism of action was recently reviewed by Joseph Larner (Diabetes 37:262, 1988).

sue levels of cAMP. Insulin also facilitates the incorporation of intracellular amino acids into protein.

Insulin is believed to influence protein synthesis at the ribosomal level in various tissues.[62] In skeletal muscles, insulin predominantly stimulates translation by increasing the rate of initiation of protein synthesis and increasing the number of ribosomes. In the liver, the predominant effect of insulin is on transcription. In cardiac muscles, insulin is believed to decrease the rate of degradation of proteins.

In the liver, there is no barrier to the transport of glucose into cells but, nevertheless, insulin influences liver metabolism, decreasing glucose output, decreasing urea production, lowering cAMP, and increasing potassium and phosphate uptake. The lowering of cAMP results in decreased activity of glycogen phosphorylase, leading to diminished glycogen breakdown and increased activity of glycogen synthetase. It appears that insulin exerts induction of specific hepatic enzymes involved in glycolysis, while inhibiting gluconeogenic enzymes. Thus insulin promotes glucose utilization through glycolysis by increasing the synthesis of glucokinase, phosphofructokinase, and pyruvate kinase. Insulin decreases the availability of glucose from gluconeogenesis by suppressing pyruvate carboxylase, phosphoenolpyruvate carboxykinase, fructose-1,6-diphosphatase, and glucose-6-phosphatase.

Insulin's effects on lipid metabolism also are important. In adipose tissue, insulin has an antilipolytic action (i.e., an effect opposing the breakdown of fatty acid triglycerides). It also decreases the supply of glycerol to the liver. Thus, at these two sites, insulin decreases the availability of precursors for the formation of triglycerides. Insulin is necessary for the activation and synthesis of lipoprotein lipases, enzymes responsible for lowering very low-density lipoprotein (VLDL) and chylomicrons in peripheral tissue. Other effects of insulin include stimulation of the synthesis of fatty acids (lipogenesis) in the liver.[63]

Diabetes mellitus is a systemic disease caused by a decrease in the secretion of insulin or in a reduced sensitivity or responsiveness to insulin by target tissue (insulin receptor activity). The disease is characterized by hyperglycemia, hyperlipidemia, and hyperaminoacedamia. Diabetes mellitus is frequently associated with the development of micro- and macrovascular diseases, neuropathy, and atherosclerosis. Various types of diabetes have been recognized, classified, and their pathophysiology discussed.[64]

The two major types of diabetes are type I, insulin-dependent diabetes mellitus (IDDM) and type II, non–insulin-dependent diabetes mellitus (NIDDM). Type I diabetes (also known as juvenile-onset diabetes) is characterized by a destruction of pancreatic β-cells, resulting in a deficiency of insulin secretion. Autoimmune complexes and viruses have been mentioned as two possible causes of β-cell destruction. Generally, in type I diabetes, receptor sensitivity to insulin is not decreased. Type II diabetes, also known as adult-onset diabetes, is characterized primarily by insulin receptor defects or postinsulin receptor defects. There is no destruction of β-cells, and insulin secretion is relatively normal. In reality, however, the two types of diabetes show a considerable overlap of clinical features.[64]

Diabetes mellitus is associated with both microangiopathy (damage to smaller vessels, e.g., the eyes and kidney) and macroangiopathy (damage to larger vessels, e.g., atherosclerosis). Hyperlipidemia (characterized by an increase in the concentration of lipoproteins such as VLDL, IDL, and LDL) has been implicated in the development of atherosclerosis and is known to occur in diabetes. Severe hyperlipidemia may lead to life-threatening attacks of acute pancreatitis. It also seems that severe hyperlipidemia causes xanthoma. Researchers also are investigating the relationship between diabetes and endogenous hyperlipidemia (hypertriglyceridemia).[63] Considering the effects of insulin on lipid metabolism, as summarized earlier, one can rationalize that in type II diabetes, in which the patient may actually have an absolute excess of insulin, in spite of the evidence of glucose tolerance tests, the effect of the excessive insulin on lipogenesis in the liver may be enough to increase the level of circulating triglycerides and of VLDL. In type I diabetes with a deficiency of insulin, the circulating level of lipids may rise because too much precursor is available, with fatty acids and carbohydrates going to the liver.

The relationship between the carbohydrate metabolic manifestations of diabetes and the development of micro- and macrovascular diseases have been extensively studied.[65,66] It is becoming increasingly clear that hyperglycemia plays a major role in the development of vascular complications of diabetes, including intercapillary glomerulosclerosis, premature atherosclerosis, retinopathy with its specific microaneurysms and retinitis proliferans, leg ulcers, and limb gangrene. First, hyperglycemia causes an increase in the activity of lysine hydroxylase and galactosyl transferase, two important enzymes involved in glycoprotein synthesis. An increase in the glycoprotein synthesis in the collagen of kidney basement membrane may lead to the development of diabetic glomerulosclerosis. Second, an increase in the uptake of glucose by non–insulin-sensitive tissues (such as nerve Schwann cells and ocular lens cells) occurs during hyperglycemia. Intracellular glucose is enzymaticallyconverted first to sorbitol and then to fructose. The buildup of these sugars inside the cells increases the osmotic pressure in ocular lens cells and Schwann cells, resulting in an increase in water uptake and impairing cell functions. Some forms of diabetic cataracts and diabetic neuropathy are believed to be caused by this pathway. Third, hyperglycemia may precipitate nonenzymatic glycosylation of a variety of proteins in the body, including hemoglobin, serum albumin, lipoprotein, fibrinogen, and basement membrane protein. Glycosylation is believed to alter the tertiary structures of proteins and possibly alter their rate of metabolism. The rate of glycosylation is a function of plasma glucose concentration and the duration of hyperglycemia. Needless to say, this mechanism might play an important role in both macro- and microvascular lesions. Finally, hyperglycemia is known to increase the rate of aggregation and agglutinization of circulating platelets. Platelets play an important role in promoting atherogenesis. The increase in the rate of platelet aggregation and agglutinization leads to the development of microemboli, which can cause transient cerebral ischemic attacks, strokes, and heart attacks.[64]

Modern concepts on the therapeutics of diabetes mellitus have been reviewed by Maurer.[67] This review emphasizes that insulin therapy does not always prevent serious complications. Even diabetics who are considered to be well under insulin therapeutic control experience wide fluctuations in blood glucose concentration, and it is hypothesized that these fluctuations eventually cause the serious complications of diabetes (e.g., kidney damage, retina degeneration, premature atherosclerosis, cataracts, neurological dysfunction, and a predisposition to gangrene).

Insulin Preparations

The various commercially available insulin preparations are listed in Table 20-10. Amorphous insulin was the first form made available for clinical use. Further purification afforded crystalline insulin that is now commonly called *regular insulin*. Insulin injection USP is made from zinc insulin crystals. For some time, regular insulin solutions have been prepared at a *p*H of 2.8 to 3.5; if the *p*H were increased above the acidic range, particles would be formed. However, more highly purified insulin can be maintained in solution over a wider range of *p*H even when unbuffered. Neutral insulin solutions have greater stability than acidic solutions; neutral insulin solutions maintain nearly full potency when stored up to 18 months at 5° and 25°C. As noted in Table 20-10, the various preparations differ in onset and duration of action. A major disadvantage of regular insulin is its short duration of action (five to seven hours), which necessitates its administration several times daily.

Many attempts have been made to prolong the duration of action of insulin; for example, the development of insulin forms possessing less water solubility than the highly soluble (in body fluids) regular insulin. Protamine insulin preparations proved to be less soluble and less readily absorbed from body tissue. Protamine zinc insulin (PZI) suspensions were even longer-acting (36 hours) than protamine insulin; these are prepared by mixing insulin, protamine, and zinc chloride with a buffered solution.

TABLE 20-10

INSULIN PREPARATIONS

Name	Particle Size (μm)	Action	Composition	pH	Duration (hr)
Insulin injection* USP		Prompt	Insulin + $ZnCl_2$	2.5–3.5	5–7
Prompt insulin zinc suspension* USP	2‡	Rapid	Insulin + $ZnCl_2$ + buffer	7.2–7.5	12
Insulin zinc suspension* USP	10–40 (70%) 2 (30%)‡	Intermediate	Insulin + $ZnCl_2$ + buffer	7.2–7.5	18–24
Extended insulin zinc suspension* USP	10–40	Long-acting	Insulin + $ZnCl_2$ + buffer	7.2–7.5	24–36
Globin zinc insulin injection* USP		Intermediate	Globin§ + $ZnCl_2$ + insulin	3.4–3.8	12–18
Protamine zinc insulin suspension† USP		Long-acting	Protamine‖ + insulin + Zn	7.1–7.4	24–36
Isophane insulin suspension* USP	30	Intermediate	Protamine# $ZnCl_2$ insulin buffer	7.1–7.4	18–24

*Clear or almost clear.
†Turbid.
‡Amorphous.
§Globin (3.6–4.0 mg/100 USP units of insulin) prepared from beef blood.
‖Protamine (1.0–1.5 mg/100 USP units of insulin) from the sperm or the mature testes of fish belonging to the genus *Oncorhynchus* or *Salmo*.
#Protamine (0.3–0.6 mg/100 USP units of insulin).

The regular insulin/PZI ratios in clinically useful preparations range from 2 : 1 to 4 : 1.

Isophane insulin suspension incorporates some of the qualities of regular insulin injection and is usually sufficiently long acting (although not as much as protamine zinc insulin) to protect the patient from one day to the next (the term *isophane* is derived from the Greek *iso* and *phane*, meaning equal and appearance, respectively). Isophane insulin is prepared by the careful control of the protamine/insulin ratio and the formation of a crystalline entity containing stoichiometric amounts of insulin and protamine. (Isophane insulin is also known as NPH; the code *N* indicates neutral *p*H, the *P* stands for protamine, and the *H* for Hegedorn, the developer of the product.) NPH insulin has a quicker onset and a shorter duration of action (28 hours) than PZI. NPH is given in single morning doses and normally exhibits greater activity during the day than at night. NPH and regular insulin can be combined conveniently and effectively for many patients with diabetes.

The posology of various insulin preparations is summarized in Table 20-10.

A major concern for PZI and NPH insulins is the potential antigenecity of protamine (obtained from fish). This concern led to the development of Lente insulins. By varying the amounts of excess zinc, by using an acetate buffer (instead of phosphate), and by adjusting the *p*H, two types of Lente insulins were prepared. At high concentrations of zinc, a microcrystalline form precipitates and is called Ultralente. Ultralente insulin is relatively insoluble, has a slower onset, and longer duration of action (comparable with PZI). At a relatively low zinc concentration, an amorphous form precipitates and is called Semilente insulin. The latter is more soluble and has a quicker onset and a shorter duration of action (comparable with that of regular insulins). A third type of insulin suspension, Lente insulin, is a

TABLE 20-11

DOSAGE AND SOURCE OF INSULIN PREPARATIONS

USP Insulin Type	Strengths and Sources	Usual Adult Dose*
Insulin injection (regular insulin, crystalline zinc insulin)	U-40 mixed, U-100 mixed: purified beef, pork; purified pork; biosynthetic human; semisynthetic human U-500: purified pork	Diabetic hyperglycemia: SC, as directed by physician 15–30 min before meals up to tid or qid
Isophane insulin suspension (NPH insulin)	U-40 mixed, U-400 mixed: beef; purified beef, pork; purified pork; biosynthetic human; semisynthetic human	SC, as directed by physician, qd 30–60 min before breakfast. An additional dose before breakfast may be necessary for some patients about 30 min before a meal or at bedtime
Isophane insulin suspension (70%) and insulin injection (30%)	U-100: purified pork; semisynthetic human	SC, as directed by physician, qd, 15–30 min before breakfast, or as directed
Insulin zinc suspension (Lente insulin)	U-40 mixed, U-100 mixed: beef; purified beef; purified pork; biosynthetic human; semisynthetic human	SC, as directed by physician, qd 30–60 min before breakfast. An additional dose may be necessary for some patients about 30 min before a meal or at bedtime
Extended insulin zinc suspension (Ultralente insulin)	U-40 mixéd, U-100 mixed: beef; purified beef	SC, as directed by physician, qd 30–60 min before breakfast
Prompt insulin zinc suspension (Semilente insulin)	U-40 mixed, U-100 mixed: beef; purified pork	SC, as directed by physician, qd 30–60 min before breakfast. An additional dose may be necessary for some patients about 30 min before a meal or at bedtime
Protamine zinc insulin suspension (PZI Insulin)	U-40 mixed, U-100 mixed: purified pork	SC, as directed by physician, qd 30–60 min before breakfast

*See USP DI for complete dosage information.

70 : 30 mixture of Ultralente and Semilente insulins. Lente insulin has a rapid onset and an intermediate duration of action (comparable with that of NPH insulin). Lente insulins are chemically incompatible with the PZI and NPH insulins because of the different buffer system used in the preparation of these insulins (an acetate buffer is used in Lente insulins and a phosphate buffer is used in PZI and NPH insulins). Dosage and sources of insulin are summarized in Table 20-11.

Additionally, regular insulin will remain fast acting when combined with NPH, but not when added to Lente. The rapid action of regular insulin is neutralized by the excess zinc present in Lente insulin.[68]

Recent advances made in the area of insulin production and delivery techniques have been reviewed by Chick[69] and Salzman.[70] Progress in alternative routes of delivery of insulin have been prompted by problems associated with conventional insulin therapy, mentioned earlier. First, various types of electromechanical devices (infusion pumps) have been developed with the aim of reducing fluctuations in blood glucose levels associated with conventional insulin therapy (subcutaneous injections). These continuous-infusion pumps are either close-loop or open-loop systems. The ultimate goal of research in this area is to develop a reliable implantable (miniature) device, for long-term use, that would eliminate the need for daily administration and monitoring of blood glucose levels. The second area of research is to study alternative routes of administration such as oral, nasal, and rectal. Preliminary results indicate that the absorption of insulin at these sites is not uniform and is unpredictable. The third approach to correcting the problems of conventional insulin therapy is to supplement the defective pancreas by transplantation with a normally functioning pancreas from an appropriate donor. The major problem with this approach is the rejection of the donor pancreas by the recipient and problems associated with the draining of exocrine enzymes. A modified procedure is to transplant only viable pancreatic islet cells or transplanting fetal or neonatal pancreas. The possibility remains, however, that in type I diabetes, the newly transplanted pancreatic β-cells could be destroyed by the same autoimmune process that caused the disease in the first place.

Glucagon

Glucagon, USP. The hyperglycemic–glycogenolytic hormone elaborated by the α-cells of the pancreas is known as glucagon. It contains 29-amino acid residues in the sequence shown. Glucagon has been isolated from the amorphous fraction of a commercial insulin sample (4% glucagon).

^{1}His–Ser–Gln–Gly–^{5}Thr–Phe–Thr–Ser–Asp–^{10}Tyr–Ser–Lys–Tyr–Leu–Asp^{15}–Ser–Arg–Arg–Ala–Gln^{20}–Asp–Phe–Val–Gln–Tyr^{25}–Leu–Met–Asn–Thr

Glucagon

Recently there has been attention focused on glucagon as a factor in the pathology of human diabetes. According to Unger, Orci, and Maugh,[71] the following observations support this implication of glucagon: an elevation in glucagon blood levels (hyperglucagonemia) has been observed in association with every type of hyperglycemia; when secretion of both glucagon and insulin are suppressed, hyperglycemia is not observed unless the glucagon levels are restored to normal by the administration of glucagon; the somatostatin-induced suppression of glucagon release in diabetic animals and humans restores blood sugar levels to normal and alleviates certain other symptoms of diabetes.

Unger, Orci, and Maugh propose that although the major role of insulin is regulation of the transfer of glucose from the blood to storage in insulin-responsive tissues (e.g., liver, fat, and muscle), the role of glucagon is regulation of the liver-mediated mobilization of stored glucose. The principal consequence of high concentrations of glucagon is liver-mediated release into the blood of abnormally high concentrations of glucose, thereby causing persistent hyperglycemia. It is therefore indicated that the presence of relative excess of glucagon is an essential factor in the development of diabetes.[71]

Glucagon's solubility is 50 μg/mL in most buffers between pH 3.5 and 8.5. It is soluble 1–10 mg/mL in the pH ranges 2.5 to 3.0 and 9.0 to 9.5. Solutions of 200 μg/mL at pH 2.5 to 3.0 are stable for at least several months at 4°C if sterile. Loss of activity by fibril formation occurs readily at high concentrations of glucagon at room temperature or above at *p*H 2.5. The isoelectric point appears to be at *p*H 7.5 to 8.5. Because it has been isolated from commercial insulin, its stability properties should be comparable with those of insulin.

As with insulin and some of the other polypeptide hormones, glucagon-sensitive receptor sites in target cells bind glucagon. This hormone–receptor interaction leads to activation of membrane adenylate cyclase, which catalyzes cAMP formation. Thus, intracellular cAMP is elevated. The mode of action of glucagon in glycogenolysis is basically the same as the mechanism of epinephrine (i.e., by stimulation of adenylate cyclase). Subsequently, the increase in cAMP results in activating the protein kinase that catalyzes phosphorylation of phosphorylase kinase → phosphophosphorylase kinase. The latter is necessary for the activation of phosphorylase to form phosphorylase a. Finally, phosphorylase a catalyzes glycogenolysis, and this is the basis for the hyperglycemic action of glucagon. Although both glucagon and epinephrine exert hyperglycemic action through cAMP, glucagon affects liver cells, and epinephrine affects both muscle and liver cells.

Fain[72] reviewed the many phenomena associated with hormones, membranes, and cyclic nucleotides, including several factors activating glycogen phosphorylase in rat liver. These factors involve not only glucagon, but also vasopressin and the catecholamines. Glucagon and β-catecholamines mediate their effects on glycogen phosphorylase through cAMP, but may also involve other factors.

Glucagon exerts other biochemical effects. Gluconeogenesis in the liver is stimulated by glucagon, and this is accompanied by enhanced urea formation. Glucagon inhibits the incorporation of amino acids into liver proteins. Fatty acid synthesis is decreased by glucagon. Cholesterol formation also is reduced. On the other hand, glucagon activates liver lipases and stimulates ketogenesis. Ultimately, the availability of fatty acids from liver triglycerides is elevated, fatty acid oxidation increases acetyl-CoA and other acyl-CoAs, and ketogenesis is promoted. As glucagon effects elevation of cAMP levels, release of glycerol and free fatty acids from adipose tissue is also increased.

Glucagon's regulatory effects on carbohydrates and fatty acid metabolism has been reviewed,[36] with particular emphasis on the enzyme systems involved.

Glucagon is therapeutically important. It is recommended for the treatment of severe hypoglycemic reactions caused by the administration of insulin to diabetic or psychiatric patients. Of course, this treatment is effective only when hepatic glycogen is available. Nausea and vomiting are the most frequently encountered reactions to glucagon.

Usual dose: parenteral, adults, 500 μg to 1 mg (0.5 to 1 unit), repeated in 20 minutes if necessary; pediatric, 25 μg/kg of body weight, repeated in 20 minutes if necessary.

GASTROINTESTINAL HORMONES

There is a formidable array of polypeptide hormones of the gastrointestinal tract that include secretin, pancreozymin–cholecystokinin, gastrin, motilin, neurotensin, vasoactive intestinal peptide, somatostatin, and others. The biosynthesis, chemistry, secretion, and actions of these hormones have been reviewed.[73]

Gastrin

Gastrin is a 17-residue polypeptide isolated from the antral mucosa. It was originally isolated in two different forms. In one of the forms, the tyrosine residue in position 12 is sulfated. Both these forms are biologically active. Cholinergic response to the presence of food in the gastrointestinal tract provides the stimulus for gastrin secretion. The lowering of *p*H in the stomach inhibits the secretion of gastrin. The effects of structural modification of gastrin on gastric acid secretion have been reviewed.[74] These studies have revealed that the four residues at the COOH-terminal end retain significant biologic activity and that the aspartate residue is the most critical for activity. The most important action of gastrin is to stimulate the secretion of gastric acid and pepsin. Other actions of gastrin include increase in the secretion of pancreatic enzymes, contraction of smooth muscles, water and electrolyte secretion by the stomach and pancreas, water and electrolyte absorption by the small

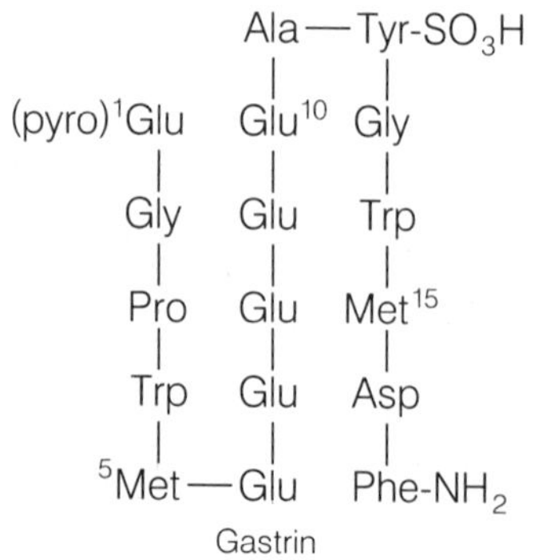

```
                  Ala—Tyr-SO3H
                   |    |
(pyro)1Glu       Glu10  Gly
      |            |    |
     Gly          Glu   Trp
      |            |    |
     Pro          Glu   Met15
      |            |    |
     Trp          Glu   Asp
      |            |    |
    5Met————————— Glu   Phe-NH2
```

Gastrin

intestine, and secretion of insulin, glucagon, and somatostatin. A synthetic pentapeptide derivative, pentagastrin, is currently employed as a gastric acid secretagogue.

Pentagastrin (Peptavlon), a physiologic gastric acid secretagogue, is the synthetic pentapeptide derivative *N-t*-butyloxycarbonyl-*β*-alanyl-L-tryptophyl-L-methionyl-L-aspartyl-L-phenylalanyl amide. It contains the COOH-terminal tetrapeptide amide (H · Try · Met · Asp · Phe · NH_2), which is considered to be the active center of the natural gastrins. Accordingly, pentagastrin appears to have the physiologic and pharmacologic properties of the gastrins, including stimulation of gastric secretion, and pepsin secretion, gastric motility, pancreatic secretion of water and bicarbonate, pancreatic enzyme secretion, biliary flow and bicarbonate output, intrinsic factor secretion, and contraction of the gallbladder.

Pentagastrin is indicated as a diagnostic agent to evaluate gastric acid secretory function, and it is useful in testing for anacidity in patients with suspected pernicious anemia, atrophic gastritis or gastric carcinoma, hypersecretion in patients with suspected duodenal ulcer or postoperative stomal ulcers, and for the diagnosis of Zollinger-Ellison tumor.

Pentagastrin is usually administered subcutaneously; the optimal dose is 6 μg/kg. Gastric acid secretion begins approximately ten minutes after administration and peak responses usually occur within 20 to 30 minutes. The usual duration of action is from 60 to 80 minutes. Pentagastrin has a relatively short plasma half-life, perhaps under ten minutes. The available data from metabolic studies indicate that pentagastrin is inactivated by the liver, kidney, and tissues of the upper intestine.

Contraindications include hypersensitivity or idiosyncrasy to pentagastrin. It should be used with caution in patients with pancreatic, hepatic, or biliary disease.

Secretin

Secretin is a 27-amino acid polypeptide that is structurally similar to glucagon. The presence of acid in the small intestine is the most important physiologic stimulus for the secretion of secretin. The primary action of secretin is on pancreatic acinar cells that regulate the secretion of water and bicarbonate. Secretin also promotes the secretion of pancreatic enzymes, to a lesser extent. Secretin inhibits the release of gastrin and, therefore, gastric acid. It also increases stomach-emptying time by reducing the contraction of the pyloric sphincter.[73]

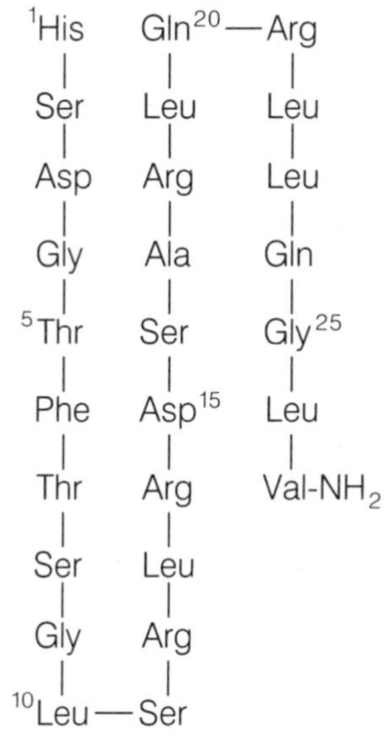

```
¹His    Gln²⁰—Arg
 |       |     |
Ser     Leu   Leu
 |       |     |
Asp     Arg   Leu
 |       |     |
Gly     Ala   Gln
 |       |     |
⁵Thr    Ser   Gly²⁵
 |       |     |
Phe     Asp¹⁵ Leu
 |       |     |
Thr     Arg   Val-NH₂
 |       |
Ser     Leu
 |       |
Gly     Arg
 |       |
¹⁰Leu — Ser
```

Secretin

Cholecystokinin–Pancreozymin

It was originally thought that cholecystokinin and pancreozymin were two different hormones. Cholecystokinin was thought to be responsible for contraction of the gallbladder, whereas pancreozymin was believed to induce secretion of pancreatic enzymes. It is now clear that both these actions are caused by a single, 33-residue polypeptide, referred to as cholecystokinin–pancreozymin (CCK–PZ). CCK–PZ is secreted in the blood, in response to the presence of food in the duodenum, especially long-chain fatty acids. The five COOH-terminal amino acid residues are identical with those in gastrin. The COOH-terminal octapeptide retains full activity of

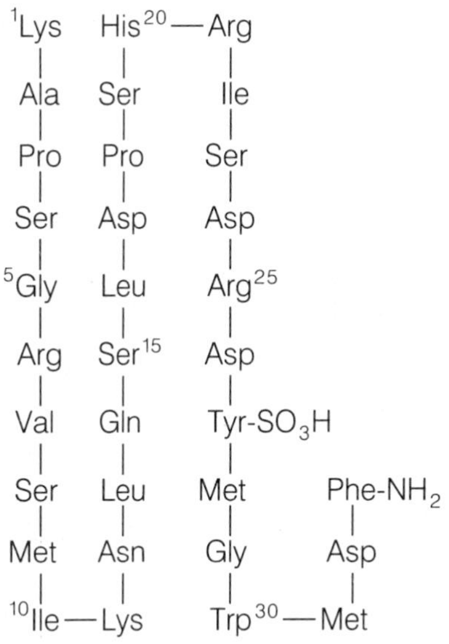

```
¹Lys    His²⁰—Arg
 |       |     |
Ala     Ser   Ile
 |       |     |
Pro     Pro   Ser
 |       |     |
Ser     Asp   Asp
 |       |     |
⁵Gly    Leu   Arg²⁵
 |       |     |
Arg     Ser¹⁵ Asp
 |       |     |
Val     Gln   Tyr-SO₃H
 |       |     |
Ser     Leu   Met        Phe-NH₂
 |       |     |          |
Met     Asn   Gly        Asp
 |       |     |          |
¹⁰Ile — Lys   Trp³⁰ —    Met
```

Cholecystokinin

the parent hormone. The octapeptide is found in the gut as well as the central nervous system. Structure–activity relationships of cholecystokinin have been reviewed.[73] The COOH-terminal octapeptide is present in significant concentrations in the central nervous system. Its possible actions here, the therapeutic implications in the treatment of Parkinson's disease and schizophrenia, and its structure–activity relationship have been reviewed.[74]

Vasoactive Intestinal Peptide

Vasoactive intestinal peptide (VIP) is widely distributed in the body and is believed to occur throughout the gastrointestinal tract. It is a 28-residue polypeptide and has structural similarities to secretin and glucagon. It causes vasodilatation and increases cardiac contractibility. VIP stimulates bicarbonate secretion, relaxes gastrointestinal and other smooth muscles, stimulates glycogenesis, inhibits gastric acid secretion, and stimulates insulin secretion. Its hormonal and neurotransmitter role has been rather recently investigated.[75]

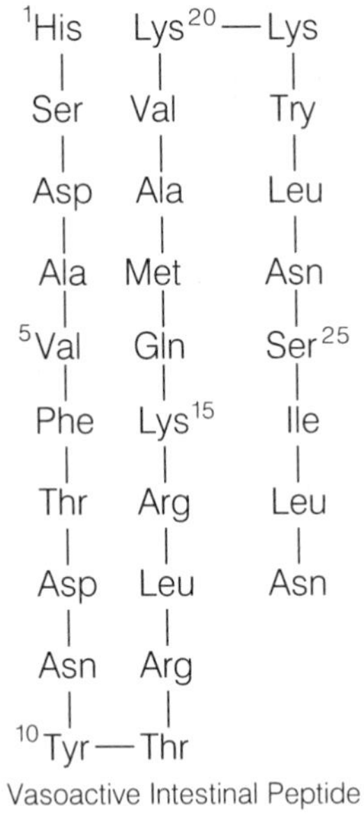

```
 1His    Lys20—Lys
  |       |      |
  Ser    Val    Try
  |       |      |
  Asp    Ala    Leu
  |       |      |
  Ala    Met    Asn
  |       |      |
 5Val    Gln    Ser25
  |       |      |
  Phe    Lys15  Ile
  |       |      |
  Thr    Arg    Leu
  |       |      |
  Asp    Leu    Asn
  |       |
  Asn    Arg
  |       |
10Tyr  — Thr
```

Vasoactive Intestinal Peptide

Gastric Inhibitory Peptide

Gastric inhibitory peptide (GIP) is a 43-amino acid polypeptide isolated from the duodenum. Secretion of GIP into the blood is stimulated by food. The primary action of GIP is inhibition of gastric acid secretion. Other actions include stimulation of insulin and glucagon secretion and stimulation of intestinal secretion.[73]

Motilin

Motilin is a 22-residue polypeptide isolated from the duodenum. Its secretion is stimulated by the presence of acid in the duodenum. Motilin inhibits gastric motor activity and delays gastric emptying.

Neurotensin

Neurotensin is a 13-amino acid peptide, first isolated from bovine hypothalamus. It has now been identified in the intestinal tract. The ileal mucosa contains 90% of the total neurotensin of the body. It is implicated as a releasing factor for several adenohypophyseal hormones. It causes vasodilatation, increases vascular permeability, and increases gastrin secretion. It decreases secretion of gastric acid and secretin.

PARATHYROID HORMONE

This hormone is a linear polypeptide containing 84-amino acid residues. Structure–activity relationship studies[76] of bovine parathyroid hormone have revealed that the biologic activity is retained by an NH_2-terminal fragment consisting of 34-amino acid residues. It regulates the concentration of calcium ion in the plasma within the normal range, in spite of variations in calcium intake, excretion, and anabolism into bone.[77] Also, for this hormone, cAMP is implicated as a second messenger. Parathyroid hormone activates adenylate cyclase in renal and skeletal cells, and this effect promotes formation of cAMP from ATP. The cAMP increases the synthesis and release of the lysosomal enzymes necessary for the mobilization of calcium from bone.

Parathyroid Injection, USP, has been employed therapeutically as an antihypocalcemic agent for the temporary control of tetany in acute hypoparathyroidism.

Calcitonin

Calcitonin (thyrocalcitonin) is a polypeptide hormone containing 32 amino acids and is secreted by parafollicular cells of the thyroid glands in response to hypocalcemia. The entire 32-residue peptide appears to be required for activity because smaller fragments are totally inactive. Common structural features of calcitonin isolated from different species are a COOH-terminal prolinamide, a disulfide bond between residues 1 and 7 at the NH_2-terminal end, and a chain length of 32 residues.[78] Calcitonin inhibits calcium resorption from bone, causing hypocalcemia, with parallel changes in plasma phosphate concentration. In general, calcitonin negates the osteolytic effects of the parathyroid hormone.

The potential therapeutic uses of calcitonin are in the treatment of hyperparathyroidism, osteoporosis and other bone disorders, hypercalcemia of malignancy, and in ideopathic hypercalcemia.

```
¹Cys-S-S-Cys — Met
 |        |      |
Gly      Thr    Leu     Gln — Thr²⁵
 |        |      |       |     |
Asn      Ser⁵   Gly¹⁰   Pro   Ala
  \      /       |       |     |
    Leu         Thr     Phe   Ile
                 |       |     |
                Tyr     Thr   Gly
                 |       |     |
                Thr     His²⁰ Val
                 |       |     |
                Gln     Phe   Gly³⁰
                 |       |     |
                Asp¹⁵   Lys   Ala
                 |       |     |
                Phe  —  Asn   Pro-NH₂
```

Calcitonin

```
¹Asp  Phe — His
 |     |     |
Arg   Pro   Leu¹⁰   α₂-Globulin
 |     |     |       |
Val   His   Val     Ser
 |     |     |       |
Tyr — Ile⁵  Ile  —  His
```

Angiotensinogen

↓ Renin

```
¹Asp  Phe — His
 |     |     |
Arg   Pro   Leu¹⁰
 |     |
Val   His
 |     |
Tyr — Ile⁵
```

Angiotensin I

↓ ACE

```
¹Asp  Phe
 |     |
Arg   Pro
 |     |
Val   His
 |     |
Tyr — Ile⁵
```

Angiotensin II

ANGIOTENSINS

The synthesis of angiotensins in the plasma is initiated by the catalytic action of renin (a peptidase elaborated by the kidneys) on angiotensinogen, an α-globulin produced by the liver and found in the plasma. The hydrolytic action of renin on angiotensinogen yields angiotensin I, a decapeptide consisting of the first ten residues of the NH_2-terminal segment of angiotensinogen. Angiotensin I has weak pharmacologic activity. Angiotensin I is converted to angiotensin II, an octapeptide, by the catalytic actions of angiotensin-converting enzyme (ACE). Angiotensin II is a highly active peptide and is hydrolyzed to angiotensin III, a heptapeptide, by an aminopeptidase. Angiotensin III retains most of the pharmacologic activity of its precursor. Further degradation of angiotensin III leads to pharmacologically inactive peptide fragments.

Angiotensin II has been shown to be the most active form; hence, it is the most investigated angiotensin for pharmacologic action and structure–activity relationships. The two primary actions of angiotensin II are vasoconstriction and stimulation of synthesis and secretion of aldosterone by the adrenal cortex. Both of these actions lead to hypertension.

Mechanisms and sites of action of angiotensin agonists and antagonists in terms of biologic activity and receptor interactions have been reviewed.[79] Additionally, compounds that inhibit ACE have found therapeutic use as antihypertensive agents (e.g., captopril). The synthesis and biologic activity of several ACE inhibitors has been reviewed.[80]

Angiotensin Amide (Hypertensin) is a synthetic polypeptide (1-L-aspariginyl-5-L-valine angiotensin octapeptide) and has twice the pressor activity of angiotensin II. It is pharmaceutically available as a lyophilized powder for injection (0.5 to 2.5 mg diluted in 500 mL of sodium chloride injection or 5% dextrose for injection) to be administered by continuous infusion. The pressor effect of angiotensin is due to an increase in peripheral resistance; it constricts resistance vessels but has little or no stimulating action on the heart and little effect on the capacitance vessels. Angiotensin has been utilized as an adjunct in various hypotensive states. It is mainly useful in controlling acute hypotension during administration of general anesthetics that sensitize the heart to the effects of catecholamines.

Erythropoetin

Erythropoetin is a 166-amino acid polypeptide. The kidney is the primary site of erythropoetin production (90%). It is elaborated in response to hypoxia in the kidneys and exerts its effects on the erythroid tissue in the bone marrow. Erythropoetin primarily increases production of red blood cells by stimulating the mitotic activity of erythroid precursor and progenitor cells.[81]

Recently, human erythropoetin has been produced by recombinant DNA technology that employed cultured mammalian cells into which the human erythropoetin gene had been inserted. The primary indication for the biosynthetic product (Epogen) is the treatment of anemia caused by reduced production of erythropoetin.

PLASMAKININS (Bradykinin, Kallidin)

These are potent vasodilators and hypotensive agents that have peptide structures. Bradykinin is a nonapeptide, whereas kallidin is a decapeptide. Kallidin is lysylbradykinin; that is, it has an additional lysine at the NH_2-terminal end of the chain. These two compounds are made available from kininogen, a blood globulin, upon hydrolysis. Trypsin, plasmin, or the proteases of certain snake venoms can catalyze the hydrolysis of kininogen.

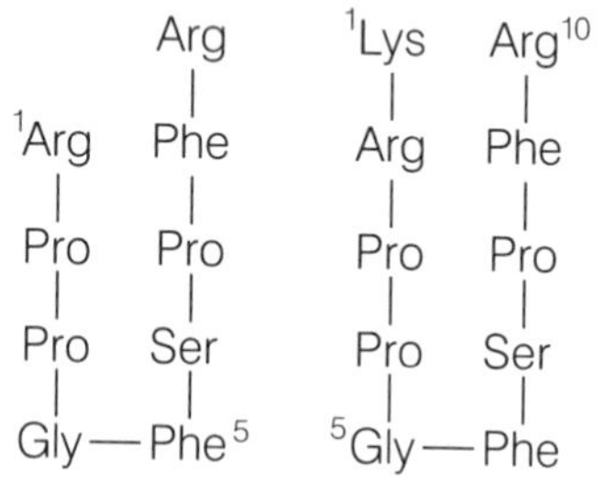

^{1}Arg—Pro—Pro—Gly—Phe^{5}—Ser—Pro—Phe—Arg

Bradykinin

^{1}Lys—Arg—Pro—Pro—^{5}Gly—Phe—Ser—Pro—Phe—Arg^{10}

Kallidin

Bradykinin is one of the most powerful vasodilators known; 0.05–0.5 $\mu g/kg$ intravenously can produce a decrease in blood pressure in all mammals yet investigated.

Although the kinins per se are not used as medicinals, kallikrein enzyme preparations that release bradykinin from the inactive precursor have been used in the treatment of Raynaud's disease, claudication, and circulatory diseases of the eyegrounds. (*Kallikreins* is the term used to designate the group of proteolytic enzymes that catalyze the hydrolysis of kininogen, forming bradykinin.)

Substance P

Substance P is a polypeptide consisting of 11-amino acid residues. Substance P has been implicated in the transmission of "painful" sensory information through the spinal cord to higher centers in the central nervous system.[74] Substance P is localized in the primary afferent sensory fibers. Other pharmacologic effects are vasodilatation, stimulation of smooth muscles, stimulation of salivary secretion, and diuresis.

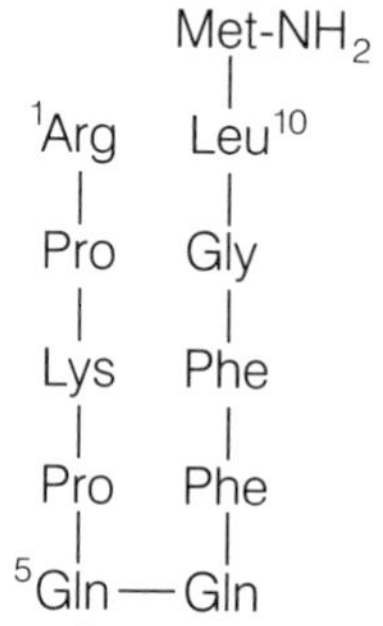

^{1}Arg—Pro—Lys—Pro—^{5}Gln—Gln—Phe—Phe—Gly—Leu^{10}—Met-NH_2

Substance P

Atrial Natriuretic Factors

Atrial natriuretic factors (ANF) are peptides elaborated by the secretory granules in the atria of mammalian hearts. Two distinct natriuretic peptides have been identified in humans and are believed to be obtained from a common precursor, atriopeptigen, a 151-amino acid peptide. A 28-amino acid peptide is the most common ANF identified. These endogenous peptides promote natriuresis and diuresis.[82]

THYROGLOBULIN

Thyroglobulin, a glycoprotein, is composed of several peptide chains; it also contains 0.5% to 1% iodine and 8% to 10% carbohydrate in the form of two types of polysaccharides. The formation of thyroglobulin is regulated by thyrotropin (TSH). Thyroglobulin has no hormonal properties. It must be hydrolyzed to release the hormonal iodothyronines: thyroxine and liothyronine (see Chap. 14).

BLOOD PROTEINS

The blood is the transport system of the organism and thus performs important distribution functions. Considering the multitude of materials transported by the blood (e.g., nutrients, oxygen, carbon dioxide, waste products of metabolism, buffer systems, antibodies, enzymes, and hormones), its chemistry is very complex. Grossly, approximately 45% consists of the formed elements that can be separated by centrifugation and, of these, only 0.2% are other than erythrocytes. The 55% of removed plasma contains approximately 8% solids of which a small portion (less than 1%) can be removed by clotting to produce defibrinated plasma, which is called serum. Serum contains inorganic and organic compounds, but the total solids are chiefly protein, mostly albumin, and the rest nearly all globulin. The plasma contains the protein fibrinogen, which is converted by coagulation to insoluble fibrin. The separated serum has an excess of the clotting agent thrombin.

Serum globulins can be separated by electrophoresis into α-, β-, and γ-globulins that contain most of the antibodies. The immunologic importance of globulins is well known. Many classes and groups of immunoglobulins are produced in response to antigens or even to a single antigen. The specificity of antibodies has been studied from various points of view, and Richards et al.[83] have reported evidence that suggests that even though immune serums appear to be highly specific for antigen binding, individual immunoglobulins may not only interact with several structurally diverse determinants, but may bind such diverse determinants to different sites within the combining region.

The importance of the blood coagulation process has been obvious for a long time. Coagulation mechanisms are well covered in several biochemistry texts;[84,85] hence, herein a brief summary suffices. The required time for blood clotting is normally five minutes, and any prolongation beyond ten minutes is considered abnormal. Thrombin, the enzyme responsible for the catalysis of fibrin formation, originates from the inactive zymogen, prothrombin; the prothrombin → thrombin transformation is dependent on calcium ions and thromboplastin. The fibrinogen → fibrin reaction catalyzed by thrombin involves proteolytic cleavage (partial hydrolysis); polymerization of the fibrin monomers from the preceding step; and actual clotting (hard clot formation). The final process forming the hard clot occurs in the presence of calcium ions and the enzyme fibrinase.

Thrombin, USP, is a sterile protein substance prepared from prothrombin of bovine origin. It is used as a topical hemostatic because of its capability of clotting blood, plasma, or a solution of fibrinogen without adding other substances. Thrombin also may initiate clotting when combined with gelatin sponge or fibrin foam.

For external use: topically to the wound, as a solution containing 100 to 2000 NIH units/mL in sodium chloride irrigation or sterile water for injection or as a dry powder.

Hemoglobin

Erythrocytes contain 32% to 55% hemoglobin, about 60% water and the rest as stroma. The last can be obtained, after hemolysis of the corpuscles by dilution, through the process of centrifuging and is found to consist of lecithin, cholesterol, inorganic salts, and a protein, stromatin. Hemolysis of the corpuscles, or laking as it sometimes is called, may be brought about by hypotonic solution, by fat solvents, by bile salts that dissolve the lecithin, by soaps or alkalies, by saponins, by immune hemolysins, and by hemolytic serums, such as those from snake venom and numerous bacterial products.

Hemoglobin (Hb) is a conjugated protein, the prosthetic group being heme (hematin) and the protein (globin) which is composed of four polypeptide chains, usually in identical pairs. The total relative molecular mass is about 66,000 including four heme molecules. The molecule has an axis of symmetry and, therefore, is composed of identical halves with an overall ellipsoid shape of the dimensions 55 × 55 × 70Å.[85]

Iron in the heme of hemoglobin (ferrohemoglobin)d, is in the ferrous state and can combine reversibly with oxygen to function as a transporter of oxygen.

$$\text{Hemoglobin} + \text{Oxygen}\ (O_2) \rightleftharpoons \text{Oxyhemoglobin}$$

In this process, the formation of a stable oxygen complex, the iron remains in the ferrous form because the heme moiety lies within a cover of hydrophobic groups of the globin. Both Hb and O_2 are magnetic, whereas HbO_2 is dimagnetic because the unpaired electrons in both molecules have become paired. When oxidized to the ferric state (methemoglobin or ferrihemoglobin), this function is lost. Carbon monoxide will combine with hemoglobin to form carboxyhemoglobin (carbonmonoxyhemoglobin) to inactivate it.

The stereochemistry of the oxygenation of hemoglobin is very complex, and it has been investigated to some extent. Some evidence from x-ray crystallographic studies reveals that the conformations of the α- and β-chains are altered when their heme moieties complex with oxygen, thus promoting the complexation with oxygen. It is assumed that hemoglobin can exist in two forms, the relative position of the subunits in each form being different. In the deoxy form, α- and β-subunits are bound to each other by ionic bonds in a compact structure that is less reactive toward oxygen than is the oxy form. Some ionic bonds are cleaved in the oxy form, relaxing the conformation. The latter conformation is more reactive to oxygen.[85,86]

REFERENCES

1. Ahmad, F., Schultz, J., Smith, E. E., and Whalen, W. J. (eds.): From Gene to Protein: Translation into Biotechnology, Vol. 19. New York, Academic Press, 1982.
2. Lehninger, A.: Principles of Biochemistry, p. 763. New York, Worth Publishers, 1982.

3. Stryer, L.: Biochemistry, 3rd ed., pp. 15–70. New York, W. H. Freeman & Co., 1988.
4. American Medical Association Department of Drugs: Drug Evaluations, 6th ed., pp. 867–870. New York, John Wiley & Sons, 1986.
5. Corey, R. B., and Pauling, L.: Proc. R. Soc. Lond. Ser. B 141:10, 1953; see also Adv. Protein Chem. p. 147, 1957.
6. Tanford, C.: The Hydrophobic Effect: Formation of Micelles and Biological Membranes, 2nd ed. New York, John Wiley & Sons, 1979.
7. McDonad, C. C., and Phillips, W. D.: J. Am. Chem. Soc. 89:6332, 1967.
8. Kato, G., and Yung, J.: Mol. Pharmacol. 7:33, 1971.
9. Elefrawi, M. E., et al.: Mol. Pharmacol. 7:104, 1971.
10. Kato, G.: Mol. Pharmacol. 8:575, 1972.
11. Kato, G.: Mol. Pharmacol. 8:582, 1972.
12. Johnson, F. H., et al.: The Kinetic Basis of Molecular Biology. New York, John Wiley & Sons, 1954.
13. Eyring, H., and Eyring, E. M.: Modern Chemical Kinetics. New York, Rheinhold, 1963.
14. Eyring, H.: Am. Chem. Soc. Natl. Meet., Dallas, April 1973; for abstract of paper, see Chem. and Eng. News, p. 17, Apr. 30, 1973.
15. Folkers, K., et al.: Biochem. Biophys. Res. Commun. 59:704, 1974.
16. Ferdinand, W.: The Enzyme Molecule. New York, John Wiley & Sons, 1976.
17. Stryer, L.: Biochemistry, 3rd ed., pp. 177–260. New York, W. H. Freeman & Co., 1988.
18. Koshland, D. E.: Sci. Am. 229:52, 1973; see also Ann. Rev. Biochem. 37:359, 1968.
19. Lowe, J. N., and Ingraham, L. L.: An Introduction to Biochemical Reaction Mechanisms. Englewood Cliffs, NJ, Prentice-Hall, 1974.
20. Hanson, K. R., and Rose, I. A.: Acc. Chem. Res. 8:1, 1975; see also Science 193:121, 1976.
21. Chem. Eng. News, p. 23, Apr. 16, 1979.
22. Ross, M. J., and Grossbard, E. B.: Annu. Rep. Med. Chem. 20:107, 1985.
23. Wallis, M., Howell, S. L., and Taylor, K. W.: The Biochemistry of the Peptide Hormone. Chichester, John Wiley & Sons, 1986.
24. Spatola, A. F.: Annu. Rep. Med. Chem. 16:199, 1981.
25. Schally, A. V., Arimura, A., and Kastin, A. J.: Science 179:341, 1973.
26. Meienhoffer, J.: Peptide and protein hormones. In Wolff, M. E. (ed.). Burger's Medicinal Chemistry, 4th ed., Part 2, p. 751. New York, John Wiley & Sons, 1979.
27. White, W. F.: Annu. Rep. Med. Chem. 8:204, 1973.
28. Otsuka, H., and Inouye, L. K.: Pharmacol. Ther. [B] 1:501, 1975.
29. Haynes, R. C., Jr., and Murad, F.: In Gilman, A. G., Goodman, L. S., Rall, T. W., and Murad, F. (eds.). Pharmacological Basis of Therapeutics, 7th ed., p. 1461. New York, Macmillan, 1985.
30. Hughes, J., et al.: Nature 258:577, 1975.
31. Jaffe, J. H., and Martin, W. R.: In Gilman, A. G., Goodman, L. S., Rall, T. W., and Murad, F. (eds.). Pharmacological Basis of Therapeutics, 7th ed., pp. 492–493, 1985.
32. Snyder, S. H.: Science 224:22, 1984.
33. Wallis, M.: J. Mol. Evol. 17:10, 1981.
34. Shome, B., and Parlow, A. F.: J. Clin. Endocrinol. Metab. 45:1112, 1977.
35. Veber, D. F., and Saperstein, R.: Annu. Rep. Med. Chem. 14:209, 1979.
36. Tulwiler, G. F., et al.: Annu. Rep. Med. Chem. 18:193, 1983.
37. Brownstein, M. J.: Science 207:373, 1980.
38. Hays, R. M.: In Gilman, A. G., Goodman, L. S., Rall, T. W., and Murad, F. (eds.). Pharmacological Basis of Therapeutics, 7th ed., p. 908. New York, Macmillan, 1985.
39. Riphagen, C. L., and Pittman, Q. J.: Fed. Proc. 45:2318, 1986.
40. American Medical Association Department of Drugs: Drug Evaluations, 6th ed., pp. 815–816. Philadelphia, W. B. Saunders, 1986.
41. Banting, F. G., and Best, C. H.: J. Lab. Clin. Med. 7:251, 1922.
42. Chang, A. Y.: Annu. Rep. Med. Chem. 9:182, 1974.
43. Rasmussen, C. R., et al.: Annu. Rep. Med. Chem. 16:173, 1981.
44. Wallis, M., Howell, S. L., and Taylor, K. W.: In The Biochemistry of Peptide Hormones, pp. 257–297. Chichester, John Wiley & Sons, 1985.
45. Steiner, D. F., et al.: Recent Prog. Horm. Res. 25:207–282, 1969.
46. Docherty, K., et al.: Proc. Natl. Acad. Sci. USA 79:4613, 1982.
47. Chan, S. J., et al.: Proc. Natl. Acad. Sci. USA 73:1964, 1976.
48. Arquilla, E. R., et al.: In Steiner, D. F., and Freinkel, N. (eds.). Handbook of Phisiology, Sect. 7. Endocrinology, Vol. 1, pp. 159–173. Washington, DC, American Physiological Society, 1972.
49. Gerich, J., et al.: Annu. Rev. Physiol. 38:353, 1976.
50. For references to Sanger's studies, see Annu. Rev. Biochem. 27:58, 1958.
51. Katsoyannis, P. G.: Science 154:1509, 1966.
52. Chance, R. E., et al.: Diabetes Care 4:147, 1981.
53. Johnson, I. S.: Diabetes Care 5(Suppl. 2):4, 1982.
54. Cosmatos, A., et al.: J. Biol. Chem. 253:6586, 1978.
55. May, J. M., et al.: J. Biol. Chem. 253:686, 1978.
56. Blundell, T. L., et al.: Crit. Rev. Biochem. 13:141, 1982.
57. Kasuga, M., et al.: Science 215:185, 1982.
58. Jarett, L., et al.: Fed. Proc. 41:2736, 1982.
59. Poderson, O.: Diabetes Care 6:301, 1983.
60. Gerich, J. E.: In Diabetes Mellitus, 9th ed., p. 53. Indianapolis, Eli Lily & Co., 1988.
61. Wallis, M., et al.: In The Biochemistry of the Polypeptide Hormones, p. 280. Chichester, John Wiley & Sons, 1985.
62. Jefferson, L. S.: Diabetes 29:487, 1980.
63. Report from the Geigy Symposium in Albuquerque, New Mexico: Diabetes Re-examined. Diabetology, Feb. 6, 1974.
64. Feldman, J. M.: In Diabetes Mellitus, 9th ed., pp. 28–42. Indianapolis, Eli Lily & Co., 1988.
65. Pirat, J.: Diabetes Care, Part 1, pp. 168–188; Part 2, pp. 252–263, 1978.
66. Raskin, P., and Rosenstock, J.: Ann. Intern. Med. 105:254, 1986.
67. Maurer, A. C.: Am. Sci. 67:422, 1979.
68. Galloway, J. A.: In Diabetes Mellitus, 9th ed., pp. 109–122. Indianapolis, Eli Lily & Co., 1988.
69. Chick, W. L.: Recent advances in insulin production and delivery. In Kozak, G. P. (ed.). Clinical Diabetes Mellitus, pp. 461–472. W. B. Saunders, 1982.
70. Salzman, R., et al.: N. Engl. J. Med. 312:1078, 1985.
71. Unger, R. J., et al.: Science 188:923, 1975.
72. Fain, J. N.: Receptors Recognition Ser. A 6:3, 1978.
73. Wallis, M., Howell, S. L., and Taylor, K. W.: In The Biochemistry of the Polypeptide Hormones, pp. 318–335. Chichester, John Wiley & Sons, 1985.
74. Emson, P. C., and Sandberg, B. E.: Annu. Rep. Med. Chem. 18:31, 1983.
75. Miller, R. J.: J. Med. Chem. 27:1239, 1984.
76. Habener, J. F., et al.: Recent Prog. Horm. Res. 33:249, 1977.
77. Cohn, D. V., and Elting, J.: Recent Prog. Horm. Res. 39:181, 1983.
78. Rodan, S. B., and Rodan, G. A.: J. Biol. Chem. 249:3068, 1974.
79. Bumpus, F. M.: Fed. Proc. 36:2128, 1977.
80. Ondetti, M. A., and Cushman, J.: J. Med. Chem. 24:355, 1981.
81. Eschbach, J. W., et al.: N. Engl. J. Med. 316:73, 1987.
82. Lappe, R. W., and Wendt, R. L.: Annu. Rep. Med. Chem. 21:273, 1986.
83. Richards, F. F., et al.: Science 187:130, 1975.
84. Harper, H. A.: Review of Physiological Chemistry, 17th ed. Los Altos, CA, Lange Medical, 1980.
85. Stryer, L. S.: Biochemistry, 3rd ed., pp. 143–176. New York, W. H. Freeman & Co., 1988.
86. Montgomery, R., et al.: Biochemistry: A Case Oriented Approach, 2nd ed., pp. 78–80. St. Louis, C. V. Mosby, 1977.

SELECTED READINGS

Boyer, P. D. (ed.): The Enzymes, 3rd ed. New York, Academic Press, 1970.

Brockerhoff, H., and Jensen, R. G.: Lipolytic Enzymes. New York, Academic Press, 1974.

Ferdinand, W.: The Enzyme Molecule. New York, John Wiley & Sons, 1976.

Galloway, J. A., Potvin, J. H., and Shuman, C. R. (eds.): Diabetes Mellitus, 9th ed. Indianapolis, Eli Lily & Co., 1988.

Gilman, A., Goodman, L. S., Rall, T. W., and Murad, F.: The Pharmacological Basis of Therapeutics, 7th ed. New York, Macmillan, 1985.

Grollman, A. P.: Inhibition of protein biosynthesis. In Brockerhoff, H., and Jensen, R. G. (eds.). Lipolytic Enzymes, pp. 231–247. New York, Academic Press, 1974.

Haschemeyer, R. H., and de Harven, E.: Electron microscopy of enzymes. Annu. Rev. Biochem. 43:279, 1974.

Jenks, W. P.: Catalysis in Chemistry and Enzymology. New York, McGraw-Hill, 1969.

Lowe, J. N., and Ingraham, L. L.: An Introduction to Biochemical Reaction Mechanisms. Englewood Cliffs, NJ, Prentice-Hall, 1974. (This book includes elementary enzymology including mechanisms of coenzyme function.)

Meinhoffer, J.: Peptide and protein hormones. In Wolff, M. E. (ed.). Burger's Medicinal Chemistry, 4th ed., Part 2, p. 751. New York, John Wiley & Sons, 1979.

Meienhofer, J.: Peptide hormones of the hypothalamus and pituitary. In Heinzelman, R. V. (ed.). Annu. Rep. Med. Chem. Vol. 10, 1975.

Mildvan, A. S.: Mechanism of enzyme action. Annu. Rev. Biochem. 43:357, 1974.

Pikes, S. J., and Parks, C. R.: The mode of action of insulin. Annu. Rev. Pharmacol. 14:365, 1974.

Rafelson, M. E., et al.: Basic Biochemistry, 4th ed., Chaps. 3, 4, 8, 11. New York, Macmillan, 1980.

Schaeffer, H. J.: Factors in the design of reversible and irreversible enzyme inhibitors. In Ariens, E. J. (ed.). Drug Design, Vol. 2, pp. 129–159. New York, Academic Press, 1971.

Stryer, L. S.: Biochemistry, 3rd ed. New York, W. H. Freeman & Co., 1988.

Tager, H. S., and Steiner, D. F.: Peptide hormones. Annu. Rev. Biochem. 43:509, 1974.

Wallis, M., Howell, S. L., and Taylor, K. W.: The Biochemistry of the Peptide Hormones. Chichester, John Wiley & Sons, 1985.

Waife, S. O. (ed.): Diabetes Mellitus, 8th ed. Indianapolis, Eli Lily & Co., 1980.

Wilson, C. A.: Hypothalamic amines and the release of gonadotrophins and other anterior pituitary hormones. In Simonds, A. B. (ed.). Adv. Drug Res. New York, Academic Press, 1974.

CHAPTER 21

Vitamins and Related Compounds

Jaime N. Delgado
Gustavo R. Ortega

Vitamins traditionally have been considered to be "accessory food factors." Generally, vitamins are among those nutrients that the human organism cannot synthesize from other dietary components. Together with certain amino acids (i.e., essential amino acids), the vitamins constitute a total of 24 organic compounds that have been characterized as dietarily essential.[1] Many vitamins function biochemically as precursors in the synthesis of coenzymes necessary in human metabolism; thus, vitamins perform essential functions. When they are not available in appropriate amounts, the consequences may lead to serious disease states.

Although there are relatively few therapeutic indications for vitamin pharmaceutical preparations, diseases caused by certain vitamin deficiencies do respond favorably to vitamin therapy. Additionally, there are products indicated for prophylactic use as dietary supplements. An optimal diet provides all of the necessary nutrients; however, in some cases of increased demands, vitamin and mineral supplementation is recommended.[2]

The medicinal chemistry of vitamins is fundamental not only to the therapeutics of nutritional problems, but also to the understanding of the biochemical actions of other medicinal agents that directly or indirectly affect the metabolic functions of vitamins and coenzymes. Accordingly, this chapter includes a brief summary of basic biochemistry of vitamins, structure–activity relationships, physicochemical properties and some stability considerations, nutritional and therapeutic applications, and brief characterizations of representative pharmaceutical products.

In 1912, Funk described a substance that was present in rice polishings and in foods that cured polyneuritis in birds and beriberi in humans. This substance was referred to as "vitamine" because it was characterized as an amine and as a vital nutritional component. After other food factors were also noted to be vital nutritional components that were not amines and did not even contain nitrogen, Drummond suggested the modification that led to the term *vitamin*. In 1913, McCollum and Davis described a lipid-soluble essential food factor in butterfat and egg yolk, and two years later reference was made to a water-soluble factor in wheat germ. Thus, the terms *fat-soluble A* and *water-soluble B* were respectively applied to these food factors. Since then, many other dietary components have been discovered to be essential nutritional components (i.e., vitamins). It is traditional to classify these compounds as either lipid-soluble or water-soluble vitamins. This classification is convenient, because members of each category possess important properties in common.

LIPID-SOLUBLE VITAMINS

The lipid-soluble vitamins include vitamins A, D, E, and K. These compounds possess other characteristics in common besides solubility. They are usually associated with the lipids of foods and are absorbed from the intestine with these dietary lipids. The lipid-soluble vitamins are stored in the liver and, thus, conserved by the organism, whereas storage of the water-soluble vitamins is usually not significant.

THE VITAMIN As

Vitamin A was first recognized as a vitamin by McCollum and Davis[3] in 1913 to 1915, but studies of the molecular mechanism of action of retinol in the visual process were not significantly productive until 1968 to 1972. The mechanism of action of vitamin A in physiologic processes, other than vision, has been very difficult to study. It has been difficult to identify a single biochemical change that is directly due to vitamin A deficiency. Nevertheless, there is convincing evidence that vitamin A performs an important function in the biosynthesis of glycoproteins. Vitamin A is involved in sugar transfer reactions in mammalian membranes.[4] Moreover, vitamin A has been demonstrated to control and direct differentiation of epithelial tissues; this has led to the suggestion that vitamin A has hormone-like properties.

Vitamin A is also required for bone growth, reproduction, embryonic development, protein synthesis, and sperm production. Investigations on the mechanism of this action have been stymied by difficulties in elucidating definitely the nature of the biochemically active form of the vitamins, whether it is all-*trans*-retinol or retinoic acid. This question remains under intensive investigation and is comprehensively reviewed by Chytil and Ong.[5]

The term *vitamin A* is currently applied to compounds possessing biologic activity similar to retinol. The term vitamin A_1 refers to all-*trans*-retinol. The term *retinoid* is applied to retinol and its naturally occurring derivatives plus synthetic analogues, which need not have vitamin A activity.

Vitamin A activity is expressed as USP units, international units (IU), retinol equivalents (RE), and β-carotene equivalents. The USP units and IU are equivalent. Each unit expresses the activity of 0.3 μg of all-*trans*-retinol, 0.334 μg of all-*trans*-retinol acetate, or 0.6 μg of β-carotene. Thus 1 mg of all-*trans*-retinol has the activity of 3333 units. One RE represents the biologic activity of 1 μg of all-*trans*-retinol or 6 μg of β-carotene.

The stereochemistry of vitamin A and related compounds is complex, and a complete stereochemical analysis is beyond the scope of this chapter. A brief summary of some stereochemical features is presented here as the basis for the characterization of the biochemical actions exerted by this vitamin. The study of the structural relationships among vitamin A and its stereoisomers has been complicated by the common use of two numbering systems, as exemplified by the vitamin A (all-*trans*-retinol) and neovitamin A (Δ^4*cis*- or 11-mono-*cis* vitamin A).

For steric reasons, the number of isomers of vitamin A most likely to occur would be limited. These are all-*trans*, 9-*cis* (Δ^3-*cis*), 13-*cis* (Δ^5-*cis*), and the 9,13-di-*cis*. A *cis* linkage at double-bond 7 or 11 encounters steric hindrance. The 11-*cis* isomer is twisted, as well as bent, at this linkage; nevertheless, this is the only isomer that is active in vision.

The biologic activity[7] of the isomers of vitamin A acetate, in terms of USP units per gram, are as follows: vitamin A, all-*trans*, 2,907,000; neovitamin

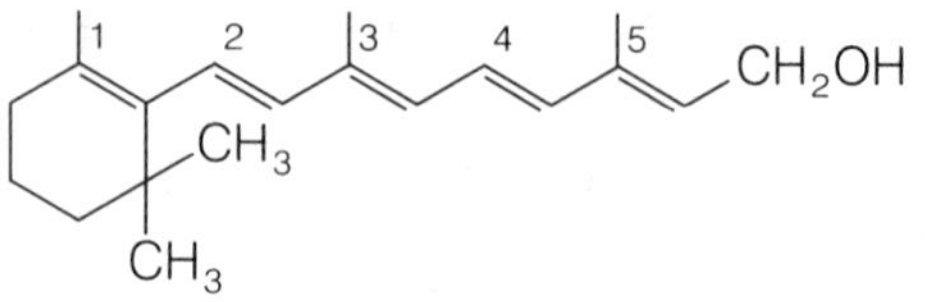

Vitamin A (Retinol)

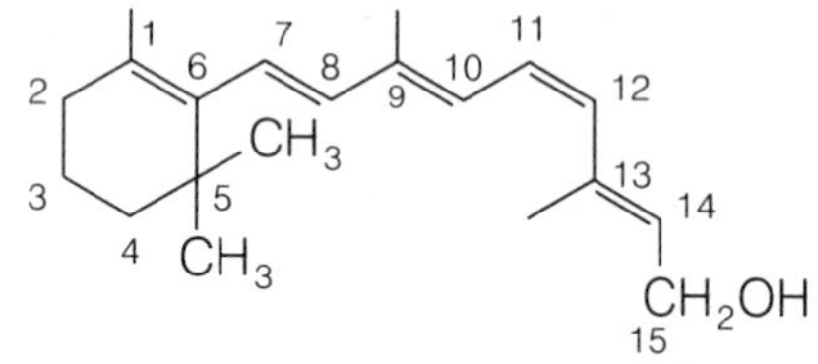

Neovitamin A

A, 2,190,111; Δ^3-*cis*, 634,000; $\Delta^{3,5}$-di-*cis*, 688,000; and $\Delta^{4,6}$-di-*cis*, 679,000. For the isomers of vitamin A aldehyde,[7] the following values have been reported: all-*trans*, 3,050,000; neo (Δ^5-*cis*), 3,120,000; Δ^3-*cis*, 637,000; $\Delta^{3,5}$-di-*cis*, 581,000; and $\Delta^{4,6}$-di-*cis*, 1,610,000.

Disregarding stereochemical variations, several compounds with structures corresponding to vitamin A, its ethers, and its esters have been prepared.[9–11] These compounds, as well as synthetic vitamin A acid, possess biologic activity.

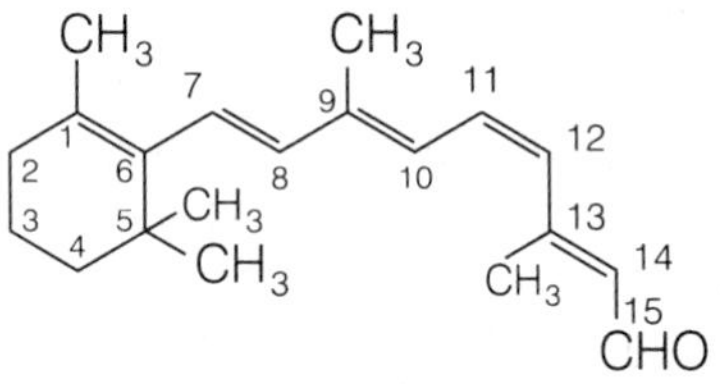

Neoretinene b (Retinal)

The chief source of natural vitamin A is fish-liver oils, which vary greatly in their content of this vitamin (Table 21-1).

Most liver oils contain vitamin A and neovitamin A in the ratio of 2 : 1. It occurs free and combined as the biologically active esters, chiefly of palmitic and some myristic and dodecanoic acids. It also is found in the livers of animals, especially those that are herbivorous. Milk and eggs are fair sources of this vitamin.

The livers of freshwater fish contain vitamin A_2 (3-dehydroretinol).

TABLE 21-1

VITAMIN A CONTENT OF SOME FISH-LIVER OILS

Source of Oil	Animal	Potency (IU/g)
Halibut, liver	*Hippoglossus hippoglossus*	60,000
Percomorph, liver	*Percomorph* fishes (mixed oils)	60,000
Shark, liver	*Galeus zygopterus*	25,500
Shark, liver	*Hypoprion brevirostris* and other varieties	16,500
Burbot, liver	*Lota maculosa*	4,880
Cod, liver	*Gadus morrhua*	850

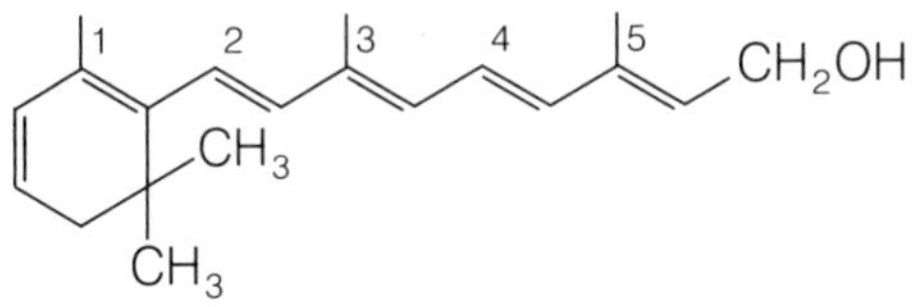

Vitamin A_2 (all-*trans*)
3-Dehydroretinol or Dehydroretinol

Physiologic doses of vitamin A are nearly completely absorbed. Even though lipids and bile aid the absorption, water-miscible forms of the vitamin are absorbed better than oily preparations. Incomplete absorption is seen following large doses of the vitamin or with disorders of fat absorption.

Retinol esters are hydrolyzed by pancreatic enzymes, absorbed by a carrier-mediated process, and reesterified. Thus, fatty acid esters enter the circulation in lymph chylomicrons. Vitamin A, primarily as the palmitate esters, is stored mainly in the liver, although the kidneys, adrenal glands, retinas, lungs, and intraperitoneal fat also have stores.

The liver releases retinol bound to an α_1-globulin referred to as retinol-binding protein (RBP). This protein, in turn, circulates in the blood complexed with a prealbumin protein that protects the RBP and retinol from metabolism and excretion. Normally, retinol esters compose less than 5% of circulating retinoids, but their concentration may increase drastically following large intake of the vitamin.

Retinol is susceptible to glucuronide conjugation, followed by enterohepatic recycling. Retinol may be oxidized to retinal and retinoic acid. Retinoic acid undergoes decarboxylation, followed by glucuronide conjugation. Normally no unchanged retinol is excreted. However, retinal, retinoic acid, and other metabolites are found in the urine and feces.

Although fish-liver oils were used for their vitamin A content, purified or concentrated forms of vitamin A are of great commercial significance. These are prepared in three ways: (1) saponification of the oil and concentration of the vitamin A in the nonsaponifiable matter by solvent extraction, the product is marketed as such; (2) molecular distillation of the nonsaponifiable matter, from which the sterols have previously been removed by freezing, giving a distillate of vitamin A containing 1 million to 2 millions IU/g; (3) subjecting the fish oil to direct molecular distillation to recover both the free vitamin A and vitamin A palmitate and myristate.

Pure crystalline vitamin A occurs as pale yellow plates or crystals. It melts at 63° to 64°C and is insoluble in water, but soluble in alcohol, the usual organic solvents, and the fixed oils. It is unstable in the presence of light and oxygen and in oxidized or readily oxidized fats and oils. It can be protected by the exclusion of air and light and by the presence of antioxidants.

Like all substances that have a polyene structure, vitamin A gives color reactions with many reagents, most of which are either strong acids or chlorides of polyvalent metals. An intense blue (Carr-Price) is obtained with vitamin A in dry chloroform solution upon the addition of a chloroform solution of antimony trichloride. This color reaction has been studied extensively and is the basis of a colorimetric assay for vitamin A.[12]

Vitamin A (all-*trans*-retinol) is biosynthesized in animals from plant pigments called *carotenoids*, which are terpenes composed of isoprenoid units; for example, β-carotene is the precursor of retinol (vitamin A). The provitamins A (e.g., β-, α-, and γ-carotenes and cryptoxanthin) are found in green parts of plants, carrots, red palm oil, butter, apricots, peaches, yellow corn, egg yolks, and other similar sources. The carotenoid pigments are utilized poorly by humans, whereas animals differ in their ability to utilize these compounds. These carotenoid pigments are provitamins A because they are converted to the active vitamin A. For example, β-carotene is absorbed intact by the intestinal mucosa, then cleaved to retinal by β-carotene-15,15-dioxygenase, which requires molecular oxygen.[13] β-Carotene can give rise to two molecules of retinal, whereas, in the other three carotenoids, only one molecule is possible by this transformation. These carotenoids have only one ring (see formula for β-carotene) at the end of the polyene chain that is identical with that found in β-carotene and is necessary for vitamin A activity. This accounts for the low activity of δ-carotene.

The conjugated double-bond systems found in vitamin A and β-carotene are necessary for activity, for when these compounds are partially or completely reduced, activity is lost. The β-ionone ring of retinol or the dehydro-β-ionone ring found in dehydroretinol (vitamin A_2) is essential for activity. Saturation results in loss of activity. The ester and methyl ethers of vitamin A have a biologic activity on a molar basis equal to vitamin A. Vitamin A acid is biologically active, but is not stored in the liver.

Absorption of dietary carotenes is dependent on absorbable fats and bile. Only 20% to 30% of it is absorbed unchanged. During absorption most carotene is metabolized to retinal.

The intestinal mucosa is the main site of β-carotene transformation[1] to retinal, but the enzyme that catalyzes the transformation also occurs in hepatic tissue. The enzyme is an iron-containing dioxygenase responsible for formation of two retinal molecules from one β-carotene molecule; subsequently, the retinal undergoes NADH or NADPH reduction to retinol.

$$\text{Retinol} \rightleftarrows \text{Retinal} \rightarrow \text{Retinoic Acid}$$

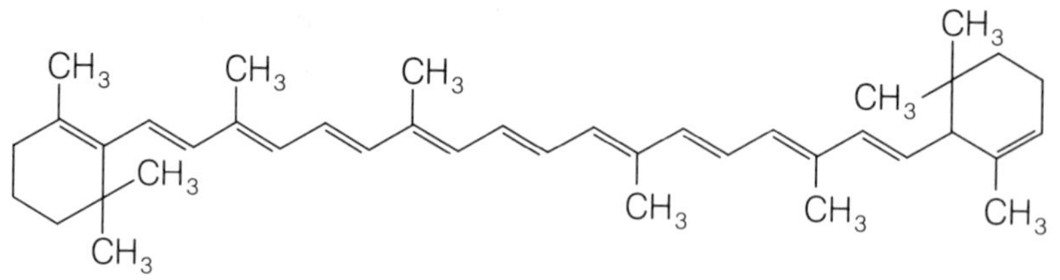

α-Carotene

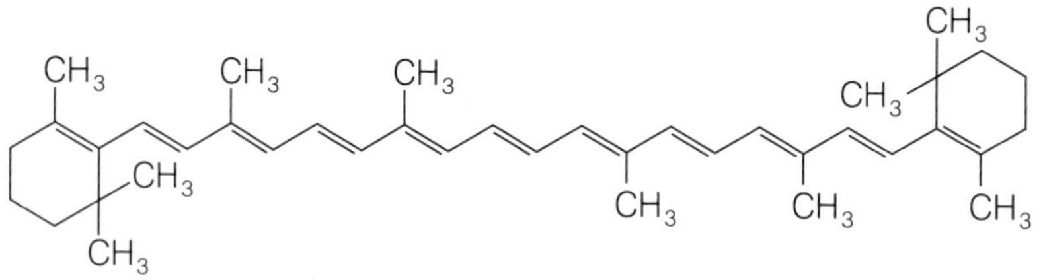

β-Carotene

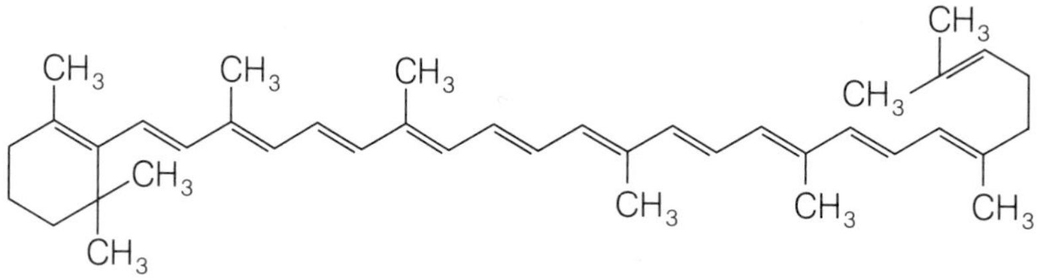

γ-Carotene

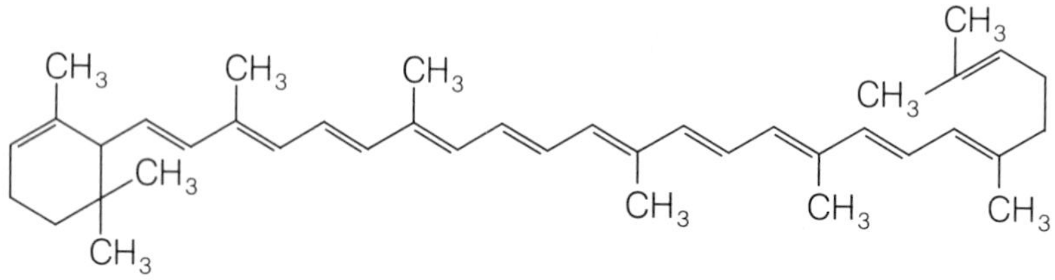

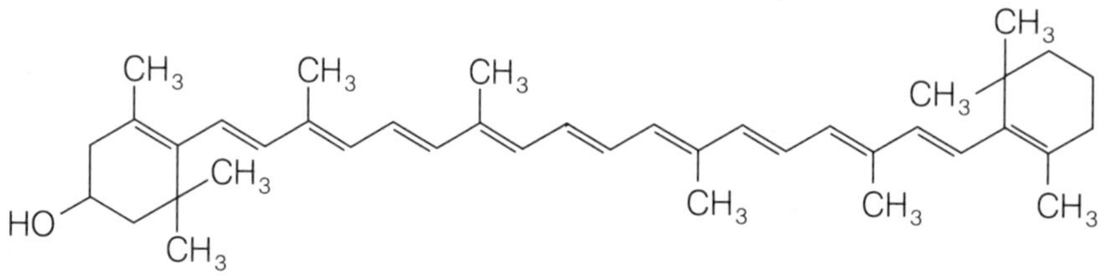

δ-Carotene

Crytoxanthin

Retinoic acid, the corresponding carboxylic acid, promotes development of bone and soft tissues and sperm production, but it does not participate in the visual process. Retinoic acid is found in the bile in the glucuronide form.

Vitamin A often is called the "growth vitamin" because a deficiency of it in the diet causes a cessation of growth in young rats. A deficiency of vitamin A is manifested chiefly by a degeneration of the mucous membranes throughout the body. This degeneration is evidenced to a greater extent in the eye than in any other part of the body and gives rise to a condition known as xerophthalmia. In the earlier stages of vitamin A deficiency, there may develop a night blindness (nyctalopia), which can be cured by vitamin A. *Night blindness* can be defined as the inability to see in dim light.

Dark adaptation or visual threshold is a more suitable description than night blindness when applied to many subclinical cases of vitamin A deficiency. The *visual threshold* at any moment is just that light intensity required to elicit a visual sensation. *Dark adaption* is the change that the visual threshold undergoes during a stay in the dark after an exposure to light. This change may be very great. After exposure of the eye to daylight, a stay of 30 minutes in the dark results in a decrease in the threshold by a factor of one million. This phenomenon is used as the basis to detect subclinical cases of vitamin A deficiencies. These tests vary in their technique, but, essentially, they measure visual dark adaptation after exposure to bright light and compare it with the normal.[14]

Advanced deficiency of vitamin A gives rise to a dryness and scaliness of the skin, accompanied by a tendency to infection. Characteristic lesions of the human skin caused by vitamin A deficiency usually occur in sexually mature persons between the ages of 16 and 30 and not in infants. These lesions appear first on the anterolateral surface of the thigh and on the posterolateral portion of the upper forearms and later spread to adjacent areas of the skin. The lesions consist of pigmented papules, up to 5 mm in diameter, at the site of the hair follicles.

Vitamin A regulates the activities of osteoblasts and osteoclasts, influencing the shape of the bones in the growing animal. The teeth also are affected. In vitamin A-deficiency states, a long overgrowth occurs. Overdoses of vitamin A in infants for prolonged periods led to irreversible changes in the bones, including retardation of growth, premature closure of the epiphyses, and differences in the lengths of the lower extremities. Thus, a close relationship exists between the functions of vitamins A and D relative to cartilage, bones, and teeth.[15]

The tocopherols exert a sparing and what appears to be a synergistic action[16] with vitamin A.

Blood levels of vitamin A decrease very slowly, and a decrease in dark adaptation was observed in only 2 of 27 volunteers (maintained on a vitamin A-free diet) after 14 months, at which time blood levels had decreased from 88 IU/100 mL of blood to 60 IU.

Vitamin A performs numerous biochemical functions; it has been demonstrated that vitamin A promotes the production of mucus by the basal cells of the epithelium, whereas in its absence keratin can be formed. Vitamin A performs a function in the biosynthesis of glycogen and some steroids, and increased quantities of coenzyme Q are found in the livers of vitamin-deficient rats. Significantly, the most well-known action of vitamin A is its function in the chemistry of vision.

Hypervitaminosis A is rarely associated with dietary intake. Vitamin A toxicity is associated with excessive supplementation. Early signs include central nervous system symptoms (fatigue, lethargy, irritability, delerium, depression, anorexia), gastrointestinal symptoms (discomfort, nausea, vomiting), and skin disorders (dryness, scaling, pruritus, erythema). Treatment involves discontinuance of the vitamin and supportive therapy.

The molecular mechanism of action of vitamin A in the visual process has been under investigation for many years. Wald in 1968 and Morton in 1972 characterized this mechanism of action. The chemistry of vision was comprehensively reviewed in *Accounts of Chemical Research* (1975) by numerous investigators. These reviews include theoretical studies of the visual chromophore, characterization of rhodopsin in synthetic systems, dynamic processes in vertebrate rod visual pigments and their membranes, and the dynamics of the visual protein opsin.[17–21]

Vitamin A (all-*trans*-retinol) undergoes isomerization to the 11-*cis* form in the liver. This transformation is catalyzed by a retinol isomerase. Subsequently, 11-*cis*-retinol interacts with a protein called retinol-binding protein (RBP) to form a complex that is transported to the retina photoreceptor cells, which contain specific receptors for the RBP–retinol complex.

The retina has been considered[20, 22] to be a double sense organ in which the rods are concerned with colorless vision at low light intensities and the cones with color vision at high light intensities. A dark-adapted, excised retina is rose-red; when it is exposed to light, its color changes to chamois, to orange, to pale yellow; finally, upon prolonged irradiation, it becomes colorless. The rods contain photosensitive visual purple (rhodopsin) that, when acted upon by light of a definite wavelength, is converted to visual yellow and initiates a series of chemical steps necessary to vision. Visual purple is a conjugated, carotenoid protein having a relative molecular mass (M_r) of about 40,000 and one prosthetic group per molecule. It contains seven hydrophobic α-helices, which are embedded in the membrane. Short hydrophilic loops interconnect the helices and are exposed to the aqueous environment on either side of the membrane. It has an absorption maximum of about 510 nm. The prosthetic group is retinene (neoretinene b or retinal), which is joined to the protein through a protonated Schiff base linkage. The function of retinene in visual purple is to provide an increased absorption coefficient in visible light and, thereby, sensitize the protein, which is denatured. This process initiates a series of physical and chemical steps necessary to vision. The protein itself differs from other proteins by having a lower energy of activation, which permits it to be denatured by a quantum of visible light. Other proteins require a quantum of ultraviolet light to be denatured. The bond between the pigment and the protein is much weaker when the protein is denatured than when it is native. The denaturation process of the protein is reversible and takes place more readily in the dark to give rise, when combined with retinene, to visual purple. The effectiveness of the spectrum in bleaching visual purple runs fairly parallel with its absorption spectrum (510 nm) and with the sensibility distribution of the eye in the spectrum at low illuminations. It has been calculated that for a human to see a barely perceptible flash of

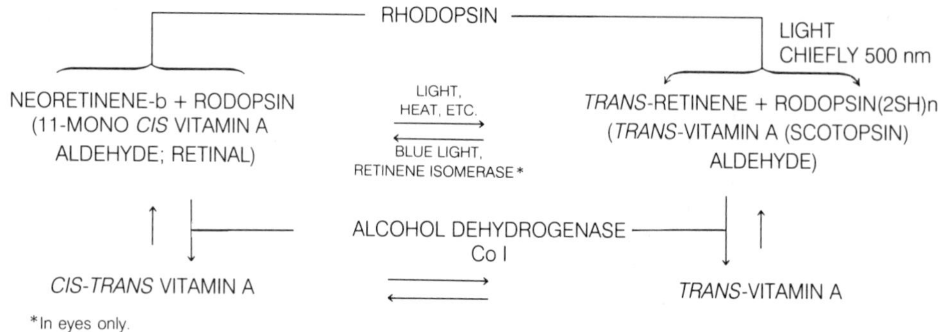

light, only one molecule of visual purple in each 5 to 14 rod cells needs to be photochemically transformed in a dark-adapted eye. The system possesses such sensitivity because of a biologic amplification system. In vivo, visual purple is constantly reformed as it is bleached by light, and under continuous illumination, an equilibrium between visual purple, visual yellow, and visual white is maintained. If an animal is placed in the dark, the regeneration of visual purple continues until a maximum concentration is obtained. Visual purple in the eyes of an intact animal may be bleached by light and regenerated in the dark an enormous number of times.

In the resting state (dark), rod and cone membranes exhibit a steady electrical current. The membrane allows sodium ions to enter freely through specific channels. A Na^+, K^+–ATPase pump maintains the ion gradient. The closing of the pores hyperpolarizes the membrane and initiates the neuronal response.

The pores are kept open by binding to cyclic-GMP. The light-induced isomerization of retinal causes a conformational change in the protein part of rhodopsin, activating the molecule. One active rhodopsin activates several hundred G-proteins called *transducin*. (G-proteins have the capability to bind with guanosine nucleotides.) Activation of transducin consists of an exchange of bound GDP for GTP. Activated transducin, in turn, activates a phosphodiesterases that hydrolyzes thousands of cGMP molecules. The decreased concentration of cGMP results in the closing of the sodium channels.

Hydrolysis of transducin-bound GTP to GDP inactivates the phosphodiesterases. However, the activated rhodopsin must also be deactivated. This is accomplished by phosphorylation of opsin by opsin kinase. Guanylate cyclase replenishes the cGMP concentration, which reopens the channels.

Visual purple occurs in all vertebrates. It is not distributed evenly over the retina. It is missing in the fovea, and in the regions outside of the fovea its concentration undoubtedly increases to a maximum in the region about 20° off center, corresponding to the high density of rods in this region. Therefore, to see an object best in the dark, one should not look directly at it.

The diagrams represent some of the changes that take place in the visual cycle involving the rhodopsin system in which the 11-mono-*cis* isomer of vitamin A is functional in the aldehyde form.[23]

NH—Lysine—Protein (+)

Light-catalyzed reaction: Rhodopsin (λmax, 506 nm) → hν, Picoseconds →

Thermal reactions:
Bathorhodopsin (λmax, 548 nm)
↓ Nanoseconds
Lumirhodopsin (λmax, 497 nm)
↓ Microseconds
Metarhodopsin-I (λmax, 478 nm)
⇅ H^+ Milliseconds
Metarhodopsin-II (λmax, 380 nm)
↓ Seconds

Bleached rhodopsin:
Opsin
+
Trans-retinal

Note: Bovine rhodopsin. λmax = wavelength maximum of each intermediate. (From Chem. Eng. News, Nov. 28, 1983.)

Only neoretinene b (retinal) (Δ^4- or 11-mono-*cis*) can combine with opsin (scotopsin) to form rhodopsin. The isomerization of *trans*-retinene may take place in the presence of blue light. However, vision continues very well in yellow, orange, and red light in which no isomerization takes place. The neoretinene b (retinal) under these circumstances is replaced by an active form of vitamin A from the bloodstream, which, in turn, obtains it from stores in the liver. The isomerization of *trans*-vitamin A in the body to *cis-trans*-vitamin A seems to keep pace with long-term processes such as growth, since vitamin A, neovitamin A, and neoretinene b (retinal) are equally active in growth tests in rats.

The sulfhydryl groups (two for each retinene molecule isomerized) exposed on the opsin initiate the transmission of impulses in the phenomenon of vision.

The assumption has been made that the mechanism of action of vitamin A might be similar to that of the steroid hormones (see the sections on steroids and steroid-hormone receptors), and this assumption has led to the characterization of two intracellular-binding proteins, one specific for retinal and the other specific for retinoic acid.[5]

The available experimental data do not provide complete evidence that these two proteins are, in fact, receptors analogous to steroid hormone receptors, but there is convincing evidence that these two proteins mediate important aspects of vitamin A function. The existence of a protein that specifically binds retinoic acid substantiates the implication of retinoic acid as a physiologic form of vitamin A.

For many years attempts have been made to demonstrate relationships between vitamin A and cancer. Recently, it has been demonstrated that dietary deficiency of vitamin A may lead to increased incidence of spontaneous and carcinogen- or virus-induced epithelial metaplasias and tumors in experimental animals and possibly in humans. It also has been observed that in several human tumors from lung, breast, skin, and stomach, cellular retinoic acid-binding protein is detectable only in the tumor and not in the normal tissue.[5]

Vitamin A, USP, is a product that contains retinol (vitamin A alcohol) or its esters from edible fatty acids, chiefly acetic and palmitic acids, the activity of which is not less than 95% of the labeled amount; 0.3 μg of vitamin A alcohol (retinol) equals 1 USP unit.

Vitamin A is indicated only for treatment of vitamin A deficiencies. Because the vitamin is prevalent in the diet, especially with supplementation of milk, this disorder is not common. It is associated with conditions resulting in the malabsorption of fats (e.g., biliary or pancreatic diseases, sprue, hepatic cirrhosis).

Pure vitamin A has the activity of 3.5 million IU/g. Moderate to massive doses of vitamin A have been used in pregnancy, lactation, acne, abortion of colds, removal of persistent follicular hyperkeratosis of the arms, persistent and abnormal warts, corns and calluses, and similar conditions. Phosphatides or the tocopherols enhance the absorption of vitamin A. Vitamin A applied topically appears to reverse the impairment of wound healing by corticoids.

Vitamin A occurs as a yellow to red, oily liquid; it is nearly odorless or has a fishy odor and is unstable to air and light. It is insoluble in water or glycerin and is soluble in absolute alcohol, vegetable oils, ether, and chloroform.

Tretinoin, USP. Retinoic acid (Retin-A). Tretinoin is a yellow to light orange crystalline powder. It is insoluble in water and slightly soluble in alcohol.

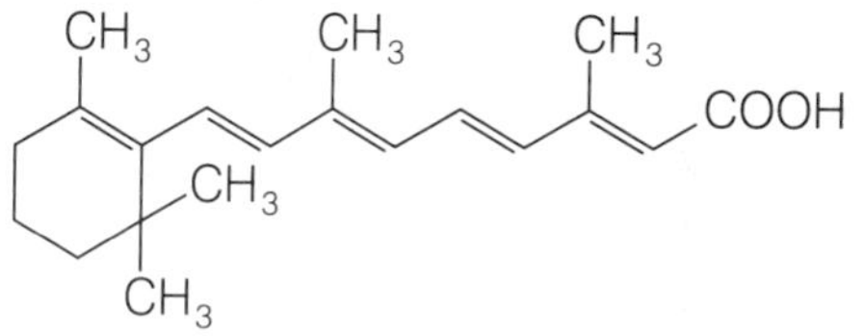

CH_3 CH_3 CH_3 COOH CH_3 CH_3

Tretinoin

Tretinoin, indicated for topical treatment of acne vulgaris, was initially used systemically. However, therapeutic doses frequently resulted in hypervitaminosis A. It appears to exert its action by decreasing the adhesion of corneocytes and by increasing the proliferation of the follicular epithelium.[24]

Tretinoin is usually applied as a 0.05% PEG-400/ethanol liquid or a 0.05% hydrophylic cream. Daily application results in inflammation, erythema, and peeling of the skin. After three to four weeks a pustular eruptions may be seen, causing the expulsion of microcomedones. Treatment may then be changed to applications every two or three days.

Because the horny layer is thinned, the skin is more susceptible to irritation by chemical or physical abuse. Thus, it is recommended that other kerolytic agents (salicylic, sulfur, resorcinol, benzoyl peroxide) be discontinued before beginning treatment with tretinoin. Sunscreens labeled SPF-15 or higher are recommended.

Unlabeled uses of tretinoin include the treatment of some forms of skin cancer, lamellar ichthyosis, Darier's disease, and photoaging.

Photoaging of the skin is mainly the result of excessive exposure to sunlight and is manifested by lax, yellow, mottled, wrinkled, leathery, rough skin. Once-daily application has been reported to aid in the early stages of photoaging.[25] Tretinoin is believed to exert this action by its function in regulating epithelial differentiation, cell division and protein synthesis.[26] However, termination of treatment results in reversal within a year.

Tretinoin is believed to exert its antineoplastic effect by promoting cellular differentiation toward normal cells.[25]

Isotretinoin, USP. 13-*cis*-Retinoic acid (Accutane). Isotretinoin is a yellow orange to orange crystalline powder. It is insoluble in water and sparingly soluble in alcohol.

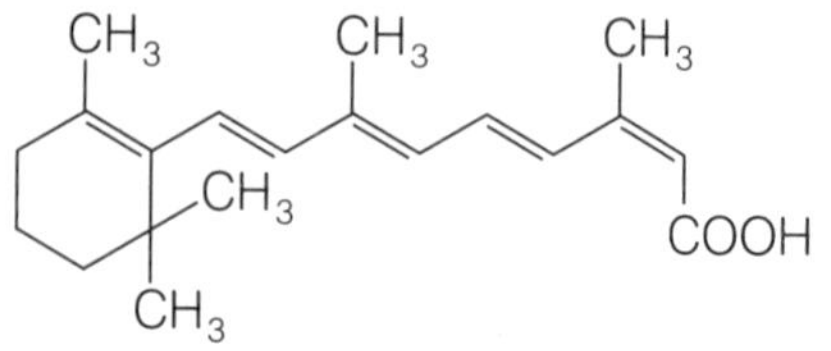

Isotretinoin

Isotretinoin is indicated for the treatment of severe recalcitrant cystic acne. Because of the risks of adverse effects, its use should be reserved for only those patients who are unresponsive to conventional acne therapies. Treatment should be individualized and modified depending on the course of the disease.

The mechanism is believed to involve inhibition of sebaceous gland function and follicular keratinization. Isotretinoin reduces sebum production, the size of the glands, and gland differentiation.

The initial dose is 0.5–1 mg/kg daily in two divided doses. Absorption is rapid, but bioavailability is low (~ 25%) because of degradation in the lumen, metabolism by the gastrointestinal mucosa and by the liver on the first pass. The chief metabolite is 4-oxoisotretinoin. Both isotretinoin and its metabolite are conjugated to the glucuronide and excreted in the urine and feces. The usual course of therapy is 15 to 20 weeks.

The adverse effects of isotretinoin are typical of chronic hypervitaminosis A. Because of the high potential to cause teratogenic effects, isotretinoin should be used with extreme caution in females of childbearing age. The manufacturer of the drug strongly recommends that patients should have pregnancy tests performed before the onset of therapy and utilize a form of birth control during therapy.

Etretinate (Tegison). Etretinate is indicated for the treatment of severe recalcitrant psoriasis. Because of its potential adverse effects, therapy should be limited to those diseases that do not respond to standard therapies. The exact mechanism of etretinate's action is unknown, but is believed to result from some of the actions common to the retinoids.

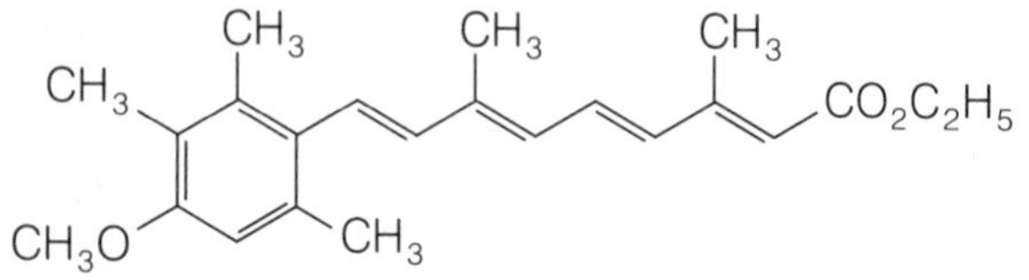

Etretinate

Oral bioavailability of etretinate is approximately 40%. Milk and lipids increase the absorption. Etretinate is significantly converted to the free acid on the first pass through the liver. The free acid (etretin) is also active. After a single dose, the half-life is 6 to 13 hours, but after long-term therapy, the half-life is 80 to 100 days. Etretinate's high lipid character results in storage in adipose tissue, from which it is released slowly. After discontinuation of therapy, etretinate can be detected for up to one year.

Etretin has the advantage of a shorter half-life, two hours after a single dose and 50 hours after multiple doses. However, it is more susceptible to conversion to 13-*cis*-etretin. Thus it appears that the ester provides metabolic stability.

The initial dosage of etretinate is 0.75–1 mg/kg in divided doses. After 8 to 16 weeks, a maintenance dosage of 0.5–0.75 mg/kg may be started. Similar to isotretinoin, extreme caution should be exercised in the administration of etretinate.

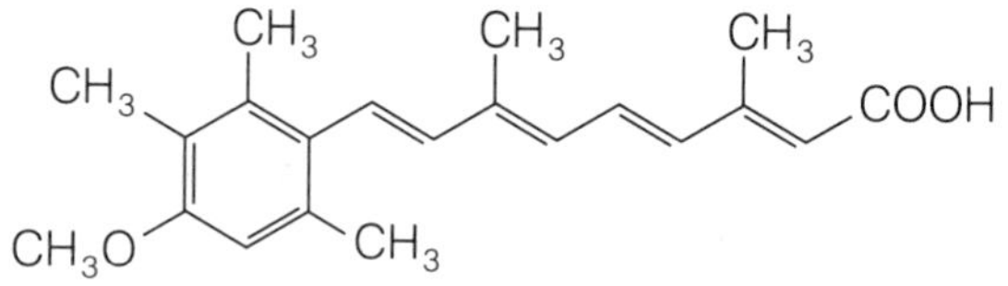

Etretin

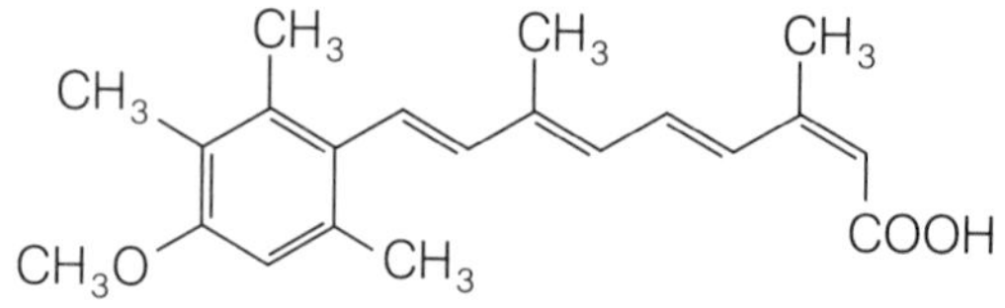

13-*cis* Etretin

β-Carotene, USP (Soletene). β-Carotene is a red or reddish brown to violet brown powder. It is insoluble in water and alcohol and sparingly soluble in vegetable oils. It is a naturally occurring carotenoid pigment found in green and yellow vegetables.

β-Carotene is indicated for the treatment of erythropoietic protoporphyria. It does not provide total protection against the sun, but patients who respond to its treatment can remain in the sun the same as normal individuals. Discontinuance of the drug results in return of the hypersensitivity. β-Carotene does not function as a sunscreen in normal patients and should not be used as such.

The dosage range is 30–300 mg/day in a single or divided doses, usually administered with food, because its absorption depends on the presence of bile and absorbable fat.

Most β-carotene is converted to retinal during absorption, but the fraction that is absorbed is distributed widely and accumulates in the skin. The metabolic pathway of β-carotene is similar to that of retinol.

Several weeks of therapy are required before sufficient amounts accumulate in the skin for it to exert its protective effects. Carotenodermia, a result of accumulation in the skin, is the major side effect. However, a "tanning" capsule containing β-carotene and canthaxanthin utilizes this effect.

Vitamin A_2

Vitamin A_2 is found in vertebrates that live, or at least begin, their lives in fresh water. Vitamin A_2 exhibits chemical, physical, and biologic properties very similar to those of vitamin A. It has the structural formula shown below.

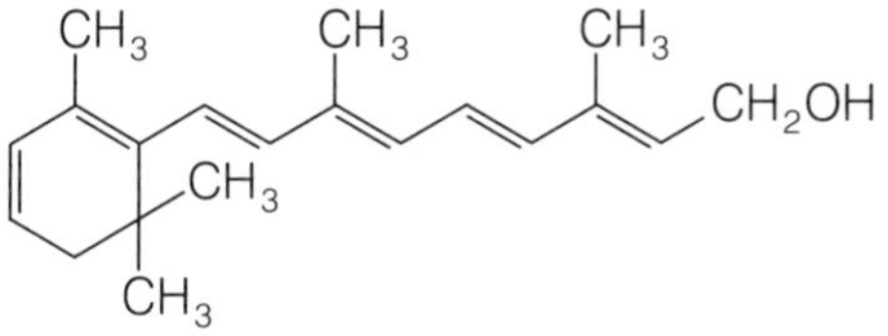

3-Dehydroretinol

Vitamin A_2 has a biologic potency of 1.3 million USP U/g, which is approximately 40% of the activity of crystalline vitamin A acetate.

THE VITAMIN Ds

The term *vitamin D* was originally applied to the material obtained by irradiation of yeast ergosterol. This material was later analyzed to be a mixture of calciferol isomers and some steroids. A 1 : 1 mixture of ergocalciferol and lumisterol$_2$ was originally designated vitamin D_1. Upon purification and further characterization, ergocalciferol (calciferol) proved to possess significant antirachitic properties and became known as vitamin D_2. Analogously, 7-dehydrocholesterol, which originates from cholesterol in human metabolism, normally undergoes transformation to cholecalciferol, vitamin D_3, upon exposure of the skin to sunlight (see the following structures).

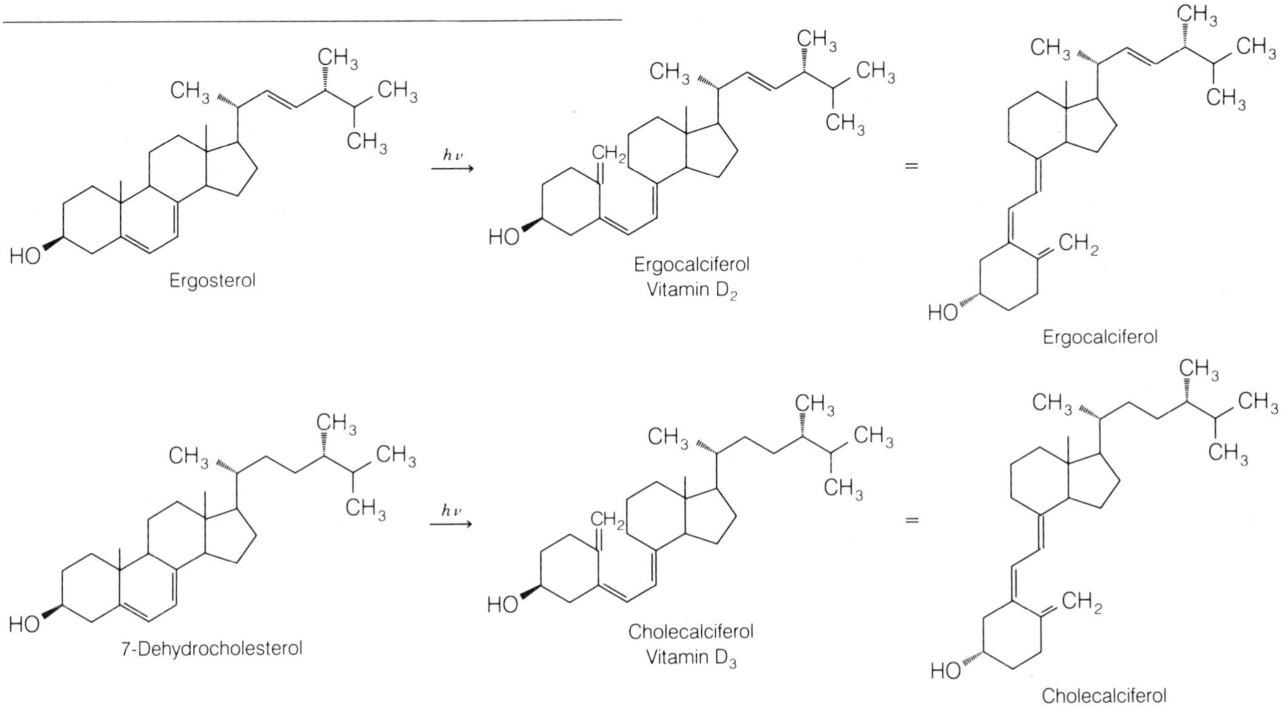

Ergosterol $\xrightarrow{h\nu}$ Ergocalciferol Vitamin D_2 = Ergocalciferol

7-Dehydrocholesterol $\xrightarrow{h\nu}$ Cholecalciferol Vitamin D_3 = Cholecalciferol

Recent advances[24] in the understanding of the biochemistry of the vitamin Ds have led to the classification of vitamin D's physiologic function as an endocrine system; vitamin D_3 (cholecalciferol) actually appears to function as a hormone precursor. According to this classification, cholecalciferol is not a vitamin because it normally can be synthesized from cholesterol in human metabolism.

Ergosterol (precursor of D_2) occurs naturally in fungi and yeast. Eggs and butter contain vitamin D_2 (ergocalciferol) or D_3 (cholecalciferol). Milk and bread are fortified by the addition of vitamin D_2. Cholecalciferol is found in fish-liver oils.

Cholecalciferol does not perform its functions directly. It must be metabolically transformed in the liver to 25-hydroxycholecalciferol (calcifediol) by a microsomal enzyme called vitamin D 25-hydroxylase which requires NAPDH and molecular oxygen. This intermediate, although possessing intrinsic activity, is further activated in renal mitochondria by the enzyme vitamin D 1-hydroxylase. This enzyme is a mixed function oxidase composed of cytochrome P-450, a flavoprotein, and ferredoxin. It, too, requires NAPDH and molecular oxygen as cofactors. This latter metabolite (calcitriol; 1,25-dihydroxycholecalciferol) is considered the biologically active form.

Subsequently, 24-hydroxylation proceeds in the kidney, and this initiates inactivation. The production of the hormone 1,25-dihydroxycholecalciferol is regulated by the body's need for calcium and phosphate ions. 1,25-Dihydroxycholecalciferol can bring about the appearance of the 24-hydroxylase system that catalyzes its metabolic inactivation. The need for calcium stimulates parathyroid hormone secretion. The parathyroid hormone, in turn, suppresses the 24-hydroxylase and stimulates the 1-hydroxylase system. When phosphate availability is below normal the 1-hydroxylase is stimulated and the 24-hydroxylase undergoes suppression.

1,25-Dihydroxycholecalciferol promotes Ca^{2+} intestinal absorption and mobilization of Ca^{2+} from bone. The mechanism of action promoting Ca^{2+} transport in the intestine involves formation of a calcium-binding protein. 1,25-Dihydroxycholecalciferol promotes availability of this protein. A calcium-dependent ATPase, Na^+, and the calcium-binding protein are necessary for the intestinal Ca^{2+}-transport process. 1,25-Dihydroxycholecalciferol also promotes intestinal phosphate absorption, mobilization of Ca^{2+} and phosphate from bone, and renal reabsorption of Ca^{2+} and phosphate.

The physiologic role of vitamin D is to maintain calcium homeostasis. Phosphate metabolism is also affected. Vitamin D accomplishes its role by enhancing the absorption of calcium and phosphate from the small intestines, promoting their mobilization from bone and by decreasing their excretion by the kidney. Also involved are the parathyroid hormone and calcitonin.

A vitamin D deficiency results in rickets in infants and children as a result of inadequate calcification of bones. In adults, osteomalacia most often occurs during pregnancy and lactation.

Hypervitaminosis D may result from large doses of the vitamin or from a hypersensitivity to the vitamin. The early symptoms are those associated with hypercalcemia, including fatigue, weakness, nausea, vomiting, vertigo, bone pain. Prolonged hypercalcemia may result in calcium deposits in the kidneys, vessels, heart, lungs, and skin. Treatment includes withdrawal of the vitamin and a low-calcium diet with an increase in fluids.

The gastrointestinal absorption of the vitamin Ds requires bile. Vitamin D_3 may be absorbed better than vitamin D_2. They enter the circulation through lymph chylomicrons. In the blood they are associated with a vitamin D-binding protein (α-globulin). The 25-hydroxylated compounds are the major circulating metabolites. These may be stored in fats and muscle for prolonged periods. The 24-hydroxy metabolites are excreted primarily in the bile.

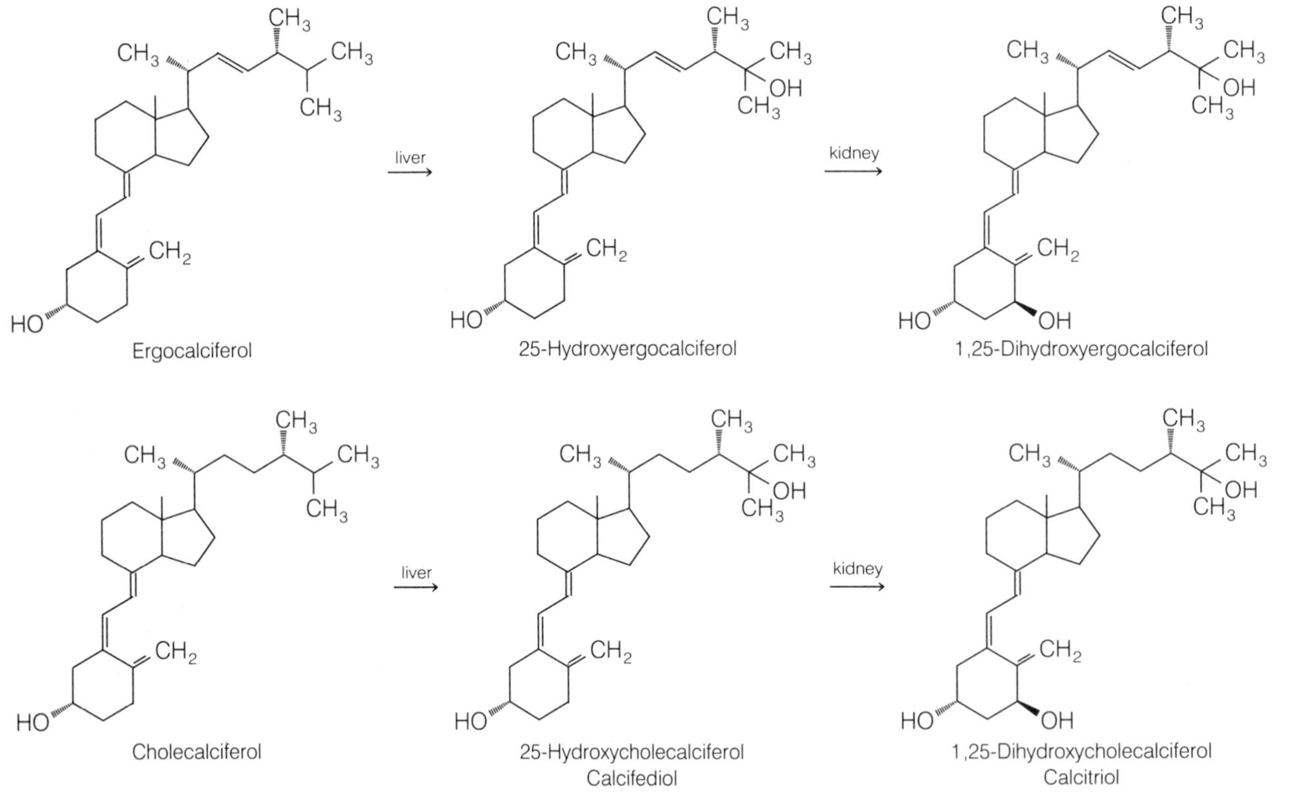

Ergocalciferol $\xrightarrow{\text{liver}}$ 25-Hydroxyergocalciferol $\xrightarrow{\text{kidney}}$ 1,25-Dihydroxyergocalciferol

Cholecalciferol $\xrightarrow{\text{liver}}$ 25-Hydroxycholecalciferol (Calcifediol) $\xrightarrow{\text{kidney}}$ 1,25-Dihydroxycholecalciferol (Calcitriol)

The vitamin Ds are important in the therapeutics of hypoparathyroidism and of vitamin D deficiency.[2] Ergocalciferol, cholecalciferol, and dihydrotachysterol are recognized by the *USP*. Although dihydrotachysterol has relatively weak antirachitic activity, it is effective and quicker acting in increasing serum Ca^{2+} concentration in parathyroid deficiency. Dihydrotachysterol has a shorter duration of action; hence, it has less potential for toxicity from hypercalcemia.

Products

Ergocalciferol, USP. Irradiated ergosta-5,7,22-trien-3β-ol; vitamin D_2; calciferol. The history and preparation of this vitamin have been described.

Vitamin D_2 is a white, odorless, crystalline compound that is soluble in fats and in the usual organic solvents, including alcohol. It is insoluble in water. Vitamin D_2 is oxidized slowly in oils by oxygen from the air, probably through the fat peroxides that are formed. Vitamin A is much less stable under the same conditions.

Cholecalciferol, USP. Activated 5,7-cholestadien-3β-ol; vitamin D_3; activated 7-dehydrocholesterol, occurs as white, odorless crystals that are soluble in fatty oils, alcohol, and many organic solvents. It is insoluble in water.

Vitamin D_3 also occurs in tuna- and halibut-liver oils. It has the same activity as vitamin D_2 in rats, but vitamin D_3 is more effective for the chick; however, both vitamins have equal activity for humans.

Vitamin D_3 exhibits stability comparable with that of vitamin D_2.

Epimerization of the —OH at C-3 in vitamin D_2 or D_3 or conversion of the —OH at C-3 to a ketone group greatly diminishes the activity, but does not completely destroy it. Ethers and esters that cannot be cleaved in the body have no vitamin D activity. Inversion of the hydrogen at C-9 in ergosterol and other 7-dehydrosterols prevents the normal course of irradiation.

Dihydrotachysterol, USP. AT 10 (Hytakerol). Dihydrotachysterol$_2$; 9,10-*seco*-5,7,22-ergastatrien-3-β-ol; Tachysterol$_2$, represented below, is a by-product from ergosterol irradiation. Reduction of tachysterol$_2$ led to dihydrotachysterol$_2$.

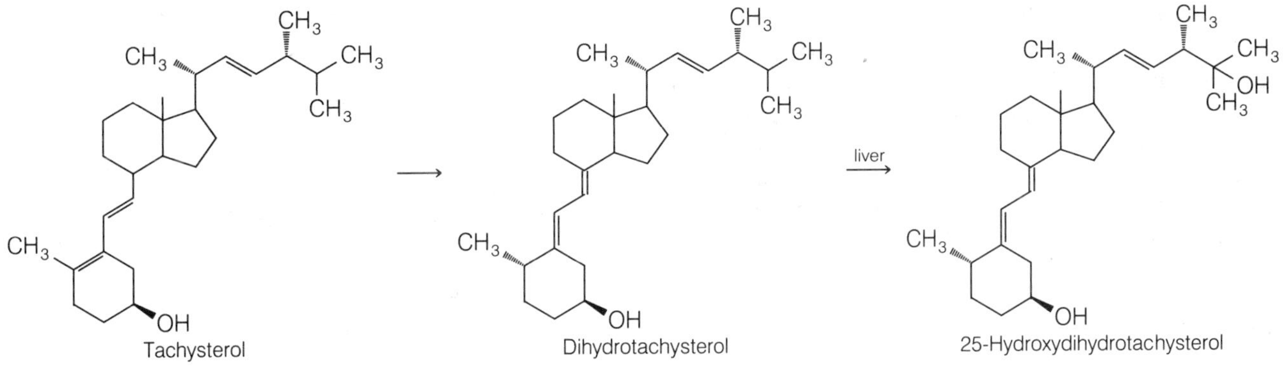

Ergocalciferol has a half-life of 24 hours (19 to 48 hours) and a duration of action of up to six months. After oral or intramuscular administration the onset of action (hypercalcemia) is 10 to 24 hours, with maximal effects seen four weeks after daily administration.

Pure vitamin D_2 protects rats from rickets in daily doses of 15 μg. However, it was soon shown that, rat unit for rat unit, vitamin D_2 or irradiated ergosterol, was not as effective as cod-liver oil for the chick. Therefore, the vitamin D of cod-liver oil appears to differ from vitamin D_2.

One microgram equals 40 USP or International units. (One mg equals 40,000 units). Because ergocalciferol is the least expensive of the vitamin D analogues, it is the preferred drug, unless the patient is unable to activate ergocalciferol.

Dihydrotachysterol occurs as colorless or white crystals or a white, crystalline, odorless powder. It is soluble in alcohol, freely soluble in chloroform, sparingly soluble in vegetable oils, and practically insoluble in water.

Dihydrotachysterol has slight antirachitic activity. It causes an increase of the calcium concentration in the blood, an effect for which tachysterol is only one-tenth as active.

In high doses, dihydrotachysterol is more effective than the other analogues for the mobilization of calcium. Thus, it is used in hypoparathyroidism.

After oral administration the onset of action is seen within hours. This fast onset of action is an advantage of this drug. Maximal activity is seen in two weeks after daily administration. Its duration of action is two weeks.

Dihydrotachysterol is activated by hepatic enzymes to its 25-hydroxylated metabolite. It does not require renal activation, for the hydroxy on ring A occupies the same position as that of the one 1-hydroxyl in the activated forms of the vitamins D.

25-Hydroxydihydrotachysterol$_3$ has weak antirachitic activity, but it is a more important bone-mobilizing agent and is more effective than dihydrotachysterol$_3$. Also, it is more effective in increasing intestinal calcium transport and bone mobilization in thyroparathyroidectomized rats. Its activity suggests that it may be the drug of choice in the treatment of hypoparathyroidism and similar bone diseases.[28]

Calcifediol, USP. 25-Hydroxycholecalciferol; 25-hydroxyvitamin D_3. Calcifediol occurs as a white powder. It is practically insoluble in water and sensitive to light and heat.

The half-life of calcifediol is 16 days (10 to 22 days). Its onset of action is seen within two to six hours, and its duration of action is 15 to 20 days.

Calcifediol is indicated for patients receiving long-term renal dialysis.

Calcitriol, USP. 1,25-Dihydroxycholecalciferol; 1,25-dihydroxyvitamin D_3. Calcitriol is the most active form of vitamin D_3. It occurs as colorless crystals that are insoluble in water.

Since calcitriol does not require activation, an increase in calcium absorption is seen within two hours of administration. Its half-life is three to eight hours and its duration of action is one to two days.

Calcitriol is indicated in patients receiving long-term renal dialysis or patients who cannot properly metabolize ergocalciferol.

VITAMIN E

Since the early 1920s it has been known that rats fed only cow's milk are not able to produce offspring. The principle from wheat germ that can rectify this deficiency in both male and female rats was named *vitamin E*. When the compound known as vitamin E was isolated in 1936, it was named tocopherol. Since then, several other closely related compounds have been discovered from natural sources, and this family of natural products took the generic name *tocopherols*.

Eight tocopherols have been isolated. Six have the 4,8,12-trimethyltridecyl saturated side chain. The other two have unsaturation in the side chain. The most well-known tocopherols include α-tocopherol (vitamin E), which has the greatest biologic activity; β-tocopherol; γ-tocopherol; and δ-tocopherol.

The base structure represented in the diagram shows that the tocopherols are various methyl-substituted tocol derivatives. β-Tocopherol is 5,8-dimethyltocol; the γ-compound is 7,8-dimethyltocol; δ-tocopherol is 8-methyltocol.

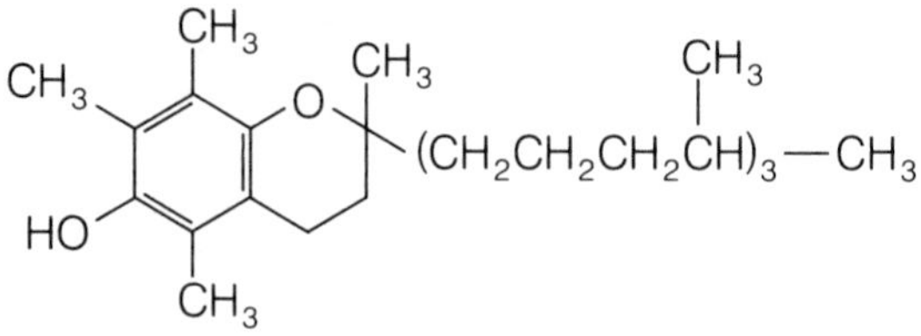

α-Tocopherol

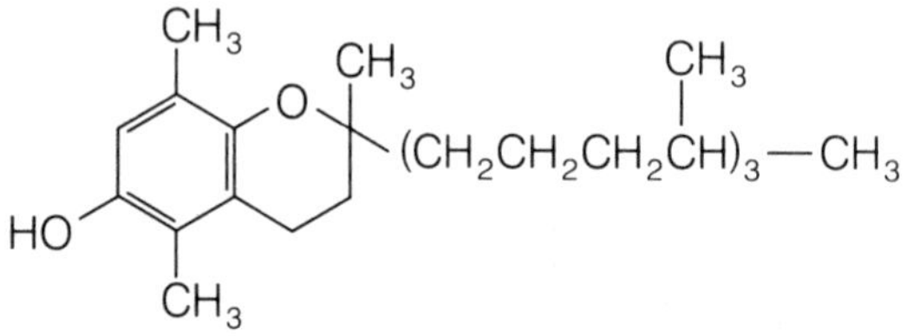

β-Tocopherol

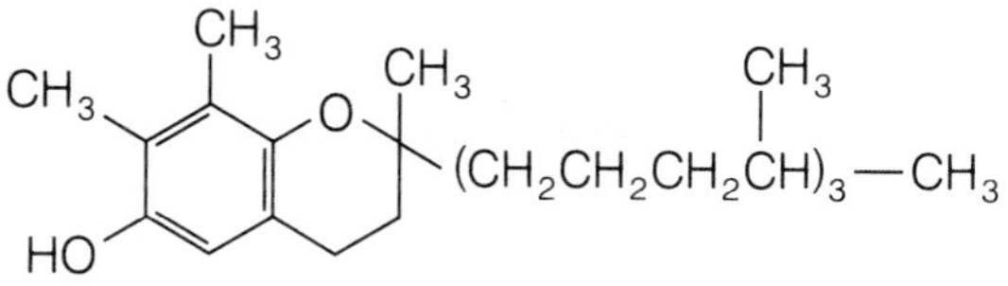

γ-Tocopherol

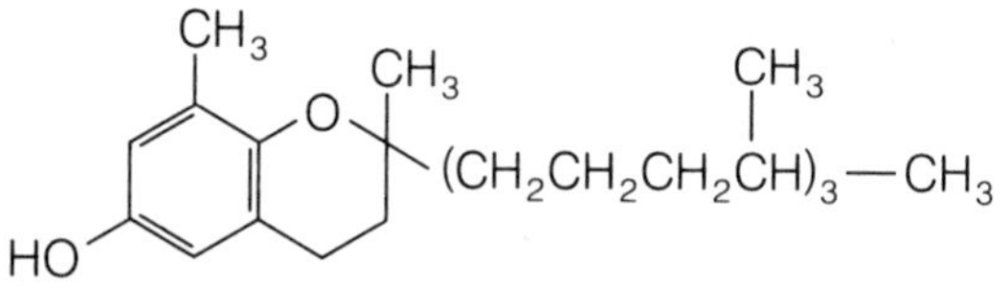

δ-Tocopherol

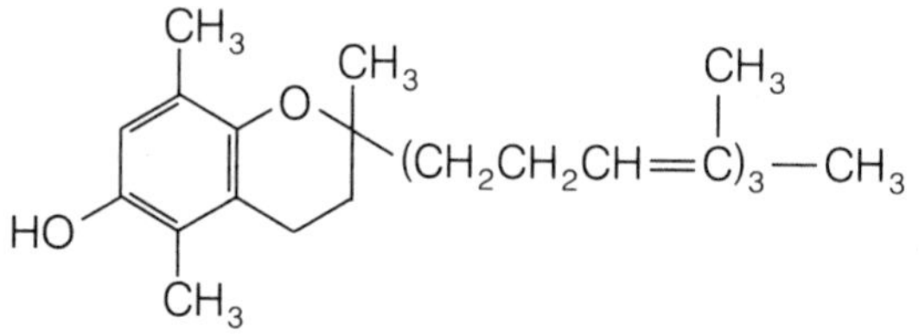

ε-Tocopherol

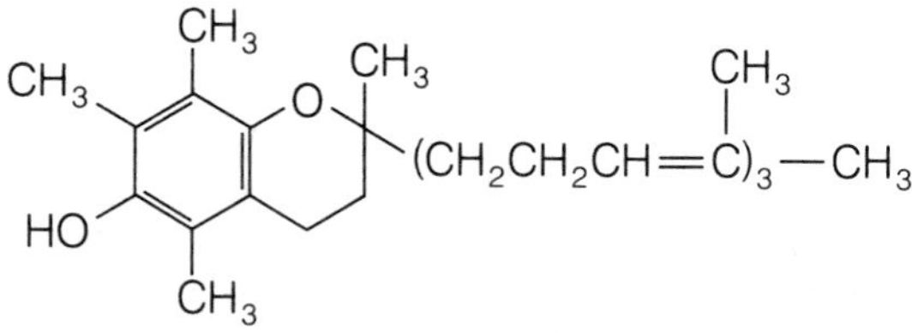

ζ_1-Tocopherol

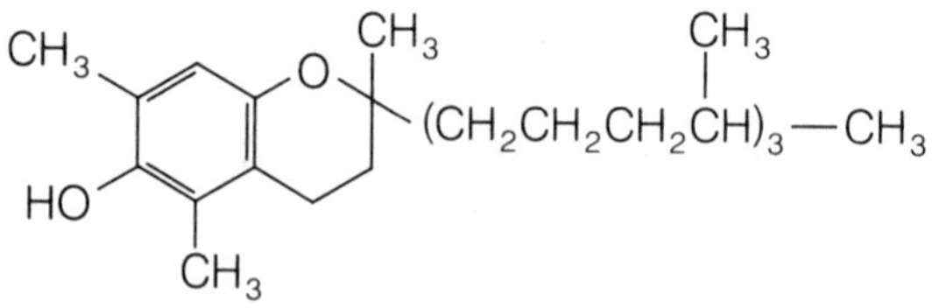

ζ_2-Tocopherol

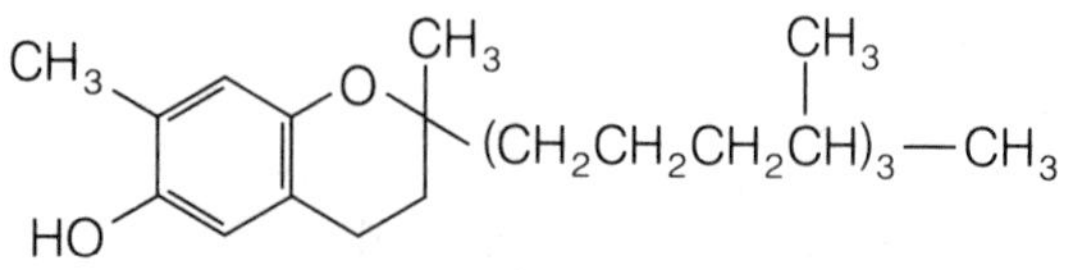

η-Tocopherol

The tocopherols are diterpenoid natural products biosynthesized from a combination of four isoprenoid units; geranylgeranyl pyrophosphate is the key intermediate that leads to these compounds.[29]

The tocopherols and their acetates are light yellow, viscous, odorless oils that have an insipid taste. They are insoluble in water and soluble in alcohol, organic solvents, and fixed oils. The acid succinate esters are white powders insoluble in water, soluble in ethanol and in vegetable oils. The tocopherols are stable in air for reasonable periods, but are oxidized slowly by air. They are oxidized readily by ferric salts, by mild oxidizing agents, and by air in the presence of alkali. They are inactivated rapidly by exposure to ultraviolet light; however, not all samples behave alike in this respect because traces of impurities apparently greatly affect the rate of oxidation. The tocopherols have antioxidant properties for fixed oils in the following decreasing order of effectiveness: δ-, γ-, β-, and α-.[30] In the process of acting as antioxidants, the tocopherols are destroyed by the accumulating fat peroxides that are decomposed by them. They are added to Light Mineral Oil NF and Mineral Oil USP because of their antioxidant property. The tocopherols can be converted to the acetates and benzoates, respectively, which are oils and are as active as the parent compounds and have the advantage of being more stable toward oxidation.

(+)-α-Tocopherol is about 1.36 times as effective as (±)-α-tocopherol in rat antisterility bioassays. β-Tocopherol is about one-half as active as α-tocopherol, and the γ- and δ-tocopherols are only 1/100 as active as α-tocopherol. The esters of the tocopherol, such as the acetate, propionate, and butyrate, are more active than the parent compound.[31] This is also true of the phosphoric acid ester of (±)-δ-tocopherol when it is administered parenterally.[32] The ethers of the tocopherols are inactive. The oxidation of the tocopherols to their corresponding quinones also leads to inactive compounds. Replacement of the methyl groups by ethyl groups leads to decreased activity. The introduction of a double bond in the 3,4-position of α-tocopherol reduces its activity by about two-thirds. Reduction of the size of the long alkyl side chain or the introduction of double bonds in this side chain markedly reduces activity.

Vitamin E activity is expressed in USP units or International units. One USP unit is equivalent to 1 IU. The relative potencies of the various commercial forms are listed in Table 21-2.

Vitamin E, USP, may consist of (+)- or (±)-α-tocopherols or their acetates or succinates, 97.0% to 100% pure. It also may be mixed tocopherols concentrate containing no less than 33% of total tocopherols of which no less than 50% is (±)- or (+)-α-tocopherol and is obtained from edible vegetable oils that may be used as diluents when needed. It also may be a 25% (±)- or (+)-α-tocopheryl acetate in concentrate, the vehicle being an edible vegetable oil.

(−)-α-Tocopherol is absorbed from the gut more rapidly than the (+)-form; however, the absorption of the mixture of (+)- and (−)-α-tocopherol was considerably higher (about 55%) than was to be expected from the data obtained after administration of the single compounds.

As doses increase, the fraction absorbed decreases. No marked differences were noted in the distribution in various tissues and the metabolic degradation of (+)- and (−)-α-tocopherols.[33] The liver is an important storage site.

The tocopherols are especially abundant in wheat germ, rice germ, corn germ, other seed germs, lettuce, soya, and cottonseed oils. All green plants contain some tocopherols, and there is some evidence that some green leafy vegetables and rose hips contain more than wheat germ. It probably is synthesized by leaves and translocated to the seeds. All four tocopherols have been found in wheat germ oil; α-, β-, and γ-tocopherols have been found in cottonseed oil. Corn oil contains predominantly γ-tocopherol and, thus, furnishes a convenient source for the isolation of this, a difficult member of the tocopherols to prepare. δ-Tocopherol is 30% of the mixed tocopherols of soya bean oil.

Most of the gastrointestinal absorption of vitamin E occurs through the mucosa and the lymphatic system. It has been demonstrated that bile performs an important function in promoting tocopherol absorption. The ester derivatives are hydrolyzed by pancreatic enzymes before absorption.

TABLE 21-2

RELATIVE POTENCIES OF VARIOUS COMMERCIAL FORMS OF VITAMIN E

Form of Vitamin E	Potency
POTENCY (USP UNITS) OF 1 MG OF THE FOLLOWING	
(±)-α-Tocopherol acid succinate	0.89
(±)-α-Tocopherol acetate	1.0
(±)-α-Tocopherol	1.1
(+)-α-Tocopherol acid succinate	1.21
(+)-α-Tocopherol acetate	1.36
(+)-α-Tocopherol	1.49
WEIGHT (MG) OF 1 USP UNIT OF THE FOLLOWING	
(±)-α-Tocopherol acid succinate	0.89
(±)-α-Tocopherol acetate	1.0
(±)-α-Tocopherol	0.91
(+)-α-Tocopherol acid succinate	0.83
(+)-α-Tocopherol acetate	0.74
(+)-α-Tocopherol	0.67

The tocopherols in lymph are associated with chylomicrons and very low-density lipoproteins (VLDL). Circulating tocopherols are also associated mainly with the blood low-density lipoproteins (LDL). The tocopherols are readily and reversibly bound to most tissues, including adipose tissue, and the vitamin is thus stored.

Vitamin E is metabolized primarily to tocopheronic acid and its γ-lactone, followed by glucuronide conjugation. These metabolites are excreted in the bile. Vitamin E may undergo some enterohepatic circulation.

For decades, there has been significant interest in investigating the biochemical functions of vitamin E, but it is still difficult to explain many of the biochemical derangements caused by vitamin E deficiency in animals. There seems to be general agreement that one of the primary metabolic functions of the vitamin is that of an antioxidant of lipids, particularly unsaturated fatty acids. This function of preventing lipid oxidation does not, however, explain all the biochemical abnormalities caused by vitamin E deficiency. Moreover, vitamin E is not the only in vivo antioxidant. Two enzyme systems, glutathione reductase and *o*-phenylenediamine peroxidase, also function in this capacity.[34]

It has been postulated that vitamin E has a role in the regulation of protein synthesis. Other actions of this vitamin have also been investigated, for example, effects on muscle creatine kinase and liver xanthine oxidase. It has been noted that vitamin E deficiency leads to an increase in the turnover of creatine kinase. There is also an increase in liver xanthine oxidase activity in vitamin E-deficient animals, and this increase is due to an increase in de novo synthesis.[34]

Although it has been difficult to establish clinical correlates of vitamin E deficiency in humans; Bieri and Farrell[34] have summarized some useful generalizations and conclusions. These workers have noted that the infant, especially the premature infant, is susceptible to tocopherol deficiency because of ineffective transfer of the vitamin from placenta to fetus, and that growth in infants requires greater availability of the vitamin. On the other hand, in adults, the tocopherol storage depots provide adequate availability that is not readily depleted, but intestinal malabsorption syndromes, when persistent, can lead to depletion of the storage depots. Children with cystic fibrosis suffer from severe vitamin E deficiency caused by malabsorption. Tropical sprue, celiac disease, gastrointestinal resections, hepatic cirrhosis, biliary obstruction, and excessive ingestion of mineral oil also may cause long-term malabsorption.

Vitamin E therapeutic indications include the clinical conditions characterized by low serum tocopherol levels, increased fragility of red blood cells to hydrogen peroxide, or conditions that require additional amounts. The latter can be exemplified by individuals who consume excessive amounts of polyunsaturated fatty acids (more than 20 g/day over normal diet).[35]

It has been claimed that vitamin E could be of therapeutic benefit in ischemic heart disease, but evidence against this claim continues to accumulate. It also has been suggested that megadoses of tocopherol be used in the treatment of peripheral vascular disease with intermittent claudication. Although some studies support this proposal, experts in the field state that further clinical studies are necessary to make a definitive recommendation. Nevertheless, it continues to be popular and controversial to consider the beneficial effects of vitamin E and other vitamins in large (mega) dietary supplements, and investigations of megavitamin E therapy for cardiovascular disease continue to appear in the literature.[34]

The eminent vitamin biochemist, R. J. Williams has emphasized that

> ...[L]ipid peroxidation, the formation of harmful peroxides, from the interaction between oxygen and highly unsaturated fats (polyunsaturates) needs to be controlled in the body. Both oxygen and the polyunsaturated lipids are essential to our existence, but if the protection against peroxidation is inadequate, serious damage to various body proteins may result. Vitamin E is thought to be the leading agent for the prevention of peroxidation and the free radical production that is associated both with it and with radiation.[36]

Williams also notes that although exact mechanisms of action of these antioxidants are not yet known,

> ...[P]roviding plenty of vitamin E and ascorbic acid—both harmless antioxidants—is indicated as a possible means of preventing premature aging, especially if one's diet is rich in polyunsaturated acids.

Considering the foregoing implication of unnecessary peroxidation of unsaturated lipids, it is interesting that rather recently it has been postulated that atherosclerosis appears to be due to a deficiency of prostacyclin, and that this deficiency is caused by inhibition of prostacyclin synthetase by lipid peroxides or by free radicals that are likely to be generated during hyperlipidemia. Although there is no direct evidence that in experimental or human atherosclerosis lipid peroxidation is the earliest sign of the disease state, lipid peroxides have been found in arteries from atherosclerotic patients and in ceroid atheromatic plaques and, at the same time, prostacyclin is hardly generated in human atheromatic plaques.[37]

THE VITAMIN Ks

The biologic activity of vitamin K was first discovered in 1929, and vitamin K was finally identified structurally in 1931 as represented in the following diagram. The term *vitamin K* was applied to the vitamin isolated from alfalfa, and a similar principle from fish meal was named vitamin K_2. Vitamin K_2 refers to a series of compounds called the menaquinones. These have a longer side chain with more unsaturation. This side chain may be composed of 1 to 13 isoprenyl units. The most common are depicted below.

Many other closely related compounds possess vitamin K activity [e.g., menadione (2-methyl-1,4-naphthoquinone) is as active as vitamin K on a molar basis]. The synthetic compounds menadione and menadiol are referred to as vitamin K_3 and K_4, respectively.

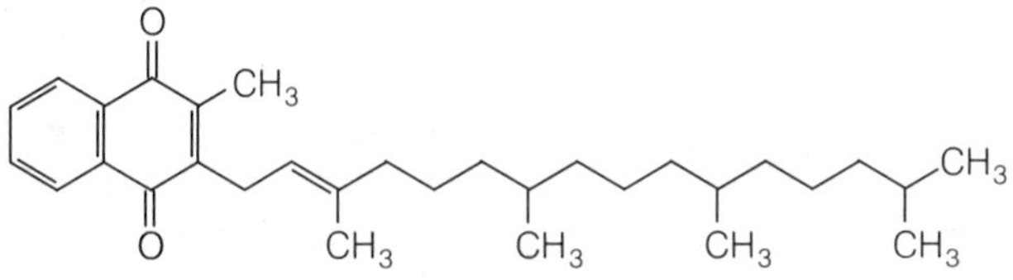

Vitamin K_1
Phytonadione
(2-Methyl-3-phytyl-1,4-naphthoquinone)

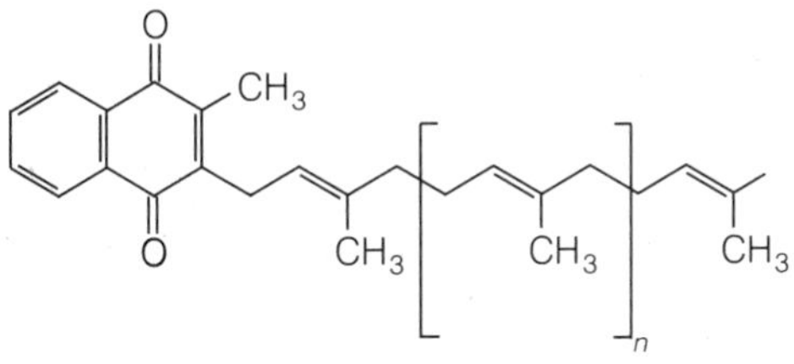

$n = 4$ = Vitamin $K_{2(30)}$
$n = 5$ = Vitamin $K_{2(35)}$

Vitamin K is a naphthoquinone derivative containing diterpenoid units biosynthesized by the intermediate, geranylpyrophosphate.[38]

Animals depend on two sources for their intake of this vitamin, dietary and bacterial synthesis. Alfalfa, chestnut leaves, and spinach are excellent sources of vitamin K_1, which also occurs in hog liver fat, hempseed, tomatoes, kale, and soy bean oil. Natural K_1 occurs as a *trans*-isomer. The synthetic commercially available form is a mixture of *cis-* and *trans*-isomers, with no more than 20% *cis*. Vitamin K_2 is synthesized by the intestinal flora, especially by gram-positive bacteria. Vitamin K_2 is not available commercially.

Numerous compounds have been tested for their antihemorrhagic activity, and significant biologic activity is manifested in compounds with the following structure when:

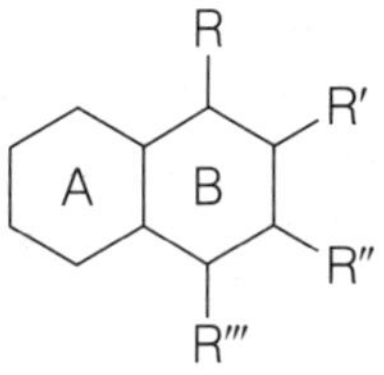

1. Ring A is aromatic or hydroaromatic.
2. Ring A is not substituted.
3. Ring B is aromatic or hydroaromatic.
4. R equals OH, CO, OR, OAc (the R in OR equals methyl or ethyl).
5. R′ equals methyl.
6. R″ equals H, sulfonic acid, dimethylamino or an alkyl group containing ten or more carbon atoms. A double bond in the β, γ-position of this alkyl group enhances potency, whereas, if the double bond is further removed, it exerts no effect. Isoprenoid groups are more effective than straight chains. In the vitamin $K_{2(30)}$-type compounds, the 6′,7′-mono-*cis* isomer is significantly less active than the all-*trans* or the 18′,19′-mono-*cis* isomer. This also was true of the vitamin $K_{2(20)}$ isoprenolog. A vitamin $K_{2(25)}$ isoprenolog was 20% more active than vitamin $K_{1(37)}$.
7. R‴ equals H, OH, NH_2, CO, OR, Ac (the R in OR equals methyl or ethyl).

Decreased antihemorrhagic activity is obtained when

1. Ring A is substituted.
2. R′ is an alkyl group larger than a methyl group.
3. R″ is a hydroxyl group.
4. R″ contains a hydroxyl group in a side chain.

It is interesting that, if ring A is benzenoid, the introduction of sulfur in place of a —CH=CH— in this ring in 2-methylnaphthoquinone permits the retention of some antihemorrhagic activity. This might indicate that, in the process of exerting vitamin K activity, the benzenoid end of the molecule must fit into a pocket carefully tailored to it. That the other end is not so closely surrounded is shown by the retention of activity on changing the alkyl group in the 2-position.

Although marked antihemorrhagic activity is found in many compounds, the possibility exists that they may be converted in the body to a vitamin K_1-type compound. The esters of the hydroquinones may be hydrolyzed, and the resulting hydroquinone may be oxidized to the quinone. The methyl tetralones, which are very active, possibly could be dehydrogenated to the methylnaphthols, which are hydroxylated, and the latter product converted to the biologically equivalent quinone. Compounds with a dihydrobenzenoid ring (such as 5,8-dihydrovitamin K_1) appear to be moderately easily dehydrogenated, whereas the corresponding tetrahydrides are resistant to such a change.

Vitamin K_1 and K_2 are absorbed by an active process in the proximal small intestines. Bile of a normal composition is necessary to facilitate the absorption. The bile component principally concerned in the absorption and transport of fat-soluble vitamin K from the digestive tract is thought to be deoxycholic acid. The molecular compound of vitamin K with deoxycholic acid was effective upon oral administration to rats with biliary fistula. Vitamin K is absorbed through the lymph in chylomicrons. It is transported to the liver, where it is concentrated, but, no significant storage occurs.

The entire metabolic pathway of Vitamin K has not been elucidated. However, the major urinary metabolites are glucuronide conjugates of carboxylic acids derived from shortening of the side chain. High fecal concentrations are probably due to bacterial synthesis.

The only known function of vitamin K in higher animals is to maintain adequate plasma levels of the protein prothrombin (factor II), and three other essential clotting factors: VII (proconvertin), IX (autoprothrombin II), and X (Stuart–Prower factor). It follows that any condition that does not permit the full utilization of the antihemorrhagic agents or the production of prothrombin would lead to an increase in the amount of time in which the blood will clot or to hemorrhagic conditions. Some of these conditions are (1) faulty absorption caused by several conditions (e.g., obstructive jaundice, biliary fistulas, intestinal polyposis, chronic ulcerative colitis, intestinal fistula, intestinal obstruction, and sprue); (2) damaged livers or primary hepatic diseases, such as atrophy, cirrhosis, or chronic hepatitis; (3) insufficient amounts of bile or abnormal bile in the intestinal tract; and (4) insufficient amounts of vitamin K.

Vitamin K-deficiency disease, which is known to lead to fatal hemorrhage, has been recognized since 1929, but the molecular mechanism of action of this vitamin continues to attract the attention of many researchers. The isolation and identification of γ-carboxyglutamic acid in bovine prothrombin in 1974 required 45 years of active research on the biochemistry of vitamin K. This discovery is considered to be a milestone because it led to better understanding of vitamin K and blood coagulation. It is already understood that vitamin K is a component of a membrane-bound carboxylase enzyme system and functions in the posttranslational carboxylation of some peptide-bound glutamate residues in several so-called vitamin K-dependent proteins that are necessary for normal blood coagulation. The vitamin K-dependent carboxylase system includes a specialized microsomal electron transport system coupled to a carbon dioxide fixation reaction. Although the reaction does not require ATP, it uses the energy from the oxidation of reduced vitamin K to execute the carboxylation of glutamic acid.[39]

Although the molecular involvement of vitamin K in blood coagulation requires further clarification, some conclusions on its mechanism can be summarized. Vitamin K undergoes reduction of the quinone ring as a preliminary step. This reduction to vitamin K hydroquinone is effected by NADH. Vitamin K hydroquinone is necessary for the carboxylation of glutamic acid to form carboxylglutamic acid, which participates in the complexation (chelation) of the necessary Ca^{2+}. It is, therefore, the general consensus that vitamin K, through its hydroquinone metabolite, participates in the formation of specific Ca^{2+}-binding sites on prothrombin and that the strong binding sites for Ca^{2+} involve adjacent γ-carboxy-lglutamic acid residues within the prothrombin structure. It is also significant that γ-carboxyglutamic acid has been found as a residue in all vitamin K-dependent clotting factors. Moreover, it is interesting that the anticoagulant coumarin derivatives and related compounds (see Chap. 14) interfere with the carboxylase system and thus interfere with the γ-carboxylation of the glutamic acid residues of the prothrombin structure.

PRODUCTS

Phytonadione, USP. 2-Methyl-3-phytyl-1,4-naphthoquinone; vitamin K_1 (Mephyton; Konakion) is described as a clear, yellow, very viscous, odorless or nearly odorless liquid.

Pure vitamin K_1 is a yellow, crystalline solid that melts at 69°C. It is insoluble in water, slightly soluble in alcohol, soluble in vegetable oils, and in the usual fat solvents. It is unstable toward light, oxidation, strong acids, and halogens. It easily can be reduced to the corresponding hydroquinone, which, in turn, can be esterfied.

The therapeutic use of vitamin K as a systemic hemostatic agent is based on the critical function that the vitamin performs in blood coagulation. Vitamin K_1 (phytonadione) is effective both in the treatment of hypoprothrombinemia caused by dietary deficiency of the vitamin or malabsorption and in bleeding caused by oral anticoagulants (e.g., coumadin derivatives). Phytonadione exerts prompt and prolonged action. It can be administered orally, subcutaneously, or intramuscularly; in emergencies it can be given by slow intravenous injection.

Vitamin K is administered in conjunction with bile salts or their derivatives in pre- and postoperative jaundiced patients to bring about and maintain a normal prothrombin level in the blood.

In the average infant, the birth values of prothrombin content are adequate, but during the first few days of life, they appear to fall rapidly, even dangerously low, and then slowly recover spontaneously. This transition period was and is critical because of the numerous sites of hemorrhagic manifestations, traumatic or spontaneous, that may prove serious if not fatal. This condition now is recognized as a type of alimentary vitamin K deficiency. The spontaneous recovery is perhaps due to the establishment of an intestinal flora capable of synthesizing vitamin K after ingestion of food. However, administration of vitamin K orally effects a prompt recovery.

Vitamin K_1 acts more rapidly (effect on prothrombin time) than menadione, within two hours after intravenous administration. However, no difference could be detected after two hours.[40]

The menadiones are much less active than vitamin K_1 in normalizing the prolonged blood-clotting times caused by dicumarol and related drugs.[40]

Vitamin K_1 is the drug of choice for humans because of its low toxicity. Its duration of action is longer than that of menadione and its derivatives. Vitamin K should not be administered to patients receiving warfarin or coumarin anticoagulants.

Vitamin K can be used to diagnose liver function accurately. The intramuscular injection of 2 mg of 2-methyl-1,4-naphthoquinone has led to response in prothrombin index in patients with jaundice of extrahepatic origin, but not in patients with jaundice of intrahepatic origin (e.g., cirrhosis).

Menadione, USP. 2-Methyl-1,4-naphthoquinone; menaphthone; Thyloquinone. Menadione can be prepared very readily by the oxidation of 2-methylnaphthalene with chromic acid. It is a bright yellow, crystalline powder and is nearly odorless. It is affected by sunlight. Menadione is practically insoluble in water; it is soluble in vegetable oils, and 1 g of it is soluble in about 60 mL of alcohol. The *NF* has a caution that menadione powder is irritating to the respiratory tract and to the skin, and an alcoholic solution has vesicant properties.

On a mole-for-mole basis, menadione is equal to vitamin K_1 in activity and can be used as a complete substitute for this vitamin. It is effective orally, intravenously, and intramuscularly. Oral absorption occurs in the distal small intestines and the colon. If given orally to patients with biliary obstruction, bile salts or their equivalent should be administered simultaneously to facilitate absorption. It can be administered intramuscularly in oil when the patient cannot tolerate an oral product, has a biliary obstruction, or when a prolonged effect is desired.

Carbon-14 labeled menadiol diacetate in small physiologic doses is converted in vivo to a vitamin $K_{2(20)}$ and the origin of the side chain probably is through mevalonic acid. This suggests that menadione may be an intermediate or a provitamin K.[40]

Menadione in oil is three times more effective than a menadione suspension in water. More of menadione than of vitamin K_1 is absorbed orally, but 38% of the former is excreted by the kidney in 24 hours, whereas only very small amounts of the latter are excreted by this route in 24 hours. In rats, menadione in part is reduced to the hydroquinone and excreted as the glucuronide 19% and the sulfate 9.3%.

Menadione Sodium Bisulfite, USP. 2-Methyl-1,4-napthoquinone sodium bisulfite; menadione bisulfite (Hykinone) is prepared by adding a solution of sodium bisulfite to menadione.

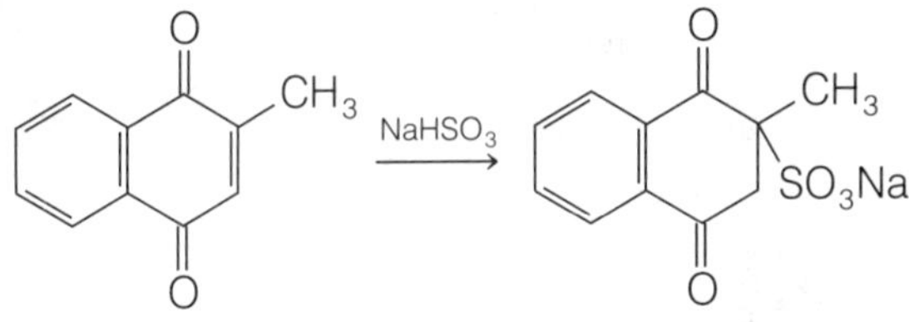

O, O, CH_3, $NaHSO_3$, SO_3Na

Menadione — Menadione sodium bisulfite

Menadione sodium bisulfite occurs as a white, crystalline, odorless powder. One gram of it dissolves in about 2 mL of water, and it is slightly soluble in alcohol. It decomposes in the presence of alkali to liberate the free quinone.

Menadiol Sodium Diphosphate, USP. Tetrasodium 2-methyl-1,4-naphthalenediol *bis*(dihydrogen phosphate); tetrasodium 2-methylnaphthohydroquinone diphosphate (Synkayvite; Kappadione) is a white hygroscopic powder, very soluble in water, giving solutions that have a *p*H of 7 to 9. It is available in ampules for use subcutaneously, intramuscularly, or intravenously and in tablets for oral administration. Unlike the other vitamin K analogues, menadiol[25] oral absorption is not dependent on the presence of bile. Once absorbed, it is converted to menadione.

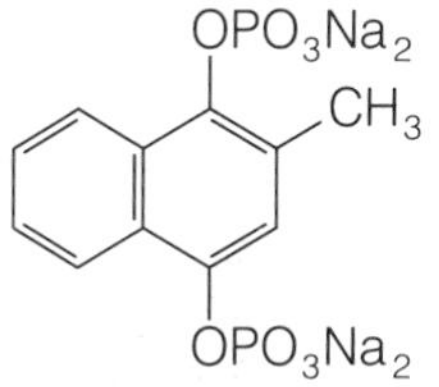

OPO_3Na_2, CH_3, OPO_3Na_2

Menadiol sodium diphosphate

Menadione bisulfite and menadiol diphosphate have been shown to produce hemolytic symptoms (reticulocytosis, increase in Heinz bodies) in newborn premature infants when given in excessive

TABLE 21-3

LIPID-SOLUBLE VITAMINS

Name Proprietary Name	***Preparations***	***Category***	***Usual Adult Dose****	***Usual Dose Range****	***Usual Pediatric Dose****
Vitamine A USP *Acon, Aquasol A, Dispatabs, Homagenets-A oral, Testavol-S, Vi-Dom-A, Vio-A, Alphalin, Anatola, Super A Vitamin*	Vitamin A capsules USP	Vitamin A (antixerophthalmic)	Prophylactic: 1.5 mg (5000 USP vitamin A units) qd; therapeutic: 3–15 mg (10,000–50,000 units) qd	Prophylactic: 1.5–2.4 mg (5000–8000 units) qd; therapeutic: 3–150 mg (10,000–500,000 units) qd	Prophylactic: the following, amounts once daily; infants up to 1 yr: 450 μg (1500 units); 1–3 yr: 600 μg (2,000 units); 3–6 yr: 750 μg (2500 units); 6–10 yr: 1.05 mg (3500 units); 10–12 yr: 1.35 mg (4500 units), 12 yr and older: see Usual Dose. Therapeutic: see Usual Dose
Ergocalciferol USP *Deltalin, Drisdol*	Ergocalciferol capsules USP Ergocalciferol solution USP Ergocalciferol tablets USP	Vitamin D (antirachitic)	Rickets, Prophylactic: 10 μg (400 USP vitamin D units) qd; therapeutic, deficiency rickets: 300 μg–1.25 mg (12,000–50,000 units) qd; refractory rickets: 1.25–25 mg (50,000–1 million units) qd; hypocalcemic tetany: 1.25–10 mg (50,000–400,000 units) qd		
Cholecalciferol USP	Decavitamin capsules USP Decavitamin tablets USP	Vitamin D (antirachitic)	1 dosage unit/day		
Dihydrotachysterol USP *Hytakerol*	Dihydrotachys-USP terol tablets	Antihypocalcemic	Initial, 800 μg–2.4 mg qd; maintenance, 200 μg/wk to 1 mg/day		
Vitamin E USP *E-Ferol, E-Ferol Succinate, Eprolin, Epsilan-M, Ecofrol, Tokols*	Vitamin E capsules USP	Vitamin E supplement		Prophylactic: from 5–30 IU of vitamin E; therapeutic: to be determined by the practitioner according to the needs of the patient	
Phytonadione USP *AquaMephyton, Konakion, Mephyton*	Phytonadione injection USP	Vitamin K (prothrombogenic)	Parenteral, 2.5–25 mg repeated in 6–8 hr, if necessary	2.5–50 mg/day	Hemorrhagic disease of the newborn, prophylactic: IM, 500 μg–1 mg; therapeutic: IM or SC 1 mg Other prothrombin deficiencies: infants, parenteral, 2 mg; older infants and children, 5–10 mg
	Phytonadion tablets USP		2.5–25 mg, repeated in 12–48 hr, if necessary	1–50 mg/day	Prothrombin deficiencies: infants, 2 mg; older infants and children, 5–10 mg
Menadione USP	Menadione injection USP Menadione tablets USP	Source of vitamin K	Oral and IM, 2 mg/day	2–10 mg	
Menadione sodium bisulfite USP *Hykinone*	Menadione sodium bisulfite injection USP	Source of vitamin K	IV and SC, 2 mg/day		
Menadiol sodium diphosphate USP *Kappadione, Synkayvite*	Menadiol sodium diphosphate injection USP Menadiol sodium diphosphate tablets USP	Source of vitamin K	Oral, IM, IV, or SC, 3–6 mg/day	5–75 mg/day	

*See USP DI for complete dosage information.

doses (more than 5–10 mg/kg). In severe cases, overt hemolytic anemia with hemoglobinuria may occur. The increased red cell breakdown may lead to hyperbilirubinemia and kernicterus.

These compounds also may interfere with bile pigment secretion. Newborns with a congenital defect of glucose-6-phosphate dehydrogenase can react with severe hemolysis, even with small doses of menadione derivatives. However, small nonhemolyzing doses can be used in the newborn, and combination with vitamin E is not considered essential.[41]

WATER-SOLUBLE VITAMINS

Although these vitamins are structurally diverse, they are characterized as a general class on the basis of water solubility to distinguish them from the lipid-soluble vitamins. This class includes the B-complex vitamins and ascorbic acid (vitamin C). The term B-*complex vitamins* usually refers to thiamine, riboflavin, pyridoxine, nicotinic acid, pantothenic acid, biotin, cyanocobalamine, and folic acid. Dietary deficiencies of any one of the B vitamins are commonly complicated by deficiencies of more than one member of the group; hence, treatment with B-complex preparations is usually indicated.

THIAMINE (VITAMIN B_1)

Thiamine was the first water-soluble vitamin to be discovered in 1926, but the complete determination of its structure and synthesis were not accomplished until 1936.

Many natural foods provide adequate amounts of this vitamin. The germ of cereals, brans, egg yolks, yeast extracts, peas, beans, and nuts usually provide enough thiamine to satisfy adult requirements. The requirement for thiamine is related directly to caloric intake, 0.2–0.3 mg/1000 calories. It is not economically practical to isolate the crystalline vitamin from natural sources on a commercial scale; hence, the commercially available thiamine is prepared synthetically.

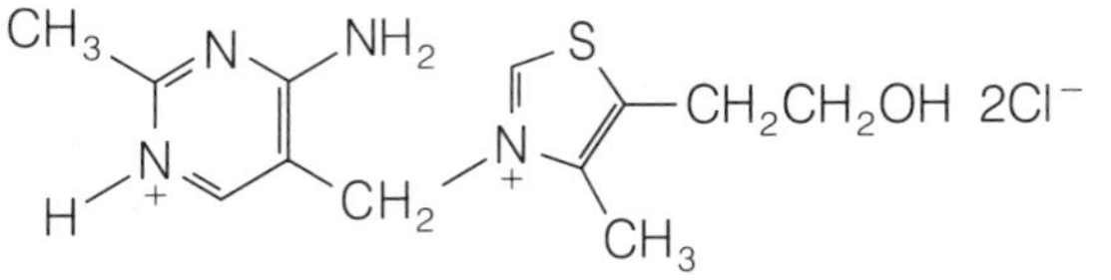

Thiamine hydrochloride

Thiamine is biologically synthesized from the pyrimidine derivative 4-amino-5-hydroxymethyl-2-methyl pyrimidine methylpyrimidine and 5-(β-hydroxyethyl)-4-methylthiazole. These two precursors are converted to phosphate derivatives under kinase catalysis, which requires ATP. The respective phosphate derivatives then interact to form thiamine phosphate. The latter reaction is catalyzed by thiamine phosphate pyrophosphorylase.

In higher mammalian organisms, thiamine is transformed to the coenzyme thiamine pyrophosphate by direct pyrophosphate transfer from ATP. This coenzyme performs important metabolic functions, for example, as cocarboxylase in the decarboxylation of α-keto acids (e.g., pyruvate to form acetyl-CoA) and in transketolases (e.g., utilization of pentoses in the hexose monophosphate shunt).

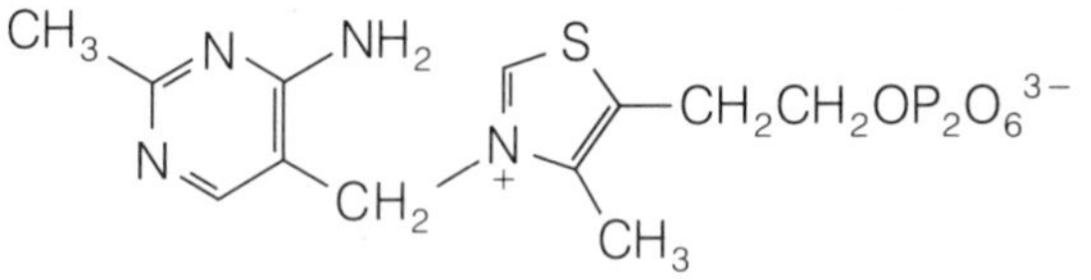

Thiamine pyrophosphate

In the decarboxylation of pyruvate, the coenzyme interacts with pyruvic acid to form so-called active aldehyde as shown below.

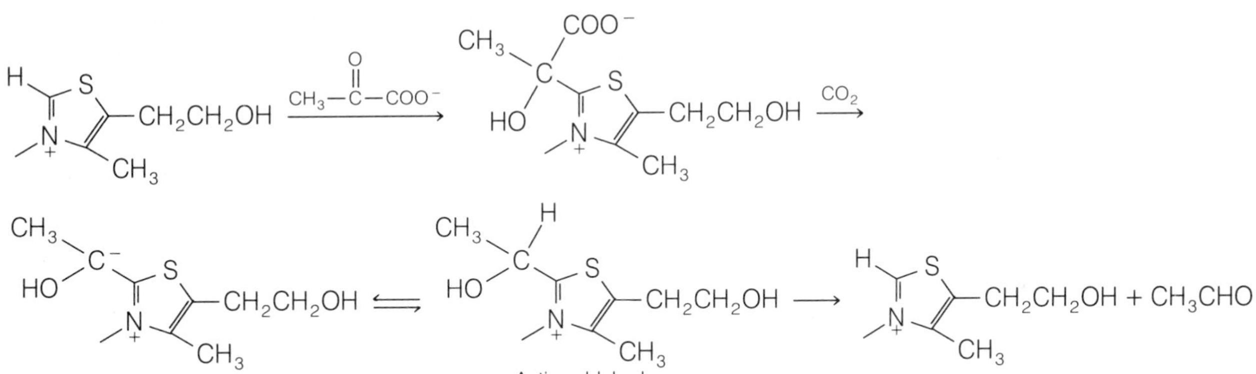

Active aldehyde

The active aldehyde intermediate then interacts with thioctic acid to form acetyl-thioctate, which is responsible for acetylating CoA-SH to form acetyl-CoA. In deficiency states, the oxidation of α-keto acids is decreased, resulting in increased pyruvate levels in the blood.

Thiamine hydrochloride is stable in acid but is unstable in aqueous solutions of pH greater than 5. Under these conditions it undergoes decomposition and inactivation. Thiamine is also susceptible to oxidation. It is readily oxidized by exposure to the atmosphere or by oxidizing reagents such as hydrogen peroxide, permanganate, or alkaline potassium ferricyanide. This oxidation forms thiochrome as represented below.

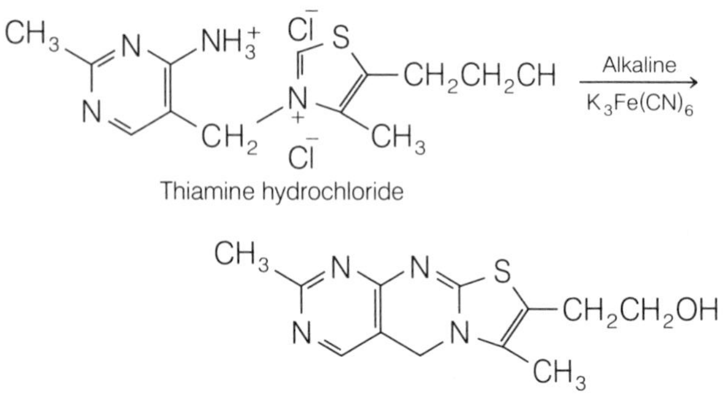

Alkaline $K_3Fe(CN)_6$

Thiamine hydrochloride

Thiochrome

Thiochrome exhibits a vivid blue fluorescence; hence, this reaction is the basis for the quantitative fluorometric assay of thiamine in the *USP*.

Oxythiamine and neopyrithiamine are antivitamines used in the study of the deficiency state. Oxythiamine is a competitive inhibitor of thiamine pyrophosphate. Neopyrithiamine inhibits the pyrophosphorylation of thiamine.

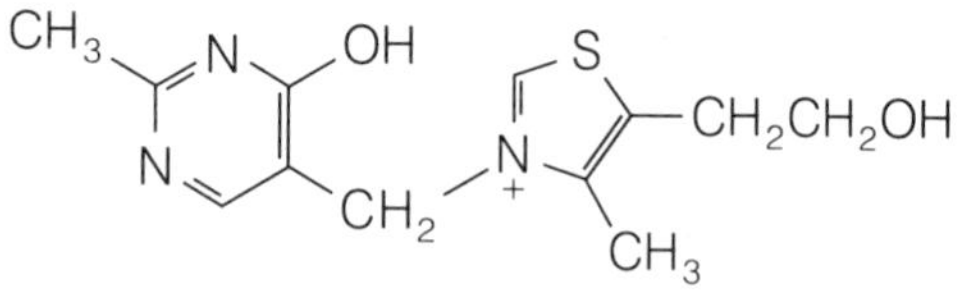

Oxythiamine

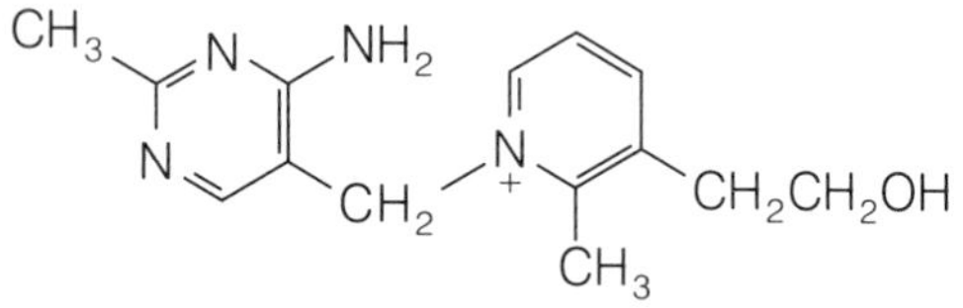

Neopyrithiamine

Thiamine Hydrochloride,* USP. Thiamine monohydrochloride; thiamine chloride; vitamin B_1 hydrochloride; vitamin B_1; aneurine hydrochloride. This occurs as small, white crystals or as a crystalline powder; it has a slight, characteristic yeastlike odor. The anhydrous product, when exposed to air, will absorb rapidly about 4% of water. One gram is soluble in 1 mL of water and in about 100 mL of alcohol. It is soluble in glycerin. An aqueous solution, 1:20, has a pH of 3. Aqueous solutions 1:100 have a pH of 2.7 to 3.4. Thiamine has pK_a values of 4.8 and 9.0.

Thiamine hydrochloride is sensitive to alkali. The addition of 3 moles of sodium hydroxide per mole of thiamine hydrochloride reacts as shown below.

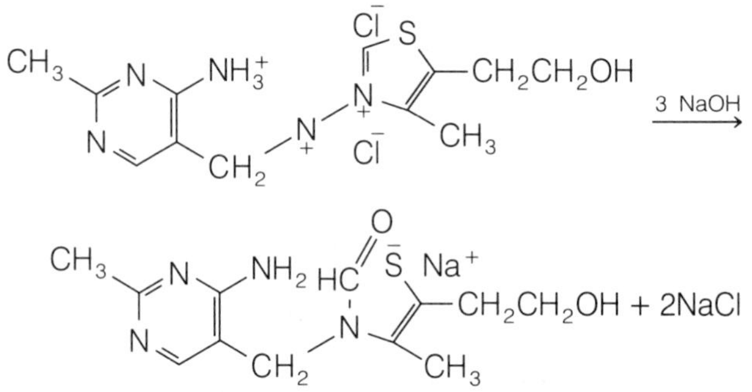

3 NaOH

+ 2NaCl

Thiamine hydrochloride absorption is a sodium-dependent process; thus, saturation limits absorption is limited to 8 to 15 mg daily. Absorption is decreased in alcoholism and cirrhosis. Food affects the rate, but not the amount, absorbed. After absorption thiamine hydrochloride is distributed to all tissue. There is limited storage of the vitamin in the body (about 30 mg).

Normally, little or no thiamine is excreted in the urine. However, if doses exceed physiologic needs after body stores are saturated, thiamine can be found in the urine as pyrimidine or as the unchanged compound.

Severe thiamine deficiency is called *beriberi*. The major organs affected are the nervous system (in dry beriberi), the cardiovascular system (in wet beriberi), and the gastrointestinal tract. Thiamine administration reverses the gastrointestinal and cardiovascular symptoms. However, the neurologic damage may be permanent if the deficiency has been severe or of long duration.

Thiamine hydrochloride is indicated in the treatment or prophylaxis of thiamine deficiencies. Dietary deficiencies are rare in the United States, but alcoholism is the most common cause of the disease. Alcoholics have poor dietary habits (deficient diet), plus alcohol interferes with the absorption of the vitamin.

Thiamine Mononitrate, USP. Thiamine nitrate; vitamin B_1 mononitrate, is a colorless compound that is soluble in water 1:35 and slightly soluble in alcohol. Two percent aqueous solutions have a pH of 6.0 to 7.1. This salt is more stable than the chloride hydrochloride in the dry state, is less hygroscopic, and is recommended for multivitamin preparations and the enrichment of flour mixes.

* The structure of thiamine hydrochloride has been revised (see Ganellin, G. R.: Chemistry and structure–activity analysis. In Roberts, G. C. K. (ed.). Drug Action at the Molecular Level. Baltimore, University Park Press, 1977).

PANTOTHENIC ACID

During the decade of the 1930s, R. J. Williams and his collaborators recognized, isolated, and synthesized pantothenic acid. Because its occurrence is so widespread, it was called *pantothenic acid* from the Greek, meaning "from everywhere."

$$HOCH_2-C(CH_3)_2-CH(OH)-C(=O)-NH-CH_2CH_2COOH$$

Pantotheinic acid

Pantothenic acid has also been called vitamin B_5. Excellent sources of the vitamin are liver, eggs, and cereals.

This vitamin is synthesized by most green plants and microorganisms. The precursors are γ-ketoisovaleric acid and β-alanine.[1] The latter originates from the decarboxylation of aspartic acid. γ-Ketoisovaleric acid is converted to ketopantoic acid by N^5,N^{10}-methylenetetrahydrofolic acid and, then, upon reduction, pantoic acid is formed. Finally, pantoic acid and β-alanine react by amide formation to form pantothenic acid.

The metabolic functions of pantothenic acid in human biochemistry are mediated through the synthesis of CoA; this vitamin is a structural component of CoA, which is necessary for many important metabolic processes. Pantothenic acid is incorporated into CoA by a series of five enzyme-catalyzed reactions. Coenzyme-A is involved in the activation of fatty acids before β-oxidation; this activation requires ATP to form the respective fatty acyl-CoA derivatives. It also participates in fatty acid β-oxidation in the final step, forming acetyl-CoA. The latter also is formed from pyruvate decarboxylation. In pyruvate decarboxylation CoA participates in collaboration with thiamine pyrophosphate and lipoic acid, two other important coenzymes. Thiamine pyrophosphate is the actual decarboxylating coenzyme that functions with lipoic acid to form acetyldihydrolipoic acid from the decarboxylation of pyruvate. Coenzyme-A then accepts the acetyl group from acetyldihydrolipoic acid to form acetyl-CoA. Acetyl-CoA participates as an acetyl donor in many processes and is also the precursor in important biosyntheses (e.g., those of fatty acids, steroids, porphyrins, and acetylcholine).

Clinical cases of pantothenic acid deficiency do not commonly develop, unless they arise in combination with deficiencies of the other B vitamins.[2] Accordingly, pantothenic acid is usually included in multivitamin preparations. The calcium salt is commonly used in pharmaceutical preparations. Panthenol, the alcohol derivative, is another form commonly used.

Products

Pantothenic Acid. Vitamin B_5 occurs as a viscous hygroscopic oil, freely soluble in water, but unstable to heat, acid, and alkali. Because of this instability, the calcium salt is used commercially.

Pantothenic acid and its derivatives have essentially no pharmacologic actions per se. Because of the ubiquitous nature of the vitamin, deficiency states usually do not develop. They have been produced by using synthetic diets devoid of the vitamin or by use of a vitamin antagonist. ω-Methylpantotheic acid has produced the following symptoms: fatigue, headache, paresthesia of the hands and feet, cardiovascular instability, and gastrointestinal problems.

$$HOCH(CH_3)-C(CH_3)_2-CH(OH)-C(=O)-NH-CH_2CH_2COOH$$

ω-Methylpantothenic acid

Pantothenic acid and its derivatives are readily absorbed and widely distributed. The highest concentrations are found in the liver, adrenal glands, heart, and kidneys. It appears pantothenic acid undergoes little if any metabolism, for the amount eliminated approximates the amount consumed. Approximately 70% of a dose is eliminated unchanged in the urine, with the remainder found in the feces.

$$\text{Adenosine-3'-}OPO_3^{2-}\text{-5'-}CH_2O-P(=O)(OH)-O-P(=O)(OH)-OCH_2-C(CH_3)_2-CH(OH)-C(=O)-NH-CH_2CH_2-C(=O)-NH-CH_2CH_2-SH$$

Coenzyme A

The only official indication of pantothenic acid is in the prevention or treatment of vitamin B deficiencies. Because a deficiency of a single B vitamin is rare, it is commonly formulated in multivitamin preparations or B-complex preparations.

Calcium Pantothenate, USP. Calcium D-pantothenate is a slightly hygroscopic, white, odorless, bitter powder that is stable in air. It is insoluble in alcohol, soluble 1 : 3 in water, and aqueous solutions have a *p*H of about 9 and $[\alpha]_D = +25°$ to $+27.5°C$. Autoclaving calcium pantothenate at 120°C for 20 minutes may cause a 10% to 30% decomposition. Some of the phosphates of pantothenic acid that occur naturally in coenzymes are quite stable to both acid and alkali, even upon heating.[43]

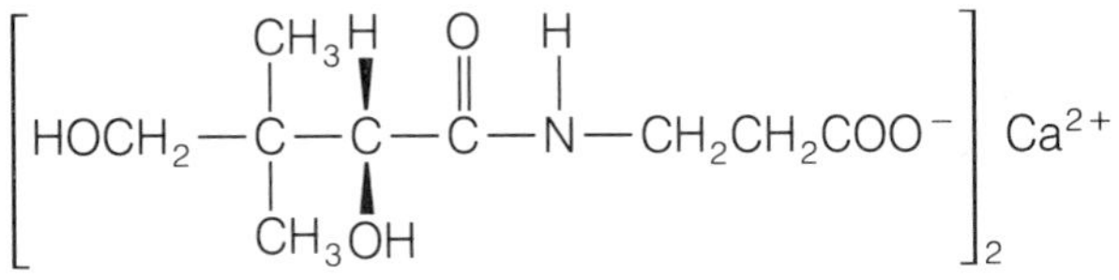

Pantothenate calcium

Racemic Calcium Pantothenate, USP, is recognized to provide a more economical source of this vitamin. Other than containing not less than 45% of the dextrorotatory biologically active form, its properties are very similar to those of Calcium Pantothenate, USP.

Panthenol, USP, the racemic alcohol analogue of pantothenic acid, exhibits both qualitatively and quantitatively the vitamin activity of pantothenic acid. It is considerably more stable than pantothenic acid in solutions with *p*H values of 3 to 5, but of about equal stability at *p*H 6 to 8. It appears to be more readily absorbed from the gut, particularly in the presence of food.

Dexpanthenol, USP, occurs as a slightly hygroscopic, viscous oil freely soluble in water and alcohol. It is the dextrorotatory alcohol derivative of pantothenic acid. Dexpanthenol is readily converted in vivo to the acid form.

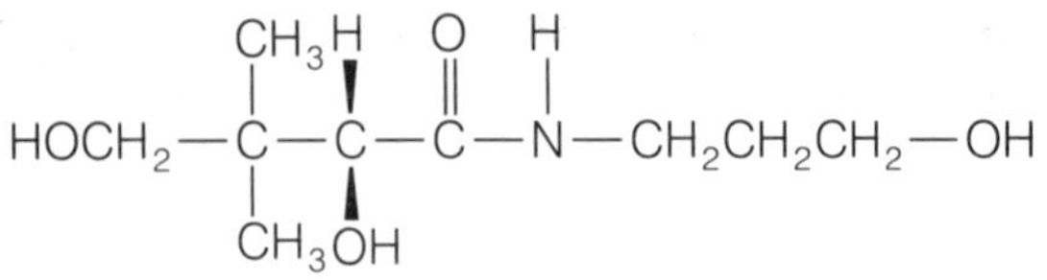

Dexpanthenol

Dexpanthenol and the racemic mixture are used in the treatment of paralytic ileus and postoperative distention. Dexpanthenol in combination with choline is used to relieve gas retention.

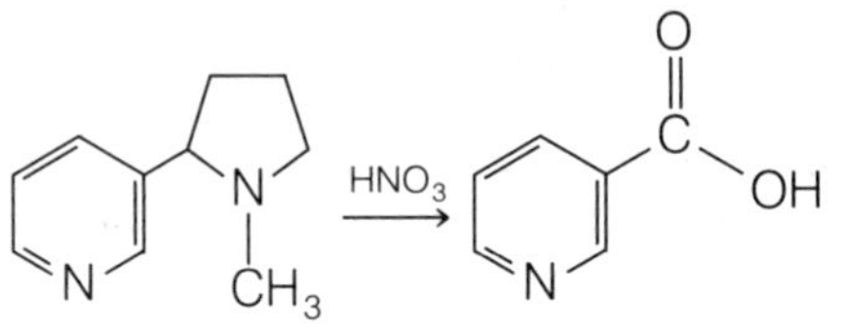

NICOTINIC ACID

Nicotinic acid (niacin) was first prepared by the oxidation of the alkaloid, nicotine, but not until 1913 was it isolated from yeast and recognized as an essential food factor (refer to the structures). In 1934–1935 nicotinamide was obtained from the hydrolysis of a coenzyme isolated from horse red blood cells. This coenzyme was later named *coenzyme II* and is now more commonly called nicotinamide-adenine dinucleotidephosphate (NADP).

Nicotinic acid $\xrightarrow{HNO_3}$ Nicotinic acid
Niacin

Generous sources of this vitamin include pork, lamb, and beef livers; hog kidneys; yeasts; pork; beef tongue; hearts; lean meats; wheat germ; peanut meal; and green peas.

Nicotinic acid can be synthesized by almost all plants and animals. Tryptophan can be metabolized to a nicotinic acid nucleotide in animals, but the efficiency of this multistep process varies from species to species. Plants and many microorganisms synthesize this vitamin through alternative routes using aspartic acid.

In the human, nicotinic acid reacts with 5-phosphoribosyl-1-pyrophosphate to form nicotinic acid mononucleotide, which then reacts with ATP to produce desamido-NAD (the intermediate dinucleotide with the nicotinic acid moiety). Finally, the latter intermediate is converted to NAD (nicotinamide-adenine dinucleotide, originally called *coenzyme I*) by transformation of the carboxyl of the nicotinic acid moiety to the amide by glutamine. This final step is catalyzed by NAD synthetase; NADP is produced from NAD by ATP under kinase catalysis.[1]

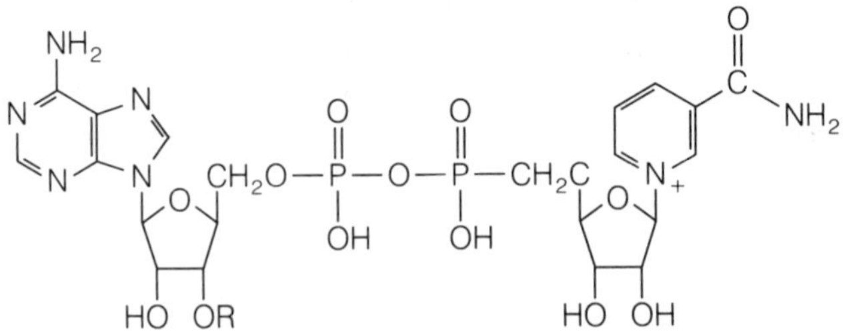

R = H = Nicotinamide adenine dinucleotide
R = PO_3^{2-} = Nicotinamide adenine dinucleotide phosphate

NAD and NADP participate as oxidizing coenzymes for many (more than 200) dehydrogenases.

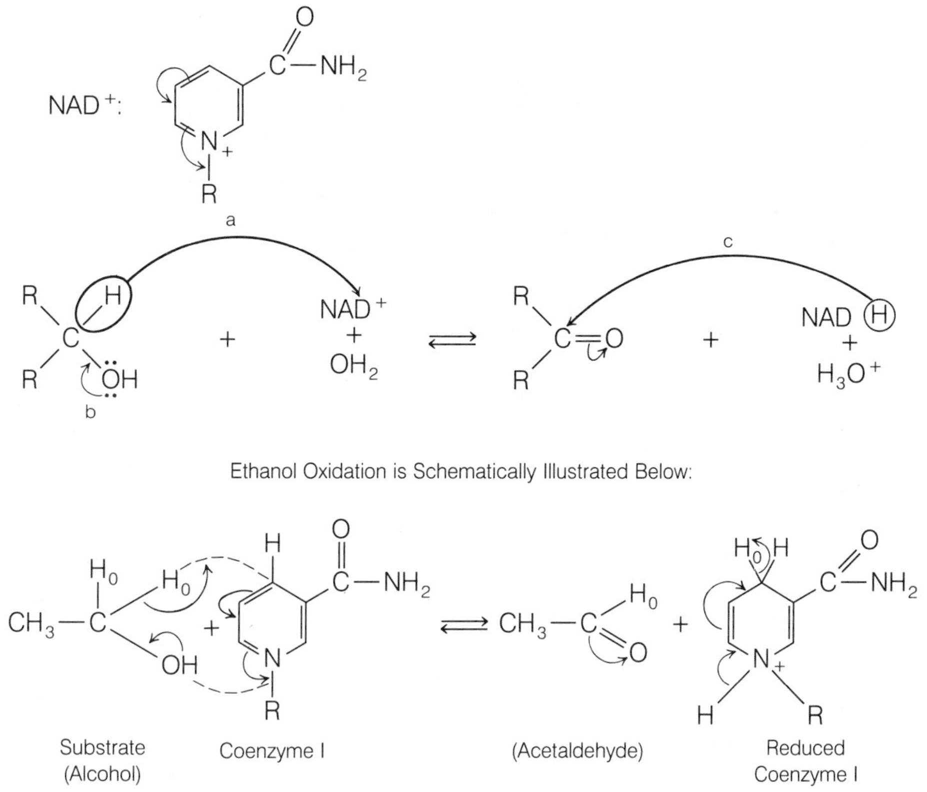

Ethanol Oxidation is Schematically Illustrated Below:

FIG. 21-1. *Generalized representation of the hydride transfer reaction.*

Some dehydrogenases require NAD, others require NADP, and some function with either. The following generalized representation (Fig. 21-1) illustrates the function of these coenzymes in metabolic oxidations and reductions. The abbreviation [NAD^+] emphasizes the electrophilicity of the pyridine C_4 moiety (which is the center of reactivity) and the substrate designated as

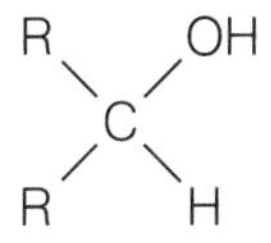

could be a primary or secondary alcohol. Arrow *a* symbolizes the function of NAD as oxidant in the hydride transfer from the substrate to the coenzyme forming NADH, reduced coenzyme. The hydroxyl of the substrate is visualized as undergoing deprotonation concertedly by either water or the pyridine nitrogen of NADH. Arrow *b* shows concerted formation of the carbonyl π-bond of the oxidation product. Arrow *c* symbolizes the reverse hydride transfer from reduced coenzyme, NADH, to the carbonyl carbon; and concertedly, as the carbonyl oxygen undergoes protonation, the reduction of the carbonyl group forms the corresponding alcohol. Thus, NAD and NADP function as hydride acceptors, whereas NADH and NADPH are hydride donors. Although the foregoing is a simplistic representation, it illustrates the dynamism of such oxidation–reduction reactions effected by these coenzymes under appropriate dehydrogenase catalysis. Alternatively, the reduced coenzymes may be utilized in ATP production through the electron transport system.

As noted earlier, nicotinic acid (also known as niacin to avoid confusion with nicotine) can be made available in nucleotide form from the amino acid tryptophan. It has been estimated that 60 mg of tryptophan equal 1 mg of nicotinic acid. Consequently; the human and other mammals can synthesize the vitamin, provided there is appropriate dietary availability of tryptophan. (Thus, it appears that nicotinic acid is not a true vitamin according to the classic definition of the term.)

Niacin, USP. Nicotinic acid; 3-pyridine carboxylic acid; vitamin B_3 occurs as white crystals or as a crystalline powder. It is odorless or it may have a slight odor. One gram of nicotinic acid dissolves in 60 mL of water. It is freely soluble in boiling water, in boiling alcohol, and in solutions of alkali hydroxides and carbonates, but is almost insoluble in ether. A 1% aqueous solution has a pH of 6. Nicotinic acid has a pK_a of 4.85.

Nicotinic acid is stable under normal storage conditions. It sublimes without decomposition.

Serious deficiency of niacin or tryptophan may lead to pellagra (from the Italian, *pelle agra*, for rough skin). The major systems affected are the gastrointestinal tract (diarrhea, enteritis, stomatitis), the skin (dermatitis), and the central nervous system (headache, dizziness, depression). Severe cases may result in delusions, hallucinations, and dementia. In the United States pellegra has become rare because flour is supplemented with nicotinic acid. Chronic alcoholism is the chief cause of pellegra and is associated with multiple vitamin deficiency. The symptoms of pellagra are completely reversed by niacin; therefore, it is indicated for the treatment and prevention of the deficiency.

Niacin, but not niacinamide, is also indicated in hyperlipidemia to lower triglycerides and cholesterol. Triglycerides, VLDL, and LDL are reduced, whereas HDL are increased. The exact mechanism is not known. The dose required (1 to 3 g three times daily) often limits the usefulness, for niacin has a direct vasodilation effect in high doses. This effect is believed to be mediated through the prostaglandins.

Niacin is readily absorbed from the gastrointestinal tract and widely distributed. At physiologic doses, little niacin is excreted unchanged. Most is excreted as *N*-methylniacin or as the glycine conjugate (nicotinuric acid). After administration of large doses, niacin can be found in the urine unchanged.

Niacinamide, USP. Nicotinamide; nicotinic acid amide. Nicotinamide is prepared by the amidation of esters of nicotinic acid or by passing ammonia gas into nicotinic acid at 320°C.

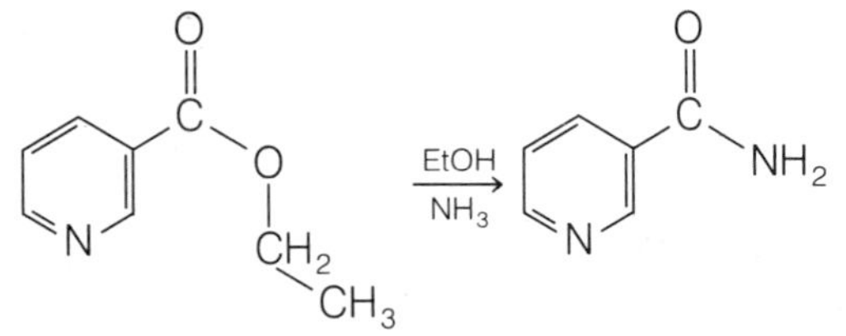

Ethyl Nicotinate $\xrightarrow[NH_3]{EtOH}$ Nicotinamide Niacinamide

Nicotinamide is a white, crystalline powder that is odorless, or nearly so, and bitter. One gram is soluble in about 1 mL of water, 1.5 mL of alcohol, and about 10 mL of glycerin. Aqueous solutions are neutral to litmus. For occurrence, action, and uses see nicotinic acid. Niacinamide has pK_a values of 0.5 and 3.35.

Similar to niacin, niacinamide is indicated in the treatment and prevention of deficiency states. Unlike niacin, niacinamide has no vasodilatory effect, which may be of therapeutic importance for compliance reasons. Niacinamide has no effect on triglycerides and lipoproteins. This product is formulated with potassium iodide and used as an iodine supplement.

Niacinamide hydrochloride is also available. It is more stable in solution and more compatible with thiamine chloride in solution.

RIBOFLAVIN (VITAMIN B_2)

Although the isolation of crystalline riboflavin was not accomplished until 1932, interest in this compound as a pigment dates back to 1881 in connection with the color in the whey of milk. In 1932, riboflavin was isolated as a coenzyme–enzyme complex from yeast by Warburg and Christian, and this complex was designated as *yellow oxidation ferment*.

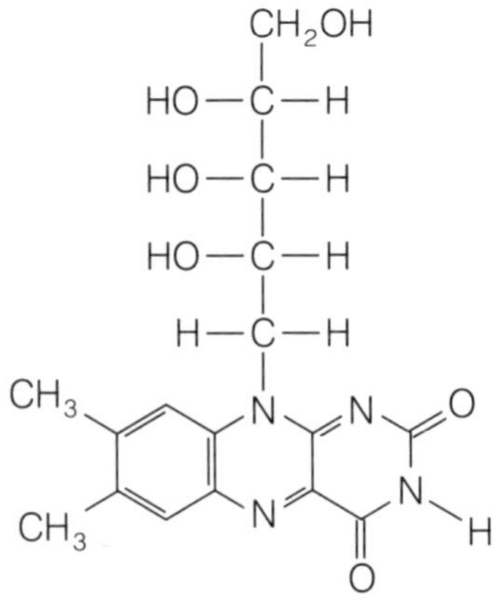

Riboflavin

Riboflavin is synthesized by all green plants and by most bacteria and fungi. Although yeast is the richest source, eggs, dairy products, legumes, and meats are the major sources in the diet. It is known that the precursor is a guanosine phosphate derivative, but the exact synthetic steps leading to the vitamin are not completely understood.

In higher mammals riboflavin is readily absorbed from the intestine and is distributed to all tissues. It is the precursor in the biosynthesis of the coenzymes flavin mononucleotide (FMN) and flavin adenine dinucleotide (FAD). The metabolic functions of this vitamin involve these two coenzymes, which participate in numerous vital oxidation–reduction processes. FMN, which is riboflavin-5′-phosphate, is produced from the vitamin and ATP under flavokinase catalytic action. This step can be inhibited by phenthiazines and the tricyclic antidepressants. FAD originates from an FMN and ATP reaction that involves reversible dinucleotide formation catalyzed by flavin nucleotide pyrophosphorylase. These coenzymes function in combination with several enzymes as coenzyme–enzyme complexes often characterized as *flavoproteins*.

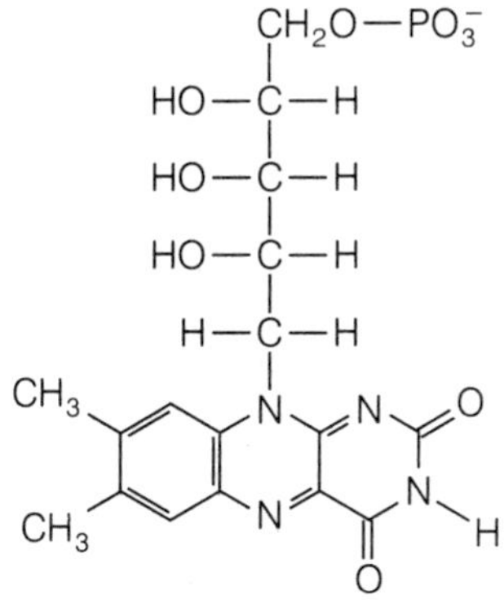

Flavin mononucleotide

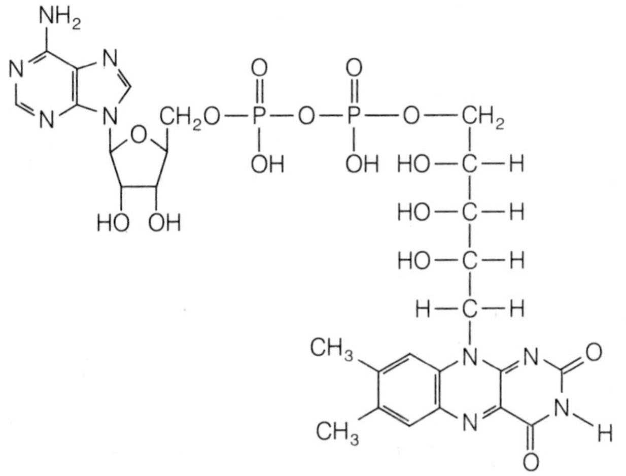

Flavin adenine dinucleotide

These flavoproteins function in aerobic or anerobic conditions as oxidases and dehydrogenases. Examples include glucose oxidase, xanthine oxidase, cytochrome reductase, and acyl-CoA dehydrogenase.

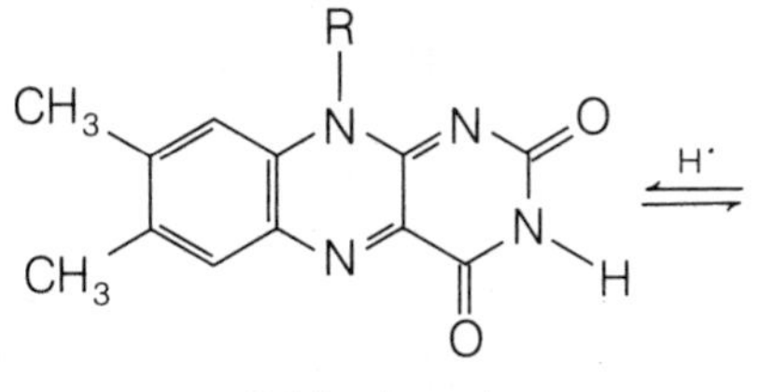

Oxidized species

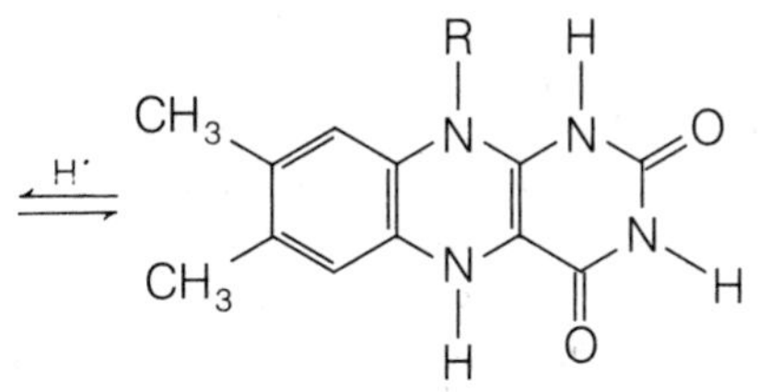

Reduced species

The riboflavin moiety of the complex is considered to be a hydrogen-transporting agent (carrier) functioning as hydrogen acceptors; the hydrogen donors may be NADH, NADPH, or some suitable substrate. The isoalloxazine rings accept two hydrides stepwise to form the dihydroriboflavin derivative.

Riboflavin, USP. Riboflavin; lactoflavin; vitamin B_2; vitamin G is a yellow to orange yellow crystalline powder with a slight odor. It is soluble in water 1 : 3000 to 1 : 20,000 mL, the variation in solubility being due to differences in internal crystalline structure, but it is more soluble in an isotonic solution of sodium chloride. A saturated aqueous solution has a pH of 6. Riboflavin has a pK_a of 10.5. It is less soluble in alcohol and insoluble in ether or chloroform. Benzyl alcohol (3%), gentistic acid (3%), urea in varying amounts, and niacinamide are used to solubilize riboflavin when relatively high concentrations of this factor are needed for parenteral solutions. Gentistic ethanol amide and sodium 3-hydroxy-2-naphthoate are also effective solubilizing agents for riboflavin.

When dry, riboflavin is not appreciably affected by diffused light; however, it deteriorates in solution in the presence of light, and this deterioration is very rapid in the presence of alkalies, producing lumiflavin. This deterioration may be retarded by buffering on the acid side. However, under acid conditions light can produce lumichrome. Neither of these decomposition products possess biologic activity.

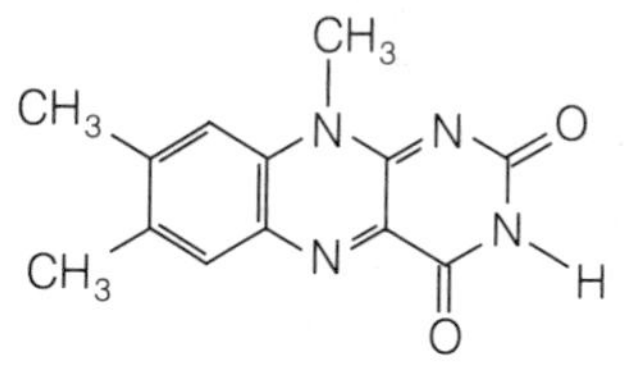

Lumiflavine

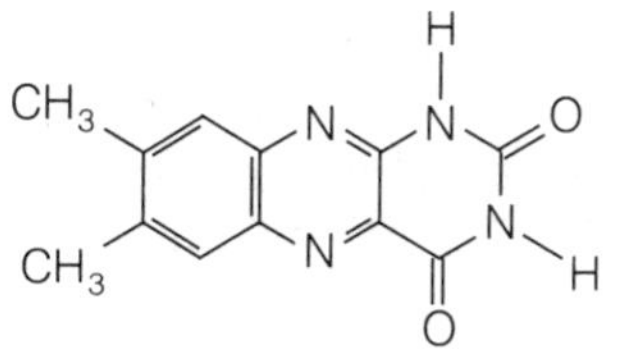

Lumichrome

The vitamin is commercially available as riboflavin, riboflavin 5-phosphate, and riboflavin 5-phosphate sodium. The phosphate esters are commercially used only in multivitamin preparations. The phosphate esters are hydrolyzed before absorption occurs. Absorption occurs through an active transport system. Riboflavin is phosphorylated by the intestinal mucosa during the absorption process. Food and bile enhance the absorption. Riboflavin is widely distributed in the body, with limited stores in the liver, spleen, heart, and kidneys. Conversion to FAD occurs primarily in the liver. FMN and FAD circulate primarily protein bound. Only small amounts (~ 9%) are excreted in the urine unchanged. Larger amounts can be found after administration of large doses.

Severe riboflavin deficiency is known as ariboflavinosis. Its major symptoms include chelosis, seborrheic dermatitis, and vascularization of the cornea. Ariboflavinosis occurs in chronic alcoholism, in combination with other vitamin deficiencies. It also has resulted for phenothiazine, tricyclic antidepressant, and probenecid therapy.

Riboflavin has no pharmacologic action and is relatively nontoxic. The only approved indication is in the treatment and prevention of ariboflavinosis.

PYRIDOXINE

In 1935, the term *vitamin B_6* was applied to the principle that cured a dermatitis in rats fed a vitamin-free diet supplemented with thiamine and riboflavin. Three years later vitamin B_6 was isolated from rice paste and yeast. In 1935, P. Gyorgy showed that "rat pellagra" was not the same as human pellagra, but that it resembled a particular disease of infancy known as "pink disease" or acrodynia. This "rat acrodynia" is characterized by a symmetric dermatosis affecting first the paws and the tips of the ears and the nose. These areas become swollen, red, and edematous, with ulcers developing frequently around the snout and on the tongue. Thickening and scaling of the ears is noted, and there is a loss of weight, with fatalities occurring in from one to three weeks after the appearance of the symptoms. Gyorgy was able to cure these conditions with a supplement obtained from yeast, which he called "vitamin B_6." In 1938, this factor was isolated from rice paste and yeast in a crystalline form in a number of laboratories. A single dose of about 100 μg produced healing in 14 days in a rat having severe vitamin B_6-deficiency symptoms.

Chemical tests, electrometric titration determinations, and absorption spectrum studies gave clues to its composition. These were substantiated by the synthesis of vitamin B_6 (1938 and 1939). This vitamin is also known as *pyridoxine* or *pyridoxol*. Two additional chemical forms, pyridoxal and pyridoxamine, have also been isolated from natural sources. These three compounds are metabolically and functionally interrelated. Currently, the term *vitamin B_6* is applied to all three forms of the vitamin.

Pyridoxine is available from whole-grain cereals, peanuts, corn, meat, poultry, and fish. However, up to 40% of the vitamin may be destroyed during cooking.

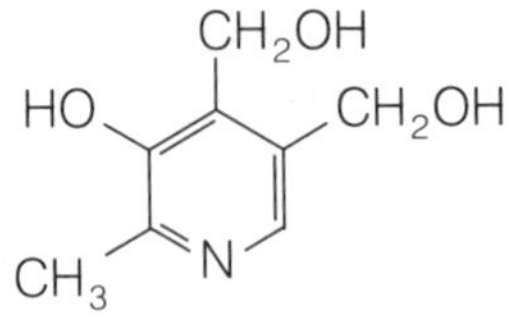

Pyridoxime

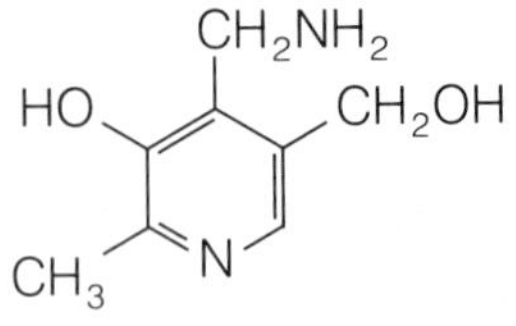

Pyridoxamine

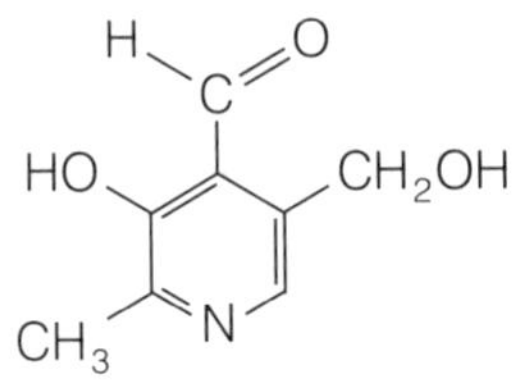

Pyridoxal

In the human, pyridoxine first undergoes hepatic phosphorylation, catalyzed by a specific kinase and, then, is oxidized to the aldehyde state by a flavoenzyme. Thus, the vitamin is the precursor necessary for the production of pyridoxal-5-phosphate, which is a coenzyme[44,45] that performs many vital functions in human metabolism. This coenzyme functions in the transaminations and decarboxylations that amino acids generally undergo; for example, it functions as a cotransaminase in the transamination of alanine to form pyruvic acid and as a codecarboxylase in the decarboxylation of dopa to form dopamine. Other biologic transformations[43,44] of amino acids in which pyridoxal can function are racemization, elimination of the α-hydrogen together with a β-substituent (i.e., OH or SH) or a γ-substituent, and probably the reversible cleavage of β-hydroxyamino acids to glycine and carbonyl compounds.

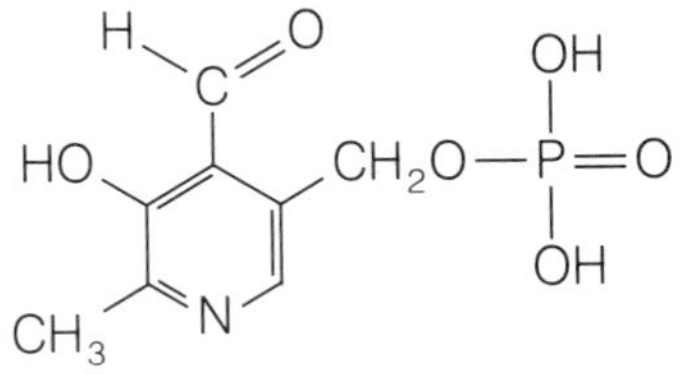

Pyridoxal-5-phosphate

An electromeric displacement of electrons from bonds a, b, or c (see diagram) would result in the release of a cation (H, R′, or COOH) and, subsequently, lead to the variety of reactions observed with pyridoxal. The extent to which one of these displacements predominates over others depends on the structure of the amino acid and the environment (*p*H, solvent, catalysts, enzymes, and such). When this mechanism applies in vivo, the pyridoxal component is linked to the enzyme through the phosphate of the hydroxymethyl group.

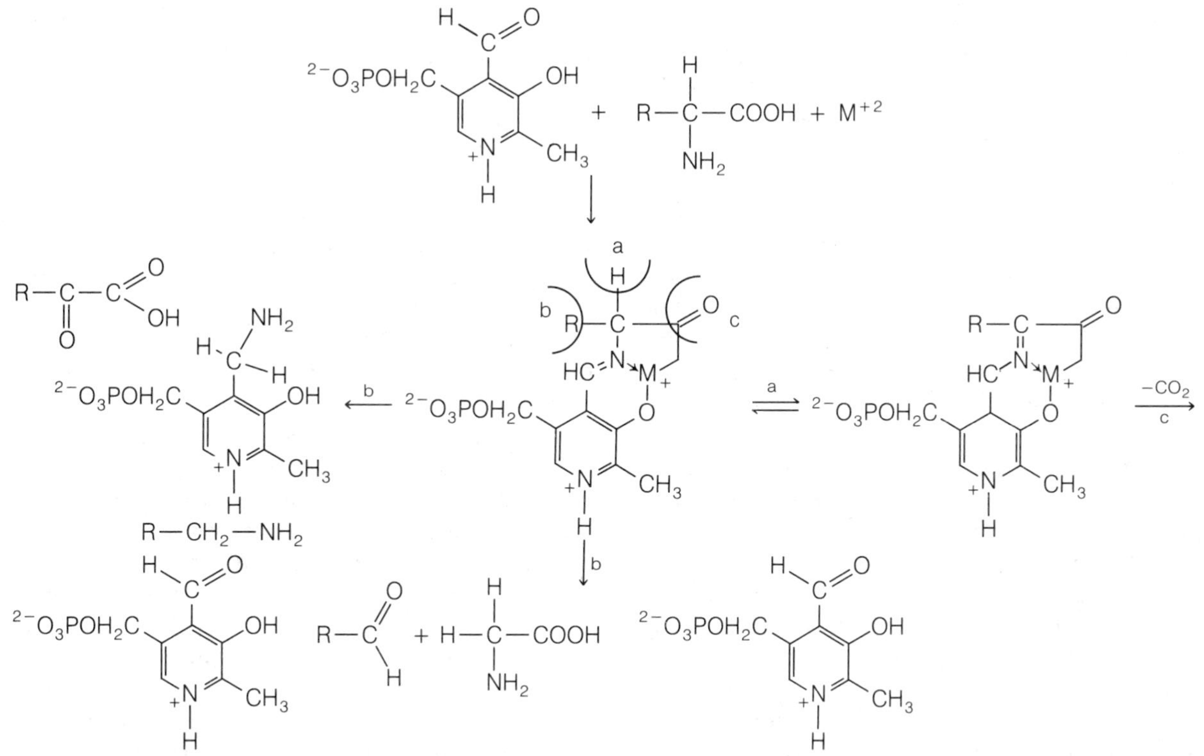

Metals such as iron and aluminum that markedly catalyze nonenzymatic transaminations in vitro probably do so by promoting the formation of the Schiff base and maintaining planarity of the conjugated system through chelate ring formation, which requires the presence of the phenolic group. This chelated metal ion also provides an additional electron-attracting group that operates in the same direction as the heterocyclic nitrogen atom (or nitro group), thereby increasing the electron displacements from the α-carbon atom as shown here.

It also should be noted that certain hydrazine derivatives, when administered therapeutically (e.g., isoniazid), can induce a deficiency of the coenzyme (pyridoxal-5-phosphate) by inactivation through the mechanism of hydrazone formation with the aldehyde functional group.

Another hydrazine derivative, hydralazine, when administered in high doses to control hypertension, can cause similar B_6 deficiency, conceivably through a similar mechanism involving hydrazone formation.

Hypochromic anemias caused by familiar-type pyridoxine dependency respond to pyridoxine therapy. Similarly, this vitamin has been useful in the treatment of hypochromic or megaloblastic anemias that are not due to iron deficiency and do not respond to other hematopoietic agents.[2]

A review by Rose[46] summarizes studies on the effects of certain hormones on vitamin B_6 nutrition in humans, on the biochemical interrelationship between steroid hormones and pyridoxal phosphate-dependent enzymes, and on the role of vitamin B_6 in regulating hypothalamus–pituitary functions. Some of these studies have important clinical implications that are noteworthy. The use of estrogen-containing oral contraceptives has been investigated as a factor leading to an abnormality of tryptophan metabolism. This abnormality resembles a dietary vitamin B_6 deficiency and responds favorably to treatment with the vitamin. For some time, there has been clinical interest in the relationship between certain hormones and vitamin B_6 function, because abnormal urinary excretions of tryptophan metabolites were observed during pregnancy and in patients with hyperthyroidism.

Estrogens and tryptophan metabolism have been studied, because estrogen administration occasionally leads to the excretion of abnormally large amounts of xanthurenic acid, a metabolic product from tryptophan. This metabolic malfunction has been related to the inhibitory effect of estrogen sulfate conjugates on another pathway of tryptophan metabolism—the transamination of the kynurenine from tryptophan. Consequently, xanthurenic acid formation appears to be abnormally increased owing to this estrogen effect or to B_6 deficiency.

In vitro studies have been conducted to determine the effect of estrogens on kynurenine aminotransferase, which catalyzes the B_6-dependent transamination of kynurenine to kynurenic acid. Some estrogen conjugates (e.g., estradiol disulfate and diethylstilbestrol sulfate) interfere with this transamination, apparently by reversible inhibition of the aminotransferase apoenzyme. It appears that the estrogen sulfate competes with pyridoxal-5-phosphate for the interaction with the apoenzyme. In contrast, free estradiol and estrone do not possess this inhibitory property.

Some women suffer from mental depression when taking estrogen-containing oral contraceptives, and this depression could be due to another malfunction in tryptophan metabolism leading to 5-hydroxytryptamine (serotonin). There is some evidence that the decarboxylation of 5-hydroxytryptophan is inhibited (in vitro) by estrogen conjugates competing with pyridoxal phosphate for the decarboxylase apoenzyme.

Other endocrine systems are interrelated. Both corticosteroids and thyroid hormones may increase the requirement for pyridoxine and affect pyridoxal-5-phosphate-dependent metabolic processes. Moreover, there appear to be associations between vitamin B_6 and anterior pituitary hormones. These associations seem to involve the hypothalamus, 5-hydroxytryptamine, and dopamine. The latter two neurotransmitters are synthesized by metabolic processes that require pyridoxal-5-phosphate.

Pyridoxine Hydrochloride, USP. 5-Hydroxy-6-methyl-3,4-pyridinedimethanol hydrochloride; vitamin B_6 hydrochloride; rat antidermatitis factor. Pyridoxine hydrochloride is a white, odorless, crystalline substance that is soluble 1 : 5 in water, 1 : 100 in alcohol, and insoluble in ether. It is relatively stable to light and air in the solid form and in acid solutions at a *p*H of not greater than 5, at which *p*H it can be autoclaved at 15 lb at 120°C for 20 to 30 minutes. Pyridoxine is unstable when irradiated in aqueous solutions at *p*H 6.8 or above. It is oxidized readily by hydrogen peroxide and other oxidizing agents. Pyridoxine is stable in mixed vitamin preparations to the same degree as riboflavin and nicotinic acid. A 1% aqueous solution has a *p*H of 3. The pK_{a_1} values for pyridoxine, pyridoxal, and pyridoxamine are 5.00, 4.22, and 3.40, respectively, and their pK_{a_2} values are 8.96, 8.68, and 8.05, respectively.

Pyridoxine is readily absorbed in the jejunum. The vitamin circulates protein bound as pyridoxal and pyridoxal phosphate. The liver is the major organ of storage (16 to 27 mg). The liver is also the organ of metabolism, oxidizing pyridoxal to 4-pyridoxic acid.

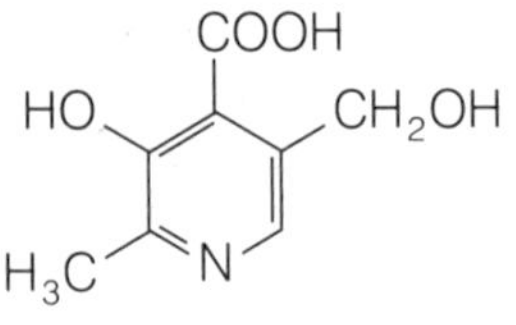

4-Pyridoxic Acid

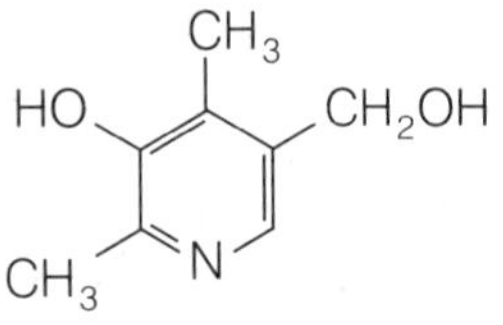

4-Desoxypyridoxol

Pyridoxime deficiencies have been studied by the use of antivitamins such as 4-desoxypyridoxal. As with the other B vitamins, dietary deficiencies are rare and are associated mainly with alcoholism. The symptoms involve the skin, nervous system, and erythropoiesis.

Because of inadequate diets, some infants suffer from severe vitamin B_6 deficiencies that can lead to epileptic-like convulsive seizures, and the convulsions can be controlled by treatment with pyridoxine.[1] It is believed that the convulsions are due to a below normal availability of the CNS neurohormone, γ-aminobutyric acid (GABA), from glutamic acid decarboxylation, which is effected by the coenzyme pyridoxal-5-phosphate.

Pyridoxime hydrochloride is indicated in the treatment and prevention of vitamin deficiency. It is also approved for concurrent administration with isoniazid and cycloserine to decrease their toxicity. Concurrent administration of pyridoxine hydrochloride with levodopa is not recommended. The decarboxylation of levodopa to dopamine in the periphery is increased by pyridoxine. This results in decreased amounts reaching the central nervous system.

THE COBALAMINS

Vitamin B_{12}, cyanocobalamin, occurs in nature as a cofactor that originally was isolated as cyanocobalamin and vitamin B_{12b} (hydroxocobalamin). In April 1948, Rickes et al.[47] isolated from clinically active liver fractions minute amounts of a red, crystalline compound that was also highly effective in promoting the growth of *Lactobacillus lactis*. This compound was called vitamin B_{12} and in single doses, as small as 3 to 6 μg, it produced positive hematologic activity in patients having addisonian pernicious anemia. Evidence indicates that its activity is comparable with that of Castle's extrinsic factor and that it can be stored in liver.

Vitamin B_{12} is found in commercial fermentation processes of antibiotics, such as those of *Streptomyces griseus*, *S. olivaceus*, *S. aureofaciens*, sewage, milorganite, and others. Some of these fermentations furnish a commercial source of vitamin B_{12}. Excellent dietary sources are meats, eggs, seafood, diary, and fermented products. However, de novo synthesis of vitamin B_{12} is restricted to microorganism. Animals depend on intestinal flora synthesis or, as in humans, consumption of animal products that have already obtained vitamin B_{12}. The only dietary plant products containing the vitamin are legumes, owing to their symbiosis with microorganisms.

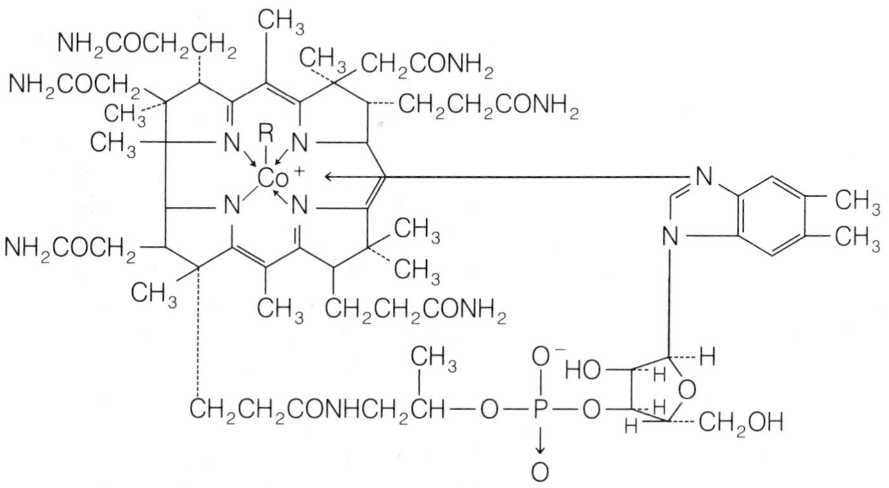

Cyanocobalamin

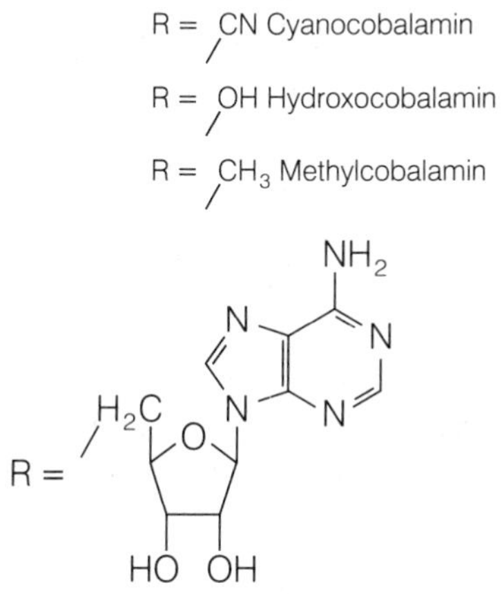

R = CN Cyanocobalamin

R = OH Hydroxocobalamin

R = CH_3 Methylcobalamin

Adenosylcobalamin

Since dietary vitamin B_{12} is protein bound, the first step in absorption is its release in the stomach. The release is enhanced by gastric pH and pancreatic proteases. The freed vitamin is immediately bound to a glycoprotein, the *intrinsic factor*, secreted by parietal cells of the gastric mucosa. The vitamin B_{12}–intrinsic factor complex is carried to the intestines where it binds with receptors in the ileum. Absorption is mainly by an active process that can be saturated by 1.5 to 3 μg of vitamin B_{12}. Excess amounts may be absorbed passively.

In the intestinal cells the complex is broken and vitamin B_{12} is absorbed into the blood, where it binds to transcobalamin II, a β-globulin, for distribution. In the liver the vitamin is converted to the active form and stored as such. Up to 90% of the vitamin (5 to 11 mg) are stored in the liver. Excretion of vitamin B_{12} is through the bile. Vitamin B_{12} undergoes extensive reabsorption.

In the biosynthesis of the coenzymes[48] derived from vitamin B_{12}, the cobalt is reduced from a trivalent to a monovalent state before the organic anionic ligands are attached to the structure. The two types of cobamides that participate as coenzymes in human metabolism are the adenosylcobamides and the methylcobamides. These coenzymes perform vital functions in methylmalonate–succinate isomerization and in methylation of homocysteine to methionine. Methylcobalamin is the major form of the coenzyme in the plasma, whereas 5-deoxyadenosylcobalamine is the major form in the liver and in other tissues. The enzyme system methylmalonyl-CoA mutase requires 5′-deoxyadenosylcobamide, and this enzyme system catalyzes the methylmalonyl-CoA transformation to succinyl-CoA, which is the major pathway of propionyl-CoA metabolism. Propionyl-CoA from lipid metabolism has to be processed through this pathway by succinyl-CoA to enter the Krebs citric acid cycle to be either converted to γ-oxaloacetate, leading to gluconeogenesis, or oxidized aerobically to CO_2, with production of ATP. The methylation of homocysteine to form methionine requires methylcobalamin, and it is catalyzed by a transmethylase that is also dependent on 5-methyltetrahydrofolic acid and reduced FAD.[48]

A deficiency, of vitamin B_{12} leads to an anemia. Pernicious anemia is due to a lack of the intrinsic factor. The symptoms involve systems that have rapidly dividing cells and the nervous system.

Herbert and Das[48] have reviewed the biochemical roles of vitamin B_{12} and folic acid in hemato- and other cell poiesis; they emphasize that vitamin B_{12} and folic acid are essential for normal growth and prolifration of all human cells. It is further noted that vitamin B_{12} also has a function in the maintenance of myelin throughout the nervous system. It is known that deficiency of either vitamin leads to megaloblastic anemia involving below-normal DNA synthesis. Interference with DNA synthesis results in the inability of cells to mature properly. These symptoms are readily reversed by vitamin B_{12} supplementation; however, the nerve damage is irreversible. It has been postulated that the myelin damage is a result of low levels of S-adenosylmethionine caused by a methionine synthetase deficiency.[49] Most vitamin B_{12} deficincies are due to malabsorption (lack of intrinsic factor, alcoholism) or increased need (pregnancy). It has been postulated that vitamin B_{12} deficiency is largely a conditioned folic acid deficiency caused by below-normal transformation of 5-methyl tetrahydrofolic acid to THF by the B_{12}-dependent homocysteine methyl transferase reaction and defective cellular uptake of 5-methyl-THF in vitamin B_{12} deficiency. This so-called *methyl-THF trap* hypothesis seems to rationalize a mechanism for the pathogenesis of megaloblastic anemia in vitamin B_{12} deficiency, and recent investigations continue to provide evidence that supports this rationalization.

Products

Cyanocobalamin, USP. Vitamin B_{12} is a cobalt-containing substance usually produced by the growth of suitable organisms or obtained from liver. It occurs as dark red crystals or an amorphous or crystalline powder. The anhydrous form is very hygroscopic and may absorb about 12% of water. One gram is soluble in about 80 mL of water. It is soluble in alcohol but insoluble in chloroform and in ether.

Vitamin B_{12} loses about 1.5% of its activity per day when stored at room temperature in the presence of ascorbic acid; whereas, vitamin B_{12b} (hydroxocobalamin) is very unstable (completely inactivated in one day). This loss in activity is accompanied by a release of cobalt and a disappearance of color. The greater stability of vitamin B_{12} is attributed to the increased strength of the bond between cobalt and the benzimidazole nitrogen by cyanide. Unusual resonance energy is imputed to the cobalt–cyanide complex, giving a positive charge to the cobalt atom and, thereby, strengthening the Co—N bond. The protective action of certain liver extracts of vitamin B_{12b} toward ascorbic acid and its sodium salt is, no doubt, due to the presence of copper and iron. Iron salts will protect vitamin B_{12b} in 0.001% concentration. Catalysis of the oxidative destruction of ascorbate by iron is well known. On exposure to air, liver extracts containing B_{12} lose most of the B_{12} activity in three months. The most favorable *p*H for a mixture of cyanocobalamin and ascorbic acid appears to be 6 to 7. Niacinamide can stabilize aqueous parenteral solutions of cyanocobalamin and folic acid at a *p*H of 6 to 6.5. However, it is unstable in B complex solution. Cyanocobalamin is stable in solutions of sorbitol and glycerin, but not in dextrose or sucrose.

Aqueous solutions of vitamin B_{12} are stable to autoclaving for 15 minutes at 121°C. It is almost completely inactivated in 95 hours by 0.015 *N* sodium hydroxide or 0.01 *N* hydrochloric acid. The optimum *p*H for the stability of cyanocobalamin is 4.5 to 5.0. Cyanocobalamin is stable in a wide variety of solvents.

Hydroxocobalamin, USP. Cobinamide dihydroxide; dihydrogen phosphate (ester) mono(inner salt); 3′-ester with 5,6-dimethyl-1-α-D-ribofuranosylbenzimidazole; vitamin B_{12b} is cyanocobalamin in which the CN group is replaced by an OH group. It occurs as dark red crystals or as a red crystalline powder that is sparingly soluble in water or alcohol and practically insoluble in the usual organic solvents.

Under the usual conditions, in the absence of cyanide ions only the hydroxo form of cobalamin is isolated from natural sources. It has good depot properties, but is less stable than cyanocobalamin.

Cyanocobalamin Co 57 Capsules, USP, contain cyanocobalamin in which some of the molecules contain radioactive cobalt (^{57}Co). Each microgram of this cyanocobalamin preparation has a specific activity of not less than 0.02 megabecquerel (0.5 μCi).

The *USP* cautions that in making dosage calculations one should correct for radioactive decay. The radioactive half-life of ^{57}Co is 270 days.

Cyanocobalamin Co 57 Solution, USP, has the same potency, dosage, and use as described under Cyanocobalamin Co 57 Capsules USP. It is a clear, colorless to pink solution that has a *p*H range of 4.0 to 5.5.

Cyanocobalamin Co 60 Capsules, USP, is the counterpart of Cyanocobalamin Co 57 Capsules in potency, dosage, and use. It differs only in its radioactive half-life, which is 5.27 years.

Cyanocobalamin Co 60 Solution, USP, has the same potency, dosage, and use as Cyanocobalamin Co 60 Capsules. It is a clear, colorless to pink solution that has a *p*H range of 4.0 to 5.5.

These four preparations must be labeled "Caution—Radioactive Material" and "Do not use after 6 months from date of standardization."

Cobalamin Concentrate, USP, derived from *Streptomyces* cultures or other cobalamin-producing microorganisms, contains 500 μg of cobalamin per gram of concentrate.

A cyanocobalamin zinc tannate complex can be used as a repository form for the slow release of cyanocobalamin when it is administered by injection.

Commercial products containing liver extract for oral and parenteral use are available. Liver extract is assayed to contain 10 to 20 μg of cyanocobalamin activity per milliliter. Crude liver extract contains 2 μg of activity per milliliter.

Vitamin B_{12} with intrinsic factor is a mixture the potency of which is expressed in terms of oral units of hematopoietic activity. One unit has no more than 15 μg of cyanocobalamin activity. The intrinsic factor is obtained from dried hog stomach, pylorus, or duodenum. One unit contains not more than 300 mg of dried product.

FOLIC ACID

In the early 1940s R. J. Williams et al.[50] reported the term *folic acid*, in referring to a vitamin occurring in leaves and foliage of spinach, from the Latin for leaf (*folium*). It has previously been called vitamin M and Vitamin B_9. Since then, folic acid has been found in whey, mushrooms, liver, yeast, bone marrow, soybeans, and fish meal, all of which serve as excellent dietary sources. The structure (see diagram) has been proved by synthesis in many laboratories (e.g., see Waller et al.[51]).

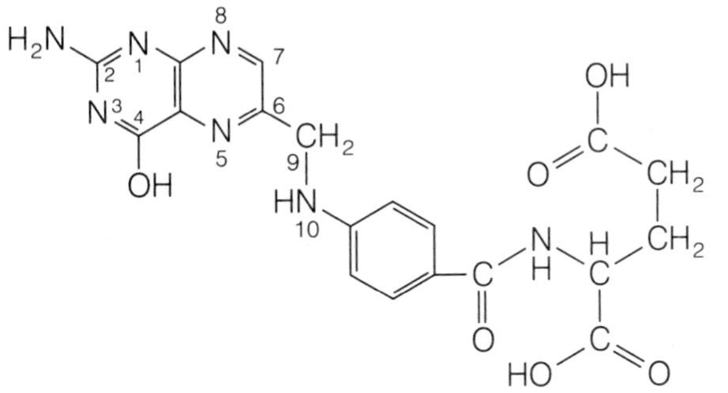

Folic Acid

Folic acid is a pteridine derivative (rings A and B constitute the pteridine heterocyclic system) synthesized by bacteria from guanosine triphosphate (GTP), *p*-aminobenzoic acid, and glutamic acid. Accordingly, the structure of folic acid is composed of three moieties: the pteridine moiety, derived from GTP; the *p*-aminobenzoic acid; and glutamic acid moieties. [It is interesting to relate that antibacterial sulfonamides (see Chap. 5) compete with *p*-aminobenzoic acid and thereby interfere with bacterial folic acid synthesis.] Of course, the human is not able to synthesize folic acid.

In the human, dietary folic acid must be reduced metabolically to tetrahydrofolic acid (THF) to exert its vital biochemical actions. This reduction, which proceeds through the intermediate dihydrofolic acid, is catalyzed by a reductase. This reductase enzyme system has been implicated as the catalyst in both reaction steps—folic acid reduction and dihydrofolic acid reduction. The coenzyme THF is converted to other cofactors by formulation of the N-10 and/or N-5 nitrogen.

These coenzymes, derived from folic acid, participate in many important reactions, including conversion of homocysteine to methionine, synthesis of glycine from serine, purine synthesis (C-2 and C-8), and histadine metabolism.

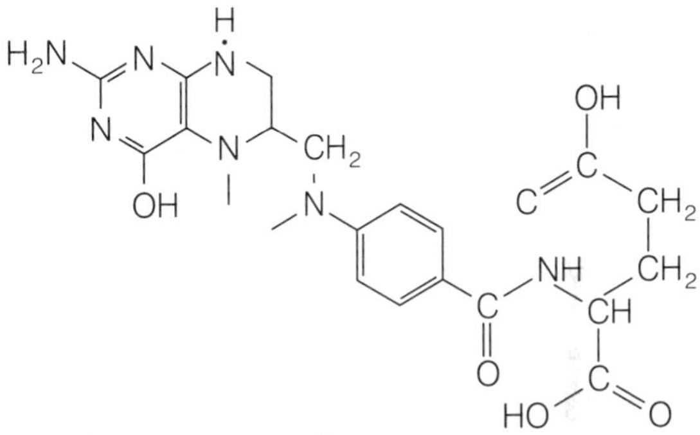

N^5	$-CH_3$	N^5-Methyltetrahydrofolic acid
N^5	$-CHO$	N^5-Formyltetrahydrofolic acid (folinic acid)
N^{10}	$-CHO$	N^{10}-Formyltetrahydrofolic acid
N^{5-10}	$=CH-$	$N^{5,10}$-Methenyltetrahydrofolic acid
N^{5-10}	$-CH_2-$	$N^{5,10}$-Methylenetetrahydrofolic acid
N^5	$-CH=NH$	N^5-Formiminotetrahydrofolic acid
N^1	$-CH_2OH$	N^5-Hydroxytetrahydrofolic acid

The most critical "one-carbon" transfer that is involved in DNA synthesis requires N^5,N^{10}-methylene-THF as the methylating coenzyme responsible

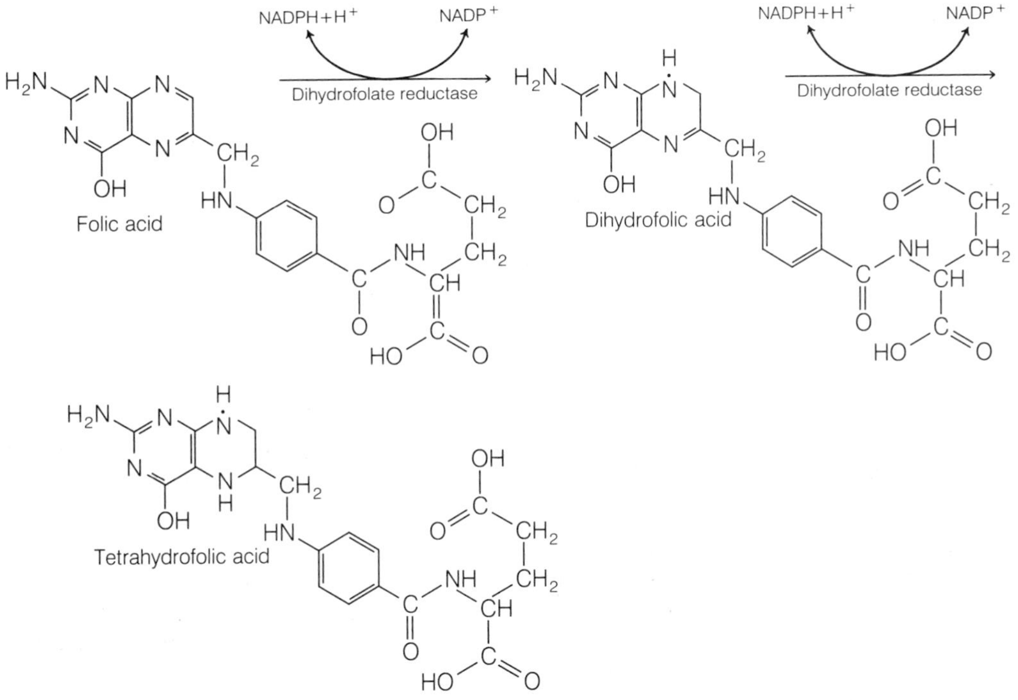

Folic acid → (Dihydrofolate reductase; $NADPH+H^+$ → $NADP^+$) → Dihydrofolic acid → (Dihydrofolate reductase; $NADPH+H^+$ → $NADP^+$) → Tetrahydrofolic acid

for converting uridylic acid to thimydilic acid. Interestingly, some folic acid antagonists useful in cancer chemotherapy (e.g., methotrexate; see Chap. 8) interfere with DNA synthesis by inhibiting this methylation step.[52]

There is a fundamental relationship between folic acid metabolism and vitamin B_{12}. The reduction of methylene-THF to 5-methyl-THF is essentially irreversible; hence, there is only one pathway for the regeneration of THF from 5-methyl-THF. The THF is regenerated by the B_{12}-dependent methyl group transfer from 5-methyl-THF to homocysteine. This biochemical interrelationship has been implicated in the etiology of megaloblastic anemia (see corresponding discussion under vitamin B_{12}^{48}).

Folic Acid, USP. *N*-[[(2-Amino-4-hydroxy-6-pteridinyl)methyl]amino]benzoylglutamic acid; pteroylglutamic acid (Folacine; Folvite). Folic acid occurs as a yellow of yellowish orange powder that is only slightly soluble in water (1 mg/100 mL). It is insoluble in the common organic solvents. The sodium salt is soluble (1 : 66) in water.

Aqueous solutions of folic acid or its sodium salt are stable to oxygen of the air, even upon prolonged standing. These solutions can be sterilized by autoclaving at a pressure of 15 lb in the usual manner. Folic acid in the dry state and in very dilute solutions is decomposed readily by sunlight or ultraviolet light. Although folic acid is unstable in acid solutions, particularly below a *p*H of 6, the presence of liver extracts has a stablizing effect at lower *p*H levels than is otherwise possible. Iron salts do not materially affect the stability of folic acid solutions.

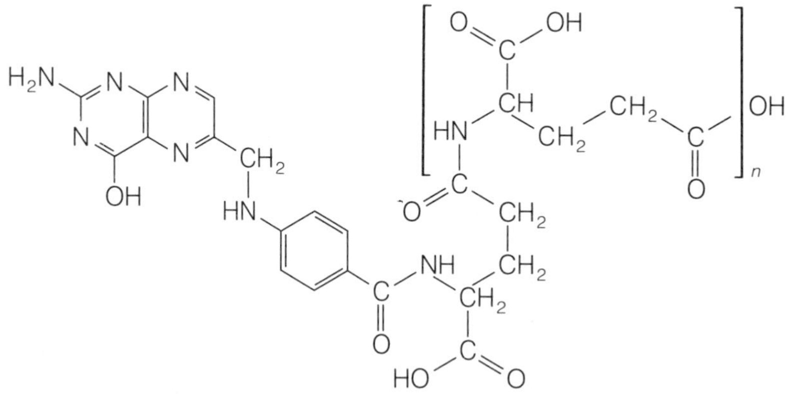

n = 1 = Pteroylglutamic acid
n = 6 = Pteroylpolyglutamic acid

The water-soluble vitamins that have a deleterious effect on folic acid are listed in their descending order of effectiveness as follows: riboflavin, thiamine hydrochloride, ascorbic acid, niacinamide, pantothenic acid, and pyridoxine. This deleterious effect may be overcome, to a considerable degree, by the inclusion of approximately 70% of sugars in the mixture.

Folic acid in foods is destroyed more readily by cooking than are the other water-soluble vitamins. These losses range from 46% in halibut to 95% in pork chops and from 69% in cauliflower to 97% in carrots.

Folic acid occurs in the diet as pteroylpolyglutamates that must be hydrolyzed to the monoglutamates before absorption. The hydrolysis is catalyzed by pteroyl-γ-glutamyl carboxypeptidase. The sites of absorption are the jejunum and upper duodenum. The mucosa in these regions possess dihydrofolate reductase, although most reduction and methylation occur in the liver. Tetrahydrofolate is distributed to all tissues, where it is stored as polyglutamates. The N^5-methyl derivative is the main transport and storage form in the body. The body stores 5 to 10 mg, approximately 50% in the liver.

The major elimination pathway for the vitamin is biliary excretion as the N^5-methyl derivative. Extensive reabsorption occurs. Only trace amounts are found in the urine. However, large doses that exceed the tubular reabsorption limit result in substantial amounts in the urine.

As with vitamin B_{12}, folic acid deficiencies are mainly the result of malabsorption or alcoholism. No neurologic abnormalities are associated with folic acid deficiency. The resulting megaloblastic anemia is indistinguishable from that caused by vitamin B_{12} because both vitamins are involved in the critical biochemical step.

Folic acid has the ability to correct the anemia caused by a vitamin B_{12} deficiency, but it has no effect on the neurologic damage. Thus, only small amounts are found in over-the-counter preparations.

Leucovorin Calcium USP. N-[4-[[(2-Amino-5-formyl-1,4,5,6,7,8-hexahydro-4-oxo-6-pteridinyl)-methyl]amino]benzoyl]-L-glutamic acid, calcium salt, calcium 5-formyl-5,6,7,8-tetrahydrofolate; calcium folinate occurs as a yellowish white or yellow, odorless, microcrystalline powder that is insoluble in alcohol and very soluble in water.

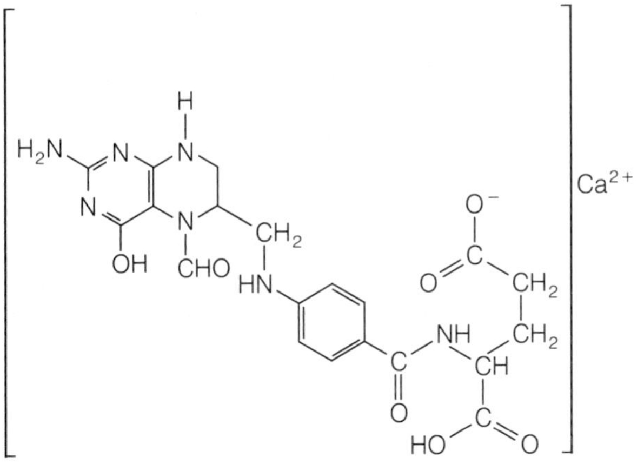

Leucovorin calcium
Calcium folinate

This product is used in chemotherapy concurrently with dihydrofolate reductase inhibitors to prevent damage to normal cells. It is not indicated for use in folic acid deficiencies.

ASCORBIC ACID

The historical significance of vitamin C was so eloquently summarized by the eminent medicinal chemist and pharmacist, the late Professor Ole Gisvold, that the following direct quotation from the seventh edition of this textbook is an appropriate introduction to the significance of ascorbic acid in medicinal chemistry and basic biochemistry:

> The disease scurvy, which now is known as a condition due to a deficiency of ascorbic acid in the diet, has considerable historical significance.[53] For example, in the war between Sweden and Russia (most likely the march of Charles XII into the Ukraine in the winter of 1708–1709) almost all of the soldiers of the Swedish army became incapacitated by scurvy. But further progress of the disease was stopped by a tea prepared from pine needles. The Iroquois Indians cured Jacques Cartier's men in the winter of 1535–1536 in Quebec by giving them a tea brewed from an evergreen tree. Many of Champlain's men died of scurvy when they wintered near the same place in 1608–1609. During the long siege of Leningrad, lack of vitamin C made itself particularly felt, and a decoction made from pine needles played an important role in the prevention of scurvy. It is somewhat common knowledge that sailors on long voyages at sea were subject to the ravages of scurvy. The British used supplies of limes to prevent this, and the sailors often were referred to as "limeys."
>
> Holst and Frolich,[54] in 1907, first demonstrated that scurvy could be produced in guinea pigs. A comparable condition cannot be produced in rats.
>
> Although Waugh and King[55] (1932) isolated crystalline vitamin C from lemon juice and showed it to be the antiscorbutic factor of lemon juice, Szent-Gyorgyi[56] had isolated the same substance from peppers in 1928, in connection with his biological oxidation–reduction studies. At the time, he failed to recognize its vitamin properties and reported it as a hexuronic acid because some of its properties resembled those of sugar acids. Hirst et al.[57] suggested that the correct formula should be one of a series of possible tautomeric isomers and offered basic proof that the formula now generally accepted is correct. The first synthesis of L-ascorbic acid (vitamin C) was announced almost simultaneously by Haworth and Reichstein,[58] in 1933. Since that time, ascorbic acid has been synthesized in a number of different ways.

This vitamin is now better known as ascorbic acid because of its acidic character and its effectiveness in the treatment and prevention of scurvy. The acidic character is due to the two enolic hydroxyls; the C-3 hydroxyl has the pK_a value of 4.1, and the C-2 hydroxyl has a pK_a of 11.6. The monobasic sodium salt is the usual salt form (e.g., Sodium Ascorbate, USP).

Ascorbic acid can be synthesized by nearly all living organisms, plants, and animals; but primates, guinea pigs, bats, and some other species are not capable of producing this vitamin. The consensus is that organisms that cannot synthesize ascorbic acid lack the liver microsomal enzyme L-gulonolactone oxidase, which catalyzes the terminal step of the biosynthetic process. Sato and Udenfriend[59] summarized studies of the biosynthesis of ascorbic acid in mammals and the biochemical and genetic basis for the incapability of some species to synthesize the vitamin. Because people are one of the few animal species that cannot synthesize ascorbic acid, the vitamin has to be available as a dietary component.

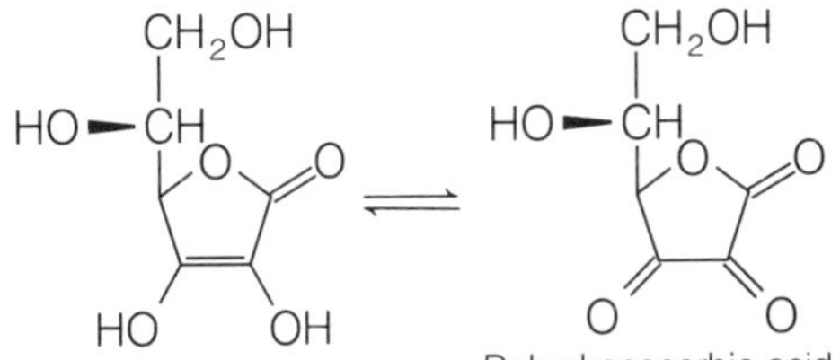

Ascorbic acid ⇌ Dehydroascorbic acid

Ascorbic acid performs important metabolic functions, as evidenced by the severe manifestations of its deficiency in humans. It has been demonstrated that this vitamin is involved in metabolic hydroxylations in numerous important metabolic processes (e.g., the synthesis of steroids and of neurotransmitters and in collagen and drug metabolism). Ascorbic acid has also been implicated as an important factor in other critical oxidation–reduction processes in human metabolism.[1]

Although it is well known that ascorbic acid is an effective reducing agent and antioxidant, the biochemical functions of this vitamin are not well understood. It is controversial to consider ascorbic acid to be an antiviral agent, but some scientists will argue that ascorbic acid is an effective cure or preventative of "common colds."[60] A recent study provides some evidence that ascorbic acid appears to help the organism recover from viral infections through an indirect mechanism on the body's immune system.[61] Ascorbic acid has also received attention as a possible anticancer agent. Recently, it has been demonstrated in cell culture studies that ascorbic acid, both alone and in combination with copper ions, is selectively toxic to melanoma cancer cells.[62]

Ascorbic Acid, USP. Vitamin C, L-ascorbic acid (Cevitamic Acid; Cebione). Ascorbic acid occurs as white or slightly yellow crystals or powder. It is odorless, and on exposure to light it gradually darkens. One gram of ascorbic acid dissolves in about 3 mL of water and in about 30 mL of alcohol. A 1% aqueous solutions has a *p*H of 2.7.

Aqueous solutions of ascorbic acid are not very stable. The ascorbic acid in such preparations undergoes oxidation, particularly under aerobic conditions. Oxidation to dehydroascorbic acid is followed by hydrolytic cleavage of the lactone. The effect of *p*H on the aerobic degradation of ascorbic acid aqueous solutions has been studied by various investigators. Rogers and Yacomeni[63] concluded that the degradation rate shows a maximum near *p*H 4 and a minimum near *p*H 5.6. It was also noted that if a preparation of ascorbic acid develops acidity on storage, and if its initial *p*H is between 5 and 5.6, the rate of degradation will increase as the *p*H decreases; hence, an initial *p*H in the range of 5.6 to 6 is recommended.

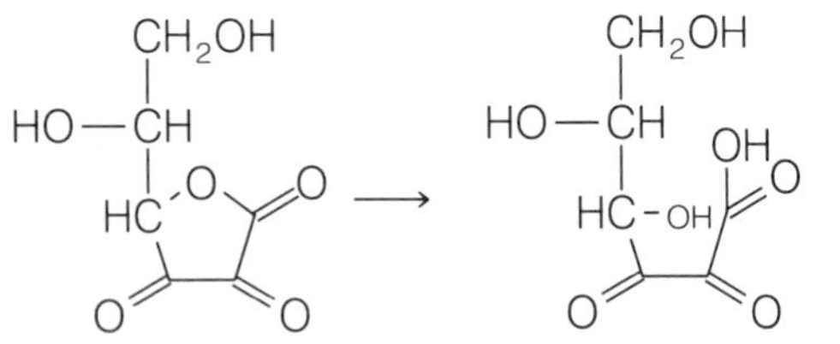

L-Ascorbic acid → 2,3-Diketo-L-gulonic acid

Dietary sources of ascorbic acid include citrus fruits, tomatoes, and potatoes. Although the sources of some commercial products are rose hips and citrus fruits, the largest amount of ascorbic acid is prepared synthetically.

Ascorbic acid is readily absorbed by an active process. Large doses can saturate this system, limiting the amounts absorbed. Once absorbed, it is distributed to all tissue. The vitamin is metabolized to oxalic acid before excretion. Ascorbic acid-2-sulfate is also a metabolite found in the urine. Large doses result in the excretion of substantial amounts of unchanged ascorbic acid. The resultant acidification of the urine is the basis for most of the vitamin's undesirable side effects.

The vitamin is indicated for the treatment and prevention of ascorbic acid deficiency. Although scurvy occurs infrequently, it is seen in the elderly, infants, alcoholics, and drug users. Ascorbic acid (but not the sodium salt) is frequently administered with methenamine to improve the effectiveness of this antibacterial agent. Because ascorbic acid increases the chelation of iron by deferoxamine, it is used in the treatment of chronic iron toxicity. It also finds usefulness as an adjunct in the treatment of methemoglobinemia.

Ascorbic Acid Injection, USP, is a sterile solution of sodium ascorbate that has a *p*H of 5.5 to 7.0. It is prepared from ascorbic acid with the aid of sodium hydroxide, sodium carbonate, or sodium bicarbonate. It may be used for intravenous injection, whereas ascorbic acid is too acidic for this purpose.

Sodium Ascorbate, USP, is a white, crystalline powder that is soluble 1 : 1.3 in water and is insoluble in alcohol.

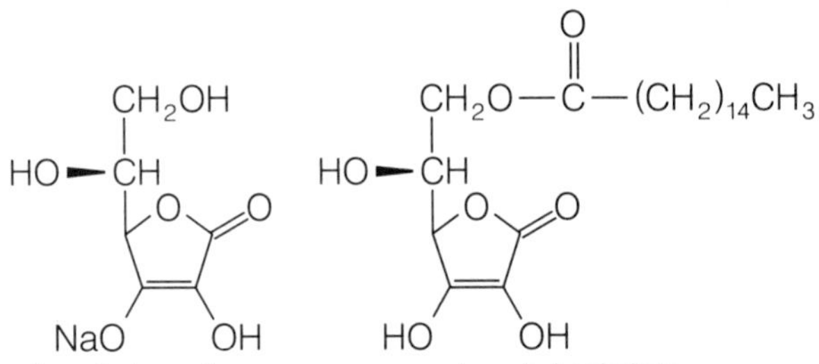

Ascorbate sodium   Ascorbyl palmitate

Ascorbyl Palmitate, NF. Ascorbic acid palmitate (ester), is the C-6 palmitic acid ester of ascorbic acid. It occurs as a white to yellowish white powder that is very slightly soluble in water and in vegetable oils. It is freely soluble in alcohol. Ascorbic acid has antioxidant properties and is a very effective synergist for the phenolic antioxidants such as propylgallate, hydroquinone, catechol, and nordihydroguaiaretic acid (NDGA) when they are used to inhibit oxidative rancidity in fats, oils, and other lipids. Long-chain, fatty acid esters of ascorbic acid are more soluble and suitable for use with lipids than is ascorbic acid.

BIOTIN

Biotin was discovered, isolated, and identified structurally in the 1930s. It has previously also been known as vitamin H. Since then, it has been noted that small amounts of biotin can be detected in almost all higher animals. The D-isomer possesses all the activity. The highest concentrations have been discovered in liver, kidney, eggs, and yeast, as a water-insoluble complex. Considerable quantities are found both free and in the complex form in vegetables, grains, and nuts. Alfalfa, string beans, spinach, and grass are fair sources of this vitamin.

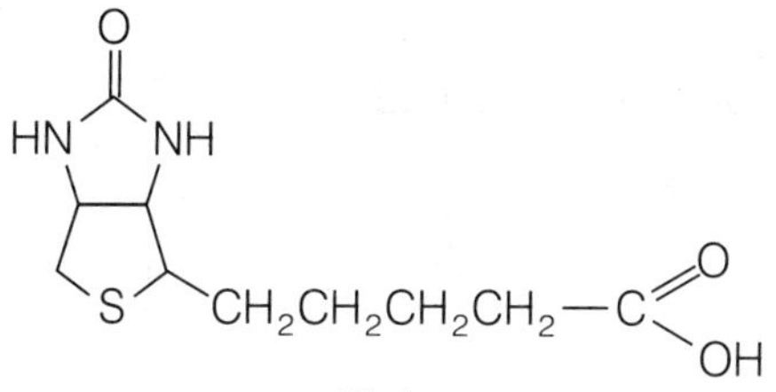

Biotin

Microorganisms synthesize biotin from the fatty acid, oleic acid. The biosynthetic process involves numerous complex reactions that remain to be better understood. The final reaction step requires formation of the sulfur heterocycle, but the source of the sulfur is not yet known.

Although this vitamin is known to perform essential metabolic functions in the human, the minimal nutritional requirement has not been established because it has been difficult to quantify the amounts of the vitamin made available by intestinal microorganisms. Nevertheless, deficiency states may develop owing to prolonged feeding of large quantities of raw egg white. Raw egg white contains avidin, a protein that complexes biotin and minimizes its absorption from the gastrointestinal tract. The symptoms of biotin deficiency include dermatitis, hyperesthesia, and glossitis.

Biotin performs vital metabolic functions in important carboxylation processes in the form of carboxybiotin, which is in combination with a carboxylase as represented below.

$$\text{Biotin–enzyme} + HCO_3^- \xrightleftharpoons{Mg^{2+}} CO_2\text{–biotin enzyme} + \text{ADP} + \text{Pi}.$$

The oxygen for ATP cleavage is derived from bicarbonate and appears in the Pi.

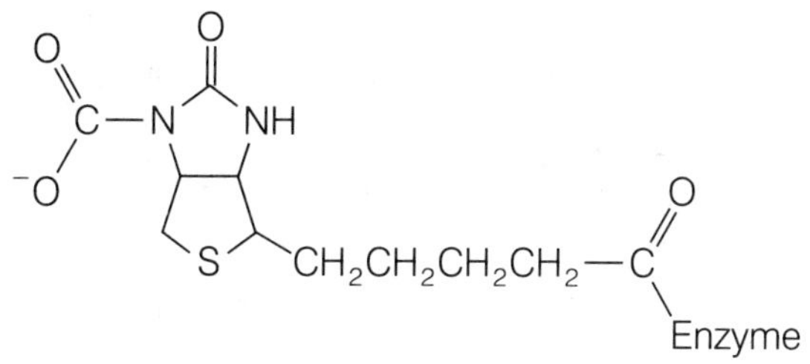

CO_2–Biotin enzyme

Purified preparations of acetyl-CoA carboxylase contained biotin [1 mole of biotin per 350,000 g of protein (enzyme)]. It catalyzed the first step in palmitate synthesis as follows:

$$CH_3COSCoA + HCO_3^- + ATP \xrightleftharpoons{Mg^{2+}} {}^-OOCCH_2COSCoA + ADP + Pi.$$

Other enzymes with which biotin appears to be intimately associated in carboxylation are β-methylcrotonyl-CoA carboxylase, propionyl-CoA carboxylase, pyruvate carboxylase, and methylmalonyloxalacetic transcarboxylase.

Biotin also is joined in an amide linkage to the ε-amino group of a lysine residue of carbamyl phosphate synthetase (CPS) to form biotin–CPS which participates with two ATP, HCO_3^-, and glutamine in the synthesis of carbamyl phosphate. This takes place stepwise as follows:

1. Biotin CPS + ATP + HCO_3^- ⇆ carbonic phosphoric anhydride biotin CPS (CPA biotin CPS) + ADP;
2. CPA biotin CPS ⇆ $^-$OOC biotin CPS + Pi;
3. $^-$OOC biotin CPS + glutamine ⇆ H_2NOC biotin CPS + ATP ⇄ biotin CPS + carbamylphosphate + ADP.

Carbamyl phosphate can participate in amino acid metabolism and some nucleic acid syntheses.

Biotin is readily absorbed from gastrointestinal tract. The body appears unable to break the fused imidazolidine and tetrahydrothiophene ring system. Biotin appears in the urine, predominately as the unchanged molecule. Only small amounts of the metabolites, biotin sulfoxide and bisnorbiotin, appear in the urine.

MISCELLANEOUS CONSIDERATIONS

Some dietary components are difficult to characterize as essential nutritional factors in human metabolism because the organism has the necessary chemistry to produce these compounds from other dietary components. (Consider vitamin D and nicotinic acid, which have been discussed earlier.) Vitamin D and nicotinic acid are, however, generally considered among the classic vitamins. Moreover, there is no clear consensus on the necessity for inositol, choline, and p-aminobenzoic acid. Nevertheless, such dietary components do perform important metabolic functions; hence, a brief characterization of these should be noted.

Inositol. 1,2,3,5-*trans*-4,6-Cyclohexanehexol; *i*-inositol; *meso*-inositol (*myo*-inositol; mouse anti-alopecia factor) is prepared from natural sources, such as corn steep liquors, and is available in limited commercial quantities. It is a white, crystalline powder that is soluble in water 1 : 6 and in dilute alcohol. It is slightly soluble in alcohol, the usual organic solvents, and in fixed oils. It is stable under normal storage conditions.

Inositol is one of nine different *cis–trans* isomers of hexahydroxycyclohexane and usually is assigned the following configuration:

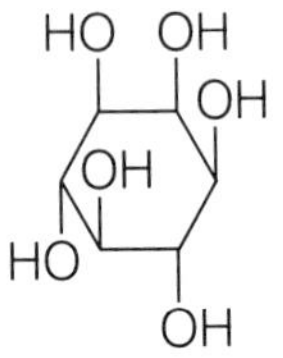

Inositol

Inositol has been found in most plants and animals tissues. It has been isolated from cereal grains, other plant parts, eggs, blood, milk, liver, brain, kidney, heart muscle, and other sources. The concentration of inositol in leaves reaches a maximum shortly before the time that the fruit ripens. Good sources of this factor are fruits, especially citrus fruits,[64] and cereal grains.

Inositol occurs free and combined in nature. In plants, it is present chiefly as the well-known phytic acid, which is inositol hexaphosphate. It is also present in the phosphatide fraction of soybean as a glycoside. In animals, much of it occurs free.[65]

Inositol in the form of phosphoinositides is almost as widely distributed as inositol, and some of these forms are more active metabolically. Phosphatidylinositol (monophosphoinositide) is the most widely distributed of the inositides and the chief fatty acid residue is stearic acid.

Recent data indicate that phosphoinositides serve as storage forms for secondary messengers. Phosphoinositides compose only a minor fraction (2% to 8%) of the lipids in cell membranes, yet they can be converted to at least three intracellular messenger molecules; arachidonic acid, inositol-1,4,5-trisphosphate (IP_3), and 1,2-diacylglyerol. The functions of arachidonic acid derivatives are discussed in Chapter 22. IP_3 functions to release intracellular calcium. The diacylglycerol acts as an essential cofactor in the activation of protein kinase C.[66]

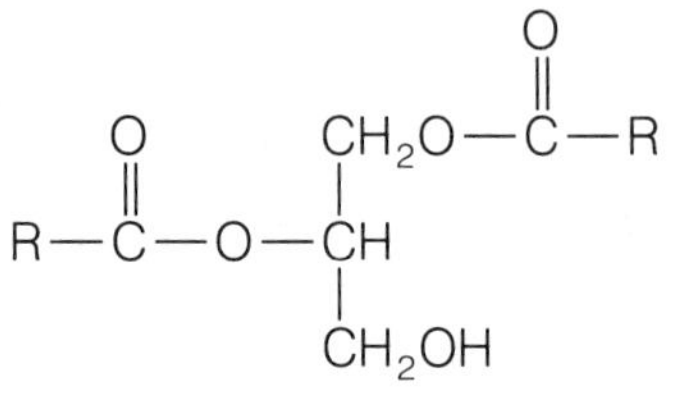

Diacylglycerol (DG or DAG)

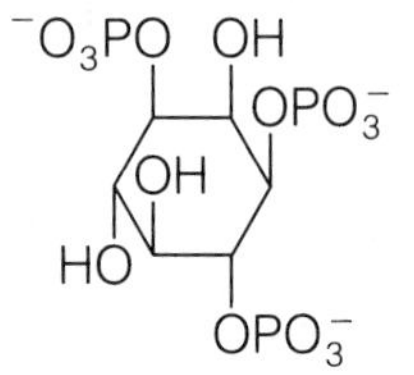

Inositol-1,4,5-trisphosphate $IP_3(1,4,5)$

The binding of many different hormones, neurotransmitters, and growth factors to the cell surface results in the activation of the polyphosphoinositide receptor system.[67] Binding to the specific receptor activates the enzyme phospholipase C through the intermediacy of a G-protein. Phospholipase C converts the phosphatidylinositol-4,5-bisphosphate (PIP_2) into inositol-1,4,5-trisphosphate (IP_3) and diacylglycerol. IP_3 releases calcium ion which, in turn, affects many cellular responses in the target cells.

IP_3 is also converted to inositol-1,3,4,5-tetrakisphosphate (IP_4). IP_4 also acts as a intracellular messenger, resulting in an influx of extracellular calcium. IP_4 is converted to $IP_3(1,3,4)$ which is dephosphorylated stepwise to inositol-1,4-diphosphate,

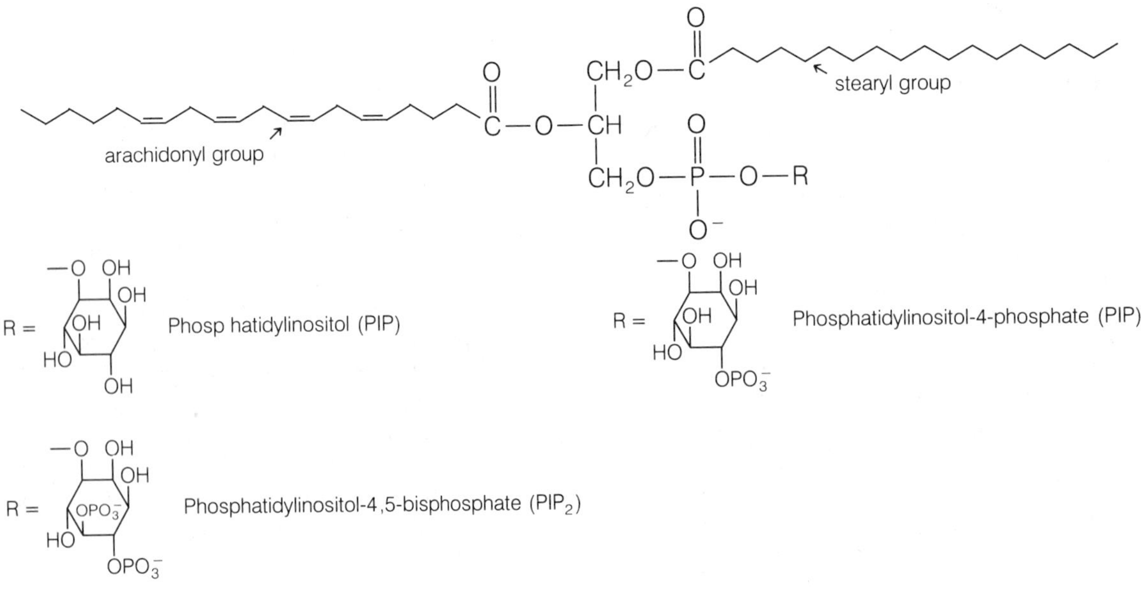

stearyl group

arachidonyl group

Phosp hatidylinositol (PIP)

Phosphatidylinositol-4-phosphate (PIP)

Phosphatidylinositol-4,5-bisphosphate (PIP_2)

inositol-1-phosphate, then inositol. The inositol is then incorporated into the diacylglycerol to form phosphatidylinositol (PI). PI is sequentially phosphorylated to form phosphatidylinositol-4-phosphate (PIP) and phosphatidylinositol-4,5-bisphosphate (PIP_2).

The diacylglycerol released has several fates. It can be converted to phosphatidylinositol (PI), as mentioned, or it can be hydrolyzed further to release the arachidonic acid component. The diacylglycerol in conjunction with calcium ion stimulates protein kinase C. By phosphorylating proteins, kinases regulate many cellular activities.

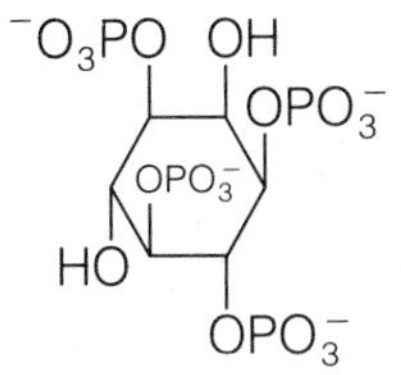

Inositol-1,3,4,5-tetrakisphosphate IP_4 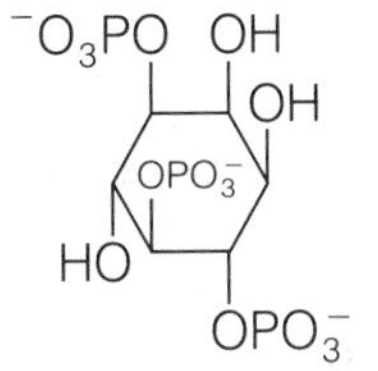 | Inositol-1,3,4-trisphosphate $IP_3(1,3,4)$ 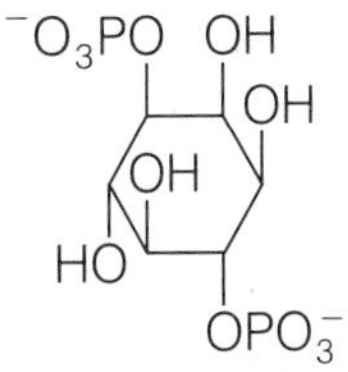 | Inositol-1,4-bisphosphate IP_2 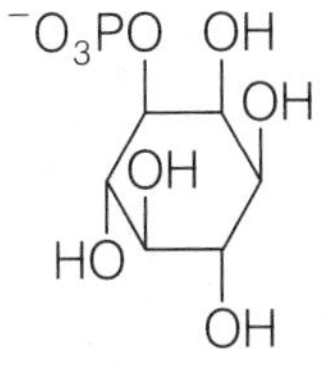 | Inositol-1-phosphate IP 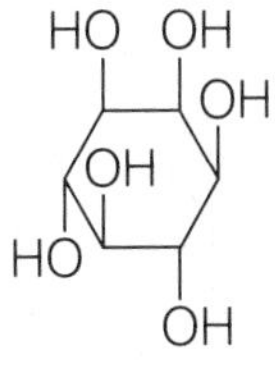 | Inositol

The cycle is completed by the phosphorylation of diacylglycerol to phosphatidic acid which, in turn, is converted to PI. The complete system remains to be fully understood. Cyclic phosphoinositol derivatives appear to also function as secondary messengers.

Inositol has been shown to be a growth factor for a wide variety of human cell lines in tissue culture. It is considered a characteristic component of seminal fluid, and the content is an index of the secretory activity of the seminal vesicles.

Evidence is accumulating to indicate that inositol will reduce elevated blood cholesterol levels. This, in turn, may prevent or mitigate cholesterol depositions in the intima of blood vessels in humans and animals and, therefore, be of value in atherosclerosis.

Inositol has also been considered as a lipotropic agent. Because the human can synthesize inositol, the need for inositol as a nutritional requirement has not been proved.[2]

Methionine, USP. An adequate diet should provide the methionine necessary for normal metabolism in the human. Methionine is considered to be an essential amino acid in people. It is the precursor in the biosynthesis of *S*-adenosylmethionine, which is an important methylating coenzyme involved in a variety of methylations, for example, in *N*-methylation of norepinephrine to form epinephrine and *O*-methylation of catecholamines catalyzed by catechol-*O*-methyltransferases. Adenosylmethionine also participates in the methylation of phosphatidylethanolamine to form phosphatidylcholine, but this pathway is not efficient enough to provide all the choline required by higher animals; hence adequate dietary availability of choline is necessary.[2]

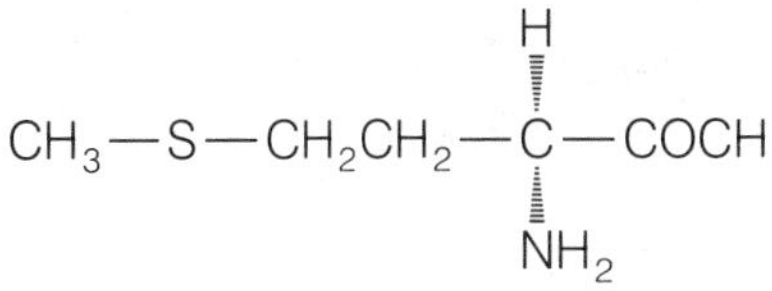

$$CH_3-S-CH_2CH_2-C(H)(NH_2)-COCH$$

Methionine

Choline is a component of many biomembranes and plasma phospholipids. Dietary sources includes eggs, fish, liver, milk, and vegetables. These sources provide choline primarily as the phospholipid lecithin. Lecithin is hydrolyzed to glycerophosphorylcholine by the intestinal mucosa before absorption. The liver liberates choline. Choline can be biosynthesized by humans; consequently, it cannot be considered a true vitamin. Biosynthesis involves methylation of ethanolamine. The methyl groups are provided by methionine or by a reaction involving vitamin B_{12} and folic acid. Therefore, deficiencies can occur only if all methyl donors are excluded from the diet.

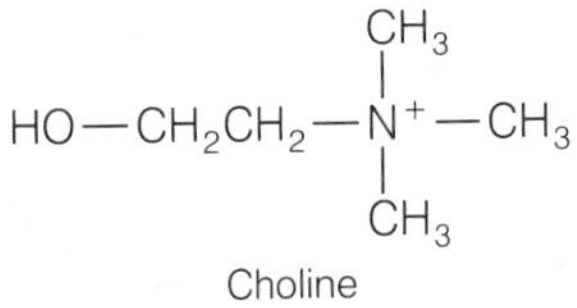

$$HO-CH_2CH_2-N^+(CH_3)_2-CH_3 \qquad {}^-O(O=)C-CH_2-N^+(CH_3)_2-CH_3$$

Choline 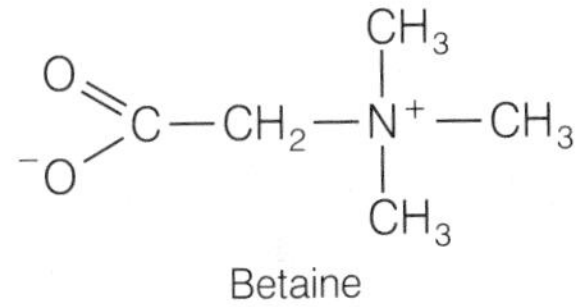 | Betaine

The therapeutic uses of choline depend on its physiologic functions. Because it is involved in the formation of plasma phospholipids, it is used as a lipotropic agent to alleviate fatty infiltration of the liver, cirrhosis. It has been used, in large doses, in certain central nervous system disorders (e.g., tardive dyskinesia, presenile dementia) because it is a precursor of acetylcholine. Choline also serves as a methyl donor in some reactions after it is converted to betaine.

***p*-Aminobenzoic Acid, USP** (PABA) has been mentioned as a biosynthetic component of folic acid in bacteria, but it is well known that higher mammalian organisms cannot synthesize folic acid from its precursors. Nevertheless, it seems that PABA performs certain metabolic functions in some animals. In the early 1950s PABA was reported to be an essential factor in the normal growth and life of the chick.

$$H_2N-C_6H_4-C(=O)OH$$

p-Aminobenzoic acid

Since these original developments in this field, various claims[69] have been made for the chromotrichial value of PABA in rats, mice, chicks, minks, and humans. The problem of nutritional achromotrichia is a complex one that may involve several vitamin or vitaminlike factors and is complicated by the synthesis and absorption from the intestinal tract of several factors produced by bacteria.

p-Aminobenzoic acid is a white, crystalline substance that occurs widely over the plant and animal kingdom. It occurs both free and combined[70] and has been isolated[71] from yeast, of which it is a natural constituent. It is soluble 1 : 170 in water, 1 : 8 in alcohol, and freely soluble in alkali.

p-Aminobenzoic acid is thought to play a role in melanin formation and to influence or catalyze tyrosine activity.[72] It inhibits oxidative destruction of epinephrine and stilbestrol, counteracts the graying of fur attributable to hydroquinone in cats and mice, exhibits antisulfanilamide activity, and counteracts the toxic effects of carbarsone and other pentavalent phenylarsonates.[33]

When given either parenterally or in the diet to experimental animals, PABA will protect them against otherwise fatal infections of epidemic or murine typhus, Rocky Mountain spotted fever, and tsutsugamushi disease.[74] These diseases have been treated clinically with most encouraging results by maintaining blood levels of 10–20 mg/100 mL for Rocky Mountain spotted fever and tsutsugamushi diseases. The mode of action of *p*-aminobenzoic acid in the treatment of these diseases appears to be rickettsiostatic rather than rickettsicidal, and the immunity mechanisms of the host finally overcome the infection.

PABA appears to function as a coenzyme in the conversion of certain precursors to purines.[75] It has been suggested as an effective sunscreen as a 5% solution in 55% to 75% ethyl alcohol on excessive sunlight-exposed areas of the skin.[76]

The historical significance of the effect of PABA on the antimicrobial action of sulfonamides and sulfones has been reviewed by Anand.[77]

REFERENCES

1. White, A., et al.: Principles of Biochemistry, 6th ed., p. 1320, 1333, 1362. New York, McGraw-Hill, 1978.
2. American Medical Association: Drug Evaluations, 3rd ed., p. 175. Littleton, MA, 1977.
3. McCollum, E. V., and Davis, M.: J. Biol. Chem. 19:245, 1914; 21:179, 1915.
4. De Luca, L. M.: Vitam. Horm. 35:1, 1977.
5. Chytil, F., and Ong, D. E.: Vitam. Horm. 36:1, 1978.
6. Robeson, C. D., et al.: J. Am. Chem. Soc. 77:4111, 1955.
7. Snell, E. E., et al.: J. Am. Chem. Soc. 77:4134, 4136, 1955.
8. Hanze, A. R., et al.: J. Am. Chem. Soc. 68:1389, 1946; Milas, N. A.: U.S. Patents 2,369,156,2,369,168, 2,382,085, 2,382,086; Isler, O., et al.: Experientia 2:31, 1946; Karrer, P., et al.: Helv. Chim. Acta 29:704, 1946; Milas, N. A., and Harrington, T. M.: J. Am. Chem. Soc. 69:2248, 1947; Oroshnik, W.: J. Am. Chem. Soc. 67:1627, 1945.
9. Milas, N. A.: Science 103:581, 1946.
10. Arens, J. F., and van Dorp, D. A.: Nature 157:190, 1946; Rec. Trav. Chim. 65:338, 1946.
11. Isler, O., et al.: Helv. Chim. Acta 30:1911, 1947.
12. Carr, F. H., and Price, E. A.: Biochem. J. 20:497, 1926.
13. Goodman, P. S., et al.: J. Biol. Chem. 242:3543, 1967.
14. Pett, L. B.: J. Lab. Clin. Med. 25:149, 1939; Hecht, S., and Mandelbaum, J.: JAMA 112:1910, 1939.
15. McLean, F., and Budy, A.: Vitam. Horm. 21:51, 1963.
16. Green, J.: Vitam. Horm. 20:485, 1962.
17. Kliger, D. S., and Menger, E. L.: Acc. Chem. Res. 8:81, 1975.
18. Hubbel, W. L.: Acc. Chem. Res. 8:85, 1975.
19. Honig, B., et al.: Acc. Chem. Res. 8:92, 1975.
20. Abrahamson, E. W.: Acc. Chem. Res. 8:101, 1975.
21. Williams, T. P.: Acc. Chem. Res. 8:107, 1975.
22. Hecht, S.: Am. Sci. 32:159, 1944.
23. White, A., et al.: Principles of Biochemistry, 6th ed., pp. 1173–1178. New York, McGraw-Hill, 1978.
24. Orfanos, C. E., Ehlert, R., and Gollnick, H.: Drugs 34:459, 1987.
25. Klingman, A. M., et al.: J. Am. Acad. Dermatol. 15:836, 1986.
26. Weiss, J. S., et al.: JAMA 259:527, 1986.
27. DeLuca, H. F.: J. Lab. Clin. Med. 87:7, 1976.
28. Suda, T., et al.: Biochemistry 9:1651, 1970.
29. Geisman, T. A., and Crout, D. H. G.: Organic Chemistry of Secondary Plant Metabolism, p. 292. San Francisco, Freeman Cooper, 1969.
30. Stern, M. A., et al.: J. Am. Chem. Soc. 69:869, 1947.
31. Demole, V., et al.: Helv. Chim. Acta 22:65, 1939.
32. Karrer, P., and Bussmann, G.: Helv. Chim. Acta 23:1137, 1940.
33. Weber, F., et al.: Biochem. Biophys. Res. Commun. 14:186, 1964.
34. Bieri, J. G., and Farrell, P. M.: Vitam. Horm. 34:31, 1976.
35. American Medical Association: Drug Evaluations, 3rd ed., pp. 186–187. Littleton, MA, 1977.
36. Williams, R. J.: Nutrition Against Disease, p. 148. New York, Pitman Publishing, 1971; see also Williams, R. J., and Kalita, D. K.: A Physician's Handbook on Orthomolecular Medicine, p. 64. New Canaan, CT, Keats Publishing, 1977; and for a general review see Williams, R. J.: Physician's Handbook of Nutritional Sciences. Springfield, IL, Charles C. Thomas, 1978.
37. Gryglewski, R. J.: Trends Pharmacol. Sci. 1:164, 1980.
38. Geisman, T. A., and Crout, D. H. G.: Organic Chemistry of Secondary Plant Metabolism, pp. 291–311. San Francisco, Freeman-Cooper, 1969.
39. Olson, R. E., and Suttie, J. W.: Vitam. Horm. 35:59, 1977.
40. See Chemistry and biochemistry of the K vitamins. Vitam. Horm. 17:531, 1959.
41. Gyorgy, P.: Vitam. Horm. 20:600, 1962.
42. White, A., et al.: Principles of Biochemistry, 6th ed., p. 1334. New York, McGraw-Hill, 1978.
43. Kig, T. E., and Strong, F. M.: Science, 112:562, 1950.
44. Braustein, A. E.: Enzymes 2:115, 1960.
45. Ikawa, K., and Snell, E.: J. Am. Chem. Soc. 76:637, 1954.

46. Rose, D. P.: Vitam. Horm. 36:53, 1978.
47. Rickes, E. L., et al.: Science 107:397, 1948; Smith: Nature 162:144, 1948; Ellis, et al.: J. Pharm. Pharmacol. 1:60, 1949.
48. Herbert, V., and Das, K. C.: Vitam. Horm. 34:1, 1976.
49. Scott, J. M., et al.: Lancet 2:234, 1981.
50. Williams, R. J., et al.: J. Am. Chem. Soc. 63:2284, 1941; 66:267, 1944.
51. Waller, C. W., et al.: J. Am. Chem. Soc. 70:19, 1948; see also Angier, R. G., et al.: Science 103, 667, 1946.
52. Montgomery, J. A.: Drugs for neoplastic disease. In Wolff, M. E. (ed.). Burger's Medicinal Chemistry, 4th ed., pp. 602–604. New York, John Wiley & Sons, 1979.
53. Schick, B.: Science 98:325, 1943.
54. Holst, A., and Frolich, T.: J. Hyg. 7:634, 1907.
55. Waugh, W. A., and King, C. C.: Science 75:357, 630, 1932; J. Biol. Chem. 97:325, 1932; Svirbely, J. L., and Szent-Gyorgyi, A.: Nature 129:576, 609, 1932; Biochem. J. 26:865, 1932; 27:279, 1933; Tillmans, J. et al.: Biochem. Z. 250:312, 1932.
56. Szent-Gyorgyi, A.: Biochem. J. 22:1387, 1928.
57. Hirst, E. L., et al.: J. Soc. Chem. Ind. 2:221, 482, 1933; Cox, E. G., and Goodwin, T. H.: J. Chem. Soc. p. 769, 1936; Nature 130:88, 1932.
58. Haworth, W. N., et al.: J. Chem. Soc. p. 1419, 1933; Reichstein, T.: Helv. Chim. Acta 16:1019, 1933.
59. Sato, P., and Udenfriend, S.: Vitam. Horm. 36:33, 1978.
60. Williams, R. J.: Physician's Handbook of Nutritional Science, p. 94, Springfield, Charles C. Thomas, 1978; Avery G. S.: Drug Treatment: Principles and Practice of Clinical Pharmacology and Therapeutics, p. 865. Littleton, MA, Publication Sciences Group, 1976.
61. Bram, S., et al.: Natue 284:629, 1980.
62. Manzella, J. P., and Roberts, N. J.: J. Immunol. 123:1940, 1979.
63. Rogers, A. R., and Yacomeni, J. A.: J. Pharm. Pharmacol. 23:2185, 1971.
64. Nelson, E. K., and Keenan, G. L.: Science 77:561, 1933.
65. Anderson, R. J., et al.: J. Biol. Chem. 125:299, 1938.
66. Majerous, P. W., et al.: Science 234:1519, 1986.
67. Nishizuka, Y.: Science 225:1365, 1984.
68. Marx, J. L.: Science 235:974, 1987.
69. Emerson, G. A.: Proc. Soc. Exp. Biol. Med. 47:448, 1941.
70. Diamond, N. S.: Science 94:420, 1941.
71. Rubbo, S. D., and Gillespie, J. M.: Nature 146:838, 1940.
72. Wisansky, W. A., et al.: J. Am. Chem. Soc. 63:1771, 1941.
73. Sandground, J. H., and Hamilton, C. R.: J. Pharmacol. Exp. Ther. 78:109, 1943.
74. Am. Prof. Pharm. 13:451, 1947.
75. Shive, W., et al.: J. Am. Chem. Soc. 69:725, 1947.
76. Avery, G. S.: Drug Treatment: Principles and Practice of Clinical Pharmacology and Therapeutics, p. 345. Littleton, MA, Publ. Sciences Group, 1976.
77. Anand, N.: Sulfonamides and sulfones. In Wolff, M. E. (ed.). Burger's Medicinal Chemistry, 4th ed., Part 2, p. 1. New York, John Wiley & Sons, 1979.

SELECTED READINGS

American Medical Association (Department of Drugs): AMA Drug Evaluations, Chaps. 13 and 14. Littleton, MA, Publishing Sciences Group, 1977.

Harper, H. A.: Review of Physiological Chemistry, 17th ed., Chap. 7. Los Altos, CA, Lange Medical, 1980.

Napoli, J. L., and DeLuca, H. F.: Blood calcium regulators. In Wolff, M. D. (ed.). Burger's Medicinal Chemistry, 4th ed., Chap. 26, p. 705. New York, John Wiley & Sons, 1979.

Sanders, H. J.: Nutrition and health. Chem. Eng. News p. 27, Mar. 26, 1979.

Suttie, J. W. (ed.): Vitamin K Metabolism and Vitamin K-Dependent Proteins. Baltimore, University Park Press, 1980.

Vitamins and Hormones, a series of annual volumes containing up-to-date reviews and current bibliographies. New York Academic Press.

White, A., et al.: Principles of Biochemistry, ed. 6, Chaps. 49, 50, 51. New York, McGraw-Hill, 1978.

CHAPTER 22

Prostaglandins and Other Eicosanoids

Thomas J. Holmes Jr.

The prostaglandins (PGA through PGH) are one group of naturally occurring 20-carbon fatty acid derivatives produced by the oxidative metabolism of 5,8,11,14-eicosatetraenoic acid, which is also called arachidonic acid. Other so-called eicosanoids produced in the complex biologic oxidation scheme called the *arachidonic acid cascade* (see Figs. 22-1 and 22-2) are thromboxane A_2(TXA_2), the leukotrienes (LKT A–D), and the highly potent antithrombotic agent, prostacyclin (PGI_2). The naming and numbering of these 20-carbon acids is included in Figures 22-1 to 22-3. Although eicosanoid-derived agents in current human clinical therapy are few, the promise of future contributions from this area is presumed to be very great. This promise stems from the fact that intermediates of arachidonic acid metabolism play an essential modulatory role in many normal and disease-related cellular processes. In fact, much of the pain, fever, swelling, nausea, and vomiting associated with "illness," in general, are probably a result of excessive prostaglandin production in damaged tissues.

HISTORY OF DISCOVERY

Early in this century (1931), it was noted by Kurzrok and Lieb that human seminal fluid could increase or decrease spontaneous muscle contractions of uterine tissue under controlled conditions.[1] This observed effect on uterine musculature was believed to be induced by an acidic vasoactive substance formed in the prostate gland, which was later (1936) termed *prostaglandin* by Von Euler.[2] Much later (1950s) it was found that the acidic extract contained not one, but several structurally related prostaglandin substances.[3] These materials were subsequently separated, purified, and characterized as the prostaglandins (PGA through PGF) varying somewhat in degree of oxygenation and dehydrogenation, and varying markedly in biologic activity (see Table 1). Specific chemical syntheses of the prostaglandins provided access to sufficient quantities of purified materials for wide-scale biologic evaluation and, also, served to confirm the structural characterization of these complex substances.[4]

Although a multitude of scientists have contributed to a refined characterization of the eicosanoid biosynthetic pathways and the biologic consequences of this cascade, the profound and persistent pioneering effort of Sune Bergstrom, Bengt Samuellson, and John R. Vane was recognized by the award of a shared Nobel Prize in Medicine in 1982. These scientists not only dedicated themselves to the chemical and biologic characterization of the eicosanoid substances, but, also, they were the first to realize the profound significance of the arachidonic acid cascade in disease processes, particularly inflammation. It was these individuals who first proved that the mechanism of anti-inflammatory action of aspirin and related nonsteroidal anti-inflammatory (NSAI) drugs was directly due to their inhibitory effect on prostaglandin formation. It has been subsequently shown that the analgesic and antipyretic effects of these NSAI agents, as well as their proulcerative and anticoagulant side effects, probably also result from their effect on eicosanoid metabolism.

A plethora of books have been published describing the role of eicosanoids in the inflammatory process, the immune system, carcinogenesis, the cardiovascular system, reproductive processes and the central nervous system (see Selected Readings). An annual update of research results in this area has been published since 1975 entitled *Advances in Prostaglandins, Thromboxanes, and Leukotrienes.*

Recent research findings in this area may appear in a variety of biochemical and clinical journals, but are the primary concern of two specific journals, *Prostaglandins* and *Prostaglandins, Leukotrienes, and Medicine*.

EICOSANOID BIOSYNTHESIS

Prostaglandins and other eicosanoids are produced by the oxidative metabolism of free arachidonic acid. Under normal circumstances, arachidonic acid is not available for metabolism, as it is present as a conjugated component of the phospholipid matrix of most cellular membranes. Release of free arachidonic acid, which may be subsequently metabolized, occurs by stimulation of phospholipase (PLA_2) enzyme activity in response to some traumatic event (e.g., tissue damage, toxin exposure, or hormonal stimulation). It is believed that the clinical anti-inflammatory effect of glucocortical steroids (i.e., hydrocortisone) is due to their ability to suppress phospholipase activity and, therefore, prevent the release of free arachidonic acid.[5] Modulation of phospholipase ac-

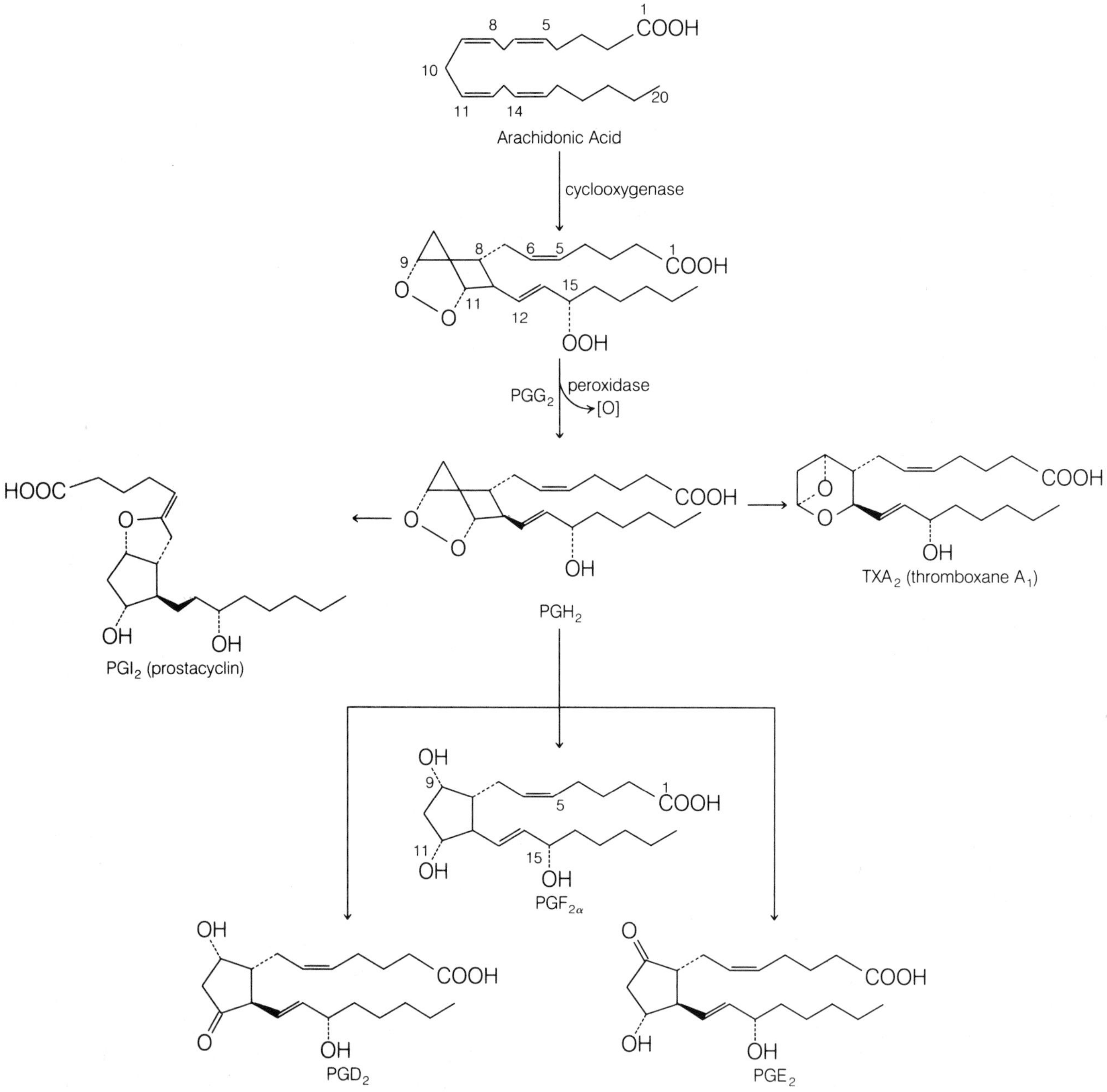

FIG. 22-1. *Cyclooxygenase pathway.*

tivity by alkali metal ions, toxins, and various therapeutic agents has recently become a major focus of biologic research because of the changes in eicosanoid production accompanying phospholipase stimulation or suppression. Although it was initially believed that the inflammatory response (swelling, redness, pain) was principally due to PGE_2, recent interest has focused on the less stable hydroperoxide intermediate PGG_2, which apparently is a more potent inducer of inflammatory symptoms.[6]

Two different routes for oxygenation of arachidonic acid have been identified: the cyclooxygenase pathway (Fig. 22-1) and the lipoxygenase pathway (Fig. 22-2). The relative significance of each of these pathways may vary in a particular tissue or disease state. The cyclooxygenase pathway, so named because of the unusual bicyclic endoperoxide (PGG_2) produced in the first step of the sequence, involves the highly stereospecific addition of two molecules of oxygen to the arachidonic acid substrate, followed by subsequent enzyme-controlled rearrangements to produce an array of oxygenated eicosanoids with diverse biologic activities (see Table 22-1). The first enzyme in this pathway, PGH-synthase, is a hemoprotein that catalyzes both the addition of oxygen (to form PGG_2) and the subsequent reduction (peroxidase activity) of the 15-position hydroperoxide to the 15-(S)-configuration alcohol (PGH_2). PGH-synthase (formerly called *cyclooxygenase* or *PG-synthetase*) has been the focus of intense investigation over the last decade because of its key role as the first enzyme in the arachidonic acid cascade. It is this enzyme that is susceptible to inhibition by NSAI agents, leading to relief of pain, fever, and inflammation.[8] And, it is this enzyme that is inhibited by the ω_3-fatty acids [eicosapentaenoic acid (EPA) and docosahexaenoic acid (OHA)] found in certain coldwater fish, leading presumably to beneficial cardiovascular effects.[9] This enzyme will metabolize 20-carbon fatty acids with one more or one less double bond than arachidonic acid, leading to prostaglandins of varied degrees of unsaturation (e.g., PGE_1 or PGE_3, for which the subscript number indicates the number of double bonds in the molecule).

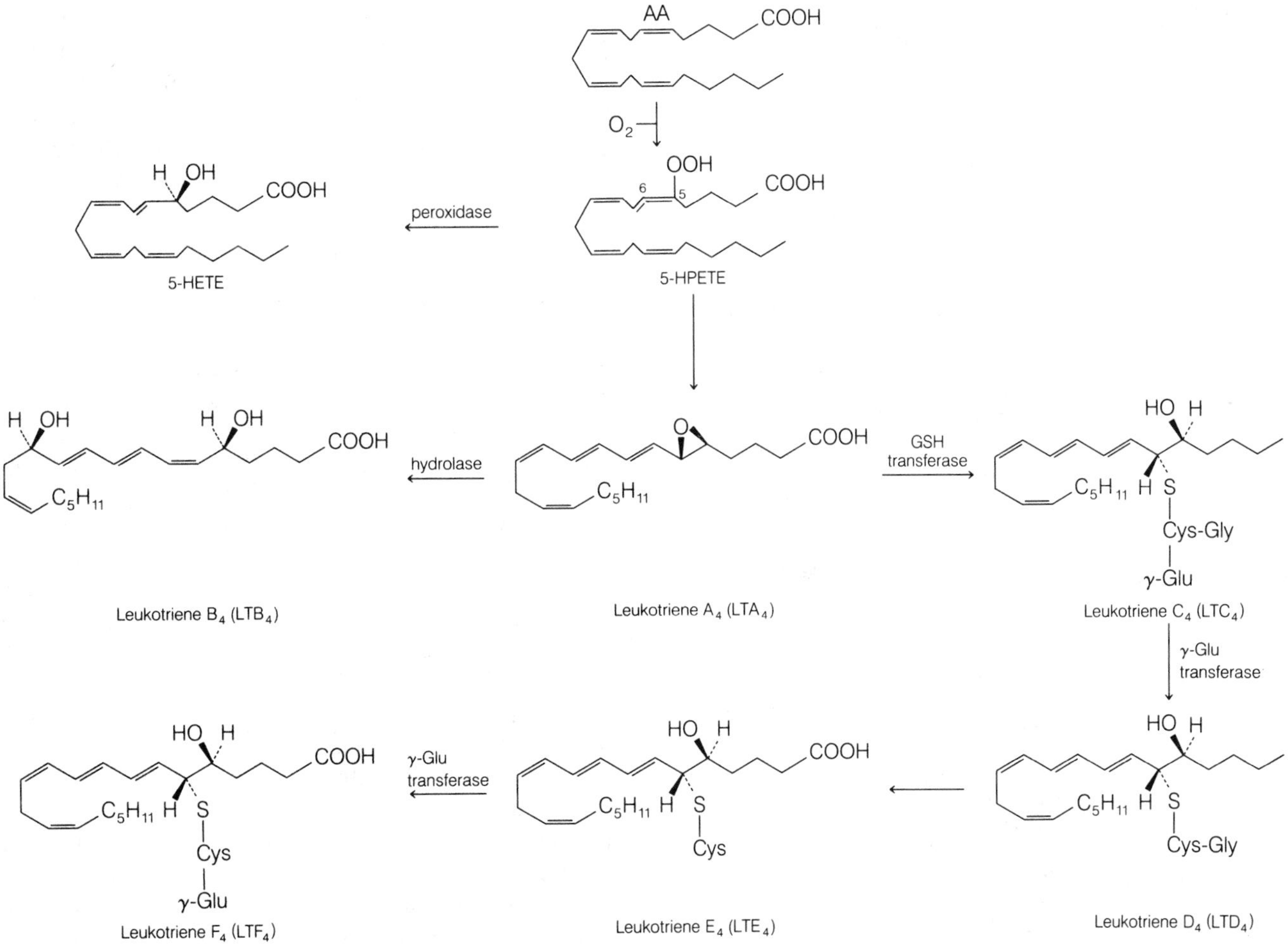

FIG. 22-2. *Lipoxygenase pathway.*

Prostaglandin H_2 serves as a branch-point substrate for specific enzymes leading to the production of the various prostaglandins, thromboxane A_2, and prostacyclin (PGI_2). Even though most tissues have the capability to produce PGH_2, the relative production of each of these derived eicosanoids is highly tissue-specific and may be subject to secondary modulation by a variety of cofactors. The characterization of enzymes involved in branches of the cyclooxygenase pathway is currently underway.

In certain cases, specific cellular or tissue response to the eicosanoids is apparently a function of available surface receptor recognition sites.[10] The variety of tissue responses observed upon eicosanoid exposure is outlined in Table 22-1. Nontissue selective inhibitors of the cyclooxygenase pathway, such as aspirin, thus, may exert a diversity of therapeutic effects, or side effects (e.g., decreased uterine muscle contraction and platelet aggregation; lowering of elevated body temperature; central and peripheral pain relief; and decreased vascular perfusion) based upon their tissue distribution.

TABLE 22-1

BIOLOGIC ACTIVITIES OBSERVED WITH THE EICOSANOIDS

Substance	*Observed Biologic Activity*
PGD_2	Weak inhibitor of platelet aggregation
PGE_1	Vasodilation Inhibitor of lipolysis Inhibitor of platelet aggregation Bronchodilatation Stimulates contraction of gastrointestinal smooth muscle
PGE_2	Stimulates hyperalgesic response Renal vasodilatation Stimulates uterine smooth-muscle contraction Protects gastrointestinal epithelia from acid degradation Reduces secretion of stomach acid Elevates thermoregulatory set-point in anterior hypothalamus
PGF_2	Stimulates breakdown of corpus luteum (luteolysis) in animals Stimulates uterine smooth-muscle contraction
PGI_2	Potent inhibitor of platelet aggregation Potent vasodilator Increases cAMP levels in platelets
TXA_2	Potent inducer of platelet aggregation Potent vasoconstrictor Decreases cAMP levels in platelets Stimulates release of ADP and serotonin from platelets
LTB_4	Increases leukocyte chemotaxis and aggregation
LTC/D_4	Slow-reacting substances of analphylaxis Potent and prolonged contraction of guinea pig ileum smooth muscle Contracts guinea pig lung parenchymal strips Bronchoconstrictive in humans Increased vascular permeability in guinea pig skin (augmented by PGEs)
5- or 12-HPETE	Vasodilatation of rat and rabbit gastric circulation Inhibits induced platelet aggregation
5- or 12-HETE	Aggregates human leukocytes Promotes leukocyte chemotaxis

The lipoxygenase pathway of arachidonic acid metabolism (Fig. 22-2) produces a variety of acyclic lipid peroxides (hydroperoxyeicosatetraenoic acids; HPETEs) and derived alcohols (hydroxyeicosatetraenoic acids; HETEs). Although the specific biologic function of each of these lipoxygenase-derived products is not completely known, they are believed to play a major role as chemotactic factors that promote cellular mobilization toward sites of tissue injury. In addition, the glutathione (GSH) conjugate, leukotriene (LKT-D_4), has been characterized as a potent, long-acting bronchoconstrictor, which is released in the lungs during severe hypersensitivity episodes (leading to its initial designation as the *slow-reacting substance of anaphylaxis*; SRSA). Because of the presumed benefit of preventing formation of LKTs in asthmatic patients, much research effort is being dedicated to the design and discovery of drugs that might selectively inhibit the lipoxygenase pathway of arachidonic acid metabolism, without affecting the cyclooxygenase pathway. It has been proposed that aspirin hypersensitivity in susceptible individuals may result from effectively "shutting down" the cyclooxygenase metabolic route, allowing only the biosynthesis of lipoxygenase pathway intermediates, including the bronchoconstrictive LKTs.[11]

DRUG ACTION MEDIATED BY EICOSANOIDS

The ubiquitous nature of the eicosanoid-producing enzymes implies their significance in a variety of essential cellular processes. Additionally, the sensitivity of these enzymes to structurally varied hydrophobic materials, particularly carboxylic acids and phenolic antioxidants, implies their susceptibility to influence by a variety of exogenously administered drugs. Because most aromatic drug molecules undergo hepatic hydroxylation, phenolic derivatives of administered drugs become readily available in vivo. Even more directly, it has been demonstrated that aromatic molecules upon in vitro incubation with microsomal PGH-synthase will become hydroxylated directly during arachidonic acid metabolism, in a process labeled *cooxidation*.[12] This cooxidative process presumably occurs during the peroxidative conversion of PGG_2 to PGH_2, which effectively makes available a nonspecific oxidizing equivalent. The cooxidation process has been implicated in the activation of polycyclic aromatic hydrocarbons to form proximate carcinogens.[13]

The only group of drugs that has been thoroughly characterized for its affect on arachidonic acid metabolism is the NSAI agents. This large group of acidic, aromatic molecules exerts a diverse spectrum of activities (previously mentioned) by inhibition of the first enzyme in the arachidonic acid cascade, PGH-synthase. Such agents as salicylic acid, phenylbutazone, naproxen, sulindac, and ibuprofen presumably act by a competitive, reversible inhibition of arachidonic acid oxygenation.[14] Aspirin and certain halogenated aromatics (including indomethacin, flurbiprofen, and meclomen) appear to inhibit PGH-synthase in a time-dependent, irreversible manner.[15] Although, for aspirin, this irreversible inhibition appears critical only for its effect on platelet aggregation and, therefore, prolongation of bleeding time.[16]

Interestingly, aspirin's primary competitor in the commercial analgetic marketplace, acetaminophen, has been shown to be a rather weak inhibitor of arachidonic acid oxygenation in vitro.[17] This, in fact, has been observed to be a characteristic of reversible noncompetitive, phenolic antioxidant inhibitors in general.[18] This determination, in concert with its lack of in vitro anti-inflammatory activity (while maintaining analgesic and antipyretic activity equivalent to the salicylates) has led to the proposal that acetaminophen is more active as an inhibitor of PGH-synthase in the brain, in which peroxide levels (which stimulate cyclooxygenase activity) are lower, than in inflamed peripheral joints, in which lipid peroxide levels are high.[14] In fact, when in vitro experimental conditions are modified to reduce the so-called peroxide tone, acetaminophen becomes as effective as aspirin in reducing arachidonic acid metabolism.[17]

DESIGN OF EICOSANOID DRUGS

The ability to successfully capitalize on the highly potent biologic effects of the various eicosanoids to develop new therapeutic agents currently seems an unfulfilled promise to medicinal chemists. Although these natural substances are highly potent effectors of various biologic functions, their use as drugs has been hampered by several factors: (1) their chemical complexity and relative instability, which has limited, to some extent, their large-scale production and formulation for clinical testing; (2) their susceptibility to rapid degradation (Fig. 22-3), which limits their effective bioactive half-life; and (3) their ability to affect diverse tissues (particularly the gastrointestinal tract, which may lead to severe nausea and vomiting) if they enter the systemic circulation even in only small amounts.

Two approaches have been employed to overcome these difficulties. First, structural analogues of particular eicosanoids have been synthesized, which are more resistant to chemical and metabolic degradation, but maintain, to a large extent, a desirable biologic activity. Although commercial production and formulation may be facilitated by this approach, biologic potency is usually reduced by several orders of magnitude. Also systemic side effects may become troublesome owing to broader distribution as a result of the increased half-life.

Primarily, structural alterations of the eicosanoids have been aimed at reducing or eliminating the very rapid metabolism of these potent substances to relatively inactive metabolites (see Fig. 22-3). Several analogues are presented in Table 22-2 to illustrate approaches that have led to potentially useful eicosanoid drugs. Methylation at the 15- or 16-position will eliminate or reduce oxidation of the essential 15-(*S*)-alcohol moiety. Esterification of the carboxylic acid function may affect formulation or absorption characteristics of the eicosanoid, whereas esterase enzymes in the bloodstream or tissues would be expected to quickly regenerate the active therapeutic agent. Somewhat surprisingly, considering the restrictive configurational requirements at the naturally asymmetric centers, a variety of hydrophobic substituents (including phenyl rings) are tolerated in the eicosanoid side chains.

A second major effect has been aimed at delivering the desired agent, either a natural eicosanoid or modified analogue, to a localized site of action by using some controlled-delivery method. The exact method of delivery may vary according to the desired site of action (e.g., uterus, stomach, lung), but has included aerosols and locally applied suppository or gel formulations.

EICOSANOIDS APPROVED FOR HUMAN CLINICAL USE

Prostaglandin $F_{2\alpha}$. Dinoprost (Prostin F2 Alpha). $PGF_{2\alpha}$ is a naturally occurring prostaglandin that is administered intra-amniotically to induce labor or abortion within the first trimester.

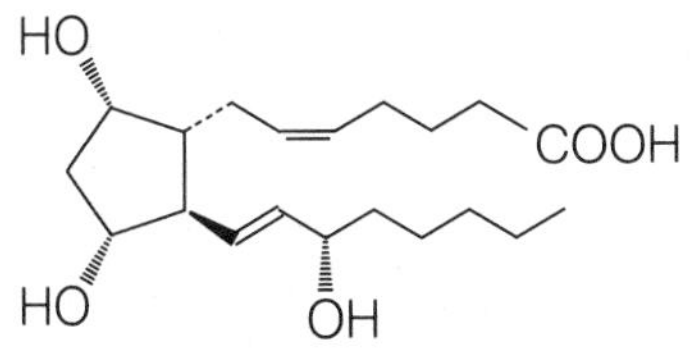

HO COOH HO OH

This product is supplied as a solution of the tromethamine salt (5 mg/mL) for direct administration.

Prostaglandin E_2. Dinoprostone (Prostin E2) PGE_2 is a naturally occurring prostaglandin that is

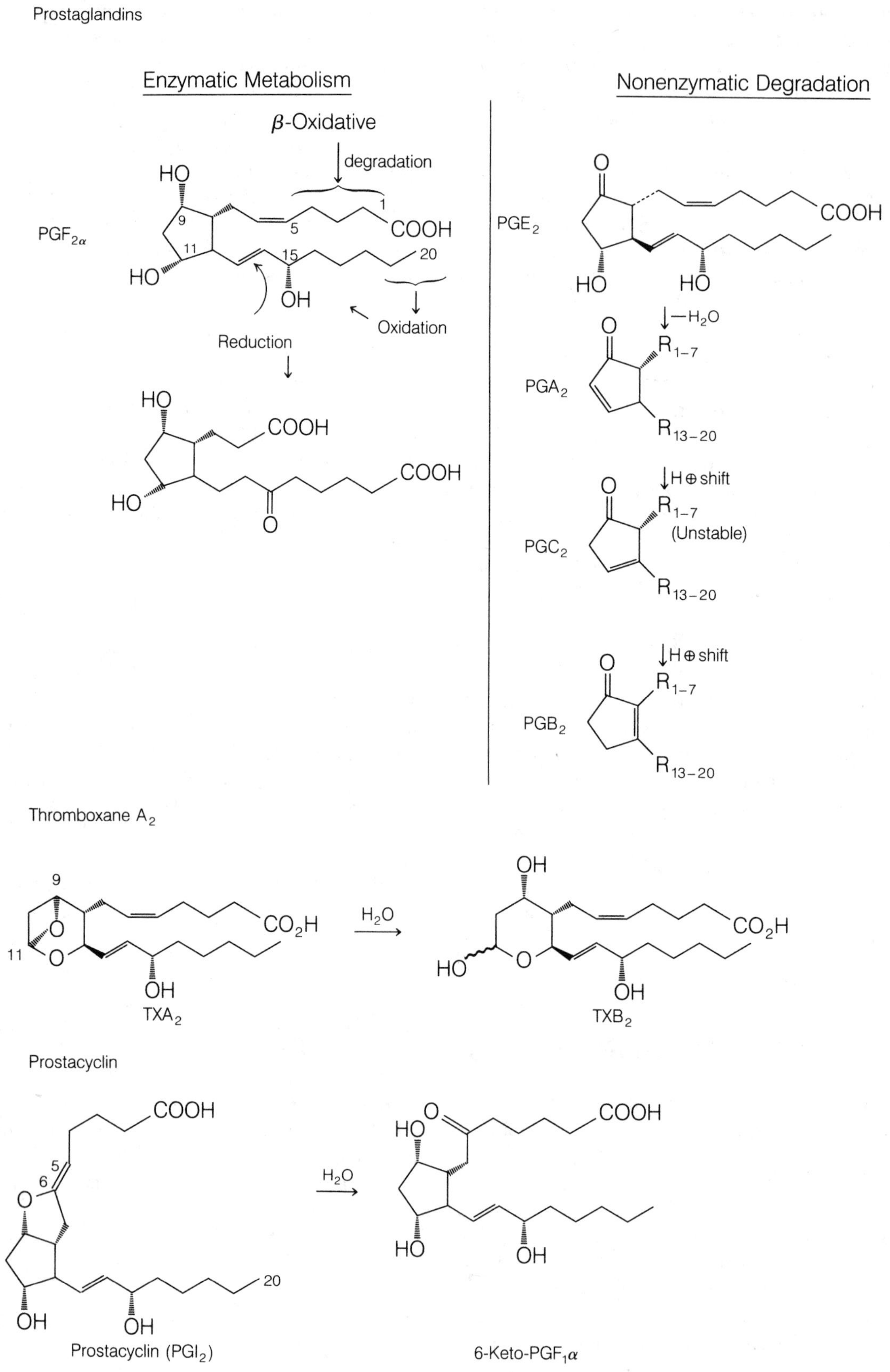

Prostaglandins

Enzymatic Metabolism

β-Oxidative degradation

$PGF_{2\alpha}$

HO HO OH COOH 1 5 9 11 15 20

Reduction

Oxidation

HO HO COOH COOH O

Nonenzymatic Degradation

PGE_2

O COOH HO HO

$-H_2O$

PGA_2

O R_{1-7} R_{13-20}

H⊕ shift

PGC_2

O R_{1-7} (Unstable) R_{13-20}

H⊕ shift

PGB_2

O R_{1-7} R_{13-20}

Thromboxane A_2

9 11 O O CO_2H OH

TXA_2

H_2O →

OH HO O CO_2H OH

TXB_2

Prostacyclin

COOH 5 6 O 20 OH OH

Prostacyclin (PGI_2)

H_2O →

O HO COOH HO OH

6-Keto-$PGF_1\alpha$

FIG. 22-3. *Eicosanoid degradation.*

TABLE 22-2

PROSTAGLANDIN ANALOGUES UNDER INVESTIGATION FOR FUTURE CLINICAL USE

Structure	*Names (Proprietary Name; Manufacturer)*	*Therapeutic Use*
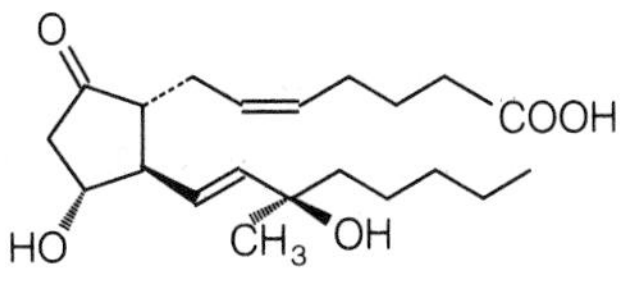	Arbaprostil (Arbocet; Upjohn); 15-(R)-methyl-PGE_2;	Gastric anti-secretory
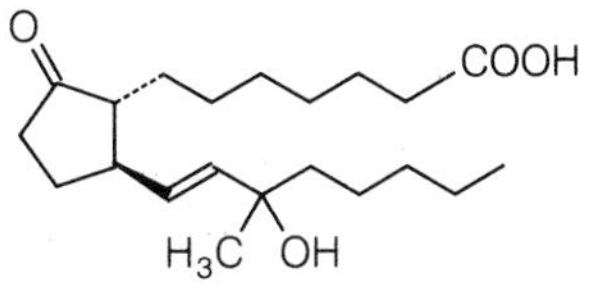	Doxaprost (Ayerst); 15-methyl-11-deoxy-PGE_1	Bronchodilatator
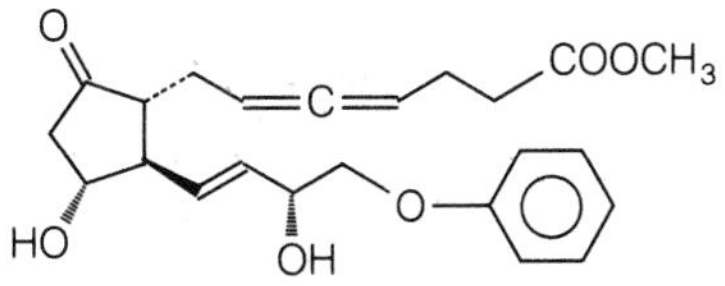	Enprostil (Syntex); PGE_2-analogue	Antisecretory; antiulcer
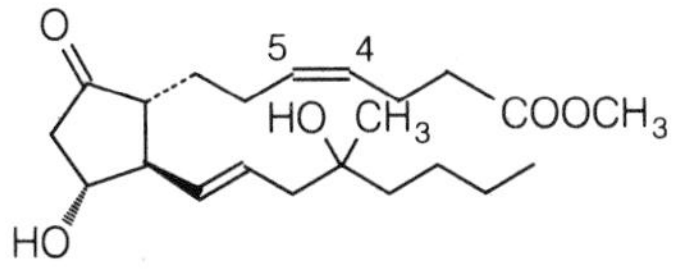	Enisoprost (Searle); 16-(R,S)-methyl-16-hydroxy-$\Delta^{4,5}$-PGE_2 methyl ester	Antiulcerative
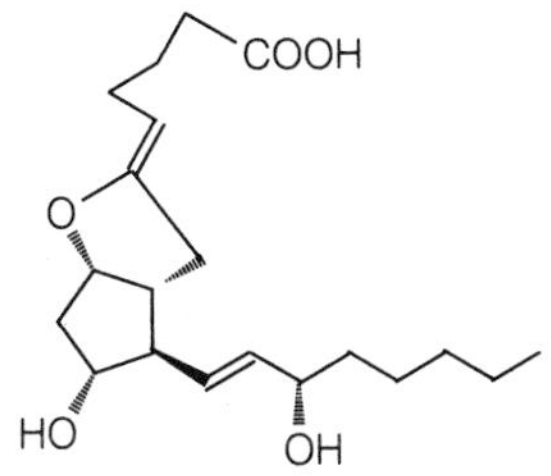	Epoprostenol (Upjohn and Burroughs-Wellcome); prostacyclin; PGI_2 Epoprostenol sodium (sodium salt of above) (Cyclo-Prostin; Upjohn / Flolan; Burroughs-Wellcome)	Platelet inhibitor; anticoagulant
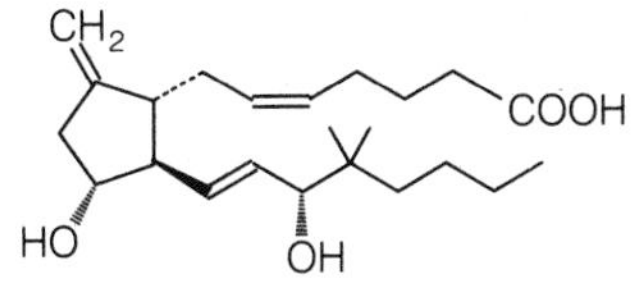	Meteneprost (Upjohn); 9-deoxo-9-methylene-16,16-dimethyl-PGE_2	Oxytocic
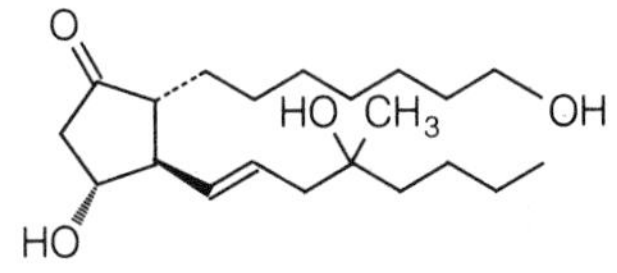	Rioprostil (OrthoMiles / Bayer Laboratories); 16-(R, S)-methyl-1,16-dihydroxy-PGE_1	Gastric anti-secretory
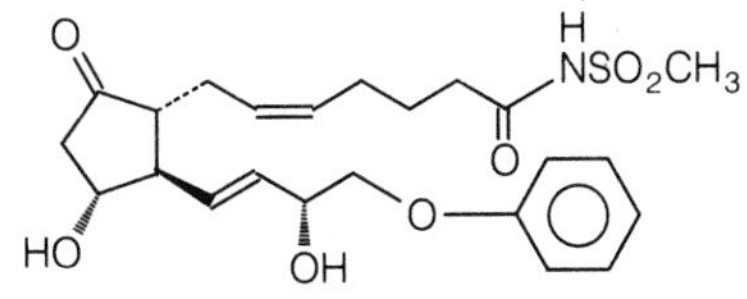	Sulprostone (Sulglandin, Pfizer/Schering A. G.); PGE_2 analogue	Oxytocic
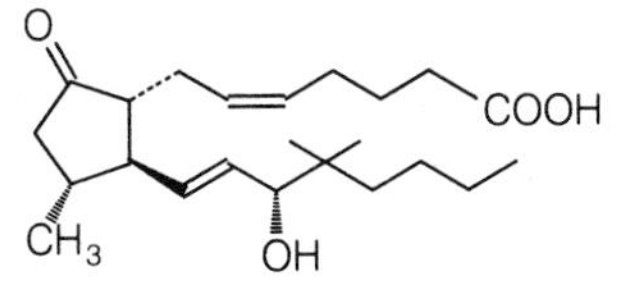	Trimoprostil (Hoffmann-LaRoche); 11-methyl-11-deoxy-16,16-dimethyl-PGE_2	Gastric antisecretory

administered in a single dose of 20 mg by vaginal suppository to induce labor or abortion.

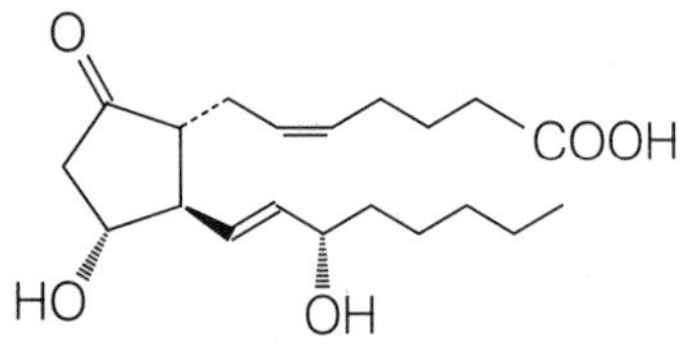

15-(S)-Methyl-PGF$_{2\alpha}$. Carboprost tromethamine (Prostin/15M) is a prostaglandin derivative which has been modified to prevent metabolic oxidation of the 15-position alcohol function.

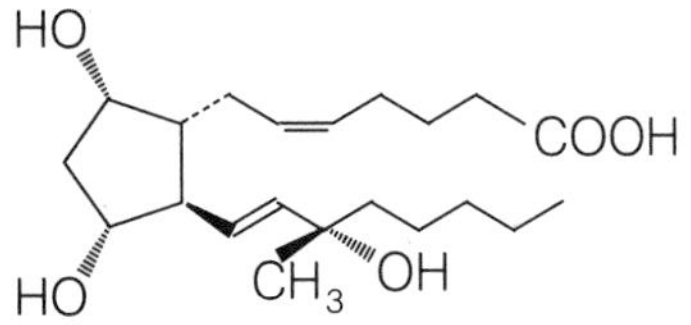

This derivative is administered in a dose of 250 μg by deep intramuscular injection to induce abortion or to ameliorate severe postpartum hemorrhage.

Prostaglandin E$_1$ USP. Alprostadil (Prostin VR Pediatric). PGE$_1$ is a naturally occurring prostaglandin that has found particular use in maintaining a patent (opened) ductus arteriosus in infants with congenital defects that restrict pulmonary or systemic blood flow.

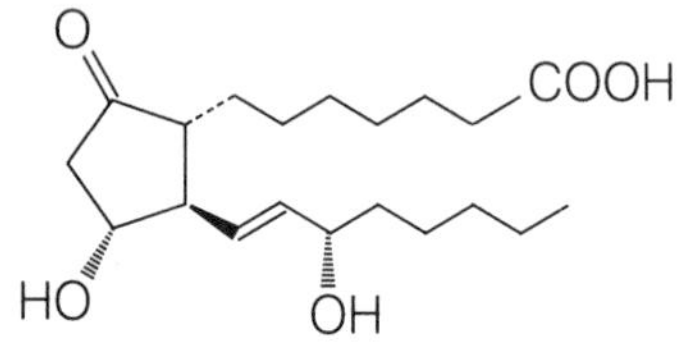

Alprostadil must be administered intravenously continually at a rate of approximately 0.1 μg/kg per minute to temporarily maintain the patency of the ductus arteriosus until corrective surgery can be performed. Up to 80% of circulating alprostadil may be metabolized in a single pass through the lungs. Because apnea has been observed in 10% to 12% of neonates with congenital heart defects, this product should be administered *only* when ventilatory assistance is immediately available. Other commonly observed side effects include decreased arterial blood pressure, which should be monitored during infusion; inhibited platelet aggregation, which might aggravate bleeding tendencies; and diarrhea.

Alprostadil is provided as a sterile solution in absolute alcohol (0.5 mg/mL) that must be diluted in saline or dextrose solution before intravenous administration.

16-(R,S)-Methyl-16-Hydroxy-PGE$_1$, Methyl Ester. Misoprostol (Cytotec) is a modified prostaglandin analog which shows potent gastric antisecretory and gastroprotective effects when administered orally.

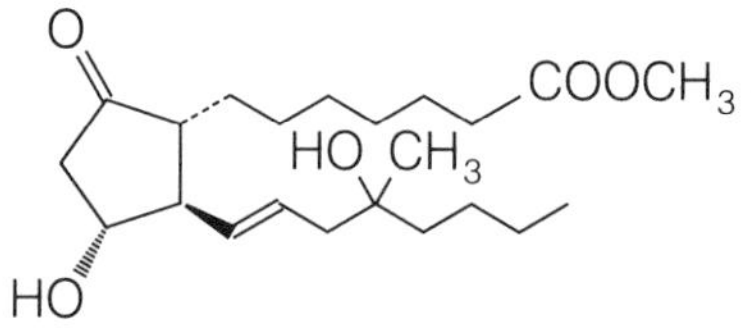

Misoprostol is administered orally in tablet form in a dose of 100–200 μg four times a day to prevent gastric ulceration in susceptible individuals who are taking nonsteroidal antiinflammatory drugs. This prostaglandin derivative should be absolutely avoided in pregnant women owing to its potential to induce abortion.

VETERINARY USES OF PROSTANOIDS

Since McCracken demonstrated that PGF$_{2\alpha}$ acts as a hormone in sheep to induce disintegration of the

TABLE 22-3

EICOSANOID PRODUCTS FOR VETERINARY USE

Proprietary Name (Manufacturer)	*Chemical Name*	*Therapeutic Use*
Equimate (ICI Corp.)	Fluprostenol; 16-[*m*-(CF$_3$)-phenoxy]-ω-tetranor-PGF$_{2\alpha}$	Induce estrus; treat equine infertility
Estrumate (ICI Corp.)	Cloprostenol; 16-*m*-chlorophenoxy-ω-tetranor-PGF$_{2\alpha}$	Synchronize estrus in cattle
Fenprostalene(Syntex)	(+)-4,5-Didehydro-16-phenoxy-ω-tetranor-PGF$_{2\alpha}$ methyl ester	Induce abortion in cattle; synchronize estrus in cows
Iliren (Hoechst-Roussel)	Tiaprost; (15*R*, *S*)-16-(3-thienyloxy)-ω-tetranor-PGF$_{2\alpha}$tromethamine salt	Induce labor; treats pyometra and persistent luteal function in cattle, sheep, pigs, and horses
Lutalyse (Upjohn)	Tromethamine salt of PGF$_{2\alpha}$	Induce and synchronize estrus in mares, cows, and sows
Synchrocept (Syntex)	Prostalene; (+)-4,5-Didehydro-15-methyl-PGF$_{2\alpha}$-methyl ester	Synchronize estrus in mares 15-methyl-PGF$_{2\alpha}$

corpus luteum (luteolysis),[19] salts of this prostaglandin and a variety of analogues have been marketed to induce or synchronize estrus in breed animals (Table 22-3). This procedure, then, allows artificial insemination of many animals during one insemination period.

EICOSANOIDS IN CLINICAL DEVELOPMENT FOR HUMAN TREATMENT

Numerous prostaglandin analogues are under investigation for the treatment of human diseases (see Table 2). Recent efforts are focused in the areas of gastroprotection as antiulcer therapy, fertility control, the development of thrombolytics (e.g., prostacyclin or thromboxane synthetase inhibitors) to treat cerebrovascular or coronary artery diseases, and the development of antiasthmatics through modulation of the lipoxygenase pathway. However, future application of eicosanoids to the treatment of hypertension or immune system disorders cannot be ruled out. Thus, although progress in this area has been slow, the further use of eicosanoids or eicosanoid analogues as therapeutic agents in the future is almost assured.

REFERENCES

1. Kurzrok, R., and Lieb, C.: Proc. Soc. Exp. Biol. N.Y. 28:268, 1931.
2. von Euler, U.S.: J. Physiol. (Lond.) 88:213, 1937.
3. Bergstrom, S., Ryhage, R., Sammuelsson, B., and Sjovall, J.: Acta Chem. Scand. 16:501, 1962.
4. Bindra, J. S., and Bindra, R.: Prostaglandin Synthesis. Academic Press, New York, 1977.
5. Hong, S. L., and Levine, L.: Proc. Natl. Acad. Sci. USA 73:1730, 1976.
6. Kuehl, F. A., Humes, J. L., Egan, R. W., et al.: Nature 265:170, 1977.
7. Miyamoto, T., Ogino, N., Yamamoto, S., and Hayaishi, O.: J. Biol. Chem. 251:2629, 1976.
8. Vane, J. R.: Nature 231:232, 1971.
9. Phillipson, B. E., Rothrock, D. W., Connor, W. E., et al.: N. Engl. J. Med. 312:1210, 1985.
10. Siegl, A. M.: In Lands, W. E. M., and Smith, W. L. (eds.). Methods Enzymol 86:179, 1982.
11. Chand, N., and Altura, B. M.: Prostaglandins Med. pp 249–256, 1981.
12. Marnett, L. J., and Eling, T. E.: Rev. Biochem. Toxicol. 5:135, 1984.
13. Robertson, I. G. C., Sivarajah, K., Eling, T. E., and Zeiger, E.: Cancer Res. 43:476, 1983.
14. Lands, W. E. M., Jr.: Trends Pharmacol. Sci. 1:78, 1981.
15. Rome, L. H., and Lands, W. E. M.: Proc. Natl. Acad. Sci. USA 72:4863, 1975.
16. Higgs, G. A., Salman, J. A., Henderson, B., and Vane, J. R.: Proc. Natl. Acad. Sci. USA 84:1417, 1987.
17. Hanel, A. M., and Lands, W. E. M.: Biochem. Pharmacol. 31:3307, 1982.
18. Kuehl, F. A., Ham, E. A., Humes, J. L., et al.: In Ramwell, P. (ed.). Prostaglandin Synthetase Inhibitors: New Clinical Applications, pp. 73–86. New York, Alan R. Liss, 1980.
19. McCracken, J. A., Carlson, J. C., Glen, M. E., et al.: Nature 238:129, 1972.

SELECTED READINGS

Bauer, R. F., Collins, P. W., and Jones, P. H.: Agents for the treatment of peptic ulcer disease. Bailey, D. M. (ed.). Annu. Rep. Med. Chem. 22:191, 1987.

Braquet, P., Garay, R. P., Frolich, J. C., and Nicosia, S. (eds.): Prostaglandins and Membrane Ion Transport (Advances in Ion Transport Regulation). New York, Raven Press, 1985.

Chakrin, L. W., and Bailey, D. M. (eds.): The Leukotrienes—Chemistry and Biology. New York, Academic Press, 1984.

Cohen, M. M. (ed.): Biological Protection with Prostaglandins, Vols. 1 and 2. Boca Raton, FL, CRC Press, 1985, 1986.

Cross, P. E., and Dickinson, R. P.: Thromboxane synthetase inhibitors and antagonists. Bailey, D. M. (ed.). Annu. Rep. Med. Chem. 22:95, 1987.

Edqvist, L. E., and Kindahl, H. (eds.): Prostaglandins in Animal Reproduction. New York, Elsevier Science Publishers, 1984.

Flower, R. J.: Eicosanoids: the Nobel Prize. Trends Pharmacol. Sci. Jan. 1983.

Greenberg, S., Kadowitz, P. J., and Burks, T. F., (eds.): Prostaglandins: Organ and Tissue-Specific Actions. New York, Marcel Dekker, 1982.

Gryglewski, R. J., Szczeklik, A., and McGiff, J. C. (eds.): Prostacyclin Clinical Trials. New York, Raven Press, 1985.

Lands, W. E. M., and Smith, W. L. (eds.): Prostaglandins and arachidonate metabolites. Methods Enzymol. Vol. 86, 1982.

Pace-Asciak, C., and Granstrom, E.: Prostaglandins and Related Substances. New York, Elsevier Science Publishers, 1983.

Piwinski, J. J., Kreutner, W., and Green, M. J.: Pulmonary and antiallergy agents. Bailey, D. M. (ed.). Ann. Rep. Med. Chem. 22:73, 1987.

Rainsford, K. D.: Anti-Inflammatory and Anti-Rheumatic Drugs, Vols. 1–3 Boca Raton, FL, CRC Press, 1985.

Roberts, S. M., and Newton, R. F. (eds.): Prostaglandins and Thromboxanes, Boston, Butterworth Scientific, 1982.

Roberts, S. M., and Scheinmann, F. (eds.): New Synthetic Routes to Prostaglandins and Thromboxanes. New York, Academic Press, 1982.

Roberts, S. M., and Scheinman, R. (eds.): Chemistry, Biochemistry, and Pharmacological Activity of Prostanoids. New York, Pergamon Press, 1978.

Robinson, H. J., and Vane, J. R. (eds.): Prostaglandin Synthetase Inhibitors. New York, Raven Press, 1974.

Scheinmann, F., and Ackroyd, J.: Leukotriene Synthesis: A New Class of Biologically Active Compounds Including SRS-A. New York, Raven Press, 1984.

Shiokawa, Y., Katori, M., and Mizushima, Y. (eds.): Perspectives in Prostaglandin Research. Excerpta Medica, Princeton, NJ, 1983.

Thaler-Dao, H., dePaulet, A. C., and Paoletti, R.: Icosanoids and Cancer. New York, Raven Press, 1984.

APPENDIX

pK_as of Drugs and Reference Compounds

Name	pK_a*	Reference
Acenocoumarol	4.7	1
Acetaminophen	9.9 (phenol)	2
Acetanilid	0.5	3
Acetarsone	3.7 (acid)	4
	7.9 (phenol)	
	9.3 (acid)	
Acetazolamide	8.8 (acetamido)	5
Acetic Acid	4.8	6
α-Acetylmethodol	8.3	7
Acetylpromazine	9.3	8
N^4-Acetylsulfadiazine	6.1	9
N^4-Acetylsulfametiazole	5.2	9
N^4-Acetylsulfamethoxypyridazine	6.9	9
N^4-Acetylsulfapyridine	8.2	9
N^4-Acetylsulfisoxazole	4.4	9
Allobarbital	7.5	10
Allopurinol	9.4	11
Allylamine	10.7	12
Allylbarbituric Acid	7.6	13
Alphaprodine	8.7	14
Alprenolol	9.6	15
Amantadine	10.8	16
Amiloride	8.7	15
p-Aminobenzoic Acid	2.4 (amine)	17
	4.9	
Aminocaproic Acid	4.4	18
	10.8	
6-Aminopenicillanic Acid	2.3 (carboxyl)	19
	4.9 (amine)	
Aminopterin	5.5 (heterocyclic ring)	20
Aminopyrine	5.0	6
Aminosalicylic Acid	1.7 (amine)	17
	3.9	
Amitriptyline	9.4	21
Ammonia	9.3	12
Amobarbital	8.0	22
Amoxicillin	2.4 (carboxyl)	23
	7.4 (amine)	
	9.6 (phenol)	
Amphetamine	9.8	24
Amphotericin B	5.7 (carboxyl)	25
	10.0 (amine)	
Ampicillin	2.7	26
	7.3 (amine)	
Aniline	4.6	6
Anisindione	4.1	27
Antazoline	10.0	28
Antifebrin	1.4	29
Antipyrine	2.2	29
Apomorphine	7.0	30
Aprobarbital	7.8	10
Arecoline	7.6	31
Arsthinol	9.5 (phenol)	4
Ascorbic Acid	4.2	18
	11.6	
Aspirin	3.5	7
Atropine	9.7	30
Barbital	7.8	32
Barbituric Acid	4.0	32
Bemegride	11.2	33
Bendroflumethiazide	8.5	34
Benzilic Acid	3.0	12
Benzocaine	2.8	30
Benzoic Acid	4.2	7
Benzphetamine	6.6	35
Benzquinamide	5.9	36
Benzylamine	9.3	12
Biscoumacetic Acid	3.1	1
	7.8 (enol)	174
Bromodiphenhydramine	8.6	37
p-Bromophenol	9.2	38
Bromothen	8.6	28
8-Bromotheophylline	5.5	13
Brucine	8.0	6
Bupivacaine	8.1	39
Butabarbital	7.9	15
Butethal	8.1	40
Butylparaben	8.4	41
Butyric Acid	4.8	12
Caffeine	14.0	42
	0.6 (amine)	
Camphoric Acid	4.7	12
Carbinoxamine	8.1	43
Carbonic Acid	6.4 (1st)	6
	10.4 (2nd)	
Cefazoline	2.3	44
Cephalexin	3.6	45
Cephaloglycin	2.5	45

*pK_a given for protonated amine.

Continued

Name	pK_a*	Reference	Name	pK_a*	Reference
Cephaloridine	3.4	46	3,5-Diiodo-L-tyrosine	2.5 (amine)	32
Cephalothin	2.4	45		6.5	
Cephradine	2.6 (carboxyl)	47		7.5 (phenol)	
	7.3 (amine)		Dimethylamine	10.7	12
Chlorambucil	5.8	48	*p*-Dimethylaminobenzoic Acid	5.1	73
Chlorcyclizine	7.8	37	*p*-Dimethylaminosalicylic Acid	3.8	73
Chlordiazepoxide	4.8	49	Dimethylbarbituric Acid	7.1	12
Chlorindione	3.6	27	Dimethylhydantoin	8.1	3
Chloroquine	8.1	50	2,4-Dinitrophenol	4.1	74
	9.9		Diperodon	8.4	75
8-Chlorotheophylline	5.3	51	Diphenhydramine	9.0	37
Chlorothiazide	6.7	52	Diphenoxylate	7.1	76
	9.5		Doxorubicin	8.2	77
Chlorpheniramine	9.2	37		10.2	
Chlorphentermine	9.6	35	Doxycycline	3.4	78
Chlorpromazine	9.2	53		7.7	
Chlorpromazine Sulfoxide	9.0	44		9.3	
Chlorpropamide	5.0	54	Doxylamine	9.2	37
Chlorprothixene	8.4	55	Droperidol	7.6	79
Chlortetracycline	3.3	56	Ephedrine	9.6	7
	7.4		Epinephrine	8.7 (phenol)	80
	9.3			9.9 (amine)	
Cinchonidine	4.2 (1st)	3	Equilenin	9.8	61
	8.4 (2nd)		Ergotamine	6.3	81
Cinchonine	4.0 (1st)	30	Erythromycin	8.8	82
	8.2 (2nd)		Erythromycin Estolate	6.9	83
Cinnamic Acid	4.5	12	17 α-Estradiol	10.7	61
Citric Acid	3.1 (1st)	57	Estriol	10.4	61
	4.8 (2nd)		Ethacrynic Acid	3.5	18
	6.4 (3rd)		Ethambutol	6.6	84
Clindamycin	7.5	58		9.5	
Clofibrate	3.0 (acid)	15	Ethanolamine	9.5	72
Clonazepam	10.5 (1-position)	59	Ethopropazine	9.6	28
	1.5 (4-position)		Ethosuximide	9.3	15
Clonidine	8.0	60	*p*-Ethoxybenzoic Acid	4.5	73
Cloxacillin	2.7	19	*p*-Ethoxysalicylic Acid	3.2	73
Cobefrin	8.5	24	Ethylamine	10.7	12
Cocaine	8.4	3	Ethylbarbituric Acid	4.4	12
Codeine	7.9	42	Ethyl Biscoumacetate	3.1	1
Colchicine	1.7	18	Ethylenediamine	6.8 (1st)	72
o-Cresol	10.3	17		9.9 (2nd)	
m-Cresol	10.1	17	Ethylparaben	8.4	41
p-Cresol	10.3	61	Ethylphenylhydantoin	8.5	85
Cyanic Acid	3.8	12	Etidocaine	7.7	39
Cyanopromazine	9.3	62	β-Eucaine	9.4	3
Cyclizine	8.2	63	Fenfluramine	9.1	35
Cyclopentamine	3.5	64	Fenoprofen	4.5	86
Cyclopentolate	7.9	65	Flucytosine	10.7 (amide)	87
Dantrolene	7.5	66		2.9 (amine)	
Debrisoquin	11.9	67	Flufenamic Acid	3.9	74
Dehydrocholic Acid	5.0	68	Flunitrazepam	1.8	59
Demeclocycline	3.3	56	*p*-Fluorobenzoic Acid	4.2	38
	7.2		Fluorouracil	8.0	88
	9.3			13.0	
Desipramine	10.2	21	Fluphenazine	8.1 (1st)	53
Dextromethorphan	8.3	43		9.9 (2nd)	
Diatrizoic Acid	3.4	69	Fluphenazine Enanthate	3.5	89
Diazepam	3.3	49		8.2	
Dibucaine	8.5	70	Flurazepam	8.2	90
Dichloroacetic Acid	1.3	42		1.9	
Dicloxacillin	2.8	26	Formic Acid	3.7	12
Dicumarol	4.4 (1st)	71	Fumaric Acid	3.0 (1st)	42
	8.0 (2nd)			4.4 (2nd)	
Diethanolamine	8.9	72	Furaltadone	5.0	91
Diethylamine	11.0	12	Furosemide	4.7	92
p-Diethylaminobenzoic Acid	6.2	73	Gallic Acid	4.2	38
p-Diethylaminosalicylic Acid	3.8	73	Glibenclamide	6.5	54
Dihydrocodeine	8.8	3	Gluconic Acid	3.6	18

*pK_a given for protonated amine.

Continued

Name	pK_a*	Reference
Glucuronic Acid	3.2	93
Glutamic Acid	4.3	12
Glutarimide	11.4	16
Glutethimide	11.8	94
Glycerophosphoric Acid	1.5 (1st)	42
	6.2 (2nd)	
Glycine	2.4	42
	9.8 (amine)	
Glycollic Acid	3.8	12
Guanethidine	11.9	67
Guanidine	13.6	3
Haloperidol	8.3	95
Heroin	7.8	14
Hexachlorophene	5.7	74
Hexetidine	8.3	96
Hexobarbital	8.3	85
Hexylcaine	9.1	70
Hippuric Acid	3.6	97
Histamine	9.9 (side chain)	98
	6.0 (imidazole)	
Homatropine	9.7	3
Hydantoin	9.1	17
Hydralazine	0.5 (ring N)	99
	6.9 (hydrazine)	
Hydrochlorothiazide	7.0	18
	9.2	
Hydrocortisone Hemisuccinate Acid	5.1	100
Hydroflumethiazide	8.9	101
	10.5	
Hydrogen Peroxide	11.3	12
Hydromorphine	7.8	3
Hydroxyamphetamine	9.6	24
p-Hydroxybenzoic Acid	4.1	38
o-Hydroxycinnamic Acid	4.7	38
m-Hydroxycinnamic Acid	4.5	38
p-Hydroxycinnamic Acid	4.4	38
Hydroxylamine	6.0	6
p-Hydroxysalicylic Acid	3.2	73
Hydroxyzine	1.8	102
Ibuprofen	5.2	103
Idoxuridine	8.3	104
Imidazole	7.0	72
Imipramine	9.5	21
Indomethacin	4.5	18
Indoprofen	5.8	105
Iodipamide	3.5	106
Iophenoxic Acid	7.5	1
Isocarboxazid	10.4	107
Isomethadone	8.1	14
Isoniazid	10.8 (pyridine)	108
	11.2 (hydazide)	
Isophthalic Acid	3.6	12
Isoproterenol	8.7 (amine)	109
	9.9 (phenol)	
Kanamycin	7.2	15
Ketamine	7.5	110
Lactic Acid	3.9	18
Leucovorin	3.1	18
	4.8	
	10.4 (phenol)	
Levallorphan Tartrate	6.9	111
Levarterenol	8.7 (phenol)	112
	9.7 (amine)	
Levodopa	2.3 (carboxyl)	113
	8.7 (amine)	
	9.7 (1st phenol)	113
	13.4 (2nd phenol)	
Levomepromazine	9.2	64
Levorphanol	8.9	14
Levulinic Acid	4.6	12
Lidocaine	7.9	114
Lincomycin	7.5	115
Liothyronine	8.4 (phenol)	116
Lorazepam	11.5	117
	1.3	
Malamic Acid	3.6	12
Maleic Acid	1.9	12
Malic Acid	3.5 (1st)	57
	5.1 (2nd)	
Malonic Acid	2.8	12
Mandelic Acid	3.8	97
Mecamylamine	11.2	7
Meclizine	3.1	118
	6.2	
Medazepam	6.2	119
Mefenamic Acid	4.3	74
Mepazine	9.3	53
Meperidine	8.7	14
Mephentermine	10.3	35
Mephenytoin	8.1	120
Mephobarbital	7.7	85
Mepivacaine	7.6	39
Mercaptopurine	7.8	121
Metaproterenol	8.8	15
Methacycline	3.5	15
	7.6	
	9.2	
Methadone	8.3	14
Methamphetamine	9.5	43
Methapyrilene	3.7	3
	8.9 (side chain)	
Methaqualone	2.5	122
Metharbital	8.2	85
Methazolamide	7.3	18
Methenamine	4.9	29
Methicillin	2.8	19
Methopromazine	9.4	62
Methotrexate	4.8	124
	5.5	
Methohexital	8.3	123
Methoxamine	9.2	64
Methoxyacetic Acid	3.5	12
o-Methoxybenzoic Acid	4.2	38
m-Methoxybenzoic Acid	4.2	38
o-Methoxycinnamic Acid	4.7	38
m-Methoxycinnamic Acid	4.5	38
p-Methoxycinnamic Acid	4.9	38
Methyclothiazide	9.4	125
Methylamine	10.6	12
1-Methylbarbituric Acid	4.4	17
Methyldopa	2.2	126
	10.6 (amine)	
	9.2 (1st phenol)	
	12.0 (2nd phenol)	
N-Methylephedrine	9.3	127
Methylergonovine	6.7	81
N-Methylglucamine	9.2	126
Methylhexylamine	10.5	64
Methylparaben	8.4	41
Methylphenidate	8.8	128
Methylprednisolone-21-phosphate	2.6	129
	6.0	
Methylpromazine	9.4	62
Methyprylon	12.0	130

*pK_a given for protonated amine.

Continued

Name	pK_a*	Reference
Methysergide	6.6	81
Metopon	8.1	14
Metoprolol	9.7	15
Metronidazole	2.6	131
Miconazole	6.9	132
Minocycline	2.8	15
	5.0	
	7.8	
	9.5	
Molindone	6.9	133
Monochloroacetic Acid	2.9	42
Morphine	8.0	32
	9.6 (phenol)	
Nafcillin	2.7	26
Nalidixic Acid	6.0 (amine)	134
	1.0	
Nalorphine	7.8	14
Naphazoline	3.9	64
1-Naphthol	9.2	38
2-Naphthol	9.4	38
Naproxen	4.2	135
Narcotine	5.9	3
Nicotine	3.1	136
	8.0	
Nicotine Methiodide	3.2	136
Nicotinic Acid	4.8	18
Nitrazepam	10.8	117
	3.2	
o-Nitrobenzoic Acid	3.2	38
m-Nitrobenzoic Acid	3.6	38
p-Nitrobenzoic Acid	3.7	38
Nitrofurantoin	7.2	91
Nitrofurazone	10.0	137
Nitromethane	11.0	12
o-Nitrophenol	7.2	38
m-Nitrophenol	8.3	38
p-Nitrophenol	7.1	38
8-Nitrotheophylline	2.1	51
Norhexobarbital	7.9	123
Norketamine	6.7	110
Norparamethadione	6.1	85
Nortrimethadione	6.2	85
Noscapine	6.2	18
Novobiocin	4.3	18
	9.1	
Ornidazole	2.6	131
Orphenadrine	8.4	15
Oxamic Acid	2.1	12
Oxazepam	1.8	138
	11.1	
Oxyphenbutazone	4.5	93
	10.0 (phenol)	
Oxytetracycline	3.3	56
	7.3	
	9.1	
Pamaquine	8.7	3
Papaverine	5.9	42
Penicillamine	1.8 (carboxyl)	18
	7.9 (amino)	
	10.5 (thiol)	
Penicillin G	2.8	42
Penicillin V	2.7	19
Penicilloic Acid	5.2	139
Pentachlorophenol	4.8	74
Pentobarbital	8.0	32
Perphenazine	7.8	21
Phenacetin	2.2	29
Phenadoxane	6.9	14
Phendimetrazine	7.6	35
Phenethicillin	2.7	19
Phenformin	11.8	140
Phenindamine	8.3	37
Phenindione	4.1	27
Pheniramine	9.3	37
Phenmetrazine	8.5	35
Phenobarbital	7.5	13
Phenol	9.9	7
Phenolsulfonphthalein	7.9	18
Phenoxyacetic Acid	3.1	12
Phentermine	10.1	35
Phenylbutazone	4.4	141
Phenylbutazone (isopropyl analog)	5.5	142
Phenylephrine	9.8 (amine)	143
	8.8 (phenol)	
Phenylethylamine	9.8	24
Phenylpropanolamine	9.4	109
Phenylpropylmethylamine	9.9	24
Phenyramidol	5.9	18
Phenytoin	8.3	144
o-Phthalamic Acid	3.8	12
Phthalic Acid	2.9	6
Phthalimide	7.4	12
Physostigmine	2.0	3
	8.1	
Picolinic Acid	5.3	38
Picric Acid	0.4	42
Pilocarpine	1.6	3
	7.1	
Piperazine	5.7	30
	10.0	
Piperidine	11.2	6
Pirbuterol	3.0 (pyridine)	145
	7.0 (pyridol)	
	10.3 (amine)	
Plasmoquin	3.5	146
	10.1	
Polymyxin B	8.9	15
Prazepam	3.0	147
Prilocaine	7.9	39
Probarbital	8.0	22
Probenecid	3.4	1
Procainamide	9.2	148
Procaine	9.0	149
Procarbazine	6.8	150
Prochlorperazine	3.6	64
	7.5	
Promazine	9.4	53
Promethazine	9.1	64
Propicillin	2.7	19
Propiomazine	6.6	151
Propionic Acid	4.9	42
Propranolol	9.5	15
i-Propylamine	10.6	12
n-Propylamine	10.6	12
Propylhexedrine	10.5	64
Propylparaben	8.4	41
Propylthiouracil	7.8	152
Pseudoephedrine	9.9	35
Pyrathiazine	8.9	64
Pyrazinamide	0.5	18
Pyridine	5.2	6
Pyridoxine	2.7	153
	5.0 (amine)	153
	9.0 (phenol)	154
Pyrilamine	4.0	3
	8.9	
Pyrimethamine	7.2	1

*pK_a given for protonated amine.

Continued

Name	pK_a*	Reference	Name	pK_a*	Reference
Pyrimethazine	9.4	53	Tetracaine	8.5	70
Pyrrobutamine	8.8	37	Tetracycline	3.3	56
Pyruvic Acid	2.5	18		7.7	
Quinacrine	8.0	3		9.5	
	10.2		Thenyldiamine	3.9	3
Quinidine	4.2	3		8.9	
	8.3		Theobromine	8.8	32
Quinine	4.2	3		0.7 (amine)	
	8.8		Theophylline	8.8	32
Reserpine	6.6	42		0.7(amine)	
Resorcinol	6.2	12	Thiamine	4.8	153
Riboflavin	1.7	3		9.0	
	10.2		Thiamylal	7.3	123
Rifampin	1.7 (C-8 phenol)	155	Thioacetic Acid	3.3	12
	7.9 (piperazine N)		Thioglycolic Acid	3.6	12
Saccharic Acid	3.0	12	Thiopental	7.5	123
Saccharin	1.6	42	Thiopropazate	3.2	64
Salicylamide	8.1	32		7.2	
Salicylic Acid	3.0	6	Thioridazine	9.5	53
	13.4 (phenol)		Thiouracil	7.5	152
Scopolamine	7.6	3	Thonzylamine	8.8	37
Secobarbital	8.0	156	L-Thyronine	9.6 (phenol)	116
Serotonin	4.9	18	L-Thyroxine	2.2 (carboxyl)	162
	9.8			6.7 (phenol)	
Sorbic Acid	4.8	12		10.1 (amine)	
Sotalol	9.8 (amine)	157	Tolazamide	5.7	15
	8.3 (sulfonamide)		Tolazoline	10.3	163
Spectinomycin	7.0	18	Tolbutamide	5.3	32
	8.7		*p*-Toluidine	5.3	7
Strychnine	2.5	3	Trichloroacetic Acid	0.9	42
	8.2		Triethanolamine	7.8	72
Succinic Acid	4.2 (1st)	57	Triethylamine	10.7	29
	5.6 (2nd)		Trifluoperazine	4.1	53
Succinimide	9.6	158		8.4	
Succinuric Acid	4.5	12	Triflupromazine	9.4	64
Sulfacetamide	5.4	159	Trimethobenzamide	8.3	164
	1.8		Trimethoprim	7.2	165
Sulfadiazine	6.5	32	Trimethylamine	9.8	12
Sulfadimethoxine	6.7	159	Tripelennamine	9.0	37
	2.0 (amine)		Triprolidine	6.5	166
Sulfadimethoxytriazine	5.0	9	Troleandomycin	6.6	18
Sulfaethidole	5.4	32	Tromethamine	8.1	72
Sulfaguanidine	2.8	32	Tropacocaine	9.7	30
Sulfamerazine	7.1	42	Tropic Acid	4.1	12
Sulfameter	6.8	18	Tropicamide	5.2	167
Sulfamethazine	7.4	32	Tropine	10.4	12
Sulfamethizole	5.4	32	Tubocurarine Chloride	7.4	168
Sulfamethoxypyridazine	7.2	9	Tyramine	9.5 (phenol)	109
Sulfanilamide	10.4	160		10.8 (amine)	
Sulfanilic Acid	3.2	6	Urea	0.2	169
Sulfaphenazole	6.5	159	Uric Acid	5.4	170
	1.9 (amine)			10.3	
Sulfapyridine	8.4	42	Valeric Acid	4.8	42
Sulfasalazine	0.6 (amine)	161	Vanillic Acid	4.5	12
	2.4 (carboxyl)		Vanillin	7.4	17
	9.7 (sulfonamide)		Vinbarbital	8.0	22
	11.8 (phenol)		Vinblastine	5.4	171
Sulfathiazole	7.1	42		7.4	
Sulfinpyrazone	2.8	93	Viomycin	8.2	18
Sulfisomidine	7.5	159		10.3	
	2.4 (amine)			12.0	
Sulfisoxazole	5.0	32	Warfarin	5.1	172
Talbutal	7.8	10	Xipamide	4.8 (phenol)	173
Tartaric Acid	3.0 (1st)	57		10.0 (sulfonamide)	
	4.3 (2nd)				

*pK_a given for protonated amine.

REFERENCES

1. Anton, A. H.: J. Pharmacol. Exp. Ther. 134:291, 1961.
2. Rhodes, H. J., et al.: J. Pharm. Sci. 64, 1387, 1975.
3. Perrin, D. D.: Dissociation Constants of Organic Bases. London, Butterworths, 1965.
4. Hiskey, C. F., and Cantwell, F. F.: J. Pharm. Sci. 57:2105, 1968.
5. Coleman, J. E.: Annu. Rev. Pharmacol. 15:238, 1975.
6. Kolthoff, I. M., and Stenger, V. A.: Volumetric Analysis, Vol. 1, 2nd ed. New York, Interscience Publishers, 1942.
7. Schanker, L. S., et al.: J. Pharmacol. Exp. Ther. 120:528, 1957.
8. Liu, S., and Hurwitz, A.: J. Colloid Interface Sci. 60:410, 1977; through Chem. Abstr. 87:44189u, 1977.
9. Scudi, J. W., and Plekss, O. J.: Proc. Soc. Exp. Biol. Med. 97:639, 1958.
10. Carstensen, J. T., et al.: J. Pharm. Sci. 53:1547, 1964.
11. Gressel, P. D., and Gallelli, S. F.: J. Pharm. Sci. 57, 335, 1968.
12. Washburn, E. W. (ed. in chief), International Critical Tables. New York, McGraw-Hill, 6:261, 1929.
13. Maulding, H. V., and Zoglio, M. A.: J. Pharm. Sci. 60:311, 1971.
14. Beckett, A. H.: J. Pharm. Pharmacol. 8:851, 1956.
15. Heel, R. C., and Avery, G. S.: in Avery, G. S. (ed.). Drug Treatment, 2nd ed., pp. 1212–1222. Sidney, ADIS, 1980.
16. Albert, A.: Selective Toxicity, 4th ed., p. 281. London, Methuen, 1968.
17. Kortüm, G. et al.: Dissociation Constants of Organic Acids. London, Butterworths, 1961.
18. Windholz, M. (ed.): The Merck Index, 9th ed. Rahway, NJ, Merck & Co., 1976.
19. Rapson, H. D. C., and Bird, A. E.: J. Pharm. Pharmacol. 15:226T, 1963.
20. Baker, R. B., and Jordan, J. H.: J. Pharm. Sci. 54:1741, 1965.
21. Green, A. L.: J. Pharm. Pharmacol. 19:10, 1967.
22. Krahl, M. E.: J. Phys. Chem. 44:449, 1940.
23. Rolinson, G. N.: J. Infect. Dis. 129:S143, 1974.
24. Leffler, E. B., et al.: J. Am. Chem. Soc. 73:2611, 1951.
25. Asher, I. M., et al.: Anal. Profiles 6:11, 1977.
26. Hou, J. P., and Poole, J. W.: J. Pharm. Sci. 58:1150, 1969.
27. Stella, V. J., and Gish, R.: J. Pharm. Sci. 68:1047, 1979.
28. Marshall, P. B.: B. J. Pharmacol. 10:270, 1955.
29. Evstratova, K. I., et al.: Farmatsiya 17:33, 1968; through Chem. Abstr. 69:9938a, 1968.
30. Kolthoff, J. M.: Biochem. Z. 162:289, 1925; through Trans. Faraday Soc. 39:338, 1945.
31. Burgen, A. S. V.: J. Pharm. Pharmacol. 16:638, 1964.
32. Ballard, B. B., and Nelson, E.: J. Pharmacol. Exp. Ther. 135:120, 1962.
33. Peinhardt, G.: Pharmazie 32:725, 1977; through Chem. Abstr. 88:141587a, 1978.
34. Agren, A., and Bäck, T.: Acta Pharm. Suec. 10:225, 1973.
35. Vree, T. B., et al.: J. Pharm. Pharmacol. 21:774, 1969.
36. Wiseman, E. H., et al.: Biochem. Pharmacol. 13:1421, 1964.
37. Lordi, N. G., and Christian, J. E.: J. Am. Pharm. Assoc. 45:300, 1956.
38. Conners, K. A., and Lipari, J. M.: J. Pharm. Sci. 65:380, 1976.
39. de Jong, R. H.: JAMA 238:1284, 1977.
40. Fincher, J. H., et al.: J. Pharm. Sci. 55:24, 1966.
41. Tammilehto, S., and Büchi, J.: Pharm. Acta Helv. 43:726, 1968.
42. Martin, A. N., et al.: Physical Pharmacy, 2nd ed., p. 194. Philadelphia, Lea & Febiger, 1969.
43. Borodkin, S., and Yunker, M. H.: J. Pharm. Sci. 59:481, 1970.
44. Nightengale, C. H. et al.: J. Pharm. Sci. 64:1907, 1975.
45. Streng, W. H.: J. Pharm. Sci. 67:667, 1978.
46. Flynn, E. H. (ed.): Cephalosporins and Penicillins, p. 316. New York, Academic Press, 1972.
47. Florey, K.: Anal. Profiles 5:36, 1976.
48. Linford, J. H.: Biochem. Pharmacol. 12:321, 1963.
49. Van der Kleijn, E.: Arch. Int. Pharmacodyn. 179:242, 1969.
50. Hong, P. D.: Anal. Profiles 5:69, 1976.
51. Meyer, M. C., and Guttman, D. E.: J. Pharm. Sci. 57:245, 1968.
52. Resetarits, D. E., and Bates, T. R.: J. Pharm. Sci. 68:126, 1979.
53. Sorby, D. L., et al.: J. Pharm. Sci. 58:788, 1966.
54. Crooks, M. J., and Brown, K. F.: J. Pharm. Pharmacol. 26:305, 1974.
55. Rudy, B. C., and Senkowski, B. Z.: Anal. Profiles 2:75, 1973.
56. Benet, L. Z., and Goyan, J. E.: J. Pharm. Sci. 55:983, 1965.
57. Pitman, I. H., et al.: J. Pharm. Sci. 57:239, 1968.
58. Taraszka, M. J.: J. Pharm. Sci. 60:946, 1971.
59. Kaplan, S. A., et al.: J. Pharm. Sci. 63:527, 1974.
60. Timmermars, P. B. M. W. M., and van Zweiten, P. A.: Arzneim. Forsch. 28:1676, 1978.
61. Hurwitz, A. R., and Liu, S. T.: J. Pharm. Sci. 66:626, 1977.
62. Hulshoff, A., and Perrin, J.: Pharm. Acta. Helv. 51:67, 1976.
63. Barlow, R. B.: Introduction to Chemical Pharmacology, 2nd ed., p. 357. New York, Wiley, 1964.
64. Chatten, L. G., and Harris, L. E.: Anal. Chem. 34:1499, 1962.
65. Wang, E. S. N., and Hammerlund, E. R.: J. Pharm. Sci. 59:1561, 1970.
66. Vallner, J. J., et al.: J. Pharm. Sci. 65:873, 1976.
67. Hengstmann, J. H., et al.: Anal. Chem. 46:35, 1974.
68. Jons, W. H., and Bates, T. R.: J. Pharm. Sci. 59:329, 1970.
69. Langecker, A. A., et al.: Arch. Exp. Pathol. Pharmacol. 222:584, 1954.
70. Truant, A. P., and Takman, B.: Anesth. Analg. 38:478, 1959.
71. Cho, M. J., et al.: J. Pharm. Sci. 60:197, 1971.
72. Bates, R. G.: Ann. N.Y. Acad. Sci. 92:341, 1961.
73. Pothisiri, P., and Carstensen, J. T.: J. Pharm. Sci. 64:1933, 1975.
74. Terrada, H., et al.: J. Med. Chem. 17:330, 1974.
75. Cohen, J. L.: Anal. Profiles 6:108, 1978.
76. Peeters, J. J.: J. Pharm. Sci. 67:129, 1978.
77. Sturgeon, R. J., and Schulman, S. G.: J. Pharm. Sci. 66:959, 1977.
78. Jaffe, J. M., et al.: J. Pharmacokinet. Biopharm. 1:281, 1973.
79. Janicki, C. A., and Gilpin, R. K.: Anal. Profiles 7:181, 1978.
80. Marten, R. B.: J. Phys. Chem. 75:2659, 1971.
81. Maulding, H. V., and Zoglio, M. A.: J. Pharm. Sci. 59:700, 1970.
82. Garrett, E. R., et al.: J. Pharm. Sci. 59:1449, 1970.
83. Mann, J. M.: Anal. Profiles 1:107, 1972.
84. Shepherd, R. G., et al.: Ann. N.Y. Acad. Sci. 135:698, 1966.
85. Butler, T. C.: J. Am. Pharm. Assoc. 44:367, 1955.
86. Ward, C. K., and Schirmer, R. E.: Anal. Profiles 6:165, 1977.
87. Waysek, E. H., and Johnson, J. H.: Anal. Profiles 5:128, 1976.
88. Rudy, B. C., and Senkowski, B. Z.: Anal. Profiles 2:234, 1973.
89. Florey, K.: Anal. Profiles 2:254, 1973.
90. Rudy, B. C., and Senkowski, B. Z.: Anal. Profiles 3:321, 1974.
91. Buzard, J. A., et al.: Am. J. Physiol. 201:492, 1961.
92. McCallister, J. B., et al.: J. Pharm. Sci. 59:1288, 1970.
93. Perel, J. M., et al.: Biochem. Pharmacol. 13:1305, 1964.
94. DeLuca, P. P., et al.: J. Pharm. Sci. 62:1321, 1973.
95. Janssen, P. A. J., et al.: J. Med. Pharm. Chem. 1:282, 1959.
96. Satzinger, G., et al.: Anal. Profiles, 7:284, 1978.
97. Parrott, E. L., and Saski, W.: Exp. Pharm. Technol. p. 255. Minneapolis, Burgess, 1965.
98. Paiva, T. B., et al.: J. Med. Chem. 13:690, 1970.
99. Naik, et al.: J. Pharm. Sci. 65:275, 1976.
100. Garrett, E. R.: J. Pharm. Sci. 51:445, 1962.
101. Smith, R. B., et al.: J. Pharm. Sci. 65:1209, 1976.
102. Florey, K.: Anal. Profiles 7:325, 1978.

103. Upjohn product information 1975.
104. Prusoff, W. H.: Pharmacol. Rev. 19:223, 1967.
105. Fucella, L. M., et al.: Clin. Pharmacol. Ther. 17:227, 1975.
106. Neudent, W., and Röpke, H.: Chem. Ber. 87:666, 1954.
107. Rudy, B. C., and Senkowski, B. Z.: Anal. Profiles 2:307, 1973.
108. Brewer, G. A.: Anal. Profiles 6:198, 1978.
109. Lewis, G. G.: B. J. Pharmacol. 9:488, 1954.
110. Cohen, M. L., and Trevor, A. J.: J. Pharmacol. Exp. Ther. 189:354, 1974.
111. Rudy, B. C., and Senkowski, B. Z.: Anal. Profiles 2:354, 1973.
112. Kappe, T., and Armstrong, M. D.: J. Med. Chem. 8:371, 1965.
113. Gorten, J. E., and Jameson, R. F.: J. Chem. Soc. (A), 2615, 1968.
114. Narahashi, T. I., et al.: J. Pharmacol. Exp. Ther. 171:32, 1970.
115. Hoerksema, H.: J. Am. Chem. Soc. 86:4223, 1964.
116. Smith, R. L.: Med. Chem. 2:477, 1964.
117. Barrett, J., et al.: J. Pharm. Pharmacol. 25:389, 1973.
118. Persson, B. A., and Schill, G.: Acta Pharm. Suec. 3:291, 1966.
119. le Pettit, G.F.: J. Pharm. Sci. 65:1095, 1976.
120. Sandoz Pharmaceuticals brochure, 1962.
121. Fox, J. J., et al.: J. Am. Chem. Soc. 80:1672, 1958.
122. Zalipsky, J. J., et al.: J. Pharm. Sci. 65:461, 1976.
123. Bush, M. T., et al.: Clin. Pharmacol. Ther. 7:375, 1966.
124. Liegler, D. G., et al.: Clin. Pharmacol. Ther. 10:849, 1969.
125. Raihle, J. A.: Anal. Profiles 5:320, 1976.
126. Balasz, L., and Pungor, E.: Mikrochim. Acta 1962:309; through Chem. Abstr. 56:13524g, 1962.
127. Halmekoski, J., and Hannikainen, H.: Acta Pharma. Suec. 3:145, 1966.
128. Siegel, S., et al.: J. Am. Pharm. Assoc. 48:431, 1959.
129. Flynn, G. L., and Lamb, D. J.: J. Pharm. Sci. 59:1436, 1970.
130. Rudy, B. C., and Senkowski, B. Z.: Anal. Profiles, 2:376, 1973.
131. Schwartz, D. E., and Jeunet, F.: Chemotherapy 22:19, 1976.
132. Peetere, J.: J. Pharm. Sci. 67:129, 1978.
133. Dudzinski, J., et al.: J. Pharm. Sci. 62:624, 1973.
134. Storoscik, R., et al.: Acta Pol. Pharm. 28:601, 1971; through Chem. Abstr. 76:158322k, 1972.
135. Chowhan, Z. T.: J. Pharm. Sci. 67:1258, 1978.
136. Barlow, R. B., and Hamilton, J. T.: B. J. Pharmacol. 18:543, 1962.
137. Sanders, H. J., et al.: Ind. Eng. Chem. 47:358, 1955.
138. Shearer, C. M., and Pilla, C. R.: Anal. Profiles 3:452, 1974.
139. Frazakenly, G. V., and Jackson, E. G.: J. Pharm. Sci. 57:335, 1968.
140. Elpern, B.: Ann. N.Y. Acad. Sci. 148:579, 1968.
141. Stella, V. J., and Pipkin, J. D.: J. Pharm. Sci. 65:1160, 1976.
142. Dayton, P. G., et al.: Fed. Proc. 18:382, 1959.
143. Riegelman, S., et al.: J. Pharm. Sci. 51:129, 1962.
144. Agarwal, S. P., and Blake, M. I.: J. Pharm. Sci. 57:1434, 1958.
145. Bansal, P. C., and Monkhouse, P. C.: J. Pharm. Sci. 66:820, 1977.
146. Christophers, S. R.: Ann. Trop. Med. 31:43, 1937.
147. Dox, T., et al.: Iyakirkin Kenkyn 9:205, 1978; through Chem. Abstr. 88:141601a, 1978.
148. Poet, R. B., and Kadin, H.: Anal. Profiles 4:360, 1975.
149. Strobel, G. B., and Bianchi, C. P.: J. Pharmacol. Exp. Ther. 172:5, 1970.
150. Rucki, R. J.: Anal. Profiles 5:415, 1976.
151. Cromble, K. B., and Cullen L. F.: Anal. Profiles 2:452, 1973.
152. Garrett, E. R., and Weber, D. J.: J. Pharm. Sci. 59:1389, 1970.
153. Carlin, H. S., and Perkins, A. J.: Am. J. Hosp. Pharm. 25:271, 1968.
154. Snell, E. E.: Vitam. Horm. 16:84, 1958.
155. Gallo, A. G., and Radaelli, P.: Anal. Profiles 5:483, 1976.
156. Knochel, J. P., et al.: J. Lab. Clin. Med. 65:361, 1965.
157. Garrett, E. R., and Schnelle, K.: J. Pharm. Sci. 60:836, 1971.
158. Conners, K. A.: Textbook of Pharmaceutical Analysis, p. 475. New York, John Wiley & Sons, 1967.
159. Suzuki, A., et al.: J. Pharm. Sci. 59:651, 1970.
160. Brueckner, A. H.: Yale J. Biol. Med. 15:813, 1943.
161. Nygard, B., et al.: Acta Pharm. Suec. 3:313, 1966.
162. Post, A., and Warren, R. J.: Anal. Profiles 5:241, 1976.
163. Shore, R. A., et al.: J. Pharmacol. Exp. Ther. 119:361, 1957.
164. Blessel, K. W., et al.: Anal. Profiles 2:564, 1973.
165. Kaplan, S. A.: J. Pharm. Sci. 59:358, 1970.
166. De Angelis, R. L., et al.: J. Pharm. Sci. 66:842, 1977.
167. Blessel, K. W., et al.: Anal. Profiles 3:577, 1974.
168. Papastephanou, C.: Anal. Profiles 7:492, 1978.
169. McLean, W. M., et al.: J. Pharm. Sci. 56:1614, 1967.
170. White, A., et al.: Principles of Biochemistry, p. 184. New York, McGraw-Hill, 1968.
171. Neuss, N., et al.: J. Am. Chem. Soc. 81:4754, 1959.
172. Hiskey, C. F., et al.: J. Pharm. Sci. 51:43, 1962.
173. Hempelmann, F. W.: Arzneim. Forsch. 27:2140, 1977.
174. Burns, J. J.: J. Am. Chem. Soc. 75:2345, 1953.

INDEX

A "t" following a number represents tabular material. Numbers followed by an "f" represent figures.

ISBN 0-397-50877-8